Introduction to Quantum Field Theory

This textbook offers a detailed and uniquely self-contained presentation of quantum and gauge field theories. Writing from a modern perspective, the author begins with a discussion of advanced dynamics and special relativity before guiding students steadily through the fundamental principles of relativistic quantum mechanics and classical field theory. This foundation is then used to develop the full theoretical framework of quantum and gauge field theories. The introductory, opening half of the book allows it to be used for a variety of courses, from advanced undergraduate to graduate level, and students lacking a formal background in more elementary topics will benefit greatly from this approach. Williams provides full derivations wherever possible and adopts a pedagogical tone without sacrificing rigor. Worked examples are included throughout the text and end-of-chapter problems help students to reinforce key concepts. A fully worked solutions manual is available online for instructors.

Anthony G. Williams is Professor of Physics at Adelaide University, Australia. He has worked extensively in the areas of hadronic physics and computational physics, studying quark and gluon substructure. For this work, he was awarded the Walter Boas Medal by the Australian Institute of Physics in 2001 and elected Fellow of the American Physical Society in 2002. In 2020, he became the deputy director of the Centre for Dark Matter Particle Physics of the Australian Research Council.

Introduction to Quantum Field Theory

Classical Mechanics to Gauge Field Theories

ANTHONY G. WILLIAMS

University of Adelaide

CAMBRIDGE
UNIVERSITY PRESS

CAMBRIDGE
UNIVERSITY PRESS

University Printing House, Cambridge CB2 8BS, United Kingdom

One Liberty Plaza, 20th Floor, New York, NY 10006, USA

477 Williamstown Road, Port Melbourne, VIC 3207, Australia

314–321, 3rd Floor, Plot 3, Splendor Forum, Jasola District Centre,
New Delhi – 110025, India

103 Penang Road, #05–06/07, Visioncrest Commercial, Singapore 238467

Cambridge University Press is part of the University of Cambridge.

It furthers the University's mission by disseminating knowledge in the pursuit of
education, learning, and research at the highest international levels of excellence.

www.cambridge.org
Information on this title: www.cambridge.org/highereducation/isbn/9781108470902
DOI: 10.1017/9781108585286

First published 2023

Printed in the United Kingdom by TJ Books Limited, Padstow Cornwall

A catalogue record for this publication is available from the British Library.

ISBN 978-1-108-47090-2 Hardback

Contents

Detailed Contents

Preface

The Standard Model of particle physics unites three of the four known forces of nature into a single, elegant theory. It is arguably the most successful physical theory devised to date. The accuracy with which the Standard Model can reproduce the measurements of precise experiments is remarkable. It describes the electromagnetic, weak and strong interactions, but it does not account for the gravitational interaction. We understand that the Standard Model is not the final word, but the path to physics beyond the Standard Model is not yet clear. We therefore set the Standard Model as our end point here.

The primary purpose of this book is to provide a single coherent framework taking us from the postulates of special relativity, Newton's laws and quantum mechanics through to the development of quantum field theory, gauge field theories and the Standard Model. The presentation is as self-contained as possible given the need to fit within a single volume. While building an understanding of quantum field theory, gauge field theories and the Standard Model is the final goal here, we have attempted to include all essential background material in a self-contained way.

There is an emphasis on showing the logically flowing development of the subject matter. In order to achieve this we have attempted to provide proofs of all key steps. Where appropriate these have been separated out from the main text in boxes. The reason for this is that the proofs sometimes require a careful mathematical discussion, and this can distract from the flow of the physical arguments. Students seeing this material for the first time can overlook the more difficult boxed proofs on a first reading while concentrating on the physics. They can be comfortable knowing that the particular result can be proved and then can come back and absorb that proof later as desired. Problem sets are provided at the end of each chapter to build familiarity and understanding of the material and practice in its application.

A word of encouragement to students: Much knowledge that is worthwhile is not easily won. If some piece of mathematics initially seems too challenging, then absorb the physical consequence of the result, move on, and come back to the maths later. Discussing with others is *always* helpful. This author remembers well when he did not understand what is written in these pages and the challenge of needing to overcome that. The hope is that enough handholds have been provided that with some effort and through discussions with others any student of quantum field theory can follow the arguments presented.

Conventions and notation: The notation and conventions that we typically follow are those of Peskin and Schroeder (1995) and Schwartz (2013). The conventions in Bjorken and Drell (1964), Bjorken and Drell (1965) and Greiner (2000) are very similar to each other and differ from our notation in the normalization choices for field operators and Dirac spinors and in their not including a factor of i in the definition of Feynman propagators. Itzykson and Zuber (1980) and Sterman (1993) use different normalization conventions again. We and the above books use the 'West coast' metric, $g^{\mu\nu} = \mathrm{diag}(1, -1, -1, -1)$, whereas Brown (1992), Srednicki (2007) and Weinberg (1995) use the 'East coast' metric, $g^{\mu\nu} = \mathrm{diag}(-1, 1, 1, 1)$. Here we denote the electric charge as q as done in

Aitchison and Hey (2013) to avoid confusion. The electron charge is then $q = -e \equiv -|e|$, where for us e will always mean the magnitude of the charge of the electron. Some texts choose e negative and some have mixed or inconsistent usage of the sign. Comparing the covariant derivative with ours, $D^\mu = \partial^\mu + iqA^\mu$, will immediately reveal the choice in each text. One needs to carefully compare equations when moving between texts, keeping these issues in mind. To know that one has the correct sign for q in D^μ one can relate it back to the Lorentz force equation in Eq. (4.3.67) or to the Maxwell equations $\partial_\mu F^{\mu\nu} = j^\nu$ as we and Aitchison and Hey (2013) have done.

Further reading: Due to the breadth of material covered in this single volume there has been a need to focus on the essentials of quantum and gauge field theories. By construction there is more than enough material for any year-long course in quantum and gauge field theories; however, there are topics and applications that could not be included here. A reader who has understood this text will be able to pick up other texts on quantum field theory and readily understand them. Both Peskin and Schroeder (1995) and Schwartz (2013) contain similar conventions and notation, so by design it is straightforward to supplement this book with examples and applications from each of these texts. Srednicki (2007), Weinberg (1995) and Weinberg (1996) provide additional applications and detail using the 'East coast' spacetime metric. For the advanced reader Weinberg's books contain insights and gems of knowledge not readily found elsewhere. A very accessible overview of quantum field theory is given in Zee (2010). There are many other very worthwhile texts and a partial list of these is given at the beginning of Chapter 6.

Organization of the Book

The first four chapters of the book lay out the foundations and structures required for the following chapters. It opens with a discussion of special relativity and Lorentz and Poincaré invariance in Chapter 1, which begins with Einstein's postulates and ends with a discussion of the representations of the Poincaré group used to classify fundamental particles. In Chapter 2 the treatment of classical point mechanics begins with Newton's laws and is developed into analytic mechanics in both its Lagrangian and Hamiltonian formulations. The normal modes of small oscillations of classical systems are treated. The analytic mechanics approach is subsequently extended to include special relativity in the special case of a single particle in an external potential. The discussion of electromagnetism in Sec. 2.7 begins from the traditional form of Maxwell's equations and moves to the relativistic formulation of electromagnetism and the consequent need to understand gauge transformations and gauge fixing. The relation of different unit systems in electromagnetism is given. Chapter 2 ends with a discussion of the Dirac-Bergmann algorithm needed to treat Hamiltonian descriptions of singular systems, which will become relevant for understanding the canonical quantization of gauge fields and fermions. The extension of classical point mechanics to classical relativistic fields is developed in Chapter 3, where the concepts of classical mechanics are extended to an infinite number of degrees of freedom. This was necessary because there is no consistent analytic mechanics formulation of a relativistic system with a finite number of degrees of freedom. We later come to understand that the quanta resulting from the quantization of the normal modes of these relativistic classical fields are the fundamental particles of the corresponding relativistic quantum field theory. A common unifying thread through all of these developments in Chapters 2 and 3 is Hamilton's principle of stationary action. Chapter 4 is devoted to the development of relativistic quantum mechanics. It begins with a review of the essential elements of quantum mechanics, including the derivation of the Feynman path integral approach to quantum mechanics. This chapter has detailed discussions of the Klein-Gordon equation, the Dirac equation, their interaction with external fields and their symmetries.

In the second part of the book we begin in Chapter 5 with a brief history and overview of particle physics to motivate the subsequent chapters. This historical perspective demonstrates the essential role of experimental discovery in driving the direction and development of the field. In Chapter 6 the formulation of free field theory is presented with explicit constructions given for scalar, charged scalar, fermion, photon and massive vector boson fields. In Chapter 7 we discuss the interaction picture, scattering cross-sections and Feynman diagrams and show how to evaluate tree-level diagrams for several theories of interest. In Chapter 8 we first consider the discrete symmetries of charge conjugation (C), parity inversion (P) and time reversal (T) and prove the CPT theorem. Then we turn to the essential elements of renormalization and the renormalization group including dimensional regularization and renormalized perturbation theory. This is followed by a discussion of spontaneous symmetry breaking, Goldstone's theorem and the Casimir effect. Finally, in Chapter 9 the extension of electromagnetism and quantum electrodynamics to nonabelian gauge theories is given. The example of quantum chromodymanics and its relation the strong interactions is then discussed and the lattice gauge theory approach to studies of nonperturbative behavior is introduced. The chapter concludes with a discussion of quantum anomalies and then finally with the construction of the Standard Model of particle physics.

How to Use This Book

This book is intended to be suitable for first-time students as well as for readers more experienced in the field. Some suggestions on how to use the material covered to build various courses are the following.

(i) For a lecture course on **Special Relativity**: Secs. 1.1, 1.2 and some or all of the material from 1.4. Some aspects of Sec. 1.3 could be included for an advanced class. Careful selections of key results from Sec. 2.6 on relativistic kinematics and Sec. 2.7.1 on the relativistic formulation of Maxwell's equations might also be considered.

(ii) For a lecture course on **Classical and/or Analytic Mechanics**: Secs. 2.1–2.7 form a sound basis. As needed, subsets of this material can be used depending on the length of the course. For an advanced class, selections of material on analytic relativistic mechanics and/or Sec. 2.9 on constrained Hamiltonian mechanics could also be included. Working through and summarizing these two advanced sections could also be assigned to students as undergraduate research or reading projects;

(iii) For a course on **Advanced Dynamics and Relativity**: Core material from Secs. 1.1, 1.2 and Secs. 2.1–2.8 form the basis of a one-semester course that I taught over many years.

(iv) For a course on **Relativistic Classical Field Theory**: Secs. 3.1–Sec. 3.3.1 are suitable for a lecture course. For a shorter course a focus on the key results and their proofs would be sufficient. For an advanced class in a longer course Sec. 3.3.2 could be included in the lectures, but this advanced material is also a candidate for a reading topic or small research project.

(v) For a course on **Relativistic Quantum Mechanics**: Secs. 4.2–4.6.5 contain all necessary course material. The review of quantum mechanics in Sec. 4.1 could be treated as assumed knowledge or some elements could be chosen for inclusion. Relevant sections of Sec. 4.1 could be assigned as reading topics in this course as well as in other courses.

(vi) For a lecture course on **Relativistic Quantum Mechanics and Particle Physics**: Material on the Klein-Gordon and Dirac equations in Secs. 4.3, 4.4 and 4.5.1 can be combined with selected elements from Chapter 5 such as the Cabibbo-Kobayashi-Maskawa (CKM) matrix, neutrino mixing, the quark model of strongly interacting particles and representations of group theory.

(vii) For a one-semester course on **Relativistic Quantum Field Theory**: Chapter 6 would form the core of such a course with an emphasis on Secs. 6.2–6.4. Technical sections such as the derivation of the functional integrals for fermions and photons could be abbreviated or omitted and the second part of Sec. 6.3.6 on the derivation of the Dirac fermion canonical anticommutation relations could be mentioned but not explicitly covered. Some of the important results in Secs. 7.1–7.6, including the Feynman rules and example tree-level cross-section calculations, could be summarized and included as course length allows.

(viii) For a one-semester course on **Gauge Field Theories**: The remainder of Chapter 7 not covered above and the core material in Chapters 8 and 9 on renormalization, gauge field theories, Goldstone's theorem, quantum chromodynamics (QCD), anomalies and the Standard Model.

(ix) For a full-year course on **Quantum and Gauge Field Theories**: Combine the material in the above two suggested courses and choose the division of material between semesters to best suit the pace of the lectures and the desired emphasis of the course.

Corrections to This Book

Despite the best efforts of all involved, there will be remaining errors in this book for which I am solely responsible. The current list of corrections along with the names of those who suggested them can be found at:

www.cambridge.org/WilliamsQFT

It would be greatly appreciated if anyone finding additional errors could please report them using the relevant corrections link provided on this website.

Acknowledgments

I offer my sincere thanks to the past and present colleagues and students who have contributed to my understanding of this material over the years. I acknowledge many useful conversations with Ross Young as well as with colleagues Rod Crewther, Paul Jackson, Derek Leinweber, Anthony Thomas, Martin White and James Zanotti. I also thank my former students Dylan Harries for help with proofreading and Ethan Carragher for both proofreading and his help preparing the solutions manual. I am also grateful to Daniel Murnane and Shanette De La Motte for assisting with the preparation of many of the figures. I thank Marc Henneaux, Don Sinclair and Kurt Sundermeyer for contributing to my understanding of the Dirac-Begmann algorithm and constrained Hamiltonian dynamics as well as Steven Avery for sharing his notes on the application of these techniques to systems with fermions. I thank Herbert Neuberger for helpful discussions regarding Gribov copies and lattice BRST.

I am enormously grateful to Jan and Ellen for their constant encouragement, love and understanding through the long hours necessary for the preparation of this book.

1 Lorentz and Poincaré Invariance

1.1 Introduction

The principles of special and general relativity are cornerstones of our understanding of the universe. The combination of gravity with quantum field theory remains an elusive goal and is outside the scope of our discussions. It will be sufficient for us to restrict our attention to special relativity, which is characterized by the Poincaré transformations. These are made up of the Lorentz transformations and spacetime translations. The Lorentz transformations are sometimes referred to as the homogeneous Lorentz transformations, since they preserve the origin, with the Poincaré transformations referred to as the inhomogeneous Lorentz transformations.

We classify fundamental particles by the representation of the Poincaré group that they transform under, which is related to their mass and intrinsic spin, e.g., spin-zero bosons, spin-half fermions and spin-one gauge bosons. Particles with integer spin (0, 1, 2, . . .) are *bosons* and obey Bose-Einstein statistics and particles with half-integer spin (1/2, 3/2, 5/2, . . .) are *fermions* and obey Fermi-Dirac statistics.

We first explore the nonrelativistic Galilean relativity that we are intuitively familiar with before we generalize to the case of special relativity.

1.1.1 Inertial Reference Frames

A *reference frame* is a three-dimensional spatial coordinate system with a clock at every spatial point. A *spacetime event* E is a point in spacetime,

$$E = (t, \mathbf{x}) = (t, x, y, z) = (t, x^1, x^2, x^3). \tag{1.1.1}$$

Note that we write the indices of the spatial coordinates $\mathbf{x} = (x^1, x^2, x^3)$ in the "up" position when discussing special relativity. The *trajectory* of a particle is then described in this reference frame by a continuous series of spacetime events; i.e., if $E = (t, \mathbf{x})$ is an element of the particle's trajectory, then the particle was at the point $\mathbf{x}$ at time t. When we discuss special relativity we will describe an event as $E = (ct, \mathbf{x})$, where c is the speed of light for reasons that will later become obvious. We refer to the set of all spacetime events as *Minkowski space*, $\mathbb{R}^{(1,3)}$, to indicate that there is one time dimension and three spatial dimensions.

When a particle or object is not subject to any external forces, we refer to it as *free*. An *inertial reference frame* is a reference frame in which a free particle or object moves in straight lines at a constant speed. An *inertial observer* is an observer who describes events in spacetime with respect to his chosen inertial reference frame. So an inertial observer describes the trajectory of a *free particle* as a straight line,

$$\mathbf{x}(t) = \mathbf{x}(0) + \mathbf{u}t, \tag{1.1.2}$$

where $\mathbf{u}$ is a constant three-vector, i.e., the constant velocity $\mathbf{u} = d\mathbf{x}/dt$.

Recall Newton's first law: "When viewed from any inertial reference frame, an object at rest remains at rest and an object moving continues to move with a constant velocity, unless acted upon by an external unbalanced force." So Newton's first law simply provides a definition of an inertial reference frame. It is in the inertial frames defined by the first law that Newton's second and third laws are formulated. In other words, Newton's laws have the same form in all inertial frames. It was assumed that all of the laws of physics have the same form in every inertial frame.

In arriving at his laws of motion Newton *assumed* that the displacement in time $\Delta t = t_2 - t_1$ between two events and all lengths would be the same in all frames. This is equivalent to Newton having assumed the laws of Galilean relativity.

1.1.2 Galilean Relativity

Consider any two observers $\mathcal{O}$ and $\mathcal{O}'$ with different reference frames such that a spacetime event E is described by observer $\mathcal{O}$ as $(t, \mathbf{x})$ and is described by observer $\mathcal{O}'$ as $(t', \mathbf{x}')$. These reference frames are not necessarily inertial.

If for every two spacetime events observers $\mathcal{O}$ and $\mathcal{O}'$ measure the same time difference and if they measure all lengths equal, then this means that

$$(t'_2 - t'_1) = (t_2 - t_1) \quad \text{and} \quad |\mathbf{x}'_2(t') - \mathbf{x}'_1(t')| = |\mathbf{x}_2(t) - \mathbf{x}_1(t)|, \tag{1.1.3}$$

where length is defined as the distance between two simultaneous events such as the locations of the end of a ruler at any given time. It then follows that the reference frames of $\mathcal{O}$ and $\mathcal{O}'$ must be related to each other by the transformation

$$t' = t + d \quad \text{and} \quad \mathbf{x}'(t') = O(t)\mathbf{x}(t) + \mathbf{b}(t), \tag{1.1.4}$$

where $O(t)$ is an orthogonal matrix at all times, $O^T(t) = O^{-1}(t)$.

Proof: Since $(t'_2 - t'_1) = (t_2 - t_1)$ for every two events we must have $t' = t + d$ where d is the difference in the settings of the clocks of the two observers.

Since the vectors $[\mathbf{x}'_2(t') - \mathbf{x}'_1(t')]$ and $[\mathbf{x}_2(t) - \mathbf{x}_1(t)]$ have the same length at all times, then they must be related by a rotation at all times; i.e., there must exist a matrix $O(t)$ that is orthogonal at all times so that

$$[\mathbf{x}'_2(t') - \mathbf{x}'_1(t')] = O(t)\,[\mathbf{x}_2(t) - \mathbf{x}_1(t)]. \tag{1.1.5}$$

In general there can be an arbitrary spatially independent function $\mathbf{b}(t)$ such that $\mathbf{x}'(t') = O(t)\mathbf{x}(t) + \mathbf{b}(t)$. The result is then proved.

If $\mathcal{O}$ and $\mathcal{O}'$ are inertial observers and if they measure the same time differences and lengths, then their reference frames are related by

$$t' = t + d \quad \text{and} \quad \mathbf{x}'(t') = O\mathbf{x}(t) - \mathbf{v}t + \mathbf{a}, \tag{1.1.6}$$

which is referred to as a *Galilean transformation*. Here d is the offset of the clock (i.e., *time translation*) of observer $\mathcal{O}'$ with respect to that of observer $\mathcal{O}$; the vector $(-\mathbf{a})$ is the displacement (i.e., *spatial translation*) of the origin of $\mathcal{O}'$ with respect to that of $\mathcal{O}$ at $t = 0$; $\mathbf{v}$ is the velocity (i.e., *velocity boost*) of $\mathcal{O}'$ with respect to $\mathcal{O}$; and O is the orthogonal matrix ($O^T = O^{-1}$) (i.e., *rotation*) that rotates the spatial axes of $\mathcal{O}'$ into the orientation of the spatial axes of $\mathcal{O}$.

Proof: If both observers are inertial, then they will each describe the trajectory of a free particle by a straight line

$$\mathbf{x}'(t') = \mathbf{x}'_0 + \mathbf{u}'t' \quad \text{and} \quad \mathbf{x}(t) = \mathbf{x}_0 + \mathbf{u}t, \tag{1.1.7}$$

where $\mathbf{x}$, $\mathbf{x}_0$, $\mathbf{u}$ and $\mathbf{u}'$ are all time-independent. Since they measure the same time differences and lengths, then from the above result we also have $t' = t + d$ and $\mathbf{x}'(t') = O(t)\mathbf{x}(t) + \mathbf{b}(t)$. It then follows that

$$\mathbf{x}'(t + d) = \mathbf{x}'_0 + \mathbf{u}'(t + d) = O(t)\mathbf{x}(t) + \mathbf{b}(t) = O(t)\left(\mathbf{x}_0 + \mathbf{u}t\right) + \mathbf{b}(t). \tag{1.1.8}$$

Acting with d/dt and using the notation $\dot{f} \equiv df/dt$ we find

$$\mathbf{u}' = \dot{O}(t)\left(\mathbf{x}_0 + \mathbf{u}t\right) + O(t)\mathbf{u} + \dot{\mathbf{b}}(t) \tag{1.1.9}$$

and acting with d/dt again gives

$$0 = \ddot{O}(t)\left(\mathbf{x}_0 + \mathbf{u}t\right) + 2\dot{O}(t)\mathbf{u} + \ddot{\mathbf{b}}(t). \tag{1.1.10}$$

This must be true for every free particle trajectory and hence must be true for every $\mathbf{u}$ and every $\mathbf{x}_0$, which means that we must have for every time t that

$$\ddot{O}(t) = \dot{O}(t) = 0 \quad \text{and so} \quad \ddot{\mathbf{b}}(t) = 0. \tag{1.1.11}$$

Then $O(t) = O$ is a time-independent orthogonal matrix and we can write $\mathbf{b}(t) = -\boldsymbol{v}t + \mathbf{a}$ for some time-independent $\boldsymbol{v}$ and $\mathbf{a}$, which proves Eq. (1.1.6).

At $t = 0$ the origin ($\mathbf{x}' = 0$) of the inertial reference frame of observer $\mathcal{O}'$ is at $\mathbf{x} = -\mathbf{a}$ in the inertial reference frame of $\mathcal{O}$ or equivalently the origin of $\mathcal{O}$ is at $\mathbf{x}' = \mathbf{a}$ in the frame of $\mathcal{O}'$. At t the origin of $\mathcal{O}'$ is at $\mathbf{x} = \boldsymbol{v}t - \mathbf{a}$, which shows that $\mathcal{O}'$ moves with a constant velocity $\boldsymbol{v}$ with respect to $\mathcal{O}$. Finally, O is the orthogonal matrix that rotates the spatial axes of $\mathcal{O}'$ to that of $\mathcal{O}$.

It was Newton's *assumption* that inertial frames are related by Galilean transformations and that all of the laws of physics are invariant under these transformations. Newton was assuming that the laws of physics were *Galilean invariant*.

An immediate consequence of Galilean invariance is the *Galilean law for the addition of velocities*, which is obtained by acting with d/dt on Eq. (1.1.6) to give

$$\mathbf{u}'(t') = O\mathbf{u}(t) - \boldsymbol{v}. \tag{1.1.12}$$

If we choose the two sets of axes to have the same orientation ($O = I$), then we recover the familiar Galilean addition of velocities $\mathbf{u}'(t) = \mathbf{u}(t) - \boldsymbol{v}$.

Einstein understood that there was an inconsistency between the assumed Galilean invariance of Newtonian mechanics and the properties of Maxwell's equations of electromagnetism. For example, if Galilean invariance applied to the propagation of light, then the velocity of light would depend on the boost $\boldsymbol{v}$; i.e., we would have $\mathbf{c}' = \mathbf{c} - \boldsymbol{v}$. In the nineteenth century it was thought that space was filled with some *luminiferous ether*[1] through which light propagated. The Earth's velocity through the ether changes as it rotates around the sun and hence light was expected to have a velocity dependent on direction. The Michelson-Morley experiment carried out in 1887 was designed to detect variations

---

[1] The term "luminiferous" means light-bearing and the original spelling of "aether" has widely been replaced by the modern English spelling "ether."

in the speed of light in perpendicular directions; however, it and subsequent experiments found no evidence of this.

1.2 Lorentz and Poincaré Transformations

Having understood Galilean relativity we can now generalize to the construction of the theory of special relativity. We first construct the Lorentz transformations and then build the Lorentz group and then extend this to the Poincaré group.

1.2.1 Postulates of Special Relativity and Their Implications

The postulates of special relativity are:

 (i) The laws of physics are the same in all inertial frames.
(ii) The speed of light is constant and is the same in all inertial frames.

Let E_1 and E_2 be any two spacetime events labeled by inertial observer $\mathcal{O}$ as $x_1^\mu = (ct_1, \mathbf{x}_1)$ and $x_2^\mu = (ct_2, \mathbf{x}_2)$, respectively. The *spacetime displacement* of these two events in the frame of $\mathcal{O}$ is defined as

$$z^\mu \equiv \Delta x^\mu = (x_2^\mu - x_1^\mu) = (c(t_2 - t_1), \mathbf{x}_2 - \mathbf{x}_1) = (c\Delta t, \Delta \mathbf{x}). \tag{1.2.1}$$

Consider the special case where event E_1 consists of a flash of light and event E_2 refers to the arrival of that flash of light. Since c is the speed of light, then $c = \Delta \mathbf{x} / \Delta t$ and so $(z^0)^2 - \mathbf{z}^2 = (c\Delta t)^2 - (\Delta \mathbf{x})^2 = 0$. Let $\mathcal{O}'$ be any other inertial observer who in his frame records these same two events as $x_1'^\mu = (ct_1', \mathbf{x}_1')$ and $x_2'^\mu = (ct_2', \mathbf{x}_2')$, respectively. From postulate (ii) it follows that $c = \Delta \mathbf{x}' / \Delta t'$ and hence that $z'^2 = (c\Delta t')^2 - (\Delta \mathbf{x}')^2 = 0$.

The Minkowski-space *metric tensor*, $g_{\mu\nu}$, is defined[2] by

$$g_{00} = -g_{11} = -g_{22} = -g_{33} = +1 \qquad \text{and} \qquad g_{\mu\nu} = 0 \text{ for } \mu \neq \nu, \tag{1.2.2}$$

which we can also write as $g_{\mu\nu} \equiv \text{diag}(1, -1, -1, -1)$. We can then further define

$$x^2 \equiv x^\mu g_{\mu\nu} x^\nu = (x^0)^2 - \mathbf{x}^2 = (ct)^2 - \mathbf{x}^2, \tag{1.2.3}$$

where we use the *Einstein summation convention* that repeated spacetime indices are understood to be summed over. In general relativity spacetime is curved and the metric tensor becomes spacetime-dependent, $g_{\mu\nu} \to g(x)_{\mu\nu}$.

For our special case consisting of the emission and reception of light at events E_1 and E_2, respectively, we say that the events have a *lightlike* displacement. Using the above notation we see that the spacetime interval in this case satisfies the condition that $z'^2 = z^2 = 0$, where $z^2 = (\Delta x)^2 = (x_2 - x_1)^2$ and $z'^2 = (\Delta x')^2 = (x_2' - x_1')^2$. If two events have a lightlike displacement in one inertial frame, then by the postulates of relativity they have a lightlike displacement in every frame. The set of all lightlike displacements makes up something that is referred to as the *light cone*.

---

[2] In general relativity and some quantum field theory textbooks $g_{\mu\nu} \equiv \text{diag}(-1, +1, +1, +1)$ is used. The more common particle physics choice is the metric tensor defined here.

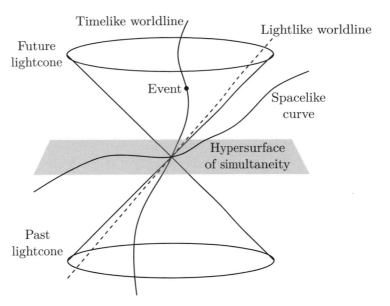

Figure 1.1 The light cone at a point on a particle worldline. In order to graphically represent the light cone, we suppress one of the three spatial dimensions.

The *spacetime interval* between spacetime events E_1 and E_2 is defined as

$$s^2 \equiv z^2 = (\Delta x)^2 = (z^0)^2 - \mathbf{z}^2 = (c\Delta t)^2 - (\Delta \mathbf{x})^2 = c^2(t_2 - t_1)^2 - (\mathbf{x}_2 - \mathbf{x}_1)^2. \qquad (1.2.4)$$

If $s^2 = 0$ the events have a lightlike displacement and they lie on the light cone.

To visualize the light cone and its usefulness in categorizing the spacetime interval, it is useful to temporarily suppress one of the spatial dimension as in Fig. 1.1. It is easy to see that all events E_2 with a lightlike displacement from event E_1 will map out a cone in the three-dimensional spacetime with the time axis as the central axis of the future and past light cones and with E_1 at the point where the light cones touch. The trajectory of a particle is a continuous sequence of spacetime events, which in special relativity is often referred to as the *worldline* of the particle.

If a particle contains event E_1, then any subsequent event E_2 on its worldline must lie inside the forward light cone of E_1. Similarly, in such a case if E_2 occurred on the worldline before E_1 then E_2 must lie inside the backward light cone. Being inside the light cone means that $s^2 > 0$ and we describe such events as *timelike* separated events. If $s^2 < 0$, then we say that the two events have a *spacelike* displacement and no particle moving at or below the speed of light could be present at both events.

We will later see that there is always an inertial reference frame where spacelike separated events are simultaneous. Simultaneous events cannot have a *causal* relationship. In addition, postulate (i) tells us that causal relationships should not depend on the observer's inertial reference frame. Thus spacelike separated events cannot be causally related; i.e., if E_1 and E_2 have a spacelike displacement, then what happens at E_1 cannot influence what happens at E_2 and vice versa.

A particle cannot be in two places at the same time in any inertial frame and so no part of its worldline can be spacelike; i.e., particles cannot travel faster than the speed of light. We will see in Sec. 2.6 that it would take an infinite amount of energy to accelerate a massive particle to the speed of light and so massive particles have worldlines that must lie entirely *within* the light cone; i.e., they have timelike worldlines. We will also see that massless particles must travel at the velocity of light and so their worldlines, like that of light, must lie entirely *on* the light cone.

Consider the special case where at $t = 0$ the clocks of the two observers are synchronized and the origins and axes of their coordinate systems coincide; i.e., at $t = 0$ the spacetime origins of $\mathcal{O}$ and $\mathcal{O}'$ are coincident, $x^\mu = x'^\mu = 0$. It must be possible to define an invertible (one-to-one and onto) transformation between their coordinate systems of the form $x^\mu \to x'^\mu = x'^\mu(x)$, where

$$dx^\mu \to dx'^\mu = \frac{\partial x'^\mu}{\partial x^\nu} dx^\nu \equiv \Lambda^\mu{}_\nu dx^\nu \quad \text{or in matrix form} \quad dx \to dx' = \Lambda\, dx, \qquad (1.2.5)$$

where we have defined $\Lambda^\mu{}_\nu \equiv \partial x'^\mu / \partial x^\nu$. The sixteen elements of Λ must be real; i.e., $\Lambda^\mu{}_\nu \in \mathbb{R}$ for all μ and $\nu = 0, 1, 2, 3$. In the matrix notation that we sometimes use x^μ is the element in row μ of a four-component column vector x, and $\Lambda^\mu{}_\nu$ is the element in row μ and column ν of the 4×4 real Lorentz transformation matrix Λ; i.e., column vector x has elements x^{row} and the matrix Λ has elements $\Lambda^{\text{row}}{}_{\text{column}}$. Note that for historical reasons we refer to x^μ as a *contravariant* four-vector. The matrices Λ with elements $\Lambda^\mu{}_\nu$ are referred to as *Lorentz transformations*.

An important and useful concept is that of the *proper time*[3] for a moving object, which is the time showing on a clock that is carried along with the object. Also important is the concept of the *proper length* of an object that is its length in its own rest frame. For a free particle we define the *comoving inertial frame* as the inertial frame in which the particle remains at rest. For an object that is accelerating we can still define an *instantaneously comoving inertial frame* as an inertial frame in which the accelerating particle is at rest at a particular instant of time.

The discussion so far allows for the possibility that the transformations $\Lambda^\mu{}_\nu$ might depend on x^μ; i.e., we might have $\Lambda^\mu{}_\nu(x)$. However from postulate (i) it follows that the Lorentz transformations $\Lambda^\mu{}_\nu$ must be spacetime independent; i.e., *any Lorentz transformation matrix Λ contains sixteen constants*. So for any x^μ we have

$$x^\mu \to x'^\mu = \Lambda^\mu{}_\nu x^\nu \qquad \text{or in matrix form} \qquad x \to x' = \Lambda x. \qquad (1.2.6)$$

Proof: Consider a free particle being observed by two inertial observers $\mathcal{O}$ and $\mathcal{O}'$. Both observers will describe the worldline as a straight line. Since by postulate (i) Minkowski space must be homogeneous, then an inertial observer must see equal increments of dx^μ in equal increments of the free particle's proper time $d\tau$; i.e., for a free particle we must have that

$$\frac{dx^\mu}{d\tau} \quad \text{and} \quad \frac{dx'^\mu}{d\tau} \quad \text{are constants} \quad \Rightarrow \quad \frac{d^2 x^\mu}{d\tau^2} = \frac{d^2 x'^\mu}{d\tau^2} = 0. \qquad (1.2.7)$$

We now consider

$$\frac{dx'^\mu}{d\tau} = \frac{\partial x'^\mu}{\partial x^\nu} \frac{dx^\nu}{d\tau} = \Lambda^\mu{}_\nu \frac{dx^\nu}{d\tau} \qquad (1.2.8)$$

and differentiate with respect to the proper time τ to give

$$\frac{d^2 x'^\mu}{d\tau^2} = \frac{\partial x'^\mu}{\partial x^\nu} \frac{d^2 x^\nu}{d\tau^2} + \frac{\partial^2 x'^\mu}{\partial x^\rho \partial x^\nu} \frac{dx^\nu}{d\tau} \frac{dx^\rho}{d\tau}. \qquad (1.2.9)$$

---

[3] The meaning of the English word *proper* in this context originates from the French word *propre* meaning *own* or *one's own*. The proper time is the time belonging to the object itself.

This must be true for all $dx^\mu/d\tau$. Then from Eq. (1.2.7) we immediately find

$$\frac{\partial^2 x'^\mu}{\partial x^\rho \partial x^\nu} = \frac{\partial}{\partial x^\rho} \Lambda^\mu{}_\nu = 0 \qquad (1.2.10)$$

and so $\Lambda^\mu{}_\nu = \partial x'^\mu/\partial x^\nu$ is independent of x; i.e., Λ is a matrix of constants. Then dividing x^μ into an infinite number of infinitesimals gives $x'^\mu = \Lambda^\mu{}_\nu x^\nu$.

It follows that the most general transformation between two inertial frames is

$$x^\mu \rightarrow x'^\mu = p(x) \equiv \Lambda^\mu{}_\nu x^\nu + a^\mu \quad \text{or in matrix form} \quad x \rightarrow x' = \Lambda x + a, \qquad (1.2.11)$$

which is referred to as a *Poincaré transformation*. In his frame observer $\mathcal{O}'$ describes the spacetime origin of $\mathcal{O}$ as being at a^μ. So a^μ is a *spacetime translation*.

Let us return to our example of the two events linked by the propagation of light as seen in two different frames; i.e., they have a lightlike displacement. Then from postulate (ii) and the above arguments we have $z^2 = z'^2 = 0$ and so

$$z'^2 = z'^T g z' = z^T \Lambda^T g \Lambda z = z^2 = z^T g z = 0, \qquad (1.2.12)$$

where $z^\mu = (z, \mathbf{z})$ with $z \equiv |\mathbf{z}|$. Then noting that $\Lambda^T g \Lambda$ is a real symmetric matrix and using the theorem given below, we see that for every Lorentz transformation Λ we must have $\Lambda^T g \Lambda = a(\Lambda)g$ for some real constant $a(\Lambda)$ independent of $\mathbf{z}$. For $\Lambda = I$ we have $I^T g I = g$ and so $a(I) = 1$. Then we have $1 = a(I) = a(\Lambda^{-1}\Lambda) = a(\Lambda^{-1})a(\Lambda)$ and so $a(\Lambda^{-1}) = 1/a(\Lambda)$. If observer $\mathcal{O}'$ is related to observer $\mathcal{O}$ by Λ, then $\mathcal{O}$ is related to $\mathcal{O}'$ by Λ^{-1}. The labeling of inertial observers is physically irrelevant, since there is no preferred inertial frame. So $a(\Lambda) = a(\Lambda^{-1}) = 1/a(\Lambda)$ and $a(\Lambda) = \pm 1$ for all Λ. Since $a(I) = 1$ and since $a(\Lambda'\Lambda) = a(\Lambda')a(\Lambda) = a(\Lambda')/a(\Lambda) = a(\Lambda'\Lambda^{-1})$ for all Λ' and Λ, then $a(\Lambda) = 1$ for all Λ. So *any* Λ must satisfy

$$\Lambda^T g \Lambda = g \qquad \text{or equivalently} \qquad \Lambda^\sigma{}_\mu g_{\sigma\tau} \Lambda^\tau{}_\nu = g_{\mu\nu}. \qquad (1.2.13)$$

This is equivalent to the statement that *the Minkowski-space metric tensor $g_{\mu\nu}$ is invariant under Lorentz transformations*.

Theorem: Let A be a real symmetric matrix, independent of $\mathbf{z}$ and define $z \equiv |\mathbf{z}|$, then if

$$\begin{pmatrix} z & \mathbf{z}^T \end{pmatrix} A \begin{pmatrix} z \\ \mathbf{z} \end{pmatrix} = 0 \qquad \text{for all} \quad \mathbf{z} \qquad (1.2.14)$$

there must exist some real constant a independent of $\mathbf{z}$ such that $A = ag$, where g is the Minkowski-space metric tensor, $g = \text{diag}(1, -1, -1, -1)$.

Proof: Since A is a real symmetric matrix, then we can write it as

$$A \equiv \begin{pmatrix} a & \mathbf{b}^T \\ \mathbf{b} & C \end{pmatrix}, \qquad (1.2.15)$$

where $a \in \mathbb{R}$, $\mathbf{b}$ is a real three-vector and C is a real symmetric 3×3 matrix. We then obtain $az^2 + 2z\mathbf{z} \cdot \mathbf{b} + \mathbf{z}^T C \mathbf{z} = 0$, which must be true for all $\mathbf{z}$. The only way this can be true for both $\pm\mathbf{z}$ is if $\mathbf{b} = 0$. Then we have $\mathbf{z}^T (aI + C) \mathbf{z} = 0$ for all $\mathbf{z}$ and since $(aI + C)$ is a real symmetric matrix, this means that we must have $C = -aI$. This then proves that $A = ag$ as required.

Since $\det(g) = \det(\Lambda^T g \Lambda) = \det(\Lambda)^2 \det(g)$ and Λ is a real matrix, then

$$\det(\Lambda) = \pm 1. \tag{1.2.16}$$

We *define* $g^{\mu\nu}$ to be the matrix elements of the matrix g^{-1}, i.e.,

$$g^{-1}g = I \qquad \text{or equivalently} \qquad g^{\mu\sigma}g_{\sigma\nu} \equiv \delta^\mu_{\ \nu}, \tag{1.2.17}$$

where the $\delta^\mu_{\ \nu}$ are the elements of the identity matrix, i.e., $\delta^\mu_{\ \nu} = 1$ if $\mu = \nu$ and zero otherwise. This is also what is done in general relativity. For the Minkowski-space metric $g = \mathrm{diag}(1, -1, -1, -1)$ that is relevant for special relativity it is clear that $g^{-1} = g$, since $g^2 = I$. So $g^{\mu\nu}$ and $g_{\mu\nu}$ are numerically equal, i.e.,

$$g^{-1} = g \qquad \text{or equivalently} \qquad g^{\mu\nu} = g_{\mu\nu}. \tag{1.2.18}$$

Acting on Eq. (1.2.13) from the left with $g^{\rho\mu}$ we find that

$$g\Lambda^T g\Lambda = gg = I \qquad \text{or equivalently} \qquad g^{\rho\mu}\Lambda^\sigma_{\ \mu}g_{\sigma\tau}\Lambda^\tau_{\ \nu} = g^{\rho\mu}g_{\mu\nu} = \delta^\rho_{\ \nu}. \tag{1.2.19}$$

Note that since $\det \Lambda \neq 0$ then Λ^{-1} exists. Since $g\Lambda^T g\Lambda = I$ then acting from the left with Λ and from the right with Λ^{-1} gives $\Lambda(g\Lambda^T g) = I$. Then $\Lambda(g\Lambda^T g) = (g\Lambda^T g)\Lambda = I$ for all Λ and so $(g\Lambda^T g)$ is the inverse of Λ,

$$\Lambda^{-1} = g\Lambda^T g \qquad \text{or equivalently} \qquad (\Lambda^{-1})^\rho_{\ \lambda} = g^{\rho\mu}\Lambda^\sigma_{\ \mu}g_{\sigma\lambda}. \tag{1.2.20}$$

Furthermore, note that since $\Lambda g\Lambda^T g = I$ and $g^2 = I$ then $\Lambda g\Lambda^T = g$ and so comparing with Eq. (1.2.13) we see that if Λ is a Lorentz transform, then so is Λ^T.

The x^μ are the contravariant four-vector spacetime coordinates. Any contravariant four-vector transforms by definition as $V^\mu \to V'^\mu = \Lambda^\mu_{\ \nu}V^\nu$. A *covariant* four-vector by definition transforms with the inverse transformation, i.e., $V_\mu \to V'_\mu = V_\nu(\Lambda^{-1})^\nu_{\ \mu}$ so that $V_\mu V^\mu = V^\mu V_\mu$ is invariant under the transformation, i.e.,

$$V'_\mu V'^\mu = V_\nu(\Lambda^{-1})^\nu_{\ \mu}\Lambda^\mu_{\ \sigma}V^\sigma = V_\nu\delta^\nu_{\ \sigma}V^\sigma = V_\nu V^\nu. \tag{1.2.21}$$

An arbitrary *tensor* with m contravariant and n covariant indices transforms as

$$T^{\mu_1\mu_2\cdots\mu_m}_{\nu_1\nu_2\ldots\nu_n} \to (T')^{\mu_1\mu_2\cdots\mu_m}_{\phantom{(T')^{\mu_1\mu_2\cdots\mu_m}}\nu_1\nu_2\ldots\nu_n} \tag{1.2.22}$$
$$= \Lambda^{\mu_1}_{\ \sigma_1}\Lambda^{\mu_2}_{\ \sigma_2}\cdots\Lambda^{\mu_m}_{\ \sigma_m}T^{\sigma_1\sigma_2\ldots\sigma_m}_{\tau_1\tau_2\ldots\tau_n}(\Lambda^{-1})^{\tau_1}_{\ \nu_1}(\Lambda^{-1})^{\tau_2}_{\ \nu_2}\cdots(\Lambda^{-1})^{\tau_n}_{\ \nu_n}$$

and we refer to such a tensor as having rank $(m + n)$ and being of type (m, n). Note that if we *contract* a contravariant and covariant index by setting them equal and summing over them, then we obtain

$$T^{\mu_1\ldots\mu_{i-1}\rho\mu_{i+1}\ldots\mu_m}_{\nu_1\ldots\nu_{j-1}\rho\nu_{j+1}\ldots\nu_n} \to (T')^{\mu_1\ldots\mu_{i-1}\rho\mu_{i+1}\ldots\mu_m}_{\nu_1\ldots\nu_{j-1}\rho\nu_{j+1}\ldots\nu_n}$$
$$= \Lambda^{\mu_1}_{\ \sigma_1}\cdots\Lambda^\rho_{\ \sigma_i}\cdots\Lambda^{\mu_m}_{\ \sigma_m}T^{\sigma_1\sigma_2\ldots\sigma_m}_{\tau_1\tau_2\ldots\tau_n}(\Lambda^{-1})^{\tau_1}_{\ \nu_1}\cdots(\Lambda^{-1})^{\tau_j}_{\ \rho}\cdots(\Lambda^{-1})^{\tau_n}_{\ \nu_n}$$
$$= \Lambda^{\mu_1}_{\ \sigma_1}\cdots\Lambda^{\mu_{i-1}}_{\ \sigma_{i-1}}\Lambda^{\mu_{i+1}}_{\ \sigma_{i+1}}\cdots\Lambda^{\mu_m}_{\ \sigma_m}T^{\sigma_1\ldots\sigma_{i-1}\lambda\sigma_{i+1}\ldots\sigma_m}_{\tau_1\ldots\tau_{j-1}\lambda\tau_{j+1}\ldots\tau_n}$$
$$\times (\Lambda^{-1})^{\tau_1}_{\ \nu_1}\cdots(\Lambda^{-1})^{\tau_{j-1}}_{\ \nu_{j-1}}(\Lambda^{-1})^{\tau_{j+1}}_{\ \nu_{j+1}}\cdots(\Lambda^{-1})^{\tau_n}_{\ \nu_n}, \tag{1.2.23}$$

where we have used $\Lambda^\rho_{\ \sigma_i}(\Lambda^{-1})^{\tau_j}_{\ \rho} = \delta^{\tau_j}_{\ \sigma_i}$. We see that if a type (m, n) tensor has a pair of spacetime indices contracted, then it becomes a type $(m - 1, n - 1)$ tensor.

We observe that $g_{\mu\nu}$ is invariant under Lorentz transformations, since

$$(\Lambda^{-1})^\sigma_{\ \mu}g_{\sigma\tau}(\Lambda^{-1})^\tau_{\ \nu} = g_{\mu\nu}. \tag{1.2.24}$$

This follows in any case since Eq. (1.2.13) is true for any Lorentz transformation.

We can use the properties of the Lorentz transformations to establish a simple relationship between the covariant and contravariant vectors. Let V^μ be a contravariant four-vector, then $(V^T g)$ is a row matrix with elements $V^\nu g_{\nu\mu}$. Under a Lorentz transformation we find that $(V^T g)$ transforms as

$$(V^T g) \to (V'^T g) = V^T \Lambda^T g = V^T g g \Lambda^T g = (V^T g) \Lambda^{-1} \qquad (1.2.25)$$

or in index notation

$$(V^\nu g_{\nu\mu}) \to (V'^\nu g_{\nu\mu}) = V^\rho \Lambda^\nu{}_\rho g_{\nu\mu} = V^\tau g_{\tau\sigma} g^{\sigma\rho} \Lambda^\nu{}_\rho g_{\nu\mu} = (V^\tau g_{\tau\sigma})(\Lambda^{-1})^\sigma{}_\mu,$$
$$\Rightarrow \quad (V^T g) V = (V^\nu g_{\nu\mu}) V^\mu = V^2. \qquad (1.2.26)$$

So for Lorentz transformations we can identify the elements of the covariant four-vector V_μ with the elements of the row vector $(gV)^T$, i.e.,

$$V_\mu \equiv (V^T g)_\mu = V^\nu g_{\nu\mu}. \qquad (1.2.27)$$

Since $g^2 = I$ it then follows that

$$V^\mu = V_\nu g^{\nu\mu} \qquad \text{or equivalently} \qquad V = ((V^T g)g)^T, \qquad (1.2.28)$$

where here the index notation is more concise than the matrix notation. These arguments extend to any spacetime indices on an arbitrary tensor. Hence we can use $g^{\mu\nu}$ to *raise* a covariant index to be a contravariant one and conversely $g_{\mu\nu}$ can be used to *lower* a contravariant index to be a covariant one, e.g.,

$$T^{\mu_1 \cdots \quad \cdots \mu_m}{}_\mu{}^\nu{}_{\nu_1 \ldots \ldots \nu_n} = g_{\mu\mu_i} g^{\nu\nu_j} T^{\mu_1 \cdots \mu_i \cdots \mu_m}{}_{\nu_1 \ldots \nu_j \ldots \nu_n}, \qquad (1.2.29)$$

where every contravariant ("up") index transforms with a Λ and every covariant ("down") index transforms with a Λ^{-1}. If a type (m, m) tensor has all of its m pairs of indices contracted, then the result is *Lorentz invariant*; e.g., if V^μ and W^μ are four-vectors, then $V^\mu W_\nu$ is a $(1, 1)$ tensor and we have Lorentz invariants,

$$V \cdot W \equiv V^\mu W_\mu = V_\mu W^\mu = V^\mu g_{\mu\nu} W^\nu = V_\mu g^{\mu\nu} W_\nu,$$
$$V^2 \equiv V \cdot V = V^\mu V_\mu = V_\mu V^\mu = V^\mu g_{\mu\nu} V^\nu = V_\mu g^{\mu\nu} V_\nu. \qquad (1.2.30)$$

If the spacetime interval between any two events E_A and E_B is Lorentz invariant and translationally invariant, then it is Poincaré invariant. This follows since if $x' = \Lambda x + a$, then

$$s'^2 = z'^2 = (x'_B - x'_A)^2 = (\Lambda x_B - \Lambda x_A)^2 = (x_B - x_A)^2 = z^2 = s^2. \qquad (1.2.31)$$

Note that *any* two spacetime displacements z'^μ and z^μ with the same interval, $z'^2 = z^2 = s^2$, are related by some Lorentz transformation by definition, $z' = \Lambda z$.

With z^0 on the y-axis and $|\mathbf{z}|$ on the x-axis constant s^2 corresponds to hyperbolic curves in (a) the forward light cone if $s^2 > 0$ and $z^0 > 0$, (b) the backward light cone if $s^2 > 0$ and $z^0 < 0$, and (c) the spacelike region if $s^2 < 0$.

If $z^2 = s^2 = d > 0$ then there is a $z'^\mu = (\pm\sqrt{d}, \mathbf{0}) = (\pm\Delta\tau, \mathbf{0})$

if $z^2 = s^2 = -d < 0$ then there is a $z'^\mu = (0, \sqrt{d}\mathbf{n}) = (0, L_0 \mathbf{n})$, $\qquad (1.2.32)$

where d is a positive real number, $\Delta\tau$ is the proper time between the events, $\mathbf{n}$ is an arbitrary unit vector, and L_0 is the proper length separating the events.

Since by definition all four-vectors transform the same way under Lorentz transformations, then we have the general results that (i) *for any timelike four-vector we can always find an inertial reference*

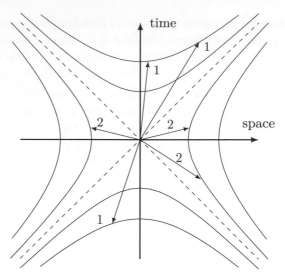

Figure 1.2 A two-dimensional representation of spacetime. Displacement vectors labeled by the same number (1 or 2) are related by Lorentz transformations.

frame where the spatial component vanishes and the time component is the proper time component, and (ii) *for any spacelike four-vector we can always find an inertial reference frame where the time component vanishes and the spatial component has its proper length* (see Fig. 1.2).

The *Lorentz transformations form a group* since they satisfy the four conditions that define a group: (i) *closure:* if Λ_1 and Λ_2 are two arbitrary Lorentz transformations, then $g = \Lambda_1^T g \Lambda_1 = \Lambda_2^T g \Lambda_2$ and hence $(\Lambda_1\Lambda_2)^T g(\Lambda_1\Lambda_2) = \Lambda_2^T \Lambda_1^T g \Lambda_1 \Lambda_2 = \Lambda_2^T g \Lambda_2 = g$ and hence $(\Lambda_1\Lambda_2)$ is also a Lorentz transformation; (ii) *associativity:* since consecutive Lorentz transformations are given by matrix multiplication, and since matrix multiplication is associative, then Lorentz transformations are also, $\Lambda_1(\Lambda_2\Lambda_3) = (\Lambda_1\Lambda_2)\Lambda_3$; (iii) *identity:* $\Lambda = I$ is a Lorentz transformation since $I^T g I = g$; and (iv) *inverse:* for every Lorentz transformation the group contains an inverse $\Lambda^{-1} = g\Lambda^T g$ as shown above. The Lorentz group is typically denoted as $O(1,3)$ and is a generalization of the rotation group $O(3)$ to four dimensions with a Minkowski-space metric. It is an example of a pseudo-orthogonal group.

Mathematical digression: Orthogonal and pseudo-orthogonal groups: Both orthogonal transformations and Lorentz transformations have the property that they are made up of real matrices A that satisfy $A^T G A = G$, where G is the relevant "metric" tensor. For the orthogonal transformations, $O(3)$, in their defining representation we have that A are real 3×3 matrices and G is the 3×3 identity matrix, i.e., with three entries of $+1$ down the diagonal such that $A^T A = I$. For the Lorentz transformations we have $G = g$, where g is given by Eq. (1.2.2) with a $+1$ and three -1 entries down the diagonal leading to the notation $O(1,3)$. The meaning of the group $O(m,n)$ in its defining representation is then straightforward to understand; it is the group of $(n+m) \times (n+m)$ real matrices A that satisfy $A^T G A = G$ with the diagonal matrix G having m entries of $+1$ and n entries of -1. If $n = 0$ then the group is the orthogonal group $O(m)$, otherwise for $n \neq 0$ we refer to the group $O(m,n)$ as *pseudo-orthogonal* or sometimes as *indefinite orthogonal*. Here we are using the "timelike" definition of the metric tensor g with $g^{00} = -g^{11} = -g^{22} = -g^{33} = +1$, which suggests writing the Lorentz transformations as

$O(1,3)$. Some texts prefer the "spacelike" definition of g with $-g^{00} = g^{11} = g^{22} = g^{33} = +1$, which would suggest writing the Lorentz transformations as the group $O(3,1)$ rather than the notation $O(1,3)$ that we use here.

It then follows that the *Poincaré transformations also form a group*, where the Lorentz transformations and the translations form two subgroups of the Poincaré group. To see that this is true, consider some Poincaré transformation p defined by $x' \equiv p(x) \equiv \Lambda x + a$. Then we have (i) *closure:* two Poincaré transformations, $x' = p_2(x)$ and $x'' = p_1(x')$, result in a Poincaré transformation, $x'' = p_1 p_2(x) \equiv p_1(p_2(x)) = p_1(x') = \Lambda_1 x' + a_1 = \Lambda_1(\Lambda_2 x + a_2) + a_1 = (\Lambda_1 \Lambda_2)x + (\Lambda_1 a_2 + a_1)$; (ii) *associativity:* $p_1(p_2 p_3(x)) = \Lambda_1(\Lambda_2 \Lambda_3 x + \Lambda_2 a_3 + a_2) + a_1 = \Lambda_1 \Lambda_2 \Lambda_3 x + \Lambda_1 \Lambda_2 a_3 + \Lambda_1 a_2 + a_1 = (\Lambda_1 \Lambda_2)(\Lambda_3 x + a_3) + (\Lambda_1 a_2 + a_1) = (p_1 p_2)(p_3(x))$; (iii) *identity:* $p(x) = x$ when $\Lambda = I$ and $a = 0$; and (iv) *inverse:* every Λ has an inverse and so $p^{-1}(x) = \Lambda^{-1}x + (-\Lambda^{-1}a)$, so that $pp^{-1}(x) = x = p^{-1}p(x)$ and so every p has an inverse.

If we set $\Lambda_1 = \Lambda_2 = \Lambda_3 = I$ in the above, then the set of translations, $t(x) \equiv x + a$, satisfies the properties of a group and this group is a subgroup of the Poincaré group. Similarly, if we set $a_1 = a_2 = a_3 = 0$ in the above, then we see that the group of Lorentz transformations, $\ell(x) \equiv \Lambda x$, forms a subgroup of the Poincaré group.

The translation group is an abelian subgroup, since all of its elements commute with each other, $t_1 t_2(x) = t_1(x + a_2) = x + a_1 + a_2 = t_2 t_1(x)$ for all t_1 and t_2. The Lorentz transformations are not an abelian group since they do not commute with themselves, $\ell_1 \ell_2(x) \neq \ell_2 \ell_1(x)$ in general; e.g., rotations do not commute with themselves and rotations and boosts do not commute. The elements of the two subgroups do not commute since $t\ell(x) = t(\Lambda x) = \Lambda x + a$, whereas $\ell t(x) = \ell(x + a) = \Lambda x + \Lambda a$. The translations are a normal (i.e., invariant) subgroup since for any Poincaré transformation $p_i(x) = \Lambda_i x + a_i$ and for any translation $t(x) = x + a$ we have $p_i t p_i^{-1}(x) = p_i t(\Lambda_i^{-1}x - \Lambda_i^{-1}a_i) = p_i(\Lambda_i^{-1}(x - a_i) + a) = x - a_i + \Lambda_i a + a_i = x + \Lambda_i a$, which is still a translation. The Lorentz transformations are not a normal subgroup of the Poincaré group, since $p_i \ell p_i^{-1}(x) = \Lambda_i \Lambda \Lambda_i^{-1}x + (I - \Lambda_i \Lambda \Lambda_i^{-1})a_i$ is not a Lorentz transformation for all Poincaré transformations p_i.

The Poincaré group is sometimes referred to as the inhomogeneous[4] Lorentz group. In fact as explained in Sec. A.7, the Poincaré group is an internal semidirect product of the normal (i.e., invariant) subgroup of spacetime translations, denoted here as $R^{1,3}$, and the subgroup of Lorentz transformations $O(1,3)$. It is denoted as the semidirect product group $R^{1,3} \rtimes O(1,3)$.

Recall that $\det(\Lambda) = \pm 1$ from Eq. (1.2.16). We can also classify Lorentz transformations by the sign of $\Lambda^0{}_0$. We see that $1 = g_{00} = \Lambda^\sigma{}_0 g_{\sigma\tau} \Lambda^\tau{}_0 = (\Lambda^0{}_0)^2 - (\Lambda^i{}_0)^2$ and so $(\Lambda^0{}_0)^2 \geq 1$. Hence either $\Lambda^0{}_0 \geq 1$ or $\Lambda^0{}_0 \leq -1$. This is illustrated in Table 1.1.

The class of proper orthochronous Lorentz transformations (i.e., rotations and boosts) themselves satisfy the requirements of a group and so form a connected subgroup of the Lorentz group. They are *proper* since they preserve orientation, $\det(\Lambda) = +1$, and they are *orthochronous* because they do not contain a time reversal (i.e., no T), meaning that $\Lambda^0{}_0 \geq +1$. This subgroup is often referred to as the *restricted Lorentz group* and is typically denoted as $SO^+(1,3)$, i.e., "S" for *special* since the determinant is $+1$ and "$+$" since it is orthochronous. In this context "proper" and "special" have the same meaning. Where it is helpful to do so, we will write $\Lambda_r \in SO^+(1,3)$ to denote an element of the restricted Lorentz group. We have then $SO^+(1,3) \subset O(1,3)$, which consists of all elements

---

[4] Lorentz transformations leave the spacetime origin unchanged and for this reason are sometimes referred to as the *homogeneous Lorentz transformations*. Translations do not leave the origin unchanged and for this reason the Poincaré transformations are sometimes referred to as the *inhomogeneous Lorentz transformations*.

Table 1.1. Classes of the Lorentz transformations

Lorentz class	$\det \Lambda$	$\Lambda^0{}_0$	Transformation
(i) Rotations and Boosts	$+1$	≥ 1	$(\Lambda_{\mathrm{r}})^\mu{}_\nu \in SO^+(1,3)$
(ii) Parity inverting	-1	≥ 1	$(P\Lambda_{\mathrm{r}})^\mu{}_\nu = P^\mu{}_\sigma (\Lambda_{\mathrm{r}})^\sigma{}_\nu;\ P^\mu{}_\nu \equiv g^{\mu\nu}$
(iii) Time reversing	-1	≤ -1	$(T\Lambda_{\mathrm{r}})^\mu{}_\nu = T^\mu{}_\sigma (\Lambda_{\mathrm{r}})^\sigma{}_\nu;\ T^\mu{}_\nu \equiv -g^{\mu\nu}$
(iv) Spacetime invert	$+1$	≤ -1	$(PT\Lambda_{\mathrm{r}})^\mu{}_\nu = (PT)^\mu{}_\sigma (\Lambda_{\mathrm{r}})^\sigma{}_\nu;\ (PT)^\mu{}_\nu \equiv -\delta^\mu{}_\nu$

Table 1.2. Classification of spacetime functions based on Lorentz transformation properties in the passive view, $(\det(\Lambda) = \pm 1)$

$s(x)$	$\rightarrow$	$s'(x') = s(x)$	Scalar
$p(x)$	$\rightarrow$	$p'(x') = \det(\Lambda) p(x)$	Pseudoscalar
$v^\mu(x)$	$\rightarrow$	$v'^\mu(x') = \Lambda^\mu{}_\nu v^\nu(x)$	Vector
$a^\mu(x)$	$\rightarrow$	$a'^\mu(x') = \det(\Lambda) \Lambda^\mu{}_\nu a^\nu(x)$	Pseudovector (or axial vector)
$t^{\mu\nu}(x)$	$\rightarrow$	$t'^{\mu\nu}(x') = \Lambda^\mu{}_\sigma \Lambda^\nu{}_\tau t^{\sigma\tau}(x)$	Second-rank tensor

Λ_{r} continuous with the identity element, I. The other three classes of Lorentz transformations in Table 1.1, labeled (ii) to (iv), do not contain the identity and so do not form a subgroup individually or in any union of the three. However, the union of the proper orthochronous subgroup with any one or more of the other three classes does form a subgroup. In fact, the proper orthochronous Lorentz transformations (the restricted Lorentz transformations) form a *normal* subgroup of the Lorentz group. Examples of classes of spacetime functions are given in Table 1.2.

Proof: Every $\Lambda \in O(1,3)$ can be written as some restricted Lorentz transformation $\Lambda_{\mathrm{r}} \in SO^+(1,3)$ combined with either a P, T or PT as shown in Table 1.1. In more mathematical language every $\Lambda \in O(1,3)$ can be written as the internal semidirect product of an element of the normal discrete subgroup $\{I, P, T, PT\}$ and $SO^+(1,3)$. Note that for any Λ we have $(P\Lambda P^{-1})^0{}_0 = P^0{}_\mu \Lambda^\mu{}_\nu P^\nu{}_0 = \Lambda^0{}_0$ and similarly for T and PT. Therefore for any $\Lambda' \in O(1,3)$ and any $\Lambda_{\mathrm{r}} \in SO^+(1,3)$ we have $\det(\Lambda' \Lambda_{\mathrm{r}} \Lambda'^{-1}) = \det \Lambda_{\mathrm{r}} = +1$ and $(\Lambda' \Lambda_{\mathrm{r}} \Lambda'^{-1})^0{}_0 = (\Lambda'_{\mathrm{r}} \Lambda_{\mathrm{r}} \Lambda'^{-1}_{\mathrm{r}})^0{}_0 \geq +1$, where Λ'_{r} is the element of $SO^+(1,3)$ corresponding to $\Lambda' \in O(1,3)$, i.e., either Λ' is Λ'_{r} itself, $P\Lambda'_{\mathrm{r}}$, $T\Lambda'_{\mathrm{r}}$ or $PT\Lambda'_{\mathrm{r}}$ as appropriate. Hence if Λ_{r} is in $SO^+(1,3)$ then so is any $\Lambda' \Lambda_{\mathrm{r}} \Lambda'^{-1}$ for any $\Lambda' \in O(1,3)$, and so $SO^+(1,3)$ is a normal subgroup of $O(1,3)$.

A parity transformation $P^\mu{}_\nu = g^{\mu\nu}$ is defined as a reversal of all three spatial dimensions. Doing a parity transformation of just two axes is equivalent to a rotation about the third spatial axis by angle π and so is a restricted Lorentz transformation. Doing a parity transformation of just one spatial axis is equivalent to a parity transformation of all three followed by a rotation of angle π about the one being inverted and so is in the parity inverting class.

1.2.2 Active and Passive Transformations

We are very often interested in transformations of physical systems. For example, transformations that leave physical systems unchanged are symmetries of the system, and such symmetries underpin

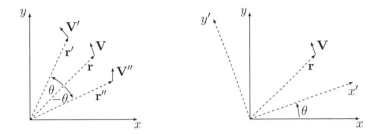

Figure 1.3 The left-hand figure illustrates active rotations by angles θ and $-\theta$, where the vector $\mathbf{V}(\mathbf{r})$ represents the physical system. The right-hand figure illustrates a passive rotation by angle θ of the coordinate system of the observer, $\mathcal{O} \to \mathcal{O}'$.

much of our understanding of modern physics. Transformations can be divided into three classes: (i) transformations of the physical system that do not transform the observer, referred to as *active transformations*; (ii) transformations of the observer that do not transform the physical system, referred to as *passive transformations*; and (iii) combinations of the two. We rarely concern ourselves with the combined transformations.

The relationship between active and passive transformations can be understood using the example of rotations in two dimensions, which is illustrated in Fig. 1.3. The observer is represented by the coordinate system, the x and y axes, and the physical system in this example is represented by vector $\mathbf{V}$ in the (x, y)-plane.

Consider the vector $\mathbf{V} \equiv (V_x, V_y)$ located at the point $\mathbf{r} \equiv (r_x, r_y)$. Under an *active* rotation by angle θ as illustrated in the left-hand side of Fig. 1.3 we have

$$\mathbf{r} \to \mathbf{r}' = R_a(\theta)\mathbf{r} \quad \text{and} \quad \mathbf{V}(\mathbf{r}) \to \mathbf{V}'(\mathbf{r}') = R_a(\theta)\mathbf{V}(\mathbf{r}), \tag{1.2.33}$$

$$R_a(\theta) = \begin{pmatrix} \cos\theta & -\sin\theta \\ \sin\theta & \cos\theta \end{pmatrix}. \tag{1.2.34}$$

After a *passive* rotation of the observer $\mathcal{O}$ by angle θ as illustrated in the right-hand side of Fig. 1.3, the observer $\mathcal{O}'$ will describe the physical system as the vector $\mathbf{V}'$ at the location $\mathbf{r}'$, where

$$\mathbf{r} \to \mathbf{r}' = R_p(\theta)\mathbf{r} \quad \text{and} \quad \mathbf{V}(\mathbf{r}) \to \mathbf{V}'(\mathbf{r}') = R_p(\theta)\mathbf{V}(\mathbf{r}) \tag{1.2.35}$$

$$R_p(\theta) = \begin{pmatrix} \cos\theta & \sin\theta \\ -\sin\theta & \cos\theta \end{pmatrix} = R_a(-\theta) = R_a^{-1}(\theta). \tag{1.2.36}$$

The relationship of $\mathbf{V}''$ to $\mathcal{O}$ is the same as the relationship $\mathbf{V}$ to $\mathcal{O}'$. So then Eqs. (1.2.33) and (1.2.35) only differ in the form of $R(\theta)$ that appears, where for active transformations we use $R(\theta) = R_a(\theta)$ and for passive transformations we use $R(\theta) = R_p(\theta)$ with $R_a(\theta) = R_p^{-1}(\theta)$.

Consider the situations where the physical system is a vector field; i.e., there is some vector $\mathbf{V}(\mathbf{r})$ defined at every point $\mathbf{r}$ in the (x, y)-plane, e.g., a three-dimensional example of a vector field is the velocity of the wind at each point in the atmosphere. We can rewrite Eqs. (1.2.33) and (1.2.35) as

$$\mathbf{V}(\mathbf{r}) \to \mathbf{V}'(\mathbf{r}) = R(\theta)\mathbf{V}(R^{-1}(\theta)\mathbf{r}), \tag{1.2.37}$$

where $R(\theta)$ is $R_a(\theta)$ or $R_p(\theta)$ as appropriate. This is the notation used for transformations in Peskin and Schroeder (1995), for example, and is physically equivalent to the above notation when discussing fields. For example, if we *actively* translate a vector field $\mathbf{V}$ by the displacement $\mathbf{a}$ then $\mathbf{r} \to \mathbf{r}' = \mathbf{r} + \mathbf{a}$ and we have

$$\mathbf{V}(\mathbf{r}) \to \mathbf{V}'(\mathbf{r} + \mathbf{a}) = \mathbf{V}(\mathbf{r}) \quad \text{or equivalently} \quad \mathbf{V}(\mathbf{r}) \to \mathbf{V}'(\mathbf{r}) = \mathbf{V}(\mathbf{r} - \mathbf{a}). \tag{1.2.38}$$

1.2.3 Lorentz Group

Consider an infinitesimal transformation from the subgroup of proper orthochronous Lorentz transformation, i.e., an element infinitesimally close to the identity. This can be written as $(\Lambda_{\mathrm{r}})^\mu{}_\nu \to \delta^\mu{}_\nu + d\omega^\mu{}_\nu$ or in matrix notation as $\Lambda \to I + d\omega$. Hence for an infinitesimal Lorentz transformation specified by $d\omega$ we can write

$$x'^\mu = \Lambda^\mu{}_\nu x^\nu = x^\mu + d\omega^\mu{}_\nu x^\nu \quad \text{or} \quad x' = \Lambda x = x + d\omega x. \tag{1.2.39}$$

We will often work in the *passive view* of transformations, where the observer is being transformed and not the system that he is observing; e.g., Λ will be the Lorentz transformation that takes observer $\mathcal{O}$ to the transformed observer $\mathcal{O}'$, where they are each observing the same physical system S. This is equivalent to the active view where the same observer $\mathcal{O}$ observes two physical systems S and S', where S' is related to S by the *inverse* transformation, e.g., Λ^{-1}. Thus shifting from the passive to the active view for any transformation is entirely seamless and is straightforwardly achieved by replacing the transformation by its inverse.

For a Poincaré transformation we must have $z'^2 = z^2$ and, since $z'^2 = z^2 + 2d\omega^\mu{}_\nu z_\mu z^\nu + \mathcal{O}(d\omega^2) = z^2 + 2d\omega_{\mu\nu} z^\mu z^\nu + \mathcal{O}(d\omega^2)$, with $d\omega_{\mu\nu} = g_{\mu\rho} d\omega^\rho{}_\nu$, this implies that $d\omega_{\mu\nu} z^\mu z^\nu$ must vanish for arbitrary z^μ, which requires that $d\omega_{\mu\nu} = -d\omega_{\nu\mu}$. Alternatively, recall from Eq. (1.2.13) that for a Lorentz transformation we must have $g = \Lambda^T g \Lambda$ and so for an infinitesimal Lorentz transformation

$$\begin{aligned} g = \Lambda^T g \Lambda &= (I - d\omega + \mathcal{O}(d\omega^2))^T g (I - d\omega + \mathcal{O}(d\omega^2)) \\ &= g - d\omega^T g - g d\omega + \mathcal{O}(d\omega^2) = g - (gd\omega)^T - g d\omega + \mathcal{O}(d\omega^2), \end{aligned} \tag{1.2.40}$$

which requires $(gd\omega)^T = -gd\omega$ and since $g_{\mu\rho} d\omega^\rho{}_\nu = d\omega_{\mu\nu}$ then $d\omega_{\mu\nu} = -d\omega_{\nu\mu}$.

Hence we see that six real constants $d\omega_{\mu\nu}$ are needed to specify a Lorentz transformation in the neighborhood of the identity. Expressed in mathematical language the manifold of $SO^+(1,3)$ has the same dimensionality as its tangent space as determined by an infinitesimal transformation; i.e., it has six real dimensions.

An arbitrary 4×4 antisymmetric real matrix can always be written as a linear combination of six linearly independent 4×4 real matrices. No linear combination of them can vanish unless all of the coefficients are zero and is directly analogous to linear independence for vectors when we consider that we can stack the columns (or rows) of a matrix to form a single longer vector.

A conventional choice for six linearly independent antisymmetric 4×4 real matrices has elements $(\delta^\mu{}_\alpha \delta^\nu{}_\beta - \delta^\nu{}_\alpha \delta^\mu{}_\beta) \equiv -i(gM^{\mu\nu})_{\alpha\beta} = -ig_{\alpha\gamma}(M^{\mu\nu})^\gamma{}_\beta$ and so

$$(M^{\mu\nu})^\alpha{}_\beta \equiv i(g^{\mu\alpha}\delta^\nu{}_\beta - g^{\nu\alpha}\delta^\mu{}_\beta) \quad \text{and} \quad (gM^{\mu\nu})_{\alpha\beta} \equiv i(\delta^\mu{}_\alpha \delta^\nu{}_\beta - \delta^\nu{}_\alpha \delta^\mu{}_\beta). \tag{1.2.41}$$

The six independent matrices are labeled by the six index pairs $\mu\nu = 01, 02, 03, 23, 31$ and 12. Clearly, we have $gM^{\mu\nu} = -gM^{\nu\mu}$ as required. Note that these matrices have an $i \equiv \sqrt{-1}$ included in the definition by convention and for future convenience. It is clear from the definition of Eq. (1.2.41) that

$$(gM^{\mu\nu})^\dagger = (gM^{\mu\nu}) \tag{1.2.42}$$

so that the $(gM^{\mu\nu})$ are Hermitian (self-adjoint) matrices (i.e., if A is Hermitian then $A^\dagger = A$, where $A^\dagger \equiv A^{*T}$). It is readily seen from the definition of Eq. (1.2.41) that these matrices have the property that for any antisymmetric $\omega_{\mu\nu}$ we have

$$\omega^{\alpha}{}_{\beta} = g^{\alpha\gamma}\omega_{\gamma\beta} = -\frac{i}{2}\omega_{\mu\nu}(M^{\mu\nu})^{\alpha}{}_{\beta}. \tag{1.2.43}$$

When we sum over all possible μ and ν we have contributions from the 12 nonzero values of $\omega_{\mu\nu}$ of which only six are independent.

Replacing $\omega_{\mu\nu}$ by $d\omega_{\mu\nu}$ we can write the infinitesimal Lorentz transformation as

$$\Lambda^{\alpha}{}_{\beta} = \delta^{\alpha}{}_{\beta} - \frac{i}{2}d\omega_{\mu\nu}\,(M^{\mu\nu})^{\alpha}{}_{\beta} + \mathcal{O}(d\omega^2) \ \text{ or } \ \Lambda = I - \frac{i}{2}d\omega_{\mu\nu}M^{\mu\nu} + \mathcal{O}(d\omega^2). \tag{1.2.44}$$

We see that $SO^+(1,3)$ is a Lie group with six independent generators $M^{\mu\nu}$ and infinitesimal parameters $d\omega_{\mu\nu}$. We can build up a finite restricted Lorentz transformation by an infinite number of infinitesimal transformations as in Eq. (A.7.3), which in this case has the form

$$\Lambda^{\alpha}{}_{\beta} = \left[\lim_{n\to\infty}\left(I - \frac{i}{2n}\omega_{\mu\nu}M^{\mu\nu}\right)^n\right]^{\alpha}{}_{\beta} = \left[\exp\left(-\frac{i}{2}\omega_{\mu\nu}M^{\mu\nu}\right)\right]^{\alpha}{}_{\beta} = [\exp(\omega)]^{\alpha}{}_{\beta}. \tag{1.2.45}$$

The exponential map in Eq. (1.2.45) is surjective (onto) but not injective (one-to-one) in the case of $SO^+(1,3)$. This means that every restricted Lorentz transformation, $\Lambda \in SO^+(1,3)$, can be written in the above form for at least one set of six real parameters $\omega_{\mu\nu} = -\omega_{\nu\mu}$, but that these parameters are not unique in general.

It is readily checked from the definition of Eq. (1.2.41) that the $M^{\mu\nu}$ matrices satisfy the commutation relations

$$[M^{\mu\nu}, M^{\rho\sigma}] = i\left(g^{\nu\rho}M^{\mu\sigma} - g^{\mu\rho}M^{\nu\sigma} - g^{\nu\sigma}M^{\mu\rho} + g^{\mu\sigma}M^{\nu\rho}\right), \tag{1.2.46}$$

which define the Lie algebra of the generators of the restricted Lorentz transformations. Comparison with Eq. (A.7.5) allows us to deduce the structure constants of $SO^+(1,3)$, but it is conventional to write the Lie algebra for the Lorentz transformations in the form of Eq. (1.2.46).

Proof: The Lorentz Lie algebra of Eq. (1.2.46) follows since

$$\begin{aligned}
[(M^{\mu\nu}), (M^{\rho\sigma})]^{\alpha}{}_{\beta} &= (M^{\mu\nu})^{\alpha}{}_{\gamma}(M^{\rho\sigma})^{\gamma}{}_{\beta} - (M^{\rho\sigma})^{\alpha}{}_{\gamma}(M^{\mu\nu})^{\gamma}{}_{\beta} \qquad (1.2.47)\\
&= (i)^2\{(g^{\mu\alpha}\delta^{\nu}{}_{\gamma} - g^{\nu\alpha}\delta^{\mu}{}_{\gamma})(g^{\rho\gamma}\delta^{\sigma}{}_{\beta} - g^{\sigma\gamma}\delta^{\rho}{}_{\beta})\\
&\quad - (g^{\rho\alpha}\delta^{\sigma}{}_{\gamma} - g^{\sigma\alpha}\delta^{\rho}{}_{\gamma})(g^{\mu\gamma}\delta^{\nu}{}_{\beta} - g^{\nu\gamma}\delta^{\mu}{}_{\beta})\}\\
&= (i)^2\{g^{\mu\alpha}g^{\rho\nu}\delta^{\sigma}{}_{\beta} - g^{\mu\alpha}g^{\sigma\nu}\delta^{\rho}{}_{\beta} - g^{\nu\alpha}g^{\rho\mu}\delta^{\sigma}{}_{\beta} + g^{\nu\alpha}g^{\sigma\mu}\delta^{\rho}{}_{\beta}\\
&\quad - g^{\rho\alpha}g^{\mu\sigma}\delta^{\nu}{}_{\beta} + g^{\rho\alpha}g^{\nu\sigma}\delta^{\mu}{}_{\beta} + g^{\sigma\alpha}g^{\mu\rho}\delta^{\nu}{}_{\beta} - g^{\sigma\alpha}g^{\nu\rho}\delta^{\mu}{}_{\beta}\}\\
&= (i)^2\{g^{\nu\rho}(g^{\mu\alpha}\delta^{\sigma}{}_{\beta} - g^{\sigma\alpha}\delta^{\mu}{}_{\beta}) - g^{\mu\rho}(g^{\nu\alpha}\delta^{\sigma}{}_{\beta} - g^{\sigma\alpha}\delta^{\nu}{}_{\beta})\\
&\quad - g^{\nu\sigma}(g^{\mu\alpha}\delta^{\rho}{}_{\beta} - g^{\rho\alpha}\delta^{\mu}{}_{\beta}) + g^{\mu\sigma}(g^{\nu\alpha}\delta^{\rho}{}_{\beta} - g^{\rho\alpha}\delta^{\nu}{}_{\beta})\}\\
&= i\{g^{\nu\rho}(M^{\mu\sigma})^{\alpha}{}_{\beta} - g^{\mu\rho}(M^{\nu\sigma})^{\alpha}{}_{\beta} - g^{\nu\sigma}(M^{\mu\rho})^{\alpha}{}_{\beta} + g^{\mu\sigma}(M^{\nu\rho})^{\alpha}{}_{\beta}\}.
\end{aligned}$$

Note that the Lorentz transformation matrices Λ are not unitary matrices and so they do not constitute a unitary representation of the Lorentz group $O(1,3)$. There are no finite-dimensional unitary representations of the Lorentz group. Hence the corresponding generators $M^{\mu\nu}$ in this representation are not Hermitian. When we consider Lorentz transformations on quantum systems we will necessarily be interested in unitary representations of the Lorentz group where the corresponding generators must be Hermitian as described in Sec. A.7.3 of the appendixes. The 4×4 matrix representation of the Lorentz transformations is referred to as the vector representation, since it acts on spacetime four-vectors. It is also referred to as the defining or standard representation of the Lorentz group.

Rotation Group

As discussed above, the rotation group $O(3)$ is a subgroup of the Lorentz group $O(1,3)$. The elements of the rotation group continuous with the identity have determinant $+1$ and form the subgroup $SO(3)$ of the rotation group called the proper rotation group, i.e., $SO(3) \subset O(3)$. The set $\{I, -I\}$, where I is the identity, forms a normal subgroup of $O(3)$ and it can be shown that $O(3)$ is the internal semidirect product of the normal subgroup $\{I, -I\}$ with the proper rotation subgroup $SO(3)$; i.e., if we combine the parity transformation with the proper rotations $SO(3)$ then we obtain $O(3)$. The continuous rotations are also a subgroup of the restricted Lorentz transformations, $SO(3) \in SO(1,3)$. Consider a Lorentz transformation, Λ_R, which is a pure rotation, then

$$x'^{\mu} = (\Lambda_R)^{\mu}{}_{\nu} \, x^{\nu} \quad \text{with} \quad \Lambda_R = \begin{pmatrix} 1 & 0 \\ 0 & R \end{pmatrix} \quad \text{giving} \quad \mathbf{x}' = R\mathbf{x}, \qquad (1.2.48)$$

where in the above R is a 3×3 real orthogonal matrix and the three-vectors $\mathbf{x}'$ and $\mathbf{x}$ are treated as three-component column vectors, e.g., where $\mathbf{x}$ has components $(x^1, x^2, x^3) = (x, y, z)$. The rotations are orthogonal transformations, where for any $R \in O(3)$ we have $R^{-1} = R^T$ and so $(\det R)^2 = \det R \det R^T = \det R \det R^{-1} = 1$ and $\det R = \pm 1$. The rotations with positive determinant are the proper rotations and those with negative determinant are the improper rotations. Under a parity transformation $P = -I$ we have $\mathbf{x}' = P\mathbf{x} = -\mathbf{x}$, which corresponds to a transformation from a right-handed to a left-handed coordinate system. We see that the rotation group is the union of the proper and improper rotations, i.e., $O(3) = \{R, PR \text{ for all } R \in SO(3)\}$, where for all $R \in SO(3)$ we have $\det R = +1$ and $\det(PR) = \det P \det R = -1$.

Note that we are defining $R^{ij} \equiv (\Lambda_R)^i{}_j$. Consider a passive transformation consisting of a rotation of the coordinate system by an angle θ counterclockwise around the z-axis. An active rotation would rotate the objects at points in space about the z-axis by an angle θ, which is equivalent to rotating the coordinate system in the opposite direction. As seen in Sec. 1.2.2 for a rotation of the system by angle θ in the positive (counterclockwise) direction we have

$$R_a(z, \theta) = \begin{pmatrix} \cos\theta & -\sin\theta & 0 \\ \sin\theta & \cos\theta & 0 \\ 0 & 0 & 1 \end{pmatrix} \quad \text{and} \quad R_p(z, \theta) = \begin{pmatrix} \cos\theta & \sin\theta & 0 \\ -\sin\theta & \cos\theta & 0 \\ 0 & 0 & 1 \end{pmatrix}, \qquad (1.2.49)$$

where R_p and R_a are for the passive and active views, respectively.

> **Mathematical digression: Active and passive view in quantum mechanics:** Many quantum mechanics textbooks adopt the active view of Messiah (1975), since that is more convenient in that context. That is, they define $R \equiv R_a$ so that the rotation of the objects in the system is $\mathbf{x}' = R_a\mathbf{x}$ and the rotated system wavefunction is $\psi'(\mathbf{x}') = \psi(\mathbf{x})$, since the rotated system in terms of rotated coordinates must have the same description as the original system in terms of the unrotated coordinates. This is equivalent to the requirement $\psi'(\mathbf{x}) = \psi(R_a^{-1}\mathbf{x})$, where $R_a^{-1} = R_p$ is the passive rotation.
>
> Using $\mathbf{L} = \mathbf{x} \times \mathbf{p}$ with $\mathbf{p} = -i\hbar\boldsymbol{\nabla}$ almost every quantum mechanics textbook includes the demonstration that $\psi'(\mathbf{x}) = \exp(-i\alpha\mathbf{n} \cdot \mathbf{L}/\hbar)\psi(\mathbf{x}) = \psi(R_a^{-1}\mathbf{x})$ for a rotation α about

the unit vector $\mathbf{n}$. Using Sec. A.2.3 it follows that $-i\hbar\boldsymbol{\nabla}\psi(\mathbf{x}) = -i\hbar\boldsymbol{\nabla}\langle\mathbf{x}|\psi\rangle = \langle\mathbf{x}|\hat{\mathbf{p}}|\psi\rangle$ and similarly $\psi'(\mathbf{x}) = \langle\mathbf{x}|\psi'\rangle = \exp(-i\alpha\mathbf{n}\cdot\mathbf{L}/\hbar)\psi(\mathbf{x}) = \langle\mathbf{x}|\exp(-i\alpha\mathbf{n}\cdot\hat{\mathbf{L}}/\hbar)|\psi\rangle = \langle R_a^{-1}\mathbf{x}|\psi\rangle = \psi(R_a^{-1}\mathbf{x})$, where $\hat{\mathbf{L}}$ is the Hilbert-space orbital angular momentum operator. An active rotation in quantum mechanics is a unitary transformation of the state with the observable operators left unchanged,

$$|\psi\rangle \to |\psi'\rangle = \hat{U}|\psi\rangle, \quad \text{where} \quad \hat{U} = \exp(-i\alpha\mathbf{n}\cdot\hat{\mathbf{L}}/\hbar). \tag{1.2.50}$$

It is useful to generalize the discussion to any unitary transformation. If we were to transform both the observer and the system together, then the expectation values must be unaffected since this is effectively just a change of orthonormal basis for the system $\langle\psi'|\hat{A}'|\psi'\rangle = \langle\psi|\hat{U}^\dagger\hat{A}'\hat{U}|\psi\rangle = \langle\psi|\hat{A}|\psi\rangle$ so that $\hat{A}' = \hat{U}\hat{A}\hat{U}^\dagger$ when $|\psi'\rangle = \hat{U}|\psi\rangle$. So the states and the observable operators transform in the same way in this sense with $\hat{U}$ acting from the left for each.

An active transformation $\hat{U}_a$ transforms the physical system by changing its state, $|\psi\rangle \to |\psi'\rangle = \hat{U}_a|\psi\rangle$, but leaves the observable operators unchanged. An equivalent passive transformation $\hat{U}_p = \hat{U}_a^{-1}$ changes the observer by transforming every physical observable operator in the *opposite* direction so that $\hat{A}' = \hat{U}_p\hat{A}\hat{U}_p^\dagger$ and leaves the state of the system unchanged. These two are physically equivalent since they leave all expectation values unchanged,

$$\textbf{Active:} \quad |\psi\rangle \to |\psi'\rangle = \hat{U}_a|\psi\rangle, \quad \textbf{Passive:} \quad \hat{A} \to \hat{A}' = \hat{U}_p\hat{A}\hat{U}_p^\dagger, \tag{1.2.51}$$

$$\textbf{Equivalent:} \quad \langle\psi'|\hat{A}|\psi'\rangle = \langle\psi|\hat{U}_a^\dagger\hat{A}\hat{U}_a|\psi\rangle = \langle\psi|\hat{U}_p\hat{A}\hat{U}_p^\dagger|\psi\rangle = \langle\psi|\hat{A}'|\psi\rangle.$$

For an active counterclockwise rotation by an angle θ about the x, y and z axes we have, respectively,

$$R_a(x,\theta) = \begin{pmatrix} 1 & 0 & 0 \\ 0 & \cos\theta & -\sin\theta \\ 0 & \sin\theta & \cos\theta \end{pmatrix}, \quad R_a(y,\theta) = \begin{pmatrix} \cos\theta & 0 & \sin\theta \\ 0 & 1 & 0 \\ -\sin\theta & 0 & \cos\theta \end{pmatrix} \quad \text{and}$$

$$R_a(z,\theta) = \begin{pmatrix} \cos\theta & -\sin\theta & 0 \\ \sin\theta & \cos\theta & 0 \\ 0 & 0 & 1 \end{pmatrix}. \tag{1.2.52}$$

Note that these results correspond to the choice of a conventional right-handed three-dimensional coordinate system for the x, y and z axes.

The group of pure rotations, $SO(3)$, is a Lie group with three generators denoted $\mathbf{J}$. We can specify an arbitrary rotation using a real three-vector $\boldsymbol{\alpha}$, where $\alpha \equiv |\boldsymbol{\alpha}|$ is the counterclockwise angle of rotation about the axis specified by the unit vector $\mathbf{n} \equiv \boldsymbol{\alpha}/\alpha$. We have $0 \le \alpha \le 2\pi$ remembering that we identify the end points 0 and 2π since the circle is a compact space. The active rotation matrix $R_a(\boldsymbol{\alpha})$ is

$$R_a(\boldsymbol{\alpha}) = \exp(-i\boldsymbol{\alpha}\cdot\mathbf{J}), \tag{1.2.53}$$

where in this vector representation of the Lorentz transformations the generators $\mathbf{J}$ are Hermitian 3×3 matrices. The negative sign for the exponent for active rotations gives the conventional sign choice for the generators of the rotation group $\mathbf{J}$.

Note that $R_a(x,\theta)$, $R_a(y,\theta)$ and $R_a(z,\theta)$ in Eq. (1.2.52) correspond to the choices $\boldsymbol{\alpha} = (\theta,0,0)$, $(0,\theta,0)$ and $(0,0,\theta)$, respectively. Using this definition of the generators $\mathbf{J} \equiv (J_x, J_y, J_z) \equiv (J^1, J^2, J^3)$ we can evaluate them using

$$J^k = i \left. \frac{\partial R_a}{\partial \alpha^k} \right|_{\alpha=0} \quad \text{for} \quad k = 1, 2, 3, \tag{1.2.54}$$

which leads to

$$J^1 = \begin{pmatrix} 0 & 0 & 0 \\ 0 & 0 & -i \\ 0 & i & 0 \end{pmatrix}, \quad J^2 = \begin{pmatrix} 0 & 0 & i \\ 0 & 0 & 0 \\ -i & 0 & 0 \end{pmatrix} \quad \text{and} \quad J^3 = \begin{pmatrix} 0 & -i & 0 \\ i & 0 & 0 \\ 0 & 0 & 0 \end{pmatrix}. \tag{1.2.55}$$

We can summarize the matrices in Eq. (1.2.55) in a very compact form as

$$\left(J^i \right)^{jk} = -i\epsilon^{ijk} = -i\epsilon_{ijk}, \tag{1.2.56}$$

where in the usual way $\epsilon^{123} = \epsilon_{123} = +1$. It is straightforward to verify from the matrices in Eq. (1.2.55) that

$$\left[J^i, J^j \right] = i\epsilon^{ijk} J^k, \tag{1.2.57}$$

which is the Lie algebra of the restricted rotation group $SO(3)$.

Relation to quantum mechanics notation: In quantum mechanics we associate the generators $\mathbf{J}$ with the angular momentum operators $\mathbf{J}^{\mathrm{QM}}$. The relationship between the two is $\mathbf{J}^{\mathrm{QM}} = \hbar \mathbf{J}$ so that the Lie algebra of the quantum operators is then

$$[(J^i)^{\mathrm{QM}}, (J^j)^{\mathrm{QM}}] = i\hbar\epsilon^{ijk}(J^k)^{\mathrm{QM}}. \tag{1.2.58}$$

We normally do not bother to differentiate the two and simply understand what is meant by $\mathbf{J}$ from the context. In any case we typically will work in later chapters in natural units where $\hbar = c = 1$. We will study this in more detail when we construct unitary representations of the Poincaré group in Sec. 1.2.4.

A number of results follow directly from the Lie algebra independent of the particular representation of the rotation group that we are considering. In particular, it follows from Eqs. (1.2.57), (A.7.16) and (A.7.17) that $\mathbf{J}^2 \equiv (J^1)^2 + (J^2)^2 + (J^3)^2 = J_x^2 + J_y^2 + J_z^2$ is a quadratic Casimir invariant of the restricted rotation group $SO(3)$, since it is readily verified from the Lie algebra and the complete antisymmetry of the structure constants $i\epsilon^{ijk}$ that

$$\left[\mathbf{J}^2, \mathbf{J} \right] = 0. \tag{1.2.59}$$

Defining $J_\pm \equiv J_x \pm iJ_y$ we can verify from the Lie algebra alone that

$$\left[\mathbf{J}^2, J_\pm \right] = 0, \quad [J_+, J_-] = 2J_z, \quad [J_z, J_\pm] = \pm J_\pm, \quad J_\pm J_\mp = \mathbf{J}^2 - J_z^2 \pm J_z. \tag{1.2.60}$$

The above relations lead to the results, proven below, that $\mathbf{J}^2$ has eigenvalues $j(j + 1)$ for some positive integer or half-integer j and that J_z has $(2j + 1)$ eigenvalues $m = j, j - 1, \ldots, -j + 1, -j$. As discussed in Sec. A.7.1, an n-dimensional representation of a group involves $n \times n$ matrices and so n is the maximum number of distinct eigenvalues for a matrix in that representation. An n-dimensional irreducible representation of the rotation group must therefore have $n = 2j + 1$ or equivalently $j = (n - 1)/2$, which directly connects the eigenvalue of $\mathbf{J}^2$ to the dimension of the group representation. For example, the vector representation that we are considering here with its 3×3 rotation matrices corresponds to $j = 1$; i.e., it is the $j = 1$ irreducible representation of the rotation group. We can use j to label different irreducible representations; i.e., the irreducible representation associated with j will be in terms of $(2j + 1) \times (2j + 1)$ matrices.

Proof: The results stated above for the eigenvalues of $\mathbf{J}^2$ and J_z are obtained readily. Since $\mathbf{J}^2$ and J_z are Hermitian and commute, then we can label their simultaneous eigenstates by their real eigenvalues, which we will temporarily label as a and b, respectively. It will be convenient to make use of quantum mechanics notation for these simultaneous eigenstates in the form $|a, b\rangle$. For an n-dimensional matrix representation of the group the $|a, b\rangle$ are to be understood as n-component column vectors rather than abstract vectors in Hilbert space. By definition we have

$$\mathbf{J}^2|a, b\rangle = a|a, b\rangle \quad \text{and} \quad J_z|a, b\rangle = b|a, b\rangle. \tag{1.2.61}$$

From the third equation of Eq. (1.2.60) we have $J_z J_+ = J_+(J_z + 1)$ and $J_z J_- = J_-(J_z - 1)$ and so

$$J_z J_+|a, b\rangle = (b+1)J_+|a, b\rangle \quad \text{and} \quad J_z J_-|a, b\rangle = (b-1)J_-|a, b\rangle, \tag{1.2.62}$$

which shows that provided $J_+|a, b\rangle \neq 0$ and $J_-|a, b\rangle \neq 0$ then J_+ and J_- raise and lower the eigenvalue of J_z by one unit, respectively.

Define $b_{\max}$ and $b_{\min}$ as the maximum and minimum eigenvalues of J_z. Then

$$J_+|a, b_{\max}\rangle = J_-|a, b_{\min}\rangle = 0, \tag{1.2.63}$$

since otherwise $b_{\max}$ and $b_{\min}$ would not be the largest and smallest eigenvalues.

We now make use of these equations together with the fourth equation of Eq. (1.2.60) and note that

$$\mathbf{J}^2|a, b_{\max}\rangle = a|a, b_{\max}\rangle = (J_-J_+ + J_z^2 + J_z)|a, b_{\max}\rangle = b_{\max}(b_{\max} + 1)|a, b_{\max}\rangle,$$
$$\mathbf{J}^2|a, b_{\min}\rangle = a|a, b_{\min}\rangle = (J_+J_- + J_z^2 - J_z)|a, b_{\min}\rangle = b_{\min}(b_{\min} - 1)|a, b_{\max}\rangle.$$

Therefore we have $a = b_{\max}(b_{\max} + 1) = b_{\min}(b_{\min} - 1)$, which requires that either $b_{\max} = -b_{\min}$ or $b_{\max} = b_{\min} - 1$. The latter is impossible since by definition $b_{\max} \geq b_{\min}$.

Hence defining $j \equiv b_{\max}$ and $m \equiv b$ we have then shown that the eigenvalues of $\mathbf{J}^2$ are $a = j(j+1)$ and the eigenvalues of J_z are $b_{\max} = j$ to $b_{\min} = -j$ in unit steps. We can only achieve this if j is either a positive integer or half-integer. There will then be $(2j+1)$ eigenvalues $j, j-1, \ldots, -j+1, -j$.

We can summarize the above arguments as

$$\mathbf{J}^2|j, m\rangle = j(j+1)|j, m\rangle \quad \text{with} \quad j = 0, \tfrac{1}{2}, 1, \tfrac{3}{2}, 2, \ldots, \tag{1.2.64}$$
$$J_z|j, m\rangle = m|j, m\rangle \quad \text{with} \quad m = -j, -j+1, \ldots, j-1, j,$$
$$J_+|j, m\rangle = |j, m+1\rangle \text{ for } m \leq j-1, \quad J_-|j, m\rangle = |j, m-1\rangle \text{ for } m \geq -j+1.$$

In terms of a matrix representation the lowest dimensional representation of $(2j+1)$ orthogonal states is in terms of $(2j+1)$ component column vectors and $(2j+1) \times (2j+1)$ dimensional complex matrices.

The irreducible representations of $SO(3)$ necessarily have $\ell \equiv j \in \mathbb{Z}$,

$$\ell = 0, 1, 2, \ldots \quad \text{with} \quad m = -\ell, -\ell+1, \ldots, -1, 0, 1, \ldots, \ell-1, \ell, \tag{1.2.65}$$

where the generators and rotations are $(2\ell + 1) \times (2\ell + 1)$ matrices, which follows since a 2π rotation in three dimensions about any axis must cause no change. We traditionally write $\mathbf{L} \equiv \mathbf{J}$

when considering integer $\ell \equiv j$, where the L then generate orbital rotations $SO(3)$ rather than rotations in any internal spin spaces.

Proof: Using quantum mechanics notation as a convenient shorthand we can consider rotations about the z-axis, for example. Since the eigenvalue equation for L_z is $L_z|j,m\rangle = \hbar m|j,m\rangle$ and since from Eqs. (1.2.50) and (1.2.53) we have $\exp(-2\pi i \hat{L}_z/\hbar) = \hat{I}$, then it follows that $\exp(-2\pi i \hat{L}_z/\hbar)|j,m\rangle = \exp(-2\pi i m)|j,m\rangle = \hat{I}|j,m\rangle = |j,m\rangle$. So m must be integer and hence j must also be integer, since $j \equiv m_{\text{max}}$.

In the vector representation that we are working with here it is readily verified from Eq. (1.2.55) that $\mathbf{J}^2 = 2I$, where $2 = j(j+1)$ with $\ell = j = 1$ and so $\mathbf{J}^2$ is already diagonal. By inspection we see that $J^3 \equiv J_z$ is not. We know from the above arguments that the eigenvalues of J_z when $j = 1$ are $m = +1, 0, -1$, which can be readily verified directly from the Hermitian matrix J_z above. The normalized eigenvectors of J_z are the column vectors $[-1, -i, 0]/\sqrt{2}$, $[0, 0, 1]$ and $[1, -i, 0]/\sqrt{2}$ corresponding to $m = +1, 0$ and -1, respectively, which is easily checked. We can diagonalize J_z by performing a unitary change of basis $J'_z = U^\dagger J_z U$, where U has the eigenvectors of J_z as its column vectors

$$U = \frac{1}{\sqrt{2}} \begin{pmatrix} -1 & 0 & 1 \\ -i & 0 & -i \\ 0 & \sqrt{2} & 0 \end{pmatrix}. \tag{1.2.66}$$

Defining $\mathbf{J}' \equiv U^\dagger \mathbf{J} U$ leads to

$$J'_x = \frac{1}{\sqrt{2}} \begin{pmatrix} 0 & 1 & 0 \\ 1 & 0 & 1 \\ 0 & 1 & 0 \end{pmatrix}, \; J'_y = \frac{1}{\sqrt{2}} \begin{pmatrix} 0 & -i & 0 \\ i & 0 & -i \\ 0 & i & 0 \end{pmatrix} \text{ and } J'_z = \begin{pmatrix} 1 & 0 & 0 \\ 0 & 0 & 0 \\ 0 & 0 & -1 \end{pmatrix}. \tag{1.2.67}$$

Note that $\mathbf{J}'^2 = U^\dagger \mathbf{J}^2 U = U^\dagger(2I)U = 2I$. The phase choices for the eigenvectors of J_z are arbitrary, but the above choice gives the conventional form for the $\mathbf{J}'$.

For Lorentz transformations in the passive view Eq. (1.2.48) leads to

$$\Lambda_R = \begin{pmatrix} 1 & 0 \\ 0 & R_p(\boldsymbol{\alpha}) \end{pmatrix} = \begin{pmatrix} 1 & 0 \\ 0 & \exp(i\boldsymbol{\alpha} \cdot \mathbf{J}) \end{pmatrix} = \exp\begin{pmatrix} 1 & 0 \\ 0 & i\boldsymbol{\alpha} \cdot \mathbf{J} \end{pmatrix}, \tag{1.2.68}$$

where the last step follows since Λ_R has a block diagonal form. Comparing this with Eq. (1.2.45) we can identify for a passive Lorentz transformation

$$-(1/2)\omega_{\mu\nu} \left(M^{\mu\nu}\right)^i{}_j = -(1/2)\omega_{k\ell} \left(M^{k\ell}\right)^i{}_j = \boldsymbol{\alpha} \cdot (\mathbf{J})^{ij}. \tag{1.2.69}$$

We can also see from Eq. (1.2.41) that $(M^{\mu\nu})^i{}_j \to (M^{k\ell})^i{}_j$ and that, e.g., $(M^{12})^i{}_j = i(g^{1i}\delta^2{}_j - g^{2i}\delta^1{}_j) = -i\epsilon^{3ij} = (J^3)^{ij}$. Equivalently in matrix notation we can readily verify from Eqs. (1.2.41) and (1.2.55) that

$$M^{12} = -M^{21} = \begin{pmatrix} 0 & 0 & 0 & 0 \\ 0 & 0 & -i & 0 \\ 0 & i & 0 & 0 \\ 0 & 0 & 0 & 0 \end{pmatrix} = \begin{pmatrix} 0 & 0 \\ 0 & J^3 \end{pmatrix} = \begin{pmatrix} 0 & 0 \\ 0 & J_z \end{pmatrix}. \tag{1.2.70}$$

Repeating this exercise for M^{23} and M^{31} we find the relationship between the Lorentz generators M and the rotational generators J, which is

$$M^{k\ell} = \epsilon^{k\ell m} \begin{pmatrix} 0 & 0 \\ 0 & J^m \end{pmatrix} \quad \text{or} \quad (J^{(4\times4)})^i \equiv \begin{pmatrix} 0 & 0 \\ 0 & J^i \end{pmatrix} = \frac{1}{2}\epsilon^{ijk}M^{jk}, \qquad (1.2.71)$$

where we have introduced a $4{\times}4$ extension of the $3{\times}3$ matrices J^i for future convenience. Whenever J^i is used in the 4×4 matrix context it is to be understood to mean $(J^{(4\times4)})^i$ and we will simply write, e.g.,

$$M^{k\ell} = \epsilon^{k\ell m} J^m \quad \text{or equivalently} \quad J^i \equiv \tfrac{1}{2}\epsilon^{ijk}M^{jk}. \qquad (1.2.72)$$

Substituting this result into Eq. (1.2.69) allows us to identify

$$\omega_{ij} = -\epsilon^{ijm}\alpha^m \quad \text{or equivalently} \quad \alpha^m = -\tfrac{1}{2}\omega_{k\ell}\epsilon^{k\ell m}. \qquad (1.2.73)$$

Since the $\mathbf{J}$ generators are Hermitian, then the pure rotations Λ_R are unitary, which follows since for any passive rotation

$$R_p(\boldsymbol{\alpha}) = \exp\left(i\boldsymbol{\alpha} \cdot \mathbf{J}\right) \quad \text{and so} \quad R_p(\boldsymbol{\alpha})^\dagger = \exp\left(-i\boldsymbol{\alpha} \cdot \mathbf{J}\right) = R_p(\boldsymbol{\alpha})^{-1}. \qquad (1.2.74)$$

These are real orthogonal matrices, $R_p(\boldsymbol{\alpha})^T = R_p(\boldsymbol{\alpha})^{-1}$, with unit determinant as we now show. Define the three-dimensional unit vector $\mathbf{n} = (n_x, n_y, n_z)$ and write

$$J_\mathbf{n} \equiv \mathbf{n} \cdot \mathbf{J} \quad \text{with} \quad \boldsymbol{\alpha} = \alpha\mathbf{n} \quad \text{and hence} \quad R_p(\boldsymbol{\alpha}) = \exp\left(i\alpha J_\mathbf{n}\right). \qquad (1.2.75)$$

It then follows that

$$\exp\left(i\alpha J_\mathbf{n}\right) = \sum_{j=0}^{\infty} \tfrac{1}{j!}\left(i\alpha\right)^j \left(J_\mathbf{n}\right)^j = I + i\sin\alpha\, J_\mathbf{n} + (\cos\alpha - 1)\left(J_\mathbf{n}\right)^2 \qquad (1.2.76)$$

and that $(iJ_\mathbf{n})$ is a real antisymmetric matrix as seen from Eq. (1.2.55). Hence

$$R_p(\boldsymbol{\alpha})^T = \exp\left(\alpha(iJ_\mathbf{n})^T\right) = \exp\left(-i\alpha J_\mathbf{n}\right) = R_p(\boldsymbol{\alpha})^{-1} \qquad (1.2.77)$$

and so every rotation $R_p(\boldsymbol{\alpha})$ is an orthogonal transformation. Recall that the generators $\mathbf{J}$ are traceless matrices and so using Eq. (A.8.4) we find

$$\det\left[R_p(\boldsymbol{\alpha})\right] = \det\left[\exp\left(i\boldsymbol{\alpha} \cdot \mathbf{J}\right)\right] = \exp\mathrm{tr}\left(i\boldsymbol{\alpha} \cdot \mathbf{J}\right) = \exp\left(i\boldsymbol{\alpha} \cdot \mathrm{tr}\mathbf{J}\right) = 1 \qquad (1.2.78)$$

as it should. We have then verified that $R_p(\boldsymbol{\alpha}) \in SO(3)$ as expected.

Both $R_p(\boldsymbol{\alpha})$ and $\mathbf{J}$ can be viewed as either 3×3 or 4×4 matrices. The 4×4 form of $\mathbf{J}$ follows from Eq. (1.2.71). The 4×4 form of $R_p(\boldsymbol{\alpha})$ has its 3×3 form in the bottom right, a one in the top left-hand corner and zeros elsewhere.

Proof of Eq. (1.2.76): We wish to show that $\exp(i\boldsymbol{\alpha} \cdot \mathbf{J}) = \exp(i\alpha J_\mathbf{n})$ leads to Eq. (1.2.76). We first note that

$$J_\mathbf{n} = \begin{pmatrix} 0 & -in_z & in_y \\ in_z & 0 & -in_x \\ -in_y & in_x & 0 \end{pmatrix}, \; (J_\mathbf{n})^2 = \begin{pmatrix} 1-n_x^2 & -n_x n_y & -n_x n_z \\ -n_y n_x & 1-n_y^2 & -n_y n_z \\ -n_z n_x & -n_z n_y & 1-n_z^2 \end{pmatrix},$$

$$(J_\mathbf{n})^3 = J_\mathbf{n} \Rightarrow (J_\mathbf{n})^{2j-1} = J_\mathbf{n}, \; (J_\mathbf{n})^{2j} = (J_\mathbf{n})^2 \text{ for } j = 1, 2, 3, \ldots. \qquad (1.2.79)$$

So $(iJ_\mathbf{n})$ is real and antisymmetric and so $\exp(i\alpha J_\mathbf{n})$ is orthogonal.

It also then follows that

$$
\begin{aligned}
e^{i\alpha J_{\mathbf{n}}} &= \sum_{j=0}^{\infty} \frac{1}{j!} t(i\alpha)^j (J_{\mathbf{n}})^j = I + J_{\mathbf{n}} \sum_{j=1,3,\dots}^{\infty} \frac{1}{j!}(i\alpha)^j + (J_{\mathbf{n}})^2 \sum_{j=2,4,\dots}^{\infty} \frac{1}{j!}(i\alpha)^j \\
&= I + \sin\alpha\,(iJ_{\mathbf{n}}) + (\cos\alpha - 1)(J_{\mathbf{n}})^2 \\
&= \begin{pmatrix}
1 + c(\alpha)(1 - n_x^2) & n_z \sin\alpha - c(\alpha)n_x n_y & -n_y \sin\alpha - c(\alpha)n_x n_z \\
-n_z \sin\alpha - c(\alpha)n_y n_x & 1 + c(\alpha)(1 - n_y^2) & n_x \sin\alpha - c(\alpha)n_y n_z \\
n_y \sin\alpha - c(\alpha)n_z n_x & -n_x \sin\alpha - c(\alpha)n_z n_y & 1 + c(\alpha)(1 - n_z^2)
\end{pmatrix},
\end{aligned}
\tag{1.2.80}
$$

where for notational brevity we have here defined $c(\alpha) \equiv (\cos\alpha - 1)$.

We recover the passive rotations in Eq. (1.2.52) in the special cases of $\mathbf{n} = (1,0,0)$, $(0,1,0)$ and $(0,0,1)$ for $R(x,\alpha)$, $R(y,\alpha)$ and $R(z,\alpha)$, respectively, and the active versions shown in Eq. (1.2.52) follow from using Eq. (1.2.53).

An arbitrary active rotation of a three-dimensional object can also be specified in terms of the three *Euler angles*, which are denoted as α, β, γ. The definition of the Euler angles is not unique, but we will adopt the most common definition of Messiah (1975), where the active rotation is specified according to the $ZY'Z''$ convention. We specify some arbitrary active rotation (i.e., rotation of the object) with

$$
R_a(\alpha, \beta, \gamma) \equiv R_a(z'', \gamma) R_a(y', \beta) R_a(z, \alpha) = e^{-i\gamma J_{z''}} e^{-i\beta J_{y'}} e^{-i\alpha J_z},
\tag{1.2.81}
$$

where this means that we first rotate the object about the z-axis by angle α, followed by a rotation around the new y-axis (y') by angle β and finally followed by a rotation around the new z-axis (z'') by angle γ. The angles α and γ vary over the range 2π and the magnitude of the range of β is π. The representation of the rotations in terms of the Euler angles is unique for almost every set of Euler angles. The $R_a(\alpha, \beta, \gamma)$ of Eq. (1.2.81) can equivalently be written as

$$
R_a(\alpha, \beta, \gamma) \equiv R_a(z, \alpha) R_a(y, \beta) R_a(z, \gamma) = e^{-i\alpha J_z} e^{-i\beta J_y} e^{-i\gamma J_z},
\tag{1.2.82}
$$

which involves rotations in the inverse order about the original unrotated axes by the same angles. The passive rotation is the inverse of this:

$$
\begin{aligned}
R_p(\alpha, \beta, \gamma) &= R_a(\alpha, \beta, \gamma)^{-1} = R(z, -\alpha) R(y', -\beta) R(z'', -\gamma) \\
&= e^{+i\alpha J_z} e^{+i\beta J_{y'}} e^{+i\gamma J_{z''}} = e^{+i\gamma J_z} e^{+i\beta J_y} e^{+i\alpha J_z}.
\end{aligned}
\tag{1.2.83}
$$

Proof of Eq. (1.2.82): This argument follows that of Messiah (1975) vol. 2. Let A be some 3×3 matrix that acts on $\mathbf{x}$ and define $\mathbf{y} = A\mathbf{x}$. Consider a rotation R such that $\mathbf{x}' = R\mathbf{x}$. We wish to establish how A transforms to A' in order that $\mathbf{y}' = A'\mathbf{x}'$ after the rotation. We have

$$
\mathbf{y}' = R\mathbf{y} = RA\mathbf{x} = (RAR^{-1})R\mathbf{x} = (RAR^{-1})\mathbf{x}'
\tag{1.2.84}
$$

for all $\mathbf{x}$ and so $A' = RAR^{-1}$. This is the analog of the quantum mechanical result given in Eq. (1.2.51).

Then $J_{y'} = (J_y)' = R_a(z, \alpha) J_y R_a(z, \alpha)^{-1} = e^{-i\alpha J_z} J_y e^{+i\alpha J_z}$, which leads to the result $e^{-i\beta J_{y'}} = e^{-i\alpha J_z} e^{-i\beta J_y} e^{+i\alpha J_z}$. Similarly, we find

$$
\begin{aligned}
e^{-i\gamma J_{z''}} = (e^{-i\gamma J_z})'' &= e^{-i\beta J_{y'}} e^{-i\alpha J_z} e^{-i\gamma J_z} e^{+i\alpha J_z} e^{+i\beta J_{y'}} \\
&= e^{-i\alpha J_z} e^{-i\beta J_y} e^{-i\gamma J_z} e^{+i\beta J_y} e^{+i\alpha J_z}
\end{aligned}
\tag{1.2.85}
$$

and so we arrive at the result

$$R_a(\alpha, \beta, \gamma) = e^{-i\gamma J_{z''}} e^{-i\beta J_{y'}} e^{-i\alpha J_z}$$
$$= e^{-i\alpha J_z} e^{-i\beta J_y} e^{-i\gamma J_z} e^{+i\beta J_y} e^{+i\alpha J_z} e^{-i\alpha J_z} e^{-i\beta J_y} e^{+i\alpha J_z} e^{-i\alpha J_z}$$
$$= e^{-i\alpha J_z} e^{-i\beta J_y} e^{-i\gamma J_z}. \tag{1.2.86}$$

Lorentz Boosts

Consider observers $\mathcal{O}$ and $\mathcal{O}'$ whose clock settings and spatial origins coincide at time $t = 0$ and whose *coordinate axes have the same orientation*, but where observer $\mathcal{O}'$ is "boosted" by velocity v in the $+x$-direction with respect to observer $\mathcal{O}$; i.e., observer $\mathcal{O}$ measures observer $\mathcal{O}'$ to have velocity $\boldsymbol{v} = (v, 0, 0)$. Some event, denoted by an abstract point $\mathcal{P}$ in spacetime, which is observed by $\mathcal{O}$ to be at $(ct, x, y, z) \equiv (x^0, x^1, x^2, x^3)$ in his coordinate frame will be observed by $\mathcal{O}'$ to be at $(ct', x', y', z') \equiv (x'^0, x'^1, x'^2, x'^3)$ in his frame, where

$$ct' = \gamma (ct - \beta x), \quad x' = \gamma (x - \beta ct), \quad y' = y, \quad z' = z \tag{1.2.87}$$

and where we have defined

$$\beta \equiv v/c \quad \text{and} \quad \gamma \equiv \frac{1}{\sqrt{1 - (v^2/c^2)}} = \frac{1}{\sqrt{1 - \beta^2}}. \tag{1.2.88}$$

If the frames and clocks of the observers are not coincident at $t = 0$ we have

$$\begin{aligned}
c\Delta t' &= \gamma (c\Delta t - \beta \Delta x), & c\Delta t &= \gamma (c\Delta t' + \beta \Delta x'), \\
\Delta x' &= \gamma (\Delta x - \beta c\Delta t), & \Delta x &= \gamma (\Delta x' + \beta c\Delta t'), \\
\Delta y' &= \Delta y, & \Delta y &= \Delta y', \\
\Delta z' &= \Delta z, & \Delta z &= \Delta z'.
\end{aligned} \tag{1.2.89}$$

The inverse transformations on the right follow from interchanging observers, $\mathcal{O} \leftrightarrow \mathcal{O}'$, and reversing the velocity, $\beta \to -\beta$. These equations are proved below.

The *proper time* is the time shown on a clock comoving with the particle. Then for a particle at rest in the frame of $\mathcal{O}'$ the proper time is $\tau = t'$. Consider some time interval, $\Delta t'$, on the clock of $\mathcal{O}'$. Since this clock is at rest in the frame of $\mathcal{O}'$ we have $\Delta x' = 0$. From the top right-hand equation in Eq. (1.2.89) we then find $c\Delta t = \gamma c\Delta t' = \gamma c\Delta \tau$ and hence we obtain the *time dilation* equation

$$\Delta t = \gamma \Delta \tau. \tag{1.2.90}$$

Then an observer measures a moving clock (with time interval $\Delta \tau$) to run slower than an equivalent clock at rest in his frame (with the corresponding time interval Δt), i.e., $\Delta \tau < \Delta t$. Every observer will record the same proper time between two events on a particle's worldline and so the proper time is Lorentz invariant.

Proof of time dilation: Consider an observer $\mathcal{O}$ and an observer $\mathcal{O}'$ moving in the $+x$-direction with velocity v with respect to $\mathcal{O}$. By assumption the speed of light, c, is the same in every frame. The observers are chosen to have the same orientation for their spatial coordinate axes in order for us to focus on boosts. In his frame, $\mathcal{O}'$ shines a light in a direction orthogonal to the x'-axis, which reflects at right angles from a mirror a distance D away. Let event E_1 be the emission of the light pulse and event E_2 be the reception of the light pulse after reflection in a mirror. The coordinates for events E_1 and E_2 are $t_1, \mathbf{x}_1$ and $t_2, \mathbf{x}_2$, respectively, in the inertial frame of $\mathcal{O}$ and $t_1', \mathbf{x}_1'$ and $t_2', \mathbf{x}_2'$, respectively, in the inertial frame of $\mathcal{O}'$. This is illustrated in Fig. 1.4. The time between events as measured by $\mathcal{O}'$ is $\Delta t' \equiv t_2' - t_1' = 2D/c$ and the spatial displacement is zero, i.e., $\Delta x' \equiv x_2' - x_1' = 0$, $\Delta y' \equiv y_2' - y_1' = 0$ and $\Delta z' \equiv z_2' - z_1' = 0$. In the frame of $\mathcal{O}$ the time between the events is $\Delta t \equiv t_2 - t_1$ and the spatial displacement between the events is $\Delta x \equiv x_2 - x_1 = v\Delta t$, $\Delta y \equiv y_2 - y_1 = 0$ and $\Delta z \equiv z_2 - z_1 = 0$. In the frame of $\mathcal{O}$ the light must travel the distance $2D$ orthogonal to the x-direction and $v\Delta t$ in the x-direction, i.e., a total distance $[(v\Delta t)^2 + (2D)^2]^{1/2}$. Then the time displacement measured by $\mathcal{O}$ is $\Delta t = [(v\Delta t)^2 + (2D)^2]^{1/2}/c = [(v\Delta t)^2 + (c\Delta t')^2]^{1/2}/c$, which we can rewrite as $(c\Delta t')^2 = (c\Delta t)^2 - (v\Delta t)^2$. This gives $\Delta t = \gamma\Delta t'$, where $\gamma \equiv [1 - (v/c)^2]^{-1/2}$, which is time dilation with the time measured in the "moving" frame of $\mathcal{O}'$ being less than that measured in the "stationary" frame of $\mathcal{O}$. Note that here $\Delta t'$ is the proper time, $\Delta t' = \Delta\tau$. So we have $\Delta t = \gamma\Delta\tau$.

Let $\Delta x = L_0$ be the length of a rod at rest in the frame of $\mathcal{O}$. The rod will be observed to be moving with velocity $\boldsymbol{v} = (-v, 0, 0)$ in the frame of $\mathcal{O}'$ and he will measure its length in his frame by observing the displacement of the two ends, $\Delta x' = L'$, at the same time in his frame, $\Delta t' = 0$. From Eq. (1.2.89) we then find $\Delta x = \gamma\Delta x' = \gamma L'$ and $\Delta t \neq 0$. However, since the rod is at rest in the frame of $\mathcal{O}$ it doesn't matter when we measure the location of the two ends and so $\Delta x = L_0$. Hence we obtain the *length contraction* equation

$$L' = L_0/\gamma; \tag{1.2.91}$$

i.e., an observer $\mathcal{O}$ measures a moving rod to be shortened, $L' < L_0$.

We also use the standard notation

$$\cosh\eta \equiv \gamma \equiv 1/\sqrt{(1 - \beta^2)}, \tag{1.2.92}$$

where η is called the *rapidity* of the boost. Using $\gamma^2(1 - \beta^2) = 1$ and the identity $\cosh^2\eta - \sinh^2\eta = 1$ gives the results

$$\sinh\eta = \beta\gamma \quad \text{and} \quad \tanh\eta = \frac{\sinh\eta}{\cosh\eta} = \beta. \tag{1.2.93}$$

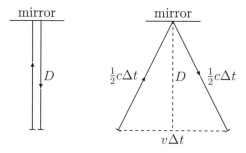

Figure 1.4   Inertial observer $\mathcal{O}'$ is moving with velocity $\boldsymbol{v} = (v, 0, 0)$ with respect to inertial observer $\mathcal{O}$. The frame of $\mathcal{O}'$ is on the left and that of $\mathcal{O}$ is on the right.

It also follows immediately that

$$e^\eta = \cosh\eta + \sinh\eta = \gamma(1+\beta) = \sqrt{\frac{1+\beta}{1-\beta}} \qquad (1.2.94)$$

and hence that

$$\eta = \frac{1}{2}\ln\left(\frac{1+\beta}{1-\beta}\right) = -\frac{1}{2}\ln\left(\frac{1-\beta}{1+\beta}\right) = \ln\left[\gamma(1+\beta)\right] = -\ln\left[\gamma(1-\beta)\right]. \qquad (1.2.95)$$

We can then summarize the passive boost in Eq. (1.2.87) by $x'^\mu = \Lambda_B{}^\mu{}_\nu x^\nu$ or in matrix form $x' = \Lambda_B x$, where

$$\Lambda_B(x,v) = \begin{pmatrix} \gamma & -\beta\gamma & 0 & 0 \\ -\beta\gamma & \gamma & 0 & 0 \\ 0 & 0 & 1 & 0 \\ 0 & 0 & 0 & 1 \end{pmatrix} = \begin{pmatrix} \cosh\eta & -\sinh\eta & 0 & 0 \\ -\sinh\eta & \cosh\eta & 0 & 0 \\ 0 & 0 & 1 & 0 \\ 0 & 0 & 0 & 1 \end{pmatrix}. \qquad (1.2.96)$$

Similarly for passive boosts in the y and z directions we have

$$\Lambda_B(y,v) = \begin{pmatrix} \gamma & 0 & -\beta\gamma & 0 \\ 0 & 1 & 0 & 0 \\ -\beta\gamma & 0 & \gamma & 0 \\ 0 & 0 & 0 & 1 \end{pmatrix} \quad \text{and} \quad \Lambda_B(z,v) = \begin{pmatrix} \gamma & 0 & 0 & -\beta\gamma \\ 0 & 1 & 0 & 0 \\ 0 & 0 & 1 & 0 \\ -\beta\gamma & 0 & 0 & \gamma \end{pmatrix} \qquad (1.2.97)$$

and their equivalent expressions in terms of $\cosh\eta$ and $\sinh\eta$.

Proof of Lorentz boost equations, Eqs. (1.2.87) and (1.2.96): To eliminate rotations and spacetime translations we choose the inertial observers $\mathcal{O}$ and $\mathcal{O}'$ to have identical orientation of spatial axes, clock settings and origins at time $t = 0$. For a boost in the x-direction there can be no change to the y and z coordinates, i.e., $y' = y$ and $z' = z$. For simplicity we can then restrict our attention to the time and x-coordinates only and represent the 4×4 Lorentz transformation $x'^\mu = \Lambda^\mu{}_\nu x^\nu$ by a 2×2 matrix as

$$\begin{pmatrix} ct' \\ x' \end{pmatrix} = \begin{pmatrix} a & b \\ d & e \end{pmatrix} \begin{pmatrix} ct \\ x \end{pmatrix} \qquad (1.2.98)$$

for some coefficients a, b, d and e. We have $\Lambda^T g \Lambda = g$ and so

$$\begin{pmatrix} a & d \\ b & e \end{pmatrix} \begin{pmatrix} 1 & 0 \\ 0 & -1 \end{pmatrix} \begin{pmatrix} a & b \\ d & e \end{pmatrix} = \begin{pmatrix} 1 & 0 \\ 0 & -1 \end{pmatrix}. \qquad (1.2.99)$$

This leads to $a^2 - d^2 = e^2 - b^2 = 1$ and $ab - de = 0$. In the limit that the boost velocity vanishes, $v \to 0$, we would have $a, e \to 1$ and $b, d \to 0$. This suggests that we try the general solution that $a = e$ and $b = d$, which we see satisfies our equations. We know from the above direct proof of time dilation that $a = \gamma$ and hence $\gamma^2 = 1/[1 - (v^2/c^2)] = a^2 = 1 + d^2$, which gives $d^2 = (v/c)^2 \gamma^2$. We then have $\gamma = a = e$ and $\pm\beta\gamma = b = d$, where $\beta \equiv v/c$. The time dilation calculation above corresponds to a passive boost from the frame of observer $\mathcal{O}$ to the frame of observer $\mathcal{O}'$ and there we saw that if $x' = 0$ then $x = vt = \beta(ct)$. From Eq. (1.2.98) we see that $ct' = \gamma ct + (\pm\beta\gamma)x$ and $x' = (\pm\gamma\beta)ct + \gamma x$ and so $x' = 0$ corresponds to $b = d = -\beta\gamma$. Finally, since $\gamma^2 - (\beta\gamma)^2 = 1$ and $\cosh^2\eta - \sinh^2\eta = 1$ for any η, then we can define the rapidity η such that $\cosh\eta \equiv \gamma = a = e$ and $\sinh\eta \equiv \beta\gamma = -b = -d$. All of our equations arising from Eq. (1.2.99) are then satisfied and we recover Eq. (1.2.96) and hence also Eq. (1.2.87) for a passive boost in the $+x$-direction.

Since the boost equations apply to any two events E_1 and E_2, then Eq. (1.2.89) will apply. Furthermore, since $\Delta x^\mu = x_2^\mu - x_1^\mu$ is spacetime translationally invariant, then Eq. (1.2.89) applies whenever there is a passive boost of $\boldsymbol{v} = (v, 0, 0)$ and the spatial axes have the same spatial origin.

There can be no length change perpendicular to the direction of motion. Consider two equal length rulers oriented perpendicular to the direction of their relative motion with their ends capable of marking the other ruler when they pass on top of each other. If one ruler was shorter it would mark the other ruler, and observers in both frames would agree on where the mark was and so which ruler was shorter. Since all inertial frames are equally valid, both observers could not see the same ruler shorter and so their perpendicular lengths must always be measured equal to that measured in the rest frame.

Note that the rapidities of *parallel* or *collinear* boosts are additive. Consider parallel passive boosts in the x-direction such that

$$\Lambda_{\mathrm{B}}(x, v_c) = \Lambda_{\mathrm{B}}(x, v_a)\Lambda_{\mathrm{B}}(x, v_b); \tag{1.2.100}$$

then v_c can be determined from η_c with

$$\eta_c = \eta_a + \eta_b. \tag{1.2.101}$$

This results follows from the identity

$$\begin{pmatrix} \cosh\eta_a & -\sinh\eta_a & 0 & 0 \\ -\sinh\eta_a & \cosh\eta_a & 0 & 0 \\ 0 & 0 & 1 & 0 \\ 0 & 0 & 0 & 1 \end{pmatrix} \begin{pmatrix} \cosh\eta_b & -\sinh\eta_b & 0 & 0 \\ -\sinh\eta_b & \cosh\eta_b & 0 & 0 \\ 0 & 0 & 1 & 0 \\ 0 & 0 & 0 & 1 \end{pmatrix}$$
$$= \begin{pmatrix} \cosh(\eta_a + \eta_b) & -\sinh(\eta_a + \eta_b) & 0 & 0 \\ -\sinh(\eta_a + \eta_b) & \cosh(\eta_a + \eta_b) & 0 & 0 \\ 0 & 0 & 1 & 0 \\ 0 & 0 & 0 & 1 \end{pmatrix}. \tag{1.2.102}$$

which is obtained from the hyperbolic relations

$$\cosh\eta_a \cosh\eta_b + \sinh\eta_a \sinh\eta_b = \cosh(\eta_a + \eta_b) \quad \text{and}$$
$$\sinh\eta_a \cosh\eta_b + \cosh\eta_a \sinh\eta_b = \sinh(\eta_a + \eta_b). \tag{1.2.103}$$

Since the choice of the direction for the x-axis is arbitrary, this result must be true for any fixed spatial direction. For sequential boosts that are not parallel, the relation between rapidities is not additive as is readily verified. Two sequential nonparallel boosts are equivalent to a single boost and a rotation and so the set of all boosts do *not* form a subgroup of the Lorentz group as they do not satisfy closure.

A particle with rest mass m moving with velocity v in the frame of observer $\mathcal{O}$ is known (see Sec. 2.6) to have energy $E = \gamma mc^2$ and momentum $\mathbf{p} = \gamma m\boldsymbol{v}$, which can be expressed in terms of the rapidity as

$$E = \gamma mc^2 = mc^2 \cosh\eta \quad \text{and} \quad |\mathbf{p}| = \gamma m|\boldsymbol{v}| = mc \sinh\eta. \tag{1.2.104}$$

It follows from this that

$$\beta = \tanh\eta = |\mathbf{p}|c/E \quad \text{and} \quad E \pm |\mathbf{p}|c = mc^2 e^{\pm\eta} \tag{1.2.105}$$

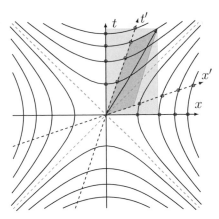

Figure 1.5 A Minkowski spacetime diagram showing the skewing of the time and spatial axes upon boosting by velocity v in the x-direction. The origins coincide at $t = t' = 0$.

and so we can express the rapidity as

$$\eta = \tanh^{-1}\left(\frac{|\mathbf{p}|c}{E}\right) = \frac{1}{2}\ln\left(\frac{E + |\mathbf{p}|c}{E - |\mathbf{p}|c}\right) = \ln\left(\sqrt{\frac{E + |\mathbf{p}|c}{E - |\mathbf{p}|c}}\right). \tag{1.2.106}$$

It is sometimes helpful to graphically represent the effect of a Lorentz boost using a two-dimensional Minkowski spacetime diagram as is done in Fig. 1.5. Consider the passive Lorentz boost of Eq. (1.2.96) from the frame of inertial observer $\mathcal{O}$ to the frame of inertial observer $\mathcal{O}'$, who is moving with velocity v in the x-direction in the frame of $\mathcal{O}$. Let us identify ct with the y-axis and x with the x-axis and retain only the nontrivial 2×2 component of the Lorentz boost in Eq. (1.2.96) and let us identify the constant a and the angle ϕ using the following definition:

$$\begin{pmatrix} ct' \\ x' \end{pmatrix} = \begin{pmatrix} \gamma & -\beta\gamma \\ -\beta\gamma & \gamma \end{pmatrix} \begin{pmatrix} ct \\ x \end{pmatrix} \equiv a \begin{pmatrix} \cos\phi & -\sin\phi \\ -\sin\phi & \cos\phi \end{pmatrix} \begin{pmatrix} ct \\ x \end{pmatrix}. \tag{1.2.107}$$

We identify $\gamma = a\cos\phi$ and $\beta\gamma = a\sin\phi$ and we must require $a^2 = a^2(\cos\phi)^2 + a^2(\sin\phi)^2 = \gamma^2 + \beta^2\gamma^2 = (1+\beta^2)/(1-\beta^2)$, which immediately gives

$$\tan\phi = \beta = \frac{v}{c} \quad \text{and} \quad a = \sqrt{\frac{1+\beta^2}{1-\beta^2}} = \sqrt{\frac{c^2+v^2}{c^2-v^2}} \geq 1. \tag{1.2.108}$$

Up to the scaling constant a this has the form of a rotation of the x-axis through an angle $+\phi$ and of the ct-axis through an angle $-\phi$. We can understand this as follows: (i) the x'-axis corresponds to $ct' = 0$ and this occurs when $ct\cos\phi - x\sin\phi = 0$ or equivalently when $ct = x\tan\phi = \beta x$, which is a straight line in the ct–x plane with slope $\beta = v/c$ and angle $\tan^{-1}\beta = \phi$; and (ii) the ct'-axis corresponds to $x' = 0$ and this occurs when $-ct\sin\phi + x\cos\phi = 0$ or equivalently when $ct = x/\tan\phi = x/\beta$, which is a straight line in the ct–x plane with slope $1/\beta = c/v$ and angle $\tan^{-1}(1/\beta) = (\pi/2) - \tan^{-1}\beta = (\pi/2) - \phi$ using $\tan[(\pi/2) - \phi] = 1/\tan\phi$.

These results are illustrated in Fig. 1.5. The hyperbolae intersect the x- and x' axes at points of equal proper length; i.e, if a point on a hyperbola intersects the x-axis at 1-meter from the origin in the frame of $\mathcal{O}$, then that hyperbola intersects the x'-axis at the 1 meter point in the frame of $\mathcal{O}'$. Similarly, the hyperbolae intersecting the ct and ct' axes do so at equal proper time intervals; i.e., each observer records the same time on their comoving clocks at these intersection points. Note that a given spacetime event is represented by a point in the plane and the event coordinates can be read off from this point in terms of (ct, x) for observer $\mathcal{O}$ or in terms of (ct', x') for observer $\mathcal{O}'$. The inset rectangle has its upper right corner at an event along the worldline of a particle, which appears just below the heavy diagonal arrow pointing to this event. Events along the worldline can be read off in both sets of coordinates. The order in which the two spacelike separated events occurred is not the same in every frame. It is also clear that no boost can ever exceed the speed of light.

In experimental particle physics it is common to use a modified version of the rapidity. Define $p_{\mathrm{L}} \equiv |\mathbf{p}_{\mathrm{L}}|$, where $\mathbf{p}_{\mathrm{L}}$ is the component of the three-momentum of a particle along the beam axis. Then the *rapidity of a particle's motion along the beam axis* is the *modified rapidity*

$$y = \tanh^{-1}\left(\frac{p_{\mathrm{L}}c}{E}\right) = \frac{1}{2}\ln\left(\frac{E + p_{\mathrm{L}}c}{E - p_{\mathrm{L}}c}\right). \tag{1.2.109}$$

Define p_{T} to be the magnitude of the component of momentum transverse to the beam, so that $\mathbf{p} = \mathbf{p}_{\mathrm{T}} + \mathbf{p}_{\mathrm{L}}$. Since the rapidities of parallel boosts are additive as shown in Eq. (1.2.102), then it follows that the difference in the modified rapidities of particles will be invariant under boosts along the beam axis; i.e., for two particles with modified rapidities y_2 and y_1 and for a boost along the beam axis of rapidity η the modified rapidities become $y_2' = y_2 + \eta$ and $y_1' = y_1 + \eta$ and so $y_2' - y_1' = y_2 - y_1$.

Let θ be the angle in the laboratory frame of the particle's motion with respect to the beam direction, $0 \leq \theta \leq \pi$. Then we have $p_{\mathrm{L}} = |\mathbf{p}|\cos\theta$, $p_{\mathrm{T}} = |\mathbf{p}|\sin\theta$ and $\tan\theta = p_{\mathrm{T}}/p_{\mathrm{L}}$. Note that $0 \leq p_{\mathrm{T}} \leq |\mathbf{p}|$ and $-|\mathbf{p}| \leq p_{\mathrm{L}} \leq |\mathbf{p}|$, where positive or negative p_{L} corresponds to forward along the beam direction ($\theta = 0$) or backward opposite to the beam direction ($\theta = \pi$), respectively. Note that y is the rapidity of the boost along the beam axis necessary to take an observer from the laboratory frame to the frame in which the particle has only momentum transverse to the beam, i.e., the frame in which the particle motion is perpendicular to the beam axis. In a fixed target experiment the "beam" is the incoming beam onto the stationary target in the laboratory frame. In a collider experiment where the two beams are directed at each other along a common beam axis we simply choose one of the two beams to define the positive "beam" direction in the definition of rapidity.

It is also common in experimental particle physics to present results in terms of the *pseudorapidity*,

$$\eta' \equiv -\ln\left[\tan\left(\frac{\theta}{2}\right)\right]. \tag{1.2.110}$$

The pseudorapidity depends only on the angle of the particle motion and not its momentum. Since $\sin^2(\theta/2) = (1 - \cos\theta)/2$ and $\cos^2(\theta/2) = (1 + \cos\theta)/2$, then

$$\eta' = -\ln\sqrt{\frac{\sin^2(\theta/2)}{\cos^2(\theta/2)}} = \frac{1}{2}\ln\left(\frac{1 + \cos\theta}{1 - \cos\theta}\right) = \frac{1}{2}\ln\left(\frac{|\mathbf{p}| + p_{\mathrm{L}}}{|\mathbf{p}| - p_{\mathrm{L}}}\right). \tag{1.2.111}$$

Note that $\eta' = 0$ corresponds to the particle moving in a direction perpendicular to the beam ($p_{\mathrm{T}} > 0$ and $p_{\mathrm{L}} = 0$), whereas $\eta' \to +\infty$ and $-\infty$ correspond to particle motion in the direction of the beam ($p_{\mathrm{T}} = 0$ and $p_{\mathrm{L}} > 0$) and opposite to the beam direction ($p_{\mathrm{T}} = 0$ and $p_{\mathrm{L}} < 0$), respectively. For massless particles or when the particle motion is highly relativistic ($E \gg m$), then we have $E \to |\mathbf{p}|c$. In that limit we see by inspection that the pseudorapidity η' and the modified rapidity y become identical. Using $\tan(\theta/2) = (1 - \cos\theta)/\sin\theta$ and $\exp(\eta') = 1/[\tan(\theta/2)]$ we can show that $p_{\mathrm{T}}\sinh\eta' = |\mathbf{p}|\cos\theta = p_{\mathrm{L}}$ and $p_{\mathrm{T}}\cosh\eta' = |\mathbf{p}|$. Then using $E/c = \sqrt{m^2c^2 + \mathbf{p}^2} = \sqrt{m^2c^2 + p_{\mathrm{T}}^2 + p_{\mathrm{L}}^2}$ we can obtain the modified rapidity y in terms of the pseudorapidity η' as

$$y = \ln\sqrt{\frac{E + p_L c}{E - p_L c}} = \ln\left(\frac{(E/c) + p_L}{\sqrt{(E/c)^2 - p_L^2}}\right) = \ln\left(\frac{\sqrt{(mc)^2 + \mathbf{p}^2} + p_L}{\sqrt{(mc)^2 + p_T^2}}\right)$$

$$= \ln\left(\frac{\sqrt{(mc)^2 + p_T^2\cosh^2\eta'} + p_T\sinh\eta'}{\sqrt{(mc)^2 + p_T^2}}\right). \tag{1.2.112}$$

We again recover $y \to \eta'$ as $m \to 0$, since $y \to \ln(\cosh\eta' + \sinh\eta') = \eta'$.

For an arbitrary boost by velocity $\boldsymbol{v} = (v_x, v_y, v_z)$ we define $\boldsymbol{\beta} = (\beta_x, \beta_y, \beta_z) = \boldsymbol{v}/c$ and $\beta \equiv |\boldsymbol{\beta}|$. Defining $\mathbf{x} \equiv (x, y, z) = (x^1, x^2, x^3)$ we can decompose $\mathbf{x}$ into components parallel and orthogonal to $\boldsymbol{v}$ using

$$\mathbf{x}_{\parallel} = \left(\frac{\mathbf{x}\cdot\boldsymbol{v}}{v}\right)\left(\frac{\boldsymbol{v}}{v}\right) = \left(\frac{\mathbf{x}\cdot\boldsymbol{\beta}}{\beta}\right)\left(\frac{\boldsymbol{\beta}}{\beta}\right) \quad\text{and}\quad \mathbf{x}_{\perp} = \mathbf{x} - \mathbf{x}_{\parallel}. \tag{1.2.113}$$

Since the orientation of the spatial axes for observer $\mathcal{O}$ is arbitrary, then from Eq. (1.2.87) it follows that a passive Lorentz boost for any velocity $\boldsymbol{v}$ has the form

$$ct' = \gamma(ct - \boldsymbol{\beta}\cdot\mathbf{x}), \quad \mathbf{x}' = \mathbf{x}_{\perp} + \gamma(\mathbf{x}_{\parallel} - \boldsymbol{\beta}ct) = \mathbf{x} + (\gamma - 1)\left(\frac{\mathbf{x}\cdot\boldsymbol{\beta}}{\beta}\frac{\boldsymbol{\beta}}{\beta}\right) - \boldsymbol{\beta}\gamma ct. \tag{1.2.114}$$

Reading off one component at a time from Eq. (1.2.114) leads to the matrix form

$$\Lambda_B(\boldsymbol{v}) = \Lambda_B(v\mathbf{n})$$

$$= \begin{pmatrix} \gamma & -\gamma\beta_x & -\gamma\beta_y & -\gamma\beta_z \\ -\gamma\beta_x & 1 + (\gamma-1)(\beta_x^2/\beta^2) & (\gamma-1)(\beta_x\beta_y/\beta^2) & (\gamma-1)(\beta_x\beta_z/\beta^2) \\ -\gamma\beta_y & (\gamma-1)(\beta_y\beta_x/\beta^2) & 1 + (\gamma-1)(\beta_y^2/\beta^2) & (\gamma-1)(\beta_y\beta_z/\beta^2) \\ -\gamma\beta_z & (\gamma-1)(\beta_z\beta_x/\beta^2) & (\gamma-1)(\beta_z\beta_y/\beta^2) & 1 + (\gamma-1)(\beta_z^2/\beta^2) \end{pmatrix}$$

$$= \begin{pmatrix} \cosh\eta & -n_x\sinh\eta & -n_y\sinh\eta & -n_z\sinh\eta \\ -n_x\sinh\eta & 1 + (\cosh\eta-1)n_x^2 & (\cosh\eta-1)n_x n_y & (\cosh\eta-1)n_x n_z \\ -n_y\sinh\eta & (\cosh\eta-1)n_y n_x & 1 + (\cosh\eta-1)n_y^2 & (\cosh\eta-1)n_y n_z \\ -n_z\sinh\eta & (\cosh\eta-1)n_z n_x & (\cosh\eta-1)n_z n_y & 1 + (\cosh\eta-1)n_z^2 \end{pmatrix}$$

$$= \begin{pmatrix} \cosh\eta & -\mathbf{n}^T\sinh\eta \\ -\mathbf{n}\sinh\eta & I_{3\times3} + \mathbf{n}\mathbf{n}^T(\cosh\eta - 1) \end{pmatrix}$$

$$= \begin{pmatrix} \gamma & -\mathbf{n}^T\beta\gamma \\ -\mathbf{n}\beta\gamma & I_{3\times3} + \mathbf{n}\mathbf{n}^T(\gamma - 1) \end{pmatrix}, \tag{1.2.115}$$

where $\mathbf{n} = (n_x, n_y, n_z) \equiv \boldsymbol{v}/v = \boldsymbol{\beta}/\beta$ is the unit vector in the boost direction. We see that this reduces to Eqs. (1.2.96) and (1.2.97) in the appropriate limits, where $\mathbf{n} = (1,0,0)$, $(0,1,0)$ and $(0,0,1)$.

We can now define the Lorentz boost generators in a way similar to what was done for the generators of rotations. However, some caution is needed since, while the rotations form a subgroup of the Lorentz group, the set of all boosts does not. However, two parallel boosts give another boost in the same direction, since parallel boosts are additive as we have seen. Thus we have closure for the set of all parallel boosts and we also have the properties of existence of the identity ($\eta = 0$), existence of the inverse (every boost has its opposite boost in the set) and associativity (since addition is associative). The addition rule for parallel boosts also implies that order of the parallel boosts is unimportant; i.e., parallel boosts commute with each other. Therefore the set of all boosts in a single direction form a one-parameter abelian Lie group that is a subgroup of the Lorentz group. There is one such subgroup for every possible spatial direction denoted by the unit vector $\mathbf{n}$, where here $\mathbf{n} = (n_x, n_y, n_z)$ denotes a unit vector in the spatial direction $\mathbf{n}$.

Therefore any passive boost by velocity $\boldsymbol{v} \equiv v\mathbf{n}$ can be written as

$$\Lambda_{\mathrm{B}}(\boldsymbol{v}) = \Lambda_{\mathrm{B}}(v\mathbf{n}) = \exp\left(i\eta K_{\mathbf{n}}\right) = \exp\left(i\eta\,\mathbf{n}\cdot\mathbf{K}\right) \qquad (1.2.116)$$

for some suitably defined generator $K_{\mathbf{n}}$. We can evaluate the generator $K_{\mathbf{n}}$ using Eq. (1.2.115):

$$K_{\mathbf{n}} = -i\left.\frac{\partial\Lambda_{\mathrm{B}}(\boldsymbol{v})}{\partial\eta}\right|_{\eta=0} = \begin{pmatrix} 0 & in_x & in_y & in_z \\ in_x & 0 & 0 & 0 \\ in_y & 0 & 0 & 0 \\ in_z & 0 & 0 & 0 \end{pmatrix}. \qquad (1.2.117)$$

In the special cases of boosts in the x, y and z directions we have, respectively, three linearly independent imaginary and anti-Hermitian matrices given by

$$K^1 = \begin{pmatrix} 0 & i & 0 & 0 \\ i & 0 & 0 & 0 \\ 0 & 0 & 0 & 0 \\ 0 & 0 & 0 & 0 \end{pmatrix}, \quad K^2 = \begin{pmatrix} 0 & 0 & i & 0 \\ 0 & 0 & 0 & 0 \\ i & 0 & 0 & 0 \\ 0 & 0 & 0 & 0 \end{pmatrix} \text{ and } K^3 = \begin{pmatrix} 0 & 0 & 0 & i \\ 0 & 0 & 0 & 0 \\ 0 & 0 & 0 & 0 \\ i & 0 & 0 & 0 \end{pmatrix}, \quad (1.2.118)$$

where we use the notation $\mathbf{K} = (K_x, K_y, K_z) = (K^1, K^2, K^3)$ and $K_{\mathbf{n}} \equiv \mathbf{n}\cdot\mathbf{K}$. It is readily checked from Eq. (1.2.41) that

$$K^i = M^{0i} \quad \text{for} \quad i = 1, 2, 3. \qquad (1.2.119)$$

Using Eqs. (1.2.116) and (1.2.45) we then identify that for a passive Lorentz boost we have

$$-(1/2)\omega_{\mu\nu}M^{\mu\nu}\big|_{\mu,\nu=0,i \text{ or } i,0} = -\omega_{0i}M^{0i} = \eta\,\mathbf{n}\cdot\mathbf{K} = \boldsymbol{\eta}\cdot\mathbf{K}, \qquad (1.2.120)$$

where $\boldsymbol{\eta} \equiv \eta\mathbf{n}$. This then gives

$$\omega_{0i} = -\eta\,n^i = -\eta^i. \qquad (1.2.121)$$

In the limit of a very small boost, $\beta \ll 1$, we have $\eta \ll 1$. Working to leading order in η we have $\sinh\eta \to \eta + \mathcal{O}(\eta^3)$ and $\cosh\eta \to 1 + \mathcal{O}(\eta^2)$. Using this in Eq. (1.2.115) we find that $\Lambda_{\mathrm{B}}(\boldsymbol{v}) \to 1 + i\eta\mathbf{n}\cdot\mathbf{K} + \mathcal{O}(\eta^2)$, which agrees with the expansion of the exponential in Eq. (1.2.116) as it must. We can also prove directly that $\exp(i\eta\mathbf{n}\cdot\mathbf{K})$ leads to Eq. (1.2.115) for any arbitrary boost velocity $\boldsymbol{v}$.

Proof of Eq. (1.2.115): We can use Eq. (1.2.116) to directly obtain the result of Eq. (1.2.115) as follows.

We begin by noting that $\exp(i\boldsymbol{\eta}\cdot\mathbf{K}) = \exp(i\eta\mathbf{n}\cdot\mathbf{K}) = \exp(i\eta K_{\mathbf{n}})$ and that

$$iK_{\mathbf{n}} = \begin{pmatrix} 0 & -n_x & -n_y & -n_z \\ -n_x & 0 & 0 & 0 \\ -n_y & 0 & 0 & 0 \\ -n_z & 0 & 0 & 0 \end{pmatrix}, \quad (iK_{\mathbf{n}})^2 = \begin{pmatrix} 1 & 0 & 0 & 0 \\ 0 & n_x^2 & n_x n_y & n_x n_z \\ 0 & n_y n_x & n_y^2 & n_y n_z \\ 0 & n_z n_x & n_z n_y & n_z^2 \end{pmatrix},$$

$$(iK_{\mathbf{n}})^3 = (iK_{\mathbf{n}}) \Rightarrow (iK_{\mathbf{n}})^{2j-1} = (iK_{\mathbf{n}}), \; (iK_{\mathbf{n}})^{2j} = (iK_{\mathbf{n}})^2 \text{ for } j = 1, 2, 3, \ldots.$$

It then follows that

$$\exp\left(i\eta K_{\mathbf{n}}\right) = \sum_{j=0}^{\infty} \frac{1}{j!}\eta^j \left(iK_{\mathbf{n}}\right)^j = I + (iK_{\mathbf{n}}) \sum_{j=1,3,\ldots}^{\infty} \frac{1}{j!}\eta^j + (iK_{\mathbf{n}})^2 \sum_{j=2,4,\ldots}^{\infty} \frac{1}{j!}\eta^j$$

$$= I + \sinh\eta\,(iK_{\mathbf{n}}) + (\cosh\eta - 1)\left(iK_{\mathbf{n}}\right)^2$$

$$= \begin{pmatrix} \cosh\eta & -n_x\sinh\eta & -n_y\sinh\eta & -n_z\sinh\eta \\ -n_x\sinh\eta & 1 + (\cosh\eta-1)n_x^2 & (\cosh\eta-1)n_x n_y & (\cosh\eta-1)n_x n_z \\ -n_y\sinh\eta & (\cosh\eta-1)n_y n_x & 1 + (\cosh\eta-1)n_y^2 & (\cosh\eta-1)n_y n_z \\ -n_z\sinh\eta & (\cosh\eta-1)n_z n_x & (\cosh\eta-1)n_z n_y & 1 + (\cosh\eta-1)n_z^2 \end{pmatrix}$$

$$= \begin{pmatrix} \cosh\eta & -\mathbf{n}^T\sinh\eta \\ -\mathbf{n}\sinh\eta & I_{3\times3} + \mathbf{n}\mathbf{n}^T(\cosh\eta - 1) \end{pmatrix}. \tag{1.2.122}$$

Relativistic Doppler Effect

Consider a source of coherent monochromatic light in its rest frame with frequency f_s, which corresponds to a period $\Delta t_s = 1/f_s$ and a wavelength $\lambda_s = c/f_s = c\Delta t_s$. Then consider an inertial observer moving away from the source with velocity v, where $v > 0$ and $v < 0$ correspond to movement away from and toward the source, respectively. In the source rest frame the distance between wavecrests is λ_s and the speed of the crests is c. We define Δt as the time in the source frame between consecutive crests arriving at the observer. If the observer is at rest ($v = 0$), then the time between two consecutive crests arriving at the observer would simply be $\Delta t = \lambda_s/c = \Delta t_s$. If the observer is not at rest with respect to the source, then the observer will have moved a distance $v\Delta t$ in the frame of the source between arrival of the crests. So the time between the crests arriving at the observer as measured in the source frame must satisfy $\Delta t = (\lambda_s + v\Delta t)/c = \Delta t_s + \beta\Delta t$, which gives $\Delta t = \Delta t_s/(1-\beta)$. Due to time dilation the moving observer's clock is running slow as observed from the source frame and so the time showing on the observer's clock between consecutive crests arriving is $\Delta t_o = \Delta t/\gamma$. This leads us to the ratio known as the *Doppler factor*,

$$\frac{\Delta t_o}{\Delta t_s} = \frac{1}{\gamma(1-\beta)} = \sqrt{\frac{1+\beta}{1-\beta}} = \frac{f_s}{f_o}. \tag{1.2.123}$$

If the observer and source are moving apart ($\beta = v/c > 0$), then $f_o < f_s$ and the observed frequency f_o is said to have a *redshift* with respect to the source frequency f_s. If they are moving together, then $f_o > f_s$ and the observed frequency is said to have a *blueshift* with respect to the source frequency. The *redshift parameter* z is defined as

$$z \equiv \frac{f_s - f_o}{f_o} = \frac{\lambda_o - \lambda_s}{\lambda_s} \quad \text{so that} \quad 1 + z = \frac{f_s}{f_o} = \frac{\lambda_o}{\lambda_s}. \tag{1.2.124}$$

Addition of Velocities

We saw earlier in Eq. (1.2.101) that collinear boosts have additive rapidities, but we would like to establish a more general relationship between velocities of particles as observed from different inertial frames. Consider a particle moving along its timelike worldline described as $\mathbf{x}(t)$ by observer $\mathcal{O}$, then at time t its velocity in the frame of $\mathcal{O}$ is $\mathbf{u} = (u_x, u_y, u_z) = d\mathbf{x}/dt$. Now consider a second observer $\mathcal{O}'$ moving in the $+x$-direction with velocity v with respect to $\mathcal{O}$. We have from Eq. (1.2.87) that

$$dt' = \gamma dt \left(1 - \frac{v}{c^2}\frac{dx}{dt}\right), \quad dx' = \gamma\left(dx - \beta cdt\right), \quad dy' = dy, \quad dz' = dz. \tag{1.2.125}$$

The velocity of the particle as observed by $\mathcal{O}'$ in his frame is $\mathbf{u}' = (u'_x, u'_y, u'_z) = d\mathbf{x}'/dt'$, which is given by

$$
\begin{aligned}
u'_x &= \frac{dx'}{dt'} = \left(\frac{dx}{dt} - v\right)\left(1 - \frac{v}{c^2}\frac{dx}{dt}\right)^{-1} = \frac{u_x - v}{1 - (vu_x/c^2)} \\
u'_y &= \frac{dy'}{dt'} = \frac{1}{\gamma}\frac{dy}{dt}\left(1 - \frac{v}{c^2}\frac{dx}{dt}\right)^{-1} = \frac{u_y}{\gamma\left[1 - (vu_x/c^2)\right]} \\
u'_z &= \frac{dz'}{dt'} = \frac{1}{\gamma}\frac{dz}{dt}\left(1 - \frac{v}{c^2}\frac{dx}{dt}\right)^{-1} = \frac{u_z}{\gamma\left[1 - (vu_x/c^2)\right]}.
\end{aligned}
\tag{1.2.126}
$$

We can use Eq. (1.2.114) to construct $\mathbf{u}'$ from $\mathbf{u}$ when observer $\mathcal{O}'$ is moving with some arbitrary velocity $\boldsymbol{v}$ with respect to observer $\mathcal{O}$. Recalling that $\boldsymbol{\beta} = \boldsymbol{v}/c$ we then have

$$dt' = \gamma dt \left(1 - \frac{\boldsymbol{\beta}}{c} \cdot \frac{d\mathbf{x}}{dt}\right) = \gamma dt \left(1 - \frac{\mathbf{u}\cdot\boldsymbol{\beta}}{c}\right) \tag{1.2.127}$$

$$d\mathbf{x}' = d\mathbf{x}_\perp + \gamma\left(d\mathbf{x}_\parallel - \boldsymbol{\beta}cdt\right) = d\mathbf{x} + (\gamma - 1)\left(\frac{d\mathbf{x}\cdot\boldsymbol{\beta}}{\beta}\frac{\boldsymbol{\beta}}{\beta}\right) - \boldsymbol{\beta}\gamma cdt,$$

which gives the general form of the velocity addition formula

$$
\begin{aligned}
\mathbf{u}' &= \frac{d\mathbf{x}'}{dt'} = \left[\mathbf{u}_\perp + \gamma\left(\mathbf{u}_\parallel - \boldsymbol{\beta}c\right)\right]\left[\gamma\left(1 - \frac{\mathbf{u}\cdot\boldsymbol{\beta}}{c}\right)\right]^{-1} \\
&= \left[\mathbf{u} + (\gamma - 1)\left(\frac{\mathbf{u}\cdot\boldsymbol{\beta}}{\beta}\frac{\boldsymbol{\beta}}{\beta}\right) - \boldsymbol{\beta}\gamma c\right]\left[\gamma\left(1 - \frac{\mathbf{u}\cdot\boldsymbol{\beta}}{c}\right)\right]^{-1}.
\end{aligned}
\tag{1.2.128}
$$

Note that $\mathbf{u}_\perp$ and $\mathbf{u}_\parallel$ are defined using Eq. (1.2.113) in the same way that we defined $\mathbf{x}_\perp$ and $\mathbf{x}_\parallel$. It is readily verified that Eq. (1.2.128) reduces to Eq. (1.2.126) in the special case where $\boldsymbol{v} = (v, 0, 0)$.

Summary of Some Important Tensor Notation

It is conventional to use Greek letters $\alpha, \beta, \gamma, \delta, \mu, \nu, \rho, \ldots$ to represent spacetime indices (e.g., $\mu \in \{0, 1, 2, 3\}$), and it is also conventional to use Latin letters $i, j, k, l, m, n, \ldots$ for spatial indices (e.g., $j \in \{1, 2, 3\} = \{x, y, z\}$).

For a contravariant four-vector V^μ we write

$$V^\mu \equiv (V^0, \mathbf{V}) = (V^0, V^1, V^2, V^3) = (V^0, V_x, V_y, V_z), \tag{1.2.129}$$

$$V_\mu = g_{\mu\nu}V^\nu \quad \text{and} \quad V^\mu = g^{\mu\nu}V_\nu.$$

This gives $V_0 = V^0$ and $V_i = -V^i$, i.e.,

$$V_\mu = (V^0, -\mathbf{V}). \tag{1.2.130}$$

The *scalar product* or "dot" product of two four-vectors is their contraction

$$U \cdot V \equiv U^\mu V_\mu = U_\mu V^\mu = U^0 V^0 - \mathbf{U} \cdot \mathbf{V} = U^\mu g_{\mu\nu}V^\nu = U_\mu g^{\mu\nu}V_\nu. \tag{1.2.131}$$

The "square" of a four-vector is the scalar product of a four-vector with itself

$$V^2 \equiv V \cdot V = V_\mu V^\mu = V_0 V^0 + V_1 V^1 + V_2 V^2 + V_3 V^3 = (V^0)^2 - \mathbf{V}^2. \tag{1.2.132}$$

We write the derivative $\partial/\partial x^\mu$ as

$$\partial_\mu \equiv \frac{\partial}{\partial x^\mu} = \left(\frac{1}{c}\frac{\partial}{\partial t}, \boldsymbol{\nabla}\right) = \left(\frac{1}{c}\frac{\partial}{\partial t}, \frac{\partial}{\partial x^1}, \frac{\partial}{\partial x^2}, \frac{\partial}{\partial x^3}\right) = \left(\frac{1}{c}\frac{\partial}{\partial t}, \frac{\partial}{\partial x}, \frac{\partial}{\partial y}, \frac{\partial}{\partial z}\right). \quad (1.2.133)$$

Note that while $\mathbf{x} = (x^1, x^2, x^3) = (x, y, z)$ we have $\boldsymbol{\nabla} = (\partial_1, \partial_2, \partial_3) = (\partial_x, \partial_y, \partial_z)$. The derivative ∂_μ must satisfy $\partial_\mu x^\nu = \delta^\nu{}_\mu$ in every frame. This means that ∂_μ must transform like a covariant four-vector,

$$\partial'_\mu = \partial_\nu (\Lambda^{-1})^\nu{}_\mu, \quad (1.2.134)$$

since only then will we have

$$\partial'_\mu x'^\nu = \partial_\rho (\Lambda^{-1})^\rho{}_\mu \Lambda^\nu{}_\sigma\, x^\sigma = (\Lambda^{-1})^\rho{}_\mu \Lambda^\nu{}_\sigma\, \partial_\rho x^\sigma = \Lambda^\nu{}_\sigma \delta^\sigma{}_\rho (\Lambda^{-1})^\rho{}_\mu = \Lambda^\nu{}_\rho (\Lambda^{-1})^\rho{}_\mu = \delta^\nu{}_\mu.$$

We have used the fact that Λ only contains constants. Since ∂_μ is a covariant four-vector, then it follows that we can write

$$\partial^\mu = g^{\mu\nu}\partial_\nu = \left(\frac{1}{c}\frac{\partial}{\partial t}, -\boldsymbol{\nabla}\right) = \frac{\partial}{\partial x_\mu}, \quad (1.2.135)$$

where we have used $x_i = -x^i$ so that $-\partial/\partial x^i = \partial/\partial x_i$. The *four-divergence* of a four-vector V^μ is Lorentz invariant and defined as

$$\partial_\mu V^\mu = \partial \cdot V = \frac{1}{c}\frac{\partial}{\partial t}V^0 + \boldsymbol{\nabla} \cdot \mathbf{V} = \partial^\mu V_\mu. \quad (1.2.136)$$

This is the relativistic extension of the usual three-dimensional divergence of a three vector $\boldsymbol{\nabla} \cdot \mathbf{V}$, which is invariant under three-dimensional rotations.

The completely antisymmetric tensor Levi-Civita is defined in Sec. A.2.1. The rank-four form used in special relativity $\epsilon^{\mu\nu\rho\sigma}$ has the properties $\epsilon^{0123} = +1$ and $\epsilon_{\mu\nu\rho\sigma} = g_{\mu\alpha}g_{\nu\beta}g_{\rho\gamma}g_{\sigma\delta}\,\epsilon^{\alpha\beta\gamma\delta}$. It follows that $\epsilon_{\mu\nu\rho\sigma} = -\epsilon^{\mu\nu\rho\sigma}$, since μ, ν, ρ, σ must all be different and so we must have the product $g_{00}g_{11}g_{22}g_{33} = -1$. Furthermore, it must transform as a rank-four *pseudotensor* under a Lorentz transformation,

$$\epsilon^{\mu\nu\rho\sigma} \to \epsilon'^{\mu\nu\rho\sigma} = \det(\Lambda)\Lambda^\mu{}_\alpha \Lambda^\nu{}_\beta \Lambda^\rho{}_\gamma \Lambda^\sigma{}_\delta\, \epsilon^{\alpha\beta\gamma\delta} = \epsilon^{\mu\nu\rho\sigma}. \quad (1.2.137)$$

Proof: From Eq. (A.8.1) we have for an $n \times n$ matrix A

$$\det A = \epsilon^{k_1 k_2 \cdots k_n} A_{1k_1} A_{2k_2} \cdots A_{nk_n}, \quad (1.2.138)$$

where the summation convention is understood. Then we have

$$\epsilon^{0123}\det\Lambda = \det\Lambda = \epsilon^{\alpha\beta\gamma\delta}\Lambda^0{}_\alpha \Lambda^1{}_\beta \Lambda^2{}_\gamma \Lambda^3{}_\delta. \quad (1.2.139)$$

Since $\det(\Lambda) = \pm 1$ then $\det(\Lambda) = 1/\det(\Lambda)$ and so we can write

$$\epsilon^{0123} = \det(\Lambda)\Lambda^0{}_\alpha \Lambda^1{}_\beta \Lambda^2{}_\gamma \Lambda^3{}_\delta\, \epsilon^{\alpha\beta\gamma\delta}. \quad (1.2.140)$$

Using $\epsilon^{0123} = -\epsilon^{1023}$, reordering two Λs and exchanging α and β we find

$$\epsilon^{1023} = \det(\Lambda)\Lambda^1{}_\alpha \Lambda^0{}_\beta \Lambda^2{}_\gamma \Lambda^3{}_\delta\, \epsilon^{\alpha\beta\gamma\delta}. \quad (1.2.141)$$

This is true for any reordering and so

$$\epsilon^{\mu\nu\rho\sigma} = \det(\Lambda)\Lambda^\mu{}_\alpha \Lambda^\nu{}_\beta \Lambda^\rho{}_\gamma \Lambda^\sigma{}_\delta\, \epsilon^{\alpha\beta\gamma\delta}. \quad (1.2.142)$$

Alternative Expression of the Lorentz Lie Algebra

We have shown that the three Hermitian generators of the rotation group, $\mathbf{J} = (J^1, J^2, J^3) = (M^{23}, M^{31}, M^{12})$, taken together with the three anti-Hermitian generators of the Lorentz boosts, $\mathbf{K} = (K^1, K^2, K^3) = (M^{01}, M^{02}, M^{03})$, make up the six generators of the restricted Lorentz group $SO^+(1,3)$. We can write these in the form

$$J^i = \tfrac{1}{2}\epsilon^{ijk}M^{jk} \text{ (or equivalently } M^{kl} = \epsilon^{klm}J^m) \quad \text{and} \quad K^i = M^{0i} \tag{1.2.143}$$

for $i = 1, 2, 3$. We are being a little careless with our notation as is the convention and it should be noted that the above equations involving $\mathbf{J}$ in the defining representation are to be understood in the 4×4 matrix sense of Eq. (1.2.71). We can therefore express the Lie algebra of the restricted Lorentz group in terms of the rotation and boost generators. It follows from the Lie algebra of the $M^{\mu\nu}$ in Eq. (1.2.46) that an alternative expression for the Lorentz Lie algebra is

$$\left[J^i, J^j\right] = i\epsilon^{ijk}J^k, \quad \left[J^i, K^j\right] = i\epsilon^{ijk}K^k, \quad \left[K^i, K^j\right] = -i\epsilon^{ijk}J^k, \tag{1.2.144}$$

where recall that $\mathbf{J}$ and $\mathbf{K}$ are Hermitian and anti-Hermitian, respectively.

Proof: The first of the three equations is just the Lie algebra of the rotation group derived earlier, i.e., Eq. (1.2.57). For the other two equations we need only Eq. (1.2.46) and Eq. (1.2.143). We have for the third equation

$$[K^i, K^j] = [M^{0i}, M^{0j}] = i(-g^{00}M^{ij}) = -iM^{ij} = -i\epsilon^{ijk}J^k. \tag{1.2.145}$$

For the second equation we see that

$$[J^i, K^j] = \tfrac{1}{2}\epsilon^{ik\ell}[M^{k\ell}, M^{0j}] = \tfrac{1}{2}\epsilon^{ik\ell}i(-g^{\ell j}M^{k0} + g^{kj}M^{\ell 0}) \tag{1.2.146}$$
$$= i\tfrac{1}{2}\epsilon^{ik\ell}(\delta^{\ell j}M^{k0} - \delta^{kj}M^{\ell 0}) = i\tfrac{1}{2}(\epsilon^{ijk}M^{0k} + \epsilon^{ij\ell}M^{0\ell}) = i\epsilon^{ijk}M^{0k} = i\epsilon^{ijk}K^k.$$

We see that the boost generators $\mathbf{K}$ do not form a closed algebra like the rotation generators $\mathbf{J}$ and this is reflected in the fact that the boosts do not satisfy closure and are therefore not a group as noted earlier. As is evident from the third equation in Eq. (1.2.144) above the difference between a boost along the x-axis followed by a boost along the y-axis and the two boosts with their order reversed involves a rotation about the z-axis. This is the origin of the Thomas precession.

As noted in Sec. A.7.2 the group of restricted Lorentz transformations $SO^+(1,3)$ must have two Casimir invariants, which can be written as

$$C_1 \equiv \mathbf{J}^2 - \mathbf{K}^2 = \tfrac{1}{2}M^{\mu\nu}M_{\mu\nu}, \quad C_2 \equiv \mathbf{J} \cdot \mathbf{K} = \mathbf{K} \cdot \mathbf{J} = -\tfrac{1}{4}\epsilon_{\mu\nu\rho\sigma}M^{\mu\nu}M^{\rho\sigma}. \tag{1.2.147}$$

Using Eq. (1.2.144) it follows that C_1 and C_2 commute with all six generators

$$\left[\mathbf{J} \cdot \mathbf{K}, J^i\right] = \left[\mathbf{J} \cdot \mathbf{K}, K^i\right] = \left[\mathbf{J}^2 - \mathbf{K}^2, J^i\right] = \left[\mathbf{J}^2 - \mathbf{K}^2, K^i\right] = 0. \tag{1.2.148}$$

Proof:
(i) $\tfrac{1}{2}M^{\mu\nu}M_{\mu\nu} = \tfrac{1}{2}M^{ij}M^{ij} - M^{0i}M^{0i} = \tfrac{1}{2}\epsilon^{ijk}J^k\epsilon^{ij\ell}J^\ell - K^iK^i = \mathbf{J}^2 - \mathbf{K}^2$.
(ii) $-\tfrac{1}{4}\epsilon_{\mu\nu\rho\sigma}M^{\mu\nu}M^{\rho\sigma}$
$$= -\tfrac{1}{4}[\epsilon_{0ijk}M^{0i}M^{jk} + \epsilon_{i0jk}M^{i0}M^{jk} + \epsilon_{ij0k}M^{ij}M^{0k} + \epsilon_{ijk0}M^{ij}M^{k0}]$$
$$= -\tfrac{1}{4}[-\epsilon^{ijk}K^iM^{jk} + \epsilon^{ijk}(-K^i)M^{jk} - \epsilon^{ijk}M^{ij}K^k + \epsilon^{ijk}M^{ij}(-K^k)]$$
$$= \tfrac{1}{2}[\mathbf{K} \cdot \mathbf{J} + \mathbf{K} \cdot \mathbf{J}] = \mathbf{J} \cdot \mathbf{K}$$

using $J^i K^i = K^i J^i$, which follows since $[J^i, K^j] = i\epsilon^{ijk}K^k$ gives $[J^i, K^i] = 0$.

(iii) $[\mathbf{J} \cdot \mathbf{K}, J^i] = J^j[K^j, J^i] + [J^j, J^i]K^j = -i\epsilon^{ijk}(J^j K^k + J^k K^j) = 0$.

(iv) $[\mathbf{J} \cdot \mathbf{K}, K^i] = J^j[K^j, K^i] + [J^j, K^i]K^j = i\epsilon^{ijk}J^j J^k + i\epsilon^{jik}K^k K^j = 0$.

(v) $[\mathbf{J}^2 - \mathbf{K}^2, J^i] = J^j[J^j, J^i] + [J^j, J^i]J^j - K^j[K^j, J^i] - [K^j, J^i]K^j$
$$= -i\epsilon^{ijk}(J^j J^k + J^k J^j) + i\epsilon^{ijk}(K^j K^k + K^k K^j) = 0.$$

(vi) $[\mathbf{J}^2 - \mathbf{K}^2, K^i] = J^j[J^j, K^i] + [J^j, K^i]J^j - K^j[K^j, K^i] - [K^j, K^i]K^j$
$$= -i\epsilon^{ijk}(J^j K^k + K^k J^j + K^j J^k + J^k K^j) = 0.$$

Summary: Any restricted (i.e., proper orthochronous) Lorentz transformation can be written in the form

$$x^\mu \to x'^\mu = \Lambda(\omega)^\mu{}_\nu x^\nu \quad \text{with} \quad \Lambda(\omega)^\mu{}_\nu = \left[\exp\left(-\tfrac{i}{2}\omega_{\alpha\beta}M^{\alpha\beta}\right)\right]^\mu{}_\nu = [\exp(\omega)]^\mu{}_\nu \quad (1.2.149)$$

for at least one choice of the six parameters $\omega^\mu{}_\nu$, where $\omega_{\mu\nu} = -\omega_{\nu\mu}$. Any Lorentz transformation can be obtained from the restricted Lorentz transformations combined with one or more of I, P or T as described in Table 1.1. The generators of the Lorentz transformations in the defining representation of the Lorentz group are the $M^{\mu\nu}$ given in Eq. (1.2.41).

Combining Eqs. (1.2.69) and (1.2.120) we see that for a passive restricted Lorentz transformation we have

$$-\tfrac{1}{2}\omega_{\mu\nu}(M^{\mu\nu}) = \boldsymbol{\alpha} \cdot \mathbf{J} + \boldsymbol{\eta} \cdot \mathbf{K}, \quad (1.2.150)$$

where $\alpha^m = -\tfrac{1}{2}\omega_{k\ell}\epsilon^{k\ell m}$ and $\eta^i = -\omega_{0i}$. So any passive restricted Lorentz transformation can be expressed in terms of a passive rotation specified by $\boldsymbol{\alpha}$ and a passive boost specified by $\boldsymbol{\eta}$. So for a passive restricted Lorentz transformation we have

$$\Lambda(\omega)^\mu{}_\nu = \Lambda_{\mathrm{p}}(\boldsymbol{\alpha}, \boldsymbol{\eta})^\mu{}_\nu = [\exp(i\boldsymbol{\alpha} \cdot \mathbf{J} + i\boldsymbol{\eta} \cdot \mathbf{K})]^\mu{}_\nu. \quad (1.2.151)$$

For an active one we have

$$\Lambda(\omega)^\mu{}_\nu = \Lambda_{\mathrm{a}}(\boldsymbol{\alpha}, \boldsymbol{\eta})^\mu{}_\nu = [\exp(-i\boldsymbol{\alpha} \cdot \mathbf{J} - i\boldsymbol{\eta} \cdot \mathbf{K})]^\mu{}_\nu, \quad (1.2.152)$$

where $\alpha^m = \tfrac{1}{2}\omega_{k\ell}\epsilon^{k\ell m}$ and $\eta^i = \omega_{0i}$.

Polar Decomposition of Restricted Lorentz Transformations

Boosts and rotations do not commute since their generators do not commute as seen in Eq. (1.2.144). However, it is always possible to express every restricted Lorentz transformation as a boost followed by a rotation and vice versa, but the $\boldsymbol{\alpha}$ and $\boldsymbol{\eta}$ are in general different depending on the order.

> **Mathematical digression: Polar decomposition:** Just as we can perform a polar decomposition for any complex number, $z \in \mathbb{C}$, as $z = re^{i\phi}$ we can write any complex square matrix as $A = UP$, where U is a unitary matrix and $P = \sqrt{A^\dagger A}$ is a positive-semidefinite Hermitian matrix. Obviously, $A^\dagger A$ is Hermitian and from Eq. (A.8.2) we know that $\det(A^\dagger A) \geq 0$ and so we can construct the positive semidefinite matrix P from the square root of the eigenvalues and the eigenvectors $A^\dagger A$ using spectral decomposition. P is unique and always exists even if A is a singular (i.e., $\det A = 0$) matrix. Then $U = AP^{-1}$ and U exists and is unique provided that A is invertible, since $U^\dagger = U^{-1} = PA^{-1}$ and U is defined if and only if $U^\dagger$ is. There is also a left or reverse polar decomposition defined by $A = P'U'$, where $P' = \sqrt{AA^\dagger}$ and $U' = P'^{-1}A = U$.

It is a useful exercise to show here how to explicitly construct the boost-rotation decomposition for any restricted Lorentz transformation using the polar decomposition of a matrix. Let Λ be any restricted passive Lorentz transformation, then

$$\Lambda = \exp(i\boldsymbol{\alpha} \cdot \mathbf{J} + i\boldsymbol{\eta} \cdot \mathbf{K}) \ \text{ and } \ \Lambda^\dagger = \exp(-i\boldsymbol{\alpha} \cdot \mathbf{J} + i\boldsymbol{\eta} \cdot \mathbf{K}) = \Lambda^T, \tag{1.2.153}$$

since Λ is real and since $\mathbf{J}^\dagger = \mathbf{J}$ and $\mathbf{K}^\dagger = -\mathbf{K}$. Any pure rotation is an orthogonal transformation $\Lambda^T = \Lambda^{-1}$ and any pure boost is a real symmetric matrix $\Lambda^T = \Lambda$ since $i\mathbf{K}$ are real symmetric. It follows from Eq. (1.2.153) that if a restricted Lorentz transformation Λ is an orthogonal matrix $\Lambda^T = \Lambda^{-1}$ then it is a pure rotation, and if it is a real symmetric matrix then it is a pure boost.

As shown earlier if Λ is a Lorentz transformation then so is Λ^T. Furthermore, since $\det(\Lambda^T) = \det\Lambda$ then if $\Lambda \in SO^+(1,3)$ then so is Λ^T and hence so are $\Lambda\Lambda^T$ and $\Lambda^T\Lambda$. Since $\Lambda\Lambda^T$ and $\Lambda^T\Lambda$ are both real symmetric then they must correspond to a pure boost; i.e., since the exponential map for the restricted Lorentz transformations is surjective (onto) then there must be at least one corresponding rapidity vector for each of them. We can then halve each rapidity and form the square root of these matrices; i.e., there must exist at least one $\boldsymbol{\eta}'$ such that

$$P \equiv \sqrt{\Lambda^\dagger\Lambda} = \sqrt{\Lambda^T\Lambda} = \exp(i\boldsymbol{\eta}' \cdot \mathbf{K}) \tag{1.2.154}$$

and similarly for $\Lambda\Lambda^T$. We obtain $\boldsymbol{\eta}'$ by forming P from Λ and then extracting $\boldsymbol{\eta}'$ using the form given in Eq. (1.2.115). Using polar decomposition we can then construct the orthogonal matrix (i.e., real Hermitian matrix)

$$O \equiv \Lambda P^{-1} = \Lambda \exp(-i\boldsymbol{\eta}' \cdot \mathbf{K}). \tag{1.2.155}$$

Clearly, we have $O^T = O^{-1}$ and $O \in SO^+(1,3)$ since Λ and P are. Again using the surjective nature of the exponential map for $SO^+(1,3)$, then there must be at least one choice of $\boldsymbol{\alpha}'$ such that

$$O = \exp(i\boldsymbol{\alpha}' \cdot \mathbf{J}), \tag{1.2.156}$$

where here we understand $\mathbf{J}$ to have the 4×4 matrix form given in Eq. (1.2.71). We can extract $\boldsymbol{\alpha}'$ by comparing this O with Eq. (1.2.68); i.e., O must correspond to a matrix with a one in the top left-hand entry and zeros elsewhere in the first row and first column since it is an orthogonal restricted Lorentz transformation and hence must be a rotation. We have thus constructed Λ as a boost followed by a rotation using the right polar decomposition

$$\Lambda = OP = \exp(i\boldsymbol{\alpha}' \cdot \mathbf{J})\exp(i\boldsymbol{\eta}' \cdot \mathbf{K}). \tag{1.2.157}$$

To express Λ as a rotation followed by a boost we simply need to perform the left polar decomposition, where $P' \equiv \sqrt{\Lambda\Lambda^T}$ is a Lorentz boost, $O' \equiv P'^{-1}\Lambda$ is a rotation and the restricted Lorentz transformations is $\Lambda = P'O'$.

1.2.4 Poincaré Group

It will be convenient here for us to expand the notation that we are using for Poincaré and Lorentz transformations and for translations to include the relevant arguments, i.e., $p(\Lambda, a)$, $\ell(\Lambda)$ and $t(a)$, respectively, such that

$$x' = p(\Lambda, a)(x) \equiv \Lambda x + a, \quad x' = \ell(\Lambda)(x) \equiv \Lambda x \ \text{ and } \ x' = t(a)(x) \equiv x + a. \tag{1.2.158}$$

Note, for example, that any Poincaré transformation can be expressed as a Lorentz transformation followed by a translation or vice versa:

$$p(\Lambda, a)(x) = t(a)\ell(\Lambda)(x) = \Lambda x + a$$
$$= \ell(\Lambda)t(\Lambda^{-1}a)(x). \tag{1.2.159}$$

It is easily seen, for example, that

$$\ell(\Lambda)t(a)\ell^{-1}(\Lambda)(x) = \Lambda(\Lambda^{-1}x + a) = x + \Lambda a = t(\Lambda a)(x) \tag{1.2.160}$$

and hence as shown earlier that

$$p(\Lambda, b)t(a)p^{-1}(\Lambda, b)(x) = t(b)\ell(\Lambda)t(a)\ell^{-1}(\Lambda)t^{-1}(b)(x) = t(b)t(\Lambda a)t^{-1}(b)(x)$$
$$= t(\Lambda a)(x), \tag{1.2.161}$$

which is still a translation and so the translations are a normal (i.e., invariant) subgroup of the Poincaré group.

Five-Dimenionsal Representation of the Poincaré Group

Using an obvious shorthand notation the Poincaré transformation $x' = t(a)\ell(\Lambda)x = \Lambda x + a$ can be written in a five-dimensional form as

$$\left(\frac{x'}{1}\right) = \left(\begin{array}{c|c} I & a \\ \hline 0 & 1 \end{array}\right)\left(\begin{array}{c|c} \Lambda & 0 \\ \hline 0 & 1 \end{array}\right)\left(\frac{x}{1}\right) = \left(\begin{array}{c|c} \Lambda & a \\ \hline 0 & 1 \end{array}\right)\left(\frac{x}{1}\right) = \left(\frac{\Lambda x + a}{1}\right). \tag{1.2.162}$$

Define $\tilde{x} \equiv (x^0, x^1, x^2, x^3, 1)$ for any spacetime point x. We can write the five-dimensional representation of the Poincaré transformation in a compact form as

$$\tilde{x}' = \tilde{t}(a)\tilde{\ell}(\Lambda)\tilde{x} \quad \text{with} \quad \tilde{t}(a) \equiv \left(\begin{array}{c|c} I & a \\ \hline 0 & 1 \end{array}\right) \quad \text{and} \quad \tilde{\ell}(\Lambda) \equiv \left(\begin{array}{c|c} \Lambda & 0 \\ \hline 0 & 1 \end{array}\right), \tag{1.2.163}$$

where $\tilde{t}(a)$ and $\tilde{\ell}(\Lambda)$ are the five-dimensional representations of the translations and the Lorentz transformations, respectively. We define the generators using Eq. (A.7.3),

$$\tilde{t}(a) = e^{-ia\cdot\tilde{P}} = \left(\begin{array}{c|c} I & a \\ \hline 0 & 1 \end{array}\right) \quad \text{and} \quad \tilde{\ell}(\Lambda) = e^{-(i/2)\omega_{\mu\nu}\tilde{M}^{\mu\nu}} = \left(\begin{array}{c|c} \Lambda & 0 \\ \hline 0 & 1 \end{array}\right), \tag{1.2.164}$$

where the generators for translations and restricted Lorentz transformations in this representation can be obtained using Eq. (A.7.2) and are

$$\tilde{P}^\mu = i\left.\frac{\partial\tilde{t}(a)}{\partial a_\mu}\right|_{a=0} = \left(\begin{array}{c|c} 0 & n^\mu \\ \hline 0 & 0 \end{array}\right) \quad \text{and} \quad \tilde{M}^{\mu\nu} = i\left.\frac{\partial\tilde{\ell}(\Lambda)}{\partial\omega_{\mu\nu}}\right|_{\omega=0} = \left(\begin{array}{c|c} M^{\mu\nu} & 0 \\ \hline 0 & 0 \end{array}\right), \tag{1.2.165}$$

where $n^0 \equiv (1,0,0,0)$, $n^1 \equiv (0,1,0,0)$, $n^2 \equiv (0,0,1,0)$ and $n^3 \equiv (0,0,0,1)$.

Obviously, the $\tilde{M}^{\mu\nu}$ will satisfy the same Lie algebra as the $M^{\mu\nu}$ given in Eq. (1.2.46). It is readily seen that $\tilde{t}(a)\tilde{t}(b) = \tilde{t}(a+b) = \tilde{t}(b)\tilde{t}(a)$ as must be the case for any representation of the translations; i.e., the translations commute in any representation, $[\tilde{t}(a), \tilde{t}(b)] = 0$. Since the $\tilde{t}$ commute for arbitrary a and b, then the generators $\tilde{P}^\mu$ must also commute with each other. Let the α and β indices take values $\alpha, \beta \in \{0, 1, 2, 3, 4\}$ with the spacetime indices μ, ν, ρ, σ having their usual range $\mu, \nu, \rho, \sigma \in \{0, 1, 2, 3\}$. Defining $g^{4\mu} = g^{\mu 4} = 0$ leads to

$$(\tilde{P}_\mu)^\alpha{}_\beta = \delta_{\beta 4}\delta^\alpha{}_\mu, \quad (\tilde{P}^\mu)^\alpha{}_\beta = \delta_{\beta 4}g^{\alpha\mu}, \quad (\tilde{M}^{\mu\nu})^\alpha{}_\beta = i(g^{\mu\alpha}\delta^\nu{}_\beta - g^{\nu\alpha}\delta^\mu{}_\beta) \tag{1.2.166}$$

from which we can readily verify that $[\tilde{P}^\mu, \tilde{M}^{\rho\sigma}] = i(g^{\mu\rho}\tilde{P}^\sigma - g^{\mu\sigma}\tilde{P}^\rho)$.

In summary, we have seen that the Poincaré group has 10 generators with six of these being the $\tilde{M}^{\mu\nu}$ and with the other four being the $\tilde{P}^\mu$. We have then found the Lie algebra of the ten generators of the Poincaré group, which is valid in any representation. We write the Poincaré Lie algebra in an arbitrary representation as

$$[P^\mu, P^\nu] = 0, \qquad [P^\mu, M^{\rho\sigma}] = i(g^{\mu\rho}P^\sigma - g^{\mu\sigma}P^\rho),$$
$$[M^{\mu\nu}, M^{\rho\sigma}] = i\left(g^{\nu\rho}M^{\mu\sigma} - g^{\mu\rho}M^{\nu\sigma} - g^{\nu\sigma}M^{\mu\rho} + g^{\mu\sigma}M^{\nu\rho}\right), \qquad (1.2.167)$$

where the form of the generators P^μ and $M^{\mu\nu}$ depends on the representation.

Representations of derivatives acting on arbitrary functions are necessarily infinite dimensional. This five-dimensional representation is only possible because we restricted our attention to the trivial extension of spacetime, $\tilde{x}$, rather than to arbitrary functions of spacetime. Eq. (1.2.166) leads to some representation-specific results: (i) $\tilde{P}^\mu\tilde{P}^\nu = 0$ for all μ and ν, and (ii) $\tilde{P}^\mu\tilde{M}^{\rho\sigma} = 0$ and $\tilde{M}^{\rho\sigma}\tilde{P}^\mu \neq 0$.

Finally, since the $\tilde{P}^\mu$ are not Hermitian matrices, then the $\tilde{t}(a)$ are not a unitary representation of the translation group. Similarly, since the $\tilde{M}^{\mu\nu}$ are not Hermitian except in the case of pure rotations, then in general the $\tilde{\ell}(\Lambda)$ are not unitary either.

Unitary Representation of the Poincaré Group

A more common approach to deriving the Poincaré Lie algebra is to study the Poincaré transformations applied to the quantum mechanical description of the motion of a single spinless (i.e., Lorentz scalar) particle. This will lead to an infinite-dimensional unitary representation of the Poincaré group with Hermitian generators. As we saw for the rotation generators the convention in quantum mechanics is to include an additional factor of $\hbar$ along with the generators $\mathbf{P}, \mathbf{J}, \mathbf{L}$ so that they correspond to the momentum, angular momentum and orbital angular momentum quantum operators, respectively. This leads to an additional factor of $\hbar$ in the Lie algebra and to a factor of $1/\hbar$ in the exponentials of these generators.

We begin with the familiar quantum mechanical definition of the momentum and orbital angular momentum operators in coordinate-space representation

$$\mathbf{P} \equiv -i\hbar\boldsymbol{\nabla} \qquad \text{and} \qquad \mathbf{J} = \mathbf{L} \equiv \mathbf{x} \times \mathbf{P} = \epsilon^{ijk}x^j P^k, \qquad (1.2.168)$$

where $\mathbf{x} \equiv (x^1, x^2, x^3) \equiv (x, y, z)$ and where the angular momentum associated with orbital motion is traditionally denoted as $\mathbf{L}$. The momentum operators $\mathbf{P}$ and the angular momentum operators $\mathbf{J} = \mathbf{L}$ are the familiar generators of three-dimensional translations and rotations, respectively, in quantum mechanics. They act on wavefunctions for a spinless particle, $\phi(ct, \mathbf{x}) = \phi(x)$. We are constructing an infinite-dimensional representation of the Poincaré group, since there are an infinite number of basis functions for the Hilbert space of square-integrable wavefunctions.

Using Eq. (1.2.71) and that the M^{ij} are antisymmetric in the indices i, j we see that in a common representation for the M, J and P we must identify

$$M^{ij} = x^i P^j - x^j P^i \quad \text{or equivalently} \quad J^i = \tfrac{1}{2}\epsilon^{ijk}M^{jk} = \epsilon^{ijk}x^j P^k. \qquad (1.2.169)$$

Using the quantum definition of the generators with the additional factor of $\hbar$ in each generator, the relativistic extension of Eqs. (1.2.168) and (1.2.169) is

$$P^\mu = (H/c, \mathbf{P}) = i\hbar\partial^\mu \qquad \text{and} \qquad M^{\mu\nu} = x^\mu P^\nu - x^\nu P^\mu, \qquad (1.2.170)$$

where we recall that $\partial/\partial x^i = \partial_i = \nabla^i$ and so $i\hbar\partial_\mu = (i\hbar\partial_0, i\hbar\boldsymbol{\nabla})$.

The Taylor expansion formula in three dimensions can be expressed in the form

$$\phi(\mathbf{x} + \mathbf{a}) = \sum_{j=1}^{\infty} \frac{1}{j!}(\mathbf{a} \cdot \nabla)^j \phi(x) = \exp(\mathbf{a} \cdot \nabla)\phi(\mathbf{x}) = \exp(i\mathbf{a} \cdot \mathbf{P}/\hbar)\phi(\mathbf{x}), \qquad (1.2.171)$$

where $\mathbf{P} = (-i\nabla)/\hbar$ is the coordinate-space representation of the momentum operator of quantum mechanics. This then generalizes to the Minkowski-space form

$$\phi(x + a) = \exp(a \cdot \partial)\phi(x) = \exp(-ia \cdot (i\hbar\partial)/\hbar)\phi(\mathbf{x}) = \exp(-ia \cdot P/\hbar)\phi(x), \qquad (1.2.172)$$

where $P_\mu \equiv i\hbar\partial_\mu = i\hbar(\partial_0, \nabla)$ is the four-momentum operator. Comparing with Eq. (1.2.38) we then see that the operator $\exp(ia \cdot P/\hbar)$ leads to an *active* spacetime translation by a^μ, $(x^\mu \to x'^\mu = x^\mu + a^\mu)$, since

$$\phi(x) \to \phi'(x) = \exp(ia \cdot P/\hbar)\phi(x) = \phi(x - a). \qquad (1.2.173)$$

Assuming doubly differentiable functions the order of derivatives is unimportant, $\partial_\mu\partial_\nu\phi(x) = \partial_\nu\partial_\mu\phi(x)$, and so the P^μ generators commute with each other, i.e., $[P^\mu, P^\nu] = 0$ as required in the Poincaré Lie algebra of Eq. (1.2.167).

Consider an infinitesimal Lorentz transformation,

$$\begin{aligned}
\exp\left(-\tfrac{i}{2}d\omega_{\mu\nu}M^{\mu\nu}/\hbar\right)\phi(x) &= \left(1 - \tfrac{i}{2}d\omega_{\mu\nu}[x^\mu P^\nu - x^\nu P^\mu]/\hbar\right)\phi(x) \\
&= \left(1 - [d\omega^\mu{}_\nu x^\nu]\partial_\mu\right)\phi(x) = \phi(x^\mu - d\omega^\mu{}_\nu x^\nu) = \phi(\Lambda(d\omega)^{-1}x),
\end{aligned} \qquad (1.2.174)$$

where we have used Eq. (1.2.39). We have used the result that

$$\phi(x + da(x)) = [1 + da(x) \cdot \partial]\phi(x) + \mathcal{O}\left((da)^2\right), \qquad (1.2.175)$$

which is the Taylor expansion result for an infinitesimal and spacetime-dependent translation $x^\mu \to x'^\mu = x^\mu + da^\mu(x)$. For Lorentz transformations we have $da(x)^\mu = d\omega^\mu{}_\nu x^\nu$. After an infinite number of the infinitesimal transformations in Eq. (1.2.174) and after using Eq. (A.8.5) and $\Lambda = \exp(\omega)$ from Eq. (1.2.45) we find

$$\phi(x) \to \phi'(x) \equiv \exp\left(-\tfrac{i}{2}\omega_{\mu\nu}M^{\mu\nu}/\hbar\right)\phi(x) = \phi(\Lambda^{-1}x). \qquad (1.2.176)$$

We can choose ω to correspond to either a passive or active Lorentz transformation as described in Eqs. (1.2.151) and (1.2.152), respectively.

Consider an infinitesimal active boost $d\eta$ in the x-direction, which according to Eq. (1.2.152) corresponds to $d\eta = d\omega_{01} = -d\omega_{10}$ with all other $d\omega_{\mu\nu}$ equal to zero. The Lorentz transformation matrix $\Lambda(d\eta)$ is given by Eq. (1.2.96) but with $\eta \to -d\eta$ for an active infinitesimal boost. Note that working to leading order in $d\eta$ we have $\cosh(d\eta) = \gamma(dv) = 1 + \mathcal{O}((dv)^2)$ and $d\eta + \mathcal{O}(d\eta^3) = \sinh(d\eta) = d\beta\gamma(dv) = d\beta = dv/c$, where we have used Eqs. (1.2.92) and (1.2.93) with $\eta \to d\eta$ and $\beta \to d\beta = dv/c$. For this active boost we find using Eq. (1.2.174) that

$$\begin{aligned}
\phi(x) \to \phi'(x) &= e^{-id\eta\mathbf{n}\cdot\mathbf{K}/\hbar}\phi(x) = e^{-id\eta K^1/\hbar}\phi(x) = (1 - (i/2)d\omega_{\mu\nu}M^{\mu\nu}/\hbar)\phi(x) \\
&= \phi(x^\mu - d\omega^\mu{}_\nu x^\nu) = \phi(x^0 - d\omega^0{}_1 x^1, x^1 - d\omega^1{}_0 x^0, x^2, x^3) \\
&= \phi(x^0 - d\eta x^1, x^1 - d\eta x^0, x^2, x^3) = \phi(\Lambda(d\eta)^{-1}x).
\end{aligned} \qquad (1.2.177)$$

From ordinary quantum mechanics we know that $[L^i, L^j] = i\hbar\epsilon^{ijk}L^k$, which follows from canonical commutation relations of nonrelativistic quantum mechanics $[x^i, P^j] = i\hbar\delta^{ij}$. The generalization of the canonical commutation relations follows from the definition of P^μ,

$$[P^\mu, x^\nu] = i\hbar g^{\mu\nu}, \qquad (1.2.178)$$

which leads to the expected result,

$$[M^{\mu\nu}, M^{\rho\sigma}] = i\hbar \left(g^{\nu\rho}M^{\mu\sigma} - g^{\mu\rho}M^{\nu\sigma} - g^{\nu\sigma}M^{\mu\rho} + g^{\mu\sigma}M^{\nu\rho}\right). \qquad (1.2.179)$$

Proof: The result follows from repeated use of the identity $P^{\mu}x^{\nu} = i\hbar g^{\mu\nu} + x^{\nu}P^{\mu}$ in the expansion of $[M^{\mu\nu}, M^{\rho\sigma}]$ to give canceling terms of the form $x^{\mu}x^{\rho}P^{\nu}P^{\sigma} = x^{\rho}x^{\mu}P^{\sigma}P^{\nu}$ and residual terms $i\hbar g^{\nu\rho}x^{\mu}P^{\sigma}$ and $-ig\hbar^{\sigma\mu}x^{\rho}P^{\nu}$.

$$\begin{aligned}
[M^{\mu\nu}, M^{\rho\sigma}] &= \{(x^{\mu}P^{\nu} - x^{\nu}P^{\mu})(x^{\rho}P^{\sigma} - x^{\sigma}P^{\rho})\} - \{\mu \leftrightarrow \rho, \nu \leftrightarrow \sigma\} \\
&= i\hbar \left(g^{\nu\rho}x^{\mu}P^{\sigma} - g^{\sigma\mu}x^{\rho}P^{\nu} - g^{\nu\sigma}x^{\mu}P^{\rho} + g^{\sigma\nu}x^{\rho}P^{\mu}\right. \\
&\qquad \left. -g^{\mu\rho}x^{\nu}P^{\sigma} + g^{\mu\rho}x^{\sigma}P^{\nu} + g^{\mu\sigma}x^{\nu}P^{\rho} - g^{\rho\nu}x^{\sigma}P^{\mu}\right) \\
&= i\hbar \left(g^{\nu\rho}M^{\mu\sigma} - g^{\mu\rho}M^{\nu\sigma} - g^{\nu\sigma}M^{\mu\rho} + g^{\mu\sigma}M^{\nu\rho}\right). \qquad (1.2.180)
\end{aligned}$$

Finally we can use Eq. (1.2.178) to obtain

$$[P^{\mu}, M^{\rho\sigma}] = P^{\mu}(x^{\rho}P^{\sigma} - x^{\sigma}P^{\rho}) - (x^{\rho}P^{\sigma} - x^{\sigma}P^{\rho})P^{\mu} = i\hbar(g^{\mu\rho}P^{\sigma} - g^{\mu\sigma}P^{\rho}). \qquad (1.2.181)$$

The Poincaré Lie algebra for the generators in this representation is then verified.

The x^{μ} and P^{μ} are Hermitian operators as we know from quantum mechanics and therefore so are the $M^{\mu\nu}$, since

$$(M^{\mu\nu})^{\dagger} = (x^{\mu}P^{\nu} - x^{\nu}P^{\mu})^{\dagger} = (P^{\nu}x^{\mu} - P^{\mu}x^{\nu}) = (x^{\mu}P^{\nu} - x^{\nu}P^{\mu}) = M^{\mu\nu}, \qquad (1.2.182)$$

where we have used again $P^{\mu}x^{\nu} = i\hbar g^{\mu\nu} + x^{\nu}P^{\mu}$. So the translations and Lorentz transformations are unitary in this representation and can be expressed as

$$U(a) \equiv t(a) = \exp(ia \cdot P/\hbar) \quad \text{and} \quad U(\Lambda) \equiv \ell(\Lambda) = \exp(-\tfrac{i}{2}\omega_{\mu\nu}M^{\mu\nu}/\hbar), \qquad (1.2.183)$$

where $U(a)^{\dagger} = U(a)^{-1}$ and $U(\Lambda)^{\dagger} = U(\Lambda)^{-1}$.

The differential operators P^{μ} and $M^{\mu\nu}$ contain a factor of $\hbar$ that is compensated for by the $1/\hbar$ in the definition of the unitary representation of the Poincaré transformations. These transformations are therefore independent of the size of $\hbar$ as they must be, since the spacetime transformations of special relativity do not depend on the scale at which quantum effects become important.

1.2.5 Representation-Independent Poincaré Lie Algebra

Note that it will be notationally convenient to use natural units $\hbar = c = 1$ for the remainder of this chapter to avoid the unnecessary clutter of factors of $\hbar$ and c. These are easily restored using dimensional arguments.

We have now shown how to obtain the Poincaré Lie algebra in a five-dimensional representation that generalizes the four-dimensional defining representation of the Lorentz Lie algebra and in an infinite-dimensional unitary representation that generalizes the Lie algebra of the rotation group in quantum mechanics. It must be possible to deduce the Poincaré Lie algebra in any arbitrary representation based only on the properties of the Poincaré group itself. We will now do this.

Since translations commute, i.e., $[t(a), t(b)] = 0$ for all a, b with $t(a) \equiv \exp(ia \cdot P)$, then it follows that the generators of the translations must commute:

$$[P^{\mu}, P^{\nu}] = 0. \qquad (1.2.184)$$

Second, since the Poincaré group has the property that

$$\ell(\Lambda)t(a)\ell(\Lambda)^{-1} = t(\Lambda a) \tag{1.2.185}$$

as shown in Eq. (1.2.160), then

$$\ell(\Lambda)\exp\left(ia \cdot P\right)\ell(\Lambda)^{-1} = \exp\left(i(\Lambda a) \cdot P\right), \tag{1.2.186}$$

$$(\Lambda a)^\mu \equiv \Lambda^\mu{}_\nu a^\nu \quad\text{or equivalently}\quad (\Lambda a)_\mu = a_\nu(\Lambda^{-1})^\nu{}_\mu \tag{1.2.187}$$

using Eq. (1.2.20). Since a is arbitrary, then both sides must agree term by term in the exponential expansion. Equating the first-order terms in a gives

$$\ell(\Lambda)(a \cdot P)\ell(\Lambda)^{-1} = (\Lambda a) \cdot P = (\Lambda a)_\mu P^\mu = a_\nu(\Lambda^{-1})^\nu{}_\mu P^\mu = a \cdot (\Lambda^{-1}P), \tag{1.2.188}$$

which leads to

$$\ell(\Lambda)P^\mu\ell(\Lambda)^{-1} = (\Lambda^{-1})^\mu{}_\nu P^\nu \quad\Rightarrow\quad \ell(\Lambda)^{-1}P^\mu\,\ell(\Lambda) = \Lambda^\mu{}_\nu P^\nu. \tag{1.2.189}$$

This is the equation that defines how the translation generators transform under Lorentz transformations. We can now use Eqs. (1.2.43) and (1.2.45) to write

$$\Lambda^\mu{}_\nu = [\exp(\omega)]^\mu{}_\nu = [\exp(-i\tfrac{1}{2}\omega_{\rho\sigma}M_d^{\rho\sigma})]^\mu{}_\nu, \quad \ell(\Lambda) = \exp(-i\tfrac{1}{2}\omega_{\rho\sigma}M^{\rho\sigma}), \tag{1.2.190}$$

where the $M_d^{\mu\nu}$ are the generators in the *defining* vector representation of the Lorentz group (i.e., the 4×4 matrices given in Eq. (1.2.41)) and the $M^{\mu\nu}$ are the generators of the arbitrary representation that we are considering. Expanding the exponentials on each side of Eq. (1.2.189) and equating terms first-order in the arbitrary $\omega_{\rho\sigma}$ leads to

$$[P^\mu, M^{\rho\sigma}] = (M_d^{\rho\sigma})^\mu{}_\nu P^\nu = i\left(g^{\mu\rho}P^\sigma - g^{\mu\sigma}P^\rho\right), \tag{1.2.191}$$

where we have used Eq. (1.2.41) for $M_d^{\rho\sigma}$.

Since the Lorentz transformations form a group, then for every Λ and Λ' there exists some Λ'' such that

$$\Lambda^\mu{}_\rho(\Lambda')^\rho{}_\sigma(\Lambda^{-1})^\sigma{}_\nu = (\Lambda'')^\mu{}_\nu. \tag{1.2.192}$$

Using Eq. (1.2.190) for Λ' and Λ'' we can write

$$\Lambda^\mu{}_\rho\left[\exp(\omega')\right]^\rho{}_\sigma(\Lambda^{-1})^\sigma{}_\nu = \sum_{j=0}^\infty \tfrac{1}{j!}\left[(\Lambda\omega'\Lambda^{-1})^j\right]^\mu{}_\nu = [\exp(\omega'')]^\mu{}_\nu, \tag{1.2.193}$$

$$\text{where}\quad \omega'' = \Lambda\omega'\Lambda^{-1} \quad\text{or}\quad (\omega'')^\mu{}_\nu = \Lambda^\mu{}_\rho\omega'^\rho{}_\sigma(\Lambda^{-1})^\sigma{}_\nu. \tag{1.2.194}$$

So ω' transforms as a Lorentz tensor to ω'' as described in Eq. (1.2.22). Then the Lorentz group property means that

$$(\omega'')^{\mu\nu} = (\Lambda\Lambda\omega')^{\mu\nu} \equiv \Lambda^\mu{}_\rho\Lambda^\nu{}_\sigma(\omega')^{\rho\sigma} \quad\text{or equivalently}$$
$$(\omega'')_{\mu\nu} = (\Lambda\Lambda\omega')_{\mu\nu} = (\omega')_{\rho\sigma}(\Lambda^{-1})^\rho{}_\mu(\Lambda^{-1})^\sigma{}_\nu, \tag{1.2.195}$$

which is a result analogous to Eq. (1.2.187).

Any representation of the Lorentz group must satisfy Eq. (1.2.192) and so

$$\ell(\Lambda)\ell(\Lambda')\ell(\Lambda)^{-1} = \ell(\Lambda'') \quad\text{with}\quad (\omega'')^{\mu\nu} = (\Lambda\Lambda\omega')^{\mu\nu}. \tag{1.2.196}$$

Then in analogy with Eq. (1.2.186) we have

$$\ell(\Lambda)\exp\left(-i\tfrac{1}{2}\omega'_{\rho\sigma}M^{\rho\sigma}\right)\ell(\Lambda)^{-1} = \exp\left(-i\tfrac{1}{2}\omega''_{\rho\sigma}M^{\rho\sigma}\right) = \exp\left(-i\tfrac{1}{2}(\Lambda\Lambda\omega')_{\rho\sigma}M^{\rho\sigma}\right).$$

Equating powers of ω' we find at first order in ω'

$$\ell(\Lambda)\left(\omega'_{\rho\sigma}M^{\rho\sigma}\right)\ell(\Lambda)^{-1} = (\Lambda\Lambda\omega')_{\rho\sigma}M^{\rho\sigma} = \omega'_{\rho\sigma}(\Lambda^{-1}\Lambda^{-1}M)^{\rho\sigma}, \tag{1.2.197}$$

which is analogous to Eq. (1.2.188). Since the ω' are arbitrary, then it follows that

$$\ell(\Lambda)M^{\mu\nu}\ell(\Lambda)^{-1} = (\Lambda^{-1})^{\mu}{}_{\rho}(\Lambda^{-1})^{\nu}{}_{\sigma}M^{\rho\sigma} \tag{1.2.198}$$

or with the replacement $\Lambda \to \Lambda^{-1}$ we can write

$$\ell(\Lambda)^{-1}M^{\mu\nu}\ell(\Lambda) = \Lambda^{\mu}{}_{\tau}\Lambda^{\nu}{}_{\lambda}M^{\tau\lambda}, \tag{1.2.199}$$

which is the analog of Eq. (1.2.189) and defines how the Lorentz transformation generators transform under Lorentz transformations in any representation. Using Eq. (1.2.190) and the fact that the ω are arbitrary we can equate terms at first-order in $\omega_{\rho\sigma}$ in the expansions of both sides of Eq. (1.2.199). This then gives

$$
\begin{aligned}
[M^{\mu\nu}, M^{\rho\sigma}] &= \left\{(M_d^{\rho\sigma})^{\mu}{}_{\tau}\delta^{\nu}{}_{\lambda} + \delta^{\mu}{}_{\tau}(M_d^{\rho\sigma})^{\nu}{}_{\lambda}\right\}M^{\tau\lambda} \\
&= i\left(g^{\rho\mu}\delta^{\sigma}{}_{\tau} - g^{\sigma\mu}\delta^{\rho}{}_{\tau}\right)M^{\tau\nu} + i\left(g^{\rho\nu}\delta^{\sigma}{}_{\lambda} - g^{\sigma\nu}\delta^{\rho}{}_{\lambda}\right)M^{\mu\lambda} \\
&= i\left(g^{\nu\rho}M^{\mu\sigma} - g^{\mu\rho}M^{\nu\sigma} - g^{\nu\sigma}M^{\mu\rho} + g^{\mu\sigma}M^{\nu\rho}\right),
\end{aligned}
\tag{1.2.200}
$$

which is the Lorentz Lie algebra. We have now obtained the full Poincaré Lie algebra of Eq. (1.2.167) in a representation-independent manner as required.

We can extend the alternative expression for the Lorentz Lie algebra given in Eq. (1.2.144) to the full Poincaré Lie algebra using Eq. (1.2.191). It is useful to temporarily restore factors of c here for future reference. We find

$$
\begin{gathered}
\left[J^i, J^j\right] = i\epsilon^{ijk}J^k, \quad \left[J^i, K^j\right] = i\epsilon^{ijk}K^k, \quad \left[K^i, K^j\right] = -i\epsilon^{ijk}J^k, \\
\left[H, P^i\right] = 0, \quad \left[P^i, P^j\right] = 0, \\
\left[H, J^i\right] = 0, \quad \left[H/c, K^i\right] = iP^i, \quad \left[P^i, J^j\right] = i\epsilon^{ijk}P^k, \quad \left[P^i, K^j\right] = i\delta^{ij}H/c,
\end{gathered}
\tag{1.2.201}
$$

where P^0 is the generator of time translations and so we identify $H/c \equiv P^0$.

Proof of Eq. (1.2.201): The first line is a restatement of Eq. (1.2.144). The second line is a restatement of $[P^\mu, P^\nu] = 0$. The third line follows from using Eq. (1.2.191) to obtain the following results:

(i) $\left[H/c, J^i\right] = \frac{1}{2}\epsilon^{ijk}\left[P^0, M^{jk}\right] = 0$;
(ii) $\left[H/c, K^i\right] = \left[P^0, M^{0i}\right] = iP^i$;
(iii) $\left[P^i, J^j\right] = \frac{1}{2}\epsilon^{j\ell k}\left[P^i, M^{\ell k}\right] = \frac{1}{2}\epsilon^{j\ell k}i\left(g^{i\ell}P^k - g^{ik}P^\ell\right) = i\epsilon^{ijk}P^k$; and
(iv) $\left[P^i, K^j\right] = \left[P^i, M^{0j}\right] = i\left(-g^{ij}\right)H/c = i\delta^{ij}H/c$.

It is straightforward to use these results to show that the Casimir invariants of the Lorentz group, $\mathbf{J} \cdot \mathbf{K}$ and $\mathbf{J}^2 - \mathbf{K}^2$, do not commute with the P^μ and so are not Casimir invariants of the Poincaré group. In order to classify the representations of the Poincaré group we need to establish what its Casimir invariants are.

Pauli-Lubanski Pseudovector (Axial Vector)

The angular momentum $\mathbf{J}$ is not a convenient quantity to discuss in a relativistic framework. The Pauli-Lubanski pseudovector provides the basis for a relativistic description of angular momentum as we will see. It is defined as

$$W_\mu \equiv -\tfrac{1}{2}\epsilon_{\mu\nu\rho\sigma}M^{\nu\rho}P^\sigma \quad \text{or} \quad W^\mu \equiv -\tfrac{1}{2}\epsilon^{\mu\nu\rho\sigma}M_{\nu\rho}P_\sigma. \tag{1.2.202}$$

The terms "pseudovector" and "axial vector" are used interchangeably in the literature. We wish to demonstrate that W_μ transforms as an axial vector.

We begin by recalling from Eq. (1.2.137) that $\epsilon^{\mu\nu\rho\sigma}$ is a rank-four pseudotensor. Under a Lorentz transformation we find

$$\begin{aligned}
W_\mu \to W'_\mu &= -\tfrac{1}{2}\epsilon'_{\mu\nu\rho\sigma}M'^{\nu\rho}P'^\sigma \\
&= -\tfrac{1}{2}\det(\Lambda)\epsilon_{\alpha\beta\gamma\delta}(\Lambda^{-1})^\alpha{}_\mu(\Lambda^{-1})^\beta{}_\nu(\Lambda^{-1})^\gamma{}_\rho(\Lambda^{-1})^\delta{}_\sigma\Lambda^\nu{}_\lambda\Lambda^\rho{}_\eta\Lambda^\sigma{}_\tau M^{\lambda\eta}P^\tau \\
&= -\tfrac{1}{2}\det(\Lambda)\epsilon_{\alpha\beta\gamma\delta}(\Lambda^{-1})^\alpha{}_\mu M^{\beta\gamma}P^\delta = \det(\Lambda)(\Lambda^{-1})^\alpha{}_\mu W_\alpha
\end{aligned} \tag{1.2.203}$$

or equivalently $W^\mu \to W'^\mu = \det(\Lambda)\Lambda^\mu{}_\alpha W^\alpha$ and so W_μ is an axial vector (pseudovector). From the Poincaré Lie algebra in Eq. (1.2.167) we have

$$[W^\mu, P^\nu] = P^\mu W_\mu = 0, \quad [W^\mu, M^{\rho\sigma}] = i(g^{\mu\rho}W^\sigma - g^{\mu\sigma}W^\rho), \tag{1.2.204}$$
$$[W_\mu, W_\nu] = i\epsilon_{\mu\nu\rho\sigma}W^\rho P^\sigma.$$

Since $P^\mu W_\mu = 0$ then W^μ has *only three* independent components.

From the Poincaré Lie algebra and the above equations it follows that both P^2 and W^2 commute with all ten Poincaré generators

$$[P^2, P^\mu] = 0, \quad [P^2, M^{\mu\nu}] = 0, \quad [W^2, P^\mu] = 0, \quad [W^2, M^{\mu\nu}] = 0 \tag{1.2.205}$$

and so they are Casimir invariants of the Poincaré group. These are the only two.

Proof: Let us prove the results in Eqs. (1.2.204) and (1.2.205) in sequence:

(i) The first result arises since

$$\begin{aligned}
[P^\mu, W_\nu] &= -\tfrac{1}{2}\epsilon_{\nu\beta\gamma\delta}[P^\mu, M^{\beta\gamma}P^\delta] = -\tfrac{1}{2}\epsilon_{\nu\beta\gamma\delta}\{M^{\beta\gamma}[P^\mu, P^\delta] + [P^\mu, M^{\beta\gamma}]P^\delta\} \\
&= -\tfrac{1}{2}\epsilon_{\nu\beta\gamma\delta}\{i(g^{\mu\beta}P^\gamma - g^{\mu\gamma}P^\beta)P^\delta\} = 0.
\end{aligned} \tag{1.2.206}$$

(iia) The indirect proof of the second result is simple: Combining the results of Eqs. (1.2.189) and (1.2.199) with the definition of W^μ and using Eq. (1.2.203), we find that $\ell(\Lambda)^{-1}W^\mu\ell(\Lambda)$ $= \det(\Lambda)\Lambda^\mu{}_\nu W^\nu$. Considering an infinitesimal Lorentz transformation, we have $\det(\Lambda) = 1$ and so the arguments that lead from Eq. (1.2.189) to Eq. (1.2.191) will again apply. Hence we obtain $[W^\mu, M^{\rho\sigma}] = i(g^{\mu\rho}W^\sigma - g^{\mu\sigma}W^\rho)$.

(iib) To prove this result directly from the Poincaré Lie algebra requires the Schouten identity, which is the statement that any quantity completely antisymmetric in its indices must vanish when the number of indices exceeds the number of values available for each index. It follows that

$$g_{\lambda\mu}\epsilon_{\nu\rho\sigma\tau} + g_{\lambda\nu}\epsilon_{\rho\sigma\tau\mu} + g_{\lambda\rho}\epsilon_{\sigma\tau\mu\nu} + g_{\lambda\sigma}\epsilon_{\tau\mu\nu\rho} + g_{\lambda\tau}\epsilon_{\mu\nu\rho\sigma} = 0, \tag{1.2.207}$$

since the left-hand side is seen to be antisymmetric under the pairwise interchange of any two of the five spacetime indices $\mu, \nu, \rho, \sigma, \tau$. We find

$$[M_{\mu\nu}, W_\rho] = -\tfrac{1}{2}\epsilon_{\rho\alpha\beta\gamma}[M_{\mu\nu}, M^{\alpha\beta}P^\gamma]$$

$$= -\tfrac{1}{2}\epsilon_{\rho\alpha\beta\gamma}g_{\mu\lambda}g_{\nu\tau}\{[M^{\lambda\tau}, M^{\alpha\beta}]P^\gamma + M^{\alpha\beta}[M^{\lambda\tau}, P^\gamma]\}$$

$$= \tfrac{1}{2i}\epsilon_{\rho\alpha\beta\gamma}g_{\mu\lambda}g_{\nu\tau}\{(g^{\tau\alpha}M^{\lambda\beta} - g^{\lambda\alpha}M^{\tau\beta} - g^{\tau\beta}M^{\lambda\alpha} - g^{\lambda\beta}M^{\tau\alpha})P^\gamma$$
$$- M^{\alpha\beta}(g^{\gamma\lambda}P^\tau - g^{\gamma\tau}P^\lambda)\}$$

$$= \tfrac{1}{2i}M^{\alpha\beta}P^\gamma[\epsilon_{\rho\nu\beta\gamma}g_{\mu\alpha} - \epsilon_{\rho\mu\beta\gamma}g_{\nu\alpha} - \epsilon_{\rho\beta\nu\gamma}g_{\mu\alpha} + \epsilon_{\rho\beta\mu\gamma}g_{\nu\alpha} - \epsilon_{\rho\alpha\beta\mu}g_{\nu\gamma} + \epsilon_{\rho\alpha\beta\nu}g_{\mu\gamma}]$$

$$= \tfrac{1}{2i}M^{\alpha\beta}P^\gamma[(g_{\mu\alpha}\epsilon_{\beta\gamma\rho\nu} + g_{\mu\beta}\epsilon_{\gamma\rho\nu\alpha} + g_{\mu\gamma}\epsilon_{\rho\nu\alpha\beta})$$
$$- (g_{\nu\alpha}\epsilon_{\beta\gamma\rho\mu} + g_{\nu\beta}\epsilon_{\gamma\rho\mu\alpha} + g_{\nu\gamma}\epsilon_{\rho\mu\alpha\beta})]$$

$$= \tfrac{1}{2i}M^{\alpha\beta}P^\gamma[(-g_{\mu\rho}\epsilon_{\nu\alpha\beta\gamma} - g_{\mu\nu}\epsilon_{\alpha\beta\gamma\rho}) - (-g_{\nu\rho}\epsilon_{\mu\alpha\beta\gamma} - g_{\nu\mu}\epsilon_{\alpha\beta\gamma\rho})]$$

$$= \tfrac{1}{2i}M^{\alpha\beta}P^\gamma[-g_{\mu\rho}\epsilon_{\nu\alpha\beta\gamma} + g_{\nu\rho}\epsilon_{\mu\alpha\beta\gamma}] = i(g_{\nu\rho}W_\mu - g_{\mu\rho}W_\nu). \tag{1.2.208}$$

In moving from the fourth to the fifth line we used the fact that only that part of $[\cdots]$ antisymmetric under $\alpha \leftrightarrow \beta$ survives given that $M^{\alpha\beta}$ is antisymmetric, which is how a term involving $g_{\mu\beta}$ on the fifth line emerged from the second of the two $g_{\mu\alpha}$ terms on the fourth line. In moving from the fifth to the sixth line we used the Schouten identity in Eq. (1.2.207) twice. A simple rearrangement of the last line of the above gives the second result in Eq. (1.2.204).

(iii) The third result follows since

$$[W_\nu, W_\nu] = -\tfrac{1}{2}\epsilon_{\mu\alpha\beta\gamma}[M^{\alpha\beta}P^\gamma, W_\nu] = -\tfrac{1}{2}\epsilon_{\mu\alpha\beta\gamma}\{M^{\alpha\beta}[P^\gamma, W_\nu] + [M^{\alpha\beta}, W_\nu]P^\gamma\}$$
$$= -\tfrac{1}{2}\epsilon_{\mu\alpha\beta\gamma}\{i(\delta_\nu^\beta W^\alpha - \delta_\nu^\alpha W^\beta)P^\gamma\} = i\epsilon_{\mu\nu\beta\gamma}W^\beta P^\gamma. \tag{1.2.209}$$

(iv) Follows immediately from the definition of W_μ.

(v) The momentum generator commutes with itself and so $[P^2, P^\mu] = 0$ and from Eq. (1.2.191) we have $[P^2, M^{\mu\nu}] = P_\rho[P^\rho, M^{\mu\nu}] + [P^\rho, M^{\mu\nu}]P_\rho = 0$; and

(vi) $[W^2, P^\mu] = W_\nu[W^\nu, P^\mu] + [W^\nu, P^\mu]W_\nu = 0$; $[W^2, M^{\mu\nu}] = W_\rho[W^\rho, M^{\mu\nu}]$ $+ [W^\rho, M^{\mu\nu}]W_\rho = i\{(W^\mu W^\nu - W^\nu W^\mu) + (W^\nu W^\mu - W^\mu W^\nu)\} = 0.$

We can rewrite the Pauli-Lubanski pseudovector in terms of the spacetime translation generators P^μ, the rotation generators $\mathbf{J}$ and the boost generators $\mathbf{K}$ as

$$W^0 = \mathbf{J} \cdot \mathbf{P} \quad \text{and} \quad \mathbf{W} = \mathbf{J}P^0 + (\mathbf{K} \times \mathbf{P}). \tag{1.2.210}$$

The commutation relations for the Pauli-Lubanski pseudovector W^μ and the six generators $\mathbf{J}$ and $\mathbf{K}$ follow from the commutation relations in Eq. (1.2.204),

$$[W^0, J^j] = 0, \ [W^0, K^j] = iW^j, \ [W^i, J^j] = i\epsilon^{ijk}W^k, \ [W^i, K^j] = iW^0\delta^{ij}. \tag{1.2.211}$$

Proof:
(i) $W^0 = -\tfrac{1}{2}\epsilon^{0jk\ell}M_{jk}P_\ell = \tfrac{1}{2}\epsilon^{0\ell jk}M^{jk}P^\ell = J^\ell P^\ell = \mathbf{J} \cdot \mathbf{P}.$

(ii) $W^i = -\tfrac{1}{2}\epsilon^{i\mu\nu\rho}M_{\mu\nu}P_\rho$
$= -\tfrac{1}{2}\epsilon^{ijk0}M_{jk}P_0 - \tfrac{1}{2}\epsilon^{i0k\ell}M_{0k}P_\ell - \tfrac{1}{2}\epsilon^{ij0\ell}M_{j0}P_\ell$
$= \tfrac{1}{2}\epsilon^{0ijk}M^{jk}P^0 + \epsilon^{0ik\ell}M^{0k}P^\ell = J^i P^0 + \epsilon^{0ik\ell}K^k P^\ell = J^i P^0 + (\mathbf{K} \times \mathbf{P})^i.$

(iii) $[W^0, J^j] = \tfrac{1}{2}\epsilon^{jk\ell}[W^0, M^{k\ell}] = 0;$ (iv) $[W^0, K^j] = [W^0, M^{0j}] = ig^{00}W^j = iW^j.$

(iv) $[W^0, K^j] = [W^0, M^{0j}] = ig^{00}W^j = iW^j.$

(v) $[W^i, J^j] = \tfrac{1}{2}\epsilon^{jk\ell}[W^i, M^{k\ell}] = \tfrac{1}{2}\epsilon^{jk\ell}i(g^{ik}W^\ell - g^{i\ell}W^k) = i\epsilon^{ijk}W^k.$

(vi) $[W^i, K^j] = [W^i, M^{0j}] = i(-g^{ij}W^0) = iW^0\delta^{ij}.$

1.3 Representations of the Lorentz Group

1.3.1 Labeling Representations of the Lorentz Group

A brief review of Lie groups and Lie algebras is provided in Sec. A.7. It can be shown that every $\Lambda \in SO^+(1,3)$ can be written as $\Lambda = \exp(\omega)$ for at least one $\omega \in \mathfrak{so}^+(1,3)$, which is a stronger form of Eq. (1.2.45). We say that the exponential map, $\exp : \mathfrak{so}^+(1,3) \to SO^+(1,3)$, is surjective (onto). The defining representation of the group $SL(2,\mathbb{C})$ is the set of complex 2×2 matrices with unit determinant. It can be shown that $SL(2,\mathbb{C})$ is a double cover of the restricted Lorentz group $SO^+(1,3)$ and so their Lie algebras are isomorphic, $\mathfrak{sl}(2,\mathbb{C}) \cong \mathfrak{so}^+(1,3)$. Since $SL(2,\mathbb{C})$ is simply connected it is the universal cover of $SO^+(1,3)$. Similarly, $SU(2)$ is a double cover of $SO(3)$ and so their Lie algebras are isomorphic, $\mathfrak{so}(3) \cong \mathfrak{su}(2)$. Since $SU(2)$ is simply connected it is the universal cover of $SO(3)$.

The Lie groups $SO^+(1,3)$ and $SO(4)$ correspond to the restricted (proper orthochronous) Lorentz transformations and the proper rotations in four dimensions, respectively. The direct product group $SU(2) \times SU(2)$ provides a double cover of $SO(4)$. So locally the Lie algebras are isomorphic and we can label the representations of $SO(4)$ using the same label used for $SU(2) \times SU(2)$, which is (j_a, j_b). Here j_a and j_b are the angular momentum representation labels with one for each copy of $SU(2)$. The Lie algebra of a direct product group is the direct sum of their Lie algebras, so the Lie algebra of both $SO(4)$ and $SU(2) \times SU(2)$ is $\mathfrak{su}(2) \oplus \mathfrak{su}(2)$.

Each of the real Lie algebras for $SO^+(1,3)$ and $SO(4)$ has six generators, which are $\mathbf{J}$, $\mathbf{K}$ and $\mathbf{J}$, $\mathbf{H} \equiv i\mathbf{K}$, respectively. So each of the complexified forms of the two groups has twelve real group parameters with the same twelve generators $\mathbf{J}$, $\mathbf{K}$, $i\mathbf{J}$ and $i\mathbf{K}$. So the two complexified Lie algebras are isomorphic and this is written as $\mathfrak{so}^+(1,3)^{\mathbb{C}} \cong \mathfrak{so}(4)^{\mathbb{C}}$ and their representations can be labeled in the same way. The realification of the twelve generators means keeping only six real group parameters and six of the twelve generators, but for $SO^+(1,3)$ and $SO(4)$ the six are chosen differently as described above. So we can specify the representations of each of $SO^+(1,3)$, $SL(2,\mathbb{C})$, $SO(4)$ and $SU(2) \times SU(2)$ using the labeling (j_a, j_b).

Define operators $\mathbf{A}$ and $\mathbf{B}$ in terms of the Lorentz group generators $\mathbf{J}$ and $\mathbf{K}$ as

$$\mathbf{A} \equiv \tfrac{1}{2}(\mathbf{J} + i\mathbf{K}) \quad \text{and} \quad \mathbf{B} \equiv \tfrac{1}{2}(\mathbf{J} - i\mathbf{K}). \tag{1.3.1}$$

It follows from the Lie algebra for $\mathbf{J}$ and $\mathbf{K}$ in Eq. (1.2.201) that

$$[A_i, A_j] = i\epsilon_{ijk}A_k, \quad [B_i, B_j] = i\epsilon_{ijk}B_k \quad \text{and} \quad [A_i, B_j] = 0, \tag{1.3.2}$$

since $\frac{1}{4}[J_i + iK_i, J_j - iK_j] = \frac{1}{4}([J_i, J_j] + i[J_i, K_j] - i[K_i, J_j] + [K_i, K_j]) = 0$ and $\frac{1}{4}[J_i \pm iK_i, J_j \pm iK_j] = \frac{1}{4}([J_i, J_j] \pm i[J_i, K_j] \pm i[H_i, J_j] - [K_i, K_j]) = i\epsilon_{ijk}\frac{1}{2}(J_k \pm iK_k)$. So $\mathbf{A}$ and $\mathbf{B}$ are generators for two different $SU(2)$ groups. We can rewrite the Casimirs for the Lorentz group given in Eq. (1.2.147) in terms of these two sets of generators using $C_1 = \mathbf{J}^2 - \mathbf{K}^2 = 2(\mathbf{A}^2 + \mathbf{B}^2)$ and $C_2 = \mathbf{J} \cdot \mathbf{K} = \mathbf{K} \cdot \mathbf{J} = -i(\mathbf{A}^2 - \mathbf{B}^2)$. So it is equally valid to view $\mathbf{A}^2$ and $\mathbf{B}^2$ as the two Casimir invariants of the Lorentz group, with their eigenvalues $j_a(j_a + 1)$ and $j_b(j_b + 1)$, respectively. Thus the notation for representations of the Lorentz group that was introduced, (j_a, j_b), is a labeling of representations by their two Casimir invariants.

The total angular momentum $\mathbf{J}$ will be the sum of the two sets of $SU(2)$ angular momentum generators, $\mathbf{J} = \mathbf{A} + \mathbf{B}$, and the usual rules for the addition of angular momenta apply in terms of Clebsch-Gordan coefficients,

$$\mathbf{J}^2|j_a j_b jm\rangle = j(j+1)|j_a j_b jm\rangle, \quad J_z|j_a j_b jm\rangle = m|j_a j_b jm\rangle, \tag{1.3.3}$$

$$|j_a j_b jm\rangle = \sum_{m_a=-j_a}^{j_a} \sum_{m_b=-j_b}^{j_b} \langle j_a j_b m_a m_b|jm\rangle |j_a j_b m_a m_b\rangle, \tag{1.3.4}$$

where $|j_a j_b m_a m_b\rangle \equiv |j_a m_a\rangle|j_b m_b\rangle$. As a change of basis the Clebsch-Gordan coefficients are the elements of a unitary matrix and they vanish unless:

$$|j_a - j_b| \le j \le j_a + j_b \quad \text{and} \quad m = m_a + m_b. \tag{1.3.5}$$

The (j_a, j_b) representation of the Lorentz group is therefore $(2j_a + 1)(2j_b + 1)$ dimensional; i.e., the representation is in terms of $(2j_a + 1)(2j_b + 1) \times (2j_a + 1)(2j_b + 1)$ matrices. The matrices can be decomposed into the different accessible values of j each of which has $(2j + 1)$ values of m, where

$$\sum_{j=|j_a-j_b|}^{j_a+j_b}(2j+1) = (2j_a+1)(2j_b+1). \tag{1.3.6}$$

Representations (j_a, j_b) important for our studies of quantum field theory include:

- $(0, 0)$: Trivial representation relevant for Lorentz scalars
- $(\frac{1}{2}, 0)$: Two-dimensional representation relevant for left-handed Weyl fermions
- $(0, \frac{1}{2})$: Two-dimensional representation relevant for right-handed Weyl fermions
- $(\frac{1}{2}, 0) \oplus (0, \frac{1}{2})$: Four-dimensional representation relevant for Dirac fermions
- $(\frac{1}{2}, \frac{1}{2})$: Four-dimensional defining representation of $SO^+(1, 3)$ for Lorentz four-vectors. One dimension for $j = \frac{1}{2} - \frac{1}{2} = 0$ and three for $j = \frac{1}{2} + \frac{1}{2} = 1$.

1.3.2 Lorentz Transformations of Weyl Spinors

The building blocks of representations for fermions are the two two-dimensional representations $(\frac{1}{2}, 0)$ and $(0, \frac{1}{2})$. For any $(j, 0)$ and $(0, j)$ we have

$$\text{Type I: } (j, 0) \quad j_a = j, \, j_b = 0 \quad \Rightarrow \mathbf{A}^2 \ne 0, \mathbf{B}^2 = 0, \mathbf{J} = i\mathbf{K}$$
$$\text{Type II: } (0, j) \quad j_a = 0, \, j_b = j \quad \Rightarrow \mathbf{A}^2 = 0, \mathbf{B}^2 \ne 0, \overline{\mathbf{J}} = -i\overline{\mathbf{K}}, \tag{1.3.7}$$

where here $\mathbf{J}, \mathbf{K}, \overline{\mathbf{J}}$ and $\overline{\mathbf{K}}$ are appropriate $(2j + 1) \times (2j + 1)$ matrix representations of $\mathbf{J}$ and $\mathbf{K}$. In the case $j = \frac{1}{2}$ relevant for fermions we have

$$\text{Type I (left-handed, denoted as } M\text{): } (\tfrac{1}{2}, 0) \quad \mathbf{J} = \tfrac{\boldsymbol{\sigma}}{2} \quad \mathbf{K} = -i\tfrac{\boldsymbol{\sigma}}{2}$$
$$\text{Type II (right-handed, denoted as } \overline{M}\text{): } (0, \tfrac{1}{2}) \quad \overline{\mathbf{J}} = \tfrac{\boldsymbol{\sigma}}{2} \quad \overline{\mathbf{K}} = +i\tfrac{\boldsymbol{\sigma}}{2}. \tag{1.3.8}$$

Since under a parity transformation $\mathbf{J} \to \mathbf{J}$ and $\mathbf{K} \to -\mathbf{K}$ and since no element of $SL(2, \mathbb{C})$ corresponds to the parity operator, then Type I and Type II are two different defining representations of $SL(2, \mathbb{C})$ that are interchanged by a parity transformation. We will see in Sec. 4.6 that Types I and II correspond to the left-handed and right-handed representations, respectively. The two-component spinors that transform under them are called the left-handed and right-handed Weyl or chiral spinors, respectively. This will be discussed in more detail in Sec. 4.6.

For every restricted Lorentz transformation there is an element of $SL(2, \mathbb{C})$ that is the exponential of a Lie algebra element. There are two different 2×2 defining representations of $SL(2, \mathbb{C})$, Type I and Type II, with elements denoted by M and $\overline{M}$, respectively. For an active rotation by angle $0 \le \alpha < 2\pi$ about the axis $\mathbf{n}$ and an active boost in the direction $\mathbf{n}_b$ with rapidity $0 \le \eta < \infty$ we have

$$M = \exp(-i\alpha\mathbf{n} \cdot \mathbf{J} - i\eta\mathbf{n}_b \cdot \mathbf{K}) = \exp\left(-\tfrac{1}{2}\alpha\mathbf{n} \cdot \boldsymbol{\sigma} - \tfrac{1}{2}\eta\mathbf{n}_b \cdot \boldsymbol{\sigma}\right),$$
$$\overline{M} = \exp(-i\alpha\mathbf{n} \cdot \overline{\mathbf{J}} - i\eta\mathbf{n}_b \cdot \overline{\mathbf{K}}) = \exp\left(-\tfrac{i}{2}\alpha\mathbf{n} \cdot \boldsymbol{\sigma} + \tfrac{1}{2}\eta\mathbf{n}_b \cdot \boldsymbol{\sigma}\right), \tag{1.3.9}$$

where $0 \leq \alpha < 2\pi$ and $0 \leq \eta < \infty$. Recall that since boosts and rotations do not commute, then Eq. (1.3.9) is not equivalent to a sequential boost of η, $\mathbf{n}_b$ followed by a rotation α, $\mathbf{n}$ or vice versa. However, we can always decompose both M and $\overline{M}$ into a boost followed by a rotation and vice versa.

1.4 Poincaré Group and the Little Group

As stated earlier, the Poincaré group is the internal semidirect product of the normal (i.e., invariant) subgroup of spacetime translations $R^{1,3}$ with the subgroup of Lorentz transformations $O(1,3)$. There is no widely accepted notation for the Poincaré group and so for $(3+1)$-dimensional Minkowski spacetime we will use the obvious notation Poincaré$(1,3)$, i.e., Poincaré$(1,3) = R^{1,3} \rtimes O(1,3)$. From the discussion of the semidirect product in Sec. A.7.1 we see that the Lorentz transformations are isomorphic to the quotient group of the Poincaré transformations with respect to spacetime translations, i.e., $O(1,3) \cong$ Poincaré$(1,3)/R^{1,3}$.

1.4.1 Intrinsic Spin and the Poincaré Group

Let us now define $L^{\mu\nu}$ as that part of $M^{\mu\nu}$ that is associated with motion in spacetime as specified by Eq. (1.2.170),

$$L^{\mu\nu} \equiv x^\mu P^\nu - x^\nu P^\mu. \tag{1.4.1}$$

The generators of the Lorentz transformation will be the sum of that part associated with the transformations in spacetime, $L^{\mu\nu}$, and that part associated with intrinsic spin transformations, $\Sigma^{\mu\nu}$. Then we have for the total Lorentz generators

$$M^{\mu\nu} \equiv L^{\mu\nu} + \Sigma^{\mu\nu}. \tag{1.4.2}$$

Since the total angular momentum generator $\mathbf{J}$ is given by $J^i = (1/2)\epsilon^{ijk}M^{jk}$, then from Eq. (1.4.2) it has both an orbital component associated with particle motion through spacetime, $\mathbf{L}$, and an intrinsic spin component, $\mathbf{S}$, where

$$\mathbf{J} = \mathbf{L} + \mathbf{S} \quad \text{with} \quad L^i \equiv \tfrac{1}{2}\epsilon^{ijk}L^{jk}, \quad S^i \equiv \tfrac{1}{2}\epsilon^{ijk}\Sigma^{jk}. \tag{1.4.3}$$

The total angular momentum, $\mathbf{J} = \mathbf{L} + \mathbf{S}$, satisfies the angular momentum Lie algebra and generates the rotations of the system. We expect that typically $\mathbf{L}$ and $\mathbf{S}$ separately satisfy the angular momentum Lie algebra and commute. It can happen that $\mathbf{L}$ is not the generator of orbital rotations and $\mathbf{S}$ is not the generator of intrinsic angular momentum rotations. We will see an example of this in Chapter 6 when we study the quantized electromagnetic field with massless photons and gauge fixing.

The total boost generator $\mathbf{K}$ of the particle is given by $K^i = M^{0i}$ and so from Eq. (1.4.2) we find that it also decomposes into an external and an intrinsic part,

$$\mathbf{K} = \mathbf{K}_{\text{ext}} + \mathbf{K}_{\text{int}} \quad \text{with} \quad K^i_{\text{ext}} \equiv L^{0i}, \quad K^i_{\text{int}} \equiv \Sigma^{0i}. \tag{1.4.4}$$

The representation of the full Lorentz group is determined by the representation of its generators $M^{\mu\nu}$. We have seen that these generators can be subdivided into the external and intrinsic components $L^{\mu\nu}$ and $\Sigma^{\mu\nu}$. Particles are classified according to the representation of the Lorentz group that they transform under.

Since $\epsilon_{\mu\nu\rho\sigma}P^\rho P^\sigma = 0$ and $L^{\mu\nu} \equiv x^\mu P^\nu - x^\nu P^\mu$, then we find that $\epsilon_{\mu\nu\rho\sigma}L^{\nu\rho}P^\sigma = 0$. So the Pauli-Lubanski pseudovector in Eq. (1.2.202) can be written as

$$W_\mu = -\tfrac{1}{2}\epsilon_{\mu\nu\rho\sigma}\left(L^{\nu\rho} + \Sigma^{\nu\rho}\right)P^\sigma = -\tfrac{1}{2}\epsilon_{\mu\nu\rho\sigma}\Sigma^{\nu\rho}P^\sigma. \tag{1.4.5}$$

Hence the only part of the Lorentz generators $M^{\mu\nu}$ that contributes to W^μ is that part associated with the intrinsic spin, $\Sigma^{\mu\nu}$. Since W^2 and P^2 are the two Casimir invariants of the Poincaré group we expect the Pauli-Lubanski pseudovector W^μ and the particle mass m to play important roles in the classification of particles.

1.4.2 The Little Group

Without necessarily restricting ourselves to a quantum mechanical context, we shall for convenience adopt quantum mechanics notation for states as we did in the discussion leading to Eq. (1.2.64). We define in the usual way $m^2 = p^2$, $p^0 > 0$ (denoted "+") and $p^0 < 0$ (denoted "−") (which we will see later correspond to particle and antiparticle states, respectively).

Consider the subspace spanned by all of the particle (or antiparticle) states specified by the four-momentum p^μ, where the particle might be massive $m^2 > 0$ (timelike) or massless $m^2 = 0$ (lightlike). There is no known physical application of the tachyonic case with $m^2 < 0$ (spacelike) and we will not consider it further here. We will refer to this subspace as the p^μ-subspace. Since $[W^\mu, P^\nu] = 0$ it follows that the subgroup formed from the generators W^μ leaves p^μ unchanged, and therefore transforms any state in this p^μ-subspace to some other state within the subspace. Then in this p^μ-subspace we can replace P^μ by p^μ and so

$$[W_\mu, W_\nu] = ip^\sigma\epsilon_{\mu\nu\rho\sigma}W^\rho, \tag{1.4.6}$$

which is a Lie algebra for the generators W^μ whose structure constants depend on p^μ. Recall also that $P^\mu W_\mu = 0$ and so only three of the four W^μ generators are independent. This Lie algebra gives rise to what Eugene Wigner called the *Little Group* of p^μ. Since the Little Group leaves p^μ unchanged, then it must act only on the internal degrees of freedom of the particle, i.e., the intrinsic spin, and takes different forms depending on the mass and momentum of the particle.

Massive Particles: $p^2 = m^2 > 0$

In the massive case, $m^2 > 0$ we have $E \equiv (\mathbf{p}^2 + m^2)^{1/2} = \gamma m$ and $\mathbf{p} = \gamma m\mathbf{v}$. Recall that we are working in natural units ($\hbar = c = 1$) throughout this section. For a state corresponding to a particle/antiparticle (corresponding to $+/-$, respectively) with rest mass m and three-momentum $\mathbf{p}$ we write

$$|\pm, m; \mathbf{p}, \sigma\rangle, \tag{1.4.7}$$

where σ is a collective label containing all other needed numbers and degeneracy labels for a particle state. For example, σ *contains the intrinsic spin labels* (s, m_s).

Recall from Eq. (1.4.3) that $S^i = \tfrac{1}{2}\epsilon^{ijk}\Sigma^{jk}$. We will see in Chapter 4 when we study spin-$\tfrac{1}{2}$ that the spin operator is appropriately identified as

$$\mathbf{S}^{\text{spin}} = \pm\mathbf{S} \quad \text{for particles and antiparticles, respectively.} \tag{1.4.8}$$

Denote a spin eigenstate as $|s, m_s\rangle$ for a particle or antiparticle with the spin quantization axis specified by the unit vector $\mathbf{n}_s$. The spin eigenvalue equations are

$$\mathbf{n}_s \equiv \text{(spin quantization axis unit vector in particle rest frame)},$$
$$(\mathbf{S}^{\text{spin}})^2 |s, m_s\rangle = \mathbf{S}^2 |s, m_s\rangle = s(s+1)|s, m_s\rangle \quad \text{and}$$
$$S_{n_s}^{\text{spin}} |s, m_s\rangle = (\mathbf{S}^{\text{spin}} \cdot \mathbf{n}_s)|s, m_s\rangle = (\pm\mathbf{S} \cdot \mathbf{n}_s)|s, m_s\rangle = m_s|s, m_s\rangle. \tag{1.4.9}$$

For a single particle of rest mass m any state representing such a particle must be an eigenstate of the Casimir invariant P^2 such that the eigenvalue is m^2 (or m^2c^2 if we temporarily restore c). Since the Casimir invariant W^2 is invariant under Poincaré transformations then we can evaluate it in any frame, such as the p^μ-subspace corresponding to $\mathbf{p} = 0$ which is the rest frame. Using the Poincaré Lie algebra in Eq. (1.2.167) we find in the rest frame

$$W^2|\pm, m; \mathbf{p} = 0, \sigma\rangle = \tfrac{1}{4}\epsilon_{\mu\nu\rho\tau}\epsilon^{\mu\alpha\beta\gamma} M^{\nu\rho} P^\tau M_{\alpha\beta} P_\gamma |\pm, m; \mathbf{p} = 0, \sigma\rangle \tag{1.4.10}$$
$$= \tfrac{1}{4}\epsilon_{\mu\nu\rho\tau}\epsilon^{\mu\alpha\beta\gamma} M^{\nu\rho}\{M_{\alpha\beta} P^\tau + i(\delta_\alpha^\tau P_\beta - \delta_\beta^\tau P_\alpha)\}P_\gamma |\pm, m; \mathbf{p} = 0, \sigma\rangle$$
$$= m^2 \tfrac{1}{4}\epsilon_{\mu\nu\rho\tau}\epsilon^{\mu\alpha\beta\gamma} M^{\nu\rho}\{M_{\alpha\beta}\delta_0^\tau + i(\delta_\alpha^\tau \delta_\beta^0 - \delta_\beta^\tau \delta_\alpha^0)\}\delta_\gamma^0 |\pm, m; \mathbf{p} = 0, \sigma\rangle$$
$$= m^2 \tfrac{1}{4}\epsilon_{ijk0}\epsilon^{imn0} M^{jk} M_{mn} |\pm, m; \mathbf{p} = 0, \sigma\rangle$$
$$= -m^2 \tfrac{1}{4}\epsilon^{ijk}\epsilon^{imn} M^{jk} M^{mn} |\pm, m; \mathbf{p} = 0, \sigma\rangle = -m^2 \mathbf{J}^2 |\pm, m; \mathbf{p} = 0, \sigma\rangle$$
$$= -m^2 \mathbf{S}^2 |\pm, m; \mathbf{p} = 0, \sigma\rangle = -m^2(\mathbf{S}^{\text{spin}})^2 |\pm, m; \mathbf{p} = 0, \sigma\rangle,$$

where we have used Eq. (1.2.72) in the final step and the fact that a particle at rest has no orbital angular momentum. Using Eq. (1.4.5) we could have also derived this equation with $M^{\mu\nu}$ replaced by $\Sigma^{\mu\nu}$ throughout; i.e., we already knew that only the intrinsic Lorentz generators contribute to the Pauli-Lubanski pseudovector.

When acting on any state of a particle/antiparticle of mass m with intrinsic spin quantum number s the Casimir invariants P^2 and W^2 take the values

$$P^2 = m^2c^2 \quad \text{and} \quad W^2 = -m^2 s(s+1), \tag{1.4.11}$$

which tell us that particle/antiparticle mass m and spin s are Poincaré invariants.

Consider the action of W^μ on a massive particle/antiparticle in its rest frame

$$W^\mu|\pm, m; \mathbf{p} = 0, \sigma\rangle = -\tfrac{1}{2}\epsilon^{\mu\nu\rho\tau} M_{\nu\rho} P_\tau |\pm, m; \mathbf{p} = 0, \sigma\rangle. \tag{1.4.12}$$

We find in the particle rest frame that

$$W^0|\pm, m; \mathbf{p} = 0, \sigma\rangle = 0 \tag{1.4.13}$$
$$W^i|\pm, m; \mathbf{p} = 0, \sigma\rangle = \pm m \tfrac{1}{2}\epsilon^{0ijk} M^{jk} |\pm, m; \mathbf{p} = 0, \sigma\rangle = \pm mS^i |\pm m; \mathbf{p} = 0, \sigma\rangle.$$

In the p^μ-subspace corresponding to the particle or antiparticle rest frame, $p^\mu = (\pm m, \mathbf{0})$, we have then arrived at the result that

$$W^\mu = (W^0, \mathbf{W}) \xrightarrow{\mathbf{p}=0} (0, \pm m\mathbf{S}) = (0, m\mathbf{S}^{\text{spin}}), \tag{1.4.14}$$

where for massive particles this equation can be taken as the *definition* of the spin operator.

A maximal set of commuting observable operators for a free massive particle/antiparticle with intrinsic spin is W^2, P^2, $\text{sgn}(P^0)$, $\mathbf{P}$ and W^3, where the eigenvalue of the operator $\text{sgn}(P^0)$ is ± 1 for particles and antiparticles, respectively. Rather than choosing $W^3 = m(S^{\text{spin}})^3$ in the rest frame we could choose $\mathbf{W} \cdot \mathbf{n}_s = m(\mathbf{S}^{\mathrm{spin}} \cdot \mathbf{n}_s)$ for spin quantization axis $\mathbf{n}_s$. In the rest frame we have $n_s^\mu = (0, \mathbf{n}_s)$ and $p^\mu = (\pm m, \mathbf{0})$. This gives

$$n_s^2 = -1 \quad \text{and} \quad p \cdot n_s = 0, \tag{1.4.15}$$

which are Lorentz-invariant equations and so are true in all inertial frames. Note that in the rest frame we have shown from the above that

$$(S^{\mathrm{spin}})^2 \equiv \frac{(-W^2)}{m^2} \xrightarrow{\mathbf{p}=0} (\mathbf{S}^{\mathrm{spin}})^2, \quad S_{n_s}^{\mathrm{spin}} \equiv \frac{(-W \cdot n_s)}{m} \xrightarrow{\mathbf{p}=0} (\mathbf{S}^{\mathrm{spin}} \cdot \mathbf{n}_s). \tag{1.4.16}$$

The operator $(S^{\mathrm{spin}})^2 \equiv (-W^2)/m^2$ is Poincaré invariant, since it is a Casimir invariant of the Poincaré group. The spin operator $S_{n_s}^{\mathrm{spin}} \equiv (-W \cdot n_s)/m$ is invariant under Lorentz boosts and rotations. Since $\mathbf{L} = \mathbf{r} \times \mathbf{p}$ is invariant under a parity transformation, then as an angular momentum spin must also be invariant. This follows from Eq. (1.4.14) since W^μ is a pseudovector. Since $\mathbf{n}_s$ simply chooses a particular linear combination of the components of $\mathbf{S}^{\mathrm{spin}}$, it is parity invariant. This means that $n^\mu(s)$ *also transforms as a pseudovector* and $S_{n_s}^{\mathrm{spin}}$ is parity invariant.

Summary for a massive particle: For a particle/antiparticle of mass m, intrinsic spin s and with the rest-frame spin quantization axis $\mathbf{n}_s$ we can write a simultaneous eigenstate of $(S^{\mathrm{spin}})^2$, P^2, $\mathrm{sgn}(P^0)$, $\mathbf{P}$ and $S_{n_s}^{\mathrm{spin}}$ as

$$|\pm, m, s; \mathbf{p}, m_s\rangle. \tag{1.4.17}$$

Note that the eigenvalues of P^0 and $\mathbf{P}$ are $\pm E$ and $\mathbf{p}$, respectively, and must satisfy $E^2 = \mathbf{p}^2 + m^2$ and so E is not independent. In summary, we then have

$$P^0|\pm, m, s; \mathbf{p}, m_s\rangle = \pm E|\pm, m, s; \mathbf{p}, m_s\rangle, \tag{1.4.18}$$
$$\mathbf{P}|\pm, m, s; \mathbf{p}, m_s\rangle = \mathbf{p}|\pm, m, s; \mathbf{p}, m_s\rangle,$$
$$P^2|\pm, m, s; \mathbf{p}, m_s\rangle = m^2|\pm, m, s; \mathbf{p}, m_s\rangle,$$
$$(S^{\mathrm{spin}})^2|\pm, m, s; \mathbf{p}, m_s\rangle = (-W^2/m^2)|\pm, m, s; \mathbf{p}, m_s\rangle = s(s+1)|\pm, m, s; \mathbf{p}, m_s\rangle,$$
$$S_{n_s}^{\mathrm{spin}}|\pm, m, s; \mathbf{p}, m_s\rangle = [(-W \cdot n)/m]|\pm, m, s; \mathbf{p}, m_s\rangle = m_s|\pm, m, s; \mathbf{p}, m_s\rangle,$$

where $\pm$ refers to particles and antiparticles, respectively.

It can be shown that the Little group corresponds to an $SU(2)$ subgroup of the Poincaré transformations associated with rotations in intrinsic spin space as expected on physical grounds. So there are $2s + 1$ values of m_s for a particle with integer or half-integer spin s as we saw in Eq. (1.2.64). We know that s is either integer or half-integer and that $m_s = -s, -s + 1, \ldots, s - 1, s$.

In the special case where we choose the rest-frame spin quantization axis to be the direction of the boost that takes the particle to its moving frame, $\mathbf{n}_s = \mathbf{n}_p$, then we have the *helicity operator*

$$S^{\mathrm{helicity}} \equiv S_{n_p}^{\mathrm{spin}} \equiv [(-W \cdot n_p)/m] \xrightarrow{\mathbf{p}=0} (\mathbf{S}^{\mathrm{spin}} \cdot \mathbf{n}_p) = (\pm \mathbf{S} \cdot \mathbf{n}_p), \quad \text{where} \tag{1.4.19}$$
$$S^{\mathrm{helicity}}|\pm, m, s; \mathbf{p}, m_s\rangle = [(-W \cdot n_p)/m]|\pm, m, s; \mathbf{p}, m_s\rangle = m_s|\pm, m, s; \mathbf{p}, m_s\rangle.$$

A massive particle with $m_s > 0$ or < 0 has the spin aligned (*positive helicity*) or anti-aligned (*negative helicity*), respectively, with the direction of motion. There are $2s + 1$ values of the helicity m_s for a massive particle. Helicity for a massive particle is invariant under continuous (proper) rotations, translations, and collinear boosts that do not reverse the direction of motion. It is not invariant under arbitrary boosts; e.g., we can boost to a frame where a massive particle is moving in the opposite direction resulting in the reversal of the helicity. Helicity reverses sign under a parity transformation and is invariant under time reversal.

Massless Particles: $p^2 = m^2 = 0$

Massless particles must always travel at the speed of light irrespective of their intrinsic spin, since $p^2 = m^2 \to 0$ implies that $|p^0| \to |\mathbf{p}|$ and so $|p^0|/|\mathbf{p}| = (\gamma mc)/(\gamma mv) = 1/\beta = c/v \to 1$, where v is the particle velocity.

For a p^μ-subspace in the limit that $m \to 0$ we have $W^2 = -m^2 s(s+1) \to 0$ unless $s \to \infty$. So for any finite-dimensional spin representation (i.e., finite spin quantum number s), we must have for any momentum eigenstate on the light cone

$$W^2|p,\sigma\rangle = P^2|p,\sigma\rangle = W \cdot P|p,\sigma\rangle = P \cdot W|p,\sigma\rangle = 0, \tag{1.4.20}$$

where σ contains all intrinsic spin information. $W \cdot P = 0$ follows from Eq. (1.2.202) and $P \cdot W = 0$ follows since $[W^\mu, P^\nu] = 0$ from Eq. (1.2.204). It should be clear from the context when W^2 refers to $W^2 = W_\mu W^\mu$ and when it refers to the y-component of $W^\mu = (W^0, W^1, W^2, W^3) = (W^0, W_x, W_y, W_z)$. Consider the massless p^μ-subspace corresponding to momentum in the z-direction $p^\mu = (\pm E, 0, 0, E) = (\pm E, E\mathbf{n}_z)$, where $\mathbf{n}_z = (0, 0, 1)$. The Pauli-Lubanski pseudovector takes the form

$$
\begin{aligned}
W^0 &= -(1/2)\epsilon^{0\nu\rho 3}M_{\nu\rho}p_3 = -\epsilon^{0123}M^{12}(-E) = E\epsilon^{123}\Sigma^{12} = ES^3 \\
W^3 &= -(1/2)\epsilon^{3\nu\rho 0}M_{\nu\rho}p_0 = -\epsilon^{3120}M^{12}(\pm E) = (\pm E)\epsilon^{123}\Sigma^{12} = (\pm E)S^3 \\
W^1 &= -(1/2)\left[\epsilon^{1\nu\rho 0}(\pm E) + \epsilon^{1\nu\rho 3}(-E)\right]M_{\nu\rho} = -\epsilon^{1230}(\pm E)M_{23} - \epsilon^{1023}(-E)M_{02} \\
&= M\Sigma^{23}(\pm E) + E\Sigma^{02} = E(\pm S^1 + K^2) \\
W^2 &= -(1/2)\left[\epsilon^{2\nu\rho 0}(\pm E) + \epsilon^{2\nu\rho 3}(-E)\right]M_{\nu\rho} = -\epsilon^{2310}(\pm E)\Sigma_{31} - \epsilon^{2013}(-E)\Sigma_{01} \\
&= \Sigma^{31}(\pm E) - E\Sigma^{01} = E(\pm S^2 - K^1),
\end{aligned} \tag{1.4.21}
$$

where recall that in W^μ only the $\Sigma^{\mu\nu}$ part of $M^{\mu\nu}$ survives. On the right-hand side we can write $\mathbf{K}_{\text{int}}$ in place of $\mathbf{K}$ since only that part survives, but to avoid cumbersome notation we do not do this. So $W^3 = \pm W^0$ for particles/antiparticles, respectively. In this subspace $P^\mu W_\mu = 0$ becomes $p^\mu W_\mu = 0$ and since we have $p^\mu = (\pm E, 0, 0, E)$ we see that this is satisfied by the above equation. We can choose W^1, W^2 and W^3 to be the three generators of the Little Group in this subspace.

In order to better understand the nature of the Little Group in the massless case, consider the linear combinations

$$F_1 \equiv \frac{\pm K^1 - S^2}{\sqrt{2}} = \pm\frac{-W^2}{E\sqrt{2}}, \quad F_2 \equiv \frac{\pm K^2 + S^1}{\sqrt{2}} = \pm\frac{W^1}{E\sqrt{2}}, \quad S^3 = \pm\frac{W^3}{E}. \tag{1.4.22}$$

These linear combinations of the Little Group generators satisfy the Lie algebra of the two-dimensional Euclidean group $E(2)$, which leave a Euclidean two-dimensional plane invariant. This consists of rotations in a two-dimensional plane about some orthogonal axis, $O(2)$, and translations in that plane, $T(2)$. Since $T(2)$ is a normal subgroup of $E(2)$, then $E(2)$ is the semidirect product of $O(2)$ extended by $T(2)$, i.e., $E(2) = T(2) \rtimes O(2)$. It then follows that $O(2)$ is the quotient group of $E(2)$ by $T(2)$, i.e., $O(2) = E(2)/T(2)$. So in the massless case the Little Group corresponds to the connected part of $E(2)$ generated by the Lie algebra, which is denoted as $SE(2)$ and which consists of proper (or pure) rotations, $SO(2)$, around an axis and the translations $T(2)$, where $SE(2) = T(2) \rtimes SO(2)$. We see that $SE(2)$ preserves orientations. It is sometimes referred to as $ISO(2)$ (Weinberg, 1995).

It can be shown that these linear combinations reproduce the Lie algebra of $E(2)$,

$$[F^i, F^j] = 0 \quad \text{and} \quad [S^3, F^i] = i\epsilon^{ij}F^j \quad \text{with} \quad \epsilon^{12} = +1 \text{ and } i, j = 1, 2, \tag{1.4.23}$$

with F^1 and F^2 playing the role of translation generators in the x and y directions in this intrinsic spin space and S^3 generating rotations around the z-axis, which is our chosen direction of motion. Had we chosen an arbitrary spatial direction $\mathbf{n}_p$ for the three-momentum $\mathbf{p} = E\mathbf{n}_p$ rather than the z-axis, then S^3 would be replaced by $\mathbf{S} \cdot \mathbf{n}_p$ for a massless particle, and F^1 and F^2 would be replaced by their translation generator equivalents in the plane orthogonal to $\mathbf{n}_p$.

Defining $\mathbf{F} \equiv (F^1, F^2)$ and $\mathbf{F}^2 = (F^1)^2 + (F^2)^2$ we see that

$$[\mathbf{F}^2, F^i] = [\mathbf{F}^2, S^3] = 0 \tag{1.4.24}$$

and so $\mathbf{F}^2$ is the Casimir of the Little Group for the massless case. Furthermore, if we define $F^\pm \equiv F^1 \pm iF^2$ then

$$[F^+, F^-] = 0, \quad [S^3, F^\pm] = \pm F^\pm \quad \text{and} \quad \mathbf{F}^2 = F^+F^- = F^-F^+. \tag{1.4.25}$$

Denoting the simultaneous eigenstate of $\mathbf{F}^2$ and S^3 as $|f; \lambda\rangle$ such that $\mathbf{F}^2|f; \lambda\rangle = f|f; \lambda\rangle$ and $S^3|f; \lambda\rangle = \lambda|f; \lambda\rangle$ we see from the above that

$$S^3 F^\pm|f; \lambda\rangle = F^\pm(S^3 \pm 1)|f; \lambda\rangle = (\lambda \pm 1)F^\pm|f; \lambda\rangle \tag{1.4.26}$$

and so F^+ and F^- raise and lower λ by one, respectively. Note that the situation here is different to that for the $SU(2)$ spin group. There the Casimir $\mathbf{S}^2$ involves all three components of $\mathbf{S}$ and as a result the raising and lowering operations terminate when $|m_s| = s$ as shown in Eq. (1.2.64). Here there is no similar mechanism to terminate the series. If the series does not terminate, then there will be an infinite number of possible helicity eigenvalues and correspondingly an infinite-dimensional representation of the Little Group. The ladder of possible helicity values is infinite unless $F^+|f; \lambda\rangle = 0$ and $F^-|f; \lambda\rangle = 0$ for some f and λ. Since there is nothing to terminate the λ ladder, then the termination can only result from the choice of f and it must then apply for all helicity eigenvalues λ; i.e., we need an f such that $F^\pm|f; \lambda\rangle = 0$ for all λ. This means that $\mathbf{F}^2|f; \lambda\rangle = F^\pm F^\mp|f; \lambda\rangle = f|f; \lambda\rangle = 0$ and hence $f = 0$ is the only possible choice if the number of possible λ values is finite. We have then arrived at $\mathbf{F}^2|\lambda\rangle = F^+|\lambda\rangle = F^-|\lambda\rangle = 0$ and $S_3|\lambda\rangle = \lambda|\lambda\rangle$, where we have defined $|\lambda\rangle \equiv |f = 0, \lambda\rangle$. This means that there is potentially only one state $|\lambda\rangle$, since the ladder operators $F^\pm$ annihilate it.

If the three-momentum of the massless particle was in an arbitrary direction $\mathbf{n}_p$ rather than the z-direction, then instead of S_3 we would have $\mathbf{S} \cdot \mathbf{n}_p$. We identify the helicity operator in the massless case as

$$S^{\text{helicity}}|\lambda\rangle \equiv (\mathbf{S}^{\text{spin}} \cdot \mathbf{n}_p)|\lambda\rangle = (\pm\mathbf{S} \cdot \mathbf{n}_p)|\lambda\rangle = \pm\lambda|\lambda\rangle, \tag{1.4.27}$$

where "$\pm$" refers to particles and antiparticles, respectively, so that a particle and its antiparticle have opposite helicity. If a particle has helicity λ, then its antiparticle has helicity $-\lambda$. *If the underlying theory is parity invariant* and since under a parity transformation $(\mathbf{S} \cdot \mathbf{n}_p) \rightarrow -(\mathbf{S} \cdot \mathbf{n}_p)$, then it follows that both particles and antiparticles will be able to be in either of two helicity states $\pm\lambda$.

Since Lorentz invariance implies invariance under $SL(2, \mathbb{C})$, then we require that any theory that is Poincaré invariant must be invariant under a rotation of 4π. This means that a helicity eigenstate $|\lambda\rangle$ should be invariant under a rotation of 4π about the momentum direction,

$$\exp[i(4\pi)(\mathbf{S}^{\text{spin}} \cdot \mathbf{n}_p)]|\lambda\rangle = e^{\pm i4\pi\lambda}|\lambda\rangle = |\lambda\rangle, \tag{1.4.28}$$

which means that λ must be half-integer or integer, $\lambda = 0, \pm\frac{1}{2}, \pm 1, \ldots$. Different values of λ correspond to different representations of the $SO(2)$ group of rotations about the direction of motion $\mathbf{n}_p$. Since this $SO(2)$ is a subgroup of the intrinsic rotations in three dimensions specified by the rotation generators $\mathbf{S}$, then it must be in the same representation. In other words we must have $\lambda = \pm s$. So the value of the helicity is determined by the representation of the $\Sigma^{\mu\nu}$, which are in turn determined by the representation of the Lorentz group that we are considering. So massless particles with spin s do not have spin in the familiar sense; they have instead the property of helicity with values $\lambda = \pm s$.

If we do not have parity invariance a massless particle may have only one value of helicity, $\lambda = +s$ or $-s$. Helicity is reversed by a parity transformation and so in the presence of parity invariance a massless particle's helicity can have two values, $\lambda = s, -s$. For example, the photon is a spin-1 particle ($s = 1$) and electromagnetism is parity invariant so the photon has two helicity states $\lambda = \pm 1$ corresponding to right and left circular polarizations of electromagnetic radiation as discussed in Sec. 2.7.2. The graviton is the massless spin-2 boson thought to mediate the gravitational interaction. Since the theory of gravity is parity invariant, then the graviton would have two helicity states, $\lambda = \pm 2$. Historically, neutrinos were thought to be massless and have only left-handed helicity, $\lambda = -\frac{1}{2}$. This corresponds to a theory of the weak interactions that is maximally parity violating. From Eq. (1.4.27) we see that the corresponding antineutrinos have right-handed helicity, $\lambda = +\frac{1}{2}$.

Summary for the massless case: For a massless particle with intrinsic spin s the helicity is $\lambda = +s$ or $-s$. The antiparticle has helicity $\lambda = -s$ or $+s$, respectively. If the relevant theory is parity invariant, then both particles and antiparticles can have either of the two helicity states; i.e., both can have $\lambda = \pm s$. We can write a simultaneous eigenstate of $(S^{\text{spin}})^2$, P^2, $\mathbf{P}$ and S^{helicity} as $|\pm, s; \mathbf{p}, \lambda\rangle$, where $\pm$ refers to particles/antiparticles, respectively, and where

$$S^{\text{helicity}}|\pm, s; \mathbf{p}, \lambda\rangle = (\mathbf{S}^{\text{spin}} \cdot \mathbf{n}_p)|\pm, s; \mathbf{p}, \lambda\rangle = \pm\lambda|\pm, s; \mathbf{p}, \lambda\rangle,$$
$$P^0|\pm, s; \mathbf{p}, \lambda\rangle = H|\pm, s; \mathbf{p}, \lambda\rangle = \pm E|\pm, s; \mathbf{p}, \lambda\rangle,$$
$$\mathbf{P}|\pm, s; \mathbf{p}, \lambda\rangle = \mathbf{p}|\pm, s; \mathbf{p}, \lambda\rangle = E\mathbf{n}_p|\pm, s; \mathbf{p}, \lambda\rangle. \tag{1.4.29}$$

Recall that for a massless particle $\mathbf{p} = E\mathbf{n}_p$ so that $E = |\mathbf{p}|$. The helicity of a massless particle is invariant under proper rotations, boosts and translations. It reverses sign under a parity transformation and is preserved under time reversal.

Summary

This chapter provides a concise statement of those aspects of Lorentz and Poincaré invariance that will be of most importance to us. Every Lorentz transformation $\Lambda \in O(1, 3)$ can be written as the internal semidirect product of an element of the normal discrete subgroup $\{I, P, T, PT\}$ and an element of the restricted Lorentz group $SO^+(1, 3)$. The Poincaré group is the internal semidirect product of the normal (i.e., invariant) subgroup of spacetime translations $R^{1,3}$ with the subgroup of Lorentz transformations $O(1, 3)$, i.e., Poincaré$(1, 3) = R^{1,3} \rtimes O(1, 3)$ with "$\rtimes$" denoting the semidirect product as discussed in Sec. A.7.1.

We saw that particles are categorized according to the representation of the Poincaré group that they transform under. Allowed values for their spin quantum number s are $s = 0, \frac{1}{2}, 1, \frac{3}{2}, \ldots$ We also found that for massive particles the spin projection quantum number for spin m_s can take on $2s + 1$ values, $m_s = -s, -s + 1, \ldots, s - 1, s$, whereas for massless particles the spin projection quantum number can take on only two values that are referred to as the helicity, $\lambda = \pm\hbar s$.

In the case of spin half particles, i.e., fermions, we saw that the appropriate form of the restricted Lorentz group was $SL(2, \mathbb{C})$, which is the double-cover of $SO^+(1, 3)$. This issue is further discussed in Sec. 4.1.4. A spin half particle only returns to its original state under an integer number of 4π

rotations, rather than 2π. However, physical observables involving fermions have the form of Lorentz covariant fermion bilinears, which transform under the defining Poincaré group, Poincaré$(1, 3) = R^{1,3} \rtimes O(1, 3)$, and so are invariant under rotations of 2π.

With regard to representations of the Poincaré group, there is an interesting theorem, the *Coleman-Mandula theorem*, that states on quite general grounds that spacetime and internal symmetries can only be combined in a trivial way. If G is a connected symmetry group of the S-matrix with some reasonable properties, then G is locally isomorphic to the direct product of an internal symmetry group and the Poincaré group; i.e., it is a trivial combination (Coleman and Mandula, 1967). Supersymmetry is one means of avoiding this theorem since it extends the Poincaré Lie algebra to a Lie superalgebra.

Problems

1.1 A muon is a more massive version of an electron and has a mass of 105.7 MeV/c^2. The dominant decay mode of the muon is to an electron, an electron antineutrino and a muon neutrino, $\mu^- \to e^- + \bar{\nu}_e + \nu_\mu$. If we have $N(t)$ unstable particles at time t, then the fraction of particles decaying per unit time is a constant; i.e., we have $dN/N = -(1/\tau)dt$ for some constant τ. This gives $dN/dt = -(1/\tau)N$, which has the solution $N(t) = N_0 e^{-t/\tau}$ where we have N_0 unstable particles at $t = 0$. The fraction of particles decaying in the interval t to $t + dt$ is $-dN/N_0 = (-dN/dt)dt/N_0 = (1/\tau)e^{-t/\tau}dt$. So the *mean lifetime* (or *lifetime*) is $\int_0^\infty t\,(-dN/dt)\,dt/N_0 = \tau \int_0^\infty xe^{-x}dx = \tau$, where $x = t/\tau$. The *half-life*, $t_{1/2}$, is the time taken for half the particles to decay, $e^{-t_{1/2}/\tau} = \frac{1}{2}$, which means that $t_{1/2} = \tau \ln 2$. The *decay rate*, Γ, is the probability per unit time that a particle will decay; i.e., $\Gamma = (-dN/dt)/N = 1/\tau$ is the inverse mean lifetime. A muon at rest has a lifetime of $\tau = 2.197 \times 10^{-6}$ s. *Cosmic rays* are high-energy particles that have traveled enormous distances from outside our solar system. *Primary cosmic rays* have been accelerated by some extreme astrophysical event and *secondary cosmic rays* result from collisions of primary cosmic rays with interstellar gas or our atmosphere. Most cosmic rays reaching our atmosphere will be stable particles such as photons, neutrinos, electrons, protons and stable atomic nuclei (mostly helium nuclei). Muon cosmic rays therefore are secondary cosmic rays produced when primary or secondary cosmic rays collide with our atmosphere. A typical height for muon production is $\sim$15 km. What is the minimum velocity that a muon moving directly downward must have so that it has a 50% chance of reaching the surface of the Earth before decaying?

1.2 The diameter of our Milky Way spiral galaxy is 100,000–180,000 light years and our solar system is approximately 25,000 light years from the galactic center. Recalling the effects of time dilation, approximately how fast would you have to travel to reach the center of the galaxy in your lifetime? Estimate how much energy would it take to accelerate your body to this speed.

1.3 Consider two events that occur at the same spatial point in the frame of some inertial observer $\mathcal{O}$. Explain why the two events occur in the same temporal order in every inertial frame connected to it by a Lorentz transformation that does not invert time. Show that the time separation between the two events is a minimum in the frame of $\mathcal{O}$. (Hint: Consider Figs. 1.2 and 1.5.)

1.4 Consider any two events that occur at the same time in the frame of an inertial observer $\mathcal{O}$. Show that by considering any Lorentz transformation there is no limit to the possible time separation of the two events and that the smallest spatial separation of the two events occurs in the frame of $\mathcal{O}$.

1.5 Two narrow light beams intersect at angle θ, where θ is the angle between the outgoing beams. The beams intersect head on when $\theta = 180°$. Using the addition of velocities formula in Eq. (1.2.126) show that for any angle θ there is always an inertial frame in which the beams intersect head on.

1.6 A ruler of rest length ℓ is at rest in the frame of inertial observer $\mathcal{O}$ and is at an angle θ with respect to the $+x$-direction. Now consider an inertial observer $\mathcal{O}'$ in an identical inertial frame except that it has been boosted by speed v in the $+x$-direction. What is the length ℓ' of the ruler and what is the angle θ' with respect to the $+x'$-direction that will be measured by $\mathcal{O}'$?

1.7 If a spaceship approaches earth at 1.5×10^8 ms^{-1} and emits a microwave frequency of 10 GHz, what frequency will an observer on Earth detect?

1.8 An inertial observer observes two spaceships moving directly toward one another. She measures one to be traveling at $0.7\,c$ in her inertial frame and the other at $0.9\,c$ in the opposite direction. What is the magnitude of the relative velocity that each spaceship measures the other to have?

1.9 A light source moves with constant velocity v in the frame of inertial observer $\mathcal{O}$. The source radiates isotropically in its rest frame. Show that in the inertial frame of $\mathcal{O}$ the light is concentrated in the direction of motion of the source, where half of the photons lie in a cone of semi-angle θ, where $\cos \theta = v/c$.

1.10 Write down the Lorentz transformation rule for an arbitrary $(3, 2)$ tensor $A^{\mu\nu\rho}{}_{\sigma\tau}$. Hence show that any double contraction of this tensor leads to a contravariant vector, i.e., to a $(1, 0)$ tensor.

1.11 A rocket of initial rest mass M_0 has a propulsion system that accelerates it by converting matter into light with negligible heat loss and directs the light in a collimated beam behind it. The propulsion system is turned on for some period of time during which the speed of the rocket is boosted by speed v. Use energy and momentum conservation to show that the final rest mass of the rocket is given by $M_v = M_0 \sqrt{(c - v)/(c + v)}$.

1.12 Two identical spaceships approach an inertial observer $\mathcal{O}$ at equal speeds but from opposite directions. A second observer $\mathcal{O}'$ traveling on one of the spaceships measures the length of the other spaceship to be only 60% of its length at rest. How fast is each space ship traveling with respect to $\mathcal{O}$?

1.13 An observer on Earth observes the hydrogen-β line of a distant galaxy shifted from its laboratory measured value of 434 nm to a value of 510 nm. How fast was the galaxy receding from the Earth at the time the light was emitted? (You may neglect the effects of the cosmological redshift due to the expansion of the universe during the time of flight of the photons.)

1.14 An inertial observer $\mathcal{O}$ is midway between two sources of light at rest in her frame with the sources 2 km apart. Each source emits a flash of light that reaches $\mathcal{O}$ simultaneously. Another inertial observer $\mathcal{O}'$ is moving parallel to the line that passes through the light sources and inertial observer $\mathcal{O}$. In her frame $\mathcal{O}'$ measures the time of each flash and finds that they are spaced 10 ns apart in time. What is the speed of $\mathcal{O}'$ with respect to $\mathcal{O}$?

1.15 A particle with integer or half-integer spin s has $2s + 1$ values of spin with respect to any arbitrary spin quantization axis $\mathbf{n}$ in the particle's rest frame. If parity is a good symmetry, a spin eigenstate for a massless particle is a helicity eigenstate with helicity eigenvalues $\pm s$. If parity is not conserved, then a single helicity eigenstate is possible with eigenvalue either $+s$ or $-s$. As succinctly as you can, summarize the key reasons for this result.

1.16 The Poincaré group is a subgroup of a larger Lie group, the *conformal group* $C(1,3)$. This fifteen-parameter group consists of the 10-parameter subgroup of *Poincaré transformations*, the one-parameter subgroup of scale transformations (*dilatations*) and the four-parameter subgroup of *special conformal transformations (SCT)*. The generators of the unitary representation of the group are $P^\mu = i\hbar\,\partial^\mu$ (translations), $M^{\mu\nu} = i\hbar\,(x^\mu \partial^\nu - x^\nu \partial^\mu) = x^\mu P^\nu - x^\nu P^\mu$ (Lorentz transformations), $D = i\hbar\,x^\mu \partial_\mu = x \cdot P$ (dilatations), and $K^\mu = i\hbar\left(2x^\mu x \cdot \partial - x^2 \partial^\mu\right) = 2x^\mu D - x^2 P^\mu$ (SCT). Using $[P^\mu, x^\nu] = i\hbar g^{\mu\nu}$ we have already seen that P^μ and $M^{\mu\nu}$ satisfy the Lie algebra of the Poincaré group. Show that the remaining commutators of the Lie algebra of the conformal group are $[D, P^\mu] = -i\hbar\,P^\mu$, $[D, K^\mu] = i\hbar\,K^\mu$, $[D, M^{\mu\nu}] = 0$, $[K^\mu, P^\nu] = -i\hbar\,2(M^{\mu\nu} + g^{\mu\nu}D)$, $[K^\mu, K^\nu] = 0$ and $[K^\mu, M^{\rho\sigma}] = i\hbar\left(g^{\mu\rho}K^\sigma - g^{\mu\sigma}K^\rho\right)$.

2

Classical Mechanics

The mathematical framework of analytical classical mechanics underpins much of the formalism used to understand modern physics including quantum field theories, gauge field theories and the Standard Model. It is important to build up the elements of the analytic approach to classical mechanics (Goldstein et al., 2013). In this chapter, unless explicitly stated otherwise, we will be working in SI units.

Nonrelativistic classical mechanics is appropriate when:

(i) velocities are much less than the speed of light (nonrelativistic motion)
(ii) the actions associated with motions of interest are much greater than $\hbar$
(iii) non-gravitational fields (e.g., electromagnetism) are sufficienty weak; and
(iv) gravitational field strength is sufficiently weak (no general relativity).

In this chapter we first introduce the Lagrangian formulation of classical mechanics. The manifestation of symmetries in this formulation is discussed and we derive Noether's theorem and show how to obtain conservation laws. The approach to studying small oscillations of arbitrarily complicated classical systems is then developed. Next we derive the Hamiltonian formalism from the Lagrangian formalism and the relationship between the Hamiltonian formalism and quantum mechanics. Dirac's canonical quantization procedure is discussed. We show that classical mechanics is the macroscopic limit of quantum mechanics. The rules of relativistic kinematics are next derived. The Lorentz covariant formulation of electromagnetism is then considered, including gauge invariance and gauge fixing. The extension of analytic mechanics to the case of a single relativistic particle in an external field is given. Finally, we study the Hamiltonian formulation of singular systems.

2.1 Lagrangian Formulation

We now introduce the underlying concepts needed to develop the Lagrangian formulation of classical mechanics.

Newton's laws: The foundation of our understanding of classical mechanics is given by Newton's laws, which can be expressed as:

(i) First law: When viewed from any inertial reference frame, an object at rest remains at rest and an object moving continues to move with a constant velocity, unless acted upon by an external unbalanced force.
(ii) Second law: The vector sum of the forces, $\mathbf{F}$, on an object is equal to the time rate of change of momentum of the object $\mathbf{F} = d\mathbf{p}/dt$.
(iii) Third law: When two bodies interact through a force, the force on the first body from the second body is simultaneous with and equal and opposite to the force on the second body from the first body.

Newton's first law serves as a definition of an inertial frame; i.e., it is a frame in which any free particle moves with a constant velocity. The first law establishes reference frames in which the other two laws are valid. Transformations between Newtonian inertial frames of reference are the transformations of Galilean relativity.

2.1.1 Euler-Lagrange Equations

For a constant mass system the second law can be written in the familiar form

$$\mathbf{F} = d\mathbf{p}/dt = m(d\mathbf{v}/dt) = m\mathbf{a}. \tag{2.1.1}$$

For a system in which the mass is changing, such as a rocket, Newton's laws must be applied to the combined system comprising the rocket and the ejected mass.

Coordinates and constraints: A *generalized coordinate* is any degree of freedom that can be used to parameterize the behavior of a classical system, e.g., $q_i(t) \in \mathbb{R}$ for $i = 1, 2, \ldots, N$ where t is the time. For n particles moving in three spatial dimensions the number of generalized coordinates is $N = 3n$, since we can choose the generalized coordinates to be the Cartesian coordinates, $\mathbf{r}_a(t) \in \mathbb{R}^3$ for $a = 1, 2, \ldots, n$. A *constraint* on the system is some condition that reduces the number of degrees of freedom by one. If we impose r constraints on n particles moving in three dimensions, then the number of degrees of freedom is $N = 3n - r$.

An important example is a *rigid body*. We can think of a rigid body as a collection of a very large number, $\mathcal{O}(N_A) \sim \mathcal{O}(10^{23})$, of atoms/molecules whose relative positions are held fixed by strong chemical forces. When this enormous number of constraints on relative positions is included we see that only six degrees of freedom remain, $N = 6$, which are the center of mass, $\mathbf{r} = (x, y, z)$, and the orientation given by the three Euler angles, $\theta_1, \theta_2, \theta_3$, where $\theta_1 \equiv \alpha \in (-\pi, \pi], \theta_2 \equiv \beta \in [0, \pi]$, and $\theta_3 \equiv \gamma \in (-\pi, \pi]$. The center of mass is defined as

$$\mathbf{r} \equiv (1/M) \sum_{a=1}^n m_a \mathbf{r}_a \quad \text{with total mass} \quad M = \sum_{a=1}^n m_a. \tag{2.1.2}$$

The six generalized coordinates are then $\vec{q} \equiv (q_1, q_2, \ldots, q_6) = (\mathbf{r}, \theta_1, \theta_2, \theta_3) = (\mathbf{r}, \boldsymbol{\theta})$.

As discussed earlier, for a rotation O in three dimensions we have $\mathbf{r}' = O\mathbf{r}$ and the requirement of the preservation of length means that

$$\mathbf{r}'^2 = \mathbf{r}'^T \mathbf{r} = (O\mathbf{r})^T (O\mathbf{r}) = \mathbf{r}^T O^T O\mathbf{r} = \mathbf{r}^T \mathbf{r} = \mathbf{r}^2 \tag{2.1.3}$$

and so $O^T O = I$. Now $\det(OO^T) = \det I = 1$ and $\det(OO^T) = \det O \det(O^T) = (\det O)^2$ so $\det O = \pm 1$ and O is nonsingular. Hence O^{-1} exists. Acting on $O^T O = I$ from the right with O^{-1} we have $O^T = O^T O O^{-1} = O^{-1}$ so that $O^T = O^{-1}$. This defines the orthogonal group in three dimensions; i.e., if $O^T = O^{-1}$ then $O \in O(3)$. If $\det O = -1$ then the orientation of the axes is not preserved, which means a parity transformation (parity inversion) and a *proper (or pure) rotation*. The orthogonal transformations with $\det O = +1$ are the proper rotations, which form the subgroup $SO(3) \subset O(3)$, where S denoting "special" means that $\det O = 1$. $O(n)$ is the orthogonal group in n dimensions, $\mathbb{R}^n$, and $SO(n)$ is the subgroup of proper rotations.

The *configuration space* or *configuration manifold*, $\mathcal{M}$, is the space of all possible configurations of the system. It can be that the configuration space varies with time, e.g., if we have time-dependent constraints. In that case we denote configuration space at time t as $\mathcal{M}_t$. For a rigid body we have $\mathcal{M} = \mathbb{R}^3 \times SO(3) \subset \mathbb{R}^6 \subset \mathbb{R}^{3n}$, where "$\times$" is the *Cartesian product* (product of two sets), which means that $\mathbb{R}^3 \times SO(3)$ consists of the set of all ordered pairs $(\mathbf{r}, \boldsymbol{\theta})$. We write $\vec{q} = (q_1, \ldots, q_6) = (\mathbf{r}, \boldsymbol{\theta}) \in \mathcal{M}$.

Holonomic and non-holonomic constraints: *Holonomic* constraints are constraints on generalized coordinates that can be written as

$$f(q_1, q_2, \ldots, q_N, t) = 0, \tag{2.1.4}$$

with no dependence on generalized velocities, $\dot{q}_1 \equiv dq_i/dt$. A constraint that cannot be written in such a form is said to be *non-holonomic*. If the holonomic constraints are time-independent then we say that the system is *scleronomous*, and if they are time-dependent we say it is *rheonomous*. A bead sliding on a frictionless wire fixed in space is an example of a *scleronomic* constraint, whereas if the wire is itself moving then this is a *rheonomic* constraint.

Holonomic constraints do no virtual work: Consider a system consisting of n spinless particles moving in three spatial directions with no constraints. In this case the system has $N = 3n$ generalized coordinates that can be taken as $(\mathbf{r}_1, \ldots, \mathbf{r}_n)$. If we now impose a set of r independent holonomic constraints,

$$f_b(\mathbf{r}_1, \mathbf{r_2}, \ldots, \mathbf{r}_n, t) = 0 \quad \text{for} \quad b = 1, 2, \ldots, r, \tag{2.1.5}$$

then the number of degrees of freedom remaining is $N = 3n - r$ with generalized coordinates $\vec{q} = (q_1, q_2, \ldots, q_N)$ at all times t. Each generalized coordinate is a function of the n particle positions and also of time,

$$q_i = q_i(\mathbf{r}_1, \mathbf{r}_2, \ldots, \mathbf{r}_n, t) \quad \text{for} \quad i = 1, 2, \ldots, N \tag{2.1.6}$$

and the time-dependent generalized coordinate manifold $\mathcal{M}_t$,

$$\vec{q} = (q_1, \ldots, q_N) \in \mathcal{M}_t \subset \mathbb{R}^{3n} \quad \text{with} \quad N = (3n - r). \tag{2.1.7}$$

As the system evolves in time then $\vec{q}(t)$ is a trajectory on $\mathcal{M}_t$, where $\mathcal{M}_t$ is itself changing with time. If the holonomic constraints are time-independent (scleronomic), then the generalized coordinate manifold $\mathcal{M}$ is time-independent.

A *virtual displacement* of the system at time t is denoted as $(\delta\mathbf{r}_1, \delta\mathbf{r}_2, \ldots, \delta\mathbf{r}_n)$ and is an infinitesimal change of the coordinates of the system that is consistent with the constraints *while holding the time fixed*; i.e., no time elapses during the coordinate change. No real displacement can take place without the passing of time. If $(\delta\mathbf{r}_1, \delta\mathbf{r}_2, \ldots, \delta\mathbf{r}_n)$ is a virtual displacement at time t, then by definition we have $(\mathbf{r}_1, \ldots, \mathbf{r}_n)$ and $(\mathbf{r}_1 + \delta\mathbf{r}_1, \ldots, \mathbf{r}_n + \delta\mathbf{r}_n) \in \mathcal{M}_t$. A *real displacement* of the system would be an infinitesimal change consistent with the constraints of the form $(d\mathbf{r}_1, \ldots, d\mathbf{r}_n, dt)$, where $(\mathbf{r}_1, \ldots, \mathbf{r}_n) \in \mathcal{M}_t$ and $(\mathbf{r}_1 + d\mathbf{r}_1, \ldots, \mathbf{r}_n + d\mathbf{r}_n) \in \mathcal{M}_{t+dt}$.

The useful property of a virtual displacement is that for every $\vec{q}$ and $\vec{q} + \delta\vec{q} \in \mathcal{M}_t$ we have a unique $(\mathbf{r}_1, \ldots, \mathbf{r}_n)$ and $(\mathbf{r}_1 + \delta\mathbf{r}_1, \ldots, \mathbf{r}_n + \delta\mathbf{r}_n) \in \mathcal{M}_t$; i.e., for every time t we have an invertible map between every $\vec{q} \in \mathcal{M}_t$ and every $\vec{r} \equiv (\mathbf{r}_1, \ldots, \mathbf{r}_n) \in \mathcal{M}_t$, but the map will be different at different times. So for every $\vec{r} \in \mathcal{M}_t$ we can write

$$\mathbf{r}_a = \mathbf{r}_a(q_1, \ldots, q_N, t) = \mathbf{r}_a(\vec{q}, t) \ \text{for} \ a = 1, \ldots, n \ \text{at every time } t. \tag{2.1.8}$$

So for virtual displacements with holonomic constraints we have

$$\delta\mathbf{r}_a = \sum_{i=1}^{N} (\partial\mathbf{r}_a/\partial q_i)\delta q_i \qquad \text{for every time } t. \tag{2.1.9}$$

For time-independent holonomic constraints this is also true for real displacements.

Virtual work, denoted as δW, is the work done under a virtual displacement by forces $(\mathbf{F}_1, \ldots, \mathbf{F}_n)$ acting on the n particles. We use a $3n$-component vector notation for the n forces and virtual displacements and so on, where, for example,

$$\vec{\mathbf{F}} \equiv (\mathbf{F}_1, \ldots, \mathbf{F}_n) \quad \text{and} \quad \delta\vec{r} \equiv (\delta\mathbf{r}_1, \ldots, \delta\mathbf{r}_n). \tag{2.1.10}$$

The $3n$-component virtual displacement vector $\delta\vec{r}$ is tangential to the constraint manifold $\mathcal{M}_t$ at fixed time t. With this notation the virtual work done by $\vec{\mathbf{F}}$ becomes

$$\delta W = \sum_{a=1}^{n} \mathbf{F}_a \cdot \delta\mathbf{r}_a = \delta W = \vec{\mathbf{F}} \cdot \delta\vec{r}. \tag{2.1.11}$$

The Cartesian velocities and accelerations of the n particles are $\mathbf{v}_a = d\mathbf{r}_a/dt$ and $\mathbf{a}_a = d\mathbf{v}_a/dt$, respectively, for $a = 1, 2, \ldots, n$. The Cartesian momenta are $\mathbf{p}_a = m_a\mathbf{v}_a$ and the total kinetic energy of the n-particle system is

$$T = \sum_{a=1}^{n} T_a = \sum_{a=1}^{n} \tfrac{1}{2}m_a\mathbf{v}_a^2 = \sum_{a=1}^{n} \tfrac{1}{2}m_a\dot{\mathbf{r}}_a^2, \tag{2.1.12}$$

where m_a is the mass of particle a. Newton's second law states that the total (i.e., net) force on particle a is the time rate of change of its momentum,

$$\mathbf{F}_a = \dot{\mathbf{p}}_a = m_a\mathbf{a}_a. \tag{2.1.13}$$

If no forces act on any of the particles, then no work done and so all particle velocities $\mathbf{v}_a$ and kinetic energies T_a are constant.

Now let us impose r holonomic constraints on this system of particles as in Eq. (2.1.5). A $3n$-vector of forces $\vec{\mathbf{F}}$ can be decomposed into a component that is orthogonal to the constraint manifold $\mathcal{M}_t$ at fixed time t and a component that is tangential to it. The component that is orthogonal to $\mathcal{M}_t$ is referred to as the constraining force, $\vec{\mathbf{F}}^{\mathrm{constr}}$. The remaining component is referred to as the applied force, $\vec{\mathbf{F}}^{\mathrm{appl}}$. So for any force $\vec{\mathbf{F}}$ at time t we have the decomposition

$$\vec{\mathbf{F}} = \vec{\mathbf{F}}^{\mathrm{appl}} + \vec{\mathbf{F}}^{\mathrm{constr}}. \tag{2.1.14}$$

Since $\delta\vec{r}$ is also tangential to $\mathcal{M}_t$, then constraint forces do no virtual work,

$$\delta W^{\mathrm{constr}} = \vec{\mathbf{F}}^{\mathrm{constr}} \cdot \delta\vec{r} = \sum_{a=1}^{n} \mathbf{F}_a^{\mathrm{constr}} \cdot \delta\mathbf{r}_a = 0. \tag{2.1.15}$$

For time-independent holonomic constraints this is also true for real displacements.

Simple example: Consider a mass m moving without friction on a plane but attached to the end of a massless string that passes through a hole in the plane. The length of the string, $\ell(t)$, can change by sliding through the hole. Consider only motion that keeps the string taut. For a fixed string length, $\ell(t) = \ell$, the motion is circular. The constraint manifold is the circle of radius ℓ and the constraint force is the centripetal force. For varying $\ell(t)$ the constraint manifold, $\mathcal{M}_t$, at any fixed time t is a circle of radius $\ell(t)$, and a virtual displacement at time t is tangential to it with an orthogonal centripetal constraining force. So no work is done by the constraint forces in a virtual displacement. A real displacement will not be tangential to the circle $\mathcal{M}_t$ since we are moving from $\mathcal{M}_t$ with radius $\ell(t)$ to $\mathcal{M}_{t+dt}$ with radius $\ell(t + dt)$ over time dt. If we shorten the string, we are doing work on the system and increasing the kinetic energy of the mass. The radial force, $\mathbf{F}$, due to the string is not orthogonal to the noncircular path of the mass and so real work is done.

D'Alembert's principle: Using the identity $\mathbf{F}_a - \dot{\mathbf{p}}_a = 0$ for $a = 1, \ldots, n$ from Newton's second law and Eq. (2.1.15), we find

$$0 = \sum_a (\mathbf{F}_a - \dot{\mathbf{p}}_a) \cdot \delta\mathbf{r}_a = \sum_a \left([\mathbf{F}_a^{\mathrm{appl}} + \mathbf{F}_a^{\mathrm{constr}}] - \dot{\mathbf{p}}_a \right) \cdot \delta\mathbf{r}_a = \sum_a \left(\mathbf{F}_a^{\mathrm{appl}} - \dot{\mathbf{p}}_a \right) \cdot \delta\mathbf{r}_a.$$

Assuming constant masses $\dot{\mathbf{p}}_a = m_a\mathbf{a}_a$ for particles $a = 1, \ldots, n$ and so we arrive at *d'Alembert's principle*, which states that

$$\sum_{a=1}^{n} \left(\mathbf{F}_a^{\mathrm{appl}} - m_a\mathbf{a}_a \right) \cdot \delta\mathbf{r}_a = 0, \tag{2.1.16}$$

where $\delta\vec{r} = (\delta\mathbf{r}_1, \delta\mathbf{r}_1, \ldots, \delta\mathbf{r}_n)$ is a virtual displacement of the system.

Lagrangian and Euler-Lagrange equations: Using Eq. (2.1.9) we see that d'Alembert's principle Eq. (2.1.16) takes the form

$$\sum_{i=1}^{N} \sum_{a=1}^{n} \left(\mathbf{F}_a^{\text{appl}} - m_a \mathbf{a}_a \right) \cdot (\partial \mathbf{r}_a / \partial q_i) \delta q_i = 0. \tag{2.1.17}$$

The variations in the generalized coordinates, δq_i, are arbitrary and so

$$Q_i - \sum_{a=1}^{n} m_a \mathbf{a}_a \cdot (\partial \mathbf{r}_a / \partial q_i) = 0 \quad \text{with} \quad Q_i \equiv \sum_{a=1}^{n} \mathbf{F}_a^{\text{appl}} \cdot (\partial \mathbf{r}_a / \partial q_i), \tag{2.1.18}$$

which defines the *generalized force*, Q_i, for the generalized coordinate q_i.

The kinetic energy in Eq. (2.1.12) has the form $T = \sum_{a=1}^{n} \frac{1}{2} m_a \dot{\mathbf{r}}_a \cdot \dot{\mathbf{r}}_a$ and so

$$\frac{\partial T}{\partial q_j} = \sum_{a=1}^{n} m_a \dot{\mathbf{r}}_a \cdot \frac{\partial \dot{\mathbf{r}}_a}{\partial q_j} \quad \text{and} \quad \frac{\partial T}{\partial \dot{q}_j} = \sum_{a=1}^{n} m_a \dot{\mathbf{r}}_a \cdot \frac{\partial \dot{\mathbf{r}}_a}{\partial \dot{q}_j}. \tag{2.1.19}$$

Since $\mathbf{r}_a = \mathbf{r}_a(\vec{q}, t)$ for $a = 1, 2, \ldots, n$ we have

$$\mathbf{v}_a = \dot{\mathbf{r}}_a(\vec{q}, t) = \frac{d}{dt} \mathbf{r}_a(\vec{q}, t) = \frac{\partial \mathbf{r}_a}{\partial t} + \sum_{i=1}^{N} \frac{\partial \mathbf{r}_a}{\partial q_i} \frac{dq_i}{dt} = \frac{\partial \mathbf{r}_a}{\partial t} + \sum_{i=1}^{N} \frac{\partial \mathbf{r}_a}{\partial q_i} \dot{q}_i. \tag{2.1.20}$$

Acting on both sides with $\partial / \partial \dot{q}_j$ we find the "cancellation of the dots" result that

$$\partial \dot{\mathbf{r}}_a / \partial \dot{q}_j = \partial \mathbf{r}_a / \partial q_j. \tag{2.1.21}$$

In arriving at the above results we treat q_i and $\dot{q}_i$ as *independent quantities* at every fixed time t, e.g., $\partial \dot{q}_i(t) / \partial q_i(t) = 0$ and $\partial q_i(t) / \partial \dot{q}_i(t) = 0$. This is appropriate since in order to specify the state of the system at time t, we must specify $2N$ quantities, i.e., the generalized coordinates $\vec{q}(t)$ and the generalized velocities $\dot{\vec{q}}(t)$.

Since $\mathbf{r}_a = \mathbf{r}_a(\vec{q}, t)$ for $a = 1, 2, \ldots, n$ we have $\partial \mathbf{r}_a / \partial \dot{q}_i = 0$. It is also clear that $\partial \mathbf{r}_a / \partial q_i$ is a function of $\vec{q}$ and t, $(\partial \mathbf{r}_a / \partial q_i)(\vec{q}, t)$, and so

$$\frac{d}{dt} \frac{\partial \mathbf{r}_a}{\partial q_j} = \frac{\partial^2 \mathbf{r}_a}{\partial t \partial q_j} + \sum_{i=1}^{N} \frac{\partial \mathbf{r}_a}{\partial q_i \partial q_j} \frac{dq_i}{dt} = \frac{\partial}{\partial q_j} \left(\frac{\partial \mathbf{r}_a}{\partial t} + \sum_{i=1}^{N} \frac{\partial \mathbf{r}_a}{\partial q_i} \dot{q}_i \right) = \frac{\partial \dot{\mathbf{r}}_a}{\partial q_j}, \tag{2.1.22}$$

where we have used Eq. (2.1.20). We are assuming that our constraint manifold $\mathcal{M}_t$ is smooth enough that $\mathbf{r}_a(q_1, q_2, \ldots, q_N, t)$ is doubly differentiable.

We can now evaluate the total time derivative of Eq. (2.1.19)

$$\frac{d}{dt} \frac{\partial T}{\partial \dot{q}_j} = \sum_{a=1}^{n} \left(m_a \ddot{\mathbf{r}}_a \cdot \frac{\partial \dot{\mathbf{r}}_a}{\partial \dot{q}_j} + m_a \dot{\mathbf{r}}_a \cdot \frac{d}{dt} \frac{\partial \dot{\mathbf{r}}_a}{\partial \dot{q}_j} \right) = \sum_{a=1}^{n} \left(m_a \ddot{\mathbf{r}}_a \cdot \frac{\partial \mathbf{r}_a}{\partial q_j} + m_a \dot{\mathbf{r}}_a \cdot \frac{d}{dt} \frac{\partial \mathbf{r}_a}{\partial q_j} \right)$$

$$= \sum_{a=1}^{n} \left(m_a \ddot{\mathbf{r}}_a \cdot \frac{\partial \mathbf{r}_a}{\partial q_j} + m_a \dot{\mathbf{r}}_a \cdot \frac{\partial \dot{\mathbf{r}}_a}{\partial q_j} \right) = \left(\sum_{a=1}^{n} m_a \ddot{\mathbf{r}}_a \cdot \frac{\partial \mathbf{r}_a}{\partial q_j} \right) + \frac{\partial T}{\partial q_j}. \tag{2.1.23}$$

Note that in moving from Eq. (2.1.12) to Eq. (2.1.23) we have not imposed any dynamics on the system and the latter equation is simply a mathematical identity that follows from Eq. (2.1.8) and the definition of kinetic energy.

Consider a system with r constraints that satisfy d'Alembert's principle as given in Eq. (2.1.16) and for which every accessible region of configuration space has $N = 3n - r$ independent generalized coordinates as in Eq. (2.1.8). This includes any physical system with r holonomic constraints. For such systems we can combine the dynamics coming from the form of Newton's second law in Eq. (2.1.18) and the mathematical identity in Eq. (2.1.23) to obtain the equations of motion

$$\frac{d}{dt}\left(\frac{\partial T}{\partial \dot{q}_j}\right) - \frac{\partial T}{\partial q_j} = \left(\sum_{a=1}^{n} m_a \ddot{\mathbf{r}}_a \cdot \frac{\partial \mathbf{r}_a}{\partial q_j}\right) = Q_j \quad \text{for} \quad j = 1, 2, \ldots, N \tag{2.1.24}$$

with the generalized forces Q_i defined in Eq. (2.1.18). These are sometimes referred to as the generalized form of the Euler-Lagrange equations.

If the virtual work done on the physical system by the applied forces is independent of the path taken in the constraint submanifold, then the system is a *conservative* system and the applied forces are referred to as *conservative forces*. Independence of the path means that the work done depends only on the initial and final points in configuration space (i.e., in the constraint submanifold), which we will refer to as $\vec{q}_a$ and $\vec{q}_b$, respectively. In systems containing friction or other dissipative forces the virtual work done will clearly depend on the path taken and so these are *not* conservative systems. The virtual work done by the applied forces along the virtual path C at some fixed time t is

$$W(\vec{q}_b, \vec{q}_a, t) = \int_C \sum_{a=1}^{n} \mathbf{F}_a^{\text{appl}} \cdot \delta \mathbf{r}_a = \int_C \sum_{i=1}^{N} \sum_{a=1}^{n} \mathbf{F}_a^{\text{appl}} \cdot \frac{\partial \mathbf{r}_a}{\partial q_i} \delta q_i = \int_C \sum_{i=1}^{N} Q_i \delta q_i \tag{2.1.25}$$

and if the virtual work is path independent (i.e., a conservative system) we can define some function $V(\vec{q}, t)$ such that

$$W(\vec{q}_b, \vec{q}_a, t) = -\left[V(\vec{q}_b, t) - V(\vec{q}_a, t)\right]. \tag{2.1.26}$$

The function $V(\vec{q}, t)$ is referred to as the *potential energy* and the negative sign is the conventional one leading to the result that the work done is $W = -\Delta V$. For an infinitesimal path $\vec{q}_b = \vec{q}_a + \delta \vec{q}$ we find from Eqs. (2.1.25) and 2.1.26

$$-\delta V = \sum_{i=1}^{N} Q_i \delta q_i \tag{2.1.27}$$

and hence we have for a conservative system

$$V = V(q_1, \ldots, q_N, t) \quad \text{and} \quad Q_i = -\partial V / \partial q_i = -\sum_{a=1}^{n} \boldsymbol{\nabla}_a V \cdot (\partial \mathbf{r}_a / \partial q_i). \tag{2.1.28}$$

Compare Eqs. (2.1.28) and (2.1.18) and observe that we have projected

$$\mathbf{F}_a^{\text{appl}} = -\boldsymbol{\nabla}_a V \quad \text{for} \quad a = 1, 2, \ldots, n \tag{2.1.29}$$

onto the set of allowed virtual displacements using the scalar product with $\partial \mathbf{r}_a / \partial q_i$.

The Lagrangian of a physical system is defined as

$$L = T - V, \tag{2.1.30}$$

where T and V are, respectively, the kinetic energy and the potential energy for the system expressed in term of the generalized coordinates, q_i, and their corresponding generalized velocities, $\dot{q}_i$. We can now combine Eq. (2.1.24) with Eqs. (2.1.28) and (2.1.30) to give the $2N$ *Euler-Lagrange equations*

$$\frac{d}{dt}\frac{\partial L}{\partial \dot{q}_i} - \frac{\partial L}{\partial q_i} = 0 \quad \text{and} \quad \dot{q}_i = \frac{dq_i}{dt} \quad \text{for} \quad i = 1, \ldots N \tag{2.1.31}$$

for the $2N$ variables $(\vec{q}, \dot{\vec{q}})$. We have used $\partial L / \partial \dot{q}_i = \partial (T - V) / \partial \dot{q}_i = \partial T / \partial \dot{q}_i$. The equations $\dot{q}_i = dq_i / dt$ are often left to be understood. The Euler-Lagrange equations apply for any physical system with holonomic constraints and conservative forces, i.e., with forces arising from a velocity-independent potential energy.

We can extend the class of physical systems that satisfy the Euler-Lagrange equations to include a very specific class of velocity-dependent potentials. Suppose there exists a velocity-dependent potential $V(\vec{q}, \dot{\vec{q}}, t)$ that satisfies

$$Q_i = -\frac{\partial V}{\partial q_i} + \frac{d}{dt}\left(\frac{\partial V}{\partial \dot{q}_i}\right). \tag{2.1.32}$$

Then we see that $L = T - V$ again leads from the equations of motion in Eq. (2.1.24) to the Euler-Lagrange equations, Eq. (2.1.31). Such systems are referred to as *monogenic* systems. For velocity-independent potentials Eq. (2.1.32) reduces to Eq. (2.1.28) and so conservative systems are a special case of monogenic systems.

Example: The Lorentz force: An example of a monogenic system with a velocity-dependent potential is a charged particle with charge q moving in electric and magnetic fields, **E** and **B**, respectively. The potential in this case is

$$V = q[\Phi - \boldsymbol{v} \cdot \mathbf{A}], \tag{2.1.33}$$

where Φ is the (scalar) electrostatic potential and **A** is the magnetic vector potential. We use SI units here. Recall that

$$\mathbf{E} = -\boldsymbol{\nabla}\Phi - \partial\mathbf{A}/\partial t \quad \text{and} \quad \mathbf{B} = \boldsymbol{\nabla} \times \mathbf{A}. \tag{2.1.34}$$

The Euler-Lagrange equations lead to the *Lorentz force*,

$$\mathbf{F} = d\mathbf{p}/dt = d(m\boldsymbol{v})/dt = q\left[\mathbf{E} + \boldsymbol{v} \times \mathbf{B}\right]. \tag{2.1.35}$$

The kinetic energy of the particle changes through the work done on it,

$$dT = \mathbf{F} \cdot d\mathbf{s} \quad \Rightarrow \quad dT/dt = \mathbf{F} \cdot \boldsymbol{v} = q\left(\mathbf{E} + \boldsymbol{v} \times \mathbf{B}\right) \cdot \boldsymbol{v} = q\mathbf{E} \cdot \boldsymbol{v}, \tag{2.1.36}$$

so that only the **E**-field can change the particle's kinetic energy.

Proof: (i) *First assume a monogenic system:* Using $\mathbf{x} = (x, y, z) = (x_1, x_2, x_3)$ as the generalized coordinates the Lagrangian is given by

$$L = T - V = \tfrac{1}{2}m\dot{\mathbf{x}}^2 - q[\Phi - \dot{\mathbf{x}} \cdot \mathbf{A}] \tag{2.1.37}$$

and so the equations of motion are

$$0 = \frac{d}{dt}\frac{\partial L}{\partial \dot{x}_i} - \frac{\partial L}{\partial x_i} = \frac{d}{dt}\left(m\dot{x}_i + qA_i\right) + q\left(\frac{\partial\Phi}{\partial x_i} - \sum_{j=1}^{3}\frac{\partial A_j}{\partial x_i}\dot{x}_j\right) \tag{2.1.38}$$

$$= m\ddot{x}_i + q\left(\frac{\partial A_i}{\partial t} + \sum_{j=1}^{3}\frac{\partial A_i}{\partial x_j}\dot{x}_j\right) + q\left(\frac{\partial\Phi}{\partial x_i} - \sum_{j=1}^{3}\frac{\partial A_j}{\partial x_i}\dot{x}_j\right)$$

$$= m\ddot{x}_i + q\left(\frac{\partial\Phi}{\partial x_i} + \frac{\partial A_i}{\partial t}\right) - q\sum_{j=1}^{3}\dot{x}_j\left(\frac{\partial A_j}{\partial x_i} - \frac{\partial A_i}{\partial x_j}\right)$$

$$= m\ddot{x}_i - qE_i - q\sum_{j=1}^{3}\dot{x}_j\left(\sum_{\ell,m=1}^{3}[\delta_{jm}\delta_{i\ell} - \delta_{j\ell}\delta_{im}]\frac{\partial A_m}{\partial x_\ell}\right)$$

$$= m\ddot{x}_i - qE_i - q\sum_{j,\ell,m,k=1}^{3}\dot{x}_j[\epsilon_{ijk}\epsilon_{klm}]\frac{\partial A_m}{\partial x_\ell}$$

$$= m\ddot{x}_i - qE_i - q\sum_{j,k=1}^{3}\dot{x}_j\epsilon_{ijk}(\boldsymbol{\nabla}\times\mathbf{A})_k = m\ddot{x}_i - qE_i - q(\dot{\mathbf{x}}\times\mathbf{B})_i,$$

which is the Lorentz force given in Eq. (2.1.35) since $\mathbf{F} = m\ddot{\mathbf{x}}$.

(ii) *Now show that the system is monogenic:* Rewrite the Lorentz force as

$$\mathbf{F} = q[-\boldsymbol{\nabla}\Phi - \tfrac{\partial\mathbf{A}}{\partial t} + \boldsymbol{v}\times(\boldsymbol{\nabla}\times\mathbf{A})] = q[-\boldsymbol{\nabla}\Phi - \tfrac{\partial\mathbf{A}}{\partial t} + \boldsymbol{\nabla}(\boldsymbol{v}\cdot\mathbf{A}) - (\boldsymbol{v}\cdot\boldsymbol{\nabla})\mathbf{A}]$$

$$= q[-\boldsymbol{\nabla}(\Phi - \boldsymbol{v}\cdot\mathbf{A}) - \tfrac{d\mathbf{A}}{dt}] = q[-\boldsymbol{\nabla}(\Phi - \boldsymbol{v}\cdot\mathbf{A}) + \tfrac{d}{dt}\boldsymbol{\nabla}_{\dot{\mathbf{x}}}(\Phi - \boldsymbol{v}\cdot\mathbf{A})]$$

$$= -\boldsymbol{\nabla}V + \tfrac{d}{dt}\boldsymbol{\nabla}_{\dot{\mathbf{x}}}V, \tag{2.1.39}$$

where we have used the following results: (a) $\mathbf{a} \times (\mathbf{b} \times \mathbf{c}) = \mathbf{b}(\mathbf{a} \cdot \mathbf{c}) - \mathbf{c}(\mathbf{a} \cdot \mathbf{b})$ for any three commuting three-vector objects; (b) $\partial \dot{x}_j / \partial x_i = 0$; (c) the definition $\boldsymbol{\nabla}_{\dot{\mathbf{x}}} \equiv (\partial/\partial \dot{x}_1, \partial/\partial \dot{x}_2, \partial/\partial \dot{x}_3) = (\partial/\partial v_1, \partial/\partial v_2, \partial/\partial v_3)$; (d) the fact that Φ and $\mathbf{A}$ are independent of velocity; and (e) that

$$\frac{d\mathbf{A}}{dt} = \frac{\partial \mathbf{A}}{\partial t} + \sum_{i=1}^3 \frac{\partial \mathbf{A}}{\partial x_i}\frac{dx_i}{dt} = \frac{\partial \mathbf{A}}{\partial t} + (\boldsymbol{v} \cdot \boldsymbol{\nabla})\mathbf{A}, \tag{2.1.40}$$

So writing Eq. (2.1.39) in component form we have shown the monogenic condition of Eq. (2.1.32), $F_i = -\frac{\partial V}{\partial x_i} + \frac{d}{dt}\left(\frac{\partial V}{\partial \dot{x}_i}\right)$.

Summary: Any monogenic system with holonomic constraints satisfies the Euler-Lagrange equations of Eq. (2.1.31) with the Lagrangian defined by Eq. (2.1.30). Conservative systems have no velocity-dependent potentials and are a special case of monogenic systems. Systems with friction and/or other dissipative forces are neither conservative nor monogenic and so do not satisfy the Euler-Lagrange equations.

Equivalent Lagrangians: Two Lagrangians are said to be equivalent if they lead to the same solutions for their equations of motion. There are two possibilities: (i) the equations of motion are the same, ensuring identical solutions; or (ii) the equations of motion are different, but the solutions are identical.

Two Lagrangians, L' and L, that differ only by multiplicative and additive constants and the total time derivative of a function depending only on generalized coordinates and time have the same equations of motion,

$$L' = aL + dF(\vec{q}, t)/dt + b. \tag{2.1.41}$$

Proof: It is clear that the constants a and b have no effect on the Euler-Lagrange equations and neither does $dF(\vec{q}, t)/dt$ since

$$
\begin{aligned}
\left(\frac{d}{dt}\frac{\partial}{\partial \dot{q}_i} - \frac{\partial}{\partial q_i}\right)\frac{d}{dt}F(\vec{q}, t) &= \frac{d}{dt}\frac{\partial}{\partial \dot{q}_i}\left[\sum_{j=1}^N \frac{\partial F}{\partial q_j}\dot{q}_j + \frac{\partial F}{\partial t}\right] - \frac{\partial}{\partial q_i}\frac{d}{dt}F(\vec{q}, t) \\
&= \frac{d}{dt}\frac{\partial F}{\partial q_i} - \frac{\partial}{\partial q_i}\frac{d}{dt}F(\vec{q}, t) \\
&= \left[\left(\sum_{j=1}^N \dot{q}_j\frac{\partial}{\partial q_j}\right) + \frac{\partial}{\partial t}\right]\frac{\partial F}{\partial q_i} - \frac{\partial}{\partial q_i}\left[\left(\sum_{j=1}^N \dot{q}_j\frac{\partial}{\partial q_j}\right) + \frac{\partial}{\partial t}\right]F(\vec{q}, t) \\
&= \left(\sum_{j=1}^N \dot{q}_j\left[\frac{\partial^2 F}{\partial q_j \partial q_i} - \frac{\partial^2 F}{\partial q_i \partial q_j}\right]\right) + \left[\frac{\partial^2 F}{\partial t \partial q_i} - \frac{\partial^2 F}{\partial q_i \partial t}\right] = 0. \tag{2.1.42}
\end{aligned}
$$

2.1.2 Hamilton's Principle

Hamilton's principle states that the classical equations of motion arise from finding a stationary point[1] of the action with respect to variations in the "path" of the system while holding the end points fixed. It is also referred to as the "principle of stationary action." The action is a functional of the path and is defined by

$$S[\vec{q}] \equiv \int_{t_i}^{t_f} dt\, L(\vec{q}, \dot{\vec{q}}, t) \tag{2.1.43}$$

[1] Sometimes this is referred to as the "principle of least action"; however, in general the classical motion corresponds to a stationary point of the action and not only a minimum. For some examples and discussion see, e.g., Gray and Taylor (2007).

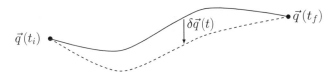

Figure 2.1   Variation of the path $\vec{q}(t)$ while holding the path end points fixed.

for the initial and final times, t_i and t_f, respectively. Specifying the path corresponds to specifying $\vec{q}(t)$ for all t from the initial to the final time, $t_i \leq t \leq t_f$. Since a classical system can only be in one state at a time, then we need only consider continuous paths that do not bend back on themselves. A variation of the path subject to fixed end points is illustrated in Fig. 2.1.

Hamilton's principle is the statement that the equations of motion for the classical system correspond to a stationary point of the action; i.e., the functional derivative of the action with respect to the path satisfies

$$\frac{\delta S[\vec{q}]}{\delta q_j(t)} = 0 \quad \text{for} \quad j = 1, 2, \ldots, N \quad \text{and} \quad t_i \leq t \leq t_f. \tag{2.1.44}$$

Functional derivative: Let S be some space of functions on some domain D. For example, consider the space of square integrable functions on $\mathbb{R}^n$, i.e., $D = \mathbb{R}^n$ and $S = \mathcal{L}^2(\mathbb{R}^n)$. A real function F with some S as its domain is referred to as a *functional*. Hence the functional F is a mapping from the space of functions S onto the real number line $\mathbb{R}$, e.g., a mapping $F : \mathcal{L}^2(\mathbb{R}^n) \to \mathbb{R}$. The notation $F[\phi]$ denotes the value of the functional F corresponding to the element ϕ in function space, i.e., $\phi \in S$ and $F[\phi] \in \mathbb{R}$.

The linear variation of F at $\overline{\phi}$ in the direction σ is defined by

$$\lim_{\epsilon \to 0} \tfrac{1}{\epsilon} \left\{ F[\overline{\phi} + \epsilon\sigma] - F[\overline{\phi}] \right\} \equiv \int_D dx\, \frac{\delta F[\phi]}{\delta \phi(x)} \bigg|_{\phi = \overline{\phi}} \sigma(x), \tag{2.1.45}$$

where, if the limit exists with $\delta F[\phi]/\delta \phi(x)$ independent of σ, then $F[\phi]$ is functionally differentiable and $\delta F[\phi]/\delta \phi(x)|_{\phi = \overline{\phi}}$ is its functional derivative with respect to ϕ at the point $\phi = \overline{\phi}$. This is a continuum version of the familiar definition of the partial derivative of some real function f of vectors $\mathbf{r} \in \mathbb{R}^N$,

$$\lim_{\epsilon \to 0} \tfrac{1}{\epsilon} \left\{ f(\overline{\mathbf{r}} + \epsilon\mathbf{s}) - f(\overline{\mathbf{r}}) \right\} \equiv \sum_{i=1}^{N} \frac{\partial f(\mathbf{r})}{\partial r_i} \bigg|_{\mathbf{r} = \overline{\mathbf{r}}} s_i. \tag{2.1.46}$$

The functional derivative is a distribution in D. A *stationary point* of the functional $F[\phi]$ is a point in function space, $\overline{\phi} \in S$, such that the gradient of F vanishes in every direction in that space,

$$\frac{\delta F[\phi]}{\delta \phi(x)} \bigg|_{\phi = \overline{\phi}} = 0. \tag{2.1.47}$$

The nth order derivative $\delta^n F[\phi]/\delta\phi(x_n)\cdots\delta\phi(x_1)$ at $\overline{\phi}$ is a functional of $\overline{\phi}$ and a function of $x_1, \ldots, x_n$. To say that F is an analytic functional in some region containing $\overline{\phi}$ is to say that in this region there is a functional Taylor expansion for $F[\phi]$ about $\overline{\phi}$ given by

$$F[\phi] = F[\overline{\phi}] + \int_D dx_1\, \frac{\delta F[\phi]}{\delta \phi(x_1)} \bigg|_{\phi = \overline{\phi}} [\phi(x_1) - \overline{\phi}(x_1)] \tag{2.1.48}$$

$$+ \tfrac{1}{2!} \int_D dx_1 dx_2\, \frac{\delta^2 F[\phi]}{\delta \phi(x_2)\delta\phi(x_1)} \bigg|_{\phi = \overline{\phi}} [\phi(x_2) - \overline{\phi}(x_2)][\phi(x_1) - \overline{\phi}(x_1)] + \cdots.$$

Functional derivatives have properties like those of partial derivatives, e.g.,

$$df = \sum_{i=1}^{N} \frac{\partial f}{\partial r_i}\bigg|_{\mathbf{r}=\bar{\mathbf{r}}} dr_i \quad \rightarrow \quad \delta F = \int_D dx\, \frac{\delta F}{\delta \phi(x)}\bigg|_{\phi=\bar{\phi}} \delta\phi(x), \tag{2.1.49}$$

$$\frac{\partial r_i}{\partial r_j} = \delta_{ij} \quad \rightarrow \quad \frac{\delta\phi(x)}{\delta\phi(y)} = \delta(x-y).$$

It is also straightforward to use the definition to show the following results:

$$\frac{\delta(d^n\phi/dx^n)(x)}{\delta\phi(y)} = \frac{d^n}{dx^n}\delta(x-y), \quad \frac{\delta[\int_D dy\, f(\phi(y))]}{\delta\phi(x)} = \frac{\partial f(\phi(x))}{\partial\phi(x)}, \tag{2.1.50}$$

$$\frac{\delta(F[\phi]G[\phi])}{\delta\phi(x)} = \frac{\delta F[\phi]}{\delta\phi(x)}G[\phi] + F[\phi]\frac{\delta G[\phi]}{\delta\phi(x)}, \quad \frac{\delta f(F[\phi])}{\delta\phi(x)} = \frac{df}{dF}\frac{\delta F[\phi]}{\delta\phi(x)}.$$

Let us define the path variation as $d\vec{q}(t) \equiv \epsilon\vec{r}(t)$ for infinitesimal ϵ so that we can use Eq. (2.1.45) to write

$$\int_{t_i}^{t_f} dt \sum_{j=1}^{N} \frac{\delta S[\vec{q}]}{\delta q_j(t)} r_j(t) = \lim_{\epsilon\to 0}\frac{1}{\epsilon}\{S[\vec{q}+\epsilon\vec{r}] - S[\vec{q}]\} \tag{2.1.51}$$

$$= \int_{t_i}^{t_f} dt \lim_{\epsilon\to 0}\frac{1}{\epsilon}\left\{L(\vec{q}+\epsilon\vec{r}, \dot{\vec{q}}+\epsilon\dot{\vec{r}}, t) - L(\vec{q}, \dot{\vec{q}}, t)\right\} = \int_{t_i}^{t_f} dt \sum_{j=1}^{N}\left\{\frac{\partial L}{\partial q_j}r_j + \frac{\partial L}{\partial \dot{q}_j}\dot{r}_j\right\}$$

$$= \int_{t_i}^{t_f} dt \sum_{j=1}^{N}\left\{\frac{\partial L}{\partial q_j}r_j - \left(\frac{d}{dt}\frac{\partial L}{\partial \dot{q}_j}\right)r_j + \frac{d}{dt}\left(\frac{\partial L}{\partial \dot{q}_j}r_j\right)\right\}$$

$$= \left[\sum_{j=1}^{N}\frac{\partial L}{\partial \dot{q}_j}r_j\right]_{t_i}^{t_f} + \int_{t_i}^{t_f} dt \sum_{j=1}^{N}\left\{\frac{\partial L}{\partial q_j} - \left(\frac{d}{dt}\frac{\partial L}{\partial \dot{q}_j}\right)\right\}r_j = \int_{t_i}^{t_f} dt \sum_{j=1}^{N}\left\{\frac{\partial L}{\partial q_j} - \frac{d}{dt}\frac{\partial L}{\partial \dot{q}_j}\right\}r_j,$$

where in the above we have made use of integration by parts and the fact that $\vec{r}(t_i) = \vec{r}(t_f) = 0$, since the path end points are required to remain fixed during the variation. Noting that $\vec{r}(t)$ is arbitrary at intermediate times, we can combine Eqs. (2.1.44) and (2.1.51) to find that Hamilton's principle leads to

$$0 = \frac{\delta S[\vec{q}]}{\delta q_j(t)} = \frac{\partial L}{\partial q_j} - \frac{d}{dt}\frac{\partial L}{\partial \dot{q}_j} \quad \text{for} \quad j = 1, 2, \ldots, N, \tag{2.1.52}$$

which are the Euler-Lagrange equations given in Eq. (2.1.31). Note that in deriving Eq. (2.1.52) we have assumed that the system has only holonomic constraints so that we are free to manipulate derivatives as we did in deriving Eq. (2.1.24).

Hamilton's principle provides an alternate proof that L and L' in Eq. (2.1.41) lead to the same Euler-Lagrange equations of motion,

$$S'[\vec{q}] = \int_{t_i}^{t_f} dt\, L' = \int_{t_i}^{t_f} dt\left[aL + \frac{d}{dt}F(\vec{q}, t) + b\right] = aS[\vec{q}] + [F(\vec{q}, t)]_{t_i}^{t_f} + b(t_f - t_i)$$

$$\Rightarrow \quad \delta S'[\vec{q}]/\delta q_j(t) = a\delta S[\vec{q}]/\delta q_j(t) = 0, \tag{2.1.53}$$

since in the application of Hamilton's principle $\vec{q}(t_f)$ and $\vec{q}(t_i)$ are held fixed.

2.1.3 Lagrange Multipliers and Constraints

To begin a discussion of the application of Lagrange multipliers to classical mechanics it is useful to first begin with the mathematical question of how to find the stationary points of a function subject to a constraint (Arfken and Weber, 1995; Goldstein et al., 2013). Consider a function F of three Cartesian coordinates $\mathbf{x} \equiv (x^1, x^2, x^3) = (x, y, z)$, i.e., $F(\mathbf{x})$. The stationary points of F in the absence of any constraints satisfy the three equations

$$\boldsymbol{\nabla} F(\mathbf{x}) = \left(\frac{\partial F}{\partial x^1}, \frac{\partial F}{\partial x^2}, \frac{\partial F}{\partial x^3} \right)(\mathbf{x}) = 0. \tag{2.1.54}$$

Then consider some constraint $G(\mathbf{x}) = 0$, which defines a two-dimensional surface in the three-dimensional space. We wish to find the stationary points of F subject to the constraint $G(\mathbf{x}) = 0$. Consider an infinitesimal variation from any point $\mathbf{x}$ on the constraint surface to a neighboring point $\mathbf{x} + d\mathbf{x}$ on the constraint surface. Then we see that the corresponding variation in G vanishes:

$$0 = dG(\mathbf{x}) = \boldsymbol{\nabla} G(\mathbf{x}) \cdot d\mathbf{x} = \sum_{i=1}^{3} \frac{\partial G}{\partial x^i} dx^i. \tag{2.1.55}$$

Since $\boldsymbol{\nabla} G(\mathbf{x}) \cdot d\mathbf{x} = 0$ for every $d\mathbf{x}$ in the constraint surface, then $\boldsymbol{\nabla} G(\mathbf{x})$ is normal to the surface. If F is stationary at some particular point $\mathbf{x}$ on the surface with respect to variations along the constraint surface, then by definition we have $dF(\mathbf{x}) = \boldsymbol{\nabla} F(\mathbf{x}) \cdot d\mathbf{x} = 0$ for every $d\mathbf{x}$ in the surface at this point. It follows then that $\boldsymbol{\nabla} F$ and $\boldsymbol{\nabla} G$ are both normal to the constraint surface at this point and so are parallel,

$$\boldsymbol{\nabla} F = -\lambda \boldsymbol{\nabla} G, \tag{2.1.56}$$

for some constant $\lambda \in \mathbb{R}$. We refer to λ as a Lagrange multiplier and the negative sign in its definition is purely a matter of convention. In Fig. 2.2 we show an illustration for the two-dimensional case, where we wish to find the stationary points of the function $F(\mathbf{x}) = F(x, y)$ subject to the constraint $G(\mathbf{x}) = G(x, y) = 0$.

Solving for this problem is then equivalent to solving for the four variables, $\mathbf{x}$ and λ, subject to the four equations

$$\boldsymbol{\nabla} \left[F(\mathbf{x}) + \lambda G(\mathbf{x}) \right] = 0 \quad \text{and} \quad G(\mathbf{x}) = 0. \tag{2.1.57}$$

We can write these four equations in a single compact form as

$$\boldsymbol{\nabla}_{x\lambda} \left[F(\mathbf{x}) + \lambda G(\mathbf{x}) \right] = 0, \tag{2.1.58}$$

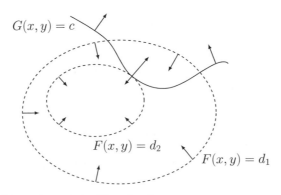

Figure 2.2   Stationary point of $F(x, y)$ subject to the constraint $G(x, y) = 0$.

where we have defined a four-component gradient operator

$$\boldsymbol{\nabla}_{x\lambda} \equiv \left(\boldsymbol{\nabla}, \frac{\partial}{\partial \lambda} \right) = \left(\frac{\partial}{\partial x^1}, \frac{\partial}{\partial x^2}, \frac{\partial}{\partial x^3}, \frac{\partial}{\partial \lambda} \right). \tag{2.1.59}$$

This generalizes to $\mathbf{x} \in \mathbb{R}^n$ and $m < n$ constraints $G_k(\mathbf{x}) = 0$ for $k = 1, \ldots, m$,

$$\boldsymbol{\nabla}_{x\lambda} \left[F(\mathbf{x}) + \sum_{\ell=1}^{m} \lambda_\ell G_\ell(\mathbf{x}) \right] = 0, \ \ \boldsymbol{\nabla}_{x\lambda} \equiv \left(\frac{\partial}{\partial x_1}, \ldots, \frac{\partial}{\partial x_n}, \frac{\partial}{\partial \lambda_1}, \ldots, \frac{\partial}{\partial \lambda_m} \right). \tag{2.1.60}$$

Let us apply this thinking to the requirement of solving for the classical motion of a system that is subject to a constraint. Consider the action for an unconstrained system of N generalized coordinates $S[\vec{q}]$ for $\vec{q}(t) = (q_1(t), \ldots, q_N(t))$ and $t_i \leq t \leq t_f$. We now subject the system to a holonomic constraint of the form

$$G(\vec{q}(t), t) = 0 \quad \text{for all} \quad t_i \leq t \leq t_f. \tag{2.1.61}$$

Let us imagine discretizing the time interval into $(m-1)$ finite time steps of size $\Delta t \equiv (t_f - t_i)/(m-1)$ such that $t_\ell = (\ell - 1)\Delta t + t_i$. We see that $t_1 = t_i$ and $t_m = t_f$. We have the correspondences $\mathbf{x} \to (\vec{q}(t_1), \ldots, \vec{q}(t_m))$, $F(\mathbf{x}) = S(\mathbf{x}) \to S[\vec{q}]$, $G_\ell(\mathbf{x}) \to G(\vec{q}(t_\ell), t_\ell)$ and $\lambda_\ell \equiv \lambda(t_\ell)\Delta t$. After applying the above formalism, and then taking the limit where the time step vanishes it is straightforward to arrive at

$$\frac{\delta S'[\vec{q}, \lambda]}{\delta \vec{q}(t)} = \frac{\delta S'[\vec{q}, \lambda]}{\delta \lambda(t)} = 0 \ \ \text{with} \ \ S'[\vec{q}, \lambda] \equiv S[\vec{q}] + \int_{t_i}^{t_f} dt \, \lambda(t) G(\vec{q}, t). \tag{2.1.62}$$

The generalization to $M < N$ holonomic constraints of the form $G_\ell(\vec{q}, t) = 0$ for $\ell = 1, \ldots, M$ is obvious:

$$\frac{\delta S'[\vec{q}, \vec{\lambda}]}{\delta \vec{q}(t)} = \frac{\delta S'[\vec{q}, \vec{\lambda}]}{\delta \vec{\lambda}(t)} = 0 \ \ \text{with} \ \ S[\vec{q}] = \int_{t_i}^{t_f} dt \, L(\vec{q}, \dot{\vec{q}}, t) \quad \text{and}$$

$$S'[\vec{q}, \vec{\lambda}] \equiv S[\vec{q}] + \int_{t_i}^{t_f} dt \sum_{\ell=1}^{M} \lambda_\ell(t) G_\ell(\vec{q}, t) \equiv \int_{t_i}^{t_f} dt \, L'(\vec{q}, \dot{\vec{q}}, \vec{\lambda}, t). \tag{2.1.63}$$

The resulting equations of motion are

$$0 = \frac{\delta S'[\vec{q}, \vec{\lambda}]}{\delta q_j(t)} = \frac{\partial L'}{\partial q_j} - \frac{d}{dt}\frac{\partial L'}{\partial \dot{q}_j} = \frac{\partial L}{\partial q_j} - \frac{d}{dt}\frac{\partial L}{\partial \dot{q}_j} + \vec{\lambda}(t) \cdot \frac{\partial \vec{G}}{\partial q_j} \ \ \text{for} \ \ j = 1, \ldots, N,$$

$$0 = \frac{\delta S'[\vec{q}, \vec{\lambda}]}{\delta \lambda_\ell(t)} = G_\ell(\vec{q}, t) \ \ \text{for} \ \ \ell = 1, \ldots, M. \tag{2.1.64}$$

The first N of these equations correspond to modified Euler-Lagrange equations with respect to the generalized coordinates $\vec{q}$, and the remaining M of these equations simply reproduce the constraint equations, $\vec{G}(\vec{q}, t) = 0$. There are $N + M$ equations that can determine the $N + M$ functions $\vec{q}(t)$ and $\vec{\lambda}(t)$, where the $\vec{q}(t)$ functions specify the classical motion of the system and the Lagrange multiplier functions $\vec{\lambda}(t)$ act to ensure that the constraints remain satisfied at all times.

Summary: If we have a monogenic and holonomic system with N generalized coordinates, then the classical equations of motion are given by the Euler-Lagrange equations of Eq. (2.1.31) and Hamilton's principle given in Eq. (2.1.52) applies; i.e., the classical equations of motion arise from finding the stationary points of the action. If such a system is subjected to an *additional* set of M holonomic constraints, $G_\ell(\vec{q}, t) = 0$, then we can introduce M real Lagrange multiplier functions, $\lambda_\ell(t)$, and define an enlarged set of $(N + M)$ generalized coordinates that we can solve for. By construction the solutions correspond to Hamilton's principle when applied to the constrained system.

Constraint forces arising from holonomic constraints: Consider a classical system of n particles moving in three dimensions in the presence of a monogenic potential, V, and a set of $r = (3n - N)$ holonomic constraints such that we have N generalized coordinates $\vec{q}$. Now consider what happens when we add an *additional* set of M holonomic constraints, $G_\ell(\vec{q}, t) = 0$ for $\ell = 1, \ldots, M$ using the method of Lagrange multipliers. Then we have from above that

$$L' = L + \sum_{k=1}^{M} \lambda_k(t) G_k(\vec{q}, t) = T - V + \sum_{k=1}^{M} \lambda_k(t) G_k(\vec{q}, t) \qquad (2.1.65)$$

and from Eq. (2.1.64) we find in place of Eq. (2.1.24) that we have

$$\frac{d}{dt}\left(\frac{\partial T}{\partial \dot{q}_j}\right) - \frac{\partial T}{\partial q_j} = Q_j + C_j \quad \text{with} \quad C_j \equiv \sum_{k=1}^{m} \lambda_k(t) \frac{\partial G_k}{\partial q_j} \qquad (2.1.66)$$

for $j = 1, 2, \ldots, N$. Note that Q_j is given by Eq. (2.1.28) and is the generalized force arising from the potential V. We see that the additional M holonomic constraints are leading to an additional conservative generalized force, C_j, for each generalized coordinate, q_j. From Eq. (2.1.18) we recognize that if we define $\mathbf{C}_a$ such that

$$C_i \equiv \sum_{a=1}^{n} \mathbf{C}_a \cdot (\partial \mathbf{r}_a / \partial q_i), \qquad (2.1.67)$$

then $\mathbf{C}_a$ is the three-dimensional force acting on particle a that arises from the additional M holonomic constraints, i.e., the additional force required to act on particle a to ensure it moves in a manner consistent with the additional M constraints.

Semi-holonomic constraints: There is a sub-class of non-holonomic constraints, *semi-holonomic* constraints, for which d'Alembert's principle is valid, i.e., that a virtual displacement does no work. Semi-holonomic constraints are linear in the generalized velocities, $G_\ell(\vec{q}, \dot{\vec{q}}, t) = A_{\ell j}(\vec{q}, t)\dot{q}_j + B_\ell(\vec{q}, t) = 0$. In addition, they must satisfy $\partial A_{\ell i}/\partial q_j = \partial A_{\ell j}/\partial q_i$ and $\partial B_\ell/\partial q_i = \partial A_{\ell i}/\partial t$ (Flannery, 2005).

2.2 Symmetries, Noether's Theorem and Conservation Laws

We now turn our attention to the deep and important connection between symmetries and conservations laws in physical systems.

Conservation laws and constants of the motion: If there exists some function of the generalized coordinates, generalized velocities and time $f(\dot{\vec{q}}(t), \vec{q}(t), t)$ such that throughout any classical trajectory we have

$$\dot{f} = df/dt = 0, \qquad (2.2.1)$$

then we say that f is a *conserved quantity* or *constant of the motion* and that $df/dt = 0$ is a *conservation law*. Conserved quantities are sometimes also referred to as *first integrals of the motion* or *integrals of the motion*.

Cyclic coordinates and generalized momentum: A generalized coordinate, q_j, is called *cyclic* or *ignorable* if it does not appear in the Lagrangian, while the corresponding generalized velocity, $\dot{q}_j$, does. The *generalized momentum* corresponding to the generalized coordinate q_j is defined as

$$p_j \equiv \partial L/\partial \dot{q}_j. \qquad (2.2.2)$$

The term "generalized momentum" is synonymous with *canonical momentum* and *conjugate momentum*. In Lagrangian mechanics we usually use the term "generalized momenta." When discussing the Hamiltonian formalism, Poisson brackets and quantum mechanics we typically use the terms "canonical momentum" or "conjugate momentum." For a cyclic coordinate, q_j, it is obvious that $\partial L/\partial q_j = 0$ and so it immediately follows from the Euler-Lagrange equations in Eq. (2.1.31) that

$$\frac{dp_j}{dt} = \frac{d}{dt}\frac{\partial L}{\partial \dot{q}_j} = 0 \tag{2.2.3}$$

and so the corresponding generalized momentum, p_j, is a conserved quantity.

Energy function: Consider the *energy function*, h, defined by

$$h(\vec{q}, \dot{\vec{q}}, t) \equiv \sum_{j=1}^{N} \dot{q}_j (\partial L/\partial \dot{q}_j) - L. \tag{2.2.4}$$

This is sometimes also referred to as *Jacobi's integral*. The energy function is so-named since h corresponds to the energy of the system in a broad class of systems called "natural systems." The total time derivative of the energy function is

$$dh/dt = -\partial L/\partial t, \tag{2.2.5}$$

which implies that the energy function is a conserved quantity if and only if the Lagrangian has no explicit dependence on time.

Proof: The total time derivative of the Lagrangian is

$$\frac{dL}{dt} = \sum_{j=1}^{N} \left[\frac{\partial L}{\partial q_i}\frac{dq_i}{dt} + \frac{\partial L}{\partial \dot{q}_i}\frac{d\dot{q}_i}{dt} \right] + \frac{\partial L}{\partial t}$$

$$= \sum_{j=1}^{N} \left[\frac{d}{dt}\left(\frac{\partial L}{\partial \dot{q}_i}\right)\dot{q}_i + \frac{\partial L}{\partial \dot{q}_i}\frac{d\dot{q}_i}{dt} \right] + \frac{\partial L}{\partial t} = \frac{d}{dt}\left[\sum_{j=1}^{N}\frac{\partial L}{\partial \dot{q}_j}\dot{q}_i\right] + \frac{\partial L}{\partial t}, \tag{2.2.6}$$

which gives $\frac{dh}{dt} = \frac{d}{dt}\left[\sum_{j=1}^{N}\frac{\partial L}{\partial \dot{q}_j}\dot{q}_j\right] - \frac{dL}{dt} = -\frac{\partial L}{\partial t}$.

Natural Lagrangians: Many systems of interest are *natural systems* with *natural Lagrangians*. These have the properties that: (i) the kinetic energy, T, is a homogeneous function of second order (or second degree) in the generalized velocities $\dot{q}_j$, i.e., $T(\vec{q}, t\dot{\vec{q}}) = t^2 T(\vec{q}, \dot{\vec{q}})$; and (ii) the potential, V, is independent of the generalized velocities $\dot{q}_j$.

Euler's homogeneous function theorem: Let $g(x, y)$ be a function that is homogenous of order n in its arguments x and y. Then $g(tx, ty) = t^n g(x, y)$ by definition, where t is some arbitrary scaling factor. Taking d/dt gives

$$nt^{n-1}g(x, y) = \frac{dg(tx, ty)}{dt} = \frac{\partial g}{\partial (tx)}\frac{d(tx)}{dt} + \frac{\partial g}{\partial (ty)}\frac{d(ty)}{dt} = x\frac{\partial g}{\partial (tx)} + y\frac{\partial g}{\partial (ty)},$$

which on choosing $t = 1$ leads to:

$$x\frac{\partial g}{\partial x} + y\frac{\partial g}{\partial y} = ng(x, y).$$

Generalizing this is straightforward. If the function $g(x_1, x_2, \ldots, x_N, z_1, z_2, \ldots)$ is homogeneous of order n in the arguments $x_1, \ldots, x_N$, then by definition $g(tx_1, \ldots, tx_N, z_1, z_2, \ldots) = t^n g(x_1, x_2, \ldots, x_N, z_1, z_2, \ldots)$ and so

$$\sum_{j=1}^{N} x_j \frac{\partial g}{\partial x_j} = ng(x_1, \ldots, x_N, z_1, z_2, \ldots). \tag{2.2.7}$$

We find for a natural Lagrangian that

$$h \equiv \sum_{j=1}^{N} \dot{q}_j (\partial L / \partial \dot{q}_j) - L = \sum_{j=1}^{N} \dot{q}_j (\partial T / \partial \dot{q}_j) - L = 2T - (T - V) = T + V \equiv E. \quad (2.2.8)$$

where we recognize that $E \equiv T + V$ is the energy of the system. We have used the properties of a natural system that $\partial V / \partial \dot{q}_j = 0$ and from Eq. (2.2.7) that $\sum_{j=1}^{N} \dot{q}_j (\partial T / \partial \dot{q}_j) = 2T$. From Eq. (2.2.5) it then follows that if we have a natural system with no explicit time-dependence ($\partial L / \partial t = 0$), then $h = E$, and $dE/dt = 0$. A simple example of a natural system with energy conservation is

$$L = T - V = \sum_{i,j=1}^{N} \tfrac{1}{2} M_{ij}(\vec{q}) \dot{q}_i \dot{q}_j - V(\vec{q}), \quad (2.2.9)$$

where M is a nonsingular real symmetric matrix and $V(\vec{q})$ is any velocity-independent and time-independent potential. The absence of an explicit time-dependence of L is equivalent to the statement that L is invariant under time translations.

A natural system with an explicit time-dependence in the Lagrangian will have $h = E$, but the energy, $E = T + V$, will vary with time. A system that is not natural but where $\partial L / \partial t = 0$ will have a conserved energy function h but $h \neq E = T + V$.

Noether's theorem for a mechanical system: Noether's theorem is named after Emmy Noether, who proved it in 1915 and published the proof in 1918. We can consider transformations of the system under transformations of the time variable and under transformations of the generalized coordinates. Regarding the first of these, we have already seen above that time translational invariance for a natural system implies $h = E$ and the conservation of energy. We now consider the implications of invariance under a transformation of the generalized coordinates.

Let q'' and $\vec{q}$ be two possible choices of the N generalized coordinates for a system. Then there must exist some invertible (i.e., nonsingular) $\vec{f}$, such that $\vec{q}' = \vec{f}(\vec{q})$. The nonsingular $N \times N$ Jacobian matrix for the transformation is $M_{ij} \equiv \partial q_i' / \partial q_j$ as in Eq. (A.2.5). We will consider only time-independent mappings here. Next consider a one-parameter family of such transformations such that $\vec{q}'(s) = \vec{f}_s(\vec{q})$, where s is a real parameter and where $s = 0$ means that $\vec{q}' = \vec{q}$, i.e.,

$$\vec{q}'(s;t) = \vec{f}_s(\vec{q}(t)), \quad \text{where} \quad s \in \mathbb{R} \quad \text{and} \quad \vec{q}'(0;t) = \vec{f}_0(\vec{q}(t)) = \vec{q}(t). \quad (2.2.10)$$

A family of transformations, $\vec{f}_s$, is by defined to be a *symmetry of the action* if $S[\vec{q}] = S[\vec{q}'(s)] = S[\vec{f}_s(\vec{q})]$ for every s and for every path $\vec{q}$ and every choice of time interval defined by t_i and t_f. Independence of the time interval implies that we must have

$$0 = \frac{d}{ds} L(\vec{q}'(s;t), \dot{\vec{q}}'(s;t), t) \Big|_{s=0}, \quad (2.2.11)$$

for every t for every path $\vec{q}$. As we know from Eq. (2.1.41) the Euler-Lagrange equations can have more symmetries than the action. Define $\vec{\eta}(\vec{q})$ as the derivative of the family of mappings at $s = 0$, i.e.,

$$\vec{\eta}(\vec{q}) \equiv \frac{d\vec{q}'}{ds} \Big|_{s=0} = \frac{d\vec{f}_s(\vec{q})}{ds} \Big|_{s=0}. \quad (2.2.12)$$

We are now able to state Noether's theorem as applied to a mechanical system.

Noether's theorem: If $\vec{f}_s$ is a symmetry of the action, then *Noether's first integral*,

$$C(\vec{q}, \dot{\vec{q}}, t) \equiv \sum_{j=1}^{N} (\partial L / \partial \dot{q}_j) \eta_j(\vec{q}) = \sum_{j=1}^{N} p_j \eta_j, \quad (2.2.13)$$

is a constant of the motion, i.e., $dC/dt = 0$.

Proof: Using Eq. (2.2.11) and the Euler-Lagrange equations we find

$$
\begin{aligned}
0 &= \frac{d}{ds} \left. L(\vec{q}\,'(s;t), \dot{\vec{q}}\,'(s;t), t) \right|_{s=0} = \sum_{j=1}^{N} \left. \left(\frac{\partial L}{\partial q_j'} \frac{dq_j'}{ds} + \frac{\partial L}{\partial \dot{q}_j'} \frac{d\dot{q}_j'}{ds} \right) \right|_{s=0} \\
&= \sum_{j=1}^{N} \left(\frac{\partial L}{\partial q_j} \eta_j + \frac{\partial L}{\partial \dot{q}_j} \frac{d\eta_j}{dt} \right) = \sum_{j=1}^{N} \left(\left(\frac{d}{dt} \frac{\partial L}{\partial \dot{q}_j} \right) \eta_j + \frac{\partial L}{\partial \dot{q}_j} \frac{d\eta_j}{dt} \right) \\
&= \frac{d}{dt} \left(\sum_{j=1}^{N} \frac{\partial L}{\partial \dot{q}_j} \eta_j \right) = \frac{dC}{dt}.
\end{aligned}
\qquad (2.2.14)
$$

To illustrate with some simple examples, again consider our system of n particles moving in three dimensions with $r = (3n - N)$ holonomic constraints, giving rise to N generalized coordinates $\vec{q}$. If the system has a monogenic potential, then $L = T - V$ and the Euler-Lagrange equations follow. Let us further restrict our attention to conservative systems as defined in Eq. (2.1.28), $V = V(\vec{q}, t)$, so that $\partial V / \partial \dot{q}_j = 0$. Under the transformation $\vec{q} \to \vec{q}\,'$ the corresponding Cartesian coordinates will also have a transformation $\mathbf{r}_a \to \mathbf{r}_a'$ where from Eq. (2.1.8) we define $\mathbf{r}_a' \equiv \mathbf{r}_a(q_1', q_2', \ldots, q_N', t)$ and both the $\mathbf{r}_a$ and $\mathbf{r}_a'$ are by construction consistent with the holonomic constraints on the system due to Eq. (2.1.8).

Using Eq. (2.1.19) and the cancellation of the dots in Eq. (2.1.21) we find

$$
p_j = \frac{\partial L}{\partial \dot{q}_j} = \frac{\partial T}{\partial \dot{q}_j} = \sum_{a=1}^{n} m_a \dot{\mathbf{r}}_a \cdot \frac{\partial \dot{\mathbf{r}}_a}{\partial \dot{q}_j} = \sum_{a=1}^{n} m_a \dot{\mathbf{r}}_a \cdot \frac{\partial \mathbf{r}_a}{\partial q_j},
\qquad (2.2.15)
$$

which we use to evaluate Noether's first integral for the family of transformations

$$
\begin{aligned}
C(\vec{q}, \dot{\vec{q}}, t) &= \sum_{j=1}^{N} p_j \eta_j = \sum_{j=1}^{N} \left(\sum_{a=1}^{n} m_a \dot{\mathbf{r}}_a \cdot \frac{\partial \mathbf{r}_a}{\partial q_j} \right) \left. \frac{dq_j'}{ds} \right|_{s=0} \\
&= \sum_{j=1}^{N} \left. \left(\sum_{a=1}^{n} m_a \dot{\mathbf{r}}_a' \cdot \frac{\partial \mathbf{r}_a'}{\partial q_j'} \frac{dq_j'}{ds} \right) \right|_{s=0} = \sum_{a=1}^{n} m_a \dot{\mathbf{r}}_a \cdot \left. \frac{d\mathbf{r}_a'}{ds} \right|_{s=0}.
\end{aligned}
\qquad (2.2.16)
$$

Examples

(i) If the Lagrangian is invariant under translations in the direction of the unit vector $\mathbf{n}$, such as V depending only on the quantities $\mathbf{n} \times \mathbf{r}_a$, then L is invariant under $\mathbf{r}_a \to \mathbf{r}_a' = \mathbf{r}_a + s\mathbf{n}$. The corresponding Noether first integral is

$$
C = \sum_a m_a \dot{\mathbf{r}}_a \cdot \mathbf{n} = \mathbf{P} \cdot \mathbf{n},
\qquad (2.2.17)
$$

where $\mathbf{P}$ is the total three-momentum of the system and the component lying in the $\mathbf{n}$-direction, $\mathbf{n} \cdot \mathbf{P}$, is a conserved quantity.

(ii) If the translational invariance applies for all directions $\mathbf{n}$, such as V depending only on relative coordinates $\mathbf{r}_a - \mathbf{r}_b$, then $\mathbf{P}$ is conserved.

(iii) If the Lagrangian is invariant under rotations of the system about some unit vector $\mathbf{n}$, then it is convenient to define the origin to lie on some point along this $\mathbf{n}$-axis. Then it follows that L is invariant under infinitesimal rotations $\mathbf{r}_a \to \mathbf{r}_a' = \mathbf{r}_a + ds(\mathbf{n} \times \mathbf{r}_a)$. For example, if V depends only on $\mathbf{r}_a \cdot \mathbf{n}$ then it follows that $\mathbf{r}_a' \cdot \mathbf{n} = \mathbf{r}_a \cdot \mathbf{n}$. The corresponding Noether first integral is

$$
C = \sum_a m_a \dot{\mathbf{r}}_a \cdot (\mathbf{n} \times \mathbf{r}_a) = \mathbf{n} \cdot \sum_a \mathbf{r}_a \times m_a \dot{\mathbf{r}}_a = \mathbf{n} \cdot \mathbf{L},
\qquad (2.2.18)
$$

where $\mathbf{L}$ is the total orbital angular momentum of the system and the component lying in the $\mathbf{n}$-direction, $\mathbf{n} \cdot \mathbf{L}$, is a conserved quantity.

(iv) If the rotational invariance about some point applies for all possible axes **n** passing through that point, then it is convenient to choose that point to be the origin. For example, if V is depends only on the quantities $\mathbf{r}_a \cdot \mathbf{r}_b$, the total orbital angular momentum, **L**, about this origin is conserved.

(v) Recall from Eq. (2.2.8) that if a natural system is invariant under time translations, then the total energy, $E = T + V$, is conserved.

2.3 Small Oscillations and Normal Modes

With complex physical systems we are often interested in small variations around equilibrium, which can be described in terms of the system's normal modes.

Consider a classical system consisting of N coupled degrees of freedom, i.e., N generalized coordinates $\vec{q} = (q_1, q_2, \ldots, q_N)$. We wish to study motions of the system that remain sufficiently close to a stable equilibrium point, i.e., a local minimum of the potential $V(\vec{q})$. Often the equilibrium point of interest is the ground state of the system (i.e., the absolute minimum of V), but the following discussion is applicable for sufficiently small oscillations around any local minimum of V.

Equilibrium: A system is said to be in *equilibrium* when all of the generalized forces in the system vanish at all times,

$$Q_i(t) = 0 \quad \text{for all } t \quad \text{and for} \quad i = 1, 2, \ldots, N, \tag{2.3.1}$$

where the generalized forces, Q_i, are defined in Eq. (2.1.18). A system is in *static equilibrium* if the generalized velocities vanish at all times, i.e., $\dot{\vec{q}}(t) = 0$ for all t. The discussion here is restricted to physical systems with the following properties:

  (i) They have a Lagrangian $L \Rightarrow$ monogenic systems with holonomic constraints.
 (ii) The system is natural and so $h = E = T + V$.
(iii) The potential V is bounded below.
 (iv) The Lagrangian has no explicit time dependence and so $dh/dt = dE/dt = 0$.
  (v) L has a Taylor series expansion in $\vec{q}$ and $\dot{\vec{q}}$ in the neighborhood of the stationary points of the potential V.
 (vi) The kinetic energy, T, is purely concave up (convex down) in the generalized velocities in the neighborhood of the stationary points of the potential.

The generalized forces satisfy $Q_i = -\partial V/\partial q_i$, so such a system is in equilibrium if the classical motion of the system is such that $\partial V/\partial q_i = 0$ for all $i = 1, \ldots, N$ at all times t; i.e., if at all times along the classical path the system is at a stationary point of the potential. There are two possibilities for equilibrium:

(a) With an isolated stationary point of V, if $T = 0$ we have a static equilibrium.

(b) The potential V has a continuous closed or open path of stationary points that necessarily all have the same value of the potential and the equations of motion of the system admit a solution which moves along this path. This is an equilibrium, but it is not a static equilibrium.

Let $\vec{q}_0$ denote a stationary point of the potential $V(\vec{q})$, i.e., by definition

$$\partial V/\partial q_i|_{\vec{q}=\vec{q}_0} = 0 \quad \text{for} \quad i = 1, \ldots, N \quad \text{or equivalently} \quad \vec{\nabla}_q V|_{\vec{q}=\vec{q}_0} = \vec{0}, \tag{2.3.2}$$

where $\vec{\nabla}_q \equiv (\partial/\partial q_1, \ldots, \partial/\partial q_N)$ is the gradient operator in the space of generalized coordinates. Define the deviation from the stationary point as

$$\vec{\eta} \equiv \vec{q} - \vec{q}_0. \tag{2.3.3}$$

and perform a Taylor expansion of the potential around $\vec{q}_0$ to give

$$V(\vec{q}) = V(\vec{q}_0) + \frac{1}{2} \sum_{i,j=1}^{N} \left. \frac{\partial^2 V}{\partial q_i \partial q_j} \right|_{\vec{q}=\vec{q}_0} \eta_i \eta_j + \mathcal{O}(\eta^3) \equiv V_0 + \frac{1}{2} \sum_{i,j=1}^{N} K_{ij} \eta_i \eta_j + \mathcal{O}(\eta^3), \tag{2.3.4}$$

where $V_0 \equiv V(\vec{q}_0)$ and the matrix K is a real symmetric matrix and so diagonalizable by an orthogonal transformation. If q_0 is a local minimum of V, then K is positive definite and hence invertible. For the physical systems being considered here the kinetic energy is purely second-order in the generalized velocities, i.e.,

$$T = \frac{1}{2} \sum_{i,j=1}^{N} M_{ij}(\vec{q}) \dot{q}_i \dot{q}_j = \frac{1}{2} \sum_{i,j=1}^{N} M_{ij}(\vec{q}) \dot{\eta}_i \dot{\eta}_j \tag{2.3.5}$$

for some $M_{ij}(\vec{q})$. The matrix $M(\vec{q})$ is real symmetric and so diagonalizable by an orthogonal transformation. Condition (vi) above means that the matrix $M(\vec{q})$ has only positive eigenvalues at the stationary points $\vec{q}_0$ of V, i.e., $M(\vec{q}_0)$ is a real, symmetric and positive-definite matrix and so $T \geq 0$ in the neighborhood of $\vec{q}_0$.

Proof: Consider n particles moving in three spatial dimensions. Since we have assumed no explicit time dependence, then the holonomic constraints will be time-independent and so Eqs. (2.1.8) and (2.1.20) become

$$\mathbf{r}_a = \mathbf{r}_a(q_1, q_2, \ldots, q_N) \quad \text{and} \quad \boldsymbol{v}_a = \dot{\mathbf{r}}_a = \sum_{i=1}^{N} \frac{\partial \mathbf{r}_a}{\partial q_i} \dot{q}_i \tag{2.3.6}$$

for $a = 1, 2, \ldots, n$. The total kinetic energy then takes the form in Eq. (2.3.5),

$$T = \sum_{a=1}^{n} \frac{1}{2} m_a \dot{\mathbf{r}}_a \cdot \dot{\mathbf{r}}_a = \sum_{a=1}^{n} \frac{1}{2} m_a \left(\sum_{i=1}^{N} \frac{\partial \mathbf{r}_a}{\partial q_i} \dot{q}_i \right) \cdot \left(\sum_{j=1}^{N} \frac{\partial \mathbf{r}_a}{\partial q_j} \dot{q}_j \right)$$

$$= \frac{1}{2} \sum_{i,j=1}^{N} \left(\sum_{a=1}^{n} m_a \frac{\partial \mathbf{r}_a}{\partial q_i} \cdot \frac{\partial \mathbf{r}_a}{\partial q_j} \right) \dot{q}_i \dot{q}_j \equiv \frac{1}{2} \sum_{i,j=1}^{N} M_{ij}(\vec{q}) \dot{q}_i \dot{q}_j, \tag{2.3.7}$$

where we have defined

$$M_{ij}(\vec{q}) \equiv \sum_{a=1}^{n} m_a \frac{\partial \mathbf{r}_a}{\partial q_i} \cdot \frac{\partial \mathbf{r}_a}{\partial q_j} = M_{ji}(\vec{q}).$$

Existence of static equilibrium: For any natural system without explicit time-dependence, if $\vec{q}_0$ is a stationary point of the potential and if at some time t_0 we have $\vec{q} \equiv \vec{q}_0$ and all of the generalized velocities are zero, then the system must be in static equilibrium; i.e., if $\vec{q}_0$ is a stationary point of V and

$$\text{if } (\vec{q}(t_0), \dot{\vec{q}}(t_0)) = (\vec{q}_0, 0), \quad \text{then } (\vec{q}(t), \dot{\vec{q}}(t)) = (\vec{q}_0, 0) \text{ for all } t. \tag{2.3.8}$$

Proofs: Consider the Euler-Lagrange equations for the systems of interest,

$$0 = \frac{d}{dt} \left(\frac{\partial L}{\partial \dot{q}_i} \right) - \frac{\partial L}{\partial q_i} = \frac{d}{dt} \left(\frac{\partial T}{\partial \dot{q}_i} \right) - \frac{\partial T}{\partial q_i} + \frac{\partial V}{\partial q_i}$$

$$= \frac{d}{dt} \left(\sum_{j=1}^{N} M_{ij} \dot{q}_j \right) - \left(\frac{1}{2} \sum_{j,k=1}^{N} \frac{\partial M_{jk}}{\partial q_i} \dot{q}_j \dot{q}_k \right) + \frac{\partial V}{\partial q_i}$$

$$= \sum_{j=1}^{N} \left(M_{ij} \ddot{q}_j + \sum_{k=1}^{N} \frac{\partial M_{ij}}{\partial q_k} \dot{q}_k \dot{q}_j \right) - \left(\frac{1}{2} \sum_{j,k=1}^{N} \frac{\partial M_{jk}}{\partial q_i} \dot{q}_j \dot{q}_k \right) + \frac{\partial V}{\partial q_i} \tag{2.3.9}$$

for $i = 1, \ldots, N$. If $\vec{q}(t_0) = \vec{q}_0$ with $\vec{q}_0$ a stationary point of V, then $\partial V/\partial q_i$ vanishes at t_0. If also $\dot{\vec{q}}(t_0) = 0$ then in an obvious matrix notation we have $0 = M\ddot{\vec{q}}$ at $t = t_0$. Since M is nonsingular, it is invertible. So acting from the left on both sides with M^{-1} gives $\ddot{\vec{q}}(t_0) = 0$. Then Eq. (2.3.8) has been proved.

Stable static equilibrium: For the systems being considered here the energy is $E = T + V$, and is conserved. So along any classical path T and V are changing but their sum is not. A point $(\vec{q}, \dot{\vec{q}}) = (\vec{q}_0, 0)$ in our $2N$-dimensional space of generalized coordinates and velocities is called a *stable static equilibrium* if there is a sufficiently small neighborhood, U, of points around it such that if the system starts from any of the points in this neighborhood, then the classical path is always contained within some other neighborhood W with $U \subseteq W$.

Existence of a stable static equilibrium: For systems satisfying conditions (i)–(vi), if $\vec{q}_0$ is a local minimum of V, then $(\vec{q}_0, 0)$ is a stable static equilibrium.

Proof: Choose a neighborhood W around $(\vec{q}_0, 0)$ such that $E = T + V \leq V_0 + \epsilon$ for some small ϵ, where $V_0 = V(\vec{q}_0)$ is as defined in Eq. (2.3.4). Choose ϵ sufficiently small so that T is concave up in the generalized velocities and V is concave up in the generalized coordinates. Then for any point $(\vec{q}(t_0), \dot{\vec{q}}(t_0)) \in W$ we will have $E \leq V_0 + \epsilon$ and since E is conserved then we must have $(\vec{q}(t), \dot{\vec{q}}(t)) \in W$ for all times t. Then $(\vec{q}_0, 0)$ is a stable static equilibrium.

Normal modes: We can now examine the behavior of a system undergoing small motion around a stable static equilibrium. Small motion means that the system has an energy $E = V_0 + \epsilon$ for some sufficiently small ϵ such that the system is trapped in the vicinity of the local minimum at $\vec{q}_0$. As we have seen, $T + (V - V_0) = \epsilon$, with $T \geq 0$ and $(V - V_0) \geq 0$. Since $M(\vec{q}_0)$ and K are positive definite matrices, then from Eqs. (2.3.4) and (2.3.5) it follows that $\eta_i|_{\max} \sim \mathcal{O}(\epsilon^{1/2})$ and $\dot{\eta}_i|_{\max} \sim \mathcal{O}(\epsilon^{1/2})$. We can omit the constant V_0 from the Taylor expansion of V, since from Eq. (2.1.41) we know the Euler-Lagrange equations will not change. So the Lagrangian appropriate for describing small motions around a static stable equilibrium is

$$L = \tfrac{1}{2} \sum_{i,j=1}^{N} M_{ij}^0 \dot{\eta}_i \dot{\eta}_j - \tfrac{1}{2} \sum_{i,j=1}^{N} K_{ij} \eta_i \eta_j, \qquad (2.3.10)$$

where $L \sim \mathcal{O}(\epsilon)$, we have neglected terms of $\mathcal{O}(\epsilon^{3/2})$ and we have defined the constant matrix $M_{ij}^0 \equiv M_{ij}(\vec{q}_0)$. The Euler-Lagrange equations are

$$\sum_{j=1}^{N} \left(M_{ij}^0 \ddot{\eta}_j + K_{ij} \eta_j \right) = 0 \quad \text{for} \quad i = 1, \ldots, N. \qquad (2.3.11)$$

If $N = 1$, this is the Lagrangian for a single harmonic oscillator, $L = \tfrac{1}{2} M^0 \dot{\eta}^2 - \tfrac{1}{2} K \eta^2$, where the angular frequency of the oscillator is $\omega = \sqrt{K/M^0}$.

In matrix notation we can write the Lagrangian of Eq. (2.3.10) as

$$L = \tfrac{1}{2} \dot{\vec{\eta}}^T M^0 \dot{\vec{\eta}} - \tfrac{1}{2} \vec{\eta}^T K \vec{\eta}, \qquad (2.3.12)$$

where both M^0 and K are real symmetric positive-definite matrices. Let O be the orthogonal matrix that diagonalizes M^0

$$D_M = O^T M^0 O, \qquad (2.3.13)$$

where the $N \times N$ diagonal matrix D_M has the positive real eigenvalues of M^0, denoted m_i, down the diagonal. The square root, $D_M^{1/2}$, of D_M exists and is the diagonal matrix with diagonal elements $m_i^{1/2}$. Similarly the inverse diagonal matrix $D_M^{-1/2}$ exists with diagonal elements $m_i^{-1/2}$. Define

$$\vec{\eta}' \equiv D_M^{1/2} O^T \vec{\eta} \quad \text{or equivalently} \quad \vec{\eta} \equiv O D_M^{-1/2} \vec{\eta}', \tag{2.3.14}$$

which allows us to write

$$L = \tfrac{1}{2} \dot{\vec{\eta}}'^{\,T} \dot{\vec{\eta}}' - \tfrac{1}{2} \vec{\eta}'^{\,T} \left(D_M^{-1/2} O^T K O D_M^{-1/2} \right) \vec{\eta}'. \tag{2.3.15}$$

The matrix in parentheses is obviously real and symmetric. Since K is positive-definite the matrix in parentheses is also positive-definite.

Proof: Let A be a nonsingular matrix. This means that for every vector $\vec{u}$ there is a unique vector $\vec{v}$ such that $\vec{u} = A\vec{v}$. This follows since if v was not unique, then $\vec{u} = A\vec{v} = A\vec{v}'$ for some $\vec{v}' \neq \vec{v}$. Then we would have $A(\vec{v} - \vec{v}') = 0$ and A would have a zero eigenvalue and hence be singular. Similarly, there is a unique vector $\vec{u}$ such that $\vec{v} = A^{-1}\vec{u}$. Thus any nonsingular matrix A is an invertible map of the entire vector space onto itself. This is true of real and complex vector spaces. If B is a real positive-definite matrix, then by definition for every vector $\vec{u}$ of nonvanishing length we have $\vec{u}^T B \vec{u} > 0$. Let $\vec{u} = A\vec{v}$; then this means that $\vec{v}^T (A^T B A) \vec{v} > 0$ for every $\vec{v}$. *Summary:* If B is real and positive-definite and if A is real and nonsingular, then $A^T B A$ is real and positive-definite. Write $A = O D_M^{-1/2}$ and note that A is nonsingular, since $\det A = \det O \det(D_M^{-1/2}) = \det O (\det D_M)^{-1/2}$ and since $\det O = \pm 1$ and $\det D_M > 0$. Then since K is real and positive-definite we have that $A^T K A = D_M^{-1/2} O^T K O D_M^{-1/2}$ is real and positive-definite, which proves the result.

Let O_ω be the orthogonal matrix that diagonalizes the real, symmetric and positive-definite matrix, $(D_M^{-1/2} O^T K O D_M^{-1/2})$, then

$$D_\omega = O_\omega^T \left(D_M^{-1/2} O^T K O D_M^{-1/2} \right) O_\omega, \tag{2.3.16}$$

where D_ω is diagonal with elements $\omega_i^2 > 0$ for $i = 1, \ldots, N$. Define the matrix

$$W \equiv O D_M^{-1/2} O_\omega \quad \text{or equivalently} \quad W^{-1} \equiv O_\omega^T D_M^{1/2} O^T. \tag{2.3.17}$$

W is not an orthogonal matrix since $D_M^{1/2}$ is not orthogonal. Then define

$$\vec{\zeta} \equiv O_\omega^T \vec{\eta}' = W^{-1} \vec{\eta} \quad \text{or equivalently} \quad \vec{\eta} \equiv O D_M^{-1/2} O_\omega \vec{\zeta} = W\vec{\zeta}, \tag{2.3.18}$$

which allows us finally to write

$$L = \tfrac{1}{2} \dot{\vec{\zeta}}^T \dot{\vec{\zeta}} - \tfrac{1}{2} \vec{\zeta}^T D_\omega \vec{\zeta} = \sum_{i=1}^N \tfrac{1}{2} \left(\dot{\zeta}_i^2 - \omega_i^2 \zeta_i^2 \right) \tag{2.3.19}$$

This derivation of Eq. (2.3.19) is equivalent to showing that

$$W^T M^0 W = I \quad \text{and} \quad W^T K W = D_\omega \quad \text{with} \quad (D_\omega)_{ij} = \omega_i^2 \delta_{ij}. \tag{2.3.20}$$

We recognize that the Lagrangian in Eq. (2.3.19) represents N independent harmonic oscillators with generalized coordinates ζ_i, generalized velocities $\dot{\zeta}_i$ and angular frequencies ω_i for $i = 1, \ldots, N$. The generalized coordinates ζ_i are *normal coordinates* and the angular frequencies of oscillation ω_i are the *normal modes* of the system. The normal modes are the resonant angular frequencies of the system for small motion about a stable static equilibrium. Each normal mode corresponds to a different type of oscillation of the system about the equilibrium point. The Euler-Lagrange equations are now isolated harmonic oscillator equations,

$$\ddot{\zeta}_i + \omega_i^2 \zeta_i^2 = 0 \tag{2.3.21}$$

and the solutions must have the standard harmonic oscillator form

$$\zeta_i(t) = C_i \cos(\omega_i t + \phi_i) = \alpha_i \cos(\omega_i t) + \beta_i \sin(\omega_i t), \tag{2.3.22}$$

where the C_i and ϕ_i are the amplitudes and phases of each of the modes and where $\alpha_i = C_i \cos \phi_i$ and $\beta_i = -C_i \sin \phi_i$ since $\cos(x+y) = \cos x \cos y - \sin x \sin y$. In terms of the original coordinates we have

$$\vec{\eta}(t) = W\vec{\zeta}(t) \quad \text{or} \quad \eta_i(t) = \sum_{j=1}^{N} W_{ij}\zeta_j(t) = \sum_{j=1}^{N} W_{ij}C_j \cos(\omega_j t + \phi_j). \tag{2.3.23}$$

The normal modes are contained entirely in W. Then we have the result that

$$W \equiv (\vec{v}_1|\vec{v}_2|\cdots|\vec{v}_N) \quad \Rightarrow \quad \vec{\eta}(t) = \sum_{j=1}^{N} C_j \vec{v}_j \cos(\omega_j t + \phi_j). \tag{2.3.24}$$

This shows that any arbitrary small-motion solution to the Euler-Lagrange equations is completely specified by the amplitudes $\vec{C} \equiv (C_1, \ldots, C_N)$ and the phases $\vec{\phi} \equiv (\phi_1, \ldots, \phi_N)$. All of the time-dependence arises from the t in the cosine as all other quantities are constants. In order to simplify the process of finding solutions for the normal modes, the following observations are helpful:

(a) The eigenvalues of the matrix $(M^0)^{-1}K$ are $\lambda_j = \omega_j^2$ and the corresponding eigenvectors are the column vectors of W, i.e., the $\vec{v}_j$.
(b) The eigenvalues λ_j are the roots of the equation $\det(K - \lambda M^0) = 0$.
(c) In general $[M^0, K] \neq 0$ and so $(M^0)^{-1}K$ is real but not symmetric and so the $\vec{v}_j$ are not orthogonal. However, the vectors $\vec{v}_i$ are orthogonal to the vectors $M^0 \vec{v}_j$ since $\vec{v}_i^T (M^0 \vec{v}_j) = \delta_{ij}$.
(d) The eigenvectors $\vec{v}_j$ can be obtained by solving $(K - \lambda_j M^0)\vec{v}_j = 0$ with $\lambda_j = \omega_j^2$ and so again there is no need to invert M^0 to solve for the $\vec{v}_j$.

Proofs

(a) From Eq. (2.3.20) we have $W^T M^0 W = I$ and so $W^T = W^{-1}(M^0)^{-1}$. Then $D_\omega = W^T K W = W^{-1}(M^0)^{-1}KW$ and so $(M^0)^{-1}KW = WD_\omega$. Using $(D_\omega)_{ij} = \omega_i^2 \delta_{ij} \Rightarrow \sum_j [(M^0)^{-1}K]_{ij}W_{jk} = \sum_j W_{ij}(D_\omega)_{jk} = \omega_k^2 W_{ik}$, then we find $[(M^0)^{-1}K]\vec{v}_k = \lambda_k \vec{v}_k$, where $\lambda_k \equiv \omega_k^2$. So λ_k and $\vec{v}_k$ are the eigenvalues and eigenvectors of the matrix $(M^0)^{-1}K$ respectively.
(b) The λ_i are the roots of $\det[(M^0)^{-1}K - \lambda I] = 0$. Since $\det M^0 \neq 0$ then $0 = (\det M^0)^{-1} \det[K - \lambda M^0]$ and so $\det[K - \lambda M^0] = 0$.
(c) Since $I = W^T M^0 W$ then $\delta_{i\ell} = (W^T)_{ij}(M^0)_{jk}W_{k\ell} = \vec{v}_i^T M^0 \vec{v}_\ell$.
(d) Since $[(M^0)^{-1}K]\vec{v}_i = \lambda_i \vec{v}_i$ and $(M^0)^{-1}$ exists then $[K - \lambda M^0]\vec{v}_i = 0$.

Summary: To study the normal modes of a system satisfying the above conditions (i)–(vi) we proceed through the following process:

(a) Identify an isolated stationary point, $\vec{q}_0$, of the potential V.
(b) Taylor expand the kinetic energy, T, in generalized velocities and the potential energy, V, in generalized coordinates about this $\vec{q}_0$ and extract M^0 and K.
(c) Solve for the eigenvalues, $\lambda_j = \omega_j^2$, and eigenvectors, $\vec{v}_j$, of $(M^0)^{-1}K$.
(d) Use Eq. (2.3.24) to describe any sufficiently small classical motion of the system in terms of the amplitudes $\vec{C}$ and the phases $\vec{\phi}$.

2.4 Hamiltonian Formulation

Lagrangian mechanics uses generalized coordinates and generalized velocities to describe the behavior of a system. An alternative approach, referred to as the Hamiltonian formalism or Hamiltonian mechanics, uses generalized coordinates and their conjugate momenta instead. The two formalisms are related by a Legendre transformation and are collectively known as *analytical mechanics*.

2.4.1 Hamiltonian and Hamilton's Equations

Consider a Lagrangian $L(\vec{q}, \dot{\vec{q}}, t)$, which is concave up in its generalized velocities. The Hessian matrix[2] for the generalized velocities for this Lagrangian is

$$(M_L)_{ij} \equiv \partial^2 L / \partial \dot{q}_i \partial \dot{q}_j. \tag{2.4.1}$$

The *Hessian matrix* M_L is real and symmetric and so can be diagonalized by an orthogonal transformation with its real eigenvalues down the diagonal. If the convex Lagrangian L is concave up in the N generalized velocities $\dot{\vec{q}}$ then all of the eigenvalues of M_L are real and positive. This allows the multi-dimensional *Legendre transform* with respect to the N generalized velocities to be defined.

The *Hamiltonian* corresponding to such a Lagrangian is then defined via the multi-dimensional Legendre transformation as

$$H(\vec{q}, \vec{p}, t) \equiv \left(\sum_{i=1}^{N} p_i \dot{q}_i \right) - L(\vec{q}, \dot{\vec{q}}(\vec{p}), t) = h(\vec{q}, \dot{\vec{q}}(\vec{p}), t), \tag{2.4.2}$$

where the generalized (i.e., canonical) momenta were defined earlier in Eq. (2.2.2),

$$p_i(t) \equiv \partial L / \partial \dot{q}_i. \tag{2.4.3}$$

If the eigenvalues have mixed signs, then the Lagrangian is not convex and the Legendre transform does not exist and so the Hamiltonian does not exist. If one of more of the eigenvalues vanishes, then we say that the system is singular (or irregular). As we will discuss in Sec. 2.9, it is still possible to obtain a Hamiltonian formulation of a singular (i.e., constrained) system. This will become important when we come to consider the quantization of gauge field theories, where we need a Hamiltonian and where gauge fixing acts as a constraint on the system.

The definition of H corresponds exactly to the energy function, h, defined in Eq. (2.2.4) when we view it as a function of the generalized momenta $\vec{p}$ rather than of the generalized velocities $\dot{\vec{q}}$. The Hamiltonian will be concave up or down in the generalized momenta if the Lagrangian is concave up or down, respectively, in the generalized velocities. Consider the Hessian matrix of the Hamiltonian defined by

$$(M_H)_{ij} \equiv \partial^2 H / \partial p_i \partial p_j. \tag{2.4.4}$$

The eigenvalues of the real symmetric matrix M_H will be all positive (or all negative) if those of M_L are all positive (or all negative). For a natural system the Hamiltonian is the energy of the system

---

[2] The Hessian or Hessian matrix of a function is the square matrix of second-order partial derivatives of the function with respect to its arguments; e.g., for a function $f(\vec{x})$ with $\vec{x} = (x_1, \ldots, x_N)$ the Hessian matrix is an $N \times N$ matrix with elements $\partial^2 f / \partial x_i \partial x_j$.

($H = h = E = T + V$) and so from Eq. (2.2.5) we see that $H = E$ is a conserved quantity if $\partial L / \partial t = 0$.

The $\vec{q}$ and $\vec{p}$ comprise the $2N$ independent variables in the Hamiltonian formalism and are referred to as the *canonical coordinates*. The set of all possible values of $\vec{q}$ and $\vec{p}$ make up the points in $2N$-dimensional *phase space*. The generalized momenta are also referred to as the *conjugate momenta*. The q_i are the canonical positions and the p_i are the canonical momenta. From this definition of p_i we see that

$$\dot{p}_i = \frac{d}{dt}\left(\frac{\partial L}{\partial \dot{q}_i}\right) = \frac{\partial L}{\partial q_i}, \tag{2.4.5}$$

where the last step follows from using the Euler-Lagrange equations of motion. The differential of the Lagrangian $L(\vec{q}, \dot{\vec{q}}, t)$ is then

$$dL = \sum_{i=1}^{N}\left(\frac{\partial L}{\partial q_i}dq_i + \frac{\partial L}{\partial \dot{q}_i}d\dot{q}_i\right) + \frac{\partial L}{\partial t}dt = \sum_{i=1}^{N}\left(\dot{p}_i dq_i + p_i d\dot{q}_i\right) + \frac{\partial L}{\partial t}dt. \tag{2.4.6}$$

The differential of the Hamiltonian $H(\vec{q}, \vec{p}, t)$ is

$$dH = \sum_{i=1}^{N}\left(\frac{\partial H}{\partial q_i}dq_i + \frac{\partial H}{\partial p_i}dp_i\right) + \frac{\partial H}{\partial t}dt. \tag{2.4.7}$$

From Eqs. (2.4.2) and (2.4.6) the differential for the Hamiltonian can be written as

$$dH = \sum_{i=1}^{N}\left(p_i d\dot{q}_i + \dot{q}_i dp_i\right) - dL = \sum_{i=1}^{N}\left(-\dot{p}_i dq_i + \dot{q}_i dp_i\right) - (\partial L/\partial t)dt. \tag{2.4.8}$$

Since dq_i, dp_i and dt are arbitrary, comparing the two expressions for dH leads to

$$\dot{q}_i = \frac{\partial H}{\partial p_i}, \quad \dot{p}_i = -\frac{\partial H}{\partial q_i} \quad \text{and} \quad \frac{\partial H}{\partial t} = -\frac{\partial L}{\partial t}. \tag{2.4.9}$$

The total time derivative of the Hamiltonian is given by

$$\frac{dH}{dt} = \sum_{j=1}^{N}\left(\frac{\partial H}{\partial q_j}\dot{q}_j + \frac{\partial H}{\partial p_j}\dot{p}_j\right) + \frac{\partial H}{\partial t} = \sum_{j=1}^{N}\left(-\dot{p}_j\dot{q}_j + \dot{q}_j\dot{p}_j\right) - \frac{\partial L}{\partial t} = -\frac{\partial L}{\partial t}, \tag{2.4.10}$$

which recovers the result of Eq. (2.2.5) since $H(\vec{q}, \vec{p}) = h(\vec{q}, \dot{\vec{q}})$.

Summary: Defining the Hamiltonian as the Legendre transform of the Lagrangian

$$H \equiv \vec{p}\cdot\dot{\vec{q}} - L \quad \text{with} \quad \vec{p} \equiv \partial L/\partial\dot{\vec{q}}, \tag{2.4.11}$$

we have arrived at *Hamilton's equations* of motion,

$$\dot{q}_i = \frac{\partial H}{\partial p_i}, \quad \dot{p}_i = -\frac{\partial H}{\partial q_i} \quad \text{and} \quad \frac{dH}{dt} = -\frac{\partial L}{\partial t}. \tag{2.4.12}$$

Example: Use the Hamiltonian formalism to derive the Lorentz force for a particle in an electromagnetic field. Recall from Eq. (2.1.37) that the Lagrangian is

$$L = T - V = \tfrac{1}{2}m\dot{\mathbf{x}}^2 - q[\Phi - \dot{\mathbf{x}}\cdot\mathbf{A}], \tag{2.4.13}$$

where $\mathbf{x}$ are the generalized coordinates here. From Eq. (2.2.2) we have

$$p_i \equiv \partial L/\partial\dot{x}_i = m\dot{x}_i + qA_i, \tag{2.4.14}$$

which gives $\dot{x}_i = (p_i - qA_i)/m$. The mechanical momentum is $\mathbf{p}^{\mathrm{mech}} = m\dot{\mathbf{x}}$ and the canonical momentum is $\mathbf{p} = \mathbf{p}^{\mathrm{mech}} + q\mathbf{A}$. The resulting Hamiltonian is

$$
\begin{aligned}
H &= \left(\sum_{i=1}^{3} \dot{x}_i p_i \right) - L \\
&= \left(\sum_{i=1}^{3} (p_i - qA_i)p_i/m \right) - \tfrac{1}{2} \sum_{i=1}^{3} (p_i - qA_i)^2/m + q \left[\Phi - \sum_{i=1}^{3} (p_i - qA_i)A_i/m \right] \\
&= \left[\sum_{i=1}^{3} (p_i - qA_i)^2/2m \right] + q\Phi = \left[(\mathbf{p} - q\mathbf{A})^2/2m \right] + q\Phi.
\end{aligned}
\tag{2.4.15}
$$

Hamilton's equations are

$$
\dot{x}_i = \frac{\partial H}{\partial p_i} = \frac{p_i - qA_i}{m}, \quad \dot{p}_i = \frac{d}{dt}(m\dot{x}_i + qA_i) = -\frac{\partial H}{\partial x_i} = -q \left(\frac{\partial \Phi}{\partial x_i} + \sum_{j=1}^{3} \frac{\partial A_j}{\partial x_i} \dot{x}_j \right),
$$

$$
\frac{dH}{dt} = \frac{d}{dt}\left(\frac{m\dot{\mathbf{x}}^2}{2} + q\Phi \right) = -\frac{\partial L}{\partial t} = q \left(\frac{\partial \Phi}{\partial t} - \dot{\mathbf{x}} \frac{\partial \mathbf{A}}{\partial t} \right).
\tag{2.4.16}
$$

The first equation is just a restatement of the definition of the canonical momentum. The second equation is identical to the first line of Eq. (2.1.38) and so again leads to the Lorentz force equation. Using $T = m\dot{\mathbf{x}}^2/2$ the third equation becomes

$$
\begin{aligned}
\frac{dT}{dt} &\equiv \frac{d(m\dot{\mathbf{x}}^2/2)}{dt} = q \left(\frac{\partial \Phi}{\partial t} - \dot{\mathbf{x}} \frac{\partial \mathbf{A}}{\partial t} \right) - q \frac{d\Phi}{dt} \\
&= q \left(\frac{\partial \Phi}{\partial t} - \dot{\mathbf{x}} \frac{\partial \mathbf{A}}{\partial t} \right) - q \left(\frac{\partial \Phi}{\partial t} + \boldsymbol{\nabla}\Phi \cdot \dot{\mathbf{x}} \right) = q \left(-\boldsymbol{\nabla}\Phi - \frac{\partial \mathbf{A}}{\partial t} \right) \cdot \dot{\mathbf{x}} = q\mathbf{E} \cdot \boldsymbol{v},
\end{aligned}
\tag{2.4.17}
$$

which reproduces Eq. (2.1.36) and in any case follows from the Lorentz force equation as was shown in obtaining Eq. (2.1.36).

Equivalent Hamiltonians: Consider the effects of modifying the Lagrangian by multiplying by a constant and adding a total time derivative as we did in Eq. (2.1.41), which as we saw leaves the Euler-Lagrange equations unchanged. Recalling that we have $F = F(\vec{q}, t)$, then the conjugate momenta become

$$
p_j' \equiv \frac{\partial L'}{\partial \dot{q}_j} = a \frac{\partial L}{\partial \dot{q}_j} + \frac{\partial}{\partial \dot{q}_j} \frac{dF}{dt} = ap_j + \frac{\partial}{\partial \dot{q}_j} \left(\frac{\partial F}{\partial t} + \sum_{k=1}^{N} \frac{\partial F}{\partial q_k} \dot{q}_k \right) = ap_j + \frac{\partial F}{\partial q_j},
\tag{2.4.18}
$$

where we have used the fact that $\partial F/\partial t$ is a function of $\vec{q}$ and t but is independent of $\dot{q}$. The Hamiltonian, H', corresponding to the Lagrangian, L', is then

$$
H'(\vec{q}, \vec{p}\,') \equiv \sum_{j=1}^{N} p_j' \dot{q}_j - L' = aH(\vec{q}, \vec{p}) + \sum_{j=1}^{N} \frac{\partial F}{\partial q_j} \dot{q}_j - \frac{dF}{dt} - b = aH(\vec{q}, \vec{p}) - \frac{\partial F}{\partial t}(\vec{q}, t) - b.
$$

So a physical system characterized by its Euler-Lagrange equations can be described by different Lagrangians, conjugate momenta and Hamiltonians.

2.4.2 Poisson Brackets

Consider any two functions on phase space, $F(\vec{q}, \vec{p}, t)$ and $G(\vec{q}, \vec{p}, t)$, that are at least doubly differentiable with respect to their arguments; i.e., the order of derivatives in a double derivative can be exchanged. Such functions are referred to as *dynamical variables*. The *Poisson bracket* of these two functions is defined as

$$\{F, G\} \equiv \sum_{j=1}^{N} \left(\frac{\partial F}{\partial q_j} \frac{\partial G}{\partial p_j} - \frac{\partial F}{\partial p_j} \frac{\partial G}{\partial q_j} \right). \tag{2.4.19}$$

If A, B and C are dynamical variables and if a and b are real constants, then the properties of the Poisson brackets can be expressed as

(i) Closure: $\{A, B\}$ is also a function on phase space.
(ii) Antisymmetry: $\{A, B\} = -\{B, A\}$.
(iii) Bilinearity: $\{aA + bB, C\} = a\{A, C\} + b\{B, C\}$.
(iv) Product rule: $\{A, BC\} = \{A, B\} C + B\{A, C\}$.
(v) Jacobi identity: $\{A, \{B, C\}\} + \{B, \{C, A\}\} + \{C, \{A, B\}\} = 0$.

These properties are the same as those of the commutator $[A, B] \equiv AB - BA$.

For any dynamical variable $A(\vec{q}, \vec{p})$ the definition of the Poisson bracket leads to

$$\{q_i, A\} = \partial A/\partial p_i \qquad \text{and} \qquad \{p_i, A\} = -\partial A/\partial q_i. \tag{2.4.20}$$

Proof of Poisson brackets properties: The first three properties, (i)–(iii), are immediately obvious from the definition. Property (iv) follows since

$$\{A, BC\} = \sum_{j=1}^{N} \left(\frac{\partial A}{\partial q_j} \frac{\partial (BC)}{\partial p_j} - \frac{\partial A}{\partial p_j} \frac{\partial (BC)}{\partial q_j} \right) \tag{2.4.21}$$

$$= \sum_{j=1}^{N} \left(\frac{\partial A}{\partial q_j} \frac{\partial B}{\partial p_j} C + \frac{\partial A}{\partial q_j} B \frac{\partial C}{\partial p_j} - \frac{\partial A}{\partial p_j} \frac{\partial B}{\partial q_j} C - \frac{\partial A}{\partial p_j} B \frac{\partial C}{\partial q_j} \right) = \{A, B\} C + B\{A, C\}.$$

The proof of the Jacobi identity of property (v) follows using:

$$\{A, \{B, C\}\} = \sum_{j,k=1}^{N} \left(\frac{\partial A}{\partial q_j} \frac{\partial}{\partial p_j} - \frac{\partial A}{\partial p_j} \frac{\partial}{\partial q_j} \right) \left(\frac{\partial B}{\partial q_k} \frac{\partial C}{\partial p_k} - \frac{\partial B}{\partial p_k} \frac{\partial C}{\partial q_k} \right). \tag{2.4.22}$$

The derivatives of products appearing in the above can be expanded as

$$\frac{\partial}{\partial p_j} \left(\frac{\partial B}{\partial q_k} \frac{\partial C}{\partial p_k} \right) = \frac{\partial^2 B}{\partial p_j \partial q_k} \frac{\partial C}{\partial p_k} + \frac{\partial B}{\partial q_k} \frac{\partial^2 C}{\partial p_j \partial p_k}, \tag{2.4.23}$$

which gives eight terms for $\{A, \{B, C\}\}$. We similarly obtain eight terms for each of $\{B, \{C, A\}\}$ and $\{C, \{A, B\}\}$. We can then identify and sum the four terms corresponding to each of $\partial^2 B/\partial q \partial p$, $\partial^2 B/\partial q \partial q$, $\partial^2 B/\partial p \partial p$ and similarly for C. Each of these six sums of four terms vanishes; e.g., for $\partial^2 B/\partial q \partial p$ we have a contribution from $\{A, \{B, C\}\}$ and $\{C, \{A, B\}\}$

$$\sum_{j,k=1}^{N} \left[\left(\frac{\partial A}{\partial q_j} \frac{\partial C}{\partial p_k} - \frac{\partial A}{\partial p_k} \frac{\partial C}{\partial q_j} \right) \frac{\partial^2 B}{\partial p_j \partial q_k} + \frac{\partial^2 B}{\partial q_j \partial p_k} \left(\frac{\partial A}{\partial p_j} \frac{\partial C}{\partial q_k} - \frac{\partial A}{\partial q_k} \frac{\partial C}{\partial p_j} \right) \right] = 0.$$

The other five sums of four terms similarly vanish.

The total time derivative for a dynamical variable $F(\vec{q}, \vec{p}, t)$ is

$$\frac{dF}{dt} = \sum_{j=1}^{N} \left(\frac{\partial F}{\partial q_j} \frac{dq_j}{dt} + \frac{\partial F}{\partial p_j} \frac{dp_j}{dt} \right) + \frac{\partial F}{\partial t} = \sum_{j=1}^{N} \left(\frac{\partial F}{\partial q_j} \frac{\partial H}{\partial p_j} - \frac{\partial F}{\partial p_j} \frac{\partial H}{\partial q_j} \right) + \frac{\partial F}{\partial t}$$

$$= \{F, H\} + \partial F/\partial t, \tag{2.4.24}$$

where we have used Hamilton's equations, which are then recovered since

$$\dot{q}_i = \{q_i, H\} = \sum_{j=1}^{N} \left(\frac{\partial q_i}{\partial q_j} \frac{\partial H}{\partial p_j} - \frac{\partial q_i}{\partial p_j} \frac{\partial H}{\partial q_j} \right) = \frac{\partial H}{\partial p_i},$$

$$\dot{p}_i = \{p_i, H\} = \sum_{j=1}^{N} \left(\frac{\partial p_i}{\partial q_j} \frac{\partial H}{\partial p_j} - \frac{\partial p_i}{\partial p_j} \frac{\partial H}{\partial q_j} \right) = -\frac{\partial H}{\partial q_i}. \tag{2.4.25}$$

If a dynamical variable F has no explicit time dependence, $F = F(\vec{q}, \vec{p})$, and if $\{F, H\} = 0$, then it is a conserved quantity, since then

$$dF/dt = \{F, H\} + \partial F/\partial t = 0. \tag{2.4.26}$$

Note that one of Hamilton's equations in Eq. (2.4.9) can be written as

$$dH/dt = \{H, H\} + \partial H/\partial t = \partial H/\partial t = -\partial L/\partial t. \tag{2.4.27}$$

It is easily seen from the definition of the Poisson bracket that

$$\{q_i, p_j\} = \delta_{ij} \quad \text{and} \quad \{q_i, q_j\} = \{p_i, p_j\} = 0, \tag{2.4.28}$$

which are sometimes referred to as the *fundamental Poisson brackets*. These relations are reminiscent of the commutation relations of nonrelativistic quantum mechanics for reasons that will soon become clear.

Let us now introduce the *symplectic notation* for Hamilton's equations, which consists of defining a $2N$-component column vector $\vec{\eta} = (\vec{q}, \vec{p})$ or, equivalently,

$$\eta_j \equiv q_j \quad \text{and} \quad \eta_{j+N} \equiv p_j \quad \text{for} \quad j = 1, 2, \ldots, N. \tag{2.4.29}$$

We also define the real antisymmetric $2N \times 2N$ matrix

$$J \equiv \begin{pmatrix} 0 & I \\ -I & 0 \end{pmatrix}, \tag{2.4.30}$$

where I is the $N \times N$ identity matrix. Note that

$$\det J = 1, \quad J^2 = -1 \quad \text{and} \quad J^T = -J = J^{-1}. \tag{2.4.31}$$

Hamilton's equations in Eq. (2.4.9) can be summarized in compact form as

$$\dot{\vec{\eta}} = J(\partial H/\partial \vec{\eta}), \tag{2.4.32}$$

where $\partial H/\partial \vec{\eta}$ is also to be understood as a $2N$-component column vector with entries $\partial H/\partial \eta_j$ for $j = 1, \ldots, 2N$. We can equally represent the column vector η as a $2N$-component vector $\vec{\eta} \equiv (\eta_1, \ldots, \eta_{2N}) \equiv (\vec{q}, \vec{p})$ and so can write $F(\vec{q}, \vec{p}, t) = F(\vec{\eta}, t)$ and $G(\vec{q}, \vec{p}, t) = G(\vec{\eta}, t)$. Define the column vectors $\partial F/\partial \vec{\eta}$ and $\partial G/\partial \vec{\eta}$ and observe that the Poisson bracket can be written in symplectic notation as

$$\{F, G\} \equiv \sum_{j=1}^{N} \left(\frac{\partial F}{\partial q_j} \frac{\partial G}{\partial p_j} - \frac{\partial F}{\partial p_j} \frac{\partial G}{\partial q_j} \right) = \left[(\partial F/\partial \vec{q})^T, (\partial F/\partial \vec{p})^T \right] \begin{pmatrix} 0 & I \\ -I & 0 \end{pmatrix} \begin{bmatrix} \partial G/\partial \vec{q} \\ \partial G/\partial \vec{p} \end{bmatrix}$$

$$= \left(\frac{\partial F}{\partial \vec{\eta}} \right)^T J \left(\frac{\partial G}{\partial \vec{\eta}} \right) = \sum_{j,k=1}^{2N} \frac{\partial F}{\partial \eta_j} J_{jk} \frac{\partial G}{\partial \eta_k}. \tag{2.4.33}$$

Selecting F and G from the η_i we reproduce the fundamental Poisson brackets of Eq. (2.4.28) in a compact notation:

$$\{\eta_i, \eta_j\} = J_{ij}. \tag{2.4.34}$$

In addition, if we choose $F = \eta_j$ and $G = H$, then using Eqs. (2.4.32) and (2.4.33) we find that $\{\eta_j, H\} = \sum_{k=1}^{2N} J_{jk}(\partial H/\partial \eta)_k = \dot{\eta}_j$. Using vector notation as shorthand we see that we can also write Hamiton's equations in the form

$$\dot{\vec{\eta}} = \{\vec{\eta}, H\} \quad \text{and} \quad dH/dt = \partial H/\partial t = -\partial L/\partial t, \tag{2.4.35}$$

which can be compared with Eqs. (2.4.9), (2.4.25) and (2.4.32).

Poisson bracket theorem: A trajectory in phase space is generated by a Hamiltonian if and only if *all* dynamical variables $F(\vec{q}, \vec{p}, t)$ and $G(\vec{q}, \vec{p}, t)$ satisfy

$$\frac{d}{dt}\{F, G\} = \{\dot{F}, G\} + \{F, \dot{G}\}. \tag{2.4.36}$$

In addition, if Eq. (2.4.36) is true for every pair of variables chosen from the $\vec{q}$ and $\vec{p}$, then it is true for all dynamical variables F and G.

Poisson's theorem (corollary): An immediate consequence of the above theorem is that if F and G are both constants of the motion, $dF/dt = dG/dt = 0$, then $\{F, G\}$ must also be a constant of the motion, $d\{F, G\}/dt = 0$.

Proof: (a) Prove that if there is a Hamiltonian, then Eq. (2.4.36) is true for all F and G: If the trajectory is determined by a Hamiltonian, H, then Eq. (2.4.24) applies for any dynamical variable. Since the Poisson bracket $\{F, G\}$ is itself a dynamical variable (assuming it is doubly differentiable), then we have

$$d\{F, G\}/dt = \{\{F, G\}, H\} + \partial\{F, G\}/\partial t \tag{2.4.37}$$

$$= -\{\{G, H\}, F\} - \{\{H, F\}, G\} + \{\partial F/\partial t, G\} + \{F, \partial G/\partial t\}$$

$$= \{\{F, H\} + \partial F/\partial t, G\} + \{F, \{G, H\} + \partial G/\partial t\} = \{\dot{F}, G\} + \{F, \dot{G}\},$$

where we have used the Jacobi identity, the linearity property of the Poisson bracket and the fact that being dynamical variables F and G are doubly differentiable so that we can exchange the order of derivatives. Thus it follows that if the system has a Hamiltonian, then Eq. (2.4.36) is true for all F and G.

(b) If Eq. (2.4.36) is true for all F and G, then it is true for $\vec{q}$ and $\vec{p}$. From Eq. (2.4.28) we have $d\{q^i, q^j\}/dt = d\{q^i, p^j\}/dt = d\{p^i, p^j\}/dt = 0$. Then if Eq. (2.4.36) is true for all $\vec{q}$ and $\vec{p}$ it follows that

$$\{\dot{q}_i, q_j\} + \{q_i, \dot{q}_j\} = 0, \quad \{\dot{q}_i, p_j\} + \{q_i, \dot{p}_j\} = 0, \quad \text{and} \quad \{\dot{p}_i, p_j\} + \{p_i, \dot{p}_j\} = 0.$$

It is readily seen from the definition of the Poisson bracket in Eq. (2.4.19) that

$$\{q_i, \dot{q}_j\} = \frac{\partial \dot{q}_j}{\partial p_i}, \{q_i, \dot{p}_j\} = \frac{\partial \dot{p}_j}{\partial p_i}, \{\dot{q}_i, p_j\} = \frac{\partial \dot{q}_i}{\partial q_j}, \{\dot{p}_i, p_j\} = \frac{\partial \dot{p}_i}{\partial q_j}, \tag{2.4.38}$$

where we have used $\partial p_i/\partial q_j = \partial q_i/\partial p_j = 0$ and $\partial p_i/\partial p_j = \partial q_i/\partial q_j = \delta_{ij}$, since the $\vec{q}$ and $\vec{p}$ are $2N$-independent variables. Hence we have

$$-\frac{\partial \dot{q}_i}{\partial p_j} + \frac{\partial \dot{q}_j}{\partial p_i} = \frac{\partial \dot{q}_i}{\partial q_j} + \frac{\partial \dot{p}_j}{\partial p_i} = -\frac{\partial \dot{p}_i}{\partial p_j} - \frac{\partial \dot{q}_j}{\partial q_i} = \frac{\partial \dot{p}_i}{\partial q_j} - \frac{\partial \dot{p}_j}{\partial q_i} = 0. \qquad (2.4.39)$$

Using the symplectic notation $\vec{\eta} \equiv (\vec{q}, \vec{p})$ and introducing $\vec{\chi} \equiv (-\dot{\vec{p}}, \dot{\vec{q}})$, or equivalently $\vec{\chi} \equiv J^T \dot{\vec{\eta}}$, we can summarize Eq. (2.4.39) as

$$\partial \chi_k / \partial \eta_\ell - \partial \chi_\ell / \partial \eta_k = 0 \qquad (2.4.40)$$

for $k, \ell = 1, 2, \ldots, 2N$. From the Poincaré lemma in Sec. A.2.6 it follows that a doubly differentiable function, $H(\vec{\eta})$, must exist such that

$$\chi_j = (J^T \dot{\eta})_j = \partial H / \partial \eta_j, \qquad (2.4.41)$$

which is Eq. (2.4.32), since $J^T = J^{-1}$. So if Eq. (2.4.36) is true for all F and G then a hamitonian exists.

(c) As we saw in showing (b) above, if Eq. (2.4.36) is true for all $\vec{q}$ and $\vec{p}$ then an H exists and hence from (a) it follows that Eq. (2.4.36) is true for all F and G.

2.4.3 Liouville Equation and Liouville's Theorem

What can be said about the time evolution of neighboring points in phase space and/or small regions of phase space?

Liouville's theorem: A region of phase space $R(t)$ at time t with $2N$-dimensional volume $V(t)$ consists of points $(\vec{q}(t), \vec{p}(t)) \in R(t)$. It will flow in time Δt to a new region $R(t + \Delta t)$ with volume $V(t + \Delta t)$ and points $(\vec{q}(t + \Delta t), \vec{p}(t + \Delta t)) \in R(t + \Delta t)$. Liouville's theorem is that

$$V(t) = V(t + \Delta t) \quad \text{for all} \quad \Delta t. \qquad (2.4.42)$$

In other words, any given region of phase space, R, moves and changes shape in general as it evolves in time, but its volume remains constant, i.e., $V(t) = V$.

Proof: An infinitesimal phase-space volume, $dV = dq_1 \ldots dq_N \, dp_1 \ldots dp_N$, is located in phase space at the point $(\vec{q}, \vec{p})$ and defined at time t. The infinitesimal volume is defined by the intervals $[q_i, q_i + dq_i]$ and $[p_i, p_i + dp_i]$ for $i = 1, \ldots, N$. After an infinitesimal time dt each point will have evolved according to Hamilton's equations, leading to a new volume $dV' = dq_1' \ldots dq_N' dp_1' \ldots dp_N'$. The evolution in the infinitesimal time step dt is

$$q_i \to q_i' = q_i + \dot{q}_i dt = q_i + \frac{\partial H}{\partial p_i} dt, \quad p_i \to p_i' = p_i + \dot{p}_i dt = p_i - \frac{\partial H}{\partial q_i} dt. \qquad (2.4.43)$$

From the discussion of Jacobians in Sec. A.2.2 it follows that

$$dR' = dq_1' \ldots dq_N' dp_1' \ldots dp_N' = \mathcal{J} dq_1 \ldots dq_N dp_1 \ldots dp_N = \mathcal{J} dR, \qquad (2.4.44)$$

where we use the notation $dR \equiv \prod_{i=1}^{N} dq_i dp_i$. The Jacobian is $\mathcal{J} = |\det M|$, where the matrix, $M \equiv I + A dt + \mathcal{O}(dt^2)$, has elements of the form

$$\frac{\partial q_i'}{\partial q_j} = \delta_{ij} + \frac{\partial^2 H}{\partial p_i \partial q_j} dt, \; \frac{\partial p_i'}{\partial p_j} = \delta_{ij} - \frac{\partial^2 H}{\partial q_i \partial p_j} dt, \; \frac{\partial q_i'}{\partial p_j} = \frac{\partial^2 H}{\partial p_i \partial p_j} dt, \; \frac{\partial p_i'}{\partial q_j} = -\frac{\partial^2 H}{\partial q_i \partial q_j} dt$$

for $i, j = 1, 2, \ldots, N$, where we have used Eq. (2.4.43). Consider $N = 1$,

$$\mathcal{J} = |\det M| = \left| \det \begin{pmatrix} 1 + \frac{\partial^2 H}{\partial p \partial q} dt & \frac{\partial^2 H}{\partial p^2} dt \\ \frac{\partial^2 H}{\partial q^2} dt & 1 - \frac{\partial^2 H}{\partial q \partial p} dt \end{pmatrix} \right| = 1 + \mathcal{O}(dt^2). \qquad (2.4.45)$$

With $M = I + A dt + \mathcal{O}(dt^2)$ we see from above that $\mathrm{tr} A = 0$ since the derivatives in q and p can be interchanged. This remains true for arbitrary N. With an infinite number of infinitesimal dt we can build up a Δt. Using Eq. (A.8.5) with $dt = \Delta t/n$ gives $M = \lim_{n \to \infty} [1 + A\Delta t/n]^n = e^{A\Delta t}$. Then using Eq. (A.8.4) gives $\det M = \exp(\mathrm{tr} \ln M) = \exp(\Delta \mathrm{tr} A t) = 1$ since $\mathrm{tr} A = 0$. Then $\mathcal{J} = |\det M| = 1$. Then we have $V = \int_R dR = \int_{R'} dR' = V'$ for every Δt.

Consider a system whose position in phase space at some initial time t is not known but is given by a phase-space probability distribution $\rho(\vec{q}, \vec{p}, t)$. We are interested in the evolution of the ρ in time due to the Hamiltonian, where the Hamiltonian may be time-dependent in general. The defining properties of a probability distribution are $\int \rho(\vec{q}, \vec{p}, t) dV = 1$ and $\rho(\vec{q}, \vec{p}, t) \geq 0$. Define the phase space velocity as $\mathbf{v}_{\mathrm{ph}} \equiv (\dot{q}_1, \ldots, \dot{q}_N, \dot{p}_1, \ldots, \dot{p}_N)$, then the probability current density is $\mathbf{j} = \rho \mathbf{v}_{\mathrm{ph}}$. The probability in some phase-space volume V is $Q = \int_V \rho dV$. The probability change is given by the probability flow through the surface S of V,

$$dQ/dt = d\left(\int_V \rho dV\right)/dt = \int_V (\partial \rho/\partial t) dV = -\int_S \mathbf{j} \cdot d\mathbf{s} = -\int_V \boldsymbol{\nabla} \cdot \mathbf{j} dV, \qquad (2.4.46)$$

where $d\mathbf{s}$ is the outward normal to the surface S and where we have used the divergence theorem in $2N$-dimensional phase space. The partial time derivative appears because the phase space points $(\vec{q}, \vec{p})$ in the fixed phase-space volume V are not changing in time. This is just the current conservation result of Eq. (4.3.22) with $2N$-dimensional phase space replacing ordinary three-dimensional coordinate space, $\mathbf{x} \to (\vec{q}, \vec{p})$. Since V is arbitrary, then we can choose it to be infinitesimal and we find the probability current conservation result that

$$0 = \partial \rho/\partial t + \boldsymbol{\nabla} \cdot \mathbf{j} = \partial \rho/\partial t + \boldsymbol{\nabla} \cdot (\rho \mathbf{v}_{\mathrm{ph}}). \qquad (2.4.47)$$

Evaluating $\boldsymbol{\nabla} \cdot \mathbf{j}$ we find

$$\boldsymbol{\nabla} \cdot \mathbf{j} = \boldsymbol{\nabla} \cdot (\rho \mathbf{v}_{\mathrm{ph}}) = \sum_{i=1}^{N} \left(\dot{q}_i \frac{\partial \rho}{\partial q_i} + \dot{p}_i \frac{\partial \rho}{\partial p_i} \right) + \rho \sum_{i=1}^{N} \left(\frac{\partial \dot{q}_i}{\partial q_i} + \frac{\partial \dot{p}_i}{\partial p_i} \right) \qquad (2.4.48)$$

$$= \sum_{i=1}^{N} \left(\frac{\partial H}{\partial p_i} \frac{\partial \rho}{\partial q_i} - \frac{\partial H}{\partial q_i} \frac{\partial \rho}{\partial p_i} \right) + \rho \sum_{i=1}^{N} \left(\frac{\partial^2 H}{\partial p_i \partial q_i} - \frac{\partial^2 H}{\partial q_i \partial p_i} \right) = \{\rho, H\}, \qquad (2.4.49)$$

which then leads to $0 = \partial \rho/\partial t + \{\rho, H\}$. Combining this with Eq. (2.4.24) leads to the **Liouville equation**,

$$0 = d\rho/dt = \partial \rho/\partial t + \{\rho, H\}, \qquad (2.4.50)$$

which is one of the foundations of statistical mechanics.

2.4.4 Canonical Transformations

Recall that in discussing Noether's theorem we considered continuous families of transformations of the generalized coordinates, i.e., $\vec{q}'(s) = \vec{f}_s(\vec{q})$ in Eq. (2.2.10). But in the symplectic notation for Hamilton's equations given in Eq. (2.4.32) we saw that we can treat the generalized coordinates, $\vec{q}$, and momenta, $\vec{p}$, in a single unified fashion. This suggests that we can consider a larger class of transformations, where the generalized coordinates and momenta can be mixed together:

$$q_i \to Q_i(\vec{q}, \vec{p}) \quad \text{and} \quad p_i \to P_i(\vec{q}, \vec{p}) \quad \text{for} \quad i = 1, 2, \dots, N. \tag{2.4.51}$$

Such transformations that preserve Hamilton's equations are referred to as *canonical transformations*. We will only consider those without explicit time-dependence. Canonical transformations that do not involve the generalized momenta are referred to as *point transformations*. These are the ones considered earlier when discussing Noether's theorem.

In symplectic notation we have $\eta \equiv (\vec{q}, \vec{p})$ and if we similarly define $\zeta = (\vec{Q}, \vec{P})$ then we can write Eq. (2.4.51) more compactly as $\eta_i \to \zeta_i(\vec{\eta})$ for $i = 1, 2, \dots, 2N$. It then follows that

$$\dot{\zeta}_i = \sum_{j=1}^{2N} \frac{\partial \zeta_i}{\partial \eta_j} \dot{\eta}_j = \sum_{j,k=1}^{2N} \frac{\partial \zeta_i}{\partial \eta_j} J_{jk} \frac{\partial H}{\partial \eta_k} = \sum_{j,k,\ell=1}^{2N} \frac{\partial \zeta_i}{\partial \eta_j} J_{jk} \frac{\partial \zeta_\ell}{\partial \eta_k} \frac{\partial H}{\partial \zeta_\ell}, \tag{2.4.52}$$

which can be written more compactly using an obvious matrix notation as

$$\dot{\vec{\zeta}} = MJM^T \frac{\partial H}{\partial \vec{\zeta}} \quad \text{with} \quad M_{jk} \equiv \frac{\partial \zeta_j}{\partial \eta_k}, \tag{2.4.53}$$

where M is the Jacobian matrix. Any change of variables that satisfies

$$MJM^T = J \quad \text{or} \quad \sum_{j,k=1}^{2N} M_{ij} J_{jk} M_{\ell k} = J_{i\ell} \quad \Rightarrow \quad \dot{\zeta} = J \frac{\partial H}{\partial \zeta}, \tag{2.4.54}$$

and so reproduces Hamilton's equations in terms of $\vec{\zeta}$. Any matrix M that satisfies $MJM^T = J$ is said to be a *symplectic matrix*. So *a change of variables is a canonical transformation if and only if it has a symplectic Jacobian, $MJM^T = J$.* The set of real $2n \times 2n$ symplectic matrices form a Lie group called $Sp(2n, \mathbb{R})$ and has a similar definition to that of the pseudo-orthogonal groups.

Invariance of the Poisson bracket

(a) The Poisson bracket is invariant under any canonical transformation.

(b) Any transformation of the type in Eq. (2.4.51) that preserves the fundamental Poisson brackets of Eq. (2.4.28) for the generalized coordinates and momenta,

$$\{Q_i, P_j\} = \delta_{ij} \quad \text{and} \quad \{Q_i, Q_j\} = \{P_i, P_j\} = 0, \tag{2.4.55}$$

is a canonical transformation.

Proof: (a) Use Eqs. (2.4.33) and (2.4.54) to write

$$\{F, G\} = \sum_{j=1}^{N} \left(\frac{\partial F}{\partial q_j} \frac{\partial G}{\partial p_j} - \frac{\partial F}{\partial p_j} \frac{\partial G}{\partial q_j} \right) = \sum_{j,k=1}^{2N} \frac{\partial F}{\partial \eta_j} J_{jk} \frac{\partial G}{\partial \eta_k} \tag{2.4.56}$$

$$= \sum_{i,j,k,\ell} \frac{\partial F}{\partial \zeta_i} M_{ij} J_{jk} M_{\ell k} \frac{\partial G}{\partial \zeta_\ell} = \sum_{i,\ell} \frac{\partial F}{\partial \zeta_i} J_{i\ell} \frac{\partial G}{\partial \zeta_\ell} = \sum_j \left(\frac{\partial F}{\partial Q_j} \frac{\partial G}{\partial P_j} - \frac{\partial F}{\partial P_j} \frac{\partial G}{\partial Q_j} \right).$$

So the Poisson brackets are invariant under canonical transformations.

(b) For any transformation $\eta \to \zeta$ with any M we can use the symplectic notation for the Poisson bracket given in Eq. (2.4.33) to write

$$\{\zeta_i, \zeta_\ell\} = \sum_{j,k=1}^{2N} \frac{\partial \zeta_i}{\partial \eta_j} J_{jk} \frac{\partial \zeta_\ell}{\partial \eta_k} = \sum_{j,k=1}^{2N} M_{ij} J_{jk} M_{\ell k} = (MJM^T)_{i\ell}. \tag{2.4.57}$$

The Poisson bracket relations of Eq. (2.4.55) in symplectic notation are written as $\{\zeta_i, \zeta_j\} = J_{ij}$. If they are valid, then it must be true that $J = MJM^T$ and so the transformation is canonical.

Consider an infinitesimal transformation of the form

$$\eta_i \to \zeta_i(\eta) = \eta_i + d\eta_i = \eta_i + d\alpha F_i(\vec{\eta}), \tag{2.4.58}$$

where $\vec{F}(\vec{\eta}) \equiv (\vec{F}_q(\vec{\eta}), \vec{F}_p(\vec{\eta}))$ is some set of $2N$ dynamical variables (i.e., $2N$ doubly differentiable functions). Note that Eq. (2.4.58) is equivalent to

$$d\eta_i/d\alpha = F_i(\vec{\eta}) \quad \text{for} \quad i = 1, 2, \ldots, 2N, \tag{2.4.59}$$

which is equivalent to

$$dq_i/d\alpha = F_{qi}(\vec{\eta}) \quad \text{and} \quad dp_i/d\alpha = F_{pi}(\vec{\eta}) \quad \text{for} \quad i = 1, 2, \ldots, N. \tag{2.4.60}$$

The Jacobian matrix for this transformation is

$$M_{ij} = \frac{\partial \zeta_i}{\partial \eta_j} = \delta_{ij} + d\alpha \frac{\partial F_i}{\partial \eta_j} \Rightarrow (MJM^T)_{i\ell} = J_{i\ell} + d\alpha \sum_{k=1}^{2N} \left(\frac{\partial F_i}{\partial \eta_k} J_{k\ell} + J_{ik} \frac{\partial F_\ell}{\partial \eta_k} \right) + \mathcal{O}(d\alpha^2).$$

For the infinitesimal transformation to be canonical the $\mathcal{O}(d\alpha)$ term must vanish. There are four cases to consider, i.e., where the indices $i\ell$ belong to the sectors qq, pp, qp and pq. These four cases, respectively, lead to the equations

$$0 = -\frac{\partial F_{qi}}{\partial p_\ell} + \frac{\partial F_{q\ell}}{\partial p_i} = \frac{\partial F_{pi}}{\partial q_\ell} - \frac{\partial F_{p\ell}}{\partial q_i} = \frac{\partial F_{qi}}{\partial q_\ell} + \frac{\partial F_{p\ell}}{\partial p_i} = -\frac{\partial F_{pi}}{\partial p_\ell} - \frac{\partial F_{q\ell}}{\partial q_i}. \tag{2.4.61}$$

Let $G = G(\vec{q}, \vec{p})$ be some dynamical variable and choose $\vec{F}(\vec{\eta})$ such that

$$F_{qi} = \frac{dq_i}{d\alpha} \equiv \frac{\partial G}{\partial p_i} \quad \text{and} \quad F_{pi} = \frac{dp_i}{d\alpha} \equiv -\frac{\partial G}{\partial q_i} \tag{2.4.62}$$

for $i = 1, 2, \ldots, N$; then we see that we have a solution for Eq. (2.4.61). The dynamical variable $G(\vec{q}, \vec{p})$ *generates* the canonical transformation. For an infinitesimal canonical transformation generated by the dynamical variable G we have

$$q_i \to Q_i = q_i + d\alpha \frac{\partial G}{\partial p_i} \quad \text{and} \quad p_i \to P_i = p_i - d\alpha \frac{\partial G}{\partial q_i}. \tag{2.4.63}$$

Note that we can write Eqs. (2.4.62) and (2.4.63) using symplectic notation as

$$\vec{F}(\vec{\eta}) = \frac{d\vec{\eta}}{d\alpha} \equiv J \frac{\partial G}{\partial \vec{\eta}} \quad \text{and} \quad \vec{\eta} \to \vec{\zeta} = \vec{\eta} + d\alpha J \frac{\partial G}{\partial \vec{\eta}}. \tag{2.4.64}$$

Observe that Eq. (2.4.62) has the same form as Hamilton's equations, with the Hamiltonian replaced by the generating function G and with time replaced by α. Thus for $\alpha \in \mathbb{R}$ we see that G generates a one-parameter family of transformations in phase space that is analogous to evolution in time under Hamilton's equations. Equivalently, we can view time evolution under a Hamiltonian as a one-parameter family of canonical transformations generated by the Hamiltonian, H.

Consider an arbitrary dynamical variable $F(\vec{q}, \vec{p}, t)$. Under an infinitesimal canonical transformation generated by some other dynamical variable $G(\vec{q}, \vec{p})$ we find

$$dF = \sum_{i=1}^{N} \left(\frac{\partial F}{\partial q_i} dq_i + \frac{\partial F}{\partial p_i} dp_i \right) = \sum_{i=1}^{N} \left(\frac{\partial F}{\partial q_i} \frac{\partial G}{\partial p_i} - \frac{\partial F}{\partial p_i} \frac{\partial G}{\partial q_i} \right) d\alpha = d\alpha \{F, G\}, \tag{2.4.65}$$

where we have used Eq. (2.4.62). Hence, under a canonical transformation generated by G every dynamical variable F satisfies

$$\frac{dF}{d\alpha} = \{F, G\}. \qquad (2.4.66)$$

Noether's theorem in the Hamiltonian context: Under an infinitesimal canonical transformation generated by G we therefore have for the Hamiltonian

$$\frac{dH}{d\alpha} = \{H, G\}. \qquad (2.4.67)$$

By definition, a *symmetry of the Hamiltonian* is a transformation that leaves the Hamiltonian unchanged. Thus the set of canonical transformations generated by G is a symmetry of the Hamiltonian if and only if

$$\{G, H\} = 0. \qquad (2.4.68)$$

Hence from Eq. (2.4.24) we see that if $G(\vec{q}, \vec{p})$ generates a symmetry of the Hamiltonian then it is conserved, since

$$\frac{dG}{dt} = 0. \qquad (2.4.69)$$

Any conserved dynamical variable $G(\vec{q}, \vec{p})$ is the generator of a one-parameter family of canonical transformations that form a symmetry of the Hamiltonian.

2.5 Relation to Quantum Mechanics

Since classical mechanics must arise as the macroscopic limit of the corresponding quantum system, then the classical and quantum descriptions of the system must be closely interconnected. In this section we explore these connections.

Poisson brackets and commutators in quantum mechanics: The classical path satisfies the Euler-Lagrange equations and is the path satisfying Hamilton's principle; i.e., it is a stationary point of the action. In the discussion of the path integral formulation of quantum mechanics in Sec. 4.1.12 following Eq. (4.1.208) we will see that macroscopic systems are those for which $S[\vec{q}] \gg \hbar$ with the classical path becoming dominant and with small quantum corrections coming from paths near to the classical path. In the *macroscopic limit*, $S[\vec{q}]/\hbar \to \infty$, these quantum corrections become vanishingly small. Since in nature $\hbar$ is fixed, then it is a conventional and very convenient misuse of language to refer to this as the limit $\hbar \to 0$ and to simply understand that this means the macroscopic limit. Another important aspect of the macroscopic limit is the extremely rapid loss of quantum coherence within and between macroscopic objects. This occurs due to the very large number of degrees of freedom in macroscopic objects and the numerous small interactions that the atoms and molecules that make up the object have with the surrounding environment.

Any path will correspond to a trajectory in phase space consisting of a series of connected points, $(\vec{q}, \vec{p})$. Denote the phase space points on the classical path as $(\vec{q}_c, \vec{p}_c)$. Consider an operator $\hat{F} \equiv F(\hat{\vec{q}}, \hat{\vec{p}}, t)$ that is a function F of $\hat{\vec{q}}$, $\hat{\vec{p}}$ and t. This is the operator analog of the classical dynamical variable $F(\vec{q}, \vec{p}, t)$.

Note that since the fluctuations around the classical path become increasingly suppressed as we take $\hbar \to 0$; then we have for the coordinate and momentum expectation values $\langle \hat{\vec{q}} \rangle \to \vec{q}_c$ and $\langle \hat{\vec{p}} \rangle \to \vec{p}_c$, respectively, and we have for the uncertainties $\sigma_{q_j} \equiv (\langle \hat{q}_j^2 \rangle - \langle \hat{q}_j \rangle^2)^{1/2} \to 0$ and

$\sigma_{p_j} \equiv (\langle \hat{p}_j^2 \rangle - \langle \hat{p}_j \rangle^2)^{1/2} \to 0$. We are using the notation $\langle \hat{A} \rangle \equiv \langle \psi | \hat{A} \psi \rangle$ where $|\psi\rangle$ is any normalized quantum state.

Consider any quantum system with N generalized coordinate operators $\hat{\vec{q}}$ and the corresponding conjugate momentum operators $\hat{\vec{p}}$. They satisfy the canonical commutation relations,

$$[\hat{q}^i, \hat{p}^j] = i\hbar \delta^{ij} \quad \text{and} \quad [\hat{q}^i, \hat{q}^j] = [\hat{p}^i, \hat{p}^j] = 0. \tag{2.5.1}$$

These lead to the *Heisenberg uncertainty relation*, also referred to as the *Heisenberg uncertainty principle*, which can be written as the uncertainty relations

$$\sigma_{q_i} \sigma_{p_j} \ge \delta_{ij} \tfrac{1}{2} \hbar. \tag{2.5.2}$$

These are a special case of the *Robertson uncertainty relation* (Robertson, 1929), where for any two Hermitian operators $\hat{A}$ and $\hat{B}$ we have

$$\sigma_A \sigma_B \ge \tfrac{1}{2} \left| \langle [\hat{A}, \hat{B}] \rangle \right|, \tag{2.5.3}$$

where $\sigma_A \equiv (\langle \hat{A}^2 \rangle - \langle \hat{A} \rangle^2)^{1/2}$ and similarly for σ_B. The Robertson uncertainty relation is a weaker form of the *Schrödinger uncertainty relation* (Schrödinger, 1930)

$$\sigma_A^2 \sigma_B^2 \ge \left| \tfrac{1}{2} \langle \{\hat{A}, \hat{B}\} \rangle - \langle \hat{A} \rangle \langle \hat{B} \rangle \right|^2 + \left| \tfrac{1}{2} \langle [\hat{A}, \hat{B}] \rangle \right|^2, \tag{2.5.4}$$

where $\{\hat{A}, \hat{B}\} \equiv \hat{A}\hat{B} + \hat{B}\hat{A}$. Dropping the positive first term on the right-hand side and taking the square root recovers the Robertson uncertainty relation.

Proof: It is then sufficient to prove the Schrödinger uncertainty relation. First define $|\chi\rangle \equiv (\hat{A} - \langle \hat{A} \rangle)|\psi\rangle$ and $|\eta\rangle \equiv (\hat{B} - \langle \hat{B} \rangle)|\psi\rangle$ for Hermitian $\hat{A}$ and $\hat{B}$, then

$$\langle \chi | \chi \rangle = \langle \psi | (\hat{A} - \langle \hat{A} \rangle)^2 \psi \rangle = \langle \hat{A}^2 \rangle - \langle \hat{A} \rangle^2 = \sigma_A^2 \quad \text{and} \quad \langle \eta | \eta \rangle = \sigma_B^2, \tag{2.5.5}$$

$$\langle \chi | \eta \rangle = \langle \psi | (\hat{A} - \langle \hat{A} \rangle)(\hat{B} - \langle \hat{B} \rangle) \psi \rangle = \langle \hat{A}\hat{B} \rangle - \langle \hat{A} \rangle \langle \hat{B} \rangle. \tag{2.5.6}$$

Also then $\langle \eta | \chi \rangle = \langle \hat{B}\hat{A} \rangle - \langle \hat{A} \rangle \langle \hat{B} \rangle$. Consider the two real quantities

$$c \equiv (1/i) \left(\langle \chi | \eta \rangle - \langle \eta | \chi \rangle \right) = (1/i) \langle [\hat{A}, \hat{B}] \rangle \in \mathbb{R},$$

$$d \equiv \langle \chi | \eta \rangle + \langle \eta | \chi \rangle = \langle \{\hat{A}, \hat{B}\} \rangle - 2 \langle \hat{A} \rangle \langle \hat{B} \rangle \in \mathbb{R}, \tag{2.5.7}$$

where we recall $\langle \chi | \eta \rangle = \langle \eta | \chi \rangle^*$. Then we have

$$|d|^2 + |c|^2 = d^2 + c^2 = (\langle \chi | \eta \rangle + \langle \eta | \chi \rangle)^2 - (\langle \chi | \eta \rangle - \langle \eta | \chi \rangle)^2 = 4 |\langle \chi | \eta \rangle|^2 \tag{2.5.8}$$

and so using the Cauchy-Schwartz inequality, $\langle \chi | \chi \rangle \langle \eta | \eta \rangle \ge |\langle \chi | \eta \rangle|^2$, we have

$$\sigma_A^2 \sigma_B^2 = \langle \chi | \chi \rangle \langle \eta | \eta \rangle \ge |\langle \chi | \eta \rangle|^2 = \tfrac{1}{4} \left(|d|^2 + |c|^2 \right), \tag{2.5.9}$$

which on rearranging gives Eq. (2.5.4) as required.

For a macroscopic action for this system the classical path will dominate and so we can expand $\hat{F}$ around this path using a Taylor expansion

$$\hat{F} = F(\vec{q}_c, \vec{p}_c, t) + (\hat{\vec{q}} - \vec{q}_c) \cdot \frac{\partial F(\vec{q}_c, \vec{p}_c, t)}{\partial \vec{q}_c} + (\hat{\vec{p}} - \vec{p}_c) \cdot \frac{\partial F(\vec{q}_c, \vec{p}_c, t)}{\partial \vec{p}_c} + \cdots. \tag{2.5.10}$$

Then for two such operators we can form their commutator using the canonical commutation relations and retain the nonvanishing $\mathcal{O}(\hbar)$ terms to give

$$[\hat{F}, \hat{G}] = \left[(\hat{\vec{q}} - \vec{q}_c) \cdot \frac{\partial F}{\partial \vec{q}_c}, (\hat{\vec{p}} - \vec{p}_c) \cdot \frac{\partial G}{\partial \vec{p}_c}\right] + \left[(\hat{\vec{p}} - \vec{p}_c) \cdot \frac{\partial F}{\partial \vec{p}_c}, (\hat{\vec{q}} - \vec{q}_c) \cdot \frac{\partial G}{\partial \vec{q}_c}\right] + \mathcal{O}(\hbar^2)$$

$$= i\hbar \left(\frac{\partial F}{\partial \vec{q}_c} \cdot \frac{\partial G}{\partial \vec{p}_c} - \frac{\partial F}{\partial \vec{p}_c} \cdot \frac{\partial G}{\partial \vec{q}_c}\right) + \mathcal{O}(\hbar^2) = i\hbar \{F, G\} + \mathcal{O}(\hbar^2). \tag{2.5.11}$$

We then have the result that

$$\lim_{\hbar \to 0} \left\langle \frac{1}{i\hbar}[\hat{F}, \hat{G}] \right\rangle = \{F, G\}, \tag{2.5.12}$$

where $F \equiv F(\vec{q}_c, \vec{p}_c, t)$ and $G \equiv G(\vec{q}_c, \vec{p}_c, t)$.

Choosing various $\hat{F}$ and $\hat{G}$ from the $\hat{\vec{q}}$ and the $\hat{\vec{p}}$ and using Eq. (2.5.12) leads again to the Poisson bracket results of Eq. (2.4.28), which we see directly arise from the canonical commutation relations in Eq. (2.5.1). It is also obvious from Eq. (2.5.12) why the mathematical properties of the Poisson bracket had to be identical to the properties of the commutator. Finally, from Eq. (4.1.24) we know that for any Heisenberg picture operator, $\hat{F}_h$, we have

$$\frac{d\hat{F}_h}{dt} = \frac{1}{i\hbar}[\hat{F}_h, \hat{H}_h] + \frac{\partial \hat{F}_h}{\partial t}, \tag{2.5.13}$$

where H_h is the Hamiltonian operator in the Heisenberg picture. Then we have

$$\frac{d\langle \hat{F} \rangle}{dt} = \frac{1}{i\hbar}\langle [\hat{F}, \hat{H}] \rangle + \left\langle \frac{\partial \hat{F}}{\partial t} \right\rangle, \tag{2.5.14}$$

which is a generalized form of the *Ehrenfest theorem* and valid in any quantum mechanics picture. Using Eq. (2.5.12) we recover Eq. (2.4.24) as $\hbar \to 0$,

$$\frac{dF}{dt} = \{F, H\} + \frac{\partial F}{\partial t}. \tag{2.5.15}$$

For the special cases of $F = q_i, p_i, H$ we recover Hamilton's equations using Eq. (2.4.25)

$$\dot{q}_i = \frac{\partial H}{\partial p_i}, \quad \dot{p}_i = -\frac{\partial H}{\partial q_i} \quad \text{and} \quad \frac{dH}{dt} = \frac{\partial H}{\partial t}. \tag{2.5.16}$$

So the Schrödinger equation leads to Hamilton's equations in the macroscopic (i.e., classical, $\hbar \to 0$) limit. The important observation is that the macroscopic limit of quantum mechanics leads inevitably to Poisson brackets and to Hamilton's classical equations of motion. This is complementary to and consistent with the classical path satisfying Hamilton's principle increasingly dominating the quantum mechanical description of the system as $\hbar \to 0$. Classical mechanics is the macroscopic limit of quantum mechanics viewed from either the Lagrangian or Hamiltonian formulation.

Operator equations of motion: We can make another connection of classical mechanics with quantum mechanics by showing that quantum operators obey the classical equations of motion. Recall from Eq. (4.1.24) that we have Ehrenfest's theorem for a Heisenberg picture operator $\hat{A}(t)$,

$$\frac{d\hat{A}(t)}{dt} = \frac{1}{i\hbar}[\hat{A}(t), \hat{H}] + \frac{\partial \hat{A}}{\partial t}. \tag{2.5.17}$$

The equal-time canonical commutation relations for the Heisenberg-picture coordinate operators $\hat{q}_1(t), \ldots, \hat{q}_N(t)$ and conjugate momenta $\hat{p}_1(t), \ldots, \hat{p}_N(t)$ are

$$[\hat{q}_i(t), \hat{q}_j(t)] = [\hat{p}_i(t), \hat{p}_j(t)] = 0 \quad \text{and} \quad [\hat{q}_i(t), \hat{p}_j(t)] = i\hbar\delta_{ij}. \tag{2.5.18}$$

The latter are just the Schrödinger picture canonical commutation relations converted to the Heisenberg picture.

The equal-time nature of the following equations is to be understood. It is easily proved by induction from the canonical commutation relations that

$$[\hat{q}_i, \hat{p}_j^n] = i\hbar\delta_{ij}\left(n\hat{p}_i^{n-1}\right) \equiv i\hbar\frac{\partial \hat{p}_j^n}{\partial \hat{p}_i}, \qquad\qquad [\hat{p}_i, \hat{q}_j^n] = -i\hbar\delta_{ij}\left(n\hat{q}_i^{n-1}\right) \equiv -i\hbar\frac{\partial \hat{q}_j^n}{\partial \hat{q}_i},$$

$$[\hat{q}_i, \hat{q}_j^n] = 0 \equiv i\hbar\frac{\partial \hat{q}_j^n}{\partial \hat{p}_i}, \qquad\qquad\qquad [\hat{p}_i, \hat{p}_j^n] = 0 \equiv -i\hbar\frac{\partial \hat{p}_j^n}{\partial \hat{q}_i}, \qquad\qquad (2.5.19)$$

where we have introduced in these equations the definition of differentiation with respect to the operators $\hat{p}_j$ and $\hat{q}_j$, respectively. If we consider any operator $\hat{A}$ that can be written as a mixed power series in the $2N$ operators $\hat{q}_1, \ldots, \hat{q}_N, \hat{p}_1, \ldots, \hat{p}_N$ then it follows from these equations that we can simply define

$$[\hat{q}_i, \hat{A}] \equiv i\hbar\,\partial\hat{A}/\partial\hat{p}_i \qquad \text{and} \qquad [\hat{p}_i, \hat{A}] \equiv -i\hbar\,\partial\hat{A}/\partial\hat{q}_i, \qquad\qquad (2.5.20)$$

where the operator derivatives behave in just the same way as their classical counterparts do. Note that these equations correspond to the Poisson bracket version of these equations in the classical context given in Eq. (2.4.20). In coordinate-space representation, where $\hat{q}_i \rightarrow q_i$ and $\hat{p}_i \rightarrow -i\hbar\partial/\partial q_i$, we can readily verify that the first and third of the above set of four equations is correct. Similarly, we can verify the second and fourth of these conveniently by working in a momentum-space basis. Furthermore, we then see that these equations apply for any function of the coordinate and momentum operators, $\hat{A} \equiv A(\hat{\vec{q}}, \hat{\vec{p}})$, with any mixed power series expansion in terms of coordinate and momentum operators.

Proof: Consider an operator $\hat{A} \equiv A(\hat{\vec{q}}, \hat{\vec{p}})$ that can be written as a power series in the coordinate and momentum operators. Using the canonical commutation relations it will always be possible to reorder operators and recombine terms in such a power series such that we can rewrite the power series in the form

$$A(\hat{\vec{q}}, \hat{\vec{p}}) = \sum c_{m_1 n_1 \cdots m_N n_N}\, \hat{q}_1^{m_1}\hat{q}_2^{m_2}\cdots\hat{q}_N^{m_N}\hat{p}_1^{n_1}\hat{p}_2^{n_2}\cdots\hat{p}_N^{n_N} \qquad (2.5.21)$$

for some set of coefficients $c_{m_1 n_1 \cdots m_N n_N} \in \mathbb{C}$. The canonical commutation relations tell us that the ordering of the various $\hat{\vec{q}}$ and $\hat{\vec{p}}$ is irrelevant except that we have each $\hat{q}_i^{m_i}$ to the left of the corresponding $p_i^{n_i}$. The opposite convention does not alter the result. For every term in this sum we have

$$[\hat{q}_i, \hat{q}_1^{m_1}\cdots\hat{q}_N^{m_N}\hat{p}_1^{n_1}\cdots\hat{p}_N^{n_N}] = \hat{q}_1^{m_1}\cdots\hat{p}_{i-1}^{n_{i-1}}[\hat{q}_i, \hat{p}_i^{n_i}]\hat{p}_{i+1}^{n_{i+1}}\cdots\hat{p}_N^{n_N}$$

$$= i\hbar n_i\hat{q}_1^{m_1}\cdots\hat{q}_N^{m_N}\hat{p}_1^{n_1}\cdots\hat{p}_i^{n_i-1}\cdots\hat{p}_N^{n_N} = i\hbar\frac{\partial}{\partial\hat{p}_i}\left(\hat{q}_1^{m_1}\cdots\hat{p}_N^{n_N}\right). \qquad (2.5.22)$$

Similarly, we follow the corresponding steps to find

$$[\hat{p}_i, \hat{q}_1^{m_1}\cdots\hat{q}_N^{m_N}\hat{p}_1^{n_1}\cdots\hat{p}_N^{n_N}] = -i\hbar\frac{\partial}{\partial\hat{p}_i}\left(\hat{q}_1^{m_1}\cdots\hat{p}_N^{n_N}\right). \qquad (2.5.23)$$

We see that Eq. (2.5.20) immediately follows.

The behavior of the operator derivatives is entirely consistent with what we would find if we replaced all operators by the classical counterparts; i.e., if $\hat{\vec{q}} \rightarrow \vec{q}$, $\hat{\vec{p}} \rightarrow \vec{p}$ and $A(\hat{\vec{q}}, \hat{\vec{p}}) \rightarrow A(\vec{q}, \vec{p})$, then we would have as expected that

$$\partial\hat{A}/\partial\hat{p}_i \rightarrow \partial A/\partial p_i \qquad \text{and} \qquad \partial\hat{A}/\partial\hat{q}_i \rightarrow \partial A/\partial q_i. \qquad (2.5.24)$$

From the Heisenberg equation of motion Eq. (2.5.17) for the cases where $\hat{O} = \hat{q}_i$ and $\hat{O} = \hat{p}_i$, respectively, we find from Eqs. (2.5.17) and (2.5.20) that

$$i\hbar\dot{\hat{q}}_i = [\hat{q}_i, \hat{H}] = i\hbar\frac{\partial\hat{H}}{\partial\hat{p}_i}, \quad i\hbar\dot{\hat{p}}_i = [\hat{p}_i, \hat{H}] = -i\hbar\frac{\partial\hat{H}}{\partial\hat{q}_i} \quad \text{and} \quad \dot{\hat{H}} = \frac{\partial\hat{H}}{\partial t}. \tag{2.5.25}$$

This finally gives Hamilton's equations in their operator form:

$$\dot{\hat{q}}_i = \frac{\partial\hat{H}}{\partial\hat{p}_i}, \quad \dot{\hat{p}}_i = -\frac{\partial\hat{H}}{\partial\hat{q}_i} \quad \text{and} \quad \dot{\hat{H}} = \frac{\partial\hat{H}}{\partial t}. \tag{2.5.26}$$

In other words, in ordinary quantum mechanics, the Heisenberg picture coordinate and conjugate momentum operators, $\hat{q}_i(t)$ and $\hat{p}_i(t)$, respectively, obey the same equations of motion as their classical counterparts; i.e., they obey Hamilton's equations and also therefore the Euler-Lagrange equations.

Comparing all of the above discussion with the Poisson bracket formalism of classical mechanics, we see that at every stage above we can move between the classical and quantum formulations by interchanging classical quantities with their quantum operator equivalents, $F \leftrightarrow \hat{F}$, and by the corresponding replacement

$$\{F, G\} \leftrightarrow \frac{1}{i\hbar}[\hat{F}, \hat{G}]. \tag{2.5.27}$$

This relationship between the classical and quantum formulations is often referred to as the *correspondence principle*. Obviously, the *expectation values* of quantum operators also obey the same equations as their classical counterparts.

It is useful to make one final observation. Since Hilbert space V_H corresponds to a complex vector space $\mathbb{C}^n$, (cf., a real vector space $\mathbb{R}^n$), then

$$\langle\psi|\hat{A}|\psi\rangle = \langle\psi|\hat{B}|\psi\rangle \quad \text{for all} \quad |\psi\rangle \in V_H \quad \Rightarrow \quad \hat{A} = \hat{B}. \tag{2.5.28}$$

Thus since the expectation value of an operator $\hat{A}$ is $\langle\hat{A}\rangle_\psi \equiv \langle\psi|\hat{A}|\psi\rangle/\langle\psi|\psi\rangle$, then it follows that equalities of expectation values for all states of the system imply equalities of the corresponding operators and vice versa.

Proof: Consider two operators $\hat{A}$ and $\hat{B}$ satisfying $\langle\psi|\hat{A}|\psi\rangle = \langle\psi|\hat{B}|\psi\rangle$ for all $|\psi\rangle \in V_H$. Let $|\psi\rangle = c_1|\psi_1\rangle + c_2|\psi_2\rangle$, where c_1, c_2, $|\psi_1\rangle$ and $|\psi_2\rangle$ are all arbitrary. It follows that $\langle\psi|(\hat{A} - \hat{B})|\psi\rangle = 0$ for all $|\psi\rangle$ and hence

$$\begin{aligned}
0 = \langle\psi|(\hat{A} - \hat{B})|\psi\rangle &= (c_1^*\langle\psi_1| + c_2^*\langle\psi_2|)(\hat{A} - \hat{B})(c_1|\psi_1\rangle + c_2|\psi_2\rangle) \\
&= c_1^*c_2\langle\psi_1|(\hat{A} - \hat{B})|\psi_2\rangle + c_1c_2^*\langle\psi_2|(\hat{A} - \hat{B})|\psi_1\rangle \\
&= \text{Re}(c_1^*c_2)(\langle\psi_1|(\hat{A} - \hat{B})|\psi_2\rangle + \langle\psi_2|(\hat{A} - \hat{B})|\psi_1\rangle) \\
&\quad + i\text{Im}(c_1^*c_2)(\langle\psi_1|(\hat{A} - \hat{B})|\psi_2\rangle - \langle\psi_2|(\hat{A} - \hat{B})|\psi_1\rangle).
\end{aligned} \tag{2.5.29}$$

Since c_1 and c_2 are arbitrary complex numbers, then $\text{Re}(c_1^*c_2)$ and $\text{Im}(c_1^*c_2)$ are arbitrary and so we must have $\langle\psi_1|(\hat{A} - \hat{B})|\psi_2\rangle = \langle\psi_2|(\hat{A} - \hat{B})|\psi_1\rangle = 0$ for arbitrary $|\psi_1\rangle$ and $|\psi_2\rangle$. This is only possible if $\hat{A} = \hat{B}$. Note that this result follows because the Hilbert space of quantum mechanics is a complex vector space. In a real vector space we would have $\text{Im}(d) = 0$, which would only imply that $\langle\psi_1|(\hat{A} - \hat{B})|\psi_2\rangle + \langle\psi_2|(\hat{A} - \hat{B})|\psi_1\rangle = 0$.

Symmetry and conserved charges in quantum mechanics: Let $\hat{G}$ be a Hermitian operator. It can then form a continuous group of unitary operators

$$\hat{U}(\alpha) \equiv e^{i\alpha\hat{G}} \quad \text{with} \quad \alpha \in \mathbb{R}, \tag{2.5.30}$$

which is a one-dimensional Lie group with $\hat{G}$ as the generator of the group. Let $\hat{F}$ be an observable operator; then we can always transform it using this transformation:

$$\hat{F}' \equiv \hat{U}(\alpha)\hat{F}\hat{U}(\alpha)^\dagger. \tag{2.5.31}$$

Under an infinitesimal transformation $d\alpha$ we have $\hat{F}' = \hat{F} + d\hat{F}$ with

$$d\hat{F} = id\alpha[\hat{G}, \hat{F}] + \mathcal{O}(d\alpha^2), \tag{2.5.32}$$

which is the quantum version of an infinitesimal canonical transformation given in Eq. (2.4.65). Assume that the Hamiltonian is invariant under this unitary transformation; then we find the quantum version of Eq. (2.4.68),

$$[\hat{G}, \hat{H}] = 0. \tag{2.5.33}$$

We see from Eq. (2.5.17) that if $\hat{G}$ has no explicit time dependence, then

$$\frac{d\hat{G}}{dt} = 0 \tag{2.5.34}$$

and so $\hat{G}$ is a conserved operator also referred to as a conserved charge operator. Note that if $\hat{U}(\alpha)\hat{H}\hat{U}(\alpha)^\dagger = \hat{H}$, then acting from the right with $\hat{U}(\alpha)$ on both sides gives $[\hat{U}(\alpha), \hat{H}] = 0$, which can only be true for arbitrary α if $[\hat{G}, \hat{H}] = 0$.

So we see that if the quantum Hamiltonian, and hence the quantum system, is invariant under a one-dimensional unitary Lie group with a generator $\hat{G}$ (with no explicit time dependence), then $\hat{G}$ is a conserved charge operator. Conversely, any conserved charge operator $\hat{G}$ (with no explicit time dependence) will generate a one-dimensional Lie group of unitary transformations that leaves the quantum Hamiltonian, and hence the quantum system, invariant.

This generalizes immediately to n-dimensional Lie groups, where we have n Hermitian generators, $\hat{G}_1, \ldots, \hat{G}_n$ (with no explicit time dependence) and where the Lie group has elements

$$\hat{U}(\vec{\alpha}) \equiv e^{i\vec{\alpha}\cdot\hat{\vec{G}}}. \tag{2.5.35}$$

Then if the system is invariant under this Lie group of transformations we must have $[\hat{U}(\vec{\alpha}), \hat{H}] = 0$ and so $\sum_{i=1}^n d\alpha_i[\hat{G}_i, \hat{H}] = 0$ for arbitrary $d\alpha_i$. This gives

$$[\hat{G}_i, \hat{H}] = 0 \quad \text{and} \quad \frac{d\hat{G}_i}{dt} = 0 \quad \text{for} \quad i = 1, \ldots, n \tag{2.5.36}$$

and we have n conserved charge operators, $\hat{G}_1, \ldots, \hat{G}_n$. Conversely if a set of n conserved charges satisfies a Lie algebra then they will generate a Lie group of unitary transformations that leave the quantum system invariant.

Example: Operator equations of motion for the Lorentz force: The nonrelativistic classical Hamiltonian describing a particle of charge q and mass m interacting with an external electromagnetic field was given in Eq. (2.4.15). It is convenient to now use the notation where the observable *mechanical momentum* associated with the motion of the particle is **p** and the canonical momentum is $\boldsymbol{\pi}$,

$$\mathbf{p} \equiv m\boldsymbol{v} = m\dot{\mathbf{x}} \quad \text{and} \quad \boldsymbol{\pi} \equiv \partial L/\partial\dot{\mathbf{x}} = \mathbf{p} + q\mathbf{A}. \tag{2.5.37}$$

The canonical momentum $\boldsymbol{\pi}$ is gauge-dependent and is therefore not an observable quantity. The corresponding quantum Hamiltonian in coordinate representation is

$$\hat{H} = \frac{(\hat{\boldsymbol{\pi}} - q\mathbf{A})^2}{2m} + q\Phi = \frac{\hat{\mathbf{p}}^2}{2m} + q\Phi, \qquad (2.5.38)$$

where the canonical momentum operator is $\hat{\boldsymbol{\pi}} = -i\hbar\boldsymbol{\nabla}$ and satisfies the canonical commutation relations and the mechanical momentum is $\hat{\mathbf{p}} = \hat{\boldsymbol{\pi}} - q\mathbf{A}$. So we have

$$[\hat{x}^i, \hat{\pi}^j] = i\hbar\delta^{ij} \quad \text{and} \quad [\hat{\pi}^i, \hat{\pi}^j] = [\hat{x}^i, \hat{x}^j] = 0 \qquad (2.5.39)$$

with $\hat{x}^i = x^i$. From Eq. (2.5.26) we know that the classical equations of motion, Eqs. (2.1.35) and (2.1.36), must follow for the Heisenberg picture operators,

$$\hat{\mathbf{F}} = \frac{d\hat{\mathbf{p}}}{dt} = q\left[\mathbf{E} + \frac{\hat{\mathbf{p}}}{m} \times \mathbf{B}\right] \quad \text{and} \quad \frac{d\hat{E}^{\mathrm{mech}}}{dt} \equiv \frac{d(\hat{\mathbf{p}}^2/2m)}{dt} = q\frac{\hat{\mathbf{p}}}{m} \cdot \mathbf{E}. \qquad (2.5.40)$$

This result is also proved directly below. In the classical limit we recover the classical nonrelativistic Lorentz force equations, Eqs. (2.1.35) and (2.1.36), as expected.

Direct proof: While Eq. (2.5.26) leads to Eq. (2.5.40) as we have shown above, it is a worthwhile exercise to obtain the operator form of the Lorentz force equation directly from the quantum Hamiltonian and the canonical commutation relations. We work in the Heisenberg picture in coordinate space representation and use the summation convention. From Eq. (4.1.24) we have

$$d\hat{\mathbf{p}}/dt = [\hat{\mathbf{p}}, \hat{H}]/i\hbar + \partial\hat{\mathbf{p}}/\partial t \quad \text{with} \quad \hat{H} = [\hat{\mathbf{p}}^2/2m] + q\Phi(t, \hat{\mathbf{x}}). \qquad (2.5.41)$$

We use relativistic notation, e.g., $\mathbf{x} = (x^1, x^2, x^3)$ and $\partial_\mu = \partial/\partial x^\mu$. We find

$$\frac{1}{i\hbar}\left[\hat{p}^i, \frac{\hat{\mathbf{p}}^2}{2m}\right] = \frac{[\hat{p}^i, \hat{p}^j]\hat{p}^j + \hat{p}^j[\hat{p}^i, \hat{p}^j]}{i\hbar 2m} = q\frac{\hat{p}^j}{m}\left(\partial_i A^j - \partial_j A^i\right) = q\left[\frac{\hat{\mathbf{p}}}{m} \times \mathbf{B}\right]^i,$$

since $\dot{x}^j\left(\partial_i A^j - \partial_j A^i\right) = (\dot{\mathbf{x}} \times \mathbf{B})^i$ from Eq. (2.1.38) and since

$$[\hat{p}^i, \hat{p}^j] = -q[\hat{\pi}^i, A^j] - q[A^i, \hat{\pi}^j] = -q[i\hbar\partial^i, A^j] - q[A^i, i\hbar\partial^j]$$
$$= -i\hbar q\left(\partial^i A^j - \partial^j A^i\right) = i\hbar q\left(\partial_i A^j - \partial_j A^i\right). \qquad (2.5.42)$$

Also $\frac{1}{i\hbar}[\hat{p}^i, q\Phi] = [\partial^i, q\Phi] = -q\partial_i\Phi$ and $\frac{\partial\hat{p}^i}{\partial t} = \frac{\partial(\hat{\pi}^i - qA^i)}{\partial t} = -q\frac{\partial A^i}{\partial t}$.
We then recover the operator form of the Lorentz force equations

$$\hat{F}^i = \frac{d\hat{p}^i}{dt} = \frac{1}{i\hbar}\left[\hat{p}^i, \frac{\hat{p}^j\hat{p}^j}{2m} + q\Phi\right] + \frac{\partial\hat{p}^i}{\partial t} = \left[\frac{\hat{p}^i}{m} \times \mathbf{B}\right]^i + q\left(-\partial_i\Phi - \frac{\partial A^i}{\partial t}\right)$$
$$= q\left[\mathbf{E} + (\hat{\mathbf{p}}/m) \times \mathbf{B}\right]^i, \qquad (2.5.43)$$

$$\frac{d\hat{p}^0}{dt} = \frac{1}{c}\frac{d\hat{E}^{\mathrm{mech}}}{dt} = \frac{1}{i\hbar}\left[\hat{p}^0, \hat{H}\right] + \frac{\partial\hat{p}^0}{\partial t} = \frac{1}{i\hbar}\left[\hat{p}^0, \frac{\hat{\mathbf{p}}^2}{2m} + q\Phi\right] - q\frac{\partial A^0}{\partial t}$$
$$= \frac{1}{i\hbar 2m}\left([\hat{p}^0, \hat{p}^j]\hat{p}^j + \hat{p}^j[\hat{p}^0, \hat{p}^j]\right) + \frac{q}{i\hbar}[\hat{p}^0, \Phi] - \frac{q}{c}\frac{\partial\Phi}{\partial t}$$
$$= \frac{q}{c}\frac{\hat{p}^j}{m}\left(-\frac{\partial A^j}{\partial t} - \partial_j\Phi\right) + \frac{q}{c}\frac{\partial\Phi}{\partial t} - \frac{q}{c}\frac{\partial\Phi}{\partial t} = \frac{q}{c}\frac{\hat{\mathbf{p}}}{m} \cdot \mathbf{E}, \qquad (2.5.44)$$

since $[\hat{p}^0, \hat{p}^j] = i\hbar\left([\partial^0, -qA^j] + [-qA^0, \partial^j]\right) = i\hbar\frac{q}{c}\left(-\frac{\partial A^j}{\partial t} - \partial_j\Phi\right)$.

Summary: The *classical limit* for a quantum system is often written as $\hbar \to 0$ for notational convenience, but what we actually mean by this is the *macroscopic limit*, i.e., $S[\vec{q}]/\hbar \gg 1$ with $\hbar$ fixed. There are three different viewpoints of how classical mechanics arises as the macroscopic limit of quantum mechanics:

(i) In the Feynman path integral formulation of quantum mechanics, to be discussed in Sec. 4.1.12, it follows that as $\hbar \to 0$ the stationary point of the action dominates and so the classical equations of motion emerge.

(ii) The correspondence of the Poisson bracket and the expectation value of the quantum commutator in Eq. (2.5.12) means that expectation values of quantum operators must obey Hamilton's equations in the macroscopic limit.

(iii) As we saw above in Eq. (2.5.26), recognizing the definition of differentiation with respect to position and momentum operators in Eq. (2.5.20) the operators themselves obey Hamilton's equations and so must their expectation values.

2.6 Relativistic Kinematics

Some relativistic extensions of classical mechanics are possible, although a fully consistent approach requires the machinery of relativistic quantum field theory.

Nonrelativistic kinematics: If all velocities are much smaller than the speed of light, then Lorentz invariance reduces to Galilean invariance, which was discussed in Sec. 1.1.2, i.e., $\beta = v/c \ll 1$ and $\gamma \simeq 1$. From Noether's theorem in Sec. 2.2 we saw that translational invariance implies three-momentum conservation in the analytical mechanics derived from Newton's laws. This also follows directly from the application of Newton's laws to the interaction of two particles. By Newton's third law the forces that particles a and b exert on each other are equal and opposite and so $\mathbf{F}_{b\,\mathrm{on}\,a} = -\mathbf{F}_{a\,\mathrm{on}\,b}$. Then by Newton's second law, $\mathbf{F} = d\mathbf{p}/dt$, it follows that $d\mathbf{p}_a/dt = -d\mathbf{p}_b/dt$. Consider the effect of the forces acting over some time interval t_1 to t_2; then $\Delta\mathbf{p}_a = \int_{t_1}^{t_2}(d\mathbf{p}_a/dt)dt = -\int_{t_1}^{t_2}(d\mathbf{p}_b/dt)dt = -\Delta\mathbf{p}_b$. The total three-momentum momentum is then conserved, $\Delta\mathbf{P} = \Delta\mathbf{p}_a + \Delta\mathbf{p}_b = 0$. The generalization to N particles is straightforward, since we can break up the system into a single particle and the set of $(N-1)$ remaining particles. This can be done particle by particle and so it follows that for N particles interacting with each other and with no external forces over any time interval $\Delta\mathbf{P} = \sum_{i=1}^{N}\Delta\mathbf{p}_i = 0$.

Consider a set of particles that are only interacting with each other and with the interactions dependent only on the distance between the particles. Then the system is translationally and rotationally invariant. As we saw in Sec. 2.2 this means that the total three-momentum $\mathbf{P}$ and the total angular momentum $\mathbf{L}$ of the system will be conserved. If the kinetic energy is given by $T = \frac{1}{2}mv^2$, then the system will be natural and since there is no explicit time dependence, then the total energy E of the system will be conserved as we saw in Sec. 2.2. If, in addition, the interactions between the particles are pure contact interactions due to elastic collisions, then all of the energy of the system must be stored as kinetic energy and so the total kinetic energy T must also be conserved.

In summary, for such a system with N_i particles in the initial state and N_f particles in the final state, we have conservation of the total momentum, total mass, total kinetic energy and total angular momentum:

$$\mathbf{P}_{\mathrm{i}} = \sum_{\ell=1}^{N_{\mathrm{i}}} \mathbf{p}_{\mathrm{i}\ell} = \sum_{\ell=1}^{N_{\mathrm{i}}} m_{\mathrm{i}\ell}\mathbf{u}_{\mathrm{i}\ell} = \sum_{k=1}^{N_{\mathrm{f}}} m_{\mathrm{f}k}\mathbf{u}_{\mathrm{f}k} = \sum_{k=1}^{N_{\mathrm{f}}} \mathbf{p}_{\mathrm{f}k} = \mathbf{P}_{\mathrm{f}} \qquad (2.6.1)$$

$$M_{\mathrm{i}} = \sum_{\ell=1}^{N_{\mathrm{i}}} m_{\mathrm{i}\ell} = \sum_{k=1}^{N_{\mathrm{f}}} m_{\mathrm{f}k} = M_{\mathrm{f}}$$

$$T_{\mathrm{i}} = \sum_{\ell=1}^{N_{\mathrm{i}}} T_{\mathrm{i}\ell} = \sum_{\ell=1}^{N_{\mathrm{i}}} \tfrac{1}{2} m_{\mathrm{i}\ell} u_{\mathrm{i}\ell}^2 = \sum_{k=1}^{N_{\mathrm{f}}} \tfrac{1}{2} m_{\mathrm{f}k} u_{\mathrm{f}k}^2 = \sum_{k=1}^{N_{\mathrm{f}}} T_{\mathrm{f}k} = T_{\mathrm{f}}$$

$$\mathbf{L}_{\mathrm{i}} = \sum_{\ell=1}^{N_{\mathrm{i}}} \mathbf{L}_{\mathrm{i}\ell} = \sum_{\ell=1}^{N_{\mathrm{i}}} (\mathbf{x}_{\mathrm{i}\ell} \times m_{\mathrm{i}\ell}\mathbf{u}_{\mathrm{i}\ell}) = \sum_{k=1}^{N_{\mathrm{f}}} (\mathbf{x}_{\mathrm{f}k} \times m_{\mathrm{f}k}\mathbf{u}_{\mathrm{f}k}) = \sum_{k=1}^{N_{\mathrm{f}}} \mathbf{L}_{\mathrm{f}k} = \mathbf{L}_{\mathrm{f}}.$$

More briefly, we can write these results as $\Delta\mathbf{P} \equiv \mathbf{P}_{\mathrm{f}} - \mathbf{P}_{\mathrm{i}} = 0$, $\Delta M \equiv M_{\mathrm{f}} - M_{\mathrm{i}} = 0$, $\Delta T \equiv T_{\mathrm{f}} - T_{\mathrm{i}} = 0$, and $\Delta\mathbf{L} \equiv \mathbf{L}_{\mathrm{f}} - \mathbf{L}_{\mathrm{i}} = 0$.

Four-vector physics: Consider a particle with rest mass m as observed by some inertial observer $\mathcal{O}$ who describes the worldline of the particle by a continuous series of events $x^\mu = (ct, \mathbf{x})$. A *massive particle* must always have a worldline within any *light cone* with its origin on the worldline, and a massless particle has a world line that is a straight line *on the light cone*. The *proper time* τ for the particle moving along its worldline is the time shown on a clock that is comoving with the particle. There is a one-to-one correspondence between the events on the worldline and the time on the comoving clock, the proper time τ. Hence observer $\mathcal{O}$ can describe the worldline as a function of proper time, $x^\mu(\tau)$, where $\mathcal{O}$ sees the particle at position $\mathbf{x}$ at time t when the comoving clock shows the time τ. Any two inertial observers obviously record the same proper time τ for any given event along the worldline of the particle. It then follows that the proper time, τ, is Poincaré invariant.

In general the particle will be experiencing forces and hence accelerations and so in general it will not be free and hence will not remain at rest in any one inertial frame. However, at any proper time τ we can always find an inertial frame in which the particle is at rest at that instant of proper time. This frame is referred to as an *instantaneously comoving frame*. If the particle velocity measured by $\mathcal{O}$ at proper time τ is $\boldsymbol{v}$, then a boost of $+\boldsymbol{v}$ from $\mathcal{O}$'s frame will obviously take us to the instantaneous comoving frame for the particle at proper time τ.

We begin with the important observations that (i) x^μ transforms as a four-vector under Lorentz transformations; (ii) τ is a Lorentz-invariant quantity; and (iii) when we describe changes to the particle worldline over an infinitesimal proper time interval $[\tau, \tau + d\tau]$ we will do so from the perspective of the comoving frame defined at time τ; i.e., the Poincaré transformation to the comoving frame is independent of $d\tau$. Hence, under a Poincaré transformation we have

$$x^\mu \to x'^\mu = \Lambda^\mu{}_\nu x^\nu + a^\mu \quad \text{and} \quad \frac{d^n x^\mu}{d\tau^n} \to \frac{d^n x'^\mu}{d\tau^n} = \Lambda^\mu{}_\nu \frac{d^n x^\nu}{d\tau^n} \qquad (2.6.2)$$

for any positive integer n; i.e., all derivatives of x^μ with respect to proper time transform as four-vectors under Lorentz transformations.

We can then define relativistic four-vector versions of familiar quantities. The four-velocity in the frame of observer $\mathcal{O}$ is defined as the rate of change of position x^μ with respect to proper time,

$$V^\mu = (V^0, \mathbf{V}) \equiv \frac{dx^\mu}{d\tau} = \frac{dt}{d\tau}\frac{dx^\mu}{dt} = \gamma\left(c, \frac{d\mathbf{x}}{dt}\right) = \gamma(c, \boldsymbol{v}), \qquad (2.6.3)$$

where we have used $dt = \gamma\, d\tau$ giving $dt/d\tau = \gamma$. If $v \equiv |\boldsymbol{v}| < c$ then $V^2 > 0$ and the four-velocity is timelike. The relativistic four-momentum is the *rest mass* m multiplied by the four-velocity,

$$p^\mu = (p^0, \mathbf{p}) \equiv mV^\mu \equiv (\gamma mc, \gamma m\boldsymbol{v}) = (E/c, \mathbf{p}), \qquad p^2 = m^2 c^2, \qquad (2.6.4)$$

where we define E as the relativistic energy and $\mathbf{p}$ as the relativistic three-momentum,

$$E \equiv cp^0 = \gamma mc^2 = \frac{mc^2}{\sqrt{1 - (v^2/c^2)}}, \qquad \mathbf{p} \equiv \frac{dx^\mu}{d\tau} = \gamma m\boldsymbol{v} = \frac{m\boldsymbol{v}}{\sqrt{1 - (v^2/c^2)}}. \qquad (2.6.5)$$

For small v/c and writing $\epsilon = (v/c)^2$ we have the expansion

$$\gamma = (1 - \epsilon)^{-1/2} = 1 + \frac{1}{2}\epsilon + \frac{3}{8}\epsilon^2 + \frac{5}{16}\epsilon^3 + \frac{35}{128}\epsilon^4 + \cdots \tag{2.6.6}$$

and so the energy can be expanded as

$$E = mc^2 + \frac{1}{2}mv^2 + \frac{3mv^4}{8c^2} + \frac{5mv^6}{16c^4} + \cdots, \tag{2.6.7}$$

where mc^2 is the rest mass energy of the particle and $(1/2)mv^2$ the nonrelativistic kinetic energy. Kinetic energy is by definition the energy associated with the motion of the particle and so we define the relativistic kinetic energy as

$$T \equiv E - mc^2 = (\gamma - 1)\,mc^2 = \frac{1}{2}mv^2 + \frac{3mv^4}{8c^2} + \frac{5mv^6}{16c^4} + \cdots. \tag{2.6.8}$$

We will show that it is the *relativistic energy and three-momentum*, E and $\mathbf{p}$, that are conserved in kinematic collisions.

The *four-acceleration* and *four-force* are similarly defined, respectively, as

$$A^\mu = (A^0, \mathbf{A}) \equiv \frac{d^2 x^\mu}{d\tau^2} = \frac{dV^\mu}{d\tau}, \qquad F^\mu = (F^0, \mathbf{F}) \equiv \frac{dp^\mu}{d\tau} = mA^\mu, \tag{2.6.9}$$

where the second equation is the relativistic version of Newton's second law. Fully contracted quantities are Lorentz invariant. Some important examples are

$$V^2 = V^\mu V_\nu = \gamma^2(c^2 - v^2) = c^2, \qquad p^2 = m^2 V^2 = m^2 c^2,$$

$$A \cdot V = \frac{dV^\mu}{d\tau}V_\mu = \frac{1}{2}\frac{d}{d\tau}V^2 = \frac{1}{2}\frac{d}{d\tau}c^2 = 0, \qquad F \cdot V = mV \cdot A = 0. \tag{2.6.10}$$

Since $p^2 = m^2 c^2 = (E/c)^2 - \mathbf{p}^2$ we have that

$$E^2 = (p^0)^2 c^2 = \mathbf{p}^2 c^2 + m^2 c^4 \qquad \text{and so} \qquad E = \pm\sqrt{\mathbf{p}^2 c^2 + m^2 c^4}, \tag{2.6.11}$$

where negative energies correspond to antiparticles as we will show in Chapter 4.

We say that a free particle is *on its mass-shell* or *on-shell* when $p^2 = m^2 c^2$, i.e., when $E = \pm\sqrt{m^2 c^4 + \mathbf{p}^2 c^2}$ where the positive/negative sign means that a particle/antiparticle is on-shell. If a particle is not on-shell we say that it is *off-shell* and we refer to it as a *virtual particle*, which is a concept that is useful in quantum field theory. Focusing for now on $E > 0$, we can write for $m > 0$ that

$$p^0 \boldsymbol{v} = (E/c)\boldsymbol{v} = \gamma mc\boldsymbol{v} = \mathbf{p}c \quad \Rightarrow \quad \mathbf{p}/p^0 = \boldsymbol{v}/c. \tag{2.6.12}$$

The definition $p^\mu = mV^\mu$ is only meaningful when $m > 0$. In the massless limit $p^2 = m^2 c^2 \to 0$ and so $p^0 = E/c = |\mathbf{p}|$. Since in the massless limit $|\mathbf{p}|/p^0 = 1$, then for consistency Eq. (2.6.12) requires that $v = c$ for all massless particles.

From Maxwell's equations we know that electromagnetic radiation travels as a wave with $c = f\lambda$, where f is the photon frequency and λ the wavelength. From the definition of Planck's constant we have that the energy of a photon is $E = hf = \hbar\omega$ with the angular frequency $\omega = 2\pi f$, which is referred to as the Planck relation. We also have the de Broglie relation between the photon momentum and wavelength, $p \equiv |\mathbf{p}| = h/\lambda$. It then follows for a photon that $p^0 = E/c = hf/c = h/\lambda = |\mathbf{p}|$.

Let B^μ and C^μ be two orthogonal four-vectors in Minkowski space, $B \cdot C = 0$; then it is straightforward to verify that

$$\text{(i)} \quad \text{If} \quad B^2 > 0 \quad \text{with} \quad B \cdot C = 0 \quad \text{then} \quad C^2 < 0. \tag{2.6.13}$$
$$\text{(ii)} \quad \text{If} \quad B^2 < 0 \quad \text{with} \quad B \cdot C = 0 \quad \text{then} \quad C^2 > 0.$$
$$\text{(iii)} \quad \text{If} \quad B^2 = 0 \quad \text{with} \quad B \cdot C = 0 \quad \text{then} \quad C^2 \leq 0.$$

So if B is timelike or spacelike and if $B \cdot C = 0$, then C is spacelike or timelike, respectively. Furthermore, if B is lightlike, then either C is spacelike ($C^2 < 0$) or C is lightlike and collinear with B.

For a massive particle $p^2 = m^2 V^2 > 0$ and so $V^2 > 0$, which means that V^μ is always timelike. From Eq. (2.6.10) we have $A \cdot V = 0$ and so the four-acceleration A^μ must always be spacelike, $A^2 < 0$, or zero, $A^\mu = (0, \mathbf{0})$. In the instantaneously comoving inertial frame we have $V^\mu = \gamma(c, \mathbf{0})$ and since $A \cdot V = 0$ in any inertial frame then we must have in the comoving frame

$$A^\mu = (0, \boldsymbol{a}_{\text{prop}}) \qquad \text{and} \qquad A^2 = -a_{\text{prop}}^2, \tag{2.6.14}$$

where $\boldsymbol{a}_{\text{prop}}$ is the three-dimensional acceleration measured in the comoving frame, i.e., the proper three-acceleration.

Let $\mathcal{O}$ be any inertial observer. Recall that $x^\mu(\tau)$ is the worldline that observer $\mathcal{O}$ measures in his frame and so the corresponding standard three-velocity and three-acceleration of the particle measured in his frame is

$$\boldsymbol{v} \equiv \frac{d\mathbf{x}}{dt} \qquad \text{and} \qquad \boldsymbol{a} \equiv \frac{d\boldsymbol{v}}{dt} = \frac{d^2\mathbf{x}}{dt^2}. \tag{2.6.15}$$

Then for the four-acceleration in the frame of $\mathcal{O}$ we have

$$A^\mu = \frac{dV^\mu}{d\tau} = \left(\gamma \frac{d}{dt}\right)(\gamma c, \gamma \boldsymbol{v}) = \left(\gamma c \frac{d\gamma}{dt}, \left[\gamma^2 \boldsymbol{a} + \gamma \boldsymbol{v} \frac{d\gamma}{dt}\right]\right)$$
$$= \gamma^2 \left(\gamma^2 \frac{\boldsymbol{a} \cdot \boldsymbol{v}}{c}, \boldsymbol{a} + \boldsymbol{v}\gamma^2 \frac{\boldsymbol{a} \cdot \boldsymbol{v}}{c^2}\right), \tag{2.6.16}$$

where we have used $dv^2/dt = d(\boldsymbol{v} \cdot \boldsymbol{v})/dt = 2\boldsymbol{a} \cdot \boldsymbol{v}$ to obtain

$$\frac{d\gamma}{dt} = \frac{d}{dt}\left(1 - \frac{v^2}{c^2}\right)^{-1/2} = -\frac{1}{2}\gamma^3 \frac{d}{dt}\left(1 - \frac{\boldsymbol{v} \cdot \boldsymbol{v}}{c^2}\right) = \gamma^3 \frac{\boldsymbol{a} \cdot \boldsymbol{v}}{c^2}. \tag{2.6.17}$$

If we boost by velocity $\boldsymbol{v}$ from the frame of observer $\mathcal{O}$ to the instantaneously comoving frame of the particle at time t, then $\boldsymbol{v}/c \to \mathbf{0}$ and $\gamma \to 1$ so that $A^\mu \to (0, \boldsymbol{a}) \equiv (0, \boldsymbol{a}_{\text{prop}})$ as expected. The four-force has the property that

$$F^\mu \equiv \frac{dp^\mu}{d\tau} = \gamma \frac{dp^\mu}{dt} = \gamma \left(\frac{1}{c}\frac{dE}{dt}, \mathbf{f}\right), \qquad \text{where} \quad \mathbf{f} \equiv \frac{d\mathbf{p}}{dt} = \frac{d}{dt}(\gamma m \boldsymbol{v}) \tag{2.6.18}$$

is the relativistic three-force. Since $F \cdot V = 0$ from Eq. (2.6.10), then

$$F \cdot V = \left(\gamma \frac{1}{c}\frac{dE}{dt}\right)(\gamma c) - (\gamma \mathbf{f}) \cdot (\gamma \boldsymbol{v}) = 0 \qquad \Rightarrow \qquad \frac{dE}{dt} = \mathbf{f} \cdot \boldsymbol{v}, \tag{2.6.19}$$

where dE/dt is the power associated with the acceleration of the particle.

Relativistic kinematics: When we construct relativistic quantum field theories then Noether's theorem will show that invariance under spacetime translations leads to the conservation of total four-momentum $\Delta P^\mu = P_f^\mu - P_i^\mu = 0$. For a system of particles interacting only through interactions with each other this can be inferred from Newton's laws and the postulates of special relativity.

If such a set of particles are all moving slowly ($v/c \ll 1$) in some inertial frame, then the nonrelativistic energy and three-momentum are conserved as discussed above, $\Delta P^\mu = P_f^\mu - P_i^\mu = 0$

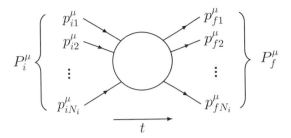

Figure 2.3   Conservation of total four-momentum in every inertial frame.

in that frame. Since the individual four-momenta of the particles transform as four-vectors, then this statement will be true in any inertial frame, when the *relative* velocities of all of the particles are small, $\Delta v/c \ll 1$. The question in an arbitrary frame is what happens if the relative velocities of the particles become very different from each other rather than being very similar. If we *assume* that there is something conserved that reduces to $\Delta P^\mu = 0$ when relative velocities are small in any inertial frame, then the only possible choice is that the conserved total quantity P^μ must be the sum over the relativistic four-momenta, p^μ, of the particles. So $\Delta P^\mu = 0$ is the Lorentz covariant generalization of nonrelativistic energy and momentum conservation. This can be written as

$$P_{\mathrm{i}}^\mu = \sum_{\ell=1}^{N_{\mathrm{i}}} p_{\mathrm{i}\ell}^\mu = \sum_{k=1}^{N_{\mathrm{f}}} p_{\mathrm{f}k}^\mu = P_{\mathrm{f}}^\mu, \tag{2.6.20}$$

where we have N_{i} and N_{f} particles in the initial and final state, respectively, as illustrated in Fig. 2.3. More explicitly, we can write

$$\mathbf{P}_{\mathrm{i}} = \sum_{\ell=1}^{N_{\mathrm{i}}} \mathbf{P}_{\mathrm{i}\ell} = \sum_{k=1}^{N_{\mathrm{f}}} \mathbf{P}_{\mathrm{f}k} = \mathbf{P}_{\mathrm{f}} \quad \text{and} \quad E_{\mathrm{i}} = \sum_{\ell=1}^{N_{\mathrm{i}}} E_{\mathrm{i}\ell} = \sum_{k=1}^{N_{\mathrm{f}}} E_{\mathrm{f}k} = E_{\mathrm{f}}. \tag{2.6.21}$$

2.7 Electromagnetism

While electromagnetism is generally regarded as part of "classical physics" rather than part of "classical mechanics," it is convenient to provide a brief introduction in this chapter for our future use.

2.7.1 Maxwell's Equations

Maxwell's equations describe electromagnetism and its interactions with matter through electric current densities, i.e, classical electrodynamics. While not strictly part of classical mechanics, electromagnetism is a cornerstone of classical physics.

As stated at the beginning of this chapter, all of our work in this chapter is in SI units unless stated explicitly otherwise. In subsequent chapters, when we move to the contexts of relativistic quantum mechanics and relativistic quantum field theory, we will follow standard practice and use Lorentz-Heaviside units. In later chapters we will move to the traditional particle physics units, which are defined as the simultaneous use of Lorentz-Heaviside units and natural units ($\hbar = c = 1$).

Newton's equations are the foundation of classical mechanics and are nonrelativistic; i.e., they are not Lorentz covariant since their form is not preserved under all Lorentz transformations. However,

Maxwell's equations are Lorentz covariant and are therefore fully consistent with special relativity. For a detailed treatment of electromagnetism and it applications see, e.g., Jackson (1975).

Maxwell's equations describe the interaction of the electric field, $\mathbf{E}$, and the magnetic field, $\mathbf{B}$, with the four-vector current density

$$j^\mu(x) = (c\rho(x), \mathbf{j}(x)) \tag{2.7.1}$$

in the vacuum and can be written in SI units as

$$\text{\textit{Faraday's law:} } \boldsymbol{\nabla} \times \mathbf{E} + \frac{\partial \mathbf{B}}{\partial t} = 0, \qquad \text{\textit{Gauss' magnetism law:} } \boldsymbol{\nabla} \cdot \mathbf{B} = 0,$$

$$\text{\textit{Gauss' law:} } \boldsymbol{\nabla} \cdot \mathbf{E} = \frac{\rho}{\epsilon_0}, \qquad \text{\textit{Ampère's law:} } \boldsymbol{\nabla} \times \mathbf{B} - \frac{1}{c^2}\frac{\partial \mathbf{E}}{\partial t} = \mu_0 \mathbf{j}, \tag{2.7.2}$$

where ϵ_0 is the vacuum permittivity, μ_0 is the vacuum permeability and

$$c = 1/\sqrt{\epsilon_0 \mu_0} \tag{2.7.3}$$

is the speed of light in the vacuum. $\rho(x)$ and $\mathbf{j}(x)$ are the electric charge density and the electric three-vector current density, respectively. In modern undergraduate physics texts it is now common to use SI units. When discussing electromagnetism there are two other frequently used systems of units, which are cgs-Gaussian units and Lorentz-Heaviside units.

Under Lorentz transformations the $\mathbf{E}$ and $\mathbf{B}$ fields mix with each other as do ρ and $\mathbf{j}$. The Lorentz covariance of Maxwell's equations above is not immediately apparent. Maxwell's equations can be expressed in terms of a *four-vector potential*

$$A^\mu(x) \equiv \left(A^0(x), \mathbf{A}(x)\right) \equiv (\Phi(x)/c, \mathbf{A}(x)) \tag{2.7.4}$$

and through doing this the Lorentz covariance of the equations can be made obvious. The term *electromagnetic field* is used to interchangeably refer either to the combined $\mathbf{E}$ and $\mathbf{B}$ fields or to the four-vector potential A^μ. We begin by observing that since there is no such thing as magnetic charge, then $\boldsymbol{\nabla} \cdot \mathbf{B} = 0$ and so it will always possible to find some three-vector function $\mathbf{A}(x)$ such that

$$\mathbf{B} = \boldsymbol{\nabla} \times \mathbf{A}, \tag{2.7.5}$$

since then $\boldsymbol{\nabla} \cdot \mathbf{B} = \boldsymbol{\nabla} \cdot (\boldsymbol{\nabla} \times \mathbf{A}) = 0$. Then we can write Faraday's law in the form

$$0 = \boldsymbol{\nabla} \times \mathbf{E} + \frac{\partial \mathbf{B}}{\partial t} = \boldsymbol{\nabla} \times \left(\mathbf{E} + \frac{\partial \mathbf{A}}{\partial t}\right). \tag{2.7.6}$$

The Poincaré lemma in Eq. (A.2.23) means that there is a field $\Phi(x)$ such that

$$\boldsymbol{\nabla}\Phi(x) = -\mathbf{E} - \frac{\partial \mathbf{A}}{\partial t}, \tag{2.7.7}$$

since then $\boldsymbol{\nabla} \times (\mathbf{E} + \partial \mathbf{A}/\partial t) = -\boldsymbol{\nabla} \times \boldsymbol{\nabla}\Phi = 0$. Hence we have

$$\mathbf{E} = -\boldsymbol{\nabla}\Phi - \frac{\partial \mathbf{A}}{\partial t}. \tag{2.7.8}$$

The *electromagnetic field strength tensor* is defined as

$$F^{\mu\nu}(x) \equiv \partial^\mu A^\nu(x) - \partial^\nu A^\mu(x) \tag{2.7.9}$$

and obviously $F^{\mu\nu} = -F^{\nu\mu}$. Since ∂^μ and A^μ both transform as four-vectors, then $F^{\mu\nu}$ transforms as a second-rank tensor. From the definitions of

$$\mathbf{E} \equiv (E_x, E_y, E_z) \equiv (E^1, E^2, E^3), \quad \mathbf{B} \equiv (B_x, B_y, B_z) \equiv (B^1, B^2, B^3) \tag{2.7.10}$$

and of $F^{\mu\nu}$ in terms of the four-vector potential A^μ it follows that

$$F^{0i} = -E^i/c \qquad \text{and} \qquad F^{ij} = -\epsilon^{ijk}B^k, \tag{2.7.11}$$

since

$$F^{0i} = \partial^0 A^i - \partial^i A^0 = \frac{1}{c}\left(\frac{\partial A^i}{\partial t} + \partial_i \Phi\right) = -\frac{E^i}{c}, \tag{2.7.12}$$

$$F^{ij} = \partial^i A^j - \partial^j A^i = -\left[\partial_i A^j - \partial_j A^i\right] = -\epsilon^{ijk}\epsilon^{mnk}\partial_m A^n = -\epsilon^{ijk}B^k.$$

In summary, we can write the matrix consisting of elements $F^{\mu\nu}$ as

$$F^{\mu\nu} = \begin{pmatrix} 0 & -E_x/c & -E_y/c & -E_z/c \\ E_x/c & 0 & -B_z & B_y \\ E_y/c & B_z & 0 & -B_x \\ E_z/c & -B_y & B_x & 0 \end{pmatrix}. \tag{2.7.13}$$

The third and fourth of Maxwell's equations in Eq. (2.7.2) in compact form are

$$\partial_\mu F^{\mu\nu} = \mu_0 j^\nu, \tag{2.7.14}$$

since

$$0 = \partial_\mu F^{\mu 0} - \mu_0 j^0 = \partial_i F^{i0} - \mu_0 \rho c = (1/c)\left[\mathbf{\nabla} \cdot \mathbf{E} - (\rho/\epsilon_0)\right], \tag{2.7.15}$$

$$0 = \partial_\mu F^{\mu j} - \mu_0 j^j = \partial_0 F^{0j} + \partial_i F^{ij} - \mu_0 j^j = -(1/c^2)(\partial E^j/\partial t) + (\mathbf{\nabla} \times \mathbf{B})^j - \mu_0 j^j.$$

The Bianchi identity is

$$\partial_\mu F_{\nu\rho} + \partial_\nu F_{\rho\mu} + \partial_\rho F_{\mu\nu} = 0 \tag{2.7.16}$$

and follows from the definition of $F_{\mu\nu}$ and from $\partial_\nu \partial_\rho A_\mu = \partial_\rho \partial_\nu A_\mu$ assuming that A_μ is doubly differentiable. Since $F_{\mu\nu} = -F_{\nu\mu}$ it is easily checked that the left-hand side of the Bianchi identity is completely antisymmetric under the pairwise interchange of the indicies μ, ν and ρ. So we can rewrite Eq. (2.7.16) as

$$\epsilon^{\mu\nu\rho\sigma}\partial_\mu F_{\rho\sigma} = 0. \tag{2.7.17}$$

Define the *dual* field strength tensor as

$$\tilde{F}^{\mu\nu} \equiv \tfrac{1}{2}\epsilon^{\mu\nu\rho\sigma}F_{\rho\sigma} \qquad \Rightarrow \qquad \tilde{F}^{\mu\nu} = -\tilde{F}^{\nu\mu}. \tag{2.7.18}$$

Since $F^{\mu\nu}$ transforms like a second-rank tensor and since from Eq. (1.2.137) we know that $\epsilon^{\mu\nu\rho\sigma}$ transforms like a rank-four pseudotensor, then $\tilde{F}^{\mu\nu}$ must transform as a second-rank pseudotensor. Recall that we use $\epsilon^{0123} = -\epsilon_{0123} = +1$ and so find

$$\tilde{F}^{0i} = -B^i \qquad \text{and} \qquad \tilde{F}^{ij} = \epsilon^{ijk}(E^k/c). \tag{2.7.19}$$

Comparing this with Eq. (2.7.13) we see that the dual transformation $F^{\mu\nu} \rightarrow \tilde{F}^{\mu\nu}$ equates to $(\mathbf{E}/c) \rightarrow \mathbf{B}$ and $\mathbf{B} \rightarrow -(\mathbf{E}/c)$,

$$\tilde{F}^{\mu\nu} = \begin{pmatrix} 0 & -B_x & -B_y & -B_z \\ B_x & 0 & E_z/c & -E_y/c \\ B_y & -E_z/c & 0 & E_x/c \\ B_z & E_y/c & -E_x/c & 0 \end{pmatrix}. \tag{2.7.20}$$

Using Eq. (2.7.17) and $\tilde{F}^{\mu\nu}$ we find a compact form of the Bianchi identity,

$$\partial_\mu \tilde{F}^{\mu\nu} = 0, \tag{2.7.21}$$

which consists of four equations. These can be written as

$$0 = \partial_\mu \tilde{F}^{\mu 0} = \partial_i \tilde{F}^{i0} = \boldsymbol{\nabla} \cdot \mathbf{B},$$
$$0 = \partial_\mu \tilde{F}^{\mu j} = \partial_0 \tilde{F}^{0j} + \partial_i \tilde{F}^{ij} = -(1/c)\left[(\partial B^j/\partial t) + (\boldsymbol{\nabla} \times \mathbf{E})^j\right], \tag{2.7.22}$$

which we recognize as the first two of Maxwell's equations in Eq. (2.7.2).

We have arrived at a Lorentz-covariant form of Maxwell's equations,

$$\partial_\mu \tilde{F}^{\mu\nu} = 0 \qquad \text{and} \qquad \partial_\mu F^{\mu\nu} = \mu_0\, j^\nu. \tag{2.7.23}$$

The first equation is a compact expression for the first two of of Maxwell's equations in Eq. (2.7.2). They contain no dynamics in the sense that they follow automatically from defining the $\mathbf{E}$ and $\mathbf{B}$ fields in terms of the four-vector potential A^μ. The second equation corresponds to the last two of Maxwell's equations and contains all of the interaction dynamics. Since $F^{\mu\nu} = -F^{\nu\mu}$ then $\partial_\mu \partial_\nu F^{\mu\nu} = 0$ and so

$$0 = \partial_\nu \partial_\mu F^{\mu\nu}(x) = \mu_0 \partial_\nu j^\nu(x). \tag{2.7.24}$$

Then the equations of motion can only have a solution provided we couple the electromagnetic field to a conserved current density $j^\mu(x)$,

$$\partial_\mu j^\mu(x) = 0. \tag{2.7.25}$$

See the discussion of conserved current density in Sec. 4.3.2.

From Eqs. (2.7.13) and (2.7.20) we easily find the results

$$F_{\mu\nu}F^{\mu\nu} = -\tilde{F}_{\mu\nu}\tilde{F}^{\mu\nu} = 2\left[\mathbf{B}^2 - (\mathbf{E}^2/c^2)\right] \qquad \text{and} \qquad \tilde{F}_{\mu\nu}F^{\mu\nu} = -(4/c)\,\mathbf{E} \cdot \mathbf{B}. \tag{2.7.26}$$

Since $F^{\mu\nu}$ is a tensor and $\tilde{F}^{\mu\nu}$ is pseudotensor, then $F_{\mu\nu}F^{\mu\nu} = -\tilde{F}_{\mu\nu}\tilde{F}^{\mu\nu}$ is a scalar and $\tilde{F}_{\mu\nu}F^{\mu\nu}$ is a pseudoscalar. It is also readily verified that

$$F^{\mu\nu} = -\tfrac{1}{2}\epsilon^{\mu\nu\rho\sigma}\tilde{F}_{\rho\sigma}. \tag{2.7.27}$$

The Poynting vector represents the magnitude and direction of the energy flux associated with the electromagnetic field and in the vacuum is given by

$$\mathbf{S} = (1/\mu_0)\,\mathbf{E} \times \mathbf{B} = c\left[\sqrt{\epsilon_0}\,\mathbf{E} \times (\mathbf{B}/\sqrt{\mu_0})\right]. \tag{2.7.28}$$

The energy density associated with the electromagnetic field in the vacuum is

$$u = \tfrac{1}{2}\left[\epsilon_0\mathbf{E}^2 + (\mathbf{B}^2/\mu_0)\right]. \tag{2.7.29}$$

The derivation of u and $\mathbf{S}$ will be given in Sec. 3.3. Since u is an energy density its dimensions are $[u] = [E]/[V] = (\mathsf{ML^2T^{-2}})/\mathsf{L^3} = \mathsf{ML^{-1}T^{-2}}$, which is the same as the dimensions of pressure since $[P] = [F]/[A] = (\mathsf{MLT^{-2}})/\mathsf{L^2} = \mathsf{ML^{-1}T^{-2}}$. From the expression for u we see that $[\sqrt{\epsilon_0}\,\mathbf{E}] = [\mathbf{B}/\sqrt{\mu_0}] = [u^{1/2}] = (\mathsf{ML^{-1}T^{-2}})^{1/2}$ and so $\mathbf{S}/c$ must have the same dimensions as u; i.e., in SI units u and $\mathbf{S}/c$ can be expressed in pascals (Pa), where $1\,\mathrm{Pa} = 1\,\mathrm{N/m^2}$.

Electromagnetism in Lorentz-Heaviside and cgs-Gaussian units: While this chapter is formulated in terms of SI units, cgs-Gaussian units and Lorentz-Heaviside units are also in common usage.

For example, in later chapters we will adopt the particle physics convention of using (i) Lorentz-Heaviside units and (ii) combining this with the use of natural units, $\hbar = c = 1$.

It is straightforward to translate Maxwell's equations from one unit system to another. Using an obvious shorthand notation, the translations are:

$$(\mathbf{E}, \Phi)^{\text{SI}} \leftrightarrow (1/\sqrt{\epsilon_0})(\mathbf{E}, \Phi)^{\text{LH}} \leftrightarrow (1/\sqrt{4\pi\epsilon_0})(\mathbf{E}, \Phi)^{\text{cgs}},$$

$$(q, \rho, \mathbf{j})^{\text{SI}} \leftrightarrow \sqrt{\epsilon_0}(q, \rho, \mathbf{j})^{\text{LH}} \leftrightarrow \sqrt{4\pi\epsilon_0}(q, \rho, \mathbf{j})^{\text{cgs}},$$

$$(\mathbf{B}, \mathbf{A})^{\text{SI}} \leftrightarrow \sqrt{\mu_0}(\mathbf{B}, \mathbf{A})^{\text{LH}} \leftrightarrow \sqrt{\mu_0/4\pi}(\mathbf{B}, \mathbf{A})^{\text{cgs}},$$

$$(j^\mu)^{\text{SI}} \leftrightarrow \sqrt{\epsilon_0}(j^\mu)^{\text{LH}} \leftrightarrow \sqrt{4\pi\epsilon_0}(j^\mu)^{\text{cgs}},$$

$$(j^0)^{\text{SI}} \equiv \rho^{\text{SI}}/c, \quad (j^0)^{\text{LH}} \equiv \rho^{\text{LH}}/c, \quad (j^0)^{\text{cgs}} \equiv \rho^{\text{cgs}}/c,$$

$$(A^\mu)^{\text{SI}} \leftrightarrow \sqrt{\mu_0}(A^\mu)^{\text{LH}} \leftrightarrow \sqrt{\mu_0/4\pi}(A^\mu)^{\text{cgs}},$$

$$(A^0)^{\text{SI}} \equiv \Phi^{\text{SI}}/c, \quad (A^0)^{\text{LH}} \equiv \Phi^{\text{LH}}, \quad (A^0)^{\text{cgs}} \equiv \Phi^{\text{cgs}},$$

$$(qA^\mu)^{\text{SI}} \leftrightarrow (qA^\mu)^{\text{LH}}/c \leftrightarrow \sqrt{1/4\pi}\,(A^\mu)^{\text{cgs}}. \tag{2.7.30}$$

The *Coulomb force* in the three systems of units is given by

$$F = \left(\frac{q_1 q_2}{4\pi\epsilon_0 r^2}\right)^{\text{SI}} = \left(\frac{q_1 q_2}{4\pi r^2}\right)^{\text{LH}} = \left(\frac{q_1 q_2}{r^2}\right)^{\text{cgs}} \tag{2.7.31}$$

and similarly for the *fine-structure constant* α we have

$$\alpha = \left(\frac{e^2}{4\pi\epsilon_0 \hbar c}\right)^{\text{SI}} = \left(\frac{e^2}{4\pi\hbar c}\right)^{\text{LH}} = \left(\frac{e^2}{\hbar c}\right)^{\text{cgs}} \simeq \frac{1}{137.036}. \tag{2.7.32}$$

For example, in both Lorentz-Heaviside and cgs-Gaussian units we have

$$\mathbf{E} = -\boldsymbol{\nabla}\Phi - \frac{1}{c}\frac{\partial \mathbf{A}}{\partial t} \quad \text{and} \quad \mathbf{B} = \boldsymbol{\nabla} \times \mathbf{A}. \tag{2.7.33}$$

In Lorentz-Heaviside units Maxwell's equations become:

Faraday's law: $\boldsymbol{\nabla} \times \mathbf{E} + \dfrac{1}{c}\dfrac{\partial \mathbf{B}}{\partial t} = 0,$ Gauss' magnetism law: $\boldsymbol{\nabla} \cdot \mathbf{B} = 0,$

Gauss' law: $\boldsymbol{\nabla} \cdot \mathbf{E} = \rho,$ Ampère's law: $\boldsymbol{\nabla} \times \mathbf{B} - \dfrac{1}{c}\dfrac{\partial \mathbf{E}}{\partial t} = \dfrac{\mathbf{j}}{c}.$
$\tag{2.7.34}$

Maxwell's equations in cgs-Gaussian units can be obtained from these equations by replacing $j^\mu \equiv (\rho/c, \mathbf{j})$ with $4\pi j^\mu$.

For both Lorentz-Heaviside units and cgs-Gaussian units we have

$$F^{\mu\nu}(x) \equiv \partial^\mu A^\nu(x) - \partial^\nu A^\mu(x), \quad \tilde{F}^{\mu\nu} \equiv \tfrac{1}{2}\epsilon^{\mu\nu\rho\sigma} F_{\rho\sigma} \quad \text{and} \quad A^\mu \equiv (\Phi, \mathbf{A}). \tag{2.7.35}$$

Note that in Lorentz-Heaviside units and cgs-Gaussian units we must define for consistency $A^0 \equiv \Phi$, whereas in SI units we have $A^0 \equiv \Phi/c$. This occurs because the above replacements for Φ and $\mathbf{A}$ differ by an overall factor of $c = 1/\sqrt{\epsilon_0\mu_0}$ when we move from SI units to either Lorentz-Heaviside units or cgs-Gaussian units. In these two unit systems it follows that

$$F^{0i} = -E^i \quad \text{and} \quad F^{ij} = -\epsilon^{ijk} B^k, \text{ since} \tag{2.7.36}$$

$$F^{0i} = \partial^0 A^i - \partial^i A^0 = \frac{1}{c}\frac{\partial A^i}{\partial t} + \partial_i \Phi = -E^i. \tag{2.7.37}$$

Hence in Lorentz-Heaviside and cgs-Gaussian units $F_{\mu\nu}$ has the form

$$F^{\mu\nu} = \begin{pmatrix} 0 & -E_x & -E_y & -E_z \\ E_x & 0 & -B_z & B_y \\ E_y & B_z & 0 & -B_x \\ E_z & -B_y & B_x & 0 \end{pmatrix}. \tag{2.7.38}$$

It is easily seen that the dual transformation $F^{\mu\nu} \to \tilde{F}^{\mu\nu}$ is now given by $\mathbf{E} \to \mathbf{B}$ and $\mathbf{B} \to -\mathbf{E}$ such that in Lorentz-Heaviside and cgs-Gaussian units we have

$$\tilde{F}^{0i} = -B^i \qquad \text{and} \qquad \tilde{F}^{ij} = \epsilon^{ijk} E^k, \tag{2.7.39}$$

which can be written as

$$\tilde{F}^{\mu\nu} = \begin{pmatrix} 0 & -B_x & -B_y & -B_z \\ B_x & 0 & E_z & -E_y \\ B_y & -E_z & 0 & E_x \\ B_z & E_y & -E_x & 0 \end{pmatrix}. \tag{2.7.40}$$

In Lorentz-Heaviside units the last two Maxwell's equations in Eq. (2.7.34) are

$$\partial_\mu F^{\mu\nu} = (1/c)j^\nu, \quad \text{since} \tag{2.7.41}$$
$$0 = \partial_\mu F^{\mu 0} - (1/c)j^0 = \partial_i F^{i0} - \rho = \mathbf{\nabla} \cdot \mathbf{E} - \rho,$$
$$0 = \partial_\mu F^{\mu j} - (1/c)j^j = -(1/c)(\partial E^j/\partial t) + (\mathbf{\nabla} \times \mathbf{B})^j - (1/c)j^j.$$

In cgs-Gaussian units we make the replacement $j^\mu \to 4\pi j^\mu$ to give $\partial_\mu F^{\mu\nu} = (4\pi/c)\, j^\nu$. In both Lorentz-Heaviside and cgs-Gaussian units we again find

$$\partial_\mu \tilde{F}^{\mu\nu} = 0, \quad \text{since} \tag{2.7.42}$$
$$0 = \partial_\mu \tilde{F}^{\mu 0} = \partial_i \tilde{F}^{i0} = \mathbf{\nabla} \cdot \mathbf{B},$$
$$0 = \partial_\mu \tilde{F}^{\mu j} = \partial_0 \tilde{F}^{0j} + \partial_i \tilde{F}^{ij} = -(1/c) \left[(\partial B^j/\partial t) + (\mathbf{\nabla} \times \mathbf{E})^j \right],$$

which we recognize as the first two of Maxwell's equations in Eq. (2.7.34).

Using the translations in Eq. (2.7.30) we find that in Lorentz-Heaviside and cgs-Gaussian units the Lorentz force takes the form

$$\mathbf{F} = q\left[\mathbf{E} + (\boldsymbol{v}/c) \times \mathbf{B}\right]. \tag{2.7.43}$$

In both Lorentz-Heaviside and cgs-Gaussian units we have

$$F_{\mu\nu} F^{\mu\nu} = -\tilde{F}_{\mu\nu} \tilde{F}^{\mu\nu} = 2\left(\mathbf{B}^2 - \mathbf{E}^2\right) \qquad \text{and} \qquad \tilde{F}_{\mu\nu} F^{\mu\nu} = -4\,\mathbf{E} \cdot \mathbf{B}. \tag{2.7.44}$$

The Poynting vector and energy density in Lorentz-Heaviside units are given by

$$\mathbf{S} = c\,\mathbf{E} \times \mathbf{B} \quad \text{and} \quad u = \tfrac{1}{2}\left(\mathbf{E}^2 + \mathbf{B}^2\right). \tag{2.7.45}$$

In cgs-Gaussian units we have

$$\mathbf{S} = (c/4\pi)\,\mathbf{E} \times \mathbf{B} \quad \text{and} \quad u = (1/8\pi)\left(\mathbf{E}^2 + \mathbf{B}^2\right). \tag{2.7.46}$$

2.7.2 Electromagnetic Waves

The propagation of free electromagnetic waves is determined by Maxwell's equations in the vacuum, i.e., $j^\mu = 0$. The curl of Faraday's law gives

$$0 = \boldsymbol{\nabla} \times \left(\boldsymbol{\nabla} \times \mathbf{E} + \frac{\partial \mathbf{B}}{\partial t} \right) = \boldsymbol{\nabla} \left(\boldsymbol{\nabla} \cdot \mathbf{E} \right) - \boldsymbol{\nabla}^2 \mathbf{E} + \frac{\partial}{\partial t} \left(\boldsymbol{\nabla} \times \mathbf{B} \right)$$

$$= -\boldsymbol{\nabla}^2 \mathbf{E} + \frac{1}{c^2} \frac{\partial^2 \mathbf{E}}{\partial t^2} = \Box \mathbf{E}, \qquad (2.7.47)$$

where we have used $\mathbf{a} \times (\mathbf{b} \times \mathbf{c}) = \mathbf{b}(\mathbf{a} \cdot \mathbf{c}) - (\mathbf{a} \cdot \mathbf{b})\mathbf{c}$, $\boldsymbol{\nabla} \cdot \mathbf{E} = 0$ and $\boldsymbol{\nabla} \times \mathbf{B} = (1/c^2)\partial \mathbf{E}/\partial t$. Similarly, the curl of Ampère's law gives

$$0 = \boldsymbol{\nabla} \times \left(\boldsymbol{\nabla} \times \mathbf{B} - \frac{1}{c^2} \frac{\partial \mathbf{E}}{\partial t} \right) = \boldsymbol{\nabla} \left(\boldsymbol{\nabla} \cdot \mathbf{B} \right) - \boldsymbol{\nabla}^2 \mathbf{B} - \frac{1}{c^2} \frac{\partial}{\partial t} \left(\boldsymbol{\nabla} \times \mathbf{E} \right)$$

$$= -\boldsymbol{\nabla}^2 \mathbf{B} + \frac{1}{c^2} \frac{\partial^2 \mathbf{B}}{\partial t^2} = \Box \mathbf{B}, \qquad (2.7.48)$$

where we have used $\boldsymbol{\nabla} \cdot \mathbf{B} = 0$ and $\boldsymbol{\nabla} \times \mathbf{E} = -\partial \mathbf{B}/\partial t$.

An equation of the form

$$\Box \psi = \frac{1}{c^2} \frac{\partial^2 \psi}{\partial t^2} - \boldsymbol{\nabla}^2 \psi = 0 \qquad (2.7.49)$$

is a wave equation for a field $\psi(x)$ where the wave propagates with the speed of light c. The plane wave solutions of this equation are specified by an amplitude $a \in \mathbb{C}$ and the wavevector, $\mathbf{k}$, and can be written as

$$\psi(x) \equiv \psi(t, \mathbf{x}) = a e^{i(\mathbf{k} \cdot \mathbf{x} - \omega t)} = a e^{ik(\hat{\mathbf{k}} \cdot \mathbf{x} - ct)} \equiv a e^{ik\kappa(x)}, \qquad (2.7.50)$$

where $c = f\lambda$, the angular frequency is $\omega = 2\pi f = ck$, f is the frequency, $T = 1/f$ is the period, λ is the wavelength, $k \equiv |\mathbf{k}| = 2\pi/\lambda$ is the magnitude of the wavevector, $\hat{\mathbf{k}} \equiv (\mathbf{k}/k)$ is the unit vector denoting the direction of motion of the plane wave and where we have defined the Lorentz scalar function

$$\kappa(x) \equiv \hat{\mathbf{k}} \cdot \mathbf{x} - ct. \qquad (2.7.51)$$

The real plane wave solutions have the form

$$\psi_R(x) \equiv \psi_R(t, \mathbf{x}) = \mathrm{Re} \left(a e^{i(\mathbf{k} \cdot \mathbf{x} - \omega t)} \right) = |a| \cos \left(\mathbf{k} \cdot \mathbf{x} - \omega t + \delta \right), \qquad (2.7.52)$$

where $a \equiv |a| \exp(i\delta)$. Note that

$$\partial_\mu \kappa(x) = (-1, \hat{\mathbf{k}}), \quad \Box \kappa = \partial_\mu \partial^\mu \kappa = 0 \quad \text{and} \quad \partial_\mu \kappa \partial^\mu \kappa = 0. \qquad (2.7.53)$$

Any doubly differentiable function of κ is a solution of the wave equation, since if

$$\psi(x) \equiv f(\kappa(x)) = f(\hat{\mathbf{k}} \cdot \mathbf{x} - ct) \qquad (2.7.54)$$

then it satisfies the wave equation

$$\Box \psi = \partial^\mu \partial_\mu f(\kappa(x)) = \partial^\mu \left(\frac{\partial f}{\partial \kappa} \partial_\mu \kappa \right) = \frac{\partial f}{\partial \kappa} \partial^\mu \partial_\mu \kappa + \frac{\partial^2 f}{\partial \kappa^2} \partial^\mu \kappa \partial_\mu \kappa = 0. \qquad (2.7.55)$$

Since $\Box \mathbf{E} = 0$ then from Eq. (2.7.52) we see that $\mathbf{E}$ has plane wave solutions,

$$\mathbf{E}(x) = \begin{pmatrix} E_x(x) \\ E_y(x) \\ E_z(x) \end{pmatrix} = \begin{pmatrix} E_x^0 \cos(\mathbf{k} \cdot \mathbf{x} - \omega t + \delta_x) \\ E_y^0 \cos(\mathbf{k} \cdot \mathbf{x} - \omega t + \delta_y) \\ E_z^0 \cos(\mathbf{k} \cdot \mathbf{x} - \omega t + \delta_z) \end{pmatrix} \tag{2.7.56}$$

and similarly for $\mathbf{B}$. To obtain Eq. (2.7.56) and the equivalent form for $\mathbf{B}$ we have only used two equations, i.e., Eqs. (2.7.47) and (2.7.48). However, there are four Maxwell's equations and so there is a need to impose two additional independent conditions in order to have a plane wave solution to Maxwell's equations. Begin with the special case $\delta \equiv \delta_x = \delta_y = \delta_z$, which gives

$$\mathbf{E}(x) = \mathbf{E}^0 \cos(\mathbf{k} \cdot \mathbf{x} - \omega t + \delta), \tag{2.7.57}$$

where $\mathbf{E}^0 \equiv (E_x^0, E_y^0, E_z^0)$ with $E_x^0, E_y^0, E_z^0 \in \mathbb{R}$. For this case it then follows from Gauss' law that we must have

$$0 = \boldsymbol{\nabla} \cdot \mathbf{E} = -(\mathbf{k} \cdot \mathbf{E}^0) \sin(\mathbf{k} \cdot \mathbf{x} - \omega t + \delta), \tag{2.7.58}$$

which requires that $\mathbf{k} \cdot \mathbf{E}^0 = 0$. This then gives the result that

$$\hat{\mathbf{k}} \cdot \mathbf{E} = 0; \tag{2.7.59}$$

i.e., $\mathbf{E}$ is orthogonal to the direction of propagation $\hat{\mathbf{k}}$. From Faraday's law we have

$$\nabla \times \mathbf{E} = -(\mathbf{k} \times \mathbf{E}^0) \sin(\mathbf{k} \cdot \mathbf{x} - \omega t + \delta) = -(\partial \mathbf{B} / \partial t), \tag{2.7.60}$$

which is satisfied if $\mathbf{B}^0 = (1/c)\hat{\mathbf{k}} \times \mathbf{E}^0$ and if the $\mathbf{B}$ field is given by

$$\mathbf{B}(x) = \mathbf{B}^0 \cos(\mathbf{k} \cdot \mathbf{x} - \omega t + \delta). \tag{2.7.61}$$

This condition can be summarized as

$$\mathbf{B} = (1/c)(\hat{\mathbf{k}} \times \mathbf{E}), \tag{2.7.62}$$

which states that (i) the $\mathbf{B}$ field is orthogonal to both $\mathbf{E}$ and to the direction of propagation and has a relative orientation to these given by the right-hand rule; and (ii) the magnitudes of the field satisfy the relation $|\mathbf{B}| = |\mathbf{E}|/c$ at all positions and at all times. This case corresponds to so-called *linear polarization* and is illustrated in Fig. 2.4 for the case of propagation in the z-direction with $\mathbf{E}$ oscillating in the y-direction and $\mathbf{B}$ oscillating in the x-direction.

How much can the condition $\delta \equiv \delta_x = \delta_y = \delta_z$ be relaxed and still have a plane wave type of solution to Maxwell's equations? It will be convenient to define our axes such that the direction of propagation is the z-direction, which we can do without loss of generality since we have shown above that Maxwell's equations are Lorentz invariant. Then $\mathbf{k} \cdot \mathbf{x} = kz$ and using Eq. (2.7.56) with Gauss' law we find

$$0 = \boldsymbol{\nabla} \cdot \mathbf{E} = \partial_x E_x + \partial_y E_y + \partial_z E_z = 0 + 0 - kE_z \sin(kz - \omega t + \delta_z); \tag{2.7.63}$$

i.e., $E_z = 0$ and so $\hat{\mathbf{k}} \cdot \mathbf{E} = 0$ is a requirement for *any* plane solution. Using $E_z = 0$ and $\partial_x \mathbf{E} = \partial_y \mathbf{E} = 0$ together with Faraday's law gives

$$\begin{aligned} (\boldsymbol{\nabla} \times \mathbf{E})_x &= -\partial_z E_y = -kE_y^0 \sin(kz - \omega t + \delta_y) = -\partial B_x / \partial t \\ (\boldsymbol{\nabla} \times \mathbf{E})_y &= \partial_z E_x = -kE_x^0 \sin(kz - \omega t + \delta_x) = -\partial B_y / \partial t \\ (\boldsymbol{\nabla} \times \mathbf{E})_z &= 0 = -\partial B_z / \partial t. \end{aligned} \tag{2.7.64}$$

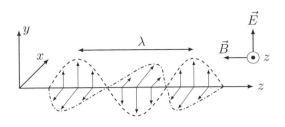

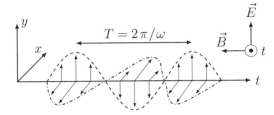

Figure 2.4 *Top:* Electromagnetic wave with linear polarization propagating in the z-direction and with $\mathbf{E}$ and $\mathbf{B}$ fields oscillating in the y-direction and the x-direction, respectively. *Bottom:* The temporal variation.

For a plane wave type solution the last of these three equations can only be satisfied if $B_z = 0$. Given that $\mathbf{B}$ must have a plane wave type form, it is readily verified that these three equations are satisfied if and only if $\mathbf{B} = (1/c)(\hat{\mathbf{k}} \times \mathbf{E})$. So this equation must be satisfied by *any* plane wave type solution to Maxwell's equations.

Summary: Defining the z-axis to be the direction of propagation, *any* plane wave solution of Maxwell's equations has an $\mathbf{E}$ field given by

$$\mathbf{E}(x) = \begin{pmatrix} E_x(x) \\ E_y(x) \\ E_z(x) \end{pmatrix} = \begin{pmatrix} E_x^0 \cos(kz - \omega t + \delta_x) \\ E_y^0 \cos(kz - \omega t + \delta_y) \\ 0 \end{pmatrix}, \tag{2.7.65}$$

i.e., $\hat{\mathbf{k}} \cdot \mathbf{E} = 0$. The magnetic field $\mathbf{B}$ is then determined by $\mathbf{B} = (1/c)(\hat{\mathbf{k}} \times \mathbf{E})$ giving

$$\mathbf{B}(x) = \begin{pmatrix} -E_y(x)/c \\ E_x(x)/c \\ 0 \end{pmatrix} = \begin{pmatrix} -(E_y^0/c) \cos(kz - \omega t + \delta_y) \\ (E_x^0/c) \cos(kz - \omega t + \delta_x) \\ 0 \end{pmatrix}. \tag{2.7.66}$$

The parameters needed to specify an electromagnetic plane wave moving in the z-direction in a vacuum are the angular frequency, $\omega = 2\pi f$, the transverse magnitudes E_x^0 and E_y^0 and the two corresponding phases δ_x and δ_y. The four quantities E_x^0, E_y^0, δ_x and δ_y determine the amplitude and polarization of the waves.

Polarization

Linear polarization: For simplicity we will continue to define the z-axis to be the direction of propagation of the plane electromagnetic wave without loss of generality. If we have $\delta \equiv \delta_x = \delta_y$ then we recover the plane wave with *linear polarization*

$$\mathbf{E}(x) = \mathbf{E}^0 \cos(kz - \omega t + \delta). \tag{2.7.67}$$

In this case the $\mathbf{E}$ field is collinear with the direction of $\mathbf{E}^0 = (E_x^0, E_y^0, 0)$ at all times and in all locations and the maximum value of the $\mathbf{E}$ field is $E^0 \equiv |\mathbf{E}^0| = [(E_x^0)^2 + (E_y^0)^2]^{1/2}$. Define θ to be the angle of $\mathbf{E}^0$ with respect to the x-axis, then $E_x^0 = E^0 \cos\theta$ and $E_y^0 = E^0 \sin\theta$. Note that this is equivalent to the superposition of two electromagnetic waves with linear polarizations in the x and y directions, respectively, and with the same frequency and phase

$$\mathbf{E}(x) = (E_x^0 \cos(kz - \omega t + \delta), 0, 0) + (0, E_y^0 \cos(kz - \omega t + \delta), 0). \tag{2.7.68}$$

The $\mathbf{B}$ field is then easily seen to be

$$\mathbf{B}(x) = \mathbf{B}^0 \cos(kz - \omega t + \delta) \tag{2.7.69}$$

with $\mathbf{B}^0 = (-E_y^0/c, E_x^0/c, 0)$.

Circular polarization: Let us next consider the case where $E^0 \equiv E_x^0 = E_y^0$ and where $\delta \equiv \delta_x$ and $\delta_y = \delta \pm \frac{\pi}{2}$, which gives

$$\mathbf{E}(x) = \begin{pmatrix} E^0 \cos(kz - \omega t + \delta) \\ \mp E^0 \sin(kz - \omega t + \delta) \\ 0 \end{pmatrix} = E^0 \begin{pmatrix} \cos(\pm[\omega t - kz - \delta]) \\ \sin(\pm[\omega t - kz - \delta]) \\ 0 \end{pmatrix}. \tag{2.7.70}$$

Consider the time-dependence of $\mathbf{E}$ at some fixed location z. Clearly, we will have $|\mathbf{E}| = E^0$. The cases $\pm[\omega t - kz - \delta]$ correspond, respectively, to right-handed/left-handed rotation of the vector $\mathbf{E}$ in the $x-y$ plane with respect to the direction of motion, z. This is referred to as *circular polarization*. Recall that positive helicity ($h = +$) is defined by the right-hand rule where the thumb is pointed in the direction of translational motion $\hat{\mathbf{k}}$ and the fingers of the right-hand wrap around in the direction of rotation. So then right-handed polarization is positive helicity and left-handed polarization is negative helicity. In optics the naming convention is opposite when describing circular polarization, since it considers the point of view of the observer looking into the direction of the plane wave. So in optics $-\hat{\mathbf{k}}$ is used as the direction of the thumb when using the right-hand rule. Then we have

$$h = + \iff \text{RH polarization} \iff \text{LH polarization (optics)},$$
$$h = - \iff \text{LH polarization} \iff \text{RH polarization (optics)}. \tag{2.7.71}$$

These concepts are illustrated in Fig. 2.5. As the wave move forwards in the positive z-direction, the intersection of the moving spiral with the $x - y$ plane traces out a counterclockwise circle as viewed from the positive z-direction. Then the figure corresponds to right-handed polarization, which is positive helicity.

Circular polarization is equivalent to the superposition of two orthogonal linear polarizations of equal amplitude and frequency with a phase difference of $\pm 90°$,

$$\mathbf{E}(x) = \left(E^0 \cos[kz - \omega t + \delta], 0, 0\right) + \left(0, E^0 \cos\left[kz - \omega t + \delta \pm (\pi/2)\right], 0\right) \tag{2.7.72}$$

corresponding to circular polarization with positive/negative helicity, respectively. In terms of complex phasor notation for a plane wave moving in the z-direction we write for the different polarizations,

$$\text{linear:} \quad \mathbf{E}(x) = E_0 \, e^{i(\mathbf{k} \cdot \mathbf{x} - \omega t + \delta)} (v_x, v_y, 0), \quad v_x, v_y \in \mathbb{R}, (v_x^2 + v_y^2)^{1/2} = 1,$$
$$\text{R circular } (h = +): \quad \mathbf{E}(x) = E_0 \, e^{i(\mathbf{k} \cdot \mathbf{x} - \omega t + \delta)} (1, i, 0)\sqrt{2},$$
$$\text{L circular } (h = -): \quad \mathbf{E}(x) = E_0 \, e^{i(\mathbf{k} \cdot \mathbf{x} - \omega t + \delta)} (1, -i, 0)/\sqrt{2}, \tag{2.7.73}$$

where $\mathbf{k} \cdot \mathbf{x} = kz$. The real parts of the phasor notation reproduce the above results. We can convert between linear polarization at $45°$ and circular polarization if we can shift the phase of either the x or y component by $90°$ with respect to the other. This is achieved in optics with a quarter-wave plate.

Elliptical polarization: In the most general case where E_x^0, E_y^0, δ_x and δ_y are arbitrary we have what is called *elliptical polarization*. For example, if we have $\delta_y = \delta_x \pm (\pi/2)$ as for circular polarization but have $E_x^0 \neq E_y^0$, then it is straightforward to see that the direction of rotation of $\mathbf{E}$ will be the same as the case for circular polarization, but instead of $\mathbf{E}$ tracing out a circle it will now trace out an ellipse with the major/minor axes given by the larger/smaller of E_x^0 and E_y^0, respectively. If we allow $E_x^0 \neq E_y^0$ and have $\delta_x < \delta_y < \delta_x + \pi$, then we will have counterclockwise rotation and an elliptical shape. If we have $\delta_x - \pi < \delta_y < \delta_x$, then we will have clockwise rotation and elliptical shape. As $\delta_y \to \delta_x \pm (\pi/2)$ the ellipse takes on a shape with x and y as the minor and major axes of the ellipse and if we also set $E_x^0 = E_y^0$ then we recover circular polarization. As $\delta_y \to \delta_x$ or $\delta_y \to \delta_x + \pi$, then the ellipse narrows and becomes a straight line and we recover linear polarization.

2.7.3 Gauge Transformations and Gauge Fixing

A transformation of the four-vector potential of the form

$$A^\mu(x) \to A'^\mu(x) = A^\mu(x) + \partial^\mu \omega(x) \tag{2.7.74}$$

is called a *gauge transformation*. The concept of gauge invariance under gauge transformations will later become the focus of much of our attention. For now, we simply note that for the case of electromagnetism it is a freedom that we have in how we map the $\mathbf{E}$ and $\mathbf{B}$ fields to the four-vector potential, A^μ. We observe that $\mathbf{E}$ and $\mathbf{B}$ are *gauge invariant* since under such a gauge transformation we have

$$\mathbf{E} \to \mathbf{E}' = -\boldsymbol{\nabla}\Phi' - \frac{\partial \mathbf{A}'}{\partial t} = -\boldsymbol{\nabla}\left(\Phi + c\partial^0 \omega\right) - \frac{\partial}{\partial t}\left(\mathbf{A} - \boldsymbol{\nabla}\omega\right)$$

$$= -\boldsymbol{\nabla}\Phi - \frac{\partial \mathbf{A}}{\partial t} - c\left(\boldsymbol{\nabla}\partial^0 - \partial^0 \boldsymbol{\nabla}\right)\omega = \mathbf{E},$$

$$\mathbf{B} \to \mathbf{B}' = \boldsymbol{\nabla} \times \mathbf{A}' = \boldsymbol{\nabla} \times (\mathbf{A} - \boldsymbol{\nabla}\omega) = \boldsymbol{\nabla} \times \mathbf{A} = \mathbf{B}, \tag{2.7.75}$$

which means that Maxwell's equations when given terms of the $\mathbf{E}$ and $\mathbf{B}$ fields are gauge invariant. It also follows from Eqs. (2.7.13) and (2.7.20) that $F^{\mu\nu}$ and $\tilde{F}^{\mu\nu}$ are gauge invariant and so the covariant forms of Maxwell's equations given in Eq. (2.7.23) are obviously gauge invariant.

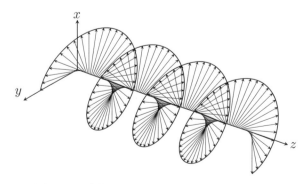

Figure 2.5   The $\mathbf{E}$ field for a right-handed circularly polarized electromagnetic wave.

It can be verified directly from the gauge transformation property of A^μ and the definition of $F^{\mu\nu}$ that $F^{\mu\nu}$ is gauge invariant, i.e.,

$$F^{\mu\nu} \to F'^{\mu\nu} = c\left(\partial^\mu A'^\nu - \partial^\nu A'^\mu\right) \tag{2.7.76}$$
$$= c\left(\partial^\mu A^\nu - \partial^\nu A^\mu + \partial^\mu \partial^\nu \omega - \partial^\nu \partial^\mu \omega\right) = F^{\mu\nu}.$$

In other words, any two four-vector potentials A^μ and A'^μ that are related by a gauge transformation lead to identical physical outcomes, where only the $\mathbf{E}$ and $\mathbf{B}$ fields have physical meaning in classical physics. However, when electromagnetism is extended into the quantum mechanical context, the use of the four-vector potential A^μ becomes essential, e.g., to understand the Aharonov-Bohm effect.

The Aharonov-Bohm effect results from the fact that a particle passing by a solenoid picks up a relative phase difference depending on which side of the solenoid it passed and this phase difference is proportional to the enclosed magnetic flux between the two paths. Since for an ideal solenoid the $\mathbf{B}$ field is only nonzero inside the solenoid, then the particle moving along either path outside feels no $\mathbf{B}$ field, but does experience an $\mathbf{A}$ field. Outside the solenoid we can have $\mathbf{B} = 0$ and $\mathbf{A} \neq 0$, since $\mathbf{B} = \nabla \times \mathbf{A}$. The Aharonov-Bohm effect is the quantum-mechanical interference of a particle with itself involving a phase difference between virtual paths that passed around the solenoid on different sides. The interference pattern is changed when the solenoid current changes. Note that while this effect requires the concept of a local interaction between the particle and the A^μ field to understand it, the final predictions can be expressed entirely in terms of $\mathbf{E}$ and $\mathbf{B}$ fields and the physical effect itself is therefore gauge invariant.

Eq. (2.7.74) shows the enormous freedom that we have to choose a four-vector potential A^μ for given $\mathbf{E}$ and $\mathbf{B}$ fields. In quantum field theory, the four-vector potential A^μ is referred to as the electromagnetic *gauge field* and electromagnetism is the simplest of a class of theories called *gauge field theories*. We can attempt to reduce this freedom we have in selecting A^μ by placing restrictions on the available choices of it. This is a process referred to as *gauge fixing*. If we could place restriction(s) on the A^μ field so that there was one and only one A^μ for any given pair of $\mathbf{E}$ and $\mathbf{B}$, then we would refer to this as an *ideal gauge fixing*.

Ideal gauge fixings are then invertible mappings from the space of gauge-fixed A^μ to the space of all possible $\mathbf{E}$ and $\mathbf{B}$ fields; i.e., the mapping is surjective (onto) and injective (one-to-one) and so it is invertible (bijective). A nonideal gauge fixing is a *partial gauge fixing* and means that there exists more than one gauge-fixed A^μ for a given pair of $\mathbf{E}$ and $\mathbf{B}$; i.e., the mapping is surjective (onto) but not injective (not one-to-one). Ideal and partial gauge fixing can be thought of as the mappings

$$\{A^\mu\}_{\text{all}} \xrightarrow{\text{partial}} \{A^\mu\}_{\text{partial}} \xrightarrow{\text{many:1}} \{\mathbf{E}, \mathbf{B}\},$$
$$\{A^\mu\}_{\text{all}} \xrightarrow{\text{ideal}} \{A^\mu\}_{\text{ideal}} \xleftarrow{\text{invertible}} \{\mathbf{E}, \mathbf{B}\}, \tag{2.7.77}$$

where, e.g., $\{A^\mu\}_{\text{all}}$ means the set of all possible gauge fields. Finding ideal gauge fixings for gauge field theories is a highly nontrivial task as we will later discuss. It is fortunate that for almost all applications where we need to fix the gauge we will find that partial gauge fixing is sufficient for our needs.

Lorenz gauge: In electromagnetism the *Lorenz gauge*,[3] also known as the *Lorenz gauge condition*, is the requirement that

$$\partial_\mu A^\mu = 0. \tag{2.7.78}$$

---

[3] The Lorenz gauge is named after Ludvig Lorenz. Historically, the name has often been confused with that of Hendrik Lorentz, after whom "Lorentz covariance" and the "Lorentz group" are named. The expression "Lorentz gauge" has therefore also been in common use for some time.

The Lorenz gauge is clearly Lorentz invariant; i.e., if A^μ is in Lorenz gauge in one inertial frame, then it is in Lorenz gauge in every inertial frame. In order to demonstrate that the Lorenz gauge condition is a gauge fixing we need to show that for any A^μ there always exists at least one ω such that the gauge-transformed A'^μ of Eq. (2.7.74) satisfies the Lorenz gauge condition, i.e., such that

$$0 = \partial_\mu A'^\mu(x) = \partial_\mu A^\mu(x) + \partial_\mu \partial^\mu \omega(x). \tag{2.7.79}$$

Define the scalar function f corresponding to the non-gauge-fixed field A^μ as

$$f(x) \equiv \partial_\mu A^\mu(x). \tag{2.7.80}$$

We see from Eq. (2.7.79) that any ω that solves the Minkowski-space version of Poisson's equation (see below)

$$\Box \omega(x) = \partial_\mu \partial^\mu \omega(x) = -f(x) \tag{2.7.81}$$

will give a gauge transformation that has A'^μ also satisfy the Lorenz gauge condition. Eq. (2.7.81) always has solutions $\omega(x)$. Let the scalar function ω_0 be any function that satisfies the Minkowski-space version of Laplace's equation

$$\Box \omega_0(x) = \partial_\mu \partial^\mu \omega_0(x) = 0. \tag{2.7.82}$$

If ω satisfies the Minkowski-space version of Poisson's equation then so will any

$$\omega'(x) \equiv \omega(x) + \omega_0(x). \tag{2.7.83}$$

We have shown that we can always find a suitable ω so that $\partial_\mu A'^\mu = 0$, which means that the Lorenz gauge condition is a valid gauge fixing. However, the gauge transformation specified by such an ω is not unique and so the Lorenz gauge is only a partial gauge fixing. The degree of freedom associated with the ambiguity of the Lorenz gauge, ω_0, is itself a solution of a massless scalar wave equation. However, as we will discuss in Sec. 6.4.4 it remains an acceptable gauge-fixing for the purposes of quantizing the electromagnetic field.

In Lorenz gauge

$$\partial_\mu F^{\mu\nu} = \partial_\mu c\left[\partial^\mu A^\nu - \partial^\nu A^\mu\right] = c\,\partial_\mu \partial^\mu A^\nu. \tag{2.7.84}$$

The second of the two Maxwell's equations in Eq. (2.7.23) can be written as

$$\Box A^\nu = \partial_\mu \partial^\mu A^\nu = (1/c^2 \epsilon_0) j^\nu = \mu_0 j^\nu. \tag{2.7.85}$$

Recall that the first of the two covariant Maxwell's equations in Eq. (2.7.23), $\partial_\mu \tilde{F}^{\mu\nu} = 0$, is a restatement of the Bianchi identity that in turn is an automatic consequence of the definition of $F^{\mu\nu}$. Thus Eq. (2.7.85) contains all of the dynamics when we work in Lorenz gauge. In the absence of any sources we have the equation for electromagnetic radiation in Lorenz gauge

$$\Box A^\nu = \partial_\mu \partial^\mu A^\nu = 0, \tag{2.7.86}$$

which is the wave equation for a massless four-vector field A^ν. We see that each of the four components of A^ν satisfies a massless wave equation.

Coulomb gauge: The *Coulomb gauge*, also known as the transverse gauge, is

$$\boldsymbol{\nabla} \cdot \mathbf{A}(x) = 0, \tag{2.7.87}$$

which is obviously not a Lorentz invariant gauge condition. If true in one inertial frame, then it will be true in any other inertial frame related by a pure rotation, by a parity transformation or by a time reversal transformation. However, it will not remain true in any inertial frame related by a boost. We first need to show that this condition can always be imposed by showing that we can always find one or more gauge transformations that lead to this condition being satisfied. We want an ω such that $A'^\mu(x) = A^\mu(x) + \partial^\mu \omega(x)$ satisfies $\boldsymbol{\nabla} \cdot \mathbf{A}'(x) = 0$. This requires that

$$0 = \boldsymbol{\nabla} \cdot \mathbf{A}' = \boldsymbol{\nabla} \cdot (\mathbf{A} - \boldsymbol{\nabla}\omega) = \boldsymbol{\nabla} \cdot \mathbf{A} - \boldsymbol{\nabla}^2 \omega. \tag{2.7.88}$$

We proceed similarly to the Lorenz gauge case and define the function g corresponding to the three-vector field $\mathbf{A}$ as

$$g(x) \equiv \boldsymbol{\nabla} \cdot \mathbf{A}. \tag{2.7.89}$$

We see from Eq. (2.7.88) that any ω that solves Poisson's equation

$$\boldsymbol{\nabla}^2 \omega(x) = g(x) \tag{2.7.90}$$

will give a gauge transformation that will give an $\mathbf{A}'$ that satisfies the Coulomb gauge condition. As we will soon discuss, Poisson's equation always has solutions. So there is always a gauge transformation that satisfies the Coulomb gauge condition.

Using Coulomb gauge, $\boldsymbol{\nabla} \cdot \mathbf{A} = 0$, in Gauss' law and Ampère's law gives

$$\boldsymbol{\nabla} \cdot \mathbf{E} = \boldsymbol{\nabla} \cdot \left(-\boldsymbol{\nabla}\Phi - \frac{\partial \mathbf{A}}{\partial t}\right) = -\boldsymbol{\nabla}^2 \Phi = \frac{\rho}{\epsilon_0}, \tag{2.7.91}$$

$$-\boldsymbol{\nabla}^2 \mathbf{A} + \frac{1}{c^2}\frac{\partial^2 \mathbf{A}}{\partial t^2} = \Box \mathbf{A} = \mu_0 \mathbf{j} - \frac{1}{c^2}\boldsymbol{\nabla}\frac{\partial \Phi}{\partial t}, \tag{2.7.92}$$

where we have used $\boldsymbol{\nabla} \times (\boldsymbol{\nabla} \times \mathbf{A}) = \boldsymbol{\nabla}(\boldsymbol{\nabla} \cdot \mathbf{A}) - \boldsymbol{\nabla}^2 \mathbf{A} = -\boldsymbol{\nabla}^2 \mathbf{A}$. We recognize that the first result is again an example of Poisson's equation

$$\boldsymbol{\nabla}^2 \Phi(x) = -\frac{\rho(x)}{\epsilon_0}. \tag{2.7.93}$$

As we show below in Eq. (2.7.103) one solution of Poisson's equation is given by

$$\Phi(x) = \frac{1}{4\pi\epsilon_0} \int \frac{\rho(t, \mathbf{x}')}{|\mathbf{x} - \mathbf{x}'|} d^3 x'. \tag{2.7.94}$$

This solution is not unique, since we can add to this any solution of Laplace's equation, $\boldsymbol{\nabla}^2 \Phi = 0$, and Poisson's equation will still be satisfied. As we prove below, the solution Φ is effectively unique in some volume V if we choose either (i) Dirichlet boundary conditions, which specify Φ everywhere on the surface S of the volume and lead to a unique Φ, or (ii) Neumann boundary conditions, which specify the normal derivative $\boldsymbol{\nabla}\Phi \cdot \mathbf{n}_s$ everywhere on the surface with $\mathbf{n}_s$ being the outward-pointing normal to the surface and which determine Φ up to an unimportant additive constant. If the charge density $\rho(x)$ vanishes at spatial infinity faster than $1/|\mathbf{x}|$, then we can choose the Dirichlet boundary condition $\Phi(x) \to 0$ as $|\mathbf{x}| \to \infty$ such that Eq. (2.7.94) is the only solution of the Poisson equation.

Mathematical digression: Divergence theorem, Laplace's equation, Poisson's equation and uniqueness

(i) Divergence theorem (also known as Gauss' theorem): Let V be some compact (i.e., bounded and closed) three-dimensional volume in $\mathbb{R}^3$ with a piecewise smooth surface S (i.e., made up of some set of differentiable surfaces). Let $\mathbf{F}(\mathbf{x})$ be a vector function that is continuously differentiable (i.e., its first derivative exists and is continuous) everywhere in V, then

$$\int_V \boldsymbol{\nabla} \cdot \mathbf{F} \, d^3x = \int_S \mathbf{F} \cdot d\mathbf{s}, \tag{2.7.95}$$

where the infinitesimal surface element $d\mathbf{s} \equiv \mathbf{n}_s ds$ is defined with $\mathbf{n}_s$ being the outward-oriented unit vector normal to the surface. A number of corollaries follow from the divergence theorem, e.g.,

$$\int_V (\boldsymbol{\nabla} \times \mathbf{F}) \, d^3x = \int_S d\mathbf{s} \times \mathbf{F} = -\int_S \mathbf{F} \times d\mathbf{s}. \tag{2.7.96}$$

This can be proven by letting $\mathbf{c}$ be any constant three-vector and noting that

$$\mathbf{c} \cdot \int_V (\boldsymbol{\nabla} \times \mathbf{F}) \, d^3x = \int_V \boldsymbol{\nabla} \cdot (\mathbf{F} \times \mathbf{c}) \, d^3x = \int_S d\mathbf{s} \cdot (\mathbf{F} \times \mathbf{c}) = \mathbf{c} \cdot \int_S d\mathbf{s} \times \mathbf{F},$$

where we have used $\mathbf{a} \cdot (\mathbf{b} \times \mathbf{c}) = \mathbf{b} \cdot (\mathbf{c} \times \mathbf{a}) = \mathbf{c} \cdot (\mathbf{a} \times \mathbf{b})$. Since the three-vector $\mathbf{c}$ is arbitrary, then Eq. (2.7.96) immediately follows.

(ii) Solution of Laplace's equation and Poisson's equation: Laplace's equation, $\boldsymbol{\nabla}^2 \Phi = 0$, is the homogeneous version of Poisson's equation, $\boldsymbol{\nabla}^2 \Phi = \rho(x)/\epsilon_0$, and so it is sufficient to study Poisson's equation (note that a homogeneous differential equation is one where if Φ is a solution, then so is $c\Phi$ for every constant c). We make the obvious definitions $\mathbf{x} \equiv (x^1, x^2, x^3) = (x, y, z)$ and $r \equiv |\mathbf{x}| = [(x^1)^2 + (x^2)^2 + (x^3)^2]^{1/2} = (x^2 + y^2 + z^2)^{1/2}$ and note that at the point $\mathbf{x}$ the vector $\mathbf{x}/r \equiv \hat{\mathbf{r}}$ is the unit vector pointing in the radial direction from the origin to $\mathbf{x}$. If the function $f(r)$ is differentiable at r, then the following identities follow (see, e.g., Arfken and Weber, 1995):

$$\boldsymbol{\nabla} r = \tfrac{\mathbf{x}}{r} = \hat{\mathbf{r}}, \quad \text{since} \quad \frac{\partial r}{\partial x^i} = \frac{\partial (\mathbf{x}^2)^{1/2}}{\partial x^i} = \tfrac{1}{2}(\mathbf{x}^2)^{-1/2} 2x^i = \tfrac{x^i}{r}; \tag{2.7.97}$$

$$\boldsymbol{\nabla} f(r) = \tfrac{\mathbf{x}}{r} \tfrac{df}{dr} = \hat{\mathbf{r}} \tfrac{df}{dr}, \quad \text{since} \quad \frac{\partial f}{\partial x^i} = \frac{\partial r}{\partial x^i} \frac{df}{dr} = \tfrac{x^i}{r} \tfrac{df}{dr};$$

$$\boldsymbol{\nabla} \cdot (\mathbf{x} f(r)) = 3f(r) + r \tfrac{df}{dr}, \text{ since } \frac{\partial (x^i f)}{\partial x^i} = \frac{\partial x^i}{\partial x^i} f(r) + x^i \frac{\partial f}{\partial x^i} = 3f(r) + r \tfrac{df}{dr};$$

$$\boldsymbol{\nabla}^2 f(r) = \boldsymbol{\nabla} \cdot \boldsymbol{\nabla} f(r) = \boldsymbol{\nabla} \cdot \left(\tfrac{\mathbf{x}}{r} \tfrac{df}{dr} \right) = 3 \left(\tfrac{1}{r} \tfrac{df}{dr} \right) + r \tfrac{d}{dr} \left(\tfrac{1}{r} \tfrac{df}{dr} \right) = \tfrac{2}{r} \tfrac{df}{dr} + \tfrac{d^2 f}{dr^2}.$$

Note that, e.g., $\boldsymbol{\nabla} f(r)$ is to be understood as being evaluated at the point $\mathbf{x}$, i.e., $\boldsymbol{\nabla} f(r)|_{\mathbf{x}}$. Any spherically symmetric function has a radial gradient, i.e., $\boldsymbol{\nabla} f(r)|_{\mathbf{x}} = \hat{\mathbf{r}}(df/dr)$. We see that for $r \neq 0$ we have

$$\boldsymbol{\nabla} \tfrac{1}{r} = -\tfrac{\mathbf{x}}{r^3} = -\tfrac{\hat{\mathbf{r}}}{r^2}, \tag{2.7.98}$$

$$\boldsymbol{\nabla} \cdot (\mathbf{x} r^n) = 3r^n + r \tfrac{dr^n}{dr} = (3+n)r^n \quad \text{so} \quad \boldsymbol{\nabla} \cdot \left(\tfrac{\mathbf{x}}{r^3} \right) = \boldsymbol{\nabla} \cdot \left(\tfrac{\hat{\mathbf{r}}}{r^2} \right) = 0,$$

$$\boldsymbol{\nabla}^2 \left(\tfrac{1}{r} \right) = \tfrac{2}{r} \tfrac{d(r^{-1})}{dr} + \tfrac{d^2(r^{-1})}{dr^2} = -\tfrac{2}{r^3} + \tfrac{2}{r^3} = 0.$$

Consider some three-dimensional sphere with its center at $r = 0$ and radius R and with V_s and S_s the volume and surface of the sphere, respectively. Then using the three-dimensional version of the divergence theorem we have

$$\int_{V_s} \boldsymbol{\nabla}^2 \left(\tfrac{1}{r} \right) d^3x = \int_{V_s} \boldsymbol{\nabla} \cdot \boldsymbol{\nabla} \left(\tfrac{1}{r} \right) d^3x = \int_{S_s} \boldsymbol{\nabla} \left(\tfrac{1}{r} \right) \cdot d\mathbf{s} = -\int_{S_s} \tfrac{\hat{\mathbf{r}}}{r^2} \cdot d\mathbf{s}$$
$$= -\int \tfrac{1}{R^2} R^2 d\Omega = -4\pi. \tag{2.7.99}$$

Note that this result is true for any radius $R > 0$. We see from Eq. (2.7.99) that $\boldsymbol{\nabla}^2(1/r) = 0$ everywhere except at $r = 0$ and so for any volume V that does not include the origin the integral vanishes, whereas for any volume that includes the origin the integral will equal -4π, i.e.,

$$\int_V \mathbf{\nabla}^2 \left(\frac{1}{r}\right) d^3 r = \left\{ \begin{array}{ll} -4\pi & \text{if } r = 0 \text{ in } V \\ 0 & \text{if } r = 0 \text{ not in } V \end{array} \right\}. \tag{2.7.100}$$

It therefore follows that

$$\mathbf{\nabla}^2 \left(\tfrac{1}{r}\right) = -4\pi \delta^3(\mathbf{x}). \tag{2.7.101}$$

Let us shift the origin by $\mathbf{x}'$ so that $r = |\mathbf{x} - \mathbf{x}'|$. Holding $\mathbf{x}'$ fixed and varying $\mathbf{x}$ gives $d(\mathbf{x} - \mathbf{x}') = d\mathbf{x}$ and so $\mathbf{\nabla}_{(\mathbf{x}-\mathbf{x}')} = \mathbf{\nabla}_{\mathbf{x}}$. Note that the subscript on the gradient is used to indicate what we are differentiating with respect to when there is a possibility of ambiguity. Then we can write

$$\mathbf{\nabla}_{\mathbf{x}}^2 \tfrac{1}{|\mathbf{x}-\mathbf{x}'|} = -4\pi \delta^3(\mathbf{x} - \mathbf{x}'). \tag{2.7.102}$$

Hence from Eq. (2.7.94) we have

$$\mathbf{\nabla}_{\mathbf{x}}^2 \Phi(x) = \frac{1}{4\pi\epsilon_0} \int \mathbf{\nabla}_{\mathbf{x}}^2 \frac{\rho(t,\mathbf{x}')}{|\mathbf{x}-\mathbf{x}'|} d^3 x' = -\frac{\rho(x)}{\epsilon_0}, \tag{2.7.103}$$

which shows that Eq. (2.7.94) is a solution of Poisson's equation in Eq. (2.7.93).

(iii) Uniqueness and boundary conditions: Finally, we consider the uniqueness of solutions of Poisson's equation. It is clear that any solution of Laplace's equation can be added to the solution of Eq. (2.7.94) and Poisson's equation will still be satisfied. Thus there is an infinite set of solutions in principle. However, with appropriate choices of boundary conditions the solution is unique. Let $\Phi_1(x)$ and $\Phi_2(x)$ be two solutions of Poisson's equation, i.e., $\mathbf{\nabla}^2 \Phi_1(x) = \mathbf{\nabla}^2 \Phi_2(x) = -\rho(x)/\epsilon_0$. Defining

$$\Sigma(x) \equiv \Phi_1(x) - \Phi_2(x) \tag{2.7.104}$$

then $\Sigma(x)$ everywhere satisfies Laplace's equation $\mathbf{\nabla}^2 \Sigma(x) = 0$ and so

$$\mathbf{\nabla} \cdot (\Sigma \mathbf{\nabla} \Sigma) = (\mathbf{\nabla}\Sigma)^2 + \Sigma \mathbf{\nabla}^2 \Sigma = (\mathbf{\nabla}\Sigma)^2 \geq 0. \tag{2.7.105}$$

Integrating over some region with volume V and surface S gives

$$0 \leq \int_V (\mathbf{\nabla}\Sigma)^2 d^3 x = \int_V \mathbf{\nabla} \cdot (\Sigma \mathbf{\nabla}\Sigma) d^3 x = \int_S \Sigma \mathbf{\nabla}\Sigma \cdot ds. \tag{2.7.106}$$

Therefore, if the surface integral on the right-hand side vanishes, we must have $\Sigma = 0$ or $\mathbf{\nabla}\Sigma = 0$ everywhere on the surface. *Dirichlet boundary conditions:* The surface integral will vanish if we have $\Sigma = \Phi_1 - \Phi_2 = 0$ everywhere on the surface S; i.e., if we specify Φ on the surface in a manner consistent with a solution of Poisson's equation, then Φ is uniquely determined everywhere. *Neumann boundary conditions:* Define the outward normal to the surface as $\mathbf{n}_s$, then $d\mathbf{s} \equiv \mathbf{n}_s ds$. The surface integral will vanish if we have $\mathbf{\nabla}\Sigma \cdot \mathbf{n}_s = \mathbf{\nabla}\Phi_1 \cdot \mathbf{n}s - \mathbf{\nabla}\Phi_2 \cdot \mathbf{n}_s = 0$ everywhere on the surface S; i.e., if we specify $\mathbf{\nabla}\Phi \cdot \mathbf{n}_s$ on the surface in a manner consistent with a solution of Poisson's equation, then $\mathbf{\nabla}\Phi$ is uniquely determined everywhere and then Φ is uniquely determined up to some irrelevant arbitrary additive constant.

Recall that we obtain Laplace's equation by choosing $\rho = 0$. All of the above arguments will still apply; i.e., Laplace's equation will have a unique solution if we specify Dirichlet boundary conditions on the surface and it will have a solution unique up to an irrelevant additive constant if we specify Neumann boundary conditions on the surface.

We can use Helmholtz's theorem (discussed below) to decompose the three-vector current density into a longitudinal and a transverse part

$$\mathbf{j}(x) = \mathbf{j}_\ell(x) + \mathbf{j}_t(x), \qquad \text{where} \qquad \mathbf{\nabla} \times \mathbf{j}_\ell(x) = \mathbf{\nabla} \cdot \mathbf{j}_t(x) = 0. \tag{2.7.107}$$

Provided that both $\rho(x)$ and $\mathbf{j}(x) \to 0$ faster than $1/|\mathbf{x}|$ as $|\mathbf{x}| \to \infty$, then we can use Helmholtz's theorem to write

$$\mathbf{j}_t(x) = \boldsymbol{\nabla}_{\mathbf{x}} \times \left(\frac{1}{4\pi} \int \frac{\boldsymbol{\nabla}_{\mathbf{x}'} \times \mathbf{j}(x')}{|\mathbf{x} - \mathbf{x}'|} \, d^3 \mathbf{x}' \right), \tag{2.7.108}$$

$$\mathbf{j}_\ell(x) = -\boldsymbol{\nabla}_{\mathbf{x}} \left(\frac{1}{4\pi} \int \frac{\boldsymbol{\nabla}_{\mathbf{x}'} \cdot \mathbf{j}(x')}{|\mathbf{x} - \mathbf{x}'|} \, d^3 \mathbf{x}' \right) = \boldsymbol{\nabla}_{\mathbf{x}} \frac{\partial}{\partial t} \left(\frac{1}{4\pi} \int \frac{\rho(t, \mathbf{x}')}{|\mathbf{x} - \mathbf{x}'|} \, d^3 \mathbf{x}' \right)$$

$$= \epsilon_0 \boldsymbol{\nabla} \frac{\partial \Phi(x)}{\partial t} = \frac{1}{\mu_0 c^2} \boldsymbol{\nabla} \frac{\partial \Phi(x)}{\partial t}, \tag{2.7.109}$$

where the volume integrals are over all of three-dimensional coordinate space, where we have used current conservation $\partial_\mu j^\mu = 0$ and where we have used the solution of Poisson's equation, Eq. (2.7.94).

The above result for $\mathbf{j}_\ell$ can be written as

$$\frac{1}{c^2} \boldsymbol{\nabla} \frac{\partial \Phi(x)}{\partial t} = \mu_0 \, \mathbf{j}_\ell(x) \tag{2.7.110}$$

and returning to Eq. (2.7.92) we see that we can write this as

$$-\boldsymbol{\nabla}^2 \mathbf{A} + \frac{1}{c^2} \frac{\partial^2 \mathbf{A}}{\partial t^2} = \mu_0 \mathbf{j} - \frac{1}{c^2} \boldsymbol{\nabla} \frac{\partial \Phi}{\partial t} = \mu_0 \, \mathbf{j}_t. \tag{2.7.111}$$

Then in Coulomb gauge: (i) the scalar potential Φ is determined entirely by the charge distribution as shown by Eq. (2.7.94), and (ii) the three-vector potential $\mathbf{A}$ is determined entirely by the transverse part of the current and satisfies a relativistic wave equation with $\mu_0 \mathbf{j}_t(x)$ as the source term. It is then clear why the Coulomb gauge is sometimes referred to as the *transverse gauge*.

Since Coulomb gauge is the condition that $\boldsymbol{\nabla} \cdot \mathbf{A} = 0$, then in a Helmholtz decomposition of $\mathbf{A}$ there is only a transverse component, $\mathbf{A} = \mathbf{A}_t$. If $\mathbf{A}$ falls off faster than $1/|\mathbf{x}|$ as $|\mathbf{x}| \to \infty$, then we have the result that

$$\mathbf{A}(x) = \boldsymbol{\nabla}_{\mathbf{x}} \times \left(\frac{1}{4\pi} \int_V \frac{\boldsymbol{\nabla}_{\mathbf{x}'} \times \mathbf{A}(t, \mathbf{x}')}{|\mathbf{x} - \mathbf{x}'|} d^3 x' \right) = \boldsymbol{\nabla}_{\mathbf{x}} \times \left(\frac{1}{4\pi} \int_V \frac{\mathbf{B}(t, \mathbf{x}')}{|\mathbf{x} - \mathbf{x}'|} d^3 x' \right). \tag{2.7.112}$$

We see that in Coulomb gauge if the current density j^μ and magnetic field $\mathbf{B}$ vanish faster than $1/|\mathbf{x}|$ at spatial infinity, then we can always choose vanishing boundary conditions for Φ and $\mathbf{A}$ and they are then completely determined by Eqs. (2.7.94) and (2.7.112), respectively. More generally, in a finite volume the surface terms of Helmholtz's theorem can be retained in Eqs. (2.7.94) and (2.7.112) and then Φ and $\mathbf{A}$ will be completely determined provided that we specify appropriate boundary conditions for them on the surface of that volume.

Mathematical digression: Green's theorem, divergence-curl uniqueness theorem and Helmholtz's theorem: There are several useful and related theorems (Arfken and Weber, 1995) that we briefly discuss below and each involves a compact volume, $V \subset \mathbb{R}^3$, with a piecewise smooth surface S.

(i) Green's theorem: Let $u(\mathbf{x})$ and $v(\mathbf{x})$ be functions that are doubly differentiable everywhere in V. The divergence theorem of Eq. (2.7.95) gives

$$\int_S u \boldsymbol{\nabla} v \cdot d\mathbf{s} = \int_V \boldsymbol{\nabla} \cdot (u \boldsymbol{\nabla} v) \, d^3 x = \int_V u \boldsymbol{\nabla}^2 v \, d^3 x + \int_V \boldsymbol{\nabla} u \cdot \boldsymbol{\nabla} v \, d^3 x. \tag{2.7.113}$$

(ii) Divergence-curl uniqueness theorem: Let $\mathbf{F}$ be a doubly differentiable vector function over a compact volume V. Let the divergence $\boldsymbol{\nabla} \cdot \mathbf{F}$ and the curl $\boldsymbol{\nabla} \times \mathbf{F}$ of the function $\mathbf{F}$ be specified throughout V and let the component normal to the surface, $F_n \equiv \mathbf{F} \cdot \mathbf{n}_s$ with $\mathbf{n}_s$ being the outward normal, also be specified. Then the vector function $\mathbf{F}$ is uniquely specified within V.

[*Proof:* Assume that there are an $\mathbf{F}_1$ and $\mathbf{F}_2$ such that $\boldsymbol{\nabla} \cdot \mathbf{F}_1 = \boldsymbol{\nabla} \cdot \mathbf{F}_2$ and $\boldsymbol{\nabla} \times \mathbf{F}_1 = \boldsymbol{\nabla} \times \mathbf{F}_2$ everywhere in V and that $\mathbf{F}_1 \cdot \mathbf{n}_s = \mathbf{F}_2 \cdot \mathbf{n}_s$ everywhere on the surface. Defining $\mathbf{G} \equiv \mathbf{F}_1 - \mathbf{F}_2$ it follows that everywhere on the surface $G_n \equiv \mathbf{G} \cdot \mathbf{n}_s = 0$ and throughout the volume $\boldsymbol{\nabla} \cdot \mathbf{G} = \boldsymbol{\nabla} \times \mathbf{G} = 0$. Since $\boldsymbol{\nabla} \times \mathbf{G} = 0$ it follows from the Poincaré lemma in Eq. (A.2.23) that there exists a function Φ such that $\mathbf{G} = \boldsymbol{\nabla}\Phi$. Since $\boldsymbol{\nabla} \cdot \mathbf{G} = 0$ then $\boldsymbol{\nabla} \cdot \boldsymbol{\nabla}\Phi = \boldsymbol{\nabla}^2 \Phi = 0$ and so Φ satisfies Laplace's equation. We can now use Green's theorem with $u = v = \Phi$ and the fact that G_n vanishes on the surface to give

$$0 = \int_S \Phi G_n ds = \int_S \Phi \mathbf{G} \cdot \mathbf{n}_s ds = \int_S \Phi \boldsymbol{\nabla}\Phi \cdot d\mathbf{s} = \int_V \boldsymbol{\nabla} \cdot (\Phi \boldsymbol{\nabla}\Phi) \, d^3 x$$
$$= \int_V \Phi \boldsymbol{\nabla}^2 \Phi d^3 x + \int_V (\boldsymbol{\nabla}\Phi)^2 \, d^3 x = \int_V (\boldsymbol{\nabla}\Phi)^2 \, d^3 x = \int_V \mathbf{G}^2 \, d^3 x. \qquad (2.7.114)$$

Since $\mathbf{G}^2 = \mathbf{G} \cdot \mathbf{G} \geq 0$ everywhere in V, then it follows that $\mathbf{G} = 0$ and so $\mathbf{F}_1 = \mathbf{F}_2$ everywhere in V, i.e., $\mathbf{F}$ is uniquely determined.]

Example application: Consider a given magnetic field $B = \boldsymbol{\nabla} \times \mathbf{A}$ in Coulomb gauge, $\boldsymbol{\nabla} \cdot \mathbf{A} = 0$, in some region V and with fixed boundary conditions for $\mathbf{A}$ on the surface S. It then follows from the theorem that $\mathbf{A}$ is uniquely determined everywhere in V. So the Coulomb gauge condition augmented with a set of boundary conditions on $\mathbf{A}$ and on Φ as discussed in Eq. (2.7.94) is an ideal gauge fixing. However, the choice of Coulomb gauge is not a Lorentz invariant constraint and this limits its utility when a relativistic treatment is required.

(iii) Helmholtz's theorem: Let $\mathbf{F}(x)$ be a doubly differentiable vector function over some compact volume $V \subset \mathbb{R}^3$. Then $\mathbf{F}$ can be decomposed into "longitudinal" (also referred to as "irrotational" or "curl-free") and "transverse" (also referred to as "rotational" or "solenoidal" or "divergence-free") components as

$$\mathbf{F}(\mathbf{x}) = \mathbf{F}_\ell(\mathbf{x}) + \mathbf{F}_t(\mathbf{x}) \quad \text{with} \quad \mathbf{F}_\ell(\mathbf{x}) = -\boldsymbol{\nabla}\Phi(\mathbf{x}), \; \mathbf{F}_t(\mathbf{x}) = \boldsymbol{\nabla} \times \mathbf{A}(\mathbf{x}), \qquad (2.7.115)$$

such that $\boldsymbol{\nabla} \times \mathbf{F}_\ell(\mathbf{x}) = \mathbf{0}$ and $\boldsymbol{\nabla} \cdot \mathbf{F}_t(\mathbf{x}) = 0$. Φ and $\mathbf{A}$ are given by

$$\Phi(\mathbf{x}) = \frac{1}{4\pi} \int_V \frac{\boldsymbol{\nabla}_{x'} \cdot \mathbf{F}(\mathbf{x}')}{|\mathbf{x}-\mathbf{x}'|} \, d^3 x' - \frac{1}{4\pi} \int_S \frac{\mathbf{F}(\mathbf{x}')}{|\mathbf{x}-\mathbf{x}'|} \cdot d\mathbf{s}' \qquad (2.7.116)$$
$$\mathbf{A}(\mathbf{x}) = \frac{1}{4\pi} \int_V \frac{\boldsymbol{\nabla}_{x'} \times \mathbf{F}(\mathbf{x}')}{|\mathbf{x}-\mathbf{x}'|} \, d^3 x' + \frac{1}{4\pi} \int_S \frac{\mathbf{F}(\mathbf{x}')}{|\mathbf{x}-\mathbf{x}'|} \times d\mathbf{s}'.$$

This theorem is referred to as *the fundamental theorem of vector calculus.*

Note that in the infinite volume limit, $V \to \mathbb{R}^3$, if $\mathbf{F}$ falls off faster than $1/|\mathbf{x}|$ as $|\mathbf{x}| \to \infty$ then the surface terms on the right-hand side vanish for each of Φ and $\mathbf{A}$. Let $\tilde{\mathbf{F}}(\mathbf{k})$, $\tilde{\Phi}(\mathbf{k})$ and $\tilde{\mathbf{A}}(\mathbf{k})$ denote the three-dimensional Fourier transforms of $\mathbf{F}(\mathbf{x})$, $\Phi(\mathbf{x})$ and $\mathbf{A}(\mathbf{x})$ (see Sec. A.2.3). Then we have

$$\tilde{\mathbf{F}}(\mathbf{k}) = \tilde{\mathbf{F}}_\ell(\mathbf{k}) + \tilde{\mathbf{F}}_t(\mathbf{k}) \quad \text{with} \quad \tilde{\mathbf{F}}_\ell(\mathbf{k}) = -\mathbf{k}\tilde{\Phi}(\mathbf{k}), \; \tilde{\mathbf{F}}_t(\mathbf{k}) = \mathbf{k} \times \tilde{\mathbf{A}}(\mathbf{k}), \qquad (2.7.117)$$

where $\tilde{\mathbf{F}}_\ell(\mathbf{k})$ points in the longitudinal direction, i.e., parallel to $\mathbf{k}$, and $\tilde{\mathbf{F}}_t(\mathbf{k})$ points in the transverse direction, i.e., orthogonal to both $\mathbf{k}$ and $\tilde{\mathbf{A}}(\mathbf{k})$.

[*Proof:* From Eq. (2.7.102) we have $\delta^3(\mathbf{x} - \mathbf{x}') = -\frac{1}{4\pi}\boldsymbol{\nabla}_\mathbf{x}^2 \frac{1}{|\mathbf{x}-\mathbf{x}'|}$ and so

$$\mathbf{F}(\mathbf{x}) = \int_V \mathbf{F}(\mathbf{x}')\delta^3(\mathbf{x} - \mathbf{x}')d^3\mathbf{x}' = -\frac{1}{4\pi}\boldsymbol{\nabla}_\mathbf{x}^2 \int_V \frac{\mathbf{F}(\mathbf{x}')}{|\mathbf{x}-\mathbf{x}'|} \, d^3\mathbf{x}'. \qquad (2.7.118)$$

Using the identity $\boldsymbol{\nabla}^2 \mathbf{v} = \boldsymbol{\nabla} \cdot \boldsymbol{\nabla}\mathbf{v} = \boldsymbol{\nabla}(\boldsymbol{\nabla} \cdot \mathbf{v}) - \boldsymbol{\nabla} \times (\boldsymbol{\nabla} \times \mathbf{v})$ we then find

$$\mathbf{F}(\mathbf{x}) = -\frac{1}{4\pi}\left[\boldsymbol{\nabla}_\mathbf{x}\left(\boldsymbol{\nabla}_\mathbf{x} \cdot \int_V \frac{\mathbf{F}(\mathbf{x}')}{|\mathbf{x}-\mathbf{x}'|}d^3\mathbf{x}'\right) - \boldsymbol{\nabla}_\mathbf{x} \times \left(\boldsymbol{\nabla}_\mathbf{x} \times \int_V \frac{\mathbf{F}(\mathbf{x}')}{|\mathbf{x}-\mathbf{x}'|}d^3\mathbf{x}'\right)\right]$$

$$= -\frac{1}{4\pi}\left[\boldsymbol{\nabla}_{\mathbf{x}}\left(\int_V \mathbf{F}(\mathbf{x}')\cdot\boldsymbol{\nabla}_{\mathbf{x}}\frac{1}{|\mathbf{x}-\mathbf{x}'|}d^3x'\right) + \boldsymbol{\nabla}_{\mathbf{x}}\times\left(\int_V \mathbf{F}(\mathbf{x}')\times\boldsymbol{\nabla}_{\mathbf{x}}\frac{1}{|\mathbf{x}-\mathbf{x}'|}d^3x'\right)\right]$$

$$= -\frac{1}{4\pi}\left[-\boldsymbol{\nabla}_{\mathbf{x}}\int_V \mathbf{F}(\mathbf{x}')\cdot\boldsymbol{\nabla}_{\mathbf{x}'}\frac{1}{|\mathbf{x}-\mathbf{x}'|}d^3x' - \boldsymbol{\nabla}_{\mathbf{x}}\times\int_V \mathbf{F}(\mathbf{x}')\times\boldsymbol{\nabla}_{\mathbf{x}'}\frac{1}{|\mathbf{x}-\mathbf{x}'|}d^3x'\right],$$

where we have used $\boldsymbol{\nabla}_{\mathbf{x}}(1/|\mathbf{x}-\mathbf{x}'|) = -\boldsymbol{\nabla}_{\mathbf{x}'}(1/|\mathbf{x}-\mathbf{x}'|)$. Using the derivative of a product formula we have the identities

$$\boldsymbol{\nabla}\cdot[f(\mathbf{x})\mathbf{v}(\mathbf{x})] = f(\mathbf{x})\boldsymbol{\nabla}\cdot\mathbf{v}(\mathbf{x}) + \mathbf{v}(\mathbf{x})\cdot\boldsymbol{\nabla}f(\mathbf{x}),$$

$$\boldsymbol{\nabla}\times[f(\mathbf{x})\mathbf{v}(\mathbf{x})] = f(\mathbf{x})\boldsymbol{\nabla}\times\mathbf{v}(\mathbf{x}) - \mathbf{v}(\mathbf{x})\times\boldsymbol{\nabla}f(\mathbf{x}).$$

Identifying $\mathbf{v}(\mathbf{x}')$ with $\mathbf{F}(\mathbf{x}')$ and $f(\mathbf{x}')$ with $1/|\mathbf{x}-\mathbf{x}'|$ we can then write

$$\mathbf{F}(\mathbf{x}) = -\frac{1}{4\pi}\left[-\boldsymbol{\nabla}_{\mathbf{x}}\left(-\int_V \frac{\boldsymbol{\nabla}_{\mathbf{x}'}\cdot\mathbf{F}(\mathbf{x}')}{|\mathbf{x}-\mathbf{x}'|}d^3x' + \int_V \boldsymbol{\nabla}_{\mathbf{x}'}\cdot\frac{\mathbf{F}(\mathbf{x}')}{|\mathbf{x}-\mathbf{x}'|}d^3x'\right)\right. \tag{2.7.119}$$

$$\left. -\boldsymbol{\nabla}_{\mathbf{x}}\times\left(\int_V \frac{\boldsymbol{\nabla}_{\mathbf{x}'}\times\mathbf{F}(\mathbf{x}')}{|\mathbf{x}-\mathbf{x}'|}d^3x' - \int_V \boldsymbol{\nabla}_{\mathbf{x}'}\times\frac{\mathbf{F}(\mathbf{x}')}{|\mathbf{x}-\mathbf{x}'|}d^3x'\right)\right]$$

$$= -\boldsymbol{\nabla}_{\mathbf{x}}\left(\frac{1}{4\pi}\int_V \frac{\boldsymbol{\nabla}_{\mathbf{x}'}\cdot\mathbf{F}(\mathbf{x}')}{|\mathbf{x}-\mathbf{x}'|}d^3x' - \frac{1}{4\pi}\int_S \frac{\mathbf{F}(\mathbf{x}')}{|\mathbf{x}-\mathbf{x}'|}\cdot d\mathbf{s}'\right)$$

$$+ \boldsymbol{\nabla}_{\mathbf{x}}\times\left(\frac{1}{4\pi}\int_V \frac{\boldsymbol{\nabla}_{\mathbf{x}'}\times\mathbf{F}(\mathbf{x}')}{|\mathbf{x}-\mathbf{x}'|}d^3x' + \frac{1}{4\pi}\int_S \frac{\mathbf{F}(\mathbf{x}')}{|\mathbf{x}-\mathbf{x}'|}\times d\mathbf{s}'\right) \equiv -\boldsymbol{\nabla}_{\mathbf{x}}\Phi(\mathbf{x}) + \boldsymbol{\nabla}_{\mathbf{x}}\times\mathbf{A}(\mathbf{x}),$$

where we have used the divergence theorem, Eq. (2.7.95) and one of its corollaries, Eq. (2.7.96).]

Summary of Coulomb gauge: We conclude that the Coulomb gauge condition $\boldsymbol{\nabla}\cdot\mathbf{A} = 0$ is an ideal gauge fixing for electromagnetism in the presence of an external current density, provided that we choose appropriate spatial boundary conditions for Φ and $\mathbf{A}$. In an infinite volume, if external charge densities and currents vanish at spatial infinity faster than $1/|\mathbf{x}|$, then these conditions together with the choice of boundary condition that $A^\mu \to 0$ at spatial infinity uniquely determine the gauge field A^μ everywhere. In particular, Gauss' law in Coulomb gauge determines $A^0(x) = \Phi(x)/c$ to have the form given in Eq. (2.7.93). In the absence of an external charge density we have $\rho(x) = j^0(x)/c = 0$ and then the above choice for A^0 becomes $A^0(x) = \Phi(x)/c = 0$. Coulomb gauge is often also referred to as the *radiation gauge* or *transverse gauge*.

Of the four components A^μ of the electromagnetic field only two components remain undetermined in this gauge; i.e., A^0 is fixed by the charge density and we also see that $\boldsymbol{\nabla}\cdot\mathbf{A} = 0$ removes one degree of freedom from $\mathbf{A}$. Since the number of degrees of freedom cannot depend on the choice of gauge, then we must conclude that the electromagnetic field has only two independent field components, i.e., two degrees of freedom at each point in space corresponding to two polarization states.

A disadvantage of Coulomb gauge is the lack of explicit Lorentz covariance in intermediate stages of calculations, which is a disincentive when relativity is an important consideration. Physical observables must be gauge invariant and so the use of a noncovariant gauge cannot lead to loss of invariance for physical observables.

2.8 Analytic Relativistic Mechanics

The purpose of this section is to explore the consequences of attempting to make relativistic extensions of classical mechanics. We consider the possibility of a relativistic extension of the mechanics of a point particle interacting with external fields.

Relativistic mechanics for a free particle: A relativistic extension of classical mechanics must be Lorentz covariant, where a *Lorentz covariant quantity*[4] is one that transforms under a particular representation of the Lorentz group. Nonrelativistically the action is a functional of the path and has no spatial indices; i.e., it is a three-dimensional scalar under rotations. Then the natural extension to the relativistic context is that the action S is a functional of the particle worldline $x^\mu(\tau)$ and that it is a Lorentz scalar; i.e., the relativistic action $S[x]$ is a Lorentz scalar that must reduce to the usual classical action in the nonrelativistic limit.

Consider a particle moving with velocity $\dot{\mathbf{x}}(t)$ at time t with respect to observer $\mathcal{O}$. The proper time τ is the time shown on a clock comoving with the particle. The proper time increment $d\tau$ for the particle is related to the time increment dt in the frame of $\mathcal{O}$ at time t in the usual way using the time-dilation equation

$$dt = \gamma(\dot{\mathbf{x}}(t))d\tau, \tag{2.8.1}$$

where the proper time interval $d\tau$ is related to the interval increment ds by

$$cd\tau = ds = \sqrt{ds^2} = \sqrt{c^2 dt^2 - d\mathbf{x}^2}. \tag{2.8.2}$$

Define the Lagrangian L in terms of the action S in the usual way,

$$S[x] \equiv S[\mathbf{x}] = \int_{t_i}^{t_f} L(\mathbf{x}, \dot{\mathbf{x}}, t)dt = \int_{\tau_i}^{\tau_f} \gamma(\dot{\mathbf{x}})L(\mathbf{x}, \dot{\mathbf{x}}, t)d\tau = \int_{[x]} \gamma(\dot{\mathbf{x}})L(\mathbf{x}, \dot{\mathbf{x}}, t)d\tau. \tag{2.8.3}$$

In Eq. (2.1.43) for the nonrelativistic case we recall that the time integral was along the particular path that $S[\vec{q}]$ was being evaluated for. Equivalently, here it is to be understood that the *time integration is along the timelike worldline* $x^\mu(\tau)$, which begins at the spacetime event $x_i^\mu = (ct_i, \mathbf{x}_i)$ and ends at the spacetime event $x_f^\mu = (ct_f, \mathbf{x}_f)$. There is a one-to-one correspondence between the uniquely defined τ and the time t for every observer and so we can define a worldline as a function of τ or in terms of t in some inertial frame. Since both S and τ are Lorentz scalars, then it follows from Eq. (2.8.3) that γL is also a Lorentz scalar.

The only scalar quantity available for a free particle is its mass m. The Lagrangian has the same units as the rest-mass energy, $[L] = [mc^2] = \mathsf{ML^2T^{-2}}$, and γ is dimensionless and so we consider $\gamma L \propto amc^2$ for some dimensionless constant a. In the usual way we write $\boldsymbol{v} = \dot{\mathbf{x}}$ and $v = |\boldsymbol{v}|$. This leads to

$$L = \frac{amc^2}{\gamma(v)} = amc^2\sqrt{1 - \frac{v^2}{c^2}} = amc^2\left(1 - \frac{1}{2}\frac{v^2}{c^2} - \frac{1}{8}\frac{v^4}{c^4} + \cdots\right). \tag{2.8.4}$$

Choosing $a = -1$, which gives the correct nonrelativistic limit,

$$L = -mc^2 + (1/2)mv^2 + \cdots = -mc^2 + T_{\text{nonrel}} + \cdots, \tag{2.8.5}$$

where the irrelevant constant $-mc^2$ on the right-hand side disappears from the Euler-Lagrange equations. So for a free relativistic particle we have

$$L = -\frac{mc^2}{\gamma(v)} = -mc^2\left(1 - \frac{v^2}{c^2}\right)^{1/2} = -mc\sqrt{c^2 - v^2}, \tag{2.8.6}$$

---

[4] The historic use of the terms "covariant four-vector" and "contravariant four-vector" (e.g., V_μ and V^μ, respectively) is potentially confusing, in that both transform in a given representation of the Lorentz group and so both are "Lorentz covariant quantities."

$$S[x] = \int_{\tau_i}^{\tau_f} \gamma L d\tau = -mc^2 \int_{[x]} \frac{1}{\gamma(v)} dt = -mc^2 \int_{[x]} d\tau \equiv -mc^2 \Delta\tau[x]. \tag{2.8.7}$$

$\Delta\tau[x]$ is the integrated proper time and is a functional of the worldline $x^\mu(\tau)$. Since we are considering massive particles we only need to consider timelike worldlines. As illustrated in Fig. 1.2 we can always boost to a frame where the initial and final events on the worldline, $x_i^\mu = (ct_i, \mathbf{x}_i)$ and $x_f^\mu = (ct_f, \mathbf{x}_f)$, satisfy $\mathbf{x}_i = \mathbf{x}_f$. In that frame, the worldline corresponding to a particle at rest has $\gamma(v) = \gamma(0) = 1$ at all points along the worldline. Any other worldline in that frame with the same end points will involve $v \neq 0$ and $\gamma(v) > 1$ along the worldline. The maximum value of the proper time interval for a worldline in that frame is given by $\Delta\tau[x] = t_f - t_i$, which corresponds to the worldline that minimizes $S[x]$. This corresponds to the worldline with the minimum integrated spatial length along the worldline in this frame, i.e., zero. In any other inertial frame this corresponds to $S[x]$ being minimized when the particle moves with a constant velocity, i.e., when the worldline is straight. For a relativistic free particle the relativistic extension of Hamilton's principle of stationary action leads to the correct physical behavior,

$$\frac{\delta S[x]}{\delta x^\mu}[\tau] = 0. \tag{2.8.8}$$

Consider the variation of the worldline as seen by any inertial observer $\mathcal{O}$. He can describe the worldline in his frame by the function $\mathbf{x}(t)$ for $t_i < t < t_f$. So the variation of the path is just as it was in the nonrelativistic Lagrangian formalism,

$$\frac{\delta S[x]}{\delta \mathbf{x}(t)} = 0, \tag{2.8.9}$$

where the action is $S[x] = \int_{t_i}^{t_f} L \, dt$ with L given in Eq. (2.8.6). Each of the mathematical steps of Eq. (2.1.51) that took us from Hamilton's principle of stationary action to the Euler-Lagrange equations remain valid and so for $\mathcal{O}$,

$$0 = \frac{\delta S[x]}{\delta x^i(t)} = \frac{\partial L}{\partial x^i} - \frac{d}{dt}\frac{\partial L}{\partial \dot{x}^i}, \tag{2.8.10}$$

which shows that the equivalence of Hamilton's principle and the Euler-Lagrange equations remain true in the relativistic context. Defining the canonical (or generalized or conjugate) momentum in the usual way gives

$$\mathbf{p} \equiv \frac{dL}{d\dot{\mathbf{x}}} = \frac{dL}{d\boldsymbol{v}} = -mc^2 \frac{d}{d\boldsymbol{v}}\left(1 - \frac{v^2}{c^2}\right)^{1/2} = \gamma m\boldsymbol{v}, \tag{2.8.11}$$

which is the relativistic three-momentum. The Euler-Lagrange equations give

$$\frac{dp^i}{dt} = \frac{d}{dt}\frac{\partial L}{\partial \dot{x}^i} = \frac{\partial L}{\partial x^i} = 0 \quad \Rightarrow \quad \frac{d\mathbf{p}}{dt} = 0 \tag{2.8.12}$$

as expected. Consider the energy function (i.e., Jacobi's integral), h, for the free relativistic particle. From Eq. (2.2.4) we have

$$h = \boldsymbol{v} \cdot \frac{\partial L}{\partial \boldsymbol{v}} - L = m\gamma v^2 + \frac{mc^2}{\gamma} = mc^2\gamma\left(\frac{v^2}{c^2} + \frac{1}{\gamma^2}\right) = \gamma mc^2 = E \tag{2.8.13}$$

and so the relativistic free particle Lagrangian corresponds to a "natural system" in the sense that the energy function equals the energy. Since the Euler-Lagrange equations are true, then the proof of Eq. (2.2.5) remains valid and so

$$\frac{dh}{dt} = -\frac{\partial L}{\partial t}. \tag{2.8.14}$$

As in the nonrelativistic case, if the system is natural and L has no explicit time dependence, then $dE/dt = dh/dt = 0$ and energy is conserved.

As discussed in Sec. 2.4.1, in order for the Legendre transform with respect to velocity to be well defined we require L to be convex with respect to velocity. As we show below, the relativistic free-particle Lagrangian is concave up with respect to velocity as it is in the nonrelativistic limit. The Hamiltonian can then be defined in the usual way as

$$H = \mathbf{p} \cdot \boldsymbol{v} - L = \frac{\mathbf{p}^2}{\gamma m} - \frac{(-mc^2)}{\gamma} = \gamma mc^2 = \left(\mathbf{p}^2 c^2 + m^2 c^4\right)^{1/2} = E. \tag{2.8.15}$$

Then $h = H = \gamma mc^2 = E$ and so it follows that

$$(\gamma mc^2)^2 = E^2 = H^2 = \mathbf{p}^2 c^2 + m^2 c^4 \tag{2.8.16}$$

as we expect. Recall from the definition of the relativistic kinetic energy that

$$E = \gamma mc^2 = mc^2 + T \quad \text{with} \quad T \equiv (\gamma - 1)mc^2. \tag{2.8.17}$$

The total energy of the free particle is the rest mass energy plus the kinetic energy.

Proof: We show that the relativistic free-particle Lagrangian L is concave up with respect to velocity. Since $\frac{\partial L}{\partial v^i} = \frac{\partial L}{\partial \dot{x}^i} = p^i = \gamma(v)m\dot{x}^i = \gamma(v)mv^i$ then

$$(M_L)^{ij} = \frac{\partial^2 L}{\partial v^i \partial v^j} = \frac{\partial}{\partial v^i}\left[\gamma(v)mv^j\right] = m\gamma(v)\delta^{ij} + mv^j \frac{\partial \gamma(v)}{\partial v^i}$$

$$= m\gamma(v)^3 \left(\frac{1}{\gamma(v)^2}\delta^{ij} + \frac{v^i v^j}{c^2}\right), \tag{2.8.18}$$

where we have used $\frac{\partial \gamma}{\partial v^i} = \left(-\frac{1}{2}\right)\gamma(v)^3 \left(-2\frac{v^i}{c^2}\right) = \gamma(v)^3 \frac{v^i}{c^2}$. Let $\mathbf{a}$ be an arbitrary nonvanishing three-vector, then

$$\mathbf{a}^T M_L \mathbf{a} = \sum_{i,j=1}^{3} a^i (M_L)^{ij} a^j = m\gamma(v)^3 \left(\frac{1}{\gamma(v)^2}\mathbf{a}^2 + \frac{(\mathbf{a} \cdot \boldsymbol{v})^2}{c^2}\right) > 0. \tag{2.8.19}$$

Then M_L is positive-definite and L is concave up with respect to the velocities.

Relativistic particle in a scalar potential: Up until this point we have only considered the case of a relativistic free particle. We want to generalize to the case of a relativistic particle interacting with external potentials. Since $\gamma L^{\text{free}} = -mc^2$ is a Lorentz scalar, then the simplest extension is to add an interaction with an external Lorentz scalar field $U(x)$ such that $\gamma L = -mc^2 - U(t, \mathbf{x})$, i.e.,

$$L = L^{\text{free}} - \frac{U(t,\mathbf{x})}{\gamma} = -\frac{mc^2}{\gamma} - \frac{U(t,\mathbf{x})}{\gamma} = -\left(1 + \frac{U}{mc^2}\right)\frac{mc^2}{\gamma} = \left(1 + \frac{U}{mc^2}\right)L^{\text{free}}. \tag{2.8.20}$$

This is a relativistic generalization of the nonrelativistic form $L = T - V$. The corresponding action is

$$S[x] = \int_{\tau_i}^{\tau_f} \left[(-mc^2) - U\right] d\tau = \int_{t_i}^{t_f} \left[\frac{(-mc^2)}{\gamma(v)} - \frac{U}{\gamma(v)}\right] dt. \tag{2.8.21}$$

In the nonrelativistic limit we can use Eq. (2.8.5) to recover $L = (1/2)mv^2 - U$ up to the irrelevant additive constant discussed above.

In the nonrelativistic limit we know that Hamilton's principle applies for systems with holonomic constraints and monogenic potentials as defined in Eq. (2.1.32). By the postulates of special relativity we know that this must be true when observed from any inertial frame. In addition, we have shown above that it is true for a relativistic free particle. We will not be imposing any non-holonomic constraints here; in fact, we will only consider the case of a particle interacting with an external field. These considerations lead us to make the only viable assumption, which is that *Hamilton's principle applies to a relativistic particle in the presence of any external Lorentz scalar interactions with an external field that corresponds to monogenic potentials in the nonrelativistic limit.* Since the mathematical steps used in the derivation of the Euler-Lagrange equations from Hamilton's principle in Eq. (2.1.51) remain valid, then *the Euler-Lagrange equations for such relativistic systems again immediately follow.*

Since $L = (1+(U/mc^2))L^{\text{free}}$ and since U is independent of velocity, then L remains concave up with respect to velocity and so the Legendre transform to the Hamiltonian is well defined. Therefore, provided that the full Lagrangian L including interactions with the potential remains concave up with respect to velocity, then *Hamilton's equations must also be valid.* Note that these consequences follow directly from the assumption of Hamilton's principle.

Since $\gamma = \gamma(\dot{\mathbf{x}})$ we observe that

$$\frac{\partial L}{\partial \mathbf{x}} = -\frac{1}{\gamma}\frac{\partial U}{\partial \mathbf{x}} \quad \text{and} \quad \frac{\partial L}{\partial \dot{\mathbf{x}}} = -\left(mc^2 + U\right)\frac{\partial}{\partial \dot{\mathbf{x}}}\frac{1}{\gamma} = \left(mc^2 + U\right)\frac{\gamma\dot{\mathbf{x}}}{c^2}. \tag{2.8.22}$$

Denote the *canonical momentum* as $\boldsymbol{\pi}$ and the purely *mechanical momentum* as $\mathbf{p} = \gamma m\boldsymbol{v}$. The canonical momentum is

$$\boldsymbol{\pi} = \frac{\partial L}{\partial \dot{\mathbf{x}}} = \frac{\partial L}{\partial \boldsymbol{v}} = \left(1 + \frac{U}{mc^2}\right)\gamma m\boldsymbol{v} = \left(1 + \frac{U}{mc^2}\right)\mathbf{p}. \tag{2.8.23}$$

Using this, the Legendre transformation to construct the Hamiltonian is

$$
\begin{aligned}
H = \boldsymbol{\pi}\cdot\boldsymbol{v} - L &= \left[1 + (U/mc^2)\right]\left(\frac{\mathbf{p}^2}{\gamma m} + \frac{mc^2}{\gamma}\right) = \left[1 + (U/mc^2)\right]\gamma mc^2\left(\frac{v^2}{c^2} + \frac{1}{\gamma^2}\right) \\
&= \left[1 + (U/mc^2)\right]\gamma mc^2 = \left[1 + (U/mc^2)\right]p^0 c = \left[1 + (U/mc^2)\right]\left(\mathbf{p}^2 c^2 + m^2 c^4\right)^{1/2} \\
&= \left(\boldsymbol{\pi}^2 c^2 + m^2 c^4\left[1 + (U/mc^2)\right]^2\right)^{1/2},
\end{aligned}
\tag{2.8.24}
$$

where $E^{\text{mech}} \equiv p^0 c = \gamma mc^2 = \left(\mathbf{p}^2 c^2 + m^2 c^4\right)^{1/2}$ is the purely *mechanical energy* carried by the particle. The four-vector form of the canonical momentum is then

$$\pi^\mu = (H/c, \boldsymbol{\pi}) = [1 + (U/mc^2)]p^\mu. \tag{2.8.25}$$

The Euler-Lagrange equations can be written down using Eq. (2.8.22) as

$$\frac{d}{dt}\left[(mc^2 + U)\frac{\gamma\dot{\mathbf{x}}}{c^2}\right] = \frac{d}{dt}\left[(mc^2 + U)\frac{\gamma\dot{\mathbf{x}}}{c^2}\right] = -\frac{1}{\gamma}\frac{\partial U}{\partial \mathbf{x}}. \tag{2.8.26}$$

Using $d/d\tau = \gamma d/dt$ and $p_\mu = mV_\mu = m\gamma(c, -\dot{\mathbf{x}})$ the equations of motion are

$$\frac{d}{d\tau}\left[\left(1 + \frac{U}{mc^2}\right)p_i\right] = \frac{d\pi_i}{d\tau} = -\frac{d\pi^i}{d\tau} = \partial_i U. \tag{2.8.27}$$

These equations are true in the frame of some inertial observer $\mathcal{O}$, but since we constructed γL to be a Lorentz scalar, then we could have chosen to work in the frame of any other inertial observer $\mathcal{O}'$ and we would have obtained the same equations written in terms of the spacetime coordinates of $\mathcal{O}'$, i.e., $x'^\mu = (ct', \mathbf{x}')$. Since boosts mix spatial and temporal components, then as we argued in Sec. 2.6 the only way that a three-vector equation can be true in every inertial frame is if it is the three-vector part of a four-vector equality. Therefore, the four-vector form of the Euler-Lagrange equations can be written as

$$d\pi_\mu/d\tau = \partial_\mu U \tag{2.8.28}$$

for $\mu = 0, 1, 2, 3$. Since $\pi^0 = H/c$ and since γ has no *explicit* time dependence, $\partial\gamma/\partial t = 0$, we see that the $\mu = 0$ case gives the result

$$\frac{dH}{d\tau} = \frac{\partial U}{\partial t} = \frac{\partial}{\partial \tau}\frac{U}{\gamma} = \frac{1}{\gamma}\frac{\partial}{\partial \tau}\frac{U}{\gamma} = -\frac{\partial L}{\partial \tau}, \tag{2.8.29}$$

which can be written as

$$\frac{dH}{dt} = \frac{\partial}{\partial t}\frac{U}{\gamma} = -\frac{\partial L}{\partial t}. \tag{2.8.30}$$

We recognize from Eq. (2.4.9) that this is one of Hamilton's equations. Note that $c\partial_0 = d/dt$ is the full time derivative and not the partial time derivative acting on explicit time dependence, $\partial/\partial t$.

Let us consider the other six Hamilton's equations in Eq. (2.4.9). Of course we must use the functional form $H = H(\mathbf{x}, \boldsymbol{\pi})$ for this purpose. First consider

$$\dot{x}^i = \frac{\partial H}{\partial \pi^i} = (1/2)\left(\boldsymbol{\pi}^2 c^2 + m^2 c^4\left[1 + (U/mc^2)\right]^2\right)^{-1/2}(2\pi^i c^2)$$

$$= \pi^i\left(\left[\boldsymbol{\pi}^2/c^2\right] + m^2\left[1 + (U/mc^2)\right]^2\right)^{-1/2}, \tag{2.8.31}$$

which has the solution $\boldsymbol{\pi} = [1 + (U/mc^2)]\gamma m\dot{\mathbf{x}} = [1 + (U/mc^2)]\mathbf{p}$ as can be verified by direct substitution into the last line of the above equation

$$\frac{[1 + (U/mc^2)]\gamma m\dot{\mathbf{x}}}{\left([1 + (U/mc^2)]^2\gamma^2 m^2(v^2/c^2) + m^2\left[1 + (U/mc^2)\right]^2\right)^{1/2}} = \sqrt{(v^2/c^2) + (1/\gamma^2)}\dot{\mathbf{x}} = \dot{\mathbf{x}}.$$

So the three Hamilton's equations in Eq. (2.8.31) have simply recovered the relation between the velocity and the canonical momentum given in Eq. (2.8.23). Evaluating the remaining three of Hamilton's equations gives

$$-\frac{d\pi_i}{dt} = \frac{d\pi^i}{dt} = \dot{\pi}^i = -\frac{\partial H}{\partial x^i} = -\frac{\partial}{\partial x^i}\left(\boldsymbol{\pi}^2 c^2 + m^2 c^4\left[1 + (U/mc^2)\right]^2\right)^{1/2} \tag{2.8.32}$$

$$= -\left(\boldsymbol{\pi}^2 c^2 + m^2 c^4\left[1 + (U/mc^2)\right]^2\right)^{-1/2} mc^2\left[1 + (U/mc^2)\right]\partial_i U$$

$$= -\left(\left[1 + (U/mc^2)\right]\gamma mc^2\right)^{-1} mc^2\left[1 + (U/mc^2)\right]\partial_i U = -(1/\gamma)\partial_i U,$$

which we recognize as $d\pi_i/d\tau = \partial_i U$. They have reproduced the three-vector part of the equations of motion, i.e., Eq. (2.8.27).

In the nonrelativistic ($\gamma \to 1$) and weak-field limit ($U \ll mc^2$) and using the expansion of γ in Eq. (2.6.6) we have for the Hamiltonian

$$H = \left[1 + (U/mc^2)\right]\gamma mc^2 \to H^{\text{nonrel}} = mc^2 + \frac{1}{2}mv^2 + U \tag{2.8.33}$$

and the four-vector form of the equations of motion in Eq. (2.8.28) becomes

$$\mu = 0: \quad \frac{dH}{d\tau} = \frac{\partial U}{\partial t} \quad \to \quad \frac{dH^{\text{nonrel}}}{dt} = \frac{\partial U}{\partial t},$$

$$\mu = 1, 2, 3: \quad \frac{d\boldsymbol{\pi}}{d\tau} = -\boldsymbol{\nabla}U \quad \to \quad \frac{d(m\boldsymbol{v})}{dt} = -\boldsymbol{\nabla}U, \tag{2.8.34}$$

which are the standard nonrelativistic results for a particle moving in a time-dependent potential U. Up to an irrelevant constant mc^2 in the nonrelativistic limit we recover the nonrelativistic Hamiltonian for particle in a potential U.

Relativistic particle in an electromagnetic field: In order to have a relativistic particle interact with an electromagnetic field, we will need to construct a Lorentz scalar potential from it first. We have the four-vector electromagnetic potential $A^\mu(x)$, but we will need another four-vector to contract this with in order to make a Lorentz scalar. One four-vector that we have available for the particle is the four-velocity of the particle, or equivalently its four-momentum $p^\mu = mV^\mu = m(\gamma c, \gamma \boldsymbol{v})$. Thus a simple choice for the relativistic Lagrangian is

$$L = L^{\text{free}} - \frac{qA^\mu V_\mu}{\gamma(v)} = -\frac{mc^2}{\gamma(v)} - \frac{qA^\mu V_\mu}{\gamma(v)} \qquad (2.8.35)$$

for some constant q, which is the electric charge of the particle. The action is

$$S[x] = \int_{\tau_i}^{\tau_f} \left[(-mc^2) - qA^\mu V_\mu \right] d\tau = \int_{t_i}^{t_f} \left[\frac{(-mc^2)}{\gamma(v)} - q\frac{A^\mu V_\mu}{\gamma(v)} \right] dt. \qquad (2.8.36)$$

Consider the gauge transformation, $A^\mu(x) \to A'^\mu(x) = A^\mu(x) + \partial^\mu \omega(x)$, then

$$S[\mathbf{x}] \to S'[\mathbf{x}] = \int_{\tau_i}^{\tau_f} \left[(-mc^2) - q\left(A^\mu + \partial^\mu \omega\right) V_\mu \right] d\tau = S[\mathbf{x}] - q\left[\omega(x_f) - \omega(x_i) \right],$$

where we have used

$$\int_{\tau_i}^{\tau_f} \partial_\mu \omega V^\mu d\tau = \int_{\tau_i}^{\tau_f} \frac{\partial \omega}{\partial x^\mu(\tau)} \frac{dx^\mu(\tau)}{d\tau} d\tau = \int_{\tau_i}^{\tau_f} \frac{d\omega(x(\tau))}{d\tau} d\tau = \left[\omega(x_f) - \omega(x_i) \right].$$

So the action is gauge invariant up to an irrelevant constant. The Lorentz invariant action chosen is the simplest gauge invariant action possible and involves the minimal (i.e., zero) number of spacetime derivatives acting on the electromagnetic field A^μ. For this reason, this choice is often referred to as *minimal coupling* and is the one that reproduces electrodynamics.

Rewrite the above Lagrangian using $V_\mu = \gamma(c, -\boldsymbol{v})$ and $A^\mu = (\Phi/c, \mathbf{A})$ to give

$$L = L^{\text{free}} - V = L^{\text{free}} - q\left(\Phi - \boldsymbol{v} \cdot \mathbf{A}\right) = -\frac{mc^2}{\gamma} - q\left(\Phi - \boldsymbol{v} \cdot \mathbf{A}\right) \qquad (2.8.37)$$

$$= -mc^2 \sqrt{1 - (v/c)^2} - q\left(\Phi - \boldsymbol{v} \cdot \mathbf{A}\right) = \frac{1}{2}m\boldsymbol{v}^2 - q\left(\Phi - \boldsymbol{v} \cdot \mathbf{A}\right) + \left[-mc^2 + \mathcal{O}(v^4/c^4) \right],$$

which is recognized as the nonrelativistic Lagrangian of Eq. (2.1.37) up to an irrelevant additive constant and relativistic corrections to the kinetic energy. So the constant q is indeed the particle's electric charge. Since the only change from Eq. (2.1.37) are the relativistic corrections to the kinetic energy term, then this introduced electromagnetic potential will be monogenic just as we showed earlier in discussing Eq. (2.1.33). Since we earlier used the nonrelativistic Lagrangian of Eq. (2.1.37) to derive the Lorentz force, $\mathbf{F} = q\left[\mathbf{E} + \mathbf{v} \times \mathbf{B}\right]$, then it follows that the nonrelativistic Lorentz force is a direct result of Lorentz invariance and minimal coupling. It seems reasonable to anticipate that the relativistic form of the Lorentz force is also true. We now have what we need to demonstrate this.

The canonical momentum is

$$\boldsymbol{\pi} = \frac{\partial L}{\partial \dot{\mathbf{x}}} = \frac{\partial L}{\partial \boldsymbol{v}} = \gamma m\boldsymbol{v} + q\mathbf{A} = \mathbf{p} + q\mathbf{A}, \qquad (2.8.38)$$

which is the relativistic version of Eq. (2.4.14), where again the electromagnetic field $\mathbf{A}$ contributes to the canonical momentum along with the mechanical momentum $\mathbf{p} = \gamma m\boldsymbol{v}$. Note that this result for the canonical momentum follows directly from the choice of the minimal coupling Lagrangian. Since we have

$$L = L^{\text{free}} - q(\Phi - \boldsymbol{v} \cdot \mathbf{A}) \quad \text{then} \quad (M_L)_{ij} = \frac{\partial^2 L}{\partial v^i \partial v^j} = \frac{\partial^2 L^{\text{free}}}{\partial v^i \partial v^j} > 0 \tag{2.8.39}$$

and the Legendre transformation to the Hamiltonian is also well defined. The Hamiltonian is constructed in the standard way,

$$H = \boldsymbol{\pi} \cdot \boldsymbol{v} - L = (\mathbf{p} + q\mathbf{A}) \cdot \boldsymbol{v} + \frac{mc^2}{\gamma} + q\left(\Phi - \boldsymbol{v} \cdot \mathbf{A}\right)$$

$$= m\gamma v^2 + \frac{mc^2}{\gamma} + q\Phi = \gamma mc^2 + q\Phi = p^0 c + q\Phi$$

$$= \left(\mathbf{p}^2 c^2 + m^2 c^4\right)^{1/2} + q\Phi = \left([\boldsymbol{\pi} - q\mathbf{A}]^2 c^2 + m^2 c^4\right)^{1/2} + q\Phi. \tag{2.8.40}$$

Combining Eqs. (2.8.38) and (2.8.40) gives the four-vector canonical momentum

$$\pi^\mu = (H/c, \boldsymbol{\pi}) = (p^0 + q[\Phi/c], \mathbf{p} + a\mathbf{A}) = p^\mu + qA^\mu. \tag{2.8.41}$$

Note that $p^2 = m^2 c^2$ means that

$$m^2 c^2 = p_\mu p^\mu = (\pi_\mu - qA_\mu)(\pi^\mu - qA^\mu) = (H - q\Phi)^2/c^2 - (\boldsymbol{\pi} - q\mathbf{A})^2, \tag{2.8.42}$$

which is simply a rearrangement of Eq. (2.8.40). Using the (v/c) expansion for γ in Eq. (2.6.6) we can expand the relativistic Hamiltonian as

$$H = \gamma mc^2 + q\Phi = mc^2 + \frac{1}{2}m\boldsymbol{v}^2 + \mathcal{O}(v^4/c^4) + q\Phi \tag{2.8.43}$$

$$= \frac{\mathbf{p}^2}{2m} + q\Phi + \left[mc^2 + \mathcal{O}(v^4/c^4)\right] = \frac{(\boldsymbol{\pi} - q\mathbf{A})^2}{2m} + q\Phi + \left[mc^2 + \mathcal{O}(v^4/c^4)\right],$$

which reduces to the nonrelativistic expression in Eq. (2.4.15) up to the relativistic corrections of $\mathcal{O}(v^4/c^4)$ and an unimportant constant (the rest mass energy mc^2). (Note that in the nonrelativistic discussion we wrote $\mathbf{p}$ rather than $\boldsymbol{\pi}$ for the canonical momentum and $m\dot{\mathbf{x}}$ rather than $\mathbf{p}$ for the mechanical momentum.)

We can construct the Euler-Lagrange equations by noting that

$$\frac{\partial L}{\partial x^i} = -\frac{q}{\gamma} V_\nu \frac{\partial A^\nu}{\partial x^i} = -\frac{q}{\gamma} V^\nu \frac{\partial A_\nu}{\partial x^i},$$

$$\frac{d}{dt}\left(\frac{\partial L}{\partial \dot{x}^i}\right) = \frac{d}{dt}\left(p^i + qA^i\right) = \frac{dp^i}{dt} + q\left(\dot{x}^j \frac{\partial A^i}{\partial x^j} + \frac{\partial A^i}{\partial t}\right) \tag{2.8.44}$$

$$= \frac{1}{\gamma}\frac{dp^i}{d\tau} - q\left(\dot{x}^j \frac{\partial A_i}{\partial x^j} + c\frac{\partial A_i}{\partial x^0}\right) = -\frac{1}{\gamma}\left[\frac{dp_i}{d\tau} + qV^\nu \frac{\partial A_i}{\partial x^\nu}\right],$$

where we have used Eq. (2.8.38). The Euler-Lagrange equations then become

$$\frac{dp_i}{d\tau} = q\left(\partial_i A_\nu - \partial_\nu A_i\right) V^\nu = qF_{i\nu} V^\nu, \tag{2.8.45}$$

where we have used $F_{\mu\nu} = \partial_\mu A_\nu - \partial_\nu A_\mu$. As previously argued, since this three-vector equation must be true in every frame and since boosts mix the temporal and spatial components of Lorentz tensors, then the four-vector form of the equation must also be true. The four-vector form of the Euler-Lagrange equations is then

$$F^\mu = dp^\mu/d\tau = qF^{\mu\nu}V_\nu, \tag{2.8.46}$$

where F^μ is the four-force and $p^\mu = mV^\mu = m\gamma(c, \boldsymbol{v})$ is the mechanical four-momentum of the particle as discussed in Sec. 2.6.

The zeroth component of the Euler-Lagrange equations gives

$$F^0 = \frac{dp^0}{d\tau} = \frac{\gamma}{c}\frac{dE^{\text{mech}}}{dt} = qF^{0j}V_j = q\left(-\frac{E^j}{c}\right)(-\gamma v^j) = \frac{\gamma}{c}q\mathbf{E}\cdot\boldsymbol{v}, \tag{2.8.47}$$

where $E^{\text{mech}} = p^0 c = \gamma mc^2 = mc^2 + T$ is the mechanical energy of the particle, i.e., its rest mass energy mc^2 plus its kinetic energy $T = (\gamma - 1)mc^2$. We then find

$$\frac{dE^{\text{mech}}}{dt} = q\mathbf{E}\cdot\boldsymbol{v} \quad \text{or equivalently} \quad dE^{\text{mech}} = q\mathbf{E}\cdot\boldsymbol{v}dt = q\mathbf{E}\cdot d\boldsymbol{s}, \tag{2.8.48}$$

which is the incremental change in energy associated with the particle undergoing a displacement $d\boldsymbol{s}$ in the presence of $\mathbf{E}$ and $\mathbf{B}$ fields. A magnetic field changes the particle's direction of motion but not its energy. The three-vector part of the Euler-Lagrange equations gives

$$F^i = \frac{dp^i}{d\tau} = \gamma\frac{dp^i}{dt} = q\left(F^{i0}V_0 + F^{ij}V_j\right) = q\left[\left(\frac{E^i}{c}\right)(\gamma c) + \left(-\epsilon^{ijk}B^k\right)\left(-\gamma v^j\right)\right]$$
$$= \gamma q\left(E^i + [\boldsymbol{v}\times\mathbf{B}]^i\right) \tag{2.8.49}$$

and so we recover the relativistic Lorentz force equation,

$$\mathbf{F} = \frac{d\mathbf{p}}{d\tau} = \frac{d(\gamma m\boldsymbol{v})}{d\tau} = \gamma(v)q\left(\mathbf{E} + \boldsymbol{v}\times\mathbf{B}\right). \tag{2.8.50}$$

Equivalently, we have the more familiar form

$$\frac{d\mathbf{p}}{dt} = \frac{d(\gamma m\boldsymbol{v})}{dt} = q\left(\mathbf{E} + \boldsymbol{v}\times\mathbf{B}\right), \tag{2.8.51}$$

where it should be noted that $\mathbf{p} = \gamma(v)m\boldsymbol{v}$ is the *relativistic* three-momentum. We recover the nonrelativistic Lorentz force equation of Eq. (2.1.35) in the limit $v \ll c$. The *relativistic Lorentz force equation* is widely exploited, e.g., in mass spectrometers and particle physics detectors at high-energy particle accelerators.

The same results follow in the Hamiltonian formalism. Consider Hamilton's equation $dH/dt = -\partial L/\partial t$. We can use this to evaluate

$$\frac{dE^{\text{mech}}}{dt} = \frac{d(\gamma mc^2)}{dt} = \frac{dH}{dt} - q\frac{d\Phi}{dt} = -\frac{\partial L}{\partial t} - q\frac{d\Phi}{dt}$$
$$= q\left(\frac{\partial\Phi}{\partial t} - \boldsymbol{v}\cdot\frac{\partial\mathbf{A}}{\partial t}\right) - q\left(\frac{\partial\Phi}{\partial t} + \boldsymbol{v}\cdot\boldsymbol{\nabla}\Phi\right) = q\mathbf{E}\cdot\boldsymbol{v}, \tag{2.8.52}$$

which has reproduced Eq. (2.8.48). Next consider the three Hamilton's equations

$$\dot{x}^i = \frac{dx^i}{dt} = \frac{\partial H(\mathbf{x}, \boldsymbol{\pi})}{\partial\pi^i} = \frac{\partial}{\partial\pi^i}\left\{\left([\boldsymbol{\pi} - q\mathbf{A}]^2 c^2 + m^2 c^4\right)^{1/2} + q\Phi\right\}$$
$$= \left([\boldsymbol{\pi} - q\mathbf{A}]^2 c^2 + m^2 c^4\right)^{-1/2}c^2\left[\pi^i - qA^i\right], \tag{2.8.53}$$

which have the solution $\mathbf{p} = \boldsymbol{\pi} - q\mathbf{A}$ as can be confirmed by direct substitution, $\dot{\mathbf{x}} = (\mathbf{p}^2 c^2 + m^2 c^4)^{-1/2} c^2 \mathbf{p} = (\gamma m c^2)^{-1} c^2 \gamma m \boldsymbol{v} = \boldsymbol{v}$. These three Hamilton's equations have simply reproduced the relationship between the canonical and mechanical momentum in Eq. (2.8.38). The last three Hamilton's equations are

$$
\begin{aligned}
\dot{\pi}^i = \frac{d\pi^i}{dt} &= -\frac{\partial H(\mathbf{x}, \boldsymbol{\pi})}{\partial x^i} = -\frac{\partial}{\partial x^i} \left\{ \left([\boldsymbol{\pi} - q\mathbf{A}]^2 c^2 + m^2 c^4 \right)^{1/2} + q\Phi \right\} \\
&= \left([\boldsymbol{\pi} - q\mathbf{A}]^2 c^2 + m^2 c^4 \right)^{-1/2} q c^2 [\boldsymbol{\pi} - q\mathbf{A}] \cdot \partial_i \mathbf{A} - q\partial_i \Phi \\
&= (\gamma m c^2)^{-1} q c^2 \mathbf{p} \cdot \partial_i \mathbf{A} + q\partial_i \Phi = q\boldsymbol{v} \cdot \partial_i \mathbf{A} - q\partial_i \Phi.
\end{aligned}
\tag{2.8.54}
$$

We can use this result to evaluate

$$
\begin{aligned}
\dot{p}^i = dp^i/dt &= \dot{\pi}^i - q\dot{A}^i = q \left[(\boldsymbol{v} \cdot \partial_i \mathbf{A} - \partial_i \Phi) - \left([\partial A^i/\partial t] + \boldsymbol{v} \cdot \boldsymbol{\nabla} A^i \right) \right] \\
&= q \left[-\partial_i \Phi - [\partial A^i/\partial t] + v^j \left(\partial_i A^j - \partial_j A^i \right) \right] = q \left(E^i + [\boldsymbol{v} \times \mathbf{B}]^i \right),
\end{aligned}
\tag{2.8.55}
$$

which has just reproduced the relativistic Lorentz force of Eq. (2.8.51). Note that in the last step we used $\partial_i = \partial/\partial x^i = \nabla^i$ and the result that

$$
\begin{aligned}
[\boldsymbol{v} \times \mathbf{B}]^i = [\boldsymbol{v} \times (\boldsymbol{\nabla} \times \mathbf{A})]^i &= \epsilon^{ijk} v^j \epsilon^{k\ell m} \nabla^\ell A^m = \epsilon^{kij} \epsilon^{k\ell m} v^j \nabla^\ell A^m \\
&= \left(\delta^{i\ell} \delta^{jm} - \delta^{im} \delta^{j\ell} \right) v^j \nabla^\ell A^m = \boldsymbol{v} \cdot \partial_i \mathbf{A} - \boldsymbol{v} \cdot \boldsymbol{\nabla} A^i.
\end{aligned}
\tag{2.8.56}
$$

2.9 Constrained Hamiltonian Systems

We have seen that singular systems do not have a well-defined Hamiltonian in the naive sense in that the Legendre transform is not well defined. This presents a significant challenge to any program of canonical quantization, such as that needed for electromagnetism. This problem motivated Dirac and others to develop a well-defined Hamiltonian approach by constructing a system with appropriate constraints (Dirac, 1950, 1958, 1982, 2001).

2.9.1 Construction of the Hamiltonian Approach

Gauge theories are of great importance in physics, and, as we will later see, they are examples of singular systems. Historically, Dirac and others approached the quantum formulation of gauge theories such as electromagnetism through the construction of generalized versions of the Hamiltonian and the associated Poisson bracket formalism. This was to be achieved through an appropriate generalization of Hamilton's equations and the Poisson bracket formalism for singular systems and the exploitation of the correspondence of the quantum commutator and the Poisson brackets in Eq. (2.5.12) up to possible ambiguities associated with operator ordering. As we will later see, a simpler and more straightforward approach to the quantization of gauge theories is through the Feynman path integral formulation to construct the path (i.e., functional) integral formulation of gauge field theories. When both formulations are sufficiently well defined, they are equivalent. The remainder of the discussion[5] in this section will focus on the construction of the classical Hamiltonian formalism for singular

---

[5] Clarifying remarks from Marc Henneaux, Don Sinclair and Kurt Sundermeyer are gratefully acknowledged.

systems, the Dirac-Bergmann algorithm, the Dirac bracket and the subsequent path to quantization. In addition to the work of Dirac referred to above, refer to Sundermeyer (1982), Henneaux and Teitelboim (1994); Rothe and Rothe (2010) and Sundermeyer (2014).

A physical system with N generalized coordinates $\vec{q} = (q_1, \ldots, q_N)$ that is holonomic and monogenic has a Lagrangian $L(\vec{q}, \dot{\vec{q}}, t)$ and its motion is determined by Hamilton's principle of stationary action, i.e., by the Euler-Lagrange equations. If the Lagrangian is convex in its generalized velocities, then we have seen that we can define the Hamiltonian $H(\vec{q}, \vec{p})$ via an N-dimensional Legendre transform and that the motion of the system is equivalently described by Hamilton's equations.

Expanding the total time derivative the Euler-Lagrange equations become

$$0 = \frac{d}{dt}\frac{\partial L}{\partial \dot{q}_i} - \frac{\partial L}{\partial q_i} = \sum_{j=1}^{N} \left(\frac{\partial^2 L}{\partial \dot{q}_i \partial \dot{q}_j} \ddot{q}_j + \frac{\partial^2 L}{\partial \dot{q}_i \partial q_j} \dot{q}_j \right) + \frac{\partial^2 L}{\partial t \partial \dot{q}_i} - \frac{\partial L}{\partial q_i}. \tag{2.9.1}$$

So we see that being able to express the N accelerations $\ddot{q}_j$ in terms of the generalized coordinates $\vec{q}$ and velocities $\dot{\vec{q}}$ requires that the real symmetric Hessian matrix $M_L(\vec{q}, \dot{\vec{q}}, t)$ be invertible; i.e., the $N \times N$ real symmetric matrix given by

$$(M_L)_{ij} \equiv \frac{\partial^2, L}{\partial \dot{q}_i \partial \dot{q}_j} \tag{2.9.2}$$

must be nonsingular, $\det M_L \neq 0$. As discussed earlier, a system with a singular Hessian matrix M_L is referred to as a *singular system*.

To proceed to a Hamiltonian formulation the conjugate momenta are defined as

$$p_j \equiv \partial L / \partial \dot{q}_j \qquad \Rightarrow \qquad (M_L)_{ij} = \partial p_j / \partial \dot{q}_i = \partial p_i / \partial \dot{q}_j, \tag{2.9.3}$$

which shows that M_L is the Jacobian matrix associated with a change of variables from the generalized velocities $\dot{\vec{q}}$ to the conjugate momenta $\vec{p}$ (see Sec. A.2.2). A singular Jacobian matrix means that the change of variables is not invertible, which means that the mapping from the $(\vec{q}, \dot{\vec{q}})$ space to the $(\vec{q}, \vec{p})$ phase space is not invertible; i.e., there is not a one-to-one correspondence between the spaces.

In particular, if the matrix M_L has rank $(N - M')$, then the nullity (i.e., dimensionality of the null space) of M_L is M'. The $2N$-dimensional space $(\vec{q}, \dot{\vec{q}})$ maps to a $(2N - M')$-dimensional submanifold within the $2N$-dimensional $(\vec{q}, \vec{p})$ phase space. We only consider systems where the rank $(N - M')$ of the matrix M_L is constant everywhere in $(\vec{q}, \dot{\vec{q}})$ space. Let this submanifold be specified by the M relations

$$\phi_r(\vec{q}, \vec{p}) = 0 \quad \text{for} \quad r = 1, 2, \ldots, M. \tag{2.9.4}$$

If all of these relations are independent, then $M = M'$, but in general $M \geq M'$.

Proof: We wish to show that we can always construct phase space constraints of the form $\phi_r(\vec{q}, \vec{p}) = 0$ for $r = 1, 2, \ldots, M$ for a singular Lagrangian L that has the $N \times N$ hessian matrix M_L with rank $(N - M')$ where $M \geq M'$.

Since $(M_L)_{ij} = \partial^2 L / \partial \dot{q}_i \partial \dot{q}_j = \partial p_i / \partial \dot{q}_j$ is a matrix with rank $(N - M')$, then there are $(N - M')$ linearly independent rows (and columns) and there are M' that are linearly dependent. The $(N - M') \times (N - M')$ submatrix formed from the independent rows and columns has rank $(N - M')$ and so is invertible (see Sec. A.8). Furthermore, since M_L is real symmetric, then so is this submatrix. There are only $(N - M')$ different $\dot{q}_j$ appearing in the derivatives of this submatrix. How we choose to label our q_j is of course irrelevant, so for the convenience of this discussion let us choose labels such that the invertible submatrix has entries $(M_L^{\mathrm{sub}})_{ij} = (M_L)_{ij} = \partial^2 L / \partial \dot{q}_i \partial \dot{q}_j = \partial p_i / \partial \dot{q}_j$ for $i, j = 1, \ldots, (N - M')$. Let us denote the velocities and momenta appearing in this *invertible* $(N - M') \times (N - M')$ submatrix as $\dot{q}_j^{\mathrm{i}}$ and p_j^{i}, respectively, for $j = 1, \ldots, (N - M')$. Similarly, we denote the *remaining* M' velocities and momenta as $\dot{q}_j^{\mathrm{r}}$ and p_j^{r}, respectively, for $j = (N - M') + 1, \ldots, N$. With this notation we then have $\dot{\vec{q}} = (\dot{\vec{q}}^{\mathrm{i}}, \dot{\vec{q}}^{\mathrm{r}})$ and $\vec{p} = (\vec{p}^{\mathrm{i}}, \vec{p}^{\mathrm{r}})$.

Note that there is an invertible change of variables

$$\dot{\vec{q}} = (\dot{\vec{q}}^{\mathrm{i}}, \dot{\vec{q}}^{\mathrm{r}}) \leftrightarrow (\vec{p}^{\mathrm{i}}, \dot{\vec{q}}^{\mathrm{r}}), \tag{2.9.5}$$

since the Jacobian matrix $J \equiv \partial(\vec{p}^{\mathrm{i}}, \dot{\vec{q}}^{\mathrm{r}}) / \partial(\dot{\vec{q}}^{\mathrm{i}}, \dot{\vec{q}}^{\mathrm{r}})$ is invertible. This follows since this Jacobian matrix has the block form

$$J = \frac{\partial(\vec{p}^{\mathrm{i}}, \dot{\vec{q}}^{\mathrm{r}})}{\partial(\dot{\vec{q}}^{\mathrm{i}}, \dot{\vec{q}}^{\mathrm{r}})} = \begin{bmatrix} [\partial p^{\mathrm{i}} / \partial \dot{q}^{\mathrm{i}}] & [\partial p^{\mathrm{i}} / \partial \dot{q}^{\mathrm{r}}] \\ [\partial \dot{q}^{\mathrm{r}} / \partial \dot{q}^{\mathrm{i}}] & [\partial \dot{q}^{\mathrm{r}} / \partial \dot{q}^{\mathrm{r}}] \end{bmatrix} = \begin{bmatrix} [M_L^{\mathrm{sub}}] & [\partial p^{\mathrm{i}} / \partial \dot{q}^{\mathrm{r}}] \\ 0 & [I_{M' \times M'}] \end{bmatrix} \tag{2.9.6}$$

and is brought to upper triangular form by the block diagonal orthogonal matrix O with diagonal blocks O^{sub} and the identity matrix $I_{M' \times M'}$, where O^{sub} is the orthogonal matrix that diagonalizes M_L^{sub} and places its eigenvalues down the diagonal. This statement is easily verified by evaluating $O^T J O$. Since the determinant of an upper triangular matrix is the product of its diagonal elements and since $\det M_L^{\mathrm{sub}} \neq 0$, then $\det J \neq 0$ and so J is invertible.

So any function $h(\vec{q}, \dot{\vec{q}})$ can be expressed as some function $f(\vec{q}, \vec{p}^{\mathrm{i}}, \dot{\vec{q}}^{\mathrm{r}})$ and vice versa. So there must exist functions f_j for $j = 1, \ldots, N$ such that

$$\dot{q}_j = f_j(\vec{q}, \vec{p}^{\mathrm{i}}, \dot{\vec{q}}^{\mathrm{r}}), \tag{2.9.7}$$

where for $j = (N - M') + 1, \ldots, N$ we must have $\dot{q}_j = f_j(\vec{q}, \vec{p}^{\mathrm{i}}, \dot{\vec{q}}^{\mathrm{r}}) = \dot{q}_j^{\mathrm{r}}$. From the definition of the momenta we then have for $j = 1, \ldots, N$ that

$$p_j = (\partial L / \partial \dot{q}_j)(\vec{q}, \dot{\vec{q}}) \equiv h_j(\vec{q}, \dot{\vec{q}}) = h_j(\vec{q}, \vec{f}(\vec{q}, \vec{p}^{\mathrm{i}}, \dot{\vec{q}}^{\mathrm{r}})) \equiv g_j(\vec{q}, \vec{p}^{\mathrm{i}}, \dot{\vec{q}}^{\mathrm{r}}), \tag{2.9.8}$$

where for $j = 1, \ldots, (N - M')$ we must have $p_j = g_j(\vec{q}, \vec{p}^{\mathrm{i}}, \dot{\vec{q}}^{\mathrm{r}}) = p_j^{\mathrm{i}}$. Now consider the M' momenta $p_j = p_j^{\mathrm{r}}$ for $j = (N - M') + 1, \ldots, N$. Assume that one of the $\vec{p}^{\mathrm{r}}$, denoted p_i^{r}, has a dependence on some $\dot{q}_k^{\mathrm{r}}$. Then we can invert Eq. (2.9.8) and solve for that $\dot{q}_k^{\mathrm{r}}$ in terms of $\vec{q}$, $\vec{p}^{\mathrm{i}}$, p_i^{r} and the remaining $(M' - 1)$ generalized velocities $\dot{q}_j^{\mathrm{r}}$ with $j \neq k$. This means that we could add this p_i^{r} to the set of $(N - M')$ momenta p_j^{i} and remove $\dot{q}_k^{\mathrm{r}}$ from the set of M' velocities $\dot{q}_j^{\mathrm{r}}$. This means that we would have an invertible mapping

$$\dot{\vec{q}} = (\dot{\vec{q}}^{\mathrm{i}}, \dot{\vec{q}}^{\mathrm{r}}) \leftrightarrow (\vec{p}^{\mathrm{i'}}, \dot{\vec{q}}^{\mathrm{r'}}), \tag{2.9.9}$$

where there are now $(N - M') + 1$ momenta in $\vec{p}^{\mathrm{i'}}$ and $(M' - 1)$ velocities in $\dot{\vec{q}}^{\mathrm{r'}}$. This would imply that there must be an $(N - M' + 1) \times (N - M' + 1)$ invertible submatrix of M_L, which would in turn imply that M_L has rank $\geq (N - M' + 1)$, which it does not. Therefore, it follows that all of the p_j^{r} must be independent of all of the $\dot{q}_j^{\mathrm{r}}$ and so we must have the M' equations

$$p_j^{\mathrm{r}} = g_j(\vec{q}, \vec{p}^{\mathrm{i}}) \quad \text{for} \quad j = (N - M') + 1, \ldots, N. \tag{2.9.10}$$

We can rewrite these equations as the M' constraint equations

$$\phi_j(\vec{q}, \vec{p}) \equiv p_j^{\text{r}} - g_j(\vec{q}, \vec{p}^{\text{i}}) = 0 \quad \text{for} \quad j = (N - M') + 1, \dots, N. \qquad (2.9.11)$$

We can always add a further $(M - M')$ dependent constraints and finally then rewrite the constraint equations in the form of Eq. (2.9.4).

Such relations represent constraints on the available phase space and are referred to as *primary constraints*, and the resulting submanifold is referred to as the *primary constraint submanifold* or sometimes as the *primary constraint surface*. The M primary constraints follow from the form of the Lagrangian density and the subsequent definitions of the conjugate momenta. One could imagine beginning with a Hamiltonian system and a set of primary constraints. Such a constrained Hamiltonian system would correspond to a singular system in the sense that if the corresponding Lagrangian could be written down then it would be that of a singular system.

We follow the standard textbook treatments and do not allow for the possibility that the constraint submanifold might have an explicit time dependence at this point, but will later argue how to generalize the results to allow for that possibility.

We want to extend the Hamiltonian formalism to apply to singular systems and to obtain a suitably modified form of Hamilton's equations that reproduce the Euler-Lagrange equations, i.e., that reproduce the solutions that follow from Hamilton's principle. Dirac and others successfully constructed such a formalism for constrained Hamiltonian dynamics. This is referred to as the *Dirac-Bergmann algorithm*.

Primary constraints: Consider usual definition of the Hamiltonian using the Legendre transformation

$$H(\vec{q}(t), \vec{p}(t)) \equiv \vec{p}(t) \cdot \dot{\vec{q}}(t) - L(\vec{q}(t), \dot{\vec{q}}(t)). \qquad (2.9.12)$$

We assume for now that the Lagrangian and hence the Hamiltonian have no explicit time dependence and will later show how to generalize the following arguments to include this. When the Lagrangian is singular, the Legendre transformation is no longer invertible and we cannot express all of the generalized velocities $\dot{\vec{q}}$ as functions of generalized coordinates $\vec{q}$ and conjugate momenta $\vec{p}$. However, we can always write $\vec{p} = \vec{p}(\vec{q}, \dot{\vec{q}})$ for a singular Lagrangian. The Hamiltonian $H(\vec{q}, \vec{p})$ remains independent of the generalized velocities, since for a variation of H we have

$$dH = d(\vec{p} \cdot \dot{\vec{q}}) - dL = \dot{\vec{q}} \cdot d\vec{p} + \vec{p} \cdot d\dot{\vec{q}} - \frac{\partial L}{\partial \vec{q}} \cdot d\vec{q} - \frac{\partial L}{\partial \dot{\vec{q}}} \cdot d\dot{\vec{q}} = \dot{\vec{q}} \cdot d\vec{p} - \frac{\partial L}{\partial \vec{q}} \cdot d\vec{q} = dH(\vec{q}, \vec{p}),$$

where we have used the definition of the conjugate momenta.

We introduce a Lagrange multiplier for each of the M constraints, i.e., $\vec{\lambda}(t) = (\lambda_1(t), \dots, \lambda_M(t))$. Using Eq. (2.1.65) the extended Lagrangian can be written as

$$L'(\vec{q}, \dot{\vec{q}}, \lambda) = L(\vec{q}, \dot{\vec{q}}) - \vec{\lambda} \cdot \vec{\phi}(\vec{q}, \vec{p}), \qquad (2.9.13)$$

where it should be noted that we have used the opposite sign for the Lagrange multipliers to that used in the Lagrangian formalism in order to conform with the choice originally made by Dirac. The sign choice for the Lagrange multipliers has no effect on the physics of the system. We now want to obtain the modified version of Hamilton's equations by generalizing the standard derivation given in Sec. 2.4.

We define the *total Hamiltonian* (Henneaux and Teitelboim, 1994; Dirac, 2001)

$$H_T(\vec{q}, \vec{p}, \vec{\lambda}) \equiv \vec{p} \cdot \dot{\vec{q}} - L'(\vec{q}, \dot{\vec{q}}) = H + \vec{\lambda} \cdot \vec{\phi}(\vec{q}, \vec{p}). \qquad (2.9.14)$$

Note that Sundermeyer (1982) refers to this as the *primary Hamiltonian*. It is just the usual Hamiltonian with the addition of each of the primary constraint functions multiplied by a corresponding Lagrange multiplier function $\lambda_r(t)$ for $r = 1, \dots, M$.

We now want to derive the appropriate form of Hamilton's equations of motion directly from Hamilton's principle, since we require that the resulting form of Hamilton's equations must be consistent with the corresponding Euler-Lagrange equations for the system. From the definition of the extended Lagrangian we can then write the extended action of Eq. (2.1.62) as

$$S'[\vec{q}, \vec{\lambda}] = \int_{t_i}^{t_f} dt \, L'(\vec{q}, \dot{\vec{q}}, \vec{\lambda}) = \int_{t_i}^{t_f} dt \left(\vec{p} \cdot \dot{\vec{q}} - H(\vec{q}, \vec{p}) - \vec{\lambda} \cdot \vec{\phi}(\vec{q}, \vec{p}) \right). \tag{2.9.15}$$

Consider a functional variation of this extended action with respect to the enlarged path $\vec{Q}(t) \equiv (\vec{q}(t), \vec{\lambda}(t))$. The classical motion of the system is determined using Hamilton's principle requiring that the action be invariant under path variations,

$$0 = \delta S' = \int_{t_i}^{t_f} dt \left(\delta \vec{p} \cdot \dot{\vec{q}} + \vec{p} \cdot \delta \dot{\vec{q}} - \delta H - \vec{\lambda} \cdot \delta \vec{\phi} - \delta \vec{\lambda} \cdot \vec{\phi} \right) \tag{2.9.16}$$

$$= \int_{t_i}^{t_f} dt \left(\delta \vec{p} \cdot \dot{\vec{q}} - \dot{\vec{p}} \cdot \delta \vec{q} - \frac{\partial H}{\partial \vec{q}} \cdot \delta \vec{q} - \frac{\partial H}{\partial \vec{p}} \cdot \delta \vec{p} - \sum_{r=1}^{M} \lambda_r \left[\frac{\partial \phi_r}{\partial \vec{q}} \cdot \delta \vec{q} + \frac{\partial \phi_r}{\partial \vec{p}} \cdot \delta \vec{p} \right] - \vec{\phi} \cdot \delta \vec{\lambda} \right)$$

$$= \int_{t_i}^{t_f} dt \left(\left[-\dot{\vec{p}} - \frac{\partial H}{\partial \vec{q}} - \sum_{r=1}^{m} \lambda_r \frac{\partial \phi_r}{\partial \vec{q}} \right] \cdot \delta \vec{q} + \left[\dot{\vec{q}} - \frac{\partial H}{\partial \vec{p}} - \sum_{r=1}^{m} \lambda_r \frac{\partial \phi_r}{\partial \vec{p}} \right] \cdot \delta \vec{p} - \vec{\phi} \cdot \delta \vec{\lambda} \right),$$

where we used integration by parts and the fact that the $\delta \vec{q}$ part of the path variations vanishes at the end points to replacement $(\vec{p} \cdot \delta \dot{\vec{q}})$ with $(-\dot{\vec{p}} \cdot \delta \vec{q})$. Since at every time t the variations in $\vec{q}, \vec{p}$ and $\vec{\lambda}$ are arbitrary and independent, we obtain the modified form of Hamilton's equations

$$\dot{q}_j = \frac{\partial H}{\partial p_j} + \vec{\lambda} \cdot \frac{\partial \vec{\phi}}{\partial p_j}, \qquad \dot{p}_j = -\frac{\partial H}{\partial q_j} - \vec{\lambda} \cdot \frac{\partial \vec{\phi}}{\partial q_j}, \qquad \phi_r = 0, \tag{2.9.17}$$

where $j = 1, \dots, N$ and $r = 1, \dots, M$. With no constraints we recover Hamilton's equations in Eq. (2.4.9). The time-dependent Lagrange multiplier functions $\vec{\lambda}(t)$ are determined by the requirement that the constraints $\vec{\phi}(\vec{q}, \vec{p}) = 0$ at all times.

The total time derivative of a dynamical variable $F(\vec{q}, \vec{p}, t)$ is

$$\frac{dF}{dt} = \sum_{j=1}^{N} \left(\frac{\partial F}{\partial q_j} \frac{dq_j}{dt} + \frac{\partial F}{\partial p_j} \frac{dp_j}{dt} \right) + \frac{\partial F}{\partial t} \tag{2.9.18}$$

$$= \sum_{j=1}^{N} \left(\frac{\partial F}{\partial q_j} \left[\frac{\partial H}{\partial p_j} + \vec{\lambda} \cdot \frac{\partial \vec{\phi}}{\partial p_j} \right] + \frac{\partial F}{\partial p_j} \left[-\frac{\partial H}{\partial q_j} - \vec{\lambda} \cdot \frac{\partial \vec{\phi}}{\partial q_j} \right] \right) + \frac{\partial F}{\partial t} = \{F, H\} + \vec{\lambda} \cdot \{F, \vec{\phi}\} + \frac{\partial F}{\partial t},$$

where we have used Eq. (2.9.17) to obtain this result.

To facilitate the discussion of Hamiltonians with constraints Dirac introduced the notion of the *weak equality* symbol "$\approx$" where

$$F(\vec{q}, \vec{p}, t) \approx G(\vec{q}, \vec{p}, t) \tag{2.9.19}$$

means that F and G are equal on the constraint submanifold at all times, i.e., that

$$F(\vec{q}, \vec{p}, t) = G(\vec{q}, \vec{p}, t) \text{ for all } (\vec{q}, \vec{p}) \text{ such that } \vec{\phi}(\vec{q}, \vec{p}) = 0; \tag{2.9.20}$$

We say that F and G are *weakly equal*. Using this notation we then have

$$\vec{\phi}(\vec{q}, \vec{p}) \approx 0;\tag{2.9.21}$$

i.e., $\vec{\phi}(\vec{q}, \vec{p})$ vanish on the constraint submanifold but are in general nonzero elsewhere. Dirac also introduced the concept of *strong equality*, denoted as $F \simeq G$ (Sundermeyer, 2014). This means that two functions and their phase space gradients are equal on the constraint submanifold at all times. Then

$$F(\vec{q}, \vec{p}, t) \simeq G(\vec{q}, \vec{p}, t)\tag{2.9.22}$$

means that

$$F(\vec{q}, \vec{p}, t) = G(\vec{q}, \vec{p}, t), \quad \left(\frac{\partial F}{\partial \vec{q}}, \frac{\partial F}{\partial \vec{p}}\right) = \left(\frac{\partial G}{\partial \vec{q}}, \frac{\partial G}{\partial \vec{p}}\right) \quad \text{when} \quad \vec{\phi}(\vec{q}, \vec{p}) = 0.\tag{2.9.23}$$

Simple *equality*, $F(\vec{q}, \vec{p}, t) = G(\vec{q}, \vec{p}, t)$, simply means that the two dynamical variables F and G are equal *everywhere* in phase space at all times. So then in increasing strength of equality we have *weak equality < strong equality < equality*.

Since the Lagrange multiplier functions $\vec{\lambda}(t)$ are independent of $\vec{q}$ and $\vec{p}$, then

$$\{F, \vec{\lambda}(t)\} = 0.\tag{2.9.24}$$

The Poisson bracket of dynamical variable $F(\vec{q}, \vec{p}, t)$ with the total Hamiltonian is

$$\{F, H_T\} = \{F, H\} + \left\{F, \vec{\lambda}\right\} \cdot \vec{\phi} + \left\{F, \vec{\phi}\right\} \cdot \vec{\lambda},\tag{2.9.25}$$

where we have used the product rule for the Poisson bracket. Since $\vec{\phi} \approx 0$ then

$$\{F, \vec{\phi} \cdot \vec{\lambda}\} \approx \{F, \vec{\phi}\} \cdot \vec{\lambda} \quad \Rightarrow \quad \{F, H_T\} \approx \{F, H\} + \left\{F, \vec{\phi}\right\} \cdot \vec{\lambda}\tag{2.9.26}$$

and so we can then rewrite Eq. (2.9.18) in the compact form

$$\frac{dF}{dt} \approx \{F, H_T\} + \frac{\partial F}{\partial t}.\tag{2.9.27}$$

In order to avoid potential ambiguity when using Poisson brackets in the treatment of singular systems, it is always to be understood that *Poisson brackets are to be evaluated before any constraints are applied.*

Secondary constraints: For consistency the time evolution of the system must preserve the constraints; i.e., we must require that the time evolution of the system does not take it off of the constraint submanifold, $\vec{\phi} \approx 0$. This requires that

$$0 \approx \dot{\phi}_r \approx \{\phi_r, H_T\} \approx \{\phi_r, H\} + \sum_{s=1}^{M}\{\phi_r, \phi_s\}\lambda_s \approx \{\phi_r, H\} + \sum_{s=1}^{M} P_{rs}\lambda_s,\tag{2.9.28}$$

where we have defined the real $M \times M$ antisymmetric matrix P with matrix elements

$$P_{rs} \equiv \{\phi_r, \phi_s\}\tag{2.9.29}$$

for $r, s = 1, \ldots, M$. Note that since P is antisymmetric, then $\det P = \det P^T = \det(-P) = (-1)^M \det P$ and so $\det P = 0$ unless the number of primary constraints M is even. A singular classical system must have either:

(a) $\det P \not\approx 0$: For this type of system we can simply invert Eq. (2.9.28) to solve for the Lagrange multipliers, which gives

$$\lambda_r \approx -\sum_{s=1}^{M}(P^{-1})_{rs}\{\phi_s, H\} \tag{2.9.30}$$

$$\Rightarrow \qquad H_T = H + \vec{\phi}\cdot\vec{\lambda} \approx H - \sum_{r,s=1}^{M}\phi_r(P^{-1})_{rs}\{\phi_s, H\}. \tag{2.9.31}$$

The time evolution, $\dot{F}(\vec{q}, \vec{p}, t) \approx \partial F/\partial t + \{F, H_T\}$, then becomes

$$\dot{F}(\vec{q}, \vec{p}, t) \approx \frac{\partial F}{\partial t} + \{F, H\} - \sum_{r,s=1}^{M}\{F, \phi_r\}(P^{-1})_{rs}\{\phi_s, H\}. \tag{2.9.32}$$

We verify that the time evolution is consistent with the constraints,

$$\dot{\phi}_r(\vec{q}, \vec{p}) \approx \{\phi_r, H\} - \sum_{s,t=1}^{M}P_{rs}(P^{-1})_{st}\{\phi_t, H\} \approx 0;\text{ or} \tag{2.9.33}$$

(b) $\det P \approx 0$: In this case some linear combinations of the Lagrange multiplier functions $\lambda_r(t)$ are left undetermined and so the number of constraints is insufficient to ensure that the motion remains on the $(2N - M')$-dimensional constraint submanifold at all times. This type of system requires the introduction of *secondary constraints* leading to associated complications.

Mathematical digression: Some properties of real antisymmetric matrices: Let P be a real and antisymmetric (i.e., skew-symmetric) $M \times M$ matrix. The following properties follow:

(i) $\det P = 0$ if M is odd, since $\det P = \det P^T = \det(-P) = (-1)^M \det P$.

(ii) P is specified by $M(M-1)/2$ real numbers, i.e., the number of entries above (or below) the diagonal, where the diagonal contains only zeros.

(iii) For any real (or complex) antisymmetric matrix P with M even it can be shown that $\det P = [\text{Pf}(P)]^2$, where $\text{Pf}(P)$ is called the Pfaffian of the matrix P and is a polynomial in the matrix elements of P that can be expressed as $\text{Pf}(P) = (1/2^M M!)\sum_{i_1,j_1,\ldots,i_M,j_M=1}^{M} \epsilon^{i_1 j_1 i_2 j_2 \ldots i_M j_M} P_{i_1 j_1} P_{i_2 j_2} \cdots P_{i_M j_M}$;

(iv) Since P has real matrix elements, then $\text{Pf}(P) \in \mathbb{R}$ so that $\det P \geq 0$.

(iv) P is a normal matrix, since $P^\dagger P = P^T P = -P^2 = PP^\dagger$.

(v) Since P is a normal matrix, then it has a complete set of orthogonal eigenvectors and is diagonalizable by a unitary transformation, $P = UDU^\dagger$, where U has the normalized eigenvectors of P as its columns.

(vi) The nonzero eigenvalues of P are pure imaginary, $\lambda = i|\lambda|$, and come in opposite sign pairs, $\pm\lambda$. [*Proof:* Since $P^\dagger P = -P^2$ then the eigenvalues λ of P satisfy $\lambda^*\lambda = -\lambda^2$ and so λ must be pure imaginary. Similar matrices have the same eigenvalues and any matrix is similar to its transpose. So if λ is an eigenvalue of P, then $-\lambda$ is an eigenvalue of $P^T = -P$. But since P^T is similar to P, then $-\lambda$ is also an eigenvalue of P.]

(vii) There exists a real orthogonal matrix O (i.e., $O^T = O^{-1}$ and $O_{ij} \in \mathbb{R}$) such that $P = O^T\Sigma O$, where Σ consists of 2×2 diagonal blocks with zeros on the diagonal and with the off-diagonal entries consisting of the pairs $\pm|\lambda|$ and with zeros everywhere else. The same 2×2 block will be repeated according to the degeneracy of the eigenvalue pair (Youla, 1961).

(viii) Define $R_P \equiv \text{rank}(P)$ and $N_P \equiv \text{nullity}(P)$, then by the rank-nullity theorem $R_P + N_P = M$. The matrix Σ has a block diagonal form with an $R_P \times R_P$ block, Σ_R, that is itself comprised of the 2×2 blocks down its diagonal and an $N_P \times N_P$ block, Σ_0, consisting entirely of zeros. By construction the eigenvectors of Σ_R span the total eigenspace pertaining to the nonzero eigenvalues and hence $\det \Sigma_R \neq 0$ and it is invertible.

We now need to generalize the discussion to include the possibility that $\det P \approx 0$. Defining $R_P \equiv \mathrm{rank}(P)$ and $N_P \equiv \mathrm{nullity}(P)$ we have $R_P + N_P = M$. As shown above we can write $\Sigma = OPO^T$, where O is an orthogonal matrix and where Σ is block diagonal with the non-null eigenspace of P corresponding to the $R_P \times R_P$ block Σ_R and the null eigenspace of P corresponding to the $N_P \times N_P$ block $\Sigma_0 = 0$. Note that since the constraints are dynamical variables, $\vec{\phi}(\vec{q}, \vec{p})$, then so are the matrices $P(\vec{q}, \vec{p})$ and $O(\vec{q}, \vec{p})$. Let $F(\vec{q}, \vec{p}, t)$ and $G(\vec{q}, \vec{p}, t)$ be any two dynamical variables, then using the product rule for the Poisson bracket we have

$$\{F\phi_r, G\} = F\{\phi_r, G\} + \{F, G\}\phi_r \approx F\{\phi_r, G\}, \tag{2.9.34}$$

since by definition the constraints satisfy $\vec{\phi}(\vec{q}, \vec{p}) \approx 0$. So for any dynamical variable G we have $O\{\vec{\phi}, G\} \approx \{O\vec{\phi}, G\}$. Making the definitions

$$\vec{\phi}' \equiv O\vec{\phi} \qquad \text{and} \qquad \vec{\lambda}' \equiv O\vec{\lambda}, \tag{2.9.35}$$

we can rewrite Eq. (2.9.28) as

$$0 \approx O\dot{\vec{\phi}} \approx O\{\vec{\phi}, H_T\} \approx O\{\vec{\phi}, H\} + OP\vec{\lambda} \approx \{\vec{\phi}', H\} + \Sigma\vec{\lambda}'. \tag{2.9.36}$$

Due to the block diagonal nature of Σ we obtain from these M equations

$$\lambda'_r \approx -\sum_{s=1}^{R_P}(\Sigma_R^{-1})_{rs}\{\phi'_s, H\} \quad \text{for } r = 1, \dots, R_P,$$
$$0 \approx \{\phi'_r, H\} \quad \text{for } r = R_P + 1, \dots, M. \tag{2.9.37}$$

The first set of R_P equations determines R_P of the transformed Lagrange multipliers, $\lambda'_1, \dots, \lambda'_{R_P}$. The second set of $N_P = (M - R_P)$ equations represents additional constraints on the system and in order to satisfy them we must further restrict the constraint submanifold. They may or may not all be independent of the primary constraints. Let $\ell' \leq N_P$ be the number of these additional constraints that are *independent* of the primary constraints. These are referred to as *secondary constraints* and we write them as

$$\chi_r(\vec{q}, \vec{p}) \approx 0 \qquad \text{for } r = 1, \dots, \ell' \leq N_P, \tag{2.9.38}$$

where here the "$\approx$" symbol now refers to the restricted constraint submanifold that satisfies both conditions $\vec{\phi}(\vec{q}, \vec{p}) = 0$ *and* $\vec{\chi}(\vec{q}, \vec{p}) = 0$. Note that primary constraints arise as a consequence of the Lagrangian being singular, whereas secondary constraints follow only when we use the equations of motion as well.

These secondary constraints must also be satisfied at all times and so we require

$$0 \approx \dot{\chi}_r \approx \{\chi_r, H_T\} \approx \{\chi_r, H\} + \sum_{s=1}^{M}\{\chi_r, \phi_s\}\lambda_s \text{ for } r = 1, \dots, \ell' \leq N_P. \tag{2.9.39}$$

These equations might then lead to ℓ'' further secondary constraints, where $\ell'' \leq \ell'$,

$$\chi_r(\vec{q}, \vec{p}) \approx 0 \quad \text{for } r = \ell' + 1, \dots, \ell' + \ell'' \tag{2.9.40}$$

and the time-independence of these constraints can lead to $\ell''' \leq \ell''$ additional constraints. This process is repeated until no new secondary constraints result. Let

$$L \equiv \ell' + \ell'' + \ell''' + \cdots \tag{2.9.41}$$

be the total number of secondary constraints generated by this process. Note that all secondary constraints are independent of the primary constraints by construction. As was the case for primary constraints, we do not insist that all secondary constraints be independent of each other. Let L' denote the number of mutually independent secondary constraints, $L' \leq L$.

Then the total number of constraints is $(M + L)$, i.e., the M primary constraints and the L secondary constraints,

$$\phi_r(\vec{q}, \vec{p}) \approx 0 \quad \text{for } r = 1, \dots, M \quad \text{and} \quad \chi_r(\vec{q}, \vec{p}) \approx 0 \quad \text{for } r = 1, \dots, L. \tag{2.9.42}$$

The restricted constraint submanifold is then $(2N - M' - L')$-dimensional, since the dimensionality is determined by the total number of independent constraints $(M' + L') \leq (M + L)$. The resulting $(M + L)$ consistency conditions are

$$0 \approx \{\phi_r, H\} + \sum_{s=1}^{m} \{\phi_r, \phi_s\}\lambda_s \quad \text{for } r = 1, \dots, M,$$
$$0 \approx \{\chi_r, H\} + \sum_{s=1}^{m} \{\chi_r, \phi_s\}\lambda_s \quad \text{for } r = 1, \dots, L, \tag{2.9.43}$$

and ensure that the time evolution of the system leaves it on the $(2N - M' - L')$-dimensional constraint submanifold in phase space.

First- and second-class constraints: Define the constraint vector, $\vec{\Phi} \approx 0$, as

$$\vec{\Phi} = \left(\Phi_1, \dots, \Phi_{(M+L)}\right) \equiv \left(\phi_1, \dots, \phi_M, \chi_1, \dots, \chi_L\right) \tag{2.9.44}$$

with $(M + L)$ components and the $(M + L) \times M$ matrix D with matrix elements

$$D_{rs} \equiv \{\Phi_r, \phi_s\}. \tag{2.9.45}$$

Then we can rewrite the consistency conditions as

$$0 \approx \{\vec{\Phi}, H\} + D\vec{\lambda}. \tag{2.9.46}$$

As noted in Sec. A.8, the row rank and column rank of any rectangular matrix are the number of linearly independent rows and columns, respectively, of the matrix and they are equal. Define $K \equiv \text{rank}(D) \leq M$, then K is the number of linearly independent combinations of the Lagrange multipliers $\vec{\lambda}$ that appear in Eq. (2.9.46).

If the Poisson bracket of a dynamical variable F with all of the constraints vanishes on the constraint submanifold,

$$\{\vec{\Phi}, F\} \approx 0, \tag{2.9.47}$$

then we define F to be a *first-class dynamical variable*. If the Poisson bracket with any one or more of the constraints does not vanish, then we define F to be a *second-class dynamical variable*. If the constraint function is a first- or second-class dynamical variable, then we refer to the constraint as a *first-* or *second-class constraint*, respectively. The classification of constraints into first and second class was introduced by Dirac and will prove to be very important in formulating the equations of motion. In fact it will be a more important distinction than the distinction between primary and secondary constraints. It is potentially confusing to have the four different definitions of "primary," "secondary," "first class" and "second class" in our discussion, but that is the terminology that was introduced historically.

We now make use of the singular value decomposition of a rectangular matrix. A real $(M+L) \times M$ matrix D has rank $K \leq M$ and we can write $D = U\Sigma V^T$, where U is an $(M + L) \times (M + L)$ real orthogonal matrix, V is an $M \times M$ real orthogonal matrix and Σ is an $(M + L) \times M$ matrix with K positive entries down the diagonal and zeros everywhere else. U and V diagonalize DD^T and $D^T D$, respectively; i.e., $U^T DD^T U = \Sigma\Sigma^T$ and $V^T D^T DV = \Sigma^T\Sigma$. We can always choose U and V such that the first K diagonal entries are the positive ones ($d_1, \dots, d_K > 0$) and the remaining

$(M - K)$ diagonal entries are the vanishing ones $(d_{(K+1)}, \ldots, d_M = 0)$ with all other entries zero, i.e., we can always arrange to have the $(M + L) \times M$ matrix Σ in the form

$$
\Sigma = \left[
\begin{array}{ccccccc}
d_1 & 0 & \cdots & 0 & 0 & \cdots & 0 \\
0 & d_2 & \cdots & \vdots & \vdots & \cdots & \vdots \\
\vdots & \cdots & \cdots & \vdots & \vdots & \cdots & \vdots \\
0 & \cdots & \cdots & d_K & 0 & \cdots & 0 \\
0 & \cdots & \cdots & 0 & 0 & \cdots & 0 \\
\vdots & \cdots & \cdots & \vdots & \vdots & \cdots & \vdots \\
0 & \cdots & \cdots & 0 & 0 & \cdots & 0 \\
\hline
0 & \cdots & \cdots & \cdots & \cdots & \cdots & 0 \\
\vdots & \cdots & \cdots & \cdots & \cdots & \cdots & \vdots \\
0 & \cdots & \cdots & \cdots & \cdots & \cdots & 0
\end{array}
\right] .
\tag{2.9.48}
$$

The horizontal line has been added to indicate the separation of the diagonal $M \times M$ component of Σ from the $L \times M$ null matrix component. Since all of the constraints $\vec{\Phi}(\vec{q}, \vec{p})$ are functions of phase space, then so are the matrices $D(\vec{q}, \vec{p})$, $\Sigma(\vec{q}, \vec{p})$, $U(\vec{q}, \vec{p})$, and $V(\vec{q}, \vec{p})$.

For any two dynamical variables F and G,

$$
\{F\Phi_r, G\} = F\{\Phi_r, G\} + \{F, G\}\Phi_r \approx F\{\Phi_r, G\}
\tag{2.9.49}
$$

since by definition the constraints satisfy $\vec{\Phi}(\vec{q}, \vec{p}) \approx 0$. Since V is an $M \times M$ real orthogonal matrix, then its rows and columns each form an orthonormal basis of $\mathbb{R}^M$. Similarly, the rows and columns of U each form an orthonormal basis of $\mathbb{R}^{(M+L)}$. Defining the rotated constraint vectors

$$
\vec{\Phi}^U \equiv U^T \vec{\Phi} \qquad \text{and} \qquad \vec{\phi}' \equiv V^T \vec{\phi},
\tag{2.9.50}
$$

we have for $r, s = 1, \ldots, M$ that

$$
d_s \delta_{rs} = \Sigma_{rs} = (U^T D V)_{rs} = \sum_{p=1}^{(M+L)} \sum_{q=1}^{M} (U^T)_{rp} \{\Phi_p, \phi_q\} V_{qs} \approx \{\Phi_r^U, \phi_s'\},
\tag{2.9.51}
$$

where on the constraint surface we can move $U^T(\vec{q}, \vec{p})$ and $V(\vec{q}, \vec{p})$ inside the Poisson bracket containing the constraints using Eq. (2.9.49). Then $\{\vec{\Phi}^U, \phi_s'\} \not\approx 0$ for $s = 1, \ldots, K$ and $\{\vec{\Phi}^U, \phi_s'\} \approx 0$ for $s = (K+1), \ldots, M$. Since $\vec{\Phi} = U\vec{\Phi}^U$ with U an orthogonal matrix and since $\vec{\Phi}^U \approx 0$, we can again use Eq. (2.9.49) to arrive at

$$
\{\vec{\Phi}, \phi_s'\} \not\approx 0 \qquad \text{for } s = 1, \ldots, K,
\tag{2.9.52}
$$

$$
\{\vec{\Phi}, \phi_s'\} \approx 0 \qquad \text{for } s = (K+1), \ldots, M.
\tag{2.9.53}
$$

We see that the ϕ_s' for $s = 1, \ldots, K$ are *second-class primary constraints*, while those for $s = (K+1), \ldots, M$ are *first-class primary constraints*.

We can now repeat the above arguments for the secondary constraints. Define the $(M + L) \times L$ matrix D^χ with matrix elements

$$
D_{rs}^\chi \equiv \{\Phi_r, \chi_s\}.
\tag{2.9.54}
$$

Define $K^\chi \equiv \mathrm{rank}(D^\chi) \leq L$. After a singular value decomposition for D^χ such that $D^\chi = U^\chi \Sigma^\chi V^\chi$ and defining $\vec{\chi}' \equiv (V^\chi)^T \vec{\chi}$ we then find

$$
\{\vec{\Phi}, \chi_s'\} \not\approx 0 \qquad \text{for } s = 1, \ldots, K^\chi,
\tag{2.9.55}
$$

$$\{\vec{\Phi}, \chi'_s\} \approx 0 \qquad \text{for } s = (K^\chi + 1), \ldots, L. \tag{2.9.56}$$

We see that the χ'_s for $s = 1, \ldots, K^\chi$ are *second-class secondary constraints*, while those for $s = (K^\chi + 1), \ldots, L$ are *first-class secondary constraints*.

Let us relabel the rotated primary and secondary constraints as

$$
\begin{aligned}
\text{first-class primary}: \quad & \vec{\phi}^{(1)} \equiv (\phi'_{(K+1)}, \ldots, \phi'_M) \\
\text{second-class primary}: \quad & \vec{\phi}^{(2)} \equiv (\phi'_1, \ldots, \phi'_K) \\
\text{first-class secondary}: \quad & \vec{\chi}^{(1)} \equiv (\chi'_{(K^\chi+1)}, \ldots, \chi'_L) \\
\text{second-class secondary}: \quad & \vec{\chi}^{(2)} \equiv (\chi'_1, \ldots, \chi'_{K^\chi})
\end{aligned}
\tag{2.9.57}
$$

and define the rotated Lagrange multipliers $\lambda'(\vec{q}, \vec{p}, t)$ using the $M \times M$ real orthogonal matrix V^T as

$$\vec{\lambda}'(t) \equiv V^T \vec{\lambda}(t). \tag{2.9.58}$$

Note that although V^T is a function of phase space, since the $\vec{\lambda}(t)$ are arbitrary functions of time in the above equation, then so are the rotated Lagrange multipiers $\lambda'(t)$. Let us further relabel these according to whether they are associated with first- or second-class primary constraints using the notation $\vec{v}$ for first-class primary and $\vec{u}$ for second-class primary, where

$$\vec{\lambda}' = (\lambda'_1, \ldots, \lambda'_M) \equiv (u_1, \ldots, u_K, v_1, \ldots, v_{(M-K)}) = (\vec{u}, \vec{v}). \tag{2.9.59}$$

Then we can write the total Hamiltonian as

$$
\begin{aligned}
H_T &= H + \vec{\lambda} \cdot \vec{\phi} = H + \vec{\lambda}' \cdot \vec{\phi}' \\
&= H + \vec{\phi}^{(1)} \cdot \vec{v} + \vec{\phi}^{(2)} \cdot \vec{u}.
\end{aligned}
\tag{2.9.60}
$$

The $(M + L)$ consistency conditions in Eq. (2.9.43) can then be rewritten as

$$
\begin{aligned}
\text{first-class primary}: \quad & 0 \approx \{\phi_r^{(1)}, H\} \tag{2.9.61} \\
\text{first-class secondary}: \quad & 0 \approx \{\chi_r^{(1)}, H\} \tag{2.9.62} \\
\text{second-class primary}: \quad & 0 \approx \{\phi_r^{(2)}, H\} + \{\phi_r^{(2)}, \vec{\phi}^{(2)}\} \cdot \vec{u} \tag{2.9.63} \\
\text{second-class secondary}: \quad & 0 \approx \{\chi_r^{(2)}, H\} + \{\chi_r^{(2)}, \vec{\phi}^{(2)}\} \cdot \vec{u}. \tag{2.9.64}
\end{aligned}
$$

Note that the Lagrange multiplier functions $\vec{v}(t)$ do not appear in these constraints and so remain undetermined; i.e., they remain arbitrary functions of time. We now wish to show that we can explicitly solve for the Lagrange multipliers $\vec{u}(t)$.

Following the notation of Sundermeyer (1982) we define the vector $\vec{\xi}(\vec{q}, \vec{p})$ of all $(K + K^\chi)$ rotated second-class constraints as

$$\vec{\xi} = (\xi_1, \ldots, \xi_{(K+K^\chi)}) \equiv (\phi_1^{(2)}, \ldots, \phi_K^{(2)}, \chi_1^{(2)}, \ldots, \chi_{K^\chi}^{(2)}) \equiv (\vec{\phi}^{(2)}, \vec{\chi}^{(2)}). \tag{2.9.65}$$

The second-class consistency conditions can be combined into a single form as

$$0 \approx \{\xi_r, H\} + \sum_{s=1}^{K} \{\xi_r, \phi_s^{(2)}\} u_s. \tag{2.9.66}$$

Let us define (Henneaux and Teitelboim, 1994)

$$H' \equiv H + \vec{\phi}^{(2)} \cdot \vec{u}, \tag{2.9.67}$$

then from Eq. (2.9.66) we see that $\{\vec{\xi}, H'\} \approx 0$. Then H' is a first-class quantity since it also has a vanishing Poisson bracket with all of the first-class constraints due to Eqs. (2.9.61) and (2.9.62). It then follows that $H_T = H' + \vec{\phi}^{(1)} \cdot \vec{u}$ is also first class since

$$\{\xi_r, H_T\} = \{\xi_r, H'\} + \{\xi_r, \vec{\phi}^{(1)} \cdot \vec{u}\} \approx \{\xi_r, \vec{\phi}^{(1)}\} \cdot \vec{u} \approx 0, \tag{2.9.68}$$

since the $\vec{\phi}^{(1)}$ are first-class constraints and so have vanishing Poisson brackets with all constraints. Define for convenience $\vec{\Phi}^{\mathrm{rot}}$ as the set of all rotated first-class and second-class constraints

$$\vec{\Phi}^{\mathrm{rot}} \equiv (\vec{\phi}^{(1)}, \vec{\chi}^{(1)}, \vec{\phi}^{(2)}, \vec{\chi}^{(2)}) = (\vec{\phi}^{(1)}, \vec{\chi}^{(1)}, \vec{\xi}); \tag{2.9.69}$$

then we have shown that H_T and H' are both first class,

$$\{\vec{\Phi}^{\mathrm{rot}}, H_T\} \approx \{\vec{\Phi}^{\mathrm{rot}}, H'\} \approx 0 \quad \Leftrightarrow \quad \{\vec{\Phi}, H_T\} \approx \{\vec{\Phi}, H'\} \approx 0. \tag{2.9.70}$$

Defining the matrix Δ with matrix elements

$$\Delta_{rs} \equiv \{\xi_r, \xi_s\}, \tag{2.9.71}$$

we observe that Δ is a $(K + K^\chi) \times (K + K^\chi)$ antisymmetric real matrix. Importantly, as we show below, Δ is also nonsingular on the constraint submanifold and hence invertible on that submanifold. Since it is nonsingular it must have maximum rank on the constraint submanifold, i.e., $\mathrm{rank}(\Delta) = (K + K^\chi)$. Since it is a real antisymmetric nonsingular matrix this means that Δ must have an even number of dimensions; i.e., *the number of second-class constraints must be even.*

Proof: We wish to prove that Δ is invertible on the constraint submanifold and that the number of second-class constraints must be even. Since Δ is a real $(K+K\chi) \times (K+K\chi)$ antisymmetric matrix it is a normal matrix ($\Delta^\dagger = \Delta^T = -\Delta \Rightarrow \Delta^\dagger \Delta = -\Delta^2 = \Delta\Delta^\dagger$) and so has a complete set of eigenvectors. If it was also singular, then it would have a zero eigenvalue and hence at least one eigenvector $\vec{R}(\vec{q}, \vec{p})$ such that

$$\sum_{s=1}^{(K+K\chi)} \Delta_{rs} R_s \approx 0. \tag{2.9.72}$$

Using the product rule for Poisson brackets and $\Delta_{rs} = \{\xi_r, \xi_s\}$ leads to

$$0 \approx \sum_s \Delta_{rs} R_s \approx \sum_{s=1}\{\xi_r, \xi_s\} R_s \approx \{\xi_r, \vec{\xi} \cdot \vec{R}\} - \{\xi_r, \vec{R}\} \cdot \vec{\xi} \approx \{\xi_r, \vec{\xi} \cdot \vec{R}\},$$

where we have used $\vec{\xi} \approx 0$. This shows that $\vec{\xi} \cdot \vec{R}$ would have a vanishing Poisson bracket with all second-class constraints. Note that $\{\phi_r^{(1)}, \vec{\xi} \cdot \vec{R}\} \approx \{\phi_r^{(1)}, \vec{\xi}\} \cdot \vec{R} \approx 0$ and $\{\chi_r^{(1)}, \vec{\xi} \cdot \vec{R}\} \approx \{\chi_r^{(1)}, \vec{\xi}\} \cdot \vec{R} \approx 0$, since $\vec{\phi}^{(1)}$ and $\vec{\chi}^{(1)}$ are first-class constraints. So if the matrix Δ was singular on the constraint submanifold then $\vec{\xi} \cdot \vec{R}$ would have a vanishing Poisson bracket with all of the constraints on the submanifold and would therefore be a first-class quantity. This would contradict the construction using the singular value decomposition that separated the first- and second-class constraints, which is not possible. Therefore, the matrix Δ must be nonsingular and hence invertible on the constraint submanifold and have maximum rank of $(K + K^\chi)$. Since Δ is a nonsingular real antisymmetric matrix, then, as shown earlier, it must have an even number of dimensions and so the number of second-class constraints $(K + K^\chi)$ must be even.

We can express Eq. (2.9.66) as

$$0 \approx \{\xi_r, H\} + \sum_{s=1}^K \Delta_{rs} u_s \equiv \{\xi_r, H\} + \sum_{s=1}^{(K+K^\chi)} \Delta_{rs} U_s, \tag{2.9.73}$$

where we have defined the $(K + K^\chi)$-component vector $\vec{U} \equiv (\vec{u}, \vec{0})$. Acting from the right with Δ^{-1} then gives

$$U_r \approx -\sum_{s=1}^{(K+K^\chi)} \Delta_{rs}^{-1} \{\xi_s, H\}, \qquad (2.9.74)$$

which can be expanded as

$$u_r \approx -\sum_{s=1}^{(K+K^\chi)} \Delta_{rs}^{-1} \{\xi_s, H\} \quad \text{for } r = 1, \ldots, K, \qquad (2.9.75)$$

$$0 \approx \sum_{s=1}^{(K+K^\chi)} \Delta_{rs}^{-1} \{\xi_s, H\} \quad \text{for } r = (K+1), \ldots, (K+K^\chi). \qquad (2.9.76)$$

So the rotated Lagrange multipliers $\vec{u}(t)$ for the second-class primary constraints have been determined.

The total Hamiltonian can now be written as

$$\begin{aligned}
H_T &= H' + \vec{\phi}^{(1)} \cdot \vec{v} = H + \vec{\phi}^{(1)} \cdot \vec{v} + \vec{\phi}^{(2)} \cdot \vec{u} \\
&\approx H + \vec{\phi}^{(1)} \cdot \vec{v} - \sum_{r=1}^{K} \sum_{s=1}^{(K+K^\chi)} \phi_r^{(2)} \Delta_{rs}^{-1} \{\xi_s, H\} \\
&\approx H + \vec{\phi}^{(1)} \cdot \vec{v} - \sum_{r,s=1}^{(K+K^\chi)} \xi_r \Delta_{rs}^{-1} \{\xi_s, H\},
\end{aligned} \qquad (2.9.77)$$

where we arrived at the last line using the result of Eq. (2.9.76). We can also write

$$H' = H + \vec{\phi}^{(2)} \cdot \vec{u} = H - \sum_{r,s=1}^{(K+K^\chi)} \xi_r \Delta_{rs}^{-1} \{\xi_s, H\}. \qquad (2.9.78)$$

Since Δ is a nonsingular real antisymmetric matrix, then so also is Δ^{-1}. Note that all constraints appear explicitly in H_T except the first-class secondary constraints. This is an issue that we will return to discuss later.

Recall that Poisson brackets are to be evaluated *before* constraints are imposed. So when H_T appears in a Poisson bracket we use the form

$$H_T = H + \vec{\phi}^{(1)} \cdot \vec{v} + \vec{\phi}^{(2)} \cdot \vec{u} \qquad (2.9.79)$$

before applying the constraints, and only *after* the Poisson bracket has been evaluated do we impose the constraints by substituting our solution for the Lagrange multipliers, $\vec{u}$ in Eq. (2.9.75).

From Eq. (2.9.27) the total time derivative of dynamical variable $F(\vec{q}, \vec{p}, t)$ is

$$\dot{F} \approx \frac{\partial F}{\partial t} + \{F, H_T\} \approx \frac{\partial F}{\partial t} + \{F, H\} + \{F, \vec{\phi}^{(1)}\} \cdot \vec{v} - \sum_{r,s} \{F, \xi_r\} \Delta_{rs}^{-1} \{\xi_s, H\}. \qquad (2.9.80)$$

As a simple check that everything has been done correctly we can verify that all constraints are preserved by the time evolution of the system. Recall that we have only considered constraints without explicit time-dependence so far and so a $\partial/\partial t$ term does not appear. For the first-class constraints, $F = \vec{\phi}^{(1)}$ and $F = \vec{\chi}^{(1)}$, the first term in Eq. (2.9.80) vanishes because of Eqs. (2.9.61) and (2.9.62) and the second and third terms vanish because $\vec{\phi}^{(1)}$ and $\vec{\chi}^{(1)}$ are first class and thus have weakly vanishing Poisson brackets with all constraints. For the second-class constraints, $F = \vec{\xi} = (\vec{\phi}^{(2)}, \vec{\chi}^{(2)})$, we have

$$\dot{\xi}_r \approx \{\xi_r, H\} + \{\xi_r, \vec{\phi}^{(1)}\} \cdot \vec{v} - \sum_{s,t} \{\xi_r, \xi_s\} \Delta_{st}^{-1} \{\xi_t, H\} \approx \{\xi_r, H\} + 0 - \{\xi_r, H\} \approx 0.$$

Thus the formalism is consistent and the rotated Lagrange multipliers $\vec{v}(t)$ remain arbitrary. Note that in the special case where all of the primary constraints are second class and where no secondary constraints are needed, then we observe that the general solution reduces to that given in Eq. (2.9.32) for $\det P \not\approx 0$ as it should.

Conditions on constraint functions: Consider our vector of primary constraint functions $\vec{\phi} = (\phi_1, \ldots, \phi_M)$ that satisfies $\vec{\phi} = 0$ on the constraint submanifold. It is assumed that this set of constraints has been expanded such that there is at least one independent constraint for each dimension of phase space that is constrained. For example, a constraint of the form $\phi(\vec{q}, \vec{p}) = p_1^2 + p_2^2 = c^2$ for some real constant c only constrains one dimension of phase space and is acceptable, whereas a constraint of the form $\phi(\vec{q}, \vec{p}) = p_1^2 + p_2^2 = 0$ should be written as two constraints because it implies that both $p_1 = 0$ and $p_2 = 0$. The latter could be written as two or more constraints in various forms such as (a) $p_1 = 0$, $p_2 = 0$; (b) $p_1^2 = 0$, $p_2^2 = 0$; (c) $p_1^2 = 0$, $p_2 = 0$; (d) $p_1 = 0$, $p_2^2 = 0$; (e) $\sqrt{|p_1|} = 0$, $p^2 = 0$; (f) $p_1 = 0$, $p_2 = 0$, $p_1^2 = 0$; (g) $p_1 = 0$, $p_2 = 0$, $ap_1 + bp_2 = 0$ for $a, b \neq 0$; and so on.

The number of *independent* primary constraints, $M' \leq M$, is the size of the smallest subset of these constraints such that if the M' independent primary constraints are zero then all M primary constraints are zero. For example, if $\phi_1, \ldots, \phi_{M'}$ are the M' independent primary constraints then, $\phi_1 = \cdots = \phi_{M'} = 0$ implies that $\phi_{M'+1} = \cdots = \phi_M = 0$. The L secondary constraints were only introduced because the primary constraints were on their own insufficient to ensure that they remain satisfied at all times; therefore, all of the *secondary constraints are by construction independent of all of the primary constraints* and follow from the equations of motion. We denote $L' < L$ as the number of independent secondary constraints; i.e., if $\chi_1, \ldots, \chi_{L'}$ are the L' independent secondary constraints, then $\chi_1 = \cdots = \chi_{L'} = 0$ implies that $\chi_{L'+1} = \cdots = \chi_L = 0$.

If not all of the constraints are independent, then we say that the constraints are *reducible* (or redundant); otherwise, we say that they *irreducible* (or nonredundant), i.e., if $M' < M$ and/or $L' < L$ then the constraints are reducible, whereas if $M' = M$ and $L' = L$ then they are irreducible.

As we have seen above we have the freedom to write the constraints that define a constraint submanifold in a variety of ways. Consider all possible combinations of the M primary constraints and choose any set $\vec{\phi}^{(c)}$ for which $\phi_1^{(c)}, \ldots, \phi_{M'}^{(c)}$ form a set of independent primary constraints such that the $M' \times 2N$ Jacobian matrix

$$J = \partial\phi_r^{(c)}/\partial(q_s, p_s) \quad \text{for} \quad r = 1, \ldots, M' \quad \text{and} \quad s = 1, \ldots, N \tag{2.9.81}$$

has maximum rank everywhere in the vicinity of the constraint submanifold, i.e., the rank M'. Similarly, for the secondary constraints we choose any combination of them for which $\chi_1^{(c)}, \ldots, \chi_{L'}^{(c)}$ form a set of independent secondary constraints and for which the $L' \times 2N$ Jacobian matrix $\partial\chi_r^{(c)}/\partial(q_s, p_s)$ for $r = 1, \ldots, L'$ and $s = 1, \ldots, N$ has maximum rank everywhere in the vicinity of the constraint submanifold, i.e., the rank L'. If the constraints have these properties we say that they satisfy the *regularity conditions* and allow us to construct the Hamiltonian formalism (Henneaux and Teitelboim, 1994). We define for later convenience $\Phi^{(c)} \equiv (\phi_1^{(c)}, \ldots, \phi_M^{(c)}, \chi_1^{(c)}, \ldots, \chi_L^{(c)})$.

Implications of the regularity conditions are that (i) the set of all independent constraint functions $\phi_1^{(c)}, \ldots, \phi_{M'}^{(c)}, \chi_1^{(c)}, \ldots, \chi_{L'}^{(c)}$ form a linearly independent set of functions in the vicinity of the constraint submanifold and so can be taken as the first $(M' + L')$ coordinates of a coordinate system in this neighborhood; (ii) the gradients of the constraints are linearly independent on the constraint submanifold; and (iii) the constraint functions are differentiable in the vicinity of the constraint submanifold. A further consequence of the regularity conditions is that if a dynamical variable vanishes on the constraint surface, $G(\vec{q}, \vec{p}) \approx 0$, then we can write

$$G(\vec{q}, \vec{p}) = \sum_{r=1}^{M+L} g_r(\vec{q}, \vec{p})\Phi_r(\vec{q}, \vec{p}) \tag{2.9.82}$$

everywhere in phase space for some set of functions $g_r(\vec{q}, \vec{p})$, where $\vec{\Phi} = (\vec{\phi}, \vec{\chi})$ is the vector consisting of all of the constraint functions.

Proof: First prove the result in the vicinity of the constraint submanifold and then extend to the whole of phase space (Henneaux and Teitelboim, 1994).

(a) *In a neighborhood of the constraint submanifold:* Define the vector of independent constraints, $\vec{y} \equiv (\phi_1^{(c)}, \ldots, \phi_{M'}^{(c)}, \chi_1^{(c)}, \ldots, \chi_{L'}^{(c)})$. It follows from the regularity conditions that sufficiently near to the constraint submanifold the components of the vector $\vec{y}$ can form the first $(M' + L')$ coordinates of a coordinate system for phase space. Let $\vec{x}$ denote the remaining $(2N - M' - L')$ coordinates. Then close to the constraint submanifold there is a one-to-one correspondence between $(\vec{q}, \vec{p})$ and $(\vec{y}, \vec{x})$ and we can write $G(\vec{y}, \vec{x})$ in place of $G(\vec{q}, \vec{p})$, where $G(0, \vec{x}) = 0$ since $\vec{y} = 0$ corresponds to the constraint submanifold. It follows that

$$G(\vec{y}, \vec{x}) = \int_0^1 \frac{d}{dt} G(t\vec{y}, \vec{x}) dt = \sum_{r=1}^{(M'+L')} y_r \int_0^1 \frac{\partial G}{\partial(t y_r)}(t\vec{y}, \vec{x}) dt$$

$$= \sum_{r=1}^{(M'+L')} g_r' y_r = \sum_{r=1}^{(M+L)} g_r' \Phi_r^{(c)} = \sum_{r=1}^{(M+L)} g_r \Phi_r, \qquad (2.9.83)$$

where $g_r' \equiv \int_0^1 \partial G/\partial(t y_r)(t\vec{y}, \vec{x}) dt$ for $r = 1, \ldots, (M' + L')$ and $g_r' \equiv 0$ for $r = (M' + L' + 1), \ldots, (M + L)$. Sufficiently close to the constraint submanifold we can always find some g_r such that the last step is true.

(b) *In all of phase space:* We start by decomposing phase space into a set of overlapping open sets (i.e., cover all of phase space with overlapping open regions) and prove the result on each of the open sets. The results on the open sets are then combined together to complete the proof.

We cover the vicinity of the constraint submanifold with the overlapping open sets denoted $O_1, O_2, \ldots$ such that each of these open sets O_k is sufficently close to the constraint submanifold that the proof in (a) above is valid, i.e., such that there is some g_{rk} such that everywhere in the open set O_k we have $G = \sum_r g_{rk} \Phi_r$. We then decompose the remainder of phase space into overlapping open sets $V_1, V_2, \ldots$ that do not intersect the constraint submanifold but that together with the open sets $O_1, O_2, \ldots$ cover all of phase space. We can choose the V_k such that on each of them at least one constraint, denoted $\Phi_{r(k)}$, does not vanish anywhere in V_k, since otherwise V_k would have to intersect the constraint surface. In the open set V_k we can then choose $g_{rk} \equiv G/\Phi_{r(k)}$ so that $G = g_{rk} \Phi_{r(k)}$ is trivially true. For example, we could choose for simplicity that V_k is the entire region of phase space on which $\Phi_k \neq 0$ so that $g_k \equiv G/\Phi_k$ and so trivially $G = g_k \Phi_k = \sum_r (\delta_{rk} g_k) \Phi_r \equiv \sum_r g_{rk} \Phi_r$.

Denote the collection of all of the open sets that cover phase space as S_α, i.e., we denote $(S_1, S_2, S_3, \ldots) \equiv (O_1, O_2, \ldots, V_1, V_2, \ldots)$. We have shown that on every one of the open sets S_α there is a set of $g_{r\alpha}$ such that $G = \sum_r g_{r\alpha} \Phi_r$, but in general $g_{r\alpha} \neq g_{r\beta}$. It remains to construct a single set of g_r that apply over all of phase space, i.e., for all S_α. The technique for doing this is referred to as the *partition of unity*, which allows for local constructions to be extended over an entire space. Let each open region S_α have a function $f_\alpha \geq 0$ that vanishes outside S_α and such that $\sum_\alpha f_\alpha = 1$ for every point $(\vec{q}, \vec{p})$ in phase space and such that only a finite number of f_α are nonzero at any point in phase space. Then we can write

$$G(\vec{q}, \vec{p}) = \sum_\alpha f_\alpha(\vec{q}, \vec{p}) G(\vec{q}, \vec{p}) = \sum_\alpha \sum_r f_\alpha(\vec{q}, \vec{p}) g_{r\alpha}(\vec{q}, \vec{p}) \Phi_r(\vec{q}, \vec{p})$$

$$\equiv \sum_r g_r(\vec{q}, \vec{p}) \Phi_r(\vec{q}, \vec{p}) \qquad (2.9.84)$$

everywhere in phase space, where we have defined $g_r \equiv \sum_\alpha f_\alpha g_{r\alpha}$. While the f_α functions making up the partition of unity are smooth (infinitely differentiable), they are not analytic everywhere and so the functions $g_r(\vec{q}, \vec{p})$ will not be analytic everywhere.

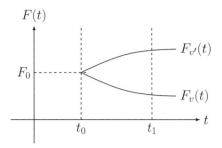

Figure 2.6   Illustration of how the time evolution of the dynamical variable $F(t) \equiv F(\vec{q}(t), \vec{p}(t), t)$ depends on the Lagrange multiplier function $v(t)$.

It is useful to consider some examples (Henneaux and Teitelboim, 1994) of constraints. Consider the examples of primary constraints listed above: the constraints in (a) are acceptable since they are irreducible (i.e., an independent set) and they satisfy the regularity conditions; those in (b)–(e) are irreducible but do not satisfy the regularity conditions and so are not acceptable constraints; those in (f) are acceptable since they are reducible and the first two form an independent set of constraints that satisfy the regularity conditions; and those in (g) are acceptable since they are reducible and any two of the three constraints form an independent set that satisfy the regularity conditions. Note that any constraint functions of the form p_i^2 or $\sqrt{|p_i|}$ lead to a violation of the regularity conditions since $\partial(p_i^2)/\partial p_i = 2p_i$ vanishes at $p_i = 0$ and $\partial(\sqrt{|p_i|})/\partial p_i$ is singular there. This is because the Jacobian matrix involving such constraints is not of constant rank everywhere in the neighborhood of the constraint surface and so they do not satisfy the regularity conditions. Additional regularity conditions required are that the rectangular matrices D and $D^\times$, defined in Eqs. (2.9.54) and (2.9.54), have constant rank K and $K^\times$, respectively, everywhere in the neighborhood of the constraint submanifold.

First-class constraints and gauge transformations: Recall that the constrained forms of Hamilton's equations of motion in Eq. (2.9.17) have been subsumed into Eq. (2.9.80), which contains the $(M - K)$ arbitrary time-dependent Lagrange multiplier functions $\vec{v}(t)$, i.e., one $v_r(t)$ for each first-class primary constraint $\phi_r^{(1)}$.

Consider the effect on the time evolution of some dynamical variable $F(\vec{q}, \vec{p}, t)$ due to a change in $\vec{v}(t)$ as shown in Fig. 2.6. Let F at $t = t_0$ be denoted as F_0, then the value of F for $t \neq t_0$ will depend on the choice of $\vec{v}(t)$. For notational brevity let us temporarily define $F(t) \equiv F(\vec{q}(t), \vec{p}(t), t)$. Then we denote $F_v(t)$ and $F_{v'}(t)$ as the results for $F(t)$ when we have chosen Lagrange multipliers $\vec{v}$ and $\vec{v}'$, respectively, where $F_0 = F_v(t_0) = F_{v'}(t_0)$. From Eq. (2.9.80) we find at $t = t_0$ that

$$\dot{F}_{v'} - \dot{F}_v = \{F_0, \vec{\phi}^{(1)}\} \cdot [\vec{v}' - \vec{v}] \,. \tag{2.9.85}$$

Since dt is infinitesimal, we can write at time $t = t_0 + dt$,

$$dF = dF_{v'} - dF_v = (\dot{F}_{v'} - \dot{F}_v)dt = \{F, \vec{\phi}^{(1)}\} \cdot [\vec{v}' - \vec{v}]\, dt \equiv \{F, \vec{\phi}^{(1)}\} \cdot d\vec{\alpha}, \tag{2.9.86}$$

where dF is the change in F over time dt due to the different choice of Lagrange multipliers $\vec{v}(t)$ and where $d\vec{\alpha} \equiv [\vec{v}' - \vec{v}]\, dt$. Equivalently, this can be expressed as

$$\partial F/\partial \alpha_r = \{F, \phi_r^{(1)}\} \quad \text{for} \quad r = 1, \ldots, (M - K). \tag{2.9.87}$$

Comparison of this equation with that for a canonical transformation in Eq. (2.4.66) shows that variation of each of the Lagrange multipliers associated with the first-class primary constraints v_r leads to the corresponding first-class constraint $\phi_r^{(1)}$ generating a canonical transformation.

Consider the above arguments with F being replaced by the coordinates $\vec{q}$ and momenta $\vec{p}$ for the system. We see that the time evolution of $\vec{q}$ and $\vec{p}$ depend on the arbitrary $\vec{v}$ and so there are an infinite number of phase space trajectories that are physically equivalent, i.e., that correspond to the same time evolution for the physical state of the system. The physical state of a classical system is uniquely specified by the values of its physical observables. Then if the dynamical variable F is a physical observable, it must be independent of the choice of Lagrange multipliers, i.e., $F_v = F_{v'}$ for all $v(t)$ and $v'(t)$.

By analogy with electromagnetism and gauge theories this invariance is referred to as *gauge invariance*. The $(M - K)$ first-class primary constraints $\vec{\phi}^{(1)}$ generate infinitesimal canonical transformations that do not change the physical state of the system and we refer to these as *gauge transformations*. The number of independent first-class primary constraints, $(M' - K)$, is the number of independent gauge transformations that are generated. A physical observable cannot depend on which physically equivalent phase-space trajectory was followed and so a physical observable must be gauge invariant; i.e., if $F_{\text{phys}}(\vec{q}, \vec{p})$ is a physically observable dynamical variable, then we must have $\{F_{\text{phys}}, \phi_r^{(1)}\} \approx 0$.

We can now generalize the above discussion and *define any canonical transformation that does not change the physical state of the system under an infinitesimal time evolution of the form of Eq. (2.9.86) as a gauge transformation.* So a physical system with a gauge symmetry will have different phase-space trajectories that correspond to the same physical time evolution of the system. If G is a generator of canonical transformations G, then for $F \to F + \alpha dF$ we must have

$$dF = \{F, G\} \cdot d\alpha \tag{2.9.88}$$

for any dynamical variable F. If we have two arbitrary functions $v(t)$ and $v'(t)$ and if in analogy with the above we define $d\alpha \equiv (v' - v)dt$ and look at the contribution of this transformation on the time evolution of F we find

$$dF = \{F, G\}(v' - v)dt, \tag{2.9.89}$$

which is the effect of deflecting the phase-space evolution onto a different phase-space trajectory. If this transformation does not change the physical state of the system, then G is a generator of gauge transformations. In this case we could make the replacement $H_T \to H_T + Gv$ in Eq. (2.9.77) and the evolution of the physical state of the system would not change, although the particular phase-space trajectory that was followed would depend on the arbitrary and physically irrelevant choice of the function $v(t)$.

Dirac observed that, for physical systems he had considered, the first-class secondary constraints $\vec{\chi}^{(1)}$ generated gauge transformations just as the first-class primary constraints did. On this basis Dirac (2001) conjectured that it was generally true that all first-class constraints are generators of gauge transformations. However, counterexamples to this *Dirac conjecture* exist and so the conjecture in general is false. However, as shown in Henneaux and Teitelboim (1994) for systems that satisfy a few additional simple conditions the Dirac conjecture can be proven. This is the case for physical systems of interest to us.

Note that the above arguments link infinitesimal gauge transformations with infinitesimal changes to a phase-space trajectory that do not change the physical state of the system. While it may be tempting to try to infer from this that physically equivalent phase-space trajectories with a finite separation in phase space are related by a finite gauge transformation, this is not true in general.[6]

Extended Hamiltonian: We will restrict our attention to systems that satisfy the conditions required for Dirac's conjecture to be true so that *all* first-class constraints are generators of gauge transformations. As discussed above we can add a Lagrange multiplier $w_r(t)$ for each of the $(L-K^\chi)$ first-class secondary constraints $\chi_r^{(1)}$ and define the *extended Hamiltonian*

$$H_E(\vec{q}, \vec{p}, \vec{v}, \vec{w}) \equiv H_T(\vec{q}, \vec{p}, \vec{v}) + \vec{\chi}^{(1)} \cdot \vec{w} \tag{2.9.90}$$

$$\approx H + \vec{\phi}^{(1)} \cdot \vec{v} + \vec{\chi}^{(1)} \cdot \vec{w} - \sum_{r,s=1}^{(K+K^\chi)} \xi_r \Delta_{rs}^{-1} \{\xi_s, H\}$$

so that for any dynamical variable $F(\vec{q}, \vec{p}, t)$ the time evolution is given by

$$\dot{F} \approx \frac{\partial F}{\partial t} + \{F, H_E\} \approx \frac{\partial F}{\partial t} + \{F, H\} + \{F, \vec{\phi}^{(1)}\} \cdot \vec{v} + \{F, \vec{\chi}^{(1)}\} \cdot \vec{w}$$

$$- \sum_{r,s=1}^{(K+K_\chi)} \{F, \xi_r\} \Delta_{rs}^{-1} \{\xi_s, H\}. \tag{2.9.91}$$

If F_{phys} is a physical observable, then it must be gauge invariant and so it follows from Eq. (2.9.88) that $\{F_{\text{phys}}, \vec{\phi}^{(1)}\} \approx 0$ and $\{F_{\text{phys}}, \vec{\chi}^{(1)}\} \approx 0$ and so

$$\dot{F}_{\text{phys}} \approx \frac{\partial F_{\text{phys}}}{\partial t} + \{F_{\text{phys}}, H_E\} \approx \frac{\partial F_{\text{phys}}}{\partial t} + \{F_{\text{phys}}, H'\}, \tag{2.9.92}$$

where H' is given in Eq. (2.9.78). The time evolution of F_{phys} is independent of the choice of $\vec{v}(t)$ and $\vec{w}(t)$ since it is a gauge-invariant quantity. If F is a gauge-dependent dynamical variable, then its time evolution will depend on the arbitrary Lagrange multiplier functions $\vec{v}(t)$ and $\vec{w}(t)$. Since from Eq. (2.9.70) we know that H_T is first class, then H_E must also be first class $\{\vec{\Phi}^{\text{rot}}, H_E\} \approx \{\vec{\Phi}^{\text{rot}}, H_T\} \approx \{\vec{\Phi}^{\text{rot}}, H'\} \approx 0$ and so H_E preserves all constraints as expected, $d\vec{\Phi}/dt \approx 0$.

Generalizing to include explicit time-dependence: If the Lagrangian has explicit time dependence, $L(\vec{q}, \dot{\vec{q}}, t)$, then the constraint submanifold is $(2N + 1)$-dimensional space $(\vec{q}, \vec{p}, t)$ and the primary and secondary constraints will be of the form $\vec{\phi}(\vec{q}, \vec{p}, t) \approx 0$ and $\vec{\chi}(\vec{q}, \vec{p}, t) \approx 0$, respectively. The constrained forms of Hamilton's equations now have the additional equation $\partial H/\partial t = -\partial L/\partial t$ in Eq. (2.4.12).

The changes to our discussion will occur when we calculate time derivatives of the constraints; e.g., the partial time derivatives of constraints will appear in the constraint consistency conditions for the primary and secondary constraints

$$0 \approx \dot{\phi}_r = \partial \phi_r/\partial t + \{\phi_r, H_T\} \quad \text{and} \quad 0 \approx \dot{\chi}_r = \partial \chi_r/\partial t + \{\chi_r, H_T\}, \tag{2.9.93}$$

(see pp. 59–60 in Sec. II.2 of Sundermeyer, 1982). The secondary constraints $\vec{\chi}$ are chosen to ensure that the possibly time-dependent primary constraints $\vec{\phi}$ are satisfied at all times. The secondary constraints are also typically time-dependent.

Working through the derivation while allowing explicit time dependence we find that Eq. (2.9.30) corresponding to the case $\det P \not\approx 0$ with $P_{rs} = \{\phi_r, \phi_s\}$ becomes

$$\lambda_r \approx -\sum_{s=1}^{M} (P^{-1})_{rs} \left(\partial \phi_s/\partial t + \{\phi_s, H\}\right). \tag{2.9.94}$$

In place of Eq. (2.9.31) we have

$$H_T = H + \vec{\phi} \cdot \vec{\lambda} \approx H - \sum_{r,s=1}^{M} \phi_r (P^{-1})_{rs} \left(\partial \phi_s/\partial t + \{\phi_s, H\}\right) \tag{2.9.95}$$

---

[6] See Sundermeyer (1982, 2014) and Pons (2004) and references therein.

and in place of Eq. (2.9.32) we have $\dot{F} \approx \partial F/\partial t + \{F, H_T\}$ so that

$$\dot{F}(\vec{q}, \vec{p}, t) \approx \partial F/\partial t + \{F, H\} - \sum_{r,s=1}^{M} \{F, \phi_r\}(P^{-1})_{rs} \left(\partial \phi_s/\partial t + \{\phi_s, H\}\right). \qquad (2.9.96)$$

Progressing further with the rederivation we find in place of Eq. (2.9.46) corresponding to $\det P \approx 0$ that

$$0 \approx \partial \vec{\Phi}/\partial t + \{\vec{\Phi}, H\} + D\vec{\lambda} \qquad (2.9.97)$$

and in place of Eq. (2.9.66) we have

$$0 \approx \partial \xi_r/\partial t + \{\xi_r, H\} + \sum_{s=1}^{K} \{\xi_r, \phi_s^{(2)}\} u_s. \qquad (2.9.98)$$

Then we observe that Eq. (2.9.77) is replaced by

$$H_T \approx H + \vec{\phi}^{(1)} \cdot \vec{v} - \sum_{r,s=1}^{(K+K^X)} \xi_r \Delta_{rs}^{-1} \left(\partial \xi_s/\partial t + \{\xi_s, H\}\right) \approx H' + \vec{\phi}^{(1)} \cdot \vec{v}, \qquad (2.9.99)$$

where

$$H' \equiv H - \sum_{r,s=1}^{(K+K^X)} \xi_r \Delta_{rs}^{-1} \left(\partial \xi_s/\partial t + \{\xi_s, H\}\right). \qquad (2.9.100)$$

Finally, it follows that Eq. (2.9.80) is replaced by

$$\dot{F} \approx \partial F/\partial t + \{F, H_T\} \approx \partial F/\partial t + \{F, H\} + \{F, \vec{\phi}^{(1)}\} \cdot \vec{v}$$
$$- \sum_{r,s} \{F, \xi_r\} \Delta_{rs}^{-1} \left(\partial \xi_s/\partial t + \{\xi_s, H\}\right). \qquad (2.9.101)$$

For systems satisfying the Dirac conjecture the addition of the first-class secondary constraints to give the extended Hamiltonian proceeds as before,

$$H_E \equiv H_T + \vec{\chi}^{(1)} \cdot \vec{w}, \qquad (2.9.102)$$

and the corresponding time evolution is again given by

$$\dot{F} \approx \partial F/\partial t + \{F, H_E\}. \qquad (2.9.103)$$

The regularity conditions on the constraints are such that if there are M and L independent primary and secondary constraints denoted $\vec{\phi}$ and, $\vec{\chi}$ respectively, then the $M \times (2N+1)$ Jacobian matrix $\partial \phi_r/\partial(q_s, p_s, t)$ and the $L \times (2N+1)$ Jacobian matrix $\partial \chi_r/\partial(q_s, p_s, t)$ must have maximum rank everywhere on the constraint submanifold at all times (i.e., rank M and rank L, respectively).

2.9.2 Summary of the Dirac-Bergmann Algorithm

The Dirac-Bergmann algorithm for a singular system may be summarized as follows.

(i) Begin with a singular Lagrangian $L(\vec{q}, \dot{\vec{q}}, t)$, which might have explicit time dependence. Construct the conjugate momenta $\vec{p} \equiv \partial L/\partial \dot{\vec{q}}$ and the standard Hamiltonian $H(\vec{q}, \vec{p}, t)$ using the Legendre transform in the usual way. Since the system is singular, this transform is not invertible and there must exist some set of primary constraints. Identify these constraints and write them in such a way that they satisfy the regularity conditions, which gives some number, M, of primary constraints $\vec{\phi}(\vec{q}, \vec{p}, t) = 0$.

(ii) Construct the $M \times M$ matrix P consisting of the Poisson brackets of the primary constraint functions, $P_{rs} \equiv \{\phi_r, \phi_s\}$.

(iii) If P is nonsingular on the primary constraint submanifold (i.e., $\det P \not\approx 0$), then invert the matrix P and solve for the Lagrange multipliers $\vec{\lambda}$ using Eq. (2.9.94). In this case there are only second-class primary constraints, i.e., no first-class primary constraints and no secondary constraints. Any dynamical variable $F(\vec{q}, \vec{p}, t)$ evolves in time according to $\dot{F} \approx \partial F/\partial t + \{F, H'\}$, where H' is given in Eq. (2.9.100). This is just Eq. (2.9.103) when there are no first-class constraints, since it follows in this case that $H_E = H_T = H'$. The algorithm is then complete in this case.

(iv) If P is singular on the primary constraint submanifold (i.e., $\det P \approx 0$), then some Lagrange multipliers are undetermined and additional constraints may be required to ensure that the time evolution of the system does not take it off the primary constraint submanifold. These are called secondary constraints. Introduce these secondary constraints to ensure that the primary constraints remain satisfied at all times. It may then be necessary to introduce further secondary constraints to ensure that these additional constraints remain satisfied at all times as the system evolves in time. This process is to be repeated until no further secondary constraints are necessary. Then rewrite these additional constraints such that they satisfy the regularity conditions and such that they are independent of the set of primary constraints (i.e., no combination of them follows from the primary constraints alone). These then are the L secondary constraints of the form $\vec{\chi}(\vec{q}, \vec{p}, t) \approx 0$.

(v) Divide the primary and secondary constraints into first and second class, where first-class constraints have a vanishing Poisson bracket with all other constraints. Constraints that are not first class are defined to be second class. This division can be performed by defining the $(M + L)$-component constraint vector $\vec{\Phi} \equiv (\phi_1, \ldots, \phi_M, \chi_1, \ldots, \chi_L)$ and then defining the $(M + L) \times M$ matrix D, where $D_{rs} \equiv \{\Phi_r, \phi_s\}$. Perform a singular value decomposition of D and construct the rotated primary constraint functions $\vec{\phi}'$, where each ϕ'_r is either first or second class. Repeat for the secondary constraints using D^χ, where $D^\chi \equiv \{\Phi_r, \chi_s\}$. This then gives the constraint functions $\vec{\phi}^{(1)}$, $\vec{\phi}^{(2)}$, $\vec{\chi}^{(1)}$ and $\vec{\chi}^{(2)}$ as given in Eq. (2.9.57). Let K and K^χ denote the number of second-class primary and second-class secondary constraints, respectively.

(vi) Let the $(K + K^\chi)$-component vector $\vec{\xi} = (\vec{\phi}^{(2)}, \vec{\chi}^{(2)})$ denote the set of all second-class constraint functions. Note that $(K + K^\chi)$ will always be an even number. Then construct the $(K + K^\chi) \times (K + K^\chi)$ matrix Δ, where $\Delta \equiv \{\xi_r, \xi_s\}$, which will always be nonsingular on the constraint submanifold. Invert Δ on the constraint submanifold.

(vii) Then construct the total Hamiltonian $H_T(\vec{q}, \vec{p}, \vec{v})$ as given in Eq. (2.9.99), where H_T depends on all of the $(K + K^\chi)$ second-class constraints and on the $(M - K)$ arbitrary Lagrange multiplier functions $\vec{v}(t)$ for the first-class primary constraints. The first class secondary constraints do not appear in H_T. Any dynamical variable $F(\vec{q}, \vec{p}, t)$ evolves in time according to Eq. (2.9.101), i.e., $\dot{F} \approx \partial F/\partial t + \{F, H_T\}$. Denote the set of all first- and second-class rotated constraints as $\vec{\Phi}^{\rm rot} = (\vec{\phi}^{(1)}, \vec{\chi}^{(1)}, \vec{\xi}) = (\vec{\phi}^{(1)}, \vec{\chi}^{(1)}, \vec{\phi}^{(2)}, \vec{\chi}^{(2)})$. In the case of time-independent constraints the total Hamiltonian H_T is first class, $\{\vec{\Phi}^{\rm rot}, H_T\} \approx 0$, which means that the constraint is true at all times, $\dot{\Phi}^{\rm rot}_j \approx 0$. In the case of time-dependent constraints we have in any case $\dot{\Phi}^{\rm rot}_j \approx \partial \Phi^{\rm rot}_j/\partial t + \{\Phi^{\rm rot}_j, H_T\} \approx 0$. This ensures that any constraint $\Phi^{\rm rot}_j$ is satisfied at all times. The dynamics of the system is then fully solved.

(viii) If the system satisfies Dirac's conjecture that all first-class constraints generate gauge transformations, then the full gauge freedom resulting from all of the first-class constraints can be made manifest by defining the extended Hamiltonian $H_E(\vec{q}, \vec{p}, \vec{v}, \vec{w})$, where $H_E \equiv H_T + \vec{\chi}^{(1)} \cdot \vec{w}$, where $\vec{w}$ are arbitrary Lagrange multiplier functions for the $(L - K^\chi)$ first-class secondary constraints $\vec{\chi}^{(1)}$. In that case the time evolution for any F is given by Eq. (2.9.103), i.e., in that case $\dot{F} \approx \partial F/\partial t + \{F, H_E\}$. For time-independent constraints the extended Hamiltonian H_E is first class, $\{\vec{\Phi}^{\text{rot}}, H_E\} \approx 0$, and so $\dot{\Phi}_j^{\text{rot}} \approx 0$, which ensures that the constraints remain true at all times. In the case of time-dependent constraints we have in any case $\dot{\Phi}_j^{\text{rot}} \approx \partial \Phi_j^{\text{rot}}/\partial t + \{\Phi_j^{\text{rot}}, H_E\} \approx 0$. The change from H_T to H_E only affects gauge-dependent dynamical variables. For a gauge-invariant dynamical variable we have $\dot{F}_{\text{phys}} \approx \partial F_{\text{phys}}/\partial t + \{F_{\text{phys}}, H_E\} \approx \partial F_{\text{phys}}/\partial t + \{F_{\text{phys}}, H_T\} \approx \partial F_{\text{phys}}/\partial t + \{F_{\text{phys}}, H'\}$; i.e., the time evolution of gauge invariant quantities is determined only by the secondary constraints and is independent of the arbitrary functions $\vec{v}$ and $\vec{w}$ that multiply the first-class constraint functions.

(ix) Define $\Delta H_E \equiv G \equiv \Delta \vec{v} \cdot \vec{\phi}^{(1)} + \Delta \vec{w} \cdot \vec{\chi}^{(1)}$ as the change to the extended Hamiltonian caused by changing the Lagrange multipliers for the first-class constraints, i.e., $\vec{v} \to \vec{v}' = \vec{v} + \Delta \vec{v}$ and $\vec{w} \to \vec{w}' = \vec{w} + \Delta \vec{w}$. Since $\dot{F} \approx \partial F/\partial t + \{F, H_E\}$, then the change to F due to the change in Lagrange multipliers is $dF_G = \{F, \Delta H_E\}dt = \{F, G\}dt$ for any dynamical functional F from the equations of motion. We then have Eq. (2.9.88), which is that G generates a canonical transformation on the constraint submanifold, which does not change physical quantities and is then by definition a gauge transformation. The set of all gauge transformations consists of all of the ways of choosing $\Delta \vec{v}$ and $\Delta \vec{w}$ in G. The set of gauge transformations has $(M-K) + (L - K_\chi)$ generators, where each first-class constraint function is a generator.

The above is for the most general case; physical systems of interest are often much simpler and as such will not require the full machinery of the algorithm outlined above. We will consider the case of electromagnetism in Sec. 3.3.2.

It is of course possible that the various primary and secondary constraints are not all consistent with each other. When this happens it simply means that there is no classically consistent theory corresponding to our initial Lagrangian. Dirac gives the simple example of $L = q$ for a single degree of freedom q. The Euler-Lagrange equations, Eq. (2.1.31), are then given by $1 = 0$, which is inconsistent. Another way to appreciate the inconsistency of this Lagrangian is to observe that Hamilton's principle has no solution, i.e., $\delta S/\delta q(t) = 1 \neq 0$, where $S = \int dt L = \int dt q(t)$. Similarly, the Lagrangian $L = a\dot{q} + V(q)$ has the Euler-Lagrange equations $dV/dq = 0$ and so the system only has solutions if $V(q)$ is a constant. So, not every Lagrangian that we can write down will correspond to a consistent classical system.

For a number of illustrative worked examples, see chapter 2 of Sundermeyer (1982) and chapter 1 of Henneaux and Teitelboim (1994). For an extensive set of exercises, see chapter 1 of Henneaux and Teitelboim (1994).

First-class constraints, gauge transformations and Lie groups: We have seen that all primary first-class constraints generate a set of canonical transformations, referred to as gauge transformations, where by definition a gauge transformation leaves the physical state of the system unchanged. All singular systems of physical interest appear to satisfy the Dirac conjecture; i.e., it seems that in systems of physical interest *all* first-class constraints, both primary and secondary, generate gauge transformations. For such systems let the $[(M-K) + (L - K^\chi)]$-component vector

$$\vec{\zeta} = (\vec{\phi}^{(1)}, \vec{\chi}^{(1)}) \tag{2.9.104}$$

denote the set of all first-class constraint functions.

Two dynamical variables F and G are first class provided that their Poisson brackets with all constraints vanish at least weakly; i.e., they must satisfy Eq. (2.9.47),

$$\{\vec{\Phi}, F\} \approx 0 \quad \text{and} \quad \{\vec{\Phi}, G\} \approx 0, \tag{2.9.105}$$

where recall that $\vec{\Phi}$ is the $(M+L)$ component vector containing all of the M primary and L secondary constraints, $\vec{\Phi} = (\vec{\phi}, \vec{\chi})$. Note that the Poisson bracket of two first-class dynamical variables $\{F, G\}$ is also first class,

$$\{\vec{\Phi}, \{F, G\}\} \approx 0. \tag{2.9.106}$$

Proof: From Eq. (2.9.82) we know that any weakly vanishing function can be written as a superposition of the constraint functions and so there must exist $f_{ij}(\vec{q}, \vec{p})$ and $g_{ij}(\vec{q}, \vec{p})$ such that

$$\{\Phi_i, F\} = \sum_j f_{ij}\Phi_j \quad \text{and} \quad \{\Phi_i, G\} = \sum_j g_{ij}\Phi_j. \tag{2.9.107}$$

From the Jacobi identity and the properties of the Poisson bracket we have

$$\{\Phi_i, \{F, G\}\} = -\{F, \{G, \Phi_i\}\} - \{G, \{\Phi_i, F\}\} = \left\{F, \sum_j g_{ij}\Phi_j\right\} - \left\{G, \sum_j f_{ij}\Phi_j\right\}$$

$$= \sum_j [g_{ij}\{F, \Phi_j\} + \{F, g_{ij}\}\Phi_j - f_{ij}\{G, \Phi_j\} - \{G, f_{ij}\}\Phi_j] \approx 0, \tag{2.9.108}$$

where we have used $\Phi_j \approx 0$ and Eq. (2.9.105).

Because the $\vec{\zeta}$ are first class then $\{\vec{\Phi}, \zeta_j\} \approx 0$ and in particular $\{\zeta_i, \zeta_j\} \approx 0$. From Eq. (2.9.82) we know that there must exist functions $g_{ijk}(\vec{q}, \vec{p})$ such that

$$\{\zeta_i, \zeta_j\} = \sum_{k=1}^{(M+L)} g_{ijk}\Phi_k. \tag{2.9.109}$$

However from Eq. (2.9.106) we know that the Poisson bracket of two first-class constraint functions is also first class, $\{\vec{\Phi}, \{\zeta_i, \zeta_j\}\} \approx 0$. So only the first-class constraint functions $\vec{\zeta}$ in $\vec{\Phi}$ can appear on the right-hand side of the above equation and again using Eq. (2.9.82) it follows that

$$\{\zeta_i, \zeta_j\} = \sum_{k=1}^{(K+K^\chi)} f_{ijk}\zeta_k \tag{2.9.110}$$

for some functions $f_{ijk}(\vec{q}, \vec{p})$. We see that at each point $(\vec{q}, \vec{p})$ in phase space the first-class constraint functions form a Lie algebra with the Poisson bracket playing the role of the Lie bracket (see Sec. A.7.2).

Define the set of operators $\hat{T}_k$ for $k = 1, \ldots, (K + K^\chi)$ that act on the space of dynamical variables as

$$\hat{T}_k F \equiv \{F, \zeta_k\}, \tag{2.9.111}$$

where $F(\vec{q}, \vec{p}, t)$ is any dynamical variable. The $\hat{T}_k$ are the generators of the canonical transformations corresponding to gauge transformations,

$$dF = \hat{T}_k F d\alpha = \{F, \zeta_k\} d\alpha. \tag{2.9.112}$$

We can calculate the commutator of two of these operators for any dynamical variable $F(\vec{q}, \vec{p}, t)$,

$$[\hat{T}_i, \hat{T}_j]F = \hat{T}_i \hat{T}_j F - \hat{T}_j \hat{T}_i F = \{\hat{T}_j F, \zeta_i\} - \{\hat{T}_i F, \zeta_j\} = \{\{F, \zeta_j\}, \zeta_i\} - \{\{F, \zeta_i\}, \zeta_j\}$$

$$= -\{\{\zeta_j, \zeta_i\}, F\} - \{\{\zeta_i, F\}, \zeta_j\} - \{\{F, \zeta_i\}, \zeta_j\} = \{F, \{\zeta_j, \zeta_i\}\} \tag{2.9.113}$$

$$= \sum_k \{F, f_{ijk}\zeta_k\} = \sum_k [f_{ijk}\{F, \zeta_k\} + \{F, f_{ijk}\}\zeta_k] = \sum_k (f_{ijk}\hat{T}_k F + \{F, f_{ijk}\}\zeta_k),$$

where we have made use of the Jacobi identity for the Poisson bracket. If the f_{ijk} are independent of the $\vec{q}$ and $\vec{p}$, then $\{F, f_{ijk}\} = 0$ and we have

$$[\hat{T}_i, \hat{T}_j] = \sum_k f_{ijk} \hat{T}_k. \tag{2.9.114}$$

So, if the f_{ijk} are constants, then we see that the generators of the gauge transformations $\hat{T}_k$ are the generators of a Lie group. If the $f_{ijk}(\vec{q}, \vec{p})$ are not constants, then since $\vec{\zeta} \approx 0$ we still find that the group property holds weakly,

$$[\hat{T}_i, \hat{T}_j] \approx \sum_k f_{ijk} \hat{T}_k; \tag{2.9.115}$$

i.e., the Lie algebra property of the generators of gauge transformations is still valid on the constraint submanifold (see sec. III.2 of Sundermeyer (1982)).

2.9.3 Gauge Fixing, the Dirac Bracket and Quantization

To remove the unphysical gauge freedom associated with the gauge transformations generated by the first-class constraints, we can impose *additional* constraints that further restrict the constraint submanifold to an *ideally gauge-fixed* submanifold.

Counting degrees of freedom: By definition on the ideally gauge-fixed submanifold there is a one-to-one correspondence between the phase-space points on this new submanifold and the physical states of the system. This definition corresponds to that of ideal gauge fixing in the context of electomagnetism given in Eq. (2.7.77). As we will later come to appreciate, finding constraints that lead to an ideally gauge-fixed submanifold is a challenging task.

Recall that the $(K + K^\chi)$ second-class constraints were independent of each other and independent of all of the first-class constraints by construction. Also recall that we defined $M' \le M$ and $L' \le L$ to be the number of independent primary and secondary constraints, respectively, and that $\vec{\zeta}$ is the set of all of the $[(M - K) + (L - K^\chi)]$ first-class constraint functions as defined in Eq. (2.9.104). The total number of *independent* first-class and second-class constraints, respectively, are

$$N_f \equiv [(M' - K) + (L' - K^\chi)] \qquad \text{and} \qquad N_s \equiv (K + K^\chi), \tag{2.9.116}$$

where we recall that N_s must be even.

For convenience, let us renumber the primary and secondary first-class constraints such that the first $(M' - K)$ and the first $(L' - K^\chi)$ of them, respectively, are independent. We can then discard the $(M - M') + (L - L')$ irrelevant redundant first-class constraints and *redefine* ζ as the N_f-component vector of *independent first-class constraint functions*

$$\vec{\zeta} = (\zeta_1, \ldots, \zeta_{N_f}) \equiv (\phi_1^{(1)} \ldots, \phi_{(M'-K)}^{(1)}, \chi_1^{(1)}, \ldots, \chi_{(L'-K^\chi)}^{(1)}). \tag{2.9.117}$$

We will assume that we are working with systems that satisfy the Dirac conjecture, such that there are then N_f generators of gauge transformations. Gauge fixing will therefore require an additional set of N_f independent constraints, $\vec{\Omega} = (\Omega_1, \ldots, \Omega_{N_f})$, to remove all gauge freedom. Allowing for the possibility of time-dependent constraints, the gauge-fixing constraints will have the form

$$\Omega_i(\vec{q}, \vec{p}, t) = 0 \quad \text{for} \quad i = 1, \ldots, N_f. \tag{2.9.118}$$

Let us now define the vector of *all independent constraint functions* including the N_f first-class ($\vec{\zeta}$), the N_f gauge fixing ($\vec{\Omega}$) and the N_s second-class ($\vec{\xi}$) constraint functions as

$$\vec{\psi} = (\psi_1, \ldots, \psi_{N_c}) \equiv (\zeta_1, \ldots, \zeta_{N_f}, \Omega_1, \ldots, \Omega_{N_f}, \xi_1, \ldots, \xi_{N_s}), \tag{2.9.119}$$

where we have defined the *total number of independent constraint functions* as

$$N_c \equiv 2N_f + N_s, \tag{2.9.120}$$

where N_c must be even.

Recall that in our initial system we had N degrees of freedom, which gave rise to a $2N$-dimensional phase space. After including all necessary constraints and then gauge-fixing to remove all unphysical path variations we arrive at a *physical submanifold* in phase space with $2N_{\text{phys}}$ dimensions, where

$$2N_{\text{phys}} \equiv 2N - N_c = 2N - 2N_f - Ns \tag{2.9.121}$$

and where N_{phys} is the number of physical degrees of freedom remaining. In the presence of time-dependent constraints, where the number of first- and second-class constraints does not change with time, the physical submanifold will have the same number of dimensions but will itself evolve in time.

Gauge fixing: Consider the real antisymmetric $N_c \times N_c$ matrix G with elements $G_{ij} \equiv \{\psi_i, \psi_j\}$ such that

$$G \equiv \begin{pmatrix} \{\psi_1, \psi_1\} & \{\psi_1, \psi_2\} & \cdots & \{\psi_1, \psi_{N_c}\} \\ \{\psi_2, \psi_1\} & \{\psi_2, \psi_2\} & \cdots & \{\psi_2, \psi_{N_c}\} \\ \vdots & \vdots & \ddots & \vdots \\ \{\psi_{N_c}, \psi_1\} & \{\psi_{N_c}, \psi_2\} & \cdots & \{\psi_{N_c}, \psi_{N_c}\} \end{pmatrix}. \tag{2.9.122}$$

Since all of the constraint functions ψ_i are independent, then no row of G can be expressed as a linear combination of the other rows and similarly for the columns. Therefore G must have maximal rank, i.e., N_c. It is therefore nonsingular,

$$\det G \not\approx 0, \tag{2.9.123}$$

and invertible. *All of the constraints ψ_i must be second class* with respect to the enlarged set of N_c constraints $\vec{\psi}$, since if any ψ_i had a vanishing Poisson bracket with all of the constraints $\vec{\psi}$, then $G_{ij} = \{\psi_i, \psi_j\} = 0$ for all $j = 1, \ldots, N_c$ and so G would not have maximal rank.

The implementation of gauge fixing has the effect of turning all of the enlarged set of constraints $\vec{\psi}$ into second-class constraints. But of course we know how to solve a system with only second-class constraints; i.e., see step (iii) of the Dirac-Bergmann algorithm. With the replacements $\vec{\xi} \to \vec{\psi}$ and $\Delta \to G$ we then have the equations of motion for our gauge-fixed system that has only second-class constraints $\vec{\psi}$

$$\dot{F} \approx \partial F/\partial t + \{F, H\} - \sum_{i,j=1}^{N_c} \{F, \psi_i\} G_{ij}^{-1} \left(\partial \psi_j/\partial t + \{\psi_j, H\}\right). \tag{2.9.124}$$

By definition any time-dependent constraint has the form $\psi_i(\vec{q}(t), \vec{p}(t), t) \approx 0$ for all times t. As a check on consistency, it follows from Eq. (2.9.124) that

$$\dot{\psi}_k \approx \partial \psi_k/\partial t + \{\psi_k, H\} - \sum_{i,j=1}^{N_c} \{\psi_k, \psi_i\} G_{ij}^{-1} \left(\partial \psi_j/\partial t + \{\psi_j, H\}\right) \approx 0, \tag{2.9.125}$$

since $\sum_i \{\psi_k, \psi_i\} G_{ij}^{-1} = \delta_{kj}$. This confirms that the equations of motion keep the system on the $(2N_{\text{phys}})$-dimensional physical submanifold at all times.

2.9.4 Dirac Bracket and Canonical Quantization

In the case of time-independent constraints we define the *Dirac bracket*

$$\{A, B\}_{\text{DB}} \equiv \{A, B\} - \sum_{i,j}^{N_c} \{A, \psi_i\} G_{ij}^{-1} \{\psi_j, B\}. \tag{2.9.126}$$

Using the properties of the Poisson bracket and the fact that G is antisymmetric, it follows that the Dirac bracket also satisfies the five properties of the Poisson bracket listed in Sec. 2.4. The properties of closure, antisymmetry, linearity and the product rule are straightforward to show. The Jacobi identity for the Dirac bracket also follows but requires some additional effort to demonstrate. A proof is provided in the appendix of Dirac (1950). See also the discussion in Dirac (1958). Note that since $G_{ij} \equiv \{\psi_i, \psi_j\}$ then $\sum_j \{\psi_i, \psi_j\} G_{jk}^{-1} = \delta_{ik}$ and so we find that the Dirac bracket of any dynamical variable F with any constraint ψ_i vanishes,

$$\{\psi_i, F\}_{\text{DB}} = \{\psi_i, F\} - \sum_{j,k=1}^{N_c} \{\psi_i, \psi_j\} G_{jk}^{-1} \{\psi_k, F\} = 0. \tag{2.9.127}$$

The Dirac bracket is the extension of the Poisson bracket to the situation where the physical system is constrained at all times to lie in the constraint submanifold. In this case the fundamental Poisson brackets of Eq. (2.4.28) are replaced by the *fundamental Dirac brackets*, which need to be evaluated and will depend on the constraints $\{q_i, p_j\}_{\text{DB}}$, $\{q_i, q_j\}_{\text{DB}}$ and $\{p_i, p_j\}_{\text{DB}}$. From above $\{\psi_i, \psi_j\}_{\text{DB}} = 0$ so that all constraints are now *first class in the context of the Dirac bracket*.

If we restrict ourselves to systems with time-independent constraints, then the equations of motion simplify to

$$\dot{F} \approx \partial F/\partial t + \{F, H\} - \sum_{i,j=1}^{N_c} \{F, \psi_i\} G_{ij}^{-1} \{\psi_j, H\} \approx \partial F/\partial t + \{F, H\}_{\text{DB}}, \tag{2.9.128}$$

where we have made use of the Dirac bracket. In particular, note from Eq. (2.9.125) that in this case the N_c constraint functions satisfy

$$\dot{\psi}_k \approx \{\psi_k, H\}_{\text{DB}} \approx 0. \tag{2.9.129}$$

For *nonsingular* systems there are no constraints and Eq. (2.9.128) becomes Eq. (2.4.26), which has the usual Possion bracket in place of the Dirac bracket and equal signs in place of "$\approx$."

In a quantum system the Hamiltonian is derived in terms of the evolution operator as given in Eq. (4.1.9), which leads to the operator equations of motion in the Heisenberg picture given in Eq. (2.5.13) for any operator $\hat{F}$,

$$\dot{\hat{F}} = \frac{\partial \hat{F}}{\partial t} + \frac{1}{i\hbar}[\hat{F}, \hat{H}]. \tag{2.9.130}$$

If the quantum system satisfies the time-independent constraints $\hat{\psi}_1, \ldots, \hat{\psi}_{N_c} = 0$ then it must be true that

$$\dot{\hat{\psi}}_k = [\hat{\psi}_k, \hat{H}] = 0 \tag{2.9.131}$$

so that the constraints remain satisfied at all times. Furthermore, any function of physical observables $\hat{F}$ must satisfy the quantum version of Eq. (2.9.127),

$$[\hat{\psi}_i, \hat{F}] = 0 \quad \text{for} \quad i = 1, \ldots, N_c, \tag{2.9.132}$$

so that the N_c constraint operators and all observables have simultaneous eigenstates. This means that the quantum system can be confined to the physical subspace V_{phys} of the Hilbert space V_H that pertains to the zero eigenvalues of the constraint operators $\hat{\psi}_i$. So any physical state $|\Psi\rangle \in V_{\text{phys}} \subset V_H$

must satisfy $\hat{\psi}_i|\Psi\rangle = 0$ for all N_c constraints and no measurement can take the state outside the subspace, if $|\Psi\rangle \in V_{\text{phys}}$ then $\hat{F}|\Psi\rangle \in V_{\text{phys}}$.

The classical system that results from taking the macroscopic limit of the quantum system will have expectation values of observable operators that obey the same equations of motion as their operator counterparts. So for every observable $\hat{F}$ and constraint $\hat{\psi}_k$ we must have

$$\lim_{\hbar \to 0}\left\langle \frac{1}{i\hbar}[\hat{F}, \hat{H}]\right\rangle \to \{F, H\}_{\text{DB}} \quad \text{and} \quad \lim_{\hbar \to 0}\left\langle \frac{1}{i\hbar}[\hat{\psi}_k, \hat{H}]\right\rangle \to \{\psi_k, H\}_{\text{DB}}. \tag{2.9.133}$$

This led Dirac to suggest that for a *singular* system one should identify

$$\lim_{\hbar \to 0}\left\langle \frac{1}{i\hbar}[\hat{A}, \hat{B}]\right\rangle = \{A, B\}_{\text{DB}} \tag{2.9.134}$$

for *any* two dynamical variables $A(\vec{q}, \vec{p}, t)$ and $B(\vec{q}, \vec{p}, t)$.

The process of attempting to construct a quantum theory from a classical theory while maintaining symmetries and other structures within the theory is referred to as *canonical quantization*. For *nonsingular* systems Dirac's proposal for canonical quantization (Dirac, 1982) was to replace the Poisson bracket by the commutator divided by $i\hbar$, i.e.,

$$\{A, B\} \to \frac{1}{i\hbar}[\hat{A}, \hat{B}]. \tag{2.9.135}$$

As we saw in Sec. 2.5 the right-hand side does indeed reduce to the left-hand side in the macroscopic limit for nonsingular systems, which justifies Dirac's proposal. For a *singular* system his proposal was to generalize this to

$$\{A, B\}_{\text{DB}} \to \frac{1}{i\hbar}[\hat{A}, \hat{B}]. \tag{2.9.136}$$

To arrive at such a result for singular systems was the primary motivation for Dirac's extensive work on the subject (Dirac, 2001). Further discussion may be found in Rothe and Rothe (2010) and the previously mentioned references.

Since quantum theory contains the classical theory as its classical limit, the classical theory cannot always uniquely determine the quantum theory through such a canonical quantization process. For example, there may be operator ordering ambiguities in the quantum theory when terms appear in the classical theory that contain combinations of coordinate and momentum variables. There are other difficulties also, such as the fact that the imposition of Dirac's proposal for all dynamical variables A and B can lead to inconsistencies when cubic and higher order polynomials of coordinates and momenta appear (Sundermeyer, 1982). Also, the fundamental Dirac brackets, where A and B are chosen from the generalized coordinates and momenta, do not have always have the simple Kronecker delta form of the fundamental Poisson brackets of Eq. (2.4.28). Global transformations in phase space to coordinates and momenta that do have the Kronecker delta form with respect to the fundamental Dirac bracket can lead to difficulties in constructing operators that satisfy the required commutation relations. In conclusion, there is no universal path to the quantization of any arbitrary classical system using Dirac's approach.

However, in simple cases of interest to us in our studies of quantum field theories, this canonical quantization program of Dirac can be carried out successfully. This introduction to singular systems has been provided so that we have the background to understand the approach to the canonical quantization of singular systems such as gauge theories. It was also useful to introduce some of the basic concepts of gauge invariance as they appear in a Hamiltonian formulation. We apply this approach to the Hamiltonian formulation of classical electromagnetism in Sec. 3.3.2. We will later see that, for theories of interest to us, the path integral formulation is equivalent and provides a more convenient and transparent method of describing quantizing systems than the Hamiltonian-based canonical approach. This is especially true for singular systems such as gauge theories.

Summary

We first applied Newton's laws to develop the machinery of the Lagrangian and Hamiltonian formalisms for nonrelativistic classical mechanics. The need to combine the postulates of special relativity with Newton's laws led us to the consequences of relativistic kinematics, including the conservation of four-momentum. For a relativistic free particle we saw that a relativistic extension of Hamilton's principle of stationary action leads to the correct relativistic free particle behavior. When this relativistic Hamilton's principle was applied to particles interacting with external fields we found appropriate equations of motion and, for example, correctly predicted that the Lorentz force equation also applies to relativistic particles moving in an electromagnetic field as we observe experimentally. Finally, we saw that the path to the Hamiltonian formulation of singular systems can be daunting, but that the construction of the Dirac-Bergmann algorithm provides a systematic method for doing this. Part of the historical motivation of Dirac and others for constructing this algorithm was to use the Hamiltonian framework to extend the canonical quantization approach to singular systems such as electromagnetism. Singular systems with first-class constraints have associated gauge transformations that leave the physical state of the system unchanged. As we will see in Chapter 3 and elsewhere, classical field theories that are invariant under gauge transformations (i.e., gauge field theories) are singular systems with first-class constraints. Although we have shown how to apply the canonical quantization approach, we will use the much simpler Lagrangian-based Feynman path integral approach to quantization for most of our applications of quantum field theory to physical systems. A summary of the path integral approach to quantum mechanics is given in Sec. 4.1.12.

Despite some successes with combining relativity with classical mechanics, there is a point beyond which we cannot proceed with this approach. As discussed in Sec. 7.10 in Goldstein et al. (2013), it can be shown that there is a "no interaction" theorem, which states that there can be "no covariant direct interaction" between a *finite* number of relativistic particles in a Lagrangian or Hamiltonian formulation of the dynamics. See, for example, discussions in Marmo et al. (1984) and references therein. Earlier foundational papers include Currie et al. (1963) and Leutwyler (1965). To proceed to describe relativistic particles acting on each other in a Lagrangian or Hamiltonian formalism, we will need to introduce the full machinery of relativistic quantum field theory. The first step in this direction is to generalize to an *infinite* number of degrees of freedom to evade the no-interaction theorem.

Problems

2.1   A cylinder of mass M, moment of inertia I about the cylindrical axis and radius R rolls on a horizontal surface without slipping.

(a) Express the no-slip constraint in differential form. Is it holonomic? Explain.
(b) Write down a Lagrangian for the system.

2.2 A block of mass m slides without friction on a larger block of mass M and is attached to a pin in this block by a massless spring with spring constant k. All motion is in one dimension. The larger block slides without friction on a flat table. Construct the Lagrangian and obtain the equations of motion. Identify any conserved quantities. Relate these to the symmetries of the system.

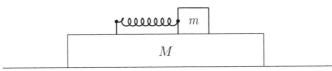

2.3 A uniform circular wire of radius R is forced to rotate about a fixed vertical diameter at constant angular velocity ω. A bead of mass m experiences gravity, is smoothly threaded on the wire and can slide without friction. Is the system holonomic? Construct a Lagrangian for the system. Show that there is just one solution for the off-axis motion of the bead where it does not oscillate. Does the system conserve energy?

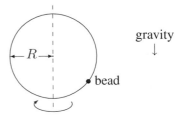

2.4 A bead of mass m is threaded without friction on a massless wire hoop of radius R that is forced to oscillate vertically in a fixed vertical plane at angular frequency ω and with amplitude a. Is the system holonomic? Construct a Hamiltonian for the system and obtain hamilton's equations. Does the system conserve energy? (Hint: Consider the external system.)

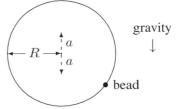

2.5 Two point masses m and M are connected by a massless rod of length ℓ and placed on a horizontal table. The mass m is also connected to a fixed point P on the table by a spring with spring constant k and zero natural length; it can slide without friction in any direction on the table. The mass M is constrained to slide along a straight line at a perpendicular distance $b < \ell$ from P. Set up a Lagrangian for the system and obtain the Euler-Lagrange equations. In the limit of small oscillations, solve for the characteristic angular frequencies and normal modes. Explain whether or not energy is conserved.

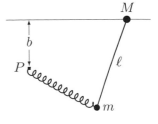

2.6 A mass-m particle moves along the x-axis subject to a potential energy $V(x) = \lambda[(x^2 - a^2)^2 + 2ax^3]$, where λ and a are positive constants.

(a) Find all equilibrium points and determine which are stable.
(b) Find the angular frequency of small oscillations about the ground state.

2.7 A uniform thin rod of length ℓ and mass m is suspended from fixed points A and B by two identical massless springs with *zero* natural length and spring constant k. At equilibrium (as shown in the figure), the angle between the springs and the vertical is ϕ and the length of the springs is a. All motion is restricted to this vertical plane.

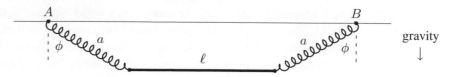

(a) Construct the Lagrangian and obtain Euler-Lagrange equations.

(b) Find the normal modes of small oscillations of the system about equilibrium and their angular frequencies.

(c) Construct the Hamiltonian, obtain Hamilton's equations and show equivalence to the Euler-Lagrange equations.

2.8 Show that the Lagrangian $L = (q_1\dot{q}_2 - q_2\dot{q}_1)^2 - a(q_1 - q_2)^4$ is invariant under the transformations

$$\begin{pmatrix} q_1 \\ q_2 \end{pmatrix} \rightarrow \begin{pmatrix} 1+s & -s \\ s & 1-s \end{pmatrix} \begin{pmatrix} q_1 \\ q_2 \end{pmatrix}$$

and construct the corresponding conserved quantity.

2.9 A system with a single degree of freedom is governed by the Lagrangian $L = e^{\beta t}(\frac{1}{2}m\dot{q}^2 - \frac{1}{2}kq^2)$.

(a) Write down the equation of motion. What sort of motion does it describe?

(b) Show that the explicit time dependence of L can be removed by introducing a new coordinate variable $Q = e^{\beta t/2}q$.

(c) Construct the first integral whose conservation is implied by this property.

2.10 A particle of mass m moves in one dimension in a potential $V(q) = -Cqe^{-\lambda q}$ where C and λ are positive constants. Write down the Lagrangian and find the point of stable equilibrium. In the small-oscillation approximation find the angular frequency of the normal mode.

2.11 The Lagrangian of a particle of mass m and charge e in a uniform, static magnetic field $\mathbf{B}$ is given by $L(\mathbf{x}, \dot{\mathbf{x}}) = \frac{1}{2}m\dot{\mathbf{x}}^2 + (e/2m)\mathbf{L} \cdot \mathbf{B}$, where $\mathbf{L} = \mathbf{x} \times m\dot{\mathbf{x}}$ is the mechanical angular momentum.

(a) Show that the Hamiltonian is $H(\mathbf{x}, \mathbf{p}) = \bar{\mathbf{p}}^2/2m$, where $\bar{\mathbf{p}} = \mathbf{p} - (e/2)\mathbf{B} \times \mathbf{x}$.

(b) Obtain the Poisson bracket result $\{\bar{p}_i, \bar{p}_j\} = e\,\epsilon_{ijk}B_k$.

(c) Show that $\bar{\mathbf{p}} \cdot \mathbf{B}$ is conserved using Poisson brackets.

(d) Are components of $\bar{\mathbf{p}}$ perpendicular to $\mathbf{B}$ conserved?

2.12 (a) Show that the relativistic Lagrangian $L = -(1/\gamma)mc^2e^{U/c^2}$ reduces to the Newtonian Lagrangian for a particle of mass m in a gravitational potential U in the slow motion, weak field limit $v \ll c, |U| \ll c^2$.

(b) Obtain the Euler-Lagrange equations of motion, and show that they can be written in the relativistic form $dV^\mu/d\tau = (g^{\mu\nu} - (1/c^2)V^\mu V^\nu)\partial_\nu U$, where V is the four-velocity and τ is the proper time of the particle. Verify that not all four-component equations are independent.

2.13 Consider a differentiable function $V(q)$ of a generalized coordinate q. Consider the Lagrangian $L = (1/12)m^2\dot{q}^4 + m\dot{q}^2V(q) - V^2(q)$. Show that this system is equivalent to a much simpler Lagrangian for a particle moving in a potential.

2.14 (a) A particle with rest mass M has kinetic energy T. Show that the magnitude of its momentum is $p = (1/c)\sqrt{T(T + 2Mc^2)}$.

    (b) A particle A decays at rest into a lighter particle B and a photon γ. Use four-vector methods to determine the energies of B and γ in terms of the rest masses m_A and m_B of A and B.

2.15 A relativistic positron e^+ with kinetic energy T annihilates a stationary electron e^- and produces two photons $e^+ + e^- \to \gamma + \gamma$.

    (a) At what angles are the photons emitted when they have equal energies?

    (b) Is it possible for the photons to have unequal energies? Explain.

3 Relativistic Classical Fields

We can generalize discrete point mechanics to the case of fields by assigning one or more degrees of freedom to each spatial point. By moving to an infinite number of degrees of freedom we can construct a consistent relativistic description in terms of Lagrangian and Hamiltonian densities and in so doing we evade the "no interaction theorem" that was discussed at the end of Chapter 2. In this chapter we first formulate the Lagrangian and Hamiltonian descriptions of relativistic classical fields, including the generalization of Poisson brackets. The role of symmetries is then examined and we see that these lead to conserved Noether currents and hence to conserved charges. Particular attention is paid to the implications of Poincaré invariance and to internal symmetries. Finally, we consider the example of the $U(1)$ gauge theory of electromagnetism. The Lagrangian formulation is relatively straightforward, but since gauge theories are singular systems then the Hamiltonian formulation requires the application of the Dirac-Bergmann algorithm developed in Sec. 2.9. We finish the chapter by explaining how Dirac's canonical quantization procedure leads to the quantum field theory formulation of electromagnetism.

3.1 Relativistic Classical Scalar Fields

For a real classical scalar field we identify one real degree of freedom with each point in three-dimensional space. The correspondence between point or discrete classical mechanics and the classical mechanics of a real scalar field $\phi(x)$ is

$$j = 1, 2, \ldots, N \to \mathbf{x} \in \mathbb{R}^3, \quad \sum_{j=1}^{N} \to \int d^3x, \quad q_j(t) \to \phi(ct, \mathbf{x}) = \phi(x)$$

$$L(\vec{q}(t), \dot{\vec{q}}(t), t) \to L(\phi(x), \partial_\mu \phi(x), x) \equiv \int d^3x \, \mathcal{L}(\phi(x), \partial_\mu \phi(x), x) \quad (3.1.1)$$

$$S[\vec{q}] = \int dt \, L \to S[\phi] = \int dt \, L = \int dt \left(\int d^3x \, \mathcal{L} \right) = (1/c) \int d^4x \, \mathcal{L},$$

where L and S are the Lagrangian and action, respectively. More carefully the correspondence can be expressed in terms of a three-dimensional lattice of spatial points $\mathbf{x}_j$ with some lattice spacing a and with $q_j(t) \equiv \phi(ct, \mathbf{x}_j)$ being the value of the field on the jth lattice point. The continuum limit $a \to 0$ can then be taken to give the right-hand side. We refer to $\mathcal{L}$ as the *Lagrangian density*. In later chapters we will adopt natural units where $\hbar = c = 1$ and so write, e.g., $S[\phi] = \int d^4x \, \mathcal{L}$.

We will be restricting our attention to local field theories where the Lagrangian density $\mathcal{L}$ is expressible as a function of a single spacetime point x, which is the reason that we can express the Lagrangian as $L = \int d^3x \, \mathcal{L}$. In addition, we will mostly be concerned with field theories that are functions of the field, its first-order derivatives and possibly an explicit spacetime dependence, e.g., through some external potential. In such cases we are then justified in writing the Lagrangian density

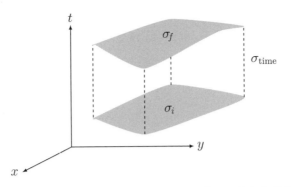

Figure 3.1   Illustration of the initial and final spacelike hypersurfaces, σ_i and σ_f, respectively, on which the field ϕ is held fixed when applying Hamilton's principle. The spacetime region R has volume V and is bounded by the hypersurface S_R.

as $\mathcal{L}(\phi(x), \partial_\mu \phi(x), x)$ as we have done above. Note that the absence of second-order and higher derivatives in the Lagrangian density is analogous to the absence of terms involving acceleration and higher-order time derivatives in the Lagrangians of nonrelativistic point mechanics. The absence of second and higher-order derivatives in the Lagrangian of point mechanics is the reason that we only considered generalized coordinates and velocities in our variational calculations leading to the Euler-Lagrange equations of Eq. (2.1.52).

A Lorentz scalar action $S[\phi]$ implies that $\mathcal{L}$ is a Lorentz scalar, since the volume element d^4x is a Lorentz scalar; i.e., the γ factors from time dilation and length contraction cancel. For example, for a boost in the z-direction from the frame of inertial observer $\mathcal{O}$ to that of inertial observer $\mathcal{O}'$ we have $dt' = \gamma dt$, $dx' = dx$, $dy' = dy$ and $dz' = dz/\gamma$. So the two infinitesimal spacetime regions denoted by d^4x' and d^4x have the same spacetime volume and contain the same spacetime events by definition. If for every possible choice of spacetime regions R and R' related by a Poincaré transformation we have

$$S'[\phi'] = (1/c) \int_{R'} d^4x' \mathcal{L}' = (1/c) \int_R d^4x \mathcal{L} = S[\phi], \tag{3.1.2}$$

then we must have

$$\mathcal{L}'\left(\phi'(x'), \partial'_\mu \phi'(x'), x'\right) = \mathcal{L}\left(\phi(x), \partial_\mu \phi(x), x\right), \tag{3.1.3}$$

where $\partial'_\mu \equiv \partial/\partial x'^\mu$. If $\mathcal{L}$ can be expressed as a function of only ϕ, $\partial^\mu \phi \partial_\mu \phi$ and scalar constants, then it will be a Lorentz scalar and $\mathcal{L}' = \mathcal{L}$.

In the case of the relativistic mechanics of a particle we saw that it was appropriate to generalize Hamilton's principle to the relativistic context; e.g., it correctly leads to the relativistic version of the Lorentz force law. Consistency requires that we use it again here as the basis for determining the classical behavior of a relativistic field. With respect to Eq. (2.1.51) we have the correspondence

$$d\vec{q}(t) \equiv \epsilon \vec{r}(t) \quad \rightarrow \quad d\phi(x) \equiv \epsilon r(x), \tag{3.1.4}$$

where the field ϕ is to be held fixed at the initial and final equal-time hypersurfaces, t_i and t_f, respectively, i.e., $r(ct_i, \mathbf{x}) = r(ct_f, \mathbf{x}) = 0$ for all relevant $\mathbf{x}$. We can replace equal-time surfaces by initial and final spacelike hypersurfaces, σ_i and σ_f, respectively, where $r(x) = 0$ for $x^\mu \in \sigma_i$ or σ_f. This is illustrated in Fig. 3.1.

Spatial boundary conditions: With regard to the spatial extent of the spacetime volume V, we will typically be interested in one of the two following scenarios:

(i) *vanishing boundary conditions:* the spatial volume is infinite with *hypersurfaces*[1] σ_{time} of the spacetime region R at spatial infinity and with all fields required to vanish at spatial infinity $|\mathbf{x}| \to \infty$ or

(ii) *periodic boundary conditions:* the spacetime volume is a finite four-dimensional rectangular box defined by t_i and t_f and the three-dimensional spatial box defined by intervals $[x_a, x_b]$, $[y_a, y_b]$ and $[z_a, z_b]$ with periodic spatial boundary conditions on the fields ϕ and their first derivatives $\partial_\mu \phi$.

In both of these scenarios the surface term arising from the use of the Minkowski-space divergence theorem vanishes as we show below.

We extend the derivation in Eq. (2.1.51) to the case of the relativistic scalar field by again making use of the appropriate generalization of Hamilton's principle given in Eq. (2.1.44), which now takes the form

$$\frac{\delta S[\phi]}{\delta \phi(x)} = 0 \quad \text{for all} \quad x^\mu \in R. \tag{3.1.5}$$

We use Eq. (2.1.45) to evaluate this functional derivative

$$
\begin{aligned}
\int_R d^4x \, \frac{\delta S[\phi]}{\delta \phi(x)} \, r(x) &= \lim_{\epsilon \to 0} \frac{1}{\epsilon} \{ S[\phi + \epsilon r] - S[\phi] \} \\
&= \frac{1}{c} \int_R d^4x \, \lim_{\epsilon \to 0} \frac{1}{\epsilon} \{ \mathcal{L}(\phi + \epsilon r, \partial_\mu \phi + \epsilon \partial_\mu r, x) - \mathcal{L}(\phi, \partial_\mu \phi, x) \} \\
&= \frac{1}{c} \int_R d^4x \, \left\{ \frac{\partial \mathcal{L}}{\partial \phi(x)} r(x) + \frac{\partial \mathcal{L}}{\partial(\partial_\mu \phi)(x)} \partial_\mu r(x) \right\} \\
&= \frac{1}{c} \int_R d^4x \, \left\{ \frac{\partial \mathcal{L}}{\partial \phi(x)} r(x) - \left(\partial_\mu \frac{\partial \mathcal{L}}{\partial(\partial_\mu \phi)(x)} \right) r(x) + \partial_\mu \left(\frac{\partial \mathcal{L}}{\partial(\partial_\mu \phi)(x)} r(x) \right) \right\} \\
&= \frac{1}{c} \int_R d^4x \, \left\{ \frac{\partial \mathcal{L}}{\partial \phi(x)} - \left(\partial_\mu \frac{\partial \mathcal{L}}{\partial(\partial_\mu \phi)(x)} \right) \right\} r(x) + \int_{S_R} \left(\frac{\partial \mathcal{L}}{\partial(\partial_\mu \phi)(x)} r(x) \right) ds_\mu \\
&= \frac{1}{c} \int_R d^4x \, \left\{ \frac{\partial \mathcal{L}}{\partial \phi(x)} - \left(\partial_\mu \frac{\partial \mathcal{L}}{\partial(\partial_\mu \phi)(x)} \right) \right\} r(x),
\end{aligned}
\tag{3.1.6}
$$

where in the above we have used the fact that the surface contribution vanishes and we have used the Minkowski-space form of the *divergence theorem*

$$\int_R \partial_\mu F^\mu \, d^4x = \int_{S_R} F^\mu ds_\mu, \tag{3.1.7}$$

where ds_μ is a four-vector infinitesimal surface element normal to the surface S_R of the spacetime region R and pointing out of the region for spacelike surfaces and into the region for timelike surfaces. The surface contribution is zero since $r(x) = 0$ on the initial and final spacelike hypersurfaces σ_i and σ_f in the application of Hamilton's principle and since we have appropriate spatial boundary conditions.

---

[1] A hypersurface in an n-dimensional space is a manifold of dimension $(n-1)$.

> **Proof:** To understand the vanishing of the surface term let us decompose it into integrals over the spacelike and timelike hypersurfaces as
>
> $$\int_{S_R} \frac{\partial \mathcal{L}}{\partial(\partial_\mu \phi)(x)} r(x) ds_\mu = \int_{\sigma_f + \sigma_i + \sigma_{\text{time}}} \frac{\partial \mathcal{L}}{\partial(\partial_\mu \phi)(x)} r(x) ds_\mu.$$
>
> The first term vanishes since the fields on the initial and final spacelike hypersurfaces are held fixed, i.e., $r(x) = 0$. In scenario (i) we have an infinite spatial volume with fields vanishing at spatial infinity, $\phi \to 0$ as $|\mathbf{x}| \to \infty$, and so provided that $\partial \mathcal{L}/\partial(\partial_\mu \phi) \to 0$ then the second term also vanishes. In scenario (ii) we have a four-dimensional rectangular box with periodic spatial boundary conditions. The second term again vanishes since the inward-oriented timelike surface integrations on the opposing timelike faces give canceling contributions.

Since $r(x)$ is arbitrary for all $x \in R$, we can combine Eqs. (3.1.5) and (3.1.6) to find that Hamilton's principle leads to the Euler-Lagrange equations

$$0 = c \frac{\delta S[\phi]}{\delta \phi(x)} = \frac{\partial \mathcal{L}}{\partial \phi(x)} - \partial_\mu \left(\frac{\partial \mathcal{L}}{\partial(\partial_\mu \phi)(x)} \right) \quad \text{for} \quad x^\mu \in R. \tag{3.1.8}$$

The generalization to a set of fields $\phi_j(x)$ for $j = 1, \ldots, N$ is straightforward,

$$S[\vec{\phi}] \equiv (1/c) \int_R d^4x \, \mathcal{L}(\vec{\phi}, \partial_\mu \vec{\phi}) \,. \tag{3.1.9}$$

The application of Hamilton's principle gives the Euler-Lagrange equations

$$0 = c \frac{\delta S[\vec{\phi}]}{\delta \phi_j(x)} = \frac{\partial \mathcal{L}}{\partial \phi_j(x)} - \partial_\mu \left(\frac{\partial \mathcal{L}}{\partial(\partial_\mu \phi_j)(x)} \right) \quad \text{for} \quad j = 1, 2, \ldots, N. \tag{3.1.10}$$

Covariant statement of Hamilton's principle in Minkowski space: In Minkowski spacetime we do not expect events with a spacelike separation to have any influence on each other. So the equations of motion in any finite region of spacetime should be independent of any particular spatial boundary conditions at spacelike infinity. An appropriate *covariant form of Hamilton's principle* is then the statement that the classical equations of motion in *any spacetime region R* arise from finding a stationary point of the action with respect to variations $\delta\vec{\phi}$ and $\delta(\partial_\mu\vec{\phi})$ that *vanish everywhere on the surface S_R of this spacetime region*. In other words, we *only consider the field fluctuations that have compact support within any arbitrary finite spacetime region R*, so that surface terms can be neglected.

Equivalent Lagrangian densities: A consequence of this covariant form of Hamilton's principle is that two Lagrangian densities that differ only by a four-divergence of the form $\partial_\mu F^\mu(\vec{\phi}, \partial_\mu\vec{\phi}, x)$ are equivalent in the sense that they will lead to the same Euler-Lagrange equations. Define

$$\mathcal{L}'(\vec{\phi}, \partial_\mu\vec{\phi}, x) \equiv \mathcal{L}(\vec{\phi}, \partial_\mu\vec{\phi}, x) + \partial_\nu F^\nu(\vec{\phi}, \partial_\mu\vec{\phi}, x) \tag{3.1.11}$$

for some four-vector function $F^\mu(\vec{\phi}, \partial_\mu\vec{\phi}, x)$. Using the Minkowski-space divergence theorem we then have for the corresponding action for the spacetime region R

$$S'[\vec{\phi}] = (1/c) \int_R d^4x \, \mathcal{L}' = (1/c) \int_R d^4x \, [\mathcal{L} + \partial_\nu F^\nu] = S[\vec{\phi}] + (1/c) \int_{S_R} F^\nu ds_\nu. \tag{3.1.12}$$

In the covariant form of Hamilton's principle we have no variations of the fields on the spacetime boundary S_R so the Euler-Lagrange equations are unchanged,

$$0 = c \frac{\delta S'[\vec{\phi}]}{\delta \phi_j(x)} = c \frac{\delta S[\vec{\phi}]}{\delta \phi_j(x)} = \frac{\partial \mathcal{L}}{\partial \phi_j(x)} - \partial_\mu \left(\frac{\partial \mathcal{L}}{\partial(\partial_\mu \phi_j)(x)} \right) \quad \text{for} \quad j = 1, \ldots, N. \tag{3.1.13}$$

Since the covariant form of Hamilton's principle allows no fluctuations of $\partial_\mu \vec{\phi}$ on the surface S_R of the spacetime region R, then F^μ can depend on the "velocities" $\partial_\mu \vec{\phi}$ as well as on $\vec{\phi}$ and x. The Euler-Lagrange equations are obviously unchanged if we vary the Lagrangian density by a multiplicative constant a and/or an overall additive constant b and so in analogy with Eq. (2.1.41) the Lagrangian density

$$\mathcal{L}'(\vec{\phi}, \partial_\mu \vec{\phi}, x) = a\mathcal{L}(\vec{\phi}, \partial_\mu \vec{\phi}, x) + \partial_\mu F^\mu(\vec{\phi}, \partial_\mu \vec{\phi}, x) + b \tag{3.1.14}$$

is equivalent to $\mathcal{L}$ in that they lead to the same Euler-Lagrange equations.

Legendre transformations for classical field theory: We can now consider how we might perform a Legendre transformation to a Hamiltonian-type formalism. There are two obvious ways that we might attempt to do this.

(i) *Traditional Hamiltonian formalism:* This approach sacrifices Lorentz covariance and Legendre transforms only the temporal derivatives, i.e., the "true" generalized velocities. This leads to the noncovariant Hamiltonian related to the energy of the field and to the zeroth component of the total four-momentum of the system. It is also related to the stress-energy tensor.

(ii) *De Donder-Weyl formalism:* The second approach is to introduce a Legendre transformation that preserves Lorentz covariance by treating temporal and spatial derivatives in a similar way, i.e., by treating them equally as types of "velocity" that transform to corresponding "momenta" under a Legendre transformation. This covariant approach is referred to as the De Donder-Weyl formalism and gives a Lorentz scalar "De Donder-Weyl Hamiltonian" density.

Both versions of the Legendre transformation are consistent in that they correctly reproduce the equations of motion as they should. The first traditional approach is the one that proves most useful in the constructions of relativistic quantum field theories since the Hamiltonian plays a key role in quantum mechanics.

Hamiltonian formalism: The Hamiltonian formulation of classical field theory is constructed in analogy with that of ordinary point mechanics beginning from the Lagrangian formulation and applying a Legendre transformation on temporal derivatives. Remembering that for every $\mathbf{x}$ there is a generalized coordinate $\phi(ct, \mathbf{x})$ then there is a corresponding conjugate momentum $\pi(ct, \mathbf{x})$ that is defined by

$$\pi(x) \equiv c\frac{\delta L}{\delta \dot{\phi}(x)} = \frac{\delta L}{\delta(\partial_0\phi)(x)} = \frac{\partial \mathcal{L}}{\partial(\partial_0\phi)(x)}. \tag{3.1.15}$$

The Legendre transform is performed as

$$H \equiv \left[\int d^3x\, \pi(x)\partial_0\phi(x)\right] - L = \int d^3x\, [\pi(x)\partial_0\phi(x) - \mathcal{L}]. \tag{3.1.16}$$

Note that $\pi(x)$ has an extra factor of c compared with the canoical momentum $p(t)$ and this cancels the extra factor of $1/c$ in $\partial_0\phi(x)$ compared with $\dot{q}(t)$. This can be written in terms of the Hamiltonian and Lagrangian densities as

$$\mathcal{H}(\phi, \boldsymbol{\nabla}\phi, \pi, x) \equiv \pi(x)\partial_0\phi(x) - \mathcal{L}(\phi, \partial_\nu\phi, x) \qquad \text{with} \qquad H \equiv \int d^3x\, \mathcal{H}. \tag{3.1.17}$$

We can now generalize the derivation of Hamilton's equations in Sec. 2.4 and so obtain Hamilton's equations of motion for a scalar field. The differential of the Lagrangian is given by $dL = \int d^3x\, (d\mathcal{L})$, where

$$dL = \frac{\partial \mathcal{L}}{\partial \phi} d\phi + \frac{\partial \mathcal{L}}{\partial(\partial_0 \phi)} d(\partial_0 \phi) + \frac{\partial \mathcal{L}}{\partial(\boldsymbol{\nabla}\phi)} \cdot d(\boldsymbol{\nabla}\phi) + \partial_\mu^{\mathrm{ex}} \mathcal{L} dx^\mu \tag{3.1.18}$$

$$= \left[\partial_0 \pi + \boldsymbol{\nabla} \cdot \left(\frac{\partial \mathcal{L}}{\partial(\boldsymbol{\nabla}\phi)}\right)\right] d\phi + \pi d(\partial_0 \phi) + \frac{\partial \mathcal{L}}{\partial(\boldsymbol{\nabla}\phi)} \cdot d(\boldsymbol{\nabla}\phi) + \partial_\mu^{\mathrm{ex}} \mathcal{L} dx^\mu,$$

where we have used the Euler-Lagrange equation in Eq. (3.1.8) and the definition of the canonical momentum density π in Eq. (3.1.15) to arrive at the second line. The partial derivatives $\partial_\mu^{\mathrm{ex}}$ are labeled with "ex" to remind us that they are acting only on any *explicit* spacetime dependence of $\mathcal{L}(\phi, \partial_\mu \phi, x)$ and $\mathcal{H}(\phi, \partial_\mu \phi, x)$; e.g., if there is no explicit x-dependence, then $\partial_\mu^{\mathrm{ex}} \mathcal{L}(\phi, \partial_\nu \phi) = 0$. The differential of the Hamiltonian is given by $dH = \int d^3x \,(d\mathcal{H})$, where

$$d\mathcal{H} = \frac{\partial \mathcal{H}}{\partial \phi} d\phi + \frac{\partial \mathcal{H}}{\partial \boldsymbol{\nabla}\phi} \cdot d(\boldsymbol{\nabla}\phi) + \frac{\partial \mathcal{H}}{\partial \pi} d\pi + \partial_\mu^{\mathrm{ex}} \mathcal{H} dx^\mu. \tag{3.1.19}$$

Using the definition of $\mathcal{H}$ in Eq. (3.1.17) we can also write the differential as

$$d\mathcal{H} = d(\partial_0 \phi \pi) - d\mathcal{L} = \pi d(\partial_0 \phi) + (\partial_0 \phi)d\pi - d\mathcal{L} \tag{3.1.20}$$

$$= -\left[\partial_0 \pi + \boldsymbol{\nabla} \cdot \left(\frac{\partial \mathcal{L}}{\partial(\boldsymbol{\nabla}\phi)}\right)\right] d\phi - \frac{\partial \mathcal{L}}{\partial(\boldsymbol{\nabla}\phi)} \cdot d(\boldsymbol{\nabla}\phi) + (\partial_0 \phi)d\pi - \partial_\mu^{\mathrm{cx}} \mathcal{L} dx^\mu,$$

where we have used the expression for $d\mathcal{L}$. Since $d\phi$, $d(\boldsymbol{\nabla}\phi)$, $d\pi$ and dx^μ are arbitrary and since $\partial\mathcal{H}/\partial(\boldsymbol{\nabla}\phi) = -\partial\mathcal{L}/\partial(\boldsymbol{\nabla}\phi)$, then Hamilton's equations of motion follow:

$$\partial_0 \phi = \frac{\partial \mathcal{H}}{\partial \pi}, \quad \partial_0 \pi = -\frac{\partial \mathcal{H}}{\partial \phi} + \boldsymbol{\nabla} \cdot \frac{\partial \mathcal{H}}{\partial(\boldsymbol{\nabla}\phi)} \quad \text{and} \quad \partial_\mu^{\mathrm{ex}} \mathcal{H} = -\partial_\mu^{\mathrm{ex}} \mathcal{L}. \tag{3.1.21}$$

Poisson brackets in classical field theory: The generalization of the Poisson bracket, Eq. (2.4.19), that is appropriate for classical field theory is (Sterman, 1993)

$$\{F, G\} \equiv \int d^3z \left(\frac{\delta F}{\delta \vec{\phi}(x^0, \mathbf{z})} \cdot \frac{\delta G}{\delta \vec{\pi}(x^0, \mathbf{z})} - \frac{\delta F}{\delta \vec{\pi}(x^0, \mathbf{z})} \cdot \frac{\delta G}{\delta \vec{\phi}(x^0, \mathbf{z})}\right), \tag{3.1.22}$$

where $\vec{\phi}(x)$ is some set of fields with conjugate momentum densities

$$\vec{\pi}(x) = \delta L / \delta(\partial_0 \vec{\phi})(x) = \partial\mathcal{L}/\partial(\partial_0 \vec{\phi})(x) \tag{3.1.23}$$

and $F[\vec{\phi}, \vec{\pi}]$ and $G[\vec{\phi}, \vec{\pi}]$ are dynamical functionals of $\vec{\phi}$ and $\vec{\pi}$ and might also have an explicit spacetime dependence. This definition of the Poisson bracket follows from the correspondence $q_j(t) \to \phi_j(ct, \mathbf{x}) = \phi_j(x^0, \mathbf{x}) = \phi_j(x)$. An example of a dynamical functional is the Hamiltonian H itself

$$H \equiv \int d^3x \, \mathcal{H}(\vec{\phi}, \boldsymbol{\nabla}\vec{\phi}, \vec{\pi}, x) = \int d^3x \left[\vec{\pi}(x) \cdot \partial_0 \vec{\phi}(x) - \mathcal{L}(\vec{\phi}, \partial_\nu \vec{\phi}, x)\right]. \tag{3.1.24}$$

Frequent use is made of the functional derivative results arising from Eq. (2.1.49),

$$\frac{\delta \phi_i(x^0, \mathbf{x})}{\delta \phi_j(x^0, \mathbf{y})} = \delta_{ij} \delta^3(\mathbf{x} - \mathbf{y}) \quad \text{and} \quad \frac{\delta \boldsymbol{\nabla}\phi_i(x^0, \mathbf{x})}{\delta \phi_j(x^0, \mathbf{y})} = \delta_{ij} \boldsymbol{\nabla}^x \delta^3(\mathbf{x} - \mathbf{y}), \tag{3.1.25}$$

where $\boldsymbol{\nabla}^x$ denotes the gradient operator with respect to x and where in the second result integration by parts is used with the assumption of fields vanishing at spatial infinity or having periodic boundary conditions in a finite volume. The fundamental Poisson brackets are easily evaluated from the above definition:

$$\{\phi(x), \phi(y)\}_{x^0=y^0} = \{\pi(x), \pi(y)\}_{x^0=y^0} = 0, \quad \{\phi(x), \pi(y)\}_{x^0=y^0} = \delta^3(\mathbf{x} - \mathbf{y}). \quad (3.1.26)$$

Proof: $\delta\phi_i(x^0, \mathbf{x})/\delta\pi_j(x^0, \mathbf{z}) = 0$ and $\delta\pi_i(x^0, \mathbf{x})/\delta\phi_j(x^0, \mathbf{z}) = 0$, since $\vec{\phi}$ and $\vec{\pi}$ are independent. The first two results follow and also the third, since

$$\{\phi_i(x), \pi_j(y)\}_{x^0=y^0} = \int d^3z \sum_k \left(\frac{\delta\phi_i(x^0,\mathbf{x})}{\delta\phi_k(x^0,\mathbf{z})} \frac{\delta\pi_j(x^0,\mathbf{y})}{\delta\pi_k(x^0,\mathbf{z})} - \frac{\delta\phi_i(x^0,\mathbf{x})}{\delta\pi_k(x^0,\mathbf{z})} \frac{\delta\pi_j(x^0,\mathbf{y})}{\delta\phi_k(x^0,\mathbf{z})} \right)$$

$$= \int d^3z \sum_k \delta_{ik}\delta^3(\mathbf{x} - \mathbf{z})\delta_{jk}\delta^3(\mathbf{y} - \mathbf{z}) = \delta_{ij}\delta^3(\mathbf{x} - \mathbf{y}). \quad (3.1.27)$$

Let $A(t) \equiv A[\vec{\phi}(t), \vec{\pi}(t)]$ be some functional of $\vec{\phi}(t, \mathbf{x})$ and $\vec{\pi}(t, \mathbf{x})$ at some fixed time t. It is clear from the definition of the Poisson bracket that we have

$$\{\phi_i(x), A(t)\}_{t=x^0} = \frac{\delta A(t)}{\delta\pi_i(x)} \quad \text{and} \quad \{\pi_i(x), A(t)\}_{t=x^0} = -\frac{\delta A(t)}{\delta\phi_i(x)}, \quad (3.1.28)$$

which is the classical field theory analog of Eq. (2.4.20). The Poisson bracket relations for Hamilton's equations in discrete mechanics are given in Eq. (2.4.25). Their classical field theory equivalent is

$$\partial_0\phi_i(x) = \{\phi_i(x), H\} = \frac{\delta H}{\delta\pi_i(x)} = \frac{\partial\mathcal{H}(x)}{\partial\pi_i(x)}$$

$$\partial_0\pi_i(x) = \{\pi_i(x), H\} = -\frac{\delta H}{\delta\phi_i(x)} = -\frac{\partial\mathcal{H}}{\partial\phi_i} + \boldsymbol{\nabla} \cdot \frac{\partial\mathcal{H}}{\partial(\boldsymbol{\nabla}\phi_i)}, \quad (3.1.29)$$

which has reproduced Hamilton's equations given in Eq. (3.1.21).

Proof: The first result immediately follows since $H \equiv \int d^3x \, \mathcal{H}(\vec{\phi}, \boldsymbol{\nabla}\vec{\phi}, \vec{\pi}, x)$,

$$\partial_0\phi_i(x) = \{\phi_i(x), H\} = \int d^3z \sum_k \delta_{ik}\delta^3(\mathbf{x} - \mathbf{z})\frac{\delta H}{\delta\pi_k(x^0,\mathbf{y})} = \frac{\delta H}{\delta\pi_i(x)} = \frac{\partial\mathcal{H}(x)}{\partial\pi_i(x)}.$$

In the Hamiltonian formalism $\vec{\phi}$ and $\boldsymbol{\nabla}\vec{\phi}$ are not independent variables for the purposes of performing variations, whereas $\vec{\phi}$ and $\vec{\pi}$ are independent. Use the notation $\delta^{\text{ex}}/\delta\phi_i(x)$ to indicate functional differentiation with respect to the explicit dependence on $\phi_i(x)$, i.e., $\delta^{\text{ex}}\phi_i(x)/\delta\phi_j(x) = \delta_{ij}\delta^3(\mathbf{x} - \mathbf{y})$ whereas $\delta^{\text{ex}}\boldsymbol{\nabla}\phi_i(x)/\delta\phi_j(y) = 0$. The second result follows since

$$\partial_0\pi_i(x) = \{\pi_i(x), H\} = -\int d^3z \sum_k \delta_{ik}\delta^3(\mathbf{x} - \mathbf{z})\frac{\delta H}{\delta\phi_k(x^0,\mathbf{y})} = -\frac{\delta H}{\delta\phi_i(x)}$$

$$= -\frac{\delta^{\text{ex}}H}{\delta\phi_i(x)} - \int d^3z \sum_k \frac{\delta^{\text{ex}}H}{\delta\boldsymbol{\nabla}\phi_k(x^0,\mathbf{z})} \cdot \frac{\delta\boldsymbol{\nabla}\phi_k(x^0,\mathbf{z})}{\delta\phi_i(x^0,\mathbf{x})} \quad (3.1.30)$$

$$= -\frac{\partial\mathcal{H}(x)}{\partial\phi_i(x)} - \int d^3z \sum_k \frac{\partial\mathcal{H}(x^0,\mathbf{z})}{\partial\boldsymbol{\nabla}\phi_k(x^0,\mathbf{z})} \cdot \delta_{ki}\boldsymbol{\nabla}^z\delta^3(\mathbf{z} - \mathbf{x}) = -\frac{\partial\mathcal{H}(x)}{\partial\phi_i(x)} + \boldsymbol{\nabla} \cdot \frac{\partial\mathcal{H}(x)}{\partial\boldsymbol{\nabla}\phi_i(x)},$$

where the gradient operator was moved using integration by parts with the surface term neglected, assuming fields vanish at spatial infinity or have periodic spatial boundary conditions.

Consider a dynamical functional $F[\vec{\phi}, \vec{\pi}]$ that might include some explicit time or spacetime dependence and not just time dependence through the field arguments $\vec{\phi}$ and $\vec{\pi}$, i.e., $F[\vec{\phi}, \vec{\pi}, x^0]$ or $F[\vec{\phi}, \vec{\pi}, x]$. Evaluating the total time derivative gives

$$\partial_0 F[\vec{\phi}, \vec{\pi}, x] = \int d^3z \sum_k \left(\frac{\delta F}{\delta \phi_k(x^0, \mathbf{z})} \partial_0 \phi_k(x^0, \mathbf{z}) + \frac{\delta F}{\delta \pi_k(x^0, \mathbf{z})} \partial_0 \pi_k(x^0, \mathbf{z}) \right) + \partial_0^{\mathrm{ex}} F$$

$$= \int d^3z \sum_k \left(\frac{\delta F}{\delta \phi_k(x^0, \mathbf{z})} \frac{\delta H}{\delta \pi_i(x^0, \mathbf{z})} - \frac{\delta F}{\delta \pi_k(x^0, \mathbf{z})} \frac{\delta H}{\delta \phi_i(x^0, \mathbf{z})} \right) + \partial_0^{\mathrm{ex}} F$$

$$= \{F, H\} + \partial_0^{\mathrm{ex}} F. \tag{3.1.31}$$

The classical field theory analog of Eq. (2.4.27) for the total time derivative of H is

$$\partial_0 H = \{H, H\} + \partial_0^{\mathrm{ex}} H = \partial_0^{\mathrm{ex}} H = -\partial_0^{\mathrm{ex}} L \quad \Rightarrow \quad \partial_0 \mathcal{H} = \partial_0^{\mathrm{ex}} \mathcal{H} = -\partial_0^{\mathrm{ex}} \mathcal{L}. \tag{3.1.32}$$

So the Hamiltonian is conserved if there is no explicit time dependence present.

De Donder-Weyl formalism: It is reasonable to ask the question of why we choose to use a noncovariant form of the Legendre transformation in the traditional Hamiltonian approach to a Lorentz covariant field theory. In the De Donder-Weyl (DW) formalism we define four generalized momentum densities in analogy with what was done in the formulation of the Hamiltonian formalism above,

$$\pi^\mu(x) \equiv \delta L / \delta(\partial_\mu \phi)(x) = \partial \mathcal{L} / \partial(\partial_\mu \phi)(x). \tag{3.1.33}$$

The temporal and spatial derivatives are being treated in the same way in order to maintain a covariant approach. The DW Hamiltonian density is then defined as the Legendre transform of the Lagrangian density with respect to all four of the generalized momentum densities,

$$\mathcal{H}^{\mathrm{DW}}(\phi, \pi^\nu, x) \equiv \pi^\mu(x) \partial_\mu \phi(x) - \mathcal{L}(\phi, \partial_\nu \phi, x). \tag{3.1.34}$$

The DW Hamiltonian density, $\mathcal{H}^{\mathrm{DW}}$, is a Lorentz scalar by construction. The Hamiltonian formalism is part way between the Lagrangian and the DW formulations in the sense that the Hamiltonian formalism does only part of the Legendre transformation that takes us from the Lagrangian formalism to the DW formalism.

It is straightforward to generalize the derivation of Hamilton's equations in Sec. 2.4 and so obtain the De Donder-Weyl equations of motion, which are the covariant analog of Hamilton's equations. The differential of the Lagrangian is given by $dL = \int d^3x\, d\mathcal{L}(\phi, \partial_\nu \phi, x)$, where

$$d\mathcal{L} = \frac{\partial \mathcal{L}}{\partial \phi} d\phi + \frac{\partial \mathcal{L}}{\partial(\partial_\mu \phi)} d(\partial_\mu \phi) + \partial_\mu^{\mathrm{ex}} \mathcal{L} dx^\mu = \partial_\mu \pi^\mu d\phi + \pi^\mu d(\partial_\mu \phi) + \partial_\mu^{\mathrm{ex}} \mathcal{L} dx^\mu. \tag{3.1.35}$$

We have made use of the Euler-Lagrange equations, Eq. (3.1.8), and the definition of the momentum densities π^μ to arrive at the final form. The differential of the DW Hamiltonian is given by $dH^{\mathrm{DW}} = \int d^3x\, (d\mathcal{H}^{\mathrm{DW}})$, where

$$d\mathcal{H}^{\mathrm{DW}}(\phi, \pi^\nu, x) = \frac{\partial \mathcal{H}^{\mathrm{DW}}}{\partial \phi} d\phi + \frac{\partial \mathcal{H}^{\mathrm{DW}}}{\partial \pi^\mu} d\pi^\mu + \partial_\mu^{\mathrm{ex}} \mathcal{H}^{\mathrm{DW}} dx^\mu. \tag{3.1.36}$$

Using the definition of $\mathcal{H}^{\mathrm{DW}}$ above we can also write the differential as

$$d\mathcal{H}^{\mathrm{DW}} = d(\pi^\mu \partial_\mu \phi) - d\mathcal{L} = \pi^\mu d(\partial_\mu \phi) + \partial_\mu \phi d\pi^\mu - d\mathcal{L}$$

$$= \pi^\mu d(\partial_\mu \phi) + \partial_\mu \phi d\pi^\mu - \partial_\mu \pi^\mu d\phi - \pi^\mu d(\partial_\mu \phi) - \partial_\mu^{\mathrm{ex}} \mathcal{L} dx^\mu$$

$$= \partial_\mu \phi d\pi^\mu - \partial_\mu \pi^\mu d\phi - \partial_\mu^{\mathrm{ex}} \mathcal{L} dx^\mu. \tag{3.1.37}$$

Since $d\phi$, $d\pi^\mu$ and dx^μ are arbitrary, then the Lorentz-covariant DW equations of motion immediately follow:

$$\partial_\mu \phi = \frac{\partial \mathcal{H}^{\mathrm{DW}}}{\partial \pi^\mu}, \quad \partial_\mu \pi^\mu = -\frac{\partial \mathcal{H}^{\mathrm{DW}}}{\partial \phi} \quad \text{and} \quad \partial_\mu^{\mathrm{ex}} \mathcal{H}^{\mathrm{DW}} = -\partial_\mu^{\mathrm{ex}} \mathcal{L}. \tag{3.1.38}$$

Since $\mathcal{H}^{\mathrm{DW}}$ is a Lorentz scalar, it cannot be associated with the energy density of the system, which must transform as the time component of a four-vector. For this reason the De Donder-Weyl formalism does not play a role in the traditional Hamiltonian-based canonical formulation of quantum field theory, which is an extension of quantum mechanics and requires a Hamiltonian that is associated with the energy of the system. We shall not consider it further here.

Free relativistic classical scalar field: The simplest Lorentz scalar Lagrangian density that is second-order in generalized velocities is that of the free scalar field,

$$\mathcal{L} = \tfrac{1}{2}(\partial_\mu \phi)^2 - \tfrac{1}{2}m^2\phi^2 = \tfrac{1}{2}\partial_0\phi^2 - \tfrac{1}{2}(\boldsymbol{\nabla}\phi)^2 - \tfrac{1}{2}m^2\phi^2. \tag{3.1.39}$$

The factors of a half are required so that the corresponding Hamiltonian gives the correct expression for the energy of the system. From Eq. (3.1.8) we find that the Euler-Lagrange equations are given by

$$0 = c\frac{\delta S[\phi]}{\delta\phi(x)} = \frac{\partial \mathcal{L}}{\partial\phi(x)} - \partial_\mu\left(\frac{\partial \mathcal{L}}{\partial(\partial_\mu\phi)(x)}\right) = -m^2\phi - \partial_\mu\partial^\mu\phi, \tag{3.1.40}$$

which gives the equation of motion

$$(\Box + m^2)\phi(x) = (\partial_\mu\partial^\mu + m^2)\phi(x) = 0. \tag{3.1.41}$$

This is referred to as the *Klein-Gordon equation* and is the relativistic wave equation for a free relativistic classical scalar field.

If we use the DW formalism we find for the generalized momenta

$$\pi^\mu(x) \equiv \frac{\partial \mathcal{L}}{\partial(\partial_\mu\phi)(x)} = \partial^\mu\phi(x) \tag{3.1.42}$$

and for the DW Hamiltonian density

$$\mathcal{H}^{\mathrm{DW}}(\phi, \pi^\nu, x) \equiv \pi^\mu(x)\partial_\mu\phi(x) - \mathcal{L}(\phi, \partial_\nu\phi, x) = \tfrac{1}{2}\pi_\mu\pi^\mu + \tfrac{1}{2}m^2\phi^2, \tag{3.1.43}$$

which gives the DW equations of motion

$$\partial_\mu\phi = \frac{\partial \mathcal{H}^{\mathrm{DW}}}{\partial\pi^\mu} = \pi^\mu \quad \text{and} \quad \partial_\mu\pi^\mu = -\frac{\partial \mathcal{H}^{\mathrm{DW}}}{\partial\phi} = -m^2\phi. \tag{3.1.44}$$

This has just reproduced the Klein-Gordon equation as expected.

If we use the Hamiltonian formalism we find for the generalized momentum

$$\pi(x) \equiv \frac{\partial \mathcal{L}}{\partial(\partial_0\phi)(x)} = \partial^0\phi(x) = \partial_0\phi(x) \tag{3.1.45}$$

and for the Hamiltonian

$$H = \int d^3x\, \mathcal{H} = \int d^3x\, [\tfrac{1}{2}\pi(x)^2 + \tfrac{1}{2}(\boldsymbol{\nabla}\phi(x))^2 + \tfrac{1}{2}m^2\phi(x)^2], \tag{3.1.46}$$

which gives the Hamiltonian equations of motion

$$\partial_0 \phi = \frac{\partial \mathcal{H}}{\partial \pi} = \pi(x) \quad \text{and} \quad \partial_0 \pi = -\frac{\partial \mathcal{H}}{\partial \phi} + \boldsymbol{\nabla} \cdot \frac{\partial \mathcal{H}}{\partial (\boldsymbol{\nabla} \phi)} = -m^2 \phi + \boldsymbol{\nabla}^2 \phi \qquad (3.1.47)$$

and so again the Klein-Gordon equation results. The Legendre transform is well defined since the Hessian matrix $M_L(x,y)|_{x^0=y^0}$ is positive definite and so the system is not singular. Recall $L = \int d^3x\, \mathcal{L}$ with $\mathcal{L}$ given in Eq. (3.1.39) and so

$$M_L(x,y)|_{x^0=y^0} = \left. \frac{\delta^2 L}{\delta(\partial_0\phi)(x)\delta(\partial_0\phi)(y)} \right|_{x^0=y^0} = \delta^3(\mathbf{x}-\mathbf{y}). \qquad (3.1.48)$$

Neglecting the term $-\frac{1}{2}(\boldsymbol{\nabla}\phi)^2$ leads to a Lagrangian density with the same form as that of a harmonic oscillator, $L = \frac{1}{2}m_{\text{osc}}\dot{q}^2 - \frac{1}{2}k_{\text{osc}}q^2$, with "mass" $m_{\text{osc}} = 1$ and with "spring constant" $k_{\text{osc}} = m^2$. Furthermore with the neglect of this term we see that the Lagrangian L is the sum of an infinite set of *uncoupled* harmonic oscillators (one for every $\mathbf{x}$). The term $-\frac{1}{2}(\boldsymbol{\nabla}\phi)^2$, which couples the harmonic oscillators, is unavoidable in a relativistic theory where one wishes the Lagrangian density to have a Lorentz-invariant form. This observation is the key element in motivating the quantization procedure leading to quantum field theory in Chapter 6.

The Klein-Gordon equation is the classical equation of motion for a free relativistic scalar field, where here "free" means that the solutions $\phi(x)$ are plane wave solutions $\exp(\pm ip \cdot x)$ with $p^2 = p_\mu p^\mu = m^2$ and their linear superpositions. The equations of motion of free classical relativistic fields are often referred to as *relativistic wave equations*; e.g., the Klein-Gordon equation is the relativistic wave equation of a real free scalar field. The plane wave solutions are the *normal modes* for this infinite system of coupled harmonic oscillators; e.g., see Sec. 2.3.

We will see in Sec. 4.3 that the Klein-Gordon equation can also be interpreted as the equation determining the behavior of a neutral spinless particle in relativistic quantum mechanics, e.g., see Bjorken and Drell (1964). The difference is entirely due to interpreting the solution to the Klein-Gordon equation either as (i) the behavior of a relativistic classical wave or (ii) the relativistic wavepacket of a spinless quantum particle.

Scalar field with a self-interaction: If we add a self-interaction $U(\phi)$,

$$\mathcal{L} = \frac{1}{2}\left(\partial_\mu\phi\right)^2 - \frac{1}{2}m^2\phi^2 - U(\phi) = \frac{1}{2}\partial_0\phi^2 - \frac{1}{2}(\boldsymbol{\nabla}\phi)^2 - \frac{1}{2}m^2\phi^2 - U(\phi), \qquad (3.1.49)$$

then the Euler-Lagrange equations become

$$\left(\partial_\mu\partial^\mu + m^2\right)\phi(x) + \partial U(\phi)/\partial\phi(x) = 0, \qquad (3.1.50)$$

e.g., if $U(\phi) = \kappa\phi^3 + \lambda\phi^4$ then $(\partial_\mu\partial^\mu + m^2)\phi(x) + 3\kappa\phi^2 + 4\lambda\phi^3 = 0$. Since $U(\phi)$ contains no derivatives of ϕ then $\pi^\mu(x)$ and $\pi(x)$ are the same as in the noninteracting case. It is straightforward to verify that both approaches reproduce the Euler-Lagrange equations as they must. The Hamiltonian is given by

$$H = \int d^3x\, [\tfrac{1}{2}\pi(x)^2 + \tfrac{1}{2}\left(\boldsymbol{\nabla}\phi(x)\right)^2 + \tfrac{1}{2}m^2\phi(x)^2 + U(\phi)] \equiv T[\pi] + V[\phi], \qquad (3.1.51)$$

where the kinetic component, $T[\pi] \equiv \int d^3x \tfrac{1}{2}\pi^2(x)$, depends only on the generalized momentum and is purely second-order in it and where the potential component, $V[\phi] \equiv \int d^3x \left[\tfrac{1}{2}\left(\boldsymbol{\nabla}\phi(x)\right)^2 + \tfrac{1}{2}m^2\phi(x)^2 + U(\phi)\right]$, depends only on the field ϕ and its *spatial* derivatives. This Hamiltonian is a straightforward generalization of the simple Hamiltonian in Eq. (4.1.198).

3.2 Noether's Theorem and Symmetries

As we have seen in Sec. 2.2, the presence of symmetries leads to conserved quantities according to Noether's theorem. In relativistic classical field theory continuous symmetries lead to conserved four-vector currents referred to as Noether currents.

3.2.1 Noether's Theorem for Classical Fields

Consider some physical system consisting of a set of N fields, $\phi_1, \ldots, \phi_N$, and with the dynamics of the system specified by the action, $S = (1/c) \int d^4x \, \mathcal{L}$. We will work in infinite Minkowski spacetime and use the covariant form of Hamilton's principle, where field fluctuations $\Delta\vec{\phi}(x)$ have an arbitrarily large but finite spatial and temporal extent; i.e., in more mathematical language we say that the $\Delta\vec{\phi}$ are functions of spacetime with compact support.

Field and coordinate transformations: Consider some arbitrary field transformation and Minkowski spacetime coordinate transformation,

$$x^\mu \to x'^\mu \quad \text{and} \quad \phi_i(x) \to \phi_i'(x'). \tag{3.2.1}$$

We will only be interested in transformations $x^\mu \to x'^\mu$ and $\phi_i(x) \to \phi_i'(x')$ that are invertible. The spacetime region will transform accordingly, $R \to R'$, and so the action can be expressed in terms of the new fields and coordinates as

$$\begin{aligned}
S[\vec{\phi}] &= (1/c) \int_R d^4x \, \mathcal{L}(\vec{\phi}, \partial_\mu\vec{\phi}, x) = (1/c) \int_{R'} d^4x' \, \mathcal{J}(x, x') \, \mathcal{L}(\vec{\phi}, \partial_\nu\vec{\phi}, x) \\
&\equiv (1/c) \int_{R'} d^4x' \, \mathcal{L}'(\vec{\phi}', \partial_\nu'\vec{\phi}', x'),
\end{aligned} \tag{3.2.2}$$

where $\partial_\nu' \equiv \partial/\partial x'^\nu$ and where $\mathcal{J}(x, x') \equiv |\det(\partial x/\partial x')|$ is the Jacobian associated with the coordinate transformation; e.g., see Sec. A.2.2. Note that in the above equation we have *defined* the transformed Lagrangian density as

$$\mathcal{L}'(\vec{\phi}', \partial_\nu'\vec{\phi}', x') \equiv \mathcal{J}(x', x) \, \mathcal{L}(\vec{\phi}, \partial_\nu\vec{\phi}, x) \tag{3.2.3}$$

so that the action for the transformed quantities can then be defined as

$$S'[\vec{\phi}'] \equiv (1/c) \int_{R'} d^4x' \, \mathcal{L}'(\vec{\phi}', \partial_\nu'\vec{\phi}', x') = (1/c) \int_R d^4x \, \mathcal{L}(\vec{\phi}, \partial_\mu\vec{\phi}, x) = S[\vec{\phi}]. \tag{3.2.4}$$

It is important to emphasize that we have done nothing here but change variables. We have not assumed any symmetry but have simply reformulated Hamilton's principle in terms of new variables while keeping the value of the action unchanged, $S'[\vec{\phi}'] = S[\vec{\phi}]$. For arbitrary transformations S' and S will almost always have different forms; i.e., they will almost always be *different functionals* of their arguments.

Derivation of Noether's theorem: We wish to consider an action $S[\vec{\phi}]$ and some coordinate and field transformation that leaves the action *form invariant*; i.e., the functionals S and S' are identical so that $S'[\vec{\phi}'] = S[\vec{\phi}'] = S[\vec{\phi}]$. Requiring that this be true for all spacetime regions R means that $\mathcal{L}$ is also form invariant, $\mathcal{L}'(\vec{\phi}', \partial_\mu'\vec{\phi}', x') = \mathcal{L}(\vec{\phi}', \partial_\mu'\vec{\phi}', x')$. Then we have

$$S[\vec{\phi}'] = (1/c) \int_{R'} d^4x' \, \mathcal{L}(\vec{\phi}', \partial_\mu'\vec{\phi}', x') = (1/c) \int_R d^4x \, \mathcal{L}(\vec{\phi}, \partial_\mu\vec{\phi}, x) = S[\vec{\phi}]. \tag{3.2.5}$$

The resulting equations of motion are then form invariant, i.e., we can replace $\mathcal{L}'$ by $\mathcal{L}$ in the transformed equations of motion,

$$0 = c\frac{\delta S'[\vec{\phi}']}{\delta\phi_j'(x')} = c\frac{\delta S[\vec{\phi}']}{\delta\phi_j'(x')} = \frac{\partial\mathcal{L}}{\partial\phi_j'(x')} - \partial_\mu'\left(\frac{\partial\mathcal{L}}{\partial(\partial_\mu\phi_j')(x')}\right) \tag{3.2.6}$$

for $j = 1, 2, \ldots, N$. Any transformation that leads to the *form invariance of the equations of motion* is defined as a *symmetry* of the action, just as it was in the discrete mechanics discussion of symmetries. Typically, only a very small subset of all possible transformations will correspond to symmetries, and as we saw in Eq. (3.1.14) not all of those will correspond to a form invariance of the action for arbitrary spacetime regions and hence form invariance of the Lagrangian density.

Consider an infinitesimal transformation of the coordinates and fields,

$$x^\mu \to x'^\mu \equiv x^\mu + \delta x^\mu(x) \equiv x^\mu + d\alpha X^\mu(x), \tag{3.2.7}$$

$$\phi_i(x) \to \phi_i'(x') \equiv \phi_i(x) + \delta\phi_i(x) \equiv \phi_i(x) + d\alpha\Phi_i(x), \tag{3.2.8}$$

where $d\alpha \ll 1$ is a real infinitesimal parameter fixing the size of the infinitesimal transformation and the $X^\mu(x)$ and the $\Phi_i(x)$ determine the nature of this spacetime and field transformations, respectively. We use the passive view of the spacetime transformation, then x'^μ and x^μ correspond to the same spacetime point P and

$$\delta\vec{\phi}(x) \equiv \vec{\phi}'(x') - \vec{\phi}(x) \equiv d\alpha\vec{\Phi}(x) \tag{3.2.9}$$

is the *local variation* of the field. This is the change in the field at the spacetime point P. The *total variation*, $\Delta\phi_i$, of the fields is defined by

$$\phi_i(x) \to \phi_i'(x) \equiv \phi_i(x) + \Delta\phi_i(x) \tag{3.2.10}$$

and includes the local field variation as well as the variation due to the coordinate transformation, since working to first order in the variation we have

$$\begin{aligned}\Delta\phi_i(x) &= \phi_i'(x) - \phi_i(x) = [\phi_i'(x) - \phi_i'(x')] + [\phi_i'(x') - \phi_i(x)]\\ &= [-\delta x^\mu(x)\partial_\mu\phi_i'(x)] + [\delta\phi_i(x)] = [-\delta x^\mu(x)\partial_\mu\phi_i(x)] + [\delta\phi_i(x)]\\ &= d\alpha\{\Phi_i(x) + [-\partial_\mu\phi_i(x)]X^\mu(x)\} \equiv d\alpha\Psi_i(x),\end{aligned} \tag{3.2.11}$$

where we have used $-\delta x^\mu(x)\partial_\mu\phi_i'(x) = -\delta x^\mu(x)\partial_\mu\phi_i(x)$ to first order. Using Eqs. (3.2.7) and (3.2.8) we have defined for notational convenience

$$\vec{\Psi}(x) \equiv \vec{\Phi}(x) - X^\mu(x)\partial_\mu\vec{\phi}(x) \quad \text{so that} \quad \Delta\vec{\phi}(x) = d\alpha\vec{\Psi}(x). \tag{3.2.12}$$

For those symmetry transformations that leave the action form invariant, we have for *any* field configuration $\vec{\phi}(x)$

$$\begin{aligned}0 = c\delta S \equiv cS[\vec{\phi}'] - cS[\vec{\phi}] &= \int_{R'} d^4x'\, \mathcal{L}(\vec{\phi}', \partial_\mu'\vec{\phi}', x') - \int_R d^4x\, \mathcal{L}(\vec{\phi}, \partial_\mu\vec{\phi}, x) \tag{3.2.13}\\ &= \left[\int_{R'} d^4x'\, \mathcal{L}(\vec{\phi}', \partial_\mu\vec{\phi}', x') - \int_R d^4x'\, \mathcal{L}(\vec{\phi}', \partial_\mu'\vec{\phi}', x')\right]\\ &\quad + \left[\int_R d^4x'\, \mathcal{L}(\vec{\phi}', \partial_\mu'\vec{\phi}', x') - \int_R d^4x\, \mathcal{L}(\vec{\phi}, \partial_\mu\vec{\phi}, x)\right].\end{aligned}$$

The purpose of the above rearrangement is to isolate the variation in spacetime and the variation in the field into the first and second terms, respectively.

Consider the final form of Eq. (3.2.13). In the first term of the second square bracket we replace the dummy variable x' by x. The second square bracket is then

$$\int_R d^4x \left[\mathcal{L}(\vec{\phi}', \partial_\mu \vec{\phi}', x) - \mathcal{L}(\vec{\phi}, \partial_\mu \vec{\phi}, x) \right] = \int_R d^4x \left[\frac{\partial \mathcal{L}}{\partial \vec{\phi}} \cdot \Delta \vec{\phi} + \frac{\partial \mathcal{L}}{\partial(\partial_\mu \vec{\phi})} \cdot \Delta(\partial_\mu \vec{\phi}) \right]$$

$$= \int_R d^4x \left[\partial_\mu \frac{\partial \mathcal{L}}{\partial(\partial_\mu \vec{\phi})} \cdot \Delta \vec{\phi} + \frac{\partial \mathcal{L}}{\partial(\partial_\mu \vec{\phi})} \cdot \partial_\mu(\Delta \vec{\phi}) \right] = \int_R d^4x \, \partial_\mu \left[\frac{\partial \mathcal{L}}{\partial(\partial_\mu \vec{\phi})} \cdot \Delta \vec{\phi} \right], \quad (3.2.14)$$

where we have used the Euler-Lagrange equations of motion given in Eq. (3.1.10) and that $\Delta(\partial_\mu \vec{\phi})(x) = \partial_\mu(\Delta \vec{\phi})(x)$ since $\Delta \vec{\phi}(x) = \vec{\phi}'(x) - \vec{\phi}(x)$ is the total variation of the fields $\vec{\phi}(x)$, i.e., since $\partial_\mu(\Delta \vec{\phi})(x) = \partial_\mu \vec{\phi}'(x) - \partial_\mu \vec{\phi}(x) = \Delta(\partial_\mu \vec{\phi})(x)$. For the first square bracket in Eq. (3.2.13) we first note that, since x' is a dummy integration variable, we can replace it by x. Defining S_R as the surface of the spacetime region R we see that the first square bracket can be written as[2]

$$\int_{R'-R} d^4x \, \mathcal{L}(\vec{\phi}', \partial_\mu \vec{\phi}', x) = \int_{S_R} [\mathcal{L}(\vec{\phi}', \partial_\mu \vec{\phi}', x) \delta x^\mu] ds_\mu = \int_R d^4x \, \partial_\mu [\mathcal{L}(\vec{\phi}, \partial_\mu \vec{\phi}, x) \delta x^\mu],$$

where in the last step we replace $\vec{\phi}'$ by $\vec{\phi}$ since we only need to work to first order in the infinitesimal transformation.

Combining the above results Eq. (3.2.13) now has the form

$$0 = \delta S \equiv S[\vec{\phi}'] - S[\vec{\phi}] = (1/c) \int_R d^4x \, \partial_\mu [\vec{\pi}^\mu(x) \cdot \Delta \vec{\phi}(x) + \mathcal{L}(\vec{\phi}, \partial_\mu \vec{\phi}, x) \delta x^\mu], \quad (3.2.15)$$

where we have used the conventional shorthand notation (used earlier in the discussion of the De Donder-Weyl formalism).

$$\vec{\pi}^\mu(x) \equiv \partial \mathcal{L} / \partial(\partial_\mu \vec{\phi})(x). \quad (3.2.16)$$

Since the spacetime region R is arbitrary we must have

$$0 = \partial_\mu \left\{ \vec{\pi}^\mu(x) \cdot \Delta \vec{\phi}(x) + \mathcal{L}(\vec{\phi}, \partial_\mu \vec{\phi}, x) \delta x^\mu \right\} \quad (3.2.17)$$

$$= d\alpha \partial_\mu \left\{ \vec{\pi}^\mu(x) \cdot \left[\vec{\Phi}(x) + [-\partial_\nu \vec{\phi}(x)] X^\nu(x) \right] + \mathcal{L}(\vec{\phi}, \partial_\mu \vec{\phi}, x) X^\mu(x) \right\}.$$

Simple form of Noether's theorem: So we have shown that if an infinitesimal transformation, characterized by X^μ and $\vec{\Phi}$ as defined in Eqs. (3.2.7) and (3.2.8), respectively, is such that it leaves the action form invariant, then there will be a conserved *Noether current* associated with this continuous symmetry of the form

$$j^\mu(x) \equiv \vec{\pi}^\mu(x) \cdot \left[\vec{\Phi}(x) - [\partial_\nu \vec{\phi}(x)] X^\nu(x) \right] + \mathcal{L}(\vec{\phi}, \partial_\mu \vec{\phi}, x) X^\mu(x), \quad (3.2.18)$$

where

$$\partial_\mu j^\mu(x) = 0. \quad (3.2.19)$$

This is the field theory analog of the Noether's theorem results for conserved quantities in discrete mechanics that were discussed in Sec. 2.2. If $X^\mu = 0$ then we say that we have an *internal symmetry*; otherwise, we say that we have a type of *spacetime symmetry*. If $\vec{\Phi} = 0$ then we have a pure

---

[2] Denote $S_{R'}$ as the surface of R'. Then for every $x^\mu \in S_R$ the corresponding spacetime point on the surface $S_{R'}$ is $x^\mu + \delta x^\mu$. Let ds_E^μ denote the Euclidean-space surface element of S. Then for any function $f(x)$ we have for infinitesimal δx^μ the usual Euclidean-space result $\int_{R'-R} f(x) d^4x = \int_{S_R} f(x) \, \delta x^\mu ds_E^\mu$. However, since $ds_E^\mu = ds_\mu$ we then have the corresponding Minkowski-space result that $\int_{R'-R} f(x) d^4x = \int_{S_R} f(x) \, \delta x^\mu ds_\mu$.

spacetime symmetry. Note that we can define and talk about a Noether current even if the action does not possess the corresponding symmetry; however, it would not be conserved in that case. For example, if the symmetry is only broken slightly then the current will be approximately conserved. This understanding of the conserved Noether current is all that we will need in most circumstances.

Generalized form of Noether's theorem: We can make a simple but important generalization of Noether's theorem by noting that the covariant form of Hamilton's principle allows for the addition of the divergence of a four-vector without changing the equations of motion as shown in the discussion surrounding Eq. (3.1.12). For some infinitesimal transformation characterized by the infinitesimal parameter $d\alpha$ we might not have exact form invariance, $\delta S = 0$, but instead we might find

$$\delta S \equiv S[\vec{\phi}'] - S[\vec{\phi}] = \frac{1}{c}\int_{S_R}(d\alpha F^\nu)ds_\nu = d\alpha\frac{1}{c}\int_R d^4x\,\partial_\nu F^\nu \qquad (3.2.20)$$

for some $F^\mu(\vec{\phi}, \partial_\mu\vec{\phi}, x)$. We see that here the action is invariant up to an overall *surface term*. Such a transformation leaves the Euler-Lagrange equations of motion unchanged based on the use of the covariant form of Hamilton's principle and so is still a symmetry transformation. In that case, since R is an arbitrary spacetime region, we see that Eq. (3.2.15) becomes

$$0 = \delta S - \frac{1}{c}\int_R d^4x\,\partial_\mu(d\alpha F^\mu) = \frac{1}{c}\int_R d^4x\,\partial_\mu\left[\vec{\pi}^\mu\cdot\Delta\vec{\phi} + \mathcal{L}\delta x^\mu - d\alpha F^\mu\right], \qquad (3.2.21)$$

which leads to the conserved generalized Noether current

$$j^\mu \equiv \vec{\pi}^\mu\cdot\left[\vec{\Phi} - [\partial_\nu\vec{\phi}]X^\nu\right] + \mathcal{L}X^\mu - F^\mu \quad \text{with} \quad \partial_\mu j^\mu = 0. \qquad (3.2.22)$$

On-shell and off-shell: In the context of quantum field theory we often use the terms *off-shell* and *on-shell*. These terms have their origin in describing whether or not a particle is *real*, which means that it is "on-mass-shell" with $p^2 = m^2c^2$, or is a *virtual* particle where $p^2 \neq m^2c^2$ as discussed in Sec. 2.6. In the context of field theory, we refer to an equation that is true without having invoked the Euler-Lagrange equations of motion as being true *off-shell*, whereas an equation that is only true after having used these equations of motion is referred to as being true *on-shell*. Off-shell equations are true for all field configurations, whereas on-shell equations are only necessarily true if the equations of motion are satisfied.

For example, the first line of Eq. (3.2.14) is true off-shell, whereas the second line is only true on-shell. This means that *the conservation of Noether currents is only true on-shell, i.e., for fields that satisfy the equations of motion.*

Since we consider variations of S over all field configurations, the quantity $\partial_\mu F^\mu$ in Eq. (3.2.21) must be a four-divergence for *arbitrary* field configurations; i.e., it must be a four-divergence off-shell as well as on-shell.

We can always define this Noether current with or without the corresponding symmetry and independent of whether we are on-shell or off-shell. However it will only be a conserved current, $\partial_\mu j^\mu(x) = 0$, if the symmetry is present and when we are on-shell since we used the equations of motion to prove its conservation.

Freedom to redefine the current: We are always free to redefine a conserved current by multiplying it by an arbitrary positive or negative real constant as convenient or as convention dictates. Any quantity that vanishes by virtue of the equations of motion can be added to a conserved Noether current and will lead to another conserved current that is equivalent on-shell, but will lead to a different current off-shell. We can always add to a current another conserved current. It is also clear that adding the four-divergence of any antisymmetric tensor, $A^{\mu\nu}$, to a conserved current amounts to a redefinition of the current; i.e., if $\partial_\mu j^\mu = 0$ then defining

$$j'^\mu \equiv j^\mu + \partial_\nu A^{\mu\nu} \quad \text{with} \quad A^{\mu\nu} = -A^{\nu\mu} \quad \text{gives} \quad \partial_\mu j'^\mu = 0. \tag{3.2.23}$$

For each conserved current

$$j^\mu(x) \equiv \left(j^0(x), \mathbf{j}(x) \right) \equiv \left(c\rho(x), \mathbf{j}(x) \right) \tag{3.2.24}$$

there is a conserved charge, which is the spatial integral of the charge density $\rho(x)$,

$$Q \equiv \int d^3x \, \rho(x) = (1/c) \int d^3x \, j^0(x). \tag{3.2.25}$$

Consider some three-dimensional region R bounded by the two-dimensional surface S_R. If the total current density passing through the surface vanishes, we then find

$$dQ/dt = \int_R d^3x \, \partial_0 j^0(x) = \int_R d^3x \, \boldsymbol{\nabla} \cdot \vec{j}(x) = \int_{S_R} d\vec{s} \cdot \vec{j}(x) = 0, \tag{3.2.26}$$

where we have used the three-dimensional form of the divergence theorem to perform the last step. Note that adding the four-divergence of an antisymmetric tensor does not change the total charge, since

$$\begin{aligned} Q' &\equiv \int d^3x \, \rho'(x) = (1/c) \int d^3x \, j'^0(x) = (1/c) \int d^3x \left[j^0(x) + \partial_\nu A^{0\nu}(x) \right] \\ &= (1/c) \int d^3x \left[j^0(x) + \partial_i A^{0i}(x) \right] = Q + \int_{S_R} ds^i A^{0i} = Q, \end{aligned} \tag{3.2.27}$$

where we have used the divergence theorem and have assumed vanishing (or periodic) boundary conditions for A^{0i} on the surface S_R.

A more general statement of this type of current redefinition is that we can always add a term Δ^μ to a current, $j^\mu \to j'^\mu = j^\mu + \Delta^\mu$, and leave both the off-shell divergence and the conserved charges unchanged, provided that the added term is conserved off-shell and that its zeroth component is a three-divergence,

$$\partial_\mu \Delta^\mu = 0 \quad \text{(off-shell)} \quad \text{and} \quad \Delta^0 = \partial_i B^i = \boldsymbol{\nabla} \cdot \mathbf{B} \tag{3.2.28}$$

for some three-vector $\mathbf{B}$. Requiring Lorentz covariance we see that this condition on Δ^0 implies that we must have $\Delta^\mu = \partial_\nu A^{\mu\nu}$ and $\partial_\mu \Delta^\mu = 0$ for some $A^{\mu\nu}$; i.e., we must have $\partial_\mu \partial_\nu A^{\mu\nu} = 0$ and this is obviously satisfied by $A^{\mu\nu} = -A^{\nu\mu}$. Both j^μ and j'^μ above have the same divergence even if they are not conserved, i.e., even if they are being considered off-shell and/or even if the symmetry is not present. When these two Noether currents are on-shell and the symmetry is present they are both conserved and lead to the same conserved charges as shown above. This technique can lead to more convenient forms for conserved currents, such as finding a symmetric form for the stress-energy tensor to be discussed in Sec. 3.2.6.

Giving currents more convenient properties using the techniques described above is sometimes referred to as *improving* the Noether current. If the improvement occurs only when the equations of motion are satisfied, then we refer to this as an *on-shell improvement*; otherwise, we say that it is an *off-shell improvement*. In this context, adding or subtracting the four-divergence of an antisymmetric tensor is said to be adding or subtracting an off-shell *improvement term*.

More concise derivation of Noether's theorem: We can arrive at a more concise derivation of Noether's theorem by making using of the Jacobian for the coordinate transformation (Sterman, 1993). Consider an infinitesimal transformation of the coordinates and fields as given in Eqs. (3.2.7) and (3.2.8). Under such an infinitesimal transformation we have for any field configuration $\vec{\phi}(x)$ that

$$\delta S[\phi] = S[\vec{\phi}'] - S[\vec{\phi}] = (1/c) \int_{R'} d^4x' \, \mathcal{L}(\vec{\phi}', \partial_\mu' \vec{\phi}', x') - (1/c) \int_R d^4x \, \mathcal{L}(\vec{\phi}, \partial_\mu \vec{\phi}, x)$$

$$= (1/c) \int_R d^4x \, |\det(\partial x'/\partial x)| \, \mathcal{L}(\vec{\phi}', \partial_\mu' \vec{\phi}', x') - (1/c) \int_R d^4x \, \mathcal{L}(\vec{\phi}, \partial_\mu \vec{\phi}, x) \qquad (3.2.29)$$

$$= (1/c) \int_R d^4x \left[(1 + \partial_\mu \delta x^\mu) \, \mathcal{L}(\vec{\phi}', \partial_\mu' \vec{\phi}', x') - \mathcal{L}(\vec{\phi}, \partial_\mu \vec{\phi}, x) \right]$$

$$= (1/c) \int_R d^4x \left[(\partial_\mu \delta x^\mu) \, \mathcal{L}(\vec{\phi}, \partial_\mu \vec{\phi}, x) + \mathcal{L}(\vec{\phi}', \partial_\mu' \vec{\phi}', x') - \mathcal{L}(\vec{\phi}, \partial_\mu \vec{\phi}, x) \right]$$

$$= (1/c) \int_R d^4x \left[(\partial_\mu \delta x^\mu) \, \mathcal{L}(\vec{\phi}, \partial_\mu \vec{\phi}, x) + \left\{ \mathcal{L}(\vec{\phi}', \partial_\mu' \vec{\phi}', x') - \mathcal{L}(\vec{\phi}', \partial_\mu \vec{\phi}', x) \right\} \right.$$

$$\left. + \left\{ \mathcal{L}(\vec{\phi}', \partial_\mu \vec{\phi}', x) - \mathcal{L}(\vec{\phi}, \partial_\mu \vec{\phi}, x) \right\} \right]$$

$$= (1/c) \int_R d^4x \left[\mathcal{L}(\partial_\mu \delta x^\mu) + \delta x^\mu (\partial_\mu \mathcal{L}) + \left\{ (\partial \mathcal{L}/\partial \vec{\phi}) \cdot \Delta \vec{\phi} + (\partial \mathcal{L}/\partial(\partial_\mu \vec{\phi})) \cdot \Delta(\partial_\mu \vec{\phi}) \right\} \right]$$

$$= (1/c) \int_R d^4x \left[\partial_\mu(\mathcal{L} \delta x^\mu) + (\partial \mathcal{L}/\partial \vec{\phi}) \cdot \Delta \vec{\phi} + \vec{\pi}^\mu \cdot \partial_\mu(\Delta \vec{\phi}) \right] \equiv (1/c) \int_R d^4x \, \delta \mathcal{L}_{\text{eff}},$$

where we have defined the *effective change in the Lagrangian density* $\delta \mathcal{L}_{\text{eff}}$ resulting from the transformation,

$$\delta \mathcal{L}_{\text{eff}} \equiv \partial_\mu(\mathcal{L} \delta x^\mu) + (\partial \mathcal{L}/\partial \vec{\phi}) \cdot \Delta \vec{\phi} + \vec{\pi}^\mu \cdot \partial_\mu(\Delta \vec{\phi}). \qquad (3.2.30)$$

We note that (i) in each step above we only retain terms to first order in the transformation; (ii) we used $\Delta(\partial_\mu \vec{\phi}) = \partial_\mu(\Delta \vec{\phi})$ since $\Delta \vec{\phi}$ is the total variation of the fields; (iii) we have made use of Eq. (3.2.7) for the coordinate transformation and Eq. (3.2.11) for the total field variation; and (iv) we have used the fact that the Jacobian $\mathcal{J}(x', x) = |\det(\partial x'/\partial x)|$ for the coordinate transformation is given by[3]

$$|\det(\partial x'/\partial x)| = |(1 + \partial_0 \delta x^0)(1 + \partial_1 \delta x^1) \cdots (1 + \partial_3 \delta x^3)| + \mathcal{O}((\delta x)^2)$$

$$= |1 + \partial_\mu \delta x^\mu| + \mathcal{O}((\delta x)^2) = 1 + \partial_\mu \delta x^\mu + \mathcal{O}((\delta x)^2), \qquad (3.2.31)$$

since $|\partial_\mu \delta x^\mu| \ll 1$ and since $\partial x'^\mu/\partial x^\nu = \delta^\mu_{\ \nu} + \partial_\nu(\delta x^\mu) + \mathcal{O}((\delta x)^2)$. Note that we have not yet used the Euler-Lagrange equations.

A transformation will be a symmetry if the action is only changed by the integral of some four-divergence $\partial_\mu F^\mu$ as given in Eq. (3.2.20). In that case we have

$$\delta S = d\alpha (1/c) \int_R d^4x \, \partial_\mu F^\mu = (1/c) \int_R d^4x \, \delta \mathcal{L}_{\text{eff}} \qquad (3.2.32)$$

for some F^μ. Since we have not yet used the Euler-Lagrange equations this result is true both on-shell and off-shell. Recall from Eqs. (3.2.7) and (3.2.11) that

$$\delta x^\mu = x'^\mu - x^\mu = d\alpha X^\mu(x), \qquad \delta \vec{\phi}(x) = \vec{\phi}'(x') - \vec{\phi}(x) = d\alpha \vec{\Phi}(x),$$

$$\Delta \vec{\phi}(x) = \delta \vec{\phi}(x) - \delta x^\mu \partial_\mu \vec{\phi}(x) = d\alpha \left\{ \vec{\Phi}(x) - X^\mu(x) \partial_\mu \vec{\phi}(x) \right\} = d\alpha \vec{\Psi}(x). \qquad (3.2.33)$$

The Lagrangian density is a Lorentz scalar field and so we can also define the *local variation of the Lagrangian density* $\delta \mathcal{L}$ as we did for the fields in Eq. (3.2.9),

$$\delta \mathcal{L} \equiv \mathcal{L}(\vec{\phi}', \partial_\mu \vec{\phi}', x') - \mathcal{L}(\vec{\phi}, \partial_\mu \vec{\phi}, x) = \delta x^\mu (\partial_\mu \mathcal{L}) + \left\{ (\partial \mathcal{L}/\partial \vec{\phi}) \cdot \Delta \vec{\phi} + \vec{\pi}^\mu \cdot \partial_\mu(\Delta \vec{\phi}) \right\}$$

$$= d\alpha \left\{ X^\mu(\partial_\mu \mathcal{L}) + \left[(\partial \mathcal{L}/\partial \vec{\phi}) \cdot \vec{\Psi} + \vec{\pi}^\mu \cdot \partial_\mu \vec{\Psi} \right] \right\}. \qquad (3.2.34)$$

---

[3] Consider some $n \times n$ matrix A with elements $A_{ij} = \delta_{ij} + \epsilon_{ij}$ where $\epsilon_{ij} \ll 1$. We wish to evaluate the determinant to $\mathcal{O}(\epsilon)$. From the definition of the determinant in Eq. (A.8.1) we see that any term in the sum containing two off-diagonal elements is $\mathcal{O}(\epsilon^2)$. So there is at most one off-diagonal element in any surviving term, i.e., $(n-1)$ diagonal elements in any relevant term. But if $(n-1)$ of the elements are diagonal, then the only term in the sum satisfying this has all elements diagonal, $A_{11} A_{22} \cdots A_{nn}$. So then $\det A = (1 + \epsilon_{11})(1 + \epsilon_{22}) \cdots (1 + \epsilon_{nn}) + \mathcal{O}(\epsilon^2) = 1 + \sum_{j=1}^n \epsilon_{jj} + \mathcal{O}(\epsilon^2)$.

Comparing Eqs. (3.2.30) and (3.2.34) shows that

$$\delta\mathcal{L}_{\text{eff}} = \mathcal{L}(\partial_\mu\delta x^\mu) + \delta\mathcal{L} = d\alpha\mathcal{L}(\partial_\mu X^\mu) + \delta\mathcal{L}. \tag{3.2.35}$$

It is sometimes convenient to write $\delta\mathcal{L}_{\text{eff}}$ after expanding out $\vec{\Psi}$,

$$\delta\mathcal{L}_{\text{eff}} = d\alpha\left\{\partial_\mu(\mathcal{L}X^\mu) + \left[(\partial\mathcal{L}/\partial\vec{\phi})\cdot\vec{\Psi} + \vec{\pi}^\mu\cdot\partial_\mu\vec{\Psi}\right]\right\} \tag{3.2.36}$$

$$= d\alpha\left\{\mathcal{L}(\partial_\mu X^\mu) + X^\mu(\partial_\mu\mathcal{L}) + \left[(\partial\mathcal{L}/\partial\vec{\phi})\cdot\vec{\Psi} + \vec{\pi}^\mu\cdot\partial_\mu\vec{\Psi}\right]\right\}$$

$$= d\alpha\left\{\mathcal{L}(\partial_\mu X^\mu) + X^\mu\left[\partial_\mu\mathcal{L} - (\partial\mathcal{L}/\partial\vec{\phi})\cdot\partial_\mu\vec{\phi} - \vec{\pi}^\nu\cdot\partial_\nu\partial_\mu\vec{\phi}\right] + (\partial\mathcal{L}/\partial\vec{\phi})\cdot\vec{\Phi}\right.$$

$$\left. + \vec{\pi}^\mu\cdot\partial_\mu\vec{\Phi} - \vec{\pi}^\nu\cdot(\partial_\nu X^\mu)(\partial_\mu\vec{\phi})\right\}$$

$$= d\alpha\left\{\mathcal{L}(\partial_\mu X^\mu) + X^\mu\left[\partial_\mu^{\text{ex}}\mathcal{L}\right] + (\partial\mathcal{L}/\partial\vec{\phi})\cdot\vec{\Phi} + \vec{\pi}^\nu\cdot\left[\partial_\nu\vec{\Phi} - (\partial_\nu X^\mu)(\partial_\mu\vec{\phi})\right]\right\},$$

where we have used

$$\partial_\mu\mathcal{L} = (\partial\mathcal{L}/\partial\vec{\phi})\cdot\partial_\mu\vec{\phi} + (\partial\mathcal{L}/\partial(\partial_\nu\vec{\phi}))\cdot\partial_\mu\partial_\nu\vec{\phi} + \partial_\mu^{\text{ex}}\mathcal{L} \tag{3.2.37}$$

for any Lagrangian density $\mathcal{L}(\vec{\phi},\partial_\mu\vec{\phi},x)$. Recall that in our notation ∂_μ^{ex} is the partial derivative with respect to the *explicit* spacetime dependence.

Using the fact that the spacetime region R is arbitrary, it follows from Eq. (3.2.29) that for a symmetry transformation we must have

$$0 = \delta\mathcal{L}_{\text{eff}} - d\alpha\left(\partial_\mu F^\mu\right) \tag{3.2.38}$$

for some F^μ. This is true off-shell, since we have not yet used the equations of motion. So *if $\delta\mathcal{L}_{\text{eff}}$ vanishes or if it can be written as a total four divergence then the corresponding transformation is a symmetry of the system.*

This gives the off-shell result

$$0 = (\delta\mathcal{L}_{\text{eff}}/d\alpha) - \partial_\mu F^\mu = \left\{\partial_\mu(\mathcal{L}X^\mu) + (\partial\mathcal{L}/\partial\vec{\phi})\cdot\vec{\Psi} + \vec{\pi}^\mu\cdot\partial_\mu\vec{\Psi}\right\} - \partial_\mu F^\mu$$

$$= \partial_\mu\left[\vec{\pi}^\mu\cdot\vec{\Psi} + \mathcal{L}X^\mu - F^\mu\right] + \Delta_{\text{off}} = \partial_\mu j^\mu + \Delta_{\text{off}}, \tag{3.2.39}$$

where we have defined the off-shell term that vanishes on-shell (for field configurations satisfying the Euler-Lagrange equations),

$$\Delta_{\text{off}} \equiv -\left\{\partial_\mu\vec{\pi}^\mu - (\partial\mathcal{L}/\partial\vec{\phi})\right\}\cdot\vec{\Psi} = -\left\{\partial_\mu\vec{\pi}^\mu - (\partial\mathcal{L}/\partial\vec{\phi})\right\}\cdot\left[\vec{\Phi} - [\partial_\nu\vec{\phi}]X^\nu\right]. \tag{3.2.40}$$

Now making use of the Euler-Lagrange equations

$$\partial_\mu\vec{\pi}^\mu(x) = \partial\mathcal{L}/\partial\vec{\phi}(x) \tag{3.2.41}$$

we have $\Delta_{\text{off}} = 0$ and so obtain the on-shell result that

$$0 = \partial_\mu\left[\vec{\pi}^\mu\cdot\vec{\Psi} + \mathcal{L}X^\mu - F^\mu\right] = \partial_\mu j^\mu, \tag{3.2.42}$$

where j^μ is the generalized Noether current in Eq. (3.2.22). So for any Noether current j^μ we have

$$\partial_\mu j^\mu = \Delta_{\text{off}} + \Delta_{\text{sv}}, \tag{3.2.43}$$

where the symmetry violating term Δ_{sv} vanishes if the symmetry is exact and where the off-shell term Δ_{off} vanishes if we are on-shell.

Alternative approach to finding the Noether current: It is also common to approach Noether's theorem from a different perspective such as that discussed in Weinberg (1995) and it is a useful exercise to relate the different approaches here. We again assume a symmetry, where the action is form invariant under the coordinate and field transformations in Eqs. (3.2.7) and (3.2.8) up to a term $d\alpha(1/c)\int d^4x\,\partial_\mu F^\mu$ for some F^μ, i.e., up to a surface term. However, let us now consider the situation where the infinitesimal parameter $d\alpha$ is replaced by an infinitesimal differentiable function of spacetime, $d\alpha \to \epsilon(x)$, so that our transformation then becomes

$$x^\mu \to x'^\mu = x^\mu + \delta x^\mu = x^\mu + \epsilon(x)X^\mu(x),$$

$$\vec{\phi}(x) \to \vec{\phi}'(x) = \vec{\phi}(x) + \Delta\vec{\phi}(x) = \vec{\phi}(x) + \epsilon(x)\left\{\vec{\Phi}(x) + [-\partial_\mu\vec{\phi}(x)]X^\mu(x)\right\}$$

$$\equiv \vec{\phi}(x) + \epsilon(x)\vec{\Psi}(x). \tag{3.2.44}$$

It is shown below that under a symmetry transformation,

$$\delta S = (1/c)\int_R d^4x\,j^\mu(\partial_\mu\epsilon) + \text{(surface terms)}, \tag{3.2.45}$$

where $j^\mu(x)$ is the conserved Noether current, Eq. (3.2.22). This can be a useful shortcut to the construction of a Noether current: If an infinitesimal transformation is a symmetry ($\delta S = 0$ up to surface terms with infinitesimal constant α), then with the replacement $\alpha \to \epsilon(x)$ and using the Euler-Lagrange equations δS can be expressed in the form of Eq. (3.2.45); and *the term multiplying* $\partial_\mu\epsilon$ *in the integrand of* δS *is the conserved Noether current,* j^μ.

Proof: If a transformation is a symmetry, then $\delta S - (1/c)\int_R d^4x\,\partial_\mu(\alpha F^\mu) = 0$ for infinitesimal constant α and $\partial_\mu j^\mu = 0$, where we are working on-shell here. It is not difficult to follow through the derivation of the Noether current that led to Eq. (3.2.21) with the replacement $\alpha \to \epsilon(x)$ and arrive at

$$\delta S - (1/c)\int_R d^4x\,\partial_\mu(\epsilon F^\mu) = (1/c)\int_R d^4x\,\partial_\mu(\epsilon j^\mu) \tag{3.2.46}$$

$$= (1/c)\int_R d^4x\,[j^\mu(\partial_\mu\epsilon) + \epsilon(\partial_\mu j^\mu)] = (1/c)\int_R d^4x\,j^\mu(\partial_\mu\epsilon),$$

since for a symmetry transformation we already know that $\partial_\mu j^\mu = 0$. Note that in the limit $\epsilon(x) \to$ constant both sides vanish. Using the divergence theorem the F^μ term becomes a surface term. So we have arrived at Eq. (3.2.45).

Examples

1. As a first simple example, consider the Lagrangian density for a massless scalar field $\mathcal{L} = \frac{1}{2}(\partial_\mu\phi)^2 = \frac{1}{2}(\partial_\mu\phi)(\partial^\mu\phi)$ under the transformation $\phi(x) \to \phi(x) + d\alpha$. We see that $\mathcal{L}$ is invariant under this transformation and that in the notation above, this transformation corresponds to $X^\mu = 0$ and $\Phi(x) = 1$. It follows that

$$j^\mu(x) \equiv \pi^\mu(x) = \partial\mathcal{L}/\partial(\partial_\mu\phi)(x) = \partial^\mu\phi(x) \tag{3.2.47}$$

is a conserved current.

2. We can consider the case of a *complex scalar field*, which can be constructed from two real scalar fields $\phi_1(x)$ and $\phi_2(x)$ as $\phi(x) \equiv 1/\sqrt{2}[\phi_1(x) + i\phi_2(x)]$. Consider a field theory specified by the free complex scalar field Lagrangian density

$$\mathcal{L} = |\partial_\mu\phi|^2 - m^2|\phi|^2 = (\partial_\mu\phi)^*(\partial^\mu\phi) - m^2\phi^*\phi$$

$$= \frac{1}{2}\left[(\partial_\mu\phi_1)^2 + (\partial_\mu\phi_2)^2 - m^2(\phi_1)^2 - m^2(\phi_2)^2\right]. \tag{3.2.48}$$

It can be shown easily from the Euler-Lagrange equations, Eqs. (3.1.10), that the field ϕ satisfies the Klein-Gordon equation, i.e., Eq. (3.1.41). One can treat either ϕ and ϕ^* as the independent fields or equivalently one can use ϕ_1 and ϕ_2. We see that $\mathcal{L}$ is invariant under the field transformation $\phi(x) \rightarrow e^{i\alpha}\phi(x)$. This transformation is referred to as a *global phase transformation*. It is an example of a *global symmetry*, where global means spacetime independent. Global phase invariance also leads to current conservation for fermions (see later). Since this transformation again leaves the spacetime coordinates unchanged, then $X^\mu = 0$. For infinitesimal α and working only to first order in that parameter for the field transformations we have $\phi \rightarrow \phi + \alpha\Delta\phi = \phi + \alpha(i\phi)$ and $\phi^* \rightarrow \phi^* + \alpha\Delta\phi^* = \phi^* + \alpha(-i\phi^*)$. Hence $\Phi(x) = i\phi(x)$ and $\Phi^*(x) = -i\phi^*(x)$. Since $\pi^\mu = \partial^\mu\phi^*$ and $\pi^{*\mu} = \partial^\mu\phi$ then using Eq. (3.2.18) we know that

$$j^\mu(x) = \pi^\mu\Phi + \pi^{*\mu}\Phi^* = i\left[\phi(\partial^\mu\phi^*) - \phi^*(\partial^\mu\phi)\right] \qquad (3.2.49)$$

is a conserved current. The complex scalar field can be viewed as a *charged scalar field* and the charged current j^μ can be, for example, the electromagnetic current associated with this field in a theory where the charged scalar field is coupled to photons; e.g., see Secs. 4.3.4 and 7.6.2.

3.2.2 Stress-Energy Tensor

For any physical system that is translationally invariant, the action is invariant under the transformation $x^\mu \rightarrow x'^\mu = x^\mu + a^\mu$ for some constant spacetime displacement a^μ. This will happen when a Lagrangian density depends only on fields and their derivatives, $\mathcal{L}(\vec{\phi}, \partial_\mu\vec{\phi}, x) \rightarrow \mathcal{L}(\vec{\phi}, \partial_\mu\vec{\phi})$. All Lagrangian densities of interest to us in field theory will have this symmetry except in cases where we are considering dynamics in the presence of external currents, fields or potentials.

For an infinitesimal translation we have four continuous one-parameter symmetries where there is one for each spacetime direction. In the notation of Eq. (3.2.7) we have $d\alpha \rightarrow da^\nu$ and $X^\mu \rightarrow X^\mu{}_\nu \equiv \delta^\mu{}_\nu$ so that

$$x^\mu \rightarrow x'^\mu = x^\mu + da^\nu\delta^\mu{}_\nu \equiv x^\mu + da^\nu X^\mu{}_\nu. \qquad (3.2.50)$$

We also have $\vec{\Phi} \rightarrow \vec{\Phi}_\nu = 0$ as there is no local variation of the fields in this transformation. Then Eq. (3.2.12) becomes

$$\vec{\Psi}_\nu = \vec{\Phi}_\nu - X^\mu{}_\nu\partial_\mu\vec{\phi} = -\partial_\nu\vec{\phi}. \qquad (3.2.51)$$

Since $\vec{\Phi}_\nu = 0$, $X^\mu{}_\nu \equiv \delta^\mu{}_\nu$ and $\partial_\rho X^\mu{}_\nu = 0$, then the effective change in the Lagrangian density given in Eq. (3.2.36) becomes in this case

$$\delta\mathcal{L}_{\text{eff}} = da^\nu \left\{ \mathcal{L}(\partial_\mu X^\mu{}_\nu) + X^\mu{}_\nu\left[\partial_\mu^{\text{ex}}\mathcal{L}\right] + (\partial\mathcal{L}/\partial\vec{\phi})\cdot\vec{\Phi}_\nu + \vec{\pi}^\nu\cdot\left[\partial_\nu\vec{\Phi}_\nu - (\partial_\nu X^\mu{}_\nu)(\partial_\mu\vec{\phi})\right] \right\}$$
$$= da^\nu X^\mu{}_\nu\left[\partial_\mu^{\text{ex}}\mathcal{L}\right] = da^\nu\partial_\nu^{\text{ex}}\mathcal{L}. \qquad (3.2.52)$$

If the Lagrangian density has no explicit spacetime dependence, $\mathcal{L}(\vec{\phi}, \partial_\mu\vec{\phi})$, then $\partial_\nu^{\text{ex}}\mathcal{L} = 0$ and so $\delta\mathcal{L}_{\text{eff}} = 0$ and the system will be symmetric. So we have confirmed the intuitively obvious statement that a Lagrangian density that has no explicit spacetime dependence is translationally invariant.

Hence from Eq. (3.2.18) we find for a translationally invariant Lagrangian density that we have four conserved currents labeled by ν, i.e., $(j^\mu)_\nu \equiv T^\mu{}_\nu$. Choosing the conventional sign and normalization for these currents we have that

$$T^\mu{}_\nu(x) = \vec{\pi}^\mu\cdot\partial_\nu\vec{\phi} - \delta^\mu{}_\nu\mathcal{L} \quad \text{with} \quad \partial_\mu T^{\mu\nu}(x) = 0, \qquad (3.2.53)$$

where the current conservation is true on-shell. We refer to $T^{\mu\nu} = T^\mu{}_\tau g^{\tau\nu}$ as the *stress-energy tensor* or sometimes as the *energy-momentum tensor*. We are of course free to choose the normalization of the stress-energy tensor, and the above is just one such choice. Note that $T^{\mu\nu}$ is not in general symmetric in its indices. We will later look at other improved forms of the stress-energy tensor. We refer to $T^{\mu\nu}$ as the *canonical stress-energy tensor*, since it is the one that arises from the standard application of Noether's theorem before any improvements are made.

Since there are four conserved currents $(j^\mu)_\nu = T^\mu{}_\nu$ for a spacetime translationally invariant system, then each will have a conserved charge of the form in Eq. (3.2.25). We are free to choose the normalization of the charges, but it is conventional to define them as

$$P_\nu \equiv \int d^3x \, (\rho)_\nu = \int d^3x \, (j^0)_\nu / c = (1/c) \int d^3x \, T^0{}_\nu. \tag{3.2.54}$$

With the definition $\vec{T}_\nu \equiv (T^1{}_\nu, T^2{}_\nu, T^3{}_\nu)$ we have

$$dP_\nu/dt = \int_R d^3x \, \partial_0 (j^0)_\nu(x) = \int d^3x \, \boldsymbol{\nabla} \cdot \vec{T}_\nu(x) = \int_{S_R} d\vec{s} \cdot \vec{T}_\nu(x) = 0 \tag{3.2.55}$$

provided that the net flux of $\vec{T}_\mu$ through the surface S_R of some spatial region R vanishes. So if $T^{\mu\nu}$ vanishes outside the region R then $dP_\nu/dt = 0$.

From Eq. (3.1.17) we see that T^{00} is the Hamiltonian density,

$$T^{00} = T^0{}_0 = \vec{\pi}^0 \cdot \partial_0 \vec{\phi} - \mathcal{L} = \vec{\pi} \cdot \partial_0 \vec{\phi} - \mathcal{L} = \mathcal{H}. \tag{3.2.56}$$

It follows that

$$P^0 = P_0 = (1/c) \int d^3x \, T^0{}_0 = \int d^3x \, (\mathcal{H}/c) = (H/c), \tag{3.2.57}$$

$$P^i = -P_i = -(1/c) \int d^3x \, T^0{}_i = -(1/c) \int d^3x \, \vec{\pi} \cdot \partial_i \vec{\phi}. \tag{3.2.58}$$

where the canonical generalized momentum densities are defined as

$$\vec{\pi} \equiv \vec{\pi}^0 = \partial \mathcal{L} / \partial(\partial_0 \vec{\phi}). \tag{3.2.59}$$

The distinction between $\vec{\pi} \equiv \vec{\pi}^0$ and our De Donder-Weyl style shorthand notation $\vec{\pi}^\mu$ should be noted. The Lorentz transformation properties of $T^{\mu\nu}$ mean that

$$P^\mu = (P^0, \mathbf{P}) = (H/c, \mathbf{P}) \tag{3.2.60}$$

must transform as a contravariant four-vector.

If the system is also natural as defined in Sec. 2.2, then the Hamiltonian will correspond to the total energy of the system, $H = E$. Any system with a Lagrangian that is second-order in the generalized velocities is a natural system, e.g., Eq. (2.2.9). We then recognize that $\mathbf{P} = (P^1, P^2, P^3)$ must be the total *physical* three momentum carried by the field. The energy density u for a natural system is then[4]

$$u = T^{00} = \mathcal{H}. \tag{3.2.61}$$

The physical three-momentum density is correspondingly given by

$$p_{\text{den}}^i = T^{0i}/c = \vec{\pi} \cdot \partial^i \vec{\phi}/c = -T^0{}_i/c = -\vec{\pi} \cdot \partial_i \vec{\phi}/c. \tag{3.2.62}$$

The total physical four-momentum is the spatial integral of the physical four-momentum density for a natural system,

$$P^\mu = \int d^3x \, p_{\text{den}}^\mu, \tag{3.2.63}$$

---

[4] Some texts normalize $T^{\mu\nu}$ such that $T^{00} = u/c^2$, which is the mass density for the field.

where we define the four-momentum density as

$$p_{\text{den}}^{\mu} = T^{0\mu}/c = (\mathcal{H}/c, \mathbf{p}_{\text{den}}) = (u/c, \mathbf{p}_{\text{den}}). \qquad (3.2.64)$$

Summary: For any translationally invariant scalar Lagrangian density the four-vector $P^{\mu} = (P^0, \mathbf{P}) = (H/c, \mathbf{P})$ is conserved. If the Lagrangian density is also second-order in the time derivative of the field, then it is a natural system and the total four-momentum of the system, $P^{\mu} \equiv (E/c, \mathbf{P})$, is conserved; i.e., the total energy E and physical three-momentum $\mathbf{P}$ of the field are conserved.

3.2.3 Angular Momentum Tensor

Consider a system that is invariant under Lorentz transformations. According to Eqs. (1.2.39) and (1.2.40) the action must be invariant under the infinitesimal transformation $x^{\mu} \to x'^{\mu} = x^{\mu} + d\omega^{\mu\nu} x_{\nu}$ with $d\omega^{\mu\nu} = -d\omega^{\nu\mu}$; i.e., there are six independent infinitesimal real parameters. This will happen when a Lagrangian density has no reference to any particular inertial reference frame; e.g., it depends only on fields and their derivatives. Using Eq. (1.2.44) for an infinitesimal Lorentz transformation

$$x^{\mu} \to x'^{\mu} = x^{\mu} + d\omega^{\mu\nu} x_{\nu} \equiv x^{\mu} + \tfrac{1}{2} d\omega^{\rho\sigma} X^{\mu}{}_{\rho\sigma}, \qquad (3.2.65)$$

where we have defined

$$X^{\mu}{}_{\rho\sigma} \equiv \left(\delta^{\mu}{}_{\rho} x_{\sigma} - \delta^{\mu}{}_{\sigma} x_{\rho} \right). \qquad (3.2.66)$$

Comparing with Eq. (3.2.7) we have $d\alpha \to \tfrac{1}{2} d\omega^{\rho\sigma}$ and $X^{\mu} \to X^{\mu}{}_{\rho\sigma}(x)$. Note that since both $d\omega^{\rho\sigma}$ and $X^{\mu}{}_{\rho\sigma}$ are antisymmetric with respect to ρ and σ exchange, the inclusion of the factor of one half is traditional so that each distinct $d\omega$ and X enter with a weighting of unity after the sums. If the components of the fields $\vec{\phi}$ are all Lorentz scalar fields ($\phi'_i(x') = \phi_i(x)$), then the Lorentz transformation will give no local variation in these fields; i.e., in the notation of Eq. (3.2.18) we have $\vec{\Phi} \to \vec{\Phi}_{\rho\sigma} = 0$. Then in the notation of Eq. (3.2.12) we have

$$\vec{\Psi}_{\rho\sigma}(x) = x_{\rho} \partial_{\sigma} \vec{\phi}(x) - x_{\sigma} \partial_{\rho} \vec{\phi}(x) \quad \text{and} \quad \Delta\vec{\phi}(x) = \tfrac{1}{2} d\omega^{\rho\sigma} \vec{\Psi}_{\rho\sigma}(x). \qquad (3.2.67)$$

If the components ϕ_i are not Lorentz scalars, then the field carries intrinsic angular momentum and we consider this generalization in Sec. 3.2.4.

Using Eq. (3.2.36) we see that the effective change in the Lagrangian density from an infinitesimal Lorentz transformation is

$$
\begin{aligned}
\delta\mathcal{L}_{\text{eff}} &= \tfrac{1}{2} d\omega^{\rho\sigma} \left\{ \mathcal{L}(\partial_{\mu} X^{\mu}{}_{\rho\sigma}) + X^{\mu}{}_{\rho\sigma}[\partial_{\mu}^{\text{ex}}\mathcal{L}] - \vec{\pi}^{\nu} \cdot (\partial_{\mu}\vec{\phi})(\partial_{\nu} X^{\mu}{}_{\rho\sigma}) \right\} \\
&= \tfrac{1}{2} d\omega^{\rho\sigma} \left\{ \mathcal{L}(g_{\rho\sigma} - g_{\sigma\rho}) + [x_{\sigma}\partial_{\rho}^{\text{ex}}\mathcal{L} - x_{\rho}\partial_{\sigma}^{\text{ex}}\mathcal{L}] - (\vec{\pi}_{\sigma} \cdot \partial_{\rho}\vec{\phi} - \vec{\pi}_{\rho} \cdot \partial_{\sigma}\vec{\phi}) \right\} \\
&= \tfrac{1}{2} d\omega^{\rho\sigma} \left\{ [x_{\sigma}\partial_{\rho}^{\text{ex}}\mathcal{L} - x_{\rho}\partial_{\sigma}^{\text{ex}}\mathcal{L}] - (\vec{\pi}_{\sigma} \cdot \partial_{\rho}\vec{\phi} - \vec{\pi}_{\rho} \cdot \partial_{\sigma}\vec{\phi}) \right\}.
\end{aligned} \qquad (3.2.68)
$$

Since the Lagrangian density is a Lorentz scalar and has no indices and since the only four-vectors available are x^{μ} and $\partial_{\mu}\vec{\phi}$, then the general functional form of the Lagrangian density consistent with Lorentz invariance for scalar fields $\vec{\phi}$ is

$$\mathcal{L}(\vec{\phi}, \partial_{\mu}\vec{\phi}, x) = f\left[\vec{\phi}^2, (\partial_{\mu}\vec{\phi} \cdot \partial^{\mu}\vec{\phi}), (x^{\mu}\partial_{\mu}\vec{\phi}) \cdot (x^{\nu}\partial_{\nu}\vec{\phi}), x^2 \right] \qquad (3.2.69)$$

for some function f. It is proved below that such an $\mathcal{L}_{\text{eff}}$ satisfies the condition $\delta\mathcal{L}_{\text{eff}} = 0$. In other words, $\delta\mathcal{L}_{\text{eff}}$ vanishes for a Lorentz-invariant Lagrangian density as expected. The third and fourth arguments of f allow for some possible explicit spacetime dependence. They would not be present in a translationally invariant system, where we would have $\mathcal{L}(\vec{\phi}, \partial_\mu \vec{\phi}) = f\left[\vec{\phi}^2, (\partial_\mu \vec{\phi} \cdot \partial^\mu \vec{\phi})\right]$.

Proof: Let us define some convenient shorthand notation

$$f_2 \equiv \frac{\partial f}{\partial(\partial_\mu \vec{\phi} \cdot \partial^\mu \vec{\phi})}, \quad f_3 \equiv \frac{\partial f}{\partial((x^\mu \partial_\mu \vec{\phi}) \cdot (x^\nu \partial_\nu \vec{\phi}))}, \quad f_4 \equiv \frac{\partial f}{\partial(x^2)}, \qquad (3.2.70)$$

where f_2, f_3, f_4 are defined to be the derivatives that act on the second, third and fourth arguments of f, respectively. Then we can write

$$\partial_\rho^{\text{ex}} \mathcal{L} = 2(\partial_\rho \vec{\phi}) \cdot (x^\nu \partial_\nu \vec{\phi}) f_3 + 2x_\rho f_4,$$
$$\vec{\pi}_\rho = \delta \mathcal{L}/\partial(\partial^\rho \vec{\phi}) = 2(\partial_\rho \vec{\phi}) f_2 + 2x_\rho (x^\nu \partial_\nu \vec{\phi}) f_3. \qquad (3.2.71)$$

Substituting into Eq. (3.2.68) we find that all terms cancel,

$$\delta \mathcal{L}_{\text{eff}} = \tfrac{1}{2} d\omega^{\rho\sigma} \left\{ x_\sigma [2(\partial_\rho \vec{\phi}) \cdot (x^\nu \partial_\nu \vec{\phi}) f_3 + 2x_\rho f_4] - x_\rho [2(\partial_\sigma \vec{\phi}) \cdot (x^\nu \partial_\nu \vec{\phi}) f_3 + 2x_\sigma f_4] \right.$$
$$\left. - [2(\partial_\sigma \vec{\phi}) f_2 + 2x_\sigma (x^\nu \partial_\nu \vec{\phi}) f_3] \cdot \partial_\rho \vec{\phi} + [2(\partial_\rho \vec{\phi}) f_2 + 2x_\rho (x^\nu \partial_\nu \vec{\phi}) f_3] \cdot \partial_\sigma \vec{\phi} \right\} = 0.$$

Using Eq. (3.2.18) we find six conserved currents labeled by $\rho\sigma$ that are antisymmetric under $\rho \leftrightarrow \sigma$, i.e., $(j^\mu)_{\rho\sigma} = \mathcal{J}^\mu{}_{\rho\sigma}$. Choosing the conventional sign and normalization for these conserved Lorentz-invariance currents we obtain

$$\mathcal{J}^\mu{}_{\rho\sigma}(x) = \left(-\vec{\pi}^\mu \cdot \partial_\nu \vec{\phi} + \delta^\mu{}_\nu \mathcal{L}\right) X^\nu{}_{\rho\sigma} = -T^\mu{}_\nu X^\nu{}_{\rho\sigma} = x_\rho T^\mu{}_\sigma - x_\sigma T^\mu{}_\rho, \qquad (3.2.72)$$

where

$$\mathcal{J}^\mu{}_{\rho\sigma} = -\mathcal{J}^\mu{}_{\sigma\rho} \quad \text{and} \quad \partial_\mu \mathcal{J}^\mu{}_{\rho\sigma} = 0 \ \text{(on-shell)} . \qquad (3.2.73)$$

We will show in Sec. 3.2.6 that in the case of Lorentz scalar fields the stress-energy tensor is symmetric, which means that we can also directly verify that $\partial_\mu \mathcal{J}^\mu{}_{\rho\sigma} = 0$ using $\partial_\mu T^{\mu\nu} = 0$ together with $T^{\mu\nu} = T^{\nu\mu}$. The six corresponding conserved charges make up the angular momentum tensor

$$M^{\rho\sigma} = (1/c) \int d^3x \, \mathcal{J}^{0\rho\sigma} = (1/c) \int d^3x \, (x^\rho T^{0\sigma} - x^\sigma T^{0\rho})$$
$$= \int d^3x \, (x^\rho p^\sigma_{\text{dens}} - x^\sigma p^\rho_{\text{dens}}). \qquad (3.2.74)$$

With the choice that we made for the normalization of the conserved currents $\mathcal{J}^\mu{}_{\rho\sigma}$ we see that the total angular momentum carried by the scalar fields $\vec{\phi}$ is given by

$$J^i = \tfrac{1}{2} \epsilon^{ijk} M^{jk} = \int d^3x \, \tfrac{1}{2} \epsilon^{ijk} \left[x^j (T^{0k}/c) - x_k (T^{0j}/c) \right]$$
$$= \int d^3x \, (\mathbf{x} \times \mathbf{p}_{\text{den}})^i \equiv L^i, \qquad (3.2.75)$$

where $\mathbf{L}$ is the total orbital angular momentum carried by the scalar fields with respect to the spatial origin $\mathbf{x} = 0$. Since a scalar field contains no intrinsic spin we have found as expected that $\mathbf{J} = \mathbf{L}$. We recognize that $\mathbf{J}$ and $M^{\mu\nu}$ are the classical field theory versions of the single particle quantum mechanics operators in Eqs. (1.2.169) and (1.2.170), respectively.

We have not yet considered the meaning of the three other components of the antisymmetric angular momentum tensor, $M^{0i} = -M^{i0}$. Let us define what is sometimes referred to as the *dynamic mass moment* of the fields, $K^i \equiv M^{0i}/c$,

$$\mathbf{K} \equiv \int d^3x \left[x^0 \mathbf{p}_{\text{dens}} - \mathbf{x}(u/c) \right] = ct\mathbf{P} - \left[\int d^3x \, \mathbf{x}(u/c) \right], \qquad (3.2.76)$$

where u is the energy density of the fields. To understand the meaning of $\mathbf{K}$ let us consider the case where we have a single free relativistic point particle with rest mass m and velocity $\boldsymbol{v}$, which corresponds to

$$\mathbf{K} = ct\mathbf{p} - \mathbf{x}(E/c) = c[t\mathbf{p} - \mathbf{x}\gamma(v)m] = -c\gamma(v)m[\mathbf{x} - \boldsymbol{v}t]. \qquad (3.2.77)$$

This result can be obtained as the classical limit of the single-particle quantum equivalent given in Eq. (1.2.170) or, alternatively, we can obtain it from the field version above by placing spatial delta-functions in u and $\mathbf{p}_{\mathrm{dens}}$. Since $dp^\mu/dt = 0$ for a particle in a Poincaré-invariant system, then

$$0 = d\mathbf{K}/dt = c\mathbf{p} - (d\mathbf{x}/dt)(E/c) = c\gamma(v)m\left[\boldsymbol{v} - (d\mathbf{x}/dt)\right], \qquad (3.2.78)$$

which has simply reproduced that $\boldsymbol{v} = d\mathbf{x}/dt$.

Returning to the case for fields, we first note that P^μ must be a four-vector from its definition and corresponds to four conserved charges, since we are assuming a Poincaré invariant system. We can always boost to the inertial frame for the fields where $\mathbf{P} = 0$ and in that frame we define the total rest mass M of the fields as

$$M \equiv (P^0/c)|_{(\mathbf{P}=0)} = \int d^3x \, (u/c^2)|_{(\mathbf{P}=0)}, \qquad (3.2.79)$$

where the total rest mass is a constant, $dM/dt = 0$, because $dP^\mu/dt = 0$. At $\mathbf{P} = 0$ we have $P^2 = M^2c^2$ and since P^μ is a four-vector then this is true in every frame. Define $\mathbf{V}$ as the analog of three-velocity $\boldsymbol{v}$ for a point particle,

$$\boldsymbol{V} \equiv \mathbf{P}/\gamma(V)M, \qquad (3.2.80)$$

where this $\mathbf{V}$ should not be confused with the four-velocity defined in Eq. (2.6.3). Under a boost of the system by velocity $\boldsymbol{v}$ we have $\mathbf{P} \neq 0$ and $P^0 = \gamma(v)Mc$, since P^μ is a four-vector. Then $P^2 = M^2c^2$ and $P^\mu = (\gamma(v)Mc, \gamma(V)M\mathbf{V})$, which for consistency requires that $\boldsymbol{v} = \mathbf{V}$. Then $\mathbf{V}$ is the boost of the system from its $\mathbf{P} = 0$ frame to the frame with total three-momentum $\mathbf{P}$. Since $dP^0/dt = 0$ in all frames, then in the rest frame we find $dM/dt = 0$. From $d\mathbf{P}/dt = 0$ it follows that $d(\gamma(V)\mathbf{V})/dt = 0$ and so $d\mathbf{V}/dt = 0$, since $\gamma(V)$ is a monotonically increasing function of V. Finally, define the *center of mass* of the fields as

$$\mathbf{X}_{\mathrm{com}} \equiv [1/\gamma(V)M] \int d^3x \left[\mathbf{x}(u/c^2)\right], \qquad (3.2.81)$$

which then allows us to write Eq. (3.2.76) as

$$\mathbf{K} = c\left[t\mathbf{P} - \gamma(V)M\mathbf{X}_{\mathrm{com}}\right], \qquad (3.2.82)$$

which has the same form as the single particle case above. For a Poincaré invariant system we then have

$$0 = d\mathbf{K}/dt = c\left[\mathbf{P} - \gamma(V)M(d\mathbf{X}_{\mathrm{com}}/dt)\right] = c\gamma(V)M\left[\boldsymbol{V} - d\mathbf{X}_{\mathrm{com}}/dt\right], \qquad (3.2.83)$$

which is the same as the single-particle form in Eq. (3.2.78) and shows that $\mathbf{V}$ is the velocity of the center of mass of the system, $\boldsymbol{V} = d\mathbf{X}_{\mathrm{com}}/dt$. So irrespective of the internal dynamics of the fields, Poincaré invariance ensures that we can define the time-independent total mass of the fields as M, the center of mass of the fields as $\mathbf{X}_{\mathrm{com}}$ and the velocity of the center of mass as $\boldsymbol{V}$. The *center-of-momentum frame* is the inertial frame in which $\boldsymbol{V} = 0$; i.e., the center of mass is at rest. Therefore, this frame is also commonly called the *center-of-mass frame*. Then we can write the spacetime coordinates of the center of mass as $X^\mu_{\mathrm{com}} = (ct, \mathbf{X}_{\mathrm{com}})$, the four-velocity as $V^\mu_{4-\mathrm{vel}} = \gamma(V)(c, \boldsymbol{V})$ and the total four-momentum of the fields as $P^\mu = (P^0, \mathbf{P}) = \gamma(V)M(c, \boldsymbol{V}) = \gamma(V)MV^\mu$ as we would for a single relativistic particle of mass M. As for a free particle the center of mass $\mathbf{X}_{\text{com}}$ moves with a constant velocity in a Poincaré invariant system.

The total orbital angular momentum about the spatial origin in Eq. (3.2.75) is

$$\mathbf{L} = \int d^3x \left[\mathbf{x} \times \mathbf{p}_{\text{dens}}\right] = \left[\mathbf{X}_{\text{com}} \times \mathbf{P}\right] + \int d^3x \left[(\mathbf{x} - \mathbf{X}_{\text{com}}) \times \mathbf{p}_{\text{dens}}\right], \qquad (3.2.84)$$

which is the sum of the orbital angular momentum of the whole system about the origin and the orbital angular momentum of the system around its center of mass. The system moves in a straight line with momentum $\mathbf{P}$ so the first term is conserved but not interesting. The center-of-momentum frame corresponds to $\mathbf{P} = 0$ and the total orbital angular momentum about the stationary center of mass is

$$\mathbf{L}_{(\mathbf{P}=0)} = \int d^3x \left[(\mathbf{x} - \mathbf{X}_{\text{com}}) \times \mathbf{p}_{\text{dens}(\mathbf{P}=0)}\right], \qquad (3.2.85)$$

which can be thought of as the "classical spin" of the whole system $\mathbf{J}_{\text{class}} \equiv \mathbf{L}_{(\mathbf{P}=0)}$.

If the fields vanish sufficiently rapidly outside some finite region and if we imagine ourselves far away from this region so that we do not resolve its internal structure, then the system of fields would appear as a single relativistic particle of mass M, classical spin $\mathbf{J}_{\text{class}}$ and velocity V. This remains true if the fields also carry intrinsic spin with $\mathbf{J}_{\text{class}} = (\mathbf{L} + \mathbf{S})_{(\mathbf{P}=0)}$, which we will now consider.

3.2.4 Intrinsic Angular Momentum

Let us now consider the case where the fields $\vec{\phi} = (\phi_1, \ldots, \phi_N)$ are not a set of N Lorentz scalar fields, but form a set of fields that transform in a nontrivial way under Lorentz transformations; i.e., $\vec{\phi}(x)$ transforms as

$$\vec{\phi}(x) \rightarrow \vec{\phi}'(x') = S\vec{\phi}(x) \quad \text{with} \quad x' = \Lambda x \qquad (3.2.86)$$

or in component form we write

$$\phi_r(x) \rightarrow \phi_r'(x') = S_{rs}\phi_s(x) \quad \text{with} \quad x'^\mu = \Lambda^\mu{}_\nu x^\nu, \qquad (3.2.87)$$

where $r, s = 1, \ldots, N$, where summation over repeated indices is to be understood, and where $S \equiv S[\Lambda]$ belongs to an $N \times N$ matrix representation of the Lorentz group. We confine our attention for now to proper orthochronous (restricted) Lorentz transformations, $\Lambda \in SO^+(1, 3)$, since we wish to study infinitesimal transformations. For example, if $\vec{\phi}(x)$ is a four-vector field, $\vec{\phi}(x) \rightarrow v^\mu(x)$ referred to in Table 1.2, then $S_{rs} \rightarrow \Lambda^\mu{}_\nu$ and the matrix $S = \Lambda$ belongs to the 4×4 defining representation of the restricted Lorentz group. When we come to consider fermions that are represented by four-component spinors in Sec. 4.4 we will construct a different 4×4 matrix representation of the Lorentz group.

Now for a physical system invariant under Lorentz transformations we can again invoke Noether's theorem as we did in constructing the angular momentum tensor. For an infinitesimal Lorentz transformation $d\omega^{\mu\nu}$ as defined in Eqs. (3.2.65) and (3.2.66) we can define $(\Sigma_{rs})_{\rho\sigma}$ using

$$S_{rs} \equiv \delta_{rs} + \tfrac{1}{2}d\omega^{\rho\sigma}\left(\Sigma_{rs}\right)_{\rho\sigma} \quad \text{with} \quad \left(\Sigma_{rs}\right)_{\rho\sigma} = -\left(\Sigma_{rs}\right)_{\sigma\rho}. \qquad (3.2.88)$$

Since the matrices S are a representation of the Lorentz group and comparing Eq. (3.2.88) with Eq. (1.2.44) we recognize that the $N \times N$ matrices $M_\Sigma^{\mu\nu} \equiv i\Sigma^{\mu\nu}$ are a representation of the generators of the Lorentz group and so must satisfy the Lorentz group Lie algebra relation in Eq. (1.2.200).

For an infinitesimal Lorentz transformation we have

$$\phi_r'(x') = S_{rs}\phi_s(x) = \phi_r(x) + \tfrac{1}{2}d\omega^{\rho\sigma}(\Sigma_{rs})_{\rho\sigma}\phi_s(x) \equiv \phi_r(x) + \tfrac{1}{2}d\omega^{\rho\sigma}\Phi_{r\rho\sigma}(x), \qquad (3.2.89)$$

where we have defined

$$\Phi_{r\rho\sigma}(x) \equiv (\Sigma_{rs})_{\rho\sigma}\phi_s(x). \tag{3.2.90}$$

Comparing with Eqs. (3.2.8) and (3.2.11) we then have

$$X^\mu \to X^\mu{}_{\rho\sigma} \equiv \left(\delta^\mu{}_\rho x_\sigma - \delta^\mu{}_\sigma x_\rho\right), \quad d\alpha \to \tfrac{1}{2}d\omega^{\rho\sigma}, \quad \Phi_i(x) \to \Phi_{r\rho\sigma}(x),$$

$$\Psi_i \to \Psi_{r\rho\sigma}(x) = (\Sigma_{rs})_{\rho\sigma}\phi_s(x) + x_\rho\partial_\sigma\phi_r(x) - x_\sigma\partial_\rho\phi_r(x). \tag{3.2.91}$$

We will also frequently use an obvious notation, where we write

$$\vec{\Phi}_{\rho\sigma} \equiv \Sigma_{\rho\sigma}\vec{\phi} \quad \text{and} \quad \vec{\Psi}_{\rho\sigma} \equiv \Sigma_{\rho\sigma}\vec{\phi} + x_\rho\partial_\sigma\vec{\phi} - x_\sigma\partial_\rho\vec{\phi}. \tag{3.2.92}$$

Using Eq. (3.2.36) the effective change in the Lagrangian density from an infinitesimal Lorentz transformation now becomes

$$\begin{aligned}
\delta\mathcal{L}_{\text{eff}} &= \tfrac{1}{2}d\omega^{\rho\sigma}\left\{\mathcal{L}(\partial_\mu X^\mu{}_{\rho\sigma}) + X^\mu{}_{\rho\sigma}[\partial_\mu^{\text{ex}}\mathcal{L}] + (\partial\mathcal{L}/\partial\vec{\phi})\cdot\vec{\Phi}_{\rho\sigma} \right. \\
&\quad \left. + \vec{\pi}^\nu\cdot\left[\partial_\nu\vec{\Phi}_{\rho\sigma} - (\partial_\mu\vec{\phi})(\partial_\nu X^\mu{}_{\rho\sigma})\right]\right\} \\
&= \tfrac{1}{2}d\omega^{\rho\sigma}\left\{[x_\sigma\partial_\rho^{\text{ex}}\mathcal{L} - x_\rho\partial_\sigma^{\text{ex}}\mathcal{L}] - (\vec{\pi}_\sigma\cdot\partial_\rho\vec{\phi} - \vec{\pi}_\rho\cdot\partial_\sigma\vec{\phi}) \right. \\
&\quad \left. + (\partial\mathcal{L}/\partial\vec{\phi})\cdot\Sigma_{\rho\sigma}\vec{\phi} + \vec{\pi}^\nu\cdot\Sigma_{\rho\sigma}\partial_\nu\vec{\phi}\right\},
\end{aligned} \tag{3.2.93}$$

where we have made use of the result from Eq. (3.2.68). If $\delta\mathcal{L}_{\text{eff}} = 0$ then the system will be Lorentz invariant. For a system that is also translationally invariant we would have $\partial_\mu^{\text{ex}}\mathcal{L} = 0$ and so we would then have Lorentz invariance if

$$(\partial\mathcal{L}/\partial\vec{\phi})\cdot\Sigma_{\rho\sigma}\vec{\phi} + \vec{\pi}^\nu\cdot\Sigma_{\rho\sigma}\partial_\nu\vec{\phi} = \vec{\pi}_\sigma\cdot\partial_\rho\vec{\phi} - \vec{\pi}_\rho\cdot\partial_\sigma\vec{\phi} = T_{\rho\sigma} - T_{\sigma\rho}, \tag{3.2.94}$$

where we have used Eq. (3.2.53). This tells us that if a Poincaré invariant system has no intrinsic angular momentum, $\Sigma_{\rho\sigma} = 0$, then the canonical stress-energy tensor is symmetric, $T^{\mu\nu} = T^{\nu\mu}$. On-shell Eq. (3.2.94) can be written as

$$\partial_\mu\left[(\partial\mathcal{L}/\partial(\partial_\mu\vec{\phi}))\cdot\Sigma_{\rho\sigma}\vec{\phi}\right] = T_{\rho\sigma} - T_{\sigma\rho}, \tag{3.2.95}$$

where we have used the Euler-Lagrange equations. This result will be exploited in Sec. 3.2.6 when we construct the Belinfante-Rosenfeld stress-energy tensor.

For a Lorentz invariant system with intrinsic angular momentum we construct the Lorentz-invariance conserved currents using Eq. (3.2.18) as we did in Eq. (3.2.72) but now including $\Phi_{r\rho\sigma}(x)$ to give

$$\begin{aligned}
\mathcal{J}^\mu{}_{\rho\sigma}(x) &= \left(-\pi_r^\mu\partial_\nu\phi_r + \delta^\mu{}_\nu\mathcal{L}\right)X^\nu{}_{\rho\sigma} + \pi_r^\mu\Phi_{r\rho\sigma} \\
&= \left(x_\rho T^\mu{}_\sigma - x_\sigma T^\mu{}_\rho\right) + \pi_r^\mu(\Sigma_{rs})_{\rho\sigma}\phi_s \equiv \left(x_\rho T^\mu{}_\sigma - x_\sigma T^\mu{}_\rho\right) + R^\mu{}_{\rho\sigma},
\end{aligned} \tag{3.2.96}$$

where we have defined

$$R^\mu{}_{\rho\sigma} \equiv \frac{\partial\mathcal{L}}{\partial(\partial_\mu\phi_r)}(\Sigma_{rs})_{\rho\sigma}\phi_s = \pi_r^\mu(\Sigma_{rs})_{\rho\sigma}\phi_s = \vec{\pi}^\mu\cdot\Sigma_{\rho\sigma}\vec{\phi}. \tag{3.2.97}$$

Again we have

$$\mathcal{J}^\mu{}_{\rho\sigma} = -\mathcal{J}^\mu{}_{\sigma\rho} \quad \text{and} \quad \partial_\mu\mathcal{J}^\mu{}_{\rho\sigma} = 0 \ \text{(on-shell)}. \tag{3.2.98}$$

The six corresponding conserved charges making up the angular momentum tensor now have the form

$$M^{\rho\sigma} = (1/c) \int d^3x \, \mathcal{J}^{0\rho\sigma} = (1/c) \int d^3x \, \left[(x^\rho T^{0\sigma} - x^\sigma T^{0\rho}) + \pi_r (\Sigma_{rs})^{\rho\sigma} \phi_s \right]$$
$$\equiv L^{\rho\sigma} + S^{\rho\sigma}, \tag{3.2.99}$$

where we recall that the canonical generalized momentum density is $\pi_r = \partial\mathcal{L}/\partial(\partial_0\phi_r)$ and where we have defined the orbital and intrinsic angular momentum components of $M^{\rho\sigma}$, respectively, as

$$L^{\rho\sigma} \equiv (1/c) \int d^3x \, \left(x^\rho T^{0\sigma} - x^\sigma T^{0\rho} \right) = \int d^3x \, \left(x^\rho p^\sigma_{\text{dens}} - x^\sigma p^\rho_{\text{dens}} \right),$$
$$\Sigma^{\rho\sigma} \equiv \int d^3x \, \left(\pi_r (\Sigma_{rs})^{\rho\sigma} \phi_s / c \right). \tag{3.2.100}$$

This decomposition of $M^{\rho\sigma}$ is the classical field theory version of the decomposition of Lorentz generators in Eq. (1.4.2). We can similarly decompose the total angular momentum (**J**) carried by the classical fields into total orbital (**L**) and total intrinsic angular momentum (**S**) components as

$$\mathbf{J} = \mathbf{L} + \mathbf{S}, \quad \text{where} \quad J^i \equiv \frac{1}{2}\epsilon^{ijk} M^{jk}, \ S^i \equiv \frac{1}{2}\epsilon^{ijk}\Sigma^{jk}, \ \text{and}$$
$$L^i \equiv \frac{1}{2}\epsilon^{ijk} L^{jk} = \int d^3x \, (\mathbf{x} \times \mathbf{p}_{\text{den}})^i. \tag{3.2.101}$$

The total intrinsic angular momentum **S** is also referred to as the total *intrinsic spin* carried by the field.

3.2.5 Internal Symmetries

Consider a symmetry where the action is invariant under a transformation of the field components between themselves at the *same spacetime point*, i.e.,

$$\phi_r(x) \to \phi'_r(x) = R_{rs}\phi_s(x). \tag{3.2.102}$$

We refer to this as an *internal symmetry*. We will typically be interested in situations where the matrix R is any arbitrary element of an $N \times N$ matrix representation of some group; e.g., the defining $N \times N$ representation of the group $SU(N)$ is one relevant example. We then say that the action is invariant under this group of internal transformations. For an infinitesimal transformation R we define λ_{rs} by

$$R_{rs} \equiv \delta_{rs} + d\alpha\lambda_{rs} \tag{3.2.103}$$

for an infinitesimal $d\alpha$ and hence

$$\phi'_r(x) = \phi_r(x) + d\alpha\lambda_{rs}\phi_s(x) \equiv \phi_r(x) + d\alpha\Phi_r(x), \tag{3.2.104}$$

which defines $\Phi_r(x)$. Clearly, in the case of internal symmetries $X^\mu(x) = 0$ and so Noether's theorem gives the conserved current

$$j^\mu(x) = \vec{\pi}^\mu(x) \cdot \vec{\Phi}(x) = \pi^\mu_r(x)\lambda_{rs}\phi_s(x) \tag{3.2.105}$$

and the conserved charge (recall $\pi_r \equiv \pi^0_r$)

$$Q = \int d^3x \, \rho(x) = \int d^3x \, j^0(x)/c = \int d^3x \, (\pi_r(x)\lambda_{rs}\phi_s(x)/c). \tag{3.2.106}$$

3.2.6 Belinfante-Rosenfeld Tensor

As observed earlier, the stress-energy tensor $T^{\mu\nu}$ is not in general symmetric; however, it is possible to construct a symmetric form using the observation of Eq. (3.2.23) that the addition of the four-divergence of an antisymmetric tensor leads to a redefinition of a conserved current. The Belinfante-Rosenfeld stress-energy tensor is an on-shell improvement of the canonical stress-energy tensor. It is this symmetric form of the stress-energy tensor that induces the curvature of spacetime in general relativity. Putting this more precisely, the Belinfante-Rosenfeld stress-energy tensor is equivalent on-shell to the symmetric Hilbert stress-energy tensor of general relativity (Belinfante, 1940; Rosenfeld, 1940; Blaschke et al., 2016).

For a Poincaré invariant system we have both translational invariance, $\partial_\mu T^{\mu\nu} = 0$, and Lorentz invariance, $\partial_\mu \mathcal{J}^\mu{}_{\rho\sigma} = 0$. It then follows from Eq. (3.2.96) that

$$
\begin{aligned}
0 = \partial_\mu \mathcal{J}^\mu{}_{\rho\sigma} &= \partial_\mu \left(\left(x_\rho T^\mu{}_\sigma - x_\sigma T^\mu{}_\rho \right) + \pi^\mu_r (\Sigma_{rs})_{\rho\sigma} \phi_s \right) \\
&= g_{\mu\rho} T^\mu{}_\sigma - g_{\mu\sigma} T^\mu{}_\rho + \partial_\mu R^\mu{}_{\rho\sigma},
\end{aligned}
\tag{3.2.107}
$$

where $R^\mu{}_{\rho\sigma}$ was defined in Eq. (3.2.97). Recall that $(\Sigma_{rs})_{\rho\sigma} = -(\Sigma_{rs})_{\sigma\rho}$ and so

$$
R^{\mu\rho\sigma} = -R^{\mu\sigma\rho}.
\tag{3.2.108}
$$

Using Eq. (3.2.107) the antisymmetric part of the stress-energy tensor is

$$
\tfrac{1}{2} \left(T^{\rho\sigma} - T^{\sigma\rho} \right) = -\tfrac{1}{2} \partial_\mu R^{\mu\rho\sigma}
\tag{3.2.109}
$$

and hence $T^{\mu\nu}$ is only nonsymmetric when the classical field has intrinsic angular momentum, $\Sigma_{rs} \neq 0$. Note that Eq. (3.2.109) is only necessarily true *on-shell* since we used current conservation in the form of Eq. (3.2.107) to obtain it; i.e., we have used the Euler-Lagrange equations of motion.

For a Poincaré invariant system we want to show that we can construct a symmetric form $\bar{T}^{\mu\nu}$ of the stress energy tensor such that on-shell we have

$$
\partial_\mu \bar{T}^{\mu\nu} = 0 \quad \text{and} \quad \bar{T}^{\mu\nu} = \bar{T}^{\nu\mu}
\tag{3.2.110}
$$

and the conserved charges are unchanged. We can exploit the fact that we are free to add the four-divergence of an antisymmetric tensor to a conserved current as shown in Eq. (3.2.23), where the two indices that need to be antisymmetric are the index contracting with the four-divergence and the index labeling the conserved current. Thus we wish to find a $K^{\mu\rho\sigma}$ such that

$$
\bar{T}^{\mu\nu} = T^{\mu\nu} + \partial_\rho K^{\rho\mu\nu} \quad \text{with} \quad K^{\rho\mu\nu} = -K^{\mu\rho\nu}.
\tag{3.2.111}
$$

Using $\bar{T}^{\mu\nu} = \bar{T}^{\nu\mu}$ and Eq. (3.2.109) we must have for consistency that

$$
\partial_\mu R^{\mu\rho\sigma} = -(T^{\rho\sigma} - T^{\sigma\rho}) = \partial_\mu K^{\mu\rho\sigma} - \partial_\mu K^{\mu\sigma\rho},
\tag{3.2.112}
$$

which is satisfied if we require

$$
R^{\mu\rho\sigma} = K^{\mu\rho\sigma} - K^{\mu\sigma\rho}.
\tag{3.2.113}
$$

Using the fact that $R^{\mu\rho\sigma} = -R^{\mu\sigma\rho}$ it is readily verified that if we define

$$
K^{\mu\rho\sigma} \equiv \tfrac{1}{2} \left(R^{\mu\rho\sigma} + R^{\rho\sigma\mu} + R^{\sigma\rho\mu} \right)
\tag{3.2.114}
$$

then all requirements are satisfied. With this choice of $K^{\mu\rho\sigma}$ we refer to the symmetrized stress-energy tensor $\bar{T}^{\mu\nu}$ as the *Belinfante-Rosenfeld tensor* or sometimes simply as the *Belinfante tensor*.

Note that the Belinfante-Rosenfeld tensor, $\bar{T}^{\mu\nu}$, is an *on-shell improvement* of the canonical stress-energy tensor, $T^{\mu\nu}$, since we have used the Euler-Lagrange equations of motion to construct it.

For the conserved charges associated with translational invariance arising from the symmetrized stress-energy tensor we have

$$\bar{P}^\nu = (1/c) \int d^3x\, \bar{T}^{0\nu} = (1/c) \int d^3x\, \left(T^{0\nu} + \partial_\mu K^{\mu 0\nu}\right) = (1/c) \int d^3x\, \left(T^{0\nu} + \partial_j K^{j0\nu}\right)$$
$$= (1/c) \int d^3x\, T^{0\nu} + (1/c) \int ds^j\, K^{j0\nu} = (1/c) \int d^3x\, T^{0\nu} = P^\mu, \tag{3.2.115}$$

where the last step is valid provided that the surface term can be neglected, i.e, if fields vanish at spatial infinity or satisfy periodic spatial boundary conditions. The total four-momentum is the same for both forms of the stress-energy tensor. Next consider the definition

$$\bar{\mathcal{J}}^\mu{}_{\rho\sigma} \equiv \left(x_\rho \bar{T}^\mu{}_\sigma - x_\sigma \bar{T}^\mu{}_\rho\right), \tag{3.2.116}$$

which is another way of writing the angular momentum tensor $\mathcal{J}^\mu{}_{\rho\sigma}$ in Eq. (3.2.96). To understand this note that

$$\bar{\mathcal{J}}^{\mu\rho\sigma} \equiv \left(x^\rho \bar{T}^{\mu\sigma} - x^\sigma \bar{T}^{\mu\rho}\right) = (x^\rho T^{\mu\sigma} - x^\sigma T^{\mu\rho}) + \left(x^\rho \partial_\lambda K^{\lambda\mu\sigma} - x^\sigma \partial_\lambda K^{\lambda\mu\rho}\right)$$
$$= (x^\rho T^{\mu\sigma} - x^\sigma T^{\mu\rho}) + \partial_\lambda \left(x^\rho K^{\lambda\mu\sigma} - x^\sigma K^{\lambda\mu\rho}\right) - (K^{\rho\mu\sigma} - K^{\sigma\mu\rho})$$
$$= (x^\rho T^{\mu\sigma} - x^\sigma T^{\mu\rho}) + \partial_\lambda \left(x^\rho K^{\lambda\mu\sigma} - x^\sigma K^{\lambda\mu\rho}\right) + R^{\mu\rho\sigma}$$
$$= \mathcal{J}^{\mu\rho\sigma} + \partial_\lambda \left(x^\rho K^{\lambda\mu\sigma} - x^\sigma K^{\lambda\mu\rho}\right), \tag{3.2.117}$$

where we have used Eqs. (3.2.96) and (3.2.113) and that $K^{\rho\mu\sigma} = -K^{\mu\rho\sigma}$. Since $\bar{\mathcal{J}}^{\mu\rho\sigma}$ and $\mathcal{J}^{\mu\rho\sigma}$ only differ by a total four-divergence and since we have antisymmetry in the first two indices of K, then it immediately follows that

$$\partial_\mu \bar{\mathcal{J}}^{\mu\rho\sigma} = 0 \quad \text{and}$$
$$\bar{M}^{\rho\sigma} = (1/c) \int d^3x\, \bar{\mathcal{J}}^{0\rho\sigma} = (1/c) \int d^3x\, \left[\mathcal{J}^{0\rho\sigma} + \partial_j(x^\rho K^{j0\sigma} - x^\sigma K^{j0\rho})\right]$$
$$= M^{\rho\sigma} + (1/c) \int ds^j\, (x^\rho K^{j0\sigma} - x^\sigma K^{j0\rho}) = M^{\rho\sigma}, \tag{3.2.118}$$

where we again assume that the field spatial boundary conditions mean that we can ignore the surface term.

3.2.7 Noether's Theorem and Poisson Brackets

Recall from Eqs. (2.4.65) to (2.4.69) that in classical mechanics any dynamical variable G generates a canonical transformation such that for every dynamical variable F we have $F \to F + dF = F + d\alpha\{F, G\}$. For the special cases of $F = q_i$ and $F = p_i$ we then have $q_i \to q_i + d\alpha\{q_i, G\} = q_i + d\alpha(\partial G/\partial p_i)$ and $p_i \to p_i + d\alpha\{p_i, G\} = p_i - d\alpha(\partial G/\partial q_i)$, respectively. We recall that canonical transformations leave all Poisson brackets invariant. Furthermore, if G is a conserved charge, then $dG/dt = \{H, G\} = 0$ and so G is the generator of a canonical transformation that is a symmetry of the system, since $dH = d\alpha\{H, G\} = 0$. This result extends to conserved Noether charges in the case of classical fields.

In a classical field theory we know from Eq. (3.2.18) that a conserved Noether current has the form

$$j^\mu(x) = \vec{\pi}^\mu(x) \cdot \vec{\Psi}(x) + \mathcal{L}(\vec{\phi}, \partial_\mu \vec{\phi}, x) X^\mu(x) \quad \text{with}$$
$$\Psi(x) = \vec{\Phi}(x) - X^\mu(x)\partial_\mu \vec{\phi}(x). \tag{3.2.119}$$

The conserved charge is $Q = \int d^3x\, j^0(x)$. The corresponding total variation of the field is $\Delta\vec{\phi}(x) = d\alpha\vec{\Psi}(x)$ as given in Eq. (3.2.12). Since Q is a conserved charge then

$$dQ/dt = \{H, Q\} = 0 \qquad (3.2.120)$$

and so Q generates a canonical transformation that is a symmetry of the system. If F is any dynamical variable then $dF = d\alpha\{F, Q\}$. For the case $F = \phi_i(x)$ the change in ϕ_i is the total variation of ϕ_i given in Eq. (3.2.11) and so $d\phi_r(x) = \Delta\phi_r(x) = d\alpha\Psi(x)$. We then have

$$d\phi_i(x) = \Delta\phi_i(x) = d\alpha\Psi(x) = d\alpha\{\phi_i(x), Q\} = d\alpha\,\delta Q/\delta\pi_i(x),$$
$$d\pi_i(x) = d\alpha\{\pi_i(x), Q\} = -d\alpha\,\delta Q/\delta\phi_i(x). \qquad (3.2.121)$$

Proof: It is a useful exercise to verify explicitly that $\Psi(x) = \delta Q/\delta\pi_i(x)$ (Greiner and Reinhardt, 1996). We work at an arbitrary fixed time t and suppress the time dependence for brevity, $\phi(x) = \phi_i(t, \mathbf{x}) \to \phi_i(\mathbf{x})$. We have

$$Q = \int d^3z\, j^0(\mathbf{z}) = \int d^3z\,(\vec{\pi}\cdot\vec{\Psi} + \mathcal{L}X^0) = \int d^3z\,(\vec{\pi}\cdot[\vec{\Phi} - X^\mu\partial_\mu\vec{\phi}] + \mathcal{L}X^0),$$

$$\frac{\delta Q}{\delta\pi_i(\mathbf{x})} = \Psi_i(\mathbf{x}) + \int d^3z\left(\vec{\pi}(\mathbf{z})\cdot\frac{\delta\vec{\Psi}(\mathbf{z})}{\delta\pi_i(\mathbf{x})} + \frac{\delta\mathcal{L}(\mathbf{z})}{\delta\pi_i(\mathbf{x})}X^0(\mathbf{z})\right) = \Psi_i(\mathbf{x}), \qquad (3.2.122)$$

where the second term vanishes as explained below. Φ_i is the local variation of the field ϕ_i and X^μ is the spacetime transformation. They are both independent of $\vec{\pi}$; e.g., see Eq. (3.2.87). For a nonsingular system the Legendre transform is invertible and so the relationship between velocities and momenta is also; i.e., there is an invertible function Π such that $\pi_i = \partial\mathcal{L}/\partial\dot{\phi}_i \equiv \Pi_i(\vec{\phi}, \dot{\vec{\phi}}, \nabla\vec{\phi})$ and so $\dot{\phi}_i = \partial\mathcal{H}/\partial\pi_i = (\Pi^{-1})_i(\vec{\phi}, \dot{\vec{\phi}}, \nabla\vec{\phi})$. Then we have

$$\vec{\pi}(\mathbf{z})\cdot\frac{\delta\vec{\Psi}(\mathbf{z})}{\delta\pi_i(\mathbf{x})} = -X^0(\mathbf{z})\,\vec{\pi}(\mathbf{z})\cdot\frac{\delta\dot{\vec{\phi}}(\mathbf{z})}{\delta\pi_i(\mathbf{x})} \quad \text{and}$$

$$\frac{\delta\mathcal{L}(\mathbf{z})}{\delta\pi_i(\mathbf{x})}X^0(\mathbf{z}) = \int d^3u\,\frac{\delta\mathcal{L}(\mathbf{z})}{\delta\dot{\vec{\phi}}(\mathbf{u})}\cdot\frac{\dot{\vec{\phi}}(\mathbf{u})}{\delta\pi_i(\mathbf{x})}X^0(\mathbf{z}) = \vec{\pi}(\mathbf{z})\cdot\frac{\dot{\vec{\phi}}(\mathbf{u})}{\delta\pi_i(\mathbf{x})}X^0(\mathbf{z}), \qquad (3.2.123)$$

which cancel when summed and the above result follows.

3.2.8 Generators of the Poincaré Group

In the case of invariance under spacetime translations considered in Sec. 3.2.2 the conserved currents were the components of the stress-energy tensor, where we found $\partial_\mu T^{\mu\nu} = 0$, and the corresponding conserved charges were the components of the total four-momentum, $P^\nu = \int d^3x\, T^{0\nu}$. We further found in Eqs. (3.2.57) and (3.2.58) that $P^0 = H = \int d^3x\,\mathcal{H}$ and that $\mathbf{P} = -\int d^3x\,\vec{\pi}\cdot\nabla\vec{\phi}$. The sign of a conserved current and charge is arbitrary, but we chose the sign that means that P^μ is the total four-momentum. The conserved charges of Lorentz invariance are the $M^{\mu\nu}$ of Eq. (3.2.99), where these have the same sign for the charges as Q above. So from Eq. (3.2.121) above and from Eqs. (3.2.51) and (3.2.92) we have

$$\{\vec{\phi}(x), P^\mu\} = -\vec{\Psi}^\nu(x) = \partial^\mu\vec{\phi}(x) \quad \text{and}$$
$$\{\vec{\phi}(x), M^{\mu\nu}\} = \vec{\Psi}^{\mu\nu}(x) = \Sigma_{\rho\sigma}\vec{\phi}(x) + x_\rho\partial_\sigma\vec{\phi}(x) - x_\sigma\partial_\rho\vec{\phi}(x). \qquad (3.2.124)$$

Recalling that the $N \times N$ matrices $M_\Sigma^{\mu\nu} = i\Sigma^{\mu\nu}$ are a representation of the generators of the Lorentz group, we can show with some effort that

$$i\{P^\mu, P^\nu\} = 0, \qquad i\{P^\mu, M^{\rho\sigma}\} = i(g^{\mu\rho}P^\sigma - g^{\mu\sigma}P^\rho),$$
$$i\{M^{\mu\nu}, M^{\rho\sigma}\} = i\left(g^{\nu\rho}M^{\mu\sigma} - g^{\mu\rho}M^{\nu\sigma} - g^{\nu\sigma}M^{\mu\rho} + g^{\mu\sigma}M^{\nu\rho}\right). \qquad (3.2.125)$$

Referring to the correspondence principle in Eq. (2.5.27), we see that the replacement of $i\{\cdot, \cdot\} \leftrightarrow [\cdot, \cdot]$ in the above reproduces the Poincaré Lie algebra in Eq. (1.2.167). So we can expect that the canonical quantization of relativistic classical fields will lead to relativistic quantum field theories in which the Poincaré Lie algebra emerges naturally. The dedicated reader may wish to prove Eq. (3.2.125). For those wanting guidance, a proof of these results can be found in chapter 2 of Greiner and Reinhardt (1996).

3.3 Classical Electromagnetic Field

As an illustration of how to extend these techniques to the case of a singular theory, we turn to the important example of a classical electromagnetic field. From this point on we will only discuss electromagnetism in terms of Lorentz-Heaviside units as discussed in Sec. 3.3. This is the usual choice in particle physics since factors of ϵ_0 and μ_0 no longer appear and only factors of c remain.

3.3.1 Lagrangian Formulation of Electromagnetism

We saw in Eqs. (2.7.41) and (2.7.42) that we can succinctly express Maxwell's equations in covariant form in Lorentz-Heaviside units as

$$\partial_\mu \tilde{F}^{\mu\nu} = 0 \qquad \text{and} \qquad \partial_\mu F^{\mu\nu} = (1/c)j^\nu, \qquad (3.3.1)$$

where the first equation is a mathematical identity that follows from the definition of $\tilde{F}^{\mu\nu}$ in Eq. (2.7.18). It reproduces the first two of Maxwell's equations in Eq. (2.7.2). The second equation contains all of the physics of the electromagnetic field interacting with a conserved four-vector current, i.e., the third and fourth of Maxwell's equations. We know from Eq. (2.7.24) that only coupling to a conserved j^μ is possible. Can we find an action $S[A]$ and Lagrangian density $\mathcal{L}$ that has this second equation as its equation of motion?

Consider the action

$$S[A] = \int dt\, L = (1/c) \int d^4x\, \mathcal{L} = (1/c) \int d^4x\, \left[-\tfrac{1}{4}F_{\mu\nu}F^{\mu\nu} - (1/c)j_\mu A^\mu\right]. \qquad (3.3.2)$$

Applying Hamilton's principle to this action gives the Euler-Lagrange equations for the four-vector field components A^μ,

$$0 = c\frac{\delta S[A]}{\delta A_\nu(x)} = \frac{\partial \mathcal{L}}{\partial A_\nu(x)} - \partial_\mu \left(\frac{\partial \mathcal{L}}{\partial (\partial_\mu A_\nu)(x)}\right) = -\frac{1}{c}j^\nu(x) + \partial_\mu F^{\mu\nu}(x), \qquad (3.3.3)$$

which follow from Eq. (3.1.8). The equations of motion are gauge invariant as described earlier; e.g., see the discussion following Eq. (2.7.74). Hence we may define

$$\mathcal{L} = -\tfrac{1}{4}F_{\mu\nu}F^{\mu\nu} - (1/c)j_\mu A^\mu \qquad (3.3.4)$$

as the Lagrangian density for the electromagnetic field interacting with a conserved four-vector current density j^μ.

Proof: This result follows since $\partial \mathcal{L}/\partial A_\nu = j^\nu/c$ and since

$$\frac{\partial \mathcal{L}}{\partial(\partial_\mu A_\nu)} = \frac{\partial\left(-\frac{1}{4}F_{\rho\sigma}F^{\rho\sigma}\right)}{\partial F_{\lambda\tau}}\frac{\partial F_{\lambda\tau}}{\partial(\partial_\mu A_\nu)} = -\frac{1}{2}F^{\lambda\tau}\left(\delta_{\mu\lambda}\delta_{\nu\tau} - \delta_{\mu\tau}\delta_{\nu\lambda}\right) = -F^{\mu\nu},$$

which gives $-\partial_\mu\left(\frac{\partial \mathcal{L}}{\partial(\partial_\mu A_\nu)}\right) = \partial_\mu F^{\mu\nu}.$ (3.3.5)

When there is no external current density, $j^\mu = 0$, we have the case of pure electromagnetism with

$$\mathcal{L} = -\tfrac{1}{4}F_{\mu\nu}F^{\mu\nu}. \tag{3.3.6}$$

Since this Lagrangian density is translationally invariant, using Eq. (3.2.53) leads to the stress-energy tensor

$$T^\mu{}_\nu(x) = \frac{\partial \mathcal{L}}{\partial(\partial_\mu A^\tau)}\partial_\nu A^\tau - \delta^\mu{}_\nu \mathcal{L} = -F^{\mu\tau}\partial_\nu A_\tau + \tfrac{1}{4}\delta^\mu{}_\nu F^{\sigma\tau}F_{\sigma\tau} \tag{3.3.7}$$

with $\partial_\mu T^{\mu\nu} = 0$. This is easily verified directly:

$$\begin{aligned}\partial_\mu T^\mu{}_\nu &= \partial_\mu\left(-F^{\mu\tau}\partial_\nu A_\tau + \tfrac{1}{4}\delta^\mu{}_\nu F^{\sigma\tau}F_{\sigma\tau}\right) = -F^{\mu\tau}\partial_\nu\partial_\mu A_\tau + \tfrac{1}{4}\partial_\nu\left(F^{\sigma\tau}F_{\sigma\tau}\right)\\ &= -\tfrac{1}{2}F^{\mu\tau}\partial_\nu F_{\mu\tau} + \tfrac{1}{2}F^{\sigma\tau}\partial_\nu F_{\sigma\tau} = 0,\end{aligned} \tag{3.3.8}$$

where we have used Eq. (3.3.1) and that $F^{\mu\tau} = \frac{1}{2}(F^{\mu\tau} - F^{\tau\mu})$.

It will prove more convenient in electromagnetism to work with the symmetrized form of the stress-energy tensor, since this can be expressed entirely in terms of gauge-invariant quantities. Using Eq. (3.2.23) we can add the four-divergence of an antisymmetric tensor and not spoil current conservation. Since $(F^{\mu\tau}A_\nu)$ is antisymmetric under the interchange of μ and τ we can define a modified conserved stress-energy tensor by choosing $K^{\mu\tau\nu} = F^{\mu\tau}A^\nu$ in Eq. (3.2.111) to give

$$\bar{T}^\mu{}_\nu \equiv T^\mu{}_\nu + \partial_\tau\left(F^{\mu\tau}A_\nu\right) = T^\mu{}_\nu + F^{\mu\tau}\partial_\tau A_\nu = -F^{\mu\tau}F_{\nu\tau} + \tfrac{1}{4}\delta^\mu{}_\nu F^{\sigma\tau}F_{\sigma\tau}, \tag{3.3.9}$$

where we have used the equations of motion $\partial_\mu F^{\mu\nu} = 0$. We then observe that

$$\bar{T}^{\mu\nu} = -F^{\mu\tau}F^\nu{}_\tau + \tfrac{1}{4}g^{\mu\nu}F^{\sigma\tau}F_{\sigma\tau}, \tag{3.3.10}$$

is a symmetric and conserved stress-energy tensor, i.e.,

$$\bar{T}^{\mu\nu} = \bar{T}^{\nu\mu} \quad \text{and} \quad \partial_\mu\bar{T}^{\mu\nu} = 0. \tag{3.3.11}$$

We recognize $\bar{T}^{\mu\nu}$ as the Belinfante-Rosenfeld tensor for electromagnetism, where we will later show that this corresponds to the construction set out in Sec. 3.2.6.

From Eq. (2.7.36) we know that $F^{\mu\nu}$ can be expressed in terms of the **E** and **B** fields alone and so this is also true of $\mathcal{L}$ and $\bar{T}^{\mu\nu}$. We find that

$$\begin{aligned}\mathcal{L} &= -\tfrac{1}{4}F^{\mu\nu}F_{\mu\nu} = -\tfrac{1}{4}\left[-2(F^{0i})^2 + (F^{ij})^2\right] = \tfrac{1}{2}\left(\mathbf{E}^2 - \mathbf{B}^2\right), \tag{3.3.12}\\ \bar{T}^{00} &= (F^{0i})^2 + \tfrac{1}{4}F^{\sigma\tau}F_{\sigma\tau} = \tfrac{1}{2}\left(\mathbf{E}^2 + \mathbf{B}^2\right) = u,\\ \bar{T}^{0i} &= -F^{0j}F^i{}_j = F^{0j}F^{ij} = (-E^j)(-\epsilon^{ijk}B^k) = (\mathbf{E}\times\mathbf{B})^i = S^i/c,\\ \bar{T}^{ij} &= -E^iE^j + \delta^{ij}\mathbf{B}^2 - B^iB^j - \tfrac{1}{4}\delta^{ij}F^{\sigma\tau}F_{\sigma\tau} = -\left(E^iE^j + B^iB^j - u\right) \equiv -\sigma^{ij},\end{aligned}$$

where sums over repeated indices are to be understood and where we have used $\epsilon^{ijk}\epsilon^{ij\ell} = 2\delta^{k\ell}$ and $\epsilon^{ik\ell}\epsilon^{jkm} = \delta^{ij}\delta^{\ell m} - \delta^{im}\delta^{j\ell}$. Note that u and $\mathbf{S}$ are the energy density and the Poynting vector, respectively, of the electromagnetic field as given in Eq. (2.7.45). σ^{ij} is referred to as the Maxwell stress tensor. We also note that $\bar{T}^{\mu\nu}$ is not only symmetric but is also gauge invariant since it can be expressed in terms of $\mathbf{E}$ and $\mathbf{B}$ fields alone. Since $F^{\mu\nu}$ is gauge invariant and A^μ is not, it is clear that the stress-energy tensor before improvement is not gauge invariant.

We can summarize the symmetrized electromagnetic stress-energy tensor as

$$\bar{T}^{\mu\nu} = \begin{pmatrix} u & S_x/c & S_y/c & S_z/c \\ S_x/c & -\sigma_{xx} & -\sigma_{xy} & -\sigma_{xz} \\ S_y/c & -\sigma_{yx} & -\sigma_{yy} & -\sigma_{yz} \\ S_z/c & -\sigma_{zx} & -\sigma_{zy} & -\sigma_{zz} \end{pmatrix}. \tag{3.3.13}$$

where $\mathbf{S} = (S^1, S^2, S^3) = (S_x, S_y, S_z)$ and we use the notation $\sigma^{11} \equiv \sigma_{xx}$, $\sigma^{12} \equiv \sigma_{xy}$ and so on. We observe that the stress-energy tensor $\bar{T}^\mu{}_\nu$ is traceless since

$$\bar{T}^\mu{}_\mu = \bar{T}^{00} - \bar{T}^{ii} = u + \sigma^{ii} = -2u + \mathbf{E}^2 + \mathbf{B}^2 = 0. \tag{3.3.14}$$

Since $\partial_\mu \bar{T}^{\mu\nu} = 0$ we have the conserved total four-momentum of the electromagnetic field $P^\mu = (E/c, \mathbf{P})$, where

$$P^0 = E/c = (1/c) \int d^3x\, u = (1/c) \int d^3x\, \bar{T}^{00} = (1/c) \int d^3x\, \tfrac{1}{2}\left(\mathbf{E}^2 + \mathbf{B}^2\right), \tag{3.3.15}$$

$$P^i = \int d^3x\, p^i_{\mathrm{dens}} = (1/c) \int d^3x\, \bar{T}^{0i} = \int d^3x\, S^i/c^2 = (1/c) \int d^3x\, (\mathbf{E} \times \mathbf{B})^i.$$

So the momentum density of the electromagnetic field is proportional to the Poynting vector, $\mathbf{p}_{\mathrm{dens}} = \mathbf{S}/c^2$. We observe that

$$P^\mu = (1/c) \int d^3x\, \bar{T}^{0\mu} = (1/c) \int d^3x \left[T^{0\mu} + \partial_j \left(F^{0j} A^\mu \right) \right] = (1/c) \int d^3x\, T^{0\mu}, \tag{3.3.16}$$

where we have used $F^{00} = 0$, the three-dimensional divergence theorem and we have assumed vanishing or periodic spatial boundary conditions so that surface terms can be neglected. So we arrive at the same P^μ using either $T^{\mu\nu}$ or $\bar{T}^{\mu\nu}$ as we knew that we should from Eq. (3.2.116).

For the electromagnetic field Eq. (3.2.87) becomes $A^\mu(x) \to A'^\mu(x') = \Lambda^\mu{}_\nu A^\nu(x)$. Comparing Eq. (3.2.88) with Eq. (1.2.44) we identify

$$S_{rs} \to \Lambda^\mu{}_\nu \quad \text{and} \quad (\Sigma^\mu{}_\nu)_{\rho\sigma} = -i(M_{\rho\sigma})^\mu{}_\nu = \delta_\rho{}^\mu g_{\sigma\nu} - \delta_\sigma{}^\mu g_{\rho\nu}, \tag{3.3.17}$$

where we have used Eq. (1.2.41). Then from Eqs. (3.2.97) and (3.2.96) we have

$$R^\mu{}_{\rho\sigma} = \frac{\partial \mathcal{L}}{\partial(\partial_\mu A_\nu)}(-iM_{\rho\sigma})_{\nu\tau}A^\tau = -F^{\mu\nu}(g_{\rho\nu}g_{\sigma\tau} - g_{\rho\tau}g_{\sigma\nu})A^\tau = F^\mu{}_\sigma A_\rho - F^\mu{}_\rho A_\sigma,$$

$$\mathcal{J}^\mu{}_{\rho\sigma} = \left(x_\rho T^\mu{}_\sigma - x_\sigma T^\mu{}_\rho \right) + R^\mu{}_{\rho\sigma} = \left(x_\rho T^\mu{}_\sigma - x_\sigma T^\mu{}_\rho \right) + F^\mu{}_\sigma A_\rho - F^\mu{}_\rho A_\sigma. \tag{3.3.18}$$

From Eq. (3.2.118) the angular momentum tensor can be obtained using

$$M^{\rho\sigma} = \bar{M}^{\rho\sigma} = \frac{1}{c} \int d^3x\, \bar{\mathcal{J}}^{0\rho\sigma} = \frac{1}{c} \int d^3x \left(x^\rho \bar{T}^{0\sigma} - x^\sigma \bar{T}^{0\rho} \right),$$

$$J^i = \frac{1}{2}\epsilon^{ijk} M^{jk} = (1/c) \int d^3x\, [\mathbf{x} \times (\mathbf{E} \times \mathbf{B})]^i, \tag{3.3.19}$$

where we have used $\bar{T}^{0i} = (\mathbf{E} \times \mathbf{B})^i$. Finally, using Eq. (3.2.114) to construct the $K^{\mu\rho\sigma}$ needed for the Belinfante-Rosenfeld tensor we find

$$K^{\mu\rho\sigma} \equiv \frac{1}{2}\left(R^{\mu\rho\sigma} + R^{\rho\sigma\mu} + R^{\sigma\rho\mu} \right) = F^{\rho\mu} A^\sigma, \tag{3.3.20}$$

which is just what we used in Eq. (3.3.9). So in defining $\bar{T}^{\mu\nu}$ in Eq. (3.3.9) we were performing the Belinfante-Rosenfeld construction. We have not as yet discussed gauge-fixing for electromagnetism in the Lagrangian formalism. This will be done when we carry out the path integral quantization of electromagnetism.

From Eq. (3.2.101) we can directly construct the intrinsic spin contribution to the total angular momentum,

$$S^i = \tfrac{1}{2}\epsilon^{ijk}\Sigma^{jk} = (1/2c)\int d^3x\,\epsilon^{ijk}(F^{0k}A^j - F^{0j}A^k) = (1/c)\int d^3x\,(\mathbf{E}\times\mathbf{A})^i, \qquad (3.3.21)$$

where the total angular momentum is $\mathbf{J} = \mathbf{L}+\mathbf{S}$. The total angular momentum of the electromagnetic field can be decomposed into orbital and spin components as

$$\mathbf{J} = (1/c)\int d^3x\,[\mathbf{x}\times(\mathbf{E}\times\mathbf{B})] = \mathbf{L} + \mathbf{S}, \qquad \text{where} \qquad (3.3.22)$$
$$\mathbf{L} = (1/c)\int d^3x\,E^i(\mathbf{x}\times\boldsymbol{\nabla})A^i \qquad \text{and} \qquad \mathbf{S} = (1/c)\int d^3x\,\mathbf{E}\times\mathbf{A}.$$

Note that the decomposition is not gauge invariant since $\mathbf{A}$ is not gauge invariant.

Proof: We wish to prove Eq. (3.3.22). Note that

$$[\mathbf{x}\times(\mathbf{E}\times\mathbf{B})]^i = \epsilon^{ijk}x^j(\mathbf{E}\times\mathbf{B})^k = \epsilon^{ijk}x^j\epsilon^{k\ell m}E^\ell B^m = \epsilon^{ijk}x^j\epsilon^{k\ell m}E^\ell\epsilon^{mnp}\partial_n A^p$$
$$= \epsilon^{ijk}x^j\epsilon^{k\ell m}E^\ell\epsilon^{mnp}\partial_n A^p = \epsilon^{ijk}x^j E^\ell[\delta^{kn}\delta^{\ell p} - \delta^{kp}\delta^{\ell n}]\partial_n A^p$$
$$= \epsilon^{ijk}x^j[E^\ell\partial_k A^\ell - E^\ell\partial_\ell A^k] = \epsilon^{ijk}x^j[E^\ell\partial_k A^\ell - \partial_\ell(E^\ell A^k)],$$

where in the last step we used $\boldsymbol{\nabla}\cdot\mathbf{E} = 0$ for a free electromagnetic field. Using integration by parts and periodic/vanishing spatial boundary conditions allows the last term to be rewritten in the required form,

$$\int d^3x\,[\mathbf{x}\times(\mathbf{E}\times\mathbf{B})]^i = \int d^3x\,\epsilon^{ijk}x^j[E^\ell\partial_k A^\ell - \partial_\ell(E^\ell A^k)] \qquad (3.3.23)$$
$$= \int d^3x\,\epsilon^{ijk}[x^j E^\ell\partial_k A^\ell + E^j A^k]$$
$$= \int d^3x\,[E^\ell(\mathbf{x}\times\boldsymbol{\nabla})^i A^\ell + (\mathbf{E}\times\mathbf{A})^i].$$

We will show in Sec. 6.4.2 that in Coulomb gauge the $\mathbf{L}$ and $\mathbf{S}$ correspond to the appropriate physical observables. Since $\boldsymbol{\nabla}\cdot\mathbf{A} = 0$ in Coulomb gauge then $\mathbf{A} = \mathbf{A}_t + \mathbf{A}_\ell = \mathbf{A}_t$ since $\mathbf{A}_\ell = 0$. Recall from Eq. (2.7.115) that $\mathbf{A}_t = \boldsymbol{\nabla}\times\mathbf{A}_A$ and $\mathbf{A}_\ell = -\boldsymbol{\nabla}\Phi_A$ for fields Φ_A and $\mathbf{A}_A$ given in Eq. (2.7.116). From Eq. (2.7.74) we have for an arbitrary gauge transformation

$$\mathbf{A} = \mathbf{A}_t + \mathbf{A}_\ell \rightarrow \mathbf{A}' = \mathbf{A}_t + \mathbf{A}_\ell - \boldsymbol{\nabla}\omega \equiv \mathbf{A}_t + \mathbf{A}'_\ell, \qquad (3.3.24)$$

where $\mathbf{A}'_\ell = -\boldsymbol{\nabla}(\Phi_A + \omega)$. Then $\mathbf{A}_t$ is gauge invariant and so the gauge-invariant physical observables, $\mathbf{L}$ and $\mathbf{S}$, are

$$\mathbf{L} = (1/c)\int d^3x\,E^i(\mathbf{x}\times\boldsymbol{\nabla})A_t^i, \qquad \mathbf{S} = (1/c)\int d^3x\,\mathbf{E}\times\mathbf{A}_t. \qquad (3.3.25)$$

3.3.2 Hamiltonian Formulation of Electromagnetism

The Hamiltonian formulation of electromagnetism is more complex than the Lagrangian formulation since electromagnetism corresponds to a singular system. While in later chapters we will work almost exclusively in the Lorentz covariant Lagrangian formalism, it is important to connect here with the more traditional canonical quantization approach that is based on the Dirac-Bergmann algorithm discussed in Sec. 2.9. Also see sec. (V.2) of Sundermeyer (1982), chapter 19 of Henneaux and Teitelboim (1994) and appendix C of Sundermeyer (2014).

Let us construct the Hamiltonian density for the electromagnetic field interacting with an external conserved current density $j^\mu(x)$. The Lagrangian density is given in Eq. (3.3.4). There are four field components, $A^\nu(x)$, so there will be four generalized momenta in the Hamiltonian formalism. The four generalized momenta are

$$\pi^\nu(x) \equiv \frac{\partial \mathcal{L}}{\partial(\partial_0 A_\nu)(x)} = -F^{0\nu}. \tag{3.3.26}$$

Note carefully that this π^μ is the canonical momentum density for each corresponding A^μ and it should not be confused with the convenient De Donders-Weyl shorthand notation of Eq. (3.2.16) that we introduced earlier when discussing Noether's theorem. For the spatial components of π^μ we have from Eq. (2.7.36) that

$$\pi^i(x) = \frac{\partial \mathcal{L}}{\partial(\partial_0 A_i)(x)} = -F^{0i} = E^i \quad \text{or equivalently} \quad \boldsymbol{\pi} = \mathbf{E}. \tag{3.3.27}$$

Cautionary note: In the appropriate way we have a raised index on π^i since we differentiated with respect to A_i with a lowered index. However, the *canonical momentum density* π^i_{can} would correspond to differentiation with respect to the raised index A^i. So for the electromagnetic field we have $\pi^i_{\mathrm{can}} = \pi_i = -\pi^i$ and so

$$\boldsymbol{\pi}_{\mathrm{can}}(x) = -\boldsymbol{\pi}(x) = -\mathbf{E}(x). \tag{3.3.28}$$

It is $\boldsymbol{\pi}_{\mathrm{can}}$ that enters Poisson brackets and canonical commutation relations in the familiar way. If expressed in terms of $\boldsymbol{\pi} = \mathbf{E}$ there will be corresponding sign change.

Due to the antisymmetry of $F^{\mu\nu}$ we see that we have the identity

$$\pi^0(x) = \frac{\partial \mathcal{L}}{\partial(\partial_0 A_0)(x)} = 0. \tag{3.3.29}$$

Comparing with Eq. (2.9.4) in Sec. 2.9 we recognize that this corresponds to a time-independent primary constraint $\pi^0(ct, \mathbf{x}) = 0$ for every $\mathbf{x}$, where each degree of freedom in our classical field is labeled by one $\mathbf{x} \in \mathbb{R}^3$. So electromagnetism corresponds to a singular classical system. It is conventional when discussing fields to refer to a result of the form $\pi^0(x) = 0$ as a single constraint, even though it is actually an infinite number of constraints, one for each $\mathbf{x}$.

It is important to be clear at this point that we are now viewing this as a system in an infinite-dimensional phase space of all field configurations, where $(\vec{A}^\mu, \vec{\pi}^\mu)$ are specified at all spatial points $\mathbf{x}$. The system evolves with time along a trajectory in phase space, where the trajectory or "path" of the system is specified by the fields (A^μ, π^μ) at all spacetime points $x^\mu = (ct, \mathbf{x}) = (x^0, \mathbf{x})$ between some initial time t_i and some final time t_f. Electromagnetism is constrained to have a trajectory that lies on a submanifold of this infinite-dimensional phase space defined by the time-independent primary constraint $\pi^0(ct, \mathbf{x}) \approx 0$. Recall that weak equality "$\approx$" means that π^0 vanishes on the primary constraint submanifold, but clearly π^0 is nonzero everywhere else in $(\vec{A}^\mu, \vec{\pi}^\mu)$ phase space. In order to discuss degrees of freedom and the dimensionality of phase space let us use the suggestive notation ∞^3 to represent the uncountable number of spatial points $\mathbf{x} \in \mathbb{R}^3$. In the usual way we can always add rigor to this notation by understanding it as the continuum and infinite-volume limits of a finite spatial lattice. With this understanding of our suggestive notation we can say that phase space has $8 \times \infty^3$ dimensions.

It is useful to note that the Hessian matrix of Eq. (2.9.2) is here given by

$$
\begin{aligned}
(M_L)^{\mu\nu}(x,y)|_{x^0=y^0} &= \frac{\delta^2 L}{\delta(\partial_0 A_\mu)(x)\delta(\partial_0 A_\nu)(y)}\bigg|_{x^0=y^0} \\
&= \frac{\partial^2 \mathcal{L}(x)}{\partial(\partial_0 A_\mu)(x)\partial(\partial_0 A_\nu)(x)}\,\delta^3(\mathbf{x}-\mathbf{y}) = -\frac{\partial F^{0\nu}(x)}{\partial(\partial_0 A_\mu)(x)}\,\delta^3(\mathbf{x}-\mathbf{y}) \\
&= \left(g^{0\mu}g^{0\nu} - g^{00}g^{\mu\nu}\right)\delta^3(\mathbf{x}-\mathbf{y}),
\end{aligned}
\tag{3.3.30}
$$

where the 4×4 matrix $(g^{0\mu}g^{0\nu} - g^{00}g^{\mu\nu})$ is diagonal with diagonal elements $(0,1,1,1)$ and therefore has rank $4-1=3$. We therefore expect one primary constraint associated with every spatial point $\mathbf{x}$, i.e., $\pi^0(ct,\mathbf{x})=0$ for all $\mathbf{x}$ as we saw above. From this point we will often say that there is, say, one constraint for simplicity with the understanding that in this context we really mean a $1\times\infty^3$ constraint.

We now need to generalize the Dirac-Bergmann algorithm described in Sec. 2.9.2 to the case of classical fields, which is a simple matter of recognizing that we have an infinite number of degrees of freedom labeled by $\mathbf{x}$. As described above we could do this in a more careful manner by constructing a three-dimensional spatial lattice in a finite volume and then taking the limits of the lattice spacing to zero followed by the spatial box size to infinity, but the outcome is the same as the straightforward approach that we use here.

Let us now follow through the steps of the Dirac-Bergmann algorithm given in Sec. 2.9.2 for the case of electromagnetism coupled to an external current density.

(i) The Lagrangian density and the conjugate momenta are given above. The standard canonical Hamiltonian density is

$$
\begin{aligned}
\mathcal{H} &= \pi^\mu \partial_0 A_\mu - \mathcal{L} = \pi^i \partial_0 A_i - \mathcal{L} = \pi^i(-\pi_i + \partial_i A_0) + \tfrac{1}{4}F^{\mu\nu}F_{\mu\nu} + (j_\mu/c)A^\mu \\
&= -\tfrac{1}{2}\pi^i\pi_i + \tfrac{1}{4}F^{ij}F_{ij} + \pi^i\partial_i A_0 + (j^0/c)A_0 + (j^i/c)A_i \\
&= \tfrac{1}{2}(\mathbf{E}^2 + \mathbf{B}^2) + \pi^i\partial_i A_0 + (j^0/c)A_0 + (j^i/c)A_i,
\end{aligned}
\tag{3.3.31}
$$

where we have used $\partial_0 A_i = F_{0i} + \partial_i A_0 = -\pi_i + \partial_i A_0$. The corresponding canonical Hamiltonian is

$$
\begin{aligned}
H &= \int d^3x\, \mathcal{H} = \int d^3x \left(-\tfrac{1}{2}\pi^i\pi_i + \tfrac{1}{4}F^{ij}F_{ij} + \pi^i\partial_i A_0 + (j^0/c)A_0 + (j^i/c)A_i\right) \\
&= \int d^3x \left(-\tfrac{1}{2}\pi^i\pi_i + \tfrac{1}{4}F^{ij}F_{ij} - (\partial_i\pi^i)A_0 + (j^0/c)A_0 + (j^i/c)A_i\right),
\end{aligned}
\tag{3.3.32}
$$

where we have used integration by parts and assumed no contribution from the spatial boundary, i.e., periodic spatial boundary conditions in a finite volume or vanishing fields at spatial infinity. We can therefore write the standard canonical Hamiltonian density as

$$
\begin{aligned}
\mathcal{H}(A^\mu,\pi^\mu,x) &= -\tfrac{1}{2}\pi^i\pi_i + \tfrac{1}{4}F^{ij}F_{ij} - (\partial_i\pi^i)A_0 + (j^0/c)A_0 + (j^i/c)A_i \\
&= -\tfrac{1}{2}\pi^i\pi_i + \tfrac{1}{2}(\mathbf{\nabla}\times\mathbf{A})^2 - (\partial_i\pi^i)A_0 + (j^0/c)A_0 + (j^i/c)A_i \\
&= \tfrac{1}{2}\left(\mathbf{E}^2 + \mathbf{B}^2\right) - (\mathbf{\nabla}\cdot\mathbf{E})A_0 + (j^0/c)A_0 + (j^i/c)A_i,
\end{aligned}
\tag{3.3.33}
$$

since $F^{ij}F_{ij} = 2\mathbf{B}^2 = 2(\mathbf{\nabla}\times\mathbf{A})^2$. There is only one primary constraint $\pi^0(x)=0$, which defines the primary constraint submanifold in the infinite-dimensional (A^μ,π^μ) functional phase space. The simplicity of the primary constraint in this case has eliminated π^0 from the Hamiltonian density.

It is straightforward to verify that Hamilton's equations based on the standard canonical Hamiltonian density reproduce Maxwell's equations given in Eq. (2.7.34). The first two of Maxwell's equations follow from expressing the $\mathbf{E}$ and $\mathbf{B}$ fields in terms of A^μ and can be summarized as $\partial_\mu \tilde{F}^{\mu\nu} = 0$ as given in Eq. (2.7.23). The second two of Maxwell's equations involve the external current and arise from Hamilton's equations for fields

$$\partial_0 A^\mu = \frac{\partial \mathcal{H}}{\partial \pi_\mu} \quad \text{and} \quad \partial_0 \pi^\mu = -\frac{\partial \mathcal{H}}{\partial A_\mu} + \boldsymbol{\nabla} \cdot \frac{\partial \mathcal{H}}{\partial(\boldsymbol{\nabla} A_\mu)}. \tag{3.3.34}$$

Proof: Since the system is singular with $\pi^0 = \partial \mathcal{L}/\partial(\partial_0 A_0) = 0$ being absent from the canonical Hamiltonian, then π_0 has no physical content. Recalling that $H = \int d^3x\, \mathcal{H}$ we are free to use integration by parts to move spatial derivatives accordingly provided that we understand that we are effectively imposing periodic spatial boundary conditions and/or requiring variations to vanish on the spatial boundary as discussed in the derivation of Noether's theorem such that boundary terms do not contribute. We then find that

$$\partial_0 A^0 = \partial \mathcal{H}/\partial \pi_0 = 0, \text{ which has no physical content,}$$

$$\partial_0 A^i = \partial \mathcal{H}/\partial \pi_i = -\pi^i + \partial^i A_0, \ \Rightarrow\ \pi^i = \partial^i A_0 - \partial_0 A^i = -F^{0i} = E^i,$$

$$\partial_0 \pi^0 = -\frac{\partial \mathcal{H}}{\partial A_0} + \boldsymbol{\nabla} \cdot \frac{\partial \mathcal{H}}{\partial(\boldsymbol{\nabla} A_0)} = \partial_i \pi^i - \frac{j^0}{c} = 0 \text{ as } \pi^0 = 0 \ \Rightarrow\ \boldsymbol{\nabla} \cdot \mathbf{E} = \rho,$$

$$\partial_0 \pi^i = -\frac{\partial \mathcal{H}}{\partial A_i} + \boldsymbol{\nabla} \cdot \frac{\partial \mathcal{H}}{\partial(\boldsymbol{\nabla} A_i)} = -\frac{j^i}{c} + \partial_j \frac{\partial \mathcal{H}}{\partial(\partial_j A_i)}$$

$$= -\frac{j^i}{c} + \partial_j \frac{\partial \left[(1/2)\epsilon^{klm}\partial_l A^m \epsilon^{knp}\partial_n A^p\right]}{\partial(\partial_j A_i)} = -\frac{j^i}{c} - \epsilon^{kji}\epsilon^{knp}\partial_j \partial_n A^p$$

$$= -j^i/c + \epsilon^{ijk}\partial_j(\boldsymbol{\nabla} \times \mathbf{A})^k = -j^i/c + [\boldsymbol{\nabla} \times (\boldsymbol{\nabla} \times \mathbf{A})]^i$$

$$= -j^i/c + [\boldsymbol{\nabla} \times \mathbf{B}]^i \ \Rightarrow\ \boldsymbol{\nabla} \times \mathbf{B} - (1/c)\partial \mathbf{E}/\partial t = \mathbf{j}/c. \tag{3.3.35}$$

So the $\partial_0 \pi^\mu$ equations reproduce the second two of Maxwell's equations.

Since Maxwell's equations involve gauge invariant (i.e., physical) quantities, they will remain unaffected by the constructions that we undertake in the Dirac-Bergmann algorithm.

(ii) In the discrete mechanics case we construct the matrix of Poisson brackets of primary constraints $\phi_r(\vec{q}, \vec{p}, t)$, i.e., $P_{r,s} \equiv \{\phi_r(t), \phi_s(t)\}$ for $r = 1, \ldots, M$. The generalization to the present case is

$$P(\mathbf{x}, \mathbf{y}) = \{\pi^0(x^0, \mathbf{x}), \pi^0(x^0, \mathbf{y})\} \equiv \{\pi^0(x), \pi^0(y)\}_{x^0 = y^0}. \tag{3.3.36}$$

From Eq. (3.1.22) we see that the Poisson bracket for electromagnetism is

$$\{F, G\} \equiv \int d^3z \left(\frac{\delta F}{\delta A^\mu(x^0, \mathbf{z})} \frac{\delta G}{\delta \pi_\mu(x^0, \mathbf{z})} - \frac{\delta F}{\delta \pi_\mu(x^0, \mathbf{z})} \frac{\delta G}{\delta A^\mu(x^0, \mathbf{z})} \right). \tag{3.3.37}$$

The fundamental Poisson brackets for electromagnetism immediately follow,

$$\{\pi^\mu(x), \pi^\nu(y)\}_{x^0=y^0} = 0, \quad \{A^\mu(x), A^\nu(y)\}_{x^0=y^0} = 0, \text{ and}$$
$$\{A^\mu(x), \pi_\nu(y)\}_{x^0=y^0} = \delta^\mu_{\ \nu}\delta^3(\mathbf{x}-\mathbf{y}) \tag{3.3.38}$$

where $\delta\pi^\mu(x)/\delta A^\nu(x^0, \mathbf{z}) = \delta A^\mu(y)/\delta\pi^\nu(x^0, \mathbf{z}) = 0$ since the field A^μ and its canonical momentum density π_μ are independent phase-space variables. Therefore $P(\mathbf{x}, \mathbf{y})$ vanishes identically,

$$P(\mathbf{x}, \mathbf{y}) = \{\pi^0(x), \pi^0(y)\}_{x^0=y^0} = 0. \tag{3.3.39}$$

Recall from the discussions of Sec. 2.9 that Poisson brackets are to be evaluated *before* constraints are applied and so we could not simply have claimed $P = 0$ because the primary constraint is $\pi^0 \approx 0$.

(iii) Since $P(\mathbf{x}, \mathbf{y}) = 0$ everywhere in functional phase space, then the functional determinant $\det P = 0$ everywhere.[5] Therefore, it is singular on the constraint submanifold, $\det P \approx 0$, and so secondary constraints will be required.

(iv) We construct the total Hamiltonian from the primary constraint according to Eq. (2.9.14), which gives

$$H_T = \int d^3x\, \mathcal{H}_T = \int d^3x \left(\mathcal{H} + v(x^0, \mathbf{x})\pi^0(x^0, \mathbf{x})\right), \tag{3.3.40}$$

where $v(x)$ is the Lagrange multiplier function. For the primary constraint to remain valid at all times we must have $\partial_0\pi^0 \approx 0$. Recalling how the Poisson bracket behaves with a Lagrange multiplier from Eq. (2.9.24), we then find using Eq. (3.1.31) and $\{\pi^0(x), F^{ij}(y)\}_{x^0=y^0} = 0$ that

$$\partial_0\pi^0(x) = \{\pi^0(x), H_T\} = \{\pi^0(x), H\} - \int d^3z \left(\{\pi^0(x), \pi^0(z)\}_{x^0=z^0} v(x^0, \mathbf{z})\right)$$
$$= \{\pi^0(x), H\} = \{\pi^0(x), \int d^3z\, A^0(z)[-\partial_i\pi^i + (j^0/c)](z)\}_{x^0=z^0} \tag{3.3.41}$$
$$= -\int d^3z\, \{\pi^0(x), A^0(z)\}_{x^0=z^0}[\partial_i\pi^i - (j^0/c)](z) = \partial_i\pi^i(x) - (j^0(x)/c).$$

For brevity we will *often suppress the "$x^0 = y^0$" notation on Poisson brackets* and leave it to be understood. We require the secondary constraint

$$\partial_i\pi^i(x) - j^0(x)/c \approx 0; \tag{3.3.42}$$

i.e., we must restrict the constraint submanifold such that $\partial_i\pi^i(x) = j^0(x)/c$ everywhere on it at all times. Our two constraints are actually $2 \times \infty^3$ constraints and leave a constraint submanifold of $6 \times \infty^3$. Since j^μ in general will depend on time, then this is a time-dependent secondary constraint. However, recall that $\pi^i = -F^{0i} = E^i$ and $\rho = j^0/c$ and so this secondary constraint is just Gauss' law given in Eq. (2.7.34),

$$\nabla \cdot \mathbf{E} = j^0/c = \rho, \tag{3.3.43}$$

and so is always true because of the equations of motion. We can also verify directly that this secondary constraint is maintained at all times by looking at the time evolution of the secondary constraint. Recall that j^μ is an external current density, then as shown below we arrive at

$$\partial_0[\partial_i\pi^i(x) - \rho(x)] = 0 \tag{3.3.44}$$

---

[5] The functional determinant of P is formed by treating $P(\mathbf{x}, \mathbf{y})$ as the matrix element with $\mathbf{x}$ as the row index and $\mathbf{y}$ as the column index. It can be defined in terms of a spatial lattice of points with the appropriate continuum and infinite volume limits taken. The limits clearly exist in this trivial case, but in general the definition of a functional determinant requires some care.

identically; i.e., it vanishes everywhere in phase space and not just on the constraint submanifold. So no additional secondary constraints are needed. In summary, we have one time-independent primary constraint $\pi^0(x) \approx 0$ and one time-dependent secondary constraint $\partial_i \pi^i - j^0 \approx 0$, where each of these is an infinite number of constraints with one for each spatial point $\mathbf{x}$;

Proof: To prove Eq. (3.3.44) we use Eqs. (2.1.50) and (3.3.38) and so

$$\delta \partial_i \pi^\mu(x^0, \mathbf{x}) / \delta A_\nu(x^0, \mathbf{y}) = 0; \quad \delta \partial_i \pi^\mu(x^0, \mathbf{x}) / \delta \pi_\nu(x^0, \mathbf{y}) = \delta^\mu_{\ \nu} \partial^x_i \delta^3(\mathbf{x} - \mathbf{y}).$$

After some thought we see that only terms in H_T containing an A^i can have a nonvanishing Poisson bracket with $\partial_i \pi^i$. Poisson brackets with the external current density j^μ vanish since it is not a dynamical variable and is independent of both A^μ or π^μ. From Eq. (3.1.31) we then find

$$\begin{aligned}
\partial_0 [\partial_i \pi^i(x) - \rho(x)] &= \{[\partial_i \pi^i - \rho](x), H_T\} + \partial^{\text{ex}}_0 [\partial_i \pi^i(x) - \rho(x)] \\
&= \{\partial_i \pi^i(x), H_T\} - \partial_0 \rho(x) \\
&= \int d^3y \{\partial_i \pi^i(x), \tfrac{1}{4} F^{jk}(y) F_{jk}(y) + (j^j(y)/c) A_j(y)\} - \partial_0 \rho(x) \\
&= \int d^3y \left[\tfrac{1}{4} \{\partial_i \pi^i(x), F^{jk}(y) F_{jk}(y)\} + \{\partial_i \pi^i(x), A_j(y)\} j^j(y)/c \right] - \partial_0 \rho(x) \\
&= \int d^3y \left[\{\partial_i \pi^i(x), A_j(y)\} j^j(y)/c \right] - \partial_0 j^0(x)/c \\
&= -\partial_i j^i(x)/c - \partial_0 j^0(x)/c = -\partial_\mu j^\mu(x)/c = 0.
\end{aligned} \tag{3.3.45}$$

In the above we have used the results that (i) the electromagnetic field can only couple to a conserved current, $\partial_\mu j^\mu = 0$; (ii) the term containing $F^{jk}(y) F_{jk}(y)$ vanishes because

$$\begin{aligned}
&\int d^3y \, \tfrac{1}{4} \{\partial_i \pi^i(x), F^{jk}(y) F_{jk}(y)\} \\
&= - \int d^3y \int d^3z \frac{\delta \partial_i \pi^i(x^0, \mathbf{x})}{\delta \pi_\mu(x^0, \mathbf{z})} \frac{\delta(F^{jk} F_{jk}/4)(x^0, \mathbf{y})}{\delta A^\mu(x^0, \mathbf{z})} \\
&= - \int d^3y \int d^3z \left[\delta^i_{\ \mu} \partial^x_i \delta^3(\mathbf{x} - \mathbf{z}) \right] \left[\frac{\partial(F^{jk} F_{jk}/4)(x^0, \mathbf{y})}{\partial(\partial_\ell A_m)(x^0, \mathbf{y})} \frac{\delta(\partial_\ell A_m)(x^0, \mathbf{y})}{\delta A^\mu(x^0, \mathbf{z})} \right] \\
&= \int d^3y \int d^3z \left(- \left[\delta^i_{\ \mu} \partial^x_i \delta^3(\mathbf{x} - \mathbf{z}) \right] \left[F^{\ell m}(x^0, \mathbf{y}) g_{m\mu} \partial^y_\ell \delta^3(\mathbf{y} - \mathbf{z}) \right] \right) \\
&= \int d^3y \int d^3z \, \partial^x_m \delta^3(\mathbf{x} - \mathbf{z}) F^{\ell m}(x^0, \mathbf{y}) \partial^y_\ell \delta^3(\mathbf{y} - \mathbf{z}) \\
&= - \int d^3y \, \partial^x_m \delta^3(\mathbf{x} - \mathbf{y}) \partial^y_\ell F^{\ell m}(x^0, \mathbf{y}) = \int d^3y \, \partial^x_m \delta^3(\mathbf{x} - \mathbf{y}) \partial^y_\ell F^{\ell m}(x^0, \mathbf{y}) \\
&= -\partial_m \partial_\ell F^{\ell m}(x^0, \mathbf{y}) = 0;
\end{aligned} \tag{3.3.46}$$

and (iii) the term containing $\{\partial_i \pi^i(x), A_j(y)\}$ satisfies

$$\begin{aligned}
\int d^3y \{\partial_i \pi^i(x), A_j(y)\} \frac{j^j(y)}{c} &= - \int d^3z \int d^3y \frac{\delta \partial_i \pi^i(x^0, \mathbf{x})}{\delta \pi_\mu(x^0, \mathbf{z})} \frac{\delta A_j(x^0, \mathbf{y})}{\delta A^\mu(x^0, \mathbf{z})} \frac{j^j(y)}{c} \\
&= - \int d^3z \int d^3y \, g^{i\mu} \partial^x_i \delta^3(\mathbf{x} - \mathbf{z}) g_{j\mu} \delta^3(\mathbf{y} - \mathbf{z}) (j^j(y)/c) \\
&= - \int d^3y \, \delta^i_{\ j} \partial^x_i \delta^3(\mathbf{x} - \mathbf{y}) (j^j(y)/c) = -\partial_i j^i(x)/c.
\end{aligned} \tag{3.3.47}$$

(v) We only have two constraints and since they have a vanishing Poisson bracket with each other, then they are both first-class constraints, i.e.,

$$\{\pi^0(x), \partial_i \pi^i(y) - \rho(y)\} = \{\pi^0(x), \partial_i \pi^i(y)\} = 0 \tag{3.3.48}$$

identically. So, we have one first-class primary constraint and one first-class secondary constraint.

(vi) There are no second-class constraints and so there is no matrix of second-class constraints, $\Delta \equiv \{\xi_r(x), \xi_s(y)\}$.

(vii) Then we construct the total Hamiltonian H_T in the classical field theory context in analogy with Eq. (2.9.99), where H_T contains one arbitrary Lagrange multiplier function $v(x)$, i.e., one for each first-class primary constraint. We have $H_T = \int d^3x \, \mathcal{H}_T$, where the total Hamiltonian density is

$$
\begin{aligned}
\mathcal{H}_T(A^\mu, \pi^\mu, v, x) &= \mathcal{H}(A^\mu, \pi^\mu, x) + v(x)\pi^0(x) \\
&= -\tfrac{1}{2}\pi^i\pi_i + \tfrac{1}{4}F^{ij}F_{ij} - (\partial_i\pi^i)A_0 + (j^0/c)A_0 + (j^i/c)A_i + v\pi^0 \\
&= \tfrac{1}{2}\left(\mathbf{E}^2 + \mathbf{B}^2\right) - (\partial_i\pi^i)A_0 + (j^0/c)A_0 + (j^i/c)A_i + v\pi^0
\end{aligned} \tag{3.3.49}
$$

and any dynamical functional $F[A^\mu, \pi^\mu, x]$ evolves in time according to Eq. (3.1.31) with H replaced by H_T,

$$
\partial_0 F = \{F, H_T\} + \partial_0^{\mathrm{ex}} F. \tag{3.3.50}
$$

The dynamics of the system is then solved but we note that it involves an arbitrary Lagrange multiplier function $v(x)$.

(viii) As we will soon see, both of the first-class constraints generate gauge transformations in electromagnetism and so we introduce an additional arbitrary Lagrange multiplier function $w(x)$ for the first-class secondary constraint and hence define the extended Hamiltonian, $H_E \equiv H_T + \int d^3x \, w(\partial_i\pi^i - \rho) \equiv \int d^3x \, \mathcal{H}_E$, where the extended Hamiltonian density of electromagnetism is

$$
\begin{aligned}
\mathcal{H}_E(A^\mu, \pi^\mu, v, w, x) &= \mathcal{H}_T(A^\mu, \pi^\mu, v, x) + w(x)[\partial_i\pi^i(x) - \rho(x)] \\
&= -\tfrac{1}{2}\pi^i\pi_i + \tfrac{1}{4}F^{ij}F_{ij} - (\partial_i\pi^i)A_0 + (j^0/c)A_0 + (j^i/c)A_i \\
&\quad + v\pi^0 + w(\partial_i\pi^i - \rho)
\end{aligned} \tag{3.3.51}
$$

and any dynamical functional $F[A^\mu, \pi^\mu, x]$ evolves in time according to Eq. (3.1.31) with H replaced by H_E,

$$
\partial_0 F = \{F, H_E\} + \partial_0^{\mathrm{ex}} F. \tag{3.3.52}
$$

The time evolution now depends in general on two arbitrary Lagrange multiplier functions, $v(x)$ and $w(x)$. Any physical observable in a classical system is uniquely determined and so physical observables must be independent of these two Lagrange multipliers. Then for any physical observable in electromagnetism we must have

$$
\partial_0 F_{\mathrm{phys}} = \{F_{\mathrm{phys}}, H_E\} + \partial_0^{\mathrm{ex}} F_{\mathrm{phys}} = \{F_{\mathrm{phys}}, H\} + \partial_0^{\mathrm{ex}} F_{\mathrm{phys}}. \tag{3.3.53}
$$

(ix) Define

$$
\Delta H_E \equiv G \equiv \int d^3x \left(\Delta v(x)\pi^0(x) + \Delta w(x)[\partial_i\pi^i(x) - \rho(x)]\right) \tag{3.3.54}
$$

as the change to the extended Hamiltonian caused by changing the Lagrange multipliers for the first-class constraints. Then $dF = \{F, \Delta H_E\}dt = \{F, G\}dt$ from the equations of motion for any F. Relabeling $d\alpha = dt$ we then have Eq. (2.9.88), which is that G generates a canonical transformation,

$$dF/d\alpha = \{F, G\} = \int d^3y \left(\Delta v(y)\{F, \pi^0(y)\} + \Delta w(y) \left\{ F, [\partial_i \pi^i(y) - \rho(y)] \right\} \right)$$
$$= \int d^3y \left(\Delta v(y)\{F, \pi^0(y)\} - \partial_i \Delta w(y) \left\{ F, \pi^i(y) \right\} \right)$$
$$= \int d^3y \; \partial_\mu \omega(y)\{F, \pi^\mu(y)\}|_{x^0=y^0} \,, \tag{3.3.55}$$

where on the last line we have explicitly indicated the understood condition $x^0 = y^0$ and where on the time slice $x^0 = ct$ we have defined

$$\partial_0 \omega(x) \equiv \Delta v(x) \quad \text{and} \quad \omega(x) \equiv \Delta w(x). \tag{3.3.56}$$

So for every generator G of a gauge transformation there is a corresponding spacetime function $\omega(x)$, where on the timeslice y^0 we can write $G[\omega]$ in the form

$$G[\omega] = \int d^3y \; [\partial_\mu \omega(y)\pi^\mu(y) - \omega(y)\rho(y)]. \tag{3.3.57}$$

Note that $\omega(x)$ need not be differentiable or even continuous since gauge transformations have no physical significance. In analogy with Eq. (2.9.111) we can associate an operator $T[\omega]$ with each generator

$$\hat{T}[\omega]F = \{F, G[\omega]\} = \int d^3y \; \partial_\mu \omega(y)\{F, \pi^\mu(y)\}|_{x^0=y^0} = dF/d\alpha, \tag{3.3.58}$$

where the operators $\hat{T}[\omega]$ form a Lie algebra, and where each element of the algebra is labeled by some ω. We make frequent use of Eq. (2.9.24), which allows us to move Lagrange multipliers and external charge densities in and out of Poisson brackets since they have no dependence on dynamical variables. We observe that the Lie algebra here is abelian, $[\hat{T}[\omega], \hat{T}[\omega']] = 0$, since for any dynamical functional F we can follow the analogous steps in Eq. (2.9.113) to find

$$\left[\hat{T}[\omega'], \hat{T}[\omega] \right] F = \hat{T}[\omega']\hat{T}[\omega]F - \hat{T}[\omega]\hat{T}[\omega']F = \{\hat{T}[\omega]F, G[\omega']\} - \{\hat{T}[\omega']F, G[\omega]\}$$
$$= \{\{F, G[\omega]\}, G[\omega']\} - \{\{F, G[\omega']\}, G[\omega]\}$$
$$= -\{\{G[\omega], G[\omega']\}, F\} - \{\{G[\omega'], F\}, G[\omega]\} - \{\{F, G[\omega']\}, G[\omega]\}$$
$$= \{F, \{G[\omega], G[\omega']\}\} = 0, \tag{3.3.59}$$

where we have used the Jacobi identity for Poisson brackets and the fundamental Poisson bracket relations $\{\pi^\mu(x), \pi^\nu(y)\}_{x^0=y^0} = 0$. So the Lie algebra of the generators of the gauge transformations of electromagnetism is abelian and the gauge transformations are an abelian group. We later recognize that this is $U(1)$. Consider the action of a gauge transformation generator on the $A^\mu(x)$ and the $\pi^\mu(x)$,

$$dA^\mu(x)/d\alpha = \hat{T}[\omega]A^\mu(x) = \int d^3y \; \partial_\nu \omega(y)\{A^\mu(x), \pi^\nu(y)\}|_{x^0=y^0} = \partial^\mu \omega(x),$$
$$d\pi^\mu(x)/d\alpha = \hat{T}[\omega]\pi^\mu(x) = \int d^3y \; \partial_\nu \omega(y)\{\pi^\mu(x), \pi^\nu(y)\}|_{x^0=y^0} = 0. \tag{3.3.60}$$

So under an infinitesimal gauge transformation the canonical momenta $\pi^\mu(x)$ are invariant and the electromagnetic field $A^\mu(x)$ transforms as

$$A^\mu(x) \rightarrow A'^\mu(x) = A^\mu(x) + d\alpha\partial^\mu \omega(x). \tag{3.3.61}$$

An infinite number of infinitesimal gauge transformations gives a finite gauge transformation; i.e., defining $d\alpha = 1/n$ then after a very large number of small gauge transformations we find

$$A^\mu(x) \rightarrow A'^\mu(x) = A^\mu(x) + \lim_{n\to\infty} \sum_{i=1}^{n} \partial^\mu[\omega(x)/n] = A^\mu(x) + \partial^\mu \omega(x). \tag{3.3.62}$$

We have then recovered the usual result for the gauge transformations of the electromagnetic field given in Eq. (2.7.74) as any consistent treatment had to do.

We are still free to choose the Lagrange multipliers to fix the gauge and we saw that this can be characterized as a choice of $\omega(x)$ to fix the gauge for each A^μ. As we concluded at the end of Sec. 2.7.3 the Coulomb gauge condition $\nabla \cdot \mathbf{A} = 0$ together with appropriate boundary conditions is an ideal gauge fixing for the electromagnetic field in the presence of an external current density. Recall that in Coulomb gauge Gauss' law leads to

$$\Phi(x) = A^0(x) = \frac{1}{4\pi} \int \frac{\rho(t, \mathbf{x}')}{|\mathbf{x} - \mathbf{x}'|} d^3 x'. \tag{3.3.63}$$

When imposing constraints we cannot assume the equations of motion and since Gauss' law is imposed as a constraint then we must also impose Eq. (3.3.63) as another gauge-fixing constraint along with the Coulomb gauge constraint.

Dirac bracket for electromagnetism: We first focus now on the case of pure electromagnetism, where there is no external current density, $j^\mu(x) = 0$. As discussed in Sec. 2.7.3 the radiation gauge is an ideal gauge fixing when there is no external current density and consists of two gauge-fixing constraints, the Coulomb gauge condition and the condition $A^0(x) = \Phi(x) = 0$; i.e., for all spacetime points x we have the two gauge-fixing constraints

$$\Omega_1(x) \equiv A^0(x) \approx 0 \quad \text{and} \quad \Omega_2(x) \equiv \nabla \cdot \mathbf{A}(x) = \partial_i A^i(x) \approx 0. \tag{3.3.64}$$

For the free electromagnetic field there are two time-independent first-class constraints per spatial point

$$\zeta_1(x) \equiv \pi^0(x) \approx 0 \quad \text{and} \quad \zeta_2(x) \equiv \nabla \cdot \boldsymbol{\pi}(x) = \partial_i \pi^i(x) \approx 0. \tag{3.3.65}$$

There are two time-independent gauge-fixing conditions per spatial point.

Following the notation of Sec. 2.9.3 we have $N_f = 2 \times \infty^3$ first-class constraints, $N_s = 0$ second-class constraints, and $N = 4 \times \infty^3$ initial degrees of freedom before any constraints are applied. So the dimension of the original unconstrained phase space is $2N = 8 \times \infty^3$ and we have the total number of constraints given by

$$N_c = 2N_f + N_s = 2(2 \times \infty^3) + 0 = 4 \times \infty^3. \tag{3.3.66}$$

Denoting the number of physical degrees of freedom as N_{phys} we find that the dimension of the physical submanifold is

$$2N_{\text{phys}} = 2N - N_c = 4 \times \infty^3. \tag{3.3.67}$$

Again following Sec. 2.9.3 we define the full set of constraints

$$\begin{aligned}
\vec{\psi}(x) &= (\psi_1(x), \psi_2(x), \psi_3(x), \psi_4(x)) \equiv (\zeta_1(x), \zeta_2(x), \Omega_1(x), \Omega_2(x)) \\
&= \left(\pi^0(x), \partial_i \pi^i(x), A^0(x), \partial_j A^j(x) \right), \tag{3.3.68}
\end{aligned}$$

which are necessarily now all second-class constraints. We then define the $N_c \times N_c$ antisymmetric matrix G with elements $G_{ij}(x, y)_{x^0 = y^0}$, where

$$G_{ij}(x, y)|_{x^0 = y^0} \equiv \{\psi_i(x), \psi_j(y)\}_{x^0 = y^0}. \tag{3.3.69}$$

It is readily verified from the definition of the Poisson brackets that

$$\{\zeta_1(x), \zeta_2(y)\}_{x^0=y^0} = 0, \quad \{\Omega_1(x), \Omega_2(y)\}_{x^0=y^0} = 0, \tag{3.3.70}$$

$$\{\zeta_1(x), \Omega_2(y)\}_{x^0=y^0} = 0, \quad \{\zeta_2(x), \Omega_1(y)\}_{x^0=y^0} = 0,$$

$$\{\zeta_1(x), \Omega_1(y)\}_{x^0=y^0} = -\delta^3(\mathbf{x}-\mathbf{y}), \quad \{\zeta_2(x), \Omega_2(y)\}_{x^0=y^0} = -\boldsymbol{\nabla}_x^2 \delta^3(\mathbf{x}-\mathbf{y}).$$

Proof: All results are straightforward except the last one. For this we find

$$\{\partial_i \pi^i(x), \partial_j A^j(y)\}_{x^0=y^0} = \int d^3z \left(-\frac{\delta \partial_i \pi^i(x)}{\delta \pi_\mu(z)} \frac{\delta \partial_j A^j(y)}{\delta A^\mu(z)} \right)_{x^0=y^0=z^0}$$

$$= \int d^3z \left(-g^{\mu i} \partial_i^x \delta^3(\mathbf{x}-\mathbf{z}) \delta^j{}_\mu \partial_j^y \delta^3(\mathbf{y}-\mathbf{z}) \right)$$

$$= \int d^3z \left(g^{ji} \partial_i^x \delta^3(\mathbf{x}-\mathbf{z}) \partial_j^z \delta^3(\mathbf{y}-\mathbf{z}) \right) \tag{3.3.71}$$

$$= \int d^3z \left(-\partial_i^x \partial_i^x \delta^3(\mathbf{x}-\mathbf{z}) \right) \delta^3(\mathbf{y}-\mathbf{z}) = -\partial_i^x \partial_i^x \delta^3(\mathbf{x}-\mathbf{y}) = -\boldsymbol{\nabla}_x^2 \delta^3(\mathbf{x}-\mathbf{y}),$$

where we have used Eq. (2.1.50), integration by parts assuming no variation is performed on the boundary so that boundary terms can be neglected, and the fact that $\partial_j^x \delta^3(\mathbf{x}-\mathbf{y}) = -\partial_j^y \delta^3(\mathbf{x}-\mathbf{y})$.

The matrix G can then be expressed as

$$G(\mathbf{x},\mathbf{y}) = \begin{pmatrix} 0 & 0 & -1 & 0 \\ 0 & 0 & 0 & -\boldsymbol{\nabla}_x^2 \\ 1 & 0 & 0 & 0 \\ 0 & \boldsymbol{\nabla}_x^2 & 0 & 0 \end{pmatrix} \delta^3(\mathbf{x}-\mathbf{y}). \tag{3.3.72}$$

Noting that G has a simple antisymmetric off-diagonal block form we can write

$$G(\mathbf{x},\mathbf{y}) \equiv \begin{pmatrix} 0 & B \\ -B & 0 \end{pmatrix} \quad \text{and} \quad G^{-1}(\mathbf{x},\mathbf{y}) = \begin{pmatrix} 0 & -B^{-1} \\ B^{-1} & 0 \end{pmatrix}. \tag{3.3.73}$$

From Eq. (2.7.102) we have $\boldsymbol{\nabla}_x^2(-1/4\pi|\mathbf{x}-\mathbf{y}|) = \delta^3(\mathbf{x}-\mathbf{y})$, which gives

$$G^{-1}(\mathbf{x},\mathbf{y}) = \begin{pmatrix} 0 & 0 & \delta^3(\mathbf{x}-\mathbf{y}) & 0 \\ 0 & 0 & 0 & \frac{-1}{4\pi|\mathbf{x}-\mathbf{y}|} \\ -\delta^3(\mathbf{x}-\mathbf{y}) & 0 & 0 & 0 \\ 0 & \frac{1}{4\pi|\mathbf{x}-\mathbf{y}|} & 0 & 0 \end{pmatrix}, \tag{3.3.74}$$

where $\int d^3z\, G(\mathbf{x},\mathbf{z}) G^{-1}(\mathbf{z},\mathbf{y}) = I \delta^3(\mathbf{x}-\mathbf{y})$ with I being the 4×4 identity matrix.

The Dirac bracket of Eq. (2.9.126) becomes

$$\{A(x^0), B(x^0)\}_{\text{DB}} \equiv \{A(x^0), B(x^0)\} \tag{3.3.75}$$

$$- \int d^3u\, d^3v \sum_{i,j=1}^4 \left(\{A(x^0), \psi_i(x^0,\mathbf{u})\} G_{ij}^{-1}(\mathbf{u},\mathbf{v}) \{\psi_j(x^0,\mathbf{v}), B(x^0)\} \right)$$

and since the constraints are time independent, then we have in analogy with Eqs. (3.1.31) and (2.9.128) that the equations of motion for any dynamical functional $F[A^\mu, \pi^\mu, x^0]$ are

$$\partial_0 F \approx \{F, H\}_{\text{DB}} + \partial_0^{\text{ex}} F, \tag{3.3.76}$$

where $H = \int d^3x\, \mathcal{H}$ is the simple canonical Hamiltonian resulting from the Legendre transformation given in Eq. (3.3.33). The Dirac bracket is sufficient to ensure that all constraints, including the gauge-fixing constraints, are satisfied at all times.

The analogs of the fundamental Poisson brackets are the fundamental Dirac brackets, which we find have the form

$$\{\pi^\mu(x), \pi^\nu(y)\}_{\text{DB}}^{x^0=y^0} = 0, \quad \{A^\mu(x), A^\nu(y)\}_{\text{DB}}^{x^0=y^0} = 0, \quad \text{and} \tag{3.3.77}$$

$$\{A^\mu(x), \pi_\nu(y)\}_{\text{DB}}^{x^0=y^0} = \left(\delta^\mu{}_\nu - g^{0\mu}\delta^0{}_\nu\right)\delta^3(\mathbf{x} - \mathbf{y}) - g^{\mu i}\delta^j{}_\nu \partial_i^x \partial_j^x \frac{1}{4\pi|\mathbf{x} - \mathbf{y}|}.$$

Proof: The fundamental Dirac brackets are straightforward to evaluate. Note that ψ_1 and ψ_2 involve only the π^μ and ψ_3 and ψ_4 involve only the A^μ. Then all A^μ will have vanishing Poisson brackets with ψ_3 and ψ_4 and all π^μ will have vanishing Poisson brackets with ψ_1 and ψ_2. Since the 2×2 diagonal blocks of G^{-1} vanish, then we must have $\{\pi^\mu(x), \pi^\nu(y)\}_{\text{DB}} = 0$ and $\{A^\mu(x), A^\nu(y)\}_{\text{DB}} = 0$.

Note that A^0 only has a non-vanishing Poisson bracket with ψ_1 and π^k only has a non-vanishing Poisson bracket with ψ_4. Since $G_{14}^{-1} = G_{41}^{-1} = 0$ then we have $\{A^0(x), \pi_k(y)\}_{\text{DB}} = 0$. Similarly, π^0 and A^k only have non-vanishing Poisson brackets with ψ_3 and ψ_2, respectively, and since $G_{23}^{-1} = G_{32}^{-1} = 0$ then we find $\{A^k(x), \pi_0(y)\}_{\text{DB}} = 0$. Rather than continue to evaluate the remaining mixed fundamental Dirac brackets on a case-by-case basis, we evaluate all of them at once:

$$\begin{aligned}
&\{A^\mu(x), \pi_\nu(y)\}_{\text{DB}} = \{A^\mu(x), \pi_\nu(y)\} \\
&\quad - \int d^3u \int d^3v \sum_{i,j=1}^{N_c} \left(\{A^\mu(x^0, \mathbf{x}), \psi_i(x^0, \mathbf{u})\} G_{ij}^{-1}(\mathbf{u}, \mathbf{v})\{\psi_j(x^0, \mathbf{v}), \pi_\nu(x^0, \mathbf{y})\}\right) \\
&= \delta^\mu{}_\nu \delta^3(\mathbf{x} - \mathbf{y}) \\
&\quad - \int d^3u \int d^3v \left(\{A^\mu(x^0, \mathbf{x}), \pi^0(x^0, \mathbf{u})\}\delta^3(\mathbf{u} - \mathbf{v})\{A^0(x^0, \mathbf{v}), \pi_\nu(x^0, \mathbf{y})\}\right. \\
&\quad\quad + \{A^\mu(x^0, \mathbf{x}), \partial_i\pi^i(x^0, \mathbf{u})\}[-1/4\pi|\mathbf{u} - \mathbf{v}|]\{\partial_j A^j(x^0, \mathbf{v}), \pi_\nu(x^0, \mathbf{y})\} \\
&\quad\quad + \{A^\mu(x^0, \mathbf{x}), A^0(x^0, \mathbf{u})\}[-\delta^3(\mathbf{u} - \mathbf{v})]\{\pi^0(x^0, \mathbf{v}), \pi_\nu(x^0, \mathbf{y})\} \\
&\quad\quad \left.+ \{A^\mu(x^0, \mathbf{x}), \partial_j A^j(x^0, \mathbf{u})\}[1/4\pi|\mathbf{u} - \mathbf{v}|]\{\partial_i\pi^i(x^0, \mathbf{v}), \pi_\nu(x^0, \mathbf{y})\}\right) \\
&= \left(\delta^\mu{}_\nu - g^{\mu 0}\delta^0{}_\nu\right)\delta^3(\mathbf{x} - \mathbf{y}) \\
&\quad - \int d^3u \int d^3v\, g^{\mu i}(\partial_i^u \delta^3(\mathbf{x} - \mathbf{u}))[-1/4\pi|\mathbf{u} - \mathbf{v}|]\delta^j{}_\nu \partial_j^v \delta^3(\mathbf{v} - \mathbf{y}) \\
&= \left(\delta^\mu{}_\nu - g^{\mu 0}\delta^0{}_\nu\right)\delta^3(\mathbf{x} - \mathbf{y}) - g^{\mu i}\delta^j{}_\nu \partial_i^x \partial_j^x [1/4\pi|\mathbf{x} - \mathbf{y}|],
\end{aligned} \tag{3.3.78}$$

where we have used integration by parts and have neglected surface terms; $\partial_i^x\delta^3(\mathbf{x} - \mathbf{y}) = -\partial_i^y\delta^3(\mathbf{x} - \mathbf{y})$; $\partial_i^x(1/|\mathbf{x} - \mathbf{y}|) = -\partial_i^y(1/|\mathbf{x} - \mathbf{y}|)$; and from Eq. (2.1.50) and the definition of the Poisson bracket for fields

$$\{\partial_i f(\mathbf{x}), g(\mathbf{y})\} = \partial_i^x\{f(\mathbf{x}), g(\mathbf{y})\}, \quad \{f(\mathbf{x}), \partial_i g(\mathbf{y})\} = \partial_i^y\{f(\mathbf{x}), g(\mathbf{y})\}. \tag{3.3.79}$$

From Eq. (3.3.77) we see that

$$\{A^0(x), \pi_\nu(y)\}_{\text{DB}}^{x^0=y^0} = 0 \quad \text{and} \quad \{A^\mu(x), \pi_0(y)\}_{\text{DB}}^{x^0=y^0} = 0 \tag{3.3.80}$$

and so the only non-vanishing fundamental Dirac brackets can be written as

$$\{A^i(x), \pi^j(y)\}_{\text{DB}}^{x^0=y^0} = g^{ij}\delta^3(\mathbf{x} - \mathbf{y}) - g^{im}g^{jn}\partial_m^x\partial_n^x\frac{1}{4\pi|\mathbf{x} - \mathbf{y}|} \tag{3.3.81}$$

$$= -\left[\delta^{ij}\delta^3(\mathbf{x} - \mathbf{y}) - \partial_x^i\partial_x^j\left(\frac{-1}{4\pi|\mathbf{x} - \mathbf{y}|}\right)\right] \equiv -\delta_{\text{tr}}^{ij}(\mathbf{x} - \mathbf{y}),$$

where we define the *transverse* and *longitudinal delta functions* as

$$\delta_{\text{tr}}^{ij}(\mathbf{x} - \mathbf{y}) \equiv \delta^{ij}\delta^3(\mathbf{x} - \mathbf{y}) - \partial_x^i\partial_x^j\left(\frac{-1}{4\pi|\mathbf{x} - \mathbf{y}|}\right) = \int\frac{d^3k}{(2\pi)^3}\,e^{i\mathbf{k}\cdot(\mathbf{x}-\mathbf{y})}\left(\delta^{ij} - \frac{k^ik^j}{\mathbf{k}^2}\right)$$

$$\delta_{\text{lo}}^{ij}(\mathbf{x} - \mathbf{y}) = \delta^{ij}\delta^3(\mathbf{x} - \mathbf{y}) - \delta_{\text{tr}}^{ij}(\mathbf{x} - \mathbf{y}). \tag{3.3.82}$$

We have used the definition of the delta function in Eq. (A.2.4) and made use of the Fourier transform relevant for the Coulomb interaction in Eq. (A.2.15). Note that $\delta_{\text{tr}}^{ij}(\mathbf{x} - \mathbf{y})$ is symmetric in i and j and in $\mathbf{x}$ and $\mathbf{y}$. We refer to this as the transverse delta function because its divergence vanishes,

$$\partial_i^x\delta_{\text{tr}}^{ij}(\mathbf{x} - \mathbf{y}) = \partial_i^y\delta_{\text{tr}}^{ij}(\mathbf{x} - \mathbf{y}) = \partial_j^x\delta_{\text{tr}}^{ij}(\mathbf{x} - \mathbf{y}) = \partial_j^y\delta_{\text{tr}}^{ij}(\mathbf{x} - \mathbf{y}) = 0. \tag{3.3.83}$$

Any three-vector function is decomposed as $\mathbf{V}(\mathbf{x}) = \mathbf{V}_t(\mathbf{x}) + \mathbf{V}_\ell(\mathbf{x})$, where

$$V_t^i(\mathbf{x}) = \int d^3y\,\delta_{\text{tr}}^{ij}(\mathbf{x} - \mathbf{y})V^j(\mathbf{y}) \quad \text{and} \quad V_\ell^i(\mathbf{x}) = \int d^3y\,\delta_{\text{lo}}^{ij}(\mathbf{x} - \mathbf{y})V^j(\mathbf{y}), \tag{3.3.84}$$

which we can relate back to Helmholtz's theorem in Eq. (2.7.107). Using Eq. (3.3.79) we see that these results ensure that

$$\{\partial_i A^i(x), \pi^j(y)\}_{\text{DB}}^{x^0=y^0} = \{A^i(x), \partial_j\pi^j(y)\}_{\text{DB}}^{x^0=y^0} = 0. \tag{3.3.85}$$

It then follows that the Dirac bracket of $A^0(x)$, $\pi^0(x)$, $\partial_i A^i(x)$ and $\partial_i\pi^i(x)$ with all quantities vanishes. These results then ensure that

$$\{A^0(x), H\}_{\text{DB}} \approx \{\pi^0(x), H\}_{\text{DB}} \approx \{\partial_i A^i(x), H\}_{\text{DB}} \approx \{\partial_i\pi^i(x), H\}_{\text{DB}} \approx 0 \tag{3.3.86}$$

and so using Eq. (3.3.76) we see that we satisfy the conditions

$$A^0 \approx 0, \quad \pi^0 \approx 0, \quad \boldsymbol{\nabla}\cdot\mathbf{A} \approx 0 \quad \text{and} \quad \boldsymbol{\nabla}\cdot\mathbf{E} \approx 0 \tag{3.3.87}$$

on the $4\times\infty^3$ physical submanifold at all times. Since we now have the Dirac bracket and since it ensures that all constraints are satisfied at all times for the free electromagnetic field, then we can take H as the standard canonical Hamiltonian of Eq. (3.3.33) with no external sources and with $\boldsymbol{\nabla}\cdot\mathbf{E} \approx 0$,

$$H = \int d^3x\left[\tfrac{1}{2}(\mathbf{E}^2 + \mathbf{B}^2) - (\boldsymbol{\nabla}\cdot\mathbf{E})A_0\right] \approx \int d^3x\,\tfrac{1}{2}[\mathbf{E}^2 + \mathbf{B}^2]. \tag{3.3.88}$$

Since A^0 and $\boldsymbol{\nabla}\cdot\mathbf{A}$ are constrained, it has now become clear that the electromagnetic field A^μ has only two field degrees of freedom rather than four. Similarly, since π^0 and $\boldsymbol{\nabla}\cdot\mathbf{E}$ are constrained, then there are only two independent field components of π^μ. In the language of Helmholtz's theorem in Eq. (2.7.115) the two degrees of freedom are the transverse components of $\mathbf{A}$ and the corresponding canonical momentum densities are the transverse components of $\boldsymbol{\pi} = \mathbf{E}$.

The generalization to include an external conserved current density is straightforward using Eq. (3.3.33) and leads to the standard canonical Hamiltonian density,

$$\mathcal{H} = \tfrac{1}{2}\left(\mathbf{E}^2 + \mathbf{B}^2\right) - [(\boldsymbol{\nabla}\cdot\mathbf{E}) - \rho]A_0 + (j^i/c)A_i, \tag{3.3.89}$$

where in place of Eqs. (3.3.64) and (3.3.65) we now have

$$\Omega_1(x) \equiv A^0(x) \approx \frac{1}{4\pi\epsilon_0}\int \frac{\rho(t,\mathbf{x}')}{|\mathbf{x}-\mathbf{x}'|}d^3x', \quad \Omega_2(x) \equiv \boldsymbol{\nabla}\cdot\mathbf{A}(x) = \partial_i A^i(x) \approx 0,$$

$$\zeta_1(x) \equiv \pi^0(x) \approx 0, \quad \zeta_2(x) \equiv \boldsymbol{\nabla}\cdot\boldsymbol{\pi}(x) - \rho(x) \approx 0. \tag{3.3.90}$$

Since we have only modified the constraints by adding external quantities that do not depend on A^μ or π^μ, then the Poisson brackets of Eq. (3.3.70) remain unchanged and then so do the Dirac brackets in Eq. (3.3.77).

It is interesting to decompose the Hamiltonian into a free part H_0 and an interacting part resulting from an external current $j^\mu = (\rho, \mathbf{j})$. In the free theory we have the Coulomb law condition $\boldsymbol{\nabla}\cdot\mathbf{E} = 0$ and so the free electric field operator is $\mathbf{E}_t$, where $\mathbf{E} = \mathbf{E}_t + \mathbf{E}_\ell$. Recall that in Coulomb gauge $\mathbf{A} = \mathbf{A}_t$ since $\boldsymbol{\nabla}\cdot\mathbf{A} = 0$ and that without an external current we have $A^0 = 0$ and since $\mathbf{E} = -\boldsymbol{\nabla}A^0 - \partial_0\mathbf{A}$ then we have $\mathbf{E}_t = -\partial_0\hat{\mathbf{A}}_t$ and $\mathbf{E}_\ell = -\boldsymbol{\nabla}A^0$, where A^0 is given above in Eq. (3.3.90) when $\rho \neq 0$. So $\mathbf{E}^2 = \mathbf{E}_t^2 + \mathbf{E}_\ell^2 = \mathbf{E}_t^2 + (\boldsymbol{\nabla}A^0)^2$ and $\boldsymbol{\nabla}\cdot\mathbf{E} = \boldsymbol{\nabla}\cdot\mathbf{E}_\ell = -\boldsymbol{\nabla}^2 A^0 = \rho$. From Eq. (3.3.89) the Hamiltonian has the form

$$H(t) = \int d^3x\,\left\{\tfrac{1}{2}\left(\mathbf{E}^2 + \mathbf{B}^2\right) - \mathbf{j}\cdot\mathbf{A}\right\} = \int d^3x\,\left\{\tfrac{1}{2}\left(\mathbf{E}_t^2 + \mathbf{B}^2\right) - \mathbf{j}\cdot\mathbf{A} + \tfrac{1}{2}\mathbf{E}_\ell^2\right\} \tag{3.3.91}$$

$$= \int d^3x\,\left\{\left[\tfrac{1}{2}\left(\mathbf{E}_t^2 + \mathbf{B}^2\right)\right] + \left[-\mathbf{j}\cdot\mathbf{A} + \tfrac{1}{2}\int d^3x'\,\frac{\rho(t,\mathbf{x})\rho(t,\mathbf{x}')}{4\pi\epsilon_0|\mathbf{x}-\mathbf{x}'|}\right]\right\} \equiv H_0 + H_I(t),$$

where using our boundary conditions and dropping a surface term we have used

$$\int d^3x\,\mathbf{E}_\ell^2 = \int d^3x\,(-\boldsymbol{\nabla}A^0)^2 = -\int d^3x\,A^0\boldsymbol{\nabla}^2 A^0 = \int d^3x\,A^0\rho. \tag{3.3.92}$$

We identify the free and interacting parts of the Hamiltonian, respectively, as

$$\hat{H}^0 \equiv \int d^3x\,\tfrac{1}{2}\left(\mathbf{E}_t^2 + \mathbf{B}^2\right), \tag{3.3.93}$$

$$H_I(t) \equiv H_C(t) - \int d^3x\,\mathbf{j}(t,\mathbf{x})\cdot\mathbf{A} \quad\text{with}\quad H_C(t) \equiv \frac{1}{2}\int d^3x'\,\frac{\rho(t,\mathbf{x})\rho(t,\mathbf{x}')}{4\pi\epsilon_0|\mathbf{x}-\mathbf{x}'|},$$

where the contribution $H_C(t)$ is the instantaneous Coulomb interaction between the charge distributions. Note that there is no faster than light interaction when we perform a full calculation. The appearance of this apparently nonlocal interaction is simply an artifact of choosing to work in Coulomb gauge.

Canonical quantization of the electromagnetic field: As we showed in Eq. (3.3.35) a naive application of Hamilton's equations reproduces the correct physical equations of motion, i.e., Maxwell's equations. However, since electromagnetism is a singular system we have the result that neither π^0 nor A^0 are dynamical variables in a Hamiltonian formulation and that a valid approach to the Hamiltonian formulation requires the use of the Dirac-Bergmann algorithm. The first two of Maxwell's equations follow from the definition of the physical $\mathbf{E}$ and $\mathbf{B}$ fields in terms of the four-vector A^μ gauge fields. The last two of Maxwell's equations are reproduced in the Dirac-Bergmann algorithm and are recovered from Eq. (3.3.53).

Applying the Dirac-Bergmann algorithm for constrained systems to the electromagnetic field in the presence of an external conserved current j^μ, we have arrived at the result that the only nonvanishing fundamental Dirac brackets are

$$\{A^i(x), \pi^j(y)\}_{\text{DB}}^{x^0=y^0} = \{A^i(x), E^j(y)\}_{\text{DB}}^{x^0=y^0} = -\delta_{ij}^{\text{tr}}(\mathbf{x} - \mathbf{y}). \tag{3.3.94}$$

According to Dirac's canonical quantization procedure of Eq. (2.9.136) the Dirac brackets of Eq. (3.3.77) imply that the fundamental equal-time commutators are

$$[\hat{\pi}^i(x), \hat{\pi}^j(y)]_{x^0=y^0} = [\hat{E}^i(x), \hat{E}^j(y)]_{x^0=y^0} = 0, \quad [\hat{A}^i(x), \hat{A}^j(y)]_{x^0=y^0} = 0,$$

$$[\hat{A}^i(x), \hat{\pi}^j(y)]_{x^0=y^0} = [\hat{A}^i(x), \hat{E}^j(y)]_{x^0=y^0} = -i\hbar\delta_{ij}^{\text{tr}}(\mathbf{x} - \mathbf{y}). \tag{3.3.95}$$

The A^0 component of the electromagnetic field is not a dynamical variable and is chosen to be a c-number (real or complex number) field as in Eq. (3.3.90) above, which obviously commutes with everything. Since the standard canonical Hamiltonian H does not contain π^0, then π^0 does not appear in the operator form of the Hamiltonian $\hat{H}$ and so it plays no further role. We notice that in terms of the canonical momentum density defined in Eq. (3.3.28) we recover the usual sign for the canonical commutation relations; i.e., we have $[\hat{A}^i(x), \hat{\pi}_{\text{can}}^j(y)]_{x^0=y^0} = i\hbar\delta_{ij}^{\text{tr}}(\mathbf{x} - \mathbf{y})$ and the only change is $\delta^{ij}\delta^3(\mathbf{x} - \mathbf{y}) \to \delta_{ij}^{\text{tr}}(\mathbf{x} - \mathbf{y})$.

The transverse delta function above ensures that

$$[\boldsymbol{\nabla} \cdot \hat{\mathbf{A}}(x), \hat{\pi}^j(y)]_{x^0=y^0} = \partial_i^x[\hat{A}^i(x), \hat{E}^j(y)]_{x^0=y^0} = -i\hbar\partial_i^x\delta_{ij}^{\text{tr}}(\mathbf{x} - \mathbf{y}) = 0,$$

$$[\hat{A}^i(x), \boldsymbol{\nabla} \cdot \hat{\mathbf{E}}(y)]_{x^0=y^0} = \partial_j^y[\hat{A}^i(x), \hat{E}^j(y)]_{x^0=y^0} = -i\hbar\partial_j^y\delta_{ij}^{\text{tr}}(\mathbf{x} - \mathbf{y}) = 0 \tag{3.3.96}$$

and so $\boldsymbol{\nabla} \cdot \hat{\mathbf{A}}$ and $[\boldsymbol{\nabla} \cdot \hat{\mathbf{E}} - \rho]$ commute with all operators and hence also with the Hamiltonian operator $\hat{H}$. So for the Hamiltonian operator $\hat{H}$ and the operators $\boldsymbol{\nabla} \cdot \hat{\mathbf{A}}$ and $[\boldsymbol{\nabla} \cdot \hat{\mathbf{E}} - \rho]$ we can always construct a set of simultaneous eigenstates. The physical Hilbert subspace will be the subspace of the total Hilbert space that pertains to the zero eigenvalues of the operators $\boldsymbol{\nabla} \cdot \hat{\mathbf{A}}$ and $[\boldsymbol{\nabla} \cdot \hat{\mathbf{E}} - \rho]$. From Eq. (3.3.88) the Hamiltonian operator for the free electromagnetic field is then

$$\hat{H} = \int d^3x \left[\tfrac{1}{2}(\hat{\mathbf{E}}^2 + \hat{\mathbf{B}}^2) - (\boldsymbol{\nabla} \cdot \hat{\mathbf{E}} - j^0)A_0 - (\mathbf{j}/c) \cdot \hat{\mathbf{A}}\right]$$

$$\approx \int d^3x \left[\tfrac{1}{2}(\hat{\mathbf{E}}^2 + \hat{\mathbf{B}}^2) - (\mathbf{j}/c) \cdot \hat{\mathbf{A}}\right], \tag{3.3.97}$$

where the second line is the form taken by $\hat{H}$ when restricted to act in the physical subspace of the Hilbert space spanned by the states $|\psi\rangle$ that satisfy $(\boldsymbol{\nabla} \cdot \hat{\mathbf{E}} - j^0)|\psi\rangle = 0$. The purpose of applying the Dirac-Bergmann algorithm has been to arrive at these fundamental equal-time commutators and constraints on the Hilbert space, which has led us to the quantization of the electromagnetic field. The Hamiltonian approach to the quantization of fields is commonly referred to as *canonical quantization of fields*. We will pick up from this point in Sec. 6.4.1.

Summary

The Lagrangian and Hamiltonian formulations of classical field theory have been obtained and we have shown how symmetries lead to conserved Noether currents. For nonsingular systems the Poisson brackets for the classical field theory combined with Dirac's canonical quantization approach then led to the corresponding quantum field theories to be discussed in Chapter 6. We have used electromagnetism as an example of how to proceed to the Hamiltonian formulation of a singular system. We have shown how obtain the corresponding Dirac brackets using the Dirac-Bergmann algorithm and then how to arrive at the canonical commutation relations for the quantized electromagnetic field.

Problems

3.1 We define a free complex scalar field as $\phi(x) = (1/\sqrt{2})[\phi_1(x) + i\phi_2(x)] \in \mathbb{C}$ in terms of two real free scalar fields $\phi_1(x), \phi_2(x) \in \mathbb{R}$ and the Lagrangian density as the sum of the two free Lagrangian densities.

    (a) Show that the Lagrangian density in terms of $\phi(x)$ and $\phi^*(x)$ is as given in Eq. (3.2.48), $\mathcal{L} = (\partial_\mu \phi)^*(\partial^\mu \phi) - m^2 \phi^* \phi$. Using the Euler-Lagrange equations for ϕ_1 and ϕ_2 show that the complex scalar field and its complex conjugate satisfy $(\partial_\mu \partial^\mu + m^2)\phi(x) = (\partial_\mu \partial^\mu + m^2)\phi^*(x) = 0$.

    (b) Using the action $S[\phi] = \int d^4 x\, \mathcal{L}$ and the Wirtinger calculus in Sec. A.2.5, show that these equations of motion follow from Hamilton's principle, $\delta S[\phi]/\delta\phi^*(x) = \delta S[\phi]/\delta\phi(x) = 0$.

    (c) Using Eq. (3.1.15) evaluate the conjugate momenta $\pi(x)$ and $\pi^*(x)$. Use these results to construct the Hamiltonian using Eq. (3.1.16) and show that $H = \int d^3 x\, \mathcal{H} = \int d^3 x\, [\pi^*(x)\pi(x) + \nabla\phi^*(x) \cdot \nabla\phi(x) + m^2 \phi^*(x)\phi(x)]$.

    (d) Obtain Hamilton's equations using Eq. (3.1.21) in terms of ϕ and ϕ^* and show that these reproduce the Klein-Gordon equations for ϕ and ϕ^*.

    (e) The complex scalar field Lagrangian in terms of ϕ_1 and ϕ_2 has a well-defined Legendre transform since the Hessian matrix M_L is positive definite. This follows from the straightforward generalization of Eq. (3.1.48),

$$(M_L)_{ij}(x,y)\big|_{x^0=y^0} = \delta^2 L/\delta\dot\phi_i(x)\delta\dot\phi_j(y)\Big|_{x^0=y^0} = \delta_{ij}\delta^3(\mathbf{x} - \mathbf{y}), \qquad (3.3.98)$$

for $i, j = 1, 2$. Show that the Wirtinger calculus version of the nonsingular Hessian matrix has the four elements

$$\delta^2 L/\delta\dot\phi(x)\delta\dot\phi(y)\big|_{x^0=y^0} = \delta^2 L/\delta\dot\phi^*(x)\delta\dot\phi^*(y)\big|_{x^0=y^0} = 0, \qquad (3.3.99)$$
$$\delta^2 L/\delta\dot\phi(x)\delta\dot\phi^*(y)\big|_{x^0=y^0} = \delta^2 L/\delta\dot\phi^*(x)\delta\dot\phi(y)\big|_{x^0=y^0} = \delta^3(\mathbf{x} - \mathbf{y}).$$

3.2 Let ϕ be a Lorentz-scalar field.

    (a) Show that the quantities $T_{\mu\nu} = \partial_\mu\phi\partial_\nu\phi - \frac{1}{2}g_{\mu\nu}\{(\partial\phi)^2 - \kappa^2\phi^2\}$ with $\lambda \in \mathbb{R}$ are components of a Lorentz $(0,2)$ tensor field.

    (b) Show that $\partial_\mu T^{\mu\nu} = 0$ if $\phi(x)$ obeys the wave equation $(\partial^2 + \kappa^2)\phi = 0$.

3.3 Consider n real equal-mass scalar fields $\phi_1, \ldots, \phi_n$ with a quartic interaction of the form $\mathcal{L}_{\text{int}} = -(\lambda/4!)(\phi_1^2 + \cdots + \phi_n^2)^2$, then

$$\mathcal{L} = \sum_{i=1}^n \mathcal{L}_i^{\text{free}} + \mathcal{L}_{\text{int}} = \sum_{i=1}^n (\tfrac{1}{2}\partial_\mu\phi_i\partial^\mu\phi_i - \tfrac{1}{2}m^2\phi_i^2) - (\lambda/4!)(\phi_1^2 + \cdots + \phi_n^2)^2$$
$$= \tfrac{1}{2}\partial_\mu\phi^T\partial^\mu\phi - \tfrac{1}{2}\phi^T\phi - (\lambda/4!)(\phi^T\phi)^2. \qquad (3.3.100)$$

(a) Obtain the Euler-Lagrange equations for this theory.

(b) Explain why this theory is invariant under $O(n)$ transformations. Derive the conserved Noether current and explain why it corresponds to the $SO(n)$ subgroup and not $O(n)$.

(c) For n complex equal-mass scalar fields write a corresponding Lagrangian density invariant under $U(n)$. Obtain the Euler-Lagrange equations. Construct the conserved Noether current.

3.4 A Lagrangian density with two complex scalar fields, ϕ and χ, is given by

$$\mathcal{L} = \partial_\mu \phi \partial^\mu \phi^* + \partial_\mu \chi \partial^\mu \chi^* - m_\phi^2 |\phi|^2 - m_\chi^2 |\chi|^2 - \lambda_\phi |\phi|^4 - \lambda_\chi |\chi|^4 \qquad (3.3.101)$$
$$- g[\chi^* \phi^2 + \chi(\phi^*)^2],$$

where $m_\phi, m_\chi, \lambda_\phi, \lambda_\chi, g$ are real constants.

(a) Using the transformations $\phi \to e^{i\alpha}\phi$ and $\chi \to e^{i\beta}\chi$ choose α, β such that $\mathcal{L}$ is invariant. Construct the Noether current j^μ.

(b) Use the Euler-Lagrange equations to verify that $\partial_\mu j^\mu(x) = 0$.

3.5 Let $F^{\mu\nu}$ be the electromagnetic tensor and consider the symmetric stress-energy tensor defined in Eq. (3.3.9), $\bar{T}^\mu{}_\nu = -F^{\mu\tau} F_{\nu\tau} + \frac{1}{4}\delta^\mu{}_\nu F^{\sigma\tau} F_{\sigma\tau}$. In the presence of an external conserved four-vector current density j^μ show that $\partial_\mu \bar{T}^{\mu\nu} = 0$ becomes $\partial_\mu \bar{T}^{\mu\nu} = (1/c)j_\mu F^{\mu\nu}$.

3.6 Let A^μ be the electromagnetic four-vector potential and define the four-vector $K_\mu = \epsilon_{\mu\nu\rho\sigma} A^\nu \partial^\rho A^\sigma$. Show that $\partial_\mu K^\mu = -(2/c)\mathbf{E} \cdot \mathbf{B}$. (Hint: Consider $\tilde{F}^{\mu\nu}$.)

3.7 A general real scalar Lagrangian density at most quadratic in the four-vector potential is $\mathcal{L} = -\frac{1}{2}(\partial_\mu A_\nu \partial^\mu A^\nu + \rho\, \partial_\mu A_\nu \partial^\nu A^\mu) + \lambda\, A_\mu A^\mu$, where $\rho, \lambda \in \mathbb{R}$.

(a) Obtain the Euler-Lagrange equations for this system.

(b) Assume plane wave solutions with the general form $A^\mu(x) = C\epsilon^\mu(\mathbf{k})e^{-ik\cdot x}$ and express the Euler-Lagrange equation in terms of $\epsilon^\mu(\mathbf{k})$ and k^μ.

(c) For $k^2 \neq 0$ we can define transverse and longitudinal projectors as $\mathcal{T}^{\mu\nu} \equiv g^{\mu\nu} - (k^\mu k^\nu/k^2)$ and $\mathcal{L}^{\mu\nu} \equiv (k^\mu k^\nu/k^2)$. Verify the properties $\mathcal{T}^{\mu\rho}\mathcal{T}_{\rho\nu} = \mathcal{T}^\mu{}_\nu$, $\mathcal{L}^{\mu\rho}\mathcal{L}_{\rho\nu} = \mathcal{L}^\mu{}_\nu$ and $\mathcal{T}^\mu{}_\nu + \mathcal{L}^\mu{}_\nu = \delta^\mu{}_\nu$. Define $\epsilon_\ell^\mu \equiv \mathcal{L}^{\mu\nu}\epsilon_\nu$ and $\epsilon_t^\mu \equiv \mathcal{T}^{\mu\nu}\epsilon_\nu$ and so verify that $k \cdot \epsilon_\ell = k \cdot \epsilon$, $k \cdot \epsilon_t = 0$ and $\epsilon_t^\mu + \epsilon_\ell^\mu = \epsilon^\mu$.

(d) Show that there is a plane wave solution to the Euler-Lagrange equations with $k^2 = m_\ell^2 \equiv 2\lambda/(1+\rho)$ that is longitudinal, $\epsilon^\mu = \epsilon_\ell^\mu$.

(e) Similarly, show that there is a plane wave solution with $k^2 = m_t^2 \equiv 2\lambda$, that is transverse.

(f) We say that the ℓ and t modes propagate with masses m_ℓ and m_t, respectively. Choosing $\rho = -1$ gives $m_\ell \to \infty$ so that the ℓ modes[6] cannot propagate, leaving only the t modes. Show that in this limit we recover the Proca Lagrangian density for a massive vector field given in Eq. (6.5.1) with m_t becoming the mass of the vector field.

3.8 Scalar electrodynamics is the gauge invariant minimal coupling of a charged scalar field and the electromagnetic field. It has the Lagrangian density $\mathcal{L} = -\frac{1}{4}F_{\mu\nu}F^{\mu\nu} + (D_\mu \phi)^*(D^\mu \phi) - m^2\phi\phi^*$, where the covariant derivative is defined as $D_\mu \equiv \partial_\mu + iqA_\mu$ and where q is the electric charge of the scalar particle.

---

[6] We avoid using the term "longitudinal modes," since this term is often used to refer to the zero helicity modes of the massive vector field.

(a) The system is invariant under global phase transformations $\phi \to e^{-i\alpha}\phi$. Use this to derive the conserved Noether current that replaces Eq. (3.2.49),

$$j^\mu = iqc\left[\phi^*(D^\mu\phi) - (D^\mu\phi)^*\phi\right] = iqc\left[\phi^*(\partial^\mu\phi) - (\partial^\mu\phi^*)\phi + 2iqA^\mu\phi^*\phi\right].$$

(b) Show that the Euler-Lagrange equations can be written as $\partial_\mu F^{\mu\nu} = j^\nu/c$ and $(D_\mu D^\mu + m^2)\phi = 0$.

(c) Show that $A^\mu(x) \to A^\mu(x) + \partial^\mu\alpha(x)$, $\phi(x) \to e^{-iq\alpha(x)}\phi(x)$ leaves the system invariant. This is an example of $U(1)$ gauge invariance.

4 Relativistic Quantum Mechanics

Relativistic quantum mechanics is the consequence of a first attempt at combining special relativity with the Schrödinger equation. It describes the relativistic quantum motion of single particles interacting with an external field. The discussion here draws from a variety of sources, including the first volume of Bjorken and Drell (1964), Itzykson and Zuber (1980), Peskin and Schroeder (1995), Weinberg (1995), Greiner (2000) and the first volume of Aitchison and Hey (2013).

This chapter begins with a review of nonrelativistic quantum mechanics including the role of symmetries, including discrete symmetries, systems of identical particles and the path integral formulation. Relativistic wavepackets are then discussed. We then discuss the Klein-Gordon equation for spin-zero bosons and the Dirac equation for spin-half fermions and its implications. We conclude with discussions of the discrete symmetries P, C and T, chirality and the Weyl representation. Relativistic quantum mechanics breaks down in the presence of strong fields and interactions, but the solutions of its wave equations underpin quantum field theory.

4.1 Review of Quantum Mechanics

To summarize the development and essentials of relativistic quantum mechanics we begin with a brief review of nonrelativistic quantum mechanics. Many of the results and techniques are essential to our understanding of quantum field theory.

4.1.1 Postulates of Quantum Mechanics

The postulates of quantum mechanics can be summarized as the following.

(i) The state of a quantum system at any time t is represented by a ray in some complex Hilbert space, V_H, where "ray" means the directional part of a vector $|\psi\rangle$ in the space; i.e., $c|\psi\rangle$ for any $c \in \mathbb{C}$ represents the same physical state so that neither the magnitude nor phase of $|\psi\rangle$ contains any physical information. If we choose $|\psi\rangle$ to be normalized, then it is still arbitrary up to a phase.

(ii) Every physical observable A of the quantum system has some corresponding Hermitian operator $\hat{A}$ that acts on the states $|\psi\rangle$ in V_H. Since $\hat{A}$ is Hermitian it has real eigenvalues and a complete orthonormal set of eigenstates $|a_i\rangle$, i.e.,

$$\hat{A}|a_i\rangle = a_i|a_i\rangle \quad \text{with} \quad a_i \in \mathbb{R}, \quad \langle a_i|a_j\rangle = \delta_{ij}, \quad \sum_i |a_i\rangle\langle a_i| = \hat{I}, \tag{4.1.1}$$

$$\hat{A} = \hat{A}\hat{I} = \hat{A}\sum_i |a_i\rangle\langle a_i| = \sum_i |a_i\rangle a_i \langle a_i| \tag{4.1.2}$$

with $\hat{I}$ the identity operator (we use countable eigenstates here).

(iii) The outcome of a measurement of the physical observable A must be one of the eigenvalues a_i of the Hermitian operator $\hat{A}$. If $|\psi\rangle$ is the physical state of the system at the time the measurement is made then immediately after the measurement the physical state of the system is given by either (a) the state $|a_i\rangle$ if the eigenvalue a_i is not degenerate or (b) the projection $\hat{\Lambda}_i|\psi\rangle$ of $|\psi\rangle$ onto the subspace spanned by the degenerate eigenvectors of a_i if it is degenerate; i.e., if $\hat{A}|a_{ir}\rangle = a_i|a_{ir}\rangle$ for $r = 1, 2, \ldots, n$, then a_i is n-fold degenerate and the projector onto the a_i subspace is $\hat{\Lambda}_i \equiv \sum_r |a_{ir}\rangle\langle a_{ir}|$.

(iv) If the physical state of the system is $|\psi\rangle$ at the time the measurement is made, then the average over a large number of identical measurements is the *expectation value* of $\hat{A}$ in the state $|\psi\rangle$

$$\langle \hat{A} \rangle_\psi \equiv \frac{\langle \psi | \hat{A} | \psi \rangle}{\langle \psi | \psi \rangle} = \frac{\sum_i a_i |c_i|^2}{\sum_i |c_i|^2} = \sum_i a_i P_i, \tag{4.1.3}$$

where the index i runs over the orthonormal basis $|a_i\rangle$, where $c_i \equiv \langle a_i | \psi \rangle$ and where the probability of measuring the outcome a_i is the probability of finding the system in the state $|a_i\rangle$ after the measurement,

$$P_i = |c_i|^2 / \sum_j |c_j|^2. \tag{4.1.4}$$

(v) The time evolution of the state of the system is determined by the time-dependent Schrödinger equation

$$i\hbar \frac{d}{dt} |\psi(t)\rangle = \hat{H}_s(t) |\psi(t)\rangle, \tag{4.1.5}$$

where $\hat{H}_s(t)$ is defined as the Schrödinger picture Hamiltonian operator for the quantum system at time t and is Hermitian.

Mathematical digression: Hilbert space: A complex Hilbert space V_H is a complete complex inner-product space. A complex inner-product space is a complex vector space with an inner product $\langle \phi | \psi \rangle$ such that there is a norm $|\psi| \equiv \sqrt{\langle \psi | \psi \rangle}$ providing a concept of "distance."

A *Cauchy sequence* is any sequence where all of the terms in the sequence eventually become arbitrarily close to one another. A Cauchy sequence of vectors $|\psi_i\rangle$ for $i = 1, 2, \ldots$ in an inner-product vector space is any chosen sequence of vectors such that for any arbitrarily small $\epsilon \in \mathbb{R}$ we can always find a sufficiently large N such that $|(\psi_i - \psi_j)| < \epsilon$ for all $i, j > N$. A *complete* inner product space is one that contains the limit of every Cauchy sequence of vectors. In simple terms, a vector space is complete provided that it contains all necessary limits, which means that we can, for example, consider such things as the calculus of variations on complete functional spaces. In a complete inner-product space we have the concept of the neighborhood of a vector, which is why such spaces are topological vector spaces with the topology determined by the inner product. Hilbert spaces are therefore also topological spaces.

We use the traditional Dirac physics ket notation for Hilbert space states $|\psi\rangle \in V_H$ and the bra notation for states in the dual space $\langle \psi |$. The bra-ket $\langle \phi | \psi \rangle$ is used to denote the inner product. A Hilbert-space operator $\hat{A}$ maps any ket $|\psi\rangle \in V_H$ to some other ket $\hat{A}|\psi\rangle \in V_H$ and so we can consider bra-ket combinations of the form $\langle \phi | \hat{A} | \psi \rangle \equiv \langle \phi | (\hat{A} | \psi \rangle) \equiv \langle \phi | \hat{A}\psi \rangle$. Mathematicians would typically use the notation $\langle A\psi, \phi \rangle$ or $(A\psi, \phi)$ for this quantity.

A finite-dimensional Hilbert space with dimension n is just the complex vector space $\mathbb{C}^n$, where the inner product is the usual complex vector "dot" or "scalar" product. In a finite-dimensional complex Hilbert space quantum states can be represented in any given orthonormal basis as complex column vectors; the dual space states are represented as complex conjugate row vectors; and operators are represented by complex matrices. For example, in matrix representation the inner product becomes $\langle\phi|\psi\rangle \rightarrow \phi^\dagger\psi = \phi^* \cdot \psi$.

All of the mathematical subtleties of Hilbert spaces arise in the case of infinite-dimensional Hilbert spaces, where the issue of completeness becomes relevant. An example of an infinite-dimensional Hilbert space is the space of square-integrable complex functions. This is the Hilbert space relevant for the quantum mechanics of the motion of a particle in a potential, which has a Hamiltonian given by Eq. (4.1.32).

Consider the origin of the time-dependent Schrödinger equation, Eq. (4.1.5), in postulate (v) at this point. When first presented to physics undergraduates it is often simply imposed without explanation; however, its origins are an obvious consequence of postulate (i) on reflection. Since the norm $|\psi| \equiv \sqrt{\langle\psi|\psi\rangle}$ of the state $|\psi\rangle$ has no physical relevance, then it need not evolve in time as the physical state of the system evolves. So the time evolution of the system can always be described by norm-preserving transformations in Hilbert space that are continuous with the identity, i.e., by unitary transformations $\hat{U}(t',t)$ such that

$$|\psi(t')\rangle = \hat{U}(t',t)|\psi(t)\rangle. \tag{4.1.6}$$

$\hat{U}(t',t)$ is the *evolution operator* and is analogous to special orthogonal transformations (proper rotations) being the length-preserving transformations in $\mathbb{R}^N$ continuous with the identity. Consider an infinitesimal time increment from t to $t + dt$,

$$|\psi(t+dt)\rangle = \hat{U}(t+dt,t)|\psi(t)\rangle. \tag{4.1.7}$$

For infinitesimal dt we obviously have that $\hat{U}(t+dt,t)$ lies in the neighborhood of the identity operator. From the discussion surrounding Eq. (A.7.2) we can always *uniquely* define a Hermitian (i.e., self-adjoint) operator $\hat{H}_s(t)$ such that

$$\hat{U}(t+dt,t) \equiv I + \left(-i\hat{H}_s(t)/\hbar\right)dt + \mathcal{O}((dt)^2) \tag{4.1.8}$$

or equivalently

$$\hat{H}_s(t) \equiv i\hbar \left.\frac{d}{dt'}\hat{U}(t',t)\right|_{t'=t}. \tag{4.1.9}$$

Using this in Eq. (4.1.7) we immediately obtain the Schrödinger equation, Eq. (4.1.5),

$$i\hbar\frac{d}{dt}|\psi(t)\rangle = i\hbar \left.\frac{d}{dt'}\hat{U}(t',t)\right|_{t'=t}|\psi(t)\rangle = H_s(t)|\psi(t)\rangle. \tag{4.1.10}$$

In summary, the Schrödinger equation is a straightforward consequence of the fact that the state of the system at any point in time is specified by a ray in Hilbert space. While we listed the Schrödinger equation as postulate (v), in principle we did not need to do this as this equation follows from postulate (i). We see that specifying the evolution operator determines the Hamiltonian operator uniquely at all times from Eq. (4.1.9). Conversely, specifying the Hamiltonian operator at all times uniquely determines the evolution operator,

$$\hat{U}(t'',t') \equiv Te^{-(i/\hbar)\int_{t'}^{t''} dt\, \hat{H}_s(t)} \quad \text{and} \quad \hat{U}^\dagger(t'',t') \equiv \overline{T}e^{+(i/\hbar)\int_{t'}^{t''} dt\, \hat{H}_s(t)}, \tag{4.1.11}$$

where T is the time-ordering operator that places operators in order of decreasing time from left to right and $\overline{T}$ is the anti-time-ordering operator that reverses the order. It follows that $\hat{U}^\dagger(t'',t')\hat{U}(t'',t') = \hat{I}$ and so $\hat{U}(t'',t')$ is unitary.

Proof: Using $|\psi(t)\rangle = \hat{U}(t,t')|\psi(t')\rangle$ in Eq. (4.1.10) gives

$$d\hat{U}(t,t')/dt = -(i/\hbar)\hat{H}_s(t)\hat{U}(t,t'). \tag{4.1.12}$$

Integrating both sides and using $\hat{U}(t',t') = \hat{I}$ gives the recurrence relation for the evolution operator,

$$\hat{U}(t'',t') = \hat{I} - (i/\hbar)\int_{t'}^{t''} dt\,\hat{H}_s(t)\hat{U}(t,t'). \tag{4.1.13}$$

Iterating this equation gives Dyson's formula for the evolution operator

$$\hat{U}(t'',t') = \hat{I} + (-i/\hbar)\int_{t'}^{t''} dt_1\hat{H}_s(t_1) + (-i/\hbar)^2\int_{t'}^{t''} dt_1\int_{t'}^{t_1} dt_2\,\hat{H}_s(t_1)\hat{H}_s(t_2)$$
$$+ (-i/\hbar)^3\int_{t'}^{t''} dt_1\int_{t'}^{t_1} dt_2\int_{t'}^{t_2} dt_3\,\hat{H}_s(t_1)\hat{H}_s(t_2)\hat{H}_s(t_3) + \cdots. \tag{4.1.14}$$

We can make all the upper limits on the integrations equal to t'' by using the time-ordering operator and compensating for the overcounting by dividing by the number of permutations of the Hamiltonian. This gives the desired result,

$$\hat{U}(t'',t') = \hat{I} + (-i/\hbar)\int_{t'}^{t''} dt_1\hat{H}_s(t_1) + (-i/\hbar)^2\frac{1}{2}\int_{t'}^{t''} dt_1\int_{t'}^{t''} dt_2\,T\left(\hat{H}_s(t_1)\hat{H}_s(t_2)\right)$$
$$+ \cdots + (-i/\hbar)^n\frac{1}{n!}\int_{t'}^{t''} dt_1\int_{t'}^{t''} dt_2\cdots\int_{t'}^{t''} dt_n\,T\left(\hat{H}_s(t_1)\hat{H}_s(t_2)\cdots\hat{H}_s(t_n)\right) + \cdots$$
$$= \sum_{n=0}^{\infty}\frac{1}{n!}(-i/\hbar)^n\int_{t'}^{t''} dt_1\int_{t'}^{t''} dt_2\cdots\int_{t'}^{t''} dt_n\,T\left(\hat{H}_s(t_1)\hat{H}_s(t_2)\cdots\hat{H}_s(t_n)\right)$$
$$\equiv T\left(\exp\left\{(-i/\hbar)\int_{t'}^{t''} dt\,\hat{H}_s(t)\right\}\right). \tag{4.1.15}$$

For consistency we must have $\hat{U}(t_3,t_2)\hat{U}(t_2,t_1) = \hat{U}(t_3,t_1)$ and so defining $\delta t \equiv (t'' - t')/(n+1)$ and $t_j = j\delta t + t'$ such that $t' = t_0$ and $t'' = t_{n+1}$ then

$$\hat{U}(t'',t') = \lim_{n\to\infty}\hat{U}(t'',t_n)\hat{U}(t_n,t_{n-1})\cdots\hat{U}(t_1,t') \tag{4.1.16}$$

$$= \lim_{n\to\infty} e^{-i\delta t\hat{H}_s(t_n)/\hbar}\cdots e^{-i\delta t\hat{H}_s(t_1)/\hbar}e^{-i\delta t\hat{H}_s(t')/\hbar} = Te^{-(i/\hbar)\int_{t'}^{t''} dt\,\hat{H}_s(t)},$$

which provides a simple explanation. The fact that $[\hat{H}_s(t_{j+1}),\hat{H}_s(t_j)] \neq 0$ is not a problem for combining the infinitesimal exponents, since corrections from the BCH formula in Eq. (A.8.6) are $\mathcal{O}(\delta t^2)$ and vanish as $\delta t \to 0$.

The set of time evolution operators $\hat{U}(t'',t')$ have group-like properties under operator multiplication provided that we require identical adjacent times in the multiplication: (i) the set is closed, $\hat{U}(t''',t'')\hat{U}(t'',t') = \hat{U}(t''',t')$; (ii) it is associative, $\hat{U}(t_a,t_b)[\hat{U}(t_b,t_c)\hat{U}(t_c,t_d)] = \hat{U}(t_a,t_d) = [\hat{U}(t_a,t_b)\hat{U}(t_b,t_c)]\hat{U}(t_c,t_d)$; (iii) it contains the identity $\hat{U}(t',t') = \hat{I}$; and (iv) every $\hat{U}(t'',t')$ has an inverse, $\hat{U}(t'',t')^{-1} = \hat{U}(t'',t')^\dagger$, which has anti-time-ordering. It is not meaningful to have a multiplication $\hat{U}(t_4,t_3)\hat{U}(t_2,t_1)$ unless $t_3 = t_2$ and so the $\hat{U}(t'',t')$ are not elements of a group in the normal sense. If the system is time-translationally invariant, $\hat{U}(t'+\Delta t,t') = \hat{U}(t+\Delta t,t) \equiv \hat{U}(\Delta t)$ for all $t',t \in \mathbb{R}$, then the $\hat{U}(\Delta t)$ do form a group; e.g., see Eq. (4.1.22).

For some observable operator, $\hat{A}$, and state of the system, $|\psi\rangle$, the Schrödinger (s) and Heisenberg (h) pictures are related by

$$\hat{A}_h(t) = \hat{U}(t,t_0)^\dagger \hat{A}_h(t_0)\hat{U}(t,t_0) = \hat{U}(t,t_0)^\dagger \hat{A}_s\hat{U}(t,t_0),$$

$$|\psi,t\rangle_s = \hat{U}(t,t_0)|\psi\rangle_h \tag{4.1.17}$$

and the two pictures are equivalent at some arbitrarily chosen time $t = t_0$, i.e., $\hat{A}_s = \hat{A}_h(t_0)$ and $|\psi,t_0\rangle_s = |\psi\rangle_h$. In the Heisenberg picture the state of the system $|\psi\rangle_h$ does not evolve in time but the operators do and hence so do their eigenstates,

$$\hat{A}_s|a\rangle_s = a|a\rangle_s, \quad \hat{A}_h(t)|a,t\rangle_h = a|a,t\rangle_h, \quad \text{where} \quad |a,t\rangle_h = \hat{U}(t,t_0)^\dagger|a\rangle_s. \tag{4.1.18}$$

Two ways of indicating time-dependence of states are used interchangeably,

$$|\psi,t\rangle \equiv |\psi(t)\rangle. \tag{4.1.19}$$

Physical observables are independent of which picture is used since the expectation values of observables are identical,

$$_s\langle\psi,t|\hat{A}_s|\psi,t\rangle_s = {}_h\langle\psi|\hat{U}(t,t_0)^\dagger \hat{A}_s\hat{U}(t,t_0)|\psi\rangle_h = {}_h\langle\psi|\hat{A}_h(t)|\psi\rangle_h, \tag{4.1.20}$$

where we assume normalized states here. The Heisenberg picture Hamiltonian is

$$\hat{H}_h(t) = \hat{U}(t,t_0)^\dagger \hat{H}_s(t)\hat{U}(t,t_0). \tag{4.1.21}$$

If the Hamiltonian has no explicit time dependence, $\hat{H}_s(t) = \hat{H}_s$, then the evolution operator takes a simpler form,

$$\hat{U}(t',t) = \hat{U}(t'-t) = e^{-i\hat{H}(t'-t)/\hbar} \quad \text{with} \quad \hat{H} \equiv H_h = \hat{H}_s. \tag{4.1.22}$$

In the general case an observable operator in the Schrödinger picture might have an *explicit* time dependence, $\hat{A}_s \to \hat{A}_s(t)$, and we might also have an explicit time-dependence in the Hamiltonian, $\hat{H}_s \to \hat{H}_s(t)$. Using Eq. (4.1.11) we then have for the Heisenberg picture forms for these

$$\hat{A}_h(t) = \hat{U}(t,t_0)^\dagger \hat{A}_s(t)\hat{U}(t,t_0), \quad \hat{H}_h(t) = \hat{U}(t,t_0)^\dagger \hat{H}_s(t)\hat{U}(t,t_0). \tag{4.1.23}$$

Using Eq. (4.1.12) and writing for temporary convenience $\hat{U} \equiv \hat{U}(t,t_0)$ we can evaluate the time derivative of $\hat{A}_h(t)$ to obtain the Heisenberg equation of motion for the observable operator $\hat{A}$ in the form

$$\begin{aligned}
\frac{d\hat{A}_h(t)}{dt} &= \frac{d\hat{U}^\dagger}{dt}\hat{A}_s(t)\hat{U} + \hat{U}^\dagger\frac{\partial\hat{A}_s(t)}{\partial t}\hat{U} + \hat{U}^\dagger\hat{A}_s(t)\frac{d\hat{U}}{dt} \\
&= \frac{1}{i\hbar}\{-\hat{U}^\dagger\hat{H}_s(t)\hat{A}_s(t)\hat{U} + \hat{U}^\dagger\hat{A}_s(t)\hat{H}_s(t)\hat{U}\} + \hat{U}^\dagger\frac{\partial\hat{A}_s(t)}{\partial t}\hat{U} \\
&= \frac{1}{i\hbar}[\hat{A}_h(t),\hat{H}_h(t)] + \frac{\partial\hat{A}_h(t)}{\partial t}, \tag{4.1.24}
\end{aligned}$$

where we *define* the explicit time-dependence of the Heisenberg picture operator as

$$\partial\hat{A}_h(t)/\partial t \equiv \hat{U}(t,t_0)^\dagger(\partial\hat{A}_s(t)/\partial t)\hat{U}(t,t_0). \tag{4.1.25}$$

The meaning of this is that an explicit time-dependence in $\hat{A}_h$ can only originate from an explicit time-dependence of $\hat{A}_s$. We have then proved Eq. (2.5.17), which is the general form of the Ehrenfest theorem at the operator level.

The question remains as to why we call this operator the "Hamiltonian" operator. In other words, what has this operator to do with the Hamiltonian of classical mechanics? We can answer this question in two steps.

(a) The definition of $\hat{H}_s(t)$ in terms of $\hat{U}(t, t')$ not only gives rise to the Schrödinger equation but also gives rise to the Heisenberg equation of motion in Eq. (4.1.24).

(b) If the Hilbert space contains time-independent (Schrödinger picture) Hermitian operators with continuous spectra that satisfy canonical commutation relations $[\hat{q}_{si}, \hat{q}_{sj}] = [\hat{p}_{si}, \hat{p}_{sj}] = 0$ and $[\hat{q}_{si}, \hat{p}_{sj}] = i\hbar\delta_{ij}$, then the Heisenberg picture operators, $\hat{\vec{q}}(t) \equiv \hat{\vec{q}}_h(t) = \hat{U}(t, t_0)^\dagger \hat{\vec{q}}_s \hat{U}(t, t_0)$ and $\hat{\vec{p}}(t) \equiv \hat{\vec{p}}_h(t) = \hat{U}(t, t_0)^\dagger \hat{\vec{p}}_s \hat{U}(t, t_0)$, satisfy the equal-time canonical commutation relations

$$[\hat{q}_i(t), \hat{p}_j(t)] = [\hat{q}_i(t), \hat{p}_j(t)] = 0 \quad \text{and} \quad [\hat{q}_i(t), \hat{p}_j(t)] = i\hbar\delta_{ij}. \qquad (4.1.26)$$

As seen in Sec. 2.5, combining the Heisenberg equations of motion in Eq. (4.1.24) with the equal-time canonical commutation relations of Eq. (4.1.26) leads to the operator equivalent of Hamilton's equations given in Eq. (2.5.26), where $\hat{\vec{q}}(t)$ and $\hat{\vec{p}}(t)$ are then recognized as generalized coordinate and momentum operators. The differentiation between coordinates and momenta comes from considering the classical limit of these operators in the system being studied. So the expectation values of quantum operators will obey the same equations of motion as their classical counterparts and the Heisenberg picture Hamiltonian $\hat{H}_h(t)$ is the quantum version of the classical Hamiltonian $H(t)$. In other words, in taking the macroscopic limit for a quantum system with the quantum Hamiltonian defined by Eq. (4.1.9) and where the generalized canonical commutation relations are true, we will arrive at a classical system with a corresponding classical version of this Hamiltonian.

So given a classical system with some classical Hamiltonian $H(\vec{q}, \vec{p}, t)$ we can immediately try to define the corresponding quantum system by defining a corresponding Hamiltonian operator $\hat{H}_h(t) \equiv H(\hat{\vec{q}}_h(t), \hat{\vec{p}}_h(t), t)$ and imposing the generalized canonical commutation relations. This is the method of *canonical quantization* proposed by Paul A. M. Dirac in his PhD thesis in 1926 and described in detail in his book (Dirac, 2001). Note that if the classical Hamiltonian contains terms that mix coordinates and momenta (e.g., $\vec{p} \cdot \vec{q}$), then there will be more than one corresponding Hamiltonian operator (since $\hat{\vec{p}} \cdot \hat{\vec{q}} \neq \hat{\vec{q}} \cdot \hat{\vec{p}}$). However, for a given quantum Hamiltonian operator the classical Hamiltonian is uniquely defined.

Dirac referred to his canonical quantization procedure for ordinary quantum mechanics as *first quantization* and he referred to its application to fields to produce quantum field theory to describe many-particle systems as *second quantization*. In first quantization language we construct Fock space by symmetrizing for bosons (or antisymmetrizing for fermions) single particle states, whereas in second quantization language we formulate Fock space in term of an occupancy number basis. The canonical quantization approach for electromagnetism is discussed in Sec. 3.3.2, where the canonical commutation relations are given in Eq. (3.3.95). The canonical quantization approach became complicated in this case because electromagnetism is a gauge theory and as such is a singular system requiring the techniques of constrained Hamiltonian systems as discussed in Secs. 2.9 and 3.3.2.

The coordinate-space and momentum-space representations of the quantum mechanics of a single particle are briefly summarized along with a discussion of Fourier transforms in Sec. A.2.3 of Appendix A. In nonrelativistic quantum mechanics we use $\int d^3x \, |\mathbf{x}\rangle\langle\mathbf{x}| = \int d^3p \, |\mathbf{p}\rangle\langle\mathbf{p}| = \hat{I}$, $\langle\mathbf{p}'|\mathbf{p}\rangle = \delta^3(\mathbf{p}' - \mathbf{p})$, $\langle\mathbf{x}'|\mathbf{x}\rangle = \delta^3(\mathbf{x}' - \mathbf{x})$. Here $|\mathbf{x}\rangle$ and $|\mathbf{p}\rangle$ are the position and momentum eigenstates, respectively, i.e., $\hat{\mathbf{x}}|\mathbf{x}\rangle = \mathbf{x}|\mathbf{x}\rangle$ and $\hat{\mathbf{p}}|\mathbf{p}\rangle = \mathbf{p}|\mathbf{p}\rangle$ with $\hat{\mathbf{x}}$ and $\hat{\mathbf{p}}$ the corresponding observable operators. The coordinate-space "matrix" elements of position and momentum operators are

$$\langle \mathbf{x}|\hat{\mathbf{x}}|\mathbf{x}'\rangle = \mathbf{x}\delta^3(\mathbf{x}' - \mathbf{x}) \quad \text{and} \quad \langle \mathbf{x}|\hat{\mathbf{p}}|\mathbf{x}'\rangle = -i\hbar\boldsymbol{\nabla}_{\mathbf{x}}\delta^3(\mathbf{x}' - \mathbf{x}). \tag{4.1.27}$$

For some Hamiltonian operator $\hat{H}(t)$ of the form

$$\hat{H}(t) \equiv H(\hat{\mathbf{x}}, \hat{\mathbf{p}}, t) \tag{4.1.28}$$

it is readily verified that the coordinate-space matrix element is

$$\langle \mathbf{x}|\hat{H}(t)|\mathbf{x}'\rangle = \langle \mathbf{x}|H(\hat{\mathbf{x}}, \hat{\mathbf{p}}, t)|\mathbf{x}'\rangle = H\left(\mathbf{x}, (-i\hbar\boldsymbol{\nabla}_{\mathbf{x}}), t\right)\delta^3(\mathbf{x}' - \mathbf{x}) \tag{4.1.29}$$

so that the coordinate-space representation of the Schrödinger equation has its familiar form

$$\begin{aligned} i\hbar\frac{\partial}{\partial t}\psi(t, \mathbf{x}) = i\hbar\frac{\partial}{\partial t}\langle \mathbf{x}|\psi(t)\rangle &= \langle \mathbf{x}|\hat{H}(t)|\psi(t)\rangle = \int d^3x'\langle \mathbf{x}|\hat{H}(t)|\mathbf{x}'\rangle\langle \mathbf{x}'|\psi(t)\rangle \\ &= H\left(\mathbf{x}, (-i\hbar\boldsymbol{\nabla}_{\mathbf{x}}), t\right)\int d^3x'\delta^3(\mathbf{x}' - \mathbf{x})\langle \mathbf{x}'|\psi(t)\rangle \\ &= H\left(\mathbf{x}, (-i\hbar\boldsymbol{\nabla}), t\right)\psi(t, \mathbf{x}). \end{aligned} \tag{4.1.30}$$

Similarly, we have

$$\langle \mathbf{x}|\hat{\mathbf{p}}|\psi(t)\rangle = (-i\hbar\boldsymbol{\nabla})\,\psi(t, \mathbf{x}). \tag{4.1.31}$$

For a Hamiltonian operator of the form $\hat{H} = \hat{\mathbf{p}}^2/2m + V(t, \hat{\mathbf{x}})$ we arrive at

$$H\left(\mathbf{x}, (-i\hbar\boldsymbol{\nabla}_{\mathbf{x}}), t\right) = -(\hbar^2/2m)\boldsymbol{\nabla}^2 + V(t, \mathbf{x}). \tag{4.1.32}$$

Up until this point we have reserved the caret symbol " ˆ " to denote abstract operators in Hilbert space that would always be combined with Dirac notation bras $\langle\psi|$ and kets $|\psi\rangle$. Despite the potential confusion it can cause, it is very helpful in coordinate-space representation to also sometimes use the caret symbol to denote the differential-operator form of operators. We will understand the definitions

$$\hat{\mathbf{p}} \equiv -i\hbar\boldsymbol{\nabla} \quad \text{and} \quad \hat{H}(t) \equiv H\left(\mathbf{x}, (-i\hbar\boldsymbol{\nabla}_{\mathbf{x}}), t\right) \tag{4.1.33}$$

whenever we are working in the coordinate-space representation and we will understand the caret symbol to denote the abstract Hilbert space operators whenever we are working with Dirac bra-ket notation. With this understanding of coordinate-space notation we can write the time-dependent Schrödinger equation in coordinate-space representation as

$$i\hbar\frac{\partial}{\partial t}\psi(t, \mathbf{x}) = \hat{H}(t)\psi(t, \mathbf{x}), \tag{4.1.34}$$

where, e.g., we might have a coordinate-space Hamiltonian operator of the form

$$\hat{H}(t) = \hat{\mathbf{p}}^2/2m + V(t, \mathbf{x}) = -(\hbar^2/2m)\boldsymbol{\nabla}^2 + V(t, \mathbf{x}). \tag{4.1.35}$$

For a classical system that is monogenic and natural we have $H = E$ and, if there is no explicit time dependence, $H(\mathbf{x}, \mathbf{p}, t) = H(\mathbf{x}, \mathbf{p})$, then the energy E is conserved. When such a system is quantized we have $H(\mathbf{x}, \mathbf{p}) \to \hat{H} \equiv H(\hat{\mathbf{x}}, \hat{\mathbf{p}})$ and the eigenvalues and eigenvectors of $\hat{H}$ are the energy eigenvalues and energy eigenvectors, respectively. Quantization in quantum mechanics has the correspondences

$$\mathbf{x} \to \hat{\mathbf{x}}, \quad \mathbf{p} \to \hat{\mathbf{p}}, \quad \text{and} \quad H(\mathbf{x}, \mathbf{p}) = E \to \hat{H} = H(\hat{\mathbf{x}}, \hat{\mathbf{p}}), \tag{4.1.36}$$

where time evolution is determined by the time-dependent Schrödinger equation.

4.1.2 Notation, Linear and Antilinear Operators

It is useful to briefly review bra-ket notation, which was introduced in Sec. A.2.7 and which is discussed in detail in Messiah (1975) (see Chapter VII for linear operators and Chapter XV for antilinear operators).

Let $\hat{A}$ be an operator mapping the Hilbert space V_H to itself, $\hat{A} : V_H \to V_H$, and $|\psi\rangle, |\phi\rangle$ be two any vectors in V_H. We define the equivalent notations

$$\hat{A}|\psi\rangle \equiv |\hat{A}\psi\rangle, \quad |c\chi + d\eta\rangle \equiv c|\chi\rangle + d|\eta\rangle \quad \text{and} \quad \langle c\chi + d\eta| \equiv c^*\langle\chi| + d^*\langle\eta| \tag{4.1.37}$$

and if $|\psi\rangle = \hat{A}|\phi\rangle = |\hat{A}\phi\rangle$ then $\langle\psi| = \langle\hat{A}\phi|$. **Note well:** *We will understand that the following bra-ket notations (inner product notations) are equivalent*:

$$\langle\psi|\hat{A}\phi\rangle \equiv \langle\psi|(\hat{A}|\phi\rangle) \equiv \langle\psi|\hat{A}|\phi\rangle = \langle\hat{A}\phi, \psi\rangle = (\hat{A}\phi, \psi), \tag{4.1.38}$$

where the last two forms on the right-hand side are used by mathematicians with the arguments of the inner product in the opposite order to that used by physicists. Mathematicians would also typically not bother with using a caret to denote an operator, but physicists often do to avoid possible confusion with an operator in Hilbert space $\hat{A}$ with its matrix representation A in some particular Hilbert space basis. An advantage of the physics ordering is that it leads more directly to a matrix representation of quantum mechanics in a finite Hilbert space. For Hermitian operators $\langle\psi|\hat{A}|\phi\rangle$ is a natural notation, but for non-Hermitian and antilinear operators the notation $\langle\psi|\hat{A}\phi\rangle$ is arguably less confusing. We refer to $|\psi\rangle \in V_H$ as a *vector* or *ket* or *state* in Hilbert space and to $\langle\psi|$ as the *bra* or *dual vector* or *covector* in the dual Hilbert space. The ket $\langle\hat{A}\psi|$ is the dual vector of the bra $|\hat{A}\psi\rangle \equiv \hat{A}|\psi\rangle$.

A *linear operator* $\hat{A}_\ell$ and an *antilinear* operator $\hat{A}_a$ acting on the Hilbert space V_H have the properties

$$|\hat{A}_\ell(c\chi + d\eta)\rangle = |c\hat{A}_\ell\chi + d\hat{A}_\ell\eta\rangle \quad \text{and} \quad |\hat{A}_a(c\chi + d\eta)\rangle = |c^*\hat{A}_a\chi + d^*\hat{A}_a\eta\rangle, \tag{4.1.39}$$

respectively, for all $|\chi\rangle, |\eta\rangle \in V_H$ and all $c, d \in \mathbb{C}$. We can also write these as

$$\hat{A}_\ell(c|\chi\rangle + d|\eta\rangle) = c\hat{A}_\ell|\chi\rangle + d\hat{A}_\ell|\eta\rangle \text{ and } \hat{A}_a(c|\chi\rangle + d|\eta\rangle) = c^*\hat{A}_a|\chi\rangle + d^*\hat{A}_a|\eta\rangle. \tag{4.1.40}$$

Note that the square of an antilinear operator is linear, since

$$|\hat{A}_a^2(c\chi + d\eta)\rangle = |\hat{A}_a(c^*\hat{A}_a\chi + d^*\hat{A}_a\eta)\rangle = |c\hat{A}_a^2\chi + d\hat{A}_a^2\eta\rangle. \tag{4.1.41}$$

Similarly, the product of any two antilinear operators $\hat{A}_{1a}\hat{A}_{2a}$ will be linear. Clearly, the product of any number of linear and antilinear operators is linear if the number of antilinear operators is even and is antilinear if the number is odd.

Two operators $\hat{A}$ and $\hat{B}$ are by definition equivalent if $\langle\psi|\hat{A}\phi\rangle = \langle\psi|\hat{B}\phi\rangle$ for all $|\psi\rangle, |\phi\rangle \in \mathbb{C}$. However, we showed earlier in the discussion of Eq. (2.5.29) that

$$\langle\psi|\hat{A}\psi\rangle = \langle\psi|\hat{B}\psi\rangle \quad \text{for all} \quad |\psi\rangle \in V_H \quad \text{if and only if} \quad \hat{A} = \hat{B}. \tag{4.1.42}$$

Another useful result is that two linear operators $\hat{A}$ and $\hat{B}$ satisfy

$$|\langle\phi|\hat{A}\psi\rangle| = |\langle\phi|\hat{B}\psi\rangle| \quad \text{for all} \quad |\phi\rangle, |\psi\rangle \in V_H \quad \text{if and only if} \quad \hat{A} = \hat{B}e^{i\alpha}, \tag{4.1.43}$$

i.e., if and only if the two operators $\hat{A}$ and $\hat{B}$ are equal up to some overall phase. This is proved as Theorem II in Chapter XV of Messiah (1975).

The *Hermitian conjugate* or *adjoint*, $\hat{A}_\ell^\dagger$, of a linear operator $\hat{A}_\ell$ acting on V_H by definition has the property that

$$\langle \hat{A}_\ell^\dagger \psi | \phi \rangle \equiv \langle \psi | \hat{A}_\ell \phi \rangle \quad \text{for all } |\psi\rangle, |\phi\rangle \in V_H. \tag{4.1.44}$$

Since $\langle \psi | \phi \rangle = \langle \phi | \psi \rangle^*$, it follows that $\langle \hat{A}_\ell \phi | \psi \rangle = \langle \phi | \hat{A}_\ell^\dagger \psi \rangle = \langle A_\ell^{\dagger\dagger} \phi | \psi \rangle$ for all $|\psi\rangle, |\phi\rangle \in V_H$ and so $\hat{A}_\ell^{\dagger\dagger} = \hat{A}_\ell$. It follows from this definition that $(c\hat{A})^\dagger = c^*\hat{A}^\dagger$, $(\hat{A} + \hat{B})^\dagger = \hat{A}^\dagger + \hat{B}^\dagger$, and $(\hat{A}\hat{B})^\dagger = \hat{B}^\dagger \hat{A}^\dagger$. Let $|\alpha\rangle, |\beta\rangle \in V_H$, then we can define a linear operator

$$\hat{A} = |\alpha\rangle\langle\beta| \quad \text{such that} \quad \langle \psi | \hat{A} \phi \rangle = \langle \psi | \alpha \rangle \langle \beta | \phi \rangle \quad \text{and} \quad \hat{A}^\dagger = |\beta\rangle\langle\alpha|. \tag{4.1.45}$$

The last result follows since if and only if $\hat{A}^\dagger = |\beta\rangle\langle\alpha|$ will we have $|\hat{A}^\dagger \psi\rangle = |\beta\rangle\langle\alpha|\psi\rangle$ and $\langle \hat{A}^\dagger \psi | \phi \rangle = \langle \alpha | \psi \rangle^* \langle \beta | \phi \rangle = \langle \psi | \hat{A} \phi \rangle$ for all $|\psi\rangle, |\phi\rangle \in V_H$ as required.

The Hermitian conjugate or adjoint of an antilinear operator, $\hat{A}_a$, needs to be defined so as to account for the antilinear behavior,

$$\langle \hat{A}_a^\dagger \psi | \phi \rangle \equiv \langle \psi | \hat{A}_a \phi \rangle^* = \langle \hat{A}_a \phi | \psi \rangle \Rightarrow \langle \phi | \hat{A}_a^\dagger \psi \rangle = \langle \hat{A}_a \phi | \psi \rangle^* = \langle \psi | \hat{A}_a \phi \rangle \tag{4.1.46}$$

for all $|\psi\rangle, |\phi\rangle \in V_H$. Writing $|\psi\rangle = |c\chi + d\eta\rangle$ it is easily verified that this definition implies that $\hat{A}_a^\dagger$ must also be antilinear. Note that $\langle \hat{A}_a \phi | \psi \rangle = \langle \hat{A}_a^\dagger \psi | \phi \rangle = \langle A_a^{\dagger\dagger} \phi | \psi \rangle$ for all $|\psi\rangle, |\phi\rangle \in V_H$ and so $\hat{A}_a^{\dagger\dagger} = \hat{A}_a$.

A *unitary operator* $\hat{U}$ is a linear operator with the property that

$$\langle \hat{U}\phi | \hat{U}\psi \rangle = \langle \phi | \psi \rangle \tag{4.1.47}$$

for every $|\phi\rangle, |\psi\rangle \in V_H$. It follows that $\langle \hat{U}\phi | \hat{U}\psi \rangle = \langle \phi | \hat{U}^\dagger \hat{U}\psi \rangle = \langle \phi | \psi \rangle$ for all $|\psi\rangle, |\phi\rangle$ so that $\hat{U}^\dagger \hat{U} = \hat{U}\hat{U}^\dagger = \hat{I}$ and hence $\hat{U}^\dagger = \hat{U}^{-1}$. Recall from Sec. A.8 that an invertible operator (or matrix) commutes with its own inverse. An *antiunitary operator* $\hat{U}_a$ is an antilinear operator with the property that

$$\langle \hat{U}_a\phi | \hat{U}_a\psi \rangle = \langle \phi | \psi \rangle^* = \langle \psi | \phi \rangle \tag{4.1.48}$$

for every $|\phi\rangle, |\psi\rangle \in V_H$. Then $\langle \psi | \phi \rangle = \langle \hat{U}_a\phi | \hat{U}_a\psi \rangle = \langle \hat{U}_a^\dagger \hat{U}_a\psi | \phi \rangle$ for all $|\psi\rangle, |\phi\rangle$ so

$$\hat{U}_a^\dagger \hat{U}_a = \hat{U}_a\hat{U}_a^\dagger = \hat{I} \quad \text{and hence} \quad \hat{U}_a^\dagger = \hat{U}_a^{-1}. \tag{4.1.49}$$

Replacing $|\phi\rangle$ with $|\hat{U}_a^\dagger \chi\rangle$ in Eq. (4.1.48) gives

$$\langle \chi | \hat{U}_a\psi \rangle = \langle \hat{U}_a^\dagger \chi | \psi \rangle^* = \langle \psi | \hat{U}_a^\dagger \chi \rangle. \tag{4.1.50}$$

Let $\hat{A}$ be a linear operator and $|\phi'\rangle = \hat{U}_a|\phi\rangle$, $|\psi'\rangle = \hat{U}_a|\psi\rangle$, then we have the result

$$\langle \phi' | \hat{A}\psi' \rangle = \langle \hat{A}^\dagger \phi' | \psi' \rangle = \langle \hat{A}^\dagger \hat{U}_a\phi | \hat{U}_a\psi \rangle = \langle \psi | \hat{U}_a^\dagger \hat{A}^\dagger \hat{U}_a\phi \rangle = \langle \psi | (\hat{U}_a^\dagger \hat{A}\hat{U}_a)^\dagger \phi \rangle. \tag{4.1.51}$$

Representations in quantum mechanics: When we talk of a representation of a quantum system we are talking of a particular choice of orthonormal basis B with basis vectors/kets $|b_i\rangle$ assuming that the Hilbert space V_H has a finite or countable basis. We have already seen how to generalize this to an uncountably infinite-dimensional Hilbert space and a basis with a continuous index when we discussed the coordinate-space representation. An orthonormal basis by definition satisfies $\langle b_i | b_j \rangle = \delta_{ij}$ and $\sum_i |b_i\rangle\langle b_i| = \hat{I}$ with $\hat{I}$ the identity operator in V_H. We typically choose an orthonormal basis consisting of the simultaneous eigenstates of one or more commuting observables. For some linear operator $\hat{A}$ we write

$$\langle \phi | \hat{A}\psi \rangle = \sum_{ij} \langle \phi | b_i \rangle \langle b_i | \hat{A} | b_j \rangle \langle b_j | \psi \rangle = \sum_{ij} d_i^* A_{ij} c_j = \vec{d}^\dagger A \vec{c}, \tag{4.1.52}$$

where $A_{ij} \equiv \langle b_i|\hat{A}|b_j\rangle \equiv \langle b_i|\hat{A}b_j\rangle$, $c_i \equiv \langle b_i|\psi\rangle$ and $d_i \equiv \langle b_i|\phi\rangle$. So the matrix A represents the linear operator $\hat{A}$ in this basis B and similarly the column vectors $\vec{c}$ and $\vec{d}$ represent the kets $|\psi\rangle$ and $|\phi\rangle$, respectively. Consider some other orthonormal basis B' with basis vectors/kets $|b_i'\rangle$. Then we have

$$\langle\phi|\hat{A}\psi\rangle = \sum_{ij}\langle\phi|b_i\rangle\langle b_i|\hat{A}|b_j\rangle\langle b_j\psi\rangle = \sum_{ijk\ell}\langle\phi|b_i\rangle\langle b_i|b_k'\rangle\langle b_k'|\hat{A}|b_\ell'\rangle\langle b_\ell'|b_j\rangle\langle b_j\psi\rangle$$
$$= \sum_{ijk\ell}d_i^*(U^\dagger)_{ik}A_{k\ell}'U_{\ell j}c_j = \vec{d}^\dagger U^\dagger A'U\vec{c} = \vec{d}'^\dagger A'\vec{c}' = \vec{d}^\dagger A\vec{c}, \quad (4.1.53)$$

where $A_{ij}' \equiv \langle b_i'|\hat{A}|b_j'\rangle$, $U_{ij} \equiv \langle b_i'|b_j\rangle$, $\vec{c}' \equiv U\vec{c}$ and $\vec{d}' \equiv U\vec{d}$. Clearly, the change of basis matrix U is unitary since $\sum_k(U^\dagger)_{ik}U_{kj} = \sum_k\langle b_i|b_k'\rangle\langle b_k'|b_j\rangle = \delta_{ij}$. As we have seen $\langle\psi|\hat{A}_1\psi\rangle = \langle\psi|\hat{A}_2\psi\rangle$ for all $|\psi\rangle$ if and only if $\hat{A}_1 = \hat{A}_2$. Now $A_1 = A_2$, or equivalently $A_1 - A_2 = 0$, in one representation means that $\vec{c}^\dagger A_1\vec{c} = \vec{c}^\dagger A_2\vec{c}$ for every $\vec{c}$, which occurs if and only if $\hat{A}_1 = \hat{A}_2$. More generally, if we have $f(A_1, A_2, A_3, \ldots) = 0$ in one representation, then in some other representation, $A_i' = U^\dagger A_i U$, we will have $f(A_1', A_2', A_3', \ldots) = 0$ if and only if $U^\dagger f(A_1, A_2, A_3, \ldots)U = f(A_1', A_2', A_3', \ldots)$. This will be true if $f(A_1', A_2', A_3', \ldots)$ contains combinations of sums, products and powers of the matrices A_i and $A_i^\dagger$. Note that $U^\dagger A^\dagger U = (U^\dagger AU)^\dagger$ and so, e.g., $U^\dagger(A_1 A_2^\dagger A_3)U = A_1'A_2'^\dagger A_3'$. In that case it follows that the equation is true in every representation and hence at the operator level, $f(\hat{A}_1, \hat{A}_2, \hat{A}_3, \ldots) = 0$. In contrast, note that $U^\dagger A^*U \neq (U^\dagger AU)^*$ and $U^\dagger A^TU \neq (U^\dagger AU)^T$ and so equations involving A_i^* and A_i^T are not representation independent. The Fourier transform discussed in Sec. A.2.3 is a change of continuous basis from the coordinate-space basis $|\mathbf{x}\rangle$ to the momentum-space basis $|\mathbf{p}\rangle$ and is a continuous version of a unitary transformation.

Complex conjugation operator (Messiah, 1975): We now turn to the definition of the complex conjugation operator, whose action depends on the basis being used. This is not surprising since a change of basis is a unitary transformation and the complex conjugation of a matrix or vector is not invariant under unitary transformations as discussed above. We *define* $\hat{K}_B$ to be the antilinear operator that leaves the basis B invariant,

$$\hat{K}_B|b_i\rangle = |b_i\rangle \quad \text{and} \quad \hat{K}_B|c\chi + d\eta\rangle = |c^*\hat{K}_B\chi + d^*\hat{K}_B\eta\rangle \quad (4.1.54)$$

for all $i = 1, 2, \ldots$ and for all $c, d \in \mathbb{C}$ and $|\chi\rangle, |\eta\rangle \in V_H$. For $|\psi\rangle$ we have

$$|\psi\rangle = \sum_i|b_i\rangle\langle b_i|\psi\rangle = \sum_i c_i|b_i\rangle \quad \text{and} \quad |\psi'\rangle \equiv \hat{K}_B|\psi\rangle = \sum_i c_i^*|b_i\rangle, \quad (4.1.55)$$

with $c_i = \langle b_i|\psi\rangle$. So in this representation $|\psi\rangle$ is represented by a column vector with elements c_i and $|\psi'\rangle = \hat{K}_B|\psi\rangle$ is represented by the complex conjugate column vector with elements c_i^*. Let $|\psi\rangle = \sum_i c_i|b_i\rangle$ and $|\phi\rangle = \sum_j d_j|b_j\rangle$, then

$$\langle\hat{K}_B\phi|\hat{K}_B\psi\rangle = \sum_{i,j}d_j c_i^*\langle b_j|b_i\rangle = \sum_i c_i^*d_i = \langle\psi|\phi\rangle \quad (4.1.56)$$

and so $\hat{K}_B$ is antiunitary, $\hat{K}_B^\dagger = \hat{K}_B^{-1}$. Note also that $\hat{K}_B^2|\psi\rangle = |\psi\rangle$ for all $|\psi\rangle$ so that $\hat{K}_B^2 = \hat{I}$. Then $\hat{K}_B$ is a self-inverse, antiunitary and Hermitian operator,

$$\hat{K}_B = \hat{K}_B^{-1} = \hat{K}_B^\dagger \quad \text{and so} \quad \hat{K}_B^2 = \hat{I}. \quad (4.1.57)$$

Recall that for any linear operator $\hat{A}$ in this representation we have

$$\hat{A} = \sum_{i,j}|b_i\rangle\langle b_i|\hat{A}|b_j\rangle\langle b_j| = \sum_{i,j}|b_i\rangle A_{ij}\langle b_j| \quad \text{with} \quad A_{ij} \equiv \langle b_i|\hat{A}b_j\rangle \quad (4.1.58)$$

so that the matrix A represents the operator $\hat{A}$ in this representation B. We observe that for the linear operator $\hat{K}_B \hat{A} \hat{K}_B$ we have

$$\hat{K}_B \hat{A} \hat{K}_B = \sum_{i,j} |b_i\rangle\langle b_i|\hat{K}_B \hat{A} \hat{K}_B|b_j\rangle\langle b_j| = \sum_{i,j} |b_i\rangle A_{ij}^* \langle b_j| \quad \text{since} \qquad (4.1.59)$$

$$\langle b_i|\hat{K}_B \hat{A} \hat{K}_B b_j\rangle = \langle b_i|\hat{K}_B \hat{A} b_j\rangle = \sum_m \langle b_i|\hat{K}_B(|b_m\rangle\langle b_m|\hat{A} b_j\rangle) = \sum_m \langle b_i|b_m\rangle A_{mj}^* = A_{ij}^*,$$

since $\hat{K}_B$ acts on everything to its right, $\hat{K}_B(|b_m\rangle A_{mj}) = |b_m\rangle A_{mj}^*$. In summary, the action of $\hat{K}_B$ is to *complex conjugate* the column vectors and matrices in the B representation of the Hilbert space V_H. Its action on column vectors and matrices in other representations arising from different basis choices will be different.

Since $\hat{K}_B$ is antiunitary and hence antilinear, then if $\hat{A}_\ell$ is any linear operator it must be that $\hat{K}_B \hat{A}_\ell$ and $\hat{A}_\ell \hat{K}_B$ are antilinear operators. Conversely, if $\hat{A}_a$ is any antilinear operator, then $\hat{K}_B \hat{A}_a$ and $\hat{A}_a \hat{K}_B$ are linear operators. If $\hat{U}$ is a unitary operator, then $\hat{K}_B \hat{U}$ and $\hat{U} \hat{K}_B$ are antiunitary. Similarly if $\hat{U}_a$ is antiunitary, then $\hat{K}_B \hat{U}_a$ and $\hat{U}_a \hat{K}_B$ are unitary. These results follow since

$$\langle \hat{K}_B \hat{U}\phi|\hat{K}_B \hat{U}\psi\rangle = \langle \hat{U}\psi|\hat{U}\phi\rangle = \langle\psi|\phi\rangle, \quad \langle \hat{U}\hat{K}_B\phi|\hat{U}\hat{K}_B\psi\rangle = \langle \hat{K}_B\phi|\hat{K}_B\psi\rangle = \langle\psi|\phi\rangle,$$

$$\langle \hat{K}_B \hat{U}_a\phi|\hat{K}_B \hat{U}_a\psi\rangle = \langle \hat{U}_a\psi|\hat{U}_a\phi\rangle = \langle\phi|\psi\rangle, \quad \langle \hat{U}_a\hat{K}_B\phi|\hat{U}_a\hat{K}_B\psi\rangle = \langle \hat{K}_B\psi|\hat{K}_B\phi\rangle = \langle\phi|\psi\rangle.$$

Since $\hat{K}_B^2 = \hat{I}$ then for *any* antilinear operator $\hat{A}_a$ and any basis choice B for the complex conjugation operator $\hat{K}_B$, we can construct a linear operator $\hat{A}_{\ell B} \equiv (\hat{A}_a \hat{K}_B)$ or $\hat{A}_{\ell B} \equiv (\hat{K}_B \hat{A}_a)$, such that we have, respectively,

$$\hat{A}_a = \hat{A}_{\ell B} \hat{K}_B \quad \text{or} \quad \hat{A}_a = \hat{K}_B \hat{A}_{\ell B}. \qquad (4.1.60)$$

Similarly, *any* antiunitary operator $\hat{U}_a$ can be written as the product of a linear unitary operator $\hat{U}_B \equiv (\hat{U}_a \hat{K}_B)$ or $\hat{U}_B \equiv (\hat{K}_B \hat{U}_a)$ and a complex conjugation operator $\hat{K}_B$, giving, respectively,

$$\hat{U}_a = \hat{U}_B \hat{K}_B \quad \text{or} \quad \hat{U}_a = \hat{K}_B \hat{U}_B. \qquad (4.1.61)$$

4.1.3 Symmetry Transformations and Wigner's Theorem

Recall from postulate (i) that the state of a quantum system is described by a ray or normalized ket, $|\psi\rangle/\sqrt{\langle\psi|\psi\rangle}$, in an appropriate Hilbert space V_H, where the norm of the vector $|\psi\rangle$ is $\|\psi\| = |\langle\psi|\psi\rangle|^{1/2}$. The outcome of the measurement of any observable $\hat{A}$ is determined by the quantities $|c_i| = |\langle a_i|\psi\rangle|$ as described in postulate (iv). More generally, any observable quantity can only ever depend on magnitudes of inner products, $|\langle\phi|\psi\rangle|$.

Any transformation that leaves $|\langle\phi|\psi\rangle|$ invariant for every $|\phi\rangle, |\psi\rangle \in V_H$ leaves the physical state of the system invariant. Such a transformation is a symmetry of the quantum system, but since it has nothing to do with the dynamics of the quantum system, it is not a symmetry in the familiar sense of Noether's theorem and conserved currents and charges. More precisely, a transformation $\hat{B} : V_H \to V_H$ such that $|\psi'\rangle = \hat{B}|\psi\rangle = |\hat{B}\psi\rangle$ is a *symmetry (in the sense of Wigner)* if

$$|\langle\phi'|\psi'\rangle| = |\langle\hat{B}\phi|\hat{B}\psi\rangle| = |\langle\phi|\psi\rangle| \qquad (4.1.62)$$

for every $|\phi\rangle, |\psi\rangle \in V_H$. It is obvious from Eqs. (4.1.47) and (4.1.48) that both unitary and antiunitary transformations are symmetries. We will later discuss how this Wigner concept of symmetry is related to a dynamical symmetry of a quantum system when time evolution is included.

Recall from the postulates of quantum mechanics in Sec. 4.1 that the state of a quantum system is determined by a ray in the corresponding Hilbert space. A ray can be represented as a unit vector in that Hilbert space and so transformations of the state of a system must be norm-preserving transformations.

Theorem: A linear operator $\hat{A}$ preserves the norm in a Hilbert space if and only if it preserves the inner product. An antilinear operator $\hat{B}$ preserves the norm if and only if it complex conjugates the inner product. In other words, for a linear operator $\hat{A}$ we have $\|\hat{A}\psi\|^2 = \|\psi\|^2$ for all $|\psi\rangle \in V_H$ if and only if $\langle \hat{A}\psi|\hat{A}\phi\rangle = \langle\psi|\phi\rangle$ for all $|\psi\rangle, |\phi\rangle \in V_H$. For an antilinear operator $\hat{B}$ we have $\|\hat{B}\psi\|^2 = \|\psi\|^2$ for all $|\psi\rangle \in V_H$ if and only if $\langle\hat{B}\psi|\hat{B}\phi\rangle = \langle\psi|\phi\rangle^* = \langle\phi|\psi\rangle$ for all $|\psi\rangle, |\phi\rangle \in V_H$. Therefore, any linear or antilinear operator that preserves the norm obviously preserves the magnitude of the inner product.

Proof: First, recall that $\|\psi\|^2 \equiv \langle\psi|\psi\rangle$. If $\langle\hat{C}\psi|\hat{C}\phi\rangle = \langle\psi|\phi\rangle$ or $\langle\phi|\psi\rangle$ for all $\psi, \phi \in V_H$, then $\|\hat{C}\psi\|^2 = \|\psi\|^2$ for all $|\psi\rangle \in V_H$. Second, we note that

$$\frac{1}{4}\left(\|\psi+\phi\|^2 - \|\psi-\phi\|^2 - i\|\psi+i\phi\|^2 + i\|\psi-i\phi\|^2\right)$$
$$= \frac{1}{4}\left(4\mathrm{Re}\langle\psi|\phi\rangle + 4\mathrm{Im}\langle\psi|\phi\rangle\right) = \langle\psi|\phi\rangle. \tag{4.1.63}$$

For $\hat{A}$ linear $\|\hat{A}(\psi \pm \phi)\| = \|(\hat{A}\psi) \pm (\hat{A}\phi)\|$ and $\|\hat{A}(\psi \pm i\phi)\| = \|(\hat{A}\psi) \pm i(\hat{A}\phi)\|$. Then if $\|\hat{A}\psi\|^2 = \|\psi\|^2$ for all $|\psi\rangle \in V_H$ we have from the above result that $\langle\hat{A}\psi|\hat{A}\phi\rangle = \langle\psi|\phi\rangle$ for all $|\psi\rangle, |\phi\rangle \in V_H$. We have $\|\hat{B}(\psi \pm \phi)\| = \|(\hat{B}\psi) \pm (\hat{B}\phi)\|$ and $\|\hat{B}(\psi \pm i\phi)\| = \|(\hat{B}\psi) \mp i(\hat{B}\phi)\|$ for an antilinear operator. Then if $\|\hat{B}\psi\|^2 = \|\psi\|^2$ for all $|\psi\rangle \in V_H$ we have from the above result that $\langle\hat{B}\psi|\hat{B}\phi\rangle = \langle\psi|\phi\rangle^*$ for all $|\psi\rangle, |\phi\rangle \in V_H$.

Wigner's theorem: Any transformation that preserves the magnitude of the inner product in a Hilbert space of two or more dimensions ($\dim V_H \geq 2$) is either a unitary or an antiunitary operator. In the one-dimensional case ($\dim V_H = 1$) there is both a unitary and antiunitary operator corresponding to any such transformation. (Proofs are given in Bargmann (1964), in Appendix A of Chapter 2 of Weinberg (1995) and as the proof of Theorem III in Chapter XV of Messiah (1975).)

Summary: By definition $\hat{B}$ is a symmetry transformation (in the sense of Wigner) if it satisfies $|\langle\hat{B}\phi|\hat{B}\psi\rangle| = |\langle\phi|\psi\rangle|$ for all $|\psi\rangle, |\phi\rangle \in V_H$. As a result of Wigner's theorem we see that $\hat{B}$ must be either unitary or antiunitary; i.e., either $\hat{B} = \hat{U}$ for some unitary $\hat{U}$ or $\hat{B} = \hat{U}_a$ for some antiunitary $\hat{U}_a$. Clearly, if $\hat{U}$ is a unitary symmetry, then so is $\hat{U}' = \hat{U}e^{i\alpha} = e^{i\alpha}\hat{U}$ and if $\hat{U}_a$ is an antiunitary symmetry then so is $\hat{U}'_a = \hat{U}_a e^{i\alpha} = e^{-i\alpha}\hat{U}_a$, where α is any arbitrary constant phase. It is easily verified from the definition of the Hermitian conjugate that

$$(\hat{A}_1\hat{A}_2\cdots\hat{A}_n)^\dagger = \hat{A}_n^\dagger\cdots\hat{A}_2^\dagger\hat{A}_1^\dagger \tag{4.1.64}$$

for any combination of linear and/or antilinear operators. Hence we have $\hat{U}'^\dagger_a = e^{-i\alpha}\hat{U}_a^\dagger = \hat{U}_a^\dagger e^{+i\alpha}$.

4.1.4 Projective Representations of Symmetry Groups

We give a brief discussion of projective representation since it is related to Wigner's theorem and the classification of representations of symmetry groups in physics. Also see Sec. 2.7 of Weinberg (1995) and Sundermeyer (2014) and references therein.

Consider a symmetry group G acting at the level of classical physics with group elements $g_1, g_2, g_3 \in G$ such that $g_1 g_2 = g_3$. If G is a symmetry at the classical level, then it is a symmetry of expectation values of the corresponding quantum system. A unitary representation of the group G has elements $\hat{U}(g)$ with $\hat{U}(g)^\dagger = \hat{U}(g)^{-1}$ such that $\hat{U}(g_1)\hat{U}(g_2) = \hat{U}(g_1g_2) = \hat{U}(g_3)$. For expectation values of a quantum system to be symmetric under G it is sufficient that

$$\hat{U}(g_1)\hat{U}(g_2) = e^{i\phi(g_1,g_2)}\hat{U}(g_3) \Rightarrow \hat{U}(g_1)\hat{U}(g_2)|\psi\rangle = \hat{U}(g_3)(e^{i\phi(g_1,g_2)}|\psi\rangle), \qquad (4.1.65)$$

where $\phi(g_1, g_2) \in \mathbb{R}$ is some phase. Both $|\psi\rangle$ and $e^{i\phi(g_1,g_2)}|\psi\rangle$ correspond to the same physical state/ray; i.e., they give the same expectation values. Any representation with this property such that $\phi \neq 0$ is called a *projective representation*. Obviously, the phase ϕ must satisfy the associativity property of a group and so for any three group elements we must have

$$[\hat{U}(g_1)\hat{U}(g_2)]\hat{U}(g_3) = \hat{U}(g_1)[\hat{U}(g_2)\hat{U}(g_3)]$$
$$\Rightarrow \quad \phi(g_1, g_2) + \phi(g_1g_2, g_3) = \phi(g_1, g_2g_3) + \phi(g_2, g_3). \qquad (4.1.66)$$

This means that finding projective representations is not a straightforward matter. If the phase has the form $\phi(g_1, g_2) = \alpha(g_3) - \alpha(g_1) - \alpha(g_2) = \alpha(g_1g_2) - \alpha(g_1) - \alpha(g_2)$ with $\alpha(g) \in \mathbb{R}$ being some phase assigned to each group element, then each side of the above phase equation becomes $\alpha(g_1g_2g_3) - \alpha(g_1) - \alpha(g_2) - \alpha(g_3)$ and so the condition is satisfied. In that case we redefine $\hat{U}'(g) \equiv \hat{U}(g)e^{i\alpha(g)}$ and so find that $\hat{U}(g_1)\hat{U}(g_2) = e^{i\phi(g_1,g_2)}\hat{U}(g_3) = e^{i[\alpha(g_3)-\alpha(g_1)-\alpha(g_2)]}\hat{U}(g_3)$, which leads to $\hat{U}'(g_1)\hat{U}'(g_2) = \hat{U}'(g_3)$. This means that we have recovered the initial group structure but now with $\phi = 0$. When the phase ϕ can be eliminated by appropriate redefinitions, then we say that we have a *linear representation* of the group and, for reasons that we do not attempt to explain here, we say that all possible phases are in the "trivial two-cocycle." When we cannot remove the phase ϕ by redefinitions, then we say that we have an *intrinsically projective* representation. Obviously, for simplicity we want to find ways of avoiding having to deal with intrinsically projective representations of symmetry groups wherever possible.

For a Lie group G the phases will impact on its Lie algebra $\mathfrak{g}$. For elements near the identity we have $\hat{U}(g) = \exp\{i\omega^a\hat{T}^a\}$, where $\hat{T}^a$ are the Hermitian generators satisfying the Lie algebra of the group in this representation. Let e be the identity element of the group, $eg = ge = g$ for all $g \in G$, then $\hat{U}(e) = \hat{I}$ is the representation of the identity. Clearly, $\hat{U}(e)\hat{U}(g) = \hat{I}\hat{U}(g) = \hat{U}(g)$ and similarly $\hat{U}(g)\hat{U}(e) = \hat{U}(g)$ and so from Eq. (4.1.65) we have $\phi(e, g) = \phi(g, e) = 0$ for all $g \Rightarrow \phi(\omega_1, 0) = \phi(0, \omega_2) = 0$. For any two elements near the identity ϕ is small and so in that case we can perform a Taylor expansion of the function ϕ and write

$$\phi(g_1, g_2) \equiv \phi(\omega_1, \omega_2) = \phi^{ab}\omega_1^a\omega_2^b + \mathcal{O}(\omega^3). \qquad (4.1.67)$$

The Lie algebra then becomes modified and takes the form

$$[\hat{T}^a, \hat{T}^b] = if^{abc}\hat{T}^c + i\Phi^{ab}\hat{I}, \qquad (4.1.68)$$

where we define $\Phi^{ab} \equiv -\phi^{ab} + \phi^{ba}$ and these are referred to as *central charges*.

Proof: First consider the standard case with $\phi = 0$. For an element g sufficiently close to the identity we can write

$$\hat{U}(g) \equiv \hat{U}(\omega) = \hat{I} + i\omega^a\hat{T}^a - \tfrac{1}{2}\omega^a\omega^b\hat{T}^{ab} + \cdots \qquad (4.1.69)$$

for some small ω^a and for Hermitian operators $\hat{T}^a$, $\hat{T}^{ab}$ and so on. Note that $\hat{T}^{ab} = \hat{T}^{ba}$. For two group elements g_1 and g_2 near the identity we have

$$\hat{U}(g_1)\hat{U}(g_2) = \hat{U}(\omega_1)\hat{U}(\omega_2) = \hat{U}(f(\omega_1, \omega_2)) = \hat{U}(g_1g_2), \qquad (4.1.70)$$

where ω_1^a, ω_2^a, $f^a(\omega_1, \omega_2)$ are small. The function $f(\omega_1, \omega_2)$ defines the group structure near the identity. Since $\hat{U}(e)\hat{U}(g) = \hat{U}(g)\hat{U}(e) = \hat{U}(g)$, then we have $\hat{U}(0)\hat{U}(\omega) = \hat{U}(f(0,\omega)) = \hat{U}(\omega)$ and $\hat{U}(\omega)\hat{U}(0) = \hat{U}(f(\omega,0)) = \hat{U}(\omega)$, which gives $f^a(0,\omega) = f^a(\omega,0) = \omega^a$. Taylor expanding $f(\omega_1, \omega_2)$ gives

$$f^a(\omega_1, \omega_2) \equiv \omega_1^a + \omega_2^a + c^{abc}\omega_1^b\omega_2^c + \cdots . \tag{4.1.71}$$

Expanding Eq. (4.1.70) and equating terms in powers of ω we find

$$(\hat{I} + i\omega_1^a\hat{T}^a - \tfrac{1}{2}\omega_1^a\omega_1^b\hat{T}^{ab} + \cdots)(\hat{I} + i\omega_2^c\hat{T}^c - \tfrac{1}{2}\omega_2^c\omega_2^d\hat{T}^{cd} + \cdots) \tag{4.1.72}$$
$$= \hat{I} + i(\omega_1^a + \omega_2^a + c^{abc}\omega_1^b\omega_2^c + \cdots)\hat{T}^a$$
$$- \tfrac{1}{2}(\omega_1^a + \omega_2^a + \cdots)(\omega_1^b + \omega_2^b + \cdots)\hat{T}^{ab} + \cdots .$$

The $\mathcal{O}(1)$, $\mathcal{O}(\omega_1)$, $\mathcal{O}(\omega_2)$, $\mathcal{O}((\omega_1)^2)$ and $\mathcal{O}((\omega_2)^2)$ terms are seen to be equal on each side of the equation. Equating the $\mathcal{O}(\omega_1\omega_2)$ terms using $\hat{T}^{ab} = \hat{T}^{ba}$ gives

$$\hat{T}^a\hat{T}^b = \hat{T}^{ab} - ic^{cab}\hat{T}^c \quad \Rightarrow \quad 0 = \hat{T}^{ab} - \hat{T}^{ba} = [\hat{T}^a, \hat{T}^b] + i(c^{cab} - c^{cba})\hat{T}^c$$
$$\Rightarrow \quad [\hat{T}^a, \hat{T}^b] = i(-c^{cab} + c^{cba})\hat{T}^c \equiv if^{abc}\hat{T}^c, \tag{4.1.73}$$

where $f^{abc} = -f^{bac}$. So we have derived the Lie algebra given in Eq. (A.7.5).

The generalization to the projective unitary representation proceeds similarly. In place of Eq. (4.1.70) we have

$$\hat{U}(\omega_1)\hat{U}(\omega_2) = e^{i\phi(\omega_1,\omega_2)}\hat{U}(f(\omega_1, \omega_2)), \tag{4.1.74}$$

which adds $i\omega_1^a\omega_2^b\phi^{ab}$ to the right-hand side of Eq. (4.1.72). This then gives

$$\hat{T}^a\hat{T}^b = \hat{T}^{ab} - ic^{cab}\hat{T}^c - i\phi^{ab}\hat{I} \tag{4.1.75}$$
$$\Rightarrow \quad [\hat{T}^a, \hat{T}^b] = i(-c^{cab} + c^{cba})\hat{T}^c + i(-\phi^{ab} + \phi^{ba})\hat{I} \equiv if^{abc}\hat{T}^c + i\Phi^{ab}\hat{I}.$$

It is sometimes possible to redefine the generators to remove the central charges Φ^{ab} and so restore the Lie algebra to its original form. For example, if the central charges have the form $\Phi^{ab} = f^{abc}r^c$, where r^c is any set of real constants, then we can redefine the generators such that

$$\hat{T}'^a \equiv \hat{T}^a + r^a\hat{I} \quad \Rightarrow \quad [\hat{T}'^a, \hat{T}'^b] = [\hat{T}^a, \hat{T}^b] = if^{abc}\hat{T}^c + i\Phi^{ab}\hat{I} = if^{abc}\hat{T}'^c. \tag{4.1.76}$$

If the central charges cannot be redefined away, we say that the Lie algebra has a nontrivial central extension. If the central charges can be redefined away, then the Lie group and its projective representation are locally the same (locally isomorphic), but the group and its projective representations may differ at the global level through the topology of the group manifold. There is a very useful theorem with a proof provided in Appendix B of Chapter 2 of Weinberg (1995), where the essential step in the proof is the use of the Poincaré lemma in a simply connected manifold.

Theorem: There are no intrinsic projective unitary representations of a Lie group G if (i) the generators of the Lie algebra of the group can be redefined in such a way as to eliminate central charges and (ii) the group is simply connected.

The first requirement ensures that the Lie agebra of the group is preserved in the representation. The second requirement means that any two elements of the group can be connected by a continuous path of elements and that the path can be continuously deformed into any other path while staying in the group manifold. The effect of this is to ensure that there are no topological obstructions to globally redefining away the phase ϕ. In mathematical language, the first requirement in the above theorem can also be expressed as the condition that there are no nontrivial central extensions of G and also as the condition that the two-dimensional cohomology of G is trivial. This brings us to another very useful theorem.

Bargmann's theorem: If the two-dimensional cohomology of a Lie group G is trivial (meaning if requirement (i) is met), then all projective unitary representations (meaning linear plus intrinsic projective representations) of G are linear unitary representations of the universal covering group $\tilde{G}$ of G. *Explanation:* Every Lie group G has a unique universal cover $\tilde{G}$, which by definition has the same Lie algebra and is simply connected. So if a Lie group G satisfies (i), then its universal cover $\tilde{G}$ satisfies (i) and (ii) and so has only linear unitary representations.

Translation: In physics we wish to know all of the ways in which a symmetry group of classical physics G might manifest itself in quantum systems and this requires us to know all of the relevant representations of that symmetry group. For a quantum system Wigner's theorem tells us that a continuous symmetry will appear as a continuous group of unitary transformations and so naturally brings us to unitary representations of some corresponding Lie group G. As we have seen above, the possible unitary representations in Hilbert space include both linear and projective unitary representations of the Lie group. If requirements (i) and (ii) are met by the Lie group G, then there are no intrinsic projective representations and we need only study the ordinary (linear) representations of G, i.e., those with $\phi(g_1, g_2) = 0$. If condition (i) is met but not condition (ii), then in general intrinsic projective representations of G will exist. However, in that case if we want to know all of the representations of G including the intrinsically projective ones, all we have to do is study the ordinary (linear) representations of the universal covering group $\tilde{G}$ of the Lie group G. A connected Lie group G is a connected manifold and connected manifolds always have a universal cover. So a universal cover $\tilde{G}$ always exists. As a cover it has the same Lie algebra and as a universal cover it is simply connected by definition and is always unique. Since the universal cover $\tilde{G}$ has the same Lie algebra as G it will also satisfy requirement (i) and being simply connected it has no intrinsic projective representations, only ordinary (linear) ones.

Summary: (1) If a symmetry group G satisfies (i) and (ii) we only need study its linear representations and (2) if G satisfies (i) but not (ii) then Bargmann's theorem applies and for every projective representation of G there is correspondence with a linear representation of its universal cover $\tilde{G}$ and so we must study those.

Examples

(a) *Projective representations of $SO(3)$:* The defining representation of the Lie group $SO(3)$ describes continuous rotations of classical objects in three dimensions as we saw in Sec. 1.2.3. $SO(3)$ is not simply connected while its double cover $SU(2)$ is. This is well known in the form of the Bali cup/plate trick also known as the belt trick or Dirac's string trick. The universal cover of $SO(3)$ is $SU(2)$ and the Lie algebra of these groups is the familiar Lie algebra of the rotation group, $\mathfrak{so}(3) \cong \mathfrak{su}(2)$. It is a double cover since $SO(3) \cong PSU(2) \equiv SU(2)/Z_2 \equiv SU(2)/\{I, -I\}$, which means that every $O \in SO(3)$ is represented twice in $SU(2)$. This Lie algebra $\mathfrak{so}(3) \cong \mathfrak{su}(2)$ allows central charges to be removed[1] and so (2) applies. So the projective (linear plus intrinsically projective) irreducible representations of $SO(3)$ correspond to the ordinary linear representations of $SU(2)$. The odd- and even-dimensional representations correspond to integer and half-integer angular momenta, respectively. The odd- and even-dimensional (matrix) representations have a projective phase $e^{i\phi}$ of $+1$ and -1, respectively, and are the integer and half-integer representations, respectively.

---

[1] As an aside, note that $Spin(3)$ is also said to be the universal cover of $SO(3)$ since $Spin(3)$ and $SU(2)$ are isomorphic, $Spin(3) \cong SU(2)$.

Consider a closed loop in $SO(3)$ given by $O_1O_2(O_1O_2)^{-1} = I$ with $O_1, O_2 \in SO(3)$. Let $U(O_i)$ be either of the two representations of O_i. We must go around the closed path in $SO(3)$ twice in $SU(2)$ to be sure of returning to the starting point since it is a double cover. Then using Eq. (4.1.65) we find

$$[\hat{U}(O_1)\hat{U}(O_2)\hat{U}^{-1}(O_1O_2)]^2 = \hat{I}, \tag{4.1.77}$$

$$\Rightarrow \quad \hat{U}(O_1)\hat{U}(O_2) = \pm\hat{U}(O_1O_2) \quad \Rightarrow \quad e^{i\phi(O_1,O_2)} = \pm 1.$$

A rotationally invariant quantum system cannot mix even- and odd-dimensional representations and so we cannot have states that are superpositions of integer and half-integer spin. This can be thought of as a *superselection rule* (Weinberg, 1995). Note that two states $|\phi_a\rangle, |\phi_b\rangle$ are separated by a *selection rule* if $\langle\phi_a|\hat{H}|\phi_b\rangle = 0$ with $\hat{H}$ the system Hamiltonian. We say that they are separated by a *superselection rule* if $\langle\phi_a|\hat{A}|\phi_b\rangle = 0$ for *all observable* operators $\hat{A}$, which means that we cannot create a quantum superposition of the two states, $c|\phi_a\rangle + d|\phi_b\rangle$.

Main lesson: To study all possible projective representations of $SO(3)$ in quantum systems we simply need to study all of the linear representations of $SU(2)$. Furthermore, since the Lie algebra is semisimple it can be shown that the irreducible projective representations of $SO(3)$ are in one-to-one correspondence with the irreducible linear representations of $SU(2)$.

(b) *Projective representations of $SO^+(1,3)$*: The restricted Lorentz group is denoted as $SO^+(1,3)$ and describes continuous Lorentz transformations of classical quantities. It is not simply connected. We know that $SL(2,\mathbb{C})$ is a simply connected double cover of $SO(1,3)$ and so it must be the universal covering group. Since $SL(2,\mathbb{C})$ is a cover, then their Lie algebras are isomorphic (the same), $\mathfrak{so}^+(1,3) \cong \mathfrak{sl}(2,\mathbb{C})$. It can be shown that this Lie algebra allows central charges to be removed and so again (II) applies. This is very similar to the $SO(3)$ and $SU(2)$ case above. So Bargmann's theorem applies and there is a one-to-one correspondence between the irreducible projective representations of $SO^+(1,3)$ and the irreducible linear representations of $SL(2,\mathbb{C})$. The construction of these representations is discussed in Sec. 1.3.1. Since $SL(2,\mathbb{C})$ is a double cover, we again must complete a closed loop twice to ensure that we return to the identity,

$$[\hat{U}(\Lambda_1)\hat{U}(\Lambda_2)\hat{U}^{-1}(\Lambda_1\Lambda_2)]^2 = \hat{I} \quad \Rightarrow \quad \hat{U}(\Lambda_1)\hat{U}(\Lambda_2) = \pm\hat{U}(\Lambda_1\Lambda_2), \tag{4.1.78}$$

which gives $e^{i\phi(\Lambda_1,\Lambda_2)} = \pm 1$ for integer/half-integer spin representations, respectively, since rotations are a subgroup of Lorentz transformations.

Main lesson: To study all possible projective representations of $SO^+(1,3)$ in quantum systems we need to study the linear representations of $SL(2,\mathbb{C})$. Since the Lie algebra is semisimple it can be shown that the irreducible projective representations of $SO^+(3)$ are in one-to-one correspondence with the irreducible linear representations of $SL(2,\mathbb{C})$.

(c) *Projective representations of the Poincaré group*: The Lie algebra of the Poincaré group allows central charges to be removed (Weinberg, 1995) and so Bargmann's theorem again applies. Interestingly, this is not the case for the Galilean group (Sundermeyer, 2014). As we saw in Chapter 1 the Poincaré group is the internal semidirect product of the normal (i.e., invariant) subgroup of spacetime translations with the subgroup of Lorentz transformations, Poincaré$(1,3) = R^{1,3} \rtimes O(1,3)$. The component continuous with the identity is $R^{1,3} \rtimes SO^+(1,3)$ and has a universal cover $R^{1,3} \rtimes SL(2,\mathbb{C})$, which again is a double cover. These representations are discussed in Sec. 1.4. Since $R^{1,3} \rtimes SL(2,\mathbb{C})$ is a double cover of the connected subgroup of the Poincaré group we again have

$$[\hat{U}(\Lambda_1,a_1)\hat{U}(\Lambda_2,a_2)\hat{U}^{-1}(\Lambda_1\Lambda_2,\Lambda_2 a_1 + a_2)]^2 = \hat{I}$$

$$\Rightarrow \quad \hat{U}(\Lambda_1)\hat{U}(\Lambda_2) = \pm\hat{U}(\Lambda_1\Lambda_2, \Lambda_2 a_1 + a_2) \tag{4.1.79}$$

and $e^{i\phi(\Lambda_1,a_1;\Lambda_2,a_2)} = \pm 1$ corresponding, respectively, to integer/half-integer projective representations of the Poincaré group.

Main lesson: To study all possible projective unitary representation of the Poincaré group we study the linear unitary representations of $R^{1,3} \rtimes SL(2,\mathbb{C})$. It can be shown that the irreducible projective unitary representations of $R^{1,3} \rtimes SO^+(1,3)$ are in one-to-one correspondence with the irreducible linear unitary representations of $R^{1,3} \rtimes SL(2,\mathbb{C})$.

4.1.5 Symmetry in Quantum Systems

We now wish to extend the above concepts to include the time evolution of a quantum system. By quantum system we mean a Hilbert space V_H with some observable operators and with a unitary time evolution operator $\hat{U}(t',t)$ as given in Eq. (4.1.6) and specified by some Hamiltonian operator $\hat{H}(t)$. Let $\hat{B} : V_H \to V_H$ be a time-independent linear transformation acting on the Hilbert space,

$$|\psi'(t)\rangle \equiv |\hat{B}\psi(t)\rangle \quad \text{for all} \quad t \in \mathbb{R}, |\psi(t)\rangle \in V_H. \tag{4.1.80}$$

We say that a dynamical quantum system has a linear symmetry $\hat{B}$ if and only if

$$\langle \hat{U}(t',t)\phi'(t)|\psi'(t)\rangle = \langle \hat{U}(t',t)\phi(t)|\psi(t)\rangle \tag{4.1.81}$$

for all $t, t' \in \mathbb{R}$ and $|\phi(t)\rangle, |\psi(t)\rangle \in V_H$. Since this must be true at $t' = t$ then we know from Wigner's theorem that $\hat{B}$ must be a unitary transformation, which we now write as $\hat{U}_s = \hat{B}$. Then Eq. (4.1.81) becomes

$$\langle \phi'(t)|\hat{U}^\dagger(t',t)\psi'(t)\rangle = \langle \phi(t)|\hat{U}_s^\dagger \hat{U}^\dagger(t',t)\hat{U}_s\psi(t)\rangle = \langle \phi(t)|\hat{U}^\dagger(t',t)\psi(t)\rangle. \tag{4.1.82}$$

So the unitary transformation $\hat{U}_s$ is a symmetry if and only if

$$\hat{U}_s^\dagger \hat{U}(t',t)\hat{U}_s = \hat{U}(t',t) \quad \text{or equivalently} \quad [\hat{U}(t',t), \hat{U}_s] = 0. \tag{4.1.83}$$

Using Eq. (4.1.11) with $\hat{H}_s(t)$ the Schrödinger picture Hamiltonian we have

$$\hat{U}(t',t) = \hat{U}_s^\dagger \hat{U}(t',t)\hat{U}_s = T\left[\exp\left\{-(i/\hbar)\int_t^{t'} dt\, \hat{U}_s^\dagger \hat{H}_s(t)\hat{U}_s\right\}\right] \tag{4.1.84}$$

for all $t, t' \in \mathbb{R}$ and so

$$\hat{H}_s(t) = \hat{U}_s^\dagger \hat{H}_s(t)\hat{U}_s \quad \text{or equivalently} \quad [\hat{H}_s(t), \hat{U}_s] = 0. \tag{4.1.85}$$

If the Hamiltonian has no explicit time dependence, then we have the familiar form for a quantum system invariant under a unitary symmetry transformation $\hat{U}_s$,

$$[\hat{H}, \hat{U}_s] = 0. \tag{4.1.86}$$

We used the above argument to connect with Wigner's theorem. We could have arrived at the same result more directly by simply observing that for a unitary symmetry $\hat{U}_s$ we require that every transformed state evolves in time in exactly the same way as the original state,

$$|\psi'(t')\rangle \equiv |\hat{U}_s\psi(t')\rangle = |\hat{U}_s\hat{U}(t',t)\psi(t)\rangle = |\hat{U}(t',t)\hat{U}_s\psi(t)\rangle = |\hat{U}(t',t)\psi'(t)\rangle,$$

which is true if and only if $[\hat{U}(t',t), \hat{U}_s] = 0$ and hence $[\hat{H}_s(t), \hat{U}_s] = 0$. We will consider antiunitary symmetries in later discussions of the time-reversal operator.

For example, under an active translation of the quantum system in coordinate space by $\mathbf{a} \in \mathbb{R}^3$ we have the unitary translation operator

$$\hat{T}(\mathbf{a}) \equiv \exp(-i\mathbf{a} \cdot \hat{\mathbf{P}}/\hbar) \quad \text{and} \quad \hat{T}^\dagger(\mathbf{a}) = \hat{T}^{-1}(\mathbf{a}) = \hat{T}(-\mathbf{a}), \qquad (4.1.87)$$

where $\hat{\mathbf{P}}$ is the total momentum operator for the system. If the quantum system is invariant under all three-dimensional translations, then we have

$$[\hat{H}, \hat{T}(\mathbf{a})] = 0 \quad \text{for all } \mathbf{a} \text{ and so} \quad [\hat{H}, \hat{\mathbf{P}}] = 0. \qquad (4.1.88)$$

Translation operators: Consider a single particle with Hilbert space position and momentum operators $\hat{\mathbf{x}}$ and $\hat{\mathbf{p}}$, respectively, in an arbitrary state $|\psi\rangle$. The coordinate-space wavefunction is $\psi(\mathbf{x}) = \langle \mathbf{x}|\psi\rangle$. From the Taylor expansion discussed in Eq. (1.2.171) and using Eq. (4.1.31) we know that

$$\langle \mathbf{x} + \mathbf{a}|\psi\rangle = \psi(\mathbf{x} + \mathbf{a}) = \exp[i\mathbf{a} \cdot (-i\hbar\boldsymbol{\nabla})/\hbar]\langle \mathbf{x}|\psi\rangle = \langle \mathbf{x}|e^{i\mathbf{a}\cdot\hat{\mathbf{p}}/\hbar}\psi\rangle = \langle e^{-i\mathbf{a}\cdot\hat{\mathbf{p}}/\hbar}\mathbf{x}|\psi\rangle$$

for every $|\psi\rangle \in V_H$, which means that

$$|\mathbf{x} + \mathbf{a}\rangle = \hat{T}(\mathbf{a})|\mathbf{x}\rangle \quad \text{with} \quad \hat{T}(\mathbf{a}) \equiv e^{-i\mathbf{a}\cdot\hat{\mathbf{p}}/\hbar}. \qquad (4.1.89)$$

Since $\hat{\mathbf{x}}$ is the position operator, then $\hat{\mathbf{x}}|\mathbf{x}\rangle = \mathbf{x}|\mathbf{x}\rangle$ and $\hat{\mathbf{x}}|\mathbf{x} + \mathbf{a}\rangle = (\mathbf{x} + \mathbf{a})|\mathbf{x} + \mathbf{a}\rangle$ and so for all $|\mathbf{x}\rangle$ we have

$$(\hat{\mathbf{x}} - \mathbf{a})|\mathbf{x} + \mathbf{a}\rangle = \mathbf{x}|\mathbf{x} + \mathbf{a}\rangle = \mathbf{x}\hat{T}(\mathbf{a})|\mathbf{x}\rangle = \hat{T}(\mathbf{a})\hat{\mathbf{x}}|\mathbf{x}\rangle = \hat{T}(\mathbf{a})\hat{\mathbf{x}}\hat{T}^\dagger(\mathbf{a})\hat{T}(\mathbf{a})|\mathbf{x}\rangle$$
$$= \hat{T}(\mathbf{a})\hat{\mathbf{x}}\hat{T}^\dagger(\mathbf{a})|\mathbf{x} + \mathbf{a}\rangle \quad \Rightarrow \quad \hat{T}(\mathbf{a})\hat{\mathbf{x}}\hat{T}^\dagger(\mathbf{a}) = \hat{\mathbf{x}} - \mathbf{a}. \qquad (4.1.90)$$

We can verify this result directly by expanding $\hat{T}(d\mathbf{a})$ to first order in $\hat{\mathbf{p}}$ for an infinitesimal translation $d\mathbf{a}$ and then using the canonical commutation relations $[\hat{x}^i, \hat{p}^j] = i\hbar\delta^{ij}$ to find $\hat{T}(d\mathbf{a})\hat{\mathbf{x}}\hat{T}^\dagger(d\mathbf{a}) = \hat{\mathbf{x}} - d\mathbf{a}$. Using Eq. (A.8.5) gives $\hat{T}(\mathbf{a})\hat{\mathbf{x}}\hat{T}^\dagger(\mathbf{a}) = \lim_{N\to\infty}[\hat{T}(\mathbf{a}/N)]^N\hat{\mathbf{x}}[\hat{T}^\dagger(\mathbf{a})/N)]^N = \hat{\mathbf{x}} + \lim_{N\to\infty} N(\hat{\mathbf{a}}/N) = \hat{\mathbf{x}} - \mathbf{a}$.

Rotation operators: Similarly under an active continuous rotation of the quantum system by counterclockwise angle $|\boldsymbol{\alpha}|$ around the axis $\mathbf{n}_\alpha \equiv \boldsymbol{\alpha}/\alpha$ we have

$$\hat{R}_a(\boldsymbol{\alpha}) \equiv \exp\left(-i\boldsymbol{\alpha} \cdot \hat{\mathbf{J}}/\hbar\right) \quad \text{and} \quad \hat{R}_a^\dagger(\boldsymbol{\alpha}) = \hat{R}_a^{-1}(\boldsymbol{\alpha}) = \hat{R}_a(-\boldsymbol{\alpha}), \qquad (4.1.91)$$

where $\hat{\mathbf{J}}$ is the total angular momentum operator for the system. If the system is invariant under continuous rotations, then

$$[\hat{H}, \hat{R}_a(\boldsymbol{\alpha})] = 0 \quad \text{for all } \boldsymbol{\alpha} \text{ and so} \quad [\hat{H}, \hat{\mathbf{J}}] = 0. \qquad (4.1.92)$$

4.1.6 Parity Operator

Let us denote the Hilbert space parity operator as $\hat{\mathcal{P}}$. Since we can construct dynamical quantum systems that are invariant under a parity transformation, then the parity transformation must be a symmetry in the sense of Wigner. So by Wigner's theorem $\hat{\mathcal{P}}$ is either unitary or antiunitary. We will soon see that $\hat{\mathcal{P}}$ must be unitary. Since all expectation values of observable operators must behave as their classical counterparts under transformations, then we must have

$$\hat{\mathbf{x}}' \equiv \hat{\mathcal{P}}\hat{\mathbf{x}}\hat{\mathcal{P}}^\dagger = -\hat{\mathbf{x}}, \quad \hat{\mathbf{p}}' \equiv \hat{\mathcal{P}}\hat{\mathbf{p}}\hat{\mathcal{P}}^\dagger = -\hat{\mathbf{p}}, \quad \hat{\mathbf{J}}' \equiv \hat{\mathcal{P}}\hat{\mathbf{J}}\hat{\mathcal{P}}^\dagger = \hat{\mathbf{J}}, \qquad (4.1.93)$$

where $\hat{\mathbf{J}}$ is any angular momentum operator. Obviously, since for orbital angular momentum we have $\hat{\mathbf{L}} = \hat{\mathbf{x}} \times \hat{\mathbf{p}}$, then $\hat{\mathbf{L}}' \equiv \mathcal{P}\hat{\mathbf{L}}\hat{\mathcal{P}}^\dagger = \mathcal{P}\hat{\mathbf{x}}\hat{\mathcal{P}}^\dagger \times \mathcal{P}\hat{\mathbf{p}}\hat{\mathcal{P}}^\dagger = \hat{\mathbf{L}}$. Then the spin operator $\hat{\mathbf{S}}$ and any other angular momentum operator must also be invariant. Since we can construct quantum systems that are invariant under a parity transformation, then $\hat{\mathcal{P}}$ must leave the canonical commutation relations and the angular momentum Lie algebra unchanged, i.e., $[\hat{x}'^i, \hat{p}'^j] = i\hbar\delta^{ij}$ and $[\hat{J}'^i, \hat{J}'^k] = i\hbar\epsilon^{ijk}J'^k$. We have

$$0 = \hat{\mathcal{P}}([\hat{x}^i, \hat{p}^j] - i\hbar\delta^{ij})\hat{\mathcal{P}}^\dagger = [\hat{x}'^i, \hat{p}'^j] - \hat{\mathcal{P}}i\hat{\mathcal{P}}^\dagger\hbar\delta^{ij} = [\hat{x}'^i, \hat{p}'^j] - i\hbar\delta^{ij}, \tag{4.1.94}$$

$$0 = \hat{\mathcal{P}}([\hat{J}^i, \hat{J}^j] - i\hbar\epsilon^{ijk}J^k)\hat{\mathcal{P}}^\dagger = [\hat{J}'^i, \hat{J}'^j] + \hat{\mathcal{P}}i\hat{\mathcal{P}}^\dagger\hbar\epsilon^{ijk}J'^k = [\hat{J}'^i, \hat{J}'^j] - i\hbar\epsilon^{ijk}J'^k,$$

if and only if $\hat{\mathcal{P}}i\hat{\mathcal{P}}^\dagger = i$. This means that $\hat{\mathcal{P}}$ must be unitary rather than antiunitary. Since two parity transformations lead to no change, then $\hat{\mathcal{P}}^2 = \hat{I}$ and $\hat{\mathcal{P}}$ is self-inverse (it is an involution). Then we have $\hat{I} = \hat{\mathcal{P}}^2 = \hat{\mathcal{P}}\hat{\mathcal{P}}^\dagger$ and so it is Hermitian, $\hat{\mathcal{P}} = \hat{\mathcal{P}}^\dagger$. Then $\hat{\mathcal{P}}$ is a unitary, Hermitian and self-inverse operator

$$\hat{\mathcal{P}} = \hat{\mathcal{P}}^{-1} = \hat{\mathcal{P}}^\dagger. \tag{4.1.95}$$

Since $\hat{\mathcal{P}}$ is Hermitian then its eigenvalues are real, $\pi_P \in \mathbb{R}$, and since it is self-inverse then its eigenvalues must satisfy $\pi_P^2 = 1$ and so $\pi_P = \pm 1$. If the quantum system is invariant under the parity operator then

$$[\hat{H}, \hat{\mathcal{P}}] = 0, \tag{4.1.96}$$

and the parity operator eigenvalue $\pi_P = \pm 1$ is a conserved quantum number. As we will discuss in Sec. 4.1.8 parity is a multiplicative quantum number, meaning that if a quantum system is made up of two subsystems, then the total Hilbert space is the tensor product of the two individial Hilbert spaces, $V_H \equiv V_{H1} \otimes V_{H2}$, and total parity π_P is the product of the parities of the subsystems,

$$\pi_P = \pi_{P1}\pi_{P2}. \tag{4.1.97}$$

The generalization to a quantum system with n subsystems immediately follows, $\pi_P = \pi_{P1}\pi_{P2}\cdots\pi_{Pn}$.

A *cyclic group* Z_n with generator $g \in Z_n$ has the properties that $g^n = e$ and $G = \{e, g, g^2, g^3, \ldots, g^{n-1}\}$, where e denotes the group identity and g^j means the jth power of the group generator g. Thus the parity operator $\hat{\mathcal{P}}$ is the generator of a cyclic group $Z_2 = \{\hat{I}, \hat{\mathcal{P}}\}$, since $\hat{\mathcal{P}}^2 = \hat{I}$ and so parity symmetry is an example of a Z_2 symmetry. All symmetries that are mathematically equivalent to parity are Z_2 symmetries and give rise to multiplicatively conserved quantum numbers.

It is proved in every elementary quantum mechanics text that the simultaneous eigenstates of the orbital angular momentum operators $\hat{\mathbf{L}}^2$ and $\hat{L}_z$ satisfy

$$\hat{\mathcal{P}}|\ell, m\rangle = (-1)^\ell|\ell, m\rangle, \tag{4.1.98}$$

where $\hat{\mathbf{L}}^2|\ell, m\rangle = \hbar^2\ell(\ell+1)|\ell, m\rangle$ and $\hat{L}_z|\ell, m\rangle = \hbar m|\ell, m\rangle$.

4.1.7 Time Reversal Operator

Let us denote the time reversal operator in Hilbert space as $\hat{\mathcal{T}}$. All expectation values of observable operators must behave classically under time reversal and so

$$\hat{\mathbf{x}}' = \hat{\mathcal{T}}\hat{\mathbf{x}}\hat{\mathcal{T}}^\dagger = \hat{\mathbf{x}}, \quad \hat{\mathbf{p}}' = \hat{\mathcal{T}}\hat{\mathbf{p}}\hat{\mathcal{T}}^\dagger = -\hat{\mathbf{p}}, \quad \hat{\mathbf{J}}' = \hat{\mathcal{T}}\hat{\mathbf{J}}\hat{\mathcal{T}}^\dagger = -\hat{\mathbf{J}}. \tag{4.1.99}$$

Since time-reversal invariance can be a symmetry of a dynamical quantum system, then it must be a symmetry in the Wigner sense and so is either a unitary or antiunitary transformation. Again we require that the canonical commutation relations and the angular momentum Lie algebra be preserved. Then we have

$$0 = \hat{\mathcal{T}}([\hat{x}^i, \hat{p}^j] - i\hbar\delta^{ij})\hat{\mathcal{T}}^\dagger = -[\hat{x}'^i, \hat{p}'^j] - \hat{\mathcal{T}}i\hat{\mathcal{T}}^\dagger\hbar\delta^{ij} = -([\hat{x}'^i, \hat{p}'^j] - i\hbar\delta^{ij}), \qquad (4.1.100)$$

$$0 = \hat{\mathcal{T}}([\hat{J}^i, \hat{J}^j] - i\hbar\epsilon^{ijk}J^k)\hat{\mathcal{T}}^\dagger = [\hat{J}'^i, \hat{J}'^j] + \hat{\mathcal{T}}i\hat{\mathcal{T}}^\dagger\hbar\epsilon^{ijk}J'^k = [\hat{J}'^i, \hat{J}'^j] - i\hbar\epsilon^{ijk}J'^k,$$

if and only if $\hat{\mathcal{T}}i\hat{\mathcal{T}}^\dagger = -i$ and so the time reversal operator $\hat{\mathcal{T}}$ must be antiunitary. Note that since $\hat{\mathcal{T}}\hat{\mathbf{J}}\hat{\mathcal{T}}^\dagger = -\hat{\mathbf{J}}$ then $\hat{\mathcal{T}}$ anticommutes with the rotation generators, $\{\hat{\mathcal{T}}, \hat{\mathbf{J}}\} = 0$. We see that $\hat{\mathcal{T}}\hat{U}(\boldsymbol{\alpha})\hat{\mathcal{T}}^\dagger = \hat{\mathcal{T}}\exp(-i\boldsymbol{\alpha}\cdot\hat{\mathbf{J}}/\hbar)\hat{\mathcal{T}}^\dagger = \hat{U}(\boldsymbol{\alpha})$ and so then $\hat{\mathcal{T}}$ commutes with the rotation operators,

$$[\hat{\mathcal{T}}, \hat{R}_a(\boldsymbol{\alpha})] = 0. \qquad (4.1.101)$$

Let $\hat{K}$ be the complex conjugation operator in the coordinate-space basis; i.e., in the notation of Eq. (4.1.54) we define $\hat{K} \equiv \hat{K}_{B_x}$ where B_x denotes the coordinate-space basis $|\mathbf{x}\rangle$ of square-integrable functions discussed in Secs. 4.1.1 and A.2.3. We have moved from Hilbert space with a countable basis to one with an uncountable basis but the arguments remain valid. Note that $\mathbf{x}^* = \mathbf{x}$, $(-i\hbar\boldsymbol{\nabla})^* = -(-i\hbar\boldsymbol{\nabla})$ and $[\mathbf{x}\times(-i\hbar\boldsymbol{\nabla})]^* = -[\mathbf{x}\times(-i\hbar\boldsymbol{\nabla})]$. This means that we have

$$\hat{K}\hat{\mathbf{x}}\hat{K}^\dagger = \hat{\mathbf{x}}, \quad \hat{K}\hat{\mathbf{p}}K^\dagger = -\hat{\mathbf{p}}, \quad \hat{K}\hat{\mathbf{L}}K^\dagger = -\hat{\mathbf{L}}. \qquad (4.1.102)$$

This follows since we have shown that these relations are true in the coordinate representation and they will remain true after a unitary transformation to any other representation. So the results are true at the operator level. Comparing with Eq. (4.1.99) we see that for a spinless particle it is sufficient to choose $\hat{\mathcal{T}} = \hat{K}$.

For a particle with intrinsic spin s the spin operators and eigenstates satisfy $\hat{S}^2|s, m_s\rangle = \hbar^2 s(s+1)|s, m_s\rangle$ and $\hat{S}^3|s, m_s\rangle = \hbar m_s|s, m_s\rangle$ for $m_s = -s, -s+1, \ldots, s-1, s$ in the usual way. In the standard representation the basis of eigenvectors of $\hat{S}^3$ is used to define the matrix representation of the Hermitian spin operators $\mathbf{S}$, with matrix elements $\mathbf{S}_{m'_s m_s} \equiv \langle s, m'_s|\hat{\mathbf{S}}|s, m_s\rangle$. It can be shown that

$$\langle s, m'_s|\hat{S}^1|s, m_s\rangle = \hbar(1/2)(\delta_{m'_s, m_s+1} + \delta_{m'_s+1, m_s})[s(s+1) - m'_s m_s]^{1/2},$$

$$\langle s, m'_s|\hat{S}^2|s, m_s\rangle = \hbar(1/2i)(\delta_{m'_s, m_s+1} - \delta_{m'_s+1, m_s})[s(s+1) - m'_s m_s]^{1/2},$$

$$\langle s, m'_s|\hat{S}^3|s, m_s\rangle = \hbar\delta_{m'_s, m_s}m_s. \qquad (4.1.103)$$

In this *standard spin basis* for any spin s we see that S^1 is a real symmetric matrix, S^2 is a pure imaginary antisymmetric matrix and S^3 is a real diagonal matrix. For example, for $s = \frac{1}{2}$ we have $\mathbf{S} = \frac{1}{2}\hbar\boldsymbol{\sigma}$ where the standard choice of Pauli spin matrices is the 2×2 matrices in Eq. (A.3.1). This standard choice for $s = 1$ is $\hbar$ times the 3×3 matrices in Eq. (1.2.67).

The Hilbert space V_H for a single particle with spin s moving in three dimensions is the tensor product of the Hilbert space for a spinless particle, $\mathcal{L}^2(\mathbb{R}^3)$, and the $(2s+1)$-dimensional Hilbert space for the intrinsic spin (denoted as V_s here). We write this as $V_H = \mathcal{L}^2(\mathbb{R}^3) \otimes V_s$. We can form a basis of this V_H from the tensor product of the coordinate-space basis with the standard spin basis and define

$$|\mathbf{x}; s, m_s\rangle \equiv |\mathbf{x}\rangle|s, m_s\rangle \qquad (4.1.104)$$

as the basis vectors. The tensor product of vector spaces is discussed in more detail in Sec. A.2.7 and below in Sec. 4.1.8.

Let us now extend the definition of $\hat{K}$ to be a complex conjugation operator with respect to the $|\mathbf{x}; s, m_s\rangle$ basis. In this basis the Hermitian spin matrices satisfy $(S^1)^* = S^1$, $(S^2)^* = -S^2$ and $(S^3)^* = S^3$, since S^2 is imaginary and S^1 and S^3 are real. So at the operator level for this $\hat{K}$ we have both Eq. (4.1.102) and

$$\hat{K}\hat{S}^1\hat{K}^\dagger = \hat{S}^1, \quad \hat{K}\hat{S}^2\hat{K}^\dagger = -\hat{S}^2, \quad \hat{K}\hat{S}^3\hat{K}^\dagger = \hat{S}^3. \tag{4.1.105}$$

We can define the unitary part $\hat{T}$ of the antiunitary time-reversal operator $\hat{\mathcal{T}}$ with respect to this $\hat{K}$ as

$$\hat{T} \equiv \hat{\mathcal{T}}\hat{K} \quad \text{with} \quad \hat{\mathcal{T}} = \hat{T}\hat{K} \quad \text{since} \quad \hat{K}^2 = \hat{I}. \tag{4.1.106}$$

(Messiah, 1975 uses the notation $\hat{\mathcal{T}} \to \hat{K}$, $\hat{K} \to K_0$ and $\hat{T} \to \hat{T}$.) The only purpose of a nontrivial choice of the unitary component $\hat{T}$ of $\hat{\mathcal{T}}$ is to deal with internal spins of particles. So $\hat{T}$ need not act on $\hat{\mathbf{x}}$ or $\hat{\mathbf{p}}$ and so can act only on the spin degrees of freedom. To recover the correct behavior of $\hat{\mathcal{T}}$ that $\hat{\mathcal{T}}\hat{\mathbf{S}}\hat{\mathcal{T}}^{-1} = -\hat{\mathbf{S}}$ we require

$$\hat{T}\hat{S}^1\hat{T}^\dagger = -\hat{S}^1, \quad \hat{T}\hat{S}^2\hat{T}^\dagger = \hat{S}^2, \quad \hat{T}\hat{S}^3\hat{T}^\dagger = -\hat{S}^3. \tag{4.1.107}$$

To reverse the signs of the x and z components of a three-vector it is sufficient to do a rotation about the y-axis by angle π. Then we can choose

$$\hat{T} \equiv \exp(-i\pi\hat{S}^2/\hbar) = \exp(-i\pi\hat{S}_y/\hbar), \tag{4.1.108}$$

where $\hat{S}^2 = \hat{S}_y$ recalling that $(\hat{S}^1, \hat{S}^2, \hat{S}^3) = (\hat{S}_x, \hat{S}_y, \hat{S}_z)$. We know in any representation the generators of the rotation group $\hat{\mathbf{J}}$ satisfy

$$\hat{U}^\dagger(\boldsymbol{\alpha})\hat{\mathbf{J}}\hat{U}(\boldsymbol{\alpha}) = \exp(+i\boldsymbol{\alpha}\cdot\hat{\mathbf{J}}/\hbar)\,\hat{\mathbf{J}}\exp(-i\boldsymbol{\alpha}\cdot\hat{\mathbf{J}}/\hbar) = R_a(\boldsymbol{\alpha})\hat{\mathbf{J}}. \tag{4.1.109}$$

The generators $\mathbf{J}$ transform as a three-vector operator as discussed below.

Mathematical digression: Vector operators: It is useful to generalize this discussion to the case of any three-vector operator $\hat{\mathbf{V}}$. These are usually referred to as a *vector operators* (Sakurai, 1994). Consider an active rotation of the system counterclockwise by angle $\alpha \equiv |\boldsymbol{\alpha}|$ about the axis $\mathbf{n}_\alpha \equiv \boldsymbol{\alpha}/\alpha$,

$$|\psi\rangle \to |\psi'\rangle = \hat{U}(\boldsymbol{\alpha})|\psi\rangle \quad \text{with} \quad \hat{U}(\boldsymbol{\alpha}) \equiv \exp(-i\boldsymbol{\alpha}\cdot\mathbf{J}/\hbar). \tag{4.1.110}$$

In order that we recover the appropriate classical limit, all expectation values of a three-vector operator $\hat{\mathbf{V}}$ must transform like a three-vector. So under an active rotation by $\boldsymbol{\alpha}$ we must have for all $|\psi\rangle$ that

$$\langle\psi|\hat{\mathbf{V}}|\psi\rangle \to \langle\psi'|\hat{\mathbf{V}}|\psi'\rangle = \langle\psi|\hat{U}^\dagger(\boldsymbol{\alpha})\hat{\mathbf{V}}\hat{U}(\boldsymbol{\alpha})|\psi\rangle = R_a(\boldsymbol{\alpha})\langle\psi|\hat{\mathbf{V}}|\psi\rangle, \tag{4.1.111}$$

where $R_a(\boldsymbol{\alpha})$ is the usual 3×3 active rotation matrix in the vector representation given in Eq. (1.2.53). So for any three-vector operator $\hat{\mathbf{V}}$ we have

$$\hat{U}^\dagger(\boldsymbol{\alpha})\hat{\mathbf{V}}\hat{U}(\boldsymbol{\alpha}) = R_a(\boldsymbol{\alpha})\hat{\mathbf{V}}, \tag{4.1.112}$$

which is a more general form of Eq. (4.1.109).

Consider an infinitesimal active rotation (using the summation convention),

$$\hat{V}^j + i(\epsilon/\hbar)[\mathbf{n}\cdot\hat{\mathbf{J}}, \hat{V}^j] = R_a(\epsilon\mathbf{n})^{jk}\hat{V}^k, \tag{4.1.113}$$

where from Eq. (1.2.56) we have $(J^i)^{jk} = -i\hbar\epsilon^{ijk}$ so that

$$R_a(\epsilon\mathbf{n})^{jk} = [\exp(-i\epsilon\mathbf{n}\cdot\mathbf{J}/\hbar)]^{jk} = \delta^{jk} - i(\epsilon/\hbar)\mathbf{n}\cdot\mathbf{J}^{jk} = \delta^{jk} - \epsilon n^i\epsilon^{ijk}.$$

Equating terms that are first order in ϵ gives $(i/\hbar)n^i[\hat{J}^i, \hat{V}^j] = -n^i\epsilon^{ijk}V^k$ for all $\mathbf{n}$ and so we must have for *any* three-vector operator $\hat{\mathbf{V}}$ that

$$[\hat{V}^i, \hat{J}^j] = i\hbar\epsilon^{ijk}V^k. \tag{4.1.114}$$

In the special case that $\hat{\mathbf{V}} = \hat{\mathbf{J}}$ we recover the angular momentum Lie algebra. Conversely, from Eq. (A.8.5) we know that we can obtain a finite transformation from an infinite number of infinitesimal ones. So if and only if an operator with three components $(\hat{V}^1, \hat{V}^2, \hat{V}^3)$ satisfies Eq. (4.1.114) will it satisfy Eq. (4.1.112) and hence be a three-vector operator that we can appropriately write as $\hat{\mathbf{V}}$.

We have now arrived at an explicit construction of the time-reversal operator

$$\hat{\mathcal{T}} = \exp(-i\pi\hat{S}_y/\hbar)\hat{K} = \hat{Y}_s\hat{K} \quad \text{with} \quad \hat{Y}_s \equiv \exp(-i\pi\hat{S}_y/\hbar), \tag{4.1.115}$$

where $\hat{\mathbf{S}}$ are the intrinsic spin operators for a particle of spin s and where $\hat{K}$ is the complex conjugation operator with respect to the $|\mathbf{x}; s, m_s\rangle$ basis. Note that in this basis S_y is a real $(2s+1)\times(2s+1)$ matrix and so $\hat{K}$ and $\hat{Y}_s$ commute, which gives

$$\hat{\mathcal{T}}^2 = (\hat{Y}_s\hat{K})^2 = \hat{Y}_s^2\hat{K}^2 = \hat{Y}_s^2 = \exp(-i2\pi\hat{S}_y/\hbar). \tag{4.1.116}$$

Consider the case of a boson; i.e., s is a nonnegative integer. Any spin state can be expressed as the superposition of the eigenstates of $\hat{S}_y$ rather than $\hat{S}_z$ and the action of $\exp(-i2\pi\hat{S}_y/\hbar)$ on every such state will leave the state unchanged so that $\hat{Y}_s^2 = \hat{I}$ for $s = 0, 1, 2, \ldots$. For integer spin a rotation of 2π in any direction returns the system to itself. We recognize by the same arguments that for $s = \frac{1}{2}, \frac{3}{2}, \frac{5}{2}, \ldots$ when acting on any $\hat{S}_y$ basis state we will have a remaining factor $\exp(-i\pi) = -1$ and so $\hat{Y}_s^2 = -\hat{I}$ for any half-integer s. This is not surprising, since we know that for a spin-$\frac{1}{2}$ particle we must rotate through 4π to return the system to itself. These arguments apply to any collection of particles. In summary, if we denote n as the number of particles in the system with half-integer spin, then

$$\hat{\mathcal{T}}^2 = \hat{Y}_s^2 = (-1)^n\hat{I} \tag{4.1.117}$$

and so $\hat{\mathcal{T}}^2$ has eigenvalues ± 1.

Since $\hat{\mathcal{T}}\hat{\mathcal{T}}^\dagger = I = (-1)^n\hat{\mathcal{T}}^2$ then if n is even or odd we have that $\hat{\mathcal{T}} = \hat{\mathcal{T}}^\dagger$ or $\hat{\mathcal{T}} = -\hat{\mathcal{T}}^\dagger$ respectively; i.e., the antiunitary operator $\hat{\mathcal{T}}$ is Hermitian or anti-Hermitian, respectively. In the case where n is even we have $\hat{\mathcal{T}}$ is Hermitian with $\hat{\mathcal{T}}^2 = \hat{I}$, which is a situation analogous to that for the parity operator $\hat{\mathcal{P}}$. So for time-reversal invariant systems containing an even number of half-integer spin particles $\hat{\mathcal{T}}$ will be an observable operator referred to as the *T-parity* operator, which has eigenvalues $\pi_T = \pm 1$. In this case $\hat{\mathcal{T}}$ is the generator of its own Z_2 symmetry group.

Kramer's theorem: In the case where n is odd we can use Eq. (4.1.50) to write for any state $|\psi\rangle$ that

$$\langle\psi|\hat{\mathcal{T}}\psi\rangle = \langle\psi|\hat{\mathcal{T}}^\dagger\psi\rangle = -\langle\psi|\hat{\mathcal{T}}\psi\rangle \quad \Rightarrow \quad \langle\psi|\hat{\mathcal{T}}\psi\rangle = 0 \tag{4.1.118}$$

and so the states $|\psi\rangle$ and $|\hat{\mathcal{T}}\psi\rangle$ are orthogonal. As we will soon discuss, if the system is time-reversal invariant, then $[\hat{H}, \hat{\mathcal{T}}] = 0$. If $|\psi\rangle$ is an energy eigenstate with energy E then $|\hat{H}\psi\rangle = E|\psi\rangle$ and so $E|\hat{\mathcal{T}}\psi\rangle = |\hat{\mathcal{T}}\hat{H}\psi\rangle = |\hat{H}\hat{\mathcal{T}}\psi\rangle$ and so $|\mathcal{T}\psi\rangle$ is an orthogonal state with the same energy E. This result for time-reversal invariant systems with half-integer spin is known as *Kramers' (degeneracy) theorem*. The degenerate orthogonal states $|\psi\rangle$ and $|\mathcal{T}\psi\rangle$ are sometimes referred to as a *Kramers' pair*. Then the subspace pertaining to any energy eigenvalue E can be spanned by an orthonormal basis made up of pairs of orthonormal energy eigenvectors $|\psi\rangle$ and $|\hat{\mathcal{T}}\psi\rangle$ and hence the subspace must be even dimensional. Any time-reversal invariant system with an odd number of half-integer particles will have energy levels that are at least doubly degenerate and will have an even degeneracy.

Examples: Hydrogen contains two fermions (or four fermions if viewed at the quark and lepton level); in either case the number is even and there is no doubling of the degeneracy of the fine structure energy levels of the electron. Deuterium has a proton and a neutron in the nucleus. The deuterium electron energy levels are slightly shifted with respect to hydrogen due to the neutron giving a small increase in the reduced mass of the system. As predicted by Kramers' theorem the hyperfine energy levels of deuterium are at least doubly degenerate. At very low energies where the internal atomic structure is not resolved, hydrogen behaves as a boson (any integer spin) and deuterium behaves as a fermion (any half-integer spin), since they contain an even and odd number of fermions, respectively. Similarly, tritium and helium-4 are bosonic and helium-3 is fermionic. Only bosonic atoms can form a Bose-Einstein condensate.

In order to have consistency with the behavior of classical physics under time reversal, the time-dependence of all expectation values of operators must be reversed. Rather than perform the passive transformation of observable operators under time-reversal, we could equally well perform an active transformation of the time evolution of the states; i.e., we define under time reversal, $t \to -t$, that

$$|\psi'(t)\rangle \equiv |\hat{\mathcal{T}}\psi(-t)\rangle \quad \text{for all } t \in \mathbb{R}. \tag{4.1.119}$$

For simplicity we flip time around $t = 0$ but we could always do this around any arbitrary time t_0. If and only if our system is time-reversal invariant, then any such $|\psi'(t)\rangle$ must evolve in time like any other state,

$$|\psi'(t')\rangle = |\hat{U}(t',t)\psi'(t)\rangle = |\hat{U}(t',t)\hat{\mathcal{T}}\psi(-t)\rangle. \tag{4.1.120}$$

Consider a time-independent $\hat{H}$ so that

$$\hat{U}(t',t) = \hat{U}(t'-t) = e^{-i\hat{H}(t'-t)/\hbar}. \tag{4.1.121}$$

Then we must have for all $|\psi(0)\rangle$ that

$$|\hat{U}(t)\hat{\mathcal{T}}\psi(0)\rangle = |\hat{U}(t)\psi'(0)\rangle = |\psi'(t)\rangle = |\hat{\mathcal{T}}\psi(-t)\rangle = |\hat{\mathcal{T}}\hat{U}(-t)\psi(0)\rangle, \tag{4.1.122}$$

where we have used Eq. (4.1.119). Then $\hat{U}(t)\hat{\mathcal{T}} = \hat{\mathcal{T}}\hat{U}(-t) = \hat{\mathcal{T}}\hat{U}^\dagger(t)$, which we can rewrite as $\hat{\mathcal{T}}^\dagger\hat{U}(t)\hat{\mathcal{T}} = \hat{U}^\dagger(t)$, and so

$$e^{+i\hat{\mathcal{T}}^\dagger\hat{H}\hat{\mathcal{T}}t/\hbar} = \hat{\mathcal{T}}^\dagger e^{-i\hat{H}t/\hbar}\hat{\mathcal{T}} = \hat{\mathcal{T}}^\dagger\hat{U}(t)\hat{\mathcal{T}} = \hat{U}^\dagger(t) = e^{+i\hat{H}t/\hbar}, \tag{4.1.123}$$

where since $\hat{\mathcal{T}}$ is antiunitary it gives $i \to -i$. So we see that for a system with a time-independent Hamiltonian we have time-reversal invariance if and only if

$$\hat{\mathcal{T}}^\dagger\hat{H}\hat{\mathcal{T}} = \hat{H} \quad \text{or equivalently} \quad [\hat{H}, \hat{\mathcal{T}}] = 0. \tag{4.1.124}$$

For a Hamiltonian to be be time-reversal invariant, $[\hat{H}, \hat{\mathcal{T}}] = 0$, it must have (i) no explicit time dependence; (ii) only real coefficients multiplying operator products; and (iii) only $\hat{\mathcal{T}}$-even operator products such as $\hat{\mathbf{p}}^2$, $\hat{\mathbf{S}}^2$, $\hat{\mathbf{L}}^2$, $\hat{\mathbf{L}} \cdot \hat{\mathbf{S}}$ and any powers of $\hat{\mathbf{x}}$. If such a system contains an odd number of half-integer spin particles, then every energy level will have even degeneracy due to Kramers' theorem.

External E and B fields: Under time-reversal, electric and magnetic fields are even and odd, respectively, $\mathbf{E} \to \mathbf{E}$ and $\mathbf{B} \to -\mathbf{B}$. If we couple a time-reversal invariant system with half-integer spin to an external **B**-field in such a way that total time-reversal invariance of the entire system is maintained, we must have done so through an internal three-vector time-reversal odd Hermitian operator $\hat{\mathbf{O}}$, e.g., a magnetic moment that is proportional to the angular momentum operator. The energy is minimized when the magnetic moment and the **B**-field align. Let H_0 be the time-reversal invariant internal Hamiltonian and let $\hat{\mathcal{T}}$ denote the *internal* time-reversal operator that does not act on the external **B**. We have $\hat{H} = \hat{H}_0 - \mathbf{B} \cdot \hat{\mathbf{O}}$, $[\hat{H}_0, \hat{\mathcal{T}}] = 0$ and $\hat{\mathcal{T}}\hat{\mathbf{O}} = -\hat{\mathbf{O}}\hat{\mathcal{T}}$ so that $[\hat{H}, \hat{\mathcal{T}}] = -\mathbf{B} \cdot [\hat{\mathbf{O}}, \hat{\mathcal{T}}] = -2\mathbf{B} \cdot \hat{\mathbf{O}}\hat{\mathcal{T}} = +2\mathbf{B} \cdot \hat{\mathcal{T}}\hat{\mathbf{O}}$. Let $[\hat{H}^0, \hat{\mathbf{O}}] = 0$ so that we can choose $|\psi\rangle$ to be a simultaneous eigenstate of $\hat{H}_0$, $\hat{\mathbf{O}}$ and $\hat{H}$. We can then write $|\hat{H}_0\psi\rangle = E_0|\psi\rangle$, $|\hat{\mathbf{O}}\psi\rangle = \mathbf{O}|\psi\rangle$ and $|\hat{H}\psi\rangle = (E_0 - \mathbf{B} \cdot \mathbf{O})|\psi\rangle$. We find that $|\hat{H}\hat{\mathcal{T}}\psi\rangle = |(\hat{H}_0 - \mathbf{B} \cdot \hat{\mathbf{O}})\hat{\mathcal{T}}\psi\rangle = |\hat{\mathcal{T}}(\hat{H}_0 + \mathbf{B} \cdot \hat{\mathbf{O}})\psi\rangle = (E_0 + \mathbf{B} \cdot \mathbf{O})|\hat{\mathcal{T}}\psi\rangle$ and so the Kramers' pair have shifted energies $E_\pm = (E_0 \pm \mathbf{B} \cdot \mathbf{O})$ and the degeneracy is lifted. In place of a **B**-field consider adding an external **E**-field with some internal three-vector time-reversal even Hermitian operator $\hat{\mathbf{D}}$ such as an electric dipole operator. An electric dipole moment occurs when there is spatial displacement of positive and negative charges. By convention an electric dipole **D** points from negative to positive while an **E**-field points from positive to negative. Then the energy is minimized when **D** and **E** align. This gives $\hat{H} = \hat{H}_0 - \mathbf{E} \cdot \hat{\mathbf{D}}$, $[\hat{H}_0, \hat{\mathbf{D}}] = 0$, $\hat{\mathcal{T}}\hat{\mathbf{D}} = \hat{\mathbf{D}}\hat{\mathcal{T}}$ and $|\hat{\mathbf{D}}\psi\rangle = \mathbf{D}|\psi\rangle$. Then we find that both $|\psi\rangle$ and $|\hat{\mathcal{T}}\psi\rangle$ have the same shifted energy $E = E_0 - \mathbf{E} \cdot \mathbf{D}$ and the Kramers' degeneracy remains.

Recalling that $|\psi'(t)\rangle \equiv |\hat{\mathcal{T}}\psi(-t)\rangle$ we note that the time-reversal analog of Eq. (4.1.81) for a time-independent Hamiltonian is

$$\langle \hat{U}(t'-t)\phi'(t)|\psi'(t)\rangle = \langle \psi(-t)|\hat{U}^\dagger(t'-t)\phi(-t)\rangle \tag{4.1.125}$$

for all $t, t' \in \mathbb{R}$ and for all $|\phi(t)\rangle, |\psi(t)\rangle \in V_H$. Choosing $t = t' = 0$ we see that this reduces to $\langle \hat{\phi}'(0)|\psi'(0)\rangle = \langle \psi(0)|\phi(0)\rangle$ and so $\hat{\mathcal{T}}$ must indeed be antiunitary according to Wigner's theorem. It also follows from Eq. (4.1.125) that

$$\langle \psi(-t)|\hat{U}^\dagger(t'-t)\phi(-t)\rangle = \langle \hat{U}(t'-t)\phi'(t)|\psi'(t)\rangle = \langle \hat{U}(t'-t)\hat{\mathcal{T}}\phi(-t)|\hat{\mathcal{T}}\psi(-t)\rangle$$
$$= \langle \hat{\mathcal{T}}(\hat{\mathcal{T}}^\dagger\hat{U}(t'-t)\hat{\mathcal{T}})\phi(-t)|\hat{\mathcal{T}}\psi(-t)\rangle = \langle \psi(-t)|(\hat{\mathcal{T}}^\dagger\hat{U}(t'-t)\hat{\mathcal{T}})\phi(-t)\rangle, \tag{4.1.126}$$

which again gives

$$\hat{\mathcal{T}}^\dagger\hat{U}(t'-t)\hat{\mathcal{T}} = \hat{U}^\dagger(t'-t) \quad \text{or equivalently} \quad [\hat{H}, \hat{\mathcal{T}}] = 0. \tag{4.1.127}$$

We can now ask about the consequences of an antiunitary symmetry that does *not* involve time reversal. Consider an antiunitary transformation $\hat{U}_a$ with $|\psi'(t)\rangle \equiv |\hat{U}_a\psi(t)\rangle$. Such an antiunitary dynamical quantum symmetry requires

$$\langle \hat{U}(t'-t)\phi'(t)|\psi'(t)\rangle = \langle \psi(t)|\hat{U}(t'-t)\phi(t)\rangle, \tag{4.1.128}$$

which gives by similar steps to those above

$$\langle \psi(t)|\hat{U}(t'-t)\phi(t)\rangle = \langle \hat{U}(t'-t)\phi'(t)|\psi'(t)\rangle = \langle \hat{U}(t'-t)\hat{U}_a\phi(t)|\hat{U}_a\psi(t)\rangle$$
$$= \langle \hat{U}_a(\hat{U}_a^\dagger\hat{U}(t'-t)\hat{U}_a)\phi(t)|\hat{U}_a\psi(t)\rangle = \langle \psi(t)|(\hat{U}_a^\dagger\hat{U}(t'-t)\hat{U}_a)\phi(t)\rangle \tag{4.1.129}$$

and hence

$$\hat{U}_a^\dagger \hat{U}(t' - t)\hat{U}_a = \hat{U}(t' - t) \quad \text{or equivalently} \quad \{\hat{H}, \hat{U}_a\} = 0. \tag{4.1.130}$$

Let $|\psi\rangle$ be an energy eigenstate $|\hat{H}\psi\rangle = E|\psi\rangle$, then since $\hat{H}\hat{U}_a = -\hat{U}_a\hat{H}$ we have $|\hat{H}\hat{U}_a\psi\rangle = -|\hat{U}_a\hat{H}\psi\rangle = -E|\hat{U}_a\psi\rangle$. So for a system with the antiunitary symmetry U_a, for every energy eigenstate $|\psi\rangle$ with energy E there is a corresponding eigenstate $|\hat{U}_a\psi\rangle$ with energy $-E$. If the energy of the nonrelativistic quantum system is not bounded above, then it would not be bounded below, which is unphysical. So we are not interested in antiunitary symmetries other than time-reversal symmetry.

Electric dipole moment (EDM): There is an interesting consequence of time-reversal invariance for the electric dipole moment (EDM) of elementary particles. If a fundamental particle has an EDM, then it has an orientation and so it could not transform as a scalar under the Poincaré group, i.e., spin $s \neq 0$. If a particle with nonzero spin s has an electric dipole moment that is not parallel or antiparallel to its spin rotation axis, then it would carry information about two spatial directions and so could not transform in the spin s representation of the Poincaré group. So an elementary particle can only have an EDM if $s \neq 0$ and if the spin axis and EDM are aligned or anti-aligned. The same is of course true of the *magnetic moment* μ of an elementary particle. Next we note that (i) under a parity transformation the spin $\mathbf{S}$ and magnetic moment μ are invariant but the EDM, $\mathbf{D}$, changes sign and (ii) under a time-reversal transformation the spin $\mathbf{S}$ and magnetic moment μ are reversed but the EDM, $\mathbf{D}$, does not change. So the existence of an elementary particle with nonzero spin $s \neq 0$ and some nonzero EDM would violate both parity and time-reversal invariance, since the EDM and spin of the particle would transform from aligned to anti-aligned or vice versa under either a P or a T transformation. So P and T symmetries are both violated. Assuming an overall CPT invariance then also implies violation of the combined CP symmetry. We already know that the weak interactions maximally violate P. Each of C, CP and T violation have also been directly observed. The Standard Model is CPT invariant and the CKM matrix discussed in Sec. 5.2.4 incorporates the observed CP violation leading to T violation. No EDM has ever been measured for an elementary particle such as an electron, nor has an EDM been measured for a compound subatomic particle such as the neutron. Only upper limits currently exist for these. The Standard Model prediction for the neutron EDM is very small and well below the currently achievable experimental sensitivity. Since the neutron is compound it could have a small contribution to an EDM by having a permanent distortion, where the centers of the positive and negative internal charge distributions are not exactly aligned.

A molecule can have a significant EDM and not imply time-reversal invariance, since it will have a spatial orientation associated with its nonspherical shape. For example, the water molecule H_2O has a bond angle of $104.45°$ between the hydrogen atoms as seen from the oxygen atom and does have an electric dipole moment associated with this shape. In contrast, the carbon dioxide CO_2 molecule is linear and the symmetric $O = C = O$ structure gives rise to no overall electric dipole moment; i.e., the left and right contributions to the nonuniform charge distributions are equal and opposite and hence cancel.

4.1.8 Additive and Multiplicative Quantum Numbers

When two or more quantum systems are combined the total Hilbert space of the combined system is the tensor product of the Hilbert space of each subsystem; e.g., if V_{H1} and V_{H2} are the Hilbert spaces of two quantum systems, then the Hilbert space of the combined systems is $V_H \equiv V_{H1} \otimes V_{H2}$.

The tensor product of vector spaces, which includes Hilbert spaces, is discussed in Sec. A.2.7 and some essential properties are summarized in Eqs. (A.2.25) and (A.2.26).

Denote the evolution operators for the systems as $\hat{U}_1(t'', t')$ and $\hat{U}_2(t'', t')$ where $\hat{H}_{s1}(t)$ and $\hat{H}_{s2}(t)$, respectively, are the corresponding time-dependent Schrödinger picture Hamiltonians given by Eq. (4.1.11). We have $[\hat{H}_{s1}(t_1), \hat{H}_{s2}(t_2)] = 0$ for all t_1, t_2. The time evolution of the combined system is

$$\hat{U}(t'', t')|\psi_1; \psi_2\rangle \equiv \hat{U}_1(t'', t')\hat{U}_2(t'', t')|\psi_1; \psi_2\rangle \quad \text{and so} \tag{4.1.131}$$

$$\hat{U}(t'', t') \equiv T\left[\exp\left\{(i/\hbar)\int_{t'}^{t''} dt\, \hat{H}_s(t)\right\}\right] = T\left[\exp\left\{(i/\hbar)\int_{t'}^{t''} dt\, [\hat{H}_{s1}(t) + \hat{H}_{s2}(t)]\right\}\right].$$

So the Hamiltonian for the combined system is

$$\hat{H}_s(t) = \hat{H}_{s1}(t) + \hat{H}_{s2}(t) \tag{4.1.132}$$

or more simply $\hat{H} = \hat{H}_1 + \hat{H}_2$ if there is no explicit time dependence.

Consider a more general situation where the evolution operator for the full system does not factorize into the product of the evolution operators for each subsystem, $U(t'', t') \neq \hat{U}_1(t'', t')\hat{U}_2(t'', t')$. We can still define the Schrödinger picture Hamiltonian for the full system $\hat{H}_s(t)$ by differentiating $U(t'', t')$ as in Eq. (4.1.9) and similarly for each of the component systems. We can then *define* the interaction between the two systems by the potential $\hat{V}(t)$, where

$$\hat{V}(t) \equiv \hat{H}_s(t) - \hat{H}_{s1}(t) - \hat{H}_{s2}(t) \tag{4.1.133}$$

and where $\hat{V}(t)$ will contain operators from both quantum subsystems. Then by construction we have for any state $|\psi(t)\rangle \in V_H = V_{H1} \otimes V_{H2}$ that

$$|\psi(t')\rangle = \hat{U}(t', t)|\psi(t)\rangle \quad \text{with} \quad U(t', t) = T\left[\exp\left\{(i/\hbar)\int_{t'}^{t''} dt\, \hat{H}_s(t)\right\}\right] \quad \text{and}$$

$$\hat{H}_s(t) = \hat{H}_{s1}(t) + \hat{H}_{s2}(t) + \hat{V}(t). \tag{4.1.134}$$

Consider the abelian unitary group of transformations generated by some time-independent Hermitian operator $\hat{Q}$ in the Hilbert space V_H,

$$|\psi'(t)\rangle = \hat{U}_Q(\alpha)|\psi(t)\rangle \equiv \exp(i\alpha\hat{Q})|\psi(t)\rangle \quad \text{for any} \quad \alpha \in \mathbb{R}. \tag{4.1.135}$$

If $\hat{U}_Q$ is a symmetry of the quantum system, then

$$[\hat{H}_s(t), \hat{Q}] = 0 \tag{4.1.136}$$

or more simply $[\hat{H}, \hat{Q}] = 0$ if the system has no explicit time-dependence. Note that here $U(1)$ is specifically a symmetry of the quantum system itself, and so the discussion of projective representations of quantum realizations of classical symmetries does not apply. In either case $\hat{Q}$ is then a conserved charge. If $\hat{Q}_1$ and $\hat{Q}_2$ are the charge operators in V_{H1} and V_{H2}, respectively, then $[\hat{Q}_1, \hat{Q}_2] = 0$. Define $|\psi_1\psi_2; t\rangle \equiv |\psi_1(t)\rangle|\psi_2(t)\rangle$, then

$$e^{i\alpha\hat{Q}}|\psi_1\psi_2; t\rangle = e^{i\alpha\hat{Q}_1}e^{i\alpha\hat{Q}_2}|\psi_1\psi_2; t\rangle \quad \text{with} \quad \hat{Q} \equiv \hat{Q}_1 + \hat{Q}_2 \tag{4.1.137}$$

and so the charge operator is *additive* and the corresponding quantum numbers are *additive quantum numbers*. If $[\hat{H}_s(t), \hat{Q}] = [\hat{H}_s(t), \hat{Q}_1] = [\hat{H}_s(t), \hat{Q}_2] = 0$, then $\hat{Q}$, $\hat{Q}_1$ and $\hat{Q}_2$ are all separately conserved. If, on the other hand, $[\hat{H}_s(t), \hat{Q}] = [\hat{H}_{s1}(t), \hat{Q}_1] = [\hat{H}_{s2}(t), \hat{Q}_2] = 0$ but $[\hat{V}(t), \hat{Q}_i] \neq 0$, then $\hat{Q}$ is conserved but $\hat{Q}_1$ and $\hat{Q}_2$ are not separately conserved.

As an illustration, assume no explicit time-dependence and let $|\psi_{q_1}(t)\rangle \in V_{H1}$ and $|\psi_{q_2}(t)\rangle \in V_{H2}$ be any Schrödinger picture eigenstates of $\hat{Q}_1$ and $\hat{Q}_2$ at time t, respectively, where $\hat{Q}_1|\psi_{q_1}(t)\rangle = q_1|\psi_{q_1}(t)\rangle$ and $\hat{Q}_2|\psi_{q_2}(t)\rangle = q_2|\psi_{q_2}(t)\rangle$. Then if $[\hat{H}, \hat{Q}_1] = [\hat{H}, \hat{Q}_2] = 0$ we have

$$\hat{Q}|\psi_{q_1}\psi_{q_2}; t\rangle = (\hat{Q}_1 + \hat{Q}_2)|\psi_{q_1}\psi_{q_2}; t\rangle = (q_1 + q_2)|\psi_{q_1}\psi_{q_2}; t\rangle \equiv q|\psi_{q_1}\psi_{q_2}; t\rangle, \qquad (4.1.138)$$

$$\hat{Q}|\psi_{q_1}\psi_{q_2}; t'\rangle = \hat{Q}e^{-i\hat{H}(t'-t)/\hbar}|\psi_{q_1}\psi_{q_2}; t\rangle = e^{-i\hat{H}(t'-t)/\hbar}\hat{Q}|\psi_{q_1}\psi_{q_2}; t\rangle = q|\psi_{q_1}\psi_{q_2}; t'\rangle,$$

where we see that $\hat{Q}|\psi_{q_1}\psi_{q_2}; t\rangle = q|\psi_{q_1}\psi_{q_2}; t\rangle$ with $q \equiv q_1 + q_2$ for all t. This is just another way to say that the total charge is conserved if $[\hat{H}, \hat{Q}] = [\hat{H}, \hat{Q}_1 + \hat{Q}_2] = 0$ but $[\hat{H}, \hat{Q}_1] \neq 0$ and $[\hat{H}, \hat{Q}_2] \neq 0$.

Since $\hat{U}_{Q1}(\alpha) = \exp(i\alpha\hat{Q}_1)$ and $\hat{U}_{Q2}(\alpha) = \exp(i\alpha\hat{Q}_2)$ we see that the unitary group operator is a *multiplicative operator*, $\hat{U}_Q(\alpha) = \hat{U}_{Q1}(\alpha)\hat{U}_{Q2}(\alpha)$, whereas the Hermitian group generator is an additive operator, $\hat{Q} = \hat{Q}_1 + \hat{Q}_2$ with a corresponding additive quantum number. It is the Hermitian operators that have observable quantum numbers and it is the behavior of these quantum numbers that is of physical interest. As we saw in the case of parity, if the transformation group elements are themselves Hermitian, then they have observable quantum numbers and these are multiplicative quantum numbers. This is the case for any parity-like transformation discussed in Sec. 4.1.6, i.e., for any Z_2 symmetry.

Any conserved charge can be used as the generator of $U(1)$ Lie group that leaves the system invariant. We call this a *global $U(1)$ symmetry* to distinguish it from a $U(1)$ gauge symmetry such as that underlying quantum electrodynamics with a spacetime-dependent phase, $\alpha \to \alpha(x)$. A quantized conserved charge leads to a compact $U(1)$, while a continuous conserved charge leads to a noncompact $U(1)$.

Compact $U(1)$: (a) If the charge is quantized such that every eigenvalue of $\hat{Q}$ is nq with $q > 0$ the magnitude of the elementary unit of charge and n an integer, $n \in \mathbb{Z}$, then the number operator for this charge is $\hat{N}_Q \equiv \hat{Q}/q$ with eigenvalues $n \in \mathbb{Z}$. Consider a charge eigenstate $|\psi_n\rangle$ with charge nq, $\hat{Q}|\psi_n\rangle = qn|\psi_n\rangle$, then

$$\hat{U}_Q(\alpha)|\psi_n\rangle \equiv e^{i\alpha\hat{Q}}|\psi_n\rangle = e^{i\alpha q\hat{N}_Q}|\psi_n\rangle = e^{i\theta\hat{N}_Q}|\psi_n\rangle = (e^{i\theta})^n|\psi_n\rangle, \qquad (4.1.139)$$

where we have defined $\theta \equiv \alpha q$. Since $e^{i(\theta+2\pi)} = e^{i\theta}$, then the range of θ is $-\pi \leq \theta < \pi$ and the corresponding range of α is $-\frac{1}{2}\alpha_0 \leq \alpha < \frac{1}{2}\alpha_0$ where $2\pi/q \equiv \alpha_0$. This means that this $U(1)$ is compact. (b) Now if $e^{i\alpha\hat{Q}} \in U(1)$ with $U(1)$ compact, then there must be some $\alpha_0 \in \mathbb{R}$ such that $e^{i\alpha\hat{Q}} = e^{i(\alpha+\alpha_0)\hat{Q}}$. This means that $\hat{I} = e^{i\alpha_0\hat{Q}} = e^{i2\pi(\alpha_0\hat{Q}/2\pi)}$ and so $\hat{N}_Q \equiv (\alpha_0/2\pi)\hat{Q}$ must have integer eigenvalues. Then $\hat{Q} = q\hat{N}_Q$ with the unit of elementary charge $q \equiv 2\pi/\alpha_0$; and so (c) the charge operator $\hat{Q}$ is quantized if and only if it generates a compact $U(1)$. As an example, consider a system rotationally invariant under rotations about the z-axis. This means that the system is invariant under transformations $e^{i\alpha\hat{Q}} = e^{-i\alpha\hat{L}_z/\hbar}$ with periodicity given by $\alpha_0 = 2\pi$ and the magnitude of the quantum of charge is $q = 1$, i.e., $Q = -\hat{L}_z/\hbar$ has eigenvalues n with $n \in \mathbb{Z}$. In some Grand Unified Theories the electromagnetic interaction emerges in the form of a compact $U(1)$ and the quantization of electric charge then arises naturally.

Non-compact $U(1)$: (a) If the eigenvalues of the charge operator are continuous, denote these as $q_c \in \mathbb{R}$. We can consider $q_c = \lim_{n\to\infty} n(q_c/n)$ with (q_c/n) playing the role of the charge quantum q with $q \to 0^+$. This corresponds to $\alpha_0 = 2\pi/q \to \infty$ and so $-\infty < \alpha < \infty$ and this $U(1)$ is non-compact; (b) if the $U(1)$ is non-compact, then $\alpha_0 \to \infty$ and so $q \to 0^+$ and the charge is continuous; and so (c) the charge operator $\hat{Q}$ is continuous if and only if it generates a non-compact $U(1)$. An example of a non-compact $U(1)$ is the translation operator in Eq. (4.1.89), where

$\hat{T} = e^{-i\mathbf{a}\cdot\hat{\mathbf{p}}/\hbar} = e^{ia(-\hat{n}\cdot\hat{\mathbf{p}}/\hbar)} = e^{ia\hat{Q}}$, where $\mathbf{a} \equiv a\hat{n}$ with $a \equiv |\mathbf{a}|$ and $\hat{n} \equiv \mathbf{a}/a$ is a unit vector. The conserved charge $\hat{Q} = -\hat{n}\cdot\hat{\mathbf{p}}/\hbar$ has continuous eigenvalues and a is unbounded, $-\infty < a < \infty$.

4.1.9 Systems of Identical Particles and Fock Space

The above arguments for a system consisting of two subsystems generalize easily to any number of subsystems. We can define the Hilbert space for n *distinguishable* particles $V_H^{(n)}$ as the tensor product of n single-particle Hilbert spaces V_H,

$$V_H^{(n)} = V_H \otimes \cdots \otimes V_H = (V_H)^{\otimes n}. \tag{4.1.140}$$

Let $|b_1\rangle \equiv \mathbf{b}_1, |b_2\rangle \equiv \mathbf{b}_2, \ldots$ be an orthonormal basis of V_H,

$$\langle b_i|b_j\rangle = \delta_{ij} \quad \text{and} \quad \sum_i |b_i\rangle\langle b_i| = \hat{I}_1, \tag{4.1.141}$$

where $\hat{I}_1$ is the identity operator in the one-particle Hilbert space V_H. We are assuming that V_H has a countable basis for simplicity here, but we can generalize to an uncountable basis such as the coordinate-space or the momentum-space basis in the usual way. Then an orthonormal basis for the Hilbert space $V_H^{(n)}$ of n distinguishable particles is

$$\begin{aligned}
|b_{i_1} b_{i_2} \cdots b_{i_n}\rangle &\equiv |b_{i_1}\rangle|b_{i_2}\rangle \cdots |b_{i_n}\rangle \equiv \mathbf{b}_{i_1} \otimes \mathbf{b}_{i_2} \otimes \cdots \otimes \mathbf{b}_{i_n}, \\
\langle b_{i_1} b_{i_2} \cdots b_{i_n}|b_{j_1} b_{j_2} \cdots b_{j_n}\rangle &= \delta_{i_1 j_1}\delta_{i_2 j_2} \cdots \delta_{i_n j_n}, \\
\sum_{i_1,\ldots,i_n} |b_{i_1} b_{i_2} \cdots b_{i_n}\rangle\langle b_{i_1} b_{i_2} \cdots b_{i_n}| &= \hat{I}_n,
\end{aligned} \tag{4.1.142}$$

where $\hat{I}_n$ is the identity operator in the Hilbert space of n distinguishable particles and where $|b_{i_1} b_{i_2} \cdots b_{i_n}\rangle$ is ordered sequentially with particle 1 first and ending with particle n. We have used both physics bra-ket notation and mathematics notation for clarity of meaning. The ordering of the one-particle basis states is relevant in the case of distinguishable particles. If the system is in the state $|\Psi\rangle$, then

$$|\Psi\rangle = \sum_{i_1,\ldots,i_n} \Psi(b_{i_1},\ldots,b_{i_n})|b_{i_1} b_{i_2} \cdots b_{i_n}\rangle, \tag{4.1.143}$$

where $\Psi(b_{i_1},\ldots,b_{i_n}) \equiv \langle b_{i_1} b_{i_2} \cdots b_{i_n}|\Psi\rangle$ and

$$P(b_{i_1},\ldots,b_{i_n}) = |\Psi(b_{i_1},\ldots,b_{i_n})|^2/\langle\Psi|\Psi\rangle \tag{4.1.144}$$

is the probability of finding particle 1 in the state $|b_{i_1}\rangle$, particle 2 in the state $|b_{i_2}\rangle$ and so on. The generalization to an n-particle state in a continuous basis such as the coordinate-space basis is straightforward, $\{|b_i\rangle : i = 1,2,\ldots\} \to \{|\mathbf{x}\rangle : \mathbf{x} \in \mathbb{R}\}$,

$$|\Psi\rangle = \int d^3x_1 d^3x_2 \cdots d^3x_n \Psi(\mathbf{x}_1, \mathbf{x}_2, \ldots, \mathbf{x}_n)|\mathbf{x}_1 \mathbf{x}_2 \cdots \mathbf{x}_n\rangle, \tag{4.1.145}$$

where $\Psi(\mathbf{x}_1, \mathbf{x}_2, \ldots, \mathbf{x}_n) \equiv \langle \mathbf{x}_1 \mathbf{x}_2 \cdots \mathbf{x}_n|\Psi\rangle$ and

$$\begin{aligned}
\langle \mathbf{x}_1 \cdots \mathbf{x}_n|\mathbf{y}_1 \cdots \mathbf{y}_n\rangle &= \delta^3(\mathbf{x}_1 - \mathbf{y}_1) \cdots \delta^3(\mathbf{x}_n - \mathbf{y}_n), \\
\int d^3x_1 \cdots d^3x_n |\mathbf{x}_1 \cdots \mathbf{x}_n\rangle\langle \mathbf{x}_1 \cdots \mathbf{x}_n| &= \hat{I}_n.
\end{aligned} \tag{4.1.146}$$

Distinguishable particles satisfy *Maxwell-Boltzmann statistics*, which is what the above construction leads to.

If we consider the case of *indistinguishable* (identical) particles, then there can be no distinction between different orderings of the arguments of the probability $P(b_{i_1},\ldots,b_{i_n})$; i.e., we can know the probability of having the n indistinguishable particles in the states labeled by $b_{1_1}, \ldots, b_{i_n}$ but it is not meaningful to consider which particle is in which state. This gives

$$P(\ldots, b_{i_k}, \ldots, b_{i_\ell}, \ldots) = P(\ldots, b_{i_\ell}, \ldots, b_{i_k}, \ldots) \quad \text{and hence}$$
$$\Psi(\ldots, b_{i_k}, \ldots, b_{i_\ell}, \ldots) = e^{i\phi} \Psi(\ldots, b_{i_\ell}, \ldots, b_{i_k}, \ldots). \tag{4.1.147}$$

If we do the same interchange twice, then we must arrive back at the same wavefunction and so we must have $e^{2i\phi} = 1$ or equivalently $e^{i\phi} = \pm 1$, so that

$$\Psi(\ldots, b_{i_k}, \ldots, b_{i_\ell}, \ldots) = \pm \Psi(\ldots, b_{i_\ell}, \ldots, b_{i_k}, \ldots) \quad \text{and hence}$$
$$|\ldots, b_{i_k}, \ldots, b_{i_\ell}, \ldots\rangle = \pm|\ldots, b_{i_\ell}, \ldots, b_{i_k}, \ldots\rangle. \tag{4.1.148}$$

If we assume that every pairwise exchange behaves the same way, then this means that we can only construct a basis for a system of n indistinguishable particles to be either symmetric or antisymmetric under every pairwise interchange of single-particle states.[2] This is sometimes referred to as the *symmetrization postulate*. By definition the symmetric case is for *bosons* and leads to *Bose-Einstein statistics* and the antisymmetric case is for *fermions* and leads to *Fermi-Dirac statistics*. Note that the antisymmetric nature of n-fermion states leads immediately to the *Pauli exclusion principle*, which states that no two indistinguishable fermions can be in the same state. In relativistic quantum field theory there are justifications for the *spin-statistics connections*, which is the proposition that integer spin particles must be bosons and half-integer spin particles must be fermions. The spin-statistics connection will be discussed in Sec. 8.1.5.

Identical bosons: For a quantum system of n identical bosons any state must be totally symmetric; i.e., the pairwise interchange of any two bosons leads to no change to the state. Let S_+ be the operator that symmetrizes an n-particle state. We define S_+ by its action on n-bosons in single-particle orthonormal basis states,

$$|b_{i_1} b_{i_2} \cdots b_{i_n}; S\rangle \equiv S_+|b_{i_1} b_{i_2} \cdots b_{i_n}\rangle \equiv \sqrt{\textstyle\prod_m n_m!/n!} \textstyle\sum_\sigma |b_{\sigma(1)} b_{\sigma(2)} \cdots b_{\sigma(n)}\rangle$$
$$= \sqrt{\textstyle\prod_m n_m!/n!} \textstyle\sum_\sigma \mathbf{b}_{\sigma(1)} \otimes \mathbf{b}_{\sigma(2)} \otimes \cdots \otimes \mathbf{b}_{\sigma(n)}, \tag{4.1.149}$$

where we sum over distinct permutations $\sigma(1), \ldots, \sigma(n)$ of $i_1, \ldots, i_n$, where n_m is the number of times b_m occurs in the n arguments $b_{i_1} b_{i_2} \cdots b_{i_n}$ and where the second line with $|b_i\rangle \equiv \mathbf{b}_i$ is the mathematics notation of the tensor product discussed in Sec. A.2.7. The square root factor is to ensure that the symmetrized states are normalized and it can be understood since the sum over all $n!$ permutations overcounts the distinct permutations by a factor of $n_m!$. Note that $\sum_m n_m = n$. By considering a few examples we can confirm that this is the correct normalization, e.g., show that $\langle b_1 b_2 b_2; S|b_1 b_2 b_2; S\rangle = \langle b_1 b_2 b_3 b_3; S|b_1 b_2 b_3 b_3; S\rangle = 1$.

Let $|\psi_{i_1}\rangle, \ldots, |\psi_{i_n}\rangle \in V_H$ be any set of n one-boson states. Then $|\psi_i\rangle = \sum_j c_{ij}|b_j\rangle$ for some set of c_{ij}. The symmetrized state $|\psi_{i_1} \cdots \psi_{i_n}; S\rangle$ has the property that

$$|\psi_{i_1} \psi_{i_2} \cdots \psi_{i_n}; S\rangle \equiv S_+|\psi_{i_1} \psi_{i_2} \cdots \psi_{i_n}\rangle = \textstyle\sum_{j_1 \ldots j_n} c_{i_1 j_1} \cdots c_{i_n j_n} S_+|b_{j_1} \cdots b_{j_n}\rangle$$
$$= \sqrt{\textstyle\prod_m n_m!/n!} \textstyle\sum_\sigma |\psi_{\sigma(1)} \cdots \psi_{\sigma(n)}\rangle, \tag{4.1.150}$$

where $\sum_\sigma$ is the sum over distinct permutations of $i_1, \ldots, i_n$. An arbitrary n boson state cannot be written as a single state $|\psi_{i_1} \psi_{i_2} \cdots \psi_{i_n}; S\rangle$ but can be written as a superposition of these, including when $|\psi_i\rangle$ are basis states.

---

[2] The relaxation of this assumption leads to parastatistics (Green, 1953). Parafermions of order n have at most n indistinguishable particles in a symmetric state; i.e., at most n particles can occupy the same state. Parabosons of order n have at most n indistinguishable particles in an antisymmetric state.

Since the ordering of the arguments is irrelevant, we can choose any arbitrary ordering rule that leads to a unique labeling of the n-boson basis states. We define an ordering prescription for the symmetrized n-boson basis states and use the notation S_0 (meaning *ordered* symmetrization) rather than S (meaning symmetrization),

$$|b_{i_1} b_{i_2} \cdots b_{i_n}; S_o\rangle \equiv |b_{i_1} b_{i_2} \cdots b_{i_n}; S\rangle \quad \text{with ordering } i_1 \leq i_2 \leq \cdots \leq i_n. \tag{4.1.151}$$

It then follows that these states $|b_{i_1} b_{i_2} \cdots b_{i_n}; S_o\rangle$ form an orthonormal basis of the Hilbert space for n identical bosons, $V_H^{(nb)} \equiv S_+((V_H)^{\otimes n})$,

$$\langle b_{j_1} \cdots b_{j_n}; S_o | b_{i_1} \cdots b_{i_n}; S_o\rangle = \delta_{i_1 j_1} \cdots \delta_{i_n j_n},$$

$$\sum_{i_1 \leq \cdots \leq i_n} |b_{i_1} b_{i_2} \cdots b_{i_n}; S_o\rangle \langle b_{i_1} b_{i_2} \cdots b_{i_n}; S_o| = \hat{I}_n, \tag{4.1.152}$$

where here $\hat{I}_n$ is the identity operator in the n-boson Hilbert space, $V_H^{(nb)}$. The ordering allows the orthonormal basis conditions to be written in a simple way, which is free of messy combinatorial factors.

Fock space for bosons is the space constructed to contain states with any number of bosons and also to contain states that are superpositions of these. This requires us to use the direct sum over these spaces for every possible number of bosons,

$$F_+(V_H) \equiv V_0 \oplus V_H \oplus S_+(V_H \otimes V_H) \oplus \cdots = \bigoplus_{n=0}^{\infty} S_+((V_H)^{\otimes n}), \tag{4.1.153}$$

where $V_H^{\otimes 1} = V_H$ and where $V_H^{\otimes 0} \equiv V_0$ is the one-dimensional Hilbert space spanned by the vacuum state $|0\rangle$, i.e., $c|0\rangle \in V_0$ for all $c \in \mathbb{C}$ and $\langle 0|0\rangle = 1$. The space $F_+(V_H)$ will always be an inner product space but it may not be complete for every choice of V_H and so may not be a Hilbert space. An inner product space that is not complete is called a pre-Hilbert space. It is always possible to construct a Hilbert space completion of a pre-Hilbert space, i.e., such that it contains the limits of all Cauchy sequences. *Fock space* is the *Hilbert space completion* of $F_+(V_H)$ and is denoted $\overline{F}_+(V_H)$. The orthonormal basis for the full boson Fock space $\overline{F}_+(V_H)$ has orthonormality and completeness relations given by

$$\langle b_{j_1} \cdots b_{j_n}; S_o | b_{i_1} \cdots b_{i_m}; S_o\rangle = \delta_{mn} \delta_{i_1 j_1} \cdots \delta_{i_n j_n},$$

$$\sum_{n=0}^{\infty} \sum_{i_1 \leq \cdots \leq i_n} |b_{i_1} b_{i_2} \cdots b_{i_n}; S_o\rangle \langle b_{i_1} b_{i_2} \cdots b_{i_n}; S_o| = \hat{I}. \tag{4.1.154}$$

Identical fermions: Conversely, for a quantum system consisting of n identical fermions any state must be totally antisymmetric, which means the pairwise interchange of any two fermions changes the sign of the state. Let $\hat{S}_-$ be the operator that antisymmetrizes an n-particle state. We define S_- by its action on n-fermions in single-particle orthonormal basis states,

$$|b_{i_1} \cdots b_{i_n}; A\rangle \equiv S_- |b_{i_1} \cdots b_{i_n}\rangle \equiv (1/\sqrt{n!}) \sum_\sigma \text{sgn}(\sigma) |b_{\sigma(1)} \cdots b_{\sigma(n)}\rangle \tag{4.1.155}$$

$$= (1/\sqrt{n!}) \sum_\sigma \text{sgn}(\sigma) \mathbf{b}_{\sigma(1)} \otimes \cdots \otimes \mathbf{b}_{\sigma(n)} = (1/\sqrt{n!}) \mathbf{b}_{i_1} \wedge \cdots \wedge \mathbf{b}_{i_n},$$

where the sum over permutations contains $n!$ terms and where $\text{sgn}(\sigma)$ means $+1$ or -1 for an even or odd number of pairwise exchanges, respectively.

Consider any set $|\psi_{i_1}\rangle, \ldots, |\psi_{i_n}\rangle \in V_H$ of n one-fermion states with $|\psi_i\rangle = \sum_j c_{ij} |b_j\rangle$ for some c_{ij}. The antisymmetrized state $|\psi_{i_1} \cdots \psi_{i_n}; A\rangle$ satisfies

$$|\psi_{i_1} \psi_{i_2} \cdots \psi_{i_n}; A\rangle \equiv S_- |\psi_{i_1} \psi_{i_2} \cdots \psi_{i_n}\rangle = \sum_{j_1 \cdots j_n} c_{i_1 j_1} \cdots c_{i_n j_n} S_- |b_{j_1} \cdots b_{j_n}\rangle$$

$$= (1/\sqrt{n!}) \sum_\sigma \text{sgn}(\sigma) |\psi_{\sigma(1)} \cdots \psi_{\sigma(n)}\rangle, \tag{4.1.156}$$

where $\sum_\sigma$ is the sum over permutations of $i_1, \ldots, i_n$. An arbitrary n fermion state can be written as a superposition of any such $|\psi_{i_1} \psi_{i_2} \cdots \psi_{i_n}; A\rangle$.

Again we can choose any ordering rule that leads to a unique labeling of the n-fermion orthonormal states. Define the *ordered* antisymmetric n-fermion state,

$$|b_{i_1} b_{i_2} \cdots b_{i_n}; A_o\rangle \equiv S_- |b_{i_1} b_{i_2} \cdots b_{i_n}\rangle \quad \text{with ordering} \quad i_1 < i_2 < \cdots < i_n. \qquad (4.1.157)$$

These states $|b_{i_1} b_{i_2} \cdots b_{i_n}; A_o\rangle$ form an orthonormal basis of the Hilbert space for n identical fermions, $V_H^{(nf)} \equiv S_-((V_H)^{\otimes n})$,

$$\langle b_{j_1} \cdots b_{j_n}; A_o | b_{i_1} \cdots b_{i_n}; A_o\rangle = \delta_{i_1 j_1} \cdots \delta_{i_n j_n},$$

$$\sum_{i_1 < \cdots < i_n} |b_{i_1} b_{i_2} \cdots b_{i_n}; A_o\rangle \langle b_{i_1} b_{i_2} \cdots b_{i_n}; A_o| = \hat{I}_n, \qquad (4.1.158)$$

where here $\hat{I}_n$ is the identity operator in the n-fermion Hilbert space, $V_H^{(nf)}$. We recognize that this is the bra-ket version of Eq. (A.2.30) and the associated discussion. If we removed the ordering, $A_o \to A$ and had no constraint on the completeness sum, then the orthonormality condition would have a sum of $n!$ products of δ-functions and the completeness sum would give $n!\hat{I}$ on the right-hand side. So the n-fermion Hilbert space is the nth exterior power of the one-fermion Hilbert space, $V_H^{(nf)} \equiv S_-((V_H)^{\otimes n}) = \bigwedge^n V_H$. See Sec. A.2.7 for a discussion of the exterior power.

For fermions Fock space contains states with any number of fermions and all superpositions of such states. This results from the direct sum over these spaces for every possible number of fermions,

$$F_-(V_H) \equiv V_0 \oplus V_H \oplus S_-(V_H \otimes V_H) \oplus \cdots = \bigoplus_{n=0}^\infty S_-((V_H)^{\otimes n})$$

$$= \bigoplus_{n=0}^\infty \bigwedge^n V_H \equiv \bigwedge V_H, \qquad (4.1.159)$$

where again $V_H^{\otimes 1} = V_H$ and where $V_H^{\otimes 0} \equiv V_0$ is the one-dimensional Hilbert space spanned by the vacuum state $|0\rangle$, i.e., $c|0\rangle \in V_0$ for all $c \in \mathbb{C}$ and $\langle 0|0\rangle = 1$. We recognize from Eq. (A.2.31) that $F_-(V_H)$ is the *exterior algebra* or *Grassmann algebra* of the single fermion Hilbert space V_H, which is denoted $\bigwedge V_H$. Fock space for fermions is denoted as $\overline{F}_-(V_H) \equiv \overline{\bigwedge V_H}$ and is the Hilbert space completion of $F_-(V_H) \equiv \bigwedge V_H$. If V_H is finite-dimensional with dimension N, then we must have $0 \leq n \leq N$. The orthonormal basis for the full fermion Fock space $\overline{F}_+(V_H)$ has orthonormality and completeness relations given by

$$\langle b_{j_1} \cdots b_{j_n}; A_o | b_{i_1} \cdots b_{i_m}; A_o\rangle = \delta_{mn} \delta_{i_1 j_1} \cdots \delta_{i_n j_n},$$

$$\sum_{n=0}^N \sum_{i_1 < \cdots < i_n} |b_{i_1} b_{i_2} \cdots b_{i_n}; A_o\rangle \langle b_{i_1} b_{i_2} \cdots b_{i_n}; A_o| = \hat{I}. \qquad (4.1.160)$$

Let $|\psi_{\alpha_1}\rangle, \ldots, |\psi_{\alpha_n}\rangle \in V_H$ be any set of n linearly independent one-fermion states. If each of these states is occupied by a fermion in an n-fermion system, the corresponding ordered ($\alpha_1 < \cdots < \alpha_n$) antisymmetrized state is

$$|\psi_{\alpha_1} \cdots \psi_{\alpha_n}; A_o\rangle = \sum_{i_1 < \cdots < i_n} |b_{i_1} \cdots b_{i_n}; A_o\rangle \langle b_{i_1} \cdots b_{i_n}; A_o | \psi_{\alpha_1} \cdots \psi_{\alpha_n}; A\rangle$$

$$= \sum_{i_1 < \cdots < i_n} \det C(i_1, \ldots, i_n) |b_{i_1} \cdots b_{i_n}; A_o\rangle$$

$$= (1/n!) \sum_{i_1, \ldots, i_n} \det C(i_1, \ldots, i_n) |b_{i_1} \cdots b_{i_n}; A\rangle \qquad (4.1.161)$$

$$= \sum_{i_1, \ldots, i_n} \Psi_{\alpha_1, \ldots, \alpha_n}(i_1, \ldots, i_n) (1/\sqrt{n!}) |b_{i_1} \cdots b_{i_n}; A\rangle,$$

where we have used Eq. (A.2.29) and have made the definitions

$$C(i_1, \ldots, i_n)_{\ell m} \equiv \langle b_{i_\ell} | \psi_{\alpha_m}\rangle \quad \text{and}$$

$$\Psi_{\alpha_1, \ldots, \alpha_n}(i_1, \ldots, i_n) \equiv (1/\sqrt{n!}) \det C(i_1, \ldots, i_n). \qquad (4.1.162)$$

If we take the one-particle basis for fermions as the coordinate-space basis and absorb the spin index into the state label α_m, then we have $C(\mathbf{x}_1,\ldots,\mathbf{x}_n)_{\ell m} = \langle \mathbf{x}_\ell | \psi_{\alpha_m} \rangle = \psi_{\alpha_m}(\mathbf{x}_\ell)$. We then arrive at

$$\Psi_{\alpha_1,\ldots,\alpha_n}(\mathbf{x}_1,\ldots,\mathbf{x}_n) \equiv \frac{1}{\sqrt{n!}} \det \begin{bmatrix} \psi_{\alpha_1}(\mathbf{x}_1) & \psi_{\alpha_1}(\mathbf{x}_2) & \cdots & \psi_{\alpha_1}(\mathbf{x}_n) \\ \psi_{\alpha_2}(\mathbf{x}_1) & \psi_{\alpha_2}(\mathbf{x}_2) & \cdots & \psi_{\alpha_2}(\mathbf{x}_n) \\ \vdots & \vdots & \ddots & \vdots \\ \psi_{\alpha_n}(\mathbf{x}_1) & \psi_{\alpha_n}(\mathbf{x}_2) & \cdots & \psi_{\alpha_n}(\mathbf{x}_n) \end{bmatrix}, \qquad (4.1.163)$$

which is referred to as the *Slater determinant*. It is a construction commonly used in descriptions of multi-electron and other multi-fermion systems and can be thought of a particular type of n-fermion wavefunction. Note that it is necessarily antisymmetric under the exchange of the one-particle state labels and the exchange of the spatial coordinates. Note that

$$(1/n!) \int d^3x_1 \cdots d^3x_n \, |\mathbf{x}_1 \cdots \mathbf{x}_n; A\rangle \langle \mathbf{x}_1 \cdots \mathbf{x}_n; A| = \hat{I}_n, \qquad (4.1.164)$$

where $\hat{I}_n$ is the identity operator in the n-fermion subspace. Consider some other set of n linearly independent one-fermion states $|\phi_{\beta_1}\rangle, \cdots, |\phi_{\beta_n}\rangle \in V_H$, then

$$\int d^3x_1 \cdots d^3x_n \, \Phi^*_{\beta_1,\ldots,\beta_n}(\mathbf{x}_1,\ldots,\mathbf{x}_n) \Psi_{\alpha_1,\ldots,\alpha_n}(\mathbf{x}_1,\ldots,\mathbf{x}_n)$$
$$= \int d^3x_1 \cdots d^3x_n \, \langle \phi_{\beta_1},\ldots,\phi_{\beta_n}; A | \mathbf{x}_1,\ldots,\mathbf{x}_n; A \rangle \langle \mathbf{x}_1,\ldots,\mathbf{x}_n; A | \psi_{\alpha_1},\ldots,\psi_{\alpha_n}; A \rangle$$
$$= \langle \phi_{\beta_1},\ldots,\phi_{\beta_n}; A | \psi_{\alpha_1},\ldots,\psi_{\alpha_n}; A \rangle. \qquad (4.1.165)$$

Not every n-fermion state can be represented by an n-fermion Slater determinant but every n-fermion state can be represented by a linear superposition of them.

4.1.10 Charge Conjugation and Antiparticles

The *charge conjugate* of a particle is referred to as its *antiparticle*. Replacing every particle by its antiparticle and vice versa while changing nothing else in a system is by definition the *charge conjugation* of that system. Charge conjugation, $\hat{C}$, does not change the position, linear momentum, orbital angular momentum or spin of a particle; i.e., it does not effect quantities that transform under Poincaré transformations. An exception to this is the case of a massless neutrino with only one helicity, such as the left-handed neutrinos in the Standard Model. In this case we need a CP transformation to replace neutrinos by antineutrinos.

The intrinsic *additive* quantum numbers of antiparticles are the opposite of those of the corresponding particle. So the total for every additive intrinsic quantum number in a system consisting of a particle and its antiparticle is zero. Fundamental particles can also carry an *intrinsic parity* in addition to the parity associated with their orbital motion. We will show in Sec. 4.5.2 that a massive spin-$\frac{1}{2}$ particle has the opposite intrinsic parity π_P to its antiparticle, where choosing all fermions to have $\pi_P = +1$ and all antifermions to have $\pi_P = -1$ leads to no inconsistencies (Weinberg, 1995). Consider any spin-$\frac{1}{2}$ fermion-antifermion pair $f_1 \bar{f}_2$ with relative orbital angular momentum quantum number L, total spin $S = 0, 1$ and total angular momentum $J = L, |L \pm 1|$. It is conventional to use the *spectroscopic notation*, $^{2S+1}L_J$, to denote such states. We know from quantum mechanics that the parity associated with the relative orbital motion is $(-1)^L$. Since parity is a multiplicative quantum number, the total parity P is the product of the two intrinsic parities and the orbital parity,

$$P(f_1 \bar{f}_2, {}^{2S+1}L_J) = (+1) \times (-1) \times (-1)^L = (-1)^{L+1}. \qquad (4.1.166)$$

The charge conjugate of the system is $\bar{f}_1 f_2$ in the same spectroscopic state and so has the same total parity since

$$P(\bar{f}_1 f_2, {}^{2S+1}L_J) = (-1) \times (+1) \times (-1)^L = (-1)^{L+1}. \qquad (4.1.167)$$

Let us here consider only couplings that conserve parity and that are invariant under charge conjugation. Consider the decay of a boson b with spin J to this $f_1 \bar{f}_2$. Charge conjugation shows that the decay of the antiboson $\bar{b}$ will be to $\bar{f}_1 f_2$, which has the same parity. So clearly the intrinsic parity of b and $\bar{b}$ must be the same. Similarly, consider any state made up of three fermions $f_1 f_2 f_3$. Under charge conjugation we arrive at the same state in terms of spins and relative motions but consisting now of $\bar{f}_1 \bar{f}_2 \bar{f}_3$. Since the product of the intrinsic parities has flipped sign, $(+1)^3 = +1 \rightarrow (-1)^3 = -1$ with nothing else changed, then the total parity must also change sign. This means that any particle with half-integer spin $s = \frac{1}{2}, \frac{3}{2}, \ldots$ (any fermion) that can decay into an odd number of spin-$\frac{1}{2}$ fermions will have an antiparticle with the opposite intrinsic parity. In summary, *bosons and antibosons have the same intrinsic parity, while fermions and antifermions have opposite intrinsic parity.*

Consider a state consisting of any n_1 spin-$\frac{1}{2}$ fermions and any n_2 spin-$\frac{1}{2}$ antifermions. Since a boson has integer spin it can only couple to such a state if $n_1 + n_2$ is even, whereas a fermion has half-integer spin and can only couple to such a state if $n_1 + n_2$ is odd. The total intrinsic parity of the state is $\pi_P = (+1)^{n_1} \times (-1)^{n_2} = (-1)^{n_2}$ and the total intrinsic parity of the charge conjugate state is $\pi_P = (-1)^{n_1} \times (+1)^{n_2} = (-1)^{n_1}$. The change in total parity is the change in total intrinsic parity, $(-1)^{n_2}/(-1)^{n_1} = (-1)^{n_2-n_1} = (-1)^{n_2+n_1}$. So if a particle can couple to a combination of spin-$\frac{1}{2}$ fermions and spin-$\frac{1}{2}$ antifermions and if parity and charge conjugation are good symmetries, then as stated above bosons and antibosons have the same intrinsic parity and fermions and antifermions have opposite intrinsic parity. While parity is not conserved by the weak interaction, it is respected by the strong and electromagnetic interactions and so intrinsic parities can be deduced from transitions induced by these forces. For example, protons are fermions and have positive parity and antiprotons have negative parity. The three pions π^0 and $\pi^\pm$ are pseudoscalar bosons and so all have negative parity. The π^0 is its own antiparticle and the π^+ and π^- are a particle-antiparticle pair.

We can regard a particle and its corresponding antiparticle as two different states of the same particle with opposite additive quantum charges q and $-q$ and identical or opposite intrinsic parities for bosons and fermions, respectively. Here q is to be understood to represent all additive quantum charges if there is more than one. For a boson b with integer spin s, additive quantum numbers q, intrinsic parity π_P we write the boson and antiboson in spin state m_s at position $\mathbf{x}$, respectively, in terms of the simultaneous eigenvalues of a maximal set of commuting observables as

$$|b(m_s, \mathbf{x})\rangle = |q, \pi_P, s, m_s, \mathbf{x}\rangle \quad \text{and} \quad |\bar{b}(m_s, \mathbf{x})\rangle = |-q, \pi_P, s, m_s, \mathbf{x}\rangle. \qquad (4.1.168)$$

Similarly for a fermion with half-integer spin s we can write the corresponding fermion and antifermion states, respectively, as

$$|f(m_s, \mathbf{x})\rangle = |q, \pi_P, s, m_s, \mathbf{x}\rangle \quad \text{and} \quad |\bar{f}(m_s, \mathbf{x})\rangle = |-q, -\pi_P, s, m_s, \mathbf{x}\rangle. \qquad (4.1.169)$$

Let $\hat{\mathcal{C}}$ denote the charge conjugation operator, then

$$\hat{\mathcal{C}}|b(m_s, \mathbf{x})\rangle = |\bar{b}(m_s, \mathbf{x})\rangle \quad \text{and} \quad \hat{\mathcal{C}}|\bar{b}(m_s, \mathbf{x})\rangle = |b(m_s, \mathbf{x})\rangle,$$

$$\hat{\mathcal{C}}|f(m_s, \mathbf{x})\rangle = |\bar{f}(m_s, \mathbf{x})\rangle \quad \text{and} \quad \hat{\mathcal{C}}|\bar{f}(m_s, \mathbf{x})\rangle = |f(m_s, \mathbf{x})\rangle. \qquad (4.1.170)$$

We require particle and antiparticle states to have the same normalization and so

$$\langle \bar{b}(m_s, \mathbf{x}) | \bar{b}(m_s, \mathbf{x}) \rangle = \langle b(m_s, \mathbf{x}) | \hat{C}^\dagger \hat{C} b(m_s, \mathbf{x}) \rangle = \langle b(m_s, \mathbf{x}) | b(m_s, \mathbf{x}) \rangle \tag{4.1.171}$$

for all $|b(m_s, \mathbf{x})\rangle$ and similarly for fermions. This means that $\hat{C}$ must be either unitary or antiunitary. We know from the arguments associated with Eq. (4.1.128) that if antiunitary, then the energy would be unbounded below and so $\hat{C}$ must be unitary, $\hat{C}^\dagger = \hat{C}^{-1}$. Clearly acting with $\hat{C}$ on a state twice returns the same state, so that $\hat{I} = \hat{C}^2$ and $\hat{C} = \hat{C}^{-1}$. This means that $\hat{C}$ is Hermitian and hence an observable,

$$\hat{C} = \hat{C}^{-1} = \hat{C}^\dagger. \tag{4.1.172}$$

The set of all $|b(m_s, \mathbf{x})\rangle$ and $|\bar{b}(m_s, \mathbf{x})\rangle$ form a basis for the one-boson Hilbert space V_H and the set of all $|f(m_s, \mathbf{x})\rangle$ and $|\bar{f}(m_s, \mathbf{x})\rangle$ form a basis for the one-fermion Hilbert space V_H. In each case V_H consists of a particle subspace spanned by $|b(m_s, \mathbf{x})\rangle$ or $|f(m_s, \mathbf{x})\rangle$ and an antiparticle subspace spanned by $|\bar{b}(m_s, \mathbf{x})\rangle$ or $|\bar{f}(m_s, \mathbf{x})\rangle$, respectively. The completions of the sums over all symmetrized and antisymmetrized tensor products of the one-boson and one-fermion Hilbert space V_H, respectively, give the boson Fock space $\overline{F}_+(V_H)$ and the fermion Fock space $\overline{F}_-(V_H)$.

Let the state $|\Psi\rangle$ be a vector in Fock space, which is an eigenstate of the observable operator $\hat{C}$ with eigenvalue $C \equiv \eta_C$. We refer to the eigenvalue as the *charge parity* or *C-parity* of the state. Both η_C and C are in common usage. Then we have

$$\hat{C}|\Psi\rangle = \eta_C|\Psi\rangle \quad \text{and} \quad \hat{C}^2|\Psi\rangle = \eta_C^2|\Psi\rangle = |\Psi\rangle \tag{4.1.173}$$

and so $C \equiv \eta_C = \pm 1$, which is a multiplicative quantum number in the same way that parity is. Clearly, only states with all additive quantum numbers equal to zero can be eigenstates of charge conjugation.

Consider a state $|\pi^+\pi^-; L\rangle$ consisting of a π^+ and a π^- with a relative orbital angular momentum L. The total parity is the product of the intrinsic parities and the orbital parity, $P = (-1)^2 \times (-1)^L = (-1)^L$. Under charge conjugation the π^+ becomes a π^- and vice versa with no other change to the state. Since they have the same intrinsic parity, then charge conjugation is equivalent to reversing both electric charges with no other change. The two pions are in an overall symmetric state in $\overline{F}_+(V_H)$ when *all* simultaneous eigenvalues are exchanged. Effectively only charge and position are exchanged since they are spinless with the same intrinsic parity. So charge conjugation is equivalent to exchanging the two particles (giving $+1$ since they are bosons) and then interchanging their positions (giving $(-1)^L$ since this is a spatial parity operation). This gives an overall factor of $\eta_C \equiv C = (+1) \times (-1)^L = (-1)^L$. We then have

$$\hat{\mathcal{P}}|\pi^+\pi^-; L\rangle = P|\pi^+\pi^-; L\rangle = (-1)^L|\pi^+\pi^-; L\rangle \quad \text{and}$$
$$\hat{C}|\pi^+\pi^-; L\rangle = C|\pi^+\pi^-; L\rangle = (-1)^L|\pi^+\pi^-; L\rangle. \tag{4.1.174}$$

Next consider a fermion-antifermion system in the state $^{2S+1}L_J$. Charge conjugation reverses the sign of the charges and the intrinsic parities but leaves the spin and coordinate-space variables unchanged. If we exchange *all* simultaneous eigenvalues of the fermion pair we get a factor of (-1) due to the antisymmetric Fock space. If we then exchange the spins of the fermion and antifermion we get a factor $(-1)^{S+1}$, since even/odd S is antisymmetric/symmetric under spin interchange and if we exchange the spatial coordinates we again get a factor of $(-1)^L$, since that is just a spatial parity transformation. Then charge conjugation gives an overall factor of $(-1) \times (-1)^{S+1} \times (-1)^L = (-1)^{L+S}$. In summary we have

$$\hat{P}|f\bar{f}; \, {}^{2S+1}L_J\rangle = P|f\bar{f}; \, {}^{2S+1}L_J\rangle = (-1)^{L+1}|f\bar{f}; \, {}^{2S+1}L_J\rangle \quad \text{and}$$

$$\hat{C}|f\bar{f}; \, {}^{2S+1}L_J\rangle = C|f\bar{f}; \, {}^{2S+1}L_J\rangle = (-1)^{L+S}|f\bar{f}; \, {}^{2S+1}L_J\rangle. \qquad (4.1.175)$$

Under charge conjugation electric charge is reversed in sign since it is an additive quantum number. So the four-vector electric current density changes sign, $j^\mu \to -j^\mu$. From the covariant form of Maxwell's equations in Eq. (2.7.23) and the definition of the electromagnetic field strength tensor $F^{\mu\nu}$ we then see that under charge conjugation we must have $A^\mu \to -A^\mu$. Operators obey the same equations as their classical counterparts and so a one-photon state created by $\hat{A}^\mu$ will also change sign under charge conjugation, it therefore follows that $C(\gamma) \equiv \eta_C(\gamma) = -1$; i.e., the photon has a C-parity of -1. A state with n photons has $C \equiv \eta_C = (-1)^n$. Since we observe the decay $\pi^0 \to \gamma + \gamma$ but not $\pi^0 \to \gamma + \gamma + \gamma$ and since $C \equiv \eta_C$ is a multiplicative quantum number, then we conclude that $C(\pi^0) \equiv \eta_C(\pi^0) = (-1)^2 = +1$. Using the notation J^{PC} we see that for a photon we have $J^{PC} = 1^{--}$ and for a pion we have $J^{PC} = 0^{-+}$.

Positronium: An example of a fermion-antifermion system is *positronium*, which is the bound state of an electron and a positron. Positronium states are resonances since they are unstable; e.g., the electron-positron pair can annihilate into two, three or more photons in the final state. Otherwise its energy levels resemble those of hydrogen. The reduced mass of positronium, $\mu_{e^+e^-} = m_{e^-}m_{e^+}/(m_{e^-} + m_{e^+}) = m_e/2$, is approximately half that of hydrogen, $\mu_H = m_e m_p/(m_e + m_p) \simeq m_e$. As for hydrogen the lowest energy states have angular momentum $L = 0$ (S-wave). The total angular momentum and spin of the $L = 0$ states can be either $J = S = 0$ or $J = S = 1$. We refer to these resonances as *para-positronium*, $|e^-e^+; \, {}^1S_0\rangle$ with ${}^{2S+1}L_J = {}^1S_0$, and *ortho-positronium*, $|e^-e^+; \, {}^3S_1\rangle$ with ${}^{2S+1}L_J = {}^3S_1$, respectively. From the above arguments positronium must have $P = (-1)^{L+1}$ and $C = (-1)^{L+S}$ and so para-positronium has $J^{PC} = 0^{-+}$ and ortho-positronium has $J^{PC} = 1^{--}$. Charge conjugation invariance then requires that para-positronium can only decay into an even number of photons and ortho-positronium can only decay into an odd number of photons. The *branching ratio* (the probability of decaying into a particular set of final state particles per decay) falls rapidly with numbers of photons in the final state, since they involve additional powers of the fine-structure constant $\alpha \simeq 1/137$. Decays into other channels, such as a neutrino-antineutrino pair, via the weak interaction are possible but have a negligible branching ratio because the weak interaction coupling is very much smaller than α.

4.1.11 Interaction Picture in Quantum Mechanics

Let us consider a time-dependent Schrödinger picture Hamiltonian of the form

$$\hat{H}_s(t) = \hat{H}_0 + \hat{H}_{\text{int}}(t), \qquad (4.1.176)$$

where H_0 is a time-independent "free" part of the Hamiltonian and $H_{\text{int}}(t)$ is a possibly time-dependent "interaction" part of the Hamiltonian. Typically, H_0 is chosen to be a Hamiltonian of free fields. We will temporarily restore $\hbar$ here. We can then define an *interaction picture*, which is a hybrid picture partway between the Schrödinger and Heisenberg pictures. From Sec. 4.1.1 we know that the free and full evolution operators are

$$\hat{U}_0(t'', t') \equiv e^{-i\hat{H}_0(t''-t')/\hbar} \quad \text{and} \quad \hat{U}(t'', t') \equiv Te^{-i\int_{t'}^{t''} dt\, \hat{H}_s(t)/\hbar}. \qquad (4.1.177)$$

The relationship between the Schrödinger (s), interaction (I) and Heisenberg (h) pictures and the definition of the interaction picture evolution operator is given by

$$s \xrightarrow{U} h \quad \text{with} \quad s \xrightarrow{U_0} I \xrightarrow{U_I} h \quad \Rightarrow \quad \hat{U}(t,t_0) \equiv \hat{U}_0(t,t_0)\hat{U}_I(t,t_0). \qquad (4.1.178)$$

This defines the interaction picture in terms of the other two pictures,

$$\hat{A}_h(t) = \hat{U}(t,t_0)^\dagger \hat{A}_s \hat{U}(t,t_0) = \hat{U}_I(t,t_0)^\dagger \hat{A}_I(t) U_I(t,t_0), \qquad (4.1.179)$$

$$|\psi\rangle_h = \hat{U}(t,t_0)^\dagger |\psi,t\rangle_s = \hat{U}_I(t,t_0)^\dagger |\psi,t\rangle_I, \qquad (4.1.180)$$

where we have defined the interaction picture operators and state of the system as

$$\hat{A}_I(t) = \hat{U}_0(t,t_0)^\dagger \hat{A}_s \hat{U}_0(t,t_0) = \hat{U}_I(t,t_0)\hat{A}_h(t)\hat{U}_I(t,t_0)^\dagger,$$

$$|\psi,t\rangle_I = \hat{U}_0(t,t_0)^\dagger |\psi,t\rangle_s = \hat{U}_I(t,t_0)|\psi\rangle_h. \qquad (4.1.181)$$

Note that the interaction picture operator $\hat{A}_I(t)$ is just the Heisenberg picture operator in the free field theory where $\hat{H}_{\text{int}}(t) = 0$. We see that we have picture-independence of expectation values as we should, since

$$_s\langle\psi,t|\hat{A}_s|\psi,t\rangle_s = {}_h\langle\psi|\hat{A}_h(t)|\psi\rangle_h = {}_I\langle\psi,t|\hat{A}_I(t)|\psi,t\rangle_I, \qquad (4.1.182)$$

where we assume normalized states here. We write the Schrödinger picture operators as time-independent but they could also be given an explicit time-dependence such as an operator involving an external source, $\hat{A}_s \to \hat{A}_s(t)$. Recall that t_0 is the time at which the Schrödinger and Heisenberg picture operators are equivalent. To define the interaction picture evolution operator between two arbitrary times, t'' and t', we consider

$$|\psi,t''\rangle_I = U_I(t'',t_0)|\psi\rangle_h = U_I(t'',t_0)U_I(t',t_0)^\dagger|\psi,t'\rangle_I$$

$$= U_0(t'',t_0)^\dagger \hat{U}(t'',t')\hat{U}_0(t',t_0)|\psi,t'\rangle_I \equiv \hat{U}_I(t'',t')|\psi,t'\rangle_I. \qquad (4.1.183)$$

So the interaction picture evolution operator for arbitrary times t'' and t' is

$$\hat{U}_I(t'',t') \equiv \hat{U}_0(t'',t_0)^\dagger \hat{U}(t'',t')\hat{U}_0(t',t_0), \qquad (4.1.184)$$

which reduces to the above definition when $t' = t_0$. We then find

$$\hat{U}_I(t''',t') = \hat{U}_I(t''',t'')\hat{U}_I(t'',t'), \qquad \hat{U}_I(t'',t') = \hat{U}_I(t',t'')^\dagger, \qquad (4.1.185)$$

using the properties of $\hat{U}_0$ and $\hat{U}$. If we set $\hat{H}_{\text{int}} = 0$ the interaction picture coincides with the Heisenberg picture. The interaction picture of Eq. (4.1.181) is the Heisenberg picture for a theory defined by the Hamiltonian $\hat{H}_0$ alone.

From Eq. (4.1.181) we have $|\psi,t\rangle_s = \hat{U}_0(t,t_0)|\psi,t\rangle_I$ and using this in the Schrödinger equation of Eq. (4.1.10) gives

$$i\hbar\frac{d}{dt}(e^{-i\hat{H}_0(t-t_0)/\hbar}|\psi,t\rangle_I) = \{\hat{H}_0 + \hat{H}_{\text{int}}(t)\}(e^{-i\hat{H}_0(t-t_0)/\hbar}|\psi,t\rangle_I) \qquad (4.1.186)$$

from which we obtain the interaction picture form of the Schrödinger equation,

$$i\hbar\frac{d}{dt}|\psi,t\rangle_I = e^{+i\hat{H}_0(t-t_0)/\hbar}\hat{H}_{\text{int}}(t)e^{-i\hat{H}_0(t-t_0)/\hbar}|\psi,t\rangle_I$$

$$= \hat{U}_0(t,t_0)^\dagger \hat{H}_{\text{int}}(t)\hat{U}_0(t-t_0)|\psi,t\rangle_I \equiv \hat{H}_I(t)|\psi,t\rangle_I, \qquad (4.1.187)$$

where following convention we define $\hat{H}_{\text{int}}(t)$ in the interaction picture as $\hat{H}_I(t)$,

$$\hat{H}_I(t) \equiv \hat{U}_0(t,t_0)^{\dagger}\hat{H}_{\text{int}}(t)\hat{U}_0(t,t_0) = (\hat{H}_{\text{int}})_I. \tag{4.1.188}$$

To avoid clashing with this conventional usage of $\hat{H}_I \equiv (\hat{H}_{\text{int}})_I$, we will use $(\hat{H})_I$ to denote the full Hamiltonian in the interaction picture following from Eq. (4.1.181),

$$(\hat{H})_I = \hat{U}_0(t,t_0)^{\dagger}\hat{H}_s\hat{U}_0(t,t_0) = \hat{U}_0(t,t_0)^{\dagger}\{\hat{H}_0 + \hat{H}_{\text{int}}\}\hat{U}_0(t,t_0) = \hat{H}_0 + \hat{H}_I. \tag{4.1.189}$$

From Eq. (4.1.187) using $|\psi,t\rangle_I = \hat{U}_I(t,t_0)|\psi\rangle_h$ we have

$$\frac{d}{dt}\hat{U}_I(t,t') = -(i/\hbar)\hat{H}_I(t)\hat{U}_I(t,t'). \tag{4.1.190}$$

From this equation we can now repeat, in the interaction picture, each of the steps that led to Eq. (4.1.15), including us arriving at the interaction picture version of the Dyson formula in Eq. (4.1.14)

$$\hat{U}_I(t'',t') = \hat{I} + (-i/\hbar)\int_{t'}^{t''} dt_1\hat{H}_I(t_1) + (-i/\hbar)^2\int_{t'}^{t''} dt_1 \int_{t'}^{t_1} dt_2\, \hat{H}_I(t_1)\hat{H}_I(t_2)$$
$$+ (-i/\hbar)^3 \int_{t'}^{t''} dt_1 \int_{t'}^{t_1} dt_2 \int_{t'}^{t_2} dt_3\, \hat{H}_I(t_1)\hat{H}_I(t_2)\hat{H}_I(t_3) + \cdots. \tag{4.1.191}$$

So we have shown that the evolution operator in the interaction picture satisfies

$$\hat{U}_I(t'',t') = U_0(t'',t_0)^{\dagger}\hat{U}(t'',t')\hat{U}_0(t',t_0) = T\left(\exp\left\{(-i/\hbar)\int_{t'}^{t''} dt\, \hat{H}_I(t)\right\}\right). \tag{4.1.192}$$

Comparing the Schrödinger equation in Eq. (4.1.12) with its interaction picture form in Eq. (4.1.190) we see that we have $\hat{U} \to \hat{U}_I$ and $\hat{H}_s \to \hat{H}_I$. Recalling that $\hat{A}_I(t) = \hat{U}_0(t,t_0)^{\dagger}\hat{A}_s\hat{U}_0(t,t_0)$ we can then follow the steps of Eq. (4.1.24) to arrive at the Heisenberg equations of motion in the interaction picture,

$$\frac{d\hat{A}_I(t)}{dt} = \frac{1}{i\hbar}[\hat{A}_I(t),\hat{H}_0] + \frac{\partial\hat{A}_I(t)}{\partial t}, \tag{4.1.193}$$
$$\partial\hat{A}_I(t)/\partial t \equiv \hat{U}_0(t,t_0)^{\dagger}(\partial\hat{A}_s(t)/\partial t)\hat{U}_0(t,t_0).$$

4.1.12 Path Integrals in Quantum Mechanics

The term "path integral" is usually used to mean functional-integral as applied to nonrelativistic quantum mechanics or quantum statistical mechanics. The possibility of a path-integral approach to nonrelativistic quantum mechanics was initially raised by Dirac and this approach was later extensively developed by Feynman (Feynman, 1948; Feynman and Hibbs, 1965). Closely related and on a sound mathematical foundation is the path-integral formulation of quantum statistical mechanics (Feynman, 1972; Wiegel, 1975). Path integrals were first introduced into statistical mechanics in a rigorous way by Wiener in the 1920s in a study of diffusion theory, and in this application they came to be known as Wiener integrals. References to many of the early contributions in this field are given in the extensive bibliography of Albeverio and Hoegh-Krohn (1976).

If we know all of the matrix elements of the evolution operator $\hat{U}(t'',t')$, with respect to some basis for a quantum system between all possible times, t', t'', then we know everything about the system. If the basis is the coordinate-space basis, then these matrix elements correspond to the *Green's function* $G(q'',q';t'',t')$ defined as

$$G(q'',q';t'',t') \equiv {}_h\langle q'',t''|q',t'\rangle_h = {}_s\langle q''|\hat{U}(t'',t')|q'\rangle_s. \tag{4.1.194}$$

Now divide $(t'' - t')$ into $(n + 1)$ equal intervals of length

$$\delta t = (t'' - t')/(n+1) \quad \text{with} \quad t_j \equiv t' + j\delta t \quad \text{for} \quad j = 0, 1, \ldots, n + 1, \tag{4.1.195}$$

where we see that $t_{j+1} = t_j + \delta t$, $t'' = t_{n+1}$ and $t' = t_0$. We omit the labels h and s for brevity from now on in this section. Noting the completeness of the coordinate-space and momentum-space bases in both the Heisenberg and Schrödinger pictures,

$$\int dq \, |q, t\rangle\langle q, t| = \int dq \, |q\rangle\langle q| = \int dp \, |p, t\rangle\langle p, t| = \int dp \, |p\rangle\langle p| = \hat{I}, \tag{4.1.196}$$

we can write

$$\langle q'', t''|q', t'\rangle = \int dq_1 \cdots \int dq_n \langle q'', t''|q_n, t_n\rangle \cdots \langle q_2, t_2|q_1, t_1\rangle\langle q_1, t_1|q', t'\rangle. \tag{4.1.197}$$

Consider a time-independent Hamiltonian of the form

$$\hat{H} \equiv H(\hat{p}, \hat{q}) = (\hat{p}^2/2m) + V(\hat{q}). \tag{4.1.198}$$

Then $\hat{U}(t'', t') = e^{-i\hat{H}(t''-t')/\hbar}$ and we can write

$$\langle q_{j+1}, t_{j+1}|q_j, t_j\rangle = \langle q_{j+1}|e^{-i\delta t \hat{H}/\hbar}|q_j\rangle = \delta(q_{j+1} - q_j) - i(\delta t/\hbar)\langle q_{j+1}|\hat{H}|q_j\rangle + O(\delta t^2).$$

We will use the conventional quantum mechanics choice of the symmetric definition of the Fourier transform given in Eq. (A.2.12), where of course the final result is independent of this choice. Using the result

$$\langle q|p\rangle = (1/\sqrt{2\pi\hbar})e^{ipq/\hbar}, \tag{4.1.199}$$

we find that

$$\langle q_{j+1}|\hat{H}|q_j\rangle = \langle q_{j+1}|H(\hat{q}, \hat{p})|q_j\rangle = \int dp_j \, \langle q_{j+1}|p_j\rangle\langle p_j|H(\hat{q}, \hat{p})|q_j\rangle \tag{4.1.200}$$
$$= \int dp_j \, H'(q_j, p_j)\langle q_{j+1}|p_j\rangle\langle p_j|q_j\rangle = \int (dp_j/2\pi\hbar)e^{ip_j(q_{j+1}-q_j)/\hbar} \, H(q_j, p_j).$$

Hence we can write

$$\langle q_{j+1}, t_{j+1}|q_j, t_j\rangle = \int (dp_j/2\pi\hbar)e^{ip_j(q_{j+1}-q_j)/\hbar} \left\{ 1 - i(\delta t/\hbar)H(q_j, p_j) + \mathcal{O}(\delta t^2) \right\}$$
$$= \int (dp_j/2\pi\hbar)e^{i(\delta t/\hbar)[p_j\{(q_{j+1}-q_j)/\delta t\} - H(q_j, p_j)]} + \mathcal{O}(\delta t^2).$$

Using this result in Eq. (4.1.197) gives

$$\langle q'', t''|q', t'\rangle = \lim_{n\to\infty} \int \prod_{i=1}^{n} dq_i \int \prod_{k=0}^{n} (dp_k/2\pi\hbar)e^{i\sum_{j=0}^{n} \delta t[p_j\{(q_{j+1}-q_j)/\delta t\} - H(q_j, p_j)]/\hbar}$$
$$\equiv \int \mathcal{D}q \, \mathcal{D}p \, e^{(i/\hbar)\int_{t'}^{t''} dt[p\dot{q} - H(q,p)]}. \tag{4.1.201}$$

The last line of Eq. (4.1.201) is shorthand notation for the line above and represents an integration over all possible *paths in phase space* with the coordinate end points $q(t'') = q''$ and $q(t') = q'$. There are no restrictions on the momentum end points. The asymmetry between the roles of p and q is due to the fact that the coordinate space Green's function is being calculated.

We will only take the continuum limit of path integrals *after* forming appropriate ratios of them that correspond to the calculation of a physical observable. This understanding means that we do not concern ourselves with the mathematical subtleties associated with the rigorous definition of a continuum path integral. For example, an exponent that is second order in the time derivative corresponds in the continuum limit to an integral over continuous but non-differentiable paths, which is to say that the set of differentiable paths is a set of measure zero. This is very much in the spirit of lattice field theory where the continuum and infinite volume limits are only taken for physical observables.

Mathematical digression: Useful Gaussian integrals: For any $\mu, \nu \in \mathbb{C}$ with $\mathrm{Re}(\mu) > 0$ we have

$$\int_{-\infty}^{\infty} d\phi \, e^{-\frac{1}{2}\mu\phi^2 + \nu\phi} = (2\pi/\mu)^{1/2} \, e^{\nu^2/2\mu}. \tag{4.1.202}$$

In particular, we have for $a, j, \epsilon \in \mathbb{R}$ and $\epsilon \to 0^+$

$$\int_{-\infty}^{\infty} d\phi \, e^{i\left(\frac{1}{2}(a+i\epsilon)\phi^2 + j\phi\right)} = \left(\frac{2\pi i}{a+i\epsilon}\right)^{1/2} \exp\left\{-\frac{i}{2}\frac{j^2}{a+i\epsilon}\right\}. \tag{4.1.203}$$

The multi-dimensional form of this is

$$\int \prod_{i=1}^{N} d\phi_i \, e^{i[\sum_{i,j} \frac{1}{2}\phi_i(A+i\epsilon I)_{ij}\phi_j + \sum_i j_i\phi_i]}$$
$$= \left(\frac{(2\pi i)^N}{\det(A+i\epsilon I)}\right)^{1/2} e^{-i\sum_{i,j}\frac{1}{2}j_i(A+i\epsilon I)_{ij}^{-1}j_j}, \tag{4.1.204}$$

where A is any real $N \times N$ symmetric matrix. The proof diagonalizes A with an orthogonal transformation, $D = OAO^T$, with the diagonal matrix D having the eigenvalues λ_i of A as its elements. Then define $\vec{\phi}' = O\vec{\phi}$ and similarly $\vec{j}'$ and note that $\prod_{i=1}^{N} d\phi_i = \prod_{i=1}^{N} d\phi_i' |\det O|^{-1} = \prod_{i=1}^{N} d\phi_i'$. This leads to a product of N one-dimensional Gaussian integrals where each a corresponds to an eigenvalue λ_i of A. The final result is arrived at after changing variables back to $\vec{j}$ and noting that $\det(A + i\epsilon) = \prod_{i=1}^{N}(\lambda_i + i\epsilon)$.

In a Euclidean formulation we have only real quantities $i(A + i\epsilon) \to -A$ and $i\vec{j} \to \vec{j}$ with A a real symmetric matrix with only positive eigenvalues, $\lambda_i > 0$. Using the same steps as in the previous proof we arrive at

$$\int \prod_{i=1}^{N} d\phi_i \, e^{-\sum_{i,j}\frac{1}{2}\phi_i A_{ij}\phi_j + \sum_i j_i\phi_i} = \left(\frac{(2\pi)^N}{\det A}\right)^{1/2} e^{\sum_{i,j}\frac{1}{2}j_i A_{ij}^{-1}j_j}. \tag{4.1.205}$$

The momentum integrations in Eq. (4.1.201) can be carried out with $H(\hat{p}, \hat{q})$ in the form of Eq. (4.1.198) since it is second-order in $\hat{p}$. From the Gaussian integrals given in Eqs. (4.1.202) and (4.1.203) and by introducing the positive real infinitesimal Feynman parameter ϵ we can make the temporary replacement $t \to t(1 - i\epsilon)$ so that $\delta t \to \delta t(1 - i\epsilon)$. Note that $\dot{q} = (q_{j+1} - q_j)/\delta t \to (q_{j+1} - q_j)/[\delta t(1 - i\epsilon)] = \dot{q}/(1 - i\epsilon)$. Hence we have

$$\int \frac{dp}{2\pi\hbar} \exp\left\{\frac{i}{\hbar}\delta t\left[p\dot{q} - \frac{p^2}{2m}\right]\right\} \to \int \frac{dp}{2\pi\hbar} \exp\left\{\frac{i}{\hbar}\delta t\left[p\dot{q} - (1 - i\epsilon)\frac{p^2}{2m}\right]\right\} \tag{4.1.206}$$
$$= \sqrt{\frac{m}{2i\pi\hbar\delta t(1-i\epsilon)}} \exp\left\{\frac{i}{\hbar}\delta t(1 + i\epsilon)\left[\frac{1}{2}m\dot{q}^2\right]\right\},$$

where we only take the limit $\epsilon \to 0^+$ at the very end of any calculation. For the Feynman infinitesimal parameter, ϵ, it is to be understood that we work only to first order in this parameter, that finite constants will be absorbed into it whenever it is convenient to do so and that it is always to be taken to zero at the end of any calculation whenever it becomes meaningful to do so. The definition in terms of the above limit is equivalent to defining the path integral in Euclidean space and then analytically continuing the result back to Minkowski space. This is referred to as a *Wick rotation*, $t \to -i\tau$, with τ being the Euclidean time and is discussed in Sec. A.4. The deformation used above, $t \to t(1 - i\epsilon)$, is an infinitesimal form of the Wick rotation. In addition, this is equivalent to specifying the Feynman boundary conditions for propagators in quantum field theory as explained in Chapter 6.

Hence using Eq. (4.1.206) in Eq. (4.1.201) we have

$$\langle q'', t'' | q', t' \rangle = \lim_{n \to \infty} \left(\frac{m}{(2 i \pi \hbar \delta t)(1 - i\epsilon)} \right)^{(n+1)/2} \int \prod_{i=1}^{n} dq_i \tag{4.1.207}$$

$$\times \exp \left\{ \frac{i}{\hbar} \delta t \sum_{j=0}^{n} \left[(1 + i\epsilon) \frac{m}{2} \left(\frac{q_{j+1} - q_j}{\delta t} \right)^2 - (1 - i\epsilon) V(q_j) \right] \right\}.$$

This is conventionally expressed in the compact form

$$\langle q'', t'' | q', t' \rangle = \int \mathcal{D}q \, \exp \left\{ \frac{i}{\hbar} S[q] \right\} = \int \mathcal{D}q \, \exp \left\{ \frac{i}{\hbar} \int_{t'}^{t''} dt \, L(q, \dot{q}) \right\}, \tag{4.1.208}$$

where $L(q, \dot{q})$ is the usual Lagrangian for the form of Hamiltonian that we are considering plus a damping term $\mathcal{O}$,

$$L(q, \dot{q}) = \tfrac{1}{2} m \dot{q}^2 - V(q) + i\epsilon(\tfrac{1}{2} m \dot{q}^2 + V(q)). \tag{4.1.209}$$

Provided that $V(q)$ is a potential bounded below we can always redefine it by a constant so that $V(q) > 0$ to make obvious that $i\epsilon(\tfrac{1}{2} m \dot{q}^2 + V(q))$ acts as a damping term in the oscillatory path integral in Eq. (4.1.208). To attach an unambiguous meaning to oscillatory formal path integrals is a nontrivial mathematical problem in general. One approach to the problem is by analytically continuing the integral from real time (i.e., Minkowski space) to imaginary time (i.e., Euclidean space), there evaluating the more easily defined nonoscillating and damped form of the path integral, and then analytically continuing the results back to real time.

For systems with Hamiltonians that are not of the form of Eq. (4.1.198) it is often possible to arrive at Eq. (4.1.208) with some effective Lagrangian L_{eff} replacing the usual Lagrangian for these systems. Examples of such Hamiltonians are those with a velocity-dependent potential and/or which are non-quadratic in momentum. See, e.g., the relevant sections of Abers and Lee (1973) and Coleman (1985). Arbitrary quantum Hamiltonians containing products of position and momentum operators lead to problems associated with the specification of operator ordering.

The action associated with any macroscopic system will be very many orders of magnitude greater than Planck's constant, i.e., $S[q]/\hbar \gg 1$. The classical path is given by the Euler-Lagrange equations, which is the one that corresponds to a stationary point of the action, i.e., the path $q(t)$ for which $\delta S / \delta q(t) = 0$ for all t.

The *Stationary Phase Approximation* is the statement that in an integration involving an oscillating exponential (or sinusoid) the integral will be dominated by the region where the exponent has a stationary point. This is related to both the *Method of Steepest Descent* and to the *Saddle-Point Method* in mathematics. In our context the stationary point corresponds to $\delta S / \delta q(t) = 0$ and hence it leads to the important fact that the classical path increasingly dominates the path integral as $S[q]/\hbar \to \infty$ and so predicts that classical mechanics is the macroscopic limit of quantum mechanics. This conclusion is based on the assumption that there is a unique classical path associated with the path end-point conditions.

Spectral Function, Partition Function and Statistical Mechanics

The energy levels in the system E_n are the eigenvalues of the Hamiltonian operator $\hat{H}$ and are related to the Green's function, Eq. (4.1.194), through the spectral function $F(t'' - t')$. We restrict ourselves to time-independent Hamiltonians in the discussion here so that $\hat{U}(t'', t') = \hat{U}(t'' - t')$. The spectral function, denoted as F here, is defined as the trace of the evolution operator

$$F(t'' - t') \equiv \text{tr}\left(e^{-i\hat{H}(t''-t')/\hbar}\right) = \sum_n \langle n|e^{-i\hat{H}(t''-t')/\hbar}|n\rangle = \sum_n e^{-iE_n(t''-t')/\hbar}$$

$$= \int dq \,_s\langle q|e^{-i\hat{H}(t''-t')/\hbar}|q\rangle_s = \int dq \,_h\langle q,t''|q,t'\rangle_h = \int dq \, G(q,q;t'',t')$$

$$= \int \mathcal{D}p\,\mathcal{D}q_{(\text{periodic})} \exp\left\{\frac{i}{\hbar}\int_{t'}^{t''} dt\,[p\dot{q} - H(p,q)]\right\}$$

$$= \int \mathcal{D}q_{(\text{periodic})} \exp\left\{\frac{i}{\hbar}\int_{t'}^{t''} dt\,L(q,\dot{q})\right\}, \tag{4.1.210}$$

where the integration is over all periodic paths in coordinate space with period $(t'' - t')$. There is no similar periodicity requirement in momentum space.

Consider the Wick rotation $t \to -i\tau$ so that $(t'' - t') \to -i(\tau'' - \tau') \equiv -i\hbar\beta$. The evolution operator then Wick rotates, $e^{-i\hat{H}(t''-t')/\hbar} \to e^{-\beta\hat{H}}$. The analogue of Eq. (4.1.201) can be obtained by following the same arguments as before

$$_s\langle q''|e^{-\beta\hat{H}}|q'\rangle_s = \,_s\langle q''|e^{-(\tau''-\tau')\hat{H}/\hbar}|q'\rangle_s = \int \mathcal{D}p\,\mathcal{D}q \exp\left\{\frac{1}{\hbar}\int_{\tau'}^{\tau''} d\tau\,[ip\dot{q} - H(p,q)]\right\}, \tag{4.1.211}$$

where the phase-space paths satisfy $q(\tau'') = q''$ and $q(\tau') = q'$. This equation can also be obtained in a simple way from Eq. (4.1.201) by making the Wick rotation $t \to -i\tau$. Note that we then have $\dot{q} \equiv (dq/dt) \to i\dot{q} \equiv i(dq/d\tau)$. Here p is unaffected by the Wick rotation and remains the canonical conjugate partner to $\hat{q}$, i.e., satisfying the standard canonical commutation relations. We are assuming throughout that $H(p,q)$ has the form of Eq. (4.1.198), i.e., that $H(p,q)$ is second-order in p, and in place of Eq. (4.1.208) we now have

$$_s\langle q''|e^{-\beta\hat{H}}|q'\rangle_s = \,_s\langle q''|e^{-(\tau''-\tau')\hat{H}/\hbar}|q'\rangle_s = \int \mathcal{D}q \exp\left\{-\frac{1}{\hbar}\int_{\tau'}^{\tau''} d\tau\,L_{\rm E}(q,\dot{q})\right\}, \tag{4.1.212}$$

where $L_{\rm E}(q,\dot{q})$ is referred to as the Euclidean Lagrangian and is given by

$$L_{\rm E}(q,\dot{q}) = \tfrac{1}{2}m\dot{q}^2 + V(q). \tag{4.1.213}$$

Eq. (4.1.212) can be obtained from Eq. (4.1.208) by the analytic continuation $t \to -i\tau$. Note that $-L_{\rm E}(q,dq/d\tau) = L(q,idq/dt)$. The Wick-rotated spectral function $F(t'' - t') \to Z(\beta)$ is recognized as the partition function of statistical mechanics,

$$Z(\beta) \equiv \text{tr}\left(e^{-\beta\hat{H}}\right) = \sum_n \langle n|e^{-\beta\hat{H}}|n\rangle = \sum_n e^{-E_n\beta} = \int dq \,_s\langle q|e^{-\beta\hat{H}}|q\rangle_s, \tag{4.1.214}$$

where $\beta \equiv (1/kT)$ is sometimes called the "inverse temperature" with T being the absolute temperature in Kelvin and k is Boltzmann's constant. So we have

$$Z(\beta) = Z((\tau'' - t')/\hbar) = \int \mathcal{D}p\,\mathcal{D}q_{(\text{periodic})} \exp\left\{\frac{1}{\hbar}\int_{\tau'}^{\tau''} d\tau\,[ip\dot{q} - H(p,q)]\right\}$$

$$= \int \mathcal{D}q_{(\text{periodic})} \exp\left\{-\frac{1}{\hbar}\int_{\tau'}^{\tau''} d\tau\,L_{\rm E}(q,\dot{q})\right\}. \tag{4.1.215}$$

Time-Ordered Products, Sources and Generating Functionals

Let us reconsider the derivation of the path integral when we have a time-dependent Schrödinger picture Hamiltonian of the form

$$\hat{H}(t) \equiv H(\hat{p},\hat{q},t) = \frac{\hat{p}^2}{2m} + V(\hat{q},t), \tag{4.1.216}$$

which is just a generalization of Eq. (4.1.198) allowing for a time-dependent potential, $V(\hat{q},t)$. By inspection we observe that all of the steps that led to Eq. (4.1.201) go through as before and so we find

$$
\begin{aligned}
{}_h\langle q'', t'' | q', t'\rangle_h &= {}_s\langle q'' | \hat{U}(t'', t') | q'\rangle_s = {}_I\langle q'', t'' | \hat{U}_I(t'', t') | q', t'\rangle_I \\
&= \int \mathcal{D}p\, \mathcal{D}q \, \exp\left\{ (i/\hbar)\int_{t'}^{t''} dt\, [p\dot{q} - H(p, q, t)] \right\} \\
&= \int \mathcal{D}q \, \exp\left\{ (i/\hbar)\int_{t'}^{t''} dt\, L(q, \dot{q}, t) \right\},
\end{aligned}
\tag{4.1.217}
$$

where $H(q, p, t) = (p^2/2m) + V(q, t)$ and $L(q, \dot{q}, t) = (p^2/2m) - V(q, t)$ are the classical Hamiltonian and classical Lagrangian for this system, respectively.

Consider now adding an external source term to the Hamiltonian operator such that $V(\hat{q}) \to V(\hat{q}, t) = V(\hat{q}) - J(t)\hat{q}$, where $J(t)$ is a real function of time and

$$
\hat{H}^J(t) \equiv H - J(t)\hat{q} = (\hat{p}^2/2m) + V(\hat{q}) - J(t)\hat{q},
\tag{4.1.218}
$$

which leads to the source-dependent evolution operator

$$
\hat{U}^J(t'', t') = T\left(\exp\left\{ -(i/\hbar)\int_{t'}^{t''} dt\, [\hat{H} - J(t)\hat{q}] \right\} \right).
\tag{4.1.219}
$$

Then we can go to an interaction picture and identify

$$
\begin{aligned}
\hat{H}(t) &\to \hat{H}^J(t), \quad \hat{H}_0 \to \hat{H}, \quad \hat{U}(t'', t') \to \hat{U}^J(t'', t'), \quad \hat{U}_0(t'', t') \to \hat{U}(t'', t'), \\
\hat{H}_{\mathrm{int}}(t) &= -J(t)\hat{q}, \quad \hat{U}_I(t'', t') = T\left(\exp\left\{ (i/\hbar)\int_{t'}^{t''} dt\, J(t)\hat{q}_I(t) \right\} \right),
\end{aligned}
\tag{4.1.220}
$$

and where we have used Eq. (4.1.192).

We make use of Eq. (4.1.217) to give the source-dependent Green's function,

$$
\begin{aligned}
{}_h\langle q'', t'' | q', t'\rangle_h^J &= {}_s\langle q'' | T\left(\exp\left\{ -(i/\hbar)\int_{t'}^{t''} dt [\hat{H} - J(t)\hat{q}] \right\} \right) | q'\rangle_s \\
&= {}_I\langle q'', t'' | T\left(\exp\left\{ (i/\hbar)\int_{t'}^{t''} dt\, J(t)\hat{q}_I(t) \right\} \right) | q', t'\rangle_I \\
&= \int \mathcal{D}p\, \mathcal{D}q \, \exp\left\{ (i/\hbar)\int_{t'}^{t''} dt\, [p\dot{q} - H(p, q) + J(t)q(t)] \right\} \\
&= \int \mathcal{D}q \, \exp\left\{ (i/\hbar)\int_{t'}^{t''} dt\, [L(q, \dot{q}) + J(t)q(t)] \right\}.
\end{aligned}
\tag{4.1.221}
$$

The spectral function of Eq. (4.1.210) becomes

$$
F^J(t'', t') = \mathrm{tr}\{\hat{U}^J(t'', t')\} = \int \mathcal{D}q_{(\mathrm{periodic})} e^{(i/\hbar)\int_{t'}^{t''} dt\, [L + J(t)q(t)]}.
\tag{4.1.222}
$$

Using the definition of the functional derivative in Eq. (2.1.45) we find

$$
(-i\hbar)^k \frac{\delta^k}{\delta J(t_1)\cdots\delta J(t_k)} \hat{U}_I(t'', t')\bigg|_{J=0} = T\left(\hat{q}_h(t_1)\cdots\hat{q}_h(t_k) \right),
\tag{4.1.223}
$$

where we take the source to zero after the derivatives are taken and where we have recognized that $\hat{H}_{\mathrm{int}} \to 0$ when $J \to 0$ and so $\hat{q}_I(t) \to \hat{q}_h(t)$. Hence using this with Eq. (4.1.221) we observe that

$$
\begin{aligned}
{}_h\langle q'', t'' | T\left[\hat{q}_h(t_1)\cdots\hat{q}_h(t_k)\right] | q', t'\rangle_h &= (-i\hbar)^k \frac{\delta^k}{\delta J(t_1)\cdots\delta J(t_k)} {}_h\langle q'', t'' | q', t'\rangle_h^J\bigg|_{J=0} \\
&= (-i\hbar)^k \frac{\delta^k}{\delta J(t_1)\cdots\delta J(t_k)} \int \mathcal{D}q \, \exp\left\{ (i/\hbar)\int_{t'}^{t''} dt\, [L(q, \dot{q}) + J(t)q] \right\}\bigg|_{J=0} \\
&= \int \mathcal{D}q\, q(t_1)\cdots q(t_k) \exp\left\{ (i/\hbar)\int_{t'}^{t''} dt\, L(q, \dot{q}) \right\},
\end{aligned}
\tag{4.1.224}
$$

where recall the understood boundary conditions $q(t'') = q''$ and $q(t') = q'$.

Let us now consider the consequences of taking the limits $t'' = T$ and $t' = -T$ and $T \to \infty$ while keeping the amount of our infinitesimal Wick rotation, ϵ, arbitrarily small but nonzero. We require that the source $J(t)$ has compact support, i.e., that it vanishes outside some arbitrarily large but finite time interval,

$$J(t) = 0 \quad \text{for all} \quad t > t_J \quad \text{and} \quad t < -t_J, \tag{4.1.225}$$

for some t_J. Let us define E_n and $|E_n\rangle_s$ for $n = 0, 1, 2, \ldots$ as the eigenvalues and normalized eigenvectors of the time-independent Schrödinger Hamiltonian, $\hat{H} \equiv H(\hat{p}, \hat{q})$. Let us denote the ground state as $|\Omega\rangle$. Hence we have

$$\hat{H}|E_n\rangle_s = E_n|E_n\rangle_s, \quad \hat{U}(t'', t') = e^{-i\hat{H}(t''-t')/\hbar} \quad \text{and} \quad |\Omega\rangle \equiv |E_0\rangle_s. \tag{4.1.226}$$

It is convenient to define the *generating functional*, $Z[J]$, as

$$Z[J] \equiv \lim_{T \to \infty(1-i\epsilon)} \frac{F^J(T, -T)}{F(T, -T)} = \lim_{T \to \infty(1-i\epsilon)} \frac{\text{tr}\{\hat{U}^J(T, -T)\}}{\text{tr}\{\hat{U}(T, -T)\}}$$

$$= \lim_{T \to \infty(1-i\epsilon)} \frac{\text{tr}\left[T\left(\exp\left\{-(i/\hbar)\int_{-T}^{T} dt\,[\hat{H} - J(t)\hat{q}]\right\}\right)\right]}{\text{tr}\left[T\left(\exp\left\{-(i/\hbar)\int_{-T}^{T} dt\,\hat{H}\right\}\right)\right]}$$

$$= \lim_{T \to \infty(1-i\epsilon)} \frac{\int \mathcal{D}q_{(\text{periodic})} \exp\left\{(i/\hbar)\int_{-T}^{T} dt\,[L(q, \dot{q}) + J(t)q(t)]\right\}}{\int \mathcal{D}q_{(\text{periodic})} \exp\left\{(i/\hbar)\int_{-T}^{T} dt\,L(q, \dot{q})\right\}}. \tag{4.1.227}$$

Some texts use the notation $W[J]$ for the generating functional, but we will follow the notation of Peskin and Schroeder (1995) and use $Z[J]$ for this. It should be noted, however, that our $Z[J]$ is normalized using the denominator so that we have $Z[0] = 1$. Using Eqs. (4.1.184) and (4.1.220) we observe that

$$F^J(T, -T) = \text{tr}\{\hat{U}^J(T, -T)\} = \sum_n {}_s\langle E_n|\hat{U}^J(T, -T)|E_n\rangle_s$$

$$= \sum_n {}_s\langle E_n|\hat{U}(T, t_0)\hat{U}_I(T, -T)\hat{U}(-T, t_0)^\dagger|E_n\rangle_s$$

$$= \sum_n e^{-iE_n(2T)/\hbar} {}_s\langle E_n|\hat{U}_I(T, -T)|E_n\rangle_s. \tag{4.1.228}$$

Taking the limit $T \to \infty(1 - i\epsilon)$ suppresses all contributions in the sum relative to that from the ground state $|\Omega\rangle = |E_0\rangle_s$,

$$\lim_{T \to \infty(1-i\epsilon)} F^J(T, -T) \to e^{-iE_0(2T)/\hbar}\langle\Omega|\hat{U}_I(T, -T)|\Omega\rangle. \tag{4.1.229}$$

Then the generating functional is given by

$$Z[J] \equiv \lim_{T \to \infty(1-i\epsilon)} \frac{\langle\Omega|\hat{U}_I(T, -T)|\Omega\rangle}{\langle\Omega|\Omega\rangle} = \lim_{T \to \infty(1-i\epsilon)} \langle\Omega|\hat{U}_I(T, -T)|\Omega\rangle, \tag{4.1.230}$$

where the latter form follows if we assume a normalized ground state, $\langle\Omega|\Omega\rangle = 1$. Using Eqs. (4.1.223) and (4.1.227) we arrive at the important result that

$$\frac{\langle\Omega|T\,[\hat{q}_h(t_1)\cdots\hat{q}_h(t_k)]\,|\Omega\rangle}{\langle\Omega|\Omega\rangle} = (-i\hbar)^k \frac{\delta^k}{\delta J(t_1)\cdots\delta J(t_k)} Z[J]\Big|_{J=0}$$

$$= \lim_{T \to \infty(1-i\epsilon)} \frac{\int \mathcal{D}q_{(\text{periodic})}\,q(t_1)\cdots q(t_k)\,e^{iS[q]/\hbar}}{\int \mathcal{D}q_{(\text{periodic})}\,e^{iS[q]/\hbar}}, \tag{4.1.231}$$

where $S[q] = \int_{-T}^{T} dt\,L(q, \dot{q})$ is the usual action with $L(q, \dot{q}) = \frac{1}{2}m\dot{q}^2 - V(q)$.

In the above we used the infinitesimal Wick rotation, which was required to define the path integral and which leads to Feynman boundary conditions for propagators, combined with an infinite time limit in order to isolate the ground-state expectation values. We could also do the full Wick rotation, $t \to -i\tau$ rather than $t \to t(1 - i\epsilon)$, and then take the Euclidean time limit $\tau'' = T_E$ and $\tau' = -T_E$ with $T_E \to \infty$ to obtain the results for the *Euclidean generating functional*,

$$\frac{\langle \Omega | T [\hat{q}_h(\tau_1) \cdots \hat{q}_h(\tau_k)] | \Omega \rangle}{\langle \Omega | \Omega \rangle} = \hbar^k \frac{\delta^k}{\delta J(t_1) \cdots \delta J(t_k)} Z_E[J] \Big|_{J=0}$$

$$= \lim_{T_E \to \infty} \frac{\int \mathcal{D}q_{(\text{periodic})} \, q(\tau_1) \cdots q(\tau_k) \, e^{-S_E[q]/\hbar}}{\int \mathcal{D}q_{(\text{periodic})} \, e^{-S_E[q]/\hbar}}, \qquad (4.1.232)$$

where $S_E[q] = \int_{-T_E}^{T_E} d\tau \, L_E(q, \dot{q})$ is the usual Euclidean action with the Euclidean Lagrangian $L_E(q, \dot{q}) = \frac{1}{2}m\dot{q}^2 + V(q)$ and $\dot{q} = dq/d\tau$. The Euclidean generating functional is given by

$$Z_E[J] = \lim_{T_E \to \infty} \frac{\int \mathcal{D}q_{(\text{periodic})} \exp\left\{ -\int_{-T_E}^{T_E} d\tau \left[L_E(q, \dot{q}) - J(\tau)q(\tau) \right] / \hbar \right\}}{\int \mathcal{D}q_{(\text{periodic})} \exp\left\{ -\int_{-T_E}^{T_E} d\tau L_E(q, \dot{q})/\hbar \right\}}. \qquad (4.1.233)$$

4.2 Wavepackets and Dispersion

Consider some real or complex plane wave with frequency f, wavelength λ, traveling in the direction of the unit vector $\mathbf{n}$. An observer $\mathcal{O}$ would describe the wave

$$\phi(t, \mathbf{x}) = a e^{i(\mathbf{k} \cdot \mathbf{x} - \omega t)} \equiv a e^{i\eta(x)}, \qquad (4.2.1)$$

where $a \in \mathbb{C}$ is the complex amplitude, $\omega = 2\pi f$ is the angular frequency, $\mathbf{k} = k\hat{\mathbf{k}} = (2\pi/\lambda)\hat{\mathbf{k}}$ is the angular wavevector, k is the angular wavenumber and

$$\eta(x) \equiv \mathbf{k} \cdot \mathbf{x} - \omega t \qquad (4.2.2)$$

is the phase of the wave at the spacetime point $x^\mu = (ct, \mathbf{x})$. If ϕ is real we would simply take the real part of the right-hand side, i.e., $\phi = \text{Re}\{a \exp[i\eta(x)]\}$.

The phase velocity is determined by the speed at which a given phase value of the wave travels in the direction of motion. Let the wave be traveling in the x-direction $\mathbf{k} = (k, 0, 0)$, then

$$kx - \omega t = \eta(t, x, y, z) = \eta(t + dt, x + dx, y, z) = (x + dx)k - (t + dt)\omega \qquad (4.2.3)$$

provided that we require $k\,dx - \omega\,dt = 0$. Hence the *phase velocity* of the wave is

$$v_p = dx/dt = \omega/k = f\lambda. \qquad (4.2.4)$$

Consider the spacetime event where $\mathcal{O}$ measures the phase of the wave at the spacetime point P that is described by him as having coordinate $x^\mu = (ct, \mathbf{x})$. Then consider observer $\mathcal{O}'$ observing the phase at the same spacetime point P described by him with coordinates $x'^\mu = (ct', \mathbf{x}')$. Clearly, both observers must observe the same phase at the same physical spacetime point P and so $\eta'(x') = \eta(x)$, which is the statement that the phase $\eta(x)$ transforms as a Lorentz scalar. Hence defining

$$k^0 \equiv \omega/c, \tag{4.2.5}$$

we can write the phase as

$$\eta(x) = \mathbf{k} \cdot \mathbf{x} - \omega t = -k^0 x^0 + \mathbf{k} \cdot \mathbf{x} = -k_\mu x^\mu = -k \cdot x. \tag{4.2.6}$$

Since $\eta(x)$ is a Lorentz scalar and x^μ is a four-vector, then

$$k^\mu = (k^0, \mathbf{k}) = (\omega/c, \mathbf{k}) \tag{4.2.7}$$

must also be a Lorentz four-vector. From the photoelectric effect we know that the energy of a photon of frequency f is $E = hf = \hbar\omega$ and so we identify

$$\hbar k^0 = \hbar\omega/c = E/c = \gamma mc = p^0. \tag{4.2.8}$$

Since p^μ and k^μ are both four-vectors, then we must have the same relationship for all four components of the four-vector and so

$$p^\mu = (p^0, \mathbf{p}) = (E/c, \mathbf{p}) = \hbar(\omega/c, \mathbf{k}) = \hbar(k^0, \mathbf{k}) = \hbar k^\mu. \tag{4.2.9}$$

It is natural to assume the same for massive particles. So combining relativity and quantum mechanics leads to the de Broglie formula for wave-particle duality

$$\mathbf{p} = \gamma m \mathbf{u} = \hbar \mathbf{k}, \tag{4.2.10}$$

where $\mathbf{u}$ is the velocity of a particle with mass m and relativistic energy E and where here $\gamma \equiv \gamma(u)$. If the particle state is described by a wavepacket that is peaked sharply around wavevector $\mathbf{k}$, then it will travel with group velocity $\mathbf{u}$, where $\mathbf{u} = \hbar\mathbf{k}/\gamma m$. The more sharply peaked it is in $\mathbf{k}$-space, then the more distributed in coordinate space it is and the smaller is the rate of dispersion of the wavepacket.

Consider a plane wave with frequency f and wavelength λ traveling in a homogeneous, but not necessarily isotropic, medium in the direction $\hat{\mathbf{k}}$,

$$\phi(x) \equiv \phi(t, \mathbf{x}) = ae^{i(\mathbf{k}\cdot\mathbf{x}-\omega t)} = ae^{ik(\hat{\mathbf{k}}\cdot\mathbf{x}-(\omega/k)t)} \equiv ae^{i\eta(x)}, \tag{4.2.11}$$

where $\omega = 2\pi f$ is the angular frequency, $k = |\mathbf{k}|$ is the angular wavenumber, $\mathbf{k} = k\hat{\mathbf{k}} = (2\pi/\lambda)\hat{\mathbf{k}}$ is the angular wavevector, $a \in \mathbb{C}$ is the complex amplitude. The plane waves are solutions of the wave equation

$$\frac{\partial^2 \phi}{\partial t^2} - \left(\frac{\omega}{k}\right)^2 \nabla^2 \phi = 0. \tag{4.2.12}$$

Writing the complex amplitude in polar coordinates as $a \equiv |a|e^{-i\eta_0}$ gives for the real part of the wavefunction

$$\begin{aligned}
\phi_R(x) &\equiv \mathrm{Re}\left(\phi(x)\right) = \mathrm{Re}(ae^{i(\mathbf{k}\cdot\mathbf{x}-\omega t)}) \\
&= |a|\,\mathrm{Re}\left[(\cos\eta_0 - i\sin\eta_0)\left\{\cos\eta(x) + i\sin\eta(x)\right\}\right] \\
&= |a|\left[\cos\eta(x)\cos\eta_0 + \sin\eta(x)\sin\eta_0\right] = |a|\cos\left[\eta(x) - \eta_0\right] \\
&= |a|\cos\left(\mathbf{k}\cdot\mathbf{x} - \omega t - \eta_0\right).
\end{aligned} \tag{4.2.13}$$

In classical mechanics only the real part of the wavefunction is physically relevant, whereas in quantum mechanics the complete complex wavefunction is required.

As explained above, the phase velocity is determined by the speed at which a given phase of the plane wave travels in the direction of motion. With the wave propagating in the $\hat{\mathbf{k}}$-direction we would then have for the phase velocity

$$\boldsymbol{v}_p = (\omega/k)\hat{\mathbf{k}} = (f\lambda)\hat{\mathbf{k}}. \tag{4.2.14}$$

An arbitrary wavefunction is a superposition of plane waves,

$$\phi(x) = \phi(t, \mathbf{x}) = \int d^3k \, a(\mathbf{k}) e^{i[\mathbf{k}\cdot\mathbf{x} - \omega(\mathbf{k})t]}, \tag{4.2.15}$$

where $a(\mathbf{k})$ is some arbitrary function of the wavevector $\mathbf{k}$. Comparing this with Eq. (A.2.10) we recognize that $a(\mathbf{k}) = \tilde{\phi}(\mathbf{k})/(2\pi)^3$, where $\tilde{\phi}(\mathbf{k})$ is the three-dimensional Fourier transform of $\phi(0, \mathbf{x})$. The function $\omega(\mathbf{k})$ is characteristic of the medium in which the wave is propagating and determines the behavior of wavepackets as they travel in the medium. A relationship between $\omega(\mathbf{k})$ and $\mathbf{k}$ for a medium is called a *dispersion relation*, for reasons that will soon become clear. In an isotropic medium ω is a function of the angular wavenumber k but is independent of the direction of wave propagation, i.e., in an isotropic medium we have $\omega(\mathbf{k}) = \omega(k)$.

For example, it is obvious that if $\omega(\mathbf{k}) = \omega(k) \propto k$, then $v_p = \omega/k$ is a constant and the phase velocity is the same for all k, i.e., for all wavelengths λ. However, in general for an isotropic medium we will have

$$\boldsymbol{v}_p(k) = f(k)\lambda\,\hat{\mathbf{k}} = [\omega(k)/k]\,\hat{\mathbf{k}} \tag{4.2.16}$$

with $\boldsymbol{v}_p$ depending on the wavenumber, $k = 2\pi/\lambda$.

Let the function $a(\mathbf{k})$ be concentrated in some small region of $\mathbf{k}$-space around some wavevector $\mathbf{k}_0$, then we will have a wavepacket whose spatial size is inversely related to the breadth of the distribution in k-space as we know from the properties of the Fourier transform (see Sec. A.2.3). Let us Taylor expand $\omega(\mathbf{k})$ around $\mathbf{k}_0$

$$\omega(\mathbf{k}) = \omega(\mathbf{k}_0) + (\mathbf{k} - \mathbf{k}_0) \cdot (\boldsymbol{\nabla}_{\mathbf{k}}\omega)(\mathbf{k}_0) + \cdots \equiv \omega_0 + (\mathbf{k} - \mathbf{k}_0) \cdot \boldsymbol{\nabla}\omega_0 + \cdots, \tag{4.2.17}$$

where we make the convenient shorthand definitions

$$\omega_0 \equiv \omega(\mathbf{k}_0) \quad \text{and} \quad \boldsymbol{\nabla}\omega_0 \equiv (\boldsymbol{\nabla}_{\mathbf{k}}\omega)(\mathbf{k}_0). \tag{4.2.18}$$

If all but the first two terms in the Taylor expansion can be neglected in the small $\mathbf{k}$-region of interest around $\mathbf{k}_0$, then we can write

$$\begin{aligned}
\phi(t, \mathbf{x}) &= \exp[it(-\omega_0 + \mathbf{k}_0 \cdot \boldsymbol{\nabla}\omega_0)] \int d^3k \, a(\mathbf{k}) \exp\left[i\mathbf{k} \cdot (\mathbf{x} - \boldsymbol{\nabla}\omega_0 t)\right] \\
&= \exp[it(-\omega_0 + \mathbf{k}_0 \cdot \boldsymbol{\nabla}\omega_0)]\phi(0, [\mathbf{x} - \boldsymbol{\nabla}\omega_0 t]).
\end{aligned} \tag{4.2.19}$$

This then shows that up to an overall time-dependent phase factor the wavepacket at time t is the same as the wavepacket at time $t = 0$ but translated by $-\boldsymbol{\nabla}\omega_0 t$. The magnitude squared of the wavefunction satisfies

$$|\phi(t, \mathbf{x})|^2 = |\phi(0, [\mathbf{x} - \boldsymbol{\nabla}\omega_0 t])|^2; \tag{4.2.20}$$

i.e., the magnitude squared of the wavefunction has a shape that is invariant and that moves with a velocity $\boldsymbol{\nabla}\omega_0$. To understand this, consider a wavepacket at $t = 0$ with a maximum of $|\phi(0, \mathbf{x})|^2$ at $\mathbf{x} = \mathbf{x}_0$. Then at time dt the maximum of the wavepacket will be at $\mathbf{x} - \boldsymbol{\nabla}\omega_0 dt = \mathbf{x}_0$, i.e., at $\mathbf{x} = \boldsymbol{\nabla}\omega_0 dt + \mathbf{x}_0$. The velocity of the wavepacket is $(\mathbf{x} - \mathbf{x}_0)/dt = \boldsymbol{\nabla}\omega_0$. Hence, if only the first two terms in the Taylor expansion of $\omega(\mathbf{k})$ are significant, then the wavepacket $\phi(t, \mathbf{x})$ propagates with its shape preserved up to a phase factor and with velocity

$$\boldsymbol{v}_g(\mathbf{k}_0) \equiv \boldsymbol{\nabla}\omega_0 = (\boldsymbol{\nabla}_{\mathbf{k}}\omega)(\mathbf{k}_0) = \left(\frac{\partial\omega}{\partial k_x}(\mathbf{k}_0), \frac{\partial\omega}{\partial k_y}(\mathbf{k}_0), \frac{\partial\omega}{\partial k_z}(\mathbf{k}_0) \right),$$

$$\phi(t,\mathbf{x}) = \exp[it(-\omega_0) + \mathbf{k}_0 \cdot \boldsymbol{v}_g(\mathbf{k}_0)]\phi(0,[\mathbf{x} - \boldsymbol{v}_g(\mathbf{k}_0)t]), \tag{4.2.21}$$

where $\boldsymbol{v}_g(\mathbf{k}_0) \equiv \boldsymbol{\nabla}\omega_0$ is referred to as the *group velocity*. A medium with such a property is called a *nondispersive medium*. In an isotropic medium the group velocity will be the same in every direction and is given by

$$v_g(k_0) = d\omega/dk|_{k=k_0}. \tag{4.2.22}$$

In a nondispersive medium the information and energy can only propagate as fast as the group velocity of the wavepacket. If the higher-order terms in the Taylor expansion cannot be neglected, then the shape of the wavepacket will not be preserved up to a phase and the wavepacket will disperse; such a medium is called a *dispersive medium*. The concept of the group velocity in a dispersive medium becomes less useful the more dispersive the medium becomes, i.e., the larger are the contributions from the quadratic and higher-order terms in the Taylor expansion of the dispersion relation $\omega(\mathbf{k})$ around $\mathbf{k}_0$.

The *signal velocity* of a wave is defined as the velocity at which it carries information. It describes how quickly information and/or energy can be communicated between two points in space. The signal velocity must always be less than or equal to the speed of light so that causality is never violated. In a nondispersive medium the signal velocity is the group velocity as we have argued above and this remains true in typical dispersive media. However, it is possible to produce highly dispersive media with unusual properties, where the group velocity is quite unrelated to the signal velocity, e.g., where the group velocity can exceed the speed of light. In such exceptional situations the concept of group velocity is no longer useful.

Example 1: Electromagnetic waves moving in the vacuum: The vacuum is isotropic and so $\omega(\mathbf{k}) = \omega(k) = E(k)/\hbar = |\mathbf{p}|c/\hbar = |\mathbf{k}|c = kc$. Then the phase velocity is $v_p = (\omega/k) = c$ and is independent of wavelength as expected since $\omega(\mathbf{k}) \propto k$. The group velocity is $v_g = d\omega/dk = c$. Hence light in the vacuum has both the phase and group velocity equal to c. Furthermore, since the Taylor expansion for $\omega(\mathbf{k})$ contains only the linear term, then any wavepacket of light in the vacuum propagates at c and maintains its shape up to a phase indefinitely.

Example 2: Electromagnetic waves in an isotropic medium: For a nonabsorptive medium the index of refraction is defined as the ratio of the phase velocity of light in the vacuum to the phase velocity of light in the medium, i.e., $n \equiv c/v_p$. In an absorptive medium the refractive index is complex, $n \to n + i\kappa$, where the damping term κ leads to exponential decay of the wavefunction as it propagates through the medium. We will focus on nonabsorptive media. When light passes from one medium to the next the frequency $f = \omega/2\pi$ of the light does not change, but the phase velocity v_p does and since $v_p = f\lambda$ then so does the wavelength $\lambda = 2\pi/k$. An equivalent definition of the refractive index is $n = \lambda_0/\lambda$, where λ_0 is the wavelength in the vacuum. This follows since $n = c/v_p = f\lambda_0/f\lambda = \lambda_0/\lambda$. In an isotropic medium we will in general have a dispersion relation specified by the function $\omega(k)$, where $\omega(k) = 2\pi f(k)$ and correspondingly $v_p(k) = f(k)\lambda = \omega(k)/k = c/n(k)$ and as always $k \equiv 2\pi/\lambda$. Note that $\omega(k) = 2\pi c/\lambda_0 = 2\pi f$ is independent of the medium, but since k is medium-dependent, then the function $\omega(k)$ is nontrivial and defines the optical properties of the medium. Since $\omega(k) = kv_p(k)$ the group velocity is

$$v_g = \frac{d\omega}{dk} = \frac{d}{dk}(kv_p) = v_p + k\frac{dv_p}{dk} = v_p - \lambda\frac{dv_p}{d\lambda} = v_p\left(1 + \frac{\lambda}{n}\frac{dn}{d\lambda}\right) = v_p\left(1 - \frac{k}{n}\frac{dn}{dk}\right), \tag{4.2.23}$$

where we have used the easily proven results that $k(dv_p/dk) = -\lambda(dv_p/d\lambda) = (v_p\lambda/n)(dn/d\lambda)$ and $(\lambda/n)(dn/d\lambda) = -(k/n)(dn/dk)$. We can also invert the relationship between ω and k and express the dispersion relation in the form $k(\omega)$. Then we can use $k(\omega) = n(\omega)\omega/c$ and write

$$v_g = \frac{d\omega}{dk} = \left[\frac{dk}{d\omega}\right]^{-1} = \left[\frac{d(n\omega/c)}{d\omega}\right]^{-1} = \frac{c}{n + \omega(dn/d\omega)} = \frac{c}{n - \lambda_0(dn/d\lambda_0)}, \qquad (4.2.24)$$

where the last result follows from $\omega(dn/d\omega) = -\lambda_0(dn/d\lambda_0)$ using $\lambda_0 = 2\pi c/\omega$. *Normal dispersion* corresponds to $dn/d\lambda < 0$ and $n > 0$ and we see from Eq. (4.2.23) that $v_g < v_p$. This is the case for air ($n \simeq 1.0003$), water ($n \simeq 1.333$) and glass ($n \simeq 1.5$), where these values correspond to the wavelength $\lambda = 589\,\text{nm}$ and are typical values for these materials at room temperature. This is the reason that the short wavelength (i.e., blue) components of a beam of white light are deflected more than the long wavelength (i.e., red) components as they pass from air through a glass prism and back to air. Otherwise, if $dn/d\lambda > 0$, then we have *anomalous dispersion* with $v_g > c$ and then it can be shown that the signal velocity is not the group velocity. If a medium has a wavelength-independent refractive index $n > 1$, then the phase and group velocities satisfy $v_p = v_g < c$ and there is no dispersion. Since the phase velocity is not the signal velocity there is no reason that the phase velocity cannot exceed c corresponding to $n < 1$, e.g., as can occur for X-rays.

Example 3: Nonrelativistic massive particle in a vacuum: In nonrelativistic quantum mechanics we have $p = mv$ and $E = p^2/2m = \hbar^2 k^2/2m$ and hence $\omega(\mathbf{k}) = \omega(k) = E(k)/\hbar = \hbar k^2/2m$. The magnitude squared of the amplitude is just the usual probability density of quantum mechanics, $|\phi(t, \mathbf{x})|^2$. Expanding around some $\mathbf{k}_0$ we have

$$\omega(\mathbf{k}) = \omega(\mathbf{k}_0) + (k - k_0)^j \left.\frac{\partial\omega}{\partial k^j}\right|_{\mathbf{k}=\mathbf{k}_0} + \frac{1}{2}(k - k_0)^i(k - k_0)^j \left.\frac{\partial^2\omega}{\partial k^i \partial k^j}\right|_{\mathbf{k}=\mathbf{k}_0} + \cdots$$

$$= \frac{\hbar k_0^2}{2m} + (\mathbf{k} - \mathbf{k}_0) \cdot \frac{\hbar \mathbf{k}_0}{m} + \mathcal{O}(|\mathbf{k} - \mathbf{k}_0|)^2, \qquad (4.2.25)$$

where no higher terms survive. For a spatially extended wavepacket consisting only of wavevectors $\mathbf{k} \simeq \mathbf{k}_0$ the third term will be very small and can be neglected to a good approximation. In such a case the dispersion of the wavepacket will be very small and the group velocity of the wavepacket will be $v_g = \hbar k_0/m = p_0/m = u_0$, where u_0 is the velocity of a nonrelativistic particle of mass m and momentum $p_0 = mu_0$. In other words, a wavepacket that is sharply peaked in momentum space disperses very slowly and moves with the same velocity as a nonrelativistic classic particle with momentum $p_0 = \hbar k_0$ as we expect. The phase velocity is given by $v_p = \omega(k_0)/k_0 = \hbar k_0/2m = (1/2)v_g$, but is not a physically relevant quantity.

Example 4: Relativistic massive particle in a vacuum: In the relativistic case we have for the energy of the particle

$$E(\mathbf{k}) = \gamma(\mathbf{k})mc^2 = (m^2c^4 + \hbar^2\mathbf{k}^2c^2)^{1/2} = (m^2c^4 + \mathbf{p}^2c^2)^{1/2} = mc^2\left[1 + (\mathbf{p}/mc)^2\right]^{1/2}$$

$$= mc^2 + p^2/2m + \cdots = mc^2 + (\hbar^2 k^2/2m) + \cdots, \qquad (4.2.26)$$

where $k \equiv |\mathbf{k}|$ and we show just the first two terms in the (p/mc) expansion. Taylor expanding this expression for $\omega(k) = E(k)/\hbar$ around some k_0 we have

$$\omega(\mathbf{k}) = \frac{E(\mathbf{k})}{\hbar} = \frac{mc^2}{\hbar}\gamma(\mathbf{k}) = \omega(k_0) + (k - k_0)^j \left.\frac{\partial\omega}{\partial k^j}\right|_{\mathbf{k}=\mathbf{k}_0} + \mathcal{O}(|\mathbf{k} - \mathbf{k}_0|)^2. \qquad (4.2.27)$$

The first term $\omega(k_0) = E(k_0)/\hbar$ is a constant and is dominated by the usual rest mass energy mc^2. Shifting the energy by a constant has no effect on the probability density, of course, since from Eq. (4.2.11) we see that if we define $\phi'(t, \mathbf{x}) = \exp(-imc^2 t)\phi(t, \mathbf{x})$ then from Eq. (4.2.21) we have

$$|\phi'(t, \mathbf{x})|^2 = |\phi(t, \mathbf{x})|^2 . \tag{4.2.28}$$

A redefinition of the energy by a constant does not change the physics of the system; i.e., changing the Hamiltonian by a constant in either classical mechanics or quantum mechanics has no effect on the dynamics of the physical system. Since a constant shift in energy is just a shift in $\omega(k)$ and since $v_p = \omega(k)/k$, then we can understand that the phase velocity is not a physically meaningful quantity. In the relativistic case we find that the phase velocity is greater than c since

$$v_p = \frac{\omega(k)}{k} = \frac{E(k)}{\hbar k} = \frac{E(k)}{p(k)} = \frac{\gamma(k)mc^2}{\gamma(k)mv} = \frac{c^2}{v(k)}, \tag{4.2.29}$$

where $\gamma(k) = [1 - (v(k)/c)^2]^{-1/2}$ and $\hbar k = p(k) = \gamma(k)mv(k)$. For our wavepacket concentrated around k_0 then we will have $v_p = c^2/u_0$, where u_0 is the velocity of a particle with relativistic energy $E_0 = \hbar\omega(k_0)$ and relativistic three-momentum $\mathbf{p}_0 = \hbar k_0 = \gamma_0 m\mathbf{u}_0$ and where $\gamma_0 \equiv \gamma(k_0)$. The group velocity is

$$v_g = \left.\frac{d\omega}{dk}\right|_{k=k_0} = \left.\frac{d(m^2c^4 + \hbar^2 k^2 c^2)^{1/2}}{\hbar dk}\right|_{k=k_0} = \frac{\hbar k_0 c^2}{E_0} = \frac{\hbar k_0}{\gamma_0 m} = \frac{\gamma_0 m u_0}{\gamma_0 m} = u_0. \tag{4.2.30}$$

Note that the wavepacket moves with the same velocity u_0 as a classical relativistic particle with energy E_0. If the wavepacket is strongly concentrated around $\mathbf{k} \simeq \mathbf{k}_0$, then the second- and higher-order derivatives in the Taylor expansion of $\omega(\mathbf{k})$ around $\mathbf{k}_0$ will in any case be small and the dispersion of the wavepacket will be small. We see then that

$$\omega(\mathbf{k}) = [\omega(k_0) - k_0 u_0] + \mathbf{k} \cdot \mathbf{u}_0 + \frac{1}{2}(k - k_0)^i (k - k_0)^j \left.\frac{\partial^2 \omega}{\partial k^i \partial k^j}\right|_{\mathbf{k}=\mathbf{k}_0} + \cdots, \tag{4.2.31}$$

since $\mathbf{k}_0$ and $\mathbf{u}_0$ are parallel. Expanding $\mathbf{k} \cdot \mathbf{x} - \omega(k)t$ around $\mathbf{k} = \mathbf{k}_0$ gives

$$\mathbf{k} \cdot \mathbf{x} - \omega(k)t = \left[\frac{E_0}{\hbar} - k_0 u_0\right]t + \mathbf{k} \cdot (\mathbf{x} - \mathbf{u}_0 t) - \frac{t}{2}(k - k_0)^i (k - k_0)^j \left.\frac{\partial^2 \omega}{\partial k^i \partial k^j}\right|_{\mathbf{k}=\mathbf{k}_0} + \cdots$$

$$= \frac{mc^2}{\hbar\gamma_0}t + \mathbf{k} \cdot (\mathbf{x} - \mathbf{u}_0 t) - \frac{t}{2}(k - k_0)^i (k - k_0)^j \left.\frac{\partial^2 \omega}{\partial k^i \partial k^j}\right|_{\mathbf{k}=\mathbf{k}_0} + \cdots, \tag{4.2.32}$$

where $\mathbf{p}_0 = \hbar k_0$, $\mathbf{u}_0 = \hbar k_0/\gamma_0 m = \mathbf{p}_0/\gamma_0 m$, and

$$E_0 - \hbar k_0 u_0 = E_0\left(1 - \frac{\hbar k_0 u_0}{E_0}\right) = E_0\left(1 - \frac{\gamma_0 m u_0^2}{\gamma_0 mc^2}\right) = E_0\left(1 - \frac{u_0^2}{c^2}\right) = \frac{E_0}{\gamma_0^2} = \frac{mc^2}{\gamma_0}. \tag{4.2.33}$$

Dissipation occurs when the second- and higher-order terms in the Taylor expansion do not vanish. The first term on the right-hand side is a time-dependent phase factor and does not affect the overall shape of the wavefunction. For a wavefunction sufficiently strongly concentrated around $\mathbf{k} \simeq \mathbf{k}_0$ the dissipative second- and higher-order derivative terms on the right-hand side will be strongly suppressed. To the extent that this is true the wavefunction will propagate with its envelope changing very slowly with time and with its velocity given by the group velocity, $\mathbf{u}_0$; i.e., there will be negligible dispersion. In that case we have from Eq. (4.2.21) that

$$\phi(t, \mathbf{x}) = e^{-i(E_0 - \mathbf{k}_0 \cdot \mathbf{u}_0)t}\phi(0, [\mathbf{x} - \mathbf{u}_0 t]) \quad \text{and} \quad |\phi(t, \mathbf{x})|^2 = |\phi(0, [\mathbf{x} - \mathbf{u}_0 t])|^2. \quad (4.2.34)$$

We saw in Example 1 above that light in a vacuum has no dispersion; i.e., wavepackets of electromagnetic radiation maintain their shape indefinitely. Hence we expect that the more relativistic the motion of a particle becomes, then the more suppressed will be the dissipation. As the group (i.e., particle) velocity $u_0 \to c$, we expect that the dissipation should vanish. We can easily prove this to be the case.

Since $\omega(k) = E(k)/\hbar = (mc^2/\hbar)\gamma(k)$ with $E(k) = (m^2 c^4 + \hbar^2 k^2 c^2)^{1/2}$ and $k^i/\gamma(k) = p^i/\hbar\gamma(k) = \gamma(k)mu^i/\hbar\gamma(k) = mu^i/\hbar$ we find

$$\frac{\partial \omega}{\partial k^i} = \frac{mc^2}{\hbar}\frac{\partial \gamma}{\partial k^i} = \frac{1}{\hbar}\frac{\partial E}{\partial k^i} = \frac{\hbar k^i c^2}{E} = \frac{\hbar k^i}{m\gamma} = u^i \implies \frac{\partial \gamma}{\partial k^i} = \left(\frac{\hbar}{mc}\right)^2 \frac{k^i}{\gamma}, \quad (4.2.35)$$

$$\frac{\partial}{\partial k^i}\frac{1}{\gamma^n} = -n\frac{1}{\gamma^{n+1}}\frac{\partial \gamma}{\partial k^i} = -n\left(\frac{\hbar}{mc}\right)^2\frac{k^i}{\gamma^{n+2}} \xrightarrow{u \to c} \mathcal{O}\left(\frac{1}{\gamma^{n+1}}\right),$$

$$\frac{\partial}{\partial k^i}\frac{k^j}{\gamma^n} = \frac{\delta^{ij}}{\gamma^n} - n\frac{k^j}{\gamma^{n+1}}\frac{\partial \gamma}{\partial k^i} = \left(\frac{\delta^{ij}}{\gamma^n} - n\frac{\hbar^2 k^i k^j}{\gamma^{n+2}m^2c^2}\right) \xrightarrow{u \to c} \mathcal{O}\left(\frac{1}{\gamma^n}\right),$$

$$\frac{\partial^2 \omega}{\partial k^i \partial k^j} = \frac{\hbar}{m}\frac{\partial}{\partial k^i}\frac{k^j}{\gamma} \xrightarrow{u \to c} \mathcal{O}\left(\frac{1}{\gamma}\right), \qquad \frac{\partial^3 \omega}{\partial k^i \partial k^j \partial k^\ell} \xrightarrow{u \to c} \mathcal{O}\left(\frac{1}{\gamma^2}\right)$$

and clearly the nth partial derivative is $\mathcal{O}(1/\gamma^{n-1})$. As $u_0 \to c$ we have $\gamma(k_0) \to \infty$ and so the second- and higher-order dissipative terms vanish and the wavefunction preserves its shape as it propagates as expected.

4.3 Klein-Gordon Equation

The simplest relativistic extension of quantum mechanics is that for a spin-zero particle, which according the spin-statistics connection must be a a boson. We have already met the Klein-Gordon equation in Sec. 3.1 in our studies of relativistic classical fields, where it was the simplest example. However, it is most commonly associated with relativistic quantum mechanics as the simplest relativistic extension of the Schrödinger equation as we will now discuss.

4.3.1 Formulation of the Klein-Gordon Equation

In a nonrelativistic quantum system consisting of a single free particle any complex coordinate-space wavefunction $\phi(t, \mathbf{x})$ can be Fourier decomposed into a superposition of plane waves in the form given in Eq. (4.2.15),

$$\phi(x) = \phi(t, \mathbf{x}) = \int d^3 k\, a(\mathbf{k})e^{i[\mathbf{k}\cdot\mathbf{x} - \omega(\mathbf{k})t]}, \quad (4.3.1)$$

where $a(\mathbf{k}) \in \mathbb{C}$ and $\omega(k) = E(k)/\hbar = \hbar k^2/2m$ as given in Example 3 above. The special case of the plane wave,

$$\phi_{\mathbf{k}}(x) = \phi_{\mathbf{k}}(t, \mathbf{x}) \equiv ae^{i[\mathbf{k}\cdot\mathbf{x} - \omega(\mathbf{k})t]}, \quad (4.3.2)$$

corresponds to the coordinate-space representation of the simultaneous eigenstate of the energy and momentum operators, $\hat{H}$ and $\hat{\mathbf{p}}$, respectively, for a free particle.

In classical mechanics changing the Hamiltonian by a constant has no effect on Hamilton's equations of motion; i.e., it has no physical consequence. For a natural classical system this corresponds to shifting the energy $H = E$ by a constant. Similarly, here shifting the energy $E(k)$ by an overall constant E_c gives

$$E(k) \rightarrow E'(k) \equiv E_c + E(k) \quad \text{and} \quad \omega(k) \rightarrow \omega'(k) \equiv (E_c/\hbar) + \omega(k). \tag{4.3.3}$$

This cannot change the physical behavior of the system since we have

$$\phi(x) \rightarrow \phi'(x) = e^{-iE_c t/\hbar}\phi(x) \tag{4.3.4}$$

and so the overall shape of wavepackets does not change,

$$|\phi'(t, \mathbf{x})|^2 = |\phi(t, \mathbf{x})|^2, \tag{4.3.5}$$

and the group velocity of wavepackets does not change,

$$v_g = \left.\frac{d\omega'}{dk}\right|_{k=k_0} = \left.\frac{d\omega}{dk}\right|_{k=k_0}. \tag{4.3.6}$$

The phase velocity does change, $v_p' = \omega'/k \neq \omega/k = v_p$, but this has no physical significance as explained above. So we are free to choose $E_c = mc^2$ so that $E(k) = mc^2 + \hbar^2 k^2/2m$. Then this $E(k)$ is the nonrelativistic limit of the relativistic energy $E(k) = \gamma(k)mc^2$. The operators $\hat{H}$ and $\hat{\mathbf{p}} = -i\hbar\boldsymbol{\nabla}$ satisfy the eigenvalue equations

$$\hat{H}\phi_{\mathbf{k}}(x) = i\hbar(\partial/\partial t)\phi_{\mathbf{k}}(x) = \hbar\omega\phi_{\mathbf{k}}(x) = E\phi_{\mathbf{k}}(x),$$
$$\hat{\mathbf{p}}\phi_{\mathbf{k}}(x) = \hbar\mathbf{k}\phi_{\mathbf{k}}(x) = \mathbf{p}\phi_{\mathbf{k}}(x). \tag{4.3.7}$$

It is natural to try to extend these considerations into the relativistic context, where we can define the contravariant four-momentum operator, $\hat{p}^\mu$, as

$$p^\mu = ((E/c), \mathbf{p}) = ((H/c), \mathbf{p}) \quad \rightarrow \quad \hat{p}^\mu = \left((\hat{H}/c), \hat{\mathbf{p}}\right). \tag{4.3.8}$$

Acting on any solution of the time-dependent Schrödinger equation, $\phi(x)$, we have

$$\hat{p}^\mu\phi(x) = \left(\frac{\hat{H}}{c}, \hat{\mathbf{p}}\right)\phi(x) = i\hbar\left(\frac{1}{c}\frac{\partial}{\partial t}, -\boldsymbol{\nabla}\right)\phi(x) = i\hbar\partial^\mu\phi(x). \tag{4.3.9}$$

A free particle of mass m satisfies[3]

$$p^2 = p^\mu p_\mu = (E/c)^2 - \mathbf{p}^2 = m^2 c^2, \tag{4.3.10}$$

where $E = \gamma mc^2$. Hence it follows that the plane wave $\phi_{\mathbf{k}}$ for a relativistic free particle is expected to satisfy the equation

$$\hat{p}^2\phi_{\mathbf{k}}(x) = \hat{p}^\mu\hat{p}_\mu\phi_{\mathbf{k}}(x) = \left(\frac{\hat{H}^2}{c^2} - \hat{\mathbf{p}}^2\right)\phi_{\mathbf{k}}(x) = -\hbar^2\partial^\mu\partial_\mu\phi_{\mathbf{k}}(x) = m^2 c^2\phi_{\mathbf{k}}(x). \tag{4.3.11}$$

---

[3] Note that when we say "mass" of the particle we *always* mean the *rest mass* of the particle, which is the standard modern convention with mass a Lorentz scalar. Some older texts define a relativistic mass as $m_{\text{rel}} \equiv \gamma m$, such that $E = \gamma mc^2 = m_{\text{rel}}c^2$ and $\mathbf{p} = \gamma m\mathbf{u} = m_{\text{rel}}\mathbf{u}$.

Thus, having incorporated the Schrödinger equation, $i\hbar\partial/\partial t\phi = \hat{H}\phi$, into the operator equivalent of the energy-mass relation for a free particle, $(\hat{H}/c)^2 - \hat{\mathbf{p}}^2 = m^2c^2$, we have arrived at the relativistic wave equation known as the *Klein-Gordon equation*

$$\left[\partial^2 + \left(\frac{mc}{\hbar}\right)^2\right]\phi = \left[\Box + \left(\frac{mc}{\hbar}\right)^2\right]\phi = 0. \tag{4.3.12}$$

Note that if we replace $(mc/\hbar)$ by m, then we arrive at the Klein-Gordon equation of Eq. (3.1.41), which was obtained when we constructed the equations of motion for a noninteracting relativistic classical scalar field. Here we have arrived at this equation as a simple relativistic extension of the Schrödinger equation, where we were attempting to interpret ϕ as a relativistic quantum wavefunction rather than as a relativistic classical field. All of the techniques that were developed for relativistic classical field theory can be applied to the complex scalar field $\phi(x)$ in relativistic quantum mechanics, which includes the definition of a Lagrangian, an action, a Hamiltonian, and the derivation of the respective equations of motion.

We see that the Klein-Gordon operator $[\partial^2 + (mc/\hbar)^2]$ is Lorentz invariant. Since the solutions of the Klein-Gordon equation, $\phi(x)$, carry no indices that transform under Lorentz transformations, then we know from the discussion of Sec. 3.2.4 that these solutions must correspond to either scalar or pseudoscalar wave functions. These will correspond to relativistic quantum mechanical descriptions of either charged scalar or charged pseodoscalar particles, respectively.

We also note that since $|\phi|^2 = \phi^*\phi$ is a Lorentz scalar we cannot identify it with a probability density as we do in ordinary nonrelativistic quantum mechanics. This is because a density is not Lorentz invariant due to volume elements being contracted in the direction of a Lorentz boost.

Solutions $\phi(x)$ that satisfy the Klein-Gordon equation include the usual plane wave solutions, $\phi(x) = \phi_{\mathbf{k}}(x)$, where

$$\omega(\mathbf{k}) = E(\mathbf{k})/\hbar = \gamma(k)mc^2/\hbar = \sqrt{(mc^2/\hbar)^2 + \mathbf{k}^2c^2} = \sqrt{m^2c^4 + \mathbf{p}^2c^2}/\hbar. \tag{4.3.13}$$

In addition, however, we observe that the Klein-Gordon equation is also satisfied by plane waves with negative frequencies (i.e., negative energies), $\omega(\mathbf{k}) \to -\omega(\mathbf{k})$, and so the plane wave solutions have the form

$$\phi_{\pm}(x) = \phi_{\pm}(t, \mathbf{x}) = a_{\pm}e^{i[\mathbf{k}\cdot\mathbf{x}\mp\omega(\mathbf{k})t]}. \tag{4.3.14}$$

The Klein-Gordon equation is a relativistic wave equation and furthermore it is a second-order wave equation in time and space, since it can be written as

$$\frac{1}{c^2}\frac{\partial^2\phi}{\partial t^2} = \left[\boldsymbol{\nabla}^2 - \left(\frac{mc}{\hbar}\right)^2\right]\phi. \tag{4.3.15}$$

A second-order equation in time requires two time boundary conditions to be specified (e.g., ϕ and $d\phi/dt$ at some $t = t_0$ or specifying ϕ at two times $t = t_1$ and t_2), whereas the Schrödinger equation is only first-order in time. The Klein-Gordon equation must therefore have an additional degree of freedom. An arbitrary solution of the Klein-Gordon equation will be a superposition of the plane wave solutions

$$\phi(x) = \int d^3k \left[a_+(\mathbf{k})e^{i[\mathbf{k}\cdot\mathbf{x}-\omega(\mathbf{k})t]} + a_-(\mathbf{k})e^{i[\mathbf{k}\cdot\mathbf{x}+\omega(\mathbf{k})t]}\right] \tag{4.3.16}$$

and so

$$\frac{\partial\phi}{\partial t}(x) = i\int d^3k\,\omega(\mathbf{k})\left[-a_+(\mathbf{k})e^{i[\mathbf{k}\cdot\mathbf{x}-\omega(\mathbf{k})t]} + a_-(\mathbf{k})e^{i[\mathbf{k}\cdot\mathbf{x}+\omega(\mathbf{k})t]}\right]. \tag{4.3.17}$$

The solutions of the nonrelativistic Schrödinger equation have $a_-(\mathbf{k}) = 0$, which means that specifying ϕ at $t = t_0$ is sufficient. However, solutions of the Klein-Gordon equation can include solutions with either $a_+(\mathbf{k}) = 0$ or $a_-(\mathbf{k}) = 0$ or neither equal to zero, which is why an extra boundary condition in time is needed.

Initially, the negative energy solutions were regarded as a problem for the Klein-Gordon equation and it was abandoned in favor of the Dirac equation. However, with the discovery of antiparticles it was eventually understood that positive and negative energy solutions actually correspond to particles and antiparticles, respectively, and that the Klein-Gordon equation is certainly a physically valid and useful equation. This is the modern *Feynman-Stueckelberg interpretation* of the negative energy solutions of the Klein-Gordon equation. In other words, what was initially considered a problem with the combination of relativity and quantum mechanics became one of the great triumphs of physics, i.e., the prediction that every particle must have a corresponding antiparticle.

We can rewrite Eq. (4.3.16) in a more compact and symmetric form. We change variables in the second term $\mathbf{k} \to -\mathbf{k}$ and note that this gives $d^3k \to -d^3k = -d^3p/\hbar^3$. We can then define $a_+(\mathbf{p}) \equiv a_+(\mathbf{k})/\hbar^3$ and $a_-^*(\mathbf{p}) \equiv -a_-(-\mathbf{k})/\hbar^3$, to give

$$\phi(x) = \int d^3p \left[a_+(\mathbf{p}) e^{-ip \cdot x/\hbar} + a_-^*(\mathbf{p}) e^{+ip \cdot x/\hbar} \right]. \tag{4.3.18}$$

The particle-like components have the conventional plane wave form, whereas the antiparticle-like components have $-p^\mu$ in the exponent rather than p^μ. We will return to this discussion after considering the conserved current associated with the Klein-Gordon equation.

4.3.2 Conserved Current

At this point it is useful to review the concept of a conserved current and its associated conserved charge. If there exists a four-vector current

$$j^\mu(x) \equiv (c\rho(x), \mathbf{j}(x)) \quad \text{such that} \quad \partial_\mu j^\mu(x) = 0, \tag{4.3.19}$$

then we say that j^μ is a conserved current density. If j^μ transforms like a Lorentz four-vector, then the current density is obviously conserved in every frame, since $\partial_\mu j^\mu$ is a Lorentz scalar. The quantity $\rho(x)$ is the associated charge density. We can also write Eq. (4.3.19) in a less covariant-looking form as

$$\frac{\partial}{\partial t} \rho(x) + \boldsymbol{\nabla} \cdot \mathbf{j}(x) = 0. \tag{4.3.20}$$

Consider some spatial volume V with surface S, then the charge inside V is

$$Q \equiv \int_V d^3x \, \rho(x). \tag{4.3.21}$$

The time-derivative of Q is

$$\frac{d}{dt} Q = \frac{d}{dt} \int_V d^3x \, \rho(x) = \int_V d^3x \, \frac{\partial}{\partial t} \rho(x) = \int_V d^3x \, \partial_0 j^0(x)$$

$$= -\int_V d^3x \, \boldsymbol{\nabla} \cdot \mathbf{j}(x) = -\int_S \mathbf{j}(x) \cdot d\mathbf{s}, \tag{4.3.22}$$

where we have used the divergence theorem (also often referred to as Gauss' theorem) for the last step. The appearance of the partial time derivative is because the spatial points in the volume V are not changing with time. It is then apparent that if there is no current density, $\mathbf{j}(x)$, through the surface S then the charge Q inside V is conserved. Furthermore, since $d\mathbf{s}$ is the outward pointing normal vector to the surface, we see that if current density flows out or in through the surface, then the total charge Q inside decreases or increases, respectively. Then if j^μ is nonzero only in some finite region and if we choose V to be all of space, then the total charge Q must be conserved.

Let us consider the real quantity $i\left(\phi^* \partial^\mu \phi - \phi \partial^\mu \phi^*\right)$, where $\phi(x)$ is a solution of the Klein-Gordon equation. Its four-divergence vanishes since

$$\partial_\mu \left(\phi^* \partial^\mu \phi - \phi \partial^\mu \phi^*\right) = \partial_\mu \phi^* \partial^\mu \phi + \phi^* \partial^2 \phi - \partial_\mu \phi \partial^\mu \phi^* - \phi \partial^2 \phi^* \tag{4.3.23}$$
$$= \phi^* \partial^2 \phi - \phi \partial^2 \phi^* = \phi^* \partial^2 \phi - (\phi^* \partial^2 \phi)^* = -(mc/\hbar)^2 \left[(\phi^* \phi) - (\phi^* \phi)^*\right] = 0,$$

where we have made use of the Klein-Gordon equation.

Initially, it was tempting to interpret $i\left(\phi^* \partial^\mu \phi - \phi \partial^\mu \phi^*\right)$ as proportional to the probability current density for the particle; however, the quantity is not positive definite and so this is not possible. It was in part this problem that motivated Dirac to formulate the Dirac equation. We show in Sec. 7.6.2 that the charge current density (in natural units) for a free charged scalar of charge q is given by

$$j^\mu = iq[\phi^* \partial^\mu \phi - (\partial^\mu \phi)^* \phi], \tag{4.3.24}$$

which is what we have also in the classical field case in Problem 3.8.

Nonrelativistic Limit: Free Schrödinger Equation

Let us consider the Klein-Gordon equation in the nonrelativistic limit in more detail. The relativistic kinetic energy T is defined by

$$T \equiv E - mc^2 = \gamma mc^2 - mc^2 = (\gamma - 1)mc^2$$
$$= \left(\frac{1}{2}\frac{v^2}{c^2} + \frac{3}{8}\frac{v^4}{c^4} + \cdots\right)mc^2 = \frac{1}{2}mv^2 + \frac{3}{8}\frac{mv^4}{c^2} + \cdots, \tag{4.3.25}$$

where we have used the result $1/\sqrt{1 - \epsilon} = 1 + (1/2)\epsilon + (3/8)\epsilon^2 + \cdots$. In the nonrelativistic limit we have $v^2 \ll c^2$ and so $T \ll mc^2$.

Define the modified wavefunction $\psi(x)$ for a particle-like solution by canceling out the dominating rest mass energy, mc^2, from the exponent in the wavefunction,

$$\psi(x) \equiv \phi(x) \exp\left((i/\hbar)mc^2 t\right). \tag{4.3.26}$$

It is natural to expect that $\psi(x)$ will be related to the nonrelativistic wavefunction, since the rest mass energy is not included in the definition of nonrelativistic energy. It follows from this definition that

$$i\hbar\frac{\partial \psi}{\partial t} = i\hbar\frac{\partial}{\partial t}\left[\phi(x) \exp\left(\frac{i}{\hbar}mc^2 t\right)\right] = (E - mc^2)\,\phi(x) \exp\left(\frac{i}{\hbar}mc^2 t\right)$$
$$= (E - mc^2)\psi(x) = T\psi(x). \tag{4.3.27}$$

In the nonrelativistic limit we have

$$|i\hbar\partial\psi/\partial t| = |T\psi| \ll |mc^2\psi| \quad \text{and} \quad \left|-\hbar^2 \partial^2\psi/\partial t^2\right| = \left|T^2\psi\right| \ll \left|m^2 c^4 \psi\right|. \tag{4.3.28}$$

We then find

$$i\hbar\frac{\partial \phi}{\partial t} = i\hbar\frac{\partial}{\partial t}\left[\psi \exp\left(-\frac{i}{\hbar}mc^2 t\right)\right] = \left(i\hbar\frac{\partial \psi}{\partial t} + mc^2\psi\right)\exp\left(-\frac{i}{\hbar}mc^2 t\right),$$
$$-\hbar^2\frac{\partial^2 \phi}{\partial t^2} = \left(-\hbar^2\frac{\partial^2 \psi}{\partial t^2} + 2mc^2 i\hbar\frac{\partial \psi}{\partial t} + m^2 c^4 \psi\right)\exp\left(-\frac{i}{\hbar}mc^2 t\right). \tag{4.3.29}$$

In the nonrelativistic limit we can safely neglect the $\hbar^2 \partial^2 \psi / \partial t^2$ term as being $\mathcal{O}((T/mc^2)^2)$ and so

$$\frac{1}{c^2}\frac{\partial^2 \phi}{\partial t^2} \to \left(\frac{-1}{\hbar^2 c^2}\right)\left(2mc^2 i\hbar \frac{\partial \psi}{\partial t} + m^2 c^4 \psi\right)\exp\left(-\frac{i}{\hbar}mc^2 t\right). \qquad (4.3.30)$$

Substituting this result into Eq. (4.3.15) gives

$$\left(\frac{-1}{\hbar^2 c^2}\right)\left(2mc^2 i\hbar \frac{\partial \psi}{\partial t} + m^2 c^4 \psi\right)\exp\left(-\frac{i}{\hbar}mc^2 t\right)$$
$$= \left[\boldsymbol{\nabla}^2 - \left(\frac{mc}{\hbar}\right)^2\right]\psi \exp\left(-\frac{i}{\hbar}mc^2 t\right). \qquad (4.3.31)$$

Canceling the terms involving the rest mass from each side and factoring out the exponentials involving the rest mass energy, we obtain the familiar Schrödinger equation for a free particle without spin,

$$i\hbar(\partial \psi / \partial t) = -(\hbar^2 / 2m)\boldsymbol{\nabla}^2 \psi. \qquad (4.3.32)$$

In the nonrelativistic limit the positive energy solutions of the Klein-Gordon equation become solutions of the usual free-particle Schrödinger equation as expected. The charge density becomes the usual probability density, $\rho(x) \to \phi(x)^*\phi(x) = \psi^*(x)\psi(x)$, and the conservation of the four-vector current j^μ in the nonrelativistic limit gives the conservation of total probability $1 \to \int d^3x\, \phi^*(x)\phi(x) = \int d^3x\, \psi^*(x)\psi(x)$. For a thorough discussion of the nonrelativistic limit of the Klein-Gordon equation see, e.g., Chapter 9 of Bjorken and Drell (1964).

Relativistic Classical Limit

The Klein-Gordon equation here specifies the evolution of a wavefunction $\phi(x)$ that attempts to describe the relativistic quantum-mechanical motion of a single free spinless (i.e., scalar) particle. In the nonrelativistic limit we should therefore recover the Schrödinger equation for a free spinless particle. This classical limit is unrelated to the Klein-Gordon equation of a relativistic classical field, where the equation has the same form but applies to a completely different physical context.

We can write the Klein-Gordon equation as

$$0 = \left[(i\hbar\partial)^2 c^2 - m^2 c^4\right]\phi = \left[\hat{p}^2 c^2 - m^2 c^4\right]\phi$$
$$= \left[\left(i\hbar\frac{\partial}{\partial t}\right)^2 - \hat{\mathbf{p}}^2 c^2 - m^2 c^4\right]\phi = \left[\hat{H}^2 - \hat{\mathbf{p}}^2 c^2 - m^2 c^4\right]\phi, \qquad (4.3.33)$$

where we have used the Schrödinger equation, $i\hbar\partial\phi/\partial t = \hat{H}\phi$. We are working in coordinate-space representation where $\hat{\mathbf{p}} = -i\hbar\boldsymbol{\nabla}$ and where the Schrödinger picture quantum state $|\phi(t)\rangle$ is specified by $\phi(x) = \langle\mathbf{x}|\phi(t)\rangle$. Recall that we require $\phi(x)$ to be square integrable at every time t so that $\int d^3x\, \rho(x)$ is finite at all times t. The Hilbert space is spanned by the set of all square integrable functions. Specifying $\phi(x)$ at any time t for every $\mathbf{x}$ specifies a ray in the Hilbert space, and the time evolution of this ray is determined by the Schrödinger equation, $i\hbar\partial\phi/\partial t = \hat{H}\phi$.

As discussed in Sec. 2.5 in the classical, i.e., macroscopic limit, we can replace operators by their macroscopic expectation values and neglect the relatively small quantum corrections. This gives

$$H^2 = \mathbf{p}^2 c^2 + m^2 c^4 \quad \Rightarrow H = \pm\left(\mathbf{p}^2 c^2 + m^2 c^4\right)^{1/2} = \gamma mc^2. \qquad (4.3.34)$$

The particle-like solution corresponds to the positive sign and leads to the relativistic free particle classical Hamiltonian in Eq. (2.8.15) and all of its consequences in Sec. 2.8. This includes the equation of motion for a free relativistic particle,

$$d\mathbf{p}/dt = 0, \quad \text{where} \quad \mathbf{p} = \gamma m \boldsymbol{v}. \tag{4.3.35}$$

4.3.3 Interaction with a Scalar Potential

We can generalize the free Klein-Gordon equation to include an interaction with a potential. However, from postulate (i) of special relativity we require the laws of physics to be the same in every frame, and so the introduction of an interaction into the Klein-Gordon equation must be done so as to make the form of the modified equations frame invariant. One way to do this is to introduce a Lorentz real scalar potential, $U(x)$, in the form

$$\left[(i\hbar\partial)^2 c^2 - \left(mc^2 + U \right)^2 \right] \phi(x) = 0 \tag{4.3.36}$$

or equivalently

$$\left[\partial^2 + (mc/\hbar)^2 \left(1 + [U/mc^2] \right)^2 \right] \phi(x) = 0, \tag{4.3.37}$$

where the solutions will no longer be simple plane waves due to the spacetime dependence of $U(x)$. Note that again we have that the four-vector current j^μ is conserved since using Eq. (4.3.37) and $U(x) \in \mathbb{R}$ gives

$$\partial_\mu j^\mu = iq\partial_\mu \left(\phi^* \partial^\mu \phi - \phi \partial^\mu \phi^* \right) = iq(\phi^* \partial^2 \phi - \phi \partial^2 \phi^*) = 0. \tag{4.3.38}$$

Nonrelativistic and weak-potential limit: Schrödinger equation with a potential: In the combined nonrelativistic and weak-potential limits, T/mc^2 and $U/mc^2 \ll 1$, we can again remove the majority of the time-dependence from $\phi(x)$ as we did above by defining

$$\psi(x) \equiv \phi(x) \exp \left((i/\hbar)mc^2 t \right). \tag{4.3.39}$$

It again follows that $(1/c^2)\partial^2 \phi/\partial t^2$ will approach the right-hand side of Eq. (4.3.30). Substituting this result into our Klein-Gordon equation with a scalar potential, Eq. (4.3.36), and again neglecting the $\hbar^2 \partial^2 \psi/\partial t^2$ term we find

$$
\begin{aligned}
\frac{1}{c^2} \frac{\partial^2 \phi}{\partial t^2} &\to \left(\frac{-1}{\hbar^2 c^2} \right) \left(2mc^2 i\hbar \frac{\partial \psi}{\partial t} + m^2 c^4 \psi \right) \exp \left(-\frac{i}{\hbar} mc^2 t \right) \\
&= \left[\boldsymbol{\nabla}^2 - \left(\frac{mc}{\hbar} \right)^2 [1 + (U/mc^2)]^2 \right] \psi \exp \left(-\frac{i}{\hbar} mc^2 t \right) \tag{4.3.40} \\
&= \left[\boldsymbol{\nabla}^2 - \left(\frac{1}{\hbar c} \right)^2 m^2 c^4 [1 + 2(U/mc^2) + (U/mc^2)^2] \right] \psi \exp \left(-\frac{i}{\hbar} mc^2 t \right). \tag{4.3.41}
\end{aligned}
$$

This gives the familiar Schrödinger equation for a spinless particle in a potential V,

$$i\hbar \frac{\partial \psi}{\partial t} = \left[-\frac{\hbar^2}{2m} \boldsymbol{\nabla}^2 + V \right] \psi, \tag{4.3.42}$$

where we have defined the usual Schrödinger equation time-dependent potential V in terms of the scalar potential U as

$$V(x) = V(t, \mathbf{x}) \equiv U(x) \left[1 + \frac{U(x)}{2mc^2}\right] \rightarrow U(x) \qquad (4.3.43)$$

after again imposing the weak potential limit $U \ll mc^2$.

For the case of the Klein-Gordon equation with a scalar potential we have explicitly shown that ordinary nonrelativistic quantum mechanics can be obtained as the weak-potential and nonrelativistic limit of relativistic quantum mechanics.

Relativistic classical limit: We can write the Klein-Gordon equation with our scalar interaction as

$$0 = \left[(i\hbar\partial)^2 c^2 - m^2 c^4 [1 + (U/mc^2)]^2\right] \phi. \qquad (4.3.44)$$

Recall from Eq. (2.4.28) that the classical canonical momenta π and coordinates $\mathbf{x}$ satisfy the fundamental Poisson bracket relations $\{x^i, \pi^j\} = \delta^{ij}$, $\{x^i, x^j\} = \{\pi^i, \pi^j\} = 0$. We reserve $\mathbf{p}$ to denote the mechanical momentum of a particle, i.e., $\mathbf{p} = \gamma m \boldsymbol{v}$. We also define $\pi^0 \equiv H/c$ as in Sec. 2.8.

Recall the correspondence between these fundamental Poisson brackets and the canonical commutation relations in Eq. (2.5.39), i.e., $[\hat{x}^i, \hat{\pi}^j] = i\hbar\delta^{ij}$, $[\hat{x}^i, \hat{x}^j] = [\hat{\pi}^i, \hat{\pi}^j] = 0$. Since $[i\hbar\partial^\mu, x^\nu] = i\hbar g^{\mu\nu}$, then we identify $\hat{\pi}^j = i\hbar\partial^j = -i\hbar\partial_j$, $\hat{\pi}^0 = \hat{H}/c = i\hbar\partial^0$, and $\hat{x}^\mu = x^\mu$ since we are working in coordinate-space representation. We can summarize these relationships as

$$[\hat{\pi}^\mu, \hat{x}^\nu] = [i\hbar\partial^\mu, x^\nu] = i\hbar g^{\mu\nu}. \qquad (4.3.45)$$

Then our Klein-Gordon equation can be written as

$$\begin{aligned}\left[\hat{\pi}^2 c^2 - m^2 c^4 [1 + (U/mc^2)]^2\right] \phi &= \left[(i\hbar\partial/\partial t)^2 - \hat{\boldsymbol{\pi}}^2 c^2 - m^2 c^4 [1 + (U/mc^2)]^2\right] \phi \\ &= \left[\hat{H}^2 - \hat{\boldsymbol{\pi}}^2 c^2 - m^2 c^4 [1 + (U/mc^2)]^2\right] \phi = 0. \qquad (4.3.46)\end{aligned}$$

In the classical (macroscopic) limit we can approximate operators by their macroscopic expectation values to give

$$H = \pm \left(\boldsymbol{\pi}^2 c^2 + m^2 c^4 [1 + (U/mc^2)]^2\right)^{1/2}. \qquad (4.3.47)$$

The particle solution corresponds to the positive sign. We recognize this particle Hamiltonian as the one in Eq. (2.8.24). We can use Hamilton's equation $\dot{x}^i = -\partial H/\partial\pi^i$ to write $\boldsymbol{\pi}$ in terms of $\dot{\mathbf{x}}$ just as we did in Eq. (2.8.31),

$$\boldsymbol{\pi} = [1 + (U/mc^2)]\gamma m\dot{\mathbf{x}} = [1 + (U/mc^2)]\mathbf{p}. \qquad (4.3.48)$$

This then gives

$$H = \pm[1 + (U/mc^2)] \left(\mathbf{p}^2 c^2 + m^2 c^4\right)^{1/2} = \pm[1 + (U/mc^2)]\gamma mc^2, \qquad (4.3.49)$$

where we use $\mathbf{p} = \gamma m \boldsymbol{v}$. Making the usual definition that $p^0 = E^{\text{mech}}/c$, where $E^{\text{mech}} = \gamma mc^2$ is the mechanical energy associated with the particle, we see that

$$[1 + (U/mc^2)](p^2 c^2 - m^2 c^4) = 0, \qquad (4.3.50)$$

which is consistent with $p^2 c^2 = m^2 c^4$ as it must be.

Given $\boldsymbol{\pi}$ we can then reconstruct the Lagrangian using $L = \dot{\mathbf{x}} \cdot \boldsymbol{\pi} - H$. We can then use the other of Hamilton's equations $\dot{\pi}^i = \partial H / \partial x^i$ and $dH/dt = -\partial L/\partial t$ to find the equations of motion for the system just as we did in Eq. (2.8.32),

$$d\pi_\mu / d\tau = \partial_\mu U. \tag{4.3.51}$$

So, we see that the classical limit of the Klein-Gordon equation with a scalar potential leads to the system of a classical relativistic particle moving in a scalar potential as discussed in Sec. 2.8. Furthermore, as we saw in that section, taking the nonrelativistic limit and neglecting the overall constant mc^2 gives $H = (1/2)mv^2 + U = \mathbf{p}^2/2m + U$, giving the classical limit of the above nonrelativistic quantum limit of the Klein-Gordon equation with a scalar potential U as it should.

4.3.4 Interaction with an Electromagnetic Field

The coupling of a Klein-Gordon particle to the electromagnetic field is determined by the principle of *minimal coupling* just as it is in relativistic mechanics; e.g., see Eq. (2.8.41). In Lorentz-Heaviside units we have

$$A^\mu = \left(A^0, \mathbf{A}\right) = (\Phi, \mathbf{A}),$$
$$\mathbf{E} = -\boldsymbol{\nabla}\Phi - (1/c)\partial\mathbf{A}/\partial t \quad \text{and} \quad \mathbf{B} = \boldsymbol{\nabla} \times \mathbf{A}. \tag{4.3.52}$$

Noting from Eq. (2.7.30) that $(qA^\mu)^{\text{SI}} = (qA^\mu)^{\text{LH}}/c$, then for a particle of charge q we have the classical replacement

$$p^\mu \to p^\mu = \pi^\mu - (q/c)A^\mu, \tag{4.3.53}$$

where π^μ is the canonical momentum and where p^μ is the mechanical momentum arising only from the motion of the particle. In coordinate-space representation in quantum mechanics π^μ always has the form $i\hbar\partial^\mu$, since this leads to the appropriate canonical commutation relations as discussed above and so we have the replacement

$$i\hbar\partial^\mu \to i\hbar D^\mu \equiv i\hbar\partial^\mu - (q/c)A^\mu \quad \text{or equivalently} \quad D^\mu \equiv \partial^\mu + i(q/\hbar c)A^\mu, \tag{4.3.54}$$

where D_μ is referred to as the *covariant derivative* for reasons to be discussed in Chapter 9. The classical-quantum correspondence for four-momentum in coordinate-space representation is

$$\text{canonical:} \ \ \pi^\mu \leftrightarrow \hat{\pi}^\mu = i\hbar\partial^\mu \qquad \text{and} \qquad \text{mechanical:} \ \ p^\mu \leftrightarrow \hat{p}^\mu = i\hbar D^\mu. \tag{4.3.55}$$

In the limit where the electromagnetic field vanishes, $A^\mu \to 0$, we have $\hat{p}^\mu = i\hbar D^\mu \to \pi^\mu = i\hbar\partial^\mu$ and we recover the free Klein-Gordon equation with $\hat{p}^\mu = i\hbar\partial^\mu$.

Minimal coupling is also referred to as *minimal substitution* and is closely connected to the concept of gauge invariance, which will be discussed in detail in later chapters. Here it is sufficient to recall that the $\mathbf{E}$ and $\mathbf{B}$ fields and the field strength tensor $F^{\mu\nu}$ are invariant under the gauge transformations

$$A_\mu(x) \to A'_\mu(x) = A_\mu(x) + \partial_\mu\omega(x). \tag{4.3.56}$$

As we know from Problem 3.8 and Sec. 7.6.2 on interaction with an electromagnetic field the Klein-Gordon equation becomes

$$\left[D^\mu D_\mu + \left(\frac{mc}{\hbar}\right)^2\right]\phi = \left[\left(\partial^\mu + i\frac{q}{\hbar c}A^\mu\right)\left(\partial_\mu + i\frac{q}{\hbar c}A_\mu\right) + \left(\frac{mc}{\hbar}\right)^2\right]\phi = 0, \tag{4.3.57}$$

which can be written as

$$\left[\partial^2 + \left(\frac{mc}{\hbar}\right)^2\right]\phi + i\frac{q}{\hbar c}\left[\partial^\mu(A_\mu\phi) + A^\mu(\partial_\mu\phi)\right] - \left(\frac{q}{\hbar c}\right)^2 A^\mu A_\mu\phi = 0. \tag{4.3.58}$$

The conserved current (in natural units) is now

$$j^\mu = iq\left[\phi^* D^\mu\phi - \phi(D^\mu\phi)^*\right]. \tag{4.3.59}$$

In coordinate-space representation the operator for the mechanical four-momentum of the particle is $\hat{p}^\mu = i\hbar D^\mu$. It is readily verified that $\partial_\mu j^\mu = 0$ since

$$\begin{aligned}
\partial_\mu\left[\phi^*(D^\mu\phi) - \phi(D^\mu\phi)^*\right] &= (\partial_\mu\phi^*)(D^\mu\phi) - (\partial_\mu\phi)(D^\mu\phi)^* + \phi^*(\partial_\mu D^\mu\phi) - \phi\partial_\mu(D^\mu\phi)^* \\
&= (D_\mu\phi)^*(D^\mu\phi) - (D_\mu\phi)(D^\mu\phi)^* + \phi^*(D_\mu D^\mu\phi) - \phi(D_\mu D^\mu\phi)^* \\
&= -\phi^*(mc/\hbar)^2\phi + \phi(mc/\hbar)^2\phi^* = 0, \tag{4.3.60}
\end{aligned}$$

where the interacting form of the Klein-Gordon equation $[D_\mu D^\mu + (mc/\hbar)^2]\phi = 0$ was used in the last step. The equality of the second and third lines follows since the four added terms involving A^μ in the third line cancel each other. When $A^\mu \to 0$ we recover the free field current density.

We can write this interacting Klein-Gordon equation as

$$\begin{aligned}
0 &= \left[(i\hbar D)^2 c^2 - m^2 c^4\right]\phi = \left[\hat{p}^2 c^2 - m^2 c^4\right]\phi \\
&= \left[\left(\hat{H}^{\text{mech}}\right)^2 - \hat{\mathbf{p}}^2 c^2 - m^2 c^4\right]\phi = \left[(i\hbar\partial - (q/c)A)^2 c^2 - m^2 c^4\right]\phi \\
&= \left[(i\hbar\partial/\partial t - q\Phi)^2 - (\hat{\boldsymbol{\pi}} - (q/c)\mathbf{A})^2 c^2 - m^2 c^4\right]\phi, \tag{4.3.61}
\end{aligned}$$

where here

$$\hat{H}^{\text{mech}} = \pm\hat{p}^0 c = \pm i\hbar D^0 c \quad\text{and}\quad \hat{\mathbf{p}} = \hat{\boldsymbol{\pi}} - (q/c)\mathbf{A} = -i\hbar\boldsymbol{\nabla} - (q/c)\mathbf{A} \tag{4.3.62}$$

are the observable operators for the mechanical energy and the mechanical three-momentum of the particle in the Schrödinger picture. Note that the above equation implies that

$$[\hat{p}^2 c^2 - m^2 c^4]\phi = 0 \tag{4.3.63}$$

as it should. At any time t we can use the eigenstates of the observable operator $\hat{\mathbf{p}}$ to define a basis for the Hilbert space for this physical system and then *any* state $|\phi\rangle$ for our system can be expressed in terms of this basis in the usual quantum mechanics way, i.e., any state $|\phi\rangle$ in the Hilbert space for our quantum system. The coordinate-space representation of an arbitrary quantum state $|\phi\rangle$ is the wavefunction $\phi(x)$.

The time evolution of our quantum system is given by the Schrödinger equation

$$i\hbar\partial\phi/\partial t = \hat{H}\phi \tag{4.3.64}$$

and we see that this combined with the interacting Klein-Gordon equation effectively defines the full Hamiltonian operator for our quantum system, $\hat{H}$. That is, since ϕ is the wavefunction corresponding to an arbitrary physical state for our quantum system, then we must have

$$\left(\hat{H} - q\Phi\right)^2 - [\hat{\boldsymbol{\pi}} - (q/c)\mathbf{A}]^2 c^2 - m^2 c^4 = 0. \tag{4.3.65}$$

This gives

$$
\begin{aligned}
\hat{H} &= \pm\left(\left[\hat{\boldsymbol{\pi}} - (q/c)\mathbf{A}\right]^2 c^2 + m^2 c^4\right)^{1/2} + q\Phi \\
&= \pm(\hat{\mathbf{p}}^2 c^2 + m^2 c^4)^{1/2} + q\Phi = \hat{H}^{\text{mech}} + q\Phi.
\end{aligned}
\tag{4.3.66}
$$

where $\hat{H}^{\text{mech}} \equiv \pm\hat{p}^0 c = \pm(\hat{\mathbf{p}}^2 c^2 + m^2 c^4)^{1/2}$. The positive sign is relevant for particles and the negative sign is associated with antiparticles.

Relativistic classical limit: We recognize that Eq. (4.3.66) with the positive sign is the operator form of the relativistic classical Hamiltonian describing the interaction of a spinless particle with an external electromagnetic field; i.e., it is the operator form of Eq. (2.8.40) expressed in Lorentz-Heaviside units. Since the canonical commutation relations remain valid in the relativistic context, the arguments in Sec. 2.5 regarding Ehrenfest's theorem and the classical limit continue to apply. So in the classical limit, the positive-energy solutions of the Klein-Gordon equation interacting with an external electromagnetic field lead to the relativistic Lorentz force equation, Eq. (2.8.51), and the equation for the rate of change of the particle's mechanical energy, Eq. (2.8.48), expressed in Lorentz-Heaviside units; i.e., we have

$$
\frac{d\mathbf{p}}{dt} = \frac{d(\gamma m\boldsymbol{v})}{dt} = q\left(\mathbf{E} + \frac{\boldsymbol{v}}{c} \times \mathbf{B}\right), \qquad \frac{dE^{\text{mech}}}{dt} = q\mathbf{E} \cdot \boldsymbol{v}.
\tag{4.3.67}
$$

Nonrelativistic quantum limit: In the nonrelativistic limit we have

$$
(\hat{\mathbf{p}}^2 c^2 + m^2 c^4)^{1/2}\phi = mc^2\left(1 + \frac{\hat{\mathbf{p}}^2}{m^2 c^2}\right)^{1/2}\phi \rightarrow \left(mc^2 + \frac{\hat{\mathbf{p}}^2}{2m}\right)\phi.
\tag{4.3.68}
$$

Again defining $\psi(x) = \phi(x)\exp(imc^2 t/\hbar)$ we find for the positive energy solutions in the nonrelativistic limit of Eq. (4.3.66) that

$$
\begin{aligned}
\hat{H}\phi &= i\hbar\frac{\partial}{\partial t}\left(\psi e^{-imc^2 t/\hbar}\right) = \left(i\hbar\frac{\partial\psi}{\partial t} + mc^2\psi\right)e^{-imc^2 t/\hbar} \\
&= \left(mc^2 + \frac{\hat{\mathbf{p}}^2}{2m} + q\Phi\right)\psi e^{-imc^2 t/\hbar}
\end{aligned}
\tag{4.3.69}
$$

and so

$$
i\hbar\frac{\partial}{\partial t}\psi = \left(\frac{\hat{\mathbf{p}}^2}{2m} + q\Phi\right)\psi = \left(\frac{\left[\hat{\boldsymbol{\pi}} - (q/c)\mathbf{A}\right]^2}{2m} + q\Phi\right)\psi = \hat{H}^{\text{nonrel}}\psi.
\tag{4.3.70}
$$

So we have recovered the nonrelativistic quantum mechanics of a particle interacting with an electromagnetic field, which in Lorentz-Heaviside units is

$$
\hat{H}^{\text{nonrel}} = \frac{\left[\hat{\boldsymbol{\pi}} - (q/c)\mathbf{A}\right]^2}{2m} + q\Phi = \frac{\hat{\mathbf{p}}^2}{2m} + q\Phi.
\tag{4.3.71}
$$

This corresponds to the Hamiltonian that was given in SI units earlier in Eq. (2.5.38).

4.4 Dirac Equation

Driven by the perception at the time that the Klein-Gordon equation was a failure in the quest to develop a relativistic form of quantum mechanics, in 1928 Dirac derived the equation named after him. This equation describes the relativistic quantum mechanics of spin-half massive particles, i.e., massive spin-half fermions. This was enormously important as it led to the prediction and first acceptance of the existence of antiparticles, predicted a magnetic moment for spin-half particles, and gave rise to an understanding of the fine structure of the hydrogen atom.

4.4.1 Formulation of the Dirac Equation

Dirac recognized that the perceived problems with the Klein-Gordon equation stemmed from the appearance of a second-order time derivative, $i\hbar(\partial^2/\partial t^2)\phi = \hat{H}^2\phi$. This gave rise to both

(a) the existence of negative energy solutions, $E < 0$
(b) negative charge density (that Dirac took to be probability density), since for $\partial_\mu j^\mu = 0$ to follow from the Klein-Gordon equation j^μ had to contain a derivative ∂^μ that allows $j^0 = \rho/c$ to become negative.

The proposal of Dirac was to look for an equation that was linear in $i\hbar\partial/\partial t$, i.e., an equation of the Schrödinger form

$$\hat{H}\psi = i\hbar\frac{\partial\psi}{\partial t} \tag{4.4.1}$$

for some $\hat{H}$, where this equation would need to be consistent with $E^2 = \mathbf{p}^2c^2 + m^2c^4$ and would need to transform covariantly under Lorentz transformations. $\hat{H}$ would need to be linear in $i\hbar c\boldsymbol{\nabla}$ in order to be covariant, since the right-hand side is linear in $i\hbar\partial/\partial t$. A simple equation of this type can be written as

$$i\hbar\frac{\partial\psi}{\partial t} = \hat{H}\psi = \left[-i\hbar c\,\boldsymbol{\alpha}\cdot\boldsymbol{\nabla} + \beta mc^2\right]\psi, \tag{4.4.2}$$

where we have defined the Hamiltonian operator as

$$\hat{H} \equiv \left[-i\hbar c\,\boldsymbol{\alpha}\cdot\boldsymbol{\nabla} + \beta mc^2\right] = \left[c\,\boldsymbol{\alpha}\cdot\hat{\boldsymbol{\pi}} + \beta mc^2\right], \tag{4.4.3}$$

where for a free particle $\hat{\mathbf{p}} = \hat{\boldsymbol{\pi}} = -i\hbar\boldsymbol{\nabla}$ and where each of $\boldsymbol{\alpha} \equiv (\alpha_1, \alpha_2, \alpha_3)$ and β are independent of $\mathbf{x}$ and t.

Note that the time evolution of an arbitrary spatial wavefunction $\psi(t, \mathbf{x})$ specified at the time t is determined by the above Schrödinger equation, $i\hbar\partial\psi/\partial t = \hat{H}\psi$. Now $\psi'(t, \mathbf{x}) \equiv \hat{H}\psi(t, x) = i\hbar\partial\psi/\partial t$ is just some other spatial wavefunction specified at time t and so $i\hbar\partial\psi'/\partial t = \hat{H}\psi'$; i.e., we must have

$$-\hbar^2\partial^2\psi/\partial t^2 = \hat{H}^2\psi \tag{4.4.4}$$

for any $\psi(t, \mathbf{x})$ that satisfies the Schrödinger equation. Since we want to recover the energy-momentum relation for our solutions,

$$E^2 = \mathbf{p}^2c^2 + m^2c^4, \tag{4.4.5}$$

we require

$$-\hbar^2\partial^2\psi/\partial t^2 = \hat{H}^2\psi = \left[\hat{\mathbf{p}}^2c^2 + m^2c^4\right]\psi = \left[-\hbar^2c^2\boldsymbol{\nabla}^2 + m^2c^4\right]\psi. \tag{4.4.6}$$

So solutions of the Dirac equation are also solutions of the Klein-Gordon equation

$$\left[\partial^2 + (mc/\hbar)^2\right]\psi = 0, \tag{4.4.7}$$

which means that we must have

$$
\begin{aligned}
\hat{H}^2 &= \left(-i\hbar c\,\boldsymbol{\alpha}\cdot\boldsymbol{\nabla} + mc^2\beta\right)^2 \\
&= \left[\sum_{i,j=1}^{3}(-i\hbar c)^2\,\alpha_i\alpha_j\nabla_i\nabla_j\right] + \left[\sum_{i=1}^{3}(-i\hbar mc^3)\,(\alpha_i\beta + \beta\alpha_i)\,\nabla_i\right] + m^2c^4\beta^2 \\
&= \left[-\hbar^2c^2\boldsymbol{\nabla}^2 + m^2c^4\right] = \left[\hat{\mathbf{p}}^2c^2 + m^2c^4\right].
\end{aligned}
\tag{4.4.8}
$$

Note that $\alpha_i\alpha_j\nabla_i\nabla_j = (1/2)(\alpha_i\alpha_j+\alpha_j\alpha_i)\nabla_i\nabla_j$ since the gradient operators commute and so only the symmetric part of $\alpha_i\alpha_j$ survives. It then follows that $\boldsymbol{\alpha}$ and β must obey the anticommutation relations of a Clifford algebra,

$$
\{\alpha_i,\alpha_j\} = 2\delta_{ij}I, \quad \{\alpha_i,\beta\} = 0, \quad \beta^2 = I \quad \text{for} \quad i,j = 1,2,3,
\tag{4.4.9}
$$

where the anticommutator is defined as

$$
\{A,B\} \equiv AB + BA.
\tag{4.4.10}
$$

Clearly, $\boldsymbol{\alpha}$ and β cannot be numbers. Dirac assumed that they were $N \times N$ matrices. **Note:** we are using the same notation for the anticommutator as we did for the Poisson bracket in classical mechanics. This is an unfortunate but common convention. However, the intended meaning of the curly brackets $\{\cdot,\cdot\}$ will always be obvious from the context.

For the Hamiltonian, $\hat{H}$, to be Hermitian $\boldsymbol{\alpha}$ and β must also be Hermitian,

$$
\boldsymbol{\alpha}^\dagger = \boldsymbol{\alpha} \quad \text{and} \quad \beta^\dagger = \beta,
\tag{4.4.11}
$$

which means that they have real eigenvalues. From the Clifford algebra we have $\alpha_i^2 = \beta^2 = I$ and so the eigenvalues of α_i^2 and β^2 are equal to 1. Since the eigenvalues of any matrix squared are the square of the eigenvalues, then the eigenvalues of the Hermitian matrices α_i and β are ± 1. We also have that $\alpha_i\beta = -\beta\alpha_i$ and so

$$
\alpha_i = -\beta\alpha_i\beta \quad \text{and} \quad \beta = -\alpha_i\beta\alpha_i.
\tag{4.4.12}
$$

Using the cyclic property of the trace we then have

$$
\begin{aligned}
\mathrm{tr}\,\alpha_i &= -\mathrm{tr}(\beta\alpha_i\beta) = -\mathrm{tr}(\beta^2\alpha_i) = -\mathrm{tr}\,\alpha_i = 0, \\
\mathrm{tr}\,\beta &= -\mathrm{tr}(\alpha_i\beta\alpha_i) = -\mathrm{tr}(\alpha_i^2\beta) = -\mathrm{tr}\,\beta = 0.
\end{aligned}
\tag{4.4.13}
$$

Recall that the trace of a matrix is the sum of its eigenvalues. Then the dimension N of the matrices must be even, since the eigenvalues are ± 1 and the matrices are traceless. If we assume that any one of the four matrices $\alpha_1,\alpha_2,\alpha_3,\beta$ could be written as a linear combination of the other three, then it is readily seen that the above Clifford algebra would be impossible to satisfy. This means that these four traceless Hermitian matrices must be linearly independent.

In the case $N = 2$ there are only three linearly independent traceless Hermitian matrices and so a representation with $N = 2$ is not possible. Any such three matrices form a basis that spans the space of 2×2 traceless Hermitian matrices when we take real linear combinations of them. This basis is typically chosen to be the Pauli matrices given in Eq. (A.3.1), $\boldsymbol{\sigma} = (\sigma^1,\sigma^2,\sigma^3)$. In nonrelatavistic quantum mechanics it is common to write the Pauli spin matrix indices in the down position, but it is convenient in the relativistic context to write them in the up position as we have done here. The set of four matrices $(I,\boldsymbol{\sigma})$ form a basis of 2×2 Hermitian matrices when we take real linear combinations of them.

Turning next to $N = 4$, it is readily checked from the properties of the Pauli spin matrices in Eq. (A.3.1) that the 4×4 matrices

$$\alpha_i = \begin{pmatrix} 0 & \sigma^i \\ \sigma^i & 0 \end{pmatrix}, \quad \beta = \begin{pmatrix} I & 0 \\ 0 & -I \end{pmatrix} \tag{4.4.14}$$

are traceless and Hermitian and satisfy the required Clifford algebra. This is the traditional *Dirac representation*, where the matrices are partitioned into two-dimensional submatrices involving the Pauli spin matrices and the 2×2 identity matrix.

In summary, we have now arrived at the *Dirac equation*

$$i\hbar \partial \psi / \partial t = \left[-i\hbar c\, \boldsymbol{\alpha} \cdot \boldsymbol{\nabla} + mc^2 \beta \right] \psi, \tag{4.4.15}$$

where the 4×4, traceless and Hermitian matrices $\boldsymbol{\alpha}$ and β satisfy the Clifford algebra of Eq. (4.4.9) and where the wavefunction $\psi(x)$ is a four-component column vector of wavefunction components

$$\psi(x) = \begin{bmatrix} \psi_1(x) \\ \psi_2(x) \\ \psi_3(x) \\ \psi_4(x) \end{bmatrix}. \tag{4.4.16}$$

We refer to a solution of the Dirac equation, $\psi(x)$, as a *Dirac spinor wavefunction*.

The term *Dirac spinor* is usually understood to be the Dirac spinor wavefunction for a particle or antiparticle moving as a plane wave after removing the plane wave exponential factor, i.e., after removing $\exp(-ip \cdot x)/\hbar$ or $\exp(+ip \cdot x/\hbar)$, respectively. The plane wave solutions of the Dirac equation will be discussed in a later section when we address the issue of its Lorentz covariance. This generalizes the concept of a two-spinor wavefunction and a two-spinor in nonrelativistic quantum mechanics, where the two-component spinor specifies the spin state of a spin-half particle. For this reason the four-component Dirac spinor is sometimes referred to as a *bi-spinor*.

In nonrelativistic quantum mechanics the spin operator is given by

$$\hat{\mathbf{S}} = (\hat{S}^1, \hat{S}^2, \hat{S}^3) = (\hat{S}_x, \hat{S}_y, \hat{S}_z) = (\hbar/2)\boldsymbol{\sigma} \tag{4.4.17}$$

and we have the angular momentum Lie algebra, $[\hat{S}^i, \hat{S}^j] = i\hbar\epsilon^{ijk}S^k$, where the summation convention is understood. The appearance of two "copies" of the Pauli spin matrices is a strong indication that the Dirac equation contains two "copies" of the spin-half algebra. The meaning of this will become clear as we proceed further, but we are immediately reminded of the $SL(2, \mathbb{C})$ representation of the restricted Lorentz group discussed in Sec. 1.3. In the nonrelativistic limit, $pc/mc^2 \ll 1$, the mass term in the Dirac equation will be dominant and so there it is natural to use the Dirac representation, where the matrix β in the mass term is diagonal.

Any four matrices related to the Dirac matrices by a similarity transformation

$$\boldsymbol{\alpha} \to \bar{\boldsymbol{\alpha}} = S\boldsymbol{\alpha}S^{-1}, \quad \beta \to \bar{\beta} = S\beta S^{-1}, \tag{4.4.18}$$

will obviously preserve the Clifford algebra of Eq. (4.4.9),

$$\{\bar{\alpha}_i, \bar{\alpha}_j\} = S\{\alpha_i, \alpha_j\}S^{-1} = \delta_{ij}SS^{-1} = \delta_{ij}, \quad \bar{\beta}^2 = S\beta^2 S^{-1} = I,$$
$$\{\bar{\alpha}_i, \bar{\beta}\} = S\{\alpha_i, \beta\}S^{-1} = 0. \tag{4.4.19}$$

If we wish to preserve the Hermitian nature of the Dirac Hamiltonian $\hat{H}$, then it follows that the nonsingular matrix S must be unitary, since

$$\bar{\boldsymbol{\alpha}}^\dagger = (S\boldsymbol{\alpha}S^{-1})^\dagger = (S^{-1})^\dagger \boldsymbol{\alpha}S^\dagger = (S^{-1})^\dagger S^{-1}\bar{\boldsymbol{\alpha}}SS^\dagger$$
$$\bar{\beta}^\dagger = (S\beta S^{-1})^\dagger = (S^{-1})^\dagger \beta S^\dagger = (S^{-1})^\dagger S^{-1}\bar{\beta}SS^\dagger \tag{4.4.20}$$

and so $\bar{\alpha}^\dagger = \bar{\alpha}$ and $\bar{\beta}^\dagger = \bar{\beta}$ if and only if $S^\dagger = S^{-1}$, i.e., if and only if S is a unitary matrix so that the similarity transformation is a unitary transformation.

4.4.2 Probability Current

The *Hermitian conjugate* or *adjoint* Dirac wavefunction is

$$\psi^\dagger = (\psi_1^*, \dots, \psi_4^*). \tag{4.4.21}$$

Taking the adjoint of the Dirac equation in Eq. (4.4.15) and remembering that α and β are Hermitian we find

$$-i\hbar \partial \psi^\dagger / \partial t = i\hbar c \nabla \psi^\dagger \cdot \alpha + \psi^\dagger \beta m c^2. \tag{4.4.22}$$

Consider

$$
\begin{aligned}
i\hbar \partial \left(\psi^\dagger \psi\right) / \partial t &= \psi^\dagger \left(i\hbar \partial \psi / \partial t\right) + \left(i\hbar \partial \psi^\dagger / \partial t\right) \psi \\
&= -i\hbar c \left(\psi^\dagger \alpha \cdot \nabla \psi + \left(\nabla \psi^\dagger\right) \cdot \alpha \psi\right) = -i\hbar c \nabla \cdot \left(\psi^\dagger \alpha \psi\right).
\end{aligned} \tag{4.4.23}
$$

Defining

$$j^\mu = (j^0, \mathbf{j}) \quad \text{with} \quad j^0 = \rho c \equiv \psi^\dagger \psi \quad \text{and} \quad \mathbf{j} \equiv c \psi^\dagger \alpha \psi, \tag{4.4.24}$$

we observe that

$$\partial_\mu j^\mu = \partial_0 j^0 + \nabla \cdot \mathbf{j} = \frac{\partial \rho}{\partial t} + \nabla \cdot \mathbf{j} = \frac{\partial}{\partial t} (\psi^\dagger \psi) + c \nabla \cdot (\psi^\dagger \alpha \psi) = 0. \tag{4.4.25}$$

Therefore we identify j^μ as the conserved probability current density, where

$$\rho = \psi^\dagger \psi = \sum_{i=1}^4 |\psi_i|^2 > 0 \tag{4.4.26}$$

is the positive-definite probability density. Using the divergence theorem we have conservation of total probability in the usual way,

$$\frac{d}{dt} \int d^3 x \, \psi^\dagger \psi = 0. \tag{4.4.27}$$

Later we will show that $j^\mu = (c\rho, \boldsymbol{j})$ is a four-vector current density as we require for the Dirac equation to be Lorentz covariant.

4.4.3 Nonrelativistic Limit and Relativistic Classical Limit

Nonrelativistic limit: As mentioned earlier it is natural to use the Dirac representation for α and β in order to consider the nonrelativistic limit of the Dirac equation. Let the four-component wavefunction $\tilde{\psi}$ be partitioned into two column spinors $\tilde{\phi}$ and $\tilde{\chi}$,

$$\psi = \begin{bmatrix} \tilde{\phi} \\ \tilde{\chi} \end{bmatrix}. \tag{4.4.28}$$

Then the Dirac equation becomes

$$i\hbar \frac{\partial}{\partial t} \begin{bmatrix} \tilde{\phi} \\ \tilde{\chi} \end{bmatrix} = -i\hbar c \begin{bmatrix} 0 & \sigma \\ \sigma & 0 \end{bmatrix} \cdot \nabla \begin{bmatrix} \tilde{\phi} \\ \tilde{\chi} \end{bmatrix} + m c^2 \begin{bmatrix} I & 0 \\ 0 & -I \end{bmatrix} \begin{bmatrix} \tilde{\phi} \\ \tilde{\chi} \end{bmatrix}, \tag{4.4.29}$$

which gives rise to two coupled first-order differential equations for $\tilde{\phi}$ and $\tilde{\chi}$,

$$i\hbar\partial\tilde{\phi}/\partial t = -i\hbar c\boldsymbol{\sigma} \cdot \boldsymbol{\nabla}\tilde{\chi} + mc^2\tilde{\phi} = c\boldsymbol{\sigma} \cdot \hat{\mathbf{p}}\tilde{\chi} + mc^2\tilde{\phi}, \tag{4.4.30}$$

$$i\hbar\partial\tilde{\chi}/\partial t = -i\hbar c\boldsymbol{\sigma} \cdot \boldsymbol{\nabla}\tilde{\phi} - mc^2\tilde{\chi} = c\boldsymbol{\sigma} \cdot \hat{\mathbf{p}}\tilde{\phi} - mc^2\tilde{\chi}. \tag{4.4.31}$$

In the rest frame, $\hat{\mathbf{p}}\psi = -i\hbar\boldsymbol{\nabla}\psi = 0$ and so the equations decouple and we find

$$\tilde{\phi} = e^{-imc^2t/\hbar} \begin{bmatrix} 1 \\ 0 \end{bmatrix}, \ e^{-imc^2t/\hbar} \begin{bmatrix} 0 \\ 1 \end{bmatrix},$$

$$\tilde{\chi} = e^{+imc^2t/\hbar} \begin{bmatrix} 1 \\ 0 \end{bmatrix}, \ e^{+imc^2t/\hbar} \begin{bmatrix} 0 \\ 1 \end{bmatrix}. \tag{4.4.32}$$

This implies a set of four linearly independent solutions for the four-spinor ψ at rest,

$$e^{-imc^2t/\hbar} \begin{bmatrix} 1 \\ 0 \\ 0 \\ 0 \end{bmatrix}, \ e^{-imc^2t/\hbar} \begin{bmatrix} 0 \\ 1 \\ 0 \\ 0 \end{bmatrix}, \ e^{+imc^2t/\hbar} \begin{bmatrix} 0 \\ 0 \\ 1 \\ 0 \end{bmatrix}, \ e^{+imc^2t/\hbar} \begin{bmatrix} 0 \\ 0 \\ 0 \\ 1 \end{bmatrix}. \tag{4.4.33}$$

Note that while the Dirac equation satisfied Dirac's desire to have a positive probability density, it has not removed the issue of negative energy states; i.e., for the first and second rest states we find $\hat{H}\psi = i\hbar\partial\psi/\partial t = mc^2\psi = E\psi$ while for the third and fourth states we find $\hat{H}\psi = i\hbar\partial\psi/\partial t = -mc^2\psi = E\psi$. Any solution for a particle in its rest frame corresponds to some linear superposition of the first two of these solutions and any solution for an antiparticle in its rest frame corresponds to some linear superposition of the last two of these solutions.

Let us consider the situation where we have a particle-like solution moving nonrelativistically, i.e., where the dominant contributions to ψ would come from the two upper components, $\tilde{\phi}$. In this limit the largest energy in the system will be the rest mass energy mc^2 and the kinetic energy will be very small compared with this. It is therefore convenient to define ϕ and χ as

$$\begin{bmatrix} \phi \\ \chi \end{bmatrix} = e^{+imc^2t/\hbar} \begin{bmatrix} \tilde{\phi} \\ \tilde{\chi} \end{bmatrix}, \tag{4.4.34}$$

which gives

$$i\hbar\partial\phi/\partial t = -i\hbar c\boldsymbol{\sigma} \cdot \boldsymbol{\nabla}\chi = c\boldsymbol{\sigma} \cdot \hat{\mathbf{p}}\chi,$$
$$i\hbar\partial\chi/\partial t = -i\hbar c\boldsymbol{\sigma} \cdot \boldsymbol{\nabla}\phi - 2mc^2\chi = c\boldsymbol{\sigma} \cdot \hat{\mathbf{p}}\phi - 2mc^2\chi. \tag{4.4.35}$$

In our limit of particle-like solutions moving nonrelativistically it follows that χ is "small" and ϕ and mc^2 are "large." So we can neglect the left-hand side of the second equation above, which gives the approximate equation

$$\chi = -i\hbar\boldsymbol{\sigma} \cdot \boldsymbol{\nabla}\phi/2mc = [\boldsymbol{\sigma} \cdot \hat{\mathbf{p}}/2mc]\phi. \tag{4.4.36}$$

Substituting this back into the first equation above gives in the particle-like nonrelativistic limit

$$i\hbar\partial\phi/\partial t = [(\boldsymbol{\sigma} \cdot \hat{\mathbf{p}})^2/2m]\phi = [\hat{\mathbf{p}}^2/2m]\phi, \tag{4.4.37}$$

where we have used $(\boldsymbol{\sigma} \cdot \mathbf{a})(\boldsymbol{\sigma} \cdot \mathbf{b}) = \mathbf{a} \cdot \mathbf{b} + i\boldsymbol{\sigma} \cdot (\mathbf{a} \times \mathbf{b})$, and where we note that $(\boldsymbol{\nabla} \times \boldsymbol{\nabla})^i\phi = \epsilon^{ijk}\partial_j\partial_k\phi = 0$. This is just the Schrödinger equation for a free particle with spin-half, where the two-component spinor is ϕ; i.e., the two components of ϕ are sufficient to describe the spin state of the spin-half particle. Note that the motion of the particle and its spin are not coupled in this limit.

Relativistic classical limit: Since by construction the solutions of the Dirac equation are also solutions of the Klein-Gordon equation as shown in Eq. (4.4.6), then in the classical limit we again arrive at that $H^2 = \mathbf{p}^2 c^2 + m^2 c^4$ and so for particle-like solutions we recover Eqs. (4.3.34) and (4.3.35); i.e., we recover the behavior of a classical relativistic free particle. So, while each of the four components of the Dirac wavefunction must obey the Dirac equation, we reproduce the classical behavior of the relativistic free particle by design.

4.4.4 Interaction with an Electromagnetic Field

As in the case of the Klein-Gordon equation, we incorporate coupling to the electromagentic field using the principle of minimal coupling as given in Eqs. (4.3.54) and (4.3.55); i.e., we have

$$A^\mu = (A^0, \mathbf{A}) = (\Phi, \mathbf{A}) \quad \text{and} \quad i\hbar\partial^\mu = \hat{\pi}^\mu = \hat{p}^\mu + (q/c)A^\mu = i\hbar D^\mu + (q/c)A^\mu. \quad (4.4.38)$$

Minimal coupling consists of recognizing that the mechanical momentum for the free particle is $\hat{p}^\mu = i\hbar\partial^\mu$ and in the interacting case is replaced by the mechanical momentum for the interacting particle, which in this case is $\hat{p}^\mu = i\hbar D^\mu$; i.e., in the Dirac equation we make the replacement $i\hbar\partial^\mu \to i\hbar D^\mu = i\hbar\partial^\mu - (q/c)A^\mu$ just as we did in the Klein-Gordon equation. This results in

$$i\hbar\partial\psi/\partial t = \hat{H}\psi = \left[c\boldsymbol{\alpha} \cdot (-i\hbar\boldsymbol{\nabla} - (q/c)\mathbf{A}) + \beta mc^2 + q\Phi\right]\psi \quad (4.4.39)$$
$$= \left[c\boldsymbol{\alpha} \cdot (\hat{\boldsymbol{\pi}} - (q/c)\mathbf{A}) + \beta mc^2 + q\Phi\right]\psi = \left[c\boldsymbol{\alpha} \cdot \hat{\mathbf{p}} + \beta mc^2 + q\Phi\right]\psi,$$

where $\hat{\boldsymbol{\pi}} = -i\hbar\boldsymbol{\nabla}$ is the *canonical three-momentum operator* and $\hat{\mathbf{p}} = \hat{\boldsymbol{\pi}} - (q/c)\mathbf{A}$ is the *mechanical three-momentum operator* in coordinate-space representation.

From the above definition of the Hamiltonian we have

$$\hat{H} - q\Phi = \left[c\boldsymbol{\alpha} \cdot (\hat{\boldsymbol{\pi}} - (q/c)\mathbf{A}) + \beta mc^2\right] \quad (4.4.40)$$

and from the properties of the Dirac matrices, $\boldsymbol{\alpha}$ and β, we find that $\hat{H}^2$ has the same form as it did in Eq. (4.3.65), i.e.,

$$(\hat{H} - q\Phi)^2 - (\hat{\boldsymbol{\pi}} - (q/c)\mathbf{A})^2 c^2 - m^2 c^4 = 0 \quad (4.4.41)$$

and Eq. (4.3.66) then follows.

Nonrelativistic and weak field limit: In terms of the two-component spinors ϕ and χ defined above, the Dirac equation now becomes

$$i\hbar\partial\phi/\partial t = [c\boldsymbol{\sigma} \cdot (-i\hbar\boldsymbol{\nabla} - (q/c)\mathbf{A})]\chi + q\Phi\phi = c\boldsymbol{\sigma} \cdot \hat{\mathbf{p}}\chi + q\Phi\phi, \quad (4.4.42)$$
$$i\hbar\partial\chi/\partial t = [c\boldsymbol{\sigma} \cdot (-i\hbar\boldsymbol{\nabla} - (q/c)\mathbf{A})]\phi - [2mc^2 - q\Phi]\chi = c\boldsymbol{\sigma} \cdot \hat{\mathbf{p}}\phi - [2mc^2 - q\Phi]\chi.$$

In the limit of particle-like solutions moving nonrelativistically χ is small and ϕ and mc^2 are large and in the weak field limit we have $q\Phi, (q/c)\mathbf{A} \ll mc^2$. In these limits we then find from the second equation that

$$\chi = [\boldsymbol{\sigma}/2mc] \cdot (-i\hbar\boldsymbol{\nabla} - (q/c)\mathbf{A})\phi = [\boldsymbol{\sigma} \cdot \hat{\mathbf{p}}/2mc]\phi. \quad (4.4.43)$$

Substituting this back into the first equation above gives the *Pauli equation*

$$i\hbar\frac{\partial\phi}{\partial t} = \left(\frac{(\boldsymbol{\sigma} \cdot \hat{\mathbf{p}})^2}{2m} + q\Phi\right)\phi = \left(\frac{(\boldsymbol{\sigma} \cdot [\hat{\boldsymbol{\pi}} - (q/c)\mathbf{A}])^2}{2m} + q\Phi\right)\phi$$
$$= \left(\frac{[\hat{\boldsymbol{\pi}} - (q/c)\mathbf{A}]^2}{2m} - \frac{\hbar q}{2mc}\boldsymbol{\sigma} \cdot \mathbf{B} + q\Phi\right)\phi, \quad (4.4.44)$$

where we have used $(\boldsymbol{\sigma} \cdot \mathbf{a})(\boldsymbol{\sigma} \cdot \mathbf{b}) = \mathbf{a} \cdot \mathbf{b} + i\boldsymbol{\sigma} \cdot (\mathbf{a} \times \mathbf{b})$ and that

$$\begin{aligned}
[\hat{\mathbf{p}} \times \hat{\mathbf{p}}]^i \phi &= \epsilon^{ijk}\hat{p}^j\hat{p}^k\phi = \epsilon^{ijk}[-i\hbar\partial_j - (q/c)A^j][-i\hbar\partial_k - (q/c)A^k]\phi \\
&= (i\hbar q/c)\phi(\epsilon^{ijk}\partial_j A^k) = (i\hbar q/c)\phi(\boldsymbol{\nabla} \times \mathbf{A}) = (i\hbar q/c)B^i\phi,
\end{aligned} \qquad (4.4.45)$$

since $\hat{\mathbf{p}} = -i\hbar\boldsymbol{\nabla} - (q/c)\mathbf{A}$ means that $[\hat{p}^i, \hat{p}^j]\phi \neq 0$.

The gyromagnetic ratio g is defined as the ratio of the magnetic field interaction arising from the spin of the particle with respect to that from the orbital motion of the particle; i.e., the total paramagnetic interaction is proportional to

$$(\hat{\mathbf{L}} + g\hat{\mathbf{s}}). \qquad (4.4.46)$$

We finally then recognize the famous success of the Dirac equation in its prediction of the gyromagnetic ratio of $g = 2$ for any elementary fermion, i.e., for any particle satisfying the Dirac equation. We refer to the quantity $a \equiv (g - 2)/2$ as the *anomalous magnetic moment*, which will be discussed in more detail in Sec. 8.7.3.

Proof

(a) $\mathbf{A} = (1/2)(\mathbf{B} \times \mathbf{x})$ is a uniform magnetic field in Coulomb gauge.

  (i) It corresponds to Coulomb gauge, since

$$2\boldsymbol{\nabla} \cdot \mathbf{A} = \boldsymbol{\nabla} \cdot (\mathbf{B} \times \mathbf{x}) = -\mathbf{B} \cdot (\boldsymbol{\nabla} \times \mathbf{x}) = -\epsilon^{ijk}B^i\delta^{jk} = 0,$$

where we have used the scalar triple product identity $\mathbf{a} \cdot (\mathbf{b} \times \mathbf{c}) = \mathbf{b} \cdot (\mathbf{c} \times \mathbf{a}) = \mathbf{c} \cdot (\mathbf{a} \times \mathbf{b})$, that $\mathbf{B}$ is independent of $\mathbf{x}$, and where we have been careful to maintain the ordering of things that do not commute.

  (ii) It is compatible with $\mathbf{B} = \boldsymbol{\nabla} \times \mathbf{A}$, since

$$\boldsymbol{\nabla} \times \mathbf{A} = \tfrac{1}{2}\boldsymbol{\nabla} \times (\mathbf{B} \times \mathbf{x}) = \tfrac{1}{2}[\mathbf{B}(\boldsymbol{\nabla} \cdot \mathbf{x}) - (\mathbf{B} \cdot \boldsymbol{\nabla})\mathbf{x}] = \tfrac{1}{2}[3\mathbf{B} - \mathbf{B}] = \mathbf{B},$$

where we have used the vector triple product identity $\mathbf{a} \times (\mathbf{b} \times \mathbf{c}) = \mathbf{b}(\mathbf{a} \cdot \mathbf{c}) - \mathbf{c}(\mathbf{a} \cdot \mathbf{b})$, that $\mathbf{B}$ is independent of $\mathbf{x}$, and where we have been careful to maintain the ordering of things that do not commute.

(b) Note that $\mathbf{A}^2 = (1/4)(\mathbf{B} \times \mathbf{x})^2$ and that

$$\begin{aligned}
[\hat{\boldsymbol{\pi}} - (q/c)\mathbf{A}]^2 \phi &= \left[-\hbar^2\boldsymbol{\nabla}^2 - 2(q/c)\mathbf{A} \cdot (-i\hbar\boldsymbol{\nabla}) + (q/c)^2\mathbf{A}^2\right]\phi + (i\hbar)(q/c)\phi(\boldsymbol{\nabla} \cdot \mathbf{A}) \\
&= \left[-\hbar^2\boldsymbol{\nabla}^2 - 2(q/c)\mathbf{A} \cdot (-i\hbar\boldsymbol{\nabla}) + (q/c)^2\mathbf{A}^2\right]\phi,
\end{aligned} \qquad (4.4.47)$$

since in Coulomb gauge $\boldsymbol{\nabla} \cdot \mathbf{A} = 0$. Using $(\mathbf{a} \times \mathbf{b}) \cdot \mathbf{c} = \mathbf{a} \cdot (\mathbf{b} \times \mathbf{c})$ we also find

$$\mathbf{A} \cdot (-i\hbar\boldsymbol{\nabla}) = (1/2)(\mathbf{B} \times \mathbf{x}) \cdot (-i\hbar\boldsymbol{\nabla}) = (1/2)\mathbf{B} \cdot [\mathbf{x} \times (-i\hbar\boldsymbol{\nabla})] = (1/2)\mathbf{B} \cdot \hat{\mathbf{L}},$$

where we have used $\hat{\mathbf{L}} = \mathbf{x} \times \hat{\boldsymbol{\pi}} = \mathbf{x} \times (-i\hbar\boldsymbol{\nabla})$. We know that we should identify $\hat{\mathbf{L}} = \mathbf{x} \times \hat{\boldsymbol{\pi}}$ and *not* $\hat{\mathbf{L}} = \mathbf{x} \times \hat{\mathbf{p}}$ since the Lie algebra of the rotation group follows from the canonical commutation relations $[\hat{x}^i, \hat{\pi}^j] = i\hbar\delta^{ij}$, which in coordinate space representation is $[x^i, -i\hbar\nabla^j] = [x^i, -i\hbar\partial_j] = i\hbar\delta^{ij}$.

Relativistic classical limit: Since we have again arrived at Eq. (4.3.66) from the Dirac equation, then we obtain Eq. (2.8.40) in the relativistic classical limit and all of the consequences of this will again follow, i.e., the relativistic Lorentz force equation and the equation for the rate of change of the mechanical energy of the particle as given in Eq. (4.3.67). The spin of an elementary particle is inherently a quantum property that has no classical (i.e., macroscopic) correspondence. However, angular momentum arising from the orbital motion of particles does have a classical (i.e., macroscopic) limit.

Dirac equation description of the hydrogen atom: The Dirac equation for an electron moving in the static Coulomb potential of a nucleus of charge Ze can be used to improve on nonrelativistic treatments of the hydrogen atom. Recall that we are working in Lorentz-Heaviside units. For the Coulomb potential we have $\mathbf{A} = 0$ and for a nuclear charge of $Ze = Z|e|$ we have $\Phi = Z\alpha/|e|r$. This leads to $\mathbf{B} = \nabla \times \mathbf{A} = 0$ and $\mathbf{E} = -\nabla\Phi = Z\alpha\hat{r}/|e|r^2$, where $\hat{r}$ is the radial unit vector and where we have used Eq. (2.7.98). This is the familiar result that a point positive or negative charge has an $\mathbf{E}$-field that is directed radially outward or inward, respectively. Since $\mathbf{A} = 0$ then we find in this special case that $\boldsymbol{\pi} = \mathbf{p} = -i\hbar\nabla$. For an electron in this Coulomb potential we have $q\Phi = -Z\alpha/r \equiv V(r)$ and the Dirac equation then becomes

$$i\hbar\partial\psi/\partial t = \hat{H}\psi = \left[c\boldsymbol{\alpha} \cdot (-i\hbar\nabla) + \beta mc^2 + V(r)\right]\psi. \tag{4.4.48}$$

It can be verified that $\hat{H}$ commutes with the total angular momentum operator, $\hat{\mathbf{J}} \equiv \hat{\mathbf{L}} + \mathbf{S}^{\mathrm{spin}} = \mathbf{r}\times(-i\hbar\nabla)+\mathbf{S}^{\mathrm{spin}}$, where the spin matrix $\mathbf{S}^{\mathrm{spin}}$ is defined in Eq. (4.4.143). So the energy eigenstates can be chosen as the simultaneous eigenstates of $\hat{H}$, $\hat{\mathbf{J}}^2$ and $\hat{J}_z$. Since the electron is a particle, as opposed to an antiparticle, $\mathbf{S}^{\mathrm{spin}} = +\mathbf{S}$.

Proof: Observe that $[\beta, \mathbf{S}^{\mathrm{spin}}] = 0$ and $[V(r), \hat{\mathbf{L}}] = [V(r), \mathbf{r} \times (-i\hbar\nabla)] = 0$ since from Eq. (2.7.97) we have $[V(r), \mathbf{r} \times \nabla] = -\mathbf{r} \times \nabla V(r) = -\mathbf{r} \times \hat{r}\frac{dV}{dr} = 0$. Trivially, we also have $[\beta, \hat{\mathbf{L}}] = [\alpha, \hat{\mathbf{L}}] = [V(r), \mathbf{S}^{\mathrm{spin}}] = [(-i\hbar\nabla), \mathbf{S}^{\mathrm{spin}}] = 0$. So only the term $c\boldsymbol{\alpha} \cdot (-i\hbar\nabla)$ requires any calculation, since βmc^2 and $V(r)$ both commute with $\hat{J}$. Next we find $[(-i\hbar\nabla)^i, \{\mathbf{r} \times (-i\hbar\nabla)^j\}] = i\hbar\epsilon^{ijk}(-i\hbar\nabla)^k$ and $[\alpha^i, S^j] = i\hbar\epsilon^{ijk}\alpha^k$. Combining these results we have

$$[\boldsymbol{\alpha} \cdot (-i\hbar\nabla), \hat{J}^j] = [\boldsymbol{\alpha} \cdot (-i\hbar\nabla), \hat{L}^j + S^j] = \alpha^i[(-i\hbar\nabla)^i, \hat{L}^j] + [\alpha^i, S^j](-i\hbar\nabla)^i$$
$$= i\hbar\epsilon^{ijk}\alpha^i(-i\hbar\nabla)^k + i\hbar\epsilon^{ijk}\alpha^k(-i\hbar\nabla)^i = 0, \tag{4.4.49}$$

which then gives the result that $[\hat{H}, \hat{\mathbf{J}}] = 0$.

The full solution of this Dirac equation requires some effort. Detailed treatments can be found in Bjorken and Drell (1964) and Greiner (2000). There is no relativistic version of the reduced mass and so the nucleus is assumed infinitely massive. This has a very small effect on energy levels as we know from the nonrelativistic limit. The energy levels of the electron are found to be

$$E_{nj} = mc^2 \left[1 + \left(\frac{Z\alpha}{\{n - (j + \frac{1}{2}) + \sqrt{(j + \frac{1}{2})^2 - (Z\alpha)^2}\}}\right)^2\right]^{-1/2}, \tag{4.4.50}$$

where $n = 1, 2, 3, \ldots$ is the principal quantum number and where the angular momentum eigenvalues have the range $j = \ell \pm \frac{1}{2}$ with $j \geq \frac{1}{2}$ and $0 \leq \ell \leq (n - 1)$. Each energy level is $(2j + 1)$-degenerate as there is no dependence of the energy on the value of $m_j = -j, \ldots, j$. Since $(j + \frac{1}{2}) \geq 1$, then we must have $Z\alpha \leq 1 \Rightarrow Z \lesssim 137$. This breakdown of the validity of the Dirac equation in the presence of strong fields is an illustration of the so-called Klein paradox that we discuss in Sec. 4.7. Recall that using the Schrödinger equation the nonrelativistic binding energy levels of hydrogen are $E_n^{\mathrm{nr-bind}} = -\mu c^2(Z\alpha)^2/2n^2$, where $\mu = mM/(m + M)$ is the reduced mass with M the mass of the nucleus and where n is the principle quantum number. Taking $M \to \infty$

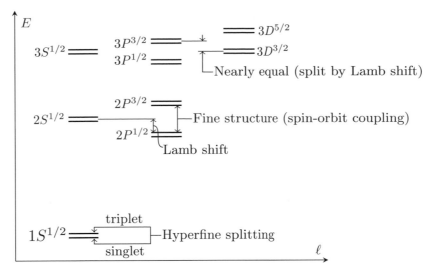

Figure 4.1   Illustration of the energy levels of hydrogen (not to scale).

we have $\mu \to m$ and so $E_n^{\text{nr-bind}} \to -mc^2(Z\alpha)^2/2n^2$ is the nonrelativistic binding energy for an infinitely heavy nucleus. Since

$$[1 + (\epsilon/\{n - \kappa + \sqrt{\kappa^2 - \epsilon^2}\})^2]^{-1/2} = [1 + (\epsilon/n)^2 + \mathcal{O}(\epsilon^4)]^{-1/2} = 1 - (\epsilon^2/2n^2) + \mathcal{O}(\epsilon^4)$$

then we find for the relativistic binding energy that

$$E_{nj}^{\text{bind}} = E_{nj} - mc^2 = -mc^2(Z\alpha)^2/2n^2 + \mathcal{O}((Z\alpha)^4) = E_n^{\text{nr-bind}} + \mathcal{O}((Z\alpha)^4), \qquad (4.4.51)$$

which shows that the Dirac equation introduces relativistic corrections at order $\mathcal{O}((Z\alpha)^4)$. This is because the electron's motion becomes increasing relativistic as it becomes more tightly bound to the nucleus with increasing Z. It is common to denote states using the corresponding spectroscopic notation of the nonrelativistic limit of the states, $n\ell_j$, where ℓ refers to the orbital quantum number $\ell = 0, 1, 2, 3, \ldots = s, p, d, f, \ldots$. The Dirac equation gives the result that energy levels E_{nj}^{bind} with the same n and j are degenerate in energy. So the $2s_{1/2}$ and the $3s_{1/2}$ levels are degenerate as are the $3p_{3/2}$ and the $3d_{3/2}$ energy levels. The $2p_{3/2}$ level is higher than the $2p_{1/2}$ and this spitting is referred to as *fine structure*. The fine-structure splitting is due to the spin-orbit coupling $\mathbf{L} \cdot \mathbf{S}$ arising from the Dirac equation. In nature there is an additional splitting referred to as the *Lamb shift*, which we understand comes from vacuum polarization effects in quantum field theory. There is in addition a *hyperfine splitting* that for a single proton nucleus splits every $n\ell_j$ energy level into two close levels. This is a result of the total angular momentum j of the electron coupling to the proton spin; i.e., it divides into two levels with $J = j \pm \frac{1}{2}$. The Dirac equation treatment above ignored the spin of the nucleus and vacuum polarization and so could not account for either the Lamb shift or the hyperfine structure of the energy levels. The hydrogen energy levels and their splittings are illustrated in Fig. 4.1.

21-cm radiation: Since the $1s^{1/2}$ level of hydrogen has no orbital motion, then the hyperfine splitting is a spin-spin interaction between the magnetic moments of the electron and proton. Two separated magnets anti-align to lower their total energy, but the electron wave function envelops the proton and so the total energy is lowest when the magnetic moments align as in a ferromagnetic material. Since the electron has a negative charge, then its magnetic moment is antiparallel to its spin. So the lower $1s^{1/2}$ hydrogen hyperfine level has the spins antiparallel and the higher level has them parallel. The frequency of a photon emitted during the spin-flip transition from parallel to antiparallel is 1420 MHz, which corresponds to a wavelength of 21.1 cm. This is in the radio part of the electromagnetic spectrum. Collisions between cold (slowly moving) hydrogen atoms in the interstellar medium can excite atoms to the higher hyperfine $1s^{1/2}$ level. The 21-cm radiation emitted during the decay to the lower level easily penetrates interstellar dust clouds that can obstruct optical observations such as near the galactic center.

4.4.5 Lorentz Covariance of the Dirac Equation

To consider the Lorentz covariance of the Dirac equation, we should restore the relativistic notation for space and time; i.e., we should express the Dirac equation in terms of $x^\mu = (ct, \mathbf{x})$ and $\partial_\mu = (\partial/c\partial t, \boldsymbol{\nabla})$. In order to achieve this we define

$$\gamma^0 \equiv \beta \quad \text{and} \quad \gamma^i \equiv \gamma^0 \alpha_i = \beta \alpha_i, \tag{4.4.52}$$

which gives the Dirac representation of the γ-matrices

$$\gamma^0 = \begin{pmatrix} I & 0 \\ 0 & -I \end{pmatrix}, \quad \gamma^i = \begin{pmatrix} 0 & \sigma^i \\ -\sigma^i & 0 \end{pmatrix}. \tag{4.4.53}$$

It is easily verified from the above and from the anticommutation relations of Eq. (4.4.9) that the Dirac γ-matrices satisfy

$$\text{tr}(\gamma^\mu) = 0 \quad \text{and} \quad \{\gamma^\mu, \gamma^\nu\} = 2g^{\mu\nu} I = 2g^{\mu\nu}, \tag{4.4.54}$$

where I is the 4×4 identity matrix that we frequently omit for brevity. The anticommutation relations correspond to the Clifford algebra $C\ell_{1,3}(\mathbb{R})$. The properties of the Dirac γ-matrices are discussed in some detail in Sec. A.3. In the Dirac representation of the γ-matrices in Eq. (4.4.53) we observe that γ^0 is Hermitian and the γ^i are anti-Hermitian,

$$(\gamma^0)^\dagger = \gamma^0 \quad \text{and} \quad (\gamma^i)^\dagger = -\gamma^i. \tag{4.4.55}$$

It then follows from Eq. (4.4.54) that

$$\gamma^0 \gamma^\mu \gamma^0 = (\gamma^\mu)^\dagger. \tag{4.4.56}$$

Consider two sets of γ-matrices that are related by a similarity transformation, $\bar{\gamma}^\mu \equiv S\gamma^\mu S^{-1}$. Both sets are traceless and satisfy the Dirac Clifford algebra, since

$$\text{tr}(\bar{\gamma}^\mu) = \text{tr}(S\gamma^\mu S^{-1}) = \text{tr}(\gamma^\mu) = 0 \quad \text{and} \quad \{\bar{\gamma}^\mu, \bar{\gamma}^\nu\} = S\{\gamma^\mu, \gamma^\nu\}S^{-1} = 2g^{\mu\nu}. \tag{4.4.57}$$

There are three commonly used representations for the Dirac γ-matrices, which are the *Dirac, chiral (or Weyl), and Majorana representations*. As discussed in Sec. A.3 these three are related by unitary transformations; i.e, they are *unitarily equivalent*, which is a special case of similarity. Let $\bar{\gamma}^\mu$ and γ^μ be two representations of the Dirac γ-matrices that are unitarily equivalent; i.e., $\bar{\gamma}^\mu = U\gamma^\mu U^\dagger$ with U a unitary 4×4 spinor matrix. If γ^μ is the Dirac representation, then

$$(\bar{\gamma}^0)^\dagger = (U\gamma^0 U^\dagger)^\dagger = U\gamma^0 U^\dagger = \bar{\gamma}^0, \quad (\bar{\gamma}^i)^\dagger = (U\gamma^i U^\dagger)^\dagger = -U\gamma^i U^\dagger = -\bar{\gamma}^i,$$
$$(\bar{\gamma}^\mu)^\dagger = (U\gamma^\mu U^\dagger)^\dagger = U(\gamma^\mu)^\dagger U^\dagger = U\gamma^0 \gamma^\mu \gamma^0 U^\dagger = \bar{\gamma}^0 \bar{\gamma}^\mu \bar{\gamma}^0. \tag{4.4.58}$$

So any representation of the γ-matrices that is unitarily equivalent to the Dirac representation will satisfy Eqs. (4.4.55) and (4.4.56). This is a special case of the more general result that if $\bar{A} = UAU^\dagger$, $\bar{B} = UBU^\dagger$ and so on with U unitary, then any equation $f(A, A^\dagger, B, B^\dagger, \dots) = 0$ will also hold for $\bar{A}$, $\bar{B}$, ... provided that f has a power series expansion. Similarly, $f(A, A^{-1}, B, B^{-1}, \dots) = 0$ is invariant under any similarity transformation, $\bar{A} = SAS^{-1}$, and $f(A, A^T, B, B^T, \dots) = 0$ is invariant under any orthogonal transformation, $\bar{A} = OAO^T$.

Multiplying Eq. (4.4.15) from the left with γ^0/c and rearranging gives

$$(i\hbar\gamma^\mu\partial_\mu - mc)\,\psi(x) = (i\hbar\slashed{\partial} - mc)\,\psi(x) = 0, \tag{4.4.59}$$

where we have introduced the commonly used *Feynman "slash" notation*,

$$\slashed{a} \equiv a_\mu\gamma^\mu. \tag{4.4.60}$$

This notation allows for compact expressions; e.g., for a Dirac particle interacting with an external electromagnetic field we find that Eq. (4.4.39) becomes

$$[i\hbar\slashed{\partial} - (q/c)\slashed{A} - mc]\,\psi(x) = (i\hbar\slashed{D} - mc)\,\psi(x) = 0, \tag{4.4.61}$$

where we recall from Eq. (4.3.54) that $D^\mu \equiv \partial^\mu + i(q/\hbar c)A^\mu$.

While we have written the Dirac equation in a form that looks covariant, we have not yet ensured that it is. In order to render it covariant we must construct a representation of the Lorentz group such that γ^μ transforms as a Lorentz contravariant four-vector with respect to these transformations. This will in turn lead to the rule for transforming the Dirac spinor wavefunction under Lorentz transformations.

Referring back to Eq. (1.2.189) we see that we need to find a 4×4 matrix representation $S(\Lambda)$ of the passive Lorentz transformations $\Lambda^\mu{}_\nu$ such that

$$S(\Lambda)^{-1}\gamma^\mu S(\Lambda) = \Lambda^\mu{}_\nu\gamma^\nu. \tag{4.4.62}$$

Then $\Lambda^\mu{}_\nu$ is the passive Lorentz transformation that transforms the observer $\mathcal{O}$ to the observer $\mathcal{O}'$. For the Dirac equation to be Lorentz covariant it must have the same form in every inertial frame; i.e., in the frame of $\mathcal{O}'$ we have

$$(i\hbar\gamma^\mu\partial'_\mu - mc)\,\psi'(x') = (i\hbar\slashed{\partial}' - mc)\,\psi'(x') = 0. \tag{4.4.63}$$

Since $S(\Lambda)$ is a representation of the Lorentz group, then $S(\Lambda^{-1}) = S(\Lambda)^{-1}$ and so

$$\begin{aligned} 0 &= (i\hbar\gamma^\mu\partial'_\mu - mc)\,\psi'(x') = (i\hbar\gamma^\mu(\Lambda^{-1})^\nu{}_\mu\partial_\nu - mc)\,\psi'(x') \\ &= (i\hbar S(\Lambda)\gamma^\nu S(\Lambda^{-1})\partial_\nu - mc)\,\psi'(x') = S(\Lambda)\,(i\hbar\gamma^\nu\partial_\nu - mc)\,S(\Lambda)^{-1}\psi'(x'). \end{aligned} \tag{4.4.64}$$

Acting on this equation from the left with $S(\Lambda)^{-1}$ we have

$$(i\hbar\slashed{\partial} - mc)\,S(\Lambda)^{-1}\psi'(x') = (i\hbar\slashed{\partial} - mc)\,\psi(x) = 0, \tag{4.4.65}$$

where Lorentz covariance requires that

$$\psi(x) \xrightarrow{\Lambda} \psi'(x') = \psi'(\Lambda x) = S(\Lambda)\psi(x). \tag{4.4.66}$$

In summary, we see that by imposing Eqs. (4.4.62) and (4.4.66) we ensure that the Dirac equation is Lorentz covariant. Note that A_μ is a covariant four-vector like ∂_μ and so $\slashed{A}$ transforms identically to $\slashed{\partial}$, which means that Eq. (4.4.61) is also Lorentz covariant. It remains to construct the representation $S(\Lambda)$. Let us begin by considering the restricted Lorentz transformations $(\Lambda_r)^\mu{}_\nu$ given in Table 1.1.

Recall that from Sec. 1.2.5 onward in Chapter 1 we used $\hbar = c = 1$ for notational simplicity. We shall show these factors explicitly here, where we have a $1/\hbar$ in exponents involving generators and the generators contain a factor of $\hbar$.

Comparing Eq. (4.4.66) with Eq. (3.2.89) we recognize that $S(\Lambda)$ is the Lorentz transformation of the intrinsic spin of the fermion and that it is $x \to x'$ that is associated with the spacetime part of the Lorentz transformation. Recalling from Eq. (1.4.2) that $M^{\mu\nu} = L^{\mu\nu} + \Sigma^{\mu\nu}$ we then realize that an appropriate notation for the generators of the restricted Lorentz transformation in $S(\Lambda)$ is $\Sigma^{\mu\nu}$. Then from Eq. (1.2.190) we have for these transformations

$$\Lambda^{\mu}{}_{\nu} = [\exp(-\tfrac{i}{2}\omega_{\rho\sigma}M_d^{\rho\sigma})/\hbar]^{\mu}{}_{\nu} \quad \text{and} \quad S(\Lambda) = \exp\left(-\tfrac{i}{2}\omega_{\rho\sigma}\Sigma^{\rho\sigma}/\hbar\right), \tag{4.4.67}$$

where the $M_d^{\mu\nu}$ are the generators in the defining vector representation of the Lorentz group (i.e., the 4×4 matrices given in Eq. (1.2.41)) and the $\Sigma^{\mu\nu}$ are the generators of the restricted Lorentz transformations that act on intrinsic spin. Since $\omega_{\mu\nu} = -\omega_{\nu\mu}$, then $\Sigma^{\mu\nu} = -\Sigma^{\nu\mu}$. Note that the $\Sigma^{\mu\nu}$ are Dirac spinor matrices with elements $(\Sigma^{\mu\nu})_{\alpha\beta}$, where α and β are Dirac spinor indices. From Eqs. (1.4.3) and (1.4.8) we recognize that the spin operators $\mathbf{S}^{\text{spin}}$ for Dirac fermions are

$$\mathbf{S}^{\text{spin}} = \pm\mathbf{S} \quad \text{with} \quad S^i = \tfrac{1}{2}\epsilon^{ijk}\Sigma^{jk}, \tag{4.4.68}$$

where $\pm$ is for particles and antiparticles, respectively.

Expanding the exponentials in Eq. (4.4.62) retaining terms to first-order in $\omega_{\mu\nu}$

$$\left(I + i\tfrac{1}{2}\omega_{\rho\sigma}\Sigma^{\rho\sigma}/\hbar + \cdots\right)\gamma^{\mu}\left(I - i\tfrac{1}{2}\omega_{\tau\lambda}\Sigma^{\tau\lambda}/\hbar + \cdots\right)$$
$$= (I - i\tfrac{1}{2}\omega_{\gamma\delta}M_d^{\gamma\delta}/\hbar + \cdots)^{\mu}{}_{\nu}\gamma^{\nu}. \tag{4.4.69}$$

Since the $\omega_{\mu\nu}$ are arbitrary, then we can equate the first-order terms to give

$$[\Sigma^{\rho\sigma}, \gamma^{\mu}] = -(M_d^{\rho\sigma})^{\mu}{}_{\nu}\gamma^{\nu} = -i(g^{\rho\mu}\delta^{\sigma}{}_{\nu} - g^{\sigma\mu}\delta^{\rho}{}_{\nu})\gamma^{\nu} = -i\left(g^{\rho\mu}\gamma^{\sigma} - g^{\sigma\mu}\gamma^{\rho}\right), \tag{4.4.70}$$

which defines the properties that we require of the generators $M^{\mu\nu}$. Note that

$$[AB, C] = ABC - CAB = A\{B, C\} - \{A, C\}B \tag{4.4.71}$$
$$\Rightarrow \quad [\gamma^{\rho}\gamma^{\sigma}, \gamma^{\mu}] = -2\left(g^{\rho\mu}\gamma^{\sigma} - g^{\sigma\mu}\gamma^{\rho}\right), \tag{4.4.72}$$

which is antisymmetric under the interchange of ρ and σ. We then have

$$[\tfrac{1}{2}\sigma^{\rho\sigma}, \gamma^{\mu}] = [\tfrac{i}{4}(\gamma^{\rho}\gamma^{\sigma} - \gamma^{\sigma}\gamma^{\rho}), \gamma^{\mu}] = -i\left(g^{\rho\mu}\gamma^{\sigma} - g^{\sigma\mu}\gamma^{\rho}\right), \tag{4.4.73}$$

where we have defined

$$\sigma^{\rho\sigma} \equiv (i/2)\left[\gamma^{\rho}, \gamma^{\sigma}\right]. \tag{4.4.74}$$

We have then arrived at

$$\Sigma^{\rho\sigma} \equiv \tfrac{1}{2}\hbar\sigma^{\rho\sigma} \quad \text{and} \quad S(\Lambda) = \exp\left(-\tfrac{i}{2}\omega_{\rho\sigma}\Sigma^{\rho\sigma}/\hbar\right) = \exp\left(-\tfrac{i}{4}\omega_{\rho\sigma}\sigma^{\rho\sigma}\right), \tag{4.4.75}$$

which is the representation of the restricted Lorentz transformations that is relevant for the Dirac equation. As we showed in Sec. 1.2.5, the generators in any representation of the Lorentz group must satisfy the Lie algebra of the Lorentz group given in Eq. (1.2.200). It is a straightforward exercise to directly verify that $\Sigma^{\mu\nu} = \tfrac{1}{2}\hbar\sigma^{\mu\nu}$ satisfies this Lie algebra. This result had to follow since from Eq. (4.4.62) we easily recover Eq. (1.2.199) with the replacements $\ell(\Lambda) \to S(\Lambda)$ and $M^{\mu\nu} \to \Sigma^{\mu\nu}$. This then leads to the Lorentz Lie algebra of Eq. (1.2.200) with $\Sigma^{\mu\nu}$ replacing $M^{\mu\nu}$. More precisely, with $\hbar$ included we arrive at Eq. (1.2.179) written in terms of $\Sigma^{\mu\nu}$.

As noted in Secs. 1.3.1 and 1.3.2 the four-dimensional representation of the restricted Lorentz group for Dirac fermions is the direct sum of two two-dimensional representations, $(\tfrac{1}{2}, 0) \oplus (0, \tfrac{1}{2})$. This is the direct sum of a Type I and Type II representation of $SL(2, \mathbb{C})$, i.e., a direct sum of the left- and right-handed Weyl representations of the restricted Lorentz group. So a four-component Dirac spinor will return to itself after a rotation of 4π rather than the familiar 2π for spatial three-vectors, since $SL(2,\mathbb{C})$ is a double cover of $SO^+(1,3)$.

It remains to construct the Dirac spinor versions of parity and time-reversal transformations to complete the construction of the Lorentz group in Table 1.1. We discuss these in Secs. 4.5.1 and 4.5.3, respectively.

4.4.6 Dirac Adjoint Spinor

From Eq. (4.4.58) we have for any representation that is unitarily equivalent to the Dirac representation that

$$\gamma^0(\gamma^\mu\gamma^\nu)^\dagger\gamma^0 = \gamma^0(\gamma^\nu)^\dagger(\gamma^\mu)^\dagger\gamma^0 = \gamma^\nu\gamma^\mu \tag{4.4.76}$$

and so recalling that $\sigma^{\mu\nu} = \frac{i}{2}[\gamma^\mu,\gamma^\nu]$ and $\Sigma^{\mu\nu} = \frac{1}{2}\hbar\sigma^{\mu\nu}$ we obtain

$$\gamma^0(\sigma^{\mu\nu})^\dagger\gamma^0 = \sigma^{\mu\nu} \quad \text{and} \quad \gamma^0(\Sigma^{\mu\nu})^\dagger\gamma^0 = \Sigma^{\mu\nu}. \tag{4.4.77}$$

We then arrive at

$$\gamma^0 S(\Lambda)^\dagger\gamma^0 = \gamma^0\left[\exp\left(-\tfrac{i}{2}\omega_{\rho\sigma}\Sigma^{\rho\sigma}/\hbar\right)\right]^\dagger\gamma^0 = \exp\left(+\tfrac{i}{2}\omega_{\rho\sigma}\gamma^0(\Sigma^{\rho\sigma}/\hbar)^\dagger\gamma^0\right)$$
$$= \exp\left(+\tfrac{i}{2}\omega_{\rho\sigma}\Sigma^{\rho\sigma}/\hbar\right) = S(\Lambda)^{-1}. \tag{4.4.78}$$

We define the *Dirac adjoint spinor* as

$$\bar{\psi}(x) \equiv \psi(x)^\dagger\gamma^0, \tag{4.4.79}$$

where $\psi^\dagger = \psi^{*T}$ is the *Hermitian adjoint* spinor. Recall that under any Lorentz transformation we have $\psi(x) \to \psi'(x') = S(\Lambda)\psi(x)$. Then for a restricted Lorentz transformation in a representation unitarily equivalent to the Dirac representation,

$$\bar{\psi}(x) \to \bar{\psi}'(x') = \psi'(x)^\dagger\gamma^0 = \psi(x)^\dagger S(\Lambda)^\dagger\gamma^0 = \bar{\psi}\gamma^0 S(\Lambda)^\dagger\gamma^0 = \bar{\psi}(x)S(\Lambda)^{-1}. \tag{4.4.80}$$

From Eqs. (4.4.62) and (4.4.66) we observe that

$$\left(i\hbar\slashed{\partial}' - mc\right)\psi'(x') = S(\Lambda)\left(i\hbar\slashed{\partial}_\nu - mc\right)\psi(x), \tag{4.4.81}$$

which is just observation that $(i\hbar\slashed{\partial} - mc)\,\psi(x)$ transforms as a Dirac spinor. Combining this with Eq. (4.4.80) we then see that we can construct the Lorentz scalar

$$\bar{\psi}'(x')\left(i\hbar\slashed{\partial}' - mc\right)\psi'(x') = \bar{\psi}(x)\left(i\hbar\slashed{\partial} - mc\right)\psi(x), \tag{4.4.82}$$

which is the Lagrangian density for a free fermion field as we will later see.

It is useful to note correspondences between the defining representation of the Lorentz group and the Dirac spinor representation of the restricted Lorentz group. For example, a contravariant four-vector field $a^\mu(x)$ and a covariant four-vector field $a_\mu(x)$ transform with Λ and Λ^{-1}, respectively, while $\psi(x)$ and $\bar{\psi}(x)$ transform with $S(\Lambda)$ and $S(\Lambda)^{-1}$, respectively. To more clearly make the comparison observe that

$$
\begin{aligned}
a^\mu(x) &\to a'^\mu(x') = \Lambda^\mu{}_\nu a^\nu(x), & \psi_\alpha(x) &\to \psi'_\alpha(x') = S(\Lambda)_{\alpha\beta}\psi_\beta(x) \\
a_\mu(x) &\to a'_\mu(x') = a_\nu(x)(\Lambda^{-1})^\nu{}_\mu, & \bar{\psi}_\alpha(x) &\to \bar{\psi}'_\alpha(x') = \bar{\psi}_\beta(x)S(\Lambda)^{-1}_{\beta\alpha} \\
a'(x') \cdot a'(x') &= a(x) \cdot a(x), & \bar{\psi}'(x')\psi'(x') &= \bar{\psi}(x)\psi(x).
\end{aligned}
\tag{4.4.83}
$$

Using $\gamma^0 S(\Lambda)^\dagger \gamma^0 = S(\Lambda)^{-1}$ we also have the correspondence

$$\Lambda^T g \Lambda = g, \qquad S(\Lambda)^\dagger \gamma^0 S(\Lambda) = \gamma^0. \tag{4.4.84}$$

4.4.7 Plane Wave Solutions

We can construct plane wave solutions of the Dirac equation by boosting the solutions for a Dirac particle at rest in Eq. (4.4.33) up to some velocity v. This is an active boost by velocity v of the Dirac wavefunction and is equivalent to a passive boost of the observer by $-v$. From Eq. (1.2.116) we have for an active boost by v

$$\Lambda \equiv \Lambda_B(-\boldsymbol{v}) = \exp(-\tfrac{i}{2}\omega_{\mu\nu}M_d^{\mu\nu}/\hbar) = \exp(-i\eta\mathbf{n}\cdot\mathbf{K}_d/\hbar) = \exp(-i\boldsymbol{\eta}\cdot\mathbf{K}_d/\hbar), \tag{4.4.85}$$

where recall from Sec. 1.2.5 that we use the subscript d to denote the defining representation of the Lorentz group given in Eq. (1.2.41). The corresponding boost generators are $\mathbf{K}_d = (K_d^1, K_d^2, K_d^3) = (M_d^{01}, M_d^{02}, M_d^{03})$ and we see that we have

$$\omega_{0i} = \eta\, n^i = \eta^i. \tag{4.4.86}$$

From Eq. (1.2.105) we have $\eta = \tanh^{-1}\beta = \tanh^{-1}(v/c)$, which is the magnitude of the rapidity of the boost, and $\boldsymbol{\eta} = \eta\mathbf{n}$, where $\mathbf{n} = \boldsymbol{v}/v$ is the direction of the active boost. This corresponds to a passive boost by η in the $-\mathbf{n}$ direction in Eq. (1.2.116).

From Eq. (4.4.66) we have for the boosted wavefunction ψ' that

$$\psi(x) \to \psi'(x') = \psi'(\Lambda x) = S(\Lambda)\psi(x) \quad\implies\quad \psi'(x) = S(\Lambda)\psi(\Lambda^{-1}x), \tag{4.4.87}$$

where $\psi(x)$ is a plane wave solution at rest from Eq. (4.4.33). In Lorentz-covariant notation we identify $\exp(\pm imc^2 t/\hbar) = \exp(\pm ip^{\mathrm{rest}}\cdot x/\hbar)$, where $p^{\mathrm{rest}\,\mu} = (mc, \mathbf{0})$, and

$$p^{\mathrm{rest}}\cdot(\Lambda^{-1}x) = p_\mu^{\mathrm{rest}}(\Lambda^{-1})^\mu{}_\nu x^\nu = (\Lambda p^{\mathrm{rest}})_\nu x^\nu = (\Lambda p^{\mathrm{rest}})\cdot x = p\cdot x, \tag{4.4.88}$$

where $p^\mu \equiv \Lambda^\mu{}_\nu p^{\mathrm{rest}\,\nu}$. From Eq. (1.2.115) we have for this Λ that

$$\begin{aligned}
\begin{bmatrix} p^0 \\ \mathbf{p} \end{bmatrix} &= [\Lambda]\begin{bmatrix} mc \\ \mathbf{0} \end{bmatrix} = \begin{bmatrix} \cosh\eta & \mathbf{n}^T\sinh\eta \\ \mathbf{n}\sinh\eta & I_{3\times3} + \mathbf{n}\mathbf{n}^T(\cosh\eta - 1) \end{bmatrix}\begin{bmatrix} mc \\ \mathbf{0} \end{bmatrix} \\
&= \begin{bmatrix} \gamma & \mathbf{n}^T\beta\gamma \\ \mathbf{n}\beta\gamma & I_{3\times3} + \mathbf{n}\mathbf{n}^T(\gamma - 1) \end{bmatrix}\begin{bmatrix} mc \\ \mathbf{0} \end{bmatrix} = \begin{bmatrix} \gamma mc \\ \gamma m\boldsymbol{v} \end{bmatrix},
\end{aligned} \tag{4.4.89}$$

where $p^\mu = (p^0, \mathbf{p}) = (\gamma mc, \gamma m\boldsymbol{v})$ is the four-momentum of a particle with mass m moving at velocity $\boldsymbol{v}$. Note that we *define* $p^0 > 0$ with

$$p^0 = E/c = \sqrt{\mathbf{p}^2 + m^2c^2} \tag{4.4.90}$$

throughout our discussions here. We have simply confirmed that the usual p^μ is a Lorentz four-vector. So under an active boost with velocity $\boldsymbol{v}$ we have obtained the result that is obvious in hindsight,

$$\exp(\pm imc^2 t/\hbar) \xrightarrow{\Lambda} \exp(\pm ip\cdot x/\hbar). \tag{4.4.91}$$

Define $u^s(0)$ and $v^s(0)$ as the positive energy (particle) and negative energy (antiparticle) spinors at rest, respectively, where s denotes the third component of the spin, i.e., $s = \uparrow$ or $\downarrow$ corresponds to $m_s = +\tfrac{1}{2}$ and $-\tfrac{1}{2}$, respectively. In the Dirac representation that we are using we have from Eq. (4.4.33),

$$u^{\uparrow}(0) = A \begin{bmatrix} 1 \\ 0 \\ 0 \\ 0 \end{bmatrix}, \quad u^{\downarrow}(0) = A \begin{bmatrix} 0 \\ 1 \\ 0 \\ 0 \end{bmatrix}, \quad v^{\downarrow}(0) = A \begin{bmatrix} 0 \\ 0 \\ 1 \\ 0 \end{bmatrix}, \quad v^{\uparrow}(0) = A \begin{bmatrix} 0 \\ 0 \\ 0 \\ 1 \end{bmatrix}, \quad (4.4.92)$$

where there are various conventions for the choice of normalization constant A but we will follow that of Peskin and Schroeder (1995). The reason for the reversal of spin for the negative energy (antiparticle) states will become clearer later. This corresponds to the *Feynman-Stueckelberg interpretation* of antiparticles with spin: *The invariant amplitude for the emission or absorption of an antiparticle with four-momentum p^{μ} and spin up/down in the rest frame corresponds to that for the absorption or emission of a particle with four-momentum $-p^{\mu}$ and spin down/up.* This arises naturally when we develop the machinery of relativistic quantum field theory.

Using Eq. (4.4.87) we see that the positive and negative energy plane wave solutions corresponding to $\psi'(x)$ are, respectively, given by

$$e^{-ip\cdot x/\hbar}u^s(p) = e^{-ip\cdot x/\hbar}S(\Lambda)u^s(0), \quad e^{+ip\cdot x/\hbar}v^s(p) = e^{+ip\cdot x/\hbar}S(\Lambda)v^s(0), \quad (4.4.93)$$

where $p^0 > 0$ and where we have defined

$$u^s(p) \equiv S(\Lambda)u^s(0), \quad v^s(p) \equiv S(\Lambda)v^s(0). \quad (4.4.94)$$

Note that later we will develop the concept of four-spin and we will distinguish between s before and after a Lorentz transformation; however, for now we do not need to add this extra notational complication as it does not affect our arguments.

Using Eqs. (1.4.4) and (4.4.85) we find have for an active boost by $\boldsymbol{v}$,

$$S(\Lambda) = \exp(-\tfrac{i}{2}\omega_{\mu\nu}\Sigma^{\mu\nu}/\hbar) = \exp(-i\eta\mathbf{n}\cdot\mathbf{K}_{\text{int}}/\hbar) = \exp(-i\boldsymbol{\eta}\cdot\mathbf{K}_{\text{int}}/\hbar), \quad (4.4.95)$$

where as above $\omega_{0i} = \eta\, n^i = \eta^i$ and where the spinor boost generators are

$$K_{\text{int}}^i/\hbar = \Sigma^{0i}/\hbar = \tfrac{1}{2}\sigma^{0i} = \tfrac{i}{4}[\gamma^0, \gamma^i] = \tfrac{i}{2}\gamma^0\gamma^i = \tfrac{i}{2}\alpha^i. \quad (4.4.96)$$

From Eq. (4.4.9) we have $\{\alpha^i, \alpha^j\} = 2\delta^{ij}I$ and so $(\mathbf{n}\cdot\boldsymbol{\alpha})^2 = \mathbf{n}\cdot\mathbf{n} = 1$, which gives

$$S(\Lambda) = \exp(-i\eta\mathbf{n}\cdot\mathbf{K}_{\text{int}}/\hbar) = \exp(\tfrac{1}{2}\eta\mathbf{n}\cdot\boldsymbol{\alpha}) = \sum_{j=0}^{\infty}(1/j!)\left(\tfrac{1}{2}\eta\right)^j(\mathbf{n}\cdot\boldsymbol{\alpha})^j \quad (4.4.97)$$

$$= \left[\sum_{j=1,3,\ldots}^{\infty}(1/j!)\left(\tfrac{1}{2}\eta\right)^j\right](\mathbf{n}\cdot\boldsymbol{\alpha}) + \left[\sum_{j=0,2,\ldots}^{\infty}(1/j!)\left(\tfrac{1}{2}\eta\right)^j\right]I$$

$$= I\cosh(\eta/2) + (\mathbf{n}\cdot\boldsymbol{\alpha})\sinh(\eta/2) = \cosh(\eta/2)\left[I + (\mathbf{n}\cdot\boldsymbol{\alpha})\tanh(\eta/2)\right]$$

$$= \sqrt{\frac{E+mc^2}{2mc^2}}\left[I + (\mathbf{n}\cdot\boldsymbol{\alpha})\frac{|\mathbf{p}|c}{E+mc^2}\right] = \frac{(p^0+mc)I + (\mathbf{p}\cdot\boldsymbol{\alpha})}{\sqrt{2mc(p^0+mc)}} = \frac{(\not{p}\gamma^0+mc)}{\sqrt{2mc(p^0+mc)}}.$$

We have used $\boldsymbol{\alpha} = \gamma^0\boldsymbol{\gamma} = -\boldsymbol{\gamma}\gamma^0$, $p^0I + p^i\alpha^i = (p^0\gamma^0 - p^i\gamma^i)\gamma^0 = \not{p}\gamma^0$ and

$$\cosh(\eta/2) = \sqrt{\tfrac{1}{2}(\cosh\eta+1)} = \sqrt{\tfrac{1}{2}(\gamma+1)} = \sqrt{\frac{E+mc^2}{2mc^2}} = \sqrt{\frac{p^0+mc}{2mc}},$$

$$\tanh(\eta/2) = \frac{\sinh\eta}{\cosh\eta+1} = \frac{\beta\gamma}{\gamma+1} = \frac{\beta\gamma mc^2}{E+mc^2} = \frac{|\mathbf{p}|c}{E+mc^2} = \frac{|\mathbf{p}|}{p^0+mc}. \quad (4.4.98)$$

This result for $S(\Lambda)$ then gives

$$u^s(p) = S(\Lambda)u^s(0) = AS(\Lambda)\begin{bmatrix}\phi_s \\ 0\end{bmatrix} = A\frac{\not{p}+mc}{\sqrt{2mc(p^0+mc)}}\begin{bmatrix}\phi_s \\ 0\end{bmatrix},$$

$$v^s(p) = S(\Lambda)v^s(0) = AS(\Lambda)\begin{bmatrix}0 \\ \chi_{-s}\end{bmatrix} = A\frac{-\not{p}+mc}{\sqrt{2mc(p^0+mc)}}\begin{bmatrix}0 \\ \chi_{-s}\end{bmatrix}, \quad (4.4.99)$$

where ϕ and χ are normal two component spinors

$$\phi_\uparrow = \chi_\uparrow = \begin{bmatrix} 1 \\ 0 \end{bmatrix} \quad \text{and} \quad \phi_\downarrow = \chi_\downarrow = \begin{bmatrix} 0 \\ 1 \end{bmatrix}. \tag{4.4.100}$$

We will often use the notation $\phi_+ = \phi_\uparrow$, $\phi_- = \phi_\downarrow$, $\chi_+ = \chi_\uparrow$, and $\chi_- = \chi_\downarrow$, Note that $v^\uparrow(p)$ and $v^\downarrow(p)$ are given in terms of $\chi_\downarrow$ and $\chi_\uparrow$, respectively, which is why $v^s(p)$ is expressed in terms of the spin-reversed two-spinor χ_{-s}.

If we choose the z-axis to be the spin quantization axis that defines ϕ_s and χ_s, then $\hat{s}_z\phi_\pm = \frac{1}{2}\hbar\sigma_z\phi_\pm = \pm\frac{1}{2}\hbar\phi_\pm = \hbar m_s\phi_\pm$ and similarly for χ_s. For ϕ_s and χ_s we make frequent use of the various notations

$$s = +, - = \uparrow, \downarrow \quad \text{to denote} \quad m_s = \tfrac{1}{2}, -\tfrac{1}{2}. \tag{4.4.101}$$

Of course we can use any direction defined by the unit three-vector $\mathbf{n}_s$ as the spin quantization axis and then $\hat{s}_z$ is replaced by

$$\hat{s}_n \equiv \hat{\mathbf{s}} \cdot \mathbf{n}_s = \tfrac{1}{2}\hbar\boldsymbol{\sigma} \cdot \mathbf{n}_s \quad \text{and} \quad \hat{s}_n\phi_\pm = \tfrac{1}{2}\hbar\boldsymbol{\sigma} \cdot \mathbf{n}_s\phi_\pm = \pm\tfrac{1}{2}\hbar\phi_\pm = \hbar m_s\phi_\pm \tag{4.4.102}$$

and similarly for $\chi_\pm$. Here $\phi_\pm$ refers to the component of spin in the $\mathbf{n}_s$ direction being spin-up or spin-down. The spinors $\phi_\pm$ form an orthonormal basis for the complex two-dimensional two-spinor space for any choice of quantization axis $\mathbf{n}_s$,

$$\phi_s^\dagger\phi_{s'} = \delta_{ss'} \quad \text{and} \quad \sum_{s=\pm} \phi_s\phi_s^\dagger = \phi_+\phi_+^\dagger + \phi_-\phi_-^\dagger = I. \tag{4.4.103}$$

Let $\phi = c_+\phi_+ + c_-\phi_-$ with $c_+, c_- \in \mathbb{C}$ be any arbitrary two-spinor, then since $(\boldsymbol{\sigma} \cdot \mathbf{n}_s)\phi_\pm = \pm\phi_\pm$ we have

$$\tfrac{1}{2}(1 \pm \boldsymbol{\sigma} \cdot \mathbf{n}_s)\phi = \tfrac{1}{2}(1 \pm \boldsymbol{\sigma} \cdot \mathbf{n}_s)(c_+\phi_+ + c_-\phi_-) = c_\pm\phi_\pm. \tag{4.4.104}$$

The same arguments apply to $\chi_\pm$. Thus the two-spinor projectors onto the one-dimensional spin-up and spin-down subspaces are, respectively, given by

$$\tfrac{1}{2}(1 + \boldsymbol{\sigma} \cdot \mathbf{n}_s) \quad \text{and} \quad \tfrac{1}{2}(1 - \boldsymbol{\sigma} \cdot \mathbf{n}_s). \tag{4.4.105}$$

Furthermore, since $\{\sigma^i, \sigma^j\} = 2\delta^{ij}$, then $\tfrac{1}{2}\{\boldsymbol{\sigma}, \boldsymbol{\sigma} \cdot \mathbf{n}_s\} = \mathbf{n}_s$ and so

$$\phi_\pm^\dagger\boldsymbol{\sigma}\phi_\pm = \tfrac{1}{2}\phi_\pm^\dagger\left[\boldsymbol{\sigma}\tfrac{1}{2}(1 \pm \boldsymbol{\sigma} \cdot \mathbf{n}_s) + \tfrac{1}{2}(1 \pm \boldsymbol{\sigma} \cdot \mathbf{n}_s)\boldsymbol{\sigma}\right]\phi_\pm$$
$$= \tfrac{1}{2}\phi_\pm^\dagger\left[\boldsymbol{\sigma} \pm \tfrac{1}{2}\{\boldsymbol{\sigma}, \boldsymbol{\sigma} \cdot \mathbf{n}_s\}\right]\phi_\pm = \tfrac{1}{2}\phi_\pm^\dagger(\boldsymbol{\sigma} \pm \mathbf{n}_s)\phi_\pm, \tag{4.4.106}$$

which gives the orientation of the spin as the expectation value of $\boldsymbol{\sigma}$,

$$\phi_\pm^\dagger\boldsymbol{\sigma}\phi_\pm = \pm\mathbf{n}_s \quad \text{and} \quad \chi_\pm^\dagger\boldsymbol{\sigma}\chi_\pm = \pm\mathbf{n}_s. \tag{4.4.107}$$

With $p^\mu = (p^0, \mathbf{p}) = (\gamma mc, \gamma m\boldsymbol{v})$ then $p^2 = m^2c^2$ and so

$$\not{p}^2 = \not{p}\not{p} = p_\mu p_\nu \gamma^\mu \gamma^\nu = \tfrac{1}{2}p_\mu p_\nu\{\gamma^\mu, \gamma^\nu\} = p_\mu p_\nu g^{\mu\nu} = p^2 = m^2c^2. \tag{4.4.108}$$

It then follows that

$$(\not{p} - mc)(\not{p} + mc) = (\not{p} + mc)(\not{p} - mc) = \not{p}^2 - m^2c^2 = 0 \tag{4.4.109}$$

and so from Eq. (4.4.99) it follows that

$$(\not{p} - mc)u^s(p) = 0 \quad \text{and} \quad (\not{p} + mc)v^s(p) = 0. \tag{4.4.110}$$

We can now verify that $\exp(-ip \cdot x/\hbar)u^s(p)$ and $\exp(+ip \cdot x/\hbar)v^s(p)$ are solutions of the Dirac equation as they were constructed to be, since

$$(i\hbar\not\partial - mc)e^{-ip\cdot x/\hbar}u^s(p) = (\not p - mc)e^{-ip\cdot x/\hbar}u^s(p) = 0 \quad \text{and}$$
$$(i\hbar\not\partial - mc)e^{+ip\cdot x/\hbar}v^s(p) = -(\not p + mc)e^{+ip\cdot x/\hbar}v^s(p) = 0. \tag{4.4.111}$$

The Dirac adjoints of $u^s(p)$ and $v^s(p)$ are given by

$$\bar{u}^s(p) = A^*[\phi_s^\dagger \ 0]\frac{(\not p^\dagger + mc)\gamma^0}{\sqrt{2mc(p^0 + mc)}} = A^*[\phi_s^\dagger \ 0]\frac{\gamma^0(\not p + mc)}{\sqrt{2mc(p^0 + mc)}} \quad \text{and}$$

$$\bar{v}^s(p) = A^*[0 \ \chi_{-s}^\dagger]\frac{(-\not p^\dagger + mc)\gamma^0}{\sqrt{2mc(p^0 + mc)}} = A^*[0 \ \chi_{-s}^\dagger]\frac{\gamma^0(-\not p + mc)}{\sqrt{2mc(p^0 + mc)}}, \tag{4.4.112}$$

where we have used $\gamma^0\gamma^\mu\gamma^0 = \gamma^{\mu\dagger}$, which is true in any representation unitarily equivalent to the Dirac representation. Using $\gamma^0 S(\Lambda)^{-1}\gamma^0 = S(\Lambda)^\dagger$ and Eq. (4.4.80) we can obtain the adjoint plane wave spinors from their rest-frame versions using

$$\bar{u}^s(p) = \bar{u}^s(0)S(\Lambda)^{-1} = u^s(0)^\dagger S(\Lambda)^\dagger\gamma^0 \quad \text{and}$$
$$\bar{v}^s(p) = \bar{v}^s(0)S(\Lambda)^{-1} = v^s(0)^\dagger S(\Lambda)^\dagger\gamma^0. \tag{4.4.113}$$

Constructing Lorentz scalars from $u^s(p)$ and $v^s(p)$ we find

$$\bar{u}^s(p)u^{s'}(p) = |A|^2 [\phi_s^\dagger \ 0]\frac{\gamma^0(\not p + mc)^2}{2mc(p^0 + mc)}\begin{bmatrix} \phi_{s'} \\ 0 \end{bmatrix} = \frac{|A|^2}{p^0 + mc}[\phi_s^\dagger \ 0]\gamma^0(\not p + mc)\begin{bmatrix} \phi_{s'} \\ 0 \end{bmatrix}$$

$$= \frac{|A|^2}{p^0 + mc}[\phi_s^\dagger \ 0]\begin{bmatrix} I & 0 \\ 0 & -I \end{bmatrix}\begin{bmatrix} p^0 + mc & \boldsymbol{\sigma}\cdot\mathbf{p} \\ -\boldsymbol{\sigma}\cdot\mathbf{p} & -p^0 + mc \end{bmatrix}\begin{bmatrix} \phi_{s'} \\ 0 \end{bmatrix}$$

$$= \frac{|A|^2}{p^0 + mc}[\phi_s^\dagger \ 0]\begin{bmatrix} (p^0 + mc)\phi_{s'} \\ \boldsymbol{\sigma}\cdot\mathbf{p}\phi_{s'} \end{bmatrix} = |A|^2\phi_s^\dagger\phi_{s'} = |A|^2\delta^{ss'},$$

$$\bar{v}^s(p)v^{s'}(p) = |A|^2 [0 \ \chi_{-s}^\dagger]\frac{\gamma^0(-\not p + mc)^2}{2mc(p^0 + mc)}\begin{bmatrix} 0 \\ \chi_{-s'} \end{bmatrix} \tag{4.4.114}$$

$$= \frac{|A|^2}{p^0 + mc}[0 \ \chi_{-s}^\dagger]\gamma^0(-\not p + mc)\begin{bmatrix} 0 \\ \chi_{-s'} \end{bmatrix}$$

$$= \frac{|A|^2}{p^0 + mc}[0 \ \chi_{-s}^\dagger]\begin{bmatrix} I & 0 \\ 0 & -I \end{bmatrix}\begin{bmatrix} -p^0 + mc & -\boldsymbol{\sigma}\cdot\mathbf{p} \\ \boldsymbol{\sigma}\cdot\mathbf{p} & p^0 + mc \end{bmatrix}\begin{bmatrix} 0 \\ \chi_{-s'} \end{bmatrix}$$

$$= \frac{|A|^2}{p^0 + mc}[0 \ \chi_{-s}^\dagger]\begin{bmatrix} -\boldsymbol{\sigma}\cdot\mathbf{p}\chi_{-s'} \\ -(p^0 + mc)\chi_{-s'} \end{bmatrix} = -|A|^2\chi_{-s}^\dagger\chi_{-s'} = -|A|^2\delta^{ss'},$$

where we have used

$$(\not p + mc)^2 = p^2 + 2mc\not p + m^2c^2 = 2mc(\not p + mc) \quad \text{and}$$
$$(-\not p + mc)^2 = p^2 - 2mc\not p + m^2c^2 = 2mc(-\not p + mc). \tag{4.4.115}$$

We follow Peskin and Schroeder (1995) and Schwartz (2013) and choose

$$A = \sqrt{2mc}, \qquad \text{which leads to} \tag{4.4.116}$$

$$u^s(p) = S(\Lambda)u^s(0) = \frac{\not p + mc}{\sqrt{(p^0 + mc)}}\begin{bmatrix} \phi_s \\ 0 \end{bmatrix}, \quad u^s(0) = \sqrt{2mc}\begin{bmatrix} \phi_s \\ 0 \end{bmatrix},$$

$$v^s(p) = S(\Lambda)v^s(0) = \frac{-\not p + mc}{\sqrt{(p^0 + mc)}}\begin{bmatrix} 0 \\ \chi_{-s} \end{bmatrix}, \quad v^s(0) = \sqrt{2mc}\begin{bmatrix} 0 \\ \chi_{-s} \end{bmatrix}, \tag{4.4.117}$$

where ϕ and χ are normal two component spinors. With this choice of normalization

$$\bar{u}^s(p)u^{s'}(p) = 2mc\delta^{ss'} \quad \text{and} \quad \bar{v}^s(p)v^{s'}(p) = -2mc\delta^{ss'}, \tag{4.4.118}$$

where the minus sign in the second equation should be noted.

It is obvious that $\bar{v}^s(p)u^{s'}(p)$ and $\bar{u}^s(p)v^{s'}(p)$ contain terms $(-\not{p}+mc)(\not{p}+mc) = 0$ and $(\not{p}+mc)(-\not{p}+mc) = 0$, respectively. Therefore we also have a type of orthogonality for the u and v plane wave spinors,

$$\bar{v}^s(p)u^{s'}(p) = \bar{u}^s(p)v^{s'}(p) = 0 \quad \text{for all} \quad s, s'. \tag{4.4.119}$$

From Eq. (4.4.116) and the fact that $\gamma^0\gamma^i\gamma^0 = -\gamma^i$ we see that

$$\gamma^0 u^s(\mathbf{p}) = u^s(p^0, -\mathbf{p}) \equiv u^s(-p) \quad \text{and} \quad \gamma^0 v^s(\mathbf{p}) = -v^s(p^0, -\mathbf{p}) \equiv -v^s(-p) \tag{4.4.120}$$

and Eq. (4.4.119) we find

$$v^{s\dagger}(p)u^{s'}(-p) = u^{s\dagger}(p)v^{s'}(-p) = 0. \tag{4.4.121}$$

We have $\not{p}u^s(p) = mc\,u^s(p)$ and $\not{p}v^s(p) = -mc\,v^s(p)$ from Eq. (4.4.110) and so

$$\bar{u}^s(p)\not{p} = u^{s\dagger}(p)\gamma^0\not{p} = u^{s\dagger}(p)\not{p}^\dagger\gamma^0 = [\not{p}u^s(p)]^\dagger\gamma^0 = mc\,\bar{u}^s(p) \quad \text{and}$$
$$\bar{v}^s(p)\not{p} = v^{s\dagger}(p)\gamma^0\not{p} = v^{s\dagger}(p)\not{p}^\dagger\gamma^0 = [\not{p}u^s(p)]^\dagger\gamma^0 = -mc\,\bar{v}^s(p). \tag{4.4.122}$$

The orthogonality condition is recovered from these relations, since

$$\bar{u}^s(p)\not{p}v^{s'}(p) = \bar{u}^s(p)[\not{p}v^{s'}(p)] = -mc\,\bar{u}^s(p)v^{s'}(p), \tag{4.4.123}$$
$$\bar{u}^s(p)\not{p}v^{s'}(p) = [\bar{u}^s(p)\not{p}]v^{s'}(p) = +mc\,\bar{u}^s(p)v^{s'}(p), \tag{4.4.124}$$

and so we must have $\bar{u}^s(p)v^{s'}(p) = 0$ and similarly for $\bar{v}^s(p)u^{s'}(p)$.

We also readily verify that

$$u^s(p)^\dagger u^{s'}(p) = v^s(p)^\dagger v^{s'}(p) = (2E/c)\delta^{ss'}. \tag{4.4.125}$$

Since the normalization condition of Eq. (4.4.118) is identically true for a massless particle, we use Eq. (4.4.125) to define the normalization for massless spinor plane wave solutions of the Dirac equation. For massive particles these two normalization conditions are of course equivalent.

Proof: We note that since $\gamma^0 = \gamma^{0\dagger}$ and $\boldsymbol{\gamma}^\dagger = -\boldsymbol{\gamma}$ then $\not{p}^\dagger = 2p^0\gamma^0 - \not{p}$. We recall also that $(-\not{p}+mc)(\not{p}+mc) = 0$. We then find

$$u^s(p)^\dagger u^{s'}(p) = 2mc\,[\phi_s^\dagger \; 0]\frac{(\not{p}^\dagger+mc)(\not{p}+mc)}{2mc(p^0+mc)}\begin{bmatrix}\phi_{s'}\\0\end{bmatrix} \tag{4.4.126}$$

$$= [\phi_s^\dagger \; 0]\frac{2p^0\gamma^0(\not{p}+mc)}{p^0+mc}\begin{bmatrix}\phi_{s'}\\0\end{bmatrix} = 2p^0[\phi_s^\dagger \; 0]\begin{bmatrix}\phi_{s'}\\0\end{bmatrix} = \frac{2E}{c}\delta^{ss'},$$

$$v^s(p)^\dagger v^{s'}(p) = 2mc\,[0 \; \chi_{-s}^\dagger]\frac{(-\not{p}^\dagger+mc)(-\not{p}+mc)}{2mc(p^0+mc)}\begin{bmatrix}0\\\chi_{-s'}\end{bmatrix} \tag{4.4.127}$$

$$= [0 \; \chi_{-s}^\dagger]\frac{(-2p^0\gamma^0)(-\not{p}+mc)}{p^0+mc}\begin{bmatrix}0\\\chi_{-s'}\end{bmatrix} = 2p^0[0 \; \chi_{-s}^\dagger]\begin{bmatrix}0\\\chi_{-s'}\end{bmatrix} = \frac{2E}{c}\delta^{ss'}.$$

4.4.8 Completeness and Projectors

We have seen above that the four plane wave spinors $u^\uparrow(p), u^\downarrow(p), v^\uparrow(p), v^\downarrow(p)$ satisfy a form of orthogonality. There is a corresponding completeness relation

$$(1/2mc) \sum_s [u^s(p)\bar{u}^s(p) - v^s(p)\bar{v}^s(p)] = I, \tag{4.4.128}$$

which follows from

$$\sum_s u^s(p)\bar{u}^s(p) = \not{p} + mc \quad \text{and} \quad \sum_s v^s(p)\bar{v}^s(p) = \not{p} - mc. \tag{4.4.129}$$

The space that is spanned by these plane wave solutions is the p^μ-subspace for Dirac particles. This is the space in which the Little Group of transformations discussed in Sec. 1.4.2 acts. Any arbitrary $\psi_\alpha(\mathbf{x})$ for $\alpha = 1, \ldots, 4$ has four complex numbers associated with every spatial point $\mathbf{x}$. The set of four plane wave spinors at every $\mathbf{p}$ form a basis of all possible four-component functions, $\psi(\mathbf{x})$, since a Fourier transform is a unitary transformation. The Dirac equation dynamics is entirely in the time evolution of such an arbitrary spinor spatial wavefunction.

Proof: Using our expressions for $u^s(p)$ and $\bar{u}^s(p)$ gives

$$\sum_s u^s(p)\bar{u}^s(p) = \frac{2mc}{2mc(p^0+mc)}(\not{p}+mc)\left(\sum_s \begin{bmatrix} \phi^s \\ 0 \end{bmatrix} [\phi_s^\dagger \ 0]\right)\gamma^0(\not{p}+mc)$$

$$= \frac{(\not{p}+mc)}{(p^0+mc)}\begin{bmatrix} I & 0 \\ 0 & 0 \end{bmatrix}\gamma^0(\not{p}+mc) = \frac{(\not{p}+mc)}{(p^0+mc)}\tfrac{1}{2}(I+\gamma^0)(\not{p}+mc)$$

$$= (p^0+mc)^{-1}\tfrac{1}{2}\left[(\not{p}+mc)^2 + (\not{p}+mc)\gamma^0(\not{p}+mc)\right]$$

$$= (p^0+mc)^{-1}(p^0+mc)(\not{p}+mc) = \not{p}+mc, \tag{4.4.130}$$

where we have used $\not{p}\gamma^0\not{p} = -\gamma^0 m^2 c^2 + 2p^0\not{p}$ and $\gamma^0\not{p}+\not{p}\gamma^0 = 2p^0$ with these both following from $\{\gamma^\mu, \gamma^\nu\} = 2g^{\mu\nu}$. Following similar steps we arrive at

$$\sum_s v^s(p)\bar{v}^s(p) = \frac{2mc}{2mc(p^0+mc)}(-\not{p}+mc)\left(\sum_s \begin{bmatrix} 0 \\ \chi_{-s} \end{bmatrix} [0 \ \chi_{-s}^\dagger]\right)\gamma^0(-\not{p}+mc)$$

$$= \frac{(-\not{p}+mc)}{(p^0+mc)}\begin{bmatrix} 0 & 0 \\ 0 & I \end{bmatrix}\gamma^0(-\not{p}+mc) = \frac{(-\not{p}+mc)}{(p^0+mc)}\tfrac{1}{2}(-I+\gamma^0)(-\not{p}+mc)$$

$$= (p^0+mc)^{-1}\tfrac{1}{2}\left[-(\not{p}+mc)^2 + (-\not{p}+mc)\gamma^0(-\not{p}+mc)\right]$$

$$= (p^0+mc)^{-1}(p^0+mc)(\not{p}-mc) = \not{p}-mc. \tag{4.4.131}$$

The orthogonality of the positive and negative energy plane wave spinors means that the *positive and negative energy projectors* are

$$P_\pm(p) = \frac{\pm\not{p}+mc}{2mc}; \tag{4.4.132}$$

i.e., from Eqs. (4.4.128) and (4.4.129) we have

$$P_+(p) + P_-(p) = \left(\frac{1}{2mc}\sum_s u^s(p)\bar{u}^s(p)\right) + \left(-\frac{1}{2mc}\sum_s v^s(p)\bar{v}^s(p)\right)$$

$$= \left(\frac{\not{p}+mc}{2mc}\right) + \left(\frac{-\not{p}+mc}{2mc}\right) = I. \tag{4.4.133}$$

These are projection operators since we can also readily check that

$$P_+(p)^2 = P_+(p), \quad P_-(p)^2 = P_-(p), \quad P_+(p)P_-(p) = P_-(p)P_+(p) = 0. \tag{4.4.134}$$

Also note that using the above results we can write the four-momentum p^μ in the form of an expectation value with respect to the plane wave spinors

$$p^\mu = \tfrac{1}{2}\bar{u}^s(p)\gamma^\mu u^s(p) = \tfrac{1}{2}\bar{v}^s(p)\gamma^\mu v^s(p). \tag{4.4.135}$$

Proof: Since $\{\gamma^\mu, \gamma^\nu\} = 2g^{\mu\nu}$ we have $\tfrac{1}{2}\{\slashed{p}, \gamma^\mu\} = p^\mu$. Using this along with $\bar{u}^s(p)u^s(p) = 2mc = -\bar{v}^s(p)v^s(p)$, $\slashed{p}u^s(p) = mcu^s(p)$, $\slashed{p}v^s(p) = -mcv^s(p)$, $\bar{u}^s(p)\slashed{p} = mc\bar{u}^s(p)$, and $\bar{v}^s(p)\slashed{p} = -mc\bar{v}^s(p)$ we observe that

$$p^\mu = \frac{p^\mu}{2mc}\bar{u}^s(p)u^s(p) = \frac{1}{2mc}\bar{u}^s(p)\tfrac{1}{2}(\slashed{p}\gamma^\mu + \gamma^\mu\slashed{p})u^s(p) = \tfrac{1}{2}\bar{u}^s(p)\gamma^\mu u^s(p), \quad \text{and}$$

$$p^\mu = \frac{-p^\mu}{2mc}\bar{v}^s(p)v^s(p) = \frac{-1}{2mc}\bar{v}^s(p)\tfrac{1}{2}(\slashed{p}\gamma^\mu + \gamma^\mu\slashed{p})v^s(p) = \tfrac{1}{2}\bar{v}^s(p)\gamma^\mu v^s(p).$$

4.4.9 Spin Vector

The spin quantization axis in the Dirac particle rest frame is represented by the unit three-vector $\mathbf{n}_s$ as discussed above. A spin eigenstate will be up or down with respect to this axis in the rest frame. So we can represent the spin state in the rest frame with the spin three-vectors $\pm\mathbf{n}_s$, which correspond to $m_s = \pm\tfrac{1}{2}$, respectively.

After the Lorentz boost Λ of Eq. (4.4.85) in the direction $\mathbf{n}_p \equiv \mathbf{p}/|\mathbf{p}|$ we have for the particle four-momentum $p^{\mu\,\mathrm{rest}} = (m, \mathbf{0}) \to p^\mu$ with p^μ given by Eq. (4.4.89). Similarly, the rest-frame four-vector $n_s^{\mu\,\mathrm{rest}} \equiv (0, \mathbf{n}_s^{\mathrm{rest}})$ after boosting becomes the *four-spin vector* $n_s^\mu = (n_s^0, \mathbf{n}_s)$, where

$$\begin{bmatrix} n_s^0 \\ \mathbf{n}_s \end{bmatrix} = \begin{bmatrix} \gamma & \mathbf{n}_p^T\beta\gamma \\ \mathbf{n}_p\beta\gamma & I_{3\times3} + \mathbf{n}_p\mathbf{n}_p^T(\gamma - 1) \end{bmatrix} \begin{bmatrix} 0 \\ \mathbf{n}_s^{\mathrm{rest}} \end{bmatrix} = \begin{bmatrix} \beta\gamma\mathbf{n}_p \cdot \mathbf{n}_s^{\mathrm{rest}} \\ \mathbf{n}_s^{\mathrm{rest}} + (\gamma - 1)(\mathbf{n}_p \cdot \mathbf{n}_s^{\mathrm{rest}})\mathbf{n}_p \end{bmatrix}, \tag{4.4.136}$$

where $\beta\gamma = \gamma v/c = |\mathbf{p}|/mc$ and $(\gamma - 1) = (p^0 - mc)/mc$. In the rest frame we have $(n_s^2)^{\mathrm{rest}} = -(\mathbf{n}_s^{\mathrm{rest}})^2 = -1$ and $(p \cdot n_s)^{\mathrm{rest}} = 0$ and so in *any* inertial reference frame,

$$n_s^2 = -1 \quad \text{and} \quad p \cdot n_s = 0, \tag{4.4.137}$$

which is Eq. (1.4.15) in the context of a spin-$\tfrac{1}{2}$ particle. From the discussion following Eq. (1.4.16) we know that $n^\mu(s)$ transforms as a pseudovector and that spin is invariant under a parity transformation.

It is convenient to introduce the γ^5 Dirac spinor matrix, which is defined as

$$\gamma^5 \equiv \gamma_5 \equiv -(i/4!)\epsilon_{\mu\nu\rho\sigma}\gamma^\mu\gamma^\nu\gamma^\rho\gamma^\sigma = -(i/4!)\epsilon^{\mu\nu\rho\sigma}\gamma_\mu\gamma_\nu\gamma_\rho\gamma_\sigma = i\gamma^0\gamma^1\gamma^2\gamma^3, \tag{4.4.138}$$

where $\epsilon^{0123} = -\epsilon_{0123} = +1$ from Sec. A.2.1. Using $\{\gamma^\mu, \gamma^\nu\} = 2g^{\mu\nu}$ we find

$$\{\gamma^\mu, \gamma^5\} = 0, \quad (\gamma^5)^2 = I, \quad [\gamma^5, \sigma^{\mu\nu}] = 0 \quad \text{and} \quad \gamma^5\sigma^{\mu\nu} = (i/2)\epsilon^{\mu\nu\rho\sigma}\sigma_{\rho\sigma}. \tag{4.4.139}$$

In representations unitarily equivalent to the Dirac representation γ^5 is Hermitian,

$$\gamma^5 = \gamma^{5\dagger}. \tag{4.4.140}$$

In the Dirac representation we find

$$\gamma^5 = \begin{pmatrix} 0 & I \\ I & 0 \end{pmatrix}. \tag{4.4.141}$$

Proof: Let us prove these results.

(i) $\{\gamma^\mu, \gamma^5\} = i(\gamma^\mu\gamma^0\gamma^1\gamma^2\gamma^3 + \gamma^0\gamma^1\gamma^2\gamma^3\gamma^\mu) = i(\gamma^\mu\gamma^0\gamma^1\gamma^2\gamma^3 - \gamma^\mu\gamma^0\gamma^1\gamma^2\gamma^3) = 0$, since any γ^μ commutes with itself and anticommutes with the other three.

(ii) $(\gamma^5)^2 = -\gamma^0\gamma^1\gamma^2\gamma^3\gamma^0\gamma^1\gamma^2\gamma^3 = -\gamma^0\gamma^1\gamma^2\gamma^0\gamma^1\gamma^2 = \gamma^0\gamma^1\gamma^0\gamma^1 = (\gamma^0)^2 = I$, where we have used $(\gamma^i)^2 = -I$ and $(\gamma^0)^2 = I$.

(iii) Since $\gamma^5\gamma^\mu\gamma^\nu = -\gamma^\mu\gamma^5\gamma^\nu = \gamma^\mu\gamma^\nu\gamma^5$ then $[\gamma^5, \gamma^\mu\gamma^\nu] = 0$ and $[\gamma^5, \sigma^{\mu\nu}] = 0$.

(iv) $\gamma^5\gamma^\lambda\gamma^\tau = -(i/4!)\epsilon^{\mu\nu\rho\sigma}\gamma_\mu\gamma_\nu\gamma_\rho\gamma_\sigma\gamma^\lambda\gamma^\tau = -(i/4!)[-\delta^{\mu\lambda}\delta^{\nu\tau}\epsilon^{\mu\nu\rho\sigma}\gamma_\rho\gamma_\sigma + \cdots] = (i/2)\epsilon^{\lambda\tau\rho\sigma}\gamma_\rho\gamma_\sigma$. We have used $\{\gamma_\mu, \gamma^\nu\} = 2\delta_\nu^\mu$ to give $\gamma_0\gamma^0 = \gamma_1\gamma^1 = \gamma_2\gamma^2 = \gamma_3\gamma^3 = I$. We also used the fact that there are four ways to match λ to one of μ, ν, ρ, σ and three ways to match τ to one of the remaining three and that each of the resulting twelve terms can be rearranged to give $(i/4!)\epsilon^{\lambda\tau\alpha\beta}\gamma_\alpha\gamma_\beta$. This then gives the result $\gamma^5\sigma^{\lambda\tau} = (i/2)\epsilon^{\lambda\tau\rho\sigma}\sigma_{\rho\sigma}$.

(v) $\gamma^{5\dagger} = (i\gamma^0\gamma^1\gamma^2\gamma^3)^\dagger = -i\gamma^{3\dagger}\gamma^{2\dagger}\gamma^{1\dagger}\gamma^{0\dagger} = i\gamma^3\gamma^2\gamma^1\gamma^0 = i\gamma^0\gamma^1\gamma^2\gamma^3 = \gamma^5$ in any representation unitarily equivalent to the Dirac representation.

(vi) Simply multiply out $i\gamma^0\gamma^1\gamma^2\gamma^3$ and use the properties of the Pauli matrices given in Sec. A.3 to obtain the above Dirac representation form for γ_5.

The γ^5 matrix transforms under Lorentz transformations as a pseudoscalar, since γ^μ transforms as a four-vector under spinor transformations and since $\epsilon_{\mu\nu\rho\sigma}$ is equal to its own Lorentz transform as a rank-four pseudotensor as given in Eq. (1.2.137),

$$\gamma^5 \xrightarrow{\Lambda} (\gamma^5)' = S(\Lambda)^{-1}\gamma^5 S(\Lambda) = -(i/4!)\epsilon_{\mu\nu\rho\sigma}[S(\Lambda)^{-1}\gamma^\mu\gamma^\nu\gamma^\rho\gamma^\sigma S(\Lambda)]$$
$$= -(i/4!)[\det(\Lambda)(\Lambda^{-1})^\alpha{}_\mu(\Lambda^{-1})^\beta{}_\nu(\Lambda^{-1})^\gamma{}_\rho(\Lambda^{-1})^\delta{}_\sigma\epsilon_{\alpha\beta\gamma\delta}][\Lambda^\mu{}_\epsilon\Lambda^\nu{}_\eta\Lambda^\rho{}_\lambda\Lambda^\sigma{}_\tau\gamma^\epsilon\gamma^\eta\gamma^\lambda\gamma^\tau]$$
$$= -(i/4!)\det(\Lambda)\epsilon_{\alpha\beta\gamma\delta}\gamma^\alpha\gamma^\beta\gamma^\gamma\gamma^\delta = \det(\Lambda)\gamma^5. \tag{4.4.142}$$

Note that $\gamma_5 \equiv \gamma^5$ and these are used interchangeably. Using Eq. (4.4.68) and since $\Sigma^{\mu\nu} = \frac{1}{2}\hbar\sigma^{\mu\nu}$ we have in the rest frame and in the Dirac representation

$$S^i = \tfrac{1}{2}\epsilon^{ijk}\Sigma^{jk} = \tfrac{1}{4}\hbar\epsilon^{ijk}\epsilon^{jk\ell}\begin{bmatrix} \sigma^\ell & 0 \\ 0 & \sigma^\ell \end{bmatrix} = \tfrac{1}{2}\hbar\begin{bmatrix} \sigma^i & 0 \\ 0 & \sigma^i \end{bmatrix},$$
$$(S^{\text{spin}})^i = \gamma^0 S^i = \tfrac{1}{2}\hbar\begin{bmatrix} \sigma^i & 0 \\ 0 & -\sigma^i \end{bmatrix} = -\tfrac{1}{2}\hbar\gamma_5\gamma^i, \tag{4.4.143}$$

where we used Eq. (A.3.10) for σ^{jk} and the fact that the upper two components of spinors at rest are for particles and the lower two are for antiparticles in this representation. Note that the spinor equations

$$\gamma^0 S^i = -\tfrac{1}{2}\hbar\gamma_5\gamma^i \quad \text{and} \quad S^i = \tfrac{1}{2}\hbar\gamma_5\gamma^0\gamma^i \tag{4.4.144}$$

are unaffected by similarity transformations and so are true in any representation.

In the rest frame with spin quantization axis $\mathbf{n}_s^{\text{rest}}$ we have then

$$S_{n_s}^{\text{spin}} \equiv \mathbf{S}^{\text{spin}} \cdot \mathbf{n}_s^{\text{rest}} = -\tfrac{1}{2}\hbar\gamma_5\boldsymbol{\gamma} \cdot \mathbf{n}_s^{\text{rest}} \quad \text{with}$$
$$S_{n_s}^{\text{spin}} u^{\pm s}(0) = \pm\tfrac{1}{2}\hbar u^{\pm s}(0) \quad \text{and} \quad S_{n_s}^{\text{spin}} v^{\pm s}(0) = \pm\tfrac{1}{2}\hbar v^{\pm s}(0). \tag{4.4.145}$$

These results are true in any representation unitarily equivalent to the Dirac representation. In any frame related by a restricted Lorentz transformation we have

$$S_{n_s}^{\text{spin}} \equiv \tfrac{1}{2}\hbar\gamma_5\,\slashed{n}_s \quad \text{with}$$
$$S_{n_s}^{\text{spin}}u^{\pm s}(p) = \pm\tfrac{1}{2}\hbar u^{\pm s}(p) \quad \text{and} \quad S_{n_s}^{\text{spin}}v^{\pm s}(p) = \pm\tfrac{1}{2}\hbar v^{\pm s}(p). \tag{4.4.146}$$

In the rest frame the spin quantization axis is $\mathbf{n}_s^{\text{rest}}$ so let us define the three vector $\mathbf{s}^{\text{rest}} \equiv \pm\mathbf{n}_s^{\text{rest}}$ for spin up and down, respectively, in the $\mathbf{n}_s^{\text{rest}}$ direction. In an arbitrary inertial frame this becomes the *spin vector*

$$s^\mu \equiv \pm n^\mu(s), \quad s^2 = -1, \quad p \cdot s = 0, \tag{4.4.147}$$

for spin up (+) and down (−) in the $\mathbf{n}_s^{\text{rest}}$ direction in the rest frame. The spacelike pseudovector $n^\mu(s)$ was discussed following Eq. (1.4.16).

For any spin vector $s^\mu = \pm n^\mu(s)$ we have the spin projector

$$\Sigma(\pm s) \equiv \tfrac{1}{2}[I \pm \gamma^5\,\slashed{n}_s] \quad \text{or equivalently} \quad \Sigma(s) \equiv \tfrac{1}{2}[I + \gamma^5\,\slashed{s}]. \tag{4.4.148}$$

Using $\slashed{n}_s^{\,2} = n_s^2 = -1$ and $\{\gamma^\mu, \gamma^5\} = 0$ we see that they are projection operators,

$$\Sigma(+s) + \Sigma(-s) = I, \quad \Sigma(+s)^2 = \Sigma(+s), \quad \Sigma(-s)^2 = \Sigma(-s),$$
$$\Sigma(+s)\Sigma(-s) = \Sigma(-s)\Sigma(+s) = 0. \tag{4.4.149}$$

In any representation unitarily equivalent to the Dirac representation we then have

$$\Sigma(s)u^s(p) = u^s(p), \quad \Sigma(s)v^s(p) = v^s(p),$$
$$\Sigma(-s)u^s(p) = \Sigma(-s)v^s(p) = 0. \tag{4.4.150}$$

We can rewrite these more compactly as

$$\Sigma(s)u^{s'}(p) = \delta_{ss'}u^s(p) \quad \text{and} \quad \Sigma(s)v^{s'}(p) = \delta_{ss'}v^s(p). \tag{4.4.151}$$

We observe that the energy and spin projection operators commute,

$$P_\pm(p)\Sigma(s) = \left(\frac{\pm\slashed{p} + mc}{2mc}\right)\left(\frac{1 + \gamma^5\,\slashed{s}}{2}\right) = \left(\frac{1 + \gamma^5\,\slashed{s}}{2}\right)\left(\frac{\pm\slashed{p} + mc}{2mc}\right) = \Sigma(s)P_\pm(p),$$

since $\slashed{p}(\gamma^5\,\slashed{s}) = (\gamma^5\,\slashed{s})\slashed{p}$. This follows from $\slashed{p}\gamma^5 = -\gamma^5\slashed{p}$ and from $\slashed{p}\slashed{s} = -\slashed{s}\slashed{p}$, where the latter is obtained from Eq. (A.3.7) and $p \cdot s = 0$. Applying projectors to the spinor completeness relation in Eqs. (4.4.128) and (4.4.129) we find

$$P_+(p)\Sigma(s) = u^s(p)\bar{u}^s(p)/2mc \quad \text{and} \quad P_-(p)\Sigma(s) = -v^s(p)\bar{v}^s(p)/2mc. \tag{4.4.152}$$

4.4.10 Covariant Interactions and Bilinears

Consider the Dirac equation with coupling to an electromagnetic potential,

$$(i\hbar\slashed{D}(x) - mc)\,\psi(x) = ([i\hbar\partial_\mu - (q/c)A_\mu(x)]\,\gamma^\mu - mc)\,\psi(x) = 0, \tag{4.4.153}$$

where according to Table 1.2 we have for the four-vector potential that $A^\mu(x) \to A'^\mu(x') = \Lambda^\mu{}_\nu A^\nu(x)$. Let us act on this Dirac equation from the left with the spinor Lorentz transformation $S(\Lambda)$; we then have from Eqs. (4.4.62) and (4.4.66) that

$$\begin{aligned}
0 &= S(\Lambda)\left(\left[i\hbar\partial_\mu - (q/c)A_\mu(x)\right]\gamma^\mu - mc\right)\psi(x) \\
&= S(\Lambda)\left(\left[i\hbar\partial_\mu - (q/c)A_\mu(x)\right]\gamma^\mu - mc\right)S(\Lambda)^{-1}S(\Lambda)\psi(x) \\
&= \left(\left[i\hbar\partial_\mu - (q/c)A_\mu(x)\right](\Lambda^{-1})^\mu{}_\nu\gamma^\nu - mc\right)\psi'(x') \\
&= \left(\left[i\hbar\partial'_\mu - (q/c)A'_\mu(x')\right]\gamma^\mu - mc\right)\psi'(x') = (i\hbar\slashed{D}' - mc)\psi'(x').
\end{aligned} \tag{4.4.154}$$

So, as expected, this equation is Lorentz covariant.

Let $V^\mu(x)$ be a pseudovector field, then according to Table 1.2 under Lorentz transformations we have $V^\mu(x) \to V'^\mu(x') = \det(\Lambda)\Lambda^\mu{}_\nu V^\nu(x)$. If $A^\mu(x)$ is replaced with a pseudovector field V^μ the resulting version of the Dirac equation would not be parity invariant. However, were we to replace $(q/c)A^\mu$ by $g\gamma^5 V^\mu(x)$ for some coupling constant $g \in \mathbb{R}$, then this would lead to $\det(\Lambda)^2 = 1$ and so the equation

$$\left(i\hbar\slashed{\partial} - mc - g\gamma_\mu\gamma^5 V^\mu(x)\right)\psi(x) = 0 \tag{4.4.155}$$

is then Lorentz covariant. Similarly, if $S(x)$ is a scalar field, $P(x)$ is a pseudoscalar field, and g is some coupling constant, then two other covariant equations are

$$(i\hbar\slashed{\partial} - mc - gS(x))\psi(x) = 0 \quad \text{and} \quad (i\hbar\slashed{\partial} - mc - g\gamma^5 P(x))\psi(x) = 0. \tag{4.4.156}$$

It is then clear how we can construct Lorentz-covariant versions of the Dirac equation that include interactions with arbitrary external fields.

We have shown that for any Lorentz transformation Λ we have

$$\begin{aligned}
\psi(x) \xrightarrow{\Lambda} \psi'(x') = S(\Lambda)\psi(x), &\qquad \bar{\psi}(x) \xrightarrow{\Lambda} \bar{\psi}'(x') = \bar{\psi}(x)S(\Lambda)^{-1}, \\
S(\Lambda)^{-1}\gamma^\mu S(\Lambda) = \Lambda^\mu{}_\nu\gamma^\nu, &\qquad S(\Lambda)^{-1}\gamma^5 S(\Lambda) = \det(\Lambda)\gamma^5.
\end{aligned} \tag{4.4.157}$$

It also follows that $\sigma^{\mu\nu} \equiv (i/2)[\gamma^\mu, \gamma^\nu]$ transforms as a second rank tensor,

$$S(\Lambda)^{-1}\sigma^{\mu\nu}S(\Lambda) = (i/2)S(\Lambda)^{-1}[\gamma^\mu, \gamma^\nu]S(\Lambda) = \Lambda^\mu{}_\rho\Lambda^\nu{}_\sigma\sigma^{\rho\sigma}. \tag{4.4.158}$$

Any spinor matrix Γ can be written as a linear superposition of the sixteen basis matrices $I, \gamma^\mu, \sigma^{\mu\nu}|_{\mu<\nu}, i\gamma^5\gamma^\mu$ and γ^5. Then any bilinear $\bar{\psi}(x)\Gamma\psi(x)$ can be written as a linear superposition of the sixteen corresponding basis bilinears. These sixteen independent fermion bilinears transform such that

$$\begin{aligned}
\text{scalar:} &\quad \bar{\psi}'(x')\psi'(x') = \bar{\psi}(x)\psi(x), \\
\text{vector:} &\quad \bar{\psi}'(x')\gamma^\mu\psi'(x') = \Lambda^\mu{}_\nu\bar{\psi}(x)\gamma^\nu\psi(x), \\
\text{second-rank tensor:} &\quad \bar{\psi}'(x')\sigma^{\mu\nu}\psi'(x') = \Lambda^\mu{}_\rho\Lambda^\nu{}_\sigma\bar{\psi}(x)\sigma^{\rho\sigma}\psi(x), \\
\text{pseudovector/axial vector:} &\quad \bar{\psi}'(x')\gamma^\mu\gamma^5\psi'(x') = \det(\Lambda)\Lambda^\mu{}_\nu\bar{\psi}(x)\gamma^\nu\gamma^5\psi(x), \\
\text{pseudoscalar:} &\quad \bar{\psi}'(x')i\gamma^5\psi'(x') = \det(\Lambda)\bar{\psi}(x)i\gamma^5\psi(x),
\end{aligned} \tag{4.4.159}$$

where the factors of i have been adjusted so that these bilinears are real in any representation unitarily equivalent to the Dirac representation. This can be readily verified from the properties of the Dirac matrices in Sec. A.3.

4.4.11 Poincaré Group and the Dirac Equation

We have seen that the generators of the Lorentz transformations for the Dirac equation are given by

$$M^{\mu\nu} = L^{\mu\nu} + \Sigma^{\mu\nu} = (x^\mu P^\nu - x^\nu P^\mu) + \tfrac{1}{2}\hbar\sigma^{\mu\nu}, \tag{4.4.160}$$

where $P^\mu = \hat{p}^\mu = i\hbar\partial^\mu$. Recall from Eqs. (1.2.72) and (1.4.3) that the total, orbital and spin angular momenta are, respectively, given by

$$J^i \equiv \tfrac{1}{2}\epsilon^{ijk}M^{jk}, \quad L^i \equiv \tfrac{1}{2}\epsilon^{ijk}L^{jk}, \quad S^i \equiv \tfrac{1}{2}\epsilon^{ijk}\Sigma^{jk} \tag{4.4.161}$$

and hence we have from Eq. (4.4.160) that $\mathbf{J} = \mathbf{L} + \mathbf{S}$.

Recall the Pauli-Lubanski pseudovector from Eq. (1.4.5),

$$W_\mu = -\tfrac{1}{2}\epsilon_{\mu\nu\rho\sigma}M^{\mu\nu}P^\sigma = -\tfrac{1}{2}\epsilon_{\mu\nu\rho\sigma}\Sigma^{\nu\rho}P^\sigma. \tag{4.4.162}$$

Following the discussion of the Little Group for massive particles in Sec. 1.4.2, consider the subspace of solutions of the Dirac equation with four-momentum p^μ such that $p^2 = m^2c^2$, i.e., the p^μ-subspace spanned by the four plane waves,

$$u^s(p)e^{-ip\cdot x/\hbar}, \ u^{-s}(p)e^{-ip\cdot x/\hbar}, \ v^s(p)e^{+ip\cdot x/\hbar} \text{ and } v^{-s}(p)e^{+ip\cdot x/\hbar}, \tag{4.4.163}$$

with the spin quantization axis given by the four-spin vector n_s^μ with $n_s^\mu \to (0, \mathbf{n}_s)$ in the rest frame. The first two plane waves span the particle subspace and the second two span the antiparticle subspace.

For any spinor wavefunction $\psi(+, p; x)$ in the particle subspace and any spinor wavefunction $\psi(-, p; x)$ in the antiparticle subspace we have, respectively,

$$\begin{aligned}
P^\mu\psi(\pm, p; x) &= \hat{p}^\mu\psi(\pm, p; x) = i\hbar\partial^\mu\psi(\pm, p; x) = (\pm p^\mu)\psi(\pm, p; x), \\
P^2\psi(\pm, p; x) &= \hat{p}^2\psi(\pm, p; x) = p^2\psi(\pm, p; x) = m^2c^2\psi(\pm, p; x), \\
W_\mu\psi(\pm, p; x) &= -\tfrac{1}{2}\epsilon_{\mu\nu\rho\sigma}(\pm p^\sigma)\Sigma^{\nu\rho}\psi(\pm, p; x).
\end{aligned} \tag{4.4.164}$$

We find that

$$\begin{aligned}
(-W \cdot n_s) &= \tfrac{1}{2}\epsilon_{\mu\nu\rho\sigma}(\pm p^\sigma)\Sigma^{\nu\rho}n_s^\mu = \tfrac{1}{4}\hbar(\pm p^\sigma)\epsilon_{\mu\nu\rho\sigma}\sigma^{\nu\rho}n_s^\mu = \tfrac{1}{4}\hbar(\pm p^\sigma)(-2i\gamma_5\sigma_{\mu\sigma})n_s^\mu \\
&= \tfrac{1}{4}\hbar(\pm p^\sigma)\gamma_5[\gamma_\mu, \gamma_\sigma]n_s^\mu = \tfrac{1}{2}\hbar(\pm p^\sigma)\gamma_5\gamma_\mu\gamma_\sigma n_s^\mu = \tfrac{1}{2}\hbar\gamma\!\!\!/ n_s(\pm p\!\!\!/) = \tfrac{1}{2}\hbar mc\gamma_5 n\!\!\!/_s,
\end{aligned} \tag{4.4.165}$$

where we have used $p \cdot n_s = 0$ and that $\pm p\!\!\!/ = mc$ when acting on the p^μ-subspace for particles and antiparticles, respectively. The last point follows since for particles $+p\!\!\!/ u^s(p) = mcu^s(p)$ and for antiparticles $-p\!\!\!/ v^s(p) = mcv^s(p)$ as shown in Eq. (4.4.110). Following Eq. (1.4.18) and keeping $\hbar$ and c explicit gives

$$\begin{aligned}
S_s^{\text{spin}} &= (-W \cdot n_s)/mc = \tfrac{1}{2}\hbar\gamma_5 n\!\!\!/_s \xrightarrow{\mathbf{P}\to 0} (\mathbf{S}^{\text{spin}} \cdot \mathbf{n}_s), \\
(S^{\text{spin}})^2 &\equiv (-W^2)/m^2c^2 \xrightarrow{\mathbf{P}\to 0} (\mathbf{S}^{\text{spin}})^2,
\end{aligned} \tag{4.4.166}$$

where the second equation follows from the same arguments that gave Eq. (1.4.10) but where now we keep $c \neq 1$ so that $p^{\text{rest}0} = \pm mc$. As discussed below Eq. (1.4.16) $n^\mu(s)$ is a pseudovector and $S_{n_s}^{\text{spin}}$ is parity invariant. We have recovered the results of Eq. (4.4.146) without starting from the rest frame this time.

It is to be understood that these equalities are valid when acting on states in the p^μ-subspace. Since S_s^{spin} and $(S^{\text{spin}})^2$ are Lorentz invariant then they have the same eigenvalues in any two frames that are related by a restricted Lorentz transformation; i.e., S_s^{spin} has eigenvalues $\hbar m_s = \pm\tfrac{1}{2}\hbar$ and $(S^{\text{spin}})^2$ has eigenvalues $\hbar^2 s(s+1) = \tfrac{3}{4}\hbar^2$ with $s = \tfrac{1}{2}$ for a Dirac fermion.

It is useful to explicitly verify that we must have $s = \tfrac{1}{2}$ for a Dirac fermion. In the rest frame we have $p^\mu = (mc, \mathbf{0})$, which gives $W_0\psi(\pm, p; x) = 0$ and

$$W_i\psi(\pm, p; x) = \mp\tfrac{1}{2}mc\epsilon_{ijk0}\Sigma^{jk}\psi(\pm, p; x) = \mp\tfrac{1}{2}mc\epsilon^{0ijk}\Sigma^{jk}\psi(\pm, p; x) \qquad (4.4.167)$$
$$= \mp mcS^i\psi(\pm, p; x) = \pm mcS_i\psi(\pm, p; x) = mcS_i^{\text{spin}}\psi(\pm, p; x),$$

which we recognize as Eq. (1.4.13). We then arrive at

$$W^2 = -m^2c^2\mathbf{S}^2 = -m^2c^2(\mathbf{S}^{\text{spin}})^2, \qquad (4.4.168)$$

where $\mathbf{S}^{\text{spin}} = \pm\mathbf{S}$ for particles and antiparticles, respectively, and where

$$S^i = \tfrac{1}{2}\epsilon^{ijk}\Sigma^{jk} = \tfrac{1}{4}\hbar\epsilon^{ijk}\sigma^{jk} = \tfrac{1}{4}\hbar\epsilon^{ijk}\epsilon^{jk\ell}\begin{bmatrix} \sigma^\ell & 0 \\ 0 & \sigma^\ell \end{bmatrix} = \tfrac{1}{2}\hbar\begin{bmatrix} \sigma^i & 0 \\ 0 & \sigma^i \end{bmatrix},$$

$$\mathbf{S}^2 = \tfrac{1}{4}\hbar^2\begin{bmatrix} \boldsymbol{\sigma}^2 & 0 \\ 0 & \boldsymbol{\sigma}^2 \end{bmatrix} = \hbar^2\tfrac{3}{4}I = \hbar^2\tfrac{1}{2}(\tfrac{1}{2}+1)I = \hbar^2 s(s+1)I, \qquad (4.4.169)$$

which confirms that $s = \tfrac{1}{2}$ as expected. For massless Dirac particles we will have $s = \tfrac{1}{2}$ and we will need to use the helicity operator discussed in Sec. 1.4.2,

$$S_n^{\text{helicity}} = (\mathbf{S}^{\text{spin}}\cdot\mathbf{n}_p). \qquad (4.4.170)$$

4.5 *P, C* and *T*: Discrete Transformations

We next need to construct the discrete symmetry transformations in the context of relativistic quantum mechanics. These are the parity inversion (P), charge conjugation (C) and time-reversal (T) operations. These results become important in the construction of the these symmetry operations in quantum field theory in Sec. 8.1.

4.5.1 Parity Transformation

The construction of the parity inversion operation is the most straighforward.

Electromagnetic field: Under a parity transformation (or parity inversion) we have $x^\mu \to x'^\mu = (ct', \mathbf{x}') = (ct, -\mathbf{x}) = P^\mu{}_\nu x^\nu$ with

$$P^\mu{}_\nu \equiv g^{\mu\nu}. \qquad (4.5.1)$$

The four-vector current j^μ transforms such that $j^0 = \rho c \to j^0$ and $\mathbf{j} \to -\mathbf{j}$. From Gauss' law, $\boldsymbol{\nabla}\cdot\mathbf{E} = \rho$, and Ampere's law, $\boldsymbol{\nabla}\times\mathbf{B}-(1/c)(\partial\mathbf{E}/\partial t) = \mathbf{j}/c$, we see that $\mathbf{E} \to -\mathbf{E}$ and $\mathbf{B} \to \mathbf{B}$. Then since $\mathbf{E} = -\boldsymbol{\nabla}\Phi-(1/c)\partial\mathbf{A}/\partial t$ and $\mathbf{B} = \boldsymbol{\nabla}\times\mathbf{A}$, we have under parity transformation that $\mathbf{A}'(x') = -\mathbf{A}(x)$ and $\Phi'(x') = \Phi(x)$; i.e., under parity transformation we have for the electromagnetic field that

$$A^\mu(x) \xrightarrow{\mathcal{P}} A'^\mu(x') = \Lambda^\mu{}_\nu A^\nu(x) \quad \text{with} \quad \Lambda^\mu{}_\nu = (P^\mu{}_\nu) = g^{\mu\nu}. \qquad (4.5.2)$$

Klein-Gordon equation: The parity transformation of a scalar field is

$$\phi(x) \xrightarrow{\mathcal{P}} \phi'(x') = \phi(x). \qquad (4.5.3)$$

Since $\Phi'(x') = \Phi(x)$, $\partial'^0 = \partial^0$ and

$$\left(-i\hbar\boldsymbol{\nabla} - \frac{q}{c}\mathbf{A}(x)\right) \xrightarrow{\mathcal{P}} \left(-i\hbar\boldsymbol{\nabla}' - \frac{q}{c}\mathbf{A}'(x')\right) = -\left(-i\hbar\boldsymbol{\nabla} - \frac{q}{c}\mathbf{A}(x)\right), \qquad (4.5.4)$$

then it is clear that Eq. (4.3.57) is invariant under a parity transformation,

$$\left[D'^{\mu} D'_{\mu} + \left(\frac{mc}{\hbar} \right)^2 \right] \phi'(x') = \left[D^{\mu} D_{\mu} + \left(\frac{mc}{\hbar} \right)^2 \right] \phi(x) = 0 \tag{4.5.5}$$

and so the Klein-Gordon equation interacting with an electromagnetic field is invariant under a parity transformation as we might have expected.

Dirac equation: Recall that in the defining representation of the Lorentz group a parity transformation is P, where $\Lambda^{\mu}{}_{\nu} = P^{\mu}{}_{\nu} \equiv g^{\mu\nu}$. Let us denote the spinor matrix representation of the parity transformation as $\mathcal{P}$, where

$$\mathcal{P} \equiv S(\Lambda)|_{\Lambda = P} = S(P). \tag{4.5.6}$$

From Eq. (4.4.62) we note that we require

$$\mathcal{P}^{-1} \gamma^{\mu} \mathcal{P} = P^{\mu}{}_{\nu} \gamma^{\nu} = g^{\mu\nu} \gamma^{\nu}. \tag{4.5.7}$$

The Clifford algebra of the Dirac matrices gives $(\gamma^0)^2 = I$ and $\gamma^0 \gamma^i \gamma^0 = -\gamma^i$, and so this equation is satisfied in any representation of the Dirac equation by

$$\mathcal{P} \equiv e^{i\phi} \gamma^0, \tag{4.5.8}$$

where $e^{i\phi}$ is an arbitrary phase. For a three-space vector two parity transformations return the vector to itself, i.e., $P^2 = I$. For fermions four parity transformations must return a Dirac spinor to itself, $\mathcal{P}^4 = I$, in analogy with the 4π rotation returning a spinor to itself and so possible phase factors are $\exp(i\phi) = \pm 1, \pm i$.

In a representation of the Dirac γ-matrices that is unitarily equivalent to the Dirac representation γ^0 is Hermitian. We can require that $\mathcal{P}$ be an observable and hence Hermitian and so then $\pi_P^{\mathrm{sp}} \equiv \exp(i\phi) = \pm 1$, where π_P^{sp} is the *intrinsic parity of the fermion species*. So we have for the parity operator for Dirac fermions

$$\mathcal{P} = \pi_P^{\mathrm{sp}} \gamma^0. \tag{4.5.9}$$

The parity of a massive particle at rest is its *intrinsic parity*. We conventionally assign $\pi_P^{\mathrm{sp}} = +1$ for all fermion species and this leads to no inconsistencies. Consider the block diagonal form of γ^0 and the plane wave solutions at rest for fermions and antifermions in Eq. (4.4.33) in the Dirac representation. These rest states are parity eigenstates with eigenvalues $\pm \pi_P^{\mathrm{sp}} = \pm 1$ for fermions/antifermions, respectively. So the *intrinsic parity* of every Dirac fermion is $\pi_P = \pi_P^{\mathrm{sp}} = +1$ and for every Dirac antifermion it is $\pi_P = -\pi_P^{\mathrm{sp}} = -1$. This is true in any representation unitarily equivalent to the Dirac representation. We find from Eq. (4.5.9) that

$$\mathcal{P} = \mathcal{P}^{-1} = \mathcal{P}^{\dagger} \quad \text{and} \quad \mathcal{P}^2 = I. \tag{4.5.10}$$

From Eq. (4.4.66) it follows that

$$\psi(x) \xrightarrow{\mathcal{P}} \psi'(x') = \psi'(ct, -\mathbf{x}) = \mathcal{P}\psi(x) = \gamma^0 \psi(x). \tag{4.5.11}$$

Note that since $\{\gamma^{\mu}, \gamma^5\} = 0$ we have

$$\mathcal{P}\gamma^5 = -\gamma^5 \mathcal{P} \tag{4.5.12}$$

and $\mathcal{P}\gamma^5 \psi(x) = -\gamma^5 \mathcal{P}\psi(x)$. So the intrinsic parities of the particle and antiparticle components of $\gamma_5 \psi(x)$ are the opposite of those of $\psi(x)$.

When acting on plane wave states $u^s(p)$ and $v^s(p)$ it is readily verified that

$$\mathcal{P}P_\pm(p)\mathcal{P}^{-1} = P_\pm(p^0, -\mathbf{p}) = P_\pm(p'), \qquad \mathcal{P}\Sigma(s)\mathcal{P}^{-1} = \Sigma(-s^0, \mathbf{s}) = \Sigma(s'), \qquad (4.5.13)$$

where $p'^\mu = (p^0, -\mathbf{p})$ and $s'^\mu = (-s^0, \mathbf{s})$ as it should since s^μ is a pseudovector. This result follows since $\mathcal{P}\gamma_5\gamma^\mu\mathcal{P}^{-1} = -g_{\mu\nu}\gamma_5\gamma^\nu$, which shows that $\gamma_5\gamma^\mu$ is a pseudovector. This shows that a parity transformation acting in the p^μ-subspace does not change the sign of the energy or the spin of the states. This is as expected since a parity transformation does not change particles into antiparticles and since spin is invariant under parity. From Eq. (4.4.159) $\bar{\psi}(x)\psi(x)$ and $i\bar{\psi}(x)\gamma_5\psi(x)$ are even and odd, respectively, under a parity transformation,

$$\bar{\psi}'(x')\psi'(x') = (\mathcal{P}\psi(x))^\dagger\gamma^0\mathcal{P}\psi(x) = \bar{\psi}(x)\psi(x),$$
$$\bar{\psi}'(x')i\gamma_5\psi'(x') = (\mathcal{P}\psi(x))^\dagger\gamma^0 i\gamma_5\mathcal{P}\psi(x) = -\bar{\psi}(x)i\gamma_5\psi(x). \qquad (4.5.14)$$

4.5.2 Charge Conjugation

Electromagnetic field: By definition *charge conjugation* reverses the sign of all quantum charges. This means that the electric charge of the particle is reversed, $q \to -q$ so that $qA^\mu \to -qA^\mu$. We can equivalently view charge conjugation as a reversal in the sign of A^μ while leaving q unchanged,

$$A^\mu(x) \xrightarrow{\mathcal{C}} A'^\mu(x) = -A^\mu(x). \qquad (4.5.15)$$

This field transformation viewpoint is convenient for studying the charge conjugation properties of the Klein-Gordon and Dirac equations.

Klein-Gordon equation: As is evident from Eq. (4.3.57) if we transform A^μ and then take the complex conjugate of this equation we arrive at

$$\left[D'^\mu D'_\mu + \left(\frac{mc}{\hbar}\right)^2\right]\phi^* = \left[\left(\partial^\mu + i\frac{q}{\hbar c}A'^\mu\right)\left(\partial_\mu + i\frac{q}{\hbar c}A'_\mu\right) + \left(\frac{mc}{\hbar}\right)^2\right]\phi^* = 0. \qquad (4.5.16)$$

So we can define the operation of charge conjugation as

$$A^\mu(x) \xrightarrow{\mathcal{C}} A'^\mu(x) = -A^\mu(x) \quad \text{and} \quad \phi(x) \xrightarrow{\mathcal{C}} \phi'(x) = \phi^*(x). \qquad (4.5.17)$$

The Klein-Gordon equation is then invariant under charge conjugation since

$$\left[D^\mu D_\mu + (mc/\hbar)^2\right]\phi(x) = 0 \quad \Longrightarrow \quad \left[D'^\mu D'_\mu + (mc/\hbar)^2\right]\phi'(x) = 0. \qquad (4.5.18)$$

Since $D^\mu\phi \to D'^\mu\phi' = (D^\mu\phi)^*$ we see that the electromagnetic current in Eq. (4.3.59) reverses sign as it should, $j^\mu = q(i\hbar/2m)[\phi^* D^\mu\phi - \phi(D^\mu\phi)^*] \to j'^\mu = -j^\mu$.

Dirac equation: Let $\psi(x)$ be some solution to the Dirac equation and $\psi_C(x)$ be the corresponding solution after charge conjugation, i.e.,

$$[i\hbar\slashed{\partial} - (q/c)\slashed{A} - mc]\psi(x) = 0 \quad \text{and} \quad [i\hbar\slashed{\partial} + (q/c)\slashed{A} - mc]\psi_C(x) = 0. \qquad (4.5.19)$$

In other words, the Dirac equation should be valid independent of whether we regard electrons as the particles and positrons as the antiparticles or vice versa.

Consider the complex conjugate of the Dirac equation

$$[-i\hbar\partial_\mu\gamma^{\mu*} - (q/c)A_\mu\gamma^{\mu*} - mc]\psi^*(x) = 0 \qquad (4.5.20)$$

and act from the left with $C\gamma^0$, where C is an invertible spinor matrix such that

$$(C\gamma^0)\left[-i\hbar\partial_\mu\gamma^{\mu*} - (q/c)A_\mu\gamma^{\mu*} - mc\right](C\gamma^0)^{-1}(C\gamma^0)\psi^*(x) = 0. \tag{4.5.21}$$

We define $\mathcal{C}$ as the charge conjugation operator, which consists of complex conjugation acting to the right followed by multiplication by the spinor matrix $C\gamma^0$. The inverse operator $\mathcal{C}^{-1}$ consists of complex conjugation acting to the right followed by multiplication by the spinor matrix $(C\gamma^0)^{-1}$. Note that the two complex conjugation operations cancel each other when acting on any quantity. The charge conjugation spinor matrix C is then defined by the requirement that it satisfy

$$\mathcal{C}\gamma^\mu\mathcal{C}^{-1} = (C\gamma^0)\gamma^{\mu*}(C\gamma^0)^{-1} = C\gamma^0\gamma^{\mu*}\gamma^0 C^{-1} = -\gamma^\mu. \tag{4.5.22}$$

Since $\gamma^5 = i\gamma^0\gamma^1\gamma^2\gamma^3$ this gives

$$\mathcal{C}\gamma^5\mathcal{C}^{-1} = (C\gamma^0)\gamma^{5*}(C\gamma^0)^{-1} = -i(C\gamma^0)\gamma^{0*}\gamma^{1*}\gamma^{2*}\gamma^{3*}(C\gamma^0)^{-1} = -\gamma^5. \tag{4.5.23}$$

Obviously, C can contain an arbitrary phase factor as was the case for the parity matrix, but we do not consider this further here. The phase freedom associated with charge conjugation is considered further in our discussion of Majorana fermions in Sec. 4.6.5. We have now arrived at our charge conjugate of the Dirac equation,

$$[i\hbar\slashed{\partial} + (q/c)\slashed{A} - mc]\psi_C(x) = 0, \tag{4.5.24}$$

where we have identified the charge-conjugated Dirac spinor wavefunction as

$$\psi(x) \xrightarrow{\mathcal{C}} \psi_C(x) \equiv \mathcal{C}\psi(x) \equiv C\gamma^0\psi^*(x) = C\bar{\psi}^T(x). \tag{4.5.25}$$

Note that in the space of Dirac spinor wavefunctions, the charge conjugation operator $\mathcal{C}$ involves complex conjugation and shows us how to transform a single-particle spinor wavefunction to the charge conjugate antiparticle spinor wavefunction. We saw in Sec. 4.1.7 that in quantum mechanics complex conjugation is an antiunitary transformation and that antiunitary symmetries other than time reversal lead to systems with energies unbounded below. This might superficially seem like a contradiction. However, in the complete quantum field theory treatment in Sec. 8.1.2 we indeed find that the charge conjugation operator is unitary.

The common representations of the γ-matrices are the Dirac, chiral and Majorana representations, which are given in Sec. A.3. They are all unitarily equivalent and so we have $\gamma^0\gamma^\mu\gamma^0 = \gamma^{\mu\dagger}$ for each of them. The Dirac and chiral representations are related by a real unitary transformation, i.e., an orthogonal transformation. Since γ^0 is real in the Dirac representation it is also real in the chiral representation, $\gamma^{0*} = \gamma^0$. Then in the Dirac and chiral representations we have

$$\gamma^0\gamma^{\mu*}\gamma^0 = (\gamma^0\gamma^\mu\gamma^0)^* = (\gamma^{\mu\dagger})^* = \gamma^{\mu T}. \tag{4.5.26}$$

In the Majorana representation the γ-matrices are pure imaginary, $\gamma^{\mu*} = -\gamma^\mu$, which means that $\gamma^{\mu\dagger} = -\gamma^{\mu T}$. We again find the above result since

$$\gamma^0\gamma^{\mu*}\gamma^0 = -\gamma^0\gamma^\mu\gamma^0 = -\gamma^{\mu\dagger} = \gamma^{\mu T}. \tag{4.5.27}$$

Substituting this result into Eq. (4.5.22) we find that for *all three* representations

$$C\gamma^0\gamma^{\mu*}\gamma^0 C^{-1} = C\gamma^{\mu T}C^{-1} = -\gamma^\mu \quad\text{or equivalently}\quad C^{-1}\gamma^\mu C = -\gamma^{\mu T}. \tag{4.5.28}$$

The conventional choice for C in the Dirac representation is

$$C = i\gamma^2\gamma^0 = -i\gamma^0\gamma^2 = \begin{pmatrix} 0 & -i\sigma^2 \\ -i\sigma^2 & 0 \end{pmatrix}. \tag{4.5.29}$$

Proof: We note that in the Dirac representation γ^2 is imaginary and $\gamma^{2T} = \gamma^2$, whereas γ^0, γ^1 and γ^3 are real so that $\gamma^{\mu\dagger} = \gamma^{\mu T}$ for $\mu = 0, 1, 3$. Also recall that $\gamma^{0\dagger} = \gamma^0$ and $\gamma^{i\dagger} = -\gamma^i$. We then have

$$C^{-1} = -i\gamma^2\gamma^0 = -C \quad \text{since} \quad (-i\gamma^2\gamma^0)(i\gamma^2\gamma^0) = \gamma^2\gamma^0\gamma^2\gamma^0 = -\gamma^2(\gamma^0)^2\gamma^2 = I,$$

$$C^{-1}\gamma^\mu C = -i\gamma^2\gamma^0\gamma^\mu i\gamma^2\gamma^0 = \gamma^2\gamma^0\gamma^\mu\gamma^2\gamma^0 = -\gamma^0\gamma^2\gamma^\mu\gamma^2\gamma^0 = -\gamma^{\mu T}, \tag{4.5.30}$$

where we have used

$$\mu \neq 2: \; -\gamma^0\gamma^2\gamma^\mu\gamma^2\gamma^0 = \gamma^0(\gamma^2)^2\gamma^\mu\gamma^0 = -\gamma^0\gamma^\mu\gamma^0 = -\gamma^{\mu\dagger} = -\gamma^{\mu T},$$

$$\mu = 2: \; -\gamma^0(\gamma^2)^3\gamma^0 = \gamma^0\gamma^2\gamma^0 = \gamma^{2\dagger} = -\gamma^2 = -\gamma^{2T}. \tag{4.5.31}$$

In the Dirac representation C is a real matrix since σ^2 is pure imaginary and hence $C = C^*$ and $C^\dagger = C^T$. This leads to

$$C = C^* \quad \text{and} \quad C = -C^{-1} = -C^\dagger = -C^T, \tag{4.5.32}$$

where $C = -C^{-1}$ was shown above and where $C = -C^T$ using Eq. (4.5.29) since $\sigma^2 = -\sigma^{2T}$. Matrix equations involving transposes are invariant under orthogonal transformations and since the Dirac and chiral representations are related by an orthogonal transformation, then in both representations (i) Eq. (4.5.28) will be satisfied by $C = i\gamma^2\gamma^0$; (ii) $C = -C^T$; and (iii) C will be real. So Eq. (4.5.32) will be true in both the Dirac and chiral representations.

In the Majorana representation we can choose $C = -i\gamma^0$, since then

$$C^{-1}\gamma^\mu C = (i\gamma^0)\gamma^\mu(-i\gamma^0) = \gamma^{\mu\dagger} = -\gamma^{\mu T}, \tag{4.5.33}$$

where we have used $\gamma^{\mu*} = -\gamma^\mu$ in this representation. Note that this C is the same explicit choice of matrix as used in the Dirac representation in Eq. (4.5.29) and so Eq. (4.5.32) again follows. This and the above results are collected and summarized in Sec. A.3. Note that two charge conjugations return the original spinor wavefunction, $\mathcal{C}^2 = I$, since in all three representations

$$(\psi_C)_C(x) = \mathcal{C}\psi_C(x) = \mathcal{C}^2\psi(x) = C\gamma^0(C\gamma^0\psi^*(x))^* = \psi(x) \tag{4.5.34}$$

following from $C\gamma^0 C^*\gamma^{0*} = (-C^{-1})\gamma^0 C^{\dagger T}\gamma^{0\dagger T} = -C^{-1}\gamma^0 C\gamma^{0T} = \gamma^{0T}\gamma^{0T} = I$.

Eqs. (4.4.132), (4.4.148), (4.5.22) and (4.5.23) give in the p^μ-subspace

$$\mathcal{C}P_\pm(p)\mathcal{C}^{-1} = P_\mp(p), \qquad \mathcal{C}\Sigma(s)\mathcal{C}^{-1} = \Sigma(s) \tag{4.5.35}$$

in any representation, since $\mathcal{C}\gamma^\mu\mathcal{C}^{-1} = -\gamma^\mu$ and $\mathcal{C}\gamma_5\gamma^\mu\mathcal{C}^{-1} = \gamma_5\gamma^\mu$. So particles are turned into antiparticles but the spin is unchanged. This is an analog of Eq. (4.5.13) for a parity transformation.

Consider a positive or negative energy plane wave with four-spin s^μ, then we must have $\psi(x) = P_\pm(p)\Sigma(s)\psi(x)$. The charge conjugate of this plane wave satisfies

$$\psi_C(x) = \mathcal{C}\psi(x) = C\bar\psi(x)^T = (C\gamma^0)\psi^*(x) = (C\gamma^0)\left[P_\pm(p)\Sigma(s)\psi(x)\right]^*$$

$$= (C\gamma^0)\left[P_\pm(p)\Sigma(s)\right]^*(C\gamma^0)^{-1}(C\gamma^0)\psi(x)^* = P_\mp(p)\Sigma(s)\psi_C(x). \tag{4.5.36}$$

This means that if $\psi(x)$ is a positive/negative energy plane wave solution with four-momentum p^μ and four-spin s^μ, then $\psi_C(x) = \mathcal{C}\psi(x) = C\gamma^0\psi^*(x)$ is a negative/positive energy plane wave solution with the same p^μ and s^μ. We note that

$$\bar\psi_C\psi_C = (C\gamma^0\psi^*)^\dagger\gamma^0(C\gamma^0\psi^*) = \psi^T\gamma^{0\dagger}C^\dagger\gamma^0 C\gamma^0\psi^* = \psi^T\gamma^0 C^{-1}\gamma^0 C\gamma^0\psi^*$$

$$= \psi^T(C\gamma^0)^{-1}\gamma^0(C\gamma^0)\psi^* = \psi^T(-\gamma^0)^*\psi^* = -(\bar\psi\psi)^*. \tag{4.5.37}$$

This means that $\psi_C(x)$ and $\psi(x)$ have the same normalization up to an irrelevant phase. These results are true in each of the Dirac, chiral and Majorana representations since Eq. (4.5.32) is true in each of these representations.

Let us explicitly verify these results for the case of the Dirac representation. For a positive-energy plane wave in the Dirac representation,

$$\psi(x) = e^{-ip\cdot x/\hbar}u^s(p) = e^{-ip\cdot x/\hbar}\frac{\not{p}+mc}{\sqrt{p^0+mc}}\begin{bmatrix}\phi_s\\0\end{bmatrix},$$

$$\psi_C(x) = (e^{-ip\cdot x/\hbar}u^s(p))_c = \mathcal{C}\psi(x) = C\bar{\psi}(x)^T = C\gamma^0\psi(x)^* = e^{ip\cdot x/\hbar}C\gamma^0(u^s(p))^*$$

$$= e^{ip\cdot x/\hbar}(C\gamma^0)\frac{\not{p}^*+mc}{\sqrt{(p^0+mc)}}(C\gamma^0)^{-1}(C\gamma^0)\begin{bmatrix}\phi_s^*\\0\end{bmatrix}$$

$$= e^{ip\cdot x/\hbar}\frac{-\not{p}+mc}{\sqrt{(p^0+mc)}}\begin{bmatrix}0\\-i\sigma^2\phi_s^*\end{bmatrix} = (-1)^{\eta_s}\left(e^{ip\cdot x/\hbar}v^s(p)\right), \tag{4.5.38}$$

where we have defined

$$\eta_s \equiv \tfrac{1}{2} - m_s. \tag{4.5.39}$$

Similarly, for a negative-energy plane wave

$$\psi(x) = e^{ip\cdot x/\hbar}v^s(p) = e^{ip\cdot x/\hbar}\frac{-\not{p}+mc}{\sqrt{p^0+mc}}\begin{bmatrix}0\\\chi_{-s}\end{bmatrix},$$

$$\psi_C(x) = (e^{ip\cdot x/\hbar}v^s(p))_c = C\bar{\psi}(x)^T = C\gamma^0\psi(x)^* = e^{-ip\cdot x/\hbar}C\gamma^0(v^s(p))^*$$

$$= e^{-ip\cdot x/\hbar}\frac{\not{p}+mc}{\sqrt{(p^0+mc)}}\begin{bmatrix}+i\sigma^2\chi_{-s}^*\\0\end{bmatrix} = (-1)^{\eta_s}\left(e^{-ip\cdot x/\hbar}u^s(p)\right). \tag{4.5.40}$$

Note that charge conjugation has turned a Dirac fermion into a Dirac antifermion and vice versa, which means that both the electric charge and the intrinsic parity have been reversed for a fermion as was discussed in Sec. 4.1.10. Factoring out the exponential factors we find from Eqs. (4.5.38) and (4.5.40)

$$C\gamma^0 u^s(p)^* = C\bar{u}^s(p)^T = (-1)^{\eta_s}v^s(p),$$

$$C\gamma^0 v^s(p)^* = C\bar{v}^s(p)^T = (-1)^{\eta_s}u^s(p) \tag{4.5.41}$$

with $\eta_S = 0, 1$ for $m_s = \pm$, respectively. This will be useful later in Sec. 8.1.2.

Proof: The unitary transformations $\pm i\sigma^2$ are active rotations about the y-axis by $\mp\pi$, since $(\sigma^2)^2 = I$ then $\exp(-i\theta\frac{1}{2}\sigma^2) = I\cos(\frac{1}{2}\theta) - i\sigma^2\sin(\frac{1}{2}\theta)$ and so $\exp(-i\{\pm\pi\}\frac{1}{2}\sigma^2) = \mp i\sigma^2$. When these unitary transformations are combined with complex conjugation we have a transformation that reverses the spin,

$$(\pm i\sigma^2)\sigma^*(\pm i\sigma^2)^\dagger = (i\sigma^2)\sigma^*(i\sigma^2)^\dagger = \sigma^2\sigma^*\sigma^2 = -\sigma, \tag{4.5.42}$$

which is readily checked for each of $\sigma^1, \sigma^2, \sigma^3$. Since $(\sigma\cdot\mathbf{n}_s)\phi_\pm = \pm\phi_\pm$, then

$$(i\sigma^2)^\dagger(\sigma\cdot\mathbf{n}_s)(i\sigma^2)\phi_\pm^* = -(\sigma^*\cdot\mathbf{n}_s)\phi_\pm^* = -((\sigma\cdot\mathbf{n}_s)\phi_\pm)^* = \mp\phi_\pm^*. \tag{4.5.43}$$

Acting from the left with $(i\sigma^2)$ gives

$$(\sigma\cdot\mathbf{n}_s)(i\sigma^2\phi_\pm^*) = \mp(i\sigma^2\phi_\pm^*), \tag{4.5.44}$$

which shows that $\phi_\pm$ and $(i\sigma^2\phi_\pm^*)$ are eigenstates of $\boldsymbol{\sigma}\cdot\mathbf{n}_s$ with opposite spins. Recall that $s = +, -$ corresponds to $m_s = +\frac{1}{2}, -\frac{1}{2}$. From Eq. (4.4.100) we find

$$-i\sigma^2\phi_s^* = (-1)^{\eta_s}\chi_{-s} \quad \text{and} \quad i\sigma^2\chi_{-s}^* = (-1)^{\eta_s}\phi_s, \tag{4.5.45}$$

where $(-1)^{\eta_s}$ removes the spin-dependent sign in the rotation, $\phi^*, \chi^* \to \chi, \phi$.

4.5.3 Time Reversal

The time-reversal transformation is more complicated than that for parity.

Electromagnetic field: We need to apply the Lorentz transformation results of Table 1.1 with due care, since these are the forms of the Lorentz transformations for the four-vector x^μ. For example, under time reversal we have

$$x^\mu \xrightarrow{\mathcal{T}} x'^\mu = (x'^0, \mathbf{x}') = (-x^0, \mathbf{x}) = T^\mu_{\ \nu}x^\nu \quad \text{with} \quad T^\mu_{\ \nu} \equiv -g^{\mu\nu}, \tag{4.5.46}$$

whereas for the four-velocity $V^\mu = dx^\mu/d\tau$ we have $V^\mu \to V'^\mu = dx'^\mu/d\tau' = (V^0, -\mathbf{V}) = \Lambda^\mu_{\ \nu}V^\nu(x) = (-T^\mu_{\ \nu})V^\nu$. The four-vector current j^μ transforms under time reversal like V^μ, since a charge density $\rho = j^0/c$ is invariant under time reversal whereas a current density $\mathbf{j}$ obviously changes sign. We have Gauss' law, $\boldsymbol{\nabla}\cdot\mathbf{E} = \rho$, and Ampère's law, $\boldsymbol{\nabla}\times\mathbf{B} - (1/c)(\partial\mathbf{E}/\partial t) = \mathbf{j}/c$, which tell us that under time reversal $\mathbf{E} \to \mathbf{E}$ and $\mathbf{B} \to -\mathbf{B}$. Then since $\mathbf{E} = -\boldsymbol{\nabla}\Phi - (1/c)\partial\mathbf{A}/\partial t$ and $\mathbf{B} = \boldsymbol{\nabla}\times\mathbf{A}$, we have under time reversal that $\mathbf{A}'(x') = -\mathbf{A}(x)$ and $\Phi'(x') = \Phi(x)$; i.e., under time reversal we have for the electromagnetic field that

$$A^\mu(x) \xrightarrow{\mathcal{T}} A'^\mu(x') = (-T^\mu_{\ \nu})A^\nu(x) = (A^0(x), -\mathbf{A}(x)) = (\Phi(x), -\mathbf{A}(x)). \tag{4.5.47}$$

Klein-Gordon equation: Under a time-reversal transformation we must have $x^\mu \to x'^\mu = T^\mu_{\ \nu}x^\nu$, $\partial^\mu \to \partial'^\mu = T^\mu_{\ \nu}\partial^\nu$, and $A^\mu(x) \to A'^\mu(x') = (-T^\mu_{\ \nu})A^\nu(x)$. Applying these transformations to the Klein-Gordon equation interacting with an electromagnetic field, Eq. (4.3.57), and taking the complex conjugate we arrive at

$$\left[\left(\partial'^\mu + i\frac{q}{\hbar c}A'^\mu(x')\right)\left(\partial'_\mu + i\frac{q}{\hbar c}A'_\mu(x')\right) + \left(\frac{mc}{\hbar}\right)^2\right]\phi^*(x') = 0. \tag{4.5.48}$$

So then we can define the operation of time reversal as

$$x^\mu \xrightarrow{\mathcal{T}} x'^\mu = T^\mu_{\ \nu}x^\nu = (-x^0, \mathbf{x}),$$

$$A^\mu(x) \xrightarrow{\mathcal{T}} A'^\mu(x') = (-T^\mu_{\ \nu})A^\nu(x) = (A^0, -\mathbf{A}),$$

$$\phi(x) \xrightarrow{\mathcal{T}} \phi'(x') = \phi^*(x) \quad \text{with} \quad T^\mu_{\ \nu} = -g^{\mu\nu}. \tag{4.5.49}$$

The Klein-Gordon equation is then invariant under time reversal since

$$\left[D^\mu D_\mu + (mc/\hbar)^2\right]\phi(x) = 0 \quad \Longrightarrow \quad \left[D'^\mu D'_\mu + (mc/\hbar)^2\right]\phi'(x') = 0. \tag{4.5.50}$$

Dirac equation: We now wish to construct the time-reversal transformation for the Dirac equation. For this purpose it is convenient to again consider a Dirac particle interacting with an external electromagnetic field, but this time with the Dirac equation in its Hamiltonian form,

$$i\hbar\frac{\partial\psi}{\partial t}(x) = \hat{H}(A, x)\psi(x) = \left[c\boldsymbol{\alpha}\cdot\left(-i\hbar\boldsymbol{\nabla} - \frac{q}{c}\mathbf{A}(x)\right) + \beta mc^2 + q\Phi(x)\right]\psi(x). \tag{4.5.51}$$

We define the time-reversed spinor as ψ_T, where under time reversal $x^\mu = (ct, \mathbf{x}) \to x'^\mu = (ct', \mathbf{x}') = (-ct, \mathbf{x})$ and

$$\psi(x) \xrightarrow{\mathcal{T}} \psi_T(x') = \psi_T(-ct, \mathbf{x}) \equiv \mathcal{T}\psi(ct, \mathbf{x}) = \mathcal{T}\psi(x). \qquad (4.5.52)$$

Just as the parity transformation operator $\mathcal{P} = \pi_P \gamma^0$ does not act on the spatial coordinates $\mathbf{x}$ or the spatial gradient operator $\boldsymbol{\nabla}$, we will construct a $\mathcal{T}$ that does not act on t or $\partial/\partial t$, so that

$$\mathcal{T}\left(\frac{\partial \psi}{\partial t}(x)\right) = \frac{\partial}{\partial t}\mathcal{T}\psi(x) = \frac{\partial}{\partial t}\psi_T(x') = -\frac{\partial \psi_T}{\partial t'}(x'). \qquad (4.5.53)$$

The time-reversed form of the above Dirac equation is then

$$\mathcal{T}\left(i\hbar\frac{\partial \psi}{\partial t}(x)\right) = -\mathcal{T}i\mathcal{T}^{-1}\hbar\frac{\partial \psi_T}{\partial t'}(x') = \mathcal{T}\hat{H}(A, x)\mathcal{T}^{-1}\psi_T(x'). \qquad (4.5.54)$$

Time-reversal invariance of the Dirac equation means that we must have

$$i\hbar\frac{\partial \psi_T}{\partial t'}(x') = \hat{H}(A', x')\psi_T(x'), \qquad (4.5.55)$$

which can be satisfied in two ways,

$$\mathcal{T}i\mathcal{T}^{-1} = \mp i \quad \text{and} \quad \mathcal{T}\hat{H}(A, x)\mathcal{T}^{-1} = \pm\hat{H}(A', x'). \qquad (4.5.56)$$

In analogy with the charge conjugation transformation $\mathcal{C}$ we attempt to construct $\mathcal{T}$ as complex conjugation acting to the right followed by the operation of some spinor matrix T, i.e., such that $\mathcal{T} = T\hat{K}$ where $\hat{K}$ is the complex conjugation operator defined in terms of the coordinate-space representation. Then we have

$$\psi(x) \xrightarrow{\mathcal{T}} \psi_T(x') = \psi_T(-ct, \mathbf{x}) = \mathcal{T}\psi(x) \equiv T\hat{K}\psi(x) = T\psi^*(x). \qquad (4.5.57)$$

This choice gives $\mathcal{T}i\mathcal{T}^{-1} = -iTT^{-1} = -i$ and hence we must have

$$\hat{H}(A', x') = \mathcal{T}\hat{H}(A, x)\mathcal{T} = T\hat{H}(A, x)^*T^{-1}. \qquad (4.5.58)$$

We see that

$$T\hat{H}(A, x)^*T^{-1} = \left[cT\boldsymbol{\alpha}^*T^{-1} \cdot (+i\hbar\boldsymbol{\nabla} - (q/c)\mathbf{A}(x)) + T\beta^*T^{-1}mc^2 + q\Phi(x)\right]$$

$$= \left[c\boldsymbol{\alpha} \cdot (-i\hbar\boldsymbol{\nabla}' - (q/c)\mathbf{A}'(x')) + \beta mc^2 + q\Phi'(x')\right] = \hat{H}(A', x'),$$

$$\Leftrightarrow \quad \mathcal{T}\boldsymbol{\alpha}\mathcal{T}^{-1} = T\boldsymbol{\alpha}^*T^{-1} = -\boldsymbol{\alpha} \quad \text{and} \quad \mathcal{T}\beta\mathcal{T}^{-1} = T\beta^*T^{-1} = \beta. \qquad (4.5.59)$$

Taking the Hermitian conjugate of these equations and noting that $\boldsymbol{\alpha}$ and β are Hermitian, it is obvious that we need $T^\dagger = T^{-1}$ and so T must be unitary.

Since $\beta = \gamma^0$ and $\alpha^i = \gamma^0\gamma^i$ the above properties can be summarized as

$$\mathcal{T}\gamma^\mu\mathcal{T}^{-1} = T\gamma^{\mu*}T^{-1} = (\gamma^0, -\boldsymbol{\gamma}) = (-T^\mu{}_\nu)\gamma^\nu = g^{\mu\nu}\gamma^\nu. \qquad (4.5.60)$$

We then note that

$$\mathcal{T}\gamma^5\mathcal{T}^{-1} = T\gamma^{5*}T^{-1} = -iT\gamma^{0*}\gamma^{1*}\gamma^{2*}\gamma^{3*}T^{-1} = i\gamma^0\gamma^1\gamma^2\gamma^3 = \gamma^5, \qquad (4.5.61)$$

$$\mathcal{T}\gamma^5\gamma^\mu\mathcal{T}^{-1} = \mathcal{T}\gamma^5\mathcal{T}^{-1}\mathcal{T}\gamma^\mu\mathcal{T}^{-1} = \gamma^5(\gamma^0, -\boldsymbol{\gamma}) = \gamma^5(-T^\mu{}_\nu)\gamma^\nu = \gamma^5 g^{\mu\nu}\gamma^\nu.$$

It then remains to show that we can find a suitable T. In both the Dirac and chiral representations γ^0, γ^1 and γ^3 are real matrices, whereas γ^2 is an imaginary matrix. This means that α_1, α_3 and β are real matrices, whereas α_2 is imaginary. It is also straightforward to show using Eq. (4.4.9) that $\alpha_1 \alpha_3$ commutes with α_1, α_3 and β but it anticommutes with α_2. It is then clear that in the Dirac and chiral representations we can choose

$$T = \pm i\alpha_1 \alpha_3 = T^\dagger = T^{-1} = -T^* \quad \text{such that} \quad T\gamma^\mu T^{-1} = \gamma^{\mu T} = \gamma_\mu^*, \qquad (4.5.62)$$

where we note that this T is unitary, Hermitian, self-inverse and pure imaginary. The sign choice for T is arbitrary. Following Bjorken and Drell (1964) we choose

$$T = -i\alpha_1 \alpha_3 = i\gamma^1 \gamma^3. \qquad (4.5.63)$$

Under a time-reversal transformation three-momentum and spin are reversed,

$$p^\mu = (p^0, \mathbf{p}) \xrightarrow{\ \mathcal{T}\ } p'^\mu = (p'^0, \mathbf{p}') = (p^0, -\mathbf{p}),$$

$$s^\mu = (s^0, \mathbf{s}) \xrightarrow{\ \mathcal{T}\ } s'^\mu = (s'^0, \mathbf{s}') = (s^0, -\mathbf{s}). \qquad (4.5.64)$$

It is readily verified from the definition of $\mathcal{T}$ above that in any representation

$$\mathcal{T}P_\pm(p)\mathcal{T}^{-1} = P_\pm(p^0, -\mathbf{p}) = P_\pm(p'), \qquad \mathcal{T}\Sigma(s)\mathcal{T}^{-1} = \Sigma(s^0, -\mathbf{s}) = \Sigma(s') \qquad (4.5.65)$$

as expected. This is the analog of Eqs. (4.5.13) and (4.5.35).

Let $\psi(x)$ be a plane wave solution of the Dirac equation with positive/negative energy, four-momentum p^μ and spin s^μ, then $\psi(x) = P_\pm(p)\Sigma(s)\psi(x)$. The time-reversed plane wave, $\psi_T(x') = \mathcal{T}\psi(x) = T\psi(x)^*$, is then

$$\psi_T(x') = \mathcal{T}\psi(x) = \mathcal{T}\left[P_\pm(p)\Sigma(s)\right]\mathcal{T}^{-1}\mathcal{T}\psi(x) = P_\pm(p')\Sigma(s')\psi_T(x') \qquad (4.5.66)$$

and so has positive/negative energy, momentum p'^μ and spin s'^μ. We also note that

$$\bar{\psi}_T(x')\psi_T(x') = (T\psi(x)^*)^\dagger \gamma^0 (T\psi(x)^*) = \psi^T(x)T^\dagger \gamma^0 T\psi^*$$
$$= \psi^T(x)T^{-1}\gamma^0 T\psi^* = \psi^T(x)\gamma^{0*}\psi^* = (\bar{\psi}(x)\psi(x))^*. \qquad (4.5.67)$$

This means that $\psi_T(x')$ and $\psi(x)$ have the same normalization up to an irrelevant phase. It will be useful later to note the result that

$$Tu^s(p) = i(-1)^{m_s + \frac{1}{2}} u^{-s}(-p)^* = e^{i\eta_+(s)} u^{-s}(-p)^*, \qquad (4.5.68)$$

$$Tv^s(p) = i(-1)^{m_s - \frac{1}{2}} v^{-s}(-p)^* = e^{i\eta_-(s)} v^{-s}(-p)^*,$$

where $\eta_+(s) \equiv \pi(m_s + 1)$ and $\eta_-(s) \equiv \pi m_s$.

Proof: In Dirac representation we have

$$T = i\gamma^1 \gamma^3 = i\begin{bmatrix} i\sigma^2 & 0 \\ 0 & i\sigma^2 \end{bmatrix} \quad \text{with} \quad i\sigma^2 = \begin{bmatrix} 0 & 1 \\ -1 & 0 \end{bmatrix} \qquad (4.5.69)$$

and also $\gamma^1\gamma^3\gamma^0 = \gamma^0\gamma^1\gamma^3 = \gamma^{0*}\gamma^1\gamma^3$, $\gamma^1\gamma^3\gamma^1 = -\gamma^1\gamma^1\gamma^3 = -\gamma^{1*}\gamma^1\gamma^3$, $\gamma^1\gamma^3\gamma^2 = \gamma^2\gamma^1\gamma^3 = -\gamma^{2*}\gamma^1\gamma^3$ and $\gamma^1\gamma^3\gamma^3 = -\gamma^3\gamma^1\gamma^3 = -\gamma^{3*}\gamma^1\gamma^3$. With $p^\mu = (p^0, \mathbf{p})$ we define $\tilde{p}^\mu \equiv (p^0, -\mathbf{p})$ and then $\gamma^1\gamma^3\not{p} = \tilde{\not{p}}^*\gamma^1\gamma^3$. In this representation it also follows from Eq. (4.4.100) that

$$i\sigma^2\phi_\pm = \mp\phi_\mp \quad \text{and} \quad i\sigma^2\chi_\pm = \mp\chi_\mp. \qquad (4.5.70)$$

Using Eq. (4.4.116) for the plane wave spinors we find

$$
Tu^{\pm}(p) = i\gamma^1\gamma^3 \frac{\not{p}+m}{\sqrt{(p^0+m)}} \begin{bmatrix} \phi_\pm \\ 0 \end{bmatrix} = i\frac{\not{p}^*+m}{\sqrt{(p^0+m)}}\gamma^1\gamma^3 \begin{bmatrix} \phi_\pm \\ 0 \end{bmatrix} \tag{4.5.71}
$$

$$
= \mp i\frac{\not{p}^*+m}{\sqrt{(p^0+m)}} \begin{bmatrix} \phi_\mp \\ 0 \end{bmatrix} = \mp i u^\mp(-p)^*,
$$

$$
Tv^{\pm}(p) = i\gamma^1\gamma^3 \frac{-\not{p}+m}{\sqrt{(p^0+m)}} \begin{bmatrix} 0 \\ \chi_\mp \end{bmatrix} = \pm i\frac{-\not{p}^*+m}{\sqrt{(p^0+m)}} \begin{bmatrix} 0 \\ \chi_\pm \end{bmatrix} = \pm i v^\mp(-p)^*.
$$

4.5.4 *CPT* Transformation

Under the combined transformations of $\mathcal{C}$, $\mathcal{P}$ and $\mathcal{T}$ we have

$$
x^\mu = (x^0, \mathbf{x}) \xrightarrow{\mathcal{CPT}} x'^\mu = (x'^0, \mathbf{x}') = (-x^0, -\mathbf{x}) = -x^\mu,
$$

$$
p^\mu = (p^0, \mathbf{p}) \xrightarrow{\mathcal{CPT}} p'^\mu = (p'^0, \mathbf{p}') = (p^0, \mathbf{p}) = p^\mu,
$$

$$
s^\mu = (s^0, \mathbf{s}) \xrightarrow{\mathcal{CPT}} s'^\mu = (s'^0, \mathbf{s}') = (-s^0, -\mathbf{s}) = -s^\mu. \tag{4.5.72}
$$

From the above results and definitions we have for *any* spinor wavefunction $\psi(x)$ in the Dirac or chiral representations that

$$
\mathcal{P}\psi(x) = e^{i\phi}\gamma^0\psi(x), \quad \mathcal{C}\psi(x) = i\gamma^2\psi^*(x), \quad \mathcal{T}\psi(x) = i\gamma^1\gamma^3\psi^*(x) \tag{4.5.73}
$$

and so as found in Bjorken and Drell (1964) we have

$$
\psi_{PCT}(x') = \psi_{PCT}(-x) = \mathcal{PCT}\psi(x) = e^{i\phi}\gamma^0(i\gamma^2[i\gamma^1\gamma^3\psi^*(x)]^*)
$$

$$
= e^{i\phi}\gamma^0\gamma^2\gamma^1\gamma^3\psi(x) = ie^{i\phi}\gamma^5\psi(x). \tag{4.5.74}
$$

Recall that we normally choose $e^{i\phi} = \pi_P^{\mathrm{sp}}$, where π_P^{sp} is the intrinsic parity of the particle in a fermion species and where it leads to no contradictions to choose $\pi_P^{\mathrm{sp}} = 1$ for every species. So the fermion has intrinsic parity $\pi_P = +\pi_P^{\mathrm{sp}} = +1$ and the antifermion has intrinsic parity $\pi_P = -\pi_P^{\mathrm{sp}} = -1$. However, we use the form $e^{i\phi}$ here to emphasize that it is in principle just some chosen phase for each type of particle. Changing the order of the $\mathcal{P}$, $\mathcal{C}$ and $\mathcal{T}$ operations only changes this result by an overall irrelevant phase,

$$
\psi_{PCT}(x') = -\psi_{PTC}(x') = \psi_{TCP}(x') = ie^{i\phi}\gamma^5\psi(x), \tag{4.5.75}
$$

$$
\psi_{CTP}(x') = -\psi_{CPT}(x') = -\psi_{TPC}(x') = ie^{-i\phi}\gamma^5\psi(x). \tag{4.5.76}
$$

In the Dirac and chiral representations we have for the ordering $\mathcal{CPT}$,

$$
\bar{\psi}_{CPT}(x')\psi_{CPT}(x') = (-ie^{-i\phi}\gamma^5\psi(x))^\dagger\gamma^0(-ie^{-i\phi}\gamma^5\psi(x))
$$

$$
= \psi(x)^\dagger\gamma^5\gamma^0\gamma^5\psi(x) = -\bar{\psi}(x)\psi(x), \tag{4.5.77}
$$

where this result follows for $\mathcal{C}$, $\mathcal{P}$ and $\mathcal{T}$ in any order. This means that $\psi_{CPT}(x')$ and $\psi(x)$ have the same normalization up to an irrelevant phase, which is what we expect since this was the case for each of the three transformations individually.

As an illustration let us consider the Dirac representation. A positive energy wavefunction is interpreted as a *particle* wavefunction,

$$\psi^{\text{part}}(p,s;x) = e^{-ip\cdot x/\hbar}u^s(p) = e^{-ip\cdot x/\hbar}\frac{\not{p}+mc}{\sqrt{p^0+mc}}\begin{bmatrix}\phi_s\\0\end{bmatrix}. \tag{4.5.78}$$

We interpret the negative-energy plane wave solutions of the Dirac equation as the *antiparticle* plane wave solutions

$$\psi^{\text{antipart}}(p,s;x) = e^{+ip\cdot x/\hbar}v^s(p) = e^{+ip\cdot x/\hbar}\frac{-\not{p}+mc}{\sqrt{p^0+mc}}\begin{bmatrix}0\\\chi_{-s}\end{bmatrix}, \tag{4.5.79}$$

where antiparticles have negative energy or equivalently can be thought of as having positive energy and moving backward in spacetime, which is the Feynman-Stueckelberg interpretation.

Consider the CPT-transformed particle and antiparticle plane waves,

$$\mathcal{CPT}\psi^{\text{part}}(p,s;x) = (-ie^{-i\phi}\gamma^5)\psi^{\text{part}}(p,s;x) = (-ie^{-i\phi}\gamma^5)e^{-ip\cdot x/\hbar}\frac{\not{p}+mc}{\sqrt{p^0+mc}}\begin{bmatrix}\phi_s\\0\end{bmatrix}$$

$$= -ie^{-i\phi}e^{-ip\cdot x/\hbar}\frac{-\not{p}+mc}{\sqrt{p^0+mc}}\begin{bmatrix}0\\\phi_s\end{bmatrix} = (-ie^{-i\phi})e^{-ip\cdot x/\hbar}v^{-s}(p) \tag{4.5.80}$$

$$= (-ie^{-i\phi})e^{+ip'\cdot x'/\hbar}v^{s'}(p') = (-ie^{-i\phi})\psi^{\text{antipart}}(p',s':x'),$$

$$\mathcal{CPT}\psi^{\text{antipart}}(p,s;x) = (-ie^{-i\phi}\gamma^5)e^{+ip\cdot x/\hbar}\frac{-\not{p}+mc}{\sqrt{p^0+mc}}\begin{bmatrix}0\\\chi_{-s}\end{bmatrix} \tag{4.5.81}$$

$$= (-ie^{-i\phi})e^{+ip\cdot x/\hbar}u^{-s}(p) = (-ie^{-i\phi})e^{-ip'\cdot x'/\hbar}u^{s'}(p') = (-ie^{-i\phi})\psi^{\text{part}}(p',s':x'),$$

where we recall that under CPT we have $x'^\mu = -x^\mu$, $p'^\mu = p^\mu$ and $s'^\mu = -s^\mu$. Up to an irrelevant phase, the CPT transformed particle plane wave becomes the corresponding antiparticle plane wave and vice versa. As we saw in Eq. (4.5.12) the parity operator and γ^5 anticommute, $\{\mathcal{P},\gamma^5\} = 0$, so that if

$$\mathcal{P}\psi^{\text{part}}(x) = \pi_P^{\text{sp}}\psi^{\text{part}}(x) \quad \text{then} \quad \mathcal{P}\psi^{\text{antipart}}(x) = (-\pi_P^{\text{sp}})\psi^{\text{antipart}}(x). \tag{4.5.82}$$

So, we confirm that particles and their antiparticles have opposite intrinsic parities.

Summary of representation-independent results: Note that in *any* representation of the Dirac equation we have

$$\mathcal{P}^{-1}\gamma^\mu\mathcal{P} = P^\mu{}_\nu\gamma^\nu = g^{\mu\nu}\gamma^\nu \quad \Rightarrow \quad \mathcal{P}\gamma^\mu\mathcal{P}^{-1} = P^\mu{}_\nu\gamma^\nu = g^{\mu\nu}\gamma^\nu,$$

$$\mathcal{C}\gamma^\mu\mathcal{C}^{-1} = (C\gamma^0)\gamma^{\mu*}(C\gamma^0)^{-1} = -\gamma^\mu,$$

$$\mathcal{T}\gamma^\mu\mathcal{T}^{-1} = T\gamma^{\mu*}T^{-1} = (-T^\mu{}_\nu)\gamma^\nu = g^{\mu\nu}\gamma^\nu, \tag{4.5.83}$$

where $\mathcal{C}i\mathcal{C}^{-1} = \mathcal{T}i\mathcal{T}^{-1} = -i$. In any representation we then find

$$\mathcal{CPT}\gamma^\mu(\mathcal{CPT})^{-1} = -\gamma^\mu \quad \text{and} \quad \mathcal{CPT}\gamma^5(\mathcal{CPT})^{-1} = \gamma^5, \tag{4.5.84}$$

since $\gamma^5 = i\gamma^0\gamma^1\gamma^2\gamma^3$. It is readily seen that applying $\mathcal{C}$, $\mathcal{P}$ and $\mathcal{T}$ in any order leads to the same results. Combining Eqs. (4.5.13), (4.5.35) and (4.5.65) we find

$$(\mathcal{CPT})P_\pm(p)(\mathcal{CPT})^{-1} = P_\mp(p) = P_\mp(p'), \tag{4.5.85}$$

$$(\mathcal{CPT})\Sigma(s)(\mathcal{CPT})^{-1} = \Sigma(-s) = \Sigma(s'),$$

where recall that under $\mathcal{CPT}$ we have $p^\mu \to p'^\mu = p^\mu$ and $s^\mu \to s'^\mu = -s^\mu$.

Recall that in any representation we can choose $\mathcal{P} = \pi_P^{\text{sp}} \gamma^0$. It follows from the above results that in any representation

$$(\mathcal{CPT})\mathcal{P}(\mathcal{CPT})^{-1} = -\mathcal{P} \quad \text{so that} \quad \{(\mathcal{CPT}), \mathcal{P}\} = 0, \tag{4.5.86}$$

which is another way to understand that particles and antiparticles have opposite intrinsic parity as shown in Eq. (4.5.82).

4.6 Chirality and Weyl and Majorana Fermions

To complete our discussions of the Dirac equation we need to explain the meaning of chirality and helicity and to define the chiral or Weyl representation of spinors. Then we consider Majorana fermions where a fermion is its own antiparticle. Finally, we introduce the Weyl spinor notation.

4.6.1 Helicity

We introduced the helicity operator in Sec. 1.4.1,

$$S^{\text{helicity}} = \mathbf{S}^{\text{spin}} \cdot \mathbf{n}_p = \mathbf{S}^{\text{spin}} \cdot \mathbf{p}/|\mathbf{p}| = \pm \mathbf{S} \cdot \mathbf{n}_p. \tag{4.6.1}$$

It appeared in Eq. (1.4.19) for massive particles and in Eq. (1.4.29) for massless particles. Recall that the spin operator is $\mathbf{S}^{\text{spin}} = +\mathbf{S}$ and $-\mathbf{S}$ for particles and antiparticles, respectively; $\mathbf{n}_p$ is the direction of motion of the particle/antiparticle; and $S^i = \frac{1}{2}\epsilon^{ijk}\Sigma^{jk}$ where $\Sigma^{\mu\nu}$ are the intrinsic spin generators in Eq. (1.4.2).

For a massive particle of spin s the helicity λ is the spin quantum number m_s when the spin quantization axis is chosen to be the direction of motion of the particle, $\mathbf{n}_s = \mathbf{n}_p$. So it can have any of $2s + 1$ values, $\lambda = m_s = -s, \ldots, s$. For a massless particle the concept of helicity is a much more fundamental one. In the absence of parity invariance the helicity of a massless particle may take only one value, either $\lambda = s$ for a right-handed particle or $\lambda = -s$ for a left-handed particle. A right-handed particle has a left-handed antiparticle and vice-versa; e.g., a massless left-handed neutrino has a corresponding massless right-handed antineutrino. In the presence of parity invariance the helicity of any massless particle or antiparticle can have two values, $\lambda = \pm s$, since a parity transformation reverses helicity.

An eigenstate of helicity for a massless particle has a positive eigenvalue when the spin is parallel to the direction of motion and is said to have a positive helicity and to be right-handed. An eigenstate with a negative eigenvalue has the spin antiparallel to the direction of motion and is said to have negative helicity and to be left-handed. The handedness is assigned by putting the thumb in the direction of motion and wrapping the fingers in the direction of the spin rotation, which selects either the right or left hand for positive or negative helicity, respectively.

We saw in Sec. 1.4.1 that the helicity of a *massive* particle is invariant under continuous (proper) rotations and translations and under collinear boosts that preserve the sign of the momentum. The helicity of a *massless* particle is invariant under any continuous Poincaré transformation, i.e., under any combination of proper rotations, boosts and translations. Recall that under a parity transformation the helicity is reversed, since $\mathbf{S} \to \mathbf{S}$ and $\mathbf{p} \to -\mathbf{p}$, and under time reversal it is conserved, since $\mathbf{S} \to -\mathbf{S}$ and $\mathbf{p} \to -\mathbf{p}$.

4.6.2 Chirality

In three spatial dimensions a parity transformation of all three spatial dimensions is the same as a parity transformation in only one spatial dimension plus a rotation. In both cases the handedness (or orientation) of the coordinate basis is flipped from right-handed to left-handed or vice versa. So *in three spatial dimensions* and in the presence of rotational invariance the following three concepts are equivalent: (i) invariance under a reversal in handedness, (ii) invariance under reflection in a mirror (parity transformation of one spatial dimension) and (iii) invariance under a parity transformation of all three spatial dimensions. The word *chiral* is derived from the Greek word for hand (kheir, or χειρ). We say that something is chiral when it is not equivalent to its mirror image, e.g., a chiral molecule in chemistry. A chiral system in three spatial dimensions necessarily violates parity.

Dirac particles are spin-$\frac{1}{2}$ fermions and in a chiral theory fermions have either right-handed (positive) or left-handed (negative) *chirality*. A spin-$\frac{1}{2}$ fermion with left-handed or right-handed chirality transforms under the left-handed (M) or right-handed ($\overline{M}$) representation of the restricted Lorentz transformations, respectively. These are, respectively, the Type I (left-handed) and Type II (right-handed) two-dimensional representations that were discussed in Sec. 1.3.2. The chirality of a particle describes which of the two representations the particle transforms under. Chiral spin-$\frac{1}{2}$ fermions are massless Dirac particles and are also commonly referred to as either *chiral fermions* or *Weyl fermions*.

Note: *The chirality of a fermion (or antifermion) is the same in all inertial frames connected by restricted Lorentz transformations. The helicity for a massive particle can be changed by a boost and so is frame dependent. Helicity and chirality are equivalent for massless fermions. For massless antifermions helicity and chirality are opposite as we will see.* A mass term in the Dirac equation mixes left and right chiralities and so a chiral fermion cannot have a Dirac mass.

Let us now show how the above is a consequence of the Dirac equation. From Eq. (4.4.110) we know that the massless plane wave solutions satisfy

$$\slashed{p}w(p) = (p^0\gamma^0 - \mathbf{p}\cdot\boldsymbol{\gamma})w(p) = 0 \quad \text{for} \quad w(p) = u^\lambda(p), v^\lambda(p), \tag{4.6.2}$$

where $\lambda = \pm s = \pm\frac{1}{2}$ for right-handed and left-handed chirality, respectively. Using $p^0 = |\mathbf{p}|$ and $(\gamma^0)^2 = I$ we can rewrite the massless Dirac equation as

$$0 = (I - \gamma^0\boldsymbol{\gamma}\cdot\mathbf{n}_p)w(p) = \left[I - (2/\hbar)\gamma_5\mathbf{S}\cdot\mathbf{n}_p\right]w(p), \tag{4.6.3}$$

where we have also used $\gamma_5^2 = I$ and $\mathbf{S} = \frac{1}{2}\hbar\gamma_5\gamma^0\boldsymbol{\gamma}$ from Eq. (4.4.144). When acting on any plane wave solution of the massless Dirac equation we have

$$S^{\text{helicity}} = \mathbf{S}^{\text{spin}}\cdot\mathbf{n}_p = \pm\mathbf{S}\cdot\mathbf{n}_p = \pm\tfrac{1}{2}\hbar\gamma_5, \tag{4.6.4}$$

where the $\pm$ refers to particle plane waves $u^\lambda(p)$ and antiparticle plane waves $v^\lambda(p)$, respectively. We can then consider γ_5 as the *chirality operator* for massless Dirac particles with eigenvalues ± 1 for right-handed and left-handed chirality, respectively.

It is then clear that the spinor matrices

$$P_R \equiv \tfrac{1}{2}\left(1 + \gamma_5\right) \quad \text{and} \quad P_L \equiv \tfrac{1}{2}\left(1 - \gamma_5\right) \tag{4.6.5}$$

are *chirality projectors* and will project particles onto right-handed and left-handed chirality states, respectively, since $\gamma_5 P_R\psi = P_R\psi$ and $\gamma_5 P_L\psi = -P_L\psi$. For antiparticles the roles of P_R and P_L are reversed. For example, when acting on a state in the p^μ-subspace for a massless particle P_R projects out the right-handed helicity components of particles and the left-handed helicity components of antiparticles. The converse is the case for P_L. We can confirm that these operators are projectors since it is easily checked that $P_R + P_L = I$, $P_R^2 = P_R$ and $P_L^2 = P_L$.

Following the discussion in Sec. 1.3.2 let us now construct the left-handed and right-handed representations of the restricted Lorentz transformations $SO^+(1, 3)$, whose elements we write as $M \in SL(2, \mathbb{C})$ and $\overline{M} \in SL(2, \mathbb{C})$ for Type I (left-handed) and Type II (right-handed), respectively. The Lorentz generators for intrinsic spin for Dirac particles are the $\Sigma^{\mu\nu} = \frac{1}{2}\hbar\sigma^{\mu\nu}$. For the internal boost generators we have $\mathbf{K}_{\text{int}}$ and for the rotation generators we have $\mathbf{S}$, which take the place of $\mathbf{K}$ and $\mathbf{J}$, respectively, in the discussions in Sec. 1.3. We have

$$K_{\text{int}}^i = \Sigma^{0i} = \tfrac{1}{2}\hbar\sigma^{0i} \quad \text{and} \quad S^i = \tfrac{1}{2}\hbar\epsilon^{ijk}\Sigma^{jk} = \tfrac{1}{4}\hbar\epsilon^{ijk}\sigma^{jk}. \tag{4.6.6}$$

4.6.3 Weyl Spinors and the Weyl Equations

It is now convenient to work in the *chiral or Weyl representation* of the Dirac matrices discussed in Sec. A.3 and we use the label χ to remind us of this. We find

$$K_{\chi\text{int}}^i = i\frac{\hbar}{2} \begin{bmatrix} -\sigma^i & 0 \\ 0 & \sigma^i \end{bmatrix} \quad \text{and} \quad S_\chi^i = \frac{\hbar}{2} \begin{bmatrix} \sigma^i & 0 \\ 0 & \sigma^i \end{bmatrix}. \tag{4.6.7}$$

Constructing the A and B of Eq. (1.3.1) we obtain

$$\mathbf{A}_\chi \equiv \frac{\mathbf{S}_\chi + i\mathbf{K}_{\chi\text{int}}}{2} = \begin{bmatrix} \frac{\hbar}{2}\sigma & 0 \\ 0 & 0 \end{bmatrix}, \quad \mathbf{B}_\chi \equiv \frac{\mathbf{S}_\chi - i\mathbf{K}_{\chi\text{int}}}{2} = \begin{bmatrix} 0 & 0 \\ 0 & \frac{\hbar}{2}\sigma \end{bmatrix}. \tag{4.6.8}$$

In the chiral representation we have for the γ_5 spinor matrix

$$\gamma_{\chi 5} = \begin{bmatrix} -I & 0 \\ 0 & I \end{bmatrix} \quad \text{and so} \quad P_{\chi R} = \begin{bmatrix} 0 & 0 \\ 0 & I \end{bmatrix}, \quad P_{\chi L} = \begin{bmatrix} I & 0 \\ 0 & 0 \end{bmatrix}. \tag{4.6.9}$$

The following results are true in this chiral representation and will be true in any representation since they are preserved by a similarity transformation,

$$AP_L = P_L A = A, \quad AP_R = P_R A = 0,$$
$$BP_R = P_R B = B, \quad BP_L = P_L B = 0. \tag{4.6.10}$$

So A generates the Lorentz transformations for left-handed chirality and B generates the Lorentz transformations for right-handed chirality. We now focus on the upper left and lower right submatrices making up $\mathbf{K}_{\chi\text{int}}$ and $\mathbf{S}_\chi$ by defining

$$\mathbf{K}_{\chi\text{int}} \equiv \begin{bmatrix} \mathbf{K}^{2\times2} & 0 \\ 0 & \overline{\mathbf{K}}^{2\times2} \end{bmatrix} \quad \text{and} \quad \mathbf{S}_\chi \equiv \begin{bmatrix} \mathbf{S}^{2\times2} & 0 \\ 0 & \overline{\mathbf{S}}^{2\times2} \end{bmatrix}. \tag{4.6.11}$$

We similarly define $\mathbf{A}^{2\times2}$, $\overline{\mathbf{A}}^{2\times2}$ and $\mathbf{B}^{2\times2}$, $\overline{\mathbf{B}}^{2\times2}$ as the diagonal blocks of $\mathbf{A}_\chi$ and $\mathbf{B}_\chi$, respectively. For the left-handed $(\frac{1}{2}, 0)$ and right-handed $(0, \frac{1}{2})$ two-dimensional representations of the intrinsic Lorentz transformations we have, respectively,

$$(\tfrac{1}{2}, 0)\colon \mathbf{A}^{2\times2} = (\hbar/2)\sigma, \ \mathbf{B}^{2\times2} = 0 \ \Rightarrow \ \mathbf{K}^{2\times2} = -i\mathbf{S}^{2\times2} = -i(\hbar/2)\sigma,$$
$$(0, \tfrac{1}{2})\colon \overline{\mathbf{A}}^{2\times2} = 0, \ \overline{\mathbf{B}}^{2\times2} = (\hbar/2)\sigma \ \Rightarrow \ \overline{\mathbf{K}}^{2\times2} = +i\overline{\mathbf{S}}^{2\times2} = +i(\hbar/2)\sigma. \tag{4.6.12}$$

Finally, then we can write an arbitrary internal Lorentz transformation as

$$S_\chi(\Lambda) = \exp[(-i\alpha\mathbf{n} \cdot \mathbf{S}_\chi - i\eta\mathbf{n}_b \cdot \mathbf{K}_{\chi\text{int}})/\hbar] = \begin{bmatrix} M & 0 \\ 0 & \overline{M} \end{bmatrix}. \tag{4.6.13}$$

With $\eta = 0$ this corresponds to an active rotation of angle α about the rotation axis $\mathbf{n}$ and if $\alpha = 0$ this corresponds to an active boost of rapidity η in the boost direction $\mathbf{n}_b$. From Eq. (4.6.13) we can now read off the left-handed (M) and right-handed ($\overline{M}$) components of the intrinsic Lorentz transformation and obtain the results in Eq. (1.3.9),

$$M(\Lambda) = \exp\left[(-i\alpha\mathbf{n} - \eta\mathbf{n}_b) \cdot \boldsymbol{\sigma}/2\right] \quad \text{and} \quad \overline{M}(\Lambda) = \exp\left[(-i\alpha\mathbf{n} + \eta\mathbf{n}_b) \cdot \boldsymbol{\sigma}/2\right], \qquad (4.6.14)$$

where the restricted Lorentz transformation Λ is specified by $(\alpha\mathbf{n}, \eta\mathbf{n}_b)$. Note that

$$\overline{M}(\Lambda) = M(\Lambda)^{-1\dagger}, \qquad (4.6.15)$$

since for any square matrix A we have $e^A e^{-A} = I$. For an invertible A we have $A^{-1\dagger} = (A^\dagger)^{-1}$, since $I = AA^{-1} = A^{-1}A \Rightarrow A^\dagger A^{-1\dagger} = (A^{-1}A)^\dagger = I = A^\dagger (A^\dagger)^{-1}$ and acting from the left with $(A^\dagger)^{-1}$ gives the result. Similarly, $A^{-1T} = (A^T)^{-1}$.

Comparing with Eq. (A.7.1) we recognize that intrinsic restricted Lorentz transformations $S_\chi(\Lambda)$ are the direct sum of the $(\frac{1}{2}, 0)$ left-handed representation of $SL(2, \mathbb{C})$ with group elements M and the $(0, \frac{1}{2})$ right-handed representation with elements $\overline{M}$. In other representations the decomposition into the direct sum remains true, of course, but the usefulness of the chiral representation is that it makes this obvious. Then in any representation we have that $S(\Lambda)$ belongs to the $(\frac{1}{2}, 0) \oplus (0, \frac{1}{2})$ representation of the restricted Lorentz transformations. This was confirmed by explicit construction in the discussion in Sec. 1.3.2.

In the chiral representation any Dirac spinor $\psi(x)$ can be written as a doublet of two-component *Weyl spinors*. We have

$$\psi(x) = \begin{bmatrix} \psi_L(x) \\ \psi_R(x) \end{bmatrix} \quad \text{with} \quad P_L\psi(x) = \begin{bmatrix} \psi_L(x) \\ 0 \end{bmatrix}, \ P_R\psi(x) = \begin{bmatrix} 0 \\ \psi_R(x) \end{bmatrix}. \qquad (4.6.16)$$

In this representation restricted Lorentz transformations in Eq. (4.4.66) are

$$\text{Type I:} \ (\tfrac{1}{2}, 0) \quad \psi_L(x) \xrightarrow{\Lambda} \psi_L'(x') = \psi_L'(\Lambda x) = M(\Lambda)\psi_L(x),$$

$$\text{Type II:} \ (0, \tfrac{1}{2}) \quad \psi_R(x) \xrightarrow{\Lambda} \psi_R'(x') = \psi_R'(\Lambda x) = \overline{M}(\Lambda)\psi_R(x), \qquad (4.6.17)$$

where we use the notation introduced in Eq. (1.3.8).

We can rewrite the Dirac equation in Eq. (4.4.59) in terms of Weyl spinors in the chiral/Weyl representation using $\sigma^\mu \equiv (I, \boldsymbol{\sigma})$ as

$$0 = (i\hbar\slashed{\partial} - mc)\psi(x) = (i\hbar\gamma_\chi^\mu \partial_\mu - mc)\begin{bmatrix} \psi_L(x) \\ \psi_R(x) \end{bmatrix} = \begin{bmatrix} mc & i\hbar\sigma\cdot\partial \\ i\hbar\bar{\sigma}\cdot\partial & -mc \end{bmatrix}\begin{bmatrix} \psi_L(x) \\ \psi_R(x) \end{bmatrix},$$
$$(4.6.18)$$

where we have used Eq. (A.3.14) for the chiral/Weyl representation of γ^μ. We note that in the massless limit this decouples into two separate equations,

$$i\hbar\bar{\sigma} \cdot \partial\psi_L(x) - mc\psi_R(x) = 0 \quad \text{and} \quad i\hbar\sigma \cdot \partial\psi_R(x) + mc\psi_L(x) = 0, \qquad (4.6.19)$$

$$\xrightarrow{m=0} \quad i\hbar\bar{\sigma} \cdot \partial\psi_L(x) = 0 \quad \text{and} \quad i\hbar\sigma \cdot \partial\psi_R(x) = 0, \qquad (4.6.20)$$

where Eq. (4.6.20) are the *Weyl equations* of motion for free massless chiral/Weyl fermions in the chiral/Weyl representation.

In this representation Eq. (4.4.62), $S(\Lambda)^{-1}\gamma^\mu S(\Lambda) = \Lambda^\mu{}_\nu \gamma^\nu$, becomes

$$\begin{bmatrix} M(\Lambda)^{-1} & 0 \\ 0 & \overline{M}(\Lambda)^{-1} \end{bmatrix}\begin{bmatrix} 0 & \sigma^\mu \\ \bar{\sigma}^\mu & 0 \end{bmatrix}\begin{bmatrix} M(\Lambda) & 0 \\ 0 & \overline{M}(\Lambda) \end{bmatrix} = \Lambda^\mu{}_\nu\begin{bmatrix} 0 & \sigma^\nu \\ \bar{\sigma}^\nu & 0 \end{bmatrix}, \qquad (4.6.21)$$

which gives

$$M(\Lambda)^{-1}\sigma^\mu \overline{M}(\Lambda) = \Lambda^\mu{}_\nu \sigma^\nu \qquad \text{and} \qquad \overline{M}(\Lambda)^{-1}\bar\sigma^\mu M(\Lambda) = \Lambda^\mu{}_\nu \bar\sigma^\nu. \qquad (4.6.22)$$

It is then easy to see that the Weyl equations are Lorentz covariant in analogy with Eq. (4.4.64). If the Weyl equation for $\psi_L(x)$ is true in the frame of $\mathcal{O}'$, then

$$\begin{aligned}
0 &= i\hbar\bar\sigma \cdot \partial' \psi_L'(x') = i\hbar\bar\sigma^\mu (\Lambda^{-1})^\nu{}_\mu \partial_\nu \psi_L'(x') \\
&= i\hbar \overline{M}(\Lambda)\bar\sigma^\nu M(\Lambda^{-1}) M(\Lambda)\partial_\nu \psi_L(x) = \overline{M}(\Lambda)i\hbar\bar\sigma \cdot \partial \psi_L(x)
\end{aligned} \qquad (4.6.23)$$

and after acting from the left with $\overline{M}(\Lambda)^{-1}$ we find $i\hbar\bar\sigma \cdot \partial \psi_L(x) = 0$ in the frame of $\mathcal{O}$. If the Weyl equation is true in one frame, then it is true in every frame related to it by a restricted Lorentz transformation. The proof of the restricted Lorentz covariance of the Weyl equation for $\psi_R(x)$ is performed similarly.

4.6.4 Plane Wave Solutions in the Chiral Representation

We have obtained solutions of the Dirac equation in the Dirac representation, but it will be useful to also construct plane wave solutions in the chiral representation. From Sec. A.3 we have for the Dirac equation in the chiral representation that

$$(\not{p}_\chi - mc)u_\chi^s(p) = \begin{bmatrix} -mcI & p\cdot\sigma \\ p\cdot\bar\sigma & -mcI \end{bmatrix} u_\chi^s(p) = 0,$$

$$(\not{p}_\chi + mc)v_\chi^s(p) = \begin{bmatrix} mcI & p\cdot\sigma \\ p\cdot\bar\sigma & mcI \end{bmatrix} v_\chi^s(p) = 0. \qquad (4.6.24)$$

We can use the unitary transformation of Eq. (A.3.13) to transform the rest frame spinors of Eq. (4.4.116) and obtain rest frame spinors in the chiral representation,

$$u_\chi^s(0) = Uu^s(0) = \sqrt{mc}\begin{bmatrix} \phi_s \\ \phi_s \end{bmatrix} \quad \text{and} \quad v_\chi^s(0) = Uv^s(0) = \sqrt{mc}\begin{bmatrix} -\chi_{-s} \\ \chi_{-s} \end{bmatrix}. \qquad (4.6.25)$$

In the rest frame we have $p\cdot\sigma = p\cdot\bar\sigma = p^0 I = mcI$ and so we readily verify that these rest-frame spinors satisfy Eq. (4.6.24). It is also easy to verify that these rest-frame spinors satisfy the normalization and orthogonality conditions given in Eqs. (4.4.116), (4.4.119) and (4.4.125) as they must since U is unitary.

To obtain $u^s(p)$ and $v^s(p)$ we must boost the rest-frame solutions in the usual way. Starting either from Eq. (4.4.97) in the chiral representation or equivalently from Eqs. (4.6.13) and (4.6.14) we have for a boost of rapidity η

$$S_\chi(\Lambda) = I\cosh(\eta/2) + (\mathbf{n}\cdot\boldsymbol{\alpha}_\chi)\sinh(\eta/2) \qquad (4.6.26)$$

$$= \begin{bmatrix} \sigma^0\cosh(\eta/2) - \mathbf{n}\cdot\boldsymbol{\sigma}\sinh(\eta/2) & 0 \\ 0 & \sigma^0\cosh(\eta/2) + \mathbf{n}\cdot\boldsymbol{\sigma}\sinh(\eta/2) \end{bmatrix},$$

where $\sigma^0 = I_{2\times2}$ and $\boldsymbol{\alpha}_\chi = \gamma_\chi^0\boldsymbol{\gamma}_\chi$. From Eqs. (1.2.92) and (1.2.93) we have

$$mc\left[\sigma^0\cosh\eta \mp \mathbf{n}\cdot\boldsymbol{\sigma}\sinh\eta\right] = \gamma mc(\sigma^0 \mp \beta\mathbf{n}\cdot\boldsymbol{\sigma}) = p^0\sigma^0 \mp \mathbf{p}\cdot\boldsymbol{\sigma} = \begin{cases} p\cdot\sigma \\ p\cdot\bar\sigma \end{cases}$$

It then follows that

$$\left.\begin{array}{c} \sqrt{p\cdot\sigma} \\ \sqrt{p\cdot\bar\sigma} \end{array}\right\} = \sqrt{mc}\left[\sigma^0\cosh(\eta/2) \mp \mathbf{n}\cdot\boldsymbol{\sigma}\sinh(\eta/2)\right], \qquad (4.6.27)$$

since $\left[\sigma^0 \cosh(\eta/2) \mp \mathbf{n} \cdot \boldsymbol{\sigma} \sinh(\eta/2)\right]^2 = \sigma^0 \cosh\eta \mp \mathbf{n} \cdot \boldsymbol{\sigma} \sinh\eta$, where we have used the hyperbolic relations in Eq. (1.2.103) and $\sigma^0 = (\mathbf{n} \cdot \boldsymbol{\sigma})^2 = I_{2 \times 2}$ from Eqs. (A.3.15) and (A.3.2). Also note that

$$(p \cdot \sigma)(p \cdot \overline{\sigma}) = (p^0)^2 (\sigma^0)^2 - (\mathbf{p} \cdot \boldsymbol{\sigma})^2 = p^2 = m^2 c^2, \qquad (4.6.28)$$

where a 2×2 identity matrix is understood on the right-hand side. Combining Eqs. (4.6.26) and (4.6.27) we have for the Lorentz boost

$$S_\chi(\Lambda) = \frac{1}{\sqrt{mc}} \left[\begin{array}{cc} \sqrt{p \cdot \sigma} & 0 \\ 0 & \sqrt{p \cdot \overline{\sigma}} \end{array} \right],$$

which gives the plane wave spinors in the chiral representation,

$$u_\chi^s(p) = S_\chi(\Lambda) u_\chi^s(0) = \left[\begin{array}{c} \sqrt{p \cdot \sigma}\,\phi_s \\ \sqrt{p \cdot \overline{\sigma}}\,\phi_s \end{array} \right], \quad v_\chi^s(p) = S_\chi(\Lambda) v_\chi^s(0) = \left[\begin{array}{c} -\sqrt{p \cdot \sigma}\,\chi_{-s} \\ \sqrt{p \cdot \overline{\sigma}}\,\chi_{-s} \end{array} \right]. \qquad (4.6.29)$$

Peskin and Schroeder (1995) and Schwartz (2013) use $v_\chi^s(p) \to -v_\chi^s(p)$ but either sign is acceptable in calculations since the phase has no physical significance.

Choose the spin quantization axis in the $+z$-direction and boost along the z-axis from the rest frame to the frame where $\mathbf{p} = (0, 0, p_3)$ with $p_3 = \pm|\mathbf{p}|$, then

$$\left. \begin{array}{c} \sqrt{p \cdot \sigma} \\ \sqrt{p \cdot \overline{\sigma}} \end{array} \right\} = \left[\begin{array}{cc} \sqrt{p^0 \mp p^3} & 0 \\ 0 & \sqrt{p^0 \pm p^3} \end{array} \right], \qquad (4.6.30)$$

$$\sqrt{p \cdot \sigma} \left[\begin{array}{c} 1 \\ 0 \end{array} \right] = \sqrt{p^0 - p^3} \left[\begin{array}{c} 1 \\ 0 \end{array} \right] \quad \text{and} \quad \sqrt{p \cdot \sigma} \left[\begin{array}{c} 0 \\ 1 \end{array} \right] = \sqrt{p^0 + p^3} \left[\begin{array}{c} 0 \\ 1 \end{array} \right], \qquad (4.6.31)$$

where for $\sqrt{p \cdot \overline{\sigma}}$ we replace $p^3 \to -p^3$. Choosing $p_3 = |\mathbf{p}|$ for a particle with spin up corresponds to positive helicity, while for spin down this will be a negative helicity state. We label these states as $u^+(p)$ and $u^-(p)$, respectively. Maintaining the spin axis in the same direction, a spin-up and spin-down antiparticle state has negative and positive helicity, respectively, with the choice $p_3 = -|\mathbf{p}|$. We label these states as $v^-(p)$ and $v^+(p)$, respectively. Summarizing these results we have

$$u_\chi^+ = \left[\begin{array}{c} \sqrt{p^0 - |\mathbf{p}|} \\ 0 \\ \sqrt{p^0 + |\mathbf{p}|} \\ 0 \end{array} \right], \; u_\chi^- = \left[\begin{array}{c} 0 \\ \sqrt{p^0 + |\mathbf{p}|} \\ 0 \\ \sqrt{p^0 - |\mathbf{p}|} \end{array} \right], \; v_\chi^+ = \left[\begin{array}{c} -\sqrt{p^0 + |\mathbf{p}|} \\ 0 \\ \sqrt{p^0 - |\mathbf{p}|} \\ 0 \end{array} \right], \; v_\chi^- = \left[\begin{array}{c} 0 \\ -\sqrt{p^0 - |\mathbf{p}|} \\ 0 \\ \sqrt{p^0 + |\mathbf{p}|} \end{array} \right]. \qquad (4.6.32)$$

For ultrarelativistic fermions $\sqrt{p^0 + |\mathbf{p}|} \to \sqrt{2p^0}$ and $\sqrt{p^0 - |\mathbf{p}|} \to 0$, which gives

$$u_\chi^+(p) \to \sqrt{\frac{2E}{c}} \left[\begin{array}{c} 0 \\ 0 \\ 1 \\ 0 \end{array} \right], \; u_\chi^-(p) \to \sqrt{\frac{2E}{c}} \left[\begin{array}{c} 0 \\ 1 \\ 0 \\ 0 \end{array} \right], \; v_\chi^+(p) \to -\sqrt{\frac{2E}{c}} \left[\begin{array}{c} 1 \\ 0 \\ 0 \\ 0 \end{array} \right], \; v_\chi^-(p) \to \sqrt{\frac{2E}{c}} \left[\begin{array}{c} 0 \\ 0 \\ 0 \\ 1 \end{array} \right]. \qquad (4.6.33)$$

In Eq. (4.6.33) $u_\chi^-(p)$ has negative helicity (and left-handed chirality) and its antiparticle state $v^+(p)$ has positive helicity (and left-handed chirality) since both are in the upper two components and so transform under $M(\Lambda)$. An example of this is the left-handed massless neutrino and its right-handed antineutrino, since for a massless particle $p^0 = |\mathbf{p}|$. Conversely, $u_\chi^+(p)$ has positive helicity (and right-handed chirality) and its antiparticle state $v^-(p)$ has negative helicity (and right-handed chirality) since both are in the lower two components and so transform under $\overline{M}(\Lambda)$.

4.6.5 Majorana Fermions

We can now consider the consequences of reducing the space containing the solutions of the Dirac equation to a smaller space. We can do this by imposing the condition that the solutions of the Dirac equation are real, $\psi^{(r)}(x)^* = \psi^{(r)}(x)$. Since the Dirac equation is $(i\gamma^\mu\partial_\mu - m)\psi(x) = 0$, then we can impose this condition if the matrices γ^μ are pure imaginary. This is satisfied by the *Majorana representation* in Sec. A.3, where the matrices $i\gamma_M^\mu$ are real. The Dirac equation can then take the purely real form referred to as the *Majorana equation*,

$$(i\gamma_M^\mu\partial_\mu - m)\psi^{(r)} = 0. \tag{4.6.34}$$

Any two fermion wavefunctions differing only by a phase ϕ are physically equivalent. So $\psi(x) = \exp(i\phi)\psi^{(r)}$ is equivalent to a real solution and satisfies the Majorana equation. Fermions with this property are referred to as *Majorana fermions* or *Majorana particles*. Since $\psi^{(r)}(x)$ has four real components, then any Majorana fermion wavefunction is specified by four real functions, $\psi_1(x), \ldots, \psi_4(x) \in \mathbb{R}$.

A Dirac equation solution equal to charge conjugate up to a phase

$$\psi(x) = e^{i\phi}\psi_C(x) \tag{4.6.35}$$

is referred to as *self-conjugate*. Recall from Eq. (4.5.25) that the charge conjugate wavefunction is given by $\psi_C(x) = C\gamma^0\psi^*(x)$. From Sec. A.3 we have in the Majorana representation $\psi_C(x) = \mathcal{C}\psi(x) = C_M\gamma_M^0\psi^*(x) = -i\psi^*(x)$ and so a real wavefunction in the Majorana representation is self-conjugate. This means that solutions to the Majorana equation are self-conjugate. As discussed in Sec. 4.1.10 an antiparticle is by definition the charge conjugate of its particle and so *Majorana fermions are their own antiparticles*. Any particle that is its own antiparticle has no additive quantum numbers; e.g., a Majorana fermion cannot have any electric charge. We can also see this directly from Eq. (4.5.19), where the particle and its charge conjugate can only satisfy the same equation if there is no coupling to the electromagnetic field, $q = 0$. A Dirac neutrino can have a conserved lepton number, whereas the existence of a massive Majorana neutrino would imply violation of lepton number by two units when two Majorana neutrinos annihilate.

Consider the Dirac equation in any representation, $(i\slashed{\partial} - mc)\chi(x) = 0$ with $\chi(x)$ a solution. Eq. (4.5.19) shows that $\chi_C(x) = C\gamma^0\chi^*(x)$ is also a solution. So we can construct a self-conjugate solution to the Dirac equation from any solution $\chi(x)$ as

$$\psi(x) \equiv \chi(x) + e^{i\phi}\chi_C(x) \text{ giving } \psi_C(x) = \chi_C(x) + e^{-i\phi}\chi(x) = e^{-i\phi}\psi(x) \tag{4.6.36}$$

with ϕ an arbitrary phase; i.e., we have $\psi(x) = e^{i\phi}\psi_C(x)$ as required.

Majorana fermions in the chiral (Weyl) representation: In the chiral representation we see from Eqs. (4.5.25) and (4.6.16) that for any spin-$\frac{1}{2}$ particle

$$\psi(x) = \begin{bmatrix} \psi_L(x) \\ \psi_R(x) \end{bmatrix}, \ \psi_C(x) = \begin{bmatrix} \psi_{CL}(x) \\ \psi_{CR}(x) \end{bmatrix} = C_\chi\gamma_\chi^0 \begin{bmatrix} \psi_L^*(x) \\ \psi_R^*(x) \end{bmatrix} = \begin{bmatrix} i\sigma^2\psi_R^*(x) \\ -i\sigma^2\psi_L^*(x) \end{bmatrix}, \tag{4.6.37}$$

where we used Eqs. (A.3.14) and (A.3.23). We observe that the charge conjugate of any left-handed spin-$\frac{1}{2}$ fermion is right-handed and that of any right-handed one is left-handed; i.e., if $\psi_R = 0$ then $\psi_{CL} = 0$ and if $\psi_L = 0$ then $\psi_{CR} = 0$.

Since a Majorana fermion is self-conjugate, $\psi(x) = e^{i\phi}\psi_C(x)$, it satisfies

$$\psi_L(x) = e^{i\phi}i\sigma^2\psi_R^*(x) \quad \text{and} \quad \psi_R(x) = -e^{i\phi}i\sigma^2\psi_L^*(x), \tag{4.6.38}$$

in the chiral representation, where $i\sigma^2$ is a real 2×2 matrix and ϕ is an arbitrary constant phase. As a check observe that $\psi_L(x) = e^{i\phi}i\sigma^2(-e^{i\phi}i\sigma^2\psi_L^*(x))^* = \psi_L(x)$ and similarly for $\psi_R(x)$. All information carried by a Majorana fermion wavefunction is contained in each of $\psi_L(x)$ and $\psi_R(x)$. From Eq. (4.6.18) we see that the Dirac equation in the chiral representation for a Majorana fermion is

$$0 = (i\gamma_\chi^\mu\partial_\mu - m)\begin{bmatrix} \psi_L(x) \\ \psi_R(x) \end{bmatrix} = \begin{bmatrix} -m & i\sigma \cdot \partial \\ i\bar{\sigma} \cdot \partial & -m \end{bmatrix}\begin{bmatrix} \psi_L(x) \\ \psi_R(x) \end{bmatrix} \tag{4.6.39}$$

and so we must have for any Majorana fermion that

$$0 = i\bar{\sigma} \cdot \partial\psi_L(x) - m\psi_R(x) = i\bar{\sigma} \cdot \partial\psi_L(x) + me^{i\phi}i\sigma^2\psi_L^*(x) \tag{4.6.40}$$

with a similar equation for $\psi_R(x)$. We note that specifying the two-component spinor $\psi_L(x)$, or alternatively $\psi_R(x)$, requires specifying two complex fields or equivalently four real fields. This is the same as the number of real fields in a real solution to the Dirac equation in Majorana representation as it should be.

4.6.6 Weyl Spinor Notation

The fermions in QED and QCD are massive leptons and massive quarks, respectively. We typically describe these theories in terms of massive Dirac fermions. In the weak interactions and in QED and QCD at extremely high energies, fermion masses are negligible and the chiral basis of Weyl fermions is convenient. In such situations helicity becomes an approximate form of chirality. This is why a chirality basis for spin is often convenient for describing high-energy processes. It is also the most covenient basis for chiral gauge theories such as the Standard Model, in which all fermions are massless before the Higgs mechanism induces mass. It is then useful to have an efficient notation for two-component Weyl (chiral) fermions. This is the *Weyl spinor notation*, which is also known as van der Waerden notation and was introduced by Bartel Leendert van der Waerden in 1929. This notation is used in discussions of supersymmetry (SUSY) (Wess and Bagger, 1992; Bailin and Love, 1994; Terning, 2006; Dreiner et al., 2010; Schwartz, 2013) and also in the twistor theory approach to quantum gravity introduced by Roger Penrose.

In this discussion we work exclusively in the chiral (Weyl) representation of the Dirac γ-matrices. Let Ψ^D be an arbitrary Dirac four-component spinor, then

$$\Psi^D = P_L\Psi^D + P_R\Psi^D = \Psi_{\text{left}}^D + \Psi_{\text{right}}^D = \begin{bmatrix} \Psi_L \\ 0 \end{bmatrix} + \begin{bmatrix} 0 \\ \Psi_R \end{bmatrix} = \begin{bmatrix} \Psi_L \\ \Psi_R \end{bmatrix}, \tag{4.6.41}$$

where P_L and P_R are defined in Eq. (4.6.5). We have Weyl spinor notation

$$\Psi_{\text{left}}^D = \begin{bmatrix} \Psi_L \\ 0 \end{bmatrix} \equiv \begin{bmatrix} \psi_\alpha \\ 0 \end{bmatrix} \quad \text{and} \quad \Psi_{\text{right}}^D = \begin{bmatrix} 0 \\ \Psi_R \end{bmatrix} \equiv \begin{bmatrix} 0 \\ \bar{\chi}^{\dot{\beta}} \end{bmatrix}, \tag{4.6.42}$$

where $\alpha, \beta = 1, 2$. Using an Einstein-type notation we write

$$\Psi^D \equiv \begin{bmatrix} \psi \\ \bar{\chi} \end{bmatrix} \equiv \begin{bmatrix} \psi_\alpha \\ \bar{\chi}^{\dot{\beta}} \end{bmatrix}, \quad \Psi_L \equiv \psi_\alpha = \begin{bmatrix} \psi_1 \\ \psi_2 \end{bmatrix} \quad \text{and} \quad \Psi_R \equiv \bar{\chi}^{\dot{\beta}} \equiv \begin{bmatrix} \bar{\chi}^{\dot{1}} \\ \bar{\chi}^{\dot{2}} \end{bmatrix}. \tag{4.6.43}$$

The left-handed two-component spinors use an undotted label and right-handed two-component spinors use a dotted label. One should be careful to avoid confusion between the "bar" notation used here in Weyl spinor notation and the standard Dirac "bar" notation $\bar{\psi} \equiv \psi^\dagger \gamma^0$ used elsewhere. From Eq. (4.6.17) we see that $\psi_\alpha \equiv \Psi_L^D$ transforms with $M(\Lambda)$ and $\bar{\chi}^{\dot{\alpha}} \equiv \Psi_R^D$ transforms with $\overline{M}(\Lambda)$.

The charge conjugate of the Dirac spinor Ψ^D in the chiral representation is

$$\Psi_C^D = C_\chi \gamma_\chi^0 \Psi^{D*} = \begin{bmatrix} i\sigma^2 \bar{\chi}^* \\ -i\sigma^2 \psi^* \end{bmatrix} \quad \text{and as required} \quad (\Psi_C^D)_C = \Psi^D, \qquad (4.6.44)$$

where we have used Eq. (4.6.37), $(\sigma^2)^* = -\sigma^2$ and $(\sigma^2)^2 = I$. Since Ψ_C^D is also a Dirac spinor by construction, then it must be that $i\sigma^2 \bar{\chi}^*$ transforms with $M(\Lambda)$ like ψ and $-i\sigma^2 \psi^*$ transforms with $\overline{M}(\Lambda)$ like $\bar{\chi}$. This can also be verified directly. We capture this fact by first defining

$$\bar{\psi}_{\dot{\alpha}} \equiv (\psi_\alpha)^* \quad \text{and} \quad \chi^\alpha \equiv (\bar{\chi}^{\dot{\alpha}})^* \qquad (4.6.45)$$

and then defining

$$\bar{\psi}^{\dot{\alpha}} \equiv \epsilon^{\dot{\alpha}\dot{\beta}} \bar{\psi}_{\dot{\beta}} \quad \text{and} \quad \chi_\alpha \equiv \epsilon_{\alpha\beta} \chi^\beta \quad \text{with}$$

$$\epsilon^{\alpha\beta} = \epsilon^{\dot{\alpha}\dot{\beta}} = -\epsilon_{\alpha\beta} = -\epsilon_{\dot{\alpha}\dot{\beta}} = i\sigma^2 = \begin{bmatrix} 0 & 1 \\ -1 & 0 \end{bmatrix}, \qquad (4.6.46)$$

which we understand to mean $\epsilon^{12} = -\epsilon^{21} = -\epsilon_{12} = \epsilon_{21} = 1$ and equivalently for the dotted indices. Note that $\epsilon^{\alpha\beta}$ and $\epsilon^{\dot{\alpha}\dot{\beta}}$ raise dotted and undotted indices, respectively, while $\epsilon_{\alpha\beta}$ and $\epsilon_{\dot{\alpha}\dot{\beta}}$ lower them. Also note that by definition they *always act from the left*, which is relevant since $\epsilon_{\alpha\beta} = -\epsilon_{\beta\alpha}$. We have chosen $\epsilon^{\alpha\beta}$ and $\epsilon_{\dot{\alpha}\dot{\beta}}$ so that

$$\epsilon_{\alpha\beta}\epsilon^{\beta\gamma} = \delta_\alpha{}^\gamma \quad \text{and} \quad \epsilon_{\dot{\alpha}\dot{\beta}}\epsilon^{\dot{\beta}\dot{\gamma}} = \delta_{\dot{\alpha}}{}^{\dot{\gamma}} \qquad (4.6.47)$$

as required for consistency, e.g., $\psi_\alpha = \epsilon_{\alpha\beta}\psi^\beta = \epsilon_{\alpha\beta}\epsilon^{\beta\gamma}\psi_\gamma = \delta_\alpha{}^\gamma\psi_\gamma$. The charge conjugate of the Dirac spinor Ψ^D is then

$$\Psi_C^D = \begin{bmatrix} i\sigma^2 \bar{\chi}^* \\ -i\sigma^2 \psi^* \end{bmatrix} = \begin{bmatrix} -\epsilon_{\alpha\beta}\chi^\beta \\ -\epsilon^{\dot{\alpha}\dot{\beta}}\bar{\psi}_{\dot{\beta}} \end{bmatrix} = \begin{bmatrix} -\chi_\alpha \\ -\bar{\psi}^{\dot{\beta}} \end{bmatrix} = \begin{bmatrix} -\chi \\ -\bar{\psi} \end{bmatrix} \xrightarrow{\text{redefine}} \begin{bmatrix} \chi \\ \bar{\psi} \end{bmatrix}, \qquad (4.6.48)$$

where it is conventional to choose the redefined sign for Ψ_C^D. We can change the sign of Ψ_C^D, which equates to choosing the sign of the charge conjugation matrix C.

From Eq. (4.6.43) we see that a left-handed Weyl spinor has $\chi = 0$ and a right-handed one has $\psi = 0$. We are free to choose the phase in Eq. (4.6.35) as $\phi = 0$ and with this choice a Majorana fermion Ψ^M satisfies $\Psi^M = \Psi_C^M$ and then $\psi = \chi$,

$$\Psi^M = \begin{bmatrix} \psi \\ \bar{\psi} \end{bmatrix} = \begin{bmatrix} \psi_\alpha \\ \bar{\psi}^{\dot{\beta}} \end{bmatrix}. \qquad (4.6.49)$$

Given any four-component Dirac spinor ψ^D we can immediately construct two Majorana spinors, $\Psi^{M_1} = \Psi_C^{M_1}$ and $\Psi^{M_2} = \Psi_C^{M_2}$, where

$$\Psi^{M_1} \equiv \tfrac{1}{2}\left(\Psi^D + \Psi_C^D\right) \quad \text{and} \quad \Psi^{M_2} \equiv -i\tfrac{1}{2}\left(\Psi^D - \Psi_C^D\right) \qquad (4.6.50)$$

such that any Dirac spinor can be decomposed in terms of them,

$$\Psi^D = \Psi^{M_1} + i\Psi^{M_2} \quad \text{and} \quad \Psi_C^D = \Psi^{M_1} - i\Psi^{M_2}. \qquad (4.6.51)$$

The Weyl notation conveniently allows for the construction of Lorentz-covariant quantities from Weyl spinors. We can rewrite Eq. (4.6.17) as

$$\psi'_\alpha(x') = M(\Lambda)_\alpha{}^\beta \psi_\beta(x) \quad \text{and} \quad \bar{\chi}'^{\dot\alpha}(x') = \overline{M}(\Lambda)^{\dot\alpha}{}_{\dot\beta} \bar{\chi}^{\dot\beta}(x). \tag{4.6.52}$$

Using Eq. (4.6.45) we find that

$$\bar{\psi}'_{\dot\alpha}(x') = [\psi'_\alpha(x')]^* = [M(\Lambda)_\alpha{}^\beta \psi_\beta(x)]^* = M(\Lambda)^*{}_{\dot\alpha}{}^{\dot\beta} \bar{\psi}_{\dot\beta}(x),$$

$$\chi'^\alpha(x') = [\bar{\chi}'^{\dot\alpha}(x')]^* = [\overline{M}(\Lambda)^{\dot\alpha}{}_{\dot\beta} \bar{\chi}^{\dot\beta}(x)]^* = \overline{M}(\Lambda)^{*\alpha}{}_\beta \chi^\beta(x), \tag{4.6.53}$$

where we have defined

$$M(\Lambda)^*{}_{\dot\alpha}{}^{\dot\beta} \equiv [M(\Lambda)_\alpha{}^\beta]^* \quad \text{and} \quad \overline{M}(\Lambda)^{*\alpha}{}_\beta \equiv [\overline{M}(\Lambda)^{\dot\alpha}{}_{\dot\beta}]^*. \tag{4.6.54}$$

Using Eq. (4.6.15) we see that $\overline{M}(\Lambda)^* = M(\Lambda)^{-1T}$ and $M(\Lambda)^* = \overline{M}(\Lambda)^{-1T}$, where we recall that $A^{-1T} = (A^T)^{-1}$. Then we can write

$$\bar{\psi}'_{\dot\alpha}(x') = (\overline{M}(\Lambda)^{-1T})_{\dot\alpha}{}^{\dot\beta} \bar{\psi}_{\dot\beta}(x) = \bar{\psi}_{\dot\beta}(x)(\overline{M}(\Lambda)^{-1})^{\dot\beta}{}_{\dot\alpha},$$

$$\chi'^\alpha(x') = (M(\Lambda)^{-1T})^\alpha{}_\beta \chi^\beta(x) = \chi^\beta(x)(M(\Lambda)^{-1})_\beta{}^\alpha. \tag{4.6.55}$$

From Eq. (4.6.46) we see that

$$\psi'_\alpha(x') = \epsilon_{\alpha\beta} \psi'^\beta(x') - \epsilon_{\alpha\beta} \overline{M}(\Lambda)^{*\beta}{}_\gamma \psi^\gamma(x) = \epsilon_{\alpha\beta} \overline{M}(\Lambda)^{*\beta}{}_\gamma \epsilon^{\gamma\delta} \psi_\delta(x),$$

$$\bar{\chi}'^{\dot\alpha}(x') = \epsilon^{\dot\alpha\dot\beta} \bar{\chi}'_{\dot\beta}(x') = \epsilon^{\dot\alpha\dot\beta} M(\Lambda)^*{}_{\dot\beta}{}^{\dot\gamma} \bar{\chi}_{\dot\gamma}(x) = \epsilon^{\dot\alpha\dot\beta} M(\Lambda)^*{}_{\dot\beta}{}^{\dot\gamma} \epsilon_{\dot\gamma\dot\delta} \bar{\chi}^{\dot\delta}(x). \tag{4.6.56}$$

In order for Eqs. (4.6.45) and (4.6.46) to be consistent we must have

$$M(\Lambda)_\alpha{}^\delta = \epsilon_{\alpha\beta} \overline{M}(\Lambda)^{*\beta}{}_\gamma \epsilon^{\gamma\delta} \quad \text{and} \quad \overline{M}(\Lambda)^{\dot\alpha}{}_{\dot\delta} = \epsilon^{\dot\alpha\dot\beta} M(\Lambda)^*{}_{\dot\beta}{}^{\dot\gamma} \epsilon_{\dot\gamma\dot\delta}, \tag{4.6.57}$$

which can be written as $M(\Lambda) = \sigma^2 \overline{M}(\Lambda)^* \sigma^2$ and $\overline{M}(\Lambda) = \sigma^2 M(\Lambda)^* \sigma^2$. Noting that $(\sigma^2)^2 = I$ and $\sigma^{2*} = -\sigma^2$, we can summarize this as

$$M(\Lambda) = \sigma^2 \overline{M}(\Lambda)^* \sigma^2 = \sigma^2 \overline{M}(\Lambda)^{-1T} \sigma^2, \tag{4.6.58}$$

where we have used Eq. (4.6.15). Note that the Pauli spin matrices satisfy

$$\sigma^2 \vec{\sigma} \sigma^2 = -\vec{\sigma}^T \tag{4.6.59}$$

and also that $\exp(A^T) = (e^A)^T$, $e^{-A} = (e^A)^{-1}$ and $\sigma^2 e^A \sigma^2 = \exp(\sigma^2 A \sigma^2)$. Combining these results with Eq. (4.6.14) we arrive at Eq. (4.6.58) and so we see that the notation and definitions are self-consistent.

The above $\bar{\psi}_{\dot\alpha}$ two-component column vector notation was used in some older texts (Wess and Bagger, 1992; Bailin and Love, 1994). In more recent discussions (Terning, 2006; Dreiner et al., 2010) the two-component row vector is used,

$$\psi^\dagger_{\dot\alpha} \equiv \bar{\psi}^T_{\dot\alpha} = (\psi_\alpha)^{*T} = (\psi_\alpha)^\dagger, \tag{4.6.60}$$

recalling that ψ_α is a two-component column vector. We can write all Weyl spinors, left or right, in terms of only left-handed Weyl spinors. To convert the above equations involving sums over components we need only make the replacements $\bar{\psi}_{\dot\alpha} \to \psi^\dagger_{\dot\alpha}$ and $\bar{\psi}^{\dot\alpha} \to \psi^{\dagger\dot\alpha}$, since in such sums there is no distinction between the components of a row vector and those of a column vector. Recalling from Eq. (4.6.15) that $\overline{M} = M^{-1\dagger}$, then we see that Eq. (4.6.52) becomes

$$\psi'_\alpha(x') = M(\Lambda)_\alpha{}^\beta \psi_\beta(x) \quad \text{and} \quad \chi'^{\dagger\dot\alpha}(x') = (M(\Lambda)^{-1\dagger})^{\dot\alpha}{}_{\dot\beta}\chi^{\dagger\dot\beta}(x) \tag{4.6.61}$$

and we see that Eq. (4.6.55) becomes

$$\psi'^\dagger_{\dot\alpha}(x') = \psi^\dagger_{\dot\beta}(x)(\overline{M}(\Lambda)^{-1})^{\dot\beta}{}_{\dot\alpha} = \psi^\dagger_{\dot\beta}(x)(M(\Lambda)^\dagger)^{\dot\beta}{}_{\dot\alpha},$$

$$\chi'^\alpha(x') = (M(\Lambda)^{-1T})^\alpha{}_\beta \chi^\beta(x) = \chi^\beta(x)(M(\Lambda)^{-1})_\beta{}^\alpha. \tag{4.6.62}$$

We can rewrite Eq. (4.6.22) as

$$M(\Lambda)^{-1}\sigma^\mu M(\Lambda)^{-1\dagger} = \Lambda^\mu{}_\nu \sigma^\nu \quad \text{and} \quad M(\Lambda)^\dagger \bar\sigma^\mu M(\Lambda) = \Lambda^\mu{}_\nu \bar\sigma^\nu. \tag{4.6.63}$$

We can assign index notation to σ^μ and $\bar\sigma^\mu$ based on these Lorentz transformation properties and compare with the above equations; i.e., we have

$$(M(\Lambda)^{-1})_\alpha{}^\beta \sigma^\mu_{\beta\dot\gamma}(M(\Lambda)^{-1\dagger})^{\dot\gamma}{}_{\dot\delta} = \Lambda^\mu{}_\nu \sigma^\nu_{\alpha\dot\delta},$$

$$(M(\Lambda)^\dagger)^{\dot\alpha}{}_{\dot\beta}\bar\sigma^{\mu\dot\beta\gamma}M(\Lambda)_\gamma{}^\delta = \Lambda^\mu{}_\nu \bar\sigma^{\nu\dot\alpha\delta}. \tag{4.6.64}$$

For any left-handed Weyl spinor ψ_α we have

$$\psi_\alpha = \epsilon_{\alpha\beta}\psi^\beta, \quad \psi^\alpha = \epsilon^{\alpha\beta}\psi_\beta, \quad \psi^\dagger_{\dot\alpha} = \epsilon_{\dot\alpha\dot\beta}\psi^{\dagger\dot\beta}, \quad \psi^{\dagger\dot\alpha} = \epsilon^{\dot\alpha\dot\beta}\psi^\dagger_{\dot\beta}, \tag{4.6.65}$$

where $\psi^{\dagger\dot\alpha}$ is a right-handed Weyl spinor. From Eq. (4.6.59) we have $\sigma^2\bar\sigma^\mu\sigma^2 = \sigma^{\mu T}$ from which it can be readily shown that

$$\sigma^\mu_{\alpha\dot\beta} = \epsilon_{\alpha\gamma}\epsilon_{\dot\beta\dot\delta}\bar\sigma^{\mu\dot\delta\gamma}, \quad \bar\sigma^{\mu\dot\alpha\beta} = \epsilon^{\dot\alpha\dot\gamma}\epsilon^{\beta\delta}\sigma^\mu_{\delta\dot\gamma},$$

$$\epsilon^{\alpha\eta}\sigma^\mu_{\eta\dot\beta} = \epsilon_{\dot\beta\dot\delta}\bar\sigma^{\mu\dot\delta\alpha}, \quad \epsilon^{\dot\alpha\dot\gamma}\sigma^\mu_{\beta\dot\gamma} = \epsilon_{\beta\eta}\bar\sigma^{\mu\dot\alpha\eta}. \tag{4.6.66}$$

The key results for constructing Lorentz-covariant quantities are Eqs. (4.6.61), (4.6.62) and (4.6.64). A convenient shorthand is possible by adopting the convention that identical undotted indices descending to the right and identical dotted indices ascending to the right are suppressed. For example, with this convention we have

$$\psi\chi \equiv \psi^\alpha \chi_\alpha, \quad \psi^\dagger\chi^\dagger \equiv \psi^\dagger_{\dot\alpha}\chi^{\dagger\dot\alpha},$$

$$\psi\sigma^\mu\chi^\dagger \equiv \psi^\alpha \sigma^\mu_{\alpha\dot\beta}\chi^{\dagger\dot\beta}, \quad \psi^\dagger\bar\sigma^\mu\chi \equiv \psi^\dagger_{\dot\alpha}\bar\sigma^{\mu\dot\alpha\beta}\chi_\beta. \tag{4.6.67}$$

With the above notation it is natural to think of $\psi^\dagger_{\dot\alpha}$ and ψ^α as row vectors and χ_α and $\chi^{\dagger\dot\alpha}$ as column vectors. We note that $\psi\chi$ and $\psi^\dagger\chi^\dagger$ are Lorentz scalars and $\psi\sigma^\mu\chi^\dagger$ and $\psi^\dagger\bar\sigma^\mu\chi$ are Lorentz four-vectors, since we have

$$\psi'\chi' = \psi'^\alpha \chi'_\alpha = \psi^\beta(M^{-1})_\beta{}^\alpha M_\alpha{}^\gamma \chi_\gamma = \psi^\beta \chi_\beta = \psi\chi,$$

$$\psi'^\dagger\chi'^\dagger = \psi'^\dagger_{\dot\alpha}\chi'^{\dagger\dot\alpha} = \psi^\dagger_{\dot\beta}(M^\dagger)^{\dot\beta}{}_{\dot\alpha}(M^{-1\dagger})^{\dot\alpha}{}_{\dot\gamma}\chi^{\dagger\dot\gamma} = \psi^\dagger_{\dot\beta}\chi^{\dagger\dot\beta} = \psi^\dagger\chi^\dagger,$$

$$\psi'\sigma^\mu\chi'^\dagger = \psi'^\alpha\sigma^\mu_{\alpha\dot\beta}\chi'^{\dagger\dot\beta} = \psi^\gamma(M^{-1})_\gamma{}^\alpha\sigma^\mu_{\alpha\dot\beta}(M^{-1\dagger})^{\dot\beta}{}_{\dot\delta}\chi^{\dagger\dot\delta} = \Lambda^\mu{}_\nu\psi\sigma^\nu\chi^\dagger,$$

$$\psi'^\dagger\bar\sigma^\mu\chi' = \psi'^\dagger_{\dot\alpha}\bar\sigma^{\mu\dot\alpha\beta}\chi'_\beta = \psi^\dagger_{\dot\gamma}(M^\dagger)^{\dot\gamma}{}_{\dot\alpha}\bar\sigma^{\mu\dot\alpha\beta}M_\beta{}^\delta \chi_\delta = \Lambda^\mu{}_\nu\psi^\dagger\bar\sigma^\nu\chi. \tag{4.6.68}$$

We observe that

$$\psi\chi = \psi^\alpha\chi_\alpha = \epsilon^{\alpha\beta}\epsilon_{\alpha\gamma}\psi_\beta\chi^\gamma = -\epsilon^{\beta\alpha}\epsilon_{\alpha\gamma}\psi_\beta\chi^\gamma = -\psi_\beta\chi^\beta = -\chi\psi, \tag{4.6.69}$$

where we have used the fact that ψ_α and χ^α as spinors are commuting objects. In a path integral formulation of quantum field theory the spinors become anticommuting C-numbers (i.e., Grassmann numbers as discussed in Sec. 6.3.2). In a canonical treatment of quantum field theory the spinors become anticommuting quantum operators. In either case, we have for *anticommuting* spinor objects that

$$\psi\chi = -\psi_\beta\chi^\beta = \chi^\beta\psi_\beta = \chi\psi. \tag{4.6.70}$$

Further details can be found in Dreiner et al. (2010).

4.7 Additional Topics

Foldy-Wouthuysen transformation: We have omitted a treatment of the Foldy-Wouthuysen transformation; discussions can be found in Bjorken and Drell (1964), Greiner (2000) and elsewhere. If all terms in the Dirac equation are small with respect to the particle mass m, then we can decouple the positive and negative energy components. This is an expansion around the nonrelativistic limit that leads to a power series in $1/m$. This is achieved using the Foldy-Wouthuysen transformation, which reduces the Dirac equation into two two-component equations. One two-component equation describes the positive-energy states and in the nonrelativistic limit becomes the Pauli equation discussed earlier. The other two-component equation describes the negative-energy states. A discussion of this and a detailed application of the Dirac equation to the hydrogen atom can be found in Chapter 4 of Bjorken and Drell (1964). One can arrive at a Foldy-Wouthuysen transformation for decoupling the positive and negative energy components of the Klein-Gordon equation in some circumstances; e.g., see Chapter 9 of Bjorken and Drell (1964).

Relationship to relativistic classical field theory: The equations of relativistic quantum mechanics have the form of relativistic wave equations with appropriate factors of $\hbar$ and c included. The techniques developed for relativistic classical field theory can be applied to the complex scalar field $\phi(x)$ in relativistic quantum mechanics. All we need do is replace m in the classical case with $(mc/\hbar)$ in the case of relativistic quantum mechanics of the complex scalar field. Maxwell's equations already describe the propagation of a photon wavepacket and so are already the wave equations describing the relativistic quantum mechanics of a photon. The derivation of the Dirac equation from the Lagrangian and Hamiltonian formalisms is complicated by the fermionic nature of Dirac particles. We defer further discussion of this until we provide a treatment of fermions in Chapter 6.

Applications of RQM: Relativistic quantum mechanics can be applied in a semi-intuitive manner to the study of elementary scattering processes as described in Bjorken and Drell (1964) and Greiner (2000) provided that the effects of particle-antiparticle pair creation can be ignored. These applications are more systematically treated using the full technology of quantum field theory that we will soon develop, so we do not attempt to summarize such applications here.

Klein paradox: Relativistic quantum mechanics does not provide a description of the relativistic quantum behavior of multiparticle systems. The "Klein paradox" is not a paradox, but a breakdown of relativistic quantum mechanics in the presence of strong and rapidly changing external fields, where reflected currents can exceed incident currents and currents can flow into forbidden regions. These effects occur when the possibility of the creation of particle-antiparticle pairs can no longer be ignored and shows the breakdown of the one-particle interpretation of relativistic quantum

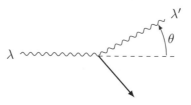

Figure 4.2   Compton scattering of a photon from an electron at rest showing the photon scattering angle θ and the recoiling electron.

mechanics. Detailed discussions of these phenomena can be found in Bjorken and Drell (1964), Greiner (2000) and many other standard texts.

Compton wavelength and particle-antiparticle creation: Arthur Compton introduced the concept of the *Compton wavelength* of a particle in his discussion of the elastic scattering of a photon by an electron, referred to as *Compton scattering* (see Fig. 4.2). A straightforward result of the application of four-momentum conservation, Eq. (2.6.20), leads to the *Compton effect*,

$$\lambda' - \lambda = (h/m_e c)(1 - \cos\theta) \equiv \lambda_C (1 - \cos\theta), \tag{4.7.1}$$

where m_e is the electron mass and where the initial electron is at rest; the initial photon has wavelength λ; the scattered photon has wavelength λ'; and the Compton wavelength for a particle of mass m is defined as the quantity

$$\lambda_C \equiv h/mc. \tag{4.7.2}$$

Proof: Let p_i^μ and p_f^μ be the initial and final four-momenta of the electron and similarly k_i^μ and k_f^μ for the photon. Then $p_i^\mu + k_i^\mu = p_f^\mu + k_f^\mu$ and $p_f^2 = m_e^2 c^2 = (p_i + k_i - k_f)^2 = m_e^2 c^2 + 2[p_i \cdot (k_i - k_f) - k_i \cdot k_f]$ using $k_i^2 = k_f^2 = 0$ for massless photons. Then we have $p_i \cdot (k_i - k_f) = k_i \cdot k_f$. In the rest frame of the initial electron $p_i^\mu = (m_e c, \mathbf{0})$, $k_i^\mu = (h/\lambda)(1, \hat{\mathbf{k}}_i)$ and $k_f^\mu = (h/\lambda')(1, \hat{\mathbf{k}}_f)$, where the photon energy $E = hf = hc\lambda$ gives $k^0 = E/c = h/\lambda$ and where $\hat{\mathbf{k}}_i$ and $\hat{\mathbf{k}}_i$ are the unit vectors in the direction of motion of the initial and final photon, respectively. We then have $m_e ch[(1/\lambda) - (1/\lambda')] = (h^2/\lambda\lambda')[1 - \hat{\mathbf{k}}_i \cdot \hat{\mathbf{k}}_f]$, which finally gives $\lambda' - \lambda = (h/m_e c)(1 - \cos\theta)$, where $\cos\theta = \hat{\mathbf{k}}_i \cdot \hat{\mathbf{k}}_f$.

Since we so often use $\hbar$ in quantum mechanics it is also common to define the *"reduced" Compton wavelength*,

$$\lambdabar \equiv \lambda_C / 2\pi = \hbar/mc, \tag{4.7.3}$$

which we refer to as "lambda-bar." If we attempt to localize a particle of mass m into a region of space of the order of its reduced Compton wavelength, $\sigma_x \sim \lambdabar_C = \hbar/mc$ then from the Heisenberg uncertainty principle in Eq. (2.5.2) we must have

$$\sigma_p \gtrsim \hbar/2\lambdabar_C = mc/2. \tag{4.7.4}$$

With further localization we find that increasing energy scales are being accessed, where there is sufficient energy to create a particle-antiparticle pair, i.e., energy scales of order $2mc^2$. At distance scales smaller than the reduced Compton wavelength the single-particle picture of relativistic quantum mechanics begins to break down. This is a similar issue to that seen in the Klein paradox. The appropriate formalism that allows for the possible creation of particle-antiparticle pairs is that of quantum field theory and it is to this that we now turn our attention.

Summary

After a review of nonrelativistic quantum mechanics we explored the behavior of relativistic wavepackets. Then we constructed the relativistic quantum mechanics of a scalar (spin-zero) boson and a spin-half fermion as the Klein-Gordon and Dirac equations, respectively, including their interaction with external scalar and electromagnetic fields. We obtained the nonrelativistic quantum limit and relativistic classical limit of these. We formulated the P, C and T transformations in this context. We studied chirality and helicity and also Majorana fermions, which are their own antiparticles. We also came to understand the breakdown of relativistic quantum mechanics in the presence of strong fields, known as the Klein paradox. Of course it is not a paradox. Rather, it is an indication that we need to develop a formalism that can describe particle annihilation and creation, i.e., quantum field theory.

Problems

4.1 Show that the symmetric n-boson basis states in Eq. (4.1.149) are normalized; i.e., show $\langle b_{i_1} \cdots b_{i_n}; S | b_{i_1} \cdots b_{i_n}; S \rangle = 1$.

4.2 For any vector operator $\hat{\mathbf{V}}$ in quantum mechanics we have $[\hat{V}^i, \hat{J}^j] = i\hbar\epsilon^{ijk}\hat{V}^k$ from Eq. (4.1.114). It will be helpful to also recall Eq. (1.2.64).
(a) Use these results to show that $\langle j_b m_b | \hat{V}_z | j_a m_a \rangle = 0$ unless $m_a = m_b$.
(b) Define $\hat{V}_\pm \equiv \hat{V}_x \pm i\hat{V}_y$ in the usual way and show that $\langle j_b m_b | \hat{V}_\pm | j_a m_a \rangle = 0$ unless $m_b = m_a \pm 1$.
(c) Evaluate $[\hat{J}_+, \hat{V}_+]$ and use this result for the special case $j = j_a = j_b$ to obtain the dependence of $\langle j(m+1) | \hat{V}_+ | jm \rangle$ on m.
(d) Extending the result of part (c), show that $\langle j m_b | \hat{\mathbf{V}} | j m_a \rangle = C_j \langle j m_b | \hat{\mathbf{J}} | j m_a \rangle$, where C_j is a constant independent of m_b and m_a.

4.3 A scalar particle with mass m and charge q interacts with a time-independent electromagnetic potential given by $qA^0 = 0$ for $z < 0$ and $qA^0 = V$ with $V > 0$ for $z \geq 0$. A particle plane wave is moving in the $+z$-direction with energy $E > 0$. Calculate the reflection (R) and transmission (T) coefficients, where these are the probability of reflection and transmission, respectively, $R + T = 1$. Define $p \equiv \sqrt{E^2 - m^2}$ and $K \equiv \sqrt{(E-V)^2 - m^2} > 0$.
(a) Show that if $|E - V| < m$, then there is no transmission, $R = 1$ and $T = 0$.
(b) Show that if $E - V > m$, then $R = |(p-K)^2/(p+K)^2| < 1$ and $T = 1 - R$ and there is partial transmission and partial reflection.
(c) Show that if $E - V < -m$, then $R = |(p+K)^2/(p-K)^2| > 1$ and $T = 1 - R$, which corresponds to a negative current being transmitted and more positive current reflected than was incident. This is the Klein paradox for the Klein-Gordon equation and shows that $|V| > 2m$ can lead to particle-antiparticle pair creation. This illustrates the breakdown of relativistic quantum mechanics in the presence of strong fields. (Hint: Use continuity of the wavefunction and its first derivative.)

4.4 Using the properties of the Dirac γ-matrices verify that the matrices $(\frac{1}{2}\sigma^{\mu\nu})$ satisfy the Lie algebra of the Lorentz transformations in Eq. (1.2.179).

4.5 Explain why the hermiticity of γ^0 and γ^5 and the antihermiticity of γ are preserved in any representation unitarily equivalent to the Dirac representation.

4.6 Provide detailed proofs of the trace identities:
(a) traces of products of γ matrices (with no $\gamma^{\mu\dagger}$) are representation independent (same for γ^μ and $\bar{\gamma}^\mu = S\gamma^\mu S^{-1}$).
(b) $\mathrm{tr}\,(I) = 4$, $\mathrm{tr}\,(\gamma^\mu\gamma^\nu) = 4g^{\mu\nu}$, $\mathrm{tr}\,(\gamma^\mu) = \mathrm{tr}\,(\text{odd \# of } \gamma^\mu\text{'s}) = \mathrm{tr}\,(\sigma^{\mu\nu}) = 0$, $\mathrm{tr}\,(\gamma^\mu\gamma^\nu\gamma^\rho\gamma^\sigma) = 4\,(g^{\mu\nu}g^{\rho\sigma} - g^{\mu\rho}g^{\nu\sigma} + g^{\mu\sigma}g^{\nu\rho})$.
(c) $\mathrm{tr}\,(\gamma^5\gamma^\mu\gamma^\nu) = \mathrm{tr}\,(\gamma^5(\text{odd \# of } \gamma^\mu\text{'s})) = 0$ and $\mathrm{tr}\,(\gamma^5\gamma^\mu\gamma^\nu\gamma^\rho\gamma^\sigma) = -4i\epsilon^{\mu\nu\rho\sigma}$.

4.7 Prove the following identities:
(a) $\gamma_\mu\gamma_\nu\gamma_\rho = g_{\mu\nu}\gamma_\rho - g_{\mu\rho}\gamma_\nu + g_{\nu\rho}\gamma_\mu + i\epsilon_{\mu\nu\rho\sigma}\gamma^\sigma\gamma^5$.
(b) $\gamma^\mu\gamma_\mu = 4$, $\gamma^\mu\gamma^5\gamma_\mu = -4\gamma^5$, $\gamma^\mu\not{k}\gamma_\mu = -2\not{k}$, $\gamma^\mu\gamma^5\not{k}\gamma_\mu = 2\gamma^5\not{k}$, $\gamma^\mu\not{k}\not{\ell}\gamma_\mu = 4k\cdot\ell$, $\gamma^\mu\not{k}\not{\ell}\not{p}\gamma_\mu = -2\not{p}\not{\ell}\not{k}$, $\gamma^\mu\not{k}\not{\ell}\not{p}\not{q}\gamma_\mu = 2\,(\not{q}\not{k}\not{\ell}\not{p} + \not{p}\not{\ell}\not{k}\not{q})$.

4.8 Prove each of the identities: $\gamma^\mu\sigma^{\rho\sigma}\gamma_\mu = 0$, $\sigma^{\mu\nu}\sigma_{\mu\nu} = 12$, $\gamma^\mu\sigma^{\rho\sigma}\gamma^\lambda\gamma_\mu = 2\gamma^\lambda\sigma^{\rho\sigma}$, $\gamma^\mu\gamma^\lambda\sigma^{\rho\sigma}\gamma_\mu = 2\sigma^{\rho\sigma}\gamma^\lambda$, $\sigma^{\mu\nu}\gamma^\rho\sigma_{\mu\nu} = 0$, $\sigma^{\mu\nu}\gamma^\rho\gamma^\sigma\sigma_{\mu\nu} = 4\,(4g^{\rho\sigma} - \gamma^\rho\gamma^\sigma)$ and $\sigma^{\mu\nu}\sigma^{\rho\sigma}\sigma_{\mu\nu} = -4\sigma^{\rho\sigma}$.

4.9 Consider the sixteen matrices $\{\Gamma_1, \Gamma_2, \ldots, \Gamma_{16}\} \equiv \{I, \gamma^\mu, \sigma^{\mu\nu}|_{\mu<\nu}, \gamma^5\gamma^\mu, i\gamma^5\}$. Show that
(a) $\gamma^0(\Gamma_I)^\dagger\gamma^0 = \Gamma_I$ in any representation unitarily equivalent to the Dirac representation;
(b) $\Gamma_I^2 = \pm 1$;
(c) $\Gamma_I\Gamma_J = \pm\Gamma_J\Gamma_I$;
(d) $\Gamma_J\Gamma_K = a_{JK}{}^L\Gamma_L$ for some L (no sum implied) with $a_{JK}{}^L = \pm 1, \pm i$;
(e) $\Gamma_I^{-1}\Gamma_J\Gamma_I = -\Gamma_J$ for $J \neq 1$ for at least one Γ_I;
(f) $\mathrm{tr}(\Gamma_J) = 0$ for $J \neq 1$;
(g) $\mathrm{tr}(\Gamma_I^{-1}\Gamma_J) = \mathrm{tr}(\Gamma_I\Gamma_J^{-1}) = 4\delta_{IJ}$;
(h) $\sum_{I=1}^{16}(\Gamma_I^{-1})_{ij}(\Gamma_I)_{k\ell} = \sum_{I=1}^{16}(\Gamma_I)_{ij}(\Gamma_I^{-1})_{k\ell} = 4\delta_{i\ell}\delta_{kj}$;
(i) if $\sum_{I=1}^{16}c_I\Gamma_I = 0$, then we must have all $c_I = 0$ and hence the Γ_J are sixteen linearly independent matrices; and
(j) using the result of part (i), explain why any complex 4×4 matrix can be written as $A = \sum_{I=1}^{16}a_I\Gamma_I$ with $a_I \equiv \frac{1}{4}\mathrm{tr}(\Gamma_I^{-1}A)$. So these sixteen matrices form a basis for the set of all 4×4 complex matrices.

4.10 (a) Prove that if $\psi(x)$ is a solution of the Dirac equation, Eq. (4.4.59), then it is also a solution of the Klein-Gordon equation, Eq. (4.3.12).
(b) Prove that ∂_μ and ∂_ν commute, whereas D_μ and D_ν do not. Evaluate $[D_\mu, D_\nu]\psi = D_\mu(D_\nu\psi) - D_\nu(D_\mu\psi)$ and so prove $[D_\mu, D_\nu] = i(q/\hbar c)F_{\mu\nu}$.
(c) Prove that if $\psi(x)$ is a solution of the Dirac equation in the presence of an electromagnetic field, Eq. (4.4.61), it is also a solution of a modified Klein-Gordon equation, $\left[D_\mu D^\mu + (q/2\hbar c)\sigma^{\mu\nu}F_{\mu\nu} + (mc/\hbar)^2\right]\psi = 0$.

4.11 Using the standard representation of the Pauli spin matrices, show that complex conjugation with a rotation of π about the y-axis reverses the orientation of the spin; i.e., show that $(i\sigma^2)\boldsymbol{\sigma}^*(i\sigma^2)^\dagger = -\boldsymbol{\sigma}$.

4.12 Prove directly from the definition of $M(\Lambda)$ and $\overline{M}(\Lambda)$ in Eq. (4.6.14) that $i\sigma^2\chi^*$ transforms with $M(\Lambda)$ and $-i\sigma^2\psi^*$ transforms with $\overline{M}(\Lambda)$ and hence verify that the charge conjugate of the Dirac spinor in the chiral representation in Eq. (4.6.44) transforms as a Dirac spinor.

Introduction to Particle Physics

The purpose of this introductory chapter is to give a context and motivation for the detailed developments to follow. It is a brief and necessarily selective overview of the understanding of particle physics phenomenology, gauge theories and the Standard Model (SM) that was applicable at the time of writing. A longer, helpful introduction to elementary particle physics can be found in Griffiths (2008). Up to date reviews of all known subatomic particles and their properties can be found in the most recent version of the *Review of Particle Physics* by the Particle Data Group (PDG).[1]

5.1 Overview of Particle Physics

The observed interactions in the universe can be understood in terms of four fundamental forces, which are listed in Table 5.1. In order of increasing strength we have: gravity, the weak interaction, electromagnetism and the strong interaction. The SM describes the last three of these but does not include gravity. The discovery of the *Higgs boson* in 2012 completed the identification of the components of the SM and led to the awarding of the 2013 Nobel Prize in Physics. At that point the SM became the most successful and precise physical theory ever devised.

Despite its impressive successes, we already know that the SM is an incomplete description of the universe. Since the SM does not describe gravity, it also does not explain the inferred *dark energy* that is leading to the accelerating expansion of the universe. The *minimal SM* has massless Dirac neutrinos. However, nonzero *neutrino masses* are necessary in order that *neutrino mixing* be possible. Neutrino mixing is necessary to explain the observed phenomenon of neutrino oscillations, where a neutrino of one flavor can turn into another neutrino flavor as the neutrino mass eigenstate propagates through space or matter. The *minimally extended SM* allows for the mixing of massive Dirac neutrinos through the *Pontecorvo-Maki-Nakagawa-Sakata (PMNS) neutrino mass matrix*, which is the neutrino analog of the quark *Cabibbo-Kobayashi-Maskawa (CKM) matrix*. This is discussed in more detail in Sec. 5.2.5. Other examples of physics Beyond the Standard Model (BSM) include (i) the existence of *dark matter*, (ii) the possibility that neutrinos are *Majorana neutrinos* and (iii) the apparent lack of sufficient CP violation in the SM to explain the magnitude of the excess of matter over antimatter in the universe.

Stable matter consists of spin-half particles and their bound states, where the spin-half particles are divided into those that experience the strong interaction (*baryons*, such as protons and neutrons) and spin-half particles that do not (*leptons*, such as electrons). Quarks have an additive quantum number called *baryon number*, B, where for quarks $B = 1/3$ and for antiquarks $B = -1/3$. So,

---

[1] See the PDG review Zyla et al. (2020). An always up-to-date version of the Review of Particle Properties can be found online at `http://pdg.lbl.gov`.

Table 5.1. Properties of the interactions

	Gravity	Electroweak		Strong	
		Weak	EM	QCD	Residual
Acts on:	Mass	Flavor	Electric Charge	Color Charge	
Particle experiencing:	All	Quarks, Leptons	Charged	Quarks, Gluons	Hadrons
Particle mediating:	Graviton	$W^+ W^-$ Z^0	γ	Gluons	Mesons
Strength (rel to EM) for:					
2 u quarks at 10^{-18} m	10^{-41}	0.8	1	25	N/A
2 u quarks at 3×10^{-17} m	10^{-41}	10^{-4}	1	60	N/A
2 protons in nucleus	10^{-36}	10^{-7}	1	N/A	20

EM, electromagnetism; QCD, quantum chromodynamics.

for example, protons and neutrons containing three quarks have $B = 1$, whereas antiprotons and antineutrons containing three antiquarks have $B = -1$. The leptons ($e^-, \nu_e, \mu^-, \nu_\mu, \tau^-, \nu_\tau$) have *lepton number* $L = 1$, whereas the antileptons ($e^+, \bar{\nu}_e, \mu^+, \bar{\nu}_\mu, \tau^+, \bar{\nu}_\tau$) have $L = -1$. The SM assumes that all fermions, including neutrinos, are Dirac fermions. We can also associate a family lepton number for each family: electron number L_e, muon number L_μ and tau number L_τ. For example, e^- and ν_e have $L_e = 1$ and e^+ and $\bar{\nu}_e$ have $L_e = -1$ and similarly for the muon and tau families. The experimental observation of neutrino mixing, which is absent in the SM, means that the SM is incomplete. We can extend the SM to include this behavior by introducing the PMNS lepton mixing matrix (to be discussed later). This leads to violation of lepton family number, while the total lepton number $L = L_e + L_\mu + L_\tau$ remains conserved (at least at the level of perturbation theory).

In order to explain the excess of matter over antimatter in the universe we need a process that allows the formation of (i) a baryonic asymmetry in the form of an excess of baryons over antibaryons, referred to as *baryogenesis*; or (ii) a leptonic asymmetry and a process to convert this into a baryonic asymmetry, referred to as *leptogenesis*; or (iii) both. Our understanding of Big Bang nucleosynthesis requires an excess of baryons over antibaryons at the level of one in a billion a few minutes after the Big Bang, whereas the requirements on leptonic asymmetry in the early universe are much weaker. Baryogenesis can occur within the SM, although the CP violation provided in the SM may not produce sufficient baryonic asymmetry to explain the observed baryon excess. For baryogenesis to occur in the SM the three *Sakharov conditions* must be met: (i) baryon number, B, violation; (ii) C and CP violation and (iii) interactions out of thermal equilibrium. In the SM lepton number, L, is conserved perturbatively and so leptogenesis is only possible in BSM theories or through nonperturbative processes. There are simple extensions of the SM that allow leptogenesis, such as the introduction of right-handed neutrinos. To produce the required baryon asymmetry there needs to be a process that converts a leptonic asymmetry into a baryon excess. A *sphaleron* is a nonperturbative solution to the electroweak field equations of the SM that can lead to violations of both B and L while preserving the difference $B - L$. It is speculated that sphalerons may have been plentiful in the early universe and could have converted an antilepton excess into a baryon excess while preserving $B - L$; i.e., if initially $B_i = 0$ and $L_i = -\Delta$ with $\Delta > 0$, then after interacting with sphalerons we would have $B_f = L_f + (B_i - L_i) = L_f - L_i > 0$ if the lepton number has increased, $L_f - L_i > 0$.

There are also questions of naturalness and fine-tuning in the SM. Why is CP violation in the strong interaction so small when it could be very large? Why is the Higgs mass at 125 GeV so light? This is referred to as the *hierarchy problem*. A model whose parameters have to be adjusted very precisely in order to agree with observation is said to be a model with *fine-tuning*. A different but related concept is that of *naturalness* in a model, which is the property that dimensionless ratios of model parameters are of order unity. If we take the view that our universe is not "special," then if two theories are capable of reproducing observations, the one that is more natural and less fine-tuned would be preferred. This can be thought of as an extension of the concept of Occam's razor. The anthropic principle is the statement that we must live in a universe that is capable of having us evolve and exist in it. This principle can counterbalance arguments relating to fine-tuning and a lack of naturalness. It is difficult to imagine life as we understand it existing in a universe without sufficient complexity, e.g., a universe that contained no atoms other than hydrogen. If the neutron-proton mass difference was $m_n - m_p > \mathcal{O}(10 \text{ MeV})$, then neutrons would decay inside nuclei and there would be nothing but hydrogen. If the proton was just a little more massive than the neutron, then protons would decay into neutrons and there would be no atoms at all.

The existence of dark matter has been inferred from its gravitational interactions with the visible universe. It appears essential for our understanding of the cosmic microwave background (CMB), the rotational velocity profiles of spiral galaxies and the mass distribution in the universe inferred from gravitational lensing. Alternatives to dark matter include modified gravity theories, such as modified Newtonian dynamics (MOND). Such theories do not yet appear capable of explaining all observations in a consistent manner and so are less favored. Dark matter has to date evaded our concerted attempts to detect it through interactions other than gravity.

The current prevailing paradigm for understanding the evolution and current structure and properties of the universe is referred to as the Lambda-CDM (or ΛCDM) model. It is sometimes called "the SM of Big Bang cosmology." Here Lambda (Λ) refers to the cosmological constant of general relativity that models dark energy. CDM refers to cold dark matter, which means that the dark matter is moving at nonrelativistic speeds ($v \ll c$). The remaining component of the ΛCDM model is the SM itself that gives rise to the visible universe.

The coupling strength of a quantum field theory depends on the energy scale of the physical process being considered. This is due to the procedure referred to as *renormalization* that is essential to our understanding of interacting quantum field theories. The SM is a quantum field theory and, in particular, belongs to a class of quantum field theories referred to as *gauge field theories* or *gauge theories*, where each force has a *gauge boson* that is the carrier of that force as shown in Table 5.2.

Table 5.2. The four fundamental forces and the gauge bosons that carry these forces				
	Gravity	Weak	Electromagnetic	Strong
Carried by:	Graviton (unobserved)	W^+, W^-, Z	Photon (γ)	Gluon
Acts on:	All	Quarks, leptons W^+, W^-, Z	Quarks, leptons, W^+, W^-	Quarks, gluons

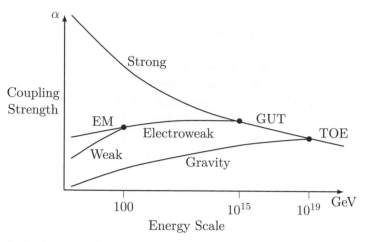

Figure 5.1 Schematic diagram indicating the merging of the strengths of the weak, electromagnetic and strong interactions. GUT, Grand Unified Theory; TOE, Theory of Everything.

The theory describing the interaction of the electromagnetic field (photons) and charged particles is known as quantum electrodynamics (QED). The original understanding of the weak interaction was in terms of the Fermi four-point interaction, where, for example, a neutron would decay into a proton and electron and an anti-electron neutrino. This process is known as β decay, since in the early days electrons were referred to as β-rays. The weak and electromagnetic interactions have their origin in a single gauge symmetry referred to as the electroweak (EW) gauge symmetry and in this sense these two interactions have been *unified*. We understand the strong interaction in terms of quantum chromodynamics (QCD).

The coupling strengths of the weak, electromagnetic and strong interactions appear to converge at an energy scale of approximately 10^{16} GeV as depicted in Fig. 5.1. This suggests that the electroweak and strong interactions unify into a single *Grand Unified Theory* (GUT) at this scale. We refer to this scale as the GUT scale, $\Lambda_{GUT} \sim 10^{15}–10^{16}$ GeV. The SM is then considered to be an *effective theory* of some GUT. As such it is accurate up to and including the electroweak symmetry breaking scale, but it breaks down as one approaches the GUT scale. In physics an effective theory is the low-energy limit of some greater theory that applies at higher energy scales; e.g., a low-energy effective theory of QCD is given in terms of interacting hadrons and on the basis of that effective theory we have built our understanding of nuclear physics.

A theory that further unifies gravity into a single underlying framework is referred to as a Theory of Everything (TOE). A fundamental theory including gravity should lead to General Relativity as part of its macroscopic limit, where it is General Relativity that determines the overall spacetime structure of the universe. The gravitational force is thought to be mediated by an unobserved particle called a graviton. This is the analog of the photon mediating the electromagnetic interaction. The weakness of the gravitational interaction means that the direct detection of a graviton would be an extremely difficult if not impossible task. General relativity equates the curvature of spacetime and gravitational forces and is essential to our understanding of the universe. The historic announcement of the detection of gravitational waves in 2016 confirmed this important prediction of the theory. Additionally, the gravitational wave pulse shapes are consistent with the in-spiraling of two merging black holes. This provides compelling evidence consistent with the strong field predictions of general relativity. The 2017 Nobel Prize in Physics was awarded to Rainer Weiss, Barry Barish and Kip

Thorne for the discovery of gravitational waves. By contrast, the SM is formulated at very small scales away from intense gravitational fields, where spacetime can be taken to be locally flat. We refer to flat spacetime as Minkowski spacetime.

It is expected that a GUT theory along with gravity would be absorbed into a larger TOE at or near the Planck energy scale, $\Lambda_P = 1.2209 \times 10^{19}$ GeV, where the Planck energy is the scale at which the quantum effects of gravity are expected to become important. It is defined in terms of the Planck mass, $\Lambda_P = m_P c^2$, where $m_P \equiv \sqrt{\hbar c/G} = 1.2209 \times 10^{19}$ GeV$/c^2 = 2.1765 \times 10^{-8}$ kg.

The gravitational potential energy between two Planck masses separated by distance d is $E = -Gm_P^2/d$. The presumed graviton is the massless spin-two gauge boson mediating the gravitational interaction just as the photon is the massless spin-one gauge boson that mediates the electromagnetic interaction. A massless gauge boson with wavelength λ has an angular wavelength $\lambda/2\pi$ and an energy $E = hc/\lambda = \hbar c/(\lambda/2\pi)$. The Planck mass is the mass for which the magnitude of the gravitational potential energy at distance d is equal to the energy of a massless gauge boson with angular wavelength d, which gives $|E| = Gm_P^2/d = \hbar c/d$ and hence $m_P = \sqrt{\hbar c/G}$. In more familiar terms, the Planck mass is approximately 10^{19} times bigger than the proton mass (938.2 MeV). The Planck mass is also approximately the mass at which a black hole's *Schwarzschild radius*,[2] $r_S = 2GM/c^2$, is equal to its Compton wavelength, $\lambda_C = h/Mc$. This equality gives a mass $M = \sqrt{\pi \hbar c/G}$ that is only $\sqrt{\pi}$ larger than the definition of m_P.

5.2 The Standard Model

It is useful to here reflect on the origins of the SM. We present the discussion in the form of a historical perspective, since the inevitable arrival at the model followed from a sequential combination of experimental and theoretical breakthroughs.[3] In the layered buildup to the model we gain valuable insights into it.

5.2.1 Development of Quantum Electrodynamics (QED)

Sylvan Schweber has written a detailed description of the history of the development of quantum electrodynamics (QED) (Schweber, 1994). We summarize some of the key developments here. After the invention of the Dirac equation to describe spin-$\frac{1}{2}$ particles in 1928 by Paul Dirac, the formulation of QED quickly followed. First attempts, including attempts to quantize the electromagnetic field, were carried out by Dirac himself. Enrico Fermi wrote down an early elegant formulation (Fermi, 1932). It soon became clear that beyond first order in perturbation theory the theory produced infinities, which brought into question the validity of the entire framework. It was recognized in the 1940s that there was disagreement between the first-order prediction and measurement for quantities such as the magnetic moment of the electron and the Lamb shift. In 1947 Hans Bethe, in a simplified nonrelativistic calculation of the Lamb shift beyond first order, assumed that the resulting infinities could be absorbed into the electron mass and charge (Bethe, 1947). This was the first application of the idea of renormalization.

---

[2] As one approaches a sufficiently dense massive object the escape velocity of any particle will increase until it reaches the speed of light. The radius at which this occurs is defined as the Schwarzschild radius. Not even light can the escape from inside this radius, leading to the concept of a *black hole*.

[3] Some interesting historical insights and details can be found in Close (2011).

The renormalization of higher-order diagrams was taken up and systematically developed independently by Sin-Itiro (also written as Shin'ichirō) Tomonaga, Julian Schwinger, Richard P. Feynman and Freeman Dyson into the covariant formulation of QED that we know today. It is noteworthy that Tomonaga was working alone in Japan during World War II and developed his results on QED independent of ongoing Western science at the time (Tomonaga, 1946). It was only after the war in 1947 that his work became more widely known and by that time both Schwinger and Feynman had independently arrived at the same results; e.g., see Schwinger (1948a,b) and Feynman (1949a,b, 1950). The 1965 Nobel Prize in Physics was awarded to Tomonaga, Schwinger and Feynman for this work.

While Tomonaga and Schwinger used operator-based methods to study QED, Feynman used his more intuitive diagram-based approach to arrive at the same results. These diagrams, now known as Feynman diagrams, resulted from Feynman's extension of his *path integral* approach to quantum mechanics to the case of QED. It is interesting that the demonstration of the path integral formulation of quantum mechanics was the basis of Feynman's PhD thesis at Princeton University in 1942 (see Sec. 4.1.12). Freeman Dyson made an essential contribution in showing that the operator-based formulations and the Feynman diagram formulation of QED were in fact equivalent (Dyson, 1949). QED is a simple example of a gauge field theory.

Each vertex in QED corresponds to a single factor of electric charge e and therefore every two vertices correspond to a power of α, where $\alpha = e^2/4\pi$. Recall that from this chapter onward we are working in natural and Lorentz-Heaviside units with $\alpha = e^2/4\pi = 137.036$, e.g., Sec. A.1. Feynman diagrams connecting a given set of initial and final states at lowest order in the fine-structure constant have no loops and if they are in one piece they are referred to as *tree diagrams*. If they consist of one or more pieces then they are referred to as *forest diagrams,* and so all tree diagrams are forest diagrams but not vice versa. They do not require renormalization. At higher order in the fine structure constant the Feynman diagrams contain loops, which give rise to infinities that require renormalization. For example, the tree diagram for electron-muon scattering is shown in Fig. 5.2 and is $\mathcal{O}(\alpha)$. The $\mathcal{O}(\alpha^2)$ diagrams for this process contain one loop and are shown in Fig. 5.3.

The external legs are removed when we calculate the scattering matrix for any process and are replaced with on-shell states containing the physical renormalized mass. We do not add loops to external lines for this reason. In a *renormalizable theory* the infinities at every order of α (at any number of loops) can be absorbed into a *finite* number of terms in the renormalized Lagrangian. The process of temporarily rendering the infinite loop integrals finite is referred to as *regularization*, where the value of the loop integral diverges as the regularization is removed. We keep *renormalized*

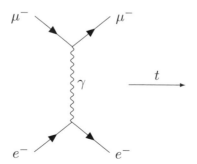

Figure 5.2 Feynman diagram for one photon exchange contribution to electron-muon scattering.

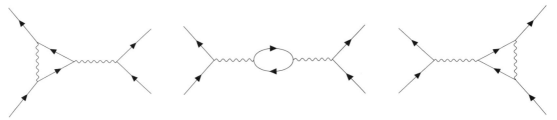

Figure 5.3 One-loop Feynman diagrams contributing to fermion-fermion scattering.

quantities finite as we remove the regularization by absorbing the resulting divergences into the regularization-dependent *bare* terms in the Lagrangian density. By working to increasingly higher levels in *perturbation theory* (higher orders in α) we find that QED gives excellent agreement with experiment.

For some QED observable f the perturbative expansion up to order N is

$$f_N(\alpha) \simeq \sum_{k=1}^{N} c_k \alpha^k, \tag{5.2.1}$$

where the Feynman rules tell us how to calculate the coefficients c_k, which become finite after renormalization. We might naively expect to recover $f(\alpha)$ exactly from the sum if we let $N \to \infty$; i.e., if the series converges then $f(\alpha) = \lim_{N \to \infty} f_N(\alpha)$.

However, there are reasons to believe that the $N \to \infty$ power series expansion has zero radius of convergence around $\alpha = 0$; i.e., it exists if $\alpha = 0$ and is divergent for all $\alpha > 0$. A quantum field theory that only exists in the limit of vanishing coupling, $\alpha \to 0$, is referred to as a *trivial theory*. Dyson gave a simple suggestive argument for triviality (Dyson, 1952). If the perturbation series was convergent for some $\alpha \in \mathbb{R}$ then it would have some finite radius of convergence R, where $f(\alpha)$ converges for all $|\alpha| < R$ with $\alpha \in \mathbb{C}$. This means that the observable with a negative fine structure constant, $f(-|\alpha|)$ with $|\alpha| < R$, would also be defined. In such a universe we would have electron-positron repulsion and electron-electron and positron-positron attraction. The energy of the universe would be lowered by having electron-positron pairs created from the vacuum with the electrons binding together and the positrons binding together with the electrons and positrons moving apart. The rest energy of the electron-positron pairs will present a small barrier between the ordinary vacuum and states with pairs created, but given sufficient time, tunneling through this barrier would occur. As additional pairs are created and separate the energy of the universe would be unbounded below and so $f(-|\alpha|)$ could not be an analytic continuation of $f(|\alpha|)$. Then $f(\alpha)$ has zero radius of convergence and by Dyson's suggestive argument QED corresponds to a trivial theory.

The perturbation series of QED is believed to be an asymptotic series. Since α is small ($\simeq 1/137$ at energy scales relevant to atomic and nuclear physics) we expect that when working at increasingly higher orders N in perturbation theory we will get increasingly good estimates of the observable f until we reach $N \sim N_{\mathrm{opt}}$. Beyond the optimal order, N_{opt}, we cease being able to improve the perturbative estimate of the observable and there is a small remaining error in the estimate of f. Due to the smallness of α this residual error is expected to be extremely small and N_{opt} is expected to be so large that it is beyond the order of any perturbative calculations that we are able to perform (Dyson, 1952). We understand this behavior as an indication that QED is a low-energy effective theory component of the SM, which in turn is a low-energy effective theory of whatever GUT theory subsumes it.

Another issue that points to QED being an effective low-energy theory of a greater theory is the fact that renormalization leads to the fine structure constant increasing logarithmically with the energy scale. The result is that at extremely high energies perturbation theory itself breaks down at an energy scale that is referred to as the *Landau pole*. This signals the onset of nonperturbative physics, which cannot be described by the asymptotic series of perturbation theory. Nonperturbative physics is described by subdominant terms with forms such as $e^{1/\alpha}$ that do not appear in the asymptotic expansion of QED perturbation theory. Landau poles are expected to appear in theories that are not asymptotically free, i.e., theories that do not have a decreasing coupling strength as the energy scale goes to infinity.

5.2.2 Development of Quantum Chromodynamics (QCD)

Before World War II the similarity of the proton and neutron masses led Werner Heisenberg and other physicists to the view that the proton and neutron were two different states of a single entity that was referred to as a *nucleon*. The nucleon was proposed to have a new property called *isospin* (or *strong isospin*), I, that was analogous to spin-$\frac{1}{2}$, where the proton and neutron had the third component of isospin as $+\frac{1}{2}$ and $-\frac{1}{2}$, respectively. Of course the masses are slightly different and so isospin is not an exact symmetry, but it is a good approximation and was very helpful in classifying the energy eigenstates of nuclei. After the Second World War the number of observed subatomic particles multiplied rapidly, first from studies of cosmic rays and subsequently from studying collisions at particle accelerators. In a very short time there were hundreds of subatomic particles and the question arose as to whether they were all elementary particles. The observation of patterns in the behavior of elements led to the construction of the periodic table, which is now understood in terms of the systematic structure of the electron orbitals in atoms. It was then logical to look at the properties of these many subatomic particles to see if there was any underlying pattern that could reveal their origin.

Strongly interacting particles with integer spin are called *mesons* and those with half-integer spin are called *baryons*. All strongly interacting particles are called *hadrons*. It was observed in 1961 that the three pions ($\pi^{\pm}, \pi^0$) appeared in an octet family of mesons and that the nucleons (p and n) appeared in an octet family of baryons. Murray Gell-Mann referred to this as the *Eightfold Way*. Gell-Mann and Yuval Ne'eman independently observed that these patterns were consistent with the representations of the Lie Group $SU(3)$. In 1962 Gell-Mann then predicted the existence, mass and quantum numbers of the Ω^- baryon in the decuplet of baryons containing the Δ baryons. This particle was discovered in 1964 at Brookhaven.

Gell-Mann and George Zweig independently understood in the period 1963–1964 that these patterns could be explained if the mesons and baryons were made up of three basic entities and their three respective antiparticles. Zweig called these entities *aces* and Gell-Mann called them *quarks*. Gell-Mann came up with a name pronounced "quork" to rhyme with "fork," but then later came upon the quote from *Finnegan's Wake*, "Three quarks for Muster Mark," where quark rhymes with "mark." For this reason the word "quark" is pronounced by some as "quark" and by others as "quork." Gell-Mann recognized that the quarks had to have fractional charges and called the three entities *up*, *down* and *strange*, which are written as u, d and s. We refer to these as different quark *flavors*. Thus the $SU(3)$ pattern of the baryons and mesons is $SU(3)$-flavor. Every hadron is either a baryon (a strongly interacting fermion) or a meson (a strongly interacting boson). The electric charge of the u quark is $+\frac{2}{3}e$ and both the d and s have electric charge $-\frac{1}{3}e$, where $e > 0$ is the magnitude of the electron charge. By contrast, leptons are defined as elementary spin-$\frac{1}{2}$ particles (fermions) that do not experience the strong interaction. The name lepton comes from the Greek word leptós (λεπτός), which means thin, fine or slender. The earliest known leptons were less massive than the hadrons.

Baryons were understood to contain three quarks and mesons were understood to contain a quark and an antiquark. The proton has electric charge $+e$ and consists of uud; the neutron has zero electric charge and consists of udd; the π^+ has electric charge $+e$ and consists of $u\bar{d}$; and so on. However, no fractional electric charge has ever been observed and the reason for this will soon become clear. Gell-Mann sometimes commented that he thought of quarks as mathematical constructs and not as physical particles. Gell-Mann received the Nobel Prize for Physics in 1969 for his contributions to our understanding of particle classification.

The spins of baryons in this quark model were understood to be the sums of the spins of the three quarks. The spin-$\frac{3}{2}$ subatomic particles Δ^{++} and Ω^- consist of uuu and sss, respectively. In each case the three spin-$\frac{1}{2}$ quarks must have the same spin to make up spin-$\frac{3}{2}$. Quarks are fermions and so must obey the Pauli exclusion principle. Wolfgang Pauli received the 1945 Nobel Prize in Physics for the development of his principle. Since the three quarks (uuu and sss) cannot be in the same state it was postulated independently by Oscar Greenberg and by Moo-Young Han and Yoichiro Nambu in 1964–1965 that the quarks must carry an additional quantum number that we now refer to as *color*, which has nothing at all to do with the color of visible light. Each quark was said to be either *red*, *green* or *blue*, where each baryon contained one quark of each color. The idea was that a baryon itself would have no net color charge; that is, it would be color neutral or more technically *color singlet*, since the three primary colors combine to be color singlet or "white" as they do on a color television. A meson contains a quark with a color and an antiquark with the corresponding anticolor. Reviews of the quark model can be found in the texts by Kokkedee (1969) and Lichtenberg (1978) and a brief but modern summary is given by the Particle Data Group (e.g., Zyla et al., 2020). A detailed discussion of the quark model and its group theory structure is beyond the scope of this text, but every student of the strong interactions should become familiar with it. Some key aspects are summarized in Sec. 5.3.

It was finally in 1973 that Harald Fritsch, Murray Gell-Mann and Heiri Leutwyler proposed the modern picture of the $SU(3)$ color group with its eight force-carrier *gluons* providing the basis of the strong interactions. This work used the nonabelian field theory developed by Chen Ning Yang and Robert Mills in 1954 (Yang and Mills, 1954). The result was the theory now referred to as QCD, where the gluons interact directly with themselves and with quarks. This contrasts with the abelian field theory of QED where the photons do not interact directly with themselves.

In the meantime studies of hadrons in high-energy collisions had led Richard Feynman in 1969 to propose that hadrons contained pointlike objects that he referred to as *partons*. Feynman had originally thought to apply his *parton model* to proton-proton scattering. Following interactions between Bjorken and Feynman this parton model was applied to the study of *deep inelastic scattering* in electron-proton collisions by James Bjorken and Emmanuel Paschos in 1969 with great success. The partons were interpreted as consisting of the three quarks providing the quantum numbers of the baryon as well as of a sea of quark-antiquark pairs. We now refer to these as the *valence quarks* and *sea quarks*, respectively. We now know that quarks, antiquarks and gluons are all partons. The success of Feynman's parton model provided very compelling evidence that quarks and gluons were actual particles and not simply mathematical constructs that help to explain the families of hadrons.

The property of the *asymptotic freedom* of QCD was discovered by David Gross, David Politzer and Frank Wilczek in 1973, for which they were awarded the Nobel Prize in Physics in 2004. Asymptotic freedom means that the interactions between quarks and gluons become weaker at high energy scales. This became the basis for the interpretation of partons as "almost free" quarks, antiquarks and gluons inside hadrons in deep inelastic scattering processes. A perturbation theory approach to QCD at high energies was then possible and the properties of perturbative QCD have since been experimentally confirmed to within a few percentage points.

If quarks and gluons are particles and if quarks have fractional charge, why has no fractional charge ever been measured? The fact that QCD is a nonabelian field theory with self-interacting gluons gives rise to asympotic freedom as already noted; i.e., the QCD coupling decreases as the energy scale increases or equivalently as the distance scale decreases. The other consequence of gluon self-interactions is that QCD coupling increases with decreasing energy scale or equivalently with increasing distance scale. At large distances the potential between two colored objects was thought to increase linearly with separation, which has since been confirmed to high precision in *lattice gauge theory* studies of QCD, referred to as *lattice QCD*. This gives rise to the property referred to as *color confinement*, where colored objects such as quarks and gluons cannot be produced as free particles, which means that they are forever trapped inside the color singlet combinations that are the hadrons. The $q\bar{q}$ combinations that make up the mesons and the qqq combinations that make up the baryons can only ever give rise to integer electric charge. At large distances the *linearly rising potential* arises from a *flux tube* of constant diameter containing gluons that forms between two oppositely colored entities, such as a quark and an antiquark. The entire object is a color singlet. The energy of the flux tube increases approximately linearly with its length because the flux tube diameter is approximately constant as the length increases. At some point the flux tube or "string" of color flux has enough energy to create a quark-antiquark pair from the vacuum, thus breaking the flux tube/string into two shorter flux tubes/strings.

In addition to the up (u), down (d) and strange (s) quark flavors initially observed we now have additionally detected charm (c), bottom (b) and top (t) quarks. Sheldon Glashow, John Iliopoulos and Luciano Maiani predicted the charm quark in 1970 on the basis of the Glashow-Iliopoulos-Maiani (GIM) mechanism, which is the mechanism that required a fourth quark to suppress flavor-changing neutral currents (FCNCs) in loop diagrams. The charm quark was detected by two teams, one led by Burton Richter and one by Samuel Ting, in 1974. The existence of the bottom and top quarks was predicted in 1973 by Makoto Kobayashi and Toshihide Maskawa as the basis of their proposed explanation of the observed CP violations in kaon decay. An experimental team led by Leon Lederman discovered the bottom quark in 1977. The very massive top quark (with approximate mass of 173 GeV/c^2) was discovered in 1995 by both the CDF and D0 (pronounced "D zero") collaborations. Kobayashi and Maskawa won the Nobel Prize in Physics in 2008 for their prediction of the bottom and top quarks. There was initially an attempt to name the t and b quarks "truth" and "beauty," but the close analogy with "up" and "down" has meant that "top" and "bottom" have emerged as the standard names.

Since quarks and gluons are confined to the interior of hadrons and since hadrons are color singlets, the question arises as to what is the nature of the strong interaction that binds protons and neutrons together inside atomic nuclei. Just as *van der Waals forces* give rise to electromagnetic forces between electrically neutral atoms and molecules, it is the QCD analog of the van der Waals forces that leads to hadron-hadron interactions. This includes the binding of nucleons to form the atomic nucleus. At long distances the hadron-hadron interaction is dominated by the exchange of the lightest color singlet hadron, the pion, π. This pion exchange gives rise to the so-called long-range Yukawa potential, $V \sim e^{-m_\pi r}/r$, between nucleons, which we derive in Sec. 7.6.3 from the Yukawa interaction. Hideki Yukawa was awarded the 1949 Nobel Prize in Physics for his prediction of mesons on the basis of studies of nuclear forces.

In high-energy processes, when a parton (a quark or gluon) is given a large amount of four-momentum and begins to move away from the interaction region, it is prevented from doing so by the large forces associated with color confinement. The result is that quark-antiquark pairs are stripped out of the vacuum between separating colored particles and the outgoing partons *hadronize into QCD jets* consisting of collimated groups of hadrons whose total four-momentum is approximately the four-momentum of the initial parton. Jets typically form tight cones of particles, which makes their identification possible. One simple picture of this hadronization process where partons evolve into jets is provided by the *Lund string model*. This exploits the simple flux tube/string breaking mechanism mentioned above. As two color-connected partons separate from each other the "string" (color flux tube) breaks into two strings and this process is repeated until there is not enough energy in the string for further breakages. An analysis of jets in high-energy collisions makes it possible to perform tests of the predictions of perturbative QCD. Models of hadronization are an essential part of the Monte Carlo event generators that compare theoretical predictions of high-energy collisions with experiment.

It remains to briefly discuss the importance of approximate *chiral symmetry* as an organizing principle for understanding the properties of the strong interaction. We have seen in Sec. 4.6 that for free massless fermions helicity and chirality are equivalent and that chirality is preserved under restricted Lorentz transformations. When massless fermions interact with gauge fields at the level of perturbation theory the chirality of the fermions is preserved; e.g., the chirality of a massless quark is unchanged when it interacts perturbatively with a photon or gluon. For example, adding an interaction with photons to the massless Dirac equation turns the partial derivative in Eqs. (4.6.18) and (4.6.20) into a covariant derivative, which does not mix the left and right chiralities. This means that chiral symmetry is an exact symmetry of massless QCD at the level of perturbation theory. In the infrared region QCD is strongly interacting and this leads to the nonperturbative formation of a *quark condensate* $\langle \bar{q}q \rangle$, which has the form of a mass term for the quarks and hence violates chiral symmetry. This process is an example of *spontaneous symmetry breaking*, where the vacuum state of the system does not respect the symmetry of the underlying Lagrangian. This spontaneous breaking of chiral symmetry in QCD is often referred to as *dynamical chiral symmetry breaking*, where the word "dynamical" reflects the fact that it is the interactions between the quarks and gluons in QCD that lead to the symmetry breaking.

When we say that a continuous system has *spontaneous symmetry breaking of a continuous symmetry*, we mean that (i) the Hamiltonian or Lagrangian defining the system has an exact continuous symmetry with the corresponding Noether currents conserved and (ii) the conserved charges arising from those conserved currents generate transformations that do not leave the vacuum or ground state of the system invariant. Put simply, the Lagrangian has a continuous symmetry that the vacuum or ground state does not.

Goldstone's theorem in relativistic quantum field theory and condensed matter physics states that, for a system with spontaneous symmetry breaking of a continuous symmetry, there will be one spin-zero massless particle for each conserved charge of the symmetry that has been broken. These spin-zero (scalar) particles are referred to as *Goldstone bosons* or *Nambu-Goldstone bosons* (NGBs). These massless Nambu-Goldstone bosons are collective excitations of the system and are the long-wavelength fluctuations of the order parameters corresponding to the broken symmetries. These collective excitations have the quantum numbers of the respective broken generators. This statement is easiest to understand in the context of condensed matter physics. In solids the spontaneous breaking of translational and rotational invariance by the ionic lattice leads to the massless collective excitations called phonons, which are massless in the sense that a phonon with vanishingly small momentum requires vanishingly small energy to create. The situation is similar with magnets, where in the ground state rotational invariance is violated by the alignment of all of the magnetic dipoles into one spatial direction. The corresponding Nambu-Goldstone boson is a magnon, which is the quantized form of a wave where the directions of the magnetic dipoles oscillate around the ground state direction. Goldstone's theorem will be discussed further in Sec. 8.8.

If the continuous symmetry that is spontaneously violated is not exact, then we say that it is *explicitly broken* or *explicitly violated* as well as spontaneously broken. In this case the currents associated with the symmetry will not be exactly conserved and the corresponding Nambu-Goldstone bosons will not be massless. However, if the explicit breaking is small, then we will have approximate current conservation and the Nambu-Goldstone bosons will be light and we refer to them as *pseudo-Goldstone bosons* or *pseudo-Nambu-Goldstone bosons* (PNGBs).

As seen from Table 5.3 the u and d are the two lightest quarks with masses $m_u \sim 2$ MeV and $m_d \sim 5$ MeV, respectively. These nonzero masses are an *explicit breaking of chiral symmetry*, but are very small compared with the proton mass of $m_p = 938$ MeV. As we see in Table 5.4 the charge and flavor content of the proton arises from the uud *valence quarks*, which have a total mass of approximately 10 MeV. The other ~ 930 MeV comes from dynamical chiral symmetry breaking in the form of *sea quarks* made up of quark-antiquark pairs and gluons. In the *constituent quark model* the sea quarks and gluons contribute a mass of approximately 330 MeV to each of the valence quarks making up baryons and most mesons.

If there were N_f different flavor quarks with the same mass, then the QCD Lagrangian would have a $U(N_f)$ unitary flavor symmetry. If these N_f quarks were all massless, then QCD would have a *flavor chiral symmetry* of the form $U(N_f)_L \times U(N_f)_R$, which means that the left-handed and

Table 5.3. The fundamental fermions (all have spin-half)

Leptons			Quarks		
Flavor	Mass (GeV/c²)	Charge (Q) e	Flavor	Mass (GeV/c²)	Charge (Q) e
ν_e electron neutrino	$< 1 \times 10^{-8}$	0	u up	0.002	2/3
e electron	0.000511	-1	d down	0.005	$-1/3$
ν_e muon neutrino	< 0.0002	0	c charm	1.3	2/3
μ muon	0.106	-1	s strange	0.1	$-1/3$
ν_τ tau neutrino	< 0.02	0	t top	173	2/3
τ tau	1.7771	-1	b bottom	4.2	$-1/3$

Table 5.4. Baryons (qqq) and antibaryons ($\bar{q}\bar{q}\bar{q}$). Baryons are fermionic hadrons. There are about 120 types of baryon.

Symbol	Name	Quark content	Electric charge	Mass (GeV/c²)	Spin
p	Proton	uud	1	0.938	1/2
$\bar{p}$	Anti-proton	$\bar{u}\bar{u}\bar{d}$	-1	0.938	1/2
n	Neutron	udd	0	0.940	1/2
Λ	Lambda	uds	0	1.116	1/2
Ω^-	Omega	sss	-1	1.672	3/2

right-handed quark components remain decoupled under restricted Lorentz transformations. Given that the c, s, b and t quarks are heavy, the QCD Lagrangian has an approximate two-flavor chiral symmetry $U(2)_L \times U(2)_R$. This group decomposes into $SU(2)_L \times SU(2)_R \times U(1)_V \times U(1)_A$, where $U(1)_V$ is the singlet vector group and $U(1)_A$ is the singlet axial group. The symmetry $U(1)_V$ corresponds to baryon number conservation, whereas $U(1)_A$ does not lead to a symmetry because of a corresponding quantum anomaly. An *anomaly* in a quantum field theory arises when the procedure of regularization necessary for renormalization unavoidably violates a symmetry.

After the spontaneous symmetry breaking due to the formation of the u and d quark condensates, the $SU(2)_L \times SU(2)_R$ flavor chiral symmetry (or $SU(2)$ flavor chiral symmetry in shorthand) is broken down to the diagonal subgroup referred to as isospin flavor symmetry $SU(2)$ (see Sec. A.7.1). Note that there are six generators of $SU(2)_L \times SU(2)_R$ and three generators of isospin $SU(2)$, since each $SU(2)$ has three generators as shown in Table A.1. For each generator of a symmetry that is broken there is one corresponding Nambu-Goldstone boson, so there are three broken generators giving rise to three Nambu-Goldstone bosons, which are the three pions $\pi^\pm, \pi^0$. As discussed following Eq. (4.6.4) the chirality operator is γ_5 and so can be expected to be present in the generators of the flavor chiral symmetry. From Eq. (4.4.159) we know that the addition of a γ_5 turns a scalar bilinear into a pseudoscalar bilinear and it is therefore not surprising that the pions are pseudoscalar Nambu-Goldstone bosons. The spin-parity quantum numbers of a pseudoscalar are conventionally written as $J^P = 0^-$, which means spin zero ($J = 0$) and odd intrinsic parity ($P = -1$).

In the limit of equal small u and d quark masses, $0 < m_u = m_d \ll m_p$, we have a small explicit chiral symmetry breaking in addition to the spontaneous chiral symmetry breaking. In that case the $SU(2)$ flavor isospin symmetry remains a good symmetry of QCD so that the pion masses arising from QCD are nonzero but are equal. For pure QCD we also have the nucleons (proton and neutron) in an isospin doublet, $m_p = m_n$, where most of the nucleon mass arises from spontaneous chiral symmetry breaking.

This remaining $SU(2)$ isospin symmetry is in turn explicitly broken by the small up-down mass difference, $m_u \neq m_d$, and by the relatively weak electromagnetic interaction, since $q_u = +\frac{2}{3}$ and $q_d = -\frac{1}{3}$. It is these small explicit sources of isospin violation that lead to the small proton-neutron mass difference, $m_n - m_p = 1.3$ MeV, and the pion mass splitting, $m_{\pi^\pm} - m_{\pi^0} = 4.5$ MeV.

While the strange quark with a mass $m_s \simeq 95 \pm 5$ MeV is significantly heavier than the up and down quarks, it is still relatively light compared with the proton and the charm, bottom and top quarks. We can then also consider the three flavors u, d and s together in an approximate symmetry given by $U(3)_L \times U(3)_R = SU(3)_L \times SU(3)_R \times U(1)_V \times U(1)_A$. However the explicit breaking of the $SU(3)_L \times SU(3)_R$ chiral symmetry (or $SU(3)$ flavor chiral symmetry in shorthand) is now greater, which is why the additional pseudo-Nambu-Goldstone bosons containing strangeness, the kaons, are heavier than the pions as we see in Table 5.5. Since $SU(3)$ has eight generators

Table 5.5. Mesons ($q\bar{q}$). Mesons are bosonic hadrons. There are about 140 types of mesons.

Symbol	Name	Quark content	Electric charge	Mass (GeV/c^2)	Spin
π^+	Pion	$u\bar{d}$	$+1$	0.140	0
K^-	Kaon	$s\bar{u}$	-1	0.494	0
ρ^+	Rho	$u\bar{d}$	$+1$	0.770	1
B^0	B-zero	$d\bar{b}$	0	5.279	0
η_c	Eta-c	$c\bar{c}$	0	2.980	0

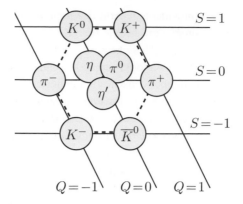

Figure 5.4   The pseudoscalar meson nonet, which consists of the eight pseudo-Nambu-Goldstone bosons of $SU(3)$ chiral symmetry mixing with the $SU(3)$ flavor singlet state. The charge and strangeness quantum number for each state is shown.

then $SU(3)_L \times SU(3)_R$ has sixteen generators, which after spontaneous chiral symmetry breaking reduced to $SU(3)$. So there are eight broken generators, leading to eight pseudoscalar pseudo-Goldstone bosons. As noted earlier, it was this fact that in part inspired Gell-man's naming of the Eightfold Way and the recognition of $SU(3)$ flavor as an organizing principle for the families of hadrons.

In Fig. 5.4 we show the nonet of pseudoscalar mesons, which arises from the octet of $SU(3)$ pseudo-Goldstone bosons mixing with the $SU(3)$ singlet state η^0. Ignoring for the moment explicit $SU(3)$ flavor violations, then the $SU(3)$ flavor octet would contain the four kaons, the three pions, and an η^0. The π^0 is the $I_3 = 0$ state of the isotriplet pions ($I = 1$). The η^0 is the isosinglet state ($I = 0$), while the η'^0 is the $SU(3)$ flavor singlet state. Since the explicit $SU(3)$ flavor breaking is much greater than $SU(2)$ isospin flavor breaking and since the π^0 contains no strange quarks, it mixes weakly with the ηs. On the other hand, the η^0 and η'^0 both contain strange quark contributions and so mix more strongly with each other to give rise to the physical η and η'. These are then the nine physical states that appear in the pseudoscalar meson nonet.

Finally, to end this brief review of QCD we note the importance of *lattice field theory* as a systematically improvable first principles approach to the study of quantum field theory in the nonperturbative regime. It puts fields on a discrete finite-volume Euclidean (imaginary time) four-dimensional spacetime lattice. It has been especially exploited in the form of *lattice QCD*, which puts quarks and gluons on a spacetime lattice. The approach requires extrapolation of the lattice to infinitesimally small lattice spacing, infinitely large volumes and light u and d quark masses in order to recover the predictions of QCD. Through the growth of computing power the approach is becoming an increasingly precise way to study many aspects of QCD and hadronic physics. For example, it has provided numerical confirmation that QCD reproduces the low-mass hadron spectrum (see Fig. 9.3), leads to confinement and confirms asymptotic freedom in the short distance regime. The field owes a great deal to Kenneth Wilson, who was a pioneer of our understanding of the *renormalization group* in quantum field theory and of lattice QCD. Wilson also did seminal work on the *operator product expansion*. He was awarded the Nobel Prize in Physics in 1982 for his work on the renormalization group and phase transitions. The renormalization group describes how the description of physical systems changes with the scale at which they are being described. This will later be seen to be essential for relating quantum field theory to physical observables and for understanding the SM.

We finish this section with a persistent puzzle associated with the QCD sector of the SM, referred to as the *strong CP problem*. We defer detailed discussion and simply describe the issue here. If QCD has no massless quarks, then there is a completely natural term that could be added to the QCD Lagrangian density that violates CP conservation, where by this we mean invariance under a CP transformation. If one quark mass vanished or if there were fewer than three generations of quarks, then the CP violation could be removed by a chiral rotation of the fields. The strength of that CP violation is characterized by an angle called the QCD θ parameter, which in principle could take any value. However, experimentally the CP violation from the strong interaction is known to be extremely small, $\theta \lesssim 10^{-9}$. For example, the θ term induces an electric dipole moment of the neutron d_n, which has a measured upper limit $d_n \lesssim 10^{-25}$ e.cm, whereas if θ were of order one we would have $d_n \sim 10^{-16}$ e.cm. This extremely small value for θ is considered unnatural and a case of fine-tuning and is referred to as the *strong CP problem*. Attempts have been made to extend QCD in order to explain the fine-tuning, where perhaps the best-known involves new light pseudoscalar particles called *axions* that arise in the theory of Peccei and Quinn (1977). Axions have been considered as a potential candidate for dark matter.

5.2.3 Development of Electroweak (EW) Theory

We begin with a brief history of the weak interactions and the electroweak theory. For an overview of the development of the electroweak theory, see Kibble (2015), and for some additional historical insights, see Close (2010) and Quigg (2015).

Antoine Becquerel was the first to discover radioactivity, and the SI unit for radioactivity, the becquerel (Bq), is named after him. Becquerel, Marie Curie[4] and Pierre Curie received the Nobel Prize in Physics in 1903 for their work on radioactivity. Ernest Rutherford observed two different types of radiation, which he referred to as α and β radiation. Becquerel discovered that the β-rays were the cathode rays (i.e., electrons) that had been studied by J. J. Thomson in 1897. The α-rays are helium nuclei and come from radioactive decays (called α decay) of nuclei resulting from the strong interaction rather than the weak interaction. James Chadwick discovered the neutron in 1932 for which he received the Nobel Prize in Physics in 1935. In 1914 Chadwick noted that the β-rays coming from the β decay of the neutron to the proton had a continuous spectrum rather than the discrete spectrum expected if a single particle was emitted, i.e., if $n \to p + e^-$. In 1930 Wolfgang Pauli suggested that another neutral particle was involved in the decay, which was named the *neutrino* ("little neutron") by Enrico Fermi. Then in 1934 Fermi proposed his theory of β decay,

$$n \to p + e^- + \bar{\nu}_e, \tag{5.2.2}$$

where a neutron decays into a proton, an electron and an anti-electron neutrino. Because this interaction has four "legs" it is often referred to as the *Fermi four-point interaction*. The strength of this interaction is characterized by the *Fermi coupling constant* G_F. Rudolph Peierls and Hans Bethe estimated that the strength of the interaction was so weak (G_F was so small) that neutrinos could pass through the Earth with very little chance of interaction. In 1956 the existence of the neutrino was confirmed using a nuclear reactor by Frederick Reines and Clyde Cowan by observing the *inverse β decay* process

$$\bar{\nu}_e + p \to n + e^+. \tag{5.2.3}$$

---

[4] Marie Curie was the first woman to receive a Nobel Prize. After receiving the Nobel Prize in Chemistry in 1911 she became the only person to win two Nobel Prizes in different sciences.

In essence Fermi's initial assumption was that the four-point interaction was the interaction of two vector currents, $V^\mu V_\mu$, in analogy with the $j^\mu A_\mu$ electromagnetic interaction of the electromagnetic fields A^μ with a charge four-vector current j^μ. One current involved the proton-neutron pair in a hadronic vector bilinear, $\bar{\psi}_n \gamma^\mu \psi_p$, and the other current involved the electron-neutrino pair in a leptonic vector bilinear, $\bar{\psi}_{\nu_e} \gamma^\mu \psi_e$, where we are referring to the fermion bilinears in Eq. (4.4.159). The interaction Hamiltonian density originally assumed by Fermi can be written as

$$\mathcal{H}_{\text{int}} = G_F J_{\text{F}}^{\dagger\mu} J_{\text{F}\mu} \quad \text{where}$$
$$J_{\text{F}}^\mu = \bar{\psi}_n \gamma^\mu \psi_p + \bar{\psi}_e \gamma^\mu \psi_{\nu_e} \quad \text{and} \quad J_{\text{F}}^{\dagger\mu} = \bar{\psi}_p \gamma^\mu \psi_n + \bar{\psi}_{\nu_e} \gamma^\mu \psi_e. \tag{5.2.4}$$

The hadron-hadron and lepton-lepton terms correspond to hadron and lepton scattering, respectively. The hadron-lepton terms give rise to the hadron-lepton transitions of the weak interactions such as β decay and inverse β decay.

Fermi was thinking in terms of quantum mechanics, whereas we will later come to understand expressions of this form in terms of quantum field theory. Note that $\mathcal{H}_{\text{int}}$ is Hermitian and that it transforms like $V^\mu V_\mu$ and so it is a Lorentz scalar. This means that it was even under a parity transformation and so was a parity-conserving operator. As a Lorentz scalar the Fermi interaction was spin independent, but it was observed by George Gamow and Edward Teller that the interaction had to be generalized to include spin-dependent interactions as well. Also by assuming such a form Fermi was implicitly assuming that parity was conserved in his four-point interaction model of the weak interactions. In the 1950s some properties of kaon decays appeared inconsistent with the assumption of parity conservation. This was historically referred to as the τ–θ puzzle, where the historically named θ^+ meson and τ^+ meson had decays $\theta^+ \to \pi^+ + \pi^0$ and $\tau^+ \to \pi^+ + \pi^+ + \pi^-$, respectively. The pion has negative intrinsic parity and parity is a multiplicative quantum number, which means that the intrinsic parities of the final states are $+1$ and -1, respectively. At low final state momenta the s-wave ($\ell = 0$) scattering will dominate and can be isolated by considering angular distributions of the final state pions. The parity associated with two particles with relative orbital angular momentum ℓ is $(-1)^\ell$. So in s-wave scattering there is no parity associated with the orbital angular momentum of the final states. Hence the θ^+ and τ^+ have positive and negative parity, respectively, which meant that these two mesons were initially assumed to be different particles. The puzzle was that the two mesons were observed to have the same mass and lifetime. We now know that the θ^+ and τ^+ mesons are in fact the same particle, the positively charged kaon, K^+, in Fig. 5.4. The θ^+ and τ^+ correspond to the two-pion and three-pion decay modes of the positively charged kaon, with the consequence that the weak interactions must violate parity.

It was this τ–θ puzzle that prompted Tsung-Dao Lee and Chen-Ning Yang in 1956 to examine the question of parity conservation. (Lee and Yang, 1956). They proposed to Chein-Shiung Wu some experiments to test this. It was resolved to test the directional properties of the emitted electrons in the β decay of cobalt 60 atoms ($^{60}_{27}\text{Co} \to {}^{60}_{28}\text{Ni} + e^- + \bar{\nu}_e + 2\gamma$). Since the electromagnetic interaction producing the photons was already known to respect parity, any violation of parity conservation must be due to the weak interaction. Wu and her colleagues performed the experiment in late 1956 and observed that parity was indeed violated. They did this by observing a nonzero correlation $\mathbf{J} \cdot \mathbf{p}$ between the (parity even) spin of the decaying cobalt nucleus $\mathbf{J}$ and the (parity odd) outgoing electron momentum $\mathbf{p}$. If parity was conserved, then the parity odd observable $\mathbf{J} \cdot \mathbf{p}$ would have to have a vanishing expectation value, but it did not. This was subsequently confirmed in studies of other decays. Lee and Yang were awarded the Nobel Prize in Physics in 1957 for their work.

In order to accommodate parity violation it was speculated that each of the hadronic and leptonic currents in the four-point interaction be a linear combination of vector ($V^\mu = \bar{\psi}\gamma^\mu\psi$) and axial vector (pseudovector) ($A^\mu = \bar{\psi}\gamma^\mu\gamma^5\psi$) bilinear currents, where the latter is not to be confused with the electromagnetic field four-vector. From a study of the various parity violating decays, Sudarsha and Marshak (1958) and subsequently Feynman and Gell-Mann (1958) concluded that the appropriate linear combination for the lepton part of the current was $V - A$,

$$\ell^\mu \equiv \bar{\psi}_e\gamma^\mu\tfrac{1}{2}(1 - \gamma^5)\psi_{\nu_e} = \bar{\psi}_e\gamma^\mu P_L\psi_{\nu_e}, \tag{5.2.5}$$

where we recall from Eq. (4.6.5) that $P_L \equiv \tfrac{1}{2}(1 - \gamma^5)$ is the left-handed chiral projection operator. Also let us recall from Eq. (4.6.33) that for massless or ultrarelativistic particles we know that P_L projects out negative helicity (left-handed helicity) for particles and positive helicity (right-handed helicity) for antiparticles. Since $P_L = P_L^2$ and $\gamma^0\gamma^\mu\gamma^5 = \gamma^5\gamma^0\gamma^\mu$ then we can write $\ell^\mu = \psi_e^\dagger P_L\gamma^0\gamma^\mu P_L\psi_{\nu_e}$. This means that the weak interaction only couples to the left-handed components of the electron and the neutrino and to the right-handed components of their respective antiparticles. Since neutrinos only interact with the rest of the SM particles through the weak interaction, then massless neutrinos must be left-handed and the corresponding massless antineutrinos must be right-handed. This extends to the muon (μ^-) and muon neutrino (ν_μ) and to the tau (τ) and tau neutrino (ν_τ).

Since the weak interactions couple *only* through left-handed couplings to the leptons there is no parity-conserving contribution to the weak interaction. We therefore say that the weak interactions are *maximally parity violating*. This was a surprising realization given that the day-to-day world that we experience appears parity invariant. This follows since our everyday world experiences are dominated by the parity-conserving forces of electromagnetism, the strong interaction and gravity.

The situation of the nucleon is more complicated and the nucleon contribution to the current can be written as

$$h^\mu \equiv \bar{\psi}_n\gamma^\mu\tfrac{1}{2}(c_V - c_A\gamma^5)\psi_p. \tag{5.2.6}$$

Ultimately, in the SM we will come to see that we replace the above with left-handed couplings to quark fields and that the chiral-symmetry breaking quark mass terms together with nonperturbative QCD effects lead to the above form in terms of nucleons. The vector current component is conserved due to baryon number conservation, which leads to the result that $c_V = 1$. Since chiral symmetry has a small explicit breaking due to the nonzero quark masses, then the axial current is only approximately conserved. This is referred to as the partial conservation of the axial current (PCAC) and results in $c_A \neq 1$. Then in the $V - A$ formulation of the Fermi four-point interaction with the Fermi coupling constant we have

$$\mathcal{H}_{\text{int}} = -\mathcal{L}_{\text{int}} = -(2\sqrt{2}G_F)J_F^{\dagger\mu}J_{F\mu} \quad \text{where}$$
$$\hat{J}_F^\mu = \bar{\psi}_n\gamma^\mu\tfrac{1}{2}(1 - c_A\gamma^5)\psi_p + \bar{\psi}_e\gamma^\mu\tfrac{1}{2}(1 - \gamma^5)\psi_{\nu_e} \quad \text{and}$$
$$\hat{J}_F^{\dagger\mu} = \bar{\psi}_p\gamma^\mu\tfrac{1}{2}(1 - c_A\gamma^5)\psi_n + \bar{\psi}_{\nu_e}\gamma^\mu\tfrac{1}{2}(1 - \gamma^5)\psi_e. \tag{5.2.7}$$

In 1963 Nicola Cabibbo noted that weak transitions involving $e - \nu_e$, $\mu - \nu_\mu$ and hadron transitions not involving changes in strangeness ($\Delta S = 0$) had similar probabilities of occurring, i.e., similar rates. He further noted that transitions involving $\Delta S = 1$ occurred at approximately $1/4$ of that rate. He introduced what we now call the *Cabibbo angle* to describe the relative admixture of the $\Delta S = 0$ and $\Delta S = 1$ components of the current. He wanted to combine the Eightfold Way of Gell-Mann with the $V - A$ coupling of the weak interaction, while preserving the idea of a universality of the

Table 5.6. Gauge bosons, the spin-one force carriers					
Unified electroweak			Strong (color)		
Name	Mass (GeV/c^2)	Charge (e)	Name	Mass (GeV/c^2)	Charge (e)
γ photon	0	0	g gluon	0	0
W^-	80.4	-1			
W^+	80.4	$+1$			
Z^0	91.187	0			

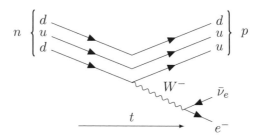

Figure 5.5   Illustration of beta decay, $n \to p + e^- + \bar{\nu}_e$, at the quark level, $d \to u + e^- + \bar{\nu}_e$.

weak interaction. We now understand the Cabibbo angle as part of the Cabibbo-Kobayashi-Maskawa (CKM) matrix that describes how three quark generations mix via the weak interaction. In terms of Fig. 5.6, the Cabibbo angle defines a 2×2 rotation matrix that mixes the d and s quarks when the third generation (the b quark) is ignored. The CKM matrix is a 3×3 unitary matrix describing the mixing of the d, s and b quarks (see Sec. 5.2.4).

Low-energy weak interactions were well described by Fermi's four-point interaction theory, but in the high-energy regime the theory broke down. For example, the predicted unbounded linear rise of neutrino cross-section with energy was inconsistent with cosmic ray data and it is unphysical. An unbounded increase in cross-section will eventually lead to the unphysical outcome that the probability of the process occurring exceeds unity. This corresponds to a *loss of unitarity* of the scattering matrix and is typical of the type of pathology that occurs when an effective theory is applied at too high energies.

Since the original intent of Fermi was to try to emulate QED in the weak interactions and since QED is a gauge theory it was natural to ask if the weak interactions could be formulated as a gauge theory. Julian Schwinger (1957) suggested that the weak and electromagnetic interactions should be combined into a single gauge theory. A variety of attempts were made at such a description, but a common problem to all of these was the fact that in order to reproduce the successful Fermi four-point interaction at low energies the exchanged gauge bosons (the W^+ and W^-) in the weak interaction had to have large masses; see Table 5.6. These masses had to be put in by hand, as gauge bosons are massless. In order for the weak and electromagnetic interactions to come from a single gauge theory with a single gauge coupling, it was clear that the $W^\pm$ bosons should have masses of order 100 GeV so as to reduce the strength of the weak interaction through the effect of the $W^\pm$ propagators at low momenta. The $W^\pm$ exchanges result in the *charged-current interaction* (see Fig. 5.5). Another issue was that the electromagnetic interaction conserved parity, while the weak interaction maximally violated it.

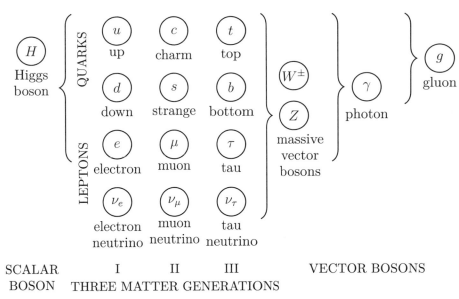

SCALAR   I   II   III   VECTOR BOSONS
BOSON   THREE MATTER GENERATIONS

Figure 5.6   Summary of the particles in the Standard Model. All matter particles are spin-half fermions, the vector bosons are spin-one and the Higgs boson has zero spin. The curly braces indicate which matter particles the different bosons interact with.

A first partial solution to the last issue was proposed by Sheldon Glashow (1961) in the form of an enlarged symmetry group, $SU(2)_L \times U(1)_Y$, that contained four vector bosons: the massive charged vector bosons, $W^{\pm} = (W_1 \mp iW_2)/\sqrt{2}$, and two neutral vector bosons, the W_3 and B, that subsequently mix together to give the massive neutral vector boson Z (also often written as Z^0) and the massless photon γ. The $W^{\pm}$ and Z are sometimes also referred to as the *intermediate vector bosons*, since they mediate the weak interactions. The "L" denotes left-handed and the "Y" denotes the *weak hypercharge* Y_W, where $Q = T_3 + \frac{1}{2}Y_W$ and where Q and T_3 denote the electric charge and the third component of *weak isospin*, respectively. The $W^{\pm}$ and Z are the carriers of the weak interaction and γ carries the electromagnetic interaction. The exchange of a Z gives rise to the *neutral-current interaction*. Subsequently, Abdus Salam and John Ward independently proposed a similar model (Salam and Ward, 1964). However, problems remained in that nonabelian theories of massive vector bosons are non-renormalizable as discussed in Sec. 8.4.1 and suffer from a loss of unitarity in high-energy scattering.

In the early 1960s there was considerable effort to understand how gauge bosons might acquire masses. This was stimulated by the interest in the Bardeen-Cooper-Schreiffer (BCS)[5] theory of superconductivity and the associated emergence of an effective photon mass as evidenced by the Meissner effect. There was a difficulty resulting from the Goldstone theorem, since naively we would expect there to be associated Nambu-Goldstone bosons coming from the spontaneous breaking of (continuous) gauge symmetries. No such particles were seen. Gerald Guralnik (1964a,b) made some initial progress on this problem. Then three groups, essentially simultaneously, published papers in that same year that explained that gauge theories did not satisfy the assumptions associated with the proof of the Goldstone theorem: François Englert and Robert Brout (Englert and Brout, 1964), Peter Higgs (1964a,b) and Gerald Guralnik, Richard Hagen and Thomas Kibble (Guralnik et al.,

---

[5] John Bardeen, Leon Cooper and John Schrieffer were awarded the 1972 Nobel Prize in Physics for their development of the BCS theory of superconductivity.

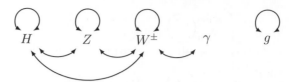

Figure 5.7 The allowed boson-boson Standard Model interactions. Standard Model fermions interact only with Standard Model bosons as shown in Fig. 5.6.

1964). The work of these authors showed that what would normally have been the Goldstone bosons coming from the broken generators of the gauge theory become the longitudinal components of the associated gauge bosons. This gives the extra degrees of freedom needed for the gauge bosons to acquire masses. However, this work was oriented toward attempts to explain the short-range behavior of the strong interactions, remembering that QCD was unknown at this time.

Finally, Steven Weinberg (1967) and Abdus Salam (1968) put together Glashow's program for the $SU(2)_L \times U(1)_Y$ framework for the electroweak interactions with the ideas of the spontaneous breaking of gauge symmetries in order to produce masses for the $W^\pm$ and Z vector bosons. The spontaneous breaking of the gauge symmetry down to the $U(1)$ of electromagnetism, $SU(2)_L \times U(1)_Y \to U(1)_{\text{em}}$, was induced using a weak isospin doublet of complex scalar fields, which is referred to as the *Higgs doublet*. From Eq. (3.2.48) we know that two complex scalar fields can be described in terms of four real scalar fields. The quartic *Higgs potential* induces a spontaneous symmetry breaking where one of the four real scalar fields takes on a nonzero vacuum expectation value $v = 246$ GeV, which is referred to as the *Higgs vacuum expectation value* (Higgs VEV). Before the symmetry breaking there were four (massless) gauge fields with $4 \times 2 = 8$ degrees of freedom and the Higgs doublet with four degrees of freedom, which gives 12 in total. After the symmetry breaking there are three massive vector bosons with 3×3 degrees of freedom, one (massless) gauge field (the photon) with two degrees of freedom, and one *Higgs boson* with one degree of freedom, which again gives a total of 12. To complete the picture Weinberg and Salam recognized that a Yukawa coupling of the Higgs doublet to massless quarks and leptons generates masses for these fermions that are proportional to the product of the Yukawa couplings and the Higgs vacuum expectation value v. The more massive the fermion, the greater is the strength of its coupling to the Higgs fields. This is why Higgs physics is closely connected with top quark physics. The SM boson-boson interactions are shown in Fig. 5.7.

The prediction of the Z immediately leads to the prediction of weak neutral currents, where neutrino-electron elastic scattering is mediated by a Z exchange. The existence of weak neutral currents was confirmed in 1973 at CERN using the Gargamelle heavy liquid bubble chamber detector (Haidt, 2015). Subsequently, Glashow, Salam and Weinberg received the 1979 Nobel Prize in Physics for their contributions to the theory of the unified weak and electromagnetic interactions and the prediction of the weak neutral current.

Weinberg and Salam expected that the electroweak theory would be renormalizable but were not able to demonstrate it. The work of Ludvig Faddeev and Victor Popov showed how to construct Feynman diagram expansions for nonabelian gauge field theories (Faddeev and Popov, 1967). Gerardus 't Hooft was a PhD student at Utrecht University under the supervision of Martinus J. G. Veltman and in 1971 he proved the renormalizability of the theory. Many years later in 1999, 't Hooft and Veltman were awarded the Nobel Prize in Physics for this achievement.

The electroweak theory predicted the masses of the $W^{\pm}$ and Z vector bosons, and they were subsequently discovered at CERN in 1983 by the UA1 and UA2 collaborations (Arnison et al., 1983; Banner et al., 1983). Carlo Rubbia and Simon van der Meer were awarded the 1984 Nobel Prize in Physics for their discovery.

The combination of the color $SU(3)_c$ of QCD, together with the $SU(2)_L \times U(1)_Y$ of the electroweak theory, gives rise to the SM with the gauge group $SU(3)_c \times SU(2)_L \times U(1)_Y$. The addition of the Higgs doublet and Higgs potential gives rise to electroweak symmetry breaking and the Higgs VEV, which in turn generates the quark and lepton masses and the masses of the $W^{\pm}$ and Z. Finally, the long-predicted Higgs boson was discovered using the Large Hadron Collider (LHC) at CERN in 2012 by the ATLAS and CMS Collaborations with a mass of approximately 125 GeV. The Nobel Prize in Physics in 2013 was awarded to François Englert and Peter Higgs for the theoretical discovery of the Higgs mechanism. Brout was deceased by this time and the maximum number of awardees for a Nobel Prize is three, which is why it was not possible to include all of those who had made such important contributions to the understanding of the Higgs mechanism, the electroweak theory, the resulting SM and the Higgs boson.

5.2.4 Quark Mixing and the CKM Matrix

The quark states that propagate freely are referred to as the quark *mass eigenstates*, whereas the quark *flavor eigenstates* are the ones that couple in the weak interactions. The unitary transformation that transforms between the mass and flavor eigenstates is the Cabibbo-Kobayashi-Maskawa (CKM) matrix (Cabibbo, 1963; Kobayashi and Maskawa, 1973; Griffiths, 2008). The CKM matrix is the extension of the work of Nicola Cabibbo, who introduced the Cabibbo angle (θ_c) to preserve the unitarity of the weak interaction. Denoting here d' and s' as the flavor eigenstates and d and s as the mass eigenstates of the down and strange quarks, the Cabibbo angle is defined such that

$$\begin{bmatrix} d' \\ s' \end{bmatrix} = \begin{bmatrix} V_{ud} & V_{us} \\ V_{cd} & V_{cs} \end{bmatrix} \begin{bmatrix} d \\ s \end{bmatrix} = \begin{bmatrix} \cos\theta_c & \sin\theta_c \\ -\sin\theta_c & \cos\theta_c \end{bmatrix} \begin{bmatrix} d \\ s \end{bmatrix}. \qquad (5.2.8)$$

We see that the Cabibbo angle describes an orthogonal (rotation) matrix, referred to as the *Cabibbo matrix*, that allows the mass and flavor eigenstates of the d and s quarks to not coincide. The physical relevance of the V_{ij} includes, but is not limited to, the fact that $|V_{ij}|^2$ is the probability that a quark of flavor i transitions to a quark of flavor j through the weak interaction. The Cabibbo matrix involves four flavors (u, d, s, c), but of course the c quark was not discovered until much later in 1974 as discussed above.

Following the discovery of CP-violation in kaon decays in 1964 it became understood that the 2×2 Cabibbo matrix could not incorporate any complex phase, which has the consequence that it could not explain CP violation. On this basis Kobayashi and Maskawa proposed a third generation of quarks, the b and t quarks, which allows a single complex CP-violating phase. Referring to Fig. 5.6 we see that the u, c and t quarks all have electric charge $+2/3$ and are referred to as the *up-type quarks* and that the d, s and b quarks all have electric charge $-1/3$ and are referred to as the *down-type quarks*. By construction the weak interaction allows the up-type and down-type quarks in the same generation to decay or mix into each other; e.g., in the first generation we have $d \to u$ in β decay and $u \to d$ in inverse β decay. The Cabibbo matrix allows mixing of the first two generations, whereas the CKM matrix allows mixing of all three generations. The intergenerational mixing could be formulated in terms of the up-type or the down-type quarks, but for historical reasons it is conventional to do the latter.

The CKM matrix is the unitary matrix V_{CKM} defined such that

$$\begin{bmatrix} d' \\ s' \\ b' \end{bmatrix} = V_{\text{CKM}} \begin{bmatrix} d \\ s \\ b \end{bmatrix} = \begin{bmatrix} V_{ud} & V_{us} & V_{ub} \\ V_{cd} & V_{cs} & V_{cb} \\ V_{td} & V_{ts} & V_{tb} \end{bmatrix} \begin{bmatrix} d \\ s \\ b \end{bmatrix}. \tag{5.2.9}$$

As we saw in Sec. 4.5.2 charge conjugation takes a fermion spinor to an antifermion spinor and this process involves complex conjugation. So the CKM matrix for antiquarks is the complex conjugate of that for quarks; i.e., in Eq. (5.2.9) we replace quarks by antiquarks and we replace V_{CKM} by V_{CKM}^*.

An $n \times n$ unitary matrix U is specified by n^2 real numbers as shown in Table A.1, since an $n \times n$ complex matrix contains $2n^2$ real numbers and $UU^\dagger = I$ is a set of n^2 constraints. In the case of n generations of quarks, we have the $n \times n$ version of Eq. (5.2.9). As discussed in Sec. 4.1.1, a normalized physical state has an overall phase ambiguity and so there are $2n$ phase ambiguities. If we multiply all $2n$ quark states with the same phase, then there is no change in the unitary $n \times n$ CKM-type matrix. This means that $2n - 1$ parameters of the $n \times n$ CKM matrix are physically irrelevant. Then the number of physically relevant, and hence measurable, parameters in an n-generation CKM matrix is $n^2 - (2n - 1) = (n - 1)^2$. An $n \times n$ orthogonal matrix has $n(n-1)/2$ real parameters as shown in Table A.1. So an $n \times n$ CKM matrix V_{CKM} can be written in terms of $n_E = n(n-1)/2$ Euler angles and $n_O = (n-1)^2 - n(n-1)/2 = (n-1)(n-2)/2$ other real parameters. So for $n = 2$ we have $n_E = 1$ and $n_O = 0$; i.e., there is only one Euler angle, the Cabibbo angle, and no other real parameters. In the $n = 3$ case we have $n_E = 3$ and $n_O = 1$. We could, for example, represent the 3×3 CKM matrix as $V_{\text{CKM}} = O_1 U_2 O_3$, where O_1 and O_3 are one-parameter orthogonal matrices and U_2 is a two-parameter unitary matrix described by an Euler angle and a phase. This is the basis for the so-called standard parameterization of the CKM matrix (Zyla et al., 2020),

$$V_{\text{CKM}} = \begin{bmatrix} 1 & 0 & 0 \\ 0 & c_{23} & s_{23} \\ 0 & -s_{23} & c_{23} \end{bmatrix} \begin{bmatrix} c_{13} & 0 & s_{13}e^{-i\delta} \\ 0 & 1 & 0 \\ -s_{13}e^{i\delta} & 0 & c_{13} \end{bmatrix} \begin{bmatrix} c_{12} & s_{12} & 0 \\ -s_{12} & c_{12} & 0 \\ 0 & 0 & 1 \end{bmatrix}$$

$$= \begin{bmatrix} c_{12}c_{13} & s_{12}c_{13} & s_{13}e^{-i\delta} \\ -s_{12}c_{23} - c_{12}s_{23}s_{13}e^{i\delta} & c_{12}c_{23} - s_{12}s_{23}s_{13}e^{i\delta} & s_{23}c_{13} \\ s_{12}s_{23} - c_{12}c_{23}s_{13}e^{i\delta} & -c_{12}s_{23} - s_{12}c_{23}s_{13}e^{i\delta} & c_{23}c_{13} \end{bmatrix}, \tag{5.2.10}$$

where $c_{ij} \equiv \cos\theta_{ij}$ and $s_{ij} \equiv \sin\theta_{ij}$; the three Euler angles are $(\theta_{12}, \theta_{23}, \theta_{13})$; and δ is the CP-violating phase. Note that if we neglect the mixing of the first and second generations to the third generation $(\theta_{13}, \theta_{23} \to 0)$, then O_3 becomes the Cabibbo matrix and $\theta_{12} \to \theta_c$. Another common parameterization is that proposed by Lincoln Wolfenstein (Wolfenstein, 1983),

$$V_{\text{CKM}} = \begin{bmatrix} 1 - (\lambda^2/2) & \lambda & A\lambda^3(\rho - i\eta) \\ -\lambda & 1 - (\lambda^2/2) & A\lambda^2 \\ A\lambda^3(1 - \rho - i\eta) & -A\lambda^2 & 1 \end{bmatrix} + \mathcal{O}(\lambda^4), \tag{5.2.11}$$

where $\lambda \equiv s_{12}$, $A\lambda^2 \equiv s_{23}$ and $A\lambda^3(\rho + i\eta) \equiv s_{13}e^{i\delta}$. The rationale for this parameterization is that we know experimentally that $\lambda \equiv s_{12} \simeq 0.226$ and $s_{13} \ll s_{23} \ll s_{12}$ and so by having $s_{13} \sim \lambda^3$, $s_{23} \sim \lambda^2$ and $s_{12} \equiv \lambda$ we can reasonably expect $A, \rho, \eta \sim \mathcal{O}(1)$. The parameters of the CKM matrix are input parameters of the SM and at this time there is no known physical explanation of the values that they take (Zyla et al., 2020).

The nine unitarity constraints on the CKM matrix are $V_{\text{CKM}}V_{\text{CKM}}^{\dagger} = I$, which implies that $V_{\text{CKM}}^{\dagger}V_{\text{CKM}} = I$ since $V_{\text{CKM}}^{\dagger} = V_{\text{CKM}}^{-1}$. In terms of matrix elements we can write these as

$$\sum_k V_{kj}V_{ki}^* = \sum_k V_{ik}V_{jk}^* = \delta_{ij}, \tag{5.2.12}$$

where the two ways of writing the nine constraints look different but the totality of their content must be the same. Naively, this appears to be eighteen constraints, but obviously only nine of the constraints are independent. For the matrix element V_{ij} we have the row index $i = 1, 2, 3 = u, c, t$ and the column index $j = 1, 2, 3 = d, s, b$. The six diagonal unitarity constraints are the so-called *weak unitarity conditions* $\sum_k |V_{ik}|^2 = \sum_k |V_{ki}|^2 = 1$, which implies that the sum of the couplings of each of the up-type quarks to the three down-type quarks is the same and vice versa.

For $i \neq j$ we have twelve nondiagonal constraints, $\sum_k V_{kj}V_{ki}^* = \sum_k V_{ik}V_{jk}^* = 0$. However, the exchange $i \leftrightarrow j$ is the same as taking the complex conjugate of the equations and so only six of these off-diagonal constraints are nontrivial. Each of these six nondiagonal constraints can be viewed as a sum of three complex numbers adding to zero, which when plotted as a sum of vectors in the complex plane corresponds to a closed triangle. Thus the unitarity of the CKM matrix leads to six nontrivial *unitarity triangles*. Useful nondiagonal unitarity constraints for the strange and bottom sectors are, respectively, $\sum_k V_{k1}V_{k2}^* = \sum_k V_{k1}V_{k3}^* = 0$, which corresponds to the unitarity triangles

$$V_{ud}V_{us}^* + V_{cd}V_{cs}^* + V_{td}V_{ts}^* = V_{ud}V_{ub}^* + V_{cd}V_{cb}^* + V_{td}V_{tb}^* = 0. \tag{5.2.13}$$

For the second of these it is conventional to divide by $V_{cd}V_{cb}^*$, since it is the best known, to give the unitarity triangle shown in Fig. 5.8

$$1 + \frac{V_{td}V_{tb}^*}{V_{cd}V_{cb}^*} + \frac{V_{ud}V_{ub}^*}{V_{cd}V_{cb}^*} \equiv c_1 + c_2 + c_3 = 0, \tag{5.2.14}$$

where the three complex numbers c_1, c_2, c_3 (vectors $\vec{c}_1, \vec{c}_2, \vec{c}_3$) that we have defined start at the origin and are oriented counterclockwise around the triangle ($\vec{c}_1 + \vec{c}_2 + \vec{c}_3 = 0$). The internal angles of the triangle are given by $\beta \equiv \phi_1 = \arg(-c_1/c_2) = \arg(-V_{cd}V_{cb}^*/V_{td}V_{tb}^*)$, $\alpha \equiv \phi_2 = \arg(-c_2/c_3) = \arg(-V_{td}V_{tb}^*/V_{ud}V_{ub}^*)$ and $\gamma \equiv \phi_3 = \arg(-c_3/c_1) = \arg(-V_{ud}V_{ub}^*/V_{cd}V_{cb}^*)$. If there is no CP violation, then all CKM parameters are real and so the area inside the triangle is zero.

The *Jarlskog invariant*, denoted J, was discovered by Cecelia Jarlskog (Jarlskog, 1985) and is defined by the relation

$$\text{Im}(V_{ij}V_{k\ell}V_{i\ell}^*V_{kj}^*) = J\sum_{m,n}\epsilon_{ikm}\epsilon_{j\ell n}, \tag{5.2.15}$$

where no summation over repeated indices i, j, k, ℓ on the left-hand side is implied. Clearly, if there is no CP violation, then all CKM matrix elements are real and so $J = 0$. From the right-hand side we note the antisymmetry under both $i \leftrightarrow k$ and $j \leftrightarrow \ell$. So up to antisymmetry there are $3 \times 3 = 9$ nonvanishing quantities, with the interesting property that as a result of unitarity all choices of i, j, k, ℓ give the same value for J. Experimentally, it is found that $J \approx 3 \times 10^{-5}$. The Jarlskog invariant can be evaluated in both the standard and Wolfenstein parameterizations,

$$J = c_{12}c_{23}c_{13}^2 s_{12}s_{23}s_{13}\sin\delta = A^2\lambda^6\eta[1 - (\lambda^2/2)] + \mathcal{O}(\lambda^{10}) \approx 3 \times 10^{-5}. \tag{5.2.16}$$

CP violation only occurs if $J \neq 0$ together with the conditions that m_u, m_c, m_t are all different and m_d, m_s, m_b are all different (Jarlskog, 1985). It is straightforward to see that the area inside each of the six unitarity triangles is the same and is equal to half of the Jarlskog invariant, i.e., triangle area $= \frac{1}{2}J$.

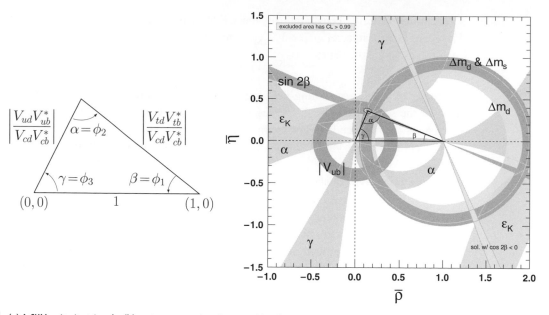

Figure 5.8   (a) A CKM unitarity triangle; (b) various constraints from precision flavor measurements (reprinted with permission from Tanabashi et al., 2018).

Proof: Consider the unitarity triangle $V_{11}V_{13}^* + V_{21}V_{23}^* + V_{31}V_{33}^* = 0$. Divide by $V_{21}V_{23}^*$ to give the rescaled unitarity triangle in Fig. 5.8. The area of the rescaled triangle is $\frac{1}{2}bh = \frac{1}{2}\text{Im}(-c_3)$ and so the area of the unscaled triangle is $\frac{1}{2}|V_{21}V_{23}^*|^2\text{Im}(-V_{11}V_{13}^*/V_{21}V_{23}^*) = -\frac{1}{2}\text{Im}(V_{11}V_{23}V_{13}^*V_{21}^*)$, where we have used the fact that if $a, b \in \mathbb{C}$ then $|b|^2\text{Im}(a/b) = \text{Im}(a|b|^2/b) = \text{Im}(ab^*)$. Using Eq. (5.2.15) we see that the area is $-\frac{1}{2}\text{Im}(V_{11}V_{23}V_{13}^*V_{21}^*) = -(-J/2) = J/2$.

Note that we could also have used $\frac{1}{2}bh = \frac{1}{2}\text{Im}(c_2)$ giving an original triangle area of $\frac{1}{2}|V_{21}V_{23}^*|^2\text{Im}(V_{31}V_{33}^*/V_{21}V_{23}^*) = \frac{1}{2}\text{Im}(V_{31}V_{23}V_{33}^*V_{21}^*)$ and again using Eq. (5.2.15) we see that the area is $\frac{1}{2}\text{Im}(V_{31}V_{23}V_{33}^*V_{21}^*) = J/2$. The same proof can be applied to each of the unitarity triangles.

The precision study of the CKM matrix elements is in an area referred to as *flavor physics*. A large number of CP-violating asymmetries have been measured as have flavor-changing neutral-current processes involving the transitions $b \to d$, $b \to s$, and $c \to u$. Violations of any of the CKM unitarity conditions would be a sign of BSM physics and so searches for such violations are of great interest. For an up-to-date summary of flavor physics measurements of the CKM matrix elements, consult the most recent Particle Data Group listings (Zyla et al., 2020).

5.2.5 Neutrino Mixing and the PMNS Matrix

The *Pontecorvo-Maki-Nakagawa-Sakata* (PMNS) matrix (Pontecorvo, 1958; Maki et al., 1962) is the unitary matrix that describes the mixing of the three known lepton generations (or flavors) and is analogous to the CKM matrix for the quark sector. It is traditional to describe the lepton generation mixing in terms of the neutrinos rather than the charged leptons in the same way that the quark generation mixing is formulated in terms of the d-type quarks rather than the u-type quarks. The PMNS matrix is also referred to as the *neutrino mixing matrix* for this reason and is the unitary matrix U_{PMNS} defined such that

$$
\begin{bmatrix} \nu_e \\ \nu_\mu \\ \nu_\tau \end{bmatrix} = U_{\text{PMNS}} \begin{bmatrix} \nu_1 \\ \nu_2 \\ \nu_3 \end{bmatrix} = \begin{bmatrix} U_{e1} & U_{e2} & U_{e3} \\ U_{\mu1} & U_{\mu2} & U_{\mu3} \\ U_{\tau1} & U_{\tau2} & U_{\tau3} \end{bmatrix} \begin{bmatrix} \nu_1 \\ \nu_2 \\ \nu_3 \end{bmatrix}, \tag{5.2.17}
$$

where ν_e, ν_μ, ν_τ are the three neutrino flavor eigenstate fields and ν_1, ν_2, ν_3 are the three neutrino mass eigenstate fields. In shorthand, we write $\nu_\ell = \sum_j U_{\ell j} \nu_j$, where $\ell = e, \mu, \tau$ and $j = 1, 2, 3$. The eigenstates of the $\hat{P}^2$ operator are the mass eigenstates that propagate freely without mixing. If $U_{\text{PMNS}} \neq I$ then it is clear that the lepton flavor eigenstates do mix as they propagate, which means that the individual lepton charges will not be conserved. In the simple extension of the SM with three mixed Dirac neutrinos U_{PMNS} is unitary, and, as for the CKM matrix, any nonunitary behavior is evidence of BSM physics, in the sense that it is beyond the simple extended SM. Since the same arguments apply regarding a 3×3 unitary matrix, we represent U_{PMNS} using both the standard parameterization in Eq. (5.2.10) and the Wolfenstein paramaterization in Eq. (5.2.11). Of course, the parameters of the quark CKM matrix and the lepton PMNS matrix are independent of each other, at least in the case of the simple extended SM. So the PMNS matrix has its own CP-violating phase and its own corresponding Jarlskog invariant that vanishes in the absence of a lepton-based CP violation.

A CP transformation turns a left-handed neutrino into a right-handed antineutrino and vice versa, since any additive charge the neutrino has (lepton number, weak isospin, weak hypercharge) is reversed by C and the helicity is reversed by P.

Experimentally, we know that there are three weakly interacting light neutrinos $m_1, m_2, m_3 \lesssim 1$ eV and in the simple extended SM these are the neutrino mass eigenstates ν_1, ν_2, ν_3. We refer to these as *active neutrinos* since they interact directly with other SM particles; i.e., they carry weak interaction charges. It is possible that there are additional neutrino species that could mix with the active neutrinos and yet not carry any weak interaction charges. We call these *sterile neutrinos*. Such additional BSM neutrinos would obviously violate lepton flavor conservation.

The PMNS matrix mixes the fields and in quantum field theory this means field operators. As we later see, an antiparticle or particle state is created by the field operator or its adjoint (Hermitian conjugate), respectively, acting on the vacuum state. The field operators and their adjoints satisfy $\hat{\nu}_\ell = \sum_j U_{\ell j} \hat{\nu}_j$ and $\hat{\nu}_\ell^\dagger = \sum_j U_{\ell j}^* \hat{\nu}_j^\dagger$. This means that the neutrino and antineutrino states, respectively, satisfy

$$
|\nu_\ell\rangle = \sum_j U_{\ell j}^* |\nu_j\rangle \quad \text{and} \quad |\bar{\nu}_\ell\rangle = \sum_j U_{\ell j} |\bar{\nu}_j\rangle. \tag{5.2.18}
$$

The possibility of Majorana neutrinos takes us beyond the Standard Model as it is usually defined, but it remains a possibility that neutrinos are Majorana particles. Majorana particles violate lepton number and have complex symmetric mass matrices. The consequence is that it is not possible to redefine the phases of n Majorana neutrinos in an $n \times n$ PMNS mixing matrix. Recalling from the CKM matrix discussion that an $n \times n$ unitary matrix has n^2 real parameters, we can still remove n phases for the n leptons leaving $n^2 - n = n(n-1)$ real parameters. We can absorb $n(n-1)/2$ into the Euler angles of an $n \times n$ orthogonal matrix leaving $n(n-1)/2$ other real parameters. This is $n(n-1)/2 - (n-1)(n-2)/2 = n-1$ additional real parameters that are chosen to be $n-1$ additional phases associated with Majorana neutrinos. For three Majorana neutrinos ($n = 3$) we have two extra CP-violating phases in addition to the one that we have for the Dirac neutrino case.

Neutrino Oscillations

Note that we work in natural units in this section. Let $\hat{Q}_i$ be the particle number operator for neutrino i with mass m_i and let $|\nu_i\rangle$ be the single particle eigenstate of that operator with no spatial properties, i.e., $\hat{Q}_j|\nu_i\rangle = \delta_{ij}|\nu_i\rangle$. So a coordinate state basis for the states of a single neutrino i can be written as the tensor product $|\nu_i, \mathbf{x}\rangle = |\mathbf{x}\rangle \otimes |\nu_i\rangle$ and a momentum state basis as $|\nu_i, \mathbf{p}\rangle = |\mathbf{p}\rangle \otimes |\nu_i\rangle$. So an arbitrary one neutrino state at time $t = 0$ can be written as $|\psi_i(0)\rangle \equiv \int d^3p\,\psi_i(\mathbf{p})|\nu_i; \mathbf{p}\rangle$ for some $\psi_i(\mathbf{p})$. At time t we have $|\psi_i(t)\rangle = \exp(-i\hat{H}t)|\psi_i(0)\rangle = \int d^3p\,\psi_i(\mathbf{p})\exp(-iE_i(p)t)|\nu_i; \mathbf{p}\rangle$, where $E_i(p) = \sqrt{\mathbf{p}^2 + m_i^2}$. In coordinate-space representation this becomes

$$\psi_i(\mathbf{x}, t)|\nu_i\rangle = \langle\mathbf{x}|\psi_i(t)\rangle = \int \frac{d^3p}{(2\pi)^{3/2}}\,\psi_i(\mathbf{p})e^{i(\mathbf{p}\cdot\mathbf{x} - E_i(p)t)}|\nu_i\rangle, \qquad (5.2.19)$$

where $\langle\mathbf{x}|\mathbf{p}\rangle = \exp(i\mathbf{p}\cdot\mathbf{x})/(2\pi)^{3/2}$. We give a brief summary of the arguments of Akhmedov and Smirnov (2009). These authors provide a wavepacket-based derivation of the neutrino oscillation formula. Also see Kayser (1981) for an earlier derivation. A full quantum field theory derivation of the neutrino oscillation formula can be found in Kobach et al. (2018). The standard oversimplified arguments also lead to the correct result but are known to be ambiguous and we do not discuss them here. References can be found in Kobach et al. (2018) and Zyla et al. (2020).

Let us choose $\psi_i(\mathbf{p}) = f_i(\mathbf{p} - \mathbf{p}_i)$ where $f_i(\mathbf{p} - \mathbf{p}_i)$ is sharply peaked around $\mathbf{p} = \mathbf{p}_i$; e.g., $f(\mathbf{p})$ could be a narrow Gaussian. Then

$$\psi_i(\mathbf{x}, t) = \int \frac{d^3p}{(2\pi)^{3/2}}\,f_i(\mathbf{p} - \mathbf{p}_i)e^{i(\mathbf{p}\cdot\mathbf{x} - E_i(p)t)} \qquad (5.2.20)$$

is a broad spatial wavepacket with mean momentum $\bar{\mathbf{p}}$ very close to $\mathbf{p}_i$. If $f_i(\mathbf{p} - \mathbf{p}_i)$ is symmetric around $\mathbf{p}_i$ then $\bar{\mathbf{p}} = \mathbf{p}_i$. For simplicity we will take this to be the case, but the arguments are easily generalized for asymmetric choices of $f(\mathbf{p})$.

In experiments we typically have $E_j \gtrsim 1$ MeV and since $m_j \lesssim 1$ eV then $|\mathbf{p}_j| \gg m_j$ and $E_j \simeq |\mathbf{p}_j|$. This means that all three neutrino mass eigenstates are typically ultrarelativistic; i.e., they are all traveling extremely close to the speed of light. As we saw in Sec. 4.2 this means that the dispersion of the wavepackets will be strongly suppressed and so it is neglected here. Retaining the first two terms in Eq. (4.2.27) we have to $\mathcal{O}(1/\gamma)$,

$$E_i(p) = E_i(p_i) + (p - p_i)^j \left.\frac{\partial E_i}{\partial p^j}\right|_{\mathbf{p}=\mathbf{p}_i} = E_i(p_i) + (\mathbf{p} - \mathbf{p}_i)\cdot\mathbf{v}_i, \qquad (5.2.21)$$

where $p = |\mathbf{p}|$, $p_i = |\mathbf{p}_i|$ and from Eq. (4.2.30) we see that $\mathbf{v}_i = p_i/E_i(p) \lesssim c = 1$ is the wavepacket group velocity. So the wavepacket moves with group velocity $\mathbf{v}_i$ very close to the speed of light and approximately preserves its wavepacket shape as it does so. This then gives

$$\psi_i(\mathbf{x}, t) = e^{-iE_i(p_i)t} \int \frac{d^3p}{(2\pi)^{3/2}}\,f_i(\mathbf{p} - \mathbf{p}_i)e^{i(\mathbf{p}\cdot\mathbf{x} - [E_i(p) - E_i(p_i)]t)}$$

$$= e^{-iE_i(p_i)t} \int \frac{d^3p}{(2\pi)^{3/2}}\,f_i(\mathbf{p} - \mathbf{p}_i)e^{i(\mathbf{p}\cdot\mathbf{x} - [\mathbf{p} - \mathbf{p}_i]\cdot\mathbf{v}_i t)} \qquad (5.2.22)$$

$$= e^{i[\mathbf{p}_i\cdot\mathbf{x} - E_i(p_i)t]}\left[\int \frac{d^3p}{(2\pi)^{3/2}}\,f_i(\mathbf{p})e^{i\mathbf{p}\cdot(\mathbf{x} - \mathbf{v}_i t)}\right] \equiv e^{i[\mathbf{p}_i\cdot\mathbf{x} - E_i(p_i)t]}g_i(\mathbf{x} - \mathbf{v}_i t),$$

where we performed a change of variable $\mathbf{p} \to \mathbf{p} + \mathbf{p}_i$ to reach the last line and where we have defined the preserved wavepacket shape factor $g_i(\mathbf{x} - \mathbf{v}_i t)$.

When a neutrino flavor eigenstate is created in a typical nuclear process it has an energy $\mathcal{O}(1\text{–}10)$ MeV and is produced in a charged current interaction. It will result in a linear superposition of the three ultrarelativistic neutrino mass eigenstates, $m_1, m_2, m_3 \lesssim 1$ eV. The production will occur over some short time interval in some small region of space, and to a very good approximation we expect the resulting wavepacket shape factors for the three ultrarelativistic mass eigenstates to be the same; i.e., we assume $g_1 = g_2 = g_3 \equiv g$. We will also neglect the tiny differences in the wavepacket group velocities and assume $\mathbf{v}_1 = \mathbf{v}_2 = \mathbf{v}_3 \equiv \mathbf{v}$ with $|\mathbf{v}|$ very close to c; e.g., if $m_j/E_j = 1/10^7 \simeq 10^{-7}$ then $\gamma(\mathbf{v}) \simeq 10^7$ and $\mathrm{v}/c \simeq 0.999999999999995$. The equal group velocity approximation is justified if we only consider propagation over distances L that are not so large that there is significant loss of overlap of the different mass wavepackets. Since the particles are extremely relativistic we neglect the effects of wavepacket dispersion as discussed in Example 4 of Sec. 4.2.

Define the common wavepacket shape for the neutrino source as $g^S(\mathbf{x} - \mathbf{v}t)$ such that its spatial midpoint is at $\mathbf{x} = 0$ at $t = 0$. The wavepacket will be finite inside some small region of space and will fall away rapidly outside that region. Then when a neutrino flavor eigenstate ν_ℓ is created in a region centered at $\mathbf{x} = 0$ at $t = 0$, we can write

$$
\begin{aligned}
|\nu_\ell(\mathbf{x}, t)\rangle &\equiv \sum_j U^*_{\ell j} \psi_j(\mathbf{x}, t)|\nu_j\rangle = \sum_j U^*_{\ell j} e^{i[\mathbf{p}_j \cdot \mathbf{x} - E_j(p_j)t]} g^S_j(\mathbf{x} - \mathbf{v}_j t)|\nu_j\rangle \\
&\simeq g^S(\mathbf{x} - \mathbf{v}t) \sum_j U^*_{\ell j} e^{i[\mathbf{p}_j \cdot \mathbf{x} - E_j(p_j)t]} |\nu_j\rangle.
\end{aligned}
\tag{5.2.23}
$$

Now consider a neutrino detector placed at location $\mathbf{L}$, where $\mathbf{L}$ is parallel to $\mathbf{v}$. If $L \equiv |\mathbf{L}|$ is not too large, the centers of the neutrino wavepackets will arrive at the detector location at approximately the same time $t = L/|\mathbf{v}| \simeq L/c$ and the wavepackets will be strongly overlapping. We can describe the detected neutrino flavor eigenstate $\nu_{\ell'}$ in terms of what we expect to be a common detector wavepacket shape, $g^D(\mathbf{x} - \mathbf{L})$, peaked at the location $\mathbf{L}$ of the detector. Then we can write for the detected neutrino flavor eigenstate

$$
|\nu_{\ell'}(\mathbf{x} - \mathbf{L})\rangle \simeq g^D(\mathbf{x} - \mathbf{L}) \sum_j U^*_{\ell' j} e^{i[\mathbf{p}_j \cdot (\mathbf{x} - \mathbf{L})]} |\nu_j\rangle.
\tag{5.2.24}
$$

The amplitude for the transition $\nu_\ell \to \nu'_\ell$ is given by

$$
\begin{aligned}
A_{\ell\ell'}(\mathbf{L}, t) &= \int d^3x \, \langle \nu_{\ell'}(\mathbf{x} - \mathbf{L})|\nu_\ell(\mathbf{x}, t)\rangle \\
&= G(\mathbf{L} - \mathbf{v}t) \sum_j U^*_{\ell j} U_{\ell' j} e^{-i[E_j(p_j)t - \mathbf{p}_j \cdot \mathbf{L}]},
\end{aligned}
\tag{5.2.25}
$$

where we have defined the effective shape factor

$$
G(\mathbf{L} - \mathbf{v}t) \equiv \int d^3x \, g^D(\mathbf{x} - \mathbf{L})^* g^S(\mathbf{x} - \mathbf{v}t)
\tag{5.2.26}
$$

and where the fact that G depends only on $(\mathbf{L} - \mathbf{v}t)$ is obvious if we do a change of integration variables $\mathbf{x} \to \mathbf{x} + \mathbf{v}t$. Let σ_{xS} and σ_{xD} be the widths of the source and detector wavepacket shape factors g^S and g^D, respectively, and let σ_x be the width of the effective shape factor G. Then we have $\sigma_x \gtrsim \max\{\sigma_{xS}, \sigma_{xD}\}$. For Gaussian wavepackets we would have exactly $\sigma_x = (\sigma^2_{xS} + \sigma^2_{xD})^{1/2}$. For simplicity we take both g^S and g^D to be even functions of their arguments with a maximum when their arguments vanish. Generalizing to asymmetric shape factors is straightforward. This means that

$G(\mathbf{L} - \mathbf{v}t)$ is also symmetric in its argument, with a maximum at $\mathbf{L} = \mathbf{v}t$. Clearly, $G(\mathbf{L} - \mathbf{v}t)$ falls off rapidly when $|\mathbf{L} - \mathbf{v}t| > \sigma_x$.

Neutrino emission and absorption times are not measured, so the probability of ν_ℓ being created at the source and $\nu_{\ell'}$ being the neutrino flavor observed in the detector at $\mathbf{L}$ is

$$P_{\ell\ell'}(\mathbf{L}) = \int_{-\infty}^{\infty} dt\, |A_{\ell\ell'}(\mathbf{L}, t)|^2 = \sum_{jk} U_{\ell j}^* U_{\ell' j} U_{\ell k} U_{\ell' k}^* I_{jk}(\mathbf{L}), \qquad (5.2.27)$$

where we have defined

$$I_{jk}(\mathbf{L}) \equiv \int_{-\infty}^{\infty} dt\, |G(\mathbf{L} - \mathbf{v}t)|^2 e^{-i\Delta\phi_{jk}(\mathbf{L}, t)},$$

$$\Delta\phi_{jk}(\mathbf{L}, t) \equiv [E_j(p_j) - E_k(p_k)]t - [p_j - p_k]L \equiv \Delta E_{jk} t - \Delta p_{jk} L. \qquad (5.2.28)$$

We readily verify that the right-hand expression for $P_{\ell\ell'}(\mathbf{L})$ is real as it must be, $P_{\ell\ell'} = P_{\ell\ell'}^*$, since complex conjugation of the argument of the sum exchanges j and k and since both are summed over. In the limit where there is no CP violation all of the elements $U_{\ell j}$ of the PMNS matrix are real and we then see that in this case $P_{\ell\ell'}(\mathbf{L}) = P_{\ell'\ell}(\mathbf{L})$.

If we consider the case of antineutrinos, then we replace all of the $U_{\ell j}$ by their complex conjugates, which is equivalent to exchanging the roles of ℓ and ℓ'. So the probability of $\bar\nu_\ell \to \bar\nu_{\ell'}$ is the same as $\nu_{\ell'} \to \nu_\ell$, i.e., $P_{\bar\ell\bar\ell'}(\mathbf{L}) = P_{\ell'\ell}(\mathbf{L})$. The result follows from our implicit assumption of CPT invariance, where a CP transformation interchanges neutrinos and antineutrinos and a time-reversal transformation T exchanges the initial and final states.

The appropriate normalization condition is

$$\int_{-\infty}^{\infty} dt\, |G(\mathbf{L} - \mathbf{v}t)|^2 = 1. \qquad (5.2.29)$$

This follows since in the limit of three degenerate neutrino masses we have $E_j(p_j) = E_k(p_k)$ and $p_j = -p_k$ by symmetry, giving $\Delta\phi_{jk}(\mathbf{L}, t) = 0$ and $I_{jk}(\mathbf{L}) = 1$. We then find $P_{\ell\ell'}(\mathbf{L}) = (\delta_{\ell\ell'})^2 = \delta_{\ell\ell'}$, where we have used the equivalent of Eq. (5.2.12) for the PMNS matrix. In other words, degenerate neutrino masses lead to no neutrino mixing. Using Eq. (5.2.12) again we have $\sum_{\ell'} U_{\ell'j} U_{\ell'k}^* = \delta_{jk}$ and so

$$\sum_{\ell'} P_{\ell\ell'}(\mathbf{L}) = \sum_j U_{\ell j}^* U_{\ell j} I_{jj}(\mathbf{L}) = \sum_j U_{\ell j}^* U_{\ell j} = 1 \qquad (5.2.30)$$

for any $\mathbf{L}$. This result is what is needed for $P_{\ell\ell'}(\mathbf{L})$ to be consistently interpreted as the probability of $\nu_\ell \to \nu_{\ell'}$.

In the ultrarelativistic limit we have

$$E_j(p_j) = \sqrt{p_j^2 + m_j^2} \simeq p_j + \frac{m_j^2}{2p_j} \simeq p_j + \frac{m_j^2}{2p} \simeq p_j + \frac{m_j^2}{2E},$$

$$\Delta E_{jk} \equiv E_j(p_j) - E_k(p_k) \simeq (p_j - p_k) + \frac{m_j^2 - m_k^2}{2p} \equiv \Delta p_{jk} + \frac{\Delta m_{jk}^2}{2p}, \qquad (5.2.31)$$

where we define p and E as the average neutrino momentum and energy, respectively. We can equally well regard p as a function of E in the mass-energy relation and rearrange to write $p(E, m^2) = \sqrt{E^2 - m^2}$ so that

$$\Delta p = \left.\frac{\partial p}{\partial E}\right|_{m^2} \Delta E + \left.\frac{\partial p}{\partial m^2}\right|_E \Delta m^2 = \frac{1}{\mathrm{v}}\Delta E - \frac{1}{2p}\Delta m^2, \qquad (5.2.32)$$

where from Eq. (4.2.22) the group velocity in natural units is $\mathrm{v} = \partial E/\partial p$. So we can write $\Delta p_{jk} = (1/\mathrm{v})\Delta E_{jk} - (1/2p)\Delta m^2_{jk}$, which is an exact expression for infinitesimal variations and reduces to the above approximate ultrarelativistic expression for ΔE_{jk} as $\mathrm{v} \to c = 1$. We then find

$$\Delta\phi_{jk}(\mathbf{L}, t) = \Delta E_{jk}t - \Delta p_{jk}L = \frac{\Delta m^2_{jk}}{2p}L - \frac{1}{\mathrm{v}}(L - \mathrm{v}t)\Delta E_{jk},$$

$$I_{jk}(\mathbf{L}) = e^{-i\Delta m^2_{jk}L/2p}\int_{-\infty}^{\infty} dt\, |G(\mathbf{L} - \mathbf{v}t)|^2 e^{i(L - \mathrm{v}t)\Delta E_{jk}/\mathrm{v}},$$

$$P_{\ell\ell'}(\mathbf{L}) = \sum_{jk} U^*_{\ell j}U_{\ell' j}U_{\ell k}U^*_{\ell' k} e^{-i\Delta m^2_{jk}L/2p}F_{jk},$$

$$F_{jk} \equiv \int_{-\infty}^{\infty} dt\, |G(\mathbf{L} - \mathbf{v}t)|^2 e^{i(L - \mathrm{v}t)\Delta E_{jk}/\mathrm{v}}, \tag{5.2.33}$$

where F_{jk} is independent of L as can be seen by a change of integration variable $t \to t + (L/\mathrm{v})$ recalling that $\mathbf{v}$ and $\mathbf{L}$ are parallel. Since the motion is ultrarelativistic we have $\mathrm{v} \simeq c = 1$ and so if $(L - \mathrm{v}t)\Delta E_{jk} \ll 1$ then the exponential in the integrand of F_{jk} becomes unity and $F_{jk} = 1$ and we recover the standard neutrino oscillation formula

$$P_{\ell\ell'}(\mathbf{L}) = \sum_{jk} U^*_{\ell j}U_{\ell' j}U_{\ell k}U^*_{\ell' k} e^{-i\Delta m^2_{jk}L/2E}, \tag{5.2.34}$$

where, as we are working to first order in the ultrarelativistic limit, we can interchange $m^2_{jk}/2E$ and $m^2_{jk}/2p$ in this equation. We can rewrite the exponential in $P_{\ell\ell'}$ as $\exp(-i2\pi L/L_{jk})$, where we have defined the jk-oscillation length as $L_{jk} \equiv 4\pi E/\Delta m^2_{jk}$. Note that ΔE_{jk} and Δp_{jk} both vanish as $\Delta m^2_{jk} \to 0$ and so to first order in Δm^2_{jk} we have $\Delta E_{jk}, \Delta p_{jk} \sim \Delta m^2_{jk}/p$, since the dimensions must balance and the only other scale available is $p \sim E$.

This standard neutrino oscillation formula requires that two conditions be met:

(i) *No decoherence condition*: The loss of overlap of the different mass wavepackets is negligible at the detector; i.e., the offset of wavepacket centers due to different group velocities is much smaller than the wavepacket size, $\Delta\mathrm{v}\, t_L = (\Delta\mathrm{v}/\mathrm{v})L \ll \sigma_x$, where $t_L \equiv L/\mathrm{v}$ is the average time taken to reach the detector. The neglected wavepacket dispersion will work to help meet this condition.

(ii) *Localization condition*: There is a negligible amount of oscillation inside the wavepacket; i.e., we want $(L - \mathrm{v}t)\Delta E_{jk} \ll 1$. Since $|G(\mathbf{L} - \mathbf{v}t)|^2$ dies off quickly when $(L - \mathrm{v}t) \gtrsim \sigma_x$ and since $\Delta E_{jk} \sim \Delta m^2_{jk}/p$, then we can rewrite the localization condition as $\sigma_x \Delta m^2_{jk}/p \ll 1$.

Example: Consider some typical numbers in a neutrino experiment. Assume that the uncertainty in the position of the neutrino creation is $\sim 2\,\mathrm{cm} = 0.02\,\mathrm{m}$. Then the wavepacket size is expected to be $\sigma_x \sim 0.02\,\mathrm{m} \simeq 10^5\,\mathrm{eV}^{-1}$, i.e., in natural units $\hbar = c = 1$ and $1 = \hbar c = 197.327\,\mathrm{MeV\,fm}$ and $1\,\mathrm{eV}^{-1} \sim 2\times10^{-7}\,\mathrm{m}$. The width of the momentum-space wavepacket is then $\sigma_p \sim 1/\sigma_x \sim 10^{-5}\,\mathrm{eV}$. Let the energy and momentum of the created neutrino wavepacket be $E \sim p \sim 10\,\mathrm{MeV}$ and note that typical neutrino mass differences satisfy $\Delta m^2 \sim 10^{-3}$–$10^{-5}\,\mathrm{eV}^2$. This gives $\Delta m^2/2p \sim 10^{-10}$–$10^{-12}\,\mathrm{eV}$. For neutrino masses $\lesssim 1\,\mathrm{eV}$ and energies $\sim 10\,\mathrm{MeV}$ we have $\gamma(\mathrm{v}) \gtrsim 10^7$, which means that $1 - \mathrm{v}/c \lesssim 10^{-14}$. If two masses differ by a factor of two, then so do their γ values to achieve the same total energy, $E = \gamma m c^2$, and then we find $\Delta\mathrm{v}/c \sim \Delta\mathrm{v}/\mathrm{v} \sim 10^{-15}$. The distance L between the neutrino source and the detector must satisfy the no decoherence condition, which requires $L(\Delta\mathrm{v}/\mathrm{v}) \ll \sigma_x$ or equivalently $L \ll \sigma_x(\mathrm{v}/\Delta\mathrm{v}) \sim 10^{13}\,\mathrm{m}$. This is twice the distance between the Sun and Pluto, so the no decoherence condition will certainly be satisfied in any terrestrial experiment. The localization condition is also comfortably satisfied: $\sigma_x(\Delta m^2/p) \lesssim 10^5 \times 10^{-10} \sim 10^{-5} \ll 1$.

Remembering that $P_{\ell\ell'}(\mathbf{L})$ was shown to be real above, we have

$$
\begin{aligned}
P_{\ell\ell'}(\mathbf{L}) &= \sum_{jk} \left\{ U_{\ell j}^* U_{\ell' j} U_{\ell k} U_{\ell' k}^* \right\} \left[\cos(\Delta m_{jk}^2 L/2E) - i\sin(\Delta m_{jk}^2 L/2E) \right] \\
&= \sum_{jk} \mathrm{Re}\left\{ U_{\ell j}^* U_{\ell' j} U_{\ell k} U_{\ell' k}^* \right\} \cos(\Delta m_{jk}^2 L/2E) \\
&\quad + \sum_{jk} \mathrm{Im}\left\{ U_{\ell j}^* U_{\ell' j} U_{\ell k} U_{\ell' k}^* \right\} \sin(\Delta m_{jk}^2 L/2E) \\
&= \sum_{jk} \mathrm{Re}\left\{ \cdots \right\} [1 - 2\sin^2(\Delta m_{jk}^2 L/4E)] + 2\sum_{j>k} \mathrm{Im}\left\{ \cdots \right\} \sin(\Delta m_{jk}^2 L/2E) \\
&= \delta_{\ell\ell'} - 4\sum_{j>k} \mathrm{Re}\left\{ U_{\ell j}^* U_{\ell' j} U_{\ell k} U_{\ell' k}^* \right\} \sin^2(\Delta m_{jk}^2 L/4E) \\
&\quad + 2\sum_{j>k} \mathrm{Im}\left\{ U_{\ell j}^* U_{\ell' j} U_{\ell k} U_{\ell' k}^* \right\} \sin(\Delta m_{jk}^2 L/2E),
\end{aligned}
\tag{5.2.35}
$$

where $\sum_{jk} \mathrm{Re}\{U_{\ell j}^* U_{\ell' j} U_{\ell k} U_{\ell' k}^*\} = \mathrm{Re}\{\sum_{jk} U_{\ell j}^* U_{\ell' j} U_{\ell k} U_{\ell' k}^*\} = \mathrm{Re}\{\delta_{\ell\ell'}^2\} = \delta_{\ell\ell'}$ using Eq. (5.2.12). We have used the observation that $\mathrm{Re}\{\cdots\}$ and $\mathrm{Im}\{\cdots\}$ are symmetric and antisymmetric, respectively, under the exchange $j \leftrightarrow k$ and similarly for $\cos(\cdots)$ and $\sin(\cdots)$. In the absence of CP violation all $U_{\ell k}$ are real and so $\mathrm{Im}\{\cdots\} = 0$ and the third term vanishes.

The above result was derived assuming that the mass eigenstates are propagating in a vacuum. We are often interested in generalizing this result to include the effects of propagation through matter. Since neutrinos interact with matter very weakly, the incoherent scattering processes that change the initial state of the neutrino (incoherent elastic, quasielastic and inelastic scattering) are negligible. However, coherent forward elastic scattering effects are much larger and give rise to an effective index of refraction for the propagating neutrinos in a way analogous to the index of refraction for light. The resulting effects on $P_{\ell\ell'}$ can be calculated in the SM. The change to neutrino oscillations caused by propagation through matter is usually referred to as the Mikheyev-Smirnov-Wolfenstein (MSW) effect. Various kinds of neutrino oscillation experiments are used to attempt to measure and constrain the PMNS matrix elements and neutrino masses. These include (i) solar neutrino experiments, where neutrinos originate from nuclear processes in the sun; (ii) atmospheric neutrinos, where neutrinos are produced in the atmosphere by incoming cosmic rays; (iii) accelerator neutrino oscillation experiments, where neutrinos are created at an accelerator and some distance away on Earth they are detected and (iv) reactor neutrino oscillation experiments, where neutrinos are produced in a nuclear reactor rather than an accelerator. Since neutrinos interact weakly, neutrino detection on Earth takes place in laboratories deep underground to avoid being overwhelmed by signals caused by incoming cosmic rays. See the most recent Particle Data Group listings for the current status of these experiments and measurements of matrix elements, neutrino masses and possible evidence for sterile neutrinos (Zyla et al., 2020).

5.2.6 Majorana Neutrinos and Double Beta Decay

In the SM neutrinos are assumed to be Dirac fermions, but there is no current experimental evidence to suggest that they are not Majorana fermions. As discussed in Sec. 4.6.5, Majorana neutrinos are their own antiparticles and their existence would lead to electroweak processes that violate lepton number conservation by two. An example of such a process is *neutrinoless double beta decay* ($0\nu\beta\beta$ decay).

Stable nuclei have just the right balance of neutron number (N) and atomic (proton) number (Z), where the element type and its chemical properties are determined by the atomic number Z, since it is the number of electrons in the neutral atom. The mass number A is the number of nucleons,

$A = Z + N$. For small Z the stable nuclei have $N \simeq Z$, but for larger Z stable nuclei have $N > Z$ to help the nuclear binding overcome the increasing Coulomb repulsion of the positively charged protons. The stable nuclei form a stability curve on an N versus Z chart of nuclei. Too far below the curve means instability due to too many protons, and too far above the curve means instability due to too many neutrons. The instability arises from Pauli blocking of the nucleon orbitals that pushes excess neutrons or excess protons to higher nuclear orbitals, leading to increasing instability the further one moves from the stability curve. Isotopes of an element have different N values for the Z that defines the element and long-lived isotopes must lie not too far from the stability curve. The standard notation for an isotope of element X is $^{A}_{Z}X$, where the element has Z protons and the particular isotope has $A - Z = N$ neutrons. Since the element name is determined by Z it is also common to simply use the notation ^{A}X to specify an isotope. Neutron-rich nuclei will typically β decay, $n \to p + e^{-} + \bar{\nu}_{e}$, and emit an electron and anti-electron neutrino, e.g., $^{60}_{27}\text{Co} \to^{60}_{28} \text{Ni} + e^{-} + \bar{\nu}_{e}$. Conversely, proton-rich nuclei will typically decay by positron emission (β plus decay) and emit a positron and an electron neutrino, e.g., $^{23}_{12}\text{Mg} \to^{23}_{11} \text{Na} + e^{+} + \nu_{e}$.

Some nuclei can decay by two simultaneous β decays, referred to as double beta decay or $\beta\beta$ decay. Since $\beta\beta$ decay is second order in the weak interaction it will be a very rare process. Some nuclei capable of $\beta\beta$ decay are energetically forbidden from single β decay, such as $^{76}_{32}\text{Ge}$, $^{100}_{42}\text{Mo}$, $^{128}_{52}\text{Te}$, $^{130}_{52}\text{Te}$ and $^{136}_{54}\text{Xe}$, and these nuclei will have extremely long lifetimes. Such nuclei have even numbers of protons and even numbers of neutrons, since the spin-pairing for the protons and for the neutrons can increase the binding energy of the nucleus, preventing single β decay from occurring.

If neutrinos are Dirac fermions, then two neutrino double beta ($2\nu\beta\beta$) decay is possible and can be searched for when single β decay is forbidden. This process is illustrated on the left in Fig. 5.9. The nuclear recoil kinetic energy will be small compared with the total lepton kinetic energy, since the leptons are much lighter than the nucleus. The total lepton kinetic energy will be equal to the binding energy difference of the initial and final nuclear state less the nuclear recoil kinetic energy. There are four leptons in the final state and so the two electrons can emerge and be detected at any relative angle. However, if neutrinos are Majorana particles, then they are their own antiparticles and neutrinoless or zero neutrino double beta ($0\nu\beta\beta$) decay is also possible. This is a lepton number-violating process and is illustrated on the right in Fig. 5.9. In the $0\nu\beta\beta$ there are only two leptons, the electrons, in the final state. Since the nuclear recoil is small they will be ejected approximately back to back with each electron kinetic energy being approximately half of the binding energy change less the small nuclear recoil kinetic energy. It is expected that the rate of ($0\nu\beta\beta$) decay will be much less than the ($2\nu\beta\beta$) rate, which makes ($0\nu\beta\beta$) detection experiments very challenging. The decay rate for light Majorana neutrinos is given by

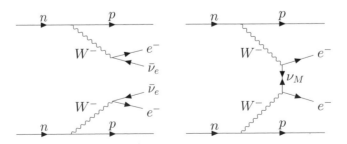

Figure 5.9   Two neutrino double β decay (left) and neutrinoless double β decay (right).

$$\Gamma = G^{0\nu}|M^{0\nu}|^2\langle m_{\beta\beta}\rangle, \tag{5.2.36}$$

where $G^{0\nu}$ is the phase space for the two electron decay, $M^{0\nu}$ is the nuclear matrix element for the decay and $\langle m_{\beta\beta}\rangle \equiv |\sum_j U_{ej}^2 m_{\nu_j}|^2$ is the effective neutrino mass. The $2\nu\beta\beta$ decay of each of the above-mentioned $\beta\beta$ nuclei has been observed with half-lives $\mathcal{O}(10^{21})$ years, but no $0\nu\beta\beta$ decay has been observed at the time of writing. For a detailed discussion of the theory of $0\nu\beta\beta$ decay, see Vergados et al. (2012), and, as always, for the current status of experimental searches, consult the most recent version of the Particle Data Group listings (Zyla et al., 2020).

5.3 Representations of *SU(N)* and the Quark Model

Even before developing the details of quantum field theory we can understand important aspects of elementary processes in particle physics from an understanding of quantum numbers and exact and partial conservation laws. This requires an understanding of how to build and decompose representations of groups.

5.3.1 Multiplets of *SU(N)* and Young Tableaux

A brief explanation of Young tableaux for $SU(N)$ is useful for discussing the quark model and more generally in discussing representations in quantum field theory. In the literature one finds the terms "multiplet" and "irreducible representation" or "irrep" used interchangeably.

Young diagrams: A *Young diagram* (sometimes referred to as a Ferrer diagram) is some finite number n of boxes arranged in rows that are left justified and stacked, where the row lengths are non-increasing as one goes down. For example, for $n = 11$ boxes we might have row lengths 4, 3, 3 and 1. We represent this Young diagram as $\lambda = (4, 3, 3, 1)$ and refer to this λ as a *partition* of the $n = 11$ boxes. Every such λ specifies a Young diagram. For $n = 11$ we can have Young diagrams such as

$$\lambda = (4, 3, 3, 1) \Rightarrow \qquad \text{and} \qquad \lambda = (4, 4, 2, 1) \Rightarrow \quad . \tag{5.3.1}$$

Young tableaux: A *standard Young tableau* of n boxes has each box labeled from 1 to n with the number in each row and each column in increasing order. A *semistandard Young tableau* has the symbols non-decreasing in rows and increasing in columns. As an illustration, for the Young diagram $\lambda = (5, 3, 1)$ we have

$$\begin{array}{|c|c|c|c|c|} \hline 1 & 2 & 4 & 7 & 8 \\ \hline 3 & 5 & 6 \\ \hline 9 \\ \hline \end{array} = \text{standard} \qquad \text{and} \qquad \begin{array}{|c|c|c|c|c|} \hline 1 & 2 & 4 & 4 & 7 \\ \hline 3 & 3 & 6 \\ \hline 8 \\ \hline \end{array} = \text{semistandard}. \tag{5.3.2}$$

So for a given Young diagram λ there are more semistandard Young tableaux than there are standard Young tableaux. *In using a Young tableau it is understood that one first symmetrizes with respect to each row and then antisymmetrizes with respect to each column.* So the same symbol can never occur twice in a column.

Mathematical digression: Symmetric group S_n**:** The symmetric group of degree n is denoted S_n and is the group whose elements are the permutations of n distinct symbols $\{X_1, X_2, \ldots, X_n\}$. The number of elements of the group (the order of the group) is the number of permutations of these symbols, $|S_n| = n!$. So S_n is a finite group. The distinct symbols could be the set of integers $\{1, 2, \ldots, n\}$ or anything else such as n basis vectors. We use the notation from Eq. (4.1.149) for a permutation σ. It is straightforward to show that the set of $n!$ permutation operations forms a group. It is also clear that we can represent the $n!$ permutations as $n!$ matrices of size $n \times n$ acting on a column vector of n symbols, where each matrix corresponds to a permutation of the n rows (or n columns) of the identity matrix. Any permutation can be written as a product of transposition of two elements. Permutations corresponding to an even (odd) number of transpositions are called even (odd) permutations.

Permutation notation: A permutation σ in S_6 can also be expressed as

$$\sigma = \begin{pmatrix} 1 & 2 & 3 & 4 & 5 & 6 \\ 6 & 3 & 2 & 1 & 5 & 4 \end{pmatrix} \quad \text{or} \quad \sigma = (164)(23)(5) \equiv (164)(23). \tag{5.3.3}$$

This notation means that in a list of six symbols we move the first to the sixth and the sixth to the fourth and the fourth to the first (164); we exchange the second and third (23); and we leave the fifth alone (5). We do not normally bother to show an unchanged symbol such as (5) and so simply write $(164)(23)$. We denote the identity permutation as (e). We refer to this as *cycle notation* since $(164), (416)$ and (641) all mean the same thing. We refer to $(\cdots)$ as a cycle. If two permutations have the same cycle structure, then they belong to the same conjugacy class (defined in Sec. A.7.1); e.g., $(164)(23)(5)$ and $(532)(64)(1)$ are in the same conjugacy class but $(164)(23)(5)$ and $(4263)(15)$ are not. Note that each conjugacy class in S_n corresponds to a partition of n; e.g., for $n = 6$ the permutations $(164)(23)(5)$ and $(4263)(15)$ correspond to the partitions $\lambda = (3, 2, 1)$ and $\lambda = (4, 2)$, respectively. In the two-line notation on the left of Eq. (5.3.3) the ordering of the columns is arbitrary. So we can make a product of two permutations by reordering the columns in the second (left) permutation and reading off the result from top right and bottom left,

$$\begin{pmatrix} 1 & 2 & 3 & 4 & 5 \\ 2 & 1 & 4 & 3 & 5 \end{pmatrix} \begin{pmatrix} 1 & 2 & 3 & 4 & 5 \\ 5 & 4 & 3 & 1 & 2 \end{pmatrix} = \begin{pmatrix} 5 & 4 & 3 & 1 & 2 \\ 5 & 3 & 4 & 2 & 1 \end{pmatrix} \begin{pmatrix} 1 & 2 & 3 & 4 & 5 \\ 5 & 4 & 3 & 1 & 2 \end{pmatrix} = \begin{pmatrix} 1 & 2 & 3 & 4 & 5 \\ 5 & 3 & 4 & 2 & 1 \end{pmatrix},$$

where the left (second) permutation is done after the first (right) permutation. In the more concise cyclic notation this is $(12)(34) \cdot (1524) = (234)(15)$.

A representation ρ of any group G is its realization as a set of complex matrices acting on some corresponding vector space V_ρ. The dimension of the representation is $\dim(V_\rho)$ and is the number of basis vectors in V_ρ. For all $g \in G$ we have a complex matrix $D^\rho(g)$ with size $\dim(V_\rho) \times \dim(V_\rho)$ that preserves the group structure. Reducible and irreducible representations are defined in the discussion associated with Eq. (A.7.1). In the representation theory of finite groups the number of inequivalent irreducible representations ρ of G is equal to the number of conjugacy classes in G.

In the case of the symmetric group, $G = S_n$, the conjugacy classes of S_n correspond to the cycle structures of the permutations, which correspond to the partitions of n. So from the above theorem we know that the number of inequivalent irreducible representations of S_n is given by the number of partitions of n, which is the number of Young diagrams with n boxes. Furthermore, for the group S_n each irreducible representation ρ is in one-to-one correspondence with a partition of n and hence to a Young diagram. The dimension of a representation ρ is given by the *hook length formula*

$$\dim(V_\rho) = n!/H_\rho, \tag{5.3.4}$$

where H_ρ is the *hook factor* and is the product of the *hook lengths* of the n boxes in the Young diagram λ_ρ. To calculate the hook length of any box in a Young diagram λ_ρ, first add the number of boxes to its right to the number of boxes below it and then add 1. Write the hook length inside every box of the Young diagram and then multiply the hook lengths together to get H_ρ. For example, for $\lambda_\rho = (5, 4, 1)$ we write in the hook lengths to give

$$\begin{array}{|c|c|c|c|c|} \hline 7 & 5 & 4 & 3 & 1 \\ \hline 5 & 3 & 2 & 1 \\ \cline{1-4} 1 \\ \cline{1-1} \end{array} \quad \Rightarrow \quad H_\rho = 7 \cdot 5 \cdot 4 \cdot 3 \cdot 1 \cdot 5 \cdot 3 \cdot 2 \cdot 1 \cdot 1 = 12{,}600. \tag{5.3.5}$$

For a finite group such as S_n the sum of the squares of the dimensions of the irreducible representations (irreps) equals the number of group elements, $\sum_i [\dim(V_{\rho_i})]^2 = |S_n| = n!$. The set of all standard Young tableaux forms a basis for the representation of S_n corresponding to the Young diagram. We will not construct these representations here. Each Young diagram corresponds to a *Young symmetrizer* that *first* symmetrizes all rows and *then* antisymmetrizes each column.

Example of S_2: There two partitions of $n = 2$ are $\lambda_{\rho_1} = (2)$ and $\lambda_{\rho_2} = (1, 1)$, which have the corresponding Young diagrams

$$\dim(V_1) = 2!/2 \cdot 1 = 1 \qquad \dim(V_2) = 2!/2! \cdot 1 = 1.$$

So S_2 has two one-dimensional irreducible representations V_1 and V_2, which are the symmetric and the antisymmetric representations, respectively.

Example of S_3: The three partitions of $n = 3$ are $\lambda_{\rho_1} = (3)$, $\lambda_{\rho_2} = (2, 1)$ and $\lambda_{\rho_3} = (1, 1, 1)$, which have the corresponding Young diagrams

$$\dim(V_{\rho_1}) = 3!/3 \cdot 2 \cdot 1 = 1 \quad \dim(V_{\rho_2}) = 3!/3 \cdot 1 \cdot 1 = 2 \quad \dim(V_{\rho_3}) = 3!/3 \cdot 2 \cdot 1 = 1.$$

$V_{\rho_1}, V_{\rho_2}, V_{\rho_3}$ are the symmetric, mixed and antisymmetric representations, respectively. As required $\sum_{i=1}^{3} [\dim(V_{\rho_i})]^2 = 1 + 4 + 1 = 6 = 3!$.

Application to $SU(N)$: All representations of $SU(N)$ can be expressed as tensor products of the $N \times N$ defining representation (lowest dimensional faithful representation), which is the group of $N \times N$ special unitary matrices U. The corresponding N-dimensional vector space is W_N and is spanned by N orthonormal basis vectors $\{|i_1\rangle, \ldots, |i_N\rangle\}$. Consider some examples: for $SU(2)_{\text{spin}}$ we have orthonormal basis vectors $|i_1\rangle = |{\uparrow}\rangle$ and $|i_2\rangle = |{\downarrow}\rangle$; for $SU(2)_{\text{isospin}}$ we have $|i_1\rangle = |u\rangle$ and $|i_2\rangle = |d\rangle$; for $SU(3)_{\text{flavor}}$ we have $|i_1\rangle = |u\rangle$, $|i_2\rangle = |d\rangle$ and $|i_3\rangle = |s\rangle$; and for $SU(3)_c$ we have $|i_1\rangle = |r\rangle$, $|i_2\rangle = |b\rangle$ and $|i_3\rangle = |g\rangle$ for red, blue and green, respectively. Here the ordering of the basis vectors was assumed to be relevant; e.g., $|uud\rangle$ is orthogonal to $|udu\rangle$. This means that we are assigning the properties to distinguishable particles. The imposition of boson or fermion statistics as discussed in Sec. 4.1.9 requires symmetry or antisymmetry, respectively, under the interchange of *all* particle quantum numbers and is imposed as needed as a final step in applications.

The n-fold tensor product of W_N is $W_N^{\otimes n} = W_N \otimes \cdots \otimes W_N$. Tensor products are discussed in Secs. 4.1.9 and A.2.7. An orthonormal basis for $W_N^{\otimes n}$ is $|i_{k_1} i_{k_2} \cdots i_{k_n}\rangle$ with $k_\ell \in \{1, 2, \ldots, N\}$ for $\ell = 1, \ldots, n$. For $SU(N)$ it is conventional to use the notation $\mathbf{N} \equiv W_N$ and so the n-fold tensor product space is typically written as $\mathbf{N}^{\otimes n} = \mathbf{N} \otimes \cdots \otimes \mathbf{N}$. For $SU(N)_{\text{flavor}}$ this gives states with n quarks of various flavors and a tensor product space of N^n dimensions. For example, for $SU(3)_{\text{flavor}}$ and $n = 3$ quarks, the tensor product space has $N^n = 3^3 = 27$ basis vectors, $|uuu\rangle$, $|uud\rangle$, $|uus\rangle$, $|udu\rangle$, $|udd\rangle$, $|uds\rangle$, $|usu\rangle$, $|usd\rangle$, $|uss\rangle$ and so on. Under an arbitrary $SU(N)$ transformation U in the single particle space $W_N = \mathbf{N}$ we have $|i_k\rangle \to \sum_{k=1}^{N} U_j{}^k |i_k\rangle$, where $U_j{}^k \equiv U_{jk}$ are the matrix elements of U.

For quantum charges such as quark flavor and color the corresponding antiquarks have antiflavors $\bar{u}, \bar{d}, \bar{s}, \bar{c}, \bar{b}, \bar{t}$ and anticolors $\bar{r}, \bar{b}, \bar{g}$. Let t^a for $a = 1, \ldots, (N^2 - 1)$ be the generators of $SU(N)$ in the $N \times N$ defining representation. The Lie algebra of the generators requires $[t^a, t^b] = i f^{abc} t^c$. The antiparticles transform in the conjugate of the defining representation (Georgi, 1999), where the generators in the conjugate representation are $\bar{t}^a = -t^{a*}$ as defined in Eq. (A.7.14). We have $[\bar{t}^a, \bar{t}^b] = i f^{abc} \bar{t}^c$ since $f^{abc} \in \mathbb{R}$. We write the N orthonormal basis for the antiparticle space $\overline{\mathbf{N}} \equiv W_{\overline{\mathbf{N}}}$ as $\{|\bar{i}_1\rangle, \ldots, |\bar{i}_N\rangle\}$; e.g., for $SU(3)_{\text{flavor}}$ we have $|\bar{i}_1\rangle = |\bar{u}\rangle$, $|\bar{i}_2\rangle = |\bar{d}\rangle$ and $|\bar{i}_3\rangle = |\bar{s}\rangle$. Note that we have $\exp\{i\vec{\omega} \cdot \vec{t}\} = (\exp\{i\vec{\omega} \cdot \vec{t}\})^* = U^*$ and so under an $SU(N)$ transformation we have $|\bar{i}_j\rangle \to \sum_{k=1}^{N} U^{j*}{}_k |\bar{i}_k\rangle$ where $U^{j*}{}_k \equiv U^*_{jk}$.

For $SU(N)$ an arbitrary state $|\Psi\rangle$ with n particles and m antiparticles is in the (n, m) tensor product space, $|\Psi\rangle \in W_{\mathbf{N}}^{\otimes n} \otimes W_{\overline{\mathbf{N}}}^{\otimes m}$, and can be written as[6]

$$|\Psi\rangle = \sum_{\text{all } k,j=1}^{N} \psi^{k_1 k_2 \cdots k_n}_{j_1 j_2 \cdots j_m} |i_{k_1} i_{k_2} \cdots i_{k_n} \bar{i}_{j_1} \bar{i}_{j_2} \cdots \bar{i}_{j_m}\rangle, \qquad (5.3.6)$$

where we have tensor coefficients $\psi^{k_1 \cdots k_n}_{j_1 \cdots j_m} \in \mathbb{C}$. We observe how the (n, m)-tensor coefficients $\psi^{k_1 \cdots k_n}_{j_1 \cdots j_m}$ transform under any $N \times N$ matrix $U \in SU(N)$,

$$|\Psi\rangle \xrightarrow{U} |\Psi'\rangle = \sum_{\text{all } r,s=1}^{N} \psi^{s_1 \cdots s_n}_{r_1 \cdots r_m} U_{s_1}{}^{k_1} \cdots U_{s_n}{}^{k_n} U^{r_1*}{}_{j_1} \cdots U^{r_m*}{}_{j_m} |i_{k_1} \cdots i_{k_n} \bar{i}_{j_1} \cdots \bar{i}_{j_m}\rangle$$

$$= \sum_{\text{all } k,j=1}^{N} \psi'^{k_1 \cdots k_n}_{j_1 \cdots j_m} |i_{k_1} \cdots i_{k_n} \bar{i}_{j_1} \cdots \bar{i}_{j_m}\rangle. \qquad (5.3.7)$$

The important observation is that any symmetrization and/or antisymmetrization in the tensor $\psi^{k_1 \cdots k_n}_{j_1 \cdots j_m}$ is preserved in the transformed tensor $\psi'^{k_1 \cdots k_n}_{j_1 \cdots j_m}$, since the order of the Us and U^*s in the transformation is arbitrary. A symmetry restriction on the tensors has the effect of only accessing a corresponding subspace of the tensor product space with the same symmetry. So the (n, m)-fold tensor product space will be the direct sum of invariant subspaces classified according to their symmetry properties. The smallest such subspaces correspond to irreducible representations of $SU(N)$ and are classified according to the symmetry property of the corresponding tensors, which are captured in the corresponding Young diagram. Each set of orthonormal basis vectors of such a minimal subspace forms a *multiplet* and each multiplet corresponds to a Young diagram and hence to an irreducible representation (irrep) of $SU(N)$. The decomposition of the tensor product space may have multiple occurrences of any given irrep/multiplet. Each standard Young tableaux for a given Young diagram corresponds to a different Young symmetrizer and hence a different realization of the irrep/multiplet in the tensor product space.

For $SU(N)$ no tensor can ever be completely antisymmetric in more than N upper or lower indices, since each index can only take one of N values. Consider an (n, m) tensor in $SU(N)$ as defined in Eq. (5.3.7) that is completely antisymmetric in any N of the particle indices $\{k_1, \ldots, k_n\}$

---

[6] Sometimes the notation $|i_k\rangle \equiv |i_k\rangle$ and $|i_j\rangle \equiv |\bar{i}_j\rangle$ is used; e.g., see chapter 10 in Georgi (1999).

or any N of the antiparticle indices $\{k_1, \ldots, k_n\}$. Such a tensor $\psi^{k_1 \cdots k_n}_{j_1 \cdots j_m}$ will be proportional to the completely antisymmetric Levi-Civita tensor in Eq. (A.2.1) with N indices. Then from the definition of the determinant in Eq. (A.8.1) we will obtain $\det U = 1$ or $\det U^* = 1$ for the transformation in the particle and antiparticle case, respectively. So these N indices transform as an $SU(N)$ singlet. For example, if $\psi^{k_1 \cdots k_n}_{j_1 \cdots j_m}$ was completely antisymmetric in the indices $k_1, \ldots, k_N$ for $n > N$, then we could define $\psi^{k_1 \cdots k_n}_{j_1 \cdots j_m} \equiv \phi^{k_{N+1} \cdots k_n}_{j_1 \cdots j_m} \epsilon^{k_1 \cdots k_N}$ and only $\phi^{k_{N+1} \cdots k_n}_{j_1 \cdots j_m}$ is relevant for defining the transformation properties and the representation. So apart from the one singlet representation, we only need consider antisymmetry in up to $N - 1$ indices for other representations.

We can view $\epsilon^{k_1 \cdots k_n}$ and $\epsilon_{j_1 \cdots j_m}$ as examples of $(n, 0)$ and $(0, m)$ tensors, respectively. Choosing $n = N$ and $m = N$ we see that both $\epsilon^{k_1 \cdots k_N}$ and $\epsilon_{j_1 \cdots j_N}$ are $SU(N)$ singlet tensors as explained above, since $\det U = \det U^* = 1$. Also note that if we contract up and down indices in a tensor then they form an $SU(N)$ singlet and so are invariant. For example, consider the contraction of the $(1, 1)$ tensor ψ^k_j to give ψ^ℓ_ℓ, where we are now adopting the *summation convention*. The contraction is a singlet since $\psi^\ell_\ell \to \psi'^\ell_\ell = \psi^s_r U_s{}^\ell U^{r*}_\ell = \psi^s_r U_{s\ell} U^*_{r\ell} = \psi^s_r U_{s\ell} U^\dagger_{\ell r} = \psi^s_r \delta_{sr} = \psi^r_r$. This is analogous to the contraction of spacetime indices, where for a $(1, 1)$ Lorentz tensor $W_\mu{}^\nu$ it follows that $W_\mu{}^\mu$ is a Lorentz scalar. This means, for example, that $\psi^{k_1 \cdots k_{p-1} \ell k_{p+1} \cdots k_n}_{j_1 \cdots j_{q-1} \ell j_{q+1} \cdots j_m}$ will transform as an $(n - 1, m - 1)$ $SU(N)$ tensor.

From Eq. (5.3.7) we have for any (n, m) tensor

$$\psi'^{k_1 \cdots k_n}_{j_1 \cdots j_m} \xrightarrow{U} \psi'^{k_1 \cdots k_n}_{j_1 \cdots j_m} = \psi^{s_1 \cdots s_n}_{j_1 \cdots j_m} U_{s_1}{}^{k_1} \cdots U_{s_n}{}^{k_n} U^{r_1*}_{j_1} \cdots U^{r_m*}_{j_m} \tag{5.3.8}$$

$$= \psi^{s_1 \cdots s_n}_{j_1 \cdots j_m} U_{s_1 k_1} \cdots U_{s_n k_n} U^*_{r_1 j_1} \cdots U^*_{r_m j_m}. \tag{5.3.9}$$

Taking the complex conjugate of this and defining $\bar{\psi}^{j_1 \cdots j_m}_{k_1 \cdots k_n} \equiv \psi^{*k_1 \cdots k_n}_{j_1 \cdots j_m}$ we see that $\bar{\psi}^{j_1 \cdots j_m}_{k_1 \cdots k_n}$ is an (m, n) tensor. This means that the complex conjugate of any (n, m) tensor is an (m, n) tensor. We say that an (n, n) tensor is self-conjugate since it and its complex conjugate are tensors in the same tensor product space. We see from Eq. (5.3.6) that $\bar{\psi}^{j_1 \cdots j_m}_{k_1 \cdots k_n}$ is the tensor associated with the bra state $\langle \Psi |$, since

$$\langle \Psi | = \langle i_{k_1} \cdots i_{k_n} \bar{i}_{j_1} \cdots \bar{i}_{j_m} | \psi^{*k_1 k_2 \cdots k_n}_{j_1 j_2 \cdots j_m} = \langle i_{k_1} \cdots i_{k_n} \bar{i}_{j_1} \cdots \bar{i}_{j_m} | \bar{\psi}^{j_1 j_2 \cdots j_m}_{k_1 k_2 \cdots k_n}. \tag{5.3.10}$$

For any two states, $| \Psi \rangle$ and $| \Phi \rangle$, in the (n, m) tensor product space the bra-ket (inner product) of the two states is

$$\langle \Phi | \Psi \rangle = \bar{\phi}^{j_1 j_2 \cdots j_m}_{k_1 k_2 \cdots k_n} \psi^{k_1 k_2 \cdots k_n}_{j_1 j_2 \cdots j_m}. \tag{5.3.11}$$

If $| \Psi \rangle$ and $| \Phi \rangle$ are vectors in an (n, m) and a (q, p) tensor product space, respectively, then their tensor product $| \Lambda \rangle \equiv | \Psi \rangle \otimes | \Phi \rangle$ is in an $(n + q, m + p)$ tensor product space with the corresponding tensor,

$$\lambda^{k_1 \cdots k_n k'_1 \cdots k'_q}_{j_1 \cdots j_m j'_1 \cdots j'_p} = [| \Psi \rangle \otimes | \Phi \rangle]^{k_1 \cdots k_n k'_1 \cdots k'_q}_{j_1 \cdots j_m j'_1 \cdots j'_p} = \psi^{k_1 \cdots k_n}_{j_1 \cdots j_m} \phi^{k'_1 \cdots k'_q}_{j'_1 \cdots j'_p}. \tag{5.3.12}$$

The goal is to express such tensor products as the sum of terms corresponding to irreducible representations of $SU(N)$.

In order to use Young tableaux and important theorems associated with them we need to consider $SU(N)$ tensors with only up (particle-like) indices or alternatively with only down (antiparticle-like) indices. Conventionally, we deal with up indices, which requires us to be able to first raise any lower indices in a tensor, then apply Young tableaux methods and finally to lower the indices again. Since the antisymmetric tensors $\epsilon^{k_1 \cdots k_N}$ and $\epsilon_{k_1 \cdots k_N}$ are $SU(N)$ invariants, then we can use these to ensure that the raising and lowering of indices is invertible. We can effectively raise the lower index of an $(n,1)$ tensor $\psi_j^{k_1\cdots k_n}$ by defining the $(n+N-1,0)$ tensor

$$\psi'^{k_1\cdots k_n \ell_1\cdots \ell_{N-1}} \equiv \epsilon^{\ell_1\cdots \ell_N} \psi_{\ell_N}^{k_1\cdots k_n}, \tag{5.3.13}$$

which is antisymmetric in $\ell_1,\ldots,\ell_{N-1}$. The identity $\epsilon_{\ell_1\cdots\ell_{N-1}j}\,\epsilon^{\ell_1\cdots\ell_N} = (N-1)!\delta_j^{\ell_N}$ with $\epsilon^{\ell_1\cdots\ell_N} = \epsilon_{\ell_1\cdots\ell_N}$ allows the index to be lowered again,

$$\frac{1}{(N-1)!}\epsilon_{\ell_1\cdots\ell_{N-1}j}\,\psi'^{k_1\cdots k_n\ell_1\cdots\ell_{N-1}} = \frac{1}{(N-1)!}\epsilon_{\ell_1\cdots\ell_{N-1}j}\,\epsilon^{\ell_1\cdots\ell_N}\psi_{\ell_N}^{k1\cdots k_n} = \psi_j^{k_1\cdots k_n}.$$

Since the raising and lowering operations are invertible, then the effective dimensions of the spaces associated with each of the tensors must be equal. This follows because of the antisymmetry in the $\ell_1,\ldots,\ell_{N-1}$ indices in $\psi'^{k_1\cdots k_n\ell_1\cdots\ell_{N-1}}$. Each lower index can be raised so that an (n,m) tensor ψ becomes an $(n+m(N-1),0)$ tensor ψ'. It is conventional to denote antisymmetrization with respect to indices $j_1,\ldots,j_p$ as $[j_1\cdots j_p]$. Consider an arbitrary (n,m) tensor that is antisymmetric in its m lower indices, $\psi_{[j_1\cdots j_m]}^{k_1\cdots k_n}$. Clearly, $m < N$ as argued above. We generalize Eq. (5.3.13) by again acting with the invariant antisymmetric tensor to give an $(n+N-m,0)$ tensor

$$\psi'^{k_1\cdots k_n\ell_1\cdots\ell_{N-m}} \equiv \epsilon^{\ell_1\cdots\ell_N}\psi_{[\ell_{N-m+1}\cdots\ell_N]}^{k_1\cdots k_n}. \tag{5.3.14}$$

The generalized identity $\epsilon_{\ell_1\cdots\ell_{N-m}j_1\cdots j_m}\,\epsilon^{\ell_1\cdots\ell_N} = m!(N-m)!\delta_{[j_{N-m+1}}^{\ell_{N-m+1}}\cdots\delta_{\ell_N]}^{j_N}$ shows that this is again invertible,

$$\frac{1}{m!(N-m)!}\epsilon_{\ell_1\cdots\ell_{N-m}j_1\cdots j_m}\,\psi'^{k_1\cdots k_n\ell_1\cdots\ell_{N-m}} = \psi_{[j_1\cdots j_m]}^{k_1\cdots k_n}.$$

Young diagrams apply to tensors with upper indices only, $(n',0)$, but any (n,m) tensor can be brought to this form for some n' with some antisymmetry built in.

Using the above results we can first raise any lower indices and then analyze the upper index tensor and its irreps using Young tableaux. For example, this is how we can study meson-type states, $|\Psi\rangle \in \mathbf{N}\otimes\overline{\mathbf{N}}$, with $(1,1)$ tensors, ψ_j^k. Consider an arbitrary $(1,1)$ state, $|\Psi\rangle = \psi_j^k|i_k\bar{i}_j\rangle$, and note that

$$|\Psi\rangle = \psi_j^k\delta^{j\ell_N}|i_k\bar{i}_{\ell_N}\rangle = \psi_j^k\frac{1}{(N-1)!}\epsilon^{\ell_1\cdots\ell_{N-1}j}\epsilon^{\ell_1\cdots\ell_N}|i_k\bar{i}_{\ell_N}\rangle$$

$$= \psi'^{k\ell_1\cdots\ell_{N-1}}\left(\frac{1}{(N-1)!}\epsilon^{\ell_1\cdots\ell_N}|i_k\bar{i}_{\ell_N}\rangle\right) \equiv \psi'^{k\ell_1\cdots\ell_{N-1}}|i_k(\bar{i})_{\ell_1\cdots\ell_{N-1}}\rangle, \tag{5.3.15}$$

where $|i_k(\bar{i})_{\ell_1\cdots\ell_{N-1}}\rangle \equiv \frac{1}{(N-1)!}\epsilon^{\ell_1\cdots\ell_N}|i_k\bar{i}_{\ell_N}\rangle$ can be thought of as "raised basis states." The Young symmetrizers act on the N indices $k,\ell_1,\ldots,\ell_{N-1}$. The Young symmetrizer can be applied to either $\psi'^{\ell_1\cdots\ell_{N-1}k}$ or $|i_k(\bar{i})_{\ell_1\cdots\ell_{N-1}}\rangle$. The Young symmetrizer for the $SU(N)$ singlet is the N component antisymmetric tensor and applying this to the raised basis states gives

$$\epsilon^{k\ell_1\cdots\ell_{N-1}}|i_k(\bar{i})_{\ell_1\cdots\ell_{N-1}}\rangle = \epsilon^{k\ell_1\cdots\ell_{N-1}}\frac{1}{(N-1)!}\epsilon^{\ell_1\cdots\ell_N}|i_k\bar{i}_{\ell_N}\rangle$$

$$= (-1)^{N-1}\delta^{k\ell_N}|i_k\bar{i}_{\ell_N}\rangle = (-1)^{N-1}|i_k\bar{i}_k\rangle, \tag{5.3.16}$$

which is a single vector spanning the one-dimensional singlet subspace. For example, for $SU(3)_{\text{flavor}}$ the normalized singlet state can be written as $\frac{1}{\sqrt{3}}|u\bar{u}+d\bar{d}+s\bar{s}\rangle$.

Some important implications for Young diagrams immediately follow from the above. Consider a $(1,0)$ tensor ψ^k in $SU(N)$ corresponding to a state $|\Psi\rangle \in \mathbf{N} \equiv \square$. Its complex conjugate is a $(0,1)$ tensor $\bar{\psi}_k \equiv \psi^{*k}$, which after raising the index becomes the tensor $\bar{\psi}'^{\ell_1\cdots\ell_{N-1}}$, which is antisymmetric in its $N-1$ indices. So the corresponding conjugate state is $|\bar{\Psi}\rangle \in \overline{\mathbf{N}}$, which corresponds to a column of $N-1$ boxes. This will be true for each column in a Young diagram.

Since we can reorder the symmetrized columns the conjugate of any Young diagram can be obtained as follows (Georgi, 1999): (i) Let $b_i \geq 1$ be the number of boxes in the ith column, then a Young diagram containing p columns is specified by $[b_1, b_2, \ldots, b_p]_{\text{col}}$; (ii) the conjugate Young diagram is given by $\overline{[b_1, b_2, \ldots, b_p]}_{\text{col}} = [N - b_p, \ldots, N - b_2, N - b_1]_{\text{col}}$. So the *conjugate* of a Young diagram is obtained by replacing every column of k boxes with a column of $N - k$ boxes and then reversing the order of the columns. Equivalently, add boxes to a Young diagram that convert it to a rectangle with N rows and p columns and then rotate the added boxes by $180°$. For example, the conjugate of the Young diagram $\lambda = (6, 4, 3, 1) = [4, 3, 3, 2, 1, 1]_{\text{col}}$ in $SU(5)$ is $\bar{\lambda} = (6, 5, 3, 2) = [4, 4, 3, 2, 2, 1]_{\text{col}}$,

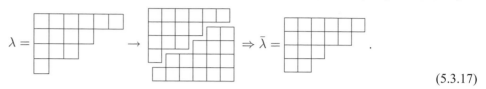

$$\tag{5.3.17}$$

A fundamental result: Each Young diagram λ_ρ for $SU(N)$ corresponds to an irrep ρ of $SU(N)$. Tensors with a symmetry corresponding to a given standard Young tableau $Y_{\rho i}$ of λ_ρ represent the ith occurrence of the irrep ρ of $SU(N)$ in the tensor product space and correspond to a minimal subspace $V_{\rho i}$ of the tensor product space. Applying the Young symmetrizer $\hat{Y}_{\rho i}$ to the basis of the tensor product space projects onto this minimal subspace $V_{\rho i}$ and the corresonding *multiplet* for this ith occurrence of the irrep ρ is an orthonormal basis for $V_{\rho i}$.

Due to the complete antisymmetry of the columns in a Young diagram, no Young diagram for $SU(N)$ can have more than N boxes in a column. Also a column with N boxes is a singlet and can be removed. So Young diagrams that are identical except for the presence of columns with N boxes correspond to the same irreducible representation. Then for $SU(N)$ the maximum number of boxes in a column in a Young diagram is $N - 1$, except the singlet that is a single column of N boxes. We can also think of the singlet as having *zero* boxes. For example, for $SU(4)$ we have

$$\square\hspace{-0.3em} = \text{singlet} \equiv \bullet \quad \text{and} \quad \square = \square . \tag{5.3.18}$$

The dimension of any irrep ρ of $SU(N)$ can be obtained from its Young diagram λ_ρ using the *factors over hooks rule* (Georgi, 1999)

$$\dim(V_\rho) = F_\rho / H_\rho. \tag{5.3.19}$$

Here H_ρ is the hook factor calculated as in Eq. (5.3.5). To calculate the F_ρ factor for $SU(N)$ put an N in the top left-hand box of the Young diagram λ_ρ. For each box to the right add 1 and for each box below subtract 1, then multiply all of the numbers in the boxes together to form F_ρ. For example, for an irrep of $SU(N)$ ($N \geq 5$) specified by the Young diagram $\lambda_\rho = (3, 2, 2, 1)$ we have

$$
\begin{array}{|c|c|c|}
\hline
N & N{+}1 & N{+}2 \\
\hline
N{-}1 & N \\
\cline{1-2}
N{-}2 & N{-}1 \\
\cline{1-2}
N{-}3 \\
\cline{1-1}
\end{array}
\Rightarrow F_\rho
\quad \text{and} \quad
\begin{array}{|c|c|c|}
\hline
6 & 4 & 1 \\
\hline
4 & 2 \\
\cline{1-2}
3 & 1 \\
\cline{1-2}
1 \\
\cline{1-1}
\end{array}
\Rightarrow H_\rho. \tag{5.3.20}
$$

Then $F_\rho = N^2(N + 1)(N + 2)(N - 1)^2(N - 2)(N - 3)$, $H_\rho = 6 \times 4 \times 4 \times 2 \times 3 = 576$ and so $\dim(V_\rho) = F_\rho / H_\rho = N^2(N + 1)(N + 2)(N - 1)^2(N - 2)(N - 3)/576$. Since a Young diagram and its conjugate correspond to an irrep and its conjugate, then the irrep and its conjugate must have the same dimension. The *adjoint irrep* has the number of generators as its dimension, which for $SU(N)$ is $N^2 - 1$. The Young diagram for the adjoint representation has a column of $N - 1$ boxes but with two boxes in the top row. Using the factors over hooks rule we confirm that $\dim(V_{\text{adjoint}}) = F/H = N(N+1)(N-1)!/N(N-2)! = (N+1)(N-1) = N^2 - 1$.

Young symmetrizers: Define $\hat{S}$ and $\hat{A}$ as permutation operations that completely symmetrize and antisymmetrize, respectively, any set of symbols that they act on. As stated above each irrep ρ has a corresponding standard Young diagram λ_ρ and each Young diagram can have various standard Young tableaux $Y_{\rho i}$ for $i = 1, 2, \ldots$ that correspond to different occurrences of the irrep ρ in the tensor product space. Each standard Young tableau $Y_{\rho i}$ corresponds to a Young symmetrizer $\hat{Y}_{\rho i}$ that acts on the tensor product basis states and projects out the corresponding irrep subspace $V_{\rho i}$ with the corresponding multiplet forming its orthonormal basis. The Young symmetrizer can be written as

$$\hat{Y}_{\rho i} = \sum_{Y_{\rho i}\,\text{columns}} \hat{A} \sum_{Y_{\rho i}\,\text{rows}} \hat{S}. \tag{5.3.21}$$

Using the cycle notation introduced in Eq. (5.3.3) we have, for example:

$$Y = \boxed{1\,2} \Rightarrow \hat{Y} = (e) + (12),\ Y - \boxed{1\,2\,3} \Rightarrow \hat{Y} = (e) + (12) + (13) + (23) + (123) + (132),$$

$$Y = \boxed{\begin{smallmatrix}1\\2\end{smallmatrix}} \Rightarrow \hat{Y} = (e) - (12),\quad Y = \boxed{\begin{smallmatrix}1\\2\\3\end{smallmatrix}} \Rightarrow \hat{Y} = (e) - (12) - (13) - (23) + (123) + (132),$$

$$Y = \boxed{\begin{smallmatrix}1&2\\3\end{smallmatrix}} \Rightarrow \hat{Y} = [(e) - (13)] \cdot [(c) + (12)] = (e) + (12) - (13) - (123),$$

$$Y = \boxed{\begin{smallmatrix}1&3\\2\end{smallmatrix}} \Rightarrow \hat{Y} = [(e) - (12)] \cdot [(e) + (13)] = (e) + (13) - (12) - (132), \tag{5.3.22}$$

where in the last two lines we used $(13) \cdot (12) = (123)$ and $(12) \cdot (13) = (132)$. Note that the last two lines correspond to the same irrep but have different standard Young tableaux and will project onto different subspaces when applied to tensor product basis states.

Labeling irreps/multiplets of *SU(N)*: It is beyond the scope of this text to provide a detailed discussion of roots, Dynkin coefficients, Dynkin diagrams, weights, weight diagrams and so on and how they are used in the construction of multiplets. Accessible treatments of these topics can be found in Lichtenberg (1978), Georgi (1999), and elsewhere. It is useful, however, to summarize a few basic concepts. The rank of $SU(N)$ is $N - 1$ and equals the maximum number of Casimir invariants of the group, $C_1, \ldots, C_{N-1}$, i.e., the maximum number of elements of the Lie algebra that commute with all group generators. This means that any irrep/multiplet ρ can be labeled by the $N - 1$ eigenvalues of the Casimir invariants of $SU(N)$, $(c_1, \ldots, c_{N-1})_C$, where the subscript C denotes "Casimir." Here $(0, \ldots, 0)_C$ denotes the singlet state. There are two simple ways to label any irrep of $SU(N)$ in terms of $N - 1$ integers.

(i) *Row lengths*: As seen above we can label any irrep/multiplet in terms of the row lengths of the corresponding Young diagram, where there are $j_r \le N - 1$ rows. We can expand λ to $N - 1$ entries by adding $(N - 1) - j_r$ zeros, $\lambda = (\lambda_1, \ldots, \lambda_{j_r}) \equiv (\lambda_1, \ldots, \lambda_{j_r}, 0, \ldots, 0)$. The singlet is written as $\lambda = (0, \ldots, 0) = \bullet$, since it is a column with N entries and such columns can be removed.

(ii) *Dynkin coefficients*: An irrep/multiplet can also be labeled by the $(N - 1)$ Dynkin coefficients, $q_1, \ldots, q_{N-1}$, in the form $(q_1, \ldots, q_{N-1})_D$, where the subscript D denotes "Dynkin." In a Young diagram the Dynkin coefficient q_1 is the number of additional boxes in row 1 as compared with row 2 and so on for the other q_i. Then $q_i = \lambda_i - \lambda_{i+1}$ for $i = 1, \ldots, N - 1$ with $\lambda_N \equiv 0$. For example, the Young diagrams in Eq. (5.3.1) in the case of $SU(5)$ can be written as $\lambda = (4, 4, 3, 1) =$

$(0, 1, 2, 1)_D$ and $\lambda = (4, 4, 2, 1) = (0, 2, 1, 1)_D$. Even though it is a column of N boxes the singlet state in $SU(N)$ is written as $(0, \ldots, 0)_D$, where we think of a column with N boxes as having no boxes. In terms of Dynkin coefficients it is easy to find the conjugate of a Young diagram, $\overline{(q_1, q_2 \ldots, q_{N-1})}_D = (q_{N-1}, \ldots, q_2, q_1)_D$. This is illustrated by the example in Eq. (5.3.17), where we see that $\overline{(2, 1, 2, 1)}_D = (1, 2, 1, 2)_D$.

Caution: Texts such as that by Georgi (1999) discuss the (n, m) irreps of $SU(3)$, where $n = q_1$ and $m = q_2$ are Dynkin coefficients. This has nothing to do with an (n, m) tensor product space. To avoid confusion we write $SU(3)$ irreps as either $(n, m)_D$ or $\lambda = (\lambda_1, \lambda_2) = (n + m, m)$.

The dimensionality of the weight diagram used to represent any $SU(N)$ multiplet is $N - 1$; e.g., for $SU(2)$, $SU(3)$ and $SU(4)$ the weight diagrams are one, two and three-dimensional, respectively. Sketching weight diagrams for $N \geq 5$ is not possible. Some weight diagrams for $SU(4)_{\text{flavor}}$ are shown in Fig. 5.10.

Irreps/multiplets of $SU(2)$**:** For $SU(2)$ we have $N - 1 = 2 - 1 = 1$ and there is only one Casimir invariant, which is $\hat{\mathbf{J}}^2$ with eigenvalues $j(j + 1)$. So the irreps of $SU(2)$ can be labeled by j as is well known. For each j the corresponding multiplet consists of the $2j + 1$ eigenstates of $\hat{J}_z$. The weight diagram for the irrep j consists of the $2j + 1$ points $-j, -j + 1, \ldots, j - 1, j$ on the real axis. The defining representation is $j = \frac{1}{2}$ or $\lambda = (1)$ or $(q_1)_D = (1)_D$ and acts in the two-dimensional space $W_2 = \mathbf{2} = \square$ spanned by the doublet $|\uparrow\rangle$ and $|\downarrow\rangle$ in the case of $SU(2)_{\text{spin}}$. To transform to $SU(2)_{\text{flavor}}$ we use $j \to I$ with I being isospin and $|\uparrow\rangle \to |u\rangle$ and $|\downarrow\rangle \to |d\rangle$. The j irrep has a $(2j + 1)$ dimensional space V_j with the eigenstates of $\hat{J}_z$ forming an orthonormal basis. The tensor product space $\mathbf{2} \otimes \mathbf{2}$ is $N^n = 2^2 = $ four-dimensional. The Young diagrams with $n = 2$ boxes are

$$
\left. \begin{array}{c} j = 0 \\ \lambda = (1, 1) \\ (q_1)_D = (0)_D \end{array} \right\} \Rightarrow \boxed{\begin{array}{c} \\ \end{array}} \Rightarrow Y = \boxed{\begin{array}{c} 1 \\ 2 \end{array}} \Rightarrow \hat{Y} = (e) - (12) \Rightarrow \tfrac{1}{\sqrt{2}} |\uparrow\downarrow - \downarrow\uparrow\rangle, \tag{5.3.23}
$$

$$
\left. \begin{array}{c} j = 1 \\ \lambda = (2) \\ (q_1)_D = (2)_D \end{array} \right\} \Rightarrow \boxed{} \Rightarrow Y = \boxed{1 \; 2} \Rightarrow \hat{Y} = (e) + (12) \Rightarrow \left\{ \begin{array}{c} |\uparrow\uparrow\rangle \\ \tfrac{1}{\sqrt{2}} |\uparrow\downarrow + \downarrow\uparrow\rangle, \\ |\downarrow\downarrow\rangle \end{array} \right.
$$

where the states have been normalized so that they are orthonormal. As a check we verify using the factors over hooks rule in Eq. (5.3.19) that $\dim(V_{j=0}) = 1$ and $\dim(V_{j=1}) = 3$, giving the singlet $\mathbf{1}$ ($j = 0$) and triplet $\mathbf{3}$ ($j = 1$), respectively. The triplet states are the orthonormal basis states of the three-dimensional space projected out by acting with $\hat{Y} = (e) + (12)$ on the four basis states $|\uparrow\uparrow\rangle, |\uparrow\downarrow\rangle, |\downarrow\uparrow\rangle$ and $|\downarrow\downarrow\rangle$. Similarly, the singlet state is the result of acting with $\hat{Y} = (e) - (12)$ on the four basis states. Non-singlet irreps are a row of $n = 2j$ boxes that symmetrize the basis states of the $(n)_D = (2j)_D$ tensor product space. The space for any $SU(2)$ multiplet/irrep j is written as $V_j = \mathbf{2j + 1}$ since $\dim(V_j) = 2j + 1$:

$$
j = 0, \; V_0 = \mathbf{1}, \; (0)_D = \boxed{\begin{array}{c} \\ \end{array}} = \bullet; \; j \geq \tfrac{1}{2}, \; V_j = \mathbf{2j + 1}, \; (n)_D = (2j)_D = \overbrace{\boxed{} \boxed{} \cdots \boxed{}}^{n \text{ boxes}},
$$

$$
\dim(V_j) = F/H = 2(2 + 1) \cdots (2 + n - 1)/n! = (n + 1)!/n! = (n + 1) = (2j + 1). \tag{5.3.24}
$$

We have the tensor product space decomposition $\mathbf{2} \otimes \mathbf{2} = \mathbf{1} \oplus \mathbf{3}$, where we have equal dimensions on each side as required, $\dim(\mathbf{2} \otimes \mathbf{2}) = 4 = \dim(\mathbf{1}) + \dim(\mathbf{3}) = 1 + 3$, where we have used $\dim(V \oplus W) = \dim(V) + \dim(W)$ and $\dim(V \otimes W) = \dim(V) \dim(W)$. In $SU(2)$ every irrep is equal to its own conjugate, where for the singlet we equate two boxes in a column with zero boxes for this purpose, $\boxed{\begin{array}{c} \\ \end{array}} = \bullet$.

Irreps of $SU(3)$: We state the procedure for constructing the weight diagram for an $(n, m)_D$ irrep/multiplet of $SU(3)$. For a proof, see Georgi (1999).

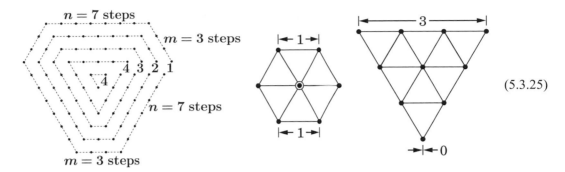

$$(5.3.25)$$

Note: The rightmost point in the left-hand diagram is located in the $x-y$ plane at $\left(\frac{n+m}{2}, \frac{n-m}{6}\sqrt{3}\right)$. We understand the left-hand figure as an explanation of how to draw the weight diagram for an arbitrary multiplet of $SU(3)$ corresponding to the Young diagram $\lambda = (n + m, m) = (n, m)_D$. It is always a hexagon unless either $n = 0$ or $m = 0$ when it becomes a triangle. An example of $(n, m)_D = (7, 3)_D = $ ▯▯▯▯▯▯▯▯▯▯ is shown. The dotted lines indicate the "layers" of the diagram, where the multiplicities of the points in each layer are indicated. The multiplicity of a point is the number of orthonormal states associated with the point. The multiplicity of the outermost layer is always one. The multiplicity increases by one for each inner layer until one arrives at the first triangular-shaped layer, at which point the multiplicity stays constant. This corresponds to multiplicities $1, 2, 3, 4, 4$ in this example. The middle example[7] is for the Young diagram $\lambda = (2, 1) = (1, 1)_D = $ ▯▯, which is an octet (**8**) since $F/H = 24/3 = 8$ from Eq. (5.3.19). Drawing the weight diagram for the multiplet we note that the central point has multiplicity two since there is no triangular layer and so we find eight states as required. The multiplet on the right is for the Young diagram $\lambda = (3) = (3, 0)_D = $ ▯▯▯ and is a decuplet (**10**) since $F/H = 60/6 = 10$. The outer layer of the weight diagram is already triangular, meaning all points in the diagram have multiplicity one and so there are 10 states in the weight diagram as required.

Coupling multiplets: tensor products of irreps: We will often be interested in coupling together irreps/multiplets of $SU(N)$, which is the tensor product of the two irrep subspaces. We wish to decompose this tensor product into a direct sum of irreps/multiplets, which is referred to as the *Clebsch-Gordan series*. We state without proof the recipe for doing this. Note that up to a reordering of the labeling of tensor-product states we can consider $A \otimes B = B \otimes A$. To couple more than two irreps couple the first two and then couple the third to each of the resulting irreps in the Clebsch-Gordan series. We follow the notation of Zyla et al. (2020).

Definition: We define a sequence of letters $a, b, c, d, \ldots$ to be an *admissible* sequence if as we move from left to right at least as many a's have occurred as bs, at least as many bs as cs and so on. So the sequences $abcd$, $aabcb$ and $aaabcbcbd$ are admissable but $abbcd$, $acbbcd$ and $aabbdc$ are not.

(i) In the tensor product $A \otimes B$ it is usually easier to choose B to contain fewer boxes. Fill the first row of B with as, the second row with bs and so on.

(ii) Then take the boxes labeled a and add them to the right of the rows of A or below the columns of A to make larger Young diagrams. Discard any diagrams that are not valid Young diagrams, i.e., if they have a column with more than N boxes or if any row is longer than any row above it. Also no

---

[7] The $(1, 1)_D$ and $(3, 0)_D$ diagrams are adapted with permission from Tanabashi et al. (2018).

column can have more than one a (since the row of as was symmetrized and the resulting columns are to be antisymmetrized). Then add the bs to the resulting diagrams in the same way again, discarding any diagram that is not a valid Young diagram or with more than one b per column. Then form a single sequence of letters by reading *right to left* in the first row, then *right to left* in the second row and so on. If the sequence for a diagram is not admissible, then discard that diagram (this avoids double counting).

(iii) Then add the cs to the surviving diagrams in the same way and repeat the procedure, then do the same for ds and so on.

(iv) The result for $A \otimes B$ will then be the direct sum of the remaining diagrams, noting that the same diagram might occur more than once. At this point columns with N boxes can be removed, except for the singlet irrep.

Example: Consider the coupling of two $SU(3)$ octets, $\mathbf{8} \otimes \mathbf{8} = \square\!\square \otimes \square\!\square$. We have $A = \mathbf{8} = \square\!\square$ and $B = \mathbf{8} = \begin{smallmatrix}a\,a\\b\end{smallmatrix}$. First add the as to $A = \square\!\square$ in all valid ways,

$$\square\,\boxed{a}\boxed{a}, \quad \square\,\boxed{a}, \quad \square^{\boxed{a}}, \quad \square\,\boxed{a}. \tag{5.3.26}$$

Then add b in all valid ways and remove diagrams with inadmissible sequences; e.g., we can't put b at the end of the first row as this leads to the inadmissible sequence baa. Form the direct sum of the remaining diagrams. Finally, remove columns of $N = 3$ boxes from any Young diagram in the direct sum except for the singlet:

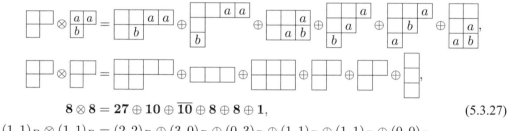

$$\mathbf{8} \otimes \mathbf{8} = \mathbf{27} \oplus \mathbf{10} \oplus \overline{\mathbf{10}} \oplus \mathbf{8} \oplus \mathbf{8} \oplus \mathbf{1}, \tag{5.3.27}$$
$$(1,1)_D \otimes (1,1)_D = (2,2)_D \oplus (3,0)_D \oplus (0,3)_D \oplus (1,1)_D \oplus (1,1)_D \oplus (0,0)_D.$$

Note that the singlet $\mathbf{1} = (0,0)_D$, the octet $\mathbf{8} = (1,1)_D$ and the 27-plet $\mathbf{27} = (2,2)_D$ are self-conjugate. Since the two $\mathbf{8}$ irreps/multiplets are different and correspond to different subspaces, it will sometimes be convenient to write them as $\mathbf{8}$ and $\mathbf{8}'$. We also observe that the total dimensions on each side are equal as expected, $\dim(\mathbf{8} \otimes \mathbf{8}) = (\dim \mathbf{8})^2 = 64 = \dim(\mathbf{27} \oplus \mathbf{10} \oplus \overline{\mathbf{10}} \oplus \mathbf{8} \oplus \mathbf{8} \oplus \mathbf{1}) = 27 + 10 + 10 + 8 + 8 + 1$.

SU(2) Examples:

$$\mathbf{2} \otimes \mathbf{2} = \mathbf{3} \oplus \mathbf{1}, \quad \square \otimes \square = \square\!\square \oplus \square;$$

$$\mathbf{3} \otimes \mathbf{3} = \mathbf{5} \oplus \mathbf{3} \oplus \mathbf{1}, \quad \square\!\square \otimes \square\!\square = \square\!\square\!\square\!\square \oplus \square\!\square \oplus \square; \tag{5.3.28}$$

$$(\mathbf{2j_1 + 1}) \otimes (\mathbf{2j_2 + 1}) = \oplus_{J=|j_1 - j_2|}^{j_1 + j_2} (\mathbf{2J + 1}).$$

SU(3) Examples:

$$\mathbf{3} \otimes \overline{\mathbf{3}} = \mathbf{8} \oplus \mathbf{1}, \quad \square \otimes \square\!\square = \square\!\square \oplus \square; \quad \mathbf{3} \otimes \mathbf{3} = \mathbf{6} \oplus \overline{\mathbf{3}}, \quad \square \otimes \square = \square\!\square \oplus \square;$$

$$\mathbf{8} \otimes \mathbf{3} = \mathbf{15} \oplus \overline{\mathbf{6}} \oplus \mathbf{3}, \quad \square\!\square \otimes \square = \square\!\square\!\square \oplus \square\!\square \oplus \square; \tag{5.3.29}$$

$$\mathbf{3} \otimes \mathbf{3} \otimes \mathbf{3} = (\mathbf{6} \oplus \bar{\mathbf{3}}) \otimes \mathbf{3} = \mathbf{10} \oplus \mathbf{8} \oplus \mathbf{8} \oplus \mathbf{1},$$

***SU(N)* Example:** The adjoint irrep in $SU(N)$ is $(\mathbf{N^2 - 1})$ since it has $N^2 - 1$ dimensions. Using the above rules we find that

$$\bar{\mathbf{N}} \otimes \mathbf{N} = (\mathbf{N^2 - 1}) \oplus \mathbf{1},$$

Clebsch-Gordan coefficients: The tensor product of the $SU(2)$ multiplets labeled by j_1 and j_2 is $(\mathbf{2j_1 + 1}) \otimes (\mathbf{2j_2 + 1})$ and has the Clebsch-Gordan series given in Eq. (5.3.28). The tensor product basis orthonormal states can be written as $|j_1 m_1 j_2 m_2\rangle$. The orthonormal basis of the $(\mathbf{2J + 1})$ and the orthonormal basis of the J irrep/multiplet in the Clebsch-Gordan series on the right-hand side can be written as $|j_1 j_2 J M\rangle$. The $SU(2)$ Clebsch-Gordan coefficients are the coefficients in the expansion of the orthonormal basis $|j_1 j_2 J M\rangle$ of the irrep/multiplet $(\mathbf{2J + 1})$ in Eq. (5.3.28) in terms of the tensor product orthonormal basis. We write the $SU(2)$ coefficients as $\langle j_1 m_1 j_2 m_2 | J M\rangle \equiv \langle j_1 m_1 j_2 m_2 | j_1 j_2 J M\rangle$, where

$$|j_1 j_2 J M\rangle = \sum_{m_1=-j_1}^{j_1} \sum_{m_2=-j_2}^{j_2} |j_1 m_1 j_2 m_2\rangle \langle j_1 m_1 j_2 m_2 | J M\rangle. \tag{5.3.30}$$

The Clebsch-Gordan coefficients for other groups are not always known. However, they are well known for $SU(2)$ (Edmonds, 2016) and $SU(3)$ (McNamee et al., 1964) and there are algorithms for calculating them for any $SU(N)$.

For $SU(3)_{\text{flavor}}$ the x and y axes of the weight diagrams correspond to I_3 and $\frac{\sqrt{3}}{2}Y$, where I_3 is the third component of isospin and Y is the hypercharge (not to be confused with the weak hypercharge Y_W). $SU(3)$ has two Casimir invariants, whereas $SU(2)$ has only one Casimir invariant, $\hat{\mathbf{J}}^2$. All states in an $SU(3)$ irrep/multiplet have the same eigenvalues for these Casimir invariants since they commute with all generators of the group. In $SU(3)_{\text{flavor}}$ it is convenient to label states in an irrep/multiplet as follows. Let ρ be an $SU(3)$ irrep/multiplet, then we can denote a state in ρ as $|\rho Y I I_3\rangle$, where Y, I, I_3 are the hypercharge, isospin and third component of isospin for the state, respectively. Consider ρ in the Clebsch-Gordan series for the tensor product of ρ_a and ρ_b, then

$$|\rho_a \rho_b \rho Y I I_3\rangle = \sum_{\substack{y_a,i_a,i_{a3} \\ y_b,i_b,i_{b3}}} |\rho_a y_a i_a i_{a3} \rho_b y_b i_b i_{b3}\rangle \langle \rho_a y_a i_a i_{a3} \rho_b y_b i_b i_{b3} | \rho Y I I_3\rangle, \tag{5.3.31}$$

where we use lowercase for the hypercharge and isospin of ρ_a and ρ_b and where $\langle \rho_a y_a i_a i_{a3} \rho_b y_b i_b i_{b3} | \rho Y I I_3\rangle \equiv \langle \rho_a y_a i_a i_{a3} \rho_b y_b i_b i_{b3} | \rho_a \rho_b \rho Y I I_3\rangle$ are the Clebsch-Gordan coefficients for $SU(3)$. The analogy with $SU(2)$ is clear. For example, if ρ_a and ρ_b are both $\mathbf{8}$ irreps/multiplets, then we are considering the $\mathbf{8} \otimes \mathbf{8}$ tensor product and the irrep/multiplet ρ is one of the $\mathbf{27}, \mathbf{10}, \overline{\mathbf{10}}, \mathbf{8}, \mathbf{8'}$ or $\mathbf{1}$ irreps/multiplets.

$\boldsymbol{SU(N) \times SU(M) \times U(1) \subset SU(N + M)}$: Let U be an $(N + M) \times (N + M)$ special unitary matrix in the defining representation of $SU(N + M)$. Similarly, let U_N and U_M be $N \times N$ and $M \times M$ matrices in the defining representations of $SU(N)$ and $SU(M)$, respectively. Define I_N and I_M as the $N \times N$ and $M \times M$ identity matrices, respectively. If M is a block diagonal matrix with blocks A and B, then $\det M = \det A \det B$. It then follows that

$$\begin{bmatrix} U_N & 0 \\ 0 & I_M \end{bmatrix}, \begin{bmatrix} I_N & 0 \\ 0 & U_M \end{bmatrix}, \begin{bmatrix} e^{i\alpha/N}I_N & 0 \\ 0 & e^{-i\alpha/M}I_M \end{bmatrix} \in SU(N+M), \qquad (5.3.32)$$

since all three types of matrix are $(N+M) \times (N+M)$ unitary matrices with unit determinant. The three types of matrices commute and they also are elements of an $(N+M) \times (N+M)$ representation of $SU(N)$, $SU(M)$ and $U(1)$, respectively. Any product of one of each of these three types of matrices is some $U \in SU(N+M)$, but not all U can be decomposed this way. This is the sense in which we can write $SU(N) \times SU(M) \times U(1) \subset SU(N+M)$. The Lie algebra must be decomposable in a similar way into three commuting subalgebras $\mathfrak{su}(2)$, $\mathfrak{su}(3)$ and a traceless diagonal matrix for $U(1)$,

$$\begin{bmatrix} \frac{1}{N}I_N & 0 \\ 0 & -\frac{1}{M}I_M \end{bmatrix}. \qquad (5.3.33)$$

The Lie algebra of the $(N+M)^2 - 1$ generators of $SU(N+M)$ contains the block diagonal forms of the $(N^2 - 1) + (M^2 - 1) + 1$ generators of the groups $SU(N)$, $SU(M)$ and $U(1)$. Comparing with $e^{i\alpha q}$ in Eq. (4.1.139) we see that under $U(1)$ the $SU(N)$ and $SU(M)$ have charges $1/N$ and $-1/M$, respectively. Note that since we can arbitrarily rescale α we could equally multiply both charges by NM and have $1/N \to M$ and $-1/M \to -N$ (Georgi, 1999). While we have argued for this decomposition in the defining representations, the properties of the Lie algebra are representation independent and so the result must be true in other representations as well. There are many ways to choose which of the $N+M$ indices are the N and which are the M. Also note that if we choose say $M = 1$, then since $SU(1)$ is the trivial group (the identity) we have the result $SU(N-1) \times U(1) \subset SU(N)$.

Example: An important early example application was the *Georgi-Glashow model* $SU(3)_c \times SU(2)_L \times U(1)_Y \subset SU(5)$, where $SU(2)_L$ refers to left-hand coupled weak isospin and here Y is understood to refer to the weak hypercharge Y_W. This was an early attempt at a Grand Unified Theory (GUT) uniting the three non-gravitational forces into a single gauge group, $SU(5)$, that was then broken down to the Standard Model gauge group $SU(3)_c \times SU(2)_L \times U(1)_Y$. In this model a scalar field analogous to the Higgs field takes on a nonzero vacuum expectation value (VEV) proportional to weak hypercharge, which leads to the spontaneous breaking of $SU(5)$ to the Standard Model. The notation

$$\mathbf{5} \to (\mathbf{3}, \mathbf{1})_{1/3} \oplus (\mathbf{1}, \mathbf{2})_{-1/2} \quad \text{or equivalently} \quad \mathbf{5} \to (\mathbf{3}, \mathbf{1})_2 \oplus (\mathbf{1}, \mathbf{2})_{-3} \qquad (5.3.34)$$

is sometimes used, which denotes that the defining representation of $SU(5)$, $\mathbf{5}$, is broken down to (1) the $\mathbf{3}$ of $SU(3)$ and the $\mathbf{1}$ of $SU(2)$ transforming, respectively, under the $U(1)$ with charge $\frac{1}{3}$ (or 2); and (ii) the $\mathbf{1}$ of $SU(3)$ and the $\mathbf{2}$ of $SU(2)$ transforming under $U(1)$ with charge $-\frac{1}{2}$ (or -3). The model leads to a prediction for a proton decay rate that is inconsistent with limits set by experiments. Proton decays have not been observed. So while the model does not appear viable, it continues to inspire the search for more complex models of this type that might be more successful.

5.3.2 Quark Model

In Table 5.7 we show the additive quantum numbers of the quarks. Antiquarks have opposite additive quantum numbers. The description of up and down quarks in terms of their isospin, rather than assigning "upness" and "down-ness" quantum numbers, is purely historical. The concept of isospin in hadronic physics was well established before the existence of quarks became known. We assign positive parity to quarks and negative parity to antiquarks as discussed in Sec. 4.5.1.

Table 5.7. Additive quantum numbers of quarks						
	d	u	s	c	b	t
B, baryon number	$+\frac{1}{3}$	$+\frac{1}{3}$	$+\frac{1}{3}$	$+\frac{1}{3}$	$+\frac{1}{3}$	$+\frac{1}{3}$
Q, electric charge	$-\frac{1}{3}$	$+\frac{2}{3}$	$-\frac{1}{3}$	$+\frac{2}{3}$	$-\frac{1}{3}$	$+\frac{2}{3}$
$I_z \equiv I_3$, third component of isospin	$-\frac{1}{2}$	$+\frac{1}{2}$	0	0	0	0
S, strangeness	0	0	-1	0	0	0
C, charm	0	0	0	$+1$	0	0
B_b, bottomness	0	0	0	0	-1	0
T, topness	0	0	0	0	0	$+1$

The charge Q and the other additive quark quantum numbers are related by the generalized Gell-Mann–Nishijima formula,

$$Q = I_z + \tfrac{1}{2}\left(B + S + C + B_b + T\right). \tag{5.3.35}$$

Conventionally, the flavor quantum number of each quark is chosen to have the same sign as its electric charge Q; e.g., the s quark has $Q = -\frac{1}{3}$ and strangeness $S = -1$. The *hypercharge* is defined as

$$Y = B + S - \tfrac{1}{3}\left(C - B_b + T\right). \tag{5.3.36}$$

By construction we have $Y = \frac{1}{3}, \frac{1}{3}, -\frac{2}{3}, 0, 0, 0$ for the d, u, s, c, b, t quarks, respectively. The u, d and s quarks form the $SU(3)_{\text{flavor}}$ triplet weight diagram in the I_z, Y plane, which is the **3**. The antiquark triplet **3̄** consists of $\bar{u}, \bar{d}, \bar{s}$. The combinations of these three quarks and antiquarks form $SU(3)_{\text{flavor}}$ irreps/multiplets. More detail of the relevant group theory, including the use of Young tableaux to determine multiplet structures of mesons and baryons, can be found in Carruthers (1966), Kokkedee (1969), Lichtenberg (1978) and Amsler (2018). Additional useful discussions are given in Hamermesh (1962), Gilmore (1974), Wybourne (1974), Choquet-Bruhat et al. (1982), Cahn (1984), Cheng and Li (1984) and Georgi (1999). Because of color confinement we will only ever be interested in QCD bound states that are singlets of $SU(3)_c$.

As discussed in Sec. 5.2.2 hadrons contain valence quarks, sea quarks and gluons. Since the additive quantum numbers of the sea quarks cancel out and since the gluons do not carry the additive quantum numbers, then the quantum numbers are determined by the valence quarks. Since we cannot tell which quark is which, then the term "valence quarks" just refers to the excess of quarks over antiquarks and conversely for valence antiquarks. Due to the effects of dynamical (spontaneous) chiral symmetry breaking the mass of hadrons is far greater than the sum of the current quark masses of the valence quarks; e.g,. the proton mass is 938 MeV and the current quark masses for uud have a combined mass $\lesssim 10$ MeV.

In the *constituent quark* model the valence quarks are taken to be dressed by the sea quarks and gluons in such a way that the valence quarks become constituent quarks with a mass that is $\simeq 330$ MeV greater than the corresponding current quark mass. Such a picture has some support from studies of quarks in lattice QCD, where the effects of dynamical (spontaneous) chiral symmetry breaking lead to an effective increase of quark mass in the low-momentum regime of approximately this amount, e.g., Bowman et al. (2005). Constituent quark masses are somewhat model dependent but typical values are $M_u = 336$ MeV, $M_d = 340$ MeV, $M_s = 486$ MeV and $M_c = 1,550$ MeV

(Griffiths, 2008). Since the electromagnetic and weak interactions have only small effects on masses, then we can expect hadrons with the same total number of u, d and s valence quarks and with a fixed number of c valence quarks to have similar masses. If we further fix the number of s quarks then the symmetry is even better.

Mesons: Mesons have zero baryon number ($B = 0$) and in the quark model are bound states of a constituent quark and antiquark, $q\bar{q}'$. Let ℓ be the relative orbital angular momentum quantum number of the q and $\bar{q}'$. Since the intrinsic parities are $+$ and $-$ for q and $\bar{q}$, respectively, then the overall parity is $P = (-1)^{\ell+1}$. As discussed in Sec. 4.1.10 in the case of a quark-antiquark pair, $q\bar{q}' \to q\bar{q}$, we can define the C-parity of the meson state, $C = (-1)^{\ell+s}$, where s is the total spin quantum number of the meson. Mesons with C-parity will not have any additive quantum numbers, such as charge or flavor. The possible total angular momenta J are $|\ell - s| \leq J \leq \ell + s$. Since each meson multiplet will contain a $q\bar{q}$ meson, then each meson multiplet can be labeled by a value for J^{PC}.

Meson multiplets with $\ell = 0$ have $P = -1$ and if $s = 0$ then they are pseudoscalar multiplets, $J^{PC} = 0^{-+}$. If $\ell = 0$ and $s = 1$ then they are a vector multiplet, $J^{PC} = 1^{--}$. If $\ell \neq 0$ then such meson states are said to have an orbital excitation. For $\ell = 1$ we can have scalars with 0^{++}, axial vectors (pseudovectors) with either 1^{+-} or 1^{++}, and tensors with 2^{++}. States that satisfy $P = (-1)^J$ are said to belong to the *natural spin-parity series* and such states must have $s = 1$ and $CP = (-1)^{2\ell+s+1} = +1$. So in the $q\bar{q}'$ quark model any meson with natural spin parity must have $CP = +1$. Experimentally, a meson is a subatomic particle that experiences the strong interaction and has integer spin. Any meson with natural parity and $CP = -1$ cannot be explained in the $q\bar{q}'$ quark model and is referred to as an *exotic meson*. Examples of exotic mesons are 0^{+-}, 1^{-+}, 2^{+-}, 3^{-+}, 4^{+-} and so on. Note that $\ell = 0$ and $s = 0$ gives $J^{PC} = 0^{-+}$ and so 0^{--} is forbidden in the $q\bar{q}'$ quark model and so also considered exotic.

It is sometimes helpful to generalize the concept of C-parity to G-parity. The G-parity operator is defined as $\hat{G} = \hat{C}e^{i\pi\hat{I}_y}$, where $\hat{C}$ is the charge conjugation operator and $\hat{I}_y$ is the y-component of the isospin operator. It is most useful when applied to mesons with only u and d valence quarks. The operator $e^{i\pi\hat{I}_y}$ is a $180°$ rotation about the y-axis in isospin space, which turns u into d and vice versa so that $I_3 \to -I_3$. The G-parity of an eigenstate of $\hat{G}$ is its corresponding eigenvalue, $G = C(-1)^I = (-1)^{I+\ell+s}$, where I is the total isospin of an isospin multiplet and C is the C-parity of the neutral member of the isospin multiplet. The pions $\pi^0, \pi^\pm$ have $G = (+1)(-1)^1 = -1$, since π^0 has $J^{PC} = 0^{-+}$ and $I = 1$. Just as an $S_z = 0$ state in a spin singlet ($S = 0$) and a spin triplet ($S = 1$) are antisymmetric and symmetric, respectively, under an interchange of spin up and spin down, an $I_3 = 0$ state in an isospin singlet (isoscalar, $I = 0$) and triplet (isovector, $I = 1$) are antisymmetric and symmetric, respectively, under an interchange of up and down quarks. This gives a factor $(-1)^I = -1$. The charged pions are also $\hat{G}$ eigenstates since taking $I_3 \to -I_3$ and then acting with $\hat{C}$ gives $\pi^\pm \to \pi^\mp \to \pi^\pm$ under G. It can be verified that $\pi^\pm$ also have $G = -1$ using the spin-flavor symmetry of their $SU(6)$ quark model states. It then follows that the G-parity of an n pion state is $(-1)^n$, since $\hat{G}|n\pi\rangle = (-1)^n|n\pi\rangle$. The $\rho^0, \rho^\pm$ are vector isovector mesons and so have $J^{PC} = 1^{--}$ and $I = 1$, which gives $G = +1$. The ω is a vector isoscalar and so has $J^{PC} = 1^{--}$ and $I = 0$, which gives $G = -1$. The ϕ is a vector meson containing $s\bar{s}$ and so is a vector isoscalar like the ω and also has $G = -1$. The approximate conservation of G-parity in the strong decays of mesons into pions means that the dominant decays to pions for the ρ, ω and ϕ are $\rho \to 2\pi$, $\omega \to 3\pi$ and $\phi \to 3\pi$.

Considering only u, d and s quarks, each allowed J^{PC} combination in the valence $q\bar{q}'$ model will have corresponding irreps/multiplets in the Clebsch-Gordan series

$$\mathbf{3} \otimes \bar{\mathbf{3}} = \mathbf{8} \oplus \mathbf{1}. \tag{5.3.37}$$

So we have an octet and singlet for each J^{PC}. The pseudoscalar (0^{-+}) octet consists of $\pi^{\pm}, \pi^0, K^0, \overline{K}^0, K^{\pm}, \eta_8$ and the singlet contains η_1. The weak and electromagnetic forces allow the mixing of $|\eta_1\rangle = \frac{1}{\sqrt{3}}|u\bar{u}+d\bar{d}+s\bar{s}\rangle$ and $|\eta_8\rangle = \frac{1}{\sqrt{6}}|u\bar{u}+d\bar{d}-2s\bar{s}\rangle$, resulting in the $|\eta\rangle$ and $|\eta'\rangle$ meson states. For the pions we have $|\pi^+\rangle = |u\bar{d}\rangle$, $|\pi^-\rangle = |\bar{u}d\rangle$ and $|\pi^0\rangle = \frac{1}{\sqrt{2}}|u\bar{u}-d\bar{d}\rangle$. The resulting nonet of particles is illustrated in Fig. 5.4. Here we are only showing the flavor content rather than the full $SU(6)$ spin-flavor content. In the vector octet we have $\rho^{\pm}, \rho^0, K^{*0}, \overline{K}^{0*}, K^{*\pm}, \omega_8$ and for the singlet we have ω_1. Here the ω_1 and ω_8 mix to give a ϕ that is almost entirely $s\bar{s}$ and an ω that is almost entirely $s\bar{s}$ free. The ρ^0 and ω are almost entirely made up of u and d valence quarks with negligible $s\bar{s}$ and $c\bar{c}$.

Extending to $SU(4)_{\text{flavor}}$ for mesons by including u, d, s and c quarks leads to

$$4 \otimes \overline{4} = \mathbf{15} \oplus \mathbf{1}. \tag{5.3.38}$$

Mixing can again occur, giving a total of sixteen meson states, which are shown in the $SU(4)$ flavor weight diagrams in Fig. 5.10 for the pseudoscalar 0^{-+} and vector 1^{--} mesons in the top left and lower left diagrams, respectively. The η_c and J/ψ are almost entirely pseudoscalar and vector $c\bar{c}$ states, respectively.

So referring to Fig. 5.10 states in a multiplet with the same C (meaning charm here rather than C-parity) and hypercharge Y but different I_z will have very similar masses. States in the same I_z–Y plane will have somewhat similar masses, whereas states in a multiplet with different values of $|C|$ (number of c and $\bar{c}$) will have very different masses. The $SU(4)$ multiplets are a convenient way to organize states according to quark content in the valence quark model but are not a good guide to symmetry since the masses of c quarks are so much greater than those of the u, d, s quarks. The $SU(3)$ multiplets (the horizontal planes) are better for that.

Baryons: Baryons have baryon number $B = 1$ and have three valence quarks. In the constituent quark model we take these to be constituent quarks. As stated above the color part of the state must be a color singlet, which means that we have complete antisymmetry under the interchange of the three color indices. Consider the limit where the constituent quark masses are equal. Since quarks are fermions then the state must be completely antisymmetric under the interchange of all quantum numbers of any two quarks and so the baryon state can be decomposed as

$$|qqq\rangle = |\text{color}\rangle_A \times |\text{space, spin, flavor}\rangle_S, \tag{5.3.39}$$

where S means symmetry under the interchange of quantum numbers and A means antisymmetry. The "space" aspect refers to the possibility of nonzero orbital angular momentum in the state. In the case of $SU(3)_{\text{flavor}}$ the baryon multiplets will belong to the multiplets on the right-hand side of

$$\mathbf{3} \otimes \mathbf{3} \otimes \mathbf{3} = \mathbf{10}_S \oplus \mathbf{8}_M \oplus \mathbf{8}_M \oplus \mathbf{1}_A, \tag{5.3.40}$$

where subscripts S, M and A indicate symmetry, mixed symmetry and antisymmetry under the interchange of quark flavors, respectively. All particles in baryon multiplets have the same spin and parity. The $SU(3)_{\text{flavor}}$ octet that contains the proton and neutron is the bottom layer of the top right $SU(4)_{\text{flavor}}$ multiplet with $J^P = \frac{1}{2}^+$ in Fig. 5.10. The $SU(3)_{\text{flavor}}$ decuplet that contains the $\Delta(1232)$ is the bottom layer of the bottom right $SU(4)_{\text{flavor}}$ multiplet with $J^P = \frac{3}{2}^+$. As for the meson case, baryons that lie along I_z lines (C and Y fixed) have very similar masses. Baryons that lie in horizontal planes in a multiplet (C fixed) have somewhat similar masses, but baryons on different horizontal sheets have very different masses.

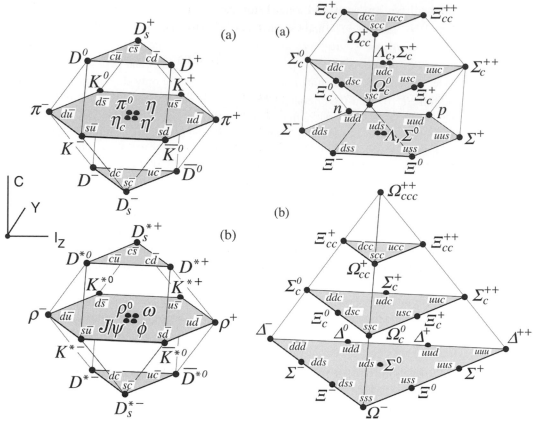

Figure 5.10 Multiplets of $SU(4)_\text{flavor}$ for pseudoscalar mesons (top left); vector mesons (bottom left); spin-$\frac{1}{2}$ baryons (top right); and spin-$\frac{3}{2}$ baryons (bottom right). Here C is the charm quantum number. (Reprinted with permission from Tanabashi et al., 2018)

$SU(6)$ **quark model:** For baryons containing only u, d and s quarks the spin and flavor quantum numbers can be combined leading to six orthonormal basis states, $|u\uparrow\rangle$, $|u\downarrow\rangle$, $|d\uparrow\rangle$, $|d\downarrow\rangle$, $|s\uparrow\rangle$, $|s\downarrow\rangle$. These are a basis for a **6** of $SU(6)_\text{spin-flavor}$. Baryons can then be assigned to multiplets in the Clebsch-Gordan series,

$$\mathbf{6} \otimes \mathbf{6} \otimes \mathbf{6} = \mathbf{56}_S \oplus \mathbf{70}_M \oplus \mathbf{70}_M \oplus \mathbf{20}_A. \tag{5.3.41}$$

The $SU(6)$ multiplets on the right-hand side can be further decomposed in terms of $SU(3)_\text{flavor}$ and $SU(2)_\text{spin}$ multiplets. Using the notation $^{2S+1}\mathbf{N}$ with S the total spin of the three quarks and N the number of states, the decompositions are

$$\mathbf{56} = {}^4\mathbf{10} \oplus {}^2\mathbf{8}, \quad \mathbf{70} = {}^2\mathbf{10} \oplus {}^4\mathbf{8} \oplus {}^2\mathbf{8} \oplus {}^2\mathbf{1}, \quad \mathbf{20} = {}^2\mathbf{8} \oplus {}^4\mathbf{1}. \tag{5.3.42}$$

Since each value of total spin S has $2S+1$ values for S_z, then $^{2S+1}\mathbf{N}$ contains $(2S+1)N$ states. The number of states on each side of the above equations are equal as they should be. From Eqs. (5.3.39) and (5.3.41) we see that the $\mathbf{56}_S$ does not need any orbital angular momentum, since with all quarks in a relative s-wave state the "space" part is symmetric as required. The $^2\mathbf{8}$ of the $\mathbf{56}_S$ then corresponds to the $J^P = \frac{1}{2}^-$ $SU(3)$ octet containing the proton and neutron. Similarly, the $^4\mathbf{10}$ of the $\mathbf{56}_S$ corresponds to the $J^P = \frac{3}{2}^+$ $SU(3)$ decuplet containing the $\Delta(1232)$. For the $\mathbf{70}_M$ and $\mathbf{20}_A$ some orbital angular momentum is necessary to ensure that $|\text{space, spin, flavor}\rangle$ is symmetric as required.

This will not be discussed further here other than to note that states with nonzero orbital angular momentum can be classified as multiplets of $SU(6) \times O(3)$. See the above references for more detail.

Summary

This chapter introduced key elements of particle physics that provide a context and motivation for the remaining chapters. It began with a brief overview of particle physics and its place in our current understanding of the universe. A discussion of the Standard Model of particle physics then followed. A historically structured approach was adopted. This began with electromagnetism and QED, then the strong interaction and QCD, and finally the weak interaction, the electroweak theory, the Higgs boson and the Standard Model. A treatment of quark mixing and the CKM matrix was provided, including unitarity triangles and the Jarlskog invariant. This was followed by a similar treatment of neutrino mixing and the PMNS matrix. An application of neutrino mixing to the study of neutrino oscillations in a vacuum was then carried out and some useful results were derived. The implications of Majorana neutrinos rather than Dirac neutrinos in the Standard Model were considered, which led to a discussion of double beta decay. The techniques for constructing representations of $SU(N)$ were developed including the use of Young tableaux. Finally, these techniques were applied to the constituent quark model and to our understanding of the quark substructure of mesons and baryons.

Problems

5.1 Show that the area of two other of the six nontrivial unitarity triangles of the CKM matrix are equal to half of the Jarlskog invariant, J.

5.2 In some circumstances the contribution of one neutrino flavor can be approximately neglected; e.g., in $\nu_\mu \to \nu_\tau$ atmospheric oscillations the role played by the ν_e is very small. Then the PMNS matrix reduces to a single 2×2 matrix of the Cabibbo form with a single mixing angle, θ. Show that in this case (in natural units) we have $P_{\ell,\ell';\ell \neq \ell'} = \sin^2(2\theta)\sin^2(\Delta m_{\ell\ell'}^2 L/4E)$.

5.3 Show that the set of permutations of n symbols S_n satisfies the properties of a group. Recall that the four properties of a group are closure, associativity, existence of the identity and the existence of inverses.

5.4 Draw the weight diagram for the $(6,3)_D$ multiplet of $SU(3)$ and indicate the multiplicities of each layer. Use this to calculate the number of states in the irrep/multiplet. Draw the corresponding Young diagram and use the factors over hooks rule to calculate the number of dimensions of the irrep/multiplet and verify that the two results agree.

5.5 Draw the Young diagram for the $(1,0,1)_D$ irrep/multiplet of $SU(4)$. Draw all standard Young tableaux for this irrep/multiplet and construct the corresponding Young symmetrizers.

5.6 Prove the $SU(2)$ results in Eq. (5.3.28): $\mathbf{2} \otimes \mathbf{2} = \mathbf{3} \oplus \mathbf{1}$; $\mathbf{3} \otimes \mathbf{3} = \mathbf{5} \oplus \mathbf{3} \oplus \mathbf{1}$; and $\mathbf{2j_1 + 1} \otimes \mathbf{2j_2 + 1} = \oplus_{J=|j_1-j_2|}^{j_1+j_2} (\mathbf{2J + 1})$.

5.7 Prove the $SU(3)$ results in Eq. (5.3.29): $\mathbf{3} \otimes \mathbf{\overline{3}} = \mathbf{8} \oplus \mathbf{1}$; $\mathbf{3} \otimes \mathbf{3} = \mathbf{6} \oplus \mathbf{\overline{3}}$; $\mathbf{8} \otimes \mathbf{3} = \mathbf{15} \oplus \mathbf{\overline{6}} \oplus \mathbf{3}$; and $\mathbf{3} \otimes \mathbf{3} \otimes \mathbf{3} = \mathbf{10} \oplus \mathbf{8} \oplus \mathbf{8} \oplus \mathbf{1}$.

5.8 Prove Eq. (5.3.38) for $SU(4)$: $\mathbf{4} \otimes \mathbf{\overline{4}} = \mathbf{15} \oplus \mathbf{1}$.

5.9　Prove Eq. (5.3.41) for $SU(6)$: $\mathbf{6} \otimes \mathbf{6} \otimes \mathbf{6} = \mathbf{56} \oplus \mathbf{70} \oplus \mathbf{70} \oplus \mathbf{20}$. Explain using Young diagrams why the $\mathbf{56}$ multiplet is symmetric, why the $\mathbf{70}$ multiplets have mixed symmetry and why the $\mathbf{20}$ is antisymmetric.

5.10　For $SU(3)_{\text{flavor}}$ recall that we defined $|i_1\rangle = |u\rangle$, $|i_2\rangle = |d\rangle$, $|i_3\rangle = |s\rangle$ and $|\bar{i}_1\rangle = |\bar{u}\rangle$, $|\bar{i}_2\rangle = |\bar{d}\rangle$, $|\bar{i}_3\rangle = |\bar{s}\rangle$. From Eq. (5.3.15) the raised basis states for the meson states in $\mathbf{3} \otimes \bar{\mathbf{3}}$ are $|i_k(\bar{i})_{\ell_1 \ell_2}\rangle \equiv \frac{1}{2}\epsilon^{\ell_1 \ell_2 \ell_3}|i_k \bar{i}_{\ell_3}\rangle$, where the Young symmetrizers $\hat{Y}$ act on the labels k, ℓ_1, ℓ_2, e.g., $(12)|i_k(\bar{i})_{\ell_1 \ell_2}\rangle = |i_{\ell_1}(\bar{i})_{k \ell_2}\rangle$ and $(123)|i_k(\bar{i})_{\ell_1 \ell_2}\rangle = |i_{\ell_3}(\bar{i})_{k \ell_1}\rangle$ and so on. The singlet Young diagram is ⬚ and the octet Young diagram is ⬚⬚. The Young symmetrizers are given in Eq. (5.3.22).

(a)　Show that applying the singlet Young symmetrizer $\hat{Y} = (e) - (12) - (13) - (23) + (123) + (132)$ to the raised basis states gives a vanishing result unless k, ℓ_1, ℓ_2 are all different. When they are all different show that $\hat{Y}$ projects out the one-dimensional space with normalized basis vector $\frac{1}{\sqrt{3}}|u\bar{u} + d\bar{d} + s\bar{s}\rangle = |\eta_1\rangle$, which recovers the result in Eq. (5.3.16) for this case.

(b)　Consider the octet Young symmetrizer $\hat{Y} = (e) + (12) - (13) - (123)$. Show that applying this when k, ℓ_1, ℓ_2 are all different leads to the states $\frac{1}{2}|u\bar{u} - d\bar{d}\rangle$, $\frac{1}{2}|d\bar{d} - s\bar{s}\rangle$ and $\frac{1}{2}|s\bar{s} - u\bar{u}\rangle$, from which we can construct two orthonormal states chosen to be $\frac{1}{\sqrt{2}}|u\bar{u} - d\bar{d}\rangle = |\pi^0\rangle$ and $\frac{1}{\sqrt{6}}|u\bar{u} + d\bar{d} - 2s\bar{s}\rangle = |\eta_8\rangle$. Since $|i_k(\bar{i})_{\ell_1 \ell_2}\rangle$ is antisymmetric in ℓ_1 and ℓ_2 then $\ell_1 \neq \ell_2$ and there are only six remaining independent possibilities $k = \ell_1 \neq \ell_2$, i.e., $k\ell_1\ell_2 = 112, 113, 221, 223, 331, 332$. Show that applying the symmetrizer $\hat{Y}$ to these leads to the orthonormal states $|u\bar{s}\rangle = |K^+\rangle$, $|u\bar{d}\rangle = |\pi^+\rangle$, $|d\bar{s}\rangle = |K^0\rangle$, $|d\bar{u}\rangle = |\pi^-\rangle$, $|s\bar{d}\rangle = |\overline{K}^0\rangle$, $|s\bar{u}\rangle = |K^-\rangle$, respectively, where we assume $J^P = 0^-$ here and so arrive at the pseudoscalar meson octet. Assuming $J^P = 1^-$ gives the vector octet.

(c)　Show that the second Young symmetrizer, $\hat{Y} = (e) + (13) - (12) - (132)$, in Eq. (5.3.22) leads to the same states and hence to the same irrep/multiplet. This is due to the antisymmetry in ℓ_1 and ℓ_2 for the mesons. The baryon octets $\mathbf{8}$ and $\mathbf{8}'$ in the Clebsch-Gordan series for $\mathbf{3} \otimes \mathbf{3} \otimes \mathbf{3}$ are different. However, for baryons we need to use the $SU(6)_{\text{spin-flavor}}$ formalism because of the need to satisfy Eq. (5.3.39).

5.11　A beam of π^- mesons is directed at a proton target and produces a variety of final states. For each of the particle states listed below propose a reaction that leads to a final state containing at least this particle. Draw a simple labeled diagram showing quark, lepton and boson lines as done, for example, in Fig. 5.5. Choose a reaction that you expect to be one of the most important ways for this final state containing the particle to be produced, keeping in mind the relative strengths of the strong, electromagnetic and weak interactions. Assume that the π^- beam has sufficient energy for each reaction to be kinematically allowed: (a) Δ^{++} resonance; (b) Z^0 boson; (c) electron neutrino, ν_e; (d) K^- meson; and (e) anti-neutrino and a charmed baryon. (Hint: The quark content of hadrons is given in Tables 5.4 and 5.5 and Fig. 5.10.)

5.12　Consider the following reactions. Which reactions are allowed? If allowed, explain what the main reaction mechanism is using a labeled diagram showing quark, lepton and boson lines. Explain whether it occurs through the strong and/or electromagnetic and/or weak interactions. If not allowed, why is it forbidden? (Hint: For the quark flavor content of hadrons, see above.) (a) $p + p \to K^+ + K^+ + \Xi^0$; (b) $\Sigma^0 \to \Lambda^0 + \gamma$; (c) $\pi^- + p \to \pi^0 + K^- + (\bar{\Sigma})^+$; (d) $\Lambda^0 \to p + \mu^- + \bar{\nu}_\mu$; (e) $\pi^+ + n \to \bar{K}^0 + K^+$; (f) $K^0 + n \to K^- + p$; (g) $\Sigma^+ \to p + \pi^0$; (h) $\nu_e + e^+ + n \to p + \pi^0$; (i) $p + \pi^+ + \pi^- \to \Delta^+ + \pi^0$; (j) $\Lambda^0 \to K^- + p$; (k) $\Sigma^- \to n + \pi^-$; (l) $\pi^0 + \nu_e \to \pi^+ + e^-$.

6 Formulation of Quantum Field Theory

This chapter introduces the key elements in the construction of free quantum field theories. There are many excellent quantum field theory texts and the reader should refer to these for additional details and insights; e.g., see Schweber (1961), Bjorken and Drell (1965), Roman (1969), Nash (1978), Itzykson and Zuber (1980), Cheng and Li (1984), Mandl and Shaw (1984), Ryder (1986), Brown (1992), Bailin and Love (1993), Sterman (1993), Peskin and Schroeder (1995), Weinberg (1995, 1996), Greiner and Reinhardt (1996), Pokorski (2000), Srednicki (2007), Zee (2010), Aitchison and Hey (2013) and Schwartz (2013) and many others. We begin with some general remarks before proceeding.

In the macroscopic limit, where the action is very large compared with $\hbar$, nonrelativistic quantum mechanics reduces to Newtonian classical mechanics. The relativistic extension of quantum mechanics is relativistic quantum mechanics, which describes the interaction of a relativistic quantum particle with an external field. When the possibility of particle creation and annihilation is also included we then arrive at quantum field theory. If we are studying physical systems such as those relevant to condensed matter physics, then we typically need not concern ourselves with relativity but do need to consider the creation and annihilation of particle-like excitations such as particle-hole pairs, Cooper pairs and phonons. In the study of gauge field theories such as QED and QCD both relativity and particle-antiparticle creation and annihilation are central to our understanding of relevant phenomena and the full machinery of relativistic quantum field theory is essential.

As we progress, we come to understand that subatomic particles are the quantized fluctuations of fields that permeate all of space. These quantum fluctuations are about a stable or perhaps metastable equilibrium state that we call the vacuum state. Our interacting quantum field theories describe how these quantized fluctuations interact with each other.

Notation and units: From this point on only natural units ($\hbar = c = 1$) will be used and combined with Lorentz-Heaviside units for electromagnetism and other gauge fields. The Einstein summation convention for repeated spacetime indices is to be understood. Similarly, summation over any repeated indices, such as Cartesian or spinor indices, is also to be assumed unless stated otherwise. Where it assists clarity, summations are sometimes explicitly written. The symbol e is the *magnitude* of the electron charge $e = |e|$; i.e., the **electron charge is** $q_e = -e \equiv -|e|$.

6.1 Lessons from Quantum Mechanics

It is useful to begin with a physical motivation for quantum field theory in terms of familiar systems before launching into the formalism. We begin by drawing an analogy with the formulation of normal modes in classical mechanics discussed in Sec. 2.3 and then generalize this to the case of a Poincaré-invariant system.

6.1.1 Quantization of Normal Modes

Consider some complicated classical system with N degrees of freedom and a stable static equilibrium as defined in Sec. 2.3. If we restrict our attention to small excitations or oscillations of the system such that we only experience the quadratic part of the kinetic and potential energies expanded around that stable static equilibrium, then the motion of the system is completely characterized by the behavior of the N normal modes of the system, with angular frequencies ω_j for $j = 1, \ldots, N$. In other words, in this regime the behavior of the system can be characterized by the motion of these N independent harmonic oscillators as given in Eq. (2.3.24).

If we sufficiently increase the size of the excitations around this equilibrium, then we will start to encounter small deviations from the quadratic behavior that can initially be characterized as small couplings between these normal modes that can be treated perturbatively. As the size of the excitations continues to increase, we will begin to probe the large-scale behavior of the system well beyond the vicinity of the small-scale normal-mode variations around the stable static equilibrium. Of course, we might not be expanding around the global energy minimum. There are various possibilities, for example: (i) the system may be oscillating around the global minimum of the energy; (ii) there might be more than one stable static equilibrium and the system may be trapped in a local energy minimum rather than the global minimum; or (iii) the system could be trapped and oscillating around a local stable static equilibrium when the system may have an energy that is not bounded below.

The normal modes in Sec. 2.3 were denoted as $\vec{\zeta}(t)$, where $\vec{\zeta} = (\zeta_1, \ldots, \zeta_N)$ and where the effective Lagrangian and effective Hamiltonian are

$$L = \sum_{j=1}^{N} \tfrac{1}{2} \left(\dot{\zeta}_j^2 - \omega_j^2 \zeta_j^2 \right), \quad H = \sum_{j=1}^{N} \tfrac{1}{2} \left(\pi_j^2 + \omega_j^2 \zeta_j^2 \right) = \sum_{j=1}^{N} \tfrac{1}{2} \left(\dot{\zeta}_j^2 + \omega_j^2 \zeta_j^2 \right). \quad (6.1.1)$$

The normal mode conjugate momenta are $\pi_j \equiv \partial L / \partial \dot{\zeta}_j = \dot{\zeta}_j$ and the normal modes satisfy the harmonic oscillator equations of motion,

$$\ddot{\zeta}_j + \omega_j^2 \zeta_j^2 = 0. \quad (6.1.2)$$

Let us now imagine the quantized version of this system. Since quantum tunneling is now possible, we require that the barrier between this minimum and any nearby minimum is sufficiently wide and/or high that quantum tunneling is negligible. Let us consider the excitation of those low-energy normal modes for which we have many energy quanta $\hbar \omega_i$ before we encounter an energy that probes the non-quadratic behavior of the excitations around the stable static equilibrium. A well-known example of such a quantum system of this type is an ionic crystal lattice whose quantized vibrational modes are referred to as *phonons*. The name derives from the fact that at long wavelengths these excitations can produce sound when they reach the surface of the crystal. Appropriate superpositions of these phonon excitations can lead to the propagation of wavepackets in the periodic crystal lattice that have a particle-like propagation. We generically refer to quantized excitations capable of exhibiting particle-like propagation as *quasiparticles*. Note that a phonon refers to a quantized collective excitation of the ionic lattice, where the corresponding wavepacket may be localized or spread through the ionic lattice.

After quantization the normal mode Hamiltonian in Eq. (6.1.1) leads to the effective theory Hamiltonian operator for this system,

$$\hat{H} = \sum_{j=1}^{N} \tfrac{1}{2} \left(\hat{\pi}_j^2 + \omega_j^2 \hat{\zeta}_j^2 \right), \quad (6.1.3)$$

where $\hat{\zeta}_j$, $\hat{\pi}_j$ and ω_j are the normal mode Hermitian coordinate operator, the Hermitian conjugate momentum operator and the angular frequency of the ith normal mode, respectively. We have shifted the energy so that the classical minimum of the energy is zero, as we did when discussing normal modes in classical mechanics. The absolute value of the Hamiltonian has no effect on the dynamics of the system and only energy changes are relevant as is obvious from Hamilton's equations.

In coordinate-space representation we have $\hat{\zeta}_j \rightarrow \zeta_j$ and $\hat{\pi}_j \rightarrow -i\hbar\partial/\partial\zeta_j$. In this or any other representation we have the equal-time canonical commutation relations for the quantized system,

$$[\hat{\zeta}_j(t), \hat{\zeta}_k(t)] = [\hat{\pi}_j(t), \hat{\pi}_k(t)] = 0 \quad \text{and} \quad [\hat{\zeta}_j(t), \hat{\pi}_k(t)] = i\hbar\delta_{jk}, \tag{6.1.4}$$

where $\hat{\zeta}_j(t)$ and $\hat{\pi}_j(t)$ are the Heisenberg picture operators.

Consider the Schrödinger picture operators,

$$\hat{\zeta}_j \equiv \hat{\zeta}_j(t_0) \quad \text{and} \quad \hat{\pi}_j \equiv \hat{\pi}_j(t_0). \tag{6.1.5}$$

The generalized coordinate operator $\hat{\zeta}_j$ for the jth normal mode has the eigenvalue equation $\hat{\zeta}_j|\zeta_j\rangle = \zeta_j|\zeta_j\rangle$ with $\zeta_j \in \mathbb{R}$ and similarly for the conjugate momentum operator, $\hat{\pi}_j|\pi_j\rangle = \pi_j|\pi_j\rangle$ with $\pi_j \in \mathbb{R}$. Since all the generalized coordinates commute and similarly the canonical momenta, we can use the simultaneous eigenstates of either as a basis for the Hilbert space of our quantized system of normal modes. We can write these simultaneous eigenstates as $|\vec{\zeta}\rangle = |\zeta_1\zeta_2\cdots\zeta_n\rangle$ and $|\vec{\pi}\rangle = |\pi_1\pi_2\cdots\pi_n\rangle$, respectively. Then we have for these two bases,

$$\langle\vec{\zeta}|\vec{\zeta}'\rangle = \delta^n(\vec{\zeta} - \vec{\zeta}') \quad \text{and} \quad \int d^n\zeta \, |\vec{\zeta}\rangle\langle\vec{\zeta}| = \hat{I},$$
$$\langle\vec{\pi}|\vec{\pi}'\rangle = \delta^n(\vec{\pi} - \vec{\pi}') \quad \text{and} \quad \int d^n\pi \, |\vec{\pi}\rangle\langle\vec{\pi}| = \hat{I}, \tag{6.1.6}$$

where obviously $\hat{\zeta}_j|\vec{\zeta}\rangle = \zeta_j|\vec{\zeta}\rangle$ and $\hat{\pi}_j|\vec{\pi}\rangle = \pi_j|\vec{\pi}\rangle$ for $j = 1, \ldots, n$. Note that Eq. (6.1.6) is valid for *any* quantum system with n generalized coordinates; it is not specific to a collection of harmonic oscillators.

Following the usual ladder operator formulation of the quantum harmonic oscillator, we define

$$\hat{a}_j \equiv \sqrt{\frac{\omega_j}{2\hbar}}\left(\hat{\zeta}_j + \frac{i}{\omega_j}\hat{\pi}_j\right) \quad \text{and} \quad \hat{a}_j^\dagger \equiv \sqrt{\frac{\omega_j}{2\hbar}}\left(\hat{\zeta}_j - \frac{i}{\omega_j}\hat{\pi}_j\right), \tag{6.1.7}$$

which are referred to as the raising and lowering operators, respectively. We can rewrite the above as

$$\hat{\zeta}_j = \sqrt{\frac{\hbar}{2\omega_j}}\left(\hat{a}_j + \hat{a}_j^\dagger\right) \quad \text{and} \quad \hat{\pi}_j = -i\sqrt{\frac{\hbar\omega_j}{2}}\left(\hat{a}_j - \hat{a}_j^\dagger\right). \tag{6.1.8}$$

Using the above the commutation relations and these definitions we obtain

$$[\hat{a}_j, \hat{a}_k] = [\hat{a}_j^\dagger, \hat{a}_k^\dagger] = 0 \quad \text{and} \quad [\hat{a}_j, \hat{a}_k^\dagger] = \delta_{jk}, \tag{6.1.9}$$

where we are adopting the standard shorthand of suppressing the identity operator, $[\hat{a}_j, \hat{a}_k^\dagger] = \delta_{jk}\hat{I}$. The first two results follow easily and the third follows since

$$[\hat{a}_j, \hat{a}_k^\dagger] = \sqrt{\frac{\omega_j}{2\hbar}}\sqrt{\frac{\omega_k}{2\hbar}}\left(\frac{1}{\omega_k}[\hat{\zeta}_j, -i\hat{\pi}_k,] + \frac{1}{\omega_j}[i\hat{\pi}_j, \hat{\zeta}_k]\right) = 1. \tag{6.1.10}$$

We can express the Hamiltonian operator as

$$\hat{H} = \sum_{j=1}^N \hbar\omega_j(\hat{a}_j\hat{a}_j^\dagger - \tfrac{1}{2}) = \sum_{j=1}^N \hbar\omega_j(\hat{a}_j^\dagger\hat{a}_j + \tfrac{1}{2}) \equiv \sum_{j=1}^N \hbar\omega_j(\hat{N}_j + \tfrac{1}{2}), \tag{6.1.11}$$

which follows since from Eq. (6.1.3) we have

$$\hat{H} = \sum_{j=1}^{N} \tfrac{1}{4}(\hbar\omega_j) \left(-\left[\hat{a}_j^{\dagger 2} - 2\hat{a}_j^{\dagger}\hat{a}_j - 1 + \hat{a}_j^2 \right] + \left[\hat{a}_j^{\dagger 2} + 2\hat{a}_j^{\dagger}\hat{a}_j + 1 + \hat{a}_j^2 \right] \right)$$
$$= \sum_{j=1}^{N} \hbar\omega_j(\hat{a}_j^{\dagger}\hat{a}_j + \tfrac{1}{2}). \tag{6.1.12}$$

In Eq. (6.1.11) we have defined the *number operator* for the jth normal mode as

$$\hat{N}_j \equiv \hat{a}_j^{\dagger}\hat{a}_j, \quad \text{where} \quad \hat{N}_j = \hat{N}_j^{\dagger} \quad \text{and} \quad [\hat{N}_j, \hat{N}_k] = [\hat{H}, \hat{N}_j] = 0. \tag{6.1.13}$$

We can form a complete orthonormal basis using the normalized simultaneous eigenstates of the number operators $\hat{N}_j$. From the commutation relations we find

$$[\hat{N}_j, \hat{a}_k^{\dagger}] = \delta_{jk}\hat{a}_j^{\dagger} \quad \text{and} \quad [\hat{N}_j, \hat{a}_k] = -\delta_{jk}\hat{a}_j. \tag{6.1.14}$$

Let c_j and $|\psi_j\rangle$ be an eigenvalue and eigenstate of $\hat{N}_j$, respectively. Then we have

$$\hat{N}_j|\psi_j\rangle = c_j|\psi_j\rangle, \; \hat{N}_j\hat{a}_j^{\dagger}|\psi_j\rangle = (c_j+1)\hat{a}_j^{\dagger}|\psi_j\rangle, \; \hat{N}_j\hat{a}_j|\psi_j\rangle = (c_j-1)\hat{a}_j|\psi_j\rangle. \tag{6.1.15}$$

We see that $\hat{a}_j^{\dagger}$ and $\hat{a}_j$ acting on an eigenstate increase and decrease the eigenvalue of $\hat{N}_j$ by 1, respectively. The eigenvalues of $\hat{H}$ must be nonnegative real numbers, since both $\hat{\pi}_j^2$ and $\hat{q}_j^2$ are the squares of Hermitian operators and so will have nonnegative real eigenvalues. So from Eq. (6.1.11) we see that there must be some eigenstate of $\hat{H}$ that cannot be lowered any further by $\hat{a}_j$; i.e., this means that there must be some state $|0_j\rangle$ such that $\hat{a}_j|0_j\rangle = 0$. This immediately leads to the result that $\hat{N}_j|0_j\rangle = \hat{a}_j^{\dagger}\hat{a}_j|0_j\rangle = 0$ and that $0, 1, 2, \ldots$ are the eigenvalues of $\hat{N}_j$.

It is then clear that the eigenvalues of each $\hat{N}_j$ are the nonnegative integers and we can write the simultaneous eigenvalues of the number operators as

$$|\vec{n}\rangle \equiv |n_1, n_2, \ldots, n_N\rangle \quad \text{where} \quad \hat{N}_j|\vec{n}\rangle = n_j|\vec{n}\rangle \quad \text{with} \quad n_j = 0, 1, 2, \ldots \tag{6.1.16}$$

for each $j = 1, 2, \ldots, N$. These simultaneous number eigenstates are then also the energy eigenstates of the system,

$$\hat{H}|\vec{n}\rangle = \sum_{j=1}^{N} \hbar\omega_j \left(\hat{N}_j + \tfrac{1}{2}\right)|\vec{n}\rangle = \sum_{j=1}^{N} \hbar\omega_j \left(n_j + \tfrac{1}{2}\right)|\vec{n}\rangle = E(\vec{n})|\vec{n}\rangle. \tag{6.1.17}$$

So we can characterize the energy levels of the system by the *occupation numbers* of the normal modes, $\vec{n} = (n_1, n_2, \ldots, n_N)$, such that

$$E(\vec{n}) = \sum_{j=1}^{N} E_j(n_j) = \sum_{j=1}^{N} \left(n_j + \tfrac{1}{2}\right)(\hbar\omega_j) \equiv E_0 + \sum_{j=1}^{N} n_j(\hbar\omega_j), \tag{6.1.18}$$

where E_0 is the quantum ground-state energy of the system with respect to the classical ground-state energy of the classical system, which is zero by construction of the Hamiltonian. We refer to E_0 as the *zero-point energy* of the quantized system,

$$E_0 \equiv \sum_{j=1}^{N} \tfrac{1}{2}\hbar\omega_j, \tag{6.1.19}$$

which is the energy when all occupation numbers vanish, $\vec{n} = (0, 0, \ldots, 0)$. Note that E_0 is the energy of the ground state of the quantized Hamiltonian *with respect to the classical equilibrium point*. Note that we cannot learn about the full physical system simply by studying the quantized excitations in the neighborhood of the equilibrium point. The occupation number of the jth normal mode tells us how many quanta of energy $\hbar\omega_j$ we have in the system. Since we can have n_j quanta with the same energy $\hbar\omega_j$ and since these excitations are indistinguishable from each other, then we see that these quanta obey Bose-Einstein statistics. This tells us, for example, that phonons are bosonic and that the ground state of the quantized ionic lattice corresponds to the absence of any phonons.

The operator that measures the total number of bosonic quasiparticles is

$$\hat{N} \equiv \sum_{j=1}^{N} \hat{N}_j \quad \text{such that} \quad \hat{N}|\vec{n}\rangle = \sum_{j=1}^{N} n_j|\vec{n}\rangle \equiv N_{\text{tot}}|\vec{n}\rangle, \tag{6.1.20}$$

where $N_{\text{tot}} \equiv n_1 + n_2 + \cdots + n_N$. Let these occupation number eigenstates $|\vec{n}\rangle$ be normalized $\langle \vec{n}|\vec{n}\rangle = 1$, then they form a complete orthonormal basis

$$\sum_{n_1,\ldots,n_N=1}^{\infty} |\vec{n}\rangle\langle \vec{n}| = \hat{I} \quad \text{and} \quad \langle \vec{n}|\vec{n}'\rangle = \delta_{\vec{n}\vec{n}'} = \delta_{n_1 n_1'} \delta_{n_2 n_2'} \cdots \delta_{n_N n_N'}. \tag{6.1.21}$$

This basis is known as the *occupancy number basis* and the Hilbert space spanned by all such basis states is the bosonic Fock space. This is a third basis for the Hilbert space of the system along with the generalized coordinate basis, $|\vec{\zeta}\rangle$, and the canonical momentum basis, $|\vec{\pi}\rangle$, discussed in Eq. (6.1.6).

The occupancy number basis doesn't attempt to assign an identity to the various quanta and so does not require the symmetrization procedures that were required to construct the bosonic basis states used in Sec. 4.1.9. We note that the bosonic Fock space is the Hilbert space of the effective theory where all non-quadratic behavior can be ignored. Fock space is the Hilbert space relevant for describing the quantum behavior of the normal modes, i.e., for describing the behavior of the noninteracting quasiparticles of the effective theory.

Now we know from above that $\hat{a}_j|\vec{n}\rangle = 0$ if $n_j = 0$. We also know that

$$\hat{a}_j^\dagger|\ldots, n_j, \ldots\rangle = c_{j+}|\ldots, n_j + 1, \ldots\rangle,$$
$$\hat{a}_j|\ldots, n_j, \ldots\rangle = c_{j-}|\ldots, n_j - 1, \ldots\rangle \quad \text{for} \quad n_j \geq 1, \tag{6.1.22}$$

for some $c_{j+}, c_{j-} \in \mathbb{C}$. Since the phase of the kets can be chosen arbitarily we can choose real coefficients and we find that

$$c_{j+} = \sqrt{n_j + 1} \quad \text{and} \quad c_{j-} = \sqrt{n_j}, \tag{6.1.23}$$

which allows us to write for any occupation number basis state

$$|\vec{n}\rangle = |n_1, n_2, \ldots, n_N\rangle = \prod_{j=1}^{N} \left((\hat{a}_j^\dagger)^{n_j}/\sqrt{n_j!}\right)|\vec{0}\rangle. \tag{6.1.24}$$

We refer to the $\hat{a}_j^\dagger$ and the $\hat{a}_j$ as *creation operators* and *annihilation operators*, respectively, since

$$\hat{a}_j^\dagger \text{ } creates \text{ a quasiparticle in the } j\text{th normal mode; and}$$

$$\hat{a}_j \text{ } annihilates \text{ a quasiparticle in the } j\text{th normal mode.} \tag{6.1.25}$$

In discussions of harmonic oscillators they are also referred to as *ladder operators*.

Proof: Temporarily writing for the sake of brevity here $|n_j\rangle = |\ldots, n_j, \ldots\rangle$ we see that the above results follow since

$$n_j + 1 = \langle n_j|(\hat{N}_j + 1)|n_j\rangle = \langle n_j|\hat{a}_j\hat{a}_j^\dagger|n_j\rangle = |c_{j+}|^2\langle n_j + 1|n_j + 1\rangle = |c_{j+}|^2,$$

$$n_j = \langle n_j|\hat{N}_j|n_j\rangle = \langle n_j|\hat{a}_j^\dagger\hat{a}_j|n_j\rangle = |c_{j-}|^2\langle n_j - 1|n_j - 1\rangle = |c_{j-}|^2. \tag{6.1.26}$$

Starting from the empty or ground state $|0\rangle$ we can build up any occupancy number basis state using $|n_j + 1\rangle = (\hat{a}_j^\dagger/\sqrt{n_j + 1})|n_j\rangle$. Doing this for every j we arrive at the above result giving any $|\vec{n}\rangle$ from the empty state $|\vec{0}\rangle$.

6.1.2 Motivation for Relativistic Quantum Field Theory

From the discussion in Sec. 4.7 we know that relativistic quantum mechanics was formulated to describe the relativistic quantum behavior of a single particle interacting with an external field, where the field is sufficiently weak and slowly varying such that particle-antiparticle creation and annihilation can be ignored and the Klein paradox so avoided. We wish to generalize our treatment to describe multiple particles and multiple antiparticles. We will need to ensure that Bose-Einstein and Fermi-Dirac statistics emerge naturally for bosons and fermions, respectively.

We also noted that relativistic quantum mechanics was formulated in terms of relativistic wave equations. We saw in Chapter 3 that relativistic wave equations are the equations of motion of relativistic classical field theories, which can be described in terms of Lagrangians and Hamiltonians for relativistic classical fields. So, apart from the judicious placement of factors of $\hbar$ and c and the entirely different interpretation of the solutions of the equations of motion, relativistic quantum mechanics and relativistic classical field theory are often considered to be essentially the same. However, the interpretation of the solutions to the equations of motion are quite different.

The above discussion of the quantum mechanics of a complicated system when we restrict our attention to small oscillations around a static stable classical equilibrium suggests that, at least for the case of bosons, we should consider the same type of approach as a means of improving on relativistic quantum mechanics. This suggests that for the study of small excitations around a stable equilibrium we should consider quantizing the normal modes of the classical field theory expanded around this equilibrium where only the quadratic parts of the Lagrangian and Hamiltonian are retained to give a purely quadratic effective theory. For scalar particles this corresponds to the Lagrangian and Hamiltonian that led to the free Klein-Gordon equation. Due to the Lorentz-covariant nature of the theory we again expect that particles and antiparticles will simultaneously and naturally emerge.

We now have the motivation that leads to the formulation of relativistic quantum field theory for free scalar particles, with its many parallels to nonrelativistic quasiparticles in quantum mechanics. Note that the Fermi-Dirac statistics of fermions will require some additional special treatment as will the quantization of gauge fields due to the fact that physical observables are gauge-invariant.

Observation: It therefore seems reasonable to contemplate the possibility that what we think of as the classical vacuum is some stable equilibrium of a larger underlying theory, where what we think of as free particles are the quasiparticles of an effective free quantum theory built on that vacuum and where the relatively weak interactions between these free particles are analogous to the small deviations from quadratic behavior discussed in the normal mode context for low-energy excitations.

Use of natural units: *As noted above, from this point on we will be using natural units* $\hbar = c = k = 1$. We may occasionally explicitly restore these constants when it is important to motivate a discussion, such as taking the classical or nonrelativistic limit. They can always be restored from expressions in natural units using dimensional arguments as and when needed.

6.2 Scalar Particles

The quantization of the real scalar field provides valuable lessons and a framework for understanding the quantization of other more complicated fields such as the gauge bosons and fermions that make up the Standard Model (SM), including QED and QCD. Since the Higgs particle of the SM is a scalar, it is also important to understand the quantization of scalars in their own right as well. A *scalar particle* is a boson (i.e., a particle that satisfies Bose-Einstein statistics) with zero spin.

6.2.1 Free Scalar Field

Recall the Lagrangian density of a free classical real scalar field from Eq. (3.1.39),

$$\mathcal{L} = \tfrac{1}{2}\left(\partial_\mu \phi\right)^2 - \tfrac{1}{2}m^2\phi^2 = \tfrac{1}{2}\partial_0\phi^2 - \tfrac{1}{2}(\boldsymbol{\nabla}\phi)^2 - \tfrac{1}{2}m^2\phi^2. \tag{6.2.1}$$

As we saw the corresponding Euler-Lagrange equation of motion is the Klein-Gordon equation given in Eq. (3.1.41),

$$\left(\partial_\mu\partial^\mu + m^2\right)\phi(x) = (\partial_0^2 - \boldsymbol{\nabla}^2 + m^2)\phi(x) = 0. \tag{6.2.2}$$

Were it not for the $\boldsymbol{\nabla}^2\phi(x)$ term these equations of motion would correspond to having one harmonic oscillator at each spatial point $\mathbf{x}$ and so we recognize that this system corresponds to an infinites number of coupled harmonic oscillators. We also recall the Hamiltonian density from Eq. (3.1.46),

$$\mathcal{H} = \tfrac{1}{2}\pi^2(x) + \tfrac{1}{2}\boldsymbol{\nabla}^2\phi(x) + \tfrac{1}{2}m^2\phi(x) \quad \text{with} \quad \pi(x) \equiv \frac{\partial\mathcal{L}}{\partial(\partial_0\phi)}(x) = \partial^0\phi(x). \tag{6.2.3}$$

The fundamental Poisson brackets for scalar field theory are given in Eq. (3.1.26). According to Dirac's canonical quantization procedure we should use the correspondence principle of Eq. (2.5.27) to arrive at the equal-time canonical commutation relations (ETCR). In the Heisenberg picture this gives

$$[\hat{\phi}(x), \hat{\phi}(y)]_{x^0=y^0} = [\hat{\pi}(x), \hat{\pi}(y)]_{x^0=y^0} = 0 \quad \text{and}$$
$$[\hat{\phi}(x), \hat{\pi}(y)]_{x^0=y^0} = i\delta^3(\mathbf{x} - \mathbf{y}). \tag{6.2.4}$$

Recall that, as discussed in Sec. 4.1.12, for a Hamiltonian with no explicit time dependence a Heisenberg picture operator, $\hat{A}(t)$, is related to its Schrödinger picture (denoted s) form $\hat{A}_s \equiv \hat{A}(t_0)$ by

$$\hat{A}(t) = e^{i\hat{H}(t-t_0)}\hat{A}_s\, e^{-i\hat{H}(t-t_0)}, \tag{6.2.5}$$

where the reference time t_0 is the arbitrary time at which the two pictures coincide.

Let us define a real combination of plane waves in spacetime,

$$\phi_{\mathbf{p}}(\mathbf{x}, t) \equiv \tfrac{1}{2}\left(c_{\mathbf{p}}e^{+ip\cdot x} + c_{\mathbf{p}}^* e^{-ip\cdot x}\right)\big|_{p^0=\omega_{\mathbf{p}}} = C_{\mathbf{p}}\cos(p \cdot x + \theta_{\mathbf{p}})|_{p^0=\omega_{\mathbf{p}}}$$
$$= C_{\mathbf{p}}\cos(\omega_{\mathbf{p}}t - \mathbf{p} \cdot \mathbf{x} + \theta_{\mathbf{p}}) \in \mathbb{R}, \tag{6.2.6}$$

where $c_{\mathbf{p}} \equiv C_{\mathbf{p}}\exp(i\theta_{\mathbf{p}}) \in \mathbb{C}$ with $C_{\mathbf{p}}, \theta_{\mathbf{p}} \in \mathbb{R}$ and where we have defined the angular frequency for a given three-momentum $\mathbf{p}$ as $\omega_{\mathbf{p}}$, where

$$E_{\mathbf{p}} \equiv \omega_{\mathbf{p}} \equiv \sqrt{|\mathbf{p}|^2 + m^2} > 0. \tag{6.2.7}$$

We observe that each $\phi_{\mathbf{p}}(\mathbf{x}, t)$ is a real solution of the Klein-Gordon equation with angular frequency $\omega_{\mathbf{p}}$,

$$\left[\frac{\partial^2}{\partial t^2} - \boldsymbol{\nabla}^2 + m^2\right]\phi_{\mathbf{p}}(\mathbf{x}, t) = \left[\frac{\partial^2}{\partial t^2} + \omega_{\mathbf{p}}^2\right]\phi_{\mathbf{p}}(\mathbf{x}, t) = 0 \quad \text{for all } \mathbf{p} \in \mathbb{R}^3, \tag{6.2.8}$$

which is the analog of the normal mode harmonic oscillator equation of motion, Eq. (6.1.2), with the correspondences

$$j \to \mathbf{p}, \quad \omega_j \to \omega_{\mathbf{p}}, \quad \zeta_j(t) \to \phi_{\mathbf{p}}(\mathbf{x}, t). \tag{6.2.9}$$

Any solution of the Klein-Gordon equation can be written as a linear superposition of these normal modes,

$$\phi(\mathbf{x}, t) = \int d^3 p \, \phi_{\mathbf{p}}(\mathbf{x}, t) = \int d^3 p \, C_{\mathbf{p}} \cos(\omega_{\mathbf{p}} t - \mathbf{p} \cdot \mathbf{x} + \theta_{\mathbf{p}}) \tag{6.2.10}$$

for some choices of the $C_{\mathbf{p}}$ and the $\theta_{\mathbf{p}}$. This is the analog of the classical mechanics small oscillation result in Eq. (2.3.24).

Comparison with Eq. (6.1.8) suggests the Schrödinger picture correspondences

$$\hat{a}_j \to \hat{a}_{\mathbf{p}} e^{i\mathbf{p} \cdot \mathbf{x}} \quad \text{and} \quad \hat{a}_j^\dagger \to \hat{a}_{\mathbf{p}}^\dagger e^{-i\mathbf{p} \cdot \mathbf{x}},$$

$$\hat{\zeta}_j = (1/\sqrt{2\omega_j})(\hat{a}_j + \hat{a}_j^\dagger) \to \hat{\phi}_{\mathbf{p}}(\mathbf{x}) = (1/\sqrt{2\omega_{\mathbf{p}}}) \left(\hat{a}_{\mathbf{p}} e^{i\mathbf{p} \cdot \mathbf{x}} + \hat{a}_{\mathbf{p}}^\dagger e^{-i\mathbf{p} \cdot \mathbf{x}} \right),$$

$$\hat{\pi}_j = -i\sqrt{\omega_j/2}(\hat{a}_j - \hat{a}_j^\dagger) \to \hat{\pi}_{\mathbf{p}}(\mathbf{x}) = -i\sqrt{\omega_{\mathbf{p}}/2} \left(\hat{a}_{\mathbf{p}} e^{i\mathbf{p} \cdot \mathbf{x}} - \hat{a}_{\mathbf{p}}^\dagger e^{-i\mathbf{p} \cdot \mathbf{x}} \right). \tag{6.2.11}$$

The properties of $\hat{a}_{\mathbf{p}}$ and its Hermitian conjugate $\hat{a}_{\mathbf{p}}^\dagger$ will be deduced from the requirements of the canonical commutation relations in Eq. (6.2.4). In the Schrödinger picture these are

$$[\hat{\phi}_s(\mathbf{x}), \hat{\phi}_s(\mathbf{y})] = [\hat{\pi}_s(\mathbf{x}), \hat{\pi}_s(\mathbf{y})] = 0 \quad \text{and} \quad [\hat{\phi}_s(\mathbf{x}), \hat{\pi}_s(\mathbf{y})] = i\delta^3(\mathbf{x} - \mathbf{y}). \tag{6.2.12}$$

The Schrödinger picture quantum field operators $\hat{\phi}_s(\mathbf{x})$ and $\hat{\pi}_s(\mathbf{x})$ are Hermitian and are given by integrating ("summing") over the quantized normal modes,

$$\hat{\phi}_s(\mathbf{x}) = \int \frac{d^3 p}{(2\pi)^3} \, \hat{\phi}_{\mathbf{p}}(\mathbf{x}) = \int \frac{d^3 p}{(2\pi)^3} \, \hat{\phi}_s(\mathbf{p}) e^{i\mathbf{p} \cdot \mathbf{x}}$$

$$\hat{\pi}_s(\mathbf{x}) = \int \frac{d^3 p}{(2\pi)^3} \, \hat{\pi}_{\mathbf{p}}(\mathbf{x}) = \int \frac{d^3 p}{(2\pi)^3} \, \hat{\pi}_s(\mathbf{p}) e^{i\mathbf{p} \cdot \mathbf{x}}, \tag{6.2.13}$$

where $\hat{\phi}_s(\mathbf{p})$ and $\hat{\pi}_s(\mathbf{p})$ are the three-dimensional Fourier transforms. We require $\hat{\phi}_s(-\mathbf{p})^\dagger = \hat{\phi}_s(\mathbf{p})$ and $\hat{\pi}_s(-\mathbf{p})^\dagger = \hat{\pi}_s(\mathbf{p})$ in order that $\hat{\phi}_s(\mathbf{x})$ and $\hat{\pi}(\mathbf{x})$ be Hermitian. We then have the Hermitian Schrödinger picture operators

$$\hat{\phi}_s(\mathbf{x}) = \int \frac{d^3 p}{(2\pi)^3} \frac{1}{\sqrt{2\omega_{\mathbf{p}}}} \left(\hat{a}_{\mathbf{p}} e^{i\mathbf{p} \cdot \mathbf{x}} + \hat{a}_{\mathbf{p}}^\dagger e^{-i\mathbf{p} \cdot \mathbf{x}} \right),$$

$$\hat{\pi}_s(\mathbf{x}) = \int \frac{d^3 p}{(2\pi)^3} (-i) \sqrt{\frac{\omega_{\mathbf{p}}}{2}} \left(\hat{a}_{\mathbf{p}} e^{i\mathbf{p} \cdot \mathbf{x}} - \hat{a}_{\mathbf{p}}^\dagger e^{-i\mathbf{p} \cdot \mathbf{x}} \right). \tag{6.2.14}$$

It is sometimes convenient to write these operators in the form

$$\hat{\phi}_s(\mathbf{x}) = \int \frac{d^3 p}{(2\pi)^3} \frac{1}{\sqrt{2\omega_{\mathbf{p}}}} \left(\hat{a}_{\mathbf{p}} + \hat{a}_{-\mathbf{p}}^\dagger \right) e^{i\mathbf{p} \cdot \mathbf{x}},$$

$$\hat{\pi}_s(\mathbf{x}) = \int \frac{d^3 p}{(2\pi)^3} (-i) \sqrt{\frac{\omega_{\mathbf{p}}}{2}} \left(\hat{a}_{\mathbf{p}} - \hat{a}_{-\mathbf{p}}^\dagger \right) e^{i\mathbf{p} \cdot \mathbf{x}}. \tag{6.2.15}$$

We can also invert the above equations to give

$$\hat{a}_{\mathbf{p}} = \int d^3 x \, \frac{1}{\sqrt{2\omega_{\mathbf{p}}}} e^{-i\mathbf{p} \cdot \mathbf{x}} \left[\omega_{\mathbf{p}} \hat{\phi}_s(\mathbf{x}) + i\hat{\pi}_s(\mathbf{x}) \right],$$

$$\hat{a}_{\mathbf{p}}^{\dagger} = \int d^3x \, \frac{1}{\sqrt{2\omega_{\mathbf{p}}}} \, e^{i\mathbf{p}\cdot\mathbf{x}} \left[\omega_{\mathbf{p}} \hat{\phi}_s(\mathbf{x}) - i\hat{\pi}_s(\mathbf{x}) \right]. \tag{6.2.16}$$

The commutation relations for the annihilation and creation operators must satisfy

$$[\hat{a}_{\mathbf{p}}, \hat{a}_{\mathbf{p}'}] = [\hat{a}_{\mathbf{p}}^{\dagger}, \hat{a}_{\mathbf{p}'}^{\dagger}] = 0 \quad \text{and} \quad [\hat{a}_{\mathbf{p}}, \hat{a}_{\mathbf{p}'}^{\dagger}] = (2\pi)^3 \delta^3(\mathbf{p} - \mathbf{p}') \tag{6.2.17}$$

in order that we recover the canonical commutation relations of Eq. (6.2.12). For example, we verify that

$$
\begin{aligned}
[\hat{\phi}_s(\mathbf{x}), \hat{\pi}_s(\mathbf{x}')] &= \int \frac{d^3p \, d^3p'}{(2\pi)^6} e^{i(\mathbf{p}\cdot\mathbf{x}+\mathbf{p}'\cdot\mathbf{x}')} \frac{(-i)}{2} \sqrt{\frac{\omega_{\mathbf{p}'}}{\omega_{\mathbf{p}}}} \left\{ -[\hat{a}_{\mathbf{p}}, \hat{a}_{-\mathbf{p}'}^{\dagger}] + [\hat{a}_{-\mathbf{p}}^{\dagger}, \hat{a}_{\mathbf{p}'}] \right\} \\
&= \int \frac{d^3p}{(2\pi)^3} i e^{i\mathbf{p}\cdot(\mathbf{x}-\mathbf{x}')} = i\delta^3(\mathbf{x} - \mathbf{x}').
\end{aligned} \tag{6.2.18}
$$

We similarly can confirm that $[\hat{\phi}_s(\mathbf{x}), \hat{\phi}_s(\mathbf{y})] = [\hat{\pi}_s(\mathbf{x}), \hat{\pi}_s(\mathbf{y})] = 0$, since the commutators in $\{\cdots\}$ have the same sign and so cancel.

Further extending the analogy with the harmonic oscillator we note that we can define the *number density operator for the mode* $\mathbf{p}$ as

$$\hat{N}_{\mathbf{p}} \equiv \frac{1}{(2\pi)^3} \hat{a}_{\mathbf{p}}^{\dagger} \hat{a}_{\mathbf{p}}, \tag{6.2.19}$$

$$\Rightarrow \quad [\hat{N}_{\mathbf{p}}, \hat{a}_{\mathbf{p}'}^{\dagger}] = \hat{a}_{\mathbf{p}}^{\dagger} \delta^3(\mathbf{p} - \mathbf{p}'), \quad [\hat{N}_{\mathbf{p}}, \hat{a}_{\mathbf{p}'}] = -\hat{a}_{\mathbf{p}} \delta^3(\mathbf{p} - \mathbf{p}'), \quad [\hat{N}_{\mathbf{p}}, \hat{N}_{\mathbf{p}'}] = 0.$$

The *number operator* includes all normal modes and is the analog of Eq. (6.1.20),

$$\hat{N} \equiv \int d^3p \, \hat{N}_{\mathbf{p}} = \int \frac{d^3p}{(2\pi)^3} \hat{a}_{\mathbf{p}}^{\dagger} \hat{a}_{\mathbf{p}}. \tag{6.2.20}$$

Since the number of particles must be Poincaré invariant, we will find that d^3p and $\hat{N}_{\mathbf{p}}$ transform oppositely under a Lorentz transformation.

Using Eq. (6.2.14) we can evaluate the Hamiltonian operator forms of $\hat{H}$ and the physical three–momentum $\hat{\mathbf{P}}$ from the free classical scalar field Hamiltonian in Eq. (3.1.46) and three-momentum in Eq. (3.2.58), respectively. Since $\hat{H}$ and $\hat{\mathbf{P}}$ for a free scalar field are conserved, there is no distinction between their Schrödinger picture and Heisenberg picture forms. Using Eq. (6.2.15) we find

$$
\begin{aligned}
\hat{H} &= \int d^3x \, \mathcal{H} = \int d^3x \left[\tfrac{1}{2}\hat{\pi}_s^2 + \tfrac{1}{2}(\boldsymbol{\nabla}\hat{\phi}_s)^2 + \tfrac{1}{2}m^2\hat{\phi}_s^2 \right] \\
&= \int d^3x \int \frac{d^3p \, d^3p'}{(2\pi)^6} e^{i(\mathbf{p}+\mathbf{p}')\cdot\mathbf{x}} \frac{1}{2} \left\{ -\frac{\sqrt{\omega_{\mathbf{p}}\omega_{\mathbf{p}'}}}{2}(\hat{a}_{\mathbf{p}} - \hat{a}_{-\mathbf{p}}^{\dagger})(\hat{a}_{\mathbf{p}'} - \hat{a}_{-\mathbf{p}'}^{\dagger}) \right. \\
&\qquad \left. + \frac{-\mathbf{p}\cdot\mathbf{p}' + m^2}{2\sqrt{\omega_{\mathbf{p}}\omega_{\mathbf{p}'}}}(\hat{a}_{\mathbf{p}} + \hat{a}_{-\mathbf{p}}^{\dagger})(\hat{a}_{\mathbf{p}'} + \hat{a}_{-\mathbf{p}'}^{\dagger}) \right\} = \int \frac{d^3p \, d^3p'}{(2\pi)^3} \delta^3(\mathbf{p} + \mathbf{p}') \frac{1}{2} \left\{ \cdots \right\} \\
&= \int \frac{d^3p}{(2\pi)^3} \frac{\omega_{\mathbf{p}}}{2} (\hat{a}_{-\mathbf{p}}^{\dagger}\hat{a}_{-\mathbf{p}} + \hat{a}_{\mathbf{p}}\hat{a}_{\mathbf{p}}^{\dagger}) = \int \frac{d^3p}{(2\pi)^3} \omega_{\mathbf{p}}(\hat{a}_{\mathbf{p}}^{\dagger}\hat{a}_{\mathbf{p}} + \tfrac{1}{2}[\hat{a}_{\mathbf{p}}, \hat{a}_{\mathbf{p}}^{\dagger}]).
\end{aligned} \tag{6.2.21}
$$

Note that the last term is the vacuum energy since

$$\langle 0|\hat{H}|0\rangle = \int \frac{d^3p}{(2\pi)^3} \frac{\omega_{\mathbf{p}}}{2} \langle 0|(\hat{a}_{-\mathbf{p}}^{\dagger}\hat{a}_{-\mathbf{p}} + \hat{a}_{\mathbf{p}}\hat{a}_{\mathbf{p}}^{\dagger})|0\rangle = \int \frac{d^3p}{(2\pi)^3} \frac{\omega_{\mathbf{p}}}{2} [\hat{a}_{\mathbf{p}}, \hat{a}_{\mathbf{p}}^{\dagger}] = \int d^3p \, \frac{\omega_{\mathbf{p}}}{2} \delta^3(\mathbf{0}),$$

where we have used $\langle 0|\hat{a}_{\mathbf{p}}\hat{a}_{\mathbf{p}}^{\dagger}|0\rangle = \langle 0|[\hat{a}_{\mathbf{p}}, \hat{a}_{\mathbf{p}}^{\dagger}]|0\rangle = [\hat{a}_{\mathbf{p}}, \hat{a}_{\mathbf{p}}^{\dagger}] = (2\pi)^3\delta^3(\mathbf{0})$. This appears troublesome since it is infinite and undefined.

Let us momentarily assume that the modes $\mathbf{p}$ are discretized by putting the system in a finite spatial region with volume V. With periodic boundary conditions we have a form of *infrared cutoff* and it leads to an infinite but countable set of momentum states; e.g., for a cubic volume with $V = L^3$ we have countable momenta $\mathbf{p} = (2\pi/L)\mathbf{n}$ with $\mathbf{n} = (n^1, n^2, n^3)$. For a dimensionless $z \in \mathbb{R}$ we can write

$$\delta(z) = (1/2\pi) \sum_{n=-\infty}^{\infty} e^{ikz} \quad \text{for } -\pi < z < \pi. \tag{6.2.22}$$

Changing variables $z = (2\pi/L)x$ and generalizing to three dimensions we arrive at

$$\delta^3(\mathbf{x}) = (1/V) \sum_{\mathbf{p}} e^{i\mathbf{p}\cdot\mathbf{x}} \quad \text{for } -(L/2) < x^j < L/2 \quad \text{and} \quad j = 1, 2, 3. \tag{6.2.23}$$

Let i be the integer that represents any counting of the discrete $\mathbf{p} = (2\pi/L)\mathbf{n}$. Note the correspondence between the infinite and finite volume relationships,

$$1 = \int \frac{d^3p}{(2\pi)^3} (2\pi)^3 \delta^3(\mathbf{p} - \mathbf{p}') = \int \frac{d^3p}{(2\pi)^3} \int d^3x \, e^{i(\mathbf{p}-\mathbf{p}')\cdot\mathbf{x}} = \int d^3x \, \delta^3(\mathbf{x}),$$

$$1 = (1/V) \sum_i V\delta_{ij} = (1/V) \sum_i \int_{-L/2}^{L/2} d^3x \, e^{i(\mathbf{p}_i - \mathbf{p}_j)\cdot\mathbf{x}} = \int_{-L/2}^{L/2} d^3x \, \delta^3(\mathbf{x}), \tag{6.2.24}$$

where we have used

$$\delta_{ij} = (1/V) \int_{-L/2}^{L/2} d^3x \, e^{i(\mathbf{p}_i - \mathbf{p}_j)\cdot\mathbf{x}}, \quad \text{since} \tag{6.2.25}$$

$$\delta_{\mathbf{nm}} = (1/V) \int_{-L/2}^{L/2} dx \, e^{i(2\pi/L)(\mathbf{n}-\mathbf{m})\cdot\mathbf{x}}.$$

We have in a finite volume $\hat{a}_{\mathbf{p}} \to \sqrt{V}\hat{a}_i$ and $\hat{a}_{\mathbf{p}}^\dagger \to \sqrt{V}\hat{a}_i^\dagger$ with $[\hat{a}_i, \hat{a}_j] = [\hat{a}_i^\dagger, \hat{a}_j^\dagger] = 0$ and $[\hat{a}_i, \hat{a}_j^\dagger] = \delta_{ij}$, which leads to correspondences in the equations above,

$$\int \frac{d^3p}{(2\pi)^3} \to \frac{1}{V} \sum_i \quad \text{and} \quad (2\pi)^3 \delta^3(\mathbf{p} - \mathbf{p}') = [\hat{a}_{\mathbf{p}}, \hat{a}_{\mathbf{p}'}^\dagger] \to V\delta_{ij} = V[\hat{a}_i, \hat{a}_j^\dagger]. \tag{6.2.26}$$

The contribution from this term in the Hamiltonian operator becomes

$$\int \frac{d^3p}{(2\pi)^3} \omega_{\mathbf{p}} \tfrac{1}{2} [\hat{a}_{\mathbf{p}}, \hat{a}_{\mathbf{p}}^\dagger] = \int d^3p \, \omega_{\mathbf{p}} \tfrac{1}{2} \delta^3(\mathbf{0}) \to \frac{1}{V} \sum_i \omega_i \tfrac{1}{2} V[\hat{a}_i, \hat{a}_i^\dagger] = \sum_i \tfrac{1}{2}\omega_i, \tag{6.2.27}$$

which is just the zero-point E_0 again as we found in Eq. (6.1.19). The infinite nature of E_0 here arises from the infinite number of normal modes. Imposing an upper limit on the magnitude of the momentum, $|\mathbf{p}| < \Lambda_{\mathrm{UV}}$, is a form of *ultraviolet cutoff* and gives rise to a sum with a finite number of terms. If we think in terms of coordinate-space integration, then we can introduce a periodic spatial lattice with lattice spacing a, where the available spatial positions labeled i are the vertices of this lattice. This is another form of ultraviolet cutoff with $\Lambda_{\mathrm{UV}} \sim 1/a$ and

$$\int d^3x \to a^3 \sum_i \quad \text{and} \quad \delta^3(\mathbf{x} - \mathbf{x}') \to (1/a^3)\delta_{ij}. \tag{6.2.28}$$

If we also introduce an infrared cutoff in the form of a finite volume, then the spatial integration becomes a sum over the finite number of lattice sites in the volume V.

Shifting the Hamiltonian by a constant has no effect on system dynamics since only relative energies are measurable. So we can simply redefine the Hamiltonian such that the quantum ground state, the *vacuum*, has vanishing energy. To do so we put things in *normal order* by the operation of *normal-ordering* of a product of creation and annihilation operators (a monomial), where all of the creation operators in the product are moved to the left of all of the annihilation operators. Normal-ordering is sometimes referred to as *Wick-ordering*. We denote normal-ordering of monomials with the "double-dot" notation, such that we have, for example,

$$\begin{aligned}
:\hat{a}^\dagger_{\mathbf{p}_1}\hat{a}_{\mathbf{p}_2}: &= :\hat{a}_{\mathbf{p}_2}\hat{a}^\dagger_{\mathbf{p}_1}: = \hat{a}^\dagger_{\mathbf{p}_1}\hat{a}_{\mathbf{p}_2}, \\
:\hat{a}_{\mathbf{p}_1}\hat{a}^\dagger_{\mathbf{p}_2}\hat{a}^\dagger_{\mathbf{p}_3}\hat{a}_{\mathbf{p}_4}\hat{a}_{\mathbf{p}_5}\hat{a}^\dagger_{\mathbf{p}_6}: &= \hat{a}^\dagger_{\mathbf{p}_2}\hat{a}^\dagger_{\mathbf{p}_3}\hat{a}^\dagger_{\mathbf{p}_6}\hat{a}_{\mathbf{p}_1}\hat{a}_{\mathbf{p}_4}\hat{a}_{\mathbf{p}_5},
\end{aligned} \tag{6.2.29}$$

where the ordering of the $\hat{a}^\dagger_{\mathbf{p}_2}\hat{a}^\dagger_{\mathbf{p}_3}\hat{a}^\dagger_{\mathbf{p}_6}$ is irrelevant since the $\hat{a}^\dagger_{\mathbf{p}}$ all commute and similarly for the ordering of $\hat{a}_{\mathbf{p}_1}\hat{a}_{\mathbf{p}_4}\hat{a}_{\mathbf{p}_5}$ since the $\hat{a}_{\mathbf{p}}$ all commute. Note that Peskin and Schroeder (1995) use N to denote normal-ordering but here we use the more traditional double-dot notation. We also define the normal-ordered identity operator to satisfy $:\hat{I}: = \hat{I}$ and $:c\hat{A}: = c:\hat{A}:$ with $c \in \mathbb{C}$ and where $\hat{A}$ is any monomial. It is often useful to use a notation where we appear to normal-order a sum of monomials (a polynomial); *however*, it must be clearly understood that this is just shorthand notation for a sum of normal-ordered monomials. That is, we are not defining normal-ordering to be a linear operation. So we *define*

$$:\hat{a}^\dagger_{\mathbf{p}_1}\hat{a}_{\mathbf{p}_2}\hat{a}_{\mathbf{p}_3} + \hat{a}_{\mathbf{p}_4}\hat{a}_{\mathbf{p}_5}\hat{a}_{\mathbf{p}_6}: \equiv :\hat{a}^\dagger_{\mathbf{p}_1}\hat{a}_{\mathbf{p}_2}\hat{a}_{\mathbf{p}_3}: + :\hat{a}_{\mathbf{p}_4}\hat{a}_{\mathbf{p}_5}\hat{a}_{\mathbf{p}_6}:, \tag{6.2.30}$$

where we strictly understand that the left-hand side is just a shorthand notation for the right-hand side. If we were to naively take normal-ordering as a linear operation, then we would arrive at a contradiction. For example, we have $:\hat{a}_{\mathbf{p}}\hat{a}^\dagger_{\mathbf{p}} - \hat{a}^\dagger_{\mathbf{p}}\hat{a}_{\mathbf{p}}: \equiv :\hat{a}_{\mathbf{p}}\hat{a}^\dagger_{\mathbf{p}}: - :\hat{a}^\dagger_{\mathbf{p}}\hat{a}_{\mathbf{p}}: = \hat{a}^\dagger_{\mathbf{p}}\hat{a}_{\mathbf{p}} - \hat{a}^\dagger_{\mathbf{p}}\hat{a}_{\mathbf{p}} = 0$, whereas if we treated normal-ordering as a linear operation then we could write $:\hat{a}_{\mathbf{p}}\hat{a}^\dagger_{\mathbf{p}} - \hat{a}^\dagger_{\mathbf{p}}\hat{a}_{\mathbf{p}}: \equiv :[\hat{a}_{\mathbf{p}}, \hat{a}^\dagger_{\mathbf{p}}]: = (2\pi)^3\delta^3(\mathbf{0}) \neq 0$, which would be a contradiction with our definition.

Let $f(\hat{\phi}, \hat{\pi})$ be any polynomial in the field operators $\hat{\phi}$ and $\hat{\pi}$, then $:f(\hat{\phi}, \hat{\pi}):$ is obtained by first expanding out f as a polynomial in $\hat{\phi}$ and $\hat{\pi}$ and then expanding those in terms of $\hat{a}_{\mathbf{p}}$ and $\hat{a}^\dagger_{\mathbf{p}}$ to arrive at a polynomial with the resulting ordering of $\hat{a}_{\mathbf{p}}$ and $\hat{a}^\dagger_{\mathbf{p}}$. Then we normal-order each monomial in that expansion (without ever using commutation relations to combine monomials). Since $\hat{a}_{\mathbf{p}}|0\rangle = 0$ and $\langle 0|\hat{a}^\dagger_{\mathbf{p}} = 0$, then this ensures that *any* normal-ordered polynomial operator, not containing a term proportional to the identity operator (i.e., $f(0,0) = 0$), will contain at least one $\hat{a}_{\mathbf{p}}$ or $\hat{a}^\dagger_{\mathbf{p}}$ and so will have a vanishing vacuum expectation value (VEV), since

$$\langle 0|:f(\hat{\phi}, \hat{\pi}):|0\rangle = 0 \quad \text{if} \quad f(0,0) = 0. \tag{6.2.31}$$

Any normal-ordered operator with at least one annihilation operator, $\hat{a}_{\mathbf{p}}$, in each monomial will annihilate the vacuum. Since $\hat{\phi}$ and $\hat{\pi}$ don't commute, then when writing down the polynomial $f(\hat{\phi}, \hat{\pi})$ we must specify the order of these operators in each monomial. Note that normal-ordering removes any operator-ordering ambiguity, since if $f_1(\hat{\phi}, \hat{\pi})$ and $f_2(\hat{\phi}, \hat{\pi})$ differ only in the ordering of non-commuting operators $\hat{\phi}$ and $\hat{\pi}$ then

$$:f_1(\hat{\phi}, \hat{\pi}): = :f_2(\hat{\phi}, \hat{\pi}):. \tag{6.2.32}$$

So then we *redefine* the Hamiltonian to be its normal-ordered form,

$$\begin{aligned}
\hat{H} &\equiv \int d^3x : \left[\tfrac{1}{2}\hat{\pi}_s^2 + \tfrac{1}{2}(\boldsymbol{\nabla}\hat{\phi}_s)^2 + \tfrac{1}{2}m^2\hat{\phi}_s^2\right] := \int \frac{d^3p}{(2\pi)^3}\frac{\omega_{\mathbf{p}}}{2} :\hat{a}^\dagger_{-\mathbf{p}}\hat{a}_{-\mathbf{p}} + \hat{a}_{\mathbf{p}}\hat{a}^\dagger_{\mathbf{p}}: \\
&= \int \frac{d^3p}{(2\pi)^3}\, \omega_{\mathbf{p}}\hat{a}^\dagger_{\mathbf{p}}\hat{a}_{\mathbf{p}} = \int d^3p\, \omega_{\mathbf{p}}\hat{N}_{\mathbf{p}} = \int d^3p\, E_{\mathbf{p}}\hat{N}_{\mathbf{p}},
\end{aligned} \tag{6.2.33}$$

since this gives the desired outcome $\hat{H}|0\rangle = 0$. Beginning with Eq. (3.2.58) we perform a straightforward calculation similar to that in Eq. (6.2.21) and obtain

$$
\begin{aligned}
\hat{\mathbf{P}} &= -\int d^3x\, \hat{\pi}_s \boldsymbol{\nabla} \hat{\phi}_s = -\int d^3x \frac{d^3p\, d^3p'}{(2\pi)^6} e^{i(\mathbf{p}+\mathbf{p}')\cdot\mathbf{x}} \frac{1}{2}\mathbf{p}' \sqrt{\frac{\omega_\mathbf{p}}{\omega_{\mathbf{p}'}}} (\hat{a}_\mathbf{p} - \hat{a}_{-\mathbf{p}}^\dagger)(\hat{a}_{\mathbf{p}'} + \hat{a}_{-\mathbf{p}'}^\dagger) \\
&= \int \frac{d^3p}{(2\pi)^3} \frac{1}{2}\mathbf{p}(\hat{a}_\mathbf{p} - \hat{a}_{-\mathbf{p}}^\dagger)(\hat{a}_{-\mathbf{p}} + \hat{a}_\mathbf{p}^\dagger) = \int \frac{d^3p}{(2\pi)^3} \frac{1}{2}\mathbf{p}(\hat{a}_\mathbf{p}^\dagger \hat{a}_\mathbf{p} + \hat{a}_\mathbf{p} \hat{a}_\mathbf{p}^\dagger) \\
&= \int \frac{d^3p}{(2\pi)^3} \mathbf{p}\left(\hat{a}_\mathbf{p}^\dagger \hat{a}_\mathbf{p} + \tfrac{1}{2}[\hat{a}_\mathbf{p}, \hat{a}_\mathbf{p}^\dagger]\right) = \int d^3p\, \mathbf{p}\hat{N}_\mathbf{p},
\end{aligned} \tag{6.2.34}
$$

where we note that $\mathbf{p}\,\hat{a}_\mathbf{p}\hat{a}_{-\mathbf{p}}$, $\mathbf{p}\,\hat{a}_{-\mathbf{p}}^\dagger \hat{a}_\mathbf{p}^\dagger$ and $\mathbf{p}\frac{1}{2}[\hat{a}_\mathbf{p}, \hat{a}_\mathbf{p}^\dagger]$ are odd with respect to $\mathbf{p}$ and hence give no contribution to the integral. So $\mathbf{p}\frac{1}{2}[\hat{a}_\mathbf{p}, \hat{a}_\mathbf{p}^\dagger]$ vanishes even without normal-ordering. So the three-momentum $\mathbf{P}$ is equal to its normal-ordered form,

$$
\hat{\mathbf{P}} \equiv -\int d^3x :\hat{\pi}\boldsymbol{\nabla}\hat{\phi}: = \int \frac{d^3p}{(2\pi)^3} \mathbf{p}\left(\hat{a}_\mathbf{p}^\dagger \hat{a}_\mathbf{p}\right) = \int d^3p\, \mathbf{p}\hat{N}_\mathbf{p}. \tag{6.2.35}
$$

We can readily check that Eq. (6.2.34) takes the same form independent of the ordering of $\hat{\pi}$ and $\boldsymbol{\nabla}\hat{\phi}$ as it should, since $\mathbf{P}$ is automatically normal-ordered.

Eqs. (6.2.33) and (6.2.35) can be combined to give the four-momentum operator,

$$
\hat{P}^\mu \equiv (\hat{H}, \hat{\mathbf{P}}) = \int d^3p\, p^\mu \hat{N}_\mathbf{p}, \tag{6.2.36}
$$

where $p^\mu = (E_\mathbf{p}, \mathbf{p}) = (\omega_\mathbf{p}, \mathbf{p})$. Since p^μ transforms as a Lorentz four-vector and since the combination of d^3p and $\hat{N}_\mathbf{p}$ is Lorentz invariant (see later discussion), then $\hat{P}^\mu$ must transform as a four-vector. It follows easily from Eq. (6.2.19) that

$$
[\hat{N}, \hat{a}_\mathbf{p}^\dagger] = \hat{a}_\mathbf{p}^\dagger, \quad [\hat{N}, \hat{a}_\mathbf{p}] = -\hat{a}_\mathbf{p}, \quad [\hat{N}, \hat{N}_\mathbf{p}] = 0, \quad [\hat{H}, \hat{N}_\mathbf{p}] = [\hat{H}, \hat{N}] = 0, \tag{6.2.37}
$$

$$
[\hat{H}, \hat{a}_\mathbf{p}^\dagger] = \omega_\mathbf{p}\hat{a}_\mathbf{p}^\dagger \equiv E_\mathbf{p}\hat{a}_\mathbf{p}^\dagger, \quad [\hat{H}, \hat{a}_\mathbf{p}] = -\omega_\mathbf{p}\hat{a}_\mathbf{p} \equiv -E_\mathbf{p}\hat{a}_\mathbf{p},
$$

$$
[\hat{\mathbf{P}}, \hat{a}_\mathbf{p}^\dagger] = \mathbf{p}\hat{a}_\mathbf{p}^\dagger, \quad [\hat{\mathbf{P}}, \hat{a}_\mathbf{p}] = -\mathbf{p}\hat{a}_\mathbf{p}.
$$

Since $[\hat{N}_\mathbf{p}, \hat{N}_{\mathbf{p}'}] = 0$, $\hat{H} = \int d^3p\, E_\mathbf{p}\hat{N}_\mathbf{p}$ and $\hat{\mathbf{P}} = \int d^3p\, \mathbf{p}\hat{N}_\mathbf{p}$, then

$$
[\hat{H}, \hat{P}^\mu] = 0, \quad [\hat{P}^\mu, \hat{a}_\mathbf{p}^\dagger] = p^\mu|_{p^0=E_\mathbf{p}} a_\mathbf{p}^\dagger \quad \text{and} \quad [\hat{P}^\mu, \hat{a}_\mathbf{p}] = -p^\mu|_{p^0=E_\mathbf{p}} a_\mathbf{p}, \tag{6.2.38}
$$

where the first equation is what we expect on the basis of translational invariance.

The operators $\hat{H}$, $\hat{\mathbf{P}}$, $\hat{N}$ and $\hat{N}_\mathbf{p}$ (for all $\mathbf{p}$) are all Hermitian and form a mutually commuting set of operators. So we can construct a set of simultaneous eigenstates of these operators that form a complete orthonormal basis of the total Hilbert space. Recall that in our notation $E_\mathbf{p}$ is *always positive* and is given by Eq. (6.2.7).

For free quantum field theories normal-ordering conveniently removes all irrelevant (i.e., unphysical) divergences. In the case of interacting field theories we need a complete regularization and renormalization program and normal-ordering is absorbed into this larger scheme. It should be noted, of course, that the divergences occur because there is an infinite number of degrees of freedom if we assume a continuous and/or infinite spacetime. If spacetime had a finite volume (an infrared cutoff) and was somehow discrete, then there would be a finite number of degrees of freedom. This is the approach referred to as lattice field theory, where calculations are done on a finite spacetime lattice and then extrapolations are made in order to recover the continuum and infinite volume limits.

Scalar Particles and the Spectrum of States

We will now follow closely the discussion of the quantization of normal modes in Sec. 6.1.1. Since this classical Hamiltonian can never be negative, then the classical energy is bounded below and the energy of every normal mode is bounded below. Every quantized normal mode makes a nonnegative energy contribution, and so the quantized Hamiltonian without normal-ordering has eigenvalues that are bounded below, which means that there must exist a normalized state $|0\rangle$ such that

$$\hat{a}_{\mathbf{p}}|0\rangle = 0, \quad \langle 0|\hat{a}_{\mathbf{p}}^{\dagger} = 0 \quad \text{for} \quad \mathbf{p} \in \mathbb{R}^3 \quad \text{with} \quad \langle 0|0\rangle = 1. \tag{6.2.39}$$

The state $|0\rangle$ is referred to as the *vacuum state*, since it immediately follows from Eqs. (6.2.19), (6.2.33) and (6.2.35) that our normal-ordered operators must satisfy

$$\hat{H}|0\rangle = \hat{\mathbf{P}}|0\rangle = \hat{N}|0\rangle = \hat{N}_{\mathbf{p}}|0\rangle = 0. \tag{6.2.40}$$

So the vacuum has no energy or physical momentum and contains no particles. Any normal-ordered operator containing at least one annihilation operator will give zero when acting on the vacuum state. We refer to this as *annihilating the vacuum*.

The state $\hat{a}_{\mathbf{p}}^{\dagger}|0\rangle$ is a simultaneous eigenstate of $\hat{H}$, $\hat{\mathbf{P}}$ and $\hat{N}$ with eigenvalues $E_{\mathbf{p}}$, $\mathbf{p}$ and 1, respectively, which follows from Eq. (6.2.37). In other words, it is a single-particle state with energy $E_{\mathbf{p}} = (|\mathbf{p}|^2 + m^2)^{1/2}$ and physical momentum $\mathbf{p}$. The state $\hat{a}_{\mathbf{p}}^{\dagger}\hat{a}_{\mathbf{q}}^{\dagger}|0\rangle$ is a two-particle state with one particle having energy $E_{\mathbf{p}} = (|\mathbf{p}|^2 + m^2)^{1/2}$ and momentum $\mathbf{p}$ and the other particle having energy $E_{\mathbf{q}} = (|\mathbf{q}|^2 + m^2)^{1/2}$ and momentum $\mathbf{q}$. Note that from the commutation relations in Eq. (6.2.17) we see that the state $\hat{a}_{\mathbf{p}}^{\dagger}\hat{a}_{\mathbf{q}}^{\dagger}|0\rangle$ is equivalent to the state $\hat{a}_{\mathbf{q}}^{\dagger}\hat{a}_{\mathbf{p}}^{\dagger}|0\rangle$. In particular, we see that $(\hat{a}_{\mathbf{p}}^{\dagger})^2|0\rangle$ is a two-particle state where both particles have the same quantum numbers; i.e., both have energy $E_{\mathbf{p}}$ and three-momentum $\mathbf{p}$. Then the quantum field theory that has been constructed satisfies *Bose-Einstein statistics* and therefore it follows that the particles being created are *bosons*.

Let us note some results that will prove useful for later calculations. Using $\hat{a}_{\mathbf{p}}|0\rangle = 0$ and Eq. (6.2.17) we see that

$$\hat{a}_{\mathbf{p}}\hat{a}_{\mathbf{q}}^{\dagger}|0\rangle = [\hat{a}_{\mathbf{p}}, \hat{a}_{\mathbf{q}}^{\dagger}]|0\rangle = (2\pi)^3\delta^3(\mathbf{p} - \mathbf{q})|0\rangle,$$

$$\langle 0|\hat{a}_{\mathbf{p}}\hat{a}_{\mathbf{q}}^{\dagger}|0\rangle = [\hat{a}_{\mathbf{p}}, \hat{a}_{\mathbf{q}}^{\dagger}] = (2\pi)^3\delta^3(\mathbf{p} - \mathbf{q}). \tag{6.2.41}$$

This result is readily extended to

$$\langle 0|\hat{a}_{\mathbf{p}_2}\hat{a}_{\mathbf{p}_1}\hat{a}_{\mathbf{q}_1}^{\dagger}\hat{a}_{\mathbf{q}_2}^{\dagger}|0\rangle = \langle 0|\hat{a}_{\mathbf{p}_2}[\hat{a}_{\mathbf{p}_1}, \hat{a}_{\mathbf{q}_1}^{\dagger}]\hat{a}_{\mathbf{q}_2}^{\dagger}|0\rangle + \langle 0|\hat{a}_{\mathbf{p}_2}\hat{a}_{\mathbf{q}_1}^{\dagger}\hat{a}_{\mathbf{p}_1}\hat{a}_{\mathbf{q}_2}^{\dagger}|0\rangle \tag{6.2.42}$$

$$= (2\pi)^6\left[\delta^3(\mathbf{p}_1 - \mathbf{q}_1)\delta^3(\mathbf{p}_2 - \mathbf{q}_2) + \delta^3(\mathbf{p}_1 - \mathbf{q}_2)\delta^3(\mathbf{p}_2 - \mathbf{q}_1)\right]$$

and in general we find

$$\langle 0|\hat{a}_{\mathbf{p}_n}\cdots\hat{a}_{\mathbf{p}_1}\hat{a}_{\mathbf{q}_1}^{\dagger}\cdots\hat{a}_{\mathbf{q}_n}^{\dagger}|0\rangle = (2\pi)^{3n}\left[\delta^3(\mathbf{p}_1 - \mathbf{q}_1)\cdots\delta^3(\mathbf{p}_n - \mathbf{q}_n)\right.$$

$$\left. + \text{ all permutations of } (\mathbf{q}_1\cdots\mathbf{q}_n)\right]. \tag{6.2.43}$$

Proof: We begin with $[A, BC] = [A, B]C + B[A, C]$ and its generalization,

$$[A, A_1A_2A_3\cdots A_n] = \sum_{k=1}^{n} A_1A_2\cdots A_{k-1}[A, A_k]A_{k+1}\cdots A_n, \tag{6.2.44}$$

which is easily proved by induction; i.e., show for $n = 1, 2$ and then assume true for n and prove for $n+1$. Using an obvious shorthand notation, where $a_j \equiv \hat{a}_{\mathbf{p}_j}$, we use the above results along with $a_j|0\rangle = 0$ and $\langle 0|a_j^{\dagger} = 0$ to give

$$\langle 0|a_n \cdots a_2 a_1 a_1^\dagger a_2^\dagger \cdots a_n^\dagger|0\rangle = \langle 0|a_n \cdots a_2 [a_1, a_1^\dagger a_2^\dagger \cdots a_n^\dagger]|0\rangle$$

$$= \sum_{k_1=1}^n [a_1, a_{k_1}^\dagger]\langle 0|a_n \cdots a_2 a_1^\dagger a_2^\dagger \cdots a_{k_1-1}^\dagger a_{k_1+1}^\dagger \cdots a_n^\dagger|0\rangle$$

$$= \sum \left(\begin{array}{c} (k_1, \cdots, k_n) = \\ \text{permutations}(1, \ldots, n) \end{array} \right) [a_1, a_{k_1}^\dagger][a_2, a_{k_2}^\dagger]\cdots[a_n, a_{k_n}^\dagger]. \qquad (6.2.45)$$

Using Eq. (6.2.41) we then find the result of Eq. (6.2.43) as required.

Summary: Each application of $\hat{a}_\mathbf{p}^\dagger$ creates an on-shell particle with energy $E_\mathbf{p} = (|\mathbf{p}|^2 + m^2)^{1/2}$ and momentum $\mathbf{p}$ and so $\hat{a}_\mathbf{p}^\dagger$ is a *creation* operator. Conversely, each application of $\hat{a}_\mathbf{p}$ removes an on-shell particle with energy $E_\mathbf{p} = (|\mathbf{p}|^2 + m^2)^{1/2}$ and momentum $\mathbf{p}$ if there is one and so $\hat{a}_\mathbf{p}$ is an *annihilation (or destruction)* operator. If there is no corresponding particle to annihilate, then $\hat{a}_\mathbf{p}$ acting on the state gives zero. Recall from Sec. 2.6 that an *on-shell* (or *on-mass-shell*) particle has its energy and three-momentum satisfy the energy-momentum relation $p^2 = E_\mathbf{p}^2 - \mathbf{p}^2 = m^2$.

Normalization of States

There are various normalization conventions for particle states in quantum field theory and each has its advantages and disadvantages. We shall throughout adopt the Lorentz-invariant normalization conventions that are used, e.g., Peskin and Schroeder (1995). The single boson state with three-momentum $\mathbf{p}$ is defined as

$$|\mathbf{p}\rangle \equiv \sqrt{2E_\mathbf{p}}\, \hat{a}_\mathbf{p}^\dagger|0\rangle. \qquad (6.2.46)$$

With this choice of normalization we find from Eqs. (6.2.17) and (6.2.39) that

$$\langle \mathbf{p}|\mathbf{q}\rangle = 2E_\mathbf{p}(2\pi)^3\delta^3(\mathbf{p}-\mathbf{q}), \qquad (6.2.47)$$

which is a Lorentz-invariant scalar product of the one-boson state vectors. The factor $\sqrt{2}$ in the normalization of Eq. (6.2.46) is not essential but it is conventional to include it since it gives a tidier form to some of the following discussion.

That Eq. (6.2.47) is Lorentz invariant can be understood as follows. Define as usual $\beta = v/c = v$ (since $c = 1$ in natural units) and $\gamma = (1 - \beta^2)^{-1/2}$. For a particle with momentum $\mathbf{p}$ and energy $E_\mathbf{p}$ since p^μ is a four-vector we find using Eq. (1.2.89) that after an *active* boost by v in, say, the $+x$-direction that $p^{1'} = \gamma(p^1 + \beta E_\mathbf{p})$, $p^{2'} = p^2$, $p^{3'} = p^3$ and $E_\mathbf{p}' = \gamma(E_\mathbf{p} + \beta p^1)$. Then from Eq. (A.2.8) we see that $E_\mathbf{p}\delta^3(\mathbf{p} - \mathbf{q})$ is Lorentz invariant, since

$$E_\mathbf{p}'\, \delta^3(\mathbf{p}' - \mathbf{q}') = E_{\mathbf{p}'}\frac{1}{|dp^{1'}/dp^1|}\, \delta^3(\mathbf{p} - \mathbf{q}) = E_\mathbf{p}\, \delta^3(\mathbf{p} - \mathbf{q}), \qquad (6.2.48)$$

where there was nothing special about the x-direction and where we have used

$$\frac{dp^{1'}}{dp^1} = \gamma\left[1 + \beta\frac{dE_\mathbf{p}}{dp^1}\right] = \gamma\left[1 + \beta\frac{p^1}{E_\mathbf{p}}\right] = \frac{E_{\mathbf{p}'}}{E_\mathbf{p}}. \qquad (6.2.49)$$

For the y and z components, $dp'^2 = dp^2$ and $dp'^3 = dp^3$. It then follows that

$$d^3p'/E_{\mathbf{p}'} = d^3p/E_\mathbf{p} \qquad (6.2.50)$$

is invariant under restricted Lorentz transformations. Using Eq. (A.2.8) and $E_\mathbf{p} > 0$ we note that for arbitrary (off-shell) p^μ

$$\delta(p^2 - m^2) = \delta((p^0)^2 - E_\mathbf{p}^2) = \sum_{p^0=\pm E_\mathbf{p}}(1/2E_\mathbf{p})\delta(p^0 \mp E_\mathbf{p}), \qquad (6.2.51)$$

which means that $\delta(p^2 - m^2)|_{p^0 \gtrless 0} = (1/2E_{\mathbf{p}})\delta(p^0 \mp E_{\mathbf{p}})$ and so

$$\frac{d^4p}{(2\pi)^3}\frac{1}{2E_{\mathbf{p}}}\delta(p^0 \mp E_{\mathbf{p}}) = \frac{d^4p}{(2\pi)^4}\,(2\pi)\delta(p^2 - m^2)\big|_{p^0 \gtrless 0}. \tag{6.2.52}$$

Then if $g(p)$ is any Lorentz-invariant function of four-momentum we find that

$$\int \frac{d^3p}{(2\pi)^3}\frac{1}{2E_{\mathbf{p}}}\,g(p)|_{p^0 = \pm E_{\mathbf{p}}} = \int \frac{d^4p}{(2\pi)^4}\,(2\pi)\,\delta(p^2 - m^2)g(p)\big|_{p^0 \gtrless 0} \tag{6.2.53}$$

is also necessarily Lorentz invariant up to time-reversal transformations (i.e., it transforms as a Lorentz scalar under restricted Lorentz transformations and parity transformations). Since p^μ is a four-vector, then under time-reversal $p^0 = \pm E_{\mathbf{p}} \to \mp E_{\mathbf{p}}$ recalling that $E_{\mathbf{p}} > 0$, which means that these quantities are not time-reversal invariant. This is just as well as we will later appreciate. We now see why the factor 2 was introduced into the normalization. The right-hand side corresponds to integration over the positive or negative branch of the particle mass-shell for $p^0 > 0$ or $p^0 < 0$, respectively. Inserting $\langle \mathbf{p}|\mathbf{q}\rangle$ in place of $g(p)|_{p^0 = \pm E_{\mathbf{p}}}$ we find using Eq. (6.2.47) that the left-hand side of the above equation is unity. This confirms that $\langle \mathbf{p}|\mathbf{q}\rangle$ is equivalent to some $g(p)|_{p^0 = \pm E_{\mathbf{p}}}$, i.e., equivalent to a Lorentz-invariant quantity evaluated on its positive or negative mass-shell.

The completeness relation for the one-particle subspace of the Fock space is then

$$\hat{I}_1 = \int \frac{d^3p}{(2\pi)^3}|\mathbf{p}\rangle\frac{1}{2E_{\mathbf{p}}}\langle\mathbf{p}|, \tag{6.2.54}$$

where $\hat{I}_1$ is the identity in the one-particle subspace and zero otherwise (i.e., it is the *projector* from the total Hilbert space of states onto the one-particle subspace). This follows since any one-particle state can be written as a superposition of momentum eigenstates $|\mathbf{p}\rangle$ and since this form of the completeness relation is that consistent with Eq. (6.2.47), where we easily confirm that $\langle \mathbf{q}|\hat{I}_1|\mathbf{q}'\rangle = \langle \mathbf{q}|\mathbf{q}'\rangle$.

Consider the action of the Schrödinger picture field operator on the vacuum

$$\hat{\phi}_s^\dagger(\mathbf{x})|0\rangle = \int \frac{d^3p}{(2\pi)^3}\frac{1}{\sqrt{2E_{\mathbf{p}}}}e^{-i\mathbf{p}\cdot\mathbf{x}}\hat{a}_{\mathbf{p}}^\dagger|0\rangle = \int \frac{d^3p}{(2\pi)^3}\frac{1}{2E_{\mathbf{p}}}e^{-i\mathbf{p}\cdot\mathbf{x}}|\mathbf{p}\rangle. \tag{6.2.55}$$

The field operator $\hat{\phi}_s(\mathbf{x})$ acting on the vacuum is a linear superposition of one-particle momentum eigenstates and is the relativistic quantum field theory analog of having a particle at the coordinate-space point $\mathbf{x}$ in ordinary quantum mechanics; i.e., it is the analog of the position eigenstate state $|\mathbf{x}\rangle$ and thus we see that $\hat{\phi}_s(\mathbf{x})$ *acting on the vacuum creates a particle at the spatial point* $\mathbf{x}$. Let us try to illustrate this point more clearly. As is apparent from the discussion in Sec. A.2.3 we can write a position eigenstate in ordinary quantum mechanics as

$$|\mathbf{x}\rangle_{\mathrm{qm}} = \int d^3p\,|\mathbf{p}\rangle_{\mathrm{qm}}\langle\mathbf{p}|\mathbf{x}\rangle_{\mathrm{qm}} = \int \frac{d^3p}{(2\pi)^3}\,e^{-i\mathbf{p}\cdot\mathbf{x}}|\mathbf{p}\rangle_{\mathrm{qm}}. \tag{6.2.56}$$

Here we are using the asymmetric form of the Fourier transform given in Eq. (A.2.9) in natural units, where $\langle\mathbf{x}'|\mathbf{x}\rangle_{\mathrm{qm}} = \delta^3(\mathbf{x}' - \mathbf{x})$, $\langle\mathbf{p}'|\mathbf{p}\rangle_{\mathrm{qm}} = (2\pi)^3\delta^3(\mathbf{p}' - \mathbf{p})$, and $\langle\mathbf{x}|\mathbf{p}\rangle_{\mathrm{qm}} = e^{i\mathbf{p}\cdot\mathbf{x}}$. In the nonrelativistic limit $E_{\mathbf{p}} \to m$ and so in that limit the integrands of the right-hand sides of Eqs. (6.2.55) and (6.2.56) have the same form up to an overall normalization constant. The analog of $\langle\mathbf{x}|\mathbf{p}\rangle_{\mathrm{qm}} = e^{i\mathbf{p}\cdot\mathbf{x}}$ is

$$\langle 0|\phi_s(\mathbf{x})|\mathbf{p}\rangle = \int \frac{d^3p'}{(2\pi)^3}\frac{1}{\sqrt{2E_{\mathbf{p}'}}}\langle 0|\left(\hat{a}_{\mathbf{p}'}e^{i\mathbf{p}'\cdot\mathbf{x}} + \hat{a}_{\mathbf{p}'}^\dagger e^{-i\mathbf{p}'\cdot\mathbf{x}}\right)\sqrt{2E_{\mathbf{p}}}\hat{a}_{\mathbf{p}}^\dagger|0\rangle$$

$$= \int \frac{d^3p'}{(2\pi)^3}\sqrt{\frac{E_{\mathbf{p}}}{E_{\mathbf{p}'}}}e^{i\mathbf{p}'\cdot\mathbf{x}}\langle 0|\hat{a}_{\mathbf{p}'}\hat{a}_{\mathbf{p}}^\dagger|0\rangle = e^{i\mathbf{p}\cdot\mathbf{x}}, \tag{6.2.57}$$

which is also quickly arrived at using the orthogonality condition of Eq. (6.2.47) and the bra version of Eq. (6.2.55),

$$\langle 0 | \phi_s(\mathbf{x}) | \mathbf{p} \rangle = \int \frac{d^3 p'}{(2\pi)^3} \frac{1}{2E_{\mathbf{p}'}} e^{i\mathbf{p}' \cdot \mathbf{x}} \langle \mathbf{p}' | \mathbf{p} \rangle = e^{i\mathbf{p} \cdot \mathbf{x}}. \tag{6.2.58}$$

Fock Space for Scalar Particles

We now proceed to establish a basis for the bosonic Fock space of our scalar particles. We could write an occupancy number basis as we did for our quantization of normal modes in Sec. 6.1.1, where $j \to \mathbf{p}$, $\omega_j \to \omega_{\mathbf{p}} \equiv E_{\mathbf{p}}$ and where we replace sums over modes by integrals over $\mathbf{p}$ and so on. However, it is convenient to work in a symmetrized basis for bosonic Fock space analogous to that used in Sec. 4.1.9.

In a momentum basis for the one-particle Hilbert space, $|b_i\rangle \to |\mathbf{p}\rangle \in V_H$, the one-particle orthonormality and completeness conditions of Eq. (4.1.141) become

$$\langle \mathbf{p} | \mathbf{q} \rangle = 2E_{\mathbf{p}} (2\pi)^3 \delta^3(\mathbf{p} - \mathbf{q}) \quad \text{and} \quad \int \frac{d^3 p}{(2\pi)^3} |\mathbf{p}\rangle \frac{1}{2E_{\mathbf{p}}} \langle \mathbf{p} | = \hat{I}_1. \tag{6.2.59}$$

The tensor product of n one-particle basis states gives an n-particle state written as $|\mathbf{p}_1 \mathbf{p}_2 \cdots \mathbf{p}_n; \mathrm{MB}\rangle$, where we can think of this as a Maxwell-Boltzmann state where the ordering of the momenta implies which one-particle Hilbert space the particle belongs to. Then we symmetrize using Eq. (4.1.149) to arrive at a symmetrized basis state for the n-boson Hilbert space,

$$|\mathbf{p}_1 \cdots \mathbf{p}_n\rangle \equiv S_+ |\mathbf{p}_1 \cdots \mathbf{p}_n; \mathrm{MB}\rangle \equiv \frac{1}{\sqrt{n!}} \sum_\sigma |\mathbf{p}_{\sigma(1)} \cdots \mathbf{p}_{\sigma(n)}; \mathrm{MB}\rangle, \tag{6.2.60}$$

where on the right-hand side we are summing over the permutations σ of the indices $1, 2, \ldots, n$. In the notation of Eq. (4.1.149) we would have written this as $|\mathbf{p}_1 \cdots \mathbf{p}_n; S\rangle$, but we will understand that $|\mathbf{p}_1 \cdots \mathbf{p}_n\rangle$ is the symmetric state. Comparing with Eq. (4.1.149) we see that we have no $(\prod_m n_m!)$ term in the normalization since for an uncountably infinite one-particle basis there is vanishingly small chance for any two bosons to have the same momentum. Put another way, in any integration over a finite region of $3n$-dimensional momentum space, that part with with any two-boson momenta equal is a set of measure zero. Rather than go through the process of constructing a Maxwell-Boltzmann state and then symmetrizing it, we see that the symmetrized state is an immediate consequence of applying n normalized one-boson creation operators, $\sqrt{2E_{\mathbf{p}}} a_{\mathbf{p}}^\dagger$, to give

$$|\mathbf{p}_1 \mathbf{p}_2 \cdots \mathbf{p}_n\rangle \equiv 2^{n/2} \sqrt{E_{\mathbf{p}_1} E_{\mathbf{p}_2} \cdots E_{\mathbf{p}_n}} \, \hat{a}_{\mathbf{p}_1}^\dagger \hat{a}_{\mathbf{p}_2}^\dagger \cdots \hat{a}_{\mathbf{p}_n}^\dagger |0\rangle, \tag{6.2.61}$$

where for $n = 1$ we recover Eq. (6.2.46) and where the symmetry of the tensor product of states follows from the fact that all of the $\hat{a}_{\mathbf{p}}^\dagger$ commute with each other.

As we did to arrive at Eq. (4.1.151) let us choose any unique ordering scheme for the momenta in $|\mathbf{p}_1 \mathbf{p}_2 \cdots \mathbf{p}_n\rangle$; e.g., we could choose to order in increasing $p_x = p^1$, which is sufficient given the chance of any two p_x being equal is vanishingly small. Then we define

$$|\mathbf{p}_1 \mathbf{p}_2 \cdots \mathbf{p}_n; S_o\rangle \equiv \left(|\mathbf{p}_1 \mathbf{p}_2 \cdots \mathbf{p}_n\rangle \text{ with ordered } \mathbf{p}_i \right). \tag{6.2.62}$$

It then follows that the set of all ordered states $|\mathbf{p}_1 \mathbf{p}_2 \cdots \mathbf{p}_n; S_o\rangle$ forms a complete basis of the Hilbert space for n identical bosons, $V_H^{(nb)} \equiv S_+((V_H)^{\otimes n})$,

$$\langle \mathbf{p}_1 \cdots \mathbf{p}_n; S_o | \mathbf{p}'_1 \cdots \mathbf{p}'_n; S_o \rangle = 2^n (2\pi)^{3n} E_{\mathbf{p}_1} \cdots E_{\mathbf{p}_n} \delta^3(\mathbf{p}_1 - \mathbf{p}'_1) \cdots \delta^3(\mathbf{p}_n - \mathbf{p}'_n),$$

$$\int \left[\frac{d^3 p_1}{2E_{\mathbf{p}_1}(2\pi)^3} \cdots \frac{d^3 p_n}{2E_{\mathbf{p}_n}(2\pi)^3} \right]_{\mathrm{ord}(\mathbf{p})} |\mathbf{p}_1 \cdots \mathbf{p}_n; S_o \rangle \langle \mathbf{p}_1 \cdots \mathbf{p}_n; S_o | = \hat{I}_n, \qquad (6.2.63)$$

where here $\hat{I}_n$ is the identity operator in the n-boson Hilbert space, $V_H^{(nb)}$, and where "ord($\mathbf{p}$)" signifies that we are only integrating over ordered momenta, e.g., $p_1^1 < p_2^1 < \cdots < p_n^1$. The ordering prescription allows the orthonormal basis conditions to be written in familiar way that is free of combinatorial factors.

In terms of the symmetric basis the above conditions become

$$\langle \mathbf{p}_1 \cdots \mathbf{p}_n | \mathbf{p}'_1 \cdots \mathbf{p}'_n \rangle = 2^n (2\pi)^{3n} E_{\mathbf{p}_1} \cdots E_{\mathbf{p}_n} \left[\delta^3(\mathbf{p}_1 - \mathbf{p}'_1) \cdots \delta^3(\mathbf{p}_n - \mathbf{p}'_n) \right.$$
$$\left. + \text{ all permutations of } (\mathbf{p}'_1, \mathbf{p}'_2, \ldots, \mathbf{p}'_n) \right],$$

$$\frac{1}{n!} \int \frac{d^3 p_1}{2E_{\mathbf{p}_1}(2\pi)^3} \cdots \frac{d^3 p_n}{2E_{\mathbf{p}_n}(2\pi)^3} |\mathbf{p}_1 \cdots \mathbf{p}_n \rangle \langle \mathbf{p}_1 \cdots \mathbf{p}_n | = \hat{I}_n, \qquad (6.2.64)$$

where the first equation follows from Eqs. (6.2.43) and (6.2.61) and where the second equation follows since the symmetric states span the space and since we easily verify that $\hat{I}_n$ satisfies $\hat{I}_n |\mathbf{p}_1 \cdots \mathbf{p}_n \rangle = |\mathbf{p}_1 \cdots \mathbf{p}_n \rangle$. We used the ordered basis $|\mathbf{p}_1 \cdots \mathbf{p}_n; S_o \rangle$ above to make contact with our earlier quantum mechanics discussion, but in quantum field theory we will normally work in the symmetric basis $|\mathbf{p}_1 \cdots \mathbf{p}_n \rangle \equiv |\mathbf{p}_1 \cdots \mathbf{p}_n; S \rangle$.

From Eq. (6.2.37) we can directly verify that

$$\hat{N}(a_{\mathbf{p}_1}^\dagger \cdots a_{\mathbf{p}_n}^\dagger)|0\rangle = a_{\mathbf{p}_1}^\dagger (\hat{N} + 1) a_{\mathbf{p}_2}^\dagger \cdots a_{\mathbf{p}_n}^\dagger |0\rangle = a_{\mathbf{p}_1}^\dagger a_{\mathbf{p}_1}^\dagger \cdots a_{\mathbf{p}_n}^\dagger (\hat{N} + n)|0\rangle$$
$$= n(a_{\mathbf{p}_1}^\dagger a_{\mathbf{p}_2}^\dagger \cdots a_{\mathbf{p}_n}^\dagger)|0\rangle. \qquad (6.2.65)$$

Then together with Eq. (6.2.61) we have as expected that

$$\hat{N}|\mathbf{p}_1 \cdots \mathbf{p}_n \rangle = n|\mathbf{p}_1 \cdots \mathbf{p}_n \rangle. \qquad (6.2.66)$$

We now have the Hilbert space of n-boson states, which is the symmetrized direct product of n one-particle Hilbert spaces, $S_+((V_H)^{\otimes n}) \equiv S_+(V_H \otimes \cdots \otimes V_H)$. As we learned in Sec. 4.1.9 the bosonic Fock space for scalar particles will then be the Hilbert space completion of the direct sum of all of the n-boson Hilbert spaces for $n = 0, 1, 2, \ldots$. The boson Fock space, $\overline{F}_+(V_H)$, is the Hilbert space completion of

$$F_+(V_H) \equiv V_0 \oplus V_H \oplus S_+(V_H \times V_H) \oplus \cdots = \bigoplus_{n=0}^{\infty} S_+((V_H)^{\otimes n}), \qquad (6.2.67)$$

where V_0 is the one-dimensional Hilbert space spanned by the vacuum state $|0\rangle$.

Any arbitrary polynomial combination of the $\hat{a}_{\mathbf{p}}^\dagger$ and $\hat{a}_{\mathbf{p}}$ operators acting on the vacuum can be reduced to a polynomial of the $\hat{a}_{\mathbf{p}}^\dagger$ only, where we use the commutation relations to bring all $\hat{a}_{\mathbf{p}}$ operators to the right where they act on the vacuum and annihilate. This leaves terms in a new polynomial acting on the vacuum containing only combinations of the $\hat{a}_{\mathbf{p}}^\dagger$, which clearly produces a vector in the Fock space. This means that Fock space contains all possible states of the system as we expect. An arbitrary state in Fock space is not an eigenvector of the number operator, but if we measure the boson number and find n bosons then the resulting state lies in the n-boson subspace corresponding to the projection operator $\hat{I}_n$.

6.2.2 Field Configuration and Momentum Density Space

Specifying an arbitrary classical field configuration ϕ at some fixed time t means specifying $\phi(x) = \phi(\mathbf{x}, t)$ for all of the uncountable infinity of spatial points $\mathbf{x}$ at this time t. This is analogous to specifying a vector in configuration space at some time t in classical mechanics, $\vec{q}(t) \in \mathcal{M}$, by choosing $q_j(t)$ for all $j = 1, \ldots, N$.

Let t_0 be the time at which the Schrödinger and Heisenberg pictures coincide in quantum mechanics as discussed in Sec. 4.1.12. The quantum mechanical Hilbert space is spanned by the simultaneous eigenstates $|\vec{q}\rangle_s$ of the mutually commuting set of the Schrödinger picture generalized coordinate operators $\hat{q}_{sj} = \hat{q}_{hj}(t_0)$ for $j = 1, \ldots, N$. Similarly, the quantum field theory analog of classical field configuration space is the space spanned by the simultaneous eigenstates $|\phi\rangle_s$ of the mutually commuting set of Hermitian Schrödinger picture operators $\hat{\phi}_s(\mathbf{x}) = \phi(\mathbf{x}, t_0)$ for all $\mathbf{x}$. We have the simple Schrödinger picture correspondence between quantum mechanics for N generalized coordinates and quantum field theory,

$$\hat{q}_{sj}|\vec{q}\rangle_s = q_j|\vec{q}\rangle_s \quad \rightarrow \quad \hat{\phi}_s(\mathbf{x})|\phi\rangle_s = \phi(\mathbf{x})|\phi\rangle_s \quad \text{for all } \mathbf{x}. \tag{6.2.68}$$

Consider the limit where we use a lattice regularization (with lattice spacing a) for three-dimensional coordinate space. The lattice sites are

$$\mathbf{x} = a\mathbf{n} \quad \text{with} \quad \mathbf{n} = (n_x, n_y, n_z) \quad \text{for} \quad n_x, n_y, n_z \in \mathbb{Z}. \tag{6.2.69}$$

Let $\mathbf{x}_1, \mathbf{x}_2, \mathbf{x}_3, \ldots$ denote the countable infinity of sites on this three-dimensional spatial lattice. We now fill space with three-dimensional cubes of edge length a that are centered on the lattice sites, where R_j labels the cube of volume a^3 centered on the lattice site $\mathbf{x}_j$. Define a cubic step function $\theta_j(\mathbf{x})$ centered at each lattice site $\mathbf{x}_j$, such that

$$\theta_j(\mathbf{x}) = 1 \quad \text{for} \quad \mathbf{x} \in R_j \quad \text{and} \quad \theta_j(\mathbf{x}) = 0 \quad \text{otherwise}, \tag{6.2.70}$$

$$(1/a^3) \int d^3x \, \theta_j(\mathbf{x})\theta_k(\mathbf{x}) = \delta_{jk} \quad \text{and} \quad \sum_{j=1}^{\infty} \theta_j(\mathbf{x})\theta_j(\mathbf{x}') = \begin{cases} 1 & \text{if } \mathbf{x}, \mathbf{x}' \in R_k \\ 0 & \text{otherwise}. \end{cases}$$

Let us now take the Schrödinger picture field operators $\hat{\phi}_s(\mathbf{x})$ and average them over each cube and so define

$$\hat{\phi}^a(\mathbf{x}_j) \equiv (1/a^3) \int d^3x \, \theta_j(\mathbf{x})\hat{\phi}_s(\mathbf{x}). \tag{6.2.71}$$

We write the eigenstates of $\hat{\phi}^a(\mathbf{x}_j)$ as $|\phi(\mathbf{x}_j)\rangle^a$ with eigenvalue $\phi(\mathbf{x}_j) \in \mathbb{R}$,

$$\hat{\phi}^a(\mathbf{x}_j)|\phi(\mathbf{x}_j)\rangle^a = \phi(\mathbf{x}_j)|\phi(\mathbf{x}_j)\rangle^a. \tag{6.2.72}$$

Clearly, from Eq. (6.2.12) we have $[\hat{\phi}^a(\mathbf{x}_j), \hat{\phi}^a(\mathbf{x}_k)] = 0$ for all $j, k = 1, 2, 3, \ldots$ and so we can construct states that are the simultaneous eigenstates of the $\hat{\phi}^a(\mathbf{x}_j)$,

$$|\phi\rangle^a \equiv |\phi(\mathbf{x}_1), \phi(\mathbf{x}_2), \phi(\mathbf{x}_3), \ldots\rangle^a \quad \text{with} \quad \hat{\phi}^a(\mathbf{x}_j)|\phi\rangle^a = \phi(\mathbf{x}_j)|\phi\rangle^a \tag{6.2.73}$$

for all lattice sites $\mathbf{x}_j$. Thus we see that with the coordinate-space lattice regularization the eigenstate for a real scalar field configuration, $|\phi\rangle^a$, is specified by one real number, $\phi(\mathbf{x}_j)$, per lattice site $\mathbf{x}_j$. Furthermore, if two configurations, ϕ and ϕ', are not identical at every lattice site $\mathbf{x}$, then they must be orthogonal. We can then choose the normalization such that we have the orthonormality conditions

$$^a\langle\phi'|\phi\rangle^a = \prod_{j=1}^{\infty} \delta(\phi'(\mathbf{x}_j) - \phi(\mathbf{x}_j)), \tag{6.2.74}$$

which is analogous to $\langle \vec{q}' | \vec{q} \rangle = \delta^n(q' - q) = \delta(q_1' - q_1)\delta(q_2' - q_2) \cdots \delta(q_n' - q_n)$ in ordinary quantum mechanics. The space of all lattice field configurations is spanned by the set of all $|\phi\rangle^a$. We have the completeness relation for this space,

$$\int \prod_{j=1}^{\infty} d\phi(\mathbf{x}_j) |\phi\rangle^a {}^a\langle\phi| = \hat{I}, \tag{6.2.75}$$

which is analogous to $\int d^n q \, |\vec{q}\rangle\langle\vec{q}| = \hat{I}$ in ordinary quantum mechanics. When the lattice regularization is removed, $a \to 0$, we can write

$$\mathbf{x}_j \to \mathbf{x}, \quad \hat{\phi}^a(\mathbf{x}_j) \to \hat{\phi}_s(\mathbf{x}), \quad |\phi\rangle^a \to |\phi\rangle_s \quad \text{and so} \quad \hat{\phi}_s(\mathbf{x})|\phi\rangle_s = \phi(\mathbf{x})|\phi\rangle_s, \tag{6.2.76}$$

which is how we can understand the meaning of Eq. (6.2.68).

In the continuum limit the lattice Schrödinger picture orthonormality and completeness relations given in Eqs. (6.2.74) and (6.2.75) and their canonical momentum density counterparts can be written as

$$_s\langle\phi'|\phi\rangle_s = \delta[\phi' - \phi]_s \quad \text{and} \quad \int \mathcal{D}\phi_s \, |\phi\rangle_s {}_s\langle\phi| = \hat{I},$$
$$_s\langle\pi'|\pi\rangle_s = \delta[\pi' - \pi]_s \quad \text{and} \quad \int \mathcal{D}\pi_s \, |\pi\rangle_s {}_s\langle\pi| = \hat{I}, \tag{6.2.77}$$

which is just the continuum analog of Eq. (6.1.6). Thus the set of all $|\phi\rangle_s$ form a basis for Fock space as does the set of all $|\pi\rangle_s$. The same is obviously true of the Heisenberg picture at any fixed time t. We thus have the quantum field theory equivalents of Eq. (4.1.196) providing completeness relations for Fock space. Note that $\delta[\phi' - \phi]_s$ and $\delta[\pi' - \pi]_s$ are functional delta functions, which vanish unless the fields are equal for every $\mathbf{x}$, and $\int \mathcal{D}\phi_s$ and $\int \mathcal{D}\pi_s$ are functional integrals, where here both are defined with respect to functions of three-space, $\phi(\mathbf{x})$ and $\pi(\mathbf{x})$, respectively. We use the subscript "s" to denote that we are concerned with functions of three-space relevant for the Schrödinger picture.

How can we understand that the Fock space described by the functional orthonormality relations in Eq. (6.2.77) corresponds to the Fock space $F^+(V_H)$ that we built up with the particle occupancy basis defined by Eqs. (6.2.64) and (6.2.67)? If we remove the $(\boldsymbol{\nabla}\hat{\phi})^2$ term in the Hamiltonian of Eq. (6.2.21), then we end up with one uncoupled harmonic oscillator at each spatial point $\mathbf{x}$ and so there is a normal mode associated with each $\mathbf{x}$. In that case it is clear that the states $|\phi\rangle$ span the Hilbert space and that Eq. (6.2.77) applies. Reintroducing the term $(\boldsymbol{\nabla}\hat{\phi})^2$ corresponds to a coupling of neighboring oscillators to produce new normal modes labeled by $E_\mathbf{p} = \omega_\mathbf{p}$, but doesn't change the Hilbert space. So the Hilbert space described by Eq. (6.2.77) is just the Fock space that we generated earlier.

6.2.3 Covariant Operator Formulation

So far the formulation of quantum field theory has been in a particular inertial frame in the Schrödinger picture. This was done to connect with the quantization of the ordinary harmonic oscillator, but it disguised the manifest Lorentz-covariant form of the quantum field theory. Let us now redress this shortcoming.

In the Heisenberg picture the field operator and its conjugate momentum operator become time-dependent. With no explicit time-dependence in $\hat{H}$ we have

$$\hat{\phi}(x) = e^{i\hat{H}(t-t_0)}\hat{\phi}_s(\mathbf{x})e^{-i\hat{H}(t-t_0)}, \quad \hat{\pi}(x) = e^{i\hat{H}(t-t_0)}\hat{\pi}_s(\mathbf{x})e^{-i\hat{H}(t-t_0)}, \tag{6.2.78}$$

where the Heisenberg and Schrödinger pictures are equivalent at the arbitrary reference time t_0. We will understand that operators are in the Heisenberg picture unless indicated otherwise. Let $\hat{\mathcal{O}}_s(\mathbf{x})$ be any time-independent Schrödinger picture operator, then its Heisenberg picture form is given by

$$\hat{O}(x) = e^{i\hat{H}(t-t_0)}\hat{O}_s(\mathbf{x})e^{-i\hat{H}(t-t_0)}. \tag{6.2.79}$$

The Heisenberg equation of motion in Eqs. (4.1.24) and (2.5.17) follows immediately from differentiating the above equation,

$$i\partial\hat{O}(x)/\partial t = [\hat{O}(x), \hat{H}]. \tag{6.2.80}$$

Note that the partial time derivative is used to indicate that it acts only on the $x^0 = t$ component of the four-vector x^μ. If we want to indicate a time derivative or a spacetime derivative with respect to *explicit* time or spacetime dependence, then we will do so with an "ex" superscript as in Eq. (3.1.18). The Schrödinger picture Hamiltonian is time-independent for the free scalar field and so the Hamiltonian is the same in either the Schrödinger or Heisenberg representation, $\hat{H} \equiv \hat{H}_s = \hat{H}(t)$, since a time-independent $\hat{H}_s$ commutes with itself at all times.

Operator Equations of Motion

The equal-time canonical commutation relations of Eq. (6.2.4) can be written as

$$[\hat{\phi}(t, \mathbf{x}), \hat{\phi}(t, \mathbf{y})] = [\hat{\pi}(t, \mathbf{x}), \hat{\pi}(t, \mathbf{y})] = 0,$$
$$[\hat{\phi}(t, \mathbf{x}), \hat{\pi}(t, \mathbf{y})] = i\delta^3(\mathbf{x} - \mathbf{y}). \tag{6.2.81}$$

We easily verify the results analogous to Eq. (2.5.19),

$$-i[\hat{\phi}(t, \mathbf{x}), \hat{\pi}(t, \mathbf{y})^n] = \delta^3(\mathbf{x} - \mathbf{y})n\hat{\pi}(t, \mathbf{x})^{n-1} \equiv \frac{\delta\hat{\pi}(t, \mathbf{y})^n}{\delta\hat{\pi}(t, \mathbf{x})},$$

$$i[\hat{\pi}(t, \mathbf{x}), \hat{\phi}(t, \mathbf{y})^n] = \delta^3(\mathbf{x} - \mathbf{y})n\hat{\phi}(t, \mathbf{x})^{n-1} \equiv \frac{\delta\hat{\phi}(t, \mathbf{y})^n}{\delta\hat{\phi}(t, \mathbf{x})}, \tag{6.2.82}$$

$$-i[\hat{\phi}(t, \mathbf{x}), \hat{\phi}(t, \mathbf{y})^n] \equiv \frac{\delta\hat{\phi}(t, \mathbf{y})^n}{\delta\hat{\pi}(t, \mathbf{x})} = 0, \qquad i[\hat{\pi}(t, \mathbf{x}), \hat{\pi}(t, \mathbf{y})^n] \equiv \frac{\delta\hat{\pi}(t, \mathbf{y})^n}{\delta\hat{\phi}(t, \mathbf{x})} = 0,$$

which define derivatives with respect to the Heisenberg picture operators $\hat{\phi}(t, \mathbf{x})$ and $\hat{\pi}(t, \mathbf{x})$ at fixed time t. Let $\hat{A}(t) \equiv A[\hat{\phi}(t), \hat{\pi}(t)]$ be an operator that is some functional of $\hat{\phi}(t, \mathbf{x})$ and $\hat{\pi}(t, \mathbf{x})$ at fixed time t. Assume that $\hat{A}(t)$ can be written as a power series in $\hat{\phi}$ and $\hat{\pi}$ at time t and integrated over all spatial arguments. Then we make definitions analogous to Eq. (2.5.20) to define functional derivatives with respect to the field and canonical momentum density operators,

$$[\hat{\phi}(t, \mathbf{x}), \hat{A}(t)] \equiv i\delta\hat{A}(t)/\delta\hat{\pi}(t, \mathbf{x}) \text{ and } [\hat{\pi}(t, \mathbf{x}), \hat{A}(t)] \equiv -i\delta\hat{A}(t)/\delta\hat{\phi}(t, \mathbf{x}). \tag{6.2.83}$$

These results hold for any order of the $\hat{\phi}$ and $\hat{\pi}$ operators in our definition of $\hat{A}(t) \equiv A[\hat{\phi}(t), \hat{\pi}(t)]$. The classical limit of Eq. (6.2.83) is the Poisson bracket form in Eq. (3.1.28) obtained from the correspondence principle in Eq. (2.5.27).

Consider the Hamiltonian of Eq. (6.2.33) in terms of Heisenberg picture fields,

$$\hat{H} \equiv \int d^3x\,\hat{\mathcal{H}}(t, \mathbf{x}) \equiv \int d^3x : \left[\tfrac{1}{2}\hat{\pi}(t, \mathbf{x})^2 + \tfrac{1}{2}\boldsymbol{\nabla}\hat{\phi}(t, \mathbf{x})^2 + \tfrac{1}{2}m^2\hat{\phi}(t, \mathbf{x})^2\right] :. \tag{6.2.84}$$

Note that $\hat{\mathcal{H}}(t, \mathbf{x}) \equiv \mathcal{H}(\hat{\phi}(t, \mathbf{x}), \boldsymbol{\nabla}\hat{\phi}(t, \mathbf{x}), \hat{\pi}(t, \mathbf{x}))$ is a function of the independent variables $\hat{\phi}$, $\boldsymbol{\nabla}\hat{\phi}$ and $\hat{\pi}$ as we saw in the classical case in Eq. (3.1.17). This is a special case in that $\hat{H}$ is time-independent. The difference between $\hat{H}$ and its non-normal-ordered form is a constant and so when we are differentiating $\hat{H}$ we can simply use its non-normal-ordered form. Using the Heisenberg equation of motion in Eqs. (6.2.80) and (6.2.83) we find the operator form of Hamilton's equations,

$$\partial_0 \hat{\phi}(t, \mathbf{x}) = \frac{\delta \hat{H}}{\delta \hat{\pi}(t, \mathbf{x})} = \frac{\partial \hat{\mathcal{H}}(t, \mathbf{x})}{\partial \hat{\pi}(t, \mathbf{x})}$$

$$\partial_0 \hat{\pi}(t, \mathbf{x}) = -\frac{\delta \hat{H}}{\delta \hat{\phi}(t, \mathbf{x})} = -\frac{\partial \hat{\mathcal{H}}(t, \mathbf{x})}{\partial \hat{\phi}(t, \mathbf{x})} + \nabla \cdot \frac{\partial \hat{\mathcal{H}}(t, \mathbf{x})}{\partial \nabla \hat{\phi}(t, \mathbf{x})}. \tag{6.2.85}$$

This is the operator equivalent of the classical field theory form of Hamilton's equations in Eq. (3.1.29). The last step in the second equation arises for the same reasons as in the classical field theory case in Eq. (3.1.30). We are assuming that the operators act on states corresponding to fields that are periodic or that vanish at spatial infinity so that surface terms can be ignored and we can use integration by parts in the usual way as we did in Chapter 3. In the free scalar field case we combine Eqs. (6.2.33) and (6.2.85) to find

$$\partial_0 \hat{\phi}(t, \mathbf{x}) = \hat{\pi}(t, \mathbf{x}) \qquad \text{and} \qquad \partial_0 \hat{\pi}(t, \mathbf{x}) = (\nabla^2 - m^2)\hat{\phi}(t, \mathbf{x}),$$

$$\Rightarrow \quad (\partial_\mu \partial^\mu + m^2)\hat{\phi}(t, \mathbf{x}) = 0, \tag{6.2.86}$$

which is just the Klein-Gordon equation of Eq. (3.1.41) in operator form.

We have shown that the Heisenberg picture operators of the scalar quantum field theory obey the same equations of motion as their classical field theory counterparts. The underlying reasons for this are (i) the fundamental Poisson brackets of the classical field theory are the classical analogs of the equal-time canonical commutation relations of the quantum field theory, and (ii) the operator derivatives defined in terms of commutators behave just as their partial derivative counterparts do. Eq. (3.1.28) is the classical field theory analog of the quantum field theory result in Eq. (6.2.83). This is just the field theory equivalent of the fact that Eq. (2.4.20) is the classical mechanics analog of the quantum mechanics result of Eq. (2.5.20).

The Hamiltonian and Lagrangian density operators have the same relationship as their classical counterparts and are related by a Legendre transformation,

$$\hat{H} \equiv \int d^3x\, \hat{\mathcal{H}} \equiv \left(\int d^3x\, \hat{\pi}(x)\dot{\hat{\phi}}(x) \right) - \hat{L} = \int d^3x\, \left(\hat{\pi}(x)\dot{\hat{\phi}}(x) - \hat{\mathcal{L}} \right), \tag{6.2.87}$$

where the canonical momentum density operator $\hat{\pi}(x)$ is defined as

$$\hat{\pi}(x) \equiv \frac{\delta \hat{L}}{\delta(\partial_0 \hat{\phi})(x)} = \frac{\partial \hat{\mathcal{L}}}{\partial(\partial_0 \hat{\phi})(x)}. \tag{6.2.88}$$

Since the operator equations of motion are the same as their classical equivalents, then the operators must also obey the Euler-Lagrange equations of Eq. (3.1.10) at the operator level for the Heisenberg picture operator $\hat{\phi}(x)$,

$$\partial_\mu \left(\frac{\delta \hat{L}}{\delta(\partial_\mu \hat{\phi})(x)} \right) - \frac{\delta \hat{L}}{\delta \hat{\phi}(x)} = 0 \quad \text{or} \quad \partial_\mu \left(\frac{\partial \hat{\mathcal{L}}}{\partial(\partial_\mu \hat{\phi})(x)} \right) - \frac{\partial \hat{\mathcal{L}}}{\partial \hat{\phi}(x)} = 0. \tag{6.2.89}$$

Conserved Current Operators

As we saw in Sec. 3.2.1, if the classical action is invariant under the infinitesimal transformations in Eqs. (3.2.7) and (3.2.8) up to an overall surface term, i.e.,

$$S[\phi] \to S[\phi'] = S[\phi] + d\alpha(1/c) \int_R d^4x\, \partial_\nu F^\nu, \tag{6.2.90}$$

then we have the corresponding conserved generalized Noether current given in Eq. (3.2.22). The conservation of this current must be encapsulated in and a consequence of the equations of motion for the system. Since the equations of motion are also obeyed at the operator level, then the conservation of the generalized Noether current must also be true at the operator level. So it must be true that

$$\partial_\mu \hat{j}^\mu = 0 \quad \text{with} \quad \hat{j}^\mu \equiv \hat{\bar{\pi}}^\mu \cdot \left[\hat{\bar{\Phi}} - [\partial_\nu \hat{\bar{\phi}}] X^\nu \right] + \hat{\mathcal{L}} X^\mu - \hat{F}^\mu, \tag{6.2.91}$$

where we have $\hat{\mathcal{L}} \equiv \mathcal{L}(\hat{\bar{\phi}}, \partial_\mu \hat{\bar{\phi}}, x)$, $\hat{F} \equiv F(\hat{\bar{\phi}}, \partial_\mu \hat{\bar{\phi}}, x)$, $\hat{\phi}'_i(x') - \hat{\phi}_i(x) \equiv d\alpha \hat{\Phi}_i(x)$ and $x'^\mu - x^\mu \equiv d\alpha X^\mu$. The corresponding conserved charge operator is

$$\hat{Q} \equiv \int d^3x \, \hat{j}^0(x) \quad \text{where} \quad \partial_0 \hat{Q} = 0. \tag{6.2.92}$$

The use of integration by parts in the various steps that lead to the above outcomes is justified since we take the Hilbert space in which our operators act to be spanned by a field configuration basis with vanishing or periodic spatial boundary conditions so that the surface terms at spatial infinity do not contribute.

We will see later that this simple picture, which is appropriate for free quantum field theories, becomes complicated by the need to regularize and renormalize interacting quantum field theories. In fact, it can happen that a classically conserved current in an interacting classical field theory has no quantum field theory analog because the regularization/renormalization procedure necessarily destroys this symmetry. The occurrence of such a nonconserved quantum current is referred to as an *anomaly* of the quantum field theory.

Free Scalar Field in the Heisenberg Picture

Note that from Eq. (6.2.37) we have

$$\hat{H} \hat{a}_{\mathbf{p}} = \hat{a}_{\mathbf{p}} \left(\hat{H} - E_{\mathbf{p}} \right) \qquad \text{and} \qquad \hat{H} \hat{a}_{\mathbf{p}}^\dagger = \hat{a}_{\mathbf{p}}^\dagger \left(\hat{H} + E_{\mathbf{p}} \right). \tag{6.2.93}$$

Hence $\hat{H}^n \hat{a}_{\mathbf{p}} = \hat{H}^{n-1} \hat{a}_{\mathbf{p}} (\hat{H} - E_{\mathbf{p}}) = \hat{a}_{\mathbf{p}} (\hat{H} - E_{\mathbf{p}})^n$ and similarly $\hat{H}^n \hat{a}_{\mathbf{p}}^\dagger = \hat{a}_{\mathbf{p}}^\dagger (\hat{H} + E_{\mathbf{p}})^n$. These results give $\exp(\hat{H}) \hat{a}_{\mathbf{p}} = \hat{a}_{\mathbf{p}} \exp(\hat{H} - E_{\mathbf{p}})$ and similarly $\exp(\hat{H}) \hat{a}_{\mathbf{p}}^\dagger = \hat{a}_{\mathbf{p}}^\dagger \exp(\hat{H} + E_{\mathbf{p}})$ from which it follows that

$$e^{i\hat{H}t} \hat{a}_{\mathbf{p}} e^{-i\hat{H}t} = \hat{a}_{\mathbf{p}} e^{-iE_{\mathbf{p}}t} \qquad \text{and} \qquad e^{i\hat{H}t} \hat{a}_{\mathbf{p}}^\dagger e^{-i\hat{H}t} = \hat{a}_{\mathbf{p}}^\dagger e^{iE_{\mathbf{p}}t}. \tag{6.2.94}$$

From Eq. (6.2.37) we see that similar results apply for $\hat{H}$ and $\hat{\mathbf{P}}$ and so we have

$$e^{i\hat{P}\cdot x} \hat{a}_{\mathbf{p}} e^{-i\hat{P}\cdot x} = \hat{a}_{\mathbf{p}} e^{-ip\cdot x} \qquad \text{and} \qquad e^{i\hat{P}\cdot x} \hat{a}_{\mathbf{p}}^\dagger e^{-i\hat{P}\cdot x} = \hat{a}_{\mathbf{p}}^\dagger e^{ip\cdot x}. \tag{6.2.95}$$

For notational simplicity and without loss of generality we choose the Schrödinger and Heisenberg picture operators to coincide at $t_0 = 0$ so that Eq. (6.2.78) becomes

$$\hat{\mathcal{O}}(x) = e^{i\hat{H}t} \hat{\mathcal{O}}_s(\mathbf{x}) e^{-i\hat{H}t}. \tag{6.2.96}$$

The Heisenberg picture field operators can then be expressed in terms of the annihilation and creation operators, since

$$\begin{aligned}
\hat{\phi}(x) = \hat{\phi}(\mathbf{x}, t) &= e^{i\hat{H}t} \hat{\phi}_s(\mathbf{x}) e^{-i\hat{H}t} = e^{i\hat{H}t} \hat{\phi}(\mathbf{x}, 0) e^{-i\hat{H}t} \\
&= e^{i\hat{H}t} \int \frac{d^3p}{(2\pi)^3} \frac{1}{\sqrt{2E_{\mathbf{p}}}} \left(\hat{a}_{\mathbf{p}} e^{i\mathbf{p}\cdot\mathbf{x}} + \hat{a}_{\mathbf{p}}^\dagger e^{-i\mathbf{p}\cdot\mathbf{x}} \right) e^{-i\hat{H}t} \\
&= \int \frac{d^3p}{(2\pi)^3} \frac{1}{\sqrt{2E_{\mathbf{p}}}} \left(\hat{a}_{\mathbf{p}} e^{-ip\cdot x} + \hat{a}_{\mathbf{p}}^\dagger e^{ip\cdot x} \right)\Big|_{p^0 = E_{\mathbf{p}}},
\end{aligned} \tag{6.2.97}$$

where the quantization of the field arises from associating an annihilation and creation operator with each normal mode of the system, i.e., with each plane wave, where the annihilation operators are associated with the positive frequencies and the creation operator with the negative frequencies. A similar result follows for $\hat{\pi}(x)$ and so the Heisenberg picture operators $\hat{\phi}(x)$ and $\hat{\pi}(x)$ can be written as

$$\hat{\phi}(x) = \int \frac{d^3p}{(2\pi)^3} \frac{1}{\sqrt{2E_\mathbf{p}}} \left(\hat{a}_\mathbf{p} e^{-ip \cdot x} + \hat{a}_\mathbf{p}^\dagger e^{ip \cdot x}\right)\big|_{p^0 = E_\mathbf{p}} \tag{6.2.98}$$

$$\hat{\pi}(x) = \int \frac{d^3p}{(2\pi)^3} (-i)\sqrt{\frac{E_\mathbf{p}}{2}} \left(\hat{a}_\mathbf{p} e^{-ip \cdot x} - \hat{a}_\mathbf{p}^\dagger e^{ip \cdot x}\right)\big|_{p^0 = E_\mathbf{p}} = \partial_0 \hat{\phi}(x) = \dot{\hat{\phi}}(x),$$

where $\pi(x) = \partial_0 \phi(x)$ from Eq. (6.2.88) as it did classically in the free field case in Eq. (3.1.45). Using Eqs. (6.2.95) and (6.2.97) it follows that we can write

$$\hat{\phi}(x) = \hat{\phi}(\mathbf{x}, t) = e^{i\hat{P} \cdot x} \hat{\phi}(0) e^{-i\hat{P} \cdot x} = e^{i(\hat{H}t - \hat{\mathbf{P}} \cdot \mathbf{x})} \hat{\phi}(0) e^{-i(\hat{H}t - \hat{\mathbf{P}} \cdot \mathbf{x})}, \tag{6.2.99}$$

$$\hat{\pi}(x) = \hat{\pi}(\mathbf{x}, t) = e^{i\hat{P} \cdot x} \hat{\pi}(0) e^{-i\hat{P} \cdot x} = e^{i(\hat{H}t - \hat{\mathbf{P}} \cdot \mathbf{x})} \hat{\pi}(0) e^{-i(\hat{H}t - \hat{\mathbf{P}} \cdot \mathbf{x})},$$

where we have defined the four-vector operator $\hat{P}^\mu \equiv (\hat{H}, \hat{\mathbf{P}})$. We see that $\hat{P}^\mu$ is the generator of translations in spacetime as is the case in ordinary quantum mechanics. Since the normal-ordering of $\hat{P}^\mu$ only changes it by a constant, then we see that Eq. (6.2.99) has the same form with or without normal-ordering since a constant contribution to the exponent on each side cancels out.

The equations for $\hat{\phi}$ and $\hat{\pi}$ can be inverted to give

$$\sqrt{2E_\mathbf{p}} \hat{a}_\mathbf{p} = 2i \int d^3x\, e^{ip \cdot x} \overleftrightarrow{\partial}_0 \hat{\phi}(x),$$

$$\sqrt{2E_\mathbf{p}} \hat{a}_\mathbf{p}^\dagger = -2i \int d^3x\, e^{-ip \cdot x} \overleftrightarrow{\partial}_0 \hat{\phi}(x), \tag{6.2.100}$$

where the right-hand sides are time-independent and where $\overleftrightarrow{\partial}_0$ is defined by[1]

$$f(x) \overleftrightarrow{\partial}_\mu g(x) \equiv \tfrac{1}{2}[f(x)(\partial_\mu g)(x) - (\partial_\mu f)(x)g(x)] \Rightarrow \overleftrightarrow{\partial}_\mu \equiv \tfrac{1}{2}[\overrightarrow{\partial}_\mu - \overleftarrow{\partial}_\mu]. \tag{6.2.101}$$

Proof: First observe that

$$\pm 2i \int d^3x\, e^{\pm ip \cdot x} \overleftrightarrow{\partial}_0 \hat{\phi}(x) = \pm 2i \int d^3x\, e^{\pm ip \cdot x} \tfrac{1}{2}[\partial_0 \hat{\phi}(x) \mp iE_\mathbf{p} \hat{\phi}(x)]$$

$$= \int d^3x\, e^{\pm ip \cdot x}[E_\mathbf{p} \hat{\phi}(x) \pm i\hat{\pi}(x)]. \tag{6.2.102}$$

Then invert Eq. (6.2.98) as done for Eq. (6.2.16) to obtain

$$\sqrt{2E_\mathbf{p}} \hat{a}_\mathbf{p} = \int d^3x\, e^{ip \cdot x}[E_\mathbf{p} \hat{\phi}(x) + i\hat{\pi}(x)],$$

$$\sqrt{2E_\mathbf{p}} \hat{a}_\mathbf{p}^\dagger = \int d^3x\, e^{-ip \cdot x}[E_\mathbf{p} \hat{\phi}(x) - i\hat{\pi}(x)]. \tag{6.2.103}$$

Then we have arrived at Eq. (6.2.100), since

$$\sqrt{2E_\mathbf{p}} \hat{a}_\mathbf{p} = \int d^3x\, e^{ip \cdot x}[E_\mathbf{p} \hat{\phi}(x) + i\hat{\pi}(x)] = 2i \int d^3x\, e^{ip \cdot x} \overleftrightarrow{\partial}_0 \hat{\phi}(x),$$

$$\sqrt{2E_\mathbf{p}} \hat{a}_\mathbf{p}^\dagger = \int d^3x\, e^{-ip \cdot x}[E_\mathbf{p} \hat{\phi}(x) - i\hat{\pi}(x)] = -2i \int d^3x\, e^{-ip \cdot x} \overleftrightarrow{\partial}_0 \hat{\phi}(x). \tag{6.2.104}$$

---

[1] Some texts do not include the $\frac{1}{2}$ in Eq. (6.2.101), e.g., Bjorken and Drell (1965) and Greiner (2000).

The action of the Heisenberg picture field operator on the vacuum,

$$\hat{\phi}(x)|0\rangle = \int \frac{d^3p}{(2\pi)^3} \frac{1}{\sqrt{2E_{\mathbf{p}}}} e^{ip\cdot x} \hat{a}_{\mathbf{p}}^{\dagger}|0\rangle = \int \frac{d^3p}{(2\pi)^3} \frac{1}{2E_{\mathbf{p}}} e^{ip\cdot x}\Big|_{p^0 = E_{\mathbf{p}}} |\mathbf{p}\rangle, \qquad (6.2.105)$$

corresponds to the creation of a particle at the spacetime point x^μ and follows from Eqs. (6.2.55) and (6.2.79)

Comparing with other texts: Most texts normalize their field operators to give the ETCR in Eq. (6.2.4). Using "BD" to denote the conventions of Bjorken and Drell (1964) and Greiner and Reinhardt (1996) we find that Eq. (6.2.98) is written as

$$\hat{a}_{\mathbf{p}} = (2\pi)^{3/2} \hat{a}(p)^{\mathrm{BD}}, \qquad (6.2.106)$$

$$\hat{\phi}(x) = \int \frac{d^3p}{(2\pi)^{3/2}} \frac{1}{\sqrt{2E_{\mathbf{p}}}} \left[\hat{a}(p)^{\mathrm{BD}} e^{-ip\cdot x} + \hat{a}^{\dagger}(p)^{\mathrm{BD}} e^{ip\cdot x} \right]\Big|_{p^0 = E_{\mathbf{p}}}.$$

6.2.4 Poincaré Covariance

Consider the set of N classical fields with intrinsic angular momentum discussed in Eq. (3.2.87) under a passive Poincaré transformation,

$$\phi_r(x) \to \phi_r'(x') = S_{rs}(\Lambda)\phi_s(x) \quad \text{with} \quad x'^\mu = \Lambda^\mu{}_\nu x^\nu + a^\mu, \qquad (6.2.107)$$

where repeated indices are summed over and where $S(\Lambda)$ is an $N \times N$ representation of the Lorentz group. We require that all of the expectation values of the field operators, $\vec{\hat{\phi}}(x)$, transform in the same manner as their classical counterparts,

$$\langle \psi | \hat{\phi}_r(x) | \psi \rangle \to \langle \psi | \hat{\phi}_r'(x') | \psi \rangle = S_{rs}(\Lambda) \langle \psi | \hat{\phi}_s(x) | \psi \rangle \quad \text{with} \quad x'^\mu = \Lambda^\mu{}_\nu x^\nu. \qquad (6.2.108)$$

This is a passive transformation in that the state of the system $|\psi\rangle$ is unchanged and the observable operator is transformed as discussed in Eq. (1.2.51). We will restrict our attention here to restricted Lorentz transformations, $\Lambda \in SO^+(1,3)$, and we will deal with parity and time-reversal transformations separately. As we saw in Eq. (2.5.28) this means that we must have at the operator level

$$\hat{\phi}_r(x) \to \hat{\phi}_r'(x') = \hat{U}_p(\Lambda, a) \hat{\phi}_r(x') \hat{U}_p(\Lambda, a)^{\dagger} = S_{rs}(\Lambda) \hat{\phi}_s(x), \qquad (6.2.109)$$

where $\hat{U}_p(\Lambda, a)$ is the appropriate unitary operator for a passive restricted Poincaré transformation, where $\Lambda \in SO^+(1,3)$ and a is any spacetime translation. The state transforms with the active unitary transformation $|\psi'\rangle = \hat{U}(\Lambda, a)|\psi\rangle$, where $\hat{U}(\Lambda, a) \equiv \hat{U}_p(\Lambda, a)^{\dagger}$. We can rearrange to write the above as

$$\hat{U}(\Lambda, a) \hat{\phi}_r(x) \hat{U}(\Lambda, a)^{\dagger} = S_{rs}^{-1}(\Lambda) \hat{\phi}_s(x') = S_{rs}^{-1}(\Lambda) \hat{\phi}_s(\Lambda x + a). \qquad (6.2.110)$$

We now restrict ourselves to the consideration of scalar fields, i.e., spinless fields. For a scalar field this reduces to

$$\hat{U}(\Lambda, a) \hat{\phi}(x) \hat{U}(\Lambda, a)^{\dagger} = \hat{\phi}(\Lambda x + a). \qquad (6.2.111)$$

Any Poincaré transformation can be written as a Lorentz transformation (Λ) followed by a translation (a) as in Eq. (1.2.11). So we can similarly write the unitary representation of any restricted Poincaré transformation as a Lorentz transformation $\hat{U}(\Lambda)$ followed by a translation $\hat{U}(a)$,

$$\hat{U}(a, \Lambda) = \hat{U}(a)\hat{U}(\Lambda). \qquad (6.2.112)$$

As the operator acts from the left on states (i.e., "kets" $|\psi\rangle$) here in $\hat{U}(a, \Lambda)$ we are doing the Lorentz transformation first and then the translation.

A unitary realization of the translations and the restricted Lorentz transformations can be specified by the 10 parameters a^μ and $\omega^{\mu\nu} = -\omega^{\nu\mu}$ and 10 Hermitian generators $\hat{P}^\mu$ and $\hat{M}^{\mu\nu} = -\hat{M}^{\nu\mu}$, where

$$\hat{U}(a) = e^{i\hat{P}\cdot a} \quad \text{and} \quad \hat{U}(\Lambda) = e^{-(i/2)\omega_{\mu\nu}\hat{M}^{\mu\nu}}. \tag{6.2.113}$$

Ignore for now that we already know the form of P^μ and let us rediscover it. For infinitesimal transformations we have to first order in da_μ and $d\omega_{\mu\nu}$,

$$\hat{U}(da, d\Lambda) = \hat{I} - (i/2)d\omega_{\mu\nu}\hat{M}^{\mu\nu} + ida_\mu\hat{P}^\mu. \tag{6.2.114}$$

Using Eqs. (1.2.174) and (1.2.175) we can write

$$\hat{\phi}(d\Lambda x + da) = (1 + da_\mu\partial^\mu + d\omega_{\mu\nu}x^\nu\partial^\mu)\hat{\phi}(x). \tag{6.2.115}$$

From Eq. (6.2.111) we have (Bjorken and Drell, 1964; Itzykson and Zuber, 1980)

$$i[da_\mu\hat{P}^\mu - \tfrac{1}{2}d\omega_{\mu\nu}\hat{M}^{\mu\nu}, \hat{\phi}(x)] = (da_\mu\partial^\mu - \tfrac{1}{2}d\omega_{\mu\nu}[x^\mu\partial^\nu - x^\nu\partial^\mu])\hat{\phi}(x), \tag{6.2.116}$$

which gives

$$i[\hat{P}^\mu, \hat{\phi}(x)] = \partial^\mu\hat{\phi}(x) \quad \text{and} \quad i[\hat{M}^{\mu\nu}, \hat{\phi}(x)] = (x^\mu\partial^\nu - x^\nu\partial^\mu)\hat{\phi}(x). \tag{6.2.117}$$

The combined actions of $\hat{U}(\Lambda)$ and $\hat{U}(a)$ on the field operator $\hat{\phi}$ have to reproduce all of the group properties of the restricted Poincaré transformations. Comparing with the discussion in Sec. 1.2.5, we identify $\ell(\Lambda) \to \hat{U}(\Lambda)$ and $t(a) \to \hat{U}(a)$ and so we see that we must have the properties that $\hat{U}(\Lambda)\hat{U}(a)U(\Lambda^{-1}) = \hat{U}(\Lambda a)$ from Eq. (1.2.185) and $\hat{U}(\Lambda)\hat{U}(\Lambda')U(\Lambda^{-1}) = \hat{U}(\Lambda'')$ from Eq. (1.2.196). Then all of the arguments in Sec. 1.2.5 will apply and so the operators $\hat{P}^\mu$ and $\hat{M}^{\mu\nu}$ are the generators of a unitary representation of the Poincaré group and satisfy the Poincaré Lie algebra of Eq. (1.2.167),

$$[\hat{M}^{\mu\nu}, \hat{M}^{\rho\sigma}] = i(g^{\nu\rho}\hat{M}^{\mu\sigma} - g^{\mu\rho}\hat{M}^{\nu\sigma} - g^{\nu\sigma}\hat{M}^{\mu\rho} + g^{\mu\sigma}\hat{M}^{\nu\rho}),$$
$$[\hat{P}^\mu, \hat{M}^{\rho\sigma}] = i(g^{\mu\rho}\hat{P}^\sigma - g^{\mu\sigma}\hat{P}^\rho), \qquad [\hat{P}^\mu, \hat{P}^\nu] = 0. \tag{6.2.118}$$

As the symbols chosen for them suggest, the appropriate forms of the 10 Poincaré generators are the operator equivalents of the four-momentum operator P^μ in Eq. (3.2.54) and the angular momentum tensor $M^{\mu\nu}$ in Eq. (3.2.74).

Now generalize to a set of N scalar field operators $\hat{\phi}_1, \ldots, \hat{\phi}_N$ in analogy with Sec. 3.2.1. From Eq. (3.2.53) the operator form of the stress-energy tensor is

$$\hat{T}^{\mu\nu}(x) = \hat{\vec{\pi}}^\mu \cdot \partial^\nu\hat{\vec{\phi}} - g^{\mu\nu}\hat{\mathcal{L}}. \tag{6.2.119}$$

From Eqs. (3.2.54) and (3.2.74) we have the four-momentum operator and angular momentum tensor operator, respectively, given by

$$\hat{P}^\mu = \int d^3x\, \hat{T}^{0\mu}(x) = (\hat{H}, \hat{\mathbf{P}}) \text{ with } \hat{H} = \int d^3x\, [\hat{\vec{\pi}} \cdot \partial^0\hat{\vec{\phi}} - \hat{\mathcal{L}}], \quad \hat{P}^i = \int d^3x\, \hat{\vec{\pi}} \cdot \partial^i\hat{\vec{\phi}},$$
$$\hat{M}^{\mu\nu} = \int d^3x\, [x^\mu\hat{T}^{0\nu}(x) - x^\nu\hat{T}^{0\mu}(x)], \tag{6.2.120}$$

where we recall that $\hat{\vec{\pi}} = \hat{\vec{\pi}}^0 = \partial\hat{\mathcal{L}}/\partial(\partial^0\hat{\vec{\phi}})$. Note that $\hat{P}^\mu$ and $\hat{M}^{\mu\nu}$ are time-independent if the Lagrangian is Poincaré invariant.

We have the result that

$$i[\hat{T}^{0\mu}(x'), \hat{\phi}_j(x)]\Big|_{x'^0=x^0} = \delta^3(\mathbf{x}' - \mathbf{x})\partial^\mu \hat{\phi}_j(x). \tag{6.2.121}$$

Proof: Using $T^{00} = \mathcal{H} = \hat{\vec{\pi}} \cdot \partial^0 \hat{\vec{\phi}} - \hat{\mathcal{L}}$ together with Eqs. (6.2.83) and (6.2.85) and using $T^{0i} = \hat{\vec{\pi}} \cdot \partial^i \hat{\vec{\phi}}$ we find that

$$i[\hat{T}^{00}(t, \mathbf{x}'), \hat{\phi}_j(t, \mathbf{x})] = i[\hat{\mathcal{H}}(t, \mathbf{x}'), \hat{\phi}_j(t, \mathbf{x})] = \delta\hat{\mathcal{H}}(t, \mathbf{x}')/\delta\hat{\pi}_j(t, \mathbf{x})$$
$$= \delta^3(\mathbf{x}' - \mathbf{x})\partial^0 \hat{\phi}_j(t, \mathbf{x}), \tag{6.2.122}$$

$$i[\hat{T}^{0i}(t, \mathbf{x}'), \hat{\phi}_j(t, \mathbf{x})] = i[\hat{\vec{\pi}}(t, \mathbf{x}'), \hat{\phi}_j(t, \mathbf{x})] \cdot \partial^i \hat{\vec{\phi}}(t, \mathbf{x}') = \delta^3(\mathbf{x}' - \mathbf{x})\partial^i \hat{\phi}_j(t, \mathbf{x}).$$

Using this we recover Eq. (6.2.117) in the form

$$i[\hat{P}^\mu, \hat{\phi}_j(x)] = \int d^3x'\ i[T^{0\mu}(x'), \hat{\phi}_j(x)]\Big|_{x'^0=x^0} = \partial^\mu \hat{\phi}_j(x),$$

$$i[\hat{M}^{\mu\nu}, \hat{\phi}_j(x)] = \int d^3x'\ (x'^\mu i[\hat{T}^{0\nu}(x'), \hat{\phi}_j(x)] - x'^\nu i[\hat{T}^{0\mu}(x'), \hat{\phi}_j(x)])\Big|_{x'^0=x^0}$$

$$= (x^\mu \partial^\nu - x^\nu \partial^\mu)\hat{\phi}_j(x). \tag{6.2.123}$$

So we have proven that the 10 generators of the Poincaré Lie algebra are indeed the four-momentum operator $\hat{P}^\mu$ and the angular momentum tensor operators $\hat{M}^{\mu\nu}$.

Generalizing to the case for a particle with spin, i.e., for a field with intrinsic angular momentum, is straightforward. From Eq. (3.2.89) we must have

$$\hat{\phi}_r(d\Lambda x + da) = (1 + \delta_{rs}da_\mu \partial^\mu + \tfrac{1}{2}d\omega_{\mu\nu}[\delta_{rs}(x^\nu \partial^\mu - x^\mu \partial^\nu) + \Sigma_{rs}^{\mu\nu}])\hat{\phi}_s(x),$$

which replaces Eq. (6.2.117) with

$$i[\hat{P}^\mu, \hat{\phi}_r(x)] = \partial^\mu \hat{\phi}_r(x), \quad i[\hat{M}^{\mu\nu}, \hat{\phi}_r(x)] = [\delta_{rs}(x^\mu \partial^\nu - x^\nu \partial^\mu) + \Sigma_{rs}^{\mu\nu}]\hat{\phi}_s(x). \tag{6.2.124}$$

Again by construction $\hat{P}^\mu$ and $\hat{M}^{\mu\nu}$ are the generators of the Poincaré transformations and satisfy the Poincaré Lie algebra. We can easily generalize Eq. (6.2.123) to the case with spin and so verify that the operator versions of the classical field theory quantities P^μ and $M^{\mu\nu}$ satisfy Eq. (6.2.124). With the addition of spin we again find that the operator versions of the classical field theory quantities are the 10 Poincaré generators, where $\hat{T}^{\mu\nu}$ and P^μ are as given above and where

$$\hat{M}^{\mu\nu} = \int d^3x\ [(x^\mu \hat{T}^{0\nu} - x^\nu \hat{T}^{0\mu}) + \hat{\pi}_r(\Sigma_{rs})^{\mu\nu}\hat{\phi}_s] \equiv \hat{L}^{\mu\nu} + \hat{\Sigma}^{\mu\nu}. \tag{6.2.125}$$

The above arguments are the quantum field theory equivalents of Eqs. (2.5.35) and (2.5.36) for the case of the unitary representation of the Poincaré Lie group.

So far in this section we have ignored normal-ordering and the above has been proven for the operators with no normal-ordering. As we show below, the normal-ordered forms of P^μ and $M^{\mu\nu}$ differ from their standard forms by (infinite) real constants. Changing an operator by a constant does not affect any of its commutation relations, $[\hat{A} + c, \hat{B} + d] = [\hat{A}, \hat{B}]$ for $c, d \in \mathbb{C}$. Shifting a Hermitian generator by a real constant has no effect on its unitary transformations, i.e., $\hat{A}^\dagger = \hat{A}$ for a unitary transformation and $\exp(i\{\hat{A} + c\})\hat{B}\exp(-i\{\hat{A} + c\}) = \exp(i\hat{A})\hat{B}\exp(-i\hat{A})$. So we can now *replace the 10 Poincaré generators P^μ and $M^{\mu\nu}$ with their normal-ordered forms* and all of the above properties remain true.

Proof: Consider first a single scalar field. Recalling that $\hat{\pi} = \hat{\pi}^0$, $\hat{T}^{00} = \hat{\mathcal{H}}$ and $[\hat{a}_{\mathbf{p}}, \hat{a}^\dagger_{\mathbf{p}'}] = (2\pi)^3 \delta^3(\mathbf{p} - \mathbf{p}')$, we can then read off from the second line of Eq. (6.2.21) that at $t = 0$ we have

$$\hat{\mathcal{H}} = \hat{T}^{00} = \hat{\pi}\,\partial^0\hat{\phi} - \mathcal{L} = \tfrac{1}{2}\hat{\pi}^2 + \tfrac{1}{2}(\boldsymbol{\nabla}\hat{\phi})^2 + \tfrac{1}{2}m^2\hat{\phi}^2$$

$$= \,:\!\hat{\mathcal{H}}\!:\, + \int \frac{d^3p\,d^3p'}{(2\pi)^6}\, e^{i(\mathbf{p}+\mathbf{p}')\cdot\mathbf{x}}\, \frac{1}{2}\left\{ \frac{\sqrt{E_{\mathbf{p}}E_{\mathbf{p}'}}}{2} + \frac{-\mathbf{p}\cdot\mathbf{p}' + m^2}{2\sqrt{E_{\mathbf{p}}E_{\mathbf{p}'}}} \right\}[\hat{a}_{\mathbf{p}}, \hat{a}^\dagger_{-\mathbf{p}'}]$$

$$= \,:\!\hat{\mathcal{H}}\!:\, + \int \frac{d^3p}{(2\pi)^3}\, \frac{1}{2}\left\{ \frac{E_{\mathbf{p}}}{2} + \frac{\mathbf{p}^2 + m^2}{2E_{\mathbf{p}}} \right\} = \,:\!\hat{\mathcal{H}}\!:\, + \int \frac{d^3p}{(2\pi)^3}\, \tfrac{1}{2}E_{\mathbf{p}}, \qquad (6.2.126)$$

where in the finite-volume case using Eq. (6.2.26) we find that the last term becomes as expected $(1/V)\sum_i \tfrac{1}{2}E_{\mathbf{p}_i}$. Similarly, using Eq. (6.2.34) at $t = 0$,

$$\hat{T}^{0i} = \mathbf{p}_{\text{dens}} = -\hat{\pi}\,\partial_i\hat{\phi} = \,:\!\hat{T}^{0i}\!:\, - \int \frac{d^3p\,d^3p'}{(2\pi)^6}\, e^{i(\mathbf{p}+\mathbf{p}')\cdot\mathbf{x}}\, \frac{1}{2}p'^i \sqrt{\frac{E_{\mathbf{p}}}{E_{\mathbf{p}'}}}[\hat{a}_{\mathbf{p}}, \hat{a}^\dagger_{-\mathbf{p}'}]$$

$$= \,:\!\hat{T}^{0i}\!:\, + \int \frac{d^3p}{(2\pi)^3}\, \tfrac{1}{2}p^i = \,:\!\hat{T}^{0i}\!:\,, \qquad (6.2.127)$$

where the last step follows since the integrand is odd in p^i. Generalizing to multiple scalar fields is straightforward. We then use these $:\!T^{0\mu}\!:$ to construct the normal-ordered $\hat{P}^\mu$ and $\hat{M}^{\mu\nu}$ for scalar fields. We see that they differ from the non-ordered fields only up to a constant, which while infinite can be regularized to a finite quantity with infrared and ultraviolet cutoffs.

In the case of a field with intrinsic spin we need only consider the additional term in Eq. (6.2.125). The $(\Sigma_{rs})^{\mu\nu}$ are constants and we easily see that

$$\hat{\pi}_r\hat{\phi}_s = \,:\!\hat{\pi}_r\hat{\phi}_s\!:\, + (-i)\int \frac{d^3p}{(2\pi)^3}. \qquad (6.2.128)$$

We then see that the same result follows for a field with intrinsic spin.

From Eq. (3.2.54) we see that the normal-ordered Poincaré generators are

$$\hat{P}^\mu = \int d^3x\, :\!\hat{T}^{0\mu}\!:\, \quad\text{and}\quad \hat{M}^{\mu\nu} = \int d^3x\, (x^\mu :\!\hat{T}^{0\nu}\!:\, - x^\nu :\!\hat{T}^{0\mu}\!:\,), \qquad (6.2.129)$$

where from this point on we will always assume that the generators are normal-ordered. The 10 normal-ordered Hermitian generators annihilate the vacuum,

$$\hat{P}^\mu|0\rangle = 0 \quad\text{and}\quad \hat{M}^{\mu\nu}|0\rangle = 0, \qquad (6.2.130)$$

which means that the vacuum is Poincaré invariant as it should be,

$$\hat{U}(\Lambda, a)|0\rangle = |0\rangle. \qquad (6.2.131)$$

Consider the one-boson momentum eigenstate state $|\mathbf{p}\rangle$ defined in Eq. (6.2.46). With the Lorentz-invariant normalization chosen for $|\mathbf{p}\rangle$ we have

$$|\mathbf{p}\rangle \to |\mathbf{p}'\rangle = \hat{U}(\Lambda)|\mathbf{p}\rangle \equiv |\Lambda\mathbf{p}\rangle, \qquad (6.2.132)$$

where we use the shorthand $p \to p' = \Lambda p$ (i.e., $p^\mu \to p'^\mu = \Lambda^\mu{}_\nu p^\nu$) and where the three-vector part of p' is written as $\mathbf{p}'$. We sometimes also make use of the convenient shorthand notation $\mathbf{p}' \equiv \Lambda\mathbf{p}$. From Eq. (6.2.46) we see that

$$\sqrt{E_{\Lambda\mathbf{p}}}\,\hat{a}^\dagger_{\Lambda\mathbf{p}}|0\rangle = \hat{U}(\Lambda)\sqrt{E_{\mathbf{p}}}\,\hat{a}^\dagger_{\mathbf{p}}|0\rangle = \hat{U}(\Lambda)\sqrt{E_{\mathbf{p}}}\,\hat{a}^\dagger_{\mathbf{p}}\hat{U}^{-1}(\Lambda)|0\rangle, \qquad (6.2.133)$$

which suggests that

$$\hat{U}(\Lambda)\sqrt{E_{\mathbf{p}}}\hat{a}_{\mathbf{p}}^{\dagger}\hat{U}^{-1}(\Lambda) = \sqrt{E_{\Lambda\mathbf{p}}}\,\hat{a}_{\Lambda\mathbf{p}}^{\dagger}. \tag{6.2.134}$$

Taking the Hermitian conjugate of this leads immediately to

$$\hat{U}(\Lambda)\sqrt{E_{\mathbf{p}}}\hat{a}_{\mathbf{p}}\hat{U}^{-1}(\Lambda) = \sqrt{E_{\Lambda\mathbf{p}}}\,\hat{a}_{\Lambda\mathbf{p}}. \tag{6.2.135}$$

We can also deduce these two results from Eq. (6.2.111) and the expression for the field operator in Eq. (6.2.97). Rewriting the latter as

$$\hat{\phi}(x) = \int \frac{d^3p}{(2\pi)^3}\frac{1}{2E_{\mathbf{p}}}\left(\sqrt{2E_{\mathbf{p}}}\hat{a}_{\mathbf{p}}e^{-ip\cdot x} + \sqrt{2E_{\mathbf{p}}}\hat{a}_{\mathbf{p}}^{\dagger}e^{ip\cdot x}\right)\Big|_{p^0=E_{\mathbf{p}}}, \tag{6.2.136}$$

we recognize from Eq. (6.2.50) that $\int d^3p/2E_{\mathbf{p}}$ is a Lorentz-invariant measure, that $\exp(\pm ip\cdot x)$ is Lorentz invariant, and that all Fourier components are linearly independent. Then $\sqrt{E_{\mathbf{p}}}\hat{a}_{\mathbf{p}}^{\dagger}$ and $\sqrt{E_{\mathbf{p}}}\hat{a}_{\mathbf{p}}$ must transform as above. Note that from Eq. (6.2.58) it follows that

$$\langle 0|\phi(x)|\mathbf{p}\rangle = e^{-ip\cdot x}\Big|_{p^0=\sqrt{\mathbf{p}^2+m^2}}. \tag{6.2.137}$$

6.2.5 Causality and Spacelike Separations

While we typically follow the notation of Peskin and Schroeder (1995) here we also incorporate notation from Bjorken and Drell (1965). Define $\Delta_+(x-y) \equiv D(x-y)$ as the vacuum expectation value of the product of two Heisenberg picture field operators at spacetime points x and y using

$$\begin{aligned}
\Delta_+(x-y) &\equiv D(x-y) \equiv \langle 0|\hat{\phi}(x)\hat{\phi}(y)|0\rangle \\
&= \int \frac{d^3p\,d^3q}{(2\pi)^6}\frac{1}{2\sqrt{E_{\mathbf{p}}E_{\mathbf{q}}}}\langle 0|\left(\hat{a}_{\mathbf{p}}e^{-ip\cdot x} + \hat{a}_{\mathbf{p}}^{\dagger}e^{ip\cdot x}\right)\Big|_{p^0=E_{\mathbf{p}}}\left(\hat{a}_{\mathbf{q}}e^{-iq\cdot y} + \hat{a}_{\mathbf{q}}^{\dagger}e^{iq\cdot y}\right)\Big|_{q^0=E_{\mathbf{q}}}|0\rangle \\
&= \int \frac{d^3p\,d^3q}{(2\pi)^6}\frac{1}{2\sqrt{E_{\mathbf{p}}E_{\mathbf{q}}}}\,e^{-i(p\cdot x-q\cdot y)}\Big|_{p^0=E_{\mathbf{p}},q^0=E_{\mathbf{q}}}\langle 0|\hat{a}_{\mathbf{p}}\hat{a}_{\mathbf{q}}^{\dagger}|0\rangle \\
&= \int \frac{d^3p}{(2\pi)^3}\frac{1}{2E_{\mathbf{p}}}\,e^{-ip\cdot(x-y)}\Big|_{p^0=E_{\mathbf{p}}},
\end{aligned} \tag{6.2.138}$$

which is Lorentz invariant up to time reversal as seen in Eq. (6.2.53). Assuming translation invariance for the system, there can only be dependence on $(x-y)$. We can similarly define

$$\Delta_-(x-y) \equiv \langle 0|\hat{\phi}(y)\hat{\phi}(x)|0\rangle = \Delta_+(y-x) = \int \frac{d^3p}{(2\pi)^3}\frac{1}{2E_{\mathbf{p}}}\,e^{+ip\cdot(x-y)}\Big|_{p^0=E_{\mathbf{p}}},$$

where the exponential factors suggest $\Delta_+(x-y)$ and $\Delta_-(x-y)$ as being associated with positive and negative energy components, respectively. Next, consider the commutator of two Heisenberg picture field operators,

$$\begin{aligned}
i\Delta(x-y) &\equiv [\hat{\phi}(x),\hat{\phi}(y)] \\
&= \int \frac{d^3p\,d^3q}{(2\pi)^6}\frac{1}{2(E_{\mathbf{p}}E_{\mathbf{q}})^{1/2}}\left[\left(\hat{a}_{\mathbf{p}}e^{-ip\cdot x} + \hat{a}_{\mathbf{p}}^{\dagger}e^{ip\cdot x}\right)\Big|_{p^0=E_{\mathbf{p}}},\left(\hat{a}_{\mathbf{q}}e^{-iq\cdot y} + \hat{a}_{\mathbf{q}}^{\dagger}e^{iq\cdot y}\right)\Big|_{q^0=E_{\mathbf{q}}}\right] \\
&= \int \frac{d^3p}{(2\pi)^3}\frac{1}{2E_{\mathbf{p}}}\left(e^{-ip\cdot(x-y)} - e^{ip\cdot(x-y)}\right)\Big|_{p^0=E_{\mathbf{p}}} = \Delta_+(x-y) - \Delta_+(y-x),
\end{aligned} \tag{6.2.139}$$

where we used the commutators in Eq. (6.2.17). Then $\Delta(x - y)$ is an odd function

$$\Delta(x - y) = -\Delta(y - x). \tag{6.2.140}$$

We can use Eq. (6.2.53) to prove that

$$i\Delta(x - y) = \int \frac{d^4p}{(2\pi)^4} 2\pi\delta(p^2 - m^2)\epsilon(p^0)e^{-ip\cdot(x-y)}, \tag{6.2.141}$$

where we have defined the odd function

$$\epsilon(p^0) \equiv \theta(p^0) - \theta(-p^0) = \begin{cases} +1 & p^0 > 0 \\ 0 & p^0 = 0 \\ -1 & p^0 < 0 \end{cases}, \tag{6.2.142}$$

where θ is the step function, i.e., $\theta(x) = 0, \frac{1}{2}, 1$ for $x < 0, x = 0, x > 0$, respectively. To see this, start with Eq. (6.2.141) and in the $p^0 < 0$ contribution change variables $p^\mu \to -p'^\mu$ and use Eq. (6.2.53) to arrive at Eq. (6.2.139).

The last line of Eq. (6.2.139) shows that the commutator $[\hat{\phi}(x), \hat{\phi}(y)]$ is a complex number and not an operator and so

$$i\Delta(x - y) = [\hat{\phi}(x), \hat{\phi}(y)] = \langle 0|[\hat{\phi}(x), \hat{\phi}(y)]|0\rangle = \Delta_+(x - y) - \Delta_+(y - x). \tag{6.2.143}$$

Consider the case where $(x-y)^\mu$ is spacelike, $(x-y)^2 < 0$. Consider the generalization of Fig. 1.2 to higher spatial dimensions with y^μ at the origin and x^μ on some spacelike hyperboloid. In two spatial dimensions we generalize this figure by recognizing that it is symmetric under rotations about the vertical time axis with an analogous behavior in three spatial dimensions that is difficult to visualize. First we perform a boost so that $x^0 - y^0 = \Delta t \to x^0 - y^0 = 0$, then we do a parity transformation in that frame so that $(0, \mathbf{x} - \mathbf{y}) \to -(0, \mathbf{x} - \mathbf{y})$ and then we do another equal boost so that $x^0 = y^0 = 0 \to x^0 - y^0 = -\Delta t$. This process will result in $(x - y)^\mu \to -(x - y)^\mu$ and did not involve time reversal. This immediately tells us that for a spacelike separation $\Delta_+(x - y) = \Delta_+(y - x)$, since $\Delta_+(x - y)$ is Lorentz invariant up to time-reversal transformations. This generalizes the result of the vanishing equal-time commutator in Eq. (6.2.81) to all spacelike separations,

$$i\Delta(x - y) \equiv [\hat{\phi}(x), \hat{\phi}(y)] = 0 \quad \text{for all} \quad (x - y)^2 < 0. \tag{6.2.144}$$

So the field operator commutes with itself whenever the spacetime points have a spacelike separation. This is the condition of causality in that the measurement of the field at point x cannot affect the measurement of the field at point y for a spacelike separation; i.e., a light signal has not had time to travel from one spacetime point to the other and so the measurements must be independent of each other. This is consistent with Fig. 3.1, where we specified an initial and final classical field configuration on spacelike hypesurfaces σ_i and σ_f, respectively.

For a free scalar field $\hat{\pi}(x) = \partial_0\hat{\phi}(x)$ and so

$$[\hat{\phi}(x), \hat{\pi}(y)]_{x^0=y^0} = [\hat{\phi}(x), \dot{\hat{\phi}}(y)]_{x^0=y^0} = (\partial/\partial y^0)i\Delta(x - y)\big|_{x^0=y^0} = i\delta^3(\mathbf{x} - \mathbf{y}),$$

$$[\hat{\pi}(x), \hat{\pi}(y)]_{x^0=y^0} = (\partial^2/\partial x^0 \partial y^0)[\hat{\phi}(x), \hat{\phi}(y)]\big|_{x^0=y^0} = 0, \tag{6.2.145}$$

where we have used the fact that we can rewrite Eq. (6.2.139) as

$$[\hat{\phi}(x), \hat{\phi}(y)] = i\Delta(x - y) = -i\int \frac{d^3p}{(2\pi)^3} \frac{1}{E_{\mathbf{p}}} e^{i\mathbf{p}\cdot(\mathbf{x}-\mathbf{y})} \sin[E_{\mathbf{p}}(x^0 - y^0)]. \tag{6.2.146}$$

So we see that we recover the ETCR of Eq. (6.2.4).

For timelike $(x - y)^\mu$ we have $(x - y)^2 > 0$ and referring again to Fig. 1.2 with y^μ at the origin we can at least picture the two spatial dimensional cases with cylindrical symmetry about the vertical time axis; we see that there is no way to move from a point on the future/past timelike hyperboloid to a point on the past/future timelike hyperboloid without including a time-reversal transformation. So we have $\Delta_+(x - y) \neq \Delta_+(y - x)$ in that case and using Eq. (6.2.143) we see that

$$i\Delta(x - y) = [\hat{\phi}(x), \hat{\phi}(y)] \neq 0 \quad \text{for} \quad (x - y)^2 > 0. \tag{6.2.147}$$

To illustrate this point, note that for any timelike $(x - y)^2 > 0$ we can boost to a frame where $\mathbf{x} - \mathbf{y} = 0$, which after defining $x^0 - y^0 \equiv t$ gives in that frame

$$\Delta_+(x - y) = \int \frac{d^3 p}{(2\pi)^3} \frac{1}{2E_\mathbf{p}} e^{-iE_\mathbf{p}t} = \int_0^\infty \frac{dp \, 4\pi p^2}{(2\pi)^3} \frac{1}{2\sqrt{p^2 + m^2}} e^{-i\sqrt{p^2+m^2}t}$$
$$= \int_{m^2}^\infty \frac{4\pi E \, dE}{(2\pi)^3} \frac{\sqrt{E^2 - m^2}}{2E} e^{-iEt} = \frac{1}{4\pi^2} \int_{m^2}^\infty dE \sqrt{E^2 - m^2} e^{-iEt}. \tag{6.2.148}$$

The general form of the odd function $\Delta(x - y)$ is given by (Bjorken and Drell, 1965)

$$\Delta(x - y) = \frac{1}{4\pi r} \frac{\partial}{\partial r} \begin{cases} J_0(m\sqrt{t^2 - r^2}), & t \geq r \\ 0, & -r < t < r \\ -J_0(m\sqrt{t^2 - r^2}), & t \leq r \end{cases}, \tag{6.2.149}$$

where $r \equiv |\mathbf{x} - \mathbf{y}|$, J_0 is the regular Bessel function and $r = t$ is the light cone.

6.2.6 Feynman Propagator for Scalar Particles

We will soon see that the time-ordered products of operators play a central role in quantum field theory. The *time-ordering operator* for bosonic field operators is

$$T\hat{A}(x)\hat{B}(y) \equiv \theta(x^0 - y^0)\hat{A}(x)\hat{B}(y) + \theta(y^0 - x^0)\hat{B}(y)\hat{A}(x)$$
$$= \begin{cases} \hat{A}(x)\hat{B}(y), & x^0 > y^0 \\ \frac{1}{2}[\hat{A}(x)\hat{B}(y) + \hat{B}(y)\hat{A}(x)], & x^0 = y^0 \\ \hat{B}(y)\hat{A}(x), & x^0 < y^0 \end{cases}. \tag{6.2.150}$$

Time ordering is not an operator in the mathematical sense, but is a shorthand notation in the same way that normal-ordering is. Referring to Fig. 1.2 we see that the time-ordering of two spacelike separated events is not Lorentz invariant, whereas for timelike separated events it is Lorentz invariant up to time-reversal transformations. This means that in order for time-ordering to be a useful concept in a relativistic context the two bosonic field operators $\hat{A}(x)$ and $\hat{B}(y)$ must commute at all spacelike separations so that the ordering for spacelike separations is irrelevant. For this reason it is common not to include the $x^0 = y^0$ possibility in the above definition. So in that case time-ordering is Lorentz invariant up to time-reversal transformations.

We define the *Feynman propagator* for a scalar field operator $\hat{\phi}(x)$ as the time-ordered product of two scalar field operators,

$$D_F(x - y) \equiv i\Delta_F^{\text{BD}}(x - y) \equiv \langle 0|T\hat{\phi}(x)\hat{\phi}(y)|0\rangle = \begin{cases} \Delta_+(x - y), & x^0 > y^0 \\ \Delta_+(y - x), & x^0 < y^0 \end{cases}, \tag{6.2.151}$$

where we have made use of the definition of $\Delta_+(x - y)$ in Eq. (6.2.138). We use the notation D_F as used by Peskin and Schroeder (1995) and Schwartz (2013) and for completeness we also show the notation used in Bjorken and Drell (1965). The Feynman propagator is an even function,

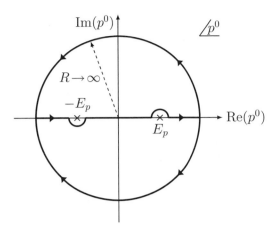

Figure 6.1 Application of Cauchy's integral theorem in the complex p^0 plane relevant for the Feynman propagator. The contour deformations are equivalent to the pole shifts $p^0 = \pm E_{\mathbf{p}} \to \pm(E_{\mathbf{p}} - i\epsilon)$ where $E_{\mathbf{p}} = \sqrt{\mathbf{p}^2 + m^2}$.

$$D_F(x - y) = D_F(y - x) \quad \text{since} \quad T\hat{\phi}(x)\hat{\phi}(y) = T\hat{\phi}(y)\hat{\phi}(x). \tag{6.2.152}$$

As we show below the Feynman propagator can be written as

$$D_F(x - y) = \int \frac{d^4p}{(2\pi)^4} \frac{i}{p^2 - m^2 + i\epsilon} e^{-ip\cdot(x-y)}, \tag{6.2.153}$$

where the presence of the infinitesimal imaginary quantity $i\epsilon$ in the denominator of the integrand tells us how to shift the poles on the real p^0 axis so as to avoid singularities in the integral. We can view an infinitesimal deformation of the poles equally validly as an infinitesimal deformation of the p^0 integration around the poles on the real axis. This deformation of the contour leads to the Feynman propagator and is illustrated in Fig. 6.1. This deformation/pole shift is variously referred to as the *Feynman prescription*, the *Feynman "$i\epsilon$" prescription* or the *Feynman boundary conditions*. The latter expression comes from the fact that the Feynman propagator gives the result that particles propagate forward in time and antiparticles propagate backward in time as we discuss below.

Referring to the discussion of Fourier transforms in Sec. A.2.3 and Eq. (A.2.9) in particular, we recognize that $(i/p^2 - m^2 + i\epsilon)$ is the asymmetric four-dimensional Fourier transform of $D_F(x)$,

$$D_F(p) \equiv \frac{i}{p^2 - m^2 + i\epsilon} = \int d^4x \, D_F(x) e^{+ip\cdot x}, \tag{6.2.154}$$

where the inverse Fourier transform is

$$D_F(x) = \int \frac{d^4p}{(2\pi)^4} D_F(p) e^{-ip\cdot x} = \int \frac{d^4p}{(2\pi)^4} \frac{i}{p^2 - m^2 + i\epsilon} e^{-ip\cdot x}. \tag{6.2.155}$$

Rather than use a notation like $\tilde{D}_F(p)$ for the Fourier transform we will just write $D_F(p)$ and understand what we mean from the argument of the function, p or x.

Proof: We wish to prove Eq. (6.2.153). We begin by recalling that $E_{\mathbf{p}} = \sqrt{\mathbf{p}^2 + m^2} > 0$ and noting that

$$(p^0 + E_{\mathbf{p}} - i\epsilon)(p^0 - E_{\mathbf{p}} + i\epsilon) = (p^0)^2 - (E_{\mathbf{p}} - i\epsilon)^2 = (p^0)^2 - E_{\mathbf{p}}^2 + i2\epsilon E_{\mathbf{p}}$$
$$= p^2 - m^2 + i\epsilon, \tag{6.2.156}$$

where we do not distinguish the positive infinitesimals ϵ and $2\epsilon E_{\mathbf{p}}$. This gives

$$\int \frac{d^4p}{(2\pi)^4} \frac{i}{p^2 - m^2 + i\epsilon} e^{-ip\cdot(x-y)}$$
$$= \int \frac{dp^0 d^3p}{(2\pi)^4} \frac{i}{(p^0 + E_{\mathbf{p}} - i\epsilon)(p^0 - E_{\mathbf{p}} + i\epsilon)} e^{-ip^0(x^0-y^0)+i\mathbf{p}\cdot(\mathbf{x}-\mathbf{y})}. \qquad (6.2.157)$$

We can apply Cauchy's integral theorem in Eq. (A.2.17) to evaluate the p^0 integral. To evaluate an integral along the real p^0 axis we choose the closed contour in the complex p^0 plane such that the semicircular section of the contour gives no contribution. Any point on the circular sections of the contour can be written as $p^0 = Re^{i\phi}$ with $R \to \infty$ and we see that in the upper and lower half-plane these points have imaginary parts given by $\mathrm{Im}(p^0) \to +\infty$ and $\mathrm{Im}(p^0) \to -\infty$, respectively. Then for $x^0 > y^0$ we see that the $\exp[-ip^0(x^0 - y^0)]$ vanishes on the lower semicircle, whereas for $x^0 < y^0$ we see that it vanishes on the upper semicircle.

So if $x^0 > y^0$ then we close the contour below and pick up the residue of the at $p^0 = +E_{\mathbf{p}}$ inside a clockwise contour. Then we find

$$\int \frac{d^4p}{(2\pi)^4} \frac{i}{(p^0 + E_{\mathbf{p}} - i\epsilon)(p^0 - E_{\mathbf{p}} + i\epsilon)} e^{-ip\cdot(x-y)} \bigg|_{x^0 > y^0}$$
$$= (-2\pi i) \int \frac{d^3p}{(2\pi)^4} \frac{i}{2E_{\mathbf{p}}} e^{-iE_{\mathbf{p}}(x^0-y^0)+i\mathbf{p}\cdot(\mathbf{x}-\mathbf{y})}$$
$$= \int \frac{d^3p}{(2\pi)^3} \frac{1}{2E_{\mathbf{p}}} e^{-ip\cdot(x-y)} \bigg|_{p^0 = E_{\mathbf{p}}} = \Delta_+(x - y), \qquad (6.2.158)$$

where the negative sign in $(-2\pi i)$ comes from contour being clockwise. Similarly, for $x^0 < y^0$ we need to close the contour above and pick up the residue of the pole at $p^0 = -E_{\mathbf{p}}$ inside a counterclockwise contour, which gives

$$\int \frac{d^4p}{(2\pi)^4} \frac{i}{(p^0 + E_{\mathbf{p}} - i\epsilon)(p^0 - E_{\mathbf{p}} + i\epsilon)} e^{-ip\cdot(x-y)} \bigg|_{x^0 < y^0}$$
$$= (2\pi i) \int \frac{d^3p}{(2\pi)^4} \frac{i}{(-2E_{\mathbf{p}})} e^{+iE_{\mathbf{p}}(x^0-y^0)+i\mathbf{p}\cdot(\mathbf{x}-\mathbf{y})}$$
$$= \int \frac{d^3p}{(2\pi)^3} \frac{1}{2E_{\mathbf{p}}} e^{-ip\cdot(y-x)} \bigg|_{p^0 = E_{\mathbf{p}}} = \Delta_+(y - x), \qquad (6.2.159)$$

where in the last step we performed a change of integration variable, $\mathbf{p} \to -\mathbf{p}$. We have then arrived at the result that

$$\int \frac{d^4p}{(2\pi)^4} \frac{i\, e^{-ip\cdot(x-y)}}{p^2 - m^2 + i\epsilon} = \theta(x^0 - y^0)\Delta_+(x - y) + \theta(y^0 - x^0)\Delta_+(y - x)$$
$$= D_F(x - y). \qquad (6.2.160)$$

We can also write this result in a general form that will be useful later:

$$\int \frac{dp^0}{2\pi} \frac{i\, e^{-ip\cdot(x-y)}}{p^2 - m^2 + i\epsilon} = \frac{1}{2E_{\mathbf{p}}} \left[\theta(x^0 - y^0)e^{-ip\cdot(x-y)} + \theta(y^0 - x^0)e^{ip\cdot(x-y)} \right]. \qquad (6.2.161)$$

In quantum mechanics we know from the time-dependent Schrödinger equation that an energy eigenstate with energy E propagates in time as e^{-iEt}. Defining $t \equiv x^0 - y^0$ we see that for the Feynman propagator if $t > 0$ then we have from Eq. (6.2.158) that $D_F \propto e^{-iE_{\mathbf{p}}t}$, whereas if $t < 0$

we have from Eq. (6.2.159) that $D_F \propto e^{+iE_\mathbf{p}t} = e^{-iE_\mathbf{p}(-t)}$. In this sense we see that particles are positive energy components that travel forward in time, whereas antiparticles can be thought of as having negative energy and traveling forward in time or as positive energy and traveling backward in time. The Feynman-Stueckelberg interpretation is that the negative energy components should be interpreted as antiparticles, which correspond to particles moving backwards in time. For example, a positron can be thought of as an electron moving backward in time, which has the effect of reversing its spin, momentum and charge.

We can also understand the Feynman prescription as arising from $i\epsilon$ prescription of Eq. (4.1.204) that allows the functional integral in Minkowski space to be defined for quantum field theory. We will discuss this in Sec. 6.2.9. It is the same as the $i\epsilon$ that was introduced in Eq. (4.1.206) to derive the path integral formulation of quantum mechanics. The $i\epsilon$ prescription can be viewed as having its origins in the Euclidean space formulation of quantum theory together with the associated *Wick rotation* back to the Minkowski space. The Wick rotation is an analytic continuation to imaginary time $t \rightarrow -i\tau \equiv -i(\hbar/kT) = -i(\hbar\beta)$ as discussed in Sec. A.4, which turns the trace of the evolution operator in quantum mechanics into the partition function of quantum mechanics as shown in Eq. (A.4.3).

We will later see that an understanding of the behavior of quantum field theories, in particular when using perturbation theory, is based on the propagation of virtual particles as they move from one interaction to the next in the Feynman diagrams that contain the physics of the theory. For the Klein-Gordon case of a scalar particle, the propagator is the inverse of the Klein-Gordon differential operator, $(\partial^2 + m^2)$, subject to the Feynman prescription.

In mathematics a Green's function for some linear differential operator $\hat{L}(x)$ is a distribution $G(x, y)$ such that $\hat{L}(x)G(x, y) = \delta(x - y)$. In general $G(x, y)$ is not unique without the specification of appropriate boundary conditions, but once specified we can think of the Green's function as the inverse of the operator subject to that boundary condition. If the operator $L(x)$ is translationally invariant, then the Green's function will have the form $G(x, y) \rightarrow G(x - y)$. The Klein-Gordon operator $L(x) = (\partial_x^2 + m^2)$ is the d'Alembertian $\Box = \partial_x^2 = \partial_\mu^x \partial_x^\mu$ plus the mass squared. It is a linear second-order differential operator and is translationally invariant. There are four choices for the boundary conditions leading to four corresponding Green's functions: the retarded propagator, the advanced propagator, the Feynman propagator and its anti-time-ordered propagator. We see that the naive quantity

$$D_{\mathrm{naive}}(x - y) = \int \frac{d^4p}{(2\pi)^4} \frac{i}{p^2 - m^2} e^{-ip\cdot(x-y)} \tag{6.2.162}$$

satisfies the equation $(\partial_x^2 + m^2)[iD_{\mathrm{naive}}(x - y)] = \delta(x - y)$, using the representation for the delta function given in Eq. (A.2.4). We identify $G(x, y) = iD(x - y)$. However, since Eq. (6.2.162) involves an integration along the real p^0 axis and since there are poles in the integrand at $p^0 = \pm E_\mathbf{p}$, where $E_\mathbf{p} \equiv \sqrt{\mathbf{p}^2 + m^2}$, then D_{naive} is ambiguous and therefore not defined. There are four ways to deform the contour around the two poles and to use Cauchy's integral formula in Sec. A.2.4 of Appendix A to give a meaning to Eq. (6.2.162). Recall that for $(x^0 - y^0) > 0$ and $(x^0 - y^0) < 0$ the semi-circular contour in the lower and upper half of the complex p^0-plane contributes nothing in the $R \rightarrow \infty$ limit. We can deform the contour above or below the poles or, equivalently, infinitesimally shift the poles below or above the real axis. There are four choices for choosing the p^0 contour to define the boundary conditions for evaluating the naive Green's function in Eq. (6.2.162):

(i) above both poles: $i/[p^2 - m^2] \rightarrow i/[(p^0 + i\epsilon)^2 - \mathbf{p}^2 - m^2]$, which vanishes for $(x^0 - y^0) < 0$ giving $D_{\mathrm{naive}}(x-y) \rightarrow D_{\mathrm{ret}}(x-y)$, where the *retarded propagator* is defined as $D_{\mathrm{ret}}(x-y) \equiv \theta(x^0 - y^0)i\Delta(x - y)$

(ii) below both poles: $i/[p^2 - m^2] \to i/[(p^0 - i\epsilon)^2 - \mathbf{p}^2 - m^2]$, which vanishes for $(x^0 - y^0) > 0$ giving $D_{\text{naive}}(x - y) \to D_{\text{adv}}(x - y)$, where the *advanced propagator* is defined as $D_{\text{adv}}(x - y) \equiv -\theta(y^0 - x^0)i\Delta(x - y)$

(iii) below the negative pole and above the positive pole: $i/[p^2 - m^2] \to i/[p^2 - m^2 + i\epsilon]$, giving $D_{\text{naive}}(x - y) \to D_F(x - y)$, where the *Feynman propagator* is defined by $D_F(x - y) \equiv \langle 0|T\hat\phi(x)\hat\phi(y)|0\rangle = \theta(x^0 - y^0)\Delta_+(x - y) + \theta(y^0 - x^0)\Delta_+(y - x)$

(iv) above the negative pole and below the positive pole: $i/[p^2 - m^2] \to i/[p^2 - m^2 - i\epsilon]$, giving $D_{\text{naive}}(x - y) \to \bar D_F(x - y)$, where the *anti-time-ordered propagator* is $-\bar D_F(x - y) \equiv \langle 0|\bar T\hat\phi(x)\hat\phi(y)|0\rangle = \theta(x^0 - y^0)\Delta_+(y - x) + \theta(y^0 - x^0)\Delta_+(x - y)$.

Using the Cauchy integral formula, Eq. (A.2.17), we can evaluate the Green's function for each case and verify the above results. We note that for the p^0 integration for the retarded propagator we have both poles inside a clockwise contour for $(x^0 - y^0) > 0$ and no poles inside a counterclockwise contour for $(x^0 - y^0) < 0$ to give

$$\int \frac{d^4p}{(2\pi)^4} \frac{i}{(p^0 + i\epsilon)^2 - \mathbf{p}^2 - m^2} e^{-ip\cdot(x-y)} \tag{6.2.163}$$

$$= \int \frac{d^4p}{(2\pi)^4} \frac{i}{(p^0 + E_{\mathbf{p}} + i\epsilon)(p^0 - E_{\mathbf{p}} + i\epsilon)} e^{-ip\cdot(x-y)}$$

$$= \theta(x^0 - y^0)(-i)\int \frac{d^3p}{(2\pi)^3} \left\{ \left.\frac{i}{2E_{\mathbf{p}}} e^{-ip\cdot(x-y)}\right|_{p^0 = E_{\mathbf{p}}} + \left.\frac{i}{-2E_{\mathbf{p}}} e^{-ip\cdot(x-y)}\right|_{p^0 = -E_{\mathbf{p}}} \right\}$$

$$= \theta(x^0 - y^0)\int \frac{d^3p}{(2\pi)^3} \frac{1}{2E_{\mathbf{p}}} \left. \left(e^{-ip\cdot(x-y)} - e^{ip\cdot(x-y)} \right)\right|_{p^0 = E_{\mathbf{p}}}$$

$$= \theta(x^0 - y^0)\langle 0|[\hat\phi(x), \hat\phi(y)]|0\rangle = \theta(x^0 - y^0)i\Delta(x - y) \equiv D_{\text{ret}}(x - y).$$

In a similar calculation for the advanced propagator the p^0 integration has no poles inside a clockwise contour for $(x^0 - y^0) > 0$ and both poles inside counterclockwise contour for $(x^0 - y^0) < 0$, which gives

$$\int \frac{d^4p}{(2\pi)^4} \frac{i}{(p^0 - i\epsilon)^2 - \mathbf{p}^2 - m^2} e^{-ip\cdot(x-y)} \tag{6.2.164}$$

$$= -\theta(y^0 - x^0)\langle 0|[\hat\phi(x), \hat\phi(y)]|0\rangle = -\theta(y^0 - x^0)i\Delta(x - y) \equiv D_{\text{adv}}(x - y).$$

The anti-time-ordering result follows from the above proof for the Feynman propagator by interchanging which pole is included inside the lower and upper contours with the result $\bar D_F(x - y)$. With anti-time-ordering the positive energy components travel backward in time and negative energy components travel forward in time.

As we saw in Eq. (6.2.144) the commutator $[\hat\phi(x), \hat\phi(y)]$ vanishes for all spacelike separations. Hence we see that D_{ret} vanishes everywhere outside the forward light cone and D_{adv} vanishes everywhere outside the backward light cone. The different prescriptions for deforming the contours correspond to different choices of boundary conditions for the Green's function $G(x - y)$. So the Feynman prescription is also referred to as the choice of Feynman boundary conditions.

From the integral expression for D_F in Eq. (6.2.153) it is easily seen that

$$(\partial_x^2 + m^2)D_F(x - y) = -i\delta^4(x - y), \tag{6.2.165}$$

where we have used the expression for the delta function in Eq. (A.2.4). It is a useful exercise to verify explicitly that $\langle 0|T\hat\phi(x)\hat\phi(y)|0\rangle = D_F(x - y)$ satisfies this equation. Evaluating the first derivative we find

$$\partial_\mu^x \langle 0|T\hat{\phi}(x)\hat{\phi}(y)|0\rangle = \partial_\mu^x \langle 0|\theta(x^0 - y^0)\hat{\phi}(x)\hat{\phi}(y) + \theta(y^0 - x^0)\hat{\phi}(y)\hat{\phi}(x)|0\rangle$$
$$= \theta(x^0 - y^0)\langle 0|\partial_\mu^x \hat{\phi}(x)\hat{\phi}(y)|0\rangle + \theta(y^0 - x^0)\langle 0|\hat{\phi}(y)\partial_\mu^x \hat{\phi}(x)|0\rangle$$
$$+ \delta^0{}_\mu \delta(x^0 - y^0)\langle 0|[\hat{\phi}(x), \hat{\phi}(y)]|0\rangle$$
$$= \langle 0|T\partial_\mu^x \hat{\phi}(x)\hat{\phi}(y)|0\rangle + \delta^0{}_\mu \delta(x^0 - y^0)\langle 0|[\hat{\phi}(x), \hat{\phi}(y)]|0\rangle = \langle 0|T\partial_\mu^x \hat{\phi}(x)\hat{\phi}(y)|0\rangle , \quad (6.2.166)$$

where we have used the fact that the equal-time commutator is zero and that

$$\partial_\mu^x \theta(x^0 - y^0) = \delta^0{}_\mu \delta(x^0 - y^0) = -\partial_\mu^x \theta(y^0 - x^0). \quad (6.2.167)$$

Taking the second derivative and adding a mass term we find as expected that

$$(\partial_x^2 + m^2)\langle 0|T\hat{\phi}(x)\hat{\phi}(y)|0\rangle$$
$$= \theta(x^0 - y^0)\langle 0|(\partial_x^2 + m^2)\hat{\phi}(x)\hat{\phi}(y)|0\rangle + \theta(y^0 - x^0)\langle 0|\hat{\phi}(y)(\partial_x^2 + m^2)\hat{\phi}(x)|0\rangle$$
$$+ \delta^0{}_\mu \delta(x^0 - y^0)\langle 0|[\partial_\mu^x \hat{\phi}(x), \hat{\phi}(y)]|0\rangle$$
$$= \langle 0|T\{(\partial_x^2 + m^2)\hat{\phi}(x)\hat{\phi}(y)\}|0\rangle + \delta(x^0 - y^0)\langle 0|[\hat{\pi}(x), \hat{\phi}(y)]|0\rangle$$
$$= \delta(x^0 - y^0)\langle 0|[\pi(x), \hat{\phi}(y)]|0\rangle = -i\delta^4(x - y), \quad (6.2.168)$$

where we have used the equation of motion $(\partial_x^2 + m^2)\hat{\phi}(x) = 0$, $\hat{\pi}(x) = \partial_0 \hat{\phi}(x)$ and $\delta(x^0 - y^0)$ $[\hat{\phi}(x), \hat{\pi}(y)] = i\delta^4(x - y)$ from the equal time canonical commutation relations in Eq. (6.2.81).

6.2.7 Charged Scalar Field

We have already considered the case of a classical complex scalar field as an example of the application of Noether's theorem in Sec. 3.2.1 with the Lagrangian given in Eq. (3.2.48). We recognized that a complex scalar field was a scalar field carrying an electric charge. The charge carried can be any kind of additive quantum number. We considered it again in relativistic quantum mechanics in the context of the Klein-Gordon equation for a complex scalar field in Sec. 4.3.1 and we studied the coupling of this charged scalar field to an external electromagnetic field in Sec. 4.3.4. We now wish to consider the charged scalar field as a quantum field theory.

From Eq. (3.2.48) we can write the normal-ordered Lagangian density operator in terms of Heisenberg picture field operators as

$$\hat{\mathcal{L}} = \tfrac{1}{2}: \left[(\partial_\mu \hat{\phi}_1)^2 + (\partial_\mu \hat{\phi}_2)^2 - m^2 \hat{\phi}_1^2 - m^2 \hat{\phi}_2^2\right] : = :\partial_\mu \hat{\phi}^\dagger \partial^\mu \hat{\phi} - m^2 \hat{\phi}^\dagger \hat{\phi}:, \quad (6.2.169)$$

where $\hat{\phi}_1(x)$ and $\hat{\phi}_2(x)$ are two independent Hermitian scalar field operators with the same mass m and where we have defined

$$\hat{\phi}(x) \equiv (1/\sqrt{2})[\hat{\phi}_1(x) + i\hat{\phi}_2(x)], \qquad \hat{\phi}^\dagger(x) \equiv (1/\sqrt{2})[\hat{\phi}_1(x) - i\hat{\phi}_2(x)]. \quad (6.2.170)$$

We have used the fact that since $\hat{\phi}_1$ and $\hat{\phi}_2$ are independent Hermitian scalar fields their operators will always commute and so we have from Eq. (6.2.81) the equal-time canonical commutation relations

$$[\hat{\phi}_i(t, \mathbf{x}), \hat{\phi}_j(t, \mathbf{y})] = [\hat{\pi}_i(t, \mathbf{x}), \hat{\pi}_j(t, \mathbf{y})] = 0,$$
$$[\hat{\phi}_i(t, \mathbf{x}), \hat{\pi}_j(t, \mathbf{y})] = i\delta_{ij}\delta^3(\mathbf{x} - \mathbf{y}), \quad (6.2.171)$$

where $\hat{\pi}_j(x) = \partial_0 \hat{\phi}_j(x)$ for free scalar fields. Since $\hat{\mathcal{L}}$ only differs from its non-normal-ordered form by a constant and since operators obey the same equations as their classical counterparts we have

$$\hat{\pi}(x) = \frac{\partial \hat{\mathcal{L}}}{\partial(\partial_0 \hat{\phi}(x))} = \partial^0 \hat{\phi}^\dagger(x), \qquad \hat{\pi}^\dagger(x) = \frac{\partial \hat{\mathcal{L}}}{\partial(\partial_0 \hat{\phi}^\dagger(x))} = \partial^0 \hat{\phi}(x). \tag{6.2.172}$$

The normal-ordered Hamiltonian density is obtained in the usual way,

$$\hat{\mathcal{H}} = :\dot{\hat{\phi}}\hat{\pi} + \hat{\pi}^\dagger \dot{\hat{\phi}}^\dagger - \hat{\mathcal{L}}: = :\hat{\pi}^\dagger \hat{\pi} + (\boldsymbol{\nabla}\hat{\phi}^\dagger)\cdot(\boldsymbol{\nabla}\hat{\phi}) + m^2\hat{\phi}^\dagger\hat{\phi}:. \tag{6.2.173}$$

The normal-ordered Hamiltonian and three-momentum operators are given by

$$\hat{H} = \int d^3x :\hat{\pi}^\dagger\hat{\pi} + (\boldsymbol{\nabla}\hat{\phi}^\dagger)\cdot(\boldsymbol{\nabla}\hat{\phi}) + m^2\hat{\phi}^\dagger\hat{\phi}:,$$
$$\hat{P} = -\int d^3x :\hat{\pi}\boldsymbol{\nabla}\hat{\phi} + \hat{\pi}^\dagger\boldsymbol{\nabla}\hat{\phi}^\dagger:. \tag{6.2.174}$$

Using $\hat{\pi}_j(x) = \partial_0\hat{\phi}_j(x)$ we then have

$$\hat{\pi}(x) \equiv (1/\sqrt{2})[\hat{\pi}_1(x) - i\hat{\pi}_2(x)], \qquad \hat{\pi}^\dagger(x) \equiv (1/\sqrt{2})[\hat{\pi}_1(x) + i\hat{\pi}_2(x)]. \tag{6.2.175}$$

We can expand in terms of annihilation operators ($\hat{a}_{1\mathbf{p}}$, $\hat{a}_{2\mathbf{p}}$) and creation operators ($\hat{a}_{1\mathbf{p}}^\dagger$, $\hat{a}_{2\mathbf{p}}^\dagger$) according to Eq. (6.2.98),

$$\hat{\phi}_j(x) = \int \frac{d^3p}{(2\pi)^3} \frac{1}{\sqrt{2E_\mathbf{p}}} \left(\hat{a}_{j\mathbf{p}}e^{-ip\cdot x} + \hat{a}_{j\mathbf{p}}^\dagger e^{ip\cdot x}\right)\bigg|_{p^0 = E_\mathbf{p}} \tag{6.2.176}$$

$$\hat{\pi}_j(x) = \int \frac{d^3p}{(2\pi)^3} (-i)\sqrt{\frac{E_\mathbf{p}}{2}} \left(\hat{a}_{j\mathbf{p}}e^{-ip\cdot x} - \hat{a}_{j\mathbf{p}}^\dagger e^{ip\cdot x}\right)\bigg|_{p^0 = E_\mathbf{p}} = \partial_0\hat{\phi}_j(x) = \dot{\hat{\phi}}_j(x)$$

for $j = 1, 2$. The $j = 1$ annihilation and creation operators satifsy Eq. (6.2.17) as do the $j = 2$ operators but all commutators involving both $j = 1$ and 2 operators vanish. For notational convenience *from this point on it is be understood that $p^0 = E_\mathbf{p} \equiv \sqrt{\mathbf{p}^2 + m^2}$ unless indicated otherwise.* It then follows that

$$\hat{\phi}(x) = \int \frac{d^3p}{(2\pi)^3} \frac{1}{\sqrt{2E_\mathbf{p}}} \left(\hat{f}_\mathbf{p}e^{-ip\cdot x} + \hat{g}_\mathbf{p}^\dagger e^{ip\cdot x}\right), \tag{6.2.177}$$

$$\hat{\phi}^\dagger(x) = \int \frac{d^3p}{(2\pi)^3} \frac{1}{\sqrt{2E_\mathbf{p}}} \left(\hat{f}_\mathbf{p}^\dagger e^{ip\cdot x} + \hat{g}_\mathbf{p}e^{-ip\cdot x}\right),$$

$$\hat{\pi}(x) = \int \frac{d^3p}{(2\pi)^3} i\sqrt{\frac{E_\mathbf{p}}{2}} \left(\hat{f}_\mathbf{p}^\dagger e^{ip\cdot x} - \hat{g}_\mathbf{p}e^{-ip\cdot x}\right) = \partial^0\hat{\phi}^\dagger(x) = \dot{\hat{\phi}}^\dagger(x),$$

$$\hat{\pi}^\dagger(x) = \int \frac{d^3p}{(2\pi)^3} (-i)\sqrt{\frac{E_\mathbf{p}}{2}} \left(\hat{f}_\mathbf{p}e^{-ip\cdot x} - \hat{g}_\mathbf{p}^\dagger e^{ip\cdot x}\right) = \partial^0\hat{\phi}(x) = \dot{\hat{\phi}}(x),$$

where we have defined new annihilation ($\hat{f}_\mathbf{p}, \hat{g}_\mathbf{p}$) and creation ($\hat{f}_\mathbf{p}^\dagger, \hat{g}_\mathbf{p}^\dagger$) operators,

$$\hat{f}_\mathbf{p} \equiv (1/\sqrt{2})[\hat{a}_{1\mathbf{p}} + i\hat{a}_{2\mathbf{p}}], \qquad \hat{f}_\mathbf{p}^\dagger \equiv (1/\sqrt{2})[\hat{a}_{1\mathbf{p}}^\dagger - i\hat{a}_{2\mathbf{p}}^\dagger],$$
$$\hat{g}_\mathbf{p} \equiv (1/\sqrt{2})[\hat{a}_{1\mathbf{p}} - i\hat{a}_{2\mathbf{p}}], \qquad \hat{g}_\mathbf{p}^\dagger \equiv (1/\sqrt{2})[\hat{a}_{1\mathbf{p}}^\dagger + i\hat{a}_{2\mathbf{p}}^\dagger]. \tag{6.2.178}$$

We readily verify from Eqs. (6.2.170) and (6.2.175) that

$$\hat{\phi}_1(x) = (1/\sqrt{2})[\hat{\phi}(x) + \hat{\phi}^\dagger(x)], \qquad \hat{\phi}_2(x) \equiv -i(1/\sqrt{2})[\hat{\phi}(x) - \hat{\phi}^\dagger(x)],$$
$$\hat{\pi}_1(x) = (1/\sqrt{2})[\hat{\pi}(x) + \hat{\pi}^\dagger(x)], \qquad \hat{\pi}_2(x) \equiv i(1/\sqrt{2})[\hat{\pi}(x) - \hat{\pi}^\dagger(x)]. \tag{6.2.179}$$

From the commutation relations for $\hat{a}_{1\mathbf{p}}, \hat{a}_{1\mathbf{p}}^\dagger, \hat{a}_{2\mathbf{p}}, \hat{a}_{2\mathbf{p}}^\dagger$ it is also straightforward to verify that the commutation relations between the new annihilation and creation operators $\hat{f}_{\mathbf{p}}, \hat{f}_{\mathbf{p}}^\dagger, \hat{g}_{\mathbf{p}}, \hat{g}_{\mathbf{p}}^\dagger$ are

$$[\hat{f}_{\mathbf{p}}, \hat{f}_{\mathbf{p}'}] = [\hat{f}_{\mathbf{p}}^\dagger, \hat{f}_{\mathbf{p}'}^\dagger] = [\hat{g}_{\mathbf{p}}, \hat{g}_{\mathbf{p}'}] = [\hat{g}_{\mathbf{p}}^\dagger, \hat{g}_{\mathbf{p}'}^\dagger] = [\hat{f}_{\mathbf{p}}, \hat{g}_{\mathbf{p}'}] = [\hat{f}_{\mathbf{p}}, \hat{g}_{\mathbf{p}'}^\dagger] = [\hat{f}_{\mathbf{p}}^\dagger, \hat{g}_{\mathbf{p}'}] = [\hat{f}_{\mathbf{p}}^\dagger, \hat{g}_{\mathbf{p}'}^\dagger] = 0,$$
$$[\hat{f}_{\mathbf{p}}, \hat{f}_{\mathbf{p}'}^\dagger] = [\hat{g}_{\mathbf{p}}, \hat{g}_{\mathbf{p}'}^\dagger] = (2\pi)^3 \delta^3(\mathbf{p} - \mathbf{p}'). \tag{6.2.180}$$

Similarly, from either Eq. (6.2.170) or Eq. (6.2.178) we can check that the only nonvanishing equal-time canonical commutation relations for $\hat{\phi}, \hat{\phi}^\dagger, \hat{\pi}, \hat{\pi}^\dagger$ are

$$[\hat{\phi}(t, \mathbf{x}), \hat{\pi}(t, \mathbf{x}')] = [\hat{\phi}^\dagger(t, \mathbf{x}), \hat{\pi}^\dagger(t, \mathbf{x}')] = i\delta^3(\mathbf{x} - \mathbf{x}'). \tag{6.2.181}$$

From the classical Noether current in Eq. (3.2.49) we know that we will have the conserved normal-ordered Noether current at the operator level,

$$\hat{j}^\mu(x) = i : \left[\hat{\phi}^\dagger(\partial^\mu \hat{\phi}) - (\partial^\mu \hat{\phi}^\dagger)\hat{\phi}\right] : \quad \text{with} \quad \partial_\mu \hat{j}^\mu(x) = 0, \tag{6.2.182}$$

where we have chosen to change the sign of the current remembering that we are always free to redefine a conserved current by multiplying it by a constant. The conserved charge is then

$$\hat{Q} = \int d^3x\, \hat{j}^0(x) = i \int d^3x : \hat{\phi}^\dagger(x)\hat{\pi}^\dagger(x) - \hat{\pi}(x)\hat{\phi}(x) := \hat{N}_f - \hat{N}_g \equiv \hat{N}, \tag{6.2.183}$$

where the number operators for the $\hat{\phi}$ and $\hat{\phi}^\dagger$ are, respectively, defined as

$$\hat{N}_f \equiv \int d^3p\, \hat{N}_{f\mathbf{p}} \equiv \int \frac{d^3p}{(2\pi)^3}\, \hat{f}_{\mathbf{p}}^\dagger \hat{f}_{\mathbf{p}}, \qquad \hat{N}_g \equiv \int d^3p\, \hat{N}_{g\mathbf{p}} \equiv \int \frac{d^3p}{(2\pi)^3}\, \hat{g}_{\mathbf{p}}^\dagger \hat{g}_{\mathbf{p}}. \tag{6.2.184}$$

We can similarly show that the normal-ordered four-momentum operator is

$$\hat{P}^\mu = (\hat{H}, \hat{\mathbf{P}}) = \int d^3p\, p^\mu \left[\hat{N}_{f\mathbf{p}} + \hat{N}_{g\mathbf{p}}\right] = \int \frac{d^3p}{(2\pi)^3}\, p^\mu \left[\hat{f}_{\mathbf{p}}^\dagger \hat{f}_{\mathbf{p}} + \hat{g}_{\mathbf{p}}^\dagger \hat{g}_{\mathbf{p}}\right]. \tag{6.2.185}$$

From the $\hat{f}$ and $\hat{g}$ commutation relations we find

$$[\hat{N}_{f\mathbf{p}}, \hat{N}_{f\mathbf{p}'}] = [\hat{N}_{f\mathbf{p}}, \hat{N}_{g\mathbf{p}'}] = [\hat{N}_{g\mathbf{p}}, \hat{N}_{g\mathbf{p}'}] = 0 \tag{6.2.186}$$

and it then follows that

$$[\hat{P}^\mu, \hat{H}] = [\hat{P}^\mu, \hat{Q}] = 0 \tag{6.2.187}$$

and so, as expected, the system is invariant under spacetime translations and the charge $\hat{Q}$ is conserved, $[\hat{H}, \hat{Q}] = 0$.

Proof: To prove Eq. (6.2.183) let us introduce the temporary shorthand notation $c_p \equiv \sqrt{E_{\mathbf{p}}}$ and

$$f_p \equiv \hat{f}_{\mathbf{p}} e^{-ip\cdot x}, \; g_p \equiv \hat{g}_{\mathbf{p}} e^{-ip\cdot x}, \; \phi \equiv \hat{\phi}(x), \; \pi \equiv \hat{\pi}(x), \; \int dp \equiv \int \frac{d^3p}{(2\pi)^3 \sqrt{2}}.$$

Then we can write

$$\phi = \int dp\,[f_p + g_p^\dagger]/c_p, \qquad \phi^\dagger = \int dp\,[f_p^\dagger + g_p]/c_p,$$
$$\pi = i\int dp\, c_p[f_p^\dagger - g_p], \qquad \pi^\dagger = -i\int dp\, c_p[f_p - g_p^\dagger]. \tag{6.2.188}$$

We find

$$\hat{Q} = i\int d^3x :\phi^\dagger \pi^\dagger - \pi\phi:$$
$$= \int d^3x \int dp \int dp' \, (c_{p'}/c_p):(f_p^\dagger + g_p)(f_{p'} - g_{p'}^\dagger) + (f_{p'}^\dagger - g_{p'})(f_p + g_p^\dagger):$$
$$= \int dp \int dp' \, (c_{p'}/c_p)\int d^3x([f_p^\dagger f_{p'} + f_{p'}^\dagger f_p] - [g_p^\dagger g_{p'} + g_{p'}^\dagger g_p]$$
$$+ [g_p f_{p'} - g_{p'} f_p] - [f_p^\dagger g_{p'}^\dagger - f_{p'}^\dagger g_p^\dagger])$$
$$= \int (d^3p/(2\pi)^3)[\hat{f}_\mathbf{p}^\dagger \hat{f}_\mathbf{p} - \hat{g}_\mathbf{p}^\dagger \hat{g}_\mathbf{p}], \tag{6.2.189}$$

where in the last step we used

$$\int d^3x \, f_p^\dagger f_{p'} = \hat{f}_\mathbf{p}^\dagger \hat{f}_{\mathbf{p}'} e^{i(E_\mathbf{p} - E_{\mathbf{p}'})t}(2\pi)^3\delta^3(\mathbf{p} - \mathbf{p}')$$
$$\int d^3x \, g_p^\dagger g_{p'} = \hat{g}_\mathbf{p}^\dagger \hat{g}_{\mathbf{p}'} e^{i(E_\mathbf{p} - E_{\mathbf{p}'})t}(2\pi)^3\delta^3(\mathbf{p} - \mathbf{p}')$$
$$\int d^3x \, g_p f_{p'} = \hat{g}_\mathbf{p} \hat{f}_{\mathbf{p}'} e^{-i(E_\mathbf{p} + E_{\mathbf{p}'})t}(2\pi)^3\delta^3(\mathbf{p} + \mathbf{p}')$$
$$\int d^3x \, f_p^\dagger g_{p'}^\dagger = \hat{f}_\mathbf{p}^\dagger \hat{g}_{\mathbf{p}'}^\dagger e^{i(E_\mathbf{p} + E_{\mathbf{p}'})t}(2\pi)^3\delta^3(\mathbf{p} + \mathbf{p}') \tag{6.2.190}$$

and that, e.g., $\int d^3p \, e^{-i2E_\mathbf{p}t}[\hat{g}_\mathbf{p}\hat{f}_{-\mathbf{p}} - \hat{g}_{-\mathbf{p}}\hat{f}_\mathbf{p}] = 0$ using a change of variables.

Starting from the expressions for $\hat{H}$ and $\hat{\mathbf{P}}$ in Eq. (6.2.174) and using the above techniques we can similarly prove the result for $\hat{P}^\mu$ in Eq. (6.2.185).

Note that since the Lagrangian density in Eq. (6.2.169) is that of two noninteracting scalar fields we know that there will be a number operator of the form of Eq. (6.2.20) associated with each,

$$\hat{N}_i = \int d^3p \, \hat{N}_{i\mathbf{p}} = \int (d^3p/(2\pi)^3)\hat{a}_{i\mathbf{p}}^\dagger a_{i\mathbf{p}}. \tag{6.2.191}$$

We can build the combined Fock space as the tensor product of the individual Fock spaces for particles of types $j = 1$ and 2. However, as we saw in Eq. (4.3.57), when we couple to the electromagnetic field it will be by coupling to the charged scalar field operators $\hat{\phi}$ and $\hat{\phi}^\dagger$. The combined Fock space basis can be formed using the eigenstates of the number operators $\hat{N}_f$ and $\hat{N}_g$.

From Eq. (6.2.177) observe that $\hat{\phi}$ annihilates a particle of type f and creates a particle of type g and conversely $\hat{\phi}^\dagger$ creates a particle of type f and annihilates a particle of type g. Since the Hamiltonian can be written as the sum of the particle 1 and particle 2 Hamiltonians, then we know that the energy of the system is bounded below. We know this in any case since before normal-ordering the Hamiltonian density has the form of a sum of terms like $\hat{A}^\dagger \hat{A}$, which from Eq. (A.8.2) we know is a positive semidefinite operator. So before normal-ordering $\hat{H}$ is bounded below and since normal-ordering only removes constants then it remains bounded below.

So there must be a vacuum state for the combined system that is the tensor product of the vacuum states,

$$|0\rangle \equiv |0\rangle_1 \otimes |0\rangle_2 = |0\rangle_f \otimes |0\rangle_g, \tag{6.2.192}$$

where $\hat{a}_{1\mathbf{p}}$ and $\hat{a}_{2\mathbf{p}}$ annihilate the vacuum. It follows from their definition that $\hat{f}_\mathbf{p}$ and $\hat{g}_\mathbf{p}$ must also annihilate this vacuum and since the $\hat{f}, \hat{f}^\dagger$ and the $\hat{g}, \hat{g}^\dagger$ operators are independent (they are mutually commuting) then the second tensor product follows also. Then we have

$$\hat{a}_{1\mathbf{p}}|0\rangle = \hat{a}_{2\mathbf{p}}|0\rangle = \hat{f}_\mathbf{p}|0\rangle = \hat{g}_\mathbf{p}|0\rangle = 0. \tag{6.2.193}$$

So we know that the eigenvalues of $\hat{N}_f$ and $\hat{N}_g$ are integers.

The natural interpretation of scalar particles with identical mass and opposite charge is as particle and antiparticle. While it is a matter of convention, let us here regard the f quanta as the particles and the g quanta as the antiparticles. With the electromagnetic coupling to the $\hat{\phi}$ field in in the form of Eq. (4.3.57), then it is natural to redefine the current to be the electric current in Eq. (4.3.59),

$$\hat{j}^{\mu}(x) = iq \colon \left[\hat{\phi}^{\dagger}(\partial^{\mu}\hat{\phi}) - (\partial^{\mu}\hat{\phi}^{\dagger})\hat{\phi} \right] \colon . \tag{6.2.194}$$

Then the charge operator becomes

$$\hat{Q} = q\hat{N}_f + (-q)\hat{N}_g = q\hat{N} \tag{6.2.195}$$

so that particles and antiparticles have electric charge $+q$ and $-q$, respectively. For any eigenstate of $\hat{N}$ the difference between the particle and antiparticle number is N and is fixed. Such a state may contain superpositions of states with different N_f and N_g but with $N = N_f - N_g$ fixed, where N is unchanged by the creation and annihilation of particle-antiparticle pairs. In analogy with Eq. (6.2.37) it is straightforward to show that

$$[\hat{N}_f, \hat{f}_{\mathbf{p}}^{\dagger}] = f_{\mathbf{p}}^{\dagger}, \quad [\hat{N}_f, \hat{f}_{\mathbf{p}}] = -f_{\mathbf{p}}, \quad [\hat{N}_g, \hat{g}_{\mathbf{p}}^{\dagger}] = g_{\mathbf{p}}^{\dagger}, \quad [\hat{N}_f, \hat{g}_{\mathbf{p}}] = -g_{\mathbf{p}},$$

$$[\hat{N}, \hat{f}_{\mathbf{p}}^{\dagger}] = f_{\mathbf{p}}^{\dagger}, \quad [\hat{N}, \hat{f}_{\mathbf{p}}] = -f_{\mathbf{p}}, \quad [\hat{N}, \hat{g}_{\mathbf{p}}^{\dagger}] = -g_{\mathbf{p}}^{\dagger}, \quad [\hat{N}, \hat{g}_{\mathbf{p}}] = g_{\mathbf{p}}. \tag{6.2.196}$$

Then we see that $\hat{\phi}^{\dagger}$ increase the total charge qN by one unit of q by either creating a particle or annihilating an antiparticle, while $\hat{\phi}$ decreases the total charge qN by one unit of q by either annihilating a particle or creating an antiparticle.

Let us now consider the propagator for the charged scalar field, i.e., the vaccum expectation value of the time-ordered product of two field operators. From Eqs. (6.2.177) and (6.2.180) we observe that

$$\langle 0|T\hat{\phi}(x)\hat{\phi}(y)|0\rangle = \langle 0|T\hat{\phi}^{\dagger}(x)\hat{\phi}^{\dagger}(y)|0\rangle = 0. \tag{6.2.197}$$

Recalling that we have $\langle 0|\hat{f}_{\mathbf{p}}\hat{f}_{\mathbf{p}'}^{\dagger}|0\rangle = \langle 0|[\hat{f}_{\mathbf{p}}, \hat{f}_{\mathbf{p}'}^{\dagger}]|0\rangle = (2\pi)^3\delta^3(\mathbf{p} - \mathbf{p}')$ and similarly for $\langle 0|\hat{g}_{\mathbf{p}}\hat{g}_{\mathbf{p}'}^{\dagger}|0\rangle$ it follows that

$$\langle 0|T\hat{\phi}(x)\hat{\phi}^{\dagger}(y)|0\rangle = \theta(x^0 - y^0)\langle 0|\hat{\phi}(x)\hat{\phi}^{\dagger}(y)|0\rangle + \theta(y^0 - x^0)\langle 0|\hat{\phi}^{\dagger}(y)\hat{\phi}(x)|0\rangle$$

$$= \int \frac{d^3p\, d^3p'}{(2\pi)^6} \frac{1}{2\sqrt{E_{\mathbf{p}}E_{\mathbf{p}'}}} \left[\theta(x^0 - y^0)\langle 0|\hat{f}_{\mathbf{p}}\hat{f}_{\mathbf{p}'}^{\dagger}|0\rangle e^{-ip\cdot x + ip'\cdot y} \right.$$

$$\left. + \theta(y^0 - x^0)\langle 0|\hat{g}_{\mathbf{p}'}\hat{g}_{\mathbf{p}}^{\dagger}|0\rangle e^{+ip\cdot x - ip'\cdot y} \right]$$

$$= \int \frac{d^3p}{(2\pi)^3} \frac{1}{2E_{\mathbf{p}}} \left[\theta(x^0 - y^0)e^{-ip\cdot(x-y)} + \theta(y^0 - x^0)e^{+ip\cdot(x-y)} \right]$$

$$= \theta(x^0 - y^0)\Delta_+(x - y) + \theta(y^0 - x^0)\Delta_+(y - x)$$

$$= D_F(x - y), \tag{6.2.198}$$

where we see that the Feynman propagator propagates a particle with positive charge forward in time or an antiparticle with negative charge backward in time. Note that if we had chosen to define $\hat{\phi}'(x) \equiv \hat{\phi}^{\dagger}(x)$, then the roles of $f_{\mathbf{p}}$ and $g_{\mathbf{p}}$ would be reversed.

6.2.8 Wick's Theorem

For a Hermitian scalar field $\hat{\phi}(x)$ we can write $\hat{\phi}(x) = \hat{\phi}^+(x) + \hat{\phi}^-(x)$, where

$$\hat{\phi}^+(x) \equiv \int \frac{d^3p}{(2\pi)^3} \frac{1}{\sqrt{2E_p}} \hat{a}_{\mathbf{p}} e^{-ip\cdot x} \quad \text{and} \quad \hat{\phi}^-(x) \equiv \int \frac{d^3p}{(2\pi)^3} \frac{1}{\sqrt{2E_p}} \hat{a}_{\mathbf{p}}^\dagger e^{ip\cdot x}, \qquad (6.2.199)$$

which are the positive energy and negative energy parts, respectively. We have

$$\hat{\phi}^+(x)|0\rangle = 0, \quad \langle 0|\hat{\phi}^-(x) = 0 \quad \Rightarrow \quad \langle 0|\hat{\phi}(x)\hat{\phi}(y)|0\rangle = \langle 0|\hat{\phi}^+(x)\hat{\phi}^-(y)|0\rangle. \qquad (6.2.200)$$

For $x^0 > y^0$ we have

$$\begin{aligned}
T\hat{\phi}(x)\hat{\phi}(y) \overset{x^0 \geq y^0}{=} \;& \hat{\phi}^+(x)\hat{\phi}^+(y) + \hat{\phi}^+(x)\hat{\phi}^-(y) + \hat{\phi}^-(x)\hat{\phi}^+(y) + \hat{\phi}^-(x)\hat{\phi}^-(y) \\
= \;& \hat{\phi}^+(x)\hat{\phi}^+(y) + \hat{\phi}^-(y)\hat{\phi}^+(x) + \hat{\phi}^-(x)\hat{\phi}^+(y) + \hat{\phi}^-(x)\hat{\phi}^-(y) + [\hat{\phi}^+(x), \hat{\phi}^-(y)] \\
= \;& :\hat{\phi}(x)\hat{\phi}(y): + [\hat{\phi}^+(x), \hat{\phi}^-(y)],
\end{aligned} \qquad (6.2.201)$$

where the normal-ordering in Eq. (6.2.29) was used. For $x^0 < y^0$ we have

$$\begin{aligned}
T\hat{\phi}(x)\hat{\phi}(y) \overset{x^0 \leq y^0}{=} \;& \hat{\phi}^+(y)\hat{\phi}^+(x) + \hat{\phi}^+(y)\hat{\phi}^-(x) + \hat{\phi}^-(y)\hat{\phi}^+(x) + \hat{\phi}^-(y)\hat{\phi}^-(x) \\
= \;& \hat{\phi}^+(x)\hat{\phi}^+(y) + \hat{\phi}^-(x)\hat{\phi}^+(y) + \hat{\phi}^-(y)\hat{\phi}^+(x) + \hat{\phi}^-(x)\hat{\phi}^-(y) + [\hat{\phi}^+(y), \hat{\phi}^-(x)] \\
=: \;& \hat{\phi}(x)\hat{\phi}(y): + [\hat{\phi}^+(y), \hat{\phi}^-(x)].
\end{aligned} \qquad (6.2.202)$$

The *contraction* or *Wick contraction* of two field operators is defined as

$$\overset{\frown}{\hat{\phi}(x)\hat{\phi}(y)} \equiv \begin{cases} [\hat{\phi}^+(x), \hat{\phi}^-(y)] & \text{for } x^0 > y^0 \\ [\hat{\phi}^+(y), \hat{\phi}^-(x)] & \text{for } x^0 < y^0 \end{cases}. \qquad (6.2.203)$$

With this definition of a contraction we can write

$$T\hat{\phi}(x)\hat{\phi}(y) = :\hat{\phi}(x)\hat{\phi}(y): + \overset{\frown}{\hat{\phi}(x)\hat{\phi}(y)}. \qquad (6.2.204)$$

Using Eq. (6.2.138) for $\Delta_+(x - y)$ we observe that

$$\begin{aligned}
[\hat{\phi}^+(x), \hat{\phi}^-(y)] &= \int \frac{d^3p}{(2\pi)^3} \frac{1}{\sqrt{2E_p}} \int \frac{d^3q}{(2\pi)^3} \frac{1}{\sqrt{2E_q}} e^{-ip\cdot x + iq\cdot y} [\hat{a}_{\mathbf{p}}, \hat{a}_{\mathbf{q}}^\dagger] \qquad (6.2.205) \\
&= \int \frac{d^3p}{(2\pi)^3} \frac{1}{2E_p} e^{-ip\cdot(x-y)} = \Delta_+(x - y) = \langle 0|\hat{\phi}(x)\hat{\phi}(y)|0\rangle.
\end{aligned}$$

Similarly, $[\hat{\phi}^+(y), \hat{\phi}^-(x)] = \langle 0|\hat{\phi}(y)\hat{\phi}(x)|0\rangle$ and so

$$\begin{aligned}
\overset{\frown}{\hat{\phi}(x)\hat{\phi}(y)} &= \theta(x^0 - y^0)\langle 0|\hat{\phi}(x)\hat{\phi}(y)|0\rangle + \theta(y^0 - x^0)\langle 0|\hat{\phi}(y)\hat{\phi}(x)|0\rangle \\
&= \langle 0|T\hat{\phi}(x)\hat{\phi}(y)|0\rangle = D_F(x - y).
\end{aligned} \qquad (6.2.206)$$

We can write the above in the form

$$\begin{aligned}
T\hat{\phi}(x)\hat{\phi}(y) &= :\hat{\phi}(x)\hat{\phi}(y): + D_F(x - y) = :\hat{\phi}(x)\hat{\phi}(y) + D_F(x - y): \\
&= :\hat{\phi}(x)\hat{\phi}(y) + \overset{\frown}{\hat{\phi}(x)\hat{\phi}(y)}:,
\end{aligned} \qquad (6.2.207)$$

since the Feynman propagator is a complex function unaffected by normal-ordering.

Wick's theorem: The generalization of the above result is referred to as *Wick's theorem* and can be written as

$$T\hat{\phi}(x_1)\hat{\phi}(x_2)\cdots\hat{\phi}(x_n) = :\{\hat{\phi}(x_1)\hat{\phi}(x_2)\cdots\hat{\phi}(x_n) + \text{all contractions}\}:. \qquad (6.2.208)$$

The meaning of this can be illustrated for the $n = 4$ case, i.e.,

$$T\hat{\phi}(x_1)\hat{\phi}(x_2)\hat{\phi}(x_3)\hat{\phi}(x_4) = :(\hat{\phi}(x_1)\hat{\phi}(x_2)\hat{\phi}(x_3)\hat{\phi}(x_4) + \overparen{\hat{\phi}(x_1)\hat{\phi}(x_2)\hat{\phi}(x_3)}\hat{\phi}(x_4)$$

$$+ \overparen{\hat{\phi}(x_1)\hat{\phi}(x_2)}\hat{\phi}(x_3)\hat{\phi}(x_4) + \hat{\phi}(x_1)\overparen{\hat{\phi}(x_2)\hat{\phi}(x_3)}\hat{\phi}(x_4) + \hat{\phi}(x_1)\hat{\phi}(x_2)\overparen{\hat{\phi}(x_3)\hat{\phi}(x_4)}$$

$$+ \hat{\phi}(x_1)\overparen{\hat{\phi}(x_2)\hat{\phi}(x_3)\hat{\phi}(x_4)} + \hat{\phi}(x_1)\hat{\phi}(x_2)\overparen{\hat{\phi}(x_3)\hat{\phi}(x_4)} + \overparen{\hat{\phi}(x_1)\hat{\phi}(x_2)}\overparen{\hat{\phi}(x_3)\hat{\phi}(x_4)}$$

$$+ \overparen{\hat{\phi}(x_1)\hat{\phi}(x_2)\hat{\phi}(x_3)\hat{\phi}(x_4)} + \hat{\phi}(x_1)\overparen{\hat{\phi}(x_2)\hat{\phi}(x_3)}\hat{\phi}(x_4)):. \qquad (6.2.209)$$

When we take the vacuum expectation value of this only the fully contracted terms will survive, since the vacuum expectation value of any normal-ordered product of operators vanishes, $\langle 0| :\hat{\phi}(x_1)\cdots\hat{\phi}(x_j): |0\rangle = 0$. We then find

$$\langle 0|T\hat{\phi}(x_1)\hat{\phi}(x_2)\hat{\phi}(x_3)\hat{\phi}(x_4)|0\rangle = D_F(x_1 - x_2)D_F(x_3 - x_4)$$
$$+ D_F(x_1 - x_3)D_F(x_2 - x_4) + D_F(x_1 - x_4)D_F(x_2 - x_3). \qquad (6.2.210)$$

Consequence of Wick's theorem: The general form of this result follows from Eq. (6.2.208) and will vanish if n is odd and if n is even has the form

$$\langle 0|T\hat{\phi}(x_1)\hat{\phi}(x_2)\cdots\hat{\phi}(x_n)|0\rangle = D_F(x_1 - x_2)D_F(x_3 - x_4)\cdots D_F(x_{n-1} - x_n)$$
$$+ \text{ all pairwise combinations of } (x_1, x_2, \ldots, x_n). \qquad (6.2.211)$$

Proof of Wick's theorem: We prove Eq. (6.2.208) by induction. It is trivially true for $n = 1$ and has already been proved for $n = 2$ in Eq. (6.2.207). We will assume it is true for n and then use this to prove it true for $n + 1$.

Since all annihilation operators commute with each other and since all creation operators commute with each other, then $: \hat{\phi}(x_1)\hat{\phi}(x_2)\cdots\hat{\phi}(x_n) :$ is independent of the ordering of the operators inside the normal-ordering for all n. It follows then that the right-hand side of Eq. (6.2.208) is independent of the ordering of the operators. Similarly, the ordering of the operators inside the time-ordering on the left-hand side is also irrelevant. Then without loss of generality we can always chose the ordering of the operators on each side of the equation to be $\hat{\phi}(x_1)\cdots\hat{\phi}(x_n)$, where $x_1^0 \geq x_2^0 \geq \cdots \geq x_n^0$. With this choice of operator ordering we have

$$T\hat{\phi}(x_1)\hat{\phi}(x_2)\cdots\hat{\phi}(x_n) = \hat{\phi}(x_1)\hat{\phi}(x_2)\cdots\hat{\phi}(x_n) \qquad (6.2.212)$$

for all n. Let us temporarily use the shorthand notation $\phi_j \equiv \hat{\phi}(x_j)$. Then the above label ordering leads to

$$T\phi_1\phi_2\cdots\phi_{n+1} = \phi_1\phi_2\cdots\phi_{n+1} = (\phi_1^+ + \phi_1^-)\phi_2\cdots\phi_{n+1} \qquad (6.2.213)$$
$$= \phi_1^-(\phi_2\cdots\phi_{n+1}) + (\phi_2\cdots\phi_{n+1})\phi_1^+ + [\phi_1^+, \phi_2\cdots\phi_{n+1}].$$

Now we note that $\phi_1^- :\phi_2\cdots\phi_{n+1}: + :\phi_2\cdots\phi_{n+1}:\phi_1^+ = :\phi_1\cdots\phi_{n+1}:$ and we make use of the assumption that Wick's theorem is true for $\phi_2\cdots\phi_{n+1}$ to observe that the sum of the first two terms in Eq. (6.2.213) satisfies

$$\phi_1^-(\phi_2\cdots\phi_{n+1}) + (\phi_2\cdots\phi_{n+1})\phi_1^+ = \phi_1^- T(\phi_2\cdots\phi_{n+1}) + T(\phi_2\cdots\phi_{n+1})\phi_1^+$$

$$= \phi_1^-\left[:\phi_2\cdots\phi_{n+1}:+:\left(\begin{array}{c}\text{all contractions}\\\text{of }\phi_2\cdots\phi_{n+1}\end{array}\right):\right]$$

$$+\left[:\phi_2\cdots\phi_{n+1}:+:\left(\begin{array}{c}\text{all contractions}\\\text{of }\phi_2\cdots\phi_{n+1}\end{array}\right):\right]\phi_1^+$$

$$=:\phi_1\phi_2\cdots\phi_{n+1}:+:\left(\begin{array}{c}\text{all contractions}\\\text{not involving }\phi_1\end{array}\right):.$$

$$(6.2.214)$$

We turn now to the third term in Eq. (6.2.213). We begin by noting that

$$[\phi_1^+,\phi_j] = [\phi_1^+,\phi_j^-] = \overbracket{\phi_1\phi_j} \tag{6.2.215}$$

for all j since $x_1^0 \geq x_j^0$. Using Eq. (6.2.44) and Wick's theorem for $n-1$ operators this generalizes to

$$[\phi_1^+,\phi_2\cdots\phi_{n+1}] = [\phi_1^+,\phi_2]\phi_3\cdots\phi_{n+1} + \cdots + \phi_2\phi_3\cdots\phi_n[\phi_1^+,\phi_{n+1}]$$

$$= \overbracket{\phi_1\phi_2}\phi_3\cdots\phi_{n+1} + \phi_2\overbracket{\phi_1\phi_3}\phi_4\cdots\phi_{n+1} + \cdots + \phi_2\cdots\phi_n\overbracket{\phi_1\phi_{n+1}}$$

$$= \overbracket{\phi_1\phi_2}T(\phi_3\cdots\phi_{n+1}) + \overbracket{\phi_1\phi_3}T(\phi_2\phi_4\cdots\phi_{n+1}) + \cdots + \overbracket{\phi_1\phi_{n+1}}T(\phi_2\phi_3\cdots\phi_n)$$

$$=:\left(\begin{array}{c}\text{all contractions}\\\text{involving }\phi_1\end{array}\right):.$$

$$(6.2.216)$$

The last two steps follow since with our labeling conventions the $(n-1)$ noncontracted operators are time-ordered. Using Wick's theorem for these gives the normal-ordered sum over all possible pairwise contractions of these $(n-1)$ operators. After multiplying with each of the corresponding single contractions involving ϕ_1 and summing over these we arrive at the normal-ordered sum over all possible contractions where ϕ_1 is always contracted over.

Combining the results of Eqs. (6.2.214) and (6.2.216) finally allows us to write Eq. (6.2.213) in the form

$$T\phi_1\phi_2\cdots\phi_{n+1} = :(\phi_1\phi_2\cdots\phi_{n+1} + \text{all possible contractions}):, \tag{6.2.217}$$

where we have arrived at this result for $n+1$ assuming the result for n and where we have restored the time-ordering operator so that the operator ordering on both sides can again be arbitrary. This completes our proof of Wick's theorem.

For a charged scalar field using Eqs. (6.2.177), (6.2.199) and (6.2.203) we have

$$\hat{\phi}^+(x) \equiv \int \frac{d^3p}{(2\pi)^3}\frac{1}{\sqrt{2E_p}}\hat{f}_{\mathbf{p}}e^{-ip\cdot x}, \quad \hat{\phi}^-(x) \equiv \int \frac{d^3p}{(2\pi)^3}\frac{1}{\sqrt{2E_p}}\hat{g}_{\mathbf{p}}^\dagger e^{ip\cdot x}, \tag{6.2.218}$$

$$(\hat{\phi}^\dagger)^+(x) \equiv \int \frac{d^3p}{(2\pi)^3}\frac{1}{\sqrt{2E_p}}\hat{g}_{\mathbf{p}}e^{-ip\cdot x}, \quad (\hat{\phi}^\dagger)^-(x) \equiv \int \frac{d^3p}{(2\pi)^3}\frac{1}{\sqrt{2E_p}}\hat{f}_{\mathbf{p}}^\dagger e^{ip\cdot x}.$$

We readily verify that

$$\overbracket{\phi(x)\phi(y)} = \overbracket{\phi^\dagger(x)\phi^\dagger(y)} = 0, \qquad \overbracket{\phi(x)\phi^\dagger(y)} = \overbracket{\phi^\dagger(x)\phi(y)} = D_F(x-y). \tag{6.2.219}$$

6.2.9 Functional Integral Formulation

In the same way that the path integral formulation of quantum mechanics was obtained in Sec. 4.1.12, we can derive a *functional integral* formulation of quantum field theory, where the integration over paths becomes replaced by an integration over all field configurations.[2] *We temporarily restore factors of $\hbar$ here* to facilitate the relationship of the discussion here to that for ordinary quantum mechanics in Sec. 4.1.12. If we know all of the matrix elements of the evolution operator, then we have completely determined the behavior of the quantum system.

Consider the quantum mechanical motion of a single degree of freedom as given in Eq. (4.1.201) generalized to N degrees of freedom,

$$\langle \vec{q}'', t''|\vec{q}', t'\rangle = {}_s\langle q''|\hat{U}(t'', t')|q'\rangle_s$$
$$= \int \mathcal{D}\vec{q}\,\mathcal{D}\vec{p}\,\exp\{i\int_{t'}^{t''} dt\,[\vec{p}\cdot\dot{\vec{q}} - H(\vec{q}, \vec{p})]/\hbar\}, \qquad (6.2.220)$$

where $\vec{q} = (q_1, \ldots, q_N)$, $\vec{p} \equiv (p_1, \ldots, p_N)$ and where we explicitly show $\hbar$ for now. If the system Hamiltonian operator $H(\hat{p}, \hat{q})$ has the simple form

$$H(\hat{\vec{q}}, \hat{\vec{p}}) = (\hat{\vec{p}}^2/2m) + V(\hat{\vec{q}}) \qquad (6.2.221)$$

that is second-order in the momentum, then we have

$$\langle \vec{q}''', t''|\vec{q}', t'\rangle = \int \mathcal{D}\vec{q}\,e^{iS[\vec{q}]/\hbar} \quad \text{with} \quad S[\vec{q}] = \int_{t'}^{t''} dt\,L(\vec{q}, \dot{\vec{q}}), \qquad (6.2.222)$$

where $L(\vec{q}, \dot{\vec{q}})$ is just the Lagrangian for the Hamiltonian in Eq. (6.2.221),

$$L(\vec{q}, \dot{\vec{q}}) = \tfrac{1}{2} m\dot{\vec{q}}^2 - V(\vec{q}) \qquad (6.2.223)$$

and where we have suppressed the boundary conditions on the action for notational convenience, $S[\vec{q}] \equiv S([\vec{q}], \vec{q}'', \vec{q}'; t'', t')$. $|\vec{q}, t\rangle$ and $|\vec{q}\rangle$ are coordinate operator eigenstates in the Heisenberg and Schrödinger pictures, respectively, i.e., $\hat{q}^k(t)|\vec{q}, t\rangle = q^k|\vec{q}, t\rangle$ and $\hat{q}^k|\vec{q}\rangle = q^k|\vec{q}\rangle$, where $|\vec{q}, t\rangle = \exp\{-i\hat{H}(t - t_0)\}|\vec{q}\rangle$.

Consider now the quantum field theory of a scalar field with a possible interaction term $U(\phi)$ with U a real function bounded below,

$$\hat{\mathcal{H}} \equiv \mathcal{H}(\hat{\phi}(x), \hat{\pi}(x)) = \tfrac{1}{2}\left\{\hat{\pi}^2(x) + (\boldsymbol{\nabla}\hat{\phi}(x))^2 + m^2\hat{\phi}^2(x)\right\} + U(\hat{\phi}), \qquad (6.2.224)$$

where we could choose, for example, $U(\hat{\phi}) = \lambda\hat{\phi}^4$. If the potential was not present, then we could use the normal-ordered Hamiltonian, which only differs from this Hamiltonian by an (infinite) constant. That constant could then be absorbed into the normalization constants described below and nothing would change. However, in the presence of an interaction we will need more than normal-ordering to obtain finite results. We will need the full renormalization procedure, to be discussed later. For now, then, we ignore the issue of normal-ordering and renormalization.

The corresponding Hamiltonian operator can be written as

$$\hat{H} \equiv H[\hat{\phi}, \hat{\pi}] \equiv \int d^3x\, \mathcal{H}(\hat{\phi}(x), \hat{\pi}(x)) = \left\{\int d^3x\, \tfrac{1}{2}\hat{\pi}^2(x)\right\} + V[\hat{\phi}],$$
$$\hat{V} \equiv V[\hat{\phi}] = \int d^3x\, \tfrac{1}{2}\{[(\boldsymbol{\nabla}\hat{\phi}(x))^2 + m^2\hat{\phi}^2(x)] + U(\hat{\phi})\}, \qquad (6.2.225)$$

where we have defined $\hat{V}$ here for temporary convenience to contain everything but the term containing the momentum. A comparison of Eq. (6.2.221) with Eq. (6.2.225) again illustrates that

---

[2] It is also common to use the term "path integral" in the context of quantum field theory, but we prefer the term "functional integral" for that purpose.

quantum field theory can be thought of as a quantum mechanical system with an infinite number of degrees of freedom.

Clearly, we would like to write down an equation analogous to Eq. (6.2.220) for our scalar quantum field theory. We use the Fock space completeness in Eq. (6.2.77) in place of the quantum mechanics completeness in Eq. (4.1.196) and we generalize the arguments of Sec. 4.1.12 using $q(t) \to \vec{q}(t) \to \phi(t, \mathbf{x})$ and $p(t) \to \vec{p}(t) \to \pi(t, \mathbf{x})$. Dividing $(t'' - t')$ into $(n+1)$ equal intervals we define

$$\delta t \equiv (t'' - t')/(n+1), \quad t_j \equiv t' + j\delta t \tag{6.2.226}$$

for $j = 0, 1, 2, \ldots, n+1$. We have $t_{j+1} = t_j + \delta t$, $t'' = t_{n+1}$ and $t' = t_0$. We have the Fock space completeness relations,

$$\int \mathcal{D}^{\mathrm{sp}}\phi \, |\phi, t\rangle\langle\phi, t| = \int \mathcal{D}^{\mathrm{sp}}\phi \, |\phi\rangle\langle\phi| = \int \mathcal{D}^{\mathrm{sp}}\pi \, |\pi, t\rangle\langle\pi, t| = \int \mathcal{D}^{\mathrm{sp}}\pi \, |\pi\rangle\langle\pi| = \hat{I} \tag{6.2.227}$$

with $|\phi, t_0\rangle = |\phi\rangle$ and $|\pi, t_0\rangle = |\pi\rangle$ and t_0 is the arbitrary time at which we choose the Heisenberg and Schrödinger pictures to coincide. Note that we have introduced the notation $\int \mathcal{D}^{\mathrm{sp}}\phi$ and $\int \mathcal{D}^{\mathrm{sp}}\pi$ to emphasize that here we are integrating over $\phi(\mathbf{x})$ and $\pi(\mathbf{x})$ as functions of space only at a fixed time. We can write

$$\langle \phi'', t'' | \phi', t' \rangle = \int \mathcal{D}^{\mathrm{sp}}\phi_1 \cdots \int \mathcal{D}^{\mathrm{sp}}\phi_n \langle \phi'', t'' | \phi_n, t_n \rangle\langle \phi_n, t_n | \phi_{n-1}, t_{n-1} \rangle$$
$$\cdots \langle \phi_2, t_2 | \phi_1, t_1 \rangle\langle \phi_1, t_1 | \phi', t' \rangle, \tag{6.2.228}$$

$$\langle \phi_{j+1}, t_{j+1} | \phi_j, t_j \rangle = \langle \phi_{j+1} | e^{-i\delta t \hat{H}/\hbar} | \phi_j \rangle = \delta[\phi_{j+1} - \phi_j] - i\frac{\delta t}{\hbar}\langle \phi_{j+1} | \hat{H} | \phi_j \rangle + O(\delta t^2).$$

Consider the symmetric Fourier transform in Sec. A.2.3 for N degrees of freedom,

$$f(\vec{q}) \equiv \langle \vec{q} | f \rangle = \int d^N p \, \langle \vec{q} | \vec{p} \rangle\langle \vec{p} | f \rangle = \int \frac{d^N p}{(2\pi\hbar)^{N/2}} e^{+i\vec{p}\cdot\vec{q}/\hbar} \tilde{f}(\vec{p})$$

$$\tilde{f}(\vec{p}) \equiv \langle \vec{p} | f \rangle = \int d^N x \, \langle \vec{p} | \vec{q} \rangle\langle \vec{q} | f \rangle = \int \frac{d^N q}{(2\pi\hbar)^{N/2}} e^{-i\vec{p}\cdot\vec{q}/\hbar} f(\vec{q}). \tag{6.2.229}$$

Let $j = 1, \ldots, N$ label the number of points $\mathbf{x}_j$ on a finite spatial lattice and make the replacements $q_j \to \phi(\mathbf{x}_j)$ and $p_j \to \pi(\mathbf{x}_j)$ and then write the formalism as if the $N \to \infty$ limit has been taken. In reality this is only a notational convenience and we will we only formally take the ultraviolet (lattice spacing to zero) and infrared (infinite volume) limits at the end of calculations when we are considering appropriate ratios of quantities where such normalizations cancel. Physical observables will always involve such ratios. Understanding this notational convenience we can use the functional notation for Fourier transforms and delta functions,

$$F[\phi] \equiv \langle \vec{\phi} | F \rangle = \int \mathcal{D}^{\mathrm{sp}}\pi \langle \phi | \pi \rangle\langle \pi | F \rangle = \frac{1}{\mathcal{N}} \int \mathcal{D}^{\mathrm{sp}}\pi \, e^{+i\int d^3x \, \pi(\mathbf{x})\phi(\mathbf{x})/\hbar} \tilde{F}[\pi],$$

$$\tilde{F}[\pi] \equiv \langle \pi | F \rangle = \int \mathcal{D}^{\mathrm{sp}}\phi \langle \pi | \phi \rangle\langle \phi | F \rangle = \frac{1}{\mathcal{N}} \int \mathcal{D}^{\mathrm{sp}}\phi \, e^{-i\int d^3x \, \pi(\mathbf{x})\phi(\mathbf{x})/\hbar} F[\phi], \tag{6.2.230}$$

$$\langle \vec{q} | \vec{q}\,' \rangle = \delta^N(\vec{q} - \vec{q}\,'), \langle \vec{p} | \vec{p}\,' \rangle = \delta^N(\vec{p} - \vec{p}\,') \to \langle \phi | \phi' \rangle = \delta[\phi - \phi'], \langle \pi | \pi' \rangle = \delta[\pi - \pi'],$$

where for any finite N the normalization factor is $\mathcal{N} = (2\pi\hbar)^{N/2}$. We also see that

$$\langle \vec{q} | \vec{p} \rangle = \frac{1}{(2\pi\hbar)^{N/2}} e^{i\vec{p}\cdot\vec{q}/\hbar} \to \langle \phi | \pi \rangle = \frac{1}{\mathcal{N}} \exp^{i\int d^3x \, \pi(\mathbf{x})\phi(\mathbf{x})/\hbar} \tag{6.2.231}$$

$$\delta^N(\vec{q}\,' - \vec{q}) = \int \frac{d^N p}{(2\pi\hbar)^N} e^{\pm i\vec{p}\cdot(\vec{q}\,'-\vec{q})/\hbar} \to \delta[\phi' - \phi] = \frac{1}{\mathcal{N}^2} \int \mathcal{D}^{\mathrm{sp}}\pi \, e^{\pm i\int d^3x \, \pi(\phi-\phi')/\hbar}$$

and similarly for $\delta[\pi - \pi']$. We then find

$$
\begin{aligned}
\langle \phi_{j+1} | H[\hat{\phi}, \hat{\pi}] | \phi_j \rangle &= \int \mathcal{D}^{\mathrm{sp}} \pi_j \langle \phi_{j+1} | (\tfrac{1}{2} \hat{\pi}^2) | \pi_j \rangle \langle \pi_j | \phi_j \rangle + \langle \phi_{j+1} | V[\hat{\phi}] | \phi_j \rangle \\
&= \left(\frac{1}{\mathcal{N}^2} \int \mathcal{D}^{\mathrm{sp}} \pi_j \, e^{i \int d^3 x \, \pi_j (\phi_{j+1} - \phi_j)/\hbar} (\tfrac{1}{2} \pi_j^2) \right) + \delta[\phi_{j+1} - \phi_j] V[\phi_j] \\
&= \frac{1}{\mathcal{N}^2} \int \mathcal{D}^{\mathrm{sp}} \pi_j \, e^{i \int d^3 x \, \pi_j (\phi_{j+1} - \phi_j)/\hbar} H[\phi_j, \pi_j]
\end{aligned} \tag{6.2.232}
$$

and Eq. (6.2.228) becomes

$$
\begin{aligned}
\langle \phi_{j+1}, t_{j+1} | \phi_j, t_j \rangle &= \frac{1}{\mathcal{N}^2} \int \mathcal{D}^{\mathrm{sp}} \pi_j \, e^{i \int d^3 x \, \pi_j (\phi_{j+1} - \phi_j)/\hbar} \left\{ 1 - i(\delta t/\hbar) H[\phi_j, \pi_j] + \mathcal{O}(\delta t^2) \right\} \\
&= \frac{1}{\mathcal{N}^2} \int \mathcal{D}^{\mathrm{sp}} \pi_j \, e^{i(\delta t/\hbar)\{ \int d^3 x \, \pi_j [(\phi_{j+1} - \phi_j)/\delta t] - H[\phi_j, \pi_j] \}} + \mathcal{O}(\delta t^2).
\end{aligned} \tag{6.2.233}
$$

Using this result in Eq. (4.1.197) gives

$$
\begin{aligned}
\langle \phi'', t'' | \phi', t' \rangle &= \lim_{n \to \infty} \frac{1}{\mathcal{N}^{2(n+1)}} \int \prod_{i=1}^{n} \mathcal{D}^{\mathrm{sp}} \phi_i \int \prod_{k=0}^{n} \mathcal{D}^{\mathrm{sp}} \pi_k \\
&\quad \times \exp \left\{ (i/\hbar) \delta t \sum_{j=0}^{n} \left[\left(\int d^3 x \, \pi_j (\phi_{j+1} - \phi_j)/\delta t \right) - H[\phi_j, \pi_j] \right] \right\} \\
&\equiv \int \mathcal{D}\pi \int \mathcal{D}\phi \, e^{(i/\hbar) \int_{t'}^{t''} dt \, d^3 x \, [\pi(x)\dot{\phi}(x) - \mathcal{H}(\phi, \pi)]},
\end{aligned} \tag{6.2.234}
$$

where the last equation is shorthand for the equation above and where here $\int \mathcal{D}\phi$ and $\int \mathcal{D}\pi$ are functional integrals over $\phi(x)$ and $\pi(x)$ as functions of four-dimensional spacetime. Note carefully the difference between $\int \mathcal{D}^{\mathrm{sp}} \phi$ and $\int \mathcal{D}\phi$.

The class of Hamiltonians that we consider here have the form

$$
\hat{H} \equiv H(\hat{\phi}, \hat{\pi}) = \int d^3 x \, \mathcal{H}(\hat{\phi}, \hat{\pi}) = \int d^3 x \, (\tfrac{1}{2} \hat{\pi}^2 + V[\hat{\phi}]). \tag{6.2.235}
$$

Most theories of interest to us can be written in this form. The momentum integrations in Eq. (6.2.234) can be carried out since $\mathcal{H}(\hat{\pi}, \hat{\phi})$ has been *restricted to be purely second-order* in $\hat{\pi}$. We introduce the positive real infinitesimal Feynman parameter ϵ and use the Gaussian integrals in Eqs. (4.1.202) and (4.1.203) to make the replacement $t \to e^{-i\epsilon} t = t(1 - i\epsilon)$ so that $\delta t \to \delta t(1 - i\epsilon)$. This is an infinitesimal Wick rotation $\theta = \epsilon$ of the type discussed in Sec. A.4. Extending Eq. (4.1.206) gives

$$
\begin{aligned}
\frac{1}{\mathcal{N}^2} \int \mathcal{D}^{\mathrm{sp}} \pi_j \, e^{(i/\hbar)\delta t \int d^3 x \left[\pi \dot{\phi} - \frac{1}{2} \pi^2 \right]} &\to \frac{1}{\mathcal{N}^2} \int \mathcal{D}^{\mathrm{sp}} \pi_j \exp \left\{ (i/\hbar)\delta t \int d^3 x \left[\pi \dot{\phi} - (1 - i\epsilon) \tfrac{1}{2} \pi^2 \right] \right\} \\
&= \frac{1}{\mathcal{N}'} \exp \left\{ (i/\hbar)\delta t (1 + i\epsilon) \int d^3 x \, \tfrac{1}{2} \dot{\phi}^2 \right\},
\end{aligned} \tag{6.2.236}
$$

where for a finite number N of spatial points $\mathcal{N}' \equiv [2i\pi\hbar\delta t(1 - i\epsilon)]^{N/2}$.

Hence using Eq. (6.2.236) in Eq. (6.2.234) we have

$$
\begin{aligned}
\langle \phi'', t'' | \phi', t' \rangle &= \lim_{n \to \infty} \frac{1}{\mathcal{N}'^{(n+1)/2}} \int \prod_{i=1}^{n} \mathcal{D}^{\mathrm{sp}} \phi_i \\
&\quad \times \exp \left\{ (i/\hbar)\delta t \left[\sum_{j=0}^{n} \left\{ \int d^3 x \, (1 + i\epsilon) \tfrac{1}{2} \left([\phi_{j+1} - \phi_j]/\delta t \right)^2 \right\} - (1 - i\epsilon) V[\phi_j] \right] \right\} \\
&\equiv \int \mathcal{D}\phi \exp \left\{ (i/\hbar) \int_{t'}^{t''} d^4 x \, \mathcal{L}(\phi, \partial_\mu \phi) \right\} = \int \mathcal{D}\phi \exp \left\{ i S[\phi]/\hbar \right\},
\end{aligned} \tag{6.2.237}
$$

where the boundary conditions on the action are understood and where

$$\begin{aligned}
\mathcal{L} &= (1 + i\epsilon)\tfrac{1}{2}\dot{\phi}^2 - (1 - i\epsilon)\left(\tfrac{1}{2}[(\boldsymbol{\nabla}\phi)^2 + m^2\phi^2] + U(\phi)\right) \\
&= \tfrac{1}{2}[\dot{\phi}^2 - (\boldsymbol{\nabla}\phi)^2 - m^2\phi^2] - U(\phi) + i\epsilon\left(\tfrac{1}{2}[\dot{\phi}^2 + (\boldsymbol{\nabla}\phi)^2 + m^2\phi^2] + U(\phi)\right) \\
&= \tfrac{1}{2}[\dot{\phi}^2 - (\boldsymbol{\nabla}\phi)^2 - m^2\phi^2] - U(\phi) + i\epsilon\mathcal{L}_E
\end{aligned} \tag{6.2.238}$$

is the standard Lagrangian density for the class of Hamiltonians that we are considering with an additional infinitesimal term arising from the Feynman parameter ϵ. We will see in Eq. (6.2.254) that $\mathcal{L}_E$ is the Euclidean Lagrangian density. Shifting a potential by a constant has no physical effect and so we can always choose $U(\phi) \geq 0$ without loss of generality so that $\mathcal{L}_E > 0$. The term multiplying $i\epsilon$ is then positive and adds an infinitesimal damping term that helps define an otherwise purely oscillatory functional integral. Field configurations with increasingly large Euclidean action will be increasingly suppressed by the damping term. For the purposes of developing a perturbation theory approach to quantum field theory it will be sufficient to only retain that part involving the mass term, $i\epsilon m^2 \phi^2$, since it leads to the Feynman boundary conditions. The remaining part acts to suppress strong fluctuations of the field ϕ in time and space but this will not be important in perturbation theory. Redefining $\epsilon m^2 \to \epsilon$ the free field Lagrangian density becomes

$$\mathcal{L} = \tfrac{1}{2}[\dot{\phi}^2 - (\boldsymbol{\nabla}\phi)^2 - (m^2 - i\epsilon)\phi^2] - U(\phi) = \tfrac{1}{2}[\partial_\mu\phi\partial^\mu\phi - (m^2 - i\epsilon)\phi^2] - U(\phi). \tag{6.2.239}$$

For the free field case this leads to the modified Klein-Gordon equation,

$$(\Box + m^2 - i\epsilon)\phi = (\partial_\mu\partial^\mu + m^2 - i\epsilon)\phi = 0. \tag{6.2.240}$$

The Green's function for the linear differential operator $(\Box + m^2 - i\epsilon)$ is the Feynman propagator of Eq. (6.2.155) with the momentum space denominator

$$p^2 - m^2 + i\epsilon. \tag{6.2.241}$$

So we see that the infinitesimal Wick rotation leads us to the Feynman boundary conditions in a natural way, $m^2 \to (m^2 - i\epsilon)$. Had we started with a Euclidean space formulation and then attempted to do an inverse Wick rotation back to Minkowski space, we would arrive at the Feynman boundary conditions as shown in Eq. (A.4.7).

Note: We will always be interested in taking ratios of regularized functional integrals, e.g., with a finite number of spacetime points or some other finite basis. In the discussion above this means taking ratios with N (number of spatial points) and n (number of intermediate time slices) finite. All of the normalization constants that we have discussed above cancel in such ratios, including the zero-point energy. Only after this cancelation do we actually take the ultraviolet and infrared limits to obtain physical observables. However, for obvious reasons it is notationally convenient to write expressions such as the last lines Eqs. (6.2.234) and (6.2.237), but we should remember what we mean by them. We refer to the $\int \mathcal{D}\phi$ and $\int \mathcal{D}\pi$ notation in the last lines as *functional integrals* over the field configurations and momentum densities, respectively. This is a more appropriate terminology than the term "path-integral" that we use for quantum mechanics; however, the latter term is sometimes also used in the context of quantum field theory.

For Hamiltonians of the form of Eq. (6.2.235) we have shown that

$$\langle\phi'', t''|\phi', t'\rangle = \langle\phi''|\hat{G}(t'' - t')|\phi'\rangle = \langle\phi''|e^{-i\hat{H}t/\hbar}|\phi'\rangle = \int\mathcal{D}\phi\, e^{iS[\phi]/\hbar}, \tag{6.2.242}$$

where the boundary conditions on the action $S[\phi]$ are to be understood; i.e., we should more fully write $S([\phi], \phi'', \phi'; t'', t')$ but the shorthand is notationally convenient. This result is the quantum field theory analog of the quantum mechanics result in Eq. (4.1.208). We write the evolution operator as $\hat{G}(t'' - t') \equiv \exp[-i\hat{H}(t'' - t')]$ and recall that $|\phi, t\rangle$ and $|\phi\rangle$ are the simultaneous eigenstates of the field operator at all points $\mathbf{x}$ on the appropriate equal-time surface in the Heisenberg and Schrödinger pictures, respectively. The path integral in quantum mechanics is an integral over all possible positions q at every time slice subject to the boundary conditions at the initial and final times, and in this sense it is an integration over all possible paths. The functional integral is an integral over all possible functions $\phi(\mathbf{x})$ at every time slice and so is an integral over all possible functions $\phi(x)$ subject to the boundary conditions at the initial and final times.

The *spectral function* is the trace of the evolution operator and so is given by

$$F(t'' - t') \equiv \mathrm{tr}\, e^{-i\hat{H}(t'' - t')/\hbar} = \int \mathcal{D}^{\mathrm{sp}} \phi \, \langle\phi| e^{-i\hat{H}(t'' - t')/\hbar} |\phi\rangle = \int \mathcal{D}^{\mathrm{sp}} \phi \, \langle\phi, t''|\phi, t'\rangle$$

$$= \int \mathcal{D}\pi \, \mathcal{D}\phi_{\text{(periodic)}} \, \exp\{(i/\hbar) \int_{t'}^{t''} d^4 x \, [\pi\dot{\phi} - \mathcal{H}(\phi, \pi)]\}$$

$$= \int \mathcal{D}\phi_{\text{(periodic)}} \, \exp\{(i/\hbar) \int_{t'}^{t''} d^4 x \, \mathcal{L}(\phi, \partial_\mu \phi)\}$$

$$= \int \mathcal{D}\phi_{\text{(periodic)}} \, \exp\{(i/\hbar) S([\phi], t'', t')\}, \tag{6.2.243}$$

where we have used the completeness of the $\phi(\mathbf{x})$ basis as defined in Eq. (6.2.77). Note that $\int \mathcal{D}\phi_{\text{(periodic)}}$ is a functional integral over the $\phi(x)$ as a function over four-space, but constrained so that $\phi(t', \mathbf{x}) = \phi(t'', \mathbf{x})$ for all $\mathbf{x} \in \mathbb{R}^3$, i.e., constrained to be the functional integral over periodic functions of the time interval.

Consider some time-independent Hamiltonian for a scalar field, then we can define the *Schrödinger functional* as

$$\mathcal{S}[\phi'', t''; \phi', t'] \equiv \langle\phi'', t''|\phi', t'\rangle = \langle\phi''| e^{-i\hat{H}(t'' - t')/\hbar} |\phi'\rangle = \langle\phi''|\hat{G}(t'', t')|\phi'\rangle. \tag{6.2.244}$$

For any Schrödinger picture state $|\Psi, t\rangle$ in Fock space we define its Schrödinger wave functional as $\Psi[\phi, t] \equiv \langle\phi|\Psi, t\rangle$. We can use the completeness relation in Eq. (6.2.227) and write the Schrödinger wave functional evolution equation

$$\Psi[\phi'', t''] \equiv \langle\phi''|\Psi, t''\rangle = \langle\phi''|\hat{G}(t'', t')|\Psi, t'\rangle = \langle\phi''| e^{-i\hat{H}(t'' - t')/\hbar} |\Psi, t'\rangle \tag{6.2.245}$$

$$= \int \mathcal{D}^{\mathrm{sp}} \phi' \, \langle\phi''| e^{-i\hat{H}(t'' - t')/\hbar} |\phi'\rangle \langle\phi'|\Psi, t'\rangle = \int \mathcal{D}^{\mathrm{sp}} \phi' \, \mathcal{S}[\phi'', t''; \phi', t'] \, \Psi[\phi', t'].$$

Note that we have formulated the above in terms of equal-time surfaces at t' and t'' and by specifying a field configuration basis on an equal-time surface in Eq. (6.2.77) because it was convenient to follow the inspiration and the familiar machinery provided by nonrelativistic quantum mechanics. However, an equal-time surface for one inertial observer will not correspond to an equal-time surface to any other inertial observer moving with respect to the first. Furthermore, the theory was constructed to be Lorentz covariant and so it must be possible to formulate the above in a more general and more covariant-looking form. Recall from Sec. 6.2.5 that the scalar field commutes with itself at all spacelike separations, i.e., $[\hat{\phi}(x), \hat{\phi}(y)] = 0$ for any two spacetime points x and y for which $(x - y)^2 < 0$. Then in place of an equal-time surface $t = t_0$, we can use any spacelike surface, σ, and construct a basis for our Hilbert space from the simultaneous eigenstates of the field operator at all points on this spacelike surface. Then for any lattice regularization of this spacelike surface we would have in place of Eq. (6.2.73)

$$|\phi_\sigma\rangle \equiv |\phi(x_1), \phi(x_2), \phi(x_3), \ldots\rangle \text{ where } x_i \in \sigma \text{ for all } i = 1, 2, 3, \ldots, \tag{6.2.246}$$

which in the limit that the lattice regularization is removed would form a basis for the Hilbert space such that

$$\langle \phi'_\sigma | \phi_\sigma \rangle = \delta^{\mathrm{sp}}[\phi'_\sigma - \phi_\sigma] \quad \text{and} \quad \int \mathcal{D}^{\mathrm{sp}}\phi_\sigma \, |\phi_\sigma\rangle\langle\phi_\sigma| = \hat{I}, \tag{6.2.247}$$

where $\delta^{\mathrm{sp}}[\phi'_\sigma - \phi_\sigma]$ vanishes unless $\phi'(x) = \phi(x)$ for every $x \in \sigma$ This generalizes the equal-time field configuration basis, Eq. (6.2.77), to a basis defined on an arbitrary spacelike surface. We can generalize the Schrödinger functional as

$$S[\phi'', \sigma''; \phi', \sigma'] \equiv \langle \phi'', \sigma''|\phi', \sigma'\rangle = \langle \phi''|\hat{G}(\sigma'', \sigma')|\phi'\rangle, \tag{6.2.248}$$

where the evolution operator between the configurations $\phi'_{\sigma'}$ and $\phi''_{\sigma''}$ on the spacelike surfaces σ' and σ'', respectively, is given by

$$\langle \phi''_{\sigma''} | \hat{G}(\sigma'', \sigma')|\phi'_{\sigma'}\rangle = \int \mathcal{D}\phi \, \exp\{i \int_{\sigma'}^{\sigma''} d^4x \, \mathcal{L}\} \tag{6.2.249}$$

for a system with a Hamiltonian of the form of Eq. (6.2.235).

6.2.10 Euclidean Space Formulation

Consider the Wick rotation of the evolution operator to Euclidean space discussed in Secs. A.4 and 4.1.12, where we analytically continue to imaginary time,

$$(t'' - t')/\hbar \to -i\beta \equiv 1/kT, \tag{6.2.250}$$

and where k and T are the Boltzmann constant and temperature, respectively. The trace of this gives the partition function,

$$F(t'' - t') \equiv \mathrm{tr}\, e^{-i\hat{H}(t''-t')/\hbar} \to \mathrm{tr}\, \exp(-\beta\hat{H}) \equiv Z(\beta). \tag{6.2.251}$$

The analytic continuation doesn't change Fock space and all completeness relations and the functional Fourier transform above are unaffected. It will be convenient to revert to natural units notation here and to again set $\hbar = 1$.

All of the steps leading to Eq. (6.2.232) remain valid, where $(t'' - t') \to -i\beta$, where we define $\delta\beta \equiv \beta/(n+1)$ and where in place of $\delta t \to \delta t(1 - i\epsilon)$ we have $\delta t \to -i\delta\beta$. We then find that Eq. (6.2.234) becomes

$$\langle \phi'', \beta|\phi', 0\rangle = \lim_{n\to\infty} \frac{1}{\mathcal{N}^{2(n+1)}} \int \prod_{i=1}^{n} \mathcal{D}^{\mathrm{sp}}\phi_i \int \prod_{k=1}^{n+1} \mathcal{D}^{\mathrm{sp}}\pi_k$$

$$\times \exp\left\{ \delta\beta \sum_{j=0}^{n} \left[\left(\int d^3x \, i\pi_j[\phi_{j+1} - \phi_j]/\delta\beta \right) - H[\phi_j, \pi_j] \right] \right\}$$

$$\equiv \int \mathcal{D}\pi \int \mathcal{D}\phi \, \exp\left\{ \int_0^\beta d^4x \left[i\pi(x)\dot{\phi}_E(x) - \mathcal{H}(\phi, \pi) \right] \right\}, \tag{6.2.252}$$

where $\dot{\phi}_E(x) \equiv \partial\phi/\partial\beta = \partial_4^E \phi$ is the Euclidean time derivative of ϕ discussed in Sec. A.4. Using the result for a Gaussian integral in Eq. (4.1.202) we then find in place of Eq. (6.2.237) that for a Hamiltonian of the form of Eq. (6.2.225)

$$\langle \phi'', \beta|\phi', 0\rangle = \lim_{n\to\infty} \frac{1}{\mathcal{N}''^{(n+1)/2}} \int \prod_{i=1}^{n} \mathcal{D}^{\mathrm{sp}}\phi_i$$

$$\times \exp\left\{ \delta\beta \left[\left\{ -\sum_{j=0}^{n} \tfrac{1}{2} \left([\phi_{j+1} - \phi_j]/\delta\beta\right)^2 \right\} - V[\phi] \right] \right\}$$

$$\equiv \int \mathcal{D}\phi \, \exp\left\{ -\int_0^\beta d^4x \, \mathcal{L}_E(\phi, \dot{\phi}_E) \right\} = \int \mathcal{D}\phi \, \exp\{-S_E[\phi]\}, \tag{6.2.253}$$

where $\mathcal{N}'' \equiv (2\pi\delta\beta)^{N/2}$, where boundary conditions on $S_E[\phi]$ are understood and where the Euclidean action and Lagrangian density are defined using Eq. (6.2.225),

$$S_E[\phi] = \int d^4x\, \mathcal{L}_E, \tag{6.2.254}$$
$$\mathcal{L}_E = \tfrac{1}{2}\dot{\phi}_E^2 + \tfrac{1}{2}(\boldsymbol{\nabla}\phi)^2 + \tfrac{1}{2}m^2\phi^2 + U(\phi) = \tfrac{1}{2}\partial_\mu^E\phi\,\partial_\mu^E\phi + \tfrac{1}{2}m^2\phi^2 + U(\phi).$$

Since $U(\phi)$ was assumed to be bounded below, then so are $\mathcal{L}_E$ and $S_E[\phi]$. In order for the Euclidean functional integral to exist we must have $S_E[\phi]$ bounded below, which requires that $U(\phi)$ be bounded below. This is why we can have an interaction like $U(\phi) = \frac{\kappa}{3!}\phi^3 + \frac{\lambda}{4!}\phi^4$ but not $U(\phi) = \frac{\kappa}{3!}\phi^3$ or any odd highest power in ϕ. Note that λ is dimensionless, while κ must have the dimension of mass (in natural units).

The partition function of statistical mechanics is defined for a countable number of energy eigenstates as $Z(\beta) \equiv \sum_i \exp(-\beta E_i) = \operatorname{tr}\exp(-\beta\hat{H})$. Here it is given by

$$Z(\beta) \equiv \operatorname{tr}\left(e^{-\beta H}\right) = \int \mathcal{D}^{\mathrm{sp}}\phi\,\langle\phi|e^{-\beta\hat{H}}|\phi\rangle = \int \mathcal{D}^{\mathrm{sp}}\phi\,\langle\phi,\beta|\phi,0\rangle \tag{6.2.255}$$
$$= \int \mathcal{D}\pi \int \mathcal{D}\phi_{(\mathrm{periodic})}\, \exp\{\textstyle\int_0^\beta d^4x\,[i\pi(x)\dot{\phi}_E(x) - \mathcal{H}(\pi,\phi)]\}$$
$$= \int \mathcal{D}\phi_{(\mathrm{periodic})}\, \exp\{-\textstyle\int_0^\beta d^4x\,\mathcal{L}_{\mathrm{E}}(\phi,\partial_\mu^E\phi)\} = \int \mathcal{D}\phi_{(\mathrm{periodic})}\, \exp\{-S_{\mathrm{E}}([\phi],\beta)\}.$$

6.2.11 Generating Functional for a Scalar Field

The quantum mechanics results of Sec. 4.1.12 carry over to the case of scalar fields in a straightforward way. Using the spatial discretization of Sec. 6.2.2 and the time discretizaton in the functional integral approach leads to a spacetime lattice regularization of the path integral, where we understand that the continuum limit is taken followed by the infinite volume limit of expressions for any physical quantity. Shifting all energies by a constant only changes $F_J(T, -T)$ in Eq. (4.1.228) by an overall phase, which cancels in the ratio $F^J(T, -T)/F(T, -T)$. So normal ordering of the free Hamiltonian does not change any of the arguments; i.e., we get the same answer for either form of $\hat{H}$ for a free field, where $\hat{H}$ is second-order in the fields.

For a free scalar field the Hamiltonian has the form

$$\hat{H} = \int d^3x\, \mathcal{H}(\hat{\phi},\hat{\pi}) = \int d^3x : \tfrac{1}{2}[\dot{\hat{\phi}}^2(x) + (\boldsymbol{\nabla}\hat{\phi}(x))^2 + m^2\hat{\phi}^2(x)]:. \tag{6.2.256}$$

Including a source term leads to the Hamiltonian

$$\hat{U}^j(t'',t') \equiv T\exp\left\{-i\textstyle\int_{t'}^{t''} dt\,\hat{H}_s(t)\right\}, \quad \text{where} \tag{6.2.257}$$
$$\hat{H}^j = \hat{H} - \int d^3x\, j(x)\hat{\phi}(x) = \int d^3x\,\{\mathcal{H} - j(x)\hat{\phi}(x)\},$$

where T is the time-ordering operator. We use this later when we consider interacting theories in Sec. 7.4. The derivation of Eq. (6.2.243) remains valid when source terms are present and so the spectral function becomes

$$F^j(t'',t') = \operatorname{tr}\{\hat{U}^j(t'',t')\} = \operatorname{tr}\left\{Te^{-i\int_{t'}^{t''} d^4x[\hat{\mathcal{H}} - j(x)\hat{\phi}(x)]}\right\} = \int \mathcal{D}\phi_{(\mathrm{periodic})}\, e^{iS[\phi,j]},$$
$$S[\phi,j] \equiv S[\phi] + \textstyle\int_{t'}^{t''} d^4x\, j(x)\phi(x) = \int_{t'}^{t''} d^4x\,[\mathcal{L} + j(x)\phi(x)], \tag{6.2.258}$$
$$\mathcal{L} = \tfrac{1}{2}[\partial_\mu\phi\,\partial^\mu\phi - m^2\phi^2] - U(\phi).$$

Following the arguments in Sec. 4.1.12 we can treat the source terms as the interaction Hamiltonian and then Eq. (4.1.220) becomes

$$\hat{H}^j = \hat{H} + \hat{H}_{\text{int}}, \quad \text{where} \quad \hat{H}_{\text{int}} = -\int d^3x \, j(x)\hat{\phi}(x),$$
$$\hat{U}_I(t'', t') = T \exp\left\{ i \int_{t'}^{t''} d^4x \, j(x)\hat{\phi}_I(x) \right\}, \tag{6.2.259}$$

where $\hat{U}_I(t'', t')$ is the evolution operator in the interaction picture. We have used Eqs. (4.1.188) and (4.1.192). From Eq. (4.1.229) we have as $T \to \infty(1 - i\epsilon)$,

$$F^j(T, -T) = \text{tr}\,\hat{U}^j(t'', t') \to e^{-iE_0(2T)} \langle \Omega|\hat{U}_I(T, -T)|\Omega\rangle \tag{6.2.260}$$
$$= e^{-iE_0(2T)} \langle \Omega|e^{i\int_{-T}^{T} d^4x \, j(x)\hat{\phi}_I(x)}|\Omega\rangle,$$

so that we isolate the ground state contribution in the spectral function. In quantum field theory we refer to the lowest energy state $|E_0\rangle \equiv |\Omega\rangle$ as the ground state or as the *vacuum state*. Unless stated otherwise, from this point onward we will always assume that it is normalized,

$$\langle \Omega|\Omega\rangle = 1. \tag{6.2.261}$$

For a free quantum field theory the ground state is the empty state, $|\Omega\rangle = |0\rangle$, where we often refer to the empty state $|0\rangle$ as the *perturbative vacuum*.

Using the notation $|0\rangle$ for the vacuum state the generating functional is

$$Z[j] \equiv \lim_{T \to \infty(1-i\epsilon)} \frac{F^j(T, -T)}{F(T, -T)} = \lim_{T \to \infty(1-i\epsilon)} \frac{\text{tr}\{\hat{U}^j(T, -T)\}}{\text{tr}\{\hat{U}(T, -T)\}} \tag{6.2.262}$$

$$= \lim_{T \to \infty(1-i\epsilon)} \frac{\text{tr}\left[Te^{-i\int_{-T}^{T} d^4x \, [\hat{\mathcal{H}} - j\hat{\phi}]}\right]}{\text{tr}\left[Te^{-i\int_{-T}^{T} d^4x \, \hat{\mathcal{H}}}\right]} = \lim_{T \to \infty(1-i\epsilon)} \langle \Omega|Te^{i\int_{-T}^{T} d^4x \, j(x)\hat{\phi}_I(x)}|\Omega\rangle$$

$$= \lim_{T \to \infty(1-i\epsilon)} \frac{\int \mathcal{D}\phi_{(\text{periodic})} \, e^{i\{S[\phi] + \int_{-T}^{T} d^4x \, j\phi\}}}{\int \mathcal{D}\phi_{(\text{periodic})} \, e^{iS[\phi]}},$$

where $S[\phi] = \int_{-T}^{T} d^4x \, \mathcal{L}$. In the noninteracting case, $U(\phi) = 0$, we have $|\Omega\rangle \to |0\rangle$ in Eq. (6.2.262). Then from Eq. (4.1.231) we see that the ground state expectation values of time-ordered products of Heisenberg picture scalar operators are given by

$$\langle \Omega|T\hat{\phi}(x_1) \cdots \hat{\phi}(x_k)|\Omega\rangle = \frac{(-i)^k \delta^k}{\delta j(x_1) \cdots \delta j(x_k)} Z[j]\Big|_{j=0} \tag{6.2.263}$$

$$= \lim_{T \to \infty(1-i\epsilon)} \frac{\int \mathcal{D}\phi_{(\text{periodic})} \, \phi(x_1) \cdots \phi(x_k) \, e^{iS[\phi]}}{\int \mathcal{D}\phi_{(\text{periodic})} \, e^{iS[\phi]}}.$$

One commonly finds that the limit $T \to \infty(1 - i\epsilon)$ and "periodic" are not explicitly indicated but are to be understood and that the denominator is not included in the definition of $Z[j]$ (Peskin and Schroeder, 1995; Schwartz, 2013). It will later be useful to define the time-ordered product before taking $j(x) = 0$

$$\langle \Omega|T\hat{\phi}(x_1) \cdots \hat{\phi}(x_k)|\Omega\rangle_j \equiv \frac{(-i)^k \delta^k}{\delta j(x_1) \cdots \delta j(x_k)} Z[j] \tag{6.2.264}$$

and so, for example, $Z[j] \equiv \langle \Omega|\Omega\rangle_j$.

After a Wick rotation of the above we arrive at the Euclidean action and generating functional, where

$$\hat{U}^j(t'',t') \to \hat{U}^j_E(t'',t') \equiv T \exp\left\{-\int_{t'}^{t''} dt\,[\hat{H} - j(t)\hat{q}]\right\}, \tag{6.2.265}$$

$$Z[j] \to Z_E[j] \equiv \lim_{T_E \to \infty} \frac{\text{tr}\{\hat{U}^j_E(T,-T)\}}{\text{tr}\{\hat{U}_E(T,-T)\}} \tag{6.2.266}$$

$$= \lim_{T_E \to \infty} \frac{\text{tr}\left[T \exp\left\{-\int_{-T_E}^{T_E} d^4x\,[\hat{\mathcal{H}} - j(x)\hat{\phi}(x)]\right\}\right]}{\text{tr}\left[T \exp\left\{-\int_{-T_E}^{T_E} d^4x\,\hat{\mathcal{H}}\right\}\right]} = \lim_{T_E \to \infty} \frac{\int \mathcal{D}\phi_{(\text{periodic})}\, e^{-S_E[\phi,j]}}{\int \mathcal{D}\phi_{(\text{periodic})}\, e^{-S_E[\phi]}},$$

where the Euclidean space point is $x_\mu = (\tau,\mathbf{x}) = (x_4,\mathbf{x}) = (x_1,x_2,x_3,x_4)$, where the Wick-rotated source-dependent evolution operator $U^j_E(t'',t')$ is exponentially decaying rather than unitary, and where the source-dependent Euclidean action is

$$S_E[\phi,j] = S_E[\phi] - \int d^4x\, j\phi = \int d^4x \left\{\tfrac{1}{2}\left[\partial_\mu\phi\partial_\mu\phi + m^2\phi^2\right] - j\phi\right\}. \tag{6.2.267}$$

The Euclidean form of Eq. (6.2.263) is then

$$\langle\Omega|T\hat{\phi}(x_1)\cdots\hat{\phi}(x_k)|\Omega\rangle_E = \frac{\delta^k}{\delta j(x_1)\cdots\delta j(x_k)} Z_E[j]\bigg|_{j=0}$$

$$= \lim_{T_E \to \infty} \frac{\int \mathcal{D}\phi_{(\text{periodic})}\,\phi(x_1)\cdots\phi(x_k)\,e^{-S_E[\phi]}}{\int \mathcal{D}\phi_{(\text{periodic})}\,e^{-S_E[\phi]}}. \tag{6.2.268}$$

The Minkowski-space vacuum expectation values of products of field operators are referred to as *Wightman functions* or *Wightman distributions*,

$$\langle\Omega|\hat{\phi}(x_1)\hat{\phi}(x_2)\cdots\hat{\phi}(x_n)|\Omega\rangle. \tag{6.2.269}$$

If one knows all of the Wightman functions for a theory, then one also knows all of the time-ordered products, which are the Green's functions. The Euclidean space analogs of the Wightman functions are called the *Schwinger functions*. The *Osterwalder-Schrader theorem* states that Schwinger functions that satisfy a property called *reflection positivity* can be analytically continued to a quantum field theory in Minkowski space, i.e., by reversing the Wick rotation. Reflection positivity is the condition that the Euclidean theory is invariant under a combination of time reversal and complex conjugation and it ensures that the Euclidean functional integral is positive. Wightman and others attempted a mathematically rigorous approach to quantum field theory known as *constructive quantum field theory*. This approach uses a set of axioms referred to as the *Wightman axioms* as the basis for the rigorous construction. In such a theory the Wightman distributions can be Wick rotated and result in Schwinger functions that satisfy reflection positivity. This interesting topic is beyond the scope of this text, but for further detail the interested reader can consult references such as Streater (1975), Jost (1979), Glimm and Jaffe (1981), Streater and Wightman (2000) and Bogolubov et al. (2012).

Free Scalar Field

Using integration by parts and the periodic boundary conditions allows the Euclidean action to be written as

$$S_E[\phi] = \int d^4x \left\{\tfrac{1}{2}\partial_\mu\phi(x)\partial_\mu\phi(x) + \tfrac{1}{2}m^2\phi(x)^2\right\}$$

$$= \int d^4x\, d^4y\, \delta^4(x-y)\left\{\tfrac{1}{2}\partial_\mu\phi(x)\partial_\mu\phi(y) + \tfrac{1}{2}m^2\phi(x)\phi(y)\right\}$$

$$= \int d^4x\, d^4y\, \tfrac{1}{2}\phi(x)\left\{\left(-\partial^x_\mu\partial^x_\mu + m^2\right)\delta^4(x-y)\right\}\phi(y). \tag{6.2.270}$$

and with the addition of a source term we have $S_E[\phi, j] = S_E[\phi] - \int d^4x\, j\phi$. The implied ultraviolet cutoff allows us to treat the first derivatives of ϕ as if they are continuous. The Euclidean scalar propagator, defined as

$$D_E(x - y) \equiv \int \frac{d^4k}{(2\pi)^4} \frac{1}{k^2 + m^2} e^{-ik\cdot(x-y)}, \qquad (6.2.271)$$

is the inverse of $\left\{ \left(-\partial_\mu^x \partial_\mu^x + m^2\right) \delta^4(x - y) \right\}$, since using integration by parts

$$
\begin{aligned}
\int d^4y &\left\{ \left(-\partial_\mu^x \partial_\mu^x + m^2\right) \delta^4(x - y) \right\} D_E(y - z) \\
&= \int d^4y\, \delta^4(x - y) \left(-\partial_\mu^y \partial_\mu^y + m^2\right) D_E(y - z) = \int d^4y\, \delta^4(x - y) \delta^4(y - z) \\
&= \delta^4(x - z).
\end{aligned}
\qquad (6.2.272)
$$

Then using Eqs. (4.1.205) and (6.2.266) we arrive at the Euclidean generating functional for a free scalar field,

$$Z_E[j] = \exp\left\{ \tfrac{1}{2} \int d^4x\, d^4y\, j(x) D_E(x - y) j(y) \right\}, \qquad (6.2.273)$$

where the periodic boundary conditions do not change the result of Eq. (4.1.205) if the sources are also periodic. Since the sources have compact support, then they vanish on the boundary. For a free field the ground state is the empty vacuum and so $|\Omega\rangle \to |0\rangle$. Then for the Euclidean two-point Schwinger function we find

$$\langle 0 | T\hat{\phi}(x)\hat{\phi}(y) | 0 \rangle_E = D_E(x - y), \qquad (6.2.274)$$

where the ground state is normalized, $\langle \Omega | \Omega \rangle = 1$, unless stated otherwise. Higher-order Schwinger functions involve sums of products of Δ_E. We observe that Green's functions with odd numbers of spacetime points are zero for a free theory.

We have for the Minkowski space action with source

$$
\begin{aligned}
S[\phi, j] &= \int d^4x \left\{ \tfrac{1}{2} \partial_\mu\phi\partial^\mu\phi - (m^2 - i\epsilon)\phi^2 + j\phi \right\} \\
&= \int d^4x\, d^4y \left[\tfrac{1}{2}\phi(x) \left\{ \left(-\partial_\mu^x \partial^{x\mu} - m^2 + i\epsilon\right) \delta^4(x - y) \right\} \phi(y) \right] + \int d^4x\, j\phi,
\end{aligned}
\qquad (6.2.275)
$$

which includes the prescription for the Feynman boundary conditions that arises from the infinitesimal Wick rotation. Using Eqs. (4.1.205) and (6.2.262) we find

$$Z[j] = \exp\left\{ -\tfrac{1}{2} \int d^4x\, d^4y\, j(x)\, D_F(x - y) j(y) \right\}. \qquad (6.2.276)$$

From Eq. (6.2.263) we evaluate the two-point function and recover Eq. (6.2.151),

$$\langle 0 | T\hat{\phi}(x)\hat{\phi}(y) | 0 \rangle = D_F(x - y). \qquad (6.2.277)$$

We will not explicitly indicate the *periodic boundary conditions in space and time* for bosons, but implicitly understand their presence.

From Eqs. (6.2.263) and (6.2.276) we see that for a free field we arrive at a result consistent with the consequence of Wick's theorem as given in Eq. (6.2.211). It is straightforward to verify that

$$
\begin{aligned}
\langle 0 | T\hat{\phi}(x_1)\cdots\hat{\phi}(x_k) | 0 \rangle &= (-i)^k \frac{\delta^k}{\delta j(t_1)\cdots\delta j(t_k)} Z[j] \Bigg|_{j=0} \\
&= \delta_{[k=\text{even}]} \left\{ D_F(x_1 - x_2) D_F(x_3 - x_4) \cdots D_F(x_{k-1} - x_k) \right. \\
&\qquad\qquad \left. + \text{all pairwise combinations of } (x_1, x_2, \ldots, x_k) \right\}.
\end{aligned}
\qquad (6.2.278)
$$

Free Charged Scalar Field

The Lagrangian density for a free charged scalar field plus a source term, $j(x)$, is

$$\mathcal{L} = \partial_\mu \phi^* \partial^\mu \phi - (m^2 - i\epsilon)\phi^* \phi + j^* \phi + \phi^* j, \tag{6.2.279}$$

where ϕ and j are now complex rather than real and where we have included the ϵ term leading to the Feynman boundary conditions. As we saw in Eq. (3.2.48) we could write this as the sum of two Lagrangian densities for real scalar fields ϕ_1 and ϕ_2 and real sources j_1 and j_2; i.e., we easily verify that

$$S[\phi, \phi^*, j, j^*] = S[\phi_1, j_1] + S[\phi_2, j_2] \tag{6.2.280}$$

with $S[\phi, j]$ is as defined above for the real scalar field in Eq. (6.2.275). In the complex plane an infinitesimal element of area is $dx dy = dz dz^*/2i$ and so we see that up to an overall irrelevant constant we can write $\mathcal{D}\phi_1 \mathcal{D}\phi_2 = \mathcal{D}\phi \mathcal{D}\phi^*$ and so

$$
\begin{aligned}
Z[j, j^*] = Z[j_1]Z[j_2] &= \lim_{T \to \infty(1-i\epsilon)} \frac{\int (\mathcal{D}\phi_1 \mathcal{D}\phi_2)\, e^{i(S[\phi_1, j_1] + S[\phi_2, j_2])}}{\int (\mathcal{D}\phi_1 \mathcal{D}\phi_2)\, e^{i(S[\phi_1] + S[\phi_2])}} \\
&= \lim_{T \to \infty(1-i\epsilon)} \frac{\int (\mathcal{D}\phi \mathcal{D}\phi^*)\, e^{iS[\phi, \phi^*, j, j^*]}}{\int (\mathcal{D}\phi \mathcal{D}\phi^*)\, e^{iS[\phi, \phi^*]}}.
\end{aligned}
\tag{6.2.281}
$$

From the definition of the Writinger derivative in Eq. (A.2.20) we see from Eq. (6.2.262) that $\delta/\delta j(x)$ and $\delta/\delta j^*(x)$ will introduce a $\hat{\phi}^\dagger(x)$ and a $\hat{\phi}(x)$, respectively, into the time-ordered product. We note that

$$
\begin{aligned}
Z[j, j^*] &= Z[j_1]Z[j_2] \\
&= \exp\left\{ -\tfrac{1}{2} \int d^4x\, d^4y\, [j_1(x)\, D_F(x-y) j_1(y) + j_2(x)\, D_F(x-y) j_2(y)] \right\} \\
&= \exp\left\{ -\int d^4x\, d^4y\, j^*(x)\, D_F(x-y) j(y) \right\},
\end{aligned}
\tag{6.2.282}
$$

since from Eq. (6.2.152) we know that D_F is an even function. It follows that

$$\langle 0|T\hat{\phi}(x)\hat{\phi}^\dagger(y)|0\rangle = (-i)^2 \left. \frac{\delta^2}{\delta j^*(x)\,\delta j(y)} Z[j, j^*]\right|_{j=0} \tag{6.2.283}$$

$$= \lim_{T \to \infty(1-i\epsilon)} \frac{\int (\mathcal{D}\phi \mathcal{D}\phi^*)\, \phi(x)\phi^*(y)\, e^{iS[\phi, \phi^*]}}{\int (\mathcal{D}\phi \mathcal{D}\phi^*)\, e^{iS[\phi, \phi^*]}} = D_F(x-y).$$

It is then readily verified that

$$
\begin{aligned}
\langle 0|T\hat{\phi}(x)\hat{\phi}^\dagger(y)|0\rangle &= \langle 0|T\hat{\phi}(y)\hat{\phi}^\dagger(x)|0\rangle = D_F(x-y), \\
\langle 0|T\hat{\phi}(x)\hat{\phi}(y)|0\rangle &= \langle 0|T\hat{\phi}^\dagger(x)\hat{\phi}^\dagger(y)|0\rangle = 0.
\end{aligned}
\tag{6.2.284}
$$

6.3 Fermions

The treatment of fermions presents challenges because in the everyday world we do not encounter anticommuting objects. So our intuition must grow and be built on mathematical constructs that we become familiar with. We first build the concept of a fermionic Fock space and introduce Grassmann variables. Then we use these to build the free fermion quantum field theory.

6.3.1 Annihilation and Creation Operators

In analogy with the case for bosons, we can choose an occupation number basis for the fermionic Fock space, $\overline{F}_-(V_H)$, that was introduced in Sec. 4.1.9. Let the single-particle Hilbert space V_H be finite-dimensional with N dimensions. Then we can choose an orthonormal basis for V_H, which we can write as $|b_i\rangle$ or $\mathbf{b}_i$ for $i = 1, \ldots, N$. An element of the occupation number basis for our fermion Fock space, $\overline{F}_-(V_H)$, has the form

$$|n_1, n_2, \ldots, n_N\rangle \quad \text{with} \quad n_i = 0, 1 \quad \text{for} \quad i = 1, \ldots, N, \tag{6.3.1}$$

where the total number of fermions in this basis state is $n = \sum_{i=1}^{N} n_i$ and clearly we must have $n \leq N$; i.e., we cannot have more fermions than there are dimensions of our one-particle Hilbert space, since by the Pauli exclusion principle no two fermions can be in the same state. So for any fermion Fock state $|\Psi\rangle \in \overline{F}_-(V_H)$

$$|\Psi\rangle = \sum_{n_1,\ldots,n_N=0}^{1} c_{n_1 n_2 \cdots n_N} |n_1, n_2, \ldots, n_N\rangle \tag{6.3.2}$$

for $c_{n_1 n_2 \cdots n_N} \in \mathbb{C}$. Denote the annihilation and creation operators for the single-fermion basis state $|b_i\rangle$ as $\hat{b}_i$ and $\hat{b}_i^\dagger$, respectively, with the properties

$$\hat{b}_i^\dagger|0\rangle = |b_i\rangle, \quad \hat{b}_i^\dagger|b_i\rangle = 0, \quad \hat{b}_i|b_i\rangle = |0\rangle, \quad \hat{b}_i|0\rangle = 0, \tag{6.3.3}$$

which we summarize in terms of the ith basis state occupation number $n_i = 0, 1$ as

$$\hat{b}_i^\dagger|n_i\rangle = (1 - n_i)|n_i + 1\rangle, \qquad \hat{b}_i|n_i\rangle = n_i|n_i - 1\rangle. \tag{6.3.4}$$

So if the state $|b_i\rangle$ is unoccupied then $\hat{b}_i^\dagger$ creates a fermion in that state and $\hat{b}_i$ acting on the state gives zero, whereas if $|b_i\rangle$ is occupied then $\hat{b}_i^\dagger$ acting on the state gives zero and $\hat{b}_i$ annihilates a fermion from that state. We also note that $\hat{N}_i = \hat{b}_i^\dagger \hat{b}_i$ is the number operator for the ith basis state and that $\{\hat{b}_i, \hat{b}_i^\dagger\}|n_i\rangle = |n_i\rangle$, since

$$\hat{N}_i|n_i\rangle = \hat{b}_i^\dagger \hat{b}_i|n_i\rangle = n_i \hat{b}_i^\dagger|n_i - 1\rangle = n_i[1 - (n_i - 1)]|n_i\rangle = n_i|n_i\rangle,$$
$$\hat{b}_i \hat{b}_i^\dagger|n_i\rangle = (1 - n_i)\hat{b}_i|n_i + 1\rangle = (1 - n_i)(n_i + 1)|n_i\rangle = (1 - n_i)|n_i\rangle,$$
$$\{\hat{b}_i, \hat{b}_i^\dagger\}|n_i\rangle = [n_i + (1 - n_i)]|n_i\rangle = |n_i\rangle, \tag{6.3.5}$$

where we have used $n_i^2 = n_i$. With respect to basis states in Eq. (4.1.158) we have

$$\hat{b}_{i_1}^\dagger \hat{b}_{i_2}^\dagger \cdots \hat{b}_{i_n}^\dagger|0\rangle = |b_{i_1} b_{i_2} \cdots b_{i_n}; A_o\rangle \quad \text{with ordering} \quad i_1 < i_2 < \cdots < i_n. \tag{6.3.6}$$

For the unordered basis in Eq. (4.1.155) we see that exchanging the positions of any two b_i entries leads to a sign change. In order to achieve this we therefore require that $\{\hat{b}_i^\dagger, \hat{b}_j^\dagger\} = 0$ for $i \neq j$. For consistency we also then require that $\{\hat{b}_i, \hat{b}_j\} = \{\hat{b}_i, \hat{b}_j^\dagger\} = 0$ for $i \neq j$. So we have arrived at the anticommutation relations for fermion annihilation and creation operators,

$$\{\hat{b}_i, \hat{b}_j^\dagger\} = \delta_{ij} \quad \text{and} \quad \{\hat{b}_i, \hat{b}_j\} = \{\hat{b}_i^\dagger, \hat{b}_j^\dagger\} = 0, \tag{6.3.7}$$

It will be useful later to note here that

$$[\hat{b}_i, \hat{b}_j^\dagger \hat{b}_k] = \hat{b}_i \hat{b}_j^\dagger \hat{b}_k - \hat{b}_j^\dagger \hat{b}_k \hat{b}_i = \hat{b}_i \hat{b}_j^\dagger \hat{b}_k + \hat{b}_j^\dagger \hat{b}_i \hat{b}_k = \{\hat{b}_i, \hat{b}_j^\dagger\}\hat{b}_k = \delta_{ij}\hat{b}_k \tag{6.3.8}$$

and similarly by symmetry $[\hat{b}_i^\dagger, \hat{b}_j \hat{b}_k^\dagger] = \delta_{ij}\hat{b}_k^\dagger$. Remembering that $n_i = 0, 1$ then we can write the fermion occupation number basis as

$$|n_1, n_2, \ldots, n_N\rangle = (\hat{b}_1^\dagger)^{n_1}(\hat{b}_2^\dagger)^{n_2} \cdots (\hat{b}_N^\dagger)^{n_N}|0\rangle, \tag{6.3.9}$$

where we understand that $(\hat{b}_i^\dagger)^0 = \hat{I}$ and $(\hat{b}_i^\dagger)^1 = \hat{b}_i^\dagger$. Since it follows from the above anticommutation relations that $[\hat{b}_i^\dagger \hat{b}_i, \hat{b}_j^\dagger] = 0$ for $i \neq j$ we then see that

$$\hat{N}_i|n_1, \ldots, n_N\rangle = n_i|n_1, \ldots, n_N\rangle, \tag{6.3.10}$$

$$\hat{N}|n_1, \ldots, n_N\rangle \equiv \sum_{i=1}^{N} \hat{N}_i|n_1, \ldots, n_N\rangle = \sum_{i=1}^{N} \hat{n}_i|n_1, \ldots, n_N\rangle \equiv n|n_1, \ldots, n_N\rangle,$$

where $\hat{N}$ is the total number operator and where there are n fermions in the state.

Consider the action of an annihilation operator $\hat{b}_j$ and a creation operator $\hat{b}_j^\dagger$ on an n-fermion ordered basis state ($i_1 < \cdots < i_n$),

$$\hat{b}_j|b_{i_1} b_{i_2} \cdots b_{i_n}; A_o\rangle = \sum_{k=1}^{n} \delta_{ji_k} (-1)^{k-1}|b_{i_1} \cdots b_{i_{k-1}} b_{i_{k+1}} \cdots \hat{b}_{i_n}; A_o\rangle,$$

$$\hat{b}_j^\dagger|b_{i_1} b_{i_2} \cdots b_{i_n}; A_o\rangle = \left[\prod_{\ell=1}^{n}(1 - \delta_{ji_\ell})\right] (-1)^k|b_{i_1} \cdots b_{i_k} b_j b_{i_{k+1}} \cdots \hat{b}_{i_n}; A_o\rangle. \tag{6.3.11}$$

The phase factor in each case comes from moving $\hat{b}_j$ or $\hat{b}_j^\dagger$ through the first $(k-1)$ or k occupied fermion states, respectively. The first result means that the right-hand side vanishes unless j corresponds to one of the occupied fermion states, say i_k; and, if so, then this state is emptied and we are left with the phase factor multiplying the $(n-1)$-fermion state. The second result means that the right-hand side vanishes if j corresponds to any already occupied state; and, if not, then it creates a particle in the state j in the ordering such that $i_k < j < i_{k+1}$ and we are left with the phase factor and the $(n+1)$-fermion state.

Proof: We have $\hat{b}_j|b_{i_1} b_{i_2} \cdots b_{i_n}; A_o\rangle = \hat{b}_j \hat{b}_{i_1}^\dagger \hat{b}_{i_2}^\dagger \cdots \hat{b}_{i_n}^\dagger|0\rangle$ and if j is not equal to any of the i_k then $\hat{b}_j$ can anticommute through to the empty state and then $\hat{b}_j|0\rangle = 0$. If $j = i_k$ then we anticommute $\hat{b}_j$ through the first $k - 1$ entries picking up the phase factor $(-1)^{k-1}$ until we have $\hat{b}_{i_1}^\dagger \cdots (\hat{b}_{i_k} \hat{b}_{i_k}^\dagger) \cdots \hat{b}_{i_n}^\dagger|0\rangle$. Then $(\hat{b}_{i_k} \hat{b}_{i_k}^\dagger)$ commutes through the other creation operators. We then use $(\hat{b}_{i_k} \hat{b}_{i_k}^\dagger)|0\rangle = |0\rangle$ and so $(\hat{b}_{i_k} \hat{b}_{i_k}^\dagger)$ then disappears. This proves the first result.

Next note that $\hat{b}_j^\dagger|b_{i_1} \cdots b_{i_n}; A_o\rangle = \hat{b}_j^\dagger \hat{b}_{i_1}^\dagger \cdots \hat{b}_{i_n}^\dagger|0\rangle$. We can anticommute $\hat{b}_j^\dagger$ through until we have either $j = i_k$ or $i_k < j < i_{k+1}$. If $j = i_k$ then we have $\hat{b}_{i_1}^\dagger \cdots (\hat{b}_{i_k}^\dagger \hat{b}_{i_k}^\dagger) \cdots \hat{b}_{i_n}^\dagger|0\rangle$ and we commute $(\hat{b}_{i_k}^\dagger \hat{b}_{i_k}^\dagger)$ through and use $(\hat{b}_{i_k}^\dagger \hat{b}_{i_k}^\dagger)|0\rangle = 0$. If $i_k < j < i_{k+1}$ then we anticommute $\hat{b}_j^\dagger$ through the first k operators giving the $(n+1)$-fermion ordered basis state.

6.3.2 Fock Space and Grassmann Algebra

As already noted following Eq. (4.1.159), the Fock space for fermions is the Hilbert space, $\bar{F}_-(V_H)$, which is the exterior or Grassmann algebra built on the single-particle Hilbert space V_H. It is common to describe the Grassmann algebra of a fermion Fock space in terms of Grassmann variables.

A brief introduction to Grassmann variables and algebra is given below. Detailed discussions can be found in Berezin (1966) and Negele and Orland (1988). Grassmann variables provide a representation of fermion Fock space and Grassmann algebra additionally provides the means to represent the operations performed on it. This leads to a representation of the quantum field theory of fermions in terms of a fermion path integral. We will use the terms *Grassmann variables* and *Grassmann generators* interchangeably, since both are in common use. Grassmann variables are also variously referred to as *Grassmann supercharges* or *Grassmann directions*. A *Grassmann number* or *supernumber* is any element of the Grassmann algebra.

Grassmann variables and Grassmann algebra: As we saw in Sec. 4.1.9, the Fock space for fermions involves the Grassmann (or exterior) algebra, named after Hermann Grassmann. Grassmann algebra can be thought of as the algebraic system whose product is the exterior product. The concepts of differentiation and integration are introduced in the Grassmann algebra, and while these may initially appear odd, once in use it becomes clear why these are natural definitions.

Let G_N be the Grassmann algebra with the N Grassmann variables or Grassmann generators $a_1, a_2, a_3, \ldots, a_N$. The generators anticommute,

$$\{a_i, a_j\} \equiv a_i a_j + a_j a_i = 0 \tag{6.3.12}$$

for all $i, j = 1, \ldots, N$. Hence, $(a_i)^2 = 0$ and $(a_i)^m = 0$ for all integers $m \geq 2$. While we can always make a representation of the Grassmann generators using matrices, it is not particularly useful to do so; e.g., for two Grassmann generators a_1 and a_2 we can construct 4×4 matrices such that $a_1^2 = a_2^2 = 0$ and $a_1 a_2 + a_2 a_1 = 0$. With more Grassmann generators we need increasingly large matrices. Consider an arbitrary function of the Grassmann generators, referred to as a *Grassmann number* or *Grassmann element* or sometimes as a *supernumber*,

$$f(a) = \sum_{\nu=0}^{N} \sum_{k_1,\ldots,k_\nu=1}^{N} f_{k_1 k_2 \cdots k_\nu} a_{k_1} a_{k_2} \cdots a_{k_\nu}, \tag{6.3.13}$$

where the coefficients $f_{k_1,\ldots,k_\nu}$ are real or complex (i.e., they are c-numbers) and antisymmetric under the pairwise interchange of any two of the indices, e.g., $f_{ijkl} = -f_{jikl} = -f_{ikjl} =$ etc. Any non-antisymmetric component of the coefficients would be projected away as $a_{k_1} a_{k_2} \cdots a_{k_\nu}$ is antisymmetric under any pairwise exchange. If we require $k_1 < k_2 < \cdots < k_N$ then the decomposition in Eq. (6.3.13) of any Grassmann number in terms of these generators is unique. The set of all distinct functions of the Grassmann generators defines the Grassmann algebra consisting of the Grassmann elements $f(a)$. There are 2^N distinct monomials, since the set of all monomials is formed using $a_1^{n_1} \cdots a_N^{i_N}$ with each $n_i = 0$ or 1. The monomials are 1; $a_1, \ldots, a_N$; $a_1 a_2, a_1 a_3, \ldots, a_{N-1} a_N$; $\ldots$; $a_1 a_2 a_3 a_4 \cdots a_N$. These 2^N distinct monomials form a basis for G_N through Eq. (6.3.13). The monomials are *even* or *odd* depending on whether they have an even or odd number of generators a_i. Those elements $f(a)$ containing only combinations of even or odd monomials are called even or odd Grassmann elements, respectively. Even elements commute with the entire Grassmann algebra just as c-numbers do and odd elements commute with even elements and anticommute with odd elements.

A left derivative operation in G_N is defined as

$$\frac{\partial}{\partial a_i} 1 \equiv 0, \quad \frac{\partial}{\partial a_i} a_j \equiv \delta_{ij}, \quad \frac{\partial}{\partial a_i} a_j \overset{i \neq j}{\equiv} -a_j \frac{\partial}{\partial a_i}, \quad \frac{\partial}{\partial a_i} \frac{\partial}{\partial a_j} = -\frac{\partial}{\partial a_j} \frac{\partial}{\partial a_i}. \tag{6.3.14}$$

The right derivative is similarly defined as

$$1 \frac{\overleftarrow{\partial}}{\partial a_i} \equiv 0, \quad a_j \frac{\overleftarrow{\partial}}{\partial a_i} \equiv \delta_{ij}, \quad a_j \frac{\overleftarrow{\partial}}{\partial a_i} \overset{i \neq j}{\equiv} -a_j \frac{\partial}{\partial a_i}, \quad \frac{\overleftarrow{\partial}}{\partial a_i} \frac{\overleftarrow{\partial}}{\partial a_j} = -\frac{\overleftarrow{\partial}}{\partial a_j} \frac{\overleftarrow{\partial}}{\partial a_i}. \tag{6.3.15}$$

So derivatives anticommute and mixed anticommutators give delta functions,

$$\{\tfrac{\partial}{\partial a_i}, \tfrac{\partial}{\partial a_j}\} = \{\tfrac{\overleftarrow{\partial}}{\partial a_i}, \tfrac{\overleftarrow{\partial}}{\partial a_j}\} = 0, \quad \{\tfrac{\partial}{\partial a_i}, a_j\} = \{a_j, \tfrac{\overleftarrow{\partial}}{\partial a_i}\} = \delta_{ij}, \tag{6.3.16}$$

where we verify these results by acting from the left on any $f(a) \in G_N$. We can also define an integration-type operation

$$\int da_i \, 1 \equiv 0, \quad \int da_i \, a_j \equiv -\int a_j \, da_i = \delta_{ij}, \quad \{a_i, da_j\} \equiv \{da_i, da_j\} \equiv 0, \qquad (6.3.17)$$

which has similarities with Grassmann differentiation.

From the above it follows that for $i \neq j$,

$$\int da_i \, a_i a_j = a_j, \quad \int da_i da_j \, a_i = -\int da_j da_i a_i = -\int da_j 1 = 0$$
$$\int da_i da_j \, a_i a_j = -\int da_i \left(\int da_j \, a_j\right) a_i = -1. \qquad (6.3.18)$$

Expanding $f(a)$ as in Eq. (6.3.13) we find a Grassmann "exact differential,"

$$\int da_i \frac{\partial}{\partial a_i} f(a) = 0. \qquad (6.3.19)$$

Integrating over all of the N Grassmann generators in G_N leads to

$$\int da_N \cdots da_1 (a_1 \cdots a_N) = 1, \quad \int da_N \cdots da_1 (a_{k_1} \cdots a_{k_N}) = \epsilon^{k_1 \cdots k_N},$$
$$\int da_N \cdots da_1 \, f(a) = \sum_{k_1,\ldots,k_N=1}^{N} \epsilon^{k_1 \cdots k_N} f_{k_1 \cdots k_N} = N! \, f_{123 \cdots N}. \qquad (6.3.20)$$

Only the term in $f(a)$ containing every generator survives.

Consider the linear change of Grassmann generators defined by

$$b_i \equiv \sum_{j=1}^{N} B_{ij} a_j, \quad \frac{\partial}{\partial b_i} \equiv \sum_{j=1}^{N} \frac{\partial}{\partial a_j} (B^{-1})_{ji}, \quad db_i \equiv \sum_{j=1}^{N} da_j (B^{-1})_{ji}, \qquad (6.3.21)$$

where B is any nonsingular $N \times N$ matrix with c-number matrix elements B_{ij}. The $b_1, \ldots, b_N$ are also Grassmann generators of G_N since it is readily verified that $\{b_i, b_j\} = \{db_i, db_j\} = \{\frac{\partial}{\partial b_i}, \frac{\partial}{\partial b_j}\} = \{b_i, db_j\} = 0$, $\frac{\partial b_i}{\partial b_j} = \delta_{ij}$ and $\int db_i \, b_j = \delta_{ij}$. The linear change of variables is just a change of basis for G_N since every $f(a) \in G_N$ can equally well be written as some $g(b) = f(a)$,

$$f(a) = \sum_{\nu=0}^{N} \sum_{k_1,\ldots,k_\nu=1}^{N} f_{k_1 \cdots k_\nu} \left(\sum_{j_1,\ldots,j_\nu=1}^{N} B_{k_1 j_1}^{-1} \cdots B_{k_\nu j_\nu}^{-1} b_{j_1} \cdots b_{j_\nu}\right)$$
$$\equiv \sum_{\nu=0}^{N} \sum_{j_1,\ldots,j_\nu=1}^{N} g_{j_1 \cdots j_\nu} b_{j_1} \cdots b_{j_\nu} \equiv g(b). \qquad (6.3.22)$$

Consider the monomial containing every new generator of G_N,

$$b_1 \cdots b_N = \sum_{k_1,\ldots,k_N=1}^{N} B_{1k_1} \cdots B_{Nk_N} a_{k_1} \cdots a_{k_N} \qquad (6.3.23)$$
$$= \sum_{k_1,\ldots,k_N=1}^{N} B_{1k_1} \cdots B_{Nk_N} \epsilon^{k_1 k_2 \cdots k_N} a_1 \cdots a_N = (\det B) a_1 \cdots a_N.$$

Similarly, we easily find that

$$db_N \cdots db_1 \equiv (\det B)^{-1} da_N \cdots da_1 \quad \Rightarrow \quad \int db_N \cdots db_1 \, b_1 \cdots b_N = 1. \qquad (6.3.24)$$

It should be noted here that the Jacobian determinant associated with the change of Grassmann variables is the inverse of the Jacobian that would result for real variables as given in Eq. (A.2.5).

Consider a set of Grassmann generators $\eta_1, \ldots, \eta_N$ in addition to $a_1 \cdots a_N$ that anticommute with all of the $a_1, \ldots, a_N$. That is,

$$\{a_i, a_j\} = \{\eta_i, \eta_j\} = \{a_i, \eta_j\} = 0 \qquad (i, j = 1, \ldots, N). \qquad (6.3.25)$$

Note that $\{(a_i + \eta_i), (a_j + \eta_j)\} = 0$. From Eq. (6.3.20) we have

$$\int da_N \cdots da_1 \, f(a + \eta) = \int da_N \cdots da_1 \, f(a). \qquad (6.3.26)$$

Consider the $2N$ generators $a_1, \bar{a}_1, a_2, \bar{a}_2, \ldots, a_N, \bar{a}_N$ of G_{2N}, where the generators a_i and $\bar{a}_i$ are independent and no relation between them is to be implied here. Let A be any $N \times N$ c-number matrix, then

$$\int (\prod_{i=1}^{N} d\bar{a}_i da_i) \exp^{-\sum_{j,k=1}^{N} \bar{a}_j A_{jk} a_k} = \det A. \qquad (6.3.27)$$

Proof: For any matrix square matrix A there is a complex matrix P such that $PAP^{-1} = J$, where J has *Jordan canonical form*. J is an upper triangular matrix with the eigenvalues of A, λ_i, down the diagonal and all other entries zero except the super-diagonal entries, which are 0 or 1. Note that

$$\sum_{j,k=1}^{N} \bar{a}_j A_{jk} a_k = \sum_{j,k=1}^{N} \bar{a}_j (P^{-1}JP)_{jk} a_k = \sum_{j,k=1}^{N} \bar{b}_j J_{jk} b_k, \qquad (6.3.28)$$

where we have defined $b_j \equiv \sum_k P_{jk} a_k$ and $\bar{b}_k \equiv \sum_k \bar{a}_j (P^{-1})_{jk} = \sum_j (P^{-1T})_{kj} \bar{a}_j$. Since $\det(P)^{-1} \det(P^{-1T})^{-1} = \det(P^{-1}) \det(P) = 1$, then from Eq. (6.3.24),

$$\prod_{i,j=1}^{N} d\bar{b}_i db_j = \prod_{i,j=1}^{N} d\bar{a}_i \, da_j \quad \text{and} \quad \prod_{i=1}^{N} d\bar{b}_i db_i = \prod_{i=1}^{N} d\bar{a}_i da_i. \qquad (6.3.29)$$

A matrix A is a *normal matrix* ($[A, A^\dagger] = 0$) if and only if it is diagonalizable by a unitary transformation U such that $UAU^\dagger = D$, where $D_{ij} = \lambda_i \delta_{ij}$. Clearly, Hermitian matrices ($A = A^\dagger$) are normal matrices. If A is normal, then we find for the left-hand side (LHS) of Eq. (6.3.27) the required result,

$$\text{LHS} = \int (\prod_{i=1}^{N} d\bar{b}_i db_i) e^{-\sum_{j=1}^{N} \bar{b}_j \lambda_j b_j} = \prod_{i=1}^{N} \int d\bar{b}_i db_i (-\bar{b}_i \lambda_i b_i) = \prod_{i=1}^{N} \lambda_i = \det(A),$$

where we have again used Eq. (6.3.20) and the fact that even Grassmann monomials commute. Returning to the general case for A we have

$$\text{LHS} = \int \prod_{i=1}^{N} d\bar{b}_i db_i e^{-\sum_{jk} \bar{b}_j J_{jk} b_k} = \int \prod_{i=1}^{N} d\bar{b}_i db_i e^{-\sum_j \bar{b}_j \lambda_j b_j} e^{-\sum_{k>j} \bar{b}_j J_{jk} b_k}$$

$$= \prod_{i=1}^{N} \int d\bar{b}_i db_i e^{-\sum_j \bar{b}_j \lambda_j b_j} = \det(A), \qquad (6.3.30)$$

which proves the result. Only unity in the expansion of the second exponential contributes, since all other terms in the expansion will contain terms $\bar{b}_j b_k$ with $k > j$ and Eq. (6.3.20) shows that such terms do not survive the integrations.

Consider now the $4N$ generators $a_1, \bar{a}_1, \ldots, a_N, \bar{a}_N$ and $\eta_1, \bar{\eta}_1, \ldots, \eta_N, \bar{\eta}_N$ of the Grassmann algebra G_{4N}. We find the result that

$$\int (\prod_{i=1}^{N} d\bar{a}_i da_i) e^{\sum_{j,k} (-\bar{a}_j A_{jk} a_k) + \sum_j (\bar{\eta}_j a_j + \bar{a}_j \eta_j)} = \det(A) e^{\sum_{i,j} \bar{\eta}_i A_{ij}^{-1} \eta_j} \qquad (6.3.31)$$

for any nonsingular complex matrix A. Using shorthand notation we have

$$-\bar{a}Aa + \bar{\eta}a + \bar{a}\eta = -(\bar{a} - \bar{\eta}A^{-1})A(a - A^{-1}\eta) + \bar{\eta}A^{-1}\eta, \qquad (6.3.32)$$

which combined with Eqs. (6.3.26) and (6.3.27) immediately proves Eq. (6.3.31). It will be useful to note that

$$(-1)\frac{\partial^2}{\partial \bar{\eta}_i \partial \eta_j} e^{\sum_{i,j=1}^{N} \bar{\eta}_i A_{ij}^{-1} \eta_j} \Big|_{\bar{\eta} = \eta = 0} = A_{ij}^{-1}, \qquad (6.3.33)$$

which generalizes to

$$\frac{(-1)^m \partial^{2m}}{\partial \bar{\eta}_{i_1} \partial \eta_{j_1} \cdots \partial \bar{\eta}_{i_m} \partial \eta_{j_m}} e^{\sum \bar{\eta}_i A_{ij}^{-1} \eta_j} \Big|_{\bar{\eta} = \eta = 0} = \sum_{k's} \epsilon^{k_1 \cdots k_m} A_{i_1 j_{k_1}}^{-1} \cdots A_{i_m j_{k_m}}^{-1}. \qquad (6.3.34)$$

Unequal numbers of $\bar{\eta}$ and η derivatives of $e^{\bar{\eta}(A^{-1})\eta}$ give zero when $\bar{\eta} = \eta = 0$.

Associate a Grassmann generator ξ_i with every basis state in the N-dimensional single-fermion Hilbert space, $|b_i\rangle$ or $\mathbf{b}_i \in V_H$, and associate a Grassmann generator ξ_i^* with every basis state in the dual single-fermion Hilbert space, $\langle b_i|$ or $\mathbf{b}_i^* \in V_H^*$. From Sec. 6.3.2 the $2N$ Grassmann generators have the properties

$$\{\xi_i, \xi_j\} = \{\xi_i^*, \xi_j^*\} = \{\xi_i, \xi_j^*\} = \xi_i^2 = \xi_i^{*2} = 0. \tag{6.3.35}$$

Consider the orthonormality conditions for the ordered n-fermion basis states in Eq. (4.1.158). Any state in fermion Fock space can be then written as

$$|\Psi\rangle = \sum_{n=0}^{N} \sum_{i_1 < \cdots < i_n = 1}^{N} c_{i_1 i_2 \cdots i_n} |b_{i_1} b_{i_2} \cdots b_{i_n}; A_o\rangle, \tag{6.3.36}$$

where $n = 0$ corresponds to the empty state contribution $c|0\rangle$. To every basis ket $|b_{i_1} b_{i_2} \cdots b_{i_n}; A_o\rangle$ we can associate the corresponding monomial $\xi_{i_1} \xi_{i_2} \cdots \xi_{i_n}$ and for every basis bra $\langle b_{i_1} b_{i_2} \cdots b_{i_n}; A_o|$ we can associate the monomial $\xi_{i_n}^* \cdots \xi_{i_2}^* \xi_{i_1}^*$. Following Berezin (1966)[3] we can choose every ket $|\Psi\rangle$ to have a one-to-one correspondence with an analytic function of Grassmann generators,

$$\Psi(\xi^*) \equiv \sum_{n=0}^{N} \sum_{i_1 < \cdots < i_n = 1}^{N} c_{i_1 i_2 \cdots i_n} \xi_{i_1}^* \xi_{i_2}^* \cdots \xi_{i_n}^*, \tag{6.3.37}$$

where $n = 0$ corresponds to the contribution $c \in \mathbb{C}$ with no generator; i.e., every state $|\Psi\rangle$ in the fermion Fock space has a corresponding unique Grassmann number associated with it, $\Psi(\xi^*)$. Under a change of basis the coefficients change. For the state Ψ and similarly for every bra $\langle\Psi|$ we have a one-to-one correspondence with

$$[\Psi(\xi^*)]^* \equiv \Psi^*(\xi) \equiv \sum_{n=0}^{N} \sum_{i_1 < \cdots < i_n = 1}^{N} c_{i_1 i_2 \cdots i_n}^* \xi_{i_n} \cdots \xi_{i_2} \xi_{i_1}. \tag{6.3.38}$$

The series terminates at $n = N$ since $\{\xi_i, \xi_j\} = 0$ means that $\xi_i^2 = 0$ and similarly $\xi_i^{*2} = 0$. Comparing the above shows the action of the "$*$" operator; i.e., it complex conjugates complex numbers such that $c \to c^*$, converts Grassmann generators such that $\xi \to \xi^*$ and $\xi^* \to \xi$ and reverses their order such that $\xi_i \xi_j \to \xi_j^* \xi_i^*$, $\xi_i^* \xi_j^* \to \xi_j \xi_i$ and $\xi_i^* \xi_j \to \xi_j^* \xi_i$. The "$*$" operator is the generalization of complex conjugation to the Grassmann algebra and it is the classical analog of the Hermitian conjugation of operators in quantum field theory. *Grassmann complex conjugation* is an involution (self-inverse operation) like complex conjugation, since

$$[[\Psi(\xi)]^*]^* = \Psi(\xi). \tag{6.3.39}$$

A Grassmann number in a Grassmann algebra with N complex generators $\xi_1, \ldots, \xi_N$ can be expressed as a finite power series in the generators $g \equiv g(\xi, \xi^*) \in G_N$. A Grassmann number g is said to be *real* if $g^* = g$, *imaginary* if $g^* = -g$ and *complex* if $g^* \neq g$. Any complex Grassmann number decomposes into its real and imaginary parts, $g = \frac{1}{2}(g + g^*) + \frac{1}{2}(g - g^*)$. We will mostly be concerned with complex Grassmann numbers and generators, where we treat ξ and ξ^* as independent variables in a Wirtinger-like way, as we do for z and z^* in Sec. A.2.5.

Let the state $|\Phi\rangle$ be defined by Eq. (6.3.36) with complex coefficients $d_{i_1 i_2 \cdots i_n}$, then the Fock space inner product $\langle\Psi|\Phi\rangle$ in Hilbert space representation is

$$\langle\Psi|\Phi\rangle = \sum_{n=0}^{N} \sum_{i_1 < \cdots < i_n = 0}^{N} c_{i_1 i_2 \cdots i_n}^* d_{i_1 i_2 \cdots i_n}. \tag{6.3.40}$$

---

[3] Note that Berezin (1966) uses typical mathematics notation for the inner product, $(\Phi, \Psi) \equiv \langle\Psi|\Phi\rangle$, uses $a^*(x)$ and $a(x)$ in place of ξ_i^* and ξ_i, uses a bar to denote complex conjugation and uses unordered summations requiring additional $1/\sqrt{n!}$ normalization factors.

We can represent this inner product in terms of the Grassmann algebra as

$$\langle \Psi | \Phi \rangle = \int \left(\prod_{i=1}^{N} d\xi_i^* d\xi_i \right) e^{-\sum_{j=1}^{N} \xi_j^* \xi_j} \Psi^*(\xi) \Phi(\xi^*). \tag{6.3.41}$$

Note that $d\xi_i^* d\xi_i$ and $\xi_i^* \xi_i$ are even Grassmann elements and so commute with all other Grassmann elements.

Proof: Note that even Grassmann monomials commute and so, e.g., we have $[\xi_i^* \xi_i, \xi_j^* \xi_j] = [d\xi_i^* d\xi_i, \xi_j^* \xi_j] = 0$. The BCH formula in Eq. (A.8.6) then gives

$$e^{-\sum_{j=1}^{N} \xi_j^* \xi_j} = e^{\sum_{j=1}^{N} \xi_j \xi_j^*} = \prod_{j=1}^{N} e^{\xi_j \xi_j^*} = \prod_{j=1}^{N} (1 + \xi_j \xi_j^*), \tag{6.3.42}$$

where the last step follows since $(\xi_j^* \xi_j)^2 = 0$. We then find the result that

$$
\begin{aligned}
&\int \left(\prod_{i=1}^{N} d\xi_i^* d\xi_i \right) e^{-\sum_{j=1}^{N} \xi_j^* \xi_j} \Psi^*(\xi) \Phi(\xi^*) \\
&= \sum_{m,n=0}^{N} \sum_{i_1 < \cdots < i_m = 1}^{N} \sum_{j_1 < \cdots < j_n = 1}^{N} c_{i_1 \cdots i_m}^* d_{j_1 \cdots j_n} \\
&\quad \times \left[\int \left(\prod_{i=1}^{N} d\xi_i^* d\xi_i \right) e^{-\sum_{j=1}^{N} \xi_j^* \xi_j} \xi_{i_m} \cdots \xi_{i_1} \xi_{j_1}^* \cdots \xi_{j_n}^* \right] \\
&= \sum_{n=0}^{N} \sum_{i_1 < \cdots < i_n = 1}^{N} c_{i_1 i_2 \cdots i_n}^* d_{i_1 i_2 \cdots i_n} = \langle \Psi | \Phi \rangle, \tag{6.3.43}
\end{aligned}
$$

where the last step can be understood as follows: The only terms in the square bracket that survive the Grassmann integration contain every ξ_k^* and every ξ_k once and only once. This means that we must have $m = n$, and, since indices are strictly increasing in the sums, then we must have $i_k = j_k$ for all $k = 1, \ldots, n$. Then the term in the square bracket can be evaluated as

$$
\begin{aligned}
&\int \left(\prod_{i=1}^{N} d\xi_i^* d\xi_i \right) e^{-\sum_{j=1}^{N} \xi_j^* \xi_j} \xi_{i_m} \cdots \xi_{i_1} \xi_{j_1}^* \cdots \xi_{j_n}^* \\
&= \delta_{mn} \delta_{i_1 j_1} \cdots \delta_{i_n j_n} \int \left(\prod_{i=1}^{N} d\xi_i^* d\xi_i \right) \left[e^{\sum_{j=1}^{N} \xi_j \xi_j^*} \xi_{i_1} \xi_{i_1}^* \cdots \xi_{i_n} \xi_{i_n}^* \right] \\
&= \delta_{mn} \delta_{i_1 j_1} \cdots \delta_{i_n j_n} \prod_{i=1}^{N} \int d\xi_i^* d\xi_i \, \xi_i \xi_i^* \\
&= \delta_{mn} \delta_{i_1 j_1} \cdots \delta_{i_n j_n}, \tag{6.3.44}
\end{aligned}
$$

where we have used $\xi_{i_n} \cdots \xi_{i_1} \xi_{i_1}^* \cdots \xi_{i_n}^* = \xi_{i_1} \xi_{i_1}^* \cdots \xi_{i_n} \xi_{i_n}^*$ and $\int d\xi^* d\xi \, \xi \xi^* = 1$.

Let $\hat{A}$ be a linear operator on fermion Fock space such that $|\Psi\rangle = \hat{A}|\Phi\rangle$, then we represent this in terms of our basis $|b_{i_1} b_{i_2} \cdots b_{i_n}; A_o\rangle$ as

$$
\begin{aligned}
|\Psi\rangle &= \sum_{m=0}^{N} \sum_{i_1 < \cdots < i_m = 1}^{N} c_{i_1 i_2 \cdots i_m} |b_{i_1} b_{i_2} \cdots b_{i_m}; A_o\rangle \tag{6.3.45} \\
&= \sum_{m,n=0}^{N} \sum_{i_1 < \cdots < i_m = 1}^{N} \sum_{j_1 < \cdots < j_n = 1}^{N} A_{i_1 \cdots i_m, j_1 \cdots j_n} d_{j_1 \cdots j_n} |b_{j_1} \cdots b_{j_n}; A_o\rangle.
\end{aligned}
$$

The properties of $\hat{A}$ are determined by the set of coefficients $A_{i_1 \cdots i_m, j_1 \cdots j_n} \in \mathbb{C}$ with ordered indices, which can be thought of as multidimensional "matrix elements" of $\hat{A}$ in this basis. Let η_i and η_i^* for $i = 1, \ldots, M$ be dummy Grassmann generators over which we integrate. Then in terms of the Grassmann algebra we have

$$\Psi(\xi^*) = \int \left(\prod_{i=1}^{N} d\eta_i^* d\eta_i \right) e^{-\sum_{j=1}^{N} \eta_j^* \eta_j} \tilde{A}(\xi^*, \eta) \Phi(\eta^*), \tag{6.3.46}$$

where $\tilde{A}$ is referred to as the *matrix form* of the operator $\hat{A}$ and is given by

$$
\begin{aligned}
\tilde{A}(\xi^*, \eta) &= \sum_{m,n=0}^{N} \sum_{i_1 < \cdots < i_m = 1}^{N} \sum_{j_1 < \cdots < j_n = 1}^{N} A_{i_1 \cdots i_m, j_1 \cdots j_n} \\
&\quad \times \xi_{i_1}^* \cdots \eta_{i_m}^* \eta_{j_n} \cdots \eta_{j_1}. \tag{6.3.47}
\end{aligned}
$$

Using Eq. (6.3.44) in Eq. (6.3.46) we immediately find that

$$\Psi(\xi^*) = \sum_{m=1}^{N} \sum_{i_1 < \cdots < i_m = 1}^{N} c_{i_1 i_2 \cdots i_m} \xi_{i_1}^* \cdots \xi_{i_m}^* \tag{6.3.48}$$
$$= \sum_{m,n=1}^{N} \sum_{i_1 < \cdots < i_m = 1}^{N} \sum_{j_1 < \cdots < j_n = 1}^{N} A_{i_1 \cdots i_m, j_1 \cdots j_n} d_{j_1 \cdots j_n} \xi_{j_1}^* \cdots \xi_{j_n}^*,$$

which is the Grassmann equivalent of Eq. (6.3.45). So if $|\Psi\rangle$ and $|\Phi\rangle$ are any two fermion Fock states and $\hat{A}$ is a linear operator, then

$$\langle\Psi|\hat{A}|\Phi\rangle = \int \left(\prod_{i=1}^{N} d\xi_i^* d\xi_i d\eta_i^* d\eta_i\right) e^{-\sum_{j=1}^{N}(\xi_j^*\xi_j + \eta_j^*\eta_j)} \Psi^*(\xi) \tilde{A}(\xi^*, \eta) \Psi(\eta^*). \tag{6.3.49}$$

Similarly, it is easily verified that the product of two linear Fock-space operators, $\hat{C} = \hat{A}\hat{B}$, is represented as

$$\tilde{C}(\xi^{*\prime}, \xi) = \int \left(\prod_{i=1}^{N} d\eta_i^* d\eta_i\right) e^{-\sum_{j=1}^{N} \eta_j^*\eta_j} \tilde{A}(\xi^{*\prime}, \eta) \tilde{B}(\eta^*, \xi). \tag{6.3.50}$$

The trace of a bounded linear operator $\hat{A}$ in Fock space is given by the sum of the diagonal elements when represented in any orthonormal basis,

$$\text{tr}\hat{A} = \sum_{n=0}^{N} \sum_{i_1 < \cdots < i_n = 1}^{N} A_{i_1 \cdots i_n, i_1 \cdots i_n}. \tag{6.3.51}$$

If the trace exists ($\text{tr}\hat{A} < \infty$) we say that the operator is *trace class*. We readily see using Eq. (6.3.44) that in terms of Grassmann algebra the trace is given by

$$\text{tr}\hat{A} = \int \left(\prod_{i=1}^{N} d\xi_i^* d\xi_i\right) e^{-\sum_{j=1}^{N} \xi_j^*\xi_j} \tilde{A}(\xi^*, -\xi). \tag{6.3.52}$$

Proof: First note from Eqs. (6.3.44) and (6.3.47) that the only terms that can survive in Eq. (6.3.52) will involve diagonal-like elements $A_{i_1 \cdots i_n, i_1 \cdots i_n}$. We also note that $\xi_{i_1}^* \cdots \xi_{i_n}^* (-\xi_{i_n}) \cdots (-\xi_{i_1}) = \xi_{i_1} \xi_{i_1}^* \cdots \xi_{i_n} \xi_{i_n}^*$. We then find

$$\int \left(\prod_{i=1}^{N} d\xi_i^* d\xi_i\right) e^{-\sum_{j=1}^{N} \xi_j^*\xi_j} \tilde{A}(\xi^*, -\xi)$$
$$= \sum_{n=0}^{N} \sum_{i_1 < \cdots < i_n = 1}^{N} A_{i_1 \cdots i_n, i_1 \cdots i_n} \int \left(\prod_{i=1}^{N} d\xi_i^* d\xi_i\right) e^{-\sum_{j=1}^{N} \xi_j^*\xi_j} \xi_{i_1} \xi_{i_1}^* \cdots \xi_{i_n} \xi_{i_n}^*$$
$$= \sum_{n=0}^{N} \sum_{i_1 < \cdots < i_n = 1}^{N} A_{i_1 \cdots i_n, i_1 \cdots i_n} = \text{tr}\hat{A}, \tag{6.3.53}$$

where we have again used Eq. (6.3.44).

The Grassmann algebra representations of the annihilation and creation operators $\hat{b}_i$ and $\hat{b}_j^\dagger$ are given by

$$(\hat{b}_i^\dagger \Psi)(\xi^*) = \xi_j^* \Psi(\xi^*) \qquad \text{and} \qquad (\hat{b}_i \Psi)(\xi^*) = (\partial/\partial\xi_j^*)\Psi(\xi^*), \tag{6.3.54}$$

where $(\hat{b}_i^\dagger \Psi)(\xi^*)$ and $(\hat{b}_i \Psi)(\xi^*)$ are the Grassmann algebra representations of $\hat{b}_i^\dagger|\Psi\rangle$ and $\hat{b}_i|\Psi\rangle$, respectively. These definitions lead to consistency with Eq. (6.3.11) including the phase factors from the anticommutators. It is also readily verified that these definitions reproduce the anticommutation relations of Eq. (6.3.7), since

$$\{\partial/\partial\xi_i^*, \xi_j^*\} = \delta_{ij} \quad \text{and} \quad \{\partial/\partial\xi_i^*, \partial/\partial\xi_j^*\} = \{\xi_i^*, \xi_j^*\} = 0. \tag{6.3.55}$$

Any bounded linear operator in Fock space can be expressed as a sum over products of annihilation and creation operators. Any such operator can always be rearranged using the anticommutation relations so that it has a normal-ordered form where all creation operators are on the left of all creation operators. We will also require that the ordering be such that $i_1 < \cdots < i_n$ for all indices as we have above. We refer to this as the normal form of the operator $\hat{A}$ and we write it as

$$\hat{A} = A(\hat{b}^\dagger, \hat{b}) \equiv \sum_{m,n=0}^{N} \sum_{i_1 < \cdots < i_m = 1}^{N} \sum_{j_1 < \cdots < j_n = 1}^{N} K_{i_1 \cdots i_m, j_1 \cdots j_n}$$
$$\times \hat{b}_{i_1}^\dagger \cdots \hat{b}_{i_m}^\dagger \hat{b}_{j_n} \cdots \hat{b}_{j_1}. \tag{6.3.56}$$

The Fock space equation $|\Psi\rangle = \hat{A}|\Phi\rangle = A(\hat{b}^\dagger, \hat{b})|\Phi\rangle$, when expressed in terms of the Grassmann algebra annihilation and creation operators, becomes

$$\Psi(\xi^*) = A(\xi^*, \partial/\partial\xi^*)\Phi(\xi^*), \tag{6.3.57}$$

where $A(\cdot, \cdot)$ is the same ordered function as in Eq. (6.3.56). It is easily verified that

$$\xi_{i_1}^* \cdots \xi_{i_m}^* (\partial/\partial\xi_{j_n}^*) \cdots (\partial/\partial\xi_{j_1}^*) e^{\sum_{j=1}^{N} \xi_j^* \eta_j} = \xi_{i_1}^* \cdots \xi_{i_m}^* \eta_{j_n} \cdots \eta_{j_1} e^{\sum_{j=1}^{N} \xi_j^* \eta_j} \tag{6.3.58}$$

and so

$$A(\xi^*, \partial/\partial\xi^*) e^{\sum_{j=1}^{N} \xi_j^* \eta_j} = A(\xi^*, \eta) e^{\sum_{j=1}^{N} \xi_j^* \eta_j}. \tag{6.3.59}$$

When we consider the identity operator, $\hat{A} = \hat{I}$, we have the result that

$$\Phi(\xi^*) = \int \left(\prod_{i=1}^{N} d\eta_i^* d\eta_i \right) e^{-\sum_{j=1}^{N} (\eta_j^* - \xi_j^*)\eta_j} \Phi(\eta^*). \tag{6.3.60}$$

Proof: To show Eq. (6.3.60) note that

$$e^{-\sum_{j=1}^{N} (\eta_j^* - \xi_j^*)\eta_j} \eta_{i_1}^* \cdots \eta_{i_n}^* = e^{-\sum_{j=1}^{N} \eta_j^* \eta_j} e^{\sum_{k=1}^{N} \xi_k^* \eta_k} \eta_{i_1}^* \cdots \eta_{i_n}^*$$
$$= e^{-\sum_{j=1}^{N} \eta_j^* \eta_j} \prod_{k=1}^{N} (1 + \xi_k^* \eta_k) \eta_{i_1}^* \cdots \eta_{i_n}^*$$
$$= e^{-\sum_{j=1}^{N} \eta_j^* \eta_j} (\xi_{i_1}^* \eta_{i_1} \eta_{i_1}^*) \cdots (\xi_{i_n}^* \eta_{i_n} \eta_{i_n}^*) + \cdots$$
$$= \xi_{i_1}^* \cdots \xi_{i_n}^* e^{-\sum_{j=1}^{N} \eta_j^* \eta_j} \eta_{i_1} \eta_{i_1}^* \cdots \eta_{i_n} \eta_{i_n}^* + \cdots, \tag{6.3.61}$$

where the terms not shown are Grassmann odd in η_i^* or η_i and hence will not survive the integration. Using Eq. (6.3.44) we then find

$$\int \left(\prod_{i=1}^{N} d\eta_i^* d\eta_i \right) e^{-\sum_{j=1}^{N} (\eta_j^* - \xi_j^*)\eta_j} \eta_{i_1}^* \cdots \eta_{i_n}^* = \xi_{i_1}^* \cdots \xi_{i_n}^*, \tag{6.3.62}$$

which when combined with Eq. (6.3.37) gives the result.

Combining the above results then gives

$$\Psi(\xi^*) = A(\xi^*, \partial/\partial\xi^*)\Phi(\xi^*) = A(\xi^*, \partial/\partial\xi^*) \int \left(\prod_{i=1}^{N} d\eta_i^* d\eta_i \right) e^{-\sum_{j=1}^{N} (\eta_j^* - \xi_j^*)\eta_j} \Phi(\eta^*)$$
$$= \int \left(\prod_{i=1}^{N} d\eta_i^* d\eta_i \right) e^{-\sum_{j=1}^{N} \eta_j^* \eta_j} [A(\xi^*, \eta) e^{\sum_{k=1}^{N} \xi_k^* \eta_k}] \Phi(\eta^*), \tag{6.3.63}$$

which has the form of Eq. (6.3.46) and hence identifies $\tilde{A}(\xi^*, \eta)$. So the Grassmann matrix form, $\tilde{A}(\xi^*, \eta)$, corresponding to the normal-ordered form of any Fock-space operator, $\hat{A} \equiv A(\hat{b}^\dagger, \hat{b})$, is determined by the function $A(\cdot, \cdot)$ and is given by

$$\tilde{A}(\xi^*, \eta) = A(\xi^*, \eta) e^{\sum_{j=1}^{N} \xi_j^* \eta_j}. \tag{6.3.64}$$

The Grassmann number operator for the ith one-fermion state is $(\xi_i^* \partial/\partial\xi_i^*)$ since

$$\hat{N}_i |b_{i_1} \cdots b_{i_n}; A_o\rangle = (\hat{b}_i^\dagger \hat{b}_i) |b_{i_1} \cdots b_{i_n}; A_o\rangle = n_i |b_{i_1} \cdots b_{i_n}; A_o\rangle$$
$$\rightarrow \quad (\xi_i^* \partial/\partial\xi_i^*) \xi_{1_1}^* \cdots \xi_{i_n}^* = n_i \xi_{1_1}^* \cdots \xi_{i_n}^*, \tag{6.3.65}$$

where $n_i = 0$ if i is not equal to any of the $i_1 < \cdots < i_n$ and $n_i = 1$ if i equals any of the $i_1 < \cdots < i_n$. Note that ξ_i^* and $\partial/\partial\xi_i^*$ are Hermitian conjugates in the Grassmann algebra representation, e.g.,

$$\langle\Psi|\hat{b}_k\Phi\rangle = \int \left(\prod_{i=1}^{N} d\xi_i^* d\xi_i\right) e^{-\sum_{j=1}^{N} \xi_j^* \xi_j} \left[\Psi(\xi^*)\right]^* \left(\partial/\partial\xi_k^*\right)\Phi(\xi^*)$$

$$= \int \left(\prod_{i=1}^{N} d\xi_i^* d\xi_i\right) e^{-\sum_{j=1}^{N} \xi_j^* \xi_j} \left[\xi_k^* \Psi(\xi^*)\right]^* \Phi(\xi^*) = \langle\hat{b}_k^\dagger\Psi|\Phi\rangle. \qquad (6.3.66)$$

Proof: Using the product rule in Eq. (6.3.14) and the exponential expansion in Eq. (6.3.42) we can write

$$\partial/\partial\xi_k^* \left(e^{-\sum_{j=1}^{N} \xi_j^* \xi_j} \Phi(\xi^*)\right) = e^{-\sum_{j=1}^{N} \xi_j^* \xi_j} \left(\partial/\partial\xi_k^* \Phi(\xi^*) - \xi_k \Phi(\xi^*)\right). \qquad (6.3.67)$$

Under integration the left-hand side of the above equation will vanish because of the exact differential result in Eq. (6.3.19) and so we can use the Grassmann version of integration by parts to replace $\partial/\partial\xi_k^* \Phi(\xi^*)$ by $\xi_k \Phi(\xi^*)$ in the integrand. Then we simply recognize from the definition of the overline operator that $[\Psi(\xi^*)]^* \xi_k \Phi(\xi^*) = [\xi_k^* \Psi(\xi^*)]^* \Phi(\xi^*)$, which proves the result.

6.3.3 Feynman Path Integral for Fermions

Let $\hat{H} \equiv H(\hat{b}^\dagger, \hat{b})$ be the fermion Fock space Hamiltonian operator in normal order and let us assume that it has no explicit time dependence. Then the evolution operator for Fock space can be written using Eq. (A.8.5) as

$$\hat{U}(t'', t') \equiv e^{-i\hat{H}(t''-t')/\hbar} = \lim_{n\to\infty} [1 - (i/\hbar)\hat{H}\delta t]^{n+1}, \qquad (6.3.68)$$

where as in Eq. (6.2.226) we divide $(t'' - t')$ into $(n + 1)$ equal intervals of length $\delta t \equiv (t'' - t')/(n + 1)$ so that $t_\ell \equiv t' + j\delta t$ for $\ell = 0, 1, 2, \ldots, n + 1$. We see that $t'' = t_{n+1}$ and $t' = t_0$. Let us for temporary convenience define $\hat{A}_\delta \equiv [1 - (i/\hbar)\hat{H}\delta t]$. Then $\hat{A}_\delta$ is in normal order since $\hat{H}$ is in normal order and so we can write

$$\langle\Psi'', t''|\Psi', t'\rangle = \langle\Psi''|\hat{U}(t'' - t')|\Psi'\rangle = \lim_{n\to\infty} \langle\Psi''|[1 - (i/\hbar)\hat{H}\delta t]^{n+1}|\Psi'\rangle$$

$$= \lim_{n\to\infty} \langle\Psi''|(\hat{A}_\delta)^{n+1}|\Psi'\rangle = \lim_{n\to\infty} \int \left(\prod_{k=0}^{n+1}\prod_{i_k=1}^{N} d\xi_{i_k}^* \xi_{i_k}\right) e^{-\sum_{\ell=0}^{n+1}\sum_{j_\ell=1}^{N} \xi_{j_\ell}^* \xi_{j_\ell}}$$

$$\times \Psi''^*(\xi_{n+1})\tilde{A}_\delta(\xi_{n+1}^*, \xi_n)\cdots\tilde{A}_\delta(\xi_2^*, \xi_1)\tilde{A}_\delta(\xi_1^*, \xi_0)\Psi'(\xi_0^*)$$

$$= \lim_{n\to\infty} \int \left(\prod_{k=0}^{n+1}\prod_{i_k=1}^{N} d\xi_{i_k}^* \xi_{i_k}\right) e^{-\sum_{\ell=0}^{n+1}\sum_{j_\ell=1}^{N} \xi_{j_\ell}^* \xi_{j_\ell}} e^{\sum_{m=1}^{n+1}\sum_{j_m=1}^{N} \xi_{j_m}^* \xi_{j_m-1}}$$

$$\times \Psi''^*(\xi_{n+1})[1 - (i/\hbar)H(\xi_{n+1}^*, \xi_n)\delta t]\cdots[1 - (i/\hbar)H(\xi_1^*, \xi_0)\delta t]\Psi'(\xi_0^*)$$

$$= \lim_{n\to\infty} \int \left(\prod_{k=0}^{n+1}\prod_{i_k=1}^{N} d\xi_{i_k}^* \xi_{i_k}\right)\Psi''^*(\xi_{n+1})e^{-\sum_{j_0=1}^{N} \xi_{j_0}^* \xi_{j_0}} \qquad (6.3.69)$$

$$\times e^{i\sum_{\ell=1}^{n+1} \delta t\{\sum_{j_\ell=1}^{N} i\hbar\xi_{j_\ell}^*(\xi_{j_\ell}-\xi_{j_\ell-1})/\delta t - H(\xi_{j_\ell}^*, \xi_{j_\ell-1})\}/\hbar}\Psi'(\xi_0^*)$$

$$\equiv \int \mathcal{D}\xi^* \mathcal{D}\xi \, \Psi''^*(\xi(t''))e^{-\sum_{k=1}^{N} \xi_k^*(t')\xi_k(t')}$$

$$\times e^{(i/\hbar)\int_{t'}^{t''} dt \, [\sum_{j=1}^{N} i\xi_j^*(t)(\hbar\partial_t)\xi_j(t) - H(\xi^*(t), \xi(t))]}\Psi'(\xi^*(t')), \qquad (6.3.70)$$

where the last line is defined by the line above it and where we have used Eqs. (6.3.41), (6.3.50) and (6.3.64) as well as the fact that

$$[1 - (i/\hbar)H(\xi_\ell^*, \xi_{\ell-1})\delta t] = \exp[-(i/\hbar)H(\xi_\ell^*, \xi_{\ell-1})\delta t] + \mathcal{O}(\delta t^2). \tag{6.3.71}$$

We have an unimportant surface-like term remaining, $\exp(-\sum_{j=1}^N \xi_j^*(t')\xi_j(t'))$. On rearranging Eq. (6.3.69), we find

$$\begin{aligned}
\langle \Psi'', t'' | \Psi', t' \rangle &= \lim_{n \to \infty} \int (\prod_{k=0}^{n+1} \prod_{i_k=1}^N d\xi_{i_k}^* \xi_{i_k}) \Psi''^*(\xi_{n+1}) e^{-\sum_{j_{n+1}=1}^N \xi_{j_{n+1}}^* \xi_{j_{n+1}}} \\
&\quad \times e^{i \sum_{\ell=1}^{n+1} \delta t \{ \sum_{j_\ell=1}^N i\hbar(\xi_{j_\ell-1}^* - \xi_{j_\ell}^*)\xi_{j_\ell-1}/\delta t - H(\xi_{j_\ell}^*, \xi_{j_\ell-1}) \}/\hbar} \Psi'(\xi_0^*) \\
&\equiv \int \mathcal{D}\xi^* \mathcal{D}\xi \; \Psi''^*(\xi(t'')) e^{-\sum_{k=1}^N \xi_k^*(t'')\xi_k(t'')} \\
&\quad \times e^{(i/\hbar) \int_{t'}^{t''} dt \, [\sum_{j=1}^N -[i\hbar \partial_t \xi_j^*(t)]\xi_j(t) - H(\xi^*(t), \xi(t))]} \Psi'(\xi^*(t')), \tag{6.3.72}
\end{aligned}$$

which shows that we can carry out integration by parts of the time derivative by replacing the surface term with another surface term, $\exp(-\sum_{j=1}^N \xi_j^*(t'')\xi_j(t''))$.

From Eqs. (6.3.49) and (6.3.70) we can immediately extract the Grassmann matrix form of the evolution operator,

$$\tilde{U}(\xi^*(t''), \xi(t')) = e^{\sum_{k=1}^N \xi_k^*(t'')\xi_k(t'')} \int \mathcal{D}\xi^* \mathcal{D}\xi \, e^{(i/\hbar) \int_{t'}^{t''} dt \, [\sum_{j=1}^N i\hbar \xi_j^* \partial_t \xi_j - H]}. \tag{6.3.73}$$

We will again be interested in the trace of the evolution operator, also known as the spectral function, which is given by

$$\begin{aligned}
\text{tr}[\hat{U}(t'', t')] &= \int \mathcal{D}\xi^*(t'') \mathcal{D}\xi(t'') \, e^{-\sum_{k=1}^N \xi^*(t'')\xi(t'')} \; \tilde{U}(\xi^*(t''), \xi(t')) \Big|_{\vec{\xi}(t') = -\vec{\xi}(t'')} \\
&= \int \mathcal{D}\xi^* \mathcal{D}\xi \; e^{(i/\hbar) \int_{t'}^{t''} dt \, [\sum_{j=1}^N i\hbar \xi_j^* \partial_t \xi_j - H(\xi^*, \xi)]} \Big|_{\vec{\xi}(t') = -\vec{\xi}(t'')}, \tag{6.3.74}
\end{aligned}$$

where we have used the formula for the trace in Eq. (6.3.52). So the trace of the evolution operator for fermions involves the integral over all Grassmann-valued paths, $\vec{\xi}(t) = (\xi_1(t), \ldots, \xi_N(t))$ for $t' \leq t \leq t''$, that are *antiperiodic in time*. Comparing Eqs. (6.3.70) and (6.3.72) we can easily verify that with the latter form we again arrive at Eq. (6.3.74) but now with $i\hbar\xi^* \partial_t \xi$ replaced by $-(i\hbar \partial_t \xi^*)\xi$. So the more symmetric form is also valid, with exponent

$$(i/\hbar) \int_{t'}^{t''} dt \, \left[\sum_{j=1}^N \tfrac{1}{2} i\hbar \{ \xi_j^* \partial_t \xi_j - (\partial_t \xi_j^*)\xi_j \} - H \right]. \tag{6.3.75}$$

Thus integration by parts for the time integral is justified for the spectral function because the antiperiodicity of the trace removes the surface terms in both cases.

6.3.4 Fock Space for Dirac Fermions

As we know from Chapter 4, antiparticles are an inevitable consequence of combining relativity with quantum mechanics. As discussed in Sec. 4.1.10, the total one-fermion Hilbert space V_H consists of a one-particle subspace spanned by one-fermion states and a one-antiparticle subspace spanned by one-antifermion states. If each subspace is N-dimensional, then the full V_H is $2N$-dimensional with annihilation and creation operators $\hat{b}_j$ and $\hat{b}_j^\dagger$ for fermions and $\hat{d}_j$ and $\hat{d}_j^\dagger$ for antifermions. Let $\hat{N}_{bj}$ and $\hat{N}_{dj}$ be the number operators measuring the number of fermions and antifermions, respectively, in the state j with energy E_j. Then the normal-ordered Hamiltonian for free particles must have the form

$$:\hat{H}: \equiv \sum_{j=1}^{N} E_j(\hat{N}_{bj} + \hat{N}_{dj}) = \sum_{j=1}^{N} E_j(\hat{b}_j^\dagger \hat{b}_j + \hat{d}_j^\dagger \hat{d}_j), \qquad (6.3.76)$$

where the annihilation operators annihilate the vacuum so that an empty state has no energy, $\hat{b}_j|0\rangle = \hat{d}_j|0\rangle = 0$. This is analogous to the discussion of the charged scalar field in Sec. 6.2.7, where $\hat{f}_{\mathbf{p}}$ and $\hat{g}_{\mathbf{p}}$ were the scalar analogs of $\hat{b}_{\mathbf{p}}^s$ and $\hat{d}_{\mathbf{p}}^s$, respectively. With the mapping

$$\hat{b}_1, \ldots, \hat{b}_N, \hat{d}_1, \ldots, \hat{d}_N \rightarrow \hat{b}_1', \ldots, \hat{b}_{2N}' \qquad (6.3.77)$$

and correspondingly for the creation operators we see that all of the above Grassmann-algebra arguments can be applied with $\hat{b} \rightarrow \hat{b}' = (\hat{b}, \hat{d})$ and $N \rightarrow 2N$. From the anticommutation relations in Eq. (6.3.7) we then obviously must have

$$\{\hat{b}_i, \hat{b}_j^\dagger\} = \{\hat{d}_i, \hat{d}_j^\dagger\} = \delta_{ij}, \quad \{\hat{b}_i, \hat{b}_j\} = \{\hat{b}_i^\dagger, \hat{b}_j^\dagger\} = \{\hat{d}_i, \hat{d}_j\} = \{\hat{d}_i^\dagger, \hat{d}_j^\dagger\} = 0,$$
$$\{\hat{b}_i, \hat{d}_j\} = \{\hat{b}_i^\dagger, \hat{d}_j\} = \{\hat{b}_i, \hat{d}_j^\dagger\} = \{\hat{b}_i^\dagger, \hat{d}_j^\dagger\} = 0. \qquad (6.3.78)$$

Let us now define our one-fermion Hilbert space V_H to be the space spanned by the plane wave solutions of the Dirac equation as discussed in Sec. 4.4. This corresponds to $j \rightarrow (s, \mathbf{p})$, where $s = \pm\frac{1}{2}$ is the spin state and $\mathbf{p}$ is the three-momentum of the state. We will follow convention and not explicitly show $\hbar$. For a theory of free relativistic fermions there must be a corresponding normal-ordered Hamiltonian operator such that

$$\hat{H} \equiv \int (d^3p/(2\pi)^3) \sum_{s=\pm\frac{1}{2}} E_{\mathbf{p}}(\hat{b}_{\mathbf{p}}^{s\dagger} \hat{b}_{\mathbf{p}}^s + \hat{d}_{\mathbf{p}}^{s\dagger} \hat{d}_{\mathbf{p}}^s), \qquad (6.3.79)$$

with $E_{\mathbf{p}} = \sqrt{\mathbf{p}^2 + m^2} > 0$ and where Eq. (6.3.78) becomes

$$\{\hat{b}_{\mathbf{p}}^s, \hat{b}_{\mathbf{p}'}^{s'\dagger}\} = \{\hat{d}_{\mathbf{p}}^s, \hat{d}_{\mathbf{p}'}^{s'\dagger}\} = \delta^{ss'}(2\pi)^3\delta^3(\mathbf{p} - \mathbf{p}'),$$
$$\{\hat{b}_{\mathbf{p}}^s, \hat{b}_{\mathbf{p}'}^{s'}\} = \{\hat{b}_{\mathbf{p}}^{s\dagger}, \hat{b}_{\mathbf{p}'}^{s'\dagger}\} = \{\hat{d}_{\mathbf{p}}^s, \hat{d}_{\mathbf{p}'}^{s'}\} = \{\hat{d}_{\mathbf{p}}^{s\dagger}, \hat{d}_{\mathbf{p}'}^{s'\dagger}\} = 0,$$
$$\{\hat{b}_{\mathbf{p}}^s, \hat{d}_{\mathbf{p}'}^{s'}\} = \{\hat{b}_{\mathbf{p}}^{s\dagger}, \hat{d}_{\mathbf{p}'}^{s'}\} = \{\hat{b}_{\mathbf{p}}^s, \hat{d}_{\mathbf{p}'}^{s'\dagger}\} = \{\hat{b}_{\mathbf{p}}^{s\dagger}, \hat{d}_{\mathbf{p}'}^{s'\dagger}\} = 0. \qquad (6.3.80)$$

With $\hat{X}_{\mathbf{p}}^s = \hat{b}_{\mathbf{p}}^s$ or $\hat{d}_{\mathbf{p}}^s$ it follows that Eq. (6.3.8) becomes

$$[\hat{b}_{\mathbf{p}}^s, \hat{b}_{\mathbf{p}'}^{s'\dagger} \hat{X}_{\mathbf{p}''}^{s''}] = \delta^{ss'}(2\pi)^3\delta^3(\mathbf{p} - \mathbf{p}')\hat{X}_{\mathbf{p}''}^{s''},$$
$$[\hat{d}_{\mathbf{p}}^s, \hat{d}_{\mathbf{p}'}^{s'\dagger} \hat{X}_{\mathbf{p}''}^{s''}] = \delta^{ss'}(2\pi)^3\delta^3(\mathbf{p} - \mathbf{p}')\hat{X}_{\mathbf{p}''}^{s''}. \qquad (6.3.81)$$

By symmetry of the commutation relations under $\hat{b}, \hat{d} \leftrightarrow \hat{b}^\dagger, \hat{d}^\dagger$ these relations are also true if we replace each operator by its Hermitian conjugate. We define the one-fermion and one-antifermion states created by $\hat{b}_{\mathbf{p}}^{s\dagger}$ and $\hat{d}_{\mathbf{p}}^{s\dagger}$, respectively, as

$$|f; \mathbf{p}, s\rangle \equiv \sqrt{2E_{\mathbf{p}}}\, b_{\mathbf{p}}^{s\dagger}|0\rangle \quad \text{and} \quad |\bar{f}; \mathbf{p}, s\rangle \equiv \sqrt{2E_{\mathbf{p}}}\, d_{\mathbf{p}}^{s\dagger}|0\rangle, \qquad (6.3.82)$$

where the normalization is chosen such that the bra-kets of such states are Lorentz invariant as in Eq. (6.2.47),

$$\langle f; \mathbf{p}, s|f; \mathbf{p}', s'\rangle = \langle \bar{f}; \mathbf{p}, s|\bar{f}; \mathbf{p}', s'\rangle = 2E_{\mathbf{p}}\delta^{ss'}(2\pi)^3\delta^3(\mathbf{p} - \mathbf{p}'),$$
$$\langle \bar{f}; \mathbf{p}, s|f; \mathbf{p}', s'\rangle = \langle f; \mathbf{p}, s|\bar{f}; \mathbf{p}', s'\rangle = 0. \qquad (6.3.83)$$

These results follow immediately from the anticommutation relations, e.g.,

$$\langle 0|\hat{b}_{\mathbf{p}}^s \hat{b}_{\mathbf{p}'}^{s'\dagger}|0\rangle = \langle 0|\{\hat{b}_{\mathbf{p}}^s, \hat{b}_{\mathbf{p}'}^{s'\dagger}\}|0\rangle = \delta^{ss'}(2\pi)^3 \delta^3(\mathbf{p} - \mathbf{p}'). \tag{6.3.84}$$

We know from Eq. (4.4.111) in Sec. 4.4.7 that $u^s(p)e^{-ip\cdot x}$ and $v^s(p)e^{ip\cdot x}$ are the plane wave solutions (normal modes) of the Dirac equation that span V_H,

$$(i\not{\partial} - m)u^s(p)e^{-ip\cdot x} = (i\not{\partial} - m)v^s(p)e^{ip\cdot x} = 0. \tag{6.3.85}$$

Define the coordinate-space Dirac (one-fermion) Hamiltonian operator as

$$h_D \equiv -i\boldsymbol{\alpha}\cdot\boldsymbol{\nabla} + \beta m. \tag{6.3.86}$$

So $u^s(p)e^{i\mathbf{p}\cdot\mathbf{x}}$ and $v^s(p)e^{-i\mathbf{p}\cdot\mathbf{x}}$ are the energy eigenfunctions of h_D with energy eigenvalues $E_{\mathbf{p}}$ and $-E_{\mathbf{p}}$, respectively, and are the normal modes of the system. As for the scalar field we construct field operators in terms of annihilation and creation operators associated with these normal modes that span V_H. In analogy with Eqs. (6.2.14) and (6.2.177) we then define the Schrödinger picture operators

$$\hat{\psi}(\mathbf{x}) \equiv \int \frac{d^3p}{(2\pi)^3} \frac{1}{\sqrt{2E_{\mathbf{p}}}} \sum_{s=\pm\frac{1}{2}} \left[\hat{b}_{\mathbf{p}}^s u^s(p)e^{i\mathbf{p}\cdot\mathbf{x}} + \hat{d}_{\mathbf{p}}^{s\dagger} v^s(p)e^{-i\mathbf{p}\cdot\mathbf{x}}\right],$$

$$\hat{\psi}^\dagger(\mathbf{x}) \equiv \int \frac{d^3p}{(2\pi)^3} \frac{1}{\sqrt{2E_{\mathbf{p}}}} \sum_{s=\pm\frac{1}{2}} \left[\hat{b}_{\mathbf{p}}^{s\dagger} u^{s\dagger}(p)e^{-i\mathbf{p}\cdot\mathbf{x}} + \hat{d}_{\mathbf{p}}^s v^{s\dagger}(p)e^{i\mathbf{p}\cdot\mathbf{x}}\right], \tag{6.3.87}$$

where the spinor index $\alpha = 1, 2, 3, 4$ is suppressed for brevity on $u^s(p)$, $v^s(p)$, $\hat{\psi}$ and its Hermitian conjugate $\hat{\psi}^\dagger$. We note that

$$\hat{\psi}_\alpha^\dagger(\mathbf{x})|0\rangle = \int (d^3p/(2\pi)^3 2E_{\mathbf{p}})\sum_s u_\alpha^{s\dagger}(p)e^{-i\mathbf{p}\cdot\mathbf{x}}|f;\mathbf{p},s\rangle \quad \text{and}$$

$$\hat{\psi}_\alpha(\mathbf{x})|0\rangle = \int (d^3p/(2\pi)^3 2E_{\mathbf{p}})\sum_s v_\alpha^s(p)e^{i\mathbf{p}\cdot\mathbf{x}}|\bar{f};\mathbf{p},s\rangle, \tag{6.3.88}$$

create the α component of a fermion and antifermion, respectively, at the point $\mathbf{x}$.

Normal-ordering for fermions should obviously involve *a minus sign for the interchange of two fermion operators*, which is consistent with the concept of an operator in normal order in the Grassmann algebra representation. Using the notation introduced for the scalar field for positive (annihilation) and negative (creation) frequency parts, we similarly label the annihilation $+$ and creation $-$ parts of the Schrödinger picture operators $\hat{\psi}(\mathbf{x})$ and $\hat{\psi}^\dagger(\mathbf{x})$ as

$$\hat{\psi}^+(\mathbf{x}) \equiv \int (d^3p/(2\pi)^3 \sqrt{2E_{\mathbf{p}}})\sum_s \hat{b}_{\mathbf{p}}^s u^s(p)e^{i\mathbf{p}\cdot\mathbf{x}},$$

$$\hat{\psi}^{\dagger+}(\mathbf{x}) \equiv \int (d^3p/(2\pi)^3 \sqrt{2E_{\mathbf{p}}})\sum_s \hat{d}_{\mathbf{p}}^s v^{s\dagger}(p)e^{i\mathbf{p}\cdot\mathbf{x}},$$

$$\hat{\psi}^-(\mathbf{x}) \equiv \int (d^3p/(2\pi)^3 \sqrt{2E_{\mathbf{p}}})\sum_s \hat{d}_{\mathbf{p}}^{s\dagger} v^s(p)e^{-i\mathbf{p}\cdot\mathbf{x}},$$

$$\hat{\psi}^{\dagger-}(\mathbf{x}) \equiv \int (d^3p/(2\pi)^3 \sqrt{2E_{\mathbf{p}}})\sum_s \hat{b}_{\mathbf{p}}^{s\dagger} u^{s\dagger}(p)e^{-i\mathbf{p}\cdot\mathbf{x}},$$

$$\hat{\psi}(\mathbf{x}) = \hat{\psi}^+(\mathbf{x}) + \hat{\psi}^-(\mathbf{x}) \quad \text{and} \quad \hat{\psi}^\dagger(\mathbf{x}) = \hat{\psi}^{\dagger-}(\mathbf{x}) + \hat{\psi}^{\dagger+}(\mathbf{x}). \tag{6.3.89}$$

Using an obvious shorthand, the normal-ordering of $\hat{\psi}^\dagger(\mathbf{x})\hat{\psi}(\mathbf{x})$ is then given by

$$:\hat{\psi}^\dagger\hat{\psi}: = \hat{\psi}^{\dagger+}\hat{\psi}^+ + \hat{\psi}^{\dagger-}\hat{\psi}^+ + \hat{\psi}^{\dagger-}\hat{\psi}^- - \hat{\psi}^-\hat{\psi}^{\dagger+}. \tag{6.3.90}$$

We have constructed the operators $\hat{\psi}$ and $\hat{\psi}^\dagger$ from the energy eigenfunctions of the Dirac Hamiltonian. We identify the Hamiltonian operator as

$$\int d^3x\, \hat{\psi}^\dagger(\mathbf{x}) h_D \hat{\psi}(\mathbf{x}) = \int d^3x\, \hat{\psi}^\dagger(\mathbf{x})(-i\boldsymbol{\alpha}\cdot\boldsymbol{\nabla} + \beta m)\hat{\psi}(\mathbf{x})$$

$$= \int d^3x\, \hat{\bar{\psi}}(\mathbf{x})(-i\boldsymbol{\gamma}\cdot\boldsymbol{\nabla} + m)\hat{\psi}(\mathbf{x})$$

$$= \int d^3x \int (d^3p\, d^3p'/(2\pi)^6 2\sqrt{E_\mathbf{p} E_{\mathbf{p}'}}) \sum_{s,s'} \left[\hat{b}^{s\dagger}_\mathbf{p} \bar{u}^s(p) e^{-i\mathbf{p}\cdot\mathbf{x}} + \hat{d}^s_\mathbf{p} \bar{v}^s(p) e^{i\mathbf{p}\cdot\mathbf{x}} \right]$$

$$\times \gamma^0 E_{\mathbf{p}'} [\hat{b}^{s'}_{\mathbf{p}'} u^{s'}(p') e^{i\mathbf{p}'\cdot\mathbf{x}} - \hat{d}^{s'\dagger}_{\mathbf{p}'} v^{s'}(p') e^{-i\mathbf{p}'\cdot\mathbf{x}}]$$

$$= \int (d^3p/(2\pi)^3 2) \sum_{ss'} [u^{s\dagger}(p) u^{s'}(p) \hat{b}^{s\dagger}_\mathbf{p} \hat{b}^{s'}_\mathbf{p} - v^{s\dagger}(p) v^{s'}(p) \hat{d}^s_\mathbf{p} \hat{d}^{s'\dagger}_\mathbf{p}]$$

$$= \int (d^3p/(2\pi)^3) \sum_s E_\mathbf{p} [\hat{b}^{s\dagger}_\mathbf{p} \hat{b}^s_\mathbf{p} - \hat{d}^s_\mathbf{p} \hat{d}^{s\dagger}_\mathbf{p}], \tag{6.3.91}$$

where we have used Eqs. (4.4.125), (4.4.121) and (4.4.111), which gives

$$[\gamma^0 E_\mathbf{p} - (-i\boldsymbol{\gamma}\cdot\boldsymbol{\nabla} + m)] e^{i\mathbf{p}\cdot\mathbf{x}} u^s(p) = [-\gamma^0 E_\mathbf{p} - (-i\boldsymbol{\gamma}\cdot\boldsymbol{\nabla} + m)] e^{-i\mathbf{p}\cdot\mathbf{x}} v^s(p) = 0.$$

So using Eq. (6.3.91) the normal-ordered Hamiltonian gives Eq. (6.3.79) as required,

$$\hat{H} \equiv H[\hat{b}^\dagger, \hat{d}^\dagger, \hat{b}, \hat{d}] \equiv \int d^3x\, \mathcal{H}(\hat{b}^\dagger, \hat{d}^\dagger, \hat{b}, \hat{d}) \equiv \int d^3x\, {:}\hat{\bar{\psi}}(\mathbf{x})(-i\boldsymbol{\gamma}\cdot\boldsymbol{\nabla} + m)\hat{\psi}(\mathbf{x}){:} \tag{6.3.92}$$

$$= \int (d^3p/(2\pi)^3) \sum_s E_\mathbf{p} {:}(\hat{b}^{s\dagger}_\mathbf{p} \hat{b}^s_\mathbf{p} - \hat{d}^s_\mathbf{p} \hat{d}^{s\dagger}_\mathbf{p}){:} = \int (d^3p/(2\pi)^3) \sum_s E_\mathbf{p} (\hat{b}^{s\dagger}_\mathbf{p} \hat{b}^s_\mathbf{p} + \hat{d}^{s\dagger}_\mathbf{p} \hat{d}^s_\mathbf{p}).$$

The effect of having normal-ordered the Hamiltonian was to add

$$\int (d^3p/(2\pi)^3) \sum_s E_\mathbf{p} \{\hat{d}^{s\dagger}_\mathbf{p}, \hat{d}^s_\mathbf{p}\} = \int d^3p \sum_s E_\mathbf{p} \delta^3(\mathbf{0}) \tag{6.3.93}$$

and so normal-ordering the Hamiltonian has removed an infinite negative constant. In the case of scalar bosons normal-ordering removed a positive infinite constant from the Hamiltonian, which was the zero-point energy in Eq. (6.2.27). In the case of fermions it is sometimes said that normal-ordering is removing the infinite sea of filled negative energy states. However, since we could just as easily have identified fermions as antifermions and vice versa, then the symmetry of the Hamiltonian under $\hat{b}^s_\mathbf{p} \leftrightarrow \hat{d}^s_\mathbf{p}$ is in fact what we should expect. Normal-ordering achieves that.

The operators $\hat{\psi}(\mathbf{x})$ and $\hat{\psi}^\dagger(\mathbf{x})$ satisfy the fermion equal-time canonical anticommutation relations

$$\{\hat{\psi}_\alpha(\mathbf{x}), \hat{\psi}^\dagger_\beta(\mathbf{y})\} = \delta_{\alpha\beta} \delta^3(\mathbf{x}-\mathbf{y}) \qquad \text{and}$$

$$\{\hat{\psi}_\alpha(\mathbf{x}), \hat{\psi}_\beta(\mathbf{y})\} = \{\hat{\psi}^\dagger_\alpha(\mathbf{x}), \hat{\psi}^\dagger_\beta(\mathbf{y})\} = 0. \tag{6.3.94}$$

Proof: The second two anticommutation relations follow easily from Eq. (6.3.80) since only anticommutators containing $\hat{b}^s_\mathbf{p}, \hat{b}^{s\dagger}_\mathbf{p}$ or $\hat{d}^s_\mathbf{p}, \hat{d}^{s\dagger}_\mathbf{p}$ pairs will be nonvanishing. The first relation arises since

$$\{\hat{\psi}_\alpha(\mathbf{x}), \hat{\psi}^\dagger_\beta(\mathbf{y})\} = \int (d^3p\, d^3p'/(2\pi)^6 2\sqrt{E_\mathbf{p} E_{\mathbf{p}'}}) \sum_{s,s'}$$

$$\times \left[\{\hat{b}^s_\mathbf{p}, \hat{b}^{s'\dagger}_{\mathbf{p}'}\} u^s_\alpha(p) u^{s'\dagger}_\beta(p') e^{i(\mathbf{p}\cdot\mathbf{x}-\mathbf{p}'\cdot\mathbf{y})} + \{\hat{d}^{s\dagger}_\mathbf{p}, \hat{d}^{s'}_{\mathbf{p}'}\} v^s_\alpha(p) v^{s'\dagger}_\beta(p') e^{-i(\mathbf{p}\cdot\mathbf{x}-\mathbf{p}'\cdot\mathbf{y})} \right]$$

$$= \int (d^3p/(2\pi)^3 2E_\mathbf{p}) \sum_s \left[u^s_\alpha(p) u^{s\dagger}_\beta(p) e^{i\mathbf{p}\cdot(\mathbf{x}-\mathbf{y})} + v^s_\alpha(p) v^{s\dagger}_\beta(p) e^{-i\mathbf{p}\cdot(\mathbf{x}-\mathbf{y})} \right]$$

$$= \int (d^3p/(2\pi)^3 2E_\mathbf{p}) \left[[(\not{p}+m)\gamma^0]_{\alpha\beta} e^{i\mathbf{p}\cdot(\mathbf{x}-\mathbf{y})} + [(\not{p}-m)\gamma^0]_{\alpha\beta} e^{-i\mathbf{p}\cdot(\mathbf{x}-\mathbf{y})} \right]$$

$$= \int (d^3p/(2\pi)^3 2E_\mathbf{p}) [2p^0 \gamma^0 \gamma^0]_{\alpha\beta} e^{i\mathbf{p}\cdot(\mathbf{x}-\mathbf{y})} = \delta_{\alpha\beta} \delta^3(\mathbf{x}-\mathbf{y}), \tag{6.3.95}$$

where we have used the spinor completeness relations in Eq. (4.4.129).

Since there is no explicit time dependence, then the evolution operator is

$$\hat{U}(t-t_0) = e^{-i\hat{H}(t-t_0)}, \tag{6.3.96}$$

where the normal-ordered Hamiltonian $\hat{H}$ is given in Eq. (6.3.92). We find that

$$e^{i\hat{H}t}\hat{b}_{\mathbf{p}}^s e^{-i\hat{H}t} = \hat{b}_{\mathbf{p}}^s e^{-iE_{\mathbf{p}}t} \quad \text{and} \quad e^{i\hat{H}t}\hat{d}_{\mathbf{p}}^s e^{-i\hat{H}t} = \hat{d}_{\mathbf{p}}^s e^{-iE_{\mathbf{p}}t}. \tag{6.3.97}$$

Proof: Using Eq. (6.3.80) and extending the result of Eq. (6.3.8) to the present context we have $[\hat{b}_{\mathbf{p}}^s, \hat{H}] = E_{\mathbf{p}}\hat{b}_{\mathbf{p}}^s$ and $[\hat{d}_{\mathbf{p}}^s, \hat{H}] = E_{\mathbf{p}}\hat{d}_{\mathbf{p}}^s$. Hence we find $[(i\hat{H}t), \hat{b}_{\mathbf{p}}^s] = (-iE_{\mathbf{p}}t)\hat{b}_{\mathbf{p}}^s$ and for j nested commutators

$$[(i\hat{H}t), [(i\hat{H}t), [\cdots [(i\hat{H}t), \hat{b}_{\mathbf{p}}^s]\cdots]]] = (-iE_{\mathbf{p}}t)^j\,\hat{b}_{\mathbf{p}}^s \tag{6.3.98}$$

and similarly for $\hat{d}_{\mathbf{p}}^s$. Then using Eq. (A.8.7) we find

$$\begin{aligned}
e^{i\hat{H}t}\hat{b}_{\mathbf{p}}^s e^{-i\hat{H}t} &= \sum_{j=0}^{\infty}(1/j!)[(i\hat{H}t), [(i\hat{H}t), [\cdots [(i\hat{H}t), \hat{b}_{\mathbf{p}}^s]\cdots]]] \\
&= \sum_{j=0}^{\infty}(1/j!)(-iE_{\mathbf{p}}t)^j\,\hat{b}_{\mathbf{p}}^s = e^{-iE_{\mathbf{p}}t}\,\hat{b}_{\mathbf{p}}^s \tag{6.3.99}
\end{aligned}$$

and similarly for $\hat{d}_{\mathbf{p}}^s$.

Then we have the Heisenberg picture operators (choosing $t_0 = 0$)

$$\hat{\psi}(x) \equiv e^{i\hat{H}t}\hat{\psi}(\mathbf{x})e^{-i\hat{H}t} = \int \frac{d^3p}{(2\pi)^3}\frac{1}{\sqrt{2E_{\mathbf{p}}}}\sum_s \left[\hat{b}_{\mathbf{p}}^s u^s(p)e^{-ip\cdot x} + \hat{d}_{\mathbf{p}}^{s\dagger} v^s(p)e^{ip\cdot x}\right], \tag{6.3.100}$$

$$\hat{\psi}^\dagger(x) \equiv e^{i\hat{H}t}\hat{\psi}^\dagger(\mathbf{x})e^{-i\hat{H}t} = \int \frac{d^3p}{(2\pi)^3}\frac{1}{\sqrt{2E_{\mathbf{p}}}}\sum_s \left[\hat{b}_{\mathbf{p}}^{s\dagger} u^{s\dagger}(p)e^{ip\cdot x} + \hat{d}_{\mathbf{p}}^s v^{s\dagger}(p)e^{-ip\cdot x}\right].$$

6.3.5 Functional Integral for Dirac Fermions

Recall that Eq. (6.3.70) results for the overlap of any two fermion Fock space states $|\Psi'\rangle$ at t' and $|\Psi''\rangle$ at t'' for a normal-ordered Hamiltonian of the form $\hat{H} \equiv H(\hat{b}^\dagger, \hat{b})$. The function $H(\cdot, \cdot)$ is what appears in the exponent of the Grassmann integration in the form $H(\xi^*, \xi)$. In the case of interest here, where V_H is the one-fermion Dirac Hilbert space, we have the correspondences

$$j \to (\mathbf{p}, s), \quad \xi_j \to \xi_{b\mathbf{p}}^s, \xi_{d\mathbf{p}}^s, \quad \xi_j^* \to \xi_{b\mathbf{p}}^{*s}, \xi_{d\mathbf{p}}^{*s}, \quad H(\xi^*, \xi) \to H[\xi_b^*, \xi_d^*, \xi_b, \xi_d],$$

$$\sum_{j=1}^N \to \int (d^3p/(2\pi)^3)\sum_s \quad \text{and} \quad \delta_{ij} \to \delta^{ss'}(2\pi^3)\delta^3(\mathbf{p} - \mathbf{p}'). \tag{6.3.101}$$

We have arrived at the fermion functional integral representation for the spectral function

$$\begin{aligned}
F(t'' - t') &= \text{tr}[\hat{U}(t'' - t')] = \int \mathcal{D}\xi_b^* \mathcal{D}\xi_d^* \mathcal{D}\xi_b \mathcal{D}\xi_d\Big|_{(\vec{\xi}_b, \vec{\xi}_d)(t') = -(\vec{\xi}_b, \vec{\xi}_d)(t'')} \\
&\quad \times e^{i\int_{t'}^{t''} dt\,[\int(d^3p/(2\pi)^3)\sum_s i\{\xi_{b\mathbf{p}}^{*s}(t)\partial_t\xi_{b\mathbf{p}}^s(t) + \xi_{d\mathbf{p}}^{*s}(t)\partial_t\xi_{d\mathbf{p}}^s(t)\} - H[\xi_b^*(t), \xi_d^*(t), \xi_b(t), \xi_d(t)]]}, \\
&= (1/N_J)\int \mathcal{D}\bar{\psi}\mathcal{D}\psi\Big|_{\psi(t') = -\psi(t'')} e^{iS[\bar{\psi}, \psi]}, \tag{6.3.102}
\end{aligned}$$

where we have arrived at the Dirac action in the exponent,

$$S[\bar{\psi}, \psi] = \int dt\, L(\bar{\psi}, \psi) = \int_{t'}^{t''} dt \int d^3x\, \mathcal{L} = \int_{t'}^{t''} dt \int d^3x\, \bar{\psi}(x)(i\slashed{\partial} - m)\psi(x) \tag{6.3.103}$$

and where the normalization factor N_J is the Jacobian determinant associated with the change of Grassmann variables (or generators). Here $\psi_\alpha(x)$ and $\bar{\psi}_\alpha(x)$ have no "^" since they are not operators. They are odd Grassmann-valued fields that are linear superpositions of the Grassmann generators given in Eq. (6.3.104), which means that $\{\psi(x)_\alpha, \psi_\beta(y)\} = \{\psi_\alpha(x), \bar{\psi}_\beta(y)\} = \{\bar{\psi}_\alpha(x), \bar{\psi}_\beta(y)\} = 0$.

The integration is over all Grassmann-valued functions of spacetime between t' and t'' that are *antiperiodic in time*.

Proof: First we define the Grassmann change of variables

$$\psi_\alpha(\mathbf{x}) \equiv \int (d^3p/(2\pi)^3\sqrt{2E_\mathbf{p}})\sum_s [\xi^s_{b\mathbf{p}}u^s_\alpha(\mathbf{p})e^{i\mathbf{p}\cdot\mathbf{x}} + \xi^{*s}_{d\mathbf{p}}v^s_\alpha(\mathbf{p})e^{-i\mathbf{p}\cdot\mathbf{x}}],$$

$$\bar{\psi}_\alpha(\mathbf{x}) \equiv \int (d^3p/(2\pi)^3\sqrt{2E_\mathbf{p}})\sum_s [\xi^{*s}_{b\mathbf{p}}\bar{u}^s_\alpha(\mathbf{p})e^{-i\mathbf{p}\cdot\mathbf{x}} + \xi^s_{d\mathbf{p}}\bar{v}^s_\alpha(\mathbf{p})e^{i\mathbf{p}\cdot\mathbf{x}}], \qquad (6.3.104)$$

which are just the Grassmann versions of Eq. (6.3.87) for $\hat{\psi}$ and $\hat{\bar{\psi}} = \psi^\dagger\gamma^0$. The $\psi_\alpha(\mathbf{x})$ and $\bar{\psi}_\alpha(\mathbf{x})$ are eight Grassmann generators for every $\mathbf{x}$. The $\xi^s_{b\mathbf{p}}$, $\xi^{*s}_{b\mathbf{p}}$, $\xi^s_{d\mathbf{p}}$ and $\xi^{*s}_{d\mathbf{p}}$ are eight Grassmann generators for every $\mathbf{p}$. Since we know the Fourier transform $\mathbf{x} \leftrightarrow \mathbf{p}$ is a unitary change of basis, then Eq. (6.3.104) is a change of basis in the Grassmann algebra as discussed in Sec. 6.3.2. With an appropriate regularization, such as putting the system in a box with an ultraviolet cut-off, this change of variable could be written in terms of some finite constant matrix B of the type $(\bar{\psi}, \psi)_i = \sum_j B_{ij}(\xi^*_b, \xi^*_d, \xi_b, \xi_d)_j$ using an obvious shorthand to connect with Eq. (6.3.21). Then using Eq. (6.3.24) we have $\mathcal{D}\bar{\psi}\mathcal{D}\psi = (\det B)^{-1}\mathcal{D}\xi^*_b\mathcal{D}\xi^*_d\mathcal{D}\xi_b\mathcal{D}\xi_d$ and we define $N_J \equiv \det B$.

Next recall that $H[\xi^*_b, \xi^*_d, \xi_b, \xi_d]$ is the normal-ordered Hamiltonian operator $\hat{H} = H[\hat{b}^\dagger, \hat{d}^\dagger, \hat{b}, \hat{d}]$ with replacements $\hat{b}^{s\dagger}_\mathbf{p}, \hat{d}^{s\dagger}_\mathbf{p}, \hat{b}^s_\mathbf{p}, \hat{d}^s_\mathbf{p} \rightarrow \xi^{*s}_{b\mathbf{p}}, \xi^{*s}_{d\mathbf{p}}, \xi^s_{b\mathbf{p}}, \xi^s_{d\mathbf{p}}$, respectively. We observe that

$$\begin{aligned} H[\xi^*_b, \xi^*_d, \xi_b, \xi_d] &= \int (d^3p/(2\pi)^3)\sum_s E_\mathbf{p}(\xi^{*s}_{b\mathbf{p}}\xi^s_{b\mathbf{p}} + \xi^{*s}_{d\mathbf{p}}\xi^s_{d\mathbf{p}}) \\ &= \int (d^3p/(2\pi)^3)\sum_s E_\mathbf{p}(\xi^{*s}_{b\mathbf{p}}\xi^s_{b\mathbf{p}} - \xi^s_{d\mathbf{p}}\xi^{*s}_{d\mathbf{p}}) \\ &= \int d^3x\, \bar{\psi}(\mathbf{x})(-i\boldsymbol{\gamma}\cdot\boldsymbol{\nabla} + m)\psi(\mathbf{x}) \equiv H[\bar{\psi}, \psi], \qquad (6.3.105) \end{aligned}$$

where $\psi(\mathbf{x})$ and $\bar{\psi}(\mathbf{x})$ are as given in Eq. (6.3.104) and where the last line follows from reversing the various steps of Eq. (6.3.91).

At time t we have $\psi(x) = \psi(t, \mathbf{x})$ and $\bar{\psi}(x) = \bar{\psi}(t, \mathbf{x})$ expressed as in Eq. (6.3.104) in terms of $\xi^s_{b\mathbf{p}}(t)$, $\xi^{*s}_{b\mathbf{p}}(t)$, $\xi^s_{d\mathbf{p}}$ and $\xi^{*s}_{d\mathbf{p}}(t)$. We observe that

$$\int d^3x\, \psi^\dagger(x)i\partial_t\psi(x) = \int (d^3p/(2\pi)^3)\sum_s i\{\xi^{*s}_{b\mathbf{p}}(t)\partial_t\xi^s_{b\mathbf{p}}(t) + \xi^{*s}_{d\mathbf{p}}(t)\partial_t\xi^s_{d\mathbf{p}}(t)\},$$

which is easily proved with the same spinor identities that we used in the proof of Eq. (6.3.91). This is left as an exercise.

The exponent in the second line of Eq. (6.3.102) is then

$$\begin{aligned} i\int_{t'}^{t''} dt\, [\int d^3x\, \psi(x)^\dagger i\partial_t\psi(x) - H[\bar{\psi}(t), \psi(t)]] \\ = i\int_{t'}^{t''} dt \int d^3x\, \bar{\psi}(x)(i\slashed{\partial} - m)\psi(x) = iS[\bar{\psi}, \psi], \qquad (6.3.106) \end{aligned}$$

where $S[\bar{\psi}, \psi]$ is referred to as the Dirac action for reasons that will soon be discussed. The result has then been proved.

So the trace of the evolution operator for fermions involves the integral over all Grassmann-valued functions that are *antiperiodic in time*. Comparing Eqs. (6.3.70) and (6.3.72) we can easily verify that with the latter form we again arrive at Eq. (6.3.102) but now with $i\hbar\bar{\psi}\partial_t\psi$ replaced by $-(i\hbar\partial_t\bar{\psi})\psi$. So, effectively, we can perform an integration by parts without a surface term for the time integral in the spectral function because the antiperiodicity of the trace in Eq. (6.3.102) removes the surface term in both cases. Assuming periodic or vanishing spatial boundary conditions we can then ignore the surface term on all spacetime coordinates and perform an integration by parts. So we effectively have in Eq. (6.3.102) that

$$S[\bar{\psi}, \psi] = \int d^4x \; \bar{\psi}(x)(i\partial\!\!\!/ - m)\psi(x) = \int d^4x \; \bar{\psi}(x)(-i\overleftarrow{\partial\!\!\!/} - m)\psi(x). \qquad (6.3.107)$$

Since we take the continuum limit at the end we recall that our continuum notation is actually a shorthand notation for finite difference methods.

Fermion Coherent States

While we have arrived at the central results for the quantum field theory of free spin-$\frac{1}{2}$ fermions without reference to coherent states, it is worthwhile to develop that additional approach here since the coherent state formulation can be applied to both fermions and bosons. A discussion of bosonic coherent states[4] is not included here but can be found in Klauder and Skagerstam (1985).

The anticommuting Grassmann generators can be further required to anticommute with the fermion annihilation and creation operators,

$$\{\xi_i, \hat{b}_j\} = \{\xi_i, \hat{b}_j^\dagger\} = \{\xi_i^*, \hat{b}_j\} = \{\xi_i^*, \hat{b}_j^\dagger\} = 0. \qquad (6.3.108)$$

We define the *fermion coherent state* associated with the Grassmann generators $\xi_1, \ldots, \xi_N$ and creation operator $\hat{b}_1^\dagger, \ldots, \hat{b}_N^\dagger$ as

$$|\xi\rangle \equiv e^{-\sum_{i=1}^N \xi_i \hat{b}_i^\dagger}|0\rangle = \left(1 - \sum_{i=1}^N \xi_i \hat{b}_i^\dagger\right)|0\rangle = \prod_{i=1}^N (1 - \xi_i \hat{b}_i^\dagger)|0\rangle. \qquad (6.3.109)$$

Fock space is a Hilbert space over the field of complex numbers, i.e., a complete inner product vector space over the field of complex numbers. This means that if V_H is a Hilbert space and $|\psi\rangle \in V_H$, then $c|\psi\rangle \in V_H$ for all $c \in \mathbb{C}$. A *super Hilbert space*, V_{SH}, is a Hilbert space over the field of supernumbers (Grassmann numbers). If $f(\xi)$ is any power series in the Grassmann generators (i.e., any Grassmann number or supernumber), then for any $|\psi\rangle \in V_H$ we have $f(\xi)|\psi\rangle \in V_{SH}$. Since the field of supernumbers contains the complex numbers, then it follows that $V_H \subset V_{SH}$. So fermion coherent states are vectors in the super Fock space, $|\xi\rangle \in V_{SH}$. However, we will always be interested in projections from the super Fock space back to Fock space. The advantage of super Fock space and the coherent state formalism is that it allows for a simultaneous treatment of bosons and fermions.

Recalling Hermitian conjugation extended to the Grassmann algebra we have

$$(e^{-\xi_i \hat{b}_i^\dagger})^\dagger = (1 - \xi_i \hat{b}_i^\dagger)^\dagger = 1 - \hat{b}_i \xi_i^* = 1 + \xi_i^* \hat{b}_i = e^{\xi_i^* \hat{b}_i} \qquad (6.3.110)$$

and so we define the coherent state bra as

$$\langle\xi| \equiv \langle 0|e^{\sum_{i=1}^N \xi_i^* \hat{b}_i} = \langle 0|\left(1 + \sum_{i=1}^N \xi_i^* \hat{b}_i\right) = \langle 0|\prod_{i=1}^N (1 + \xi_i^* \hat{b}_i). \qquad (6.3.111)$$

Fermion coherent states have the properties

$$\hat{b}_j|\xi\rangle = \xi_j|\xi\rangle, \quad \langle\xi|\hat{b}_j^\dagger = \langle\xi|\xi_j^*, \quad \hat{b}_j^\dagger|\xi\rangle = -\frac{d}{d\xi_j}|\xi\rangle, \quad \langle\xi|\hat{b}_j = \frac{d}{d\xi_j^*}\langle\xi|,$$

$$\langle\xi|\eta\rangle = e^{\sum_{j=1}^N \xi_j^* \eta_j}. \qquad (6.3.112)$$

The coherent state is again an eigenstate of the annihilation operator.

---

[4] Roy J. Glauber was a co-winner of the 2005 Nobel Prize in Physics for his work on the quantum theory of optical coherence. His work makes extensive use of coherent states.

Proof: We see that

$$\hat{b}_j|\xi\rangle = \hat{b}_j \prod_{i=1}^{N}(1 - \xi_i \hat{b}_i^\dagger)|0\rangle = \prod_{i=1, i\neq j}^{N}(1 - \xi_i \hat{b}_i^\dagger)\xi_j \hat{b}_j \hat{b}_j^\dagger|0\rangle$$

$$= \prod_{i=1, i\neq j}^{N}(1 - \xi_i \hat{b}_i^\dagger)\xi_j|0\rangle = \xi_j \prod_{i=1}^{N}(1 - \xi_i \hat{b}_i^\dagger)|0\rangle = \xi_j|\xi\rangle,$$

$$\hat{b}_j^\dagger|\xi\rangle = \hat{b}_j^\dagger \prod_{i=1}^{N}(1 - \xi_i \hat{b}_i^\dagger)|0\rangle = \prod_{i=1, i\neq j}^{N}(1 - \xi_i \hat{b}_i^\dagger)\hat{b}_j^\dagger|0\rangle$$

$$= -\frac{d}{d\xi_j}\prod_{i=1}^{N}(1 - \xi_i \hat{b}_i^\dagger)|0\rangle = -\frac{d}{d\xi_j}|\xi\rangle. \tag{6.3.113}$$

The proofs for $\langle\xi|\hat{b}_j^\dagger$ and $\langle\xi|\hat{b}_j$ are similar and left as an exercise. We see that

$$\langle\xi|\eta\rangle = \langle 0|\prod_{i=1}^{N}(1 + \xi_i^* \hat{b}_i)\prod_{j=1}^{N}(1 - \eta_j \hat{b}_j^\dagger)|0\rangle = \langle 0|\prod_{i=1}^{N}(1 + \xi_i^* \hat{b}_i)(1 - \eta_i \hat{b}_i^\dagger)|0\rangle$$

$$= \langle 0|\prod_{i=1}^{N}(1 + \xi_i^* \eta_i \hat{b}_i \hat{b}_i^\dagger)|0\rangle = \prod_{i=1}^{N}(1 + \xi_i^* \eta_i) = e^{\sum_{i=1}^{N} \xi_i^* \eta_i}. \tag{6.3.114}$$

Let $|\Psi\rangle$ be any state in fermion Fock space as given in Eq. (6.3.36), then we have

$$\langle\xi|\Psi\rangle = \sum_{n=0}^{N}\sum_{i_1<\cdots<i_n=1}^{N} c_{i_1 i_2\cdots i_n}\langle\xi|b_{i_1}b_{i_2}\cdots b_{i_n}; A_o\rangle$$

$$= \sum_{n=0}^{N}\sum_{i_1<\cdots<i_n=1}^{N} c_{i_1 i_2\cdots i_n}\langle\xi|b_{i_1}^\dagger b_{i_2}^\dagger \cdots b_{i_n}^\dagger|0\rangle$$

$$= \sum_{n=0}^{N}\sum_{i_1<\cdots<i_n=1}^{N} c_{i_1 i_2\cdots i_n}\xi_{i_1}^*\xi_{i_2}^*\cdots\xi_{i_n}^* = \Psi(\xi^*), \tag{6.3.115}$$

where we have used Eqs. (6.3.6), (6.3.112) and (6.3.37). We observe that

$$\langle\xi|b_{i_1}b_{i_2}\cdots b_{i_n}; A_o\rangle = \xi_{i_1}^*\cdots\xi_{i_n}^* \quad \text{and} \quad \langle b_{i_1}b_{i_2}\cdots b_{i_n}; A_o|\xi\rangle = \xi_{i_n}\cdots\xi_{i_1}. \tag{6.3.116}$$

Let $\hat{A}$ be any linear operator in Fock space. Using the orthonormal basis given in Eq. (4.1.160) we have

$$\hat{A} = \sum_{m,n=0}^{N}\sum_{i_1<\cdots<i_m=1}^{N}\sum_{j_1<\cdots<j_n=1}^{N} A_{i_1\cdots i_m, j_1\cdots j_n}|b_{i_1}\cdots b_{i_m}; A_o\rangle\langle b_{j_1}\cdots b_{j_n}; A_o|,$$

$$A_{i_1\cdots i_m, j_1\cdots j_n} \equiv \langle b_{i_1}\cdots b_{i_m}; A_o|\hat{A}|b_{j_1}\cdots b_{j_n}; A_o\rangle. \tag{6.3.117}$$

Using similar steps to those in Eq. (6.3.115) we easily see that

$$\langle\Psi|\xi\rangle = \Psi^*(\xi) \quad \text{and} \quad \langle\xi|\hat{A}|\eta\rangle = \tilde{A}(\xi^*, \eta), \tag{6.3.118}$$

where we have used Eqs. (6.3.38) and (6.3.47). The resolution of the identity in terms of coherent states is given by

$$\hat{I} = \int \left(\prod_{i=1}^{N} d\xi_i^* d\xi_i\right) e^{-\sum_{j=1}^{N} \xi_j^* \xi_j} |\xi\rangle\langle\xi|. \tag{6.3.119}$$

Proof: Consider the matrix element of the above between any two orthonormal basis states in Fock space,

$$\langle b_{i_1}\cdots b_{i_m}; A_o|\hat{I}|b_{j_1}\cdots b_{j_n}; A_o\rangle$$

$$= \int \left(\prod_{i=1}^{N} d\xi_i^* d\xi_i\right) e^{-\sum_{j=1}^{N} \xi_j^* \xi_j} \langle b_{i_1}\cdots b_{i_m}; A_o|\xi\rangle\langle\xi|b_{j_1}\cdots b_{j_n}; A_o\rangle$$

$$= \int \left(\prod_{i=1}^{N} d\xi_i^* d\xi_i\right) e^{-\sum_{j=1}^{N} \xi_j^* \xi_j} \xi_{i_m}\cdots\xi_{i_1}\xi_{j_1}^*\cdots\xi_{j_n}^*$$

$$= \delta_{mn}\delta_{i_1 j_1}\cdots\delta_{i_n j_n} = \langle b_{i_1}\cdots b_{i_m}; A_o|b_{j_1}\cdots b_{j_n}; A_o\rangle, \tag{6.3.120}$$

where we have used Eq. (6.3.44). Then for any two states in Fock space we have $\langle\Psi|\hat{I}|\Phi\rangle = \langle\Psi|\Phi\rangle$ and so the operator $\hat{I}$ above is the identity operator.

It is now clear that the fermion coherent state formalism will reproduce all of the quantities and relationships that we derived above starting from Eq. (6.3.37). It is a useful exercise for the interested reader to complete this task. A coherent state formalism can also be constructed for bosons analogous to what was done for fermions. These coherent states live in what is called Bargmann-Fock space. We will not pursue this further but further discussions can be found in Itzykson and Zuber (1980) and Klauder (2011).

6.3.6 Canonical Quantization of Dirac Fermions

The canonical quantization of fermions requires some care. Often a simple canonical argument is given that suffices for a basic introduction. It can also be done more rigorously with some effort. We present both here for completeness.

Simple Canonical Argument

We begin with a quick and simple argument for canonical quantization similar to that presented, for example, in Bjorken and Drell (1965). Consider the Dirac action

$$S[\bar{\psi}, \psi] = \int dt \, L = \int d^4x \, \mathcal{L} = \int d^4x \, \bar{\psi}(x)(i\slashed{\partial} - m)\psi(x), \tag{6.3.121}$$

where $\psi(x)$ is a four-component column vector with components $\psi_\alpha(x) \in \mathbb{C}$, where $\psi^\dagger(x)$ is the complex conjugate row vector with components $\psi_\alpha^*(x)$, and where $\bar{\psi}_\alpha(x) \equiv \psi^\dagger(x)\gamma^0$. Note that here we are treating $\psi_\alpha(x)$ as a complex field and *not* as the odd Grassman-valued field in Eq. (6.3.104). We will return to this important distinction later. Using the Wirtinger calculus of Sec. A.2.5 we have the Wirtinger derivatives $\delta/\delta\psi_\alpha(x)$ and $\delta/\delta\psi_\alpha^*(x)$. The Dirac action $S[\bar{\psi}, \psi]$ is a real function of the complex fields $\psi_\alpha(x)$ and we will have a stationary point of the Dirac action when $\delta S/\delta\psi_\alpha(x) = 0$ or equivalently when $\delta S/\delta\psi_\alpha^*(x) = 0$. Note that the latter condition can also be expressed as $\delta S/\delta\bar{\psi}_\alpha(x) = 0$. So, similarly to the complex scalar field case, we can express Hamilton's principle for the Dirac action as

$$\frac{\delta S[\bar{\psi}, \psi]}{\delta \bar{\psi}_\alpha(x)} = \frac{\delta S[\bar{\psi}, \psi]}{\delta \psi_\alpha(x)} = 0, \tag{6.3.122}$$

which leads to the recovery of the Dirac equation as the equations of motion,

$$(i\slashed{\partial} - m)\psi(x) = \bar{\psi}(x)(-i\overleftarrow{\slashed{\partial}} - m) = 0. \tag{6.3.123}$$

> **Proof:** Recall from Sec. 3.1 that we only consider field fluctuations with compact support when deriving the equations of motion so that surface terms can be neglected. Using integration by parts we then find
>
> $$0 = \delta S[\bar{\psi}, \psi]/\delta\psi(x) = \delta/\delta\psi(x)\int d^4x \, \bar{\psi}(x)(-i\overleftarrow{\slashed{\partial}} - m)\psi(x) = \bar{\psi}(x)(-i\overleftarrow{\slashed{\partial}} - m),$$
> $$0 = \delta S[\bar{\psi}, \psi]/\delta\bar{\psi}(x) = (i\slashed{\partial} - m)\psi(x). \tag{6.3.124}$$
>
> From Sec. A.2.5 we know that these two equations must be equivalent. This is seen by taking the complex conjugate transpose of the equations of motion $\bar{\psi}(x)(-i\overleftarrow{\slashed{\partial}} - m) = 0$ and acting from the left with γ^0,
>
> $$0 = \gamma^0(i\partial_\mu\gamma^{\mu\dagger} - m)[\bar{\psi}(x)]^\dagger = (i\partial_\mu\gamma^0\gamma^{\mu\dagger}\gamma^0 - m)\gamma^0[\bar{\psi}(x)]^\dagger$$
> $$= (i\slashed{\partial} - m)[\bar{\psi}(x)\gamma^0]^\dagger = (i\slashed{\partial} - m)\psi(x), \tag{6.3.125}$$

where we have used $\gamma^{0\dagger} = \gamma^0$, $(\gamma^0)^2 = 1$ and $\gamma^0\gamma^\mu\gamma^0 = \gamma^{\mu\dagger}$. Note that $\delta/\delta\psi_\alpha^\dagger = [\gamma^0(\delta/\delta\bar\psi)]_\alpha$, which follows since then

$$\delta\psi_\alpha^\dagger(\mathbf{x})/\delta\psi_\beta^\dagger(\mathbf{y}) = \sum_{\gamma,\delta}\gamma_{\beta\delta}^0\gamma_{\gamma\alpha}^0[\delta\bar\psi_\gamma(\mathbf{x})/\delta\bar\psi_\delta(\mathbf{y})] = \delta_{\alpha\beta}\delta^3(\mathbf{x}-\mathbf{y}). \qquad (6.3.126)$$

Of course, we can also obtain these results from the Lagrangian density using the Euler-Lagrange equations.

The canonical momentum densities conjugate to ψ_α and $\bar\psi_\alpha$ are, respectively,

$$\pi_\alpha(x) = \delta L/\delta\dot\psi_\alpha(x) = \partial\mathcal{L}/\partial\dot\psi_\alpha(x) = i\psi_\alpha^\dagger(x) = i(\bar\psi(x)\gamma^0)_\alpha,$$
$$\bar\pi_\alpha(x) = \delta L/\delta\dot{\bar\psi}_\alpha(x) = \partial\mathcal{L}/\partial\dot{\bar\psi}_\alpha(x) = 0. \qquad (6.3.127)$$

Note that our discussion could be framed in terms of $\psi^\dagger$ rather than $\bar\psi$, but the latter is more convenient for our purposes since we will often be constructing the fermion bilinears of Eq. (4.4.159). It is clear that the Hessian associated with a Legendre transform of the Dirac action is singular since all second derivatives with respect to velocities $\dot\psi_\alpha$ and $\dot{\bar\psi}_\alpha$ vanish. Let us temporarily ignore this problem and construct the naive Hamiltonian anyway,

$$H = \int d^3x \sum_\alpha[\dot\psi_\alpha\pi_\alpha + \dot{\bar\psi}_\alpha\bar\pi_\alpha] - L = \int d^3x\,[\sum_\alpha(\dot\psi_\alpha\pi_\alpha + \dot{\bar\psi}_\alpha\bar\pi_\alpha) - \mathcal{L}]$$
$$= \int d^3x\,[i\bar\psi\gamma^0\partial_0\psi - \bar\psi(i\slashed\partial - m)\psi] = \int d^3x\,\bar\psi[-i\boldsymbol\gamma\cdot\boldsymbol\nabla + m]\psi. \qquad (6.3.128)$$

This is the classical form of the Dirac Hamiltonian in Eq. (6.3.91).

Recall from Sec. 3.2.2 that the invariance of a classical field theory under arbitrary spacetime displacements will occur when the action has no explicit spacetime dependence. Such a symmetry leads to the existence of a stress-energy tensor, $T^\mu{}_\nu$, with four conserved currents. The Dirac action is spacetime-translation invariant and so we find the stress-energy tensor using Eq. (3.2.53),

$$T^\mu{}_\nu = \pi^\mu\,\partial_\nu\psi + \bar\pi^\mu\,\partial_\nu\bar\psi - \delta^\mu{}_\nu\mathcal{L} = i\bar\psi\gamma^\mu\partial_\nu\psi - \delta^\mu{}_\nu\bar\psi(i\slashed\partial - m)\psi, \qquad (6.3.129)$$

which is true on-shell or off-shell. We have suppressed the summation over the spinor indices for brevity and we have defined in the usual way

$$\pi_\alpha^\mu \equiv \partial\mathcal{L}/\partial(\partial_\mu\psi_\alpha) = i(\bar\psi\gamma^\mu)_\alpha \quad \text{and} \quad \bar\pi_\alpha^\mu \equiv \partial\mathcal{L}/\partial(\partial_\mu\bar\psi_\alpha) = 0. \qquad (6.3.130)$$

Recall that Noether currents are only conserved on-shell (when the equations of motion are satisfied). Since $(i\slashed\partial - m)\psi = 0$ on-shell, then $\mathcal{L} = 0$ on-shell leading to

$$T^\mu{}_\nu = \pi^\mu\,\partial_\nu\psi + \bar\pi^\mu\,\partial_\nu\bar\psi = i\bar\psi\gamma^\mu\partial_\nu\psi \quad \text{and} \quad \partial_\mu T^\mu{}_\nu = 0. \qquad (6.3.131)$$

From Eq. (3.2.54) we see that the four conserved charges are the components of the total four-momentum,

$$P^\nu = (H, \mathbf{P}) = \int d^3x\,T^{0\nu} = \int d^3x\,i\bar\psi\gamma^0\partial^\nu\psi = \int d^3x\,i\psi^\dagger\partial^\nu\psi. \qquad (6.3.132)$$

Since $(i\slashed\partial - m)\psi = 0$ on-shell then we can write the on-shell Hamiltonian as

$$H = \int d^3x\,i\psi^\dagger\partial_0\psi = \int d^3x\,\psi^\dagger(-i\boldsymbol\alpha\cdot\boldsymbol\nabla + \beta m)\psi = \int d^3x\,\bar\psi[-i\boldsymbol\gamma\cdot\boldsymbol\nabla + m]\psi. \qquad (6.3.133)$$

The total conserved three-momentum is given by

$$\mathbf{P} = \int d^3x\,\psi^\dagger(-i\boldsymbol\nabla)\psi. \qquad (6.3.134)$$

Similarly, since the Dirac action is Lorentz invariant by construction and since Dirac spinors carry intrinsic angular momentum (they are spin-$\frac{1}{2}$), then the results of Sec. 3.2.4 will apply. There will be six conserved charges making up the angular momentum tensor, $M^{\mu\nu}$. From Eq. (3.2.99) we have

$$
\begin{aligned}
M^{\rho\sigma} &= \int d^3x \left[\left(x^\rho T^{0\sigma} - x^\sigma T^{0\rho} \right) + \pi_\alpha (\Sigma_{\alpha\beta})^{\rho\sigma} \psi_\beta + \bar{\pi}_\alpha (\Sigma_{\alpha\beta})^{\rho\sigma} \bar{\psi}_\beta \right] \\
&= \int d^3x \left[\left(x^\rho i\psi^\dagger \partial^\sigma \psi - x^\sigma i\psi^\dagger \partial^\rho \psi \right) + i\psi_\alpha^\dagger (\Sigma_{\alpha\beta})^{\rho\sigma} \psi_\beta \right] \\
&= \int d^3x\, i\psi^\dagger \left(x^\rho \partial^\sigma - x^\sigma \partial^\rho + \tfrac{1}{4}[\gamma^\rho, \gamma^\sigma] \right) \psi = \int d^3x\, \psi^\dagger \left(x^\rho i\partial^\sigma - x^\sigma i\partial^\rho + \Sigma_{\mathrm{Dirac}}^{\mu\nu} \right) \psi,
\end{aligned} \qquad (6.3.135)
$$

where we sum over repeated spinor indices and where a comparison of Eqs. (3.2.88) and (4.4.75) shows that

$$
i\Sigma^{\rho\sigma} \equiv \Sigma_{\mathrm{Dirac}}^{\rho\sigma} = \tfrac{1}{2}\sigma^{\rho\sigma} = \tfrac{i}{4}[\gamma^\rho, \gamma^\sigma]. \qquad (6.3.136)
$$

Note: The $\Sigma^{\mu\nu}$ used here corresponds to the form used in Chapter 3, whereas we now write the $\Sigma^{\mu\nu}$ of Sec. 4.4 as $\Sigma_{\mathrm{Dirac}}^{\mu\nu}$ to avoid confusion. Recall from Eq. (1.2.72) that $J^i = \frac{1}{2}\epsilon^{ijk}M^{jk}$ and that from Eq. (4.4.143) we have $S^i = \frac{1}{2}\epsilon^{ijk}\Sigma_{\mathrm{Dirac}}^{jk}$ so in the Dirac representation

$$
\begin{aligned}
\mathbf{J} = (M^{23}, M^{31}, M^{12}) &= \int d^3x\, \psi^\dagger \left(\mathbf{x} \times (-i\boldsymbol{\nabla}) + \mathbf{S} \right) \psi \\
&= \int d^3x\, \psi^\dagger \left(\mathbf{x} \times (-i\boldsymbol{\nabla}) + \tfrac{1}{2} \begin{bmatrix} \boldsymbol{\sigma} & 0 \\ 0 & \boldsymbol{\sigma} \end{bmatrix} \right) \psi,
\end{aligned} \qquad (6.3.137)
$$

which in the nonrelativistic limit just reduces to the familiar form of $\mathbf{J} = \mathbf{L} + \mathbf{S}$ of nonrelativistic quantum mechanics. Also note that we can write

$$
M^{\rho\sigma} = \int d^3x\, \psi^\dagger M^{\rho\sigma}|_{\mathrm{Dirac}} \psi, \qquad (6.3.138)
$$

where $M^{\rho\sigma}|_{\mathrm{Dirac}}$ is the operator given in Eq. (4.4.160).

Consider a naive application of Dirac's canonical quantization procedure as we did for the scalar field arriving at the canonical commutation relations in Eq. (6.2.4). *Note:* We will use the notation $\{\cdot, \cdot\}_{\mathrm{PB}}$ to denote the Poisson bracket when needed to distinguish it from the anticommutator, $\{\cdot, \cdot\}$. Treating $\psi_\alpha(x)$ as a complex field using the Wirtinger calculus of Sec. A.2.5 we would arrive at the fundamental Poisson brackets $\{\psi_\alpha(\mathbf{x}), \pi_\beta(\mathbf{x})\}_{\mathrm{PB}} = \delta_{\alpha\beta}\delta^3(\mathbf{x} - \mathbf{y})$ and $\{\psi_\alpha(\mathbf{x}), \psi_\beta(\mathbf{x})\}_{\mathrm{PB}} = \{\pi_\alpha(\mathbf{x}), \pi_\beta(\mathbf{x})\}_{\mathrm{PB}} = 0$, where we recall that we have $\pi_\alpha(x) = i\psi_\alpha^\dagger(x)$. Then applying the correspondence principle of Eq. (2.5.27) would give $[\hat{\psi}_\alpha(\mathbf{x}), \hat{\psi}_\beta^\dagger(\mathbf{y})] = \delta_{\alpha\beta}\delta^3(\mathbf{x} - \mathbf{y})$ and $[\hat{\psi}_\alpha(\mathbf{x}), \hat{\psi}_\beta(\mathbf{y})] = [\hat{\psi}_\alpha^\dagger(\mathbf{x}), \hat{\psi}_\beta^\dagger(\mathbf{y})] = 0$. This is the same as the canonical anticommutation relations in Eq. (6.3.94) that we derived using normal modes and fermion Fock space, *except* that here we have commutators rather than anticommutators. At this point the simple approach is to say that in order to recover the required fermion antisymmetry, captured in the anticommutation relations of Eq. (6.3.80), we must replace commutators with anticommutators, which then results in Eq. (6.3.94).

The above naive canonical quantization procedure is inadequate for two reasons:
(i) we have not accounted for the fact that we have a singular Lagrangian, and
(ii) the "classical limit" of a fermion quantum field theory is a Grassmann-valued field, not a complex field, as we have seen.

More Careful Canonical Treatment

We present a concise summary of the key results and arguments so that the interested reader can fill in further details for themself. In Sec. A.2.5 we showed how Wirtinger calculus allows us to extend the constructions of classical mechanics from real variables to complex variables and we applied that approach to the complex/charged scalar field and naively to Dirac fermion fields in the simple canonical argument above. As shown in Chapters 6 and 7 of Henneaux and Teitelboim (1994) we can extend the constructions of classical mechanics to the case of Grassmann-valued variables. See also the discussion in Das (2008). Since this leads to a generalization of the Poisson bracket we can then generalize the Dirac-Bergmann algorithm for singular systems as discussed in Sec. 2.9.1. The extension of classical mechanics to apply over a Grassmann algebra unifies the approach to bosons and fermions. The classical limit ($\hbar \to 0$) for bosons leads to c-number (real or complex number) variables, whereas for fermions it leads to Grassmann numbers.

Classical mechanics over a Grassmann algebra: We will work only with Grassmann left-derivatives (meaning acting from the left) for simplicity, but the presentation could be repeated with the right-derivative as well. The properties of a Grassmann algebra are discussed in previous sections of this chapter. Consider a Grassmann algebra, G_N, with N generators $\xi_1, \ldots, \xi_N$. Any element $g \in G_N$ is referred to as a Grassmann number or Grassmann element and can be written as

$$g = \sum_{n=0}^{N} \sum_{i_1 < \cdots < i_n = 1}^{N} c_{i_1 i_2 \cdots i_n} \xi_{i_1} \xi_{i_2} \cdots \xi_{i_n}$$
$$= c_0 + \sum_{i=1}^{N} c_i \xi_i + \sum_{j<k=1}^{N} c_{ij} \xi_i \xi_j + \cdots, \qquad (6.3.139)$$

where $c_{i_1 i_2 \cdots i_n} \in \mathbb{C}$. An *even* Grassmann number contains only even monomials and an *odd* Grassmann number contains only odd monomials. An even Grassmann number commutes with all Grassmann numbers and an odd Grassmann number anticommutes with an odd Grassmann number. So even Grassmann numbers are bosonic in nature, whereas odd Grassmann numbers are fermionic.

In a classical system built on a Grassmann algebra we denote a bosonic generalized coordinate as $q(t)$ and a fermionic one as $\theta(t)$, where

$$q(t) = \sum_{n,\text{even}=0}^{N} \sum_{i_1 < \cdots < i_n = 1}^{N} c_{i_1 i_2 \cdots i_n}(t) \xi_{i_1} \xi_{i_2} \cdots \xi_{i_n}$$
$$\theta(t) = \sum_{n,\text{odd}=1}^{N} \sum_{i_1 < \cdots < i_n = 1}^{N} d_{i_1 i_2 \cdots i_n}(t) \xi_{i_1} \xi_{i_2} \cdots \xi_{i_n}. \qquad (6.3.140)$$

A space containing both bosonic and fermionic degrees of freedom is called a *superspace*. We are uninterested in generalized coordinates that are neither even nor odd, i.e., that are neither bosonic nor fermionic. All time dependence is contained in the complex coefficients and so time derivatives are easily understood. For now we follow Henneaux and Teitelboim (1994) and assume that the Grassmann generators are real, $\xi_j^* = \xi_j$, and that all coefficients $c_{i_1 i_2 \cdots i_n}$ and $d_{i_1 i_2 \cdots i_n}$ are real. This means that the $q(t)$ and $\theta(t)$ are real. Recall that in our notation ξ^* is the Grassmann complex conjugate of ξ. We can easily extend the results to the complex case by introducing another N Grassmann generators $\xi_1^*, \ldots, \xi_N^*$, by using the Wirtinger calculus for $q(t)$ and $q^*(t)$ and by the addition of $\theta^*(t)$ in terms of the ξ^*.

Having reached this point we have no further need to explicitly refer to the Grassmann generators and we will talk only in terms of the generalized coordinates. The dynamics of the generalized coordinates of a classical system in Chapter 2 were independent of the choice of an underlying Cartesian coordinate system. In the same sense the dynamics of the generalized coordinates here do not depend on our choice of Grassmann generators. All that matters is how the Lagrangian for the system is specified in terms of the generalized coordinates. If we have N_b bosonic generalized coordinates $q_1(t), \ldots, q_{N_b}(t)$ and N_f fermionic generalized coordinates $\theta_1(t), \ldots, \theta_{N_f}(t)$, then we have the commutation and anticommutation relations

$$[q_i(t), q_j(t)] = [q_i(t), \theta_\alpha(t)] = 0 \quad \text{and} \quad \{\theta_\alpha(t), \theta_\beta(t)\} = 0. \qquad (6.3.141)$$

In order to construct our classical system in terms of a Lagrangian we also need to introduce the corresponding independent bosonic and fermionic velocities, $\dot{q}_i(t)$ and $\dot{\theta}_\alpha(t)$. All coordinates and velocities commute except those involving two odd quantities, which will anticommute,

$\{\theta_\alpha(t), \theta_\beta(t)\} = \{\theta_\alpha(t), \dot{\theta}_\beta(t)\} = \{\dot{\theta}_\alpha(t), \dot{\theta}_\beta(t)\} = 0$. An action and Lagrangian for the system are then specified in the usual way as

$$S[\vec{q}, \vec{\theta}] = \int_{t_i}^{t_f} dt\, L(\vec{q}, \dot{\vec{q}}, \vec{\theta}, \dot{\vec{\theta}}, t). \tag{6.3.142}$$

We *define* the derivatives with respect to coordinates and velocities based on a power series expansion in terms of these quantities, rather than in terms of any infinitesimal limit process as we did for quantum operators. So for a function f expressed as an arbitrary power series of coordinates and velocities we have

$$df = \sum_{i=1}^{N_b} \left(dq_i \frac{\partial f}{\partial q_i} + d\dot{q}_i \frac{\partial f}{\partial \dot{q}_i} \right) + \sum_{\alpha=1}^{N_f} \left(d\theta_\alpha \frac{\partial f}{\partial \theta_\alpha} + d\dot{\theta}_\alpha \frac{\partial f}{\partial \dot{\theta}_\alpha} \right), \tag{6.3.143}$$

where $\partial f/\partial q_i$ and $\partial f/\partial \dot{q}_i$ have their usual meanings when acting on a power series and where $\partial f/\partial \theta_\alpha$ and $\partial f/\partial \dot{\theta}_\alpha$ are the usual left Grassmann derivatives. For example, $(\partial f/\partial q_i)(q_j q_i^n q_k) = nq_j q_i^{n-1} q_k$ and $(\partial f/\partial \theta_\alpha)(\theta_\beta \theta_\alpha \theta_\gamma) = -\theta_\beta \theta_\gamma$. The odd Grassmann differentials $d\theta_\alpha$ are as defined in Eq. (6.3.17). Grassmann-odd differentials are always placed to the left of the corresponding Grassmann left derivatives, since then $d\theta_\alpha(\partial/\partial\theta_\alpha)(\theta_\beta\theta_\alpha\theta_\gamma) = \theta_\beta d\theta_\alpha \theta_\gamma$ because $\{d\theta_\alpha, \theta_\beta\} = 0$.

We briefly summarize and generalize the classical mechanics arguments of Chapter 2. We need to take care to track minus signs coming from the ordering of the fermionic (odd) quantities. Hamilton's principle is written as

$$\delta S/\delta \vec{q}(t) = \delta S/\delta \vec{\theta}(t) = 0. \tag{6.3.144}$$

Using the fact that the total differential for the Lagrangian is

$$dL = \sum_{j=1}^{N_b} \left[dq_j \frac{\partial L}{\partial q_j} + d\dot{q}_j \frac{\partial L}{\partial \dot{q}_j} \right] + \sum_{\alpha=1}^{N_f} \left[d\theta_\alpha \frac{\partial L}{\partial \theta_\alpha} + d\dot{\theta}_\alpha \frac{\partial L}{\partial \dot{\theta}_\alpha} \right] + dt \frac{\partial L}{\partial t}, \tag{6.3.145}$$

then we can generalize the arguments of Eq. (2.1.51) at fixed time t to obtain the Euler-Lagrange equations of Eq. (2.1.52) in the form

$$0 = \frac{\partial L}{\partial q_j} - \frac{d}{dt}\frac{\partial L}{\partial \dot{q}_j} = \frac{\partial L}{\partial \theta_\alpha} - \frac{d}{dt}\frac{\partial L}{\partial \dot{\theta}_\alpha} \tag{6.3.146}$$

for $j = 1, \ldots, N_b$ and $\alpha = 1, \ldots, N_f$. We define the conjugate generalized momenta in the standard way as

$$p_j \equiv \partial L/\partial \dot{q}_j \quad \text{and} \quad \pi_\alpha \equiv \partial L/\partial \dot{\theta}_\alpha. \tag{6.3.147}$$

Using the Euler-Lagrange equations we have

$$\dot{p}_j = \partial L/\partial q_j \quad \text{and} \quad \dot{\pi}_\alpha = \partial L/\partial \theta_\alpha \quad \text{and so} \tag{6.3.148}$$

$$dL = \sum_{j=1}^{N_b} [dq_j \dot{p}_j + d\dot{q}_j p_j] + \sum_{\alpha=1}^{N_f} \left[d\theta_\alpha \dot{\pi}_\alpha + d\dot{\theta}_\alpha \pi_\alpha \right] + dt(\partial L/\partial t). \tag{6.3.149}$$

We assume for now that the Lagrangian is nonsingular; i.e., the generalized momenta are invertible functions of the generalized velocities. We can then define the Hamiltonian

$$H = \sum_{j=1}^{N_b} \dot{q}_j p_j + \sum_{\alpha=1}^{N_f} \dot{\theta}_\alpha \pi_\alpha - L. \tag{6.3.150}$$

The total differential for the Hamiltonian, $H(\vec{q}, \vec{p}, \vec{\theta}, \vec{\pi}, t)$, can be written both as

$$dH = \sum_{j=1}^{N_b} \left[dq_j \frac{\partial H}{\partial q_j} + dp_j \frac{\partial H}{\partial p_j} \right] + \sum_{\alpha=1}^{N_f} \left[d\theta_\alpha \frac{\partial H}{\partial \theta_\alpha} + d\pi_\alpha \frac{\partial H}{\partial \pi_\alpha} \right] + dt \frac{\partial H}{\partial t}$$

and from Eqs. (6.3.149) and (6.3.150) as

$$dH = \sum_{j=1}^{N_b}[d\dot{q}_j p_j + dp_j \dot{q}_j] + \sum_{\alpha=1}^{N_f}[d\dot{\theta}_\alpha \pi_\alpha - d\pi_\alpha \dot{\theta}_\alpha] - dL$$

$$= \sum_{j=1}^{N_b}[-dq_j \dot{p}_j + dp_j \dot{q}_j] + \sum_{\alpha=1}^{N_f}[-d\theta_\alpha \dot{\pi}_\alpha - d\pi_\alpha \dot{\theta}_\alpha] - dt(\partial L/\partial t). \qquad (6.3.151)$$

Since dq_i, dp_i, $d\theta_\alpha$, $d\pi_\alpha$ and dt are independent differentials, then we must have

$$\dot{q}_j = \frac{\partial H}{\partial p_j}, \quad \dot{p}_j = -\frac{\partial H}{\partial q_j}, \quad \dot{\theta}_\alpha = -\frac{\partial H}{\partial \pi_\alpha}, \quad \dot{\pi}_\alpha = -\frac{\partial H}{\partial \theta_\alpha}, \quad \frac{\partial H}{\partial t} = -\frac{\partial L}{\partial t}, \qquad (6.3.152)$$

where the sign of the $\dot{\theta}_\alpha$ equation should be noted. This is the generalized form of Hamilton's equations. Note that we can decompose any dynamical variable, $F(\vec{q}, \vec{p}, \vec{\theta}, \vec{\pi}, t)$, into even and odd Grassmann parts, $F = F_{\text{even}} + F_{\text{odd}}$. We will only be interested in situations where the action, S, is even, which leads to an even Lagrangian, L. Then π_α is odd and so the Hamiltonian, H, is even.

Similar to Eq. (6.3.143), for any dynamical variable F we have

$$dF = \sum_{i=1}^{N_b}\left(dq_i \frac{\partial F}{\partial q_i} + dp_i \frac{\partial F}{\partial p_i}\right) + \sum_{\alpha=1}^{N_f}\left(d\theta_\alpha \frac{\partial F}{\partial \theta_\alpha} + d\pi_\alpha \frac{\partial F}{\partial \pi_\alpha}\right) + \frac{\partial F}{\partial t}, \qquad (6.3.153)$$

which gives using Hamilton's equations

$$\dot{F} \equiv \frac{dF}{dt} = \sum_{i=1}^{N_b}\left(\frac{dq_i}{dt}\frac{\partial F}{\partial q_i} + \frac{dp_i}{dt}\frac{\partial F}{\partial p_i}\right) + \sum_{\alpha=1}^{N_f}\left(\frac{d\theta_\alpha}{dt}\frac{\partial F}{\partial \theta_\alpha} + \frac{d\pi_\alpha}{dt}\frac{\partial F}{\partial \pi_\alpha}\right) + \frac{\partial F}{\partial t}$$

$$= \sum_{i=1}^{N_b}\left(\frac{\partial H}{\partial p_i}\frac{\partial F}{\partial q_i} - \frac{\partial H}{\partial q_i}\frac{\partial F}{\partial p_i}\right) - \sum_{\alpha=1}^{N_f}\left(\frac{\partial H}{\partial \pi_\alpha}\frac{\partial F}{\partial \theta_\alpha} + \frac{\partial H}{\partial \theta_\alpha}\frac{\partial F}{\partial \pi_\alpha}\right) + \frac{\partial F}{\partial t}$$

$$\equiv \{F, H\}_{\text{PB}} + \frac{\partial F}{\partial t}, \qquad (6.3.154)$$

where the last line is the *definition* of the Poisson bracket of F with the Hamiltonian, H. Recall that H is even, so for any *even* dynamical variable E and any even or odd dynamical variable F the Poisson bracket $\{F, E\}_{\text{PB}}$ is also defined as above with H replaced by E. Let E and O denote even and odd dynamical variables, respectively. The Poisson brackets must have the same symmetry properties as the commutators/anticommmutators for the coordinates in Eq. (6.3.141). So we require

$$\{E_1, E_2\}_{\text{PB}} = -\{E_2, E_1\}_{\text{PB}}, \quad \{E, O\}_{\text{PB}} = -\{O, E\}_{\text{PB}}, \qquad (6.3.155)$$

$$\{O_1, O_2\}_{\text{PB}} = \{O_2, O_1\}_{\text{PB}} \text{ and } \{O_1 O_2, O_3\}_{\text{PB}} = O_1\{O_2, O_3\}_{\text{PB}} - \{O_1, O_3\}_{\text{PB}}O_2,$$

where the last is to match the result $[O_1 O_2, O_3] = O_1\{O_2, O_3\} - \{O_1, O_3\}O_2$ since $O_1 O_2$ is even. As shown in Henneaux and Teitelboim (1994) the generalization of the Poisson bracket that has all of these properties for any two (even or odd) dynamical variables $F(\vec{q}, \vec{p}, \vec{\theta}, \vec{\pi})$ and $G(\vec{q}, \vec{p}, \vec{\theta}, \vec{\pi})$ is

$$\{F, G\}_{\text{PB}} = \sum_{j=1}^{N_b}\left[\frac{\partial F}{\partial q_j}\frac{\partial G}{\partial p_j} - \frac{\partial F}{\partial p_j}\frac{\partial G}{\partial q_j}\right] + (-1)^{\epsilon_F}\sum_{\alpha=1}^{N_f}\left[\frac{\partial F}{\partial \theta_\alpha}\frac{\partial G}{\partial \pi_\alpha} + \frac{\partial F}{\partial \pi_\alpha}\frac{\partial G}{\partial \theta_\alpha}\right], \qquad (6.3.156)$$

where ϵ_F is the Grassmann parity of F, i.e., $\epsilon_F = 0, 1$ for F even/odd, respectively. If F is neither even nor odd, then we decompose $F = F_{\text{even}} + F_{\text{odd}}$ and use $\{F, G\}_{\text{PB}} = \{F_{\text{even}}, G\}_{\text{PB}} + \{F_{\text{odd}}, G\}_{\text{PB}}$. So we only need deal with dynamical variables that are either odd or even. The following results can be shown:

$$\{F, G\}_{\text{PB}} = -(-1)^{\epsilon_F \epsilon_G}\{G, F\}_{\text{PB}}, \quad \{F, G\}_{\text{PB}}^* = -\{\bar{G}^*, \bar{F}^*\}_{\text{PB}}, \quad \epsilon_{\{F, G\}_{\text{PB}}} = \epsilon_F + \epsilon_G,$$

$$\{F, G_1 G_2\}_{\text{PB}} = \{F, G_1\}_{\text{PB}}G_2 + (-1)^{\epsilon_F \epsilon_{G_1}}G_1\{F, G_2\}_{\text{PB}}. \qquad (6.3.157)$$

The fundamental Poisson brackets that generalize the standard fundamental Poisson brackets of Eq. (2.4.28) can then be evaluated from Eq. (6.3.156) and are

$$\{q_i, p_j\}_{\text{PB}} = -\{p_j, q_i\}_{\text{PB}} = \delta_{ij}, \quad \{\theta_\alpha, \pi_\beta\}_{\text{PB}} = \{\pi_\beta, \theta_\alpha\}_{\text{PB}} = -\delta_{\alpha\beta},$$
$$\{q_i, q_j\}_{\text{PB}} = \{p_i, p_j\}_{\text{PB}} = \{\theta_\alpha, \theta_\beta\}_{\text{PB}} = \{\pi_\alpha, \pi_\beta\}_{\text{PB}} = 0,$$
$$\{q_i, \theta_\alpha\}_{\text{PB}} = \{p_i, \theta_\alpha\}_{\text{PB}} = \{q_i, \pi_\alpha\}_{\text{PB}} = \{p_i, \pi_\alpha\}_{\text{PB}} = 0. \tag{6.3.158}$$

Having constructed the generalized Poisson bracket then canonical quantization follows from applying the correspondence principle of Eq. (2.5.27), but where now

$$\{F, G\}_{\text{PB}} \leftrightarrow \frac{1}{i} \begin{cases} [\hat{F}, \hat{G}] & \text{if } F \text{ and/or } G \text{ are even} \\ \{\hat{F}, \hat{G}\} & \text{if } F \text{ and } G \text{ are odd} \end{cases}. \tag{6.3.159}$$

Canonical quantization of Dirac fermions: We can now improve the previous simple canonical quantization argument by combining classical mechanics over a Grassmann algebra with the Dirac-Bergmann algorithm to manage the fact that the Dirac Lagrangian is singular. We now briefly discuss the derivation of the Dirac fermion quantum field theory canonical commutation relations and Hamiltonian.[5]

We start with the Dirac action written in term of Grassmann-valued fields, $\psi(x)$ and $\bar{\psi}(x)$, given in Eq. (6.3.103),

$$S[\bar{\psi}, \psi] = \int dt\, L = \int d^4x\, \mathcal{L} = \int d^4x\, \bar{\psi}(i\slashed{\partial} - m)\psi. \tag{6.3.160}$$

The Grassmann quantities ψ_α and $\psi_\alpha^\dagger = (\psi_\alpha)^*$ are related by Grassmann complex conjugation, but we can treat ψ and $\psi^\dagger$ as independent Grassmann variables in the Wirtinger sense as previously discussed. It will be convenient to deal with the Grassmann quantities ψ and $\bar{\psi} = \psi^\dagger \gamma^0$ in analogy with what we did earlier. The Euler-Lagrange equations result from the application of Hamilton's principle for this action as given in Eq. (6.3.124) but now with the left Grassmann derivatives,

$$0 = \delta S[\bar{\psi}, \psi]/\delta\psi(x) = \delta/\delta\psi(x)\int d^4x\, \bar{\psi}(x)(-i\overleftarrow{\slashed{\partial}} - m)\psi(x) = -\bar{\psi}(x)(-i\overleftarrow{\slashed{\partial}} - m),$$
$$0 = \delta S[\bar{\psi}, \psi]/\delta\bar{\psi}(x) = (i\slashed{\partial} - m)\psi(x), \tag{6.3.161}$$

where we have used integration by parts in the first equation. So we recover the Dirac equation in the Grassmann context,

$$(i\slashed{\partial} - m)\psi(x) = 0 \quad \text{and} \quad \bar{\psi}(x)(-i\overleftarrow{\slashed{\partial}} - m) = 0. \tag{6.3.162}$$

We define conjugate momenta in terms of the Dirac Lagrangian using the left Grassmann derivative to give

$$\pi_\alpha^\mu = \partial\mathcal{L}/\partial(\partial_\mu\psi_\alpha) = -i(\bar{\psi}\gamma^\mu)_\alpha \quad \text{and} \quad \bar{\pi}_\alpha^\mu = \partial\mathcal{L}/\partial(\partial_\mu\bar{\psi})_\alpha = 0,$$
$$\pi_\alpha = \partial\mathcal{L}/\partial\dot{\psi}_\alpha = -i(\bar{\psi}\gamma^0)_\alpha = -i\psi_\alpha^\dagger \quad \text{and} \quad \bar{\pi}_\alpha = \partial\mathcal{L}/\partial\dot{\bar{\psi}}_\alpha = 0, \tag{6.3.163}$$

which differ in sign from Eq. (6.3.127) since we have Grassmann quantities.

Note that the conjugate momenta π_α and $\bar{\pi}_\alpha$ for $\alpha = 1, \ldots, 4$ are not invertible functions of their respective velocities $\dot{\psi}_\alpha$ and $\dot{\bar{\psi}}_\alpha$ and so these eight equations must be imposed on the system as primary constraints,

$$\phi_1 \equiv \pi + i\bar{\psi}\gamma^0 \approx 0 \quad \text{and} \quad \phi_2 \equiv \bar{\pi} \approx 0. \tag{6.3.164}$$

[5] Steven G. Avery is gratefully acknowledged for sharing his summary of these arguments.

Recall from Sec. 2.9 that "$\approx$" is the weak equality symbol meaning that the equality is true on the constraint submanifold. See the summary of the Dirac-Bergmann algorithm in Sec. 2.9.2 and also its application to the electromagnetic field in Sec. 3.3.2. The standard Hamiltonian from Eq. (6.3.150) is then

$$
\begin{aligned}
H &= \int d^3x \sum_\alpha (\dot{\psi}_\alpha \pi_\alpha + \dot{\bar\psi}_\alpha \bar\pi_\alpha) - L \approx \int d^3x \sum_\alpha (\dot{\psi}_\alpha \pi_\alpha) - L \\
&\approx \int d^3x \sum_\alpha \dot{\psi}_\alpha (-i\bar\psi\gamma^0)_\alpha - L \approx \int d^3x \, (i\bar\psi\gamma^0\dot\psi - \mathcal{L}) \\
&\approx \int d^3x \, (i\bar\psi\gamma^0\dot\psi - i\bar\psi\gamma^0\dot\psi - i\bar\psi\boldsymbol{\gamma}\cdot\boldsymbol{\nabla}\psi + m\bar\psi\psi) \approx \int d^3x \, \bar\psi(-i\boldsymbol{\gamma}\cdot\boldsymbol{\nabla} + m)\psi,
\end{aligned}
\tag{6.3.165}
$$

which is the same as found earlier in Eq. (6.3.128). Using the Poisson bracket in Eq. (6.3.156) with $\theta_\alpha \to \big(\psi_\alpha(\mathbf{z}), \bar\psi_\alpha(\mathbf{z})\big)$ and $\pi_\alpha \to (\pi_\alpha(\mathbf{z}), \bar\pi_\alpha(\mathbf{z}))$, we arrive at

$$
\begin{aligned}
\{F, G\}_{\text{PB}} = (-1)^{\epsilon_F} \int d^3z \sum_{\gamma=1}^4 \Bigg[&\frac{\delta F}{\delta\psi_\gamma(\mathbf{z})} \frac{\partial G}{\partial\pi_\gamma(\mathbf{z})} + \frac{\delta F}{\delta\pi_\gamma(\mathbf{z})} \frac{\delta G}{\delta\psi_\gamma(\mathbf{z})} \\
&+ \frac{\delta F}{\delta\bar\psi_\gamma(\mathbf{z})} \frac{\partial G}{\partial\bar\pi_\gamma(\mathbf{z})} + \frac{\delta F}{\delta\bar\pi_\gamma(\mathbf{z})} \frac{\delta G}{\delta\bar\psi_\gamma(\mathbf{z})} \Bigg].
\end{aligned}
\tag{6.3.166}
$$

We immediately see that only nonvanishing fundamental Poisson brackets are

$$
\begin{aligned}
\{\psi_\alpha(\mathbf{x}), \pi_\beta(\mathbf{y})\}_{\text{PB}} &= \{\pi_\beta(\mathbf{y}), \psi_\alpha(\mathbf{x})\}_{\text{PB}} = -\delta_{\alpha\beta}\delta^3(\mathbf{x} - \mathbf{y}), \\
\{\bar\psi_\alpha(\mathbf{x}), \bar\pi_\beta(\mathbf{y})\}_{\text{PB}} &= \{\bar\pi_\beta(\mathbf{y}), \bar\psi_\alpha(\mathbf{x})\}_{\text{PB}} = -\delta_{\alpha\beta}\delta^3(\mathbf{x} - \mathbf{y}).
\end{aligned}
\tag{6.3.167}
$$

We wish to evaluate the Poisson brackets of the primary constraints and so find the elements of the infinite-dimensional "matrix"

$$
P_{i\alpha, j\beta}(\mathbf{x}, \mathbf{y}) \equiv \{\phi_{i\alpha}(\mathbf{x}), \phi_{j\beta}(\mathbf{y})\}_{\text{PB}}
\tag{6.3.168}
$$

for $i, j = 1, 2$. We find that

$$
\begin{aligned}
P_{1\alpha, 2\beta}(\mathbf{x}, \mathbf{y}) &= \{\phi_{i\alpha}(\mathbf{x}), \phi_\beta(\mathbf{y})\}_{\text{PB}} = \{\pi_\alpha(\mathbf{x}) + i(\bar\psi(\mathbf{x})\gamma^0)_\alpha, \bar\pi_\beta(\mathbf{y})\}_{\text{PB}} \\
&= i \sum_\delta \{\bar\psi(\mathbf{x})_\delta, \bar\pi_\beta(\mathbf{y})\}_{\text{PB}} \gamma^0_{\delta\alpha} = -i\gamma^0_{\beta\alpha}\delta^3(\mathbf{x} - \mathbf{y}),
\end{aligned}
\tag{6.3.169}
$$

and similarly $P_{2\alpha, 1\beta}(\mathbf{x}, \mathbf{y}) = -i\gamma^0_{\alpha\beta}\delta^3(\mathbf{x} - \mathbf{y})$ with all other matrix elements zero. Since the spatial dependence of P is the spatial identity matrix, $\delta^3(\mathbf{x} - \mathbf{y})$, let us factor this out and define

$$
P_{1\alpha, 2\beta}(\mathbf{x}, \mathbf{y}) \equiv M_{1\alpha, 2\beta}\delta^3(\mathbf{x}, \mathbf{y}),
\tag{6.3.170}
$$

where the 8×8 matrix M is given by

$$
M = -i \begin{bmatrix} 0 & \gamma^{0T} \\ \gamma^0 & 0 \end{bmatrix} \quad \text{and so} \quad M^{-1} = i \begin{bmatrix} 0 & \gamma^0 \\ \gamma^{0T} & 0 \end{bmatrix}.
\tag{6.3.171}
$$

A block matrix M with four equal size blocks A, B, C, D and $[C, D] = 0$ has $\det M = \det(AD - BC)$, which gives $\det M = -i\det(-\gamma^{0T}\gamma^0) = -i(-1)^4(\det\gamma^0)^2 = -i$. We have $P^{-1}(\mathbf{x}, \mathbf{y}) = M^{-1}\delta^3(\mathbf{x} - \mathbf{y})$, since $\int d^3z \, P^{-1}(\mathbf{x}, \mathbf{z}) P(\mathbf{z}, \mathbf{y}) = I_{8\times8}\delta^3(\mathbf{x} - \mathbf{y})$.

Since P is invertible the summary of the Dirac-Bergmann algorithm in Sec. 2.9.2 then tells us that there are only these two second-class primary constraints; i.e., there are no secondary constraints and there are no first-class constraints requiring gauge-fixing such as that used for electromagnetism in Sec. 3.3.2. Furthermore, we can construct the so-called total Hamiltonian of Eq. (2.9.31),

$$
H_T \equiv H + \sum_{i=1}^2 \sum_\alpha \int d^3x \, \phi_{i\alpha}(\mathbf{x})\lambda_{j\alpha}(\mathbf{x}),
\tag{6.3.172}
$$

where $\lambda_{1\alpha}(\mathbf{x})$ and $\lambda_{2\alpha}(\mathbf{x})$ are the Lagrange multipliers for the constraints $\phi_{1\alpha}(\mathbf{x})$ and $\phi_{2\alpha}(\mathbf{x})$ in Eq. (6.3.164). The Lagrange multipliers are given by Eq. (2.9.30), which here take the form

$$\lambda_{i\alpha}(\mathbf{x}) = -\sum_{k=1}^{2}\int d^3z\sum_{\gamma=1}^{4}P_{i\alpha,k\gamma}^{-1}(\mathbf{x},\mathbf{z})\{\phi_{k\gamma}(\mathbf{z}),H\}_{\mathrm{PB}}$$
$$= -\sum_k\sum_\gamma M_{i\alpha,k\gamma}^{-1}\{\phi_{k\gamma}(\mathbf{x}),H\}_{\mathrm{PB}}. \tag{6.3.173}$$

Remembering that we are dealing with Grassmann-valued fields we next find that

$$\{\phi_1(\mathbf{x}),H\}_{\mathrm{PB}} = \{\pi(\mathbf{x}) + i\bar{\psi}(\mathbf{x})\gamma^0, \int d^3z\,\bar{\psi}(\mathbf{z})(-i\boldsymbol{\gamma}\cdot\boldsymbol{\nabla}+m)\psi(\mathbf{z})\}_{\mathrm{PB}} \tag{6.3.174}$$
$$= \{\pi(\mathbf{x}), \int d^3z\,\bar{\psi}(\mathbf{z})(i\boldsymbol{\gamma}\cdot\overleftarrow{\boldsymbol{\nabla}}+m)\psi(\mathbf{z})\}_{\mathrm{PB}} = \bar{\psi}(\mathbf{x})(i\boldsymbol{\gamma}\cdot\overleftarrow{\boldsymbol{\nabla}}+m),$$
$$\{\phi_2(\mathbf{x}),H\}_{\mathrm{PB}} = \{\bar{\pi}(\mathbf{x}), \int d^3z\,\bar{\psi}(\mathbf{z})(-i\boldsymbol{\gamma}\cdot\boldsymbol{\nabla}+m)\psi(\mathbf{z})\}_{\mathrm{PB}} = (i\boldsymbol{\gamma}\cdot\boldsymbol{\nabla}-m)\psi(\mathbf{x}),$$

which then gives the results that

$$\lambda_1(\mathbf{x}) = -(-i)\gamma^0\{\phi_2(\mathbf{x}),H\}_{\mathrm{PB}} = i\gamma^0(i\boldsymbol{\gamma}\cdot\boldsymbol{\nabla}-m)\psi(\mathbf{x}),$$
$$\lambda_2(\mathbf{x}) = -(-i)\{\phi_1(\mathbf{x}),H\}_{\mathrm{PB}}\gamma^0 = i\bar{\psi}(\mathbf{x})(i\boldsymbol{\gamma}\cdot\overleftarrow{\boldsymbol{\nabla}}+m)\gamma^0. \tag{6.3.175}$$

This gives an explicit form for the total Hamiltonian H_T. However, we are primarily interested in constructing the Dirac bracket, $\{F,G\}_{\mathrm{DB}}$, of Eq. (2.9.126). We only need the standard Hamiltonian H to specify the dynamics of the constrained system,

$$\dot{F} = \partial F/\partial t + \{F,H\}_{\mathrm{DB}}. \tag{6.3.176}$$

The Dirac bracket replaces the usual Poisson bracket in the correspondence principle of canonical quantization for a singular system; i.e., the Grassmann form of $\{F,G\}_{\mathrm{DB}}$ replaces $\{F,G\}_{\mathrm{PB}}$ in Eq. (6.3.159) for a singular classical system over a Grassmann algebra. Since we have only primary second-class constraints, then the matrix of all second-class constraints G just becomes the matrix P, so that the Dirac bracket is given by

$$\{A,B\}_{\mathrm{DB}} = \{A,B\}_{\mathrm{PB}} - \sum_{i,j,\gamma\delta}\int d^3u\,d^3v\,\{A,\phi_{i\gamma}(\mathbf{u})\}_{\mathrm{PB}}P_{i\gamma,j\delta}^{-1}(\mathbf{u},\mathbf{v})\{\phi_{j\delta}(\mathbf{v}),B\}_{\mathrm{PB}}$$
$$= \{A,B\}_{\mathrm{PB}} - \sum_{i,j,\gamma\delta}\int d^3u\,\{A,\phi_{i\gamma}(\mathbf{u})\}_{\mathrm{PB}}M_{i\gamma,j\delta}^{-1}\{\phi_{j\delta}(\mathbf{u}),B\}_{\mathrm{PB}}$$
$$= \{A,B\}_{\mathrm{PB}} - i\sum_{\gamma\delta}\int d^3u\,\Big[\{A,\phi_{1\gamma}(\mathbf{u})\}_{\mathrm{PB}}\gamma_{\gamma\delta}^0\{\phi_{2\delta}(\mathbf{u}),B\}_{\mathrm{PB}}$$
$$+ \{A,\phi_{2\gamma}(\mathbf{u})\}_{\mathrm{PB}}\gamma_{\gamma\delta}^{0T}\{\phi_{1\delta}(\mathbf{u}),B\}_{\mathrm{PB}}\Big]. \tag{6.3.177}$$

Using the fundamental Poisson brackets of Eq. (6.3.167) we can evaluate the fundamental Dirac brackets. Since $\phi_1 \equiv \pi + i\bar{\psi}\gamma^0$ and $\phi_2 \equiv \bar{\pi}$, then the fundamental Dirac brackets will only differ from the fundamental Poisson brackets for the pair $\bar{\psi}$ and $\bar{\pi}$ and the pair $\bar{\psi}$ and ψ. We find that

$$\{\bar{\pi}_\alpha(\mathbf{x}),\bar{\psi}_\beta(\mathbf{y})\}_{\mathrm{DB}} = \{\bar{\pi}_\alpha(\mathbf{x}),\bar{\psi}_\beta(\mathbf{y})\}_{\mathrm{PB}}$$
$$- i\sum_{\gamma,\delta}\int d^3u\,\{\bar{\pi}_\alpha(\mathbf{x}),(i\bar{\psi}(\mathbf{u})\gamma^0)_\gamma\}_{\mathrm{PB}}\gamma_{\gamma\delta}^0\{\bar{\pi}_\delta(\mathbf{u}),\bar{\psi}_\beta(\mathbf{y})\}_{\mathrm{PB}}$$
$$= \{\bar{\pi}_\alpha(\mathbf{x}),\bar{\psi}_\beta(\mathbf{y})\}_{\mathrm{PB}} + \delta_{\alpha\beta}\delta^3(\mathbf{x}-\mathbf{y}) = 0, \tag{6.3.178}$$
$$\{\psi_\alpha(\mathbf{x}),\bar{\psi}_\beta(\mathbf{y})\}_{\mathrm{DB}} = -i\sum_{\gamma,\delta}\int d^3u\,\{\psi_\alpha(\mathbf{x}),\pi_\gamma(\mathbf{u})\}_{\mathrm{PB}}\gamma_{\gamma\delta}^0\{\bar{\pi}_\delta(\mathbf{u}),\bar{\psi}_\beta(\mathbf{y})\}_{\mathrm{PB}}$$
$$= -i\gamma_{\alpha\beta}^0\delta^3(\mathbf{x}-\mathbf{y}). \tag{6.3.179}$$

Note that Eq. (6.3.178) is what we must have in a consistent Hamiltonian dynamics where $\bar{\pi}$ vanishes on the constraint submanifold, i.e., since $\bar{\pi} \approx 0$. For two odd variables $\{O_1,O_2\}_{\mathrm{PB}} = \{O_2,O_1\}_{\mathrm{PB}}$ and so the matrix P in Eq. (6.3.168) will be symmetric. Hence the fundamental Dirac brackets will also be symmetric.

The only nonvanishing fundamental Dirac brackets are then

$$\{\psi_\alpha(\mathbf{x}), \pi_\beta(\mathbf{y})\}_{\mathrm{DB}} = \{\pi_\beta(\mathbf{y}), \psi_\alpha(\mathbf{x})\}_{\mathrm{DB}} = -\delta_{\alpha\beta}\delta^3(\mathbf{x}-\mathbf{y}),$$
$$\{\psi_\alpha(\mathbf{x}), \bar{\psi}_\beta(\mathbf{y})\}_{\mathrm{DB}} = \{\bar{\psi}_\beta(\mathbf{y}), \psi_\alpha(\mathbf{x})\}_{\mathrm{DB}} = -i\gamma^0_{\alpha\beta}\delta^3(\mathbf{x}-\mathbf{y}), \tag{6.3.180}$$

which are equivalent since we have the constraint $\pi \approx -i\bar{\psi}\gamma^0$. We started with the four independent Grassmann-valued spinor fields ψ, $\bar{\psi}$, π and $\bar{\pi}$ describing our phase space. These reduce to two after incorporating the constraints $\bar{\pi} \approx 0$ and $\pi \approx -i\bar{\psi}\gamma^0 = -i\psi^\dagger$ that define the constraint submanifold. The Dirac brackets define the dynamics of the system in such a way that time evolution of the system keeps it on the constraint submanifold at all times. We readily see from the above that $\dot{\phi}_1 = \dot{\phi}_2 = 0$. We will use ψ and $\psi^\dagger$ as our independent Grassmann-valued spinor fields. As we have shown, these have the fundamental Dirac brackets,

$$\{\psi_\alpha(\mathbf{x}), \psi^\dagger_\beta(\mathbf{y})\}_{\mathrm{DB}} = \{\psi^\dagger_\beta(\mathbf{x}), \psi_\alpha(\mathbf{y})\}_{\mathrm{DB}} = -i\delta_{\alpha\beta}\delta^3(\mathbf{x}-\mathbf{y}),$$
$$\{\psi_\alpha(\mathbf{x}), \psi_\beta(\mathbf{y})\}_{\mathrm{DB}} = \{\psi^\dagger_\beta(\mathbf{x}), \psi^\dagger_\alpha(\mathbf{y})\}_{\mathrm{DB}} = 0. \tag{6.3.181}$$

Using the correspondence principle of Eq. (6.3.159) we finally arrive at the canonical quantization procedure for Dirac fermions. So we have again found the canonical anticommutation relations of Eq. (6.3.94) noting that $\{\cdots, \cdots\}$ means the anticommutator here,

$$\{\hat{\psi}_\alpha(\mathbf{x}), \hat{\psi}^\dagger_\beta(\mathbf{y})\} = \delta_{\alpha\beta}\delta^3(\mathbf{x}-\mathbf{y}),$$
$$\{\hat{\psi}_\alpha(\mathbf{x}), \hat{\psi}_\beta(\mathbf{y})\} = \{\hat{\psi}^\dagger_\beta(\mathbf{x}), \hat{\psi}^\dagger_\alpha(\mathbf{y})\} = 0. \tag{6.3.182}$$

The Hamiltonian operator (before normal-ordering) is given by

$$\hat{H} = \int d^3x\, \hat{\bar{\psi}}(-i\boldsymbol{\gamma}\cdot\boldsymbol{\nabla}+m)\hat{\psi} \tag{6.3.183}$$

and the time evolution of any operator $\hat{F}$ is given by

$$d\hat{F}/dt = \partial\hat{F}/\partial t - i[\hat{F}, \hat{H}]. \tag{6.3.184}$$

We can repeat the derivation of Noether's theorem for classical field theory over a Grassmann algebra by taking some care with ordering of Grassmann-odd quantities. This includes having Grassmann-odd differentials on the left since we are using the left Grassmann derivative. This means that for Grassmann-odd fields we must keep velocities and space-time derivatives of fields to the left of momenta as we did in the Legendre transform defining the Hamiltonian above.

Following through the derivation of Noether's theorem in Sec. 3.2.1 it is not difficult to see that in the Grassmann algebra case we arrive at the same Noether current given in Eq. (3.2.18) with the appropriate ordering,

$$j^\mu(x) \equiv \left[\vec{\Phi}(x) - [\partial_\nu\vec{\phi}(x)]X^\nu(x)\right]\cdot\vec{\pi}^\mu(x) + \mathcal{L}(\vec{\phi}, \partial_\mu\vec{\phi}, x)X^\mu(x), \tag{6.3.185}$$

where $\partial_\mu j^\mu(x) = 0$ on-shell and where $X^\mu(x)$ and $\Phi(x)$ are defined in Eqs. (3.2.7) and (3.2.9), respectively. The conservation of the ordered stress-energy tensor follows from spacetime translational invariance and can be arrived at as before,

$$T^\mu{}_\nu(x) = \partial_\nu\vec{\phi}\cdot\vec{\pi}^\mu - \delta^\mu{}_\nu\mathcal{L} \quad \text{with} \quad \partial_\mu T^{\mu\nu}(x) = 0 \tag{6.3.186}$$

with the components of the total four-momentum, $P^\mu = \int d^3x\, T^{0\mu}$, as the conserved charges. Similarly, following through the derivation of the angular momentum tensor in the presence of intrinsic spin in Sec. 3.2.4 we again arrive at Eqs. (3.2.72)–(3.2.74) for the conserved Noether current, $\mathcal{T}^{\mu}_{\ \rho\sigma}$, and conserved charges, $M^{\mu\nu}$.

Recalling that spinor summations are to be understood and being careful with ordering, we arrive at the following on-shell results for Dirac fermions,

$$
\begin{aligned}
T^{\mu\nu} &= \partial^{\nu}\psi\,\pi^{\mu} + \partial^{\nu}\bar{\psi}\,\bar{\pi}^{\mu} - g^{\mu\nu}\mathcal{L} = \partial^{\nu}\psi(-i\bar{\psi}\gamma^{\mu}) = i\bar{\psi}\gamma^{\mu}\partial^{\nu}\psi, \\
P^{\mu} &= \int d^3x\, T^{0\mu} = \int d^3x\,[\partial^{\mu}\psi\pi + \partial^{\mu}\bar{\psi}\bar{\pi} - g^{0\mu}\mathcal{L}] = \int d^3x\, \partial^{\mu}\psi\pi = \int d^3x\, i\psi^{\dagger}\partial^{\mu}\psi, \\
\mathcal{J}^{\mu\rho\sigma} &= x^{\rho}T^{\mu\sigma} - x^{\sigma}T^{\mu\rho} + (\Sigma^{\rho\sigma}\psi)\pi^{\mu} + (\Sigma^{\rho\sigma}\bar{\psi})\bar{\pi}^{\mu} \\
&= i\bar{\psi}\gamma^{\mu}(x^{\rho}\partial^{\sigma} - x^{\sigma}\partial^{\rho} + \Sigma^{\rho\sigma})\psi = i\bar{\psi}\gamma^{\mu}(x^{\rho}\partial^{\sigma} - x^{\sigma}\partial^{\rho} + \tfrac{1}{4}[\gamma^{\rho},\gamma^{\sigma}])\psi, \\
M^{\rho\sigma} &= \int d^3x\, \mathcal{J}^{0\rho\sigma} = \int d^3x\,[x^{\rho}T^{0\sigma} - x^{\sigma}T^{0\rho} + (\Sigma^{\rho\sigma}\psi)\pi + (\Sigma^{\rho\sigma}\bar{\psi})\bar{\pi}] \\
&= \int d^3x\, i\psi^{\dagger}\,(x^{\rho}\partial^{\sigma} - x^{\sigma}\partial^{\rho} + \tfrac{1}{4}[\gamma^{\rho},\gamma^{\sigma}])\,\psi = \int d^3x\, \psi^{\dagger}\,(x^{\rho}i\partial^{\sigma} - x^{\sigma}i\partial^{\rho} + \Sigma^{\mu\nu}_{\text{Dirac}})\,\psi.
\end{aligned}
$$

$$(6.3.187)$$

We have imposed the constraints to arrive at the right-hand sides of these equation. These are the same as the results obtained in the previous simple canonical treatment but now ψ and $\psi^{\dagger}$ are Grassmann-valued fields. We note that the above quantities are all Grassmann-even and so commute with all Grassman numbers. Using Eq. (6.3.187) and the fundamental Dirac brackets of Eq. (6.3.180) we find

$$
\begin{aligned}
\{\psi(x), P^{\mu}\}_{\text{DB}} &= \partial^{\mu}\psi(x)\,, \\
\{\psi(x), M^{\mu\nu}\}_{\text{DB}} &= x^{\mu}\partial^{\nu}\psi(x) - x^{\nu}\partial^{\mu}\psi(x) + \Sigma^{\mu\nu}\psi(x)\,.
\end{aligned}
$$

$$(6.3.188)$$

where we recall that for Dirac fermions $\Sigma^{\mu\nu} = \tfrac{1}{4}[\gamma^{\mu},\gamma^{\nu}] = \tfrac{1}{i}\Sigma^{\mu\nu}_{\text{Dirac}}$ and where this result corresponds to Eq. (3.2.124). Similarly, we can show that

$$
\begin{aligned}
i\{P^{\mu}, P^{\nu}\}_{\text{DB}} &= 0, \qquad i\{P^{\mu}, M^{\rho\sigma}\}_{\text{DB}} = i(g^{\mu\rho}P^{\sigma} - g^{\mu\sigma}P^{\rho}), \\
i\{M^{\mu\nu}, M^{\rho\sigma}\}_{\text{DB}} &= i\,(g^{\nu\rho}M^{\mu\sigma} - g^{\mu\rho}M^{\nu\sigma} - g^{\nu\sigma}M^{\mu\rho} + g^{\mu\sigma}M^{\nu\rho}),
\end{aligned}
$$

$$(6.3.189)$$

which is Eq. (3.2.125) emerging in the Dirac fermion context. From the correspondence principle of Eq. (6.3.159) we then understand that the Poincaré Lie algebra will again emerge in the quantum field theory for Dirac fermions.

In summary, we have extended classical mechanics to apply over a Grassmann algebra and we have shown that a consistent application of the canonical quantization procedure leads to the same quantum field theory for Dirac fermions that we arrived at earlier using Fock space arguments in Sec. 6.3.4. The point of this discussion has been to explain why we expect the operators of the Dirac quantum field theory to obey the same equations of motion and conservation laws as they do classically (classical in the Grassmann sense) and to satisfy the Poincaré Lie algebra, as was the case for scalar field theory in Sec. 6.2.3.

6.3.7 Quantum Field Theory for Dirac Fermions

We now have everything that we need to describe the quantum field theory for free Dirac fermions. In the massless limit Dirac fermions can be decomposed into independent left-handed and right-handed chiral fermions as discussed in Sec. 4.6.2. By restricting the space of solutions of the Dirac equation to self-conjugate solutions we arrive at a different form of massive fermion, a Majorana fermion, as discussed in Sec. 4.6.5. A Dirac fermion necessarily has both a particle and antiparticle, whereas a Majorana fermion is its own antiparticle. From this point on, when we discuss fermions we will mean Dirac fermions unless specified otherwise.

The canonical quantization procedure of Sec. 6.3.6 has led us to the Heisenberg picture fermion equal time anticommutation relations (ETACR),

$$\{\hat{\psi}_\alpha(t,\vec{x}), \hat{\psi}_\beta^\dagger(t,\vec{y})\} = \delta_{\alpha\beta}\delta^3(\vec{x} - \vec{y}) \tag{6.3.190}$$

$$\{\hat{\psi}_\alpha(t,\vec{x}), \hat{\psi}_\beta(t,\vec{y})\} = \{\hat{\psi}_\alpha^\dagger(t,\vec{x}), \hat{\psi}_\beta^\dagger(t,\vec{y})\} = 0,$$

which follow from Eqs. (6.3.182) and (6.3.100). After normal-ordering the Hamiltonian operator of Eq. (6.3.183) we are led directly to Eq. (6.3.92), which we repeat here for convenience,

$$\hat{H} = \int d^3x \, :i\hat{\bar{\psi}}\gamma^0\partial_0\hat{\psi} - \hat{\mathcal{L}}: = \int d^3x \, :i\hat{\bar{\psi}}\gamma^0\partial_0\hat{\psi} - \hat{\bar{\psi}}(i\not{\partial} - m)\hat{\psi}:$$

$$= \int d^3x \, :\hat{\bar{\psi}}(-i\boldsymbol{\gamma}\cdot\boldsymbol{\nabla} + m)\hat{\psi}: = \int (d^3p/(2\pi)^3)\sum_s E_{\mathbf{p}}(\hat{b}_{\mathbf{p}}^{s\dagger}\hat{b}_{\mathbf{p}}^s + \hat{d}_{\mathbf{p}}^{s\dagger}\hat{d}_{\mathbf{p}}^s), \tag{6.3.191}$$

where it is to be understood that

$$i\hat{\bar{\psi}}\gamma^\mu\partial_\mu\hat{\psi} \equiv i\hat{\bar{\psi}}\gamma^\mu\tfrac{1}{2}[\vec{\partial}_\mu - \overleftarrow{\partial}_\mu]\hat{\psi} \equiv i\hat{\bar{\psi}}\gamma^\mu\overleftrightarrow{\partial}_\mu\hat{\psi} \quad \text{so that}$$

$$(i\hat{\bar{\psi}}\gamma^\mu\overleftrightarrow{\partial}_\mu\hat{\psi})^\dagger = i\hat{\bar{\psi}}\gamma^\mu\overleftrightarrow{\partial}_\mu\hat{\psi} \tag{6.3.192}$$

and $\hat{H}$ and $\hat{L}$ are Hermitian. Also recall from Eq. (6.3.184) that

$$\hat{F}/dt = \partial\hat{F}/\partial t - i[\hat{F}, \hat{H}]. \tag{6.3.193}$$

From the canonical quantization procedure we know that the field operators $\hat{\psi}(x)$ must satisfy the classical equations of motion. Then we understand why it was appropriate to write $\hat{\psi}(x)$ in terms of plane wave solutions of the Dirac equation in Eq. (6.3.100),

$$\hat{\psi}(x) = \int \frac{d^3p}{(2\pi)^3}\frac{1}{\sqrt{2E_{\mathbf{p}}}}\sum_s \left[\hat{b}_{\mathbf{p}}^s u^s(p)e^{-ip\cdot x} + \hat{d}_{\mathbf{p}}^{s\dagger}v^s(p)e^{ip\cdot x}\right], \tag{6.3.194}$$

since this automatically satisfies the Dirac equation,

$$(i\not{\partial} - m)\hat{\psi}(x) = 0 \quad \text{and equivalently} \quad \hat{\bar{\psi}}(x)(-i\overleftarrow{\not{\partial}} - m) = 0. \tag{6.3.195}$$

These are the same as the Dirac equations satisfied by the Grassmann-valued fields in Eq. (6.3.162) and so, as for the scalar field case, we see that the quantum field theory operators obey the same equations of motion as their classical counterparts. As we went to some trouble to show, this is because the Grassmann extension of the usual canonical quantization procedure could be successfully applied.

To express the annihilation and creation operators in terms of the fermion field operators it is convenient to define the shorthand notation

$$u_{ps}(x) \equiv u^s(p)e^{-ip\cdot x} \quad \text{and} \quad v_{ps}(x) \equiv v^s(p)e^{+ip\cdot x}. \tag{6.3.196}$$

It is readily shown using Eq. (4.4.125) that

$$\sqrt{2E_{\mathbf{p}}}\hat{b}_{\mathbf{p}}^{s\dagger} = \int d^3x\,\hat{\psi}^\dagger(x)u_{ps}(x), \quad \sqrt{2E_{\mathbf{p}}}\hat{d}_{\mathbf{p}}^{s\dagger} = \int d^3x\,v_{ps}^\dagger(x)\hat{\psi}(x), \tag{6.3.197}$$

$$\sqrt{2E_{\mathbf{p}}}\hat{b}_{\mathbf{p}}^s = \int d^3x\,u_{ps}^\dagger(x)\hat{\psi}(x), \quad \sqrt{2E_{\mathbf{p}}}\hat{d}_{\mathbf{p}}^s = \int d^3x\,\hat{\psi}^\dagger(x)v_{ps}(x).$$

This is analogous to Eq. (6.2.100) for scalar fields.

Comparing with other texts: Many modern texts normalize their field operators to give the ETACR in Eq. (6.3.190). Using BD to denote the conventions of Bjorken and Drell (1964) and Greiner and Reinhardt (1996) we write Eq. (6.3.194) as

$$u^s(p) = \sqrt{2m}\, u(p,s)^{\mathrm{BD}}, \qquad\qquad v^s(p) = \sqrt{2m}\, v(p,s)^{\mathrm{BD}},$$

(6.3.198)

$$\hat{b}_{\mathbf{p}}^s = (2\pi)^{3/2}\hat{b}(p,s)^{\mathrm{BD}}, \qquad\qquad \hat{d}_{\mathbf{p}}^s = (2\pi)^{3/2}\hat{d}(p,s)^{\mathrm{BD}},$$

$$\hat{\psi}(x) = \int \frac{d^3p}{(2\pi)^{3/2}}\sqrt{\frac{m}{E_{\mathbf{p}}}}\sum_s \left[\hat{b}(p,s)^{\mathrm{BD}} u(p,s)^{\mathrm{BD}} e^{-ip\cdot x} + \hat{d}^\dagger(p,s)^{\mathrm{BD}} v(p,s)^{\mathrm{BD}} e^{ip\cdot x}\right].$$

From Eq. (6.3.187) we can write down the normal ordered forms of the generators of the Poincaré group,

$$\hat{P}^\mu = \int d^3x\, :i\hat{\psi}^\dagger(x)\partial^\mu\hat{\psi}(x):,$$

$$\hat{M}^{\mu\nu} = \int d^3x\, :i\hat{\psi}^\dagger\left(x^\mu\partial^\nu - x^\nu\partial^\mu + \tfrac{1}{4}[\gamma^\mu,\gamma^\nu]\right)\hat{\psi}:,$$

(6.3.199)

which satisfy the Poincaré Lie algebra of Eq. (1.2.167). We expect this to be the case for the operators before normal-ordering because of the canonical quantization procedure and the Dirac bracket results of Eq. (6.3.189). Since normal-ordering only changes $\hat{P}^\mu$ and $\hat{M}^{\mu\nu}$ by at most an overall constant, the Poincaré Lie algebra will be true for these operators with or without normal-ordering. We can also directly prove the Poincaré Lie algebra for $\hat{P}^\mu$ and $\hat{M}^{\mu\nu}$ using the ETACR above.

Recall that the ETACR along with Eq. (6.3.194) are equivalent to the anticommutation relations of Eq. (6.3.80),

$$\{\hat{b}_{\mathbf{p}}^s, \hat{b}_{\mathbf{p}'}^{s'\dagger}\} = \{\hat{d}_{\mathbf{p}}^s, \hat{d}_{\mathbf{p}'}^{s'\dagger}\} = \delta^{ss'}(2\pi)^3\delta^3(\mathbf{p}-\mathbf{p}')$$

(6.3.200)

with all other anticommutators of $\hat{b}_{\mathbf{p}}^s$, $\hat{b}_{\mathbf{p}}^{s\dagger}$, $\hat{d}_{\mathbf{p}}^s$ and $\hat{d}_{\mathbf{p}}^{s\dagger}$ vanishing. Also recall that for the vacuum state $|0\rangle$ we have have

$$\hat{b}_{\mathbf{p}}^s|0\rangle = \hat{d}_{\mathbf{p}}^s|0\rangle = 0.$$

(6.3.201)

Since the vacuum state should be Poincaré invariant we want $\hat{P}^\mu|0\rangle = \hat{M}^{\mu\nu}|0\rangle = 0$, which follows if we normal-order these Poincaré generators. Since we do not want normal-ordering to change the Poincaré Lie algebra, then it should change these generators by at most an infinite constant. The generators involve sums of products of pairs of annihilation and creation operators such as in the Hamiltonian above. We note for example that

$$\hat{b}_{\mathbf{p}}^s\hat{b}_{\mathbf{p}'}^{s'\dagger} = -\hat{b}_{\mathbf{p}'}^{s'\dagger}\hat{b}_{\mathbf{p}}^s + \{\hat{b}_{\mathbf{p}}^s, \hat{b}_{\mathbf{p}'}^{s'\dagger}\},$$

(6.3.202)

where the second term on the right-hand side is either zero or a real infinite constant. So we understand why normal-ordering for fermions has an *additional negative sign* in its definition for fermion-odd quantities,

$$:\hat{b}_{\mathbf{p}}^s\hat{b}_{\mathbf{p}'}^{s'\dagger}: = -\hat{b}_{\mathbf{p}'}^{s'\dagger}\hat{b}_{\mathbf{p}}^s, \qquad\qquad :\hat{b}_{\mathbf{p}}^s\hat{d}_{\mathbf{p}'}^{s'\dagger}: = -\hat{d}_{\mathbf{p}'}^{s'\dagger}\hat{b}_{\mathbf{p}}^s,$$

$$:\hat{d}_{\mathbf{p}}^s\hat{d}_{\mathbf{p}'}^{s'\dagger}: = -\hat{d}_{\mathbf{p}'}^{s'\dagger}\hat{d}_{\mathbf{p}}^s, \qquad\qquad :\hat{d}_{\mathbf{p}}^s\hat{b}_{\mathbf{p}'}^{s'\dagger}: = -\hat{b}_{\mathbf{p}'}^{s'\dagger}\hat{d}_{\mathbf{p}}^s.$$

(6.3.203)

Using similar steps to those in Eq. (6.3.91) we can show that before normal-ordering we have $\hat{P} = \int d^3x\, \hat{\psi}^\dagger(x)(-i\boldsymbol{\nabla})\hat{\psi}(x) = \int (d^3p/(2\pi)^3)\Sigma_s\mathbf{p}(\hat{b}_{\mathbf{p}}^{s\dagger}\hat{b}_{\mathbf{p}}^s - \hat{d}_{\mathbf{p}}^s\hat{d}_{\mathbf{p}}^{s\dagger})$, which after normal-ordering becomes

$$\hat{\mathbf{P}} = \int d^3x\, :\hat{\psi}^\dagger(x)(-i\boldsymbol{\nabla})\hat{\psi}(x): = \int \frac{d^3p}{(2\pi)^3}\sum_s \mathbf{p}(\hat{b}_{\mathbf{p}}^{s\dagger}\hat{b}_{\mathbf{p}}^s + \hat{d}_{\mathbf{p}}^{s\dagger}\hat{d}_{\mathbf{p}}^s) \quad\text{and so}$$

$$\hat{P}^\mu = (\hat{H}, \hat{\mathbf{P}}) = \int \frac{d^3p}{(2\pi)^3}\sum_s p^\mu(\hat{b}_{\mathbf{p}}^{s\dagger}\hat{b}_{\mathbf{p}}^s + \hat{d}_{\mathbf{p}}^{s\dagger}\hat{d}_{\mathbf{p}}^s),$$

(6.3.204)

where we recall that $p^0 \equiv E_{\mathbf{p}} \equiv \sqrt{\mathbf{p}^2 + m^2}$.

The Dirac action in terms of the Grassmann-valued fields given in Eq. (6.3.103) is invariant under a global phase transformation like the complex scalar field in Eq. (3.2.48). If $\psi(x) \to e^{i\alpha}\psi(x)$ then $\bar{\psi}(x) \to e^{-i\alpha}\bar{\psi}(x)$. Then under an infinitesimal change we have $\psi_\beta(x) \to \psi_\beta(x) + \alpha(i\psi(x))$ and $\bar{\psi}_\beta(x) \to \psi_\beta(x) - \alpha(i\bar{\psi}(x))$ and so in terms of Eqs. (3.2.7) and (3.2.8) we have $\Phi_\beta(x) = i\psi_\beta(x)$, $\bar{\Phi}_\beta(x) = -i\psi_\beta(x)$ and $X^\mu(x) = 0$. The corresponding Grassmann Noether current, Eq. (6.3.185), is then

$$j^\mu = a\sum_\beta [\Phi_\beta(x)\pi_\beta^\mu(x) + \bar{\Phi}_\beta(x)\bar{\pi}_\beta^\mu(x)] = a\sum_\beta (i\psi_\beta)(-i[\bar{\psi}\gamma^\mu]_\beta) = -a\bar{\psi}\gamma^\mu\psi,$$

where a is an arbitrary constant to be chosen and where from Eq. (6.3.163) we have $\pi_\beta^\mu = -i(\bar{\psi}\gamma^\mu)_\beta$ and $\bar{\pi}_\beta^\mu = 0$. The conventional choice of constant is $a = -1$ so that we arrive at the conserved Noether current density,

$$j^\mu = \bar{\psi}\gamma^\mu\psi. \tag{6.3.205}$$

So we expect the operator $\hat{j}^\mu = \hat{\bar{\psi}}(x)\gamma^\mu\hat{\psi}(x)$ to be a conserved operator current density, which can be readily verified using the Dirac equation in Eq. (6.3.195),

$$\partial_\mu \hat{j}^\mu = (\partial\hat{\bar{\psi}})\gamma^\mu\hat{\psi} + \hat{\bar{\psi}}\slashed{\partial}\hat{\psi} = im\hat{\bar{\psi}}\hat{\psi} - im\hat{\bar{\psi}}\hat{\psi} = 0. \tag{6.3.206}$$

Since normal-ordering will change this current by at most an infinite constant, then the normal-ordered current density operator will also be conserved and so we define

$$\hat{j}^\mu(x) \equiv :\hat{\bar{\psi}}(x)\gamma^\mu\hat{\psi}(x): \qquad \text{with} \qquad \partial_\mu\hat{j}^\mu = 0. \tag{6.3.207}$$

The corresponding conserved charge operator is

$$\hat{Q} = \int d^3x\, \hat{j}^0(x) = \int d^3x\, :\hat{\psi}^\dagger(x)\hat{\psi}(x):$$
$$= \int \frac{d^3p}{(2\pi)^3} \sum_s (\hat{b}_{\mathbf{p}}^{s\dagger}\hat{b}_{\mathbf{p}}^s - \hat{d}_{\mathbf{p}}^{s\dagger}\hat{d}_{\mathbf{p}}^s) \equiv N^+ - N^- = \hat{N}, \tag{6.3.208}$$

where the fermion and antifermion number operators are $\hat{N}^+$ and $\hat{N}^-$, respectively, and where their definitions are obvious from the above. The total fermion number operator $\hat{N}$ measures the number of fermions minus the number of antifermions. We arrive at this result using similar steps to those in Eq. (6.3.91) to first show that $\int d^3x\, \hat{\psi}^\dagger(x)\hat{\psi}(x) = \int [d^3p/(2\pi)^3] \sum_s (\hat{b}_{\mathbf{p}}^{s\dagger}\hat{b}_{\mathbf{p}}^s + \hat{d}_{\mathbf{p}}^s\hat{d}_{\mathbf{p}}^{s\dagger})$, which after fermion normal-ordering gives the above result.

It is useful to observe that the normal-ordered fermion current density can be written as a commutator (Bjorken and Drell, 1965; Greiner and Reinhardt, 1996)

$$\hat{j}^\mu \equiv :\hat{\bar{\psi}}\gamma^\mu\hat{\psi}: = \tfrac{1}{2}[\hat{\bar{\psi}}_\alpha, (\gamma^\mu\hat{\psi})_\alpha] = \tfrac{1}{2}[\hat{\bar{\psi}}, \gamma^\mu\hat{\psi}], \tag{6.3.209}$$

where the commutator form is referred to as the *symmetrized current operator*.

Proof: Using the plane wave expansion for the fermion field operator in Eq. (6.3.194) and writing for brevity $u_{\mathbf{p}}^s \equiv u^s(p)$ and so on we have

$$\tfrac{1}{2}[\hat{\bar{\psi}}_\alpha, (\gamma^\mu\hat{\psi})_\alpha] = \gamma_{\alpha\beta}^\mu \tfrac{1}{2}(\hat{\bar{\psi}}_\alpha\hat{\psi}_\beta - \hat{\psi}_\beta\hat{\bar{\psi}}_\alpha) = \tfrac{1}{2}\int [d^3p'\, d^3p/(2\pi)^6 2\sqrt{E_{\mathbf{p}'}E_{\mathbf{p}}'}] \sum_{s',s}$$
$$\times \Big\{ (\hat{b}_{\mathbf{p}'}^{s'\dagger}\hat{b}_{\mathbf{p}}^s - \hat{b}_{\mathbf{p}}^s\hat{b}_{\mathbf{p}'}^{s'\dagger})\bar{u}_{\mathbf{p}'}^{s'}\gamma^\mu u_{\mathbf{p}}^s e^{+i(p'-p)\cdot x} + (\hat{d}_{\mathbf{p}'}^{s'}\hat{d}_{\mathbf{p}}^{s\dagger} - \hat{d}_{\mathbf{p}}^{s\dagger}\hat{d}_{\mathbf{p}'}^{s'})\bar{v}_{\mathbf{p}'}^{s'}\gamma^\mu v_{\mathbf{p}}^s e^{-i(p'-p)\cdot x}$$
$$+ (\hat{b}_{\mathbf{p}'}^{s'\dagger}\hat{d}_{\mathbf{p}}^{s\dagger} - \hat{d}_{\mathbf{p}}^{s\dagger}\hat{b}_{\mathbf{p}'}^{s'\dagger})\bar{u}_{\mathbf{p}'}^{s'}\gamma^\mu v_{\mathbf{p}}^s e^{i(p'+p)\cdot x} + (\hat{d}_{\mathbf{p}'}^{s'}\hat{b}_{\mathbf{p}}^s - \hat{b}_{\mathbf{p}}^s\hat{d}_{\mathbf{p}'}^{s'})\bar{v}_{\mathbf{p}'}^{s'}\gamma^\mu u_{\mathbf{p}}^s e^{-i(p'+p)\cdot x} \Big\}$$

$$
\begin{aligned}
&= \int [d^3p' \, d^3p/(2\pi)^6 2\sqrt{E_{\mathbf{p}'} E_{\mathbf{p}}'}] \sum_{s',s} \Big\{ \hat{b}_{\mathbf{p}'}^{s'\dagger} \hat{b}_{\mathbf{p}}^s \bar{u}_{\mathbf{p}'}^{s'} \gamma^\mu u_{\mathbf{p}}^s e^{+i(p'-p)\cdot x} \\
&\quad - \hat{d}_{\mathbf{p}}^{s\dagger} \hat{d}_{\mathbf{p}'}^{s'} \bar{v}_{\mathbf{p}'}^{s'} \gamma^\mu v_{\mathbf{p}}^s e^{-i(p'-p)\cdot x} + \hat{b}_{\mathbf{p}'}^{s'\dagger} \hat{d}_{\mathbf{p}}^{s\dagger} \bar{u}_{\mathbf{p}'}^{s'} \gamma^\mu v_{\mathbf{p}}^s e^{i(p'+p)\cdot x} \\
&\quad + \hat{d}_{\mathbf{p}'}^{s'} \hat{b}_{\mathbf{p}}^s \bar{v}_{\mathbf{p}'}^{s'} \gamma^\mu u_{\mathbf{p}}^s e^{-i(p'+p)\cdot x} + Y^\mu \Big\} = \, :\hat{\bar{\psi}}\gamma^\mu \hat{\psi}:.
\end{aligned} \tag{6.3.210}
$$

We have used the anticommutation relations of Eq. (6.3.200) to arrive at the expression containing Y^μ defined below. Using Eq. (4.4.135) we find

$$
Y^\mu \equiv \delta^{ss'} \delta^3(\mathbf{p}' - \mathbf{p}) \tfrac{1}{2}(-\bar{u}_{\mathbf{p}}^s \gamma^\mu u_{\mathbf{p}}^s + \bar{v}_{\mathbf{p}}^s \gamma^\mu v_{\mathbf{p}}^s) = \delta^{ss'} \delta^3(\mathbf{p}' - \mathbf{p})(-p^\mu + p^\mu) = 0.
$$

Let us recall, rather than rewrite, a number of important earlier results: the anticommutation relations of the annihilation operators ($\hat{b}_{\mathbf{p}}^s$ and $\hat{d}_{\mathbf{p}}^s$) and creation operators ($\hat{b}_{\mathbf{p}}^{s\dagger}$ and $\hat{d}_{\mathbf{p}}^{s\dagger}$) in Eq. (6.3.80); the normalization and orthonormality of one-fermion and one-antifermion momentum eigenstates, $|f; \mathbf{p}, s\rangle$ and $|\bar{f}; \mathbf{p}, s\rangle$, respectively, in Eqs. (6.3.82) and (6.3.83); and the Heisenberg fermion field operators $\hat{\psi}(x)$ and $\hat{\psi}^\dagger(x)$ in terms of the annihilation and creation operators in Eq. (6.3.100).

From Eq. (6.3.100) it immediately follows that the fermion field operators obey the Heisenberg equations of motion in Eq. (6.3.193),

$$
i\frac{\partial}{\partial t}\hat{\psi}_\alpha(x) = [\hat{\psi}_\alpha(x), \hat{H}] \qquad \text{and} \qquad i\frac{\partial}{\partial t}\hat{\psi}_\alpha^\dagger(x) = [\hat{\psi}_\alpha^\dagger(x), \hat{H}]. \tag{6.3.211}
$$

Using the definitions $\hat{P}^\mu$ and $\hat{M}^{\mu\nu}$ and the ETACR in Eq. (6.3.190) we find

$$
\begin{aligned}
&[\hat{\psi}(x), \hat{P}^\mu] = i\partial^\mu \hat{\psi}(x) \qquad \text{and} \\
&[\hat{\psi}(x), \hat{M}^{\mu\nu}] = i[x^\mu \partial^\nu \hat{\psi}(x) - x^\nu \partial^\mu \hat{\psi}(x) + \Sigma^{\mu\nu}\hat{\psi}(x)],
\end{aligned} \tag{6.3.212}
$$

which as expected also follow from the application of the correspondence principle of Eq. (6.3.159) to the Dirac bracket results in Eq. (6.3.189). Using these results we can then further show that

$$
\begin{aligned}
&[\hat{P}^\mu, \hat{P}^\nu] = 0, \qquad [\hat{P}^\mu, \hat{M}^{\rho\sigma}] = i(g^{\mu\rho}\hat{P}^\sigma - g^{\mu\sigma}\hat{P}^\rho), \\
&[\hat{M}^{\mu\nu}, \hat{M}^{\rho\sigma}] = i(g^{\nu\rho}\hat{M}^{\mu\sigma} - g^{\mu\rho}\hat{M}^{\nu\sigma} - g^{\nu\sigma}\hat{M}^{\mu\rho} + g^{\mu\sigma}\hat{M}^{\nu\rho}),
\end{aligned} \tag{6.3.213}
$$

which correpond to the Dirac bracket results in Eq. (6.3.189). It also follows that the operators $\hat{P}^\mu$ and $\hat{M}^{\mu\nu}$ are Hermitian before and after normal-ordering. Then

$$
\hat{P}^{\mu\dagger} = \hat{P}^\mu \qquad \text{and} \qquad \hat{M}^{\mu\nu\dagger} = \hat{M}^{\mu\nu} \tag{6.3.214}
$$

and as anticipated the operators $\hat{P}^\mu$ and $\hat{M}^{\mu\nu}$ are the Hermitian generators of a unitary representation of the Poincaré group. The proof is left as Problem 6.9. Any Poincaré transformation can be represented as a Lorentz transformation, $\hat{U}(\Lambda)$, followed by a translation, $\hat{U}(a)$, or vice versa. From Eq. (6.2.110) we know that the fermion field operator must transform under a Poincaré transformation as

$$
\hat{U}(\Lambda, a)\hat{\psi}(x)\hat{U}(\Lambda, a)^\dagger = S^{-1}(\Lambda)\hat{\psi}(\Lambda x + a), \tag{6.3.215}
$$

where $S(\Lambda)$ was defined in Eq. (4.4.67) in terms of what we are now referring to as $\Sigma^{\mu\nu}|_{\text{Dirac}} = \frac{1}{2}\sigma^{\mu\nu} = (i/4)[\gamma^\mu, \gamma^\nu] = i\Sigma^{\mu\nu}$, where we now understand $\Sigma^{\mu\nu}$ to refer to the quantity appearing in Eq. (3.2.89). Recall from Eq. (6.2.113) that

$$\hat{U}(a) = e^{i\hat{P}\cdot a} \quad \text{and} \quad \hat{U}(\Lambda) = e^{-(i/2)\omega_{\mu\nu}\hat{M}^{\mu\nu}}. \tag{6.3.216}$$

We have defined the theory in terms of the ETACR, which raises the question of the covariance of the theory. So let us consider the anticommutators at arbitrary times x^0 and y^0. Using Eq. (6.3.100) we find for the anticommutator of $\hat{\psi}$ and $\hat{\bar{\psi}}$,

$$\begin{aligned}
\{\hat{\psi}_\alpha(x), \hat{\bar{\psi}}_\beta(y)\} &= \int \frac{d^3p\, d^3p'}{(2\pi)^6} \frac{1}{2\sqrt{E_{\mathbf{p}}E_{\mathbf{p}'}}} \sum_{s,s'} \left[\{\hat{b}^s_{\mathbf{p}}, \hat{b}^{s'\dagger}_{\mathbf{p}'}\} u^s_\alpha(p) \bar{u}^{s'}_\beta(p') e^{i(p'\cdot y - p\cdot x)} \right. \\
&\qquad\qquad\qquad \left. + \{\hat{d}^{s\dagger}_{\mathbf{p}}, \hat{d}^{s'}_{\mathbf{p}'}\} v^s_\alpha(p) \bar{v}^{s'}_\beta(p') e^{-(p'\cdot y - p\cdot x)} \right] \\
&= \int \frac{d^3p}{(2\pi)^3} \frac{1}{2E_{\mathbf{p}}} \left[(\not{p} + m)_{\alpha\beta} e^{-ip\cdot(x-y)} + (\not{p} - m)_{\alpha\beta} e^{ip\cdot(x-y)} \right] \\
&= (i\not{\partial}_x + m)_{\alpha\beta}\, i\Delta(x-y), \\
\{\hat{\psi}_\alpha(x), \hat{\psi}_\beta(y)\} &= \{\hat{\bar{\psi}}_\alpha(x), \hat{\bar{\psi}}_\beta(y)\} = 0, \tag{6.3.217}
\end{aligned}$$

where we have used the anticommutation relations in Eq. (6.3.80) and the spinor identities in Eq. (4.4.129). Note that from Eq. (6.2.146) we have $i\Delta|_{x^0=y^0} = 0$ and $\partial^x_\mu i\Delta|_{x^0=y^0} = -i\delta^0_\mu \delta^3(\mathbf{x} - \mathbf{y})$ and so from the above equations we see that we recover the ETACR of Eq. (6.3.190) in the limit $(x^0 - y^0) \to 0$. Since $\Delta(x-y)$ vanishes for spacelike separations, then

$$\{\hat{\psi}_\alpha(x), \hat{\bar{\psi}}_\beta(y)\} = (i\not{\partial}_x + m)_{\alpha\beta}\, i\Delta(x-y) = 0 \qquad \text{for} \quad (x-y)^2 < 0. \tag{6.3.218}$$

While the fermion field operators do not commute at spacelike separations, all physical observables are constructed from bilinears at a single spacetime point such as those in Eq. (4.4.159). The commutator of any two fermion bilinears vanishes at spacelike separations as expected from causality. This follows from Eq. (6.3.218),

$$\begin{aligned}
[\hat{\bar{\psi}}_\alpha(x)\hat{\psi}_\beta(x), \hat{\bar{\psi}}_\gamma(y)\hat{\psi}_\delta(y)] &= \hat{\bar{\psi}}_\alpha(x)\{\hat{\psi}_\beta(x), \hat{\bar{\psi}}_\gamma(y)\}\hat{\psi}_\delta(y) - \{\hat{\bar{\psi}}_\alpha(x), \hat{\bar{\psi}}_\gamma(y)\}\hat{\psi}_\beta(x)\hat{\psi}_\delta(y) \\
&\quad + \hat{\bar{\psi}}_\gamma(y)\hat{\bar{\psi}}_\alpha(x)\{\hat{\psi}_\beta(x), \hat{\psi}_\delta(y)\} - \hat{\bar{\psi}}_\gamma(y)\{\hat{\bar{\psi}}_\alpha(x), \hat{\psi}_\delta(y)\}\hat{\psi}_\beta(x) \\
&= \hat{\bar{\psi}}_\alpha(x)\{\hat{\psi}_\beta(x), \hat{\bar{\psi}}_\gamma(y)\}\hat{\psi}_\delta(y) - \hat{\bar{\psi}}_\gamma(y)\{\hat{\bar{\psi}}_\alpha(x), \hat{\psi}_\delta(y)\}\hat{\psi}_\beta(x) \\
&= 0 \qquad\qquad \text{for} \quad (x-y)^2 < 0. \tag{6.3.219}
\end{aligned}$$

Feynman Propagator for Fermions

The states corresponding to having a fermion and an antifermion at the point $x \equiv (x^0, \mathbf{x}) \equiv (t, \mathbf{x})$ are, respectively, given by

$$\hat{\psi}^\dagger(x)|0\rangle \quad \text{and} \quad \hat{\psi}(x)|0\rangle, \tag{6.3.220}$$

which means that the operators $\hat{\psi}^\dagger(x)$ and $\hat{\psi}(x)$ when acting on the vacuum create a fermion and an antifermion, respectively. To create a fermion at the point $x \equiv (x^0, \mathbf{x})$ and to annihilate it at the point $y \equiv (y^0, \mathbf{y})$ at a later time (where $y^0 > x^0$), we have the corresponding amplitude

$$\langle 0|\hat{\psi}_\beta(y)\hat{\psi}^\dagger_\alpha(x)|0\rangle\, \theta(y^0 - x^0). \tag{6.3.221}$$

To create an antifermion at the point y and to annihilate it at the point x at a later time ($x^0 > y^0$) we use the opposite ordering for the operators giving the amplitude

$$\langle 0|\hat{\psi}_\alpha^\dagger(x)\hat{\psi}_\beta(y)|0\rangle \, \theta(x^0 - y^0). \tag{6.3.222}$$

The amplitudes increase the fermion number at x and decrease it at y by one.

We define $S_F(x - y)$ in terms of the difference of these two amplitudes as

$$[S_F(y - x)\gamma^0]_{\beta\alpha} \equiv \langle 0|\hat{\psi}_\beta(y)\hat{\psi}_\alpha^\dagger(x)|0\rangle\theta(y^0 - x^0) - \langle 0|\hat{\psi}_\alpha^\dagger(x)\hat{\psi}_\beta(y)|0\rangle\theta(x^0 - y^0),$$

where it then follows that

$$(i\slashed{\partial}_y - m)_{\alpha\gamma}[S_F(y - x)\gamma^0]_{\gamma\beta} = i\gamma^0_{\alpha\gamma}\langle 0|\{\hat{\psi}_\gamma(y), \hat{\psi}_\beta^\dagger(x)\}|0\rangle\delta(x^0 - y^0)$$

$$= i\gamma^0_{\alpha\beta}\delta^4(y - x), \tag{6.3.223}$$

where we have used $(i\slashed{\partial}_y - m)\hat{\psi}(y) = 0$ and $\partial_0^y\theta(y^0 - x^0) = \delta(y^0 - x^0) = \delta(x^0 - y^0)$. Acting from the right with γ^0 we can then write

$$(i\slashed{\partial}_y - m)S_F(y - x) = i\delta^4(y - x), \tag{6.3.224}$$

where the presence of the Dirac spinor 4×4 spinor identity matrix is to be understood in the usual way. This shows that $S_F(x - y)$ is a Green's function. Following similar steps to those used in Eq. (6.3.217) we find

$$\langle 0|\hat{\psi}_\beta(y)\hat{\bar{\psi}}_\alpha(x)|0\rangle = \{\hat{\psi}_\beta^+(y), \hat{\bar{\psi}}_\alpha^-(x)\} = \int \frac{d^3p}{(2\pi)^3}\frac{1}{2E_\mathbf{p}}\sum_s u_\beta^s(\mathbf{p})\bar{u}_\alpha^s(\mathbf{p})e^{ip\cdot(x-y)}$$

$$= (i\slashed{\partial}_y + m)_{\beta\alpha}\int \frac{d^3p}{(2\pi)^3}\frac{1}{2E_\mathbf{p}}e^{-ip\cdot(y-x)} = (i\slashed{\partial}_y + m)_{\beta\alpha}\Delta_+(y - x),$$

$$\langle 0|\hat{\bar{\psi}}_\alpha(x)\hat{\psi}_\beta(y)|0\rangle = \{\hat{\bar{\psi}}_\alpha^+(x), \hat{\psi}_\beta^-(y)\} = \int \frac{d^3p}{(2\pi)^3}\frac{1}{2E_\mathbf{p}}\sum_s v_\beta^s(\mathbf{p})\bar{v}_\alpha^s(\mathbf{p})e^{-ip\cdot(x-y)}$$

$$= -(i\slashed{\partial}_y + m)_{\beta\alpha}\int \frac{d^3p}{(2\pi)^3}\frac{1}{2E_\mathbf{p}}e^{-ip\cdot(x-y)} = -(i\slashed{\partial}_y + m)_{\beta\alpha}\Delta_+(x - y), \tag{6.3.225}$$

where we have used $\langle 0|\hat{b}_\mathbf{p}^s\hat{b}_{\mathbf{p}'}^{s'\dagger}|0\rangle = \langle 0|\{\hat{b}_\mathbf{p}^s, \hat{b}_{\mathbf{p}'}^{s'\dagger}\}|0\rangle = \delta^{ss'}(2\pi)^3\delta^3(\mathbf{p} - \mathbf{p}')$ and similarly $\langle 0|\hat{d}_\mathbf{p}^s\hat{d}_{\mathbf{p}'}^{s'\dagger}|0\rangle = \delta^{ss'}(2\pi)^3\delta^3(\mathbf{p} - \mathbf{p}')$ and the definition of $\Delta_+(x - y)$ in Eq. (6.2.138). From its definition above it follows that

$$S_F(y - x)_{\beta\alpha} \equiv \langle 0|\hat{\psi}_\beta(y)\hat{\bar{\psi}}_\alpha(x)|0\rangle\theta(y^0 - x^0) - \langle 0|\hat{\bar{\psi}}_\alpha(x)\hat{\psi}_\beta(y)|0\rangle\theta(x^0 - y^0)$$

$$= (i\slashed{\partial}_y + m)_{\beta\alpha}[\Delta_+(y - x)\theta(y^0 - x^0) + \Delta_+(x - y)\theta(x^0 - y^0)] = (i\slashed{\partial}_y + m)_{\beta\alpha}D_F(y - x)$$

$$= \int \frac{d^4p}{(2\pi)^4}\frac{i(\slashed{p} + m)_{\beta\alpha}}{p^2 - m^2 + i\epsilon}e^{-ip\cdot(y-x)}, \tag{6.3.226}$$

where we have used Eq. (6.2.151) for $D_F(x - y)$.

It is natural to define the *time-ordering operator for fermions* with a negative sign for pair-wise interchanges; i.e., for fermion fields $\hat{A}(x)$ and $\hat{B}(y)$ we define

$$T[\hat{A}(x)\hat{B}(y)] \equiv \hat{A}(x)\hat{B}(y)\theta(x^0 - y^0) - \hat{B}(y)\hat{A}(x)\theta(y^0 - x^0) = -T[\hat{B}(y)\hat{A}(x)]. \tag{6.3.227}$$

This extra negative sign appeared in the definition of S_F above and was important because it led to the Feynman boundary conditions appearing in the expression for S_F. Finally, then, we have arrived at the Feynman propagator for fermions,

$$S_F(x - y) \equiv i S_F^{\text{BD}}(x - y) = \langle 0 | T[\hat{\psi}(x)\hat{\bar{\psi}}(y)] | 0 \rangle = (i \partial\!\!\!/_x + m) D_F(x - y) \tag{6.3.228}$$

$$= \int \frac{d^4 p}{(2\pi)^4} \frac{i(\not{p} + m)}{p^2 - m^2 + i\epsilon} e^{-ip \cdot (x - y)} \equiv \int \frac{d^4 p}{(2\pi)^4} S_F(p) e^{-ip \cdot (x - y)},$$

where we again indicate the Bjorken and Drell (1965) notation for the propagator that differs by a factor of i from ours (Peskin and Schroeder, 1995). Note that $S_F(p)$ is the Fourier transform of $S_F(x)$. We write the free fermion propagator as

$$S_F(x - y) = \int \frac{d^4 p}{(2\pi)^4} \frac{i}{\not{p} - m + i\epsilon} e^{-ip \cdot (x - y)}, \qquad \text{since} \tag{6.3.229}$$

$$\frac{1}{\not{p} - m + i\epsilon} = (\not{p} - m + i\epsilon)^{-1} (\not{p} + m - i\epsilon)^{-1} (\not{p} + m - i\epsilon)$$

$$= [(\not{p} + m - i\epsilon)(\not{p} - m + i\epsilon)]^{-1} (\not{p} + m - i\epsilon) = \frac{\not{p} + m - i\epsilon}{p^2 - m^2 + 2im\epsilon} \to \frac{\not{p} + m}{p^2 - m^2 + i\epsilon}.$$

The Fourier transform of $S_F(x - y)$ is

$$S_F(p) \equiv \frac{i}{\not{p} - m + i\epsilon} = \frac{i(\not{p} + m)}{p^2 - m^2 + i\epsilon} = \frac{i \sum_s u^s(p) \bar{u}^s(p)}{p^2 - m^2 + i\epsilon} \qquad \text{such that}$$

$$S_F(x) = \int \frac{d^4 p}{(2\pi)^4} S_F(p) e^{-ip \cdot x} \qquad \text{and} \qquad S_F(p) = \int d^4 x \, S_F(x) e^{+ip \cdot x}, \tag{6.3.230}$$

where we have used $\not{p} + m = \sum_s u^s(p) \bar{u}^s(p)$ from Eq. (4.4.129).

6.3.8　Generating Functional for Dirac Fermions

Recall from Eq. (6.3.102) that the fermion spectral function is given by

$$F(t'' - t') = (1/N_J) \int \mathcal{D}\bar{\psi} \mathcal{D}\psi \, e^{iS[\bar{\psi}, \psi]}, \qquad \text{where}$$

$$S[\bar{\psi}, \psi] = \int_{t'}^{t''} dt \int d^3 x \, \bar{\psi}(x)(i\partial\!\!\!/ - m)\psi(x) \tag{6.3.231}$$

and where for notational simplicity we now understand, but no longer explicitly indicate, that the Grassmann-valued functions are *antiperiodic in time* and *periodic or vanishing in space*. We further understand from Eq. (6.3.107) that in this spectral function we can use integration by parts in the action and ignore the surface terms.

The normal-ordered free Dirac Hamiltonian is as given in Eq. (6.2.225),

$$\hat{H} = \int d^3 x \, \mathcal{H} = \int d^3 x : \hat{\bar{\psi}}(\mathbf{x})(-i\boldsymbol{\gamma} \cdot \boldsymbol{\nabla} + m)\hat{\psi}(\mathbf{x}) :, \tag{6.3.232}$$

which differs by a constant from the form without normal-ordering. We can define the evolution operator with Grassmann-valued sources $\bar{\eta}, \eta$ included as

$$\hat{U}^{\bar{\eta}\eta}(t'', t') = T \exp \left\{ -i \int_{t'}^{t''} dt \, \hat{H}^{\bar{\eta}\eta} \right\}, \tag{6.3.233}$$

$$\hat{H}^{\bar{\eta}\eta} = \hat{H} - \int d^3 x \, [\bar{\eta}(x)\hat{\psi}(x) - \hat{\bar{\psi}}(x)\eta(x)] = \int d^3 x \, [\hat{\mathcal{H}} - \bar{\eta}(x)\hat{\psi}(x) - \hat{\bar{\psi}}(x)\eta(x)].$$

Multiplying a Hilbert space operator by a Grassmann number means that $H^{\bar{\eta}\eta}$ is an operator in a super-Hilbert space rather than a Hilbert space as discussed in Sec. 6.3.6. We always set the sources η and $\bar{\eta}$ to zero after differentiating and so the resulting identities remain valid in Hilbert space.

The derivation of Eq. (6.3.102) is readily extended to include the source terms using Eq. (6.3.31) and we find for the spectral function with sources that

$$F^{\bar{\eta}\eta}(t'', t') = \text{tr}\{\hat{U}^{\bar{\eta}\eta}(t'', t')\} = \int \mathcal{D}\bar{\psi}\mathcal{D}\psi \, e^{iS[\bar{\psi}, \psi, \bar{\eta}, \eta]},$$

$$S[\bar{\psi}, \psi, \bar{\eta}, \eta] \equiv S[\bar{\psi}, \psi] + \int d^4x \, [\bar{\eta}\psi + \bar{\psi}\eta] = \int_{t'}^{t''} d^4x \, [\mathcal{L} + \bar{\eta}\psi + \bar{\psi}\eta]. \tag{6.3.234}$$

Following the arguments in Sec. 4.1.12 we can again treat the source terms as the interaction Hamiltonian and then Eq. (4.1.220) becomes

$$\hat{H}^{\bar{\eta}\eta} = \hat{H} + \hat{H}_{\text{int}}, \quad \text{where} \quad \hat{H}_{\text{int}} = -\int_{t'}^{t''} d^4x \, [\bar{\eta}\hat{\psi} + \hat{\bar{\psi}}\eta],$$

$$\hat{U}_I(t'', t') = T \exp\left\{i\int_{t'}^{t''} d^4x \, [\bar{\eta}\hat{\psi}_I + \hat{\bar{\psi}}_I\eta]\right\}, \tag{6.3.235}$$

where $\hat{U}_I(t'', t')$ is the evolution operator in the interaction picture. We have again used Eqs. (4.1.188) and (4.1.192). From Eq. (4.1.229) we have as $T \to \infty(1 - i\epsilon)$,

$$F^{\bar{\eta}\eta}(T, -T) \to e^{-iE_0(2T)} \langle\Omega|\hat{U}_I(T, -T)|\Omega\rangle = e^{-iE_0(2T)} \langle\Omega|e^{i\int_{-T}^{T} d^4x \, [\bar{\eta}\hat{\psi}_I + \hat{\bar{\psi}}_I\eta]}|\Omega\rangle,$$

so that we again isolate the ground state contribution. The generating functional is then analogous to the noninteracting version of Eq. (6.2.262) and is given by

$$Z[\bar{\eta}, \eta] \equiv \lim_{T \to \infty(1-i\epsilon)} \frac{F^{\bar{\eta}\eta}(T, -T)}{F(T, -T)} = \lim_{T \to \infty(1-i\epsilon)} \frac{\text{tr}\{\hat{U}^{\bar{\eta}\eta}(T, -T)\}}{\text{tr}\{\hat{U}(T, -T)\}} \tag{6.3.236}$$

$$= \lim_{T \to \infty(1-i\epsilon)} \frac{\text{tr}\left[Te^{-i\int_{-T}^{T} d^4x \, [\hat{\mathcal{H}} - \bar{\eta}\hat{\psi} - \hat{\bar{\psi}}\eta]}\right]}{\text{tr}\left[Te^{-i\int_{-T}^{T} d^4x \, \hat{\mathcal{H}}}\right]} = \lim_{T \to \infty(1-i\epsilon)} \frac{\langle 0|Te^{i\int_{-T}^{T} d^4x \, [\bar{\eta}\hat{\psi}_I + \hat{\bar{\psi}}_I\eta]}|0\rangle}{\langle 0|0\rangle}$$

$$= \lim_{T \to \infty(1-i\epsilon)} \frac{\int \mathcal{D}\bar{\psi}\mathcal{D}\psi \, e^{iS[\bar{\psi}, \psi, \bar{\eta}, \eta]}}{\int \mathcal{D}\bar{\psi}\mathcal{D}\psi \, e^{iS[\bar{\psi}, \psi]}},$$

where here $S[\bar{\psi}, \psi, \bar{\eta}, \eta] = \int_{-T}^{T} d^4x \, [\mathcal{L} + \bar{\eta}\psi + \bar{\psi}\eta]$ and $S[\bar{\psi}, \psi] = \int_{-T}^{T} d^4x \, \mathcal{L}$. Recall that the fermion fields are antiperiodic in time with vanishing or periodic spatial boundary conditions.

Putting everything together and using Grassmann functional derivatives we then find a result analogous to that for the scalar field,

$$\langle\Omega|T\hat{\psi}(x_1) \cdots \hat{\psi}(x_{k_1})\hat{\bar{\psi}}(x_{k_1+1}) \cdots \hat{\bar{\psi}}(x_{k_1+2})|\Omega\rangle \tag{6.3.237}$$

$$= (-i)^{k_1}i^{k_2} \frac{\delta^{k_1+k_2}}{\delta\bar{\eta}(x_1) \cdots \delta\bar{\eta}(x_{k_1})\delta\eta(x_{k_1+1}) \cdots \delta\eta(x_{k_1+k_2})} Z[\bar{\eta}, \eta]\bigg|_{\bar{\eta}, \eta=0}$$

$$= \lim_{T \to \infty(1-i\epsilon)} \frac{\int \mathcal{D}\bar{\psi}\mathcal{D}\psi \, \psi(x_1) \cdots \psi(x_{k_1})\bar{\psi}(x_{k_1+1}) \cdots \bar{\psi}(x_{k_1+k_2}) \, e^{iS[\bar{\psi}, \psi]}}{\int \mathcal{D}\bar{\psi}\mathcal{D}\psi \, e^{iS[\bar{\psi}, \psi]}}.$$

For a free Dirac fermion the ground state vacuum is the empty pertubative vacuum, $|\Omega\rangle = |0\rangle$, and the normal-ordered Hamiltonian is given in Eq. (6.3.232). As we have shown, the corresponding Lagrangian density is

$$\mathcal{L} = \bar{\psi}(i\slashed{\partial} - m)\psi. \tag{6.3.238}$$

Referring back to Eq. (6.3.69) we see that the infinitesimal Wick rotation $\delta \to \delta t(1 - i\epsilon)$ leads to no change to the $\partial_t = \partial_0$ term but can be viewed as changing $H \to H(1 - i\epsilon)$. Consulting Eq. (6.3.102) shows that the action becomes

$$S[\bar{\psi}, \psi] = \int d^4x \, \bar{\psi}\left(i\gamma^0\partial_0 - (1 - i\epsilon)[i\boldsymbol{\gamma}\cdot\boldsymbol{\nabla} + m]\right)\psi. \tag{6.3.239}$$

For notational convenience we will ignore the $\mathcal{O}(\epsilon)$ term for now, but recall it when we want to establish the Feynman boundary conditions at the end. Using integration by parts and neglecting the surface term, since the action is to appear in the trace of the evolution operator, we can rewrite the action as

$$S[\bar{\psi}, \psi] = \int d^4x \, d^4y \, \bar{\psi}(x) \left[(i\partial\!\!\!/^x - m)\delta^4(x - y) \right] \psi(y). \tag{6.3.240}$$

Using Eqs. (6.3.236) and (6.3.31) we obtain the Dirac fermion generating functional

$$Z[\bar{\eta}, \eta] \equiv \frac{\int \mathcal{D}\bar{\psi}\mathcal{D}\psi \, e^{iS[\bar{\psi}, \psi] + i \int d^4x \, [\bar{\eta}(x)\psi(x) + \bar{\psi}(x)\eta(x)]}}{\int \mathcal{D}\bar{\psi}\mathcal{D}\psi \, e^{iS[\bar{\psi}, \psi]}}$$

$$= \exp\left\{ -\int d^4x \, d^4y \, \bar{\eta}(x)S_F(x - y)\eta(y) \right\}, \tag{6.3.241}$$

where S_F is the fermion propagator in Eq. (6.3.228). This follows from identifying

$$M(x, y) \equiv (i\partial\!\!\!/^x - m)\delta^4(x - y) \qquad \text{and so} \tag{6.3.242}$$

$$M^{-1}(x, y) = -iS_F(x - y) \equiv \int \frac{d^4p}{(2\pi)^4} \frac{(p\!\!\!/ + m)}{p^2 - m^2 + i\epsilon} e^{-ip\cdot(x-y)}, \tag{6.3.243}$$

where we used integration by parts and neglected surface terms,

$$\int d^4z \, [(i\partial\!\!\!/^x - m)\delta^4(x - z)]S_F(z - y) = \int d^4z \, [(-i\partial\!\!\!/^z - m)\delta^{(4)}(x - z)]S_F(z - y)$$

$$= \int d^4z \, \delta^{(4)}(x - z)(i\partial\!\!\!/^z - m)S_F(z - y) = i\delta^{(4)}(x - y). \tag{6.3.244}$$

We have also used Eq. (6.3.224). We have not yet justified the $i\epsilon$ term in S_F. Let us now reestablish the $\mathcal{O}(\epsilon)$ term from Eq. (6.3.239) leading to

$$M(x, y) \equiv (i\gamma^0\partial_0 - (1 - i\epsilon)[i\boldsymbol{\gamma} \cdot \boldsymbol{\nabla} + m])\delta^4(x - y) \qquad \text{and so}$$

$$M^{-1}(x, y) = \int \frac{d^4p}{(2\pi)^4} \frac{(\gamma^0 p_0 + (1 - i\epsilon)[-\boldsymbol{\gamma} \cdot \mathbf{p} + m])}{(p^0)^2 - (1 - i\epsilon)^2[\mathbf{p}^2 + m^2]} e^{-ip\cdot(x-y)}$$

$$\to \int \frac{d^4p}{(2\pi)^4} \frac{(p\!\!\!/ + m)}{p^2 - m^2 + i\epsilon} e^{-ip\cdot(x-y)} = -iS_F(x - y), \tag{6.3.245}$$

where the last step follows since the ϵ term in the numerator has no effect, since $(1 - i\epsilon)^2 = (1 - 2i\epsilon) + \mathcal{O}(\epsilon^2)$ and since in the denominator we can make the replacement $i2\epsilon(\mathbf{p}^2 + m^2) \to i\epsilon$. So we recover the Feynman fermion propagator in Eq. (6.3.241) for the generating functional $Z[\bar{\eta}, \eta]$. We have again arrived at Eq. (6.3.228), but this time from the functional integral approach,

$$\langle 0|T\hat{\psi}(x)\hat{\bar{\psi}}(y)|0\rangle = \frac{\delta^2}{\delta\bar{\eta}(x)\delta\eta(y)} Z[\bar{\eta}, \eta]\Big|_{\bar{\eta}=\eta=0} = \frac{\int \mathcal{D}\bar{\psi}\mathcal{D}\psi \, \psi(x)\bar{\psi}(x)e^{iS[\bar{\psi}, \psi]}}{\int \mathcal{D}\bar{\psi}\mathcal{D}\psi \, e^{iS[\bar{\psi}, \psi]}}$$

$$= S_F(x - y). \tag{6.3.246}$$

Wick's Theorem for Fermions

Using Eqs. (6.3.237) and (6.3.241) it is relatively straightforward to show the fermion analog of the result for free scalar bosons in Eq. (6.2.278),

$$\langle 0|T\hat{\psi}(x_1)\hat{\psi}(x_2)\cdots\hat{\psi}(x_n)\hat{\bar{\psi}}(y_1)\hat{\bar{\psi}}(y_1)\cdots\hat{\bar{\psi}}(y_m)|0\rangle$$

$$= (-i)^m i^n \frac{\delta^{m+n}}{\delta\bar{\eta}(x_1)\cdots\delta\bar{\eta}(x_m)\delta\eta(y_1)\cdots\delta\eta(y_n)} Z[\bar{\eta}, \eta]\Big|_{\bar{\eta}=\eta=0}$$

$$= \delta_{mn}\Sigma_{k_1,\ldots,k_m} \epsilon^{k_1,\ldots,k_m} S_F(x_1 - y_{k_1})S_F(x_2 - y_{k_2})\cdots S_F(x_m - y_{k_m}). \tag{6.3.247}$$

Note that this result will always be zero unless the number of ψs and $\bar{\psi}$s are equal, i.e., unless the numbers of $\bar{\eta}$ and η derivatives are equal. The antisymmetric tensor $\epsilon^{k_1,\ldots,k_m}$ ensures that we pick up a negative sign for each pairwise exchange of fermion operators. Comparing the above expression with Eq. (A.8.1) we see that it is the determinant of an $m \times m$ matrix with matrix elements $A_{ij} \equiv S_F(x_i - y_j)$. This is reminiscent of the Slater determinant in Eq. (4.1.163).

We could also arrive at this result by proving Wick's theorem for fermions using the fermion definition of normal-ordering and time-ordering with their negative signs for the pairwise interchange of fermion operators; i.e., see Eqs. (6.3.90) and (6.3.227). For example, if $x_3^0 > x_1^0 > x_4^0 > x_2^0$ then

$$T\hat{\psi}(x_1)\hat{\bar{\psi}}(x_2)\hat{\psi}(x_3)\hat{\psi}(x_4) = (-1)^3 \hat{\psi}(x_3)\hat{\psi}(x_1)\hat{\psi}(x_4)\hat{\bar{\psi}}(x_2). \tag{6.3.248}$$

We define the *fermion (Wick) contraction* in terms of the time-ordered and normal-ordered product of pairs of fermion operators in analogy with the scalar field in Eq. (6.2.204),

$$T\hat{A}(x)\hat{B}(y) \equiv \;:\hat{A}(x)\hat{B}(y): + \overline{\hat{A}(x)\hat{B}(y)}, \tag{6.3.249}$$

where $\hat{A}$, $\hat{B}$ are each one of $\hat{\psi}$ or $\hat{\bar{\psi}}$. We further require it to have the same symmetry properties under the pairwise exchange of fermion operators such that

$$T\hat{A}(x)\hat{C}(z)\hat{B}(y) = -T\hat{A}(x)\hat{B}(y)\hat{C}(z), \qquad :\hat{A}(x)\hat{C}(z)\hat{B}(y): = -:\hat{A}(x)\hat{B}(y)\hat{C}(z):,$$
$$\overline{\hat{A}(x)\hat{C}(z)\hat{B}(y)} = -\overline{\hat{A}(x)\hat{B}(y)}C(z). \tag{6.3.250}$$

Repeating the steps analogous to Eqs. (6.2.201) and (6.2.202) we find

$$\overline{\hat{\psi}(x)\hat{\psi}(y)} = \overline{\hat{\bar{\psi}}(x)\hat{\bar{\psi}}(y)} = 0, \qquad \overline{\hat{\psi}(x)\hat{\bar{\psi}}(y)} = -\overline{\hat{\bar{\psi}}(y)\hat{\psi}(x)} = 0,$$
$$\overline{\hat{\psi}(x)\hat{\bar{\psi}}(y)} \equiv \left\{ \begin{array}{ll} \{\hat{\psi}^+(x), \hat{\bar{\psi}}^-(y)\} & \text{for } x^0 > y^0 \\ -\{\hat{\bar{\psi}}^+(y), \hat{\psi}^-(x)\} & \text{for } x^0 < y^0 \end{array} \right\} = S_F(x-y), \tag{6.3.251}$$

where we have used

$$\langle 0|\hat{\psi}(x), \hat{\bar{\psi}}(y)|0\rangle = \langle 0|\{\hat{\psi}^+(x), \hat{\bar{\psi}}^-(y)\}|0\rangle = \{\hat{\psi}^+(x), \hat{\bar{\psi}}^-(y)\},$$
$$\langle 0|\hat{\bar{\psi}}(y), \hat{\psi}(x)|0\rangle = \langle 0|\{\hat{\bar{\psi}}^+(y), \hat{\psi}^-(x)\}|0\rangle = \{\hat{\bar{\psi}}^+(y), \hat{\psi}^-(x)\} \tag{6.3.252}$$

and $S_F(x - y) = \langle 0|T\hat{\psi}(x)\hat{\bar{\psi}}(y)|0\rangle$. Since the appropriate antisymmetry is now built in we can use Eq. (6.3.251) to obtain Wick's theorem for fermions using similar steps to that used to prove the scalar boson case. We find

$$T\hat{\psi}(x_1)\hat{\bar{\psi}}(x_2)\hat{\bar{\psi}}(x_3) \cdots \hat{\psi}(x_k)$$
$$= \;:\left(\hat{\psi}(x_1)\hat{\bar{\psi}}(x_2)\hat{\bar{\psi}}(x_3) \cdots \hat{\psi}(x_k) + \text{ all signed contractions}\right): . \tag{6.3.253}$$

As in the scalar boson case when we take the vacuum expectation value of these time-ordered products we find that only fully contracted terms survive and we again arrive at the result in Eq. (6.3.247).

6.4 Photons

Next we consider the quantization of the electromagnetic field. We begin with the canonical approach to its quantization. We discussed classical electromagnetism in Sec. 2.7 and again in the context of a relativistic classical field theory in Sec. 3.3. We showed that Maxwell's equations were the Euler-Lagrange equations for the Lagrangian of electromagnetism in Sec. 3.3.1.

6.4.1 Canonical Quantization of the Electromagnetic Field

In Sec. 3.3.2 we developed the Hamiltonian formalism, which presented a challenge since electromagnetism is a singular classical system. This required the application in Sec. 3.3.2 of the Dirac-Bergmann algorithm that was developed in Sec. 2.9.

These considerations resulted in the canonical quantization procedure for the electromagnetic field in the Coulomb/radiation gauge. The results were given in Eqs. (3.3.95)–(3.3.97). We gather and summarize these results here and so converge with traditional treatments such as that in Bjorken and Drell (1965). Recalling that $\hat{\pi}_{\mathrm{can}} = -\hat{\pi} = -\mathbf{E}$ we have the equal-time canonical commutation relations (ETCR) for the electromagnetic field in the Heisenberg picture,

$$[\hat{E}^i(x), \hat{E}^j(y)]_{x^0=y^0} = [\hat{A}^i(x), \hat{A}^j(y)]_{x^0=y^0} = 0,$$
$$[\hat{A}^j(y), \hat{E}^i(x)]_{x^0=y^0} = -i\delta_{ij}^{\mathrm{tr}}(\mathbf{x} - \mathbf{y}), \tag{6.4.1}$$

where the transverse delta function is defined as

$$\delta_{ij}^{\mathrm{tr}}(\mathbf{x} - \mathbf{y}) \equiv \int [d^3k/(2\pi)^3]\, e^{i\mathbf{k}\cdot(\mathbf{x}-\mathbf{y})} \left[\delta_{ij} - (k^i k^j / \mathbf{k}^2)\right]. \tag{6.4.2}$$

The Hamiltonian operator for the system is given in Eq. (3.3.97),

$$\hat{H} = \int d^3x \left[\tfrac{1}{2}(\hat{\mathbf{E}}^2 + \hat{\mathbf{B}}^2) - (\mathbf{\nabla} \cdot \hat{\mathbf{E}} - j^0)A_0 - \mathbf{j} \cdot \hat{\mathbf{A}}\right], \tag{6.4.3}$$

where $\hat{\mathbf{B}} = \mathbf{\nabla} \times \mathbf{A}$ and where the term containing $\mathbf{\nabla} \cdot \hat{\mathbf{E}} - j^0$ vanishes when acting on states in the physical subspace; i.e., the physical subspace is the null space of the operator $\mathbf{\nabla} \cdot \hat{\mathbf{E}} - j^0$. The coordinate-like operators are the components of $\mathbf{A}$ and the conjugate momentum-like operators are the component of $\hat{\mathbf{E}}$. As discussed following Eq. (3.3.95), A^0 is not an operator but is a c-number field given in Eqs. (2.7.94) and (3.3.90). As a result of the ETCR and the properties of the transverse delta function we have from Eq. (3.3.96) that

$$[\mathbf{\nabla} \cdot \hat{\mathbf{A}}(x), \hat{E}^j(y)]_{x^0=y^0} = -i\partial_i^x \delta_{ij}^{\mathrm{tr}}(\mathbf{x} - \mathbf{y}) = 0, \quad [\mathbf{\nabla} \cdot \hat{\mathbf{A}}(x), \hat{A}^j(y)]_{x^0=y^0} = 0,$$
$$[\hat{A}^i(x), \mathbf{\nabla} \cdot \hat{\mathbf{E}}(y)]_{x^0=y^0} = -i\partial_j^y \delta_{ij}^{\mathrm{tr}}(\mathbf{x} - \mathbf{y}) = 0, \quad [\hat{E}^i(x), \mathbf{\nabla} \cdot \hat{\mathbf{E}}(y)]_{x^0=y^0} = 0. \tag{6.4.4}$$

Then operators expressed in terms of $\hat{\mathbf{A}}$ and $\hat{\mathbf{E}}$, such as $\hat{H}$, will commute with $\mathbf{\nabla} \cdot \hat{\mathbf{A}}$ and $(\mathbf{\nabla} \cdot \hat{\mathbf{E}} - j^0)$. So $\mathbf{\nabla} \cdot \hat{\mathbf{A}}$, $(\mathbf{\nabla} \cdot \hat{\mathbf{E}} - j^0)$ and $\hat{H}$ can be chosen to have simultaneous eigenstates. The evolution operator is $\hat{U}(t'' - t') = e^{-i\hat{H}(t''-t')}$ and so the constraint operators $\mathbf{\nabla} \cdot \hat{\mathbf{A}}$ and $(\mathbf{\nabla} \cdot \hat{\mathbf{E}} - j^0)$ commute with it. We will frequently use $\rho \equiv j^0$.

Let V_{Full} be the Hilbert space spanned by the eigenstates of the operators $\hat{\mathbf{A}}(\mathbf{x})$ and $\hat{\mathbf{E}}(\mathbf{x})$ for all $\mathbf{x} \in \mathbb{R}^3$. We define the physical Hilbert space, $V_{\mathrm{Phys}} \subset V_{\mathrm{Full}}$, as the subspace of V_{Full} that is the null space of $(\mathbf{\nabla} \cdot \mathbf{E} - j^0)$. We can further define the gauge-fixed Hilbert space, $V_{\mathrm{gf}} \subset V_{\mathrm{Phys}}$, as the subspace of V_{Phys} that is the nullspace of $\mathbf{\nabla} \cdot \mathbf{A}$. As discussed in Sec. 2.7.3 the use of Coulomb gauge with A^0 given in Eq. (3.3.90) is an ideal gauge-fixing for the electromagnetic field, which means that every state $|\psi\rangle \in V_{\text{gf}}$ is a unique physical state and by definition

$$\text{if} \quad |\psi\rangle \in V_{\text{gf}} \quad \text{then} \quad (\boldsymbol{\nabla} \cdot \mathbf{E} - \rho)|\psi\rangle = \boldsymbol{\nabla} \cdot \mathbf{A}|\psi\rangle = 0. \tag{6.4.5}$$

Recall that in the Heisenberg picture we have for any operator $\hat{F}$,

$$\partial_0 \hat{F} = \dot{\hat{F}} = (\partial \hat{F}/\partial t) - i[\hat{F}, \hat{H}] \quad \text{since} \quad \hat{F}(t'') = \hat{U}(t'', t')^\dagger \hat{F}(t')\hat{U}(t'', t'). \tag{6.4.6}$$

For an operator without explicit time dependence we have $\dot{\hat{F}} = -i[\hat{F}, \hat{H}]$. By construction of the canonical quantization procedure we expect that operators should obey the same equations of motion as their classical counterparts. We can explicitly verify that we recover Maxwell's equations in Eq. (2.7.34) for the operators when they are constrained to act on states within the gauge-fixed subspace,

$$\begin{aligned}
&\text{Faraday's law: } \boldsymbol{\nabla} \times \hat{\mathbf{E}} = -\dot{\hat{\mathbf{B}}}, &&\text{Gauss' magnetism law: } \boldsymbol{\nabla} \cdot \hat{\mathbf{B}} = 0, \\
&\text{Gauss' law: } \boldsymbol{\nabla} \cdot \hat{\mathbf{E}} = \rho, &&\text{Ampere's law: } \boldsymbol{\nabla} \times \hat{\mathbf{B}} - \mathbf{j} = \dot{\hat{\mathbf{E}}}.
\end{aligned} \tag{6.4.7}$$

Proof: As discussed above, Gauss' law $\boldsymbol{\nabla} \cdot \hat{\mathbf{E}} = \rho$ is a physical constraint that defines the physical subspace, V_{Phys}. Since $\hat{\mathbf{B}} = \boldsymbol{\nabla} \times \hat{\mathbf{A}}$ then Gauss' magnetism law follows from the vector calculus identity $\boldsymbol{\nabla} \cdot (\boldsymbol{\nabla} \times \mathbf{v}) = 0$ for any vector quantity $\mathbf{v}$. In the physical and gauge-fixed subspaces, V_{Phys}, we have

$$\hat{H} = \int d^3x \left[\tfrac{1}{2}(\hat{\mathbf{E}}^2 + \hat{\mathbf{B}}^2) - \mathbf{j} \cdot \hat{\mathbf{A}} \right], \tag{6.4.8}$$

$$\begin{aligned}
\Rightarrow \quad \dot{\hat{B}}^i(t, \mathbf{x}) &= -i[\hat{B}^i(t, \mathbf{x}), \hat{H}] = -i\tfrac{1}{2}\int d^3y\, [\hat{B}^i(t, \mathbf{x}), \hat{\mathbf{E}}^2(t, \mathbf{y})] \\
&= -i\tfrac{1}{2}\int d^3y\, \epsilon^{ijk}\partial_j^x[\hat{A}^k(t, \mathbf{x}), \hat{\mathbf{E}}^2(t, \mathbf{y})] = -\int d^3y\, \epsilon^{ijk}\partial_j^x \hat{E}^\ell(t, \mathbf{y})\delta_{\ell k}^{\text{tr}}(\mathbf{y} - \mathbf{x}) \\
&= -\int d^3y\, \hat{E}^\ell(t, \mathbf{y})\int[d^3k/(2\pi)^3]e^{i\mathbf{k}\cdot(\mathbf{y}-\mathbf{x})}\epsilon^{ijk}(-ik^j)[\delta_{\ell k} - (k^\ell k^j/\mathbf{k}^2)] \\
&= -\int d^3y\, \hat{E}^\ell(t, \mathbf{y})\epsilon^{ijk}(-\partial_j^y)\delta^3(\mathbf{y} - \mathbf{x}) = -\epsilon^{ijk}\partial_j^x \hat{E}^k(t, \mathbf{x}) = -(\boldsymbol{\nabla} \times \hat{\mathbf{E}})^i,
\end{aligned} \tag{6.4.9}$$

which gives Faraday's law. A very similar calculation leads to Ampere's law,

$$\begin{aligned}
\dot{\hat{E}}^i(t, \mathbf{x}) &= -i[\hat{E}^i(t, \mathbf{x}), \hat{H}] = -i\int d^3y \left([\hat{E}^i(t, \mathbf{x}), \tfrac{1}{2}\hat{\mathbf{B}}^2(t, \mathbf{y}) - \mathbf{j} \cdot \hat{\mathbf{A}}(t, \mathbf{y})] \right) \\
&= -i\int d^3y\, (\epsilon^{j\ell k}\hat{B}^j(t, \mathbf{y})\partial_\ell^y - j^k)[\hat{E}^i(t, \mathbf{x}), \hat{A}^k(t, \mathbf{y})] \\
&= -i\int d^3y\, (\epsilon^{j\ell k}\hat{B}^j(t, \mathbf{y})\partial_\ell^y - j^k)i\delta_{ik}^{\text{tr}}(\mathbf{x} - \mathbf{y}) \\
&= \int d^3y\, \epsilon^{j\ell k}\hat{B}^j(t, \mathbf{y})\partial_\ell^y \delta_{ik}^{\text{tr}}(\mathbf{x} - \mathbf{y}) - j_t^i(t, \mathbf{x}) = (\boldsymbol{\nabla} \times \hat{\mathbf{B}})^i - j_t^i(t, \mathbf{x}). \tag{6.4.10}
\end{aligned}$$

As we saw in Eq. (2.7.111), in Coulomb gauge only the transverse component, $\mathbf{j}_t$, of the three-vector current density contributes to the equations of motion determining $\mathbf{A}$. Another way to see this is that since in the gauge-fixed subspace we have $\boldsymbol{\nabla} \cdot \hat{\mathbf{A}} = 0$, then $\hat{\mathbf{A}}$ is transverse and so only $\mathbf{j}_t$ survives in $\mathbf{j} \cdot \hat{\mathbf{A}}$. So as expected we have also recovered Ampere's law.

6.4.2 Fock Space for Photons

Recall from Sec. 2.7 that the electromagnetic field has two polarization states and that electromagnetic waves by definition travel at the speed of light. Photons are the particles associated with the quantization of the electromagnetic field. We know that we can describe electromagnetism in terms of the four-vector field A^μ and we know from Sec. 1.3.1 that four-vectors transform under the $\left(\frac{1}{2}, \frac{1}{2}\right)$ representation of the Lorentz group, which has one $j = 0$ degree of freedom and three corresponding to $j = 1$. We also know from Sec. 1.4.2 that massless particles can only have two spin states, which are helicity states, $\lambda = \pm s$. So photons with spin $s = 1$ admit only two helicity states $\lambda = \pm s = \pm 1$, which are the circular polarization states. The four degrees of freedom in A^μ are reduced to two with one removed by the need to impose Gauss' law and the other removed by the need for gauge-fixing. By the same reasoning a massive vector particle must correspond to a vector field and have spin $s = 1$, but as we know from Sec. 1.4.2 it will have three spin states $m_s = 0, \pm 1$. This means that one of four degrees of freedom of A^μ should be eliminated when we formulate the classical field theory for a massive vector field. This is what we will find in Sec. 6.5 when we study the Proca equation.

We will continue to work in Coulomb gauge in this discussion. We know that photons are bosons and so photon Fock space must be constructed in terms of bosonic annihilation and creation operators. In this sense the only change to the scalar field case is the two polarization states. From Eq. (6.2.17) we can immediately write down the commutation relations required to construct photon Fock space,

$$[\hat{a}_{\mathbf{k}}^\lambda, \hat{a}_{\mathbf{k}'}^{\lambda'}] = [\hat{a}_{\mathbf{k}}^{\lambda\dagger}, \hat{a}_{\mathbf{k}'}^{\lambda'\dagger}] = 0 \quad \text{and} \quad [\hat{a}_{\mathbf{k}}^\lambda, \hat{a}_{\mathbf{k}'}^{\lambda'\dagger}] = (2\pi)^3 \delta^{\lambda\lambda'} \delta^3(\mathbf{k} - \mathbf{k}'). \tag{6.4.11}$$

Recall that we take $A^0 = 0$ for a free electromagnetic field, then $F^{0i} = -E^i = -\partial_0 A^i$. Since $\partial_0 A^i = -F^{0i}$ are the generalized velocities, we see from the expression for the free photon Lagrangian in the first line of Eq. (3.3.12) that $\mathcal{L}$ is a natural Lagrangian density as defined in Sec. 2.2. So for a free electromagnetic field the Hamiltonian operator will be the energy operator as expected. Since $\mathcal{L}$ has no explicit time-dependence, then $\hat{H}$ will have no explicit time-dependence and the energy will be conserved. Since we have a natural system we generalize Eq. (6.2.33) and identify the normal-ordered Hamiltonian as the total energy operator,

$$\hat{H} = \int \frac{d^3k}{(2\pi)^3} \sum_{\lambda=1}^2 \omega_{\mathbf{k}} \hat{a}_{\mathbf{k}}^{\lambda\dagger} \hat{a}_{\mathbf{k}}^\lambda = \int d^3k \sum_{\lambda=1}^2 \omega_{\mathbf{k}} \hat{N}_{\mathbf{k}}^\lambda = \int d^3k \sum_{\lambda=1}^2 E_{\mathbf{k}} \hat{N}_{\mathbf{k}},$$

$$\hat{N}_{\mathbf{k}}^\lambda \equiv \frac{1}{(2\pi)^3} \hat{a}_{\mathbf{k}}^{\lambda\dagger} \hat{a}_{\mathbf{k}}^\lambda \quad \text{and} \quad \hat{N} \equiv \int d^3k \sum_{\lambda=1}^2 \hat{N}_{\mathbf{k}}^\lambda. \tag{6.4.12}$$

Here $\omega_{\mathbf{k}} \equiv E_{\mathbf{k}} = \sqrt{\mathbf{k}^2}$ is the energy of a photon with three-momentum $\mathbf{k}$ and polarization $\lambda = 1, 2$ and $\hat{N}_{\mathbf{k}}^\lambda$ is corresponding occupation number density. The operator $\hat{N}$ measures the total number of photons. Since $\hat{a}_{\mathbf{k}}^{\lambda\dagger} |0\rangle = 0$, the free vacuum state $|0\rangle$ has zero energy and zero photons.

In Coulomb gauge $\nabla \cdot \mathbf{A} = 0$ and $A^0 = 0$ for a free field, which is called the radiation gauge (Bjorken and Drell, 1965). Any real classical three-vector field with the property $\nabla \cdot \mathbf{A} = 0$ can be expanded in terms of plane waves as

$$\mathbf{A}(\mathbf{x}) = \int d^3k \sum_{\lambda=1}^2 \boldsymbol{\epsilon}(k, \lambda)(c_{\mathbf{k}}^\lambda e^{i\mathbf{k}\cdot\mathbf{x}} + c_{\mathbf{k}}^{*\lambda} e^{-i\mathbf{k}\cdot\mathbf{x}}), \tag{6.4.13}$$

where $\boldsymbol{\epsilon}(k, \lambda) \in \mathbb{R}$ are polarization vectors and $c_{\mathbf{k}}^\lambda \in \mathbb{C}$ are arbitrary coefficients. The polarization three-vectors are orthogonal to $\mathbf{k}$ and orthonormal to each other,

$$\boldsymbol{\epsilon}(k, \lambda) \cdot \mathbf{k} = 0 \qquad \text{and} \qquad \boldsymbol{\epsilon}(k, \lambda) \cdot \boldsymbol{\epsilon}(k, \lambda') = \delta_{\lambda\lambda'}. \tag{6.4.14}$$

For example, if the unit vector $\hat{\mathbf{k}} \equiv \mathbf{k}/|\mathbf{k}|$ is aligned with the z direction, then $\boldsymbol{\epsilon}(k, 1)$ and $\boldsymbol{\epsilon}(k, 2)$ can define the x and y directions, respectively, which implies that

$$\hat{\mathbf{k}} \times \boldsymbol{\epsilon}(k, 1) = \boldsymbol{\epsilon}(k, 2) \quad \text{and} \quad \hat{\mathbf{k}} \times \boldsymbol{\epsilon}(k, 2) = -\boldsymbol{\epsilon}(k, 1), \tag{6.4.15}$$

$$\Rightarrow \quad \hat{\mathbf{k}} \times \boldsymbol{\epsilon}(k, \lambda) = (-1)^{1-\lambda} \boldsymbol{\epsilon}(k, \bar{\lambda}) \quad \text{where} \quad \lambda = 1, 2 \Leftrightarrow \bar{\lambda} \equiv 2, 1.$$

Orthogonality to $\mathbf{k}$ ensures $\boldsymbol{\nabla} \cdot \mathbf{A} = 0$ and a three vector with a constraint has two remaining degrees of freedom. Since we define the xyz axes to obey the right-hand rule we do not want the reversal of a photon's momentum to change the handedness of our polarization axes. For this reason we define (Bjorken and Drell, 1965)

$$\epsilon(-k, 1) \equiv -\epsilon(k, 1) \qquad \text{and} \qquad \epsilon(-k, 2) \equiv +\epsilon(k, 2), \tag{6.4.16}$$

$$\Rightarrow \quad \epsilon(-k, \lambda) = (-1)^\lambda \epsilon(k, \lambda) \qquad \text{and} \qquad \epsilon(k, \lambda) \cdot \epsilon(-k, \lambda') = (-1)^\lambda \delta_{\lambda\lambda'}.$$

The vectors $\epsilon(k, 1)$ and $\epsilon(k, 2)$ form a complete orthonormal basis of the two-dimensional space orthogonal to $\mathbf{k}$. The projector $P_{\mathbf{k}_\perp}$ onto this space is

$$P^{ij}_{\mathbf{k}_\perp} \equiv \sum_{\lambda=1}^{2} \epsilon^i(k, \lambda) \epsilon^j(k, \lambda) = \delta^{ij} - k^i k^j / \mathbf{k}^2, \tag{6.4.17}$$

where $P_{\mathbf{k}_\perp}$ is a 3×3 matrix and where $P_{\mathbf{k}_\parallel} = I - P_{\mathbf{k}_\perp}$ is the longitudinal projector with matrix elements $P^{ij}_{\mathbf{k}_\parallel} = k^i k^j / \mathbf{k}^2$. The polarization of electromagnetic waves was discussed in Sec. 2.7.2 and here we are working in the linear polarization basis.

Using the above and in analogy with what was done for the scalar field, for a Hermitian operator $\hat{A}(\mathbf{x})$ satisfying $\boldsymbol{\nabla} \cdot \hat{\mathbf{A}} = 0$ we can write

$$\hat{\mathbf{A}}(\mathbf{x}) = \int \frac{d^3k}{(2\pi)^3} \frac{1}{\sqrt{2\omega_{\mathbf{k}}}} \sum_{\lambda=1}^{2} \epsilon(k, \lambda) (\hat{a}^\lambda_{\mathbf{k}} e^{i\mathbf{k}\cdot\mathbf{x}} + \hat{a}^{\lambda\dagger}_{\mathbf{k}} e^{-i\mathbf{k}\cdot\mathbf{x}}), \tag{6.4.18}$$

where the operators $\hat{a}^\lambda_{\mathbf{k}}$ and $\hat{a}^{\lambda\dagger}_{\mathbf{k}}$ are the Fock space annihilation and creation operators, respectively, in Eq. (6.4.11) above. From Eq. (2.7.92) we have for the free field classical equation of motion in Coulomb gauge that $\Box \mathbf{A}(t, \mathbf{x}) = 0$. The Heisenberg picture operators must satisfy the same equation of motion and so we define

$$\hat{\mathbf{A}}(x) = \int \frac{d^3k}{(2\pi)^3} \frac{1}{\sqrt{2\omega_{\mathbf{k}}}} \sum_{\lambda=1}^{2} \epsilon(k, \lambda) (\hat{a}^\lambda_{\mathbf{k}} e^{-ik\cdot x} + \hat{a}^{\lambda\dagger}_{\mathbf{k}} e^{ik\cdot x}), \tag{6.4.19}$$

where $k^0 \equiv \omega_k = E_{\mathbf{k}} = \sqrt{\mathbf{k}^2}$. In radiation gauge we have $A^0 = 0$ and from Eqs. (2.7.8) and (3.3.28) we then have $\hat{\boldsymbol{\pi}}_{\text{can}} = -\hat{\boldsymbol{\pi}} = -\hat{\mathbf{E}} = \partial_0 \hat{\mathbf{A}}$,

$$\hat{\boldsymbol{\pi}}_{\text{can}}(x) = -\hat{\mathbf{E}}(x) = \int \frac{d^3k}{(2\pi)^3} (-i) \sqrt{\frac{\omega_{\mathbf{k}}}{2}} \sum_{\lambda=1}^{2} \epsilon(k, \lambda) (\hat{a}^\lambda_{\mathbf{k}} e^{-ik\cdot x} - \hat{a}^{\lambda\dagger}_{\mathbf{k}} e^{ik\cdot x}). \tag{6.4.20}$$

These closely resemble the corresponding scalar boson equations in Eq. (6.2.98), except that we now have the polarization vectors included. Generalizing the proofs of Eqs. (6.2.16) and (6.2.100) using Eq. (6.4.14) gives the results

$$\sqrt{2\omega_{\mathbf{k}}} \hat{a}^\lambda_{\mathbf{k}} = \int d^3x \, e^{ik\cdot x} \epsilon(k, \lambda) \cdot [\omega_{\mathbf{k}} \hat{\mathbf{A}}(x) + i\hat{\boldsymbol{\pi}}_{\text{can}}(x)] = 2i \int d^3x \, e^{ik\cdot x} \overset{\leftrightarrow}{\partial_0} \epsilon(k, \lambda) \cdot \hat{\mathbf{A}}(x),$$

$$\sqrt{2\omega_{\mathbf{k}}} \hat{a}^{\lambda\dagger}_{\mathbf{k}} = -2i \int d^3x \, e^{-ik\cdot x} \overset{\leftrightarrow}{\partial_0} \epsilon(k, \lambda) \cdot \hat{\mathbf{A}}(x). \tag{6.4.21}$$

We use Eq. (6.4.16) to write the analog of the expressions in Eqs. (6.2.15),

$$\mathbf{A}(\mathbf{x}) = \int \frac{d^3k}{(2\pi)^3} \frac{1}{\sqrt{2\omega_{\mathbf{k}}}} \sum_{\lambda=1}^{2} \epsilon(k, \lambda) \left[\hat{a}^\lambda_{\mathbf{k}} + (-1)^\lambda \hat{a}^{\lambda\dagger}_{-\mathbf{k}} \right] e^{i\mathbf{k}\cdot\mathbf{x}},$$

$$\mathbf{B}(\mathbf{x}) = \boldsymbol{\nabla} \times \mathbf{A}(\mathbf{x}) = \int \frac{d^3k}{(2\pi)^3} \frac{1}{\sqrt{2\omega_{\mathbf{k}}}} \sum_{\lambda=1}^{2} i\mathbf{k} \times \epsilon(k, \lambda) \left[\hat{a}^\lambda_{\mathbf{k}} + (-1)^\lambda \hat{a}^{\lambda\dagger}_{-\mathbf{k}} \right] e^{i\mathbf{k}\cdot\mathbf{x}},$$

$$\hat{\mathbf{E}}(\mathbf{x}) = -\hat{\boldsymbol{\pi}}_{\text{can}}(\mathbf{x}) = \int \frac{d^3k}{(2\pi)^3} i\sqrt{\frac{\omega_{\mathbf{k}}}{2}} \sum_{\lambda=1}^{2} \epsilon(k, \lambda) \left[\hat{a}^\lambda_{\mathbf{k}} - (-1)^\lambda \hat{a}^{\lambda\dagger}_{-\mathbf{k}} \right] e^{i\mathbf{k}\cdot\mathbf{x}}. \tag{6.4.22}$$

The generalization of Eq. (6.2.18) then becomes

$$
\begin{aligned}
[\hat{A}^i(\mathbf{x}), \hat{\pi}_{\text{can}}^j(\mathbf{x}')] &= \int \frac{d^3k\, d^3k'}{(2\pi)^6} e^{i(\mathbf{k}\cdot\mathbf{x}+\mathbf{k}'\cdot\mathbf{x}')} \frac{(-i)}{2} \sqrt{\frac{\omega_{\mathbf{k}'}}{\omega_{\mathbf{k}}}} \sum_{\lambda,\lambda'} \epsilon^i(k,\lambda)\epsilon^j(k',\lambda') \\
&\quad \times \left\{ -(-1)^{\lambda'}[\hat{a}_{\mathbf{k}}^\lambda, \hat{a}_{-\mathbf{k}'}^{\lambda'\dagger}] + (-1)^\lambda[\hat{a}_{-\mathbf{k}}^{\lambda\dagger}, \hat{a}_{\mathbf{k}'}^{\lambda'}] \right\} \\
&= i \int \frac{d^3k}{(2\pi)^3} e^{i\mathbf{k}\cdot(\mathbf{x}-\mathbf{x}')} \sum_\lambda \epsilon^i(k,\lambda)\epsilon^j(-k,\lambda)(-1)^\lambda \\
&= i \int \frac{d^3k}{(2\pi)^3} e^{i\mathbf{k}\cdot(\mathbf{x}-\mathbf{x}')} [\delta^{ij} - (k^i k^j/\mathbf{k}^2)] = i\delta_{\text{tr}}^{ij}(\mathbf{x}-\mathbf{x}').
\end{aligned}
\tag{6.4.23}
$$

We also find that $[\hat{A}^i(\mathbf{x}), \hat{A}^j(\mathbf{x}')] = [\hat{\pi}_{\text{can}}^i(\mathbf{x}), \hat{\pi}_{\text{can}}^j(\mathbf{x}')] = 0$, since in each case the two commutators in $\{\cdots\}$ have the same sign and so cancel. We have then recovered the ETCR in Eq. (6.4.1), which justifies the choice of $\hat{\mathbf{A}}$ made in Eq. (6.4.18). Since $\mathbf{k}\cdot\epsilon(k,\lambda) = 0$ we also obviously have

$$
\boldsymbol{\nabla}\cdot\hat{\mathbf{A}}(x) = \boldsymbol{\nabla}\cdot\hat{\mathbf{E}}(x) = 0.
\tag{6.4.24}
$$

Using steps similar to those in Eq. (6.2.21) we can evaluate the Hamiltonian. Note that the spatial integral will give $\int [d^3x/(2\pi)^3] e^{i(\mathbf{k}+\mathbf{k}')\cdot\mathbf{x}} = \delta^3(\mathbf{k}+\mathbf{k}')$ so that $\mathbf{k} = -\mathbf{k}'$. Also since $(\mathbf{A}\times\mathbf{B})\cdot(\mathbf{C}\times\mathbf{D}) = (\mathbf{A}\cdot\mathbf{C})(\mathbf{C}\cdot\mathbf{D}) - (\mathbf{B}\cdot\mathbf{C})(\mathbf{A}\cdot\mathbf{D})$, then we have $[i\mathbf{k}\times\epsilon(k,\lambda)]\cdot[-i\mathbf{k}\times\epsilon(-k,\lambda')] = \mathbf{k}^2\epsilon(k,\lambda)\cdot\epsilon(-k,\lambda')$ in the $\mathbf{B}^2$ contribution. Also recall that $\mathbf{k}^2 = \omega_{\mathbf{k}}^2$. So similar to Eq. (6.2.21) we arrive at

$$
\begin{aligned}
\hat{H} &= \int d^3x\, \tfrac{1}{2}(\hat{\mathbf{E}}^2 + \hat{\mathbf{B}}^2) \\
&= \int \frac{d^3k}{(2\pi)^3} \frac{1}{2} \sum_{\lambda,\lambda'} \epsilon(k,\lambda)\cdot\epsilon(-k,\lambda') \left\{ -\frac{\omega_{\mathbf{k}}}{2}\left[\hat{a}_{\mathbf{k}}^\lambda - (-1)^\lambda\hat{a}_{-\mathbf{k}}^{\lambda\dagger}\right]\left[\hat{a}_{-\mathbf{k}}^{\lambda'} - (-1)^{\lambda'}\hat{a}_{\mathbf{k}}^{\lambda'\dagger}\right] \right. \\
&\qquad \left. +\frac{\omega_{\mathbf{k}}}{2}\left[\hat{a}_{\mathbf{k}}^\lambda + (-1)^\lambda\hat{a}_{-\mathbf{k}}^{\lambda\dagger}\right]\left[\hat{a}_{-\mathbf{k}}^{\lambda'} + (-1)^{\lambda'}\hat{a}_{\mathbf{k}}^{\lambda'\dagger}\right] \right\} \\
&= \int \frac{d^3k}{(2\pi)^3} \frac{\omega_{\mathbf{k}}}{2} \sum_{\lambda,\lambda'} \epsilon(k,\lambda)\cdot\epsilon(-k,\lambda') \left\{ a_{\mathbf{k}}^\lambda(-1)^{\lambda'}\hat{a}_{\mathbf{k}}^{\lambda'\dagger} + (-1)^\lambda\hat{a}_{-\mathbf{k}}^{\lambda\dagger}a_{-\mathbf{k}}^{\lambda'} \right\} \\
&= \int \frac{d^3k}{(2\pi)^3} \frac{\omega_{\mathbf{k}}}{2} \sum_\lambda \left\{ a_{\mathbf{k}}^\lambda\hat{a}_{\mathbf{k}}^{\lambda\dagger} + \hat{a}_{-\mathbf{k}}^{\lambda\dagger}a_{-\mathbf{k}}^\lambda \right\} = \int \frac{d^3k}{(2\pi)^3} \omega_{\mathbf{k}} \sum_\lambda \left\{ \hat{a}_{\mathbf{k}}^{\lambda\dagger}a_{\mathbf{k}}^\lambda + \tfrac{1}{2}[\hat{a}_{\mathbf{k}}^\lambda, \hat{a}_{\mathbf{k}}^{\lambda\dagger}] \right\}.
\end{aligned}
\tag{6.4.25}
$$

After normal-ordering the Hamiltonian we then obtain Eq. (6.4.12) as required,

$$
\hat{H} = \int d^3x\, {:}\hat{\pi}^i\partial_0\hat{A}_i - \mathcal{L}{:} = \int d^3x\, \tfrac{1}{2}{:}\hat{\mathbf{E}}^2 + \hat{\mathbf{B}}^2{:} = \int [d^3k/(2\pi)^3]\omega_{\mathbf{k}} \sum_\lambda \hat{a}_{\mathbf{k}}^{\lambda\dagger}a_{\mathbf{k}}^\lambda.
\tag{6.4.26}
$$

We can evaluate the photon three-momentum operator using Eq. (3.3.15). The spatial integration gives $\delta^3(\mathbf{k}+\mathbf{k}')$ and since $\mathbf{A}\times(\mathbf{B}\times\mathbf{C}) = (\mathbf{A}\cdot\mathbf{C})\mathbf{B} - (\mathbf{A}\cdot\mathbf{B})\mathbf{C}$ we have $\epsilon(k,\lambda)\times[\mathbf{k}\times\epsilon(-k,\lambda')] = \mathbf{k}\,\epsilon(k,\lambda)\cdot\epsilon(-k,\lambda')$. This gives

$$
\begin{aligned}
\hat{\mathbf{P}} &= \int d^3x\, \hat{\mathbf{E}}\times\hat{\mathbf{B}} \\
&= \int \frac{d^3k}{(2\pi)^3} \frac{1}{2} \sum_{\lambda,\lambda'} \mathbf{k}\,\epsilon(k,\lambda)\cdot\epsilon(-k,\lambda') \left\{ \left[\hat{a}_{\mathbf{k}}^\lambda - (-1)^\lambda\hat{a}_{-\mathbf{k}}^{\lambda\dagger}\right]\left[\hat{a}_{-\mathbf{k}}^{\lambda'} + (-1)^{\lambda'}\hat{a}_{\mathbf{k}}^{\lambda'\dagger}\right] \right\} \\
&= \int \frac{d^3k}{(2\pi)^3} \frac{\mathbf{k}}{2} \sum_\lambda \left\{ a_{\mathbf{k}}^\lambda\hat{a}_{\mathbf{k}}^{\lambda\dagger} - \hat{a}_{-\mathbf{k}}^{\lambda\dagger}a_{-\mathbf{k}}^\lambda \right\} = \int \frac{d^3k}{(2\pi)^3} \frac{\mathbf{k}}{2} \sum_\lambda \left\{ a_{\mathbf{k}}^\lambda\hat{a}_{\mathbf{k}}^{\lambda\dagger} + \hat{a}_{\mathbf{k}}^{\lambda\dagger}a_{\mathbf{k}}^\lambda \right\} \\
&= \int \frac{d^3k}{(2\pi)^3} \mathbf{k} \sum_\lambda \left\{ \hat{a}_{\mathbf{k}}^{\lambda\dagger}a_{\mathbf{k}}^\lambda + \tfrac{1}{2}[\hat{a}_{\mathbf{k}}^\lambda, \hat{a}_{\mathbf{k}}^{\lambda\dagger}] \right\},
\end{aligned}
\tag{6.4.27}
$$

where in passing from the second to the third line the terms proportional to $\hat{a}_{\mathbf{k}}^{\lambda}\hat{a}_{-\mathbf{k}}^{\lambda'}$ and $\hat{a}_{-\mathbf{k}}^{\lambda\dagger}\hat{a}_{\mathbf{k}}^{\lambda'\dagger}$ are overall odd under $\mathbf{k} \to -\mathbf{k}$ and so vanish in the integral. In the usual way we define the momentum operator to be its normal-ordered form,

$$\hat{\mathbf{P}} \equiv \int d^3x : \hat{\mathbf{E}} \times \hat{\mathbf{B}} : = \int \frac{d^3k}{(2\pi)^3}\mathbf{k}\sum_{\lambda} \hat{a}_{\mathbf{k}}^{\lambda\dagger}\hat{a}_{\mathbf{k}}^{\lambda}. \tag{6.4.28}$$

Using $k^0 = \omega_{\mathbf{k}}$ we can summarize these results for the four-momentum operator as

$$\hat{P}^\mu = \int \frac{d^3k}{(2\pi)^3}k^\mu\sum_{\lambda=1}^{2} \hat{a}_{\mathbf{k}}^{\lambda\dagger}\hat{a}_{\mathbf{k}}^{\lambda} = \int d^3k \, k^\mu \sum_{\lambda=1}^{2} \hat{N}_{\mathbf{k}}^{\lambda} \quad \text{with} \quad \hat{P}^\mu|0\rangle = 0. \tag{6.4.29}$$

From Eq. (3.3.22) we identify the normal-ordered total angular momentum, orbital angular momentum and spin angular momentum operators, respectively, as

$$\hat{\mathbf{J}} = \int d^3x\,[\mathbf{x} \times (:\hat{\mathbf{E}} \times \hat{\mathbf{B}}:)] = \hat{\mathbf{L}} + \hat{\mathbf{S}}, \quad \text{where}$$

$$\hat{\mathbf{L}} = \int d^3x : \hat{E}^i(\mathbf{x} \times \boldsymbol{\nabla})\hat{A}^i : \quad \text{and} \quad \hat{\mathbf{S}} = \int d^3x : \hat{\mathbf{E}} \times \hat{\mathbf{A}}:. \tag{6.4.30}$$

We are in Coulomb gauge and from above $\hat{\mathbf{J}}$ is gauge invariant, while $\hat{\mathbf{L}}$ and $\hat{\mathbf{S}}$ are not. Evaluating the spin operator $\hat{\mathbf{S}}$ before normal-ordering gives

$$\hat{\mathbf{S}} = \int \frac{d^3k}{(2\pi)^3}\frac{i}{2}\sum_{\lambda,\lambda'}\boldsymbol{\epsilon}(k,\lambda) \times \boldsymbol{\epsilon}(-k,\lambda')\left\{\left[\hat{a}_{\mathbf{k}}^{\lambda} - (-1)^\lambda\hat{a}_{-\mathbf{k}}^{\lambda\dagger}\right]\left[\hat{a}_{-\mathbf{k}}^{\lambda'} + (-1)^{\lambda'}\hat{a}_{\mathbf{k}}^{\lambda'\dagger}\right]\right\} \tag{6.4.31}$$

$$= \int \frac{d^3k}{(2\pi)^3}\frac{i}{2}\sum_{\lambda,\lambda'}\boldsymbol{\epsilon}(k,\lambda) \times \boldsymbol{\epsilon}(k,\lambda')\left\{a_{\mathbf{k}}^{\lambda}\hat{a}_{\mathbf{k}}^{\lambda'\dagger} - \hat{a}_{-\mathbf{k}}^{\lambda\dagger}a_{-\mathbf{k}}^{\lambda'}\right\} = \int \frac{d^3k}{(2\pi)^3}i\hat{\mathbf{k}}\left\{\hat{a}_{\mathbf{k}}^{1}a_{\mathbf{k}}^{2\dagger} - \hat{a}_{\mathbf{k}}^{1\dagger}\hat{a}_{\mathbf{k}}^{2}\right\},$$

where in passing from the first to the second line the terms proportional to $\hat{a}_{\mathbf{k}}^{\lambda}\hat{a}_{-\mathbf{k}}^{\lambda'}$ and $\hat{a}_{-\mathbf{k}}^{\lambda\dagger}\hat{a}_{\mathbf{k}}^{\lambda'\dagger}$ are overall odd under $\mathbf{k} \to -\mathbf{k}$ and so vanish in the integral. We have also used $\boldsymbol{\epsilon}(k,\lambda) \times \boldsymbol{\epsilon}(k,\lambda') = (-1)^{\lambda-1}(1 - \delta_{\lambda\lambda'})\hat{k}$, which we can understand by regarding the unit vectors $\boldsymbol{\epsilon}(k,1)$, $\boldsymbol{\epsilon}(k,2)$ and $\hat{k}$ to be the unit vectors in the x, y, z directions, respectively, in the conventional right-handed coordinate system.

We can use a linear polarization basis ($\lambda = 1, 2$) or a helicity/circular polarization basis ($h = \pm$) to describe the electromagnetic field. The linear polarization basis, $\boldsymbol{\epsilon}(k,1)$ and $\boldsymbol{\epsilon}(k,2)$, is real. The helicity/circular polarization basis, $\boldsymbol{\epsilon}(k,+) \equiv \boldsymbol{\epsilon}(k,R)$ and $\boldsymbol{\epsilon}(k,-) \equiv \boldsymbol{\epsilon}(k,L)$, is complex since from Eq. (2.7.73) it follows that

$$\boldsymbol{\epsilon}(k,+) = [\boldsymbol{\epsilon}(k,1) + i\boldsymbol{\epsilon}(k,2)]/\sqrt{2}, \quad \boldsymbol{\epsilon}(k,-) = [\boldsymbol{\epsilon}(k,1) - i\boldsymbol{\epsilon}(k,2)]/\sqrt{2}, \tag{6.4.32}$$

$$\boldsymbol{\epsilon}(k,1) = [\boldsymbol{\epsilon}(k,+) + \boldsymbol{\epsilon}(k,-)]/\sqrt{2}, \quad \boldsymbol{\epsilon}(k,2) = -i[\boldsymbol{\epsilon}(k,+) - \boldsymbol{\epsilon}(k,-)]/\sqrt{2},$$

$$\boldsymbol{\epsilon}(k,h) \cdot \mathbf{k} = 0, \quad \boldsymbol{\epsilon}^*(k,h) \cdot \boldsymbol{\epsilon}(k,h') = \delta_{hh'}, \quad \boldsymbol{\epsilon}^*(k,h) = \boldsymbol{\epsilon}(k,-h) = -\boldsymbol{\epsilon}(-k,h).$$

The expression for the photon field operator must be independent of the polarization basis chosen and so from Eq. (6.4.19) we can write

$$\hat{\mathbf{A}}(x) = \int \frac{d^3k}{(2\pi)^3}\frac{1}{\sqrt{2\omega_{\mathbf{k}}}}\sum_{\lambda=1}^{2}\left[\boldsymbol{\epsilon}(k,\lambda)\hat{a}_{\mathbf{k}}^{\lambda}e^{-ik\cdot x} + \boldsymbol{\epsilon}(k,\lambda)\hat{a}_{\mathbf{k}}^{\lambda\dagger}e^{ik\cdot x}\right]$$

$$= \int \frac{d^3k}{(2\pi)^3}\frac{1}{\sqrt{2\omega_{\mathbf{k}}}}\sum_{h=\pm}\left[\boldsymbol{\epsilon}(k,h)\hat{a}_{\mathbf{k}}^{h}e^{-ik\cdot x} + \boldsymbol{\epsilon}(k,h)^*\hat{a}_{\mathbf{k}}^{h\dagger}e^{ik\cdot x}\right], \tag{6.4.33}$$

where for consistency we must have $\sum_{\lambda}\boldsymbol{\epsilon}(k,\lambda)\hat{a}_{\mathbf{k}}^{\lambda} = \sum_{h}\boldsymbol{\epsilon}(k,h)\hat{a}_{\mathbf{k}}^{h}$. It follows that

$$\hat{a}_{\mathbf{k}}^{+} \equiv \hat{a}_{\mathbf{k}}^{L} = [\hat{a}_{\mathbf{k}}^{1} - i\hat{a}_{\mathbf{k}}^{2}]/\sqrt{2}, \quad \hat{a}_{\mathbf{k}}^{-} \equiv \hat{a}_{\mathbf{k}}^{R} = [\hat{a}_{\mathbf{k}}^{1} + i\hat{a}_{\mathbf{k}}^{2}]/\sqrt{2},$$

$$\hat{a}_{\mathbf{k}}^{1} = [\hat{a}_{\mathbf{k}}^{+} + \hat{a}_{\mathbf{k}}^{-}]/\sqrt{2}, \quad \hat{a}_{\mathbf{k}}^{2} = i[\hat{a}_{\mathbf{k}}^{+} - \hat{a}_{\mathbf{k}}^{-}]/\sqrt{2}, \tag{6.4.34}$$

where we take the Hermitian conjugate to obtain the results for $\hat{a}_{\mathbf{k}}^{\pm\dagger}$. It is readily verified for the helicity basis $h = \pm$ that as expected

$$[\hat{a}_{\mathbf{k}}^{h}, \hat{a}_{\mathbf{k}'}^{h'}] = [\hat{a}_{\mathbf{k}}^{h\dagger}, \hat{a}_{\mathbf{k}'}^{h'\dagger}] = 0 \quad \text{and} \quad [\hat{a}_{\mathbf{k}}^{h}, \hat{a}_{\mathbf{k}'}^{h'\dagger}] = (2\pi)^3 \delta^{hh'} \delta^3(\mathbf{k} - \mathbf{k}'). \tag{6.4.35}$$

In analogy with Eq. (6.2.46) we can define the single photon state with three-momentum $\mathbf{k}$ and circular polarization $h = \pm 1$ as

$$|\mathbf{k}, h\rangle = \sqrt{2\omega_{\mathbf{k}}} \hat{a}_{\mathbf{k}}^{h\dagger} |0\rangle . \tag{6.4.36}$$

Following the arguments leading to Eq. (6.2.58) the quantum field theory analog of the single photon plane wave wavefunction is

$$\langle 0|\hat{\mathbf{A}}(x)|\mathbf{k}, h\rangle = \boldsymbol{\epsilon}(k, h) e^{-ik\cdot x}. \tag{6.4.37}$$

Using the commutation relations in Eq. (6.4.34) we find that

$$\hat{a}_{\mathbf{k}}^{1} a_{\mathbf{k}}^{2\dagger} - \hat{a}_{\mathbf{k}}^{1\dagger} \hat{a}_{\mathbf{k}}^{2} = -(i/2) \left\{ \hat{a}_{\mathbf{k}}^{+} \hat{a}_{\mathbf{k}}^{+\dagger} + \hat{a}_{\mathbf{k}}^{+\dagger} \hat{a}_{\mathbf{k}}^{+} - \hat{a}_{\mathbf{k}}^{-} \hat{a}_{\mathbf{k}}^{-\dagger} - \hat{a}_{\mathbf{k}}^{-\dagger} \hat{a}_{\mathbf{k}}^{-} \right\}, \tag{6.4.38}$$

So for the normal-ordered spin operator we find

$$\hat{\mathbf{S}} = \int d^3x \; :\mathbf{E} \times \mathbf{A}: = \int \frac{d^3k}{(2\pi)^3} \hat{\mathbf{k}} (\hat{a}_{\mathbf{k}}^{+\dagger} a_{\mathbf{k}}^{+} - \hat{a}_{\mathbf{k}}^{-\dagger} a_{\mathbf{k}}^{-}) = \int d^3k \, \hat{\mathbf{k}} (\hat{N}_{\mathbf{k}}^{+} - \hat{N}_{\mathbf{k}}^{-}), \tag{6.4.39}$$

where $\hat{N}_{\mathbf{k}}^{\pm}$ are the number density operators for photons with helicity $h = \pm 1$ and three-momentum $\mathbf{k}$ similar to the scalar field in Eq. (6.2.19). This is a classically intuitive definition of the total spin operator that we might have been tempted to write down from a Fock space point of view. A single photon state with linear polarization in the $\lambda = 1, 2$ direction is denoted $|\mathbf{k}, \lambda\rangle$ and satisfies

$$|\mathbf{k}, \lambda\rangle \equiv \sqrt{2\omega_{\mathbf{k}}} \hat{a}_{\mathbf{k}}^{\lambda\dagger} |0\rangle, \quad \langle \mathbf{k}, \lambda | \hat{\mathbf{S}} | \mathbf{k}, \lambda \rangle = 0, \tag{6.4.40}$$

where the result on the right-hand side is easily understood using

$$\langle \mathbf{k}, 1 | \hat{a}_{\mathbf{k}}^{2\dagger} \hat{a}_{\mathbf{k}}^{1} - \hat{a}_{\mathbf{k}}^{1\dagger} \hat{a}_{\mathbf{k}}^{2} | \mathbf{k}, 1 \rangle = 0 \quad \text{since} \quad \hat{a}_{\mathbf{k}}^{2} |\mathbf{k}, 1\rangle = 0, \tag{6.4.41}$$

$$\langle \mathbf{k}, 2 | \hat{a}_{\mathbf{k}}^{2\dagger} \hat{a}_{\mathbf{k}}^{1} - \hat{a}_{\mathbf{k}}^{1\dagger} \hat{a}_{\mathbf{k}}^{2} | \mathbf{k}, 2 \rangle \propto \langle 0 | a_{\mathbf{k}}^{1} - \hat{a}_{\mathbf{k}}^{1\dagger} | 0 \rangle = 0 \quad \text{since} \quad \langle 0 | \hat{a}_{\mathbf{k}}^{1} | 0 \rangle = 0.$$

Any circular polarization state is a superposition of $|\mathbf{k}, 1\rangle$ and $|\mathbf{k}, 2\rangle$. Any linearly polarized photon state is a superposition of equal and opposite circular polarizations/helicities and so has zero total spin angular momentum.

The Coulomb gauge result for $\hat{\mathbf{S}}$ is clearly a physical observable since $\hat{N}_{\mathbf{k}}^{\pm}$ are. Since $\hat{\mathbf{J}}$ is gauge invariant and therefore physical, this means that $\hat{\mathbf{L}}$ in Coulomb gauge is also the appropriate physical observable. Since $\hat{\mathbf{A}}$ is purely transverse in Coulomb gauge, then we have arrived at the gauge-invariant expressions for $\hat{\mathbf{L}}$ and $\hat{\mathbf{S}}$ that we anticipated at the classical level in Eq. (3.3.25),

$$\hat{\mathbf{L}} = \int d^3x \; :\hat{E}^i (\mathbf{x} \times \boldsymbol{\nabla}) \hat{A}_t^i:, \qquad \hat{\mathbf{S}} = \int d^3x \; :\hat{\mathbf{E}} \times \hat{\mathbf{A}}_t:. \tag{6.4.42}$$

Note: A word of caution is needed here. The Poincaré Lie algebra survives the canonical quantization procedure and so the total angular momentum operators $\hat{\mathbf{J}}$ are the generators of the rotation group and so must satisfy the usual Lie algebra,

$$[\hat{J}^i, \hat{J}^j] = i\epsilon^{ijk} \hat{J}^k. \tag{6.4.43}$$

We have defined the decomposition $\hat{\mathbf{J}} = \hat{\mathbf{L}} + \hat{\mathbf{S}}$ and both observable operators $\hat{\mathbf{L}}$ and $\hat{\mathbf{S}}$ transform as three-vector operators under rotations generated by $\hat{\mathbf{J}}$ in the usual way and so satisfy Eq. (4.1.114).

However, neither $\hat{\mathbf{L}}$ nor $\hat{\mathbf{S}}$ separately obey the angular momentum Lie algebra, which means that $\hat{\mathbf{L}}$ cannot be the generator of spatial rotations and $\hat{\mathbf{S}}$ cannot be the generator of internal spin rotations. Furthermore, $\hat{\mathbf{L}}$ and $\hat{\mathbf{S}}$ do not commute with each other and so are not simultaneously observable. Since the number density operators commute with each other, $[\hat{N}_{\mathbf{k}}^{\lambda}, \hat{N}_{\mathbf{k}'}^{\lambda'}] = 0$, then from Eq. (6.4.39) it follows that the spin operator components all commute,

$$[\hat{S}^i, \hat{S}^j] = 0. \tag{6.4.44}$$

This means that all three components of $\hat{\mathbf{S}}$ are simultaneously observable. So what we have written as $\mathbf{L}$ and $\mathbf{S}$ in the decomposition of $\hat{\mathbf{J}}$ are not the orbital angular momentum and spin angular momentum operators in the familiar sense. The operator $\hat{\mathbf{S}}$ generates transformations of the polarization of the electromagnetic field, while maintaining the fact that the field is purely transverse. For further discussion of the details of the angular momentum decomposition of photons, see van Enk and Nienhuis (1994) and Barnett et al. (2016) and references therein.

Where is the harmonic oscillator hiding in electromagnetism? Recall that the harmonic oscillator form for the Hamiltonian for the scalar bosonic field shown in Eqs. (6.2.3) and (6.2.21) was what led us to the bosonic Fock space. We therefore expect that there is a harmonic oscillator form for the electromagnetic field Hamiltonian. Recall that in Coulomb gauge the coordinate-like quantity is $\mathbf{A}_t(x)$, the momentum-like quantity is $\boldsymbol{\pi}_{\mathrm{can}} = -\boldsymbol{\pi} = -\mathbf{E} = \partial_0 \mathbf{A}_t$ and that any Coulomb gauge field configuration at fixed time can be written as

$$\mathbf{A}_t(\mathbf{x}) = \int \frac{d^3 k}{(2\pi)^3} \frac{1}{\sqrt{2\omega_{\mathbf{k}}}} \sum_{\lambda=1}^{2} \boldsymbol{\epsilon}(k, \lambda)(c_{\mathbf{k}}^{\lambda} e^{i\mathbf{k}\cdot\mathbf{x}} + c_{\mathbf{k}}^{\lambda*} e^{-i\mathbf{k}\cdot\mathbf{x}}). \tag{6.4.45}$$

The classical Hamiltonian is then seen to have a harmonic oscillator form,

$$\begin{aligned}
H &= \int d^3 x \, \tfrac{1}{2}\left[\mathbf{E}^2 + \mathbf{B}^2\right] = \int d^3 x \, \tfrac{1}{2}\left[(\partial_0 \mathbf{A}_t)^2 + (\boldsymbol{\nabla} \times \mathbf{A}_t)^2\right] \\
&= \sum_{i=1}^{3} \int d^3 x \, \tfrac{1}{2}\left[(\partial_0 A_t^i)^2 + (\boldsymbol{\nabla} A_t^i)^2\right],
\end{aligned} \tag{6.4.46}$$

which is the scalar field form in Eq. (6.2.3) except that we have three fields and the mass is zero. Only two independent fields remain after gauge-fixing. The last step is understood as follows: expanding out $\int d^3 x \, (\boldsymbol{\nabla} \times \mathbf{A}_t)^2$ and performing the spatial integral can only lead to terms with $\mathbf{k}' = \pm\mathbf{k}$. We also have an overall factor

$$[\mathbf{k} \times \boldsymbol{\epsilon}(k, \lambda)] \cdot [\mathbf{k}' \times \boldsymbol{\epsilon}(k', \lambda')] = (\mathbf{k} \cdot \mathbf{k}')\boldsymbol{\epsilon}(k, \lambda) \cdot \boldsymbol{\epsilon}(k', \lambda') - \mathbf{k}' \cdot \boldsymbol{\epsilon}(k, \lambda)\mathbf{k} \cdot \boldsymbol{\epsilon}(k', \lambda')$$

$$\rightarrow (\mathbf{k} \cdot \mathbf{k}')\boldsymbol{\epsilon}(k, \lambda) \cdot \boldsymbol{\epsilon}(k', \lambda') = \sum_{i=1}^{3}[\mathbf{k}\epsilon^i(k, \lambda)] \cdot [\mathbf{k}'\epsilon^i(k', \lambda')], \tag{6.4.47}$$

where to arrive at the second line we have used the fact that $\mathbf{k}' = \pm\mathbf{k}$ with Eq. (6.4.14), $\mathbf{k} \cdot \boldsymbol{\epsilon}(k, \lambda) = 0$. Substituting this result back into our full expanded form we find $\int d^3 x \, (\boldsymbol{\nabla} \times \mathbf{A}_t)^2 = \sum_i \int d^3 x \, (\boldsymbol{\nabla} A_t^i)^2$ as claimed.

Photon Propagator in Coulomb Gauge

Consider the time-ordered photon propagator in Coulomb gauge, where as for the scalar field we define it as the vacuum expectation value of the time-ordered product of two coordinate-like field operators,

$$\begin{aligned}
D_V^{\mu\nu}(x - y) &\equiv iD_V^{\mathrm{BD}\mu\nu}(x - y) \equiv \langle 0|T\hat{A}^{\mu}(x)\hat{A}^{\nu}(y)|0\rangle_{\mathrm{Coul}} \\
&= \langle 0|\hat{A}^{\mu}(x)\hat{A}^{\nu}(y)|0\rangle_{\mathrm{Coul}}\theta(x^0 - y^0) + \langle 0|\hat{A}^{\nu}(y)\hat{A}^{\mu}(x)|0\rangle_{\mathrm{Coul}}\theta(y^0 - x^0),
\end{aligned} \tag{6.4.48}$$

where BD denotes the Bjorken and Drell (1965) convention. We add the label "Coul" to the time-ordered vacuum expectation value simply to remind us that Coulomb gauge leads to a frame dependence and so we cannot expect $\langle 0|T\hat{A}^{\mu}(x)\hat{A}^{\nu}(y)|0\rangle_{\text{Coul}}$ to be Lorentz covariant. In Coulomb/radiation gauge in the quantization inertial frame for a free photon we define $\hat{A}^0(x) = 0$ and extend Eq. (6.4.19) and write

$$\hat{A}^{\mu}(x) = \int \frac{d^3 k}{(2\pi)^3} \frac{1}{\sqrt{2\omega_{\mathbf{k}}}} \sum_{\lambda=1}^{2} \left(\epsilon^{\mu}(k,\lambda)\hat{a}_{\mathbf{k}}^{\lambda}e^{-ik\cdot x} + \epsilon^{\mu}(k,\lambda)^*\hat{a}_{\mathbf{k}}^{\lambda\dagger}e^{ik\cdot x} \right), \tag{6.4.49}$$

where $\lambda = 1, 2$ or $\lambda = \pm 1$ for the linear polarization and helicity bases, respectively. The polarization four vectors in the quantization frame are

$$\epsilon^{\mu}(k,\lambda) \equiv (0, \boldsymbol{\epsilon}(k,\lambda)) \quad \text{for} \quad \lambda = 1, 2 \text{ or } \pm 1. \tag{6.4.50}$$

We can then evaluate this propagator in the usual way and find

$$D_V^{\mu\nu}(x-y) = \int \frac{d^3 k}{2\omega_{\mathbf{k}}(2\pi)^3} \sum_{\lambda} \epsilon^{\mu}(k,\lambda)\epsilon^{\nu}(k,\lambda)^* [\theta(x^0-y^0)e^{-ik\cdot(x-y)} + \theta(y^0-x^0)e^{ik\cdot(x-y)}]$$

$$= \int \frac{d^4 k}{(2\pi)^4} \frac{i}{k^2+i\epsilon} \sum_{\lambda=1}^{2} \epsilon^{\mu}(k,\lambda)\epsilon^{\nu}(k,\lambda)^* e^{-ik\cdot(x-y)}, \tag{6.4.51}$$

where $\sum_{\lambda} \epsilon^{\mu}(k,\lambda)\epsilon^{\nu}(k,\lambda)^*$ in the quantization inertial frame is independent of k^0 so that we were able to use Eq. (6.2.161).

We want to obtain the propagator as seen by any inertial observer and not just an observer in the quantization frame in which we defined the Hamiltonian and fixed to Coulomb gauge. Let the photon four-momentum in the quantization inertial frame be $k^{\mu} = (|\mathbf{k}|, \mathbf{k})$. Define $n^{\mu} \equiv (1, 0, 0, 0)$ in the quantization frame and in any frame

$$\tilde{k}^{\mu} \equiv \frac{k^{\mu} - (k \cdot n)n^{\mu}}{\sqrt{(k \cdot n)^2 - k^2}}. \tag{6.4.52}$$

In the quantization frame $\tilde{k}^{\mu} = (0, \hat{\mathbf{k}}) = (0, \mathbf{k}/|\mathbf{k}|)$. So we have a set of four orthonormal four-vectors in this frame in the sense that

$$\tilde{k}^2 = \tilde{k}_{\mu}\tilde{k}^{\mu} = \epsilon(k,1)_{\mu}\epsilon(k,1)^{\mu} = \epsilon(k,2)_{\mu}\epsilon(k,2)^{\mu} = -1, \quad n^2 = +1, \tag{6.4.53}$$

$$n \cdot \tilde{k} = n \cdot \epsilon(k,\lambda) = \tilde{k} \cdot \epsilon(k,\lambda) = \epsilon(k,1) \cdot \epsilon(k,2) = 0 \quad \text{for} \quad \lambda = 1, 2.$$

In the quantization inertial frame we have $\mathbf{k} \cdot \boldsymbol{\epsilon}(k,\lambda) = 0$, $\boldsymbol{\epsilon}(k,\lambda') \cdot \boldsymbol{\epsilon}(k,\lambda)^* = \delta_{\lambda'\lambda}$ and $\epsilon^{\mu} \equiv (0, \boldsymbol{\epsilon})$ and so we can write

$$k_{\mu}\epsilon^{\mu}(k,\lambda) = k_{\mu}\epsilon^{\mu}(k,\lambda)^* = 0, \quad \epsilon_{\mu}(k,\lambda')^*\epsilon^{\mu}(k,\lambda) = -\delta_{\lambda'\lambda}, \quad \lambda = 1, 2 \text{ or } \pm 1. \tag{6.4.54}$$

Since these orthonormality conditions have a Lorentz-invariant form, then we can move from the quantization frame to that of any other inertial observer by Lorentz transforming all four-vector quantities. The four four-vectors $\tilde{k}^{\mu}$, n^{μ}, $\epsilon^{\mu}(k,1)$ and $\epsilon^{\mu}(k,2)$ or $\tilde{k}^{\mu}$, n^{μ}, $\epsilon^{\mu}(k,+)$ and $\epsilon^{\mu}(k,-)$ then form a basis for the space of all momentum four-vectors and we have a completeness relation,

$$\delta^{\mu}{}_{\nu} = -\sum_{\lambda=1,2 \text{ or } \pm} \epsilon^{\mu}(k,\lambda)\epsilon_{\nu}(k,\lambda)^* - \tilde{k}^{\mu}\tilde{k}_{\nu} + n^{\mu}n_{\nu}, \tag{6.4.55}$$

which is verified by writing an arbitrary four-vector as a superposition of the four basis vectors, acting on that four-vector with the right-hand side of the above and then showing that we recover the original four-vector. Acting on this completeness relation with $g^{\nu\sigma}$, changing $\sigma \to \nu$ and then rearranging gives

$$\sum_\lambda \epsilon^\mu(k,\lambda)\epsilon^\nu(k,\lambda)^* = -g^{\mu\nu} - \tilde{k}^\mu \tilde{k}^\nu + n^\mu n^\nu \tag{6.4.56}$$
$$= -g^{\mu\nu} - \frac{k^\mu k^\nu - (k\cdot n)(k^\mu n^\nu + n^\mu k^\nu)}{(k\cdot n)^2 - k^2} - \frac{k^2 n^\mu n^\nu}{(k\cdot n)^2 - k^2},$$

where for on-shell photons the third term is absent since $k^2 = 0$. Then the Coulomb gauge time-ordered propagator for any inertial observer is

$$D_V^{\mu\nu}(x-y) \equiv \int \frac{d^4k}{(2\pi)^4}\, D_V^{\mu\nu}(k)\, e^{-ik\cdot(x-y)} \tag{6.4.57}$$
$$= \int \frac{d^4k}{(2\pi)^4}\, \frac{i\, e^{-ik\cdot(x-y)}}{k^2 + i\epsilon}\left[-g^{\mu\nu} - \frac{k^\mu k^\nu - (k\cdot n)(k^\mu n^\nu + n^\mu k^\nu)}{(k\cdot n)^2 - k^2} - \frac{k^2 n^\mu n^\nu}{(k\cdot n)^2 - k^2}\right],$$

In the quantization inertial frame we set $\hat{A}^0 = 0$ and so $D_V^{00} = D_V^{0j} = D_V^{j0} = 0$, which we can verify for the right-hand side of Eq. (6.4.57) using $n^\mu = (1, \mathbf{0})$. In the quantization frame in Coulomb gauge we find

$$D_V^{ij}(x-y) = \int \frac{d^4k}{(2\pi)^4}\, \frac{i\, e^{-ik\cdot(x-y)}}{k^2 + i\epsilon}\left[\delta^{ij} - \frac{k^i k^j}{\mathbf{k}^2}\right], \qquad D_V^{00} = D_V^{0j} = D_V^{j0} = 0. \tag{6.4.58}$$

Note that $\partial_i^x D_V^{ij}(x-y) = \partial_j^x D_V^{ij}(x-y) = 0$. For this reason the Coulomb gauge propagator is often referred to as the *transverse propagator*.

The transverse propagator is written with three terms. The first term is the Feynman gauge photon propagator that we will soon discuss. The second term is proportional to k^μ and/or k^ν ($-i\partial^\mu$ and/or $-i\partial^\nu$ in coordinate space) and vanishes if acting on conserved currents. In the quantization frame the third term is

$$\int \frac{d^4k}{(2\pi)^4}\, \frac{i\, e^{-ik\cdot(x-y)}}{k^2 + i\epsilon}\, \frac{-k^2 n^\mu n^\nu}{(k\cdot n)^2 - k^2} \to \int \frac{d^4k}{(2\pi)^4}\, \frac{-ig^{\mu 0}g^{\nu 0}e^{-ik\cdot(x-y)}}{\mathbf{k}^2} = \frac{-ig^{\mu 0}g^{\nu 0}\delta(x^0 - y^0)}{4\pi|\mathbf{x}-\mathbf{y}|},$$

where we have used Eq. (A.2.15). This contribution cancels the contribution from the Coulomb interaction in the Hamiltonian of Eq. (3.3.91) when we calculate physical amplitudes (Bjorken and Drell, 1965; Weinberg, 1995). We show this in Sec. 7.5.3. So in practice one can ignore the Coulomb term in the propagator and the Coulomb term in the Hamiltonian since they cancel. Then the relevant Coulomb gauge propagator is the Feynman propagator with no Coulomb term. This can be obtained from the generating functional as will be shown in Sec. 6.4.4.

6.4.3 Functional Integral for Photons

If we ignore the Gauss' law constraint and the gauge-fixing constraint, then we would have the full Hilbert space, V_{Full}, with coordinate-like operators $\hat{\mathbf{A}}$, momentum-like operators $\hat{\mathbf{E}}$ and the Hamiltonian given in Eq. (6.4.3). Without the Gauss' law or Coulomb constraints we will have $\delta_{ij}^{\text{tr}}(\mathbf{x}-\mathbf{y}) \to \delta_{ij}(\mathbf{x}-\mathbf{y})$ in the ETCR given in Eq. (6.4.3) and we can then make the replacements $\hat{\phi} \to \hat{\mathbf{A}}$ and $\hat{\pi} \to \hat{\mathbf{E}}$. This would allow us to follow analogous steps that lead from Eq. (6.2.228) to Eq. (6.2.234) and then on to Eq. (6.2.243) to obtain a spectral function $F(t'', t')$. The constraints require us to restrict these functional integrations over $\mathbf{A}$ and $\mathbf{E}$ to be only over those configurations that span the gauge-fixed subspace $V_{\text{gf}} \subset V_{\text{Phys}} \subset V_{\text{Full}}$.

So let us first consider how to enforce the Gauss' law constraint. To restrict the full Hilbert space to the physical Hilbert space, V_{Phys}, we replace the identity operator in V_{Full} with the transverse projection operator onto V_{Phys} denoted as $\hat{P}_E$,

$$\hat{I} = \int \mathcal{D}^{\text{sp}} \mathbf{E} \, |\mathbf{E}\rangle\langle\mathbf{E}| \to \hat{P}_E \equiv \int \mathcal{D}^{\text{sp}} \mathbf{E}_{\text{Phys}} \, |\mathbf{E}_{\text{Phys}}\rangle\langle\mathbf{E}_{\text{Phys}}|$$

$$= (1/\mathcal{N}_E) \int \mathcal{D}^{\text{sp}} \mathbf{E} \, |\mathbf{E}\rangle\delta[\boldsymbol{\nabla} \cdot \mathbf{E} - \rho]\langle\mathbf{E}|, \qquad (6.4.59)$$

where we have adapted Eq. (6.2.227) to the present case and where $\mathcal{N}_E$ is an irrelevant normalization constant since $\delta[\boldsymbol{\nabla} \cdot \mathbf{E} - \rho]/\delta\mathbf{E}(\mathbf{x})$ is independent of $\mathbf{E}$. We understand that the functional integral and hence $\mathcal{N}_E$ are appropriately regularized. The quantity $\mathcal{N}_E$ will cancel in ratios before the regularization is removed.

To be clear, here $\int \mathcal{D}^{\text{sp}} \mathbf{E}$ is the integral over all three-vector fields $\mathbf{E}(\mathbf{x})$, whereas $\int \mathcal{D}^{\text{sp}} \mathbf{E}_{\text{Phys}}$ only integrates over those fields $\mathbf{E}_{\text{Phys}}(\mathbf{x})$ that satisfy Gauss' law. Since the canonical momentum density is $\boldsymbol{\pi}_{\text{can}} = -\boldsymbol{\pi} = -\mathbf{E}$ then $\vec{q} \cdot \vec{p} \to -\int d^3x \, \mathbf{A} \cdot \mathbf{E}$ in the extension of Eq. (6.2.231). We then have

$$\langle\mathbf{A}|\mathbf{E}\rangle = (1/\mathcal{N})e^{-i\int d^3x \, \mathbf{E}(\mathbf{x}) \cdot \mathbf{A}(\mathbf{x})}, \quad \delta[\mathbf{A}' - \mathbf{A}] = (1/\mathcal{N}^2)\int \mathcal{D}^{\text{sp}} \mathbf{E} \, e^{\pm i\int d^3x \, \mathbf{E} \cdot (\mathbf{A} - \mathbf{A}')} \quad (6.4.60)$$

and the delta function can be expressed as

$$\delta[\boldsymbol{\nabla} \cdot \mathbf{E} - \rho] = (1/\mathcal{N}') \int \mathcal{D}^{\text{sp}} A^0 \, e^{+i\int d^3x \, \delta t A^0 (\nabla \cdot \mathbf{E} - \rho)}, \qquad (6.4.61)$$

where the overall normalization $\mathcal{N}'$ is unimportant and cancels in ratios. We have introduced a functional integral over a dummy function $f(\mathbf{x}) = \delta t A^0(\mathbf{x})$ and have absorbed the δt in the integration variable into the normalization $\mathcal{N}'$. The reason for introducing the notations δt and A^0 will soon become clear.

With the above replacements, the insertion of the projection operator Eq. (6.2.232) and using the notation $\rho_j(\mathbf{x}) \equiv \rho(t_j, \mathbf{x})$ we find

$$\langle\mathbf{A}_{j+1}|\hat{H}\hat{P}_E|\mathbf{A}_j\rangle = (1/\mathcal{N}_E) \int \mathcal{D}^{\text{sp}} \mathbf{E}_j \langle\mathbf{A}_{j+1}|\hat{H}|\mathbf{E}_j\rangle\delta[\boldsymbol{\nabla} \cdot \mathbf{E}_j - \rho_j]\langle\mathbf{E}_j|\mathbf{A}_j\rangle \qquad (6.4.62)$$

$$= \frac{1}{\mathcal{N}_E \mathcal{N}'} \int \mathcal{D}^{\text{sp}} \mathbf{E}_j \mathcal{D}^{\text{sp}} A^0_j \, e^{i\int d^3x \, \delta t A^0_j (\nabla \cdot \mathbf{E}_j - \rho_j)}\langle\mathbf{A}_{j+1}|\mathbf{E}_j\rangle\langle\mathbf{E}_j|\mathbf{A}_j\rangle\frac{\langle\mathbf{A}_{j+1}|\hat{H}_{\text{Phys}}|\mathbf{E}_j\rangle}{\langle\mathbf{A}_{j+1}|\mathbf{E}_j\rangle}$$

$$\equiv (1/\mathcal{N}'') \int \mathcal{D}^{\text{sp}} \mathbf{E}_j \mathcal{D}^{\text{sp}} A^0_j \, e^{i\int d^3x \, \{\delta t A^0_j (\nabla \cdot \mathbf{E}_j - \rho_j) - \mathbf{E}_j \cdot (\mathbf{A}_{j+1} - \mathbf{A}_j)\}} H(\mathbf{E}_j, \mathbf{A}_{j+1}),$$

where $\mathcal{N}'' \equiv \mathcal{N}_E \mathcal{N}' \mathcal{N}^2$ and where

$$\hat{H}_{\text{Phys}} = \int d^3x \left[\tfrac{1}{2}(\hat{\mathbf{E}}^2 + \hat{\mathbf{B}}^2) - \mathbf{j} \cdot \hat{\mathbf{A}}\right], \qquad (6.4.63)$$

$$H(\mathbf{E}_j, \mathbf{A}_{j+1}) \equiv \frac{\langle\mathbf{A}_{j+1}|\hat{H}_{\text{Phys}}|\mathbf{E}_j\rangle}{\langle\mathbf{A}_{j+1}|\mathbf{E}_j\rangle} = \int d^3x \left[\tfrac{1}{2}(\mathbf{E}_j^2 + \mathbf{B}_{j+1}^2) - \mathbf{j} \cdot \mathbf{A}_{j+1}\right]. \qquad (6.4.64)$$

It then follows that Eq. (6.2.234) becomes

$$\langle\mathbf{A}_{j+1}, t_{j+1}|\mathbf{A}_j, t_j\rangle = \langle\mathbf{A}_{j+1}|[1 - i\delta t\hat{H}]\hat{P}_E|\mathbf{A}_j\rangle + \mathcal{O}(\delta t^2) \qquad (6.4.65)$$

$$= (1/\mathcal{N}'') \int \mathcal{D}^{\text{sp}} \mathbf{E}_j \mathcal{D}^{\text{sp}} A^0_j e^{i\int d^3x \{\delta t A^0_j (\nabla \cdot \mathbf{E}_j - \rho_j) - \mathbf{E}_j \cdot (\mathbf{A}_{j+1} - \mathbf{A}_j)\}}$$

$$\times [1 - i\delta t H(\mathbf{E}_j, \mathbf{B}_{j+1})] + \mathcal{O}(\delta t^2)$$

$$= (1/\mathcal{N}'') \int \mathcal{D}^{\text{sp}} \mathbf{E}_j \mathcal{D}^{\text{sp}} A^0_j e^{i\delta t \int d^3x \{A^0_j (\nabla \cdot \mathbf{E}_j - \rho_j) - \mathbf{E}_j \cdot [(\mathbf{A}_{j+1} - \mathbf{A}_j)/\delta t] - \mathcal{H}(\mathbf{E}_j, \mathbf{A}_{j+1})\}} + \mathcal{O}(\delta t^2)$$

and Eq. (6.2.234) becomes

$$\langle \mathbf{A}'', t'' | \mathbf{A}', t' \rangle = \lim_{n \to \infty} (1/\mathcal{N}'')^{(n+1)} \int \prod_{i=1}^{n} \mathcal{D}^{\mathrm{sp}} \mathbf{A}_i \int \prod_{k=0}^{n} \mathcal{D}^{\mathrm{sp}} \mathbf{E}_k \mathcal{D}^{\mathrm{sp}} A_{0k}$$

$$\times \exp \left\{ i \delta t \sum_{j=0}^{n} \int d^3 x [A_j^0 (\nabla \cdot \mathbf{E}_j - \rho_j) - \mathbf{E}_j \cdot [(\mathbf{A}_{j+1} - \mathbf{A}_j)/\delta t] - \mathcal{H}(\mathbf{E}_j, \mathbf{A}_{j+1})] \right\}$$

$$\equiv \int \mathcal{D}\mathbf{E}\mathcal{D}\mathbf{A}\mathcal{D}A^0 \exp \left\{ i \int_{t'}^{t''} d^4 x \left[-\mathbf{E} \cdot \partial^0 \mathbf{A} + A^0 (\nabla \cdot \mathbf{E} - \rho) - \mathcal{H}(\mathbf{E}, \mathbf{A}) \right] \right\}$$

$$= \int \mathcal{D}\mathbf{E}\mathcal{D}\mathbf{A}\mathcal{D}A^0 \exp \left\{ i \int_{t'}^{t''} d^4 x \left[-\mathbf{E} \cdot (\partial^0 \mathbf{A} + \nabla A^0) - A^0 j^0 - \tfrac{1}{2} \mathbf{E}^2 - \tfrac{1}{2} \mathbf{B}^2 + \mathbf{j} \cdot \mathbf{A} \right] \right\}$$

$$\equiv \int \mathcal{D}A^\mu \exp \left\{ i \int_{t'}^{t''} d^4 x \left[\tfrac{1}{2} (\partial^0 \mathbf{A} + \nabla A^0)^2 - \tfrac{1}{2} \mathbf{B}^2 - j^\mu A_\mu \right] \right\}$$

$$\equiv \int \mathcal{D}A^\mu \exp \left\{ i \int_{t'}^{t''} d^4 x \left[-\tfrac{1}{4} F^{\mu\nu} F_{\mu\nu} - j^\mu A_\mu \right] \right\} = \int \mathcal{D}A^\mu \, e^{iS[A,j]}, \qquad (6.4.66)$$

where $S[A, j]$ is the action for electromagnetism with an external current density,

$$S[A, j] \equiv \int d^4 x \, \mathcal{L} = \int_{t'}^{t''} d^4 x \left[-\tfrac{1}{4} F^{\mu\nu} F_{\mu\nu} - j^\mu A_\mu \right] \qquad (6.4.67)$$

and where the boundary conditions $\mathbf{A}''$ at t'' and $\mathbf{A}'$ at t' are understood. To do the Gaussian functional integration over $\mathbf{E}$ we used Eq. (4.1.204) and we understand that the infinitesimal Wick rotation is present as always but do not explicitly indicate it here for brevity. Note that in the above we used integration by parts and again dropped the boundary term. We also used the result that

$$\tfrac{1}{2} (\partial^0 \mathbf{A} + \nabla A^0)^2 - \tfrac{1}{2} \mathbf{B}^2 = \tfrac{1}{2} [(\partial^0 A^i - \partial^i A^0)^2 - \mathbf{B}^2] = \tfrac{1}{2} [(F^{0i})^2 - \mathbf{B}^2]$$

$$= -\tfrac{1}{4} F^{\mu\nu} F_{\mu\nu} = \mathcal{L}, \qquad (6.4.68)$$

where we have used Eqs. (2.7.36) and (3.3.12). The purpose of introducing A^0 was to implement Gauss' law, which has led us to a functional integral over the coordinate-like fields involving the Lorentz-invariant action. However, we are not yet done as we still have not implemented the Coulomb gauge condition or any other gauge condition for that matter. This is the next challenge that we must face.

6.4.4 Gauge-Fixing

As we saw in Sec. 2.7.3 the description of electromagnetism in terms of the four-vector potential A^μ requires gauge-fixing. We first discuss Coulomb gauge-fixing before considering gauge-fixing in a general context. Then we consider the issue of Gribov copies and why we can neglect them in electromagentism and finally we consider the photon propagator in a variety of gauges.

Coulomb Gauge-Fixing

The construction of the trace of the evolution operator will lead to the use of bosonic boundary conditions, where we have periodic temporal boundary conditions and the usual periodic or vanishing spatial boundary conditions. We saw that $F^{\mu\nu}$ is gauge invariant in Eq. (2.7.76) and we saw that the electromagnetic field can only couple to conserved currents in Eq. (2.7.24). As a consequence the action $S[A, j]$ is gauge invariant since under a gauge transformation $A^\mu \to A^{\mu\prime} = A^\mu + \partial^\mu \omega$ we have

$$S[A, j] \to S[A', j] = S[A'] - \int d^4 x \, j^\mu A'_\mu = S[A] - \int d^4 x \, j^\mu A_\mu = S[A, j]. \qquad (6.4.69)$$

We have used $\int d^4 x \, j^\mu \partial_\mu \omega = -\int d^4 x \, (\partial_\mu j^\mu) \omega = 0$, which results from integration by parts, the Minkowski-space divergence theorem in Eq. (3.1.7), the bosonic boundary conditions to eliminate the surface term and also current conservation.

Let us now implement gauge-fixing to Coulomb gauge. This requires us to restrict the physical Hilbert space, V_{Phys}, to the gauge-fixed Hilbert space, V_{gf}, by including an additional projection operator,

$$\hat{I} = \int \mathcal{D}^{\text{sp}} \mathbf{A} \, |\mathbf{A}\rangle\langle\mathbf{A}| \to \hat{P}_A \equiv \int \mathcal{D}^{\text{sp}} \mathbf{A}_{\text{gf}} \, |\mathbf{A}_{\text{gf}}\rangle\langle\mathbf{A}_{\text{gf}}| = \frac{1}{\mathcal{N}_A} \int \mathcal{D}^{\text{sp}} \mathbf{A} \, |\mathbf{A}\rangle \delta[\boldsymbol{\nabla} \cdot \mathbf{A}]\langle\mathbf{A}|, \quad (6.4.70)$$

where $|\mathbf{A}_{\text{gf}}\rangle \in V_{\text{gf}}$ and where like $\mathcal{N}_E$ in Eq. (6.4.59) it follows that $\mathcal{N}_A$ is an irrelevant constant that cancels in ratios.

Adding the Coulomb gauge projector $\hat{P}_A$ along with $\hat{P}_E$ into Eq. (6.4.65) and following through the same steps leads to the gauge fixed form of Eq. (6.4.66),

$$\langle\mathbf{A}''_{\text{gf}}, t''|\mathbf{A}'_{\text{gf}}, t'\rangle = (1/\mathcal{N}'_A) \int \mathcal{D}A^\mu \, \delta[\boldsymbol{\nabla} \cdot \mathbf{A}] e^{iS[A,j]} = \int \mathcal{D}A^\mu_{\text{gf}} \, e^{iS[A_{\text{gf}},j]}, \quad (6.4.71)$$

where the delta functionals over spatial functions become delta functionals over spacetime functions, $\prod_{i=0}^{n} \delta[\boldsymbol{\nabla} \cdot \mathbf{A}](t_i) \to \delta[\boldsymbol{\nabla} \cdot \mathbf{A}]$ and also $\prod_{i=0}^{n} \mathcal{N}_A \to \mathcal{N}'_A$. It follows from the above derivation that in Coulomb gauge we have

$$\int \mathcal{D}A^\mu_{\text{gf}} = \int \mathcal{D}A^0 \mathcal{D}\mathbf{A}_{\text{gf}}. \quad (6.4.72)$$

Applying the arguments of Sec. 6.2.11 leads to an expression for the generating functional for the photon field. The evolution operator in the physical subspace is determined by the physical Hamiltonian, $\hat{H}_{\text{Phys}}$, given in Eq. (6.4.63),

$$\hat{U}^j(t'', t') = T \exp\{-i \int_{t'}^{t''} dt \, \hat{H}_{\text{Phys}}\} = T \exp\{-i \int_{t'}^{t''} dt \, \hat{H} - \int d^3x \, \mathbf{j} \cdot \hat{\mathbf{A}}\}$$
$$= T \exp\{-i \int_{t'}^{t''} d^4x \, [\hat{\mathcal{H}} - \mathbf{j} \cdot \hat{\mathbf{A}}]\}, \quad (6.4.73)$$

where $\hat{H} = \int d^3x \, \mathcal{H} = \int d^3x \, \frac{1}{2}(\mathbf{E}^2 + \mathbf{B}^2)$. Dividing the above into free and interacting parts as we do in Eq. (3.3.91) gives

$$\hat{U}^j(t'', t') = T \exp\{-i \int_{t'}^{t''} dt \, \hat{H}_{\text{Phys}}\} = T \exp\{-i \int_{t'}^{t''} dt \, \hat{H}_0 + \hat{H}_I\}$$
$$= T \exp\{-i \int_{t'}^{t''} d^4x \, \hat{\mathcal{H}}_0 + [\hat{\mathcal{H}}_C - \mathbf{j} \cdot \hat{\mathbf{A}}]\}, \quad (6.4.74)$$

where the free and Coulomb interaction Hamiltonian densities are, respectively,

$$\mathcal{H}_0(t, \mathbf{x}) \equiv \frac{1}{2}(\hat{\mathbf{E}}_t^2 + \hat{\mathbf{B}}^2)$$
$$\mathcal{H}_C(t, \mathbf{x}) \equiv \frac{1}{2} j_0(t, \mathbf{x}) A^0(t, \mathbf{x}) = \frac{1}{2} \int d^3x' \, j^0(t, \mathbf{x}) j^0(t, \mathbf{x}')/4\pi|\mathbf{x} - \mathbf{x}'|, \quad (6.4.75)$$

where here we understand $A^0(t, \mathbf{x}) \equiv \int d^3x' \, j^0(t, \mathbf{x}')/4\pi|\mathbf{x} - \mathbf{x}'|$ to mean the solution of Poisson's equation given in Eq. (2.7.94). The spectral function in V_{gf} is then

$$F^j(t'', t') = \text{tr}\{\hat{U}^j(t'', t')\} = \int \mathcal{D}\mathbf{A}_{\text{gf}} \, \langle\mathbf{A}_{\text{gf}}|T \exp^{-i \int_{t'}^{t''} d^4x \, [\hat{\mathcal{H}}_0 + \mathcal{H}_C - \mathbf{j} \cdot \hat{\mathbf{A}}]} |\mathbf{A}_{\text{gf}}\rangle$$
$$= \int \mathcal{D}\mathbf{A}_{\text{gf}} \, \langle\mathbf{A}_{\text{gf}}, t''|\mathbf{A}_{\text{gf}}, t'\rangle$$
$$= \int \mathcal{D}A^\mu_{\text{periodic}} \, \delta[\boldsymbol{\nabla} \cdot \mathbf{A}] \, e^{iS[A,j]}, \quad (6.4.76)$$

where $S[A, j]$ is given in Eq. (6.4.67) and where we neglect irrelevant multiplicative constants. The generating functional is defined by

$$Z[j] \equiv \lim_{T \to \infty(1-i\epsilon)} \frac{F^j(T, -T)}{F(T, -T)} = \lim_{T \to \infty(1-i\epsilon)} \frac{\text{tr}\{\hat{U}^j(T, -T)\}}{\text{tr}\{\hat{U}(T, -T)\}}$$
$$= \lim_{T \to \infty(1-i\epsilon)} \frac{\int \mathcal{D}A^\mu_{\text{periodic}} \, \delta[\boldsymbol{\nabla} \cdot \mathbf{A}] \, e^{i\{S[A] - \int_{-T}^{T} d^4x \, j_\mu A^\mu\}}}{\int \mathcal{D}A^\mu_{\text{periodic}} \, \delta[\boldsymbol{\nabla} \cdot \mathbf{A}] \, e^{iS[A]}}, \quad (6.4.77)$$

Before we can explicitly evaluate the generating function $Z[j]$ we need to deal with the gauge-fixing delta functional. Rather than continue to focus only on Coulomb gauge it will be convenient to generalize to other choices of gauge. Coulomb gauge was the most convenient for implementing the canonical quantization procedure, but having come this far other gauge choices will better meet our subsequent needs.

However, before leaving Coulomb gauge let us return to consider the Coulomb gauge photon propagator. We *define* the Coulomb gauge Feynman propagator, $D_F(x-y)$, in terms of the generating functional. Working in the quantization inertial frame where $n^\mu = (1, \mathbf{0})$ and using Eqs. (6.4.74), (6.4.76) and (6.4.77) gives

$$D_F^{\mu\nu}(x-y) \equiv (i)^2 \frac{\delta^2}{\delta j_\mu(x)\delta j_\nu(y)} Z[j] \Big|_{j=0} \equiv \langle 0|T\hat{A}^\mu(x)\hat{A}^\nu(y)|0\rangle_{\text{Coul}} + \frac{ig^{\mu 0}g^{\nu 0}\delta(x^0-y^0)}{4\pi|\mathbf{x}-\mathbf{x}'|}$$

$$= D_V^{\mu\nu}(x-y) + \frac{ig^{\mu 0}g^{\nu 0}\delta(x^0-y^0)}{4\pi|\mathbf{x}-\mathbf{x}'|} \tag{6.4.78}$$

$$= \lim_{T\to\infty(1-i\epsilon)} \frac{\int \mathcal{D}A_{\text{periodic}}^\mu \, A^\mu(x)A^\nu(y)\delta[\boldsymbol{\nabla}\cdot\mathbf{A}]\, e^{i\{S[A]-\int_{-T}^T d^4x\, j_\mu A^\mu\}}}{\int \mathcal{D}A_{\text{periodic}}^\mu \, \delta[\boldsymbol{\nabla}\cdot\mathbf{A}]\, e^{iS[A]}}.$$

Proof: Only the first line above requires explanation. We have used

$$(i)^2 \frac{\delta^2}{\delta j_\mu(x)\delta j_\nu(y)} T \exp^{-i\int_{t'}^{t''} d^4x\,[\hat{\mathcal{H}}_0+\mathcal{H}_C-\mathbf{j}\cdot\hat{\mathbf{A}}]} \Big|_{j=0} \tag{6.4.79}$$

$$= T\left[\hat{A}^\mu(x)\hat{A}^\nu(y) + \{ig^{\mu 0}g^{\nu 0}\delta(x^0-y^0)/4\pi|\mathbf{x}-\mathbf{x}'|\}\right]\exp^{-i\int_{t'}^{t''} d^4x\,\hat{\mathcal{H}}_0},$$

where we recall that $\hat{A}^0 = 0$ when $j^\mu = 0$. Taking the trace of this and the limit $T \to \infty(1-i\epsilon)$ produces the vacuum expectation value of this quantity, which in the Heisenberg picture becomes

$$\langle 0|T[\hat{A}^\mu(x)\hat{A}^\nu(y)]|0\rangle + \{ig^{\mu 0}g^{\nu 0}\delta(x^0-y^0)/4\pi|\mathbf{x}-\mathbf{x}'|\}. \tag{6.4.80}$$

It will later be useful to note that

$$D_F^{ij}(x-y) = D_V^{ij}(x-y). \tag{6.4.81}$$

Cautionary note: As mentioned earlier, since Coulomb gauge quantization was formulated in the quantization inertial frame we do not have a manifestly covariant formulation; i.e., we have frame dependence through n^μ. So $\langle 0|T\hat{A}^\mu(x)\hat{A}^\nu(y)|0\rangle_{\text{Coul}}$ is not Lorentz covariant even though time-ordering is covariant (excluding time-reversal transformations). In Sec. 6.4.5 we show that a covariant canonical quantization procedure leads to a covariant version, $D_F(x-y) = \langle 0|T\hat{A}^\mu(x)\hat{A}^\nu(y)|0\rangle$.

From the discussion of $D_V^{\mu\nu}$ in Sec. 6.4.2 we observe that the Coulomb term in Eq. (6.4.78) cancels the Coulomb term (the third term) in $D_V^{\mu\nu}$ in Eq. (6.4.57). So the Coulomb gauge Feynman propagator in an arbitrary inertial frame is

$$D_F^{\mu\nu}(x-y) \equiv (i)^2 \frac{\delta^2}{\delta j_\mu(x)\delta j_\nu(y)} Z[j] \Big|_{j=0} = \int \frac{d^4k}{(2\pi)^4} D_F^{\mu\nu}(k)\, e^{-ik\cdot(x-y)} \tag{6.4.82}$$

$$= \int \frac{d^4k}{(2\pi)^4} \frac{i\, e^{-ik\cdot(x-y)}}{k^2+i\epsilon}\left[-g^{\mu\nu} - \frac{k^\mu k^\nu - (k\cdot n)(k^\mu n^\nu + k^\nu n^\mu)}{(k\cdot n)^2 - k^2}\right],$$

where $n^\mu = (1, \mathbf{0})$ in the quantization inertial frame.

It is $D_F^{\mu\nu}$ that will occur whenever we differentiate the generating functional $Z[j]$ and $D_F^{\mu\nu}$ that will appear in the calculation of physical observables as we will see in Sec. 7.5.3. In the free case an odd number j of derivatives vanish after taking $j = 0$ and even numbers of derivatives will just produce permutations of products of $D_F^{\mu\nu}$ as in the scalar boson case in Eq. (6.2.278).

General Gauge-Fixing

The expression in Eq. (6.4.77) for the Coulomb-gauge photon-generating functional is a central one. We now wish to understand how to extend it to other gauge choices. Any gauge transformation of a gauge field configuration, $A^\mu \to A'^\mu = A^\mu + \partial^\mu \omega$, leads to physically equivalent configurations since the same $\mathbf{E}$ and $\mathbf{B}$ fields result as we saw in Eq. (2.7.75). Recall that we have the bosonic boundary conditions discussed previously, where we require that all fields vanish at spatial infinity faster than $1/r^2$ whenever needed or that we have periodic spatial boundary conditions so that surface terms in integration by parts can be ignored. When calculating the generating functional we also must have periodic temporal boundary conditions as we have seen, which means that there we can also ignore temporal surface terms when doing integration by parts.

Given some ideal gauge-fixing condition we can always make the decomposition

$$A^\mu = A^\mu_{\mathrm{gf}} + \partial^\mu \omega, \tag{6.4.83}$$

which is a generalization of the Helmholtz decomposition. In Coulomb gauge, where A^μ_{gf} satisfies $\boldsymbol{\nabla} \cdot \mathbf{A}_{\mathrm{gf}} = 0$, it leads to the familiar $\mathbf{A} = \mathbf{A}_t + \mathbf{A}_\ell$ and no gauge-fixing of A^0. Choose any gauge field configuration A^μ and then consider the set of all gauge transformations of it,

$$A_\mu \xrightarrow{g(\omega)} A'_\mu \equiv A^\omega_\mu = A_\mu + \partial_\mu \omega. \tag{6.4.84}$$

As illustrated in Fig. 6.2(a) the set of all such A'^μ form the *gauge orbit*. We see that every A^μ lies on one and only one gauge orbit and so gauge field configuration space is densely filled with gauge orbits. We typically illustrate gauge orbits as closed because we often deal with compact gauge groups. In Fig. 6.2(b) we illustrate an ideal gauge-fixing condition, $F[A] = 0$, which defines some submanifold within the space of all gauge configurations where the ideal gauge-fixing manifold intersects each gauge orbit once and only once. We can also consider some other non-ideal or partial gauge-fixing condition, $F_{\mathrm{ni}}[A] = 0$, which intersects gauge orbits more than once. These multiple intersection points are referred to as *Gribov copies*[6] or Gribov ambiguities and will be discussed later

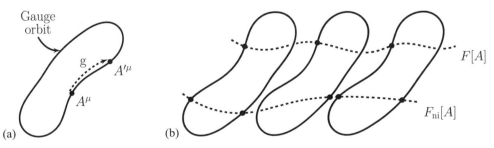

Figure 6.2   (a) A gauge transformation g of a configuration around its gauge orbit; (b) an ideal gauge-fixing condition $F[A] = 0$ intersecting every orbit once and only once and a non-ideal (partial) gauge-fixing condition $F_{\mathrm{ni}}[A] = 0$ with multiple intersections.

[6] Named after Vladimir Gribov.

when we consider nonabelian gauge theories such as quantum chromodynamics. We first encountered the issue of ideal and non-ideal gauge-fixing in Sec. 2.7.3. The gauge orbit that contains the $A^\mu = 0$ gauge field configuration is referred to as the *trivial orbit* and corresponds to $\mathbf{E} = \mathbf{B} = 0$. So in the weak $\mathbf{E}$ and $\mathbf{B}$ field limit we are staying close to the trivial orbit.

Assume an ideal gauge-fixing condition such that if $F[A](x) = 0$, then $A^\mu = A_{\rm gf}^\mu$ is uniquely determined. Then we can use $A_{\rm gf}^\mu$ to uniquely label the gauge orbits. Let us temporarily restrict $\omega(x)$ such that it vanishes as $x^\mu \to \pm\infty$, which ensures that if $\partial_\mu\omega = 0$ everywhere then $\omega(x) = 0$ everywhere. Then defining $A_\mu^\omega = A_{\rm gf\mu} + \partial_\mu\omega$ means that if $A_\mu^\omega = A_{\rm gf\mu}$, then $\omega(x) = 0$. Using Eq. (A.2.7) we can then write

$$1 = \int \mathcal{D}\omega\, \delta[\omega] = \int\mathcal{D}\omega\, |\det M_F[A^\omega]|\, \delta[F[A^\omega]], \qquad (6.4.85)$$
$$\text{where} \quad M_F[A^\omega](x,y) \equiv \delta F[A^\omega](x)/\delta\omega(y).$$

Relaxing the restriction on ω gives an irrelevant (regularized) infinite constant in place of unity on the left-hand side of Eq. (6.4.85) Inserting this constant factor $\int\mathcal{D}\omega\,\delta[\omega]$ into the functional integral before gauge-fixing gives

$$\int \mathcal{D}A^\mu\, e^{iS[A,j]} \propto \int\mathcal{D}\omega\int\mathcal{D}A^\mu\, |\det M_F[A^\omega]|\,\delta[F[A^\omega]]\, e^{iS[A,j]} \qquad (6.4.86)$$
$$= [\textstyle\int\mathcal{D}\omega]\int\mathcal{D}A^\mu\,|\det M_F[A]|\,\delta[F[A]]\,e^{iS[A,j]}$$
$$\equiv \mathcal{N}_\omega\int\mathcal{D}A^\mu\,|\det M_F[A]|\,\delta[F[A]]\,e^{iS[A,j]},$$

where the second line is obtained by the inverse gauge transformation $A_\mu^\omega \to A_\mu$ and using the fact that $\int\mathcal{D}A^\mu$ and $S[A,j]$ are gauge invariant and so unchanged. The third line follows since there is no longer any ω-dependence and so $[\int\mathcal{D}\omega] \equiv \mathcal{N}_\omega$ just contributes another irrelevant (regularized) infinite constant. Discarding $\mathcal{N}_\omega$ removes the infinity associated with the gauge-invariant functional integral and allows us to write in place of Eq. (6.4.76)

$$F^j(t'',t') = \int\mathcal{D}A_{\rm periodic}^\mu\, |\det M_F[A]|\,\delta[F[A]]\,e^{iS[A,j]}. \qquad (6.4.87)$$

In the case of Coulomb gauge we have $F[A] = \boldsymbol{\nabla}\cdot\mathbf{A}$ and $M_F[A]$ is independent of A_μ. Neglecting the (regularized) infinite constant $|\det M_F[A]|$ we recover Eq. (6.4.76). This confirms that Eq. (6.4.87) is the appropriate generalization of Coulomb gauge.

Let us assume that there exists a family of ideal gauge-fixings of the form

$$F_c[A](x) \equiv f(A)(x) - c(x) = 0, \qquad (6.4.88)$$
$$M_{F_c}[A](x,y) = \frac{\delta F_c[A](x)}{\delta\omega(y)} = \frac{\delta[f(A)-c](x)}{\delta\omega(y)} = \frac{\delta f(A)(x)}{\delta\omega(y)} \equiv M_f[A](x,y),$$

where $c(x)$ is an arbitrary real function of spacetime. We refer to M_F and M_f as the *Faddeev-Popov matrix* and to $\det M_F$ and $\det M_f$ as the *Faddeev-Popov determinant*. We can define a new type of "gauge" depending on a *gauge parameter* ξ by forming a Gaussian-like weighted average over the function $c(x)$. Understanding that $T \to (1-i\epsilon)\infty$ in the usual way, the oscillating integral is defined through Gaussian damping. Ignoring the irrelevant constant normalization associated with the integral $\int\mathcal{D}c\, e^{-i(1/2\xi)\int d^4x\, c^2(x)}$ we have

$$F^j(t'',t') = \int\mathcal{D}c\int\mathcal{D}A_{\rm periodic}^\mu\,|\det M_f[A]|\,\delta[f(A)-c]\,e^{iS[A,j]-i(1/2\xi)\int d^4x\, c^2(x)}$$
$$= \int\mathcal{D}A_{\rm periodic}^\mu\,|\det M_f[A]|\,e^{i\{S[A,j]-(1/2\xi)\int d^4x\, f(A)^2\}}. \qquad (6.4.89)$$

The gauge parameter ξ is the width of the Gaussian distribution. In the limit of zero width, $\xi \to 0$, we approach the exact constraint $f(A) = 0$.

Arbitrary Covariant Gauge (R_ξ Gauge) Propagator

Ignoring for now the issue that the Lorenz gauge is not an ideal gauge-fixing, we generalize it to $\partial_\mu A^\mu - c(x) = 0$, i.e., $f(A) = \partial_\mu A^\mu$. This gauge-fixing is referred to as *arbitrary covariant gauges* or as R_ξ *gauges* and has an arbitrary real gauge parameter ξ. The Faddeev-Popov matrix M_f is independent of the gauge field,

$$M_f(x,y) = \delta(\partial_\mu A^\mu - c)(x)/\delta\omega(y) = \delta(\partial^2\omega(x))/\delta\omega(y) = \partial_x^2\delta^4(x-y), \tag{6.4.90}$$

which means that $|\det M_f[A]|$ is another irrelevant (regularized) infinite constant that can be discarded. So we can write the spectral function as

$$F^j(t'',t') = \int \mathcal{D}A^\mu_{\text{periodic}}\, e^{i\{S_\xi[A] - \int d^4x\, j_\mu A^\mu\}}, \quad \text{where}$$

$$S_\xi[A] \equiv \int d^4x\, \{-(1/4)F_{\mu\nu}F^{\mu\nu} - (1/2\xi)(\partial_\mu A^\mu)^2\}. \tag{6.4.91}$$

The generating functional can then be expressed in the simple form

$$Z[j] = \lim_{T\to\infty(1-i\epsilon)} \frac{\int \mathcal{D}A^\mu_{\text{periodic}}\, e^{i\{S_\xi[A] - \int_{-T}^T d^4x\, j_\mu A^\mu\}}}{\int \mathcal{D}A^\mu_{\text{periodic}}\, e^{iS_\xi[A]}}. \tag{6.4.92}$$

Common choices of the R_ξ gauge parameter are

$$\text{Landau gauge: } \xi \to 0; \quad \text{Feynman gauge: } \xi = 1; \quad \text{Yennie gauge: } \xi = 3. \tag{6.4.93}$$

Note that Landau gauge corresponds to the limit of zero width of the Gaussian distribution over $c(x)$ and selects the Lorenz gauge exactly, $\partial_\mu A^\mu = 0$.

Using the periodic boundary conditions to eliminate surface terms and using steps like those in Eq. (6.2.270), the gauge-fixed action S_ξ can be expressed as

$$\begin{aligned}
S_\xi[A] &= \int d^4x\, \{-\tfrac{1}{2}(\partial_\mu A_\nu \partial^\mu A^\nu - \partial_\mu A_\nu \partial^\nu A^\mu) - \tfrac{1}{2\xi}(\partial_\mu A^\mu)^2\} \\
&= \int d^4x\, d^4y\, \delta^4(x-y)\tfrac{1}{2}\{-\partial_\mu^x A_\nu(x)\partial_y^\mu A^\nu(y) + \partial_x^\mu A_\nu(x)\partial_y^\nu A_\mu(y) \\
&\qquad\qquad - \tfrac{1}{\xi}\partial_x^\mu A_\mu(x)\partial_y^\nu A_\nu(y)\} \\
&= \int d^4x\, d^4y\, \tfrac{1}{2}A_\mu(x)\big[\{g^{\mu\nu}\partial_x^2 - [1-\tfrac{1}{\xi}]\partial_x^\mu\partial_x^\nu\}\delta^4(x-y)\big]A_\nu(y) \\
&\equiv \int d^4x\, d^4y\, \tfrac{1}{2}A_\mu(x)K^{\mu\nu}(x,y)A_\nu(y). \tag{6.4.94}
\end{aligned}$$

It is not difficult to show that $K^{\mu\nu}(x,y) = K^{\nu\mu}(y,x)$ and so K is a real symmetric matrix with row index (μ,x) and column index (ν,y). From Eq. (4.1.204) we then obtain the photon-generating functional in an R_ξ gauge,

$$Z[j] = \exp\left\{-\tfrac{1}{2}\int d^4x\, d^4y\, j_\mu(x)D_F^{\mu\nu}(x-y)j_\nu(y)\right\}, \tag{6.4.95}$$

$$D_F^{\mu\nu}(x-y) \equiv \int \frac{d^4k}{(2\pi)^4}\frac{i}{k^2+i\epsilon}\left(-g^{\mu\nu} + (1-\xi)\frac{k^\mu k^\nu}{k^2+i\epsilon}\right)e^{-ik\cdot(x-y)}$$

$$\equiv \int \frac{d^4k}{(2\pi)^4}D_F^{\mu\nu}(k)e^{-ik\cdot(x-y)}, \tag{6.4.96}$$

where $D_F^{\mu\nu}(x-y)$ is the covariant *Feynman propagator* for the photon. We see that $D_F^{\mu\nu}(x-y) = i(K^{-1})^{\mu\nu}(x,y)$ since

$$\{g^{\mu\nu}\partial_x^2 - [1-\tfrac{1}{\xi}]\partial_x^\mu\partial_x^\nu\}D_{F\nu\rho}(x-y) = i\delta^\mu{}_\rho\delta^4(x-y), \quad \text{which gives}$$

$$\int d^4y\, K^{\mu\nu}(x,y)D_{F\nu\rho}(y-z) = i\delta^\mu{}_\rho\delta^4(x-z). \tag{6.4.97}$$

Proof: The above follows since

$$\{g^{\mu\nu}\partial_x^2 - [1 - \tfrac{1}{\xi}]\partial_x^\mu\partial_x^\nu\}D_{F\nu\rho}(x-y)$$

$$= \int \frac{d^4k}{(2\pi)^4}[-g^{\mu\nu}k^2 + k^\mu k^\nu(1 - \tfrac{1}{\xi})][g_{\nu\rho} - \frac{k_\nu k_\rho}{k^2}(1-\xi)]\frac{-i}{k^2+i\epsilon}e^{-ik\cdot(x-y)}$$

$$= \int \frac{d^4k}{(2\pi)^4}[-\delta^\mu_{\ \rho}k^2]\frac{-i}{k^2+i\epsilon}e^{-ik\cdot(x-y)} = i\delta^\mu_{\ \rho}\delta^4(x-y), \qquad (6.4.98)$$

$$\int d^4y\, K^{\mu\nu}(x,y)D_{F\nu\rho}(y-z) = \int d^4y\,\{g^{\mu\nu}\partial_x^2 - [1 - \tfrac{1}{\xi}]\partial_x^\mu\partial_x^\nu\}\delta^4(x-y)D_F^{\nu\rho}(y-z)$$

$$= \int d^4y\,\delta^4(x-y)\{g^{\mu\nu}\partial_y^2 - [1 - \tfrac{1}{\xi}]\partial_y^\mu\partial_y^\nu\}D_F^{\nu\rho}(y-z) = i\delta^\mu_{\ \rho}\delta^4(x-z), \qquad (6.4.99)$$

where we neglect surface terms when integrating by parts since we need only consider field configurations with periodic/vanishing boundary conditions.

To understand the appearance of $k^2 + i\epsilon$ that leads to the Feynman boundary condition, we can first formulate the theory in Euclidean space and then Wick-rotate the result back to Minkowski space using the discussion in Sec. A.4. A number k of functional derivatives with respect to the source gives

$$\frac{(i)^k\,\delta^k}{\delta j_{\mu_1}(x_1)\cdots\delta j_{\mu_k}(x_k)}\,Z[j]\Big|_{j=0} = \lim_{T\to\infty(1-i\epsilon)}\frac{\int \mathcal{D}A^\mu_{\text{periodic}}\,A^{\mu_1}(x_1)\cdots A^{\mu_k}(x_k)\,e^{iS_\xi[A]}}{\int \mathcal{D}A^\mu_{\text{periodic}}\,e^{iS_\xi[A]}}.$$
$$(6.4.100)$$

From Eq. (6.4.95) the covariant Feynman propagator for the photon is

$$D_F^{\mu\nu}(x-y) = (i)^2\frac{\delta^2}{\delta j_\mu(x)\delta j_\nu(y)}\,Z[j]\Big|_{j=0}. \qquad (6.4.101)$$

Since physical quantities must be independent of gauge choice, then the ξ-dependence must vanish in any physical observable. The covariant Gupta-Bleuler operator formalism leads to the same result for the photon propagator as we show in Sec. 6.4.5.

We see that the Landau gauge has a transverse photon propagator and that the Feynman gauge has the simplest form. The Yennie gauge has convenient properties in the infrared that lead to convenient cancellations when calculating low-energy phenomena. Note that the limit $\xi \to \infty$ corresponds to a Gaussian with infinite width and hence no gauge-fixing. In this limit the propagator $D_{\mu\nu}$ in Eq. (6.4.96) and the generating functional $Z[j]$ in Eq. (6.4.95) become undefined as expected (Itzykson and Zuber, 1980; Greiner and Reinhardt, 1996).

Gribov Copies Can Be Ignored in Electromagnetism

We now need to consider the issue that Lorenz gauge is not an ideal gauge-fixing condition and that Gribov copies are therefore expected in the R_ξ gauges. We wish to show that this does not affect the above results in the case of electromagnetism. Let us define A^μ_{gf} to be the result of an *ideal* gauge-fixing $F[A] = 0$. Then the A_{gf} label the gauge orbits. For some other non-ideal gauge-fixing $F_{\text{ni}}[A] = 0$, there are multiple A_μ that satisfy that condition, which are the Gribov copies. Let the gauge orbit labeled by A^μ_{gf} have $N_G[A_{\text{gf}}]$ Gribov copies, where the copies are $A^{\omega_k}_\mu = A_{\text{gf}\mu} + \partial_\mu\omega_k$ for which $F_{\text{ni}}[A^{\omega_k}] = 0$. Then Eq. (6.4.85) becomes

$$\int \mathcal{D}\omega\,|\det M_F[A^\omega]|\,\delta[F[A^\omega]] = \sum_{k=1}^{N_G[A_{\text{gf}}]}\int \mathcal{D}\omega\,\delta[\omega - \omega_k] = N_G[A_{\text{gf}}]. \qquad (6.4.102)$$

This means that inserting the left-hand side of Eq. (6.4.102) into $\int \mathcal{D}A^\mu \, e^{iS[A,j]}$ as we did in Eq. (6.4.86) will lead to the insertion of a gauge-orbit dependent term $N_G[A_{\mathrm{gf}}]$ into the integrand in $F^j(t'', t')$ in Eq. (6.4.89). Then each gauge orbit would contribute proportionally to the number of Gribov copies that it has, which would change the theory and so be unacceptable.

In R_ξ gauges $F_{\mathrm{ni}}[A](x) = \partial^\mu A_\mu(x) - c(x) = 0$ and the Gribov copies correspond to the multiple solutions of $\partial^2 \omega(x) = 0$. This is because $0 = \partial^\mu A_\mu^{\omega_1}(x) - c(x)$ and $0 = \partial^\mu A_\mu^{\omega_2}(x) - c(x)$ means that $0 = \partial^\mu(\partial_\mu \omega_2 - \partial_\mu \omega_1) = \partial^2(\omega_2 - \omega_1)$. Since $\partial^2 \omega(x) = 0$ is independent of A_μ, then the number of solutions is the same for every orbit, $N_G = N_G[A]$, and we can again discard an irrelevant (regularized) infinite constant. So the Gribov copies associated with electromagnetism in the R_ξ gauge can be ignored and the generating functional is unaffected. This will not be the case for nonabelian gauge theories such as quantum chromodynamics as we will see in Sec. 9.2.1. This will be true in electromagnetism for any gauge-fixing choice with $F[A] = f(A) - c = 0$ where $f(A)$ is first-order in A^μ, such as the Coulomb gauge, R_ξ-gauges or the axial gauge that we discuss next.

Axial Gauge Propagator

The *axial gauge* is the condition $n \cdot A = n_\mu A^\mu = 0$, where n^μ is a fixed four-vector such that $n^2 = \pm 1$. Typical choices are $n^\mu = (0,0,0,1)$ and $n^\mu = (1,0,0,0)$. The latter is called *temporal gauge*. In axial gauge the Faddeev-Popov matrix becomes

$$M_f(x,y) = \delta(n_\mu A^\mu - c)(x)/\delta\omega(y) = \delta(n_\mu \partial^\mu \omega(x))/\delta\omega(y) = n \cdot \partial_x \delta^4(x-y),$$

$$|\det M_f[A]| = |\det n \cdot \partial|, \tag{6.4.103}$$

which is independent of the gauge field configuration. The number of Gribov copies is the number is solutions of $n_\mu \partial^\mu \omega = 0$ and is independent of A_μ. Using Eq. (6.4.89) and the steps in Eq. (6.4.94) the gauge-fixed action for $f(A) - c = n \cdot A - c = 0$ is

$$S_\xi[A] \equiv \int d^4x \, \{-\tfrac{1}{4} F_{\mu\nu} F^{\mu\nu} - \tfrac{1}{2\xi}(n_\mu A^\mu)^2\} \equiv \int d^4x \, d^4y \, \tfrac{1}{2} A_\mu(x) K^{\mu\nu}(x,y) A_\nu(y)$$

$$K^{\mu\nu}(x,y) \equiv \{g^{\mu\nu}\partial_x^2 - \partial_x^\mu \partial_x^\nu - \tfrac{1}{\xi} n^\mu n^\nu\} \delta^4(x-y). \tag{6.4.104}$$

The photon propagator in axial gauge is the inverse of this $K^{\mu\nu}(x,y)$, which is

$$D_F^{\mu\nu}(x-y) = \int \frac{d^4k}{(2\pi)^4} \frac{-i}{k^2 + i\epsilon} \left(g^{\mu\nu} + \frac{(n^2 + \xi k^2)k^\mu k^\nu}{(k \cdot n)^2} - \frac{k^\mu n^\nu + k^\nu n^\mu}{k \cdot n} \right) e^{-ik \cdot (x-y)}, \tag{6.4.105}$$

Proof: The proof follows the same steps as that of Eq. (6.4.97). In momentum space we have $K^{\mu\nu}(k) = -g^{\mu\nu} k^2 + k^\mu k^\nu - \tfrac{1}{\xi} n^\mu n^\nu$. The only essential new step in the proof is the demonstration that

$$[-g^{\mu\nu} k^2 + k^\mu k^\nu - \tfrac{1}{\xi} n^\mu n^\nu] \left[g_{\nu\rho} + \frac{(n^2 + \xi k^2)k_\nu k_\rho}{(k \cdot n)^2} - \frac{k_\nu n_\rho + k_\rho n_\nu}{(k \cdot n)} \right] \frac{(-i)}{k^2 + i\epsilon} = i\delta^\mu_{\ \rho}.$$

This is a straightforward task but requires careful bookkeeping of the 15 terms that result in order to verify the cancellation of fourteen of them.

Eq. (6.4.105) has undesirable large-k behavior since the ξ-dependent part of the integrand is $\mathcal{O}((k)^0)$. For this reason we choose the Landau-like limit for axial gauge, where the Gaussian width $\xi \to 0$

with the axial gauge condition then imposed exactly, $n \cdot A = 0$. The axial-gauge photon propagator is then

$$D_F^{\mu\nu}(x-y) = \int \frac{d^4k}{(2\pi)^4} \frac{-i}{k^2 + i\epsilon} \left(g^{\mu\nu} + \frac{n^2 k^\mu k^\nu}{(k \cdot n)^2} - \frac{k^\mu n^\nu + k^\nu n^\mu}{k \cdot n} \right) e^{-ik \cdot (x-y)}. \tag{6.4.106}$$

A gauge choice that is closely related to the axial gauge is the *light-cone gauge*,

$$A^+(x) \equiv A^0(x) + A^3(x) = 0. \tag{6.4.107}$$

The lack of manifest Lorentz covariance of the axial and light-cone gauges presents significant challenges. So here we will typically choose to work in the Lorentz-covariant R_ξ gauges.

Coulomb Gauge Propagator

We can rewrite the quantization-frame Coulomb gauge condition, $\mathbf{\nabla} \cdot \mathbf{A} = 0$, in an arbitrary frame as $f(A) = [\partial - (n \cdot \partial)n] \cdot A = 0$, where $n^\mu = (1, 0, 0, 0)$ in the quantization frame. As we did in the R_ξ and axial gauges we can generalize this condition to $f(A) - c = [\partial - (n \cdot \partial)n] \cdot A - c = 0$ and then perform a Gaussian weighted average over the function $c(x)$ with width ξ. The arguments leading to Eq. (6.4.89) remain valid. The Faddeev-Popov matrix M_f is again independent of A^μ since $f(A)$ is first order in A. So we arrive at the gauge-fixed action

$$S_\xi[A] \equiv \int d^4x \left\{ -(1/4)F_{\mu\nu}F^{\mu\nu} - (1/2\xi)([\partial - (n \cdot \partial)n] \cdot A)^2 \right\} \tag{6.4.108}$$
$$= \int d^4x \, d^4y \, \tfrac{1}{2} A_\mu(x) K^{\mu\nu}(x,y) A_\nu(y), \qquad \text{where}$$
$$K^{\mu\nu}(x,y) \equiv \left\{ g^{\mu\nu}\partial_x^2 - \partial_x^\mu \partial_x^\nu - \tfrac{1}{\xi}[\partial - (n \cdot \partial)n]^\mu [\partial - (n \cdot \partial)n]^\nu \right\} \delta^4(x-y).$$

In momentum space $K^{\mu\nu}(k) = -g^{\mu\nu}k^2 + k^\mu k^\nu - \frac{1}{\xi}n(k)^\mu n(k)^\nu$, where we have defined $n(k)^\mu \equiv -i[k^\mu - (n \cdot k)n^\mu]$ since $i\partial^\mu \to k^\mu$. Replacing n^μ with this $n(k)^\mu$ in the axial gauge result in Eq. (6.4.105) gives the generalized Coulomb gauge propagator,

$$D_F^{\mu\nu}(x-y) = \int \frac{d^4k}{(2\pi)^4} \frac{ie^{-ik \cdot (x-y)}}{k^2 + i\epsilon} \left(-g^{\mu\nu} - \frac{[n(k)^2 + \xi k^2]k^\mu k^\nu}{[k \cdot n(k)]^2} + \frac{k^\mu n(k)^\nu + k^\nu n(k)^\mu}{k \cdot n(k)} \right).$$

Taking the Coulomb gauge limit, $\xi \to 0$, and using $n^2 = 1$ leads to

$$D_F^{\mu\nu}(x-y) = \int \frac{d^4k}{(2\pi)^4} \frac{ie^{-ik \cdot (x-y)}}{k^2 + i\epsilon} \left(-g^{\mu\nu} - \frac{k^\mu k^\nu - (k \cdot n)(k^\mu n^\nu + k^\nu n^\mu)}{(k \cdot n)^2 - k^2} \right), \tag{6.4.109}$$

reproducing the Coulomb gauge Feynman propagator in Eqs. (6.4.78) and 6.4.82).

6.4.5 Covariant Canonical Quantization for Photons

To close our discussion of the free photon field we note that a covariant formulation of the canonical quantization of the electromagnetic field is possible. This approach was developed by Gupta (1950) and Bleuler (1950). This leads to additional complications, including the appearance of negative norm states and care is required to define the physical Hilbert space to contain only positive norm states. Discussions of the Gupta-Bleuler formalism are given in Bogolyubov and Shirkov (1959), Schweber (1961), Itzykson and Zuber (1980) and Greiner and Reinhardt (1996). The key to the Gupta-Bleuler approach is to add a term to the Langrangian that keeps the classical system nonsingular with this addition becoming irrelevant once the full Hilbert space, V, is restricted to the physical Hilbert space, $V \to V_{\text{Phys}}$.

We start with the classical field theory of a modified Maxwell Lagrangian density,[7]

$$\mathcal{L}_\xi = \mathcal{L}_{\text{em}} - \tfrac{1}{2\xi}(\partial_\mu A^\mu)^2 = -\tfrac{1}{4}F_{\mu\nu}F^{\mu\nu} - \tfrac{1}{2\xi}(\partial_\mu A^\mu)^2, \tag{6.4.110}$$

where ξ is a free parameter. We recognize that this is the same as the Lagrangian density that we use in the functional integral in the arbitrary covariant gauges in Eq. (6.4.91), where ξ is the R_ξ gauge parameter. Generalizing the derivation of Eq. (3.3.3) gives the equations of motion

$$\partial_\mu F^{\mu\nu} + \tfrac{1}{\xi}\partial^\nu(\partial_\mu A^\mu) = \Box A^\nu - (1 - \tfrac{1}{\xi})\partial^\nu(\partial_\mu A^\mu) = 0. \tag{6.4.111}$$

It is convenient at this point to make the choice $\xi = 1$, which corresponds to Feynman gauge, which we recognize from Eq. (6.4.91). This gives the equation of motion $\Box A^\mu(x) = 0$, which is four massless Klein-Gordon equations and

$$\mathcal{L}_{\xi=1} = \mathcal{L}_{\text{em}} - \tfrac{1}{2}(\partial_\mu A^\mu)^2 = -\tfrac{1}{2}\partial_\mu A_\nu \partial^\mu A^\nu + \tfrac{1}{2}\partial_\mu A_\nu \partial^\nu A^\mu - \tfrac{1}{2}(\partial_\mu A^\mu)^2 \tag{6.4.112}$$

$$= -\tfrac{1}{2}\partial_\mu A_\nu \partial^\mu A^\nu + \tfrac{1}{2}\partial_\mu[A_\nu(\partial^\nu A^\mu) - (\partial_\nu A^\nu)A^\mu]. \tag{6.4.113}$$

As known from Eq. (3.1.11) a four-divergence in a Lagrangian density does not change the equations of motion and can be removed. Omitting this gives

$$\mathcal{L}_{\xi=1} \to \mathcal{L}' = -\tfrac{1}{2}\partial_\mu A_\nu \partial^\mu A^\nu. \tag{6.4.114}$$

The Euler-Lagrange equations for $\mathcal{L}'$ are again $\Box A^\mu = 0$ as expected. The canonical momentum densities corresponding to the A^μ are

$$\pi_\mu = \frac{\partial \mathcal{L}'}{\partial(\partial_0 A^\mu)} = -\partial^0 A_\mu = -\dot{A}_\mu \tag{6.4.115}$$

and none vanishes so the system is nonsingular. From Eq. (3.3.31) the corresponding Hamiltonian density is

$$\mathcal{H}' = \pi^\mu \partial_0 A_\mu - \mathcal{L}' = -\pi^\mu \pi_\mu + \tfrac{1}{2}\partial_\mu A_\nu \partial^\mu A^\nu = -\tfrac{1}{2}\pi^\mu \pi_\mu + \tfrac{1}{2}\partial_k A_\nu \partial^k A^\nu \tag{6.4.116}$$

$$= \tfrac{1}{2}\sum_{k=1}^3 \left[(\dot{A}^k)^2 + (\boldsymbol{\nabla}A^k)^2\right] - \tfrac{1}{2}\left[(\dot{A}^0)^2 + (\boldsymbol{\nabla}A^0)^2\right].$$

This has the form of four harmonic oscillator Hamiltonian densities except that the one for A^0 has the wrong sign and so the classical energy is unbounded below. This problem is resolved below when the Lorenz gauge condition is imposed. The canonical energy-momentum tensor of Eq. (3.3.7) becomes

$$T^\mu{}_\nu(x) = \frac{\partial \mathcal{L}'}{\partial(\partial_\mu A^\tau)}\partial_\nu A^\tau - \delta^\mu{}_\nu \mathcal{L}' = -\partial^\mu A_\tau \partial_\nu A^\tau + \delta^\mu{}_\nu \tfrac{1}{2}\partial_\sigma A_\tau \partial^\sigma A^\tau \tag{6.4.117}$$

with $\partial_\mu T^{\mu\nu} = -(\Box A_\tau)\partial^\nu A^\tau - (\partial^\mu A_\tau)(\partial_\mu \partial^\nu A^\tau) + (\partial^\nu \partial_\sigma A^\tau)(\partial^\sigma A^\tau) = 0$ as expected since $\Box A^\mu = 0$. Note that $T^{\mu\nu}$ is already symmetric and needs no improvement. The Hamiltonian density $\mathcal{H}' = T^{00}$ can be reexpressed as

$$\mathcal{H}'' = \tfrac{1}{2}(\partial^0 \mathbf{A} + \boldsymbol{\nabla}A^0)^2 + \tfrac{1}{2}(\boldsymbol{\nabla} \times \mathbf{A})^2 = \tfrac{1}{2}(\mathbf{E}^2 + \mathbf{B}^2), \tag{6.4.118}$$

where $\mathcal{H}'$ and $\mathcal{H}''$ are equivalent *provided that* (i) we impose the Lorenz gauge condition $\partial_\mu A^\mu = \dot{A}^0 + \boldsymbol{\nabla} \cdot \mathbf{A} = 0$; (ii) we use the equation of motion $\Box A^\mu = 0$; and (iii) we neglect terms that are a total three-divergence since they will not contribute to the spatial integration in $H = \int d^3x\, \mathcal{H}''$ assuming vanishing spatial boundary conditions. The proof is left as an exercise here but a proof can also be found in Greiner and Reinhardt (1996). The imposition of the Lorenz gauge condition

---

[7] Note that $\lambda \equiv 1/\xi$ and $\zeta = 1/\xi$ are used in Itzykson and Zuber (1980) and Greiner and Reinhardt (1996), respectively.

has ensured that $\mathcal{H}''$ is positive and is the usual energy density of the electromagnetic field. The momentum density is $p^i_{\text{dens}} = T^{0i} = -\partial^0 A^\tau \partial^i A_\tau$ so that

$$\mathbf{p}_{\text{dens}} = \dot{A}^\mu \boldsymbol{\nabla} A_\mu = \dot{A}^0 \boldsymbol{\nabla} A^0 - \sum_{j=1}^{3} \dot{A}^j \boldsymbol{\nabla} A^j, \tag{6.4.119}$$

which we will later see becomes the usual photon momentum density after imposing Lorenz gauge in the quantum theory.

Recalling that π_μ are the canonical momenta for the A^μ, then the fundamental Poisson brackets in Eqs. (3.1.26) and (3.3.38) apply,

$$\{A^\mu(x), A^\nu(y)\}_{x^0=y^0} = \{\pi^\mu(x), \pi^\nu(y)\}_{x^0=y^0} = 0 \quad \text{and}$$
$$\{A^\mu(x), \pi_\nu(y)\}_{x^0=y^0} = \delta^\mu{}_\nu \delta^3(\mathbf{x} - \mathbf{y}). \tag{6.4.120}$$

Since the system is nonsingular the simple canonical quantization approach in Eq. (2.9.135) is appropriate and so

$$[\hat{A}^\mu(x), \hat{A}^\nu(y)]_{x^0=y^0} = [\hat{\pi}^\mu(x), \hat{\pi}^\nu(y)]_{x^0=y^0} = 0 \quad \text{and}$$
$$[\hat{A}^\mu(x), \hat{\pi}^\nu(y)]_{x^0=y^0} = ig^{\mu\nu} \delta^3(\mathbf{x} - \mathbf{y}). \tag{6.4.121}$$

Taking spatial derivatives of $[\hat{A}^\mu(x), \hat{A}^\nu(y)]_{x^0=y^0} = 0$ shows that spatial derivatives commute at equal times, $[\partial_i \hat{A}^\mu(x), \hat{A}^\nu(y)]_{x^0=y^0} = [\partial_i \hat{A}^\mu(x), \partial_j \hat{A}^\nu(y)]_{x^0=y^0} = 0$. The same applies for the $\hat{\pi}^\mu$. Since $\hat{\pi}^\mu = -\partial_0 \hat{A}^\mu$, then Eq. (6.4.121) becomes

$$[\hat{A}^\mu(x), \hat{A}^\nu(y)]_{x^0=y^0} = [\dot{\hat{A}}^\mu(x), \dot{\hat{A}}^\nu(y)]_{x^0=y^0} = 0 \quad \text{and}$$
$$[\hat{A}^\mu(x), \dot{\hat{A}}^\nu(y)]_{x^0=y^0} = -ig^{\mu\nu} \delta^3(\mathbf{x} - \mathbf{y}), \tag{6.4.122}$$

where we note that $[\hat{A}^0(x), \dot{\hat{A}}^0(y)]_{x^0=y^0} = -i\delta^3(\mathbf{x} - \mathbf{y})$ appears to have the "wrong" sign compared with that for the $\hat{A}^i$ and the scalar field in Eq. (6.2.4).

Imposing Lorenz gauge would cancel the added term in Eq. (6.4.110). When we quantized in Coulomb gauge we required Gauss' law $\boldsymbol{\nabla} \cdot \hat{\mathbf{E}} |\Psi\rangle = 0$ and the Coulomb gauge condition $\boldsymbol{\nabla} \cdot \hat{\mathbf{A}} |\Psi\rangle = 0$ for all $|\Psi\rangle \in V_{\text{Phys}}$; i.e., we required the operator identities $\boldsymbol{\nabla} \cdot \hat{\mathbf{E}} = 0$ and $\boldsymbol{\nabla} \cdot \hat{\mathbf{A}} = 0$ in the physical Hilbert space. However, we cannot do the same with the Lorenz gauge-fixing condition, since that leads to conflict with Eq. (6.4.122); i.e., we *cannot* set $\partial_\mu \hat{A}^\mu = 0$ to define the gauge-fixed Hilbert space V_{Lor} since[

$$[\partial_\mu \hat{A}^\mu(x), \hat{A}^\nu(y)]_{x^0=y^0} = [\dot{\hat{A}}^0(x), \hat{A}^\nu(y)]_{x^0=y^0} = ig^{0\nu} \delta^3(\mathbf{x} - \mathbf{y}) \neq 0,$$

where we have used $[\boldsymbol{\nabla} \cdot \hat{\mathbf{A}}(x), \hat{A}^\nu(y)]_{x^0=y^0} = \boldsymbol{\nabla} \cdot [\hat{\mathbf{A}}(x), \hat{A}^\nu(y)]_{x^0=y^0} = 0$. The Gupta-Bleuler approach exploits the linear nature of the Lorenz gauge-fixing condition to impose a weaker condition on the space of physical states,

$$\partial_\mu \hat{A}_\mu^{(+)}(x) |\Psi\rangle = 0 \quad \text{and} \quad \langle \Psi | \partial^\mu \hat{A}_\mu^{(-)}(x) \quad \text{for all} \quad |\Psi\rangle \in V_{\text{Phys}}, \tag{6.4.123}$$

where $\hat{A}_\mu \equiv \hat{A}_\mu^{(+)} + \hat{A}_\mu^{(-)}$ with $\hat{A}_\mu^{(+)}$ and $\hat{A}_\mu^{(-)}$ corresponding to the annihilation (positive frequency) and creation (negative frequency) parts of $\hat{A}_\mu$, respectively. Even though $\partial_\mu \hat{A}^\mu(x) |\Psi\rangle \neq 0$ we easily verify that the matrix element of $\partial_\mu \hat{A}^\mu$ between any two physical states satisfies the Lorenz condition,

$$\langle \Psi' | \partial_\mu \hat{A}^\mu(x) |\Psi\rangle = 0 \quad \text{for all} \quad |\Psi\rangle, |\Psi'\rangle \in V_{\text{Phys}}. \tag{6.4.124}$$

Define $\epsilon^\mu(k, 0) \equiv n^\mu$ and $\epsilon^\mu(k, 3) \equiv \tilde{k}^\mu$, where n^μ and $\tilde{k}^\mu$ are given in Eq. (6.4.52). It is readily verified using Eq. (6.4.55) that orthonormality-like conditions hold,

$$\epsilon(k, \lambda')^* \cdot \epsilon(k, \lambda) = g_{\lambda'\lambda} \quad \text{and} \quad \sum_{\lambda=0}^{3} g_{\lambda\lambda} \epsilon(k, \lambda)^* \cdot \epsilon(k, \lambda) = g_{\mu\nu}. \tag{6.4.125}$$

Here we allow for the possibility that $\lambda = 1, 2$ might refer to the helicity basis where $\epsilon^\mu(k, 1)$ and $\epsilon^\mu(k, 1)$ are complex. The four polarization vectors $\epsilon^\mu(k, 0), \ldots, \epsilon^\mu(k, 3)$ form a basis of four-vectors; i.e., we have simply written Eq. (6.4.55) in a different notation. In any inertial frame for on-shell photons ($k^2 = 0$) we must have

$$k \cdot \epsilon(k, 1) = k \cdot \epsilon(k, 2) = 0 \quad \text{and} \quad k \cdot \epsilon(k, 0) = -k \cdot \epsilon(k, 3) = n \cdot k, \tag{6.4.126}$$

which can be verified using the quantization frame. It is conventional to refer to $\epsilon^\mu(k, 1)$ and $\epsilon^\mu(k, 2)$ as *transverse polarization* vectors and $\epsilon^\mu(k, 0)$ and $\epsilon^\mu(k, 3)$ as *scalar* and *longitudinal polarization* vectors, respectively. Expand the field operator in terms of all possible normal modes *before* imposing the Lorenz gauge condition in Eq. (6.4.123),

$$\hat{A}^\mu(x) = \int \frac{d^3k}{(2\pi)^3} \frac{1}{\sqrt{2\omega_{\mathbf{k}}}} \sum_{\lambda=0}^3 \left(\epsilon^\mu(k, \lambda) \hat{a}_{\mathbf{k}}^\lambda e^{-ik \cdot x} + \epsilon^\mu(k, \lambda)^* \hat{a}_{\mathbf{k}}^{\lambda\dagger} e^{ik \cdot x} \right). \tag{6.4.127}$$

The Lorenz-like gauge condition requires that for any physical state $|\Psi\rangle$,

$$\partial_\mu \hat{A}_\mu^{(+)}(x)|\Psi\rangle = -i \int \frac{d^3k}{(2\pi)^3} \frac{1}{\sqrt{2\omega_{\mathbf{k}}}} e^{-ik \cdot x} \sum_{\lambda=0}^3 k \cdot \epsilon(k, \lambda) \hat{a}_{\mathbf{k}}^\lambda |\Psi\rangle = 0. \tag{6.4.128}$$

Using Eq. (6.4.126) we see that $\sum_{\lambda=0}^3 k \cdot \epsilon(k, \lambda) \hat{a}_{\mathbf{k}}^\lambda = (n \cdot k)(\hat{a}_{\mathbf{k}}^0 - \hat{a}_{\mathbf{k}}^3)$ and since Eq. (6.4.128) is required to be true for every x^μ, then

$$\hat{L}_{\mathbf{k}}|\Psi\rangle \equiv (\hat{a}_{\mathbf{k}}^0 - \hat{a}_{\mathbf{k}}^3)|\Psi\rangle = 0. \tag{6.4.129}$$

The Hermitian conjugate is $\langle\Psi|\hat{L}_{\mathbf{k}}^\dagger = \langle\Psi|(\hat{a}_{\mathbf{k}}^{0\dagger} - \hat{a}_{\mathbf{k}}^{3\dagger}) = 0$. We can write this as

$$\hat{a}_{\mathbf{k}}^0|\Psi\rangle = \hat{a}_{\mathbf{k}}^3|\Psi\rangle \quad \text{and also} \quad \langle\Psi|\hat{a}_{\mathbf{k}}^{0\dagger} = \langle\Psi|\hat{a}_{\mathbf{k}}^{3\dagger} \tag{6.4.130}$$

for every three-momentum $\mathbf{k}$ and every gauge-fixed state $|\Psi\rangle \in V_{\text{Lor}}$. Recall from Eq. (2.7.83) that Lorenz gauge is not an ideal gauge-fixing and the same will be true in the Hilbert space with the Lorenz-like gauge-fixing of Eq. (6.4.123), V_{Lor}. We will later show that there is a whole class of gauge equivalent states in V_{Lor} that correspond to the same physical state; i.e., we have $V_{\text{Phys}} \subset V_{\text{Lor}}$.

Let us follow the steps analogous to those for the scalar field in Eqs. (6.2.15)–(6.2.18). We first evaluate $\hat{\pi}^\mu(x) = -\partial^0 \hat{A}^\mu(x)$ to obtain the analog of Eq. (6.2.15) using Eq. (6.4.127). Then use Eq. (6.4.125) and $g_{\lambda'\lambda} = g_{\lambda\lambda}\delta_{\lambda'\lambda}$ to obtain

$$\hat{a}_{\mathbf{k}}^\lambda = g_{\lambda\lambda} \int d^3x \frac{1}{\sqrt{2\omega_{\mathbf{k}}}} e^{ik \cdot x} \epsilon^\mu(k, \lambda)^* \left[\omega_{\mathbf{k}} \hat{A}_\mu(x) + i\dot{\hat{A}}_\mu(x) \right]. \tag{6.4.131}$$

The Hermitian conjugate gives $\hat{a}_{\mathbf{k}}^{\lambda\dagger}$. It is then relatively straightforward to use Eq. (6.4.121) to show that

$$[\hat{a}_{\mathbf{k}}^\lambda, \hat{a}_{\mathbf{k}'}^{\lambda'}] = [\hat{a}_{\mathbf{k}}^{\lambda\dagger}, \hat{a}_{\mathbf{k}'}^{\lambda'\dagger}] = 0 \quad \text{and} \quad [\hat{a}_{\mathbf{k}}^\lambda, \hat{a}_{\mathbf{k}'}^{\lambda'\dagger}] = -g_{\lambda\lambda'}(2\pi)^3\delta^3(\mathbf{k} - \mathbf{k}'), \tag{6.4.132}$$

where for the last result we used $g_{\lambda\lambda}g_{\lambda'\lambda'}\epsilon(\mathbf{k}, \lambda)^\mu \epsilon(\mathbf{k}, \lambda)^*_\mu = g_{\lambda\lambda}g_{\lambda'\lambda'}g_{\lambda\lambda'} = g_{\lambda\lambda'}$. For $\lambda = 1, 2, 3$ these commutation relations are the same as those that we will obtain for the massive vector field. However, the $\lambda = 0$ case has the opposite, unnatural sign. Using $\mathcal{H}'$ in Eq. (6.4.116) we can define the normal-ordered Hamiltonian,

$$\hat{H} \equiv \tfrac{1}{2} \int d^3x \,{:}\hat{\pi}^\mu \hat{\pi}_\mu + \boldsymbol{\nabla}A^\mu \cdot \boldsymbol{\nabla}\hat{A}_\mu{:} = \tfrac{1}{2} \int d^3x \,{:} \sum_{i=1}^3 [(\dot{\hat{A}}^i)^2 + (\boldsymbol{\nabla}\hat{A}^i)^2] - [(\dot{\hat{A}}^0)^2 + (\boldsymbol{\nabla}\hat{A}^0)^2]{:}$$

$$= \int \frac{d^3k}{(2\pi)^3} E_{\mathbf{k}} \sum_{\lambda=0}^3 (-g_{\lambda\lambda}) \hat{a}_{\mathbf{k}}^{\lambda\dagger} \hat{a}_{\mathbf{k}}^\lambda = \int \frac{d^3k}{(2\pi)^3} E_{\mathbf{k}} \left[\left(\sum_{\lambda=1}^3 \hat{a}_{\mathbf{k}}^{\lambda\dagger} \hat{a}_{\mathbf{k}}^\lambda \right) - \hat{a}_{\mathbf{k}}^{0\dagger} \hat{a}_{\mathbf{k}}^0 \right], \tag{6.4.133}$$

where we obtain this result using $E_{\mathbf{k}} \equiv \omega_{\mathbf{k}} = |\mathbf{k}|^2$ and Eqs. (6.4.132) and (6.4.125). Using Eq. (6.4.119) we similarly find

$$\hat{\mathbf{P}} = \int d^3x\, \hat{\mathbf{p}}_{\text{dens}} \equiv \int d^3x : \dot{\hat{A}}^\mu \boldsymbol{\nabla} \hat{A}_\mu : = -\int d^3x : (\textstyle\sum_{\lambda=1}^3 \dot{\hat{A}}^i \boldsymbol{\nabla} \hat{A}_i) - \dot{\hat{A}}^0 \boldsymbol{\nabla} \hat{A}_0 : \qquad (6.4.134)$$
$$= \int \frac{d^3k}{(2\pi)^3}\, \mathbf{k} \textstyle\sum_{\lambda=0}^3 (-g_{\lambda\lambda}) \hat{a}_{\mathbf{k}}^{\lambda\dagger} \hat{a}_{\mathbf{k}}^\lambda = \int \frac{d^3k}{(2\pi)^3}\, \mathbf{k} \left[\left(\textstyle\sum_{\lambda=1}^3 \hat{a}_{\mathbf{k}}^{\lambda\dagger} \hat{a}_{\mathbf{k}}^\lambda \right) - \hat{a}_{\mathbf{k}}^{0\dagger} \hat{a}_{\mathbf{k}}^0 \right].$$

These results can be summarized as

$$\hat{P}^\mu = \int d^3k\, k^\mu \textstyle\sum_{\lambda=0}^3 \hat{N}_{\mathbf{k}}^\lambda \quad \text{with} \quad \hat{N}_{\mathbf{k}}^\lambda \equiv \int \frac{d^3k}{(2\pi)^3}\, \mathbf{k} \textstyle\sum_{\lambda=0}^3 (-g_{\lambda\lambda}) \hat{a}_{\mathbf{k}}^{\lambda\dagger} \hat{a}_{\mathbf{k}}^\lambda, \qquad (6.4.135)$$

which is identical to our earlier result for $\hat{P}^\mu$ in Eq. (6.4.29) except that now we have scalar and longitudinal polarization contributions, $\lambda = 0$ and 3, in addition to the physical transverse polarizations, $\lambda = 1$ and 2. However, using Eq. (6.4.130) it is clear that $\langle \Psi' | \hat{a}_{\mathbf{k}}^{0\dagger} \hat{a}_{\mathbf{k}}^0 | \Psi \rangle = \langle \Psi' | \hat{a}_{\mathbf{k}}^{3\dagger} \hat{a}_{\mathbf{k}}^3 | \Psi \rangle$ for any two physical states and so the scalar and longitudinal contributions then cancel. So only transverse polarizations ($\lambda = 1, 2$) contribute to expectation values and matrix elements,

$$\langle \Psi' | \hat{P}^\mu | \Psi \rangle = \int d^3k\, k^\mu \textstyle\sum_{\lambda=1}^2 \langle \Psi' | \hat{N}_{\mathbf{k}}^\lambda | \Psi \rangle \quad \text{for all} \quad |\Psi\rangle, |\Psi'\rangle \in V_{\text{Lor}}, \qquad (6.4.136)$$

which effectively recovers Eq. (6.4.29) obtained in Coulomb gauge.

The full Hilbert space V is the Fock space created using the creation operators $\hat{a}_{\mathbf{k}}^0, \ldots, \hat{a}_{\mathbf{k}}^3$ in the usual way. The gauge-fixed subspace $V_{\text{Lor}} \subset V$ consists of states $|\Psi\rangle \in V_{\text{Lor}}$ such that $\hat{L}_{\mathbf{k}} |\Psi\rangle = (\hat{a}_{\mathbf{k}}^0 - \hat{a}_{\mathbf{k}}^3) |\Psi\rangle = 0$ for all $\mathbf{k}$. Let $|\Phi_T\rangle$ be a pure transverse state constructed using only combinations of the $\hat{a}_{\mathbf{k}}^{1\dagger}$ and $\hat{a}_{\mathbf{k}}^{2\dagger}$ creation operators. So $|\Phi_T\rangle$ is in a smaller purely transverse subspace, which is the physical subspace, $|\Phi_t\rangle \in V_{\text{Phys}} \subset V_{\text{Lor}} \subset V$, where V_{Phys} is spanned by basis vectors

$$|0\rangle, \quad \textstyle\prod_{i=1}^{n_1} \hat{a}_{\mathbf{k}_i}^{1\dagger} |0\rangle, \quad \textstyle\prod_{j=1}^{n_2} \hat{a}_{\mathbf{k}_i'}^{2\dagger} |0\rangle, \quad \textstyle\prod_{i=1}^{n_1} \textstyle\prod_{j=1}^{n_2} \hat{a}_{\mathbf{k}_i}^{1\dagger} \hat{a}_{\mathbf{k}_j'}^{2\dagger} |0\rangle \qquad (6.4.137)$$

for $n_1, n_2 = 1, 2, 3, \ldots$ and all possible momenta. The states in V_{Phys} involve only the observable transverse modes and so all $|\Psi_T\rangle \in V_{\text{Phys}}$ are physically distinct. A gauge choice that restricted states to V_{Phys} would be an ideal gauge-fixing.

We have $\hat{L}_{\mathbf{k}} |\Phi_T\rangle = 0$ since for annihilation operators $\hat{a}^\lambda |0\rangle = 0$. The commutation relations in Eq. (6.4.132) lead to $[\hat{L}_{\mathbf{k}}, \hat{L}_{\mathbf{k}'}] = [\hat{L}_{\mathbf{k}}^\dagger, \hat{L}_{\mathbf{k}'}^\dagger] = [\hat{L}_{\mathbf{k}}, \hat{L}_{\mathbf{k}'}^\dagger] = 0$, since

$$[\hat{L}_{\mathbf{k}}, \hat{L}_{\mathbf{k}'}^\dagger] = [\hat{a}_{\mathbf{k}}^0 - \hat{a}_{\mathbf{k}'}^3, \hat{a}_{\mathbf{k}}^{0\dagger} - \hat{a}_{\mathbf{k}'}^{3\dagger}] = [\hat{a}_{\mathbf{k}}^0, \hat{a}_{\mathbf{k}'}^{0\dagger}] + [\hat{a}_{\mathbf{k}}^3, \hat{a}_{\mathbf{k}'}^{3\dagger}] = 0 \qquad (6.4.138)$$

and so on. For arbitrary functions $g_1(\mathbf{k}), g_2(\mathbf{k}, \mathbf{k}'), \ldots$ define

$$|\Psi\rangle \equiv \hat{G} |\Phi_T\rangle \equiv \left[1 + \int d^3k\, g_1(\mathbf{k}) \hat{L}_{\mathbf{k}}^\dagger + \int d^3k\, d^3k'\, g_2(\mathbf{k}, \mathbf{k}') \hat{L}_{\mathbf{k}}^\dagger \hat{L}_{\mathbf{k}'}^\dagger + \cdots \right] |\Phi_T\rangle, \qquad (6.4.139)$$

which satisfies $\hat{L}_{\mathbf{k}} |\Psi\rangle = 0$ and so $|\Psi\rangle \in V_{\text{Lor}}$. It follows that $[\hat{G}, \hat{L}_{\mathbf{k}}] = [\hat{G}, \hat{L}_{\mathbf{k}}^\dagger] = [\hat{G}, \hat{G}'] = [\hat{G}^\dagger, \hat{G}'^\dagger] = [\hat{G}^\dagger, \hat{G}'] = 0$. Any such state $|\Psi\rangle$ is in the gauge-fixed subspace V_{Lor} since $\hat{L}_{\mathbf{k}} \hat{G} |\Phi_T\rangle = \hat{G} \hat{L}_{\mathbf{k}} |\Phi_T\rangle = 0$ for all $\mathbf{k}$. Any $\hat{G}$ that contains only the operators $\hat{a}_{\mathbf{k}}^{0\dagger}$ and $\hat{a}_{\mathbf{k}}^{3\dagger}$ and satisfies $[\hat{G}, \hat{L}_{\mathbf{k}}] = 0$ must have the form given in Eq. (6.4.139). (*Proof:* Any function of $\hat{a}_{\mathbf{k}}^{0\dagger}$ and $\hat{a}_{\mathbf{k}}^{3\dagger}$ can be written as a function of $\hat{L}_{\mathbf{k}}^\dagger = \hat{a}_{\mathbf{k}}^{0\dagger} - \hat{a}_{\mathbf{k}}^{3\dagger}$ and $\hat{a}_{\mathbf{k}}^{0\dagger} + \hat{a}_{\mathbf{k}}^{3\dagger}$. Any factor of $\hat{a}_{\mathbf{k}}^{0\dagger} + \hat{a}_{\mathbf{k}}^{3\dagger}$ in $\hat{G}$ would lead to the result $\hat{L}_{\mathbf{k}} |\Psi\rangle = \hat{L}_{\mathbf{k}} \hat{G} |\Phi_T\rangle \neq 0$, since $\hat{L}_{\mathbf{k}}$ and $\hat{a}_{\mathbf{k}}^{0\dagger} + \hat{a}_{\mathbf{k}}^{3\dagger}$ do not commute.) Then any $|\Psi\rangle \in V_{\text{Lor}}$ can be written as $\hat{G} |\Phi_T\rangle$ for some such $\hat{G}$ and some $|\Phi_T\rangle$. So there is a whole class of states $|\Psi\rangle \in V_{\text{Lor}}$ that correspond to the same pure transverse physical state $|\Phi_T\rangle$. The Lorenz-like gauge is not an ideal gauge-fixing as we expected. Let $|\Psi\rangle, |\Psi'\rangle \in V_{\text{Lor}}$ have pure transverse parts $|\Phi_T\rangle, |\Phi'_T\rangle \in V_{\text{Lor}}$, respectively. Then *only the pure transverse parts can contribute to physical amplitudes* since

$$\langle \Psi' | \Psi \rangle = \langle \Phi'_T | \hat{G}^{\dagger'} \hat{G} | \Phi_T \rangle = \langle \Phi'_T | \hat{G} \hat{G}^{\dagger'} | \Phi_T \rangle = \langle \Phi'_T | \Phi_T \rangle, \qquad (6.4.140)$$

where $\hat{G}^{\dagger'} | \Phi_T \rangle = | \Phi_T \rangle$ and $\langle \Phi'_T | \hat{G} = \langle \Phi'_T |$ since $\hat{L}_{\mathbf{k}} | \Phi_T \rangle = 0$ and $\langle 0 | \hat{L}^{\dagger}_{\mathbf{k}} = 0$, respectively. An observable operator $\hat{O}$ is gauge invariant, which means that if $|\Psi\rangle$ satisfies some gauge condition, then so must $\hat{O} | \Psi \rangle$. Hence if $|\Psi\rangle \in V_{\text{Lor}}$, then $\hat{O} | \Psi \rangle \in V_{\text{Lor}}$. So the scalar and longitudinal photons are unobservable if we work in V_{Lor}. A gauge-fixing that achieved the constraint $V \to V_{\text{Phys}}$ would be an ideal gauge-fixing, whereas the Lorenz-like constraint that we have implemented has led to the partial gauge-fixing $V \to V_{\text{Lor}}$ that will be sufficient for perturbation theory.

We have not yet explained the earlier reference to negative norm states in the full Hilbert space in the Gupta-Bleuler approach. We define one-photon states in the usual way by generalizing Eq. (6.2.46) to give for $\lambda = 0, 1, 2, 3$

$$| \mathbf{k}, \lambda \rangle = \sqrt{2E_{\mathbf{k}}} \hat{a}^{\lambda}_{\mathbf{k}} | 0 \rangle. \qquad (6.4.141)$$

Then using the commutation relations in Eq. (6.4.132) and $\langle 0 | 0 \rangle = 1$ we find

$$\langle \mathbf{k}', \lambda' | \mathbf{k}, \lambda \rangle = \sqrt{4E_{\mathbf{k}} E_{\mathbf{k}'}} \langle 0 | \hat{a}^{\lambda'}_{\mathbf{k}'} \hat{a}^{\lambda \dagger}_{\mathbf{k}} | 0 \rangle = \sqrt{4E_{\mathbf{k}} E_{\mathbf{k}'}} \langle 0 | [\hat{a}^{\lambda'}_{\mathbf{k}'}, \hat{a}^{\lambda \dagger}_{\mathbf{k}}] | 0 \rangle \qquad (6.4.142)$$
$$= (-g_{\lambda' \lambda}) 2E_{\mathbf{k}} (2\pi)^3 \delta^3(\mathbf{k}' - \mathbf{k}).$$

So the states $|\mathbf{k}, 1\rangle, |\mathbf{k}, 2\rangle, |\mathbf{k}, 3\rangle$ have positive norm while the state $|\mathbf{k}, 0\rangle$ has negative norm. As shown, only the purely transverse part of any state contributes to physical amplitudes. Gauge-fixed states have positive norm, $\langle \Psi | \Psi \rangle \geq 0$ for all $|\Psi\rangle \in V_{\text{Lor}}$.

We have the conventional-looking mode expansion for the field operator $\hat{A}^{\mu}(x)$ in Eq. (6.4.127) and for the equal-time field commutators in Eq. (6.4.122) with a factor $(-g^{\mu\nu})$. This in turn leads to conventional-looking commutation relations for annihilation and creation operators with a factor $(-g_{\lambda\lambda'})$. When inserting a complete set of states between two operators it should be with respect to the complete indefinite norm Hilbert space V to preserve locality (Itzykson and Zuber, 1980). Physical matrix elements of operators should be between states in the physical Hilbert space V_{Phys}. So the trace of any physical observable is a trace over states in V_{Phys}, e.g., the spectral function $\text{tr}\, \hat{U}(t'' - t') = \text{tr}(T \exp\{-i \int_{t'}^{t''} dt \hat{H}\})$ where

$$\hat{H} = \tfrac{1}{2} \int d^3x : \sum_{i=1}^{3} [(\dot{\hat{A}}^i)^2 + (\boldsymbol{\nabla} \hat{A}^i)^2] - [(\dot{\hat{A}}^0)^2 + (\boldsymbol{\nabla} \hat{A}^0)^2] : \qquad (6.4.143)$$

from Eq. (6.4.133). The energy is unbounded below in V but is bounded below in V_{Phys} by the vacuum state $|0\rangle \in V_{\text{Phys}}$.

Provided that we keep in mind these points we can reproduce generalized forms of many of the results obtained for the scalar boson field in a straightforward way (Itzykson and Zuber, 1980). For example, using Eq. (6.4.132) we find in analogy with Eqs. (6.2.138) and (6.2.205) that

$$\langle 0 | \hat{A}^{\mu}(x) \hat{A}^{\nu}(y) | 0 \rangle = -g^{\mu\nu} \Delta_+(x - y) = [\hat{A}^{(+)\mu}(x), \hat{A}^{(-)\nu}(y)], \qquad (6.4.144)$$

where we have used Eqs. (6.4.125) and (6.4.132). From Eq. (6.2.160) we find that $D_F(x - y) = \theta(x^0 - y^0) \Delta_+(x - y) + \theta(y^0 - x^0) \Delta_+(y - x)$ and so

$$\langle 0 | T \hat{A}^{\mu}(x) \hat{A}^{\nu}(y) | 0 \rangle = -g^{\mu\nu} D_F(x - y) = D_F^{\mu\nu}(x - y) = \overbrace{\hat{A}^{\mu}(x) \hat{A}^{\nu}(y)}, \qquad (6.4.145)$$

where $D_F^{\mu\nu}$ is in Feynman gauge as expected and where the contraction of two fields is as defined in Eq. (6.2.203). The extra Lorenz-gauge-like condition was only needed to remove the additional degree of freedom that we added. The derivation used for Eq. (6.2.202) again follows and so does the analog of Eq. (6.2.204), which gives

$$T\hat{A}^\mu(x)\hat{A}^\nu(y) = \,:\hat{A}^\mu(x)\hat{A}^\nu(y): + \overline{\hat{A}^\mu(x)\hat{A}^\nu(y)} = \,:\hat{A}^\mu(x)\hat{A}^\nu(y): + D_F^{\mu\nu}(x-y). \quad (6.4.146)$$

The proof of Wick's theorem now applied to the $\hat{A}^\mu(x)$ operators can then proceed as before. The final result is the extension of Eq. (6.2.211) for Feynman gauge,

$$\langle 0|T\hat{A}^{\mu_1}(x_1)\hat{A}^{\mu_2}(x_2)\cdots\hat{A}^{\mu_n}(x_n)|0\rangle = D_F^{\mu_1\mu_2}(x_1-x_2)\cdots D_F^{\mu_{n-1}\mu_n}(x_{n-1}-x_n)$$

$$+ \text{ all pairwise combinations of } \{(\mu_1,x_1),(\mu_2,x_2),\ldots,(\mu_n,x_n)\} \quad (6.4.147)$$

$$= \left.\frac{(i)^k\,\delta^k\,Z_{\xi=1}[j]}{\delta j_{\mu_1}(x_1)\cdots\delta j_{\mu_k}(x_n)}\right|_{j=0} = \lim_{T\to\infty(1-i\epsilon)}\frac{\int \mathcal{D}A_{\text{periodic}}^\mu\,A^{\mu_1}(x_1)\cdots A^{\mu_k}(x_n)\,e^{iS_{\xi=1}[A]}}{\int \mathcal{D}A_{\text{periodic}}^\mu\,e^{iS_{\xi=1}[A]}},$$

which vanishes if n is odd. Since the trace is over the physical states, then the four-vector fields at $\pm T$, $A^\mu(\pm T,\mathbf{x})$, are equal and in the physical space, whereas at all intermediate times the $A^\mu(t,\mathbf{x})$ are in the indefinite metric space. This detail does not affect perturbation theory arguments.

Alternate Derivation of the Covariant Gauge Feynman Propagator

Let us return to $\mathcal{L}_\xi$ in Eq. (6.4.110) and now keep ξ an arbitrary real number. Generalizing the derivation of Eq. (3.3.3) gives the equations of motion

$$\partial_\mu F^{\mu\nu} + \tfrac{1}{\xi}\partial^\nu(\partial_\mu A^\mu) = \Box A^\nu - (1-\tfrac{1}{\xi})\partial^\nu(\partial_\mu A^\mu) \quad (6.4.148)$$

$$= [\Box g^{\nu\mu} - (1-\tfrac{1}{\xi})\partial^\nu\partial^\mu]A_\mu = 0,$$

which reproduces Eq. (6.4.111). In place of Eq. (3.3.26) the four generalized momenta are now

$$\pi^\nu(x) \equiv \frac{\partial\mathcal{L}_\xi}{\partial(\partial_0 A_\nu)(x)} = -F^{0\nu} - \tfrac{1}{\xi}g^{0\nu}(\partial_\mu A^\mu). \quad (6.4.149)$$

Recalling that π_μ are the canonical momenta for the A^μ, then the fundamental Poisson brackets are as in Eq. (3.3.38),

$$\{A^\mu(x),A^\nu(y)\}_{x^0=y^0} = \{\pi^\mu(x),\pi^\nu(y)\}_{x^0=y^0} = 0 \quad \text{and}$$

$$\{A^\mu(x),\pi_\nu(y)\}_{x^0=y^0} = \delta^\mu{}_\nu\delta^3(\mathbf{x}-\mathbf{y}). \quad (6.4.150)$$

Since the system is nonsingular the simple canonical quantization approach in Eq. (2.9.135) is appropriate and this gives

$$[\hat{A}^\mu(x),\hat{A}^\nu(y)]_{x^0=y^0} = [\hat{\pi}^\mu(x),\hat{\pi}^\nu(y)]_{x^0=y^0} = 0 \quad \text{and}$$

$$[\hat{A}^\mu(x),\hat{\pi}^\nu(y)]_{x^0=y^0} = ig^{\mu\nu}\delta^3(\mathbf{x}-\mathbf{y}). \quad (6.4.151)$$

Taking spatial derivatives of $[\hat{A}^\mu(x),\hat{A}^\nu(y)]_{x^0=y^0} = 0$ shows that spatial derivatives commute at equal times, $[\partial_i\hat{A}^\mu(x),\hat{A}^\nu(y)]_{x^0=y^0} = [\partial_i\hat{A}^\mu(x),\partial_j\hat{A}^\nu(y)]_{x^0=y^0} = 0$. The same applies for the $\hat{\pi}^\mu$. We have $\hat{\pi}^0 = -\tfrac{1}{\xi}\partial_\rho\hat{A}^\rho$ and $\hat{\pi}^i = -\partial^0\hat{A}^i + \partial^i\hat{A}^0$. With some effort we can verify the following commutation relations,

$$[\hat{A}^\mu(x),\hat{A}^\nu(y)]_{x^0=y^0} = [\dot{\hat{A}}^0(x),\dot{\hat{A}}^0(y)]_{x^0=y^0} = [\dot{\hat{A}}^i(x),\dot{\hat{A}}^j(y)]_{x^0=y^0} = 0, \quad (6.4.152)$$

$$[\dot{\hat{A}}^\mu(x),\hat{A}^\nu(y)]_{x^0=y^0} = ig^{\mu\nu}[1+(\xi-1)g_{\mu 0}]\delta^3(\mathbf{x}-\mathbf{y}),$$

$$[\dot{\hat{A}}^0(x),\hat{A}_j(y)]_{x^0=y^0} = i(1-\xi)\frac{\partial}{\partial x^j}\delta^3(\mathbf{x}-\mathbf{y}).$$

Sketch proof: We use an obvious shorthand here. Making use of the commuting of spatial derivatives we use $\hat{\pi}^0$ to access $-\frac{1}{\xi}\dot{\hat{A}}^0$ and $\hat{\pi}^i$ to access $-\dot{\hat{A}}^i$.

(i) Prove the second line on a case-by-case basis: Consider $[\hat{A}^0, \hat{\pi}^0]$ and $[\hat{A}^0, \hat{\pi}^j]$ and so verify the result for $[\dot{\hat{A}}^\mu, \hat{A}^0]$. Then consider $[\hat{A}^j, \hat{\pi}^0]$ and $[\hat{A}^j, \hat{\pi}^k]$ and verify the result for $[\dot{\hat{A}}^\mu, \hat{A}^j]$.

(ii) Then prove the second and third results on the first line: First note from the second line that $[\dot{\hat{A}}^\mu, \hat{A}^\nu] = 0$ if $\mu \neq \nu$ and hence also $[\dot{\hat{A}}^\mu, \partial^j \hat{A}^\nu] = 0$ if $\mu \neq \nu$. We find the results by first considering $[\hat{\pi}^0, \hat{\pi}^0]$ and then $[\hat{\pi}^i, \hat{\pi}^j]$. (iii) The last line follows from $0 = [\hat{\pi}^0, \hat{\pi}_j]$: First use this to show that $[\dot{\hat{A}}^0, \dot{\hat{A}}_j] = -[\partial_k \hat{A}^k, \dot{\hat{A}}_j] + [\dot{\hat{A}}^0, \partial_j \hat{A}^0]$. Evaluate $[\hat{A}^k, \dot{\hat{A}}_j]$ and $[\dot{\hat{A}}^0, \hat{A}^0]$ use the second line and add spatial derivatives to obtain the last line.

The equations of motion are true by construction, $[\Box g^{\nu\mu} - (1 - \frac{1}{\xi})\partial^\nu \partial^\mu]\hat{A}_\mu = 0$, which follows from the canonical quantization formalism. We know from Eq. (6.4.97) that the Green's function satisfying the Feynman boundary conditions for the equation of motion in Eq. (6.4.111) is the covariant gauge Feynman propagator, $D_F^{\mu\nu}(x-y)$, given in Eq. (6.4.96),

$$[\Box_x g^{\mu\rho} - (1 - \tfrac{1}{\xi})\partial_x^\mu \partial_x^\rho]D_{F\rho\nu}(x-y) = i\delta^\mu{}_\nu \delta^4(x-y). \tag{6.4.153}$$

It follows that

$$D_F^{\mu\nu}(x-y) = \langle 0|T\hat{A}^\mu(x)\hat{A}^\nu(y)|0\rangle \tag{6.4.154}$$

$$-\int \frac{d^4k}{(2\pi)^4} \frac{-i[g^{\mu\nu} - (1-\xi)\frac{k^\mu k^\nu}{k^2+i\epsilon}]}{k^2 + i\epsilon} e^{-ik\cdot(x-y)} = \int \frac{d^4k}{(2\pi)^4} D_F^{\mu\nu}(k)e^{-ik\cdot(x-y)}.$$

Proof: We first need to show that

$$[\Box_x g^{\sigma\mu} - (1 - \tfrac{1}{\xi})\partial_x^\sigma \partial_x^\mu]\langle 0|T\hat{A}_\mu(x)\hat{A}_\nu(y)|0\rangle = i\delta^\sigma{}_\nu \delta^4(x-y) \tag{6.4.155}$$

so that we can make the identification $D_F^{\mu\nu}(x-y) = \langle 0|T\hat{A}^\mu(x)\hat{A}^\nu(y)|0\rangle$. There is a problem if we attempt to derive this result using annihilation and creation operators using a mode expansion for A^μ, since for $\xi \neq 1$ the simple commutation relations in Eq. (6.4.132) do not result from Eq. (6.4.152). So Fock space cannot be constructed in the usual way (Greiner and Reinhardt, 1996). However, the result follows from Eq. (6.4.152) directly. Note that:

(i) $\partial_x^0 T\hat{A}_\mu(x)\hat{A}_\nu(y) = \partial_0^x\left(\theta(x^0 - y^0)\hat{A}_\mu(x)\hat{A}_\nu(y) + \theta(y^0 - x^0)\hat{A}_\nu(y)\hat{A}_\mu(x)\right)$

$\qquad = \delta(x^0 - y^0)[\hat{A}_\mu(x), \hat{A}_\nu(y)] + \theta(x^0 - y^0)\dot{\hat{A}}_\mu(x)\hat{A}_\nu(y) + \theta(y^0 - x^0)\hat{A}_\nu(y)\dot{\hat{A}}_\mu(x)$

$\qquad = \theta(x^0 - y^0)\dot{\hat{A}}_\mu(x)\hat{A}_\nu(y) + \theta(y^0 - x^0)\hat{A}_\nu(y)\dot{\hat{A}}_\mu(x)$

(ii) $\partial_x^0 \partial_x^0 T\hat{A}_\mu(x)\hat{A}_\nu(y) = \delta(x^0 - y^0)[\dot{\hat{A}}_\mu(x), \hat{A}_\nu(y)] + \theta(x^0 - y^0)\ddot{\hat{A}}_\mu(x)\hat{A}_\nu(y)$

$\qquad\qquad + \theta(y^0 - x^0)\hat{A}_\nu(y)\ddot{\hat{A}}_\mu(x)$

(iii) $\partial_x^\rho T\hat{A}_\mu(x)\hat{A}_\nu(y) = \theta(x^0 - y^0)\partial_x^\rho \hat{A}_\mu(x)\hat{A}_\nu(y) + \theta(y^0 - x^0)\hat{A}_\nu(y)\partial_x^\rho \hat{A}_\mu(x)$

(iv) $\partial_x^\sigma \partial_x^\rho T\hat{A}_\mu(x)\hat{A}_\nu(y) = g^{\sigma 0}\delta(x^0 - y^0)g^{\rho 0}[\dot{\hat{A}}_\mu(x), \hat{A}_\nu(y)]$

$\qquad\qquad + \theta(x^0 - y^0)\partial_x^\sigma \partial_x^\rho \hat{A}_\mu(x)\hat{A}_\nu(y) + \theta(y^0 - x^0)\hat{A}_\nu(y)\partial_x^\sigma \partial_x^\rho \hat{A}_\mu(x),$

where we used $[\partial_x^j \hat{A}_\mu(x), \hat{A}_\nu(y)]_{x^0=y^0} = \partial_x^j [\hat{A}_j(x), \hat{A}_\nu(y)]_{x^0=y^0} = 0$ so that $[\partial_x^\rho \hat{A}_\mu(x),$ $\hat{A}_\nu(y)]_{x^0=y^0} = g^{\rho 0}[\hat{A}_\mu(x), \hat{A}_\nu(y)]_{x^0=y^0}$ in the last result. Combining these results and using $[\Box g^{\sigma\mu} - (1 - \frac{1}{\xi})\partial^\sigma \partial^\mu]\hat{A}_\mu(x) = 0$ we find

$$[\Box_x g^{\sigma\mu} - (1 - \tfrac{1}{\xi})\partial_x^\sigma \partial_x^\mu] T \hat{A}_\mu(x) \hat{A}_\nu(y) \tag{6.4.156}$$

$$= \delta(x^0 - y^0)([\dot{\hat{A}}^\sigma(x), \hat{A}_\nu(y)] + (\tfrac{1}{\xi} - 1)g^{\sigma 0}[\dot{\hat{A}}_0(x), \hat{A}_\nu(y)]) = i\delta^\sigma_{\ \nu}\delta^4(x - y),$$

where we have used $[\dot{\hat{A}}_\mu(x), \hat{A}_\nu(y)]_{x^0=y^0} = ig_{\mu\nu}[1 + (\xi - 1)g^{\mu 0}]\delta^3(\mathbf{x} - \mathbf{y})$ to evaluate the middle part of the above equation,

$$i\delta^4(x - y)[\delta^\sigma_{\ \nu} + \delta^\sigma_{\ \nu}(\xi - 1)g_{\sigma 0} + (\tfrac{1}{\xi} - 1)g^{\sigma 0}g_{0\nu}(1 + \xi - 1)] = i\delta^\sigma_{\ \nu}\delta^4(x - y).$$

Time-ordering is Lorentz covariant (with no time reversal) and since we have an entirely covariant formulation, then the left side of Eq. (6.4.156) must be covariant. So the noncovariant terms on the right had to cancel.

We can relate the derivation of the covariant gauge Feynman propagator in Eq. (6.4.154) to the functional integral derivation leading to Eq. (6.4.96). Since physical quantities must be independent of gauge we can extend the above discussion to the more general gauge choice $\partial_\mu A^\mu(x) - c(x) = 0$, where $c(x)$ is some real scalar field. Then we generalize the Lagrangian in Eq. (6.4.110) to

$$\mathcal{L}_\xi = \mathcal{L}_{em} - \tfrac{1}{2\xi}[\partial_\mu A^\mu - c(x)]^2 = -\tfrac{1}{4}F_{\mu\nu}F^{\mu\nu} - \tfrac{1}{2\xi}(\partial_\mu A^\mu - c)^2. \tag{6.4.157}$$

It is straightforward to see that now $\pi^\nu = -F^{0\nu} - \tfrac{1}{\xi}g^{\nu 0}(\partial_\mu A^\mu - c)$ and the equations of motion become $\Box A^\nu - (1 - \tfrac{1}{\xi})\partial^\nu \partial_\mu A^\mu - \tfrac{1}{\xi}\partial^\nu c = 0$. Since the system remains non singular, then canonical quantization based on the Poisson brackets of Eq. (6.4.120) is again sufficient and leads to Eq. (6.4.151). Since $\hat{\pi}^\nu(x) = -\hat{F}^{0\nu} - \tfrac{1}{\xi}g^{\nu 0}(\partial_\mu \hat{A}^\mu - c)$ with $c(x)$ remaining a real scalar function, then $c(x)$ disappears from the commutation relations and we again arrive at Eq. (6.4.151). The operator equations of motion are $[\Box g^{\nu\mu} - (1 - \tfrac{1}{\xi})\partial^\nu \partial^\mu]\hat{A}_\mu - \tfrac{1}{\xi}\partial^\nu c = 0$. Since no physical observable can depend on the gauge choice we can always do a Gaussian-weighted average over all functions $c(x)$ centered around $c = 0$ just as we did in the functional approach. Here the width of the Gaussian is arbitrary, but we are free to choose it to be ξ. The Gaussian average of $\partial^\nu c$ vanishes, resulting in the operator equation of motion $\Box A^\nu - (1 - \tfrac{1}{\xi})\partial^\nu \partial_\mu A^\mu = 0$ as before. Then the proof of Eqs. (6.4.156) and (6.4.153) follow as before and we see the analogy between the functional and covariant canonical approaches. This raises the question of whether we could derive the covariant gauge functional integral directly from the covariant canonical quantization. We do not attempt this here. First, the Gupta-Bleuler constraint is not a simple operator identity but a constraint on the physical space in terms of creation and annihilation operators. Second, the imposition of constraints in a Hamiltonian derivation of a functional integral can be more complicated than was found in Coulomb gauge, e.g., if second-class constraints are present (Senjanovic, 1976).

6.5 Massive Vector Bosons

Massive vector bosons are relevant as they arise in the Standard Model following the Higgs mechanism and as vector meson bound states in QCD such as the ρ and ω mesons. Interacting theories with elementary massive vector bosons lead to violations of unitarity, which are unphysical. While they do not appear as fundamental particles, it will be still useful to develop the formalism to treat such particles (Greiner and Reinhardt, 1996) since they appear as effective degrees of freedom after the Higgs mechanism in the Standard Model.

6.5.1 Classical Massive Vector Field

To construct the classical Lagrangian density for a massive vector field we simply add a mass term to the electromagnetism Lagrangian density,

$$\mathcal{L} = -\tfrac{1}{4} F_{\mu\nu} F^{\mu\nu} + \tfrac{1}{2} m^2 A_\mu A^\mu - j_\mu A^\mu, \quad \text{with} \quad F^{\mu\nu} \equiv \partial^\mu A^\nu - \partial^\nu A^\mu, \tag{6.5.1}$$

where the sign is such that for the spatial components we have $\tfrac{1}{2} m^2 \mathbf{A}^2$ as for the scalar field. We see that $\mathcal{L}$ is no longer invariant under a gauge transformation, $A^\mu \to A^\mu + \partial^\mu \omega$, since obviously the mass term is not gauge invariant. As for the scalar boson the field $A^\mu(x)$ is real, which corresponds to a neutral massive vector field. For a charged/complex massive vector field we have in analogy with the charged scalar in Eq. (3.2.48),

$$\mathcal{L} = -\tfrac{1}{2} F_{\mu\nu}^* F^{\mu\nu} + m^2 A_\mu^* A^\mu - j_\mu^* A^\mu - j^\mu A_\mu^*. \tag{6.5.2}$$

For simplicity we deal here with the neutral case, but the extension of the following to the charged case is straightforward following similar steps used in going from the neutral to the charged scalar field. It is useful to continue to use $\mathbf{E}$ and $\mathbf{B}$ notation purely for notational convenience. They are *not* electromagnetic fields. We define

$$E^i \equiv F^{i0} \quad \text{and} \quad \mathbf{B} \equiv \mathbf{\nabla} \times \mathbf{A}. \tag{6.5.3}$$

The Euler-Lagrange equations of motion are obtained by modifying the derivation in Sec. 3.3.1 to include a mass term and we arrive at

$$\partial_\mu F^{\mu\nu} + m^2 A^\nu = j^\nu. \tag{6.5.4}$$

In terms of the field A^μ this has the form

$$\Box A^\nu - \partial^\nu (\partial_\mu A^\mu) + m^2 A^\nu = j^\nu, \tag{6.5.5}$$

which is commonly referred to as the *Proca equation*. Taking the four-divergence of the Proca equation gives

$$\partial_\mu A^\mu = \partial_\mu j^\mu / m^2. \tag{6.5.6}$$

By definition in our discussion here $m \in \mathbb{R}$ and so $m^2 > 0$. We will only consider the case of coupling to a conserved four-vector current, $\partial_\mu j^\mu = 0$. In electromagnetism it was only possible to couple to a conserved current, but here it is a simplifying choice sufficient for our needs. So the equations of motion mean that we must automatically satisfy the Lorenz-gauge type transversality condition,

$$\partial_\mu A^\mu = 0 \tag{6.5.7}$$

and the equation of motion for each component A^μ is a Klein-Gordon equation

$$(\Box + m^2) A^\mu = (\partial^2 + m^2) A^\mu = j^\mu, \tag{6.5.8}$$

subject to the constraint in Eq. (6.5.7). Only three of the fields are independent. If we specifiy $\mathbf{A}$ and $\mathbf{E}$, then A^0 is automatically known from the Proca equation,

$$\partial_i F^{i0} + m^2 A^0 - j^0 = \mathbf{\nabla} \cdot \mathbf{E} + m^2 A^0 - j^0 = 0 \quad \Rightarrow \quad A^0 = \frac{-\mathbf{\nabla} \cdot \mathbf{E} + j^0}{m^2}. \tag{6.5.9}$$

For a free field we construct the canonical stress-energy tensor in the usual way and find in place of Eq. (3.3.7),

$$T^{\mu\nu} = -F^{\mu\tau} \partial^\nu A_\tau + \tfrac{1}{4} g^{\mu\nu} F^{\sigma\tau} F_{\sigma\tau} - \tfrac{1}{2} m^2 g^{\mu\nu} A_\sigma A^\sigma = T_{\text{em}}^{\mu\nu} - \tfrac{1}{2} m^2 g^{\mu\nu} A_\sigma A^\sigma, \tag{6.5.10}$$

where the mass term is absent in the electromagnetic case. We easily see that $R^\mu{}_{\rho\sigma} = F^\mu{}_\sigma A_\rho - F^\mu{}_\rho A_\sigma$ from Eq. (3.3.18) is unchanged and so we again find that $K^{\mu\rho\sigma} = F^{\rho\mu} A^\sigma$ and that the improvement term is $\partial_\tau K^{\tau\mu\nu}$ as in Eq. (3.3.9). Then from Eqs. (3.2.111) and (3.3.9) the symmetric stress-energy tensor is given by

$$\begin{aligned} \bar{T}^{\mu\nu} &= T^{\mu\nu} + \partial_\tau (F^{\mu\tau} A^\nu) = T^{\mu\nu} + (\partial_\tau F^{\mu\tau}) A^\nu + F^{\mu\tau} \partial_\tau A^\nu \\ &= -F^{\mu\tau} F^\nu{}_\tau + \tfrac{1}{4} g^{\mu\nu} F^{\sigma\tau} F_{\sigma\tau} - \tfrac{1}{2} m^2 g^{\mu\nu} A_\sigma A^\sigma + m^2 A^\mu A^\nu \\ &= \bar{T}_{\text{em}}^{\mu\nu} - \tfrac{1}{2} m^2 g^{\mu\nu} A_\sigma A^\sigma + m^2 A^\mu A^\nu, \end{aligned} \tag{6.5.11}$$

where we have used the equations of motion, Eq. (6.5.4). In place of Eq. (3.3.15) we find for the momentum four-vector,

$$\begin{aligned} P^0 &= E = \int d^3x\, \bar{T}^{00} = \int d^3x\, \{\tfrac{1}{2}(\mathbf{E}^2 + \mathbf{B}^2) + \tfrac{1}{2} m^2 [(A^0)^2 + \mathbf{A}^2]\}, \\ P^i &= \int d^3x\, \bar{T}^{0i} = \int d^3x\, \{(\mathbf{E} \times \mathbf{B})^i + m^2 A^0 A^i\}. \end{aligned} \tag{6.5.12}$$

The Noether current associated with angular momentum can be written as either $\mathcal{J}^{\mu\rho\sigma}$ or $\bar{\mathcal{J}}^{\mu\rho\sigma}$ and both lead to the same conserved charges,

$$\begin{aligned} M^{\rho\sigma} &= \bar{M}^{\rho\sigma} = \int d^3x\, \bar{\mathcal{J}}^{0\rho\sigma} = \int d^3x\, \left(x^\rho \bar{T}^{0\sigma} - x^\sigma \bar{T}^{0\rho}\right) \\ &= M_{\text{em}}^{\rho\sigma} + \int d^3x\, m^2 \left[A^0(x^\rho A^\sigma - x^\sigma A^\rho) - \tfrac{1}{2} A^2(x^\rho g^{0\sigma} - x^\sigma g^{0\rho})\right], \\ J^i &= \tfrac{1}{2} \epsilon^{ijk} M^{jk} = J_{\text{em}}^i + \int d^3x\, m^2 A^0 (\mathbf{x} \times \mathbf{A})^i \\ &= \int d^3x\, \{[\mathbf{x} \times (\mathbf{E} \times \mathbf{B})]^i + m^2 A^0 (\mathbf{x} \times \mathbf{A})^i\}. \end{aligned} \tag{6.5.13}$$
$$\tag{6.5.14}$$

The massive vector field transforms under Lorentz transformations the same as the electromagnetic field and so Eq. (3.3.17) applies also here, which we can use in Eqs. (3.2.100) and (3.2.101). The total spin $S^i = \tfrac{1}{2} \epsilon^{ijk} \Sigma^{jk}$ is then the same as that for electromagnetism,

$$\mathbf{S} = \int d^3x\, \mathbf{E} \times \mathbf{A}. \tag{6.5.15}$$

The extra mass-dependent term adds to the orbital angular momentum of electromagnetism to give

$$\mathbf{L} = \int d^3x\, \{E^i (\mathbf{x} \times \mathbf{\nabla}) A^i + m^2 A^0 (\mathbf{x} \times \mathbf{A})\}. \tag{6.5.16}$$

In order to carry out the canonical quantization procedure for massive vector bosons, the arguments of Sec. 3.3.2 for electromagnetism can be followed and modified as appropriate. Canonical momentum densities are

$$\pi^\nu \equiv \frac{\partial \mathcal{L}}{\partial(\partial_0 A_\nu)} = -F^{0\nu}, \quad \pi^0 = 0, \quad \pi^i = -F^{0i} = E^i, \quad \boldsymbol{\pi}_{\text{can}} = -\boldsymbol{\pi} = -\mathbf{E}. \tag{6.5.17}$$

Since $\pi^0 = 0$, then we have a primary constraint on our system and only three independent degrees of freedom, $\mathbf{A}$. Since the Proca Lagrangian corresponds to a singular system, that means that we must use the machinery of the Dirac-Bergmann algorithm and use it to construct the Dirac bracket:

(i) The standard canonical Hamiltonian density is

$$\mathcal{H} = \pi^\mu \partial_0 A_\mu - \mathcal{L} = \pi^i(-\pi_i + \partial_i A_0) + \tfrac{1}{4} F^{\mu\nu} F_{\mu\nu} - \tfrac{1}{2} m^2 A_\mu A^\mu + j_\mu A^\mu$$
$$= \tfrac{1}{2}(\mathbf{E}^2 + \mathbf{B}^2) - (\partial_i \pi^i) A_0 - \tfrac{1}{2} m^2 A_\mu A^\mu + j^0 A_0 + j^i A_i, \qquad (6.5.18)$$

where we have used $\partial_0 A_i = F_{0i} + \partial_i A_0 = -\pi_i + \partial_i A_0$ and integrated by parts and neglected surface terms to make the replacement $\pi^i \partial_i A_0 \to -(\partial_i \pi^i) A^0$.

(ii) Since π^0 does not appear in $\mathcal{H}$, there is no Hamiltonian dynamics to ensure that it vanishes. So we have one primary constraint as for electromagnetism,

$$\pi^0(x) \approx 0. \qquad (6.5.19)$$

The fundamental Poisson brackets are also the same and given in Eq. (3.3.38),

$$\{\pi^\mu(x), \pi^\nu(y)\}_{x^0=y^0} = 0, \quad \{A^\mu(x), A^\nu(y)\}_{x^0=y^0} = 0, \quad \text{and}$$
$$\{A^\mu(x), \pi_\nu(y)\}_{x^0=y^0} = \delta^\mu{}_\nu \delta^3(\mathbf{x} - \mathbf{y}). \qquad (6.5.20)$$

The Poisson bracket of the constraint with itself vanishes identically,

$$P(\mathbf{x}, \mathbf{y}) = \{\pi^0(x), \pi^0(y)\}_{x^0=y^0} = 0. \qquad (6.5.21)$$

(iii) Since $P(\mathbf{x}, \mathbf{y}) = 0$ everywhere in functional phase space, then the functional determinant $\det P$ vanishes everywhere including on the constraint submanifold. Since $\det P \approx 0$, then one or more secondary constraints will be required.

(iv) We use the total Hamiltonian as in Eq. (3.3.40),

$$H_T = \int d^3x\, \mathcal{H}_T = \int d^3x\, \left[\mathcal{H} + v(x)\pi^0(x)\right] = H_T^{\text{em}} - \tfrac{1}{2} m^2 \int d^3x\, A^2(x), \qquad (6.5.22)$$

to find the time variation of the primary constraint dynamical variable π^0,

$$\partial_0 \pi^0(x) = \{\pi^0(x), H_T\} = \{\pi^0(x), H\}$$
$$= \int d^3z\, \{\pi^0(x), A^0(z)\}_{x^0=z^0} [-\partial_i \pi^i(z) - m^2 A^0(z) + j^0(z)]$$
$$= \partial_i \pi^i(x) + m^2 A^0(x) - j^0(x). \qquad (6.5.23)$$

So if the system is on the constraint submanifold at time x^0, $\pi(x^0, \mathbf{x}) \approx 0$, then for the system to stay on the constraint submanifold at all times we must impose the secondary constraint

$$\partial_i \pi^i(x) + m^2 A^0(x) - j^0(x) = \boldsymbol{\nabla} \cdot \mathbf{E}(x) + m^2 A^0(x) - j^0(x) \approx 0, \qquad (6.5.24)$$

which we recognize as Eq. (6.5.9). This secondary constraint must be true at all times, so we require $\partial_0[\partial_i \pi^i(x) + m^2 A^0(x) - j^0(x)] \approx 0$, where

$$\partial_0[\partial_i \pi^i(x) + m^2 A^0(x) - j^0(x)] = \{[\cdots], H_T\} + \partial_0^{\text{ex}}[\cdots]$$
$$= \{[\cdots], H_T^{\text{em}}\} + \{[\cdots], -\tfrac{1}{2} m^2 \int d^3y\, A^2(y)\} + \partial_0 j^0$$
$$= \left[\{\partial_i \pi^i(x) - j^0(x), H_T^{\text{em}}\} + \partial_0 j^0\right] + m^2 \{A^0(x), H_T^{\text{em}}\} + \{[\cdots], H_T - H_T^{\text{em}}\}$$
$$= m^2 \{A^0(x), H_T^{\text{em}}\} - \tfrac{1}{2} m^2 \int d^3y\, \{[\cdots], A^2(y)\}_{x^0=y^0}$$
$$= m^2 \int d^3y\, \left(v(y)\{A^0(x), \pi^0(y)\} - \tfrac{1}{2}\{\partial_i \pi^i(x), A^2(y)\}\right)_{x^0=y^0}$$
$$= m^2[v(x) + \partial_i A^i(x)] \approx 0, \qquad (6.5.25)$$

where the first term in square brackets in the third line above vanishes for a conserved current as we saw in Eq. (3.3.45). We are then free to use the Lagrange multiplier for its intended purpose and *choose* $v(x) = -\partial_i A^i(x)$ so that the secondary constraint is true at all times on the constraint submanifold. Then no further constraints are required. We have no further need of the total Hamiltonian H_T and proceed to construct the Dirac brackets.

(v) We have one primary and one secondary constraint and these are both second-class constraints since their Poisson bracket is nonzero,

$$\psi_1 \equiv \pi^0, \quad \psi_2 \equiv \partial_i \pi^i(x) + m^2 A^0(x) - j^0(x),$$

$$G_{12}(x,y)_{x^0=y^0} \equiv \{\psi_1(x), \psi_2(y)\}_{x^0=y^0} = \{\pi^0(x), \partial_i \pi^i(y) + m^2 A^0(y) - j^0(y)\}_{x^0=y^0}$$

$$= m^2 \{\pi^0(x), A^0(y)\}_{x^0=y^0} = -m^2 \delta^3(\mathbf{x} - \mathbf{y}). \tag{6.5.26}$$

Note that $G_{11} = G_{22} = 0$ and $G_{12} = -G_{21}$ and so the 2×2 matrix G is known. Since there are no first-class constraints, then there is no need for any gauge-fixing.

The inverse of the functional matrix G is

$$G^{-1}(x,y)_{x^0=y^0} \equiv G^{-1}(\mathbf{x},\mathbf{y}) = \begin{pmatrix} 0 & 1/m^2 \\ -1/m^2 & 0 \end{pmatrix} \delta^3(\mathbf{x} - \mathbf{y}). \tag{6.5.27}$$

We now have all that we need to construct the Dirac bracket using Eq. (2.9.126), which has the same overall form as Eq. (3.3.75),

$$\{C(x^0), D(x^0)\}_{\text{DB}} \equiv \{C(x^0), D(x^0)\} \tag{6.5.28}$$

$$- \int d^3u \, d^3v \sum_{i,j=1}^2 \left(\{C(x^0), \psi_i(u)\} G_{ij}^{-1}(u,v) \{\psi_j(x), D(x^0)\} \right)_{x^0=u^0=v^0}.$$

Using this gives the fundamental Dirac brackets (recall that $\mathbf{E} = \boldsymbol{\pi}$),

$$\{\pi^\mu(x), \pi^\nu(y)\}_{\text{DB}}^{x^0=y^0} = \{\pi^0(y), A^\mu(x)\}_{\text{DB}}^{x^0=y^0} = \{A^i(x), A^j(y)\}_{\text{DB}}^{x^0=y^0} = 0, \tag{6.5.29}$$

$$\{A^0(x), \pi^i(y)\}_{\text{DB}}^{x^0=y^0} = \{A^0(x), A^0(y)\}_{\text{DB}}^{x^0=y^0} = 0,$$

$$\{A^i(x), \pi^j(y)\}_{\text{DB}}^{x^0=y^0} = -\delta^{ij} \delta^3(\mathbf{x} - \mathbf{y}),$$

$$\{A^0(x), A^j(y)\}_{\text{DB}}^{x^0=y^0} = -(1/m^2)\partial_j^x \delta^3(\mathbf{x} - \mathbf{y}),$$

where from Eq. (2.9.127) we know that $\psi_1 = \pi^0$ and $\psi_2 = \partial_i \pi^i(x) + m^2 A^0(x) - j^0(x)$ must have a vanishing Dirac bracket with every dynamical variable.

From Eq. (2.9.128) we see that the time evolution of all dynamical variables is determined by the canonical Hamiltonian H and the Dirac bracket,

$$\partial_0 F = \dot{F} \approx \{F, H\}_{\text{DB}} + \partial_0^{\text{ex}} F, \tag{6.5.30}$$

which ensures that the system remains on the constraint submanifold. Recall that the evaluation of Poisson and Dirac brackets should occur before the constraints are implemented and so in Eq. (6.5.30) we should use

$$H = \int d^3x \left\{ \tfrac{1}{2}[\mathbf{E}^2 + (\boldsymbol{\nabla} \times \mathbf{A})^2] - (\boldsymbol{\nabla} \cdot \mathbf{E})A_0 - \tfrac{1}{2}m^2 A_\mu A^\mu + j^0 A_0 - \mathbf{j} \cdot \mathbf{A} \right\}. \tag{6.5.31}$$

The equations of motion follow from Eqs. (6.5.29)–(6.5.31),

$$\partial_0 A^i(x) = \dot{A}^i(x) = \{A^i(x), H\}_{\text{DB}}$$

$$= \int d^3y \{A^i(x), \tfrac{1}{2}\mathbf{E}^2(y) - (\boldsymbol{\nabla} \cdot \mathbf{E})(y)A^0(y)\}_{\text{DB}}^{x^0=y^0}$$

$$= -E^i(x) - \partial_i A^0(x) = -E^i(x) + (1/m^2)\partial_i[(\boldsymbol{\nabla} \cdot \mathbf{E})(x) - j^0(x)], \tag{6.5.32}$$

$$\partial_0 E^i(x) = \dot{E}^i(x) = \{E^i(x), H\}_{\mathrm{DB}}$$
$$= \int d^3 y \{E^i(x), \tfrac{1}{2}(\boldsymbol{\nabla} \times \mathbf{A})^2(y) + \tfrac{1}{2}m^2 \mathbf{A}^2(y) - \mathbf{j} \cdot \mathbf{A}(y)\}_{\mathrm{DB}}^{x^0 = y^0}$$
$$= -\boldsymbol{\nabla}^2 A^i(x) + \partial_i(\boldsymbol{\nabla} \cdot \mathbf{A}) + m^2 A^i(x) - j^i(x), \tag{6.5.33}$$

where the first equation has just reproduced $F^{i0} = E^i = \partial^i A^0 - \partial^0 A^i$ with A^0 given in Eq. (6.5.9) and the second equation has just reproduced the Proca equation given in Eq. (6.5.5). The latter is readily seen by expanding out both the second equation and the Proca equation in terms of ∂_0, ∂_i, A^0 and A^i. This confirms that we have correctly constructed the constrained Hamiltonian formalism using the Dirac-Bergmann algorithm.

6.5.2 Normal Modes of the Massive Vector Field

From Eq. (6.5.8) we have $(\Box + m^2)A^\mu = 0$ and so we know that any solution of the Proca equation can be written as a superposition of plane waves. As we know from our studies of the Little Group in Sec. 1.4.2, a massive spin-one particle has three spin states. This is why the Proca equation has three independent degrees of freedom, rather than the two independent degrees of freedom of electromagnetism. In the rest frame of the massive particle we could choose three orthonormal three vectors $\epsilon^\mu(k, \lambda)$ such as $\boldsymbol{\epsilon}(k, 1) = (1, 0, 0)$, $\boldsymbol{\epsilon}(k, 2) = (0, 1, 0)$ and $\boldsymbol{\epsilon}(k, 3) = (0, 0, 1)$. However, the conventional and convenient choice valid in any frame is two normalized spacelike vectors corresponding to circular polarization around the direction of motion and a third spacelike vector with its three-momentum in the direction of motion chosen such that $k \cdot \epsilon(k, \lambda) = 0$. For a massive vector boson moving in the z-direction we have $k^\mu = (E_\mathbf{k}, 0, 0, |\mathbf{k}|)$. This choice of polarization four-vectors is

$$\epsilon^\mu(k, \lambda) \equiv (0, \boldsymbol{\epsilon}(k, \lambda)) \text{ for } \lambda = 1, 2 \text{ with } \boldsymbol{\epsilon}(k, 1) = (1, 0, 0), \ \boldsymbol{\epsilon}(k, 2) = (0, 1, 0),$$
$$\epsilon^\mu(k, 3) \equiv (|\mathbf{k}|/m, 0, 0, E_\mathbf{k}/m), \tag{6.5.34}$$

where for a helicity basis we can define $\epsilon(k, 1)$ and $\epsilon(k, 2)$ with the circular polarization three-vectors given in Eqs. (2.7.73) and (6.4.32). We also define

$$\epsilon^\mu(k, 0) \equiv \hat{k}^\mu \equiv k^\mu/m \tag{6.5.35}$$

for any on-shell momentum k^μ. It then follows that

$$\epsilon(k, 1)^* \cdot \epsilon(k, 1) = \epsilon(k, 2)^* \cdot \epsilon(k, 2) = \epsilon(k, 3)^* \cdot \epsilon(k, 3) = -1, \quad \hat{k}^2 = +1, \tag{6.5.36}$$
$$\hat{k} \cdot \epsilon(k, \lambda) = \hat{k} \cdot \epsilon(k, \lambda)^* = 0, \quad \epsilon(k, \lambda)^* \cdot \epsilon(k, \lambda') = -\delta_{\lambda\lambda'} \quad \text{for} \quad \lambda, \lambda' = 1, \dots, 3.$$

So $\epsilon^\mu(k, 0)$, $\epsilon^\mu(k, 1)$, $\epsilon^\mu(k, 2)$ and $\epsilon^\mu(k, 3)$ form an orthonormal-like basis of four-vectors with orthonormality and completeness relations, respectively, given by

$$\epsilon(k, \lambda')^* \cdot \epsilon(k, \lambda) = g_{\lambda'\lambda} \quad \text{for} \quad \lambda, \lambda' = 0, \dots, 3,$$
$$\delta^\mu_{\ \nu} = -\sum_{\lambda=1}^3 \epsilon^\mu(k, \lambda)\epsilon_\nu(k, \lambda)^* + \hat{k}^\mu \hat{k}_\nu \equiv \sum_{\lambda=0}^3 g_{\lambda\lambda'}\epsilon^\mu(k, \lambda)\epsilon_\nu(k, \lambda)^*. \tag{6.5.37}$$

Completeness is checked by writing an arbitrary four-vector as a superposition of the four basis vectors, acting on that four-vector with the right-hand side of the above and then showing that we recover the original four-vector. Rearranging gives

$$\sum_{\lambda=1}^3 \epsilon^\mu(k, \lambda)\epsilon^\nu(k, \lambda)^* = -g^{\mu\nu} + \hat{k}^\mu \hat{k}^\nu = -g^{\mu\nu} + (k^\mu k^\nu/m^2). \tag{6.5.38}$$

Any real solution to the free Proca equation can be written as

$$A^\mu(x) = \int \frac{d^3k}{(2\pi)^3} \frac{1}{\sqrt{2\omega_\mathbf{k}}} \sum_{\lambda=1}^3 \left[\epsilon^\mu(k,\lambda) c_\mathbf{k}^\lambda e^{-ik\cdot x} + \epsilon^\mu(k,\lambda)^* c_\mathbf{k}^{\lambda*} e^{ik\cdot x} \right], \qquad (6.5.39)$$

which we readily see satisfies both $(\Box + m^2)A^\mu(x) = 0$ since $(k^2 + m^2) = 0$ and $\partial_\mu A^\mu = 0$ since $k_\mu \epsilon^\mu(k,\lambda) = 0$ for $\lambda = 1, 2, 3$. The normal modes of the massive vector field are $\mathrm{Re}[\epsilon^\mu(k,\lambda)e^{-ik\cdot x}]$.

6.5.3 Quantization of the Massive Vector Field

We proceed to canonical quantization based on Eq. (2.9.136) in the usual way. We replace $\mathbf{A}$ and $\mathbf{E}$ with their Hermitian operator forms $\hat{\mathbf{A}}$ and $\hat{\mathbf{E}}$ and every Dirac bracket is replaced by a commutator according to $i\{A, B\}_{\mathrm{DB}} \to [\hat{A}, \hat{B}]$. In the operator form we implement the constraints and write the Hamiltonian in normal-ordered form,

$$\hat{H} = \int d^3x : \frac{1}{2} \left[\hat{\mathbf{E}}^2 + (\boldsymbol{\nabla} \times \hat{\mathbf{A}})^2 + m^2\hat{\mathbf{A}}^2 + \frac{(\boldsymbol{\nabla} \cdot \hat{\mathbf{E}})^2}{m^2} + \frac{(j^0)^2}{m^2} \right] - j^0 \frac{\boldsymbol{\nabla} \cdot \hat{\mathbf{E}}}{m^2} - \mathbf{j} \cdot \hat{\mathbf{A}} :$$

$$= \int d^3x : \tfrac{1}{2}(\hat{\mathbf{E}}^2 + \hat{\mathbf{B}}^2) + \tfrac{1}{2}m^2\hat{\mathbf{A}}^2 + \tfrac{1}{2}(-\boldsymbol{\nabla} \cdot \hat{\mathbf{E}} + j^0)^2/m^2 - \mathbf{j} \cdot \hat{\mathbf{A}} :. \qquad (6.5.40)$$

The canonical Hamiltonian is specified only in terms of the three coordinate-like $\mathbf{A}$ and the three canonical momentum-like $\boldsymbol{\pi}_{\mathrm{can}} = -\boldsymbol{\pi} = -\mathbf{E}$ plus source terms. The variable π^0 does not appear anywhere and can be ignored. The fundamental Dirac brackets involving only $\mathbf{A}$ and/or $\mathbf{E}$ become the operator commutators

$$[\hat{E}^i(x), \hat{E}^j(y)]_{x^0=y^0} = [\hat{A}^i(x), \hat{A}^j(y)]_{x^0=y^0} = 0,$$
$$[\hat{A}^i(x), \hat{E}^j(y)]_{x^0=y^0} = -i\delta^{ij}\delta^3(\mathbf{x} - \mathbf{y}). \qquad (6.5.41)$$

We note that the constraint

$$\hat{A}^0 = (-\boldsymbol{\nabla} \cdot \hat{\mathbf{E}} + j^0)/m^2 \qquad (6.5.42)$$

together with the above commutators automatically ensures that

$$[\hat{A}^0(x), \hat{E}^j(y)]_{x^0=y^0} = [-\boldsymbol{\nabla} \cdot \mathbf{E}(x), \hat{E}^j(y)]_{x^0=y^0}/m^2 = 0, \qquad (6.5.43)$$
$$[\hat{A}^0(x), \hat{A}^j(y)]_{x^0=y^0} = [-\boldsymbol{\nabla} \cdot \mathbf{E}(x), \hat{A}^j(y)]_{x^0=y^0}/m^2 = -(1/m^2)\partial_j^x \delta^3(\mathbf{x} - \mathbf{y}),$$

which shows that all of the Dirac brackets in Eq. (6.5.29) are reflected in the corresponding operator commutators. So the only commutators that we need are those in Eq. (6.5.41), and their corresponding fundamental Dirac brackets are equal to their fundamental Poisson brackets forms in Eq. (6.5.20).

Recall from Eq. (3.2.14) and the discussion of on-shell and off-shell that Noether currents and charges are constructed after implementing the Euler-Lagrange equations. So P^μ and $M^{\rho\sigma}$ can be written only in terms of $\mathbf{A}$ and $\mathbf{E}$. To arrive at the Poisson bracket Poincaré Lie agebra in Eq. (3.2.125) we only need the fundamental Poisson brackets involving $\mathbf{A}$ and/or $\mathbf{E}$ and we note that these are the same as the corresponding Dirac brackets. Then the Poisson brackets leading to the Poincaré Lie algebra correspond to the commutators in Eq. (6.5.41). This means that the operators $\hat{P}^\mu$ and $\hat{M}^{\rho\sigma}$ are given only in terms of $\mathbf{A}$ and $\mathbf{E}$ and that the commutators in Eq. (6.5.41) must then lead to these operators being the generators of the Poincaré transformations in the quantum field theory and satisfying the operator Poincaré lie algebra in Eq. (6.2.118). Since the normal-ordered operators differ only by constants from their non-ordered form the Poincaré Lie algebra is unaffected.

Setting the external source to zero and implementing the constraints we obtain the conserved charges P^μ, which we then normal-order to give

$$\hat{P}^0 = \hat{H} = \int d^3x\, \hat{\bar{T}}^{00} = \int d^3x : \tfrac{1}{2}[\hat{\mathbf{E}}^2 + \hat{\mathbf{B}}^2 + m^2\hat{\mathbf{A}}^2 + (1/m^2)(\boldsymbol{\nabla}\cdot\hat{\mathbf{E}})^2]:,$$

$$\hat{P}^i = \int d^3x\, \hat{\bar{T}}^{0i} = \int d^3x :(\hat{\mathbf{E}}\times\hat{\mathbf{B}})^i - (\boldsymbol{\nabla}\cdot\hat{\mathbf{E}})\hat{A}^i:, \qquad (6.5.44)$$

where as usual $\hat{\mathbf{B}} = \boldsymbol{\nabla}\times\hat{\mathbf{A}}$. We can repeat this with $\hat{M}^{\rho\sigma}$ and so obtain the normal-ordered operators $\hat{\mathbf{J}}$, $\hat{\mathbf{L}}$ and $\hat{\mathbf{S}}$ in terms of $\hat{\mathbf{A}}$ and $\hat{\mathbf{E}}$ corresponding to their classical forms in Eqs. (6.5.14)–(6.5.16).

From Eq. (6.5.39) we know that Hermitian operators that satisfy the free Proca equation can be written as

$$\hat{A}^\mu(x) = \int \frac{d^3k}{(2\pi)^3}\frac{1}{\sqrt{2\omega_{\mathbf{k}}}}\sum_{\lambda=1}^3 \left[\epsilon^\mu(k,\lambda)\hat{a}_{\mathbf{k}}^\lambda e^{-ik\cdot x} + \epsilon^\mu(k,\lambda)^*\hat{a}_{\mathbf{k}}^{\lambda\dagger}e^{ik\cdot x}\right], \qquad (6.5.45)$$

for some operators $\hat{a}_{\mathbf{k}}^\lambda$. By construction $\hat{A}^\mu(x)$ satisfies the free Proca equation,

$$\partial_\mu \hat{A}^\mu = 0 \quad \text{and} \quad (\Box + m^2)\hat{A}^\mu(x) = 0, \qquad (6.5.46)$$

since $k\cdot\epsilon(k,\lambda) = 0$ for $\lambda = 1,2,3$ and $-k^2 + m^2 = 0$. In the case of a charged/complex massive vector field we would follow the generalizations in Sec. 6.2.7 and begin with

$$\hat{A}^\mu(x) = \int \frac{d^3k}{(2\pi)^3}\frac{1}{\sqrt{2\omega_{\mathbf{k}}}}\sum_{\lambda=1}^3 \left[\epsilon^\mu(k,\lambda)\hat{f}_{\mathbf{k}}^\lambda e^{-ik\cdot x} + \epsilon^\mu(k,\lambda)^*\hat{g}_{\mathbf{k}}^{\lambda\dagger}e^{ik\cdot x}\right]. \qquad (6.5.47)$$

We will continue to consider the neutral massive vector field here for simplicity. Extension to the case of a charged massive vector field is left as an exercise.

Constructing the $\hat{\mathbf{E}}$-field operator we have

$$\hat{E}^i = F^{i0} = -\partial_0\hat{A}^i + \partial_i\hat{A}^0$$

$$= -\int \frac{d^3k}{(2\pi)^3}\frac{1}{\sqrt{2\omega_{\mathbf{k}}}}\sum_{\lambda=1}^3 \left[\{i\omega_{\mathbf{k}}\epsilon^i(k,\lambda) - ik^i\epsilon^0(k,\lambda)\}\hat{a}_{\mathbf{k}}^\lambda e^{-ik\cdot x} + \{\cdots\}^*\hat{a}_{\mathbf{k}}^{\lambda\dagger}e^{ik\cdot x}\right]$$

$$= -\int \frac{d^3k}{(2\pi)^3}\frac{1}{\sqrt{2\omega_{\mathbf{k}}}}\sum_{\lambda=1}^3 (i\omega_{\mathbf{k}})\left[\tilde{\epsilon}^i(k,\lambda)\hat{a}_{\mathbf{k}}^\lambda e^{-ik\cdot x} - \tilde{\epsilon}^i(k,\lambda)^*\hat{a}_{\mathbf{k}}^{\lambda\dagger}e^{ik\cdot x}\right], \qquad (6.5.48)$$

where we have defined

$$\tilde{\boldsymbol{\epsilon}}(k,\lambda) \equiv \boldsymbol{\epsilon}(k,\lambda) - (\mathbf{k}/\omega_{\mathbf{k}})\epsilon^0(k,\lambda) = \boldsymbol{\epsilon}(k,\lambda) - \mathbf{k}\frac{\mathbf{k}\cdot\boldsymbol{\epsilon}(k,\lambda)}{\omega_{\mathbf{k}}^2} \qquad (6.5.49)$$

and where $k\cdot\epsilon(k,\lambda) = \omega_{\mathbf{k}}\epsilon^0(k,\lambda) - \mathbf{k}\cdot\boldsymbol{\epsilon}(k,\lambda) = 0$ gives

$$\epsilon^0(k,\lambda) = \mathbf{k}\cdot\boldsymbol{\epsilon}(k,\lambda)/\omega_{\mathbf{k}}. \qquad (6.5.50)$$

We also observe that

$$\mathbf{k}\cdot\tilde{\boldsymbol{\epsilon}}(k,\lambda) = [1 - (\mathbf{k}^2/\omega_{\mathbf{k}}^2)]\mathbf{k}\cdot\boldsymbol{\epsilon}(k,\lambda) = (m^2/\omega_{\mathbf{k}}^2)\mathbf{k}\cdot\boldsymbol{\epsilon}(k,\lambda) = (m^2/\omega_{\mathbf{k}})\epsilon^0(k,\lambda),$$

which recovers the expected free-field constraint of Eq. (6.5.9) in operator form,

$$\hat{A}^0(x) = -\boldsymbol{\nabla}\cdot\hat{\mathbf{E}}(x)/m^2. \qquad (6.5.51)$$

From Eq. (6.5.36) and the above equations we find for $\lambda,\lambda' = 1,2,3$ that

$$\tilde{\boldsymbol{\epsilon}}(k,\lambda)^*\cdot\boldsymbol{\epsilon}(k,\lambda') = \boldsymbol{\epsilon}(k,\lambda)^*\cdot\boldsymbol{\epsilon}(k,\lambda') - \epsilon^0(k,\lambda)^*(\mathbf{k}/\omega_{\mathbf{k}})\cdot\boldsymbol{\epsilon}(k,\lambda')$$

$$= \boldsymbol{\epsilon}(k,\lambda)^*\cdot\boldsymbol{\epsilon}(k,\lambda') - \epsilon^0(k,\lambda)^*\epsilon^0(k,\lambda') = -\epsilon(k,\lambda)^*\cdot\epsilon(k,\lambda') = \delta_{\lambda\lambda'}. \qquad (6.5.52)$$

Using this result it is not difficult to show that

$$\hat{a}_{\mathbf{k}}^{\lambda} = \int d^3x \frac{e^{-ik\cdot\mathbf{x}}}{\sqrt{2\omega_{\mathbf{k}}}} \left[\omega_{\mathbf{k}} \tilde{\epsilon}(k,\lambda)^* \cdot \hat{\mathbf{A}}(\mathbf{x}) - i\epsilon(k,\lambda)^* \cdot \hat{\mathbf{E}}(\mathbf{x}) \right], \tag{6.5.53}$$

$$\hat{a}_{\mathbf{k}}^{\lambda\dagger} = \int d^3x \frac{e^{ik\cdot\mathbf{x}}}{\sqrt{2\omega_{\mathbf{k}}}} \left[\omega_{\mathbf{k}} \tilde{\epsilon}(k,\lambda) \cdot \hat{\mathbf{A}}(\mathbf{x}) + i\epsilon(k,\lambda) \cdot \hat{\mathbf{E}}(\mathbf{x}) \right].$$

Using the ETCR in Eq. (6.5.41) we recover the Fock space annihilation and creation operator commutation relations for $\lambda, \lambda' = 1, 2, 3$,

$$[\hat{a}_{\mathbf{k}}^{\lambda}, \hat{a}_{\mathbf{k}'}^{\lambda'}] = [\hat{a}_{\mathbf{k}}^{\lambda\dagger}, \hat{a}_{\mathbf{k}'}^{\lambda'\dagger}] = 0 \quad \text{and} \quad [\hat{a}_{\mathbf{k}}^{\lambda}, \hat{a}_{\mathbf{k}'}^{\lambda'\dagger}] = (2\pi)^3 \delta_{\lambda\lambda'} \delta^3(\mathbf{k}-\mathbf{k}'). \tag{6.5.54}$$

Using the definitions of the Hamiltonian and three-momentum operators in Eq. (6.5.44) and after some lengthy but straightforward calculations we find

$$\hat{H} = \int \frac{d^3k}{(2\pi)^3} \sum_{\lambda=1}^3 \omega_{\mathbf{k}} \hat{a}_{\mathbf{k}}^{\lambda\dagger} \hat{a}_{\mathbf{k}}^{\lambda} = \int d^3k \sum_{\lambda=1}^3 \omega_{\mathbf{k}} \hat{N}_{\mathbf{k}}^{\lambda} = \int d^3k \sum_{\lambda=1}^3 E_{\mathbf{k}} \hat{N}_{\mathbf{k}}^{\lambda},$$

$$\hat{\mathbf{P}} = \int \frac{d^3k}{(2\pi)^3} \sum_{\lambda=1}^3 \mathbf{k} \, \hat{a}_{\mathbf{k}}^{\lambda\dagger} \hat{a}_{\mathbf{k}}^{\lambda} = \int d^3k \sum_{\lambda=1}^3 \mathbf{k} \hat{N}_{\mathbf{k}}^{\lambda}. \tag{6.5.55}$$

These are the same as the Fock space expressions for the photon case and the proofs are similar, except that there are three spin states for each massive vector boson rather than two. Further detail outlining some of the steps in the proofs of these results can be found in Greiner and Reinhardt (1996). These Fock space expressions had to emerge since we know from Eq. (6.5.46) that we have three independent fields with each satisfying the same Klein-Gordon equation. This is how the underlying harmonic oscillator behavior emerges in this case.

Feynman Propagator for Massive Vector Bosons

We define the naive propagator $\Delta_V^{\mu\nu}(x-y)$ for the massive vector boson as the time-ordered product of field operators in the familiar way and write it as

$$\Delta_V^{\mu\nu}(x-y) \equiv \langle 0|T\hat{A}^\mu(x)\hat{A}^\nu(y)|0\rangle_{\text{NC}}$$

$$= \langle 0|\hat{A}^\mu(x)\hat{A}^\nu(y)|0\rangle_{\text{NC}}\theta(x^0-y^0) + \langle 0|\hat{A}^\nu(y)\hat{A}^\mu(x)|0\rangle_{\text{NC}}\theta(y^0-x^0)$$

$$= \int \frac{d^3k\,d^3k'}{(2\pi)^6} \frac{1}{2\sqrt{\omega_{\mathbf{k}}\omega_{\mathbf{k}'}}} \sum_{\lambda,\lambda'=1}^3 \left[\epsilon^\mu(k,\lambda)\epsilon^{\nu*}(k',\lambda')e^{-ik\cdot x+ik'\cdot y}\langle 0|\hat{a}_{\mathbf{k}}^{\lambda}\hat{a}_{\mathbf{k}'}^{\lambda'\dagger}|0\rangle\theta(x^0-y^0) \right.$$

$$\left. +\epsilon^\nu(k',\lambda')\epsilon^{\mu*}(k,\lambda)e^{+ik\cdot x-ik'\cdot y}\langle 0|\hat{a}_{\mathbf{k}'}^{\lambda'}\hat{a}_{\mathbf{k}}^{\lambda\dagger}|0\rangle\theta(y^0-x^0) \right]$$

$$= \int \frac{d^3k}{(2\pi)^3} \frac{1}{2\omega_{\mathbf{k}}} \sum_{\lambda=1}^3 \left[\epsilon^\mu(k,\lambda)\epsilon^{\nu*}(k,\lambda)e^{-ik\cdot(x-y)}\theta(x^0-y^0) \right.$$

$$\left. +\epsilon^\nu(k,\lambda)\epsilon^{\mu*}(k,\lambda)e^{ik\cdot(x-y)}\theta(y^0-x^0) \right]$$

$$= \int \frac{d^3k}{(2\pi)^3} \frac{1}{2\omega_{\mathbf{k}}} \left(-g^{\mu\nu} + \frac{k^\mu k^\nu}{m^2} \right) \left[e^{-ik\cdot(x-y)}\theta(x^0-y^0) + e^{ik\cdot(x-y)}\theta(y^0-x^0) \right]$$

$$= -\left(g^{\mu\nu} + \frac{\partial_x^\mu \partial_x^\nu}{m^2} \right) D_F(x-y) - i\frac{g^{\mu 0}g^{\nu 0}}{m^2}\delta^4(x-y)$$

$$= \int \frac{d^4x}{(2\pi)^4} i \left\{ \frac{[-g^{\mu\nu}+(k^\mu k^\nu/m^2)]}{k^2-m^2+i\epsilon} - \frac{g^{\mu 0}g^{\nu 0}}{m^2} \right\} e^{-ik\cdot(x-y)}$$

$$\equiv \int \frac{d^4x}{(2\pi)^4} \Delta_V^{\mu\nu}(k)\, e^{-ik\cdot(x-y)}, \tag{6.5.56}$$

where $D_F(x-y)$ is the scalar boson Feynman propagator given in Eq. (6.2.153), $\Delta_V^{\mu\nu}(k)$ is the momentum-space massive vector boson propagator, and the odd-looking noncovariant term containing $g^{\mu 0}g^{\nu 0}/m^2$ is explained below. We refer to this as a contact term since it is proportional to $\delta^4(x-y)$. As we found in the case of the photon, because we quantized with respect to $\hat{\mathbf{A}}$ and $\hat{\mathbf{E}}$ we have chosen a quantization frame and so $\langle 0|T\hat{A}^\mu(x)\hat{A}^\nu(y)|0\rangle_{\rm NC}$ is not manifestly covariant. That is why we use the annotation "NC" to remind us of this. We arrive at the usual Feynman boundary conditions and the addition of the $i\epsilon$ by Wick-rotating back from a Euclidean formulation. Let us write the covariant part of $\Delta_V^{\mu\nu}$ as the covariant Feynman propagator,

$$
\Delta_F^{\mu\nu}(x-y) \equiv \int \frac{d^4x}{(2\pi)^4}\, i\left\{\frac{[-g^{\mu\nu}+(k^\mu k^\nu/m^2)]}{k^2-m^2+i\epsilon}\right\}e^{-ik\cdot(x-y)}
$$

$$
\equiv \int \frac{d^4x}{(2\pi)^4}\Delta_F^{\mu\nu}(k)e^{-ik\cdot(x-y)},\quad\text{where}
$$

$$
\Delta_V^{\mu\nu}(x-y) = \Delta_F^{\mu\nu}(x-y) - i\frac{g^{\mu 0}g^{\nu 0}}{m^2}\delta^4(x-y). \tag{6.5.57}
$$

Proof of Eq. (6.5.56): We make use of Eq. (6.2.161) in passing from the three-dimensional integral with time step functions to the four-dimensional integral. We would like to remove $k^\mu k^\nu/m^2$ from inside the $\int d^3k$ by acting on the exponentials $e^{\pm ik\cdot(x-y)}$ with $i\partial_x^\mu i\partial_x^\nu$, but if either $\mu=0$ or $\nu=0$ then we have a time derivative that also acts on the time step functions. We find that

$$
i\partial_x^\mu i\partial_x^\nu[e^{-ik\cdot(x-y)}\theta(x^0-y^0)+e^{ik\cdot(x-y)}\theta(y^0-x^0)]
$$
$$
= k^\mu k^\nu[e^{-ik\cdot(x-y)}\theta(x^0-y^0)+e^{ik\cdot(x-y)}\theta(y^0-x^0)]
$$
$$
+ i[g^{\mu 0}k^\nu+k^\mu g^{\nu 0}]\delta(x^0-y^0)(e^{-ik\cdot(x-y)}+e^{ik\cdot(x-y)})
$$
$$
- g^{\mu 0}g^{\nu 0}\delta'(x^0-y^0)(e^{-ik\cdot(x-y)}-e^{ik\cdot(x-y)})
$$
$$
\xrightarrow{\text{no odd }\mathbf{k}} k^\mu k^\nu[e^{-ik\cdot(x-y)}\theta(x^0-y^0)+e^{ik\cdot(x-y)}\theta(y^0-x^0)]
$$
$$
+ ig^{\mu 0}g^{\nu 0}k^0\delta(x^0-y^0)(e^{-ik\cdot(x-y)}+e^{ik\cdot(x-y)}). \tag{6.5.58}
$$

To arrive at this last form the $g^{\mu 0}k^i$ and $k^i g^{\nu 0}$ terms were omitted since they are odd in $\mathbf{k}$ and so vanish in the integral over $\mathbf{k}$. For the derivative of the delta function, $\delta'(x^0-y^0)\equiv\partial_0\delta(x^0-y^0)$, we used the result that

$$
\int dx^0\delta'(x^0-y^0)g(x^0) = -\int dx^0\delta(x^0-y^0)\partial_0 g(x^0) \tag{6.5.59}
$$

and so made the replacement $\delta'(x^0-y^0)(\cdots)\to-\delta(x^0-y^0)\partial_0(\cdots)$. The second and third terms in the mid-part of the equation then partially cancel and the remainder is the second term in the last part of the equation. So to replace $k^\mu k^\nu$ by $i\partial_x^\mu i\partial_x^\nu$ we add the negative of the last term above to compensate for the time derivatives acting on the time-ordering step functions,

$$
\int\frac{d^3k}{(2\pi)^3}\frac{1}{2\omega_{\mathbf{k}}}\frac{-ig^{\mu 0}g^{\nu 0}\omega_{\mathbf{k}}}{m^2}\delta(x^0-y^0)(e^{-ik\cdot(x-y)}+e^{ik\cdot(x-y)}) = \frac{-ig^{\mu 0}g^{\nu 0}}{m^2}\delta^4(x-y),
$$

which then proves Eq. (6.5.56).

Why does the noncovariant contact term appear in what must be a covariant theory, since it is based on a Poincaré-covariant Lagrangian density and action? A consistent calculation of physical observables must lead to covariant outcomes, but how does this occur? In the canonical quantization procedure to construct the Hamiltonian we have $\hat{\mathcal{H}}_{\text{int}} \neq -\hat{\mathcal{L}}_{\text{int}}$. This difference is what is needed to give a covariant evolution operator, $T \exp\{-i \int_{t'}^{t''} d^4x\, \mathcal{H}\}$. Comparing Eqs. (6.5.1) and (6.5.40) and recalling that for the free theory $\hat{A}^0 = -\boldsymbol{\nabla} \cdot \hat{\mathbf{E}}/m^2$ we see that $\hat{\mathcal{L}}_{\text{int}} = -j_\mu \hat{A}^\mu$ and $\hat{\mathcal{H}}_{\text{int}} = j_\mu \hat{A}^\mu + (j^0)^2/2m^2$. As we will soon see, the noncovariant $(j^0)^2/2m^2$ term is exactly what is needed to give Lorentz covariance of the evolution operator.

6.5.4 Functional Integral for Massive Vector Bosons

To obtain the functional integral formulation of a massive vector boson we can follow many of the same steps as for electromagnetism, but we do not need to deal with the issue of gauge-fixing. We start with the canonical commutation relations given in Eq. (6.5.41), where we note that the three coordinate-like fields are $\hat{\mathbf{A}}(x)$ and the corresponding canonically conjugate momentum-like fields are $\hat{\boldsymbol{\pi}}_{\text{can}}(x) = -\hat{\boldsymbol{\pi}}(x) = -\hat{\mathbf{E}}(x)$. We have the usual results

$$\hat{I} = \int \mathcal{D}^{\text{sp}}\mathbf{E}\, |\mathbf{E}\rangle\langle\mathbf{E}|, \quad \hat{I} = \int \mathcal{D}^{\text{sp}}\mathbf{A}\, |\mathbf{A}\rangle\langle\mathbf{A}|, \tag{6.5.60}$$

$$\langle\mathbf{A}|\mathbf{E}\rangle = (1/\mathcal{N})\exp^{-i\int d^3x\, \mathbf{E}(\mathbf{x})\cdot\mathbf{A}(\mathbf{x})}, \quad \delta[\mathbf{A}'-\mathbf{A}] = (1/\mathcal{N}^2)\int \mathcal{D}^{\text{sp}}\mathbf{E}\, e^{\pm i\int d^3x\, \mathbf{E}\cdot(\mathbf{A}'-\mathbf{A})},$$

where time is fixed in all of these expressions and $\mathcal{N}$ is an irrelevant overall constant. To simplify notation for the remainder of this discussion we will simply *omit irrelevant normalization constants* when they appear, but understand that they are there. We divide the time interval $[t', t'']$ in our usual way

$$\delta t \equiv (t'' - t')/(n+1), \quad t_j \equiv t' + j\delta t \tag{6.5.61}$$

for $j = 0, 1, \ldots, n+1$ with $t_{j+1} = t_j + \delta t$, $t'' = t_{n+1}$ and $t' = t_0$. The matrix element of the Hamiltonian operator in Eq. (6.5.40) without normal-ordering between the field operators at adjacent times t_j and t_{j+1} is

$$\langle\mathbf{A}_{j+1}|\hat{H}|\mathbf{A}_j\rangle = \int \mathcal{D}^{\text{sp}}\mathbf{E}_j \langle\mathbf{A}_{j+1}|\hat{H}|\mathbf{E}_j\rangle\langle\mathbf{E}_j|\mathbf{A}_j\rangle$$

$$= \int \mathcal{D}^{\text{sp}}\mathbf{E}_j\, e^{-i\int d^3x\, \mathbf{E}_j\cdot(\mathbf{A}_{j+1}-\mathbf{A}_j)} H(\mathbf{E}_j, \mathbf{A}_{j+1}), \tag{6.5.62}$$

where $H(\mathbf{E}_j, \mathbf{A}_{j+1})$ is the Hamiltonian of Eq. (6.5.40) without normal-ordering and with $\hat{\mathbf{E}} \to \mathbf{E}_j$ and $\hat{\mathbf{A}} \to \mathbf{A}_{j+1}$. Then in the usual way we find

$$\langle\mathbf{A}'', t''|\mathbf{A}', t'\rangle = \lim_{n\to\infty} \int \prod_{i=1}^n \mathcal{D}^{\text{sp}}\mathbf{A}_i \int \prod_{k=0}^n \mathcal{D}^{\text{sp}}\mathbf{E}_k \tag{6.5.63}$$

$$\times \exp\left\{i\delta t \sum_{j=0}^n \int d^3x[-\mathbf{E}_j \cdot [(\mathbf{A}_{j+1}-\mathbf{A}_j)/\delta t] - \mathcal{H}(\mathbf{E}_j, \mathbf{A}_{j+1})]\right\}$$

$$\equiv \int \mathcal{D}\mathbf{E}\mathcal{D}\mathbf{A} \exp\left\{i\int_{t'}^{t''} d^4x \left[-\mathbf{E}\cdot\dot{\mathbf{A}} - \mathcal{H}(\mathbf{E}, \mathbf{A})\right]\right\}$$

$$= \int \mathcal{D}\mathbf{E}\mathcal{D}\mathbf{A} \exp\left\{i\int_{t'}^{t''} d^4x \left[-\mathbf{E}\cdot\dot{\mathbf{A}} - \tfrac{1}{2}(\mathbf{E}^2+\mathbf{B}^2) - \tfrac{1}{2}m^2\mathbf{A}^2\right.\right.$$

$$\left.\left. - \tfrac{1}{2}(-\boldsymbol{\nabla}\cdot\mathbf{E})^2/m^2 + \tfrac{1}{2}(j^0)^2/m^2 + j^0\boldsymbol{\nabla}\cdot\mathbf{E}/m^2 + \mathbf{j}\cdot\mathbf{A}\right]\right\}$$

$$= \int \mathcal{D}\mathbf{E}\mathcal{D}\mathbf{A} \exp\left\{i\int_{t'}^{t''} d^4x \left[-\mathbf{E}\cdot\dot{\mathbf{A}} - \tfrac{1}{2}(\mathbf{E}^2+\mathbf{B}^2) - \tfrac{1}{2}m^2\mathbf{A}^2\right.\right.$$

$$\left.\left. - \tfrac{1}{2}(-\boldsymbol{\nabla}\cdot\mathbf{E}+j^0)^2/m^2 + \mathbf{j}\cdot\mathbf{A}\right]\right\}.$$

Up to an overall irrelevant normalization we can write unity as an arbitrary Gaussian integral, which we choose here to be

$$1 = \int \mathcal{D}A^0 \, \exp\left\{ i\tfrac{1}{2}m^2 \int d^4x \, [A^0 + (\boldsymbol{\nabla} \cdot \mathbf{E} - j^0)/m^2]^2 \right\}, \tag{6.5.64}$$

where A^0 is our dummy integration variable. Inserting this in the above gives

$$\begin{aligned}
\langle \mathbf{A}'', t'' | \mathbf{A}', t' \rangle^j &= \langle \mathbf{A}'' | \hat{U}^j(t'', t') | \mathbf{A}' \rangle \\
&= \int \mathcal{D}\mathbf{E}\mathcal{D}\mathbf{A} \exp\left\{ i \int_{t'}^{t''} d^4x \left[-\mathbf{E} \cdot \dot{\mathbf{A}} - \mathcal{H}(\mathbf{E}, \mathbf{A}) \right] \right\} \\
&= \int \mathcal{D}\mathbf{E}\mathcal{D}\mathbf{A}\mathcal{D}A^0 \exp\left\{ i \int_{t'}^{t''} d^4x \left[\tfrac{1}{2}m^2[A^0 + (\boldsymbol{\nabla} \cdot \mathbf{E} - j^0)/m^2]^2 - \mathbf{E} \cdot \dot{\mathbf{A}} - \mathcal{H}(\mathbf{E}, \mathbf{A}) \right] \right\} \\
&= \int \mathcal{D}\mathbf{E}\mathcal{D}A^\mu \exp\left\{ i \int_{t'}^{t''} d^4x \left[-\mathbf{E} \cdot (\partial^0 \mathbf{A} + \boldsymbol{\nabla}A^0) - \tfrac{1}{2}\mathbf{E}^2 - \tfrac{1}{2}\mathbf{B}^2 + \tfrac{1}{2}m^2 A^\mu A_\mu - j^\mu A_\mu \right] \right\} \\
&\equiv \int \mathcal{D}A^\mu \exp\left\{ i \int_{t'}^{t''} d^4x \left[-\tfrac{1}{4}F^{\mu\nu}F_{\mu\nu} + \tfrac{1}{2}m^2 A^\mu A_\mu - j^\mu A_\mu \right] \right\} = \int \mathcal{D}A^\mu \, e^{iS[A,j]},
\end{aligned} \tag{6.5.65}$$

where we have used a spatial integration by parts, where the $\mathbf{E}$ functional integral is the same as in Eq. (6.4.66) and where we recognize from Eq. (6.5.1) that $S[A, j]$ is the usual Proca action in the presence of a conserved current source j^μ,

$$\begin{aligned}
S[A, j] &\equiv S[A] - \int d^4x \, j^\mu A_\mu = \int d^4x \, \mathcal{L} \\
&= \int_{t'}^{t''} d^4x \left[-\tfrac{1}{4}F^{\mu\nu}F_{\mu\nu} + \tfrac{1}{2}m^2 A^\mu A_\mu - j^\mu A_\mu \right]
\end{aligned} \tag{6.5.66}$$

with boundary conditions $\mathbf{A}'$ and $\mathbf{A}''$ at t' and t'', respectively. We can also obtain this result in Euclidean space (Kashiwa, 1981) and then Wick-rotate back to Minkowski space. In Euclidean space we have $i\tfrac{1}{2} \int d^4x \, (A^0)^2 \to -\tfrac{1}{2} \int d^4x_E \, (A_4)^2$, which shows that Eq. (6.5.64) is a Gaussian integral. This leads to $m^2 \to m^2 - i\epsilon$ in the denominator of the Feynman propagator, which helps to define the Minkowski-space functional integral and leads to the Feynman boundary conditions.

As an aside, recall that the physical Hilbert space is spanned by the states $|\mathbf{A}\rangle$ and also by the states $|\mathbf{E}\rangle$ and that when we write the operator $\hat{A}^0$ for a *free* massive vector field we have defined it such that

$$\hat{A}^0 \equiv -\boldsymbol{\nabla} \cdot \hat{\mathbf{E}}/m^2. \tag{6.5.67}$$

From Eq. (6.5.65) we obtain the spectral function for the massive vector field with a source,

$$\begin{aligned}
F^j(t'', t') = \text{tr}\{\hat{U}^j(t'', t')\} &= \int \mathcal{D}^{\text{sp}}\mathbf{A} \, \langle \mathbf{A} | Te^{-i\int_{t'}^{t''} dt \, \hat{H}} | \mathbf{A} \rangle = \int \mathcal{D}^{\text{sp}}\mathbf{A} \, \langle \mathbf{A}, t'' | \mathbf{A}, t' \rangle^j \\
&= \int \mathcal{D}A^\mu_{\text{periodic}} \exp(i\{S[A] - \int_{t'}^{t''} d^4x \, j^\mu A_\mu\}),
\end{aligned} \tag{6.5.68}$$

where $\int \mathcal{D}^{\text{sp}}\mathbf{A}$ means integrating over spatial functions $\mathbf{A}(\mathbf{x})$ at a fixed time rather than $\int \mathcal{D}\mathbf{A}$, which means integrating over all spacetime functions $\mathbf{A}(x)$ between t' and t''. Without loss of generality we can require that $A^0(x)$ is also periodic in time along with the spatial components $\mathbf{A}(x)$. In the above $\hat{H} = H(\hat{\mathbf{E}}, \hat{\mathbf{A}})$ is as defined in Eq. (6.5.44) but here is understood to not be normal-ordered since we are using the functional integral representation. If we use normal-ordering $\hat{H}$ only changes by an irrelevant constant since it is second-order in the fields. This only changes $F^j(t'', t')$ by an irrelevant overall phase that cancels in ratios. So whether we use normal-ordering or not in the discussion here is unimportant.

Note that only because of the appearance of the $(j^0)^2/2m^2$ term in the Hamiltonian do we arrive at the Lorentz-covariant form of the evolution operator in Eq. (6.5.68) in terms of the action $S[A]$ in

Eq. (6.5.66). So Lorentz covariance of the time evolution of the quantum system has resulted from carefully following the canonical quantization procedure using the Dirac-Bergmann algorithm.

The generating functional is defined in the usual way and is given by

$$
Z[j] \equiv \lim_{T \to \infty(1-i\epsilon)} \frac{F^j(T, -T)}{F(T, -T)} = \lim_{T \to \infty(1-i\epsilon)} \frac{\int \mathcal{D} A^\mu_{\text{periodic}} \, e^{i\{S[A] - \int_{-T}^{T} d^4 x \, j_\mu A^\mu\}}}{\int \mathcal{D} A^\mu_{\text{periodic}} \, e^{iS[A]}} \tag{6.5.69}
$$

$$
= \lim_{T \to \infty(1-i\epsilon)} \frac{\int \mathcal{D}\mathbf{A} \, \langle \mathbf{A} | T e^{-i\int_{-T}^{T} d^4 x \{\hat{\mathcal{H}}_0 + j_\mu \hat{A}^\mu + [(j^0)^2/2m^2]\}} | \mathbf{A} \rangle}{\int \mathcal{D}\mathbf{A} \, \langle \mathbf{A} | T e^{-i\int_{-T}^{T} d^4 x \, \hat{\mathcal{H}}_0} | \mathbf{A} \rangle}. \tag{6.5.70}
$$

So differentiating with the source current j^μ on the functional integral expression in the first line will bring down classical fields A^μ into the integrand in the usual way. Differentiating the operator form in the second line with respect to j^μ will bring down $\hat{A}^\mu$ operators into the time-ordered product in the usual way but will also act on the j^0 in the exponential factor when $\mu = 0$. Since the functional integral form is clearly Poincaré invariant, then we note that the contributions from the $(j^0)^2$ term are exactly what is required to cancel the noncovariant nature of the time-ordered product in the case of the massive vector field.

We can evaluate the generating functional using the functional integral expression in a way similar to what was done in the case of electromagnetism. We rewrite the source-free action in Eq. (6.5.66) using periodic boundary conditions as

$$
\begin{aligned}
S[A] &= \int_{t'}^{t''} d^4 x \left[-\tfrac{1}{4} F^{\mu\nu} F_{\mu\nu} + \tfrac{1}{2} m^2 A^\mu A_\mu \right] \\
&= \int d^4 x \left\{ -\tfrac{1}{2}(\partial_\mu A_\nu \partial^\mu A^\nu - \partial_\mu A_\nu \partial^\nu A^\mu) + \tfrac{1}{2} m^2 A^\mu A_\mu \right\} \\
&= \int d^4 x \, d^4 y \, \tfrac{1}{2} A_\mu(x) \left[\{ g^{\mu\nu}(\partial_x^2 + m^2) - \partial_x^\mu \partial_x^\nu \} \delta^4(x - y) \right] A_\nu(y) \\
&\equiv \int d^4 x \, d^4 y \, \tfrac{1}{2} A_\mu(x) K^{\mu\nu}(x, y) A_\nu(y),
\end{aligned} \tag{6.5.71}
$$

where we easily verify that $K^{\mu\nu}(x, y) = K^{\nu\mu}(y, x)$. From Eq. (4.1.204) we then obtain the massive vector boson–generating functional,

$$
Z[j] = \exp \left\{ -\tfrac{1}{2} \int d^4 x \, d^4 y \, j_\mu(x) \Delta_F^{\mu\nu}(x - y) j_\nu(y) \right\}, \quad \text{where} \tag{6.5.72}
$$

$$
\Delta_F^{\mu\nu}(x - y) = \int \frac{d^4 k}{(2\pi)^4} \frac{-i}{k^2 - m^2 + i\epsilon} \left(g^{\mu\nu} - \frac{k^\mu k^\nu}{m^2} \right) e^{-ik \cdot (x - y)} \tag{6.5.73}
$$

$$
\equiv \int \frac{d^4 k}{(2\pi)^4} \Delta_F^{\mu\nu}(k) e^{-ik \cdot (x - y)}. \tag{6.5.74}
$$

Similar to photons in Eq. (6.4.97) we have $\Delta_F^{\mu\nu}(x - y) = i(K^{-1})^{\mu\nu}(x, y)$, since

$$
\{ g^{\mu\nu}(\partial_x^2 + m^2) - \partial_x^\mu \partial_x^\nu \} \Delta_{F\nu\rho}(x - y) = i\delta^\mu{}_\rho \delta^4(x - y), \quad \text{which gives}
$$
$$
\int d^4 y \, K^{\mu\nu}(x, y) \Delta_{F\nu\rho}(y - z) = i\delta^\mu{}_\rho \delta^4(x - z). \tag{6.5.75}
$$

Proof: It is easiest to show this in momentum space. Note that:

$[g^{\mu\nu}(-k^2 + m^2) + k^\mu k^\nu] k_\nu k_\rho / m^2 = k^\mu k_\rho,$

$[g^{\mu\nu}(-k^2 + m^2) + k^\mu k^\nu] g_{\nu\rho} = \delta^\mu{}_\rho(-k^2 + m^2) + k^\mu k_\rho,$

$[g^{\mu\nu}(-k^2 + m^2) + k^\mu k^\nu][-g_{\nu\rho} + (k_\nu k_\rho / m^2)] = \delta^\mu{}_\rho[k^2 - m^2],$ and so finally

$[g^{\mu\nu}(-k^2 + m^2) + k^\mu k^\nu] \Delta_{F\nu\rho}(k) = i\delta^\mu{}_\rho$ and Eq. (6.5.75) then follows.

The covariant part, $\Delta_F^{\mu\nu}$, of the noncovariant massive vector boson propagator, $\Delta_V^{\mu\nu}$, has appeared naturally in the generating functional. We have then arrived at

$$\Delta_F^{\mu\nu}(x-y) = (i)^2 \frac{\delta^2}{\delta j_\mu(x)\delta j_\nu(y)} Z[j]\bigg|_{j=0} = \lim_{T\to\infty(1-i\epsilon)} \frac{\int \mathcal{D}A^\mu_{\text{periodic}}\, A^\mu(x)A^\nu(y)\, e^{iS[A]}}{\int \mathcal{D}A^\mu_{\text{periodic}}\, e^{iS[A]}}$$

$$= \langle 0|T\hat{A}^\mu(x)\hat{A}^\nu(y)|0\rangle_{\text{NC}} + i\frac{g^{\mu 0}g^{\nu 0}}{m^2}\delta^4(x-y)$$

$$= \Delta_V^{\mu\nu}(x-y) + i\frac{g^{\mu 0}g^{\nu 0}}{m^2}\delta^4(x-y), \tag{6.5.76}$$

where the first line follows from Eqs. (6.5.69) and (6.5.72), the second line follows from Eq. (6.5.70) and where we recall that the time-ordered product is in the Heisenberg picture. Note that we have recovered the relationship between $\Delta_V^{\mu\nu}$ and $\Delta_F^{\mu\nu}$ in Eq. (6.5.57), which is an indication that we have correctly formulated the functional integral. The noncovariant contribution in $\Delta_V^{\mu\nu}$ is canceled by so-called *normal-dependent terms* that appear in the canonical quantization approach to the construction of the perturbation series.

6.5.5 Covariant Canonical Quantization for Massive Vector Bosons

The Gupta-Bleuler covariant quantization method used for photons can be generalized and the massless photon and massive vector boson cases can each be obtained as limiting cases of this. We add a mass term to Eq. (6.4.110) (Itzykson and Zuber, 1980) and also coupling to an external current j^μ,

$$\mathcal{L} = -\tfrac{1}{4}F_{\mu\nu}F^{\mu\nu} + \tfrac{1}{2}m^2 A_\mu A^\mu - \tfrac{1}{2\xi}(\partial_\mu A^\mu)^2 - j_\mu A^\mu. \tag{6.5.77}$$

The limit $\xi \to \infty$ restores the Proca Lagrangian of Eq. (6.5.1) and taking $m \to 0$ leads to Eq. (6.4.110) for photons. We again assume that we have only coupling to conserved currents, $\partial_\mu j^\mu = 0$. The equations of motion are now

$$(\Box + m^2)A^\nu - (1 - \tfrac{1}{\xi})\partial^\nu(\partial_\mu A^\mu) = j^\nu \tag{6.5.78}$$

in place of Eq. (6.4.148). The conjugate momenta are the same as in Eq. (6.4.149), $\pi^\nu = -F^{0\nu} - \tfrac{1}{\xi}g^{\nu 0}(\partial_\mu A^\mu)$.

Since the system is nonsingular and the conjugate momenta π^μ are as for the photon case, then the Poisson brackets and equal-time canonical commutation relations are as obtained before in Eqs. (6.4.120), (6.4.151) and (6.4.152). The field operators must satisfy the equation of motion and so $[(\Box + m^2)g^{\nu\mu} - (1 - \tfrac{1}{\xi})\partial^\nu\partial^\mu]\hat{A}_\mu = 0$. All the steps of the proof of Eq. (6.4.156) then go through as before and the noncovariant contributions again cancel in this fully covariant approach. So the only change is the m^2 term in the equation of motion. It therefore follows that

$$[(\Box + m^2)g^{\sigma\mu} - (1 - \tfrac{1}{\xi})\partial^\sigma\partial^\mu]\, T\hat{A}_\mu(x)\hat{A}_\nu(y) = i\delta^\sigma_{\ \nu}\delta^4(x-y). \tag{6.5.79}$$

Acting on both sides of the equation of motion in Eq. (6.5.78) with ∂_ν gives

$$(\Box + \xi m^2)(\partial_\nu A^\nu) \equiv (\Box + \mu^2)(\partial_\nu A^\nu) = 0 \quad \text{with} \quad \mu^2 \equiv \xi m^2. \tag{6.5.80}$$

From this point on we will restrict to $\xi \geq 0$. Then when $\xi \neq 0$ both m^2 and μ^2 act as squared masses and $(\partial \cdot A)$ satisfies the Klein-Gordon equation $(\Box + \mu^2)(\partial \cdot A) = 0$. In this covariant formulation using Feynman boundary conditions we then have

$$\langle 0|T\hat{A}^\mu(x)\hat{A}^\nu(y)|0\rangle = \int \frac{d^4k}{(2\pi)^4} i \left(\frac{-g^{\mu\nu} + k^\mu k^\nu/m^2}{k^2 - m^2 + i\epsilon} - \frac{k^\mu k^\nu/m^2}{k^2 - \mu^2 + i\epsilon} \right) e^{-ik\cdot(x-y)}. \quad (6.5.81)$$

Proof: We wish to use Eq. (6.5.79) to show Eq. (6.5.81). Working in momentum space we want to show that

$[\{g^{\mu\nu}(-k^2 + m^2) + k^\mu k^\nu\} + \{-\frac{1}{\xi}k^\mu k^\nu\}](C + D) \equiv [A + B](C + D) = \delta^\mu{}_\rho,$

where $[A + B]$ is an obvious shorthand notation and similarly $(C + D)$ is shorthand notation for the term with parentheses in Eq. (6.5.81). From the proof of Eq. (6.5.75) we already have $AC = \delta^\mu{}_\rho$ and so we need to show that $AD + BC + BD = 0$. We find that $AD = k^\mu k_\rho/k^2 - \xi m^2 + i\epsilon$, $BC = \frac{1}{\xi}k^\mu k_\rho/m^2$, $BD = -\frac{1}{\xi}k^2(k^\mu k_\rho/m^2)/k^2 - \xi m^2 + i\epsilon$. Multiply by $(k^2 - \xi m^2 + i\epsilon)$ and sum:

$(AD + BC + BD)(k^2 - \xi m^2 + i\epsilon) = \{k^\mu k_\rho\} + \{\frac{1}{\xi}k^2(k^\mu k_\rho/m^2) - k^\mu k_\rho\} + \{-\frac{1}{\xi}k^2(k^\mu k_\rho/m^2)\} = 0.$

It is not difficult to verify that

$$\frac{-g^{\mu\nu} + (k^\mu k^\nu/m^2)}{k^2 - m^2 + i\epsilon} - \frac{k^\mu k^\nu/m^2}{k^2 - \mu^2 + i\epsilon} = \frac{1}{k^2 - m^2 + i\epsilon}\left[-g^{\mu\nu} + \frac{(1-\xi)k^\mu k^\nu}{(k^2 - \xi m^2 + i\epsilon)}\right]. \quad (6.5.82)$$

Taking the limit $m^2 \to 0$ we recover the covariant (R_ξ) gauge photon propagator Eq. (6.4.154) and we understand now why the double pole term had the form $1/(k^2 + i\epsilon)^2$ (Itzykson and Zuber, 1980). However, it is conventional to not always be careful about explicitly writing this to simplify notation. In the limit $\xi \to \infty$ we recover the covariant Proca Feynman propagator $\Delta_F^{\mu\nu}(x - y)$ as we would expect.

Summary

In this chapter we formulated free quantum field theories corresponding to spin-zero, spin-half and massless and massive spin-one representations of the Poincaré group. The integer spin cases are relativistic bosonic systems and the half-integer spin case corresponds to relativistic fermions. The Hilbert spaces appropriate for free quantum field theories are bosonic and fermionic Fock spaces as appropriate.

The lessons that we learned from our understanding of normal modes and the harmonic oscillator of quantum mechanics underpin our understanding of the spin-zero and massless and massive spin-one bosonic systems. The harmonic oscillator normal modes that we quantize are the plane wave solutions of the corresponding relativistic wave equations, which are the Klein-Gordon equation, Maxwell's equations and the Proca equation, respectively. All of the Hamiltonians consequently have the simple harmonic oscillator form in Fock space in terms of annihilation and creation operators for these modes. For each bosonic system we also arrive at the Fock space result using the canonical quantization procedure of Dirac, which for the singular spin-one systems requires the Dirac-Bergmann algorithm to formulate an appropriate Hamiltonian in terms of the physical degrees of freedom. In the case of electromagnetism, as a gauge theory, arriving at the physical degrees of freedom also requires a gauge-fixing procedure, where there is freedom in how we choose the gauge. In each case we can either start from the construction of a bosonic Fock space built on relativistic normal modes or we can start from Dirac's canonical quantization procedure of the corresponding classical field theory, where we arrive at the same free bosonic quantum field theory either way. This happens because, despite the sometimes complicated formalism, once we arrive at a Hamiltonian in terms of physical degrees of freedom in each case it has an underlying harmonic oscillator form. The boson propagators were first obtained from the operator formalism and then were obtained from the generating functional using the functional integral formulation. We also showed that a covariant canonical quantization approach is possible using the Gupta-Bleuler formalism and saw that we again arrive at the same Feynman propagators for the photon and massive vector boson.

In the case of fermions we began with the construction of a fermion Fock space and introduced Grassmann algebra to represent the states in a fermion Fock space and also to represent operations on that space. We then chose the normal modes labeling the fermion annihilation and creation operators to be the plane wave solutions of the Dirac equation. The resulting fermion field operators then satisfy the Dirac equation by construction and they satisfy canonical equal-time anticommutation relations because of the anticommutation relations obeyed by the fermion annihilation and creation operators. By extending the definition of classical mechanics to apply to a Grassmann algebra and by extending the principle of canonical quantization to both bosons and fermions, we directly arrived at the quantum field theory of Dirac fermions in a canonical quantization approach. The fermion propagator was obtained in the operator formalism and then was obtained from a fermion-generating functional using a Grassmann-based definition of a functional integral.

In bosonic and/or fermionic Fock space we can move from any state to any other state using operators constructed solely from annihilation and creation operators. This is because any state can be written as a superposition of basis states constructed using products of creation operators. We can state this more formally as the fundamental theorem that *any* operator in Fock space can be represented as sums of products of annihilation and creation operators. A proof of this theorem is given in Sec. 4.2 of Weinberg (1995). We can change the order of the annihilation and creation operators in any sum of products using the commutation relations to arrive at some new sum of products. So we can express any operator as a sum of products of annihilation and creation operators in normal order.

We understood that generating functionals are ratios of functional integrals, where the limit of the removal of the regularization of the functional integrals is only to be carried out as the last step when calculating physical observables. Since the free theories that we have considered have observable operators second-order in the field operators, then normal-ordering is sufficient to remove infinities such as the zero-point energy. The most important results of this chapter are the expressions for the generating functionals for the different free fields and the surrounding discussion for each. These results are $Z[j]$ in Eq. (6.2.276) and $Z[j, j^*]$ in Eq. (6.2.282) for the real and complex scalar fields, respectively; $Z[\bar{\eta}, \eta]$ in Eq. (6.3.241) for the free fermion field; $Z[j]$ in Eq. (6.4.95) for the photon in the R_ξ gauges; and $Z[j]$ in Eq. (6.5.72) for the free massive vector boson. We now must study the implications of introducing interactions into and between our quantum field theories.

Problems

6.1 Substitute Eq. (6.2.14) for $\hat{\phi}_s(\mathbf{x})$ and $\hat{\pi}_s(\mathbf{x})$ into Eq. (6.2.16) and hence show that the expressions given for the annihilation and creation operators, $\hat{a}_{\mathbf{p}}$ and $\hat{a}_{\mathbf{p}}^\dagger$, are correct.

6.2 Show that the commutation relations of $\hat{a}_{\mathbf{p}}$ and $\hat{a}_{\mathbf{p}}^\dagger$ in Eq. (6.2.17) reproduce each of the Schrödinger picture canonical commutation relations in Eq. (6.2.12).

6.3   Obtain the result for the total three-momentum operator $\mathbf{P}$ for a free scalar field in terms of annihilation and creation operators as given in Eq. (6.2.34).

6.4   Explain why the expressions for the n-boson identity operator $\hat{I}_n$ in Eqs. (6.2.63) and (6.2.64) are identical and verify that $\hat{I}_n|\mathbf{p}_1\cdots\mathbf{p}_n\rangle = |\mathbf{p}_1\cdots\mathbf{p}_n\rangle$ and $\hat{I}_n|\mathbf{p}_1\cdots\mathbf{p}_n;S_o\rangle = |\mathbf{p}_1\cdots\mathbf{p}_n;S_o\rangle$ as required.

6.5   Use Eq. (6.2.129) and symmetry to explain why $\hat{\mathbf{P}}$ and $\hat{M}^{ij}$ do not need normal-ordering. Show that for a scalar boson field

$$\hat{L}^{ij} = -i\int \frac{d^3p}{(2\pi)^3}\,\hat{a}_{\mathbf{p}}^{\dagger}\left(p^i\frac{\partial}{\partial p^j} - p^j\frac{\partial}{\partial p^i}\right)\hat{a}_{\mathbf{p}}\quad\text{and hence}$$

$$\hat{L}^i = \tfrac{1}{2}\epsilon^{ijk}\hat{L}^{jk} = i\epsilon^{ijk}\int\frac{d^3p}{(2\pi)^3}\,\hat{a}_{\mathbf{p}}^{\dagger}p^k\frac{\partial}{\partial p^j}\hat{a}_{\mathbf{p}},\qquad\qquad(6.5.83)$$

where $\hat{L}^{\mu\nu}$ is the coordinate component of $\hat{M}^{\mu\nu}$ and $\hat{\mathbf{L}}$ is the orbital angular momentum operator. Show that the orbital angular momentum of a single boson at rest is zero, i.e., $\hat{\mathbf{L}}|\mathbf{p}=\mathbf{0}\rangle = 0$. (Hint: Consider integration by parts to move the momentum derivative.)

6.6   In Eq. (6.2.168) we showed, using its vacuum expectation value form, that $D_F(x-y)$ satisfied $(\partial_x^2 + m^2)D_F(x-y) = -i\delta(x-y)$. Using their vacuum expectation value forms, similarly show that this is true for each of $D_{\mathrm{ret}}(x-y)$, $D_{\mathrm{adv}}(x-y)$ and $\bar{D}_F(x-y)$.

6.7   Prove that (a) the only nonzero commutation relations for the annihilation and creation operators of a charged scalar field $(\hat{f}_{\mathbf{p}}, \hat{f}_{\mathbf{p}}^{\dagger}, \hat{g}_{\mathbf{p}}, \hat{g}_{\mathbf{p}}^{\dagger})$ are given by Eq. (6.2.180), i.e., $[\hat{f}_{\mathbf{p}}, \hat{f}_{\mathbf{p}'}^{\dagger}] = [\hat{g}_{\mathbf{p}}, \hat{g}_{\mathbf{p}'}^{\dagger}] = (2\pi)^3\delta^3(\mathbf{p}-\mathbf{p}')$, and (b) the only nonzero equal-time canonical commutation relations for the complex field operator $\hat{\phi}$ and its canonical momentum density operator $\hat{\pi}$ are given by Eq. (6.2.181), i.e., $[\hat{\phi}(t,\mathbf{x}), \hat{\pi}(t,\mathbf{x}')] = [\hat{\phi}^{\dagger}(t,\mathbf{x}), \hat{\pi}^{\dagger}(t,\mathbf{x}')] = i\delta^3(\mathbf{x}-\mathbf{x}')$.

6.8   Prove Eq. (6.3.197) for writing the fermion creation and annihilation operators in terms of the field operators.

6.9   Show that the fermion operators $\hat{P}^{\mu}$ and $\hat{M}^{\mu\nu}$ are the generators of the Poincaré group by proving Eqs. (6.3.214), (6.3.212) and (6.3.213). Show that $\hat{P}^{\mu}$ and $\hat{M}^{\mu\nu}$ are Hermitian. (Hint: First prove before normal-ordering and then prove that normal-ordering only involves adding a real constant. While each step is relatively straightforward, this proof requires care and effort.)

6.10 In the covariant quantization approach:

(a) Prove Eq. (6.4.148); i.e., the addition of $-\frac{1}{2\xi}(\partial_\mu A^\mu)^2$ to the Maxwell langrangian gives the equations of motion $\Box A^\nu - (1-\frac{1}{\xi})\partial^\nu(\partial_\mu A^\mu) = 0$.

(b) Complete the details of the sketch proof and show how to obtain the covariant canonical commutation relations for $\hat{A}^\mu$ and $\dot{\hat{A}}^\mu$ in Eq. (6.4.152).

6.11 Generalize the arguments for the neutral vector field and obtain some of the key results for the charged vector field case in analogy with the extension of the neutral scalar field to the charged scalar field. (Hint: A line-by-line derivation is not necessary, but explain the rationale for the summary of results.)

6.12 For a massive vector boson field:

(a) Prove Eq. (6.5.51), which is the result that $\hat{A}^0(x) = -\boldsymbol{\nabla} \cdot \hat{\mathbf{E}}(x)/m^2$ for a free massive vector boson field.

(b) Prove Eq. (6.5.54), which are the commutation relations for the annihilation and creation operators.

6.13 (a) Define $\{V_1, V_2, \ldots, V_{16}\} \equiv \frac{1}{2}\{I, \gamma^0, i\gamma^j, i\sigma^{0j}, \sigma^{jk}|_{j<k}, i\gamma^5\gamma^0, \gamma^5\gamma^j, \gamma^5\}$. Show that these sixteen matrices are Hermitian, $V_J = V_J^\dagger$.

(b) Using the γ-matrix anticommutation relations show that (i) $V_J^2 = \frac{1}{4}I$; (ii) $V_J V_K = \pm V_K V_J$; and (iii) $V_K V_J V_K = -\frac{1}{4}V_J$ for every $J \neq 1$ for at least one K ($K = 16$ for $J = 2, \ldots, 5$ and $12, \ldots, 15$; K is any of $2, \ldots, 5$ for $J = 16$; and K is at least one of $2, \ldots, 5$ for $J = 6, \ldots, 11$).

(c) Show for every V_J and V_K that $V_J V_K = \frac{1}{2}c_{JKL} V_L$ where $c_{JKL} = \pm 1, \pm i$ for only one V_L and where *no sum* over L is implied. Note that for fixed J each K leads to a different L and for fixed K each J leads to a different L. This is referred to as the *rearrangement lemma*.

(d) By considering $\mathrm{tr}(V_K V_J V_K)$ explain why $\mathrm{tr}V_J = 0$ for every $J \neq 1$.

(e) Explain why trace identities lead to "orthonormality", $\mathrm{tr}(V_J^\dagger V_K) = \mathrm{tr}(V_J V_K) = \delta_{JK}$.

(f) Show that $c_{JKL} = 2\mathrm{tr}(V_J V_K V_L)$.

(g) Define the 16 vectors $(X_J)_r \equiv (V_J)_{ij}$ with $r \equiv j + 4(i-1) = 1, \ldots, 16$. Hence show that $\mathrm{tr}(V_J^\dagger V_K) = \sum_r (X_J)_r^*(X_K)_r = X_J^\dagger X_K = \delta_{JK}$ so that the X_J are 16 orthonormal sixteen component complex vectors and hence are an orthonormal basis of $\mathbb{C}^{16}$. Show that the completeness relation for this orthonormal basis $\sum_J (X_J)_r (X_J^*)_s = \delta_{rs}$ can be written as $\sum_J (V_J)_{ij}(V_J^*)_{k\ell} = \sum_J (V_J)_{ij}(V_J^\dagger)_{\ell k} = \sum_J (V_J)_{ij}(V_J)_{\ell k} = \delta_{ik}\delta_{j\ell}$.

(h) Any 4×4 complex matrix can be regarded as a vector in $\mathbb{C}^{16}$. Completeness of the X_J then means completeness of the matrices V_J and so for any matrix A we can write $A \equiv \sum_J c_J V_J$. Show that $c_J = \mathrm{tr}(V_K A)$.

(i) Explain why a matrix that commutes with all four γ-matrices must commute with all matrices and so must be a multiple of the identity. This is sometimes referred to *Schur's lemma*.

(j) Show $(V_J)_{ij}(V_K)_{k\ell} = \sum_{m,n}(V_J)_{im}(V_K)_{kn}\delta_{mj}\delta_{n\ell} = \sum_L (V_J V_L)_{i\ell}(V_K V_L)_{kj}$ using completeness. Then show using the rearrangement lemma that $(V_J)_{ij}(V_K)_{k\ell} = (1/4)\sum_L c_{JLM}c_{KLN}(V_M)_{i\ell}(V_N)_{kj}$. Explain why we can write for some coefficients $C_{JKMN} \in \mathbb{C}$,

$$(V_J)_{ij}(V_K)_{k\ell} \equiv \sum_{M,N} C_{JKMN}(V_M)_{i\ell}(V_N)_{kj}, \qquad (6.5.84)$$

which is referred to as the general form of the *Fierz identity*. Explain why $C_{JKMN} = 0$ if $J = K$ unless $M = N$ and vice versa.

(k) Using this result and orthonormality show that $C_{JKMN} = \mathrm{tr}(V_M V_J V_N V_K)$.

(l) Rewrite the 16 basis matrices as $2V_J \to \Gamma_{\mathcal{A}}^a$, where a labels the appropriate combination of spacetime indices and where $\mathcal{A} = S, V, T, A, P$ denote scalar, vector, tensor, axial vector and pseudoscalar, respectively. Specifically, we define $\Gamma_S^1 \equiv I, \Gamma_V^1, \ldots, \Gamma_V^4 \equiv \gamma^\mu$, $\Gamma_T^1, \ldots, \Gamma_T^6 \equiv \sigma^{\mu\nu}|_{\mu<\nu}, \Gamma_A^1, \ldots, \Gamma_A^4 \equiv i\gamma^5\gamma^\nu$, and $\Gamma_P^1 \equiv \gamma^5$. Use the Einstein summation convention for the collective spacetime index a. Summations over the index $\mathcal{A}$ will always be explicitly written. We define $\Gamma_{\mathcal{A}a} \equiv g_{ab}\Gamma_{\mathcal{A}}^b$, where g_{ab} represents the appropriate number of $g_{\mu\nu}$ for the particular $\mathcal{A}$. For example, we have $\Gamma_{S1} \equiv I, \Gamma_{V1}, \ldots, \Gamma_{V4} \equiv \gamma_\mu$, $\Gamma_{T1}, \ldots, \Gamma_{T6} \equiv \sigma_{\mu\nu}|_{\mu<\nu}, \Gamma_{A1}, \ldots, \Gamma_{A4} \equiv i\gamma^5\gamma_\mu$ and $\Gamma_{P1} \equiv \gamma^5$. With this notation show that the above results can be rewritten as:

$\text{tr}(\Gamma_A^a \Gamma_{Bb}) = 4\delta_{AB}\delta_b^a$; $A \equiv \sum_A c_{Aa}\Gamma_A^a$ for some $c_{Aa} \in \mathbb{C}$;

$\sum_A (\Gamma_A^a)_{ij} (\Gamma_{Aa})_{k\ell} = \sum_A (\Gamma_{Aa})_{ij} (\Gamma_A^a)_{k\ell} = 4\delta_{i\ell}\delta_{kj}$;

$\Gamma_A^a \Gamma_B^b = \rho_{ABC}^{abc}\Gamma_{Cc}$ with $\rho_{ABC}^{abc} = \frac{1}{4}\text{tr}(\Gamma_A^a\Gamma_B^b\Gamma_C^c)$;

$(\Gamma_A^a)_{ij} (\Gamma_B^b)_{kl} = \sum_{C,D} C_{ABCD}^{abcd} (\Gamma_{Cc})_{i\ell} (\Gamma_{Dd})_{kj}$ (Fierz transformation);

$C_{ABCD}^{abcd} = \frac{1}{16}\text{tr}(\Gamma_C^c\Gamma_A^a\Gamma_D^d\Gamma_B^b)$, where C_{ABCD}^{abcd} vanishes if $(a,A) = (b,B)$ unless $(c,C) = (d,D)$ and vice versa.

(m) The standard Fierz transformation has $(a,A) = (b,B)$ and hence $(c,C) = (d,D)$ and is a special case. Recall that we sum over repeated spacetime indices. Show that $(\Gamma_A^a)_{ij} (\Gamma_{Aa})_{kl} = \sum_C C_{AC} (\Gamma_C^c)_{i\ell} (\Gamma_{Cc})_{kj}$ with $C_{AB} = \frac{1}{16N_B}\text{tr}(\Gamma_{Bb}\Gamma_A^a\Gamma_B^b\Gamma_{Aa})$ and $N_B \equiv 1, 4, 6, 4, 1$ for $B = S, V, T, A, P$, respectively.

(n) The contraction identities can be written as $\Gamma_A^a\Gamma_B^b\Gamma_{Aa} \equiv f_{AB}\Gamma_B^b$, which defines the coefficients f_{AB}. Show that $C_{AB} = \frac{1}{4}f_{AB}$, which can be written in matrix form as

$$C = \frac{1}{4}f = \frac{1}{4}\begin{bmatrix} 1 & 1 & 1 & 1 & 1 \\ 4 & -2 & 0 & 2 & -4 \\ 6 & 0 & -2 & 0 & 6 \\ 4 & 2 & 0 & -2 & -4 \\ 1 & -1 & 1 & -1 & 1 \end{bmatrix}, \tag{6.5.85}$$

where the row (A) and column (B) indices occur in the order S, V, T, A, P. Explicitly verify at least six of these matrix elements.

(o) Adopt the notation of Itzykson and Zuber (1980) to write the standard Fierz identities in a more familiar form. Consider four arbitrary spinors $\psi_1, \ldots, \psi_4$ and five Lorentz scalar quadrilinear spinor combinations,

$$s(1,2;3,4) \equiv (\bar{\psi}_1\Gamma_S\psi_2)(\bar{\psi}_3\Gamma_S\psi_4) = (\bar{\psi}_1\psi_2)(\bar{\psi}_3\psi_4)$$

$$v(1,2;3,4) \equiv (\bar{\psi}_1\Gamma_V^a\psi_2)(\bar{\psi}_3\Gamma_{Va}\psi_4) = (\bar{\psi}_1\gamma^\mu\psi_2)(\bar{\psi}_3\gamma_\mu\psi_4)$$

$$t(1,2;3,4) \equiv (\bar{\psi}_1\Gamma_T^a\psi_2)(\bar{\psi}_3\Gamma_{Ta}\psi_4) = (\bar{\psi}_1\sigma^{\mu\nu}|_{\mu<\nu}\psi_2)(\bar{\psi}_3\sigma_{\mu\nu}|_{\mu<\nu}\psi_4)$$

$$= \tfrac{1}{2}(\bar{\psi}_1\sigma^{\mu\nu}\psi_2)(\bar{\psi}_3\sigma_{\mu\nu}\psi_4)$$

$$a(1,2;3,4) \equiv (\bar{\psi}_1\Gamma_A^a\psi_2)(\bar{\psi}_3\Gamma_{Aa}\psi_4) = (\bar{\psi}_1(i\gamma^5\gamma^\mu)\psi_2)(\bar{\psi}_3(i\gamma^5\gamma_\mu)\psi_4)$$

$$= (\bar{\psi}_1\gamma^5\gamma^\mu\psi_2)(\bar{\psi}_3\gamma_\mu\gamma^5\psi_4)$$

$$p(1,2;3,4) \equiv (\bar{\psi}_1\Gamma_P\psi_2)(\bar{\psi}_3\Gamma_P\psi_4) = (\bar{\psi}_1\gamma^5\psi_2)(\bar{\psi}_3\gamma^5\psi_4). \tag{6.5.86}$$

Define Ψ to be the column vector with elements Ψ_A being s, v, t, a, p for $A = S, V, T, A, P$, respectively, i.e., $\Psi(1,2;3,4) \equiv [s,v,t,a,p](1,2;3,4)$. Show that the standard Fierz identity can be compactly summarized as

$$\Psi(1,4;3,2) = C\Psi(1,2;3,4) \tag{6.5.87}$$

for arbitrary spinors $\psi_1, \psi_2, \psi_3, \psi_4$. So the action of the matrix C is to exchange the second and fourth spinors (or equivalently the first and third). We have $\Psi(1,2;3,4) = C\Psi(1,4;3,2) = C^2\Psi(1;2;3,4)$ and so $C^2 = I$. Explicitly verify that this is true of the matrix C. Note that if instead of spinors we have anticommuting fermion operators then the required exchange of two fermion operators in the reordering gives an overall minus sign, i.e., $\hat{\Psi}(1,2;3,4) = -C\hat{\Psi}(1,4;3,2)$, where the caret on the Ψ denotes that it is an operator quadilinear made up of four fermion operators.

7 Interacting Quantum Field Theories

Now that we have shown how to construct free field theories it is necessary to learn how to introduce interactions between the particles in our theories. Once we have done that we need to build an understanding of how to perform observations of them using scattering theory and to calculate cross-sections using Feynman rules and Feynman diagrams.

7.1 Physical Spectrum of States

If a state of an interacting system contains only a single particle of mass m or a single bound state of mass M, then these single-particle states can be chosen to be exact eigenstates of the full interacting Hamiltonian operator $\hat{H} = \hat{P}^0$ and the total three-momentum operator $\hat{\mathbf{P}}$. The eigenstates will correspond to plane wave states in terms of these masses. Note that these masses arise when the interactions of the theory are included and differ from the free particle mass m_0 of the noninteracting theory. Multiparticle asymptotic states approach exact eigenstates of $\hat{P}^\mu = (\hat{H}, \hat{\mathbf{P}})$ when the interactions between particles fall off sufficiently rapidly with distance. We can make an asymptotic state by creating single particles in arbitrary but finite wavepackets with different group velocities and considering $t \to \pm\infty$. The arbitrary wavepackets will have an infinite separation between them and will not be interacting if the forces between them fall off fast enough. The wavepackets can then be taken to be very large and the momentum spread in the wavepacket to become very small. So the asymptotic states can arbitrarily closely approach a collection of noninteracting physical particles in approximate plane wave states and they will be approximate eigenstates of $\hat{P}^\mu$. We can then discuss the energies and momenta of multiparticle asymptotic states in the same way that we would for a free theory, but now in terms of the physical single-particle mass m and including the possibility of one or more bound states with physical masses M_1, M_2 and so on. For the purpose of building asymptotic states we consider bound states to be different types of particles. In general a bound state might have nonzero total angular momentum, which is then the spin of this corresponding particle. This overall picture of the spectrum of asymptotic states is illustrated in Fig. 7.1.

We can then build a mixed Fock space in terms of the full vacuum $|\Omega\rangle$, physical singe-particle states and the different bound states. This is a Fock space constructed from free single particles of mass m and free bound states of masses M_1, M_2 and so on. Denote this Fock space as V_{Fock} and the full Hilbert space of the interacting theory as V_{full}. Let $|\mathbf{p}\rangle$ denote a single particle or a stable bound state in an on-shell momentum eigenstate in the full theory. Then clearly for all $|\mathbf{p}\rangle$ we have

$$|\Omega\rangle, |\mathbf{p}\rangle \in V_{\text{Fock}} \quad \text{and} \quad |\Omega\rangle, |\mathbf{p}\rangle \in V_{\text{full}}. \tag{7.1.1}$$

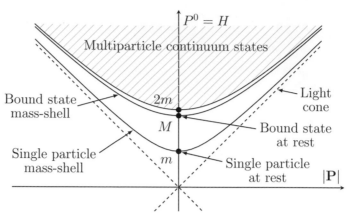

Figure 7.1 Energy ($H = P^0$) and three-momentum (**P**) eigenstates of physical asymptotic states. A continuum of states begins above $2m$ and bound states can also occur.

Consider some Heisenberg picture state in the full Hilbert space, $|\alpha\rangle \in V_{\text{full}}$ with the Heisenberg and Schrödinger pictures coinciding at $t = t_0$. Using the evolution operator $\hat{U}(t'', t') = e^{-i\hat{H}(t''-t')}$ we have for the Schrödinger picture state $|\alpha, t\rangle_s$,

$$|\alpha, t''\rangle_s = \hat{U}(t'', t_0)|\alpha\rangle = \hat{U}(t'', t')|\alpha, t'\rangle_s \quad \text{and} \quad |\alpha\rangle = |\alpha, t_0\rangle_s. \tag{7.1.2}$$

In an interacting theory two particles will not stay a finite distance apart forever unless they are in a bound state, but in that case we treat the bound state as a separate type of single particle in our mixed Fock space as discussed above. If the *cluster decomposition principle* holds in the full theory, then for every state $|\alpha, t\rangle_s \in V_{\text{full}}$ with finite spatial extent at some finite time t there will be a state $|\alpha, t\rangle_s^F \in V_{\text{Fock}}$ such that

$$|\alpha, t\rangle_s \in V_{\text{full}} \xrightarrow{\ t\to\pm\infty\ } |\alpha, t\rangle_s^F \in V_{\text{Fock}}, \tag{7.1.3}$$

where $|\alpha, t\rangle_s^F$ evolves in time as a set of free particles (including stable bound states) with their physical masses, m, M_1, M_2 and so on. Since the particles are becoming infinitely far apart with the forces between them becoming negligible, then they approach the limit of a collection of free particles and evolve as such. The point is that *if the cluster decomposition principle holds, then any state with finite spatial extent in the interacting theory evolves as $t \to \pm\infty$ into a state that asymptotically approaches a state in the asymptotic mixed Fock space.* It is not being implied that the asymptotic mixed Fock space V_{Fock} is equal (or isomorphic) to the full Hilbert space V_{full}, but a subspace of V_{full} spanned by states with large wavepacket separations ($> d$) will asymptotically approach the corresponding subspace in V_{Fock} as the limit $d \to \infty$ is taken.

> **Cluster decomposition principle:** In the above discussion we are implicitly assuming that when stable particles and/or stable bound states have spacelike separations approaching infinity, then we can neglect their influence on each other. This is a familiar property of physical systems. Were it not the case, then to describe the behavior of any system we would have to describe the behavior of the whole universe. This property is necessary in order to be able to define an *isolated system*. The cluster decomposition principle is sometimes called the cluster decomposition theorem. Consider a system consisting of two clusters of particles whose

spacelike separation is characterized by the distance d. As $d \to \infty$ the propagator for the system of particles factorizes into a product of two propagators with one propagator per cluster when the principle is valid. In the case of axiomatic quantum field theory, Streater and Wightman (2000) prove the cluster decomposition principle as the factorization of the Wightman functions in their Theorem 3-4 for theories with and without a mass gap. In such theories they show that the interaction between two clusters falls off faster than any power of $1/d$ if the theory has a mass gap, i.e., if the theory has no massless particles. If the theory does not have a mass gap, then it has one or more massless particles and the interaction may fall off as slowly as $1/d^2$. In realistic theories, such as gauge theories, the Wightman axioms are not satisfied and the cluster decomposition principle has to be considered on a case-by-case basis (Strocchi, 1978). In a theory with a Yukawa interaction, which is the exchange of a massive boson, the interactions fall off exponentially with distance and the cluster decomposition principle is satisfied. In quantum electrodynamics (QED) the massless photon leads to the Coulomb force, which falls off as $1/d^2$ and is a marginal case. However, in real QED scattering processes charged particles are typically shielded by other matter at very large separations that lead to electric neutrality and so in practice the cluster decomposition holds in that case. At low energies the description of the scattering of charged particles from a Coulomb potential requires the use of Coulomb wavefunctions (Messiah, 1975). In the case of quantum chromodynamics (QCD), the cluster decomposition must fail for quarks, gluons and any colored states since such states are never observed in nature (Strocchi, 1976). This property of QCD is referred to as color confinement as discussed in Sec. 5.2.2 and is observed in nonperturbative studies of QCD using the techniques of lattice gauge theory. A discussion of the general form of the cluster decomposition property in terms of the factorization of the S-matrix can be found in Weinberg (1995).

Fock space for higher spin states: We saw how to construct free field theories and their Fock spaces for spins $s = 0, \frac{1}{2}, 1$ in Chapter 6. However, we have not constructed a Fock spaces for particles with *higher spin* ($s \geq \frac{3}{2}$). In his discussion of general causal fields Weinberg has shown how to construct field theories for arbitrary irreducible representations, (j_1, j_2), of the restricted Lorentz group, from which free field theories of arbitrary spin can be constructed (Weinberg, 1995). As discussed there, the usual problems reported for the construction of field theories with $s \geq \frac{3}{2}$ only arise when they are coupled to external sources or are otherwise interacting. These difficulties are not relevant for free particles in a Fock space and so we can construct a Fock space for free bound states of arbitrary spin. We only consider interacting quantum field theories involving *fundamental* particles with spins $s = 0, \frac{1}{2}, 1$, since the formulation of consistent interacting quantum field theories for fundamental particles of higher spin ($s \geq \frac{3}{2}$) remains an unsolved problem.

As discussed in Chapter 6 we should use symmetrized or antisymmetrized states for bosons and fermions, respectively. For simplicity we temporarily put this issue aside, since it is a simple matter to impose this requirement at the end of our discussion.

Consider Fock space states $|\alpha, t\rangle_s^F$ that consist of the free particles in momentum eigenstates, $|\alpha, t\rangle_s^F = |\mathbf{p}_1 \cdots \mathbf{p}_m, t\rangle_s^F$. In that case we have by definition

$$|\mathbf{p}_1 \cdots \mathbf{p}_m, t\rangle_s^F = e^{-iE_\alpha(t-t_0)}|\mathbf{p}_1 \cdots \mathbf{p}_m\rangle^F, \qquad (7.1.4)$$

where $E_\alpha \equiv E_{\mathbf{p}_1} + \cdots + E_{\mathbf{p}_m}$ is the total energy of the free particles and where $|\mathbf{p}_1 \cdots \mathbf{p}_m\rangle^F$ is the Heisenberg picture state in the Fock space.

The *scattering matrix* or *S-matrix* is defined through its matrix elements

$$S_{\beta\alpha} \equiv \langle\beta| \lim_{T\to\infty} \hat{U}(T,-T)|\alpha\rangle = \langle\beta|\hat{S}|\alpha\rangle, \tag{7.1.5}$$

where $\hat{U}(t'',t')$ is the evolution operator in Eq. (4.1.11) and where we have defined the S-matrix operator, $\hat{S} \equiv \lim_{T\to\infty} \hat{U}(T,-T)$. Here $|\alpha\rangle$ and $|\beta\rangle$ are Schrödinger picture states defined at any common time, i.e., $|\alpha\rangle \equiv |\alpha,t\rangle_s$ and $|\beta\rangle \equiv |\beta,t\rangle_s$ for any time t. To study scattering we choose the common time as either $t \to \pm\infty$ so that we can use Eq. (7.1.3). In that case we are free to choose $|\alpha,t\rangle$ and $|\beta,t\rangle$ at fixed common time t as any two states in the noninteracting Fock space. For scattering we will typically choose $|\alpha\rangle \equiv |\alpha,t\rangle = |\mathbf{p}_1 \cdots \mathbf{p}_m\rangle^F$ and $|\beta\rangle \equiv |\beta,t\rangle = |\mathbf{k}_1 \cdots \mathbf{k}_n\rangle^F$ with an m particle initial state and an n particle final state. These states are Heisenberg picture states in the noninteracting field theory for particles of physical mass m.

It will later be convenient to omit the Fock-space label "F" and understand what is meant by the states $|\mathbf{p}_1 \cdots \mathbf{p}_m\rangle$ and $|\mathbf{k}_1 \cdots \mathbf{k}_n\rangle$. We will return to explore the S-matrix further in Sec. 7.3. It would be more appropriate to use finite localized wavepackets that can become plane waves after taking $T \to \infty$. This will be discussed in Sec. 7.5.1.

The subspace of this asymptotic mixed Fock space containing only one type of scalar particle of mass m and no bound states is built from the full vacuum $|\Omega\rangle$ and the one-particle momentum eigenstates $|\mathbf{p}\rangle$ of mass m. The states in this subspace consist of states made from particles of mass m that contain all of the particle self-interactions but do not include interactions *between* these physical particles. The effect of the self-interactions has been to replace the *bare mass* m_0 of the noninteracting theory with the single-particle physical mass m.

We can define *asymptotic single-particle annihilation and creation operators*, $\hat{a}_{\mathrm{asp}}$ and $\hat{a}^\dagger_{\mathrm{asp}}$, respectively, as the operators that move between the basis vectors of this Fock subspace of free particles of mass m in the usual way. Since we are defining a Fock space they must satisfy Eq. (6.2.17),

$$[\hat{a}_{\mathrm{asp}}, \hat{a}_{\mathrm{asp}'}] = [\hat{a}^\dagger_{\mathrm{asp}}, \hat{a}^\dagger_{\mathrm{asp}'}] = 0 \quad \text{and} \quad [\hat{a}_{\mathrm{asp}}, \hat{a}^\dagger_{\mathrm{asp}'}] = (2\pi)^3 \delta^3(\mathbf{p}-\mathbf{p}'). \tag{7.1.6}$$

We can then use these operators as we did in the free field case in Eq. (6.2.97) to define an asymptotic field $\hat{\phi}_{\mathrm{as}}(x)$,

$$\hat{\phi}_{\mathrm{as}}(x) = \int \frac{d^3p}{(2\pi)^3} \frac{1}{\sqrt{2E_\mathbf{p}}} \left(\hat{a}_{\mathrm{asp}} e^{-ip\cdot x} + \hat{a}^\dagger_{\mathrm{asp}} e^{ip\cdot x} \right)\big|_{p^0=E_\mathbf{p}}. \tag{7.1.7}$$

We can then proceed with all of the previous free scalar field theory construction provided that we understand that these are valid when the interactions between our particles of mass m can be neglected. For example, it follows that

$$\langle\Omega|T\hat{\phi}_{\mathrm{as}}(x)\hat{\phi}_{\mathrm{as}}(y)|\Omega\rangle = \int \frac{d^4p}{(2\pi)^4} \frac{i}{p^2 - m^2 + i\epsilon} e^{-ip\cdot(x-y)} \equiv D_0(m^2;x-y), \tag{7.1.8}$$

where D_0 denotes the free form of the scalar boson propagator. We will reserve the notation D_F for the full nonperturbative propagator of the interacting theory.

Operator equations of motion: Provided that we arrive at interacting quantum field theories through the canonical quantization approach, then the operators of the interacting quantum field theory will obey the classical equations of motion. This includes the equal-time canonical commutation and anticommutation relations for the field operators and their canonical momentum operators, which follow from the Poisson bracket or Dirac bracket as appropriate for the field theory. We can specify operators at some initial time, say, $t = t_0$, when the Schrödinger and Heisenberg pictures coincide. The fields will evolve with the full Hamiltonian operator for the interacting theory, $\hat{H}$. Thus according to Eqs. (4.1.23) and (6.2.78) we have

$$\hat{\phi}(x) = \hat{U}(t,t_0)^\dagger \hat{\phi}(t_0,\mathbf{x})\hat{U}(t,t_0) = \hat{U}(t,t_0)^\dagger \hat{\phi}_s(\mathbf{x})\hat{U}(t,t_0), \tag{7.1.9}$$

$$\hat{\pi}(x) = \hat{U}(t,t_0)^\dagger \hat{\pi}(t_0,\mathbf{x})\hat{U}(t,t_0) = \hat{U}(t,t_0)^\dagger \hat{\pi}_s(\mathbf{x})\hat{U}(t,t_0),$$

where the ETCR are those given in Eq. (6.2.12), $[\hat{\phi}_s(\mathbf{x}),\hat{\pi}_s(\mathbf{x}')] = i\delta^3(\mathbf{x}-\mathbf{x}')$ and $[\hat{\phi}_s(\mathbf{x}),\hat{\phi}_s(\mathbf{x}')] = [\hat{\pi}_s(\mathbf{x}),\hat{\pi}_s(\mathbf{x}')] = 0$. In an interacting theory we can have $\hat{\pi}(x) \neq d\hat{\phi}(x)/dt$, such as in scalar QED, for example. From Eq. (7.1.9) it follows that the ETCR are true for the interacting Heisenberg picture fields at all times.

7.2 Källén-Lehmann Spectral Representation

Recall that physical single-particle states are exact eigenstates of the full Hamiltonian and have physical mass m rather than the bare m_0. We only need to make use of the asymptotic regime when there is more than one particle or bound state present. The relationship between the physical and bare masses depends on the interaction and is something that we will consider in more detail later.

Just as we did in the free case in Eq. (6.2.54) we can write the exact projection operator onto one-particle states of the full theory as

$$\hat{I}_1 = \int \frac{d^3p}{(2\pi)^3}|\mathbf{p}\rangle\frac{1}{2E_\mathbf{p}}\langle\mathbf{p}|, \tag{7.2.1}$$

where $|\mathbf{p}\rangle$ denotes exact eigenstates of $\hat{P}^\mu$ with $p^2 = m^2$. It remains to consider the other exact eigenstates of $\hat{P}^\mu$ in the interacting theory. By construction our theories are Poincaré invariant and so we know that we can classify states according to their eigenvalues with respect to the Casimir operators $\hat{P}^2$ and $\hat{W}^2$; their transformation properties under C, P and T, and any conserved charges that they might carry.

An interacting Hermitian scalar field operator $\hat{\phi}(x)$ can only produce states from the physical vacuum $|\Omega\rangle$ with scalar quantum numbers. So while the interacting theory might admit bound states with nonzero angular momentum, $\hat{\phi}(x)|\Omega\rangle$ will not contain any contribution from such states. Let us label all of the scalar states in the interacting theory, other than the vacuum state, using the label κ such that $|\kappa;\mathbf{p}\rangle$ is an eigenstate of $\hat{P}^\mu = (\hat{H},\hat{\mathbf{P}})$ with eigenvalues p^μ with $p^0 = E_\mathbf{p} = \sqrt{\mathbf{p}^2 + m_\kappa^2}$. Then we can write the projection operator onto the subspace of all scalar states as

$$\hat{I}_{\text{scalar}} = |\Omega\rangle\langle\Omega| + \sum_\kappa \int \frac{d^3p}{(2\pi)^3}|\kappa;\mathbf{p}\rangle\frac{1}{2E_\mathbf{p}}\langle\kappa;\mathbf{p}|, \tag{7.2.2}$$

where we understand that $\sum_\kappa$ may have both sum and integral components to it.

Consider the two-point Green's function with $x^0 > y^0$,

$$\langle\Omega|\hat{\phi}(x)\hat{\phi}(y)|\Omega\rangle = \langle\Omega|\hat{\phi}(x)\hat{I}\hat{\phi}(y)|\Omega\rangle = \langle\Omega|\hat{\phi}(x)\hat{I}_{\text{scalar}}\hat{\phi}(y)|\Omega\rangle \tag{7.2.3}$$

$$= \langle\Omega|\hat{\phi}(x)|\Omega\rangle\langle\Omega|\hat{\phi}(y)|\Omega\rangle + \sum_\kappa \int \frac{d^3p}{(2\pi)^3}\langle\Omega|\hat{\phi}(x)|\kappa;\mathbf{p}\rangle\frac{1}{2E_\mathbf{p}}\left.\langle\kappa;\mathbf{p}|\hat{\phi}(y)|\Omega\rangle\right|_{E_\mathbf{p}\equiv\sqrt{p^2+m_\kappa^2}}$$

$$= \langle\Omega|\hat{\phi}(0)|\Omega\rangle^2 + \sum_\kappa |\langle\Omega|\hat{\phi}(0)|\kappa;\mathbf{p}=0\rangle|^2 \int \frac{d^3p}{(2\pi)^3}\frac{1}{2E_\mathbf{p}}\left. e^{-ip\cdot(x-y)}\right|_{E_\mathbf{p}\equiv\sqrt{p^2+m_\kappa^2}}.$$

We have used Eqs. (6.2.111) and (6.2.113) to write

$$\hat{\phi}(x + a) = e^{i\hat{P}\cdot a}\hat{\phi}(x)e^{-i\hat{P}\cdot a} \quad \Rightarrow \quad \hat{\phi}(x) = e^{i\hat{P}\cdot x}\hat{\phi}(0)e^{-i\hat{P}\cdot x}, \tag{7.2.4}$$

which using $\hat{P}^\mu|\kappa;\mathbf{p}\rangle = p^\mu|\kappa;\mathbf{p}\rangle$ with $p^0 = E_\mathbf{p} = \mathbf{p}^2 + m_\kappa^2 > 0$ gives

$$\langle\Omega|\hat{\phi}(x)|\kappa;\mathbf{p}\rangle = \langle\Omega|\hat{\phi}(0)|\kappa;\mathbf{p} = 0\rangle e^{-ip\cdot x}\Big|_{p^0 = E_\mathbf{p} \equiv \sqrt{\mathbf{p}^2 + m_\kappa^2}}. \tag{7.2.5}$$

Only a neutral scalar field has the quantum numbers of the vacuum (no conserved charges and $J^{PC} = 0^{++}$) and so only a neutral scalar field can have a nonzero field vacuum expectation value (VEV), $\langle\Omega|\hat{\phi}(x)|\Omega\rangle \neq 0$. Since we are considering translationally invariant theories we can only ever have $\langle\Omega|\hat{\phi}(x)|\Omega\rangle = v$ with v a constant. A nonzero VEV was previously briefly discussed in terms of the Higgs boson in Sec. 5.2.3. In the case of a nonzero VEV we can always shift the field by defining $\hat{\phi}'(x) \equiv \hat{\phi}(x) - v$ such that $\langle\Omega|\hat{\phi}'(x)|\Omega\rangle = 0$. Shifting the neutral scalar field by a constant does not change $\frac{1}{2}\partial_\mu\hat{\phi}\partial^\mu\hat{\phi}$, but there will be consequences for any mass and interaction terms in the Lagrangian density as we will later see in the Standard Model. For now we consider only the case with no VEV, $v = 0$.

In that case we can use Eq. (6.2.158) and write

$$\langle\Omega|\hat{\phi}(x)\hat{\phi}(y)|\Omega\rangle\Big|_{x^0 > y^0} = \sum_\kappa |\langle\Omega|\hat{\phi}(0)|\kappa;\mathbf{p} = 0\rangle|^2 \Delta_+(m_\kappa^2; x - y), \tag{7.2.6}$$

where $\Delta_+(m_\kappa^2; x - y)$ is just $\Delta_+(x - y)$ with m^2 replaced by m_κ^2. It follows that

$$\langle\Omega|T\hat{\phi}(x)\hat{\phi}(y)|\Omega\rangle$$
$$= \sum_\kappa |\langle\Omega|\hat{\phi}(0)|\kappa;\mathbf{p} = 0\rangle|^2 \left[\theta(x^0 - y^0)\Delta_+(m_\kappa^2; x - y) + \theta(y^0 - x^0)\Delta_+(m_\kappa^2; y - x)\right]$$
$$= \sum_\kappa |\langle\Omega|\hat{\phi}(0)|\kappa;\mathbf{p} = 0\rangle|^2 D_0(m_\kappa^2; x - y) = \int_0^\infty ds\, \rho(s)D_0(s; x - y)$$
$$= \int \frac{d^4p}{(2\pi)^4} \int_0^\infty ds\, \frac{i\rho(s)}{p^2 - s + i\epsilon}e^{-ip\cdot(x-y)}, \tag{7.2.7}$$

where we have used Eq. (6.2.160), where $D_0(s; x - y)$ is the free propagator with mass-squared s and where the *spectral density function* has been defined as

$$\rho(s) \equiv \sum_\kappa |\langle\Omega|\hat{\phi}(0)|\kappa;\mathbf{p} = 0\rangle|^2 \delta(s - m_\kappa^2) > 0. \tag{7.2.8}$$

We can then define the full propagator of the interacting theory as

$$D_F(x - y) \equiv \langle\Omega|T\hat{\phi}(x)\hat{\phi}(y)|\Omega\rangle \tag{7.2.9}$$

$$= \int \frac{d^4p}{(2\pi)^4} \left[\int_0^\infty ds\, \frac{i\rho(s)}{p^2 - s + i\epsilon}\right] e^{-ip\cdot(x-y)} \equiv \int \frac{d^4p}{(2\pi)^4} D_F(p)e^{-ip\cdot(x-y)},$$

where the last line defines the momentum-space full propagator $D_F(p)$. This expression of the two-point function is referred as the *spectral representation* or as the *Källén-Lehmann representation* (Källén, 1952; Lehmann, 1954). From Eq. (7.2.8) we see that $\rho(s) > 0$. A typical shape for the spectral function is given in Fig. 7.2, which shows the one-particle contribution, the bound states with scalar quantum numbers and the scalar multiparticle contributions. The contribution from scalar n-particle states occurs in the region $s \geq (nm)^2$. The corresponding analytic structure of the two-point Green's function is shown in Fig. 7.3. We can write

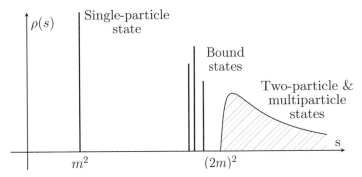

Figure 7.2   Typical form of the spectral function $\rho(s)$ for an interacting scalar theory. Shown here is a pole for the single-particle state at $s = m^2$, several bound states in the region $m^2 < s < (2m)^2$ and a continuum of states for $s \geq (2m)^2$.

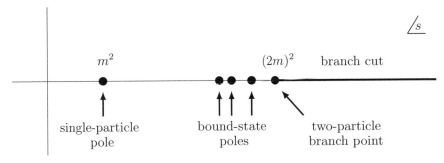

Figure 7.3   Singularity structure of the two-point Green's function in the complex s-plane. There are single-particle and bound-state poles and n-particle branch cuts starting at branch points $s = (nm)^2$. The two-particle branch cut is shown and starts at $s = (2m)^2$.

$$\rho(s) = Z\delta(s - m^2) + [\text{bound state } \delta\text{-functions}] \tag{7.2.10}$$
$$+ [\text{2-particle part for } s \geq (2m)^2] + [\text{3-particle part for } s \geq (3m)^2] + \cdots.$$

The magnitude of the one-particle contribution is determined by $Z > 0$, which is referred to as as the *wave-function or field-strength renormalization*. There may be three-particle bound states lying below the three-particle threshold and above the two-particle threshold and so on for bound states of n particles. Our definition of $\rho(s)$ differs from that in Peskin and Schroeder (1995) by a factor of 2π. Note that

$$D_F(p) = Z\frac{i}{p^2 - m^2 + i\epsilon} + \int_{s>m^2}^{\infty} ds\, \frac{i\rho(s)}{p^2 - s + i\epsilon}, \tag{7.2.11}$$

$$Z = \lim_{p^2 \to m^2 + i\epsilon} \frac{(p^2 - m^2 + i\epsilon)}{i} D_F(p) \quad \text{and} \quad \lim_{p^2 \to m^2 + i\epsilon} D_F(p) \to \frac{iZ}{p^2 - m^2 + i\epsilon}.$$

A charged scalar field cannot have a VEV and so $\langle\Omega|\hat{\phi}(x)|\Omega\rangle = \langle\Omega|\hat{\phi}(x)^\dagger|\Omega\rangle = 0$. Following steps analogous to those in Eqs. (7.2.3)–(7.2.6) and using $\langle\kappa; \mathbf{p} = 0|\hat{\phi}^\dagger(0)|\Omega\rangle = \langle\Omega|\hat{\phi}(0)|\kappa; \mathbf{p} = 0\rangle^*$ and a similar result with $\phi \to \phi^\dagger$ gives

$$\langle\Omega|\hat{\phi}(x)\hat{\phi}^\dagger(y)|\Omega\rangle\Big|_{x^0 > y^0} = \sum_\kappa |\langle\Omega|\hat{\phi}(0)|\kappa; \mathbf{p} = 0\rangle|^2 \Delta_+(m_\kappa^2; x - y),$$

$$\langle\Omega|\hat{\phi}^\dagger(y)\hat{\phi}(x)|\Omega\rangle\Big|_{y^0 > x^0} = \sum_\kappa |\langle\Omega|\hat{\phi}(0)|\kappa; \mathbf{p} = 0\rangle|^2 \Delta_+(m_\kappa^2; y - x), \tag{7.2.12}$$

and so the two-point Green's function is as for the neutral scalar case,

$$\langle \Omega | T \hat{\phi}(x) \hat{\phi}^\dagger(y) | \Omega \rangle$$
$$= \sum_\kappa |\langle \Omega | \hat{\phi}(0) | \kappa; \mathbf{p} = 0 \rangle|^2 \left[\theta(x^0 - y^0) \Delta_+(m_\kappa^2; x - y) + \theta(y^0 - x^0) \Delta_+(m_\kappa^2; y - x) \right]$$
$$= \int_0^\infty ds \, \rho(s) D_0(s; x - y) = D_F(x - y). \tag{7.2.13}$$

An interacting scalar field without derivative interactions has $\hat{\pi}(x) = \dot{\hat{\phi}}(x)$. In that case we can show under certain assumptions discussed below that

$$0 \leq Z < 1. \tag{7.2.14}$$

Proof: In the absence of a VEV the two-point Wightman function is given by

$$\langle \Omega | \hat{\phi}(x) \hat{\phi}(y) | \Omega \rangle = \int_0^\infty ds \, \rho(s) \int \frac{d^3 p}{(2\pi)^3} \frac{1}{2E_\mathbf{p}} \, e^{-ip \cdot (x-y)} \bigg|_{E_\mathbf{p} \equiv \sqrt{p^2 + s}}$$
$$= \int_0^\infty ds \, \rho(s) \int \frac{d^4 p}{(2\pi)^4} (2\pi) \, \delta(p^2 - s) e^{-ip \cdot (x-y)} \bigg|_{p^0 > 0} = \int_0^\infty ds \, \rho(s) \Delta_+(s; x - y),$$

where we have used Eqs. (7.2.3), (7.2.8), (6.2.158) along with (6.2.52) to show the four-dimensional form in the second line. Using Eq. (6.2.139) the vacuum expectation value of the commutator is then

$$\langle \Omega | [\hat{\phi}(x), \hat{\phi}(y)] | \Omega \rangle = \int_0^\infty ds \, \rho(s) [\Delta_+(s; x - y) - \Delta_+(s; y - x)]$$
$$= \int_0^\infty ds \, \rho(s) i \Delta(s; x - y). \tag{7.2.15}$$

From Eq. (6.2.145) we have

$$i\delta^3(\mathbf{x} - \mathbf{y}) = \langle \Omega | i\delta^3(\mathbf{x} - \mathbf{y}) | \Omega \rangle = \langle \Omega | [\hat{\phi}(x), \hat{\pi}(y)] | \Omega \rangle_{x^0 = y^0} \tag{7.2.16}$$
$$= \langle \Omega | [\hat{\phi}(x), \dot{\hat{\phi}}(y)] | \Omega \rangle_{x^0 = y^0} = \frac{\partial}{\partial y_0} \langle \Omega | [\hat{\phi}(x), \hat{\phi}(y)] | \Omega \rangle |_{x^0 = y^0}$$
$$= \int_0^\infty ds \, \rho(s) \frac{\partial}{\partial y_0} i\Delta(s; x - y) |_{x^0 = y^0} = \int_0^\infty ds \, \rho(s) i\delta^3(\mathbf{x} - \mathbf{y}),$$

which shows that $1 = \int_0^\infty ds \, \rho(s) = Z + \int_{>m^2}^\infty ds \, \rho(s)$. Since $\rho(s) \geq 0$ and since in an interacting theory there is a nonzero contribution from the continuum we arrive at Eq. (7.2.14), $0 \leq Z < 1$. The proof has assumed Lorentz invariance and that $\int_{>m^2}^\infty ds \, \rho(s)$ can be defined.

7.3 Scattering Cross-Sections and Decay Rates

The scattering matrix or S-matrix was defined earlier in Eq. (7.1.5). It is the amplitude for an initial state $|i\rangle \in V_{\text{full}}$ in the infinite past to scatter into a final state $|f\rangle \in V_{\text{full}}$ in the infinite future,

$$S_{fi} \equiv \langle f | \lim_{T \to \infty} \hat{U}(T, -T) | i \rangle = \langle f | \hat{S} | i \rangle, \tag{7.3.1}$$

where $\hat{U}(t'', t') = T e^{-i\hat{H}(t'' - t')}$ is the evolution operator, $\hat{S} = \lim_{T \to \infty} \hat{U}(T, -T)$ is the S-matrix operator and $|i\rangle$ and $|f\rangle$ are Schrödinger-picture states defined at any common time t. They can be chosen as Heisenberg-picture states for a time-independent Hamiltonian $\hat{H}$ as discussed in Sec. 7.1.

In general the i and f labels are continuous and so the S-matrix is not a matrix in the usual sense. In a time translationally invariant quantum system it does not matter how we take the infinite future $t'' \to \infty$ or infinite past $t' \to -\infty$ limits and so in that case there is no loss of generality in taking them symmetrically as done here with $t'' = T$ and $t' = -T$ and $T \to \infty$. It is conventional to separate the S-matrix operator $\hat{S}$ into a part where no transition takes place, $\hat{I}$, and a part where transitions occur, $i\hat{T}$,

$$\hat{S} \equiv \hat{I} + i\hat{T} \quad \text{and} \quad S_{fi} = \delta_{fi} + iT_{fi}. \tag{7.3.2}$$

The factor of i in the $\hat{T}$ operator definition is conventional. The T_{fi} are the elements of the T-matrix, which is also referred to as the *transition matrix* or *transfer matrix*. We understand that $\delta_{fi} \equiv \langle f|i\rangle$ and assume unit normalization for our states.

In scattering experiments we are interested in considering those $|i\rangle, |f\rangle \in V_{\text{full}}$ comprised of stable particles or bound states that are in finite wavepackets that are arbitrarily far apart in space so that the interactions between them are negligible. We assume that the cluster decomposition principle applies. Let the characteristic size of the wavepackets be σ and the characteristic separation of the wavepackets be d. Provided that we first take $d \to \infty$ we can then take σ arbitrarily large so that the wavepackets can become approximate plane waves with very narrowly peaked momenta. With these understandings we can regard such initial and final states as being approximate plane waves in the mixed Fock space, V_{Fock}. In particular in this limit we can identify such $|i\rangle$ and $|f\rangle$ with Heisenberg-picture plane wave states in the mixed Fock space as discussed in Sec. 7.1, where these were written as $|i\rangle = |\mathbf{p}_1 \cdots \mathbf{p}_m\rangle$ and $|f\rangle = |\mathbf{k}_1 \cdots \mathbf{k}_n\rangle$.

Now consider an initial state consisting of two particles with masses m_A and m_B and three-momenta $\mathbf{p}_A$ and $\mathbf{p}_B$ so that $|i\rangle = |\mathbf{p}_A\mathbf{p}_B\rangle$ and a final state consisting of n particles with masses $m_1, \ldots, m_n$ and three-momenta $\mathbf{p}_1, \ldots, \mathbf{p}_n$ so that $|f\rangle = |\mathbf{p}_1 \cdots \mathbf{p}_n\rangle$ In a Poincaré-invariant system we have four-momentum conservation and we can factor out a momentum-conserving delta function,

$$T_{fi} = \langle f|\hat{T}|i\rangle \equiv (2\pi)^4 \delta^4(p_A + p_B - \textstyle\sum_{i=1}^n p_i)\mathcal{M}_{fi}, \tag{7.3.3}$$

where $i = (\mathbf{p}_A, \mathbf{p}_B)$ and $f = (\mathbf{p}_1, \ldots, \mathbf{p}_n)$ and $\mathcal{M}_{fi}$ is defined as the *invariant amplitude* or *invariant matrix element*. We initially consider scalar particles for simplicity.

Using a simple shorthand we see that in all cases where $f \neq i$ we have

$$|\langle f|\hat{S}|i\rangle|^2 \stackrel{f \neq i}{=} |\langle f|\hat{T}|i\rangle|^2 = (2\pi)^8 \delta^4(0)\delta^4(p_A + p_B - \textstyle\sum_{j=1}^n p_j)|\mathcal{M}_{fi}|^2$$
$$= (V\Delta T)(2\pi)^4 \delta^4(p_A + p_B - \textstyle\sum_{j=1}^n p_j)|\mathcal{M}_{fi}|^2, \tag{7.3.4}$$

where we have used the result that in any regularization of the spacetime volume we have the correspondence $(2\pi)^4\delta^4(0) \to (V\Delta T)$, where V is the finite spatial volume and ΔT is a finite time interval. This is a straightforward generalization of the three-dimensional result given in Eq. (6.2.26), $(2\pi)^3\delta^3(0) \to V$.

7.3.1 Cross-Section

Classical cross-section: We can measure the *classical* cross-section A_X of a classical three-dimensional object as seen from any given direction by using a uniform beam of classical particles arriving from that direction and measuring how many particles are scattered by the object.

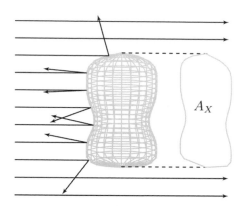

A_X

The classical cross-section A_X of a solid object in a given direction can be measured using a uniform beam of particles. A_X is the area of the beam that is scattered.

This is illustrated in Fig. 7.4. Let Φ denote the flux of the incident beam, which is defined as the number density of the beam ρ multiplied by the beam velocity v,

$$\Phi \equiv \rho v = \text{number of particles/unit time/unit area.} \tag{7.3.5}$$

Φ is the number of beam particles passing though a unit area orthogonal to the beam per second. The current of incident particles on an area A orthogonal to the beam is $I_{\text{inc}} = \Phi A = \rho v A$. The number of particles scattered is N_{sc}. The number scattered per unit time by the target is the scattered current, I_{sc}, which is also called the *scattering rate*, R, so that we have the following equivalences:

$$R \equiv \frac{dN_{\text{sc}}}{dt} \equiv I_{\text{sc}}. \tag{7.3.6}$$

The *total cross-section* σ is the scattered current divided by the beam flux,

$$\sigma \equiv \frac{\text{\# of scattered particles/sec}}{\text{\# of incident particles/sec/unit area}} = \frac{I_{\text{sc}}}{\Phi}. \tag{7.3.7}$$

In classical physics every beam particle that hits the solid object will scatter and so in that case we have $I_{\text{sc}} = \Phi A_X$ and so

$$\sigma = \frac{I_{\text{sc}}}{\Phi} = \frac{\Phi A_X}{\Phi} = A_X, \tag{7.3.8}$$

which is why we refer to σ as a cross-section and why it has the units of area. *The total cross-section σ is the area of the beam that is scattered.* It is also *the scattering area of the target seen by each particle in the beam.*

We can also use the above definition of the cross-section σ when a classical charged particle scatters from the electromagnetic field of a single target. Recall that the momentum p of the beam particles determines their de Broglie wavelength ($\lambda = 2\pi/p$) and any target much smaller than λ will appear as an approximate point particle to the beam and will typically behave as such. We are using the word "particle" in this generic sense in our discussion of the single target here. The case of a target particle with nonzero total charge requires care because the Coulomb potential falls off relatively slowly as $1/r$. For a neutral target particle such as an atom or molecule the interaction can occur through higher moments of the electromagnetic field, such as an electric dipole moment, and the falloff is much more rapid. We can divide the beam particles into those that undergo negligible

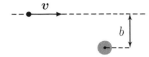

The impact parameter b of a particle approaching a target.

scattering and those with significant scattering. The boundary is determined by our ability to detect beam particles with very *forward scattering*. Forward scattering refers to incoming particles that have zero or very small angles of deflection with respect to the beam in the forward direction.

Clearly, there is no concept of A_X being the physical cross-section of a solid in this case but σ is still the effective area of the beam that is scattered. Imagine a spherical region with the target particle at its center and with some arbitrary large radius r. Consider the infinitesimal current of particles $dI_{sc}(\theta, \phi)$ scattering into some infinitesimal solid angle $d\Omega = \sin\theta d\theta d\phi$. We can write this as

$$dI_{sc}(\theta, \phi) = \frac{dI_{sc}}{d\Omega}(\theta, \phi)d\Omega, \tag{7.3.9}$$

where $dI_{sc}/d\Omega$ is the number of scattered particles per unit time per unit of solid angle at (θ, ϕ) on the sphere. We define the *differential cross-section* as

$$\frac{d\sigma}{d\Omega} \equiv \frac{\#\text{ of scattered particles/sec/unit solid angle}}{\#\text{ of incident particles/sec/unit area}} = \frac{1}{\Phi}\frac{dI_{sc}(\theta, \phi)}{d\Omega}. \tag{7.3.10}$$

We can refer to $d\sigma$ as the differential of the cross-section. Similarly, the variation of the cross-section with any combination of kinematic variables is said to be a differential cross-section, e.g., $d\sigma/d\theta, d\sigma/dE, d^2\sigma/dxdy$ for any kinematic variables x and y and so on. We obviously have

$$\sigma = \int d\Omega \frac{d\sigma}{d\Omega} = \int_0^\pi d\theta \int_0^{2\pi} \sin\theta d\phi \frac{d\sigma}{d\Omega}(\theta, \phi). \tag{7.3.11}$$

There is a subtlety associated with interactions that do not strictly vanish outside some finite radius R, which occurs with any field-mediated interaction. In the classical case every incident particle in the beam will be deflected at least very slightly by the interaction. So if the uniform beam has a finite cross-sectional area A_{beam}, then since every particle in the beam is scattered slightly we would have $\sigma = A_{\text{beam}}$. For this reason we do not typically consider the total cross-section σ but instead study scattering by considering the differential cross-section $d\sigma/d\Omega$ away from forward scattering. Experimentally, we cannot place detectors in the path of the beam either behind or in front of the target and so we can never in practice have a detector system that makes up a true "4π *detector*," also known as a *hermetic detector*. A real detector system will cover some fraction of 4π solid angle and will consist of one or more single detectors. Each single detector will occupy some solid angle $\Delta\Omega$ with $\Delta\sigma = \int_{\Delta\Omega}[d\sigma/d\Omega]d\Omega$ being the area of the beam that is scattered into the solid angle $\Delta\Omega$.

The *impact parameter* illustrated in Fig. 7.5 of a particle approaching a target is defined as the distance of closest approach of the incoming particle trajectory to the center of the target. It is the length of a line orthogonal to the trajectory that goes from the trajectory to the center of the target.

Simple cross-section in quantum mechanics: Consider a quantum particle in a wavepacket that well approximates a plane wave coming from the z-direction and scattering off some target. In the quantum case for a finite-range potential not every particle will be scattered because of

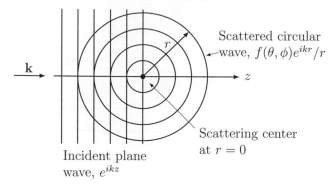

Figure 7.6   An incoming approximate plane wave scattered into an outgoing spherical wave.

quantum tunneling effects. For a short-range energy-conserving scattering in nonrelativistic quantum mechanics the total wavefunction as $r \to \infty$ can be written as

$$\psi(t, \mathbf{x}) \overset{r \to \infty}{\longrightarrow} c \left\{ e^{ikz} + f(\theta, \phi) \frac{e^{ikr}}{r} \right\} e^{iE_{\mathbf{k}}t}, \tag{7.3.12}$$

where $\mathbf{k} = (0, 0, k)$ and $E_{\mathbf{k}}$ are the particle three-momentum and energy, respectively, and where $r \equiv |\mathbf{x}|$ with the target located at $\mathbf{r} \equiv \mathbf{x} = 0$. A short-range interaction falls off faster than the Coulomb interaction; i.e., the potential falls off faster than $1/r$. The incident plane wave component is e^{ikz} and the scattered component is $f(\theta, \phi)e^{ikr}/r$ as shown in Fig. 7.6. There are three distinct distance scales here. First, there is the effective range over which the target can significantly influence the scattering particle, $r \lesssim r_{\text{target}}$. Second, there is the scale that characterizes the finite transverse extent of our incoming plane wave, ℓ_{tr}. Last there is the distance from the target at which we make our observations, $r \gtrsim r_{\text{observ}}$. We are interested in the typical situation, where $r_{\text{targ}} \ll \ell_{\text{tr}} \ll r_{\text{observ}}$. In other words the incoming particle wavepacket presents as an approximate plane wave to the target, but is concentrated within a narrow cylindrical shape passing through the target with respect to an observer a macroscopic distance away. In practice there will be many such particles making up a beam. Note that there is only interference between a scattered wave and the approximate plane wave when $\theta \simeq 0, \pi$, i.e., in the very forward or backward directions. For now we neglect the finite length of a real wavepacket, but it is straightforward to generalize to include that as well.

The associated conserved current for a spinless particle is the conserved Klein-Gordon current, which gives as $r \gg 1/k$ in the plane wave limit,

$$\mathbf{j}(t, \mathbf{x}) = -(i/2m)(\psi^* \boldsymbol{\nabla} \psi - \psi \boldsymbol{\nabla} \psi^*) \tag{7.3.13}$$

$$= \frac{|c|^2}{m} \text{Re} \left[-i \left\{ e^{-ikz} + f^*(\theta, \phi) \frac{e^{-ikr}}{r} \right\} \boldsymbol{\nabla} \left\{ e^{ikz} + f(\theta, \phi) \frac{e^{ikr}}{r} \right\} \right]$$

$$= |c|^2 \text{Re} \left[\frac{k}{m} \left\{ e^{-ikz} + f^*(\theta, \phi) \frac{e^{-ikr}}{r} \right\} \left\{ e^{ikz} \hat{\mathbf{e}}_z + f(\theta, \phi) \hat{\mathbf{e}}_r \frac{e^{ikr}}{r} + \mathcal{O}(1/r^2) \right\} \right],$$

where we have used

$$\boldsymbol{\nabla} = \hat{\mathbf{e}}_r \frac{\partial}{\partial r} + \frac{1}{r} \hat{\mathbf{e}}_\theta \frac{\partial}{\partial \theta} + \frac{1}{r \sin \theta} \hat{\mathbf{e}}_\phi \frac{\partial}{\partial \phi} \tag{7.3.14}$$

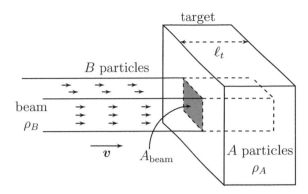

A uniform beam of B particles impinging on a fixed target of A particles.

and where we do not explicitly write the $\hat{\mathbf{e}}_\phi$ and $\hat{\mathbf{e}}_\theta$ terms as they are higher order in $1/r$. Then we have

$$\mathbf{j} \xrightarrow{r\to\infty} |c|^2 \left[\left\{ \frac{\mathbf{k}}{m} \right\} + \left\{ \hat{\mathbf{e}}_r \frac{k}{m} |f(\theta,\phi)|^2 \frac{1}{r^2} \right\} + \left\{ \frac{k}{m} \text{Re}\, \mathcal{O}(e^{\pm ik(r-z)}/r) \right\} \right] \tag{7.3.15}$$

$$\equiv \mathbf{j}_{\text{inc}} + \mathbf{j}_{\text{sc}} + \mathbf{j}_{\text{int}} \tag{7.3.16}$$

where the three current density contributions are identified by $\{\cdots\}$ and are the incident, scattered and interference terms, respectively. Since the approximate plane wave has finite transverse size $\simeq \ell_{\text{tr}}$, then $\mathbf{j}_{\text{inc}}$ and $\mathbf{j}_{\text{int}}$ are concentrated within the narrow cylindrical shape containing the approximate plane wave, i.e., $\theta \simeq 0, \pi$ as $r \to \infty$. Outside that region only the scattered current density, $\mathbf{j}_{\text{sc}}$, is seen.

The incoming flux is $\Phi = |\mathbf{j}_{\text{inc}}| = |c|^2(k/m)$ and the scattered current into the solid angle $d\Omega$ is $dI_{\text{sc}} = |\mathbf{j}_{\text{sc}}|r^2 d\Omega = |c|^2(k/m)|f(\theta,\phi)|^2 d\Omega$ and so the differential cross-section and total cross-section are, respectively, given by

$$\frac{d\sigma}{d\Omega} = \frac{1}{\Phi}\frac{dI_{\text{sc}}}{d\Omega} = |f(\theta,\phi)|^2 \quad \text{and} \quad \sigma = \int d\Omega\, |f(\theta,\phi)|^2. \tag{7.3.17}$$

Fixed target scattering experiments: Consider a target consisting of particles of type A with density ρ_A and target thickness ℓ_t. We have a uniform beam of area A_{beam} of particles of type B with density ρ_B moving with velocity v orthogonal to the target. The target area is larger than the beam area and all of the beam overlaps the target. The laboratory frame is the frame of the fixed target.

The beam flux in the laboratory frame is $\Phi_B = \rho_B v$. The current incident on the target is $I_B = \Phi_B A_{\text{beam}}$ and is the number of B particles incident on the target per unit time. If the target and beam densities and the target thickness are not too large, then the beam can be assumed to be unattenuated as it moves through the target and the interaction of the beam and target will be the sum of the interactions between single-beam particles B and single-target particles A. Define σ_{AB} and $d\sigma_{AB}/d\Omega$ as the single-target particle cross-section and differential cross-section in the laboratory frame, respectively. Note that σ_{AB} *is the effective scattering area of each A particle as seen by each B particle and vice versa.*

As illustrated in Fig. 7.7, the number of target particles A in the path of the beam will be $N_A = \rho_A A_{\text{beam}} \ell_t$. So the total scattered current is N_A times the scattered current per target particle,

$I_{\rm sc}^A = \Phi_B \sigma_{AB}$, where we have used Eq. (7.3.7). The total scattering rate of the beam on the target in the laboratory frame is then

$$R \equiv \frac{dN_{\rm sc}}{dt} \equiv I_{\rm sc} = N_A I_{\rm sc}^A = N_A \Phi_B \sigma_{AB} = \rho_A \rho_B v A_{\rm beam} \ell_t \sigma_{AB} \equiv \mathcal{L} \sigma_{AB}, \qquad (7.3.18)$$

where we have defined the *luminosity*, $\mathcal{L}$, as

$$\mathcal{L} \equiv N_A \Phi_B = \rho_A \rho_B v A_{\rm beam} \ell_t = \rho_A \rho_B v V_{\rm int} = \Phi_B \rho_A V_{\rm int}, \qquad (7.3.19)$$

where $V_{\rm int} \equiv A_{\rm beam} \ell_t$ is the volume of the region in which the interactions are taking place. Luminosity is the number of scattering opportunities for A and B particles per unit area per unit time. The *integrated luminosity*, $L \equiv \int_{t_0}^{t_0+T} dt\,\mathcal{L}$, is the luminosity integrated over a time interval $[t_0, t_0 + T]$ and corresponds to the number of scattering opportunities per unit area. The total number of scattering events and the number of scattering events per unit of solid angle over time T are

$$N_{\rm sc} = L\,\sigma_{AB} \quad \text{and} \quad \frac{dN_{\rm sc}}{d\Omega} = L\,\frac{d\sigma_{AB}}{d\Omega} \quad \text{with} \quad L \equiv \int_{t_0}^{t_0+T} dt\,\mathcal{L}. \qquad (7.3.20)$$

In particle and nuclear physics cross-sections are usually expressed using the unit of a *barn* denoted by the symbol b, where $1\,{\rm b} = 10^{-28}\,{\rm m}^2 = 10^{-24}\,{\rm cm}^2 = 100\,{\rm fm}^2$, where $1\,{\rm fm} = 10^{-15}\,{\rm m}$. A barn is the typical cross-sectional area of a nucleus.[1] For example, the radius of a uranium nucleus is $r \simeq 7.5\,{\rm fm}$, giving a classical cross-sectional area of approximately $\pi r^2 \simeq 180\,{\rm fm}^2 = 1.8 \times 10^{-28}\,{\rm m}^2 = 1.8\,{\rm b}$. In natural units $1 = \hbar c = 197.327\,{\rm MeV\,fm}$ and $1\,{\rm GeV}^{-2} = 0.38938 \times 10^{-28}\,{\rm m}^2 = 0.3894\,{\rm mb}$. So in natural units $1\,{\rm mb} = 2.5682\,{\rm GeV}^{-2}$.

If particles A and B stay intact and do not become excited states of themselves and do not produce or become any other particles, then we say that they have undergone *elastic scattering*, $AB \to AB$. In this case the kinetic energy of each particle is conserved in the center of momentum frame but the particle directions can change. If A and/or B become excited states of themselves and/or if one or both become other particles and/or if additional particles are produced, then we say they have undergone *inelastic scattering* In this case kinetic energy is converted during the collision. The interactions $e^-p \to e^-p$, $e^-e^+ \to e^-e^+$, $e^-\,{\rm Au} \to e^-\,{\rm Au}$ and $pp \to pp$ are examples of elastic scattering, where Au denotes a gold nucleus or at very low energies it could be understood as a gold atom. Examples of inelastic interactions are $e^+e^- \to \mu^+\mu^-$, $e^-p \to e^-p\mu^+\mu^-$, $pp \to pp\pi^0$ and $e^-\,{\rm Au} \to e^-\,{\rm Au}^*$, where ${\rm Au}^*$ denotes an excited state of the gold atom or the gold nucleus as appropriate. If a relatively small amount of kinetic energy is converted during the scattering, we say that this is *quasielastic scattering*. When two particles scatter inelastically but where the two products of the scattering experience only a small deflection from the direction of the incoming particles we say that they have undergone *diffractive scattering*. For example, if two protons inelastically scatter with one or both protons breaking up but staying tightly clustered together with little deflections, then we refer to this as *single-diffractive dissociation* or *double-diffractive dissociation*, respectively. The strong interaction contribution to diffractive scattering is often associated with the virtual exchange of a pseudoparticle called a *pomeron*.

Particles A and B with sufficient total energy will be able to scatter into a variety of final states. These final states can be classified into *scattering channels*, which are labeled by the combination of particles that make up the final state. Here we are considering a particle and its excited states to be different final state particles for the purpose of this classification. Let us denote X as the set of all

---

[1] During the development of the atom bomb during the Second World War US physicists were scattering neutrons from uranium nuclei, which were described as being "as big as a barn." They started using "barn" to mean $10^{-28}\,{\rm m}^2$, which then was adopted as a standard definition.

accessible combinations of final state particles such that $AB \to X$ includes all possible scatterings. The total cross-section for AB is then $\sigma_{AB} \equiv \sigma(AB \to X)$.

We can also specify a part of the total cross-section with the requirement that one or more particles must appear in the final state. For example, we might specify that the final state contains particle C but we do not restrict what other particles might be present. The cross-section for this is written as $\sigma(AB \to CX)$ or also frequently as $\sigma(AB \to C + X)$, where here X is the set of all possible combinations of particles that can appear along with C. We refer to $\sigma(AB \to CX)$ as the *production cross-section* for particle C. This is also referred to as an *inclusive cross-section* in that it includes all accessible particle combinations along with particle C. An *exclusive cross-section* refers to a process where the scattering channel is completely specified with all final state particles identified, $AB \to C_1 C_2 \cdots C_n$. Examples of production/inclusive cross-sections are $\sigma(pp \to W^+ X)$, $\sigma(pp \to \pi^+ X)$, $\sigma(pp \to \mu^+ \mu^- X)$ and $\sigma(\gamma p \to \Delta^+ X)$. Examples of exclusive cross-sections are $\sigma(e^+ e^- \to \mu^+ \mu^-)$, $\sigma(pp \to pp\pi^0)$, $\sigma(pp \to pn\pi^+)$ and $\sigma(pp \to p\Delta^+)$. The associated scattering processes are referred to as *inclusive and exclusive processes*, respectively.

In particle physics the term *fiducial cross-section* is typically used to mean the cross-section that will be detectable after allowing for the finite solid angle covered by the detectors and the imperfect detector efficiencies and may include essential kinematic cuts. Fiducial cross-sections will necessarily exclude the very forward and backward regions near the path of the beam. When an unstable particle is produced in a collision it may have many decay channels; the probability of the particle decaying into a particular channel is referred to as the *branching ratio* or *branching fraction*. With a very large number of decays of identical unstable particles it is the fraction formed by the ratio of the number of times a decay is into a particular decay channel to the total number of decays.

Colliding beam experiments: Almost all colliding beam experiments are head-on collisions of bunches of particles or very nearly head-on collisions. When the number of bunches becomes too large, then it is advantageous to have the colliding beams crossing at a very small angle to each other so that only one bunch from each beam are colliding at a time.

Consider two colliding beams of particles with particles of type A in one beam and particles of type B in the other beam. The beams consist of bunches of particles that for simplicity we take as rectangular cuboids with uniform particle number densities ρ_A and ρ_B, bunch lengths ℓ_A and ℓ_B and the same beam area orthogonal to the direction of motion, A_{beam}. Of course, each particle is also in a sufficiently broad quantum wavepacket that it has a relatively small uncertainty in its momentum and hence its velocity. The particle velocities in the laboratory frame are collinear but have opposite direction so that $|\boldsymbol{v}_A - \boldsymbol{v}_B| = v_A + v_B$. This is illustrated in Fig. 7.8, where we show two particle bunches approaching each other in the central region of our detector system called the collision region. Ideally, our system of detectors covers as much of the 4π solid angle surrounding the collision region as possible. They will typically have multiple components with

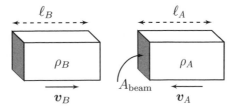

Figure 7.8   Two colliding packets of particles of type A and type B with the same beam area and collinear velocities $\boldsymbol{v}_A$ and $\boldsymbol{v}_B$ in the laboratory frame, where $|\boldsymbol{v}_A - \boldsymbol{v}_B| = v_A + v_B$.

each component specialized in detecting certain types of particles and measuring their momentum, energy and sometimes even polarization or spin orientation. The colliding beams experiment is set up so that the bunch velocities and bunch spacings result in the bunches crossing each other in the collision region with some frequency f_{coll}. The time between bunch collisions in the collision region is then $\Delta t = 1/f_{\text{coll}}$.

Let us proceed in several steps to arrive at an expression for the luminosity of our colliding beams. First, let us consider one bunch of A particles as the target and imagine for now that the bunch of B particles is arbitrarily long so that it forms a continuous beam. In this case we can adapt the fixed target result in Eq. (7.3.19) above with the replacements $v \to v_A + v_B$ and $\ell_t \to \ell_A$ to give the result, $\mathcal{L}_{\text{cont B}} = \rho_A \rho_B (v_A + v_B) A_{\text{beam}} \ell_A$. Whenever a bunch of B particles enters the collision region it aways sees an identical bunch of A particles. So all we need to do now is account for the fact that our continuous B beam is actually chopped up and is only "turned on" for a fraction of the time. The time taken for a B bunch of length ℓ_B to pass through the front face of the colliding A bunch in the collision region is $\Delta t_B = \ell_B/(v_A + v_B)$. The time between collisions is $\Delta t = 1/f_{\text{coll}}$ and so the fraction of time that the B beam is on is $\Delta t_B/\Delta t$, which gives a luminosity of $\mathcal{L} = (\Delta t_B/\Delta t)\mathcal{L}_{\text{cont B}}$. The luminosity of our colliding beams is then

$$\mathcal{L} = [\ell_B/(v_A + v_B)\Delta t]\rho_A \rho_B (v_A + v_B) A_{\text{beam}} \ell_A$$
$$= \frac{\rho_A \rho_B A_{\text{beam}} \ell_A \ell_B}{\Delta t} = \frac{N_A N_B}{A_{\text{beam}} \Delta t}, \tag{7.3.21}$$

where N_A and N_B are the number of particles in the A and B bunches, respectively. The $A \leftrightarrow B$ symmetry of the above result is as expected. In the case of a circular collider where each bunch completes f_{circ} circuits per unit time and with N_{bunch} being the number of bunches of each type around the circuit, then the time between collisions is $\Delta t = 1/f_{\text{coll}} = 1/N_{\text{bunch}} f_{\text{circ}}$ and so in that case

$$\mathcal{L} = \frac{N_A N_B f_{\text{coll}}}{A_{\text{beam}}} = \frac{N_A N_B f_{\text{circ}} N_{\text{bunch}}}{A_{\text{beam}}}. \tag{7.3.22}$$

In the case of more realistic bunch crossings there is a small angle between the beams, which introduces a geometric luminosity reduction factor, $F_{\text{red}} \lesssim 1$. Assuming that both beams have same gaussian shapes it can be shown that we have the replacement $A_{\text{beam}} \to 4\pi\sigma_x\sigma_y$. So a more realistic expression for the luminosity is

$$\mathcal{L} = \frac{N_A N_B}{4\pi\sigma_x\sigma_y \Delta t} F_{\text{red}} = \frac{N_A N_B f_{\text{coll}}}{4\pi\sigma_x\sigma_y} F_{\text{red}} = \frac{N_A N_B f_{\text{circ}} N_{\text{bunch}}}{4\pi\sigma_x\sigma_y} F_{\text{red}}, \tag{7.3.23}$$

where σ_x^2 and σ_y^2 are the gaussian variances of the beam in the transverse directions where we define the beams to be in the $\pm z$-direction. The variance σ_z^2 along the beam direction does not appear in the final result in the same way that ℓ_A and ℓ_B did not appear above. Non-gaussian transverse beam bunch shapes introduce additional small corrections to the above results. As an example, let us now consider proton-proton collisions at the Large Hadron Collider (LHC) at CERN.[2]

Example: Proton-proton collisions at the LHC: The LHC has four experimental regions (ATLAS, CMS, LHCb, ALICE as of the year 2020) distributed around the circular 26.7-km tunnel. The following numbers are typical of those used in the LHC operation at that time. In each region the counter-rotating proton beams intersect at a collision point at a very small crossing angle of

---

[2] CERN (derived from Conseil Européenne pour la Recherche Nucléaire) is the European Organization for Nuclear Research. It was established in 1954 and is based in Geneva. The circular 26.7-km Large Hadron Collider (LHC) tunnel spans the Franco-Swiss border.

$285\,\mu\text{rad} = 2.85 \times 10^{-4}\,\text{rad}$, which gives a luminosity reduction factor of $F_{\text{red}} = 0.835$. The counter-rotating proton bunches each contain $N_p = 1.15 \times 10^{11}$ protons. The circulation frequency is $f_{\text{circ}} = 11.245\,\text{kHz}$ and the number of bunches in each beam is $N_{\text{bunch}} = 2808$, which gives an average collision frequency of $f_{\text{coll}} = f_{\text{circ}} N_{\text{bunch}} = 11{,}245 \times 2808\,\text{Hz} = 31.6\,\text{MHz}$. The average time between bunch collisions is then $\Delta t = 1/f_{\text{coll}} = 31.7\,\text{ns}$. The transverse size of the squeezed beams at the intersection points is given by $\sigma_x = \sigma_y = 16.7\,\mu\text{m} = 1.67 \times 10^{-3}\,\text{cm}$. The bunch length is specified in terms of the gaussian variance in the beam direction, $\sigma_z = 7.7\,\text{cm}$. So the LHC luminosity at each experimental intersection point is

$$\mathcal{L} = \frac{N_p^2 f_{\text{coll}} F_{\text{red}}}{4\pi\sigma_x\sigma_y} \simeq \frac{(1.15 \times 10^{11})^2 (3.16 \times 10^7)(0.835)}{4\pi(1.67 \times 10^{-3})^2}\,\text{cm}^{-2}\text{s}^{-1}$$
$$\simeq 1.0 \times 10^{34}\,\text{cm}^{-2}\text{s}^{-1}. \tag{7.3.24}$$

For example, the 2018 run of the LHC produced an integrated luminosity of ~ 65 inverse femtobarns, $L = \int_{2018} dt\,\mathcal{L} \simeq 65\,\text{fb}^{-1} = 6.5 \times 10^{40}\,\text{cm}^{-2}$ with a run of approximately 180 days. Note that $1\,\text{fb} = 10^{-15}\,\text{b} = 10^{-43}\,\text{m}^2 = 10^{-39}\,\text{cm}^2$. At the above design luminosity it would take $T \simeq (6.5 \times 10^{40}/10^{34})\,\text{s} = 6.5 \times 10^6\,\text{s}$ or approximately 75 days to achieve $L \simeq 65\,\text{fb}^{-1}$. However, the average luminosity is significantly less than the design luminosity, since once the beam has been "filled" with $N_p = 1.15 \times 10^{11}$ protons per bunch it is then left to run with the number of protons per bunch slowly and exponentially decreasing. After approximately 10 hours, or sometimes less, the beam is "refilled" and the peak luminosity is again reached. The loss of protons in the bunches is due to collisions in the interaction regions as well as other various loss mechanisms as the bunches circulate, e.g., collisions with residual gas in the vacuum tube. Also not every bunch hits another bunch at the intersection regions since there are occasional gaps in the bunch spacing to allow for beam abort to the beam dump. The time between collisions is often quoted as 25 ns, which is the typical time spacing. However, the bunch spacing is not uniform and the value 31.7 ns used above is the average time between bunch collisions. The LHC achieved its design luminosity of $10^{34}\,\text{cm}^{-2}\text{s}^{-1}$ in June 2016 and after various improvements were made in 2017 the LHC achieved double its design luminosity.

The energy of each proton is 6.5 TeV, giving a combined center of momentum energy for each proton-proton collision of 13 TeV. The Lorentz factor is $\gamma = E_p/m_p c^2 = 6.5 \times 10^3/0.938 = 6{,}930$, giving $v/c = \sqrt{1 - \gamma^{-2}} = 0.999999990$. We verify the correct tunnel length $L_{\text{tunnel}} = v/f_{\text{circ}} = (3.0 \times 10^8/11{,}245)\,\text{m} = 26.7\,\text{km}$. The average distance between bunches is $L_{\text{tunnel}}/N_{\text{bunch}} = 26{,}700/2808 = 9.50\,\text{m}$.

The total cross-section for proton-proton scattering at 6.5 TeV on 6.5 TeV is approximately $\sigma_{\text{total}} \simeq 111\,\text{mb}$. This consists of (Antchev et al., 2019) (i) 80 mb inelastic and (ii) 31 mb elastic. The inelastic scattering cross-section is further broken down into diffractive (forward and/or soft) inelastic scattering and hard (significant transverse components) inelastic scattering with the diffractive contribution 25–30% (i.e., $\sim 27.5\%$) of the hard contribution (ATLAS Collaboration et al., 2012). The detectors in use at the LHC do not cover very forward or very backward angles to avoid issues with the beam and so they are only sensitive to particles scattering at sufficiently high angles away from the beam axis, where a significant amount of transverse energy has been transferred. Such detectable events are almost always hard and inelastic and so have a detectable cross-section $\sigma_{\text{inel}} \simeq (1 - 0.275) \times 80\,\text{mb} \simeq 58\,\text{mb} = 58 \times 10^{-27}\,\text{cm}^2$. Then the number of detectable events per second will be

$$R_{\text{det}} = \frac{dN_{\text{det}}}{dt} \simeq \mathcal{L}\,\sigma_{\text{inel}} = 10^{34} \times (58 \times 10^{-27})\,\text{s}^{-1} = 5.8 \times 10^8\,\text{s}^{-1}. \tag{7.3.25}$$

Since $f_{\text{coll}} = 31.6\,\text{MHz}$ is the number of bunch crossings per unit time, then the number of detectable collisions per bunch crossing is

$$N_{\text{det/bunch}} = R_{\text{det}}/f_{\text{coll}} = 5.8 \times 10^8/31.6 \times 10^6 \simeq 18. \tag{7.3.26}$$

So there are approximately 20 detectable inelastic proton-proton scattering events per bunch crossing. The actual number fluctuates per bunch crossing according to a Poisson distribution.

Let us consider Higgs production. The Higgs mass is now known to be $m_H \simeq 125\,\text{GeV}$. During the search for the Higgs boson its mass was unknown so theoretical predictions for Higgs production and decay were presented as a function of the Higgs mass. The total cross-section for Higgs production and subsequent Higgs decay to two photons in the ATLAS detector, given the fiducial coverage of the detector, is

$$\sigma_{\text{fid}}(pp \to H + X, H \to \gamma\gamma) \simeq 65\,\text{fb} = 65 \times 10^{-39}\,\text{cm}^2, \tag{7.3.27}$$

where this is inclusive of all production processes resulting from the pp collision. It includes the effects of the finite acceptance of the detector for diphoton events and the efficiency of the detectors for observing photons. So the detectable rate for diphoton production arising from Higgs decay in pp collisions in ATLAS at the LHC at *peak* luminosity of $10^{34}\,\text{cm}^{-2}\text{s}^{-1}$ is

$$\begin{aligned} R(pp \to H + X, H \to \gamma\gamma) &= \mathcal{L}\,\sigma_{\text{fid}}(pp \to H + X, H \to \gamma\gamma) \\ &= 10^{34} \times (65 \times 10^{-39})\,\text{s}^{-1} = 6.5 \times 10^{-4}\,\text{s}^{-1}. \end{aligned} \tag{7.3.28}$$

Using the actual integrated luminosity for the 2018 run, $L^{2018} \simeq 65\,\text{fb}^{-1}$, and the fiducial cross-section for $H \to \gamma\gamma$ for the ATLAS detector, we can then estimate the number of observed $H \to \gamma\gamma$ events in 2018 in ATLAS as

$$N_{H\to\gamma\gamma}^{2018} = L^{2018}\,\sigma_{\text{fid}}(pp \to H + X, H \to \gamma\gamma) = (65\,\text{fb}^{-1}) \times (65\,\text{fb}) \simeq 4200.$$

An upgrade of the LHC to the High Lumiosity LHC (HL-LHC) is achievable by increasing the number of protons per bunch and by improvements to beam optics. The result is expected to be an increase in proton-proton collisions per bunch crossing from $\sim$20 to $\sim$140–200, which is a phenomenon referred to as *pileup*. Disentangling the large number of events per bunch crossing presents challenges, typically in the form of softer hadronic physics events complicating the detection of hard events of interest.

7.3.2 Relating the Cross-Section to the S-Matrix

Assume that all of space is filled with A and B particles with number densities ρ_A and ρ_B, respectively, and where A and B particles are in some kind of wavepackets with group velocities $\boldsymbol{v}_A$ and $\boldsymbol{v}_B$ that are collinear but with opposite direction. For simplicity we will assume that our particles are scalar bosons for now. Then as above we define $v_{AB} \equiv |\boldsymbol{v}_A - \boldsymbol{v}_B| = (v_A + v_B)$. Consider some spatial volume V. We can adapt Eq. (7.3.18) to this case by the replacements $v \to v_{AB}$ and $V_{\text{int}} \to V$. Let ΔT be some long time period, then the number of AB scatterings in this volume over this time is

$$N_{\text{sc}} = R\Delta T = \rho_A \rho_B v_{AB}(V\Delta T)\sigma_{AB}. \tag{7.3.29}$$

The number of ways to choose AB pairs in volume V is $N_A N_B = \rho_A \rho_B V^2$. So the average probability of any single AB pair scattering in volume V and time ΔT is

$$P_{AB} \equiv N_{\text{sc}}/N_A N_B = (\Delta T/V) v_{AB} \sigma_{AB}. \tag{7.3.30}$$

Now consider the limit where the two colliding wavepackets are taken to be plane waves. Then there is only one kind of initial state i with the two plane waves filling space and approaching each other from opposite directions. So the initial state is $i = (\mathbf{p}_A, \mathbf{p}_B)$, where $\mathbf{p}_A = (0, 0, p_A^z)$ and $\mathbf{p}_B = (0, 0, p_B^z)$ are the three-momenta of the A and B plane waves, respectively. We have defined the z-axis to be the direction of relative motion. Since $E_A = \gamma_A m_A$ and $\mathbf{p}_A = \gamma_A m_A \mathbf{v}_A$, then $\mathbf{v}_A = \mathbf{p}_A/E_A$ and similary $\mathbf{v}_B = \mathbf{p}_B/E_B$. We have the following equalities,

$$|\mathbf{v}_A - \mathbf{v}_B| = |v_A^z - v_B^z| = v_A + v_B = \left| \frac{\mathbf{p}_A}{E_A} - \frac{\mathbf{p}_B}{E_B} \right|, \tag{7.3.31}$$

where $v_A = |\mathbf{v}_A| = |v_A^z|$ and $v_B = |\mathbf{v}_B| = |v_B^z|$. In this limit the probability of a single AB pair scattering ($f \neq i$) in volume V over time $\Delta T = T - (-T) = 2T$ is

$$P_{AB} = \sum_f \frac{|\langle f|\hat{S}|i\rangle|^2}{\langle i|i\rangle\langle f|f\rangle} = \sum_f \frac{|\langle f|\hat{T}|i\rangle|^2}{\langle i|i\rangle\langle f|f\rangle}, \tag{7.3.32}$$

where here $|i\rangle \neq |f\rangle$ and we are allowing for the initial and final states $|i\rangle$ and $|f\rangle$ to not be normalized. We take these states to be

$$|i\rangle \to |\mathbf{p}_A \mathbf{p}_B\rangle \quad \text{and} \quad |f\rangle \to |\mathbf{p}_1 \mathbf{p}_2 \cdots \mathbf{p}_n\rangle, \tag{7.3.33}$$

which are the product of one-particle states normalized according to the one-particle state orthonormality and completeness relations in Eq. (6.2.59). Note that

$$\langle \mathbf{p}|\mathbf{p}\rangle = 2E_{\mathbf{p}}(2\pi)^3 \delta^3(\mathbf{0}) \to 2E_{\mathbf{p}} V, \tag{7.3.34}$$

where again we use Eq. (6.2.26) to identify $(2\pi)^3 \delta^3(\mathbf{0}) \to V$ in a finite spatial volume. Recall from Eq. (6.2.48) that $E_{\mathbf{p}} \delta^3(\mathbf{p} - \mathbf{q})$ is Lorentz invariant, which here corresponds to the Lorentz invariance of $E_{\mathbf{p}} V$. This then gives

$$\begin{aligned}
\langle i|i\rangle &= \langle \mathbf{p}_A \mathbf{p}_B | \mathbf{p}_A \mathbf{p}_B \rangle \to 4 E_{\mathbf{p}_A} E_{\mathbf{p}_B} V^2, \\
\langle f|f\rangle &= \langle \mathbf{p}_1 \mathbf{p}_2 \cdots \mathbf{p}_n | \mathbf{p}_1 \mathbf{p}_2 \cdots \mathbf{p}_n \rangle \to \prod_{j=1}^n (2E_{\mathbf{p}_j} V), \\
\sum_f &\to \sum_{n=1,2,\ldots} \int \prod_{j=1}^n [V/(2\pi)^3] d^3 p_j,
\end{aligned} \tag{7.3.35}$$

where we have used Eq. (6.2.26) to rewrite a sum over discrete momenta in the form $\sum_i \to (V/2\pi) \int d^3 p$. In summing over final states we sum over all possible particle number and types and integrate over all momenta. The T-matrix is

$$\langle \mathbf{p}_1 \cdots \mathbf{p}_n | \hat{T} | \mathbf{p}_A \mathbf{p}_B \rangle = (2\pi)^4 \delta^4(p_A + p_B - \textstyle\sum_{i=1}^n p_i) \mathcal{M}_{p_A p_B \to p_1 \cdots p_n}, \tag{7.3.36}$$

where $i = (\mathbf{p}_A, \mathbf{p}_B)$ and $f = (\mathbf{p}_1, \ldots, \mathbf{p}_n)$. We have from Eq. (7.3.4) that

$$|\langle \mathbf{p}_1 \cdots \mathbf{p}_n | \hat{T} | \mathbf{p}_A \mathbf{p}_B \rangle|^2 = (V \Delta T)(2\pi)^4 \delta^4(p_A + p_B - \textstyle\sum_{j=1}^n p_j) |\mathcal{M}_{fi}|^2, \tag{7.3.37}$$

where we use the shorthand $\mathcal{M}_{fi} \equiv \mathcal{M}_{p_A p_B \to p_1 \cdots p_n}$. Putting this all together, the probability in Eq. (7.3.32) of a single pair of A and B particles to scatter becomes

$$P_{AB} = \sum_{n=1,2,\ldots} \int \left[\prod_{k=1}^{n} \frac{V d^3 p_k}{(2\pi)^3} \right] \frac{(V\Delta T)(2\pi)^4 \delta^4(p_A + p_B - \sum_{j=1}^{n} p_j)|\mathcal{M}_{fi}|^2}{4 E_{\mathbf{p}_A} E_{\mathbf{p}_B} V^2 \prod_{j=1}^{n}(2 E_{\mathbf{p}_j} V)} \qquad (7.3.38)$$

$$= \frac{\Delta T}{V} \sum_{n=1,2,\ldots} \int \left[\prod_{k=1}^{n} \frac{d^3 p_k}{(2\pi)^3 2 E_{\mathbf{p}_j}} \right] \frac{(2\pi)^4 \delta^4(p_A + p_B - \sum_{j=1}^{n} p_j)|\mathcal{M}_{fi}|^2}{4 E_{\mathbf{p}_A} E_{\mathbf{p}_B}}.$$

Recall from Eq. (7.3.30) that $P_{AB} = N_{\text{sc}}/N_A N_B$ and so is clearly Lorentz invariant. From Eq. (7.3.34) we have that $E_{\mathbf{p}} V$ is Lorentz invariant. Similarly, recall from Eq. (7.3.4) that for the Lorentz-invariant four-dimensional δ-function we have $(2\pi)^4 \delta^4(0) \to V\Delta T$. Since P_{AB} and $V\Delta T/E_{\mathbf{p}_A} E_{\mathbf{p}_B}$ and the integration measures are Lorentz invariant, then it must be that $|\mathcal{M}_{fi}|^2$ is Lorentz invariant, which is why we refer to $\mathcal{M}_{fi}$ as the invariant amplitude or invariant matrix element.

Using Eq. (7.3.30) we find for the total scattering cross-section ($f \neq i$) that

$$\sigma \equiv \sigma_{AB} = \sum_{n=1,2,\ldots} \left[\int \prod_{k=1}^{n} \frac{d^3 p_k}{(2\pi)^3 2 E_{\mathbf{p}_j}} \right] \frac{(2\pi)^4 \delta^4(p_A + p_B - \sum_{j=1}^{n} p_j)|\mathcal{M}_{fi}|^2}{|v_A - v_B| 4 E_{\mathbf{p}_A} E_{\mathbf{p}_B}}. \qquad (7.3.39)$$

Note that all of the factors of V and $\Delta T = 2T$ have canceled out and so we can now take the limits $V, T \to \infty$.

If we consider the contribution to the total cross-section arising from the scattering into n particles with infinitesimal momentum intervals $d^3 p_1 \cdots d^3 p_n$ we have the differential cross-section

$$d\sigma = \frac{|\mathcal{M}_{fi}|^2}{4 E_{\mathbf{p}_A} E_{\mathbf{p}_B}|v_A - v_B|} d\Pi_n^{\text{LIPS}}, \qquad (7.3.40)$$

where we have defined an infinitesimal element of the n-body *Lorentz-invariant phase space (LIPS)* as

$$d\Pi_n^{\text{LIPS}} \equiv (2\pi)^4 \delta^4(p_A + p_B - \sum_{j=1}^{n} p_j) \prod_{k=1}^{n} [d^3 p_k/(2\pi)^3 2 E_{\mathbf{p}_k}]. \qquad (7.3.41)$$

We know that $d\Pi_n^{\text{LIPS}}$ is invariant under Lorentz transformations since the delta function preserves total four-momentum in every inertial frame and since $d^3 p/E_{\mathbf{p}}$ is invariant as we showed in Eq. (6.2.50).

The cross-section σ has the units of area and since our constructions have respected Lorentz invariance, then σ must transform under Lorentz transformations as an area. Let us understand how this comes about. Let us first define the *Lorentz-invariant Møller flux factor* as

$$F \equiv \sqrt{(p_A \cdot p_B)^2 - m_A^2 m_B^2}. \qquad (7.3.42)$$

Had we chosen to work in an inertial frame where the plane waves were not collinear, then the flux of one on the other would still be proportional to their relative velocity; i.e., we would still have a factor $|v_A - v_B|$ and normalization factors $E_{\mathbf{p}_A}$ and $E_{\mathbf{p}_A}$ but with v_A and v_B not collinear. Then using Eq. (7.3.31) we note that

$$(p_A \cdot p_B)^2 - (E_{\mathbf{p}_A} E_{\mathbf{p}_B}|v_A - v_B|)^2 = (E_{\mathbf{p}_A} E_{\mathbf{p}_B} - \mathbf{p}_A \cdot \mathbf{p}_B)^2 - (E_{\mathbf{p}_B}\mathbf{p}_A - E_{\mathbf{p}_A}\mathbf{p}_B)^2$$

$$= E_{\mathbf{p}_A}^2 E_{\mathbf{p}_B}^2 + (\mathbf{p}_A \cdot \mathbf{p}_B)^2 - E_{\mathbf{p}_B}^2 \mathbf{p}_A^2 - E_{\mathbf{p}_A}^2 \mathbf{p}_B^2 = E_{\mathbf{p}_A}^2 m_B^2 + (\mathbf{p}_A \cdot \mathbf{p}_B)^2 - E_{\mathbf{p}_B}^2 \mathbf{p}_A^2$$

$$= E_{\mathbf{p}_A}^2 m_B^2 + \mathbf{p}_A^2 \mathbf{p}_B^2 - E_{\mathbf{p}_B}^2 \mathbf{p}_A^2 - (\mathbf{p}_A \times \mathbf{p}_B)^2 = E_{\mathbf{p}_A}^2 m_B^2 - \mathbf{p}_A^2 m_B^2 - (\mathbf{p}_A \times \mathbf{p}_B)^2$$

$$= m_A^2 m_B^2 - (\mathbf{p}_A \times \mathbf{p}_B)^2, \qquad (7.3.43)$$

where we have used $(\mathbf{p}_A \cdot \mathbf{p}_B)^2 = \mathbf{p}_A^2 \mathbf{p}_B^2 - (\mathbf{p}_A \times \mathbf{p}_B)^2$. We can then define

$$F' \equiv E_{\mathbf{p}_A} E_{\mathbf{p}_B} |v_A - v_B| = \sqrt{(p_A \cdot p_B)^2 - m_A^2 m_B^2 + (\mathbf{p}_A \times \mathbf{p}_B)^2} = \sqrt{F^2 + (\mathbf{p}_A \times \mathbf{p}_B)^2},$$

$$d\sigma = \frac{|\mathcal{M}_{fi}|^2}{4F'} d\Pi_n^{\mathrm{LIPS}} = \frac{|\mathcal{M}_{fi}|^2}{4\sqrt{F^2 + (\mathbf{p}_A \times \mathbf{p}_B)^2}} d\Pi_n^{\mathrm{LIPS}}, \tag{7.3.44}$$

where $F' \equiv \sqrt{F^2 + (\mathbf{p}_A \times \mathbf{p}_B)^2}$ is the flux factor. A *collinear inertial frame* is one in which $\mathbf{p}_A$ and $\mathbf{p}_B$ are collinear. Examples include the center-of-momentum frame and the fixed target frame. Note that $1/F'$ has units of area and reduces to the Lorentz-invariant inverse Møller flux factor in any collinear inertial frame,

$$F' = E_{\mathbf{p}_A} E_{\mathbf{p}_B} |v_A - v_B| \xrightarrow{\text{collinear}} F = \sqrt{(p_A \cdot p_B)^2 - m_A^2 m_B^2}. \tag{7.3.45}$$

So under all Lorentz transformations that maintain collinearity of the two beams it follows that $F' = F$ is invariant. Since $|\mathcal{M}_{fi}|^2$ and $d\Pi_n^{\mathrm{LIPS}}$ are Lorentz invariant, then $d\sigma$ transforms as an infinitesimal cross-sectional area orthogonal to the colliding beams in a collinear inertial frame as it should.

Asymptotic States and Wavepackets

In our discussion to this point we have treated the wavepackets of particles A and B in the plane wave limit, where they filled all of space. A more realistic physical picture is provided in Fig. 7.9, where the ovals represent the quantum wavepackets of the particles (Peskin and Schroeder, 1995). Consider a single A wavepacket in either a fixed target ($v_A = 0$) or in a collinear colliding beam ($v_A \neq 0$, v_A antiparallel to v_B). Over a long period of time ΔT the A wavepacket will experience B wavepackets from all possible values of the impact parameter vectors, $\mathbf{b} \in \mathbb{R}^2$, that lie in a plane orthogonal to the direction of relative motion. The roles of the A and B particles can also be reversed.

We can then construct a wavefunction for a free particle A in terms of a superposition of plane waves and we can define a state $|\phi_A, t\rangle$ using Eq. (4.2.15) as

$$\psi_A(t, \mathbf{x}) = \int \frac{d^3 k}{(2\pi)^3} \phi_A(\mathbf{k}) e^{-ik \cdot x}, \quad |\phi_A, t\rangle \equiv \int \frac{d^3 k}{(2\pi)^3} \frac{1}{\sqrt{2E_{\mathbf{k}}}} \phi_A(\mathbf{k}) e^{-iE_{\mathbf{k}} t} |\mathbf{k}\rangle. \tag{7.3.46}$$

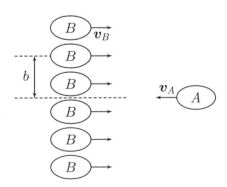

Figure 7.9   The beam of particle B wavepackets incident on a single particle A wavepacket as seen from the laboratory frame. b is the wavepacket impact parameter.

Note that $|\phi_A, t\rangle$ is a superposition of one-particle momentum eigenstates. The normalization of $|\phi_A, t\rangle$ follows from normalizing the wavefunction,

$$\int d^3x \, |\psi_A(t, \mathbf{x})|^2 = \int \frac{d^3k}{(2\pi)^3} |\phi_A(\mathbf{k})|^2 = \langle \phi_A, t | \phi_A, t \rangle = 1, \tag{7.3.47}$$

where we have used $\langle \mathbf{p} | \mathbf{q} \rangle = 2E_{\mathbf{p}}(2\pi)^3 \delta^3(\mathbf{p} - \mathbf{q})$ from Eq. (6.2.59). For two such states we find

$$\langle \phi_A, t | \phi_B, t \rangle = \int \frac{d^3k}{(2\pi)^3} \phi_A^*(\mathbf{k}) \phi_B(\mathbf{k}) e^{i(E_A - E_B)t} = \int d^3x \, \psi_A^*(t, \mathbf{x}) \psi_B(t, \mathbf{x}), \tag{7.3.48}$$

where we now write $E_A \equiv E_A(\mathbf{k}) = \sqrt{\mathbf{k}^2 + m_A^2}$ and $E_B \equiv E_B(\mathbf{k}) = \sqrt{\mathbf{k}^2 + m_B^2}$ in place of $E_{\mathbf{k}}$ since we may have $m_A \neq m_B$. So $\phi_A(\mathbf{k})$ and $\phi_A(\mathbf{k})e^{-iE_At}$ are the three-dimensional asymmetric Fourier transforms of $\psi_A(0, \mathbf{x})$ and $\psi_A(t, \mathbf{x})$, respectively. Recall that we are interested in the limit where the wavepackets are narrow in momentum space and sharply peaked around $\mathbf{p}_A$ and $\mathbf{p}_B$, respectively. As we saw in Eq. (4.2.30), the group velocity for the narrow momentum-space wavepackets ϕ_A and ϕ_B is the same as the velocity of particles with momenta $\mathbf{p}_A$ and $\mathbf{p}_B$. Let $\boldsymbol{v}_A$ be the group velocity for particle A with mass m_A and peak three-momentum $\mathbf{p}_A$ and define the corresponding peak energy as $E_{\mathbf{p}_A} = \sqrt{\mathbf{p}_A^2 + m_A^2}$ and similarly for particle B. The effects of the small dispersion of our wavepackets do not affect the following arguments and so we will not include them for simplicity here. From Eq. (4.2.34) we have up to small dispersion corrections

$$\psi_A(x) = \psi_A(t, \mathbf{x}) \simeq e^{-i(E_{\mathbf{p}_A} - \mathbf{p}_A \cdot \boldsymbol{v}_A)t} \psi_A(0, [\mathbf{x} - \boldsymbol{v}_A t]) \tag{7.3.49}$$

and similarly for $\psi_B(x)$. Let $\psi_A(0, \mathbf{x})$ and $\psi_B(0, \mathbf{x})$ be wavefunctions centered at $\mathbf{x} = 0$. Consider the situation where $\boldsymbol{v}_A \neq \boldsymbol{v}_B$, then the wavepackets $\psi_A(t, \mathbf{x})$ and $\psi_B(t, \mathbf{x})$ are colliding at $t = 0$ since they both have their centers at $\mathbf{x} = 0$ at that time. As we take $t \to \pm\infty$ with $\boldsymbol{v}_A \neq \boldsymbol{v}_B$ the wavepackets become infinitely separated and they have no overlap.

If the cluster decomposition principle holds, then as $t \to \pm\infty$ a two-particle state in the full Hilbert space will consist of the tensor product of the two one-particle states, which is then a two-particle state in the noninteracting Fock space, V_{Fock}. So for $t \to \pm\infty$ we can write the two-particle state as the Fock-space state

$$|\phi_A, \phi_B, t\rangle \equiv \int \frac{d^3k_A d^3k_B}{(2\pi)^6 2\sqrt{E_A E_B}} \phi_A(\mathbf{k}_A) \phi_B(\mathbf{k}_B) e^{-i(E_A + E_B)t} |\mathbf{k}_A \mathbf{k}_B\rangle. \tag{7.3.50}$$

Since we define the initial state and the final state at any arbitrarily large positive or negative *fixed time* t_{fix}, then we can redefine $\phi_A(\mathbf{k}_A)e^{-iE_A t_{\text{fix}}} \to \phi_A(\mathbf{k})$ and $\phi_B(\mathbf{k}_A)e^{-iE_B t_{\text{fix}}} \to \phi_B(\mathbf{k})$, giving a simple form for our two-particle initial state,

$$|i\rangle = |\phi_A \phi_B\rangle \equiv \int \frac{d^3k_A d^3k_B}{(2\pi)^6 2\sqrt{E_A E_B}} \phi_A(\mathbf{k}_A) \phi_B(\mathbf{k}_B) |\mathbf{k}_A \mathbf{k}_B\rangle, \tag{7.3.51}$$

where it follows from above that $|i\rangle$ is normalized, $\langle i|i\rangle = 1$. Choosing $\boldsymbol{v}_A$ antiparallel to $\boldsymbol{v}_B$, which means $\mathbf{p}_A$ antiparallel to $\mathbf{p}_B$, we have a direct head-on collision of the A and B wavepackets with zero impact parameter, $b = 0$.

If instead of a head-on collision $\psi_B(x)$ is translated in space by a vector $\mathbf{b}$, then $\psi_B(t, \mathbf{x} - \mathbf{b}) \to \psi_B(t, \mathbf{x})$. This means that we have the replacement $e^{i\mathbf{k} \cdot \mathbf{x}} \to e^{i\mathbf{k} \cdot (\mathbf{x} - \mathbf{b})}$,

$$\psi_B(t, \mathbf{x}) = \int \frac{d^3k}{(2\pi)^3} \phi_B(\mathbf{k}) e^{-i\mathbf{k} \cdot \mathbf{b}} e^{-ik \cdot x}. \tag{7.3.52}$$

We see that the translation is achieved by the replacement $\phi_B(\mathbf{k}) \to \phi_B(\mathbf{k})e^{-i\mathbf{k}\cdot\mathbf{b}}$. In this case our two-particle Heisenberg-picture initial state in Fock space becomes

$$|i\rangle = |\phi_A\phi_B; \mathbf{b}\rangle \equiv \int \frac{d^3k_A d^3k_B}{(2\pi)^6 2\sqrt{E_A E_B}} \phi_A(\mathbf{k_A})\phi_B(\mathbf{k_B})e^{-i\mathbf{k}_B\cdot\mathbf{b}}|\mathbf{k}_A\mathbf{k}_B\rangle. \tag{7.3.53}$$

Again choosing $\boldsymbol{v}_A$ antiparallel to $\boldsymbol{v}_B$ we have a collision of the A and B wavepackets with impact parameter b as illustrated in Fig. 7.9. The translation can be absorbed through a redefinition, $\phi'_B(\mathbf{k}) \equiv \phi_B(\mathbf{k})e^{-i\mathbf{k}\cdot\mathbf{b}}$. With this understanding we recognize that Eq. (7.3.51) is a general form for any two-particle asymptotic state in Fock space with arbitrary relative translation of the wavepackets at $t = 0$.

Generalizing this to n outgoing and well-separated momentum-space wavepackets in the distant future we can write any normalized corresponding asymptotic final state as the Fock-space state

$$|f\rangle \equiv |\phi_1\phi_2\cdots\phi_n\rangle \equiv \int \prod_{j=1}^{n} \left[\frac{d^3k_j}{(2\pi)^3} \frac{1}{\sqrt{2E_j}} \phi_j(\mathbf{k_j}) \right] |\mathbf{k}_1\mathbf{k}_2\cdots\mathbf{k}_n\rangle. \tag{7.3.54}$$

Now consider a very large spatial volume, which we are free to choose as $V = AL = A(v_{AB}\Delta T)$ for later convenience. We have a large area A orthogonal to the direction of the incoming wavepackets and length $L = \Delta T v_{AB}$, where $\Delta T = 2T$ as above and is also very large. With our choice of spatial volume we have from Eq. (7.3.30),

$$\sigma_{AB} = (V/\Delta T)P_{AB}/v_{AB} = (V/L)P_{AB} = AP_{AB}. \tag{7.3.55}$$

In any experiment we observe the outcomes of many scattering events and averaging over these will lead to averages over impact parameters. So we must replace P_{AB} and σ_{AB} with averages over the impact parameter vector,

$$P_{AB} = (1/A)\int d^2b\, P_{AB;\mathbf{b}} \quad \text{and} \quad \sigma_{AB} = (1/A)\int d^2b\, \sigma_{AB;\mathbf{b}}, \tag{7.3.56}$$

which gives the result that

$$\sigma \equiv \sigma_{AB} = AP_{AB} = \int d^2b\, P_{AB;\mathbf{b}}. \tag{7.3.57}$$

As noted above our initial state $|i\rangle = |\phi_A\phi_B; \mathbf{b}\rangle$ is normalized and so

$$P_{AB;\mathbf{b}} = \sum_f \frac{|\langle f|\hat{T}|\phi_A\phi_B; \mathbf{b}\rangle|^2}{\langle f|f\rangle}, \tag{7.3.58}$$

where since we are "summing" over all asymptotic final states in the Fock space we do not need to build any outgoing wavepackets and can instead "sum" over the n-particle Fock-space plane wave states in the Fock space. This is discussed further in Sec. 7.5.1. In place of Eq. (7.3.36) we then have

$$\langle \mathbf{p}_1\cdots\mathbf{p}_n|\hat{T}|\phi_A\phi_B; \mathbf{b}\rangle = \int \frac{d^3k_A d^3k_B}{(2\pi)^6 2\sqrt{E_A E_B}} \phi_A(\mathbf{k_A})\phi_B(\mathbf{k_B})e^{-i\mathbf{k}_B\cdot\mathbf{b}}$$
$$\times \langle \mathbf{p}_1\cdots\mathbf{p}_n|\hat{T}|\mathbf{k}_A\mathbf{k}_B\rangle, \tag{7.3.59}$$

where $\langle \mathbf{p}_1\cdots\mathbf{p}_n|\hat{T}|\mathbf{k}_A\mathbf{k}_B\rangle$ is given in Eq. (7.3.36). Putting things together leads to

$$\sigma \equiv \sigma_{AB} = \int d^2b \sum_{n=1,2,\ldots} \int \left[\prod_{k=1}^{n} \frac{d^3p_k}{(2\pi)^3 2E_{\mathbf{p}_j}} \right] \left[\prod_{\ell=A,B} \frac{d^3k_\ell d^3k'_\ell}{(2\pi)^6 2\sqrt{E_{\mathbf{k}_\ell}E_{\mathbf{k}'_\ell}}} \right] \tag{7.3.60}$$
$$\times e^{-i(\mathbf{k}_B - \mathbf{k}'_B)\cdot\mathbf{b}}(2\pi)^4\delta^4(k_A + k_B - \textstyle\sum_{r=1}^n p_r)(2\pi)^4\delta^4(k'_A + k'_B - \textstyle\sum_{s=1}^n p_s)$$
$$\times \phi(\mathbf{k}_A)\phi(\mathbf{k}_B)\phi(\mathbf{k}'_A)^*\phi(\mathbf{k}'_B)^* \mathcal{M}_{k_a k_b \to p_1\cdots p_n}\mathcal{M}^*_{k'_a k'_b \to p_1\cdots p_n}.$$

We find $\int d^2b\, e^{-i(\mathbf{k}_B - \mathbf{k}'_B)\cdot\mathbf{b}} = (2\pi)^2 \delta^2(\mathbf{k}_B^\perp - \mathbf{k}_B^\perp)$ after integrating over the impact parameter vector $\mathbf{b}$, where we note that $d^3k = d^2k^\perp dk^z$. We can combine this two-dimensional δ-function with the second four-dimensional δ-function in Eq. (7.3.60) above to eliminate all six integrals over $\mathbf{k}'_A$ and $\mathbf{k}'_B$. We find

$$
\begin{aligned}
&\int d^2b \int d^3k'_A d^3k'_B\, e^{-i(\mathbf{k}_B - \mathbf{k}'_B)\cdot\mathbf{b}} \delta^4(k'_A + k'_B - \textstyle\sum_s p_s)\\
&= (2\pi)^2 \int d^3k'_A d^3k'_B\, \delta^2(\mathbf{k}_B^\perp - \mathbf{k}_B^\perp)\delta^3(\mathbf{k}'_A + \mathbf{k}'_B - \textstyle\sum_s \mathbf{p}_s)\delta(E'_A + E'_B - \textstyle\sum_s E_s)\\
&= (2\pi)^2 \int d^3k'_B\, \delta^2(\mathbf{k}_B^\perp - \mathbf{k}_B^\perp)\delta(E'_A + E'_B - \textstyle\sum_s E_s)\big|_{\mathbf{k}'_A = \sum_s \mathbf{p}_s - \mathbf{k}'_B}\\
&= (2\pi)^2 \int d^3k'_B\, \delta^2(\mathbf{k}_B^\perp - \mathbf{k}_B^\perp)\left|\frac{k'^z_A}{E'_A} - \frac{k'^z_B}{E'_B}\right|^{-1}\delta(k'^z_B + k'^z_A - \textstyle\sum_s p^z_s)\big|_{\mathbf{k}'_A = \sum_s \mathbf{p}_s - \mathbf{k}'_B}\\
&= (2\pi)^2 \left|\frac{k'^z_A}{E'_A} - \frac{k'^z_B}{E'_B}\right|^{-1}\Bigg|_{\mathbf{k}'_A = \sum_s \mathbf{p}_s - \mathbf{k}'_B,\, E'_A = \sum_s E_s - E'_B,\, \mathbf{k}_B^\perp = \mathbf{k}_B^\perp}
\end{aligned}
\tag{7.3.61}
$$

where $k'^z_B = g(\mathbf{k}'_B)$ is the condition on k'^z_B that results in $E'_A + E'_B = \sum_s E_s$, where we have used Eq. (A.2.8) to arrive at

$$
\begin{aligned}
&\delta(E'_A + E'_B - \textstyle\sum_s E_s)\big|_{\mathbf{k}'_A = \sum_s \mathbf{p}_s - \mathbf{k}'_B}\\
&= \delta\left(\sqrt{(\textstyle\sum_s \mathbf{p}_s - \mathbf{k}'_B)^2 + m_A^2} + \sqrt{\mathbf{k}_B'^2 + m_B^2} - \textstyle\sum_s E_s\right)\\
&\equiv \delta(f(k'^z_B)) = |df/dk'^z_B|^{-1}\, \delta(k'^z_B + k'^z_A - \textstyle\sum_s p^z_s)\big|_{\mathbf{k}'_A = \sum_s \mathbf{p}_s - \mathbf{k}'_B}\\
&= \left|\frac{k'^z_A}{E'_A} - \frac{k'^z_B}{E'_B}\right|^{-1}\delta(k'^z_B + k'^z_A - \textstyle\sum_s p^z_s)\big|_{\mathbf{k}'_A = \sum_s \mathbf{p}_s - \mathbf{k}'_B}
\end{aligned}
\tag{7.3.62}
$$

and where last line follows since

$$
|df/dk'^z_B| = \left|\frac{-(\sum_s p^z_s - k'^z_B)}{\sqrt{(\sum_s \mathbf{p}_s - \mathbf{k}'_B)^2 + m_A^2}} + \frac{k'^z_B}{\sqrt{\mathbf{k}_B'^2 + m_B^2}}\right| = \left|-\frac{k'^z_A}{E'_A} + \frac{k'^z_B}{E'_B}\right|^{-1}\Bigg|_{\mathbf{k}'_A = \sum_s \mathbf{p}_s - \mathbf{k}'_B}.
$$

We wish to find the solution for k'^z_B of the equation $f(k'^z_B) = 0$,

$$
\textstyle\sum_s E_s = E'_A + E'_B = \sqrt{(\textstyle\sum_s \mathbf{p}_s - \mathbf{k}'_B)^2 + m_A^2} + \sqrt{(\mathbf{k}'_B)^2 + m_B^2},
\tag{7.3.63}
$$

where since the ϕ functions are sharply peaked we only require one solution. We need $k'^z_B > 0$ for the B wavepackets moving in the $+z$ direction.

The four-momentum δ-functions ensure that $k^\mu_A + k^\mu_B = \sum_s p^\mu = k'^\mu_A + k'^\mu_B$ and we also have $\mathbf{k}_B'^\perp = \mathbf{k}_B^\perp$. As we saw above these give us six contraints on the $\mathbf{k}'_A$ and $\mathbf{k}'_B$ and so these must be expressible in terms of $\mathbf{k}_A$, $\mathbf{k}_B$, m_A and m_B. The on-mass-shell conditions $k_A^2 = k_A'^2 = m_A^2$, $k_B^2 = k_B'^2 = m_A^2$ are understood and specify E_A, E_B, E'_A and E'_B in terms of their respective three-momenta. Since $\mathbf{k}_A^\perp + \mathbf{k}_B^\perp = \mathbf{k}_A'^\perp + \mathbf{k}_B'^\perp = \mathbf{k}_A'^\perp + \mathbf{k}_B^\perp$, then we must have $\mathbf{k}_A'^\perp = \mathbf{k}_A^\perp$. Using this we can write $E'_A + E'_B = E_A + E_B$ as

$$
\begin{aligned}
&\sqrt{(k'^z_A)^2 + (\mathbf{k}_A^\perp)^2 + m_A^2} + \sqrt{(k'^z_B)^2 + (\mathbf{k}_B^\perp)^2 + m_B^2}\\
&= \sqrt{(k^z_A)^2 + (\mathbf{k}_A^\perp)^2 + m_A^2} + \sqrt{(k^z_B)^2 + (\mathbf{k}_B^\perp)^2 + m_B^2}.
\end{aligned}
\tag{7.3.64}
$$

Combining this with $k'^z_A + k'^z_B = k^z_A + k^z_B$ we have two equations in the four unknowns k^z_A, k^z_B, k'^z_A and k'^z_B and the obvious solution to these equations is $k'^z_A = k^z_A$ and $k'^z_B = k^z_B$. We have arrived at the result that $\mathbf{k}'_A = \mathbf{k}_A$ and $\mathbf{k}'_B = \mathbf{k}_B$. So Eq. (7.3.60) has now become

$$\sigma \equiv \sigma_{AB} = \sum_{n=1,2,\dots} \int \left[\prod_{k=1}^{n} \frac{d^3 p_k}{(2\pi)^3 2E_{\mathbf{p}_j}} \right] \left[\frac{d^3 k_A d^3 k_B}{(2\pi)^6 4E_A E_B} \right] \tag{7.3.65}$$

$$\times (2\pi)^4 \delta^4(k_A + k_B - \sum_{r=1}^{n} p_r) |\phi(\mathbf{k}_A)|^2 |\phi(\mathbf{k}_B)|^2 \left| \frac{k_A^z}{E_A} - \frac{k_B^z}{E_B} \right|^{-1} |\mathcal{M}_{k_a k_b \to p_1 \cdots p_n}|^2.$$

Since we are interested in the limit where $\phi(\mathbf{k}_A)$ and $\phi(\mathbf{k}_B)$ are very sharply peaked functions centered around $\mathbf{p}_A$ and $\mathbf{p}_B$, respectively, then functions that are relatively smooth in $\mathbf{k}_A$ and $\mathbf{k}_B$ can have their arguments replaced by the peak values and be moved outside the integrals. Since we have collinear beams here then $v_A = p_A^z/E_{\mathbf{p}_A}$ and $v_B = p_B^z/E_{\mathbf{p}_B}$, which gives

$$\sigma = \sum_{n=1,2,\dots} \left[\int \prod_{k=1}^{n} \frac{d^3 p_k}{(2\pi)^3 2E_{\mathbf{p}_j}} \right] \frac{|\mathcal{M}_{p_a p_b \to p_1 \cdots p_n}|^2}{|\mathbf{v}_A - \mathbf{v}_B| 4E_{\mathbf{p}_A} E_{\mathbf{p}_B}} \left[\int \frac{d^3 k_A d^3 k_B}{(2\pi)^6} \right]$$

$$\times (2\pi)^4 \delta^4(k_A + k_B - \sum_{r=1}^{n} p_r) |\phi(\mathbf{k}_A)|^2 |\phi(\mathbf{k}_B)|^2. \tag{7.3.66}$$

In a real experiment, where we sample the differential cross-section with physical detectors, our measurements of final state momenta will have greater uncertainties Δp_r^μ than the widths of the incoming momentum-space wavepackets. Then the $\mathbf{k}_A$ and $\mathbf{k}_B$ that are not strongly suppressed by the narrow ϕ functions will appear to be consistent with four-momentum conservation. Hence we can use the peak values in the δ-function and move it outside the integrations, which leaves two normalization integrals of the form of Eq. (7.3.47). Finally, we recover the cross-section in Eq. (7.3.39) arrived at earlier using the simple plane wave approximation,

$$\sigma \equiv \sigma_{AB} = \sum_{n=1,2,\dots} \left[\int \prod_{k=1}^{n} \frac{d^3 p_k}{(2\pi)^3 2E_{\mathbf{p}_j}} \right] \frac{(2\pi)^4 \delta^4(p_A + p_B - \sum_{j=1}^{n} p_j) |\mathcal{M}_{fi}|^2}{|\mathbf{v}_A - \mathbf{v}_B| 4E_{\mathbf{p}_A} E_{\mathbf{p}_B}}. \tag{7.3.67}$$

The total cross-section cannot be measured and instead we integrate the differential cross-section $d\sigma$ over the range of the detectors in some channel, $AB \to X_n$, where

$$d\sigma_{AB \to X_n} \equiv \left[\prod_{k=1}^{n} \frac{d^3 p_k}{(2\pi)^3 2E_{\mathbf{p}_j}} \right] \frac{(2\pi)^4 \delta^4(p_A + p_B - \sum_{j=1}^{n} p_j) |\mathcal{M}_{fi}|^2}{|\mathbf{v}_A - \mathbf{v}_B| 4E_{\mathbf{p}_A} E_{\mathbf{p}_B}} \tag{7.3.68}$$

$$= \frac{|\mathcal{M}_{fi}|^2}{|\mathbf{v}_A - \mathbf{v}_B| 4E_{\mathbf{p}_A} E_{\mathbf{p}_B}} d\Pi_n^{\mathrm{LIPS}} = \frac{|\mathcal{M}_{fi}|^2}{4F'} d\Pi_n^{\mathrm{LIPS}},$$

where the Lorentz-invariant phase space differential, $d\Pi^{\mathrm{LIPS}}$, is defined in Eq. (7.3.41).

7.3.3 Particle Decay Rates

Recall from Problem 1.1 that the decay rate Γ_A of particles of type A is the number of A decays per unit time divided by the number of A particles present,

$$\Gamma_A = -\frac{(dN_A/dt)}{N_A} \equiv \frac{1}{\tau_A}, \tag{7.3.69}$$

where τ_A is the lifetime or mean lifetime of particles of type A as discussed in Problem 1.1 . This equation has the solution $N_A(t) = N_A(0)e^{-t/\tau_A}$. The half-life of particles of type A, $t_{1/2}^A$, is the time taken for half the particles in a collection of A particles to decay and $t_{1/2}^A = \tau_A \ln 2$.

We now wish to relate the S-matrix to the decay rate. The initial state now only contains a single particle,

$$|i\rangle = |\mathbf{p}_A\rangle \quad \text{and} \quad \langle i|i\rangle = \langle \mathbf{p}_A|\mathbf{p}_A\rangle \to 2E_{\mathbf{p}_A} V, \tag{7.3.70}$$

and the final state might contain multiple particles as in the case of scattering in Eq. (7.3.35). The probability to transition from an initial state i to any possible final state f in volume V and time ΔT is again given by Eq. (7.3.32) and the subsequent equations all follow, where the only change is the nature of initial state $|i\rangle$ given above. Then in place of Eq. (7.3.38) we now have

$$P_A = \Delta T \sum_{n=1,2,\dots} \int \left[\prod_{k=1}^{n} \frac{d^3 p_k}{(2\pi)^3 2E_{\mathbf{p}_j}} \right] \frac{(2\pi)^4 \delta^4(p_A - \sum_{j=1}^{n} p_j)|\mathcal{M}_{fi}|^2}{2E_{\mathbf{p}_A}}. \tag{7.3.71}$$

The probability per unit time for particle A to decay in volume V is the decay rate,

$$\Gamma_{\mathbf{p}_A} = \sum_{n=1,2,\dots} \int \left[\prod_{k=1}^{n} \frac{d^3 p_k}{(2\pi)^3 2E_{\mathbf{p}_j}} \right] \frac{(2\pi)^4 \delta^4(p_A - \sum_{j=1}^{n} p_j)|\mathcal{M}_{fi}|^2}{2E_{\mathbf{p}_A}}, \tag{7.3.72}$$

where V and ΔT no longer appear and so we can take $V, \Delta T \to \infty$. For a particle A decaying at rest we have $E_A = \gamma_A m_A \to m_A$ in the denominator We recognize the Lorentz factor γ_A is just the time dilation factor, which means that the decay rate slows the faster the A particle is going in the laboratory frame. We normally define lifetime and half-life in the particle rest frame, then

$$\tau_A \equiv \tau_A(0) = \tau_A(v_A)/\gamma_A \quad \text{with} \quad \gamma_A = (1 - v_A^2)^{-1/2}. \tag{7.3.73}$$

We define the differential decay rate into the n-particle channel as

$$d\Gamma_{p_A} = \frac{|\mathcal{M}_{fi}|^2}{2E_{p_A}} d\Pi_n^{\text{LIPS}} \tag{7.3.74}$$

with $d\Pi_n^{\text{LIPS}}$ from Eq. (7.3.41). The decay rate for a particle at rest is $\Gamma_A = \Gamma_{\mathbf{p}_A=0}$.

In the case of scattering we assumed that the two particles in the initial state and all of the particles in the final state were so far apart that we could ignore their interactions and treat them as free particles in the $t \to \pm\infty$ limits, respectively. We are inherently assuming that these separated particles are stable and do not decay. In the case of a single particle decaying we need to be more subtle in our thinking. Since particle A decays, then it is not an eigenstate of the full Hamiltonian and so the state $|i\rangle = |\mathbf{p}_A\rangle$ had to be prepared at some time. Let us define the preparation time as $t = 0$. Imagine some large finite time T such that we consider the decay of particle A in the time interval $[0, T]$ so that here $\Delta T = T$. Eq. (7.3.32) still applies for our prepared state $|i\rangle$ with the understanding that now $\hat{S} = \hat{U}(T, 0)$ rather than $\hat{U}(T, -T)$. The subsequent arguments all follow and lead to the result for Γ given above. We could arrive at the equation for Γ by adapting the scattering wavepacket arguments, where the wavepacket for the unstable particle or state A is produced by some process at time $t = 0$.

7.3.4 Two-Body Scattering ($2 \to 2$) and Mandelstam Variables

The Mandestam variables s, t and u for two-body scattering are illustrated in Fig. 7.10 and are given by

$$s = (p_1 + p_2)^2, \qquad t = (p_1 - p_3)^2, \qquad u = (p_1 - p_4)^2. \tag{7.3.75}$$

The usefulness of these parameters is that when the interaction is mediated or dominated by the exchange of a single particle (shown as a dashed line with four-momentum q^μ), then s, t and u are the four-momentum squared, q^2, of the exchanged particle in each case. We refer to these exchanges as being in the *s-channel*, *t-channel* and *u-channel*, respectively. Recall the derivation in Sec. 2.6 of

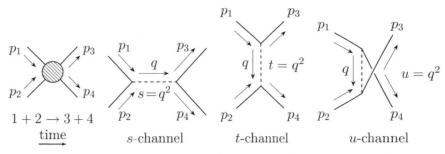

Figure 7.10 Two-body scattering and Mandelstam variables: $s = (p_1 + p_2)^2$, $t = (p_1 - p_3)^2$ and $u = (p_1 - p_4)^2$.

the conservation of four-momentum, which gives $p_1^\mu + p_2^\mu - p_3^\mu - p_4^\mu = 0$. Using this we have for the s-channel $q^\mu = (p_1^\mu + p_2^\mu) = (p_3^\mu + p_4^\mu)$ and $s = q^2$, for the t-channel we have $q^\mu = (p_1^\mu - p_3^\mu) = (p_4^\mu - p_2^\mu)$ and $t = q^2$, and for the u-channel we have $q^\mu = (p_1^\mu - p_4^\mu) = (p_3^\mu - p_2^\mu)$ and $u = q^2$. Note that $s = p_1^2 + p_s^2 + 2p_1 \cdot p_2$, $t = p_1^2 + p_3^2 - 2p_1 \cdot p_3$ and $u = p_1^2 + p_4^2 - 2p_1 \cdot p_4$. Summing these we find

$$
\begin{aligned}
s + t + u &= m_1^2 + m_2^2 + m_3^2 + m_4^2 + 2p_1^2 + 2p_1 \cdot p_2 - 2p_1 \cdot p_3 - 2p_1 \cdot p_4 \\
&= m_1^2 + m_2^2 + m_3^2 + m_4^2 + 2p_1 \cdot (p_1 + p_2 - p_3 - p_4) \\
&= m_1^2 + m_2^2 + m_3^2 + m_4^2.
\end{aligned} \tag{7.3.76}
$$

For two particles scattering into a two-particle final state, $1 + 2 \to 3 + 4$, we have in any collinear inertial frame,

$$
d\sigma = \frac{|\mathcal{M}_{fi}|^2}{4E_1 E_2 |\boldsymbol{v}_1 - \boldsymbol{v}_2|} d\Pi_2^{\text{LIPS}} = \frac{|\mathcal{M}_{fi}|^2}{4F} d\Pi_2^{\text{LIPS}},
$$
$$
d\Pi_2^{\text{LIPS}} \equiv (2\pi)^4 \delta^4(p_1 + p_2 - p_3 - p_4)[d^3 p_3/(2\pi)^3 2E_{\mathbf{p}_3}][d^3 p_4/(2\pi)^3 2E_{\mathbf{p}_4}]. \tag{7.3.77}
$$

CM frame: Consider two-body scattering in the *center-of-momentum (CM)* frame,[3] which is an inertial frame in which the total three-momentum vanishes. We can always perform a Lorentz boost to the CM frame. The two-body CM energy is given by $\sqrt{s}$, since in the CM frame $\mathbf{p}_1 = -\mathbf{p}_2$ and so

$$
s = (p_1 + p_2)^2 = (E_1 + E_2)^2 + (\mathbf{p}_1 + \mathbf{p}_2)^2 = (E_1 + E_2)^2 = E_{\text{cm}}^2. \tag{7.3.78}
$$

Four-momentum conservation also gives $\mathbf{p}_3 = -\mathbf{p}_4$ and $\sqrt{s} = E_3 + E_4 = E_{\text{cm}}$. Defining $p_3 \equiv |\mathbf{p}_3|$ the integral over two-particle phase space in the CM frame is

$$
\begin{aligned}
\int d\Pi_2^{\text{LIPS}} &= \int (2\pi)^4 \delta^4(p_1 + p_2 - p_3 - p_4)[d^3 p_3/(2\pi)^3 2E_{\mathbf{p}_3}][d^3 p_4/(2\pi)^3 2E_{\mathbf{p}_4}] \\
&= \int (2\pi)\delta(E_{\mathbf{p}_1} + E_{\mathbf{p}_2} - E_{\mathbf{p}_3} - E_{\mathbf{p}_4})[d^3 p_3/(2\pi)^3 2E_{\mathbf{p}_3} 2E_{\mathbf{p}_4}]\big|_{\mathbf{p}_3 + \mathbf{p}_4 = \mathbf{p}_1 + \mathbf{p}_2 = 0} \\
&= \int \frac{d\Omega_{\mathbf{p}_3} \, dp_3 \, p_3^2}{(2\pi)^3 2E_{\mathbf{p}_3} 2E_{\mathbf{p}_4}} (2\pi)\delta(E_{\mathbf{p}_1} + E_{\mathbf{p}_2} - E_{\mathbf{p}_3} - E_{\mathbf{p}_4})\big|_{\mathbf{p}_3 + \mathbf{p}_4 = \mathbf{p}_1 + \mathbf{p}_2 = 0} \\
&= \int \frac{d\Omega_{\mathbf{p}_3} \, p_3^2}{16\pi^2 E_{\mathbf{p}_3} E_{\mathbf{p}_4}} \left| \frac{p_3}{E_{\mathbf{p}_3}} + \frac{p_3}{E_{\mathbf{p}_4}} \right|^{-1} \Bigg|_{\mathbf{p}_3 + \mathbf{p}_4 = \mathbf{p}_1 + \mathbf{p}_2 = 0} = \int \frac{d\Omega_{\mathbf{p}_3}}{16\pi^2} \frac{|\mathbf{p}_3|}{\sqrt{s}} \Bigg|_{\mathbf{p}_3 + \mathbf{p}_4 = \mathbf{p}_1 + \mathbf{p}_2 = 0},
\end{aligned} \tag{7.3.79}
$$

---

[3] This is also sometimes referred to as the "center-of-mass frame." Strictly speaking, the center-of-mass for a system is a *point* defined classically in Eqs. (2.1.2) and (3.2.81). Center-of-momentum refers to an *inertial frame* in which all relativistic three-momenta sum to zero.

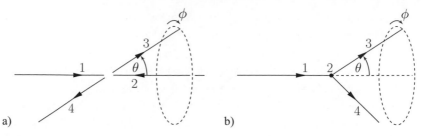

Figure 7.11 Two-body scattering: (a) in the center-of-momentum (CM) frame; (b) in the fixed-target frame.

where $\mathbf{p}_1$ and $\mathbf{p}_2$ are fixed, where we have used Eq. (7.3.62) with $\mathbf{p}_4 = -\mathbf{p}_3$ and where $\sqrt{s} = E_{\mathbf{p}_3} + E_{\mathbf{p}_4} = E_{\mathbf{p}_1} + E_{\mathbf{p}_2} = E_{\text{cm}}$. Defining

$$p_i^{\text{cm}} \equiv |\mathbf{p}_1| = |\mathbf{p}_2| \quad \text{and} \quad p_f^{\text{cm}} \equiv |\mathbf{p}_3| = |\mathbf{p}_4|, \tag{7.3.80}$$

the CM-invariant Møller flux factor can be written as

$$
\begin{aligned}
F &= E_{\mathbf{p}_1} E_{\mathbf{p}_2} |\boldsymbol{v}_1 - \boldsymbol{v}_2| = E_{\mathbf{p}_1} E_{\mathbf{p}_2} [(p_i^{\text{cm}}/E_{\mathbf{p}_1}) + (p_i^{\text{cm}}/E_{\mathbf{p}_2})] \\
&= p_i^{\text{cm}} (E_{\mathbf{p}_2} + E_{\mathbf{p}_1}) = p_i^{\text{cm}} \sqrt{s} = p_i^{\text{cm}} E_{\text{cm}}.
\end{aligned} \tag{7.3.81}
$$

Since outgoing particles 3 and 4 move away from the collision region back-to-back we only need to specify the direction of one and so we write $d\Omega \equiv d\Omega_{\mathbf{p}_3}$ as the corresponding infinitesimal solid angle. From Eq. (7.3.77) the CM differential cross-section is then given by (Thomson, 2013)

$$\left(\frac{d\sigma}{d\Omega} \right)_{\text{cm}} = \frac{|\mathcal{M}_{fi}|^2}{4F} d\Pi_2^{\text{LIPS}} = \frac{|\mathcal{M}_{fi}|^2}{4p_i^{\text{cm}}\sqrt{s}} \frac{p_f^{\text{cm}}}{16\pi^2\sqrt{s}} = \frac{1}{64\pi^2 s} \frac{p_f^{\text{cm}}}{p_i^{\text{cm}}} |\mathcal{M}_{fi}|^2. \tag{7.3.82}$$

Note that $d\Omega = d\theta \sin\theta \, d\phi = d\phi \, d(\cos\theta)$ with θ and ϕ defined in Fig. 7.11. In the case of spinless particles we will have azimuthal (cylindrical) symmetry around the collinear beam direction, which means ϕ-independence. In that case we can integrate over ϕ and pick up a factor of 2π to give

$$\left(\frac{d\sigma}{d(\cos\theta)} \right)_{\text{cm}} = \frac{1}{32\pi s} \frac{p_f^{\text{cm}}}{p_i^{\text{cm}}} |\mathcal{M}_{fi}|^2 \xrightarrow{\text{all masses equal}} \frac{|\mathcal{M}_{fi}|^2}{32\pi s}, \tag{7.3.83}$$

where the special case $m_1 = m_2 = m_3 = m_4$ leads to $p_i^{\text{cm}} = p_f^{\text{cm}}$ because then every particle must have energy $E_{\text{cm}}/2$. Since $d\Omega$ is not Lorentz invariant it is preferable to use a collinear-invariant form. Let the momentum of particle 1 define the z-axis and assuming no ϕ-dependance we are free to take $\phi = 0$. Using $p_j \equiv |\mathbf{p}_j|$ for the jth particle, we can then write in any collinear frame

$$p_1^\mu = (E_{\mathbf{p}_1}, 0, 0, p_1), \quad p_3^\mu = (E_{\mathbf{p}_3}, p_3 \sin\theta, 0, p_3 \cos\theta) \quad \text{and}$$
$$p_1 \cdot p_3 = E_{\mathbf{p}_1} E_{\mathbf{p}_3} - p_1 p_3 \cos\theta \tag{7.3.84}$$

and so for the Mandelstam variable t we have

$$t = (p_1 - p_3)^2 = m_1^2 + m_3^2 - 2p_1 \cdot p_3 = m_1^2 + m_3^2 - 2E_{\mathbf{p}_1} E_{\mathbf{p}_3} + 2p_1 p_3 \cos\theta. \tag{7.3.85}$$

Consider an arbitrary collinear frame and an experiment with the initial energies, $E_{\mathbf{p}_1}$ and $E_{\mathbf{p}_2}$, and initial momenta, $\mathbf{p}_1$ and $\mathbf{p}_2$, fixed. From four-momentum conservation then $E_{\mathbf{p}_3} + E_{\mathbf{p}_4}$ and $\mathbf{p}_3 + \mathbf{p}_4$ are fixed. In any given scattering channel the final state masses m_3 and m_4 are known. So if we fix $E_{\mathbf{p}_3}$ we know $E_{\mathbf{p}_4}$ from energy conservation and then we know p_3 and p_4 from the on-shell mass-energy relation. In any collinear frame there will only be one combination of angles $\theta \equiv \theta_3$ and θ_4 that satisfy the two conditions that $\mathbf{p}_3^\parallel + \mathbf{p}_4^\parallel = \mathbf{p}_1 + \mathbf{p}_2$ and the vanishing transverse momentum condition $\mathbf{p}_3^{\perp} + \mathbf{p}_4^{\perp} = 0$. In other words, if we have ϕ-indepenence, then specifying $E_{\mathbf{p}_3}$ determines θ, where $0 \leq \theta \leq \pi$, and the full final state is known. Conversely, specifying θ fixes the ratio $p_3^{\perp}/p_3^{\parallel}$ and leads to $p_3^{\perp}/p_3^{\parallel} = -p_4^{\perp}/p_3^{\parallel}$. Conservation of collinear momentum then determines $p_4^{\parallel}$ as a function of $p_3^{\parallel}$. So $\mathbf{p}_3$ and $\mathbf{p}_4$ are both known functions of $p_3^{\parallel}$ and hence so are $E_{\mathbf{p}_3}$ and $E_{\mathbf{p}_4}$. We can then solve for $p_3^{\parallel}$ using energy conservation and the full final state is known. So if we have azimuthal symmetry about the beam and given the two final state particle masses, then the final state is determined by a single real number, such as θ or $\cos\theta$ or $E_{\mathbf{p}_3}$ or t. If we choose, say, $\cos\theta$, then $E_{\mathbf{p}_3} = E_{\mathbf{p}_3}(\cos\theta)$ and $t = t(\cos\theta)$.

In the special case of the CM frame any back-to-back $\mathbf{p}_3$ and $\mathbf{p}_4$ that satisfy four-momentum conservation can be rotated through any angle about the collision point and will remain a solution. This means that in the CM frame $E_{\mathbf{p}_3}$ and hence p_3 are independent of θ. Hence in the CM frame we have from Eq. (7.3.85),

$$dt = 2p_1 p_3 d(\cos\theta) = 2p_i^{\mathrm{cm}} p_f^{\mathrm{cm}} d(\cos\theta). \tag{7.3.86}$$

Then using Eq. (7.3.83) we find in the CM frame that

$$\frac{d\sigma}{dt} = \frac{d\sigma}{d(\cos\theta)} \frac{d(\cos\theta)}{dl} = \frac{1}{64\pi s} \frac{1}{(p_i^{\mathrm{cm}})^2} |\mathcal{M}_{fi}|^2 = \frac{1}{64\pi s} \frac{1}{|\mathbf{p}_1^{\mathrm{cm}}|^2} |\mathcal{M}_{fi}|^2, \tag{7.3.87}$$

where the right-hand side is to be evaluated in the CM frame. However, since $d\sigma$ is invariant under any collinear boost and since t is Lorentz invariant, then the left-hand side, $d\sigma/dt$, is the same in *every collinear inertial frame*.

Fixed-target frame: Two-body scattering in the fixed-target frame is also illustrated in Fig. 7.11. Again assuming azimuthal symmetry and using $\phi = 0$ gives

$$p_1^{\mu} = (E_{\mathbf{p}_1}, 0, 0, p_1), \quad p_2^{\mu} = (m_2, 0, 0, 0), \quad p_3^{\mu} = (E_{\mathbf{p}_3}, p_3\sin\theta, 0, p_3\cos\theta), \tag{7.3.88}$$

with $p_4^{\mu} = (E_{\mathbf{p}_4}, p_4\sin\theta_4, 0, p_4\cos\theta_4)$ determined by four-momentum conservation. Since we are not in the CM frame $E_{\mathbf{p}_3}$ and p_3 will now depend on $\cos\theta$ and so we can write $E_{\mathbf{p}_3}(\cos\theta)$ and $p_3(\cos\theta)$. We can then write t as both

$$t = (p_1 - p_3)^2 = m_1^2 + m_3^2 - 2p_1 \cdot p_3 = m_1^2 + m_3^2 - 2E_{\mathbf{p}_1}E_{\mathbf{p}_3} + 2p_1 p_3\cos\theta \text{ and}$$
$$t = (p_4 - p_2)^2 = m_2^2 + m_4^2 - 2p_2 \cdot p_4 = m_2^2 + m_4^2 - 2m_2 E_4$$
$$= m_2^2 + m_4^2 - 2m_2(E_{\mathbf{p}_1} + m_2 - E_{\mathbf{p}_3}). \tag{7.3.89}$$

Since $E_{\mathbf{p}_1}$ is fixed, from the second expression we find

$$\frac{dt}{d(\cos\theta)} = 2m_2 \frac{dE_{\mathbf{p}_3}}{d(\cos\theta)}. \tag{7.3.90}$$

Acting on $E_{\mathbf{p}_3}^2 - p_3^2 = m_3^2$ with $d/d(\cos\theta)$ gives

$$2E_{\mathbf{p}_3} \frac{E_{\mathbf{p}_3}}{d(\cos\theta)} = 2p_3 \frac{dp_3}{d(\cos\theta)} \quad \Rightarrow \quad \frac{dp_3}{d(\cos\theta)} = \frac{E_{\mathbf{p}_3}}{p_3} \frac{dE_{\mathbf{p}_3}}{d(\cos\theta)}. \tag{7.3.91}$$

Equating the right-hand sides of the above two expressions for t gives

$$m_1^2 + m_3^2 - 2E_{\mathbf{p}_1}E_{\mathbf{p}_3} + 2p_1 p_3\cos\theta = m_2^2 + m_4^2 - 2m_2(E_{\mathbf{p}_1} + m_2 - E_{\mathbf{p}_3}) \tag{7.3.92}$$

and acting on this with $d/d(\cos\theta)$ then leads to

$$-2E_{\mathbf{p}_1} \frac{dE_{\mathbf{p}_3}}{d(\cos\theta)} + 2p_1 p_3 + 2p_1\cos\theta \frac{dp_3}{d(\cos\theta)} = 2m_2 \frac{dE_{\mathbf{p}_3}}{d(\cos\theta)}. \tag{7.3.93}$$

Using Eq. (7.3.91) in Eq. (7.3.93) we can solve for $dE_{\mathbf{p}_3}/d(\cos\theta)$. This leads to

$$\frac{dt}{d(\cos\theta)} = 2m_2\frac{dE_{\mathbf{p}_3}}{d(\cos\theta)} = \frac{2m_2 p_1 p_3^2}{p_3(m_2 + E_{\mathbf{p}_1}) - p_1 E_{\mathbf{p}_3}\cos\theta}.$$

The differential cross-section in the fixed-target frame is then

$$\begin{aligned}
\frac{d\sigma}{d\Omega} &= \frac{1}{2\pi}\frac{d\sigma}{d(\cos\theta)} = \frac{d\sigma}{dt}\frac{dt}{d\Omega} = \frac{d\sigma}{dt}\frac{1}{2\pi}\frac{dt}{d(\cos\theta)} = \frac{d\sigma}{dt}\frac{m_2}{\pi}\frac{dE_{\mathbf{p}_3}}{d(\cos\theta)} \\
&= \frac{1}{64\pi s}\frac{1}{(p_i^{\text{cm}})^2}|\mathcal{M}_{fi}|^2\frac{m_2}{\pi}\frac{p_1 p_3^2}{p_3(m_2 + E_{\mathbf{p}_1}) - p_1 E_{\mathbf{p}_3}\cos\theta} \\
&= \frac{1}{64\pi^2}\frac{1}{p_1 m_2}\frac{p_3^2}{p_3(m_2 + E_{\mathbf{p}_1}) - p_1 E_{\mathbf{p}_3}\cos\theta}|\mathcal{M}_{fi}|^2,
\end{aligned} \tag{7.3.94}$$

where we have used Eqs. (7.3.42) and (7.3.81) to arrive at

$$s(p_i^{\text{cm}})^2 = F^2 = (p_1\cdot p_2)^2 - m_1^2 m_2^2 = E_{\mathbf{p}_1}^2 m_2^2 - m_1^2 m_2^2 = |\mathbf{p}_1|^2 m_2^2 = p_1^2 m_2^2. \tag{7.3.95}$$

Recall that for convenience here we understand that $p_1^2 = |\mathbf{p}_1|^2$ and $p_3^2 = |\mathbf{p}_3|^2$ with $p_1 \equiv |\mathbf{p}_1|$ and $p_3 \equiv |\mathbf{p}_3|$, whereas $p_1\cdot p_2 = p_{1\mu}p_2^\mu$ as usual. Note that $d\sigma/dt$ is the same in every collinear frame. The frame dependence arises from $dt/d\Omega$, which depends on the collinear inertial frame used.

7.3.5 Unitarity of the S-Matrix and the Optical Theorem

Consider some orthonormal basis, $\{|b_1\rangle, |b_2\rangle, \ldots\} \in V_{\text{full}}$, of the full Hilbert space V_{full} for our interacting system, where $\hat{U}(t'', t') = T\exp\{-i\int_{t'}^{t''} dt\, H_s(t)\}$ is the evolution operator, becoming $\hat{U}(t'', t') = e^{-i\hat{H}(t''-t')}$ for a time-independent Hamiltonian as we assume here. For now we will select the initial and final states for S to be from this complete orthonormal set of basis states, i.e., consider S_{jk} with $|j\rangle, |k\rangle \in \{|b_1\rangle, |b_2\rangle, \ldots\}$. Since the Hamiltonian is Hermitian, $\hat{H} = \hat{H}^\dagger$, the evolution operator is unitary, $\hat{U}(t'', t')^\dagger = \hat{U}(t', t'') = \hat{U}(t'', t')^{-1}$. Then the S-matrix is unitary, $S^\dagger = S^{-1}$, since using completeness of the basis we have

$$\begin{aligned}
(S^\dagger S)_{jk} &= \sum_\ell S_{j\ell}^\dagger S_{\ell k} = \lim_{T\to\infty}\sum_\ell\langle j|\hat{U}(T, -T)^\dagger|\ell\rangle\langle\ell|\hat{U}(T, -T)|k\rangle \\
&= \lim_{T\to\infty}\langle j|\hat{U}(T, -T)^\dagger\hat{U}(T, -T)|k\rangle = \langle j|k\rangle = \delta_{jk}.
\end{aligned} \tag{7.3.96}$$

where $j, k, \ell \in \{|b_1\rangle, |b_2\rangle, \ldots\}$. Note that

$$-i(T - T^\dagger) = T^\dagger T \quad\Rightarrow\quad -i(T_{jk} - T_{kj}^*) = \sum_\ell T_{\ell j}^* T_{\ell k}, \tag{7.3.97}$$

since $I = S^\dagger S = (I - iT^\dagger)(I - iT) = I + i(T - T^\dagger) + T^\dagger T$. Choosing $j = k$ gives

$$2\operatorname{Im} T_{jj} = \sum_\ell |T_{\ell j}|^2. \tag{7.3.98}$$

This result is valid for any choice of orthonormal basis for our full Hilbert space. We know how to generalize these arguments for an uncountable continuous basis.

Let us now select the initial and final states $|i\rangle$ and $|f\rangle$ to be asymptotic plane wave states as discussed in Sec. 7.1 and the beginning of Sec. 7.3. Then using Eq. (7.3.3) we obtain

$$\begin{aligned}
-i(2\pi)^4\delta^4(P_f - P_i)(\mathcal{M}_{fi} - \mathcal{M}_{if}^*) &= \sum_\ell(2\pi)^8\delta^4(P_f - P_\ell)\mathcal{M}_{f\ell}^\dagger\delta^4(P_\ell - P_i)\mathcal{M}_{\ell i} \\
&= (2\pi)^4\delta^4(P_f - P_i)\sum_\ell(2\pi)^4\delta^4(P_\ell - P_i)\mathcal{M}_{\ell f}^*\mathcal{M}_{\ell i}.
\end{aligned} \tag{7.3.99}$$

Choosing $i = f$ gives

$$2\operatorname{Im}\mathcal{M}_{ii} = \sum_{\ell}(2\pi)^4\delta^4(P_\ell - P_i)\mathcal{M}_{i\ell}^*\mathcal{M}_{\ell i} = \sum_{\ell}(2\pi)^4\delta^4(P_\ell - P_i)|\mathcal{M}_{\ell i}|^2. \qquad (7.3.100)$$

We temporarily neglect the boson and fermion symmetrization discussed in Sec. 4.1.9 as it is straightforward to apply this later. Then the orthonormality and completeness relations for our asymptotic states can be written as

$$\langle \mathbf{p}_1\cdots\mathbf{p}_n|\mathbf{q}_1\cdots\mathbf{q}_m\rangle = \delta_{nm}(2\pi)^{3n}(2E_{\mathbf{p}_1})\cdots(2E_{\mathbf{p}_n})\delta^3(\mathbf{p}_1 - \mathbf{q}_1)\cdots\delta^3(\mathbf{p}_n - \mathbf{q}_n),$$

$$\hat{I} = \sum_{\ell}|\ell\rangle\langle\ell| \quad\rightarrow\quad \hat{I} = \sum_n \int\left[\prod_{k=1}^n \frac{d^3 p_k}{(2\pi)^3 2E_{\mathbf{p}_k}}\right]|\mathbf{p}_1\cdots\mathbf{p}_n\rangle\langle\mathbf{p}_1\cdots\mathbf{p}_n|. \qquad (7.3.101)$$

In writing the completeness relation in terms of the asymptotic states, we are using the fact that $\mathcal{M}$ is only nonvanishing if the initial state leads to a scattering. The expression for $\hat{I}$ on the right-hand side above is the compleness relation for V_{Fock}. Combining Eqs. (7.3.100) and (7.3.101) and choosing $|i\rangle = |\mathbf{p}_A\mathbf{p}_B\rangle$ we arrive at

$$2\operatorname{Im}\mathcal{M}_{p_A p_B \to p_A p_B} = \sum_n \int d\Pi_n^{\text{LIPS}}|\mathcal{M}_{p_A p_B \to p_1\cdots p_n}|^2, \qquad (7.3.102)$$

where $d\Pi_n^{\text{LIPS}}$ is defined in Eq. (7.3.101) and where the second line is the graphical representation of the first. From Eq. (7.3.44) we have for the channel $AB \to n$ particles in any collinear frame

$$\sigma_{AB\to n} = \int d\Pi_n^{\text{LIPS}}|\mathcal{M}_{fi}|^2/4F. \qquad (7.3.103)$$

Then the total cross-section for $AB \to X$ in any collinear frame is given by

$$\sigma_{\text{tot}} \equiv \sigma_{AB\to X} = \sum_n \sigma_{AB\to n} = \sum_n \int d\Pi_n^{\text{LIPS}}|\mathcal{M}_{p_A p_B \to p_A p_B}|^2/4F$$
$$= \operatorname{Im}\mathcal{M}_{p_A p_B \to p_A p_B}/2F = \operatorname{Im}\mathcal{M}_{p_A p_B \to p_A p_B}/2p_i^{\text{cm}}E_{\text{cm}}, \qquad (7.3.104)$$

which is referred to as the *optical theorem*. Recall from Eq. (7.3.81) that in the CM frame we have $F = p_i^{\text{cm}}\sqrt{s} = p_i^{\text{cm}}E_{\text{cm}}$, where in this frame $|\mathbf{p}_A| = |\mathbf{p}_B| \equiv p_i^{\text{cm}}$. In quantum field theory we often also refer to Eq. (7.3.102) as the optical theorem. The optical theorem has a long history in optics and quantum mechanics and a concise summary of this along with references can be found in Newton (1976). Its derivation in terms of partial waves is included in most quantum mechanics courses, where it takes the form $\sigma_{\text{tot}} = (4\pi/k)\operatorname{Im}f(\theta,\phi)$ with $f(\theta,\phi)$ defined in Eq. (7.3.12).

7.4 Interaction Picture and Feynman Diagrams

The discussion of the interaction picture in quantum mechanics in Sec. 4.1.11 applies immediately to quantum field theory, since we will have a Schrödinger picture Hamiltonian operator, $\hat{H}_s$, with a free part, $\hat{H}_0$, and an interaction part, $\hat{H}_{\text{int}}$ in a Hilbert space, with $\hat{H}_s = \hat{H}_0 + \hat{H}_{\text{int}}$. As above, $\hat{H}_0$

will always be be time-independent and will be chosen to be the sum of one or more free field theory Hamiltonians. In this section we will see how the interaction picture leads to Feynman diagrams.

7.4.1 Interaction Picture

The interaction part $\hat{H}_{\text{int}}$ can be a self-interaction of a field, an interaction between two or more fields, an interaction with an external source or a combination of these. When the interaction does not include any external sources, then $\hat{H}_{\text{int}}$ will be time-independent and so $\hat{H}_s$ will be time-independent. If one or more external sources are present, then there will be explicit time-dependence and $\hat{H}_{\text{int}}(t)$ and hence $\hat{H}_s(t)$ will depend on time. Examples of coupling to external sources were given in Chapter 6. We typically define the source to strictly vanish outside an arbitrarily large but finite time interval, $j(x) = 0$ for $|t| > T_j$ for some arbitrarily large time T_j.

Recall from Eqs. (4.1.176), (4.1.177) and (4.1.192) that for a time-dependent Schrödinger picture Hamiltonian we have

$$\hat{H}_s(t) = \hat{H}_0 + \hat{H}_{\text{int}}(t), \ \hat{U}_0(t'',t') \equiv e^{-i\hat{H}_0(t''-t')}, \ \hat{U}(t'',t') \equiv Te^{-i\int_{t'}^{t''} dt\,\hat{H}_s(t)},$$

$$\hat{U}_I(t'',t') = U_0(t'',t_0)^\dagger \hat{U}(t'',t')\hat{U}_0(t',t_0) = Te^{-i\int_{t'}^{t''} dt\,\hat{H}_I(t)},$$

$$\hat{H}_I(t) \equiv \hat{U}_0(t,t_0)^\dagger \hat{H}_{\text{int}}(t)\hat{U}_0(t,t_0). \tag{7.4.1}$$

As in Chapter 6, unless otherwise indicated field operators will be in the Heisenberg picture, $\hat{\phi}(x) \equiv \hat{\phi}_h(x)$. From Eqs. (4.1.179) and (4.1.192) it then follows that

$$\hat{\phi}(x) = \hat{U}(t,t_0)^\dagger \hat{\phi}_s(\mathbf{x})\hat{U}(t,t_0) = \hat{U}_I(t,t_0)^\dagger \hat{\phi}_I(x)\hat{U}_I(t,t_0), \tag{7.4.2}$$

where $\hat{U}_I(t'',t')$ has the properties given in Eq. (4.1.185). Note that some other texts (Itzykson and Zuber, 1980; Peskin and Schroeder, 1995) use the notation $\hat{U}(t'',t')$ in place of $\hat{U}_I(t'',t')$ but we prefer to make explicit that we mean the interaction picture evolution operator. The ground state of the free (non-interacting) Hamiltonian $\hat{H}_0$ is defined as $|0\rangle$ and, since we will always be taking $\hat{H}_0$ as normal-ordered, then

$$\hat{H}_0|0\rangle = 0. \tag{7.4.3}$$

The state $|0\rangle$ is referred to as the *perturbative vacuum*.

Consider the case with no external source, $j(x) = 0$, then H_{int} is time-independent and so is the Hamiltonian $\hat{H} = \hat{H}_h = \hat{H}_s = \hat{H}_0 + \hat{H}_{\text{int}}$. The ground state of this $\hat{H}$ is the *full* or *nonperturbative vacuum* state, denoted as $|\Omega\rangle$. For simplicity we assume enumerable and ordered energy eigenstates ($E_{n+1} \geq E_n$ for $n = 0, 1, 2, \ldots$),

$$\hat{H}|\Omega\rangle = E_0|\Omega\rangle \equiv E_0|E_0\rangle \quad \text{and} \quad \hat{H}|E_n\rangle = E_n|E_n\rangle. \tag{7.4.4}$$

The full energy eigenstates form an orthonormal basis for the Hilbert space,

$$\langle E_n|E_m\rangle = \delta_{nm} \quad \text{and} \quad \hat{I} = \sum_{n=0}^{\infty} |E_n\rangle\langle E_n|. \tag{7.4.5}$$

Let us consider the infinite time limit in conjunction with the $i\epsilon$ prescription that we saw was associated with the Feynman boundary conditions in Sec. 6.2.6 and that we used throughout Chapter 6. It then follows that

$$e^{-i\hat{H}T}|0\rangle = \sum_{n=0}^{\infty} e^{-i\hat{E}_nT}|E_n\rangle\langle E_n|0\rangle \xrightarrow{T\to\infty(1-i\epsilon)} e^{-iE_0T}|\Omega\rangle\langle\Omega|0\rangle, \tag{7.4.6}$$

since the contributions from all $n \geq 1$ will be exponentially suppressed relative to that from $n = 0$. Making the very reasonable assumption that $\langle 0|\Omega\rangle \neq 0$ and inverting this equation we have

$$|\Omega\rangle = \lim_{T\to\infty(1-i\epsilon)} e^{-i\hat{H}T}|0\rangle / e^{-iE_0 T}\langle\Omega|0\rangle. \tag{7.4.7}$$

This result remains true if we make the replacement $T \to T+t_0 = t_0-(-T)$ and so using $\hat{H}_0|0\rangle = 0$ we find that

$$|\Omega\rangle = \lim_{T\to\infty(1-i\epsilon)} e^{-i\hat{H}(t_0-(-T))}|0\rangle / e^{-iE_0(t_0-(-T))}\langle\Omega|0\rangle \tag{7.4.8}$$

$$= \lim_{T\to\infty(1-i\epsilon)} \frac{e^{-i\hat{H}(t_0-(-T))}e^{-i\hat{H}_0(-T-t_0)}|0\rangle}{e^{-iE_0(t_0-(-T))}\langle\Omega|0\rangle} = \lim_{T\to\infty(1-i\epsilon)} \frac{\hat{U}_I(t_0,-T)|0\rangle}{e^{-iE_0(t_0-(-T))}\langle\Omega|0\rangle}.$$

So we see that, up to a constant, we can obtain $|\Omega\rangle$ by evolving $|0\rangle$ into the infinite past using $\hat{U}_I(0,-T)$ and the $i\epsilon$ prescription. Similarly, we arrive at

$$\langle\Omega| = \lim_{T\to\infty(1-i\epsilon)} \frac{\langle 0|\hat{U}_I(T,t_0)}{e^{-iE_0(T-t_0)}\langle 0|\Omega\rangle}. \tag{7.4.9}$$

Combining these results gives

$$\frac{\langle 0|\hat{U}_I(T,t_0)\hat{U}_I(t_0,-T)|0\rangle}{e^{-iE_0(2T)}|\langle 0|\Omega\rangle|^2} = \frac{\langle 0|\hat{U}_I(T,-T)|0\rangle}{e^{-iE_0(2T)}|\langle 0|\Omega\rangle|^2} \xrightarrow{T\to\infty(1-i\epsilon)} \langle\Omega|\Omega\rangle = 1, \tag{7.4.10}$$

where $\langle\Omega|\Omega\rangle = 1$ since the energy eigenstates, including the vacuum, were chosen to be orthonormal. Combining these results along with Eq. (7.4.2) consider the two-point vacuum expectation value of Heisenberg picture operators with $x^0 > y^0$,

$$\langle\Omega|\hat{\phi}(x)\hat{\phi}(y)|\Omega\rangle = \langle\Omega|\hat{U}_I(x^0,t_0)^\dagger\hat{\phi}_I(x)\hat{U}_I(x^0,t_0)\hat{U}_I(y^0,t_0)^\dagger\hat{\phi}_I(y)\hat{U}_I(y^0,t_0)|\Omega\rangle$$

$$= \langle\Omega|\hat{U}_I(x^0,t_0)^\dagger\hat{\phi}_I(x)\hat{U}_I(x^0,y_0)\hat{\phi}_I(y)\hat{U}_I(y^0,t_0)|\Omega\rangle$$

$$= \lim_{T\to\infty(1-i\epsilon)} \frac{\langle 0|\hat{U}_I(T,x^0)\hat{\phi}_I(x)\hat{U}_I(x^0,y_0)\hat{\phi}_I(y)\hat{U}_I(t^0,-T)|0\rangle}{e^{-iE_0(2T)}|\langle 0|\Omega\rangle|^2}$$

$$= \lim_{T\to\infty(1-i\epsilon)} \frac{\langle 0|\hat{U}_I(T,x^0)\hat{\phi}_I(x)\hat{U}_I(x^0,y_0)\hat{\phi}_I(y)\hat{U}_I(t^0,-T)|0\rangle}{\langle 0|\hat{U}_I(T,-T)|0\rangle}.$$

$$= \lim_{T\to\infty(1-i\epsilon)} \frac{\langle 0|T\{\hat{\phi}_I(x)\hat{\phi}_I(y)e^{-i\int_{-T}^{T} dt\,\hat{H}_I(t)}\}|0\rangle}{\langle 0|T\{e^{-i\int_{-T}^{T} dt\,\hat{H}_I(t)}\}|0\rangle}. \tag{7.4.11}$$

In the second to last step we used Eq. (7.4.10) to replace the denominator and in the last step we used the properties of time-ordering. The above result holds for all $x^0 > y^0$ and clearly generalizes to any number of Heisenberg picture operators.

We are considering here a normal-ordered Hamiltonian of the form of Eq. (6.2.256),

$$\hat{H} = \hat{H}_0 + \hat{H}_{\rm int} = \int d^3x : \tfrac{1}{2}[\dot{\hat{\phi}}^2(x) + (\boldsymbol{\nabla}\hat{\phi}(x))^2 + m^2\hat{\phi}^2(x)]: + U(\hat{\phi}), \tag{7.4.12}$$

where $\hat{H}_0$ is normal-ordered but $\hat{H}_{\rm int}$ is not and where $U(\phi)$ contains no derivatives. We have arrived at the result

$$\langle\Omega|T\hat{\phi}(x_1)\cdots\hat{\phi}(x_n)|\Omega\rangle = \lim_{T\to\infty(1-i\epsilon)} \frac{\langle 0|T\{\hat{\phi}_I(x_1)\cdots\hat{\phi}_I(x_n)e^{-i\int_{-T}^{T} dt\,\hat{H}_I(t)}\}|0\rangle}{\langle 0|T\{e^{-i\int_{-T}^{T} dt\,\hat{H}_I(t)}\}|0\rangle},$$

$$\hat{H}_I(t) = \hat{U}_0(t,t_0)^\dagger\hat{H}_{\rm int}(t)\hat{U}_0(t,t_0) = \hat{U}_0(t,t_0)^\dagger[\int d^3x\,U(\hat{\phi})]\hat{U}_0(t,t_0)$$

$$= \int d^3x\,U(\hat{\phi}_I), \tag{7.4.13}$$

where recall that the interaction picture operators are the Heisenberg picture operators of the free field theory. We then have from Eq. (6.2.151)

$$\langle 0|T\hat{\phi}_I(x)\hat{\phi}_I(y)|0\rangle = D_F(x-y) \tag{7.4.14}$$

and Wick's theorem in Eq. (6.2.208) applies,

$$T\hat{\phi}_I(x_1)\cdots\hat{\phi}_I(x_n) = :\{\hat{\phi}_I(x_1)\cdots\hat{\phi}_I(x_n) + \text{all contractions}\}:, \tag{7.4.15}$$

which leads to Eq. (6.2.211)

$$\langle 0|T\hat{\phi}_I(x_1)\cdots\hat{\phi}_I(x_n)|0\rangle = \delta_{[n=\text{even}]}\{D_F(x_1-x_2)D_F(x_3-x_4)\cdots D_F(x_{n-1}-x_n) \\ + \text{ all pairwise combinations of } (x_1,x_2,\ldots,x_n)\}. \tag{7.4.16}$$

As in Eq. (6.2.257) we can now add in a time-dependent external source term, which gives in the Schrödinger picture

$$\hat{H}_s(t) = \hat{H} - \int d^3x\, j(x)\hat{\phi}_s(x) \tag{7.4.17}$$

and obtain a modified evolution operator, $\hat{U}^j(t'',t') = Te^{-i\int_{t'}^{t''} d^4x\, \hat{\mathcal{H}}_s}$. We will always be interested in the $j(x) \to 0$ limit and so the trace is to be taken over the exact eigenstates of the full Hamiltonian $\hat{H}$. Incorporating this result along with Eq. (6.2.262) gives for the generating functional

$$Z[j] \equiv \lim_{T\to\infty(1-i\epsilon)} \frac{\text{tr}\{\hat{U}^j(T,-T)\}}{\text{tr}\{\hat{U}(T,-T)\}} = \lim_{T\to\infty(1-i\epsilon)} \frac{\text{tr}\left[Te^{-i\int_{-T}^{T} d^4x\, [\hat{\mathcal{H}}-j\hat{\phi}]}\right]}{\text{tr}\left[Te^{-i\int_{-T}^{T} d^4x\, \hat{\mathcal{H}}}\right]}$$

$$= \lim_{T\to\infty(1-i\epsilon)} \frac{\langle\Omega|Te^{i\int_{-T}^{T} d^4x\, j(x)\hat{\phi}_I(x)}|\Omega\rangle}{\langle\Omega|\Omega\rangle} = \lim_{T\to\infty(1-i\epsilon)} \frac{\langle 0|Te^{-i\int_{-T}^{T} d^4x\, [\hat{\mathcal{H}}_I-j\hat{\phi}_I]}|0\rangle}{\langle 0|Te^{-i\int_{-T}^{T} d^4x\, \hat{\mathcal{H}}_I}|0\rangle}$$

$$= \lim_{T\to\infty(1-i\epsilon)} \frac{\int \mathcal{D}\phi_{\text{(periodic)}}\, e^{i\{S[\phi]+\int_{-T}^{T} d^4x\, j\phi\}}}{\int \mathcal{D}\phi_{\text{(periodic)}}\, e^{iS[\phi]}}, \tag{7.4.18}$$

where we have used Eq. (6.2.260). Then from Eq. (7.4.13) we obtain the time-ordered vacuum expectation values in terms of the functional integral in Eq. (6.2.263),

$$\langle\Omega|T\hat{\phi}(x_1)\cdots\hat{\phi}(x_n)|\Omega\rangle = \frac{(-i)^n\delta^k}{\delta j(x_1)\cdots\delta j(x_n)}Z[j]\Bigg|_{j=0} \tag{7.4.19}$$

$$= \lim_{T\to\infty(1-i\epsilon)} \frac{\int \mathcal{D}\phi_{\text{(periodic)}}\,\phi(x_1)\cdots\phi(x_n)\, e^{iS[\phi]}}{\int \mathcal{D}\phi_{\text{(periodic)}}\, e^{iS[\phi]}},$$

which has the form of Eq. (6.2.263). In the free field limit the interaction picture reduces to the Heisenberg picture and Eq. (7.4.13) reduces to Eq. (6.2.278).

As always infrared and ultraviolet regularization of the numerator and denominator of Eq. (7.4.19) is to be understood. The restoration of Poincaré invariance is only required for physical observables after renormalization when regularizations can be safely removed. In this way we do not concern ourselves with *Haag's theorem*, which is essentially the statement that the interaction picture does not exist in an interacting, relativistic, quantum field theory (Haag, 1955, 1958). The proof of Haag's theorem relies on translational invariance, which we do not have in the intermediate stages of our calculations. In this way the interaction picture is simply seen as an intermediate mathematical device for doing calculations and it is not required that it be defined in the absence of regularization.

Define the free and interacting parts of the action and Lagrangian density in an obvious way,

$$S[\phi] = S_0[\phi] + S_{\text{int}}[\phi] = \int d^4x [\mathcal{L}_0(\phi, \partial\phi) + \mathcal{L}_{\text{int}}(\phi)]. \tag{7.4.20}$$

By expanding the exponential term by term we can verify that

$$\int \mathcal{D}\phi \, e^{i\{S[\phi] + \int d^4x \, j\phi\}} = e^{iS_{\text{int}}[-i\frac{\delta}{\delta j}]} \int \mathcal{D}\phi \, e^{i\{S_0[\phi] + \int d^4x \, j\phi\}}, \tag{7.4.21}$$

where, for example, we might have $\mathcal{L}_{\text{int}} = -\frac{\lambda}{4!}\phi^4$. We have collected these results here for ease of comparison and convenience. We will later see how they are generalized to various interacting field theories containing combinations of scalar bosons, fermions, gauge fields and massive vector bosons.

7.4.2 Feynman Diagrams

The time-ordered vacuum expectation values of a field theory are commonly referred to as its Green's functions, as we saw in Chapter 6, and as its (time-ordered) correlation functions. We wish to understand how to perform a perturbative expansion of these in an interacting field theory using the technique of Feynman diagrams. As an illustration we consider the case of a scalar field theory with a quartic self-interaction, which has the Hamiltonian

$$\hat{H} = \hat{H}_0 + \hat{H}_{\text{int}} = \hat{H}_0 + \int d^3x \, \frac{\lambda}{4!}\hat{\phi}^4, \tag{7.4.22}$$

where $\hat{H}_0$ is the normal-ordered Klein-Gordon Hamiltonian in Eq. (6.2.33) and the $\hat{\phi}^4$ interaction is not normal-ordered. The inclusion of the 4! in the coupling term is a convenience that simplifies later combinatorics. Both Eqs. (7.4.13) and (7.4.19) lead to the same result, which is that these interacting field theory Green's functions can be expanded as sums of products of free field Green's functions in a perturbation series in the coupling λ. The perturbation series comes from expanding the exponentials $e^{-i\int_{-T}^{T} dt \hat{H}_I(t)}$ in Eq. (7.4.13) and $e^{iS[\phi]}$ in Eq. (7.4.19).

Let us first consider the numerator of Eq. (7.4.13),

$$\langle 0|T\{\hat{\phi}_I(x_1)\cdots\hat{\phi}_I(x_n)e^{-i\int_{-T}^{T} dt\,\hat{H}_I(t)}\}|0\rangle$$

$$= \langle 0|T\{\hat{\phi}_I(x_1)\cdots\hat{\phi}_I(x_n)[1 + \{-i\int_{-T}^{T} dt\,\hat{H}_I(t)\} + \frac{1}{2!}\{-i\int_{-T}^{T} dt\,\hat{H}_I(t)\}^2 + \cdots]\}|0\rangle$$

$$= \langle 0|T\{\hat{\phi}_I(x_1)\cdots\hat{\phi}_I(x_n)[1 + (\frac{-i\lambda}{4!})\int_{-T}^{T} d^4x\,\hat{\phi}_I(z)^4$$

$$+ \frac{1}{2!}(\frac{-i\lambda}{4!})^2\int_{-T}^{T} d^4z_1 d^4z_2\,\hat{\phi}_I(z_1)^4\hat{\phi}_I(z_2)^4 + \cdots]\}|0\rangle, \tag{7.4.23}$$

which is a perturbation series in λ with the $\mathcal{O}(\lambda^m)$ term given by

$$\langle 0|T\{\hat{\phi}_I(x_1)\cdots\hat{\phi}_I(x_n)\frac{1}{m!}(\frac{-i\lambda}{4!})^m\int_{-T}^{T} d^4z_1\cdots d^4z_m\hat{\phi}_I(z_1)^4\cdots\hat{\phi}_I(z_m)^4\}|0\rangle. \tag{7.4.24}$$

We evaluate such terms using Wick's theorem in the form of Eq. (6.2.211). For example, for the four-point Green's function the $\mathcal{O}(\lambda^0)$ term in the numerator of Eq. (7.4.13) is just the free field result given in Eq. (6.2.210). This results in the sum of three different contractions containing two free propagators each. We can represent the three contractions by introducing the concept of Feynman diagrams, where each free propagator $D_F(x - y)$ is represented by a line,

$$\langle 0|T\{\hat{\phi}_I(x_1)\cdots\hat{\phi}_I(x_4)\}|0\rangle = \qquad + \qquad + \qquad . \tag{7.4.25}$$

For the $\mathcal{O}(\lambda^1)$ term, $\langle 0|T\{\hat{\phi}_I(x_1)\cdots\hat{\phi}_I(x_4)(\frac{-i\lambda}{4!})\int d^4z\hat{\phi}_I(z)^4\}|0\rangle$, we have seven *connected* Feynman diagrams

$$(7.4.26)$$

and three *disconnected Feynman diagrams*

$$(7.4.27)$$

Each of the 10 diagrams contains four free propagators and one four-vertex.

We say that a diagram is made up of *disconnected subdiagrams* if we can draw an imaginary curve through the diagram that cuts no lines and divides the diagram into two or more pieces. When a piece of a diagram cannot be further divided in such a way we refer to it as a disconnected subdiagram. Of the 10 $\mathcal{O}(\lambda)$ diagrams above, the first has no disconnected subdiagrams, the next six each contain two disconnected subdiagrams and each of the last three contain three disconnected subdiagrams. Any disconnected subdiagram that contains no external points of the Green's function is referred to as a *vacuum subdiagram* or as *vacuum bubble*. Only the last three of the 10 diagrams contain a vacuum subdiagram and these have the form of two loops connected by a four-vertex, i.e., a figure-eight. *By definition a connected Feynman diagram contains no vacuum subdiagrams. A Feynman diagram that is not connected is by definition disconnected.* This is why we have referred to the first seven diagrams as connected and the last three as disconnected.

The first of the 10 $\mathcal{O}(\lambda)$ diagrams corresponds to contractions of the type

$$\hat{\phi}_I(x_1)\hat{\phi}_I(x_2)\hat{\phi}_I(x_3)\hat{\phi}_I(x_4)\hat{\phi}_I(z)\hat{\phi}_I(z)\hat{\phi}_I(z)\hat{\phi}_I(z), \qquad (7.4.28)$$

where there are four ways to contract $\hat{\phi}_I(x_1)$ with a $\hat{\phi}_I(z)$, three ways to contract $\hat{\phi}_I(x_2)$ with a remaining $\hat{\phi}_I(z)$ and two ways to contract $\hat{\phi}_I(x_3)$ with a remaining $\hat{\phi}_I(z)$. So there are $4! = 12$ equivalent contractions corresponding to the first diagram, which cancels the $1/4!$ in the coupling so that the first diagram gives

$$(4!)(\tfrac{-i\lambda}{4!})\int d^4z\, D_F(x_1 - z)D_F(x_2 - z)D_F(x_3 - z)D_F(x_4 - z). \qquad (7.4.29)$$

The second of the 10 $\mathcal{O}(\lambda)$ diagrams contains contractions of the form

$$\hat{\phi}_I(x_1)\hat{\phi}_I(x_2)\hat{\phi}_I(x_3)\hat{\phi}_I(x_4)\hat{\phi}_I(z)\hat{\phi}_I(z)\hat{\phi}_I(z)\hat{\phi}_I(z), \qquad (7.4.30)$$

where there are four ways to contract $\hat{\phi}_I(x_1)$ with a $\hat{\phi}_I(z)$ and three ways to contract $\hat{\phi}_I(x_2)$ with a $\hat{\phi}_I(z)$ and no choices remain. So there are $4 \times 3 = 12$ equivalent contractions and the contribution from the second diagram is then

$$(4 \times 3)(\tfrac{-i\lambda}{4!})\int d^4z\, D_F(x_1 - z)D_F(x_2 - z)D_F(z - z)D_F(x_3 - x_4). \qquad (7.4.31)$$

To similarly write down the contributions from the other eight diagrams is then a straightforward matter. To proceed to higher orders observe that there is one four-vertex for every power of λ in the full perturbative expansion and while contractions rapidly become more complicated the way to perform the calculations should now be clear. A systematic means of counting equivalent contractions is needed.

As an example, consider an $\mathcal{O}(\lambda^2)$ contribution to the numerator of the two-point Green's function given by the contraction

$$\langle 0|T[\tfrac{1}{2!}(\tfrac{-i\lambda}{4!})^2 \int d^4z_1 d^4z_2 \tag{7.4.32}$$

$$\times \hat{\phi}_I(x_1)\hat{\phi}_I(x_2)\hat{\phi}_I(z_1)\hat{\phi}_I(z_1)\hat{\phi}_I(z_1)\hat{\phi}_I(z_1)\hat{\phi}_I(z_2)\hat{\phi}_I(z_2)\hat{\phi}_I(z_2)\hat{\phi}_I(z_2)]|0\rangle$$

$$= \tfrac{1}{2!}(\tfrac{-i\lambda}{4!})^2 \int d^4z_1 d^4z_2\, D_F(x_1-z_1)D_F(x_2-z_1)[D_F(z_1-z_2)]^2 D_F(z_2-z_2).$$

However, there are many contractions that lead to the result on the right-hand side:

$$\underbrace{2!}_{\substack{\text{permutions} \\ \text{of } z_1 \text{ and } z_2 \\ \text{vertices}}} \times \underbrace{4!}_{\substack{\text{ways to contract} \\ \text{into } z_1 \text{ vertex}}} \times \underbrace{(4 \times 3)}_{\substack{\text{ways to contract} \\ \text{into } z_2 \text{ vertex}}} \times \underbrace{\tfrac{1}{2!}}_{\substack{\text{corrects overcounting} \\ \text{of multiple contractions} \\ \text{between same two vertices}}} \tag{7.4.33}$$

The first 2! comes since we can clearly exchange z_1 and z_2 to obtain additional contractions that lead to the same right-hand side because z_1 and z_2 are dummy integration variables. Then once the roles of the z_1 and z_2 are fixed there are 4! ways of performing the contractions into the z_1 vertex. Because the z_2 vertex has a contraction with itself then there are only 4×3 ways of contracting into the z_2 vertex. Noting that there are two contractions between z_1 and z_2 then this counting scheme has overcounted the contractions by the number of permutations of the two contractions, i.e., by 2!. To convince yourself of this overcounting it is a simple but somewhat tedious matter to draw the relevant contractions and count them. The sum of all equivalent contractions is represented by a single Feynman diagram, which in the case of contractions equivalent to Eq. (7.4.32) is

$$x_1 \underline{\qquad\qquad} x_2.$$

We have 4×3 in the third term rather than 4! due to the self-contraction at the z_2 vertex, which corresponds to the top loop in the diagram. The final factor of two is due to the two identical propagators connecting z_1 and z_2 in the bottom loop of the diagram. So instead of the naive $2! \times (4!)^2$ contractions for the diagram we have $2! \times (4!)^2/S$ where the factor $S = 2 \times 2! = 4$ is called the *symmetry factor* of the diagram. An $\mathcal{O}(\lambda^3)$ example is given in Peskin and Schroeder (1995).

The above discussion generalizes to $\mathcal{O}(\lambda^n)$ contributions, which have n vertices. Any Feynman diagram with n vertices will have a weighting given by

$$\underbrace{n!}_{\substack{\text{permutations} \\ \text{of } n \text{ vertices}}} \times \underbrace{(4!)^n}_{\substack{\text{naive contractions} \\ \text{into the } n \text{ vertices}}} \times \underbrace{\tfrac{1}{S}}_{\substack{\text{symmetry factor to} \\ \text{correct naive overcounting}}} \tag{7.4.34}$$

We note that the $n! \times (4!)^n$ cancels against the factors in front of the n-integrals.

Coordinate-space Feynman rules: So the value of any given Feynman diagram with n vertices then simplifies to the following *Feynman rules*:

(i) Each line $x_1 \underline{\qquad} x_2$ corresponds to a propagator, $D_F(x_1 - x_2)$.

(ii) Each vertex z_j corresponds to $(-i\lambda) \int d^4z_j$.

(iii) Associate unity with each external point, $x_1, \ldots, x_n$.

(iv) For each Feynman diagram divide by its symmetry factor S.

The motivation for adding 4! into the self-coupling term $(\lambda/4!)\hat{\phi}^4$ is now clear.

Calculating Symmetry Factors

We need to understand how to calculate the symmetry factor S of any Feynman diagram. The naive counting of equivalent contractions for any given n-vertex Feynman diagram is $N^{\text{naive}} \equiv n!(4!)^n$. Let N be the number of actual contractions of this diagram, then the symmetry factor for the diagram is by definition

$$S \equiv \frac{N^{\text{naive}}}{N} = \frac{n!(4!)^n}{N}. \qquad (7.4.35)$$

Every symmetry possessed by any given Feynman diagram under the exchange of propagators and vertices represents a lost opportunity for a distinct contraction. So the symmetry factor S is the number of symmetries that the diagram possesses under propagator and vertex exchanges. In mathematical language the number of symmetries is the order of its group of automorphisms, where an automorphism is a transformation that leaves an object unchanged; i.e., it is a symmetry. Let us build our understanding with some examples.

To determine S we can proceed as follows: (i) Use a dot to represent each of the n external points and label the dots; (ii) draw every vertex with four lines emerging and number the lines never using the same number twice; (iii) then systematically build the diagram of interest by connecting lines to external points and joining pairs of lines to form a single line joining two points; (iv) at each step make a specific choice for the next connection that is made but count the number of possible options available for each connection as the choice is made; (v) multiply the number of options at each step together to obtain N; (vi) calculate $S = n!(4!)^n/N$; and then (vii) confirm that S is the number of symmetries of the diagram as a check.

Examples: The following Feynman diagram examples show that counting symmetries is the easiest way to calculate S, but we can always resort to brute force.

a) Join 1 to 2 (3 choices), join 3 to 4 (1 choice), which gives $N = 3 \times 1 = 3$ and so $S = 1!(4!)^1/N = 4!/3 = 8$. Symmetry check: Flip each loop about the horizontal (2×2), flip diagram about a vertical line through vertex (2) and so $S = 2^3 = 8$.

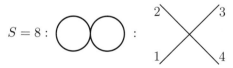

b) Join x to 1 (4 choices), join y to 4 (3 choices), join 2 to 3 (1 choice), which gives $N = 4 \times 3 = 12$ and so $S = 1!(4!)^1/N = 4!/12 = 2$. Symmetry check: Flip loop about a vertical axis (2) and so $S = 2$.

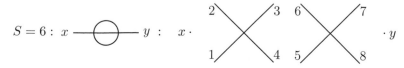

c) Join x to 1 ($2 \times 4 = 8$ choices), join y to 8 ($1 \times 4 = 4$ choices), join 2 to 7 (3 choices), join 3 to 6 (2 choices), join 4 to 5 (1 choice), which gives $N = 8 \times 4 \times 3 \times 2 = 8 \times 4!$ and so $S = 2!(4!)^2/N = 2!4!/8 = 6$. Symmetry check: $3! = 6$ ways to exchange lines in loop and so $S = 6$.

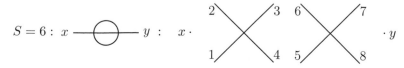

d) Join x to 1 (8 choices), join y to 4 (3 choices), join 2 to 7 (4 choices), join 3 to 6 (3 choices), join 5 to 8 (1 choice), which gives $N = 8 \times 3 \times 4 \times 3 = 12 \times 4!$ and so $S = 2!(4!)^2/N = 2!4!/12 = 4$. Symmetry check: Flip top loop about vertical axis (2) and exchange lines in bottom loop (2) so $S = 2^2 = 4$.

$S = 4$:

e) Join 1 to 5 (4 choices), join 2 to 6 (3 choices), join 3 to 7 (2 choices), join 4 to 8 (1 choice), which gives $N = 4!$ and $S = 2!(4!)^2/N = 2 \times 4! = 48$. Symmetry check: There are 4! permutations of lines, 2! vertex exchanges and so $S = 48$.

$S = 48$:

f) Join x to 1 (12 choices), join y to 12 (8 choices), join 2 to 3 (3 choices, either join 2-3 or 2-4 or 3-4), join 4 to 9 (3 choices), join 10 to 7 (4 choices), join 11 to 6 (3 choices), join 5 to 8 (1 choice), giving $N = 12 \times 8 \times 3^3 \times 4 = 18 \times (4!)^2$ and $S = 3!(4!)^3/N = 8$. Symmetry check: Single loop flip about vertical (2), top loop flip about vertical (2), exchange lines in bottom loop (2) and so $S = 2^3 = 8$.

$S = 8$:

g) Join x to 1 (12 choices), join y to 4 (3 choices), join 3 to 6 (8 choices), join 2 to 11 (4 choices), join 7 to 10 (3 choices), join 5 to 9 (2 choices), join 8 to 12 (1 choice), giving $N = 12 \times 4 \times 8 \times 3^2 \times 2 = 12 \times (4!)^2$ and $S = 3!(4!)^3/N = 3!4!/12 = 12$. Symmetry check: There are 3! ways of exchanging lines in the top loop and 2! ways to exchange vertices (flip about vertical line) and so $S = 3! \times 2 = 12$.

$S = 12$:

Factoring Vacuum Subdiagrams

A Feynman diagram will in general be a product of pieces and hence is typically a disconnected diagram. We wish to prove the important result that *only connected Feynman diagrams contribute to Green's functions*.

Consider the set of all vacuum subdiagrams

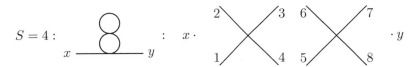

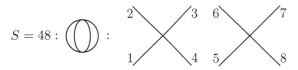

It will be convenient to label the subdiagrams as V_i and to also understand V_i to mean the value of the subdiagram when evaluated using the Feynman rules above. The intended usage will be clear from the context. We can similarly label all of the connected Feynman diagrams with n external legs as C_α for $\alpha = 1, 2, \ldots$ and understand that this can also mean the value of the diagram.

The value of a Feynman diagram containing n_i copies of V_i is

$$\underbrace{V_i V_i \cdots V_i}_{n_i \text{ copies}} = \frac{1}{n_i!} V_i^{n_i}. \tag{7.4.36}$$

For a Feynman diagram containing a connected contribution C_α, n_i copies of V_i and n_j copies of V_j,

$$C_\alpha \underbrace{V_i V_i \cdots V_i}_{n_i \text{ copies}} \underbrace{V_j V_j \cdots V_j}_{n_j \text{ copies}} = C_\alpha \frac{1}{n_i!} V_i^{n_i} \frac{1}{n_j!} V_j^{n_j}, \tag{7.4.37}$$

and similarly when any number of mixed vacuum subdiagrams are included.

Proof: (a) Eq. (7.4.36): Let m be the number of vertices in V_i. Then to calculate the symmetry factor we first need to count the number N of contractions of $n_i m$ vertices into the n_i sets of m vertices. There are $(n_i m)!/m![(n_i - 1)m]!$ ways to choose the first m vertices. There are $[(n_i - 1)m]!/m![(n_i - 2)m]!$ ways to choose the next m vertices. So naively the number of choices is

$$\frac{(n_i m)!}{m![(n_i - 1)m]!} \frac{[(n_i - 1)m]!}{m![(n_i - 2)m]!} \cdots \frac{(2m)!}{m!} = \frac{(n_i m)!}{(m!)^{n_i}}. \tag{7.4.38}$$

However, we have overlooked something important. We have counted the number of choices as if the n_i sets of vertices are distinguishable, which they are not. Since any of the $n_i!$ permutations of the n_i sets is one single choice, then the correct number of ways to choose n_i sets of m is reduced by a factor $1/n_i!$,

$$\# \text{ choices of } n_i \text{ sets of } m \text{ vertices} = (n_i m)!/n_i!(m!)^{n_i}. \tag{7.4.39}$$

To see this clearly, count by hand the number of ways of choosing $n_i = 3$ sets of $m = 2$ vertices from $n_i m = 6$ vertices. One finds 15 choices and we confirm that $(n_i m)!/n_i!(m!)^{n_i} = 6!/3!(2!)^3 = 15$.

Let N_i be the number of ways to contract the m vertices into the vacuum subdiagram V_i. Then the symmetry factor for V_i is $S_i = m!(4!)^m/N_i$. The total number of contractions N_tot to form the n_i copies of V_i is the number of ways to choose the n_i sets multiplied by the number of ways to contract every set,

$$N_\text{tot} = [(n_i m)!/n_i!(m!)^{n_i}] \times N_i^{n_i}. \tag{7.4.40}$$

Using Eq. (7.4.35) gives $S_i = m!(4!)^m/N_i$ and gives the total symmetry factor

$$S_\text{tot} = (n_i m)!(4!)^{n_i m}/N_\text{tot} = n_i! S_i^{n_i}. \tag{7.4.41}$$

Then the value of n_i copies of V_i in a Feynman diagram contributes $1/n_i!$ of the value of n_i products of the value of V_i, which proves Eq. (7.4.36).

(b) Eq. (7.4.37) with $n_j = 0$: Let r be the number of vertices in C_α. The symmetry factor for the diagram C_α is $S_\alpha = r!(4!)^r/N_\alpha$, where N_α is the number of contractions of r vertices and n

external points to form C_α. To form $C_\alpha V_i \cdots V_i$ we perform contractions of $(r + n_i m)$ vertices. We see that there are $(r + n_i m)!/r!(n_i m)!$ ways to choose the first r vertices to form C_α. Then as above there are $(n_i m)!/n_i!(m!)^{n_i}$ ways to choose the remaining n_i sets of m vertices to form the V_i. So the total number of contractions is

$$N_{\text{tot}} = [(r + n_i m)!/r!(n_i m)!][(n_i m)!/n_i!(m!)^{n_i}]N_\alpha N_i^{n_i}$$
$$= (1/n_i!)[(r + n_i m)!/r!(m!)^{n_i}]N_\alpha N_i^{n_i}. \tag{7.4.42}$$

Then Eq. (7.4.37) with $n_j = 0$ follows since the total symmetry factor is

$$S_{\text{tot}} = (r + n_i m)!(4!)^{r + n_i m}/N_{\text{tot}} = n_i! S_\alpha S_i^{n_i}. \tag{7.4.43}$$

(c) Eq. (7.4.37): It should now be obvious that for $n_j \geq 1$ with V_j containing m_j vertices the total number of contractions is

$$N_{\text{tot}} = (1/n_i!)(1/n_j!)[(r + n_i m + n_j m_j)!/r!(m!)^{n_i}(m_j!)^{n_j}]N_\alpha N_i^{n_i} N_j^{n_j}.$$

We find $S_{\text{tot}} = n_i! n_j! S_\alpha S_i^{n_i} S_j^{n+j}$ and so Eq. (7.4.37) follows. The extension of this to include any number of different subdiagrams is straightforward.

The denominator of Eq. (7.4.13) is

$$D \equiv \langle 0|T\{e^{-i \int_{-T}^{T} dt\, \hat{H}_I(t)}\}|0\rangle \tag{7.4.44}$$
$$= \langle 0|T[1 + \{-i \int_{-T}^{T} dt\, \hat{H}_I(t)\} + \tfrac{1}{2!}\{-i \int_{-T}^{T} dt\, \hat{H}_I(t)\}^2 + \cdots]|0\rangle$$
$$= \langle 0|T[1 + (\tfrac{-i\lambda}{4!}) \int d^4z\, \hat{\phi}_I(z)^4 + \tfrac{1}{2!}(\tfrac{-i\lambda}{4!})^2 \int d^4z_1 d^4z_2\, \hat{\phi}_I(z_1)^4 \hat{\phi}_I(z_2)^4 + \cdots]|0\rangle$$
$$\equiv 1 + A_1 + A_2 + \dots,$$

where we have introduced the notation A_n for the $\mathcal{O}(\lambda^n)$ contribution to the denominator. We see that

$$A_1 = \;\;\text{⬒}\;\; = V_1 \quad \text{and} \quad A_2 = \;\;\text{⬒}\;\; + \;\;\text{⬭}\;\; + \;\;\text{⬒⬒}\;\; = V_2 + V_3 + \tfrac{1}{2!}V_1^2.$$

We can similarly write down by inspection the sum of all combinations of the V_i that have a total of n vertices to give the $\mathcal{O}(\lambda^n)$ term, A_n. Let D_1 be the sum of all contributions to D that contain no factors of V_1. Then

$$D = D_1 + D_1 V_1 + D_1 \tfrac{1}{2!}V_1^2 + D_1 \tfrac{1}{3!}V_1^3 = \cdots = D_1 e^{V_1}. \tag{7.4.45}$$

We can then define D_{12} as the sum of all contributions to D that contain no V_1 or V_2 and find $D = D_{12}e^{V_1 + V_2}$. Proceeding in this way through every V_i we arrive at

$$D \equiv \langle 0|T\{e^{-i \int_{-T}^{T} dt\, \hat{H}_I(t)}\}|0\rangle = 1 + A_1 + A_2 + \cdots = e^{\sum_i V_i}. \tag{7.4.46}$$

The numerator of Eq. (7.4.13) is

$$N \equiv \langle 0|T\{\hat{\phi}_I(x_1) \cdots \hat{\phi}_I(x_n)e^{-i \int_{-T}^{T} dt\, \hat{H}_I(t)}\}|0\rangle \tag{7.4.47}$$
$$= \langle 0|T\{\hat{\phi}_I(x_1) \cdots \hat{\phi}_I(x_n)[1 + \{-i \int_{-T}^{T} dt\, \hat{H}_I(t)\} + \cdots]\}|0\rangle$$
$$= \langle 0|T\{\hat{\phi}_I(x_1) \cdots \hat{\phi}_I(x_n)[1 + (\tfrac{-i\lambda}{4!}) \int d^4z\, \hat{\phi}_I(z)^4 + \cdots]\}|0\rangle$$
$$\equiv (\textstyle\sum_\alpha C_\alpha)[1 + A_1 + A_2 + \cdots] = (\textstyle\sum_\alpha C_\alpha)e^{\sum_i V_i},$$

where $\sum_\alpha C_\alpha$ is the sum of all connected Feynman diagrams. So we see that the vacuum subdiagrams have been factored out of the numerator and have been cancelled by the denominator. So we have arrived at the remarkable result that *the n-point Green's function is the sum of all connected Feynman diagrams*,

$$\langle \Omega | T \hat{\phi}(x_1) \cdots \hat{\phi}(x_n) | \Omega \rangle = \sum_\alpha C_\alpha = \begin{bmatrix} \text{sum of all connected Feynman} \\ \text{diagrams with } n \text{ external points} \end{bmatrix} \quad (7.4.48)$$

$$= \frac{(-i)^k \, \delta^k}{\delta j(x_1) \cdots \delta j(x_k)} Z[j] \Bigg|_{j=0},$$

where $Z[j]$ is given in Eq. (7.4.18). So we see that our $Z[J]$ is the *generating functional for Green's functions*, which are the sum of connected Feynman diagrams. Recall from Eq. (6.2.262) that the numerator of our $Z[j]$ is the $Z[j]$ in some other texts; e.g., in Peskin and Schroeder (1995) $Z^{\mathrm{PS}}[j] = \int \mathcal{D}\phi \, e^{iS[\phi,j]}$ is the generator of *all* Feynman diagrams and $\ln Z^{\mathrm{PS}}[j]$ is the generator of connected Feynman diagrams. For the four-point function the $\mathcal{O}(\lambda^0)$ connected contributions are those in Eq. (7.4.25) and the $\mathcal{O}(\lambda^1)$ connected contributions are those in Eq. (7.4.26). As we have just seen disconnected contributions such as those in Eq. (7.4.27) do not contribute.

7.4.3 Feynman Rules in Momentum Space

The Feynman rules above tell us how to evaluate Feynman diagrams in terms of coordinate-space propagators, momentum integrals and coupling constants. Recall from Eq. (6.2.155) that the scalar boson propagator is

$$D_F(x - y) = \langle 0 | T \hat{\phi}(x) \hat{\phi}(y) | 0 \rangle = \int \frac{d^4 p}{(2\pi)^4} \frac{i}{p^2 - m^2 + i\epsilon} e^{-ip \cdot (x-y)}. \quad (7.4.49)$$

It is straightforward to see that we arrive at an identical expression for the value of any Feynman diagram by using the momentum-space version of these rules,

$$(7.4.50)$$

Momentum-space Feynman rules: First, arbitrarily associate a momentum with every propagator on a Feynman diagram as shown in Eq. (7.4.50) and then:

(i) For each propagator, assign: $\dfrac{\quad p \quad}{} = \dfrac{i}{p^2 - m^2 + i\epsilon}.$

(ii) For each vertex, assign: $= -i\lambda (2\pi)^4 \delta^4(p_1 + p_2 + p_3 + p_4).$

(iii) For each external point x, assign: $\overset{x}{\underset{p}{\longleftarrow}} = e^{-ip \cdot x}.$

(iv) Integrate over all four-momenta using: $\int \dfrac{d^4 p}{(2\pi)^4}.$

(v) Divide the value of the Feynman diagram by its symmetry factor S.

For example, from the coordinate-space Feynman rules for the first diagram in Eq. (7.4.50) we note that $S = 2$ and then obtain

$$\frac{1}{2}(-i\lambda) \int d^4z \, D_F(x-z) D_F(z-z) D_F(z-y) \tag{7.4.51}$$

$$= \frac{1}{2}(-i\lambda) \int d^4z \, \frac{d^4p_1}{(2\pi)^4} \frac{d^4p_2}{(2\pi)^4} \frac{d^4p_3}{(2\pi)^4} \frac{i^3 e^{-ip_1 \cdot (x-z) - ip_3 \cdot (z-z) - ip_2 \cdot (z-y)}}{(p_1^2 - m^2 + i\epsilon)(p_2^2 - m^2 + i\epsilon)(p_3^2 - m^2 + i\epsilon)}$$

$$= \frac{1}{2} \int \frac{d^4p_1}{(2\pi)^4} \frac{d^4p_2}{(2\pi)^4} \frac{d^4p_3}{(2\pi)^4} [-i\lambda(2\pi)^4 \delta^4(-p_1 + p_2 - p_3 + p_3)] e^{-ip_1 \cdot x + ip_2 \cdot y}$$

$$\times \frac{i}{p_1^2 - m^2 + i\epsilon} \frac{i}{p_2^2 - m^2 + i\epsilon} \frac{i}{p_3^2 - m^2 + i\epsilon},$$

where the last expression is just what we find by directly using the momentum-space Feynman rules. To obtain the momentum space version of this one-loop Feynman diagram contribution remove both $\int \frac{d^4p_1}{(2\pi)^4} e^{-ip_1 \cdot x}$ and $\int \frac{d^4p_2}{(2\pi)^4} e^{-ip_2 \cdot y}$. If we amputate the external propagators we are then left with

$$(2\pi)^4 \delta^4(p_2 - p_1)(-i\lambda) \frac{1}{2} \int \frac{d^4p_3}{(2\pi)^4} \frac{i}{p_3^2 - m^2 + i\epsilon}, \tag{7.4.52}$$

which will require regularization and renormalization to make sense of.

Feynman Diagrams from Functional Integrals

Denote the free Lagrangian density in Eq. (6.2.258) as $\mathcal{L}_0$ so that

$$\mathcal{L} = \mathcal{L}_0 + \mathcal{L}_{\text{int}} = \tfrac{1}{2}[\partial_\mu \phi \partial^\mu \phi - m^2 \phi^2] - \tfrac{\lambda}{4!}\phi^4, \tag{7.4.53}$$

where $\mathcal{L}_{\text{int}} = -\mathcal{H}_{\text{int}} = -\frac{\lambda}{4!}\phi^4$. The free generating functional in Eq. (6.2.262) is

$$Z_0[j] = \lim_{T \to \infty(1-i\epsilon)} \frac{\int \mathcal{D}\phi_{(\text{periodic})} \, e^{i\int_{-T}^{T} d^4x \, (\mathcal{L}_0 + j\phi)}}{\int \mathcal{D}\phi_{(\text{periodic})} \, e^{i\int_{-T}^{T} d^4x \, \mathcal{L}_0}} \equiv \frac{\int \mathcal{D}\phi \, e^{i\int d^4x \, (\mathcal{L}_0 + j\phi)}}{\int \mathcal{D}\phi \, e^{i\int d^4x \, \mathcal{L}_0}}$$

$$= \exp\left\{-\tfrac{1}{2} \int d^4x \, d^4y \, j(x) \, D_F(x-y) j(y)\right\}, \tag{7.4.54}$$

where we have used the result in Eq. (6.2.276). On the right-hand side we have suppressed the periodicity and the $T \to \infty(1-i\epsilon)$ limit for brevity but understand that they are implied. Similarly for the interacting theory we write the full generating functional using this shorthand as

$$Z[j] = \frac{\int \mathcal{D}\phi \, e^{i\int d^4x \, (\mathcal{L}_0 + \mathcal{L}_{\text{int}} + j\phi)}}{\int \mathcal{D}\phi \, e^{i\int d^4x \, (\mathcal{L}_0 + \mathcal{L}_{\text{int}})}} = \frac{\int \mathcal{D}\phi \, e^{\int d^4y \, (\frac{-i\lambda}{4!})\phi^4} \, e^{i\int d^4x \, (\mathcal{L}_0 + j\phi)}}{\int \mathcal{D}\phi \, e^{\int d^4y \, (\frac{-i\lambda}{4!})\phi^4} \, e^{i\int d^4x \, \mathcal{L}_0}}. \tag{7.4.55}$$

Define for convenience

$$N_0 \equiv \int \mathcal{D}\phi \, e^{i\int d^4x \, \mathcal{L}_0} \qquad \text{then} \qquad \int \mathcal{D}\phi \, e^{i\int d^4x \, (\mathcal{L}_0 + j\phi)} = N_0 Z_0[j] \tag{7.4.56}$$

and we can write the numerator of Eq. (7.4.55) as

$$\int \mathcal{D}\phi \, e^{i\{S[\phi] + \int d^4x \, j\phi\}} = \int \mathcal{D}\phi \, e^{i\int d^4x \, (\mathcal{L}_0 + \mathcal{L}_{\text{int}} + j\phi)} = \int \mathcal{D}\phi \, e^{\int d^4y \, (\frac{-i\lambda}{4!})\phi^4} \, e^{i\int d^4x \, (\mathcal{L}_0 + j\phi)}$$

$$= \exp\left\{i \int d^4y \, \frac{1}{i^4} \left(\frac{-i\lambda}{4!}\right) \frac{\delta^4}{\delta j(y)^4}\right\} \int \mathcal{D}\phi \, e^{i\int d^4x \, (\mathcal{L}_0 + j\phi)}$$

$$= \sum_{n=0}^{\infty} \frac{i^n}{n!} \left\{\int d^4y \, \left(\frac{-i\lambda}{4!}\right) \frac{\delta^4}{\delta j(y)^4}\right\}^n \int \mathcal{D}\phi \, e^{i\int d^4x \, (\mathcal{L}_0 + j\phi)}$$

$$= \sum_{n=0}^{\infty} \frac{i^n}{n!} \left\{\int d^4y \, \left(\frac{-i\lambda}{4!}\right) \frac{\delta^4}{\delta j(y)^4}\right\}^n N_0 e^{-\int d^4x \, d^4y \, j(x) D_F(x-y) j(y)}. \tag{7.4.57}$$

Taking $j(x) = 0$ of this expression gives the denominator of Eq. (7.4.55). It is then not difficult to see that the denominator gives N_0 multiplied by the sum of all of the Feynman vacuum subdiagrams, since there are no external points. Acting on the numerator with

$$(-i)^n \frac{\delta^k}{\delta j(x_1) \cdots \delta j(x_n)} \tag{7.4.58}$$

and then taking $j(x) = 0$ similarly gives N_0 times the sum of all n-point diagrams. Taking the ratio then cancels the N_0 constant and as seen earlier all of the vacuum subdiagrams also cancel. We have then confirmed as expected

$$\langle \Omega | T \hat{\phi}(x_1) \cdots \hat{\phi}(x_n) | \Omega \rangle = \frac{\int \mathcal{D}\phi \, \phi(x_1) \cdots \phi(x_n) \, e^{iS[\phi]}}{\int \mathcal{D}\phi \, e^{iS[\phi]}} = \frac{(i)^n \delta^k}{\delta j(x_1) \cdots \delta j(x_n)} Z[j] \Big|_{j=0}$$

$$= \sum_\alpha C_\alpha = \begin{bmatrix} \text{sum of all connected Feynman} \\ \text{diagrams with } n \text{ external points} \end{bmatrix}. \tag{7.4.59}$$

7.5 Calculating Invariant Amplitudes

The Lehmann-Symanzik-Zimmermann (LSZ) formalism was developed by Harry Lehmann, Kurt Symanzik and Wolfhart Zimmermann (Lehmann et al., 1955) to calculate the invariant amplitudes that are in turn needed to calculate scattering cross-sections. We will follow traditional arguments similar to those in Bjorken and Drell (1965) and Greiner and Reinhardt (1996). An alternative and arguably more intuitive approach is given in Peskin and Schroeder (1995).

7.5.1 LSZ Reduction Formula for Scalars

Instead of inserting the full identity operator into Eq. (7.2.3) let us insert the one-particle projection operator given in Eq. (7.2.1) to isolate the one-particle contribution. This equation becomes

$$\langle \Omega | \hat{\phi}(x) \hat{\phi}(y) | \Omega \rangle^{1-\text{particle}} = \langle \Omega | \hat{\phi}(x) \hat{I}_1 \hat{\phi}(y) | \Omega \rangle \tag{7.5.1}$$

$$= \int \frac{d^3 p}{(2\pi)^3} \langle \Omega | \hat{\phi}(x) | \mathbf{p} \rangle \frac{1}{2 E_\mathbf{p}} \langle \mathbf{p} | \hat{\phi}(y) | \Omega \rangle \Big|_{E_\mathbf{p} \equiv \sqrt{p^2 + m^2}}$$

$$= |\langle \Omega | \hat{\phi}(0) | \mathbf{p} = 0 \rangle|^2 \int \frac{d^3 p}{(2\pi)^3} \frac{1}{2 E_\mathbf{p}} e^{-ip \cdot (x-y)} \Big|_{E_\mathbf{p} \equiv \sqrt{p^2 + m^2}}$$

$$= |\langle \Omega | \hat{\phi}(0) | \mathbf{p} = 0 \rangle|^2 \Delta_+(m^2; x - y).$$

Then in place of Eq. (7.2.7) we find

$$\langle \Omega | T \hat{\phi}(x) \hat{\phi}(y) | \Omega \rangle^{1-\text{particle}} = |\langle \Omega | \hat{\phi}(0) | \mathbf{p} = 0 \rangle|^2 D_0(m^2; x - y). \tag{7.5.2}$$

Comparing Eq. (7.2.7) with (7.2.10) and using Eq. (7.2.5) shows that

$$Z = |\langle \Omega | \hat{\phi}(0) | \mathbf{p} = 0 \rangle|^2 = |\langle \Omega | \hat{\phi}(0) | \mathbf{p} \rangle|^2 \quad \text{and}$$

$$\langle \Omega | T \hat{\phi}(x) \hat{\phi}(y) | \Omega \rangle^{1-\text{particle}} = Z D_0(m^2; x - y) = Z \langle \Omega | T \hat{\phi}_{\text{as}}(x) \hat{\phi}_{\text{as}}(y) | \Omega \rangle, \tag{7.5.3}$$

where $\hat{\phi}_{\mathrm{as}}(x)$ is the asymptotic field operator and we have used Eq. (7.1.8). From Eq. (6.2.46) for a free field we must have for the asymptotic field case that

$$|\mathbf{p}\rangle \equiv \sqrt{2E_\mathbf{p}}\,\hat{a}^\dagger_{\mathrm{as}\mathbf{p}}|\Omega\rangle \tag{7.5.4}$$

and from Eq. (6.2.137) we have

$$\langle\Omega|\hat{\phi}_{\mathrm{as}}(x)|\mathbf{p}\rangle = e^{-ip\cdot x}. \tag{7.5.5}$$

Similarly, since the single-particle state $|\mathbf{p}\rangle$ is an exact eigenstate of $\hat{P}^\mu$ and $\hat{\phi}$ is a scalar field, we must have $\langle\Omega|\hat{\phi}(x)|\mathbf{p}\rangle \propto e^{-ip\cdot x}$. Combining with Eq. (7.5.3) gives

$$\langle\Omega|\hat{\phi}(x)|\mathbf{p}\rangle = \sqrt{Z}e^{-ip\cdot x} = \sqrt{Z}\langle\Omega|\hat{\phi}_{\mathrm{as}}(x)|\mathbf{p}\rangle. \tag{7.5.6}$$

It is natural to choose the phases of $\langle\Omega|\hat{\phi}(0)|\mathbf{p}=0\rangle$ and $\langle\Omega|\hat{\phi}_{\mathrm{as}}(0)|\mathbf{p}=0\rangle$ to be the same. This is possible since only magnitudes of matrix elements are observable. When constrained to operate within the one-particle subspace of the full Hilbert space, $\hat{\phi}(x)$ must be equivalent to $\hat{\phi}_{\mathrm{as}}$ up to this constant factor $\sqrt{Z}$,

$$\hat{I}_1\hat{\phi}(x)\hat{I}_1 = \sqrt{Z}\hat{\phi}_{\mathrm{as}}(x). \tag{7.5.7}$$

since $\langle\Omega|\hat{\phi}(x)|\Omega\rangle = \langle\Omega|\hat{\phi}_{\mathrm{as}}(x)|\Omega\rangle = 0$ and $\langle\mathbf{p}|\hat{\phi}(x)|\mathbf{k}\rangle = \langle\mathbf{p}|\hat{\phi}_{\mathrm{as}}(x)|\mathbf{k}\rangle = 0$. Let $|\alpha\rangle$ and $|\beta\rangle$ be any states in the Fock space built on the physical single-particle states, then from Eq. (7.5.7) we have

$$\langle\beta|\hat{\phi}(x)|\alpha\rangle = \sqrt{Z}\langle\beta|\hat{\phi}_{\mathrm{as}}(x)|\alpha\rangle. \tag{7.5.8}$$

This is referred to as the *asymptotic condition* (Lehmann et al., 1955).

In and Out States

Consider the scattering between asymptotic states where we use arbitrary wavepacket states in the initial and the final states using the arguments leading to Eqs. (7.3.53) and (7.3.54). The S-matrix is then

$$S_{fi} = \lim_{\sigma\to\infty}\langle f|\lim_{T\to\infty}\hat{U}(T,-T)|i\rangle = \langle f|\hat{S}|i\rangle, \tag{7.5.9}$$

where $\hat{S} \equiv \lim_{T\to\infty}\hat{U}(T,-T)$. It is to be understood that we work with an arbitrarily large T and that the $T\to\infty$ limit is only to be taken at the end of calculations. For m particles scattering to n particles we have

$$|i\rangle \equiv |f_{\mathbf{k}_1}f_{\mathbf{k}_2}\cdots f_{\mathbf{k}_m}\rangle \equiv \int\prod_{j=1}^m\left[\frac{d^3k_j'}{(2\pi)^3}f_{\mathbf{k}_j}(\mathbf{k}_j')\right]|\mathbf{k}_1'\mathbf{k}_2'\cdots\mathbf{k}_m'\rangle, \tag{7.5.10}$$

$$|f\rangle \equiv |f_{\mathbf{p}_1}f_{\mathbf{p}_2}\cdots f_{\mathbf{p}_n}\rangle \equiv \int\prod_{\ell=1}^m\left[\frac{d^3p_\ell'}{(2\pi)^3}f_{\mathbf{p}_\ell}(\mathbf{p}_\ell')\right]|\mathbf{p}_1'\mathbf{p}_2'\cdots\mathbf{p}_n'\rangle$$

and where $\hat{U}(t'',t')$ is the evolution operator given in Eq. (4.1.11). For notational brevity we have defined $f_\mathbf{k}(\mathbf{k}') \equiv \phi(\mathbf{k}')/\sqrt{2E_\mathbf{k}'}$, where $\phi(\mathbf{k}')$ is a momentum wavefunction strongly peaked at $\mathbf{k}' = \mathbf{k}$. This can be seen by comparing the above equations with Eq. (7.3.54). We can define $t = 0$ to be the approximate collision time when the centers of the initial wavepackets are at their closest approach and approximately coincident. The spatial distribution of the wavepackets at $t = 0$ defines an effective interaction region for the scattering event. One typically expects harder scattering to occur when the wavepacket centers are closer together at $t = 0$ and softer scattering when they are further apart. If the incoming wavepackets are small compared with the typical interaction region, then Fig. 7.9 and the associated analysis is relevant. However, we saw that averaging over impact parameter gives the same results as the large wavepacket (plane wave) limit for the initial wavepackets. When we detect the final state particles in a scattering experiment we cannot know which combination of final state wavepackets led to this outcome. Any basis for the asymptotic final state Fock space is equally valid and choosing this basis to be approximate plane waves is obviously the most convenient. So we can always take the limit of large spatial wavepackets, $\sigma \to \infty$, for both initial and final states. As they approach plane waves the details of the interaction time and interaction region become unimportant. Taking the large wavepacket limit, $\sigma \to \infty$, after the infinite time limit, $T \to \infty$, ensures that the asymptotic condition is always satisfied.

To relate our discussion to traditional notation (Bjorken and Drell, 1965; Itzykson and Zuber, 1980; Greiner and Reinhardt, 1996) we can choose to define *in states* and *out states*, respectively, as

$$|i \, \text{in}\rangle \equiv |i\rangle = |f_{\mathbf{k}_1} f_{\mathbf{k}_2} \cdots f_{\mathbf{k}_m}\rangle, \tag{7.5.11}$$

$$\langle f \, \text{out}| \equiv \langle f|\hat{S} = \langle f| \lim_{T \to \infty} \hat{U}(T, -T) = \langle f_{\mathbf{p}_1} f_{\mathbf{p}_2} \cdots f_{\mathbf{p}_n}| \lim_{T \to \infty} \hat{U}(T, -T).$$

Note that we could alternatively have chosen $\langle f \, \text{out}| \equiv \langle f|$ and $|i \, \text{in}\rangle \equiv \hat{S}|i\rangle$ and the end result would not change. With this notation we can express the S-matrix as

$$S_{fi} = \lim_{\sigma \to \infty} \lim_{T \to \infty} \langle f|\hat{U}(T, -T)|i\rangle = \lim_{\sigma \to \infty} \langle f|\hat{S}|i\rangle = \lim_{\sigma \to \infty} \langle f \, \text{out}|i \, \text{in}\rangle. \tag{7.5.12}$$

Since from Eq. (4.1.11) the evolution operator $\hat{U}(t'', t')$ is unitary, the in and out states are related by

$$|\alpha \, \text{out}\rangle = \hat{S}^\dagger |\alpha \, \text{in}\rangle \qquad \text{and} \qquad |\alpha \, \text{in}\rangle = \hat{S}|\alpha \, \text{out}\rangle. \tag{7.5.13}$$

We can also define in and out creation operators such that they satisfy

$$\sqrt{2E_{\mathbf{k}}}\hat{a}^\dagger_{\mathbf{k}\text{in}}|\Omega\rangle = |\mathbf{k} \, \text{in}\rangle \qquad \text{and} \qquad \langle\Omega|\sqrt{2E_{\mathbf{k}}}\hat{a}_{\mathbf{k}\text{out}} = \langle\mathbf{k} \, \text{out}|. \tag{7.5.14}$$

Taking the Hermitian conjugate gives the in and out annihilation operators, $\hat{a}_{\mathbf{k}\text{in}}$ and $\hat{a}_{\mathbf{k}\text{out}}$. We easily verify that Eq. (7.5.14) follows from the definitions

$$\hat{a}^\dagger_{\mathbf{k}\text{in}} \equiv \hat{a}^\dagger_{\text{a}\mathbf{k}} \qquad \text{and} \qquad \hat{a}^\dagger_{\mathbf{k}\text{out}} \equiv \hat{S}^\dagger \hat{a}^\dagger_{\text{a}\mathbf{k}} \hat{S} = \hat{S}^\dagger \hat{a}^\dagger_{\mathbf{k}\text{in}} \hat{S}. \tag{7.5.15}$$

By definition the vacuum state $|\Omega\rangle$ is stable. We can always adjust the phase of $\hat{S}$ by shifting the Hamiltonian by a constant, which means that we can always choose

$$\hat{S}|\Omega\rangle = \lim_{T \to \infty} \hat{U}^\dagger(T, -T)|\Omega\rangle = |\Omega\rangle; \tag{7.5.16}$$

e.g., for a time-independent Hamiltonian we can choose $\hat{H}|\Omega\rangle = 0$. Then it follows that $|\Omega\rangle = |\Omega \, \text{in}\rangle = |\Omega \, \text{out}\rangle$. For our purposes it will be sufficient to assume a time-independent $\hat{H}$ so that $\hat{U}(t'', t') = e^{-i\hat{H}(t''-t')}$. The inclusion of a time-dependent source that vanishes outside a finite time interval is discussed in Bjorken and Drell (1965) and Greiner and Reinhardt (1996) in terms of the *Yang-Feldman equations* (Yang and Feldman, 1950).

Clearly, $\hat{a}_{\mathbf{k}\text{in}}$ and $\hat{a}^\dagger_{\mathbf{k}\text{in}}$ satisfy the commutation relations in Eq. (7.1.6). Since $\hat{S}$ is unitary, then so do $\hat{a}_{\mathbf{k}\text{out}}$ and $\hat{a}^\dagger_{\mathbf{k}\text{out}}$. It follows that

$$\hat{a}_{\mathbf{k}\text{in}} = \hat{S}\hat{a}_{\mathbf{k}\text{out}}\hat{S}^\dagger \quad \text{and} \quad \hat{a}^\dagger_{\mathbf{k}\text{in}} = \hat{S}\hat{a}^\dagger_{\mathbf{k}\text{out}}\hat{S}^\dagger. \tag{7.5.17}$$

Using Eq. (7.5.15) we can define $\hat{\phi}_{\text{in}}(x) = \hat{\phi}_{\text{as}}(x)$. At finite T the evolution operator is $\hat{S} = \hat{U}(T, -T)$ and in that case we can define $\phi_{\text{out}}(x)$ according to

$$\hat{\phi}_{\text{out}}(t + T, \mathbf{x}) \equiv \hat{S}^\dagger \hat{\phi}_{\text{as}}(t - T, \mathbf{x})\hat{S} = \hat{S}^\dagger \hat{\phi}_{\text{in}}(t - T, \mathbf{x})\hat{S}. \tag{7.5.18}$$

The asymptotic condition in Eq. (7.5.8) can then be expressed in the form

$$\langle f|\hat{\phi}(x)|i\rangle \xrightarrow{x^0 = -T \to -\infty} \sqrt{Z}\langle f|\hat{\phi}_{\text{in}}(x)|i\rangle, \tag{7.5.19}$$

$$\langle f|\hat{\phi}(x)|i\rangle \xrightarrow{x^0 = T \to \infty} \sqrt{Z}\langle f|\hat{\phi}_{\text{out}}(x)|i\rangle,$$

since with $x^0 = \pm T \to \pm\infty$ we know that $\hat{\phi}(x)$ is operating in the Fock space, V_{Fock}, containing $|i\rangle$ and $|f\rangle$.

From Eq. (6.2.100) we have for the asymptotic annihilation operator[4]

$$\sqrt{2E_{\mathbf{k}}}\hat{a}^\dagger_{\text{ask}} = -2i \int d^3x\, e^{-ik \cdot x} \overset{\leftrightarrow}{\partial}_0 \hat{\phi}_{\text{as}}(x), \tag{7.5.20}$$

where $\overset{\leftrightarrow}{\partial}_0$ is given in Eq. (6.2.101) and where the right-hand side is time-independent. Then it follows that we similarly have the time-independent quantities,

$$\sqrt{2E_{\mathbf{k}}}\hat{a}^\dagger_{\mathbf{k}\,\text{in}} = \sqrt{2E_{\mathbf{k}}}\hat{a}^\dagger_{\text{ask}} = -2i \int d^3x\, e^{-ik \cdot x} \overset{\leftrightarrow}{\partial}_0 \hat{\phi}_{\text{in}}(x)|_{x^0 = -T}, \tag{7.5.21}$$

$$\sqrt{2E_{\mathbf{k}}}\hat{a}^\dagger_{\mathbf{k}\,\text{out}} = \sqrt{2E_{\mathbf{k}}}\hat{S}^\dagger \hat{a}^\dagger_{\text{ask}}\hat{S} = -2i \int d^3x\, e^{-ik \cdot x} \overset{\leftrightarrow}{\partial}_0 \hat{\phi}_{\text{out}}(x)|_{x^0 = T},$$

where the second result follows from the first using Eq. (7.5.18).

Reduction Formula

While we have been careful to construct our arguments in terms of physically sensible wavepackets, let us initially take the plane wave limits of Eqs. (7.5.9) and (7.5.10) and show at the end that the same result follows using the wavepackets. We do this to keep the presentation as simple as possible. We then have $|i\rangle \to |\mathbf{k}_1\mathbf{k}_2 \cdots \mathbf{k}_m\rangle$ and $|f\rangle \to |\mathbf{p}_1\mathbf{p}_2 \cdots \mathbf{p}_n\rangle$ so that

$$S_{fi} = \langle f|\hat{S}|i\rangle = \langle \mathbf{p}_1\mathbf{p}_2 \cdots \mathbf{p}_n|\hat{S}|\mathbf{k}_1\mathbf{k}_2 \cdots \mathbf{k}_m\rangle = \langle f\,\text{out}|i\,\text{in}\rangle. \tag{7.5.22}$$

Recall Eq. (7.5.14) and consider the S-matrix element

$$S_{fi} = \langle f|\hat{S}|i\rangle = \langle f\,\text{out}|i\,\text{in}\rangle = \sqrt{2E_{\mathbf{k}_1}}\langle f\,\text{out}|\hat{a}^\dagger_{\mathbf{k}_1\,\text{in}}|(i - \mathbf{k}_1)\,\text{in}\rangle \tag{7.5.23}$$

$$= \sqrt{2E_{\mathbf{k}_1}}\left\{ \langle f\,\text{out}|\hat{a}^\dagger_{\mathbf{k}_1\,\text{out}}|(i - \mathbf{k}_1)\,\text{in}\rangle + \langle f\,\text{out}|[\hat{a}^\dagger_{\mathbf{k}_1\,\text{in}} - \hat{a}^\dagger_{\mathbf{k}_1\,\text{out}}]|(i - \mathbf{k}_1)\,\text{in}\rangle \right\},$$

where $|(i - \mathbf{k}_1)\rangle \equiv |\mathbf{k}_2\mathbf{k}_3 \cdots \mathbf{k}_m\rangle$ is the initial state $|i\rangle$ with the boson with momentum $\mathbf{k}_1$ removed. Consider the first term in the last line of the above equation. If $\mathbf{k}_1$ is not equal to any of the $\mathbf{p}_j$, then this term will vanish. If $\mathbf{k}_1 = \mathbf{p}_j$ for some j, then a boson has not scattered and has passed straight through the interaction region. This contributes to the forward scattering, which is not of physical interest. So we can neglect this first term. So in practice we are only concerned with the second term in Eq. (7.5.23).

---

[4] Note: $\overset{\leftrightarrow}{\partial}_0$ in Bjorken and Drell (1965), Itzykson and Zuber (1980) and Greiner and Reinhardt (1996) is twice our $\overset{\leftrightarrow}{\partial}_0$.

Using Eqs. (7.5.21) and (7.5.19) gives for this second term

$$\sqrt{2E_{\mathbf{k}_1}} \langle f \, \text{out}|\hat{a}^\dagger_{\mathbf{k}_1\text{in}} - \hat{a}^\dagger_{\mathbf{k}_1\text{out}}|(i-\mathbf{k}_1)\,\text{in}\rangle \qquad (7.5.24)$$

$$= \lim_{T\to\infty} -2i \int d^3x\, e^{-ik_1\cdot x}\overleftrightarrow{\partial}_0 \langle f\,\text{out}|[\hat{\phi}_{\text{in}}(x)|_{x^0=-T} - \hat{\phi}_{\text{out}}(x)|_{x^0=T}]|(i-\mathbf{k}_1)\,\text{in}\rangle$$

$$= (i/\sqrt{Z})\lim_{T\to\infty} -2\int d^3x\, e^{-ik_1\cdot x}\overleftrightarrow{\partial}_0 \langle f\,\text{out}|[\hat{\phi}(x)|_{x^0=-T} - \hat{\phi}(x)|_{x^0=T}]|(i-\mathbf{k}_1)\,\text{in}\rangle$$

$$= (i/\sqrt{Z})\lim_{T\to\infty} 2\int_{-T}^{T} d^4x\, \partial_0\left[e^{-ik_1\cdot x}\overleftrightarrow{\partial}_0 \langle f\,\text{out}|\hat{\phi}(x)|(i-\mathbf{k}_1)\,\text{in}\rangle\right]$$

$$= (i/\sqrt{Z})\lim_{T\to\infty} \int_{-T}^{T} d^4x [e^{-ik_1\cdot x}\partial_0^2 - (\partial_0^2 e^{-ik_1\cdot x})]\langle f\,\text{out}|\hat{\phi}(x)|(i-\mathbf{k}_1)\,\text{in}\rangle$$

$$= (i\sqrt{Z})\lim_{T\to\infty} \int_{-T}^{T} d^4x\, e^{-ik_1\cdot x}(\overrightarrow{\Box}+m^2)\langle f\,\text{out}|\hat{\phi}(x)|(i-\mathbf{k}_1)\,\text{in}\rangle$$

since $\int_{-T}^{T} dt\, \partial^0 g(t) = g(T)-g(-T)$, $2\partial_0(g\overleftrightarrow{\partial}_0 h) = \partial_0(g\partial_0 h - h\partial_0 g) = g\partial_0^2 h - h\partial_0^2 g$, $\partial_0^2 e^{-ik_1\cdot x} = (\nabla^2 - m^2)e^{-ik_1\cdot x}$ and integration by parts and vanishing/periodic spatial boundary conditions shifts ∇^2 onto $\hat{\phi}(x)$. Since, strictly speaking, we should be acting on localized wavepackets, this is certainly justified. Neglecting forward-scattering contributions gives

$$S_{fi} = \langle f|\hat{S}|i\rangle = \lim_{T\to\infty} \langle \mathbf{p}_1\cdots\mathbf{p}_n|\hat{U}(T,-T)|\mathbf{k}_1\cdots\mathbf{k}_m\rangle \qquad (7.5.25)$$

$$= (i/\sqrt{Z})\lim_{T\to\infty} \int_{-T}^{T} d^4x\, e^{-ik_1\cdot x}(\overrightarrow{\Box}_x + m^2)\langle f\,\text{out}|\hat{\phi}(x)|(i-\mathbf{k}_1)\,\text{in}\rangle.$$

Now let us repeat this exercise by removing $\mathbf{p}_1$ from the out state in the above,

$$\langle f\,\text{out}|\hat{\phi}(x)|(i-\mathbf{k}_1)\,\text{in}\rangle = \sqrt{2E_{\mathbf{p}_1}}\langle (f-\mathbf{p}_1)\,\text{out}|\hat{a}_{\mathbf{p}_1\text{out}}\hat{\phi}(x)|(i-\mathbf{k}_1)\,\text{in}\rangle \qquad (7.5.26)$$

$$= \sqrt{2E_{\mathbf{p}_1}}\left[\langle (f-\mathbf{p}_1)\,\text{out}|\hat{\phi}(x)\hat{a}_{\mathbf{p}_1\text{in}}|(i-\mathbf{k}_1)\,\text{in}\rangle \right.$$
$$\left. + \langle (f-\mathbf{p}_1)\,\text{out}|[\hat{a}_{\mathbf{p}_1\text{out}}\hat{\phi}(x) - \hat{\phi}(x)\hat{a}_{\mathbf{p}_1\text{in}}]|(i-\mathbf{k}_1)\,\text{in}\rangle\right].$$

The first term will only contribute if $\mathbf{p}_1$ is equal to one of $\mathbf{k}_2,\ldots,\mathbf{k}_m$ and so again only contributes to forward scattering. Focusing on the second term and using the Hermitian conjugate of Eq. (7.5.20) we find using similar steps,

$$\sqrt{2E_{\mathbf{p}_1}}\langle (f-\mathbf{p}_1)\,\text{out}|[\hat{a}_{\mathbf{p}_1\text{out}}\hat{\phi}(x) - \hat{\phi}(x)\hat{a}_{\mathbf{p}_1\text{in}}]|(i-\mathbf{k}_1)\,\text{in}\rangle \qquad (7.5.27)$$

$$= (i/\sqrt{Z})\lim_{T\to\infty} 2\int d^3y\, e^{ip_1\cdot y}\overleftrightarrow{\partial}_0^y \langle (f-\mathbf{p}_1)\,\text{out}|\, [\hat{\phi}(y)_{y^0=T}\hat{\phi}(x)$$
$$- \hat{\phi}(x)\hat{\phi}(y)_{y^0=-T}]\,|(i-\mathbf{k}_1)\,\text{in}\rangle$$

$$= (i/\sqrt{Z})\lim_{T\to\infty} 2\int d^3y\, e^{ip_1\cdot y}\overleftrightarrow{\partial}_0^y \langle (f-\mathbf{p}_1)\,\text{out}|[T\hat{\phi}(y)\hat{\phi}(x)]_{y^0=T}$$
$$- [T\hat{\phi}(y)\hat{\phi}(x)]_{y^0=-T}|(i-\mathbf{k}_1)\,\text{in}\rangle$$

$$= (i/\sqrt{Z})\lim_{T\to\infty} 2\int_{-T}^{T} d^4y\, \partial_0^y\left[e^{ip_1\cdot y}\overleftrightarrow{\partial}_0^y \langle (f-\mathbf{p}_1)\,\text{out}|T\hat{\phi}(y)\hat{\phi}(x)|(i-\mathbf{k}_1)\,\text{in}\rangle\right]$$

$$= (i/\sqrt{Z})\lim_{T\to\infty} \int_{-T}^{T} d^4y\, e^{ip_1\cdot y}(\overrightarrow{\Box}_y + m^2)\langle (f-\mathbf{p}_1)\,\text{out}|T\hat{\phi}(y)\hat{\phi}(x)|(i-\mathbf{k}_1)\,\text{in}\rangle.$$

Note the natural appearance of time-ordering. Since even the interacting fields must satisfy causality, then $[\hat{\phi}(x),\hat{\phi}(y)] = 0$ for spacelike separations and so Eq. (6.2.166) remains valid. This allows us to move ∂_0^y in and out of the time-ordering operator without concern.

This process can be repeated until all particles from the in and out states have been removed, leaving the time-ordered vacuum expectation value of $(m + n)$ field operators. The limit $T \to \infty(1 - i\epsilon)$ is to be understood and is what ensures the Feynman boundary conditions in the usual way. When all $\mathbf{p}_i$ are different from all $\mathbf{k}_j$ there is no forward scattering. In that case the S-matrix is given by the *Lehmann-Symanzik-Zimmermann (LSZ) reduction formula* (Lehmann et al., 1955)

$$S_{fi} = \langle f|\hat{S}|i\rangle = \lim_{T \to \infty(1-i\epsilon)} \langle \mathbf{p}_1 \cdots \mathbf{p}_n|\hat{U}(T, -T)|\mathbf{k}_1 \cdots \mathbf{k}_m\rangle = \langle f\,\text{out}|i\,\text{in}\rangle \qquad (7.5.28)$$

$$= \left(i/\sqrt{Z}\right)^{m+n} \int d^4y_1 \cdots d^4y_n \, d^4x_1 \cdots d^4x_m$$

$$\times \, e^{ip_1 \cdot y_1}(\overrightarrow{\Box}_{y_1} + m^2 - i\epsilon) \cdots e^{ip_n \cdot y_n}(\overrightarrow{\Box}_{y_n} + m^2 - i\epsilon)$$

$$\times \, \langle\Omega|T\hat{\phi}(y_1) \cdots \hat{\phi}(y_n)\hat{\phi}(x_1) \cdots \hat{\phi}(x_m)|\Omega\rangle$$

$$\times \, (\overleftarrow{\Box}_{x_1} + m^2 - i\epsilon)e^{-ik_1 \cdot x_1} \cdots (\overleftarrow{\Box}_{x_m} + m^2 - i\epsilon)e^{-ik_m \cdot x_m}.$$

Wavepacket states: We now repeat the above analysis using the wavepacket states in Eq. (7.5.10) rather than plane waves used in Eq. (7.5.22). Then we have

$$S_{fi} = \langle f|\hat{S}|i\rangle = \langle f_{\mathbf{p}_1} f_{\mathbf{p}_2} \cdots f_{\mathbf{p}_n}|\hat{S}|f_{\mathbf{k}_1} f_{\mathbf{k}_2} \cdots f_{\mathbf{k}_m}\rangle = \langle f\,\text{out}|i\,\text{in}\rangle, \qquad (7.5.29)$$

where recall that we have defined $f_{\mathbf{k}}(\mathbf{k}') \equiv \phi(\mathbf{k}')/\sqrt{2E'_{\mathbf{k}}}$, with $\phi(\mathbf{k}')$ a momentum wavefunction strongly peaked at $\mathbf{k}' = \mathbf{k}$ as described in the discussion leading to Eq. (7.3.54). In place of Eq. (7.5.5) we now have

$$\langle\Omega|\hat{\phi}(x)|f_{\mathbf{p}}\rangle = \sqrt{Z} \int \frac{d^3k'}{(2\pi)^3} f_{\mathbf{p}}(\mathbf{k}')\langle\Omega|\hat{\phi}_{\text{as}}(x)|\mathbf{k}'\rangle = \sqrt{Z} \int \frac{d^3k'}{(2\pi)^3} f_{\mathbf{p}}(\mathbf{k}')e^{-ik' \cdot x} \qquad (7.5.30)$$

$$\equiv \sqrt{Z} f_{\mathbf{p}}(x),$$

which justifies Eq. (7.5.19). There is a slight subtlety here, which is that $f_{\mathbf{p}}(0, \mathbf{x})$ is the asymmetric Fourier transform of $f_{\mathbf{p}}(\mathbf{k}) = \phi(\mathbf{k})/\sqrt{2E_{\mathbf{k}}}$, whereas $\psi(0, \mathbf{x})$ in Eq. (7.3.46) is that for $\phi(\mathbf{k})$. Nonetheless, $f_{\mathbf{p}}(x)$ is a normalizable wavepacket solution of the Klein-Gordon equation with spatial extent that we denote as σ and with $f_{\mathbf{p}}(\mathbf{k})$ sharply peaked at $\mathbf{k} = \mathbf{p}$ so that Eq. (7.3.49) applies to it. It follows that

$$\langle f\,\text{out}|i\,\text{in}\rangle = \langle f_{\mathbf{p}_1} \cdots f_{\mathbf{p}_n}\,\text{out}|f_{\mathbf{k}_1} \cdots f_{\mathbf{k}_m}\,\text{in}\rangle \qquad (7.5.31)$$

$$= \int \frac{d^3k'_1}{(2\pi)^3} f_{\mathbf{k}_1}(\mathbf{k}'_1)\langle f_{\mathbf{p}_1} \cdots f_{\mathbf{p}_n}\,\text{out}|\mathbf{k}'_1 f_{\mathbf{k}_2} \cdots f_{\mathbf{k}_m}\,\text{in}\rangle$$

$$= \int \frac{d^3k'_1}{(2\pi)^3} f_{\mathbf{k}_1}(\mathbf{k}'_1)\sqrt{2E_{\mathbf{k}'_1}}\langle f_{\mathbf{p}_1} \cdots f_{\mathbf{p}_n}\,\text{out}|\hat{a}^\dagger_{\mathbf{k}'_1\,\text{in}}|f_{\mathbf{k}_2} \cdots f_{\mathbf{k}_m}\,\text{in}\rangle$$

$$= \int \frac{d^3k'_1}{(2\pi)^3} f_{\mathbf{k}_1}(\mathbf{k}'_1)(-2i) \int d^3x\, e^{-ik'_1 \cdot x} \overleftrightarrow{\partial}_0 \langle f_{\mathbf{p}_1} \cdots f_{\mathbf{p}_n}\,\text{out}|[\hat{\phi}_{\text{in}}(x)|_{x^0=-T}]|f_{\mathbf{k}_2} \cdots f_{\mathbf{k}_m}\,\text{in}\rangle$$

$$= (-2i) \int d^3x\, f_{\mathbf{k}_1}(x) \overleftrightarrow{\partial}_0 \langle f_{\mathbf{p}_1} \cdots f_{\mathbf{p}_n}\,\text{out}|[\hat{\phi}_{\text{in}}(x)|_{x^0=-T}]|f_{\mathbf{k}_2} \cdots f_{\mathbf{k}_m}\,\text{in}\rangle$$

$$= (-2i)\sqrt{Z} \int d^3x\, f_{\mathbf{k}_1}(x) \overleftrightarrow{\partial}_0 \langle f_{\mathbf{p}_1} \cdots f_{\mathbf{p}_n}\,\text{out}|[\hat{\phi}(x)|_{x^0=-T}]|f_{\mathbf{k}_2} \cdots f_{\mathbf{k}_m}\,\text{in}\rangle,$$

which has had the effect of replacing $e^{-ik_1 \cdot x}$ with $f_{\mathbf{k}_1}(x)$. Keeping only the relevant second term in Eq. (7.5.23), it and Eq. (7.5.24) become

$$S_{fi} = (i/\sqrt{Z}) \lim_{T \to \infty} (-2) \int d^3x \, f_{\mathbf{k}_1}(x) \overset{\leftrightarrow}{\partial}_0 \tag{7.5.32}$$

$$\times \langle f_{\mathbf{p}_1} \cdots f_{\mathbf{p}_n} \, \mathrm{out} | [\hat{\phi}(x)|_{x^0=-T} - \hat{\phi}(x)|_{x^0=T}] | f_{\mathbf{k}_2} \cdots f_{\mathbf{k}_m} \, \mathrm{in} \rangle$$

$$= (i/\sqrt{Z}) \lim_{T \to \infty} 2 \int_{-T}^{T} d^4x \, \partial_0 \left[f_{\mathbf{k}_1}(x) \overset{\leftrightarrow}{\partial}_0 \langle f_{\mathbf{p}_1} \cdots f_{\mathbf{p}_n} \, \mathrm{out} | \hat{\phi}(x) | f_{\mathbf{k}_2} \cdots f_{\mathbf{k}_m} \, \mathrm{in} \rangle \right]$$

$$= (i/\sqrt{Z}) \lim_{T \to \infty} \int_{-T}^{T} d^4x [f_{\mathbf{k}_1}(x)\partial_0^2 - (\partial_0^2 f_{\mathbf{k}_1}(x))] \langle f \, \mathrm{out} | \hat{\phi}(x) | (i - \mathbf{k}_1) \, \mathrm{in} \rangle$$

$$= (i\sqrt{Z}) \lim_{T \to \infty} \int_{-T}^{T} d^4x \, f_{\mathbf{k}_1}(x) (\overset{\rightarrow}{\Box} + m^2) \langle f \, \mathrm{out} | \hat{\phi}(x) | (i - \mathbf{k}_1) \, \mathrm{in} \rangle,$$

where we used that $f_{\mathbf{k}}(x)$ is a solution of the Klein-Gordon equation. Each $e^{-ik_j \cdot x}$ is replaced by $f_{\mathbf{k}_j}(x)$ for each in state. Similarly, for each out state $e^{+ip_j \cdot x}$ is replaced by $f_{\mathbf{p}_j}^*(x)$. We have then arrived at the wavepacket form of the LSZ result,

$$S_{fi} = \langle f | \hat{S} | i \rangle = \lim_{\sigma \to \infty} \lim_{T \to \infty(1-\epsilon)} \langle f_{\mathbf{p}_1} \cdots f_{\mathbf{p}_n} | \hat{U}(T, -T) | f_{\mathbf{k}_1} \cdots f_{\mathbf{k}_m} \rangle \tag{7.5.33}$$

$$= \lim_{\sigma \to \infty} \lim_{T \to \infty(1-\epsilon)} \left(i/\sqrt{Z} \right)^{m+n} \int_{-T}^{T} d^4y_1 \cdots d^4y_n \, d^4x_1 \cdots d^4x_m$$

$$\times f_{\mathbf{p}_1}^*(y_1)(\overset{\rightarrow}{\Box}_{y_1} + m^2 - i\epsilon) \cdots f_{\mathbf{p}_n}^*(y_n)(\overset{\rightarrow}{\Box}_{y_n} + m^2 - i\epsilon)$$

$$\times \langle \Omega | T\hat{\phi}(y_1) \cdots \hat{\phi}(y_n)\hat{\phi}(x_1) \cdots \hat{\phi}(x_m) | \Omega \rangle$$

$$\times (\overset{\leftarrow}{\Box}_{x_i} + m^2 - i\epsilon) f_{\mathbf{k}_i}(x_i) \cdots (\overset{\leftarrow}{\Box}_{x_m} + m^2 - i\epsilon) f_{\mathbf{k}_m}(x_m).$$

We see that there is no obstruction to first taking the $T \to \infty(1 - \epsilon)$ and then taking the plane wave limit, $f_{\mathbf{p}}(\mathbf{k}) \to (2\pi)^3 \delta^3(\mathbf{p} - \mathbf{k})$, which leads to $f_{\mathbf{k}}(x) \to e^{-ik \cdot x}$. So as anticipated, the LSZ result in Eq. (7.5.28) follows from the wavepacket treatment.

Define $G^{(n)}(p_1, \ldots, p_n)$ as the four-dimensional asymmetric Fourier transform of the n-point Green's function $G^{(n)}(x_1, \ldots, x_n)$,

$$G^{(n)}(p_1, \ldots, p_n) = \int \prod_{i=1}^{n} d^4x_i \, e^{ip_i \cdot x_i} G^{(n)}(x_1, \ldots, x_n) \tag{7.5.34}$$

$$= \int \prod_{i=1}^{n} d^4x_i \, e^{ip_i \cdot x_i} \langle \Omega | T\hat{\phi}(x_1) \cdots \hat{\phi}(x_n) | \Omega \rangle$$

$$G(x_1, \ldots, x_n) = \int \prod_{i=1}^{n} \frac{d^4p_i}{(2\pi)^4} e^{-ip_i \cdot x_i} G(p_1, \ldots, p_n),$$

where all momenta are outgoing. $(\Box + m^2 - i\epsilon)$ becomes $-(p^2 - m^2 + i\epsilon)$ in momentum space. Note that in a spacetime translationally invariant theory $G^{(n)}(p_1, \ldots, p_n)$ will contain the four-momentum conserving factor $\delta^4(\sum_i p_i)$. Using Eq. (7.2.11) the LSZ reduction formula in momentum space is given by

$$S_{fi} = \left(\frac{-i}{\sqrt{Z}} \right)^{m+n} \left[\left\{ \prod_{j=1}^{n} (p_j^2 - m^2 + i\epsilon) \right\} G^{(m+n)} \left\{ \prod_{i=1}^{m} (k_i^2 - m^2 + i\epsilon) \right\} \right]_{\text{on-shell}} \tag{7.5.35}$$

$$= \sqrt{Z}^{m+n} \left[\left\{ \prod_{j=1}^{n} D_F^{-1}(p_j) \right\} G^{(m+n)}(p_1, \ldots, p_n, -k_1, \ldots, -k_m) \left\{ \prod_{i=1}^{m} D_F^{-1}(k_i) \right\} \right]_{\text{on-shell}},$$

where forward scattering contributions have been ignored. The negative signs on the incoming particle momenta are because the momenta in Eq. (7.5.34) are outgoing. We now impose the bosonic requirement that we must symmetrize the initial momentum states and also symmetrize the final momentum states in the usual way; i.e., we symmetrize over the various p_j and also over the various k_i.

Alternative approach: Other derivations of the LSZ reduction formula are possible, but reduce to the same argument. In Bjorken and Drell (1965), Greiner and Reinhardt (1996) and elsewhere, wavepacket solutions of the Klein-Gordon equation are used to smear the operators rather than the states, which amounts to replacing $e^{-ik \cdot x}$ in Eq. (7.5.21) with wavepacket solutions $f_{\mathbf{k}}(x)$ to create smeared creation operators. For example, comparing the first and third lines of Eq. (7.5.31) it becomes clear that

$$\hat{a}_{\text{in}}^{\dagger}(f_{\mathbf{k}})|\Omega\rangle = |f_{\mathbf{k}} \text{ in}\rangle, \quad \text{with} \quad \hat{a}_{\text{in}}^{\dagger}(f_{\mathbf{k}}) \equiv \int \frac{d^3 k'}{(2\pi)^3} f_{\mathbf{k}}(\mathbf{k}') \sqrt{2E_{\mathbf{k}'}}\, \hat{a}_{\mathbf{k}' \text{in}}^{\dagger}, \tag{7.5.36}$$

which shows that $\hat{a}_{\text{in}}^{\dagger}(f_{\mathbf{k}})$ is the creation operator for the wavepacket in state $|f_{\mathbf{k}} \text{ in}\rangle$. From the smeared annihilation and creation operators we can define smeared in and out field operators. Rather than have our point-like field operators with wavepacket states one could have smeared operators with plane wave states. So the operator smearing approach is equivalent to the wavepacket approach in this way, which is why Eq. (16.81) in Bjorken and Drell (1965) and Eq. (9.85) in Greiner and Reinhardt (1996) are identical to our Eq. (7.5.33).

Intuitive argument: The above derivation of the LSZ formalism for scalars largely follows traditional arguments. While logically laid out, it is demanding to work through. Peskin and Schroeder (1995) offer a more intuitive approach. An S-matrix element can be obtained by Fourier transforming an $(m+n)$-point Green's function and projecting out the one-particle parts of $G^{(m+n)}$ in momentum space by acting on it with inverse on-shell propagators. All but the one-particle pole parts of $G^{(m+n)}$ will vanish since the inverse propagators vanish at the poles.

The LSZ reduction formula in Eqs. (7.5.28) and (7.5.35) and its generalizations are key results that will underpin the calculations of scattering amplitudes and cross-sections in quantum field theory. For additional details regarding the LSZ reduction formula, see some of the original papers (Yang and Feldman, 1950; Lehmann et al., 1955) and also discussions in Bjorken and Drell (1965), Itzykson and Zuber (1980), Peskin and Schroeder (1995), Greiner and Reinhardt (1996) and other texts. The LSZ reduction formula is then combined with Eq. (7.4.13), which expresses the Green's functions of the full theory in terms of the interaction picture, where the Green's functions are to be evaluated using Eq. (7.4.48) and the rules for evaluating Feynman diagrams developed in Sec. 7.4.2. A major issue that remains to be addressed is the program of renormalization needed to manage the apparent infinities arising from loops in Feynman diagrams.

Amputated and One-Particle Irreducible Amplitudes

An n-point Green's function $G^{(n)}(x_1, \ldots, x_n)$ is shown on the left-hand side of Eq. (7.5.37).

$$\tag{7.5.37}$$

As we know from Eq. (7.4.48) we can evaluate $G^{(n)}(x_1, \ldots, x_n)$ as the sum of all connected Feynman diagrams with n external points. Note that propagators in the Feynman diagrams are the *bare Feynman propagators*, denoted $D_0(x - y)$, that contain the bare mass m_0 and not the physical mass m,

$$D_0(x - y) \equiv \int \frac{d^4p}{(2\pi)^4} \frac{i}{p^2 - m_0^2 + i\epsilon} e^{-ip \cdot (x-y)} \equiv \int \frac{d^4p}{(2\pi)^4} D_0(p) e^{-ip \cdot (x-y)}. \qquad (7.5.38)$$

The Fourier transform $G^{(n)}(p_1, \ldots, p_n)$ can similarly be obtained as the sum of these Feynman diagrams evaluated in momentum space. A comparison of Eqs. (7.5.34) and (7.5.35) shows that our definition of $G^{(n)}(p_1, \ldots, p_n)$ corresponds to all momenta $p_1, \ldots, p_n$ flowing outward.

Consider any arbitrarily complicated connected n-point Feynman diagram contributing to $G^{(n)}$. Choose any external point, say, x_1, and follow the connected diagram inward until reaching the furthest point of the *last* bare Feynman propagator that when cut would disconnect x_1 from the rest of the diagram. Then repeat this process for each of x_2 to x_n. The n-point diagram that lies inside each of these n cut points has no external bare Feynman propagators and is referred to as an *amputated Feynman diagram*. The *sum of all connected amputated Feynman diagrams* gives the amputated Green's function, $G_{\text{amp}}^{(n)}$, which is shown on the right-hand side of Eq. (7.5.37). Now we need to establish what is meant by the legs with shaded circles in this equality.

Again take the arbitrary Feynman diagram and consider the part of it between x_1 and the place where the cut would be that separates off the amputated part of the diagram. Now imagine holding the rest of the diagram fixed but adding every possible connected contribution between x_1 and the cut, which is represented as the line with the shaded blob. We know from Eq. (7.4.48) that the sum of all connected two-point Feynman diagrams is just the two-point Green's function,

$$D_F(x - y) = \langle \Omega | T \hat{\phi}(x) \hat{\phi}(y) | \Omega \rangle = x - \!\!\bullet\!\!- y. \qquad (7.5.39)$$

This then explains the equality of the countably infinite sum of Feynman diagrams on the left and right sides of Eq. (7.5.37). Every connected n-point Feynman diagram appears once and only once on each side of the figure. From Eq. (7.5.37) we have

$$G^{(n)}(p_1, \ldots, p_n) = D_F(p_1) \cdots D_F(p_n) \, G_{\text{amp}}^{(n)}(p_1, \ldots, p_n), \qquad (7.5.40)$$

where all momenta flow outward. Comparing with Eq. (7.5.35) we find the *LSZ reduction formula for scalars*,

$$S_{fi} = \langle \mathbf{p}_1 \cdots \mathbf{p}_n | \hat{S} | \mathbf{k}_1 \cdots \mathbf{k}_m \rangle \qquad (7.5.41)$$

$$= \sqrt{Z}^{m+n} \, G_{\text{amp}}^{(m+n)}(p_1, \ldots, p_n, -k_1, \ldots, -k_m) \Big|_{\text{on-shell}} + \begin{bmatrix} \text{forward scattering} \\ \text{contributions} \end{bmatrix}.$$

It will later be seen that as part of the renormalization program the field operator $\hat{\phi} \equiv \hat{\phi}_0$ is regarded as a bare field field operator, which is related to the renormalized field operator $\hat{\phi}_r$ by $\hat{\phi}_0 = \sqrt{Z} \hat{\phi}_r$. This is equivalent to redefining the field operator such that the residue of the single-particle pole is unity,

$$D_{r,F}(x - y) \equiv \langle \Omega | T \hat{\phi}_r(x) \hat{\phi}_r(y) | \Omega \rangle, \quad D_{r,F}(p) = \frac{i}{p^2 - m^2 + i\epsilon} + \int_{s > m^2}^{\infty} ds \, \frac{1}{Z} \frac{i\rho(s)}{p^2 - s + i\epsilon}.$$

Similarly for the renormalized Green's function we have

$$G_r^{(n)}(x_1, \ldots, x_n) \equiv \langle \Omega | \hat{\phi}_r(x_1) \cdots \hat{\phi}_r(x_n) | \Omega \rangle = (1/\sqrt{Z})^n G^{(n)}(x_1, \ldots, x_n). \qquad (7.5.42)$$

Comparing with Eq. (7.5.35) we observe that S_{fi} arises from amputating the renormalized propagators from the renormalized Green's function. Define the renormalized amputated Green's function as

$$G_{r,\text{amp}}^{(n)}(p_1, \ldots, p_n) \equiv D_{r,F}(p_1)^{-1} \cdots D_{r,F}(p_n)^{-1} G_r^{(n)}(p_1, \ldots, p_n). \tag{7.5.43}$$

Then using Eqs. (7.3.2) and (7.3.3) we have

$$S_{fi} = \delta_{fi} + iT_{fi} = \delta_{fi} + i(2\pi)^4 \delta^4 \left(\sum_i p_i - \sum_j k_j \right) \mathcal{M}_{fi} \tag{7.5.44}$$

$$= \langle \mathbf{p}_1 \cdots \mathbf{p}_n | \hat{S} | \mathbf{k}_1 \cdots \mathbf{k}_m \rangle$$

$$= G_{r,\text{amp}}^{(m+n)}(p_1, \ldots, p_n, -k_1, \ldots, -k_m) \Big|_{\text{on-shell}} + \begin{bmatrix} \text{forward scattering} \\ \text{contributions} \end{bmatrix},$$

where $G_{r,\text{amp}}^{(m+n)}$ is $G_r^{(m+n)}$ with the renormalized propagators $D_{r,F}(p)$ amputated. We have not yet explicitly imposed symmetrization of the initial and final states, but since the boson operators commute in the time-ordered product the symmetry is built in already and so no further symmetrization is needed.

Also note that we want to integrate over all distinct final quantum states. Since a quantum state with n identical particles is unchanged by any of the $n!$ permutations of the identical particles, then integrating over the n-body Lorentz-invariant phase space $d\Pi_n^{\text{LIPS}}$ would overcount states by $n!$. *For this reason we must divide by $n!$ and use $(1/n!)d\Pi_n^{\text{LIPS}}$ instead when we have n identical final-state particles.* We will typically be interested in $2 \to n$ particle scattering or $1 \to n$ particle decay.

In the renormalization program the existence of loops in Feynman diagrams will require renormalization of the theory in terms of a wavefunction renormalization Z, a bare mass m_0 and a bare coupling λ_0. As mentioned earlier, a Feynman diagram in one piece with no loops is referred to as a *tree diagram* or a *tree-level diagram*. A diagram with no loops is a *forest diagram* and a forest diagram with only one piece is a tree diagram. A forest diagram with two or more disjoint pieces only contributes to forward scattering, which is why we only need to consider tree diagrams in scattering if we have no loops. If we work only at tree-level, no renormalization is required. So, for calculations of cross-sections at tree-level we use $Z \to 1$, $m_0 \to m$ and $\lambda_0 \to \lambda$ and then $G_{r,\text{ampl}}^{(m+n)} = G_{\text{ampl}}^{(m+n)}$. For ϕ^4 theory if we consider two-particle ($m = 2$) to two-particle ($n = 2$) scattering at tree-level we only have one diagram contributing. Using momentum space Feynman rules we find

$$G_{\text{amp},r}^{(2+2)}(p_1, p_2, -k_1, -k_2) = p_1 \begin{matrix} p_2 \\ \times \\ k_1 \end{matrix} k_2 = -i\lambda(2\pi)^4 \delta^4(p_1 + p_2 - k_1 - k_2). \tag{7.5.45}$$

The symmetrization of the initial and final states does not lead to a topologically distinct diagram and so can be ignored. Then the invariant amplitude is

$$\mathcal{M}_{fi} = \mathcal{M}_{k_1 k_2 \to p_1 p_2} = -\lambda. \tag{7.5.46}$$

As discussed earlier, since we have two identical bosons in the final state, then we must include a factor of $\frac{1}{2!} = \frac{1}{2}$ along with $d\Pi_2^{\text{LIPS}}$. Then using Eq. (7.3.40) the differential cross-section at tree-level is

$$d\sigma = \frac{|\mathcal{M}_{fi}|^2}{4E_{\mathbf{k}_1} E_{\mathbf{k}_2} |\boldsymbol{v}_1 - \boldsymbol{v}_2|} \frac{1}{2} d\Pi_2^{\text{LIPS}} \xrightarrow{\text{tree-level}} \frac{\lambda^2}{4E_{\mathbf{k}_1} E_{\mathbf{k}_2} |\boldsymbol{v}_1 - \boldsymbol{v}_2|} \frac{1}{2} d\Pi_2^{\text{LIPS}}, \tag{7.5.47}$$

where $d\Pi_2^{\mathrm{LIPS}}$ is defined in Eq. (7.3.41) and where $\boldsymbol{v}_1$ and $\boldsymbol{v}_2$ are the velocities of the incoming scalar particles corresponding to momenta $\mathbf{k}_1$ and $\mathbf{k}_2$, respectively.

We can apply a form of the above amputation process to the two-point function itself. Start from y in the figure in Eq. (7.5.39) and identify the furthest point on the last bare propagator, *excluding* the bare propagator connected to x, that divides the diagram in two with a cut of a *single* bare propagator. The location of the cut is between the light and dark shaded circles in the figure in the top line of Eq. (7.5.49). The location of the cut *defines* the light shaded circle we write as $-i\Pi(p^2)$. The bare scalar boson propagator is

$$D_0(p) = \underset{\longrightarrow}{\overset{p}{}} = \frac{i}{p^2 - m_0^2 + i\epsilon}. \tag{7.5.48}$$

Whenever there might be confusion about the presence or absence of a bare Feynman propagator in a diagram we will use a "dot" on a line to designate D_0. Using this notation and working in momentum space it then follows that

$$D_F(p) = \underset{}{\bigcirc} = \underset{}{\bullet} + \underset{}{\bullet\bigcirc\bigcirc} = D_0(p) + D_0(p)[-i\Pi(p^2)]D_F(p)$$

$$= \sum(\text{connected two-point diagrams}), \tag{7.5.49}$$

$$-i\Pi(p^2) \equiv \underset{}{\bigcirc} = \underset{}{\bigcirc} + \underset{}{\oplus} + \underset{}{8} + \cdots = \sum(\text{1PI two-point diagrams}).$$

Feynman diagrams that cannot be divided into two parts by the cut of any single bare propagator are referred to as *one-particle irreducible (1PI)* or *proper* diagrams. A diagram that is 1PI cannot have external bare propagators. As we have defined it, $-i\Pi(p^2)$ is the sum of all 1PI two-point Feynman diagrams and so $-i\Pi(p^2)$ is a *1PI amplitude*. For reasons that will soon become clear, we refer to $\Pi(p^2)$ as the *scalar self-energy*. We can understand the equation of the first line of Eq. (7.5.49) by recognizing that every diagram in the full propagator appears once and only once on each side of the equation; i.e., every diagram on the left can be uniquely identified with a corresponding diagram on the right and vice versa.

Acting on Eq. (7.5.49) with $D_0^{-1}(p)$ and $D_F^{-1}(p)$ from the left and right, respectively, we find

$$D_0^{-1}(p) = D_F^{-1}(p) - i\Pi(p^2) \quad \Rightarrow \quad D_F^{-1}(p) = D_0^{-1}(p) + i\Pi(p^2). \tag{7.5.50}$$

Combining this with Eq. (7.5.64) it follows that

$$D_F(p) = \underset{}{\bigcirc} = \int_0^\infty ds\, \frac{i\rho(s)}{p^2 - s + i\epsilon} = \frac{i}{p^2 - m_0^2 - \Pi(p^2) + i\epsilon}. \tag{7.5.51}$$

Comparing with Eq. (7.2.11) we see that $D_F(p^2)$ has a pole at $p^2 = m^2$ by definition and so $m^2 - m_0^2 - \Pi(m^2) = 0$, which gives for the physical squared mass

$$m^2 = m_0^2 + \Pi(m^2). \tag{7.5.52}$$

We now see that the self-energy, $\Pi(p^2)$, is that contribution from the self-interaction of a particle that changes its propagator from its bare form to its physical form. Recursively expanding Eq. (7.5.49) it follows that in momentum space

$$D_F = D_0 + D_0[-i\Pi]D_0 + D_0[-i\Pi]D_0[-i\Pi]D_0 + \cdots. \tag{7.5.53}$$

It is convenient to define a *renormalized scalar self-energy* $\Pi_r(p^2)$ such that

$$\left.\frac{d\Pi_r(p^2)}{dp^2}\right|_{p^2=m^2} = 0 \quad \text{and hence} \quad D_F(p) = \frac{iZ}{p^2 - m^2 - \Pi_r(p^2) + i\epsilon}, \tag{7.5.54}$$

which has a pole at $p^2 = m^2$. We determine Z from $\Pi(p^2)$ using

$$\frac{1}{Z} = \left[1 - \frac{d\Pi}{dp^2} \right]_{p^2 = m^2}. \tag{7.5.55}$$

Proof: Equating Eqs. (7.5.51) and (7.5.54) we find

$$Z[p^2 - m_0^2 - \Pi(p^2)] = p^2 - m^2 - \Pi_r(p^2), \tag{7.5.56}$$

$$\Rightarrow \quad \Pi_r(p^2) = Z\Pi(p^2) - (Z-1)(p^2 - m_0^2) - (m^2 - m_0^2). \tag{7.5.57}$$

The renormalization factor Z is determined by $d\Pi_r(p^2)/dp^2 = 0$ at the pole, $p^2 = m^2$, which gives $[Z d\Pi/dp^2 - (Z-1)]_{p^2 = m^2} = 0$ and hence Eq. (7.5.55).

7.5.2 LSZ for Fermions

The above arguments generalize to the case of fermions. Detailed discussions can be found in Bjorken and Drell (1965) and Greiner and Reinhardt (1996) but note the minor difference in normalization convention that was explained in Eq. (6.3.198).

Källen-Lehmann representation for fermions: As for scalars there is a bare mass m_0 for the free fermion in the fermion free particle Fock space. When interactions are present, the physical spectrum will again have a form like that in Fig. 7.1, where there is a single fermion state with mass m, possible bound state masses M_i and a continuum that starts at the threshold mass. If only a single species of fermion is present, then the continuum threshhold is $m_{\text{th}} = 2m$.

If the cluster decomposition principle holds in the interacting theory, then as $t \to \pm\infty$ the states become asymptotic states. If no bound states are present in the asymptotic states, then these states behave as Fock space states in a free fermion theory with physical mass m built on the interacting theory vacuum $|\Omega\rangle$. We can create separate Fock spaces for bound states as discussed earlier in the scalar case. We can construct the asymptotic fermion fields in this Fock space in the usual way,

$$\hat{\psi}_{\text{as}}(x) = \int \frac{d^3p}{(2\pi)^3} \frac{1}{\sqrt{2E_{\mathbf{p}}}} \sum_s \left[\hat{b}_{\text{as}\mathbf{p}}^s u^s(p) e^{-ip \cdot x} + \hat{d}_{\text{as}\mathbf{p}}^{s\dagger} v^s(p) e^{ip \cdot x} \right], \tag{7.5.58}$$

where the asymptotic annihilation and creation operators satisfy the usual Fock space anticommutation relations in Eq. (6.3.80),

$$\{\hat{b}_{\text{as}\mathbf{p}}^s, \hat{b}_{\text{as}\mathbf{p}'}^{s'\dagger}\} = \{\hat{d}_{\text{as}\mathbf{p}}^s, \hat{d}_{\text{as}\mathbf{p}'}^{s'\dagger}\} = \delta^{ss'}(2\pi)^3 \delta^3(\mathbf{p} - \mathbf{p}') \tag{7.5.59}$$

with all other anticommutators vanishing. The ETACR relations are then as for the free field in Eq. (6.3.190),

$$\{\hat{\psi}_{\text{as}\alpha}(t, \vec{x}), \hat{\psi}_{\text{as}\beta}^\dagger(t, \vec{y})\} = \delta_{\alpha\beta} \delta^3(\vec{x} - \vec{y}) \tag{7.5.60}$$

$$\{\hat{\psi}_{\text{as}\alpha}(t, \vec{x}), \hat{\psi}_{\text{as}\beta}(t, \vec{y})\} = \{\hat{\psi}_{\text{as}\alpha}^\dagger(t, \vec{x}), \hat{\psi}_{\text{as}\beta}^\dagger(t, \vec{y})\} = 0.$$

Eqs. (6.3.82) and (6.3.83) for fermion (f) and antifermion ($\bar{f}$) states of mass m then follow. Summarizing these using a more compact notation, we have

$$|\mathbf{p}s\rangle \equiv |f; \mathbf{p}, s\rangle \equiv \sqrt{2E_{\mathbf{p}}}\, b_{\text{as}\mathbf{p}}^{s\dagger} |\Omega\rangle, \quad |\bar{\mathbf{p}}\bar{s}\rangle \equiv |\bar{f}; \bar{\mathbf{p}}, \bar{s}\rangle \equiv \sqrt{2E_{\bar{\mathbf{p}}}}\, d_{\text{as}\bar{\mathbf{p}}}^{\bar{s}\dagger} |\Omega\rangle,$$

$$\langle \mathbf{p}s | \mathbf{p}'s'\rangle = 2E_{\mathbf{p}} \delta^{ss'}(2\pi)^3 \delta^3(\mathbf{p} - \mathbf{p}'), \quad \langle \bar{\mathbf{p}}\bar{s} | \bar{\mathbf{p}}'\bar{s}'\rangle = 2E_{\bar{\mathbf{p}}} \delta^{\bar{s}\bar{s}'}(2\pi)^3 \delta^3(\bar{\mathbf{p}} - \bar{\mathbf{p}}'),$$

$$\langle \mathbf{p}s | \bar{\mathbf{p}}'\bar{s}' \rangle = \langle \bar{\mathbf{p}}\bar{s} | \mathbf{p}'s' \rangle = 0. \tag{7.5.61}$$

From Eq. (6.3.100) we then have

$$\langle \Omega | \hat{\psi}_{\text{as}}(x) | \mathbf{p}s \rangle = u^s(p)e^{-ip\cdot x}, \quad \langle \Omega | \hat{\bar{\psi}}_{\text{as}}(x) | \bar{\mathbf{p}}\bar{s} \rangle = \bar{v}^{\bar{s}}(\bar{p})e^{-i\bar{p}\cdot x}, \tag{7.5.62}$$

$$\langle \mathbf{p}s | \hat{\bar{\psi}}_{\text{as}}(x) | \Omega \rangle = \bar{u}^s(p)e^{ip\cdot x}, \quad \langle \bar{\mathbf{p}}\bar{s} | \hat{\psi}_{\text{as}}(x) | \Omega \rangle = v^s(p)e^{i\bar{p}\cdot x}.$$

The fermion Feynman propagator for the asymptotic fields is the free propagator with mass m,

$$S_0(m^2; x - y) = \langle \Omega | T[\hat{\psi}_{\text{as}}(x)\hat{\bar{\psi}}_{\text{as}}(y)] | \Omega \rangle = (i\slashed{\partial}_x + m)D_0(m^2; x - y) \tag{7.5.63}$$

$$= \int \frac{d^4p}{(2\pi)^4} \frac{i(\slashed{p} + m)}{p^2 - m^2 + i\epsilon} e^{-ip\cdot(x-y)} \equiv \int \frac{d^4p}{(2\pi)^4} S_0(m^2; p)e^{-ip\cdot(x-y)}.$$

The full fermion propagator for the interacting theory is

$$S_F(x - y) \equiv \langle \Omega | T[\hat{\psi}(x)\hat{\bar{\psi}}(y)] | \Omega \rangle \tag{7.5.64}$$

$$= \int \frac{d^4p}{(2\pi)^4} \left[\int_0^\infty ds \frac{i\{\rho_1(s)\slashed{p} + \rho_2(s)\}}{p^2 - s + i\epsilon} \right] e^{-ip\cdot(x-y)} \equiv \int \frac{d^4p}{(2\pi)^4} S_F(p)e^{-ip\cdot(x-y)},$$

where $\rho_1(s), \rho_2(s) \in \mathbb{R}$, $\rho_1(s) \geq 0$ and $s\rho_1(s) - \rho_2(s) \geq 0$. For a proof of this result, see Sec. 16.8 of Bjorken and Drell (1965). This simple form results from the assumptions of CPT invariance and also P invariance. The weak interactions are maximally parity violating, but it is usually sufficient to treat these using perturbation theory. So we typically do not need to write the full nonperturbative fermion propagator in its more general form, $\rho_1(s)\slashed{p} + \rho_2(s) \to \rho_1(s)\slashed{p} + \rho_2(s) + \tilde{\rho}_1(s)\slashed{p}\gamma_5 + \tilde{\rho}_2(s)\gamma^5$.

There will be a fermion version of Fig. 7.3 and Eq. (7.2.10), where the one-particle contribution to the spectral representation is by definition at $s = m^2$ with

$$[\rho_1(s)\slashed{p} + \rho_2(s)]_{1-\text{part}} = [\rho_1(m^2)\slashed{p} + \rho_2(m^2)]\,\delta(s - m^2) = Z_2\,[\slashed{p} + m]\,\delta(s - m^2),$$

$$\text{where} \quad Z_2 \equiv \rho_1(m^2) \quad \text{and} \quad m \equiv \rho_2(m^2)/Z_2. \tag{7.5.65}$$

As in Eq. (7.2.11) the full propagator is

$$S_F(p) = Z_2 \frac{i\{\slashed{p} + m\}}{p^2 - m^2 + i\epsilon} + \int_{s>m^2}^\infty ds \frac{i\{\rho_1(s)\slashed{p} + \rho_2(s)\}}{p^2 - s + i\epsilon}. \tag{7.5.66}$$

Similar to Eq. (7.5.20) we invert the relationship between the asymptotic field operators and their annihilation and creation operators using Eq. (6.3.197),

$$\sqrt{2E_{\mathbf{p}}}\hat{b}^{s\dagger}_{\text{as}\mathbf{p}} = \int d^3x \; \hat{\psi}^\dagger_{\text{as}}(x)u_{ps}(x), \quad \sqrt{2E_{\mathbf{p}}}\hat{d}^{s\dagger}_{\text{as}\mathbf{p}} = \int d^3x \; v^\dagger_{ps}(x)\hat{\psi}_{\text{as}}(x), \tag{7.5.67}$$

$$\sqrt{2E_{\mathbf{p}}}\hat{b}^s_{\text{as}\mathbf{p}} = \int d^3x \; u^\dagger_{ps}(x)\hat{\psi}_{\text{as}}(x), \quad \sqrt{2E_{\mathbf{p}}}\hat{d}^s_{\text{as}\mathbf{p}} = \int d^3x \; \hat{\psi}^\dagger_{\text{as}}(x)v_{ps}(x)$$

$$u_{ps}(x) \equiv u^s(p)e^{-ip\cdot x} \quad \text{and} \quad v_{ps}(x) \equiv v^s(p)e^{+ip\cdot x}.$$

The neutral scalar boson arguments for the scattering matrix, S_{fi}, extend to the fermion case in a straightforward way leading to fermion versions of Eqs. (7.5.9)–(7.5.17). The only differences are that initial and final states can contain both fermions and antifermions, that each fermion and antifermion state has both a three-momentum and spin label and that now we have $\hat{b}, \hat{b}^\dagger$ for fermions and $\hat{d}, \hat{d}^\dagger$ for antifermions in place of $\hat{a}, \hat{a}^\dagger$. Eqs. (7.5.17) and (7.5.18) become

$$\hat{b}^s_{\text{kin}} = \hat{S}\hat{b}^s_{\text{kout}}\hat{S}^\dagger \quad \text{and} \quad \hat{b}^{s\dagger}_{\text{kin}} = \hat{S}\hat{b}^{s\dagger}_{\text{kout}}\hat{S}^\dagger, \tag{7.5.68}$$

$$\hat{d}^s_{\text{kin}} = \hat{S}\hat{d}^s_{\text{kout}}\hat{S}^\dagger \quad \text{and} \quad \hat{d}^{s\dagger}_{\text{kin}} = \hat{S}\hat{d}^{s\dagger}_{\text{kout}}\hat{S}^\dagger,$$

$$\hat{\psi}_{\text{out}}(t + T, \mathbf{x}) \equiv \hat{S}^\dagger \hat{\psi}_{\text{as}}(t - T, \mathbf{x})\hat{S} = \hat{S}^\dagger \hat{\psi}_{\text{in}}(t - T, \mathbf{x})\hat{S},$$

$$\hat{\psi}_{\text{out}}^\dagger(t + T, \mathbf{x}) \equiv \hat{S}^\dagger \hat{\psi}_{\text{as}}^\dagger(t - T, \mathbf{x})\hat{S} = \hat{S}^\dagger \hat{\psi}_{\text{in}}^\dagger(t - T, \mathbf{x})\hat{S},$$

where we make the arbitrary choice to identify $\hat{b}_{\mathbf{k}\text{in}}^s \equiv \hat{b}_{\text{ask}}^s$ and $\hat{d}_{\mathbf{k}\text{in}}^s \equiv \hat{d}_{\text{ask}}^s$ as we did for scalars in Eq. (7.5.15). The fermion asymptotic conditions are then

$$\langle f|\hat{\psi}(x)|i\rangle \xrightarrow{x^0 = -T \to -\infty} \sqrt{Z_2}\langle f|\hat{\psi}_{\text{in}}(x)|i\rangle, \qquad (7.5.69)$$

$$\langle f|\hat{\psi}(x)|i\rangle \xrightarrow{x^0 = T \to \infty} \sqrt{Z_2}\langle f|\hat{\psi}_{\text{out}}(x)|i\rangle,$$

with the same asymptotic conditions for $\hat{\psi}^\dagger(x)$. The understanding is that $|i\rangle$ and $|f\rangle$ contain spatially well-separated wavepacket solutions to the Dirac equation. In terms of in and out operators Eq. (7.5.67) becomes

$$\sqrt{2E_{\mathbf{p}}}\hat{b}_{\mathbf{p}\text{in}}^{s\dagger} = \int d^3x\, \hat{\psi}_{\text{in}}^\dagger(x)u_{ps}(x)|_{x^0 = -T}, \quad \sqrt{2E_{\mathbf{p}}}\hat{d}_{\mathbf{p}\text{in}}^{s\dagger} = \int d^3x\, v_{ps}^\dagger(x)\hat{\psi}_{\text{in}}(x)|_{x^0 = -T},$$

$$\sqrt{2E_{\mathbf{p}}}\hat{b}_{\mathbf{p}\text{out}}^s = \int d^3x\, u_{ps}^\dagger(x)\hat{\psi}_{\text{out}}(x)|_{x^0 = T}, \quad \sqrt{2E_{\mathbf{p}}}\hat{d}_{\mathbf{p}\text{out}}^s = \int d^3x\, \hat{\psi}_{\text{out}}^\dagger(x)v_{ps}(x)|_{x^0 = T}.$$

Define fermion and antifermion wavepacket solutions of the Dirac equation as done for scalars in Eq. (7.5.10),

$$|g_{\mathbf{p}}^s\rangle \equiv \int \frac{d^3p'}{(2\pi)^3} g_{\mathbf{p}}(\mathbf{p}')|\mathbf{p}'s\rangle \quad \text{and} \quad |\bar{g}_{\bar{\mathbf{p}}}^{\bar{s}}\rangle \equiv \int \frac{d^3\bar{p}'}{(2\pi)^3} \bar{g}_{\bar{\mathbf{p}}}(\bar{\mathbf{p}}')|\bar{\mathbf{p}}'\bar{s}\rangle, \qquad (7.5.70)$$

where we are using Eq. (7.5.61). Then from Eq. (7.5.62) we find

$$\langle\Omega|\hat{\psi}_{\text{as}}(x)|g_{\mathbf{p}}^s\rangle = \int \frac{d^3p'}{(2\pi)^3} g_{\mathbf{p}}(\mathbf{p}')u_{ps}(x) \equiv U_{ps}(x), \qquad (7.5.71)$$

$$\langle\Omega|\hat{\psi}_{\text{as}}^\dagger(x)|\bar{g}_{\bar{\mathbf{p}}}^{\bar{s}}\rangle = \int \frac{d^3\bar{p}'}{(2\pi)^3} \bar{g}_{\bar{\mathbf{p}}}(\bar{\mathbf{p}}')v_{\bar{p}\bar{s}}^\dagger(x) \equiv V_{\bar{p}\bar{s}}^\dagger(x),$$

where $u_{ps}(x)$ and $v_{ps}(x)$ are the plane wave solutions in Eq. (7.5.67). The wavepackets in coordinate space are $U_{ps}(x)$ and $V_{\bar{p}\bar{s}}(x)$ and satisfy the Dirac equation. The plane wave limit is $g_{\mathbf{p}}(\mathbf{p}') \xrightarrow{\sigma \to \infty} (2\pi)^3\delta^3(\mathbf{p}' - \mathbf{p})$ and $\bar{g}_{\bar{\mathbf{p}}}(\bar{\mathbf{p}}') \xrightarrow{\sigma \to \infty} (2\pi)^3\delta^3(\bar{\mathbf{p}}' - \bar{\mathbf{p}})$, which leads to $U_{ps}(x) \xrightarrow{\sigma \to \infty} u_{ps}(x)$ and $V_{\bar{p}\bar{s}}(x) \xrightarrow{\sigma \to \infty} v_{ps}(x)$. Define the smeared in and out operators as

$$\hat{b}_{\text{in}}^{s\dagger}(g_{\mathbf{k}}) \equiv \int \frac{d^3k'}{(2\pi)^3} g_{\mathbf{k}}(\mathbf{k}')\sqrt{2E_{\mathbf{k}'}}\, \hat{b}_{\mathbf{k}'\text{in}}^{s\dagger}, \quad \hat{d}_{\text{in}}^{s\dagger}(\bar{g}_{\mathbf{k}}) \equiv \int \frac{d^3k'}{(2\pi)^3} \bar{g}_{\mathbf{k}}(\mathbf{k}')\sqrt{2E_{\mathbf{k}'}}\, \hat{d}_{\mathbf{k}'\text{in}}^{s\dagger}, \qquad (7.5.72)$$

$$\hat{b}_{\text{out}}^s(g_{\mathbf{k}}) \equiv \int \frac{d^3k'}{(2\pi)^3} g_{\mathbf{k}}^*(\mathbf{k}')\sqrt{2E_{\mathbf{k}'}}\, \hat{b}_{\mathbf{k}'\text{out}}^{s\dagger}, \quad \hat{d}_{\text{out}}^s(\bar{g}_{\mathbf{k}}) \equiv \int \frac{d^3k'}{(2\pi)^3} \bar{g}_{\mathbf{k}}^*(\mathbf{k}')\sqrt{2E_{\mathbf{k}'}}\, \hat{d}_{\mathbf{k}'\text{out}}^{s\dagger}.$$

These operators create the in and out wavepacket states

$$|g_{\mathbf{k}}^s \text{ in}\rangle = \hat{b}_{\text{in}}^{s\dagger}(g_{\mathbf{k}})|\Omega\rangle, \quad |\bar{g}_{\mathbf{k}}^{\bar{s}} \text{ in}\rangle = \hat{d}_{\text{in}}^{\bar{s}\dagger}(g_{\bar{\mathbf{k}}})|\Omega\rangle, \qquad (7.5.73)$$

$$\langle g_{\mathbf{k}}^s \text{ out}| = \langle\Omega|\hat{b}_{\text{in}}^s(g_{\mathbf{k}}), \quad \langle\bar{g}_{\mathbf{k}}^{\bar{s}} \text{ out}| = \langle\Omega|\hat{b}_{\text{in}}^{\bar{s}}(\bar{g}_{\bar{\mathbf{k}}}).$$

In terms of these new operators we now have in place of Eq. (7.5.67),

$$\hat{b}_{\text{in}}^{s\dagger}(g_{\mathbf{p}}) = \int d^3x\, \hat{\psi}_{\text{in}}^\dagger(x)U_{ps}(x)|_{x^0 = -T}, \quad \hat{d}_{\text{in}}^{s\dagger}(\bar{g}_{\mathbf{p}}) = \int d^3x\, V_{ps}^\dagger(x)\hat{\psi}_{\text{in}}(x)|_{x^0 = -T}, \qquad (7.5.74)$$

$$\hat{b}_{\text{out}}^s(g_{\mathbf{p}}) = \int d^3x\, U_{ps}^\dagger(x)\hat{\psi}_{\text{out}}(x)|_{x^0 = T}, \quad \hat{d}_{\text{out}}^s(\bar{g}_{\mathbf{p}}) = \int d^3x\, \hat{\psi}_{\text{out}}^\dagger(x)V_{ps}(x)|_{x^0 = T},$$

where recall that factors of $\sqrt{2E_{\mathbf{p}}}$ were absorbed into these operators in Eq. (7.5.72).

Let $|i\rangle$ be an initial state consisting of m fermions and $\bar{m}$ antifermions and $|f\rangle$ be a final state of n fermions and $\bar{n}$ antifermions all in wavepacket states,

$$|i\rangle = |g^{s_1}_{\mathbf{k}_1} \cdots g^{s_m}_{\mathbf{k}_m} ; \bar{g}^{\bar{s}_1}_{\bar{\mathbf{k}}_1} \cdots \bar{g}^{\bar{s}_{\bar{m}}}_{\bar{\mathbf{k}}_{\bar{m}}}\rangle \xrightarrow{\sigma \to \infty} |\mathbf{k}_1 s_1 \cdots \mathbf{k}_m s_m; \bar{\mathbf{k}}_1 \bar{s}_1 \cdots \bar{\mathbf{k}}_{\bar{m}} \bar{s}_{\bar{m}}\rangle, \qquad (7.5.75)$$

$$\langle f| = \langle \bar{g}^{\bar{r}_{\bar{n}}}_{\bar{\mathbf{p}}_{\bar{n}}} \cdots \bar{g}^{\bar{r}_1}_{\bar{\mathbf{p}}_1} ; g^{r_n}_{\mathbf{p}_n} \cdots g^{r_1}_{\mathbf{p}_1}| \xrightarrow{\sigma \to \infty} \langle \bar{\mathbf{p}}_{\bar{n}} \bar{r}_{\bar{n}} \cdots \bar{\mathbf{p}}_1 \bar{r}_1; \mathbf{p}_n r_n \cdots \mathbf{p}_1 r_1|,$$

$$|i \, \text{in}\rangle \equiv |i\rangle, \qquad \langle f \, \text{out}| \equiv \lim_{T \to \infty} \langle f|\hat{U}(T, -T), \qquad (7.5.76)$$

where $|i \, \text{in}\rangle$ and $\langle f \, \text{out}|$ contain spatially well-separated wavepackets so that the asymptotic condition in Eq. (7.5.69) is valid. The fermion version of Eq. (7.5.28) is

$$S_{fi} = \langle f|\hat{S}|i\rangle = \lim_{\sigma \to \infty} \lim_{T \to \infty (1 - i\epsilon)} \langle f|\hat{U}(T, -T)|i\rangle$$

$$= \left(\frac{-i}{\sqrt{Z_2}} \right)^{m+n} \left(\frac{i}{\sqrt{Z_2}} \right)^{\bar{m}+\bar{n}} \int d^4 x_1 \cdots d^4 x_m \, d^4 \bar{x}_1 \cdots d^4 \bar{x}_{\bar{m}} \, d^4 y_1 \cdots d^4 y_n \, d^4 \bar{y}_1 \cdots d^4 \bar{y}_{\bar{n}}$$

$$\times \bar{v}_{\bar{k}_{\bar{m}} \bar{s}_{\bar{m}}}(\bar{x}_{\bar{m}})(i\overrightarrow{\not{\partial}}_{\bar{x}_{\bar{m}}} - m + i\epsilon) \cdots \bar{v}_{\bar{k}_1 \bar{s}_1}(\bar{x}_1)(i\overrightarrow{\not{\partial}}_{\bar{x}_1} - m + i\epsilon)$$

$$\times \bar{u}_{p_n r_n}(y_n)(i\overrightarrow{\not{\partial}}_{y_n} - m + i\epsilon) \cdots \bar{u}_{p_1 r_1}(y_1)(i\overrightarrow{\not{\partial}}_{y_1} - m + i\epsilon)$$

$$\times \langle \Omega|T\hat{\bar{\psi}}(\bar{y}_{\bar{n}}) \cdots \hat{\bar{\psi}}(\bar{y}_1)\hat{\psi}(y_n) \cdots \hat{\psi}(y_1)\hat{\bar{\psi}}(x_1) \cdots \hat{\bar{\psi}}(x_m)\hat{\psi}(\bar{x}_1) \cdots \hat{\psi}(\bar{x}_{\bar{m}})|\Omega\rangle$$

$$\times (-i\overleftarrow{\not{\partial}}_{x_1} - m + i\epsilon)u_{k_1 s_1}(x_1) \cdots (-i\overleftarrow{\not{\partial}}_{x_m} - m + i\epsilon)u_{k_m s_m}(x_m)$$

$$\times (-i\overleftarrow{\not{\partial}}_{\bar{y}_1} - m + i\epsilon)v_{\bar{p}_1 \bar{r}_1}(\bar{y}_1) \cdots (-i\overleftarrow{\not{\partial}}_{\bar{y}_{\bar{n}}} - m + i\epsilon)v_{\bar{p}_{\bar{n}} \bar{r}_{\bar{n}}}(\bar{y}_{\bar{n}}), \qquad (7.5.77)$$

where we have neglected forward-scattering contributions and where spinor summations linking quantities with the same spacetime point are to be understood.

Proof: The proof of Eq. (7.5.77) is similar to that for the boson case. Considering a fermion in the initial state we have in place of Eq. (7.5.23),

$$S_{fi} = \langle f|\hat{S}|i\rangle = \langle f \, \text{out}|i \, \text{in}\rangle = \langle f \, \text{out}|\hat{b}^{s_1 \dagger}_{\text{in}}(g_{\mathbf{k}_1})|(i - \mathbf{k}_1 s_1) \, \text{in}\rangle \qquad (7.5.78)$$

$$= \left\{ \langle f \, \text{out}|\hat{b}^{s_1 \dagger}_{\text{out}}(g_{\mathbf{k}_1})|(i - \mathbf{k}_1) \, \text{in}\rangle + \langle f \, \text{out}|[\hat{b}^{s_1 \dagger}_{\text{in}}(g_{\mathbf{k}_1}) - \hat{b}^{s_1 \dagger}_{\text{out}}(g_{\mathbf{k}_1})]|(i - \mathbf{k}_1 s_1) \, \text{in}\rangle \right\}.$$

In the plane wave limit the first term vanishes unless $\mathbf{k}_1$ is equal to one of the $\mathbf{p}_j$, which is a forward-scattering contribution and so is of no interest.

Using Eqs. (7.5.67) and (7.5.69) the second term becomes

$$\langle f \, \text{out}|[\hat{b}^{s_1 \dagger}_{\text{in}}(g_{\mathbf{k}_1}) - \hat{b}^{s_1 \dagger}_{\text{out}}(g_{\mathbf{k}_1})]|(i - \mathbf{k}_1 s_1) \, \text{in}\rangle \qquad (7.5.79)$$

$$= \int d^3 x \, \langle f \, \text{out}|[\hat{\psi}^\dagger_{\text{in}}(x)U_{k_1 s_1}(x)|_{x^0 = -T} - \hat{\psi}^\dagger_{\text{out}}(x)U_{k_1 s_1}(x)|_{x^0 = T}]|(i - \mathbf{k}_1 s_1) \, \text{in}\rangle$$

$$= (1/\sqrt{Z_2}) \int d^3 x \, \langle f \, \text{out}|[\hat{\psi}^\dagger(x)U_{k_1 s_1}(x)|_{x^0 = -T}$$
$$- \hat{\psi}^\dagger(x)U_{k_1 s_1}(x)|_{x^0 = T}]|(i - \mathbf{k}_1 s_1) \, \text{in}\rangle$$

$$= -(1/\sqrt{Z_2}) \int_{-T}^T d^4 x \, \partial_0^x \langle f \, \text{out}|\hat{\psi}^\dagger(x)U_{k_1 s_1}(x)|(i - \mathbf{k}_1 s_1) \, \text{in}\rangle$$

$$= -(i/\sqrt{Z_2}) \int_{-T}^T d^4 x \, \langle f \, \text{out}|\hat{\bar{\psi}}(x)|(i - \mathbf{k}_1 s_1) \, \text{in}\rangle(-i\overleftarrow{\not{\partial}}^x - m)U_{k_1 s_1}(x). \qquad (7.5.80)$$

The last step needs explanation. Since the wavepacket is a solution of the Dirac equation, then $(i\not{\partial} - m)U_{ks}(x) = 0$ and after multiplying by $-i\gamma^0$ we find $\partial_0 U_{ks}(x) = i\gamma^0(i\gamma^k \partial_k - m)U_{ks}(x)$. This leads to

$$\int d^4x\, \partial_0^x[\hat{\psi}^\dagger(x)U_{k_1s_1}(x)] = \int d^4x[\{\partial_0^x\hat{\psi}^\dagger(x)\}U_{k_1s_1}(x) + \hat{\psi}^\dagger(x)\{\partial_0^x U_{k_1s_1}(x)\}]$$

$$= \int d^4x[\{\partial_0^x\hat{\psi}^\dagger(x)\}U_{k_1s_1}(x) + \hat{\psi}^\dagger(x)\{i\gamma^0(i\gamma^k\partial_k^x - m)U_{k_1s_1}(x)\}]$$

$$= i\int d^4x\, \hat{\bar{\psi}}(x)(-i\overleftarrow{\partial}^x - m)U_{k_1s_1}(x), \qquad (7.5.81)$$

where we used vanishing/periodic spatial boundary conditions to do an integration by parts to have ∂_k^x act on $\hat{\bar{\psi}}(x)$. The above steps can be repeated for the other three cases to give the four results

$$S_{fi} = \frac{-i}{\sqrt{Z_2}}\int_{-T}^{T} d^4x_1\, \langle f\,\mathrm{out}|\hat{\bar{\psi}}(x_1)|(i-\mathbf{k}_1s_1)\,\mathrm{in}\rangle(-i\overleftarrow{\partial}^{x_1} - m)U_{k_1s_1}(x_1) \qquad (7.5.82)$$

$$= \frac{-i}{\sqrt{Z_2}}\int_{-T}^{T} d^4y_1\, \bar{U}_{p_1r_1}(y_1)(i\overrightarrow{\partial}^{y_1} - m)\langle(f-\mathbf{p}_1r_1)\,\mathrm{out}|\hat{\psi}(x)|i\,\mathrm{in}\rangle$$

$$= \frac{i}{\sqrt{Z_2}}\int_{-T}^{T} d^4\bar{x}_1\, \bar{V}_{\bar{k}_1\bar{s}_1}(\bar{x}_1)(i\overrightarrow{\partial}^{\bar{x}_1} - m)\langle f\,\mathrm{out}|\hat{\psi}(\bar{x}_1)|(i-\bar{\mathbf{k}}_1\bar{s}_1)\,\mathrm{in}\rangle$$

$$= \frac{i}{\sqrt{Z_2}}\int_{-T}^{T} d^4\bar{y}_1\, \langle(f-\bar{\mathbf{p}}_1\bar{r}_1)\,\mathrm{out}|\hat{\bar{\psi}}(\bar{y}_1)|i\,\mathrm{in}\rangle(-i\overleftarrow{\partial}^{y_1} - m)V_{\bar{p}_1\bar{r}_1}(\bar{y}_1).$$

Following the same steps used in Eqs. (7.5.26) and 7.5.27 we can combine any two of the above results to arrive at a time-ordered pair of operators. Iterating this process leads to Eq. (7.5.77) after we remove all particles and antiparticles from the initial and final states. We remove the particles and antiparticles from $|i\rangle$ and $|f\rangle$ in the same order in which they appear in Eq. (7.5.75) in order to arrive at Eq. (7.5.77). Doing so in any other order will only change S_{fi} by a physically irrelevant overall sign that has no effect on $|\mathcal{M}_{fi}|^2$. Recall that we should really always be taking the limit $T \to \infty(1 - i\epsilon)$ leading to the Feynman boundary conditions and the $i\epsilon$ terms shown in Eq. (7.5.77). After this we can safely take the plane wave limit $\sigma \to \infty$ so that $U \to u$ and $V \to v$.

Analogous to the scalar case, define $G(x_1, \ldots, x_n; y_1, \ldots, y_{n'})$ as the fermion Green's function,

$$G(x_1, \ldots, x_n; y_1, \ldots, y_{n'}) \equiv \langle\Omega|T\hat{\psi}(y_1)\cdots\hat{\psi}(y_{n'})\hat{\bar{\psi}}(x_1)\cdots\hat{\bar{\psi}}(x_n)|\Omega\rangle. \qquad (7.5.83)$$

Analogous to Eq. (7.5.34) we define the momentum space Green's function,

$$G(p_1, \ldots, p_n; q_1, \ldots, q_{n'}) \equiv \int \prod_{i=1}^{n} d^4x_i e^{ip_i\cdot x_i} \prod_{j=1}^{n'} d^4y_j e^{iq_j\cdot y_j} \qquad (7.5.84)$$

$$\times\, G(x_1, \ldots, x_n; y_1, \ldots, y_{n'}),$$

$$G(x_1, \ldots, y_{n'}) = \int \prod_{i=1}^{n} \frac{d^4p_i}{(2\pi)^4} e^{-ip_i\cdot x_i} \prod_{j=1}^{n'} \frac{d^4q_j}{(2\pi)^4} e^{-iq_j\cdot y_j} G(p_1, \ldots, q_{n'}).$$

Recall that with fermion time-ordering the interchange of any two fermion operators gives a sign change. However, the sign of the invariant amplitude is irrelevant. We will use the symbol $\pm$ to mean "up to a sign" and freely change the order of time-ordered fermion operators. So we write the Green's function in Eq. (7.5.77) as

$$G(x_1, \ldots, x_m, \bar{y}_1, \ldots, \bar{y}_{\bar{n}}; y_1, \ldots, y_n, \bar{x}_1, \ldots, \bar{x}_{\bar{m}}) = \pm\langle\Omega|T\hat{\bar{\psi}}(\bar{y}_{\bar{n}})\cdots\hat{\psi}(\bar{x}_{\bar{m}})|\Omega\rangle.$$

with the Fourier transform

$$G(k_1, \ldots, k_m, \bar{p}_1, \ldots, \bar{p}_{\bar{n}}; p_1, \ldots, p_n, \bar{k}_1, \ldots, \bar{k}_{\bar{m}}), \qquad (7.5.85)$$

where all momenta are outgoing. Comparing the sign of the exponential in the Fourier transform in Eq. (7.5.84) with the signs of the exponentials in Eq. (7.5.77) we obtain the momentum space expression for the scattering matrix,

$$
\begin{aligned}
S_{fi} = \pm \left(\frac{-i}{\sqrt{Z_2}}\right)^{m+n} \left(\frac{i}{\sqrt{Z_2}}\right)^{\bar{m}+\bar{n}} & \left[\bar{v}^{\bar{s}_{\bar{m}}}(\bar{k}_{\bar{m}})(-\bar{\not{k}}_{\bar{m}} - m + i\epsilon) \cdots \bar{v}^{\bar{s}_1}(\bar{k}_1)(-\bar{\not{k}}_1 - m + i\epsilon) \right. \\
& \times \bar{u}^{r_n}(p_n)(\not{p}_n - m + i\epsilon) \cdots \bar{u}^{r_1}(p_1)(\not{p}_1 - m + i\epsilon) \\
& \times G(-k_1, \ldots, -k_m, \bar{p}_{\bar{n}}, \ldots, \bar{p}_{\bar{n}}; p_1, \ldots, p_n, -\bar{k}_1, \ldots, -\bar{k}_{\bar{m}}) \\
& \times (\not{k}_1 - m + i\epsilon)u_{k_1 s_1}(x_1) \cdots (\not{k}_m - m + i\epsilon)u^{s_m}(k_m) \\
& \left. \times (-\bar{\not{p}}_1 - m + i\epsilon)v^{\bar{r}_1}(\bar{p}_1) \cdots (-\bar{\not{p}}_{\bar{n}} - m + i\epsilon)v^{\bar{r}_{\bar{n}}}(\bar{p}_{\bar{n}})\right]_{\text{on-shell}},
\end{aligned}
$$

$$(7.5.86)$$
$$(7.5.87)$$

where forward-scattering contributions are neglected and where "on-shell" means that we take $p^2 \to m^2 - i\epsilon$ for all external momenta. How do we understand the momentum signs here? For the fermion we see that $u^{s_j}(k_j)$ is the source of incoming momentum k_j^μ and $\bar{u}^{r_j}(p_j)$ is the sink of outgoing momentum p_j^μ. For the antifermion, which propagates backward in time, we can view $v^{\bar{r}_j}(\bar{p}_j)$ as the source of incoming momentum $-\bar{p}_j$ and $\bar{v}^{\bar{s}_j}(\bar{k}_j)$ as the sink of the outgoing momentum $-\bar{k}_j$.

Using Eqs. (7.5.66) and (6.3.230) we find that

$$
\not{p} - m + i\epsilon = \left(\frac{\not{p} + m}{p^2 - m^2 + i\epsilon}\right)^{-1} = iZ_2 \lim_{p^2 \to m^2 - i\epsilon} S_F(p)^{-1}. \tag{7.5.88}
$$

Then similar to Eq. (7.5.41) for the scalar case we have

$$
\begin{aligned}
S_{fi} = \pm(-1)^{\bar{m}+\bar{n}}\sqrt{Z_2}^{m+\bar{m}+n+\bar{n}} & \bar{v}^{\bar{s}_{\bar{m}}}(\bar{k}_{\bar{m}}) \cdots \bar{v}^{\bar{s}_1}(\bar{k}_1)\bar{u}^{r_n}(p_n) \cdots \bar{u}^{r_1}(p_1) \\
& \times \left[S_F(-\bar{\not{k}}_{\bar{m}})^{-1} \cdots S_F(-\bar{\not{k}}_1)^{-1}S_F(\not{p}_n)^{-1} \cdots S_F(\not{p}_1)^{-1}\right. \\
& \times G(-k_1, \ldots, -k_m, \bar{p}_{\bar{n}}, \ldots, \bar{p}_{\bar{n}}; p_1, \ldots, p_n, -\bar{k}_1, \ldots, -\bar{k}_{\bar{m}}) \\
& \left. \times S_F(\not{k}_1)^{-1} \cdots S_F(\not{k}_m)^{-1}S_F(-\bar{\not{p}}_1)^{-1} \cdots S_F(-\bar{\not{p}}_{\bar{n}})^{-1}\right]_{\text{on-shell}} \\
& \times u_{k_1 s_1}(x_1) \cdots u^{s_m}(k_m)v^{\bar{r}_1}(\bar{p}_1) \cdots v^{\bar{r}_{\bar{n}}}(\bar{p}_{\bar{n}}) \\
= (-1)^{\bar{m}+\bar{n}}\sqrt{Z_2}^{m+\bar{m}+n+\bar{n}} & \bar{v}^{\bar{s}_{\bar{m}}}(\bar{k}_{\bar{m}}) \cdots \bar{v}^{\bar{s}_1}(\bar{k}_1)\bar{u}^{r_n}(p_n) \cdots \bar{u}^{r_1}(p_1) \\
& \times G_{\text{amp}}(\bar{p}_{\bar{n}}, \ldots, -\bar{k}_{\bar{m}})\big|_{\text{on-shell}} \, u^{s_1}(k_1) \cdots u^{s_m}(k_m)v^{\bar{r}_1}(\bar{p}_1) \cdots v^{\bar{r}_{\bar{n}}}(\bar{p}_{\bar{n}}),
\end{aligned}
$$

$$(7.5.89)$$

where $G_{\text{amp}}(-k_1, \ldots, -k_m, \bar{p}_{\bar{n}}, \ldots, \bar{p}_{\bar{n}}; p_1, \ldots, p_n, -\bar{k}_1, \ldots, -\bar{k}_{\bar{m}})\big|_{\text{on-shell}}$ is the on-shell amputated Green's function with all momenta outgoing. The incoming momenta are the k_j^μ and $\bar{k}_j^\mu$ and the outgoing momenta are the p_j^μ and the $\bar{p}_j^\mu$.

We can again regard the field operator as the bare field operator and define the renormalized field operator $\hat{\psi}_r$ as $\hat{\psi}(x) \equiv \hat{\psi}_0(x) \equiv \sqrt{Z_2}\hat{\psi}_r(x)$. So from Eq. (7.5.66) the renormalized propagator has its perturbative form at the mass pole,

$$
S_{r,F}(x-y) \equiv \langle\Omega|T\hat{\psi}_r(x)\hat{\bar{\psi}}_r(y)|\Omega\rangle, \quad S_{r,F}(p) = \frac{i\{\not{p}+m\}}{p^2 - m^2 + i\epsilon} + \int_{s>m^2}^\infty ds \, \frac{i\{\rho_1(s)\not{p} + \rho_2(s)\}}{Z_2(p^2 - s + i\epsilon)},
$$

since $S_{r,F}(x-y) = (1/Z_2)S_F(x-y)$. For a renormalized Green's function,

$$
G_r(x_1, \ldots, x_n; y_1, \ldots, y_{n'}) = (1/\sqrt{Z_2})^{n+\bar{n}}G(x_1, \ldots, x_n; y_1, \ldots, y_{n'}). \tag{7.5.90}
$$

The renormalized amputated Green's function is obtained by acting on a renormalized Green's function G_r with inverse renormalized propagators $S_{r,F}$,

$$
\sqrt{Z_2}^{n+n'}G_{\text{amp}}(p_1, \ldots, p_n; q_1, \ldots, q_{n'}) = G_{r,\text{amp}}(p_1, \ldots, p_n; q_1, \ldots, q_{n'}). \tag{7.5.91}
$$

Then we have up to an irrelevant sign the *LSZ reduction formula for fermions*

$$S_{fi} = \pm\, \bar{v}^{\bar{s}_{\bar{m}}}(\bar{k}_{\bar{m}}) \cdots \bar{v}^{\bar{s}_1}(\bar{k}_1) \bar{u}^{r_n}(p_n) \cdots \bar{u}^{r_1}(p_1)\, G_{r,\mathrm{amp}}(\bar{p}_{\bar{n}}, \ldots, -\bar{k}_{\bar{m}})\big|_{\text{on-shell}}$$
$$\times\, u^{s_1}(k_1) \cdots u^{s_m}(k_m) v^{\bar{r}_1}(\bar{p}_1) \cdots v^{\bar{r}_{\bar{n}}}(\bar{p}_{\bar{n}}) + \begin{bmatrix} \text{forward scattering} \\ \text{contributions} \end{bmatrix}. \tag{7.5.92}$$

The antisymmetrization for the initial and final states automatically follows from the antisymmetry of operator exchange in the fermion time-ordered product and does not need to be separately imposed.

Again use Eq. (7.5.44) to relate S_{fi} and $\mathcal{M}_{fi}$ and ignore forward-scattering contributions and an irrelevant overall sign. Then similar to the scalar case we have for the invariant amplitude $\mathcal{M}_{fi}$ for a scattering process,

$$i(2\pi)^4 \delta^4\big(\textstyle\sum_i p_i - \sum_j k_j\big) \mathcal{M}_{fi} = \bar{v}^{\bar{s}_{\bar{m}}}(\bar{k}_{\bar{m}}) \cdots \bar{v}^{\bar{s}_1}(\bar{k}_1) \bar{u}^{r_n}(p_n) \cdots \bar{u}^{r_1}(p_1) \tag{7.5.93}$$
$$\times\, G_{r,\mathrm{amp}}(\bar{p}_{\bar{n}}, \ldots, \bar{p}_1, p_n, \ldots, p_1; -k_1, \ldots, -k_m, -\bar{k}_1, \ldots, -\bar{k}_{\bar{m}})\big|_{\text{on-shell}}$$
$$\times\, u^{s_1}(k_1) \cdots u^{s_m}(k_m) v^{\bar{r}_1}(\bar{p}_1) \cdots v^{\bar{r}_{\bar{n}}}(\bar{p}_{\bar{n}}).$$

So we obtain $\mathcal{M}_{fi}$ by evaluating the renormalized amputated Green's function and contracting all spinor indices with the appropriate u and v spinors. We can then use this to evaluate $2 \to n$ scattering cross-sections and $1 \to n$ particle decay rates.

The bare fermion propagator with the spinor indices showing is

$$S_0(p)_{\beta\alpha} = \underset{\alpha \quad\;\; \beta}{\xrightarrow{\hspace{0.6cm}p\hspace{0.6cm}}} = \frac{(i\!\not{p} + m_0)_{\beta\alpha}}{p^2 - m_0^2 + i\epsilon} = \left(\frac{i}{\not{p} - m_0 + i\epsilon}\right)_{\beta\alpha}. \tag{7.5.94}$$

Note that if the momentum arrow has the opposite direction to the arrow on the fermion line, then this corresponds to $S_0(-p)$. Comparing with Eq. (7.5.89) we observe that *for an incoming or outgoing fermion with momentum p* the fermion line arrow and the momentum arrow are aligned and pointing forward in time and we use $S_F(p)$; *for an incoming or outgoing antifermion with momentum p* the momentum arrow points forward in time and the fermion line arrow points backward in time and so we use $S_F(-p)$. From Eq. (6.3.226) we know that in $S_F(y-x)_{\beta\alpha}$ a fermion flows as $(x, \alpha) \to (y, \beta)$, which is the direction of the arrow on the fermion propagator. An antifermion flows $(y, \beta) \to (x, \alpha)$ opposite to the fermion arrow. Since $e^{-ip\cdot(y-x)}$ is a positive energy plane wave moving in the $\mathbf{x} \to \mathbf{y}$ direction with momentum $\mathbf{p}$, then from Eq. (6.3.226) we know that $S_F(p)_{\beta\alpha}$ is a fermion propagator with the arrow pointing $\alpha \to \beta$.

As for the scalar case in Eq. (7.5.49) the amputation process can be applied to the fermion propagator itself to separate the 1PI part. We again use a dot on a bare propagator line to indicate its presence when there is a possible ambiguity. Using this notation and working in momentum space it then follows that

$$S_F(p) = \text{\raisebox{-0.3em}{\includegraphics{diagram}}} = S_0(p) + S_0(p)[-i\Sigma(p)]S_F(p)$$
$$= \sum (\text{connected two-point diagrams}), \tag{7.5.95}$$

$$-i\Sigma(p) \equiv \text{\raisebox{-0.3em}{\includegraphics{diagram}}} = \sum (\text{1PI two-point diagrams}),$$

where $\Sigma(p)$ is called the *fermion self-energy* for reasons that become clear below.

Acting on Eq. (7.5.95) with $S_0^{-1}(p)$ from the left and $S_F^{-1}(p)$ from the right gives

$$S_0^{-1}(p) = S_F^{-1}(p) - i\Sigma(p) \quad \Rightarrow \quad S_F^{-1}(p) = S_0^{-1}(p) + i\Sigma(p). \tag{7.5.96}$$

Combining this with Eq. (7.2.9) it follows that

$$S_F(p) = \text{---}\!\!\bigcirc\!\!\text{---} = \int_0^\infty ds\, \frac{i\{\rho_1(s)\slashed{p} + \rho_2(s)\}}{p^2 - s + i\epsilon} = \frac{i}{\slashed{p} - m_0 - \Sigma(p) + i\epsilon}. \tag{7.5.97}$$

As for the scalar case, the fermion self-energy, $\Sigma(p)$, is the contribution from the self-interaction of a particle that changes its propagator from its bare form to its physical form. Recursively expanding Eq. (7.5.49) we find in momentum space

$$S_F = S_0 + S_0[-i\Sigma]S_0 + S_0[-i\Sigma]S_0[-i\Sigma]S_0 + \cdots. \tag{7.5.98}$$

7.5.3 LSZ for Photons

We begin with the Coulomb gauge discussion in Sec. 6.4, where we only need consider the spatial components of the photon operator, $\hat{\mathbf{A}}$. Compared with the scalar boson case the primary differences are that we have three field operators and polarization vectors are present in the plane wave expansion. From Eq. (6.4.33) we have for the asymptotic photon operators

$$\hat{\mathbf{A}}_{\text{as}}(x) = \int \frac{d^3k}{(2\pi)^3} \frac{1}{\sqrt{2E_{\mathbf{k}}}} \sum_{\lambda=1}^2 \left[\boldsymbol{\epsilon}(k,\lambda)\hat{a}_{\mathbf{k}\text{as}}^\lambda e^{-ik\cdot x} + \boldsymbol{\epsilon}(k,\lambda)^* \hat{a}_{\mathbf{k}\text{as}}^{\lambda\dagger} e^{ik\cdot x} \right], \tag{7.5.99}$$

where $E_{\mathbf{k}} = \omega_{\mathbf{k}} = \sqrt{\mathbf{k}^2}$. Recall that if we take λ to be the linear polarization label, $\lambda = 1, 2$, then $\boldsymbol{\epsilon}(k, \lambda)$ is real, whereas if we take λ to be the helicity label, $\lambda = \pm 1$, then $\boldsymbol{\epsilon}(k, \lambda)$ is complex. We have in either case from Eq. (6.4.32) that

$$\boldsymbol{\epsilon}(k,\lambda)^* \cdot \boldsymbol{\epsilon}(k,\lambda') = \delta_{\lambda\lambda'}, \qquad \epsilon^\mu(k,\lambda)^* \epsilon_\mu(k,\lambda') = -\delta_{\lambda\lambda'}, \tag{7.5.100}$$

where the first equation is true in the quantization inertial frame and the second is true in any inertial frame. The fact that the photon mass is zero introduces a subtlety since there is no gap between the single physical photon mass and the mass of a physical multiphoton state. This infrared problem is handled carefully in practical calculations, but for now we imagine assigning multiphoton states an arbitrarily small threshold mass that is subsequently taken to zero, $m_{\text{th}} \to 0^+$. We work directly in the naive plane wave limit here, since as seen for scalar bosons and fermions this leads to the same result as a careful wavepacket treatment.

Recall that Eq. (6.4.58) is the free photon propagator defined by the time-ordered product in the quantization frame in Coulomb gauge. The full photon propagator in momentum space has the spectral representation for the spatial components,

$$D_V^{ij}(k) = D_F^{ij}(k) = \left(Z_3 \frac{i}{k^2 + i\epsilon} + \int_{m_{\text{th}}^2 \to 0^+}^\infty ds\, \frac{i\rho(s)}{k^2 - s + i\epsilon} \right) \left[\delta^{ij} - \frac{k^i k^j}{\mathbf{k}^2} \right], \tag{7.5.101}$$

where $0 \le Z_3, \rho(s)$. A proof of this is given in Bjorken and Drell (1965). In place of Eq. (7.5.8) we have the asymptotic condition

$$\langle\beta|\hat{\mathbf{A}}(x)|\alpha\rangle = \sqrt{Z_3}\langle\beta|\hat{\mathbf{A}}_{\text{as}}(x)|\alpha\rangle, \tag{7.5.102}$$

where $|\alpha\rangle$ and $|\beta\rangle$ are arbitrary asymptotic states.

In place of Eqs. (7.5.4)–(7.5.6) and Eqs. (7.5.61)–(7.5.62) we have

$$|\mathbf{k}\lambda\rangle \equiv |\mathbf{k}, \lambda\rangle \equiv \sqrt{2E_{\mathbf{k}}}\, \hat{a}_{\mathbf{k}\text{as}}^{\lambda\dagger}|\Omega\rangle, \quad \langle\mathbf{k}\lambda|\mathbf{k}'\lambda'\rangle = 2E_{\mathbf{k}}\delta^{\lambda\lambda'}(2\pi)^3\delta^3(\mathbf{k} - \mathbf{k}'),$$

$$\langle\Omega|\hat{\mathbf{A}}(x)|\mathbf{k}\lambda\rangle = \sqrt{Z_3}\langle\Omega|\hat{\mathbf{A}}_{\text{as}}(x)|\mathbf{k}\lambda\rangle = \sqrt{Z_3}\,\boldsymbol{\epsilon}(k,\lambda)e^{-ik\cdot x}, \tag{7.5.103}$$

where λ can be either a linear polarization or a helicity index. Eq. (6.4.21) becomes

$$\sqrt{2\omega_{\mathbf{k}}}\hat{a}_{\mathbf{k}}^{\lambda} = 2i \int d^3x \, e^{ik\cdot x} \overset{\leftrightarrow}{\partial}_0 \boldsymbol{\epsilon}(k,\lambda)^* \cdot \hat{\mathbf{A}}_{\mathrm{as}}(x),$$

$$\sqrt{2\omega_{\mathbf{k}}}\hat{a}_{\mathbf{k}}^{\lambda\dagger} = -2i \int d^3x \, e^{-ik\cdot x} \overset{\leftrightarrow}{\partial}_0 \boldsymbol{\epsilon}(k,\lambda) \cdot \hat{\mathbf{A}}_{\mathrm{as}}(x). \tag{7.5.104}$$

It is now clear how to generalize the LSZ derivation for the scalar boson to the photon in Coulomb gauge. Let the initial and final states be

$$|i\rangle = |\mathbf{k}_1\lambda_1 \cdots \mathbf{k}_m\lambda_m\rangle, \quad |f\rangle = |\mathbf{p}_1\kappa_1 \cdots \mathbf{p}_n\kappa_n\rangle. \tag{7.5.105}$$

In place of Eq. (7.5.25) it then follows that

$$S_{fi} = \langle f|\hat{S}|i\rangle = \lim_{T\to\infty} \langle \mathbf{p}_1\kappa_1 \cdots \mathbf{p}_n\kappa_n|\hat{U}(T,-T)|\mathbf{k}_1\lambda_1 \cdots \mathbf{k}_m\lambda_m\rangle \tag{7.5.106}$$

$$= (i/\sqrt{Z_3}) \lim_{T\to\infty} \int_{-T}^{T} d^4x \, e^{-ik_1\cdot x} \overset{\rightarrow}{\Box}_x \boldsymbol{\epsilon}(k_1,\lambda_1) \cdot \langle f \, \mathrm{out}|\hat{\mathbf{A}}(x)|(i-\mathbf{k}_1)\,\mathrm{in}\rangle,$$

since the plane waves satisfy the massless Klein-Gordon equation, $\Box \, e^{\pm ik\cdot x} = 0$. As usual, forward-scattering contributions are neglected. Note that removing a photon from the in state involved $\hat{a}_{\mathbf{k}}^{\lambda\dagger}$ and the use of Eq. (7.5.104) and so introduced an $\boldsymbol{\epsilon}$. Removing a photon from the out state will involve an $\hat{a}_{\mathbf{k}}^{\lambda}$ and will introduce an $\boldsymbol{\epsilon}^*$. Time-ordering will again appear naturally. So the Coulomb gauge photon result in the quantization inertial frame is just the scalar result with minor changes: $\hat{\phi}(x) \to \hat{\mathbf{A}}(x)$; $|\mathbf{p}\rangle \to |\mathbf{p}\lambda\rangle$; $Z \to Z_3$; $\Box + m^2 \to \Box$; for every initial state photon an $\boldsymbol{\epsilon}(\mathbf{k}_i,\lambda_i)$ is contracted in a scalar product with its corresponding $\hat{\mathbf{A}}$; and for every final state photon an $\boldsymbol{\epsilon}(\mathbf{k}_i,\lambda_i)^*$ is contracted with its corresponding $\hat{\mathbf{A}}$. Then in place of the scalar boson result in Eq. (7.5.28) we have the result for photons,

$$S_{fi} = \langle f|\hat{S}|i\rangle = \lim_{T\to\infty(1-i\epsilon)} \langle \mathbf{p}_1\kappa_1 \cdots \mathbf{p}_n\kappa_n|\hat{U}(T,-T)|\mathbf{k}_1\lambda_1 \cdots \mathbf{k}_m\lambda_m\rangle = \langle f \, \mathrm{out}|i \, \mathrm{in}\rangle$$

$$= \left(i/\sqrt{Z_3}\right)^{m+n} \int d^4y_1 \cdots d^4y_n \, d^4x_1 \cdots d^4x_m$$

$$\times e^{ip_1\cdot y_1} \epsilon^{j_1}(p_1,\kappa_1)^*(\overset{\rightarrow}{\Box}_{y_1} - i\epsilon) \cdots e^{ip_n\cdot y_n} \epsilon^{j_n}(p_n,\kappa_n)^*(\overset{\rightarrow}{\Box}_{y_n} - i\epsilon)$$

$$\times \langle\Omega|T\hat{A}^{j_1}(y_1) \cdots \hat{A}^{j_n}(y_n)\hat{A}^{i_1}(x_1) \cdots \hat{A}^{i_m}(x_m)|\Omega\rangle$$

$$\times (\overset{\leftarrow}{\Box}_{x_1} - i\epsilon)\epsilon^{i_1}(k_1,\lambda_1)e^{-ik_1\cdot x_1} \cdots (\overset{\leftarrow}{\Box}_{x_m} - i\epsilon)\epsilon^{i_m}(k_m,\lambda_m)e^{-ik_m\cdot x_m}. \tag{7.5.107}$$

Recall that in the quantization frame in Coulomb gauge we have $\epsilon^\mu = (0,\boldsymbol{\epsilon})$ and $\hat{A}^\mu = (0,\hat{\mathbf{A}})$. Define the Fourier transform of the Green's function in the usual way,

$$G^{(n)}(p_1,\ldots,p_n)^{\mu_1\cdots\mu_n} \equiv \int \prod_{i=1}^{n} d^4x_i \, e^{ip_i\cdot x_i}\langle\Omega|T\hat{A}^{\mu_1}(x_1)\cdots\hat{A}^{\mu_n}(x_n)|\Omega\rangle, \tag{7.5.108}$$

where all momenta are outgoing. Neglecting forward scattering, using covariant notation and noting that in momentum space $(\Box - i\epsilon) \to -(p^2 + i\epsilon)$ we find

$$S_{fi} = \left(-i/\sqrt{Z_3}\right)^{m+n} \left[\left\{\prod_{j=1}^{n} \epsilon_{\nu_j}(p_j,\kappa_j)^*(p_j^2 + i\epsilon)\right\}\right. \tag{7.5.109}$$

$$\left.\times G^{(m+n)}(p_1,\ldots,p_n,-k_1,\ldots,-k_m)^{\nu_1\cdots\nu_n\mu_1\cdots\mu_m}\left\{\prod_{i=1}^{m}(k_i^2 + i\epsilon)\epsilon_{\mu_i}(k_i,\lambda_i)\right\}\right]_{\mathrm{on\text{-}shell}}.$$

This result is gauge and Lorentz invariant, where $G^{(m+n)}$ is the *covariant Green's function* obtained from the covariant generating functional as discussed below. Furthermore, Z_3 and $\rho(s)$ in Eq. (7.5.101) are gauge-invariant quantities.

Proof: *Gauge invariance:* The effect of $(p_j^2 + i\epsilon)$ and $(k_i^2 + i\epsilon)$ is to truncate the external photon propagators from $G^{(m+n)}$. Let us understand how this occurs. In the case of Coulomb gauge we have from Eqs. (6.4.53) and (6.4.54) that in any frame $n \cdot k = 0$ and $k \cdot \epsilon = k \cdot \epsilon^* = 0$. This means that $\epsilon_{\nu_j}(p_j, \kappa_j)^* (p_j^2 + i\epsilon)$ and $(k_i^2 + i\epsilon)\epsilon_{\mu_i}(k_i, \lambda_i)$ will have the effect of truncating external Coulomb-gauge propagators from $G^{(m+n)}$, since only the $-g^{\mu\nu}/k^2 + i\epsilon$ part of the Coulomb-gauge photon propagator in Eq. (6.4.82) survives. As we will show when discussing the Ward identity in Eq. (8.3.35), the truncated invariant amplitude vanishes when contracted with the relevant photon four-momentum k^μ due to current conservation. So terms such as $k^\mu k^\nu$, $k^\mu n^\nu$ and $k^\nu n^\mu$ in the external photon propagators will vanish, since any term containing a k^μ and/or a k^ν will either be contracted with an ϵ^μ, $\epsilon^{*\mu}$ or the truncated scattering amplitude and the term will then vanish. This means that the external photon propagator can be in *any* of the Coulomb, covariant (R_ξ) or axial gauges and it will still be truncated in Eq. (7.5.109). So the photon propagator truncation is gauge-independent.

What about the choice of gauge for internal propagators? As we show in Eq. (8.3.36) as a result of the Ward identity we similarly find that any term in an internal photon propagator containing a k^μ or k^ν will vanish and so we can also use any gauge for the internal propagators in the truncated invariant scattering amplitude.

Lorentz invariance: Replacing all internal and external propagators with any covariant-gauge photon propagators must lead to a Lorentz invariant result for the scattering amplitude. Since we have shown that the scattering amplitude is gauge invariant, then it must also be Lorentz invariant in any gauge.

Gauge invariance of Z_3 and $\rho(s)$: Finally let us return to the Coulomb-gauge propagator in the quantization frame, where we have

$$D_V^{ij}(x - y) = D_F^{ij}(x - y) = \langle \Omega | T \hat{A}^i(x) \hat{A}^j(y) | \Omega \rangle . \tag{7.5.110}$$

Extracting the gauge-invariant part using the $\epsilon(k, \lambda)$ (Bjorken and Drell, 1965) leads to ,

$$\epsilon^i(k, \lambda')^* D_V^{ij}(k)\epsilon^j(k, \lambda) = \epsilon^i(k, \lambda')^* D_F^{ij}(k)\epsilon^j(k, \lambda) \tag{7.5.111}$$

$$= \epsilon^i(k, \lambda')^* (\cdots) \left[\delta^{ij} - \frac{k^i k^j}{\mathbf{k}^2} \right] \epsilon^j(k, \lambda) = (\cdots) \delta_{\lambda'\lambda}$$

$$= \left(Z_3 \frac{i}{k^2 + i\epsilon} + \int_{m_{\text{th}}^2 \to 0^+}^\infty ds \, \frac{i\rho(s)}{k^2 - s + i\epsilon} \right) \delta_{\lambda'\lambda} ,$$

where we have also used Eq. (7.5.101). So the last line is a gauge-invariant distribution in k^2. *It then follows that Z_3 and $\rho(s)$ are gauge-invariant quantities.* [*Proof:* For Z_3 this is obvious. If some $\rho(s)$ and $\rho'(s)$ have the same result for the integral for every k^2, then we must have $\rho(s) = \rho'(s)$. So $\rho(s)$ is unique and gauge invariant since the integral is gauge invariant for every k^2.]

We expand on these issues below.

Summary: It has taken some effort to arrive at a simple and elegant conclusion. It is the covariant Green's functions that appear in the calculation of physical observables in Eq. (7.5.109). The covariant Green's functions are given by

$$G^{(n)}(x_1, \ldots, x_n) \tag{7.5.112}$$

$$= \frac{(i)^k \, \delta^k \, Z[j]}{\delta j_{\mu_1}(x_1) \cdots \delta j_{\mu_k}(x_n)}\bigg|_{j=0} = \lim_{T \to \infty(1-i\epsilon)} \frac{\int \mathcal{D}A^\mu_{\text{periodic}} A^{\mu_1}(x_1) \cdots A^{\mu_k}(x_n) \, e^{iS_\xi[A]}}{\int \mathcal{D}A^\mu_{\text{periodic}} \, e^{iS_\xi[A]}},$$

where S_ξ and $Z[j]$ are the action and covariant generating functional for that gauge choice. For arbitrary covariant gauges, S_ξ and $Z[j]$ are given in Eqs. (6.4.91) and (6.4.92), respectively. In the Coulomb gauge and axial gauges $Z[j]$ has the same form but with S_ξ given in Eqs. (6.4.108) and (6.4.104), respectively. The two point Green's functions are the Feynman propagators, $G^{(2)}(x, y) = D_F(x - y)$. For the covariant ($R_\xi$) gauges, the Coulomb gauge and axial gauges, these are given in Eqs. (6.4.96), (6.4.109) and (6.4.106), respectively.

The covariant Green's function and the vacuum expectation value of the time-ordered product of field operators were shown in Eq. (6.4.147) to be equal in the case of the Gupta-Bleuler approach formulated in the Feynman gauge,

$$G^{(n)}_{\xi=1}(x_1, \ldots, x_n) = \langle 0|T \hat{A}^{\mu_1}(x_1) \hat{A}^{\mu_2}(x_2) \cdots \hat{A}^{\mu_n}(x_n)|0 \rangle_{\text{GB}}, \tag{7.5.113}$$

where GB denotes the Gupta-Bleuler formalism. Eq. (7.5.109) can also be obtained in the covariant quantization approach of Gupta-Bleuler (Itzykson and Zuber, 1980). In other gauges the covariant Green's function is not in general the time-ordered product of photon field operators as seen in Coulomb gauge in Eq. (6.4.78).

Consider the full photon Feynman propagator $D_F^{\mu\nu}$ in any gauge and its expansion in terms of Feynman diagrams in terms of Feynman propagators. We write the bare photon propagator in that gauge as $D_0^{\mu\nu}$. We can separate out the 1PI part of the full propagator as we have done before and write it as $i\Pi^{\mu\nu}$. This then gives

$$D_F^{\mu\nu}(q) \equiv \text{〜⬤〜} = \text{〜〜⬤〜〜} + \text{〜⬤〜⬤〜} = D_0^{\mu\nu}(q) + D_0^{\mu\sigma}(q)[i\Pi_{\sigma\tau}(q)]D_F^{\tau\nu}(q)$$

$$= \sum(\text{connected two-point diagrams}), \tag{7.5.114}$$

$$i\Pi^{\mu\nu}(p) \equiv \text{〜⬤〜} = \sum(\text{1PI two-point diagrams}),$$

where we refer to $\Pi^{\mu\nu}(q)$ as the *photon polarization tensor*. Iterating the above equation in momentum space gives

$$D_F^{\mu\nu} = D_0^{\mu\nu} + D_0^{\mu\sigma}[i\Pi_{\sigma\tau}]D_0^{\tau\nu} + D_0^{\mu\sigma}[i\Pi_{\sigma\rho}]D_0^{\rho\lambda}[i\Pi_{\lambda\tau}]D_0^{\tau\nu} + \cdots. \tag{7.5.115}$$

The result of the photon coupling to a conserved current leads to the *Ward-Takahashi identities* that will be discussed in Sec. 8.3.2. A special case of these is the Ward identity given in Eq. (8.3.35), which leads to the result that the polarization tensor is purely transverse,

$$q_\mu \Pi^{\mu\nu}(q) = 0. \tag{7.5.116}$$

It then follows that we can define the scalar quantity $\Pi(q^2)$ such that[5]

$$\Pi^{\mu\nu}(q) = \left[-g^{\mu\nu}q^2 + q^\mu q^\nu\right]\Pi(q^2). \tag{7.5.117}$$

We will work almost exclusively in covariant gauges, where $\Pi(q^2)$ can only be a function of q^2. However, since we show below that $\Pi(q^2)$ is gauge invariant this will be true in any gauge. We can write $i\Pi^{\mu\nu}(q) = -iq^2\Pi(q^2)\left[g^{\mu\nu} - (q^\mu q^\nu/q^2)\right]$, where $\left[g^{\mu\nu} - (q^\mu q^\nu/q^2)\right]$ is the transverse

---

[5] Our sign for $\Pi(q^2)$ is opposite to that used by Peskin and Schroeder (1995).

projector. Acting on $D^{\mu\nu}(q^2)$ in a covariant gauge from both the left and right with the transverse projector $[g^{\mu\nu} - (q^\mu q^\nu/q^2)]$ we can define the scalar function $d(q^2)$ as

$$D_{F,t}^{\mu\nu}(q) \equiv \frac{-id(q^2)}{q^2 + i\epsilon}\left[g^{\mu\nu} - (q^\mu q^\nu/q^2)\right]. \tag{7.5.118}$$

Performing this same projection in the right-hand side of Eq. (7.5.114) gives

$$D_{F,t}^{\mu\nu}(q) = \frac{-i}{q^2 + i\epsilon}\left\{1 + (-iq^2)\Pi(q^2)\frac{-id(q^2)}{q^2 + i\epsilon}\right\}\left[g^{\mu\nu} - (q^\mu q^\nu/q^2)\right],$$

which shows that $d(q^2) = 1 - \Pi(q^2)d(q^2)$ and so $d(q^2) = 1/[1 + \Pi(q^2)]$. Comparing with Eq. (6.4.96) and using Eq. (7.5.101) we have in an arbitrary covariant gauge,

$$D_F^{\mu\nu}(q) = \frac{i}{q^2 + i\epsilon}\left[\frac{-g^{\mu\nu} + (q^\mu q^\nu/q^2)}{1 + \Pi(q^2)} - \xi\frac{q^\mu q^\nu}{q^2}\right]. \tag{7.5.119}$$

In the absence of other massless particles in the theory $\Pi(q^2)$ cannot have any singularity at $q^2 = 0$. In the limit that the multiphoton threshold is taken to zero, $m_{\rm th} \to 0^+$, we can expect a multiphoton cut reaching down to $q^2 = 0$ but not a singularity. Then $\Pi(q^2)$ is finite at $q^2 = 0$ and so the one-photon singularity in the gauge-independent part of $D_F^{\mu\nu}$ remains at $q^2 = 0$ and the photon remains massless. The transverse part of the propagator is "dressed" by a factor $1/[1 + \Pi(q^2)]$, while the non-transverse part remains unaffected and maintains its free form.

From Eq. (7.5.111) we have the gauge-invariant quantity

$$\epsilon^i(q,\lambda')^* D_F^{ij}(q)\epsilon^j(q,\lambda) = \frac{i\delta_{\lambda'\lambda}}{q^2 + i\epsilon}\frac{1}{1 + \Pi(q^2)} = \left(Z_3\frac{i}{q^2 + i\epsilon} + \int_{m_{\rm th}^2 \to 0^+}^{\infty} ds\,\frac{i\rho(s)}{q^2 + s + i\epsilon}\right)\delta_{\lambda'\lambda}.$$

This shows that $\Pi(q^2)$ must be independent of gauge choice and also that

$$Z_3 = \frac{1}{1 + \Pi(0)} \quad \text{and} \quad \frac{1}{1 + \Pi(q^2)}\bigg|_{q^2 \neq 0} = \int_{m_{\rm th}^2 \to 0^+}^{\infty} ds\,\frac{(q^2 + i\epsilon)\rho(s)}{q^2 + s + i\epsilon}\bigg|_{q^2 \neq 0}, \tag{7.5.120}$$

$$D_F^{\mu\nu}(q) = \left(Z_3\frac{i}{q^2 + i\epsilon} + \int_{m_{\rm th}^2 \to 0^+}^{\infty} ds\,\frac{i\rho(s)}{q^2 - s + i\epsilon}\right)\left[-g^{\mu\nu} + (q^\mu q^\nu/q^2)\right] - \frac{i\xi(q^\mu q^\nu/q^2)}{q^2 + i\epsilon}.$$

For a QED-invariant amplitude $\mathcal{M}$ with an external photon with polarization $\epsilon_\mu(k,\lambda)$ we define $M^\mu(k)$ such that $\mathcal{M}(k) \equiv \epsilon_\mu(k,\lambda)M^\mu(k)$. According to the Ward identity in Eq. (8.3.35) it follows that

$$k_\mu M^\mu(k) = 0, \tag{7.5.121}$$

for any k^μ. We discuss this further in Sec. 8.3.2. It then follows in an arbitrary covariant gauge that

$$\left[\epsilon_{\nu_j}(p_j,\kappa_j)^*(p_j^2 + i\epsilon)G^{(m+n)}(\cdots)^{\nu_1\cdots\nu_n\mu_1\cdots\mu_n}\right]_{\text{on-shell}} \tag{7.5.122}$$

$$= -iZ_3\left[\epsilon^{\sigma_j}(p_j,\kappa_j)^*(D_F^{-1})_{\sigma_j\nu_j}(p_j)\,G^{(m+n)}(\cdots)^{\nu_1\cdots\nu_n\mu_1\cdots\mu_n}\right]_{\text{on-shell}},$$

$$(D_F^{-1})^{\mu\nu}(q) \equiv -i\left\{(-g^{\mu\nu}q^2 + q^\mu q^\nu)[1 + \Pi(q^2)] + (i/\xi)q^\mu q^\nu\right\}, \tag{7.5.123}$$

where D_F^{-1} is the inverse photon propagator, $(D_F^{-1})_{\mu\nu}(q)\,D_F^{\nu\rho}(q) = \delta_\mu{}^\rho$, and where the $q^\mu q^\nu$ terms in the inverse propagator vanish when acting on $G^{(m+n)}$ because of the Ward identity. The same argument applies for the $\epsilon_{\mu_i}(k_i,\lambda_i)$ contributions.

Define the amputated Green's function as we did earlier in Eq. (7.5.37), where we identify and remove the full Feynman propagators $D_F^{\mu\nu}$ from the diagram. Finally, we have the *LSZ reduction formula for photons* in arbitrary covariant gauge,

$$
\begin{aligned}
S_{fi} &= (-\sqrt{Z_3})^{m+n}\, \epsilon_{\nu_1}(p_1,\kappa_1)^* \cdots \epsilon_{\nu_n}(p_n,\kappa_n)^* G_{\mathrm{amp}}^{(m+n)}(p_1,\ldots,-k_m)^{\nu_1\cdots\mu_m} \\
&\quad \times \epsilon_{\mu_1}(k_1,\lambda_1)\cdots\epsilon_{\mu_m}(k_m,\lambda_m), \\
&= (-1)^{m+n}\epsilon_{\nu_1}(p_1,\kappa_1)^* \cdots \epsilon_{\nu_n}(p_n,\kappa_n)^* G_{r,\mathrm{amp}}^{(m+n)}(p_1,\ldots,-k_m)^{\nu_1\cdots\mu_m} \\
&\quad \times \epsilon_{\mu_1}(k_1,\lambda_1)\cdots\epsilon_{\mu_m}(k_m,\lambda_m),
\end{aligned}
\tag{7.5.124}
$$

where $G_{\mathrm{amp}}^{(m+n)}$ and $G_{r,\mathrm{amp}}^{(m+n)}$ are the amputated and renormalized amputated covariant Green's functions, respectively. We can again use Eq. (7.5.44) to relate S_{fi} and $\mathcal{M}_{fi}$ and ignore forward-scattering contributions and an irrelevant overall sign. The invariant amplitude $\mathcal{M}_{fi}$ in arbitrary covariant gauge is

$$
\begin{aligned}
i(2\pi)^4\delta^4(\textstyle\sum_i p_i - \sum_j k_j)\mathcal{M}_{fi} &= \epsilon_{\nu_1}(p_1,\kappa_1)^* \cdots \epsilon_{\nu_n}(p_n,\kappa_n)^* \\
\times\, G_{r,\mathrm{amp}}^{(m+n)}(p_1,\ldots,p_n,-k_1,\ldots,-k_m)^{\nu_1\cdots\nu_n\mu_1\cdots\mu_m} &\,\epsilon_{\mu_1}(k_1,\lambda_1)\cdots\epsilon_{\mu_m}(k_m,\lambda_m).
\end{aligned}
\tag{7.5.125}
$$

Note that the $q^\mu q^\nu$ terms would vanish even without invoking the Ward identity since from Eq. (6.4.54) we have $\epsilon^\sigma(p,\kappa)^* p_\sigma p_\nu = 0$ on shell. Extracting all of the $D_F^{\mu\nu}$ from $G^{(m+n)}$ to leave $G_{\mathrm{amp}}^{(m+n)}$ leads to terms in the on-shell limit of the form

$$
\lim_{p^2\to-i\epsilon}\left\{\epsilon_\nu(p,\kappa)^*(p^2+i\epsilon)D_F^{\nu\rho}(p)\right\}
\tag{7.5.126}
$$

$$
= \lim_{p^2\to-i\epsilon}\left\{\epsilon_\nu(p,\kappa)^* i g^{\nu\rho}/[1+\Pi(p^2)]\right\} = -iZ_3\epsilon^\rho(p,\kappa)^*,
\tag{7.5.127}
$$

which is another way to understand the above result. As discussed above, if we consider Coulomb or axial gauge with $D_F^{\mu\nu}(p)$ given in Eq. (6.4.82) or Eq. (6.4.105) respectively, then we would again arrive at this result. So the LSZ result given in the form of Eqs. (7.5.124) and (7.5.125) also applies in Coulomb gauge or axial gauge. A formal proof of gauge invariance for QED scattering amplitudes will be given along with a discussion of the Ward and Ward-Takahashi identities in Sec. 8.3.2.

It is left as an exercise to generalize these arguments for the photon to the case of the massive Proca vector boson. The essential differences are that we need not be concerned with gauge invariance, there is a nonzero mass and there are three polarizations rather than two. Again working in the quantization frame it is not difficult to arrive at Eq. (7.5.109) with $p^2 + i\epsilon \to p^2 - m^2 + i\epsilon$ for each momentum, with the three polarizations, and with $Z_3 \to Z_B$ for some appropriate Z_B. Using $k\cdot\epsilon(p,\lambda) = 0$ from Eq. (6.5.36), the Ward identity and the massive vector boson Feynman propagator $D_F^{\mu\nu}$ in Eq. (6.5.72) we again find Eq. (7.5.125) with the corresponding amputated covariant Green's function.

7.6 Feynman Rules

We have seen in Eqs. (7.3.68) and (7.3.74) that differential cross-sections and differential decay rates are calculated in terms of $|\mathcal{M}_{fi}|^2$, where the $\mathcal{M}_{fi}$ is the invariant amplitude for the process. From the LSZ reduction formulae in Eqs. (7.5.44), (7.5.93) and (7.5.125) we know how to obtain $\mathcal{M}_{fi}$ from the renormalized, amputated and covariant Green's function, $G_{r,\mathrm{amp}}$, for the interacting theory. We obtain these renormalized amputated Green's functions by adding renormalization constants into the Lagrangian as discussed leading up to Eq. (7.5.46). There we explained that at tree-level all renormalization constants can be set to unity. We will discuss renormalization and renormalization constants in detail in Chapter 8. Recall that the amputated covariant Green's functions G_{amp} are obtained from the covariant Green's functions G by amputating the external dressed propagators.

The connected covariant Green's functions G are obtained from the covariant generating functional $Z[j]$ for the full theory by differentiating with respect to sources and then setting the sources to zero as seen in Eqs. (6.2.263), (6.3.237) and (7.5.112). The functional arguments leading to Eq. (7.4.59) also apply to the fermion, photon and massive vector boson cases and so we can use this generalization to obtain the Feynman diagram-based perturbative expansion for the covariant Green's function G as we did in the scalar case. The purpose of this section is to collect together the Feynman rules that show how to construct the connected and amputated Feynman diagrams that contribute to the amputated covariant Green's function and hence to the $\mathcal{M}_{fi}$ to the desired order of perturbation theory. Note that Feynman diagrams that consist of two or more disjoint parts will contribute to forward scattering and so do not contribute to the scattering amplitude $\mathcal{M}_{fi}$.

7.6.1 External States and Internal Lines

We begin with the Feynman rules for $i\mathcal{M}_{fi}$ for the external states, internal lines corresponding to bare Feynman propagators, and symmetry factors.

Neglect renormalization for now: It is simplest and conventional to first introduce and discuss Feynman rules while ignoring the issue of renormalization. Recall that for tree-level calculations this is sufficient. Beyond the lowest order in perturbation theory we will encounter loops, and the renormalization procedures to be discussed in Sec. 8.4 will become essential for the calculation of physical amplitudes.

Assign relative signs for Feynman diagrams: The set of all contractions leads to the set of all topologically distinct Feynman diagrams, so *all topologically distinct diagrams need to be included. If two diagrams are topologically identical up to the exchange of two external boson lines or two external fermion lines, then there is a relative plus sign or relative minus sign between the diagrams, respectively.* This is true independent of whether the lines are both in the initial state, both in the final state or one of each. The argument is similar to that given below for a fermion loop. This comes from the need to rearrange the field operators in the Green's function in the LSZ formalism to bring together the operators to be contracted into lines. We will consider examples of this in Sec. 7.6.3.

Identical particles in the final state: For n identical final-state particles include a factor of $1/n!$ in $d\Pi_n^{\mathrm{LIPS}}$ in Eq. (7.3.41) so as not to overcount final states.

Divide diagrams by their symmetry factor: Calculate the symmetry factor S for each Feynman diagram according to the principles discussed in Sec. 7.4.2. *Divide the contribution of each Feynman diagram by its symmetry factor S.*

Add a minus sign for fermion loops: *When a Feynman diagram contains a fermion loop it comes with a minus sign and the loop contribution has the form*

$$(-1)\mathrm{tr}[S_0\Gamma S_0\Gamma\cdots\Gamma S_0] \tag{7.6.1}$$

with the Γs being appropriate spinor matrices and with the bare propagators ordered from right to left in the direction of the fermion arrow in the loop. If the theory has more than one type of fermion interaction, then the Γs may be mixed. To understand this result note that any fermion interaction that we will consider will be of the form $\bar{\psi}(x)\Gamma\psi(x) = \bar{\psi}_\alpha(x)\Gamma_{\alpha\beta}\psi_\beta(x)$ for some spinor matrices Γ. A fermion loop will occur when a group of interaction terms $\bar{\psi}\Gamma\psi$ are fully contracted within themselves; i.e., there is no contraction with a fermion operator associated with an external fermion or antifermion. For example, consider a loop contraction of three interaction terms, remembering that we are in the interaction picture,

$$\langle 0|T\cdots\hat{\bar{\psi}}_{\alpha_1}(x_1)\Gamma_{\alpha_1\beta_1}\hat{\psi}_{\beta_1}(x_1)\,\hat{\bar{\psi}}_{\alpha_2}(x_2)\Gamma_{\alpha_2\beta_2}\hat{\psi}_{\beta_2}(x_2)\,\hat{\bar{\psi}}_{\alpha_3}(x_3)\Gamma_{\alpha_3\beta_3}\hat{\psi}_{\beta_3}(x_3)\cdots|0\rangle$$

$$= -\langle 0|T\cdots\hat{\psi}_{\beta_3}(x_3)\hat{\bar{\psi}}_{\alpha_1}(x_1)\Gamma_{\alpha_1\beta_1}\hat{\psi}_{\beta_1}(x_1)\,\hat{\bar{\psi}}_{\alpha_2}(x_2)\Gamma_{\alpha_2\beta_2}\hat{\psi}_{\beta_2}(x_2)\,\hat{\bar{\psi}}_{\alpha_3}(x_3)\Gamma_{\alpha_3\beta_3}\cdots|0\rangle$$

$$= (-1)\mathrm{tr}\left[S_0(x_3-x_1)\Gamma S_0(x_1-x_2)\Gamma S_0(x_2-x_3)\right]\langle 0|T\cdots|0\rangle, \qquad (7.6.2)$$

with the fermion arrow in the orientation $x_3 \to x_2 \to x_1 \to x_3$. The minus sign arises since the fermion operators anticommute in the time-ordered product. We have used Eq. (7.5.94). Since all topologically distinct attachments to the loop must be summed over, only one orientation of the fermion loop should be included.

Initial and final states: For clarity we here indicate the end of a line connected to the rest of the Feynman diagram with a bullet (•), and here time is flowing from left to right. As we have seen from the LSZ formalism *all external propagators are amputated and replaced in Feynman diagrams as follows:*

Since the charged scalar field is just the superposition of two real scalar fields, the external state is the same as for the neutral scalar. The arrow on the charged scalar indicates the flow of the charge. For the massive vector boson we use the same symbol as for the photon line, $\sim\!\sim\!\sim\!\sim$, and $\epsilon^\mu(p,\lambda)$ has three polarization states rather than two. For the charged massive vector boson we add an arrow as for the charged scalar, $\sim\!\sim\!\blacktriangleright\!\sim\!\sim$, and since it is the superposition of two real massive vectors it again has $\epsilon^\mu(p,\lambda)$ with three polarizations. Some texts omit the arrow on charged massive vector boson lines.

Internal Feynman propagators: Feynman diagrams are obtained from the covariant generating functional and so contain the bare/undressed Feynman propagators, which are

$$\text{(neutral, charged massive vector)} \quad \text{\textasciitilde\textasciitilde}, \text{\textasciitilde\textasciitilde}\blacktriangleright\text{\textasciitilde} \quad = \frac{i}{p^2 - m^2 + i\epsilon}\left[-g^{\mu\nu} + \frac{p^\mu p^\nu}{m^2}\right] = \Delta_0^{\mu\nu}(p).$$

Summary of momentum-space Feynman rules: To calculate the contributions from a Feynman diagram we perform the following steps:

(i) *Each line for an external particle is amputated and replaced with the corresponding quantity according to* Eq. (7.6.3).

(ii) *Each internal line is replaced with a Feynman propagator according to* Eq. (7.6.4).

(iii) *Charges must be conserved at every vertex.*

(iv) *Divide by the symmetry factor S for the diagram.*

(v) *Assign a relative minus sign to the diagram if it corresponds to another diagram that is topologically equivalent up to the exchange of two identical external fermion lines (see Sec. 7.6.3 for examples of this).*

(vi) *Each vertex contains a momentum-conserving factor $V_F(2\pi)^4\delta^4(\sum_j p_j)$ with V_F some vertex structure depending on the detailed nature of the interaction, e.g., $V_F = -i\lambda$ for the $(\lambda/4!)\phi^4$ interaction as we saw in Sec. 7.4.3.*

(vii) *Integrate over all internal momenta with $\int d^4p/(2\pi)^4$ (we will need renormalization to make sense of integrations over loop momenta).*

(viii) *The result will be a contribution to $i(2\pi)^4\delta^4(\sum p_f - \sum p_i)\mathcal{M}_{fi}$.*

Simpler approach used in practice: It is not difficult see a simpler and equivalent approach to steps (vi)–(viii):

(vi) *Each vertex contains a vertex structure V_F for the theory, where there may be more than one V_F defining the interactions of the theory. We typically just refer to V_F as a Feynman diagram vertex.*

(vii) *Integrate over only the undetermined momenta with $\int d^4p/(2\pi)^4$ (we will need renormalization to make sense of integrations over loop momenta).*

(viii) *The result will be a contribution to $i\mathcal{M}_{fi}$.*

The number of undetermined momenta is equal to the number of loops. To see why we have a contribution to $i\mathcal{M}_{fi}$ rather than $i(2\pi)^4\delta^4(\sum p_f - \sum p_i)\mathcal{M}_{fi}$, refer to the discussion leading to Eq. (7.4.52). The overall $(2\pi)^4\delta^4(\sum p_f - \sum p_i)$ does *not* occur in the simple shortcut. *From now on we use this simple shortcut to evaluate Feynman diagrams and we will describe interaction vertices in this context.*

Polarizations: When calculating cross-sections we will be considering a beam on a target or two colliding beams. If the beam and target or the two beams are both unpolarized, then they will consist of initial state particles with an equal admixture of the possible polarizations. The final state particles can scatter into any allowed polarization state. In such a case we should *average the cross-section over initial polarizations and sum over final polarizations.*

7.6.2 Examples of Interacting Theories

It now remains to consider examples of simple interacting quantum field theories and to explain the Feynman rules for the interaction vertices for each of them.

Scalar field with quartic self-interaction: We have already discussed ϕ^4 theory in Sec. 7.4.2. Using KG to denote Klein-Gordon, the system is specified by

$$\mathcal{L} = \mathcal{L}_0 + \mathcal{L}_{\text{int}}^{\phi^4} = \mathcal{L}_{\text{KG}} + \mathcal{L}_{\text{int}}^{\phi^4} = \tfrac{1}{2}\left[\partial_\mu \phi \partial^\mu \phi - m^2\phi^2\right] - (\lambda/4!)\phi^4, \tag{7.6.5}$$

$$H = H_0 + H_{\text{int}}^{\phi^4} = H_{\text{KG}} + H_{\text{int}}^{\phi^4} = H_{\text{KG}} + \int d^3x(\lambda/4!)\phi^4. \tag{7.6.6}$$

Since the interaction term contains no time derivatives, there is a simple relationship between the Hamiltonian and Lagrangian formulations, $\mathcal{L}_{\text{int}}^{\phi^4} = -\mathcal{H}_{\text{int}}^{\phi^4} = -(\lambda/4!)/\phi^4$. There are no complications in the canonical quantization procedure, so the construction of the evolution operator and spectral function is straightforward. We can generate the spectral function for the interacting theory with a source from that for the corresponding Klein-Gordon spectral function,

$$F^j(t'', t') = \text{tr}\{\hat{U}^j(t'', t')\} = \text{tr}\left\{Te^{-i\int_{t'}^{t''} d^4x[\hat{\mathcal{H}} - j\hat{\phi}]}\right\} \tag{7.6.7}$$

$$= e^{-i\int_{t'}^{t''} d^4x\, \mathcal{H}_{\text{int}}^{\phi^4}\left(-i\frac{\delta}{\delta j(x)}\right)}\text{tr}\left\{Te^{-i\int_{t'}^{t''} d^4x[\hat{\mathcal{H}}_{\text{KG}} - j\hat{\phi}]}\right\} = \int \mathcal{D}\phi\, e^{i\int_{t'}^{t''} d^4x[\mathcal{L} + j\phi]}.$$

The generating functional was given earlier in Eq. (7.4.18),

$$Z[j] \equiv \lim_{T\to\infty(1-i\epsilon)} \frac{\text{tr}\{\hat{U}^j(T,-T)\}}{\text{tr}\{\hat{U}(T,-T)\}} = \lim_{T\to\infty(1-i\epsilon)} \frac{\langle 0|Te^{-i\int_{-T}^{T} d^4x\,[\hat{\mathcal{H}}_I - j\hat{\phi}_I]}|0\rangle}{\langle 0|Te^{-i\int_{-T}^{T} d^4x\,\hat{\mathcal{H}}_I}|0\rangle}$$

$$= \frac{\int \mathcal{D}\phi\, e^{i\{S[\phi] + \int d^4x\, j\phi\}}}{\int \mathcal{D}\phi\, e^{iS[\phi]}} = \frac{e^{i\int d^4x\, \mathcal{L}_{\text{int}}^{\phi^4}\left(\frac{-i\delta}{\delta j(x)}\right)}\int \mathcal{D}\phi\, e^{i\int d^4x[\mathcal{L}_{\text{KG}} - j\phi]}}{e^{i\int d^4x\, \mathcal{L}_{\text{int}}^{\phi^4}\left(\frac{-i\delta}{\delta j(x)}\right)}\int \mathcal{D}\phi\, e^{i\int d^4x\, \mathcal{L}_{\text{KG}}}}, \tag{7.6.8}$$

where $\hat{\mathcal{H}}_I = \hat{U}_0(t,t_0)^\dagger \hat{\mathcal{H}}_{\text{int}}(t)\hat{U}_0(t,t_0) = \mathcal{H}_{\text{int}}(\hat{\phi}_I) = (\lambda/4!)\hat{\phi}_I^4$ and where for the sake of brevity from here on we omit writing $\lim_{T\to\infty(1-i\epsilon)}$ and periodic/antiperiodic for the functional integral ratios. We obtain time-ordered vacuum expectation values in terms of the functional integral in Eq. (6.2.263),

$$\langle \Omega|T\hat{\phi}(x_1)\cdots\hat{\phi}(x_n)|\Omega\rangle = \frac{(-i)^n\delta^k}{\delta j(x_1)\cdots\delta j(x_n)}Z[j]\bigg|_{j=0}. \tag{7.6.9}$$

In Eq. (7.4.57) we saw that Eq. (7.6.8) leads to the Feynman diagram vertex

$$i\mathcal{L}_{\text{int}}^{\phi^4} = -i\frac{\lambda}{4!}\phi^4 \qquad\qquad \Rightarrow \qquad\qquad \diagup\!\!\!\!\!\diagdown = -i\lambda. \tag{7.6.10}$$

Important note: If an interaction has $n!$ equivalent permutations of the fields, then we divide by $n!$ to keep the Feynman diagram vertex simple; e.g., for an interaction term $\mathcal{L}_{\text{int}}^{\phi n} = -(\lambda/n!)\phi^n$ we have a vertex $(-i\lambda)$ but now with n legs.

Yukawa interaction: The simplest interacting theory containing fermions is a *Yukawa theory* with a Yukawa interaction between a scalar/pseudoscalar boson and a fermion. The interaction contains no time derivatives and so the Lagrangian and Hamiltonian densities are again simply related. For a scalar or pseudoscalar boson, respectively, we have

$$\mathcal{L}_{\text{int}}^{\text{Yuk}}(\phi, \bar{\psi}, \psi) = -\mathcal{H}_{\text{int}}^{\text{Yuk}}(\phi, \bar{\psi}, \psi) \equiv -g\bar{\psi}\phi\psi \text{ or } -g\bar{\psi}i\gamma_5\phi\psi, \tag{7.6.11}$$

$$\mathcal{L} = \mathcal{L}_0 + \mathcal{L}_{\text{int}}^{\text{Yuk}} = \mathcal{L}_{\text{KG}} + \mathcal{L}_{\text{Dirac}} + \mathcal{L}_{\text{int}}^{\text{Yuk}},$$

$$\mathcal{H} = \mathcal{H}_0 + \mathcal{H}_{\text{int}} = \mathcal{H}_0 + \mathcal{H}_{\text{int}}^{\text{Yuk}} = \mathcal{H}_{\text{KG}} + \mathcal{H}_{\text{Dirac}} + \mathcal{H}_{\text{int}}^{\text{Yuk}}. \tag{7.6.12}$$

Let m be the fermion mass and m_ϕ be the scalar mass. The canonical quantization procedure is then a straightforward combination of those for the scalar and fermion fields. Denoting the evolution operator with sources as $\hat{U}^{j\bar{\eta}\eta}(t'', t')$, the spectral function for the Yukawa theory with sources is

$$F^{j\bar{\eta}\eta}(t'', t') = \text{tr}\{\hat{U}^{j\bar{\eta}\eta}(t'', t')\} = \text{tr}\left\{ T e^{-i\int_{t'}^{t''} d^4x[\hat{\mathcal{H}} - j\hat{\phi} - \bar{\eta}\hat{\psi} - \hat{\bar{\psi}}\eta]} \right\} \tag{7.6.13}$$

$$= e^{-i\int d^4x\, \mathcal{H}_{\text{int}}^{\text{Yuk}}\left(\frac{-i\delta}{\delta j(x)}, \frac{-i\delta}{\delta\bar{\eta}(x)}, \frac{i\delta}{\delta\eta(x)}\right)} \text{tr}\left\{ T e^{-i\int_{t'}^{t''} d^4x[\hat{\mathcal{H}}_{\text{KG}} + \mathcal{H}_{\text{Dirac}} - j\hat{\phi} - \bar{\eta}\hat{\psi} - \hat{\bar{\psi}}\eta]} \right\}$$

$$= e^{i\int d^4x\, \mathcal{L}_{\text{int}}^{\text{Yuk}}\left(\frac{-i\delta}{\delta j(x)}, \frac{-i\delta}{\delta\bar{\eta}(x)}, \frac{i\delta}{\delta\eta(x)}\right)} \int \mathcal{D}\phi \mathcal{D}\bar{\psi} \mathcal{D}\psi\, e^{i\int_{t'}^{t''} d^4x[\mathcal{L}_{\text{KG}} + \mathcal{L}_{\text{Dirac}} + j\phi + \bar{\eta}\psi + \bar{\psi}\eta]}$$

$$= \int \mathcal{D}\phi \mathcal{D}\psi \mathcal{D}\bar{\psi}\, e^{i\int_{t'}^{t''} d^4x[\mathcal{L} + j\phi + \bar{\eta}\psi + \bar{\psi}\eta]} = \int \mathcal{D}\phi \mathcal{D}\psi \mathcal{D}\bar{\psi}\, e^{iS[\phi,\bar{\psi},\psi] + i\int_{t'}^{t''} d^4x[j\phi + \bar{\eta}\psi + \bar{\psi}\eta]}.$$

The generating functional for the Yukawa theory is

$$Z[j, \bar{\eta}, \eta] \equiv \lim_{T \to \infty(1-i\epsilon)} \frac{\text{tr}\{\hat{U}^{j\bar{\eta}\eta}(T, -T)\}}{\text{tr}\{\hat{U}(T, -T)\}} = \frac{\int \mathcal{D}\phi \mathcal{D}\bar{\psi} \mathcal{D}\psi\, e^{iS[\phi,\bar{\psi},\psi] + i\int d^4x[j\phi + \bar{\eta}\psi + \bar{\psi}\eta]}}{\int \mathcal{D}\phi \mathcal{D}\bar{\psi} \mathcal{D}\psi\, e^{iS[\phi,\bar{\psi},\psi]}}. \tag{7.6.14}$$

From Eq. (7.6.13) it follows that the Feynman diagram vertex is

$$i\mathcal{L}_{\text{int}}^{\text{Yuk}} = -ig\bar{\psi}\phi\psi,\ g\bar{\psi}\gamma_5\phi\psi \qquad \Rightarrow \qquad = -ig,\ g\gamma_5, \tag{7.6.15}$$

where we understand that the vertex with no γ_5 contains a 4×4 spinor identity matrix.

Quantum electrodynamics (QED): In Sec. 4.4.4 we discussed the interaction of a Dirac fermion with an electromagnetic field in the context of relativistic quantum mechanics, and minimal coupling was the simplest choice for such an interaction. We also used the principle of minimal coupling to derive the Lorentz force for a relativistic classical particle and for a spinless relativistic quantum particle in Secs. 2.8 and (4.3.4), respectively. We further understand from Eq. (2.7.25) that the electromagnetic field can only couple to conserved currents. When that current is the fermion conserved current, $j_{\text{Dirac}}^\mu(x) = \bar{\psi}(x)\gamma^\mu\psi(x)$, we arrive at QED after we quantize. The classical Lagrangian density for electrodynamics is given by

$$\mathcal{L} = \mathcal{L}_0 + \mathcal{L}_{\text{int}}^{\text{QED}} = \mathcal{L}_{\text{Dirac}} + \mathcal{L}_{\text{Maxwell}} + \mathcal{L}_{\text{int}}^{\text{QED}} \tag{7.6.16}$$

$$= [\bar{\psi}(i\slashed{\partial} - m)\psi] + [-\tfrac{1}{4}F_{\mu\nu}F^{\mu\nu}] + [-q_c\bar{\psi}\slashed{A}\psi] = \bar{\psi}(i\slashed{D} - m)\psi - \tfrac{1}{4}F_{\mu\nu}F^{\mu\nu},$$

$$\mathcal{L}_{\text{int}}^{\text{QED}}(A^\mu, \bar{\psi}, \psi) \equiv -q_c A_\mu j_{\text{Dirac}}^\mu = -q_c A_\mu \bar{\psi}\gamma^\mu\psi = -q_c\bar{\psi}\slashed{A}\psi, \qquad D_\mu \equiv \partial_\mu + iq_c A_\mu,$$

where q_c is the electric charge of the fermion and D_μ is the covariant derivative introduced earlier in Eq. (4.3.54). We use q_c for electric charge from this point on to avoid confusion with four-momentum transfer q^μ, e.g., $q_c = -e \equiv -|e|$ for an electron. At the classical level we should really treat the fermion fields as Grassmann-valued fields as we did in Eqs. (6.3.160) and (6.3.161). However, we saw that doing this leads to the same equations of motion as the simpler approach where we treated the fermion fields as c-number (complex) fields in Eqs. (6.3.121) and (6.3.123). In either case the Euler-Lagrange equations resulting from this action are

$$(i\slashed{D} - m)\psi(x) = 0, \qquad \partial_\mu F^{\mu\nu} = q_c j_{\text{Dirac}}^\nu(x) = q_c\bar{\psi}(x)\gamma^\nu\psi(x). \tag{7.6.17}$$

The physical behavior of the theory should only depend on the $\mathbf{E}$ and $\mathbf{B}$ fields and not on the particular choice of A^μ; i.e., it should not depend on the choice of gauge for the four-vector field A^μ. The Dirac action is invariant under a global phase transformation, $\psi(x) \to e^{i\alpha}\psi(x)$, which by Noether's theorem leads to conservation of the fermion current j_{Dirac}^μ as we saw in Eq. (6.3.205). The fermion current j^μ_{Dirac} is invariant under a *local phase transformation*, $\psi(x) \to e^{i\alpha(x)}\psi(x)$. A local phase transformation of the fermion field in the QED Lagrangian density is equivalent to a change of gauge for the four-vector potential. This means that the QED Lagrangian density is invariant under the combined transformations characterized by $\alpha(x)$,

$$\psi(x) \to \psi'(x) = e^{i\alpha(x)}\psi(x), \quad \bar\psi(x) \to \bar\psi'(x) = e^{-i\alpha(x)}\bar\psi(x), \qquad (7.6.18)$$
$$A_\mu(x) \to A'_\mu(x) = A_\mu(x) - (1/q_c)\partial_\mu\alpha(x).$$

This illustrates how to produce a *gauge theory* from a non-gauge theory: (i) Turn a global phase invariance into a local one; (ii) introduce gauge fields for which the non-invariant term(s) are equivalent to a change of gauge; (iii) the resulting theory is invariant under the combined transformations. This is achieved using the minimal coupling prescription where the partial derivative, ∂_μ, is replaced with the covariant derivative, D_μ. The combined transformations are referred to as *gauge transformations* for this *gauge theory*, which by construction is *gauge invariant*.

Proof: We know from Eqs. (2.7.13) and (2.7.76) that $F^{\mu\nu}$ is gauge invariant,

$F_{\mu\nu} \to F'_{\mu\nu} = \partial_\mu A'_\nu - \partial_\nu A'_\mu = \partial_\mu A_\nu - \partial_\nu A_\mu - \frac{1}{q}(\partial_\mu\partial_\nu - \partial_\nu\partial_\mu)\alpha = F_{\mu\nu}$. Then, it follows that $-\frac{1}{4}F_{\mu\nu}F^{\mu\nu} \to -\frac{1}{4}F'_{\mu\nu}F'^{\mu\nu} = -\frac{1}{4}F_{\mu\nu}F^{\mu\nu}$. Trivially, we find that $\bar\psi m\psi \to \bar\psi' m\psi' = \bar\psi m\psi$. Finally, note that

$$\bar\psi(i\slashed{D})\psi = \bar\psi(i\partial_\mu - q_c A_\mu)\gamma^\mu\psi \to \bar\psi'(i\slashed{D}')\psi' = \bar\psi e^{-i\alpha}(i\partial_\mu - q_c A_\mu + \partial_\mu\alpha)\gamma^\mu e^{i\alpha}\psi$$
$$= \bar\psi e^{-i\alpha}e^{i\alpha}(i\partial_\mu - q_c A_\mu + \partial_\mu\alpha - \partial_\mu\alpha)\gamma^\mu\psi = \bar\psi(i\slashed{D})\psi.$$

We cancel the non-invariant term, $-\partial_\mu\alpha$, with a term from the change of gauge, $\partial_\mu\alpha$. Then $\mathcal{L} \to \mathcal{L}' = \mathcal{L}$ and QED is gauge invariant as stated.

Since the interaction contains no time derivatives, then

$$\mathcal{H}^{\mathrm{QED}}_{\mathrm{int}}(A^\mu, \bar\psi, \psi) = -\mathcal{L}^{\mathrm{QED}}_{\mathrm{int}}(A^\mu, \bar\psi, \psi) = q_c\bar\psi\slashed{A}\psi = q_c A_\mu j^\mu_{\mathrm{Dirac}}, \qquad (7.6.19)$$
$$\mathcal{H} = \mathcal{H}_0 + \mathcal{H}^{\mathrm{QED}}_{\mathrm{int}} = \mathcal{H}_{\mathrm{Maxwell}} + \mathcal{H}_{\mathrm{Dirac}} + \mathcal{H}^{\mathrm{QED}}_{\mathrm{int}}.$$

The canonical quantization procedure is then a straightforward combination of those for the free Maxwell and Dirac fields. The fermion and photon field operators are independent and commute, which follows from the canonical quantization prescription given in Eq. (6.3.159). It is to be understood that the appropriate gauge fixing terms are included.

When we work at the operator level it is simplest to understand that we are working in the covariant Gupta-Bleuler formalism for the photon described in Sec. 6.4.5 with $A^\mu \to \hat{A}^\mu$. This corresponds to the choice of Feynman gauge. If we restrict our attention to only the functional integral formalism, then we can work conveniently in any R_ξ gauge, rather than just Feynman gauge. Denoting the evolution operator with sources as $\hat{U}^{j\bar\eta\eta}(t'', t')$ we obtain the spectral function for QED,

$$F^{j\bar\eta\eta}(t'', t') = \mathrm{tr}\{\hat{U}^{j\bar\eta\eta}(t'', t')\} = \mathrm{tr}\left\{Te^{-i\int_{t'}^{t''}d^4x[\mathcal{H}+j\cdot\hat{A}-\bar\eta\hat\psi-\hat{\bar\psi}\eta]}\right\} \qquad (7.6.20)$$
$$= e^{-i\int d^4x\,\mathcal{H}^{\mathrm{QED}}_{\mathrm{int}}\left(\frac{i\delta}{\delta j_\mu(x)}, \frac{-i\delta}{\delta\bar\eta(x)}, \frac{i\delta}{\delta\eta(x)}\right)}\mathrm{tr}\left\{Te^{-i\int_{t'}^{t''}d^4x[\hat{\mathcal{H}}_{\mathrm{Maxwell}}+\hat{\mathcal{H}}_{\mathrm{Dirac}}+j\cdot\hat{A}-\bar\eta\hat\psi-\hat{\bar\psi}\eta]}\right\}$$
$$= e^{i\int d^4x\,\mathcal{L}^{\mathrm{QED}}_{\mathrm{int}}\left(\frac{i\delta}{\delta j_\mu(x)}, \frac{-i\delta}{\delta\bar\eta(x)}, \frac{i\delta}{\delta\eta(x)}\right)}\int\mathcal{D}A^\mu\mathcal{D}\bar\psi\mathcal{D}\psi\,e^{i\int_{t'}^{t''}d^4x[\hat{\mathcal{L}}_{\mathrm{Maxwell}}+\mathcal{L}_{\mathrm{Dirac}}+\bar\eta\psi+\bar\psi\eta-j\cdot A]}$$
$$= \int\mathcal{D}A^\mu\mathcal{D}\bar\psi\mathcal{D}\psi\,e^{i\int_{t'}^{t''}d^4x[\mathcal{L}+\bar\eta\psi+\bar\psi\eta-j\cdot A]} = \int\mathcal{D}A^\mu\mathcal{D}\bar\psi\mathcal{D}\psi\,e^{iS[A,\bar\psi,\psi]+i\int_{t'}^{t''}d^4x[\bar\eta\psi+\bar\psi\eta-j\cdot A]}.$$

The generating functional for QED is then

$$Z[j, \bar{\eta}, \eta] \equiv \lim_{T \to \infty(1-i\epsilon)} \frac{\text{tr}\{\hat{U}^{j\bar{\eta}\eta}(T, -T)\}}{\text{tr}\{\hat{U}(T, -T)\}} = \frac{\int \mathcal{D}A^\mu \mathcal{D}\bar{\psi} \mathcal{D}\psi \, e^{iS[A, \bar{\psi}, \psi] + i\int d^4x[\bar{\eta}\psi + \bar{\psi}\eta - j\cdot A]}}{\int \mathcal{D}A^\mu \mathcal{D}\bar{\psi} \mathcal{D}\psi \, e^{iS[A, \bar{\psi}, \psi]}}.$$

$$(7.6.21)$$

From Eq. (7.6.20) it follows that the Feynman diagram vertex is

$$i\mathcal{L}_{\text{int}}^{\text{QED}} = -iq_c \bar{\psi}\slashed{A}\psi \equiv -iq_c A_\mu j_{\text{Dirac}}^\mu = -iq_c \bar{\psi} A_\mu \gamma^\mu \psi \quad \Rightarrow \quad \text{\raisebox{-0.5em}{\includegraphics}} = -iq_c\gamma^\mu, \quad (7.6.22)$$

where for an electron $q_c = -e \equiv -|e|$ and for u and d quarks we have $q_c = +\frac{2}{3}|e|$ and $q_c = -\frac{1}{3}|e|$, respectively.

Scalar quantum electrodynamics (scalar QED): A complex/charged Klein-Gordon field (cKG) has global phase invariance that leads to a conserved current and hence to a conserved charge. It was discussed earlier as a classical field theory in Eq. (3.2.48), in the context of relativistic quantum mechanics in Sec. 4.3.2, and as a charged scalar quantum field in Sec. 6.2.7. We produce a gauge field theory from this by making the phase transformation local, by introducing the gauge field A^μ and by replacing ∂_μ with D_μ. This leads to the gauge theory known as scalar electrodynamics and after quantizing as scalar QED. The Lagrangian density is

$$\mathcal{L} = -\tfrac{1}{4}F_{\mu\nu}F^{\mu\nu} + (D_\mu\phi)^*(D^\mu\phi) - m^2\phi^*\phi = \mathcal{L}_{\text{Maxwell}} + (D_\mu\phi)^*(D^\mu\phi) - m^2\phi^*\phi$$

$$\equiv \mathcal{L}_{\text{Maxwell}} + \mathcal{L}_{\text{cKG}} + \mathcal{L}_{\text{int}}^{\text{sQED}} \equiv \mathcal{L}_0 + \mathcal{L}_{\text{int}}^{\text{sQED}} = \mathcal{L}_0 + \left[q_c A_\mu j_{\text{KG}}^\mu + q_c^2 A^2 \phi^*\phi\right] \quad (7.6.23)$$

$$= \left[-\tfrac{1}{4}F_{\mu\nu}F^{\mu\nu}\right] + \left[\partial_\mu\phi^*\partial^\mu\phi - m^2\phi^*\phi\right] + \left[q_c A_\mu i\left\{\phi(\partial^\mu\phi^*) - \phi^*\partial^\mu\phi\right\} + q_c^2 A^2 \phi^*\phi\right]$$

and is invariant under the gauge transformation

$$\phi(x) \to \phi'(x) = e^{i\alpha(x)}\phi(x), \quad \phi^*(x) \to \phi'^*(x) = e^{-i\alpha(x)}\phi^*(x), \quad (7.6.24)$$

$$A_\mu(x) \to A'_\mu(x) = A_\mu(x) - (1/q_c)\partial_\mu\alpha(x).$$

The Euler-Lagrange equations are given by Eqs. (7.6.2) and (3.3.3) as we showed in Problem 3.8,

$$(D_\mu D^\mu + m^2)\phi = 0, \quad \partial_\mu F^{\mu\nu} = j^\nu, \quad \text{where} \quad j^\mu \equiv iq_c\left[\phi^* D^\mu\phi - (D^\mu\phi)^*\phi\right]. \quad (7.6.25)$$

The canonical momenta for the charged scalar field are

$$\pi^* = \frac{\partial\mathcal{L}}{\partial(\partial_0\phi^*)} = D^0\phi = (\partial^0 + iq_c A^0)\phi, \quad \pi = \frac{\partial\mathcal{L}}{\partial(\partial_0\phi)} = (D^0\phi)^* = (\partial^0 - iq_c A^0)\phi^*. \quad (7.6.26)$$

Since $\mathcal{L}_{\text{int}}^{\text{sQED}}$ contains no time derivatives of A^μ, then the usual Maxwell Hamiltonian will result within the full Hamiltonian density, which is given by

$$\mathcal{H} = \mathcal{H}_{\text{Maxwell}} + \pi\dot{\phi} + \pi^*\dot{\phi}^* - (D_\mu\phi)^*(D^\mu\phi) + m^2\phi^*\phi \quad (7.6.27)$$

$$= \mathcal{H}_{\text{Maxwell}} + [\pi^*\pi + \boldsymbol{\nabla}\phi^* \cdot \boldsymbol{\nabla}\phi + m^2\phi^*\phi] + [-\mathcal{L}_{\text{int}}^{\text{sQED}} - q_c^2(A^0)^2\phi^*\phi]$$

$$= \mathcal{H}_{\text{Maxwell}} + \mathcal{H}_{\text{cKG}} + \mathcal{H}_{\text{int}}^{\text{sQED}} \equiv \mathcal{H}_0 + \mathcal{H}_{\text{int}}^{\text{sQED}},$$

where from Eq. (6.2.173) we have classically $\mathcal{H}_{\text{cKG}} = \pi^*\pi + \boldsymbol{\nabla}\phi^* \cdot \boldsymbol{\nabla}\phi + m^2\phi^*\phi$. Note that $\mathcal{H}_{\text{Maxwell}}$ is the photon Hamiltonian density appropriate to the choice of gauge such as the Gupta-Bleuler form corresponding to Eq. (6.4.116).

Proof: To prove Eq. (7.6.27) we use the result that

$$\pi\dot{\phi} + \pi^*\dot{\phi}^* - (D_\mu\phi)^*(D^\mu\phi) + m^2\phi^*\phi \tag{7.6.28}$$

$$= \pi(\pi^* - iq_cA^0\phi) + \pi^*(\pi + iq_cA^0\phi^*) - \pi^*\pi - (D_k\phi)^*(D^k\phi) + m^2\phi^*\phi$$

$$= [\pi^*\pi + \boldsymbol{\nabla}\phi^* \cdot \boldsymbol{\nabla}\phi + m^2\phi^*\phi] + iq_cA^0(\pi^*\phi^* - \pi\phi) + iq_cA_k(\phi^*\partial^k\phi - \{\partial^k\phi^*\}\phi)$$

$$- q_c^2A_kA^k\phi^*\phi$$

$$= \mathcal{H}_{\text{cKG}} + iq_cA_\mu(\phi^*\partial^\mu\phi - \phi\partial^\mu\phi^*) - q_c^2A_\mu A^\mu\phi^*\phi - q_c^2(A^0)^2\phi^*\phi$$

$$= \mathcal{H}_{\text{cKG}} - \mathcal{L}_{\text{int}}^{\text{sQED}} - q_c^2(A^0)^2\phi^*\phi \equiv \mathcal{H}_{\text{cKG}} + \mathcal{H}_{\text{int}}^{\text{sQED}}.$$

Note that the derivative interaction involving ϕ leads to $\mathcal{H}_{\text{int}}^{\text{sQED}} \neq -\mathcal{L}_{\text{int}}^{\text{sQED}}$ and

$$\mathcal{H}_{\text{int}}^{\text{sQED}} = -\mathcal{L}_{\text{int}}^{\text{sQED}} - q^2(A^0)^2\phi^*\phi, \tag{7.6.29}$$

which contains a noncovariant contribution involving A^0. This noncovariant term is exactly what is needed to make the spectral function covariant, which it must be in a correct Hamiltonian formulation of a covariant Lagrangian. This term compensates for the effect of derivative operators in the interaction acting on the time-ordering operator as we have seen before. Working in the Gupta-Bleuler formalism for the photon the interacting system is nonsingular and canonical quantization then proceeds in the usual way with $A^\mu \to \hat{A}^{\mu\dagger}$. For the charged scalar field we have $\phi \to \hat{\phi}$ and $\phi^* \to \hat{\phi}^\dagger$. This leads to the Hamiltonian density operator $\hat{\mathcal{H}}$. Denoting the evolution operator with sources as $\hat{U}^{jJ^*J}(t'',t')$ we obtain the spectral function for scalar QED,

$$F^{jJ^*J}(t'',t') = \text{tr}\{\hat{U}^{jJ^*J}(t'',t')\} = \text{tr}\left\{Te^{-i\int_{t'}^{t''} d^4x[\hat{\mathcal{H}}+j\cdot\hat{A}-J^*\hat{\phi}-J\hat{\phi}^*]}\right\} \tag{7.6.30}$$

$$= \text{tr}\left\{Te^{-i\int_{t'}^{t''} d^4x[\hat{\mathcal{H}}_{\text{Maxwell}}+\hat{\mathcal{H}}_{\text{cKG}}+\hat{\mathcal{H}}_{\text{int}}^{\text{sQED}}+j\cdot\hat{A}-J^*\hat{\phi}-J\hat{\phi}^*]}\right\}$$

$$= e^{i\int d^4x \,\mathcal{L}_{\text{int}}^{\text{sQED}}\left(\frac{i\delta}{\delta j_\mu(x)}, \frac{-i\delta}{\delta J^*(x)}, \frac{-i\delta}{\delta J(x)}\right)} \int \mathcal{D}A^\mu\mathcal{D}\phi\mathcal{D}\phi^* \, e^{i\int_{t'}^{t''} d^4x[\mathcal{L}_{\text{Maxwell}}+\mathcal{L}_{\text{cKG}}+J^*\phi+J\phi^*-j\cdot A]}$$

$$= \int \mathcal{D}A^\mu\mathcal{D}\phi\mathcal{D}\phi^* \, e^{i\int_{t'}^{t''} d^4x[\mathcal{L}+J^*\phi+J\phi^*-j\cdot A]}$$

$$= \int \mathcal{D}A^\mu\mathcal{D}\phi\mathcal{D}\phi^* \, e^{iS[A,\phi^*,\phi]+i\int_{t'}^{t''} d^4x[J^*\phi+J\phi^*-j\cdot A]}.$$

Passing from the second to the third line in the above equation uses the noncovariant term in $\mathcal{H}_{\text{int}}^{\text{sQED}}$ to compensate for the effect of moving the time derivatives in $\mathcal{L}_{\text{int}}^{\text{sQED}}$ through the time-ordering operator, which can be verified with a little effort. An argument in terms of the interaction picture is given in Itzykson and Zuber (1980). We can be satisfied that the above result must be true since the trace of the evolution operator must be covariant and gauge invariant if everything is done correctly and since we must reproduce the noninteracting result if $q = 0$. The generating functional for scalar QED is then

$$Z[j,J^*,J] \equiv \lim_{T\to\infty(1-i\epsilon)} \frac{\text{tr}\{\hat{U}^{j\bar{\eta}\eta}(T,-T)\}}{\text{tr}\{\hat{U}(T,-T)\}} = \frac{\int \mathcal{D}A^\mu\mathcal{D}\phi\mathcal{D}\phi^* \, e^{iS[A,\phi^*,\phi]+i\int d^4x[J^*\phi+J\phi^*-j\cdot A]}}{\int \mathcal{D}A^\mu\mathcal{D}\bar{\phi}\mathcal{D}\psi \, e^{iS[A,\phi^*,\phi]}}. \tag{7.6.31}$$

From Eq. (7.6.30) we see that the vertices in Feynman diagrams arise from

$$i\mathcal{L}_{\text{int}}^{\text{sQED}} = (-iq_c)A_\mu\left\{\phi^*i\partial^\mu\phi - \phi i\partial^\mu\phi^*\right\} + iq_c^2g_{\mu\nu}A^\mu A^\nu\phi^*\phi \tag{7.6.32}$$

and the two terms, respectively, correspond to the Feynman diagram vertices

$$\text{(left diagram)} = -iq_c(p'^\mu + p^\mu) \qquad \text{and} \qquad \text{(right diagram)} = 2iq_c^2 g_{\mu\nu}, \qquad (7.6.33)$$

for a scalar boson with electric charge q_c. To understand the three-point vertex on the left consider the case of a boson of charge q_c with incoming momentum p^μ and outgoing momentum p'^μ, i.e., boson to boson. The operator $\hat{\phi}$ in the vertex annihilates an initial state boson of charge q_c and $\hat{\phi}^\dagger$ annihilates the boson in the outgoing state. The issue of moving ∂^μ through the time-ordering operator is taken care of in the above formalism and so we ignore this for the present purposes. Equivalently, we can focus only on the spatial components and then understand that the final result must be covariant. Referring to Eq. (6.2.177) we see that in the above case the term $\hat{\phi}^\dagger i\partial^\mu \hat{\phi}$ will generate p^μ through the $\hat{f}_{\mathbf{p}} e^{-ip\cdot x}$ part of $\hat{\phi}$ and the term $\hat{\phi} i\partial^\mu \hat{\phi}^\dagger$ will generate $-p'^\mu$ through the $\hat{f}_{\mathbf{p}}^\dagger e^{ip\cdot x}$ part of $\hat{\phi}^\dagger$. It is left as an exercise to show that the other three cases all correspond to the same vertex, i.e., antiboson to antiboson, boson-antiboson annihilation, and boson-antiboson creation. Based on the $(1/n!)\phi^n$ discussion above, the four-point vertex on the right is straightforward to understand since the second term in the interaction involves $A^2 = g_{\mu\nu} A^\mu A^\nu$ rather than $(1/2!)A^2$ and so gives a factor of $2! = 2$ in this Feynman diagram vertex. This diagram is referred to as a *seagull diagram*, since with some imagination it resembles an approaching flying seagull.

7.6.3 Example Tree-Level Results

For the case of a scalar field with quartic self-interaction we have seen earlier in Sec. 7.5.1 that the differential cross-section for $\phi\phi \to \phi\phi$ scattering is given by Eq. (7.5.47). We now consider other examples of tree-level caculations.

Yukawa Interaction

Consider the invariant amplitude arising from the scattering of two *distinguishable* fermions, $f_1 f_2 \to f_1 f_2$, with the masses m_1 and m_2 through the exchange of a virtual (off-shell) scalar boson of mass m_ϕ and with coupling constant g. We label the spinors as $u_1^s(p)$ and $u_2^s(p)$ and so on to denote that they correspond to masses m_1 and m_2, respectively. Using the Feynman rules at tree level we have only the t-channel diagram contribution

$$i\mathcal{M}_{f_1 f_2 \to f_1 f_2} = \qquad \xrightarrow{\text{time}} \qquad (7.6.34)$$

$$= (-ig)\bar{u}_1^{s'}(p')u_1^s(p)\frac{i}{(p'-p)^2 - m_\phi^2}(-ig)\bar{u}_2^{r'}(k')u_2^r(k),$$

where *time is from left to right* and we omit the $i\epsilon$ since $(p'-p)$ is spacelike.

In the case of identical fermions there are two contributing tree-level diagrams with the final (or initial) states exchanged and a relative minus sign between them. We have $p^\mu + k^\mu = p'^\mu + k'^\mu$ from four-momentum conservations, and the Mandelstam variables defined in Fig. 7.10 are

$$s = (p+k)^2 = (p'+k')^2, \quad t = (p'-p)^2 = (k'-k)^2, \quad u = (k'-p)^2 = (p'-k)^2. \quad (7.6.35)$$

There are tree-level t-channel and u-channel contributions,

$$i\mathcal{M}_{ff\to ff} = \text{(diagrams)} \tag{7.6.36}$$

$$= i(-ig)^2 \left\{ \frac{\bar{u}^{s'}(p')u^s(p)\bar{u}^{r'}(k')u^r(k)}{t - m_\phi^2} - \frac{\bar{u}^{s'}(p')u^r(k)\bar{u}^{r'}(k')u^s(p)}{u - m_\phi^2} \right\},$$

with a relative minus sign. They are related by $p', s' \leftrightarrow k', r'$ (or equivalently $p, s \leftrightarrow k, r$). Exchanging both initial and final states would not lead to a topologically distinct diagram.

In the case of fermion-antifermion scattering at tree level we have a t-channel and an s-channel contribution to the invariant amplitude,

$$i\mathcal{M}_{f\bar{f}\to f\bar{f}} = \text{(diagrams)} \tag{7.6.37}$$

$$= i(-ig)^2 \left\{ \frac{\bar{u}^{s'}(p')u^s(p)\bar{v}^r(k)v^{r'}(k')}{t - m_\phi^2} - \frac{\bar{u}^{s'}(p')v^{r'}(k')\bar{v}^r(k)u^s(p)}{s - m_\phi^2} \right\}.$$

The relative minus sign between the two diagrams arises since the relative time orientation of the diagrams does not change the argument about contractions being between adjacent operators in the time-ordered product in the LSZ formalism. Rotating the right-hand diagram counterclockwise by $90°$ the diagrams are the same under the interchange $p', s' \leftrightarrow -k, -r$ or $p, s \leftrightarrow -k', -r'$ and are antisymmetric.

If the antifermion is not the same species as the particle, $f_1\bar{f}_2 \to f_1\bar{f}_2$, then only the t-channel contribution occurs. In the t-channel the virtual boson is spacelike, whereas in the s-channel it is timelike. If we consider the annihilation and creation of particle-antiparticle pairs of different species, $f_1\bar{f}_1 \to f_2\bar{f}_2$, then only the s-channel can contribute unless the exchanged scalar particle carries species charges.

Yukawa potential: Consider the tree-level Yukawa interaction between two distinguishable fermions of masses m_1 and m_2 in the nonrelativistic limit. Keeping terms of $\mathcal{O}(|\mathbf{p}|/m)$ we have

$$p^\mu \to (m_1, \mathbf{p}),\ p'^\mu \to (m_1, \mathbf{p}'),\ k^\mu \to (m_2, \mathbf{k}),\ k'^\mu \to (m_s, \mathbf{k}'), \tag{7.6.38}$$

$$\Rightarrow t = (p' - p)^2 = (k' - k)^2 = -|\mathbf{p}' - \mathbf{p}|^2 + \mathcal{O}(\mathbf{p}^4),\quad \mathbf{p}' - \mathbf{p} = -(\mathbf{k}' - \mathbf{k}).$$

Using Eq. (7.6.34) we need only keep $\mathcal{O}((|\mathbf{p}|/m)^0)$ in spinors so $u^s(p) \to u^s(0)$. From Eq. (4.4.118) we have then $\bar{u}_1^{s'}(p')u_1^s(p) \to \bar{u}_1^{s'}(0)u_1^s(0) = 2m_1\phi_{s'}^\dagger\phi_s = 2m_1\delta^{ss'}$ and $\bar{u}_2^{r'}(k')u_2^r(k) \to \bar{u}_2^{r'}(0)u_2^r(0) = 2m_2\phi_{r'}^\dagger\phi_r = 2m_2\delta^{rr'}$, which leads to

$$i\mathcal{M} \to \frac{ig^2}{|\mathbf{p}' - \mathbf{p}|^2 + m_\phi^2} 4m_1 m_2 \delta^{ss'}\delta^{rr'}. \tag{7.6.39}$$

We see that the spin of the fermions is unchanged by the interaction in this limit. The nonrelativistic limit of the Yukawa interaction leads to the attractive spin-independent Yukawa potential between two nonrelativistically moving fermions,

$$V(r) = -\frac{g^2}{4\pi r} e^{-m_\phi r}, \tag{7.6.40}$$

which is negative, and so the Yukawa potential is attractive between two fermions.

Proof: *S-matrix in quantum mechanics:* We have $\langle \mathbf{p}'|\mathbf{p}\rangle = (2\pi)^3\delta^3(\mathbf{p}' - \mathbf{p})$ in quantum mechanics and so $\langle \mathbf{x}|\mathbf{p}\rangle = e^{i\mathbf{p}\cdot\mathbf{x}}$. Given $\hat{H} = \hat{H}_0 + \hat{V}$, then the nonrelativistic scattering matrix to first order in the interaction potential $\hat{V}$ is

$$S_{fi} = \delta_{fi} - i2\pi\delta(E_f - E_i)\langle\phi_f|\hat{V}|\phi_i\rangle,$$

which is the Born approximation. Here $|\phi_i\rangle$ and $|\phi_f\rangle$ are eigenstates of $\hat{H}_0$ and $\delta_{fi} \equiv \langle\phi_f|\phi_i\rangle$. For distinguishable two-particle states $|\phi\rangle = |\mathbf{p}, \mathbf{k}\rangle$ and with a translationally invariant two-particle potential $\hat{V}$ we have

$$\begin{aligned}
\langle\phi_f|\hat{V}|\phi_i\rangle &= \langle\mathbf{p}', \mathbf{k}'|\hat{V}|\mathbf{p}, \mathbf{k}\rangle = \int d^3x_1 d^3x_2 V(\mathbf{x}_2 - \mathbf{x}_1)e^{i(\mathbf{p}-\mathbf{p}')\cdot\mathbf{x}_1}e^{i(\mathbf{k}-\mathbf{k}')\cdot\mathbf{x}_2}\\
&= \int d^3X\, e^{i(\mathbf{p}+\mathbf{k}-\mathbf{p}'-\mathbf{k}')\cdot\mathbf{X}}\int d^3r\, V(\mathbf{r})\exp\left\{-i\left(\frac{m_2(\mathbf{p}-\mathbf{p}')}{m_1+m_2} - \frac{m_1(\mathbf{k}-\mathbf{k}')}{m_1+m_2}\right)\cdot\mathbf{r}\right\}\\
&= (2\pi)^3\delta^3(\mathbf{p}+\mathbf{k}-\mathbf{p}'-\mathbf{k}')\int d^3r\, V(\mathbf{r})e^{-i(\mathbf{k}'-\mathbf{k})\cdot\mathbf{r}}\\
&= (2\pi)^3\delta^3(\mathbf{p}+\mathbf{k}-\mathbf{p}'-\mathbf{k}')V(\mathbf{q}),
\end{aligned}$$

where $\mathbf{q} \equiv \mathbf{k}' - \mathbf{k} = -(\mathbf{p}' - \mathbf{p})$, $V(\mathbf{q})$ is the asymmetric Fourier transform of $V(\mathbf{r})$ and the center-of-mass and relative coordinates are, respectively,

$$\mathbf{X} \equiv \frac{m_1\mathbf{x}_1 + m_2\mathbf{x}_2}{m_1+m_2}, \quad \mathbf{r} \equiv \mathbf{x}_2 - \mathbf{x}_1.$$

We have used the fact that the Jacobian for the change of variable in the integrations is unity. The two-particle scattering matrix is then

$$S_{fi} = \delta_{fi} + i(2\pi)^4\delta^4(p + k - p' - k'))[-V(\mathbf{q})].$$

Finally, we obtain $V(\mathbf{r})$ as the inverse asymmetric Fourier transform

$$V(\mathbf{r}) = \int (d^3q/(2\pi)^3)\, e^{i\mathbf{q}\cdot\mathbf{x}}V(\mathbf{q}).$$

Relation to S-matrix in QFT: Recall that the normalization of our fermion momentum states is given by Eq. (6.3.83),

$$\langle f; \mathbf{p}, s|f; \mathbf{p}', s'\rangle = 2E_{\mathbf{p}}\delta^{ss'}(2\pi)^3\delta^3(\mathbf{p} - \mathbf{p}')$$

and similarly for antifermion states $|\bar{f}; \mathbf{p}, s\rangle$. The difference with quantum mechanics normalization is $\sqrt{2E}$ for each $|\mathbf{p}\rangle$ and nonrelativistically $\sqrt{2E} \to \sqrt{2m}$. This gives an overall additional factor of $4m_1m_2$ in our nonrelativistic QFT two-particle scattering matrix, $S_{fi}^{\text{QFT}} \to 4m_1m_2 S_{fi}^{\text{QM}}$. In QFT we have

$$S_{fi} = \delta_{fi} + iT_{fi} = \delta_{fi} + i(2\pi)^4\delta^4(p_A + p_B - \sum_{i=1}^n p_i)\mathcal{M}_{fi}.$$

Putting things together we then identify in the nonrelativistic limit

$$\frac{\mathcal{M}}{4m_1m_2} = \frac{g^2}{|\mathbf{p}' - \mathbf{p}|^2 + m_\phi^2} = \frac{g^2}{\mathbf{q}^2 + m_\phi^2} = -V(\mathbf{q}) = -V(|\mathbf{q}|), \tag{7.6.41}$$

where we omit the factor $\delta^{ss'}\delta^{rr'}$ since both cases are spin preserving. Then

$$V(r) = \int \frac{d^3q}{(2\pi)^3}e^{i\mathbf{q}\cdot\mathbf{x}}V(\mathbf{q}) = -g^2\int \frac{d^3q}{(2\pi)^3}e^{i\mathbf{q}\cdot\mathbf{x}}\frac{1}{\mathbf{q}^2 + m_\phi^2} = -\frac{g^2}{4\pi r}e^{-m_\phi r}, \tag{7.6.42}$$

where we have used Eq. (A.2.14).

If we replace one of the two fermions by an antifermion, then we find: (i) a minus sign from Eq. (4.4.118) since $\bar{v}_2^r(k)v_2^{r'}(k') \to \bar{v}_2^r(0)v_2^{r'}(0) = -2m_2\delta^{rr'}$, and (ii) a second minus sign because ignoring the time orientation of the diagram we have exchanged two identical fermion lines as we did in Eq. (7.6.37). The two minus signs cancel and so the fermion-antifermion Yukawa potential is also attractive. Finally, if we also replace the other fermion with an antifermion, then there are two additional sign changes that again cancel. So the antifermion-antifermion Yukawa potential is again attractive. As discussed in Sec. 5.2.2 Yukawa used the typical range of the strong interaction of $\simeq 1$ fm to predict the existence and approximate mass ($m_\pi \sim 200$ MeV) of the pion. The pion is a pseudoscalar and so the relevant Yukawa interaction is $-g\bar{\psi}i\gamma_5\phi\psi$. The γ^5 will mix large and small components of the spinors in the Dirac representation, giving rise to a leading p-wave nature of the pion-nucleon interaction since the pseudoscalar pion parity will couple to odd ℓ in a partial wave expansion for the scattering. However, the essentials of the above argument survive. A discussion of the pion-nucleon interaction in terms of the isospin formalism can be found in Chapter 10 of Bjorken and Drell (1964).

Quantum Electrodynamics

Consider the invariant amplitude arising from the scattering of two distinguishable charged fermions, $f_1 f_2 \to f_1 f_2$, with charges q_1, q_2 and masses m_1, m_2 through the exchange of a virtual (off-shell) photon in an arbitrary covariant gauge, e.g., electron-muon scattering, $e^- \mu^- \to e^- \mu^-$. Similar to the case of the Yukawa interaction the Feynman rules give at tree level a t-channel contribution to $f_1 f_2 \to f_1 f_2$,

$$i\mathcal{M}_{f_1 f_2 \to f_1 f_2} = \qquad \qquad \text{time} \longrightarrow \qquad \qquad (7.6.43)$$

$$= (-iq_1)\bar{u}_1^{s'}(p')\gamma^\mu u_1^s(p)\frac{i\left[-g_{\mu\nu} + (1-\xi)(q_\mu q_\nu/q^2)\right]}{q^2}(-iq_2)\bar{u}_2^{r'}(k')\gamma^\nu u_2^r(k),$$

where the four-momentum transfer is $q^\mu \equiv (p'-p)^\mu = -(k'-k)^\mu$ and $t = q^2 = q_\mu q^\mu$. Current conservation for a free Dirac fermion is $\partial_\mu \hat{j}_{\text{Dirac}}^\mu(x) = \partial_\mu{:}\hat{\bar{\psi}}(x)\gamma^\mu\hat{\psi}(x){:} = 0$, where we recall that Noether currents are conserved when the equations of motion are satisfied, i.e., when the field operators are on-shell. For plane wave solutions, $u^s(p)$ and $v^s(p)$, current conservation gives

$$(p'-p)_\mu \bar{u}^{s'}(p')\gamma^\mu u^s(p) = (p'-p)_\mu \bar{v}^{r'}(p)\gamma^\mu v^s(p') = 0 \qquad (7.6.44)$$

$$(p'+p)_\mu \bar{v}^{s'}(p')\gamma^\mu u^s(p) = (p'+p)_\mu \bar{u}^{r'}(p)\gamma^\mu v^s(p') = 0.$$

Proof: There are two proofs of this result.

(i) *Direct proof:* From Eqs. (4.4.110) and (4.4.122) we have $(\slashed{p}-m)u^s(p) = (\slashed{p}+m)v^s(p) = 0$ and $\bar{u}^s(p)(\slashed{p}-m) = \bar{v}^s(p)(\slashed{p}+m) = 0$. Use the shorthand notation $u \equiv u^s(p)$, $u' \equiv u^{s'}(p')$ and so on. Then $\slashed{p}u = mu$, $\slashed{p}'u' = mu'$, $\slashed{p}v = -mv$, $\slashed{p}'v' = -mv'$, $\bar{u}\slashed{p} = \bar{u}m$, $\bar{u}'\slashed{p}' = \bar{u}'m$, $\bar{v}\slashed{p} = -m\bar{v}$ and $\bar{v}'\slashed{p}' = -m\bar{v}'$. Then we note that $(p'-p)_\mu\bar{u}'\gamma^\mu u = \bar{u}'(\slashed{p}'-\slashed{p})u = (m-m)\bar{u}'u = 0$. Similarly, $(p'-p)_\mu\bar{v}\gamma^\mu v' = (-m+m)\bar{v}v' = 0$, $(p'+p)_\mu\bar{u}\gamma^\mu v' = (-m+m)\bar{u}v' = 0$, and $(p'+p)_\mu\bar{v}\gamma^\mu u' = (m-m)\bar{v}u' = 0$.

(ii) *Proof from current conservation:* Acting with ∂_μ on $\hat{j}_{\text{Dirac}}^\mu(x)$ in Eq. (6.3.210) we see that current conservation requires that the terms containing $\hat{b}^\dagger\hat{b}$, $\hat{d}^\dagger\hat{d}$, $\hat{b}^\dagger\hat{d}^\dagger$, and $\hat{d}\hat{b}$ must each vanish and Eq. (7.6.44) then follows.

This shows that the $q^\mu q^\nu$ terms in $\mathcal{M}_{f_1 f_2 \to f_1 f_2}$ vanish and so the result is gauge invariant as expected. So we can choose any convenient ξ. For all fermion calculations below it is easily verified that Eq. (7.6.44) ensures gauge invariance. We present each result in Feynman gauge, $\xi = 1$, since it is the most convenient.

In the case of identical fermions scattering, $ff \rightarrow ff$, we find at tree level,

$$iM_{ff \rightarrow ff} = \quad \text{} \tag{7.6.45}$$

$$= i(-iq_c)^2 \left\{ \frac{\bar{u}^{s'}(p')\gamma^\mu u^s(p)[-g_{\mu\nu}]\bar{u}^{r'}(k')\gamma^\mu u^r(k)}{t} - \frac{\bar{u}^{s'}(p')\gamma^\mu u^r(k)[-g_{\mu\nu}]\bar{u}^{r'}(k')\gamma^\nu u^s(p)}{u} \right\},$$

where both the t and u channels contribute and the relative sign is due to the interchange of two final state fermions. An important example is $e^-e^- \rightarrow e^-e^-$, which is referred to as *Møller scattering*.

For fermion-antifermion scattering at tree level we again have a t-channel and an s-channel contribution to the invariant amplitude as we did in the Yukawa case,

$$iM_{f\bar{f} \rightarrow f\bar{f}} = \quad \text{} \tag{7.6.46}$$

$$= i(-iq_c)^2 \left\{ \frac{\bar{u}^{s'}(p')\gamma^\mu u^s(p)[-g_{\mu\nu}]\bar{v}^r(k)\gamma^\nu v^{r'}(k')}{t} - \frac{\bar{u}^{s'}(p')\gamma^\mu v^{r'}(k')[-g_{\mu\nu}]\bar{v}^r(k)\gamma^\nu u^s(p)}{s} \right\},$$

where the relative sign occurs for the same reason as it did in Eq. (7.6.37). An example of this is electron-positron scattering, $e^-e^+ \rightarrow e^-e^+$, which is referred to as *Bhabha scattering*. If the antifermion is not the same species as the fermion, $f_1\bar{f}_2 \rightarrow f_1\bar{f}_2$, then only the t-channel contribution occurs, e.g., $e^-\mu^+ \rightarrow e^-\mu^+$. If we consider the annihilation and creation of particle-antiparticle pairs of different species, $f_1\bar{f}_1 \rightarrow f_2\bar{f}_2$, then only the s-channel can contribute, e.g., $e^-e^+ \rightarrow \mu^-\mu^+$.

For fermion-antifermion annihilation into photons, $f\bar{f} \rightarrow \gamma\gamma$, we find at tree level,

$$iM_{f\bar{f} \rightarrow \gamma\gamma} = \quad \text{} \tag{7.6.47}$$

$$= i(-iq_c)^2 \epsilon^\mu(p', \lambda')^* \epsilon^\nu(k', \kappa')^* \bar{v}^r(k) \left\{ \frac{\gamma_\nu[(\not{p} - \not{p}') + m]\gamma_\mu}{t - m^2} + \frac{\gamma_\mu[(\not{p} - \not{k}') + m]\gamma_\nu}{u - m^2} \right\} u^s(p),$$

where λ', κ' are the polarizations of the photons with momenta p', k', respectively. Spinor objects are ordered right to left in the direction of the fermion arrow.

In the case of fermion-photon scattering, $f\gamma \rightarrow f\gamma$, we find at tree level,

$$iM_{f\gamma \rightarrow f\gamma} = \quad \text{} \tag{7.6.48}$$

$$= i(-iq_c)^2 \epsilon^\nu(k', \kappa')^* \bar{u}^{s'}(p') \left\{ \frac{\gamma_\nu[(\not{p} + \not{k}) + m]\gamma_\mu}{s - m^2} + \frac{\gamma_\mu[(\not{p} - \not{k'}) + m]\gamma_\nu}{u - m^2} \right\} u^s(p)\epsilon^\mu(k, \lambda).$$

An important example is electron-photon scattering, $e^-\gamma \to e^-\gamma$, which is *Compton scattering*. The kinematics of this was briefly discussed in Sec. 4.7.

Bremsstrahlung (German for "braking radiation") is the QED process in which a charged particle emits a photon due to its interaction with an external classical electromagnetic field, A^μ_{cl}. If the external field is macroscopic, then it is a reasonable approximation to treat it as a classical field. For example, the external electromagnetic field could be that in a particle accelerator or it could be the Coulomb field of a very heavy or fixed nucleus. The QED Lagrangian density in this case has the replacement $A^\mu \to A^\mu + A^\mu_{\text{ext}}$, where $A^\mu_{\text{ext}}(x)$ is the classical background electromagnetic field and where the corresponding background classical **E** and **B** fields are obtained in the usual way from this. The lowest-order Bremsstrahlung process corresponds to Eq. (7.6.48) but where for the initial photon we make the replacement $\epsilon^\mu(k, \lambda) \to A^\mu_{\text{ext}}(k)$ with $A^\mu_{\text{ext}}(k)$ being the Fourier transform of $A^\mu_{\text{ext}}(x)$. In the case of a fixed nucleus with charge Ze we have $\mathbf{A}_{\text{ext}} = 0$ and $A^0_{\text{ext}} = -Ze/4\pi|\mathbf{x}|$, which is obtained from Eq. (2.7.94) in the case of a point nucleus. The detailed calculation is given in Itzykson and Zuber (1980) and is meaningful provided $Ze/(v/c) \ll 1$, where v is the initial charged particle velocity with respect to the nucleus. The result of this calculation is the *Bethe-Heitler cross-section*, which is a little too complicated to derive here. However, it is noteworthy that this cross-section has a $1/\omega$ divergence as the emitted photon energy $k'^0 \equiv \omega$ vanishes, $\omega \to 0$. This is an example of an *infrared divergence* that results due to the massless nature of the photon. We can regulate the divergence by adding a small photon mass term m_γ and then taking $m_\gamma \to 0$ at the end of the calculation. It is not possible to measure the presence or absence of an arbitrarily soft photon ($\omega \to 0$) with any real detector. That means that we need to consider all processes differing only by the number of initial or final state photons that can contribute terms of the same order in e to the cross-section. When we do so we find that the infrared divergences cancel in these sums over photon number. For further discussion and details, see, for example, Chapter 20 of Schwartz (2013), Chapter 6 of Peskin and Schroeder (1995) and Chapter 13 of Weinberg (1995).

There is no tree-level contribution to *light-by-light scattering*, $\gamma\gamma \to \gamma\gamma$, which is a classically forbidden process. The lowest-order diagram consists of a single fermion loop with two initial state photons and two final state photons attached. There are four vertices and so we have $\mathcal{M}_{\gamma\gamma \to \gamma\gamma} \propto \alpha^2$, where $\alpha \equiv e^2/4\pi \simeq 1/137$ is the fine structure constant. So we have for the differential cross-section of light by light in a vacuum, $d\sigma_{\gamma\gamma \to \gamma\gamma} \propto |\mathcal{M}_{\gamma\gamma \to \gamma\gamma}|^2 \propto \alpha^4$, which leads to a small cross-section. The first observations were in the form of Delbrück scattering, which is the contribution to the scattering of light from the Coulomb field of a heavy nucleus, e.g., lead with atomic number $Z = 82$. In this case the four couplings are to a fermion loop in the Coulomb field of the nucleus, with a coupling constant e for the incoming and outgoing scattered photon and an effective coupling constant Ze from each of the two vertices arising from the Coulomb field of the nucleus. This effectively leads to $\mathcal{M} \propto Z^2\alpha^2$ and an effective enhancement of the light-by-light contribution to the photon scattering cross-section. That there must be at least four electromagnetic vertices in the loop in Delbrück scattering rather than three is due to Furry's theorem, which will be discussed later, but is essentially the result that only fermion loops with an even number of photon lines attached are nonvanishing due to the C invariance of QED.

Coulomb potential: Consider the interaction between two different fermions of masses m_1 and m_2 and charges q_1 and q_2, respectively, in the nonrelativistic limit. We follow the arguments used to obtain the Yukawa potential in Eq. (7.6.40). At leading order in $(|\mathbf{p}|/m)$ Eq. (7.6.38) again applies. At leading order in $(|\mathbf{p}|/m)$ for the amplitude we can take $u^s(p) \to u^s(0)$ and $v^s(p) \to v^s(0)$. From Eq. (4.4.118) we then have $\bar{u}_1^{s'}(p')\gamma^\mu u_1^s(p) \to \bar{u}_1^{s'}(0)\gamma^0 u_1^s(0) = 2m_1\phi_{s'}^\dagger\phi_s = 2m_1\delta^{ss'}$ and similarly $\bar{u}_2^{r'}(k')\gamma^\mu u_2^r(k) \to 2m_2\delta^{rr'}$. These results are easily understood in the Dirac representation since the three γ matrices mix upper and lower components, while γ^0 does not, but fully contracted spinor quantities such as $\bar{u}^{s'}(p')\Gamma u^s(p)$ are representation independent. With $-g^{00} = -1$ from the photon propagator we arrive at

$$i\mathcal{M} \to \frac{-iq_1q_2}{|\mathbf{p}' - \mathbf{p}|^2}4m_1m_2\delta^{ss'}\delta^{rr'} = \frac{-iq_1q_2}{|\mathbf{q}|^2}4m_1m_2\delta^{ss'}\delta^{rr'}. \tag{7.6.49}$$

Then $\mathcal{M}/4m_1m_2 = -q_1q_2/|\mathbf{q}|^2 = -V(\mathbf{q})$, where we again ignore the spin delta functions as we did in Eq. (7.6.41). The Coulomb potential between two nonrelativistic fermions is then given by

$$V(\mathbf{q}) = \frac{q_1q_2}{|\mathbf{q}|^2} \quad \Rightarrow \quad V(r) = \int \frac{d^3q}{(2\pi)^3}e^{i\mathbf{q}\cdot\mathbf{x}}V(\mathbf{q}) = \frac{q_1q_2}{4\pi r} = \frac{Z_1Z_2\alpha}{r}, \tag{7.6.50}$$

where $Z_1 \equiv q_1/e$ and $Z_2 \equiv q_2/e$ are the charges in units of $e \equiv |e|$. So fermions with the same sign charges repel and those with opposite sign charges attract. If a fermion is replaced by an antifermion, then $\bar{v}_1^{s'}(p')\gamma^\mu v_1^s(p) \to \bar{v}_1^{s'}(0)\gamma^0 v_1^s(0) = -2m_1\chi_{s'}^\dagger\chi_s = -2m_1\delta^{ss'}$. The extra minus sign changes the sign of the potential and equates to changing the sign of the charge, $q_1 \to -q_1$. So, as expected, fermions and antifermions have opposite electric charges and the Coulomb potential between a fermion-antifermion pair is attractive. For two antifermions the extra signs cancel and the potential is the same as for two fermions.

Scalar Quantum Electrodynamics

In the case of identical scalars scattering, $\phi\phi \to \phi\phi$, we find at tree level,

$$i\mathcal{M}_{\phi\phi\to\phi\phi} = \quad\quad\quad\quad\quad\quad\quad\quad\quad\quad \tag{7.6.51}$$

$$= i(-iq_c)^2\left\{\frac{(p'+p)^\mu[-g_{\mu\nu}](k'+k)^\nu}{t} + \frac{(k'+p)^\mu[-g_{\mu\nu}](p'+k)^\nu}{u}\right\},$$

where the relative plus sign between the t and u channels is due to the bosonic nature of the scalars. This is the scalar QED version of Møller scattering. The photon four-momentum is $q^\mu = (p'-p)^\mu = -(k'-k)^\mu$ in the first diagram and $q^2 = t$. Note that $q\cdot(p'+p) = (p'-p)\cdot(p'+p) = p'^2 - p^2$ and $q\cdot(k'+k) = -(k'-k)\cdot(k'+k) = -k'^2 + k^2$ and these vanish for on-shell states since $p'^2 = p^2 = k'^2 = k^2 = m^2$. In the second diagram we have $q^\mu = (k'-p)^\mu = -(p'-k)$ and $q^2 = u$ and the same result follows. So the amplitude is gauge invariant and we can use the Feynman gauge for the photon propagator in scalar QED as we did in QED. As for QED this is related to current conservation of the free particle, since in the mode expansion of the operator form of the free current,

$\hat{j}^{\mu}_{\mathrm{KG}} = i[\hat{\phi}^{\dagger}\partial^{\mu}\hat{\phi} - (\partial^{\mu}\hat{\phi})^{\dagger}\hat{\phi}]$, the on-shell current is conserved, $(p'-p)\cdot(p'+p) = m_{\phi}^2 - m_{\phi}^2 = 0$. So in both scalar QED and QED, we see that gauge invariance of the tree-level amplitudes follows from current conservation.

In the case of scalar-antiscalar scattering, $\phi\phi^* \to \phi\phi^*$, the tree-level result is

$$i\mathcal{M}_{\phi\phi^*\to\phi\phi^*} = \quad\raisebox{-1em}{}\quad + \quad\raisebox{-1em}{} \tag{7.6.52}$$

$$= i(-iq_c)^2 \left\{ \frac{(p'+p)^{\mu}[-g_{\mu\nu}](-k'-k)^{\nu}}{t} + \frac{(p'-k')^{\mu}[-g_{\mu\nu}](p-k)^{\nu}}{s} \right\}.$$

For scalar-antiscalar annihilation into photons, $\phi\phi^* \to \gamma\gamma$, we have at tree level,

$$i\mathcal{M}_{\phi\phi^*\to\gamma\gamma} = \quad\raisebox{-1em}{}\quad + \quad + \quad \tag{7.6.53}$$

$$= i(-iq_c)^2 \epsilon^{\mu}(p',\lambda')^* \epsilon^{\nu}(k',\kappa')^* \left\{ \frac{(2p-p')_{\mu}(k'-2k)_{\nu}}{t-m_{\phi}^2} + \frac{(2p-k')_{\mu}(p'-2k)_{\nu}}{u-m_{\phi}^2} - 2g_{\mu\nu} \right\}.$$

In the case of scalar-photon scattering, $\phi\gamma \to \phi\gamma$, we find at tree level,

$$i\mathcal{M}_{\phi\gamma\to\phi\gamma} = \quad + \quad + \quad \tag{7.6.54}$$

$$= i(-iq_c)^2 \epsilon^{\nu}(k',\kappa')^* \left\{ \frac{(2p+k)_{\mu}(2p'+k')_{\nu}}{s-m_{\phi}^2} + \frac{(2p-k')_{\nu}(2p'-k)_{\mu}}{u-m_{\phi}^2} - 2g_{\mu\nu} \right\} \epsilon^{\mu}(k,\kappa),$$

Coulomb potential in scalar QED: If we consider the Coulomb interaction between two distinguishable charged scalars, we need only consider the t-channel contribution to Møller scattering in Eq. (7.6.51). In the nonrelativistic limit we have $t = q^2 = (p'-p)^2 \to -|\mathbf{p}'-\mathbf{p}|^2 = -|\mathbf{q}|^2$ and $(p'+p)\cdot(k'+k) =\to (2m_{\phi})^2 = 4m_{\phi}^2$. Then $i\mathcal{M}_{\phi\phi\to\phi\phi} \to -iq_c^2 4m_{\phi}^2/|\mathbf{q}|^2$. Our QFT normalization for scalar momentum eigenstates contains a factor $\sqrt{2E_{\mathbf{p}}}$ as in the fermion case and so to compare with the nonrelativistic formalism we divide by $4m_{\phi}^2$ as done in Eq. (7.6.41). This results in $V(\mathbf{q}) = V(|\mathbf{q}|) = -\mathcal{M}_{\phi\phi\to\phi\phi}/4m_{\phi}^2 = q_c^2/|\mathbf{q}|^2$ and so $V(r) = q_c^2/4\pi r$, which is just the repulsive Coulomb potential found in the fermion case in Eq. (7.6.40). If one scalar boson is replaced by its antiparticle, then the potential changes sign and becomes attractive, which follows since $(k'-k)^{\mu} \to (-k'-k)^{\mu}$ in the t-channel as we saw in $\mathcal{M}_{\phi\phi^*\to\phi\phi^*}$ and the s-channel contribution vanishes in the nonrelativistic limit since $(p'-k')\cdot(p-k)/s \to 0$. In the case of two antibosons there are two sign changes, which leads to repulsion. So the nonrelativistic limit of scalar QED again leads to the familiar Coulomb interaction.

7.6.4 Substitution Rules and Crossing Symmetry

An important consequence of the Feynman rules for calculating invariant amplitudes is that if two Feynman diagrams differ only in having one external line moved from the initial to the final state or vice versa, then the two amplitudes are related by the *substitution rules*:

$$\text{scalar in } (p) \longleftrightarrow \text{scalar out } (p' = -p), \tag{7.6.55}$$

$$\text{charged scalar in } (p) \longleftrightarrow \text{charged antiscalar out } (p' = -p),$$

$$\text{photon in } (k, h), \ \epsilon^\mu(k, h) \longleftrightarrow \text{photon out } (k' = -k, h' = -h), \ \epsilon^\mu(k', h')^*,$$

$$\text{fermion in } (p, s), \ u^s(p) \longleftrightarrow \text{antifermion out } (p' = -p, s' = -s), \ v^{s'}(p'),$$

$$\text{antifermion in } (p, s), \ \bar{v}^s(p) \longleftrightarrow \text{fermion out } (p' = -p, s' = -s), \ \bar{u}^{s'}(p'),$$

where h is the photon helicity with $h = +1, -1$ for right and left circular polarization, respectively. For exchanges of fermions with antifermions the relative sign of the diagram is assigned based on the rules above. Substitution rules are reversible and can also be sequentially applied. Two reactions related by one or more applications of the substitution rule are referred to as *crossed reactions*. Provided we only ever consider theories with fermion number conservation we will always be considering dual fermion exchanges, f in and f out goes to $\bar{f}$ out and $\bar{f}$ in. Crossed reactions can have useful relationships between their invariant amplitudes, $\mathcal{M}$, which are referred to as either *crossing symmetry* or crossing relations. A discussion of this in relation to causality and analyticity can be found in Omnès and Froissart (1963), Roman (1969), Itzykson and Zuber (1980). Since p^0 and p'^0 are always positive for physical particles, then using $p' = -p$ to go between invariant amplitudes means that there is an assumed analytic continuation in the p^0 plane. In two-body scattering of scalar particles there are three physical regions of s, t, u corresponding to the reactions labeled I, II and III, where I is $\phi(p_1)\phi(p_2) \rightarrow \phi(p_3)\phi(p_4)$, II is $\phi(p_1)\bar{\phi}(-p_3) \rightarrow \bar{\phi}(-p_2)\phi(p_4)$, and III is $\phi(p_1)\bar{\phi}(-p_4) \rightarrow \bar{\phi}(-p_2)\phi(p_3)$. Referring to these reactions we see that $\sqrt{s}$, $\sqrt{t}$ and $\sqrt{u}$ each becomes the center of mass energy, respectively. This can be understood by rearranging each of the t-channel and u-channel diagrams in Fig. 7.10 to resemble the s-channel diagram. The crossing symmetry for the invariant amplitudes is given by $\mathcal{M}_{\text{I}}(p_1, p_2; p_3, p_4) = \mathcal{M}_{\text{II}}(p_1, -p_3; -p_2, p_4) = \mathcal{M}_{\text{III}}(p_1, -p_4; p_3, -p_2)$, where p_1, p_2, p_3, p_4 can only be physical for at most one amplitude at a time. Using these momentum replacements we can rewrite these crossing symmetry relations in terms of s, t, u arguments as

$$\mathcal{M}_{\text{I}}(s, t, u) = \mathcal{M}_{\text{II}}(t, s, u) = \mathcal{M}_{\text{III}}(u, t, s), \tag{7.6.56}$$

which is meaningful if we can analytically continue between the three physics regions in the Mandelstam plane. This is shown in Fig. 7.12 for the equal mass case. The charge conjugate and time-reversed processes do not lead to new s, t, u regions. In the case of particles with spin such as fermions and photons the situation becomes more complicated.

Crossing symmetry plays an important role in the S-matrix approach to the strong interaction. This theoretical approach was popular in the 1960s and was championed by Geoffrey Chew and colleagues at the University of California, Berkeley. S-matrix theory was an attempt to formulate a theory of the strong interaction using the analytic properties of the scattering matrix to calculate the interactions of bound states without assuming the existence of any underlying point-particle description. This led to the concept of *bootstrap theory* where the properties of each subatomic particle arise solely as a result of its relationship to all other subatomic particles. Chew and Steven Frautschi observed that the mesons appear to fall into families known as straight-line *Regge trajectories*. In these trajectories

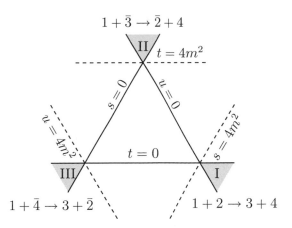

$$1 + \bar{3} \rightarrow \bar{2} + 4$$

II $\quad t = 4m^2$

$s = 0$ $\quad$ $u = 0$

$u = 4m^2$ $\quad$ $s = 4m^2$

$t = 0$

III $\quad$ I

$$1 + \bar{4} \rightarrow 3 + \bar{2} \qquad\qquad 1 + 2 \rightarrow 3 + 4$$

Figure 7.12   The Mandelstam plane. The physical regions are shaded and labeled as I, II and III.

the square of the mass of a meson appears approximately linearly proportional to its spin. Since this is a pattern seen in bound states in quantum mechanics, it was viewed by many at the time that this was evidence that hadrons were bound states of elementary particles. The concepts of crossing symmetry, analyticity, causality and dispersion relations are all connected in the S-matrix approach (Chew, 1962; Frautschi, 1963; Omnès and Froissart, 1963; Barone and Predazzi, 2002). With the advent of QCD the S-matrix approach has historically become of less interest, although it continues to provide many useful lessons.

7.6.5 Examples of Calculations of Cross-Sections

Now that we have experience in using the Feynman rules to calculate tree-level invariant amplitudes $\mathcal{M}$, it is useful to calculate some examples of differential cross-sections using them.

Differential cross-section for $\phi^+\phi^+ \rightarrow \phi^+\phi^+$ in scalar QED: The scattering of two identical scalar bosons with charge $q_c = Ze \equiv Z|e|$ has both the t-channel and u-channel contributions in Eq. (7.6.51). Using the Mandelstam variables in Eq. (7.6.35) we find

$$(p' + p) \cdot (k' + k) = -(p' + p)^2 = -2m_\phi^2 - 2p \cdot p' = -2m_\phi^2 + 2p \cdot (p + k' + k) = 2p \cdot (k' + k)$$
$$= (2m_\phi^2 + 2k \cdot p) - (2m_\phi^2 - 2k' \cdot p) = s - u \tag{7.6.57}$$

and similarly $(k' + p) \cdot (p' + k) = -(p' + k)^2 = 2k \cdot (p + k') = s - t$. Combining these results with Eq. (7.3.82) and using $p_i^{\rm cm} = p_f^{\rm cm}$ we find for the differential cross-section in the center-of-momentum frame

$$\left(\frac{d\sigma}{d\Omega}\right)_{\rm cm} = \frac{1}{64\pi^2 s}|\mathcal{M}_{fi}|^2 = \frac{q_c^2}{64\pi^2 s}\left|\frac{s-u}{t} + \frac{s-t}{u}\right|^2 = \frac{Z^2\alpha^2}{4s}\left|\frac{s-u}{t} + \frac{s-t}{u}\right|^2. \tag{7.6.58}$$

Polarizations, tricks and trace identities for fermions: Recall the discussions in Sec. 7.6.1. Where we do not have initially polarized beams or target, we must average the cross-section over initial polarizations. If we do not measure the polarization of any of the final state particles, then we must sum the cross-section over final state polarizations. For two initial state fermions with spins r, s and two final state fermions with spins r', s' this means that we can define the unpolarized squared modulus of the invariant amplitude as

$$|\mathcal{M}|^2 \xrightarrow{\text{unpolarized}} \overline{|\mathcal{M}|^2} = \sum_{r',s'} \left(\tfrac{1}{4}\sum_{r,s}\right)|\mathcal{M}|^2. \tag{7.6.59}$$

The following types of identities are useful when evaluating $|\mathcal{M}|^2 = \mathcal{M}\mathcal{M}^*$,

$$[\bar{\psi}\gamma^\mu\chi]^* = [\psi^\dagger\gamma^0\gamma^\mu\chi]^\dagger = \chi^\dagger\gamma^{\mu\dagger}\gamma^0\psi = \bar{\chi}\gamma^\mu\psi, \text{ and similarly } [\bar{\psi}\chi]^* = \bar{\chi}\psi,$$

$$[\bar{\psi}i\gamma^5\chi]^* = \bar{\chi}i\gamma_5\psi, \quad [\bar{\psi}\gamma^\mu\gamma^5\chi]^* = \bar{\chi}\gamma^\mu\gamma^5\psi, \quad [\bar{\psi}\sigma^{\mu\nu}\chi]^* = \bar{\chi}\sigma^{\mu\nu}\psi. \tag{7.6.60}$$

The different forms arise alone or in combination depending on the interactions in the theory and correspond to the five bilinear forms in Eq. (4.4.159). Let Γ be any spinor matrix and define $\overline{\Gamma} \equiv \gamma^0\Gamma^\dagger\gamma^0$ then $\overline{\Gamma_1\Gamma_2} = \overline{\Gamma_2}\,\overline{\Gamma_1}$ and $[\bar{\psi}\Gamma\chi]^* = \bar{\chi}\overline{\Gamma}\psi$. We find that $\overline{\gamma^\mu} = \gamma^\mu$, $\overline{\sigma^{\mu\nu}} = \sigma^{\mu\nu}$, $\overline{\gamma^\mu\gamma^5} = \gamma^\mu\gamma^5$ and $\overline{\gamma^5} = -\gamma^5$. Using the spinor relations $\sum_s u_\alpha^s(p)\bar{u}_\beta^s(p) = (\slashed{p}+m)_{\alpha\beta}$ and $\sum_s v_\alpha^s(p)\bar{v}_\beta^s(p) = (\slashed{p}-m)_{\alpha\beta}$ from Eq. (4.4.129) we arrive at the *Casimir trick*,

$$\sum_{s_1,s_2}[\bar{u}_1^{s_1}(p_1)\Gamma u_2^{s_2}(p_2)][\bar{u}_1^{s_1}(p_1)\Gamma' u_2^{s_2}(p_2)]^* = \mathrm{tr}[\Gamma(\slashed{p}_2+m_2)\overline{\Gamma}'(\slashed{p}_1+m_1)]. \tag{7.6.61}$$

Observe that if $u_1 \to v_1$ then $m_1 \to -m_1$ and if $u_2 \to v_2$ then $m_2 \to -m_2$.

> **Proof:** Using $[\bar{\psi}\Gamma'\chi]^* = \bar{\chi}\overline{\Gamma}'\psi$ the left-hand side becomes
>
> $$\sum_{s_1,s_2} \bar{u}_{1\alpha}^{s_1}(p_1)\Gamma_{\alpha\beta}u_{2\beta}^{s_2}(p_2)\bar{u}_{2\gamma}^{s_2}(p_2)\overline{\Gamma}'_{\gamma\delta}u_{1\delta}^{s_1}(p_1) \tag{7.6.62}$$
>
> $$= \sum_{s_1,s_2} u_{1\delta}^{s_1}(p_1)\bar{u}_{1\alpha}^{s_1}(p_1)\Gamma_{\alpha\beta}u_{2\beta}^{s_2}(p_2)\bar{u}_{2\gamma}^{s_2}(p_2)\overline{\Gamma}'_{\gamma\delta}$$
>
> $$= \Gamma_{\alpha\beta}(\slashed{p}_2+m_2)_{\beta\gamma}\overline{\Gamma}'_{\gamma\delta}(\slashed{p}_1+m_1)_{\delta\alpha} = \mathrm{tr}[\Gamma(\slashed{p}_2+m_2)\overline{\Gamma}'(\slashed{p}_1+m_1)].$$

We now recognize the importance of the trace identities and contractions given in Sec. A.3. They are needed to evaluate fermion loop contributions to Feynman diagrams as we saw in Eq. (7.6.1) and they are needed to evaluate differential cross-sections using the Casimir trick. For calculations of the trace of a large number of γ-matrices there are a variety of computer programs publicly available. In the case where we wish to select one or more of the initial or final spins, then we can do so by including the relevant Γ spin projection operators as given in Eq. (4.4.148). We could choose this to be a helicity projection, where helicity is defined in Eq. (4.6.1). In the limit of massless fermions or when the momenta are very large we would insert the chirality projection operators in Eq. (4.6.5).

Differential cross-section for $e^+e^- \to \mu^+\mu^-$: At tree level the invariant amplitude corresponds to the right-hand diagram in Eq. (7.6.46) with the relative minus sign irrelevant. Both e^- and μ^- are the particle rather than the antiparticle by convention and both have charge $q_c = -e = -|e|$ and their masses are m_e and m_μ, respectively. Recalling that $s = (p+k)^2 = (p'+k')^2$ the result is

$$i\mathcal{M} = ie^2 \frac{\bar{u}^{s'}(p')\gamma^\mu v^{r'}(k')\bar{v}^r(k)\gamma_\mu u^s(p)}{s}, \tag{7.6.63}$$

where $\bar{u}^{s'}(p')$, $v^{r'}(k')$ are the final state muon, antimuon, respectively, and where $u^s(p)$, $\bar{v}^r(k)$ are the initial state electron, positron, respectively. This leads to

$$|\mathcal{M}|^2 = \frac{e^4}{s^2}[\bar{u}^{s'}(p')\gamma^\mu v^{r'}(k')\bar{v}^r(k)\gamma_\mu u^s(p)][\bar{u}^{s'}(p')\gamma^\nu v^{r'}(k')\bar{v}^r(k)\gamma_\nu u^s(p)]^*. \tag{7.6.64}$$

If we are only interested in unpolarized scattering, then we use Eq. (7.6.61) to obtain

$$\overline{|\mathcal{M}|^2} = \frac{1}{4}\sum_{r,s,r',s'}|\mathcal{M}|^2 = \frac{e^4}{4s^2}\mathrm{tr}[\gamma^\mu(\slashed{k}'-m_\mu)\gamma^\nu(\slashed{p}'+m_\mu)]\,\mathrm{tr}[\gamma_\mu(\slashed{p}+m_e)\gamma_\nu(\slashed{k}-m_e)]$$

$$= \frac{2e^4}{s^2}\left[t^2 + u^2 + 4s(m_e^2 + m_\mu^2) - 2(m_e^2 + m_\mu^2)^2\right]. \tag{7.6.65}$$

Proof: The right-hand side of the first line follows from Eq. (7.6.61). We use the trace identities in Sec. A.3 to find

$$\text{tr}[\gamma_\mu(\not{p} + m_e)\gamma_\nu(\not{k} - m_e)] = p^\alpha k^\beta \text{tr}[\gamma_\mu \gamma_\alpha \gamma_\nu \gamma_\beta] - m_e^2 \text{tr}[\gamma_\mu \gamma_\nu]$$
$$= 4(p_\mu k_\nu + p_\nu k_\mu - g_{\mu\nu} p \cdot k) - 4m_e^2 g_{\mu\nu}, \tag{7.6.66}$$

$$\text{tr}[\gamma^\mu(\not{k}' - m_\mu)\gamma^\nu(\not{p}' + m_\mu)] = 4(k'^\mu p'^\nu + k'^\nu p'^\mu - g^{\mu\nu} p' \cdot k') - 4m_\mu^2 g^{\mu\nu}, \tag{7.6.67}$$

$$\overline{|\mathcal{M}|^2} = (4e^4/s^2)[k'^\mu p'^\nu + k'^\nu p'^\mu - g^{\mu\nu}(p' \cdot k' + m_\mu^2)][p_\mu k_\nu + p_\nu k_\mu - g_{\mu\nu}(p \cdot k + m_e^2)]$$
$$= (8e^4/s^2)[p \cdot p'k \cdot k' + p \cdot k'k \cdot p' + m_\mu^2 p \cdot k + m_e^2 p' \cdot k' + 2m_e^2 m_\mu^2]. \tag{7.6.68}$$

The Mandelstam variables satisfy

$$s = 2m_e^2 + 2p \cdot k = 2m_\mu^2 + 2p' \cdot k', \quad t = m_e^2 + m_\mu^2 - 2p \cdot p' = m_e^2 + m_\mu^2 - 2k \cdot k',$$
$$u = m_e^2 + m_\mu^2 - 2p \cdot k' = m_e^2 + m_\mu^2 - 2k \cdot p', \tag{7.6.69}$$

which leads to $p \cdot p' = k \cdot k' = \frac{1}{2}(t - m_e^2 - m_\mu^2)$, $p \cdot k' = k \cdot p' = \frac{1}{2}(u - m_e^2 - m_\mu^2)$, $p \cdot k = \frac{1}{2}s - m_e^2$ and $p' \cdot k' = \frac{1}{2}s - m_\mu^2$. Substituting into Eq. (7.6.66) we find

$$\overline{|\mathcal{M}|^2} = (2e^4/s^2)\left[t^2 + u^2 + 2(m_e^2 + m_\mu^2)^2 + 2(m_e^2 + m_\mu^2)(s - t - u)\right]$$
$$= (2e^4/s^2)\left[t^2 + u^2 + 4s(m_e^2 + m_\mu^2) - 2(m_e^2 + m_\mu^2)^2\right], \tag{7.6.70}$$

where we have used Eq. (7.3.76) to give $-t - u = s - 2(m_e^2 + m_\mu^2)$.

The unpolarized differential cross-section in the CM frame is then obtained from Eq. (7.3.82), where in the CM frame we have $p_i^{\text{cm}} = |\mathbf{p}| = |\mathbf{k}|$, $p_f^{\text{cm}} = |\mathbf{p}'| = |\mathbf{k}'|$, $p^0 = p'^0 = k^0 = k'^0 = \frac{1}{2}E_{\text{cm}} = \frac{1}{2}\sqrt{s} \equiv E$, $\mathbf{p} = -\mathbf{k}$, $\mathbf{p}' = -\mathbf{k}'$ and $\mathbf{p}' \cdot \mathbf{p} = |\mathbf{p}'||\mathbf{p}|\cos\theta$. Note that θ is the angle between the initial e^- and the final μ^- in the CM frame. Comparing with the left-hand side of Fig. 7.11 we have $p_1 = p$, $p_2 = k$, $p_3 = p'$ and $p_4 = k'$ and the result for $e^+e^- \to \mu^+\mu^-$ is then

$$\left(\frac{d\sigma}{d\Omega}\right)_{\text{cm}} = \frac{1}{64\pi^2 s}\frac{p_f^{\text{cm}}}{p_i^{\text{cm}}}\overline{|\mathcal{M}|^2} = \frac{\alpha^2}{2s^3}\frac{|\mathbf{p}'|}{|\mathbf{p}|}\left[t^2 + u^2 + 4s(m_e^2 + m_\mu^2) - 2(m_e^2 + m_\mu^2)^2\right]$$
$$= \frac{\alpha^2}{16E^6}\frac{|\mathbf{p}'|}{|\mathbf{p}|}\left[E^4 + |\mathbf{p}|^2|\mathbf{p}'|^2\cos^2\theta + E^2(m_e^2 + m_\mu^2)\right], \tag{7.6.71}$$

where the last line follows from Eq. (7.6.69) since

$$s = 4E^2, \quad t = m_e^2 + m_\mu^2 - 2E^2 + 2\mathbf{p} \cdot \mathbf{p}', \quad u = m_e^2 + m_\mu^2 - 2E^2 - 2\mathbf{p} \cdot \mathbf{p}'. \tag{7.6.72}$$

If $E \gg m_e$ then we can neglect the electron mass and take $\mathbf{p} \to E$, which simplifies the expression. In the ultrarelativistic limit of $E \gg m_\mu$ we find for $e^+e^- \to \mu^+\mu^-$

$$\left(\frac{d\sigma}{d\Omega}\right)_{\text{cm}} = \frac{\alpha^2}{16E^2}\left[1 + \cos^2\theta\right] = \frac{\alpha^2}{4s}\left[1 + \cos^2\theta\right] \Rightarrow \sigma_{e^+e^- \to \mu^+\mu^-} = \frac{4\pi\alpha^2}{3s}, \tag{7.6.73}$$

since $\int d\Omega\,(1 + \cos^2\theta) = 2\pi\int_{-1}^{+1} d\cos\theta\,(1 + \cos^2\theta) = 2\pi[x + \frac{1}{3}x^3]_{-1}^{+1} = 16\pi/3$.

Differential cross-section for $e^- \mu^- \to e^- \mu^-$**:** Only the first diagram in Eq. (7.6.45) will contribute since e^- and μ^- are distinguishable, which gives

$$i\mathcal{M} = ie^2 \, \frac{\bar{u}^{s'}(p')\gamma^\mu u^s(p)\bar{u}^{r'}(k')\gamma_\mu u^r(k)}{t}. \tag{7.6.74}$$

Using Eq. (7.6.61) for the unpolarized cross-section the analog of Eq. (7.6.65) is

$$
\begin{aligned}
\overline{|\mathcal{M}|^2} &= \frac{1}{4}\sum_{r,s,r',s'}|\mathcal{M}|^2 = \frac{e^4}{4t^2}\,\mathrm{tr}[\gamma^\mu(\slashed{k}+m_\mu)\gamma^\nu(\slashed{k}'+m_\mu)]\,\mathrm{tr}[\gamma_\mu(\slashed{p}+m_e)\gamma_\nu(\slashed{p}'+m_e)] \\
&= \frac{8e^4}{t^2}\left[p\cdot k' p'\cdot k + p\cdot k p'\cdot k' - m_\mu^2 p\cdot p' - m_e^2 k\cdot k' + 2m_e^2 m_\mu^2\right] \\
&= \frac{2e^4}{t^2}\left[u^2 + s^2 + 4t(m_e^2+m_\mu^2) - 2(m_e^2+m_\mu^2)^2\right],
\end{aligned} \tag{7.6.75}
$$

where we have p, p' for e^- and k, k' for μ^-. The first line above is equivalent to the first line of Eq. (7.6.65) with the replacements $(p, k; p', k') \to (p, -p'; k', -k)$. This is an example of crossing symmetry and can be deduced from rotating the relevant Feynman diagrams. With these replacements in Eq. (7.6.68) we arrive at the second line above. It also follows that

$$s = (p+k)^2 \to (p-p')^2 = t, \ t = (p'-p)^2 \to (k'-p)^2 = u, \ u = (k'-p)^2 \to (k+p)^2 = s,$$

which when applied to Eq. (7.6.65) lead to the third line above. While we can perform the explicit calculations to obtain the results in Eq. (7.6.75) we obtain them more easily using the crossing symmetry replacements in Eq. (7.6.65).

The kinematics for reactions related by crossing symmetry are of course different as illustrated in Fig. 7.12. Since the initial and final state particles are the same, then using four-momentum conservation we arrive at the following CM frame kinematics when we neglect the electron mass for simplicity ($m_e \to 0$),

$$
\begin{aligned}
&p \equiv |\mathbf{p}| = |\mathbf{p}'|, \quad E^2 = p^2 + m_\mu^2 \\
&\mathbf{p}\cdot\mathbf{p}' = p^2\cos\theta, \quad E_{\mathrm{cm}} = E + p.
\end{aligned} \tag{7.6.76}
$$

Note that $p_i^{\mathrm{cm}} = p_f^{\mathrm{cm}} = p$ in this notation. We find $p\cdot k = p'\cdot k' = p(E+p)$, $p'\cdot k = p\cdot k' = p(E+p\cos\theta)$, $p\cdot p' = p^2(1-\cos\theta)$, $k\cdot k' = E^2 - p^2\cos\theta$, $s = (E+p)^2$ and $t = -2p^2(1-\cos\theta)$. Substituting into the second line of Eq. (7.6.75) gives

$$\overline{|\mathcal{M}|^2} = \frac{2e^4}{p^2(1-\cos\theta)^2}\left[(E+p)^2 + (E+p\cos\theta)^2 - m_\mu^2(1-\cos\theta)\right]. \tag{7.6.77}$$

So the tree-level differential cross-section for $e^-\mu^- \to e^-\mu^-$ in the limit $m_e \to 0$ is

$$\left(\frac{d\sigma}{d\Omega}\right)_{\mathrm{cm}} = \frac{1}{64\pi^2 s}\frac{p_f^{\mathrm{cm}}}{p_i^{\mathrm{cm}}}\overline{|\mathcal{M}|^2} = \frac{\alpha^2\left[(E+p)^2 + (E+p\cos\theta)^2 - m_\mu^2(1-\cos\theta)\right]}{2p^2(E+p)^2(1-\cos\theta)^2}. \tag{7.6.78}$$

For scattering in the forward direction, $\theta \to 0$, the tree-level differential cross-section diverges like $1/\theta^4$. The divergence as we approach forward scattering is expected based on our introductory discussion of the cross-section. For example, Rutherford scattering is the scattering of charged particles by the Coulomb interaction. As we approach the forward scattering limit we will find a diverging cross-section since every particle in the beam will be deflected at least infinitesimally by the infinite-range potential. The above forward divergence has the same origin.

Differential cross-section for $e^- \mu^+ \to e^- \mu^+$**:** Since μ^- like e^- is the particle, then we should treat μ^+ as an antiparticle. This gives the tree-level result

$$i\mathcal{M} = ie^2 \, \frac{\bar{u}^{s'}(p')\gamma^\mu u^s(p)\bar{v}^r(k)\gamma_\mu v^{r'}(k')}{t}, \tag{7.6.79}$$

$$\overline{|\mathcal{M}|^2} = \frac{1}{4}\sum_{r,s,r',s'}|\mathcal{M}|^2 = \frac{e^4}{4t^2}\,\mathrm{tr}[\gamma^\mu(\slashed{k}' - m_\mu)\gamma^\nu(\slashed{k} - m_\mu)]\,\mathrm{tr}[\gamma_\mu(\slashed{p} + m_e)\gamma_\nu(\slashed{p}' + m_e)]$$

$$= \frac{2e^4}{t^2}\left[u^2 + s^2 + 4t(m_e^2 + m_\mu^2) - 2(m_e^2 + m_\mu^2)^2\right], \tag{7.6.80}$$

which is identical to the result for μ^-. This follows from Eq. (7.6.75) and crossing symmetry. To understand this, convert the $\mathrm{tr}[\cdots]\mathrm{tr}[\cdots]$ term in Eq. (7.6.75) into the corresponding term in Eq. (7.6.80) with the replacement $k \leftrightarrow -k'$. This gives

$$s = (p + k)^2 \to (p - k')^2 = u, \ t = (p' - p)^2 \to (p' - p)^2 = t, \ u = (k' - p)^2 \to (-k - p)^2 = s$$

and the result then follows. One might ask why replacing a repulsive interaction with an attractive one leaves the tree-level cross-section invariant. The reason is that for a given wavepacket B in Fig. 7.9 we will obtain the same angle of deflection behavior from the wavepacket A at lowest order in α if we reverse both the sign of the interaction and also the sign of the impact parameter $b \to -b$. To calculate the cross-section we integrate over b and so changing the sign of the interaction does not change the tree-level cross-section. At higher order in α interference terms will lead to differences between $e^- \mu^- \to e^- \mu^-$ and $e^- \mu^+ \to e^- \mu^+$.

Compton scattering $\gamma e^- \to \gamma e^-$**:** The tree-level Feynman diagrams and the invariant amplitude for Compton scattering are shown in Eq. (7.6.48). Since $p^2 = p'^2 = m_e^2$ and $k^2 = k'^2 = 0$, we have for the denominators that $s - m_e^2 = (p+k)^2 - m_e^2 = 2p \cdot k$ and $u - m_e^2 = (p-k')^2 - m_e^2 = -2p \cdot k'$. For the numerators we have $(\slashed{p} + m_e)\gamma_\mu u^s(p) = [2p_\mu - \gamma_\mu(\slashed{p} - m_e)]u^s(p) = 2p_\mu u^s(p)$, using $\{\gamma^\mu, \gamma^\mu\} = 2g^{\mu\nu}$ and $(\slashed{p} - m_e)u^s(p) = 0$. Then the invariant amplitude becomes

$$i\mathcal{M} = -ie^2 \epsilon^\nu(k', \kappa')^* \bar{u}^{s'}(p')\left\{\frac{\gamma_\nu \slashed{k}\gamma_\mu + 2\gamma_\nu p_\mu}{2p \cdot k} + \frac{-\gamma_\mu \slashed{k}'\gamma_\nu + 2\gamma_\mu p_\nu}{-2p \cdot k'}\right\} u^s(p)\epsilon^\mu(k, \lambda).$$

The *Ward identity* is the result that if $i\mathcal{M}$ is a QED invariant amplitude and if

$$i\mathcal{M} = i\epsilon^\mu(k, \kappa)^* \mathcal{M}_\mu(k) \quad \text{or} \quad i\mathcal{M} = i\epsilon^\mu(k, \kappa)\mathcal{M}_\mu(k), \quad \text{then} \quad k^\mu \mathcal{M}_\mu(k) = 0, \tag{7.6.81}$$

where here $\epsilon^\mu(k, \kappa)$ is the polarization vector for an on-shell photon, $k^2 = 0$. We will not prove this identity until Sec. 8.3.2, but it is useful here. Write the Compton scattering invariant amplitude as $i\mathcal{M} \equiv i\epsilon^\nu(k', \kappa')^* \epsilon^\mu(k, \lambda)\mathcal{M}_{\nu\mu}(k', k)$, which defines

$$\mathcal{M}_{\nu\mu}(k', k) = -e^2 \bar{u}^{s'}(p')\left\{\frac{\gamma_\nu \slashed{k}\gamma_\mu + 2\gamma_\nu p_\mu}{2p \cdot k} + \frac{-\gamma_\mu \slashed{k}'\gamma_\nu + 2\gamma_\mu p_\nu}{-2p \cdot k'}\right\} u^s(p). \tag{7.6.82}$$

The Ward identity leads to $k'^\nu \mathcal{M}_{\nu\mu}(k', k) = k^\mu \mathcal{M}_{\nu\mu}(k', k) = 0$. When we calculate the unpolarized cross-section we will need to evaluate

$$\sum_{\kappa',\lambda}|\mathcal{M}|^2 = \sum_{\kappa',\lambda}\epsilon^\nu(k', \kappa')^* \epsilon^\mu(k, \lambda)\mathcal{M}_{\nu\mu}(k', k)\epsilon^\sigma(k', \kappa')\epsilon^\rho(k, \lambda)^* \mathcal{M}_{\sigma\rho}(k', k)^*$$

$$= (-g^{\nu\sigma})(-g^{\rho\mu})\mathcal{M}_{\nu\mu}(k', k)\mathcal{M}_{\sigma\rho}(k', k)^* = \mathcal{M}_{\nu\mu}(k', k)\mathcal{M}^{\nu\mu}(k', k)^*, \tag{7.6.83}$$

where we see that only the $-g^{\mu\nu}$ part of Eq. (6.4.56) survives for on-shell photons because of the Ward identity. For temporary notational convenience we define $\Gamma(k',k,p)$ using $\mathcal{M}_{\nu\mu}(k',k) \equiv -e^2 \bar{u}^{s'}(p')\Gamma_{\nu\mu}(k',k,p)u^s(p)$. Averaging over initial and summing over final photon polarizations and electrons spin gives

$$\overline{|\mathcal{M}|^2} = \tfrac{1}{4}\sum_{\kappa',\lambda}\sum_{s,s'}|\mathcal{M}|^2 = \tfrac{1}{4}\sum_{s,s'}\mathcal{M}_{\nu\mu}(k',k)\mathcal{M}^{\nu\mu}(k',k)^* \tag{7.6.84}$$

$$= (e^4/4)\sum_{s,s'}[\bar{u}^{s'}(p')\Gamma_{\nu\mu}(k',k,p)u^s(p)][\bar{u}^{s'}(p')\Gamma^{\nu\mu}(k',k,p)u^s(p)]^*$$

$$= (e^4/4)\mathrm{tr}[\Gamma_{\nu\mu}(k',k,p)(\not{p}+m_e)\overline{\Gamma^{\nu\mu}(k',k,p)}(\not{p}'+m_e)]$$

$$= \frac{e^4}{4}\left[\frac{A_1}{(2p\cdot k)^2} + \frac{A_2}{(2p\cdot k)(2p\cdot k')} + \frac{A_3}{(2p\cdot k')(2p\cdot k)} + \frac{A_4}{(2p\cdot k')^2}\right], \tag{7.6.85}$$

where we have used the Casimir trick in Eq. (7.6.61), where we note that $\overline{\gamma^\mu} = \gamma^\mu$ and $\overline{\gamma_\nu\not{k}\gamma_\mu} = \gamma_\mu\not{k}\gamma_\nu$, and where we have defined

$$A_1 \equiv \mathrm{tr}[(\gamma_\nu\not{k}\gamma_\mu + 2\gamma_\nu p_\mu)(\not{p}+m_e)(\gamma^\mu\not{k}\gamma^\nu + 2\gamma^\nu p^\mu)(\not{p}'+m_e)], \tag{7.6.86}$$

$$A_2 \equiv -\mathrm{tr}[(\gamma_\nu\not{k}\gamma_\mu + 2\gamma_\nu p_\mu)(\not{p}+m_e)(-\gamma^\nu\not{k}'\gamma^\mu + 2\gamma^\mu p^\nu)(\not{p}'+m_e)],$$

$$A_3 \equiv -\mathrm{tr}[(-\gamma_\mu\not{k}'\gamma_\nu + 2\gamma_\mu p_\nu)(\not{p}+m_e)(\gamma^\mu\not{k}\gamma^\nu + 2\gamma^\nu p^\mu)(\not{p}'+m_e)],$$

$$A_4 \equiv \mathrm{tr}[(-\gamma_\mu\not{k}'\gamma_\nu + 2\gamma_\mu p_\nu)(\not{p}+m_e)(-\gamma^\nu\not{k}'\gamma^\mu + 2\gamma^\mu p^\nu)(\not{p}'+m_e)].$$

We see that $A_1 \leftrightarrow A_4$ when $k^\mu \leftrightarrow -k'^\mu$, since μ and ν are dummy indices. Using Eq. (A.3.29) we reverse the gamma matrices in a trace to show that $A_2 = A_3$.

We can use the trace and contraction identities in Sec. A.3 to evaluate the traces, remembering that only traces with even numbers of γ-matrices are nonzero. Of the 16 terms in A_1 only eight are nonzero. The first term is

$$\mathrm{tr}[\gamma_\nu\not{k}\gamma_\mu\not{p}\gamma^\mu\not{k}\gamma^\nu\not{p}'] = \mathrm{tr}[\gamma_\nu\not{k}(-2\not{p})\not{k}\gamma^\nu\not{p}'] = -2\mathrm{tr}[-2\not{k}\not{p}\not{k}\not{p}'] \tag{7.6.87}$$

$$= 4\mathrm{tr}[\not{k}\not{p}\not{k}\not{p}'] = 16[(k\cdot p)(k\cdot p') - k^2(p\cdot p') + (k\cdot p')(k\cdot p)] = 32(k\cdot p)(k\cdot p'),$$

since $k^2 = 0$. Using similar steps and $p^2 = p'^2 = 0$ we eventually arrive at the result

$$A_1 = 16[4m_e^4 - 2m_e^2 p\cdot p' + 4m_e^2 k\cdot p - 2m_e^2 k\cdot p' + 2(k\cdot p)(k\cdot p')] \tag{7.6.88}$$

$$= 16[2m_e^4 + m_e^2 t + 2m_e^2(s-m_e^2) + m_e^2(u-m_e^2) - \tfrac{1}{2}(s-m_e^2)(u-m_e^2)]$$

$$= 16[2m_e^4 + m_e^2(s-m_e^2) - \tfrac{1}{2}(s-m_e^2)(u-m_e^2)],$$

where $s + t + u = 2m_\gamma^2 + 2m_e^2 = 2m_e^2$ with the Mandelstam variables given by

$$s = (p+k)^2 = (p'+k')^2 = m_e^2 + 2p\cdot k = m_e^2 + 2p'\cdot k', \tag{7.6.89}$$

$$t = (p'-p)^2 = (k'-k)^2 = 2m_e^2 - 2p'\cdot p = -2k'\cdot k,$$

$$u = (k'-p)^2 = (p'-k)^2 = m_e^2 - 2p\cdot k' = m_e^2 - 2p'\cdot k,$$

and where we note that $p\cdot k = p'\cdot k'$, $p\cdot k' = p'\cdot k$ and $k'\cdot k = p'\cdot p - m_e^2$. If we take $k^\mu \to -k'^\mu$ then t is unchanged but $s \leftrightarrow u$ and so we find

$$A_4 = 16[4m_e^4 - 2m_e^2 p\cdot p' - 4m_e^2 k'\cdot p + 2m_e^2 k'\cdot p' + 2(k'\cdot p)(k'\cdot p')] \tag{7.6.90}$$

$$= 16[2m_e^4 + m_e^2(u-m_e^2) - \tfrac{1}{2}(s-m_e^2)(u-m_e^2)]. \tag{7.6.91}$$

We obtain A_2 by again evaluating eight traces and combining these results gives

$$A_2 = A_3 = -16m_e^2[2m_e^2 + (p\cdot k - p'\cdot k)] = -8m_e^2[4m_e^2 + (s-m_e) + (u-m_e^2)]. \tag{7.6.92}$$

Using the expressions for $A_1, \ldots, A_4$ and after some algebra we arrive at

$$
\overline{\mathcal{M}^2} = 2e^4 \left\{ \left[\frac{2m_e^2}{(s-m_e^2)} + \frac{2m_e^2}{(u-m_e^2)} \right]^2 + \frac{4m_e^2}{(s-m_e^2)} + \frac{4m_e^2}{(u-m_e^2)} - \left[\frac{u-m_e^2}{s-m_e^2} + \frac{s-m_e^2}{u-m_e^2} \right] \right\}
$$
$$
= 2e^4 \left\{ m_e^4 \left[\frac{1}{p \cdot k} - \frac{1}{p \cdot k'} \right]^2 + 2m_e^2 \left[\frac{1}{p \cdot k} - \frac{1}{p \cdot k'} \right] + \left[\frac{p \cdot k'}{p \cdot k} + \frac{p \cdot k}{p \cdot k'} \right] \right\}. \qquad (7.6.93)
$$

We typically consider Compton scattering in the fixed-target or laboratory frame as shown in Fig. 4.2. For the unpolarized cross-section calculation we can use the differential cross-section result in the fixed-target frame in Eq. (7.3.94) and identify $m_1 = m_3 = m_\gamma = 0$, $m_2 = m_4 = m_e$, $p_1^\mu = k^\mu$, $p_2^\mu = p^\mu$, $p_3^\mu = k'^\mu$ and $p_4^\mu = p'^\mu$. Now write $\omega \equiv k^0 = |\mathbf{k}| = 1/\lambda$ and $\omega' \equiv k'^0 = |\mathbf{k}'| = 1/\lambda'$. This leads to

$$
\frac{d\sigma}{d(\cos\theta)} = \frac{1}{32\pi} \frac{1}{\omega m_e} \frac{\omega'}{m_e + \omega(1-\cos\theta)} \overline{|\mathcal{M}|^2} = \frac{1}{32\pi} \frac{\omega'^2}{m_e^2 \omega^2} \overline{|\mathcal{M}|^2}, \qquad (7.6.94)
$$

where to arrive at the last result we have used the Compton effect formula in Eq. (4.7.1) after rearranging it into the form

$$
(1/\omega') - (1/\omega) = (1-\cos\theta)/m_e \quad \Rightarrow \quad m_e\omega/\omega' = m_e + \omega(1-\cos\theta). \qquad (7.6.95)
$$

Recognizing that $p \cdot k = m_e\omega$ and $p \cdot k' = m_e\omega'$ we arrive at

$$
\overline{\mathcal{M}^2} = 2e^4 \left\{ m_e^2 \left[\frac{1}{\omega} - \frac{1}{\omega'} \right]^2 + 2m_e \left[\frac{1}{\omega} - \frac{1}{\omega'} \right] + \left[\frac{\omega'}{\omega} + \frac{\omega}{\omega'} \right] \right\} = 2e^4 \left\{ -\sin^2\theta + \left[\frac{\omega'}{\omega} + \frac{\omega}{\omega'} \right] \right\},
$$

where the final form follows from using Eq. (7.6.95). We then find the *Klein-Nishina formula* for unpolarized photon-electron scattering (Klein and Nishina, 1929),

$$
\frac{d\sigma}{d(\cos\theta)} = 2\pi \frac{d\sigma}{d\Omega} = \frac{\pi\alpha^2}{m_e^2} \frac{\omega'^2}{\omega^2} \left\{ -\sin^2\theta + \left[\frac{\omega'}{\omega} + \frac{\omega}{\omega'} \right] \right\}. \qquad (7.6.96)
$$

Note from Eq. (7.6.95) that $\omega' = \omega/[1 + (\omega/m_e)(1-\cos\theta)]$. Then for soft incident photons such that $E_\gamma = \omega \ll m_e$ we have $\omega' \to \omega$ and we obtain

$$
\frac{d\sigma}{d(\cos\theta)} = 2\pi \frac{d\sigma}{d\Omega} \to \frac{\pi\alpha^2}{m_e^2} \left\{ 1 + \cos^2\theta \right\} \quad \Rightarrow \quad \sigma = \frac{8\pi\alpha^2}{3m_e^2}, \qquad (7.6.97)
$$

since $\int_{-1}^{+1} d(\cos\theta)[1 + \cos^2\theta] = 8/3$. In this limit the Klein-Nishina formula has reduced to the Thomson cross-section for an e.m. wave scattering from an electron.

Cross-section for $e^+e^- \to \gamma\gamma$: The tree-level Feynman diagrams and invariant amplitude for this are given in Eq. (7.6.47). For the unpolarized cross-section the simple crossing-symmetry relations can be used, subject to an overall sign change due to the presence of a fermion that we will discuss later. In this sense the diagrams in Eq. (7.6.47) can be obtained from those in Eq. (7.6.48) by the replacements $p \to p$, $k \to -p'$, $p' \to -k$ and $k' \to k'$. We make these substitutions directly into the second line of Eq. (7.6.93) to obtain the needed result. For the Mandelstam variables the crossing symmetry corresponds to $s \leftrightarrow t$, $u \leftrightarrow u$, where now s, t, u are

$$
s = 2m_e^2 + 2p \cdot k = 2p' \cdot k', \quad t = m_e^2 - 2p' \cdot p = m_e^2 - 2k' \cdot k, \quad u = m_e^2 - 2p' \cdot k = m_e^2 - 2k' \cdot p.
$$

Substituting these into the first line of Eq. (7.6.93) after $s \leftrightarrow t$ obviously leads again to the same result as the direct momentum substitution into the second line. After introducing an overall sign change the result is

$$\overline{\mathcal{M}^2} = -2e^4 \left\{ \left[\frac{2m_e^2}{(t - m_e^2)} + \frac{2m_e^2}{(u - m_e^2)} \right]^2 + \frac{4m_e^2}{(t - m_e^2)} + \frac{4m_e^2}{(u - m_e^2)} - \left[\frac{u - m_e^2}{t - m_e^2} + \frac{t - m_e^2}{u - m_e^2} \right] \right\}$$

$$= 2e^4 \left\{ -m_e^4 \left[\frac{1}{p \cdot p'} + \frac{1}{p \cdot k'} \right]^2 + 2m_e^2 \left[\frac{1}{p \cdot p'} + \frac{1}{p \cdot k'} \right] + \left[\frac{p \cdot k'}{p \cdot p'} + \frac{p \cdot p'}{p \cdot k'} \right] \right\}, \quad (7.6.98)$$

where we have used $p' \cdot p = k' \cdot k$ and $p' \cdot k = k' \cdot p$. *Why did we need the overall sign change?* The simple crossing relations in Eq. (7.6.56) are valid for scalars, but for particles with spin there can be complications. In going from Compton scattering to pair annihilation we have replaced a fermion leg with an antifermion leg. This leads to a reversal of a fermion momentum and an associated $m_e \to -m_e$ as discussed when we considered the Casimir trick in Eq. (7.6.61). This leads to the replacement $\not{p} + m_e \to -\not{p} - m_e$ in the trace and hence an overall sign change in addition to the simple crossing relations. We can verify Eq. (7.6.98) directly without reference to crossing symmetry by evaluating the relevant versions of $A_1, \ldots, A_4$. In arriving at Eq. (7.6.80) from Eq. (7.6.75) we replaced two fermion legs with two antifermion legs and so the two minus signs canceled and as we saw the simple crossing relations were sufficient. *Simple rule: Introduce one minus sign for each replacement fermion $\leftrightarrow$ antifermion when using crossing symmetry relations for unpolarized fermions.*

The kinematics for $e^+ e^- \to \gamma\gamma$ in the CM frame are given by

$$p \equiv |\mathbf{p}|, \quad E = |\mathbf{p}'| = \sqrt{|\mathbf{p}|^2 + m_e^2}$$
$$\mathbf{p} \cdot \mathbf{p}' = pE \cos\theta, \quad E_{\text{cm}} = 2E. \quad (7.6.99)$$

We have $p_i^{\text{cm}} = |\mathbf{p}| = p$, $p_f^{\text{cm}} = |\mathbf{p}'| = E$, $p \cdot k = E^2 + p^2$, $p' \cdot k' = 2E^2$, $p' \cdot k = p \cdot k' = E(E + p \cos\theta)$, $p \cdot p' = k \cdot k' = E(E - p \cos\theta)$ and $s = 4E^2$. After some algebra we find that the tree-level differential cross-section in the CM frame for $e^+ e^- \to \gamma\gamma$ is

$$\left(\frac{d\sigma}{d\Omega} \right)_{\text{cm}} = \frac{1}{2\pi} \left(\frac{d\sigma}{d(\cos\theta)} \right)_{\text{cm}} = \frac{1}{64\pi^2 s} \frac{p_f^{\text{cm}}}{p_i^{\text{cm}}} \overline{|\mathcal{M}|^2} \quad (7.6.100)$$

$$= \frac{\alpha^2}{s} \frac{E}{p} \left\{ \frac{-2m_e^4}{(m_e^2 + p^2 \sin^2\theta)^2} + \frac{2m_e^2}{m_e^2 + p^2 \sin^2\theta} + \frac{E^2 + p^2 \cos^2\theta}{m_e^2 + p^2 \sin^2\theta} \right\}, \quad (7.6.101)$$

where we have used $E^2 - p^2 \cos^2\theta = m_e^2 + p^2 - p^2 \cos^2\theta = m_e^2 + p^2 \sin^2\theta$. In the high-energy limit ($E \gg m_e$) we have $(d\sigma/d(\cos\theta))_{\text{cm}} = (2\pi\alpha^2/s)(1 + \cos^2\theta)/\sin^2\theta$.

7.6.6   Unstable Particles

For a stable particle all of the standard arguments apply and lead to the LSZ formalism and to decay rate $\Gamma = 0$ from Eq. (7.3.69) and $\Pi(p^2) \in \mathbb{R}$ from Eqs. (7.2.8) and (7.5.51). For a complex scalar self-energy, $\Pi(p^2) = \Pi^R(p^2) + \Pi^I(p^2)$, where $\Pi^R(p^2) \equiv \text{Re}\,\Pi(p^2)$ and where $\Pi^I(p^2) \equiv i\text{Im}\,\Pi(p^2)$

is pure imaginary. If we initially neglect $\Pi^I(p^2)$ then we can use $\Pi^R(p^2)$ in place of $\Pi(p^2)$ and we can define, in analogy with Eqs. (7.5.51) and (7.5.54),

$$D_F^R(p) \equiv \frac{i}{p^2 - m_0^2 - \Pi^R(p^2) + i\epsilon} = \frac{iZ}{p^2 - m^2 - \Pi_r^R(p^2) + i\epsilon} \qquad (7.6.102)$$

in place of the $D_F(p)$ given in Eq. (7.5.51). Here we have defined the real mass m, the corresponding renormalization factor Z and the renormalized real self-energy $\Pi_r^R(p^2)$ using the analogs of Eqs. (7.5.52), (7.5.55) and (7.5.57), respectively. With the neglect of $\Pi^I(p^2)$ we can define stable asymptotic states of mass m in the usual way and the LSZ formalism will involve amputating $D_F^R(p)$ propagators.

The full propagator includes all possible insertions of $\Pi^I(p^2)$ in analogy with Eq. (7.5.49), $D_F(p) = D_F^R(p) + D_F^R(p)[-i\Pi^I(p^2)]D_F(p)$. This leads to the result $D_F^{-1}(p) = (D_F^R)^{-1}(p) + i\Pi^I(p^2)$ and so we have,

$$D_F(p) = \frac{i}{p^2 - m_0^2 - \Pi(p^2) + i\epsilon} = \frac{iZ}{p^2 - m^2 - \Pi_r^R(p^2) - iZ\mathrm{Im}\,\Pi(p^2)}, \qquad (7.6.103)$$

where $\Pi^I(p^2) = i\mathrm{Im}\,\Pi(p^2)$ and where $\Pi_r^R(p^2)$ is defined in terms of $\Pi^R(p^2)$ in analogy with Eq. (7.5.57). Recall that $\Pi_r^R(p^2)$ vanishes quadratically as $p^2 \to m^2$. We now see that there is no pole at $p^2 = m^2$ but $D_F(p)$ is enhanced there if $|\Pi^I(p^2)| = |\mathrm{Im}\,\Pi(p^2)| \ll m^2$. We refer to $p^2 \simeq m^2$ as the *resonance region*. Sufficiently close to the resonance region we have $\Pi_r^R(p^2) \simeq 0$ and

$$D_F(p) \xrightarrow{p^2 \simeq m^2} \frac{iZ}{p^2 - m^2 - iZ\mathrm{Im}\,\Pi(p^2)}. \qquad (7.6.104)$$

Consider the derivation of the LSZ formalism for the special case $S_{fi} = \langle \mathbf{p}|\hat{S}|\mathbf{k}\rangle$; the first term in the second line of Eq. (7.5.23) has the form

$$\sqrt{2E_\mathbf{k}}\langle \mathbf{p}\,\text{out}|\hat{a}^\dagger_{\mathbf{k}\,\text{out}}|\Omega\rangle = \sqrt{4E_\mathbf{k}E_\mathbf{p}}\langle\Omega|\hat{a}_{\mathbf{p}\,\text{out}}\hat{a}^\dagger_{\mathbf{k}\,\text{out}}|\Omega\rangle = (2\pi)^3\delta^3(\mathbf{p} - \mathbf{k}), \qquad (7.6.105)$$

which is a purely real contribution to forward scattering. Similarly, the first term in the second line of Eq. (7.5.26) will be a purely real contribution to forward scattering. Then any imaginary contribution to S_{fi} could only arise from the LSZ contribution given in Eq. (7.5.35). To first order in $\Pi^I(p^2)$ the contribution to the two-point function is $G_{\Pi^I}^{(2)}(p, -k) = (2\pi)^4\delta^4(p - k)D_F^R(p)[-i\Pi^I(p^2)]D_F^R(p)$, where we are using the analog of Eq. (7.5.53). Using the LSZ formalism the relevant part of S_{fi} with $m = n = 1$ and first-order in $\Pi^I(p^2)$ is

$$\begin{aligned} iT_{pk}^{\Pi^I} &= i(2\pi)^4\delta^4(p - k)\mathcal{M}_{pk}^{\Pi^I} = Z[(D_F^R)^{-1}(p)G_{\Pi^I}^{(2)}(p, -k)(D_F^R)^{-1}(k)]_{\mathrm{p}^2 = \mathrm{m}^2} \\ &= -iZ(2\pi)^4\delta^4(p - k)\Pi^I(m^2), \end{aligned} \qquad (7.6.106)$$

where we have used Eqs. (7.3.2), (7.3.3), (7.5.34), (7.5.35) and (7.5.39). This gives

$$\mathcal{M}_{pp} = -Z\Pi^I(m^2) \quad \Rightarrow \quad \mathrm{Im}\,\mathcal{M}_{pp} = -Z\,\mathrm{Im}\,\Pi(m^2). \qquad (7.6.107)$$

For a single particle in the initial and final state we have $|\mathbf{p}\rangle = |i\rangle = |f\rangle$ and the optical theorem result in Eq. (7.3.102) gives

$$2\,\mathrm{Im}\,\mathcal{M}_{pp} = \sum_{X_n} \int d\Pi_{X_n}^{\mathrm{LIPS}}|\mathcal{M}_{p \to X_n}|^2, \qquad (7.6.108)$$

where $d\Pi_{X_n}^{\mathrm{LIPS}}$ is defined in Eq. (7.3.41) and where X_n is shorthand for any set of n particles the particle can decay into. Rewriting Eq. (7.3.72) for the decay rate of the particle at rest with this notation gives

$$\Gamma = \sum_{X_n} \int d\Pi^{\text{LIPS}}_{X_n} \frac{|\mathcal{M}_{p \to X_n}|^2}{2m}\bigg|_{\mathbf{p}=0} = \frac{\text{Im}\,\mathcal{M}_{pp}}{m} = -\frac{Z\,\text{Im}\,\Pi(m^2)}{m}, \tag{7.6.109}$$

since $\mathcal{M}_{pp}$ is Lorentz invariant. So a particle is stable if and only if $\mathcal{M}_{pp} \in \mathbb{R}$ and is unstable if and only if $\text{Im}\,\mathcal{M}_{pp} \propto \text{Im}\,\Pi(m^2) \neq 0$. A particle is unstable when access to a collection of on-shell particles is kinematically allowed, $\mathcal{M}_{p \to X_n} \neq 0$. In the resonance region the particle propagator has the approximate form of Eq. (7.6.104),

$$D_F(p) \xrightarrow{p^2 \simeq m^2} \frac{iZ}{p^2 - m^2 + im\Gamma} \propto \frac{1}{p^2 - m^2 + im\Gamma}, \tag{7.6.110}$$

which is the *relativistic Breit-Wigner distribution* for a resonance. It is valid near the resonance mass region for a narrow resonance, $\Gamma \ll m$. The mass m is sometimes called the *real pole mass* or the *Breit-Wigner mass*. The Breit-Wigner form corresponds to a pole in the complex $s = p^2$ plane at $m^2 - im\Gamma$. When plotted as a function of $s = p^2 \in \mathbb{R}$ the full width at half-maximum of the Breit-Wigner distribution is given by $2m\Gamma$. We sometimes refer to Γ as the *resonance width* for this reason. So a resonance with a narrow *width* has a small decay rate and the particle becomes stable as the width vanishes. It is also common to describe the resonance in terms of the complex pole parameters, m_c and Γ_c, where the pole in the complex s-plane is $\sqrt{s} = m_c - i\Gamma_c/2$ by definition. Since the pole is at $s - m^2 + im\Gamma = 0 = s - (m_c - i\Gamma_c/2)^2$ then $m^2 = (m_c^2 - \frac{1}{4}\Gamma_c^2)$ and $m\Gamma = m_c\Gamma_c$.

Consider two particles scattering with the s-channel quantum numbers corresponding to some accessible unstable state with Breit-Wigner mass m and width Γ. If there are no nearby particle production thresholds or other resonances, then in the resonance region the differential cross-section will have the form

$$d\sigma \propto \left|\frac{1}{s - m^2 + im\Gamma}\right|^2 = \frac{1}{(s - m^2)^2 + m^2\Gamma^2}, \tag{7.6.111}$$

which has a characteristic Breit-Wigner resonance bump at $s \simeq m^2$. As expected $\Gamma \to 0$ is the stable particle limit since

$$\lim_{\Gamma \to 0} \frac{m\Gamma}{\pi} \frac{1}{(s - m^2)^2 + m^2\Gamma^2} = \delta(s - m^2). \tag{7.6.112}$$

When a resonance is not sufficiently narrow, $\Gamma \not\ll m$, the various approximations that we have made lose validity. In that case the resonance behavior is characterized by a momentum-dependent width, $\Gamma \to \Gamma(s)$. For further discussion of this and related issues, see Sec. 6.3 of Brown (1992) and also Zyla et al. (2020) and references therein.

Summary

We introduced the spectrum of asymptotic states and the corresponding Källén-Lehmann spectral representation for an interacting scalar field in Eq. (7.2.9). The S-matrix and the invariant amplitude for particle scattering were defined in Eqs. (7.3.2) and (7.3.3) respectively. The cross-section and differential cross-section for particle scattering were defined in Eqs. (7.3.7) and (7.3.10). The connection between the modulus-squared of the invariant amplitude and the differential cross-section was derived and given in Eq. (7.3.40). A similar result for the differential decay rate of an unstable particle was given in Eq. (7.3.74). The Mandelstam variables for two-body scattering were defined in Eq. (7.3.75) in terms of Fig. 7.10. The unitarity of the S-matrix and the optical theorem were then discussed in Sec. 7.3.5.

The interaction picture, familiar from quantum mechanics, was developed in the context of an interacting quantum field theory. We formulated Green's functions in terms of a Feynman diagram expansion and showed how to calculate symmetry factors for these diagrams. For an interacting scalar field theory we obtained Eq. (7.4.59), which shows the relationship between the vacuum expectation value of the time-ordered product of field operators, the infinite sum of connected Feynman diagrams and the generating functional defined in terms of the functional integral.

The LSZ formalism was developed for each of interacting scalar, fermion and photon fields in Sec. 7.5. The invariant amplitude was obtained in terms of the amputated Green's functions for each of these in Eqs. (7.5.44), (7.5.93) and (7.5.125), respectively. In Sec. 7.6 we presented a summary of the Feynman rules for calculating invariant amplitudes with Feynman diagrams and we considered some examples of interacting theories, including a scalar field with a quartic self-interaction, the Yukawa interaction of a fermion and a scalar (or pseudoscalar), quantum electrodynamics and scalar quantum electrodynamics. Example derivations of some tree-level invariant amplitudes were presented in Sec. 7.6.3. In the nonrelativistic limit the Yukawa and Coulomb potentials were shown to result from the Yukawa interaction and from (fermion/scalar) elcctrodynamics, respectively. Crossing symmetry was then introduced and used to discuss relations between some sample cross-sections calculated at tree level in Sec. 7.6.5. Finally, the Breit-Wigner form in Eq. (7.6.110) for the propagator of an unstable particle was obtained in Sec. 7.6.6.

Problems

7.1 Consider the proton-proton collisions in the Large Hadron Collider as discussed in Sec. 7.3.1. Assume that: improved beam optics gave an improved luminosity reduction factor $F_{\mathrm{red}} = 0.88$ and a transverse beam size of 15 μm, average time spacing between bunches reduced to 24 ns, and the initial protons per bunch increased to 1.3×10^{11}. Calculate the following:
(a) the number of proton bunches in each beam,
(b) the peak luminosity,
(c) the peak rate of detectable events,
(d) the peak number of detectable events per bunch crossing and
(e) the approximate peak observed rate for diphoton events coming from Higgs decays.

7.2 Sketch three examples of connected four-point ($n = 4$) Feynman diagrams in ϕ^3 theory contributing to the left-hand side of Eq. (7.5.37). Show that each diagram can be uniquely identified as a contribution to the right-hand side.

7.3 Sketch three examples of connected two-point ($n = 2$) Feynman diagrams with 10 vertices in ϕ^4 theory that contribute to $D_F(p)$. Show that these can be uniquely identified with contributions to the right-hand side of Eq. (7.5.49).

7.4   Consider a Yukawa theory with the Lagrangian density

$$\mathcal{L} = \tfrac{1}{2}\partial_\mu\phi\partial^\mu\phi - \tfrac{1}{2}m^2\phi^2 + \bar\psi(i\slashed{\partial} - M)\psi - g\bar\psi(1 + ir\gamma^5)\phi\psi, \qquad (7.6.113)$$

where $g, r \in \mathbb{R}$, ϕ is a scalar field and ψ is the field for fermion f.

(a) Does this theory respect parity? Explain.
(b) Write down the Feynman rules for this theory.
(c) Calculate the tree-level invariant amplitude for $f\bar{f} \to f\bar{f}$.
(d) Sketch but do not evaluate all distinct Feynman diagrams contributing to the invariant amplitude for $\phi\phi \to \phi\phi$ at lowest nonvanishing order in g.

7.5   By drawing on the key steps in the derivation of the LSZ formalism for scalars and fermions, outline the key steps in the derivation of the LSZ formalism for charged scalar bosons. There is no need to build explicit wavepackets to satisfy the asymptotic condition, which can be assumed.

7.6   Write down the version of the photon LSZ formula of Eq. (7.5.125) that is appropriate for massive vector Proca bosons. Justify your answer by describing the key steps needed to arrive at this result.

7.7   Consider a theory with three interacting scalar fields Φ, ϕ_a and ϕ_b with masses M, m and m, respectively, with $3m > M > 2m$ and with the Lagrangian density

$$\mathcal{L} = \tfrac{1}{2}\partial_\mu\Phi\partial^\mu\Phi - \tfrac{1}{2}M^2\Phi^2 + \tfrac{1}{2}\partial_\mu\phi_a\partial^\mu\phi_a - \tfrac{1}{2}m^2\phi_a^2 + \tfrac{1}{2}\partial_\mu\phi_b\partial^\mu\phi_b - \tfrac{1}{2}m^2\phi_b^2$$
$$- (\lambda/4!)\Phi^4 + \tfrac{1}{2}g_{aa}\Phi\phi_a^2 + g_{ab}\Phi\phi_a\phi_b. \qquad (7.6.114)$$

(a) What decays of the heavy scalar Φ are possible in a tree-level calculation? Sketch the lowest order Feynman diagram for the process $\Phi \to \phi_b\phi_b$.
(b) Can ϕ_a or ϕ_b decay? Explain.
(c) Evaluate the ratio of partial decay rates $\Gamma(\Phi \to \phi_a\phi_b)/\Gamma(\Phi \to \phi_a\phi_a)$ at tree level.
(d) Evaluate the total tree-level decay rate for a Φ at rest, $\Gamma(\Phi \to \text{anything})$.

7.8   Consider a physical system described by the Lagrangian density

$$\mathcal{L} = \tfrac{1}{2}[\partial_\mu\phi\partial^\mu\phi + \partial_\mu\chi\partial^\mu\chi - m_\phi^2\phi^2 - m_\chi^2\chi^2] - \tfrac{1}{2}g\chi^2\phi - (\lambda/4!)\chi^4, \qquad (7.6.115)$$

where $\phi, \chi, m_\phi, m_\chi, g$ and λ are real.

(a) Calculate $d\sigma/d\cos\theta$ in the center-of-momentum (CM) frame at tree level for $\chi\phi \to \chi\phi$ in terms of CM energy and scattering angle.
(b) Repeat this calculation for $\chi\chi \to \chi\chi$ at tree level in the CM frame. Express your answer in terms of the Mandelstam variables s, t and u.

7.9   The tree-level invariant amplitude for the equivalent of Compton scattering in scalar QED was calculated in Eq. (7.6.54). Use this to calculate the unpolarized differential cross-section $d\sigma/d\Omega$ in the fixed-target frame (initial ϕ at rest). (Hint: Use $\sum_\kappa \epsilon^\mu(k, \kappa)\epsilon^\mu(k, \kappa)^* \to -g^{\mu\nu}$ as in Eq. (7.6.83).)

7.10 Consider a gauge theory of charged scalars and charged fermions interacting with photons, which is a combined form of QED and scalar QED,

$$\mathcal{L} = \mathcal{L}_{\text{Dirac}} + \mathcal{L}_{\text{KG}} + \mathcal{L}_{\text{Maxwell}} + \mathcal{L}_{\text{int}}^{\text{QED}} + \mathcal{L}_{\text{int}}^{\text{sQED}}. \tag{7.6.116}$$

(a) Calculate the invariant amplitude for the process $e^- e^+ \to \phi \phi^*$ to lowest order in α, where the ϕ charge is $q_c = +|e|$.

(b) Calculate $d\sigma/d\Omega$ for this process in the CM frame. Compare with the result for $e^+ e^- \to \mu^+ \mu^-$ in Eq. (7.6.71).

8 Symmetries and Renormalization

In this chapter we discuss the issues of symmetries and renormalization in interacting quantum field theories. We begin with a study of the discrete symmetries P, C and T and the CPT theorem. This is followed by a discussion of the spin-statistics connection. Continuous symmetries are then considered along with Schwinger-Dyson equations. The systematics of regularization and renormalization of loops in Feynman diagrams are then developed and applied to QED as an illustration. The spontaneous breaking of continuous symmetries is then explored, which leads to Goldstone's theorem. Finally, we turn to a derivation of the Casimir effect that describes the attractive force between two parallel plates arising solely from quantum fluctuations in the vacuum.

8.1 Discrete Symmetries: P, C and T

The P, C and T transformations studied in relativistic quantum mechanics in Sec. 4.5 were operations on the space of single-particle relativistic wave functions, which admitted both single-particle and single-antiparticle contributions. They showed us how to transform solutions of the one-particle relativistic wave equations under these transformations. In Sec. 4.1 we studied P, C and T in the context of nonrelativistic quantum mechanics. Here we are interested in defining these transformations in the Hilbert spaces of relativistic quantum field theory, which for free fields are the relativistic Fock spaces that we constructed in Chapter 6 for scalar bosons, fermions, photons and massive vector bosons. We know from Wigner's theorem in Sec. 4.1.3 that P, C and T in a Hilbert space must be either unitary or antiunitary. We saw that an antiunitary operator effectively reverses time as discussed in Sec. 4.1.7. So P and C must be unitary and T must be antiunitary.

8.1.1 Parity

A parity transformation is a discrete Lorentz transformation as discussed in Chapter 1 and it has previously been considered in the context of quantum mechanics in Sec. 4.1.6 and relativistic quantum mechanics in Sec. 4.5.1. We now wish to study parity in the context of relativistic quantum field theory with the unitary parity operator $\hat{\mathcal{P}}$. However, in addition to the usual positive and negative parity of classical quantities, $\pi_P = \pm 1$, there is the possibility of additional arbitrary phases in quantum systems as seen in Eq. (4.1.65) when discussing projective representations of symmetries. This is also true of charge conjugation $\hat{\mathcal{C}}$ and time reversal $\hat{\mathcal{T}}$.

From the above and from Sec. 4.1.6 it follows that the parity transformation will be effected by a unitary operator $\hat{\mathcal{P}}$, where in the passive view we have

$$\hat{\phi}(x) \xrightarrow{\mathcal{P}} \hat{\phi}'(x') = \hat{\mathcal{P}}\hat{\phi}(t, -\mathbf{x})\hat{\mathcal{P}}^{-1} = \zeta_P^{\phi}\hat{\phi}(x),$$

$$\hat{V}^\mu(x) \xrightarrow{\mathcal{P}} \hat{V}'^\mu(x') = \hat{\mathcal{P}}\hat{V}^\mu(t,-\mathbf{x})\hat{\mathcal{P}}^{-1} = \zeta_P^V(-1)^{(\mu)}V^\mu(x),$$

$$\hat{\psi}(x) \xrightarrow{\mathcal{P}} \hat{\psi}'(x') = \hat{\mathcal{P}}\hat{\psi}(t,-\mathbf{x})\hat{\mathcal{P}}^{-1} = \zeta_P^\psi\gamma^0\psi(x), \tag{8.1.1}$$

where the various $\zeta_P^i \in \mathbb{C}$ are phases, $|\zeta_P^\phi| = |\zeta_P^V| = |\zeta_P^\psi| = 1$ and where we have introduced the convenient shorthand notation used in Peskin and Schroeder (1995),

$$(-1)^{(\mu)} \equiv \begin{cases} +1 & \text{for } \mu = 0 \\ -1 & \text{for } \mu = 1,2,3. \end{cases} \tag{8.1.2}$$

These transformations preserve the equations of motion for the field operators, such as the Klein-Gordon equation for the scalar case and the Dirac equation for the fermion case. The above equations can be summarized as

$$\hat{\mathcal{P}}\hat{\phi}(x)\hat{\mathcal{P}}^{-1} = \zeta_P^\phi\hat{\phi}(t,-\mathbf{x}), \quad \hat{\mathcal{P}}\hat{V}^\mu(x)\hat{\mathcal{P}}^{-1} = \zeta_P^V(-1)^{(\mu)}\hat{V}^\mu(t,-\mathbf{x}),$$

$$\hat{\mathcal{P}}\hat{\psi}(x)\hat{\mathcal{P}}^{-1} = \zeta_P^\psi\gamma^0\hat{\psi}(t,-\mathbf{x}). \tag{8.1.3}$$

Note that a unitary operator is linear and so acts only on Hilbert space operators and states, but not any any *c*-number quantities. Since $\hat{\mathcal{P}}$ is unitary,

$$\hat{\mathcal{P}}^{-1} = \hat{\mathcal{P}}^\dagger \quad \text{and so} \quad \hat{\mathcal{P}}\hat{\psi}^\dagger(x)\hat{\mathcal{P}}^{-1} = \zeta_P^{\psi*}\hat{\psi}^\dagger(t,-\mathbf{x})\gamma^0. \tag{8.1.4}$$

For any Hermitian boson field operator $\hat{\phi}(x) = \hat{\phi}^\dagger(x)$ we have

$$\hat{\mathcal{P}}\hat{\phi}(x)\hat{\mathcal{P}}^{-1} = \hat{\mathcal{P}}\hat{\phi}^\dagger(x)\hat{\mathcal{P}}^{-1} = (\hat{\mathcal{P}}\hat{\phi}(x)\hat{\mathcal{P}}^{-1})^\dagger = \zeta_P^{\phi*}\hat{\phi}^\dagger(t,-\mathbf{x}) = \zeta_P^{\phi*}\hat{\phi}(t,-\mathbf{x}) \tag{8.1.5}$$

and so $\zeta_P^\phi = \zeta_P^{\phi*} = \pm1$ for a Hermitian boson field.

Note that we are free to choose the phases ζ_P^i to be real for any field and doing so means that $\hat{\mathcal{P}}^2\hat{\phi}(x)(\hat{\mathcal{P}}^{-1})^2 = \hat{\phi}(x)$ and similarly for all fields. As we will see below this means that $\hat{\mathcal{P}}^2$ leaves all annihilation and creation operators unchanged and since $\hat{\mathcal{P}}^2|0\rangle = |0\rangle$ then it leaves all basis states and hence all states unchanged. This means that $\hat{\mathcal{P}}^2$ is equal to the identity operator,

$$\hat{\mathcal{P}}^2 = \hat{I} \quad \Rightarrow \quad \hat{\mathcal{P}} = \hat{\mathcal{P}}^{-1} = \hat{\mathcal{P}}^\dagger, \tag{8.1.6}$$

and so $\hat{\mathcal{P}}$ is unitary and Hermitian. Since $\hat{\mathcal{P}}$ is Hermitian, the parity operator $\hat{\mathcal{P}}$ is an observable operator. This means that the intrinsic parities of states created by boson and fermion operators are observables. The intrinsic parity for boson operators is $\pi_P^\phi = \zeta_P^\phi = \pm1$ and $\pi_P^V = \zeta_P^V(-1)^{(\mu)} = \pm(-1)^{(\mu)}$. As stated in Sec. 4.5.1 defining the intrinsic parity of a fermion as $\pi_P^\psi = \zeta_P^\psi = +1$ leads to no inconsistencies. This can be understood since in any calculation of an amplitude fermion operators occur in pairs. Scalar and pseudoscalar fields have $\pi_P^\phi = \pm1$, respectively. For a spin-one field we have $\pi_P^V = \zeta_P^V(-1)^{(\mu)} = \pm(-1)^{(\mu)}$ for a vector and pseudovector, respectively.

Intrinsic parity is a multiplicative quantum number; i.e., for any boson field operators $\hat{O}^i(x)$ we have $\mathcal{P}\hat{O}^i(x)\mathcal{P}^{-1} = \zeta_P^i\hat{O}^i(t,-\mathbf{x})$ and so

$$\mathcal{P}\hat{O}^1\cdots\hat{O}^k\mathcal{P}^{-1} = \mathcal{P}\hat{O}^1\mathcal{P}^{-1}\cdots\mathcal{P}\hat{O}^k\mathcal{P}^{-1} = (\zeta_P^1\cdots\zeta_P^k)\hat{O}_1\cdots\hat{O}_k, \tag{8.1.7}$$

where for brevity we have suppressed the *t* and **x** arguments. Note that fermion field operators always appear in the form of bosonic bilinears, $\hat{O}^i = \hat{\bar{\psi}}^i\Gamma\hat{\psi}^i$ with Γ some Dirac matrix, and so the above still applies.

In ordinary quantum mechanics under a parity transformation $\hat{\mathcal{P}}\hat{\mathbf{x}}\hat{\mathcal{P}}^{-1} = -\hat{\mathbf{x}}$, which leads to $\hat{\mathcal{P}}|\mathbf{x}\rangle = |-\mathbf{x}\rangle$ since $\hat{\mathbf{x}}(\hat{\mathcal{P}}|\mathbf{x}\rangle) = \hat{\mathcal{P}}(-\hat{\mathbf{x}})\hat{\mathcal{P}}^{-1}\hat{\mathcal{P}}|\mathbf{x}\rangle = -\hat{\mathcal{P}}\hat{\mathbf{x}}|\mathbf{x}\rangle = -\mathbf{x}(\hat{\mathcal{P}}|\mathbf{x}\rangle)$. We also have

$Y_{\ell m}(\Omega) \to (-1)^\ell Y_{\ell m}(\Omega)$ and so we find $\hat{\mathcal{P}}^{-1}|\ell, m\rangle = (-1)^\ell|\ell, m\rangle$, since $Y_{\ell m}(\Omega) = \langle \mathbf{x}|\ell, m\rangle \to \langle -\mathbf{x}|\ell, m\rangle = \langle \mathbf{x}|\hat{\mathcal{P}}^{-1}|\ell, m\rangle = (-1)^\ell Y_{\ell m}(\Omega) = (-1)^\ell\langle \mathbf{x}|\ell, m\rangle$ for all $|\mathbf{x}\rangle = |r, \theta, \phi\rangle$. Let $|1; \ell, m_\ell\rangle$ be a single scalar boson angular momentum eigenstate, then we must have $\hat{\mathcal{P}}^{-1}|1; \ell, m\rangle = \hat{\mathcal{P}}|1; \ell, m\rangle = (-1)^\ell|1; \ell, m\rangle$. Using Eq. (6.2.55) the wavefunction for this state in terms of the Schrödinger picture operator is

$$\langle 0|\hat{\phi}^\dagger(\mathbf{x})|1; \ell, m\rangle = \langle 0|\hat{\phi}(\mathbf{x})|1; \ell, m\rangle, \tag{8.1.8}$$

since $\hat{\phi} = \hat{\phi}^\dagger$. The vacuum is defined to have even parity, $\hat{\mathcal{P}}|0\rangle = |0\rangle$, and so under a parity transformation

$$\langle 0|\hat{\phi}(\mathbf{x})|1; \ell, m\rangle \xrightarrow{\mathcal{P}} \langle 0|\hat{\phi}(-\mathbf{x})|1; \ell, m\rangle = \pi_P^\phi\langle 0|\hat{\mathcal{P}}\hat{\phi}(\mathbf{x})\hat{\mathcal{P}}^{-1}|1; \ell, m\rangle$$
$$= \pi_P^\phi(-1)^\ell\langle 0|\hat{\phi}(\mathbf{x})|1; \ell, m\rangle, \tag{8.1.9}$$

which means that the total parity of the wavefunction is the product of the intrinsic parity and the orbital angular momentum parity giving $\pi_P^\phi(-1)^\ell$. For example, a pseudoscalar boson in a state with even or odd ℓ has total parity that is odd (-1) or even $(+1)$, respectively. So parity is a multiplicative quantum number.

From Eqs. (6.2.97) and (8.1.3) we have for a neutral (Hermitian) free scalar field

$$\hat{\mathcal{P}}\hat{\phi}(x)\hat{\mathcal{P}}^{-1} = \int[d^3p/(2\pi)^3]\,(1/\sqrt{2E_\mathbf{p}})(\hat{\mathcal{P}}\hat{a}_\mathbf{p}\hat{\mathcal{P}}^{-1}e^{-ip\cdot x} + \hat{\mathcal{P}}\hat{a}_\mathbf{p}^\dagger\hat{\mathcal{P}}^{-1}e^{ip\cdot x}) \tag{8.1.10}$$
$$= \zeta_P^\phi\hat{\phi}(t, -\mathbf{x}) = \zeta_P^\phi\int[d^3p/(2\pi)^3]\,(1/\sqrt{2E_\mathbf{p}})(\hat{a}_{-\mathbf{p}}e^{-ip\cdot x} + \hat{a}_{-\mathbf{p}}^\dagger e^{ip\cdot x}),$$

where we changed integration variable $\mathbf{p} \to -\mathbf{p}$ on the right-hand side. Since the plane waves are linearly independent, then we must have

$$\hat{\mathcal{P}}\hat{a}_\mathbf{p}\hat{\mathcal{P}}^{-1} = \zeta_P^\phi\hat{a}_{-\mathbf{p}} \qquad \text{and} \qquad \hat{\mathcal{P}}\hat{a}_\mathbf{p}^\dagger\hat{\mathcal{P}}^{-1} = \zeta_P^\phi\hat{a}_{-\mathbf{p}}^\dagger, \tag{8.1.11}$$

where $\zeta_P^\phi = \pm 1$ for a scalar and pseudoscalar boson, respectively. The canonical commutation relations are unaffected by the parity transformation as they should be in a Poincaré-invariant formulation,

$$[\hat{\mathcal{P}}\hat{a}_\mathbf{p}\hat{\mathcal{P}}^{-1}, \hat{\mathcal{P}}\hat{a}_{\mathbf{p}'}^\dagger\hat{\mathcal{P}}^{-1}] = (\zeta_P^\phi)^2[\hat{a}_{-\mathbf{p}}, \hat{a}_{-\mathbf{p}'}^\dagger] = (2\pi)^3\delta(\mathbf{p} - \mathbf{p}'), \tag{8.1.12}$$

where the ETCR of Eq. (6.2.4) are then also unchanged. The parity operator for a neutral scalar and pseudoscalar can be written as (Bjorken and Drell, 1965)

$$\hat{\mathcal{P}} = \exp\left\{\pm i\,(\pi/2)\int[d^3p/(2\pi)^3]\left[\hat{a}_\mathbf{p}^\dagger\hat{a}_\mathbf{p} - \zeta_P^\phi\hat{a}_\mathbf{p}^\dagger\hat{a}_{-\mathbf{p}}\right]\right\}, \tag{8.1.13}$$

where $\zeta_P^\phi = \pm 1$ and either sign choice for the exponent is valid and both lead to the same properties for the operator $\hat{\mathcal{P}}$. We see that $\hat{\mathcal{P}} = \hat{\mathcal{P}}^\dagger = \hat{\mathcal{P}}^{-1}$ and $\hat{\mathcal{P}}|0\rangle = |0\rangle$ as expected. The parity operator is similar for free neutral (Hermitian) vector bosons, such as photons and free neutral massive vector and axial vector bosons.

Proof: We write $\hat{\mathcal{P}} \equiv \exp\{i\hat{P}\}$ with the definition here that

$$\hat{P} \equiv -(\pi/2)\int[d^3p/(2\pi)^3][\hat{a}_\mathbf{p}^\dagger\hat{a}_\mathbf{p} - \zeta_P^\phi\hat{a}_\mathbf{p}^\dagger\hat{a}_{-\mathbf{p}}]. \tag{8.1.14}$$

Using the commutation relations in Eq. (6.2.17) we find

$$[\hat{P}, \hat{a}_\mathbf{p}] = \pi\tfrac{1}{2}(\hat{a}_\mathbf{p} - \zeta_P^\phi\hat{a}_{-\mathbf{p}}), \tag{8.1.15}$$
$$[\hat{P}, [\hat{P}, \hat{a}_\mathbf{p}]] = \pi\tfrac{1}{2}([\hat{P}, \hat{a}_\mathbf{p}] - \zeta_P^\phi[\hat{P}, \hat{a}_{-\mathbf{p}}]) = \pi^2\tfrac{1}{2}(\hat{a}_\mathbf{p} - \zeta_P^\phi\hat{a}_{-\mathbf{p}}) = \pi[\hat{P}, \hat{a}_\mathbf{p}]$$

and similarly for additional commutations with $\hat{P}$. We have used $(\zeta_P^\phi)^2 = 1$. From Eq. (A.8.7) we have

$$
\begin{aligned}
\hat{\mathcal{P}}\hat{a}_{\mathbf{p}}\hat{\mathcal{P}}^{-1} &= \hat{a}_{\mathbf{p}} + i[\hat{P},\hat{a}_{\mathbf{p}}] + (i^2/2!)[\hat{P},[\hat{P},\hat{a}_{\mathbf{p}}]] + (i^3/3!)[\hat{P},[\hat{P},[\hat{P},\hat{a}_{\mathbf{p}}]]] + \cdots \\
&= \hat{a}_{\mathbf{p}} + [\hat{P},\hat{a}_{\mathbf{p}}](i + i^2\pi/2! + i^3\pi^2/3! + \cdots) \\
&= \hat{a}_{\mathbf{p}} - ([\hat{P},\hat{a}_{\mathbf{p}}]/\pi) + ([\hat{P},\hat{a}_{\mathbf{p}}]/\pi)(1 + i\pi + (i\pi)^2/2! + (i\pi)^3/3! + \cdots) \\
&= \hat{a}_{\mathbf{p}} - \tfrac{1}{2}(\hat{a}_{\mathbf{p}} - \zeta_P^\phi \hat{a}_{-\mathbf{p}}) + \tfrac{1}{2}(\hat{a}_{\mathbf{p}} - \zeta_P^\phi \hat{a}_{-\mathbf{p}})e^{i\pi} \\
&= \tfrac{1}{2}(\hat{a}_{\mathbf{p}} + \zeta_P^\phi \hat{a}_{-\mathbf{p}}) - \tfrac{1}{2}(\hat{a}_{\mathbf{p}} - \zeta_P^\phi \hat{a}_{-\mathbf{p}}) = \zeta_P^\phi \hat{a}_{-\mathbf{p}}.
\end{aligned}
\tag{8.1.16}
$$

Note that changing the sign of $\hat{P}$ is achieved by the replacement $\pi \to -\pi$ and this leads to $e^{i\pi} \to e^{-i\pi}$ above and so leads to the same result. Since $\hat{\mathcal{P}}$ is unitary taking the Hermitian conjugate gives $\hat{\mathcal{P}}\hat{a}_{\mathbf{p}}^\dagger\hat{\mathcal{P}}^{-1} = \zeta_P^\phi \hat{a}_{-\mathbf{p}}^\dagger$.

For a charged (non-Hermitian) scalar field we have

$$
\hat{\mathcal{P}}\hat{\phi}(x)\hat{\mathcal{P}}^{-1} = \zeta_P^\phi \hat{\phi}(t,-\mathbf{x}) \quad \text{and} \quad \hat{\mathcal{P}}\hat{\phi}^\dagger(x)\hat{\mathcal{P}}^{-1} = \zeta_P^{\phi*}\hat{\phi}^\dagger(t,-\mathbf{x}),
\tag{8.1.17}
$$

where ζ_P^ϕ can be any complex phase. From Eq. (6.2.177) we have for a free field

$$
\begin{aligned}
\hat{\mathcal{P}}\hat{f}_{\mathbf{p}}\hat{\mathcal{P}}^{-1} &= \zeta_P^\phi \hat{f}_{-\mathbf{p}} \quad \text{and} \quad \hat{\mathcal{P}}\hat{g}_{\mathbf{p}}^\dagger\hat{\mathcal{P}}^{-1} = \zeta_P^\phi \hat{g}_{-\mathbf{p}}^\dagger, \\
\hat{\mathcal{P}}\hat{f}_{\mathbf{p}}^\dagger\hat{\mathcal{P}}^{-1} &= \zeta_P^{\phi*}\hat{f}_{-\mathbf{p}}^\dagger \quad \text{and} \quad \hat{\mathcal{P}}\hat{g}_{\mathbf{p}}\hat{\mathcal{P}}^{-1} = \zeta_P^{\phi*}\hat{g}_{-\mathbf{p}},
\end{aligned}
\tag{8.1.18}
$$

where the Hermitian conjugate of the first line gives the second line. An explicit form for the parity operator in this case is

$$
\hat{\mathcal{P}} = \exp\left\{-i\int\frac{d^3p}{(2\pi)^3}\left[\pm\frac{\pi}{2}(\hat{f}_{\mathbf{p}}^\dagger\hat{f}_{-\mathbf{p}} - \hat{g}_{\mathbf{p}}^\dagger\hat{g}_{-\mathbf{p}}) + \left(\phi_P \mp \frac{\pi}{2}\right)(\hat{f}_{\mathbf{p}}^\dagger\hat{f}_{\mathbf{p}} - \hat{g}_{\mathbf{p}}^\dagger\hat{g}_{\mathbf{p}})\right]\right\},
\tag{8.1.19}
$$

where $\zeta_P^\phi \equiv e^{i\phi_P} = \pm 1$ for scalar and pseudoscalar, respectively, and where both sign choices are valid. This again gives $\hat{\mathcal{P}} = \hat{\mathcal{P}}^\dagger = \hat{\mathcal{P}}^{-1}$ and $\hat{\mathcal{P}}|0\rangle = |0\rangle$.

Proof: Define $\hat{\mathcal{P}}_1 \equiv \exp\{i\hat{P}_1\}$ and $\hat{\mathcal{P}}_2 \equiv \exp\{i\hat{P}_2\}$, where

$$
\hat{P}_1 \equiv -\alpha\int[d^3p/(2\pi)^3][\hat{f}_{\mathbf{p}}^\dagger\hat{f}_{-\mathbf{p}} - \hat{g}_{\mathbf{p}}^\dagger\hat{g}_{-\mathbf{p}}], \quad \hat{P}_2 \equiv -\beta\int[d^3p/(2\pi)^3][\hat{f}_{\mathbf{p}}^\dagger\hat{f}_{\mathbf{p}} - \hat{g}_{\mathbf{p}}^\dagger\hat{g}_{\mathbf{p}}]
$$

and we see that $\hat{P}_1 = \hat{P}_1^\dagger$ and $\hat{P}_2 = \hat{P}_2^\dagger$. This then gives using Eq. (A.8.7)

$$
[\hat{P}_1,\hat{f}_{\mathbf{p}}] = \alpha\hat{f}_{-\mathbf{p}}, \; [\hat{P}_1,[\hat{P}_1,\hat{f}_{\mathbf{p}}]] = \alpha^2\hat{f}_{\mathbf{p}}, \; [\hat{P}_1,\hat{g}_{\mathbf{p}}^\dagger] = \alpha\hat{g}_{-\mathbf{p}}^\dagger, \; [\hat{P}_1,[\hat{P}_1,\hat{g}_{\mathbf{p}}^\dagger]] = \alpha^2\hat{g}_{\mathbf{p}}^\dagger,
$$

$$
\begin{aligned}
\hat{\mathcal{P}}_1\hat{f}_{\mathbf{p}}\hat{\mathcal{P}}_1^{-1} &= \hat{f}_{\mathbf{p}} + i[\hat{P}_1,\hat{f}_{\mathbf{p}}] + (i^2/2!)[\hat{P}_1,[\hat{P}_1,\hat{f}_{\mathbf{p}}]] + (i^3/3!)[\hat{P}_1,[\hat{P}_1,[\hat{P}_1,\hat{f}_{\mathbf{p}}]]] + \cdots \\
&= \hat{f}_{\mathbf{p}}\cos\alpha + i\hat{f}_{-\mathbf{p}}\sin\alpha
\end{aligned}
\tag{8.1.20}
$$

and similarly $\hat{\mathcal{P}}_1\hat{g}_{\mathbf{p}}^\dagger\hat{\mathcal{P}}_1^{-1} = \hat{g}_{\mathbf{p}}^\dagger\cos\alpha + i\hat{g}_{-\mathbf{p}}^\dagger\sin\alpha$. Then choosing

$$
\alpha = \pm\pi/2 \quad \Rightarrow \quad \hat{\mathcal{P}}_1\hat{f}_{\mathbf{p}}\hat{\mathcal{P}}_1^{-1} = \pm i\hat{f}_{-\mathbf{p}} \quad \text{and} \quad \hat{\mathcal{P}}_1\hat{g}_{\mathbf{p}}^\dagger\hat{\mathcal{P}}_1^{-1} = \pm i\hat{g}_{-\mathbf{p}}^\dagger.
\tag{8.1.21}
$$

We also have using Eq. (A.8.7)

$$
[\hat{P}_2,\hat{f}_{\mathbf{p}}] = \beta\hat{f}_{\mathbf{p}}, \; [\hat{P}_2,[\hat{P}_2,\hat{f}_{\mathbf{p}}]] = \beta^2\hat{f}_{\mathbf{p}}, \; [\hat{P}_2,\hat{g}_{\mathbf{p}}^\dagger] = \beta\hat{g}_{\mathbf{p}}^\dagger, \; [\hat{P}_2,[\hat{P}_2,\hat{g}_{\mathbf{p}}^\dagger]] = \beta^2\hat{g}_{\mathbf{p}}^\dagger,
$$

$$
\hat{\mathcal{P}}_2\hat{f}_{\mathbf{p}}\hat{\mathcal{P}}_2^{-1} = \hat{f}_{\mathbf{p}} + i[\hat{P}_2,\hat{f}_{\mathbf{p}}] + (i^2/2!)[\hat{P}_2,[\hat{P}_2,\hat{f}_{\mathbf{p}}]] + \cdots = \hat{f}_{\mathbf{p}}e^{i\beta}
\tag{8.1.22}
$$

and similarly $\hat{\mathcal{P}}_2 \hat{g}_{\mathbf{p}}^\dagger \hat{\mathcal{P}}_2^{-1} = \hat{g}_{\mathbf{p}}^\dagger e^{i\beta}$. It follows from Eq. (6.2.19) that $[\hat{N}, \hat{a}_{\mathbf{p}}^\dagger \hat{a}_{\mathbf{p}'}] = \hat{a}_{\mathbf{p}}^\dagger [\hat{N}, \hat{a}_{\mathbf{p}'}] + [\hat{N}, \hat{a}_{\mathbf{p}}^\dagger] \hat{a}_{\mathbf{p}'} = -\hat{a}_{\mathbf{p}}^\dagger \hat{a}_{\mathbf{p}'} + \hat{a}_{\mathbf{p}}^\dagger \hat{a}_{\mathbf{p}'} = 0$. Similarly, we find $[\hat{P}_1, \hat{P}_2] = 0$. The BCH formula in Eq. (A.8.6) then leads to $[\hat{\mathcal{P}}_1, \hat{\mathcal{P}}_2] = 0$ and

$$\hat{\mathcal{P}} \equiv \hat{\mathcal{P}}_1 \hat{\mathcal{P}}_2 = e^{i(\hat{P}_1 + \hat{P}_2)}$$

$$= \exp\{-i \int [d^3 p/(2\pi)^3][\alpha(\hat{f}_{\mathbf{p}}^\dagger \hat{f}_{-\mathbf{p}} - \hat{g}_{\mathbf{p}}^\dagger \hat{g}_{-\mathbf{p}}) + \beta(\hat{f}_{\mathbf{p}}^\dagger \hat{f}_{\mathbf{p}} - \hat{g}_{\mathbf{p}}^\dagger \hat{g}_{\mathbf{p}})]\}. \qquad (8.1.23)$$

For the choices $\alpha = \pm\pi/2$ we have from Eqs. (8.1.21) and (8.1.22) that $\hat{\mathcal{P}} \hat{f}_{\mathbf{p}} \hat{\mathcal{P}}^{-1} = \pm i e^{i\beta} \hat{f}_{-\mathbf{p}} = \zeta_P^\phi \hat{f}_{-\mathbf{p}}$ and so we have $\zeta_P^\phi \equiv e^{i\phi_P} = \pm i e^{i\beta} = e^{i(\beta \pm \pi/2)}$ which means that $\beta = \phi_P \mp \pi/2$. Note that we could further generalize by making a different choice of α for each of the $\hat{f}$ and $\hat{g}$ terms, but we have not explicitly indicated this. These freedoms allow different expressions for $\hat{\mathcal{P}}$ while achieving the same purpose. Note that (i) ignoring the $\hat{g}$-terms, replacing $\hat{f}$ with $\hat{a}$ and choosing $\alpha = +\pi/2$, we recover Eq. (8.1.13) since $\phi_P = 0$ and π for scalar and pseudoscalar, respectively; and (ii) choosing $\alpha = -\pi/2$ and $+\pi/2$ for the $\hat{f}$ and $\hat{g}$ terms, respectively, gives the result obtained in Greiner and Reinhardt (1996).

The four photon field operators are Hermitian and so the $\zeta_P^{A^\mu}$ are real. As we know from Eq. (4.5.2) the photon field transforms as a vector and so from Eq. (8.1.3) we must have $\zeta_P^A = +1$,

$$\hat{\mathcal{P}} \hat{A}^\mu(x) \hat{\mathcal{P}}^{-1} = \zeta_P^A (-1)^{(\mu)} \hat{A}^\mu(t, -\mathbf{x}) = (-1)^{(\mu)} \hat{A}^\mu(t, -\mathbf{x}). \qquad (8.1.24)$$

This follows since we want $j_\mu \hat{A}^\mu$ to transform as a scalar under parity and since j^μ transforms as a vector, $j^\mu(t, \mathbf{x}) \xrightarrow{\mathcal{P}} (-1)^{(\mu)} j^\mu(t, -\mathbf{x}) = P^\mu{}_\nu j^\nu(t, -\mathbf{x})$ where recall from Table 1.1 that $P^\mu{}_\nu \equiv g^{\mu\nu}$. It is easiest to work in Coulomb gauge in the quantization inertial frame, where the degrees of freedom are represented entirely by the operator $\hat{\mathbf{A}}(x)$ given by Eqs. (6.4.49) and (6.4.50). So we require

$$\hat{\mathcal{P}} \hat{\mathbf{A}}(t, \mathbf{x}) \hat{\mathcal{P}}^{-1} = -\hat{\mathbf{A}}(t, -\mathbf{x}). \qquad (8.1.25)$$

Note $A^0(t, \mathbf{x}) \to A^0(t, -\mathbf{x})$ since in the presence of an external conserved current

$$A^0(x) = (1/4\pi) \int d^3 x' \, j^0(t, \mathbf{x}')|\mathbf{x} - \mathbf{x}'|. \qquad (8.1.26)$$

We have the requirement that

$$\hat{\mathcal{P}} \hat{a}_{\mathbf{k}}^\lambda \hat{\mathcal{P}}^{-1} = -(-1)^\lambda a_{-\mathbf{k}}^\lambda \quad \text{and} \quad \hat{\mathcal{P}} \hat{a}_{\mathbf{k}}^{\lambda\dagger} \hat{\mathcal{P}}^{-1} = -(-1)^\lambda a_{-\mathbf{k}}^{\lambda\dagger}, \qquad (8.1.27)$$

where the minus sign comes directly from that in Eq. (8.1.25) and where changing integration variable $\mathbf{k} \to -\mathbf{k}$ restores the exponential leading to $a_{\mathbf{k}}^\lambda \to a_{-\mathbf{k}}^\lambda$ and $\epsilon(k, \lambda) \to \epsilon(-k, \lambda) = (-1)^\lambda \epsilon(k, \lambda)$. Greiner and Reinhardt (1996) have an additional minus sign in Eq. (8.1.27) since they insert an additional overall minus sign into the Bjorken and Drell (1965) definition of $\epsilon(-k, \lambda) \equiv (-1)^\lambda \epsilon(k, \lambda)$ that we use here. In terms of the helicity basis using Eq. (6.4.33) we must have

$$\hat{\mathcal{P}} \hat{a}_{\mathbf{k}}^h \hat{\mathcal{P}}^{-1} = \hat{a}_{-\mathbf{k}}^{-h} \quad \text{and} \quad \hat{\mathcal{P}}(\hat{a}_{\mathbf{k}}^h)^\dagger \hat{\mathcal{P}}^{-1} = (\hat{a}_{-\mathbf{k}}^{-h})^\dagger, \qquad (8.1.28)$$

which follows since $\epsilon(-k, h) = -\epsilon(k, -h)$ from Eq. (6.4.32). Using Eq. (8.1.13) and replacing ζ_P^ϕ by $-(-1)^\lambda$ and $+1$ we can write down the photon parity operator in each basis, respectively, as

$$\hat{\mathcal{P}} = \exp\{\pm i(\pi/2) \int [d^3 k/(2\pi)^3] \sum_{\lambda=1}^2 [\hat{a}_{\mathbf{k}}^{\dagger\lambda} \hat{a}_{\mathbf{k}}^\lambda + (-1)^\lambda \hat{a}_{\mathbf{k}}^{\dagger\lambda} \hat{a}_{-\mathbf{k}}^\lambda]\}$$

$$= \exp\{\pm i(\pi/2) \int [d^3 k/(2\pi)^3] \sum_{h=\pm} [\hat{a}_{\mathbf{k}}^{\dagger h} \hat{a}_{\mathbf{k}}^h - \hat{a}_{\mathbf{k}}^{\dagger h} \hat{a}_{-\mathbf{k}}^{-h}]\}, \qquad (8.1.29)$$

where the above proof needs only minor modifications to arrive at these results. To obtain $\hat{P}$ in inertial frames other than the quantization intertial frame we perform a Lorentz transformation of the operator in the usual way, $\mathcal{P} \rightarrow \mathcal{P}' = \hat{U}(\Lambda)\hat{P}\hat{U}(\Lambda)^\dagger$. We do not attempt to construct the photon parity operator in other gauges, since it is enough for our purposes to know that it must exist. It is also clear that a change of inertial frame can lead to a change of gauge; e.g., the Coulomb/radiation gauge is not a Lorentz-invariant gauge condition. Note that Bjorken and Drell (1965) invert this thinking and talk of modifying the Lorentz transformation for the photon by a gauge term such that the Coulomb gauge condition is valid in every frame. We prefer the former view where the Lorentz transformation has its canonical form.

For a neutral massive vector and massive axial vector in the quantization inertial frame we have

$$\hat{P} = \exp\{\pm i(\pi/2) \int [d^3k/(2\pi)^3] \sum_{\lambda=1}^3 [\hat{a}_{\mathbf{k}}^{\dagger\lambda}\hat{a}_{\mathbf{k}}^{\lambda} + \zeta_P^V(-1)^\lambda \hat{a}_{\mathbf{k}}^{\dagger\lambda}\hat{a}_{-\mathbf{k}}^{\lambda}]\}, \tag{8.1.30}$$

where we sum over the three polarizations rather than two and where $\zeta_P^V = \pm 1$ is for vector and axial vector, respectively. For example, for a free massive vector field from Eq. (6.5.51) we have $\hat{A}^0 = -\boldsymbol{\nabla} \cdot \hat{\mathbf{E}}/m^2$, which gives $\hat{P}\hat{A}^0(t, -\mathbf{x})\hat{P}^{-1} = \hat{A}^0(t, -\mathbf{x})$ since $\boldsymbol{\nabla} \rightarrow -\boldsymbol{\nabla}$ and from Eq. (6.5.48) we have $\hat{P}\hat{\mathbf{E}}(t, \mathbf{x})\hat{P}^{-1} = -\hat{\mathbf{E}}(t, -\mathbf{x})$. For charged vector and axial vector fields we can perform a construction similar to that for the charged scalar field and there will again be an arbitrary phase for each.

For free fermions we use Eqs. (6.3.194) and (8.1.1) and after a change of integration variable $\mathbf{p} \rightarrow -\mathbf{p}$ to restore the exponential we are left with

$$u^s(p)\mathcal{P}\hat{b}_{\mathbf{p}}^s\mathcal{P}^{-1} = \zeta_P^\psi \gamma^0 u^s(-p)\hat{b}_{-\mathbf{p}}^s = \zeta_P^\psi u^s(p)\hat{b}_{-\mathbf{p}}^s,$$
$$v^s(p)\mathcal{P}\hat{d}_{\mathbf{p}}^{s\dagger}\mathcal{P}^{-1} = \zeta_P^\psi \gamma^0 v^s(-p)\hat{d}_{-\mathbf{p}}^{s\dagger} = -\zeta_P^\psi v^s(p)\hat{d}_{-\mathbf{p}}^{s\dagger}, \tag{8.1.31}$$

where we have used $u^s(p) = \gamma^0 u^s(-p)$ and $v^s(p) = -\gamma^0 v^s(-p)$ from Eq. (4.4.120) and $(\gamma^0)^2 = I$. Fermions only ever appear in physically relevant quantities in the form of bilinears and from Eq. (8.1.4) it is clear that the phase only ever enters bilinears in the form $\zeta_P^{\psi*}\zeta_P^\psi = |\zeta_P^\psi|^2 = 1$. So ζ_P^ψ is not relevant and it is conventional to assign it the value $\zeta_P^\psi = 1$. Then we can rewrite the above result as

$$\mathcal{P}\hat{b}_{\mathbf{p}}^s\mathcal{P}^{-1} = \hat{b}_{-\mathbf{p}}^s \quad \text{and} \quad \mathcal{P}\hat{d}_{\mathbf{p}}^{s\dagger}\mathcal{P}^{-1} = -\hat{d}_{-\mathbf{p}}^{s\dagger}, \tag{8.1.32}$$

where taking the Hermitian conjugate of each shows that $\hat{b}_{\mathbf{p}}^{s\dagger}$ and $\hat{d}_{\mathbf{p}}^s$ transform as $\hat{b}_{\mathbf{p}}^s$ and $\hat{d}_{\mathbf{p}}^{s\dagger}$, respectively. The annihilation and creation operator anticommutation relations in Eq. (6.3.80) are invariant under the above parity transformation as expected. We observe that the intrinsic parities of a fermion and an antifermion are $+1$ and -1 respectively, since $\hat{b}_{\mathbf{p}}^{s\dagger}$ creates a fermion and $\hat{d}_{\mathbf{p}}^{s\dagger}$ creates an antifermion. Since $\hat{P}$ is a unitary operator, then

$$\hat{P}\hat{\psi}^\dagger(x)\hat{P}^{-1} = (\hat{P}\hat{\psi}(x)\hat{P}^{-1})^\dagger = \hat{\psi}^\dagger(t, -\mathbf{x})\gamma^0 \Rightarrow \hat{P}\hat{\bar{\psi}}(x)\hat{P}^{-1} = \hat{\bar{\psi}}(t, -\mathbf{x})\gamma^0 \tag{8.1.33}$$

in any representation of the Dirac matrices unitarily equivalent to the Dirac representation. Since $\hat{P}$ is a Hilbert space operator with no Dirac indices, then for any arbitrary bilinear $\hat{\bar{\psi}}(x)\Gamma\hat{\psi}(x)$ with Γ being some combination of Dirac matrices,

$$\hat{P}\hat{\bar{\psi}}(x)\Gamma\hat{\psi}(x)\hat{P}^{-1} = \hat{P}\hat{\bar{\psi}}(x)\hat{P}^{-1}\Gamma\hat{P}\hat{\psi}(x)\hat{P}^{-1} = \hat{\bar{\psi}}(t, -\mathbf{x})\gamma^0\Gamma\gamma^0\hat{\psi}(t, -\mathbf{x}). \tag{8.1.34}$$

Since $\gamma^0 I \gamma^0 = I$, $\gamma^0 \gamma^\mu \gamma^0 = (-1)^{(\mu)}\gamma^\mu$, $\gamma^0 \sigma^{\mu\nu}\gamma^0 = (-1)^{(\mu)}(-1)^{(\nu)}\sigma^{\mu\nu}$, $\gamma^0\gamma^\mu\gamma^5\gamma^0 = -(-1)^{(\mu)}\gamma^\mu\gamma^5$, and $\gamma^0 i\gamma^5\gamma^0 = -i\gamma^5$, then we see that the operator forms of the 16 independent fermion bilinears transform as expected under parity. The explicit expression for the fermion parity operator is (Bjorken and Drell, 1965)

$$\hat{\mathcal{P}} = \exp\{\pm i(\pi/2) \int [d^3p/(2\pi)^3] \sum_s [\hat{b}_{\mathbf{p}}^{\dagger s} \hat{b}_{\mathbf{p}}^s - \hat{b}_{\mathbf{p}}^{\dagger s} \hat{b}_{-\mathbf{p}}^s + \hat{d}_{\mathbf{p}}^{\dagger s} \hat{d}_{\mathbf{p}}^s + \hat{d}_{\mathbf{p}}^{\dagger s} \hat{d}_{-\mathbf{p}}^s]\} \tag{8.1.35}$$

and is obtained following similar steps to those used above for bosons. Again we see that $\hat{\mathcal{P}} = \hat{\mathcal{P}}^\dagger = \hat{\mathcal{P}}^{-1}$ and $\hat{\mathcal{P}}|0\rangle = |0\rangle$.

Proof: We write $\hat{\mathcal{P}} \equiv \exp\{i\hat{P}\}$ with the definition

$$\hat{P} \equiv \pm(\pi/2) \int [d^3p/(2\pi)^3] \sum_s [\hat{b}_{\mathbf{p}}^{\dagger s} \hat{b}_{\mathbf{p}}^s - \hat{b}_{\mathbf{p}}^{\dagger s} \hat{b}_{-\mathbf{p}}^s + \hat{d}_{\mathbf{p}}^{\dagger s} \hat{d}_{\mathbf{p}}^s + \hat{d}_{\mathbf{p}}^{\dagger s} \hat{d}_{-\mathbf{p}}^s]. \tag{8.1.36}$$

Using the commutation relations in Eq. (6.3.80) we find Eq. (6.3.81) and so

$$[\hat{P}, \hat{b}_{\mathbf{p}}^s] = \mp\pi \tfrac{1}{2}(\hat{b}_{\mathbf{p}}^s - \hat{b}_{-\mathbf{p}}^s), \quad [\hat{P}, \hat{d}_{\mathbf{p}}^s] = \mp\pi \tfrac{1}{2}(\hat{d}_{\mathbf{p}}^s + \hat{d}_{-\mathbf{p}}^s), \tag{8.1.37}$$

$$[\hat{P}, [\hat{P}, \hat{b}_{\mathbf{p}}^s]] = \mp\pi \tfrac{1}{2}([\hat{P}, \hat{b}_{\mathbf{p}}^s] - [\hat{P}, \hat{b}_{-\mathbf{p}}^s]) = \pi^2 \tfrac{1}{2}(\hat{b}_{\mathbf{p}}^s - \hat{b}_{-\mathbf{p}}^s) = \mp\pi[\hat{P}, \hat{b}_{\mathbf{p}}^s]$$

and similarly $[\hat{P}, [\hat{P}, \hat{d}_{\mathbf{p}}^s]] = \mp\pi[\hat{P}, \hat{d}_{\mathbf{p}}^s]$. Using Eq. (A.8.7) again we have

$$
\begin{aligned}
\hat{\mathcal{P}} \hat{b}_{\mathbf{p}}^s \hat{\mathcal{P}}^{-1} &= \hat{b}_{\mathbf{p}}^s + i[\hat{P}, \hat{b}_{\mathbf{p}}^s] + (i^2/2!)[\hat{P}, [\hat{P}, \hat{b}_{\mathbf{p}}^s]] + (i^3/3!)[\hat{P}, [\hat{P}, [\hat{P}, \hat{b}_{\mathbf{p}}^s]]] + \cdots \\
&= \hat{b}_{\mathbf{p}}^s + [\hat{P}, \hat{b}_{\mathbf{p}}^s][i + i^2(\mp\pi)/2! + i^3(\mp\pi)^2/3! + \cdots] \\
&= \hat{b}_{\mathbf{p}}^s + \tfrac{1}{2}(\hat{b}_{\mathbf{p}}^s - \hat{b}_{-\mathbf{p}}^s)[i(\mp\pi) + (\mp i\pi)^2/2! + (\mp i\pi)^3/3! + \cdots] \\
&= \hat{b}_{\mathbf{p}}^s - \tfrac{1}{2}(\hat{b}_{\mathbf{p}}^s - \hat{b}_{-\mathbf{p}}^s) + \tfrac{1}{2}(\hat{b}_{\mathbf{p}}^s - \hat{b}_{-\mathbf{p}}^s)e^{\mp i\pi} = \hat{b}_{-\mathbf{p}}^s \tag{8.1.38}
\end{aligned}
$$

and also $\hat{\mathcal{P}} \hat{d}_{\mathbf{p}}^s \hat{\mathcal{P}}^{-1} = \hat{d}_{\mathbf{p}}^s - \tfrac{1}{2}(\hat{d}_{\mathbf{p}}^s + \hat{d}_{-\mathbf{p}}^s) + \tfrac{1}{2}(\hat{d}_{\mathbf{p}}^s + \hat{d}_{-\mathbf{p}}^s)e^{\mp i\pi} = -\hat{d}_{-\mathbf{p}}^s$. Hermitian conjugation gives $\hat{\mathcal{P}} \hat{b}_{\mathbf{p}}^{s\dagger} \hat{\mathcal{P}}^{-1} = \hat{b}_{-\mathbf{p}}^{s\dagger}$ and $\hat{\mathcal{P}} \hat{d}_{\mathbf{p}}^{s\dagger} \hat{\mathcal{P}}^{-1} = -\hat{d}_{-\mathbf{p}}^{s\dagger}$. Finally, let $\hat{P}_\pm$ refer to the two sign choices for $\hat{P}$. Note that $\hat{\mathcal{P}} = e^{i\hat{P}_+} = e^{i\hat{P}_-}$ and that $(e^{i\hat{P}_+})^\dagger = e^{i\hat{P}_-}$ and that $e^{i\hat{P}_+}e^{i\hat{P}_-} = \hat{I}$. Then $\hat{\mathcal{P}} = \hat{\mathcal{P}}^\dagger = \hat{\mathcal{P}}^{-1}$ and $\hat{\mathcal{P}}|0\rangle = |0\rangle$.

As already noted the exponents are normal-ordered in each case above, then

$$\hat{\mathcal{P}}|0\rangle = |0\rangle, \tag{8.1.39}$$

which means that the vacuum is an even parity eigenstate of the parity operator as previously stated. The above discussion has been for the case of free field theories.

For each of the free quantum field theories that we constructed in Chapter 6 we can readily verify that the normal-ordered operators behave as they should under a parity transformation. For example, we find that in each case

$$\hat{\mathcal{P}} \hat{H} \hat{\mathcal{P}}^{-1} = \hat{H} \quad \hat{\mathcal{P}} \hat{\mathbf{P}} \hat{\mathcal{P}}^{-1} = -\hat{\mathbf{P}} \quad \text{and} \quad \hat{\mathcal{P}} \hat{\mathbf{J}} \hat{\mathcal{P}}^{-1} = \hat{\mathbf{J}}. \tag{8.1.40}$$

Recall from Eq. (6.2.32) that normal-ordering removes any ordering ambiguity associated with combinations of non-commuting field and conjugate momentum operators. This is true for both bosons and fermions, where we recall that with fermion normal-ordering we have an additional minus sign for each exchange of annihilation and/or creation operators as discussed in Sec. 6.3.4. As we saw in Sec. 4.1.5 as a result of Wigner's theorem any symmetry transformation $\hat{U}$ in a quantum system is either a unitary or antiunitary transformation. This means that $\hat{U}^\dagger = \hat{U}^{-1}$. The antiunitary operator is a combination of a unitary operator and a complex conjugation operator. Also recall from the discussion around Eq. (4.1.130) that antiunitary symmetries other than time reversal lead to unphysical quantum systems with energies not bounded below. As we have seen for the P transformation and as we will see for the C and T transformations, an annihilation operator remains an annihilation operator and a creation operator remains a creation operator. So normal-ordering and any of the P, C and T symmetry transformations are commuting operations in the following sense. If $\hat{U}\hat{a}_i\hat{U}^{-1} = \hat{a}_i'$ then $\hat{U}\hat{a}_i^\dagger\hat{U}^{-1} = \hat{a}_i'^\dagger$ and

$$\hat{U}{:}\hat{a}_1\hat{a}_2^\dagger\hat{a}_3{:}\hat{U}^{-1} = \hat{U}\hat{a}_2^\dagger\hat{a}_1\hat{a}_3\hat{U}^{-1} = \hat{a}_2'^\dagger\hat{a}_1'\hat{a}_3' = {:}\hat{a}_1'\hat{a}_2'^\dagger\hat{a}_3'{:}. \tag{8.1.41}$$

The conserved current density operator for a charged scalar field in Eq. (6.2.182) transforms as a vector under a parity transformation as it should,

$$\hat{\mathcal{P}}\hat{j}^\mu(x)\hat{\mathcal{P}}^{-1} = i{:}\left[\hat{\phi}^\dagger(t, -\mathbf{x})(-1)^{(\mu)}(\partial^\mu\hat{\phi})(t, -\mathbf{x}) - (-1)^{(\mu)}(\partial^\mu\hat{\phi}^\dagger)(t, -\mathbf{x})\hat{\phi}(t, -\mathbf{x})\right]{:}$$
$$= (-1)^{(\mu)}\hat{j}^\mu(t, -\mathbf{x}). \tag{8.1.42}$$

Similarly, using Eq. (8.1.34) the normal-ordered four-vector current for a fermion in Eq. (6.3.207) transforms as a vector under parity since

$$\hat{\mathcal{P}}\hat{j}^\mu(x)\hat{\mathcal{P}}^{-1} = {:}\hat{\bar{\psi}}(t, -\mathbf{x})\gamma^0\gamma^\mu\gamma^0\hat{\psi}(t, -\mathbf{x}){:} = (-1)^{(\mu)}\hat{j}^\mu(t, -\mathbf{x}). \tag{8.1.43}$$

From Eq. (6.3.137) we have the normal-ordered angular momentum operator for Dirac fermions,

$$\hat{\mathbf{J}} = \int d^3x\,{:}\hat{\psi}^\dagger(x)\left(\mathbf{x} \times (-i\boldsymbol{\nabla}) + \mathbf{S}\right)\hat{\psi}(x){:}, \tag{8.1.44}$$

where $S^i = \frac{1}{2}\epsilon^{ijk}\Sigma_{\text{Dirac}}^{jk} = \frac{1}{4}\epsilon^{ijk}\sigma^{jk}$. Under a parity transformation we find

$$\hat{\mathbf{J}} \xrightarrow{P} \hat{\mathbf{J}}' = \hat{\mathcal{P}}\hat{\mathbf{J}}\hat{\mathcal{P}}^{-1} = \int d^3x\,{:}\hat{\psi}^\dagger(t, -\mathbf{x})\gamma^0\left(\mathbf{x} \times (-i\boldsymbol{\nabla}) + \mathbf{S}\right)\gamma^0\hat{\psi}(t, -\mathbf{x}){:}$$
$$= \int d^3x\,{:}\hat{\psi}^\dagger(t, -\mathbf{x})\left(\mathbf{x} \times (-i\boldsymbol{\nabla}) + \mathbf{S}\right)\hat{\psi}(t, -\mathbf{x}){:} = \hat{\mathbf{J}}, \tag{8.1.45}$$

where we used $\gamma^0 S^i\gamma^0 = \frac{1}{4}\epsilon^{ijk}\gamma^0\sigma^{jk}\gamma^0 = \frac{1}{4}\epsilon^{ijk}\sigma^{jk} = S^i$ since $\gamma^0\sigma^{jk}\gamma^0 = \sigma^{jk\dagger} = \sigma^{jk}$ and where in the last step we performed a change of integration variable $\mathbf{x} \to -\mathbf{x}$. So the angular momentum operator transforms as an axial vector as it should.

In an interacting field theory we have a total Hamiltonian operator $\hat{H}(t)$ that contains the free field terms plus the interaction terms. From Eq. (4.1.23) we know that in the Heisenberg picture all field operators evolve like those in Eq. (7.1.9). We can still express $\hat{\phi}_s(\mathbf{x})$ in a three-dimensional mode expansion as done in Eq. (6.2.15) but now $\omega_\mathbf{p}$ is some new function of $\mathbf{p}$ and $\hat{a}_\mathbf{p}$ and $\hat{a}_\mathbf{p}^\dagger$ can no longer be associated with single-particle annihilation and creation. Since we must satisfy the ETCR then $\hat{\phi}_s(\mathbf{x})$ and $\hat{\pi}_s(\mathbf{x})$ must act in the same interacting Hilbert space (or subspace if other fields are present) and so we can also express $\hat{\pi}_s(\mathbf{x})$ as an Hermitian linear combination of the same modes, i.e., in terms of the same $\hat{a}_\mathbf{p}$ and $\hat{a}_\mathbf{p}^\dagger$. We define the function $\omega_\mathbf{p}$ in the interacting theory to have the form that allows us to write the mode expansions for $\hat{\phi}_s(\mathbf{x})$ and $\hat{\pi}_s(\mathbf{x})$ in the form of Eq. (6.2.14). For the free field we use $\omega_\mathbf{p} = \sqrt{\mathbf{p}^2 + m^2}$ since that is the choice that leads to $\hat{\pi}(x) = d\hat{\phi}(x)/dt$ as appropriate for that case. From Eq. (6.2.14) we can write $\hat{a}_\mathbf{p}$ and $\hat{a}_\mathbf{p}^\dagger$ in the form of Eq. (6.2.16) with this interacting form of $\omega_\mathbf{p}$ and we then again arrive at Eq. (6.2.17) for the interacting case, i.e., $[\hat{a}_\mathbf{p}, \hat{a}_{\mathbf{p}'}] = [\hat{a}_\mathbf{p}^\dagger, \hat{a}_{\mathbf{p}'}^\dagger] = 0$ and $[\hat{a}_\mathbf{p}, \hat{a}_{\mathbf{p}'}^\dagger] = (2\pi)^3\delta^3(\mathbf{p} - \mathbf{p}')$. So at $t = t_0$ all of the above arguments apply and we can label the resulting $\hat{\mathcal{P}}$ as $\hat{\mathcal{P}}_s \equiv \hat{\mathcal{P}}(t_0)$, which evolves according to

$$\hat{\mathcal{P}}(t) = \hat{U}(t, t_0)^\dagger\hat{\mathcal{P}}_s\hat{U}(t, t_0). \tag{8.1.46}$$

If $[\hat{H}(t), \hat{P}_s] = 0$ then parity is a conserved quantity and $\hat{\mathcal{P}}(t) = \hat{\mathcal{P}}_s$. We will typically be interested in Hamiltonians without any explicit time dependence, $\hat{H}(t) = \hat{H}$, where we have

$\hat{U}(t, t_0) = \exp\{-i\hat{H}(t - t_0)\}$. The presence of an external source such as $j(x)$ or $j^\mu(x)$ will obviously introduce time dependence, $\hat{H} \to \hat{H}(t)$. Similar arguments apply for other fields and also for the $\hat{C}$ and $\hat{\mathcal{T}}$ operations. So our arguments for free fields also apply for interacting fields (Bjorken and Drell, 1965).

8.1.2 Charge Conjugation

Charge conjugation is the replacement of particles by antiparticles and vice versa. It is a discrete symmetry unrelated to spacetime and so is not a Poincaré transformation. It has previously been discussed in terms of quantum mechanics in Sec. 4.1.10 and relativistic quantum mechanics in Sec. 4.5.2. Here we mean charge in a very general sense where it is any quantum number carried in opposite amounts by particles and antiparticles.

As for the parity operator the charge conjugation operator $\hat{C}$ must be unitary, $\hat{C}^{-1} = \hat{C}^\dagger$, since it can be a symmetry of a quantum system and it cannot be antiunitary as explained earlier. A neutral scalar boson is its own antiparticle and so charge conjugation can only change the Hermitian scalar field operator, $\hat{\phi}(x) = \hat{\phi}^\dagger(x)$, by at most a phase,

$$\hat{C}\hat{\phi}(x)\hat{C}^{-1} = \zeta_C^\phi\hat{\phi}(x) \quad \Rightarrow \quad \hat{C}\hat{a}_{\mathbf{p}}(x)\hat{C}^{-1} = \zeta_C^\phi\hat{a}_{\mathbf{p}} \quad \text{and} \quad \hat{C}\hat{a}_{\mathbf{p}}^\dagger\hat{C}^{-1} = \zeta_C^\phi\hat{a}_{\mathbf{p}}^\dagger. \tag{8.1.47}$$

Taking the Hermitian conjugates of these shows that $\zeta_C^\phi = \pm 1$ since $\zeta_C^\phi \in \mathbb{R}$. For $\zeta_C^\phi = \pm 1$ we say that the field has even or odd C-parity, respectively. Using Eq. (8.1.22) we can construct $\hat{C}$ as $\hat{C} = \hat{P}_2 = e^{i\hat{P}_2}$ where $\hat{P}_2 \equiv -\beta \int [d^3p/(2\pi)^3]\hat{a}_{\mathbf{p}}^\dagger\hat{a}_{\mathbf{p}}$ with $\beta = \arg\zeta_C^\phi = 0, \pm\pi$ for an even and odd C-parity operator, respectively, so that the C-parity discussed in Sec. 4.1.10 is $\eta_C^\phi = \zeta_C^\phi = e^{i\beta} = \pm 1$.

For a charged scalar field the operator $\hat{C}$ will be a unitary operator that Hermitian conjugates the field operator up to a phase,

$$\hat{C}\hat{\phi}(x)\hat{C}^{-1} = \zeta_C^\phi\hat{\phi}^\dagger(x) \quad \text{and} \quad \hat{C}\hat{\phi}^\dagger(x)\hat{C}^{-1} = \zeta_C^{\phi*}\hat{\phi}(x), \tag{8.1.48}$$

where ζ_C^ϕ is an arbitrary complex phase. Using Eq. (6.2.177) and matching the operators multiplying the exponentials $e^{\pm ip\cdot x}$ we find

$$\hat{C}\hat{f}_{\mathbf{p}}\hat{C}^{-1} = \zeta_C^\phi\hat{g}_{\mathbf{p}} \quad \text{and} \quad \hat{C}\hat{g}_{\mathbf{p}}^\dagger\hat{C}^{-1} = \zeta_C^\phi\hat{f}_{\mathbf{p}}^\dagger,$$
$$\hat{C}\hat{f}_{\mathbf{p}}^\dagger\hat{C}^{-1} = \zeta_C^{\phi*}\hat{g}_{\mathbf{p}}^\dagger \quad \text{and} \quad \hat{C}\hat{g}_{\mathbf{p}}\hat{C}^{-1} = \zeta_C^{\phi*}\hat{f}_{\mathbf{p}}. \tag{8.1.49}$$

An explicit form for the charge conjugation operator can be obtained in a manner very similar to that used for the parity operator in the case that $\zeta_C^\phi = \pm 1$,

$$\hat{C} = \exp\left\{-i\frac{\pi}{2}\int\frac{d^3p}{(2\pi)^3}\left[\pm(\hat{f}_{\mathbf{p}}^\dagger\hat{g}_{\mathbf{p}} + \hat{g}_{\mathbf{p}}^\dagger\hat{f}_{\mathbf{p}}) \mp \zeta_C^\phi(\hat{f}_{\mathbf{p}}^\dagger\hat{f}_{\mathbf{p}} + \hat{g}_{\mathbf{p}}^\dagger\hat{g}_{\mathbf{p}})\right]\right\}, \tag{8.1.50}$$

where either sign choice in the above expression is valid. This gives $\hat{C} = \hat{C}^\dagger = \hat{C}^{-1}$. The result is generalized to the case of an arbitrary complex phase below.

Proof: Define $\hat{C}_1 \equiv \exp\{i\hat{C}_1\}$ and $\hat{C}_2 \equiv \exp\{i\hat{C}_2\}$, where

$$\hat{C}_1 \equiv -\alpha \int [d^3p/(2\pi)^3][\hat{f}_{\mathbf{p}}^\dagger\hat{g}_{\mathbf{p}} + \hat{g}_{\mathbf{p}}^\dagger\hat{f}_{\mathbf{p}}], \quad \hat{C}_2 \equiv -\beta \int [d^3p/(2\pi)^3][\hat{f}_{\mathbf{p}}^\dagger\hat{f}_{\mathbf{p}} + \hat{g}_{\mathbf{p}}^\dagger\hat{g}_{\mathbf{p}}],$$

where we note that $\hat{C}_1 = \hat{C}_1^\dagger$ and $\hat{C}_2 = \hat{C}_2^\dagger$. This then gives using Eq. (A.8.7)

$$[\hat{C}_1, \hat{f}_{\mathbf{p}}] = \alpha\hat{g}_{\mathbf{p}}, \ [\hat{C}_1, \hat{g}_{\mathbf{p}}^\dagger] = -\alpha\hat{f}_{\mathbf{p}}^\dagger, \ [\hat{C}_1, \hat{f}_{\mathbf{p}}^\dagger] = -\alpha\hat{g}_{\mathbf{p}}^\dagger, \ [\hat{C}_1, \hat{g}_{\mathbf{p}}] = \alpha\hat{f}_{\mathbf{p}},$$

$$[\hat{C}_1, [\hat{C}_1, \hat{f}_{\mathbf{p}}]] = \alpha^2 \hat{f}_{\mathbf{p}}, \quad [\hat{C}_1, [\hat{C}_1, \hat{g}_{\mathbf{p}}^\dagger]] = \alpha^2 \hat{g}_{\mathbf{p}}^\dagger,$$

$$\hat{C}_1 \hat{f}_{\mathbf{p}} \hat{C}_1^{-1} = \hat{f}_{\mathbf{p}} + i[\hat{C}_1, \hat{f}_{\mathbf{p}}] + (i^2/2!)[\hat{C}_1, [\hat{C}_1, \hat{f}_{\mathbf{p}}]] + (i^3/3!)[\hat{C}_1, [\hat{C}_1, [\hat{C}_1, \hat{f}_{\mathbf{p}}]]] + \cdots$$

$$= \hat{f}_{\mathbf{p}} \cos\alpha + i\hat{g}_{\mathbf{p}} \sin\alpha \tag{8.1.51}$$

and similarly $\hat{C}_1 \hat{g}_{\mathbf{p}}^\dagger \hat{C}_1^{-1} = \hat{g}_{\mathbf{p}}^\dagger \cos\alpha - i\hat{f}_{\mathbf{p}}^\dagger \sin\alpha$. Then choosing

$$\alpha = \pm\pi/2 \quad \Rightarrow \quad \hat{C}_1 \hat{f}_{\mathbf{p}} \hat{C}_1^{-1} = \pm i\hat{g}_{\mathbf{p}} \quad \text{and} \quad \hat{C}_1 \hat{g}_{\mathbf{p}}^\dagger \hat{C}_1^{-1} = \mp i\hat{f}_{\mathbf{p}}^\dagger. \tag{8.1.52}$$

We also have using Eq. (A.8.7)

$$[\hat{C}_2, \hat{f}_{\mathbf{p}}] = \beta\hat{f}_{\mathbf{p}}, \quad [\hat{C}_2, [\hat{C}_2, \hat{f}_{\mathbf{p}}]] = \beta^2 \hat{f}_{\mathbf{p}}, \quad [\hat{C}_2, \hat{g}_{\mathbf{p}}^\dagger] = -\beta\hat{g}_{\mathbf{p}}^\dagger, \quad [\hat{C}_2, [\hat{C}_2, \hat{g}_{\mathbf{p}}^\dagger]] = \beta^2 \hat{g}_{\mathbf{p}}^\dagger,$$

$$\hat{C}_2 \hat{f}_{\mathbf{p}} \hat{C}_2^{-1} = \hat{f}_{\mathbf{p}} + i[\hat{C}_2, \hat{f}_{\mathbf{p}}] + (i^2/2!)[\hat{C}_2, [\hat{C}_2, \hat{f}_{\mathbf{p}}]] + \cdots = \hat{f}_{\mathbf{p}} e^{i\beta} \tag{8.1.53}$$

and similarly $\hat{C}_2 \hat{g}_{\mathbf{p}}^\dagger \hat{C}_2^{-1} = \hat{g}_{\mathbf{p}}^\dagger e^{-i\beta}$. It is not difficult to show that $[\hat{C}_1, \hat{C}_2] = 0$ and so using the BCH formula in Eq. (A.8.6) we have $[\hat{\mathcal{C}}_1, \hat{\mathcal{C}}_2] = 0$ and

$$\hat{\mathcal{C}} \equiv \hat{\mathcal{C}}_1 \hat{\mathcal{C}}_2 = e^{i(\hat{C}_1 + \hat{C}_2)}$$

$$= \exp\{-i \int [d^3p/(2\pi)^3][\alpha(\hat{f}_{\mathbf{p}}^\dagger \hat{g}_{\mathbf{p}} + \hat{g}_{\mathbf{p}}^\dagger \hat{f}_{\mathbf{p}}) + \beta(\hat{f}_{\mathbf{p}}^\dagger \hat{f}_{\mathbf{p}} + \hat{g}_{\mathbf{p}}^\dagger \hat{g}_{\mathbf{p}})]\}, \tag{8.1.54}$$

where $\alpha = \pm\pi/2$. For the case of $\hat{f}_{\mathbf{p}}$ we require $\zeta_C^\phi = \pm ie^{i\beta} = e^{i(\beta\pm\pi/2)}$ and for the case of $\hat{g}_{\mathbf{p}}^\dagger$ we require $\zeta_C^\phi = \mp ie^{-i\beta} = e^{-i(\beta\pm\pi/2)}$, which means that $\zeta_C^\phi = e^{i(\beta\pm\pi/2)} = e^{-i(\beta\pm\pi/2)}$. So we must have $\beta = (n \mp \frac{1}{2})\pi$ for $n \in \mathbb{Z}$, where for n even/odd we have $\zeta_C^\phi = \pm 1$, respectively. Then for $\zeta_C^\phi = \pm 1$ we can choose $\beta = \mp\zeta_C^\phi(\pi/2)$. If the $\pm$ notation is confusing one should work through the two cases of $\alpha = \pm\pi/2$ separately. The choice $\alpha = -\pi/2$ reproduces the $\zeta_C^\phi \in \mathbb{R}$ result in Greiner and Reinhardt (1996).

Generalization to arbitrary phase: For $\zeta_C^\phi \notin \mathbb{R}$ the situation becomes more complicated because of the different signs for $\hat{f}_{\mathbf{p}}$ and $\hat{g}_{\mathbf{p}}^\dagger$ in Eq. (8.1.52). Replace ζ_C^ϕ in Eq. (8.1.50) with $+1$ and label the unitary result $\hat{\mathcal{C}}'$, then

$$\hat{\mathcal{C}}' \hat{f}_{\mathbf{p}} \hat{\mathcal{C}}'^{-1} = \hat{g}_{\mathbf{p}} \quad \text{and} \quad \hat{\mathcal{C}}' \hat{g}_{\mathbf{p}}^\dagger \hat{\mathcal{C}}'^{-1} = \hat{f}_{\mathbf{p}}^\dagger \tag{8.1.55}$$

and similarly for $\hat{f}_{\mathbf{p}}^\dagger$ and $\hat{g}_{\mathbf{p}}$. Define $\hat{\mathcal{C}}_3 \equiv \exp(i\hat{C}_3)$ with

$$\hat{C}_3 \equiv -\gamma \int [d^3p/(2\pi)^3][\hat{f}_{\mathbf{p}}^\dagger \hat{f}_{\mathbf{p}} - \hat{g}_{\mathbf{p}}^\dagger \hat{g}_{\mathbf{p}}] = \hat{C}_3^\dagger$$

$$\Rightarrow \quad \hat{\mathcal{C}}_3 \hat{f}_{\mathbf{p}} \hat{\mathcal{C}}_3^{-1} = e^{i\gamma} \hat{f}_{\mathbf{p}} \quad \text{and} \quad \hat{\mathcal{C}}_3 \hat{g}_{\mathbf{p}}^\dagger \hat{\mathcal{C}}_3^{-1} = e^{i\gamma} \hat{g}_{\mathbf{p}}^\dagger \tag{8.1.56}$$

with $\hat{\mathcal{C}}_3$ unitary. Defining $\zeta_C^\phi \equiv e^{-i\gamma}$ and $\hat{\mathcal{C}} \equiv \hat{\mathcal{C}}_3 \hat{\mathcal{C}}'$ then $\hat{\mathcal{C}}$ is unitary and

$$\hat{\mathcal{C}} \hat{f}_{\mathbf{p}} \hat{\mathcal{C}}^{-1} = \hat{\mathcal{C}}_3 \hat{g}_{\mathbf{p}} \hat{\mathcal{C}}_3^{-1} = \zeta_C^\phi \hat{g}_{\mathbf{p}} \quad \text{and} \quad \hat{\mathcal{C}} \hat{g}_{\mathbf{p}}^\dagger \hat{\mathcal{C}}^{-1} = \hat{\mathcal{C}}_3 \hat{f}_{\mathbf{p}}^\dagger \hat{\mathcal{C}}_3^{-1} = \zeta_C^\phi \hat{f}_{\mathbf{p}}^\dagger. \tag{8.1.57}$$

Since the exponents are normal-ordered, then the vacuum has positive C-parity,

$$\hat{\mathcal{C}}|0\rangle = |0\rangle. \tag{8.1.58}$$

Recall that the relevant Hilbert space V_H is a Fock space and that this has a basis constructed from creation operators acting on the vacuum. Note that

$$\hat{\mathcal{C}}^2 \hat{f}_{\mathbf{p}}^\dagger (\hat{\mathcal{C}}^{-1})^2 = \zeta_C^{\phi*} \hat{\mathcal{C}} \hat{g}_{\mathbf{p}}^\dagger \hat{\mathcal{C}}^{-1} = \hat{f}_{\mathbf{p}}^\dagger \quad \text{and} \quad \hat{\mathcal{C}}^2 \hat{g}_{\mathbf{p}}^\dagger (\hat{\mathcal{C}}^{-1})^2 = \zeta_C^\phi \hat{\mathcal{C}} \hat{f}_{\mathbf{p}}^\dagger \hat{\mathcal{C}}^{-1} = \hat{g}_{\mathbf{p}}^\dagger. \tag{8.1.59}$$

This means that $\hat{C}^2$ leaves every basis state invariant since

$$\hat{C}^2(\hat{f}_{\mathbf{p}_1}^\dagger \cdots \hat{f}_{\mathbf{p}_i}^\dagger \hat{g}_{\mathbf{p}_{i+1}}^\dagger \cdots \hat{g}_{\mathbf{p}_j}^\dagger)|0\rangle = \hat{C}^2(\hat{f}_{\mathbf{p}_1}^\dagger \cdots \hat{g}_{\mathbf{p}_j}^\dagger)(\hat{C}^{-1})^2|0\rangle = (\hat{f}_{\mathbf{p}_1}^\dagger \cdots \hat{g}_{\mathbf{p}_j}^\dagger)|0\rangle. \qquad (8.1.60)$$

Then for all $|\psi\rangle \in V_H$ we have $\hat{C}^2|\psi\rangle = |\psi\rangle$ and so $\hat{C}^2$ is the identity operator,

$$\hat{C}^2 = \hat{I} \quad \Rightarrow \quad \hat{C} = \hat{C}^{-1} = \hat{C}^\dagger, \qquad (8.1.61)$$

and so $\hat{C}$ is unitary and Hermitian. Since $\hat{C}$ is Hermitian it is an observable with real eigenvalues and since $\hat{C}^2 = \hat{I}$ these eigenvalues are ± 1 corresponding to even and odd C-parity, respectively. Note that $\hat{\phi}^\dagger(x)|0\rangle$ and $\hat{\phi}(x)|0\rangle$ produce charged particle and antiparticle states, respectively. Since $\hat{C}\hat{\phi}^\dagger(x)|0\rangle = \zeta_C^{\phi*}\hat{\phi}(x)|0\rangle$ and $\hat{C}\hat{\phi}(x)|0\rangle = \zeta_C^\phi\hat{\phi}^\dagger(x)|0\rangle$, then charged states cannot be C-parity eigenstates and only neutral particles can. Clearly, C-parity will be a multiplicative intrinsic quantum number in the same way that intrinsic parity is.

Proof: There is a direct proof of Eq. (8.1.61). (i) For $\zeta_C^\phi \in \mathbb{R}$ we see from Eq. (8.1.50) that $\hat{C} \to \hat{C}^\dagger$ simply changes the arbitrary sign choice in the exponent and so we have $\hat{C} = \hat{C}^\dagger = \hat{C}^{-1}$. (ii) The arbitrary complex phase case is only slightly more complicated. We know that $[\hat{C}_1, \hat{C}_2] = 0$ and similarly $[\hat{C}_2, \hat{C}_3] = 0$ but $[\hat{C}_1, \hat{C}_3] \neq 0$ and so $[\hat{C}', \hat{C}_3] \neq 0$. From the real phase result we know $\hat{C}' = \hat{C}'^\dagger = \hat{C}'^{-1}$ and then $\hat{C}^\dagger = (\hat{C}_3\hat{C}')^\dagger = \hat{C}'^\dagger\hat{C}_3^\dagger = \hat{C}'\hat{C}_3^\dagger$, where $\hat{C}_3^\dagger = \hat{C}_3^{-1}$ is $\hat{C}_3$ with $\gamma \to -\gamma$. From Eq. (8.1.55) we see that $\hat{C}'$ leads to the exchange $\hat{f} \leftrightarrow \hat{g}$ and this exchange in $\hat{C}_3$ and $\hat{C}_3^\dagger$ reverses the sign of γ and so $\hat{C}'\hat{C}_3^\dagger\hat{C}' = \hat{C}_3$. This gives $\hat{C}^\dagger = \hat{C}'\hat{C}_3^\dagger = \hat{C}_3\hat{C}' = \hat{C}$ and so $\hat{C} = \hat{C}^\dagger = \hat{C}^{-1}$ as concluded above.

The annihilation and creation operator commutation relations of Eq. (6.2.180) are readily verified to be invariant under charge conjugation since $\hat{f} \leftrightarrow \hat{g}$ and phases cancel. Similarly, since $\hat{\pi} = \partial_0\hat{\phi}^\dagger$ and $\hat{\pi}^\dagger = \partial_0\hat{\phi}$, then the ETCR of Eq. (6.2.181) are invariant since $\hat{\phi} \leftrightarrow \hat{\phi}^\dagger$, $\hat{\pi} \leftrightarrow \hat{\pi}^\dagger$ with phases canceling.

From Eq. (8.1.48) it follows that any combination of field and momentum operators with $\hat{\phi}, \hat{\pi}$ appearing symmetrically with $\hat{\phi}^\dagger, \hat{\pi}^\dagger$ will be invariant under charge conjugation. In situations where $\pi, \hat{\pi}^\dagger$ appear together in the same term as $\hat{\phi}, \hat{\phi}^\dagger$ operator-ordering issues may arise since they do not commute, but as previously discussed normal-ordering removes these ambiguities and symmetrizes the term. For example, $\hat{\mathbf{P}}$ in Eq. (6.2.174) is independent of how we order $\hat{\pi}, \hat{\pi}^\dagger$ and $\boldsymbol{\nabla}\hat{\phi}, \boldsymbol{\nabla}\hat{\phi}^\dagger$ due to the normal ordering. It follows from Eq. (6.2.174) that

$$\hat{C}\hat{H}\hat{C}^{-1} = \hat{H} \quad \text{and} \quad \hat{C}\hat{\mathbf{P}}\hat{C}^{-1} = \hat{\mathbf{P}}, \qquad (8.1.62)$$

which we can also see immediately from their expression in terms of normal-ordered annihilation and creation operators in Eq. (6.2.185). Similarly, it is easily shown that the angular momentum operator is invariant, $\hat{C}\hat{\mathbf{L}}\hat{C}^{-1} = \hat{\mathbf{L}}$, and that the conserved current operator reverses sign as expected,

$$\hat{j}^\mu \xrightarrow{C} j'^\mu = \hat{C}j^\mu\hat{C}^{-1} = i\hat{C} :[\hat{\phi}^\dagger(\partial^\mu\hat{\phi}) - (\partial^\mu\hat{\phi}^\dagger)\hat{\phi}]:\hat{C}^{-1} = i:\hat{C}[\hat{\phi}^\dagger(\partial^\mu\hat{\phi}) - (\partial^\mu\hat{\phi}^\dagger)\hat{\phi}]\hat{C}^{-1}:$$
$$= i:[\hat{\phi}(\partial^\mu\hat{\phi}^\dagger) - (\partial^\mu\hat{\phi})\hat{\phi}^\dagger]: = -i:[\hat{\phi}^\dagger(\partial^\mu\hat{\phi}) - (\partial^\mu\hat{\phi}^\dagger)\hat{\phi}]: = -\hat{j}^\mu. \qquad (8.1.63)$$

It is obvious from Eq. (8.1.49) that the order of charge conjugation and normal ordering can be exchanged with no effect. Note that in moving from the first to the second line we exchanged the order of the field and field derivatives and used the fact that normal-ordering for bosons involves no sign change. For $\hat{\mathbf{j}}$ this was trivial since $\hat{\phi}$ and $\boldsymbol{\nabla}\hat{\phi}$ commute as do their Hermitian conjugates.

However, $\hat{j}^0 = i{:}\hat{\phi}^\dagger\hat{\pi}^\dagger - \hat{\pi}\hat{\phi}{:}$ and so $\hat{j}'^0 = i{:}\hat{\phi}\hat{\pi} - \hat{\pi}^\dagger\hat{\phi}^\dagger{:}$ and equating $j'^0 = -\hat{j}^0$ is only possible because the normal-ordering allows the rearrangement, $\hat{\phi}\hat{\pi} \to \hat{\pi}\hat{\phi}$ and $\hat{\pi}^\dagger\hat{\phi}^\dagger \to \hat{\phi}^\dagger\hat{\pi}^\dagger$.

From Eq. (4.5.15) we know that the electromagnetic field must transform as

$$\hat{A}^\mu(x) \xrightarrow{\mathcal{C}} \hat{A}'^\mu(x) = \hat{\mathcal{C}}\hat{A}^\mu(x)\hat{\mathcal{C}}^{-1} = -\zeta_C^A \hat{A}^\mu(x) = -\hat{A}^\mu(x), \qquad (8.1.64)$$

which follows since under charge conjugation a charge current density changes sign, $j^\mu(x) \xrightarrow{\mathcal{C}} -j^\mu(x)$ and since under charge conjugation $j_\mu \hat{A}^\mu$ should be invariant. For an arbitrary vector field we write

$$\hat{\mathcal{C}}\hat{V}^\mu(x)\hat{\mathcal{C}}^{-1} = -\zeta_C^V \hat{V}^\mu(x), \qquad (8.1.65)$$

where for a neutral vector field $\eta_C^V = -\zeta_C^V = \pm 1$ for a even and odd *C*-parity, respectively. It will be convenient when we come to discuss *CPT* transformations to define ζ_C^V in this way. The photon has odd *C*-parity as discussed in Sec. 4.1.10, $\eta_C^A = -\zeta_C^A = -1$. In Coulomb gauge in the quantization inertial frame,

$$\hat{\mathcal{C}}\hat{\mathbf{A}}(t, \mathbf{x})\hat{\mathcal{C}}^{-1} = -\hat{\mathbf{A}}(t, -\mathbf{x}). \qquad (8.1.66)$$

This requires

$$\hat{\mathcal{C}}\hat{a}_{\mathbf{k}}^\lambda\hat{\mathcal{C}}^{-1} = -\hat{a}_{\mathbf{k}}^\lambda \quad \text{and} \quad \hat{\mathcal{C}}\hat{a}_{\mathbf{k}}^{\lambda\dagger}\hat{\mathcal{C}}^{-1} = -\hat{a}_{\mathbf{k}}^{\lambda\dagger}. \qquad (8.1.67)$$

This is similar to equation Eq. (8.1.25) except that there is no need to invert the sign of $\mathbf{x}$. We arrive at the explicit form (Bjorken and Drell, 1965)

$$\hat{\mathcal{C}} = \exp\{\pm i\,\pi \int [d^3k/(2\pi)^3] \sum_{\lambda=1}^2 \hat{a}_{\mathbf{k}}^{\lambda\dagger}\hat{a}_{\mathbf{k}}^\lambda\}, \qquad (8.1.68)$$

which also satisfies Eq. (8.1.61).

Proof: We write $\hat{\mathcal{C}} \equiv \exp\{i\hat{C}\}$ with the definition here that

$$\hat{C} \equiv \pm\pi \int [d^3k/(2\pi)^3] \sum_{\lambda=1}^2 \hat{a}_{\mathbf{k}}^{\lambda\dagger}\hat{a}_{\mathbf{k}}^\lambda. \qquad (8.1.69)$$

Using the commutation relations in Eq. (6.4.11) we find

$$[\hat{C}, \hat{a}_{\mathbf{k}}^\lambda] = \mp\pi\hat{a}_{\mathbf{k}}^\lambda, \quad [\hat{C}, [\hat{C}, \hat{a}_{\mathbf{k}}^\lambda]] = (\mp\pi)^2\hat{a}_{\mathbf{k}}^\lambda, \qquad (8.1.70)$$

and similarly for additional commutations with $\hat{C}$. Using Eq. (A.8.7) gives

$$\begin{aligned}
\hat{\mathcal{C}}\hat{a}_{\mathbf{k}}^\lambda\hat{\mathcal{C}}^{-1} &= \hat{a}_{\mathbf{k}}^\lambda + i[\hat{C}, \hat{a}_{\mathbf{k}}^\lambda] + (i^2/2!)[\hat{C}, [\hat{C}, \hat{a}_{\mathbf{k}}^\lambda]] + (i^3/3!)[\hat{C}, [\hat{C}, [\hat{C}, \hat{a}_{\mathbf{k}}^\lambda]]] + \cdots \\
&= \hat{a}_{\mathbf{k}}^\lambda[1 + (\mp i\pi) + (\mp i\pi)^2/2! + (\mp i\pi)^3/3! + \cdots] = \hat{a}_{\mathbf{k}}^\lambda e^{\mp i\pi} = -\hat{a}_{\mathbf{k}}^\lambda. \quad (8.1.71)
\end{aligned}$$

Since $\hat{C} = \hat{C}^\dagger$ then $\hat{\mathcal{C}}$ is unitary. We also see that $\hat{\mathcal{C}}^2\hat{a}_{\mathbf{k}}^\lambda(\hat{\mathcal{C}}^{-1})^2 = \hat{a}_{\mathbf{k}}^\lambda$ and $\hat{\mathcal{C}}|0\rangle = |0\rangle$ and so $\hat{\mathcal{C}}^2 = \hat{I}$ in all of Fock space $\Rightarrow \hat{\mathcal{C}} = \hat{\mathcal{C}}^{-1} = \hat{\mathcal{C}}^\dagger$ as claimed.

Since an *n*-photon state will involve *n* photon creation operators and since

$$\hat{\mathcal{C}}[\hat{a}_{\mathbf{k}_1}^{\lambda_1} \cdots \hat{a}_{\mathbf{k}_n}^{\lambda_n}]|0\rangle = \hat{\mathcal{C}}[\hat{a}_{\mathbf{k}_1}^{\lambda_1} \cdots \hat{a}_{\mathbf{k}_n}^{\lambda_n}]\hat{\mathcal{C}}^{-1}|0\rangle = (-1)^n\hat{a}_{\mathbf{k}_1}^{\lambda_1} \cdots \hat{a}_{\mathbf{k}_n}^{\lambda_n}|0\rangle, \qquad (8.1.72)$$

then any *n*-photon state is a *C*-parity eigenstate with *C*-parity $\eta_C = (-1)^n$. This has important consequences for theories that are invariant under charge conjugation such as quantum electrodynamics. In such a theory any even or odd *C*-parity eigenstate can only decay to even or odd number of photons, respectively, which is an example of a selection rule. This also leads to an on-shell version of Furry's theorem, in the sense that in quantum electrodynamics a state with an even and odd number of photons can only transition into states with an even and odd number of photons, respectively; i.e., we cannot transition between states with even and odd numbers of photons. As we will see later *Furry's theorem* in electrodynamics has a stronger form and says that this applies even to virtual photons. More precisely, it is the statement that loops in Feynman graphs with an odd number of photons attached vanish in quantum electrodynamics.

In the case of fermions recall that for the Dirac equation to preserve its form under charge conjugation we required the spinor to transform as in Eq. (4.5.25). For this to be true at the operator level we require a unitary operator $\hat{\mathcal{C}}$ such that

$$\hat{\psi}(x) \xrightarrow{\mathcal{C}} \hat{\psi}'(x) = \hat{\mathcal{C}}\hat{\psi}(x)\hat{\mathcal{C}}^{-1} = C\hat{\bar{\psi}}^T(x), \tag{8.1.73}$$

where the spinor charge conjugation matrix in the Dirac represnntation is given in Eq. (4.5.29), $C = i\gamma^2\gamma^0 = -i\gamma^0\gamma^2$, and has the properties that $C^{-1}\gamma^\mu C = -\gamma^{\mu T}$ and $C = C^* = -C^{-1} = -C^\dagger = -C^T$. We could have introduced an arbitrary phase ζ_C^ψ in this transformation as we did in the case of parity in Eq. (8.1.1), but since only fermion bilinears are relevant we can simply choose the fermion phase to be unity without loss of generality, $\zeta_C^\psi = 1$. In the Dirac representation we also have $\gamma^{0T} = \gamma^0$ and $\gamma^{2T} = \gamma^2$. It follows from above that

$$\hat{\mathcal{C}}\hat{\psi}^\dagger\hat{\mathcal{C}}^{-1} = [\hat{\bar{\psi}}^T]^\dagger C^\dagger = [\gamma^{0T}\hat{\psi}^{\dagger T}]^\dagger C^\dagger = \hat{\psi}^T\gamma^{0T}C = \hat{\psi}^T\gamma^0 C = -\hat{\psi}^T C^{-1}\gamma^0$$

$$\Rightarrow \quad \hat{\mathcal{C}}\hat{\bar{\psi}}\hat{\mathcal{C}}^{-1} = -\hat{\psi}^T C^{-1}. \tag{8.1.74}$$

Using this with the plane wave expansion of Eq. (6.3.194) and identifying the quantities multiplying the exponential factors $e^{\pm ip\cdot x}$ we obtain

$$\hat{\mathcal{C}}\hat{b}_{\mathbf{p}}^s\hat{\mathcal{C}}^{-1}u^s(p) = \hat{d}_{\mathbf{p}}^s C\bar{v}^s(p)^T = (-1)^{\eta_s}\hat{d}_{\mathbf{p}}^s u^s(p),$$

$$\hat{\mathcal{C}}\hat{d}_{\mathbf{p}}^{s\dagger}\hat{\mathcal{C}}^{-1}v^s(p) = \hat{b}_{\mathbf{p}}^{s\dagger} C\bar{u}^s(p)^T = (-1)^{\eta_s}\hat{b}_{\mathbf{p}}^{s\dagger} v^s(p), \tag{8.1.75}$$

where in the last step in each equation above we have used Eq. (4.5.41) with $\eta_s = 0, 1$ for spin up/down, respectively. It then follows that

$$\hat{\mathcal{C}}\hat{b}_{\mathbf{p}}^s\hat{\mathcal{C}}^{-1} = (-1)^{\eta_s}\hat{d}_{\mathbf{p}}^s \quad \text{and} \quad \hat{\mathcal{C}}\hat{d}_{\mathbf{p}}^{s\dagger}\hat{\mathcal{C}}^{-1} = (-1)^{\eta_s}\hat{b}_{\mathbf{p}}^{s\dagger} \tag{8.1.76}$$

and similarly for $\hat{b}_{\mathbf{p}}^{s\dagger}$ and $\hat{d}_{\mathbf{p}}^s$, respectively. The annihilation and creation operator anticommutation relations in Eq. (6.3.80) are invariant under the above charge conjugation transformation as expected. Using $(-1)^{\eta_s} = e^{i\pi\eta_s}$ an explicit form for the unitary operator $\hat{\mathcal{C}}$ is (Bjorken and Drell, 1965)

$$\hat{\mathcal{C}} \equiv \hat{\mathcal{C}}_2\hat{\mathcal{C}}_1, \quad \text{where here we define}$$

$$\hat{\mathcal{C}}_1 \equiv \exp\{\pm i\pi \int [d^3p/(2\pi)^3] \sum_s \eta_s[\hat{b}_{\mathbf{p}}^{\dagger s}\hat{b}_{\mathbf{p}}^s - \hat{d}_{\mathbf{p}}^{\dagger s}\hat{d}_{\mathbf{p}}^s]\},$$

$$\hat{\mathcal{C}}_2 \equiv \exp\{\pm i(\pi/2) \int [d^3p/(2\pi)^3] \sum_s[\hat{b}_{\mathbf{p}}^{\dagger s}\hat{b}_{\mathbf{p}}^s - \hat{b}_{\mathbf{p}}^{\dagger s}\hat{d}_{\mathbf{p}}^s + \hat{d}_{\mathbf{p}}^{\dagger s}\hat{d}_{\mathbf{p}}^s - \hat{d}_{\mathbf{p}}^{\dagger s}\hat{b}_{\mathbf{p}}^s]\}. \tag{8.1.77}$$

This operator is also self-inverse, $\hat{\mathcal{C}}^2 = \hat{I}$ and so we again have Eq. (8.1.61).

Proof: Define $\hat{C}_2$ from $\hat{\mathcal{C}}_2 \equiv \exp\{i\hat{C}_2\}$. Using Eq. (6.3.81) gives

$$[\hat{C}_2, \hat{b}_{\mathbf{p}}^s] = \mp\pi\tfrac{1}{2}(\hat{b}_{\mathbf{p}}^s - \hat{d}_{\mathbf{p}}^s), \quad [\hat{C}_2, \hat{d}_{\mathbf{p}}^s] = \mp\pi\tfrac{1}{2}(\hat{d}_{\mathbf{p}}^s - \hat{b}_{\mathbf{p}}^s), \tag{8.1.78}$$

$$[\hat{C}_2, [\hat{C}_2, \hat{b}_{\mathbf{p}}^s]] = \mp\pi\tfrac{1}{2}([\hat{C}_2, \hat{b}_{\mathbf{p}}^s] - [\hat{C}_2, \hat{d}_{\mathbf{p}}^s]) = \pi^2\tfrac{1}{2}(\hat{b}_{\mathbf{p}}^s - \hat{d}_{\mathbf{p}}^s) = \mp\pi[\hat{C}_2, \hat{b}_{\mathbf{p}}^s]$$

and similarly $[\hat{C}_2, [\hat{C}_2, \hat{d}_{\mathbf{p}}^s]] = \mp \pi [\hat{C}_2, \hat{d}_{\mathbf{p}}^s]$. Then following Eq. (8.1.38) we obtain

$$\hat{C}_2 \hat{b}_{\mathbf{p}}^s \hat{C}_2^{-1} = d_{\mathbf{p}}^s \quad \text{and} \quad \hat{C}_2 \hat{d}_{\mathbf{p}}^s \hat{C}_2^{-1} = b_{\mathbf{p}}^s. \tag{8.1.79}$$

Similarly, defining $\hat{C}_1$ from $\hat{C}_1 \equiv \exp\{i\hat{C}_1\}$ and using Eq. (A.8.7) gives

$$[\hat{C}_1, \hat{b}_{\mathbf{p}}^s] = \mp \pi \eta_s \hat{b}_{\mathbf{p}}^s \quad \text{and} \quad [\hat{C}_1, \hat{d}_{\mathbf{p}}^s] = \pm \pi \eta_s \hat{d}_{\mathbf{p}}^s, \tag{8.1.80}$$

$$\hat{C}_1 \hat{b}_{\mathbf{p}}^s \hat{C}_1^{-1} = \hat{b}_{\mathbf{p}}^s + i[\hat{C}_1, \hat{b}_{\mathbf{p}}^s] + (i^2/2!)[\hat{C}_1, [\hat{C}_1, \hat{b}_{\mathbf{p}}^s]] + (i^3/3!)[\hat{C}_1, [\hat{C}_1, [\hat{C}_1, \hat{b}_{\mathbf{p}}^s]]] + \cdots$$
$$= \hat{b}_{\mathbf{p}}^s[1 + (\mp i\pi\eta_s) + (\mp i\pi\eta_s)^2/2! + (\mp i\pi\eta_s)^3/3! + \cdots] = \hat{b}_{\mathbf{p}}^s e^{\mp i\pi\eta_s} = (-1)^{\eta_s} \hat{b}_{\mathbf{p}}^s$$

and similarly $\hat{C}_1 \hat{d}_{\mathbf{p}}^s \hat{C}_1^{-1} = \hat{d}_{\mathbf{p}}^s e^{\pm i\pi\eta_s} = (-1)^{\eta_s} \hat{d}_{\mathbf{p}}^s$. Note that $\hat{C}_1^\dagger = \hat{C}_1$ and $\hat{C}_2^\dagger = \hat{C}_2$ and so $\hat{C}_1$ and $\hat{C}_2$ are both unitary. Then $\hat{C} = \hat{C}_2\hat{C}_1$ is unitary since $\hat{C}\hat{C}^\dagger = \hat{C}_2\hat{C}_1\hat{C}_1^\dagger\hat{C}_2^\dagger = \hat{I}$. Normal-ordered exponents give $\hat{C}|0\rangle = |0\rangle$. It is also clear that $\hat{C}^2 \hat{b}_{\mathbf{p}}^s (\hat{C}^{-1})^2 = \hat{b}_{\mathbf{p}}^s$ and $\hat{C}^2 \hat{d}_{\mathbf{p}}^s (\hat{C}^{-1})^2 = \hat{d}_{\mathbf{p}}^s$ and so $\hat{C}^2 = \hat{I} \Rightarrow \hat{C} = \hat{C}^{-1} = \hat{C}^\dagger$.

The normal-ordered four-vector current density for a fermion in Eq. (6.3.207) changes sign under charge conjugation as it should, since

$$\hat{C}\hat{j}^\mu(x)\hat{C}^{-1} = \hat{C} :\hat{\bar{\psi}}(x)\gamma^\mu\hat{\psi}(x): \hat{C}^{-1} = :\hat{C}\hat{\bar{\psi}}(x)\hat{C}^{-1}\gamma^\mu\hat{C}\hat{\psi}(x)\hat{C}^{-1}: \tag{8.1.81}$$
$$= :-\hat{\psi}(x)^T C^{-1} \gamma^\mu C \hat{\bar{\psi}}(x)^T: = :\hat{\psi}(x)^T \gamma^{\mu T} \hat{\bar{\psi}}(x)^T:$$
$$= :\hat{\psi}_\alpha(x)\gamma_{\beta\alpha}^\mu \hat{\bar{\psi}}_\beta(x): = -:\hat{\bar{\psi}}_\beta(x)\gamma_{\beta\alpha}^\mu \hat{\psi}_\alpha(x): = -\hat{j}^\mu(x),$$

where recall that a unitary operator passes through the normal-ordering, that $\hat{C}$ acts only on annihilation and creation operators, that $C^{-1}\gamma^\mu C = -\gamma^{\mu T}$ and that interchanging fermion operators in fermion normal ordering adds a negative sign. Similarly, using the symmetrized current in Eq. (6.3.209), $\hat{j}^\mu = \frac{1}{2}[\hat{\bar{\psi}}, \gamma^\mu\hat{\psi}]$, we have

$$\hat{C}\hat{j}^\mu\hat{C}^{-1} = \hat{C} \tfrac{1}{2}[\hat{\bar{\psi}}_\alpha, (\gamma^\mu\hat{\psi})_\alpha]\hat{C}^{-1} = \tfrac{1}{2}\gamma_{\alpha\beta}^\mu\hat{C}[\hat{\bar{\psi}}_\alpha, \hat{\psi}_\beta]\hat{C}^{-1} = \tfrac{1}{2}\gamma_{\alpha\beta}^\mu[-(\hat{\psi}^T C^{-1})_\alpha, (C\hat{\bar{\psi}}^T)_\beta]$$
$$= -\tfrac{1}{2}C_{\gamma\alpha}^{-1}\gamma_{\alpha\beta}^\mu C_{\beta\delta}[\hat{\psi}_\gamma, \hat{\bar{\psi}}_\delta] = -\tfrac{1}{2}\gamma_{\delta\gamma}^\mu[\hat{\bar{\psi}}_\delta, \hat{\psi}_\gamma] = -\tfrac{1}{2}[\hat{\bar{\psi}}_\delta, (\gamma^\mu\hat{\psi})_\delta] = -j^\mu. \tag{8.1.82}$$

8.1.3 Time Reversal

In classical mechanics time-reversal symmetry means that given some trajectory in phase space for a system $\vec{q}(t), \vec{p}(t)$, then a trajectory $\vec{q}\,'(t) = \vec{q}(-t), \vec{p}\,'(t) = -\vec{p}(-t)$ is also a valid classical trajectory in that system. We earlier studied time-reversal in the context of quantum mechanics in Sec. 4.1.7 and we saw that the time-reversal operator $\hat{T}$ must be antiunitary. We also know from Eq. (4.1.61) that we can always choose an arbitrary basis for Fock space and define a complex conjugation operator $\hat{K}$ for that basis and a corresponding unitary operator $\hat{U} \equiv \hat{T}\hat{K}$ such that

$$\hat{T} = \hat{U}\hat{K}. \tag{8.1.83}$$

From Secs. 4.1.7 and 4.5.3 we require that time-reversal transformations be effected by an antiunitary operator $\hat{T}$, such that in the passive view we have

$$\hat{\phi}(x) \xrightarrow{T} \hat{\phi}'(x') = \hat{T}\hat{\phi}(-t, \mathbf{x})\hat{T}^{-1} = \zeta_T^\phi \hat{\phi}(x),$$
$$\hat{V}^\mu(x) \xrightarrow{T} \hat{V}'^\mu(x') = \hat{T}\hat{V}^\mu(-t, \mathbf{x})\hat{T}^{-1} = -\zeta_T^V (-1)^{(\mu)}\hat{V}^\mu(x),$$

$$\hat{\psi}(x) \xrightarrow{\mathcal{T}} \hat{\psi}'(x') = \hat{\mathcal{T}}\hat{\psi}(-t, \mathbf{x})\hat{\mathcal{T}}^{-1} = \zeta_T^\psi T\hat{\psi}(x), \tag{8.1.84}$$

where the $\zeta_T^i \in \mathbb{C}$ are arbitrary phases, $|\zeta_T^\phi| = |\zeta_T^V| = |\zeta_T^\psi| = 1$ and where from Eqs. (4.5.62) and (4.5.63) the spinor matrix T in the Dirac representation is

$$T = i\gamma^1\gamma^3 = T^\dagger = T^{-1} = -T^* \quad \text{such that} \quad T\gamma^\mu T^{-1} = \gamma^{\mu T} = \gamma_\mu^*. \tag{8.1.85}$$

We again set the fermion phase to unity since only fermion bilinears are relevant, $\zeta_T^\psi = 1$. The above transformations preserve the equations of motion for the operators under time reversal. They can be summarized as

$$\hat{\mathcal{T}}\hat{\phi}(x)\hat{\mathcal{T}}^{-1} = \zeta_T^\phi\hat{\phi}(-t, \mathbf{x}), \quad \hat{\mathcal{T}}\hat{V}^\mu(x)\hat{\mathcal{T}}^{-1} = \zeta_T^V(-1)^{(\mu)}\hat{V}^\mu(-t, \mathbf{x}),$$
$$\hat{\mathcal{T}}\hat{\psi}(x)\hat{\mathcal{T}}^{-1} = T\hat{\psi}(-t, \mathbf{x}). \tag{8.1.86}$$

Note that a unitary operator is linear and so acts only on Hilbert space operators and states, but not on c-number quantities. For a fermion field we have

$$\hat{\mathcal{T}}\hat{\psi}^\dagger(x)\hat{\mathcal{T}}^{-1} = [\hat{\mathcal{T}}\hat{\psi}(x)\hat{\mathcal{T}}^{-1}]^\dagger = [T\hat{\psi}(-t, \mathbf{x})]^\dagger = \hat{\psi}^\dagger(-t, \mathbf{x})T^{-1}, \tag{8.1.87}$$
$$\hat{\mathcal{T}}\hat{\bar{\psi}}(x)\hat{\mathcal{T}}^{-1} = \hat{\mathcal{T}}\hat{\psi}^\dagger(x)\hat{\mathcal{T}}^{-1}\gamma^{0*} = \hat{\psi}^\dagger(-t, \mathbf{x})T^{-1}\gamma^{0*} = \hat{\bar{\psi}}(-t, \mathbf{x})T^{-1},$$

since $T^{-1}\gamma^{0*} = \gamma^0 T^{-1}$. For any Hermitian boson field operator $\hat{O}^i(x) = \hat{O}^{i\dagger}(x)$,

$$\hat{\mathcal{T}}\hat{O}^{i\dagger}(x)\hat{\mathcal{T}}^{-1} = [\hat{\mathcal{T}}\hat{O}^i(x)\hat{\mathcal{T}}^{-1}]^\dagger = \zeta_T^{i*}\hat{O}^{i\dagger}(t, -\mathbf{x}) = \zeta_T^{i*}\hat{O}^i(-t, \mathbf{x}) \tag{8.1.88}$$

and so $\zeta_T^i = \zeta_T^{i*} = \pm 1$. Since $\hat{T}$ is antiunitary it cannot correspond to a Hermitian observable. So, for a neutral (Hermitian) boson field operator $\hat{O}^i$, the real phase ζ_T^i is *not* an observable like parity or C-parity. It then can be arbitrarily assigned either sign, $\zeta_T^i = \pm 1$. For charged (non-Hermitian) boson field operators $\hat{O}^i$ the phase ζ_T^i is completely arbitrary. These observations will be important to remember when we come to discuss the CPT theorem in Sec. 8.1.4.

For a charged scalar field the antiunitary operator $\hat{\mathcal{T}}$ has the properties that

$$\hat{\mathcal{T}}\hat{\phi}(x)\hat{\mathcal{T}}^{-1} = \zeta_T^\phi\hat{\phi}(-t, \mathbf{x}) \quad \text{and} \quad \hat{\mathcal{T}}\hat{\phi}^\dagger(x)\hat{\mathcal{T}}^{-1} = \zeta_T^{\phi*}\hat{\phi}^\dagger(-t, \mathbf{x}), \tag{8.1.89}$$

where ζ_T^ϕ is an arbitrary complex phase. We can use the creation operators $\hat{f}_\mathbf{p}$ and $\hat{g}_\mathbf{p}$ in the plane wave expansion of Eq. (6.2.177) to construct a basis for Fock space. It is then convenient to define the complex conjugation operator, $\hat{K} = \hat{K}^\dagger = \hat{K}^{-1}$, to be the one that leaves this plane wave Fock space basis invariant. This is equivalent to the statement that the complex conjugation operator leaves the vacuum invariant and commutes with all of these creation operators,

$$\hat{K}|0\rangle = |0\rangle, \quad [\hat{K}, \hat{f}_\mathbf{p}^\dagger] = [\hat{K}, \hat{g}_\mathbf{p}^\dagger] = 0 \quad \Rightarrow \quad [\hat{K}, \hat{f}_\mathbf{p}] = [\hat{K}, \hat{g}_\mathbf{p}] = 0. \tag{8.1.90}$$

The Hermitian conjugate of the commutators with the creation operators gives the commutators with the annihilation operators on the right of the above equation. Then we have, for example, $\hat{K}\hat{f}_\mathbf{p}e^{-ip\cdot x}\hat{K}^{-1} = \hat{f}_\mathbf{p}e^{ip\cdot x}\hat{K}\hat{K}^{-1} = \hat{f}_\mathbf{p}e^{ip\cdot x}$. Using this in Eqs. (6.2.177) and (8.1.89) gives

$$\hat{\mathcal{T}}\hat{\phi}(x)\hat{\mathcal{T}}^{-1} = \hat{U}\hat{K}\hat{\phi}(x)\hat{K}^{-1}\hat{U}^{-1} = \hat{U}\hat{\phi}(-x)\hat{U}^{-1} = \zeta_T^\phi\hat{\phi}(-t, \mathbf{x}). \tag{8.1.91}$$

So the action of $\hat{U}$ on a scalar field is the same as a parity transformation $\hat{U} = \hat{\mathcal{P}}$, with the phase ζ_T^ϕ replacing ζ_P^ϕ. For $\hat{U}$ for a neutral scalar field we use Eq. (8.1.13) with $\zeta_T^\phi = \pm 1$ and for a charged scalar field we use Eq. (8.1.19) with arbitrary phase $\zeta_T^\phi = e^{i\phi_T}$.

For the electromagnetic field under time reversal we have $\mathbf{E} \to \mathbf{E}$ and $\mathbf{B} \to -\mathbf{B}$ because charge distributions are unchanged but currents are reversed, which means that $A^\mu(x) \to (-1)^{(\mu)}A(-t, \mathbf{x})$.

This is equivalent to the requirement that $j_\mu A^\mu$ is invariant under time reversal. Working in Coulomb gauge in the quantization inertial frame we then require at the operator level that

$$\hat{\mathcal{T}}\hat{\mathbf{A}}(x)\hat{\mathcal{T}}^{-1} = \hat{\mathcal{U}}\hat{K}\hat{\mathbf{A}}(x)\hat{K}^{-1}\hat{\mathcal{U}}^{-1} = \hat{\mathcal{U}}\hat{\mathbf{A}}(-x)\hat{\mathcal{U}}^{-1} = -\hat{\mathbf{A}}(-t,\mathbf{x}). \tag{8.1.92}$$

The required action of $\hat{\mathcal{U}}$ is again to perform a parity transformation and so we can use $\hat{\mathcal{P}}$ of Eq. (8.1.29) as the $\hat{\mathcal{U}}$ here.

For the fermion case in Eqs. (8.1.86) and (8.1.87) things are a little more complicated. First note that two time reversals for a fermion field operator lead to a change of sign for the operator,

$$\hat{\mathcal{T}}^2\hat{\psi}(x)\hat{\mathcal{T}}^{-2} = \hat{\mathcal{T}}T\hat{\psi}(-t,\mathbf{x})\hat{\mathcal{T}}^{-1} = T^*T\hat{\psi}(x) = -\hat{\psi}(x), \tag{8.1.93}$$

where recall from Eq. (4.5.62) that $T = T^\dagger = T^{-1} = -T^*$. So $\hat{\psi}(x)$ returns to itself after four time reversals, which is analogous to a fermion having to be rotated through 2π twice to return to itself or more generally to go around a closed path twice in the space of Lorentz transformations as discussed in Sec. 4.1.4. Consider Eqs. (6.3.194) and (8.1.86) and note that, for example,

$$\hat{\mathcal{T}}\hat{b}^s_{\mathbf{p}}u^s(p)e^{-ip\cdot x}\hat{\mathcal{T}}^{-1} = \hat{\mathcal{U}}\hat{K}\hat{b}^s_{\mathbf{p}}u^s(p)e^{-ip\cdot x}\hat{K}^{-1}\hat{\mathcal{U}}^{-1} = \hat{\mathcal{U}}\hat{b}^s_{\mathbf{p}}\hat{\mathcal{U}}^{-1}u^s(p)^*e^{+ip\cdot x},$$

$$\hat{\mathcal{T}}\hat{d}^{s\dagger}_{\mathbf{p}}u^s(p)e^{ip\cdot x}\hat{\mathcal{T}}^{-1} = \hat{\mathcal{U}}\hat{d}^{s\dagger}_{\mathbf{p}}\hat{\mathcal{U}}^{-1}v^s(p)^*e^{-ip\cdot x}. \tag{8.1.94}$$

Then we require that

$$\int \left[d^3p/(2\pi)^3\sqrt{2E_{\mathbf{p}}}\right]\sum_s\left[\hat{\mathcal{U}}\hat{b}^s_{\mathbf{p}}\hat{\mathcal{U}}^{-1}u^s(p)^*e^{ip\cdot x} + \hat{\mathcal{U}}\hat{d}^{s\dagger}_{\mathbf{p}}\hat{\mathcal{U}}^{-1}v^s(p)^*e^{-ip\cdot x}\right] \tag{8.1.95}$$

$$= \int \left[d^3p/(2\pi)^3\sqrt{2E_{\mathbf{p}}}\right]\sum_s\left[\hat{b}^s_{\mathbf{p}}Tu^s(p)e^{i(p^0t+\mathbf{p}\cdot\mathbf{x})} + \hat{d}^{s\dagger}_{\mathbf{p}}Tv^s(p)e^{-i(p^0t+\mathbf{p}\cdot\mathbf{x})}\right]$$

$$= \int \left[d^3p/(2\pi)^3\sqrt{2E_{\mathbf{p}}}\right]\sum_s\left[e^{i\eta_+(-s)}\hat{b}^{-s}_{-\mathbf{p}}u^s(p)^*e^{ip\cdot x} + e^{i\eta_-(-s)}\hat{d}^{-s\dagger}_{-\mathbf{p}}Tv^s(p)^*e^{-ip\cdot x}\right],$$

where in the last line we changed variables $(s,\mathbf{p})\to(-s,-\mathbf{p})$ and used Eq. (4.5.68),

$$Tu^s(p) = e^{i\eta_+(s)}u^{-s}(-p)^* \quad\text{and}\quad Tv^s(p) = e^{i\eta_-(s)}v^{-s}(-p)^*, \tag{8.1.96}$$

$$\eta_+(s) \equiv \pi(m_s+1) \quad\text{and}\quad \eta_-(s) \equiv \pi m_s.$$

Applying T twice to the above equations gives

$$T^2u^s(p) = e^{i\eta_+(s)}Tu^{-s}(-p)^* = e^{i\eta_+(s)}[-Tu^{-s}(-p)]^* = -e^{i[\eta_+(s)-\eta_+(-s)]}u^s(p) = u^s(p),$$

$$T^2v^s(p) = -e^{i[\eta_-(s)-\eta_-(-s)]}v^s(p) = v^s(p), \tag{8.1.97}$$

which is as expected since $T^2 = I$ and follows since

$$\eta_\pm(s) - \eta_\pm(-s) = \pi. \tag{8.1.98}$$

Comparing the first and third lines of Eq. (8.1.95) leads to the requirements

$$\hat{\mathcal{U}}\hat{b}^s_{\mathbf{p}}\hat{\mathcal{U}}^{-1} = e^{i\eta_+(-s)}\hat{b}^{-s}_{-\mathbf{p}} = -e^{i\eta_+(s)}\hat{b}^{-s}_{-\mathbf{p}},$$

$$\hat{\mathcal{U}}\hat{d}^{s\dagger}_{\mathbf{p}}\hat{\mathcal{U}}^{-1} = e^{i\eta_-(-s)}\hat{d}^{-s\dagger}_{-\mathbf{p}} = -e^{i\eta_-(s)}\hat{d}^{-s\dagger}_{-\mathbf{p}}, \tag{8.1.99}$$

which can be satisfied by (Bjorken and Drell, 1965)

$$\hat{\mathcal{U}} \equiv \hat{\mathcal{U}}_2\hat{\mathcal{U}}_1, \quad\text{where} \tag{8.1.100}$$

$$\hat{\mathcal{U}}_1\hat{b}^s_{\mathbf{p}}\hat{\mathcal{U}}_1^{-1} = e^{i\eta_+(s)}\hat{b}^s_{\mathbf{p}} \quad\text{and}\quad \hat{\mathcal{U}}_1\hat{d}^{s\dagger}_{\mathbf{p}}\hat{\mathcal{U}}_1^{-1} = e^{i\eta_-(s)}\hat{d}^{s\dagger}_{\mathbf{p}},$$

$$\hat{\mathcal{U}}_2\hat{b}^s_{\mathbf{p}}\hat{\mathcal{U}}_2^{-1} = -\hat{b}^{-s}_{-\mathbf{p}} \quad\text{and}\quad \hat{\mathcal{U}}_2\hat{d}^{s\dagger}_{\mathbf{p}}\hat{\mathcal{U}}_2^{-1} = -\hat{d}^{-s\dagger}_{-\mathbf{p}}. \tag{8.1.101}$$

Explicit forms for unitary operators $\hat{\mathcal{U}}_1$ and $\hat{\mathcal{U}}_2$ are

$$\hat{\mathcal{U}}_1 = \exp\{\pm i\pi \int [d^3p/(2\pi)^3] \sum_s [\eta_+(s)\hat{b}_\mathbf{p}^{\dagger s}\hat{b}_\mathbf{p}^s - \eta_-(s)\hat{d}_\mathbf{p}^{\dagger s}\hat{d}_\mathbf{p}^s]\}, \tag{8.1.102}$$

$$\hat{\mathcal{U}}_2 = \exp\{\pm i(\pi/2) \int [d^3p/(2\pi)^3] \sum_s [\hat{b}_\mathbf{p}^{\dagger s}\hat{b}_\mathbf{p}^s + \hat{b}_\mathbf{p}^{\dagger s}\hat{b}_{-\mathbf{p}}^{-s} + \hat{d}_\mathbf{p}^{\dagger s}\hat{d}_\mathbf{p}^s + \hat{d}_\mathbf{p}^{\dagger s}\hat{d}_{-\mathbf{p}}^{-s}]\}.$$

> **Proof:** We construct $\hat{\mathcal{U}}_1$ based on the form of $\hat{\mathcal{C}}_1$ for charge conjugation in Eq. (8.1.77). There is no obstruction to us using different phases for the $\hat{b}_\mathbf{p}^{\dagger s}\hat{b}_\mathbf{p}^s$ and $\hat{d}_\mathbf{p}^{\dagger s}\hat{d}_\mathbf{p}^s$ terms and so we arrive at $\hat{\mathcal{U}}_1$ above. The relative minus sign arises since it is $\hat{d}_\mathbf{p}^{s\dagger}$ that requires the phase $e^{i\eta_-(s)}$. Similarly, we can modify the parity operator $\hat{\mathcal{P}}$ of Eq. (8.1.35) by the replacements $\hat{b}_{-\mathbf{p}}^s, \hat{d}_{-\mathbf{p}}^s \to \hat{b}_{-\mathbf{p}}^{-s}, \hat{d}_{-\mathbf{p}}^{-s}$ and changing the sign of the term producing $\hat{b}_{-\mathbf{p}}^{-s}$ so that we obtain the minus sign in $\hat{\mathcal{U}}_2\hat{b}_\mathbf{p}^s\hat{\mathcal{U}}_2^{-1} = -\hat{b}_{-\mathbf{p}}^{-s}$. This gives the expression for $\hat{\mathcal{U}}_2$ above.

For the free charged scalar field it follows from Eqs. (6.2.174) and (8.1.89) that

$$\hat{\mathcal{T}}\hat{H}\hat{\mathcal{T}}^{-1} = \hat{H} \quad \text{and} \quad \hat{\mathcal{T}}\hat{\mathbf{P}}\hat{\mathcal{T}}^{-1} = -\hat{\mathbf{P}}, \tag{8.1.103}$$

since $\hat{\pi} = \partial_0 \hat{\phi}^\dagger$ and $\hat{\pi}^\dagger = \partial_0 \hat{\phi}$ and $\partial_0 \xrightarrow{\mathcal{T}} -\partial_0$ then we find $\hat{\mathcal{T}}\hat{\pi}\hat{\mathcal{T}}^{-1} = -\hat{\pi}$ and $\hat{\mathcal{T}}\hat{\pi}^\dagger\hat{\mathcal{T}}^{-1} = -\hat{\pi}^\dagger$. This can also be seen directly from their expression in terms of normal-ordered annihilation and creation operators in Eq. (6.2.185). Similarly, from Eqs. (3.2.62) and (3.2.75) we note that

$$\hat{\mathcal{T}}\hat{\mathbf{L}}\hat{\mathcal{T}}^{-1} = -\hat{\mathbf{L}}. \tag{8.1.104}$$

The conserved current operator also behaves as expected under time reversal,

$$\begin{aligned} \hat{\mathcal{T}}j^\mu(x)\hat{\mathcal{T}}^{-1} &= \hat{\mathcal{T}}\, i{:}[\hat{\phi}^\dagger(\partial^\mu\hat{\phi}) - (\partial^\mu\hat{\phi}^\dagger)\hat{\phi}]{:}\,\hat{\mathcal{T}}^{-1} \\ &= -i\,{:}\hat{\mathcal{T}}[\hat{\phi}^\dagger(\partial^\mu\hat{\phi}) - (\partial^\mu\hat{\phi}^\dagger)\hat{\phi}]\hat{\mathcal{T}}^{-1}{:} = (-1)^{(\mu)}\hat{j}^\mu(-t,\mathbf{x}). \end{aligned} \tag{8.1.105}$$

The normal-ordered four-vector current density for a fermion in Eq. (6.3.207) behaves identically, since

$$\begin{aligned} \hat{\mathcal{T}}j^\mu(x)\hat{\mathcal{T}}^{-1} &= \hat{\mathcal{T}}\,{:}\hat{\bar{\psi}}(x)\gamma^\mu\hat{\psi}(x){:}\,\hat{\mathcal{T}}^{-1} = {:}\hat{\mathcal{T}}\,\hat{\bar{\psi}}\hat{\mathcal{T}}^{-1}\hat{\mathcal{T}}\gamma^\mu\hat{\mathcal{T}}^{-1}\hat{\mathcal{T}}\hat{\psi}\,\hat{\mathcal{T}}^{-1}{:} \\ &= {:}\hat{\bar{\psi}}T^{-1}\gamma^{\mu*}T\hat{\psi}{:} = {:}\hat{\bar{\psi}}\gamma_\mu\hat{\psi}{:} = \hat{j}_\mu(-t,\mathbf{x}) = (-1)^{(\mu)}\hat{j}^\mu(-t,\mathbf{x}), \end{aligned} \tag{8.1.106}$$

where we used Eq. (8.1.85) to find $T\gamma^{\mu*}T^{-1} = \gamma_\mu$. Using Eqs. (6.3.191) it is straightforward to verify for fermions that for the normal-ordered four-momentum operators $\hat{P}^\mu = (\hat{H}, \hat{\mathbf{P}})$ in Eq. (6.3.199)

$$\begin{aligned} \hat{\mathcal{T}}^{-1}\hat{P}^\mu\hat{\mathcal{T}}^{-1} &= \int d^3x\,{:}\hat{\mathcal{T}}i\hat{\psi}^\dagger(x)\partial^\mu\hat{\psi}(x)\hat{\mathcal{T}}^{-1}{:} = -i\int d^3x\,{:}\hat{\psi}^\dagger T^{-1}(-\partial_\mu)T\hat{\psi}{:} \\ &= i\int d^3x\,{:}\hat{\psi}^\dagger\partial_\mu\hat{\psi}{:} = \hat{P}_\mu = (-1)^{(\mu)}\hat{P}^\mu. \end{aligned} \tag{8.1.107}$$

This result is also readily obtained from the expressions for $\hat{P}^\mu$ in terms of annihilation and creation operators in Eq. (6.3.204). The fermion angular momentum operators are given in Eq. (8.1.44) and reverse sign under time reversal as expected,

$$\begin{aligned} \hat{\mathcal{T}}\hat{\mathbf{J}}\hat{\mathcal{T}}^{-1} &= \int d^3x\,{:}\hat{\mathcal{T}}[\hat{\psi}^\dagger(x)\,(\mathbf{x}\times(-i\boldsymbol{\nabla}) + \mathbf{S})\,\hat{\psi}(x)]\hat{\mathcal{T}}^{-1}{:} \\ &= \int d^3x\,{:}\hat{\psi}^\dagger(x)\,(\mathbf{x}\times(i\boldsymbol{\nabla}) + T^{-1}\mathbf{S}^*T)\,\hat{\psi}(x){:} = -\hat{\mathbf{J}}, \end{aligned} \tag{8.1.108}$$

since $T^{-1}\sigma^{jk*}T = (-i/2)T^{-1}[\gamma^{j*}, \gamma^{k*}]T = -(i/2)[\gamma^j, \gamma^k] = -\sigma^{jk}$ and this then leads to $T^{-1}S^{i*}T = \frac{1}{4}\epsilon^{ijk}T^{-1}\sigma^{jk*}T = -\frac{1}{4}\epsilon^{ijk}\sigma^{jk} = -S^i$.

Using Eq. (8.1.84) we have for a charged scalar field with $x' = (-t, \mathbf{x})$,

$$\hat{\phi}(x) \xrightarrow{\ \mathcal{T}\ } \hat{\phi}'(x') = \hat{\mathcal{T}}\hat{\phi}(-t, \mathbf{x})\hat{\mathcal{T}}^{-1} = \zeta_T^\phi \hat{\phi}(x), \tag{8.1.109}$$

$$\partial_0 \hat{\phi}(x) \xrightarrow{\ \mathcal{T}\ } \partial_0' \hat{\phi}'(x') = -\partial_0 \hat{\mathcal{T}}\hat{\phi}(-t, \mathbf{x})\hat{\mathcal{T}}^{-1} = -\zeta_T^\phi \partial_0 \hat{\phi}(x),$$

$$[\hat{\phi}(t, \mathbf{x}), \partial_0 \hat{\phi}^\dagger(t, \mathbf{y})] \xrightarrow{\ \mathcal{T}\ } [\hat{\phi}'(t', \mathbf{x}), \partial_0' \hat{\phi}'^\dagger(t', \mathbf{y})] = \hat{\mathcal{T}}[\hat{\phi}(-t, \mathbf{x}), -\partial_0 \hat{\phi}^\dagger(-t, \mathbf{y})]\hat{\mathcal{T}}^{-1}$$

$$= -[\hat{\phi}(t, \mathbf{x}), \partial_0 \hat{\phi}^\dagger(t, \mathbf{y})] = \hat{\mathcal{T}}i\delta^3(\mathbf{x} - \mathbf{y})\hat{\mathcal{T}}^{-1} = -i\delta^3(\mathbf{x} - \mathbf{y}).$$

We see that the nonvanishing equal-time canonical commutation relations for a charged scalar field in Eq. (6.2.181) change sign due to the presence of the time derivative. We have arrived at $[\hat{\phi}'(t', \mathbf{x}), \partial_0' \hat{\phi}'^\dagger(t', \mathbf{y})] = -i\delta^3(\mathbf{x} - \mathbf{y})$ for the time-reversed system, which is to be compared with $[\hat{\phi}(t, \mathbf{x}), \partial_0 \hat{\phi}^\dagger(t, \mathbf{y})] = i\delta^3(\mathbf{x} - \mathbf{y})$ that has also reemerged. Note that if time reversal was not antiunitary, then we would have arrived at an inconsistency with the wrong sign emerging for the standard canonical commutation relations. This is discussed in Messiah (1975), volume II, and Sakurai (1994), where time reversal must complex conjugate all coefficients and so be antiunitary in order to avoid inconsistencies. For example, consider the commutation relations for angular momentum. Under time reversal $\hat{\mathbf{J}} \xrightarrow{\ \mathcal{T}\ } \hat{\mathbf{J}}' = -\hat{\mathbf{J}}$ and $[\hat{J}^i, \hat{J}^j] = i\epsilon^{ijk}\hat{J}^k$ becomes

$$[\hat{J}'^i, \hat{J}'^j] = \hat{\mathcal{T}}[\hat{J}^i, \hat{J}^j]\hat{\mathcal{T}}^{-1} = \hat{\mathcal{T}}i\epsilon^{ijk}\hat{J}^k \hat{\mathcal{T}}^{-1} = -i\epsilon^{ijk}\hat{J}'^k \tag{8.1.110}$$

and ensures consistency since $[(-\hat{J}^i), (-\hat{J}^j)] = -i\epsilon^{ijk}(-\hat{J}^k)$.

In the fermion case we have from Eqs. (6.3.190), (8.1.86) and (8.1.87)

$$\{\hat{\psi}_\alpha(t, \vec{x}), \hat{\psi}_\beta^\dagger(t, \vec{y})\} \xrightarrow{\ \mathcal{T}\ } \{\hat{\psi}_\alpha'(t', \vec{x}), \hat{\psi}_\beta'^\dagger(t', \vec{y})\} = \hat{\mathcal{T}}\{\hat{\psi}_\alpha(-t, \vec{x}), \hat{\psi}_\beta^\dagger(-t, \vec{y})\}\hat{\mathcal{T}}^{-1}$$

$$= T_{\alpha\gamma}T_{\delta\beta}^{-1}\{\hat{\psi}_\gamma(t, \vec{x}), \hat{\psi}_\delta^\dagger(t, \vec{y})\} = T_{\alpha\gamma}T_{\delta\beta}^{-1}\delta_{\gamma\delta}\delta^3(\mathbf{x} - \mathbf{y}) = \delta_{\alpha\beta}\delta^3(\mathbf{x} - \mathbf{y}), \tag{8.1.111}$$

which shows that the fermion equal-time anticommutation relations are unchanged as expected since they contain no complex coefficients. Let $|\chi(\mathbf{p}, s)\rangle$ be a one-fermion state in Fock space with four-momentum p^μ and spin s, then

$$|\chi(\mathbf{p}, s)\rangle = \hat{b}_{\mathbf{p}}^{s\dagger}|0\rangle \quad \text{and} \tag{8.1.112}$$

$$\hat{\mathcal{T}}|\chi(\mathbf{p}, s)\rangle = \hat{\mathcal{T}}\hat{b}_{\mathbf{p}}^{s\dagger}|0\rangle = \hat{\mathcal{T}}\hat{b}_{\mathbf{p}}^{s\dagger}\hat{\mathcal{T}}^{-1}|0\rangle = -e^{-i\eta_+(s)}\hat{b}_{-\mathbf{p}}^{-s\dagger}|0\rangle = -e^{-i\eta_+(s)}|\chi(-\mathbf{p}, -s)\rangle.$$

Since the quantum field theory equivalent of a fermion at the point $\mathbf{x}$ at time t is $\hat{\psi}_\alpha(t, \mathbf{x})|0\rangle$ then the corresponding equivalent of the wavefunction for this state is (Bjorken and Drell, 1965)

$$\psi_{\alpha \mathbf{p}s}(t, \mathbf{x}) = \langle 0|\hat{\psi}_\alpha(t, \mathbf{x})|\chi(\mathbf{p}, s)\rangle = \langle 0|\hat{\psi}_\alpha(t, \mathbf{x})\chi(\mathbf{p}, s)\rangle = \langle \hat{K}\hat{\psi}_\alpha(t, \mathbf{x})\chi(\mathbf{p}, s)|\hat{K}0\rangle$$

$$= \langle \hat{K}\hat{\psi}_\alpha(t, \mathbf{x})\chi(\mathbf{p}, s)|\hat{U}^\dagger\hat{U}\hat{K}0\rangle = \langle \hat{U}\hat{K}\hat{\psi}_\alpha(t, \mathbf{x})\chi(\mathbf{p}, s)|\hat{U}\hat{K}0\rangle = \langle \hat{\mathcal{T}}\hat{\psi}_\alpha(t, \mathbf{x})\chi(\mathbf{p}, s)|\hat{\mathcal{T}}0\rangle$$

$$= \langle \hat{\mathcal{T}}\hat{\psi}_\alpha(t, \mathbf{x})\chi(\mathbf{p}, s)|0\rangle = \langle 0|\hat{\mathcal{T}}\hat{\psi}_\alpha(t, \mathbf{x})\chi(\mathbf{p}, s)\rangle^* = \langle 0|\hat{\mathcal{T}}\hat{\psi}_\alpha(t, \mathbf{x})\hat{\mathcal{T}}^{-1}\hat{\mathcal{T}}|\chi(\mathbf{p}, s)\rangle^*$$

$$= -e^{-i\eta_+(s)}T_{\alpha\beta}\langle 0|\hat{\psi}_\beta(-t, \mathbf{x})|\chi(-\mathbf{p}, -s)\rangle^* = -e^{-i\eta_+(s)}T_{\alpha\beta}\psi_{\beta, -\mathbf{p}, -s}^*(-t, \mathbf{x}). \tag{8.1.113}$$

where we have used Eq. (4.1.56). We can identify $\psi_T(-t, \mathbf{x}) \equiv e^{i\eta_+(s)}\psi_{\alpha, -\mathbf{p}, -s}(-t, \mathbf{x})$ as the time reversed wavefunction and so the above can be summarized as

$$\psi(t, \mathbf{x}) = -T\psi_T^*(-t, \mathbf{x}) \quad \text{or equivalently} \quad \psi_T(-t, \mathbf{x}) = T\psi^*(t, \mathbf{x}), \tag{8.1.114}$$

where we recall that $T = T^{-1} = -T^*$. We see that this has reproduced the one-particle relativistic quantum mechanics result of Eq. (4.5.57). We find a similar result in the antifermion case and for arbitrary wavepackets made from superpositions of these plane wave wavefunctions.

8.1.4 The *CPT* Theorem

All of the free theories that we have studied so far are invariant under each of the P, C and T transformations. Interacting quantum field theories will have interaction terms that may or may not respect these symmetries. Violations of the P, C and T symmetries have all been observed in nature. In our everyday experience of the macroscopic world the effects of these symmetry-violating interactions are relatively small and we can typically ignore them, but they are a vitally important part of our understanding of the subatomic world. For example, without the parity-violating weak interactions two protons could not combine to form a deuterium nucleus (neutron plus proton) and there would be no fusion process in stars. To describe nature the Standard Model and any future enhancements of it must then violate each of P, C and T. However, the Standard Model is invariant under a combined CPT transformation and no violation of the combined symmetry CPT has been observed in nature to date. There is a very powerful proof that any "reasonable" theory that we can write down will be CPT invariant. It is this CPT theorem that we will now discuss. Note that the order of the letters is unimportant and, while the alphabetic ordering of CPT is used here, PCT and other permutations of the three letters are also in common use.

In the case of fermions the arbitrary phases $\zeta_P, \zeta_C, \zeta_T$ that we could have included in our discussion of P, C and T transformations were set to $+1$ since they disappear in all fermion bilinears and so were irrelevant. In the case of bosons we saw that the most restrictive case for the phases was that of a neutral (Hermitian) boson field. In that case both the intrinsic parity $\pi_P = \pm 1$ and C-parity $\eta_C = \pm 1$ are observable and so are fixed. For a neutral (Hermitian) boson field the time-reversal phase is real, $\zeta_T = \pm 1$, but is arbitrary since $\hat{\mathcal{T}}$ is antiunitary and so not an observable operator. Then for *any* field operator we can choose the phases such that

$$\zeta_C^i, \zeta_P^i, \zeta_T^i \in \mathbb{R} \quad \text{and} \quad \zeta_C^i \zeta_P^i \zeta_T^i = 1. \tag{8.1.115}$$

We *define* $\hat{\Theta}$ to be the combined C, P and T transformations such that

$$\Theta \equiv \hat{C}\hat{P}\hat{\mathcal{T}} \quad \text{with} \quad \zeta_C, \zeta_P, \zeta_T = \pm 1 \quad \text{and} \quad \zeta_C \zeta_P \zeta_T = 1, \tag{8.1.116}$$

where since $\hat{C}$ and $\hat{P}$ are unitary and $\hat{\mathcal{T}}$ is antiunitary then $\hat{\Theta}$ is antiunitary. It is this $\hat{\Theta}$ that generates the CPT transformation discussed in the literature such as in Weinberg (1995) and Streater and Wightman (2000).

Under this CPT transformation $\hat{\Theta}$ we have from our previous discussions

$$\hat{\Theta}\hat{\phi}(x)\hat{\Theta}^{-1} = \zeta_C^\phi \zeta_P^\phi \zeta_T^\phi \hat{\phi}^\dagger(-x) = \hat{\phi}^\dagger(-x), \tag{8.1.117}$$

$$\hat{\Theta}\hat{V}^\mu(x)\hat{\Theta}^{-1} = [-\zeta_C^V][\zeta_P^V(-1)^{(\mu)}][\zeta_T^V(-1)^{(\mu)}]\hat{V}^{\mu\dagger}(-x) = -V^{\mu\dagger}(-x),$$

$$\hat{\Theta}\hat{\psi}_\alpha(x)\hat{\Theta}^{-1} = \zeta_C^\psi \zeta_P^\psi \zeta_T^\psi [T\gamma^0 C\hat{\bar{\psi}}^T(-x)]_\alpha = [T\gamma^0 C\gamma^{0T}]_{\alpha\beta}\hat{\psi}_\beta^\dagger(-x) = -i\gamma_{\alpha\beta}^5 \hat{\psi}_\beta^\dagger(-x),$$

where recall that we are working in the Dirac representation here and

$$T\gamma^0 C\gamma^{0T} = T\gamma^0 C\gamma^0 = (i\gamma^1\gamma^3)\gamma^0(i\gamma^2\gamma^0)\gamma^0 = -i\gamma^5. \tag{8.1.118}$$

The order of CPT is unimportant for scalar and vector fields. The fermion result agrees with Greiner and Reinhardt (1996), who considered CPT, but differs by a sign from the result in Bjorken and Drell (1965), who considered PCT. For fermion bilinears this difference disappears and the order of the C, P and T transformations becomes irrelevant. Using Eq. (8.1.117) and $\{\gamma^5, \gamma^\mu\} = 0$ gives

$$\hat{\Theta}\hat{\psi}(x)\hat{\Theta}^{-1} = i\gamma^0\gamma^5\hat{\bar{\psi}}^T(-x) \quad \text{and} \quad \hat{\Theta}\hat{\bar{\psi}}(x)\hat{\Theta}^{-1} = \hat{\psi}^T(-x)i\gamma^5\gamma^0, \tag{8.1.119}$$

since in Dirac representation $\gamma^0 = \gamma^{0*} = \gamma^{0\dagger} = (\gamma^0)^{-1}$, $\gamma^5 = \gamma^{5*} = \gamma^{5\dagger} = (\gamma^5)^{-1}$, and

$$\hat{\Theta}\hat{\bar{\psi}}_\alpha(x)\hat{\Theta}^{-1} = \hat{\Theta}\hat{\psi}_\beta^\dagger(x)\hat{\Theta}^{-1}\gamma^0_{\beta\alpha} = [\hat{\Theta}\hat{\psi}_\beta(x)\hat{\Theta}^{-1}]^\dagger\gamma^0_{\beta\alpha} = [-i\gamma^5\hat{\psi}^{\dagger T}(-x)]^\dagger_\beta\gamma^0_{\beta\alpha}$$
$$= [\hat{\psi}^T(-x)i\gamma^5]_\beta\gamma^0_{\beta\alpha} = [\hat{\psi}^T(-x)i\gamma^5\gamma^0]_\alpha. \tag{8.1.120}$$

Due to the $\hat{\mathcal{P}}$ and $\hat{\mathcal{T}}$ components of $\hat{\Theta}$ we have

$$\hat{\Theta}\partial_\mu\hat{\Theta}^{-1} = -\partial_\mu. \tag{8.1.121}$$

Let $\hat{\Phi}_{\mu_1\cdots\mu_n}(x)$ be any normal-ordered local product of complex numbers, scalar and vector field operators and derivatives. We have arrived at the important result (Weinberg, 1995; Greiner and Reinhardt, 1996)

$$\hat{\Theta}\hat{\Phi}_{\mu_1\cdots\mu_n}(x)\hat{\Theta}^{-1} = (-1)^n\hat{\Phi}^\dagger_{\mu_1\cdots\mu_n}(-x), \tag{8.1.122}$$

where recall that since $\hat{\Theta}$ is antiunitary it complex conjugates any coefficients in $\hat{\Phi}_{\mu_1\cdots\mu_n}(x)$ as does taking the Hermitian conjugate on the right-hand side. Note that the partial derivative ∂_μ is understood to be unaffected here when the Hermitian conjugate is taken.

Proof: From Eqs. (8.1.117) and (8.1.121) we understand that under the CPT transformations there is one factor of -1 for each uncontracted spacetime index. The only issue requiring further explanation is the need for normal-ordering of $\hat{\Phi}_{\mu_1\cdots\mu_n}(x)$. If it was not normal-ordered, then after CPT the operators would be Hermitian conjugated in their original order, whereas $\hat{\Phi}^\dagger_{\mu_1\cdots\mu_n}(x)$ has the Hermitian conjugate operators in reverse order. Distinguishable boson fields $\hat{\phi}_a$ and $\hat{\phi}_b$ commute; however, ϕ_a and $\phi_a^\dagger$ do not and so we cannot reorder as we would like. Normal-ordering fixes this problem as it did in Eq. (8.1.63) when we considered the charged boson normal-ordered current under charge conjugation. For example,

$$\hat{\Theta}\hat{\phi}_a^\dagger(x)\partial_\mu\hat{\phi}_a(x)\hat{\Theta}^{-1} = -\phi_a(x)\partial_\mu\hat{\phi}_a^\dagger(x) \neq -\left(\hat{\phi}_a^\dagger(x)\partial_\mu\hat{\phi}_a(x)\right)^\dagger = -\partial_\mu\hat{\phi}_a^\dagger(x)\phi_a(x),$$

whereas if we impose normal-ordering then

$$\hat{\Theta}{:}\hat{\phi}_a^\dagger(x)\partial_\mu\hat{\phi}_a(x){:}\hat{\Theta}^{-1} = {:}\hat{\Theta}\hat{\phi}_a^\dagger(x)\partial_\mu\hat{\phi}_a(x)\hat{\Theta}^{-1}{:} = -{:}\hat{\phi}_a(x)\partial_\mu\hat{\phi}_a^\dagger(x){:} \tag{8.1.123}$$
$$= -{:}\partial_\mu\hat{\phi}_a^\dagger(x)\phi_a(x){:} = -{:}\left(\hat{\phi}_a^\dagger(x)\partial_\mu\hat{\phi}_a(x)\right)^\dagger{:} = -\left({:}\hat{\phi}_a^\dagger(x)\partial_\mu\hat{\phi}_a(x){:}\right)^\dagger.$$

We have used the facts that (i) since none of C, P or T exchanges the roles of annihilation and creation operators, then neither does CPT and so we can move the $\hat{\Theta}$ operators in and out of the normal-ordering as desired with the understanding that normal-ordering is not acting on $\hat{\Theta}$ itself; and (ii) we can move the Hermitian conjugate through normal-ordering as well since, e.g., $({:}\hat{a}_1^\dagger\hat{a}_2\hat{a}_3^\dagger{:})^\dagger = (\hat{a}_1^\dagger\hat{a}_3^\dagger\hat{a}_2)^\dagger = \hat{a}_2^\dagger\hat{a}_3\hat{a}_1 = {:}\hat{a}_2^\dagger\hat{a}_3\hat{a}_1{:} = {:}\hat{a}_3\hat{a}_2^\dagger\hat{a}_1{:} = {:}(\hat{a}_1^\dagger\hat{a}_2\hat{a}_3^\dagger)^\dagger{:}$.

Note: In moving from the first to the second line in Eq. (8.1.123) we have used the *spin-statistics connection for bosons*, which implies that boson fields commute within a normal-ordered product.

Next we need to consider the effect of this CPT transformation on normal-ordered fermion bilinears. Let Γ denote any one of the spinor matrices that appear in the fermion bilinears in Eq. (4.4.159),

$$\Gamma \in \{I, i\gamma^5, \gamma^\mu, \gamma^\mu\gamma^5, \sigma^{\mu\nu}\}, \tag{8.1.124}$$

where these matrices span the space of all spinor matrices and have the property

$$[\hat{\bar{\psi}}_a(x)\Gamma\hat{\psi}_b(x)]^\dagger = \hat{\bar{\psi}}_b(x)\Gamma\hat{\psi}_a(x), \tag{8.1.125}$$

since $\gamma^0\Gamma\gamma^0 = \Gamma^\dagger$. Consider the normal-ordered fermion bilinear $\hat{\bar{\psi}}_a(x)\Gamma\hat{\psi}_b(x)$ under this CPT transformation,

$$\begin{aligned}
\hat{\Theta}:\hat{\bar{\psi}}_a(x)\Gamma\hat{\psi}_b(x):\hat{\Theta}^{-1} &= :\hat{\Theta}\hat{\bar{\psi}}_a(x)\Gamma\hat{\psi}_b(x)\hat{\Theta}^{-1}: = :\hat{\Theta}\hat{\bar{\psi}}_a(x)\hat{\Theta}^{-1}\hat{\Theta}\Gamma\hat{\Theta}^{-1}\hat{\Theta}\hat{\psi}_b(x)\hat{\Theta}^{-1}: \\
&= -:\hat{\psi}_a^T(-x)\gamma^5\gamma^0\Gamma^*\gamma^0\gamma^5\hat{\bar{\psi}}_b^T(-x): = +:\hat{\bar{\psi}}_b(-x)[\gamma^5\gamma^0\Gamma^*\gamma^0\gamma^5]^T\hat{\psi}_a(-x): \\
&= (-1)^{n_\Gamma}:\hat{\bar{\psi}}_b(-x)\Gamma\hat{\psi}_a(-x): = (-1)^{n_\Gamma}:\hat{\bar{\psi}}_a(-x)\Gamma\hat{\psi}_b(-x):^\dagger, \tag{8.1.126}
\end{aligned}$$

where n_Γ is the number of spacetime indices in Γ. To arrive at this result we have used Eq. (8.1.125), the fact that all fermion operators anticommute inside normal-ordering and

$$[\gamma^5\gamma^0\Gamma^*\gamma^0\gamma^5]^T = \gamma^5\gamma^0\Gamma^\dagger\gamma^0\gamma^5 = \gamma^5\Gamma\gamma^5 = (-1)^{n_\Gamma}\Gamma, \tag{8.1.127}$$

since $\gamma^5\gamma^\mu\gamma^5 = -\gamma^\mu$, $\gamma^5(\gamma^\mu\gamma^5)\gamma^5 = -\gamma^\mu\gamma^5$. Note that this result depends on us using the *spin-statistics connections for fermions* as well as normal-ordering so that exchanging the order of $\hat{\psi}_a$ and $\hat{\psi}_b$ gives a minus sign and nothing more even when $a = b$. This was just what we discussed earlier when considering the charge conjugation of the fermion four-vector current density in Eq. (8.1.81). We have $n_\Gamma = 0$ if $\Gamma = I$ or $i\gamma_5$, $n_\Gamma = 1$ if $\Gamma = \gamma^\mu$ or $\gamma^\mu\gamma_5$ and $n_\Gamma = 2$ if $\Gamma = \sigma^{\mu\nu}$.

Let $\hat{\Psi}_{\mu_1\cdots\mu_n}(x)$ be any normal-ordered local product of complex numbers, fermion bilinear operators $\hat{\bar{\psi}}_b(x)\Gamma\hat{\psi}_a(x)$ and derivatives. Then the above arguments lead to

$$\hat{\Theta}\hat{\Psi}_{\mu_1\cdots\mu_n}(x)\hat{\Theta}^{-1} = (-1)^n\hat{\Psi}_{\mu_1\cdots\mu_n}^\dagger(-x), \tag{8.1.128}$$

where again recall that since Θ is antiunitary it complex conjugates any coefficients in $\hat{\Phi}_{\mu_1\cdots\mu_n}(x)$ as does taking the Hermitian conjugate on the right-hand side.

Boson operators are defined to commute with fermion operators. This leads to no inconsistencies and is in any case what we expect based on the behavior of even and odd Grassmann numbers. It is then straightforward to extend the notion of normal-ordering to include products of bosons and fermion bilinears. Let $\hat{F}_{\mu_1\cdots\mu_n}$ be any local normal-ordered product of complex numbers, boson operators, fermion bilinears and derivatives with n uncontracted spacetime indices, then we have

$$\hat{\Theta}\hat{F}_{\mu_1\cdots\mu_n}(x)\hat{\Theta}^{-1} = (-1)^n\hat{F}_{\mu_1\cdots\mu_n}^\dagger(-x). \tag{8.1.129}$$

By construction of our quantum field theories $\hat{F}_{\mu_1\cdots\mu_n}(x)$ transforms as a rank n tensor under all restricted (proper orthochronous) Lorentz transformations. So if $n = 0$ then $\hat{F}$ is invariant under restricted Lorentz transformations and $\hat{\Theta}\hat{F}(x)\hat{\Theta}^{-1} = \hat{F}^\dagger(-x)$. If $\hat{F}$ is also Hermitian then $\hat{\Theta}\hat{F}(x)\hat{\Theta}^{-1} = \hat{F}(-x)$. Let $\hat{\mathcal{L}}$ be any Hermitian sum of local normal-ordered products of complex numbers, boson operators, fermion bilinears and derivatives, then $\hat{\mathcal{L}}$ is invariant under the CPT transformation since

$$\hat{\Theta}\hat{\mathcal{L}}(x)\hat{\Theta}^{-1} = \hat{\mathcal{L}}(-x). \tag{8.1.130}$$

These results have brought us to a very important theorem.

CPT theorem: Any local quantum field theory described by a Hermitian, normal-ordered Lagrangian density $\hat{\mathcal{L}}(x)$ that is invariant under restricted Lorentz transformations and whose operators obey the spin-statistics connection will be invariant under the CPT transformation $\hat{\Theta}$; i.e., it will satisfy Eq. (8.1.130).

The Hamiltonian density operator is obtained from the Lagrangian density operator in a similar way for both boson and fermion fields. Let $\hat{\mathcal{L}}_{\mathrm{nno}}$ and $\hat{\mathcal{H}}_{\mathrm{nno}}$ denote the *non-normal-ordered* (nno) Lagrangian and Hamiltonian densities, respectively. Let r be an index that runs over any massive boson fields present. We have a relationship of the general form

$$\hat{\mathcal{H}}_{\mathrm{nno}} = \sum_r [\dot{\hat{\phi}}_r \hat{\pi}_r + \hat{\pi}_r^\dagger \dot{\hat{\phi}}_r^\dagger] - \hat{\mathcal{L}}_{\mathrm{nno}} \quad \text{with} \quad \hat{\pi}_r = \frac{\partial \hat{\mathcal{L}}_{\mathrm{nno}}}{\partial(\partial_0 \hat{\phi}_r)}, \quad \hat{\pi}_r^\dagger = \frac{\partial \hat{\mathcal{L}}_{\mathrm{nno}}}{\partial(\partial_0 \hat{\phi}_r^\dagger)}. \tag{8.1.131}$$

So after normal-ordering we have

$$\hat{\mathcal{H}} = \sum_r :\dot{\hat{\phi}}_r \hat{\pi}_r + \hat{\pi}_r^\dagger \dot{\hat{\phi}}_r^\dagger: - \hat{\mathcal{L}}. \tag{8.1.132}$$

From Eq. (8.1.117) we have for any boson field $\hat{\Theta}\hat{\phi}_r(x)\hat{\Theta}^{-1} = \hat{\phi}_r^\dagger(-x)$ and so $\hat{\Theta}\hat{\phi}_r^\dagger(x)\hat{\Theta}^{-1} = \hat{\phi}_r(-x)$, $\hat{\Theta}\dot{\hat{\phi}}_r(x)\hat{\Theta}^{-1} = -\dot{\hat{\phi}}_r^\dagger(-x)$ and $\hat{\Theta}\dot{\hat{\phi}}_r^\dagger(x)\hat{\Theta}^{-1} = -\dot{\hat{\phi}}_r(-x)$ since $\hat{\Theta}\partial_0\hat{\Theta}^{-1} = -\partial_0$. Since $\hat{\pi}_r$ is $\hat{\mathcal{L}}_{\mathrm{nno}}$ with a factor of $\partial_0\hat{\phi}_r$ removed, then the effect is to remove a phase of -1. After normal-ordering, $\hat{\mathcal{L}}$ is invariant and the phases of (-1) for $\hat{\pi}_r$ and $\partial_0\hat{\phi}_r$ cancel. Since boson operators commute inside normal-ordering,

$$\hat{\Theta}:\dot{\hat{\phi}}_r(x)\hat{\pi}_r(x) + \hat{\pi}_r^\dagger(x)\dot{\hat{\phi}}_r^\dagger(x):\hat{\Theta}^{-1} = :\dot{\hat{\phi}}_r(-x)\hat{\pi}_r(-x) + \hat{\pi}_r^\dagger(-x)\dot{\hat{\phi}}_r^\dagger(-x):, \tag{8.1.133}$$

which then shows for massive bosons that

$$\hat{\Theta}\hat{\mathcal{H}}(x)\hat{\Theta}^{-1} = \hat{\mathcal{H}}(-x). \tag{8.1.134}$$

Similarly, for spin-half fermions from Eq. (6.3.191) we have

$$\hat{\Theta}\hat{\mathcal{H}}(x)\hat{\Theta}^{-1} = \hat{\Theta}[:i\bar{\hat{\psi}}(x)\gamma^0 \overleftrightarrow{\partial}_0 \hat{\psi}(x): - \hat{\mathcal{L}}(x)]\hat{\Theta}^{-1} = :i\bar{\hat{\psi}}(-x)\gamma^0 \overleftrightarrow{\partial}_0 \hat{\psi}(-x): - \hat{\mathcal{L}}(-x)$$
$$= \hat{\mathcal{H}}(-x), \tag{8.1.135}$$

since $:i\bar{\hat{\psi}}(x)\gamma^0 \overleftrightarrow{\partial}_0 \hat{\psi}(x):$ is a normal-ordered Hermitian fermion bilinear with the sign change for γ^0 balanced by the sign change for ∂_0 so that Eq. (8.1.130) applies to it. Using the same arguments as in the massive boson case above and Eq. (6.4.26), we have for the electromagnetic field in Coulomb gauge

$$\hat{\Theta}\hat{\mathcal{H}}(x)\hat{\Theta}^{-1} = \hat{\Theta}[\tfrac{1}{2}:\hat{\mathbf{E}}^2(x) + \hat{\mathbf{B}}^2(x):]\hat{\Theta}^{-1} = \hat{\Theta}[:\hat{\pi}^i(x)\partial_0\hat{A}_i(x): - \hat{\mathcal{L}}(x)]\hat{\Theta}^{-1} \tag{8.1.136}$$
$$= :\hat{\pi}^i(-x)\partial_0\hat{A}_i(-x): - \hat{\mathcal{L}}(-x) = \tfrac{1}{2}:\hat{\mathbf{E}}^2(-x) + \hat{\mathbf{B}}^2(-x): = \hat{\mathcal{H}}(-x),$$

where $\pi^i = E^i$. For the Coulomb gauge condition we find as expected

$$\hat{\Theta}\boldsymbol{\nabla}\cdot\hat{\mathbf{E}}(x)\hat{\Theta}^{-1} = \hat{\Theta}\partial_i\hat{\pi}^i(x)\hat{\Theta}^{-1} = \partial_i\hat{\pi}^i(-x) = \boldsymbol{\nabla}\cdot\hat{\mathbf{E}}(-x) = 0. \tag{8.1.137}$$

This shows that the normal-ordered Hamiltonian density operator is CPT invariant if and only if the normal-ordered Lagrangian density is,

$$\hat{\Theta}\hat{\mathcal{L}}(x)\hat{\Theta}^{-1} = \hat{\mathcal{L}}(-x) \qquad \Longleftrightarrow \qquad \hat{\Theta}\hat{\mathcal{H}}(x)\hat{\Theta}^{-1} = \hat{\mathcal{H}}(-x). \tag{8.1.138}$$

We have then arrived after some effort at a proof of the important CPT theorem. We took the journey in the traditional manner so that we could learn about P, C and T along the way. Note that we have assumed the usual spin-statistics connection in our derivation, where we picked up positive or negative signs when interchanging any two boson or any two fermion operators, respectively, inside the normal-ordering. We shall discuss this further in the next section.

For a CPT-invariant theory the universe is indistinguishable from its "CPT-mirror" image, where time and space are reversed and every particle is replaced by its antiparticle. As a result of this *particles and their antiparticles must have opposite conserved charges and the same masses*. In addition, *particle and antiparticle lifetimes are equal to first order in the C-violating interaction*.

Proof: A conserved charge, such as electric charge, is associated with a conserved Hermitian normal-ordered four-vector current density operator $\hat{j}^\mu(x)$ and a conserved charge operator, $\hat{Q} = \int d^3x\, \hat{j}^0(x)$ with $d\hat{Q}/dt = 0$. From Eq. (8.1.129) we have $\hat{j}^\mu(x) \xrightarrow{\Theta} -\hat{j}^\mu(-x)$, which leads to $\hat{Q} \xrightarrow{\Theta} -\hat{Q}$.

The equality of masses is something that we already know from the discussions in Sec. 1.4. The Casimir invariants W^2 and P^2 are formed from the 10 generators of the Poincaré group, P^μ and $M^{\mu\nu}$. These are the generators of the continuous subgroup of the Poincaré group consisting of the restricted Lorentz transformations and translations and have no relation to $\hat{\mathcal{P}}, \hat{\mathcal{T}}$ or $\hat{\mathcal{C}}$. The two Casimir invariants of this continuous subgroup are W^2 and P^2 and using this we were able to construct the possible massive particle/antiparticle states in Eq. (1.4.18) and the massless particle/antiparticle states in Eq. (1.4.29). From these results we see that the masses of particles and antiparticles are always the same independent of their transformation properties under P, C or T.

We also give the traditional explicit proof for the case of massive particles. Let us use the notation $|a, m_s\rangle \equiv |a, m_a; \mathbf{p} = 0, s, m_s\rangle$ to denote a particle of type a at rest with mass m_a, spin s_a and third component of spin m_s. For a CPT-invariant theory with a time-independent $\hat{\mathcal{H}}$ it follows from Eq. (8.1.138) that $\hat{\Theta}\hat{H}\hat{\Theta}^{-1} = \hat{H}$. The action of a CPT operator on a state is to produce a state with reversed sign of all conserved charges, reversed spin and with the particle replaced by an antiparticle where there is an overall arbitrary phase factor. This means

$$\hat{\Theta}|a, m_s\rangle = e^{i\phi}|\bar{a}, -m_s\rangle. \tag{8.1.139}$$

Then using $\hat{\Theta}\hat{H}\hat{\Theta}^{-1} = \hat{H}$ we have for the mass of particle a,

$$m_a = \langle a, m_s|\hat{H}|a, m_s\rangle = \langle a, m_s|\Theta^{-1}\hat{\Theta}\hat{H}\hat{\Theta}^{-1}\hat{\Theta}|a, m_s\rangle$$
$$= \langle a, m_s|\Theta^{-1}\hat{H}\hat{\Theta}|a, m_s\rangle = \langle \bar{a}, -m_s|\hat{H}|\bar{a}, -m_s\rangle = m_{\bar{a}}, \tag{8.1.140}$$

where we have used Eq. (4.1.51).

The proof of the equality of lifetimes to first order is more complicated and is omitted here. An early proof is given in Matthews and Salam (1959) and a more recent discussion can be found in Greiner and Reinhardt (1996). The partial widths of various decay channels can be different for particles and antiparticles, but the total widths must be equal to first order.

Since CPT invariance applies to such a broad class of theories, it is sometimes considered to almost have the status of a fundamental physical law. For example, even though the Standard Model violates each of the C, P, and T symmetries, it is still CPT invariant. If a CPT-invariant theory has a violation of CP, then it necessarily violates T as well and vice versa. T violation is sometimes spoken of as a form of CP violation for this reason. A thorough discussion of C, P, and T, with a primary focus on CP violation and its experimental consequences, can be found in Bigi and Sanda (2009).[1]

---

[1] In their Tables 4.1 and 4.2 these authors choose pseudoscalars to have a phase -1 under CPT rather than our choice of $+1$. Since i changes sign under CPT, then to relate our pseudoscalar bosons and fermion bilinears to theirs, use $(\text{pseudo})_{\text{us}} = i(\text{pseudo}_{\text{them}})$.

CPT was more often referred to as PCT in early work. It was first discussed and proved by Luders (1954) and in 1955 by Pauli (1988). Jost (1957) used a more axiomatic approach to quantum field theory to prove the theorem. It was also demonstrated independently by Schwinger (1951), although at the time of its publication this work was not generally recognized as a proof of the CPT theorem. A rigorous and detailed proof using the assumptions of axiomatic quantum field theory was given in 1964 by Streater and Wightman (2000) resulting in the CPT theorem in their Eq. 4-7. It uses a complexified form of the restricted Lorentz group to prove a number of theorems regarding vacuum expectation values of products of operators, which then lead to the CPT theorem. While efforts to construct interacting quantum field theories in four dimensions that satisfy the Wightman axioms have not been successful, we still have the above derivation to meet our needs.

8.1.5 Spin-Statistics Connection

The *spin-statistics connection*, also referred to as the *spin-statistics theorem*, asserts that all particles with integer $(0, 1, 2, \ldots)$ spin satisfy Bose-Einstein statistics and all particles with odd-half-integer $(\frac{1}{2}, \frac{3}{2}, \ldots)$ spin satisfy Fermi-Dirac statistics.

The construction of Fock space requires commutation relations for bosonic annihilation and creation operators and anticommutation relations for fermionic annihilation and creation operators. These operators are associated with the plane wave solutions of a relativistic wave equation, e.g., the plane-wave solutions of the Klein-Gordon equation, Maxwell's equations, the Proca equation or the Dirac equation. The annihilation and creation operators and the plane waves are combined to define local field operators that also satisfy the relativistic wave equation by construction. Each relativistic wave equation transforms in some representation of the Lorentz group and the spin-statistics connection is that if the spin of the representation of the wave equation is integer or odd-half-integer, then we are to use commutation relations or anticommutation relations, respectively, for the corresponding Fock space. The question is: Why do we need to do this?

As we have seen, with the above process using the spin-statistics connection we find free theories for scalar $(s = 0)$ and vector $(s = 1)$ bosons and for Dirac fermions $(s = \frac{1}{2})$ with energies bounded below, with appropriate causality requirements satisfied and with conserved currents being conserved at the operator level after normal-ordering as they should be. We also used the spin-statistics connection in the above proof of CPT invariance, which suggests some type of relationship between CPT invariance and the spin-statistics connection.

So, what happens if we use the *wrong* spin-statistics connection? Consider a scalar field given by Eq. (6.2.98)

$$\hat{\phi}(x) = \int \frac{d^3p}{(2\pi)^3} \frac{1}{\sqrt{2E_{\mathbf{p}}}} \left(\hat{a}_{\mathbf{p}} e^{-ip \cdot x} + \hat{a}_{\mathbf{p}}^\dagger e^{ip \cdot x} \right) \big|_{p^0 = E_{\mathbf{p}}}, \tag{8.1.141}$$

but where in place of commutators in Eq. (6.2.17) we generalize to

$$[\hat{a}_{\mathbf{p}}, \hat{a}_{\mathbf{p}'}]_{\mp} = [\hat{a}_{\mathbf{p}}^\dagger, \hat{a}_{\mathbf{p}'}^\dagger]_{\mp} = 0 \quad \text{and} \quad [\hat{a}_{\mathbf{p}}, \hat{a}_{\mathbf{p}'}^\dagger]_{\mp} = (2\pi)^3 \delta^3(\mathbf{p} - \mathbf{p}') \tag{8.1.142}$$

with $\mp$ denoting bosonic commutators and fermionic anticommutators, respectively, $[\hat{A}, \hat{B}]_{-} \equiv [\hat{A}, \hat{B}]$ and $[\hat{A}, \hat{B}]_{+} \equiv \{\hat{A}, \hat{B}\}$. Using Eq. (8.1.142) we obtain

$$
\begin{aligned}
[\hat{\phi}(x), \hat{\phi}(y)]_{\mp} &= \int \frac{d^3p\, d^3q}{(2\pi)^6} \frac{1}{2(E_{\mathbf{p}} E_{\mathbf{q}})^{1/2}} \left[\left(\hat{a}_{\mathbf{p}} e^{-ip \cdot x} + \hat{a}_{\mathbf{p}}^\dagger e^{ip \cdot x} \right), \left(\hat{a}_{\mathbf{q}} e^{-iq \cdot y} + \hat{a}_{\mathbf{q}}^\dagger e^{iq \cdot y} \right) \right]_{\mp} \\
&= \int \frac{d^3p}{(2\pi)^3} \frac{1}{2E_{\mathbf{p}}} \left(e^{-ip \cdot (x-y)} \mp e^{ip \cdot (x-y)} \right),
\end{aligned} \tag{8.1.143}
$$

where we understand that $p^0 = E_{\mathbf{p}}$ and $q^0 = E_{\mathbf{q}}$ as usual. If we choose commutation relations for the scalar field, we find that the $[\hat{\phi}(x), \hat{\phi}(y)] \equiv [\hat{\phi}(x), \hat{\phi}(y)]_-$ vanishes for all spacelike $(x - y)$ as given in Eq. (6.2.144). If we choose anticommutation relations, then $\{\hat{\phi}(x), \hat{\phi}(y)\} \equiv [\hat{\phi}(x), \hat{\phi}(y)]_+$ does not vanish for spacelike separations and we do not know how to make sense of this from the point of view of causality.

Proof: Note that $[\hat{a}_{\mathbf{p}}, \hat{a}_{\mathbf{q}}]_{\mp} = [\hat{a}_{\mathbf{p}}^\dagger, \hat{a}_{\mathbf{q}}^\dagger]_{\mp} = 0$ when we use commutation and anticommutation relations, respectively, and so

$$[(\hat{a}_{\mathbf{p}}e^{-ip\cdot x} + \hat{a}_{\mathbf{p}}^\dagger e^{ip\cdot x}), (\hat{a}_{\mathbf{q}}e^{-iq\cdot y} + \hat{a}_{\mathbf{q}}^\dagger e^{iq\cdot y})]_{\mp} \qquad (8.1.144)$$
$$= [\hat{a}_{\mathbf{p}}, \hat{a}_{\mathbf{q}}^\dagger]_{\mp}e^{-ip\cdot x + iq\cdot y} + [\hat{a}_{\mathbf{p}}^\dagger, \hat{a}_{\mathbf{q}}]_{\mp}e^{ip\cdot x - iq\cdot y}$$
$$= (2\pi)^3 \delta^3(\mathbf{p} - \mathbf{q})(e^{-ip\cdot(x-y)} \mp e^{-ip\cdot(x-y)}).$$

Next we consider fermions with the fermionic anticommutation relations for annihilation and creation operators in Eq. (6.3.78) replaced with corresponding bosonic commutators, $\{\hat{A}, \hat{B}\} \to [\hat{A}, \hat{B}]$. Before normal-ordering the Dirac Hamiltonian can still be calculated as in Eq. (6.3.91) to give the same result,

$$\hat{H}_{\mathrm{nno}} = \int \frac{d^3p}{(2\pi)^3} \sum_s E_{\mathbf{p}}[\hat{b}_{\mathbf{p}}^{s\dagger}\hat{b}_{\mathbf{p}}^s - \hat{d}_{\mathbf{p}}^s\hat{d}_{\mathbf{p}}^{s\dagger}]. \qquad (8.1.145)$$

With bosonic commutation relations for the annihilation and creation operators we do *not* have a sign change when exchanging two of these operators inside normal-ordering. So after normal-ordering the Dirac Hamiltonian we would obtain

$$\hat{H} = \int \frac{d^3p}{(2\pi)^3} \sum_s E_{\mathbf{p}} : \hat{b}_{\mathbf{p}}^{s\dagger}\hat{b}_{\mathbf{p}}^s - \hat{d}_{\mathbf{p}}^s\hat{d}_{\mathbf{p}}^{s\dagger} := \int \frac{d^3p}{(2\pi)^3} \sum_s E_{\mathbf{p}}[\hat{b}_{\mathbf{p}}^{s\dagger}\hat{b}_{\mathbf{p}}^s - \hat{d}_{\mathbf{p}}^{s\dagger}\hat{d}_{\mathbf{p}}^s], \qquad (8.1.146)$$

which makes no sense since the energy would be unbounded below. We could lower the energy indefinitely by increasing the number of antifermions, since $\hat{d}_{\mathbf{p}}^{s\dagger}\hat{d}_{\mathbf{p}}^s$ is the number operator for antifermions with three-momentum $\mathbf{p}$ and spin m_s. Similarly, we would find for the charge operator in place of Eq. (6.3.208) that $\hat{Q} = \hat{N}^+ + \hat{N}^-$ and the system could only have positive charge.

With our two examples above we have illustrated the conclusions of Pauli (1940), which were that (i) if integer spin systems are quantized with Fermi-Dirac statistics they will violate causality, and (ii) if half-integer spin systems are quantized with Bose-Einstein statistics they will have an energy unbounded below. In $3+1$ Minkowski space no consistent quantum field theory with the wrong spin-statistics connection has been constructed, where "consistent" includes the requirements that it be a local quantum field theory, invariant under restricted Lorentz transformations, satisfying causality and with an energy bounded below.

It is common to consider the spin-statistics connection as a fundamental requirement of quantum field theories, and many important historical figures have constructed a variety of arguments and proofs for this that are too numerous to detail here. An extensive review of these can be found in the book by Duck and Sudarshan (1998a) with a summary in a shorter review by the same authors (Duck and Sudarshan, 1998b). An additional valuable perspective is provided in the review of this book by Wightman (1999). In the context of axiomatic quantum field theory the spin-statistics connection was proved in Streater and Wightman (2000) and stated in their theorem 4-10. A proof using a construction of general causal fields and the Lorentz invariance of the S-matrix is given in Sec. 5.7

of Weinberg (1995). The prevailing view is that an elegant and elementary argument for the spin-statistics connection remains to be found. As observed by Richard Feynman, this suggests that there is an underlying principle that we do not yet fully understand (volume III of Feynman (1963–1965)].

8.2 Generating Functionals and the Effective Action

Functional techniques can provide a powerful tool for understanding the relationships between infinite sums of Feynman diagrams of different classes. The effective action is especially useful for understanding spontaneous symmetry breaking in quantum field theories.

8.2.1 Generating Functional for Connected Green's Functions

Consider the generating functional for an interacting scalar theory with source $J(x)$,

$$Z[J] = \frac{\int \mathcal{D}\phi \, e^{i \int d^4 x \, (\mathcal{L} + J\phi)}}{\int \mathcal{D}\phi \, e^{i \int d^4 x \, \mathcal{L}}} \equiv e^{-iE[J]} \equiv e^{iW[J]} \equiv e^{G_c[J]}, \tag{8.2.1}$$

where the ϕ^4 version of this was given in Eq. (7.4.55) and where we have defined $E[J] = -W[J] = iG_c[j]$ in terms of $Z[J]$. We again defer consideration of regularization and renormalization of loops here for simplicity and will return to address this in Sec. 8.4. In the literature both $E[J]$ and $W[J]$ are in common use in discussions and Itzykson and Zuber (1980) and Sterman (1993) use $G_c[J]$. We will typically follow Peskin and Schroeder (1995) and use $E[J]$ but sometimes $G_c[J] = -iE[J]$ will be a more convenient quantity when discussing Green's functions.

Recall that by definition a connected Feynman diagram contains no vacuum subdiagrams as illustrated in Eq. (7.4.26). However, a connected Feynman diagram might consist of one or more disjoint pieces, where in Eq. (7.4.26) all diagrams but the left-hand one are disjoint. If every two lines in a Feynman diagram are connected through any continuous sequence of lines and vertices, then the diagram has only one piece and is not disjoint. By definition a *connected Green's function contains only connected and non-disjoint Feynman diagrams*. These two uses of the word "connected" are potentially confusing so in summary note that (i) *a connected Feynman diagram has no vacuum subdiagrams*, and (ii) *a connected Green's function contains only connected and non-disjoint (i.e., single-piece) Feynman diagrams*.

Write an n-point connected Green's function as $G_c^{(n)}(x_1, \ldots, x_n)$, where the ordinary Green's function is $G^{(n)}(x_1, \ldots, x_n) = \langle \Omega | T\hat{\phi}(x_1) \cdots \hat{\phi}(x_n) | \Omega \rangle$. As $Z[J]$ is the generating functional of the $G^{(n)}(x_1, \ldots, x_n)$ in Eq. (6.2.263) it is shown below that $G_c[J]$ *is the generating functional for the connected Green's functions* such that

$$G_c^{(n)}(x_1, \ldots, x_n) = \frac{(-i)^n \delta^n}{\delta J(x_1) \cdots \delta J(x_n)} G_c[J] \bigg|_{J=0}. \tag{8.2.2}$$

The result that $G_c[J] = iW[J] = -iE[J] = \ln Z[J]$ is the generator of the connected Green's functions is referred to as the *linked cluster theorem*. The relationship between the Green's functions and generating functionals is

$$Z[J] = 1 + \sum_{n=1}^{\infty} i^n \frac{1}{n!} \int d^4 x_1 \cdots d^4 x_n J(x_1) \cdots J(x_n) \, G^{(n)}(x_1, \ldots, x_n), \tag{8.2.3}$$

$$G_c[J] = \sum_{k=1}^{\infty} i^k \frac{1}{k!} \int d^4 x_1 \cdots d^4 x_k J(x_1) \cdots J(x_k) \, G_c^{(k)}(x_1, \ldots, x_k), \tag{8.2.4}$$

where we have $G_c[0] = -iE[0] = 0$ since $Z[0] = 1$. An explicit demonstration of this result for the two- and three-point connected Green's functions is given in Peskin and Schroeder (1995). Additional discussion can be found in Itzykson and Zuber (1980), Cheng and Li (1984), Weinberg (1996) and elsewhere.

Proof: (a) *Detailed version*: Note that the scalar Green's functions are symmetric under permutations of the spacetime indices $x_1, \ldots, x_n$. Any Green's function $G^{(n)}(x_1, \ldots, x_n)$ can be written as the sum of all possible products of connected Green's functions with a total of n legs. Then we can write

$$\frac{i^n}{n!} \int d^4x_1 \cdots d^4x_n J(x_1) \cdots J(x_n) G^{(n)}(x_1, \ldots, x_n) \qquad (8.2.5)$$

$$= \frac{i^n}{n!} \sum_{q_1 m_1 + \cdots + q_p m_p = n} \int d^4x_1 \cdots d^4x_n J(x_1) \cdots J(x_n)$$

$$\times G_c^{(m_1)}(x_1, \ldots, x_{m_1}) \cdots G_c^{(m_p)}(x_{n-m_p+1}, \ldots, x_n),$$

where $G_c^{(m_1)}$ first appears q_1 times followed by $G_c^{(m_2)}$ appearing q_2 times and so on. The $1/n!$ divides the total spacetime volume such that the symmetric product $J(x_1) \cdots J(x_n)$ effectively occurs only once in the integrals in Eq. (8.2.5). The integration over the symmetric J products ensures that any unique assignment of the spacetime indices to the connected Green's functions is sufficient. Let us use a convenient shorthand notation, where in the expression

$$[i^{m_j} \int d^4x_1 \cdots d^4x_{m_j} J(x_1) \cdots J(x_{m_j}) G_c^{(m_j)}(x_1, \ldots, x_{m_j})]^{q_j}, \qquad (8.2.6)$$

we understand that each of the q_j factors has a different set of dummy spacetime coordinates $x_1, \ldots, x_{m_j}$ that we integrate over. In order to ensure that every combination of the symmetric product $J(x_1) \cdots J(x_{q_j m_j})$ occurs only once in the $q_j m_j$ different spacetime integrals in the above expression we must normalize this expression as

$$\frac{1}{q_j!} [\frac{i^{m_j}}{m_j!} \int d^4x_1 \cdots d^4x_{m_j} J(x_1) \cdots J(x_{m_j}) G_c^{(m_j)}(x_1, \ldots, x_{m_j})]^{q_j} \equiv \frac{1}{q_j!} \left[C_{m_j}[J] \right]^{q_j},$$

where comparing with Eq. (8.2.4) we see that $G_c[J] = \sum_{k=1}^{\infty} C_k[J]$. Using $i^n = \prod_{j=1}^{p} (i^{m_j})^{q_j}$ since $q_1 m_1 + \cdots + q_p m_p = n$, multiplying the $j = 1, \ldots, p$ forms of the above expression together and summing with $\sum_{q_1 m_1 + \cdots + q_p m_p = n}$ we have an equivalent form for Eq. (8.2.5). All we have done is to rearrange the way the spacetime integrals are expressed and to adjust the spacetime volume accordingly. So the generating functional is

$$Z[J] = 1 + \sum_n i^n \frac{1}{n!} \int d^4x_1 \cdots d^4x_n J(x_1) \cdots J(x_n) G^{(n)}(x_1, \ldots, x_n) \qquad (8.2.7)$$

$$= 1 + \sum_n i^n \frac{1}{n!} \int d^4x_1 \cdots d^4x_n J(x_1) \cdots J(x_n) \sum_{q_1 m_1 + \cdots + q_p m_p = n}$$

$$\times G_c^{(m_1)}(x_1, \ldots, x_{m_1}) \cdots G_c^{(m_p)}(x_{n-m_p+1}, \ldots, x_n)$$

$$= 1 + \sum_n \sum_{q_1 m_1 + \cdots + q_p m_p = n} \prod_{j=1}^{p} \frac{1}{q_j!} (C_{m_j}[J])^{q_j}$$

$$= e^{C_1[J]} e^{C_2[J]} e^{C_3[J]} \cdots = e^{\sum_{k=1}^{\infty} C_k[J]} = e^{G_c[J]},$$

which proves the result. To proceed from the second last to the last line we proceed term by term; i.e., every term has a form like $\frac{1}{2!}(C_2)^2 \frac{1}{5!}(C_4)^5 C_5 \frac{1}{3!}(C_8)^3 \cdots$ and occurs once and only once in each line.

(b) *Simple version*: Write $Z[J] = \sum_{k=0}^{\infty} D_k[J]$, where $D_0[J] = 1$, where $D_k[0] = 0$ for $k \geq 1$ and where each $D_k[J]$ is sum of all connected (but possibly disjoint) Feynman diagrams with k legs attached to the external source field J. Let $C_i[J]$ with $i = 1, 2, \ldots$ be the set of all possible connected and non-disjoint Feynman diagram pieces with i legs attached to source J. Then $G_c[J] = \sum_{i=1}^{\infty} C_i[J]$ with $C_i[0] = 0$ for all i. Every connected Feynman diagram contributing to $D_k[J]$ can be expressed as the product of its connected non-disjoint pieces. If some C_i occurs m_i times in a connected Feynman diagram, then it must be divided by the symmetry factor $1/m_i!$ according to the Feynman rules and so occurs as $(C_i[J])^{m_i}/m_i!$. It follows that $D_k[J] = \sum_{q_1 m_1 + \cdots + q_p m_p = k} \prod_{j=1}^{p} \frac{1}{q_j!}(C_{m_j}[J])^{q_j}$. The sum over k for all $D_k[J]$ will contain every possible product of the different $(C_i[J])^{m_i}/m_i!$ once and only once and so $Z[J] = \sum_{k=0}^{\infty} D_k[J]$ becomes

$$Z[J] = \prod_{k=1}^{\infty} \left(1 + C_k[J] + \tfrac{1}{2!}C_k[J]^2 + \cdots\right) = \prod_{k=1}^{\infty} e^{C_k[J]} = e^{\sum_{k=1}^{\infty} C_k[J]} = e^{G_c[J]}.$$

8.2.2 The Effective Action

Define the classical field configuration $\phi_c(x)$ as

$$\phi_c(x) \equiv -\frac{\delta E[J]}{\delta J(x)} = -i\frac{\delta G_c[J]}{\delta J(x)} = -i\frac{\delta \ln Z[J]}{\delta J(x)} = \frac{-i}{Z[J]}\frac{\delta Z[J]}{\delta J(x)} = \frac{\langle\Omega|\hat{\phi}(x)|\Omega\rangle_J}{\langle\Omega|\Omega\rangle_J}, \qquad (8.2.8)$$

where the subscript "c" on ϕ implies "classical" rather than connected and where we have used Eq. (6.2.264). Note that $\phi_c(x)$ is a functional of the source $J(x)$ and while we should write it as $\phi_c(x, [J])$ it is notationally convenient to leave this J-dependence to be understood. We regard the classical field $\phi_c(x)$ as the vacuum expectation value of the scalar field operator in the presence of the source $J(x)$.

For a translationally invariant theory when $J = 0$ we expect $\phi_c(x)$ to be a constant field ϕ_0 and the one-point connected Green's function to be x-independent,

$$\phi_0 \equiv \phi_c(x)|_{J=0} = \langle\Omega|\hat{\phi}(x)|\Omega\rangle = -i\left.\frac{\delta G_c[J]}{\delta J(x)}\right|_{J=0} = G_c^{(1)}. \qquad (8.2.9)$$

For the Lagrangian density $\mathcal{L} = \frac{1}{2}\partial_\mu\phi\partial^\mu\phi - U(\phi)$ with a symmetric potential U such that $U(\phi) = U(-\phi)$ then either $\phi_0 = 0$ and the symmetry is preserved or $\phi_0 \neq 0$ and we have spontaneous symmetry breaking (SSB) as discussed in Sec. 5.2.2. For the example $U(\phi) = \frac{1}{2}\mu^2\phi^2 + \frac{\lambda}{4!}\phi^4$ we see that $U(\phi)$ is convex and symmetric about $\phi = 0$ with the classical ground state at $\phi_0^{\mathrm{cl}} = 0$. SSB could only occur through quantum effects that effectively reverse the sign of the quadratic μ^2 term leading to some $\phi_0 \neq \phi_0^{\mathrm{cl}}$. In the case $U(\phi) = -\frac{1}{2}\mu^2\phi^2 + \frac{\lambda}{4!}\phi^4$, we have a double-well potential and SSB occurs already at the classical level with two symmetric degenerate minima $\phi_0^{\mathrm{cl}} = \pm v$ for some $v \in \mathbb{R}$. If U is not symmetric about $\phi = 0$, such as $U(\phi) = \pm\frac{1}{2}\mu^2\phi^2 + \frac{\kappa}{3!}\phi^3 + \frac{\lambda}{4!}\phi^4$, then the classical minimum will be at some $\phi_0^{\mathrm{cl}} \neq 0$. In that case typically we might also have $\phi_0 \neq \phi_0^{\mathrm{cl}}$, but this does not indicate SSB since there was no initial symmetry to break. The cubic term is a source of *explicit symmetry breaking* for the $\phi \to -\phi$ symmetry. Note that $\phi_0 \neq 0$ can only occur when the field ϕ has the quantum numbers of the vacuum, $J^{PC} = 0^{++}$. If this is not the case the vacuum expectation value (VEV) of the field, $\langle\Omega|\hat{\phi}(x)|\Omega\rangle$, must vanish. Vacuum expectation values can also occur for composite operators with vacuum quantum numbers such as the quark condensate in the spontaneous breaking of chiral symmetry in QCD. In the case where $U(\phi)$ contains an explicit cubic term ϕ^3, then there are contributions to the one-point function in the form of so-called *tadpole diagrams*. An example at one loop is a propagator leading to a ϕ^3 vertex with a closed loop:

$$ = \text{one-loop tadpole diagram from a } \phi^3 \text{ interaction.} \tag{8.2.10}$$

We will typically be interested in theories where at the classical level $\phi = 0$ is the symmetric classical field configuration and so in such situations a nonzero VEV implies SSB.

It is useful to define the *effective action* $\Gamma[\phi_c]$ in terms of $E[J]$ where the functional dependence on J is replaced by a functional dependence on $\phi_c(x)$,

$$\Gamma[\phi_c] \equiv -E[J] - \int d^4y \, J(y)\phi_c(y) = W[J] - \int d^4y \, J(y)\phi_c(y). \tag{8.2.11}$$

$\Gamma[\phi_c]$ is sometimes called the quantum effective action (Weinberg, 1996). This transformation is analogous to a functional form of the Legendre transform of classical mechanics, although compared with that we have a relative minus sign in how we have chosen to relate J and ϕ_c. By construction $\Gamma[\phi_c]$ is independent of J since we easily verify using Eq. (8.2.8) that $\delta\Gamma[\phi_c]/\delta J(x) = 0$. The inverse of the transform that defines $\Gamma[\phi_c]$ can be written as

$$E[J] \equiv -\Gamma[\phi_c] - \int d^4y \, J(y)\phi_c(y) \quad \text{with} \quad J(x) = -\frac{\delta\Gamma[\phi_c]}{\delta\phi_c(x)}, \tag{8.2.12}$$

where we observe that $E[J]$ is independent of ϕ_c since $\delta E[J]/\delta\phi_c(x) = 0$.

As usual the Euclidean space version is more readily defined mathematically. It corresponds to a statistical mechanics treatment, where the spectral function becomes the partition function. The Legendre-Fenchel or convex transform is a generalization of the Legendre transform used for nonconvex functions. It can be used to convexify a nonconvex function by applying a transform and then an inverse transform. This *convexification* is more easily achieved by drawing a straight line between two points on a curve so as to replace any nonconvex part of the curve. This is related to the Maxwell construction for thermodynamic systems with a first-order phase transition, where states in equilibrium can be in a simultaneous admixture of two phases. This Maxwell construction is equivalent to replacing the corresponding van der Waals Helmholtz free energy by its convexified form.

Expanding $\mathcal{L}$ and temporarily restoring factors of $\hbar$ we are defining

$$Z[J] = e^{\frac{i}{\hbar}W[J]} = \frac{\int \mathcal{D}\phi \, e^{\frac{i}{\hbar}\int d^4x \left[\frac{1}{2}\partial_\mu\phi\partial^\mu\phi - U(\phi) + J\phi\right]}}{\int \mathcal{D}\phi \, e^{\frac{i}{\hbar}\int d^4x \left[\frac{1}{2}\partial_\mu\phi\partial^\mu\phi - U(\phi)\right]}} = \frac{\int \mathcal{D}\phi \, e^{\frac{i}{\hbar}S[\phi,J]}}{\int \mathcal{D}\phi \, e^{\frac{i}{\hbar}S[\phi]}}, \tag{8.2.13}$$

so that the effective action $W[J] = -E[J]$ has the units of $\hbar$, i.e., the same units as the classical action, $S[\phi] = \frac{1}{2}\partial_\mu\phi\partial^\mu\phi - U(\phi)$. Then $\Gamma[\phi_c]$ also has the dimensions of the action. Setting the external source to zero gives

$$\left.\frac{\delta\Gamma[\phi_c]}{\delta\phi_c(x)}\right|_{J=0} = 0, \tag{8.2.14}$$

which is similar to the equations of motion in classical field theory, $\delta S[\phi]/\delta\phi(x) = 0$. This is one motivation for referring to $\Gamma[\phi_c]$ as the effective action. For additional discussion of this, see Chapter 16 of Weinberg (1996).

Effective action as the generator of the 1PI Green's functions: Differentiate Eq. (8.2.8) with respect to $\phi_c(y)$ to give

$$\delta^4(x-y) = \frac{\delta\phi_c(x)}{\delta\phi_c(y)} = -i\frac{\delta}{\delta\phi_c(y)}\frac{\delta G_c[J]}{\delta J(x)} = -i\int d^4z\,\frac{\delta J(z)}{\delta\phi_c(y)}\frac{\delta^2 G_c[J]}{\delta J(z)J(x)} \tag{8.2.15}$$

$$= i\int d^4z\,\frac{\delta^2\Gamma[\phi_c]}{\delta\phi_c(y)\delta\phi_c(z)}\frac{\delta^2 G_c[J]}{\delta J(z)J(x)} = \frac{1}{i}\int d^4z\,\frac{\delta^2\Gamma[\phi_c]}{\delta\phi_c(y)\delta\phi_c(z)}\frac{(-i)^2\delta^2 G_c[J]}{\delta J(z)J(x)},$$

where we have used Eq. (8.2.12). For now let us restrict our attention to cases where there is no VEV, $\phi_0 \equiv \phi_c(x)|_{J=0} = 0$. We can expand $\Gamma[\phi_c]$ as a functional Taylor series about $\phi_c = 0$,

$$\Gamma[\phi_c] \equiv \sum_{k=1}^{\infty}\int d^4x_1\cdots d^4x_k\,\Gamma^{(k)}(x_1,\ldots,x_k)\phi_c(x_1)\cdots\phi_c(x_k). \tag{8.2.16}$$

It follows that $\delta^2\Gamma[\phi_c]/\delta\phi_c(y)\delta\phi_c(z)$ is the inverse of $i\delta^2 G_c[J]/\delta J(x)J(z)$ from Eq. (8.2.15). Then setting $J=0$ we have from Eqs. (8.2.2) and (8.2.16) that

$$\delta^4(x-y) = (1/i)\int d^4z\,\Gamma^{(2)}(y,z)G_c^{(2)}(z,x). \tag{8.2.17}$$

For a translationally invariant theory we have $G_c^{(2)}(x_1,x_2) = G_c^{(2)}(x_1-x_2)$ and $\Gamma^{(2)}(x_1,x_2) = \Gamma^{(2)}(x_1-x_2)$. No VEV means that $G_c^{(1)} = 0$ and so we must have $G_c^{(2)}(x_1,x_2) = G^{(2)}(x_1-x_2) = D_F(x_1-x_2)$. Then Eq. (8.2.17) becomes

$$\delta^4(x-y) = (1/i)\int d^4z\,\Gamma^{(2)}(y-z)D_F(z-x) \tag{8.2.18}$$

$$\Rightarrow \quad \Gamma^{(2)}(p) = i[D_F(p)]^{-1} = p^2 - m_0^2 - \Pi(p^2),$$

where we have used Eq. (7.5.51) and where $\Gamma^{(2)}(p)$ and $D_F(p)$ are the usual asymmetric Fourier transforms of $\Gamma^{(2)}(x-y)$ and $D_F(x-y)$ as given in Eq. (6.2.155). Recall from Eq. (7.5.49) that $-i\Pi(p^2)$ is a one-particle irreducible (1PI) amplitude and so $i\Gamma^{(2)}(p)$ is a 1PI amplitude plus an inverse bare propagator, $p^2 - m_0^2$.

We can view Eq. (8.2.15) as a "master equation" in the sense that by taking additional derivatives $\delta/\delta\phi_c(x)$ and setting $J=0$ we can generate additional relations to arbitrarily high order. Taking one additional derivative, $\delta/\delta\phi_c(u)$, removes the δ-function and gives after a simple calculation

$$0 = \int d^4z\left[\frac{1}{i}\Gamma^{(3)}(u,y,z)G_c^{(2)}(z,x) + \int d^4u\,G_c^{(3)}(v,z,x)\frac{1}{i}\Gamma^{(2)}(y,z)\frac{1}{i}\Gamma^{(2)}(u,v)\right]$$

$$= \int d^4z\left[\frac{1}{i}\Gamma^{(3)}(u,y,z)D_F(z-x) + \int d^4u\,G_c^{(3)}(v,z,x)D_F^{-1}(y-z)D_F^{-1}(u-v)\right] \tag{8.2.19}$$

or $0 = \frac{1}{i}\Gamma^{(3)}_{uyz}G_{czx}^{(2)} + G_{cvzx}^{(3)}\frac{1}{i}\Gamma^{(2)}_{yz}\frac{1}{i}\Gamma^{(2)}_{uv} = \frac{1}{i}\Gamma^{(3)}_{uyz}D_{Fzx} + G_{cvzx}^{(3)}D_{Fyz}^{-1}D_{Fuv}^{-1}$ in an obvious shorthand. Using the symmetry of the spacetime labels and Eq. (8.2.17) we find

$$G_c^{(3)}(x,y,z) = \int d^4x'd^4y'd^4z'\,i\Gamma^{(3)}(x',y',z')D_F(x'-x)D_F(y'-y)D_F(z'-z)$$

$$= \int d^4x'd^4y'd^4z'\,G_{c,\mathrm{amp}}^{(3)}(x',y',z')D_F(x'-x)D_F(y'-y)D_F(z'-z), \tag{8.2.20}$$

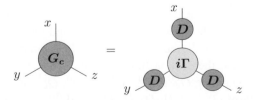

and so $\Gamma^{(3)}(x_1,x_2,x_3) = G_{c,\mathrm{amp}}^{(3)}(x_1,x_2,x_3)$, where $G_{c,\mathrm{amp}}^{(3)}(y_1,y_2,y_3)$ is the connected amputated three-point function, which is just the 1PI three-point function. We show below that $\Gamma^{(n)}(x_1,\ldots,x_n)$

is the 1PI n-point function for all $n \geq 3$ and so we say that $\Gamma[\phi_c]$ *is the generating functional for the 1PI amplitudes*. For the four-point connected Green's function we find, for example:

Outline of proof: Consider the shorthand version of the first line of Eq. (8.2.19),
(i) $0 = \frac{1}{i}\Gamma^{(3)}_{uyz}G^{(2)}_{czx} + G^{(3)}_{cvzx}\frac{1}{i}\Gamma^{(2)}_{yz}\frac{1}{i}\Gamma^{(2)}_{uv}$.

We are being a little careless with notation here since we do not yet take $J = 0$ as we wish to perform additional derivatives before taking $J = 0$, which also means $\phi_c = 0$. What we are writing in this proof as $G^{(n)}_{cx_1 \cdots x_n}$ and $\Gamma^{(n)}_{x_1 \cdots x_n}$ should be understood before and after taking $J = 0$ as shorthand for

$$(-i)^n \delta^n G_c[J]/\delta J(x_1) \cdots \delta J(x_n) \xrightarrow{J=0} G^{(n)}_c(x_1, \ldots, x_n) \equiv G^{(n)}_{cx_1 \cdots x_n},$$
$$\delta^n \Gamma[\phi_c]/\delta\phi_c(x_1) \cdots \delta\phi_c(x_n) \xrightarrow{J=0} \Gamma^{(n)}(x_1, \ldots, x_n) \equiv \Gamma^{(n)}_{x_1 \cdots x_n}.$$

In this notation $\delta_{xy} = \frac{1}{i}\Gamma^{(2)}_{yz}G^{(2)}_{czx}$ represents Eq. (8.2.15) before taking $J = 0$ and Eq. (8.2.17) afterward. Next note that the effect of $\delta/\delta\phi_c(x) \equiv \delta/\delta\phi_{cx}$ in this same shorthand is
(ii) $\dfrac{\delta}{\delta\phi_{cx}}\Gamma^{(n)}_{x_1 \cdots x_n} = \Gamma^{(n+1)}_{x_1 \cdots x_n x}$ and $\dfrac{\delta}{\delta\phi_{cx}}G^{(n)}_{cx_1 \cdots x_n} = G^{(n+1)}_{cx_1 \cdots x_n v}\frac{1}{i}\Gamma^{(2)}_{vx}$.
Then acting on line (i) above with $\delta/\delta\phi_c(w) \equiv \delta/\delta\phi_{cw}$ gives
(iii) $0 = \frac{1}{i}\Gamma^{(4)}_{uyzw}G^{(2)}_{czx} + \frac{1}{i}\Gamma^{(3)}_{uyz}G^{(3)}_{czxv}\frac{1}{i}\Gamma^{(2)}_{vw} + G^{(4)}_{cvzxs}\frac{1}{i}\Gamma^{(2)}_{sw}\frac{1}{i}\Gamma^{(2)}_{yz}\frac{1}{i}\Gamma^{(2)}_{uv}$
$\qquad + G^{(3)}_{cvzx}\frac{1}{i}\Gamma^{(3)}_{yzw}\frac{1}{i}\Gamma^{(2)}_{uv} + G^{(3)}_{cvzx}\frac{1}{i}\Gamma^{(2)}_{yz}\frac{1}{i}\Gamma^{(3)}_{uvw}$.

Since the inverse of $G^{(2)}_c$ is $\frac{1}{i}\Gamma^{(2)}$, then we can multiply the above equation by $G^{(2)}_{cuu'}, G^{(2)}_{cyy'}, G^{(2)}_{cww'}$ and replace $x \to x'$ to give

$$0 = \frac{1}{i}\Gamma^{(4)}_{uyzw}G^{(2)}_{cuu'}G^{(2)}_{cyy'}G^{(2)}_{czx'}G^{(2)}_{cww'} + G^{(4)}_{cu'y'x'w'} + \frac{1}{i}\Gamma^{(3)}_{uyz}G^{(3)}_{czx'w'}G^{(2)}_{cuu'}G^{(2)}_{cyy'}$$
$$+ G^{(3)}_{cu'zx'}\frac{1}{i}\Gamma^{(3)}_{yzw}G^{(2)}_{cyy'}G^{(2)}_{cww'} + G^{(3)}_{cvy'x'}\frac{1}{i}\Gamma^{(3)}_{uvw}G^{(2)}_{cuu'}G^{(2)}_{cww'}.$$

Next insert $G^{(3)}_{cx'y'z'} = i\Gamma^{(3)}_{xyz}G^{(2)}_{cxx'}G^{(2)}_{cyy'}G^{(2)}_{czz'}$ and rearrange to give

$$G^{(4)}_{cu'y'x'w'} = i\Gamma^{(4)}_{uyzw}G^{(2)}_{cuu'}G^{(2)}_{cyy'}G^{(2)}_{czx'}G^{(2)}_{cww'} + [G^{(2)}_{cuu'}G^{(2)}_{cyy'}i\Gamma^{(3)}_{uyz}G^{(2)}_{czs}i\Gamma^{(3)}_{srt}G^{(2)}_{crx'}G^{(2)}_{ctw'}$$
$$+ (u \leftrightarrow w) + (y \leftrightarrow w)].$$

Relabeling and tidying up the indices and taking $J = 0$ we have arrived at

$$G^{(4)}_{cwxyz} = D_{Fww'}D_{Fxx'}D_{Fyy'}D_{Fzz'}i\Gamma^{(4)}_{w'x'y'z'}$$
$$+ [D_{Fww'}D_{Fxx'}D_{Fyy'}D_{Fzz'}i\Gamma^{(3)}_{w'x'u}D_{Fuv}i\Gamma^{(3)}_{vy'z'} + (x \leftrightarrow y) + (x \leftrightarrow z)],$$

which is represented graphically above. Any four-point connected non-disjoint diagram that is one-particle reducible occurs once and only once in the square bracket. So the remaining amplitude, $i\Gamma^{(4)}$, must contain only 1PI diagrams.

We can then act on line (iii) with $\delta/\delta\phi_c$ to obtain an expression relating $G^{(5)}_c$ and $i\Gamma^{(5)}$ and containing $G^{(4)}_c$, $i\Gamma^{(4)}$, $G^{(3)}_c$, $i\Gamma^{(3)}$, $G^{(2)}_c$ and $i\Gamma^{(2)}$. Rearranging this gives an equation of the form $G^{(5)}_c = D_F D_F D_F D_F D_F i\Gamma^{(5)} + [\cdots]$, where again $[\cdots]$ contains "dressed" tree diagrams consisting of D_F and the 1PI diagrams $i\Gamma^{(4)}$ and $i\Gamma^{(3)}$. Any connected one-particle

> reducible Feynman diagram occurs once and only once in $[\cdots]$ as before. It is not difficult to see that this pattern of increasingly complicated "dressed" tree diagrams continues indefinitely leading to the conclusion that the $i\Gamma^{(n)}$ are the 1PI n-point amplitudes.

The *background field method* provides a useful means to calculate the effective action of a quantum field theory by expanding a quantum field around a classical background field. The Green's functions are then evaluated as a function of the background field. An advantage of the approach is that gauge invariance can be manifestly preserved when applied to gauge theories. An explicit demonstration of this is the derivation of the one-loop contribution to the β-function of nonabelian gauge theories is given in Peskin and Schroeder (1995).

8.2.3 Effective Potential

Recall that we expect the $\phi_0 \equiv \phi_c(x)_{J=0}$ to be a constant field as discussed in Eq. (8.2.9). We also know that $\Gamma[\phi_c]$ has the dimensions of an action and so we can define the *effective potential* $V_{\text{eff}}(\phi_c)$ of an *arbitrary constant field* ϕ_c using

$$\Gamma[\phi_c] \equiv - \int d^4x \, V_{\text{eff}}(\phi_c) = -(VT)V_{\text{eff}}(\phi_c), \tag{8.2.21}$$

where VT is the spacetime volume that we are working in and where we will take $VT \to \infty$ in the usual way. Based on the arguments of statistical mechanics mentioned earlier, the effective potential should be a convex function and should we encounter a nonconvex form in a theoretical calculation we must use the convexified version as illustrated in Fig. 8.1. Label the two minima as ϕ_1 and ϕ_2 with $\phi_1 < \phi_2$. For values of ϕ_c between the minima, $\phi_1 < \phi_c < \phi_2$, when in equilibrium the system will simultaneously exist in two phases with V_1 the total volume of coordinate space in the ϕ_1 phase and $V_2 = V - V_1$ the total volume in the ϕ_2 phase. This is analogous to the first-order liquid-vapor phase transition. The spatial average over the total volume $V = V_1 + V_2$ of the effective potential is $V_{\text{eff}} = rV_{\text{eff}}(\phi_1)+(1-r)V_{\text{eff}}(\phi_2)$ where r is the ratio $r \equiv V_1/V$. Since our definition of V_{eff} assumed a constant ϕ_c it is not surprising that we can incorrectly arrive at a nonconvex form; however, we now know how to convexify V_{eff} and resolve this issue. Define ϕ_{min} as the minimum value of the convex V_{eff}. It will be the unique solution of $dV_{\text{eff}}/d\phi_c = 0$ and we expect after quantization that $\phi_0 \equiv \langle\Omega|\hat{\phi}(x)|\Omega\rangle \simeq \phi_{\text{min}}$.

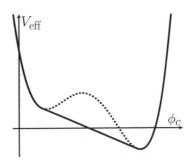

Figure 8.1   A possible form for a nonconvex effective potential with the nonconvex portion shown as a dashed curve. The solid line is the actual V_{eff} and is the convexified form.

8.2.4　Loop Expansion

An expansion of Feynman diagrams in terms of the number of independent loops can be identified with an expansion in powers of $\hbar$, which means that such a *loop expansion* is equivalent to an $\hbar$-*expansion*. Let us understand this statement.

The number of *independent loops*, L, in a Feynman diagram is by definition the number of undetermined internal momenta once we apply the momentum-space Feynman rules discussed in Sec. 7.6.1. A planar Feynman diagram is one that can be drawn on a two-dimensional plane with no lines crossing. In such a case L is equal to a naive counting of closed loops. For a non-planar diagram care must be taken to establish the number of independent internal momenta in order to find L. For a Feynman diagram with $I \geq 1$ internal lines and $V \geq 0$ vertices we arrive at

$$L = I - (V - 1) = I - V + 1. \tag{8.2.22}$$

This follows since every vertex has a four-momentum conserving δ-function, where one of these results in the four-momentum conservations of the E external legs. That leaves $(V - 1)$ δ-functions to constrain the momenta of the I internal propagators, which means that $L = I - (V - 1)$ internal momenta remain independent.

Rewrite Eq. (8.2.13) with the free and interacting terms in $\mathcal{L}$ separated,

$$Z[J] = \frac{\int \mathcal{D}\phi\, e^{\frac{i}{\hbar} \int d^4 x\, [\mathcal{L}_0 + \mathcal{L}_{\mathrm{int}} + J\phi]}}{\int \mathcal{D}\phi\, e^{\frac{i}{\hbar} \int d^4 x\, [\mathcal{L}_0 + \mathcal{L}_{\mathrm{int}}]}} = \frac{\int \mathcal{D}\phi\, e^{\frac{i}{\hbar} \int d^4 x\, [\frac{1}{2}(\partial_\mu \phi \partial^\mu \phi - m^2 \phi^2) + \mathcal{L}_{\mathrm{int}} + J\phi]}}{\int \mathcal{D}\phi\, e^{\frac{i}{\hbar} \int d^4 x\, [\frac{1}{2}(\partial_\mu \phi \partial^\mu \phi - m^2 \phi^2) + \mathcal{L}_{\mathrm{int}}]}}. \tag{8.2.23}$$

There is also a factor of $\hbar$ in the mass term since we know from Eq. (4.3.15) that m is $mc/\hbar$ when we do not use natural units, but this simply defines m and is not relevant to our argument. A propagator in a Feynman diagram is the inverse of the differential operator in the free part of $\mathcal{L}$ and so $D_0(p) \propto \hbar$. Each vertex in a Feynman diagram arises from an occurence of $\mathcal{L}_{\mathrm{int}}$ and so has a factor of $1/\hbar$. This leads to the result that for any Feynman diagram there will be an associated factor

$$\hbar^{E+I-V} \xrightarrow{I \geq 1} \hbar^{E+L-1}, \tag{8.2.24}$$

since there are $E + I$ free propagators and V vertices in the diagram and where we have used Eq. (8.2.22). A Feynman diagram with external legs removed has $E = 0$ and so has a factor $\hbar^{L-1}$. These arguments are very general. For example, using the Feynman rules in Sec. 7.6 we see that the above results apply also to theories that include fermions, photons and vector bosons.

It is also possible to establish a relationship between a loop expansion and a perturbative expansion in powers of the coupling constant in a theory-dependent form. For example, consider $\mathcal{L}_{\mathrm{int}} = -(\lambda/n!)\phi^n(x)$. Every vertex connects to n lines, every external line connects to one vertex and every internal line connects to two vertices and so for $I \geq 1$ we find

$$nV = E + 2I \quad \Rightarrow \quad V = [E + 2(L - 1)]/(n - 2), \tag{8.2.25}$$

where we have again used Eq. (8.2.22). A diagram with V vertices is order V in perturbation theory, i.e., $\propto \lambda^V$. So if we keep the number of external legs E fixed and change the order of perturbation theory by ΔV, then the number of independent loops changes by ΔL, where $\Delta V = 2\Delta L/(n - 2)$; i.e., $\lambda^V \to \lambda^{V+\Delta V}$ means that $\hbar^{E+L-1} \to \hbar^{E+L+\Delta L-1}$. If a theory has more than one kind of vertex, such as $\mathcal{L}_{\mathrm{int}} = -(\kappa/3!)\phi^3 - (\lambda/4!)\phi^4$, then Eq. (8.2.22) remains valid and we simply generalize to $3V_3 + 4V_4 = E + 2I$ where there are V_3 ϕ^3-vertices and V_4 ϕ^4-vertices.

Consider some simple examples in ϕ^n theories and check the above formulae:

$$: n = 4, E = 2, I = 1, L = 1, V = 1, \mathcal{O}(\lambda), \mathcal{O}(\hbar^2) \qquad (8.2.26)$$

$$: n = 4, E = 2, I = 3, L = 2, V = 2, \mathcal{O}(\lambda^2), \mathcal{O}(\hbar^3)$$

$$: n = 4, E = 2, I = 5, L = 3, V = 3, \mathcal{O}(\lambda^3), \mathcal{O}(\hbar^4)$$

$$: n = 3, E = 4, I = 10, L = 3, V = 8, \mathcal{O}(\kappa^8), \mathcal{O}(\hbar^6).$$

The last example is a non-planar diagram. Note that there are only three independent loop momenta, not four, which is more easily deduced using $I - V + 1 = 3 = L$.

8.3 Schwinger-Dyson Equations

The classical equations of motion come from Hamilton's principle, which are the classical configurations that lead to a stationary action. We can arrive at a form of "equations of motion" for quantum field theories by considering variations of the fields in the functional integral formalism. This leads to a form of equations of motion for Green's functions, which are referred to as *Schwinger-Dyson equations* or sometimes as *Dyson-Schwinger equations*.

8.3.1 Derivation of Schwinger-Dyson Equations

Consider an integrable function $g(\vec{x})$, i.e., $g(\vec{x}) \to 0$ fast enough as $|\vec{x}| \to \pm\infty$ that the integral exists. Under a change of variable $\vec{x} \to \vec{x}' = \vec{x} + \vec{\epsilon}$ for some constant $\vec{\epsilon}$ we have $d^n x = d^n x'$ since the Jacobian is unity. It then follows that

$$\int_{-\infty}^{\infty} d^n x \, g(\vec{x}) = \int_{-\infty}^{\infty} d^n x' \, g(\vec{x}') = \int_{-\infty}^{\infty} d^n x \, g(\vec{x} + \vec{\epsilon}) = \int_{-\infty}^{\infty} d^n x \left[g(\vec{x}) + \vec{\epsilon} \cdot \frac{dg}{d\vec{x}} + \mathcal{O}(\epsilon^2) \right].$$

Since $\vec{\epsilon}$ is arbitrary then $\int_{-\infty}^{\infty} dx \, dg(x)/dx_i = 0$ for all $i = 1, \ldots, n$. So for a convergent integral the integral of a total derivative of the integrand vanishes. If a function $s(x)$ satisfies $s(x) \to +\infty$ as $x \to \pm\infty$ and if $f(x)$ is less than exponentially divergent as $x \to \pm\infty$, then we can choose $g(x) = f(x)e^{-s(x)}$. We could also choose $g(x) = f(x)e^{i[s(x)+i\epsilon x^2]}$ with ϵ giving an exponential damping term analogous to the Feynman parameter. Functional extensions of these examples are

$$\int \mathcal{D}\phi \, \frac{\delta}{\delta\phi(x)} \left\{ F[\phi]e^{-S_E[\phi]} \right\} = 0 \quad \text{and} \quad \int \mathcal{D}\phi \, \frac{\delta}{\delta\phi(x)} \left\{ F[\phi]e^{iS[\phi]} \right\} = 0, \qquad (8.3.1)$$

where the first is a Euclidean functional integral with the Euclidean action $S_E[\phi]$ and the second is a Minkowski space version with the Feynman boundary conditions providing the damping term in the Minkowski-space action $S[\phi]$. The Minkowski space form of Eq. (8.3.1) can be written as

$$\int \mathcal{D}\phi \left\{ \frac{\delta F[\phi]}{\delta\phi(x)} + iF[\phi]\frac{\delta S[\phi]}{\delta\phi(x)} \right\} e^{iS[\phi]} = 0. \qquad (8.3.2)$$

Note that the term $\{\cdots\}$ will in general contain constants, the field ϕ and spacetime derivatives of the field such as $\partial^\mu \phi$, $\partial^2 \phi$ and so on. We are free to move all constants and all spacetime derivatives outside the functional integral, which leaves the functional integral with terms of the form $\int \mathcal{D}\phi\, \phi(x_1)\cdots\phi(x_n)e^{iS[\phi]}$. Recall from Eq. (6.2.263) that $\int \mathcal{D}\phi\, \phi(x_1)\cdots\phi(x_n)e^{iS[\phi]}/\int \mathcal{D}\phi\, e^{iS[\phi]} = \langle\Omega|T\hat{\phi}(x_1)\cdots\hat{\phi}(x_n)|\Omega\rangle$. This means that for different choices of $F[\phi]$ we can use Eq. (8.3.2) to generate a series of equations for the Green's functions of the theory, which can be thought of as the quantum field theory extension of the classical equations of motion since they involve $\delta S[\phi]/\delta\phi(x)$. This infinite set of equations are the *Schwinger-Dyson equations* and they can be expressed in various forms as we shall see.

Consider $F[\phi] = \phi(x_1)\cdots\phi(x_n)$ in the case of a free scalar field theory, where $\delta S/\delta\phi(x) = -(\partial_x^2 + m^2)\phi(x) = -(\Box + m^2)\phi(x)$ from Eq. (3.1.40). Then we have

$$0 = \sum_{i=1}^n \delta^4(x - x_i)\int \mathcal{D}\phi\, \phi(x_1)\cdots\phi(x_{i-1})\phi(x_{i+1})\cdots\phi(x_n)e^{iS[\phi]} \tag{8.3.3}$$

$$- i(\Box_x + m^2)\int \mathcal{D}\phi\, \phi(x_1)\cdots\cdots\phi(x_n)\phi(x)e^{iS[\phi]}, \tag{8.3.4}$$

which leads to Schwinger-Dyson equations of the form

$$(\Box_x + m^2)\langle\Omega|T\hat{\phi}(x)\hat{\phi}(x_1)\cdots\hat{\phi}(x_n)|\Omega\rangle$$
$$= -i\sum_{i=1}^n \delta^4(x - x_i)\langle\Omega|T\hat{\phi}(x_1)\cdots\hat{\phi}(x_{i-1})\hat{\phi}(x_{i+1})\cdots\hat{\phi}(x_n)|\Omega\rangle. \tag{8.3.5}$$

In the special case of $n = 1$ we see that we have simply reproduced Eq. (6.2.168), $(\Box_x + m^2)\langle\Omega|T\hat{\phi}(x)\hat{\phi}(x_1)|\Omega\rangle = -i\delta^4(x - x_1)$. Since the Heisenberg picture operators must obey the classical equations of motion, $(\Box_x + m^2)\hat{\phi}(x) = 0$, then the right-hand side of this equation contains the contact terms that arise when we move the operator $(\Box_x + m^2)$ inside the time-ordered product as we saw in deriving Eq. (6.2.168). Let us add a polynomial interaction such as $\mathcal{L}_{\text{int}} = -\frac{\kappa}{3!}\phi^3 - \frac{\lambda}{4!}\phi^4$, then $\delta S_{\text{int}}/\delta\phi(x) = \partial\mathcal{L}_{\text{int}}/\partial\phi(x) = -\frac{\kappa}{2!}\phi^2(x) - \frac{\lambda}{3!}\phi^3(x) \equiv \mathcal{L}'_{\text{int}}(\phi(x))$ and we have

$$(\Box_x + m^2)\langle\Omega|T\hat{\phi}(x)\hat{\phi}(x_1)\cdots\hat{\phi}(x_n)|\Omega\rangle - \langle\Omega|T\mathcal{L}'_{\text{int}}(\hat{\phi}(x))\hat{\phi}(x)\hat{\phi}(x_1)\cdots\hat{\phi}(x_n)|\Omega\rangle$$
$$= -i\sum_{i=1}^n \delta^4(x - x_i)\langle\Omega|T\hat{\phi}(x_1)\cdots\hat{\phi}(x_{i-1})\hat{\phi}(x_{i+1})\cdots\hat{\phi}(x_n)|\Omega\rangle. \tag{8.3.6}$$

We can move a *single* spacetime derivative through a time-ordered product of fields since

$$\partial_\mu^x\langle\Omega|T\hat{\phi}(x)\hat{\phi}(x_1)\cdots\hat{\phi}(x_n)|\Omega\rangle = \langle\Omega|T\partial_\mu^x\hat{\phi}(x)\hat{\phi}(x_1)\cdots\hat{\phi}(x_n)|\Omega\rangle, \tag{8.3.7}$$

which follows as a generalization of Eq. (6.2.166). The time-ordering involves sums of products of $\theta(x_i^0 - x_j^0)$ and whenever the derivative acts on some $\theta(x_i^0 - x_j^0)$ there will be an otherwise identical term in the sum with $\theta(x_j^0 - x_i^0)$ instead. Then we use $\langle\Omega|[\hat{\phi}(x_i), \hat{\phi}(x_j)]|\Omega\rangle = 0$ when $x_i^0 = x_j^0$ from the equal-time commutators to eliminate those two terms. We proceed until all such pairs of terms have canceled. Acting with a second derivative generates contact terms as we saw in Eq. (6.2.168) and above.

Analogous results apply for photons, vector bosons and fermions. The vanishing of the integral of a total derivative of the integrand in the case of fermions is a result of Eq. (6.3.19), $\int da_i \frac{\partial}{\partial a_i} f(a) = 0$ and its generalizations.

Conserved Currents and Schwinger-Dyson Equations

Consider an arbitrary field variation but with the requirement that it have a unit Jacobian for simplicity, $\mathcal{D}\phi \to \mathcal{D}\phi' = \mathcal{D}\phi$, which will occur for any unitary transformation of the fields. We only ever consider ratios of functional integrals since that is how we regulate and define them. So it is sufficient that the Jacobian be independent of the fields being integrated over, $\mathcal{D}\phi \to \mathcal{D}\phi' \propto \mathcal{D}\phi$, since any ϕ-independent term will cancel in the ratio. In such cases we will simply write $\mathcal{D}\phi' = \mathcal{D}\phi$ as well as if the Jacobian were unity. It follows that

$$\int \mathcal{D}\phi\, F[\phi]e^{iS[\phi]} = \int \mathcal{D}\phi'\, F[\phi']e^{iS[\phi']} = \int \mathcal{D}\phi\, F[\phi']e^{iS[\phi']} \tag{8.3.8}$$
$$= \int \mathcal{D}\phi\, \left\{ F[\phi]e^{iS[\phi]} + \delta F[\phi]e^{iS[\phi]} + i\delta S[\phi]F[\phi]e^{iS[\phi]} + \mathcal{O}(\delta^2) \right\}.$$

Let the infinitesimal parameter be spacetime dependent (i.e., local), $\alpha \to \epsilon(x)$, so that we can use Eq. (3.2.45) for δS. Equating the $\mathcal{O}(\delta) = \mathcal{O}(\epsilon)$ terms gives

$$0 = \int \mathcal{D}\phi\, \left\{ \delta F[\phi] + i\delta S[\phi]F[\phi] \right\} e^{iS[\phi]} = \int \mathcal{D}\phi\, \left\{ \delta F[\phi] + i[\int d^4x\, (\partial_\mu \epsilon)j^\mu]F[\phi] \right\} e^{iS[\phi]}$$
$$= \int d^4x\, \epsilon(x) \int \mathcal{D}\phi\, \left\{ \delta F[\phi]/\delta\epsilon(x) - i\partial_\mu j^\mu(x)F[\phi] \right\} e^{iS[\phi]}, \tag{8.3.9}$$

where we have used integration by parts to move ∂_μ since we can choose $\epsilon(x)$ to have compact support (vanish outside an arbitrary finite region). Note that we have $\delta F[\phi] = \int d^4x\, \epsilon(x)\delta F[\phi]/\delta\epsilon(x) + \mathcal{O}(\epsilon^2)$. Since $\epsilon(x)$ is an arbitrary infinitesimal function with compact support, then finally we have

$$\partial_\mu \int \mathcal{D}\phi\, j^\mu(x)F[\phi]e^{iS[\phi]} = (-i) \int \mathcal{D}\phi\, \delta F[\phi]/\delta\epsilon(x)\, e^{iS[\phi]}, \tag{8.3.10}$$

where the order of ∂_μ and $\int \mathcal{D}\phi$ can be freely exchanged. These types of Schwinger-Dyson equations involving symmetries and conserved currents lead to some important and useful results.

Examples

1. Global phase invariance for the charged scalar field leads to the conserved current $j^\mu(x) = i\left[\phi^*(\partial^\mu\phi) - (\partial^\mu\phi^*)\phi\right]$ as we saw in Eq. (3.2.49). The Lagrangian density is invariant under $\phi(x) \to \phi'(x) = e^{i\alpha}\phi(x)$ and $\phi^*(x) \to \phi'^*(x) = e^{-i\alpha}\phi^*(x)$ and so no surface terms arise. As above, replace $\alpha \to \epsilon(x)$ so that we can arrive at the Noether current using Eq. (3.2.45). From $\mathcal{L} = \partial_\mu\phi\partial^\mu\phi^* - m^2\phi\phi^*$ we find to $\mathcal{O}(\epsilon)$ that $\delta\mathcal{L} = (\partial_\mu\epsilon)[i(\phi\partial^\mu\phi^* - \phi^*\partial^\mu\phi)] = (\partial_\mu\epsilon)j^\mu$ and so $\delta S = \int d^4x\, (\partial_\mu\epsilon)j^\mu(x)$. The sign of j^μ depends on the sign of ϵ and needs to be checked each time to get the correct sign for $j^\mu(x)$ in Eq. (8.3.10). For the charged scalar field we have

$$\frac{\delta F[\phi^*,\phi]}{\delta\epsilon(x)} = \frac{\delta F}{\delta\phi(x)}\frac{\partial\phi(x)}{\partial\epsilon(x)} + \frac{\delta F}{\delta\phi^*(x)}\frac{\partial\phi^*(x)}{\partial\epsilon(x)} = i\left\{ \frac{\delta F}{\delta\phi(x)}\phi(x) - \frac{\delta F}{\delta\phi^*(x)}\phi^*(x) \right\},$$

since $\phi \to \phi + i\epsilon\phi$ and $\phi^* \to \phi^* - i\epsilon\phi^*$. Choosing $F[\phi] = \phi(x_1)\phi^*(y_1)$ we find

$$\partial_\mu \int \mathcal{D}\phi^*\mathcal{D}\phi\, j^\mu(x)\phi(x_1)\phi^*(y_1)e^{iS[\phi^*,\phi]} \tag{8.3.11}$$
$$= (-i) \int \mathcal{D}\phi^*\mathcal{D}\phi\, i\left\{ \delta^4(x-x_1) - \delta^4(x-y_1) \right\} \phi(x_1)\phi^*(y_1)\, e^{iS[\phi^*,\phi]}.$$

We see that the above equation can be written as

$$\partial_\mu\langle\Omega|T\hat{j}^\mu(x)\hat{\phi}(x_1)\hat{\phi}^*(y_1)|\Omega\rangle \tag{8.3.12}$$
$$= \left\{ \delta^4(x-x_1) - \delta^4(x-y_1) \right\}\langle\Omega|T\hat{\phi}(x_1)\hat{\phi}^*(y_1)|\Omega\rangle.$$

Since we are taking the divergence of the current, normal-ordering is not relevant. Choosing $F[\phi^*,\phi] = \phi(x_1)\phi^*(y_1)\cdots\phi(x_n)\phi^*(y_n)$ this easily generalizes to

$$\partial_\mu\langle\Omega|T\hat{j}^\mu(x)\hat{\phi}(x_1)\hat{\phi}^*(y_1)\cdots\hat{\phi}(x_n)\hat{\phi}^*(y_n)|\Omega\rangle \tag{8.3.13}$$
$$= \sum_{i=1}^n \left\{ \delta^4(x-x_i) - \delta^4(x-y_i) \right\}\langle\Omega|T\hat{\phi}(x_1)\hat{\phi}^*(y_1)\cdots\hat{\phi}(x_n)\hat{\phi}^*(y_n)|\Omega\rangle,$$

where the right-hand side is made up of the contact terms that would vanish by current conservation if ∂_μ were acting inside the time-ordered product.

2. For a free fermion field the Lagrangian density, $\mathcal{L} = \bar{\psi}(i\slashed{\partial} - m)\psi$, is invariant under a global phase transformation, $\psi(x) \to e^{-i\alpha}\psi(x)$, $\bar{\psi}(x) \to e^{i\alpha}\bar{\psi}(x)$. This leads to a conserved vector current, $j^\mu(x) = \bar{\psi}(x)\gamma^\mu\psi(x)$, as we know from Eqs. (4.4.25) and (6.3.207). Again making the phase transformation local (spacetime dependent), $\alpha \to \epsilon(x)$, and using $i\partial_\mu[e^{-i\epsilon}\psi] = e^{-i\epsilon}[i\partial_\mu\psi + (\partial_\mu\epsilon)\psi]$ we find to $\mathcal{O}(\epsilon)$ that $\delta\mathcal{L} = (\partial_\mu\epsilon)\bar{\psi}\gamma^\mu\psi = (\partial_\mu\epsilon)j^\mu$ so that $\delta S = \int d^4x\,(\partial_\mu\epsilon)j^\mu$. For the fermion field Eq. (8.3.10) becomes

$$\partial_\mu \int \mathcal{D}\bar{\psi}\mathcal{D}\psi\, j^\mu(x)F[\bar{\psi},\psi]e^{iS[\bar{\psi},\psi]} = (-i)\int \mathcal{D}\bar{\psi}\mathcal{D}\psi\, \delta F[\bar{\psi},\psi]/\delta\epsilon(x)\, e^{iS[\bar{\psi},\psi]}. \qquad (8.3.14)$$

To express these results in terms of time-ordered products of the current operator and other fermion operators we use Eq. (6.3.237) in place of Eq. (6.2.263). If $F[\bar{\psi},\psi] = \psi(x_1)\bar{\psi}(y_1)$ then $\delta F[\bar{\psi},\psi] = \{-i\epsilon(x_1) + i\epsilon(y_1)\}\psi(x_1)\bar{\psi}(y_1)$ and so

$$\partial_\mu \int \mathcal{D}\bar{\psi}\mathcal{D}\psi\, j^\mu(x)\psi(x_1)\bar{\psi}(x_2)e^{iS[\bar{\psi},\psi]} \qquad (8.3.15)$$
$$= (-i)\int \mathcal{D}\bar{\psi}\mathcal{D}\psi\, \left\{-i\delta^4(x-x_1) + i\delta^4(x-y_1)\right\}\psi(x_1)\bar{\psi}(y_1)\, e^{iS[\bar{\psi},\psi]}.$$

Using Eq. (6.3.237) leads to a result similar to that for the charged scalar field,

$$\partial_\mu\langle\Omega|T\hat{j}^\mu(x)\hat{\psi}(x_1)\hat{\bar{\psi}}(y_1)|\Omega\rangle \qquad (8.3.16)$$
$$= \left\{-\delta^4(x-x_1) + \delta^4(x-y_1)\right\}\langle\Omega|T\hat{\psi}(x_1)\hat{\bar{\psi}}(y_1)|\Omega\rangle.$$

Choosing $F[\bar{\psi},\psi] = \psi(x_1)\bar{\psi}(y_1)\cdots\psi(x_n)\bar{\psi}(y_n)$ this generalizes to

$$\partial_\mu\langle\Omega|T\hat{j}^\mu(x)\hat{\psi}(x_1)\hat{\bar{\psi}}(y_1)\cdots\hat{\psi}(x_n)\hat{\bar{\psi}}(y_n)|\Omega\rangle \qquad (8.3.17)$$
$$= \sum_{i=1}^n\{-\delta^4(x-x_i) + \delta^4(x-y_i)\}\langle\Omega|T\hat{\psi}(x_1)\hat{\bar{\psi}}(y_1)\cdots\hat{\psi}(x_n)\hat{\bar{\psi}}(y_n)|\Omega\rangle,$$

noting that $\hat{\psi}\hat{\bar{\psi}}$ pairs commute with other such pairs in a time-ordered product.

3. Now consider the case of an internal symmetry as discussed in Sec. 3.2.5. Here the n scalar fields transform between themselves as $\phi_r(x) \to \phi'_r(x) = R_{rs}\phi_s(x)$ with summation understood. We have $R_{rs} = \delta_{rs} + \alpha\lambda_{rs}$ for an infinitesimal transformation. Generalizing to a local infinitesimal parameter in the usual way, $\alpha \to \epsilon(x)$, we have $\phi'_r(x) = \phi_r(x) + \epsilon(x)\lambda_{rs}\phi_s(x) \equiv \phi_r(x) + \epsilon(x)\Phi_r(x)$. We know from the derivation of the Noether current for scalar fields that the conserved current $j^\mu(x) = \vec{\pi}^\mu(x) \cdot \vec{\Phi}(x) = (\partial\mathcal{L}/\partial(\partial_\mu\vec{\phi})) \cdot \vec{\Phi}$ from Eq. (3.2.105) has the correct sign so that $\delta S = \int d^4x\,(\partial_\mu\epsilon)j^\mu(x)$. In this case Eq. (8.3.10) becomes

$$\partial_\mu \int \mathcal{D}\vec{\phi}\, j^\mu(x)F[\vec{\phi}]e^{iS[\vec{\phi}]} = (-i)\int \mathcal{D}\vec{\phi}\, \delta F[\vec{\phi}]/\delta\epsilon(x)\, e^{iS[\vec{\phi}]}. \qquad (8.3.18)$$

Choose $F[\vec{\phi}] = \phi_r(x_1)\phi_s(x_2)$, then since $\delta\phi_r(x_i)/\delta\epsilon(x) = \delta^4(x-x_i)\Phi_r(x)$ we find

$$\partial_\mu\langle\Omega|T\hat{j}^\mu(x)\hat{\phi}_r(x_1)\hat{\phi}_s(y_1)|\Omega\rangle \qquad (8.3.19)$$
$$= (-i)\{\delta^4(x-x_1)\langle\Omega|T\hat{\Phi}_r(x_1)\hat{\phi}_s(y_1)|\Omega\rangle + \delta^4(x-y_1)\langle\Omega|T\hat{\phi}_r(x_1)\hat{\Phi}_s(y_1)|\Omega\rangle\}.$$

8.3.2 Ward and Ward-Takahashi Identities

Let us extend Example 2 above to QED, which is to say that we wish to understand the implications of the global phase invariance of the fermion part of the QED action, $\mathcal{L} = \bar{\psi}(i\slashed{D} - m)\psi - \frac{1}{4}F_{\mu\nu}F^{\mu\nu}$. Recall that $D_\mu \equiv \partial_\mu + iq_c A_\mu$ with q_c the fermion electric charge. Even though we are going to allow the phase transformation to be local as we did above, we are not performing a gauge transformation of the form of Eq. (7.6.18) since the photon field is not changed. It is convenient to *now include the fermion charge q_c in the definition of the current j^μ_{Dirac}* as opposed to Eq. (7.6.22). To do that we consider the transformation $\psi \to \psi' = e^{-iq_c\epsilon}\psi$, $\bar\psi \to \bar\psi' = e^{iq_c\epsilon}\bar\psi$ and $A^\mu \to A'^\mu = A^\mu$. To $\mathcal{O}(\epsilon)$ this gives $\delta\mathcal{L} = (\partial_\mu\epsilon)q_c\bar\psi\gamma^\mu\psi = (\partial_\mu\epsilon)j^\mu$. Then $\delta S = \int d^4x\,(\partial_\mu\epsilon)j^\mu$ with now $j^\mu = q_c\bar\psi\gamma^\mu\psi$ and so Eq. (8.3.14) in QED becomes

$$\partial_\mu \int \mathcal{D}\bar\psi \mathcal{D}\psi \mathcal{D}A^\mu\, j^\mu(x)F[\bar\psi,\psi,A]e^{iS[\bar\psi,\psi,A]} \tag{8.3.20}$$
$$= (-i)\int \mathcal{D}\bar\psi\mathcal{D}\psi\mathcal{D}A^\mu\, \delta F[\bar\psi,\psi,A]/\delta\epsilon(x)\, e^{iS[\bar\psi,\psi,A]}.$$

With $F[\bar\psi,\psi,A] = \psi(x_1)\bar\psi(y_1)\cdots\psi(x_n)\bar\psi(y_n)A^{\mu_1}(z_1)\cdots A^{\mu_m}(z_m)$ it then follows that Eq. (8.3.17) becomes (Itzykson and Zuber, 1980)

$$\partial^x_\mu\langle\Omega|T\hat{j}^\mu(x)\hat\psi(x_1)\hat{\bar\psi}(y_1)\cdots\hat\psi(x_n)\hat{\bar\psi}(y_n)\hat{A}^{\mu_1}(z_1)\cdots\hat{A}^{\mu_m}(z_m)|\Omega\rangle \tag{8.3.21}$$
$$= q_c\sum_{i=1}^n\{-\delta^4(x-x_i)+\delta^4(x-y_i)\}$$
$$\times\langle\Omega|T\hat\psi(x_1)\hat{\bar\psi}(y_1)\cdots\hat\psi(x_n)\hat{\bar\psi}(y_n)\hat{A}^{\mu_1}(z_1)\cdots\hat{A}^{\mu_m}(z_m)|\Omega\rangle.$$

Since the operators must obey the classical equations of motion, then $\partial_\mu\hat{j}^\mu(x) = 0$ and so the right-hand side of the above equation again arises from the contact terms that occur when moving ∂_0 through the time-ordering. Generalizing the arguments that led to Eq. (6.2.166) we can then conclude that

$$[\hat{j}^0(x),\hat{A}^\mu(x')]\delta(x^0-x'^0) = 0,\quad [\hat{j}^0(x),\hat\psi(x')]\delta(x^0-x'^0) = -q_c\hat\psi(x)\delta^4(x-x'),$$
$$[\hat{j}^0(x),\hat{\bar\psi}(x')]\delta(x^0-x'^0) = q_c\hat{\bar\psi}(x)\delta^4(x-x'), \tag{8.3.22}$$

which we can also verify directly from the equal-time commutation and anticommutation relations for QED in Eqs. (6.3.182) and (6.4.122) and using the mixed commutators $[\hat{A}^\mu(t,\mathbf{x}),\hat\psi(t,\mathbf{x}')] = [\hat{A}^\mu(t,\mathbf{x}),\hat{\bar\psi}(t,\mathbf{x}')] = 0$. The latter follow since the photon and fermion fields are independent as discussed in the QED part of Sec. 7.6.2. These equal-time commutation and anticommutation relations remain valid in the interacting quantum field theory just as the fundamental Poisson brackets do in interacting classical theories because we arrived at them using canonical quantization.

The meaning of these Schwinger-Dyson equations is more easily understood in momentum space. We can perform an asymmetric Fourier transform in the usual way, where we recall from Eqs. (7.5.108) and (7.5.85) that $\int d^4x e^{ip\cdot x}$ leads to an outgoing momentum p. Let us choose outgoing momenta p'_i for the $\psi(x_i)$, incoming momentum q for j^μ, incoming momenta p_i for the $\bar\psi(y_i)$ and outgoing momenta k_i for the $A^{\mu_i}(z_i)$. These choices are arbitrary. The momentum-space amplitude for the left-hand side of Eq. (8.3.21) is then defined as

$$M^\mu(q;p_1,\ldots,p_n;p'_1,\ldots,p'_n;k_1,\ldots,k_m) \equiv \int d^4x(\textstyle\prod_{i=1}^n d^4x_i d^4y_i)(\prod_{j=1}^m d^4z_j)$$
$$\times e^{-iq\cdot x}e^{i\sum_{k=1}^n(p'_k\cdot x_k-p_k\cdot y_k)}e^{i\sum_{\ell=1}^m k_\ell\cdot z_\ell}\langle\Omega|Tj^\mu(x)\hat\psi(x_1)\cdots\hat{A}^{\mu_m}(z_m)|\Omega\rangle. \tag{8.3.23}$$

We also define a similar amplitude except that we omit the current operator $\hat{j}^\mu(x)$,

$$M_0(p_1,\ldots,p_n;p'_1,\ldots,p'_n;k_1,\ldots,k_m) \equiv \int(\textstyle\prod_{i=1}^n d^4x_i d^4y_i)(\prod_{j=1}^m d^4z_j)$$
$$\times e^{i\sum_{k=1}^n(p'_k\cdot x_k-p_k\cdot y_k)}e^{i\sum_{\ell=1}^m k_\ell\cdot z_\ell}\langle\Omega|T\hat\psi(x_1)\cdots\hat{A}^{\mu_m}(z_m)|\Omega\rangle. \tag{8.3.24}$$

The Fourier transform of the right-hand side of Eq. (8.3.21) involves quantities like

$$q_c \int d^4x\, d^4x_i\, d^4y_i\, e^{-iq\cdot x} e^{i(p_i'\cdot x_i - p_i\cdot y_i)} \delta^4(x - x_i)\langle\Omega|T\hat{\psi}(x_1)\cdots \hat{A}^{\mu_m}(z_m)|\Omega\rangle \quad (8.3.25)$$

$$= q_c \int d^4x_i\, d^4y_i\, e^{i[(p_i'-q)\cdot x_i - p_i\cdot y_i]}\langle\Omega|T\hat{\psi}(x_1)\cdots \hat{A}^{\mu_m}(z_m)|\Omega\rangle. \quad (8.3.26)$$

The Fourier transform of Eq. (8.3.21) can then be written as

$$iq_\mu M^\mu(q; p_1, \ldots, p_n; p_1', \ldots, p_n'; k_1, \ldots, k_m) \quad (8.3.27)$$
$$= q_c \sum_{i=1}^n \{M_0(p_1, \ldots, (p_i + q), \ldots, p_n; p_1', \ldots, p_n'; k_1, \ldots, k_m)$$
$$- M_0(p_1, \ldots, p_n; p_1', \ldots, (p_i' - q), \ldots, p_n'; k_1, \ldots, k_m)\}.$$

This infinite set of Schwinger-Dyson equations is collectively referred to as the *Ward-Takahashi identities*. Since we took four-dimensional Fourier transforms all four-momenta can be off-shell or on-shell and are arbitrary up to the conservation of four-momentum that follows from the spacetime translational invariance of the Green's functions, $\sum_{i=1}^n (p_i'^\mu - p_i^\mu) + \sum_{j=1}^m k_i^\mu - q^\mu = 0$.

The Ward-Takahashi identity for QED is not a result of gauge invariance; rather, it follows from the global phase invariance of the fermion part of the action and the resulting conserved vector current $j^\mu(x)$. This is the sense in which we observed in Eq. (2.7.25) that the photon can only couple to a conserved current. We can write down a similar result for scalar QED that builds on Example 1 using Eq. (8.3.13). The forms of Eqs. (8.3.13) and (8.3.17) agree up to an overall sign difference and so

$$-iq_\mu N^\mu(q; p_1, \ldots, p_n; p_1', \ldots, p_n'; k_1, \ldots, k_m) \quad (8.3.28)$$
$$= q_c \sum_{i=1}^n \{N_0(p_1, \ldots, (p_i + q), \ldots, p_n; p_1', \ldots, p_n'; k_1, \ldots, k_m)$$
$$- N_0(p_1, \ldots, p_n; p_1', \ldots, (p_i' - q), \ldots, p_n'; k_1, \ldots, k_m)\},$$

where $M^\mu \to N^\mu$ and $M_0 \to N_0$ with the replacements $\psi \to \phi$ and $\bar{\psi} \to \phi^*$. This sign difference is related to the fact that $\mathcal{L}_{\text{int}}^{\text{QED}} = -q_c A_\mu j_{\text{Dirac}}^\mu$ in Eq. (7.6.16), whereas $\mathcal{L}_{\text{int}}^{\text{sQED}} = +q_c A_\mu j_{\text{KG}}^\mu + q_c^2 A^2 \phi^* \phi$ in Eq. (7.6.23). In scalar QED the additional interaction $q_c^2 A^2 \phi^* \phi$ leads to complications in applying the Ward-Takahashi identities in scalar QED (Schwartz, 2013). Since gauge invariance was not invoked then there is nothing special about photons coupling to the conserved current $j^\mu(x)$ for fermions. For example, the Ward-Takahashi identities are still valid if the photons are replaced by massive vector (Proca) bosons coupled to conserved currents.

Ward Identities

We want to explore the implications of the QED Ward-Takahashi identity for the calculation of QED scattering amplitudes. Consider some QED Green's function

$$\langle\Omega|T\hat{\psi}(x_1)\hat{\bar{\psi}}(y_1)\cdots \hat{\psi}(x_n)\hat{\bar{\psi}}(y_n)\hat{A}^{\mu_1}(z_1)\cdots \hat{A}^{\mu_m}(z_m)|\Omega\rangle, \quad (8.3.29)$$

where we understand that we are using the Gupta-Bleuler formalism for photons to avoid complications and so that we can treat the photon operators much as we would scalar boson operators as discussed in Sec. 6.4.5. For example, this leads to the simple equal-time canonical commutation relations in Eq. (6.4.122). Recall that in this formalism we added a "Feynman-like" gauge-fixing term to the Lagrangian density so that the classical equations of motion are $\Box A^\mu(x) = 0$ for a free photon field or $\Box A^\mu = j_{\text{Dirac}}^\mu = q_c \bar{\psi}\gamma^\mu \psi$ for QED. This was complemented by the requirement

$\langle \Psi | \partial_\mu \hat{A}^\mu(x) | \Psi \rangle = 0$, which defines the physical subspace, $|\Psi\rangle \in V_{\text{Phys}}$. Heisenberg picture operators satisfy the classical equations of motion and so in QED we have $\Box_x \hat{A}^\mu(x) = \hat{j}^\mu_{\text{Dirac}}(x)$.

Next recall that the LSZ formalism for photons is given in Eq. (7.5.107) and so we need to calculate

$$\Box_{z_1} \cdots \Box_{z_m} \langle \Omega | T \hat{\psi}(x_1) \hat{\bar{\psi}}(y_1) \cdots \hat{\psi}(x_n) \hat{\bar{\psi}}(y_n) \hat{A}^{\mu_1}(z_1) \cdots \hat{A}^{\mu_m}(z_m) | \Omega \rangle. \tag{8.3.30}$$

It is not difficult to show that the form of the Schwinger-Dyson equation analogous to Eq. (8.3.6) for QED is

$$\Box_z \langle \Omega | T \cdots \hat{A}^\mu(z) \hat{A}^{\mu_1}(z_1) \cdots \hat{A}^{\mu_m}(z_m) | \Omega \rangle = \langle \Omega | T \cdots \hat{j}^\mu(z) \hat{A}^{\mu_1}(z_1) \cdots \hat{A}^{\mu_m}(z_m) | \Omega \rangle$$
$$+ i \sum_{i=1}^m \delta^4(z - z_i) g^{\mu \mu_i} \langle \Omega | T \cdots \hat{A}^{\mu_2}(z_2) \cdots \hat{A}^{\mu_{i-1}}(z_{i-1}) \hat{A}^{\mu_{i+1}}(z_{i+1}) \cdots \hat{A}^{\mu_m}(z_m) | \Omega \rangle.$$

From Eq. (6.4.153) we have $\Box_x g^{\mu\rho} D_{0\rho\nu}(x-y) = i \delta^\mu_{\ \nu} \delta^4(x-y)$ since the Gupta-Bleuler propagator corresponds to Feynman gauge. This is just the familiar result that $\Box_x D_0(x-y) = -i\delta^4(x-y)$, since $D_0^{\mu\nu}(x-y) = -g^{\mu\nu} D_0(x-y)$. So in an obvious shorthand we can rewrite the above as

$$\Box_z \langle \cdots \hat{A}^\nu(z) \cdots \rangle = \langle \cdots \hat{j}^\nu(z) \cdots \rangle \tag{8.3.31}$$
$$+ \Box_z \sum_{i=1}^m D_0^{\nu\mu_i}(z - z_1) \langle \cdots \hat{A}^{\mu_{i-1}}(z_{i-1}) \hat{A}^{\mu_{i+1}}(z_{i+1}) \cdots \rangle.$$

The second term contains a disconnected photon propagator, $D_0^{\nu\mu_i}(z-z_1)$, that connects two external points and so the second term does not contribute to scattering amplitudes and can be neglected. Applying this to Eq. (8.3.30) we see that up to terms that do not contribute to scattering we have effectively

$$\Box_{z_1} \cdots \Box_{z_m} \langle \Omega | T \cdots \hat{A}^{\mu_1}(z_1) \cdots \hat{A}^{\mu_m}(z_m) | \Omega \rangle = \langle \Omega | T \cdots \hat{j}^{\mu_1}(z_1) \cdots \hat{j}^{\mu_m}(z_m) | \Omega \rangle. \tag{8.3.32}$$

So *to amputate the external photon propagators in a QED scattering amplitude we replace the photon operators in the Green's function with the conserved current operators.*

Consider some invariant scattering amplitude $\mathcal{M}(q; \dots)$ for QED with an external photon with momentum q^μ and polarization vector $\epsilon^\mu(q)$. From the covariant form of the LSZ formula in Eq. (7.5.107) appropriate for the Gupta-Bleuler formalism, we know from Eq. (8.3.32) that the scattering amplitude in coordinate space will contain $\Box_z \langle \Omega | T \hat{A}^\mu(z) \cdots | \Omega \rangle = \langle \Omega | T \hat{j}^\mu(z) \cdots | \Omega \rangle$ up to irrelevant terms. Then we can write $\mathcal{M}(q; \dots) \equiv \epsilon_\mu(q) M^\mu(q; \dots)|_{\text{on-shell}}$, where $M^\mu(q; \dots)$ is the Fourier transform of some $\langle \Omega | T \hat{j}^\mu(z) \cdots | \Omega \rangle$ with unconstrained momenta; i.e., they are not required to be on-shell. Define $M_0(\cdots)$ as M^μ with the external photon removed and recall the Ward-Takahashi identity in Eq. (8.3.27), then

$$iq_\mu M^\mu(q; \dots) = q_c \sum_{i=1}^n \left\{ M_0(\dots, (p_i + q), \dots) - M_0(\dots, (p_i' - q), \dots) \right\}. \tag{8.3.33}$$

We have $\mathcal{M}(q, \dots) \equiv \epsilon_\mu(q) M^\mu(q, \dots)|_{\text{on-shell}}$ with on-shell meaning $p_i^2 = p_i'^2 = m^2$ and $q^2 = k_i^2 = 0$. Before being evaluated on-shell $\mathcal{M}$ had external propagators amputated. For fermions we multiply by $\not{p}_i - m + i\epsilon$, which was effectively multiplying by $p_i^2 - m^2 + i\epsilon$. The poles in M_0 on the right-hand side are at $(p_i + q)^2 - m^2 + i\epsilon = 0$ and $(p_i' - q)^2 - m^2 + i\epsilon$ and since $[p^2 - m^2 + i\epsilon] / [(p \pm q)^2 - m^2 + i\epsilon] \to 0$ as we take $p^2 \to m^2 - i\epsilon$, then the right-hand side must vanish when the fermions on the left-hand side are on-shell. Note that there are isolated points in momentum space where this argument does not hold, $\pm 2p \cdot q + q^2 = 0$, but since $\mathcal{M}$ is a physical observable the p_i and p_i' dependence of M^μ should be continuous. So $q_\mu M^\mu$ vanishes on-shell since there is one fermion in each term on the right that is off-shell.

Summary: If $\mathcal{M}$ is an invariant scattering amplitude with an external incoming photon of momentum q^μ and polarization vector $\epsilon^\mu(q)$, then M^μ is defined by

$$\mathcal{M}(q; p_1, .., p_n; p_1', .., p_n'; k_1, .., k_m) \tag{8.3.34}$$
$$\equiv \epsilon_\mu(q) M^\mu(q; p_1, .., p_n; p_1', .., p_n'; k_1, .., k_m)|_{\text{on-shell}},$$

where there are n fermion lines entering, n fermion lines leaving and $m + 1$ photon lines in total. For an outgoing photon we use $\epsilon_\mu(q)^*$ in place of $\epsilon_\mu(q)$. We are defining $M^\mu(\cdots)$ above as a function of arbitrary momenta, where the $(2n + m + 1)$ momenta are not required to be on-shell. The *Ward identities* are the result that

$$q_\mu M^\mu(q; p_1, .., p_n; p_1', .., p_n'; k_1, .., k_m)|_{\text{all fermions on-shell}} = 0 \tag{8.3.35}$$

and do not require any of the $(m + 1)$ photons to be on-shell. So it does not matter what choice of covariant gauge we use for the external photons, since a $q^\mu q^\nu / q^2$ term will never contribute and any covariant gauge choice is equivalent to Feynman gauge. Since this is true for any value of q_c, then *these results are true order-by-order in perturbation theory*; i.e., contributions from different powers of charge q_c must satisfy these results individually.

Proof of gauge invariance: What about the choice of gauge for internal photon lines? Consider the order q_c^n contribution to an on-shell scattering amplitude $\mathcal{M}$ and denote it as $\mathcal{M}^{(n)}$. Let the number of external fermion lines and external photon lines, E_f and E_γ, respectively, be fixed. E_f must be even and each vertex brings one power of q_c. External photons involve one vertex and internal photons involve two vertices, which gives $n = 2I_\gamma + E_\gamma$. So the number of internal photon propagators $D_0^{\mu\nu}$ is $I_\gamma = (n - E_\gamma)/2$. Then we can write

$$\mathcal{M}^{(n)} = q_c^n \left[\prod_{i=1}^{E_\gamma} \epsilon_{\mu_i}(k_i) \right] \left[\prod_{j=1}^{I_\gamma} \int \frac{d^4 q_j}{(2\pi)^4} D_{0\nu_j \rho_j}(q_j) \right] \tag{8.3.36}$$
$$\times \mathcal{M}^{\nu_1 \rho_1 \cdots \nu_{I_\gamma} \rho_{I_\gamma} \mu_1 \cdots \mu_{E_\gamma}}(k_1, \ldots, k_{E_\gamma}, q_1, \ldots, q_{I_\gamma}, p_1, \ldots, p_{E_f/2}, p_1', \ldots, p_{E_f/2}'),$$

where for simplicity we do not distinguish between ϵ and ϵ^* for external photons here and where $\mathcal{M}^{\nu_1 \rho_1 \cdots}$ is the fermion-amputated Fourier transform of $\langle \Omega | T \cdots | \Omega \rangle$, where the dots represents $n = 2I_\gamma + E_\gamma$ current operators $\hat{j}^\mu$ and $E_f/2$ of each of $\hat{\psi}$ and $\hat{\bar{\psi}}$ operators. We have used Eq. (8.3.32) and Feynman gauge for $D_0^{\mu\nu}$ to arrive at this result, but having done so we see that because of the Ward identity in Eq. (8.3.35) we could use any covariant gauge for the internal propagators $D_0^{\mu\nu}$. In summary, invariant scattering amplitudes $\mathcal{M}$ are independent of the covariant gauge parameter ξ. *So QED amplitudes are gauge invariant and this follows from fermion current conservation.*

Diagrammatic proof: A proof can also be given in terms of Feynman diagrams. This is based on the observation that a fermion line in a Feynman diagram either flows continuously through the diagram or is an internal loop in the diagram. Consider the set of all Feynman diagrams contributing to some amplitude M_0. To obtain the corresponding M^μ we must attach an additional photon with incoming momentum q at every possible point for each diagram in M_0 and then amputate the propagator, which is equivalent to inserting a current operator $\hat{j}^\mu(x)$. For a fermion line moving through the diagram this is represented as

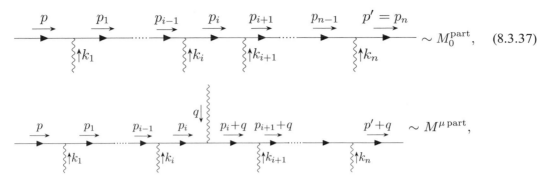

$$\sim M_0^{\text{part}}, \quad (8.3.37)$$

$$\sim M^{\mu\,\text{part}},$$

where we must sum over all insertion points for the photon with momentum q. The spinor identity $-iq_c\slashed{q} = -iq_c[(\slashed{p}_i + \slashed{q} - m_0) - (\slashed{p}_i - m_0)]$ is the tree-level Ward-Takahashi identity and gives $S_0(p_i + q)(-iq_c\slashed{q})S_0(p_i) = q_c[S_0(p_i) - S_0(p_i + q)]$. Contributions to $k_\mu M^\mu$ cancel in the sum over all insertions in the second line of Eq. (8.3.37) except that we are left with the difference of two terms of the form of the first line of Eq. (8.3.37), $(p \to p + q) - (p' \to p' - q)$. This is just what we need to give the right-hand side of Eq. (8.3.27). So what of the closed loops? Using the above result and closing the loop we can integrate over the loop momentum. Provided that we can do a shift of the loop momentum ℓ such that $\ell \to \ell + q$, the two terms then cancel as a contribution to $k_\mu M^\mu$. This then leads to the Ward-Takahashi identity in Eq. (8.3.27). Details can be found in Bjorken and Drell (1965), Peskin and Schroeder (1995) and elsewhere. There are two subtleties: (i) The two external fermion lines are dressed by all of the self-energy contribution not shown here and these give rise to the physical mass m on these external lines, but this does not affect our argument about the cancellation of the internal contributions; and (ii) acceptable regularizations cannot violate our ability to shift loop momenta or this will violate our proof that loops do not contribute. So it is not sufficient to simply impose a naive momentum cutoff as a form of regulator.

Consider the sum of all Feynman diagrams for a QED amplitude at any given order of perturbation theory, $\mathcal{O}(\alpha^n)$. Now add one additional internal photon line to give a diagram of order $\mathcal{O}(\alpha^{n+1})$. Fix one end of the new photon line somewhere in the diagram. The other end of the photon line must then be attached at all possible locations and the sum of these diagrams must appear in the α^{n+1} amplitude. Similarly, external photons must couple to all possible points in a diagram. So the above arguments then tell us that *the $k^\mu k^\nu / k^2$ part of the propagator will give a vanishing contribution when we sum over diagrams and so we can neglect it.* This is why the choice of gauge for the internal photon lines is irrelevant and we can choose any gauge, such as Feynman gauge where $D_0^{\mu\nu} = -ig^{\mu\nu}/k^2 + i\epsilon$. Note that if we couple a massive vector (Proca) boson to fermions with no self-coupling of the vector bosons, then current conservation will lead to the same conclusion for the $k^\mu k^\nu / m^2$ contribution to the vector boson propagator and we can neglect it.

Simple argument for special case of QED Ward identities: It is common to find a brief argument leading to a weaker form of the Ward identities. The argument considers the replacement $\epsilon^\mu(k, \lambda) \to \tilde{\epsilon}^\mu(k, \lambda) = \epsilon^\mu(k, \lambda) + c(k)k^\mu$, where $c(k)$ is some real function. Assuming an on-shell photon, $k^2 = 0$, we see that the orthonormality conditions in Eq. (6.4.54) remain valid for $\tilde{\epsilon}^\mu$ but it cannot be written as a linear superposition of $\epsilon^\mu(k, 1)$ and $\epsilon^\mu(k, 2)$. It is sometimes argued that this replacement is allowed by Lorentz transformations that are in the Little Group ($\Lambda^\mu{}_\nu k^\nu = k^\mu$) and an explicit construction is given in Schwartz (2013). It is also sometimes argued (Aitchison and Hey, 2013) that the incomplete gauge fixing provided by the Lorenz gauge ($\partial_\mu A^\mu = 0$) allows residual gauge transformations that lead to this replacement. Consider a more general form of this replacement, $\epsilon^\mu(k, \lambda) \to \tilde{\epsilon}^\mu(k, \lambda) = \epsilon'^\mu(k, \lambda) + c(k)k^\mu$, where the $\epsilon'^\mu(k, \lambda)$ is any linear combination of $\epsilon^\mu(k, 1)$ and $\epsilon^\mu(k, 2)$. From Eq. (8.3.34) we have for an invariant amplitude with an external photon, $\mathcal{M} = \epsilon^\mu M^\mu|_{\text{on-shell}}$. This must be both Lorentz and gauge invariant in any sensible theory, which means that

$$\mathcal{M} = \epsilon^\mu M^\mu|_{\text{on-shell}} = [\epsilon'^\mu + c(k)k^\mu]M'^\mu|_{\text{on-shell}} = \epsilon'^\mu M'^\mu|_{\text{on-shell}} = \mathcal{M}'.$$

The above arguments suggest that in sensible theories $k_\mu M'^\mu|_{\text{on-shell}} = 0$, where on-shell includes $k^2 = 0$. This is a special case of the Ward identity in Eq. (8.3.35) proved in Sec. 8.3.2, which only required the fermions to be on-shell.

8.4 Renormalization

We have previously referred to the fact that loops that occur in Feynman diagrams are divergent and that a renormalization procedure is needed to make sense of them. A preliminary discussion of some aspects of renormalization was given at the end of Sec. 7.5.1. It is now time to attempt a more systematic treatment.

8.4.1 Superficial Degree of Divergence

A Feynman diagram with L independent loops will involve L loop-momentum integrals and so the diagram will contain $\int d^4k_1 \cdots d^4k_L$. In the case of a 1PI Feynman diagram we have $L \geq 1$ and every internal propagator is part of a loop. Consider an arbitrary 1PI Feynman diagram containing scalars, fermions, photons, Proca massive vector bosons and interactions that might contain derivatives. Let us define some useful quantities for such a diagram:

$E_\phi, E_f, E_\gamma, E_V \equiv$ number of external scalar, fermion, photon, vector boson lines

$I_\phi, I_f, I_\gamma, I_V \equiv$ number of internal scalar, fermion, photon, vector boson lines

$L \equiv$ number of independent loops; $V_i \equiv$ number of vertices of interaction type i

$n_i \equiv$ number of derivatives in a vertex of interaction type i

$k \in \{\phi, f, \gamma, V\}$ with c_k defined by $c_f = 1$, $c_\phi = c_\gamma = c_V = 2$

Φ_k is defined by $\Phi_\phi = \phi$, $\Phi_f = \psi$, $\Phi_\gamma = A^\mu$, $\Phi_V = V^\mu$

n_{ik} is the number of fields of type k in the vertex of interaction type i.   (8.4.1)

Scalar boson propagators come with $1/(k^2 - m_\phi^2)$, fermions with $1/(\slashed{k} - m_f)$, photons with $1/k^2$, vector bosons with $1/(k^2 - m_V^2)$ giving the respective dimensions c_k and each derivative gives some momentum k. Consider a 1PI Feynman diagram in momentum space after removing the $(2\pi)^4\delta^4(\sum p_j)$ arising from translational invariance. In place of Eq. (8.2.22) we now have by the same arguments

$$L = \sum_k I_k - \sum_i V_i + 1.$$ (8.4.2)

The generalization of Eq. (8.2.25) is

$$2I_k + E_k = \sum_i V_i n_{ik},$$ (8.4.3)

since the right-hand side is the total number of connections between vertices and Φ_k and all internal Φ_k lines use two connections and all external Φ_k lines use one.

We know from the Feynman rules that a 1PI Feynman diagram, $i\Gamma_{\text{diag}}$, contributing to its $i\Gamma$ has an ultraviolet contribution to the loop integrals of the form

$$i\Gamma_{\text{diag}}(p_1, p_2, \ldots) \propto \int \frac{d^4k_1 \cdots d^4k_L\, k^{\sum_i V_i n_i}}{(k_\phi^2)^{I_\phi}(k_f)^{I_f}(k_\gamma^2)^{I_\gamma}(k_V^2)^{I_V}} \propto \int \frac{dk\, k^{4L-1}\, k^{\sum_i V_i n_i}}{k^{\sum_k I_k c_k}}, \qquad (8.4.4)$$

where the k, k_ϕ, k_f, k_γ, k_V momenta are sums and differences of at least one of the independent loop momenta $k_1, \ldots, k_L$ as well as the various fixed external momenta $p_1, p_2, \ldots$ and masses. There are $E_\phi + E_f + E_\gamma + E_V$ external momenta, $p_1, p_2, \ldots$, and they and masses become irrelevant in the ultraviolet limit when all internal momenta are simultaneously large. Differentiating the Feynman diagram contribution with respect to any of the p_i improves the convergence of the integral by decreasing the power of k in the numerator or increasing the power of k in the denominator. After a sufficient number of such derivatives the remaining integral is finite. So in a Taylor expansion of any Feynman diagram contribution $\Gamma_{\text{diag}}(p_1, p_2, \ldots)$ around $p_1, p_2, \ldots = 0$ only a finite number of the lowest-order terms will be divergent.

From Eq. (6.4.96) we see that the photon propagator goes as $1/k^2$ since we can always use Feynman gauge, $-ig^{\mu\nu}/k^2$. For the vector boson propagator we can use $-ig^{\mu\nu}/k^2$ since from Eq. (6.5.74) the propagator is $-i[g^{\mu\nu} - \frac{k^\mu k^\nu}{m^2}]/k^2$ and since the $k^\mu k^\nu/m^2$ term in the numerator does not contribute if the vector boson couples only to conserved currents that do not involve itself (Weinberg, 1995). *Abelian gauge theories coupling to conserved currents are renormalizable when we add explicit gauge-symmetry breaking masses to the gauge bosons for this reason*; e.g., the Proca Lagrangian is renormalizable. However, *in a nonabelian gauge theory if we add explicit masses for the gauge bosons, then the theory becomes nonrenormalizable*. This was why it was necessary to find a renormalizable theory that allows the generation of gauge-boson masses to successfully describe the weak interactions. This is what is achieved by the Higgs mechanism in the Standard Model.

Working in natural units the action is dimensionless and so all terms in the Lagrangian density have dimension $[\mathcal{L}] = \mathsf{M}^4$ since $[d^4x] = \mathsf{M}^{-4}$. From the free Lagrangian densities we know that the field dimensions are $[\Phi_k] = \mathsf{M}^{\Delta_k}$, where

$$\Delta_\phi = \Delta_\gamma = \Delta_V = 1 \quad \text{and} \quad \Delta_f = \frac{3}{2}, \qquad (8.4.5)$$

since $[\partial] = \mathsf{M}^{\Delta_\partial} = \mathsf{M}$. It will be useful to observe the relationship that

$$\Delta_k = 2 - \tfrac{1}{2}c_k. \qquad (8.4.6)$$

Let λ_i be the coupling constant for vertex of type i. Then this vertex has the form

$$\mathcal{L}_{\text{int},i} = -\lambda_i (\partial)^{n_i} \prod_k (\Phi_k)^{n_{ik}}, \qquad (8.4.7)$$

where since $[\mathcal{L}_{\text{int},i}] = \mathsf{M}^4$ the coupling constant dimension must be

$$[\lambda_i] = \mathsf{M}^{4-n_i-\sum_k n_{ik}\Delta_k} \equiv \mathsf{M}^{\delta_i} \quad \text{with} \quad \delta_i \equiv 4 - n_i - \sum_k n_{ik}\Delta_k, \qquad (8.4.8)$$

where δ_i is the mass dimension of the coupling constant λ_i. For ϕ^4 theory we have $\mathcal{L}_{\text{int}} = -\frac{\lambda}{4!}\phi^4$ and so $\frac{\lambda}{4!}$ is dimensionless since $\delta_i = 4 - n_i - \sum_k n_{ik}\Delta_k = 4 - 0 - 4 = 0$. Similarly for QED the charge $q_c = e \equiv |e|$ is dimensionless since $\mathcal{L}_{\text{int}} = -q_c\bar{\psi}\slashed{A}\psi$ and $\delta_i = 4 - n_i - \sum_k n_{ik}\Delta_k = 4 - 0 - [2(\frac{3}{2}) + 1] = 0$.

The *superficial degree of divergence* D is defined as the power of k in the numerator (including dk) less the power of k in the denominator of Eq. (8.4.4),

$$D \equiv 4L + \sum_i V_i n_i - \sum_k I_k c_k = \sum_k I_k (4 - c_k) + \sum_i V_i (n_i - 4) + 4 \qquad (8.4.9)$$
$$= 4 - \sum_k E_k \Delta_k - \sum_i V_i \delta_i,$$

where Eq. (8.4.2) was used to eliminate L, Eq. (8.4.3) was used to eliminate I_k and we have used Eqs. (8.4.6) and (8.4.8) to obtain the last line. Naively Eq. (8.4.4) might be as divergent as $\int dk \, k^{D-1}$. Introducing an ultraviolet (UV) cutoff Λ to regulate the integral, it then becomes $\int_0^\Lambda dk \, k^{D-1} \propto \Lambda^D$. Note that D is the mass dimension of this integral, which we write as $[\int_0^\Lambda dk \, k^{D-1}] = \mathsf{M}^D$. Then for $D \leq -1$ we expect the integral to have no divergence, for $D = 0$ we expect a $\ln \Lambda$ divergence, and for $D > 0$ we expect a Λ^D divergence. However, this is only a guide and 1PI diagrams may be more or less divergent than naively expected. For example, symmetries such as Ward identities might lead to divergence less than expected from D and diagrams with divergent subdiagrams (see below) might be more divergent than expected. One-loop diagrams have no subdiagrams and so will have a divergence no worse than predicted by D.

From the Feynman rules and the above arguments we know that ultraviolet behavior of a 1PI Feynman diagram has the form of coupling constants multiplied by the loop integrals up to dimensionless constants that we omit here,

$$i\Gamma_{\text{diag}} = \prod_i (\lambda_i)^{V_i} \int_0^\Lambda dk \, k^{D-1}, \qquad (8.4.10)$$

which means that it has dimensions

$$[i\Gamma_{\text{diag}}] = [\prod_i (\lambda_i)^{V_i}][\int_0^\Lambda dk \, k^{D-1}] = \mathsf{M}^{D + \sum_i V_i \delta_i} = \mathsf{M}^{4 - \sum_k E_k \Delta_k}. \qquad (8.4.11)$$

This is independent of the internal details of the Feynman diagram, since all diagrams $i\Gamma_{\text{diag}}$ contributing to $i\Gamma = \sum i\Gamma_{\text{diag}}$ must have the same dimension.

We can also arrive at the dimensionality of $i\Gamma$ and the result for D directly using dimensional analysis: The 1PI amplitude $i\Gamma$ has the dimensions of an amputated Greens's function in momentum space after factoring out an overall momentum-conserving δ-function. The Green's function has E_k external lines for field Φ_k with $\sum_k E_k$ the total number of external lines. The dimensionality of the connected Green's function in coordinate space is determined by the field operators and is $\sum_k E_k \Delta_k$. Each Fourier transform of an operator lowers its dimension by four due to $[d^4 x] = \mathsf{M}^{-4}$, which gives a dimensionality of $\sum_k E_k (\Delta_k - 4)$. Factoring out the δ-function with $[\delta^4(\sum_j p_j)] = \mathsf{M}^{-4}$ leads to the dimensionality $\mathsf{M}^{4 + \sum_k E_k (\Delta_k - 4)}$ for the momentum-space Green's function. Each external propagator has dimensionality $\mathsf{M}^{-c_k} = \mathsf{M}^{2\Delta_k - 4}$ and so after all amputations we have

$$[i\Gamma] = \mathsf{M}^{4 + \sum_k E_k (\Delta_k - 4) - \sum_k E_k (2\Delta_k - 4)} = \mathsf{M}^{4 - \sum_k E_k \Delta_k}, \qquad (8.4.12)$$

which has reproduced the right-hand side of Eq. (8.4.11) since we must have $[i\Gamma_{\text{diag}}] = [i\Gamma]$. Recall that $[i\Gamma_{\text{diag}}] = [\prod_i (\lambda_i)^{V_i}][\int_0^\Lambda dk \, k^{D-1}] = \mathsf{M}^{D + \sum_i V_i \delta_i}$ in Eq. (8.4.11) on dimensional grounds. So we again find $D = 4 - \sum_k E_k \Delta_k - \sum_i V_i \delta_i$ using only dimensional arguments.

From Eq. (8.4.5) we see that all fields considered have a positive mass dimension, $\Delta_k > 0$. If we also have a nonnegative coupling constant mass dimension, $\delta_i \geq 0$, then $D \leq 4 - \sum_k E_k \Delta_k$ and so all superficially divergent 1PI diagrams ($D \geq 0$) will have an upper limit on the number of external legs, $\sum_k E_k \Delta_k \leq 4$.

It is useful to make the following definitions of classes of theories:

(i) *super-renormalizable theory* (all $\delta_i > 0$): This contains only a finite number of superficially divergent diagrams, since both the number of external legs (the E_k) and the number of vertices (the V_i) will have upper limits.

(ii) *renormalizable theory* (all $\delta_i \geq 0$): For those interactions with $\delta_i = 0$ there is no limit on the number of vertices in diagrams that are superficially divergent, but such diagrams will have an upper limit on the number of external legs.

(iii) *non-renormalizable theory* (one or more $\delta_i < 0$): In this case, adding vertices (increasing V_i) of the type with $\delta_i < 0$ decreases D and so at high enough order in the associated λ_i all diagrams will be superficially divergent.

Similarly, we can define classes of interactions, $\mathcal{L}_{\text{int},i}$:

(i) *super-renormalizable or relevant interaction* ($\delta_i > 0$)
(ii) *renormalizable or marginal interaction* ($\delta_i = 0$)
(iii) *nonrenormalizable or irrelevant interaction* ($\delta_i < 0$).

8.4.2 Superficial Divergences in QED

From Eq. (8.4.2) we find $L = I_f + I_\gamma - V + 1$. Using Eq. (8.4.3) the number of vertices can be expressed as

$$V = E_\gamma + 2I_\gamma = \tfrac{1}{2}E_f + I_f \tag{8.4.13}$$

and from Eq. (8.4.9) the superficial degree of divergence is

$$D = 4 - \tfrac{3}{2}E_f - E_\gamma. \tag{8.4.14}$$

From Eq. (8.4.8) we know that the charge $q = e \equiv |e|$ is dimensionless, $\delta = 0$, which means that by definition QED is a renormalizable theory and its one interaction is a marginal interaction. The only 1PI amplitudes that are superficially divergent in QED have $D = 4 - \tfrac{3}{2}E_f - E_\gamma > 0$.

These 1PI or proper amplitudes are the following:

(a) the sum of 1PI vacuum subdiagrams with $D = 4$, not relevant for scattering
(b) the one-photon proper tadpole with $D = 4 - 1 = 3$, which vanishes
(c) the electron self-energy $-i\Sigma(p)$ of Eq. (7.5.95) with $D = 4 - 3 = 1$
(d) the photon self-energy or vacuum polarization $i\Pi^{\mu\nu}(q)$ with $D = 4 - 2 = 2$
(e) the electron-photon proper vertex $\Gamma^\mu(p',p)$ with $D = 4 - 3 - 1 = 0$
(f) the three-photon proper vertex with $D = 4 - 3 = 1$, which vanishes
(g) the four-photon proper vertex with $D = 4 - 4 = 0$.

Graphically, these 1PI amplitudes are represented as:

Note that the proper tadpole (b) must vanish since the photon does not have the quantum numbers of the vacuum, $J^{PC} = 1^{--} \neq 0^{++}$, and so $\langle\Omega|\hat{A}^\mu(x)|\Omega\rangle = 0$. Recall from Eq. (8.1.64) that the photon field changes sign under charge conjugation, $\hat{\mathcal{C}}\hat{A}^\mu(x)\hat{\mathcal{C}}^{-1} = -\hat{A}^\mu(x)$, and the vacuum is invariant under it, $\hat{\mathcal{C}}|\Omega\rangle = |\Omega\rangle$. Then any Green's function with only an odd number of photon operators must vanish, since

$$\langle\Omega|T\hat{A}^{\mu_1}\cdots\hat{A}^{\mu_{2n+1}}|\Omega\rangle = \langle\Omega|\hat{\mathcal{C}}^{-1}\hat{\mathcal{C}}T\hat{A}^{\mu_1}\cdots\hat{A}^{\mu_{2n+1}}\hat{\mathcal{C}}^{-1}\hat{\mathcal{C}}|\Omega\rangle \qquad (8.4.15)$$
$$= \langle\Omega|T\hat{\mathcal{C}}\hat{A}^{\mu_1}\cdots\hat{A}^{\mu_{2n+1}}\hat{\mathcal{C}}^{-1}|\Omega\rangle = (-1)^{2n+1}\langle\Omega|T\hat{A}^{\mu_1}\cdots\hat{A}^{\mu_{2n+1}}|\Omega\rangle = 0.$$

The result that all amplitudes with only an odd number of external photon lines must vanish is known as *Furry's theorem*. For this reason the three-photon proper vertex in (f) must vanish. There is a closely related result, which is that *any QED Feynman diagram containing a fermion loop with an odd number of vertices will not contribute to any amplitude and can be neglected.*

Proof: From Eq. (4.5.33) we know that $C^{-1}\gamma^\mu C = -\gamma^{\mu T}$ and so it follows that $C^{-1}S_F(p)C = S_F(-p)^T$ and $C^{-1}S_F(x_1,x_2)C = S_F(x_2,x_1)^T$. Working in coordinate space for clarity, the Feynman rules tell us that any such loop with orientation $x_1, x_2, \ldots, x_n, x_1$ will contain a trace of the form

$$\text{tr}[\gamma^{\mu_1}S_F(x_1,x_n)\gamma^{\mu_n}S_F(x_n,x_{n-1})\cdots\gamma^{\mu_2}S_F(x_2,x_1)]$$
$$= \text{tr}[C^{-1}\gamma^{\mu_1}S_F(x_1,x_n)\gamma^{\mu_n}S_F(x_n,x_{n-1})\cdots\gamma^{\mu_2}S_F(x_2,x_1)C]$$
$$= (-1)^n\,\text{tr}[\gamma^{\mu_1 T}S_F(x_n,x_1)^T\gamma^{\mu_n T}S_F(x_{n-1},x_n)^T\cdots\gamma^{\mu_2 T}S_F(x_1,x_2)^T]$$
$$= (-1)^n\,\text{tr}[\gamma^{\mu_1}S_F(x_1,x_2)\gamma^{\mu_2}S_F(x_2,x_3)\cdots\gamma^{\mu_n}S_F(x_n,x_1)],$$

which is $(-1)^n$ times the opposite orientation of the fermion loop, i.e., the loop direction $x_1, x_n, x_{n-1}, \ldots, x_2, x_1$. The two loop orientations are topologically distinct and so both must occur in any amplitude. If n is odd then they will cancel. If n is even they are identical (Itzykson and Zuber, 1980).

Next consider the four-photon proper vertex in diagram (g). This must have the form $M^{\mu_1\mu_2\mu_3\mu_4}(k_1, k_2, k_3, k_4)$ and is a function of the external momenta $k_1^\mu, \ldots, k_4^\mu$. There are only three independent external momenta after factoring out the momentum-conserving delta function as $(2\pi)^4\delta^4(\sum_{j=1}^4 k_j)$. The lowest-order Feynman diagram contributing to this 1PI amplitude is the fermion box diagram,

$\equiv M(1234); \quad M^{\mu_1\mu_2\mu_3\mu_4} = 2[M(1234) + M(1324) + M(1432)], \qquad (8.4.16)$

where for a single loop orientation there are the three distinct ways of attaching the photons to the loop and these are added because the photon is a boson. The other orientation is included by doubling this contribution since the number of legs is even as discussed above. The 1PI amplitude $M^{\mu_1\mu_2\mu_3\mu_4}(k_1, k_2, k_3, k_4)$ is invariant under the pairwise interchange of any two photons as it should be. From the Ward identity in Eq. (8.3.35) we must have $k_1^{\mu_1}M_{\mu_1\mu_2\mu_3\mu_4} = \cdots = k_4^{\mu_4}M_{\mu_1\mu_2\mu_3\mu_4} = 0$. The only available tensors are $g^{\mu\nu}$ and $k_1^{\mu_1}, \ldots, k_4^{\mu_4}$. The only quantities that vanish in a scalar product with $k_{1\mu_1}$ have the form $(g^{\mu_1\mu_2}k_1^{\mu_3} - g^{\mu_1\mu_3}k_1^{\mu_2})$, $(g^{\mu_1\mu_2}k_1^{\mu_4} - g^{\mu_1\mu_4}k_1^{\mu_2})$ and $(g^{\mu_1\mu_3}k_1^{\mu_4} - g^{\mu_1\mu_4}k_1^{\mu_3})$ and any linear combination of these. Similar statements apply for $k_2^{\mu_2}$, $k_3^{\mu_3}$ and $k_4^{\mu_4}$. We need one such factor for each of the four vertices constructed to have the correct symmetry. The result is that $M^{\mu_1\mu_2\mu_3\mu_4} = C[\mathcal{O}(k^4)]^{\mu_1\mu_2\mu_3\mu_4}$ where C is a function of Lorentz invariants and where $[C] = \mathsf{M}^{D-4} = \mathsf{M}^{-4}$. So current conservation has led to the Ward identity and this has given an *effective* superficial degree of divergence $D_{\mathrm{eff}} = -4$ for the loop integrals in C. The four-photon amplitude is finite up to issues that might arise from divergent subdiagrams.

Only the superficially divergent diagrams (c), (d) and (e) remain and these are referred to as *primitively divergent* diagrams. The renormalization program for QED consists of showing that all divergences can be absorbed into four bare constants that depend on the regularization. As the regularization is removed all divergences are absorbed into these bare constants, leaving all physical observables unchanged.

Divergent subdiagrams: The superficial degree of divergence D was calculated based on the contribution to a Feynman diagram when all loop momenta go to infinity together. A *subdiagram* is any part of a Feynman diagram. A *divergent subdiagram* is a subdiagram that diverges when only loop momenta in the subdiagram go to infinity. For example, some Feynman diagrams contributing to Compton scattering, $\gamma e^- \to \gamma e^-$, are shown in Eq. (8.4.17). Diagram (a) has no divergent subdiagrams and has convergent loop momenta, whereas the subdiagrams shown in dashed boxes in diagrams (b) and (c) are divergent. We see that they diverge when the internal loop momentum in the dashed box goes to infinity:

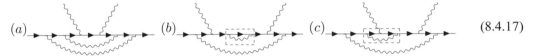

$$(8.4.17)$$

An important theorem by Weinberg and refined by others (Weinberg, 1960; Hahn and Zimmermann, 1968; Zimmermann, 1968) states that *a Feynman diagram integral converges only if the superficial divergence D of the whole diagram as well as of every subdiagram is negative.* The dashed boxes in Eq. (8.4.17) are examples of defining a subdiagram. If all momenta flowing into or out of a box are held fixed and if the subdiagram inside has nonnegative D, then it is a divergent subdiagram.

We will later see that we can introduce wavefunction and mass renormalization as well coupling constant renormalization to deal with the three primitively divergent diagrams either in the form of multiplicative renormalization constants or as counterterms. We can understand that these counterterms cancel the appearance of divergent subdiagrams that have the form of the three primitively divergent diagrams at the one-loop level. The one-loop form of these is

$$(8.4.18)$$

Overlapping divergences: Difficulties with this simple picture of the removal of divergences occur when one goes beyond one-loop divergent subdiagrams and has *overlapping* or *nested* divergences. This can occur when we add additional internal photon lines to the one-loop electron and photon self-energy diagrams:

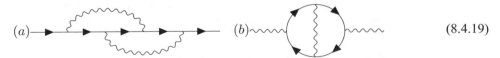

$$(8.4.19)$$

The central fermion propagator in diagram (a) and the vertical photon propagator in diagram (b) have two loop momenta running through them. In both examples the two divergent integrals are not independent. We need to show that the renormalization of fields, masses and couplings is sufficient to remove overlapping divergences.

BPHZ (Bogliubov-Parasiuk-Hepp-Zimmermann) theorem: *All divergences in a renormalizable theory are removed by the renormalization of the primitively divergent diagrams through the renormalization of fields, masses and coupling constants.*

There is an additional assumption in this theorem, which is that for all infinities to be removed when renormalizing the theory one typically must include all renormalizable interactions consistent with the symmetries of the theory. For example, in a Yukawa theory with interaction $g\phi\bar{\psi}\psi$ a fermion loop with four ϕ external lines will lead to a logarithmic divergence and so a ϕ^4 interaction must be included so that the associated renormalization of the coupling in the $(\lambda/4!)\phi^4$ interaction can absorb this divergence (Weinberg, 1995). The BPHZ theorem is the result of a systematic procedure for removing the ultraviolet divergences. The technical analysis was developed by Bogliubov and Parasiuk (1957) and completed by Hepp (1966). Subsequently, this work was refined in Zimmermann (1969) and also by Zimmermen in his lectures in Deser et al. (1970). The extension to the case of massless particles was developed in Lowenstein and Zimmermann (1975) and Lowenstein (1976). This procedure removed all superficial divergences from the full diagram as well as all of its subdiagrams and was shown to be equivalent to a renormalization of the fields, coupling constants and masses appearing in the Lagrangian of the renormalizable theory. Using the above theorem by Weinberg it then follows that any Feynman diagram integral converges and so the theory is finite. In other words, a renormalizable quantum field theory as defined in Sec. 8.4.1 is rendered finite at all orders of perturbation theory by renormalizing the fields, masses and coupling constants.

The proof of the BPHZ theorem is long and technical and will not be given here. The general proof involves the concept of *forests* (not to be confused with forest diagrams defined in Sec. 5.2.1). Consider some Feynman diagram and imagine drawing one box around any divergent 1PI subdiagram in every possible way, where the whole diagram may be one such divergent 1PI subdiagram. Next, add an additional box in every possible way. Then keep adding additional boxes in every possible way until no more can be added. Each of the above boxed diagrams is referred to as a *forest* and includes the case where zero boxes are drawn. A forest may be (i) disjoint, where no boxes have overlapping areas; (ii) nested, where one box lies entirely inside another box; or (iii) overlapping, where two boxes partially overlap. A complicated forest may have combinations of the above. This definition of forests allowed Zimmerman to prove the recursive forest formula and obtain the renormalized value of a Feynman diagram (Zimmermann, 1969; Deser et al., 1970). The BPHZ theorem uses a Taylor expansion of the Feynman diagram in terms of the external momenta around zero. A detailed treatment of the Weinberg and BPHZ theorems in the case of QED is given in Bjorken and Drell (1965). A pedagogical discussion for a ϕ^3 theory can be found in Das (2008) and an illustration for a ϕ^4 theory at the two-loop level is given in Peskin and Schroeder (1995). For a detailed discussion of this and other issues regarding renormalization, see Collins (1984).

8.5 Renormalized QED

In the following we will discuss quantum electrodynamics in d spacetime dimensions with N_f flavors of fermion. For notational convenience we will no longer use the subscript F to denote a full Feynman propagator and so write $D^{\mu\nu}(k)$ for the full photon propagator and $S^f(p)$ for the full fermion propagator of flavor f. Aspects of the following discussion can also be found in Bjorken and Drell (1965) and Itzykson and Zuber (1980). Let us denote the *renormalized* action in R_ξ gauge as

$$
\begin{aligned}
\tilde{S}[\bar\psi, \psi, A_\mu] &= \int d^d x \left[\sum_{f=1}^{N_f} \bar\psi_0^f \left(i\slashed\partial - m_0^f - q_0^f \slashed A_0 \right) \psi_0^f - \tfrac{1}{4} F_{0\mu\nu} F_0^{\mu\nu} - \tfrac{1}{2\xi_0} (\partial_\mu A_0^\mu)^2 \right] \\
&= \int d^d x \left[\sum_f \left\{ Z_2^f \bar\psi^f (i\slashed\partial - m_0^f)\psi^f - Z_2^f \sqrt{Z_3} q_0^f \bar\psi^f \slashed A \psi^f \right\} - Z_3 \tfrac{1}{4} F_{\mu\nu} F^{\mu\nu} - Z_3 \tfrac{1}{2\xi_0} (\partial_\mu A^\mu)^2 \right] \\
&= \int d^d x \left[\sum_f \left\{ Z_2^f \bar\psi^f (i\slashed\partial - m_0^f)\psi^f - Z_1^f q^f \bar\psi^f \slashed A \psi^f \right\} - Z_3 \tfrac{1}{4} F_{\mu\nu} F^{\mu\nu} - \tfrac{1}{2\xi} (\partial_\mu A^\mu)^2 \right] \\
&= \int d^d x \left[\sum_f \bar\psi^f \left(i\slashed\partial - m^f - q^f \slashed A \right) \psi^f - \tfrac{1}{4} F_{\mu\nu} F^{\mu\nu} - \tfrac{1}{2\xi} (\partial_\mu A^\mu)^2 \right. \\
&\quad \left. + \sum_f \left\{ (Z_2^f - 1)\bar\psi^f i\slashed\partial \psi^f - (Z_2^f m_0^f - m^f)\bar\psi^f \psi^f - (Z_1^f - 1)q^f \bar\psi^f \slashed A \psi^f \right\} - (Z_3 - 1)\tfrac{1}{4} F_{\mu\nu} F^{\mu\nu} \right] \\
&= \int d^d x \left[\sum_f \bar\psi^f \left(i\slashed\partial - m^f - q^f \slashed A \right) \psi^f - \tfrac{1}{4} F_{\mu\nu} F^{\mu\nu} - \tfrac{1}{2\xi} (\partial_\mu A^\mu)^2 \right. \\
&\quad \left. + \sum_f \left\{ \bar\psi^f \left(i\delta_2^f \slashed\partial - \delta_m^f \right) \psi^f - \delta_1^f q^f \bar\psi^f \slashed A \psi^f \right\} - \delta_3 \tfrac{1}{4} F_{\mu\nu} F^{\mu\nu} \right],
\end{aligned} \tag{8.5.1}
$$

where the bare fields, masses, charges and gauge parameter are $\psi_0^f, A_0^\mu, m_0^f, q_0^f$ and ξ_0, respectively, and their renormalized versions are ψ^f, A^μ, m^f, q^f and ξ. We have the multiplicative renormalization constants: (i) Z_2^f is the wavefunction renormalization constant for flavor f; (ii) Z_3 is the wavefunction renormalization constant for the photon; and (iii) Z_1^f is the charge renormalization constant for flavor f. Comparing the first, second and third lines of Eq. (8.5.6) we see that we have defined

$$
\psi_0^f \equiv \sqrt{Z_2^f}\, \psi^f, \quad A_0^\mu \equiv \sqrt{Z_3}\, A^\mu, \quad q_0^f \equiv \frac{Z_1^f}{Z_2^f \sqrt{Z_3}} q^f, \quad \xi_0 \equiv Z_3 \xi. \tag{8.5.2}
$$

Comparing the last two equalities we see that we have the counterterm coefficients

$$
Z_2^f \equiv 1 + \delta_2^f, \quad Z_3 \equiv 1 + \delta_3, \quad Z_2^f m_0^f \equiv m^f + \delta m^f, \quad Z_1^f \equiv 1 + \delta_1^f. \tag{8.5.3}
$$

We adopt notation consistent with Peskin and Schroeder (1995) here and note that Schwartz (2013) uses a different definition of δ_m^f, which we denote as $\delta_m'^f$. In renormalized perturbation theory the relation between them is defined as $\delta_m^f \equiv (\delta_m'^f + \delta_2^f) m^f$. We can also define the mass renormalization constant Z_m^f in terms of the above quantities,

$$
m_0^f \equiv Z_m^f m^f \quad \Rightarrow \quad Z_m^f = [1 + (\delta_m^f / m^f)]/Z_2^f = 1 + \delta_m'^f + \mathcal{O}[(\delta_2^f)^2]. \tag{8.5.4}
$$

Note that when we do renormalized perturbation theory if we work to lowest nontrivial order then we can neglect $\mathcal{O}(\delta^2)$ terms and simply use $Z_m^f = 1 + \delta_m'^f$.

Note that the second to last line of Eq. (8.5.1) contains only renormalized quantities. The terms in the last line are referred to as the *counterterms*, which are adjusted to absorb divergences as regulators are removed and to leave physical observables unchanged. Note that $\xi_0 = Z_3 \xi$ is the statement that there is no gauge-parameter counterterm, which follows because only the transverse part of the photon propagator is dressed by interactions according to the Ward identity in Eq. (7.5.116), $q_\mu \Pi^{\mu\nu} = 0$. Also we will later show that

$$
Z_1^f = Z_2^f \quad \Rightarrow \quad q^f = \sqrt{Z_3}\, q_0^f, \tag{8.5.5}
$$

which shows that the only change to bare charges is due to vacuum polarization of the photon. So all flavor charges are renormalized by the same factor, $\sqrt{Z_3}$, which is why the charges of the electron, muon and tauon can all remain the same in the presence of renormalization. In this sense fermion *charge universality* is expected in theories where Z_1^f/Z_2^f is independent of flavor f. Sometimes in the literature $m_B^f \equiv Z_2^f m_0^f$ and $q_B^f \equiv Z_1^f q^f$ are referred to as the bare mass and bare charge (Collins, 1984).

We continue to defer a discussion of the details of regularization until Sec. 8.6 so that we do not become overwhelmed with detail at this point. It will be sufficient here to let Λ denote the regularization parameter that controls ultraviolet divergences, where $\Lambda \to \infty$ corresponds to the removal of the regularization. As mentioned earlier we require regularization to allow shifts of loop momenta inside loop integrals, so that Λ cannot be a simple ultraviolet cutoff. We can also introduce a renormalization scale μ and renormalization conditions at that scale, where the renormalized masses and charges can be specified at that scale with those renormalization conditions. So we write $m^f(\mu)$ and $q^f(\mu)$ since in general masses and charges change with the renormalization scale. For example, the fine structure constant at low energies is $\alpha(\mu \approx 0) \sim 1/137.04$, while at a scale of the Z boson mass it decreases to $\alpha(\mu \approx 90\,\mathrm{GeV}) \sim 1/127.5$. In a renormalizable theory we can hold the $m^f(\mu)$ and $q^f(\mu)$ fixed while taking $\Lambda \to \infty$, with all Λ-dependence absorbed into the renormalization constants. After taking this limit all physical properties of the theory are determined and independent of the regulator Λ. So we have the dependencies $Z_1^f(\mu,\Lambda)$, $Z_2^f(\mu,\Lambda)$, $Z_3(\mu,\Lambda)$, $m_0(\Lambda)$, $q_0^f(\Lambda)$ and $\xi_0(\Lambda)$ all varying with Λ such that $m^f(\mu)$, $q^f(\mu)$ and $\xi(\mu)$ remain constant as $\Lambda \to \infty$. In QED it is common to use the *on-shell renormalization scheme*, where masses and coupling are defined with all particles on their mass shell. We will discuss this later in this section. With this choice m^f is the physical fermion mass, which is also referred to as the *pole mass*, since it is the location of the pole in the renormalized fermion propagator.

8.5.1 QED Schwinger-Dyson Equations with Bare Fields

In order to avoid cumbersome notation in this section we will *temporarily* drop the subscript "0" that denotes bare fields and write $\bar\psi_0^f, \psi_0^f, A_0^\mu$ as $\bar\psi^f, \psi^f, A^\mu$. This was also done in our discussion of the LSZ formalism, where, for example, $S(p)$ in Eq. (7.5.95) and $D^{\mu\nu}(q)$ in Eq. (7.5.114) were the *bare-field propagators*. As always $S_0(p)$ and $D_0^{\mu\nu}$ are the perturbative bare propagators. With this understanding it follows that the generating functional for the *bare-field Green's functions* is

$$Z[\bar\eta, \eta, j] = e^{G_c[\bar\eta, \eta, j]} = \int \mathcal{D}A^\mu \mathcal{D}\bar\psi \mathcal{D}\psi \, e^{i\tilde S[\bar\psi, \psi, A_\mu] + i\int d^dx \, [\sum_f (\bar\psi^f \eta^f + \bar\eta^f \psi^f) + j\cdot A]},$$

$$S[\bar\psi, \psi, A] = \int d^dx \left[\sum_{f=1}^{N_f} \bar\psi^f (i\slashed\partial - m_0^f - q_0^f \slashed A)\psi^f - \tfrac14 F_{\mu\nu}F^{\mu\nu} - \tfrac{1}{2\xi_0}(\partial_\mu A^\mu)^2 \right], \qquad (8.5.6)$$

where j^μ, $\bar\eta^f$ and η^f are arbitrary source fields for the photon, fermions and antifermions, respectively, and where from Eq. (8.2.1) we recognize that $G_c[j, \bar\eta, \eta]$ is the generator of connected Green's functions. We have defined Z so that $Z[0,0,0] = 1$ and $G_c[0,0,0] = 0$. Note that if j^μ were to be interpreted as an external current we would have $-j \cdot A$ but since j is an arbitrary source for simplicity we use $+j \cdot A$. It is straightforward to reintroduce the renormalized fields once the derivations are complete by introducing the field renormalizations Z_2 and Z_3. Since we assume an appropriate regularization scheme with Λ kept finite in these derivations, then all quantities are finite. Once we have restored renormalization constants we can then safely take $\Lambda \to \infty$.

As seen in Sec. 8.3.1 in Eqs. (8.3.1) and (8.3.2) Schwinger-Dyson equations follow because the functional integral of a total functional derivative is zero. So we find

$$0 = \int \mathcal{D}A^\mu \mathcal{D}\bar{\psi}\mathcal{D}\psi \, \frac{\delta}{\delta A_\mu(x)} e^{iS[\bar{\psi},\psi,A_\mu]+i\int d^dx \, [\sum_f(\bar{\psi}^f\eta^f+\bar{\eta}^f\psi^f)+j\cdot A]}$$

$$= \int \mathcal{D}A^\mu \mathcal{D}\bar{\psi}\mathcal{D}\psi \left\{ \frac{\delta S}{\delta A_\mu(x)} + j^\mu(x) \right\} e^{iS[\bar{\psi},\psi,A_\mu]+i\int d^dx \, [\sum_f(\bar{\psi}^f\eta^f+\bar{\eta}^f\psi^f)+j\cdot A]}$$

$$= \left\{ \frac{\delta S}{\delta A_\mu(x)} \left[\frac{i\delta}{\delta\eta}, \frac{-i\delta}{\delta\bar{\eta}}, \frac{-i\delta}{\delta j} \right] + j^\mu(x) \right\} Z[\bar{\eta},\eta,j] \, . \tag{8.5.7}$$

The derivative in the second line of Eq. (8.5.6) gives

$$\frac{\delta S}{\delta A_\mu(x)} = \left[\Box g_{\mu\nu} - (1-\tfrac{1}{\xi_0})\partial_\mu\partial_\nu \right] A^\nu - \sum_f q_0^f \, \bar{\psi}^f \gamma_\mu \psi^f \, . \tag{8.5.8}$$

Note that (a) $\delta G_c/\delta X = \delta \ln Z/\delta X = (1/Z)(\delta Z/\delta X)$ from Eq. (8.2.1). This leads to (b) $\delta^2 G_c/\delta X \delta Y = \delta^2 \ln Z/\delta X \delta Y = -(\delta G_c/\delta X)(\delta G_c/\delta Y) + (1/Z)(\delta^2 Z/\delta X \delta Y)$. Rearranging then gives (c) $(1/Z)(\delta^2 Z/\delta X \delta Y) = (\delta G_c/\delta X)(\delta G_c/\delta Y) + \delta^2 G_c/\delta X \delta Y$. Using Eq. (8.5.8) in Eq. (8.5.7), dividing it by Z and using these results we find

$$-j_\mu(x) = \left[\Box g_{\mu\nu} - \left(1-\tfrac{1}{\xi_0}\right)\partial_\mu\partial_\nu \right] \frac{\delta G_c}{i\delta j_\nu(x)} \tag{8.5.9}$$

$$- \sum_f q_0^f \left(\frac{\delta G_c}{\delta\eta^f(x)} \gamma_\mu \frac{\delta G_c}{\delta\bar{\eta}^f(x)} + \frac{\delta}{\delta\eta^f(x)} \left[\gamma_\mu \frac{\delta G_c}{\delta\bar{\eta}^f(x)} \right] \right) \, .$$

Here we used (a) to effectively replace A^ν in Eq. (8.5.8) with $\delta G_x/i\delta j_\nu$ in Eq. (8.5.9) and we used (c) to replace $\bar{\psi}^f \gamma_\mu \psi^f$ in Eq. (8.5.8) with the term $(\cdots)$ in the second line of Eq. (8.5.9). This equation can be thought of as a compact form of the quantum field theory version of Maxwell's equations. Performing a Legendre transformation gives the generating functional for the 1PI Green's functions as found in Sec. 8.2.2,

$$i\Gamma[\bar{\psi},\psi,A] = G_c[\bar{\eta},\eta,j] - i\int d^dx \left[\bar{\psi}^f\eta^f + \bar{\eta}^f\psi^f + j_\mu A^\mu \right] \, . \tag{8.5.10}$$

It would be better to add a subscript "c" on the classical fields as we did in Sec. 8.2.1 and write them as $\bar{\psi}_c^f$, ψ_c^f and A_c^μ to avoid confusion with fields in the functional integral, but for brevity we leave this to be understood. From Eq. (8.5.10) we have

$$A_\mu(x) = \frac{\delta G_c}{i\delta j^\mu(x)}, \qquad \psi^f(x) = \frac{\delta G_c}{i\delta\bar{\eta}^f(x)}, \qquad \bar{\psi}^f(x) = -\frac{\delta G_c}{i\delta\eta^f(x)}, \tag{8.5.11}$$

$$j_\mu(x) = -\frac{\delta\Gamma}{\delta A^\mu(x)}, \qquad \eta^f(x) = -\frac{\delta\Gamma}{\delta\bar{\psi}^f(x)}, \qquad \bar{\eta}^f(x) = \frac{\delta\Gamma}{\delta\psi^f(x)}. \tag{8.5.12}$$

From Eq. (8.5.11) we see that we now have expressions for $\bar{\psi}$, ψ and A_μ in terms of $\bar{\eta}$, η and j_μ and vice versa, i.e., $\bar{\psi}_\alpha^f(x) = \bar{\psi}_\alpha^f[\bar{\eta},\eta,j_\mu](x) = i\delta G_c[\bar{\eta},\eta,j_\mu]/\delta\eta_\alpha^f(x)$ with the spinor index α now explicitly shown. These "classical fields" A^μ, $\bar{\psi}^f$ and ψ^f vanish when the sources $j^\mu = \bar{\eta}^f = \eta^f = 0$. Also the sources vanish when the classical fields are set to zero. So first derivatives of G_c in Eq. (8.5.11) vanish when sources are set to zero and first derivatives of Γ in Eq. (8.5.12) vanish when the classical fields are set to zero. Using Eq. (8.5.11) and $\delta\eta_\beta^g(x)/\delta\eta_\alpha^f(y) = \delta_{\alpha\beta}\delta_{fg}\delta^d(x-y)$ we find in analogy with Eq. (8.2.15),

$$i\int d^dz \sum_h \frac{\delta^2 G_c}{\delta\eta_\alpha^f(x)\bar{\eta}_\gamma^h(z)} \frac{\delta^2\Gamma}{\delta\psi_\gamma^h(z)\bar{\psi}_\beta^g(y)} \bigg|_{\substack{\eta=\bar{\eta}=0 \\ \psi=\bar{\psi}=0}} = \delta_{\alpha\beta}\delta_{fg}\delta^d(x-y), \tag{8.5.13}$$

where f, g, h are flavor labels. When the fermion sources $(\bar{\eta}, \eta)$ vanish the first derivatives of G_c in Eq. (8.5.9) vanish and so it can be written as

$$
\frac{\delta\Gamma}{\delta A^\mu(x)}\bigg|_{\psi=\bar{\psi}=0} = \left[\Box g_{\mu\nu} - (1 - \tfrac{1}{\xi_0})\partial_\mu\partial_\nu\right]A^\nu(x) + \sum_f q_0^f \mathrm{tr}\left[\gamma_\mu S^f(x, x, [A_\mu])\right],
$$

$$
S^f(x, y, [A]) \equiv \frac{\delta^2 G_c}{\delta\bar{\eta}^f(x)\delta\eta^f(y)}\bigg|_{\bar{\eta}=\eta=0} = \left(\frac{\delta^2 i\Gamma}{\delta\bar{\psi}^f(x)\delta\psi^f(y)}\bigg|_{\psi=\bar{\psi}=0}\right)^{-1}, \tag{8.5.14}
$$

where we have made the identification based on Eq. (6.3.246) that $S^f(x, y, [A_\mu])$ is the propagator of the fermion of flavor f in an external electromagnetic field A_μ. Note that based on Eq. (8.5.14) we can equally well view this as a functional of the photon source field, $S^f(x, y, [A]) \equiv S^f(x, y, [j])$. This identification follows from Eq. (8.5.13) and the fact that the second derivative of G_c is the full two-point fermion Green's function, i.e., the fermion propagator in the presence of a background field, $S^f(x, y, [A_\mu])$. The full fermion Green's function in a vacuum is $S^f(x, y) = S^f(x, y, [A = 0])$.

To obtain the Schwinger-Dyson equation for the photon polarization tensor it remains only to act with $i\delta/\delta A_\nu(y)$ on Eq. (8.5.14) and set $J_\mu(x) = 0$. The proper photon-fermion vertex for flavor f is Γ_μ^f and is defined by

$$
-iq_0^f \Gamma_\mu^f(x; y, z) \equiv \frac{\delta}{\delta A^\mu(x)}\frac{\delta^2 i\Gamma}{\delta\bar{\psi}^f(y)\delta\psi^f(z)}\bigg|_{\substack{A=0\\\psi=\bar{\psi}=0}}, \tag{8.5.15}
$$

where at tree level we have $\Gamma_\mu^f(x; y, z) \to \gamma^\mu$. This follows since the tree-level contribution to the 1PI vertex is $-iq_0^f\gamma^\mu$ as we know from Eq. (7.6.22). In analogy with Eq. (8.2.15) we can show that the second derivative of $i\Gamma$ with respect to A gives the inverse photon propagator, $(D^{-1})^{\mu\nu}(x, y)$. Thus from Eq. (8.5.14) we obtain the Schwinger-Dyson equation for the inverse photon propagator

$$
D_{\mu\nu}^{-1}(x, y) = \frac{\delta^2 i\Gamma}{\delta A^\mu(x)\delta A^\nu(y)}\bigg|_{\substack{A=0\\\psi=\bar{\psi}=0}} = i\left[\Box g_{\mu\nu} - \left(1 - \frac{1}{\xi_0}\right)\partial_\mu\partial_\nu\right]\delta^d(x - y) - i\Pi_{\mu\nu}(x, y)
$$

$$
= D_{0\mu\nu}^{-1}(x - y) - i\Pi_{\mu\nu}(x, y), \tag{8.5.16}
$$

where we have defined the photon polarization tensor $\Pi_{\mu\nu}$ such that

$$
i\Pi_{\mu\nu}(x, y) = -i\sum_f q_0^f \mathrm{tr}\left[\gamma_\mu \frac{\delta S^f(x, x, [A_\mu])}{\delta A^\nu(y)}\right] \tag{8.5.17}
$$

$$
= (-1)\sum_f \int d^d z_1\, d^d z_2\, \mathrm{tr}\left[(-iq_0^f)\gamma_\mu S^f(x, z_1)(-iq_0^f)\Gamma_\nu^f(y; z_1, z_2)S^f(z_2, x)\right].
$$

To understand the last step define $(M^f)^{-1} \equiv S^f$ and use Eqs. (8.5.14) and (8.5.15) to obtain $\delta(M^f)^{-1}/\delta A = -(M^f)^{-1}[\delta M^f/\delta A](M^f)^{-1} = -S^f[-iq_0^f\Gamma^f]S^f$, where we use a simple matrix notation here and where we have used Eq. (A.8.9).

Using translational invariance and working in momentum space we can write Eq. (8.5.16) as $D^{-1}(q) = D_{0\mu\nu}^{-1}(q) - i\Pi_{\mu\nu}(q)$. Acting from the left with $D_{0\mu\nu}$ and from the right with $D_{\mu\nu}$ we find

$$
D^{\mu\nu}(q) = D_0^{\mu\nu}(q) + D_0^{\mu\rho}(q)i\Pi_{\rho\sigma}(q)D^{\sigma\nu}(q), \tag{8.5.18}
$$

$$
i\Pi^{\mu\nu}(q) = (-1)\sum_f \int \frac{d^d\ell}{(2\pi)^d}\mathrm{tr}[(-iq_0^f)\gamma^\mu S^f(\ell)(-iq_0^f)\Gamma^{f\nu}(\ell, \ell + q)S^f(\ell + q)],
$$

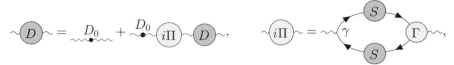

where the two Schwinger-Dyson equations for the photon are represented diagrammatically in the last line. While we labored to arrive at these equations starting from the generating functional, it is much easier to write these diagrammatic equations down by hand by simply recognizing that every Feynman diagram contributing on one side of the equation contributes once and only once to the other side. We use the standard Feynman rules for evaluating these diagrams but with perturbative quantities replaced with their full nonperturbative forms where indicated in the diagram. The factor of (-1) in $i\Pi^{\mu\nu}$ is due to the fermion loop in the usual way. As anticipated, consistency tells us that $\Gamma_\mu^f = \gamma_\mu + \mathcal{O}(\alpha)$. Similarly, $D^{\mu\nu} = D_0^{\mu\nu} + \mathcal{O}(\alpha) = i[-g^{\mu\nu} + (1-\xi)(k^\mu k^\mu/k^2)]/k^2 + \mathcal{O}(\alpha)$ and $S^f = S_0^f + \mathcal{O}(\alpha) = (i/\not{p} - m_0) + \mathcal{O}(\alpha)$.

From the Ward identities in Eq. (8.3.35) we must have $q_\mu \Pi^{\mu\nu}(q) = q_\nu \Pi^{\mu\nu}(q) = 0$. The Ward-Takahashi identities in Eq. (8.3.27) tell us that the vertex satisfies

$$q_\mu \Gamma^{f\mu}(p+q,p) = (S^f)^{-1}(p+q) - (S^f)^{-1}(p), \qquad (8.5.19)$$

where factors are correct since at tree level $q_\mu \gamma^\mu = (\not{p} + \not{q} - m_0) - (\not{p} - m_0)$. Eq. (8.5.19) is often referred to as the *Ward-Takashi identity* and was derived by Takahashi directly from the field equations (Takahashi, 1957). It generalized the Ward identity in the left-hand side of Eq. (8.5.36) (Ward, 1950). Then we find $S^f(\ell)q_\nu \Gamma^{f\nu}(\ell, q+\ell)S^f(q+\ell) = S^f(\ell+q) - S^f(\ell)$ and inserting into Eq. (8.5.18) gives

$$q_\nu \Pi^{\mu\nu}(q) \propto \int d^d\ell \, \mathrm{tr}[\gamma^\mu \{S^f(\ell+q) - S^f(\ell)\}] = 0. \qquad (8.5.20)$$

Since $D^{\mu\nu}$ and $D_0^{\mu\nu}$ are symmetric, then $\Pi^{\mu\nu}$ must be symmetric and so again $q_\mu \Pi^{\mu\nu}(q) = q_\nu \Pi^{\mu\nu}(q) = 0$. This result depended on us being able to shift the loop momentum $\ell \to \ell + q$. So using Eq. (7.5.117), $\Pi^{\mu\nu}(q) \equiv \left[-g^{\mu\nu}q^2 + q^\mu q^\nu\right]\Pi(q^2)$, to define $\Pi(q^2)$ we again arrive at Eq. (7.5.119),

$$D^{\mu\nu}(q) = \frac{i}{q^2 + i\epsilon}\left[\frac{-g^{\mu\nu} + (q^\mu q^\nu/q^2)}{1 + \Pi(q^2)} - \xi_0 \frac{q^\mu q^\nu}{q^2}\right]. \qquad (8.5.21)$$

The photon remains massless because $\Pi^{\mu\nu}$ is purely transverse. This occurs since the photon can only couple to conserved currents leading to the Ward-Takahashi identity.

Following a procedure similar to that above (Itzykson and Zuber, 1980) we can derive an integral equation for the fermion propagator using

$$
\begin{aligned}
0 &= \int \mathcal{D}A^\mu \mathcal{D}\bar\psi \mathcal{D}\psi \, \frac{\delta}{\delta\bar\psi(x)} e^{iS[\bar\psi,\psi,A_\mu] + i\int d^dx\,[\sum_f(\bar\psi^f\eta^f + \bar\eta^f\psi^f) + j\cdot A]} \\
&= \left\{\frac{\delta S}{\delta\bar\psi^f(x)}\left[\frac{i\delta}{\delta\eta}, \frac{-i\delta}{\delta\bar\eta}, \frac{-i\delta}{\delta j}\right] + \eta^f(x)\right\} Z[\bar\eta, \eta, j] \\
&= \left\{\left(i\not\partial - m_0^f - q_0^f \gamma^\mu \frac{\delta}{i\delta j^\mu(x)}\right)\frac{\delta}{i\delta\bar\eta^f(x)} + \eta^f(x)\right\} Z[\bar\eta, \eta, j]. \qquad (8.5.22)
\end{aligned}
$$

Acting on this equation with $\delta/i\delta\eta^f(y)$, using Eq. (6.3.246) and setting all fermion sources to zero $(\bar\eta = \eta = 0)$ gives

$$0 = \left(i\not\partial - m_0^f - q_0^f \gamma^\mu \frac{\delta}{i\delta j^\mu(x)}\right) S^f(x, y, [j])Z[0, 0, j] - i\delta^d(x-y)Z[0, 0, j].$$

Rearranging this and setting the photon source to zero ($j = 0$) leads to

$$i\delta^d(x - y) = (i\not\partial - m_0^f)S^f(x - y) - q_0^f\gamma^\mu \frac{\delta}{i\delta j^\mu(x)} S^f(x, y, [j])\Big|_{j=0}, \tag{8.5.23}$$

where we have used the fact the $\delta Z[0, 0, j]/\delta j^\mu(x)|_{j=0} = 0$. Next observe that

$$\frac{\delta S^f(x, y, [j])}{i\delta j^\mu(z)}\Big|_{j=0} = \frac{\delta^3}{i\delta j^\mu(z)\delta\bar\eta^f(x)\delta\eta^f(y)} Z[\bar\eta, \eta, j]\Big|_{\bar\eta=\eta=j=0} \tag{8.5.24}$$

$$= \langle\Omega|T\hat A_\mu(z)\hat{\bar\psi}^f(x)\hat\psi^f(y)|\Omega\rangle$$

$$= \int d^d u\, d^d v\, d^d w\, D_{\mu\nu}(z - u)S^f(x - v)(-iq_0^f)\Gamma^{f\nu}(u; v, w)S^f(w - y), \tag{8.5.25}$$

where $(-iq_0^f)\Gamma^{f\nu}(u; v, w)$ is the 1PI three-point amplitude obtained by amputating the external legs. Substituting this into Eq. (8.5.23) and taking the Fourier transform leads to

$$i = (\not p - m_0^f)S^f(p) - q_0^f \int \frac{d^d\ell}{(2\pi)^d} \gamma^\mu S^f(\ell + p)D_{\mu\nu}(\ell)(-iq_0^f)\Gamma^{f\nu}(\ell + p, p)S^f(p).$$

Dividing both sides by i and right-multiplying by $(S^f)^{-1}(p)$ gives

$$(S^f)^{-1}(p) = (S_0^f)^{-1}(p) - \int \frac{d^d\ell}{(2\pi)^d} (-iq_0^f)\gamma^\mu S^f(\ell + p)D_{\mu\nu}(\ell)(-iq_0^f)\Gamma^{f\nu}(\ell + p, p)$$

$$\equiv (S_0^f)^{-1}(p) + i\Sigma^f(p), \tag{8.5.26}$$

where we have identified the fermion self-energy for flavor f using Eq. (7.5.96). Combining this with Eq. (7.5.95) leads to the fermion Schwinger-Dyson equations,

$$S^f(p) = S_0^f(p) + S_0^f(p)[-i\Sigma^f(p)]S^f(p), \tag{8.5.27}$$

$$-i\Sigma(p) = \int \frac{d^d\ell}{(2\pi)^d} (-iq_0^f)\gamma^\mu S^f(\ell + p)D_{\mu\nu}(\ell)(-iq_0^f)\Gamma^{f\nu}(\ell + p, p),$$

As was the case for the photon Schwinger-Dyson equation it is far simpler to arrive at these results by simply equating the contributing Feynman diagrams on each side of the diagrams and then write the equations down by inspection. Observe that the two-point amplitudes in Eqs. (8.5.18) and (8.5.27) would form a closed system except that they depend on the three-point amplitude $\Gamma^{f\mu}$. This is the pattern with Schwinger-Dyson equations, where an n-point amplitude depends on $(n + 1)$-point and higher amplitudes.

The Schwinger-Dyson equation for the 1PI (proper) photon-fermion vertices can be similarly derived (Itzykson and Zuber, 1980). However, we now understand that we can obtain the appropriate equations based on equating Feynman diagrams. This leads to the diagrammatic equality in Eq. (8.5.28). The corresponding integral equation in Eq. (8.5.29) can then be then written down by inspection from the first line of Eq. (8.5.28) using the usual Feynman rule conventions.

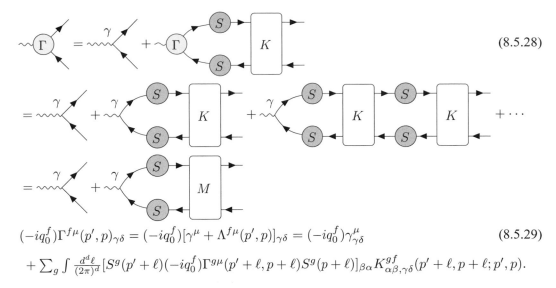

$$(-iq_0^f)\Gamma^{f\mu}(p',p)_{\gamma\delta} = (-iq_0^f)[\gamma^\mu + \Lambda^{f\mu}(p',p)]_{\gamma\delta} = (-iq_0^f)\gamma^\mu_{\gamma\delta} \tag{8.5.29}$$

$$+ \sum_g \int \frac{d^d\ell}{(2\pi)^d} [S^g(p'+\ell)(-iq_0^f)\Gamma^{g\mu}(p'+\ell,p+\ell)S^g(p+\ell)]_{\beta\alpha} K^{gf}_{\alpha\beta,\gamma\delta}(p'+\ell,p+\ell;p',p).$$

The fermion-antifermion *scattering kernel*, K^{fg}, is two-particle irreducible (2PI) with respect to the fermion-antifermion lines and does not contain any fermion-antifermion single-photon annihilation contributions (no intermediate single-photon state) since these would not be 1PI contributions to $\Gamma^{f\mu}$ with respect to the photon line. Annihilations into multiphoton intermediate states is possible and can lead to flavor mixing in K^{fg}, i.e., $f \neq g$. The fermion-antifermion scattering amplitude M^{fg} is 1PI with respect to the fermion lines and does not contain any fermion-antifermion single-photon annihilation contributions. Using an obvious shorthand notation it follows that $M = K + K(S)^2 K + K(S)^2 K(S)^2 K + \cdots = K + K(S)^2 M$. There is also an integral equation for M^{fg} (or K^{fg}), which can be written as an integral equation (Bjorken and Drell, 1965) involving higher-point functions.

8.5.2 Renormalized QED Green's Functions

To obtain the renormalized Green's functions we recall that the fields in the section above were actually the bare fields and so to write things in terms of the renormalized fields we use $\bar{\psi}^f \to \bar{\psi}_0^f = (Z_2^f)^{1/2}\bar{\psi}^f$, $\psi^f \to \psi_0^f = (Z_2^f)^{1/2}\psi^f$ and $A^\mu \to A_0^\mu = (Z_3^f)^{1/2}A^\mu$. The renormalized Green's functions will be marked with a tilde as done in Bjorken and Drell (1965) to avoid confusion and are the time-ordered products of the renormalized fields. So we immediately have

$$S^f(p) = Z_2^f \tilde{S}^f(p) \quad \text{and} \quad D^{\mu\nu}(q) = Z_3 \tilde{D}^{\mu\nu}(q). \tag{8.5.30}$$

Recall that the bare amputated three-point function was $(-iq_0^f)\Gamma^{f\mu}(p',p)$. Before amputation the three-point function is $\langle\Omega|T\hat{A}^\mu(z)\hat{\psi}(x)\bar{\hat{\psi}}(y)|\Omega\rangle$, which brings a factor $Z_2^f Z_3^{1/2}$. After amputating three propagators to give the 1PI vertex we are dividing by $(Z_2^f)^2 Z_3$ giving an overall factor $(Z_2^f)^{-1}Z_3^{-1/2}$. Following Eq. (8.5.15) we *define* the renormalized 1PI three-point function as $(-iq^f)\tilde{\Gamma}^{f\mu}(p',p)$ and so

$$(-iq_0^f)\Gamma^{f\mu}(p',p) = (Z_2^f)^{-1}Z_3^{-1/2}[(-iq^f)\tilde{\Gamma}^{f\mu}(p',p)] = (Z_1^f)^{-1}(-iq_0^f)\tilde{\Gamma}^{f\mu}(p',p),$$

where we have used $q_0^f = Z_1^f (Z_2^f)^{-1} Z_3^{-1/2} q^f$. It then follows that

$$\Gamma^{f\mu}(p', p) = (Z_1^f)^{-1} \tilde{\Gamma}^{f\mu}(p', p). \tag{8.5.31}$$

These arguments extend to arbitrary renormalized Green's functions.

The renormalization constants and bare parameters are determined by imposing the renormalization point boundary conditions. In the *on-shell renormalization* scheme we define the renormalized fermion self-energy $-i\tilde{\Sigma}(p)$, the renormalized vacuum polarization $i\tilde{\Pi}(q^2)$ and the renormalized vertex dressing $\tilde{\Lambda}^{f\mu}(p', p)$ in terms of the renormalized propagators and vertices as

$$\tilde{S}^f(p) = \frac{i}{\not{p} - m^f - \tilde{\Sigma}^f(p)}, \quad \tilde{D}^{\mu\nu}(q) = \frac{i[-g^{\mu\nu} + \frac{q^\mu q^\nu}{q^2 + i\epsilon}]}{(q^2 + i\epsilon)[1 + \tilde{\Pi}(q^2)]} - \frac{i\xi q^\mu q^\nu}{(q^2 + i\epsilon)^2}, \tag{8.5.32}$$

$$\tilde{\Gamma}^{f\mu}(p', p) = \gamma^\mu + \tilde{\Lambda}^{f\mu}(p', p),$$

where m^f are the physical masses of the fermions as defined in the LSZ formalism. We refer to the m^f as the *pole masses*, since they correspond to the poles in the renormalized fermion propagators. In the on-shell renormalization scheme the fermion and photon propagators take their perturbative form as we approach their mass shells and similarly for the proper vertex when we take the soft photon limit,

$$\tilde{\Sigma}^f(p)\Big|_{\not{p}=m^f} = \frac{d\tilde{\Sigma}^f(p)}{d\not{p}}\Big|_{\not{p}=m^f} = \tilde{\Pi}(q^2)\Big|_{q^2=0} = \tilde{\Lambda}^{f\mu}(p, p)\Big|_{\not{p}=m^f} = 0. \tag{8.5.33}$$

Proof: Omit the fermion species label f for simplicity here. We wish to understand the first, second and fourth conditions. The formal equation $\not{p} = m$ cannot be taken literally since $\not{p}$ can never be proportional to the identity matrix. Since $p^2 = \not{p}^2$ and since $\not{p}$ commutes with itself, then $\tilde{\Sigma}(p)$ and $\tilde{\Lambda}^\mu(p, p)$ are only functions of $\not{p}$, i.e., $\tilde{\Sigma}(p) \equiv \tilde{\Sigma}(\not{p})$ and similarly $\tilde{\Lambda}^\mu(p, p) \equiv \tilde{\Lambda}^\mu(\not{p})$.

Let $f(\not{p})$ be some function of $\not{p}$. We can expand $f(\not{p})$ as a power series in $(\not{p} - m)$ for any constant m, $f(\not{p}) = g(\not{p} - m)$. (This is possible since (i) we can write $f(\not{p}) = f_d(p^2)(\not{p} - m) + f_s(p^2) = h_d(p^2 - m^2)(\not{p} - m) + h_s(p^2 - m^2)$; (ii) since $(p^2 - m^2) = (\not{p} + m)(\not{p} - m) = (\not{p} - m + 2m)(\not{p} - m) = (\not{p} - m)^2 + 2m(\not{p} - m)$ then any $h(p^2 - m^2) = k(\not{p} - m)$ for some function $k(\cdots)$; and (iii) it follows that $f(\not{p}) = k_d(\not{p} - m)(\not{p} - m) + k_s(\not{p} - m) \equiv g(\not{p} - m)$.)

So writing $\tilde{\Sigma}(\not{p}) = g(\not{p} - m) = \sum_{n=0}^\infty c_n(\not{p} - m)^n$ and having $c_0 = c_1 = 0$ ensures that the fermion propagator has its perturbative form at the mass pole, which is what we require. We understand then that $\not{p} = m$ *really means replace $\not{p}$ everywhere with m*. With this understanding we see that $\tilde{\Sigma}(p)|_{\not{p}=m} = c_0$ and $(d\tilde{\Sigma}(p)/d\not{p})|_{\not{p}=m} = c_1$. These arguments extend to the vertex dressing $\tilde{\Lambda}^\mu(p, p) = f_a(p^2)\gamma^\mu + f_b(p^2)p^\mu \equiv g_a(\not{p} - m)\gamma^\mu + g_b(\not{p} - m)p^\mu$, where $\tilde{\Lambda}^\mu(p, p)|_{\not{p}=m} = 0$ means that $c_0^a = c_0^b = 0$ in the power series for g_a and g_b.

The conditions on $\tilde{\Sigma}^f$ determine both Z_2^f and m_0^f, where Z_2^f is adjusted to give unit normalization of the propagator and m_0^f is adjusted to ensure the physical mass m^f. The condition on $\tilde{\Pi}$ determines Z_3 and the Ward identity ensures that the photon remains massless. The condition on the $\tilde{\Gamma}^{f\mu}$ determines the Z_1^f. The relationship between q_0^f and q^f is then fixed and so we can choose the q_0^f to obtain the desired physical value for the electric charges q^f. From the Ward-Takahashi identity for the bare fields in Eq. (8.5.19) it follows that

$$(Z_2^f/Z_1^f)q_\mu\tilde{\Gamma}^{f\mu}(p+q,p) = (\tilde{S}^f)^{-1}(p+q) - (\tilde{S}^f)^{-1}(p). \tag{8.5.34}$$

This must be true for all momenta and so taking $(p+q)^2 = p^2 = m^{f2}$ we have $(Z_2^f/Z_1^f)q_\mu\tilde{\Gamma}^{f\mu}(p+q,p) = (\not{p}+\not{q}-m^f) - (\not{p}-m^f) = \not{q}$. Now taking limit $q^\mu \to 0$ gives $(Z_2^f/Z_1^f)q_\mu\tilde{\Gamma}^{f\mu}(p+q,p) \to (Z_2^f/Z_1^f)\not{q} = \not{q}$ and so for each flavor f we have

$$Z_1^f = Z_2^f \quad \Rightarrow \quad q_\mu\tilde{\Gamma}^{f\mu}(p+q,p) = (\tilde{S}^f)^{-1}(p+q) - (\tilde{S}^f)^{-1}(p). \tag{8.5.35}$$

So in QED charge renormalization arises from vacuum polarization corrections alone, $q_0^f = (Z_3)^{-1/2}q^f$, and *charge renormalization is universal*. For example, if a bare electron and muon have the same charge, then even though their masses are different their charges are the same. From Eq. (8.5.35) it follows that

$$\tilde{\Gamma}^{f\mu}(p,p) = \frac{\partial(\tilde{S}^f)^{-1}(p)}{\partial p_\mu} \quad \text{and} \quad \tilde{S}^f(p+q) = q_\mu\tilde{\Gamma}^{f\mu}(p+q,p)\Big|_{\not{p}=m^f}, \tag{8.5.36}$$

where the left-hand equation is the *original Ward identity* (Ward, 1950; Bjorken and Drell, 1965). The right-hand equation shows that if we know the proper vertex then we can obtain the fermion propagator. In the right-hand equation we have used $\tilde{S}^f(p)|_{\not{p}=m^f} = 0$, where $\not{p} = m^f$ means replacing any occurrence of $\not{p}$ by m^f.

Recall that $S^f = Z_2^f\tilde{S}^f$, $D^{\mu\nu} = Z_3\tilde{D}^{\mu\nu}$, $\Gamma^{f\mu} = (1/Z_1^f)\tilde{\Gamma}^{f\mu}$ and $q_0^f = (Z_3)^{-1/2}q^f$, which means, for example, that $(q_0^f)^2 D^{\mu\nu} = (q^f)^2\tilde{D}^{\mu\nu}$. Note that since $Z_1^f = Z_2^f$ we can and will use these interchangeably. From this it follows that we can write

$$\tilde{D}_{\mu\nu}^{-1}(q) = Z_3[D_{0\mu\nu}^{-1}(q) - i\Pi_{\mu\nu}(q)] \equiv Z_3 D_{0\mu\nu}^{-1}(q) - i\Pi'_{\mu\nu}(q), \tag{8.5.37}$$

$$(\tilde{S}^f)^{-1}(p) = Z_2^f[(S_0^f)^{-1}(p) + i\Sigma^f(p)] \equiv Z_2^f(S_0^f)^{-1}(p) + i\Sigma^{f\prime}(p), \tag{8.5.38}$$

$$\tilde{\Gamma}^{f\mu}(p',p) = Z_1^f[\gamma^\mu + \Lambda^{f\mu}(p',p)] \equiv Z_1^f\gamma^\mu + \Lambda'^{f\mu}(p',p), \tag{8.5.39}$$

where we have defined $\Pi'_{\mu\nu}$, Σ'^f and $\Lambda^{f\mu}(p',p)$ for notational convenience. Then Eqs. (8.5.18) and (8.5.27) can be written in the form

$$\tilde{D}^{\mu\nu}(q) = Z_3^{-1}\{D_0^{\mu\nu}(q) + D_0^{\mu\rho}(q)i\Pi'_{\rho\sigma}(q)\tilde{D}^{\sigma\nu}(q)\}, \tag{8.5.40}$$

$$i\Pi'^{\mu\nu}(q) = (-1)Z_1^f\sum_f\int\frac{d^d\ell}{(2\pi)^d}\text{tr}[(-iq^f)\gamma^\mu\tilde{S}^f(\ell)(-iq^f)\tilde{\Gamma}^{f\nu}(\ell,\ell+q)\tilde{S}^f(\ell+q)],$$

$$\tilde{S}^f(p) = (Z_2^f)^{-1}\{S_0^f(p) + S_0^f(p)[-i\Sigma'^f(p)]\tilde{S}^f(p)\}, \tag{8.5.41}$$

$$-i\Sigma'(p) = Z_1^f\int\frac{d^d\ell}{(2\pi)^d}(-iq^f)\gamma^\mu\tilde{S}^f(\ell+p)\tilde{D}_{\mu\nu}(\ell)(-iq^f)\tilde{\Gamma}^{f\nu}(\ell+p,p),$$

$$\Lambda'^{f\mu}(p',p) = \sum_g\int\frac{d^d\ell}{(2\pi)^d}[\tilde{S}^g(p'+\ell)\tilde{\Gamma}^{g\mu}(p'+\ell,p+\ell)\tilde{S}^g(p+\ell)]_{\beta\alpha}$$
$$\times \tilde{K}^{gf}_{\alpha\beta,\gamma\delta}(p'+\ell,p+\ell;p',p).$$

To understand the last equation note that $K^{gf} = (Z_2^g)^{-1}\tilde{K}^{gf}(Z_2^f)^{-1}$, since the scattering kernel is a four-point function bringing $Z_2^g Z_2^f$ followed by four fermion amputations bringing $(Z_2^g Z_2^f)^{-2}$ leaving an overall factor of $(Z_2^g Z_2^f)^{-1}$. This means that $\Lambda^{f\mu} \sim \Gamma^g S^g S^g K^{gf} = (Z_1^g)^{-1}\tilde{\Gamma}^g(Z_2^g)^2\tilde{S}^g\tilde{S}^g(Z_2^g)^{-1}\tilde{K}^{gf}(Z_2^f)^{-1} = (Z_1^f)^{-1}\tilde{\Gamma}^g\tilde{S}^g\tilde{S}^g\tilde{K}^{gf}$, where we have used $Z_1^f = Z_2^f$. Then $\Lambda'^{f\mu} = Z_1^f\Lambda^{f\mu} \sim \tilde{\Gamma}^g\tilde{S}^g\tilde{S}^g\tilde{K}^{gf}$.

Introducing Π', Σ'^f and $\Lambda'^{f\mu}$ has allowed easily remembered forms for these quantities and for Eqs. (8.5.37), (8.5.38) and (8.5.39). They are the same as the bare Schwinger-Dyson equations but with all unrenormalized dressed quantities now renormalized and with any perturbative bare quantity replacements $\gamma^\mu \to Z_1^f\gamma^\mu$, $D_0^{\mu\nu} \to Z_3^{-1}D_0^{\mu\nu}$ and $S_0^f \to (Z_2^f)^{-1}S_0^f$.

Define $\Pi'(q^2) \equiv Z_3 \Pi(q^2)$ so that $\Pi'_{\mu\nu}(q) = [-g_{\mu\nu}q^2 - q_\mu q_\nu]\Pi'(q^2) = Z_3 \Pi_{\mu\nu}(q)$, where we recall from Eq. (7.5.117) that $\Pi_{\mu\nu}(q) = [-g_{\mu\nu}q^2 - q_\mu q_\nu]\Pi(q^2)$. Focusing for now only on the transverse propagator (using Landau gauge) we see that

$$\tilde{D}^{\mu\nu}(q) = \frac{D^{\mu\nu}(q)}{Z_3} = \frac{i[-g^{\mu\nu} + \cdots]}{q^2 Z_3[1 + \Pi(q^2)]} = \frac{i[-g^{\mu\nu} + \cdots]}{q^2[Z_3 + \Pi'(q^2)]} = \frac{i[-g^{\mu\nu} + \cdots]}{q^2[1 + \tilde{\Pi}(q^2)]}. \tag{8.5.42}$$

So the condition $\tilde{\Pi}(0) = 0$ means that $Z_3[1 + \Pi(0)] = Z_3 + \Pi'(0) = 1$ and so as we found earlier in Eq. (7.5.120),

$$Z_3 = \frac{1}{1 + \Pi(0)} = 1 - \Pi'(0) \quad \text{and} \quad \tilde{\Pi}(q^2) = Z_3 + \Pi'(q^2) - 1 = \Pi'(q^2) - \Pi'(0). \tag{8.5.43}$$

For the fermions we have $i(\tilde{S}^f)^{-1}(p) = \not{p} - m^f - \tilde{\Sigma}^f(p) = Z_2^f(\not{p} - m_0^f) - \Sigma^{f'}(p)$ from Eqs. (8.5.32) and (8.5.38). From the on-shell renormalization conditions in Eq. (8.5.33) it follows that

$$Z_2^f = 1 + \frac{d\Sigma^{f'}(p)}{d\not{p}}\bigg|_{\not{p}=m^f}, \quad m_0^f = m^f - \frac{1}{Z_2^f}\Sigma^{f'}(p)\big|_{\not{p}=m^f}. \tag{8.5.44}$$

We can define $A^f(p^2)\not{p} - B^f(p^2) \equiv \not{p} - m^f - \tilde{\Sigma}^f(p)$ so that

$$\tilde{S}^f(p) \equiv \frac{i}{A^f(p^2)\not{p} - B^f(p^2)} \equiv \frac{iZ^f(p^2)}{\not{p} - M^f(p^2)}, \tag{8.5.45}$$

where $Z^f(p^2) = 1/A^f(p^2)$, $M^f(p^2) = B^f(p^2)/A^f(p^2)$. The on-shell renormalization conditions are *not* simply $Z^f((m^f)^2) = 1$ and $M^f((m^f)^2) = B^f((m^f)^2) = m^f$. Since $df(p^2)/d\not{p} = 2\not{p}(df/dp^2)$ the p^2-dependence of A and B lead to somewhat complicated forms: $0 = \tilde{\Sigma}^f(p)|_{\not{p}=m^f} = B^f((m^f)^2) - m^f A^f((m^f)^2)$ and we similarly find $0 = (d\tilde{\Sigma}^f/d\not{p})|_{\not{p}=m^f} = 1 - A^f((m^f)^2) - 2m_f[m^f(dA^f/dp^2) - (dB^f/dp^2)]_{p^2=(m^f)^2}$.

We have $\tilde{\Gamma}^{f\mu} = Z_1^f \gamma^\mu + \Lambda'^{f\mu}(p', p) = \gamma^\mu + \tilde{\Lambda}^{f\mu}(p', p)$ for the fermion-photon 1PI vertex from Eqs. (8.5.33) and (8.5.39). Then $\tilde{\Lambda}_\mu^f(p', p) = \Lambda_\mu^{f'}(p', p) + (Z_1^f - 1)\gamma_\mu$ and from the vertex boundary condition $\tilde{\Lambda}^{f\mu}(p, p)|_{\not{p}=m^f} = 0$ in Eq. (8.5.33) we find

$$(1 - Z_1^f)\gamma_\mu = \Lambda_\mu^{f'}(p, p)\big|_{\not{p}=m^f} \Rightarrow \tilde{\Lambda}_\mu^f(p', p) = \Lambda_\mu^{f'}(p', p) - \Lambda_\mu^{f'}(p, p)\big|_{\not{p}=m^f}. \tag{8.5.46}$$

Summary: In this discussion of QED with f fermion flavors we have derived the Schwinger-Dyson equations for the photon and fermion propagators and for the 1PI (proper) photon-fermion vertices. We first obtained the bare nonperturbative form of these as $D^{\mu\nu}(p)$, $S^f(p)$ and $\Lambda^{f\mu}(p', p)$. The deviation from the bare perturbative quantities $D_0^{\mu\nu}$, $S_0^f(p)$, and γ^μ was captured in the bare vacuum polarization $\Pi^{\mu\nu}(q^2)$, the bare fermion self-energies $\Sigma^f(p)$ and bare 1PI vertex dressing $\Lambda^{f\mu}(p', p)$. These quantities were all specified in terms of the bare fermion masses m_0^f and the bare electric charges q_0^f and are finite and dependent on some large but finite regularization parameter Λ. We recognized that we could write these equations down by inspection using the diagrammatic approach. We also saw that the Ward identity leads to $q_\mu \Pi^{\mu\nu}(q) = 0$ so that $\Pi^{\mu\nu}(q) = [-g^{\mu\nu}q^2 - q^\mu q^\nu]\Pi(q^2)$ and that the Ward-Takashi identity leads to $q_\mu \Gamma^{f\mu}(p + q, p) = (S^f)^{-1}(p + q) - (S^f)^{-1}(p)$.

We renormalized the theory by introducing a multiplicative renormalization constant for each of the primitively divergent diagrams, $Z_1(\Lambda)$, $Z_2(\Lambda)$ and $Z_3(\Lambda)$. Since we have multiple flavors these become the $2f + 1$ renormalization constants $Z_1^f(\Lambda)$, $Z_2^f(\Lambda)$ and $Z_3(\Lambda)$. The Ward-Takahashi identity then leads to $Z_1^f(\Lambda) = Z_2^f(\Lambda)$. We wish to hold the physical masses m^f and charges q^f fixed while we remove the regularization $\Lambda \to \infty$. Since QED is renormalizable in $d = 4$ dimensions, then we can always choose $Z_1^f(\Lambda)$, $Z_2^f(\Lambda)$ and $Z_3(\Lambda)$, $m_0^f(\Lambda)$, $q_0^f(\Lambda)$ such that m^f and q^f remain fixed as $\Lambda \to \infty$; the renormalized Green's functions $\tilde{D}^{\mu\nu}(q)$, $\tilde{S}^f(p)$ have their perturbative form on-shell; and $\tilde{\Gamma}^{f\mu}(p',p)$ and all physical amplitudes become independent of Λ in the limit $\Lambda \to \infty$.

Since the Schwinger-Dyson equations are an infinite tower of coupled integral equations, it is not feasible to attempt to solve them in a rigorous way. However, our purpose was to demonstrate the general procedure of nonperturbative renormalization and to show that renormalization is not dependent on perturbation theory. This observation is crucial for the implementation of lattice gauge theory, for example. Schwinger-Dyson equations have been exploited in various approximation schemes to attempt to probe the nonperturbative behavior of theories such as QED and QCD (Roberts and Williams, 1994; Alkofer and von Smekal, 2001). Comparison of such studies with lattice QCD calculations can provide some useful insights.

8.5.3 Renormalization Group

How can we specify a different *renormalization scheme* while leaving the physical behavior of the theory unchanged?

In the on-shell renormalization scheme used above the propagators and three-point vertex were required to have their perturbative form in the neighborhood of the mass shell, which required the on-shell renormalization boundary conditions in Eq. (8.5.33). In a *momentum subtraction* (MOM) renormalization scheme we simply fix $\tilde{S}^f$, $\tilde{D}^{\mu\nu}$ and $\tilde{\Gamma}^{f\mu}$ at chosen momentum points so that they appear to have a perturbative form at those specific momenta. This is much simpler than having to differentiate and set $\not{p} = m^f$ as done in on-shell renormalization. Two examples are (i) the symmetric MOM scheme ($\overline{\text{MOM}}$) with $p'^2 = p^2 = q^2 = -\mu^2$ for arbitrary μ, which implies that $2p \cdot q = \mu^2$; and (ii) an asymmetric MOM scheme ($\widetilde{\text{MOM}}$) with $p'^2 = q^2 = -\mu^2$ and $p^2 = 0$, which implies that $2p \cdot q = 0$.

Consider the $\overline{\text{MOM}}$ scheme, where in place of Eqs. (8.5.32) and (8.5.33) we have

$$\tilde{S}^f(\mu;p) = \frac{i}{\not{p} - m^f(\mu) - \tilde{\Sigma}^f(\mu;p)} = \frac{iZ^f(\mu;p^2)}{\not{p} - M^f(p^2)}, \quad \tilde{\Gamma}^{f\rho}(\mu;p',p) = \gamma^\rho + \tilde{\Lambda}^{f\rho}(\mu;p',p),$$

$$\tilde{D}^{\rho\sigma}(\mu;q) = \frac{i[-g^{\rho\sigma} + \frac{q^\rho q^\sigma}{q^2 + i\epsilon}]}{(q^2 + i\epsilon)[1 + \tilde{\Pi}(\mu;q^2)]} - \frac{i\xi(\mu)q^\rho q^\sigma}{(q^2 + i\epsilon)^2} \tag{8.5.47}$$

$$\tilde{\Sigma}^f(\mu;p)\Big|_{p^2 = -\mu^2} = 0, \quad \tilde{\Pi}(\mu;q^2)\Big|_{q^2 = -\mu^2} = 0, \quad \tilde{\Lambda}^{f\mu}(\mu;p',p)\Big|_{p'^2 = p^2 = q^2 = -\mu^2} = 0.$$

Eq. (8.5.34) must still apply and at $p'^2 = p^2 = q^2 = -\mu^2$ the above boundary conditions lead to $[Z_2(\mu,\Lambda)/Z_1(\mu,\Lambda)]q_\mu\gamma^\mu = [\not{p}' - m^f(\mu)] - [\not{p} - m^f(\mu)] = \not{q}$ so that $Z_1(\mu,\Lambda) = Z_2(\mu,\Lambda)$ and the Ward-Takahasi identity in Eq. (8.5.35) again follows in the $\overline{\text{MOM}}$ scheme. It is possible to choose renormalization schemes where $Z_1 \neq Z_2$ but there is little reason to do so. Recall that S^f, $D^{\mu\nu}$ and $\Gamma^{f\mu}$ are bare dressed quantities dependent on the regularization paramater Λ and so we can write

$$S^f(p) \equiv S_{\text{bare}}^f(\Lambda;p) = Z_2^f(\Lambda)\tilde{S}^f(p) = Z_2^f(\mu;\Lambda)\tilde{S}^f(\mu;p), \tag{8.5.48}$$

$$D^{\rho\sigma}(q) \equiv D_{\text{bare}}^{\rho\sigma}(\Lambda;q) = Z_3(\Lambda)\tilde{D}^{\rho\sigma}(q) = Z_3(\mu;\Lambda)\tilde{D}^{\rho\sigma}(\mu;q),$$

$$\Gamma^{f\rho}(p',p) \equiv \Gamma_{\text{bare}}^{f\rho}(\Lambda;p',p) = [Z_1^f(\Lambda)]^{-1}\tilde{\Gamma}^{f\rho}(p',p) = [Z_1^f(\mu;\Lambda)]^{-1}\tilde{\Gamma}^{f\rho}(\mu;p',p),$$

$$q_0^f = [Z_1^f(\Lambda)/Z_2^f(\Lambda)\sqrt{Z_3(\Lambda)}]q^f = [Z_3(\Lambda)]^{-1/2}q^f = [Z_3(\mu,\Lambda)]^{-1/2}q^f(\mu).$$

From Eq. (8.5.45) we have $Z_2^f(\Lambda)Z(p^2) = Z_2^f(\mu,\Lambda)Z^f(\mu;p^2)$ and $M(p^2) = M(\mu;p^2)$ for all *renormalization points* or *renormalization scales* μ. So the fermion mass function $M^f(p^2)$ is μ-independent and the wavefunction renormalization function $Z(\mu;p^2)$ carries all of the μ-dependence. It follows that $m^f(\mu) = M^f(p^2)|_{p^2=-\mu^2}$. For the photon propagator we have $Z_3(\Lambda)/[1+\tilde\Pi(q^2)] = Z_3(\mu,\Lambda)/[1+\tilde\Pi(\mu;q^2)]$. The renormalized mass and charge at the renormalization point μ are $m^f(\mu)$ and $q^f(\mu)$. The renormalized charge at renormalization point μ is related to the standard on-shell definition of the charge as $q^f(\mu) = \lim_{\Lambda\to\infty}[Z_3(\mu,\Lambda)/Z_3(\Lambda)]^{1/2}q^f$, where we recall that all renormalized quantities are finite and Λ-independent as $\Lambda\to\infty$.

It is convenient to define

$$Z_1^f(\mu',\mu) \equiv \lim_{\Lambda\to\infty}[Z_1^f(\mu',\Lambda)/Z_1^f(\mu,\Lambda)],\ \ Z_2^f(\mu',\mu) \equiv \lim_{\Lambda\to\infty}[Z_2^f(\mu',\Lambda)/Z_2^f(\mu,\Lambda)],$$

$$Z_3(\mu',\mu) \equiv \lim_{\Lambda\to\infty}[Z_3(\mu',\Lambda)/Z_3(\mu,\Lambda)],\ \ Z_m^f(\mu',\mu) \equiv [M^f(-\mu'^2)/M^f(-\mu^2)]. \tag{8.5.49}$$

Then, for example, comparing two different renormalization points we have

$$\tilde S^f(\mu';p) = [Z_2^f(\mu',\mu)]^{-1}\tilde S^f(\mu;p),\ \ \ \tilde D^{\rho\sigma}(\mu';q) = [Z_3(\mu',\mu)]^{-1}\tilde D^{\rho\sigma}(\mu;q),$$

$$q^f(\mu') = [Z_3(\mu',\mu)]^{1/2}q^f(\mu'),\ \ \ m^f(\mu') = Z_m^f(\mu',\mu)\,m^f(\mu). \tag{8.5.50}$$

Strictly speaking, we should distinguish between renormalized quantities with arbitrarily large but finite Λ and those with $\Lambda\to\infty$, but since in a renormalizable theory the limit exists we do not bother to explicitly indicate this subtlety.

Let $Z(\mu',\mu)$ be any of the quantities in Eq. (8.5.49). The set of $Z(\mu',\mu)$ for all μ',μ *almost* form a group in a sense, but they are not a group. Consider the μ' subset of these consisting of all Zs where at least one argument is μ'. We can define a restricted operation of multiplication on such a subset, $Z(\mu'',\mu) = Z(\mu'',\mu')Z(\mu',\mu)$, but not between arbitrary elements of the full set. The result of the multiplication, $Z(\mu'',\mu)$, lies in the full set but not the μ' subset and so the subset is not closed under the restricted operation. So neither the full set nor the subset can form a group under such an operation, since we only satisfy a form of *partial closure*. However, the restricted operation leads to other group-like properties, such as *associativity*: $Z(\mu''',\mu'')[Z(\mu'',\mu')Z(\mu',\mu)] = [Z(\mu''',\mu'')Z(\mu'',\mu')]Z(\mu',\mu)$; *identity*: $Z(\mu,\mu) = 1$; and *inverse*: $[Z(\mu',\mu)]^{-1} = Z(\mu,\mu')$ since $Z(\mu,\mu')Z(\mu',\mu) = Z(\mu,\mu) = 1$. Note that if we had the situation where $Z(\mu+\Delta\mu,\mu) = Z(\Delta\mu)$ for all μ, then we could multiply any two elements of the set together and satisfy closure and we would then have a group. An analog of this is the evolution operator for a time-dependent Hamiltonian where $\hat U(t+\Delta t,t) \neq \hat U(\Delta t)$ plays the role of $Z(\mu+\Delta\mu,\mu) \neq Z(\Delta\mu)$. With the above in mind we now understand why the set of transformations between different renormalization scales is referred to as the *renormalization group* (RG), even though strictly speaking it is not a group.

For notational simplicity we consider a single fermion flavor with bare mass m_0 and bare charge q_0. Consider an arbitrary QED bare Green's function, $G_{\rm bare}^{(n_\psi,n_\gamma)}$, involving the bare fields A_0^ν, $\bar\psi_0$ and ψ_0 with a total of n_ψ fermion operators and n_γ photon operators. Since we have $A_0^\nu = \sqrt{Z_3(\mu,\Lambda)}A^\nu$, $\bar\psi_0 = \sqrt{Z_2(\mu,\Lambda)}\bar\psi$ and $\psi_0 = \sqrt{Z_2(\mu,\Lambda)}\psi$ the bare fields are functions of Λ and the renormalized fields are functions of μ. The bare and renormalized Green's functions are related by

$$G_{{\rm bare}\,\nu_1,\ldots,\nu_{n_\gamma}}^{(n_\psi,n_\gamma)}(\cdots) = \langle\Omega|T\hat A_{0\nu_1}(z_1)\cdots\hat A_{0\nu_{n_\gamma}}(z_{n_\gamma})\hat\psi_0(x_1)\cdots\hat{\bar\psi}_0(x_{n_\psi})|\Omega\rangle$$

$$= [Z_2(\mu,\Lambda)]^{n_\psi/2}[Z_3(\mu,\Lambda)]^{n_\gamma/2}\langle\Omega|T\hat A_{\nu_1}(z_1)\cdots\hat A_{\nu_{n_\gamma}}(z_{n_\gamma})\hat\psi(x_1)\cdots\hat{\bar\psi}(x_{n_\psi})|\Omega\rangle$$

$$= [Z_2(\mu,\Lambda)]^{n_\psi/2}[Z_3(\mu,\Lambda)]^{n_\gamma/2}G_{\nu_1,\ldots,\nu_{n_\gamma}}^{(n_\psi,n_\gamma)}(\cdots). \tag{8.5.51}$$

For brevity we now suppress spacetime indices and coordinates. Note that $G_{\text{bare}}^{(n_\psi, n_\gamma)}$ depends only on the bare quantities Λ, $q_0(\Lambda)$, $m_0(\Lambda)$ and $\xi_0(\Lambda)$. The renormalized Green's function $G^{(n_\psi, n_\gamma)}$ is independent of Λ since QED is renormalizable but depends on the renormalization scale μ and on $q(\mu)$, $m(\mu)$ and $\xi(\mu)$. For simplicity we work in Feynman gauge here so that $\xi_0(\Lambda) = Z_3(\mu, \Lambda)\xi(\mu) = 0$. So we can write

$$G_{\text{bare}}^{(n_\psi, n_\gamma)}(\Lambda, q_0(\Lambda), m_0(\Lambda)) = [Z_2(\mu, \Lambda)]^{n_\psi/2}[Z_3(\mu, \Lambda)]^{n_\gamma/2} G^{(n_\psi, n_\gamma)}(\mu, q(\mu), m(\mu)).$$

Recall that we remove the regulator, $\Lambda \to \infty$, while holding all renormalized quantities fixed at any chosen renormalization point, μ. Similarly, we vary the renormalization point μ while holding all bare quantities fixed. So by construction

$$0 = \mu \frac{\partial}{\partial \mu} G_{\text{bare}}^{(n_\psi, n_\gamma)} = \mu \frac{\partial}{\partial \mu} \left([Z_2(\mu, \Lambda)]^{n_\psi/2}[Z_3(\mu, \Lambda)]^{n_\gamma/2} G^{(n_\psi, n_\gamma)}(\mu, q(\mu), m(\mu)) \right)$$

$$= Z_2^{n_\psi/2} Z_3^{n_\gamma/2} \left(\mu \frac{\partial}{\partial \mu} + \frac{n_\gamma}{2}\gamma_3 + \frac{n_\psi}{2}\gamma_2 + \beta \frac{\partial}{\partial q} + \gamma_m m \frac{\partial}{\partial m} \right) G^{(n_\psi, n_\gamma)}, \qquad (8.5.52)$$

where we have defined

$$\beta \equiv \mu \frac{\partial q}{\partial \mu}, \quad \gamma_3 \equiv \frac{\mu}{Z_3} \frac{\partial Z_3}{\partial \mu}, \quad \gamma_2 \equiv \frac{\mu}{Z_2} \frac{\partial Z_2}{\partial \mu}, \quad \gamma_m \equiv \frac{\mu}{m} \frac{\partial m}{\partial \mu}. \qquad (8.5.53)$$

We have arrived at the *Callan-Symanzik equation* (Callan, 1970; Symanzik, 1970),

$$\left(\mu \frac{\partial}{\partial \mu} + \frac{n_\gamma}{2}\gamma_3 + \frac{n_\psi}{2}\gamma_2 + \beta \frac{\partial}{\partial q} + \gamma_m m \frac{\partial}{\partial m} \right) G^{(n_\psi, n_\gamma)} = 0. \qquad (8.5.54)$$

We refer to β as the *β-function* and to γ_m as the *anomalous mass dimension*. Equations such as the Callan-Symanzik equation are often referred to collectively as *renormalization group equations*.

8.6 Regularization

We have been careful to note throughout that some suitable infrared and ultraviolet regularization has been assumed so that we were dealing with finite quantities at intermediate steps. The regularization is only to be removed when evaluating physical amplitudes at the end of calculations. It has been implicitly assumed that symmetries of the classical system would either be preserved by the regularization or be recovered when the regularization was removed and so be symmetries of the regularized and renormalized quantum field theory. This is not always the case, however. If for *every* regulator a symmetry is broken and not restored when the regulator is removed, then this is referred to as an *anomaly* of the theory. One important example of this is the chiral anomaly of Adler, Bell and Jackiw, which is discussed in Sec. 9.3 (Adler, 1969; Bell and Jackiw, 1969).

8.6.1 Regularization Methods

There are a variety of regularization methods with different levels of usefulness. We discuss some of these below. Every known regularization method violates one or more symmetries and it is essential to choose a method that preserves the most important ones for the theory of interest. We summarize some of these below and we consider dimensional regularization on its own in the next section since it will be most important to us.

Momentum cutoff: A simple momentum cutoff, Λ, violates both Lorentz invariance and gauge invariance. Violations of gauge invariance remain even after $\Lambda \to \infty$. For example, we find that $q_\mu \Pi^{\mu\nu}(q)|_{\Lambda \to \infty} \neq 0$. To illustrate why this is, substitute $S(p) \to \not{p} + m/p^2 - m^2 + i\epsilon$ into Eq. (8.5.20) with a cutoff to give $q_\nu \Pi^{\mu\nu}(q) \propto \int^\Lambda d^4\ell \, \text{tr}[\gamma^\mu \{ S^f(\ell + q) - S^f(\ell) \}]$. We can evaluate perturbation theory integrals in Euclidean space since the Wick rotation gives no contribution from curves C_1 and C_3 for ℓ^0 in Fig. A.1 of Sec. A.4. From Eqs. (A.4.6) and (A.6.2),

$$k^2 \to -k_E^2, \quad d^4\ell \to i d^4\ell_E \quad \text{and} \quad d^4\ell_E = (2\pi^2)\ell_E^3 d\ell_E = \pi^2 \ell_E^2 d\ell_E^2. \tag{8.6.1}$$

Evaluate the trace and retain leading terms as $\ell^2 \to \infty$ after analytic continuation to Euclidean space. This leads to

$$q_\nu \Pi^{\mu\nu}(q) \sim \int^\Lambda d^4\ell \, \text{tr}[\gamma^\mu \not{q} - \gamma^\mu \not{\ell}(2q \cdot \ell/\ell^2)]/\ell^2 \sim q^\mu \int^{\Lambda^2} d\ell_E^2 \sim q^\mu \Lambda^2, \tag{8.6.2}$$

which does not vanish as $\Lambda \to \infty$. We have not bothered to indicate the rotation of q^μ and γ^μ to Euclidean space and back. This result remains true if we attempt to use some smooth cutoff such as a Gaussian function since any kind of cutoff destroys the momentum-translational invariance of the loop integral. So momentum cutoffs are not appropriate regulators in gauge theories. However, they can be useful in some effective theories that allow the consistent application of a cutoff.

Consider the single loop contribution to the scalar 1PI self-energy $\Pi(p^2)$ in $(\lambda/4!)\phi^4$ theory in Eq. (7.5.49). Using the Feynman rules in Sec. 7.6.1 we have $-i\lambda$ for the vertex from Eq. (7.6.10), a symmetry factor of $S = 2$ from a vertical flip of the loop, a scalar propagator $i/\ell^2 - m^2 + i\epsilon$ and a loop integral $\int d^4\ell/(2\pi)^4$. Define the $\mathcal{O}(\lambda^1)$ contribution to $-i\Pi(p^2)$ as $-i\Pi_1(p^2)$. With a cutoff it is

$$-i\Pi_1(p^2) = \quad \raisebox{-1em}{\includegraphics{}} \quad = -i\frac{\lambda}{2} \int^\Lambda \frac{d^4\ell}{(2\pi)^4} \frac{i}{\ell^2 - m^2 + i\epsilon}, \tag{8.6.3}$$

where the result is independent of p^2, there are no external legs on a 1PI diagram, and the subscript 1 denotes $\mathcal{O}(\lambda^1)$. Rotating to Euclidean space and using rotational invariance, $d^4\ell \to i d^4\ell_E = i\pi^2 \ell_E^2 d\ell_E^2$ and $k^2 \to -k_E^2$ we find

$$-i\Pi_1 = -i\lambda\frac{1}{2}[\pi^2/(2\pi)^4] \int^{\Lambda^2} d\ell_E^2 \, [\ell_E^2/(\ell_E^2 + m^2)] \tag{8.6.4}$$

$$= -i(\lambda/32\pi^2) \int^{\Lambda^2} d\ell_E^2 \, [1 - m^2/(\ell_E^2 + m^2)] = -i(\lambda/32\pi^2)[\Lambda^2 - m^2 \ln(\{\Lambda^2 + m^2\}/m^2)]$$

$$= -i(\lambda/32\pi^2)[\Lambda^2 - m^2 \ln(\Lambda^2/m^2) + \mathcal{O}(1/\Lambda^2)].$$

We see that Π_1 is both quadratically and logarithmically divergent as $\Lambda \to \infty$.

Derivative method: We earlier noted that each derivative of a divergent loop integral with respect to external momentum increases the power of loop momentum in the denominator and improves convergence of the integral. This technique simply throws away any infinite constant of integration in a nonsystematic way and so is not a useful regulator in quantitative calculations.

Pauli-Villars regularization: The essence of Paul-Villars regularization (Pauli and Villars, 1949) is to subtract from any loop integral the same loop integral with a much larger mass M in the propagators. This suppresses the loop integral at large loop momenta, $\ell^2 \gg M^2$, where masses and external momenta are unimportant.

For example, at one loop in $(\kappa/3!)\phi^3$ theory with mass m we have an $\mathcal{O}(\kappa^2)$ 1PI contribution to the ϕ self-energy, $-i\Pi(p^2)$, given by

$$-i\Pi_2(p^2) = \underset{p}{\overleftarrow{\quad}}\overbrace{\underset{\ell+p}{\bigcirc}}^{\ell}\underset{p}{\overleftarrow{\quad}} \equiv (-i\kappa)^2 \tfrac{1}{2} I(p^2), \quad \text{where} \tag{8.6.5}$$

$$I(p^2) \equiv \int \frac{d^4\ell}{(2\pi)^4} \left[\frac{i}{\ell^2 - m^2 + i\epsilon} \frac{i}{(p+\ell)^2 - m^2 + i\epsilon} - \frac{i}{\ell^2 - M^2 + i\epsilon} \frac{i}{(p+\ell)^2 - M^2 + i\epsilon} \right].$$

There is one $(-i\kappa)$ for each vertex, a symmetry factor $S = 2$ for 2! ways to exchange lines in the loop and no external propagators. Use $\ell^0 \to i\ell_4^E$, $\ell^2 \to -\ell_E^2$ and $d^4\ell_E = 2\pi^2\ell_E^2 d\ell_E^2 = \pi^2\ell_E^2 d\ell_E^2$. At large ℓ_E^2 we have $I(0) = (i/16\pi^2)(-1)^2 \int d\ell_E^2 \, \ell_E^2 [\mathcal{O}(1/\ell_E^4) - \mathcal{O}(1/\ell_E^4)]$ and so the large ℓ_E^2 contribution to I is $\int d\ell_E^2 \, \mathcal{O}(1/\ell_E^4)$ and so converges. Using a table of integrals we evaluate I at $p^\mu = 0$,

$$I(0) = -(i/16\pi^2)(-1)^2 \int d\ell_E^2 \, \ell_E^2 \left\{ [1/(\ell_E^2 + m^2)^2] - [1/(\ell_E^2 + M^2)^2] \right\} \tag{8.6.6}$$
$$= -(i/16\pi^2)\ln(M^2/m^2).$$

Here M is now playing the role of the UV regulator written as Λ in Sec. 8.5 and so we take $M \to \infty$ when calculating physical amplitudes. One can arrive at Eq. (8.6.5) by adding to the theory a *ghost particle* partner for the ϕ particle that is a scalar with the same couplings but obeying fermionic statistics and with a large mass M; i.e., the ghost particle has the wrong spin-statistics connection. Similarly for a fermion particle ψ we use a corresponding spin-half spinor field with large mass M_ψ satisfying bosonic statistics. The sign difference can be understood from Eq. (7.6.2) as bosons commute in time-ordered products, whereas fermions anticommute.

If we attempt to use Pauli-Villars while keeping the cutoff in Eq. (8.6.3) we find $-i\Pi_1^{PV}(p^2) = -i\Pi_1(p^2)_m - i\Pi_1(p^2)_M \propto [M^2 \ln(\Lambda^2/M^2) - m^2 \ln(\Lambda^2/m^2) + \mathcal{O}(1/\Lambda^2)]$, which is still logarithmically divergent in Λ and so we cannot take $\Lambda \to \infty$. So for this case a second Pauli-Villars subtraction is required to remove the Λ logarithmic divergence, which complicates matters.

Pauli-Villars regularization was widely used in the early days of quantum field theory. For an application to QED see, for example, Bjorken and Drell (1965), Das (2008) and Schwartz (2013). It preserves the momentum translational invariance of momentum integrations, while also being an intuitively satisfying means of controlling large-momentum behavior. However, there are a number of shortcomings that limit its modern use. For diagrams with multiple loops one ghost per particle is insufficient and additional ghost particles are required. While it can be useful in abelian gauge theories where a massive Proca vector boson coupled to conserved currents is consistent, the method breaks down in the case of nonabelian gauge theories where massive vector bosons cannot be consistently described. The method also fails in the case of chiral gauge theories where fermions are massless.

Schwinger parameterization and proper time regularization: The Laplace transform of a function $f(s)$ for $s \geq 0$ is $F(u) \equiv \int_0^\infty ds \, f(s) e^{-su}$, where $u \in \mathbb{C}$. This transform is frequently used in studying differential equations. For the special case of $f(s) = s^n$ for $n = 0, 1, 2, \ldots$ we have $F(u) = \Gamma(n+1)/u^{n+1} = n!/u^{n+1}$. It was observed by Julian Schwinger that if $1/u \equiv i/A$ was a propagator, then

$$(i/A)^{n+1} = [1/\Gamma(n+1)] \int_0^\infty ds \, s^n e^{isA} = (1/n!) \int_0^\infty ds \, s^n e^{isA}, \tag{8.6.7}$$

which is the *Schwinger parameterization*. Since Feynman boundary conditions give $A = C + i\epsilon$ with $C \in \mathbb{R}$, then the integral converges. Also note that

$$(1/A_1 A_2) = -\int_0^\infty ds_1 ds_2 \, e^{i(s_1 A_1 + s_2 A_2)} = -\int_0^1 dx \int_0^\infty d\lambda \, \lambda e^{i\lambda[A_1 + (A_2 - A_1)x]} \tag{8.6.8}$$
$$= \int_0^1 dx \, 1/[A_1 + (A_2 - A_1)x]^2 = \int_0^1 dx \, 1/[(1-x)A_1 + xA_2]^2,$$

where $\lambda = s_1 + s_2$, $x \equiv s_2/(s_1 + s_2) = s_2/\lambda$ and $s_1 = (1 - x)\lambda$. Using Eq. (A.2.6) we find the Jacobian is $|\lambda|$ giving $ds_1 ds_2 = \lambda d\lambda dx$ in Eq. (8.6.8). We also used $-\int_0^\infty d\lambda\, \lambda e^{-\lambda a} = \frac{d}{da}\int_0^\infty d\lambda\, e^{-\lambda a} = \frac{d}{da}[-\frac{1}{a}e^{-\lambda a}]_0^\infty = -\frac{1}{a^2}$. Eq. (8.6.8) reproduces the Feynman parameterization result in Eq. (A.5.1). Another useful relation is

$$\frac{1}{A^m B^n} = \frac{\Gamma(m+n)}{\Gamma(m)\Gamma(n)} \int_0^\infty ds\, \frac{s^{n-1}}{(A+Bs)^{m+n}}. \tag{8.6.9}$$

The variable s in a loose sense is the proper time for the motion of a particle and so is called the *Schwinger proper time*; see, for example, Schwartz (2013).

Consider the ϕ^4 one-loop self-energy in Eq. (8.6.3) and perform a Schwinger parameterization on the Euclidean integral using $1/A = \int_0^\infty ds\, e^{-sA}$ to give

$$-i\Pi_1(p^2) = [-i(\lambda/32\pi^2)] \int d\ell_E^2 \, [\ell_E^2/\ell_E^2 + m^2] = [\cdots] \int d\ell_E^2 \, \ell_E^2 \int_0^\infty ds\, e^{-s(\ell_E^2 + m^2)}$$

$$= [\cdots] \int_0^\infty ds\, e^{-sm^2} \int d\ell_E^2 \, \ell_E^2 e^{-s\ell_E^2} = [-i(\lambda/32\pi^2)] \int_0^\infty ds\, (1/s^2) e^{-sm^2}. \tag{8.6.10}$$

Observe that the ultraviolet divergence at large ℓ_E^2 is now replaced by a divergence at small proper time, $s \to 0$. The integral can be regulated by introducing a suppression at small s of the form $\exp(-1/\lambda^2 s)$ and this is an example of *proper time regularization*. Evaluating the integral using a table of integrals we find

$$-i\Pi_1(p^2) = [-i(\lambda/32\pi^2)] \int_0^\infty ds\, (1/s^2) e^{-sm^2 - (1/\Lambda^2 s)} \tag{8.6.11}$$

$$= -i(\lambda/32\pi^2)[\Lambda^2 - m^2 \ln(\Lambda^2/m^2) + \mathcal{O}(1/\Lambda^2)], \tag{8.6.12}$$

which is the same as found with the cutoff method in Eq. (8.6.4). For details of the proof of this result, see Das (2008).

Lattice regularization: Euclidean spacetime is put onto a finite four-dimensional lattice. It has been used with considerable success to study the nonpertubative behavior of quantum field theories and QCD. It is a first principles approach to nonperturbative studies of quantum field theory in that it is systematically improvable. Lattice gauge theory is a gauge invariant form of cutoff regularization. Gauge invariance is the most important symmetry to maintain in any regularization of gauge theories. The lattice violates rotational and translational invariance in Euclidean space since it uses a finite spacetime volume; however, these symmetries are recovered in the continuum and infinite volume limits. The treatment of chiral symmetry on the lattice requires some care. A brief introduction to lattice QCD is presented in Sec. 9.2.5.

8.6.2 Dimensional Regularization

This is the most useful and the commonly used regularization for perturbative studies of gauge theories[2] and consists of considering spacetime dimension $d = 4 - \epsilon$. Dimensional regularization was developed by 't Hooft and Veltman to enable the regularization and subsequent renormalization of nonabelian gauge theories ('t Hooft and Veltman, 1972; 't Hooft, 1973), where Pauli-Villars regularization was known to be unsuccessful. Such theories are also commonly referred to as *Yang-Mills theories*. The method works in evaluating Feynman diagrams with overlapping divergences

---

[2] In principle dimensional regularization can also be used in nonpertubative studies although this is not commonly done, e.g., see Gusynin et al. (1999).

and respects momentum translational invariance of momentum integrals, i.e., invariance under shifts in the momentum integration variable. It also respects unitarity and causality. A summary of dimensional regularization is given in Sec. A.6. Complications can arise when considering implicitly four-dimensional quantities such as $\epsilon^{\mu\nu\rho\sigma}$ and γ^5, which are related since $\gamma^5 = -(i/4!)\epsilon_{\mu\nu\rho\sigma}\gamma^\mu\gamma^\nu\gamma^\rho\gamma^\sigma$. However, this is only an issue in theories with anomalies. This is discussed further in Sec. 9.3.

In $d < 4$ dimensions logarithmically divergent integrals such as Eq. (8.6.5) are convergent without the Pauli-Villars regulator or any form of cutoff. We define

$$d \equiv 4 - \epsilon, \tag{8.6.13}$$

so that $\epsilon \equiv 4 - d$. Beware that the choice $2\epsilon = 4 - d$ is also common and used in some texts. We have $\Lambda \sim 1/\epsilon$ as the ultraviolet regulator, where $\epsilon \to 0^+$ corresponds to $\Lambda \to \infty$ and is taken at the end of calculations. Because we can make shifts in the momentum integral, then Eq. (8.5.20) remains valid and $q_\nu \Pi^{\mu\nu} = 0$ as required.

In natural units the action $S[\phi] = \int d^d x\, \mathcal{L}$ for any theory must be dimensionless in d dimensions since we exponentiate it in the path integral. Since $[d^d x] = \mathsf{M}^{-d}$ then we require $[\mathcal{L}] = \mathsf{M}^d$. Since $[\partial_\mu] = \mathsf{M}$ then we can deduce the dimension of any field from its kinetic term. For example, for scalars we must have $[\partial_\mu \phi \partial^\mu \phi] = \mathsf{M}^d$ and so $[\phi] = \mathsf{M}^{(d-2)/2}$. For a fermion, since we must have $[\bar{\psi}\,\partial\!\!\!/\,\psi] = \mathsf{M}^d$, then $[\psi] = \mathsf{M}^{(d-1)/2}$. Since particle masses m appear once for each ∂_μ in the kinetic terms of theories, then $[m] = [\partial_\mu] = \mathsf{M}$. In ϕ^4 theory in d-dimensions the interaction term has dimension $[(\lambda/4!)\phi^4] = \mathsf{M}^d$ and so $[\lambda] = \mathsf{M}^{d-2(d-2)} = \mathsf{M}^{4-d}$. However, since we prefer to keep the coupling dimensionless as it is in $d = 4$ we introduce an *arbitrary mass scale*, μ, with $[\mu] = \mathsf{M}$ and make the replacement $\lambda \to \mu^{4-d}\lambda = \mu^\epsilon \lambda$ so that now $[\lambda] = \mathsf{M}^0$. In summary, for ϕ^4 theory in d-dimensions the Lagrangian density is

$$\mathcal{L} = \tfrac{1}{2}\partial_\mu \phi \partial^\mu - \tfrac{1}{2}m^2\phi^2 - (\mu^{4-d}\lambda/4!)\phi^4, \tag{8.6.14}$$

where λ is dimensionless. In the limit $\epsilon \to 0^+$ we have $\mu^{4-d} = \mu^\epsilon \to 1$ and so μ and ϵ both become irrelevant in the $\epsilon \to 0^+$ limit for physical quantities in a renormalizable theory. In dimensional regularization the arbitrary mass scale μ can be used as the renormalization scale μ.

Similarly in d-dimensions we find for the QED Lagrangian density of Eq. (7.6.16) $[A^\mu] = \mathsf{M}^{(d-2)/2}$, $[\psi] = \mathsf{M}^{(d-1)/2}$, $[m] = \mathsf{M}$ and $[q_c] = \mathsf{M}^{d-(d-1)-(d-2)/2} = \mathsf{M}^{(4-d)/2} = \mathsf{M}^{\epsilon/2}$. Using an arbitrary mass scale μ to keep q_c dimensionless in natural units we replace $q_c \to \mu^{(4-d)/2}q_c = \mu^{\epsilon/2}q_c$. The d-dimensional Lagrangian density is

$$\mathcal{L} = \bar{\psi}(i\partial\!\!\!/ - m)\psi - \tfrac{1}{4}(\partial_\mu A_\nu - \partial_\nu A_\mu)(\partial^\mu A^\nu - \partial^\nu A^\mu) - \mu^{(4-d)/2}q_c\bar{\psi}A\!\!\!/\psi \tag{8.6.15}$$
$$= \bar{\psi}(i D\!\!\!\!/ - m)\psi - \tfrac{1}{4}F_{\mu\nu}F^{\mu\nu},$$

where in d-dimensions the covariant derivative is $D^\mu = \partial^\mu + i\mu^{(4-d)/2}q_c A^\mu$. In Yukawa theory we similarly have $g \to \mu^{(4-d)/2}g = \mu^{\epsilon/2}g$.

In $(\kappa/3!)\phi^3$ theory we have $[(\kappa/3!)\phi^3] = \mathsf{M}^d$ and so $[\kappa] = \mathsf{M}^{d-3(d-2)/2} = \mathsf{M}^{3-(d/2)}$ and so we make the replacement $\kappa \to \mu^{3-(d/2)}\kappa$ so that $[\kappa] = \mathsf{M}^0$. In place of Eq. (8.6.5) we have for the one-loop scalar self-energy

$$-i\Pi_2(p^2) = (-i\mu^{3-(d/2)}\kappa)^2 \tfrac{1}{2} \int \frac{d^d\ell}{(2\pi)^d} \frac{i}{\ell^2 - m^2 + i\epsilon} \frac{i}{(p+\ell)^2 - m^2 + i\epsilon}. \tag{8.6.16}$$

Again for simplicity choose $p^\mu = 0$, which leads to

$$-i\Pi_2(0) = \frac{\mu^{6-d}\kappa^2}{2} \int \frac{d^d\ell}{(2\pi)^d} \frac{1}{(\ell^2 - m^2 + i\epsilon)^2} = \frac{\mu^{6-d}\kappa^2}{2} \left[\frac{i(-1)^2}{(4\pi)^{d/2}}\right] \frac{\Gamma(2 - \frac{d}{2})}{\Gamma(2)} \left(\frac{1}{m^2}\right)^{2-\frac{d}{2}}$$

$$= \frac{i\mu^2\kappa^2}{2} \frac{\Gamma(2 - \frac{d}{2})}{(4\pi)^{d/2}} \left(\frac{\mu^2}{m^2}\right)^{2-\frac{d}{2}} = \frac{i\mu^2\kappa^2}{32\pi^2} \left[\frac{2}{\epsilon} - \ln\left(\frac{m^2}{\mu^2}\right) - \gamma + \ln(4\pi) + \mathcal{O}(\epsilon)\right]$$

$$= \frac{i\mu^2\kappa^2}{32\pi^2} \left[\frac{2}{\epsilon} - \ln\left(\frac{m^2}{4\pi e^{-\gamma}\mu^2}\right) + \mathcal{O}(\epsilon)\right] = \frac{i\mu^2\kappa^2}{32\pi^2} \left[\frac{2}{\epsilon} - \ln\left(\frac{m^2}{\tilde{\mu}^2}\right) + \mathcal{O}(\epsilon)\right], \quad (8.6.17)$$

where we used Eqs. (A.6.4) and (A.6.17) and we defined

$$\tilde{\mu}^2 \equiv 4\pi e^{-\gamma}\mu^2 \quad (8.6.18)$$

because this combination occurs frequently. The integral is finite but diverges as $\Lambda \sim 1/\epsilon$ as $\epsilon \to 0^+$. We note the emergence of a logarithm in the finite part.

Similarly evaluating the ϕ^4 one-loop self-energy in Eq. (8.6.3) using dimensional regularization and Eq. (A.6.4) gives (Das, 2008)

$$-i\Pi_1(p^2) = -i\mu^{4-d}\lambda\frac{1}{2}\int^\Lambda \frac{d^d\ell}{(2\pi)^d}\frac{i}{\ell^2 - m^2 + i\epsilon} = \frac{\mu^{4-d}\lambda}{2}\int^\Lambda \frac{d^d\ell}{(2\pi)^d}\frac{1}{\ell^2 - m^2 + i\epsilon}$$

$$= \frac{\mu^{4-d}\lambda}{2}\left[\frac{-i}{(4\pi)^{d/2}}\right]\frac{\Gamma(1 - \frac{d}{2})}{\Gamma(1)}\left(\frac{1}{m^2}\right)^{1-\frac{d}{2}} \simeq \frac{-im^2\lambda}{32\pi^2}(4\pi)^{\epsilon/2}\left[-\frac{2}{\epsilon} + (\gamma - 1)\right]\frac{\mu^2}{m^2}$$

$$\simeq \frac{im^2\lambda}{32\pi^2}\left(1 + \frac{\epsilon}{2}\ln 4\pi\right)\left[\frac{2}{\epsilon} - (\gamma - 1)\right]\left[1 + \frac{\epsilon}{2}\ln\left(\frac{\mu^2}{m^2}\right)\right] \quad (8.6.19)$$

$$= \frac{im^2\lambda}{32\pi^2}\left[\frac{2}{\epsilon} - \ln\left(\frac{m^2}{4\pi e^{-(\gamma-1)}\mu^2}\right) + \mathcal{O}(\epsilon)\right] = \frac{im^2\lambda}{32\pi^2}\left[\frac{2}{\epsilon} - \ln\left(\frac{m^2}{\tilde{\mu}^2 e}\right) + \mathcal{O}(\epsilon)\right],$$

where $\Gamma(1 - \frac{d}{2}) = \Gamma(-1 + \frac{\epsilon}{2}) = \Gamma(\frac{\epsilon}{2})/(-1 + \frac{\epsilon}{2}) \simeq -(\frac{2}{\epsilon} - \gamma)(1 + \frac{\epsilon}{2}) \simeq -\frac{2}{\epsilon} + (\gamma - 1)$ from Eq. (A.6.16). The integral is finite for nonzero ϵ and so has been regularized. Observe the similarity of Eqs. (8.6.17) and (8.6.19). Note that the tadpole contribution $\Pi_1(p^2)$ is independent of p^2.

8.7 Renormalized Perturbation Theory

In this section we want to develop the essential tools and techniques for performing perturbative calculations with a focus on using dimensional regularization. We consider the examples of scalar ϕ^4 theory and QED. While we focused primarily on multiplicative renormalization for the nonperturbative treatment of QED in Sec. 8.5, it is conventional in perturbation theory to discuss renormalization using counterterms while working order by order in powers of the coupling constant.

8.7.1 Renormalized Perturbation Theory for ϕ^4

We can define the renormalized ϕ^4 Lagrangian density in a manner similar to what was done for QED in Eq. (8.5.1). We have

$$\mathcal{L} = \tfrac{1}{2}\partial_\mu\phi_0\partial^\mu\phi_0 - \tfrac{1}{2}m_0^2\phi_0^2 - (\lambda_0/4!)\phi_0^4 = \tfrac{1}{2}Z\partial_\mu\phi\partial^\mu\phi - \tfrac{1}{2}Zm_0\phi^2 - (\lambda_0/4!)Z^2\phi^4$$

$$= \left[\tfrac{1}{2}\partial_\mu\phi\partial^\mu\phi - \tfrac{1}{2}m^2\phi^2 - (\mu^\epsilon\lambda/4!)\phi^4\right] + \left\{\tfrac{1}{2}\delta_z\partial_\mu\phi\partial^\mu\phi - \tfrac{1}{2}\delta_m\phi^2 - (\mu^\epsilon\delta_\lambda/4!)\phi^4\right\}, \quad (8.7.1)$$

where m and λ are the renormalized mass and coupling, respectively, in the renormalized part of $\mathcal{L}$ indicated by $[\cdots]$. In nonperturbative studies one typically starts with the first line of Eq. (8.7.1), whereas in perturbation theory we use the second line. The counterterms are in $\{\cdots\}$ with the counterterm coefficients defined as

$$\delta_z \equiv Z - 1, \quad \delta_m \equiv Z m_0^2 - m^2, \quad \mu^\epsilon \delta_\lambda \equiv Z^2 \lambda_0 - \mu^\epsilon \lambda. \tag{8.7.2}$$

The effect of adding the counterterms is to introduce new contributions to the Feynman rules in addition to Eqs. (7.6.4) and (7.6.10). These are read off from the Lagrangian density and indicated with a crossed circle,

$$\longrightarrow\!\!\otimes\!\!\longrightarrow \; = \; i(p^2 \delta_z - \delta_m), \qquad\qquad \bigotimes \; = \; -i\mu^\epsilon \delta_\lambda. \tag{8.7.3}$$

These diagrams are to be understood and used as new two-point and three-point interaction vertices. The obvious question arises as to why we do not simply modify the scalar Feynman propagator to

$$\frac{i}{p^2 - m^2 + i\epsilon} \;\rightarrow\; \frac{i}{(1 + \delta Z)p^2 - (1 + \delta m)m^2 + i\epsilon} \tag{8.7.4}$$

rather than treat $i(p^2 \delta_z - \delta_m)$ as a new two-point vertex. The reason is that $\delta_z, \delta_m \sim \mathcal{O}(\lambda)$ and $\delta\lambda \sim \mathcal{O}(\lambda^2)$ and we wish to do a systematic perturbative expansion in powers of the coupling constant λ. Each of δ_z, δ_m and δ_λ will be a power series expansion in λ with each term removing divergences at the corresponding order. The right-hand side of Eq. (8.7.4) generates all powers of λ in a polynomial expansion,

$$\text{RHS} = \frac{i}{p^2 - m^2 + i\epsilon} + \frac{i}{p^2 - m^2 + i\epsilon}[i(p^2 \delta_z - \delta_m)]\frac{i}{p^2 - m^2 + i\epsilon} + \cdots,$$

where we have expanded around $\lambda = 0$. We have made use of the geometric series $1/[a - r] = (1/a)/[1 - (r/a)] = (1/a)[1 + (r/a) + (r/a)^2 + \cdots]$ for $|r/a| < 1$ with $a = -i(p^2 - m + i\epsilon)$ and $r = -i(-p^2 \delta_z + \delta_m)$. We see that $i(p^2 \delta_z - \delta_m)$ naturally emerges as the two-point interaction vertex.

In the language of Secs. 8.5.2 and 8.5.3 the renormalized quantities m and λ depend on the renormalization scheme and the renormalization point μ, which we can imply by writing $m(\mu)$ and $\lambda(\mu)$. The counterterms $\delta_z(\mu, \Lambda)$ and $\delta_m(\mu, \Lambda)$ depend on the renormalization scheme, the renormalization point μ and the regularization parameter Λ, where $\Lambda \sim 1/\epsilon$ for dimensional regularization. The advantage of the counterterm approach to perturbation theory is that we now use perturbation theory based on renormalized m and λ and the form of $\mathcal{L}$ in the second line of Eq. (8.7.1) to calculate amplitudes. These are a function of m, λ, δ_z, δ_m, δ_λ and $\Lambda \sim 1/\epsilon$. There are no longer any explicit Z factors as these have been absorbed into the counterterms. Write the renormalized propagator as $D(p)$, the renormalized two-point 1PI amplitude as $-i\Pi(p^2)$ and the renormalized 1PI vertex as $-i\lambda\Gamma(p_1, \ldots, p_4)$. With our conventions the p_i are outgoing momenta and it is understood that $\sum_i p_i = 0$ so that *only three momenta are independent*. We have

$$D(p) = \frac{i}{p^2 - m^2 - \Pi(p^2) + i\epsilon} \; = \; \bigcirc\!\!\!-, \qquad -i\Pi(p^2) \; = \; \bigcirc\!\!\!-, \tag{8.7.5}$$

$$-i\lambda\Gamma(p_1, \ldots, p_4) \; = \; \bigotimes .$$

The four-point vertex has only three independent momenta denoted p'', p', p, since $p_1 + p_2 = p_3 + p_4$ for incoming momenta p_1, p_2 and outgoing momenta p_3, p_4. For the propagator Eqs. (7.5.51)–(7.5.55) still apply but now with the replacements $m_0 \to m$, $D_0(p) \to i/p^2 - m^2 + i\epsilon$, $\Pi_r(p^2) \to \Pi(p^2)$ and $Z \to 1$ since we will require $\Pi(p^2) = 0$ and $d\Pi/dp^2 = 0$ at the renormalization point. Similarly, we require for the vertex $\Gamma(\cdots) = 1$ at the renormalization point.

Consider the on-shell renormalization scheme, where m is the physical mass of the ϕ particle and λ is the coupling at zero momentum transfer, where all four external momenta are on shell $p_1^2 = \cdots = p_4^2 = m^2$. By symmetry $\Gamma(p_1, \ldots, p_4)$ is invariant under permutations of the momenta. In terms of the Mandelstam variables in Fig. 7.10 this renormalization point is $s = 4m^2$, $t = u = 0$, where s is the square of the sum of any two incoming momenta and t and u are the squared differences of each of the outgoing momenta with one of the incoming momenta.

The on-shell scheme is then specified by the *on-shell renormalization conditions*,

$$\left. \frac{d\Pi(p^2)}{dp^2} \right|_{p^2=m^2} = 0, \quad \Pi(m^2) = 0, \quad \Gamma(\cdots)|_{s=4m^2,\, t=u=0} = 1. \tag{8.7.6}$$

The value of m here is the *pole mass*, and the renormalization conditions ensure that pole mass occurs at $p^2 = m^2$ and that the residue of the pole is unity. More precisely, the pole mass is the location of the smallest p^2 singularity in $D(p)$ in the on-shell renormalization scheme. The pole mass is the physical mass of the particle as shown in the Kállén-Lehmann representation in Fig. 7.2. The subtractions leading to Eq. (8.7.5) remove all divergences and so $\Pi(p^2)$ and $D(p)$ remain finite and approach well-defined finite limits as $\Lambda \sim 1/\epsilon \to \infty$. The resulting $D(p)$ is the $D_F(p)$ of Eqs. (7.2.11) and (7.5.54) but with $Z = 1$. Since the difference is vanishingly small, we typically do not bother to distinguish between $D(p)$ with $\Lambda \sim 1/\epsilon \gg 1$ and with $\Lambda \sim 1/\epsilon \to \infty$ and understand which is intended from the context.

Since λ is an arbitrary parameter, every order of perturbation theory must be independently free of divergences. The counterterm parameters will themselves be power series expansions in λ and so will provide independent contributions order by order in λ to cancel divergences. Calculations of contributions to amplitudes at $\mathcal{O}(\lambda^n)$ become rapidly more complex as n increases.

Propagator renormalization: The Feynman diagrams contributing to the 1PI two-point amplitude $-i\Pi(p^2)$ were given earlier in Eq. (7.5.49); however, we must now add the appropriate counterterm Feynman diagrams using the new vertices in Eq. (8.7.3). Recall that δ_z and δ_m can in principle contribute at $\mathcal{O}(\lambda)$ and above and that δ_λ can contribute at $\mathcal{O}(\lambda^2)$ and above, since any vertex contribution beyond tree level must contain at lease two vertices. Explicitly showing all diagrams that can contribute to $\mathcal{O}(\lambda^2)$ we have

$$-i\Pi(p^2) = \tag{8.7.7}$$

If we only keep terms of $\mathcal{O}(\lambda)$ then $-i\Pi(p^2) = $, where we understand that we only imply the $\mathcal{O}(\lambda)$ contribution from the counterterm here. The other four diagrams and the λ^2 term in contribute at $\mathcal{O}(\lambda^2)$. Working to $\mathcal{O}(\lambda)$ we use the calculation of the first diagram in Eq. (8.6.19) and write

$$-i\Pi(p^2) = -i\Pi_1(p^2) + i(p^2\delta_z - \delta_m) + \mathcal{O}(\lambda^2). \tag{8.7.8}$$

Since $-i\Pi(p^2)$ is from a contact interaction it is p^2-independent. The requirement that $d\Pi(p^2)/dp^2|_{p^2=m^2} = 0$ leads to $\delta_z = d\Pi_1(p^2)/dp^2|_{p^2=m^2} = 0 + \mathcal{O}(\lambda^2)$. The requirement that $\Pi(m^2) = 0$ leads to $\delta_m = -\Pi_1(m^2) + \mathcal{O}(\lambda^2)$. In summary then,

$$\delta_z = 0 + \mathcal{O}(\lambda^2), \tag{8.7.9}$$

$$\delta_m = -\frac{\nu^{4-d}\lambda\Gamma(1-\frac{d}{2})}{2(4\pi)^{d/2}(m^2)^{1-\frac{d}{2}}} + \mathcal{O}(\lambda^2) = \frac{m^2\lambda}{32\pi^2}\left[\frac{2}{\epsilon} - \ln\left(\frac{m^2}{\tilde{\nu}^2 e}\right) + \mathcal{O}(\epsilon)\right] + \mathcal{O}(\lambda^2).$$

where we now use ν rather than μ to denote the dimensional regularization mass scale in the on-shell renormalization scheme, since here it is not a renormalization scale. It will disappear when we take the limit $\epsilon \to 0$ in renormalized quantities. When we later introduce the MS and $\overline{\text{MS}}$ renormalization schemes this mass scale is used to denote the renormalization scale and so we will use μ then. We have $\Pi(p^2) = 0 + \mathcal{O}(\lambda^2)$ and so $D(p) = (i/p^2 - m^2 + i\epsilon) + \mathcal{O}(\lambda^2)$, which means that changes to the propagator in ϕ^4 theory start at $\mathcal{O}(\lambda^2)$, which implies two-loop contributions.

Turning now to $\mathcal{O}(\lambda^2)$ contributions it follows from the above arguments that the sum of the second diagram in parentheses and the last counterterm diagram cancel at $\mathcal{O}(\lambda^2)$ and so do not contribute at that order, $\begin{matrix}\includegraphics\end{matrix} + \begin{matrix}\includegraphics\end{matrix} = \mathcal{O}(\lambda^3)$. This will then be a general property of the theory where in moving from $\mathcal{O}(\lambda^n)$ to $\mathcal{O}(\lambda^{n+1})$, the addition of a self-energy loop on a propagator will be canceled by the diagram with the loop replaced by the counterterm. So in ϕ^4 theory we do not need to bother with such diagrams, which is a welcome simplification and so we have

$$-i\Pi(p^2) = \bigcirc + \ominus + \mathcal{O}(\lambda^3) + \otimes + \bigcirc, \tag{8.7.10}$$

where the $\mathcal{O}(\lambda^2)$ term of $\otimes$ is determined by Eq. (8.7.6). However, to evaluate the last diagram we first need the vertex counterterm $\times$ in Eq. (8.7.3) at $\mathcal{O}(\lambda^2)$. To build physical intuition note that the $\mathcal{O}(\lambda^2)$ contribution from the right-hand diagram $\bigcirc$ cancels divergences when only one of the loop momenta in $\ominus$ becomes infinite, whereas the $\mathcal{O}(\lambda^2)$ contribution from $\otimes$ cancels divergences when both loop momenta become infinite.

Vertex renormalization: The lowest-order contribution to the vertex renormalization is at $\mathcal{O}(\lambda^2)$. We now redefine p_1 and p_2 as *incoming momenta* and p_3 and p_4 as *outgoing momenta* so that we can use the Mandelstam variables defined in Sec. 7.3.4. We are not in general assuming $p_1, \ldots, p_4$ are on-shell here. For off-shell momenta s, t and u still have their usual meanings. This then leads to

$$\begin{matrix} p_1 \\ \\ p_2 \end{matrix} \begin{matrix}\includegraphics\end{matrix} \begin{matrix} p_3 \\ \\ p_4 \end{matrix} \xrightarrow{\text{1PI part}} -i\lambda\Gamma(-p_1,-p_2,p_3,p_4) = \begin{matrix}\includegraphics\end{matrix} \tag{8.7.11}$$

$$= \begin{matrix}\includegraphics\end{matrix} + \left(\begin{matrix}\includegraphics\end{matrix}\overset{\widehat{\ell+p}}{\underset{\ell}{}} + [s \to t] + [s \to u]\right) + \mathcal{O}(\lambda^3) + \begin{matrix}\includegraphics\end{matrix}.$$

The $\mathcal{O}(\lambda^2)$ contribution is the sum of the terms in parentheses $(\cdots)$ and the $\mathcal{O}(\lambda^2)$ contribution from the vertex counterterm $\times$. The renormalization point is specified by $s = (p_1 + p_2)^2 = m^2$, $t = (p_1 - p_3)^2 = 0$ and $u = (p_1 - p_4)^2 = 0$ with all momenta on-shell. In the first diagram in $(\cdots)$ we see that $s = p^2$. The second diagram in $(\cdots)$ is the first rotated clockwise by $90°$ and the third is the same as the second but the two outgoing lines (or incoming lines) are crossed. The effect is that the p in the loop satisfies $s = p^2$, $t = p^2$ and $u = p^2$, respectively, for the three diagrams. So we only need evaluate one diagram as a function of p^2 and we know the result for all three. Using Feynman rules the result for any one of the three diagrams in $d = 4 - \epsilon$ dimensions with $\lambda \to \nu^\epsilon \lambda$ for some arbitrary ν we have

$$
\begin{aligned}
(-i\nu^\epsilon\lambda)\Gamma_2(p^2) &\equiv \includegraphics{loop} = \frac{(-i\nu^\epsilon\lambda)^2}{2} \int \frac{d^d\ell}{(2\pi)^d} \frac{i}{\ell^2 - m^2 + i\epsilon} \frac{i}{(\ell + p)^2 - m^2 + i\epsilon} \\
&= -\frac{(-i\nu^\epsilon\lambda)^2}{2} \int_0^1 dx \int \frac{d^d\ell}{(2\pi)^d} \frac{1}{\{(1-x)[\ell^2 - m^2 + i\epsilon] + x[(\ell + p)^2 - m^2 + i\epsilon]\}^2} \quad (8.7.12) \\
&= -\frac{(-i\nu^\epsilon\lambda)^2}{2} \int_0^1 dx \int \frac{d^d\ell}{(2\pi)^d} \frac{1}{\{\ell^2 + 2xp\cdot\ell + xp^2 - m^2 + i\epsilon\}^2} \\
&= -\frac{(-i\nu^\epsilon\lambda)^2}{2} \int_0^1 dx \left[i\frac{1}{(4\pi)^{d/2}} \right] \frac{\Gamma(2 - \frac{d}{2})}{\Gamma(2)} \left(\frac{1}{(m^2 - xp^2) + (xp)^2} \right)^{2 - \frac{d}{2}} \\
&= \frac{i\nu^\epsilon\lambda^2}{2} \int_0^1 dx \frac{\Gamma(2 - \frac{d}{2})}{(4\pi)^{d/2}} \left(\frac{\nu^2}{m^2 - x(1-x)p^2} \right)^{2 - \frac{d}{2}} \\
&= \frac{i\nu^\epsilon\lambda^2}{32\pi^2} \int_0^1 dx \left\{ \frac{2}{\epsilon} - \ln\left[\frac{m^2 - x(1-x)p^2}{\tilde{\nu}^2} \right] + \mathcal{O}(\epsilon) \right\},
\end{aligned}
$$

where we have used (i) a symmetry factor $S = 2$ from a horizontal flip of the loop, (ii) the Feynman parameterization in Eq. (A.5.1), (iii) the dimensional regularization result of Eq. (A.6.11), (iv) the ϵ expansion in Eq. (A.6.17) and (v) the definition $\tilde{\nu}^2$ in Eq. (8.6.18).

Up to $\mathcal{O}(\lambda^2)$ the contribution to the 1PI vertex, $-i\nu^\epsilon\lambda\Gamma(\cdots)$, is then

$$
-i\nu^\epsilon\lambda\Gamma(\cdots) = -i\nu^\epsilon\lambda \left\{ 1 + [\Gamma_2(s) + \Gamma_2(t) + \Gamma_2(u)] \right\} + \mathcal{O}(\lambda^3) + (-i\nu^\epsilon\delta_\lambda). \quad (8.7.13)
$$

The renormalization point requirement is that $\Gamma(\cdots)|_{s=4m^2,\, t=u=0} = 1$ and so

$$
\delta_\lambda = \frac{\lambda^2}{32\pi^2} \int_0^1 dx \left\{ \frac{6}{\epsilon} - 3\gamma + 3\ln(4\pi) - \ln\left[\frac{m^2 - x(1-x)4m^2}{\nu^2} \right] - 2\ln\frac{m^2}{\nu^2} + \mathcal{O}(\epsilon) \right\} + \mathcal{O}(\lambda^3).
$$

This is what is needed to calculate the last diagram in Eq. (8.7.10), but we will not complete that task here. All $1/\epsilon$ divergences then cancel and taking $\epsilon \to 0$ gives the finite result for the 1PI renormalized vertex

$$
\begin{aligned}
-i\lambda\Gamma(\cdots) = -i\lambda - \frac{i\lambda^2}{32\pi^2} \int_0^1 dx \left\{ \ln\left[\frac{m^2 - x(1-x)s}{m^2 - x(1-x)4m^2} \right] \right. & \\
\left. + \ln\left[\frac{m^2 - x(1-x)t}{m^2} \right] + \ln\left[\frac{m^2 - x(1-x)u}{m^2} \right] \right\} + \mathcal{O}(\lambda^3), &
\end{aligned} \quad (8.7.14)
$$

which satisfies the renormalization point condition. If the external legs are all on-shell and since m is the physical mass and $Z = 1$, then in the LSZ formalism the 1PI vertex in ϕ^4 theory is the invariant scattering amplitude for two-body scattering,

$$i\mathcal{M}(s, t, u) = -i\lambda\Gamma(\cdots)|_{p_1^2 = \cdots = p_4^2 = m^2}. \tag{8.7.15}$$

So Eq. (8.7.14) gives the $\mathcal{O}(\lambda^2)$ one-loop correction to the tree-level invariant amplitude in Eq. (7.5.46), which we recall was $i\mathcal{M}^{\text{tree}} = -i\lambda$.

We will not pursue the renormalization program for perturbative ϕ^4 further here, but the essential elements have been laid out. An elegant discussion of ϕ^4 theory at $\mathcal{O}(\lambda^3)$ involving two-loop diagrams can be found in Peskin and Schroeder (1995). There it is shown that the two-loop contribution to the 1PI four-point function, $-i\Gamma(\cdots)$, is finite and that the local counterterms in Eq. (8.7.3) remove the overlapping divergences that occur. This is as expected from the BPHZ theorem.

8.7.2 Renormalized Perturbative Yukawa Theory

The Yukawa theory interaction is $i\mathcal{L}_{\text{int}}^{\text{Yuk}} = -ig\bar{\psi}\phi\psi$ from Eq. (7.6.15) and with scalar and fermion physical masses m and m_f, respectively. The counterterms are

$$\text{—}\!\otimes\!\text{—} = i(p^2\delta_z - \delta_m), \quad \blacktriangleleft\!\otimes\!\blacktriangleleft = i(\not{p}\delta_2 - \delta_{m_f}), \quad \text{—}\!\otimes\!\diagup = -i\nu^{\epsilon/2}\delta_g, \tag{8.7.16}$$

where these follow from the Yukawa theory equivalent of Eq. (8.5.1) for QED. Let us consider the $\mathcal{O}(g)$ one-loop correction to the 1PI two-point amplitude,

$$-i\Pi(p^2) = \text{—}\bigcirc\text{—} + \mathcal{O}(g^3) + \text{—}\!\otimes\!\text{—}. \tag{8.7.17}$$

The fermion loop gives a factor (-1), which leads to an $\mathcal{O}(g^2)$ contribution

$$-i\Pi(p^2) = -(-i\nu^{\epsilon/2}g)^2 \int \frac{d^d\ell}{(2\pi)^d} \text{tr}\left[\frac{i(\not{p} + \not{\ell} + m_f)i(\not{\ell} + m_f)}{[(p+\ell)^2 - m_f^2 + i\epsilon][\ell^2 - m_f^2 + i\epsilon]}\right] + i(p^2\delta_z - \delta_m)$$

$$= -\nu^\epsilon g^2 \int \frac{d^d\ell}{(2\pi)^d} \frac{4[(p+\ell)\cdot\ell + m_f^2]}{[(p+\ell)^2 - m_f^2 + i\epsilon][\ell^2 - m_f^2 + i\epsilon]} + i(p^2\delta_z - \delta_m)$$

$$= -4\nu^\epsilon g^2 \int_0^1 dx \int \frac{d^d\ell}{(2\pi)^d} \frac{(p+\ell)\cdot\ell + m_f^2}{\{\ell^2 + 2xp\cdot\ell + xp^2 - m_f^2 + i\epsilon\}^2} + i(p^2\delta_z - \delta_m)$$

$$= -4\nu^\epsilon g^2 \int_0^1 dx \int \frac{d^d\ell'}{(2\pi)^d} \frac{\ell'^2 - x(1-x)p^2 + m_f^2}{\{\ell'^2 + x(1-x)p^2 - m_f^2 + i\epsilon\}^2} + i(p^2\delta_z - \delta_m)$$

$$= -4\nu^\epsilon g^2 \left[\frac{i}{(4\pi)^{d/2}}\right] \int_0^1 dx \frac{-\frac{d}{2}\Gamma(1 - \frac{d}{2}) + \Gamma(2 - \frac{d}{2})}{\{m_f^2 - x(1-x)p^2\}^{1-(d/2)}} + i(p^2\delta_z - \delta_m)$$

$$= \frac{4i\nu^\epsilon g^2}{(4\pi)^{d/2}} \int_0^1 dx \frac{(d-1)\Gamma(2 - \frac{d}{2})}{\{m_f^2 - x(1-x)p^2\}^{1-(d/2)}(1 - \frac{d}{2})} + i(p^2\delta_z - \delta_m), \tag{8.7.18}$$

where we have used (i) the d-dimensional trace identities in Eq. (A.6.20), (ii) the Feynman parameterization in Eq. (A.5.1) with $(p + \ell)$ in A_1, (iii) a change of variables $\ell' = \ell + xp$ to have the denominator a function of ℓ'^2 only, (iv) that odd integrals vanish as discussed following

Eq. (A.6.10) so that $\int d^d \ell'(p \cdot \ell') f(\ell'^2) = 0$, (v) Eqs. (A.6.4) and (A.6.5) to evaluate the ℓ' integral and (vi) Eq. (A.6.16) to obtain the result $\Gamma(1 - \frac{d}{2}) = \Gamma(2 - \frac{d}{2})/(1 - \frac{d}{2})$. Differentiate Eq. (8.7.18) with respect to p^2, set $p^2 = m_f^2$ and use the on-shell renormalization condition $d\Pi/dp^2|_{p^2=m^2} = 0$, to find

$$\delta_z = -\frac{4g^2}{(4\pi)^{d/2}} \int_0^1 dx \, \frac{x(1-x)(d-1)\Gamma(2 - \frac{d}{2})(\nu^2)^{2-(d/2)}}{\{m_f^2 - x(1-x)m^2\}^{2-(d/2)}} \tag{8.7.19}$$

$$= -\frac{3g^2}{4\pi^2} \int_0^1 dx \, x(1-x) \left\{ \frac{2}{\epsilon} - \frac{2}{3} - \ln\left[\frac{m_f^2 - x(1-x)m^2}{\tilde{\nu}^2}\right] + \mathcal{O}(\epsilon) \right\},$$

where we used Eq. (A.6.17), $(d-1)[\frac{2}{\epsilon} + \cdots] = (3-\epsilon)[\frac{2}{\epsilon} + \cdots] = 3[\frac{2}{\epsilon} - \frac{2}{3} + \cdots]$, $\nu^\epsilon = (\nu^2)^{2-(d/2)}$ and $\tilde{\nu}^2$ as defined by Eq. (8.6.18). We can now determine δ_m by requiring that $\Pi(p^2)|_{p^2=m^2} = 0$. The $\mathcal{O}(g^2)$ correction to the renormalized scalar propagator can then be obtained and ν disappears as $\epsilon \to 0$. The completion of the calculation is left as an exercise.

Referring to the discussion in Sec. A.6 below Eq. (A.6.12), observe that δ_z is associated with a logarithmic divergence $\sim \Gamma(2 - \frac{d}{2})$ and δ_m is associated with a quadratic divergence $\sim \Gamma(1 - \frac{d}{2})$. This is the usual pattern for a scalar field, although in ϕ^4 theory we saw from Eq. (8.7.12) that the mass renormalization was associated with a logarithmic divergence $\sim \Gamma(2 - \frac{d}{2})$. This observation of a quadratic mass divergence for scalar fields is the basis of the so-called hierarchy problem for the Higgs mass. If gravity at the Planck scale or some extension of the Standard Model at the GUT scale provides the mechanism for the origin of the Higgs field, then the renormalization of the quadratic divergence of the Higgs mass to give a small physical value relative to these scales seems to be fine-tuned or unnatural.

8.7.3 Renormalized Perturbative QED

Consider a single fermion species here. From Eq. (8.5.1) we can read off the counterterm contributions to the Feynman rules as $i\mathcal{L}$ in the usual way,

$$\vcenter{\hbox{\includegraphics{photon}}} = -i(q^2 g^{\mu\nu} - q^\mu q^\nu)\delta_3, \quad \vcenter{\hbox{\includegraphics{fermion}}} = i(\slashed{p}\,\delta_2 - \delta_m), \quad \vcenter{\hbox{\includegraphics{vertex}}} = -i\nu^{\epsilon/2} q_c \gamma^\mu \delta_1,$$

$$\tag{8.7.20}$$

where q_c is the renormalized charge of the fermion at the on-shell renormalization point. The on-shell renormalization conditions of Eq. (8.5.33) now have the form

$$\Sigma(p)|_{\slashed{p}=m} = \frac{d\Sigma(p)}{d\slashed{p}}\bigg|_{\slashed{p}=m} = \Pi(q^2)|_{q^2=0} = \Lambda^\mu(p,p)|_{\slashed{p}=m} = 0. \tag{8.7.21}$$

The renormalized propagators and vertex are

$$S(p) = \frac{i}{\slashed{p} - m - \Sigma(p)}, \quad D^{\mu\nu}(q) = \frac{i[-g^{\mu\nu} + \frac{q^\mu q^\nu}{q^2 + i\epsilon}]}{(q^2 + i\epsilon)[1 + \Pi(q^2)]} - \frac{i\xi q^\mu q^\nu}{(q^2 + i\epsilon)^2}, \tag{8.7.22}$$

$$\Gamma^\mu(p',p) = \gamma^\mu + \Lambda^\mu(p',p).$$

To do calculations of the one-loop $\mathcal{O}(q_c^2)$ contributions to these renormalized quantities we use dimensional regularization. To control infrared divergences we temporarily introduce a photon mass, m_γ, and we take $m_\gamma \to 0$ at the end of calculations. A discussion of both Pauli-Villars and dimensional regularization for these calculations can be found in Peskin and Schroeder (1995) and

Schwartz (2013). Since gauge invariance is valid for any value of the charge, q_c, then physical amplitudes must be gauge invariant order by order in perturbation theory. We work in Feynman gauge here.

Renormalized fermion self-energy:[3] This is $-i\Sigma(p) = -i\Sigma_2(p) + \mathcal{O}(q_c^4)$, where the $\mathcal{O}(q_c^2)$ contribution is calculated similarly to Eq. (8.7.18),

$$-i\Sigma_2(p) = \text{[diagram]} + \text{[diagram]} \tag{8.7.23}$$

$$= (-i\nu^{\epsilon/2}q_c)^2 \int \frac{d^d\ell}{(2\pi)^d} \frac{\gamma^\nu i(\slashed{p}+\slashed{\ell}+m)\gamma^\mu}{(p+\ell)^2 - m^2 + i\epsilon} \frac{i(-g_{\nu\mu})}{\ell^2 - m_\gamma^2 + i\epsilon} + i(\slashed{p}\delta_2 - \delta_m)$$

$$= -\nu^\epsilon q_c^2 \int \frac{d^d\ell}{(2\pi)^d} \frac{-(2-\epsilon)(\slashed{p}+\slashed{\ell}) + (4-\epsilon)m}{[(p+\ell)^2 - m^2 + i\epsilon][\ell^2 - m_\gamma^2 + i\epsilon]} + i(\slashed{p}\delta_2 - \delta_m)$$

$$= -\nu^\epsilon q_c^2 \int_0^1 dx \int \frac{d^d\ell}{(2\pi)^d} \frac{-(2-\epsilon)(\slashed{p}+\slashed{\ell}) + (4-\epsilon)m}{\{\ell^2 + 2xp\cdot\ell + xp^2 - xm^2 + (1-x)m_\gamma^2 + i\epsilon\}^2} + i(\slashed{p}\delta_2 - \delta_m)$$

$$= -\nu^\epsilon q_c^2 \int_0^1 dx \int \frac{d^d\ell'}{(2\pi)^d} \frac{-(2-\epsilon)[(1-x)\slashed{p} + 2\slashed{\ell}'] + (4-\epsilon)m}{\{\ell'^2 + x(1-x)p^2 - xm^2 - (1-x)m_\gamma^2 + i\epsilon\}^2} + i(\slashed{p}\delta_2 - \delta_m)$$

$$= -\nu^\epsilon q_c^2 \left[\frac{i}{(4\pi)^{d/2}}\right] \int_0^1 dx \frac{\Gamma(2-\frac{d}{2})\{(4-\epsilon)m - (2-\epsilon)(1-x)\slashed{p}\}}{\{xm^2 + (1-x)m_\gamma^2 - x(1-x)p^2\}^{2-(d/2)}} + i(\slashed{p}\delta_2 - \delta_m)$$

$$= -\frac{iq_c^2}{(4\pi)^{d/2}} \int_0^1 dx \frac{\Gamma(2-\frac{d}{2})\{(4-\epsilon)m - (2-\epsilon)x\slashed{p}\}(\nu^2)^{2-(d/2)}}{\{(1-x)m^2 + xm_\gamma^2 - x(1-x)p^2\}^{2-(d/2)}} + i(\slashed{p}\delta_2 - \delta_m)$$

$$= -\frac{iq_c^2}{(4\pi)^2} \int_0^1 dx \left[\left\{\frac{2}{\epsilon} - \ln\frac{\Delta_\Sigma}{\tilde{\nu}^2}\right\}\{4m - 2x\slashed{p}\} - 2(m - x\slashed{p})\right] + i(\slashed{p}\delta_2 - \delta_m),$$

where we defined for brevity $\Delta_\Sigma \equiv (1-x)m^2 + xm_\gamma^2 - x(1-x)p^2$ and we have used $-\gamma - \ln(\Delta_\Sigma/\nu^2) + \ln(4\pi) = -\ln(\Delta_\Sigma/\tilde{\nu}^2)$. The spinor matrices are ordered right to left in the direction of the arrow (γ^μ is left, γ^ν is right) and the loop momentum ℓ is counterclockwise. We have used (i) the contractions $\gamma_\mu\gamma^\mu = d = (4-\epsilon)$ and $\gamma_\mu\slashed{k}\gamma^\mu = -(2-\epsilon)\slashed{k}$ from Eq. (A.6.19), (ii) the Feynman parameterization in Eq. (A.5.1) with A_1 containing $(p+\ell)$, (iii) a change of variable $\ell' = \ell + xp$, (iv) that odd ℓ' integrals vanish, (v) Eq. (A.6.4) to do the ℓ' integral, (vi) that Eq. (A.5.1) is symmetric under the exchange $x \leftrightarrow (1-x)$ and this simplifes the last line and finally (vii) Eq. (A.6.17). Recall $p^2 = \slashed{p}^2$ in the discussion of the formal equation $\slashed{p} = m$ in the proof of Eq. (8.5.33). The on-shell renormalization conditions in Eq. (8.7.21) require $d\Sigma_2(p)/d\slashed{p}|_{\slashed{p}=m} = 0$ and $\Sigma_2(p)|_{\slashed{p}=m} = 0$. Using $2 - \frac{d}{2} = \frac{\epsilon}{2}$ and neglecting $\mathcal{O}(\epsilon)$ terms gives to $\mathcal{O}(\alpha^2)$,

$$\delta_2 = \frac{q_c^2}{(4\pi)^{d/2}} \int_0^1 dx \frac{\Gamma(2-\frac{d}{2})(\nu^2)^{2-(d/2)}}{\{(1-x)^2m^2 + xm_\gamma^2\}^{2-(d/2)}} \tag{8.7.24}$$

$$\times \left[-(2-\epsilon)x + \frac{\epsilon}{2}\frac{2m^2x(1-x)[2(2-x) - \epsilon(1-x)]}{(1-x)^2m^2 + xm_\gamma^2}\right]$$

$$= -\frac{q_c^2}{4\pi}\frac{1}{2\pi} \int_0^1 dx \left[x\left\{\frac{2}{\epsilon} - \ln\frac{(1-x)^2m^2 + xm_\gamma^2}{\tilde{\nu}^2} - 1 - \frac{m^2(1-x)[2(2-x)]}{(1-x)^2m^2 + xm_\gamma^2}\right\}\right],$$

---

[3] Our sign for $\Sigma(p)$ is that of Peskin and Schroeder (1995) but Schwartz (2013) uses the opposite sign. For the photon scalar self-energy $\Pi(q^2)$ the converse is true. Here δ_2 and δ_m are as in Peskin and Schroeder (1995), whereas Schwartz (2013) defines δ_m^{Sch} as $\delta_m = (\delta_m^{\text{Sch}} + \delta_2)m$.

$$\delta_m = m\delta_2 - \frac{q_c^2}{(4\pi)^{d/2}} \int_0^1 dx \, \frac{\Gamma(2-\frac{d}{2})(\nu^2)^{2-(d/2)} m[2(2-x) - \epsilon(1-x)]}{\{(1-x)^2 m^2 + x m_\gamma^2\}^{2-(d/2)}} \tag{8.7.25}$$

$$= m\delta_2 - \frac{q_c^2}{4\pi} \frac{m}{2\pi} \int_0^1 dx \left[\left\{ \frac{2}{\epsilon} - \ln \frac{(1-x)^2 m^2 + x m_\gamma^2}{\tilde{\nu}^2} \right\} (2-x) - (1-x) \right].$$

Renormalized vacuum polarization: The renormalized vacuum polarization is $i\Pi^{\mu\nu}(q)(p) = i\Pi_2^{\mu\nu}(q)(p) + \mathcal{O}(q_c^4)$, where $\Pi^{\mu\nu}(q) = \left[-g^{\mu\nu} q^2 + q^\mu q^\nu \right] \Pi(q^2)$ from Eq. (7.5.117). The calculation of the $\mathcal{O}(q_c^2)$ contribution is

$$i\Pi_2^{\mu\nu}(q) = \raisebox{-1ex}{\includegraphics[height=3ex]{placeholder}} \tag{8.7.26}$$

$$= (-1)(-i\nu^{\epsilon/2} q_c)^2 \int \frac{d^d\ell}{(2\pi)^d} \, \mathrm{tr} \left[\frac{i(\slashed{\ell}+m)\gamma^\mu i(\slashed{\ell}+\slashed{q}+m)\gamma^\nu}{[(\ell+q)^2 - m^2 + i\epsilon][\ell^2 - m^2 + i\epsilon]} \right] - i(q^2 g^{\mu\nu} - q^\mu q^\nu)\delta_3$$

$$= -4\nu^\epsilon q_c^2 \int \frac{d^d\ell}{(2\pi)^d} \frac{\ell^\mu(\ell+q)^\nu + \ell^\nu(\ell+q)^\mu + g^{\mu\nu}[m^2 - \ell\cdot(\ell+q)]}{[(\ell+q)^2 - m^2 + i\epsilon][\ell^2 - m^2 + i\epsilon]} - i(q^2 g^{\mu\nu} - q^\mu q^\nu)\delta_3$$

$$= -4\nu^\epsilon q_c^2 \int_0^1 dx \int \frac{d^d\ell}{(2\pi)^d} \frac{\ell^\mu(\ell+q)^\nu + \ell^\nu(\ell+q)^\mu + g^{\mu\nu}[m^2 - \ell\cdot(\ell+q)]}{\{\ell^2 + 2xq\cdot\ell + xq^2 - m^2 + i\epsilon\}^2} - i(q^2 g^{\mu\nu} - q^\mu q^\nu)\delta_3$$

$$= -4\nu^\epsilon q_c^2 \int_0^1 dx \int \frac{d^d\ell'}{(2\pi)^d} \frac{2\ell'^\mu\ell'^\nu - g^{\mu\nu}\ell'^2 - 2x(1-x)q^\mu q^\nu + g^{\mu\nu}[m^2 + x(1-x)q^2] + \mathcal{O}(\ell')}{\{\ell'^2 + x(1-x)q^2 - m^2 + i\epsilon\}^2}$$
$$\quad - i(q^2 g^{\mu\nu} - q^\mu q^\nu)\delta_3$$

$$= \frac{4i\nu^\epsilon q_c^2}{(4\pi)^{d/2}} \int_0^1 dx \frac{\Gamma(2-\frac{d}{2})[a^2 g^{\mu\nu} + 2x(1-x)q^\mu q^\nu - g^{\mu\nu}(2m^2 - a^2)]}{\{a^2\}^{2-(d/2)}} - i(q^2 g^{\mu\nu} - q^\mu q^\nu)\delta_3$$

$$= \frac{4i\nu^\epsilon q_c^2}{(4\pi)^{d/2}} \int_0^1 dx \frac{\Gamma(2-\frac{d}{2})2x(1-x)[-q^2 g^{\mu\nu} + q^\mu q^\nu]}{\{m^2 - x(1-x)q^2\}^{2-(d/2)}} - i(q^2 g^{\mu\nu} - q^\mu q^\nu)\delta_3$$

$$= -i[q^2 g^{\mu\nu} - q^\mu q^\nu] \left[\frac{8q_c^2}{(4\pi)^{d/2}} \int_0^1 dx \frac{\Gamma(2-\frac{d}{2})\nu^{2-(d/2)} x(1-x)}{\{m^2 - x(1-x)q^2\}^{2-(d/2)}} + \delta_3 \right]$$

$$= -i[q^2 g^{\mu\nu} - q^\mu q^\nu] \left[\frac{2}{\pi} \frac{q_c^2}{4\pi} \int_0^1 dx \, x(1-x) \left\{ \frac{2}{\epsilon} - \ln\left(\frac{m^2 - x(1-x)q^2}{\tilde{\nu}^2} \right) \right\} + \delta_3 \right]$$

$$\equiv -i[q^2 g^{\mu\nu} - q^\mu q^\nu]\Pi_2(q^2),$$

where we neglected $\mathcal{O}(\epsilon)$ terms and we have a (-1) from the fermion loop. Here q is from left to right, the loop momentum ℓ flows clockwise, the top and bottom fermion momenta are $\ell + q$ and ℓ, respectively, and γ^ν is on the left and γ^μ is on the right. We have used (i) the trace identities in Eqs. (A.3.24) and (A.3.25) since they are unchanged in dimensional regularization; (ii) Eq. (A.5.1); (iii) $\ell' = \ell + xq$; (iv) odd ℓ' integrals vanish; (v) Eqs. (A.6.4), (A.6.5) and (A.6.6) with $a^2 \equiv m^2 - x(1-x)q^2$; (vi) $(1-\frac{d}{2})\Gamma(1-\frac{d}{2}) = \Gamma(2-\frac{d}{2})$ from Eq. (A.6.16); and (vii) Eq. (A.6.17). As expected the Ward identity of Eqs. (8.3.35) and (7.5.116) is satisfied.

Since $\Pi(q^2) = \Pi_2(q^2) + \mathcal{O}(q_c^4)$ and since we require $\Pi(0) = 0$ then this leads to

$$\delta_3 = -\frac{q_c^2}{4\pi}\frac{2}{\pi}\int_0^1 dx\, x(1-x)\left\{\frac{2}{\epsilon} - \ln\left(\frac{m^2}{\tilde{\nu}^2}\right)\right\} = -\frac{q_c^2}{4\pi}\frac{1}{3\pi}\left\{\frac{2}{\epsilon} - \ln\left(\frac{m^2}{\tilde{\nu}^2}\right)\right\}, \qquad (8.7.27)$$

since $\int_0^1 dx\, x(1-x) = 1/6$. Taking $\epsilon \to 0$ we find that

$$\Pi(q^2) = -\frac{2}{\pi}\frac{q_c^2}{4\pi}\int_0^1 dx\, x(1-x)\,\ln\left(\frac{m^2 - x(1-x)q^2}{m^2}\right) + \mathcal{O}(q_c^4). \qquad (8.7.28)$$

For $q^2 < 0$, which is the case for spacelike virtual one-photon exchanges in the t or u channels, $\Pi(q^2)$ is real and well defined. For $q^2 > 0$, which is the case for a timelike photon in the s channel, the same is true provided that $m^2 - x(1-x)q^2$ remains positive. However, since the maximum value of $x(1-x)$ is $1/4$, then argument of the logarithm can become negative when $q^2 \geq 4m^2$. We see that there is a logarithmic branch cut on the negative real axis starting at $q^2 = 4m^2$, which is associated with the fact that such a photon can decay into an on-shell fermion-antifermion pair.

In the nonrelativistic limit the Coulomb potential calculation in Eq. (7.6.49) is modified by $V(\mathbf{q}) = q_1q_2/|\mathbf{q}|^2 \to q_1q_2/|\mathbf{q}|^2[1 + \Pi(-|\mathbf{q}|^2)]$. For an electron and proton $q_1q_2 = -e^2$ so using $\alpha = e^2/4\pi$ and taking the Fourier transform leads to

$$V(r) = \int \frac{d^3q}{(2\pi)^3} e^{i\mathbf{q}\cdot\mathbf{x}} V(\mathbf{q}) = -\frac{\alpha}{r}\left(1 + \frac{e^2}{6\pi^2}\int_1^\infty dx\, e^{-2mrx}\frac{2x^2+1}{2x^4}\sqrt{x^2-1}\right)$$

$$\xrightarrow{r\gg 1/m} -\frac{\alpha}{r}\left(1 + \frac{\alpha}{4\sqrt{\pi}}\frac{e^{-2mr}}{(mr)^{3/2}}\right), \qquad (8.7.29)$$

which is the referred to as the *Uehling potential*. For a derivation and further discussion, see Uehling (1935), Peskin and Schroeder (1995), Greiner and Rheinhardt (2009) and Schwartz (2013). The Uehling potential arises from vacuum polarization and contributes to the Lamb shift. Next consider the limit where $Q^2 \equiv -q^2 \gg m^2$,

$$\Pi(q^2) = -\frac{2}{\pi}\frac{q_c^2}{4\pi}\int_0^1 dx\, x(1-x)\left\{\ln[x(1-x)] + \ln\left(\frac{-q^2}{m^2}\right) + \mathcal{O}(m^2/q^2)\right\} + \mathcal{O}(q_c^4)$$

$$= -\frac{1}{3\pi}\frac{q_c^2}{4\pi}\left\{\ln\left(\frac{-q^2}{m^2}\right) - \frac{5}{3} + \mathcal{O}\left(\frac{m^2}{-q^2}\right)\right\} + \mathcal{O}(q_c^4), \qquad (8.7.30)$$

where we have used $\int_0^1 dx\, x(1-x) = 1/6$ and $\int_0^1 dx\, x(1-x)\ln[x(1-x)] = -5/18$ from Eq. (A.5.7). The one-loop contribution for the photon gives a $Q^2 = -q^2$ dependence to the effective fine structure constant, so that for an electron we have

$$\alpha_{\text{eff}}(Q^2) \equiv \frac{\alpha}{1 + \Pi(-Q^2)} \simeq \frac{\alpha}{1 - (\alpha/3\pi)\ln[(Q^2/m^2)e^{-5/3}]}. \qquad (8.7.31)$$

The effective coupling increases at shorter distances, which corresponds to larger Q^2. This is understood as the $e^+ - e^-$ pairs forming virtual dipoles with the positron ends attracted to the negative bare electron charge and the electron ends repelled. The effect of this *vacuum polarization* shields the bare electron charge and so at shorter distances we probe closer to the bare electron and a larger effective charge.

Renormalized vertex dressing: Using dimensional regularization and defining $q \equiv (p' - p)$ the vertex dressing is $\Lambda^\mu(p', p) = \Lambda^\mu_2(p', p) = +\mathcal{O}(q^4_c)$, where

$$(-i\nu^{\epsilon/2}q_c)\Lambda^\mu_2(p', p) = \quad \text{[diagram]} \quad (8.7.32)$$

$$= (-i\nu^{\epsilon/2}q_c)^3 \int \frac{d^d\ell}{(2\pi)^d} \frac{-ig_{\nu\alpha}}{(\ell - p)^2 - m^2_\gamma + i\epsilon} \frac{\gamma^\nu i(\slashed{\ell}' + m)\gamma^\mu i(\slashed{\ell} + m)\gamma^\alpha}{[\ell'^2 - m^2 + i\epsilon][\ell^2 - m^2 + i\epsilon]} + (-i\nu^{\epsilon/2}q_c)\gamma^\mu\delta_1$$

$$= 2(\nu^{\epsilon/2}q_c)^3 \int \frac{d^d\ell}{(2\pi)^d} \frac{\{(\slashed{\ell}\gamma^\mu\slashed{\ell}' - \frac{\epsilon}{2}\slashed{\ell}'\gamma^\mu\slashed{\ell}) - 2m(\ell'^\mu + \ell^\mu) + m^2\gamma^\mu\}}{[(\ell - p)^2 - m^2_\gamma + i\epsilon][\ell'^2 - m^2 + i\epsilon][\ell^2 - m^2 + i\epsilon]} + (-i\nu^{\epsilon/2}q_c)\gamma^\mu\delta_1$$

$$= 4(\nu^{\epsilon/2}q_c)^3 \int_0^1 [dxdydz\,\delta(1 - x - y - z)] \int \frac{d^d\ell}{(2\pi)^d} \frac{\{\cdots\}}{[xA_1 + yA_2 + zA_3]^3} - i\nu^{\epsilon/2}q_c\gamma^\mu\delta_1$$

$$= 4(\nu^{\epsilon/2}q_c)^3 \int_0^1 [\cdots] \int \frac{d^dk}{(2\pi)^d} \frac{(1 - \frac{\epsilon}{2})\slashed{k}\gamma^\mu\slashed{k} + N^\mu}{[k^2 - \Delta + i\epsilon]^3} - i\nu^{\epsilon/2}q_c\gamma^\mu\delta_1$$

$$= 4(\nu^{\epsilon/2}q_c)^3 \int_0^1 [\cdots] \frac{i(-1)^3}{(4\pi)^{d/2}} \left\{ \frac{\Gamma(2 - \frac{d}{2})(2 - \epsilon)^2\gamma^\mu}{8\Delta^{2-(d/2)}} + \frac{\Gamma(3 - \frac{d}{2})N^\mu}{2\Delta^{3-(d/2)}} \right\} - i\nu^{\epsilon/2}q_c\gamma^\mu\delta_1$$

$$= -4i\nu^{\epsilon/2}q^3_c \int_0^1 [\cdots] \frac{\Gamma(2 - \frac{d}{2})}{(4\pi)^{d/2}} \left[\frac{\nu^2}{\Delta}\right]^{2-(d/2)} \left\{ \frac{(2 - \epsilon)^2\gamma^\mu}{8} + \frac{(2 - \frac{d}{2})N^\mu}{2\Delta} \right\} - i\nu^{\epsilon/2}q_c\gamma^\mu\delta_1$$

$$= \frac{-4i\nu^{\epsilon/2}q^3_c}{(4\pi)^2} \int_0^1 [\cdots] \left[\left\{ \frac{2}{\epsilon} - \ln\frac{\Delta}{\nu^2} - \gamma + \ln(4\pi) - 2 \right\} \frac{\gamma^\mu}{2} + \frac{N^\mu}{2\Delta} + \mathcal{O}(\epsilon) \right] - i\nu^{\epsilon/2}q_c\gamma^\mu\delta_1$$

$$= -i\nu^{\epsilon/2}q_c \left\{ \frac{q^2_c}{4\pi} \frac{1}{2\pi} \int_0^1 [\cdots] \left[\left\{ \frac{2}{\epsilon} - \ln\frac{\Delta}{\tilde{\nu}^2} - 2 \right\} \gamma^\mu + \frac{N^\mu}{\Delta} + \mathcal{O}(\epsilon) \right] + \gamma^\mu\delta_1 \right\},$$

where we have made use of (i) the contraction identities in Eq. (A.6.19) with $\mathcal{O}(\epsilon)$ terms retained for only the divergent ℓ^2 numerator contributions; (ii) the Feynman parameterization $1/A_1 A_2 A_3 = 2 \int dxdydz\,\delta(1 - x - y - z)/[xA_1 + yA_2 + zA_3]^3$ in Eq. (A.5.6) with A_1, A_2, A_3 related to the $\ell, \ell', \ell - p$ terms respectively; (iii) that we have $D \equiv xA_1 + yA_2 + zA_3 = \ell^2 + 2\ell \cdot (yq - zp) + yq^2 + zp^2 - (x + y)m^2 - zm^2_\gamma + i\epsilon$; (iv) a shift in the loop momentum $k = \ell + yq - zp$ and the definition

$$\Delta \equiv -2yzq \cdot p - y(1 - y)q^2 - z(1 - z)p^2 + (1 - z)m^2 + zm^2_\gamma \qquad (8.7.33)$$

leading to $D = k^2 - \Delta + i\epsilon$ after using $(x + y) = (1 - z)$; (v) that terms odd in k in the numerator vanish giving the numerator $\slashed{k}\gamma^\mu\slashed{k} + N(p, q)$, where we have defined

$$N^\mu \equiv (-y\slashed{q} + z\slashed{p})\gamma^\mu[(1 - y)\slashed{q} + z\slashed{p}] - 2m(1 - 2y)q^\mu - 4mzp^\mu + m^2\gamma^\mu; \qquad (8.7.34)$$

(vi) $\slashed{k}\gamma^\mu\slashed{k} = k_\rho k_\sigma\gamma^\rho\gamma^\mu\gamma^\sigma$ and Eqs. (A.6.4) and (A.6.6) for the k integral and then $\gamma^\nu\gamma^\mu\gamma_\nu = -(2 - \epsilon)\gamma^\mu$ and $\Gamma(3 - \frac{d}{2}) = (2 - \frac{d}{2})\Gamma(2 - \frac{d}{2})$ from Eq. (A.6.16); and finally (vii) Eq. (A.6.17) again.

For on-shell fermions $(p + q)^2 = p^2 = m^2$, which gives $0 = 2p \cdot q + q^2$ and leads to

$$\Delta_0(q^2) \equiv \Delta_{(p+q)^2=p^2=m^2} = yzq^2 - y(1 - y)q^2 - z(1 - z)m^2 + (1 - z)m^2 + zm^2_\gamma$$
$$= -xyq^2 + (1 - z)^2m^2 + zm^2_\gamma. \qquad (8.7.35)$$

Evaluating the vertex between fermion spinors to form an amplitude means that we can use $\bar{u}^{s'}(p')\not{p}' = \bar{u}^{s'}(p')m$ and $\not{p}u^s(p) = mu^s(p)$" to evaluate δ_1. Because of momentum conservation only two of p', p, q are independent and for on-shell fermions we want to identify the q-dependence. We rewrite $N^\mu(p,q)$ in terms of $(p'+p)^\mu$ and $q^\mu \equiv p'^\mu - p^\mu$ by replacing p^μ in N^μ with $\frac{1}{2}[(p'+p)^\mu - q^\mu]$. We define $N_0^\mu(p,q)$ as the simplified version of $N^\mu(p,q)$ that comes from evaluating it between two on-shell spinors, which means that $\bar{u}^{s'}(p')N^\mu(p,q)u^s(p) = \bar{u}^{s'}(p')N_0^\mu(p,q)u^s(p)$. Note that N^μ is a four-vector and there are only three independent four-vectors available, γ^μ, q^μ and $(p'+p)^\mu$. There is only one independent non-constant Lorentz scalar available for on-shell fermions, which can be $2p' \cdot p$, $(p'+p)^2 = 2m^2 + 2p' \cdot p$ or $q^2 = 2m^2 - 2p' \cdot p$. Choosing q^2 for this purpose it follows that we can write

$$N_0^\mu(p,q) \equiv N^\mu|_{p'=p=m} = \gamma^\mu f(q^2) + (p'+p)^\mu g(q^2) + q^\mu h(q^2) \qquad (8.7.36)$$
$$= \gamma^\mu[(1-x)(1-y)q^2 + (1-2z-z^2)m^2] + (p'+p)^\mu[mz(z-1)] + q^\mu[m(z-2)(x-y)]$$
$$\to \gamma^\mu[(1-x)(1-y)q^2 + (1-2z-z^2)m^2] + (p'+p)^\mu[mz(z-1)],$$

where the functions $f(q^2)$, $g(q^2)$ and $h(q^2)$ in the second line are obtained from Eq. (8.7.34) after some lengthy algebra. The proof is left as an exercise. The third line follows since Δ_0 is symmetric in x and y whereas the q^μ term is odd and vanishes in the integration over x and y. This is related to the Ward identity.

Since QED is renormalizable, then we must be able to satisfy the renormalization condition in Eq. (8.7.21). This means that if $\not{p}' = \not{p} = m$ and $q = 0$, then $\Lambda^\mu = 0$ with a suitable choice of δ_1. Since δ_1 can only cancel the γ^μ component, then any other contribution must vanish at $q = 0$. We can understand this using the Gordon identity in Eq. (A.3.44), which is valid because we have made the replacements $\not{p}' = \not{p} = m$. So using $(p'^\mu + p^\mu) \to 2m\gamma^\mu - i\sigma^{\mu\nu}q_\nu$ we find the required form,

$$N_0^\mu(p,q) = \gamma^\mu[f(q^2) + 2mg(q^2)] - i\sigma^{\mu\nu}q_\nu g(q^2) \equiv \gamma^\mu N_\gamma(q^2) + \frac{i\sigma^{\mu\nu}q_\nu}{2m}N_\sigma(q^2)$$
$$= \gamma^\mu[(1-x)(1-y)q^2 + (1-4z+z^2)m^2] + \frac{i\sigma^{\mu\nu}q_\nu}{2m}[2m^2 z(1-z)]. \qquad (8.7.37)$$

To obtain δ_1 using Eq. (8.7.21) set $\not{p}' = \not{p} = m$, $q^\mu = 0$. This gives

$$\delta_1 = -\frac{q_c^2}{4\pi}\frac{1}{2\pi}\int_0^1 [\cdots]\left[\left\{\frac{2}{\epsilon} - \ln\frac{\Delta_0(0)}{\tilde{\nu}^2} - 2\right\} + \frac{N_\gamma(0)}{\Delta_0(0)}\right] \qquad (8.7.38)$$
$$= -\frac{q_c^2}{4\pi}\frac{1}{2\pi}\int_0^1 dz(1-z)\left[\frac{2}{\epsilon} - 2 - \ln\frac{(1-z)^2 m^2 + zm_\gamma^2}{\tilde{\nu}^2} + \frac{(1-4z+z^2)m^2}{(1-z)^2 m^2 + zm_\gamma^2}\right],$$

where we have neglected $\mathcal{O}(\epsilon)$ terms and have used $\int_0^1 dxdydz\,\delta(1-x-y-z)f(z) = \int_0^1 dz(1-z)f(z)$. Finally, then we arrive at the renormalized vertex

$$\Gamma^\mu(p',p) = \gamma^\mu + \Lambda^\mu(p',p) = \gamma^\mu + \frac{q_c^2}{4\pi}\frac{1}{2\pi}\int_0^1 dxdydz\,\delta(1-x-y-z) \qquad (8.7.39)$$
$$\times \left[\ln\frac{\Delta_0(0)}{\Delta}\gamma^\mu + \frac{N^\mu}{\Delta} - \frac{N_\gamma(0)}{\Delta_0(0)}\right] + \mathcal{O}(q_c^4),$$

where we have taken $\epsilon \to 0$ and the mass scale ν has disappeared as expected. Since the only purpose of the photon mass, m_γ, is to control infrared divergences it can be set to zero in the numerators and in the logarithm.

For the invariant amplitude of a fermion scattering from a virtual photon we have

$$iM^\mu = -iq_c\bar{u}^{s'}(p')\,\Gamma^\mu(p',p)|_{p'^2=p^2=m^2}\,u^s(p) \qquad (8.7.40)$$
$$= -iq_c\bar{u}^{s'}(p')\,\Gamma^\mu(p',p)|_{\not{p}'=\not{p}=m}\,u^s(p),$$

where the second line follows since when $\not{p}$ and $\not{p}'$ act on $u^s(p)$ and $\bar{u}^{s'}(p')$ from the right and left, respectively, they give m. Note that iM^μ is simply that part of an invariant amplitude that couples to a single virtual photon of momentum $q = p' - p$. It is conventional to define *form factors* $F_1(q^2)$ and $F_2(q^2)$ using

$$\Gamma^\mu(p',p)|_{\not{p}'=\not{p}=m} \equiv \gamma^\mu F_1(q^2) + \frac{i\sigma^{\mu\nu}q_\nu}{2m}F_2(q^2), \qquad (8.7.41)$$

where the on-shell renormalization conditions in Eq. (8.7.21) require that $F_1(0) = 1$. We use $\Delta|_{\not{p}'=\not{p}=m} = \Delta_0(q^2)$ and $N^\mu|_{\not{p}'=\not{p}=m} = N_0^\mu(p,q)$ in the form of Eq. (8.7.37) to obtain the one-loop results for the form factors from Eq. (8.7.39),

$$F_1(q^2) = 1 + \frac{q_c^2}{4\pi}\frac{1}{2\pi}\int_0^1 dx\,dy\,dz\,\delta(1-x-y-z)\left[\ln\frac{\Delta_0(0)}{\Delta_0(q^2)} + \frac{N_\gamma(q^2)}{\Delta_0(q^2)} - \frac{N_\gamma(0)}{\Delta_0(0)}\right] + \mathcal{O}(q_c^4),$$

$$F_2(q^2) = \frac{q_c^2}{4\pi}\frac{1}{2\pi}\int_0^1 dx\,dy\,dz\,\delta(1-x-y-z)\frac{N_\sigma}{\Delta_0(q^2)} + \mathcal{O}(q_c^4), \qquad (8.7.42)$$

where N_σ is defined in Eq. (8.7.37). The Ward identity in Eq. (8.3.35), $q_\mu M^\mu = 0$, is satisfied since $q_\mu \sigma^{\mu\nu} q_\nu = 0$ and $\bar{u}^{s'}(p')\not{q}u^s(p) = \bar{u}^{s'}(p')(\not{p}' - \not{p})u^s(p) = 0$.

Observe that $F_1(0) = 1$ as required. For the electron, muon and tauon we have $q_c^2/4\pi = \alpha$ and so for these elementary particles

$$F_2(q^2) = \frac{\alpha}{2\pi}\int_0^1 dx\,dy\,dz\,\delta(1-x-y-z)\frac{2m^2 z(1-z)}{(1-z)^2 m^2 + zm_\gamma^2 - xyq^2} + \mathcal{O}(q_c^4). \qquad (8.7.43)$$

For a non-timelike virtual photon we have $Q^2 \equiv -q^2 \geq 0$ and so the denominator has no singularity, $(1-z)^2 m^2 + zm_\gamma^2 + xyQ^2 > 0$. There is no infrared divergence in $F_2(q^2)$, since with $q^2 = 0$ and taking the photon mass $m_\gamma = 0$ we find

$$F_2(0) = (\alpha/2\pi)\int_0^1 dx\,dy\,dz\,\delta(1-x-y-z)\frac{2z}{1-z} + \mathcal{O}(q_c^4) = \frac{\alpha}{2\pi} + \mathcal{O}(q_c^4), \qquad (8.7.44)$$

since $\int_0^1 dx\,dy\,dz\,2z/(1-z) = \int_0^1 dz\int_0^{1-z} dy\,2z/(1-z) = \int_0^1 dz\,2z = [z^2]_0^1 = 1$. This result was obtained by Julian Schwinger in 1948 and as we show below is the $\mathcal{O}(\alpha)$ contribution to the anomalous magnetic moment of the electron (Schwinger, 1948a).

Verifying that $Z_1 = Z_2$: Eq. (8.5.35) tells us that $Z_1 = Z_2$ at every order in α and so we must have $\delta_1 = \delta_2$ to every order since $Z_1 = 1 + \delta_1$ and $Z_2 = 1 + \delta_2$. Since $\int_0^1 dx\,cx = \int_0^1 dx(1-x)c = c/2$ for any constant c, then in $\delta_2 - \delta_1$ we can cancel common constant terms in the integrands and rearrange to find

$$\delta_1 - \delta_2 = \frac{q_c^2}{4\pi}\frac{1}{2\pi}\int_0^1 dz\left\{(1-z)\left[1+\ln\frac{(1-z)^2 m^2 + zm_\gamma^2}{\nu^2} - \frac{(1-4z+z^2)m^2}{(1-z)^2 m^2 + zm_\gamma^2}\right]\right.$$

$$\left. -z\left[\ln\frac{(1-z)^2 m^2 + zm_\gamma^2}{\nu^2} + \frac{2m^2(1-z)(2-z)}{(1-z)^2 m^2 + zm_\gamma^2}\right]\right\} \qquad (8.7.45)$$

$$= \frac{q_c^2}{4\pi}\frac{1}{2\pi}\int_0^1 dz\left\{(1-2z)\ln\frac{(1-z)^2 m^2 + zm_\gamma^2}{\nu^2} + (1-z) - \frac{(1-z)(1-z^2)m^2}{(1-z)^2 m^2 + zm_\gamma^2}\right\} = 0$$

using Eq. (8.7.46) below. As expected $Z_1 = Z_2$ at $\mathcal{O}(\alpha)$. To finish, note that $\int_0^1 dz\,(1-2z)\ln(\nu^2) = 0$ and define $u(z) \equiv z(1-z)$ and $v(z) \equiv \ln\{(1-z)^2 m^2 + zm_\gamma^2\}$. Then using integration by parts $\int_0^1 dz\,u'v = [uv]_0^1 - \int_0^1 dz\,uv'$ we find

$$\int_0^1 dz\,(1-2z)\ln\frac{(1-z)^2m^2+zm_\gamma^2}{\nu^2}=\int_0^1 dz\,(1-2z)\ln\{(1-z)^2m^2+zm_\gamma^2]\} \qquad (8.7.46)$$

$$=\left[z(1-z)\ln\{\cdots\}\right]_0^1-\int_0^1 dz\,z(1-z)\frac{(-2)(1-z)m^2+m_\gamma^2}{(1-z)^2m^2+zm_\gamma^2}$$

$$=\int_0^1 dz\,z(1-z)\frac{2(1-z)m^2-m_\gamma^2}{(1-z)^2m^2+zm_\gamma^2}=\int_0^1 dz\left[-(1-z)+\frac{(1-z)(1-z^2)m^2}{(1-z)^2m^2+zm_\gamma^2}\right],$$

where it is easiest to verify the last line working right to left.

Scattering from external E and B fields: We are also interested in scattering of a single fermion from external **E** and **B** fields. For scattering from an external potential $A_\mu^{\mathrm{ext}}(x)$ we need to add an additional interaction term so that Eq. (7.6.19) becomes $\mathcal{H}_{\mathrm{int}}^{\mathrm{QED}}=-\mathcal{L}_{\mathrm{int}}^{\mathrm{QED}}=q_c(A_\mu+A_\mu^{\mathrm{ext}})j_{\mathrm{Dirac}}^\mu$. If $A_\mu^{\mathrm{ext}}(x)$ depends on t and $\mathbf{x}$, then spacetime translational invariance is lost and total four-momentum is not conserved. Then we cannot extract a $\delta^4(P_f-P_i)$ from the T-matrix $T_{fi}=\langle f;p',s'|\hat{T}|f;p,s\rangle$ to define an invariant amplitude $\mathcal{M}$. However if A_μ^{ext} is time-independent, then total energy will be conserved and we can define an amplitude $\mathcal{M}'$ using

$$T_{fi}=\langle f;p',s'|\hat{T}|f;p,s\rangle\equiv 2\pi\delta(E_f-E_i)\mathcal{M}'_{fi}, \qquad (8.7.47)$$

which is the replacement $(2\pi)^3\delta^3(\mathbf{P}_f-\mathbf{P}_i)\mathcal{M}\to\mathcal{M}'$ in Eq. (7.3.3). Following through the arguments of Sec. 7.3.2 in this case we have, for example, in place of Eq. (7.3.37) that $|T_{fi}|^2=\Delta T(2\pi)\delta(E_f-E_i)|\mathcal{M}'_{fi}|^2$. From there it is not difficult to see that the differential cross-section of Eq. (7.3.40) now becomes

$$d\sigma=\frac{d^3p'}{(2\pi)^3 2E_{p'}}\frac{(2\pi)\delta(E_{\mathbf{p}'}-E_{\mathbf{p}})|\mathcal{M}'_{fi}|^2}{|\boldsymbol{v}|2E_{\mathbf{p}}}, \qquad (8.7.48)$$

where $\mathbf{p}=\gamma m\boldsymbol{v}$ and $i\to f$ means $\mathbf{p}\to\mathbf{p}'$. A single interaction with $A_\mu^{\mathrm{ext}}(x)$ gives

$$iT_{fi}=i(2\pi)\delta(E_f-E_i)\mathcal{M}'_{fi}=iM^\mu A_\mu^{\mathrm{ext}}(q) \qquad (8.7.49)$$

$$=-iq_c\bar{u}^{s'}(p')\left[\gamma^\mu F_1(q^2)+\frac{i\sigma^{\mu\nu}q_\nu}{2m}F_2(q^2)\right]u^s(p)\,A_\mu^{\mathrm{ext}}(q),$$

where we use the four-diemnsional Fourier transform $A_\mu^{\mathrm{ext}}(q)=\int d^4x\,e^{iq\cdot x}A_\mu^{\mathrm{ext}}(x)$. In the nonrelativistic limit this is the Born approximation. For the static potential $A_\mu^{\mathrm{ext}}(\mathbf{x})$ we can also define

$$A_\mu^{\mathrm{ext}}(q)\equiv 2\pi\delta(q^0)\tilde{A}_\mu(\mathbf{q}), \qquad (8.7.50)$$

where $\tilde{A}^\mu(\mathbf{q})$ is the three-dimensional Fourier transform of $A_\mu^{\mathrm{ext}}(\mathbf{x})$ and where $\delta(q^0)=\delta(E_{\mathbf{p}'}-E_{\mathbf{p}})$. Then we have

$$i\mathcal{M}'_{fi}=-iq_c\bar{u}^{s'}(p')\left[\gamma^\mu F_1(q^2)+\frac{i\sigma^{\mu\nu}q_\nu}{2m}F_2(q^2)\right]u^s(p)\,\tilde{A}_\mu(\mathbf{q}). \qquad (8.7.51)$$

We wish to consider the nonrelativistic limit by keeping the lowest nonvanishing power of momenta. It is convenient (Peskin and Schroeder, 1995) to work in the chiral representation using Eqs. (4.6.29), (4.6.27) and (1.2.104), which gives

$$u^s(p)=\begin{bmatrix}\sqrt{p\cdot\sigma}\phi_s\\\sqrt{p\cdot\overline{\sigma}}\phi_s\end{bmatrix}\quad\text{with}\quad\left.\begin{matrix}\sqrt{p\cdot\sigma}\\\sqrt{p\cdot\overline{\sigma}}\end{matrix}\right\}=\sqrt{m}\left(1\mp\frac{\mathbf{p}\cdot\boldsymbol{\sigma}}{2m}\right)+\mathcal{O}(\mathbf{p}^2). \qquad (8.7.52)$$

This leads to results valid in any representation,

$$\bar{u}^{s'}(p')\gamma^0 u^s(p) = \bar{u}^{s'\dagger}(p')u^s(p) = 2m\phi_{s'}^\dagger \phi_s + \mathcal{O}(\mathbf{p}^2), \tag{8.7.53}$$

$$\bar{u}^{s'}(p')\gamma^i u^s(p) = 2m\phi_{s'}^\dagger \left[\frac{(p'+p)^i}{2m} - \frac{i\epsilon^{ijk}q^j\sigma^k}{2m}\right]\phi_s + \mathcal{O}(\mathbf{p}^2),$$

$$\bar{u}^{s'}(p')\sigma^{0j}q_j u^s(p) = \mathcal{O}(\mathbf{p}^2), \quad \bar{u}^{s'}(p')\sigma^{ij}q_j u^s(p) = 2m\phi_{s'}^\dagger[-\epsilon^{ijk}q^j\sigma^k]\phi_s + \mathcal{O}(\mathbf{p}^2),$$

where we used $\sigma^i\sigma^j = \delta^{ij} + i\epsilon^{ijk}\sigma^k$, where $\mathcal{O}(\mathbf{p}^2)$ means $\mathcal{O}(\mathbf{p}^2)$, $\mathcal{O}(\mathbf{p'}^2)$ or $\mathcal{O}(\mathbf{p'}\cdot\mathbf{p})$ and where we recall that $q^j = -q_j$.

E field only: For a static electric field we have $A_\mu^{\text{ext}} = (\Phi(\mathbf{x}), 0)$ and $\mathbf{E}(\mathbf{x}) = -\boldsymbol{\nabla}\Phi(\mathbf{x})$. If $\mathbf{E}(\mathbf{x})$ is weak enough that $\Phi(\mathbf{x})$ only varies significantly over macroscopic distances, then $\tilde{\Phi}(\mathbf{q})$ will only be nonvanishing when $|\mathbf{q}| \simeq 0$. Conversely, $F_1(q^2)$ and $F_2(q^2)$ correspond to microscopic distance scales and so are very broad functions in momentum space, which means it is an extremely good approximation to replace them with $F_1(0)$ and $F_2(0)$ in this case. Then Eq. (8.7.51) becomes

$$i\mathcal{M}'_{fi} = -iq_c\bar{u}^{s'}(p')\left[\gamma^0 F_1(0) + \frac{i\sigma^{0j}q_j}{2m}F_2(0)\right]u^s(p)\,\tilde{\Phi}(\mathbf{q}) \tag{8.7.54}$$

$$\rightarrow -iq_c 2m F_1(0)\tilde{\Phi}(\mathbf{q})\phi_{s'}^\dagger \phi_s = -iq_c 2m\tilde{\Phi}(\mathbf{q})\delta_{s's}, \tag{8.7.55}$$

since $F_1(0) = 1$. The nonrelativistic limit of a particle scattering in a spin-independent static potential is $-\mathcal{M}'_{fi}/2m \rightarrow V(\mathbf{q})\delta_{s's}$ where $V(\mathbf{q}) = \int d^3x\, V(\mathbf{x})e^{-i\mathbf{q}\cdot\mathbf{x}}$ and $V(\mathbf{x})$ is the static nonrelativistic potential. We have then arrived at

$$V(\mathbf{x}) = q_c\Phi(\mathbf{x}), \tag{8.7.56}$$

which is the usual electrostatic potential experienced by a particle with charge q_c.

Proof: We adapt the proof of the Yukawa potential in Eq. (7.6.40). For a single particle the normalization difference is $2m$ rather than $4m_1 m_2$ and so $S_{fi}^{\text{QFT}} \rightarrow 2m S_{fi}^{\text{QM}}$. Here there is no spatial translational invariance. We have
$S_{fi}^{\text{QM}} = \delta_{fi}^{\text{QM}} - i(2\pi)\delta(E_f - E_i)\langle\phi_f|\hat{V}|\phi_i\rangle; S_{fi}^{\text{QFT}} = \delta_{fi}^{\text{QFT}} + i(2\pi)\delta(E_f - E_i)\mathcal{M}'_{fi}$.
It then follows that: $-\mathcal{M}'_{fi}/2m = \langle\phi_f|\hat{V}|\phi_i\rangle = \delta_{s's}\langle\mathbf{p}'|\hat{V}|\mathbf{p}\rangle = \delta_{s's}\int d^3x\, V(\mathbf{x})e^{-i(\mathbf{p}'-\mathbf{p})\cdot\mathbf{x}} = \delta_{s's}V(\mathbf{q})$.

B field only: Next consider a static magnetic field with $A^{\text{ext}\mu}(x) = (0, \mathbf{A}(\mathbf{x}))$ where $\mathbf{B} = \boldsymbol{\nabla}\times\mathbf{A}$. Again we assume $\mathbf{B}$ is sufficiently weak that $\mathbf{A}(\mathbf{x})$ varies only on macroscopic scales and again we only need consider $F_1(0)$ and $F_2(0)$. Defining $\tilde{\mathbf{A}}(\mathbf{q}) \equiv \int d^3x\, \mathbf{A}(\mathbf{x})e^{-i\mathbf{q}\cdot\mathbf{x}}$ and $\tilde{\mathbf{B}}(\mathbf{q}) \equiv \int d^3x\, \mathbf{B}(\mathbf{x})e^{-i\mathbf{q}\cdot\mathbf{x}}$ then Eq. (8.7.51) becomes

$$i\mathcal{M}'_{fi} = -iq_c\bar{u}^{s'}(p')\left[\gamma^i F_1(0) + \frac{i\sigma^{ij}q_j}{2m}F_2(0)\right]u^s(p)\,\tilde{A}_i(\mathbf{q}) \tag{8.7.57}$$

$$= -iq_c 2m\left\{-F_1(0)\frac{(p'+p)^i}{2m}\delta_{s's}\tilde{A}^i(\mathbf{q}) - [F_1(0) + F_2(0)]\frac{\phi_{s'}^\dagger \sigma^k \phi_s}{2m}\tilde{B}^k(\mathbf{q})\right\},$$

where we have used $B^i(\mathbf{x}) = \epsilon^{ijk}\nabla_j A^k(\mathbf{x}) = \int[d^3q/(2\pi)^3]i\epsilon^{ijk}q^k A^k(\mathbf{q})e^{i\mathbf{q}\cdot\mathbf{x}}$ so that $\tilde{B}^i(\mathbf{q}) = i\epsilon^{ijk}q^j A^k(\mathbf{q})$. Write the spin operator for a fermion as $\mathbf{S} = \boldsymbol{\sigma}/2$. From the arguments above we recognize that $-\mathcal{M}_{fi}/2m = \phi_{s'}^\dagger V(\mathbf{x}, \mathbf{S})\phi_s$, where $V(\mathbf{x}, \mathbf{S})$ is the spin-dependent potential for the nonrelativistic interaction with a magnetic field $\mathbf{B}$. We have found that this is given by

$$V(\mathbf{x}, \mathbf{S}) = -q_c \frac{\mathbf{p}' + \mathbf{p}}{2m} \cdot \mathbf{A}(\mathbf{x}) - \boldsymbol{\mu} \cdot \mathbf{B} \quad \text{with} \quad \boldsymbol{\mu} \equiv g(q_c/2m)\mathbf{S}, \tag{8.7.58}$$

where $g \equiv 2[F_1(0) + F_2(0)] = 2 + 2F_2(0) = 2 + \mathcal{O}(\alpha)$ is the *Landé g-factor* for the fermion. The *anomalous magnetic moment* is defined in terms of g as

$$a \equiv (g - 2)/2 = F_2(0). \tag{8.7.59}$$

For the electron, muon and tauon we have $a = F_2(0) = (\alpha/2\pi) + \mathcal{O}(q_c^4)$ from Eq. (8.7.44), where $\alpha/2\pi \approx 0.00116140973$. The experimentally measured values of a for the electron and muon are very close to this value and are approximately $a_e^{\mathrm{exp}} \simeq 0.00115965218$ and $a_\mu^{\mathrm{exp}} \simeq 0.001165918$. Calculations in QED up to $\mathcal{O}(\alpha^5)$ have been carried out for both the electron (Aoyama et al., 2012b) and the muon (Aoyama et al., 2012a). For the electron the agreement with experiment is up to 10 significant figures, which is one of the most accurate physical predictions ever made. The Standard Model prediction for an anomalous magnetic moment is

$$a^{\mathrm{SM}} = a^{\mathrm{QED}} + a^{\mathrm{EW}} + a^{\mathrm{Hadron}}, \tag{8.7.60}$$

where the QED and electroweak contributions, a^{QED} and a^{EW}, can be calculated with perturbation theory but the QCD contribution a^{Hadron} has nonperturbative QCD contributions. For the electron both a_e^{EW} and a_e^{Hadron} are very small even at 10 significant figures. For the larger-mass muon both a_μ^{EW} and a_μ^{Hadron} can be significant and must be calculated to compare a_μ^{SM} with a_μ^{exp}. Any tension between a^{SM} and a^{exp} is a possible indication of physics beyond the Standard Model and so this is a subject of considerable interest. New physics contributions are typically stronger for the muon relative to the electron by a factor $\sim (m_\mu/m_e)^2$.

Both E and B fields: In the presence of macroscopic static $\mathbf{E}$ and $\mathbf{B}$ fields the Born approximation is not suitable. Instead, at any instant of time in the scattering the momentum transfer will be $\mathbf{q} \ll \mathbf{p}$ and so $(\mathbf{p}' + \mathbf{p})/2m \to \boldsymbol{v}$, where $\boldsymbol{v}$ is the instantaneous velocity of the nonrelativistic particle. Then the total spin-independent potential becomes $V^{\mathrm{total}} = q_c(\Phi - \boldsymbol{v} \cdot \mathbf{A})$, which is just the potential in Eq. (2.1.33) in the nonrelativistic derivation of the Lorentz force from the Lagrangian formulation. Of course, the Lorentz force is also true relativistically as we saw in relativistic single-particle classical mechanics in Eq. (2.8.50). The relativistic quantum mechanics limit ignores quantum field theory effects, $F_1(0) = 1$ and $F_2(0) = 0$, and then we see that Eq. (8.7.51) reduces to the QED interaction in the Dirac equation. The spin-independent behavior of the Dirac equation is the same as that of the Klein-Gordon equation and we know from Eq. (4.3.67) that the relativistic classical limit of that leads to the relativistic Lorentz force equation.

8.7.4 Minimal Subtraction Renormalization Schemes

The imposition of on-shell renormalization conditions to produce finite renormalized Green's functions is a demanding exercise even at $\mathcal{O}(\alpha)$. It is often convenient to adopt a simpler approach. In dimensional regularization the *minimal subtraction scheme* or *MS renormalization scheme* consists of subtracting *only* the divergent part of the amplitude, which means defining δ_2, δ_m, δ_3 and δ_1 to consist of only the $(2/\epsilon)$ contribution. In the *modified minimal subtraction scheme* or $\overline{MS}$ *(MS-bar) renormalization scheme* we also subtract off the γ and $\ln(4\pi)$ terms that come with $(2/\epsilon)$ because of Eq. (A.6.17). From Eqs. (8.7.24), (8.7.25), (8.7.27) and (8.7.38) we find

$$\delta_1^{\text{MS}} = \delta_2^{\text{MS}} = -\frac{q_c^2}{16\pi^2}\frac{2}{\epsilon}, \ \delta_m^{\text{MS}} = m\left(\delta_2^{\text{MS}} - \frac{3q_c^2}{16\pi^2}\frac{2}{\epsilon}\right) = -\frac{q_c^2}{4\pi^2}\frac{2}{\epsilon}, \ \delta_3^{\text{MS}} = -\frac{q_c^2}{12\pi^2}\frac{2}{\epsilon},$$

$$\delta_1^{\overline{\text{MS}}} = \delta_2^{\overline{\text{MS}}} = -\frac{q_c^2}{16\pi^2}\left(\frac{2}{\epsilon} + \ln(4\pi e^{-\gamma})\right), \ \delta_m^{\overline{\text{MS}}} = -m\frac{q_c^2}{4\pi^2}\left(\frac{2}{\epsilon} + \ln(4\pi e^{-\gamma})\right),$$

$$\delta_3^{\overline{\text{MS}}} = -\frac{q_c^2}{12\pi^2}\left(\frac{2}{\epsilon} + \ln(4\pi e^{-\gamma})\right), \tag{8.7.61}$$

where we have used $[-\gamma + \ln(4\pi)] = \ln(4\pi e^{-\gamma})$. These results were calculated in Feynman gauge. The counterterms $\delta_1 = \delta_2$ are gauge-dependent but since $\Pi(q^2)$ is gauge-independent then so is δ_3. The additional simplicity of the $\overline{\text{MS}}$ scheme leads to its more frequent use. In these schemes we will write μ rather than ν as the dimensional regularization scale since μ is now the renormalization scale.

At fixed regularization parameter $\Lambda \sim 1/\epsilon$ if we hold the bare quantities fixed in the first line of Eq. (8.5.1), then the last line of this equation is also fixed. If we change the renormalization prescription, then we change the definitions of δ_1, δ_2, δ_3 and δ_m, which means that we are also changing the definitions of m, q_c, ξ and the renormalized fields to compensate for that change. In MS and $\overline{\text{MS}}$ at fixed ϵ we have introduced a renormalization scale μ through the use of dimensional regularization and so renormalized quantities will in general depend on μ. The total μ dependence of any physical observable must vanish. The MS and $\overline{\text{MS}}$ renormalization schemes are examples of *mass-independent renormalization schemes*, since the counterterms have no dependence on any particle masses. This means that the β-function and other renormalization group parameters are mass-independent as we will see below.

How to use the MS and $\overline{\text{MS}}$ renormalization schemes: Calculate with renormalized perturbation theory using the Feynman rules and counterterms as for the on-shell renormalization scheme, except that now there are renormalization-scale dependent masses, charges and gauge parameters, $m(\mu)$, $q_c(\mu)$ and $\xi(\mu)$. So it is $m(\mu)$ and $q_c(\mu)$ that appear in Eq. (8.7.61). The physical (pole) mass m and the physically measurable charge q_c are different from $m(\mu)$ and $q_c(\mu)$. Fixing $m(\mu)$ and $q_c(\mu)$ at some μ in some scheme in some gauge determines them at every μ and determines the physical values of m and q_c. The counterterm Feynman rules, such as those in Eq. (8.7.20), are unchanged except that the values assigned to δ_1, δ_2, δ_3 and δ_m depend on the scheme. Physical observables must be both μ and gauge independent and these requirements provide valuable checks on calculations. Recall from Eq. (8.5.4) that $Z_m = [1 + (\delta m/m)]/Z_2$ and so to $\mathcal{O}(q_c^2)$ we have

$$Z_m^{\overline{\text{MS}}} = 1 + \frac{\delta_m^{\overline{\text{MS}}}}{m(\mu)} - \delta_2^{\overline{\text{MS}}} = 1 + \delta_m'^{\overline{\text{MS}}} = 1 - \frac{3q_c(\mu)^2}{(4\pi)^2}\left[\frac{2}{\epsilon} + \ln(4\pi e^{-\gamma})\right]. \tag{8.7.62}$$

It is straightforward to obtain the renormalized $\Sigma(p)$, $\Pi(q^2)$ and $\Lambda^\mu(p', p)$ at $\mathcal{O}(q_c^2)$ in the MS scheme from Eqs. (8.7.23), (8.7.26) and (8.7.32), respectively, by setting $\delta_1 = \delta_2 = \delta_3 = \delta_m = 0$, removing the $(2/\epsilon)$ terms and writing μ in place of ν. To obtain the results in the $\overline{\text{MS}}$ scheme from the results in the MS scheme we replace $\tilde{\mu}^2$ in the logarithms with μ^2. This corresponds to the removal of $[(2/\epsilon) - \gamma + \ln(4\pi)]$ in place of $(2/\epsilon)$. For example, the vacuum polarization in Eq. (8.7.26) leads to

$$\Pi^{\overline{\text{MS}}}(q^2) = -\frac{2}{\pi}\frac{q_c^2}{4\pi}\int_0^1 dx\, x(1-x)\ln\left(\frac{m^2 - x(1-x)q^2}{\mu^2}\right) + \mathcal{O}(q_c^4). \tag{8.7.63}$$

Interpretation of μ as a physical scale: Renormalized perturbation theory will converge most quickly in powers of the coupling q_c when the loop corrections are as small as possible. Consider the one-loop result for $\Pi(q^2)$ above for $Q^2 = -q^2 \gtrsim m^2$. The effects of the loop correction are smallest when $\mu^2 \sim Q^2$. This pattern continues at higher loops with higher powers of q_c^2. So *we should choose μ to be similar to the characteristic momentum scale relevant to the physical process* so that we have optimal convergence at a given order in the renormalized perturbation theory. In this way we associate μ with the characteristic scale Q^2 in a physical process. So, while the on-shell scheme directly connects with physical mass and charge, when Q^2 is large perturbation theory will converge best in schemes like the MS or $\overline{\text{MS}}$ with the renormalization scale μ chosen such that $\mu^2 \sim Q^2$.

Adoption of the $\overline{\text{MS}}$ renormalization scheme: *From this point onward we always work in the $\overline{\text{MS}}$ renormalization scheme unless stated otherwise.* We will no longer explicitly use the $\overline{\text{MS}}$ label on quantities but leave it to be understood.

8.7.5 Running Coupling and Running Mass in QED

From the arguments of Sec. 8.6.2 we have $[A_0^\mu] = \mathsf{M}^{(d-2)/2}$, $[\psi_0] = \mathsf{M}^{(d-1)/2}$, $[m] = \mathsf{M}$ and $[q_{0c}] = \mathsf{M}^{(4-d)/2} = \mathsf{M}^{\epsilon/2}$. Keeping the renormalized charge $q_c(\mu)$ dimensionless in natural units means that the renormalized charge in Eq. (8.5.1) is now replaced by $\mu^{\epsilon/2} q_c(\mu)$. Note that $Z_1(\mu, \epsilon)$, $Z_2(\mu, \epsilon)$, $Z_3(\mu, \epsilon)$ and $Z_m(\mu, \epsilon)$ remain dimensionless. Using $Z_1(\mu, \epsilon) = Z_2(\mu, \epsilon)$ in Eq. (8.5.2) we have

$$q_{0c}(\epsilon) = Z_q(\mu, \epsilon) \mu^{\epsilon/2} q_c(\mu) \quad \text{with} \tag{8.7.64}$$

$$Z_q(\mu, \epsilon) \equiv \frac{Z_1(\mu, \epsilon)}{Z_2(\mu, \epsilon) \sqrt{Z_3(\mu, \epsilon)}} = Z_3(\mu, \epsilon)^{-1/2} = 1 - \tfrac{1}{2} \delta_3 + \mathcal{O}(q_c^4),$$

where $Z_q \equiv Z_3^{-1/2}$ is defined here for notational convenience and δ_3 is given in the last line of Eq. (8.7.61). The bare charge is independent of μ and so,

$$0 = \mu \frac{\partial}{\partial \mu} q_{0c} = \mu \frac{\partial}{\partial \mu} [Z_q \mu^{\epsilon/2} q_c] = \mu^{\epsilon/2} Z_q q_c \left[\frac{\epsilon}{2} + \frac{\mu}{Z_q} \frac{\partial Z_q}{\partial \mu} + \frac{\mu}{q_c} \frac{dq_c}{d\mu} \right] \tag{8.7.65}$$

$$\Rightarrow \quad \beta(q_c) \equiv \mu \frac{dq_c}{d\mu} = -\frac{\epsilon}{2} q_c - q_c \frac{\mu}{Z_q} \frac{\partial Z_q}{\partial \mu} = -\frac{\epsilon}{2} q_c - \frac{q_c}{Z_q} \frac{\partial Z_q}{\partial q_c} \mu \frac{dq_c}{d\mu},$$

where $q_c(\mu)$ depends only on μ, where $Z_q \equiv Z_3^{-1/2}$ depends on μ only through $q_c(\mu)$ and where we have defined $\beta(q_c)$, which is referred to as the QED β-function. To facilitate a later comparison with QCD in Sec. 9.2.3 it is convenient to define

$$\beta_0 \equiv -4/3, \tag{8.7.66}$$

where the reason for this will soon become clear. Note that $Z_q = 1 + \mathcal{O}(q_c^2)$, $\partial Z_q / \partial q_c = -\frac{1}{2} (\partial \delta_3 / \partial q_c)$ and to leading order $\mu(dq_c/d\mu) = -(\epsilon/2) q_c$. This gives

$$\beta(q_c) = -\frac{\epsilon}{2} q_c - \frac{q_c}{Z_q} \left(-\tfrac{1}{2} \right) \left[\frac{-2q_c}{12\pi^2} \frac{2}{\epsilon} \right] \left[-\frac{\epsilon}{2} q_c + \mathcal{O}(q_c^3) \right] = -\frac{\epsilon}{2} q_c + \frac{q_c^3}{12\pi^2} + \mathcal{O}(q_c^5) \tag{8.7.67}$$

$$= -\frac{\epsilon}{2} q_c - \frac{\beta_0 q_c^3}{(4\pi)^2} + \mathcal{O}(q_c^5),$$

where we can now take $\epsilon \to 0$. Using $\beta(q_c) = \mu(\partial q_c/\partial \mu) = -\beta_0 q_c^3/(4\pi)^2 + \mathcal{O}(q_c^5)$ we can rearrange and integrate from μ_a to μ_b to find at one loop

$$\int_{q_c(\mu_a)}^{q_c(\mu_b)} \frac{dq_c}{\beta(q_c)} = \int_{\mu_a}^{\mu_b} \frac{d\mu}{\mu} \quad \Rightarrow \quad \frac{(4\pi)^2}{2\beta_0}\left[\frac{1}{q_c(\mu_b)^2} - \frac{1}{q_c(\mu_a)^2}\right] = \ln(\mu_b/\mu_a). \tag{8.7.68}$$

Choosing $q_c = -e \equiv -|e|$ we rearrange this using $\ln a = \frac{1}{2}\ln a^2$ to give

$$\alpha(\mu_b) = \frac{\alpha(\mu_a)}{1 + \frac{\beta_0}{4\pi}\alpha(\mu_a)\ln(\frac{\mu_b^2}{\mu_a^2})} = \alpha(\mu_a)\left[1 - \frac{\beta_0\alpha(\mu_a)}{4\pi}\ln\left(\frac{\mu_b^2}{\mu_a^2}\right) + \mathcal{O}(\alpha^2)\right], \tag{8.7.69}$$

where $\alpha(\mu) = e(\mu)^2/4\pi$ is the *running coupling* of QED. Observe that $\alpha(\mu_b)/\alpha(\mu_a)$ increases when μ_b/μ_a increases and so the running coupling increases with renormalization scale in QED. There is a singularity in Eq. (8.7.69) when $\ln(\mu_b^2/\mu_a^2) = 3\pi/\alpha(\mu_a)$. Choosing $\mu_a = m_e = 0.511$ MeV and the corresponding on-shell value $\alpha(m_e) \simeq 1/137$ we find that the singularity occurs at $\mu_b = \Lambda_{\mathrm{QED}}$, where

$$\ln\frac{\Lambda_{\mathrm{QED}}^2}{m_e^2} \equiv -\frac{4\pi}{\beta_0\alpha(m_e)} = \frac{3\pi}{\alpha(m_e)} \simeq 1{,}291 \quad \Rightarrow \quad \Lambda_{\mathrm{QED}} \simeq 10^{286} \text{ eV}. \tag{8.7.70}$$

Λ_{QED} is referred to as the *Landau pole*. The above expressions are only meaningful when $\mu < \Lambda_{\mathrm{QED}}$, since otherwise $\alpha(\mu) = g^2(\mu)/4\pi$ would become negative. Obviously, our leading order approximation will begin to break down when $\mu \gg m_e$, but if the QED β function remains finite and positive at all scales a singularity will eventually occur. The scale 10^{286} eV is so high that there is no practical difficulty associated with the Landau pole; however, its existence does suggest that QED on its own is a trivial theory as discussed earlier in Sec. 5.2.1.

We can also define a β-function in terms of α as a power series expansion in α. Following the conventions of Schwartz (2013) for the coefficients β_i we write

$$\beta(\alpha) \equiv \mu\frac{\partial\alpha}{\partial\mu} = \frac{2e}{4\pi}\mu\frac{\partial e}{\partial\mu} = \frac{2e}{4\pi}\left[\frac{e^3}{12\pi^2} + \mathcal{O}(e^5)\right] \equiv -2\alpha\left[\beta_0\frac{\alpha}{4\pi} + \beta_1\left(\frac{\alpha}{4\pi}\right)^2 + \cdots\right], \tag{8.7.71}$$

where comparing with Eq. (8.7.67) we see that this β_0 is indeed the $\beta_0 = -4/3$ as defined earlier. We can rewrite Eq. (8.7.69) as

$$\begin{aligned}
\alpha(\mu_b) &= \frac{1}{[\beta_0/4\pi]\{[4\pi/\beta_0\alpha(\mu_a)] - \ln(\mu_a^2) + \ln(\mu_b^2)\}} \\
&= \frac{4\pi}{\beta_0\{\ln(\mu_b^2) - \ln(\Lambda_{\mathrm{QED}}^2)\}} = \frac{4\pi}{\beta_0\ln(\mu_b^2/\Lambda_{\mathrm{QED}}^2)} = \frac{2\pi}{\beta_0\ln(\mu_b/\Lambda_{\mathrm{QED}})},
\end{aligned} \tag{8.7.72}$$

where the first line shows that $[4\pi/\beta_0\alpha(\mu_a)] - \ln(\mu_a^2)$ must be independent of μ_a and so it can be replaced with $[4\pi/\beta_0\alpha(m_e)] - \ln(m_e^2) = -\ln(\Lambda_{\mathrm{QED}})$. This is a very powerful one-loop result that shows that (i) since $\beta_0 = -4/3 < 0$ and $\mu < \Lambda_{QED}$ then $\alpha(\mu)$ increases with increasing renormalization scale, and (ii) the coupling at scale μ is determined by the location of the Landau pole, Λ_{QED}. Since $\alpha(\mu) = e(\mu)^2/4\pi \geq 0$ and $\beta_0 < 0$, then Eq. (8.7.72) can only be valid when $\mu < \Lambda_{\mathrm{QED}}$, which means that $\ln(\mu/\Lambda_{\mathrm{QED}}) < 0$ for all relevant μ. The effective replacement of the dimensionless coupling α with a dimensionful scale Λ_{QED} is sometimes referred to as *dimensional transmutation*. Using this formula we find that the value of the fine-structure constant at the Z-boson mass is approximately $\alpha_{\mathrm{naive}}(M_Z) \simeq 1/134$. However, at high-momentum scales we should also include the photon coupling to more massive charged particles including the μ and τ leptons, the quarks and the $W^{\pm}$ bosons. The measured value at M_Z is $\alpha(M_Z) \simeq 1/128$.

The anomalous mass dimension $\gamma_m(\mu)$ is defined in Eq. (8.5.53) and the bare mass $m_0(\epsilon) = Z_m(\mu, \epsilon) m(\mu)$ is independent of μ. It then follows that

$$0 = \mu \frac{\partial m_0}{\partial \mu} = \mu \frac{\partial (Z_m m)}{\partial \mu} = Z_m m \left[\frac{\mu}{m} \frac{\partial m}{\partial \mu} + \frac{\mu}{Z_m} \frac{\partial Z_m}{\partial \mu} \right] \Rightarrow \gamma_m = -\frac{\mu}{Z_m} \frac{\partial Z_m}{\partial \mu}. \tag{8.7.73}$$

To use this result note that $Z_m(\mu, \epsilon)$ depends on μ only through $q_c(\mu)$ so that $\partial Z_m/\partial \mu = (\partial Z_m/\partial q_c)(\partial q_c/\partial \mu)$. Then using $\partial Z_m/\partial q_c = \partial \delta'_m/\partial q_c$ with $Z_m = 1 + \delta'_m$ in Eq. (8.7.62) and $\beta(q_c) = \mu(dq_c/d\mu) = -(\epsilon/2)q_c + \mathcal{O}(q_c^3)$ we then find to $\mathcal{O}(q_c^2)$,

$$\gamma_m = -\frac{\mu}{Z_m} \frac{\partial Z_m}{\partial q_c} \frac{dq_c}{d\mu} = -\frac{3q_c}{8\pi^2} \frac{2}{\epsilon} \frac{\epsilon}{2} q_c = -\frac{6q_c^2}{(4\pi)^2} = -\frac{3(\frac{q_c}{e})^2 \alpha}{2\pi} = \frac{-3(\frac{q_c}{e})^2}{\beta_0 \ln(\frac{\mu}{\Lambda_{\mathrm{QED}}})}, \tag{8.7.74}$$

where Eq. (8.7.72) was used in the last step. The definition $\gamma_m = (\mu/m)(\partial m/\partial \mu)$ can be rearranged as $dm/m = d \ln m = \gamma_m d\mu/\mu = \gamma_m d \ln \mu = \gamma_m d \ln(\mu/\Lambda_{\mathrm{QED}})$ since $d \ln(\mu/\Lambda_{\mathrm{QED}})/d \ln \mu = 1$. Again choosing $q_c = -e = -|e|$ and integrating this equation from μ_a to μ_b we find

$$\int_{m(\mu_a)}^{m(\mu_b)} d \ln m = \ln \left[\frac{m(\mu_b)}{m(\mu_a)} \right] = \int_{\mu_a}^{\mu_b} d \ln \left[\frac{\mu}{\Lambda_{\mathrm{QED}}} \right] \gamma_m(\mu) = \int_{\mu_a}^{\mu_b} d \ln \left[\frac{\mu}{\Lambda_{\mathrm{QED}}} \right] \frac{(-3)}{\beta_0 \ln(\mu/\Lambda_{\mathrm{QED}})}$$

$$= (-3/\beta_0) \ln[\ln(\mu_b/\Lambda_{\mathrm{QED}})/\ln(\mu_a/\Lambda_{\mathrm{QED}})]$$

$$\Rightarrow \frac{m(\mu_b)}{m(\mu_a)} = \left[\frac{\ln(\mu_b^2/\Lambda_{\mathrm{QED}}^2)}{\ln(\mu_a^2/\Lambda_{\mathrm{QED}}^2)} \right]^{-3/\beta_0} = \left[\frac{\alpha(\mu_a)}{\alpha(\mu_b)} \right]^{-3/\beta_0} = \left[\frac{\alpha(\mu_a)}{\alpha(\mu_b)} \right]^{9/4}, \tag{8.7.75}$$

where we used Eq. (8.7.72). We refer to this $m(\mu)$ as the QED *running mass* and we see that at one-loop in QED the running mass logarithmically decreases with renormalization scale. This has occurred because $\gamma_m < 0$ in QED. Since Λ_{QED} is so large this growth is very slow. To verify that we recover γ_m in Eq. (8.7.74) from this solution, define C such that $m(\mu) = C \ln(\mu/\Lambda_{\mathrm{QED}})^{-3/\beta_0}$. Then we confirm that

$$\gamma_m(\mu) = \frac{\mu}{m} \frac{dm(\mu)}{d\mu} = \mu \left[\frac{-3}{\beta_0} \right] \frac{1}{\ln(\frac{\mu}{\Lambda_{\mathrm{QED}}})} \frac{d \ln(\frac{\mu}{\Lambda_{\mathrm{QED}}})}{d\mu} = \frac{-3}{\beta_0 \ln(\frac{\mu}{\Lambda_{\mathrm{QED}}})}. \tag{8.7.76}$$

8.7.6 Renormalization Group Flow and Fixed Points

Assume that we know the β-function $\beta(g)$ of some theory with coupling $g(\mu)$. If the β-function has one or more zeros, then interesting things occur. Consider, for example, the β-function shown in Fig. 8.2, which has zeros at $g = 0, g_1, g_2$. Recall that $\beta(g) = (1/\mu)(dg/d\mu) = dg/d \ln \mu$. So when $\beta(g) > 0$ then $g(\mu)$ increases with $\ln \mu$ and when $\beta(g) < 0$ then $g(\mu)$ decreases with $\ln \mu$. So as μ increases $g(\mu)$ will flow in the directions of the arrows in Fig. 8.2.

Let g_z be any one of the zeros of $\beta(g)$, then $\beta(g_z) = 0$ and for g sufficiently close to g_z we will have for some constant B_z that

$$\beta(g) \simeq -B_z(g - g_z) = \frac{dg}{d \ln \mu} \quad \Rightarrow \quad g(\mu_b) \simeq g_z + [g(\mu_a) - g(z)] \left[\frac{\mu_b}{\mu_a} \right]^{-B_z}, \tag{8.7.77}$$

where the right-hand side follows from integrating $dg/(g - g_z) = -B_z d \ln \mu$ between μ_a and μ_b. Note that $-B_z$ is the slope of $\beta(g)$ at the zero crossing g_z. There are three cases for a zero crossing: (i) if $B_z > 0$ then for fixed μ_a as $\mu_b \to \infty$ we have $g(\mu_b) \to g_z$ and we say that g_z is an *ultraviolet*

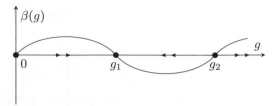

Figure 8.2   Illustration of a β-function with three zeros, $\beta(0) = \beta(g_1) = \beta(g_2) = 0$. The arrowheads indicate the direction of the flow of the coupling $g(\mu)$ as μ increases.

fixed point; (ii) if $B_z < 0$ then for fixed μ_a as $\mu_b \to 0$ we have $g(\mu_b) \to g_z$ and we say that g_z is an *infrared fixed point*; and (iii) if $B_z = 0$ then there is no flow at that coupling. Referring to Fig. 8.2 we then understand that $g = g_1$ is an ultraviolet fixed point and $g = 0$ and $g = g_2$ are infrared fixed points.

For QED we have $\beta(g) = g^3/12\pi + \mathcal{O}(g^5)$ from Eq. (8.7.67) and so $\beta(g) > 0$ near $g = 0$. Conversely, QCD has $\beta(g) = -(g^3/16\pi^2)[11 - \frac{2}{3}n_f]$ as we will see in Eq. (9.2.30), where n_f is the number of quark flavors. Then for QCD $\beta(g) \propto -g^3$ and so $\beta(g) < 0$ near $g = 0$. The slope $-B_z$ at $g = 0$ vanishes, since in a free theory there is no interaction to induce a coupling flow, $(d\beta/dg)(0) = 0$. So we expect $\beta(g) = Cg^3$ for some constant C near $g = 0$ and in this sense $g = 0$ is a trivial fixed point. Nonetheless, the direction-of-flow arguments remain true and $g = 0$ is an infrared fixed point in QED and an ultraviolet fixed point in QCD. We say that QCD is asymptotically free as discussed in Sec. 5.2.2, since $g \to 0$ as $\mu \to \infty$.

8.8 Spontaneous Symmetry Breaking

The concept of spontaneous symmetry breaking was briefly discussed in Secs. 5.2 and 8.2.2. It occurs when the ground state of a system does not respect the symmetry of the underlying Lagrangian. The example in Eq. (6.2.225) was a classical scalar field ϕ with the Lagrangian density and Hamiltonian given by

$$\mathcal{L} = \tfrac{1}{2}\partial_\mu\phi\partial^\mu\phi - U(\phi) = \tfrac{1}{2}\partial_\mu\phi\partial^\mu\phi + \tfrac{1}{2}\mu^2\phi^2 - (\lambda/4!)\phi^4 \tag{8.8.1}$$
$$H = T[\pi] + V[\phi] = \int d^3x \tfrac{1}{2}\pi^2 + V[\phi] = \int d^3x[\tfrac{1}{2}\pi^2 + \tfrac{1}{2}(\boldsymbol{\nabla}\phi)^2 - \tfrac{1}{2}\mu^2\phi^2 + (\lambda/4!)\phi^4],$$

where $\pi = \partial_0\phi$ and $\mu^2 > 0$. Note the discrete symmetry under $\phi \to -\phi$. The ground state of the system is a constant field ϕ_0 that mimimizes $V[\phi]$ and $U(\phi)$,

$$\phi_0 = \pm v = \pm(\sqrt{6/\lambda})\mu. \tag{8.8.2}$$

A spatially infinite system requires an infinite amount of energy to change ground state. So we write the Lagrangian for the system expanded around one of the ϕ_0. For the case $\phi_0 = v = (\sqrt{6/\lambda})\mu$ we then write

$$\sigma \equiv \phi - \phi_0 \quad \Rightarrow \quad \mathcal{L} = \tfrac{1}{2}\partial_\mu\sigma\partial^\mu\sigma - \tfrac{1}{2}(2\mu^2)\sigma^2 - (\sqrt{\lambda/6})\mu\sigma^3 - (\lambda/4!)\sigma^4 \tag{8.8.3}$$
$$\equiv \tfrac{1}{2}\partial_\mu\sigma\partial^\mu\sigma - \tfrac{1}{2}m_\sigma^2\sigma^2 - (\kappa_\sigma/3!)\sigma^3 - (\lambda/4!)\sigma^4,$$

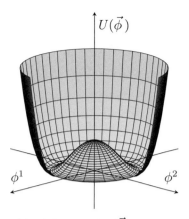

Figure 8.3   Mexican hat potential for the linear sigma model with $N = 2$ and $U(\vec{\phi}_0) = 0$.

where an irrelevant constant term has been omitted and where $m_\sigma \equiv \mu\sqrt{2}$ and $\kappa_\sigma/3! \equiv \mu\sqrt{\lambda/6}$. The $\phi \to -\phi$ symmetry is no longer manifest but remains encoded in the detailed relationship of the coefficients m_σ, κ_σ and λ.

Linear sigma model: This simple model has a set of N real scalar fields, $\vec{\phi} = (\phi^1, \ldots, \phi^N)$. This is referred to as the linear sigma model and can be written as

$$\mathcal{L} = \tfrac{1}{2}\partial_\mu\vec{\phi} \cdot \partial^\mu\vec{\phi} - U(\vec{\phi}) = \tfrac{1}{2}\partial_\mu\vec{\phi} \cdot \partial^\mu\vec{\phi} + \tfrac{1}{2}\mu^2\vec{\phi}^2 - (\lambda/4)(\vec{\phi}^2)^2. \tag{8.8.4}$$

The system has a continuous symmetry under rotations of the vector of fields $\vec{\phi}$, where $\vec{\phi}' = O\vec{\phi}$ with O an $N \times N$ orthogonal matrix, $O^T = O^{-1}$. An illustration of the Mexican hat shaped potential $U(\vec{\phi})$ for $N = 2$ is shown in Fig. 8.3. We observe that there is a continuously degenerate set of ground states $\vec{\phi}_0$ that satisfy

$$|\vec{\phi}_0| = \mu/\sqrt{\lambda} \equiv v, \tag{8.8.5}$$

where v is the magnitude of the vacuum expectation value (VEV). Let us choose the orientation of $\vec{\phi}$ such that $\vec{\phi}_0 = (0, \ldots, 0, v)$ so that $\phi_0^i = 0$ for $i = 1, \ldots, N-1$ and $\phi_0^N = v$. In Fig. 8.3 this is the choice $\vec{\phi}_0 = (\phi_0^1, \phi_0^2) = (0, v)$. Then define the shifted fields $(\vec{\pi}, \sigma)$ and rewrite the Lagrangian density in terms of them,

$$(\vec{\pi}, \sigma) = (\pi^1, \ldots, \pi^{N-1}, \sigma) \equiv \vec{\phi} - \vec{\phi}_0 = \vec{\phi} - (0, \ldots, 0, v),$$

$$\mathcal{L} = \tfrac{1}{2}(\partial_\mu\vec{\pi})^2 + \tfrac{1}{2}(\partial_\mu\sigma)^2 - \tfrac{1}{2}(2\mu^2)\sigma^2 - \mu\sqrt{\lambda}\sigma^3 - \mu\sqrt{\lambda}\sigma\vec{\pi}^2 - \tfrac{\lambda}{4}\sigma^4 - \tfrac{\lambda}{2}\sigma^2\vec{\pi}^2 - \tfrac{\lambda}{4}(\vec{\pi}^2)^2$$

$$= \tfrac{1}{2}(\partial_\mu\vec{\pi})^2 + \tfrac{1}{2}(\partial_\mu\sigma)^2 - \tfrac{1}{2}m_\sigma^2\sigma^2 - \lambda v\sigma^3 - \lambda v\sigma\vec{\pi}^2 - \tfrac{\lambda}{4}\sigma^4 - \tfrac{\lambda}{2}\sigma^2\vec{\pi}^2 - \tfrac{\lambda}{4}(\vec{\pi}^2)^2, \tag{8.8.6}$$

where $m_\sigma \equiv \sqrt{2}\mu$ is the σ mass. The $N-1$ $\vec{\pi}$-fields are massless and are referred to as *Goldstone bosons*. The $\vec{\pi}$ fields are not to be confused with conjugate momenta. The motivation for using $\vec{\pi}$ is that in QCD the pions are the pseudo-Goldstone bosons associated with spontaneous chiral symmetry breaking. In the $N = 2$ case the π field corresponds to fluctuations in the flat direction orthogonal to the ϕ^2 axis at the point $(0, v)$ and σ field corresponds to fluctuations in the ϕ^2 direction at this point, where $\partial^2 U/\partial(\phi^2)^2|_{\vec{\phi}=(0,v)} = 2\mu^2 = m_\sigma^2$. It is now straightforward to quantize the theory about the spontaneously broken vacuum. We find the Lagrangian with counterterms for renormalized perturbation theory about the unbroken vacuum by replacing $m^2 \to -\mu^2$, $4! \to 4$ and $\delta_m \to \delta_\mu$ in Eq. (8.7.1), which leads to

$$\begin{aligned}
\mathcal{L} &= \left[\tfrac{1}{2}(\partial_\mu\vec{\phi})^2 + \tfrac{1}{2}\mu^2\vec{\phi}^2 - \tfrac{\lambda}{4}(\vec{\phi}^2)^2\right] + \left\{\tfrac{1}{2}\delta_z(\partial_\mu\vec{\phi})^2 - \tfrac{1}{2}\delta_\mu\vec{\phi}^2 - (\delta_\lambda/4)(\vec{\phi}^2)^2\right\} \\
&= \left[\tfrac{1}{2}(\partial_\mu\vec{\pi})^2 + \tfrac{1}{2}(\partial_\mu\sigma)^2 - \tfrac{1}{2}(2\mu^2)\sigma^2 - \lambda v\sigma^3 - \lambda v\sigma\vec{\pi}^2 - \tfrac{\lambda}{4}\sigma^4 - \tfrac{\lambda}{2}\sigma^2\vec{\pi}^2 - \tfrac{\lambda}{4}(\vec{\pi}^2)^2\right] \\
&\quad + \left\{\tfrac{1}{2}\delta_z(\partial_\mu\vec{\pi})^2 - \tfrac{1}{2}(\delta_\mu + \delta_\lambda v^2)\vec{\pi}^2 + \tfrac{1}{2}\delta_z(\partial_\mu\sigma)^2 - \tfrac{1}{2}(\delta_\mu + 3\delta_\lambda v^2)\sigma^2\right. \\
&\quad \left. -(\delta_\mu + \delta_\lambda v^2)v\sigma - \delta_\lambda v\sigma\vec{\pi}^2 - \delta_\lambda v\sigma^3 - \tfrac{1}{4}\delta_\lambda\sigma^4 - \tfrac{1}{2}\delta_\lambda\sigma^2\vec{\pi}^2 - \tfrac{1}{4}\delta_\lambda(\vec{\pi}^2)^2\right\}.
\end{aligned} \tag{8.8.7}$$

These counterterms for the renormalized perturbation theory about the shifted vacuum follow from the replacement $\vec{\phi} \to (\vec{\pi}, \sigma + v)$ in the counterterms for the unshifted Lagrangian density. The Feynman rules for the renormalized perturbation theory of the shifted theory follow using Eq. (8.8.7) just as in Sec. 8.7.1.

All 10 counterterms in $\{\cdots\}$ depend on only three counterterm parameters δ_z, δ_μ and δ_λ because of the hidden symmetry encoded in the renormalized parameters of the shifted theory. The first two counterterms in $\{\cdots\}$ give the $N-1$ propagator counterterms for the $\vec{\pi}$ and the next two do the same for σ. The fifth counterterm renormalizes the tadpole diagrams for the σ, where σ tadpole diagrams can lead to a vacuum expectation value for σ as we saw in Eq. (8.2.10). The remaining five renormalize the three- and four-point vertices. Since there are three counterterm parameters, we need to specify three renormalization conditions. The most convenient choice for these is (i) adjust the fifth counterterm to exactly cancel the σ tadople diagrams so that $\langle\Omega|\hat{\sigma}|\Omega\rangle = 0$, so that quantum effects do not shift the vacuum away from its classical value and $\langle\Omega|\hat{\phi}^N|\Omega\rangle = v = \mu/\sqrt{\lambda}$ remains valid in the quantum theory; (ii) the σ propagator has its perturbative form at the sigma mass, $p^2 = m_\sigma^2$; and (iii) choose the $\sigma\sigma \to \sigma\sigma$ four-point amplitude at $s = 4m_\sigma^2$, $t = u = 0$, which fixes λ at that scale. Since these conditions are defined at the renormalized mass m_σ they do not fix it and it will differ from its classical value of $\sqrt{2}\mu$ due to the fourth counterterm and residual quantum effects. However, it remains finite because the theory is renormalizable. The renormalized parameters of the theory are then v, λ and m_σ. The one-loop contribution to the mass of the $\vec{\pi}$ particles cancels exactly. This remains true at all orders of perturbation theory and the Goldstone bosons remain massless in the quantum theory.

Goldstone's theorem: More generally it can be shown that *the spontaneous breaking of each continuous global symmetry leads to the existence of a massless particle referred to as a Goldstone boson*. This result is known as *Goldstone's theorem* and was proved in Goldstone (1961) and Goldstone et al. (1962).

Consider the broken $O(N)$ symmetry of the linear sigma model. As summarized in Table A.1 this group has $N(N-1)/2$ generators, since an $N \times N$ orthogonal matrix has this many independent real parameters. (*Proof:* An $N \times N$ matrix has $N^2 - N = N(N-1)$ off-diagonal elements $\Rightarrow$ $N + \tfrac{1}{2}N(N-1) = \tfrac{1}{2}N(N+1)$ independent elements in a symmetric matrix. Since both sides are symmetric in $O^T O = I$ there are $\tfrac{1}{2}N(N+1)$ constraints on N^2 elements $\Rightarrow$ the number of independent elements in O is $N^2 - \tfrac{1}{2}N(N+1) = \tfrac{1}{2}N(N-1)$.) After shifting the fields, the $N-1$ fields $\vec{\pi}$ have an $O(N-1) \subset O(N)$ rotational invariance with $\tfrac{1}{2}(N-1)(N-2)$ generators, meaning that the number of broken symmetries is $\tfrac{1}{2}[N(N-1) - (N-1)(N-2)] = N-1$, which is the number of Goldstone bosons. The number of broken generators is the number of Goldstone bosons. A detailed discussion of the nonlinear sigma model using renormalized perturbation theory can be found in Peskin and Schroeder (1995).

Goldstone bosons remain massless after quantization: Consider a set of N scalar fields $\vec{\phi}$ with a continuous global symmetry G such that for $R \in G$ the global transformation $\vec{\phi} \to \phi' = R\vec{\phi}$ leaves the Lagrangian density $\mathcal{L}$ invariant. R is in the $N \times N$ matrix representation of the group G. This was referred to as an internal symmetry in Sec. 3.2.5, where we discussed Noether's theorem.

The corresponding conserved Noether current is given in Eq. (3.2.105). The linear sigma model is an explicit example of this with $G = O(N)$ and $\vec{\phi}' = O\vec{\phi}$. Let $\mathcal{L}$ have classical ground state $\vec{\phi}_0$ that does not respect the symmetry G. Then there must be a continuous degeneracy of ground states consisting of the field configurations $\vec{\phi}_0' = R\vec{\phi}_0$. To isolate the mass matrix for the theory we choose *constant fields* ϕ^i_{con} so that all $\partial_\mu \vec{\phi}$ terms in $\mathcal{L}$ vanish and then $\mathcal{L} \to -U(\vec{\phi})$, which is the potential density shown in Fig. 8.3 in the $O(N)$ case. The classical mass matrix for the shifted vacuum $\vec{\phi}_0$ is

$$(M^2_{\text{class}})_{ij} \equiv \left. \frac{\partial^2 U(\vec{\phi}_{\text{con}})}{\partial \phi^i_{\text{con}} \partial \phi^j_{\text{con}}} \right|_{\vec{\phi}_{\text{con}} = \vec{\phi}_0}. \tag{8.8.8}$$

The eigenvalues of the matrix M^2_{class} are the squared classical masses, $m^2_1, \ldots, m^2_N$. Since $\vec{\phi}_0$ is a ground state then all $m^2_i \geq 0$. The number of vanishing masses is the number of zero eigenvalues of M^2_{class} and is the number of Goldstone bosons. It is also the number of directions in $\vec{\phi}$ space at $\vec{\phi}_0$ that leave $U(\vec{\phi})$ invariant.

Consider the effective action $\Gamma[\vec{\phi}_c]$ defined in Eq. (8.2.11). Recall that ϕ_0 is constant and is the minimum of $-\Gamma[\phi_c]$. For constant fields $\Gamma[\vec{\phi}_{\text{con}}] = -(VT)V_{\text{eff}}(\vec{\phi}_{\text{con}})$. Referring to Eq. (8.2.1) and generalizing to N fields we see that if $\mathcal{L}$ is invariant under $\vec{\phi} \to R\vec{\phi}$ then $Z[\vec{J}]$ and $W[\vec{J}]$ are invariant under $\vec{J} \to R^T\vec{J}$ since $\vec{J}$ only enters through $\vec{J} \cdot \vec{\phi}$. Similarly, since we have the definition $\Gamma[\vec{\phi}_c] = W[\vec{J}] - \int d^4y\, \vec{J}(j) \cdot \vec{\phi}_c(y)$ then $\Gamma[\phi_c]$ must be invariant under $\phi_c(y) \to R\phi_c(y)$. So the effective action $\Gamma[\phi_c]$ has the same internal symmetries as the action $S[\vec{\phi}] = \int d^4x\, \mathcal{L}$. Here we have a vacuum expectation value $\vec{\phi}_0$ and Eq. (8.2.16) should be rewritten as a power series about $\vec{\phi}_0$ but otherwise the arguments are unchanged. Renormalization counterterms do not spoil internal symmetries so we can choose a counterterm to cancel tadpoles such that the classical value of $\vec{\phi}_0$ remains as the vacuum expectation value, $\phi^i_0 = \langle \Omega | \hat{\phi}^i(x) | \Omega \rangle$. Then the N-field generalization of Eq. (8.2.18) is

$$\Gamma^{(2)}_{ij}(x - y) = \left. \frac{\delta^2 \Gamma[\phi_c]}{\delta \phi^i(x) \delta \phi^j(y)} \right|_{\vec{\phi}_c = \vec{\phi}_0} \quad \Rightarrow \quad \Gamma^{(2)}_{ij}(p) = p^2 - (M^2)_{ij}(p^2), \tag{8.8.9}$$

where $\Gamma^{(2)}_{ij}(p) = \int d^4x\, e^{ip\cdot(x-y)}\Gamma^{(2)}_{ij}(x - y)$. If we restrict ourselves to constant fields then Eq. (8.8.9) must transform like Eq. (8.8.8) and it will remain invariant in the same directions in $\vec{\phi}$ space. Constant fields also isolate $p = 0$ so that

$$\left. \frac{\partial^2 V_{\text{eff}}(\vec{\phi}_{\text{con}})}{\partial \phi^i_{\text{con}} \partial \phi^j_{\text{con}}} \right|_{\vec{\phi}_{\text{con}} = \vec{\phi}_0} = -\Gamma^{(2)}_{ij}(0) = (M^2)_{ij}(0). \tag{8.8.10}$$

So the directions in $\vec{\phi}$ space that leave U invariant must be the same directions that leave V_{eff} invariant. These correspond to the zero eigenvalues of the mass matrix, which are the Goldstone bosons in the full renormalized theory; i.e., if $m^i(0) = 0$ for some i then $D^i(p^2) = i/[p^2 - m^i(p^2)]$ has a pole at $p^2 = 0$. This shows that Goldstone bosons remain massless after quantization and renormalization.

If we restrict ourselves to building Hilbert space using field configurations that vanish outside a finite volume, then it is clear that any state in the Hilbert space built on one vacuum will be orthogonal to a state in the Hilbert space built on another degenerate vacuum. It would require an infinite amount of kinetic energy to move from one degenerate vacuum to another in infinite coordinate space. For additional discussion of this, see Weinberg (1996). In the presence of a small explicit symmetry breaking the Goldstone bosons acquire a small mass and become pseudo-Goldstone bosons or pseudo-Nambu-Goldstone bosons as discussed in Sec. 5.2.2.

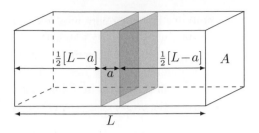

Two parallel perfectly conducting plates are placed a distance a apart in a box composed of six perfectly conducting faces. The box has length L and end area A.

8.9 Casimir Effect

We have not yet considered the simplest example of an ultraviolet divergence, which occurs when calculating the Casimir effect. This effect was first discovered by Hendrik Casimir (Casimir, 1948; Casimir and Polder, 1948) and was later generalized to more realistic situations by Lifshitz and collaborators (Dzyaloshinskii et al., 1961). Consider the diagram in Fig. 8.4. Casimir showed that there is an attractive force per unit area between the two plates that in the limit $L, A \to \infty$ is given by

$$\frac{F}{A} = \frac{\hbar c \pi^2}{240\, a^4} \simeq \frac{1.300}{a(\mu)^4} \times 10^{-7} \text{ N cm}^{-2} = \frac{0.01300}{a(\mu)^4} \text{ dyne cm}^{-2}, \qquad (8.9.1)$$

where $a(\mu)$ is the separation in microns (1 μm $= 10^{-6}$ m). While this effect was predicted in 1948 it was not verified experimentally until 1997 (Lamoreaux, 1997). It has since been confirmed in later experiments. We restore $\hbar$ and c in this section.

As we know from Eq. (6.4.12) the zero-point energy E_0 for the electromagnetic field is set to zero by the normal-ordering prescription for renormalization. Since there are an infinite number of angular frequency modes ω in a quantum field theory without both ultraviolet and infrared regularization, then $E_0 = \frac{1}{2} \sum_n \hbar \omega_n \to \infty$ as regularization is removed. First consider the zero-point energy E_0^a for the modes between the two plates in Fig. 8.4. Since we will be taking $A \to \infty$ we can write for $n = 0, 1, 2, \ldots$ that $\omega_n^a(k_x, k_y) = c\sqrt{k_x^2 + k_y^2 + (n^2\pi^2/a^2)} = c\sqrt{k^2 + (n^2\pi^2/a^2)}$. Only transverse modes contribute to the energy. There are two polarizations for each transverse mode with $n \geq 1$ but only one mode for $n = 0$.[4] We then we have

$$E_0^a = (\tfrac{1}{2}\hbar) \int [A dk_x dk_y/(2\pi)^2] \left\{ \omega_0^a(k_x, k_y) + 2\sum_{n=1}^{\infty} \omega_n^a(k_x, k_y) \right\} \qquad (8.9.2)$$

$$= (A\hbar/2\pi) \int_0^\infty k dk \sum_{n=(0)1}^{\infty} \omega_n^a(k^2), \qquad (8.9.3)$$

where we have used the two-dimensional equivalent of Eq. (6.2.26) to "sum" over the $x - y$ modes and in the second line we have used polar coordinates in the $k_x - k_y$ plane. The notation $\sum_{n=(0)1}^{\infty}$ is defined to mean that the $n = 0$ contribution has a factor of $\frac{1}{2}$. We are interested in the difference between the total energy of the zero modes with the two central plates compared with that when the box is empty,

$$\delta E \equiv [E_0^a + 2E_0^{(L-a)/2}] - E_0^L. \qquad (8.9.4)$$

---

[4] If $k_z = 0$ then waves move in the $x - y$ plane with no variation in the z-direction. The linear polarization parallel to the plates vanishes on the plates and so vanishes everywhere between.

When $L \to \infty$ the sum in E_0^L becomes an integral, $\sum_{n=(0)1}^{\infty} \to \int_0^{\infty} dn = (L/\pi) \int_0^{\infty} dk_z$, where we have defined $k_z \equiv n\pi/L$. Then we have for very large L

$$E_0^L \to (L/\pi)(A\hbar c/2\pi) \int_0^{\infty} kdk \int_0^{\infty} dk_z \sqrt{k^2 + k_z^2} \qquad (8.9.5)$$

$$\Rightarrow 2E_0^{(L-a)/2} - E_0^L \to -(a/\pi)(A\hbar c/2\pi) \int_0^{\infty} kdk \int_0^{\infty} dk_z \sqrt{k^2 + k_z^2},$$

$$\Rightarrow \delta E \to (A\hbar c/2\pi) \int_0^{\infty} kdk \left\{ \sum_{n=(0)1}^{\infty} \sqrt{k^2 + (n^2\pi^2/a^2)} - (a/\pi) \int_0^{\infty} dk_z \sqrt{k^2 + k_z^2} \right\}$$

$$= (A\hbar c\pi^2/4a^3) \left\{ \sum_{n=(0)1}^{\infty} \left[\int_0^{\infty} du\sqrt{u+n^2} \right] - \int_0^{\infty} dn \left[\int_0^{\infty} du\sqrt{u+n^2} \right] \right\},$$

where we obtain the last line by defining $u \equiv a^2 k^2/\pi^2$ and redefining the continuum n as $n \equiv (a/\pi)k_z$ giving $kdk = \frac{1}{2}dk^2 = (\pi^2/2a^2)du$ and $(a/\pi)dk_z = dn$.

The two terms in the last line of Eq. (8.9.5) are ultraviolet divergent. Following Casimir (Casimir, 1948; Itzykson and Zuber, 1980) we introduce a smooth function $f(k)$ such that for some large ultraviolet k_m we have $f(k) = 1$ for $k/k_m \lesssim 1$ and $f(k) \to 0$ sufficiently rapidly as $k/k_m \to \infty$ so that quantities are well defined. This is physically required since sufficiently high frequencies will simply pass through any realistic conducting plates. The final result is independent of the regularization used as demonstrated in detail in Schwartz (2013). All that matters is that high frequencies are cutoff in some way as they would be in any real experiment. Note that $\mathbf{k} = (k_x, k_y, k_z)$ and so $\mathbf{k}^2 = k^2 + k_z^2 = (\pi/a)^2(u + n^2)$, which means that $f(k) = f(\frac{\pi}{a}\sqrt{u+n^2})$. The energy change due to the presence of the plates is then

$$\delta E = (A\hbar c\pi^2/4a^3) \left\{ \sum_{n=(0)1}^{\infty} F(n) - \int_0^{\infty} dn\, F(n) \right\}, \quad \text{where} \qquad (8.9.6)$$

$$F(n) \equiv \int_0^{\infty} du\sqrt{u+n^2}\, f(\tfrac{\pi}{a}\sqrt{u+n^2}). \qquad (8.9.7)$$

The Euler-Maclaurin formula (page 16 in Abramowitz and Stegun, 1972) leads to

$$\sum_{n=(0)1}^{\infty} F(n) - \int_0^{\infty} dn\, F(n) = -\tfrac{1}{12}F'(0) + \tfrac{1}{720}F'''(0) + \cdots, \qquad (8.9.8)$$

where to arrive at this result we have used that $F(n)$ and all derivatives of $F(n)$ vanish as $n \to \infty$ because $f(n) \to 0$ as $n \to \infty$. To evaluate derivatives it is convenient to define $w = u + n^2$ so that $F(n) = \int_{n^2}^{\infty} dw\, \sqrt{w}\, f(\frac{\pi}{a}w)$. Then, for example, it follows that $F'(n) = 2n(dF/dn^2) = -2n^2 f(\frac{\pi}{a}n^2)$, $F''(n) = -4nf - 2n^2 f'$ and similarly $F'''(n) = -4f - 8nf' - 2n^2 f''$. By construction all derivatives of $f(k)$ vanish at $k = 0$ and so it follows that $F^{(4)}(0) = 0$ and similarly all higher derivatives of F vanish at $n = 0$. Then since $F'(0) = 0$ and $F'''(0) = -4$ the right-hand side of Eq. (8.9.8) is $\frac{1}{720}F'''(0) = -\frac{1}{180}$, which leads to the Casimir force given above, since

$$\delta E = -A\hbar c\pi^2/720a^3 \quad \Rightarrow \quad F/A = -d(\delta E/A)/da = \hbar c\pi^2/240a^4. \qquad (8.9.9)$$

Since δE decreases when a decreases this is an attractive force between the plates.

Summary

In this chapter we obtained the parity (P), charge conjugation (C) and time-reversal (T) transformations in quantum field theory and proved the *CPT* theorem. The generating functional for connected Green's functions was then derived along with the effective action and effective potential useful for understanding spontaneous symmetry breaking. The general form of the Schwinger-Dyson equations was then established and used to derive the Ward and Ward-Takahashi identities for QED. We then discussed the systematics of renormalizability and ultraviolet divergences in Feynman diagrams. The renormalization of QED was then considered and the renormalization group was introduced.

Regularization was then discussed and we adopted dimensional regularization for the purposes of perturbation theory. Examples of the regularization and renormalization of several field theories were then considered along with choices of renormalization schemes. The running mass and running coupling of QED were derived and spontaneous symmetry breaking in quantum field theory was then treated. To complete the chapter we derived and discussed the Casimir effect.

Problems

8.1 Use the equal-time canonical commutation and anticommutation relations to prove Eq. (8.3.22). Then for either a boson or fermion operator $\hat{X}(y)$ generalize Eq. (6.2.166) and show that $\partial^x_\mu \langle \Omega | T \hat{j}^\mu(x) \hat{X}(y) | \Omega \rangle = \delta(x^0 - y^0) \langle \Omega | [\hat{j}^0(x), \hat{X}(y)] | \Omega \rangle$. Then further generalize this result to verify the QED Ward identity in Eq. (8.3.21).

8.2 Using the Feynman rules for a $\phi\bar{\psi}\psi$ Yukawa theory, write down the one-loop contribution to the ϕ^4 proper vertex. Show that this one-loop contribution is divergent and explain why we must include a ϕ^4 vertex in Yukawa theory.

8.3 Prove Eq. (8.7.36).

8.4 Prove the results in Eq. (8.7.53) in the chiral representation or any other representation that you care to use.

8.5 Complete the detailed steps to arrive at the scattering of a charged fermion from a static electromagnetic field given in Eq. (8.7.48).

8.6 Consider the pseudoscalar Yukawa theory in Eq. (7.6.11) with a ϕ^4 interaction,

$$\mathcal{L} = \tfrac{1}{2}(\partial_\mu \phi \partial^\mu \phi - m_\phi^2 \phi^2) + \bar{\psi}(i\slashed{\partial} - m)\psi - \tfrac{\lambda}{4!}\phi^4 - ig\phi\bar{\psi}\gamma^5\psi. \tag{8.9.10}$$

(Hint: This problem requires care and effort.)

(a) Calculate the one-loop self-energy for the scalar and for the fermion and write down the $\overline{\text{MS}}$ counterterms.

(b) Calculate the $\overline{\text{MS}}$ one-loop β-functions for both couplings, $\beta_\lambda(g, \lambda)$ and $\beta_g(g, \lambda)$.

8.7 Calculate a one-loop contribution to the $\gamma\gamma\gamma$ 1PI vertex in QED. Show that it vanishes as expected from Furry's theorem and so needs no renormalization. (Hint: Eq. (A.3.29) may prove useful.).

8.8 In the discussion of the Casimir effect we placed the parallel plates in the middle of the box as shown in Fig. 8.4. Show that the Casimir effect depends only on the plate separation a and not on plate location as $L, A \to \infty$.

Nonabelian Gauge Theories

In this final chapter we first provide an introduction to nonabelian gauge theories and their quantization. We then focus on QCD and its properties including the Feynman rules, the running coupling, asymptotic freedom, the running of quark masses and BRST invariance. Lattice QCD is then briefly discussed. We then turn to quantum anomalies with a focus in the example of the Adler-Bell-Jackiw anomaly. This is followed by an introduction to the electroweak sector, electroweak symmetry breaking and finally to the Standard Model itself.

9.1 Nonabelian Gauge Theories

We can extend the $U(1)$ abelian gauge theories of QED or scalar QED to the case of nonabelian gauge theories. Consider a set of n fields $\vec{\psi} = (\psi_1, \ldots, \psi_n)$ that could be boson or fermion, which transform according the $n \times n$ unitary representation of some Lie group G (see Sec. A.7) as $\vec{\psi}' = U(g)\vec{\psi}$ with $g \in G$. If $U(g)$ is independent of spacetime and if the Lagrangian density $\mathcal{L}$ is invariant under these transformations, then the system is invariant under the global symmetry G. If we require that $\mathcal{L}$ be invariant under *local* transformations $U(g(x))$, then we have a gauge theory. If G is a nonabelian Lie group, then it is a nonabelian gauge theory.

9.1.1 Formulation of Nonabelian Gauge Theories

Write the generators of the Lie algebra $\mathfrak{g}$ of the Lie group G in an $n \times n$ unitary representation as T^a for $a = 1, 2, \ldots, d(G)$, where $d(G)$ is the number of generators of G. We will understand that $\mathfrak{g}$ and G can refer to either their abstract form or their $n \times n$ representation depending on context. With this understanding we can write $U \in G$. For any $U \in G$ sufficiently close to the identity we can write $U(\vec{\alpha}) = \exp(i\alpha^a T^a)$, where the summation convention is understood and $\vec{\alpha} \in \mathbb{R}^{d(G)}$. Since the $U(\vec{\alpha})$ are unitary, then the T^a are Hermitian, $T^{a\dagger} = T^a$.

Consider a set of n fermions $\psi = (\psi_1, \ldots, \psi_n)$, where ψ is a column vector of the n fields. Then $\mathcal{L} = \bar{\psi}_i(i\slashed{\partial} - m_i)\psi_i$ is invariant under the global transformation $\psi_i(x) \rightarrow \psi_i'(x) = U_{ij}(\vec{\alpha})\psi_j(x) = (e^{i\alpha^a T^a})_{ij}\psi_j(x)$, where we again use the summation convention. Invariance under local transformations $U(\vec{\alpha}) = e^{i\alpha^a(x)T^a}$ means that like QED in Eq. (7.6.18) we must introduce $d(G)$ gauge fields $A_\mu^a(x)$ such that we define the covariant derivative acting on the fermion fields,[1]

$$D_\mu \equiv \partial_\mu - igA_\mu, \quad \text{or in component form} \quad (D_\mu)_{ij} \equiv \partial_\mu \delta_{ij} - igA_\mu^a T_{ij}^a, \quad (9.1.1)$$

$$\text{with} \quad (A_\mu)_{ij}(x) \equiv A_\mu^a(x)T_{ij}^a \quad \text{and} \quad \psi_i'(x) = U(\vec{\alpha}(x))_{ij}\psi_j(x) = (e^{i\alpha^a(x)T^a})_{ij}\psi_j(x).$$

---

[1] We follow Peskin and Schroeder (1995) and Schwartz (2013) where the sign of g in D_μ is chosen to be opposite to that of q_c in D_μ in QED. Some texts use $\mathbf{A}_\mu$ to denote the $n \times n$ matrix nature of $A_\mu \equiv A_\mu^a T^a$, but we do not do this.

Since it carries the label a, $A_\mu^a(x)$ fields transform under the adjoint representation of G. We suppress the $n \times n$ identity matrices and understand that ∂_μ means $I\partial_\mu$ in $D_\mu = \partial_\mu - igA_\mu$. Since ψ is shorthand for the column vector $\psi = (\psi_1, \ldots, \psi_n)$, then $\bar{\psi}$ is shorthand for the row vector $\bar{\psi} = (\bar{\psi}_1, \ldots, \bar{\psi}_n)$. The fermion part of the Lagrangian density can be written in several equivalent ways using these shorthand notations,

$$\mathcal{L} = \bar{\psi}_i(i\slashed{\partial}\delta_{ij} + g\slashed{A}^aT_{ij}^a - m\delta_{ij})\psi_j = \bar{\psi}(i\slashed{\partial} + g\slashed{A} - m)\psi \tag{9.1.2}$$
$$= \bar{\psi}_i(i\slashed{D}_{ij} - m\delta_{ij})\psi_j = \bar{\psi}(i\slashed{D} - m)\psi \quad \text{with} \quad \psi' = U\psi, \quad \bar{\psi}' = \bar{\psi}U^\dagger.$$

Note that $A_\mu(x)$ takes its value in the Lie algebra, i.e., $A_\mu(x) = A_\mu^a(x)T^a \in \mathfrak{g}$ for all x. For $\mathcal{L}$ to be invariant under $U \in G$ we must have for all ψ that

$$\mathcal{L} = \bar{\psi}'(i\slashed{D}' - m)\psi' = \bar{\psi}(i\slashed{D} - m)\psi \quad \Rightarrow \quad D_\mu'\psi' = UD_\mu\psi \quad \Rightarrow \quad U^\dagger D_\mu'U\psi = D_\mu\psi,$$
$$\Rightarrow \quad U^\dagger(\partial_\mu - igA_\mu')U\psi = (\partial_\mu - igA_\mu)\psi \quad \Rightarrow \quad A_\mu' = UA_\mu U^\dagger - (i/g)(\partial_\mu U)U^\dagger, \tag{9.1.3}$$

which expressed in terms of $A_\mu^a(x)$ and expanded in powers of $\vec{\alpha}(x)$ becomes

$$A_\mu^a(x) \to A_\mu'^a(x) = A_\mu^a(x) + (1/g)\partial_\mu\alpha^a(x) - f^{abc}\alpha^b(x)A_\mu^c(x) + \mathcal{O}(\vec{\alpha}^2). \tag{9.1.4}$$

The gauge transformation of $A^\mu \to A_\mu'$ in Eq. (9.1.3) must be true for all $U \in G$ not just for those U close to the identity with small $\vec{\alpha}(x)$.

It remains to construct the gauge-invariant component of $\mathcal{L}$ describing the gauge fields in the absence of fermions. Recall the covariant derivative in QED from Eq. (7.6.16), $D_\mu = \partial_\mu + iq_cA_\mu$, where $A_\mu(x)$ is not a matrix. We find $[D_\mu, D_\nu]\psi = iq_c([\partial_\mu, A_\nu] - [\partial_\nu, A_\mu])\psi = iq_cF_{\mu\nu}\psi$, which means that $[D_\mu, D_\nu] = -iq_cF_{\mu\nu}$ and is just a function not a differential operator. In the nonabelian case we find $[D_\mu, D_\nu]\psi = (-ig[\partial_\mu, A_\nu] + ig[\partial_\nu, A_\mu] - g^2[A_\mu, A_\nu])\psi$ and so it is natural to define

$$F_{\mu\nu} \equiv (i/g)[D_\mu, D_\nu] = \partial_\mu A_\nu - \partial_\nu A_\mu - ig[A_\mu, A_\nu] \tag{9.1.5}$$
$$\Rightarrow \quad F_{\mu\nu}^a = \partial_\mu A_\nu^a - \partial_\nu A_\mu^a + gf^{abc}A_\mu^b A_\nu^c, \tag{9.1.6}$$

where we have used $[A_\mu, A_\nu] = A_\mu^b A_\nu^c[T^b, T^c] = if^{bca}T^a A_\mu^b A_\nu^c = if^{abc}T^a A_\mu^b A_\nu^c$. In QED $F_{\mu\nu}$ is gauge invariant, but this is not the case in a nonabelian theory since

$$F_{\mu\nu} \to F_{\mu\nu}' = F_{\mu\nu}'^a T^a = (i/g)[D_\mu', D_\nu'] = (i/g)U[D_\mu, D_\nu]U^\dagger = UF_{\mu\nu}U^\dagger$$
$$= F_{\mu\nu}^a e^{i\alpha^b T^b}T^a e^{-i\alpha^c T^c} = (F_{\mu\nu}^a - f^{abc}\alpha^b F_{\mu\nu}^c)T^a + \mathcal{O}(\vec{\alpha}^2)$$
$$\Rightarrow \quad F_{\mu\nu}^a \to F_{\mu\nu}'^a = F_{\mu\nu}^a - f^{abc}\alpha^b F_{\mu\nu}^c + \mathcal{O}(\vec{\alpha}^2), \tag{9.1.7}$$

where we have used $D_\mu' = UD_\mu U^\dagger$ and that f^{abc} is completely antisymmetric. Since $\text{tr}(F'^{\mu\nu}F_{\mu\nu}') = \text{tr}(UF^{\mu\nu}F_{\mu\nu}U^\dagger) = \text{tr}(F^{\mu\nu}F_{\mu\nu})$, then this quantity is gauge invariant. Also note that $\text{tr}(F^{\mu\nu}F_{\mu\nu}) = F^{a\mu\nu}F_{\mu\nu}^b\text{tr}(T^aT^b) = C(r)F^{a\mu\nu}F_{\mu\nu}^a$, where $C(r)$ for the group G in the representation r is defined in Eq. (A.7.12). So a simple extension of the abelian $\mathcal{L} = -\frac{1}{4}F^{\mu\nu}F_{\mu\nu}$ to the nonabelian case is to define for the gauge fields

$$\mathcal{L} = -\frac{1}{4C(r)}\text{tr}(F^{\mu\nu}F_{\mu\nu}) = -\frac{1}{4}F^{a\mu\nu}F_{\mu\nu}^a, \tag{9.1.8}$$

giving cubic and quartic interactions of the A_μ^a fields. We are interested in $SU(N)$ theories with the T^a in the defining (fundamental) representation, where $C(r) \to C(N) = T_F = \frac{1}{2}$ from Eq. (A.7.26). This leads to the *Yang-Mills* Lagrangian density,

$$\mathcal{L} = -\frac{1}{2}\text{tr}(F^{\mu\nu}F_{\mu\nu}) = -\frac{1}{4}F^{a\mu\nu}F_{\mu\nu}^a. \tag{9.1.9}$$

One other quantity that is also gauge invariant is $\text{tr}(\tilde{F}^{\mu\nu}F_{\mu\nu})$, where $\tilde{F}_{\mu\nu} = \tilde{F}^a_{\mu\nu}T^a$ and $\tilde{F}^{a\mu\nu} \equiv \frac{1}{2}\epsilon^{\mu\nu\alpha\beta}F^a_{\alpha\beta}$ is the dual field strength tensor for the generator T^a as defined in Eq. (2.7.18). This follows since $\tilde{F}'^{\mu\nu} = U\tilde{F}^{\mu\nu}U^\dagger$.

Since the gauge fields carry the index a, then they transform under the adjoint representation. The appropriate covariant derivative to use in that case is its adjoint representation given in Eq. (A.7.15), $T^a_{ij} \to t^a_{bc}(A) = -if^{abc}$. This leads to

$$(D_\mu A_\nu)^a \equiv D^{ab}_\mu A^b_\nu = [\partial_\mu\delta^{ab} - gf^{abc}A^c_\mu]A^b_\nu = \partial_\mu A^a_\nu + gf^{abc}A^b_\mu A^c_\nu, \qquad (9.1.10)$$
$$\Rightarrow \quad A^a_\mu(x) \to A'^a_\mu(x) = A^a_\mu(x) + (1/g)D^{ab}_\mu\alpha^b(x) + \mathcal{O}(\vec{\alpha}^2),$$

where the gauge transformation of A^a_μ in Eq. (9.1.4) is now expressed in terms of D^{ab}_μ. The representation of the covariant derivative is defined by what it acts on.

The arguments about renormalizibility in Sec. 8.4 again apply. So the most general gauge-invariant renormalizable Lagrangian density for $SU(N)$ is

$$\mathcal{L} = -\frac{1}{4}F^{a\mu\nu}F^a_{\mu\nu} + \bar{\psi}(i\slashed{D} - m)\psi + c_\theta\tilde{F}^{a\mu\nu}F^a_{\mu\nu}, \qquad (9.1.11)$$

where c_θ is an arbitrary real constant. The term c_θ is referred to as the *theta term* in QCD. It can be written as a total four-divergence and so does not contribute to dynamics probed by field fluctuations with compact support such as perturbation theory. However, since QCD contains spontaneous (also called dynamical) chiral symmetry breaking, the vacuum is nonperturbative and the theta term can contribute in that case due to the appearance of large gauge transformations that can change the topological charge (winding number) of the gauge field. Since it could take any value, the fact that c_θ is found to be extremely small in QCD appears to have no simple explanation. A nonzero value of c_θ leads to CP violation in the strong interactions and for this reason the smallness of c_θ is referred to as the *strong CP problem*.

If a group can be expressed as the direct product of two subgroups such as $G \cong H \times K$, then ψ will be a vector in the tensor product space as discussed in Secs. A.2.7, 4.1.9 and 5.3. Consider the $m \times m$ matrix representation of H with $j = 1, \ldots, m$ and the $n \times n$ matrix representation of K with $k = 1, \ldots, n$. The fields will be labeled as $\psi_{jk}(x)$ and the unitary matrices U_H and U_K will act on the j and k indices, respectively, and so commute with each other. Here the notation "$\otimes$" means the combination of one matrix acting on the j indices and another matrix acting on the k indices. Then we can write $U = U_K \otimes U_H$. The covariant derivative for $G \cong H \times K$ is then a straightforward extension of the previous case,

$$D_\mu = \partial_\mu - ig_H A_\mu \otimes I_H - ig_K I_H \otimes B_\mu, \qquad (9.1.12)$$

where I_H and I_K are $m \times m$ and $n \times n$ identity matrices and where $A_\mu(x) = T^a_H A^a_\mu(x)$ and $B_\mu(x) = T^b_K B^b_\mu(x)$. Here T^b_H for $a = 1, \ldots, d(H)$ and T^b_K for $b = 1, \ldots, d(K)$ are the generators in the appropriate representations for groups H and K, respectively. So there are $d(H)$ gauge fields for H with coupling constant g_H and $d(K)$ gauge fields for K with coupling constant g_K. There is no reason for these couplings to be identical, and, as we will soon discuss, they will evolve differently with renormalization scale, μ, if H and K are different groups. Since we can write[2] $U(N) \cong U(1) \times SU(N)$ then $U(N)$ will not be respected by renormalization since the $U(1)$ and $SU(N)$ subgroups will evolve differently with renormalization scale even if the couplings were chosen equal at some scale. In addition, $U(1)$ symmetries of this type are sometimes broken by anomalies (see Sec. 9.3). So it will always be important to factor out any $U(1)$ subgroups and to

---

[2] In the $n \times n$ representation the more formal result is $U(n) \cong [U(1) \times SU(n)]/\mathbb{Z}_n$ where $\mathbb{Z}_n$ are the nth roots of unity, since if $S \in SU(n)$, then $\det(e^{i\theta}S) = 1$ if $e^{in\theta} = 1$. If we define the elements of $U(1)$ to be identity matrices with only one diagonal element multiplied by $e^{i\theta}$, then it is meaningful to write $U(n) \cong U(1) \times SU(n)$ in any representation.

separately consider semisimple nonabelian groups, e.g., $SU(N)$. The unitary representations of semisimple nonabelian groups have Hermitian traceless generators T^a. Other than $U(1)$, we are almost always interested in gauge theories with compact semisimple Lie groups.

9.1.2 Wilson Lines and Wilson Loops

One can view gauge theory from the perspective of differential geometry (Itzykson and Zuber, 1980; Cheng and Li, 1984; Peskin and Schroeder, 1995). The boson and/or fermion ψ fields at different spacetime points can have arbitrary phases in a gauge theory and so the usual spatial derivatives of them using ∂_μ have no well-defined meaning. Consider an $n \times n$ unitary representation of G. A choice of $U(x)$ for every x specifies a point in the gauge group manifold, where this manifold is the tensor product of all the copies of G that we write as $\prod_x G_x$. So every point in the gauge group manifold, $\prod_x G_x$, specifies a gauge transformation.

We cannot directly compare $\psi(x)$ and $\psi(y)$ since they are only defined up to gauge transformations, $\psi(x) \to \psi'(x) = U(x)\psi(x)$ and $\psi(y) \to \psi'(y) = U(y)\psi(y)$. This ambiguity can be removed by specifying a rule that tells us how to compare ψ at any two different spacetime points. This rule has the effect of defining the curvature of the vector-field manifold containing the vectors $\psi(x) = (\psi_1(x), \ldots, \psi_n(x))$. We will see that this rule can be characterized by a gauge field A_μ. In this sense $A_\mu(x)$ forms a *connection* analogous to the Christoffel connection in general relativity.

We proceed by defining a parallel transport operator, which is a bilocal $n \times n$ matrix that we write as $U(y, x)$. In gauge field theory $U(y, x)$ is more commonly referred to as a *Wilson line*. **Note:** *Do not confuse the Wilson line from x to y, $U(y, x)$, with the gauge transformation $U(x)$ at x.*[3] For a given choice of the Wilson line $U(y, x)$ we can compare $\psi(x)$ and $\psi(y)$ by first "parallel transporting" $\psi(x)$ to y. For example, we perform a subtraction of $\psi(y)$ and $\psi(x)$ as $\psi(y) - U(y, x)\psi(x)$. This difference must be a vector on the vector-field manifold at y. Then under a gauge transformation $U(x)$ the Wilson line $U(y, x)$ must have the property that

$$\psi(y) - U(y, x)\psi(x) \xrightarrow{U(x)} \psi'(y) - U'(y, x)\psi'(x) = U(y)\left[\psi(y) - U(y, x)\psi(x)\right],$$
$$\Rightarrow \quad U'(y, x)U(x) = U(y)U(y, x) \quad \Rightarrow \quad U'(y, x) = U(y)U(y, x)U(x)^\dagger, \qquad (9.1.13)$$

since $\psi'(x) = U(x)\psi(x)$. We need to ensure that $U(y, x)$ has this property. In a gauge field theory the gauge field will have an action that suppresses "rough" gauge field configurations and in any case the gauge field configuration will be smoothed by whatever ultraviolet regulator we impose. Then it is sufficient for us to consider Wilson lines that are differentiable with respect to their arguments so that $\partial_\mu^x U(y, x)$ and $\partial_\mu^y U(y, x)$ exist. Consider the case where $y^\mu = x^\mu + dx^\mu = x^\mu + \epsilon n^\mu$, then

$$U(x + dx, x) = I + dx^\mu \partial_\mu^y U(y, x)|_{y=x} + \mathcal{O}(\epsilon^2) \equiv I + igdx^\mu A_\mu(x) + \mathcal{O}(\epsilon^2), \qquad (9.1.14)$$

which is the definition of the gauge field $A_\mu(x)$ in terms of the Wilson line $U(y, x)$. Since the purpose of $U(x, y)$ is to perform a gauge transformation as we move from x to y then $U(y, x) \in G$. This means that $A_\mu(x)$ lies in the tangent space of the group and so takes its values in the Lie algebra of G. This means that we can define $A_\mu^a(x)$ such that $A_\mu(x) = A_\mu^a(x)T^a$ where the T^a are the Hermitian $n \times n$ generators of G. Subtracting $\psi(x)$ from $\psi(x + dx)$ and working to first order in dx^μ leads to

$$\psi(x + dx) - U(x + dx, x)\psi(x) = dx^\mu[\partial_\mu - igA_\mu(x)]\psi(x) \equiv dx^\mu D_\mu\psi(x), \qquad (9.1.15)$$

---

[3] $U(y, x)$ is written as $W(y, x)$ in Schwartz (2013) and avoids confusion. We follow Peskin and Schroeder (1995) and use $U(y, x)$ since it is the more conventional choice in lattice gauge theory.

where we have defined the covariant derivative, $D_\mu \equiv \partial_\mu - igA_\mu$. Comparing Eqs. (9.1.13) and (9.1.14) we can deduce how the gauge field must transform under a gauge transformation, $A_\mu(x) \to A'_\mu(x)$. To first order in dx^μ we have

$$U'(x+dx) = U(x+dx)U(x+dx,x)U(x)^\dagger = [U(x) + dx^\mu \partial_\mu U(x)][I + igdx^\mu A_\mu(x)]U(x)^\dagger$$
$$= I + igdx^\mu \left\{ U(x)A_\mu(x)U(x)^\dagger - (i/g)[\partial_\mu U(x)]U(x)^\dagger \right\} \equiv I + igdx^\mu A'_\mu(x),$$
$$\Rightarrow \quad A_\mu(x) \xrightarrow{U(x)} A'_\mu(x) = U(x)A_\mu(x)U(x)^\dagger - (i/g)[\partial_\mu U(x)]U(x)^\dagger, \tag{9.1.16}$$

which reproduces Eq. (9.1.3) and its consequences, e.g., $D'_\mu \psi'(x) = U(x)D_\mu \psi(x)$.

Consider some path in spacetime from x to y, and divide it up into N intervals of the form $z_0 = x, z_1, \ldots, z_{N-1}, z_N = y$ such that $z_i^\mu - z_{i-1}^\mu = dz_i^\mu = \epsilon n_i^\mu(z_i)$. Then taking $N \to \infty$ such that $\epsilon \to 0$ we find using a generalization of Eq. (A.8.5) that

$$U(y,x) = \lim_{N\to\infty} \prod_{i=1}^N [I + igdz_i^\mu A_\mu(z_i)] = P \exp\{ig \int_x^y dz^\mu A_\mu(z)\}$$
$$= P \exp\{ig \int_0^1 dt\, A_\mu(z(t))[dz^\mu/dt]\}, \tag{9.1.17}$$

where $z^\mu(t)$ for $0 \le t \le 1$ is the path $x \to y$ and where the path-ordering operator P is analogous to the time-ordering operator T and is necessary because the $A_\mu(z)$ do not commute. While the form in the second line of Eq. (9.1.17) makes the path dependence explicit, the first line remains a convenient shorthand. We confirm that the Wilson line is a unitary matrix, since $U(x,y)U(y,x) = I$ and since

$$U(x,y) = P \exp\{ig \int_y^x dz^\mu A_\mu(x)\} = \overline{P} \exp\{-ig \int_x^y dz^\mu A_\mu(x)\} = U(y,x)^\dagger, \tag{9.1.18}$$

where $\overline{P}$ denotes anti-path-ordering. The Wilson line $U(x,y)$ depends on the gauge field $A_\mu(x)$ and on the path. Let C be a path $z^\mu(t)$ that is a closed loop, where $x^\mu = z^\mu(0) = z^\mu(1) = y^\mu$. Let Σ denote a surface that has the loop C as its boundary. The corresponding Wilson line is called a *Wilson loop*. The nonabelian Stokes theorem for a Wilson loop is the result that

$$U^C \equiv \mathrm{tr}U(x,x) = \mathrm{tr}P \exp\{ig \int_0^1 A_\mu(z(t))[dz^\mu(t)/dt]dt\}$$
$$= \mathrm{tr}P_t \exp\{ig \int_0^1 dt \int_0^1 ds\, F_{\mu\nu}(z(s,t))[dz^\mu(s,t)/ds][dz^\nu(s,t)/dt]\}, \tag{9.1.19}$$

where path ordering with respect to t is denoted P_t and is sufficient to remove ordering ambiguities. A special case of this occurs in the construction of lattice gauge theory where Σ is a lattice plaquette. For a discussion and proofs of the nonabelian Stokes theorem, see Hirayama and Ueno (2000) and references therein.

9.1.3 Quantization of Nonabelian Gauge Theories

Here we do not construct the nonabelian Hamiltonian using the Dirac-Bergmann algorithm as we did for electromagnetism in Sec. 3.3.2. However, a brief summary of this procedure is given in Itzykson and Zuber (1980) and Weinberg (1996). Detailed discussions of this approach can also be found in Henneaux and Teitelboim (1994). We proceed directly to the functional integral approach to quantization, which follows from adding the appropriate cubic and quartic interactions for the gauge fields and imposing an appropriate gauge fixing. This will be illustrated using QCD, but the approach generalizes in a straightforward way to $SU(N)$ for any N in any $n \times n$ representation and with any number of additional fermion species of different "flavor."

9.2 Quantum Chromodynamics

The QCD Lagrangian density has quarks (the fermions) in the defining (fundamental) 3×3 representation of $SU(3)$ with three quark colors red (r), green (g) and blue (b) labeling the quark color column vector. The color gauge fields A_μ^a are the eight gluons. The quarks come in $n_f = 6$ quark flavors as shown in Table 5.3. In order of increasing quark mass these are up (u), down (d), strange (s), charm (c), bottom (b), and top (t). So the QCD Lagrangian density is

$$\mathcal{L} = -\tfrac{1}{4} F^{a\mu\nu} F_{\mu\nu}^a + \bar{\psi}^f (i\slashed{D} - m^f)\psi^f, \tag{9.2.1}$$

where sums over repeated color and flavor indices, $\sum_{a=1}^8$ and $\sum_{f=1}^{n_f}$, respectively, are to be understood. For detailed discussions devoted to QCD, see, for example, Ynduráin (1983) and Muta (1987). The color gauge symmetry $SU(3)$ implies a global $SU(3)$ color symmetry and so there will be eight conserved color charges, Q^a. However, these charges are not gauge invariant and so are not observable.

9.2.1 QCD Functional Integral

The extension of the functional integral for an abelian gauge field to the nonabelian case is the generalization of Eq. (6.4.89) to

$$F^j(t'', t') = \int \mathcal{D}c \int \mathcal{D}A_{\text{periodic}}^\mu \; |\det M_f[A]| \; \delta[f(A) - c] \; e^{iS[A,j] - i(1/2\xi)\int d^4 x\, c^2(x)}$$
$$= \int \mathcal{D}A_{\text{periodic}}^\mu \; |\det M_f[A]| \; e^{i\{S[A,j] - (1/2\xi)\int d^4 x\, f(A)^2\}}, \tag{9.2.2}$$

$$\text{where} \quad S[A,j] \equiv \int d^4 x \left\{ -\tfrac{1}{4} F_{\mu\nu}^a F^{a\mu\nu} - j^{a\mu} A_\mu^a \right\} = S[A] - \int d^4 x\, j^{a\mu} A_\mu^a,$$
$$S_\xi[A,j] = S_\xi[A] - \int d^4 x\, j^{a\mu} A_\mu^a = S[A,j] - (1/2\xi)\int d^4 x\, f(A)^2\}. \tag{9.2.3}$$

Generalizing Eq. (6.4.88) to gluon fields, the Faddeev-Popov matrix is given by $M_f^{ab}(x,y) = \delta f^a(A)(x)/\delta\omega^b(y)$, where $\omega^a(x)$ replaces $\alpha^a(x)$ in Eq. (9.1.4). Then

$$A_\mu'^a(x) = A_\mu^a(x) + (1/g)D_\mu^{ab}\omega^b(x) + \mathcal{O}(\vec{\alpha}^2) \tag{9.2.4}$$

$$\Rightarrow \quad M_f^{ab}(x,y) = \frac{\partial f(A^a)(x)}{\partial A_\mu^a(x)} \frac{\delta A_\mu^a(x)}{\delta\omega^b(y)} = \frac{1}{g}\frac{\partial f(A^a)(x)}{\partial A_\mu^a(x)} D_\mu^{ab}(x)\delta^4(x-y), \tag{9.2.5}$$

where $f(A)$ is chosen linear in A_μ^a. Evaluating M_f for familiar choices, we find

$$R_\xi \text{ gauges: } f(A^a) = \partial^\mu A_\mu^a \Rightarrow \quad M_f^{ab}(x,y) = (1/g)(\partial^\mu D_\mu^{ab})(x)\delta^4(x-y), \tag{9.2.6}$$

$$\text{axial } \xi \text{ gauges: } f(A^a) = n^\mu A_\mu^a \Rightarrow \quad M_f^{ab}(x,y) = (1/g)(n^\mu D_\mu^{ab})(x)\delta^4(x-y). \tag{9.2.7}$$

If we limit our considerations to perturbation theory, then the gluon fields are not excited to have strong fluctuations away from the action minimum at $S[A] = 0$. The gauge fields that are gauge-equivalent to $A_\mu = 0$ form the trivial orbit and have $S[A] = 0$. Only gauge orbits that lie close to the trivial gauge orbit, meaning those with small $S[A]$, are accessed in perturbation theory. For the gauge-fixed action $S_\xi[A]$ the associated Gribov copies lie around each gauge orbit and on the trivial orbit these are all the solutions of $S_\xi[A] = 0$. The number of Gribov copies for every orbit in the neighborhood of the trivial orbit will be the same and for every Gribov copy in this neighborhood the Faddeev-Popov determinant is positive, $\det M_f[A] > 0$. As we will later discuss when we introduce the Haar measure for gauge fields, each Gribov copy contributes equally. This means that we are simply overcounting by the number of Gribov copies on the trivial orbit and this cancels in ratios just as it did in the abelian case.

So for the purposes of perturbation theory we can effectively ignore Gribov copies and omit the modulus on the determinant and write the key result,

$$1 = \int \mathcal{D}\omega \, \det M_F[A]\delta[F[A]] = \int \mathcal{D}\omega \, \det(\delta F[A](x)/\delta\omega)\delta[F[A]], \qquad (9.2.8)$$

which replaces the form for an assumed ideal gauge fixing in Eq. (6.4.85). The equality in Eq. (9.2.8) is adopted as the starting point for discussions in many textbooks. It forms the basis of what is referred to as the *Faddeev-Popov method* and the *Faddeev-Popov ansatz*. With the choice $F[A] = f(A) - c$ we use Eq. (6.4.88) and $M_F[A] \to M_f[A] = \delta f(A)/\delta\omega$ and integrate over the Gaussian distribution of $c(x)$ to absorb the delta functional $\delta[f[A] - c]$. Then in place of Eq. (9.2.2) we have

$$F^j(t'', t') = \int \mathcal{D}A^\mu_{\text{periodic}} \, \det M_f[A] \, e^{i\{S[A,j] - (1/2\xi)\int d^4x \, f(A)^2\}}. \qquad (9.2.9)$$

The next step is to recognize from Eq. (6.3.27) that we can represent the determinant $\det M_f[A]$ using an appropriate integration over Grassmann-valued fields,

$$\det M_f[A] \propto \det(igM_f[A]) = \int \mathcal{D}\bar{c}\mathcal{D}c \, e^{-ig\int d^4x d^4y \, \bar{c}^a(y)M_f^{ab}(x,y)c^b(x)}, \qquad (9.2.10)$$

where will use the fact that the overall normalization of the determinant cancels in ratios. The Grassmann-valued fields $c^a(x)$ and $\bar{c}^a(x)$ correspond to fermion fields, but they transform as Lorentz scalars and carry a color index. As we know from Sec. 8.1.5 these fields have the wrong spin-statistics connection and so cannot be physical particles. For this reason they are referred to as *Faddeev-Popov ghosts*, where c^a is the ghost and $\bar{c}^a$ is the antighost. **Note:** The field $c(x)$ used in the Gaussian averaging is unrelated to the ghost and antighost fields, $c^a(x)$ and $\bar{c}^a(x)$.

Faddeev-Popov formulation of QCD: For brevity we now omit the "periodic" label on $\int \mathcal{D}A^\mu$. In place of Eq. (7.6.21) we have now arrived at the Faddeev-Popov QCD generating functional in an arbitrary covariant gauge,

$$Z[j, \bar{\eta}, \eta] = \frac{\int \mathcal{D}A^\mu \mathcal{D}\bar{\psi}\mathcal{D}\psi\mathcal{D}\bar{c}\mathcal{D}c \, e^{iS_\xi[A,\bar{\psi},\psi,\bar{c},c] - i\int d^4x \, [\bar{\eta}\psi + \bar{\psi}\eta - j\cdot A]}}{\int \mathcal{D}A^\mu \mathcal{D}\bar{\psi}\mathcal{D}\psi\mathcal{D}\bar{c}\mathcal{D}c \, e^{iS_\xi[A,\bar{\psi},\psi,\bar{c},c]}}, \qquad (9.2.11)$$

where $\quad S_\xi[A, \bar{\psi}, \psi, \bar{c}, c] = \int d^4x \, \mathcal{L}$,

$$\mathcal{L} \equiv -\tfrac{1}{4}F^{a\mu\nu}F^a_{\mu\nu} + \bar{\psi}^f(i\slashed{D} - m^f)\psi^f - \tfrac{1}{2\xi}(\partial^\mu A^a_\mu)^2 + \bar{c}^a(-\partial^\mu D^{ab}_\mu)c^b \qquad (9.2.12)$$

$$= \left[-\tfrac{1}{4}(\partial_\mu A^a_\nu - \partial_\nu A^a_\mu)^2 - \tfrac{1}{2\xi}(\partial^\mu A^a_\mu)^2 + \bar{\psi}^f(i\slashed{\partial} - m^f)\psi^f + \bar{c}^a(-\partial^\mu \partial_\mu)c^a \right]$$

$$+ \left\{ -gf^{abc}(\partial_\mu A^a_\nu)A^b_\mu A^c_\nu - \tfrac{1}{4}g^2(f^{eab}A^a_\mu A^b_\nu)(f^{ecd}A^c_\mu A^d_\nu) + g\bar{\psi}^f_i \slashed{A}^a T^a_{ij}\psi^f_j \right.$$

$$\left. + gf^{abc}(\partial_\mu \bar{c}^a)A^b_\mu c^c \right\},$$

where we have used Eq. (9.2.6) for the R_ξ-covariant gauges. The terms in $[\cdots]$ are the free field parts of the Faddeev-Popov Lagrangian density $\mathcal{L}$ and from these terms we read off the Feynman propagators for each particle. The terms in $\{\cdots\}$ are the interaction terms and using the techniques from Sec. 7.6.2 we can deduce the corresponding interaction vertices. The resulting propagators and vertices are (gluon propagator) $\quad b,\nu \,\text{〰}\, a,\mu \;=\; \dfrac{i\delta^{ab}}{p^2+i\epsilon}\left[-g^{\mu\nu}+(1-\xi)\dfrac{p^\mu p^\nu}{p^2}\right]$

(quark propagator) $\quad \beta,j \xrightarrow{\;p\;} \alpha,i \;=\; \dfrac{i\delta_{ij}(\slashed{p}+m^f)_{\alpha\beta}}{p^2-(m^f)^2+i\epsilon}$ (9.2.13)

(ghost propagator) $\quad b \cdots\!\xrightarrow{\;p\;}\!\cdots a \;=\; \dfrac{i\delta^{ab}}{p^2+i\epsilon}$

(quark-gluon vertex) $\quad = ig\gamma^\mu T^a_{ij}$

(triple-gluon vertex) $\quad = gf^{abc}\left[g^{\mu\nu}(k-p)^\rho + g^{\nu\rho}(p-q)^\mu + g^{\rho\mu}(q-k)^\nu\right]$

(four-gluon vertex)
$$
\begin{aligned}
= \; -ig^2\big[& f^{abe}f^{cde}(g^{\mu\rho}g^{\nu\sigma}-g^{\mu\sigma}g^{\nu\rho}) \\
& + f^{ace}f^{bde}(g^{\mu\nu}g^{\rho\sigma}-g^{\mu\sigma}g^{\nu\rho}) \\
& + f^{ade}f^{bce}(g^{\mu\nu}g^{\rho\sigma}-g^{\mu\rho}g^{\nu\sigma})\big]
\end{aligned}
$$

(ghost-gluon vertex) $\quad = -gf^{abc}p^\mu.$

For convenience we reproduce Eq. (A.7.26) here:

$$T_F \equiv C(N) = \tfrac{1}{2}, \quad C_F \equiv C_2(N) = (N^2-1)/2N, \quad C_A \equiv C_2(A) = C(A) = N,$$
$$\sum_a (T^aT^a)_{ij} = C_F\delta_{ij}, \quad \mathrm{tr}(T^aT^b) = T^a_{ij}T^b_{ji} = T_F\delta^{ab}, \quad f^{acd}f^{bcd} = C_A\delta^{ab}. \quad (9.2.14)$$

Here T^a are understood to be the traceless Hermitian matrix generators in the $N \times N$ defining (fundamental) representation, $T^a = t_a(N)$. For QCD we have $N = 3$, $T_F = \tfrac{1}{2}$, $C_A = 3$ and $C_F = \tfrac{4}{3}$. These results are frequently used in calculations.

For the axial ξ gauges with arbitrary ξ we replace $\partial^\mu D^{ab}_\mu$ with $n^\mu D^{ab}_\mu$. In the limit $\xi \to 0$ we arrive at the exact axial-gauge constraint $n^\mu A^a_\mu = 0$. In that case $n^\mu D^{ab}_\mu = n^\mu \partial_\mu \delta^{ab}$, which means that the Faddeev-Popov ghosts decouple and we can simply ignore them. For this reason we say that axial gauge is a *ghost-free gauge*. It is for this reason that axial gauge is the choice of gauge used in the Hamiltonian formulation and the canonical quantization procedure as done in Itzykson and Zuber (1980) and Weinberg (1996).[4] Since we want to avoid involving time in the constraints in the Hamiltonian formulation the axial-gauge choice $A^a_3 = 0$ is commonly used. For practical calculations of physical processes in perturbation theory the maintenance of Lorentz covariance is

---

[4] The axial gauge condition $A^a_3 = 0$ alone is not an ideal gauge fixing. Any $\omega^a(t,x,y)$ leaves A^a_3 invariant since $\partial\omega^a/\partial z = 0$. It is ideal if we also impose $A^a_\mu(\mathbf{x}) \to \mathbf{0}$ as $|\mathbf{x}| \to \infty$ as for Coulomb gauge, since then $\omega^a(x) = 0$. This condition excludes any global gauge transformations.

often preferred, despite the cost of having to include the Faddeev-Popov ghosts. Complications can arise in axial gauges at higher orders due to the presence of the unphysical $1/(k \cdot n)$ singularities.

9.2.2 Renormalization in QCD

The renormalized version of the QCD lagangian density in Eq. (9.2.11) is a generalization of that for QED given in Eqs. (8.5.1)–(8.5.4),

$$\mathcal{L} = -Z_3 \tfrac{1}{4}(\partial_\mu A_\nu^a - \partial_\nu A_\mu^a)^2 - \tfrac{Z_3}{2\xi}(\partial^\mu A_\mu^a)^2 + Z_2^f \bar{\psi}^f(i\slashed{\partial} - m_0^f)\psi^f + Z_{3c}\bar{c}^a(-\partial^\mu \partial_\mu)c^a$$
$$- Z_{A3}gf^{abc}(\partial_\mu A_\nu^a)A_\mu^b A_\nu^c - Z_{A4}\tfrac{1}{4}g^2(f^{eab}A_\mu^a A_\nu^b)(f^{ecd}A_\mu^c A_\nu^d) + Z_1^f g\bar{\psi}_i^f \slashed{A}^a T_{ij}^a \psi_j^f$$
$$+ Z_{1c}gf^{abc}(\partial_\mu \bar{c}^a)A_\mu^b c^c \tag{9.2.15}$$
$$\equiv \mathcal{L}_{\mathrm{ren}} + \mathcal{L}_{\mathrm{ct}},$$

where we have the renormalization constants

$$Z_2 \equiv 1 + \delta_2, \quad Z_3 \equiv 1 + \delta_3, \quad Z_{3c} \equiv 1 + \delta_{3c}, \quad Z_{A3} \equiv 1 + \delta_{A3}, \quad Z_{A4} \equiv 1 + \delta_{A4},$$
$$Z_1 \equiv 1 + \delta_1, \quad Z_{1c} \equiv 1 + \delta_{1c}, \quad Z_2 m_0^f \equiv m^f + \delta_m^f \equiv Z_2 Z_m^f m^f, \tag{9.2.16}$$
$$Z_1 g = Z_2 \sqrt{Z_3}\, g_0, \; Z_{1c}g = Z_{3c}\sqrt{Z_3}\, g_0, \; Z_{A3}g = (Z_3)^{3/2} g_0, \; Z_{A4}g^2 = (Z_3)^2 g_0^2.$$

The relations in the last line follow from the field redefinitions analogous to those for QED in Eq. (8.5.2), $\psi_{0i}^f \equiv (Z_2^f)^{1/2}\psi_i^f$, $A_{0\mu}^a \equiv (Z_3)^{1/2}A_\mu^a$ and $c_0^a \equiv (Z_{3c})^{1/2}c^a$. They follow from the fact that QCD is renormalizable. The renormalized $\mathcal{L}$, $\mathcal{L}_{\mathrm{ren}}$, is obtained by setting all Z factors to unity, replacing m_0^f with m^f and arriving at the counterterm Lagrangian density, $\mathcal{L}_{\mathrm{ren}}$, with the replacements $Z_i \to \delta_i$ except that $Z_2 m_0^f \to \delta_m^f$. Recall from Eq. (8.5.4) that $m_0^f/m^f \equiv Z_m^f = [1 + (\delta_m^f/m^f)]/Z_2 = 1 + \delta_m'^f + \mathcal{O}(\delta^2)$, where $\delta_m'^f \equiv (\delta_m^f/m^f) - \delta_2$. Note that the last line of Eq. (9.2.16) can be expressed as relationships between the renormalization constants,

$$Z_1^f/Z_2^f = Z_{1c}/Z_{3c} = Z_{A3}/Z_3 = \sqrt{Z_{A4}/Z_3} = (g_0/g)\sqrt{Z_3}. \tag{9.2.17}$$

Working to $\mathcal{O}(g^2)$ for each δ and discarding $\mathcal{O}(\delta^2)$ terms leads to

$$\delta_1^f - \delta_2^f = \delta_{1c} - \delta_{3c} = \delta_{A3} - \delta_3 = \tfrac{1}{2}(\delta_{A4} - \delta_3), \tag{9.2.18}$$

which is a useful check on the $\mathcal{O}(g^2)$ calculations. The renormalizability of nonabelian gauge theories was proved by Gerard 't Hooft in 't Hooft (1971). See also discussions in Collins (1984). Given that the theory is renormalizable, it must be true that $Z_1^f/Z_2^f = (g_0/g)\sqrt{Z_3}$ and so Z_1^f/Z_2^f is independent of flavor. This is the QCD equivalent of charge universality in QED and means that all quark flavors continue to couple equally to gluons after renormalization.

    The calculation of one-loop renormalized perturbation theory proceeds very much as it did in QED, except there are now more diagrams to evaluate. If we are only interested in the renormalization group behavior, then we only need to evaluate the counterterms. In order to do that in the MS scheme we only need to identify the divergent $1/\epsilon$ contributions. We obtain the $\overline{\mathrm{MS}}$ result from that by the replacement $(2/\epsilon) \to (2/\epsilon) + \ln(4\pi e^{-\gamma})$. There are some useful shortcuts for doing this (Schwartz, 2013). For example, the product of two propagators in a loop integral can always be rewritten using the Feynman parameterization in Eq. (8.7.61) as $1/(\ell^2 + \cdots)^2$, where ℓ is the loop momentum. Then using Eqs. (A.6.4) and (A.6.17) we find that

$$\int \frac{d^4\ell}{(2\pi)^4} \frac{1}{\ell^2 + \cdots} = \frac{i}{(4\pi)^2}\left[\frac{2}{\epsilon} + \ln(4\pi e^{-\gamma}) + \cdots\right]. \tag{9.2.19}$$

The unspecified terms are unimportant for identifying the counterterm.

Gluon self-energy: Gluon self-energy diagrams at $\mathcal{O}(g^2)$ are

$$(9.2.20)$$

Quark self-energy: Quark self-energy diagrams at $\mathcal{O}(g^2)$ are

$$(9.2.21)$$

Ghost self-energy: Ghost self-energy diagrams at $\mathcal{O}(g^2)$ are

$$(9.2.22)$$

Quark-gluon vertex: Contributions to the quark-gluon vertex at $\mathcal{O}(g^2)$ are

$$(9.2.23)$$

The drawing of the diagrams for the $\mathcal{O}(g^2)$ corrections to the ghost-gluon, three-gluon and four-gluon vertices is left as an exercise.

Consider the quark loop contribution to the gluon vacuum polarization. This calculation using the Feynman rules in Eq. (9.2.13) is the same as that for the photon in QED except that we have n_f fermions, $-q_c \to g$ and the quark-gluon vertices bring a T^a and T^b. Using Eq. (8.7.26), $g \to \mu^{\epsilon/2} g$ in dimensional regularization and $\text{tr}[T^a T^b] = T_F \delta^{ab}$ we find

$$\text{} \equiv i\Pi^{ab\mu\nu}_{2,q\text{-loop}}(q) \tag{9.2.24}$$

$$= (-1)\text{tr}[T^a T^b](-i\mu^{\epsilon/2}g)^2 \sum_{f=1}^{n_f} \int \frac{d^d\ell}{(2\pi)^d} \text{tr}\left[\frac{i(\slashed{\ell} - m^f)\gamma^\mu i(\slashed{\ell} + \slashed{q} - m^f)\gamma^\nu}{[(\ell+q)^2 - m^{f2} + i\epsilon][\ell^2 - m^{f2} + i\epsilon]}\right]$$

$$= i[-q^2 g^{\mu\nu} + q^\mu q^\nu]\delta^{ab} T_F \frac{2}{\pi}\frac{g^2}{4\pi}\sum_{f=1}^{n_f} \int_0^1 dx\, x(1-x)\left\{\frac{2}{\epsilon} - \ln\left(\frac{m^{f2} - x(1-x)q^2}{\tilde{\mu}^2}\right)\right\},$$

where the counterterm $\text{} = -i\delta^{ab}(q^2 g^{\mu\nu} - q^\mu q^\nu)\delta_3$ is to be included when all diagrams are added together to form $\Pi^{ab\mu\nu}_2(q)$ as shown in Eq. (9.2.20). If we replace $\tilde{\mu}$ with μ in the logarithm, then $(2/\epsilon) \to (2/\epsilon) + \ln(4\pi e^{-\gamma})$ and we can read off the quark loop contribution to δ_3 in $\overline{\text{MS}}$. Since $\int_0^1 dx\, x(1-x) = 1/6$ we find

$$\delta_{3,q} = -\frac{g^2}{4\pi}\frac{n_f T_F}{3\pi}\left[\frac{2}{\epsilon} + \ln(4\pi e^{-\gamma})\right] = \left\{\left[\frac{2}{\epsilon} + \ln(4\pi e^{-\gamma})\right]\frac{g^2}{2(4\pi)^2}\right\}\left[-\frac{8}{3}n_f T_F\right], \tag{9.2.25}$$

which we could also read off directly from Eq. (8.7.27). We recognize this as the quark loop contribution to the full δ_3 given in Eq. (9.2.28). The evaluation of the remaining three diagrams in Eq. (9.2.20) is here left as an exercise, but details can be found elsewhere (Peskin and Schroeder, 1995; Schwartz, 2013).

We can similarly evaluate the quark self-energy diagram $\text{} = -i(\Sigma_{2,g})_{ij}(p)$ in Eq. (9.2.21) in Feynman gauge by adding color factors to the QED result in Eq. (8.7.23). There is one δ^{ba} from the gluon propagator, one $\delta_{\ell k}$ from the internal quark propagator and a T^a_{kj} and $T^b_{i\ell}$ from the quark-gluon vertices, which gives a color factor $T^b_{i\ell}\delta^{ba}\delta_{\ell k}T^a_{kj} = (T^a T^a)_{ij} = C_F \delta_{ij}$, where the fermion arrow direction is from T^a to T^b and the summation convention is understood. From Eqs. (8.7.24) and (8.7.25) we can then read off the result that in Feynman gauge,

$$\delta_2 = \{\cdots\}\,[-2C_F] \quad \text{and} \quad \delta'_m = (\delta^f_m/m^f) - \delta_2 = \{\cdots\}\,[-6C_F], \tag{9.2.26}$$

where $\{\cdots\}$ is the corresponding term in Eq. (9.2.25). Note that in $\overline{\text{MS}}$ (and also in MS) δ_2 and δ'_m have no dependence on the renormalized quark mass, m^f, and so are flavor independent. We see these Feynman gauge results for δ_2 and δ'_m contained in Eq. (9.2.28). Note from Eq. (9.2.28) that δ'_m is gauge invariant.

We can evaluate the gluon-loop contribution diagram for the quark-gluon vertex $\equiv$ $-ig(\Lambda^{\mu a}_{2,1g})_{ij}(p',p)$ by adding color factors to Eq. (8.7.32). Writing the color matrices right to left in the direction of the fermion arrow, $T^c_{i\ell}T^a_{\ell k}T^b_{kj}\delta^{bc}$, gives

$$(T^c T^a T^b)\delta^{bc} = (T^b T^a T^b) = T^b T^b T^a + T^b[T^a, T^b] = C_F T^a + T^b i f^{abc}T^c \tag{9.2.27}$$
$$= C_F T^a + \tfrac{1}{2}i f^{abc}[T^b, T^c] = C_F T^a - \tfrac{1}{2}f^{abc}f^{bcd}T^d = (C_F - \tfrac{1}{2}C_A)T^a.$$

Inserting this into Eq. (8.7.38) and using $\int_0^1 dz(1-z) = \tfrac{1}{2}$ we can extract the $\overline{\text{MS}}$ part of this contribution to δ_1, which gives $\delta_{1,1g} = \{\cdots\}\,[-2C_F + C_A]$. An evaluation of $\equiv -ig(\Lambda^{\mu a}_{2,2g})_{ij}(p',p)$ leads to $\delta_{1,2g} = \{\cdots\}\,[-3C_A]$, which gives $\delta_1 = \delta_{1,1g} + \delta_{1,2g} = \{\cdots\}\,[-2C_F - 2C_A]$, i.e., Eq. (9.2.28). in Feynman gauge.

Evaluating all one-loop diagrams in arbitrary covariant (R_ξ) gauge we find that the $\overline{\text{MS}}$ counterterms for $SU(N)$ Yang-Mills theory with n_f flavors of fermions are

$$\delta_1 = \left\{\left[\tfrac{2}{\epsilon} + \ln(4\pi e^{-\gamma})\right]\frac{g^2}{2(4\pi)^2}\right\}[-2C_F - 2C_A + 2(1-\xi)C_F + \tfrac{1}{2}(1-\xi)C_A],$$
$$\delta_2 = \{\cdots\}\,[-2C_F + 2(1-\xi)C_F], \quad \delta_3 = \{\cdots\}\left[\tfrac{10}{3}C_A - \tfrac{8}{3}n_f T_F + (1-\xi)C_A\right],$$
$$\delta'_m = \{\cdots\}\,[-6C_F], \quad \delta_{A3} = \{\cdots\}\left[\tfrac{4}{3}C_A - \tfrac{8}{3}n_f T_F + \tfrac{3}{2}(1-\xi)C_A\right],$$
$$\delta_{A4} = \{\cdots\}\left[-\tfrac{2}{3}C_A - \tfrac{8}{3}n_f T_F + 2(1-\xi)C_A\right], \quad \delta_{1c} = \{\cdots\}\,[-C_A + (1-\xi)C_A],$$
$$\delta_{3c} = \{\cdots\}\,[C_A + \tfrac{1}{2}(1-\xi)C_A], \tag{9.2.28}$$

where the above form is convenient for identifying Feynman gauge ($\xi = 1$) (Ynduráin, 1983; Muta, 1987; Schwartz, 2013). Feynman gauge can often be simpler to use in calculations. Using the above we verify Eq. (9.2.18) and find that

$$\delta^f_1 - \delta^f_2 = \delta_{1c} - \delta_{3c} = \delta_{A3} - \delta_3 = \tfrac{1}{2}(\delta_{A4} - \delta_3) = \{\cdots\}\,[-\tfrac{1}{2}(3+\xi)C_A]. \tag{9.2.29}$$

In $\overline{\text{MS}}$ (and MS) at one loop there is no flavor dependence, $\delta^f_1 = \delta_1$ and $\delta^f_2 = \delta_2$.

9.2.3 Running Coupling and Running Quark Mass

Similar to the QED treatment in Sec. 8.7.5 we can obtain one-loop results for the running coupling and the running mass of quarks. The essential inputs to this are the QCD β-function, $\beta(\alpha_s)$, and the quark anomalous mass dimension, $\gamma_m(\mu)$.

Consider the quark gluon vertex, since the bare coupling is independent of μ

$$0 = \mu\frac{\partial g_0}{\partial\mu} = \mu\frac{\partial}{\partial\mu}\left[g\frac{\mu^{\epsilon/2}Z_1}{Z_2\sqrt{Z_3}}\right] \Rightarrow \beta(g) \equiv \mu\frac{dg}{d\mu} = -g\left[\frac{\epsilon}{2} + \mu\frac{\partial}{\partial\mu}(\delta_1 - \delta_2 - \tfrac{1}{2}\delta_3) + \cdots\right],$$

where we neglect terms $\mathcal{O}(\epsilon\delta)$ and $\mathcal{O}(\delta^2)$. The counterterms depend on μ only through $g(\mu)$ and so neglecting terms $\mathcal{O}(\epsilon g^3)$ and $\mathcal{O}(g^5)$ we have

$$\beta(g) = -g\left[\frac{\epsilon}{2} + \mu\frac{\partial g}{\partial\mu}\frac{\partial}{\partial g}(\delta_1 - \delta_2 - \tfrac{1}{2}\delta_3) + \cdots\right] = -g\left[\frac{\epsilon}{2} - g\frac{\epsilon}{2}\frac{\partial}{\partial g}(\delta_1 - \delta_2 - \tfrac{1}{2}\delta_3) + \cdots\right]$$

$$= -\frac{\epsilon}{2}g - \frac{g^3}{16\pi^2}[\tfrac{11}{3}C_A - \tfrac{4}{3}n_f T_F] + \cdots = -\frac{\epsilon}{2}g - \frac{g^3}{16\pi^2}[11 - \tfrac{2}{3}n_f] + \cdots, \qquad (9.2.30)$$

where we have used the $\overline{\text{MS}}$ counterterms above and have specialized to $SU(3)_{\text{color}}$ in the last step. Observe that the gauge parameter ξ has disappeared in $\beta(g)$.

Define the coefficients of the expansion of the QCD β-function as we did for QED,

$$\beta(\alpha_s) \equiv \mu\frac{\partial\alpha_s}{\partial\mu} = \frac{2g}{4\pi}\mu\frac{dg}{d\mu} = \frac{2g}{4\pi}\beta(g) \equiv -2\alpha_s\left[\frac{\epsilon}{2} + \beta_0\frac{\alpha_s}{4\pi} + \beta_1\left(\frac{\alpha_s}{4\pi}\right)^2 + \beta_2\left(\frac{\alpha_s}{4\pi}\right)^3 + \cdots\right],$$

$$\Rightarrow \beta_0 = [\tfrac{11}{3}C_A - \tfrac{4}{3}n_f T_F] = 11 - \tfrac{2}{3}n_f. \qquad (9.2.31)$$

In QED with $\epsilon \to 0$ from Eq. (8.7.67) we have $\beta(q_c) = -\beta_0^{\text{QED}}q_c^3/(4\pi)^2$, where $\beta_0^{\text{QED}} = -4/3$ and correspondingly from Eq. (9.2.30) we have $\beta(g) = -\beta_0 g^3/(4\pi)^2$. So the arguments leading to Eq. (8.7.69) again apply with the QCD version of β_0,

$$\alpha_s(\mu_b) = \frac{\alpha_s(\mu_a)}{1 + \frac{\beta_0}{4\pi}\alpha_s(\mu_a)\ln(\frac{\mu_b^2}{\mu_a^2})} = \alpha_s(\mu_a)\left[1 - \frac{\beta_0}{4\pi}\alpha_s(\mu_a)\ln\left(\frac{\mu_b^2}{\mu_a^2}\right) + \mathcal{O}(\alpha_s^2)\right]. \qquad (9.2.32)$$

There is a singularity when $\ln(\mu_b^2) = \ln(\mu_a^2) - [4\pi/\beta_0\alpha_s(\mu_a)]$, which must be true for all μ_a since the left-hand side is independent of μ_a. So we can then define $\ln(\Lambda_{\text{QCD}}^2) \equiv \ln(\mu_a^2) - [4\pi/\beta_0\alpha_s(\mu_a)]$ and write the QCD analog of Eq. (8.7.72),

$$\alpha_s(\mu) = \frac{4\pi}{\beta_0\ln(\mu^2/\Lambda_{\text{QCD}}^2)} = \frac{2\pi}{\beta_0\ln(\mu/\Lambda_{\text{QCD}})}. \qquad (9.2.33)$$

Λ_{QCD} is the QCD equivalent of the Landau pole in QED, but importantly since $\beta_0 > 0$ the above equation is only meaningful when $\mu > \Lambda_{\text{QCD}}$. So Λ_{QCD} is a lower limit to μ in one-loop perturbation theory in QCD, whereas Λ_{QED} was an upper limit on μ in one-loop perturbation theory in QED. Since $\beta_0 > 0$, as μ increases then $\alpha_s(\mu)$ decreases, which is the property of asymptotic freedom discussed in Sec. 5.2.2. Note that in QCD we have $n_f = 6$ and so $\beta_0 = 11 - \tfrac{2}{3}n_f > 0$. If we had $n_f \geq 17$ quark flavors, then QCD would cease being asymptotically free.

Recall that $Z_m = 1 + \delta_m'$ with δ_m' given in Eq. (9.2.28). In the case of QCD we have $\beta(g) = -(\epsilon/2)g + \cdots$ and $\partial Z_m/\partial g = \partial\delta_m'/\partial g = -(2/\epsilon)[8g/(4\pi)^2] + \cdots$, where we used $C_F = \tfrac{4}{3}$. As we did for QED using Eqs. (8.7.73) and (8.7.74) it is now straightforward to evaluate the quark anomalous mass dimension,

$$\gamma_m = -\frac{\mu}{Z_m}\frac{\partial Z_m}{\partial g}\frac{dg}{d\mu} = -\frac{8g}{(4\pi)^2}\frac{2}{\epsilon}\frac{\epsilon}{2}g = -\frac{8g^2}{(4\pi)^2} = \frac{-4}{\beta_0\ln(\frac{\mu}{\Lambda_{\text{QCD}}})}. \qquad (9.2.34)$$

Repeating the steps leading to Eq. (8.7.75) the quark running mass $m(\mu)$ satisfies

$$\frac{m(\mu_b)}{m(\mu_a)} = \left[\frac{\ln(\mu_b^2/\Lambda_{\text{QCD}}^2)}{\ln(\mu_a^2/\Lambda_{\text{QCD}}^2)}\right]^{-\frac{4}{\beta_0}} = \left[\frac{\ln(\mu_a^2/\Lambda_{\text{QCD}}^2)}{\ln(\mu_b^2/\Lambda_{\text{QCD}}^2)}\right]^{\frac{12}{33-2n_f}} = \left[\frac{\alpha_s(\mu_a)}{\alpha_s(\mu_b)}\right]^{\frac{12}{33-2n_f}}, \qquad (9.2.35)$$

since $\beta_0 = 11 - \tfrac{2}{3}n_f$. So the quark mass logarithmically decreases with renormalization scale as expected since $\gamma_m < 0$. Measuring $\alpha_s(\mu)$ at some scale then determines Λ_{QCD} and $\alpha_s(\mu)$

at all scales. To verify that we recover γ_m in Eq. (9.2.34) from this solution, define C such that $m(\mu) = C \ln(\mu/\Lambda_{\text{QCD}})^{-4/\beta_0}$. Then we confirm that

$$\gamma_m = \frac{\mu}{m}\frac{dm(\mu)}{d\mu} = \mu\left[\frac{-4}{\beta_0}\right]\frac{1}{\ln(\frac{\mu}{\Lambda_{\text{QCD}}})}\frac{d\ln(\frac{\mu}{\Lambda_{\text{QCD}}})}{d\mu} = \frac{-4}{\beta_0\ln(\frac{\mu}{\Lambda_{\text{QCD}}})}. \tag{9.2.36}$$

Decoupling of heavy particles: Appelquist-Carrazone theorem: In the above the focus was on the QCD counterterms in the mass-independent $\overline{\text{MS}}$ scheme, which leads to the mass-independent β-function and renormalization group equations. In those discussions flavor dependence due to different quark masses did not arise; however, quark mass dependence remains in the renormalized Green's functions and amplitudes. For example, consider the quark loop contribution to the gluon vacuum polarization function $\Pi(q^2)$, where $\Pi^{\mu\nu}(q) = [-g^{\mu\nu}q^2 + q^\mu q^\nu]\Pi(q^2)$. From Eq. (9.2.24) the $\mathcal{O}(g^2)$ quark loop contribution from n_f quarks in $\overline{\text{MS}}$ is

$$\Pi_{n_f}(q^2) = -T_F\frac{g^2}{2\pi^2}\sum_{f=1}^{n_f}\int_0^1 dx\, x(1-x)\ln\left(\frac{m^{f2} - x(1-x)q^2}{\mu^2}\right). \tag{9.2.37}$$

Consider the spacelike region, $Q^2 = -q^2 > 0$. The decoupling theorem is most transparent if we work in a momentum subtraction (MOM) scheme as discussed in Sec. 8.5.3, where $\Pi^{\text{MOM}}(q^2) = \Pi(q^2) - \Pi(-\mu^2)$ so that $\Pi^{\text{MOM}}(-\mu^2) = 0$. The quark loop contribution is $\Pi_{n_f}^{\text{MOM}}(q^2) = \Pi_{n_f}(q^2) - \Pi_{n_f}(-\mu^2)$, which gives

$$\Pi_{n_f}^{\text{MOM}}(q^2) = -T_F\frac{g^2}{2\pi^2}\sum_{f=1}^{n_f}\int_0^1 dx\, x(1-x)\ln\left(\frac{m^{f2} + x(1-x)Q^2}{m^{f2} + x(1-x)\mu^2}\right). \tag{9.2.38}$$

Assume some quark of flavor f' has a very large mass such that $m^{f'2} \equiv M^2 \gg \mu^2, Q^2$. Observing that $\ln[1 + x(1-x)(Q^2/M^2)] = x(1-x)(Q^2/M^2) + \mathcal{O}(Q^4/M^4)$ and similarly $\ln[1 + x(1-x)(\mu^2/M^2)] = x(1-x)(\mu^2/M^2) + \mathcal{O}(\mu^4/M^4)$, then this flavor quark only gives a contribution $\mathcal{O}(Q^2/M^2, \mu^2/M^2)$. So if any $m^{f2} \gg \mu^2, Q^2$ then that very massive ("heavy") quark can be ignored up to small corrections, $\mathcal{O}(Q^2/m^{f2}, \mu^2/m^{f2})$, and we can work with only the remaining $(n_f - 1)$ flavors.

This is an illustration of the *Appelquist-Carrazone theorem* also known as the *decoupling theorem* (Symanzik, 1970, 1973; Appelquist and Carazzone, 1975; Ovrut and Schnitzer, 1980; Ynduráin, 1983; Grządkowski et al., 1984), which can be summarized as: *The effects of very massive ("heavy") particles either can be absorbed into a renormalization of coupling constants of the theory or will appear as small corrections suppressed by inverse powers of the heavy mass, provided that the remaining low-energy theory without the heavy particles is renormalizable.*

There are six quarks with wide-ranging masses; we are interested in physical processes that span all of these scales and need a simple way to manage this issue. A convenient approach is to define n_f to be dependent on the momentum scale Q^2, where we can use, for example, the interpolation formula (Ynduráin, 1983),

$$n_f(Q^2) = \sum_{f=1}^{n_f}[1 - 4(m^{f2}/Q^2)]^{1/2}[1 + 2(m^{f2}/Q^2)]\theta(Q^2 - 4m^{f2}). \tag{9.2.39}$$

Other interpolation formulae are possible, but the details are not very important as QCD calculations are relatively insensitive to how this is done.

QCD running coupling constant: In Fig. 9.1 we show the running coupling constant $\alpha_s(\mu)$, which is also written as $\alpha_s(Q^2) \equiv \alpha_s(\mu)|_{Q^2=\mu^2}$. The theoretical uncertainty is indicated by the solid lines and the experimental determinations are shown for a variety of experiments. The degree

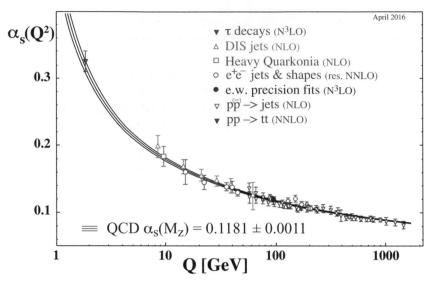

Figure 9.1 Measurements of α_s with $Q^2 = \mu^2$ and compared with the theoretical prediction from QCD (solid lines). NLO = next-to-leading order; NNLO = next-to-next-to leading order; res. NNLO = NNLO matched with resummed next-to-leading logs; N$_3$LO = next-to-NNLO. Image reprinted with permission (Tanabashi et al., 2018).

of QCD renormalized perturbation theory used to extract α_s in the various experiments is shown in parentheses. For further detail refer to Zyla et al. (2020) and references therein.

Evidence for three colors: We arrived at the $e^+e^- \to \mu^+\mu^-$ cross-section in the ultrarelativistic limit in Eq. (7.6.73). Extending to the case of a fermion of charge q_c^f and mass m^f we have $\sigma_{e^+e^- \to \bar{f}f} = (q_c^f)^2 4\pi\alpha^2/3s = (q_c^f/e)^2 \sigma_{e^+e^- \to \mu^+\mu^-}$ provided that $s \gg (m^f)^2$. Consider the case $e^+e^- \to$ hadrons. As s increases and we cross the $\bar{q}^f q^f$ quark production threshold we access new hadronic final states containing these quarks. With increasing s the final state QCD interactions have little effect and we expect an approximate plateau up to the next threshold. Since there are three quark colors, $N_c = 3$, for each flavor f we expect to see

$$R \equiv \frac{\sigma_{e^+e^- \to \text{ hadrons}}}{\sigma_{e^+e^- \to \mu^+\mu^-}} \simeq \sum_{f \text{ for } s>(m^f)^2} N_c \frac{(q_c^f)^2}{e^2}. \tag{9.2.40}$$

The factor $N_c = 3$ is essential in order to be consistent with experiments.

9.2.4 BRST Invariance

The Faddeev-Popov Lagrangian density in Eq. (9.2.12) has an important symmetry referred to as *BRST invariance* or BRST symmetry, which is named after C. Becchi, A. Rouet, R. Stora and I. V. Tyutin (Becchi et al., 1975, 1976; Tyutin, 1975). For additional discussion and perspective, see Itzykson and Zuber (1980), Ynduráin (1983), Muta (1987), Peskin and Schroeder (1995), Weinberg (1996) and Schwartz (2013). This symmetry is valid in the presence of the gauge-fixing and ghost terms.

Let us construct this invariance by imposing a suitable generalized gauge transformation. Rewrite the Faddeev-Popov Lagrangian density in Eq. (9.2.12) as

$$\mathcal{L} = \mathcal{L}_{\text{gi}} - \tfrac{1}{2\xi}(\partial^\mu A_\mu^a)^2 + \bar{c}^a(-\partial^\mu D_\mu^{ab})c^b, \tag{9.2.41}$$

where $\mathcal{L}_{\text{gi}}$ contains the gauge-invariant parts of $\mathcal{L}$ given in Eq. (9.2.1). Recall that $\mathcal{L}$ appears in the exponent of the functional integral, where A_μ^a are real fields and quarks and ghosts are Grassmann fields. Begin by introducing a new constant Grassman number θ that anticommutes with itself and the Grassmann fields. Define a gauge transformation of the quark and gluon fields, $\alpha^a(x) = \theta c^a(x)$, where $c^a(x)$ are the ghost fields. Note that $\alpha^a(x)$ is an even Grassman function and so commutes with everything. Since $\theta^2 = 0$ we do not need infinitesimal gauge transformations in order to keep only the linear terms in θ. Then Eqs. (9.1.1) and (9.1.10) become

$$\psi_i^f \to \psi_i'^f = \psi_i^f + i\alpha^a T_{ij}^a \psi_j^f = \psi_i^f + i\theta c^a T_{ij}^a \psi_j^f, \qquad (9.2.42)$$
$$A_\mu^a \to A_\mu'^a = A_\mu^a + (1/g)D_\mu^{ab}\alpha^b = A_\mu^a + (1/g)\theta D_\mu^{ab}c^b.$$

Since $\mathcal{L}_{\text{gi}}$ is gauge invariant, then it is unchanged by these gauge transformations, $\mathcal{L}_{\text{gi}} \to \mathcal{L}_{\text{gi}}$. It follows from the above that

$$D_\mu^{ab} \to D_\mu^{ab} - \theta f^{abc} D_\mu^{cd} c^d, \quad (\partial^\mu A_\mu^a)^2 \to (\partial^\mu A_\mu^a)^2 + (2/g)\theta(\partial^\nu A_\nu^a)\partial^\mu D_\mu^{ab} c^b, \qquad (9.2.43)$$

which leads to the preliminary result that

$$\mathcal{L} \xrightarrow{\text{prelim}} \mathcal{L} + (1/\xi g)\theta(\partial^\nu A_\nu^a)(-\partial^\mu D_\mu^{ab} c^b) - \theta\bar{c}^a(-\partial^\mu f^{abc} D_\mu^{cd} c^d)c^b. \qquad (9.2.44)$$

Since the only purpose of the ghosts and antighosts is to produce the Faddeev-Popov determinant, then there is no relationship between c^a and $\bar{c}^a$ and they can transform differently. We can then choose these two transformations to cancel the two terms on the right-hand side of Eq. (9.2.44). The necessary transformations are

$$\bar{c}^a \to \bar{c}'^a = \bar{c}^a - (1/\xi g)\theta(\partial^\mu A_\mu^a) \quad \text{and} \quad c^a \to c'^a = c^a - \tfrac{1}{2}\theta f^{abc} c^b c^c. \qquad (9.2.45)$$

The $\bar{c}^a$ transformation obviously cancels the $(1/\xi g)$ term in Eq. (9.2.44). The final term on the right-hand side of Eq. (9.2.44) arose because $D_\mu^{ab} c^b$ in Eq. (9.2.41) was not invariant since D_μ^{ab} is not invariant. The above transformation of c^a is chosen specifically so that $D_\mu^{ab} c^b$ becomes invariant, which then leaves $\mathcal{L}$ invariant.

Summary: The BRST transformations are defined in Eqs. (9.2.42) and (9.2.45) and by construction leave the Faddeev-Popov Lagrangian $\mathcal{L}$ invariant.

Proof: It only remains to show that $D_\mu^{ab} c^b$ is invariant. We find that

$$D_\mu^{ab} c^b \to D_\mu^{ab} c^b - \theta f^{abc}(D_\mu^{cd} c^d)c^b - \tfrac{1}{2}\theta f^{bcd} D_\mu^{ab}(c^c c^d) = D_\mu^{ab} c^b, \quad \text{since}$$
$$\tfrac{1}{2} f^{bcd} D_\mu^{ab}(c^c c^d) = \tfrac{1}{2} f^{acd}[(\partial_\mu c^c)c^d + c^c(\partial_\mu c^d)] - \tfrac{1}{2} g f^{bcd} f^{abe} A_\mu^e c^c c^d \qquad (9.2.46)$$
$$= f^{acd}(\partial_\mu c^c)c^d + \tfrac{1}{2} g(f^{dab} f^{cbe} + f^{acb} f^{dbe}) A_\mu^e c^c c^d$$
$$= f^{acd}(\partial_\mu c^c)c^d + g f^{dab} f^{cbe} A_\mu^e c^c c^d = -f^{abc}(D_\mu^{cd} c^d)c^b,$$

where the Jacobi identity in Eq. (A.7.7) has been used and summed indices have been relabeled as needed to arrive at the final form.

Since θ is a constant Grassmann number, then BRST is a global invariance. The global phase invariance of QED leads to current conservation, which in turn leads to the QED Ward-Takahashi identities. Similarly, BRST invariance in a nonabelian theory leads to a generalized form of the Ward-Takahashi identities referred to as the *Slavnov-Taylor identities* (Taylor, 1971; Slavnov, 1972). One such identity is that the self-energy for gluons is transverse, $q_\mu \Pi^{\mu\nu}(q) = 0$, as it is for photons in

QED (Ynduráin, 1983). The first discussion of these identities in a restricted form was by 't Hooft in his proof of the renormalizability of nonabelian gauge theories ('t Hooft, 1971). The complete form of these identities was subsequently developed in the above references by Slavnov and Taylor.

9.2.5 Lattice QCD

In our discussions of quantum field theory and functional integrals we have always had the implicit understanding that we had both ultraviolet and infrared regulators that would be removed once we had formed a physical observable. We have also understood that the Wick rotation from Euclidean space to Minkowski space is the origin of the Feynman boundary conditions and the damping of oscillatory Minkowski space functional integrals. In Sec. A.4 the relation between Euclidean and Minkowski-space quantities was discussed and the Wick rotation explained. In Sec. 6.2.11 we showed the relationship between the Euclidean and Minkowski-space generating functionals for a scalar field. The next step is to realize the formulation of quantum field theory in Euclidean space on a spacetime lattice, which will allow us to move beyond perturbation theory and to study the nonperturbative behavior of quantum field theories. The running coupling of QCD shown in Fig. 9.1 demonstrates that at short distance scales QCD is asymptotically free and that it is strong coupling at long distance scales. Lattice gauge theory is a very valuable tool for understanding the low-energy behavior of QCD from first principles. A number of texts provide a pedagogical introduction to this extensive subject, including Montvay and Münster (1994), DeGrand and DeTar (2006), Rothe (2012) and Knechtli et al. (2017).

Let Γ be a hypercubic lattice of Euclidean spacetime points. Define such a lattice as $\Gamma = \{x : x = \sum_{\mu=1}^{4} an_\mu \hat{\mu}, \, n_\mu \in \mathbb{Z}, \, 1 \leq n_\mu \leq N\}$, where $a \in \mathbb{R}$ is the lattice spacing, $\hat{\mu}$ are four orthonormal unit vectors and $L = aN$ is the lattice length in each direction. We often use a longer time direction, $L^4 \to L^3 T$ with $T > L$ and $N_t > N$, to extract large Euclidean time behavior. Anisotropic lattices with $a_t < a$ and $T = N_t a_t$ are sometimes used. A lattice site $x \in \Gamma$ is specified by (n_1, n_2, n_3, n_4). The infrared regulator is $1/L$ and the ultraviolet regulator is $\Lambda = 1/a$.

Scalar fields on the lattice: The Euclidean action for a real scalar field was given in Eq. (6.2.270). On the lattice the partial derivative is $\partial_\mu \to [\phi(x + a\hat{\mu}) - \phi(x)]/a$ and the volume integral is $\int d^4x \to a^4 \sum_{x \in \Gamma}$. Using $\Box = \partial_\mu \partial_\mu$ we can write the lattice version of $\Box \phi(x)$ in the form $\Box \phi(x) \to (1/a^2)\hat{\Box}\phi(x) = (1/a^2) \sum_\mu [\phi(x + a\hat{\mu}) + \phi(x - a\hat{\mu}) - 2\phi(x)]$. The lattice action for a free scalar field is then

$$S_E[\phi] = \int d^4x \, \tfrac{1}{2} \left\{ \partial_\mu \phi(x) \partial_\mu \phi(x) + m_0^2 \phi(x)^2 \right\} = \int d^4x \, \tfrac{1}{2} \phi(x) \left\{ -\Box + m_0^2 \right\} \phi(x)$$

$$\xrightarrow{\text{lattice}} S_{\text{latt}}[\phi] = a^4 \sum_{x \in \Gamma} \tfrac{1}{2} \{ \phi(x)[-(1/a^2)\hat{\Box}\phi(x)] + \phi(x)m_0^2\phi(x) \} \qquad (9.2.47)$$

$$= \sum_{x,y \in \Gamma} \phi(x) M(x,y) \phi(y),$$

$$M(x,y) \equiv a^4 \tfrac{1}{2} \left\{ -\sum_\mu \tfrac{1}{a^2} [\delta_{(x+a\hat{\mu}),y} + \delta_{(x-a\hat{\mu}),y} - 2\delta_{x,y}] + m_0^2 \delta_{x,y} \right\}.$$

Recall that the spectral function is the trace of the evolution operator and was given in Eq. (6.2.243), $F(t'' - t') = \text{tr} \, e^{-i\hat{H}(t''-t')/\hbar} = \int \mathcal{D}\phi \, \exp\{(i/\hbar)S([\phi], t'', t')\}$ with the ϕ field *periodic in space and time*. Also recall that with the addition of a source term $j\phi$ the spectral function leads to the generating functional $Z[j]$ for the scalar field as given in Eq. (6.2.262). By taking derivatives of $Z[j]$ with respect to j and setting $j = 0$ we can bring any quantity of interest, $O[\phi]$, into the time-ordered vacuum expectation value. Using Eq. (6.2.263) we then arrive at

$$\langle\Omega|T\{O[\hat{\phi}]\}|\Omega\rangle = \lim_{T\to\infty(1-i\epsilon)}\frac{\mathrm{tr}[T\{O[\hat{\phi}]e^{-i\hat{H}T}\}]}{\mathrm{tr}[e^{-i\hat{H}T}]} = \lim_{T\to\infty(1-i\epsilon)}\frac{\int\mathcal{D}\phi\,O[\phi]e^{iS[\phi]}}{\int\mathcal{D}\phi\,e^{iS[\phi]}},$$

$$\langle\Omega|T\{O[\hat{\phi}]\}|\Omega\rangle_E = \lim_{T\to\infty}\frac{\mathrm{tr}[T\{O[\hat{\phi}]e^{-\hat{H}T}\}]}{\mathrm{tr}[e^{-\hat{H}T}]} = \lim_{T\to\infty}\frac{\int\mathcal{D}\phi\,O[\phi]e^{-S_E[\phi]}}{\int\mathcal{D}\phi\,e^{-S_E[\phi]}}, \qquad (9.2.48)$$

where the second form is the Euclidean space version involving the partition function $\mathrm{tr}[e^{-\hat{H}T}]$. Since $S_E[\phi] \geq 0$, then $P[\phi] \equiv e^{-S_E[\phi]}/\int\mathcal{D}\phi\,e^{-S_E[\phi]}$ is a probability density functional in configuration space with $0 \leq P[\phi] \leq 1$ and $\int\mathcal{D}\phi\,P[\phi] = 1$. It is sufficient for $S_E[\phi]$ to be bounded below since adding a constant to $S_E[\phi]$ does not change $P[\phi]$. Then it follows that $\langle\Omega|T\{O[\hat{\phi}]\}|\Omega\rangle_E = \lim_{T\to\infty}\int\mathcal{D}\phi\,O[\phi]P[\phi]$. Observe that $P[\phi']/P[\phi] = e^{-S_E[\phi']}/e^{-S_E[\phi]}$.

Monte Carlo methods: Consider an *ensemble of lattice field configurations* $\phi_i(x)$ for $i = 1,\ldots,N_{\mathrm{config}}$ that have been sampled with the relative probability $P[\phi_i]/P[\phi_j] = e^{-S_{\mathrm{latt}}[\phi_i]}/e^{-S_{\mathrm{latt}}[\phi_j]}$. We calculate $\langle\Omega|T\{O[\hat{\phi}]\}|\Omega\rangle_E$ by taking the average over the ensemble $\{\phi_1,\ldots,\phi_{N_{\mathrm{config}}}\}$ with appropriate limits,

$$\langle\Omega|T\{O[\hat{\phi}]\}|\Omega\rangle_E = \lim_{L\to\infty}\ \lim_{a\to0,\,L\,\mathrm{fixed}}\ \lim_{N_{\mathrm{config}}\to\infty}\frac{\sum_{i=1}^{N_{\mathrm{config}}}O[\phi_i]}{N_{\mathrm{config}}}. \qquad (9.2.49)$$

This equation is the basis of the lattice approach to quantum field theories. As required if $O[\phi] = 1$ we find $\langle\Omega|\Omega\rangle = 1$. We are using *importance sampling* to perform a multidimensional *Monte Carlo integration*. If $O[\phi]$ is sufficiently smooth, the method can be very efficient even for modest N_{config}.

To form an ensemble of lattice configurations $\{\phi_1,\ldots,\phi_{N_{\mathrm{config}}}\}$ first generate a long sequence of lattice configurations $\{c_1,c_2,\ldots\}$ referred to as a *Markov chain* using an updating rule that has $P(c \to c')$ as the transition probability to move from configuration c in the chain to the next configuration c'. Let $P(c \to c')$ have the properties of (i) *ergodicity:* any configuration can be accessed with a long enough Markov chain; and (ii) *detailed balance:* $P(c \to c')/P(c' \to c) = e^{-S_{\mathrm{latt}}[c']}/e^{-S_{\mathrm{latt}}[c]}$. Configurations in the Markov chain that are separated by a very large number of configuration update steps will be uncorrelated. Taking a selection of N_{config} uncorrelated configurations from such a Markov chain produces an ensemble with the required properties. A simple way to generate such a Markov chain is to use the *Metropolis algorithm* (also referred to as the Metropolis-Hastings algorithm). There are more sophisticated algorithms with improved efficiency such as the hybrid Monte Carlo (HMC) algorithm. Various algorithms are discussed in Rothe (2012).

Fermions on the lattice: For a free fermion the Euclidean action is given by $S_E[\bar{\psi},\psi] = \int d^4x\,\bar{\psi}[\gamma_\mu\partial_\mu + m_0]\psi$. where we understand that here the γ-matrices are the Euclidean matrices in Eq. (A.4.4). On the lattice this can be written as

$$S_E[\bar{\psi},\psi] \to S_{\mathrm{latt}}^F = a^4\sum_{x\in\Gamma}\{\bar{\psi}(x)\sum_\mu\gamma_\mu\frac{1}{2a}[\psi(x+a\hat{\mu}) - \psi(x-a\hat{\mu})] + m_0\bar{\psi}(x)\psi(x)\}$$

$$= \sum_{x,y\in\Gamma}\bar{\psi}_\alpha(x)M_{\alpha\beta}(x,y)\psi_\beta(y),$$

$$M_{\alpha,\beta}(x,y) \equiv a^4\left\{\sum_\mu\frac{1}{2a}(\gamma_\mu)_{\alpha\beta}[\delta_{(x+a\hat{\mu}),y} - \delta_{(x-a\hat{\mu}),y}] + m_0\delta_{xy}\delta_{\alpha\beta}\right\}. \qquad (9.2.50)$$

From Eq. (6.3.245) we recognize that the fermion lattice propagator is $M_{\alpha\beta}^{-1}(x,y)$. In the infinite volume limit the finite lattice spacing restricts available momenta to lie within the Brillouin zone (BZ), $-\pi/a \leq k_\mu < \pi/a$. For a scalar particle the inverse propagator has the form $\tilde{k}_\mu^2 + m_0^2$, where $\tilde{k}_\mu = (2/a)\sin(k_\mu a/2)$. So $\tilde{k}_\mu$ and k_μ have a one-to-one correspondence in the BZ. For simple fermions, however, we have $\tilde{k}_\mu = (1/2)\sin(k_\mu a)$ and there are two values of $\tilde{k}_\mu$ for every k_μ in the BZ, where one is at small k_μ and one at large k_μ. The latter are called "fermion doublers." This is the essence of the fermion doubling problem and is closely related to the issue of chiral symmetry. Wilson's solution (Wilson, 1974) was to add a term higher order in a, which becomes irrelevant as $a \to 0$,

$$S^W_{\text{latt}}[\bar{\psi}, \psi] = S^F_{\text{latt}}[\bar{\psi}, \psi] - (r/2)a^5 \sum_{x \in \Gamma} \bar{\psi}(x)\Box\psi(x), \tag{9.2.51}$$

where recall $\Box = \hat{\Box}/a^2$. This leads to a Wilson fermion form of $M_{\alpha\beta}(x, y)$,

$$M^W_{\alpha\beta}(x, y) \equiv a^4\{(m_0 + 4\tfrac{r}{a})\delta_{xy}\delta_{\alpha\beta} - \sum_\mu \tfrac{1}{2a}[(r - \gamma_\mu)_{\alpha\beta}\delta_{(x+a\hat{\mu}),y} \tag{9.2.52}$$

$$+ (r + \gamma_\mu)_{\alpha\beta}\delta_{(x-a\hat{\mu}),y}]\}.$$

The Wilson term pushes the fermion doublers to very high momenta at the cost of violating chiral symmetry, since r gives a mass-like term even when $m_0 = 0$.

Then evaluating the fermion functional integral using Eq. (6.3.31) we have

$$\int \mathcal{D}\bar{\psi}\mathcal{D}\psi\, e^{-S_E[\bar{\psi},\psi]} \to \int (\prod_{x \in \Gamma} \prod_{\alpha=1}^4 d\bar{\psi}_{x\alpha}d\psi_{x\alpha})\, e^{-S_{\text{latt}}[\bar{\psi},\psi]} = \det M, \tag{9.2.53}$$

where recall that this is the Euclidean version of the fermion spectral function and so we require the Grassmann fields to be *antiperiodic in time and periodic in space* as follows from Eq. (6.3.74). Defining the finite difference derivative over two lattice spacings will prove convenient when we introduce coupling to gauge fields. There are subtleties with fermion lattice actions such as the fermion doubling problem and computational challenges in reaching the physical pion mass due to the violation of chiral symmetry on the lattice. However, these can be managed as discussed in the above references. Other common fermion actions include staggered (Kogut-Susskind), domain wall, twisted mass and overlap.

Gauge fields on the lattice and the Haar measure: It remains to implement gauge fields on the lattice. While scalar bosons and fermions "live on the lattice sites," $x \in \Gamma$, the gauge fields "live on the links" between the lattice sites in the form of parallel transport operators, $U_\mu(x) \equiv U(x + a\hat{\mu}, x)$ for all $x \in \Gamma$ and $\mu = 1, \ldots, 4$. The $U_\mu(x)$ are referred to as *link variables* or simply *links*. For QCD we have $U_\mu(x) \in SU(3)$. Each $SU(3)$ matrix requires eight real numbers to specify it, each site $x \in \Gamma$ has four links $U_\mu(x)$ in the positive μ directions and there are $N^3 \times N_t$ sites. To specify the full lattice link configuration U requires specifying every link, which means $8 \times 4 \times N^3 \times N_t$ real numbers. The integration over lattice link configurations U is an integration over a $(32 \times N^3 \times N_t)$-dimensional compact manifold, since each link field takes its values in $SU(3)$. The integration over the link configurations makes use of as the *Haar measure*, which assigns an invariant volume to subsets of group elements and so defines an integration over the gauge group. The integral is defined as

$$\int \mathcal{D}U \equiv \prod_{x \in \Gamma} \prod_{\mu=1}^4 dU_\mu(x), \qquad \text{with the properties:} \tag{9.2.54}$$

(i) $\int \mathcal{D}U\, F[U] = \int \mathcal{D}U\, F[UU_0] = \int \mathcal{D}U\, F[U_0U]$;   (ii) $\int \mathcal{D}U\, 1 = 1$;   (9.2.55)

(iii) $\int \mathcal{D}U\, \{cF[U] + dG[U]\} = c \int \mathcal{D}U\, F[U] + d \int \mathcal{D}U\, G[U]$;   (iv) $\int \mathcal{D}U\, U = 0$.

The first property with arbitrary U_0 is the analog of translational invariance around the compact group manifold. The second is a normalization choice. The third is the property of linearity. Property (i) tells us that $\int \mathcal{D}U\, U = \int \mathcal{D}U\, U_1 U U_2 = U_1(\int \mathcal{D}U\, U)U_2$ for all U_1, U_2 and so property (iv) follows. The simplest way to implement the Haar measure is to randomly generate an ensemble of link configurations, $\{U_1, \ldots, U_{N_{\text{config}}}\}$, and then use $\int \mathcal{D}U\, F[U] = \lim_{N_{\text{config}} \to \infty}(\sum_{i=1}^{N_{\text{config}}} F[U_i])/N_{\text{config}}$. This definition reproduces the Haar measure. However, we are almost always interested in evaluating $\int \mathcal{D}U\, F[U]e^{-S[U]}$, where $S[U] \in \mathbb{R}$ and is bounded below. In practice we use a Markov chain to generate an ensemble of link configurations $\{U_1, \ldots, U_{N_{\text{config}}}\}$ with relative probability $P[U']/P[U] = e^{-S[U']}/e^{-S[U]}$ so that

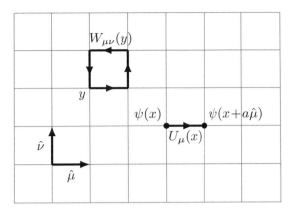

Figure 9.2   Lattice unit vectors $\hat{\mu}$ and $\hat{\nu}$, lattice links $U_\mu(x)$ and lattice plaquettes $W_{\mu\nu}(y)$.

$$\int \mathcal{D}U \, F[U] e^{-S[U]} / \int \mathcal{D}U \, e^{-S[U]} = \lim_{N_{\text{config}} \to \infty} \left(\sum_{i=1}^{N_{\text{config}}} F[U_i] \right) / N_{\text{config}}. \qquad (9.2.56)$$

The links $U_\mu(x)$ are related to the lattice version of the gauge fields $A_\nu(x)$ through the definitions

$$U_\mu(x) \equiv U(x + a\hat{\mu}, x) \equiv e^{iag_0 A_\mu(x)}, \quad U_\mu^\dagger(x) = U(x, x + a\hat{\mu}) \equiv e^{-iag_0 A_\mu(x)}, \qquad (9.2.57)$$

which are the lattice versions of Eqs. (9.1.14) and (9.1.17). For the purposes of constructing perturbation theory recall that we used the Faddeev-Popov method to fix the gauge and only considered small field fluctuations around $A_\mu = 0$. In the continuum limit and the weak coupling limit $dU_\mu(x) \to iag_0 dA_\mu(x)$.

A *plaquette* is defined as the smallest Wilson loop that can be made with lattice links and is denoted as $W_{\mu\nu}(y)$ for all $y \in \Gamma$ and $\mu, \nu = 1, \ldots, 4$, where

$$W_{\mu\nu}(y) = U_\nu^\dagger(y) U_\mu^\dagger(y + a\hat{\nu}) U_\nu(y + a\hat{\mu}) U_\mu(y) \quad \Rightarrow \quad W_{\mu\nu}^\dagger(y) = W_{\nu\mu}(y). \qquad (9.2.58)$$

This is illustrated in Fig. 9.2. The lattice field strength tensor $F_{\mu\nu}(y)$ is defined as

$$W_{\mu\nu}(y) \equiv \exp\{ia^2 g_0 F_{\mu\nu}(y)\} \quad \text{for all} \quad y \in \Gamma. \qquad (9.2.59)$$

We show below that this leads to the result that

$$F_{\mu\nu}(y) = \frac{1}{a}\{[A_\nu(y + a\hat{\mu}) - A_\nu(y)] - [A_\mu(y + a\hat{\nu}) - A_\mu(y)]\} - ig_0[A_\mu(y), A_\nu(y)] + \mathcal{O}(a)$$
$$\xrightarrow{a \to 0} \partial_\mu A_\nu(y) - \partial_\nu A_\mu(y) - ig_0[A_\mu(y), A_\nu(y)] = F_{\mu\nu}(y)|_{\text{continuum}}, \qquad (9.2.60)$$

which shows that $F_{\mu\nu}$ takes its continuum form as $a \to 0$. We can also understand this result as a special case of the nonabelian Stokes theorem in Eq. (9.1.19).

Proof: Define for convenience

$$a\Delta_\nu A_\mu(y) \equiv A_\mu(y + a\hat{\nu}) - A_\mu(y), \quad a\Delta_\mu A_\nu(y) \equiv A_\nu(y + a\hat{\mu}) - A_\nu(y).$$

For brevity then also define $A = -iag_0 A_\nu(y)$, $B = -iag_0[A_\mu(y) + a\Delta_\nu A_\mu(y)]$, $C = iag_0[A_\nu(y) + a\Delta_\mu A_\nu(y)]$ and $D = iag_0 A_\mu(y)$. From Eq. (A.8.6) we have the BCH formula, $e^A e^B = e^{A+B+\frac{1}{2}[A,B]+\mathcal{O}(a^3)}$, and using this three times gives

$$W_{\mu\nu}(y) = e^A e^B e^C e^D = e^{A+B+\frac{1}{2}[A,B]+\mathcal{O}(a^3)} e^{C+D+\frac{1}{2}[C,D]+\mathcal{O}(a^3)} \qquad (9.2.61)$$
$$= e^{A+B+\frac{1}{2}[A,B]+C+D+\frac{1}{2}[C,D]+\frac{1}{2}[A+B,C+D]+\mathcal{O}(a^3)}.$$

It is straightforward to show $A + B + C + D = ia^2 g_0 (\Delta_\mu A_\mu - \Delta_\nu A_\mu) + \mathcal{O}(a^3)$, $\frac{1}{2}[A, B] + \frac{1}{2}[C, D] = a^2 g_0^2 [A_\mu, A_\nu] + \mathcal{O}(a^3)$ and $\frac{1}{2}[A + B, C + D] = \mathcal{O}(a^3)$. Then $W_{\mu\nu} = e^{ia^2 g_0 \{\Delta_\mu A_\mu - \Delta_\nu A_\mu - ig_0 [A_\mu, A_\nu] + \mathcal{O}(a)\}}$, which proves Eq. (9.2.60).

Recall from Eq. (9.1.9) that for any Yang-Mills theory the pure gauge field component of the action in Euclidean space is $S_{\mathrm{YM}}[A] = \int d^4 x \, \frac{1}{2} \mathrm{tr}(F_{\mu\nu} F_{\mu\nu})$, where sums over repeated indices are understood. Then we find

$$\sum_{x \in \Gamma} \sum_{\mu,\nu} \mathrm{tr}[I - W_{\mu\nu}(x)] = \sum_{x \in \Gamma} \sum_{\mu,\nu} \mathrm{tr}[I - \exp\{ia^2 g_0 F_{\mu\nu}(x)\}] \qquad (9.2.62)$$
$$= \sum_{x \in \Gamma} \sum_{\mu,\nu} \mathrm{tr}[-ia^2 g_0 F_{\mu\nu} + \tfrac{1}{2} a^4 g_0^2 F_{\mu\nu}(x) F_{\mu\nu}(x) + \mathcal{O}(a^6)]$$
$$= g_0^2 \, a^4 \sum_{x \in \Gamma} \tfrac{1}{2} \mathrm{tr}[\textstyle\sum_{\mu,\nu} F_{\mu\nu}(x) F_{\mu\nu}(x)] + \mathcal{O}(a^6),$$

where the term linear in $F_{\mu\nu}$ vanishes because it is traceless and also because it is antisymmetric in μ and ν. The lattice version of the Yang-Mills action is defined as

$$S_{\mathrm{latt}}[U] \equiv \frac{1}{g_0^2} \sum_{x \in \Gamma} \sum_{\mu,\nu} \mathrm{Re} \, \mathrm{tr}[I - W_{\mu\nu}] = \frac{2}{g_0^2} \sum_P \mathrm{Re} \, \mathrm{tr}[I - W_{\mu\nu}] \qquad (9.2.63)$$
$$= \beta \sum_P \left(1 - \tfrac{1}{n} \mathrm{Re} \, \mathrm{tr}[W_{\mu\nu}]\right) \xrightarrow{a \to 0} \int d^4 x \, \tfrac{1}{2} \mathrm{tr}(F_{\mu\nu} F_{\mu\nu}) = S_{\mathrm{YM}}[A],$$

where taking the real part ensures that no small imaginary contributions survive, where $n = 3$ is the dimension of the T^a matrices, where $\beta \equiv 2n/g_0^2 = 6/g_0^2$ and where $\sum_P = \sum_{x \in \Gamma} \sum_{\mu < \nu} = \frac{1}{2} \sum_{x \in \Gamma} \sum_{\mu,\nu}$ is the sum over distinct plaquettes. The constraint $\mu < \nu$ ensures that plaquettes occur once with a single orientation. We refer to $S_{\mathrm{latt}}[U]$ as the *Wilson gauge action* (Wilson, 1974).

In order to construct a lattice gauge theory with matter fields we need to ensure gauge invariance of the action. That means that when $\bar{\psi}(x)$ and $\psi(y)$ appear in a term they need to be parallel transported to the same point first so that only combinations of the form $\bar{\psi}_\alpha(x) U(x, y) \psi_\beta(y)$ occur. For example, it follows that $\bar{\psi}_\alpha(x) \psi_\beta(x - a\hat{\mu}) \to \bar{\psi}_\alpha(x) U(x, x - a\hat{\mu}) \psi_\beta(x - a\hat{\mu}) = \bar{\psi}_\alpha(x) U_\mu(x - a\hat{\mu}) \psi_\beta(x - a\hat{\mu})$ and also $\bar{\psi}_\alpha(x) \psi_\beta(x + a\hat{\mu}) \to \bar{\psi}_\alpha(x) U(x, x + a\hat{\mu}) \psi_\beta(x + a\hat{\mu}) = \bar{\psi}_\alpha(x) U_\mu^\dagger(x) \psi_\beta(x + a\hat{\mu})$. Then we arrive at the simplest QCD lattice action,

$$S_{\mathrm{latt}}^{\mathrm{QCD}} = S_{\mathrm{latt}}[U] + S_{\mathrm{latt}}^W[\bar{\psi}, \psi, U] \qquad (9.2.64)$$
$$= \beta \sum_P \left(1 - \tfrac{1}{n} \mathrm{Re} \, \mathrm{tr}[W_{\mu\nu}]\right) + a^4 \sum_{x \in \Gamma} (m_0 + 4\tfrac{r}{a}) \bar{\psi}(x) \psi(x)$$
$$- a^4 \sum_{x \in \Gamma, \mu} \bar{\psi}(x) \tfrac{1}{2a} [(r - \gamma_\mu) U_\mu^\dagger(x) \psi(x + a\hat{\mu}) + (r + \gamma_\mu) U_\mu(x - a\hat{\mu}) \psi(x - a\hat{\mu})],$$

where the QCD Euclidean action is recovered as $a \to 0$, $S_{\mathrm{latt}}^{\mathrm{QCD}} \xrightarrow{a \to 0} S_E^{\mathrm{QCD}}$.

Proof: Given Eq. (9.2.63) it remains to show $S_{\mathrm{latt}}^W[\bar{\psi}, \psi, U] \to \int d^4 x \, \bar{\psi}(\slashed{D} - m_0)\psi$ as $a \to 0$. Remembering that $\frac{r}{a} \hat{\square} = ar\square$ we see that the Wilson term vanishes like a and is irrelevant. So we can ignore all r terms. The m_0 term is clearly correct. Finally, using Eq. (9.2.57) and expanding the exponential gives

$$\bar{\psi}(x) \tfrac{1}{2a} \gamma_\mu [U_\mu^\dagger(x) \psi(x + a\hat{\mu}) - U_\mu(x - a\hat{\mu}) \psi(x - a\hat{\mu})] \qquad (9.2.65)$$
$$= \bar{\psi}(x) \tfrac{1}{2a} \gamma_\mu [\{1 - iag_0 A_\mu(x)\} \psi(x + a\hat{\mu}) - \{1 + iag_0 A_\mu(x - a\hat{\mu})\} \psi(x - a\hat{\mu})] + \mathcal{O}(a)$$
$$= \bar{\psi}(x) \gamma_\mu [\partial_\mu - ig_0 A_\mu(x)] \psi(x) + \mathcal{O}(a) = \bar{\psi}(x) \slashed{D} \psi(x) + \mathcal{O}(a).$$

For brevity now write the lattice QCD action as

$$S[\bar{\psi}, \psi, U] \equiv S[U] + \sum_{x,y \in \Gamma} \bar{\psi}_\alpha(x) M_{\alpha\beta}(x, y) \psi_\beta(y), \qquad (9.2.66)$$

where $S[U]$ is the gluon action such as Eq. (9.2.63) and where $M[U]$ is the fermion matrix; e.g., for Wilson fermions $M[U]$ is M^W in Eq. (9.2.52) with the addition of the links $U_\mu^\dagger(x)$ and $U_\mu(x-a\hat\mu)$ as shown above in Eq. (9.2.64). Approaching the continuum with a fermion action that respects chiral symmetry $\Rightarrow \det M > 0$. However, with Wilson fermions with very light quark masses one can sometimes encounter exceptional configurations that cause a zero crossing of the determinant. The use of improved fermion actions resolves this issue so that $\det M > 0$.

The fermion integrations are carried out using Eqs. (6.3.31) and (6.3.34) and Wick's theorem in Eq. (6.3.247). Consider the lattice estimate of some Euclidean Green's function, $\langle \Omega | T(O[\hat A, \hat{\bar\psi}, \hat\psi]) | \Omega \rangle$, where $O[\hat A, \hat{\bar\psi}, \hat\psi]$ is chosen to be a gauge invariant quantity. This leads to a corresponding $O[U, \hat{\bar\psi}, \hat\psi]$ in the path integral,

$$\langle \Omega | T(O[\hat A, \hat{\bar\psi}, \hat\psi]) | \Omega \rangle = \frac{\int \mathcal{D}U (\prod_f \mathcal{D}\bar\psi^f \mathcal{D}\psi^f) O[U, \bar\psi, \psi] e^{-S[\bar\psi, \psi, U]}}{\int \mathcal{D}U (\prod_f \mathcal{D}\bar\psi^f \mathcal{D}\psi^f) e^{-S[\bar\psi, \psi, U]}} \tag{9.2.67}$$

$$= \frac{\int \mathcal{D}U \, O[U, M^{-1}[U]] \{\prod_f \det M^f[U]\} e^{-S[U]}}{\int \mathcal{D}U \{\prod_f \det M^f[U]\} e^{-S[U]}} = \lim_{N_{\mathrm{config}} \to \infty} \frac{\sum_{i=1}^{N_{\mathrm{config}}} O[U, M^{-1}[U]]}{N_{\mathrm{config}}},$$

where f is the quark flavor. Here we use the relative probability distribution

$$P[U']/P[U] = \{\textstyle\prod_f \det(M^f[U'])\} e^{-S[U']} / \{\textstyle\prod_f \det(M^f[U])\} e^{-S[U]} \tag{9.2.68}$$

to generate the ensemble of lattice configurations. Here we are implicitly assuming pairs of mass-degenerate fermions so that $\prod_f \det(M^f[U']) > 0$, since $\det M^f[U] \in \mathbb{R}$. Since the u and d masses are nearly degenerate this is a good approximation. To include the more massive s quark the pseudofermion approach provides a very good approximation method (Knechtli et al., 2017; Rothe, 2012). This is the generalized form of Eq. (9.2.49) and (9.2.56). Early calculations of lattice QCD used the *quenched approximation*, which used the approximation $\det(M^f[U]) = 1$ to strongly reduce the cost of calculations. With modern computers this approximation is no longer necessary. Here the continuum and infinite volume limits are understood. The $M^{-1}[U]$ are the quark propagators in the background gauge field U and all possible contractions of the fermion fields must be performed to construct $O[U, M^{-1}[U]]$.

We have not yet addressed the issue of gauge fixing. In any realistic lattice QCD calculation N_{config} will be finite and extremely tiny compared with the number of real numbers allowed by machine precision on a modern computer. For two configurations in an ensemble to be on the same gauge orbit they must *at the very least* have the same value for the gauge action to machine precision, $S[U_i] = S[U_j]$. This will never happen in practice. Obviously, for any ensemble one can directly check this. So we will never in practice have Gribov copies in the ensemble and every gauge orbit is sampled once or not at all. This is equivalent to sampling from any ideal gauge fixing because of property (i) of the Haar measure in Eq. (9.2.55). *So in the extraction of physical observables from the lattice we can ignore gauge fixing.*

Gauge fixing on the lattice: It can be useful to implement gauge fixing on the lattice, such as for Fourier accelerating lattice algorithms (Davies et al., 1988). Any gauge-dependent quantity will vanish with the Haar measure due to property (i). To consider such quantities we must fix the gauge. Issues arise when using axial gauges on the lattice, including the breaking of translational invariance. A common choice is the lattice Landau gauge (i.e., Lorenz gauge), which results from finding local minima of the lattice version of the functional $G[A] = \int d^4x \, \mathrm{tr}(A_\mu)^2$ with respect to arbitrary gauge transformations. The Lorenz gauge condition, $\partial_\mu A_\mu = 0$, is satisfied by these local minima. The Hessian of this gauge-fixing functional is $-\partial D_\mu^{ab}[A]$, which we recognize is also the

Faddeev-Popov operator (Zwanziger, 1992). Since for local minima the Hessian is a positive definite matrix, this means that all eigenvalues of the Faddeev-Popov operator are positive and its determinant is positive. These local minima form the *Gribov region*, Ω. The set of all *global* minima is said to form the *fundamental modular region* (FMR), Λ. Clearly, $\Lambda \subset \Omega$ and so Gribov copies remain in Ω. Using local mimima is sufficient for many purposes such as the Fourier acceleration of lattice algorithms. To sample from the FMR, Λ, means finding the global minima of $G[A]$ on each sampled gauge orbit. This is an NP-hard problem analogous to finding the ground state of a spin glass.

One can choose to neglect the effects of Gribov copies and attempt to study gauge-dependent quantities such as quark and gluon propagators and the quark-gluon vertex in the Gribov region. This can lead to useful insights and appears to recover perturbative QCD behavior at large momenta. See, for example, Bowman et al. (2007), Pawlowski et al. (2010) and references therein. Since QCD is asymptotically free at high momenta we expect contributions there to be dominated by configurations with small action, which are those near the trivial orbit. As commented earlier, because of property (i) of the Haar measure all Gribov copies on the trivial orbit can equally well be associated with $A_\mu = 0$ and so all must contribute identically to perturbation theory. So we simply overcount in perturbation theory by the number of Gribov copies on the trivial orbit and that number cancels when we take ratios. This is how we can understand QCD perturbation theory in the context of the Haar measure of lattice gauge theory, where $dU_\mu(x) = iag\,dA_\mu(x) + \mathcal{O}(a^2 g_0^2)$. *For perturbative QCD we can safely ignore Gribov copies.* Finally, the possible nonperturbative lattice formulation of BRST invariance has also been considered. The simplest implementation of this was not successful since the Gribov copies occur in canceling pairs around the gauge orbits (Neuberger, 1987).

Hadron masses from lattice QCD: Consider two composite and gauge-invariant operators, $\hat{\chi}_1^h(x)$ and $\hat{\chi}_2^h(x)$, constructed from quark and gluon fields (links) with the same quantum numbers corresponding to some hadron, h. Write the energy eigenstates with the quantum numbers of h as $|E_0^h\rangle, |E_1^h\rangle, \ldots$ and define states

$$|\chi_1^h\rangle \equiv \sum_{\mathbf{x}} \hat{\chi}_1^{h\dagger}(t_0,\mathbf{x})|\Omega\rangle = \sum_{i=0}^{\infty} c_{1i}|E_i^h\rangle, \quad |\chi_2^h\rangle \equiv \sum_{\mathbf{x}} \hat{\chi}_2^{h\dagger}(t_0,\mathbf{x})|\Omega\rangle = \sum_{j=0}^{\infty} c_{2j}|E_j^h\rangle,$$

where we sum over spatial points since we will move to the lattice. The corresponding energy eigenvalues are E_i^h so that $\hat{H}|E_i^h\rangle = E_i^h|E_i^h\rangle$. Consider the Euclidean Green's function constructed from these operators at times t and t_0 with $t > t_0$,

$$\begin{aligned}
G(t - t_0) &\equiv \langle\Omega|T\left[\sum_{\mathbf{x}} \hat{\chi}_2^h(t,\mathbf{x})\hat{\chi}_1^{h\dagger}(t_0,0)\right]|\Omega\rangle = \tfrac{1}{N_{\mathbf{x}}}\langle\Omega|\sum_{\mathbf{y},\mathbf{z}} \hat{\chi}_2^h(t,\mathbf{z})\hat{\chi}_1^{h\dagger}(t_0,\mathbf{y})|\Omega\rangle \\
&= \tfrac{1}{N_{\mathbf{x}}}\langle\Omega|\sum_{\mathbf{y},\mathbf{z}} \hat{\chi}_2^h(t_0,\mathbf{z})e^{-(t-t_0)\hat{H}}\hat{\chi}_1^{h\dagger}(t_0,\mathbf{y})|\Omega\rangle \\
&= \tfrac{1}{N_{\mathbf{x}}}\langle\chi_2^h|e^{-(t-t_0)\hat{H}}|\chi_1^h\rangle = \tfrac{1}{N_{\mathbf{x}}}\sum_{i=0}^{\infty} c_{2i}^* c_{1i} e^{-(t-t_0)E_i^h},
\end{aligned} \tag{9.2.69}$$

where $N_{\mathbf{x}}$ is the number of spatial points. Since $\sum_{\mathbf{x}} \hat{\chi}_i^{h\dagger}(t,\mathbf{x})$ has no spatial dependence, then it cannot impart any momentum to the vacuum $|\Omega\rangle$ and so the E_i^h accessed here all correspond to the hadron having zero three-momentum, $\mathbf{P} = 0$. Then here $E_i^h = M_i^h$, where the M_0^h is the mass of the hadron h in its ground state and the M_i^h for $i \geq 1$ are the masses of its excited states.

So after generating an ensemble of lattice gauge configurations $\{U_1, U_2, \ldots\}$ and performing all contractions of fermion field operators and replacing them with fermion propagators $M^{-1}[U]$ we obtain the relevant form of $O[U, M^{-1}[U]](t-t_0)$ from the $T[\cdots](t)$ in Eq. (9.2.69). We then evaluate $G(t - t_0)$ from Eq. (9.2.67) and obtain the ground state mass of the hadron h using

$$M_0^h = \lim_{t',t \gg t_0} \frac{-1}{t' - t} \ln\left[\frac{G(t' - t_0)}{G(t - t_0)}\right]. \tag{9.2.70}$$

In Fig. 9.3 we show the good agreement of light hadron masses extracted from lattice QCD and their experimentally measured values. This is the most fundamental prediction of lattice QCD

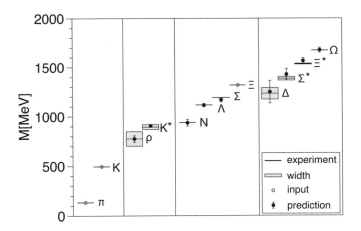

Figure 9.3 $N_f = 2 + 1$ lattice QCD predictions for light hadrons according to Dürr et al. (2008). The three open circles are inputs. Experimental numbers are from Amsler et al. (2008). Reprinted with permission from Fodor and Hoelbling (2012).

and provides compelling evidence that QCD is the correct theory of the strong interactions. Lattice QCD calculations have now found their way into a wide range of studies of the strong interaction and its implications, which include, for example, (i) decay constants, (ii) form factors, (iii) determination of the $\overline{\text{MS}}$ strong coupling constant and quark masses in perturbative QCD applications using lattice perturbation theory as a bridge, (iv) calculation of the hadronic contribution to light-by-light scattering needed for a precise calculation of the muon magnetic moment and (v) studies of the strong interaction at finite temperature (finite lattice extent in the Euclidean time direction $T = 1/kT_{\text{temp}}$). Studies of QCD at finite density (finite chemical potential μ) remain at a rudimentary level because at $\mu \neq 0$ the fermion determinant suffers from a sign problem and Eq. (9.2.68) fails.

Modern lattice QCD studies use a variety of *improved actions*, where irrelevant operators, like the Wilson term for fermions, are added to improve convergence to the continuum limit and chiral behavior. QED has been added into lattice QCD calculations to make them more realistic. Lattice gauge theory has an important role to play in studies of theories beyond the Standard Model. Approaching the continuum and infinite volume limits with the small quark masses needed to obtain physical pion masses is a computationally challenging task. As of 2020 high-performance computers are approaching exaflop capability (10^{18} floating-point operations per second) but to extend the range and maturity of lattice calculations will require ever-faster computers with ever-increasing bandwidth to memory. For an up-to-date review of the status of lattice QCD, its applications and an extensive list of references, consult the most recent Review of Particle Physics (Zyla et al., 2020).

9.3 Anomalies

An anomaly in a quantum field theory arises when the procedure of regularization necessary for renormalization unavoidably violates a symmetry of the unrenormalized theory. It is the process of quantization itself that breaks the classical symmetry; e.g., a classical theory with no scale such as pure Yang-Mills theory has scale invariance violated by quantization, which is referred to as a *scale anomaly*. Anomalies that break global symmetries are not a problem, since they do not render the theory inconsistent. However, a *gauge anomaly* leading to nonconservation of a gauge current in a gauge theory does lead to an inconsistent theory. This important topic of anomalies could fill a chapter on its own. We will need to limit ourselves to a brief introduction and a summary of some key points. For some pedagogical lectures on anomalies and many references, see Harvey (2003) and Bilal (2008). Also see the original papers of Adler (1969) and Bell and Jackiw (1969) and discussions in Itzykson and Zuber (1980), Cheng and Li (1984), Huang (1992), Peskin and Schroeder (1995), Weinberg (1996) and Schwartz (2013).

ABJ anomaly: The *ABJ anomaly* is named after its discoverers, Stephen Adler, John Bell and Roman Jackiw. It is the anomalous breaking of a global chiral symmetry and so is a *chiral anomaly*. Consider QED and the three-point functions,

$$T_{\mu\nu\rho}(k_1, k_2, q) \equiv i \int d^4x_1 d^4x_2 \langle \Omega | T[\hat{J}_\mu(x_1)\hat{J}_\nu(x_2)\hat{J}_\rho^5(0)]|\Omega\rangle e^{i(k_1 \cdot x_1 + k_2 \cdot x_2)}, \qquad (9.3.1)$$

$$T_{\mu\nu}(k_1, k_2, q) \equiv i \int d^4x_1 d^4x_2 \langle \Omega | T[\hat{J}_\mu(x_1)\hat{J}_\nu(x_2)\hat{P}(0)]|\Omega\rangle e^{i(k_1 \cdot x_1 + k_2 \cdot x_2)},$$

where from Eq. (8.3.23) k_1^μ and k_2^μ are outgoing momenta and we have defined

$$\hat{J}_\mu(x) \equiv \hat{\bar\psi}(x)\gamma_\mu\hat\psi(x), \quad \hat{J}_\mu^5(x) \equiv \hat{\bar\psi}(x)\gamma_\mu\gamma^5\hat\psi(x), \quad \hat{P}(x) \equiv \hat{\bar\psi}(x)\gamma^5\hat\psi(x). \qquad (9.3.2)$$

$\hat{J}_\mu(x)$ is the conserved electromagnetic current operator leading to the QED Ward identities, $\hat{J}_\mu^5(x)$ is the axial current operator and $\hat{P}(x) \equiv \hat{J}^5(x)$ is the pseudoscalar density operator. Temporarily ignoring issues of regularization and renormalization, the QED equations of motion lead to

$$\partial^\mu \hat{J}_\mu(x) = 0 \quad \text{and} \quad \partial^\mu \hat{J}_\mu^5(x) = 2im\hat{P}(x), \qquad (9.3.3)$$

where m is the fermion mass. As expected $\hat{J}^\mu(x)$ is conserved. It also appears that $\hat{J}_\mu^5(x)$ is conserved in the chiral (massless fermion) limit. The second equation is referred to as the *partial conservation of axial current (PCAC)*. Since there are no external fermion lines, we know from the Ward identity in Eq. (8.3.35) that

$$k_1^\mu T_{\mu\nu} = k_2^\nu T_{\mu\nu} = 0 \qquad \text{and} \qquad k_1^\mu T_{\mu\nu\rho} = k_2^\nu T_{\mu\nu\rho} = 0. \qquad (9.3.4)$$

We can also see this in another way since Eq. (6.2.166) gives

$$\partial_x^\mu \langle \Omega | T[\hat{J}_\mu(x)\hat{O}(y)]|\Omega\rangle = \langle \Omega | T[\partial_x^\mu \hat{J}_\mu(x)\hat{O}(y)]|\Omega\rangle + [\hat{J}_0(x), \hat{O}(y)]\delta(x_0 - y_0) \qquad (9.3.5)$$

$$= [\hat{J}_0(x), \hat{O}(y)]\delta(x_0 - y_0),$$

where we have used Eq. (6.2.166). For $\hat{O}(y) = \hat{J}_\mu(y), \hat{J}_\mu^5(y)$ and $\hat{P}(y)$ and products of these we have from Eq. (8.3.22) that $[\hat{J}_0(x), \hat{O}(y)]\delta(x^0 - y^0) = 0$ and so again arrive at Eq. (9.3.4). It is tempting to assume that in the chiral ($m = 0$) limit $\partial^\mu \hat{J}_\mu^5(x) = 0$ and that there would be a corresponding axial Ward identity. This could be expected to lead to $q^\rho T_{\mu\nu\rho} = 0$ for $m = 0$ and for nonzero m this would become $-q^\rho T_{\mu\nu\rho} = 2mT_{\mu\nu}$, where $q_\mu \equiv k_1^\mu + k_2^\mu$ is an incoming momentum. However, these results for $\hat{J}_\mu^5(x)$ and $q^\rho T_{\mu\nu\rho}$ are incorrect. This can be seen by studying the one-loop triangle diagrams consisting of the fermion propagators forming a triangle with either J_μ, J_ν, J_ρ^5 at the vertices or J_μ, J_ν, P.

When one evaluates the triangle diagram for $T_{\mu\nu\rho}$ one meets a linear divergence that must be regularized. This does not occur with three J_μ vertices in QED because of Furry's theorem. It is not possible to regularize this in such a way that the Ward identity results in Eq. (9.3.4) and the PCAC identity $-q^\rho T_{\mu\nu\rho} = 2mT_{\mu\nu}$ are simultaneously satisfied. Since we know that the Ward identities are true, then the PCAC relation must be invalid in any regularized theory (Cheng and Li, 1984; Huang,

1992). At higher loops this problem does not occur and so it arises entirely at the one-loop level. We say that the chiral anomaly is *one-loop exact*. It can be shown diagramatically that the correct result for the divergence of the axial current in the case of fermions of charge $q_c = e$ is (Adler, 1969; Bell and Jackiw, 1969)

$$\partial^\mu \hat{J}^5_\mu(x) = 2im\hat{P}(x) - (\alpha/2\pi)\hat{\tilde{F}}^{\mu\nu}(x)\hat{F}_{\mu\nu}(x), \qquad (9.3.6)$$

where $\tilde{F}^{\mu\nu}$ is the dual tensor. The result is often quoted with $m = 0$. For renormalized field operators use e and m and for bare field operators use e_0 and m_0.

Transforming the functional measure: A satisfying way of understanding the ABJ anomaly is through studying the functional measure (Fujikawa, 1979). If $\psi(x) \to \psi'(x) = \Delta(x)\psi(x)$ then $\bar{\psi}(x) \to \bar{\psi}'(x) = \bar{\psi}(x)\gamma^0\Delta^\dagger(x)\gamma^0 \equiv \bar{\psi}(x)\bar{\Delta}(x)$, where $\Delta(x)$ is some spinor matrix. It will be convenient to borrow notation from quantum mechanics so that the following manipulations of spacetime variables are easier to follow. We use the notation $|x\rangle$ such that $\langle x|y\rangle = \delta^4(x-y)$. Define the spacetime coordinate operator $\hat{x}$ so that $\hat{x}|x\rangle = x|x\rangle$. Define functional matrices $\mathcal{U}$ and $\bar{\mathcal{U}}$ where $\mathcal{U}(y,x) \equiv \langle y|\hat{\mathcal{U}}|x\rangle \equiv \langle y|\Delta(\hat{x})|x\rangle = \Delta(x)\langle y|x\rangle = \Delta(x)\delta^4(x-y)$ and similarly for $\bar{\mathcal{U}}$ in terms of $\bar{\Delta}(x)$. We use Det and Tr to denote the functional determinant and trace, respectively, that act on functional matrices such as $\mathcal{U}$ and $\bar{\mathcal{U}}$. It then follows that $\mathcal{D}\psi \to \mathcal{D}\psi' = [\mathrm{Det}\,\mathcal{U}]^{-1}\mathcal{D}\psi \equiv \mathcal{J}^{-1}\mathcal{D}\psi$ and similarly we also have $\mathcal{D}\bar{\psi} \to \mathcal{D}\bar{\psi}' = [\mathrm{Det}\,\bar{\mathcal{U}}]^{-1}\mathcal{D}\bar{\psi} \equiv \bar{\mathcal{J}}^{-1}\bar{\psi}$, where we have used Eq. (6.3.24). This leads to $\mathcal{D}\bar{\psi}\mathcal{D}\psi \to \mathcal{D}\bar{\psi}'\mathcal{D}\psi' = \mathcal{D}\bar{\psi}\mathcal{D}\psi\,(\mathcal{J}\bar{\mathcal{J}})^{-1}$. Using Eq. (A.8.4) we can write

$$\begin{aligned}
\mathcal{J} &= \mathrm{Det}\,\mathcal{U} = \exp\{\mathrm{Tr}\ln\mathcal{U}\} = \exp\{\textstyle\int d^4x\, \langle x|\mathrm{tr}\ln\hat{\mathcal{U}}|x\rangle\} \\
&= \exp\{\textstyle\int d^4x\, \delta^4(x-x)\mathrm{tr}\ln\Delta(x)\} = \exp\{\textstyle\int d^4x\, \delta^4(0)\mathrm{tr}\ln\Delta(x)\}
\end{aligned} \qquad (9.3.7)$$

and similarly for $\bar{\mathcal{J}}$ in terms of $\bar{\Delta}(x)$. Note that $\delta^4(0)$ is an ultraviolet divergence since if we suppress high momenta then $\delta^4(x)$ becomes finite at $x = 0$.

In the derivation of the QED vector Ward identities we used $\Delta(x) = e^{-iq_c\epsilon(x)I}$ with I the spinor identity matrix, which gives $\mathcal{J} = \exp\{-iq_c\int d^4x\, \delta^4(0)4\epsilon(x)\}$. Also $\bar{\Delta}(x) = \gamma^0\Delta^\dagger(x)\gamma^0 = e^{iq_c\epsilon(x)I}$ and so $\bar{\mathcal{J}} = \exp\{iq_c\int d^4x\, \delta^4(0)4\epsilon(x)\}$. Any reasonable regularization of $\delta^4(x)$ will then lead to $\mathcal{J}\bar{\mathcal{J}} = 1$. So $\mathcal{D}\bar{\psi}'\mathcal{D}\psi' = \mathcal{D}\bar{\psi}\mathcal{D}\psi$ as we had earlier assumed and the standard Ward identities emerge as expected.

A chiral transformation has the form $e^{i\epsilon\gamma^5}$ with ϵ a constant and for massless fermions this leaves the QED Lagrangian invariant. To try to derive the conserved current we allow $\epsilon \to \epsilon(x)$ in the usual way which gives $\Delta(x) = e^{i\epsilon(x)\gamma^5}$. This leads to $\mathcal{J} = \exp\{\int d^4x\, \langle x|\mathrm{tr}\ln\hat{\mathcal{U}}|x\rangle\} = \exp\{i\int d^4x\, \delta^4(0)\epsilon(x)\mathrm{tr}\gamma^5\}$, which is undefined since $\mathrm{tr}\gamma^5 = 0$ and $\delta^4(0) \to \infty$. In order to have a well-defined Jacobian we need to regularize $\hat{\mathcal{U}}$ such that (i) it is not local in $\hat{x}$ so that $\delta^4(0)$ does not emerge; (ii) it is spinor-matrix valued so that we do not arrive at $\mathrm{tr}\gamma^5 = 0$; (iii) it does not spoil gauge invariance; and (iv) it has an ultraviolet regularization parameter Λ such that the regularization is removed as $\Lambda \to \infty$. For this purpose we use some smooth function $f(\hat{\slashed{D}}^2/\Lambda^2)$, where $f(z)$ has the properties that $f(0) = 1$, $f(z) \to 0$ as $z \to \infty$, $zf'(z)|_{z=0} = 0$ and $zf'(z) \to 0$ as $z \to \infty$. We will choose for simplicity $f(z) = e^{-z}$ but the final result does not depend on this choice of $f(z)$ (Bilal, 2008). We use the notation $i\hat{\slashed{D}} = i\hat{\slashed{\partial}} - q_c\slashed{A}$. We also define $|p\rangle$ such that $\hat{p}^\mu|p\rangle = p^\mu|p\rangle$ and $\langle p|x\rangle = e^{ip\cdot x}/(2\pi)^2$ and have completeness in the form $\hat{I} = \int d^4p\, |p\rangle\langle p|$, which are four-dimensional extensions of quantum mechanics notation. Then

$$\slashed{D}^2 = \tfrac{1}{2}\{\gamma^\mu,\gamma^\nu\}D_\mu D_\nu + \tfrac{1}{2}[\gamma^\mu,\gamma^\nu]D_\mu D_\nu = D^2 + \tfrac{1}{4}[\gamma^\mu,\gamma^\nu][D_\mu,D_\nu] \qquad (9.3.8)$$

$$= D^2 + \tfrac{1}{2}q_c\sigma^{\mu\nu}F_{\mu\nu}, \qquad (9.3.9)$$

where we have used $[D_\mu, D_\nu] = iq_c F_{\mu\nu}$ from Eq. (9.1.5) and $\sigma^{\mu\nu} = \frac{i}{2}[\gamma^\mu, \gamma^\nu]$ from Eq. (A.3.5). We regularize using $\gamma^5 \to \gamma^5 f(\slashed{D}^2/\Lambda^2) = \gamma^5 e^{-\slashed{D}^2/\Lambda^2}$ so that

$$\text{Tr}\ln\mathcal{U} = \lim_{\Lambda\to\infty}\int d^4x\,\langle x|\text{tr}\ln\hat{\mathcal{U}}|x\rangle = \lim_{\Lambda\to\infty}\int d^4x\,\langle x|i\epsilon(\hat{x})\text{tr}[\gamma^5 e^{-\slashed{D}^2/\Lambda^2}]|x\rangle \tag{9.3.10}$$
$$= \lim_{\Lambda\to\infty}\int d^4x\,i\epsilon(x)\langle x|\text{tr}[\gamma^5\exp\{-[\hat{D}^2 + \tfrac{1}{2}q_c\sigma^{\mu\nu}F_{\mu\nu}(\hat{x})]/\Lambda^2\}]|x\rangle.$$

From Eq. (A.3.30) we know that a trace with γ^5 vanishes unless there are an even number other γ-matrices and at least four of these. So the term of lowest order in $1/\Lambda^2$ is quadratic in $F_{\mu\nu}$. Higher powers of F are suppressed relative to this by powers $\mathcal{O}(1/\Lambda^2)$. Commutators of D^2 with quadratic $F_{\mu\nu}$ terms will also be higher order in $1/\Lambda^2$ and so become irrelevant. The gauge action suppresses the high-freqency contributions to A_μ and so in the large momentum region we can effectively make the replacement $D_\mu \to \partial_\mu$. Since we can ignore the commutators we can factor the exponential into two parts,

$$\text{Tr}\ln\mathcal{U} = \lim_{\Lambda\to\infty}\int d^4x\,i\epsilon(x)\text{tr}[\gamma^5 e^{\{-\frac{1}{2}q_c\sigma^{\mu\nu}F_{\mu\nu}(x)\}/\Lambda^2}]\langle x|e^{-\hat{D}^2/\Lambda^2}|x\rangle. \tag{9.3.11}$$

As explained above only the term in the expansion of the exponential quadratic in $F_{\mu\nu}$ both survives the trace and remains relevant as $\Lambda \to \infty$. This leads to

$$\text{tr}[\gamma^5 e^{\{-\frac{1}{2}q_c\sigma^{\mu\nu}F_{\mu\nu}(x)\}/\Lambda^2}] = \text{tr}[\gamma^5 \tfrac{1}{2!}\{-\tfrac{1}{2}q_c\sigma^{\mu\nu}F_{\mu\nu}(x)\}^2/\Lambda^4] + \mathcal{O}(1/\Lambda^6) \tag{9.3.12}$$
$$= \tfrac{1}{8}q_c^2 F_{\mu\nu}(x)F_{\rho\sigma}(x)\text{tr}[\gamma^5\sigma^{\mu\nu}\sigma^{\rho\sigma}]/\Lambda^4 + \mathcal{O}(1/\Lambda^6)$$
$$= \tfrac{i}{2}q_c^2\epsilon^{\mu\nu\rho\sigma}F_{\mu\nu}F_{\rho\sigma}/\Lambda^4 + \mathcal{O}(1/\Lambda^6) = iq_c^2\tilde{F}_{\mu\nu}F^{\mu\nu}/\Lambda^4 + \mathcal{O}(1/\Lambda^6).$$

Since effectively $(D_\mu - \partial_\mu)/\Lambda \to 0$ as $\Lambda \to \infty$ then

$$\langle x|e^{-\hat{D}^2/\Lambda^2}|x\rangle = \langle x|e^{-\hat{\partial}^2/\Lambda^2}e^{-(\hat{D}^2-\hat{\partial}^2+\mathcal{O}([\hat{D}^2,\hat{\partial}^2]))/\Lambda^2}|x\rangle \tag{9.3.13}$$
$$\xrightarrow{\Lambda\to\infty}\langle x|e^{-\hat{\partial}^2/\Lambda^2}|x\rangle = \lim_{x\to y}\int d^4p\,\langle y|e^{-\hat{\partial}^2/\Lambda^2}|p\rangle\langle p|x\rangle$$
$$= \lim_{x\to y}\int[d^4p/(2\pi)^4]e^{ip\cdot(x-y)}e^{p^2/\Lambda^2} = i\int[d^4p_E/(2\pi)^4]e^{-p_E^2/\Lambda^2} = i\Lambda^4/16\pi^2,$$

where the integral was evaluated in Euclidean space after a Wick rotation in the usual way. Then Eq. (9.3.11) becomes

$$\text{Tr}\ln\mathcal{U} = -i\int d^4x\,\epsilon(x)\frac{q_c^2}{32\pi^2}\epsilon^{\mu\nu\rho\sigma}\tilde{F}_{\mu\nu}F_{\rho\sigma} = -i\int d^4x\,\epsilon(x)\frac{q_c^2}{16\pi^2}\tilde{F}_{\mu\nu}F^{\mu\nu}. \tag{9.3.14}$$

Since $\mathcal{J} = \bar{\mathcal{J}} = \exp\{\text{Tr}\ln\mathcal{U}\}$ then $\mathcal{D}\bar{\psi}'\mathcal{D}\psi' = \mathcal{D}\bar{\psi}\mathcal{D}\psi(\mathcal{J}\bar{\mathcal{J}})^{-1}$, which then leads to $\mathcal{D}\bar{\psi}'\mathcal{D}\psi' = \mathcal{D}\bar{\psi}\mathcal{D}\psi\exp\{i\int d^4x\,\epsilon(x)(q_c^2/8\pi^2)\tilde{F}_{\mu\nu}F^{\mu\nu}\}$.

Let us rederive the conserved vector current using $\psi(x) \to \psi'(x) = e^{i\epsilon(x)}\psi(x)$. This leads to $\bar{\psi}(x)i\slashed{\partial}\psi(x) \to \bar{\psi}'(x)i\slashed{\partial}\psi'(x) = \bar{\psi}(x)i\slashed{\partial}\psi(x) - J^\mu(x)\partial_\mu\epsilon(x)$ with other terms invariant. We have defined $J^\mu(x) \equiv \bar{\psi}(x)\gamma^\mu\psi(x)$. Since we know that the functional measure is invariant under this transformation, then we must have

$$\int\mathcal{D}\bar{\psi}\mathcal{D}\psi\mathcal{D}A\,e^{i\int d^4x\,\mathcal{L}_{\text{QED}}} = \int\mathcal{D}\bar{\psi}'\mathcal{D}\psi'\mathcal{D}A\,e^{i\int d^4x\,\mathcal{L}'_{\text{QED}}} = \int\mathcal{D}\bar{\psi}\mathcal{D}\psi\mathcal{D}A\,e^{i\int d^4x\,\mathcal{L}'_{\text{QED}}}$$
$$= \int\mathcal{D}\bar{\psi}\mathcal{D}\psi\mathcal{D}A\exp\{i\int d^4x\,[\mathcal{L}_{\text{QED}} - J^\mu(x)\partial_\mu\epsilon(x)]\}$$
$$= \int\mathcal{D}\bar{\psi}\mathcal{D}\psi\mathcal{D}A\exp\{i\int d^4x\,[\mathcal{L}_{\text{QED}} + \epsilon(x)\partial^\mu J_\mu(x)]\}, \tag{9.3.15}$$

where the boundary conditions on spectral functions mean as always that the surface terms do not contribute in the integration by parts. Since $\epsilon(x)$ is arbitary, then we must have $\partial_\mu J^\mu(x) = 0$ and so the vector current $J^\mu(x)$ must be conserved.

Consider the local chiral transformation $\psi(x) \to \psi'(x) = e^{i\gamma^5 \epsilon(x)}\psi(x)$ and work to $\mathcal{O}(\epsilon)$. Then $\bar\psi(x)\psi(x) \to \bar\psi'(x)\psi'(x) = \bar\psi(x)\psi(x) + 2iP(x)$ with $P(x) \equiv \bar\psi(x)\gamma^5\psi(x)$ and we also have $\bar\psi(x)\slashed\partial\psi(x) \to \bar\psi'(x)\slashed\partial\psi'(x) = \bar\psi(x)\slashed\partial\psi(x) + \epsilon(x)\partial^\mu J_\mu^5(x)$, where we have made the definition $J_\mu^5(x) \equiv \bar\psi(x)\gamma_\mu\gamma^5\psi(x)$. Under this transformation both the Lagrangian density and the functional measure change,

$$\int \mathcal{D}\bar\psi \mathcal{D}\psi \mathcal{D}A\, e^{i\int d^4x\, \mathcal{L}_{\mathrm{QED}}} = \int \mathcal{D}\bar\psi' \mathcal{D}\psi' \mathcal{D}A\, e^{i\int d^4x\, \mathcal{L}'_{\mathrm{QED}}} \tag{9.3.16}$$
$$= \int \mathcal{D}\bar\psi \mathcal{D}\psi \mathcal{D}A\, \exp\{i\int d^4x\, \mathcal{L}'_{\mathrm{QED}} + \epsilon(x)(q_c^2/8\pi^2)\tilde F_{\mu\nu}F^{\mu\nu}\}$$
$$= \int \mathcal{D}\bar\psi \mathcal{D}\psi \mathcal{D}A\, \exp\{i\int d^4x\, [\mathcal{L}_{\mathrm{QED}} + \epsilon(x)\{\partial^\mu J_\mu^5(x) - 2imP(x) + (q_c^2/8\pi^2)\tilde F_{\mu\nu}F^{\mu\nu}\}]\}.$$

Since $\epsilon(x)$ is arbitrary, then the term multiplying it must vanish and so

$$\partial^\mu J_\mu^5(x) = 2imP(x) - (q_c^2/8\pi^2)\tilde F_{\mu\nu}(x)F^{\mu\nu}(x). \tag{9.3.17}$$

This proves the result in Eq. (9.3.6) since it must also be true at the operator level in the usual way. The pseudoscalar contribution from $m \neq 0$ is an explicit breaking of chiral symmetry, whereas the $\tilde F F$ contribution is the quantum anomaly due to the noninvariance of the measure used to quantize the theory. The derivation of the ABJ anomaly from the measure is nonpertubative and so is valid to all orders in perturbation theory. Since this is also the one-loop result we say that the ABJ anomaly is one-loop exact.

9.4 Introduction to the Standard Model

The elements of the Standard Model (SM) were previously discussed from a historical perspective in Sec. 5.2. The SM Lagrangian density can be decomposed into parts for convenience and in the following we will construct each part in turn,

$$\mathcal{L}_{\mathrm{SM}} = \mathcal{L}_{\mathrm{gauge}} + \mathcal{L}_{\mathrm{Higgs}} + \mathcal{L}_{\mathrm{kin\text{-}ferm}} + \mathcal{L}_{\mathrm{Yukawa}}. \tag{9.4.1}$$

9.4.1 Electroweak Symmetry Breaking

As discussed in Chapter 5, the SM is based on the direct product group $SU(3)_c \times SU(2)_L \times U(1)_Y$, where $SU(3)_c$ leads to QCD. We refer to $SU(2)_L$ as the *weak isospin* gauge group and $U(1)_Y$ as the *weak hypercharge* gauge group. The Higgs mechanism causes electroweak symmetry breaking, $SU(2)_L \times U(1)_Y \to U(1)_{\mathrm{em}}$, where $U(1)_{\mathrm{em}}$ is the usual QED gauge group.

The *Higgs multiplet* or *Higgs doublet*[5] Φ is a weak-isospin doublet of Higgs complex scalar fields with weak hypercharge $Y = \frac{1}{2}$, where for brevity here we write Y rather than the notation Y_W used in Chapter 5. Without quarks or leptons,

$$\mathcal{L}_{\mathrm{gauge}+\Phi} = -\tfrac{1}{4}(W_{\mu\nu}^a)^2 - \tfrac{1}{4}(B_{\mu\nu})^2 + (D_\mu\Phi)^\dagger(D^\mu\Phi) + m^2\Phi^\dagger\Phi - \lambda(\Phi^\dagger\Phi)^2, \tag{9.4.2}$$

where W_μ^a for $a = 1, 2, 3$ are the three $SU(2)_L$ gauge bosons, B_μ is the $U(1)_Y$ gauge boson, $B_{\mu\nu} \equiv \partial_\mu B_\nu - \partial_\nu B_\mu$ is the B field-strength tensor, and $W_{\mu\nu}^a$ are the $SU(2)_L$ field strength tensors analogous to those in QCD for $SU(3)_c$. Note that the mass-like Higgs term, $+m^2\Phi^\dagger\Phi$, in $\mathcal{L}$ has the opposite sign to that of a normal mass term. The sign of the Higgs quartic coupling constant λ ensures that $V(\Phi) \equiv -m^2\Phi^\dagger\Phi + \lambda(\Phi^\dagger\Phi)^2$ is bounded below. The Higgs potential $V(\Phi)$ is chosen to have a Mexican hat shape for the doublet similar to that in Fig. 8.3 so that spontaneous symmetry

---

[5] We use Φ for the Higgs doublet like Peskin and Schroeder (1995) and h for the Higgs boson like Schwartz (2013).

breaking will occur. Using Eq. (9.1.12) we write the covariant derivative for the electroweak direct product group as[6]

$$D_\mu = \partial_\mu + ig W_\mu^a T^a + ig' Y B_\mu, \tag{9.4.3}$$

where g and g' are the coupling constants for the $SU(2)_L$ and $U(1)_Y$ gauge groups, respectively. Recall from Eqs. (9.1.1) and (9.1.10) that the representation of the covariant derivative depends on what it acts on. Since the Higgs is a weak isodoublet, then the representation of the $SU(2)_L$ generators that act on it is $T^a \equiv \tau^a \equiv \frac{1}{2}\sigma^a$, where $\sigma^1, \sigma^2, \sigma^3$ are the Pauli spin matrices. Then we have

$$D_\mu \Phi = \partial_\mu \Phi + ig W_\mu^a T^a \Phi + ig' Y B_\mu \Phi, \tag{9.4.4}$$

and here $Y = \frac{1}{2}$ because D_μ is acting on the Higgs doublet. We will later see that fermion electric charge is given by $Q = (T^3 + Y)$ in units of $e \equiv |e|$.[7]

From Sec. 8.8 we know that $V(\Phi) = -m^2|\Phi|^2 + \lambda(|\Phi|^2)^2$ will lead to spontaneous symmetry breaking. The Higgs isodoublet contains two complex scalar fields or equivalently four real scalar fields. Temporarily put all fields to zero except one real field. Stationary points of V then satisfy $dV/d\Phi = -2m^2\Phi + 4\lambda\Phi^3 = 0$ and so are at $\Phi = 0$ for the unstable vacuum and at $\Phi = m/\sqrt{2\lambda} \equiv v/\sqrt{2}$ corresponding to the spontaneously broken vacuum. We refer to $v \equiv m/\sqrt{\lambda}$ as the *Higgs vacuum expectation value* or the *Higgs VEV*. The values of the Higgs weak isodoublet corresponding to the bottom of the Higgs Mexican hat are

$$\Phi = e^{2i\pi^a T^a/v} \begin{bmatrix} 0 \\ (v/\sqrt{2}) + (h/\sqrt{2}) \end{bmatrix} \quad \text{with} \quad v \equiv \frac{m}{\sqrt{\lambda}}. \tag{9.4.5}$$

The $e^{2i\pi^a T^a/v}$ generates the "rotations" around the bottom of the Mexican hat potential such that $|\Phi| = v/\sqrt{2}$ and h is the real *Higgs boson* field. The $\sqrt{2}$ changes the normalization of the kinetic term for a complex scalar field to that of a real scalar field, e.g., see Eq. (3.2.48). The $\pi^a(x)$ fields are the massless real Goldstone boson fields arising from the symmetry breaking. The factor of two is included with the $\pi^a(x)$ so that they are like pions in chiral models, where $T^a = \sigma^a/2 \to T^a = \sigma^a$. We recognize that $\exp(\cdots)$ is just an $SU(2)_L$ gauge transformation. We are free to choose the gauge so that it cancels the π^a contribution. So, we can take the simplifying step of choosing $\pi^a = 0$, which is referred to as the choice of *unitary gauge*. To understand this in a simpler context, see discussions of the *Abelian Higgs model* in Cheng and Li (1984), Peskin and Schroeder (1995), Schwartz (2013) and elsewhere. The Abelian Higgs model has a $U(1)$ gauge field A_μ interacting with a charged scalar field ϕ, where the charged scalar field has a Higgs-like potential, $V(\phi) = -m^2|\phi|^2 + \lambda(|\phi|^2)^2$. We say that *the gauge boson has "eaten" the Goldstone boson* through the choice of unitary gauge in the Higgs mechanism.

The mass terms for the gauge bosons arise from $v \neq 0$. To find these mass terms we set $h = 0$ and discard terms containing derivatives. Using Eq. (9.4.4) we find

---

[6] Since we will connect with QED we use the sign for the covariant derivative couplings consistent with Eq. (7.6.16) leading to the appropriate sign for the Lorentz force (Aitchison and Hey, 2013).

[7] We follow Peskin and Schroeder (1995) and Schwartz (2013) and use the "half-scale" definition of Y but it is also common to use $Y' = 2Y$ with $Q = (T^3 + \frac{1}{2}Y')$.

$$D_\mu \Phi \to i\frac{gv}{2\sqrt{2}} \begin{bmatrix} \frac{g'}{g}B_\mu + W_\mu^3 & W_\mu^1 - iW_\mu^2 \\ W_\mu^1 + iW_\mu^2 & \frac{g'}{g}B_\mu - W_\mu^3 \end{bmatrix} \begin{bmatrix} 0 \\ 1 \end{bmatrix} \tag{9.4.6}$$

$$|D_\mu \Phi|^2 \to \frac{g^2 v^2}{8} \begin{bmatrix} 0 & 1 \end{bmatrix} \begin{bmatrix} \frac{g'}{g}B_\mu + W_\mu^3 & W_\mu^1 - iW_\mu^2 \\ W_\mu^1 + iW_\mu^2 & \frac{g'}{g}B_\mu - W_\mu^3 \end{bmatrix} \begin{bmatrix} \frac{g'}{g}B_\mu + W_\mu^3 & W_\mu^1 - iW_\mu^2 \\ W_\mu^1 + iW_\mu^2 & \frac{g'}{g}B_\mu - W_\mu^3 \end{bmatrix} \begin{bmatrix} 0 \\ 1 \end{bmatrix}$$

$$= \frac{g^2 v^2}{8} \left[W_\mu^1 W^{1\mu} + W_\mu^2 W^{2\mu} + \left(\frac{g'}{g}B_\mu - W_\mu^3\right)\left(\frac{g'}{g}B^\mu - W^{3\mu}\right) \right]. \tag{9.4.7}$$

It remains to diagonalize the term containing B_μ and W_μ^3, which is achieved by rotating with the *Weinberg angle*, θ_w, where

$$\tan\theta_w \equiv g'/g, \tag{9.4.8}$$

$$\Rightarrow \left\{ \begin{array}{l} A_\mu = \cos\theta_w B_\mu + \sin\theta_w W_\mu^3 \\ Z_\mu = -\sin\theta_w B_\mu + \cos\theta_w W_\mu^3 \end{array} \right\} \Leftrightarrow \left\{ \begin{array}{l} B_\mu = \cos\theta_w A_\mu - \sin\theta_w Z_\mu \\ W_\mu^3 = \sin\theta_w A_\mu + \cos\theta_w Z_\mu \end{array} \right\}.$$

With these definitions we see that $\frac{g'}{g}B_\mu - W_\mu^3 = Z_\mu/\cos\theta_w$ and we can read off the tree-level masses from Eq. (9.4.7). The photon field A_μ remains massless, $m_\gamma = 0$. For the W_μ^1 and W_μ^2 we have $\frac{1}{2}m_W^2 \equiv g^2 v^2/8$ and so $m_W \equiv gv/2$. It follows that

$$m_Z = \frac{m_W}{\cos\theta_w} = \frac{gv}{2\cos\theta_w}, \quad \cos\theta_w = \frac{g}{\sqrt{g^2 + g'^2}}, \quad \sin\theta_w = \frac{g'}{\sqrt{g^2 + g'^2}}. \tag{9.4.9}$$

We define the charged W-bosons as $W^\pm \equiv \frac{1}{\sqrt{2}}(W_\mu^1 \mp iW_\mu^2)$ and we also define $T^\pm \equiv (T^1 \pm iT^2)$ so that in the defining (fundamental) representation of $SU(2)_L$ we have $T^\pm = \frac{1}{2}(\sigma^1 \pm \sigma^2) = \sigma^{\pm}$.[8] Using the result that $W^1 T^1 + W^2 T^2 = \frac{1}{\sqrt{2}}[W^+ T^+ + W^- T^-]$ and Eq. (9.4.8) we can rewrite the covariant derivative in Eq. (9.4.3) as

$$D_\mu = \partial_\mu + i\frac{g}{\sqrt{2}}[W_\mu^+ T^+ + W_\mu^- T^-] + \frac{iZ_\mu[g^2 T^3 - g'^2 Y]}{\sqrt{g^2 + g'^2}} + \frac{igg' A_\mu[T^3 + Y]}{\sqrt{g^2 + g'^2}}, \tag{9.4.10}$$

where $T^\pm$ and T^3 are in the appropriate representation of $SU(2)_L$. Comparing the A_μ component of the above D_μ with Eq. (7.6.16) we identify the electric charge as

$$q_c \to eQ \equiv e[T^3 + Y] \Rightarrow Q \equiv [T^3 + Y], \quad e \equiv |e| = \frac{gg'}{\sqrt{g^2 + g'^2}} = g\sin\theta_w. \tag{9.4.11}$$

We can equivalently write the covariant derivative as

$$D_\mu = \partial_\mu + i\frac{g}{\sqrt{2}}[W_\mu^+ T^+ + W_\mu^- T^-] + i\frac{g}{\cos\theta_w}Z_\mu[T^3 - \sin^2\theta_w Q] + ieA_\mu Q. \tag{9.4.12}$$

Since the upper and lower components of the Higgs doublet have $T^3 = \pm\frac{1}{2}$, respectively, and both have $Y = \frac{1}{2}$ then $\Phi = (\Phi^+, \Phi^0)$ and so after electroweak symmetry breaking h and v are neutral scalars as expected.

We must now expand $\mathcal{L}_{\text{gauge}}$ in Eq. (9.4.2) using the unitary gauge Higgs doublet

$$\Phi = \begin{bmatrix} 0 \\ \frac{v+h}{\sqrt{2}} \end{bmatrix} \Rightarrow \left\{ \begin{array}{l} V(\Phi) = -m^2|\Phi|^2 + \lambda|\Phi|^4 = \lambda[|\Phi|^2 - \frac{v^2}{2}]^2 - \frac{\lambda v^4}{4} \\ = \lambda(h^2 v^2 + h^3 v + \frac{h^4}{4} - \frac{v^4}{4}) \equiv V(h). \end{array} \right. \tag{9.4.13}$$

---

[8] This $T^\pm$ is the Peskin and Schroeder (1995) definition. Schwartz (2013) uses $\tau^\pm = \frac{1}{\sqrt{2}}[\tau^1 \pm \tau^2]$.

Reading off the h^2 term in $V(h)$ we define $\frac{1}{2}m_h^2 h^2 \equiv \lambda v^2 h^2$ and so the Higgs mass at tree level is $m_h = v\sqrt{2\lambda} = \sqrt{2}m$. The constant term $-\lambda v^4/4$ is irrelevant. The Higgs boson h has cubic and quartic self-interactions. We obtain $\mathcal{L}_{\text{gauge}}$ in terms of the fields h, $W_\mu^\pm$, Z_μ and A_μ using the covariant derivative in Eq. (9.4.12) and replacing B_μ and W_μ^a in terms of $W_\mu^\pm$, Z_μ and A_μ. This leads to the field derivatives and all of their mutual couplings. Define $F_{\mu\nu} \equiv \partial_\mu A_\nu - \partial_\nu A_\mu$ and similarly $Z_{\mu\nu} \equiv \partial_\mu Z_\nu - \partial_\nu Z_\mu$. We find $\mathcal{L}_{\text{gauge}+\Phi} = \mathcal{L}_{\text{gauge}} + \mathcal{L}_{\text{Higgs}}$, where

$$
\begin{aligned}
\mathcal{L}_{\text{gauge}} = &-\tfrac{1}{4}F_{\mu\nu}F^{\mu\nu} - \tfrac{1}{4}Z_{\mu\nu}Z^{\mu\nu} + \tfrac{1}{2}m_Z^2 Z_\mu Z^\mu - \tfrac{1}{2}(\partial_\mu W_\nu^+ - \partial_\nu W_\mu^+)(\partial^\mu W^{\nu-} - \partial^\nu W^{\mu-}) \\
&+ m_W^2 W_\mu^+ W^{\mu-} + ie\cot\theta_w\left[\partial^\mu Z^\nu(W_\mu^+ W_\nu^- - W_\nu^+ W_\mu^-)\right. \\
&\left. - Z^\mu(\partial_\mu W_\nu^+ - \partial_\nu W_\mu^+)W^{\nu-} + Z^\mu(\partial_\mu W_\nu^- - \partial_\nu W_\mu^-)W^{\nu+}\right] + ie[Z\to A] \\
&+ [e^2/2\sin^2\theta_w]\{W_\mu^+ W^{\mu+} W_\nu^- W^{\nu-} - (W_\mu^+ W^{\mu-})^2\} \\
&+ e^2\cot^2\theta_w[Z^\mu W_\mu^+ Z^\nu W_\nu^- - Z_\mu Z^\mu W_\nu^+ W^{\nu-}] + e^2[Z\to A] \\
&+ e^2\cot\theta_w[A^\mu W_\mu^+ Z^\nu W_\nu^- + A^\mu W_\mu^- Z^\nu W_\nu^+ - 2A_\mu Z^\mu W_\nu^+ W^{\nu-}],
\end{aligned}
\tag{9.4.14}
$$

$$
\begin{aligned}
\mathcal{L}_{\text{Higgs}} = &\tfrac{1}{2}(\partial_\mu h\partial^\mu h - m_h^2 h^2) - \left[\frac{gm_h^2}{4m_W}\right]h^3 - \left[\frac{g^2 m_h^2}{32m_W^2}\right]h^4 \\
&+ \left[\frac{2}{v}\right]h(m_W^2 W_\mu^+ W^{\mu-} + \tfrac{1}{2}m_Z^2 Z_\mu Z^\mu) + \left[\frac{1}{v^2}\right]h^2(m_W^2 W_\mu^+ W^{\mu-} + \tfrac{1}{2}m_Z^2 Z_\mu Z^\mu).
\end{aligned}
\tag{9.4.15}
$$

The coefficients for h^3 and h^4 in Eq. (9.4.15) are just complicated ways of writing λv and $\lambda/4$ from Eq. (9.4.13), respectively, in terms of physical masses. The Feynman rules are constructed in the usual way from these Lagrangian densities.

9.4.2 Quarks and Leptons

As discussed in Sec. 5.2.3 the weak interactions are maximally parity violating. We saw in Sec. 4.6.2 that in the absence of a mass term the left and right chiral fermions transform independently under Lorentz transformations. Similarly, vector interactions do not mix the chiralities of fermions since $\slashed{\partial}$ and $\slashed{A}$ have the same spinor structure. Then we can create a *chiral gauge theory* using the chirality projectors in Eq. (4.6.5), $P_R = \frac{1}{2}(1+\gamma^5)$ and $P_L = \frac{1}{2}(1-\gamma^5)$. Consider $A_\mu = A_\mu^a t^a$ with gauge fields A_μ^a and group generators t^a, then we can write, for example,

$$
\mathcal{L}_{\text{chiral}} = \bar{\psi}i(\slashed{\partial} + ig\slashed{A}_\mu P_L)\psi = \bar{\psi}_L i(\slashed{\partial} + ig\slashed{A})\psi_L + \bar{\psi}_R i\slashed{\partial}\psi_R.
\tag{9.4.16}
$$

We have ψ_L in the same group representation as t^a and the representation of the free ψ_R is unimportant. The term with ψ_L corresponds to a left-handed chiral gauge theory and the second term is a noninteracting right-handed fermion that we can simply neglect. So left-handed and right-handed chiral fermions can transform under different representations of a gauge group or even under different gauge groups.

In the SM we begin with massless fermions with only the left-handed leptons and quarks coupling to $SU(2)_L$. The quark and lepton masses will arise through electroweak symmetry breaking and are proportional to the Higgs VEV. Specifically, we take left-handed fermions in weak-isospin doublets ($T^3 = \pm\frac{1}{2}$) and right-handed fermions as weak-isospin singlets ($T^3 = 0$). We then set the weak hypercharge Y such that each particle has the observed charge, $Q = (T^3 + Y)$. There are three generations of $SU(2)_L$ weak isodoublets for each of the leptons and quarks,

$$
(L^1, L^2, L^3) \equiv \begin{bmatrix} \nu_{eL} \\ e_L \end{bmatrix}, \begin{bmatrix} \nu_{\mu L} \\ \mu_L \end{bmatrix}, \begin{bmatrix} \nu_{\tau L} \\ \tau_L \end{bmatrix}, \quad (Q^1, Q^2, Q^3) \equiv \begin{bmatrix} u_L \\ d_L \end{bmatrix}, \begin{bmatrix} c_L \\ s_L \end{bmatrix}, \begin{bmatrix} b_L \\ t_L \end{bmatrix}.
\tag{9.4.17}
$$

These are left-handed chiral fermions transforming in the $(\frac{1}{2}, 0)$ representation of the Lorentz group as seen in Eq. (4.6.17). Upper doublet components have $T^3 = +\frac{1}{2}$ and lower doublet components have $T^3 = -\frac{1}{2}$. Assigning $Y = Y_L \equiv -\frac{1}{2}$ for the L^i and $Y = Y_Q \equiv +\frac{1}{6}$ for the Q^i leads to the correct fermion electric charges, $Q = (T^3 + Y)$: (i) $Q = 0$ for $\nu_{eL}, \nu_{\mu L}, \nu_{\tau L}$; (ii) $Q = -1$ for e_L, μ_L and τ_L; (iii) $Q = +\frac{2}{3}$ for u_L, c_L, b_L; and (iv) $Q = -\frac{1}{3}$ for d_L, s_L, t_L. For the right-handed fermions we have

$$(\nu_R^1, \nu_R^2, \nu_R^3) \equiv (\nu_{eR}, \nu_{\mu R}, \nu_{\tau L}), \qquad (e_R^1, e_R^2, e_R^3) \equiv (e_R, \mu_R, \tau_R), \qquad (9.4.18)$$

$$(u_R^1, u_R^2, u_R^3) \equiv (u_R, c_R, b_R), \qquad (d_R^1, d_R^2, d_R^3) \equiv (d_R, s_R, t_R), \qquad (9.4.19)$$

where since $T^3 = 0$ we simply set $Y = Q$ such that $Y = Y_\nu \equiv 0$ for the three ν_R^i, $Y = Y_e \equiv -1$ for the three e_R^i, $Y = Y_u \equiv +\frac{2}{3}$ for the three u_R^i, and $Y = Y_d \equiv -\frac{1}{3}$ for the three d_R^i. The right-handed chiral fermions transform under the $(0, \frac{1}{2})$ representation of the Lorentz group. We then use these values for Y and the appropriate $SU(2)_L$ representation to form the covariant derivative in Eq. (9.4.12).

Fermion kinetic terms: The fermion kinetic and gauge-coupling terms in the SM Lagrangian density are

$$\begin{aligned}
\mathcal{L}_{\text{kin-ferm}} &= \bar{L}^i i \slashed{D} L^i + \bar{Q}^i i \slashed{D} Q^i + \bar{u}_R^i i \slashed{D} u_R^i + \bar{d}_R^i i \slashed{D} d_R^i + \bar{e}_R^i i \slashed{D} e_r^i + \bar{\nu}_R^i i \slashed{D} \nu_R^i \\
&= i \bar{L}^i (\slashed{\partial} + i g \slashed{W}^a T^a + i g' Y_L \slashed{B}) L^i + i \bar{Q}^i (\slashed{\partial} + i g \slashed{W}^a T^a + i g' Y_Q \slashed{B}) Q^i \quad (9.4.20) \\
&\quad + i \bar{u}_R^i (\slashed{\partial} + i g' Y_u \slashed{B}) u_R^i + i \bar{d}_R^i (\slashed{\partial} + i g' Y_d \slashed{B}) d_R^i + i \bar{e}_R^i (\slashed{\partial} + i g' Y_e \slashed{B}) e_R^i + i \bar{\nu}_R^i \slashed{\partial} \nu_R^i.
\end{aligned}$$

The left-handed and right-handed quarks also experience $SU(3)_c$ covariant derivatives for QCD but we do not show this for brevity and leave it to be understood. We see from the final term that the ν_R^i do not interact and so the *right-handed neutrinos ν_R^i disappear from the theory*. We can use the covariant derivative from Eq. (9.4.10) and the above quantum number assignments to arrive at $\mathcal{L}_{\text{kin-ferm}}$ in terms of physical fields. For example, consider the part of $\mathcal{L}_{\text{kin-ferm}}$ involving the charged leptons coupling to the photon. Using $Y_L = -\frac{1}{2}$ and $Y_e = -1$ we find

$$\begin{aligned}
\mathcal{L}_{\text{kin-ferm}} &= \cdots - e \left\{ (-\tfrac{1}{2} + Y_L) \bar{e}_L^i \slashed{A} e_L^i + Y_e \bar{e}_R^i \slashed{A} e_R^i \right\} + \cdots \quad (9.4.21) \\
&= \cdots - [\bar{e} q_c \slashed{A} e + \bar{\mu} q_c \slashed{A} \mu + \bar{\tau} q_c \slashed{A} \tau] + \cdots,
\end{aligned}$$

where $q_c = -e \equiv -|e|$ for the e, μ, τ and we recover the usual QED interactions. Using shorthand matrix notation[9] such as $u_L \equiv (u_L^1, u_L^2, u_L^3) = (u_L, c_L, b_L)$ the complete form in terms of the physical vector boson fields is

$$\begin{aligned}
\mathcal{L}_{\text{kin-ferm}} &= \bar{L} i \slashed{\partial} L + \bar{Q} i \slashed{\partial} Q + \bar{u}_R i \slashed{\partial} u_R + \bar{d}_R i \slashed{\partial} d_R + \bar{e}_R i \slashed{\partial} e_R \quad (9.4.22) \\
&\quad - g \left[W_\mu^+ J^{\mu+} + W_\mu^- J^{\mu-} + Z_\mu J_Z^\mu \right] - e A_\mu J_{\text{em}}^\mu,
\end{aligned}$$

where we have defined the currents

$$J_W^{\mu+} = \frac{1}{\sqrt{2}} [\bar{\nu}_L \gamma^\mu e_L + \bar{u}_L \gamma^\mu d_L], \quad J_W^{\mu-} = \frac{1}{\sqrt{2}} [\bar{e}_L \gamma^\mu \nu_L + \bar{d}_L \gamma^\mu u_L], \quad (9.4.23)$$

$$\begin{aligned}
J_Z^\mu &= (1/\cos\theta_w) \big[\bar{\nu}_L \gamma^\mu (\tfrac{1}{2}) \nu_L + \bar{e}_L \gamma^\mu (-\tfrac{1}{2} + \sin^2\theta_w) e_L + \bar{e}_R \gamma^\mu (\sin^2\theta_w) e_R \\
&\quad + \bar{u}_L \gamma^\mu (\tfrac{1}{2} - \tfrac{2}{3}\sin^2\theta_w) u_L + \bar{u}_R \gamma^\mu (-\tfrac{2}{3}\sin^2\theta_w) u_R + \bar{d}_L \gamma^\mu (-\tfrac{1}{2} + \tfrac{1}{3}\sin^2\theta_w) d_L \\
&\quad + \bar{d}_R \gamma^\mu (\tfrac{1}{3}\sin^2\theta_w) d_R \big],
\end{aligned}$$

$$J_{\text{em}}^\mu = \bar{u} \gamma^\mu (+\tfrac{2}{3}) u + \bar{d} \gamma^\mu (-\tfrac{1}{3}) d + \bar{e} \gamma^\mu (-1) e.$$

---

[9] Whether notation such as u_L refers to a three-generation column vector or just the first-generation field u_L is to be understood from the context.

Yukawa couplings of the fermions and Higgs: In the chiral representation $m\bar\psi\psi = m(\bar\psi_L^\dagger \psi_R + \bar\psi_R^\dagger \psi_L)$. If ψ_L and ψ_R are in different representations of a gauge group, then any mass term violates gauge invariance. So in the SM such mass terms are not permitted. The Higgs doublet Φ has $Y = \frac{1}{2}$, transforms under the defining representation of $SU(2)$, and has only a lower component with $T^3 = -\frac{1}{2}$. Under a $U(1)_Y$ transformation Φ^* will have the opposite weak hypercharge to Φ. From Eq. (4.5.45) we know that for a spin-half system $i\sigma\phi_{-1/2}^* = \chi_{1/2}$. It follows that

$$\tilde\Phi \equiv i\sigma^2 \Phi^* \tag{9.4.24}$$

has $Y = -\frac{1}{2}$, also transforms under the defining representation of $SU(2)$ and has only an upper component with $T^3 = +\frac{1}{2}$. It then follows that combinations such as $\bar{Q}^i \Phi$, $\bar{Q}^i \tilde\Phi$, $\bar{L}^i \Phi$, $\bar{L}^i \tilde\Phi$ and their Hermitian conjugates (h.c.) are $SU(2)_L$ singlets. In order to make $SU(3)_c \times SU(2)_L \times U(1)_Y$ invariant terms we combine these with right-handed fermions so that we have total $Y = 0$ and $SU(3)_c$ singlets. The latter means that we can only combine left-handed and right-handed quarks and left-handed and right-handed leptons. We then arrive at the Yukawa-type interactions,

$$\mathcal{L}_{\text{Yukawa}} = -\lambda_d^{ij} \bar{Q}^i \Phi d_R^j - \lambda_u^{ij} \bar{Q}^i \tilde\Phi u_R^j - \lambda_e^{ij} \bar{L}^i \Phi e_R^j + \text{h.c.} \tag{9.4.25}$$

$$= -\lambda_d^{ij} \bar{d}_L^i \frac{1}{\sqrt{2}}(v+h) d_R^j - \lambda_u^{ij} \bar{u}_L^i \frac{1}{\sqrt{2}}(v+h) u_R^j - \lambda_e^{ij} \bar{L}^i \frac{1}{\sqrt{2}}(v+h) e_R^j + \text{h.c.}$$

$$= -\frac{1}{\sqrt{2}}(v+h)[\bar{d}_L \lambda_d d_R + \bar{u}_L \lambda_u u_R + \bar{e}_L \lambda_e e_R], \tag{9.4.26}$$

where the last line uses a matrix notation with three-component column vectors d_L, d_R, u_L, u_R, e_L and e_R and arbitrary 3×3 complex matrices λ_d, λ_u and λ_e. Note that since there are no right-handed neutrinos there is no neutrino Yukawa term. We verify that each has $Y = 0$; e.g., using $Y_Q = -Y_{\bar{Q}} = +\frac{1}{6}$, $Y_\Phi = -Y_{\tilde\Phi} = \frac{1}{2}$, $Y_L = -Y_{\bar{L}} = -\frac{1}{2}$, $Y_d = -\frac{1}{3}$, $Y_u = +\frac{2}{3}$ and $Y_e = -1$ we find that $Y_{\bar{Q}^i \Phi d_R^j} = -\frac{1}{6} + \frac{1}{2} - \frac{1}{3} = 0$, $Y_{\bar{Q}^i \tilde\Phi u_R^j} = -\frac{1}{6} - \frac{1}{2} + \frac{2}{3} = 0$ and $Y_{\bar{e}_L^i \Phi e_R^j} = \frac{1}{2} + \frac{1}{2} - 1 = 0$. Total $Y = 0$ follows since each term is an $SU(2)_L$ singlet with zero total charge, $Q = 0$.

Under a CP transformation each term is exchanged with its Hermitian conjugate with the λ matrices unchanged, which means that if the SM had exact CP invariance then the λ matrices would be real. We can bring the Yukawa matrices λ_d, λ_u and λ_e into real diagonal form with nonnegative elements using the singular-value decomposition of an arbitrary complex matrix, which generalizes the real case in Eq. (2.9.48). For any complex square matrix λ we can find unitary matrices U and W such that $\lambda = U D W^\dagger$ where D is diagonal with diagonal elements D_{ii} being the non-negative eigenvalues of $\lambda^\dagger \lambda$ and $\lambda \lambda^\dagger$. So we can write for each λ matrix,

$$\lambda_u = U_u D_u W_u^\dagger, \quad \lambda_d = U_d D_d W_d^\dagger, \quad \lambda_u = U_e D_e W_e^\dagger. \tag{9.4.27}$$

Since the left-handed and right-handed fermions are independent fields we can absorb the W matrices into change of basis for the right-handed fermion fields and similarly for the U matrices with the left-handed fermion fields,

$$d_R \to W_d d_R, \quad u_R \to W_u u_R, \quad e_R \to W_e e_R, \tag{9.4.28}$$

$$d_L \to U_d d_L, \quad u_L \to U_u u_L, \quad e_L \to U_e e_L. \tag{9.4.29}$$

So the U and W unitary matrices no longer appear in $\mathcal{L}_{\text{Yukawa}}$ and we simply replace each matrix λ with its corresponding diagonal matrix D,

$$\mathcal{L}_{\text{Yukawa}} = -\tfrac{1}{\sqrt{2}}(v+h)[\bar{d}_L D_d d_R + \bar{u}_L D_u u_R + \bar{e}_L D_e e_R] \tag{9.4.30}$$

$$= -m_d^i(1+\tfrac{h}{v})d_L^i d_R^i - m_u^i(1+\tfrac{h}{v})u_L^i u_R^i - m_e^i(1+\tfrac{h}{v})e_L^i e_R^i + \text{h.c.},$$

$$m_d^i \equiv \tfrac{1}{\sqrt{2}}v D_d^{ii}, \quad m_d^i \equiv \tfrac{1}{\sqrt{2}}v D_d^{ii}, \quad m_d^i \equiv \tfrac{1}{\sqrt{2}}v D_d^{ii}. \tag{9.4.31}$$

We see that m_d^i, m_u^i and m_e^i are the fermion masses and that we have a Yukawa coupling of the fermions with the Higgs boson h proportional to the fermion mass.

The fermion change of basis diagonalized $\mathcal{L}_{\text{Yukawa}}$ but now we need to consider the effect on $\mathcal{L}_{\text{kin-ferm}}$. Consider Eq. (9.4.22) and the electroweak currents given Eq. (9.4.23). For terms involving $\bar{\psi}\cdots\psi$, where $\psi = d_L, d_R, u_L, u_R, e_L, e_R$ we have $W^\dagger W = U^\dagger U = I$ and the change of basis has no effect. The only terms not of this form are the charged currents $J_W^{\mu+}$ and $J_W^{\mu-}$, which become

$$J_W^{\mu+} = \tfrac{1}{\sqrt{2}}\left[\bar{\nu}_L \gamma^\mu e_L + \bar{u}_L \gamma^\mu V d_L\right], \quad J_W^{\mu-} = \tfrac{1}{\sqrt{2}}\left[\bar{e}_L \gamma^\mu \nu_L + \bar{d}_L \gamma^\mu V^\dagger u_L\right], \tag{9.4.32}$$

where $V \equiv V_{\text{CKM}} \equiv U_u^\dagger U_d$ and so $V^\dagger \equiv V_{\text{CKM}}^\dagger = U_d^\dagger U_u$. We recognize that $V \equiv V_{\text{CKM}}$ is the Cabibbo-Kobayashi-Masjawa (CKM) matrix discussed earlier in Sec. 5.2.4.

There are some remaining subtleties. (i) As we saw in our discussion of the ABJ anomaly in Sec. 9.3 the global chiral rotations that we have used to express $\mathcal{L}_{\text{SM}}$ in its mass basis can lead to chiral anomalies. Using the fact that the anomaly terms can be written as total derivatives as we saw in Eq. (9.1.11), it can be shown that for the SM the terms arising from $SU(2)_L$ and $U(1)_Y$ lead to no effects on observables (Peskin and Schroeder, 1995). For $SU(3)_c$ the generated term has the form of the QCD θ term and so can be absorbed into that. There are then two potential sources of CP violation in the SM in the mass basis, which are a possible CP-violating phase in the CKM matrix and a nonzero value for θ. That θ appears to be so small is referred to as the strong CP problem discussed earlier. (ii) The SM has no gauge anomalies and this is a consquence of detailed cancellations due to the quantum number assignments in the SM. Had the quantum numbers been chosen randomly, then the gauge anomalies would not cancel and the SM would not be a consistent theory. (iii) As we have seen, the neutrinos remain massless in the SM but they can be given Dirac or Majorana masses in a variety of ways leading to the neutrino PMNS matrix and its consequences, discussed in Sec. 5.2.5.

Summary

In this final chapter we extended our discussions to nonabelian gauge theories. We generalized the covariant derivative to the nonabelian case and constructed an appropriate Lagrangian density. We considered parallel transport using Wilson lines and introduced the gauge-invariant Wilson loop. The quantization of QCD was used as an example and renormalized perturbative QCD was used to extract the QCD running quark mass and the QCD running coupling constant. The decoupling of heavy quarks using the Appelquist-Carrazone theorem was discussed as was the experimental evidence for three colors. The BRST invariance of perturbative QCD was demonstrated and it was noted that this leads to the Slavnov-Taylor identities as generalizations of the Ward-Takahashi identities. We then presented a brief summary of lattice QCD and illustrated how calculations are performed using the extraction of hadron masses as an example. Quantum anomalies were then introduced and the ABJ chiral anomaly was considered as an explicit example. We ended with a derivation of the Standard Model and connected this back to our discussions in Chapter 5. We have now reached the limit of what can be covered in this text. However, the reader is now well positioned to pursue additional content in texts focusing on the Standard Model and beyond.

Problems

9.1 (a) Show that Noether's theorem for global $SU(3)_{\text{color}}$ symmetry in QCD leads to the conserved color current $j_\mu^a = \bar\psi_i \gamma_\mu T_{ij}^a \psi_j - f^{abc} A_\nu^b F_{\mu\nu}^c$ and hence that $dQ^a/dt = 0$ where $Q^a = \int d^3x\, j_0^a$ in the usual way.

   (b) Show that the quark component of the color current $j_{q\mu}^a \equiv \bar\psi_i \gamma_\mu T_{ij}^a \psi_j$ satisfies $D^{\mu bc} j_{q\mu}^c = 0$ and hence that $\partial^\mu j_{q\mu}^a \neq 0$. Explain why QED-like Ward-Takahashi and Ward identities do not apply in QCD.

9.2 Calculate the $\overline{\text{MS}}$ counterterm for the quark-gluon vertex in an arbitrary covariant gauge and obtain the result for δ_1 in Eq. (9.2.28).

9.3 Draw the diagrams contributing to the $\mathcal{O}(g^2)$ (one-loop) corrections to the tree-level ghost-gluon, three-gluon and four-gluon vertices. Write down the loop integrals for the ghost-gluon vertex.

9.4 Define the BRST transformation of some field ϕ as $Q\phi$, where under the BRST transformation $\phi \to \phi + \delta\phi$ with $\delta\phi \equiv \theta Q\phi$. For example, from Eq. (9.2.42) we see that $QA_\mu^a = (1/g)D_\mu^{ab} c^b$. Show that the BRST variations are nilpotent, which is the statement that $Q^2\phi = 0$ when acting on all fields ϕ.

9.5 Starting from Eq. (9.4.2), complete the derivation of the Lagrangian densities $\mathcal{L}_{\text{gauge}}$ and $\mathcal{L}_{\text{Higgs}}$ in terms of the physical fields ($W_\mu^\pm$, Z_μ, A_μ and h) given in Eqs. (9.4.14) and (9.4.15), respectively.

9.6 Show that the fermion kinetic contribution $\mathcal{L}_{\text{kin-ferm}}$ in Eq. (9.4.20) can be rewritten in terms of the physical vector boson fields as Eq. (9.4.22) with the electroweak currents $J_W^{\mu\pm}$, J_Z^μ and J_{em}^μ given in Eq. (9.4.23).

9.7 (a) Use the definition of the Fermi coupling constant in Eq. (5.2.7) and the low-momentum limit of the tree-level contribution to the scattering $e^- \bar\nu_e \to \mu^- \bar\nu_\mu$ to show that $G_F/\sqrt{2} = g^2/8M_W^2 = 1/2v^2$. Since we know from many experimental measurements that $G_F \simeq 1.167 \times 10^{-5}\ \text{GeV}^{-2}$ then $v = (\sqrt{2}G_F)^{-1/2} \simeq 246\ \text{GeV}$.

   (b) Calculate the total cross-section for $e^+ e^- \to Zh$ at tree level in the CM frame (the electron-Higgs coupling is very small and can be neglected).

   (c) Estimate the total cross-section for $e^+ e^- \to$ hadrons at tree level in the CM frame, including both the intermediate γ and Z contributions to $\mathcal{M}$ and their interference in the total cross-section, $\sigma(e^+ e^- \to \text{hadrons})$.

9.8 (a) Calculate to lowest order the decay rate $h \to b\bar b$.

   (b) Calculate the decay rate for $h \to gg$. (Hint: This problem requires a one-loop calculation and some care and effort.)

   (c) Draw the lowest order diagrams contributing to the decay $h \to \gamma\gamma$.

9.9 Consider an $SU(3)$ nonabelian gauge theory with coupling to eight real scalar fields in the adjoint representation, ϕ^a for $a = 1, \ldots, 8$. The covariant derivative is then $D_\mu^{ab}\phi^b = \partial_\mu\phi^a + gf^{abc}A_\mu^b\phi^c$. Define $\Phi = \phi^a T^a$, where T^a are the eight $SU(3)$ generators. Write down the Lagrangian density for the theory. Assume that some potential is added that causes Φ to have a vacuum expectation value, $\Phi_0 = \langle\Omega|\hat{\Phi}|\Omega\rangle \neq 0$. Since Φ_0 is a traceless Hermitian matrix we can diagonalize it. Assume that after doing so we found $\Phi_0 \to \Phi_0^D = vD$ with $v > 0$ and with D being the diagonal matrix $D = \text{diag}(\frac{1}{2}, \frac{1}{2}, -1)$. By referring to Eq. (5.3.33) explain why this corresponds to the symmetry breaking $SU(3) \to SU(2) \times U(1)$ with $SU(3)$ and $SU(2)$ in their defining 3×3 and 2×2 representations, respectively. Explain why the "hypercharge operator" Y must satisfy $Y \propto D$. How many gauge bosons acquire a mass and how many remain massless? Calculate the mass of the massive gauge bosons at tree level.

Appendix

A.1 Physical Constants

$$c = 2.99792458 \times 10^8 \,\text{m/s (exact value)} \qquad \hbar = 6.582119 \times 10^{-22} \,\text{MeV s}$$

$$\hbar \equiv h/2\pi = 1.0545718 \times 10^{-34} \,\text{J s} \qquad \hbar c = 197.32698 \,\text{MeV fm},$$

$$e \equiv |e| = 1.6021766 \times 10^{-19} \,\text{C} \qquad \alpha \equiv e^2/4\pi\epsilon_0 \hbar c = 1/137.035999$$

$$G_F/(\hbar c)^3 = 1.16638 \times 10^{-5} \,\text{GeV}^{-2} \qquad \alpha_s(m_{Z^0}) \simeq 0.1182(12)$$

$$\sin^2(\theta_w)_{m_{Z^0}} \simeq 0.231 \qquad m_e = 0.510999 \,\text{MeV}/c^2$$

$$m_p,\, m_n = 938.2721,\, 939.5654 \,\text{MeV}/c^2 \qquad m_{\text{deuteron}} = 1875.613 \,\text{MeV}/c^2$$

$$m_\mu = 105.6584 \,\text{MeV}/c^2 \qquad m_\tau = 1777 \,\text{MeV}/c^2$$

$$m_{\pi^0},\, m_{\pi^\pm} = 134.98,\, 139.57 \,\text{MeV}/c^2 \qquad m_{W^\pm},\, m_{Z^0} = 80.4,\, 91.19 \,\text{GeV}/c^2$$

$$(\hbar c)^2 = 0.3893794 \,\text{GeV}^2 \,\text{mbarn} \qquad 1\,\text{b} \equiv 1\,\text{barn} \equiv 10^{-28} \,\text{m}^2 = 10^{-24} \,\text{cm}^2$$

$$1\,\text{Å} \equiv 1\,\text{angstrom} \equiv 10^{-10} \,\text{m} \qquad 1\,\text{fm} \equiv 1\,\text{fermi} \equiv 10^{-15} \,\text{m}$$

$$1\,\text{eV} \equiv 1.6021766 \times 10^{-19} \,\text{J} \qquad 1\,\text{eV}/c^2 \equiv 1.7826619 \times 10^{-36} \,\text{kg}$$

$$0°\,\text{C} \equiv 273.15 \,\text{K} \qquad 1\,\text{G} \equiv 10^{-4} \,\text{T}$$

$$1\,\text{erg} \equiv 10^{-7} \,\text{J} \qquad 1/kT\,(\text{at } 300\,\text{K}) = 38.682 \,\text{eV}^{-1}$$

$$\text{(A.1.1)}$$

$$\mu_B \equiv \text{Bohr magneton} \equiv e\hbar/2m_e = 9.2740 \times 10^{-24} \,\text{J T}^{-1} = 5.78838^{-11} \,\text{MeV T}^{-1}$$

$$\mu_N \equiv \text{nuclear magneton} \equiv e\hbar/2m_p = 3.152451^{-14} \,\text{MeV T}^{-1}$$

$$N_A \equiv \text{Avogadro's number} = 6.022141 \times 10^{23} \,\text{mol}^{-1}$$

$$k \equiv \text{Boltzmann constant} = 1.38065 \times 10^{-23} \,\text{J K}^{-1} = 8.6173 \times 10^{-5} \,\text{eV K}^{-1}$$

A.2 Notation and Useful Results

In order to simplify comparison with other treatments, the notation and conventions adopted in this text will typically be those of Peskin and Schroeder (1995) and/or Schwartz (2013). For the avoidance of doubt, e is here *always* defined as the *magnitude* of the electron charge, $e \equiv |e|$.

A.2.1 Levi-Civita Tensor

The completely antisymmetric Levi-Civita tensor of rank n is antisymmetric under the interchange of *any two* of its n indices

$$\epsilon^{i_1 i_2 \cdots i_j \cdots i_k \cdots i_n} = -\epsilon^{i_1 i_2 \cdots i_k \cdots i_j \cdots i_n}, \quad \epsilon^{123\cdots n} = +1. \tag{A.2.1}$$

We equate up and down three-vector indices but must use $g_{\mu\nu}$ to relate up and down spacetime indices,

$$\epsilon_{ijk} \equiv \epsilon^{ijk}, \quad \epsilon_{\mu\nu\rho\sigma} = g_{\mu\mu'} g_{\nu\nu'} g_{\rho\rho'} g_{\sigma\sigma'} \epsilon^{\mu'\nu'\rho'\sigma'}, \quad \epsilon^{0123} = -\epsilon_{0123} = +1. \tag{A.2.2}$$

Note that this sign convention agrees with Jackson (1975), Itzykson and Zuber (1980), and Peskin and Schroeder (1995), but is opposite in sign to that used by Bjorken and Drell; i.e., see Section 7.2 of Bjorken and Drell (1964). Some useful results for contractions of the rank-four antisymmetric tensor are

$$\epsilon^{\mu\nu\rho\sigma}\epsilon_{\mu\nu\rho\sigma} = -4!, \; \epsilon^{\mu\nu\rho\alpha}\epsilon_{\mu\nu\rho\beta} = -3!\delta^\alpha{}_\beta, \; \epsilon^{\mu\nu\alpha\beta}\epsilon_{\mu\nu\gamma\delta} = -2(\delta^\alpha{}_\gamma \delta^\beta{}_\delta - \delta^\alpha{}_\delta \delta^\beta{}_\gamma). \tag{A.2.3}$$

A.2.2 Dirac Delta Function and Jacobians

$$\int d^n x \, \delta^n(x - x_0) f(x) = f(x_0), \qquad \delta^n(k - k_0) \equiv \int \frac{d^n x}{(2\pi)^n} \, e^{\pm i(k - k_0)\cdot x}. \tag{A.2.4}$$
$$\delta(x) = d\theta(x)/dx, \qquad\qquad\qquad \theta(x) = 0 \text{ for } x < 0, \; 1 \text{ for } x > 0.$$

Consider an invertible change of variable $\vec{x} \to \vec{y}$, then $d^n x = d^n y \, \det[M(x, y)]$, where the *Jacobian matrix* is given by

$$M(\vec{x}, \vec{y}) \equiv \begin{pmatrix} \partial x^1/\partial y^1 & \cdots & \partial x^1/\partial y^n \\ \vdots & \ddots & \vdots \\ \partial x^n/\partial y^1 & \cdots & \partial x^n/\partial y^n \end{pmatrix}. \tag{A.2.5}$$

Let some region R in x-space with volume V_R map to region R' in y-space, then

$$V_x = \int_R d^n x = \int_{R'} d^n y \, \mathcal{J}(\vec{x}, \vec{y}), \tag{A.2.6}$$

where $\mathcal{J}(\vec{x}, \vec{y}) \equiv |\det M(\vec{x}, \vec{y})|$ is the *Jacobian* and where the orientation of the y-integration is defined so that $V_{R'} = \int_{R'} d^n y > 0$ is the volume of R' in y-space. The modulus of the determinant allows us to always choose the orientation of our integrations so that the volume is positive.

Since for $\vec{x}_0 \in R$ we have $1 = \int_R d^n x \, \delta^n(\vec{x} - \vec{x}_0)$, then

$$1 = \int_R d^n x \, \delta^n(\vec{x} - \vec{x}_0) = \int_{R'} d^n y \mathcal{J}(\vec{x}, \vec{y}) \, \delta^n(\vec{x}(\vec{y}) - \vec{x}_0(\vec{y})) = \int_{R'} d^n y \, \delta^n(\vec{y} - \vec{y}_0), \tag{A.2.7}$$

and hence we must have for $\vec{y}_0 \equiv \vec{y}(\vec{x}_0)$ that $\delta^n(\vec{y} - \vec{y}_0) = \mathcal{J}(\vec{x}, \vec{y}) \, \delta^n(\vec{x} - \vec{x}_0)$. This result extends to a noninvertible change of variable $\vec{y} = \vec{y}(\vec{x})$ provided that it is invertible in the neighborhood of all points $\vec{x}_0^{(1)}, \ldots, \vec{x}_0^{(N)} \in R$ that satisfy the equation $\vec{y}(\vec{x}_0^{(k)}) = \vec{y}_0$. In one dimension this result has the form

$$\delta(y - y_0) = \sum_{k=1}^N \left. (|dx/dy|) \right|_{x=x_0^{(k)}} \delta(x - x_0^{(k)}). \tag{A.2.8}$$

A.2.3 Fourier Transforms

In quantum field theory we use the asymmetric definition of the Fourier transform,

$$f(x) = \int \frac{d^n k}{(2\pi)^n} e^{-ik\cdot x} \tilde{f}(k), \qquad\qquad \tilde{f}(k) = \int d^n x \, e^{+ik\cdot x} f(x), \qquad\qquad \text{(A.2.9)}$$

where the signs of the exponentials are conventional for four spacetime dimensions. In three spatial dimensions it is conventional in a field theory context to use

$$f(\mathbf{x}) = \int \frac{d^3 k}{(2\pi)^3} e^{+i\mathbf{k}\cdot\mathbf{x}} \tilde{f}(\mathbf{k}), \qquad\qquad \tilde{f}(\mathbf{k}) = \int d^3 x \, e^{-i\mathbf{k}\cdot\mathbf{x}} f(\mathbf{x}). \qquad\qquad \text{(A.2.10)}$$

The coordinate-space represention of a plane wave moving in the $+\mathbf{k}$-direction is given by its conventional form, i.e., $\sim \exp(+i\mathbf{k}\cdot\mathbf{x})$. Since $k\cdot x \equiv k^0 x^0 - \mathbf{k}\cdot\mathbf{x}$ we see the motivation for the sign convention for the exponents in Eq. (A.2.9).

In nonrelativistic quantum mechanics we use the symmetric Fourier transform,

$$f(\mathbf{x}) \equiv \langle \mathbf{x}|f\rangle = \int d^3 k \, \langle \mathbf{x}|\mathbf{k}\rangle\langle \mathbf{k}|f\rangle = \int \frac{d^3 k}{(2\pi)^{3/2}} e^{+i\mathbf{k}\cdot\mathbf{x}} \tilde{f}(\mathbf{k}), \qquad\qquad \text{(A.2.11)}$$

$$\tilde{f}(\mathbf{k}) \equiv \langle \mathbf{k}|f\rangle = \int d^3 x \, \langle \mathbf{k}|\mathbf{x}\rangle\langle \mathbf{x}|f\rangle = \int \frac{d^3 x}{(2\pi)^{3/2}} e^{-i\mathbf{k}\cdot\mathbf{x}} f(\mathbf{x}). \qquad\qquad \text{(A.2.12)}$$

The quantum mechanics completeness and orthogonality normalization conventions are $\int d^3 x \, |\mathbf{x}\rangle\langle \mathbf{x}| = \int d^3 k \, |\mathbf{k}\rangle\langle \mathbf{k}| = \hat{I}$, $\langle \mathbf{k}'|\mathbf{k}\rangle = \delta^3(\mathbf{k}' - \mathbf{k})$ and $\langle \mathbf{x}'|\mathbf{x}\rangle = \delta^3(\mathbf{x}' - \mathbf{x})$. Since $\langle \mathbf{x}'|\mathbf{x}\rangle = \delta^3(\mathbf{x}' - \mathbf{x}) = \int d^3 k \, \langle \mathbf{x}|\mathbf{k}\rangle\langle \mathbf{k}|\mathbf{x}\rangle = \int d^3 k \, |\langle \mathbf{x}|\mathbf{k}\rangle|^2$ and using Eq. (A.2.4), i.e., $\delta^3(\mathbf{x}' - \mathbf{x}) = \int d^3 k/(2\pi)^3 \exp[\pm i\mathbf{k}\cdot(\mathbf{x}' - \mathbf{x})]$, it follows that $\langle \mathbf{x}|\mathbf{k}\rangle = 1/(2\pi)^{3/2} e^{i\mathbf{k}\cdot\mathbf{x}}$ where we make the conventional sign choice for the exponent. Factors of $\hbar$ are restored using $\mathbf{p} = \hbar\mathbf{k}$ instead of $\mathbf{k}$.

From Eq. (4.1.202) it immediately follows that the Fourier transform of a Gaussian function is a Gaussian function, since

$$\int \frac{dk}{2\pi} e^{ik(x-x_0)} e^{-\frac{1}{2}ak^2} = \frac{1}{\sqrt{2\pi a}} e^{-\frac{1}{2}[(x-x_0)^2/a]} \xrightarrow{a\to 0^+} \delta(x - x_0). \qquad\qquad \text{(A.2.13)}$$

An important three-dimensional Fourier transform leads to the Yukawa potential in the nonrelativistic limit of a massive particle exchange,

$$\int \frac{d^3 k}{(2\pi)^3} e^{i\mathbf{k}\cdot\mathbf{x}} \frac{1}{\mathbf{k}^2 + m^2} = \frac{1}{4\pi|\mathbf{x}|} e^{-mx}. \qquad\qquad \text{(A.2.14)}$$

In the case of photon exchange, $m \to 0$, we recover the Coulomb potential,

$$\int \frac{d^3 k}{(2\pi)^3} e^{i\mathbf{k}\cdot\mathbf{x}} \frac{1}{\mathbf{k}^2} = \frac{1}{4\pi|\mathbf{x}|} \quad\Rightarrow\quad \nabla^2\left(\frac{-1}{4\pi|\mathbf{x}|}\right) = \int \frac{d^3 k}{(2\pi)^3} e^{i\mathbf{k}\cdot\mathbf{x}} = \delta^3(\mathbf{x}). \qquad\qquad \text{(A.2.15)}$$

A.2.4 Cauchy's Integral Theorem

$$f'(z_0) \equiv \frac{df}{dz}(z_0) \equiv \lim_{z\to z_0} \frac{f(z) - f(z_0)}{z - z_0}, \qquad\qquad \frac{\partial f}{\partial x}(z_0) = -i\frac{\partial f}{\partial y}(z_0), \qquad\qquad \text{(A.2.16)}$$

$$f(a) = \frac{1}{2\pi i} \oint_c dz \frac{f(z)}{z - a}. \qquad\qquad \text{(A.2.17)}$$

$c \equiv$ closed counterclockwise contour in $\mathbb{C}$ containing a. Define $f^{(n)}(z) \equiv \frac{d^n f}{dz^n}(z)$.

$$f^{(n)}(a) \equiv \frac{d^n}{dz'^n} f(z')\Big|_{z'=a} = \frac{d^n}{dz'^n} \frac{1}{2\pi i} \oint_c dz \frac{f(z)}{z - z'}\Big|_{z'=a} = \frac{n!}{2\pi i} \oint_c dz \frac{f(z)}{(z-a)^{n+1}}. \qquad\qquad \text{(A.2.18)}$$

Using $\frac{1}{(z-a)(z-b)} = \frac{1}{a-b}\left(\frac{1}{z-a} - \frac{1}{z-b}\right)$ leads to

$$\frac{1}{2\pi i} \oint_c dz \frac{f(z)}{(z-a)(z-b)} = \frac{1}{2\pi i} \oint_c dz \frac{f(z)}{a-b}\left(\frac{1}{z-a} - \frac{1}{z-b}\right) = \frac{f(a) - f(b)}{a-b}, \frac{f(a)}{a-b}, \frac{-f(b)}{a-b}, 0 \qquad \text{(A.2.19)}$$

for a and b inside contour c, only a in c, only b in c and neither in c, respectively.

A.2.5 Wirtinger Calculus

We briefly summarize the definition and properties of Wirtinger calculus (Henrici, 1993; Schreier and Scharf, 2010). Let $z = x + iy, z^* = x - iy \in \mathbb{C}$ with $x, y \in \mathbb{R}$.

$$\text{Wirtinger} \atop \text{derivatives}: \qquad \frac{\partial}{\partial z} \equiv \frac{1}{2}\left(\frac{\partial}{\partial x} - i\frac{\partial}{\partial y}\right), \qquad \frac{\partial}{\partial z^*} \equiv \frac{1}{2}\left(\frac{\partial}{\partial x} + i\frac{\partial}{\partial y}\right), \qquad \left(\frac{\partial}{\partial z}\right)^* = \frac{\partial}{\partial z^*}. \tag{A.2.20}$$

Define $f(z, z^*) \equiv g(x, y)$ noting that $x = \frac{z+z^*}{2}, y = \frac{z-z^*}{2i}$, then

$$df = (\partial g/\partial x)dx + (\partial g/\partial y)dy = (\partial f/\partial z)dz + (\partial f/\partial z^*)dz^*. \tag{A.2.21}$$

The proofs of the following results are left as an exercise:

$\partial(cz)/\partial z = \partial(cz^*)/\partial z^* = c$ and $\partial(cz^*)/\partial z = \partial(cz)/\partial z^* = 0$ for any $c \in \mathbb{C}$,

$\partial(zz^*)/\partial z = z^*$ and $\partial(zz^*)/\partial z^* = z$,

$(\partial/\partial z)[f_1(z)f_2(z)] = (\partial f_1/\partial z)f_2(z) + f_1(z)(\partial f_2/\partial z)$,

$(\partial/\partial z^*)[f_1(z)f_2(z)] = (\partial f_1/\partial z^*)f_2(z) + f_1(z)(\partial f_2/\partial z^*)$.

For a function of a function of z, $f(z) = w(v(z)) = (w \circ v)(z)$, we find that

$(\partial f/\partial z) = (\partial w/\partial v')|_{v'=v(z)}(\partial v/\partial z) + (\partial w/\partial v'^*)|_{v'^*=v^*(z)}(\partial v^*/\partial z)$,

$(\partial f/\partial z^*) = (\partial w/\partial v')|_{v'=v(z)}(\partial v/\partial z^*) + (\partial w/\partial v'^*)|_{v'^*=v^*(z)}(\partial v^*/\partial z^*)$,

$[\partial f/dz]^* = \partial f^*/dz^*$ and $[\partial f/dz^*]^* = \partial f^*/dz$.

So the product rule and chain rule apply as for normal derivatives. The extension to multiple complex variables, $z_1, \ldots, z_N \in \mathbb{C}$, is straightforward using the fact that $\partial z_i/\partial z_j = \partial z_i^*/\partial z_j^* = \delta_{ij}$ and $\partial z_i/\partial z_j^* = \partial z_i^*/\partial z_j = 0$.

A.2.6 Exactness, Conservative Vector Fields and Integrating Factors

Let $\vec{V} \equiv \vec{V}(\vec{q})$ be a function on an N-dimensional manifold with coordinates $\vec{q}$. A differential $\vec{V} \cdot d\vec{q}$ is said to be *exact* if there exists a function $\phi(\vec{q})$ such that

$$d\phi = \vec{V} \cdot d\vec{q} \quad \text{or equivalently} \quad V_j = \frac{\partial \phi}{\partial q_j}. \tag{A.2.22}$$

If and only if the condition

$$\frac{\partial V_i}{\partial q_j} - \frac{\partial V_j}{\partial q_i} = 0 \quad \text{for all} \quad i, j = 1, 2, \ldots, N \tag{A.2.23}$$

holds throughout a simply connected (no holes) region does there exist a doubly differentiable function $\phi(\vec{q})$ such that $V_i = \partial\phi/\partial q_i$ in that region (i.e., such that $\vec{V} \cdot d\vec{q}$ is an exact differential). This is a special case of the *Poincaré lemma*. For example, if $\nabla \times \mathbf{V}(\mathbf{x}) = 0$ for every $\mathbf{x} \in \mathbb{R}^3$, Eq. (A.2.23) applies and there is a $\phi(\mathbf{x})$ such that $\mathbf{V} = \nabla\phi$. We say that such a $\mathbf{V}$ is a *conservative vector field* and that ϕ is its *scalar potential*. Stokes' theorem implies that the integral of a conservative vector field along any path depends only on the end points,

$$\int_p \mathbf{V} \cdot d\mathbf{x} = \phi(\mathbf{x}_f) - \phi(\mathbf{x}_i). \tag{A.2.24}$$

A function $\mu(\vec{q})$ is defined to be an *integrating factor* for some differential $\vec{V}(\vec{q}) \cdot d\vec{q}$ if the modified differential $\vec{V}'(\vec{q}) \cdot d\vec{q} \equiv \mu(\vec{q})\vec{V}(\vec{q}) \cdot d\vec{q}$ is exact.

A.2.7 Tensor and Exterior Products

Tensor product of vector spaces: Let V and W be vector spaces over the field F, where typically here $F = \mathbb{C}$. The *tensor product* of these two vector spaces is denoted as $V \otimes W$. The tensor product of two vectors, $\mathbf{v} \otimes \mathbf{w}$, behaves "as a product should" and is defined by the tensor properties:

(i) $\lambda(\mathbf{v} \otimes \mathbf{w}) = (\lambda \mathbf{v}) \otimes \mathbf{w} = \mathbf{v} \otimes (\lambda \mathbf{w})$

(ii) $(\mathbf{v}_1 + \mathbf{v}_2) \otimes \mathbf{w} = \mathbf{v}_1 \otimes \mathbf{w} + \mathbf{v}_2 \otimes \mathbf{w}$

(iii) $\mathbf{v} \otimes (\mathbf{w}_1 + \mathbf{w}_2) = \mathbf{v} \otimes \mathbf{w}_1 + \mathbf{v} \otimes \mathbf{w}_2$

for all $\mathbf{v}, \mathbf{v}_1, \mathbf{v}_2 \in V$, for all $\mathbf{w}, \mathbf{w}_1, \mathbf{w}_2 \in W$, and for all $\lambda \in F$. The tensor product of vector spaces is defined as the set of all possible linear combinations of tensors $\mathbf{v} \otimes \mathbf{w}$. It is easily seen that $V \otimes W$ is a vector space. Note that physicists typically write the tensor product in terms of kets and we will understand the following equivalent notations for elements of $V \otimes W$,

$$|\mathbf{v}\mathbf{w}\rangle \equiv |\mathbf{v}, \mathbf{w}\rangle \equiv |\mathbf{v}; \mathbf{w}\rangle \equiv |\mathbf{v}\rangle|\mathbf{w}\rangle \equiv \mathbf{v} \otimes \mathbf{w} \in V \otimes W. \tag{A.2.25}$$

The inner (or scalar) product on $V \otimes W$ is

$$\text{math notation:} \quad (\mathbf{v}_1 \otimes \mathbf{w}_1, \mathbf{v}_2 \otimes \mathbf{w}_2) \equiv (\mathbf{v}_1, \mathbf{v}_2)(\mathbf{w}_1, \mathbf{w}_2),$$

$$\text{bra-ket notation:} \quad \langle \mathbf{v}_1 \otimes \mathbf{w}_1 | \mathbf{v}_2 \otimes \mathbf{w}_2 \rangle \equiv \langle \mathbf{v}_2; \mathbf{w}_2 | \mathbf{v}_1; \mathbf{w}_1 \rangle \equiv \langle \mathbf{v}_2 | \mathbf{v}_1 \rangle \langle \mathbf{w}_2 | \mathbf{w}_1 \rangle. \tag{A.2.26}$$

The tensor products of vector spaces, in particular, Hilbert spaces, play a central role throughout quantum mechanics and quantum field theory.

Exterior or wedge product: This is important for building Hilbert spaces for fermions. The exterior product of k copies of the vector space V is referred to as the *kth exterior power* of V and is denoted as $\bigwedge^k V$. This space is a vector space spanned by the set of all completely antisymmetric tensors of the form $\mathbf{v}_1 \wedge \cdots \wedge \mathbf{v}_k$ for $\mathbf{v}_1, \ldots, \mathbf{v}_k \in V$. We refer to "$\wedge$" as the *exterior product* or *wedge product*. The properties of $\mathbf{v}_1 \wedge \cdots \wedge \mathbf{v}_k$ are:

(i) it changes sign under the pairwise interchange of any two $\mathbf{v}_i$ and $\mathbf{v}_j$

(ii) $\mathbf{v}_1 \wedge \cdots \wedge (\lambda \mathbf{v}_i) \wedge \cdots \wedge \mathbf{v}_k = \lambda(\mathbf{v}_1 \wedge \cdots \wedge \mathbf{v}_i \wedge \cdots \wedge \mathbf{v}_k)$

(iii) $\mathbf{v}_1 \wedge \cdots \wedge (\mathbf{v}_i + \mathbf{v}_i') \wedge \cdots \wedge \mathbf{v}_k = (\mathbf{v}_1 \wedge \cdots \wedge \mathbf{v}_i \wedge \cdots \wedge \mathbf{v}_k) + (\mathbf{v}_1 \wedge \cdots \wedge \mathbf{v}_i' \wedge \cdots \wedge \mathbf{v}_k)$

(iv) $(\mathbf{v}_1 \wedge \cdots \wedge \mathbf{v}_j) \wedge (\mathbf{u}_1 \wedge \cdots \wedge \mathbf{u}_k) = \mathbf{v}_1 \wedge \cdots \wedge \mathbf{v}_j \wedge \mathbf{u}_1 \wedge \cdots \wedge \mathbf{u}_k$.

Let $\sigma(1), \ldots, \sigma(k)$ denote any permutation of the indices $1, \ldots, k$. The complete antisymmetry of the exterior product gives

$$\mathbf{v}_{\sigma(1)} \wedge \cdots \wedge \mathbf{v}_{\sigma(k)} = \epsilon^{\sigma(1) \cdots \sigma(k)} \mathbf{v}_1 \wedge \cdots \wedge \mathbf{v}_k = \text{sgn}(\sigma) \mathbf{v}_1 \wedge \cdots \wedge \mathbf{v}_k.$$

If any two of the vectors are the same, then $\mathbf{v}_{i_1} \wedge \cdots \wedge \mathbf{v}_{i_k} = 0$. Note that

$$\text{sgn}(\sigma) = \epsilon^{\sigma(1)\sigma(2)\cdots\sigma(k)}$$

is the sign of the permutation of the indices $1, 2, \ldots, k$ with $\text{sgn}(\sigma) = +1$ or -1 for even or odd permutations, respectively. Since $\text{sgn}(\sigma) = 1/\text{sgn}(\sigma)$, then we also have

$$\mathbf{v}_1 \wedge \cdots \wedge \mathbf{v}_k = \text{sgn}(\sigma) \mathbf{v}_{\sigma(1)} \wedge \cdots \wedge \mathbf{v}_{\sigma(k)}$$

for any given permutation σ. The k vectors $\mathbf{v}_1, \ldots, \mathbf{v}_k \in V$ span a k-dimensional subspace of V if they are a linearly independent set. Let $\mathbf{e}_1, \ldots, \mathbf{e}_k$ be any basis spanning a k-dimensional subspace containing these vectors, then we can write $\mathbf{v}_i = \sum_{j=1}^k C_{ij} \mathbf{e}_j$ and

$$\mathbf{v}_1 \wedge \mathbf{v}_2 \wedge \cdots \wedge \mathbf{v}_k = \sum_{j_1,\ldots,j_k=1}^{k} C_{1j_1} C_{2j_2} \cdots C_{kj_k} \mathbf{e}_{j_1} \wedge \mathbf{e}_{j_2} \wedge \cdots \wedge \mathbf{e}_{j_k} \tag{A.2.27}$$

$$= \sum_{j_1,\ldots,j_k=1}^{k} \epsilon^{j_1 j_2 \cdots j_k} C_{1j_1} C_{2j_2} \cdots C_{kj_k} \mathbf{e}_1 \wedge \mathbf{e}_2 \wedge \cdots \wedge \mathbf{e}_k$$

$$= \det C \, \mathbf{e}_1 \wedge \mathbf{e}_2 \wedge \cdots \wedge \mathbf{e}_k = \sum_{\sigma} \mathrm{sgn}(\sigma) \, \mathbf{v}_{\sigma(1)} \otimes \cdots \otimes \mathbf{v}_{\sigma(k)},$$

where we have used Eq. (A.8.1) for the determinant and where we sum over permutations σ. If the $\mathbf{v}_i$ are not linearly independent, then the $k \times k$ matrix C will not have rank k so that $\det C = 0$ and $\mathbf{v}_1 \wedge \mathbf{v}_2 \wedge \cdots \wedge \mathbf{v}_k = 0$ as it should.

Since all permutations of the $\mathbf{e}_i$ in $\mathbf{e}_1 \wedge \mathbf{e}_2 \wedge \cdots \wedge \mathbf{e}_k$ are equal up to a sign, then linearly independent exterior products of the basis vectors contain different basis vectors. The number of ways to choose k basis vectors out of the n basis vectors of V is $\binom{n}{k} = n!/k!(n-k)!$ and so this is the number of linearly independent $\mathbf{e}_{j_1} \wedge \mathbf{e}_{j_2} \wedge \cdots \wedge \mathbf{e}_{j_k}$. Then $\dim \bigwedge^k = n!/k!(n-k)!$, e.g., $\dim \bigwedge^1 V = n$, $\dim \bigwedge^2 V = n(n-1)/2$, and $\dim \bigwedge^n V = 1$. For $k > n$ all exterior products $\mathbf{v}_1 \wedge \mathbf{v}_2 \wedge \cdots \wedge \mathbf{v}_k$ must vanish, since the maximum number of linearly independent vectors in V is n. If $V = \mathbb{R}^n$ then an n-dimensional parallelepiped with edges $\mathbf{c}_1, \ldots, \mathbf{c}_n$ has volume $V_{\mathrm{vol}} = |\det C|$, where C is the matrix with elements c_{ij} and where $\mathbf{c}_i = \sum_{j=1}^{n} c_{ij} \mathbf{e}_j$ with $\{\mathbf{e}_1, \ldots, \mathbf{e}_n\}$ an orthonormal basis. Then $\mathbf{c}_1 \wedge \cdots \wedge \mathbf{c}_n = \det C \, \mathbf{e}_1 \wedge \cdots \wedge \mathbf{e}_n$. For this reason $\mathbf{c}_1 \wedge \cdots \wedge \mathbf{c}_n$ is sometimes referred to as the *volume tensor*.

We can embed the $\bigwedge^k V$ into the tensor product space $V^{\otimes n}$. For $k = n$ it is the antisymmetric one-dimensional subspace of $V^{\otimes n}$. For any $k \leq n$ we can identify

$$\mathbf{v}_1 \wedge \cdots \wedge \mathbf{v}_k = \sum_{\sigma} \mathrm{sgn}(\sigma) \, \mathbf{v}_{\sigma(1)} \otimes \cdots \otimes \mathbf{v}_{\sigma(k)}, \tag{A.2.28}$$

where we are summing over all permutations σ. This reproduces the required properties. Note that the normalization on the right-hand side is a matter of convention. We have made one simple and common choice. The other common choice is to include a factor $1/k!$ on the right-hand side. The normalization choice is relevant when evaluating inner products and defining an orthonormal basis. The inner product for tensor product spaces was given above. From this it follows that

$$(1/k!)\langle \mathbf{v}_1 \wedge \cdots \wedge \mathbf{v}_k | \mathbf{w}_1 \wedge \cdots \wedge \mathbf{w}_k \rangle = \det C \quad \text{with} \quad C_{ij} \equiv \langle \mathbf{v}_i | \mathbf{w}_j \rangle \tag{A.2.29}$$

and with C being a $k \times k$ matrix. Then the set of all $(1/\sqrt{k!})\mathbf{e}_{i_1} \wedge \cdots \wedge \mathbf{e}_{i_k}$ with the ordering such that $i_1 \leq i_2 \leq \cdots \leq i_k$ form an orthonormal basis of $\bigwedge^k V$, since

$$(1/k!)\langle \mathbf{e}_{i_1} \wedge \cdots \wedge \mathbf{e}_{i_k} | \mathbf{e}_{j_1} \wedge \cdots \wedge \mathbf{e}_{j_k} \rangle = \delta_{i_1 j_1} \cdots \delta_{i_k j_k} \tag{A.2.30}$$

and since this set spans $\bigwedge^k V$. The direct sum of all exterior products of V is

$$F_-(V) = \bigoplus_{k=0}^{\infty} \bigwedge^k V \tag{A.2.31}$$

and is called the *exterior algebra* or *Grassmann algebra* of the vector space V. The product operation of the algebra is the exterior product.

A.3 Dirac Algebra

Pauli spin matrices:
$$\sigma^1 = \begin{pmatrix} 0 & 1 \\ 1 & 0 \end{pmatrix}, \quad \sigma^2 = \begin{pmatrix} 0 & -i \\ i & 0 \end{pmatrix}, \quad \sigma^3 = \begin{pmatrix} 1 & 0 \\ 0 & -1 \end{pmatrix}, \tag{A.3.1}$$

$$\mathrm{tr}\sigma^i = 0, \quad (\sigma^i)^\dagger = \sigma^i, \quad [\sigma^i, \sigma^j] = 2i\epsilon^{ijk}\sigma^k, \quad \{\sigma^i, \sigma^j\} = 2\delta^{ij}, \tag{A.3.2}$$

$$\sigma^i\sigma^j = \delta^{ij} + i\epsilon^{ijk}\sigma^k, \quad (\boldsymbol{\sigma}\cdot\mathbf{a})(\boldsymbol{\sigma}\cdot\mathbf{b}) = \mathbf{a}\cdot\mathbf{b} + i\boldsymbol{\sigma}\cdot(\mathbf{a}\times\mathbf{b}), \quad \sigma^\pm \equiv (1/2)(\sigma^1 \pm i\sigma^2).$$

Dirac γ-matrices:
$$\{\gamma^\mu, \gamma^\nu\} \equiv \gamma^\mu\gamma^\nu + \gamma^\nu\gamma^\mu = 2g^{\mu\nu} \Rightarrow (\gamma^0)^2 = -(\gamma^i)^2 = I, \tag{A.3.3}$$

$$\gamma^0\gamma^i = -\gamma^i\gamma^0, \quad \gamma^0\gamma^i\gamma^0 = -\gamma^i, \quad \gamma^i\gamma^0\gamma^i = \gamma^0 \Rightarrow \mathrm{tr}\gamma^i = \mathrm{tr}\gamma^0 = 0, \tag{A.3.4}$$

$$\gamma^5 \equiv \gamma_5 \equiv i\gamma^0\gamma^1\gamma^2\gamma^3 = \frac{-i}{4!}\epsilon_{\mu\nu\rho\sigma}\gamma^\mu\gamma^\nu\gamma^\rho\gamma^\sigma, \quad \sigma^{\mu\nu} \equiv \frac{i}{2}[\gamma^\mu, \gamma^\nu], \tag{A.3.5}$$

$$\{\gamma^\mu, \gamma^5\} = 0, \quad (\gamma^5)^2 = I, \quad \gamma^\mu\gamma^\nu = \tfrac{1}{2}\{\gamma^\mu, \gamma^\nu\} + \tfrac{1}{2}[\gamma^\mu, \gamma^\nu] = g^{\mu\nu} - i\sigma^{\mu\nu}, \tag{A.3.6}$$

$$\not{k}\not{\ell} = k\cdot\ell - i\sigma_{\mu\nu}k^\mu\ell^\nu, \quad [\gamma^5, \sigma^{\mu\nu}] = 0 \quad \text{and} \quad \gamma^5\sigma^{\mu\nu} = \frac{i}{2}\epsilon^{\mu\nu\rho\sigma}\sigma_{\rho\sigma}. \tag{A.3.7}$$

Dirac representation:
$$\gamma^{0\dagger} = \gamma^0, \quad \gamma^{i\dagger} = -\gamma^i, \quad \gamma^{\mu\dagger} = \gamma_\mu = (\gamma^\mu)^{-1}, \tag{A.3.8}$$

$$\gamma^0 = \begin{pmatrix} I & 0 \\ 0 & -I \end{pmatrix}, \quad \boldsymbol{\gamma} = \begin{pmatrix} 0 & \boldsymbol{\sigma} \\ -\boldsymbol{\sigma} & 0 \end{pmatrix}, \quad \gamma^5 = \begin{pmatrix} 0 & I \\ I & 0 \end{pmatrix}, \tag{A.3.9}$$

$$\sigma^{0i} = \begin{pmatrix} 0 & i\sigma^i \\ i\sigma^i & 0 \end{pmatrix}, \quad \sigma^{ij} = \epsilon_{ijk}\begin{pmatrix} \sigma^k & 0 \\ 0 & \sigma^k \end{pmatrix}, \tag{A.3.10}$$

$$\gamma^{5\dagger} = \gamma^5, \quad \gamma^0\gamma^\mu\gamma^0 = \gamma^{\mu\dagger}, \quad \gamma^0\gamma^5\gamma^0 = -\gamma^{5\dagger} = -\gamma^5, \quad \gamma^0\sigma^{\mu\nu}\gamma^0 = (\sigma^{\mu\nu})^\dagger, \tag{A.3.11}$$

$$\gamma^0\left(\gamma^5\gamma^\mu\right)\gamma^0 = \left(\gamma^5\gamma^\mu\right)^\dagger. \tag{A.3.12}$$

Similar and unitarily equivalent matrices: Let S be nonsingular ($\det S \neq 0$) and U be unitary ($U^\dagger = U^{-1}$). Then $A' \equiv SAS^{-1}$ is *similar* to A and $\bar{A} \equiv UAU^{-1}$ is *unitarily equivalent* to A. $\bar{A}$ is a special case of an A'. Any representation corresponds to a choice of basis in the vector space and so any change of basis is a similarity transformation. All algebraic relations containing only γ-matrices (with no $\gamma_\mu^\dagger$ or $\gamma^{\mu T}$) are true in any representation. Those involving both the γ_μ and $\gamma_\mu^\dagger$ remain true in unitarily equivalent representations. Both the chiral (or Weyl) (γ_χ^μ) and the Majorana (γ_M^μ) representations are unitarily equivalent to the Dirac (γ_D^μ) representation, so Eqs. (A.3.8), (A.3.11) and (A.3.12) remain valid.

Chiral representation:
$$\gamma_\chi^\mu = U\gamma_D^\mu U^\dagger \quad \text{with} \quad U = \frac{1}{\sqrt{2}}\begin{pmatrix} I & -I \\ I & I \end{pmatrix}, \quad U^\dagger = U^T, \tag{A.3.13}$$

$$\gamma_\chi^0 = \begin{pmatrix} 0 & I \\ I & 0 \end{pmatrix}, \quad \boldsymbol{\gamma}_\chi = \begin{pmatrix} 0 & \boldsymbol{\sigma} \\ -\boldsymbol{\sigma} & 0 \end{pmatrix}, \quad \gamma_\chi^5 = \begin{pmatrix} -I & 0 \\ 0 & I \end{pmatrix}, \tag{A.3.14}$$

$$\sigma^\mu \equiv (I, \boldsymbol{\sigma}), \quad \bar{\sigma}^\mu \equiv (I, -\boldsymbol{\sigma}), \quad \mathrm{tr}(\sigma^\mu\bar{\sigma}^\nu) = 2g^{\mu\nu}, \quad \gamma_\chi^\mu = \begin{pmatrix} 0 & \sigma^\mu \\ \bar{\sigma}^\mu & 0 \end{pmatrix}, \tag{A.3.15}$$

$$\sigma_\chi^{0i} = i\begin{pmatrix} -\sigma^i & 0 \\ 0 & \sigma^i \end{pmatrix}, \quad \sigma_\chi^{ij} = \epsilon_{ijk}\begin{pmatrix} \sigma^k & 0 \\ 0 & \sigma^k \end{pmatrix}. \tag{A.3.16}$$

Majorana repre- $\quad\gamma_M^\mu = U\gamma_D^\mu U^\dagger \quad$ with $\quad U = U^\dagger = \dfrac{1}{\sqrt{2}}\begin{pmatrix} I & \sigma^2 \\ \sigma^2 & -I \end{pmatrix},$ (A.3.17)
sentation:

$$\gamma_M^0 = \begin{pmatrix} 0 & \sigma^2 \\ \sigma^2 & 0 \end{pmatrix},\ \gamma_M^1 = \begin{pmatrix} i\sigma^3 & 0 \\ 0 & i\sigma^3 \end{pmatrix},\ \gamma_M^2 = \begin{pmatrix} 0 & -\sigma^2 \\ \sigma^2 & 0 \end{pmatrix},\ \gamma_M^3 = \begin{pmatrix} -i\sigma^1 & 0 \\ 0 & -i\sigma^1 \end{pmatrix},$$

$$\gamma_M^{\mu*} = -\gamma_M^\mu,\quad \gamma_M^5 = \begin{pmatrix} \sigma^2 & 0 \\ 0 & -\sigma^2 \end{pmatrix}.$$ (A.3.18)

Charge conjugation $\quad (C\gamma^0)\gamma^{\mu*}(C\gamma^0)^{-1} \equiv -\gamma^\mu \quad$ from Eq. (4.5.22), (A.3.19)
matrix:

For $\gamma_D^\mu, \gamma_\chi^\mu, \gamma_M^\mu$: $\quad C^{-1}\gamma^\mu C = -\gamma^{\mu T},\quad C^{-1}\gamma^5 C = \gamma^{5T},$ (A.3.20)

$$C^{-1}\sigma^{\mu\nu}C = -\sigma^{\mu\nu T},\quad C^{-1}\left(\gamma^5\gamma^\mu\right)C = \left(\gamma^5\gamma^\mu\right)^T,$$ (A.3.21)

$$C^{-1} = C^T = C^\dagger = -C,\quad C^2 = -I,\quad C^\dagger C = CC^\dagger = I,$$ (A.3.22)

$$C_D = C_M = \begin{pmatrix} 0 & -i\sigma^2 \\ -i\sigma^2 & 0 \end{pmatrix},\quad C_\chi = \begin{pmatrix} i\sigma^2 & 0 \\ 0 & -i\sigma^2 \end{pmatrix}.$$ (A.3.23)

Trace identities: Identities with only γ-matrices (no $\gamma_\mu^\dagger$ or γ_μ^T) are representation independent, since $\mathrm{tr}(\gamma'^{\mu_1}\cdots\gamma'^{\mu_n}) = \mathrm{tr}(S\gamma^{\mu_1}S^{-1}\cdots S\gamma^{\mu_n}S^{-1}) = \mathrm{tr}(\gamma^{\mu_1}\cdots\gamma^{\mu_n})$.

$$\mathrm{tr}\,(I) = 4,\quad \mathrm{tr}\,(\gamma^\mu\gamma^\nu) = 4g^{\mu\nu},\quad \mathrm{tr}\,(\gamma^\mu) = \mathrm{tr}\,(\text{odd \# of } \gamma^{\mu\prime}\text{s}) = \mathrm{tr}\,(\sigma^{\mu\nu}) = 0.$$ (A.3.24)

Traces with four or more γ-matrices:

$$\mathrm{tr}\,(\gamma^\mu\gamma^\nu\gamma^\rho\gamma^\sigma) = 4\left(g^{\mu\nu}g^{\rho\sigma} - g^{\mu\rho}g^{\nu\sigma} + g^{\mu\sigma}g^{\nu\rho}\right),$$ (A.3.25)

$$\mathrm{tr}\,(\gamma^\mu\gamma^\nu\sigma^{\rho\sigma}) = 4i\left(-g^{\mu\rho}g^{\nu\sigma} + g^{\mu\sigma}g^{\nu\rho}\right),\ \mathrm{tr}\,(\sigma^{\mu\nu}\sigma^{\rho\sigma}) = 4\left(g^{\mu\rho}g^{\nu\sigma} - g^{\nu\rho}g^{\mu\sigma}\right),$$ (A.3.26)

$$\mathrm{tr}\,(\gamma^{\mu_1}\cdots\gamma^{\mu_{2N}}) = g^{\mu_1\mu_2}\mathrm{tr}\,(\gamma^{\mu_3}\gamma^{\mu_4}\cdots\gamma^{\mu_{2N}}) - g^{\mu_1\mu_3}\mathrm{tr}\,(\gamma^{\mu_2}\gamma^{\mu_4}\cdots\gamma^{\mu_{2N}})$$ (A.3.27)

$$+ g^{\mu_1\mu_4}\mathrm{tr}\,(\gamma^{\mu_2}\gamma^{\mu_3}\gamma^{\mu_5}\cdots\gamma^{\mu_{2N}})\cdots + g^{\mu_1\mu_{2N}}\mathrm{tr}\,(\gamma^{\mu_2}\gamma^{\mu_3}\cdots\gamma^{\mu_{2N-1}})$$

$$= \sum_{1=i_1<i_2<\cdots<i_N;i_k<j_k}^{N} 4\epsilon^{i_1 j_1\cdots i_N j_N}g^{\mu_{i_1}\mu_{j_1}}\cdots g^{\mu_{i_N}\mu_{j_N}},$$

$$\mathrm{tr}\,(\sigma^{\mu_1\mu_2}\sigma^{\mu_3\mu_4}\sigma^{\mu_5\mu_6}) = -4i\,[g^{\mu_1\mu_3}(g^{\mu_2\mu_5}g^{\mu_4\mu_6} - g^{\mu_2\mu_6}g^{\mu_4\mu_5})$$ (A.3.28)

$$+ g^{\mu_1\mu_4}(g^{\mu_2\mu_6}g^{\mu_3\mu_5} - g^{\mu_2\mu_5}g^{\mu_3\mu_6}) + g^{\mu_1\mu_5}(g^{\mu_2\mu_4}g^{\mu_3\mu_6} - g^{\mu_2\mu_3}g^{\mu_4\mu_6})$$

$$+ g^{\mu_1\mu_6}(g^{\mu_2\mu_3}g^{\mu_4\mu_5} - g^{\mu_2\mu_4}g^{\mu_3\mu_5})]$$

$$\mathrm{tr}\,(\gamma^{\mu_1}\gamma^{\mu_2}\cdots\gamma^{\mu_N}) = \mathrm{tr}\,(\gamma^{\mu_N}\cdots\gamma^{\mu_2}\gamma^{\mu_1}).$$ (A.3.29)

Explanation of the last result: If $\{\gamma^\mu,\gamma^\nu\} = 2g^{\mu\nu}I$ then $\{\gamma^{\mu T},\gamma^{\nu T}\} = 2g^{\mu\nu}I$ and so γ^μ and $\gamma^{\mu T}$ are both representations. Since $\mathrm{tr}\,(\gamma^{\mu_1}\gamma^{\mu_2}\cdots\gamma^{\mu_N})$ can be evaluated using only $\{\gamma^\mu,\gamma^\nu\} = 2g^{\mu\nu}I$, then it is representation independent. Since $\mathrm{tr}A = \mathrm{tr}A^T$ then $\mathrm{tr}\,(\gamma^{\mu_1}\gamma^{\mu_2}\cdots\gamma^{\mu_N}) = \mathrm{tr}\,(\gamma^{\mu_1 T}\gamma^{\mu_2 T}\cdots\gamma^{\mu_N T}) = \mathrm{tr}\,(\gamma^{\mu_N}\cdots\gamma^{\mu_2}\gamma^{\mu_1})$.

Traces involving γ_5:

$$\mathrm{tr}\,(\gamma^5) = \mathrm{tr}\,(\gamma^5\gamma^\mu) = \mathrm{tr}\,(\gamma^5\gamma^\mu\gamma^\nu) = \mathrm{tr}\,(\gamma^5(\text{odd \# of } \gamma^{\mu\prime}\text{s})) = 0,$$ (A.3.30)

$$\mathrm{tr}\,(\gamma^5\gamma^\mu\gamma^\nu\gamma^\rho\gamma^\sigma) = -4i\epsilon^{\mu\nu\rho\sigma}.$$

Traces with an even number of γ^5-matrices reduce to a trace with none since γ^5 anticommutes with all γ-matrices and $(\gamma^5)^2 = I$. A trace with an odd number of γ^5 matrices reduces to a trace with one. To reduce numbers of γ-matrices use

$$\gamma_\mu\gamma_\nu\gamma_\rho = g_{\mu\nu}\gamma_\rho - g_{\mu\rho}\gamma_\nu + g_{\nu\rho}\gamma_\mu + i\epsilon_{\mu\nu\rho\sigma}\gamma^\sigma\gamma^5.$$ (A.3.31)

Contraction identities

$$\gamma^\mu \gamma_\mu = 4I, \quad \gamma^\mu \gamma^5 \gamma_\mu = -4\gamma^5, \quad \gamma^\mu \not{k} \gamma_\mu = -2\not{k}, \quad \gamma^\mu \gamma^5 \not{k} \gamma_\mu = 2\gamma^5 \not{k}, \tag{A.3.32}$$

$$\gamma^\mu \not{k} \not{\ell} \gamma_\mu = 4k \cdot \ell, \quad \gamma^\mu \not{k} \not{\ell} \not{p} \gamma_\mu = -2\not{p} \not{\ell} \not{k}, \quad \gamma^\mu \not{k} \not{\ell} \not{p} \not{q} \gamma_\mu = 2 \left(\not{q} \not{k} \not{\ell} \not{p} + \not{p} \not{\ell} \not{k} \not{q} \right), \tag{A.3.33}$$

$$\gamma^\mu \sigma^{\rho\sigma} \gamma_\mu = 0, \quad \sigma^{\mu\nu} \sigma_{\mu\nu} = 12, \quad \gamma^\mu \sigma^{\rho\sigma} \gamma^\lambda \gamma_\mu = 2\gamma^\lambda \sigma^{\rho\sigma}, \quad \gamma^\mu \gamma^\lambda \sigma^{\rho\sigma} \gamma_\mu = 2\sigma^{\rho\sigma} \gamma^\lambda, \tag{A.3.34}$$

$$\sigma^{\mu\nu} \gamma^\rho \sigma_{\mu\nu} = 0, \quad \sigma^{\mu\nu} \gamma^\rho \gamma^\sigma \sigma_{\mu\nu} = 4 \left(4g^{\rho\sigma} - \gamma^\rho \gamma^\sigma \right), \quad \sigma^{\mu\nu} \sigma^{\rho\sigma} \sigma_{\mu\nu} = -4\sigma^{\rho\sigma}. \tag{A.3.35}$$

Dirac spinor identities

$$(\not{p} - m)u^s(p) = 0, \quad (\not{p} + m)v^s(p) = 0, \quad p^2 = m^2, \quad p^\mu = (p^0, \mathbf{p}) = (E_{\mathbf{p}}, \mathbf{p}), \tag{A.3.36}$$

$$\bar{u}^s(p) \equiv u^{\dagger s}(p)\gamma^0, \quad \bar{v}^s(p) \equiv v^{\dagger s}(p)\gamma^0, \quad \bar{u}^s(p)(\not{p} - m) = 0, \quad \bar{v}^s(p)(\not{p} + m) = 0. \tag{A.3.37}$$

Spinor normalization is that of Peskin and Schroeder (1995) and Schwartz (2013),

$$\bar{u}^r(p)u^s(p) = 2m\delta^{rs}, \quad \bar{v}^r(p)v^s(p) = -2m\delta^{rs}, \quad \bar{u}^r(p)v^s(p) = \bar{v}^r(p)u^s(p) = 0, \tag{A.3.38}$$

$$\bar{u}^s(p)\gamma^0 = u^{\dagger s}(p) = \bar{u}^s(\tilde{p}), \quad \bar{v}^s(p)\gamma^0 = v^{\dagger s}(p) = -\bar{v}^s(\tilde{p}) \text{ with } \tilde{p}^\mu \equiv (E_{\mathbf{p}}, -\mathbf{p}), \tag{A.3.39}$$

$$\bar{u}^r(p)\gamma^0 u^s(p) = u^{r\dagger}(p)u^s(p) = \bar{u}^r(\tilde{p})u^s(p) = 2E_{\mathbf{p}}\delta^{rs}, \tag{A.3.40}$$

$$\bar{v}^r(p)\gamma^0 v^s(p) = v^{r\dagger}(p)v^s(p) = -\bar{v}^r(\tilde{p})v^s(p) = 2E_{\mathbf{p}}\delta^{rs}, \tag{A.3.41}$$

$$u^{r\dagger}(\tilde{p})v^s(p) = v^{r\dagger}(\tilde{p})u^s(p) = 0, \tag{A.3.42}$$

where $\gamma^0 \widetilde{\not{p}} \gamma^0 = \not{p}$ leads to $\bar{u}^s(p)\gamma^0 = u^{\dagger s}(p) = \bar{u}^s(\tilde{p})$ and $\bar{v}^s(p)\gamma^0 = v^{\dagger s}(p) = -\bar{v}^s(\tilde{p})$. Sums over fermion polarizations give

$$\sum_s u^s(p)\bar{u}^s(p) = \not{p} + m, \quad \sum_s v^s(p)\bar{v}^s(p) = \not{p} - m. \tag{A.3.43}$$

Defining $q^\mu \equiv p'^\mu - p^\mu$ we have the *Gordon identity* (or Gordon decomposition),

$$\bar{u}^{s'}(p')\gamma^\mu u^s(p) = \bar{u}^{s'}(p') \left[\frac{p'^\mu + p^\mu}{2m} + \frac{i\sigma^{\mu\nu}q_\nu}{2m} \right] u^s(p). \tag{A.3.44}$$

Proof: Note that $\{\gamma^\mu, \gamma^\nu\} = 2g^{\mu\nu}$ and so $[\gamma^\mu, \gamma^\nu] = 2g^{\mu\nu} - 2\gamma^\nu \gamma^\mu$, which can be rewritten as $i\sigma^{\mu\nu} = \gamma^\nu \gamma^\mu - g^{\mu\nu}$. Noting the antisymmetric nature of $\sigma^{\mu\nu}$ we also have $i\sigma^{\mu\nu} = -i\sigma^{\nu\mu} = g^{\nu\mu} - \gamma^\mu \gamma^\nu$. Using these results gives $\bar{u}^{s'}(p')i\sigma^{\mu\nu}(p'_\nu - p_\nu)u^s(p) = \bar{u}^{s'}(p')[(\gamma^\nu \gamma^\mu - g^{\mu\nu})p'_\nu - (g^{\nu\mu} - \gamma^\mu \gamma^\nu)p_\nu]u^s(p) = \bar{u}^{s'}(p')[\not{p}'\gamma^\mu + \gamma^\mu \not{p} - (p'^\mu + p^\mu)]u^s(p) = \bar{u}^{s'}(p')[2m\gamma^\mu - (p'^\mu + p^\mu)]u^s(p)$. Re-arranging this equation gives the Gordon identity.

A.4 Euclidean Space Conventions

It is often convenient to analytically continue quantum field theory to Euclidean space, e.g., lattice gauge theory and the evaluation of momentum loop integrals in perturbation theory. The analytic continuation to Euclidean space is

$$x^0 \to -ix_4^E, \qquad \mathbf{x} \equiv (x^1, x^2, x^3) = (x_1^E, x_2^E, x_3^E) \equiv \mathbf{x}^E,$$

$$x^2 = x \cdot x = (x^0)^2 - \mathbf{x}^2 \to -(x^E)^2 = -x^E \cdot x^E = -[(\mathbf{x}^E)^2 + (x_4^E)^2], \tag{A.4.1}$$

and is referred to as the *Wick rotation*. It is a rotation in time in the sense that we rotate x^0 by $90°$ *clockwise* from $\theta = 0$ to $\theta = \pi/2$ in the complex time plane,

$$x^0 \equiv e^{i0}r \rightarrow e^{-i\theta}r \rightarrow e^{-i\pi/2}r = -ir \equiv -ix_4. \tag{A.4.2}$$

The three-vector part involves no rotation and equates Euclidean and Minkowski three-vector quantities. Only the time components of quantities are Wick rotated. The Wick rotation converts the trace of the evolution operator into the partition function of statistical mechanics, i.e., $t/\hbar \rightarrow -i\hbar\beta$ with $\beta \equiv 1/kT$,

$$\mathrm{tr}\, e^{-i\hat{H}t/\hbar} \xrightarrow{t \rightarrow -i\hbar\beta} \mathrm{tr}\, e^{-\beta\hat{H}} = \textstyle\sum_n e^{-\beta E_n} \,, \tag{A.4.3}$$

where here $\hat{H}$ is the Hamiltonian and E_n are the corresponding energy eigenvalues. The Euclidean Dirac matrices can be chosen to be Hermitian and satisfy

$$\gamma_4^E = \gamma^0, \quad \gamma_j^E = -i\gamma^j, \quad \{\gamma_\mu^E, \gamma_\nu^E\} = 2\,\delta_{\mu\nu},$$
$$\gamma_5^E = -\gamma_1^E\,\gamma_2^E\,\gamma_3^E\,\gamma_4^E = \gamma^5 = \gamma_5 = i\,\gamma^0\,\gamma^1\,\gamma^2\,\gamma^3, \tag{A.4.4}$$

so that, for example, $\mathrm{tr}[\gamma_5^E\,\gamma_\lambda^E\,\gamma_\mu^E\,\gamma_\nu^E\,\gamma_\rho^E] = -4\,\epsilon^{\lambda\mu\nu\rho}$ with $\epsilon^{1234} = +1$. Note that

$$\partial_0 \equiv \frac{\partial}{\partial x^0} \rightarrow i\partial_4^E \equiv i\frac{\partial}{\partial x_4^E} \,, \qquad \partial_i \equiv \frac{\partial}{\partial x^i} \equiv \partial_i^E \equiv \frac{\partial}{\partial x_i^E} \tag{A.4.5}$$

and since $k_\mu \leftrightarrow i\partial_\mu$ then the appropriate transcription for momentum is

$$k^0 = k_0 \rightarrow ik_4^E \,, \qquad k^i = -k_i \equiv -k_i^E \,,$$
$$k^2 = k \cdot k = (k^0)^2 - \mathbf{k}^2 \rightarrow -(k^E)^2 = -k^E \cdot k^E, \tag{A.4.6}$$

where we are rotating k^0 *counterclockwise* by $90°$. Since the covariant derivative involves a linear combination of ∂_μ and A_μ we transform A_μ like ∂_μ.

The rotation of x^0 and p^0 are done together and in opposite directions so that during the Wick rotation we have $p^0x^0 \rightarrow (e^{i\theta}p)(e^{-i\theta}x) = px \rightarrow (ip^0)(-ix^0) = p_4x_4$. Then $p \cdot x \in \mathbb{R}$ is fixed throughout the Wick rotation and a Fourier transform remains so throughout. The Feynman prescription for the contour integration is shown in Fig. 6.1. For the Feynman propagator $D_F(x)$ we can deform the integration along the real p^0 axis to one along the imaginary p^0 axis (the p_4 axis), where for each $p \equiv |p^0|$ we hold $p \cdot x$ fixed. The lengths of contours C_1 and C_3 in Fig. A.1 increase

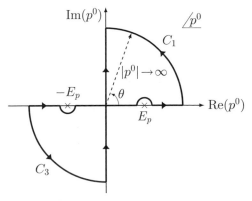

Figure A.1   Wick rotation is allowed if there are no poles or cuts in the first or third quadrant of the complex p^0 plane and if the arcs C_1 and C_3 have a vanishing contribution.

linearly with p while the integrand falls off like $1/p^2$ and so as $p \to \infty$ their contribution vanishes. So the Cauchy integral theorem means that the integral along the real axis has become an integration along the imaginary axis (the p_4 axis) and

$$D_F(x) = \int \frac{d^4p}{(2\pi)^4} \frac{ie^{-ip\cdot x}}{p^2 - m^2 + i\epsilon} \to \int \frac{d^4p^E}{(2\pi)^4} \frac{e^{-ip^E\cdot x^E}}{p^{E2} + m^2} = D_E(x^E). \tag{A.4.7}$$

The Wick-rotated Feynman propagator is the Euclidean propagator $D_E(x^E)$. We can arrive at this with simple transcriptions in Eqs. (A.4.1) and (A.4.6); however, this is only valid since (i) there were no poles or cuts in the first and third quadrants, and (ii) contours C_1 and C_3 at infinity give no contribution. The naive substitution is adequate for pertubative calculations but not nonperturbatively.

A.5 Feynman Parameterization

To simplify the evaluation of loop integrals in perturbation theory there are a number of useful identities as we will see in Sec. A.6. The simplest of these is

$$\frac{1}{A_1 A_2} = \int_0^1 dx_1 dx_2 \frac{\delta(x_1 + x_2 - 1)}{[x_1 A_1 + x_2 A_2]^2} = \int_0^1 dx \frac{1}{[xA_1 + (1-x)A_2]^2}. \tag{A.5.1}$$

The parameters x_1, x_2 and x are all referred to as *Feynman parameters*. The proof of this result is straightforward from right to left using

$$\int_0^1 dx \frac{1}{[ax + b]^2} = \frac{1}{a} \int_b^{a+b} dy \frac{1}{y^2} = \frac{1}{a} \left[-\frac{1}{y}\right]_b^{a+b} = \frac{1}{(a+b)b}, \tag{A.5.2}$$

where we have used the change of variable $y = ax + b$ and where we identify $a = (A_1 - A_2)$ and $b = A_2$. After differentiating Eq. (A.5.1) $n_1 - 1$ times with respect to A_1 and $n_2 - 1$ times with respect to A_2 we immediately obtain

$$\frac{1}{A_1^{n_1} A_2^{n_2}} = \frac{\Gamma(n_1 + n_2)}{\Gamma(n_1)\Gamma(n_2)} \int_0^1 dx_1 dx_2 \, \delta(x_1 + x_2 - 1) \frac{x_1^{n_1-1} x_2^{n_2-1}}{[x_1 A_1 + x_2 A_2]^{n_1+n_2}}, \tag{A.5.3}$$

since

$$\frac{d^{n_1+n_2-2}}{dA_1^{n_1-1} dA_2^{n_2-1}} \frac{1}{A_1 A_2} = (-1)^{n_1+n_2-2} \frac{(n_1-1)!(n_2-1)!}{A_1^{n_1} A_2^{n_2}}, \tag{A.5.4}$$

$$\frac{d^{n_1+n_2-2}}{dA_1^{n_1-1} dA_2^{n_2-1}} \frac{1}{[x_1 A_1 + x_2 A_2]^2} = (-1)^{n_1+n_2-2} \frac{(n_1+n_2-1)! \, x_1^{n_1-1} x_2^{n_2-1}}{[x_1 A_1 + x_2 A_2]^{n_1+n_2}}.$$

Here we have used the usual gamma function, $\Gamma(\alpha)$, where $\Gamma(n) = (n-1)!$ for integer n. Since both sides of Eq. (A.5.3) are well defined when we extend n_1 and n_2 to noninteger or complex powers (they are each analytic functions of these powers), it then follows that this result extends to

$$\frac{1}{A_1^{\alpha_1} A_2^{\alpha_2}} = \frac{\Gamma(\alpha_1 + \alpha_2)}{\Gamma(\alpha_1)\Gamma(\alpha_2)} \int_0^1 dx_1 dx_2 \, \delta(x_1 + x_2 - 1) \frac{x_1^{\alpha_1-1} x_2^{\alpha_2-1}}{[x_1 A_1 + x_2 A_2]^{\alpha_1+\alpha_2}}, \tag{A.5.5}$$

where $\alpha_1, \alpha_2 \in \mathbb{C}$. This result can be further generalized to

$$\frac{1}{A_1^{\alpha_1} A_2^{\alpha_2} \cdots A_n^{\alpha_n}} = \frac{\Gamma(\alpha_1 + \cdots + \alpha_n)}{\Gamma(\alpha_1) \cdots \Gamma(\alpha_n)} \int_0^1 \frac{\left(\prod_{i=1}^n dx_i \, x_i^{\alpha_i - 1}\right) \delta\left(1 - \sum_{i=1}^n x_i\right)}{[x_1 A_1 + \cdots + x_n A_n]^{\alpha_1 + \alpha_2 + \cdots + \alpha_n}} \tag{A.5.6}$$

for complex powers $\alpha_1, \alpha_2, \ldots, \alpha_n \in \mathbb{C}$.

It can be useful to have the results

$$\int dx \, x^m \ln x = x^{m+1} \left(\frac{\ln x}{m+1} - \frac{1}{(m+1)^2}\right) \quad \text{for} \quad m \neq -1,$$

$$\Rightarrow \quad \int_0^1 dx \, x^m \ln x = \int_0^1 dx \, (1-x)^m \ln(1-x) = -\frac{1}{(m+1)^2}. \tag{A.5.7}$$

A.6 Dimensional Regularization

Integrals that are logarithmically divergent in four dimensions are finite in $d = 4 - \epsilon$ with $\epsilon > 0$. In d-dimensions we have

$$g^{\mu\nu} g_{\mu\nu} = \delta^{\mu}{}_{\mu} = d. \tag{A.6.1}$$

Here we understand that a^2 means $a^2 - i\epsilon_F$, where ϵ_F leads to the Feynman boundary conditions and is not to be confused with the ϵ in $d = 4 - \epsilon$. It is safest to Wick rotate to Euclidean space, analytically continue to d-dimensions and then continue back to Minkowski space.

Solid angle and volume in d-dimensions: Consider some d-dimensional sphere with radius R in d-dimensonal Euclidean space. The volume of the sphere, V_d, and the solid angle, Ω_d, are defined according to

$$\int_{\text{sphere,R}} d^d \ell = \int d\Omega_d \int_0^R \ell^{d-1} d\ell = V_d \quad \text{with} \quad \int d\Omega_d = \Omega_d. \tag{A.6.2}$$

It follows that

$$V_d = \Omega_d \int_0^R d\ell \, \ell^{d-1} = \Omega_d \frac{1}{d} R^d \quad \text{with} \quad \Omega_d = \frac{2\pi^{d/2}}{\Gamma(d/2)}, \tag{A.6.3}$$

since $\pi^{d/2} = (\int_{-\infty}^{\infty} dx \, e^{-x^2})^d = \int d^d \ell \, e^{-\vec{\ell}^2} = \int d\Omega_d \int d\ell \, \ell^{d-1} e^{-\ell^2} = \Omega_d \frac{1}{2} \Gamma(\frac{d}{2})$. Using $\Gamma(n) = (n-1)!$ and $\Gamma(\frac{1}{2} + n) = (2n)! \sqrt{\pi}/4^n n!$ gives familiar results: $\Omega_2 = 2\pi$, $\Omega_3 = 4\pi$ and $\Omega_4 = 2\pi^2$ for a circle (S^1), sphere (S^2) and three-sphere (S^3), respectively. The corresponding volumes are $V_2 = \pi R^2$, $V_3 = \frac{4}{3}\pi R^3$ and $V_4 = \frac{1}{2}\pi^2 R^4$.

Evaluating loop integrals in d-dimensions: The following Minkowski-space results are useful,

$$\int \frac{d^d \ell}{(2\pi)^d} \frac{1}{(\ell^2 - a^2)^\alpha} = \left[\frac{i(-1)^\alpha}{(4\pi)^{d/2}}\right] \frac{\Gamma(\alpha - \frac{d}{2})}{\Gamma(\alpha)} \left(\frac{1}{a^2}\right)^{\alpha - \frac{d}{2}} \tag{A.6.4}$$

$$\int \frac{d^d \ell}{(2\pi)^d} \frac{\ell^2}{(\ell^2 - a^2)^\alpha} = -\left[\frac{i(-1)^\alpha}{(4\pi)^{d/2}}\right] \frac{d}{2} \frac{\Gamma(\alpha - \frac{d}{2} - 1)}{\Gamma(\alpha)} \left(\frac{1}{a^2}\right)^{\alpha - \frac{d}{2} - 1} \tag{A.6.5}$$

$$\int \frac{d^d \ell}{(2\pi)^d} \frac{\ell^\mu \ell^\nu}{(\ell^2 - a^2)^\alpha} = -\left[\frac{i(-1)^\alpha}{(4\pi)^{d/2}}\right] \frac{g^{\mu\nu}}{2} \frac{\Gamma(\alpha - \frac{d}{2} - 1)}{\Gamma(\alpha)} \left(\frac{1}{a^2}\right)^{\alpha - \frac{d}{2} - 1} \tag{A.6.6}$$

$$\int \frac{d^d\ell}{(2\pi)^d} \frac{(\ell^2)^2}{(\ell^2 - a^2)^\alpha} = \left[\frac{i(-1)^\alpha}{(4\pi)^{d/2}}\right] \frac{d(d+2)}{4} \frac{\Gamma(\alpha - \frac{d}{2} - 2)}{\Gamma(\alpha)} \left(\frac{1}{a^2}\right)^{\alpha - \frac{d}{2} - 2} \tag{A.6.7}$$

$$\int \frac{d^d\ell}{(2\pi)^d} \frac{\ell^\mu \ell^\nu \ell^\rho \ell^\sigma}{(\ell^2 - a^2)^\alpha} = \left[\frac{i(-1)^\alpha}{(4\pi)^{d/2}}\right] \frac{1}{4} \frac{\Gamma(\alpha - \frac{d}{2} - 2)}{\Gamma(\alpha)} \left(\frac{1}{a^2}\right)^{\alpha - \frac{d}{2} - 2}$$
$$\times \left[g^{\mu\nu}g^{\rho\sigma} + g^{\mu\rho}g^{\nu\sigma} + g^{\mu\sigma}g^{\nu\rho}\right]. \tag{A.6.8}$$

These results are most easily shown by Wick rotating to Euclidean space, where $d^d\ell \to id^d\ell^E$, $(\ell^2 - a^2) \to -(\ell^{E2} + a^2)$ as explained in Sec. A.4. Useful formulae for arriving at the above results for convergent integrals are

$$\int d^d\ell\, \ell^\mu \ell^\nu f(\ell^2) = \int d^d\ell\, \tfrac{1}{d} \ell^2 g^{\mu\nu} f(\ell^2), \tag{A.6.9}$$

$$\int d^d\ell\, \ell^\mu \ell^\nu \ell^\rho \ell^\sigma f(\ell^2) = \int d^4\ell\, \tfrac{1}{(d+2)d} \left(\ell^2\right)^2 [g^{\mu\nu}g^{\rho\sigma} + g^{\mu\nu}g^{\rho\sigma} + g^{\mu\nu}g^{\rho\sigma}]f(\ell^2). \tag{A.6.10}$$

Note that for an odd number of momenta $\ell^\mu \ell^\nu \cdots$ such integrals vanish, while for any even number, $2n$, we must have a completely symmetrized sum of n factors of the metric tensor $g^{\mu\nu}$, where the appropriate normalization can be deduced.

Proof: Note that $\int_{-\infty}^{\infty} dx\, x g(x^2 + b^2) = 0$ vanishes by symmetry if the integral exists. So in Euclidean space in d-dimensions $\int d^d\ell^E\, \ell_\mu^E g(\ell^{E2}) = 0$ and similarly for any odd number of ℓ_μ^E. Then $\int d^d\ell^E\, \ell_\mu^E \ell_\nu^E g(\ell^{E2}) = \delta_{\mu\nu} C$ for some C, since to avoid the previous argument we need $\mu = \nu$. Set $\mu = \nu$, sum over μ and use $\delta_{\mu\mu} = d$ (summation implied) to find $\int d^d\ell^E\, \ell^{E2} g(\ell^{E2}) = dC$. Rearranging leads to $\int d^d\ell^E\, \ell_\mu^E \ell_\nu^E g(\ell^{E2}) = \delta_{\mu\nu} C = \delta_{\mu\nu} \int d^d\ell^E\, \tfrac{1}{d}\ell^{E2} g(\ell^{E2})$. We are only interested in g being a product of propagator denominators so that the Wick rotation is a simple transcription $g(\ell^{E2}) = g(-\ell^2) \equiv f(\ell^2)$. This then leads to Eq. (A.6.9).

For the same reason $\int d^d\ell^E\, \ell_\mu^E \ell_\nu^E \ell_\rho^E \ell_\sigma^E g(\ell^{E2}) = [\delta_{\mu\nu}\delta_{\rho\sigma} + \delta_{\mu\rho}\delta_{\nu\sigma} + \delta_{\mu\sigma}\delta_{\nu\rho}]C$ for some constant C, where the indices on the right must be arranged to reflect the symmetry on the left. Setting $\mu = \nu$ and summing over μ and then setting $\rho = \sigma$ and summing over ρ leads to $\ell_\mu^E \ell_\nu^E \ell_\rho^E \ell_\sigma^E \to \ell^{E2} \ell_\rho^E \ell_\sigma^E \to (\ell^{E2})^2$ and also $[\delta_{\mu\nu}\delta_{\rho\sigma} + \delta_{\mu\rho}\delta_{\nu\sigma} + \delta_{\mu\sigma}\delta_{\nu\rho}] \to [d\delta_{\rho\sigma} + \delta_{\rho\sigma} + \delta_{\rho\sigma}] = (d+2)\delta_{\rho\sigma} \to (d+2)d$. Then $\int d^d\ell^E\, (\ell^{E2})^2 g(\ell^{E2}) = (d+2)dC$. In Minkowski space we arrive at Eq. (A.6.10).

The above results can be used to build up other results of interest,

$$\int \frac{d^d\ell}{(2\pi)^d} \frac{1}{(\ell^2 + 2k\cdot\ell - a^2)^\alpha} = \left[i\frac{(-1)^\alpha}{(4\pi)^{d/2}}\right] \frac{\Gamma(\alpha - \frac{d}{2})}{\Gamma(\alpha)} \left(\frac{1}{a^2 + k^2}\right)^{\alpha - \frac{d}{2}}, \tag{A.6.11}$$

$$\int \frac{d^d\ell}{(2\pi)^d} \frac{\ell^\mu}{(\ell^2 + 2k\cdot\ell - a^2)^\alpha} = \left[i\frac{(-1)^\alpha}{(4\pi)^{d/2}}\right] (-k^\mu) \frac{\Gamma(\alpha - \frac{d}{2})}{\Gamma(\alpha)} \left(\frac{1}{a^2 + k^2}\right)^{\alpha - \frac{d}{2}}.$$

These follow from $\ell^2 + 2k\cdot\ell - a^2 = (\ell + k)^2 - (k^2 + a^2) = \ell'^2 - (k^2 + a^2)$ with $\ell' = \ell + k$ and then changing variables, $\ell' \to \ell$, and using Eq. (A.6.4).

Note that if $a^2 = 0$ and $\alpha < d/2$, then the right-hand side of Eq. (A.6.4) is zero. For consistency in dimensional regularization we *define*

$$\int \frac{d^d\ell}{(2\pi)^d} \frac{1}{(\ell^2)^\alpha} \equiv 0 \quad \text{for all } \alpha, \tag{A.6.12}$$

which can be understood since such integrals contain no quantity with a scale and so they must vanish. This is an indirect consequence of requiring that the momentum integral be invariant under shifts in momentum; i.e., there is no cut-off scale of any kind and so no value can be assigned on the right-hand side other than zero. Observe from Eq. (A.6.4), for example, that if $\alpha = 2$ then $\int d^d\ell[1/(\ell^2 - a^2)^2] \sim \Gamma(2 - \frac{d}{2})$ and so the *logarithmic divergence* ($\sim \ln \Lambda$) as $d \to 4$ is associated with the pole in $\Gamma(z)$ at $z = 0$ since $\Gamma(2 - \frac{d}{2}) \to \Gamma(0)$. Similarly, if $\alpha = 1$ then we have $\int d^d\ell[1/(\ell^2 - a^2)] \sim \Gamma(1 - \frac{d}{2})$ and so a *quadratic divergence* ($\sim \Lambda^2$) in $d = 4$ is associated with a pole at $z = -1$ since $\Gamma(1 - \frac{d}{2}) \to \Gamma(-1)$.

In perturbation theory the leading divergent piece is identified by expanding around $d = 4 - \epsilon$ dimensions. In particular, for dimensionless a^2 we have

$$\left(1/a^2\right)^{\epsilon/2} = 1 - (\epsilon/2)\ln(a^2) + \mathcal{O}(\epsilon^2) \quad \text{with} \quad (\epsilon/2) = 2 - (d/2), \tag{A.6.13}$$

which follows since $x^\epsilon = e^{\epsilon \ln x} = 1 + \epsilon \ln x + \mathcal{O}(\epsilon^2)$. Note that a^2 means $a^2 - i\epsilon_F$ if needed. It is necessary to know how to expand the Γ-function around its poles at $z = 0, -1, -2, \ldots$. The expansion around $z = -n$ is

$$\Gamma(-n + \epsilon) = \frac{(-1)^n}{n!}\left[\frac{1}{\epsilon} - \gamma + \sum_{k=1}^{n}\frac{1}{k} + \mathcal{O}(\epsilon)\right], \tag{A.6.14}$$

which for the special case of an expansion around $z = 0$ gives

$$\Gamma(\epsilon) = (1/\epsilon) - \gamma + \mathcal{O}(\epsilon). \tag{A.6.15}$$

Here γ is the *Euler-Mascheroni constant*, which has the value $\gamma \simeq 0.5772156649$. For nonpositive z an analytic continuation from the positive complex plane can be achieved by using the recursion formula,

$$\Gamma(z) = \frac{\Gamma(z + n + 1)}{z(z + 1)\cdots(z + n)}, \tag{A.6.16}$$

where n is chosen so that $z + n$ is positive. As we saw above there are frequently occurring combinations with forms such as

$$\frac{\Gamma(2 - \frac{d}{2})}{(4\pi)^{d/2}}\left[\frac{1}{a^2}\right]^{2 - \frac{d}{2}} = \frac{1}{(4\pi)^2}\left[\frac{2}{\epsilon} - \ln(a^2) - \gamma + \ln(4\pi) + \mathcal{O}(\epsilon)\right]. \tag{A.6.17}$$

Using steps similar to those in the proof below this result can be generalized.

Proof: First note that Eq. (A.6.15) gives
$\Gamma(2 - \frac{d}{2}) = \Gamma(\frac{\epsilon}{2}) = [\frac{2}{\epsilon} - \gamma + \mathcal{O}(\epsilon)]$.
Then using Eq. (A.6.13) it follows that
$1/(4\pi)^{d/2} = 1/(4\pi)^{2 - (\epsilon/2)} = \{1/(4\pi)^2\}[1 + \frac{\epsilon}{2}\ln(4\pi) + \mathcal{O}(\epsilon^2)]$,
$(1/a^2)^{2 - (d/2)} = (1/a^2)^{\epsilon/2} = [1 - \frac{\epsilon}{2}\ln(a^2) + \mathcal{O}(\epsilon^2)]$.
Multiplying the three expansions together gives Eq. (A.6.17).

Dirac algebra in d-dimensions: The extension of Dirac matrices to $d = 4 - \epsilon$ dimensions is a less obvious procedure.

First, briefly consider the generalization of the *Dirac algebra* to n spacetime dimensions where there is one time dimension and $(n - 1)$ spatial dimensions. For the metric tensor $g^{00} = -g^{11} = \cdots = -g^{(n-1)(n-1)} = +1$ with all other entries zero. Hence, we see that $\text{tr}(g) = -(n - 2)$,

$g^{\mu\sigma}g_{\sigma\nu} = \delta^{\mu}{}_{\nu}$, and $\delta^{\mu}{}_{\mu} = n$ and there will be n Dirac γ-matrices satisfying $\{\gamma^{\mu}, \gamma^{\nu}\} = 2g^{\mu\nu}$ with $\gamma^{0} = \gamma^{0\dagger}$ and $\gamma^{i} = -\gamma^{i\dagger}$. The contraction identities in Eq. (A.3.32) change, since they will obviously vary with the dimension of spacetime. We find for these, for example:

$$\gamma^{\mu}\gamma_{\mu} = nI, \quad \gamma^{\mu}\not{k}\gamma_{\mu} = -(n-2)\not{k}, \quad \gamma^{\mu}\not{k}\not{\ell}\gamma_{\mu} = 4k\cdot\ell - (4-n)\not{k}\not{\ell},$$
$$\gamma^{\mu}\not{k}\not{\ell}\not{p}\gamma_{\mu} = -2\not{p}\not{\ell}\not{k} + (4-n)\not{k}\not{\ell}\not{p}. \tag{A.6.18}$$

As the number of linearly independent γ-matrices increases with increasing spacetime dimensions n, the dimensionality of these matrices will also in general need to increase. So trace identities will also vary with n, since $\mathrm{tr}I = n_{\mathrm{matrix}}$, where the representation in n spacetime dimensions consists of $n_{\mathrm{matrix}} \times n_{\mathrm{matrix}}$ matrices.

When extending from integer, $n \in \mathbb{Z}$, to noninteger spacetime dimensions, $d \in \mathbb{R}$, we define the contraction identities with n replaced by $d = 4 - \epsilon$,

$$\gamma^{\mu}\gamma_{\mu} = (4-\epsilon)I, \quad \gamma^{\mu}\not{k}\gamma_{\mu} = -(2-\epsilon)\not{k}, \quad \gamma^{\mu}\not{k}\not{\ell}\gamma_{\mu} = 4k\cdot\ell - \epsilon\not{k}\not{\ell},$$
$$\gamma^{\mu}\not{k}\not{\ell}\not{p}\gamma_{\mu} = -2\not{p}\not{\ell}\not{k} + \epsilon\not{k}\not{\ell}\not{p}. \tag{A.6.19}$$

The trace identities can be generalized to $\mathrm{tr}I = f(d)$ and $\mathrm{tr}(\gamma^{\mu}\gamma^{\nu}) = f(d)g^{\mu\nu}$, where $f(d)$ is some arbitrary smooth function of d (Itzykson and Zuber, 1980) such that $f(n) = n_{\mathrm{matrix}}$. We will almost exclusively be interested in analytically continuing from $n = 4$ dimensions to $d = 4 - \epsilon$ dimensions. Without loss of generality it is convenient to choose $f(d) = 4$ for $d = 4 - \epsilon$ so that, e.g.,

$$\mathrm{tr}I = 4, \quad \mathrm{tr}\gamma^{\mu} = 0 \quad \text{and} \quad \mathrm{tr}(\gamma^{\mu}\gamma^{\nu}) = 4g^{\mu\nu}. \tag{A.6.20}$$

It is important to use $g^{\mu\nu}g_{\nu\mu} = \delta^{\mu}{}_{\mu} = d$ before taking $\epsilon \to 0$. These choices have the advantage that *the trace identities in $d = 4 - \epsilon$ dimensions are the same as they are in four-dimensions*; i.e., the identities in Eqs. (A.3.24)–(A.3.29) still apply.

The definition of γ^{5} is intrinsically four-dimensional due to its definition in terms of the four-dimensional antisymmetric tensor $\epsilon^{\mu\nu\rho\sigma}$ in Eq. (A.3.5). A variety of recipes for defining γ^{5} in d-dimensions exist (see, e.g., Chanowitz et al., 1979; Fujii et al., 1981, and references therein). However, common practice (Ynduráin, 1983; Muta, 1987) is to simply assume that in d-dimensions there exists some Hermitian matrix γ^{5} that satisfies

$$\{\gamma^{\mu}, \gamma^{5}\} = 0, \quad \gamma^{5} = \gamma^{5\dagger}, \quad (\gamma^{5})^{2} = I. \tag{A.6.21}$$

This is an adequate definition of γ_{5} in d-dimensions in anomaly-free theories. The original proposal of 't Hooft and Veltman ('t Hooft and Veltman, 1972) was to define $\gamma^{5} \equiv i\gamma^{0}\gamma^{1}\gamma^{2}\gamma^{3}$ in any dimension such that $\{\gamma_{5}, \gamma^{\mu}\} = 0$ for $\mu = 0, 1, 2, 3$ but $[\gamma_{5}, \gamma^{\mu}] = 0$ otherwise. To understand what this means in practice, see, for example, the discussion of the *Adler-Bell-Jackiw anomaly* in Peskin and Schroeder (1995). This 't Hooft-Veltman definition of γ_{5} successfully reproduces the correct form of the Adler-Bell-Jackiw anomaly equation, which is also referred to as the *axial anomaly* in QED and is necessary to understand the decay of the neutral pion into two photons, $\pi^{0} \to \gamma\gamma$. It is an example of a *chiral anomaly*, which is the anomalous nonconservation of a chiral current, where in turn an *anomaly* in quantum field theory is said to occur when a symmetry of the classical theory is not a symmetry of *any* regularization of the quantum field theory.

For further discussion of the above issues and for additional references, see, for example, Ynduráin (1983), Collins (1984), Itzykson and Zuber (1980), Muta (1987), Peskin and Schroeder (1995) and Schwartz (2013).

A.7 Group Theory and Lie Groups

We provide a brief summary of some of the results needed for our discussions.

A.7.1 Elements of Group Theory

Definition of a group: A set of elements $\{g_1, g_2, g_3, \ldots\}$ forms a *group*, G, when a binary group operation (or group multiplication), $g_i g_j \equiv g_i \circ g_j$, has the following properties:

(i) *Closure:* If $g_i, g_j \in G$, then $g_i g_j \in G$, i.e., $g_i g_j = g_k$ for some $g_k \in G$.

(ii) *Associativity:* $g_i(g_j g_k) = (g_i g_j) g_k$ for all $g_i, g_j, g_k \in G$.

(iii) *Identity:* There exists some e (or we will often write I) $\in G$, called the identity, such that $e g_i = g_i e = g_i$ for every $g_i \in G$.

(iv) *Inverse:* For every $g_i \in G$ there exists some $g_i^{-1} \in G$, called the inverse of g_i, such that $g_i g_i^{-1} = g_i^{-1} g_i = e$.

We have used i, j to label group elements but the group index can also be continuous for a continuous group, e.g., the rotation group. If $g_i g_j = g_j g_i$ then $[g_i, g_j] = 0$ and we say that the group is *abelian*. Otherwise, we say that it is *nonabelian*. The group of translations is abelian, while the group of rotations is nonabelian.

Representations of a group: A *representation* of a group G is a specific realization D of the abstract group in terms of matrices, where there is a matrix $D(g)$ for every $g \in G$ and where the group operation corresponds to matrix multiplication. Groups must contain the inverse of every element and so a matrix representation is a mapping of the abstract group onto a set of invertible matrices such that (i) if $g, g', g'' \in G$ and if $gg' = g''$ then $D(g)D(g') = D(g'')$, and (ii) $D(g^{-1}) = D(g)^{-1}$ for every g. An $n \times n$ matrix representation D of the group G is said to have *dimension n* and is necessarily a subgroup of the set of $n \times n$ invertible complex matrices known as the general linear group, $GL(\mathbb{C}^n)$, i.e., $D(g) \subset GL(n, \mathbb{C}) \equiv GL(\mathbb{C}^n)$.

A *reducible* representation $D(g)$ has a nontrivial proper invariant subspace; i.e., a vector in the subspace remains in the subspace when acted upon by $D(g)$ for every $g \in G$. Otherwise, we refer to $D(g)$ as an *irreducible representation*. If every $D(g)$ can be brought into block-diagonal form by some constant similarity transformation S, then we have a *completely reducible representation*, i.e., if there is an S such that

$$SD(g)S^{-1} = \begin{pmatrix} D_1(g) & 0 & 0 & \cdots \\ 0 & D_2(g) & 0 & \cdots \\ 0 & 0 & D_3(g) & \cdots \\ \vdots & \vdots & \vdots & \ddots \end{pmatrix} \tag{A.7.1}$$

for all $g \in G$. A completely reducible representation is decomposable into a direct sum of matrices, $D(g) = D_1(g) \oplus D_2(g) \oplus D_3(g) \oplus \cdots$ with dimensions such that $\dim[D] = n = \dim[D_1] + \dim[D_2] + \cdots$. Each of the submatrices $D_i(g)$ is itself a lower-dimensional representation of the group. Note that $\dim[D_i]$ is the dimension of the vector space V_i in which the $D_i(g)$ act, where the $D_i(g)$ are $\dim[D_i] \times \dim[D_i]$ matrices.

We say that the representation is *faithful* if every matrix is mapped to by at most one element $g \in G$; i.e., every $g \in G$ maps to a unique $D(g)$ and no information is lost. A *trivial representation* for a group is a mapping of the entire group to the identity matrix so all information is lost. The irreducible trivial representation of a group is unity, i.e., the 1×1 identity matrix. The *defining* (or *standard*) representation is the lowest-dimensional faithful representation. This typically means an invertible correspondence between the g and the $D(g)$; i.e., it is typically the lowest-dimensional representation that is isomorphic to the group. For example, for $SU(N)$ the defining representation is the set of $N \times N$ special unitary matrices. A *fundamental* representation is an irreducible finite-dimensional representation of a semisimple Lie group or Lie algebra whose highest weight is a fundamental weight (not all of these terms will be defined here). The defining representation of a classical Lie group is a fundamental representation, but in general there is more than one fundamental representation, e.g., for $SO(n)$, $SU(n)$ and $Sp(n)$. We physicists sometimes use the phrase *fundamental representation* with regard to some classical Lie group, when what we mean is the *defining representation*. In quantum field theory continuous symmetry transformations are represented by unitary transformations in Hilbert space. Hence we will often be interested in *unitary representations of groups* where $D(g)^{\dagger} = D(g)^{-1}$.

A *subset* of a set is any selection of elements from the set; a *proper subset* is a subset that is not the whole set itself; a *subgroup* S of G is a nonempty subset of the elements of G, $S \subset G$, that also forms a group; a *proper subgroup* is a subgroup that is not the whole group itself; a *normal subgroup (or invariant subgroup)* N of G is a subgroup of G with the property that for every $n \in N$ and every $g \in G$ we have $gng^{-1} \in N$. Some simple results follow: (i) The intersection of two subgroups is a subgroup; (ii) the intersection of two normal subgroups is a normal subgroup. A *simple group* is a group that does not contain any nontrivial normal subgroups. ("Nontrivial" means not the whole group itself and not the identity on its own.) A *semisimple* group has no nontrivial normal abelian subgroups. Hence a simple group is necessarily semisimple but not vice versa.

The *center* $Z(G)$ of a group G is the set of elements that commute with all other elements of the group. The elements of $Z(G)$ form a normal abelian subgroup of G. A group and a topology defined on the set of elements of the group form a *topological group* if the group operation and the inverse operation are continuous; i.e., (i) if $g', g'' \in G$ and g' and g'' are in the same neighborhood, then gg' and gg'' are in the same neighborhood, for all $g \in G$; and (ii) if g' and g'' are in the same neighborhood, then so are their inverses g'^{-1} and g''^{-1}. A *continuous group* is a topological group whose elements in any given neighborhood can be labeled by continuous real parameters, e.g., the Euler angles for rotations in three dimensions. A *compact group* is a topological group with a compact manifold; i.e., the group parameters take their values in a compact space (closed and bounded space), e.g., the proper rotations $SO(3)$ is a compact group, while the translation group is not.

A mapping $G \to G'$ for two groups G and G' is a group *isomorphism* if it is an invertible mapping that preserves the group multiplication rule, i.e., if $g_1, g_2, g_3 \in G$ and the corresponding mapped elements after the mapping are $g_1', g_2', g_3' \in G'$ and if for every $g_1 g_2 = g_3$ we have $g_1' g_2' = g_3'$ then the mapping is isomorphic. We say that G and G' are isomorphic to each other and this is often represented by the symbol "$\cong$" i.e., $G \cong G'$. An isomorphism of G with itself is called an *automorphism*. The set of all automorphisms of a group themselves form a group. Since by definition the automorphisms preserve all of the relations within the group, then the larger the group of automorphisms is the more symmetry the group must possess. A group *homomorphism* is a structure-preserving mapping $f : G \to G'$ where more than one element of G may map to a single element of G' and not every element of G' may be mapped to. An isomorphism is then a special case of a homomorphism, i.e., when there is a one-to-one correspondence so that the homomorphism is invertible.

The *conjugacy class* containing g_i is the set of elements gg_ig^{-1} for all $g \in G$. Clearly, every $g_i \in G$ belongs to a conjugacy class and no $g \in G$ can belong to more than one conjugacy class. (*Proof:* Consider two distinct conjugacy classes defined, respectively, by $g_i, g_j \in G$. If g' belongs simultaneously to both classes, then there must exist some g_a and g_b such that $g' = g_ag_ig_a^{-1} = g_bg_jg_b^{-1}$. This implies that $g_i = (g_a^{-1}g_b)g_j(g_a^{-1}g_b)^{-1}$ and hence that g_i and g_j are in the same conjugacy class, which contradicts our starting assumption. Hence no g' can belong to two different conjugacy classes as claimed.) So any group G can be decomposed into its distinct conjugacy classes and the number of these is referred to as the *class number*.

A *covering group* C of some topological group G is a topological group that is locally isomorphic to the group G, but there may be more than one element of C for each $g \in G$. For example, the group $SL(2, \mathbb{C})$ is a cover of the group of restricted Lorentz transformations $SO^+(1,3)$ with cover index 2 (see Sec. 1.3); i.e., $SL(2, \mathbb{C})$ is a *double cover* of $SO^+(1,3)$. Similarly, $SU(2)$ is a double cover of $SO(3)$.

Consider two groups $G = \{g_1, g_2, \ldots\}$ and $H = \{h_1, h_2, \ldots\}$, then the *direct-product group* is denoted as $P = G \times H$ and consists of elements (g_i, h_j) for all i, j with the corresponding group multiplication law $(g_i, h_j)(g_k, h_\ell) = (g_ig_k, h_jh_\ell)$. Examples of direct product groups are $SU(2) \times U(1)$ and $SU(3) \times SU(3)$. The two components G and H are each normal subgroups of the larger group $G \times H$. G and H are isomorphic to the quotient groups $G \cong P/H$ and $H \cong P/G$.

Let D_1 be a representation of group G in vector space V_1 with dimension $\dim[D_1]$ so that the $D_1(g)$ are $\dim[D_1] \times \dim[D_1]$ matrices and similarly D_2 and V_2 for the group H. Representations of the direct product group $P = G \times H$ are formed by the tensor products of representations of G and H. We can write this as $D_1 \otimes D_2$, where $\dim[D_1 \otimes D_2] = \dim[D_1]\dim[D_2]$ so that $\dim[D_1]\dim[D_2] \times \dim[D_1]\dim[D_2]$ matrices act in the vector space formed by the tensor product of the two vector spaces, $V_1 \otimes V_2$. We can also consider the direct product of a group with itself, $G \times G$. The tensor product of two irreducible representations of G is irreducible when viewed as a representation of the product group $G \times G$. An example of interest to us discussed in Sec. 1.3.1 is $SU(2) \times SU(2)$, which can be used to label representations of the restricted Lorentz group $SO^+(1,3)$. Representations of $SU(n)$ obtained by decomposing direct products of $SU(n)$ with itself into direct sums of irreducible representations (irreps) of $SU(n)$ are important and are discussed in Sec. 5.3.1.

Consider any group G with a normal subgroup N and another subgroup H. If every $g \in G$ can be written as hn (or nh) with $n \in N$ and $h \in H$ and the only common element of N and H is the identity e, $N \cap H = \{e\}$, then we say that G is *semidirect product* of N and H and we write this as $G = N \rtimes H$. An example is the Poincaré group that is the internal semidirect product of the translations, an abelian normal subgroup and the subgroup of the Lorentz transformations. Another example of this is the Euclidean group $E(n)$ in n-dimensions made up of the translations, an abelian normal subgroup and the orthogonal transformations $O(n)$. For example, a rotation, translation, inverse rotation is still a translation, which means that the translations are a normal subgroup of $E(n)$, whereas a translation, rotation, inverse translation is almost never a rotation and so the rotations are not a normal subgroup of $E(n)$.

A.7.2 Lie Groups

A *manifold* is a space that locally resembles Euclidean space. In a *differentiable manifold* we can locally use calculus. A Lie group is both a continuous group and a differentiable manifold, where the points in the manifold are the group elements. The elements of a *continuous group* in any given neighborhood can be labeled by continuous real parameters, e.g., the Euler angles for rotations in three dimensions. Lie groups arise naturally from continuous symmetries.

A region of some manifold is *path connected* if every two points on the region can be connected by a continuous path that stays in the region. The region is said to be *simply connected* if every path connecting any two points can be continuously deformed into any other path while staying in the space (the region has no holes).

More specifically, a continuous group is a *Lie group* if it is infinitely differentiable with respect to its real parameters. Lie groups are therefore also real smooth manifolds. More precisely, this is the definition of a *real Lie group* that is typically what is understood when Lie groups are discussed and includes, e.g., $U(n)$, $SU(n)$, $O(n)$ and $SO(1,3)$. We can also extend our considerations to *complex Lie groups* with complex group parameters that give rise to a complex group manifold. Examples of complex Lie groups that will be of particular importance to us are $GL(n, \mathbb{C})$ and $SL(2, \mathbb{C})$. We will be restricting our attention to Lie groups that admit a matrix representation, since this covers all Lie groups of relevance to us.

For a Lie group with n real parameters, $\vec{\omega} \in \mathbb{R}^n$, we have for sufficiently small $\vec{\omega}$,

$$g(\vec{\omega}) = I + i\vec{\omega} \cdot \vec{T} + \mathcal{O}(\omega^2) = I + i\omega^a T^a + \mathcal{O}(\omega^2) \quad \text{with} \quad T^a \equiv -i \left. \frac{\partial g}{\partial \omega^a} \right|_{\vec{\omega}=\vec{0}}, \qquad (A.7.2)$$

where the summation convention over a is understood here and where the operators T^a are referred to as the *generators* of the Lie group. The T^a are necessarily linearly independent since each infinitesimal $\vec{\omega}$ uniquely specifies a group element near the identity. The number of parameters n of the group is referred to as the dimension of the group $d(G) \equiv n$.

We are interested in matrix representations of Lie groups, where the abstract generators T^a are represented by some $d \times d$ matrices that we also write as T^a for now. Later we use t_r^a for the matrices representing the abstract T^a in the irreducible representation r. Then using Eqs. (A.7.2) and (A.8.5) we have

$$g(\vec{\omega}) = \lim_{N \to \infty} g(\vec{\omega}/N)^N = \lim_{N \to \infty} [I + i(1/N)\vec{\omega} \cdot \vec{T}]^N = e^{i\vec{\omega} \cdot \vec{T}} \qquad (A.7.3)$$

provided $\vec{\omega}$ is sufficiently close to the origin in $\mathbb{R}^n$. The topology of a Lie group will not be the trivial topology of $\mathbb{R}^n$. So clearly we cannot have a one-to-one correspondence between all possible points $\vec{\omega}$ in $\mathbb{R}^n$ and all elements g of the Lie group. However, there will be a *finite region* (or *patch*) of parameter space near the identity where this one-to-one correspondence is satisfied and where we can uniquely identify $g(\vec{\omega}) = \exp(i\vec{\omega} \cdot \vec{T})$ as written in Eq. (A.7.3).

Let $\hat{\omega}_1$ and $\hat{\omega}_2$ be two unit vectors in parameter space. Using the Baker-Campbell-Hausdorff formula, Eq. (A.8.6), the product of two infinitesimal group elements is

$$g(\epsilon_1 \hat{\omega}_1) g(\epsilon_2 \omega_2) = \exp\{i(\epsilon_1 \hat{\omega}_1 + \epsilon_2 \hat{\omega}_2) \cdot \vec{T} - \tfrac{1}{2}\epsilon_1 \epsilon_2 \omega_1^a \omega_2^b [T^a, T^b] + \cdots\} \qquad (A.7.4)$$
$$= \exp\{i\epsilon_3 \hat{\omega}_3 \cdot \vec{T}\} = g(\epsilon_3 \hat{\omega}_3)$$

by closure for some $\epsilon_3 \hat{\omega}_3$. Since $\epsilon_1 \omega_1^a$ and $\epsilon_2 \omega_2^b$ are arbitrary, then we must have

$$[T^a, T^b] = i f^{abc} T^c \qquad (A.7.5)$$

for some constants $f^{abc} \in \mathbb{C}$, where *we use the summation convention over repeated indices*. For a detailed proof, see Sec. 4.1.4. The *Lie algebra* of the Lie group consists of the vector space of matrices made from the linear combinations of the generators, $\vec{\omega} \cdot \vec{T}$, along with an operation defined by the Lie bracket, which here is Eq. (A.7.5). Clearly, $f^{abc} = -f^{bac}$ and we refer to the constants

f^{abc} as the *structure constants* of the Lie algebra. From Eq. (A.7.2) we see that the Lie algebra is the tangent space for the Lie group at its identity element.

Note that for finite-dimensional matrices $\mathrm{tr}AB = \mathrm{tr}BA$ and so $\mathrm{tr}[A,B] = 0$. Then it follows that $\mathrm{tr}[T^a, T^b] = if^{abc}\mathrm{tr}T^c = 0$. For a semisimple Lie algebra it can be shown that there is an $(f^{-1})^{abc}$ such that $(f^{-1})^{abd}f^{abc} = \delta^{dc}$ and so $\mathrm{tr}T^d = 0$. So *a semisimple Lie algebra has only traceless generators*. Consider a connected Lie group whose Lie algebra has I as a generator. This is not a semisimple Lie algebra. The group will have $U(1)$ as a normal (invariant) abelian subgroup, with elements $e^{i\omega I} \in U(1)$, and so by the above definition will not be a semisimple Lie group. Examples of Lie groups with semisimple Lie algebras are given in Table A.1. We can always add $U(1)$ as a normal subgroup to a semisimple Lie group using the direct product as needed, e.g., $U(3) = U(1) \times SU(3)$.

It is easily verified that the Lie algebra and the associativity property of the generators lead to the *Jacobi identity*,

$$[T^a, [T^b, T^c]] + [T^b, [T^c, T^a]] + [T^c, [T^a, T^b]] = 0, \tag{A.7.6}$$

which in turn leads to

$$f^{ade}f^{bcd} + f^{bde}f^{cad} + f^{cde}f^{abd} = 0. \tag{A.7.7}$$

More generally in the abstract treatment of Lie algebras the binary operation $[A, B]$ is referred to as the *Lie bracket*. There are various equivalent ways to specify the Lie bracket and one such way is (i) *bilinearity*: $[aX + bY, Z] = a[X, Z] + b[Y, Z]$ and $[Z, aX + bY] = a[Z, X] + b[Z, Y]$; (ii) *alternativity*: $[X, X] = 0$; and (iii) the *Jacobi identity*: $[X, [Y, Z]] + [Z, [X, Y]] + [Y, [Z, X]]$. This leads to $[X, Y] = -[Y, X]$ since $0 = [X + Y, X + Y] = [X, Y] + [Y, X]$. The cross-product of three-dimensional vectors, the commutator of operators/matrices and the Poisson bracket in Sec. 2.4.2 each have these properties and so are examples of the Lie bracket.

Distinct Lie groups can have the same Lie algebra and so can be locally isomorphic. For example, $SU(2)$ and $SO(3)$ have the same Lie algebra as do $SL(2, \mathbb{C})$ and $SO^+(1, 3)$, but their Lie groups have a different topological or global character. One Lie group can be the cover of another Lie group. A *universal cover* of a Lie group is a simply connected cover. *Every Lie group has a unique universal cover*.

For a Lie group G it is conventional to denote the corresponding Lie algebra with the same name using the lowercase gothic font $\mathfrak{g}$ for the group G. For example, the Lie algebras for $SU(n)$, $O(n)$ and $SO(n)$ are denoted as $\mathfrak{su}(n)$, $\mathfrak{o}(n)$ and $\mathfrak{so}(n)$, respectively. The Lie algebra is the vector space $\mathfrak{g}$ together with Lie bracket as the binary operation on the space. The direct sum $\mathfrak{g}$ of two Lie algebras $\mathfrak{a}$ and $\mathfrak{b}$ is defined such that if $x \in \mathfrak{a}$ and $y \in \mathfrak{b}$ then $z \equiv (x, y) \in \mathfrak{g}$, where the Lie bracket operation is defined such that $[z, z'] = [(x, y), (x', y')] \equiv ([x, x'], [y, y'])$. In a matrix representation we can write every z in block diagonal form like the right-hand side of Eq. (A.7.1) with one block containing x and one block containing y.

The direct product of two Lie groups corresponds to the direct sum of their Lie algebras; i.e., if A and B are two Lie groups, then their direct product $G = A \times B$ has a Lie algebra $\mathfrak{g}$ that is the direct sum of their respective Lie algebras; i.e., $\mathfrak{g} = \mathfrak{a} \oplus \mathfrak{b}$.

A *connected Lie group* is path connected over the whole manifold in that any two elements of the group can be continuously deformed onto each other. A *simply connected Lie group* has a simply connected manifold, where every path can be continuously deformed into any other path (no holes).

A subset $\mathfrak{g}'$ of a Lie algebra $\mathfrak{g}$ is defined to be a *Lie subalgebra* if $[X', Y'] \in \mathfrak{g}'$ for all $X', Y' \in \mathfrak{g}'$; i.e., the subset is closed under the binary operation of the Lie bracket. An *ideal* or an *invariant Lie subalgebra* is a Lie subalgebra for which $[X, X'] \in \mathfrak{g}'$ for all $X \in \mathfrak{g}$ and $X' \in \mathfrak{g}'$; i.e., an element of the Lie subalgebra stays in the subalgebra when acted on by an arbitrary element of the algebra. These definitions are analogs of those for a subgroup and a normal (i.e., invariant) subgroup.

A *simple Lie group* is defined as a connected nonabelian Lie group with no nontrivial connected normal subgroups. A *simple Lie algebra* is a nonabelian Lie algebra whose only ideals are zero and the whole algebra itself. A semisimple Lie algebra was defined above. The *Lie correspondence* is the result that a connected Lie group will be simple if its Lie algebra is simple. Note that a "simple Lie group" is not the same as a "simple group" as defined earlier in the section, since a simple Lie group may contain normal subgroups, whereas a simple group does not.

A connected Lie group that does not contain any nontrivial connected abelian normal subgroups is referred to as a *semisimple Lie group*. It can be shown that a connected Lie group is semisimple if and only if its Lie algebra is semisimple. A simple Lie group is then necessarily also semisimple, but the converse is not true.

A *Casimir invariant* commutes with every generator of the group and so commutes with every element of the Lie algebra. Both $SO(3)$ and $SU(2)$ have one Casimir invariant and $SO(4)$ and $SU(3)$ each have two. Since the restricted Lorentz group $SO^+(1,3)$ and $SO(4)$ have the same complexified forms (see below), then their Lie algebras will each have two Casimir invariants. In any irreducible representation of a Lie group the Casimir invariants are matrices proportional to the identity matrix. The constants of proportionality classify the representations of the Lie algebra and hence the Lie group.

Up to this point we have been focusing our attention on *real Lie algebras* and their corresponding *real Lie groups*, where the Lie algebra consists only of real linear combinations of the generators, $\vec{T}$, i.e., consists of $\vec{\omega} \cdot \vec{T}$ with $\vec{\omega} \in \mathbb{R}^n$ defining a real smooth group manifold of n-dimensions. A *complex Lie algebra* and its corresponding *complex Lie group* has $X = \vec{\omega} \cdot \vec{T} \to X = \vec{c} \cdot \vec{T}$ where $\vec{c} \in \mathbb{C}^n$. Examples of complex Lie groups are $SL(2, \mathbb{C})$ and $GL(n, \mathbb{C})$. Any real Lie algebra has a *complexification* by generalizing its group parameter space from $\mathbb{R}^n$ to $\mathbb{C}^n$, i.e., by doubling the number of real group parameters. It then follows that any complex Lie algebra can always be viewed as a real Lie algebra of twice the dimension. Examples of interest to us are the complexifications $SU(2) \to SL(2, \mathbb{C})$, $SO^+(1,3) \to SL(2, \mathbb{C}) \times SL(2, \mathbb{C})/Z_2$, $SO(4) \to SL(2, \mathbb{C}) \times SL(2, \mathbb{C})/Z_2$ and $SU(2) \times SU(2) \to SL(2, \mathbb{C}) \times SL(2, \mathbb{C})$. Since $SO^+(1,3) = SL(2, \mathbb{C})/Z_2 = SL(2, \mathbb{C})/\{I, -I\}$ then it follows that under complexification we also have $SL(2, \mathbb{C}) \to SL(2, \mathbb{C}) \times SL(2, \mathbb{C})$. So the complexified forms of $SO^+(1,3)$, $SO(4)$, $SL(2, \mathbb{C})$ and $SU(2) \times SU(2)$ all have the same Lie algebra.

Weyl's unitarian trick: The classification of the irreducible representations of real Lie groups is most easily done by considering the representations of the Lie algebra of their complexified forms. This is *Weyl's unitarian trick*. Complex representations of semisimple (i.e., a direct sum of simple) Lie algebras and their corresponding Lie groups are fully reducible (i.e., a direct sum of irreducible representations).

For any compact connected Lie group, G, any $g \in G$ can be written as $g = e^{iX}$ for at least one element of the Lie algebra, $X \in \mathfrak{g}$, but there may be more than one such X. Table A.1 shows that this is true for $U(n)$, $SU(n)$ and $SO(n)$ but not for $O(n)$ or $SL(2, \mathbb{C})$, for example. It is also the case for $SO^+(1,3)$.

A.7.3 Unitary Representations of Lie Groups

In quantum mechanics and quantum field theory we are especially interested in unitary representations of symmetry groups as explained in Secs. 4.1.3–4.1.8. A (matrix) representation r of abstract

Lie group[a]	Rep[b]	Compact	Connected	Simply connected	Lie algebra[c]	Dimension[d] $(\mathbb{R}$ or $\mathbb{C})^e$
$U(n)$	$\mathbb{R}$	Y	Y	N	$\mathfrak{u}(n)$	n^2
$SU(n)^f$	$\mathbb{R}$	Y	Y	Y	$\mathfrak{su}(n), s, ss$	$n^2 - 1$
$O(n)$	$\mathbb{R}$	Y	N	N	$\mathfrak{so}(n), s, ss$	$n(n-1)/2$
$SO(n)$	$\mathbb{R}$	Y	Y	N	$\mathfrak{so}(n), s, ss$	$n(n-1)/2$
$GL(n,\mathbb{R})$	$\mathbb{R}$	N	N	N	$M(n,\mathbb{R})$	n^2
$SL(n,\mathbb{R})$	$\mathbb{R}$	N	Y	N	$\mathfrak{sl}(n,\mathbb{R}), s, ss$	$n^2 - 1$
$GL(n,\mathbb{C})$	$\mathbb{C}$	N	Y	N	$M(n,\mathbb{C})$	n^2
$SL(n,\mathbb{C})$	$\mathbb{C}$	N	Y	Y	$\mathfrak{sl}(n,\mathbb{C}), s, ss$	$n^2 - 1$

Table A.1. Some common Lie groups

[a] A connected Lie group is semisimple if and only if its Lie algebra is semisimple (ss).

[b] Real ($\mathbb{R}$) and complex ($\mathbb{C}$) refer to the representation for the group.

[c] $s \equiv$ "simple" and $ss \equiv$ "semisimple" refers to the properties of the Lie algebra.

[d] The dimension n is the number of real or complex numbers required to specify an element of the real or complex Lie algebra, respectively.

[e] For a complex Lie group the real dimension is twice the complex dimension.

[f] Note that the entry for $SU(n)$ is for $n \geq 2$.

operators T^a will be denoted as $t^a_r \equiv D_{(r)}(T^a)$ and as we have shown for a semisimple Lie algebra we must have $\operatorname{tr} t^a_r = 0$ and

$$[t^a_r, t^b_r] = i f^{abc} t^c_r. \tag{A.7.8}$$

The group elements will be represented by matrices $D_r(g)$. For unitary representations $D_r(g)^\dagger = D(g)^{-1}$ and since sufficiently close to the identity we have

$$D_r(g(\vec{\omega})) = e^{i\vec{\omega} \cdot \vec{t}_r}, \tag{A.7.9}$$

then the t^a_r are Hermitian, $t^a_r = t^{a\dagger}_r$. Taking the Hermitian conjugate of Eq. (A.7.8) shows that the structure constants must be real,

$$f^{abc} \in \mathbb{R}. \tag{A.7.10}$$

If the Lie algebra is semisimple and the Lie group is also connected, then the Lie group is semisimple and so has no abelian subgroup, i.e., no $U(1)$ subgroup.

Taking the trace of two generators gives

$$\operatorname{tr}(t^a_r t^b_r) \equiv M^{ab}, \tag{A.7.11}$$

where the matrix M is real, symmetric and positive definite. We can always choose the generators t^a_r of a Lie algebra such that for any arbitrary $C(r) > 0$ we have

$$\operatorname{tr}(t^a_r t^b_r) = C(r)\delta^{ab}. \tag{A.7.12}$$

The notation $T(r) \equiv C(r)$ is also in common use (Schwartz, 2013).

Proof: M is symmetric since $\operatorname{tr} AB = \operatorname{tr} BA$. Hermitian A and B give $(\operatorname{tr} AB)^* = \operatorname{tr}(AB)^\dagger = \operatorname{tr}(B^\dagger A^\dagger) = \operatorname{tr} BA = \operatorname{tr} AB$ and so M is real. Since M is real symmetric, then it diagonalizes with an orthogonal transformation $OMO^T = D$ with $D^{ab} = d^a \delta^{ab}$ for some $d^a \in \mathbb{R}$.

Define $t_r'^a \equiv O^{ab}t_t^b$ using the summation convention so that $D^{ab} = d^a\delta^{ab} = (OMO^T)^{ab} = O^{ac}\text{tr}(t_r^c t_r^d)(O^T)^{db} = \text{tr}(t_r'^a t_t'^b)$. Then $d^a = \text{tr}[(t_r'^a)^2] > 0$ since for any Hermitian A we have $\text{tr}A^2 = \text{tr}(A^\dagger A) = \sum_i |\lambda_i|^2$ with λ_i the eigenvalues of A. Generators are not zero matrices, $t_r^a \neq 0$, and so $d_a = \text{tr}[(t_r^a)^2] > 0$. Next define $t_r''^a \equiv t_r'^a/\sqrt{d^a}$ so that $\text{tr}(t_r''^a t_r''^b) = \delta^{ab}$. Choosing any arbitrary normalization $C(r) > 0$ we can finally redefine our t_r^a as $t_r^a \equiv \sqrt{C(r)}t_r''^a$ so that $\text{tr}(t_r^a t_r^b) = C(r)\delta^{ab}$.

Using Eq. (A.7.8) and Eq. (A.7.12)) we find for the structure constants

$$f^{abc} = -\frac{i}{C(r)}\text{tr}\left\{\left[t_r^a, t_r^b\right]t_r^c\right\} = -\frac{i}{C(r)}\text{tr}\left\{t_r^a t_r^b t_r^c - t_r^b t_r^a t_r^c\right\}, \quad (A.7.13)$$

since $if^{abc} = if^{abc}\text{tr}[(t_r^c)^2]/C(r) = \text{tr}([t_r^a, t_r^b]t_r^c)/C(r)$. The cyclic nature of the trace in Eq. (A.7.13) means that the f^{abc} (i) are invariant under cyclic permutations of a, b, c; (ii) vanish if any of a, b, c are equal; and (iii) change sign under the pairwise interchange of a, b, c. So the f^{abc} are completely antisymmetric.

For any irreducible representation r there is a *conjugate representation* $\bar{r}$,

$$t_{\bar{r}}^a \equiv -(t_r^a)^* = -(t_r^a)^T, \quad (A.7.14)$$

since $if^{abc}(-t_r^{c*}) = (if^{abc}t_r^c)^* = [t_r^a, t_r^b]^* = [t_r^{a*}, t_r^{b*}] = [-t_r^{a*}, -t_r^{b*}]$. In the special case where $t_{\bar{r}}^a = t_r^a$ we say that the representation is *real*. Define

$$(t_A^a)_{bc} \equiv -if^{abc}. \quad (A.7.15)$$

Using the Jacobi identity in Eq. (A.7.7) it is readily verified that the t_A^a matrices also satisfy Eq. (A.7.8) and so are also a representation of the Lie algebra, which is referred to as the *adjoint representation*, $r = A$.

The quadratic *Casimir invariant* is defined as

$$\mathbf{t}_r^2 \equiv \mathbf{t}_r \cdot \mathbf{t}_r \equiv \sum_a t_r^a t_r^a \equiv t_r^a t_r^a. \quad (A.7.16)$$

It is an invariant because it commutes with all of the group generators,

$$[t_r^a, \mathbf{t}_r^2] = [t_r^a, t_r^b]t_r^b + t_r^b[t_r^a, t_r^b] = if^{abc}\left\{t_r^b, t_r^c\right\} = 0, \quad (A.7.17)$$

which follows since f^{abc} is antisymmetric. Schur's lemma states that any element of an irreducible representation of a Lie algebra that commutes with all generators is a multiple of the identity. So there is a real constant $C_2(r)$ such that

$$\mathbf{t}_r^2 = C_2(r)I, \quad (A.7.18)$$

where I is the identity matrix for the representation. $C_2(r)$ is real since $\mathbf{t}_r^2$ is Hermitian. A familiar example of a Casimir invariant is $\mathbf{J}^2 = J_x^2 + J_y^2 + J_z^2$ for the Lie algebra of the rotation group. In this case we have $C_2(r) = C_2(j) = j(j+1)$, where we use $r = j$ to label the irreducible representation of $SU(2)$.

Considering Eq. (A.7.12) and then taking the trace with respect to the generator indices a and b and comparing with the trace of Eq. (A.7.18) we find

$$d(G)C(r) = d(r)C_2(r), \quad (A.7.19)$$

where $d(r)$ is the *dimension of the irreducible representation* r (i.e., the matrices in r are $d(r) \times d(r)$ matrices) and where $d(G)$ is the *dimension of the group* G (i.e., there are $d(G)$ generators in the Lie group). Thus once a representation normalization is fixed through choosing $C(r)$ in Eq. (A.7.12) we can obtain $C_2(r)$ from Eq. (A.7.19).

Example of SU(N): The special unitary groups $SU(N)$ play an important role in particle physics and gauge theories. These are unitary Lie groups with unit determinant ($U^\dagger U = UU^\dagger = I$ and $\det U = 1$). It is a simple Lie group with a simple Lie algebra and it is also compact and simply connected. The lowest-dimensional faithful representation ($r = N$) is given in terms of $N \times N$ complex matrices, i.e., $d(r) = d(N) = N$. This is the defining or standard representation that is frequently referred to as the "fundamental representation." We can write Eq. (A.7.18) in representation r with Casimir $C_2(r)$ as

$$\sum_a (t_r^a t_r^a)_{ij} = C_2(r)\delta_{ij}. \tag{A.7.20}$$

The conventional normalization for $SU(2)$ in the defining representation, $r = 2$ and $d(2) = 2$, has the three 2×2 generators, $t_2^a = \frac{1}{2}\sigma^a$. This gives $C(2) = \frac{1}{2}$, since

$$\mathrm{tr}[t_2^a t_2^b] = C(2)\delta^{ab} = \tfrac{1}{2}\delta^{ab}. \tag{A.7.21}$$

We will generalize to $SU(N)$ with the first three generators in the defining representation, $r = N$, always being simple generalizations of the $SU(2)$ generators. So we adopt Eq. (A.7.21) as the normalization convention for the defining representation of $SU(N)$ in terms of $d(N) \times d(N) = N \times N$ matrices as

$$\mathrm{tr}[t_N^a t_N^b] = C(N)\delta^{ab} = \tfrac{1}{2}\delta^{ab}. \tag{A.7.22}$$

There are $N^2 - 1$ linearly independent traceless Hermitian $N \times N$ matrices and so $N^2 - 1$ generators in $SU(N)$, i.e., $d(G) \equiv d(SU(N)) = N^2 - 1$. Using Eqs. (A.7.18) and (A.7.19) we find the quadratic Casimir invariant in the defining representation

$$\sum_a (t_r^a t_r^a)_{ij} = C_2(r)C_2(N) = \frac{d(SU(N))C(N)}{d(N)} = \frac{N^2 - 1}{2N}. \tag{A.7.23}$$

The same result will clearly hold for the conjugate representation, $r = \bar{N}$, since $d(\bar{N}) = d(N) = N$ and $C(\bar{N}) = C(N) = \frac{1}{2}$. Using Eqs. (A.7.22) and (A.7.8) it follows since the generators are traceless that

$$t_N^a t_N^b = \tfrac{1}{2}\mathrm{tr}[t_N^a, t_N^b] + \tfrac{1}{2}\{t_N^a, t_N^b\} = \tfrac{1}{2N}\delta^{ab} + \tfrac{1}{2}d^{abc}t_N^c + \tfrac{1}{2}if^{abc}t_N^c, \tag{A.7.24}$$

where $\mathrm{tr}(t_N^a t_N^b) = \frac{1}{2}\delta^{ab}$ and where it is readily verified that $d^{abc} = 2\mathrm{tr}(t_N^a\{t_N^b, t_N^c\})$ is totally symmetric. For $SU(2)$ we see that $d^{abc} = 0$.

For the adjoint representation recall that the generators are $(t_A^a)_{bc} = -if^{abc}$ and so these are $(N^2 - 1) \times (N^2 - 1)$ traceless Hermitian matrices. So the adjoint representation ($r = A$) of $SU(N)$ has dimension $d(r) = d(A) = N^2 - 1$, which is the same as the group dimension $d(G) = d(SU(N)) = N^2 - 1$. This gives the result $C_2(A) = d(SU(N))C(A)/d(A) = C(A)$. Using Eq. (A.7.12) we obtain

$$C(A)\delta^{ab} = \mathrm{tr}(t_A^a t_A^b) = -f^{acd}f^{bdc} = f^{acd}f^{bcd} = N\delta^{ab}, \tag{A.7.25}$$

which can be explicitly checked for $SU(2)$ and $SU(3)$. The proof for all N is not given here. So for $SU(N)$ we have $C_2(A) = C(A) = N$.

Key SU(N) results: We adopt the common notations T_F, C_F and C_A and recall that $t_F^a \equiv t^a(N)$ is the defining (fundamental) representation:

$$T_F \equiv C(N) = \tfrac{1}{2}, \quad C_F \equiv C_2(N) = \tfrac{N^2-1}{2N}, \quad C_A \equiv C_2(A) = C(A) = N,$$

$$\sum_a (t_F^a t_F^a)_{ij} = C_F \delta_{ij}, \quad \text{tr}(t_F^a t_F^b) = t_{Fij}^a t_{Fji}^b = T_F \delta^{ab}, \quad f^{acd} f^{bcd} = C_A \delta^{ab}. \quad (A.7.26)$$

Other notation: In Sec. 5.3.1 we used $\mathbf{N}$ and $\overline{\mathbf{N}}$ to denote the defining ($r = N$) and conjugate ($r = \overline{N}$) representations, respectively, which is the conventional notation in that context. The tensor products of irreducible representations of $SU(N)$ and their decomposition into the direct sum of irreducible representations was discussed in Sec. 5.3.1. For example, we have $\overline{\mathbf{N}} \otimes \mathbf{N} = (\mathbf{N^2 - 1}) \oplus \mathbf{1}$ from Eq. (5.3.30), where $(\mathbf{N^2 - 1})$ is another notation for the adjoint representation $r = A$ and $\mathbf{1}$ is the trivial representation $d(r) = 1$. So we have $\dim(A) = \dim(\mathbf{N^2 - 1}) = N^2 - 1$.

Results for $SU(3)$: The standard basis for the fundamental representation of $SU(3)$ is $t_F^a \equiv \lambda^a / 2$ for $a = 1, \ldots, 8$, where λ^a are the eight Gell-Mann matrices

$$\lambda^1 = \begin{pmatrix} 0 & 1 & 0 \\ 1 & 0 & 0 \\ 0 & 0 & 0 \end{pmatrix} = \begin{pmatrix} \sigma^1 & 0 \\ & 0 & 0 \end{pmatrix}, \quad \lambda^2 = \begin{pmatrix} 0 & -i & 0 \\ i & 0 & 0 \\ 0 & 0 & 0 \end{pmatrix} = \begin{pmatrix} \sigma^2 & 0 \\ & 0 & 0 \end{pmatrix},$$

$$\lambda^3 = \begin{pmatrix} 1 & 0 & 0 \\ 0 & -1 & 0 \\ 0 & 0 & 0 \end{pmatrix} = \begin{pmatrix} \sigma^3 & 0 \\ & 0 & 0 \end{pmatrix}, \quad \lambda^4 = \begin{pmatrix} 0 & 0 & 1 \\ 0 & 0 & 0 \\ 1 & 0 & 0 \end{pmatrix}, \quad \lambda^5 = \begin{pmatrix} 0 & 0 & -i \\ 0 & 0 & 0 \\ i & 0 & 0 \end{pmatrix},$$

$$\lambda^6 = \begin{pmatrix} 0 & 0 & 0 \\ 0 & 0 & 1 \\ 0 & 1 & 0 \end{pmatrix}, \quad \lambda^7 = \begin{pmatrix} 0 & 0 & 0 \\ 0 & 0 & -i \\ 0 & i & 0 \end{pmatrix}, \quad \lambda^8 = \frac{1}{\sqrt{3}} \begin{pmatrix} 1 & 0 & 0 \\ 0 & 1 & 0 \\ 0 & 0 & -2 \end{pmatrix}. \quad (A.7.27)$$

We see that the Gell-Mann matrices λ^a are 3×3 generalizations of the Pauli spin matrices σ^i. The following results can be shown (Ynduráin, 1983):

$$f^{123} = 1, \ f^{147} = f^{246} = f^{257} = f^{345} = 1/2, \ f^{156} = f^{367} = -1/2, \ f^{458} = f^{678} = \sqrt{3}/2,$$

$$\{t^a, t^b\} \equiv \sum_c d^{abc} t^c + \tfrac{1}{3} \delta^{ab}, \quad (A.7.28)$$

$$d^{118} = d^{228} = d^{338} = \frac{1}{\sqrt{3}}, \quad d^{448} = d^{558} = d^{668} = d^{778} = -\frac{1}{2\sqrt{3}}, \quad (A.7.29)$$

$$d^{888} = -\frac{1}{\sqrt{3}}, \quad d^{146} = d^{157} = d^{247} = d^{256} = d^{344} = d^{355} = \tfrac{1}{2}, \quad d^{366} = d^{377} = -\tfrac{1}{2},$$

$$\text{tr}(t_F^a t_F^b t_F^c) = \tfrac{i}{4} f^{abc} + \tfrac{1}{4} d^{abc}, \quad \sum_{abc} (d^{abc})^2 = \tfrac{40}{3}, \quad (A.7.30)$$

$$\sum_{abc} (f^{abc})^2 = 24, \quad \sum_{mka} \epsilon_{imk} (t_F^a)_{jm} (t_F^a)_{k\ell} = -\tfrac{2}{3} \epsilon_{ijl}. \quad (A.7.31)$$

The f^{abc} are totally antisymmetric, the d^{abc} are totally symmetric and values not listed are zero.

A.8 Results for Matrices

Similarity transformation: $A' = SAS^{-1}$, where S is any invertible matrix. S^{-1} exists if and only if S is nonsingular, i.e., if and only if $\det S \neq 0$.

Transpose: Every matrix A is similar to its transpose A^T.

Hermitian conjugate or adjoint of a matrix: $A^\dagger = A^{*T}$.

Normal matrix: $A^\dagger A = AA^\dagger$; Matrix A is diagonalizable by a unitary transformation if and only if it is normal. Any normal A has a complete set of eigenvectors.

Hermitian or self-adjoint matrix: $A^\dagger = A$.

Determinant: For any $n \times n$ complex matrix A

$$\det A = \sum_{k_1,k_2,\ldots,k_n=1}^{n} \epsilon^{k_1 k_2 \cdots k_n} A_{1k_1} A_{2k_2} \cdots A_{nk_n}, \tag{A.8.1}$$

where $\epsilon^{k_1 k_2 \cdots k_n}$ is the rank-n Levi-Civita (antisymmetric) tensor with $\epsilon^{12 \cdots n} = +1$.

Rank of a matrix: The row and column spaces of a rectangular $m \times n$ matrix A are the spaces spanned by the m rows and n columns of A, respectively. The dimensions of the spaces are called the row rank and column rank, respectively. By comparing A and A^T it can be shown that the row and column ranks are always equal and so we write this as $\mathrm{rank}(A)$. Clearly, $\mathrm{rank}(A) \leq \min(m,n)$. For a square $n \times n$ matrix A it follows that $\det A = 0$ unless A has maximum rank, $\mathrm{rank}(A) = n$.

Determinant properties: For any $n \times n$ complex matrix A

$$\det A = \det A^T, \quad \det(A^*) = (\det A)^*, \quad \det(A^\dagger) = (\det A)^*, \quad \det(A^\dagger A) \geq 0,$$

$$\det(AB) = \det A \det B, \quad \det A^{-1} = 1/\det A \ \text{ if } \ \det A \neq 0, \tag{A.8.2}$$

$$\epsilon^{\ell_1 \ell_2 \cdots \ell_n} \det(A) = \sum_{k_1,k_2,\ldots,k_n=1}^{n} \epsilon^{k_1 k_2 \cdots k_n} A_{\ell_1 k_1} A_{\ell_2 k_2} \cdots A_{\ell_n k_n}, \tag{A.8.3}$$

$$\det(A) = \prod_{i=1}^{\ell} (\lambda_i)^{k_i}, \quad \mathrm{tr}(A) = \sum_{i=1}^{\ell} k_i \lambda_i, \quad \det(A) = \exp(\mathrm{tr}\ln A), \tag{A.8.4}$$

$$\exp A \equiv \lim_{N\to\infty} [1 + (A/N)]^N = \sum_{k=1}^{\infty} (1/k!) A^k, \tag{A.8.5}$$

where the ℓ distinct eigenvalues of A are $\lambda_1, \lambda_2, \ldots, \lambda_\ell$ with corresponding algebraic multiplicities $k_1, k_2, \ldots, k_\ell$ and where $\sum_{i=1}^{\ell} k_i = n$.

Inverse of a matrix: If $AB = I$ then $\det A \det B = 1 \Rightarrow \det A, \det B \neq 0$ and so $(BA - I)B = 0$ $\Rightarrow BA - I = 0 \Rightarrow AB = BA = I \Rightarrow A = B^{-1}$.

Baker-Campbell-Hausdorff formula and related results

$$e^A e^B = \exp\left(A + B + \tfrac{1}{2}[A,B] + \tfrac{1}{12}[A,[A,B]] - \tfrac{1}{12}[B,[A,B]] + \cdots\right). \tag{A.8.6}$$

$$e^A B e^{-A} = B + [A,B] + (1/2!)[A,[A,B]] + (1/3!)[A,[A,[A,B]]] + \cdots$$
$$= \sum_{j=0}^{\infty} (1/j!)[A, \ldots, [A, [A, [A, B]]] \cdots]. \tag{A.8.7}$$

$$e^A e^B e^{-A} = e^{(e^A B e^{-A})} = e^{(B+[A,B]+\frac{1}{2!}[A,[A,B]]+\frac{1}{3!}[A,[A,[A,B]]]+\cdots)}$$
$$= \exp\left(\sum_{j=1}^{\infty} (1/j!)[A, \ldots, [A, [A, [A, B]]] \cdots]\right). \tag{A.8.8}$$

For the derivative of the inverse of a matrix,

$$dM^{-1}(a)/da = -M^{-1}(a)[dM(a)/da]M^{-1}(a) \tag{A.8.9}$$

(**Proof:** Let matrix $M(a)$ depend on a variable a. Denote $M' = dM/da$. Then use $0 = I' = (MM^{-1})' = M'M^{-1} + M(M^{-1})' \Rightarrow (M^{-1})' = -M^{-1}M'M^{-1}$.)

References

Abers, Ernest S., and Lee, Benjamin W. 1973. Gauge theories. *Physics Reports*, **9**(1), 1.

Abramowitz, Milton, and Stegun, Irene A. 1972. *Handbook of Mathematical Functions with Formulas, Graphs, and Mathematical Tables*. 10th ed. New York: Dover.

Adler, Stephen L. 1969. Axial-vector vertex in spinor electrodynamics. *Phys. Rev.*, **177**(Jan), 2426–2438.

Aitchison, I. J. R., and Hey, A. J. G. 2013. *Gauge Theories in Particle Physics: 2 Volume Set: A Practical Introduction*. Boca Raton, FL: CRC Press.

Akhmedov, E. Kh., and Smirnov, A. Yu. 2009. Paradoxes of neutrino oscillations. *Physics of Atomic Nuclei*, **72**(8), 1363–1381.

Albeverio, S. A., and Hoegh-Krohn, R. J. 1976. *Mathematical Theory of Feynman Path Integrals*. Lecture Notes in Mathematics, vol. 523. Berlin: Springer-Verlag.

Alkofer, Reinhard, and von Smekal, Lorenz. 2001. The infrared behaviour of QCD Green's functions: Confinement, dynamical symmetry breaking, and hadrons as relativistic bound states. *Physics Reports*, **353**(5), 281–465.

Amsler, C., et al. 2008. The Review of Particle Physics. *Physics Letters B*, **667**, 1.

Amsler, Claude. 2018. *The Quark Structure of Hadrons: An Introduction to the Phenomenology and Spectroscopy*. Lecture Notes in Physics, vol. 949. Cham: Springer International Publishing.

Antchev, G., et al. 2019. First measurement of elastic, inelastic and total cross-section at $\sqrt{s} = 13$ TeV by TOTEM and overview of cross-section data at LHC energies. *Eur. Phys. J.*, **C79**(2), 103.

Aoyama, Tatsumi, Hayakawa, Masashi, Kinoshita, Toichiro, and Nio, Makiko. 2012a. Complete tenth-order QED contribution to the muon $g-2$. *Phys. Rev. Lett.*, **109**(Sep), 111808.

Aoyama, Tatsumi, Hayakawa, Masashi, Kinoshita, Toichiro, and Nio, Makiko. 2012b. Tenth-order QED contribution to the electron $g-2$ and an improved value of the fine structure constant. *Phys. Rev. Lett.*, **109**(Sep), 111807.

Appelquist, Thomas, and Carazzone, J. 1975. Infrared singularities and massive fields. *Phys. Rev. D*, **11**(May), 2856–2861.

Arfken, G. B, and Weber, H. J. 1995. *Mathematical Methods for Physicists*. San Diego, CA: Academic Press.

Arnison, G., et al. 1983. Experimental observation of isolated large transverse energy electrons with associated missing energy at s = 540 GeV. *Physics Letters B*, **122**(1), 103–116.

ATLAS Collaboration, Aad, G., et al. 2012. Rapidity gap cross sections measured with the ATLAS detector in pp collisions at $\sqrt{s} = 7$ TeV. *The European Physical Journal C*, **72**(3), 1926.

Bailin, D., and Love, A. 1994. *Supersymmetric Gauge Field Theory and String Theory*. Boca Raton, FL: CRC Press.

Bailin, David, and Love, Alexander. 1993. *Introduction to Gauge Field Theory*. Revised ed. Graduate Student Series in Physics. New York: Taylor & Francis Group.

Banner, M., et al. 1983. Observation of single isolated electrons of high transverse momentum in events with missing transverse energy at the CERN pp collider. *Physics Letters B*, **122**(5), 476–485.

Bargmann, V. 1964. Note on Wigner's theorem on symmetry operations. *Journal of Mathematical Physics*, **5**(7), 862–868.

Barnett, Stephen M., Allen, L., Cameron, Robert P., Gilson, Claire R., Padgett, Miles J., Speirits, Fiona C., and Yao, Alison M. 2016. On the natures of the spin and orbital parts of optical angular momentum. *Journal of Optics*, **18**(6), 064004.

Barone, Vincenzo, and Predazzi, Enrico. 2002. *The Relativistic S-Matrix*. Berlin: Springer.

Becchi, C., Rouet, A., and Stora, R. 1975. Renormalization of the abelian Higgs-Kibble model. *Communications in Mathematical Physics*, **42**(2), 127–162.

Becchi, C., Rouet, A., and Stora, R. 1976. Renormalization of gauge theories. *Annals of Physics*, **98**(2), 287–321.

Belinfante, Frederik J. 1940. On the current and the density of the electric charge, the energy, the linear momentum and the angular momentum of arbitrary fields. *Physica*, **7**(5), 449–474.

Bell, J. S., and Jackiw, R. 1969. A PCAC puzzle: $\pi^0 \to \gamma\gamma$ in the σ-model. *Il Nuovo Cimento A (1965–1970)*, **60**(1), 47–61.

Berezin, F. A. 1966. *The Method of Second Quantization by F. A. Berezin*. Translated by Nobumichi Mugibayashi and Alan Jeffrey. Pure and Applied Physics vol. 24. New York: Academic Press.

Bethe, H. A. 1947. The electromagnetic shift of energy levels. *Phys. Rev.*, **72**(Aug), 339–341.

Bigi, I. I., and Sanda, A. I. 2009. *CP Violation*. 2nd ed. Cambridge Monographs on Particle Physics, Nuclear Physics and Cosmology. Cambridge: Cambridge University Press.

Bilal, Adel. 2008. Lectures on anomalies, arXiv:0802.0634.

Bjorken, J. D., and Drell, S. D. 1964. *Relativistic Quantum Mechanics*. International Series in Pure and Applied Physics. New York: McGraw-Hill.

Bjorken, J. D., and Drell, S. D. 1965. *Relativistic Quantum Fields*. International Series in Pure and Applied Physics. New York: McGraw-Hill.

Blaschke, Daniel N., Gieres, François, Reboud, Méril, and Schweda, Manfred. 2016. The energy–momentum tensor(s) in classical gauge theories. *Nuclear Physics B*, **912**, 192–223. Mathematical Foundations of Quantum Field Theory: A Volume dedicated to the Memory of Raymond Stora.

Bleuler, Konrad. 1950. Eine neue Methode zur Behandlung der longitudenalen und skalaren Photonen. *Hlvetica Physica Acta*, **23**(5), 567–586.

Bogliubov, N. N., and Parasiuk, O. S. 1957. On the multiplication of the causal function in the quantum theory of fields. *Acta mathematica*, **97**, 227–266.

Bogolubov, N. N., Logunov, A. A., Oksak, A. I., and Todorov, I. 2012. *General Principles of Quantum Field Theory*. Mathematical Physics and Applied Mathematics. Dordrecht: Springer Netherlands.

Bogolyubov, N. N., and Shirkov, D. V. 1959. Introduction to the theory of quantized fields. *Intersci. Monogr. Phys. Astron.*, **3**, 1–720.

Bowman, Patrick O., Heller, Urs M., Leinweber, Derek B., Parappilly, Maria B., Williams, Anthony G., and Zhang, Jian-bo. 2005. Unquenched quark propagator in Landau gauge. *Phys. Rev. D*, **71**, 054507.

Bowman, Patrick O., Heller, Urs M., Leinweber, Derek B., Parappilly, Maria B., Sternbeck, André, von Smekal, Lorenz, Williams, Anthony G., and Zhang, Jianbo. 2007. Scaling behavior and positivity violation of the gluon propagator in full QCD. *Phys. Rev. D*, **76**(Nov), 094505.

Brown, L. S. 1992. *Quantum Field Theory*. Cambridge: Cambridge University Press.

Cabibbo, Nicola. 1963. Unitary symmetry and leptonic decays. *Phys. Rev. Lett.*, **10**(Jun), 531–533.

Cahn, R. N. 1984. *Semi-Simple Lie Algebras and Their Representations*. Menlo Park, CA: Benjamin/Cummings.

Callan, Curtis G. 1970. Broken scale invariance in scalar field theory. *Phys. Rev. D*, **2**(Oct), 1541–1547.

Carruthers, P. 1966. *Introduction to Unitary Symmetry*. New York: Interscience.

Casimir, H. B. G. 1948. On the attraction between two perfectly conducting plates. *Proceedings of the Royal Netherlands Academy of Arts and Sciences*, **51**, 793–795.

Casimir, H. B. G., and Polder, D. 1948. The influence of retardation on the London–Van der Waals forces. *Phys. Rev.*, **73**(Feb), 360–372.

Chanowitz, M., Furman, M., and Hinchliffe, I. 1979. The axial current in dimensional regularization. *Nuclear Physics B*, **159**(1), 225–243.

Cheng, T. P., and Li, L. F. 1984. *Gauge Theory of Elementary Particle Physics*. Oxford Science Publications. Oxford: Oxford University Press.

Chew, Geoffrey F. 1962. *S-Matrix Theory of the Strong Interactions*. Frontiers in Physics. New York: W. A. Benjamin.

Choquet-Bruhat, Y., DeWitt-Morette, C., and Dillard-Bleick, M. 1982. *Analysis, Manifolds, and Physics*. Revised ed. Amsterdam: North-Holland.

Close, F. E. 2010. *Neutrino*. Oxford: Oxford University Press.

Close, Frank E. 2011. *The Infinity Puzzle*. Oxford: Oxford University Press.

Coleman, Sidney. 1985. *Aspects of Symmetry: Selected Erice Lectures*. Cambridge University Press.

Coleman, Sidney, and Mandula, Jeffrey. 1967. All possible symmetries of the *S* matrix. *Phys. Rev.*, **159**(Jul), 1251–1256.

Collins, J. C. 1984. *Renormalization*. Cambridge University Press.

Currie, D. G., Jordan, T. F., and Sudarshan, E. C. G. 1963. Relativistic invariance and Hamiltonian theories of interacting particles. *Rev. Mod. Phys.*, **35**(Apr), 350–375.

Das, Ashok. 2008. *Lectures on Quantum Field Theory*. World Scientific Publishing Company.

Davies, C. T. H., Batrouni, G. G., Katz, G. R., Kronfeld, A. S., Lepage, G. P., Wilson, K. G., Rossi, P., and Svetitsky, B. 1988. Fourier acceleration in lattice gauge theories. I. Landau gauge fixing. *Physical Review D, Particles and Fields*, **37**(6), 1581–1588.

DeGrand, Thomas, and DeTar, Carleton. 2006. *Lattice Methods for Quantum Chromodynamics*. Singapore: World Scientific Publishing Company.

Deser, S., Grisaru, M., and Pendleton, H. 1970. *Lectures on Elementary Particles and Quantum Field Theory*. Cambridge, MA: MIT Press.

Dirac, P. A. M. 1950. Generalized Hamiltonian dynamics. *Canadian Journal of Mathematics*, **2**, 129–148.

Dirac, P. A. M. 1958. Generalized Hamiltonian dynamics. *Proceedings of the Royal Society of London. Series A. Mathematical and Physical Sciences*, **246**(1246), 326–332.

Dirac, P. A. M. 1982. *Principles of Quantum Mechanics*. Oxford: Oxford University Press.

Dirac, P. A. M. 2001. *Lectures on Quantum Mechanics: An Unabridged Reprint of the Work Originally Published by the Belfer Graduate School of Science, Yeshiva University, New York, 1964*. New York: Dover.

Dreiner, Herbi K., Haber, Howard E., and Martin, Stephen P. 2010. Two-component spinor techniques and Feynman rules for quantum field theory and supersymmetry. *Physics Reports*, **494**(1), 1–196.

Duck, Ian, and Sudarshan, E. C. G. 1998a. *Pauli and the Spin-Statistics Theorem*. Singapore: World Scientific Publishing Company.

Duck, Ian, and Sudarshan, E. C. G. 1998b. Toward an understanding of the spin-statistics theorem. *American Journal of Physics*, **66**(4), 284–303.

Dürr, S., Fodor, Z., Frison, J., Hoelbling, C., Hoffmann, R., Katz, S. D., Krieg, S., Kurth, T., Lellouch, L., Lippert, T., Szabo, K. K., and Vulvert, G. 2008. Ab initio determination of light hadron masses. *Science*, **322**(5905), 1224.

Dyson, F. J. 1949. The radiation theories of Tomonaga, Schwinger, and Feynman. *Phys. Rev.*, **75**(Feb), 486–502.

Dyson, F. J. 1952. Divergence of perturbation theory in quantum electrodynamics. *Phys. Rev.*, **85**(Feb), 631–632.

Dzyaloshinskii, I. E., Lifshitz, E. M., and Pitaevskii, Lev P. 1961. General theory of Van der Waals' forces. *Soviet Physics Uspekhi*, **4**(2), 153–176.

Edmonds, A. R. 2016. *Angular Momentum in Quantum Mechanics*. Investigations in Physics. Princeton, NJ: Princeton University Press.

Englert, F., and Brout, R. 1964. Broken symmetry and the mass of gauge vector mesons. *Phys. Rev. Lett.*, **13**(Aug), 321–323.

Faddeev, L. D., and Popov, V. N. 1967. Feynman diagrams for the Yang-Mills field. *Physics Letters B*, **25**(1), 29 – 30.

Fermi, Enrico. 1932. Quantum theory of radiation. *Rev. Mod. Phys.*, **4**(Jan), 87–132.

Feynman, R. P. 1948. Space-time approach to non-relativistic quantum mechanics. *Rev. Mod. Phys.*, **20**(Apr), 367–387.

Feynman, R. P. 1949a. Space-time approach to quantum electrodynamics. *Phys. Rev.*, **76**(Sep), 769–789.

Feynman, R. P. 1949b. The theory of positrons. *Phys. Rev.*, **76**(Sep), 749–759.

Feynman, R. P. 1950. Mathematical formulation of the quantum theory of electromagnetic interaction. *Phys. Rev.*, **80**(Nov), 440–457.

Feynman, R. P. 1972. *Statistical Mechanics*. Reading, MA: W. A. Benjamin.

Feynman, R. P., and Gell-Mann, M. 1958. Theory of the Fermi interaction. *Physical Review*, **109**(1), 193–198.

Feynman, R. P., and Hibbs, A. R. 1965. *Quantum Mechanics and Path Integrals*. New York: McGraw-Hill.

Feynman, Richard P. 1963–1965. *The Feynman Lectures on Physics*. Reading, MA: Addison-Wesley.

Flannery, M. R. 2005. The enigma of nonholonomic constraints. *American Journal of Physics*, **73**(3), 265–272.

Fodor, Zoltan, and Hoelbling, Christian. 2012. Light hadron masses from lattice QCD. *Rev. Mod. Phys.*, **84**(Apr), 449–495.

Frautschi, Steven C. 1963. *Regge Poles and S-Matrix Theory*. New York: W. A. Benjamin.

Fujii, Yasunori, Ohta, Nobuyoshi, and Taniguchi, Hiroshi. 1981. On the definitions of γ_5 in continuous dimensions. *Nuclear Physics B*, **177**(2), 297–324.

Fujikawa, Kazuo. 1979. Path-integral measure for gauge-invariant fermion theories. *Phys. Rev. Lett.*, **42**(Apr), 1195–1198.

Georgi, H. 1999. *Lie Algebras in Particle Physics*. Reading, MA: Perseus.

Gilmore, R. 1974. *Lie Groups, Lie Algebra and Some of Their Applications*. New York: Wiley-Interscience.

Glashow, Sheldon L. 1961. Partial-symmetries of weak interactions. *Nuclear Physics*, **22**(4), 579–588.

Glimm, J., and Jaffe, A. 1981. *Quantum Physics: A Functional Integral Point of View*. New York: Springer–Verlag.

Goldstein, H., Safko, J. L., and Poole, C. P., Jr. 2013. *Classical Mechanics*. 3rd ed. San Francisco, CA: Pearson.

Goldstone, J. 1961. Field theories with "Superconductor" solutions. *Il Nuovo Cimento (1955–1965)*, **19**(1), 154–164.

Goldstone, Jeffrey, Salam, Abdus, and Weinberg, Steven. 1962. Broken symmetries. *Phys. Rev.*, **127**(Aug), 965–970.

Gray, C. G., and Taylor, Edwin F. 2007. When action is not least. *American Journal of Physics*, **75**(5), 434–458.

Green, H. S. 1953. A generalized method of field quantization. *Phys. Rev.*, **90**(Apr), 270–273.

Greiner, W. 2000. *Relativistic Quantum Mechanics: Wave Equations*. 3rd ed. Berlin: Springer-Verlag.

Greiner, W., and Rheinhardt, J. 2009. *Quantum Electrodynamics*. 4th ed. Berlin: Springer-Verlag.

Greiner, Walter, and Reinhardt, Joachim. 1996. *Field Quantization*. Berlin: Springer-Verlag.

Griffiths, D. 2008. *Introduction to Elementary Particles*. 2nd revised ed. KGaA, Weinheim: Wiley-VCH Verlag GmbH & Co.

Grządkowski, B., Krawczyk, P., and Pokorski, S. 1984. Natural relations and Appelquist-Carazzone decoupling theorem. *Phys. Rev. D*, **29**(Apr), 1476–1487.

Gupta, Suraj N. 1950. Theory of longitudinal photons in quantum electrodynamics. *Proceedings of the Physical Society A*, **63**(July), 681–691.

Guralnik, G. S. 1964a. Photon as a symmetry-breaking solution to field theory. I. *Phys. Rev.*, **136**(Dec), B1404–B1416.

Guralnik, G. S. 1964b. Photon as a symmetry-breaking solution to field theory. II. *Phys. Rev.*, **136**(Dec), B1417–B1422.

Guralnik, G. S., Hagen, C. R., and Kibble, T. W. B. 1964. Global conservation laws and massless particles. *Phys. Rev. Lett.*, **13**(Nov), 585–587.

Gusynin, V. P., Schreiber, A. W., Sizer, T., and Williams, A. G. 1999. Chiral symmetry breaking in dimensionally regularized nonperturbative quenched QED. *Phys. Rev. D*, **60**(Aug), 065007.

Haag, R. 1955. On quantum field theories. *Matematisk-fysiske Meddelelser*, **29**, 1–37.

Haag, R. 1958. Quantum field theories with composite particles and asymptotic conditions. *Physical Review*, **112**(2), 669–673.

Hahn, Yukap, and Zimmermann, Wolfhart. 1968. An elementary proof of Dyson's power counting theorem. *Communications in Mathematical Physics*, **10**(4), 330–342.

Haidt, Dieter. 2015. The discovery of weak neutral currents. *Adv. Ser. Dir. High Energy Phys.*, **23**, 165–183.

Hamermesh, M. 1962. *Group Theory and Its Application to Physical Problems*. Reading, MA: Addison-Wesley.

Harvey, J. A. 2003. TASI 2003 lectures on anomalies, arXiv:hep-th/0509097.

Henneaux, M., and Teitelboim, C. 1994. *Quantization of Gauge Systems*. Princeton, NJ: Princeton University Press.

Henrici, P. 1993. *Applied and Computational Complex Analysis, vol. 3: Discrete Fourier Analysis, Cauchy Integrals, Construction of Conformal Maps, Univalent Functions*. Applied and Computational Complex Analysis. Hoboken, NJ: Wiley.

Hepp, Klaus. 1966. Proof of the Bogoliubov-Parasiuk theorem on renormalization. *Communications in Mathematical Physics*, **2**(1), 301–326.

Higgs, Peter W. 1964a. Broken symmetries and the masses of gauge bosons. *Phys. Rev. Lett.*, **13**(Oct), 508–509.

Higgs, P. W. 1964b. Broken symmetries, massless particles and gauge fields. *Physics Letters*, **12**(2), 132–133.

Hirayama, Minoru, and Ueno, Masataka. 2000. Non-abelian Stokes theorem for Wilson loops associated with general gauge groups. *Progress of Theoretical Physics*, **103**(1), 151–159.

Huang, Kerson. 1992. *Quarks, Leptons and Gauge Fields*. 2nd ed. Singapore: World Scientific.

Itzykson, C., and Zuber, J. B. 1980. *Quantum Field Theory*. New York: McGraw-Hill.

Jackson, J. D. 1975. *Classical Electrodynamics*. 2nd ed. Wiley.

Jarlskog, C. 1985. Commutator of the quark mass matrices in the standard electroweak model and a measure of maximal CP nonconservation. *Phys. Rev. Lett.*, **55**(Sep), 1039–1042.

Jost, R. 1979. *The General Theory of Quantized Fields*. Lectures in Applied Mathematics; Proceedings of the Summer Seminar. American Mathematical Society.

Jost, Res. 1957. Eine Bemerkung zum PCT Theorem. *Helv. Phys. Acta*, **30**, 409.

Källén, G. 1952. On the definition of the renormalization constants in quantum electrodynamics. *Helv. Phys. Acta*, **25**(4), 417–435.

Kashiwa, Tarō. 1981. Euclidean path-integral representation of the vector field. I: The massive vector case. *Progress of Theoretical Physics*, **66**(5), 1858–1878.

Kayser, Boris. 1981. On the quantum mechanics of neutrino oscillation. *Phys. Rev. D*, **24**(Jul), 110–116.

Kibble, T. W. B. 2015. History of electroweak symmetry breaking. *Journal of Physics: Conference Series*, **626**(1), 012001.

Klauder, J., and Skagerstam, B. 1985. *Coherent States*. Singapore: World Scientific Publishing Company.

Klauder, John R. 2011. *A Modern Approach to Functional Intergration: Applied and Numerical Harmonic Analysis*. Basel: Birkhäuser.

Klein, O., and Nishina, Y. 1929. Über die Streuung von Strahlung durch freie Elektronen nach der neuen relativistischen Quantendynamik von Dirac. *Zeitschrift für Physik*, **52**(11), 853–868.

Knechtli, Francesco, Günther, Michael, and Peardon, Michael. 2017. *Lattice Quantum Chromodynamics: Practical Essentials*. SpringerBriefs in Physics. Springer.

Kobach, Andrew, Manohar, Aneesh V., and McGreevy, John. 2018. Neutrino oscillation measurements computed in quantum field theory. *Physics Letters B*, **783**, 59–75.

Kobayashi, Makoto, and Maskawa, Toshihide. 1973. CP-violation in the renormalizable theory of weak interaction. *Progress of Theoretical Physics*, **49**(2), 652–657.

Kokkedee, J. J. J. 1969. *The Quark Model*. New York: W. A. Benjamin.

Lamoreaux, S. K. 1997. Demonstration of the Casimir force in the 0.6 to 6μm range. *Phys. Rev. Lett.*, **78**(Jan), 5–8.

Lee, T. D., and Yang, C. N. 1956. Question of parity conservation in weak interactions. *Phys. Rev.*, **104**(Oct), 254–258.

Lehmann, H. 1954. Uber Eigenschaften von Ausbreitungsfunktionen und Renormierungskonstanten quantisierter Felder. *Il Nuovo Cimento (1943–1954)*, **11**(4), 342–357.

Lehmann, H., Symanzik, K., and Zimmerman, W. 1955. Zur Formulierung quantisierter Feldtheorien. *Nuovo Cimento*, **1**(1), 205–225.

Leutwyler, H. 1965. A no-interaction theorem in classical relativistic Hamiltonian particle mechanics. *Il Nuovo Cimento (1955–1965)*, **37**(2), 556–567.

Lichtenberg, D. B. 1978. *Unitary Symmetry and Elementary Particles*. 2nd ed. New York: Academic Press.

Lowenstein, J. H. 1976. Convergence theorems for renormalized Feynman integrals with zero-mass propagators. *Communications in Mathematical Physics*, **47**(1), 53–68.

Lowenstein, J. H., and Zimmermann, W. 1975. The power counting theorem for Feynman integrals with massless propagators. *Communications in Mathematical Physics*, **44**(1), 73–86.

Luders, Gerhart. 1954. On the equivalence of invariance under time reversal and under particle-antiparticle conjugation for relativistic field theories. *Kong. Dan. Vid. Sel. Mat. Fys. Med.*, **28N5**(5), 1–17.

Maki, Ziro, Nakagawa, Masami, and Sakata, Shoichi. 1962. Remarks on the unified model of elementary particles. *Progress of Theoretical Physics*, **28**(5), 870–880.

Mandl, F., and Shaw, G. 1984. *Quantum Field Theory*. New York: Wiley-Interscience.

Marmo, G., Mukunda, N., and Sudarshan, E. C. G. 1984. Relativistic particle dynamics: Lagrangian proof of the no-interaction theorem. *Phys. Rev. D*, **30**(Nov), 2110–2116.

Matthews, P. T., and Salam, Abdus. 1959. Relativistic theory of unstable particles. II. *Phys. Rev.*, **115**(Aug), 1079–1084.

McNamee, P., J., S., and Chilton, Frank. 1964. Tables of Clebsch-Gordan coefficients of SU_3. *Rev. Mod. Phys.*, **36**(Oct), 1005–1024.

Messiah, A. 1975. *Quantum Mechanics*, 2 vols. Amsterdam: North Holland.

Montvay, Istvan, and Münster, Gernot. 1994. *Quantum Fields on a Lattice*. Cambridge Monographs on Mathematical Physics. Cambridge: Cambridge University Press.

Muta, T. 1987. *Foundations of Quantum Chromodynamics*. Singapore: World Scientific.

Nash, C. 1978. *Relativistic Quantum Fields*. London: Academic Press.

Negele, John W., and Orland, Henri. 1988. *Quantum Many-Particle Systems*. Frontiers in physics vol. 68. Redwood City, CA: Addison-Wesley.

Neuberger, Herbert. 1987. Nonperturbative BRS invariance and the Gribov problem. *Physics Letters B*, **183**(3), 337–340.

Newton, Roger G. 1976. Optical theorem and beyond. *American Journal of Physics*, **44**(7), 639–642.

Omnès, R., and Froissart, M. 1963. *Mandelstaam Theory and Regge Poles*. Frontiers in Physics. New York: W. A. Benjamin.

Ovrut, Burt A., and Schnitzer, Howard J. 1980. Decoupling theorems for effective field theories. *Phys. Rev. D*, **22**(Nov), 2518–2533.

Pauli, W. 1940. The connection between spin and statistics. *Phys. Rev.*, **58**(Oct), 716–722.

Pauli, W., and Villars, F. 1949. On the invariant regularization in relativistic quantum theory. *Rev. Mod. Phys.*, **21**(Jul), 434–444.

Pauli, Wolfgang. 1988. Exclusion principle, Lorentz group and reflection of space-time and charge. Pages 459–479 of Wolfgang Pauli, ed., *Wolfgang Pauli*. Springer.

Pawlowski, Jan M., Spielmann, Daniel, and Stamatescu, Ion-Olimpiu. 2010. Lattice Landau gauge with stochastic quantisation. *Nuclear Physics B*, **830**(1), 291–314.

Peccei, R. D., and Quinn, H. R. 1977. CP conservation in the presence of pseudoparticles. *Physical Review Letters*, **38**, 1440–1443.

Peskin, M. E., and Schroeder, D. V. 1995. *An Introduction to Quantum Field Theory*. Reading, MA: Addison-Wesley.

Pokorski, Stefan. 2000. *Gauge Field Theories*. 2nd ed. Cambridge Monographs on Mathematical Physics. Cambridge: Cambridge University Press.

Pons, Josep. 2004. On Dirac's incomplete analysis of gauge transformations. *Stud. Hist. Philos. Mod. Phys.*, **36**, 491–518.

Pontecorvo, Brun. 1958. Inverse beta processes and nonconservation of lepton charge. *Soviet Physics JETP*, **7**, 172–173.

Quigg, Chris. 2015. Electroweak symmetry breaking in historical perspective. *Annual Review of Nuclear and Particle Science*, **65**(1), 25–42.

Roberts, Craig D., and Williams, Anthony G. 1994. Dyson-Schwinger equations and their application to hadronic physics. *Prog. Part. Nucl. Phys.*, **33**, 477–575.

Robertson, H. P. 1929. The uncertainty principle. *Phys. Rev.*, **34**(Jul), 163–164.

Roman, Paul. 1969. *Introduction to Quantum Field Theory*. New York: John Wiley & Sons.

Rosenfeld, Léon. 1940. Sur le tenseur dímpulsion-énergie. *Mém. Acad. Roy. Belg. Sci.*, **18**, 1–30.

Rothe, Heinz J. 2012. *Lattice Gauge Theories: An Introduction*. 4th ed. World Scientific Lecture Notes in Physics, vol. 82. Singapore: World Scientific.

Rothe, Heinz J., and Rothe, Klaus D. 2010. *Classical and Quantum Dynamics of Constrained Hamiltonian Systems*. World Scientific Lecture Notes in Physics. Singapore: World Scientific.

Ryder, L. H. 1986. *Quantum Field Theory*. Cambridge: Cambridge University Press.

Sakurai, J. J. 1994. *Modern Quantum Mechanics*. Revised ed. New York: Addison-Wesley.

Salam, A., and Ward, J. C. 1964. Electromagnetic and weak interactions. *Physics Letters*, **13**(2), 168–171.

Salam, Abdus. 1968. Weak and electromagnetic interactions. Pages 367–377 of: N. Svartholm, ed., *8th Nobel Symposium Lerum, Sweden, May 19–25, 1968*, vol. C680519. Stockholm: Almqvist & Wiksell.

Schreier, Peter J., and Scharf, Louis L. 2010. Complex differential calculus (Wirtinger calculus). Pages 277–286 of *Statistical Signal Processing of Complex-Valued Data: The Theory of Improper and Noncircular Signals*. Cambridge: Cambridge University Press.

Schrödinger, E. 1930. Zum Heisenbergschen Unschärfeprinzip. *Sitzungsberichte der Preussischen Akademie der Wissenschaften, Physikalisch-mathematische Klasse*, **14**, 296–303.

Schwartz, M. D. 2013. *Quantum Field Theory and the Standard Model*. New York: Cambridge University Press.

Schweber, Silvan. 1994. *QED and the Men Who Did It: Dyson, Feynman, Schwinger, and Tomonaga*. Princeton Series in Physics. Princeton, NJ: Princeton University Press.

Schweber, S. S. 1961. *An Introduction to Relativistic Quantum Field Theory*. New York: Harper & Row.

Schwinger, Julian. 1948a. On quantum-electrodynamics and the magnetic moment of the electron. *Phys. Rev.*, **73**(Feb), 416–417.

Schwinger, Julian. 1948b. Quantum electrodynamics. I. A covariant formulation. *Phys. Rev.*, **74**(Nov), 1439–1461.

Schwinger, Julian. 1951. The theory of quantized fields. I. *Phys. Rev.*, **82**(Jun), 914–927.

Schwinger, Julian. 1957. A theory of the fundamental interactions. *Annals of Physics*, **2**(5), 407–434.

Senjanovic, P. 1976. Path integral quantization of field theories with second class constraints. *Annals Phys.*, **100**, 227–261. [Erratum: Annals Phys. 209, 248 (1991)].

Slavnov, A. A. 1972. Ward identities in gauge theories. *Theoretical and Mathematical Physics*, **10**(2), 99–104.

Srednicki, Mark. 2007. *Quantum Field Theory*. Cambridge: Cambridge University Press.

Sterman, G. 1993. *An Introduction to Quantum Field Theory*. Cambridge: Cambridge University Press.

Streater, R. F. 1975. Outline of axiomatic relativistic quantum field theory. *Reports on Progress in Physics*, **38**(7), 771–846.

Streater, R. F., and Wightman, A. S. 2000. *PCT, Spin and Statistics, and All That*. Landmarks in Physics. Princeton, NJ: Princeton University Press.

Strocchi, F. 1976. Locality, charges and quark confinement. *Physics Letters B*, **62**(1), 60–62.

Strocchi, F. 1978. Local and covariant gauge quantum field theories. Cluster property, superselection rules, and the infrared problem. *Phys. Rev. D*, **17**(Apr), 2010–2021.

Sudarsha, E. C. G., and Marshak, R. E. 1958. Chirality invariance and the universal Fermi interaction. *Physical Review*, **109**(5), 1860–1862.

Sundermeyer, K. 1982. *Constrained Dynamics*. New York: Springer-Verlag.

Sundermeyer, K. 2014. *Symmetries in Fundamental Physics*. New York: Springer.

Symanzik, K. 1970. Small distance behaviour in field theory and power counting. *Communications in Mathematical Physics*, **18**(3), 227–246.

Symanzik, K. 1973. Infrared singularities and small-distance-behaviour analysis. *Communications in Mathematical Physics*, **34**(1), 7–36.

't Hooft, G. 1971. Renormalizable Lagrangians for massive Yang-Mills fields. *Nuclear Physics B*, **35**(1), 167–188.

't Hooft, G. 1973. Dimensional regularization and the renormalization group. *Nuclear Physics B*, **61**, 455–468.

't Hooft, G., and Veltman, M. 1972. Regularization and renormalization of gauge fields. *Nuclear Physics B*, **44**(1), 189–213.

Takahashi, Y. 1957. On the generalized ward identity. *Il Nuovo Cimento (1955–1965)*, **6**(2), 371–375.

Tanabashi, M., et al. 2018. The Review of Particle Physics. *Phy. Rev. D*, **98**, 030001.

Taylor, J. C. 1971. Ward identities and charge renormalization of the Yang-Mills field. *Nuclear Physics B*, **33**(2), 436–444.

Terning, John. 2006. *Modern Supersymmetry: Dynamics and Duality*. International Series of Monographs on Physics, no. 132. Oxford: Oxford Science Publications.

Thomson, Mark. 2013. *Modern Particle Physics*. New York: Cambridge University Press.

Tomonaga, S. 1946. On a relativistically invariant formulation of the quantum theory of wave fields. *Progress of Theoretical Physics*, **1**(2), 27–42.

Tyutin, L. V. 1975. Lebedev preprint FIAN.

Uehling, E. A. 1935. Polarization effects in the positron theory. *Phys. Rev.*, **48**(Jul), 55–63.

van Enk, S. J., and Nienhuis, G. 1994. Spin and orbital angular momentum of photons. *Europhysics Letters (EPL)*, **25**(7), 497–501.

Vergados, J. D., Ejiri, H., and Šimkovic, F. 2012. Theory of neutrinoless double-beta decay. *Reports on Progress in Physics*, **75**(10), 106301.

Ward, J. C. 1950. An identity in quantum electrodynamics. *Phys. Rev.*, **78**(Apr), 182–182.

Weinberg, S. 1995. *Quantum Theory of Fields, vol. 1: Foundations*. Cambridge: Cambridge University Press.

Weinberg, S. 1996. *Quantum Theory of Fields, vol. 2: Modern Applications*. Cambridge: Cambridge University Press.

Weinberg, Steven. 1960. High-energy behavior in quantum field theory. *Physical Review*, **118**(3), 838–849.

Weinberg, Steven. 1967. A model of leptons. *Phys. Rev. Lett.*, **19**(Nov), 1264–1266.

Wess, Julius, and Bagger, Jonathan. 1992. *Supersymmetry and Supergravity*. 2nd revised ed. Princeton, NJ: Princeton University Press.

Wiegel, F. W. 1975. Path integral methods in statistical mechanics. *Physics Reports*, **16**(2), 57–114.

Wightman, Arthur S. 1999. Review of *Pauli and the Spin-Statistics Theorem* by I. Duck and E. C. G. Sudarshan. *American Journal of Physics*, **67**(8), 742–746.

Wilson, Kenneth G. 1974. Confinement of quarks. *Phys. Rev. D*, **10**(Oct), 2445–2459.

Wolfenstein, Lincoln. 1983. Parametrization of the Kobayashi-Maskawa matrix. *Phys. Rev. Lett.*, **51**(Nov), 1945–1947.

Wybourne, B. 1974. *Classical Groups for Physicists*. New York: Wiley-Interscience.

Yang, C. N., and Feldman, David. 1950. The S-matrix in the Heisenberg representation. *Phys. Rev.*, **79**(Sep), 972–978.

Yang, C. N., and Mills, R. L. 1954. Conservation of isotopic spin and isotopic gauge invariance. *Phys. Rev.*, **96**(Oct), 191–195.

Ynduráin, F. J. 1983. *Quantum Chromodynamics: An Introduction to the Theory of Quark and Gluons*. Berlin: Springer–Verlag.

Youla, D. C. 1961. A normal form for a matrix under the unitary congruence group. *Canadian Journal of Mathematics*, **13**, 693–704.

Zee, A. 2010. *Quantum Field Theory in a Nutshell*. 2nd ed. Princeton, NJ: Princeton University Press.

Zimmermann, W. 1969. Convergence of Bogoliubov's method of renormalization in momentum space. *Communications in Mathematical Physics*, **15**(3), 208–234.

Zimmermann, Wolfhart. 1968. The power counting theorem for Minkowski metric. *Communications in Mathematical Physics*, **11**(1), 1–8.

Zwanziger, Daniel. 1992. Critical limit of lattice gauge theory. *Nuclear Physics B*, **378**(3), 525–590.

Zyla, P. A., et al. 2020. Review of Particle Physics. *PTEP*, **2020**(8), 083C01.

Index